26th Polymer Processing Society Annual Meeting 2010 (PPS-26)

Banff, Canada
4 - 8 July 2010

Volume 1 of 2

ISBN: 978-1-5108-3973-1

Printed from e-media with permission by:

Curran Associates, Inc.
57 Morehouse Lane
Red Hook, NY 12571

Some format issues inherent in the e-media version may also appear in this print version.

Printed by Curran Associates, Inc. (2017)

For permission requests, please contact Polymer Processing Society
at the address below.

Polymer Processing Society
810 E. 10th St.
Lawrence, KS 66044
USA

Phone: +1-800-627-0326
+1-785-865-9403
Fax: +1-785-843-6153

pps@allenpress.com

Additional copies of this publication are available from:

Curran Associates, Inc.
57 Morehouse Lane
Red Hook, NY 12571 USA
Phone: 845-758-0400
Fax: 845-758-2633
Email: curran@proceedings.com
Web: www.proceedings.com

TABLE OF CONTENTS

VOLUME 1

PLENARY SESSIONS

S01 – BIOMEDICAL MATERIALS

S02 – POLYMERS FOR ENERGY: OIL & GAS, FUEL CELLS AND SOLAR CELLS

S03 - RECYCLING AND REUSE OF POLYMERS

G01 – BLOW MOLDING, THERMOFORMING AND ROTOMOLDING

G02 - COMPOSITES

G03 – POLYMER NANOCOMPOSITES

G04 – EXTRUSION AND EXTRUSION PROCESSES

G05 – FIBER, FILMS AND MEMBRANES

G06 – INJECTION MOLDING AND MOLDS

G07 – MIXING AND COMPOUNDING

VOLUME 2

G08 – MORPHOLOGY AND STRUCTURE DEVELOPMENT

G09 – POLYMER BLENDS AND ALLOYS

G10 – POLYMER FOAMS

G11 – POLYMERS FROM RENEWABLE RESOURCES

G12 – PROCESS MODELING AND SIMULATION

G13 – PROCESS MONITORING, CONTROL AND SENSORS

G14 – REACTIVE PROCESSING

G15 – RHEOLOGY AND RHEOMETRY

G16 – RUBBER PROCESSING

I01 – CHRIS MACOSKO SESSION (ONLY BY INVITATION)

I02 – JOHN VLACHOPOULOS SESSION (ONLY BY INVITATION

I03 – JAMES L. WHITE SESSION (ONLY BY INVITATION)

Chairman's Message

On behalf of the Polymer Processing Society (PPS), it is my great pleasure to welcome you to the Jewel of Canada, Banff, Alberta for the 26th Annual Meeting of the PPS in 2010. This is actually the 25th Anniversary of the PPS because the inaugural meeting happened in 1985 in Akron, Ohio.

One of PPS founders and giant in polymer processing research, Dr. James L. White, passed away in 2009 and we are having a special session on Sunday July 4 to honor his contributions to the field, and more importantly, contributions to our society. We are also honoring two other individuals, Professor Chris Macosko and Professor John Vlachopoulos for their scientific contributions to polymer processing. Dr. Macosko is being honored on the occasion of his 65th Birthday and Dr. Vlachopoulos is being honored for his longtime contribution to PPS particularly in the area of PPS membership. There are two Morand Lambla Award winners in 2010 who will also be honored: Dr. Marianna Kontopoulou (Canada) and Dr. Wei Yu (China).

Moraine Lake, Banff

More than six hundred contributions will be given, with more than three hundred being oral presentations (over fifty keynotes and ten plenaries). We chose the spectacular mountain location of Banff, Canada for the 26th Annual Meeting of the Polymer Processing Society, July 4-8, 2010 purposefully because the inspiring mountain venue will enable creative and energizing discussion. The 2010 Polymer Processing Society Conference brings together over 500 international polymer processing researchers to enhance scientific understanding and technical innovation in polymer processing by providing a discussion forum for the field's world-wide community of engineers and scientists. The meeting promises to be an exciting blend of advanced materials like biopolymers, polymer nanomaterials, and bio-medical polymers, in addition to simulation and novel processing methods.

There will be time for social interactions, discussion, and networking and these are very important elements of the annual meeting. I would like to thank the PPS-26 attendees, presenters, sponsors, symposium organizers and especially the hard working organizing committee of PPS-26. I would like to give special thanks to the University of Alberta and the University of Calgary for their support and commitment to the conference.

Banff National Park is Canada's oldest national park, established in 1885 in the Rocky Mountains and encompasses 6,641 square kilometres with numerous glaciers, ice fields, evergreen forest, and alpine landscapes. There will be opportunities for visits to beautiful Lake Louise, Columbia Icefield and other touristic sites before, during and after the meeting. You will experience Canada's Beauty in one of the most extraordinary settings in the world. We hope that you will be inspired by Canada's nature during PPS-26 and remember it for a lifetime!

Sincerely,

U. Sundararaj, Chair of PPS-26

INVITED LECTURE – DONALD R. PAUL

PPS-26

26TH ANNUAL MEETING
POLYMER PROCESSING
SOCIETY
July 4-8, 2010 | Banff, Canada

PPS-26

Polymer-Nanoparticle Materials

D. R. Paul
Department of Chemical Engineering and Texas Materials Institute
University of Texas at Austin
Austin, TX,
USA

There has been much interest in incorporation of nanoparticles into polymer formulations focusing especially on both processing and performance issues. High aspect ratio particles like clay platelets or nanotubes have been of interest for mechanical reinforcement, flame retardance, and other functions. In most cases, when rigid fillers are added to a polymer, toughness is reduced even in the case of nanofillers. As a result, rubber toughening in combination with nanoparticle reinforcement has been of some interest. More recently, it has been realized that nanoparticles, especially clay platelets, can significantly alter the morphology of polymer blends; typically dispersed phase particle size is greatly reduced probably as a result of a reduction in the rate of phase coalesence. Since the properties of blends depend on the size of the dispersed polymer phase, many questions, and possibly opportunities, arise when nanoparticles are incorporated into multiphase polymer blends. These issues will be explored in this presentation.

MELTING OF POLYMER BLENDS AND CONCOMITANT MORPHOLOGY DEVELOPMENT IN SINGLE SCREW EXTRUDERS

S. M. Cunha, A. Gaspar-Cunha and J. A. Covas*

I3N / IPC – Institute for Polymers and Composites, Department of Polymer Engineering, University of Minho, Guimarães, Portugal - denebe@dep.uminho.pt, agc@dep.uminho.pt, jcovas@dep.uminho.pt

Abstract - The current understanding of the melting stage in single screw extruders results from pioneering research efforts that were initiated in the fifties and continued for more than thirty years. Most of these theoretical and experimental studies used homopolymers as model systems, whereas in industrial practice there has been a considerable evolution in terms of the complexity of the materials being extruded. This work reports an attempt to monitor the melting sequence and the morphology development of immiscible physical and chemically compatibilized PA6/PP blends. A hybrid melting mechanism, incorporating elements of the Tadmor and of the Dispersive melting mechanisms seems to develop; the early stages of morphology development seem to be similar to those observed in the Haake mixer and Twin-screw extruder.

Introduction

Phenomenological studies of melting in single screw extruders were initiated in the fifties by Maddock (1959). Tadmor *et al.* (1966) developed the first mathematical description of Maddock's experimental observations. In his model, Tadmor assumes the existence of an elastic solid bed, formed by the compactation of the solid pellets, that is separated from the inner hot barrel wall by a thin melt film. The relative velocity between solids and barrel pushes this film into a melt pool, which grows near to the screw pushing flight. Simultaneously, the melt pool pushes the solid bed against the screw trailing flight, exposing more solids for melting at the solids-melt film interface. This sequence of events induces the progressive decrease of the solid bed width.

A rather different melting mechanism was observed by Zhu and Chen (1991) when using specific screw geometries, where dispersion of the solid particles in molten polymer was achieved by forcing the solid bed to break-up, followed by melting of these particles. Rauwendaal (1993) developed the first model of this dispersive melting mechanism, while Huang and Peng (1993) and Potente and Pape (2001) suggested different mathematical approaches to describe the same phenomena.

The above studies were performed using homopolymers, while complex systems are routinely processed nowadays. In the case of polymer blends, it has been shown that the melting stage is crucial to morphology development and, consequentially, to the final blend properties (Machado *et al.* (2006)).

A morphology evolution mechanism for the initial polymer blending stages was proposed by Scott and Macosko (1991) for batch mixers and confirmed by Sundararaj *et al.* (1992) for twin-screw extruders. However, and despite of some efforts (for example, Gosh and Tyagi (2002)), the morphology development in the case of single screw extruders is not clear.

This work reports a study of melting and morphology development of immiscible and compatibilized PA6/PP blends in a single screw extruder. The ultimate goal of the work is to correlate operating conditions and screw geometry with the morphologies that induce optimal properties.

Experimental

Materials

Polyamide 6 / Polypropylene (physical blends, PA6/PP and compatibilized *in situ* with Maleic anhydride grafted Polypropylene, PA6/PP/PP-g-MA), with different compositions, were selected as model systems. The specific grades (see Table 1) were chosen so that the viscosity ratio for the physical blends was approximately one. The melting temperature (from DSC thermograms) of PP and PA6 is 166°C and 226°C, respectively.

Table 1- Raw materials

Material	Trade name	Supplier
PP	ISPLEN 030 G1E	REPSOL
PP-g-MA	OREVAC CA 100	ARKEMA
PA6	Akulon F 130	DSM

Equipment

The blends were prepared in a prototype modular single screw extruder, with a diameter of 30 mm, L/D = 30 and a compression ratio of 2.5, fitted with material sampling devices and pressure transducers along the barrel. A hydraulic contraption is available to extract quickly the screw in order to perform Maddock-type experiments. The screw profile can be modified using the same modularity approach as that of co-rotating twin screw extruders.

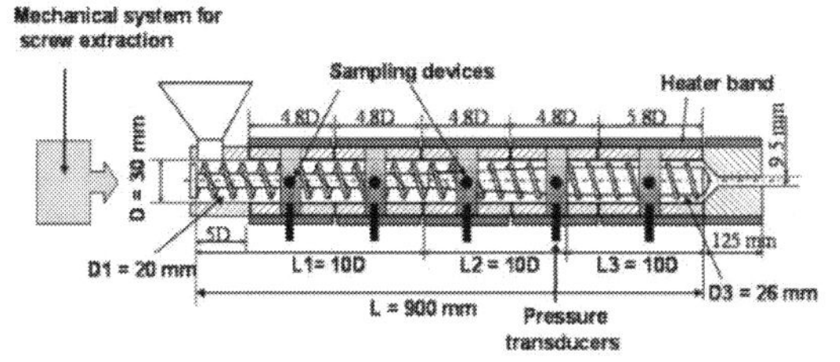

Figure 1- Extruder layout

Procedure

The blend components were pre-mixed in a drum mixer for 10 minutes before extrusion. After reaching steady state, the screw rotation was interrupted, the die was quickly removed and the screw was extracted from the barrel. This operation took approximately 1 minute. The polymer helix was removed from the screw and further cooled with compressed air. Cross-sections were cut from the polymer helices at regular down-channel intervals, immersed in epoxy resin for 24h, polished and immersed for one hour in a dyeing bath in order to tint PA6, thus facilitating its identification. The melting sequence was studied using a stereoscopic magnifying glass, while the morphology evolution was observed by SEM (after freezing in liquid nitrogen and fracture). Samples were also submitted to selective chemical dissolution by formic acid or hot p-xylene to extract PA6 and PP, respectively.

Results and Discussion

A comprehensive study of melting of polymers blends must encompass various observation scales. In terms of melt development, Figure 2 shows that it is more complex than in the case of homopolymers.

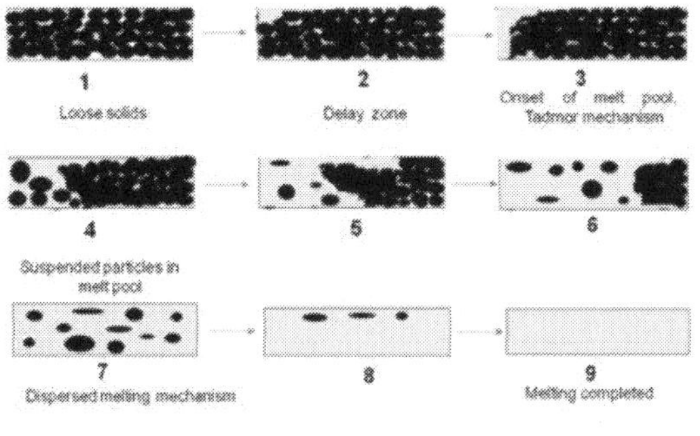

Figure 2- Hybrid melting mechanism of polymer blends.

If the early melting stages seem analogous to the conventional observations for homopolymers, from stage 4 onwards both the Tadmor and the dispersive melting sequences seem to develop concurrently. This evolution is valid for both the physical and compatibilized blends, although the duration of the intermediate steps depends on composition and operating conditions.
Monitoring the contribution of each blend component to the overall melting performance requires not only its detection, but also the quantification of its melting evolution. Figure 3 illustrates the role of each polymer in the melting sequence, taking into consideration the blend composition.

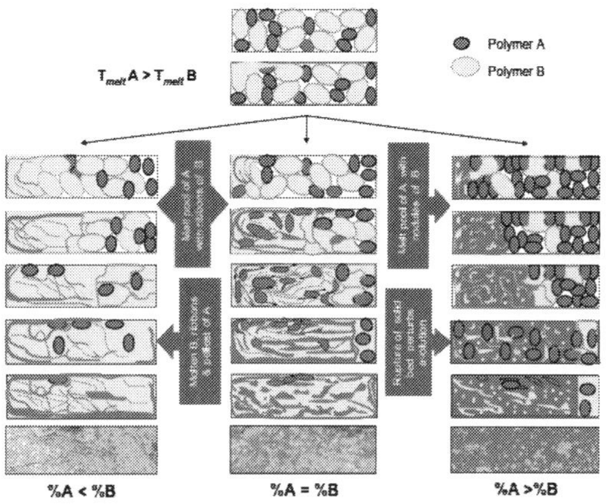

Figure 3- Melting mechanism of polymer blends

Not only the onset, but also rate of melting of each polymer is different. Therefore, it makes sense to investigate the corresponding morphological development. Figure 4 presents the type of data generated for this purpose. In this case, PA6 was extracted.

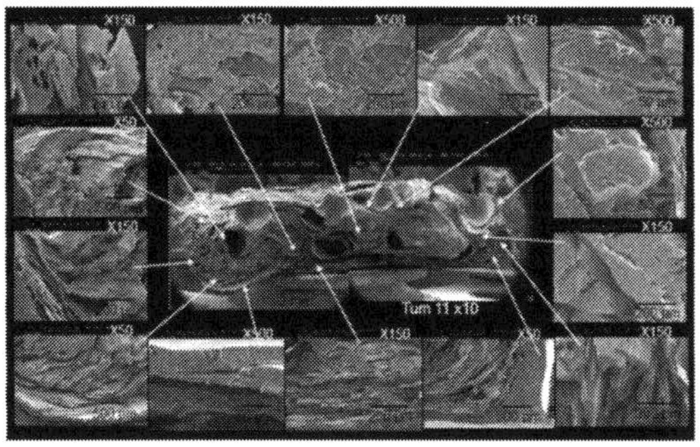

Figure 4- SEM micrographs at screw turn 11 for the PA6/PP 50/50 w/w blend.

The various micrographs across the channel depict different morphologies. Since the cross-channel flow pattern in single screw extruder channels is well known, it is possible to recreate the global morphology evolution. Anyway, the individual photos of Figure 4 show structures that were reported by Scott and Macosko (1991).

Conclusions

The melting mechanism of polymer blends is more complex than that reported for homopolymers. A hybrid melting sequence, incorporating elements of the Tadmor and Dispersive melting mechanisms, seems to develop. The duration of the sequential steps is affected by blend composition and operating conditions. The blend components melt at different rates that depend again on blend composition, operating conditions, and location of the pellets in the channel. Morphological structures similar those reported for polymer blends processed in other types of equipment were observed upon melting. A rapid morphology evolution is preferentially observed at the solid pellets/molten material interface.

Acknowledgements

The authors are grateful to Portuguese Fundação para a Ciência e Tecnologia for supporting this work under grant SFRH/BD/19997/2004, to DSM, the Netherlands, for supplying PA6 and ARKEMA, Spain, for supplying PP-g-MA.

References

Gosh, A. K., Tyagi, S., "Morphology development during blending of immiscible polymers in Screw Extruders", Polym. Eng. Sci., **42**, 1309-1321 (2002)

Huang H. X., Peng Y. C., "Theoretical modeling of dispersive melting mechanism of Polymers in an Extruder", Adv. Polym. tech. papers, **4**, 343-352 (1993)

Maddock B. H., SPE Journal, **15**,383-389 (1959)

Machado A. V., Yquel V., Covas J. A., Flat J., Ghamri N., Wollny A., " The effect of compatibilization route of PA/PO blends on the physico-chemical phenomena developing along a twin-screw extruder", Macromo. Symp., **233**, 86-94 (2006)

Potente H. and Pape J., "The melting behaviour of solid particles surrounded by melt under shear conditions", In 17th Annual Meeting of the PPS, (2001)

Rauwendaal C., "Dispersed solids melting theory", SPE ANTEC Tech. Papers, 2232-2237 (1993)

Scott C.E., Macosko C.W., "Model experiments concerning morphology development during the initial stages of polymer blending" Polymer Bulletin, **26**, 341-348 (1991)

Sundararaj U., Macosko C. W., Rolando R. J.,

Tadmor Z., Marshall D.I., Klein I.. "Fundamentals of plasticating extrusion. I. A theoretical model for melting", Polym. Eng. Sci., **6**, 185-190 (1966)

Zhu F., Chen L., "Studies on the theory of single screw plasticating extrusion. Part I: A new experimental method to extrusion", Polym. Eng. Sci., **31**, 1113-1116 (1991)

INVITED LECTURE – MUSA KAMAL

PPS-26

26TH ANNUAL MEETING
POLYMER PROCESSING SOCIETY
July 4-8, 2010 | Banff, Canada

PPS-26

Micro and Nano Structure Development in Polymer Processing

Musa R. Kamal
Department of Chemical Engineering
McGill University, Montreal,
Canada

Recently, there has been growing interest in producing materials and products that have dimensions at the micro or nano scale and in the transcription of micro patterns, as in micro injection molding and micro injection-compression molding, respectively. Other materials, such as nanocomposites, require the control of the nanostructure of the product. In fact, it is gradually becoming evident that control of performance of polymeric products, even at the macro scale, requires the micro or nano structuring of the material. The above is especially true for products based on semi-crystalline polymers, polymer blends and polymeric composites. This goes significantly beyond the characterization or observation of the micro or nano structure, and requires the ability to predict, manipulate and control these characteristics. We shall discuss some of the requirements posed by micro and nano structuring and the issues that they raise in the field of polymer processing.

26TH ANNUAL MEETING
POLYMER PROCESSING SOCIETY
July 4-8, 2010 | Banff, Canada
PPS-26

On the performance of static mixers

Han E. H. Meijer , Mrityunjay K. Singh, and Patrick D. Anderson
Materials Technology, Eindhoven University of Technology,
Eindhoven, The Netherlands.

The performance of various static mixers, like the Multiflux mixer, the Kenics mixer, the Ross Low-Pressure Drop (LPD) and Low-Low-Pressure Drop (LLPD) mixer, the standard Sulzer SMX mixer, and the recently developed new designs of the SMX, known as SMX(n), and variations thereof, the SMXn, and a new splitting serpentine channel known as the Dentincx mixer, is compared using as criteria (i) energy consumption, measured in terms of the dimensionless pressure drop, (ii) compactness, measured as the dimensionless length, and (iii) quality of the resulting uniformity in striation thicknesses, as a measure for perfectness the way the baker's transformation is performed. All analyses are based on applying the Mapping Method as the most powerful tool present today to analyze distributive mixing and to optimize mixing protocols and mixer designs. As a quantitative mixing measure the cross section averaged, flux-weighted, discrete, intensity of segregation is used, as is follows directly from the Mapping results. Mixers with simple geometries give the lowest pressure drop, at the cost of long lengths with as the winner the Kenics mixer. Most compact are the new designs of the Sulzer geometries, the SMX(n) and the SMXn. The Multiflux performs best in mimicking the optimal baker's transformation but the much more easy to fabricate Dentincx mixer comes close.

INVITED LECTURE – JOHN VLACHOPOULOS

26TH ANNUAL MEETING
POLYMER PROCESSING SOCIETY
PPS-26 July 4-8, 2010 | Banff, Canada PPS-26

Romancing the Rheological Design of Extrusion Dies

John Vlachopoulos
CAPPA-D, Chemical Engineering,
McMaster University
Hamilton, ON
Canada

Extrusion dies are used for the production of film, sheet, profiles, tubing and pipes, wire and cable coating and other products. The first requirement of extrusion die design is balancing of the flow, which means same local area-average velocity at the die exit. Other problems relate to local wall shear rates and stresses, residence times, temperature rise due by viscous dissipation, regions of stagnating flow and the rheological behavior of polymer melts. Despite the recent advances in numerical methods and rheology, trial and error procedures continue to be used by many designers and manufacturers throughout the world. When computer flow simulation packages are employed they seldom go beyond the generalized Newtonian fluid for the description of material behavior. Inverse methods, which have been developed by some researchers, had little, if any, impact on extrusion die design. The reasons for this unsatisfactory state of the technology, as well as ways to improve it, will be discussed. Several examples will be presented illustrating how rational approaches in extrusion die design and optimization can be implemented on the factory floor by using computer flow simulation software and rheological characterization of polymer melts.

POLYMERIC NANOCOMPOSITES: DYNAMICS IN CONFINED SPACE

Masami Okamoto[1]*

[1] *Advanced Polymeric Nanostructured Materials Engineering, Graduate School of Engineering, Toyota Technological Institute, 2-12-1, Hisakata Tempaku, Nagoya, Aichi, Japan – okamoto@toyota-ti.ac.jp*

Abstract - The nanocomposites and their self-assembly behavior have recently been approached to produce nanoscale polymeric materials. Additionally, these nanocomposites have been proposed as model systems to examine polymer structure and dynamics in confined environments. The existence of percolated organo-clay network structure, which is ignored in many literatures as always, is responsible for the soft glassy dynamics in various nanocomposite melt rheology. This mesoscopic structure is intrinsically disordered metastable state and out-of-equilibrium, and offers an apt analogy to soft colloidal gels. The mobility and relaxation of macromolecular chains are retarded by the geometric confinement of the organo-clay networks. This review provides details concerning the mesoscale network structure in nanocomposite melts in the light of rheometry with a combination of scattering experiment and electron microscopy. We also present the nonisothermal disorder-order phase transition (chain packing) kinetics and disorder transition (chain melting) behaviors including the conformational changes of the chain segment of the alkyl chains in confined space (nano-gallery space of the layered fillers). Knowledge of such investigations shall be useful in assessing how does the phase transitions appear to be fundamentally different from those observed in bulk.

Introduction

Over the last few years, the utility of inorganic nanoscale particles as filler to enhance the polymer performance has been established. Of particular interest is recently developed nanocomposite technology consisting of a polymer and organically modified layered filler (organo-clay) because they often exhibit remarkably improved mechanical and various other materials properties as compared with those of virgin polymer or conventional composite (micro/macro-composites) [1, 2]. These concurrent property improvements are well beyond what can be generally achieved through the micro/macro-composites preparation.

The synthetic strategy and molecular design was first explored by Toyota group with nylon 6 as the matrix polymer [1]. This new class of material is now being introduced in structural applications, such as gas barrier film and other load-bearing applications [1]. Polymer/clay nano-composites (PCNs) and their self-assembly behaviors have recently been approached to produce nanoscale polymeric materials. Additionally, these nanocomposites have been proposed as model systems to examine polymer structure and dynamics in confined environments [3-5].

In order to understand the processability of these materials, i.e., the final stage of any polymeric material, one must understand the detailed rheological behavior of these materials in the molten state. Understanding the rheological properties of PNC melts are not only important in gaining a fundamental knowledge of the processability, but is also helpful in understanding the structure-property relationships in these materials. Although rheological measurements is indirect probe, they are a well-established approach to probe the interaction between nano-filler and polymer matrix and the time-dependent structure development. In addition, more clear nanoscale and mesoscale structure development of the systems could be provided when combined with X-ray/light scattering experiments and electron microscopy.

The original mesoscale structure in PNCs consists of randomly oriented exfoliated layers or tactoids of layers. This randomly distributed nano-filler forms a "organo-clay network" structure, which is mediated by polymer chains and organo-clay-organo-clay interactions, responsible for the linear viscoelastic response observed in PCN melts. This meso-structure, which is intrinsically disordered metastable state and out-of-equilibrium, and offers an apt analogy to soft colloidal glasses and gels, was extensively discussed [6, 7].

In this paper, we discuss meso-structure development in PNC melts with the primary focus on flow behavior and an analogy to soft glassy dynamics. We also present the nonisothermal order transition (chain packing) and disorder transition (chain melting) behavior including the conformational changes of the chain segment of the intercalants in one-dimensional confined space.

Analogy to soft colloids

Ren et al [8] discussed the analogy to the dynamics of soft colloidal glasses, exhibiting yield stress, thixotropy, and slow stress recovery under deformation [9]. A glass is homogenious on interparticle structure with elasticity derived from caging effect, which is characterized by long-range electrostatic repulsion and short-range attraction . In the colloidal solid, the storage modulus and complex viscosity show logarithmic dependence time like PCN melts. For example, in an aqueous Laponite dispersion, the structure re organization responsible for the rheological behavior depend upon the ionic strength of the dispersion, which tunes the balance of attractive and repulsive forces between clay domains.

On the other hand, a gel also shows the soft glassy dynamics, which is associated with metastable, structural heterogeneity on mesoscale length, caused by a percolation infinite (volume spanning) network. PCNs have a structural hierarchy without long-range repulsive interaction, leaving short-range attractive forces (van der Waals force) between polymer and clay because of the surface modification of clay by cationic intercalant. For this reason, PCN melts was classified as an attractive colloidal gels [6]. This picture supports other experimental results on transient start-up viscosity under shear flow [10].

Reversibility of network formation process

The central discussion in PCN melts is the structure with elasticity. The idea is an organo-clay network with a few wt%. An important question is how long the build of the solid-like network continues during annealing, i.e., reversibility of the network formation process.

In an aqueous dilute suspensions of smectite swelling clays such as hectorite and montmorillonite (MMT), the existence of a long-range structual correlation (~100 nm) was proposed *via* in-situ small-angle neutron scattering (SANS) experiment of the smectite suspension under quiescent state [11]. This is a worthy hypothesis for new interpretation of the network formation and complex rheological properties. Such magnitude of the structural length in the clay suspensions is suitable for Rayleigh scattering experiment on the basis of the different polarizability between clay particles and medium.

Okamoto et al. constructed rheo-optical device, small-angle light scattering apparatus under shear flow (Rheo-SALS), which enables us to perform time-resolved measurements of light intensity scattered from the internal structure developed under shear flow [12]. They examined long-range correlation and its evolution in the organo-Hectorite/styrene (3.5/96.5 vol/vol) suspensions (dense condition: volume fraction $\phi \geq 0.01$ [13]) under shear flow. In this region, the power law relation for both zero-ω limiting value of modulus and fluidizing stress (σ_{fl}) were given by

$$G'(\omega \to 0) \sim \sigma_{fl} \propto \phi^{4\pm0.2} \quad (1)$$

The ratio of $G''(\omega)/G'(\omega)(=\tan\delta)$ exhibits independence of φ and remains constant value of about 0.1. This behavior relates to self-similarity of the linked network structure in gelation region reported by Winter [14].

Nanocomposite structure

Fig. 1 shows the results of TEM bright field images of nylon 6-based nanocomposite (N6C3.7: loading amount of MMT= 3.7wt%), in which white entities are the cross section of the discrete lamella and the dark area is the nylon 6 matrix due to the staining with 12 tungstophosphoric acid at 80 °C for 2 h. The black strip inside the white part is an individual (exfoliated) MMT particle. The silicate layers are more homogeneously and finely dispersed. The figure shows larger view permitting the observation of discrete nano-layers. The disorder and delaminated layer structure are observed in the TEM image. We estimated the form factors obtained from TEM images, i.e., average value of the particle length (L), thickness (D), of the dispersed particles and the correlation length (ξ) between the particles. The details of the evaluation were described in our previous paper [15]. For N6C3.7, D, L and ξ are in the range of 1-2 nm, 90±10 nm and is 25±10 nm, respectively. The ξ value (≅ 25 nm) corresponds to the random coil size of nylon 6 in N6C3.7. The root-mean-square radius of gyration $<s^2>^{1/2}$ is around 9 nm, which is calculated by $<s^2>^{1/2} = 4.0 \times 10^{-2} M_w^{1/2}$. The chain mobility of the nylon 6 presumably is decreased in translational motion. The network structure in N6C3.7 can be assessed by the relation $\phi_{MMT} > (L/D)^{-1}$, where $(L/D)^{-1}$ is the (aspect ratio)$^{-1}$, confirming the formation of percolated network structure.

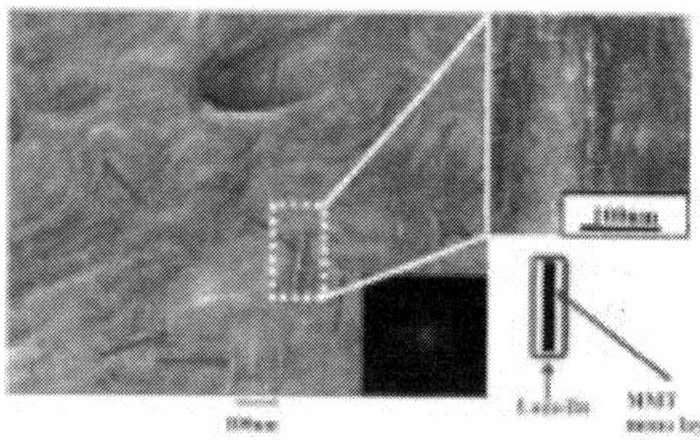

Fig. 1. Bright field TEM images of N6C3.7. The insert is a computed FFT spectrum of the micrograph.

Fig. 2 shows the the temperature dependence of relaxation rate (ω_{rel}), crystallization *half-time* ($t_{1/2}$) calculated from SALS, and rotational relaxation rate ($1/t_d$). The estimated time scale for rotational relaxation rates is about 0.02-0.04 s^{-1} in the range of temperature between 225 and 275 °C. These values are much smaller than the relaxation time scale ($1/\omega_{rel}$) (those are in fact 0.066-0.25 s^{-1}), suggesting the major driving force for the structural evolution (relaxation of the mesoscale network) is not simple Brownian relaxation of the MMT particles.

In the range of T_c between 200 and 210 °C, a high relaxation rate of the network is observed, that seems to be overwhelming the retarding effect. That meams at high T_c region, the network formation is faster and resulted in the *shish-kebab* type of structure in the network. On the other hand, the T_c range between 160-190 °C, the crystallization with large degree of supercooling provides the disturbed crystal growth and hinder the stable growth of the γ–form due to the lack of time for crystalline structure development [16].

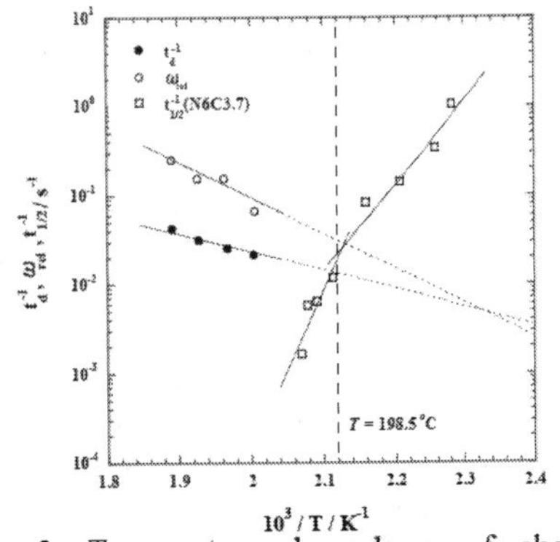

Fig. 2. Temperature dependence of characteristic relaxation rates (ω_{rel} and $1/t_d$) and crystallization rate ($1/t_{1/2}$) of N6C3.7.

Phase transition in nanoconfined space

Table 1 show the characteristic parameters of the organic modified layered fillers (OMLFs) used in this study. They are layered titanate (HTO) intercalated with ocatadecyl tri-methyl ammonium cation (C_{18}TM) and montmorillonite (MMT) intercalated with di-octadecyl di-methyl ammonium cation (DC_{18}DM). The HTO have high surface charge density compared to MMT [2] and the average distance between the exchange sites was found to be 0.888 and 1.188 nm for HTO and MMT respectively.

Table 1- Characteristic parameters of fillers

Chemical formula	HTO	MMT
	$H_{1.07}Ti_{1.73}O_{3.95}\cdot 0.5H_2O$	$Na_{0.33}(Al_{1.67}Mg_{0.33})Si_4O_{10}(OH)_2$
Particle size /nm	~100-200	~100-200
CEC/meq/100g a)	~200	~90
Surface charge Density/ e^-/nm^2 b)	1.26	0.708

a) Methylene blue absorption
b) Surface charge density = ξ/a.b (ξ:layer charge; a, b: cell parameters)

The surface charge density of HTO and MMT were calculated to analyze the structure of intercalant in nanoconfined space. The values were 1.26 and 0.708 for HTO and MMT respectively. Fig. 3 shows the typical WAXD patterns of HTO-C_{18}TM, C_{18}TM and HTO (pristine HTO). The peak HTO-C_{18}TM showed between 2.96-3.16 degrees (2θ) represents the interlayer opening ($d_{(001)}$). The interlayer opening was calculated by subtraction of layer thickness value of 0.374nm from d_{001} value (2.80-2.99 nm) of HTO. When the intercalant is coupled with an all-trans conformation, the alkyl chains are oriented at an angle of 70-90 degrees with respect to the HTO surface. This causes the formation of an inter-digitated layer with a peak that shifts the diffraction angle towards the lower side. The large interlayer opening due to the inter-chain distances implies poor chain packing in HTO when intercalated with C_{18}TM. The reflections at 2Θ =22.01°, 24.10° and 41.97°corresponds to $(110)_{in}$, $(200)_{in}$ and $(201)_{in}$ represents the C_{18}TM crystallite, are observed as small peaks in HTO-C_{18}TM. The peak occurring from $(110)_{in}$ plane (parallel to the alkyl chains) is shifted towards the higher reflection angle. The large inter-chain distances may be due to the formation of good chain packing in the confined space.

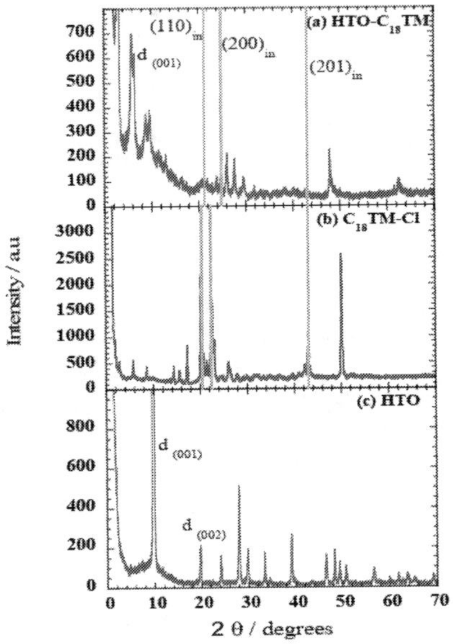

Fig. 3. Typical WAXD profiles of (a) HTO-C_{18}TM, (b) C_{18}TM-Cl, and (c) HT OLL (HTO-H_3O^+).

FTIR spectroscopy was used to probe the conformational change and lateral chain-chain interaction in HTO-C_{18}TM. The frequency, width and intensity of the asymmetric ($\nu_{as}(CH_2)$) and symmetric ($\nu_s(CH_2)$) methylene stretching bands near 2920 and 2851 cm^{-1}, respectively, are sensitive to the gauche/trans conformer ratio.

Figs. 4 and 5 show the peak wavenumber and the half maximum full-width (FWHM) of the characteristic band ($\nu_{as}(CH_2)$) in HTO-C_{18}TM and bulk C_{18}TM respectively. The conformational change from trans form to gauche form improved the molecular mobility of C_{18}TM in the interspacing of HTO layers in the case of heating. The alkyl chains are tethered to the silicate surface on one end and the translational freedom of the chains is limited due to the ion-dipole interaction between the head group of $N^+(CH_3)_2$ and silicate surface. On cooling, the variation is a reversible process with a hysteresis. The formation of gauche form is enhanced and the poor chain packing takesplace in the ordering of the alkyl chains in HTO-C_{18}TM as compared to the bulk C_{18}TM. There is also a possibility of pressure drop (Δp) into the nano-galleries [17] on cooling and thus affecting the mobility of the alkyl chains. For $C_{18}TM$, the sharp changes in peak maxima and FWHM were seen around the T_m of the crystallized alkyl chains in the bulk.

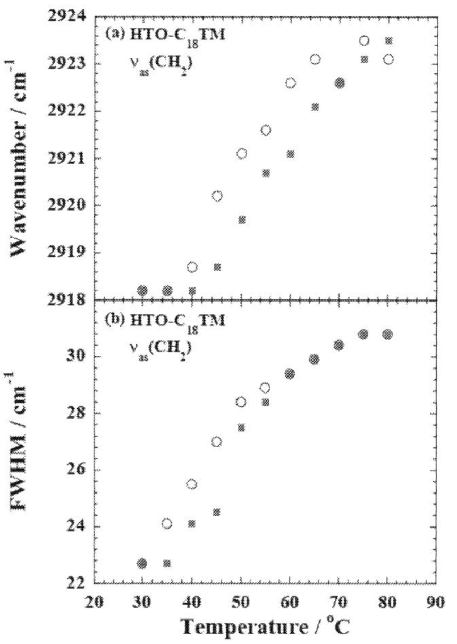

Fig. 4. Temperature variations of (a) frequency shifts and (b) FWHM for characteristic band ($\nu_{as}(CH_2)$) upon cyclic thermal treatment for HTO-$C_{18}TM$.

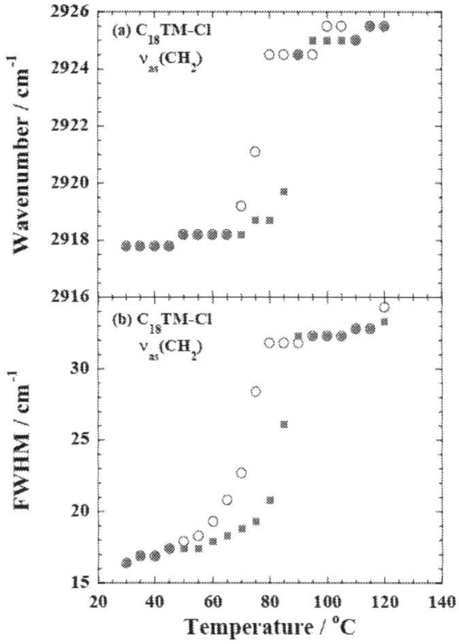

Fig. 5. Temperature variations of (a) frequency shifts and (b) FWHM for characteristic band (ν_{as} (CH_2)) upon cyclic thermal treatment for $C_{18}TM$.

On the other hand, for HTO-$C_{18}TM$, the confinement effects lead to a more steady increase up to 80 °C. The hysteresis phenomena was observed for both HTO-$C_{18}TM$ and $C_{18}TM$, i.e. the peak maxima in the cooling process exhibited higher frequency when compared to the heating process owing to the difference between T_m and T_c, confirming dependence of chain mobility with temperature.

TMDSC data from the nonisothermal cooling process was used to understand the order-disorder transition kinetics of alkyl chains in HTO-$C_{18}TM$. $1/t_{1/2}$ is defined as a measure of the overall order-disorder transition (crystallization) rate and plotted against cooling rate as shown in Fig. 6.

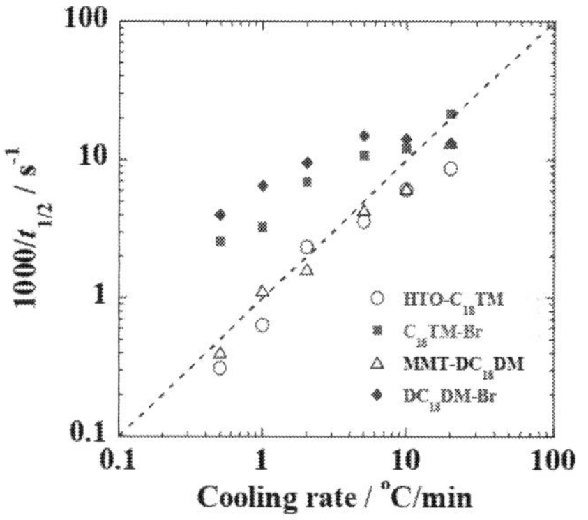

Fig. 6. Cooling rate dependence of overall order-disorder transition rate ($1/t_{1/2}$). The dashed line indicates the linear relation between x- and y- axis.

The confined molecules ($C_{18}TM$) in one or two dimensional order provided significant contribution to the enhancement of the nonisothermal phase transition (chain packing) at higher cooling rate (~ 5.0-20.0 °C/min) despite a slow downward deviation (~2-20.0 °C/min). Similar trend was observed in the case of $DC_{18}DM$ too. The interesting feature is that, the order transition rate increased throughout the experiment in the case of HTO-$C_{18}TM$ and MMT-$DC_{18}DM$.

Ozawa-Flynn-Wall method based on [18] Doyle's approximation can be used to find the effective activation energy E_a for nonisothermal ordering (crystallization) is expressed as follows:

$$\text{Cooling rate} \sim \exp(1.052 E_a / RT_c)$$

where E_a is the activation energy for the nonisothermal phase transition and RT_c is the thermal energy.

For both the cases (HTO-$C_{18}TM$ and MMT-$DC_{18}DM$) examined, the plots conformed to two straight lines in the different T_c regime. The slope that reflects the activation energy of the nonisothermal process was reduced when the cooling rate was beyond 5.0 °C/min. In the polymer melt crystallization, the E_a value decreased upon increasing the crystallization temperature throughout both the glass and melt crystallization regions. The increase of order transition rate for the both fillers may be due to the low activation energy. This feature is similar to the result shown in Fig. 7.The promoted chain packing in different T_c ranges of HTO-C_{18}TM and MMT-C_{18}TM (cooling rate > 5.0 °C/min) showed much lower energy barrier of (E_a= 330, 110 kJ/mol) when compared to C_{18}TM and DC_{18}-DM (E_a= 400, 250 kJ/mol), respectively. This indicates a possible change in the disorder-order phase transition mechanism. A significant decrease in E_a is correlated with the nonisothermal ordering associated with the arrangement of the alkyl chains and its molecular mobility in the nanogalleries (HTO-C_{18}TM & MMT-DC_{18}DM).

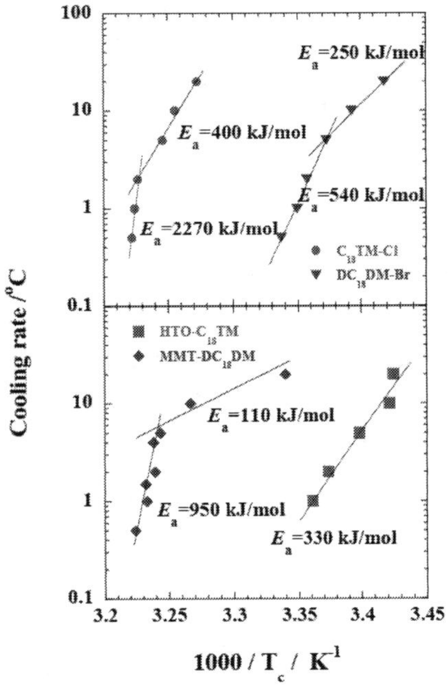

Fig. 7. Relation of Cooling rate and 1000/T_c. Activation energy (E_a) was calculated from the slope of dashed line with Ozawa-Flynn-Wall method.

For HTO-C_{18}TM, E_a was found to be constant (330 kJ/mol), while for C_{18}TM-Cl, the E_a (2270 to 400 kJ/mol) showed a decreasing trend. In the case of MMT-DC_{18}DM, a significant change in E_a (= 950 to 110 kJ/mol) was found when compared with that of DC_{18}DM-Br (E_a= 540 to 250 kJ/mol). Thus it can be concluded the order of non isothermal ordering associated with alkyl chain packing of HTO-C_{18}TM is better than MMT-DC_{18}DM. This may be due to the high surface charge density of HTO than MMT.

Via order-disorder transition kinetic, it was found that the ordering in the confined space was more obvious as opposed to the bulk crystallization. For C_{18}TM, the overall crystallization rates exhibited large value one order higher in magnitude when compared to HTO-C_{18}TM. However, these values were strongly affected by the increase of the cooling rate and later achieved a constant value. The confined molecules (C_{18}TMs) in one or two dimensional order on HTO exhibited significant contribution to the enhancement of the nonisothermal ordering (chain packing) for a higher cooling rate when compared to the DC_{18}DMs on MMT, can be attributed to the high surface charge density of HTO over MMT.

Acknowledgement

This work was supported by the FR-Project (2009-2010).

References

1. Okada, A., Usuki, A., "Twenty Years of Polymer-Clay Nanocomposites", Macromol. Mater. Eng., **291**, 1449-1476 (2006)
2. Hussain, F., Hojjati, M., Okamoto, M., Gorga, R. E., "Polymer–Matrix Nanocomposites, Processing, Manufacturing, and Application: An Overview", J. Composite Mater., **40**, 1511-1575 (2006)
3. Vaia, R. A., Giannelis, E. P., "Polymer Melt Intercalation in Organically-Modified Layered Silicates: Model Predictions and Experiment" Macromolecules, **30**, 8000-8009 (1997)
4. Rao, Y., Pochan, J. M., "Mechanics of Polymer–Clay Nanocomposites", Macromolecules, **40**, 290-296 (2007)
5. Manias, E., Kuppa, V., "The Origins of Fast Segmental Dynamics in 2 nm Thin Confined Polymer Films", Eur. Phys. J., **E8**, 193-199 (2002)
6. Treece, M. A., Oberhauser, J. P., "Soft Glassy Dynamics in Polypropylene–Clay Nanocomposites", Macromolecules, **40**, 571-582 (2007)
7. Galgali, G., Ramesh, C., Lele, A., "A Rheological Study on the Kinetics of Hybrid Formation in Polypropylene Nanocomposites", Macromolecules, **34**, 852-858 (2001)
8. Ren, J., Casanueva, B. F., Mitchell, C. A., Krishnamoorti, R., "Disorientation Kinetics of Aligned Polymer Layered Silicate Nanocomposites", Macromolecules, **36**, 4188-4194 (2003)
9. Tanaka, H., Meunier, J., Bonn, D., "Nonergodic States of Charged Colloidal Suspensions: Repulsive and Attractive Glasses and Gels", Phys Rev., **E 69**, 031404-1-6 (2004)
10. Treece, M. A., Oberhauser, J. P., "Ubiquity of Soft Glassy Dynamics in Polypropylene–clay Nanocomposites", Polymer, **48**, 1083-1095 (2007)
11. Mourchid, A., Delville, A., Lambard, J., Lecolier, E., Levitz, P., "Phase Diagram of Colloidal Dispersions of Anisotropic Charged Particles: Equilibrium Properties, Structure, and Rheology of Laponite Suspensions", Langmuir, **11**, 1942-1950 (1995)
12. Okamoto, M., Sato, H., Taguchi, H., Kotaka, T., "Shear-Induced Aggregation Behavior in Lipophilized Smectite Clay/Styrene Suspension", Nippon Rheology Gakkaishi, **28**, 199-200 (2000)
13. Okamoto, M., Taguchi, H., Sato, H., Kotaka, T., Tatayama, H., "Dispersed Structure and

Rheology of Lipophilized-Smectite/Toluene Suspensions", Langmuir, **16**, 4055-4058 (2000)
14. Winter, H. H., Chambon, F., "Analysis of Linear Viscoelasticity of a Crosslinking Polymer at the Gel Point" J. Rheol., **30**, 367-382 (1986)
15. Sinha Ray, S., Yamada, K., Okamoto, M., Ogami, A., Ueda, K., "New Polylactide/Layered Silicate Nanocomposites.3. High-Performance Biodegradable Materials", Chem Mater., **15**, 1456-1465 (2003)
16. Katoh, Y., Okamoto, M., "Crystallization Controlled by Layered Silicate in Nylon 6-Cay Nano-composite", Polymer, **50**, 4718-4726 (2009).
17. Yoshida, O., Okamoto, M., "Direct Melt Intercalation of Polylactide Chains into Nano-Galleries: Interlayer Expansion and Nano-Composite Structure", Macromol. Rapid Commun., **27**, 751-757 (2006)
18. Ozawa, T., "Kinetic Analysis of Derivative Curves in Thermal Analysis" J. Therm. Anal. Calorim., **2**, 301-324 (1970)

PPS-26

26TH ANNUAL MEETING
POLYMER PROCESSING SOCIETY
July 4-8, 2010 | Banff, Canada

PPS-26

Morphology and Properties of Exfoliated Graphene/Thermoplastic Nanocomposites

Christopher Macosko
Department of Chemical Engineering and Materials Science
University of Minnesota, MN
USA

Graphite has attracted large attention as a reinforcement for polymers due to its ability to modify electrical conductivity, mechanical and gas barrier properties of host polymers and its potentially lower cost than carbon nanotubes. If graphite can be exfoliated into atomically thin graphene sheets, it is possible to achieve the highest property enhancements at the lowest loading. However, small spacing and strong van der Waals forces between graphene layers make exfoliation of graphite via conventional composite manufacturing strategies challenging. Recently, two different approaches to obtain exfoliated graphite prior to blending were reported: thermal treatment (FGS) (Schniepp et al., 2006) and chemical modification (iGO) (Stankovich et al., 2006) of graphite oxide. We describe and evaluate these exfoliation approaches and the methods used to produce graphene reinforced thermoplastics, particularly polyester, polycarbonate, polyethylene and polyurethane (TPU) nanocomposites. Three different dispersion methods - melt blending, solution mixing and in-situ polymerization – are compared. Characterization of dispersion quality is illustrated with TEM, rheology and in electrical conductivity, tensile modulus and gas barrier property improvements.

Synthesis and exfoliation of isocyanate-treated graphene oxide nanoplatelets, Stankovich, S.; Piner, R. D.; Nguyen, S. T.; Ruoff, R. S. *Carbon* **2006**, *44*, 3342-3347.

Functionalized Single Graphene Sheets Derived from Splitting Graphite Oxide, Schniepp, H. C.; Li, J.-L.; McAllister, M. J.; Sai, H.; Herrera-Alonso, M.; Adamson, D. H.; Prud'homme, R. K.; Car, R.; Saville, D. A.; Aksay, I. A. *J. Phys. Chem. B* **2006**, *110*, 8535-8539.

Morphology and Properties of Polyester/Exfoliated Graphite Nanocomposites, H. Kim and C.W. Macosko, *Macromolecules,* **2008**, 41, 3317-3327.

INVITED LECTURE – CHRISTOPHER MACOSKO

Processing-property Relationships of Polycarbonate/Graphene Composites, H. Kim and C. W. Macosko *Polymer* **2009,** 50, 3797–3809.

Morphology and Properties of Exfoliated Graphite/Thermoplastic Polyurethane Nanocomposties, H. Kim and C. W. Macosko *SPE ANTEC* **2009**.

DESIGN, FABRICATION AND APPLICATIONS OF POLYMER MICRO/NANOFLUIDIC BIOCHIPS

L. James Lee

Department of Chemical and Biomolecular Engineering, The Ohio State University, USA - lee.31@osu.edu

Abstract - Micro/Nanotechnology is initiated from the electronics industry. In recent years, it has been extended to micro-electro-mechanic system (M/NEMS) for producing miniature devices based on silicon and semi-conductor materials. However, the use of these hard materials alone is inappropriate for many biomedical devices. Soft polymeric materials possess many attractive properties such as high toughness and recyclability. Some possess excellent biocompatibility, are biodegradable, and can provide various biofunctionalities. I will first give a brief overview of major activities in our center on micro/nanomanufacturing of polymeric materials and micro/nanofluidics, followed with several example biochips. An enzyme immunoassay chip will be discussed as an example for a low-cost and mass-producible lab-on-a-chip platform for molecular and biological analyses. The platform is a microfluidic CD for Enzyme-Linked Immunosorbent Assays (ELISA) that reduces cost, accelerates results, and improves reliability of analyses for food borne contaminants, cancer diagnoses and environmental contamination. The second example is a microfluidic-electrospray system to produce multifunctional nanoparticles containing gene/drug for in vivo applications. The third example is a nanochannel based cell electroporation device capable of dose and location controlled in vitro drug/gene delivery to individual cells.

Polymer CD-ELISA Microfluidic Biochip

Capillary valves are one of the most important components for the realization of a fully integrated microfluidic system, such as lab-on-a-chip (LOC). This is because the capillary valve is simple and can be implemented easily for applications in various microfluidic devices. Capillary valving has the advantage of not requiring any moving parts and external actuation. For pure water or buffer solutions, the capillary valve works well because a proper hydrophobic polymer surface can be chosen to provide a desirable contact angle. However, proteins exist in the solutions used in most biomedical applications. The phenomena of protein adsorption onto plastic substrates have been widely observed. Due to protein adsorption, the surface of the capillary valve gradually becomes hydrophilic and the solution would wick through, leading to the failure of the valving function. One way to overcome this problem is to implement a superhydrophobic coating in the valve areas. We developed polyaniline nanograss along with fluorine coating to create nanopatterned capillary valves of a Compact Disk (CD)-based Enzyme-Linked Immunosorbent Assay (ELISA) chip. "Hierarchical roughness" of nanosized papillae makes the valve surface superhydrophobic and much less affected by protein binding. This polyaniline nanoconstruct provides a reliable and low-cost approach to control fluid transport on microfluidic biochips as shown in Fig. 1.

Polymer Microfluidic-Electrospray Chip

Nanomedicine is a rapidly developing new

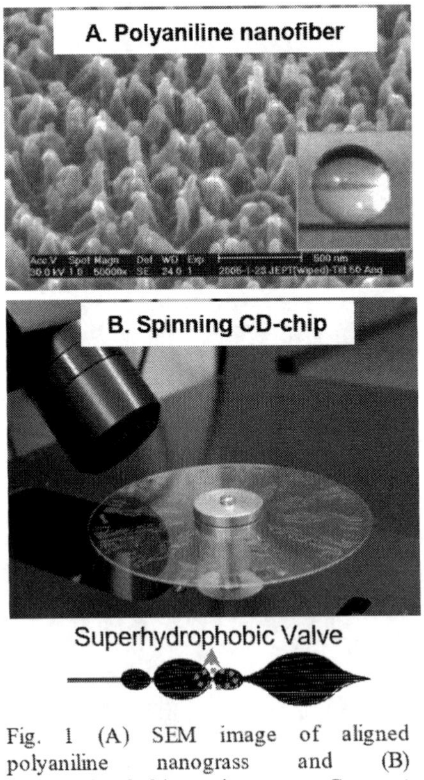

Fig. 1 (A) SEM image of aligned polyaniline nanograss and (B) Superhydrophobic valves on Compact Disk-based chip treated with polyaniline

field that utilizes nanoparticle gene/drug delivery systems for therapeutic applications. The size, size distribution, surface charge, morphology, and encapsulation efficiency of the nanoparticle delivery systems play an extremely important role in the success and efficacy of gene/drug delivery *in vitro* and *in vivo*. Although the conventional preparation method, bulk mixing, is straightforward, the formation of the nanoparticles is not well controlled. Here, we present a novel method to produce multifunctional nanoparticles by the combination of microfluidics and electrospray. In this method, microfludics is used to bring all components in a well controlled manner, and the preformed liquid mixture is dispersed into tiny droplets by electrospray. This technique can be used to produce various multifunctional nanoparticles As shown in Fig. 2, polyplex nanoparticles are produced using bulk mixing method with the energy transfer pair of Qdot605 and Cy5 labeled nucleic acids. Upon excitation at 488 nm, QD-FRET-mediated Cy5 emission indicates a compact and intact nanoparticle. The loss of Cy5 emission indicates a disrupted and deconjugated nanoparticle.

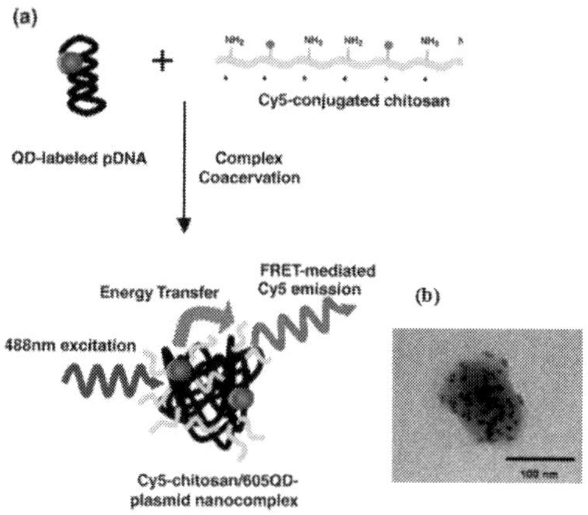

Fig. 2 Formation (a) and structure (b, TEM image) of polyplex nanoparticles.

Different from the bulk mixing method, nanoparticles can be produced using microfluidic electrospray in a controlled way. As shown in Fig. 3, Qdot605 and Cy5-ODN are brought together through the two inlets of the microfluidic device and are well mixed at the outlet, as shown in the upper center picture. The mixture is fed through the inner needle. A lipid mixture is fed though the outer needle. With a positive high voltage applied, a Taylor-cone jet forms at the end of the needle. The jet breaks into tiny droplets.

Compared with bulk mixing, microfluidic electrospray is able to produce more monodispersed lipoplex nanoparticles. The nanoparticles produced by microfluidic electrospray have unilamellar structure, while those produced by bulk mixing have multilamellar structure, which might be due to the fast evaporation of the solvent in the electrospray process. The structure difference might lead to different intracellular behaviors.

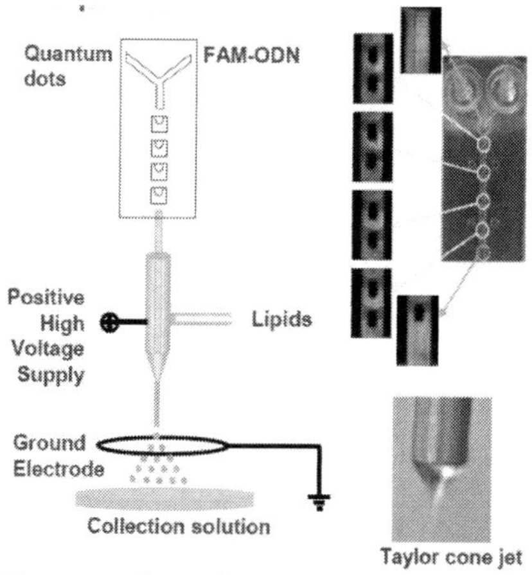

Fig. 3 Schematic diagram of the microfluidic-electrospray method.

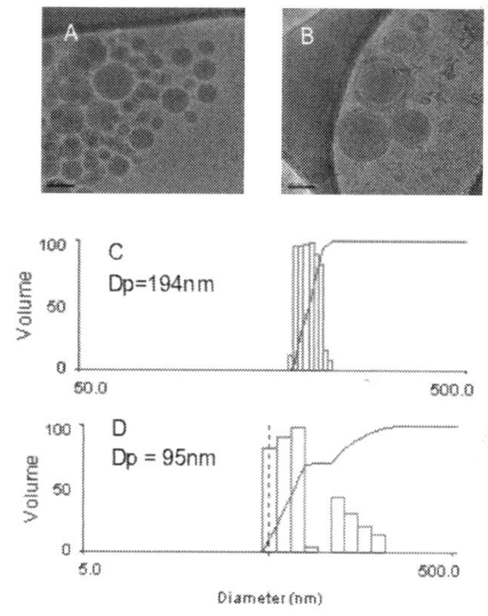

Fig. 4 Structure and size distribution of multifunctional lipoplex nanoparticles produced by microfluidic electrospray (A and C) and bulk mixing (B and D).

Nanoelectroporation

Gene therapy for the treatment of chronic diseases involves the introduction of large plasmid or linear genes and small nucleic acid molecules such as oligodeoxynucleotides (ODN), small interfering RNA (siRNA), and microRNA (miR) into the intra-cellular space. Most of the current gene delivery methods, including viral and non-viral nanoparticles, electroporation (EP), gene gun, etc., are based on bulk stochastic processes in which 10^5-10^6 cells are transfected randomly by a large amount of gene/drugs. With these methods, it is impossible to control the injected dose, particularly for multiple gene/drug delivery. New technologies at the nanoscale that can control the dosage and type of drug/gene, reduce the concentration of gene, and provide high cell viability are clearly required for successful gene delivery [1-2]. We have fabricated a low-cost nano-electroporation device for high precision and benign gene delivery. The proof-of-concept has been successfully demonstrated using localized delivery of PI dye, ODN, plasmid-DNA (GFP) and Q-dots [3].

References:

[1] Salem *et al.* "Multifunctional nanorods for gene delivery" Nature Materials **2**, 668 (2003).
[2] Guignet *et al.* "Suspended-drop electroporation for high-throughput delivery of biomolecules into cells" Nature Methods **5**, 393 (2008).
[3] Guan, J. *et al.* "Large Laterally Ordered Nanochannel Arrays from DNA Combing and Imprinting" Adv. Mater. 2010 (accepted).

26TH ANNUAL MEETING
POLYMER PROCESSING SOCIETY
PPS-26
July 4-8, 2010 | Banff, Canada

Microcellular Foaming – Art vs Science

Chul B. Park and Siu N. Leung
Microcellular Plastics Manufacturing Laboratory,
Department of Mechanical & Industrial Engineering,
University of Toronto, Toronto
Canada

Cellular and microcellular plastics have significant market potentials in a wide spectrum of applications, ranging from daily household commodities to advanced engineering or biomedical applications.Although the foaming technology has been an art, extensive research has revealed its scientific aspects, and has led to its continuous development. Various studies have aimed to control cell morphology to achieve desirable foam properties such as acoustical and thermal insulation abilities, optical properties, as well as mechanical properties while reducing material costs. This is a non-trivial task because polymeric foaming processes involve delicate thermodynamic, kinetic, and rheological mechanisms. Furthermore, environmental concerns urge the replacement of ozone-depleting gases (e.g., CFCs or HCFCs) by non-ozone-depleting ones (e.g., CO2 and N2) for foaming. The uses of these viable alternatives pose additional challenges in foam fabrication because of their low solubility and high diffusivity in polymer melt. In this context, it is critical to elucidate the fundamental mechanisms that govern cell nucleation and growth processes during foaming in order to drive the development in this technology. This paper presents a series of theoretical investigations, simulations, and experimental studies to identify the mechanisms of cell formation, growth, coarsening, and coalescence of thermoplastics foams with the presence of inorganic fillers, organic dispersed phase, crystalline phase, or nanoparticles in the polymer matrices. This work also reveals the roles of various additives or dispersed phases to control cell formation and cell growth as well as to suppress cell ripening and coalescence.

MODELING DRUG RELEASE FROM SOLID MATRICES BASED ON POLYMERIC HYDROGELS

I. Galdi[1], G. Lamberti[1], G. Titomanlio[1*], A.A. Barba[2], M. d'Amore[2]

Università degli Studi di Salerno, Via Ponte don Melillo, 84084 Fisciano (SA) Italy
[1] Dipartimento di Ingegneria Chimica e Alimentare, Tel +39 089964220 (G. Titomanlio), Fax +39 089964057, e-mails: igaldi@unisa.it, glamberti@unisa.it, gtitomanlio@unisa.it
[2] Dipartimento di Scienze Farmaceutiche, e-mails: aabarba@unisa.it, mdamore@unisa.it

The drug release from matrices made of hydrogels is a complex process which can be divided into the following steps: water diffusion into the matrix from a liquid external phase, the polymer plasticization and swelling by effect of water, drug diffusion into the gel toward the outer medium, and the polymer erosion at the external interface. Such a complex behavior, to our knowledge, was not described by a full model up to now.
In this work, the full complex process of drug release from polymer (HPMC) based matrices systems was investigated and described. The phenomena, previously observed and quantified experimentally in our laboratory (water up-take, polymer swelling and erosion, and drug diffusion trough the polymer and release in the outer medium), were described by transient mass balances in presence of variable coefficient pseudo-diffusion, and the resulting set of differential equations were solved by using finite element methods.
The resulting code was found able to describe quantitatively the main features of the observed phenomena, including the tablet size increase and the tablet shape deformation as result of water up-take and polymer erosion.

Introduction

The drug release from solid matrices systems, made of polymeric bio-composites (polymer plus drug), is a basic concept for studies on controlled drug release. The most interesting class of polymers in this application is given by hydrogels, also pH-sensitive ones. Matrices based on hydrogels, once swallowed (during the in-vitro tests, once immersed in the solvent mimicking the body fluid), start to absorb water from the surrounding (water up-take) and the water causes a number of phenomena: the hydrogel swelling, the polymer plasticization (lowering of the glass transition temperature), the diffusion coefficient increase, the erosion phenomenon (due to polymer disentanglement). As a result, the drug previously embedded in the matrix can be released (by diffusion into the matrix and then by dissolution in the outer solvent) (Siepmann and Peppas, 2001).

The design of these systems requires a deep knowledge of the phenomena involved during tablet hydration/dissolution in the stomach and in the gastro-intestinal tract. These phenomena can be described in terms of mathematical equations (modeling) and these last can be solved by properly designed software (numerical code). The mathematical modeling of drug release can indeed significantly facilitate the development of new dosage systems and the optimization of existing pharmaceutical products. Certainly, the identification of the physical model and its mathematical description, in a way adequate to describe the real behavior of a pharmaceutical system, can be used to simulate the effect of the device design parameters on the release kinetics. The availability of the right model allows to predict *a priori* the system formulation parameters to obtain the desired drug release kinetics (Siepmann and Siepmann, 2008).

Aim of this work is to point out a model and the related numerical code, which has to be able to describe all the phenomena observed during drug release from tablets made of swellable/erodible polymer matrices.

Experimental

Materials

Hydroxypropyl methylcellulose (HPMC, Methocel K15M Premium Grade) was kindly offered by Colorcon (UK) and theophylline (TP, CAS no. 58-55-9) was purchased from Sigma Aldrich (Italy). Both the materials were used as obtained, in form of fine powders with maximum particle size lower than 500 μm, and the mean volume diameter for the two materials are of the order of 100 μm (manufacturer data). In the following the water will be identified as the component 1; the drug, TP, as the component 2; and the polymer, HPMC, as the component 3.

Methods

Matrices preparation and hydration

Mixtures of HPMC and TP 50:50 w/w blends were obtained by mixing the powders in a mortar. The cylindrical matrices (0.35 g, 13 mm diameter, 2.0 mm thickness) were prepared by compressing the powder in a tabletting machine (Specac PN3000), equipped with flat-faced punches, diameter 13 mm, with a compression force of 50 kN (by a Carver Press) kept for 5 minutes.

The "radial" test
The tablets, clamped between two glass slides to allow the water uptake only by the lateral surfaces, were immersed in distilled water kept at 37°C for given time intervals. After the immersion, the samples were removed from the bath, and then the partially hydrated tablets were cut by annular punches of different size, obtaining several annuli, and one core disc. The different amounts of partially hydrated polymer in different annuli were carefully weighted, dried, weighted once more. In this way it was possible to determine the average concentration of water in each annulus, i.e. as function of tablet radius. Then, the dried samples were totally dissolved and the drug contents were assayed by a spectrometer, $\lambda = 275$ nm. Thus, the average mass fractions of water and drug in each annulus, and in the core, were measured [Barba et al. 2009a, 2009b].

The "overall" test
The tablets were immersed in a USP dissolution tester type II containing distilled water kept at 37°C. In this test, the entire tablet's surface is in contact with the medium. The tablets were placed in a suitable sample holder to avoid the sticking of the matrices on the bottom of the vessel. After given immersion intervals, three hydrated samples were removed, weighted, dried, weighted once more and at last they were fully dissolved, to allow the drug content assaying. By this way the water absorbed, as well as the polymer and drug residual into the tablet were determined. One matrix for each immersion time were cut and photographed. Size (diameter and thickness) and shape of the hydrated matrices were obtained from image analysis.

Modeling

Balance equations
The transport of water and drug in the matrix can be view as two pseudo-diffusion phenomena, which can be described by two transient mass balances ($k = 1$ for the water and 2 for the drug):

$$\rho\frac{\partial\omega_k}{\partial t} = \vec{\nabla}\cdot\left(\rho D_k \vec{\nabla}\omega_k\right) \tag{1}$$

In eq. (1), the matrix density is ρ, ω_k are the water and drug mass fractions, D_k are the pseudo-diffusion coefficients.
The initial conditions for integration are given by eq. (2), in which Ω is the integration domain (i.e. the matrix) and $\omega_{k,0}$ are the initial homogeneous mass fraction of water ($k = 1$) and drug ($k = 2$).

$$@\, t=0 \quad \forall\vec{x}\in\Omega \quad \omega_k\left(t=0,\vec{x}\right)=\omega_{k,0} \tag{2}$$

The boundary conditions are defined on the moving boundary $\Gamma(t)$, and they are given by eq. (3).

$$@\, \vec{x}\in\Gamma(t) \quad \forall t>0 \quad \omega_k\left(t>0,\vec{x}\in\Gamma(t)\right)=\omega_{k,eq} \tag{3}$$

In eq. (3), the $\omega_{k,eq}$ are the equilibrium values for water ($k = 1$) and drug ($k = 2$) mass fraction.

Constitutive equations
To solve equation (1), the pseudo-diffusion coefficients, D_k (for $k = 1, 2$), have to be evaluated. In polymeric systems subjected to swelling, the diffusion coefficients are not constant, being low in the dry polymer and increasing as the water content increases (into the gel). They can be modeled according to Siepmann *et al.*:

$$D_k(\omega_1) = D_k^* \cdot \exp\left[-\beta_k \cdot \left(1-\frac{\omega_1}{\omega_{1,eq}}\right)\right] \tag{4}$$

where $D_k^*/\exp(\beta_k)$ is the value of the pseudo-diffusion coefficient in the dry matrix ($\omega_1 = 0$), and D_k^* is the value of the pseudo-diffusion coefficient in the fully swollen matrix ($\omega_1 = \omega_{1,eq}$).
The density of the partially hydrated matrix can be calculated from the simplest mixing rule which can be written for the specific volume:

$$\frac{1}{\rho} = \frac{\omega_1}{\rho_1} + \frac{\omega_2}{\rho_2} + \frac{1-\omega_1-\omega_2}{\rho_3} \tag{5}$$

where ρ_1, ρ_2 and ρ_3 are the water, the drug and the polymer densities, respectively.

Modeling of the swelling and of the erosion
The water up-take causes the matrix swelling, and the polymer disentanglement at the matrix surface causes the matrix erosion. In turn, these two phenomena, swelling and erosion, causes the matrix surface to be a moving boundary. Mathematically, there is a trace of this statement in eq. (3), where the surface has been described by the term $\Gamma(t)$. Therefore, there is the need for modeling the two phenomena, with the aim of obtaining the function $\Gamma(t)$.
Thus, the movement of a surface element is due to the swelling phenomenon (which causes the increase of the matrix size) and to the erosion phenomenon (which causes the decrease of the matrix size). In term of element velocity, v, the governing equation is:

$$v = v_{swe} + v_{eros} \tag{6}$$

In which v_{swe} is the size-increase velocity due to the swelling (a positive value) and v_{eros} is the size-decrease velocity due to the erosion (a negative value).
According with the Camera-Roda and Sarti model, the flow of the solvent, j_1 in the present case, 1 being the water, in a swelling system can be obtained by summing up two contribution, one due to the diffusion

and one due to the swelling phenomenon itself. The first one, $j_{1,diff}$, is thus described by the Fick law of diffusion, the second one, $j_{1,swe}$, should account for a diffusion term plus a relaxation term:

$$j_1 = j_{1,diff} + j_{1,swe} = = (\rho D_1 \nabla\omega_1) + \left(\rho D_{swe}\nabla\omega_1 + \rho\tau \frac{\partial \omega_1}{\partial t}\right) \qquad (7)$$

Neglecting the relaxation term (i.e. assuming the time scale of the relaxation is lower than the time scale of the diffusion, $\tau \rightarrow 0$), and hypothesizing that the swelling diffusion coefficient is proportional to the diffusion coefficient itself (i.e. $D_{swe} = k_{swe}D_1$, in which k_{swe} is a proper constant, to be evaluated), the water flow due to the swelling is given by:

$$j_{1,swe} = \rho k_{swe} D_1 \nabla \omega_1 \qquad (8)$$

A water balance on a surface element of thickness δ, which is subject to the swelling phenomenon due to the water uptake, gives:

$$v_{swe} = \frac{d\delta}{dt} = -\frac{j_{1,swe}}{\rho} = -\frac{k_{swe} j_{1,diff}}{\rho} \qquad (9)$$

Locally on the surface, the diffusion normal flux is negative (inward), thus the velocity given by eq. (9) is a positive value (causing the size increase).
The boundary movement velocity due to the erosion phenomenon is calculated as proportional to the local surface area, since the erosion is a surface phenomenon:

$$v_{eros} = -k_{eros} A_E(t) \qquad (10)$$

In eq. (10) $A_E(t)$ is the erosion surface area (which is time-dependent) and k_{eros} is a proper proportionality constant.

Code solving
The FEM software used in this work to implement the simulations is COMSOL Multiphysics® 3.4 (Copyright © 1994–2007 by COMSOL AB, Tegnérgatan 23 SE-111 40 Stockholm). This software allows transforming conventional models for each kind of physical model into multi-physics models, which solve coupled physics phenomena simultaneously (Comsol, 2007). The development and implementation of the simulations have been carried out with the help of a workstation based on the processor Intel® Core™2 Duo E8500, with a clock rate of 3.16 GHz and a RAM of 3 Gb, 800 MHz.

Results and Discussion

Experiments were carried out by immersion of tablets for several time intervals, following the protocol outlined in the section named "radial" test, and the data in terms of drug, polymer and water masses were summarized in Figure 1 (symbols). Therefore, the time evolutions for these masses were obtained; these latter were used to validate the model predictions (curves) and, indirectly, the correct evaluation of each single phenomenon observed. Indeed, the water up-take due to the water diffusion causes the increase in water mass, the erosion causes the decrease in polymer mass, the drug diffusion causes the decrease in drug mass. The inspection of Figure 1 confirms that each single phenomenon is correctly quantified by the models.

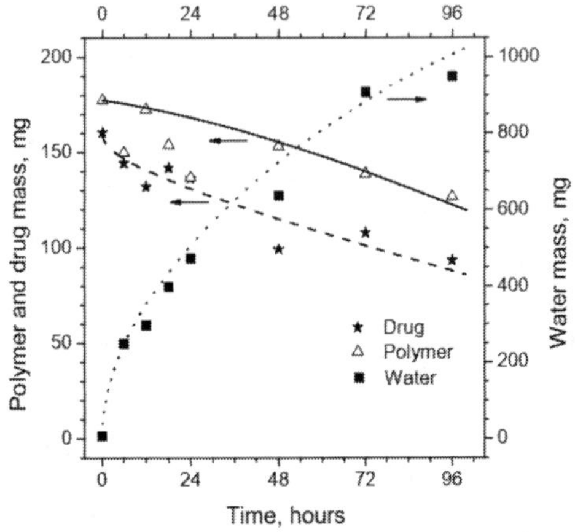

Figure 1. Time evolutions of polymer, drug and water mass in the HPMC-TP 1:1 matrices subjected to the lateral hydration ("radial" test). Symbols, experimental data (△, polymer, ★, drug, ■, water); curves, model predictions (continuous, polymer; dashed, drug; dotted, water).

The parameters optimized were D_1^* and D_2^* as well as k_{swe} and k_{eros}. All the parameters but these four were taken from literature or from previous estimations. It is worth to note that the set of experimental data is large, and it comes from different measurements (mass of water and polymers along the tablet radius, concentration of the drug in the dissolution medium).

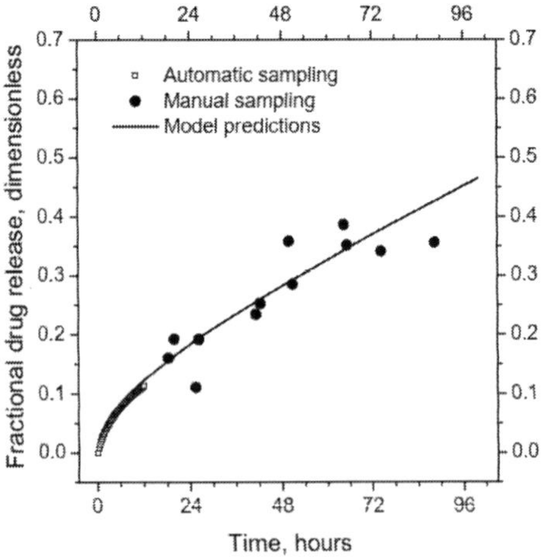

Figure 2. Fractional release of the drug for the HPMC-TP 1:1 matrices subjected to the lateral hydration ("radial" test). Symbols, experimental data (□, automatic sampling, ●, manual sampling); curve, model predictions.

Of course, the correct evaluation of each phenomenon makes the model able to predict the main result of the release test, i.e. the fractional drug release evolution with time. In Figure 2 the experimental data of fractional release during the "radial" test are reported.
The first period (up to 12 hours) were followed by automated sampling in the USP Type II apparatus. The longer immersion times, with consequently higher drug concentrations, have been monitored manually (because of both the long time and of the high drug concentration in the medium). The model calculations were carried out along all the observation period, and they nicely describe the experimental data. It is worth to note that usually, this is the only kind of experimental data available studying the release from solid matrices using the USP Type II apparatus, and thus the tuning of the model parameters was carried out on the basis of such a limited number of experimental data.

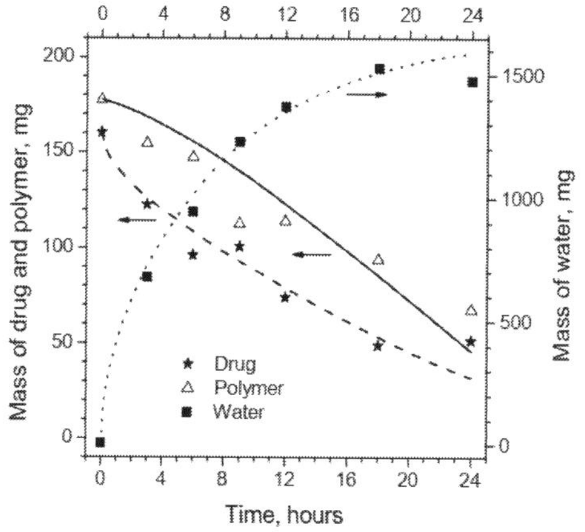

Figure 3. Time evolutions of polymer, drug and water mass in the HPMC-TP 1:1 matrices subjected to the full hydration. Symbols, experimental data (△, polymer, ★, drug, ■, water); curves, model predictions (continuous, polymer; dashed, drug; dotted, water).

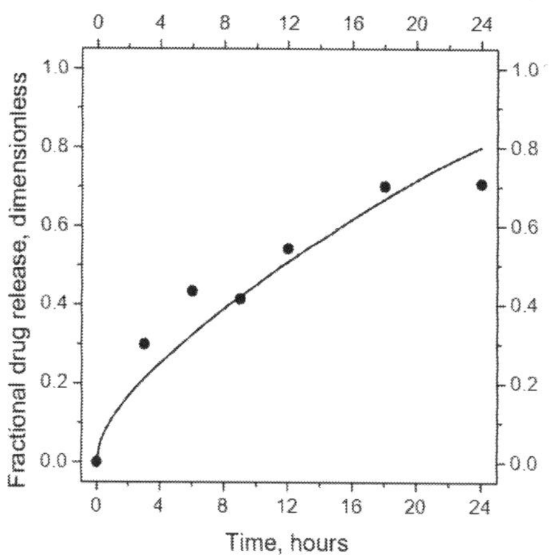

Figure 4. Fractional release of the drug for the HPMC-TP 1:1 matrices subjected to the full hydration ("overall" test). Symbols, experimental data; curve, model predictions.

In Figure 3 data similar to those presented in Figure 1 were reported. In this case, they were obtained during the "overall" test, i.e. when the entire tablet's surface is in contact with the dissolution medium. Even in this case, the model reveals itself able to capture all the phenomena observed. Indeed, the model curves nicely fit the experimental data. Only the two parameters k_{swe} and k_{eros} were further optimized. Figure 4 reports the fractional drug release evolution with time, both experimental (symbols) and model (curve) results. Of course, without any further optimization, the model was able to predict the kinetic of the drug release.

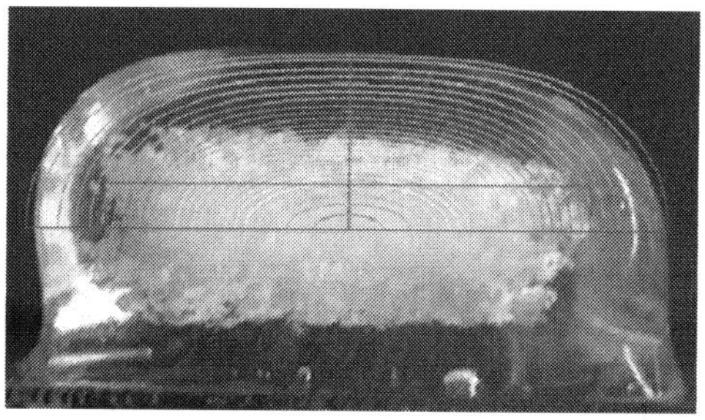

Figure 5. Shape and hydration levels for the HPMC-TP 1:1 matrices subjected to the hydration. Snapshot of cut matrices (experimental) and superimposed simulated profiles (model) after 12 hours of hydration. The initial, non-swollen half-matrices shape is reported as a rectangle (diameter 13 mm, semi-thickness 1.1 mm).

The most important result, to our knowledge never obtained before, is the ability of the code to predict even the shape assumed by the swelling/eroding matrices during the hydration. Once the code has been tuned (i.e. each single parameter has been determined by fitting the experimental data from Figure 3), the shape of the matrices as predicted by the code was compared with the real one, as observed by cutting and photographing the hydrated matrices. Figure 5 shows an example, after 12 hours of hydration. The code predictions fits perfectly the experimental data, therefore the code has been proven not only descriptive of all the phenomena observed, but also predictive of some peculiar features of the process.

Conclusions

The large sets of experimental data obtained during previous research activities within our group allowed the quantification of all the phenomena (the swelling and the erosion, the mass evolutions, the mass fraction profiles).
In this work, we pointed out and proposed a model which was found able to reproduce all the observed phenomena, even the shape assumed by the tablet during the hydration. The model is thus fully descriptive, i.e. it is able to quantitatively reproduce the full behavior experimentally observed.

Acknowledgements

The authors gratefully acknowledge the financial support of Fondazione Cassa di Risparmio Salernitana – Salerno.

References

Siepmann J., Peppas N.A., "Modeling of drug release from delivery systems based on hydroxypropyl methylcellulose (HPMC)", Adv. Drug Deliv. Rev., **48**, 139-157 (2001)

Siepmann J., Siepmann F., "Mathematical modeling of drug delivery", Int. J. Pharm. **364**, 328-343 (2008)

Barba A.A., d'Amore M., Cascone S., Chirico S., Lamberti G., Titomanlio G., "On the behavior of HPMC/theophylline matrices for controlled drug delivery", Journal of Pharmaceutical Sciences, **98,** 4100-4110 (2009)

Barba A.A., Chirico S., d'Amore M., Lamberti G., Titomanlio G., "Swelling of cellulose derivative (HPMC) matrix systems for drug delivery", Carbohydrate Polymers, **78**, 469-474 (2009)

COMSOL Multiphysics User's Guide, © COPYRIGHT 1994–2007 by COMSOL AB. All rights reserved (2007)

CONTROLLING BURST AND FINAL DRUG RELEASE TIMES FROM POROUS POLYLACTIDE DEVICES DERVIED FROM CO-CONTINUOUS POLYMER BLENDS

Basil D. Favis* and Zhenyu Xiang

CREPEC, Department of Chemical Engineering,

École Polytechnique de Montréal, P.O. Box 6079 Station Centre-ville, Montréal, (QC) Canada H3C 3A7

Abstract –This work demonstrates that it is possible to exercise a wide range of control over both the initial burst release and the final drug release times from porous polylactide (PLA) devices derived from co-continuous polymer blends. Two strategies are used, a layer-by-layer polyelectrolyte surface deposition approach on the porous PLA surface and the application of a partially closed-cell protocol (Figure 1). A PLA porous substrate with a pore-size of 1.5 μm, derived from a blend of PLA and polystyrene (PS) via selective solvent extraction of the PS phase, is used as the drug delivery device. The surface area and pore dimensions are examined via BET nitrogen adsorption and image analysis. Porous PLA substrates with 0, 3 and 5 layers of polyelectrolyte and with open areas of 100%, 12 % and 2% are studied both separately and in combination. In vitro release tests are performed to study the release profile of bovine serum albumin (BSA) from the devices via UV spectrophotometry. It is shown that, while both are important, surface modification is more dominant in controlling the release rate than the partially closed cell approach (Figure 2). When a 5 polyelectrolyte layer surface modification of the PLA and a partially closed cell approach (2 % open area) are combined, in the L5C sample, the synergy is dramatic with a 5 times reduction in the first two hour burst release amount and a total release time which is extended by 123 times as compared to the 100% open cell, surface unmodified, reference sample. The L5C sample ultimately releases 89% of the total BSA loaded demonstrating the high level of interconnectivity of the micro channels in the porous PLA. Figure 3 shows that the mechanism of release in this system is clearly diffusion controlled with well defined concentration gradients, as measured by X-ray Photoelectron Spectroscopy (XPS), observed in the direction of release for both the 100% open cell system (L0) and the surface modified, partially closed sample with one end opened (L5E). These effects point towards a diffusion mechanism combined with a sorption/desorption interaction of the BSA with the modifed PLA surface.

REFERENCES

1. Zhu, Y.; Gao, C.; He, T.; Liu, X.; Shen, J. *Biomacromolecules* **2003**, *4*, 446-452.
2. Zacharia, N.S.; DeLongchamp, D.M.; Modestino, M.; Hammond, P.T. *Macromolecules* **2007**, *40*, 1598-1603.
3. Tang, Z.; Wang, Y.; Podsiadlo, P.; Kotov, N.A. *Adv. Mat*er. **2006**, *18*, 3203-3224.
4. Heuberger, R.; Sukhorukov, G.; Voros, J.; Textor, M.; Mohwald, H. *Adv. Func. Mat*, **2005**, *26*, 357-366.
5. Rahman, N.A.; Mathiowitz, E. *J. Controlled Release*, **2004**, *94*, 163-175.
6. Matsumoto, A.; Matsukawa, Y.; Suzuki, T.; Yoshino, H. *J. Controlled Release*, **2005**, *106*, 172-180.
7. Whang, K.; Thomas, H.; Healy K.E.; Nuber, G. *Polymer* **1995**, *36*, 837-842.
8. Li, W.; Laurencin, C.T.; Tuan, R.S.; Ko, F.K. *J. Biomedical Mater. Res.* **2002**, *60*, 613-621.
9. Xie, B.; Parkhill, R.L.; Warren, W.L.; Smay, J.E. *Adv. Func. Mater.* **2006**, *16*, 1685-1693.
10. Mikos, A.G.; Thorsen, A.J.; Czerwonka, L.A.; Bao, Y.; Langer, R.; Winslow, D.N.; Vacanti, J.P. *Polymer* **1994**, *26*, 1068-1077.
11. Sarazin, P.; Favis, B.D. *Biomacromolecules* **2003**, *4*, 1669.
12. Yuan, Z.; Favis, B.D. *Biomaterials* **2004**, *25*, 2161-2170.
13. Favis, B.D.; Sarazin, P.; Li, J.; Yuan, Z. U.S. Patent Application, 10/552,357 and WIPO (PCT) WO 2004/087797.
14. Sarazin, P.; Favis, B.D. *Polymer* **2005**, *46*, 5966.
15. Roy, X.; Sarazin, P.; Favis., B.D. *Adv. Mater*, **2006**, *18*, 1015-1019.
16. Roy, X.; Sarazin, P.; Favis., B.D. WIPO (PCT) WO 2007/053955.
17. Salehi, P.; Sarazin, P.; Favis, B.D. *Biomacromolecules* **2008**, *9*, 1131-1138.
18. Alem, H.; Blindeau, F.; Glinel, K.; Demiustrier-Champagne, S.; Jonas, A.M. *Macromolecules* **2007**, *40*, 3366-3372.
19. Sarazin, P.; Virgilio, N.; Favis, B.D. *J. Appl. Polym. Sci*, **2006**, *100*, 1039-1047.

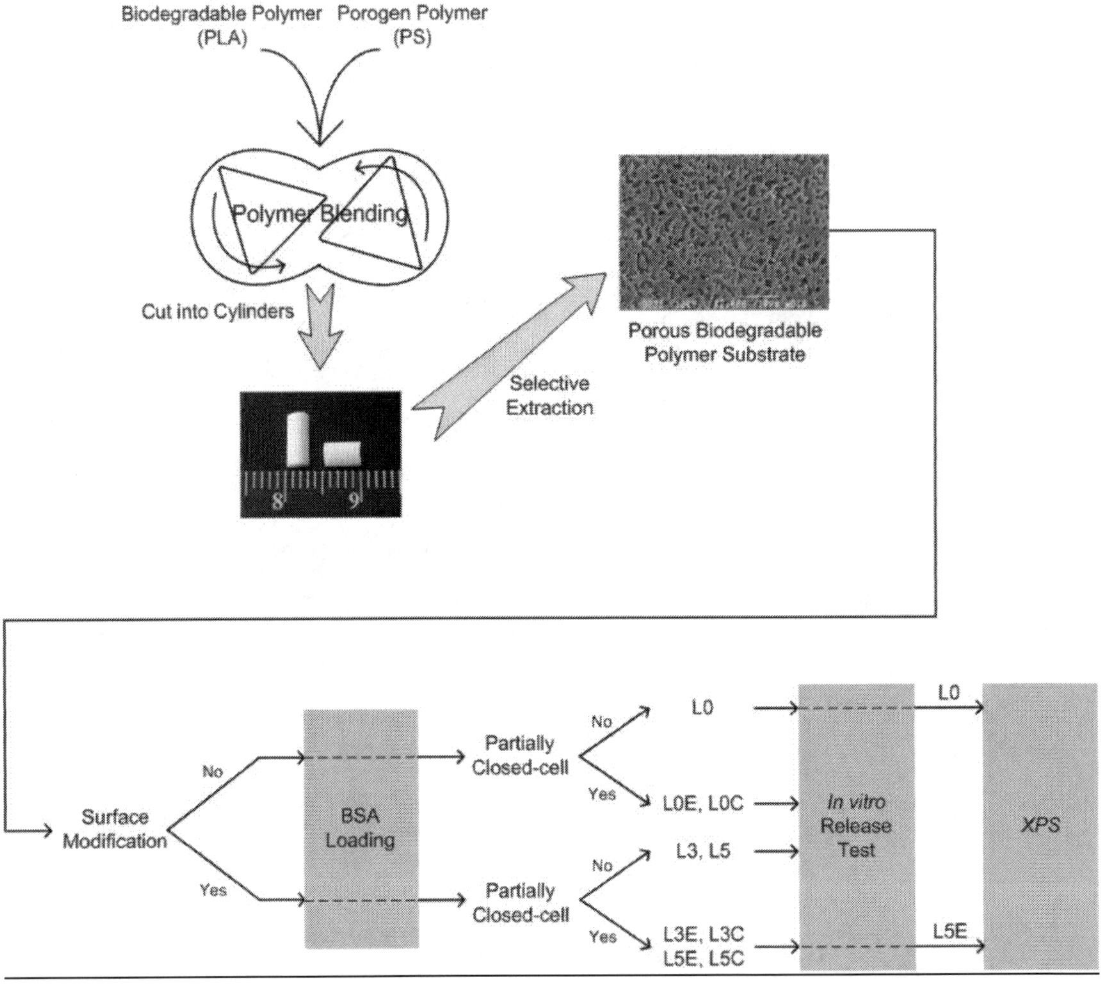

Figure 1. Experimental protocol. L0, L3 and L5 stand for 0, 3 and 5 layers of polyelectrolyte surface deposition, respectively.

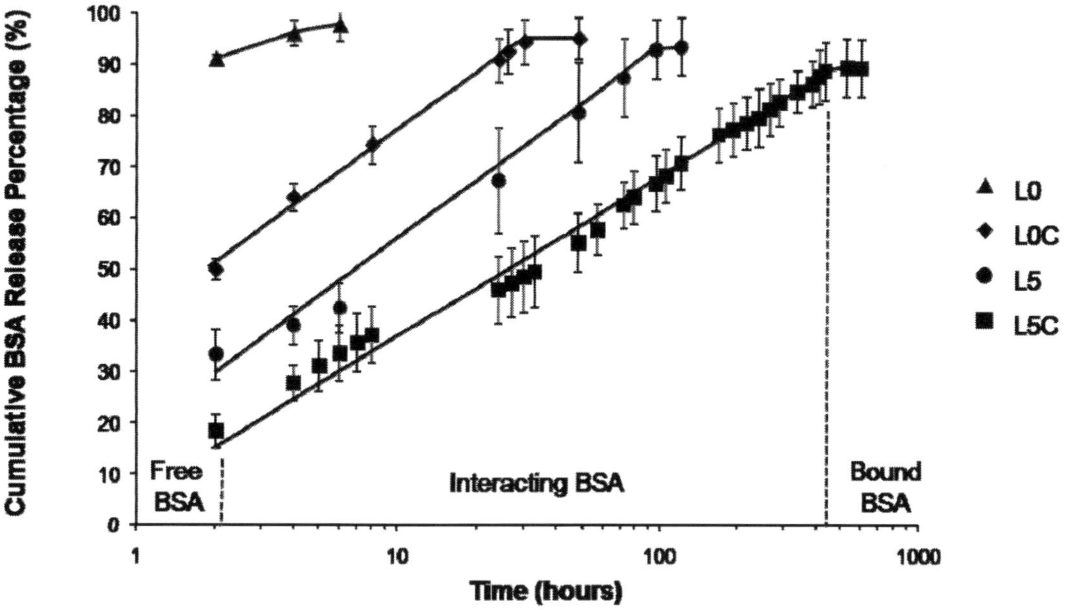

Figure 2. Effect of polyelectrolyte layers and partially closed-cell strategy on BSA release. Three BSA regions are shown for the L5C sample: free, interacting and bound BSA.

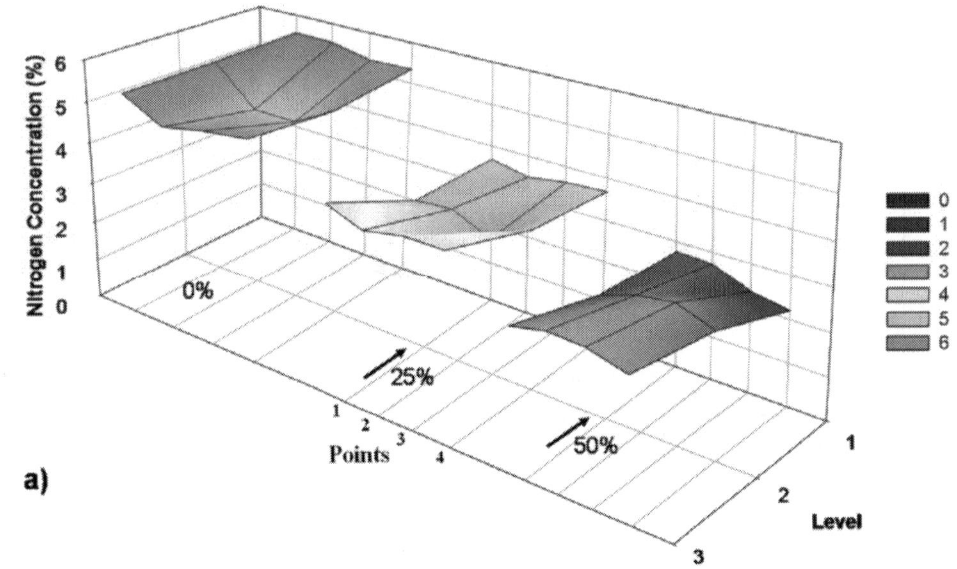

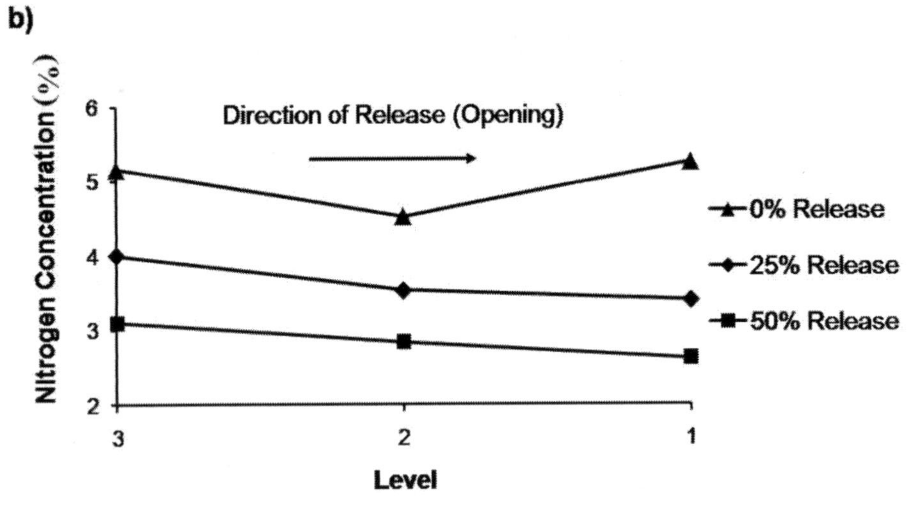

Figure 3. BSA distribution, as measured by XPS, inside the one-end opened porous template with 5 layers of polyelectrolyte surface deposition (L5E sample)

a) BSA distribution of closed-cell with one end PLA substrate, preloaded with 5 layers of polyelectrolyte. Arrows show the direction of the opening.
b) Nitrogen concentration from sample bottom to opening when 0%, 25% and 50% cumulative BSA is released.

Table 1. Characteristic properties of the materials

	M_w	Density (g/cm^3)		Glass Temp (°C)	Supplier
		20 °C	200 °C		
PLA	78,000	1.23	1.09	55	NatureWorks
PS	192,000	1.04	0.97	105	Dow Chemical
PDADMAC	100,000~200,000				Sigma-Aldrich
PSS	70,000				Sigma-Aldrich
BSA	64,000				Sigma-Aldrich

Affects of size diameter of biodegradation of Poly(ε-carprolactone) electrospun fibers mats

*Pornpen Siridamrong**

(1) chulalongkorm university - bangkok - thailand

Poly(ε-caprolactone)(PCL) is a semi-crystalline linear hydrophobic polymer. It is used as surgical material in medical and it has been reported to have biomedical application in controlled drug delivery and implants for orthopedic surgery.PCL has low glass-transition temperature (~600C) and low melting (600C). The affect parameters of PCL biodegradation for tissue engineering are depending on surface treatment, and size diameter of fiber mats. In this research, 15-30wt% of PCL fibers were electrospun from PCL solution in a mixed solvent of dichloromethane(DCM) and N,N-dimethylformamide(DMF) at different applied concentration, flow rate, potentials and deposition distances. The relationship between electrospnning parameters and fiber diameter had been investigated. The average diameters are in the range of 200, 500nanometer and 1 micrometer that were allowed to degrade in Pseudomonas cepacia lipase (PS), pH 7.4 phosphate buffer at 370C. The degradation process was followed by a total organic carbon (TOC) measurement at every 24 hrs until fiber mats completely decomposed. Fiber morphology and mechanical property of obtained fiber mats were observed and measured using scanning electron microscope (SEM), tensile strength and differential scanning calorimeter (DSC) that used to monitor property change during degradation such as surface morphology and thermal property. The aim of this study is to develop a facile and effocoent process to provide a size diameter difference of electrospun PCL scaffold with degradation. The change in size diameter of PCL scaffold has been investigated for biodegradation time by enzyme. The results indicated that both processing parameters affected fiber morphology which in turned affected mechanical strength and degradation of the fiber mats.

Drug Release from a Biomaterial

A. Beygi[1], G. Mir-Mohammad-Sadeghi[1], M. Jafari-Nodoushan[2]

[1]Department of Polymer Engineering,Amirkabir University Technology, Tehran, Iran – ahmad.beygi@aut.ac.ir; gsadeghi@aut.ac.ir
2 Novel Drug Delivery Systems Department, Iran Polymer & Petrochemical, Institute, Tehran, Iran – m.jafari@ippi.ac.ir

Biomaterial drug release matrices is a good choice for in vivo tissue regeneration. However, the ability to predict and control the rate of release from delivery systems is still a challenge. A model was simulated that a drug is released into a region containing damaged nerve ends. The biomaterial holding the drug has a cylindrical shape and serves a dual purpose: It acts as a guide to help regenerating nerve cells connect, and it also stimulates the healing process through targeted drug release

Introduction
Biomaterial matrices for drug release are useful for in vivo tissue regeneration. Delivery of proteins from materials can be controlled by modifying the affinity of the material for the protein drug, the number of protein binding sites, or the degradation rate of the material [1,2]. The scaffolds should also deliver appropriate cues for promoting nerve regeneration in a controlled, localized manner[3]. The following study describes the release of a drug from a biomaterial matrix to damaged cell tissue. Specifically, a nerve guide delivers a regenerating drug to damaged nerve ends. The model examines detailed drug-release kinetics with rate expressions handling drug dissociation/association reactions as well as matrix degradation by enzyme catalysis. The enzyme reaction is described by Michaelis-Menten kinetics. With this simulation it is easy to investigate design parameters governing the rate of drug release such as drug-to-biomaterial affinity, biomaterial degradation, drug loading, and of course the influence of geometry and composition of the biomaterial matrix.

Experimental
Model Definition
In this model a drug is released into a region containing damaged nerve ends. The biomaterial holding the drug has a cylindrical shape and serves a dual purpose: It acts as a guide to help regenerating nerve cells connect, and it also stimulates the healing process through targeted drug release.
The model simulates transient chemical reactions and species transport in a 2D geometry with axial symmetry.
Fig.1 shows the full 3D geometry as well as the 2D modeling domain, reduced by axial symmetry and a mirror plane. The subdomains are:

- The nerve-cell tissue (Ω1)
- The biomaterial matrix (Ω2)
- The surrounding medium (Ω3)

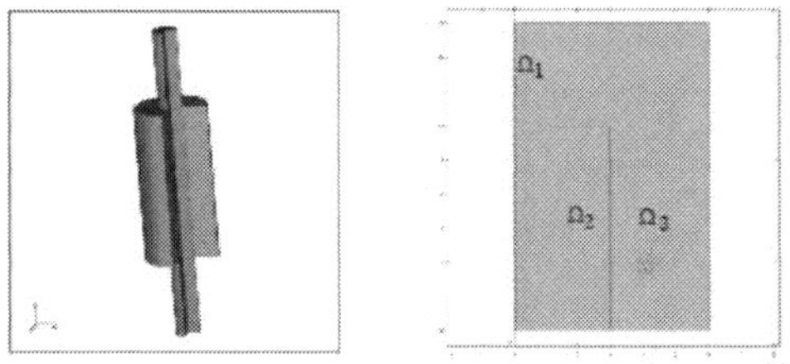

Fig.1. The full 3D geometry (left) and the equivalent modeling domain reduced to 2D by axial symmetry (right). The subdomains are: the nerve-cell tissue (Ω1), the biomaterial matrix (Ω2), and the surrounding medium (Ω3).

In the biomaterial matrix, a drug molecule, d, binds to a peptide, p, which in turn is anchored to the matrix,m. Matrix-bound species are labeled mpd and mp, respectively, the latter referring to a species where no drug is bound to the peptide. Species mpd and mp are active only in subdomain Ω2.
Two mechanisms release the drug from the matrix. First, the drug can simply dissociate from the matrix site mp. Second, matrix degradation by an enzyme, e, originating from the cell-tissue domain, leads to release of the drug-peptide species, pd, from which the drug subsequently dissociates. The unbound species p, d, pd, and eare active in the entire model domain. Fig.2. illustrates the complete reaction scheme.

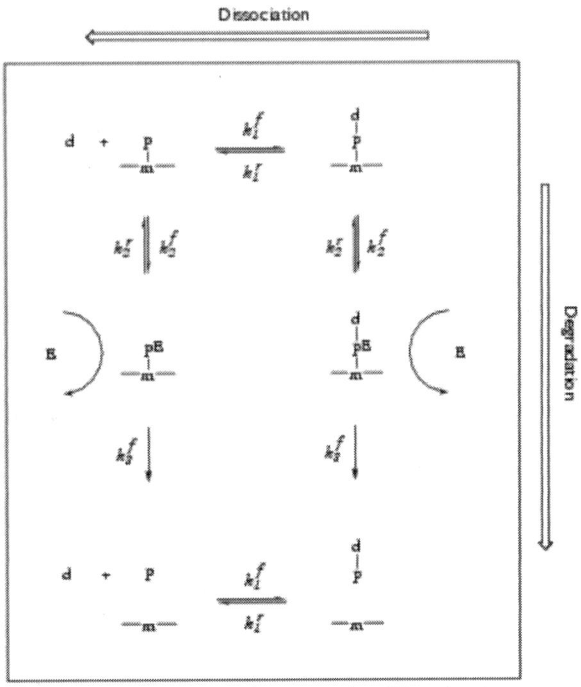

Fig.2. Reaction scheme describing drug dissociation/association reactions (horizontal) and matrix-degradation reactions (vertical).

The simulation makes use of two diffusion application modes. The time-dependent mass balance per species is described by:

(1) $$\frac{\partial c_i}{\partial t} + \nabla \cdot (-D_{ik} \nabla c_i) = R_{ik} \quad \text{in } \Omega_k$$

whereDik (m2/s) is the diffusion coefficient for species i in the medium of subdomain Ωk. Further, Rik(mol/(m3· s)) is the rate expression for species i in the medium of subdomain Ωk. In the matrix subdomain, all the reactions described in Fig.2 are possible, leading to the following rate expressions (Equ. 2-6):

(2) $$R_{d2} = -k_1^f c_d (c_{mp} + c_p) + k_1^r (c_{mpd} + c_{pd})$$

(3) $$R_{p2} = -k_1^f c_d c_p + k_1^r c_{pd} + R_{MMmp}$$

(4) $$R_{pd2} = k_1^f c_d c_p - k_1^r c_{pd} + R_{MMmpd}$$

(5) $$R_{mpd2} = k_1^f c_d c_{mp} - k_1^r c_{mpd} - R_{MMmpd}$$

(6) $$R_{mp2} = -k_1^f c_d c_{mp} + k_1^r c_{mpd} - R_{MMmp}$$

The rate terms RMMmp and RMMmpd refer to the Michaelis-Menten kinetics describing the enzyme catalyzed degradation of the matrix(Equ. 7-10):

(7) $$R_{MMmp} = \frac{V_{max} c_{mp}}{K_M + c_{mp}}$$

(8) $$R_{MMmpd} = \frac{V_{max} c_{mpd}}{K_M + c_{mpd}}$$

with

(9) $$V_{max} = k_3^f c_e$$

(10) $$K_M = \frac{k_3^f + k_2^r}{k_2^f}$$

RMMmp describes the disappearance of mp sites and the production of p species. RMMmpd describes the disappearance of mpd sites and the production of pd species. In the cell region (Ω1) and in the surrounding medium (Ω3) only dissociation/association reactions occur, leading to the rate expressions(Equ. 11,12)

(11) $$R_{d1} = R_{d3} = R_{p1} = R_{p3} = -k_1^f c_d c_p + k_1^r c_{pd}$$

(12) $$R_{pd1} = R_{pd3} = k_1^f c_d c_p - k_1^r c_{pd}$$

The boundary condition is axial symmetry along the rotational axis and insulation/symmetry elsewhere. Values for diffusion coefficients and rate constants come from the literature[4].

Results and Discussion

Fig.3 shows the concentration transients of the reacting species in a perfectly mixed (space-independent) system.

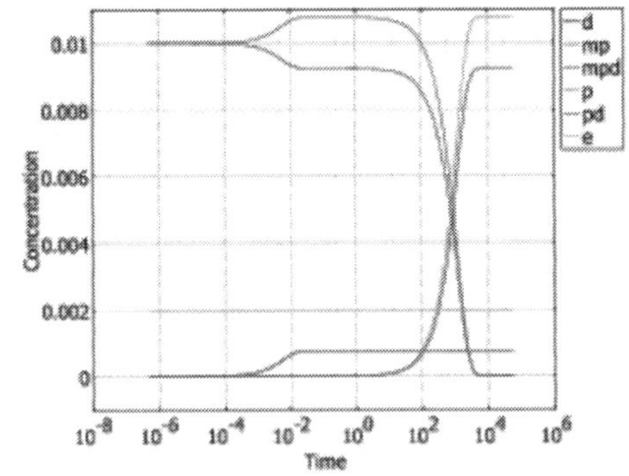

Fig.3. Concentrations of all reacting species (mol/ m3) as functions of time (s).

The effect of enzyme degradation is clearly visible, with matrix-bound peptide species (mp and mpd) decreasing and free peptide species (p and pd) increasing with time. The matrix is completely degraded after approximately 5000 seconds. As the drug and peptide species have the same association/dissociation kinetics, no matter whether the peptide is free or matrix-bound, the steady-state concentration of drug is constant during the degradation process.

Solving the space-dependent mass balances of Equation 1 results in concentration distributions of all participating species as functions of time.Fig.4. shows the concentration of drug across the modeling domain.

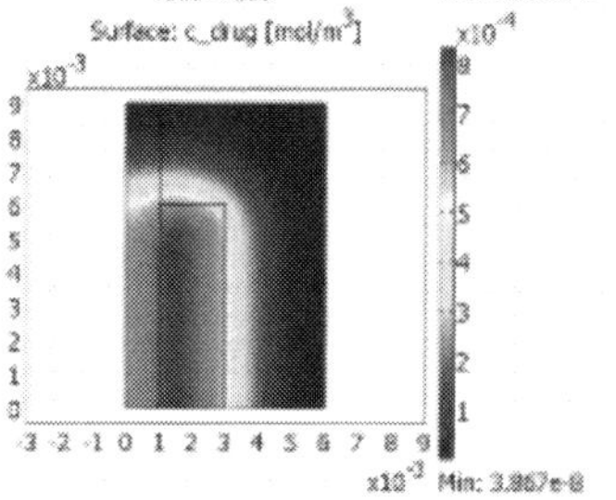

Fig.4. Drug concentration at the end of the simulation (7200 s).

You can visualize the effect of matrix degradation on the release of the drug by plotting the total drug concentration over a cross section of the matrix as a function of time.

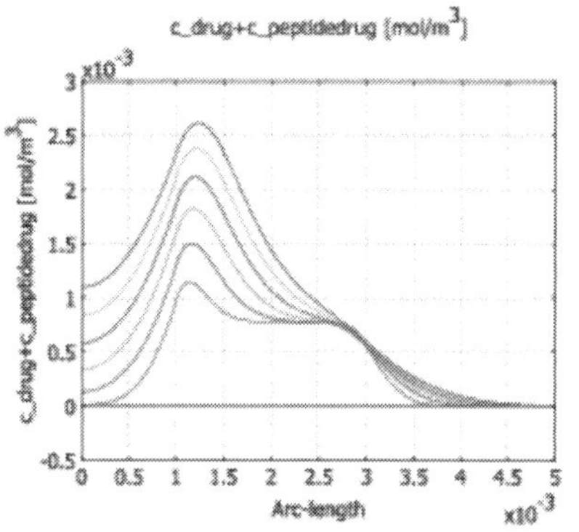

Fig.5. Concentration profiles describing the total drug concentration (cd + cpd) across the modeling domain. Profiles were collected at times up to 3600 s.

As noted earlier, the enzyme originates from the nerve-cell subdomain. From Fig.5. it is clear that matrix degradation has a directing effect on the drug release toward the damaged cell region.

To visualize the degradation front passing through the biomaterial geometry, plot the total matrix site concentration (cmp+ cmpd) across the matrix subdomain (Fig.6.).

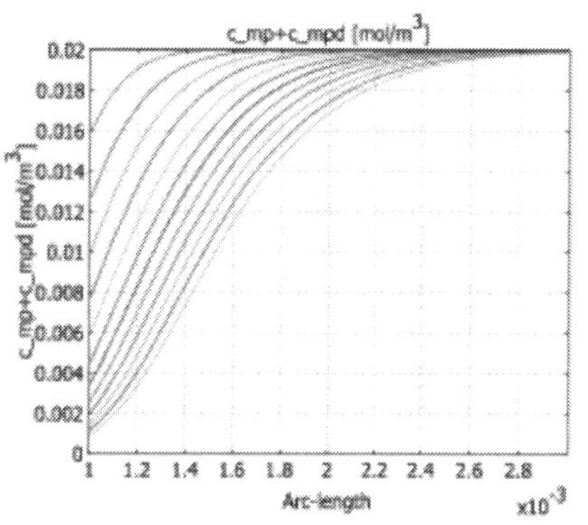

Fig.6. Concentration profiles describing the total matrix site concentration (cmp+ cmpd) with profiles collected at times up to 7200 s.

The detailed reaction/transport description in this model allows for the investigation of many design parameters relevant to bioengineering.

Conclusion

A model was developed that simulated release profile of a drug delivers from a biomaterial matrix. This simulation presents the effect of matrix degradation on drug release as a function of time and geometry. moreover, it is straightforward to study the influence of the drug/peptide affinity by varying the rate constants kf1and kr1, or the influence of drug loading by varying the cmp :cmpd ratio.

References

1. Edelman E, Mathiowitz E, Langer R, Klagsbrun M. Controlled and modulated release of basic fibroblast growth factor, Biomaterials,12,612–26 **(1991).**

2. Schroeder-Tefft J, Bentz H, Estridge T, Collagen and heparin matrices for growth factor delivery, J Control Release, 49, 291–8 **(1997).**

3. Stephanie M. Willerth, Shelly E. Sakiyama-Elbert, Approaches to neural tissue engineering using scaffolds for drug delivery, Advanced Drug Delivery Reviews, 59, 325–338 **(2007)**

4. Maxwell D.J., Hicks B.C., Parsons S., Sakiyama-Elbert S.E., Development of rationally designed affinity-based drug delivery, ActaBiomaterialia 1, 101-103 **(2005).**

EXTRUSION OF PHARMACEUTICAL POLYMER FORMULATIONS

A.L. Kelly[1,2*], A. Paradkar[1,2], T. Gough[1,2], R. Dhumal[1] and P.D.Coates[2]

[1]Centre for Pharmaceutical Engineering Science, University of Bradford, UK – A.L.Kelly@Bradford.ac.uk, A.Paradkar1@Bradford.ac.uk, T.Gough@Bradford.ac.uk, R.S.Dhumal@Bradford.ac.uk
[2] Polymer IRC, School of Engineering, Design and Technology, University of Bradford, UK – P.D.Coates@Bradford.ac.uk

Polymeric materials are widely used in pharmaceutical applications, for example as excipients (non active ingredients) in tablets or capsules. Increasingly, extrusion is being used to create polymer-drug blends or composites. The active pharmaceutical ingredient (API) can be incorporated as a particulate dispersion or a molecular solution, depending upon the thermal and solubility properties of the constituents. Selection of extrusion conditions such as screw configuration, temperature, throughput and screw speed has a significant effect on the properties of the compound. The aim of this study was to develop a solid dispersion (SD) by melt extrusion for enhanced solubility of the poorly soluble antidiabetic drug glipizide (GPZ) and to understand the effect of extrusion conditions on the properties of the formulation.

Introduction

More than 40% of newly discovered compounds and marketed drugs are either poorly soluble or water insoluble (Dubin, 2005). Low solubility and/or dissolution rate in the gastrointestinal tract compromise oral bioavailability. Therefore, delivery of poorly soluble drugs is one of the major challenges confronting formulation scientists in pharmaceutical industries. Preparation of solid dispersions is the most popular approach (Serajuddin, 1999), which involves dispersion of one or more active ingredients in an inert carrier or matrix in the solid state. SDs are prepared either by dissolving actives in organic solvents followed by subsequent evaporation of solvents by different means (Dhumal et al., 2008) or by melt extrusion (Crowley et al., 2007).

For pharmaceutical systems, several research groups have recently demonstrated that melt processing of pharmaceutical powders is a viable method to prepare granules, SDs, sustained release tablets, transdermal drug delivery systems and implants (Repka et al., 2007). Apart from being solvent free, extrusion is a continuous, single-step, easily controlled and readily scalable technique. Closed process units also prevent cross-contamination and readily address the regulatory demand for QbD (Quality by Design) and PAT (Process Analytical Technology). High mixing efficiency and short processing time make extrusion suitable for pharmaceuticals. The technology has been successfully commercialized for development of an anti-HIV formulation Kaletra (which is a combination of lopinavir and ritonavir) (Crowley et al., 2007).

In the present work we have developed SD of Glipizide, an antidiabetic drug with poor water solubility by hot melt extrusion technology in order to understand issues in processing of SDs and to develop a systematic interdisciplinary approach to implement proof of concept to scale-up.

Experimental

Glipizide (GPZ) and Polyethylene oxide (POLYOX WSR N-80) were blended in powdered forms in a range of blend ratios, as shown in table 1. A 16mm co-rotating twin screw extruder (Pharmalab, Thermo Scientific, UK, shown in Figure 1) was used with a screw configuration having length to diameter ratio of 40:1, consisting of forward feeding elements, a mixing section and a discharge section. The feeding section comprised transport elements with a double helix to convey the material along the screw length. The mixing section consisted of bi-lobal kneading blocks of 4mm thickness arranged to effectively blend the materials. The mixing section (28 screw diameters from the feed end) consisted of 18 mixing paddles arranged with increasing staggering angle (9 elements at 30°, 5 elements at 60° and 4 elements at 90°).

Table 1: Batch codes with the ratio of drug to polymer and processing temperatures at three levels (3^2).

Batch Code	Drug: Polymer ratio (w/w)	Process temperature (°C)
SD 2100	1:2 (-1)	100 (-1)
SD 2120	1:2 (-1)	120 (0)
SD 2140	1:2 (-1)	140 (+1)
SD 4100	1:4 (0)	100 (-1)
SD 4120	1:4 (0)	120 (0)
SD 4140	1:4 (0)	140 (+1)
SD 6100	1:6 (1)	100 (-1)
SD 6120	1:6 (1)	120 (0)
SD 6140	1:6 (1)	140 (+1)

The extruder was equipped with a Brabender gravimetric twin-screw powder feeder. Extruder set temperatures shown in Table 1 (TGA experiments showed that no decomposition of the polymer or drug occurred at these temperatures). Once the temperature was stabilised, extrusion was carried out at a powder feed rate of 10 g/min and a screw speed of 100 rpm.

Extrudates were collected after cooling at ambient temperature and subsequently pelletised. All samples were stored in desiccators at room temperature.

DSC profiles were generated in the range of 25–250 °C using TA instruments Q2000 DSC with RCS90 cooling unit. The crystallinity of samples from individual batches was assessed using X-ray powder diffraction (XRPD) using a Bruker D8 diffractometer (wavelength of X-rays 0.154 nm Cu source, voltage 40 kV, and filament emission 40 mA). Dissolution studies were performed using USP 26 type II dissolution test apparatus (Copley Scientific, Nottingham, UK). Samples were placed in a dissolution vessel containing 900 ml 0.1 N HCl, maintained at 37 ± 0.5 °C and stirred at 100 rpm. Samples were collected periodically and replaced with a fresh dissolution medium.

Results and Discussion

All of the batches listed in the Table 1 could be processed without any difficulty, indicating the suitability of PEO for melt extrusion at the selected parameters. However, the downstream processing of these batches needed adjustment in order to cut the extruded strands into pellets. It was not possible to produce pellets with a similar particle size distribution from all batches. This is important for ease of downstream processing and also dictates the in-vitro and in-vivo performance of the final product. Figure 1 shows images of extrudates processed at 100°C with different amounts of polymer. For batches processed at 100°C and 120°C, as the level of polymer increased, the diameter of pelletised extrudate increased, with a narrower particle size distribution. However, for batches extruded at 140°C, the reverse trend was observed with respect to the particle size distribution. Diameter of the drawn extrudate was observed to vary with polymer content and processing temperature due to changes in rheology of the drug-polymer composite.

Figure 1 – Photographs of extruded solid dispersions with different blend ratios, processed at 100°C

SD16100 SD14100 SD12100

The DSC thermogram of GPZ was typical of a crystalline anhydrous substance with a sharp melting endothermic peak at 209 °C. PEO is a crystalline polymer with a melting point of 62 °C. A physical mixture of GPZ with PEO in 1:4 ratio also showed an endotherm at 62 °C corresponding to the melting of PEO but with a small GPZ melting peak with only 10 % heat of fusion of that of pure crystalline drug, indicating good miscibility of drug in PEO melt. The molten PEO starts dissolving the drug as its mobility increases. Similar findings were reported for nifedipine in PEO (Li et al, 2006). All physical mixtures were found to exhibit similar thermograms.

The drug melting peak completely disappeared in all batches after extrusion, indicating a loss of drug crystallinity in molten PEO. The melting point of PEO was also found to decrease after extrusion with GPZ from 62 °C for pure PEO and physical mixture to 60 °C for all the extruded samples. This may be due to molecular interactions between PEO and GPZ. Heat of fusion was also found to be decrease after extrusion indicating a slight decrease in the crystallinity of PEO following extrusion.

Figure 2 – DSC thermograms of GPZ, PEO, physical mixture and solid dispersions

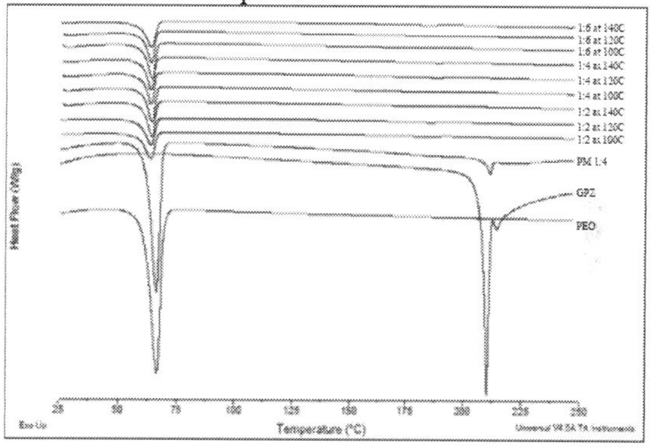

The XRPD patterns of GPZ, physical mixtures and the SDs are presented in Figure 3. GPZ showed sharp diffraction peaks confirming the crystalline nature of the pure drug. The crystalline nature of PEO was confirmed by the characteristic dominant peaks at 2-theta angles of 19.3 and 23.4 degrees. Typical diffraction peaks of GPZ and PEO but with reduced intensity in physical mixtures indicated the presence of free crystalline drug and PEO. Following extrusion the intensity of PEO peaks was significantly reduced. However, the diffraction peaks corresponding to GPZ were totally absent indicating a complete loss of GPZ crystallinity after extrusion at all the studied ratios and processing temperatures. Both DSC and XRPD results clearly demonstrate the loss of GPZ crystallinity, indicating the ability of PEO melt to effectively dissolve GPZ and form SD.

Figure 3 – XRPD patterns of GPZ, physical mixture and solid dispersions.

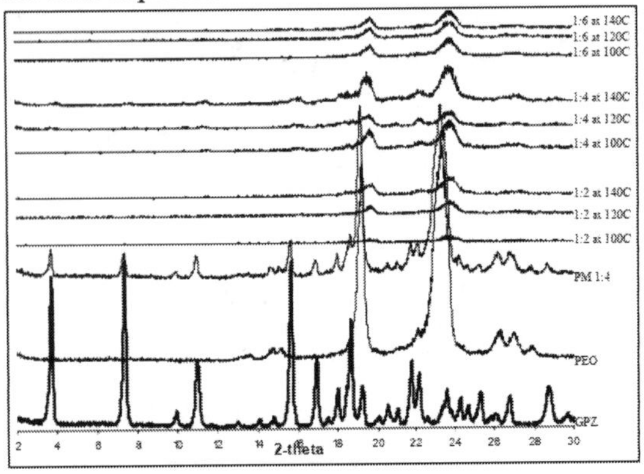

An example of the GPZ dissolution profiles obtained from pure GPZ, physical mixture and different SDs at processing temperatures are shown in Figure 4. GPZ is known to show pH dependent solubility with higher solubilities at higher pH (Mehramizi et al., 2007). Therefore, dissolution testing was carried out in 0.1 N HCl to discriminate the dissolution performance of different formulations. While pure drug and physical mixture dissolved slowly over the test period of 90 minutes, all SDs reached the plateau concentration in 15 minutes or less. This enhancement in the rate and extent of dissolution of SDs can be attributed to several mechanisms including amorphisation and increased wetability due to dispersion in the hydrophilic polymer.

Figure 4 – Dissolution profiles of glipizide and SD 4 at different temperatures

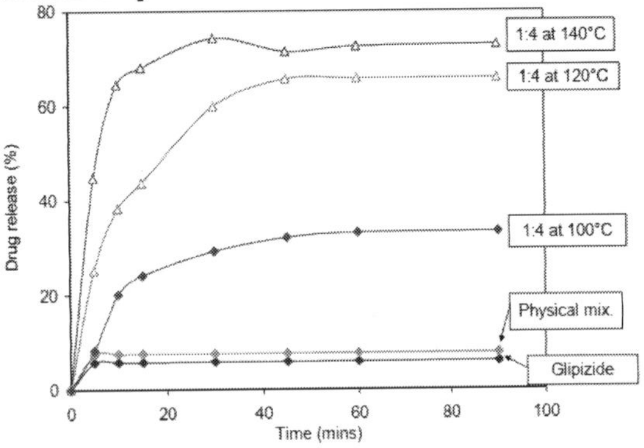

Conclusions

A solid dispersion system of glipizide and PEO was developed by melt extrusion. DSC and XRPD confirmed the loss of drug crystallinity at all the polymer ratios and processing temperatures studied. The PEO maintained a semi-crystalline structure, although some decrease in crystallinity was observed. All batches exhibited improved drug dissolution compared to the pure drug and physical mixtures indicating the potential of the system for use in solubility enhancement of poorly soluble compounds.

Acknowledgements

The authors would like to acknowledge Dow Chemical Company, Midland, MI for the gift samples of POLYOX WSR N-80.

References

Crowley MM., et al., Pharmaceutical applications of hot-melt extrusion: part I. Drug Dev. Ind. Pharm. **33**, 909-926 (2007)

Dhumal R.S., et al., Preparation of amorphous cefuroxime axetil nanoparticles by sonoprecipitation for enhancement of bioavailability. Eur. J. Pharm. Biopharm. **70**, 109-15 (2008)

Dubin C. H., Formulation Frustrations. Drug Del. Tech., **5**, 30–34 (2005)

Mehramizi A., et al., Solid carriers for improved solubility of glipizide in osmotically controlled oral drug delivery system. Drug Dev. Ind. Pharm. 33, 812-823 (2007)

Repka M.A., et al., Pharmaceutical applications of hot-melt extrusion: Part II. Drug Dev. Ind. Pharm. **33**, 1043-57 (2007)

Serajuddin A.T., Solid dispersion of poorly water-soluble drugs: early promises, subsequent problems, and recent breakthroughs. J. Pharm Sci. **88**, 1058-66 (1999)

LONG-TERM FATIGUE AND DYNAMIC CREEP PERFORMANCE OF DENDRITIC POLYISOBUTYLENE-BASED NANOCOMPOSITES FOR BIOMEDICAL DEVICES

C. Götz[1], G.T. Lim[2], F. Fischer[1], F. Wolff-Fabris[1], J.E. Puskas[2] and V. Altstädt[1,*]

[1] *Department of Polymer Engineering, University of Bayreuth, FAN A, Universitaetsstrasse 30, Bayreuth, D-95447, Germany - christian.goetz@uni-bayreuth.de; frank.fischer@uni-bayreuth.de; altstaedt@uni-bayreuth.de*
[2] *Department of Polymer Science, The University of Akron, Akron, OH 44325-3909, USA – gtlim@uakron.edu; jpuskas@uakron.edu*

Abstract
Polymers are lately becoming very important biomaterials for new biomedical devices, therapeutic drug delivery and tissue engineering. Depending on the application, biopolymers need to maintain a specific level of material performance against fatigue degradation and fracture during their service life. Linear poly(styrene-*b*-isobutylene-*b*-styrene) (SIBS) block copolymer belongs to a unique class of biocompatible rubbers, currently in clinical practice as a coating on drug eluting stents. However, due to relatively high creep it is not suitable for load bearing applications. This work intends to look into a new generation of SIBS, i.e., dendritic SIBS copolymer with a branched polyisobutylene core with poly(isobutylene-*b*-methylstyrene) end blocks, which was further reinforced with 37.5 wt% carbon black (CB). Electron micrographs revealed that the CB filler was well dispersed in the matrix to have a pronounced influence on the polymer phase morphology of the resulting nanocomposite. This led to superior tensile properties relative to the neat polymer and silicone rubber. Using a hysteresis measurement methodology to mimic the dynamic loading experienced by a soft tissue implant, the long-term dynamic creep and fatigue response of these nanocomposites were also found to be better compared to the neat polymer and silicone rubber.

Introduction

Thermoplastic elastomers (TPEs) are steadily becoming more relevant for engineering science, as their mechanical properties can be tailored to balance the stiffness and strength of thermoplastics, and the ductility and toughness of elastomers. Like their thermoplastic counterparts, TPEs can be processed with industrial techniques, such as extrusion and injection molding without any need of post-curing procedures, and therefore carry the "green" label of recyclability [Holden et al., 2004]. In general, TPEs can be grouped into five basic types, (a) styrenic block copolymers (e.g. Kraton® - Kraton Polymers), (b) thermoplastic polyurethane elastomer (e.g. Pellethane® - Dow Chemicals), (c) thermoplastic polyester elastomer (e.g. Hytrel® - DuPont), (d) thermoplastic amide elastomer (e.g. Pebax® - Arkema) and finally (e) thermoplastic olefin elastomer (e.g. rubber-modified polypropylene). These five types of TPEs have their own unique properties to be tailored for usage in specific fields of application, e.g. in biomedical devices [Ehrenstein, 1999; Holden et al., 2004].

In particular, styrenic block copolymers are distinguished by the presence of a soft elastomeric phase with a low glass transition temperature (T_g) (below room temperature) and another hard styrene phase with a much higher T_g (~ 100 °C). For styrenic TPEs, the overall material performance is determined by the polymer phases and their proportion [Holden et al., 2004]. Poly(styrene-*b*-isoprene-*b*-styrene), poly(styrene-*b*-butadiene-*b*-styrene) (SBS) are commercially developed styrenic TPE polymers under various trade names.

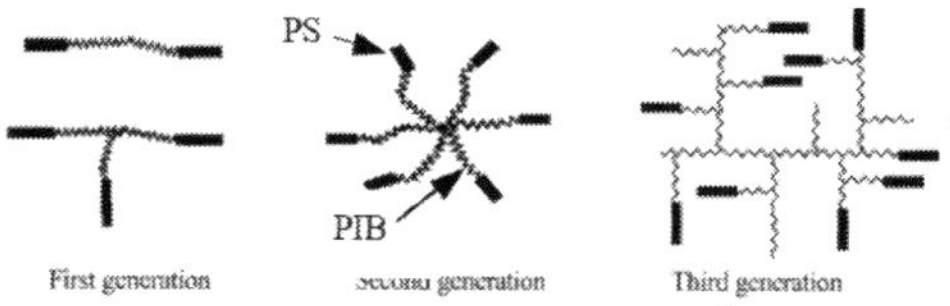

Figure 1 – Architectures of (a) the 1st generation (L_SIBS), (b) the 2nd generation and (c) the 3rd generation (D_IBS) of SIBS polymers [Puskas et al., 2009a].

This first generation of SIBS (linear triblock L_SIBS) (see Fig. 1) has shown its potentials and applicability as a biomaterial due to properties such as biocompatibility and a superior combination of mechanical properties when compared to silicone rubber [El Fray et al., 2006; Holden et al., 2004]. The biocompatibility of L_SIBS has been well established to earn the approval of the US Food and Drug Administration for use as a polymer coating (Translute®) on implanted coronary stents [FDA, 2004]. L_SIBS is also marketed by Kaneka Co. for general applications (SIBSTAR®).

Subsequently, newer generations of this material had evolved with various architectures, such as the second generation of multi-arm star polymer [Jacob and Kennedy, 1999; Storey and Shoemake, 1999], and more recently, the third generation with an arborescent or tree-like branched structure [Puskas et al., 2004; 2005], as shown in Fig. 1.

Due to the phase separation between polyisobutylene (PIB) and polystyrene (PS), and the lower surface energy of the rubbery PIB domain, Puskas and co-workers demonstrated by XPS that there is a thin (~ 10 nm) layer of PIB over the surface of PIB-based block copolymers [Puskas and Kwon, 2006]. This layer of PIB is believed to bring about an excellent biocompatibility to both L_SIBS and D_SIBS polymers. From a comprehensive review by Pinchuk and his co-workers, there are deficiencies of L_SIBS such as relatively low creep resistance, and the tendency of stress cracking in the presence of organic solvents [Pinchuk et al., 2008]. This is likely to limit the application of linear SIBS in biomedical devices and requires material reinforcement. In D_SIBS, the more sophisticated branched architecture poses entropic restriction on chain entanglement to give the polymer a different physical and mechanical response to external forces. The synthesis technology was scaled up to the pilot plant (pound/batch) level [Puskas and Kwon, 2006]. Recently, the fourth-generation of dendritic SIBS block copolymers were synthesized, having short copolymer end blocks that phase separated even when those end blocks had low T_g (< -50 °C). This technology was also scaled up and proved to be more robust than earlier processes [Puskas et al., 2009b].

The synthesis and morphological characterization of SIBS copolymers with different architectures is well documented in the literature [Gyor et al., 1994; Jacob et al., 1996; Kaszas, 1992; Kennedy et al., 1990; Koshimura and Sato, 1992; Puskas et al., 1998; 2002; Puskas and Kaszas, 1996; Storey and Showmake, 1999; Storey et al., 2000]; however, relatively less is known about the structure-property relationships of the newly developed dendritic SIBS materials with copolymer end blocks. As part of our continued collaboration in characterizing these new materials, this work focuses on the quasi-static mechanical and long-term dynamic fatigue characterization of a block copolymer with a branched PIB midblocks attached with poly(isobutylene-co-p-methylstyrene) end blocks, and its composite reinforced with carbon black (CB).

Fillers, such as CB and silica particles, and layered silicates or mineral clay, are widely added to polymers for reinforcement or as processing aids. For the production of rubber tires, carbon black particles are commonly introduced as a color pigment, a reinforcement, and to conduct heat and static charges away from the tires to the ground. It was the seminal work of Kojima and his co-workers in their successful development of nylon-clay nanocomposites at the Toyota research laboratory [Kojima et al., 1993] to demonstrate the synergistic benefit of using nanoclay and carbon nanotubes to modify the performance of the final materials. This nanofiller technology has opened up new possibilities for polymer formulation and established an entirely new area of research in polymer science and engineering. However, to fully harness this synergistic effect, there needs to be a good dispersion of these nanofillers uniformly throughout the polymeric matrix. Once a good filler-matrix surface interaction can be established, polymer nanocomposites can offer lighter materials with improved properties at a lower filler loading (and hence lower cost) over conventional composites with larger than nanometer size reinforcing fillers [Sengupta et al., 2007]. However, rubber nanocomposites have not lived up to the high expectations, and commercially still regular carbon black is used for rubber reinforcement.

This work investigated the quasi-static mechanical properties of a neat and a reinforced dendritic block copolymer and compared with medical grade silicone rubber. Furthermore, the dynamic fatigue behavior and morphology of the materials were characterized to give insights into the structure-property relationships of these innovative biomaterials for soft tissue applications.

2. Experimental

2.1 Materials

The polymer sample with a branched PIB midblocks flanked with poly(isobutylene-co-p-methylstyrene) end blocks was synthesized and characterized as reported [Puskas et al., 2009b]. For simplicity's sake, it will be denoted D_IB-MS10, where the number refers to the overall wt% PMS content in the polymer. The PMS content in the end blocks was determined to be 40.7 wt% [Puskas et al., 2009b]. The number-average molecular weight of the polymer (M_n) was measured to be 291,600 g/mol, and the polydispersity index of 1.94 represents the ratio of weight-average to number-average molecular weight.

After polymerization, the D_IB-MS10 block copolymer was mixed with 37.5 wt% N234 carbon black (Cabot Corporation) in a Brabender mixer using a 78.8 % fill factor as reported [Puskas et al., 2009b]. Table 1 shows the characteristics of the CB used in this study.

Table 1 – Characteristics of CB in the current study.

CB Filler	Mean diameter [nm]	Heterogeneity Index	EMSA* [m^2/g]
N234	20	1.57	124

*EMSA: Estimated Mean Surface Area (Specific)

Medical-grade silicone rubber, MED 4050, reinforced with silicon dioxide (silica) particles was used as a benchmark material, courtesy of NuSil Technology.

2.2 Sample preparation

D_IB-MS10 and its CB nanocomposite (D_IB-MS10_CB) were compression molded using a hydraulic hot press (Webber PW10) into 1 mm thick flat sheets, while silicone rubber was used as supplied (1.5 mm thickness). From these sheets, microdumbbells were cut with a cutting tool to the dimensions following the specifications defined by DIN ISO 53504. Before characterization, the microdumbbells were annealed for 24 h in water at 37 °C, followed by 48 h of drying in a vacuum oven at 23 °C before being stored in a desiccator.

2.3 Transmission Electron Microscopy (TEM)

Thin sections of D_IB-MS10 and D_IB-MS10_CB samples were prepared using a cryomicrotome, and were then stained for 30 minutes using an aqueous solution of ruthenium tetroxide (RuO_4 - 0.5 wt%) at room temperature. Staining occurs in the hard PMS phase through the chemical reaction between RuO_4 and the benzene ring [Sawyer and Grubb, 1996].

2.4 Quasi-Static Tensile Testing

Quasi-static tensile tests were carried out with a Zwick Z2.5 universal test equipment with a load cell of 0.5 kN, using a cross-head speed of 500 mm/min and a grip distance of 26.5 mm. Measurements were made using automatic and self-tightening clamps to prevent slipping of the microdumbbells from the clamps. Tensile tests were performed on the materials under the standard laboratory conditions to evaluate ultimate tensile stress (σ_{ult}), elongation at break (ε_{max}) and the stresses at specific elongation according ASTM D412-06a. All tensile results were averaged from 5 samples.

2.5 Fatigue testing (SILT)

To conduct the hysteresis measurements, a servo-hydraulic test machine with a proper digital controller (Instron 8400/8800) and a special software package were used for the evaluation of the hysteresis loops. To apply the required loadings, a 50 N load cell and a 10 kN servo cylinder were adopted. During the hysteresis measurements, the resulting strain in the material was measured using the displacement between the specimen fixtures. Using this setup and the software package, the applied loadings and the material response can continuously be digitalized for data collection and analysis, as already described in previous publications [El Fray and Altstädt, 2003; Götz et al., 2009].

In Fig. 2, a scheme of a measured hysteresis loop is shown with the division of the loop for equal strain [Ehrenstein, 1993; Renz et al., 1988]. Using this approach, it is possible to determine strain, stiffness and energy related properties, as well as, to study non-linear viscoelastic material behavior. In the hysteresis testing, changes in material stiffness related parameters, e.g. dynamic stiffness and material damping can be determined and used as damage criteria to characterize the fatigue resistance of the material.

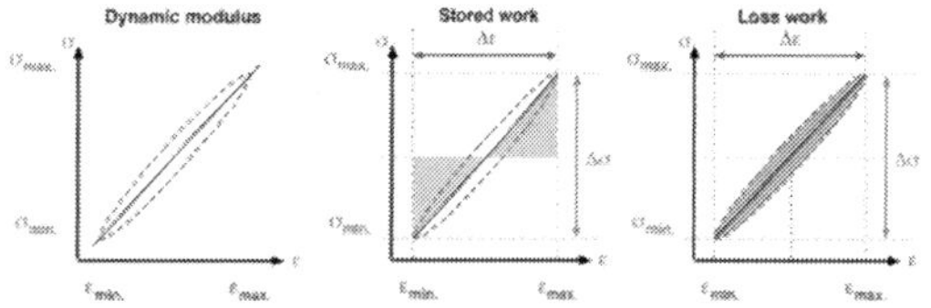

Figure 2 – Evaluation of dynamic modulus E_{dyn}, and stored and loss work to characterize fatigue behavior from a hysteresis loop [Ehrenstein, 1993; Renz et al., 1988].

The stepwise increasing load test (SILT) was used for the hysteresis methodology to accelerate the fatigue failure of a material [Ehrenstein, 1993; Götz et al., 2009; Renz et al., 1988;]. Based on the ultimate tensile stress (σ_{ult}) as determined from quasi-static tensile testing, the initial dynamic load in SILT was selected to be 1 % of σ_{ult}. The small step size was chosen because our preliminary fatigue studies showed "pinched" hysteresis loops at high strains [Puskas et al., 2009a].

As shown in Fig. 3, the dynamic load was then increased by 1 % after every 1,000 cycles, while maintaining a constant load ratio (R) of 0.1 where R is the minimum stress divided by the maximum stress ($R = \sigma_{min}/\sigma_{max}$). This selected R value was used to maintain a tensile state of stress in the material during testing.

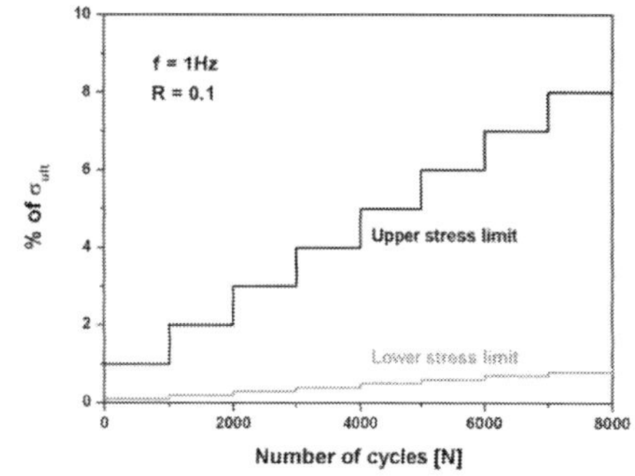

Figure 3 – Loading pattern as a function of the number of cycles (N) for SILT.

In the present study, a constant frequency of 1 Hz was used for every hysteretic cycle to limit the strain rate, prevent hysteretic heating of the sample and other effects on the results so that they are solely influenced by the applied dynamic load. From the SILT, the critical stress level (σ_L) of the tested material is determined based on the criterion when the dynamic modulus (E_{dyn}) - the slope of the hysteresis loop (see Fig. 2) - decreases by 5 % or more within a single stress level.

3. Results and Discussion

3.1 Phase Morphology

Fig. 4(a) and (b) represents the phase separated microstructure of D_IB-MS10 and its composite. This phase separation is responsible for the rubber-like behavior of the investigated block copolymers [Holden et al., 2004]. Due to the staining with RuO_4 [Sawyer and Grubb, 1996], the darker regions correspond to the PMS-rich discreet phases and the lighter regions correspond to the continuous PIB phase of the material.

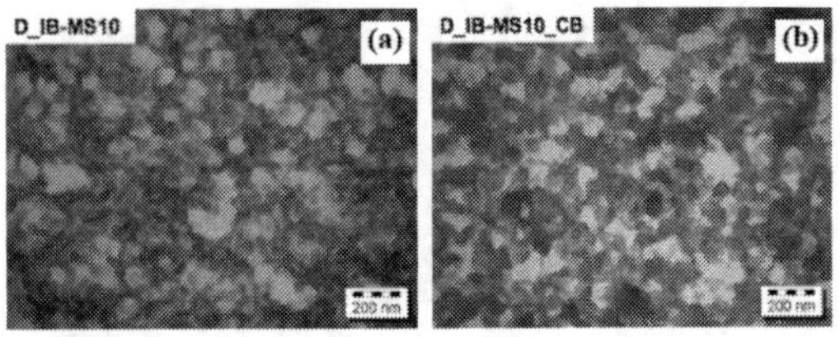

Figure 4 – TEM images of microtomed and stained (a) D_IB-MS10 and (b) D_IB-MS10_CB.

In Fig. 4(a), the D_IB-MS10 with 10 wt% overall PMS content shows spherical PMS-rich domains distributed in the continuous PIB soft phase. Interestingly from Fig. 4(b), there are no signs of CB agglomeration either in the PMS phases or in the PIB soft phase. This leads to the conclusion that the carbon black is well-dispersed, and the (reinforced) copolymer can be classified as a true nanocomposite. Furthermore, one can propose that the excellent CB dispersion developed a secondary network structure for the reinforcement of the neat polymer. The D_IB-MS10_CB nanocomposite still shows spherical discreet phases. However, Fig. 4 clearly indicates that adding CB to the neat D_IB-MS10 copolymer leads to a drastic morphological change, with an increase of the PMS-rich domain size as compared to the neat D_IB-MS10 (see Table 3). An explanation for this domain size increase is likely the affinity of CB to PMS over PIB, which is known to have poor filler interaction with CB. The excellent dispersion of the CB and its effect on the phase separation is responsible for the drastic enhancement of the tensile properties, to be discussed in the next section.

Table 3 – Measured domain size of hard styrene phase.

Material	Domain Size [nm]
D_IB-MS10	41 ± 6
D_IB-MS10_CB	79 ± 32

3.2 Tensile Performance

Fig. 5 shows the representative stress-strain plots of D_IB-MS10 and its D_IB-MS10_CB nanocomposite. Both materials exhibit typical rubber-like behavior where an initial elastic part of the stress–strain curve is followed by a flat stress plateau. During this flat plateau, polymer chains start to align in the loading direction as they are released from the physical entanglement [Ehrenstein, 1999; Holden et al., 2004]. With this chain alignment, the polymer chains recover their load-carrying capability to explain the strain-hardening observed thereafter. Also, it is known that IBS polymers can crystallize at large stretch ratios to strengthen the material [Kaszas, 1992]. At high strains, these synergistic effects of strain hardening and strain-induced crystallization were observed by the loss of transparency in the neat D_IB-MS10 polymer.

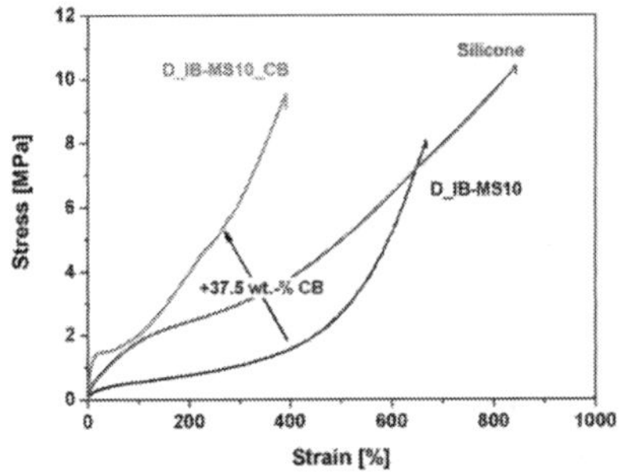

Figure 5 – Stress – strain tensile plots of D_IB-MS10 and D_IB-MS10_CB microdumbbells tested at 500 mm/min.

Table 4 shows the measured moduli values as indicated by the stress (σ) at specific strains (100 %, 200 % and 300 %), the ultimate tensile stress (σ_{ult}) and the elongation at break (ε_{max}) for the materials considered in this study. From Table 4 and Fig. 5, the D_IB-MS10 polymer has lower moduli and tensile strength and lesser ductility than silicone rubber.

Table 4 – Quasi-static tensile properties of different materials using microdumbbells tested at 500 mm/min.

Material	σ at 100 % [MPa]	σ at 200 % [MPa]	σ at 300 % [MPa]	σ_{ult} [MPa]	ε_{max} [%]
D_IB-MS10	0.5 ± 0.0	0.7 ± 0.1	1.1 ± 0.2	8.2 ± 0.4	647 ± 21
D_IB-MS10_CB	2.4 ± 0.4	4.6 ± 0.6	7.5 ± 1.1	10.2 ± 1.4	370 ± 31
Silicone	1.8 ± 0.1	2.4 ± 0.0	3.0 ± 0.1	10.2 ± 0.2	845 ± 36

From Table 4, the addition of CB filler to D_IB-MS10 significantly increased the moduli and σ_{ult}, but lowered ε_{max} as compared to the neat polymer and silicone. While the above impacts of the introduction of CB filler on the tensile behavior are not unexpected [Mai and Yu, 2006], it is important to relate them to the morphological changes that CB filler had caused in D_IB-MS10. The first morphological change is the creation of a new "secondary reinforcement network" by the dispersion (local) and distribution (global) of CB filler in the polymer matrix. To ensure an increase in both modulus and strength, it is essential that CB filler is well dispersed and distributed, and has a good surface interaction with the polymer matrix. Agglomeration of the filler particles and poor adhesion to the polymer matrix can lead to the premature failure of the composite material. It is evident from Fig. 4 and 5 that the excellent dispersion of the CB filler leads to the reinforcement of the D_IB-MS10 copolymer.

The second morphological change is unique to the segregated nature of the D_IB-MS10 copolymer. It is apparent in the TEM images of Fig. 4 that the CB filler promotes a better segregation between the PIB and PMS-rich phases and increase the domain size of the discreet spherical domains, as quantified in Table 4. This improved phase segregation has a pronounced impact to enhance the stress transfer between the PIB and PMS phases when the material is deformed. Furthermore, the good interaction between the CB filler and PMS phases ensures the integrity of the PMS phases to resist failure at high stress levels. Last but not least, it is very likely that the CB filler also influences the onset and extent of strain-induced crystallization at large elongation.

3.3 Dynamic Fatigue Tensile Testing

Fig. 6 displays the dynamic moduli E_{dyn} profile during the SILT testing of the neat D_IBS polymer and the corresponding nanocomposite as well as that of silicone rubber. The critical load levels for each material were obtained and presented in Table 5.

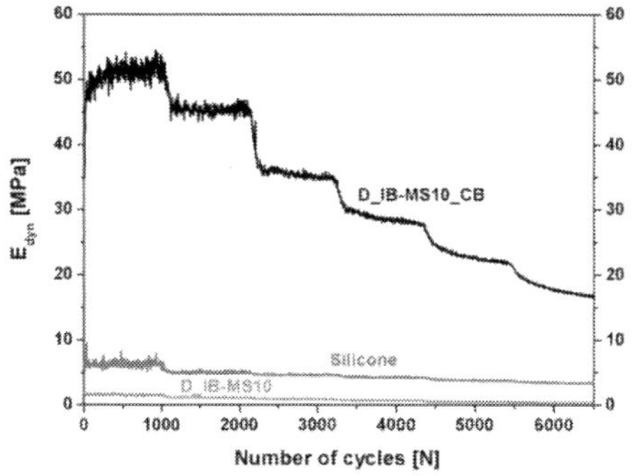

Figure 6 – Change in E_{dyn} during SILT.

Table 5 – Critical loading levels and critical stresses σ_c for various materials using SILT.

Material	Critical loading level	Critical stress σ_c [MPa]
D_IB-MS10	2	0.16
D_IB-MS10_CB	4	0.41
Silicone rubber	5	0.51

Fig. 6 indicates that the addition of carbon black increases the E_{dyn} of D_IB-MS10 from around 2.5 MPa to almost 50 MPa. The higher stiffness of CB and the formation of the secondary CB network (see Section 3.2) increased the dynamic moduli of the nanocomposite, similarly to the quasi-static tensile results. Furthermore, the larger PMS domain size (see Table 3) led to a stiffer material. Fig. 7 shows the change of the mid-strain, indicative of creep, during SILT in D_IB-MS10, its nanocomposite and silicone rubber. One can see that D_IB-MS10 has a higher overall deformation than silicone rubber due to its lower modulus (see Fig. 5) and "physically" crosslinked structure. Being chemically crosslinked and reinforced with silica, silicone rubber has smaller mid-strain compared to the neat D_IBS copolymer (126 % versus 11%). However, when CB filler was added to D_IB-MS10 the mid-strain was reduced significantly to 10 %, which is even lower than that of the silicone rubber.

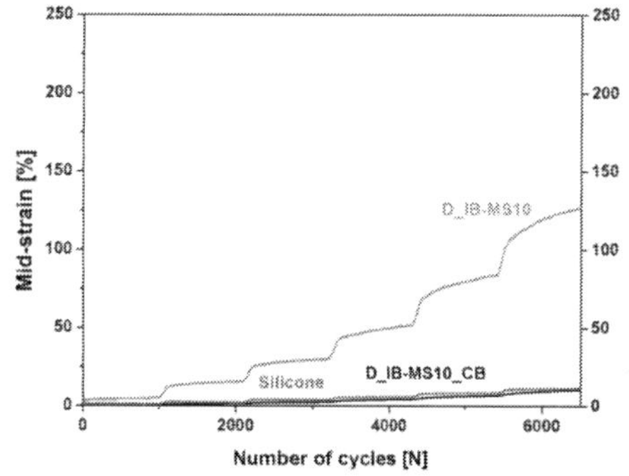

Figure 7 – Variation of mid-strain over the number of cycles using SILT.

It is discussed in the literature that mainly the soft phase is responsible for the creep behavior of TPEs [Ehrenstein, 1999; Grellmann and Seidler, 2001; Ryan and Hamely, 1997]. Therefore in the D_IB-MS10_CB composite, it can be postulated that the applied dynamic load can be sustained through the continuous PIB phase, as well as by the secondary network of the CB filler and the good filler-matrix interaction in the discreet PMS-rich phases.

Based on the SILT results a loading pattern for future single loading tests (SLT) was determined by taking the average of the critical stresses of the nanocomposites and will be used in all future SLT measurements for suitable comparison of the long-term fatigue and creep behavior of the materials under investigation.

4. Conclusions

This study investigated the morphology and quasi-static and dynamic mechanical properties of a new class of styrenic thermoplastic elastomers. The neat polymer (D_IB-MS10) and its carbon black (CB) reinforced nanocomposite (D_IB-MS10_CB) were considered in this work with a focus to examine the influence of CB on the material properties. As these materials are intended to be biomaterials, it is of great interest to understand their structure-property relationships and long-term dynamic creep response. Characterization approaches such as TEM, tensile testing and hysteresis measurements (SILT) were employed to determine the morphological, quasi-static and dynamic behavior of these materials. From these evaluations, the properties of both materials were compared to a medical-grade silica-reinforced silicone rubber as a benchmark material for biomedical devices. Morphological studies revealed that the size of discreet hard phases of D_IB-MS10 increased with the addition of 37.5 wt% of CB filler. Furthermore, the CB was well dispersed mainly in the hard phase, which appeared to form a secondary network that helped to reinforce this block copolymer. Indeed together with the good filler-matrix interaction, the use of CB filler led to an improvement in quasi-static tensile properties of D_IB-MS10_CB. Also when the CB filler was introduced, the mid-strain value of D_IB-MS10 in the step-wise increasing load tests was significantly reduced to be comparable to that of chemically crosslinked and silica-reinforced silicone rubber. This suggests that the creep performance of D_IB-MS10_CB will be similar to a chemically crosslinked reinforced rubber.

5. Acknowledgements

The authors are grateful to the German Science Foundation (DFG) under Al 474/12-1 and National Science Foundation under NSF Grant DMR-0509687 for financial support. In addition, the authors would like to acknowledge Mrs. C. Kunert for providing the TEM images.

6. References

[ASTM D412-06a]
ASTM International, ASTM D412-06a: Standard test method for vulcanized rubbers and thermoplastic elastomers – Tension, 2006.

[DIN ISO 53504]
DIN ISO 53504, Determination of tensile stress/strain properties of rubber, 2005.

[Ehrenstein, 1993]
Ehrenstein, G.W.: Hysteresis Messverfahren, Universität Erlangen-Nürnberg Lehrstuhl für Kunststofftechnik, Erlangen (1993).

[Ehrenstein, 1999]
Ehrenstein G.W.: Polymer-Werkstoffe, Carl HanserVerlag, München (1999).

[El Fray and Altstädt, 2003]
El Fray, M., Altstädt, V., "Fatigue behaviour of multiblock thermoplastic elastomers. 2. Dynamic creep of poly(aliphatic/aromatic-ester) copolymers", Polymer 44, 4635-4650 (2003).

[El Fray et al., 2006]
El Fray, M., Prowans, P., Puskas, J. E., Altstädt, V., "Biocompatibility and fatigue properties of polystyrene-polyisobutylene-polystyrene, an emerging thermoplastic elastomeric biomaterial" Biomacromolecules 7, 844-850 (2006).

[FDA, 2004]
U.S. Food and Drug Administration, Publication P030025: TAXUS Express2TM Paclitaxel-Eluting Coronary Stent System (Monorail and Over the Wire). Washington, DC: U.S. Government Printing Office (2004).

[Götz et al., 2009]
Götz, C., Handge, U.A., Piatek, M., El Fray, M., Altstädt V., "Influence of e-beam irradiation on the dynamic creep and fatigue properties of poly(aliphatic-aromatic-ester) copolymers for biomedical applications", Polymer 50, 5499-5507 (2009).

[Grellmann and Seidler, 2001]
Grellmann, W., Seidler, S.: Deformation and fracture behavior of polymers, Springer Verlag, Heidelberg (2001).

[Gyor et al., 1994]
Gyor, M., Fodor, Z., Wang, H.C., Faust, "Polyisobutylene-based thermoplastic elastomers. I. Synthesis and characterization of Polystyrene-Polyisobutylene-Polystyrene triblock copolymers", J. Macromol. Sci. A, A31(12), 2055-2065 (1994).

[Holden et al., 2004]
Holden, G., Kricheldorf, H.R., Quirk, R.P.: Thermoplastic Elastomers, 4th Edition, Hanser Publishers, Munich (2004).

[Jacob et al., 1996]
Jacob, S., Majoros, I., Kennedy, J.P., "New stars: eight Polyisobutylene arms emanating from a calixarene core", Macromolecules 29, 27, 8631-8641 (1996).

[Jacob and Kennedy, 1999]
Jacob, S., Kennedy, J.P., "Synthesis, Characterization and Properties of Octa-Arm Polyisobutylene-Based Star Polymers", Adv. Polym. Sci., 146, 1-38 (1999).

[Kaszas, 1992]
Kaszas, G. Polym. Mater. Sci. Eng. 67, 325-327 (1992).

[Kennedy et al., 1990]
US 4,946,899 (1990) Kennedy, J.P., Puskas, J.E., Kaszas, G., Hager, W.G.

[Kojima, 1993]
Kojima, Y., Usuki, A., Kawasumi, M., Okada, A., Fukushima, Y., Karauchi, T., Kamigaito, O. J., "Mechanical properties of nylon 6-clay hybrid", Mater. Res., 8, 1185-1189 (1993).

[Koshimura and Sato, 1992]
Koshimura, K., Sato, H. "Application study of styrene-isobutylene-styrene block copolymer as a new thermoplastic elastomer",Polym. Bull. 29, 705-711 (1992).

[Mai and Yu, 2006]
Mai, Y.-W., Yu, Z.-Z.: Polymer nanocomposites, Woodhead Publishing Limited, Cambridge (2006).

[Pinchuk et al., 2008]
Pinchuk, L., Wilson, P.J., Barry, J.J., Schoephoerster, R.T., Parel, J.-M., Kennedy, J.P., "Medical applications of poly(styrene-block-isobutylene-block styrene) (''SIBS'')", Biomaterials 29, 448-460 (2008).

[Puskas and Kaszas, 1996]
Puskas, J.E., Kaszas G., "Polyisobutylene-based thermoplastic elastomers: a review", Rubber Reviews. Rubber Chem. Technol. 69 3, 462-475 (1996).

[Puskas et al., 1998]
Puskas, J.E., Pattern, W., Wetmore, P.M., Krukonis, V., "Synthesis and characteritation of novel multiarm-star Polyisobutylene-Polystyrene thermoplastic elastomers", Rubber Chem. Technol. 72, 559-568 (1998).

[Puskas et al., 2002]
Puskas, J.E., Antony, P., Kwon, Y., Kovar, M., Norton, P.R., "Study of the surface morphology of polyisobutylene-based block copolymers by atomic force microscopy", J. Macromol. Sci., Macromol. Symp. 183, 191-198 (2002).

[Puskas et al., 2004]
U.S. Patent 6,747,098 (2004) Puskas, J.E., Paulo, C., Antony, P.

[Puskas et al., 2005]
Puskas, J.E., Kwon, Y., Antony, P., Bhowmick, A.K., "Synthesis and characterization of novel dendritic (arborescent, hyperbranched) polyisobutylene–polystyrene block copolymers", J. Polym. Sci. Pol. Chem., 43, 1811 (2005).

[Puskas and Kwon, 2006]
Puskas, J.E., Kwon, Y., "Biomacromolecular engineering: design, synthesis and characterization. One-pot synthesis of block copolymers of arborescent polyisobutylene and polystyrene", Polym. Adv. Technol., 17, 615-620 (2006).

[Puskas et al., 2009a]
Puskas, J.E., Dos Santos, L.M., Fischer, F., Goetz, C., El Fray, M., Altstädt, V., Tomkins, M., "Fatigue testing of implantable specimens: Effect of sample size and branching on the dynamic fatigue properties of polyisobutylene-based biomaterials", Polymer, 50, 591-597 (2009).

[Puskas et al., 2009b]
Puskas, J.E., Dos Santos, L.M., Kaszas, G., Kulbaba, K., "Novel thermoplastic elastomers based on arborescent (dendritic) polyisobutylene with short copolymer end sequences", J. Polym. Sci. Pol. Chem., 47, 1148-1158 (2009).

[Ryan and Hamely, 1997]
Ryan, A.J., Hamely, I.W. The Physics of Glassy Polymers, Haward RN, Young RJ (Eds.) London: Chapman and Hall, 1997.

[Renz et al., 1988]
Renz, R., Altstädt, V., Ehrenstein, G.W., "Hysteresis Measurements for Characterizing the Dynamic Fatigue of R-SMC", J. Reinf. Plast. Comp. 1988, 7, 413-434.

[Sengupta et al., 2007]
Sengupta, R., Chakraborty, S., Bandyopadhyay, S., Dasgupta, S., Mukhopadhyay, R., Auddy, K., Deuri, A.S., ") A short review on rubber/clay nanocomposites with emphasis on mechanical properties", Polym. Eng. Sci., 47, 1956-1974 (2007).

[Sawyer and Grubb, 1996]
Sawyer, L.C., Grubb, D.T: Polymer Microscopy, 2nd Edition, Chapman & Hall, London (1996).

[Storey and Showmake, 1999]
Storey, R.F., Showmake, K.A., "Poly(styrene-*b*-isobutylene) multiarm star-block copolymers", J. Polym. Sci., Part A: Polym. Chem. 37 11, 1629-1641 (1999).

[Storey et al., 2000]
Storey, R.F., Baugh, D.W., "Poly(styrene-*b*-isobutylene-*b*-styrene) block copolymers and ionomers therefrom: morphology as determined by small-angle X-ray scattering and transmission electron microscopy", Polymer 41, 3205-3211 (2000).

NaOH TREATED MICRO/NANO FIBROUSE POLYETHYLENE TEREPHTHALATE MATS FOR BIOMEDICAL APPLICATION

Hadjizadeh Afra [(1), (2)], Bureau Martin N [(2)], Ajji Abdellah [(1)*]

[(1)] Department of Chemical Engineering, Ecole Polytechnique of Montreal, Quebec, Canada, [(2)] Industrial Materials Institute - National Research Council Canada, Quebec, Canada

Abstract - Recently micro/nano-fibrous polyethylene terephthalate (PET) nonwovens have attracted attention for their potential in biomedical applications. NaOH treatment has been used as a simple and cost effective method for their surface functionalization; however, their physical and mechanical properties have been poorly studied. This study reports the physical and tensile characterization of NaOH treated micro/nano-fibrous PET mats. PET webs with an average fiber diameter of 6 µm and 10 µm were produced by melt blowing process and the fibrous PET mates were fabricated using consolidation of a desired number of melt blown PET web layers. PET mats with a fiber diameter of 0.7 and 1.5 µm were also produced by using electro-spinning. The mats were treated using NaOH 1N at 65°C in various times. The morphology and tensile properties of the mats was evaluated using SEM and micromechanical tester, respectively. Mats with different fiber diameter showed different fiber surfaces features, tensile properties, weight loss, porosity, thickness and shrinkage, depending on the treatment time. These results demonstrated that in NaOH treatment of PET matrices, the selected treatment condition should be compatible with the matrices structure and also the properties required for their end use, in order to control variation imposed in the matrices properties.

1. Introduction

Porous fibrous materials have found a wide variety of applications in different areas including biotechnology and biomedical application (Andrews, Feugier et al. 2008; Robin Ng 2009). PET had been one of the potential candidates in fabricating those structures, due to their structural and mechanical properties. In addition, this biostable polymer has been long used for biomedical applications such as vascular grafts.

In particular tissue engineering applications, the contact guidance effect of fibers on cells migrations (Hadjizadeh and Doillon 2010) and the typical surface roughness, created by microfibers (Flemming, Murphy et al. 1999; Palmaz, Benson et al. 1999), modulate cell behavior. Therefore, the capability of PET for being processed into fibers makes it a high potential candidate for the biomedical applications.

Nonwoven fibrous PET matrices with various fiber diameter ranges can be produced using fiber spinning techniques including melt spinning, electro-spinning and melt blowing. In the melt blowing process, PET nonwovens with fiber diameter range below 50 µm can be produced. Such a range in fiber diameter appears to be of interest for biomedical applications. Moreover, this linear aromatic semi-crystalline polyester has been processed to nanofibers in recent years by using electro-spinning for the aim of being applied in many applications including tissue engineering. PET surface can be modified with NaOH (Robin Ng 2009). A simple NaOH treatment method was developed for functionalization of nonwoven fibrous matrices of polyethylene terephthalate (PET) since long time ago, mainly for biological applications to make the PET fabric surface cell adhesive.

With regard to the afore mentioned discussions, an interesting concept is to apply NaOH treatment method on fibrous mats produced by melt blowing or electro spinning composed of microfibers and nanofibers with a large surface area to alter their surface properties in terms of surface features, wetability and functionality. However, the effect of this treatment on the physical and mechanical properties of the matrices is unknown and should be studied which is the objective of this study.

2. Materials and methods

Polyethylene terephthalate webs with two different average fiber diameters (AFD) of 6 ± 2.5 µm and 10 ± 4 µm were produced by melt blowing process. A number of these webs were consolidated to prepare fibrous mats (Mat-A1, AFD of 6 ± 2.5 µm and Mat-B1, AFD of 10 ± 4 µm) using a thermal treatment. Electro-spun PET mats (Mat-A2 and Mat-B2) with two different AFD (0.7 ± 0.04 and 1.8 ± 0.2 µm, respectively) were also produced. The matrices were treated using NaOH 1N at 65°C for various durations (ranging from 20 min to 24 h). Then, in addition to the physical properties such as weight loss, thickness, porosity and shrinkage, the morphology and tensile properties of mats were also evaluated using scanning electron microscopy and micromechanical tester, respectively.

3. Result and discussion

It was observed that the hydrolysis rate of all the samples increased with treatment time and the weight loss of fibers with small diameters was higher than that of the fibers with large diameters (Table 1 and 2). This surface degradation created pores on the surface of the fibers (Fig. 1) and altered the physical (Table 1 and 2) and mechanical properties (Fig. 2) of the mats which depended on the treatment time and also fiber diameter. In general, by increasing treatment duration, weight loss, porosity and shrinkage increased; while thickness decreased. As a result of treatment, pores appeared on the surface of individual fibers, and tensile stress decreased. Mats with different fiber diameters showed different physical and mechanical properties.

Table 1: Physical properties of melt-blown PET mats

Sample	Treatment time (min)	Weight loss (%)	Thickness (μm)	Porosity (%)	Shrinkage (%)
Mat-A1	0	-	156.4 ± 24.7	76.1 ± 4.4	-
	20	3.0 ± 1.2	176.7 ± 30.3	78.8 ± 4.1	6.8 ± 9.3
	40	11.2 ± 6.4	194.1 ± 62.75	80.3 ± 3.4	15.1 ± 9.2
	60	18.0 ± 7.4	197.5 ± 66.5	79.9 ± 3.2	22.8 ± 13.2
	120	42.0 ± 5.1	241.0 ± 56.5	85.2 ± 2.3	37.6 ± 15.3
	240	67.4 ± 3.5	227.0 ± 53.8	88.5 ± 1.2	55.3 ± 11.2
	1440	95.6 ± 3.9	-	-	-
Mat-B1	0	-	208.1 ± 18.4	83.6 ± 2.0	
	20	3.4 ± 3.1	202.1 ± 16.7	82.7 ± 1.3	2.8 ± 3.0
	40	7.0 ± 3.34	228.6 ± 65.7	84.7 ± 2.7	8.7 ± 4.8
	60	14.4 ± 4.0	264.3 ± 56.1	85.9 ± 2.8	14.4 ± 5.7
	120	31.4 ± 12.0	334.5 ± 78.7	87.9 ± 1.6	33.8 ± 8.8
	240	46.0 ± 2.6	307.4 ± 57.8	89.6 ± 2.1	37.8 ± 10.9
	1440	91.0 ± 2	-	-	-

Table 2: Physical properties of electro-spun PET mats

Sample	Treatment time (min)	Thickness (μm)	Weight loss (%)	Porosity	Shrinkage (%)
Mat-A2	0	100 ± 10	-	81.5 ± 1.7	-
	20	165 ± 13.2	23.0 ± 5.3	87.1 ± 4.5	46.4 ± 3.3
	40	84 ± 12.5	57.3 ± 9.22	82.2 ± 10.4	68.0 ± 6.0
	60	-	70.4 ± 6.0	-	-
	120	-	96.0 ± 5.4	-	-
	240	-	100.0 ± 0.0	-	-
Mat-B2	0	100 ± 10	-	83.6 ± 1.6	-
	20	124.6 ± 7.8	15.2 ± 2.7	88.37 ± 1.7	22.0 ± 3.1
	40	82.4 ± 8.9	41.6 ± 4.0	87.5 ± 3.2	25.0 ± 3.6
	60	45.7 ± 5.2	61.0 ± 5.1	85.1 ± 5.2	35 ± 2.6
	120	-	82.0 ± 1.0	-	-
	240	-	97.5 ± 5	-	-

Figure 1: Representative FESEM images of untreated and NaOH treated PET mats

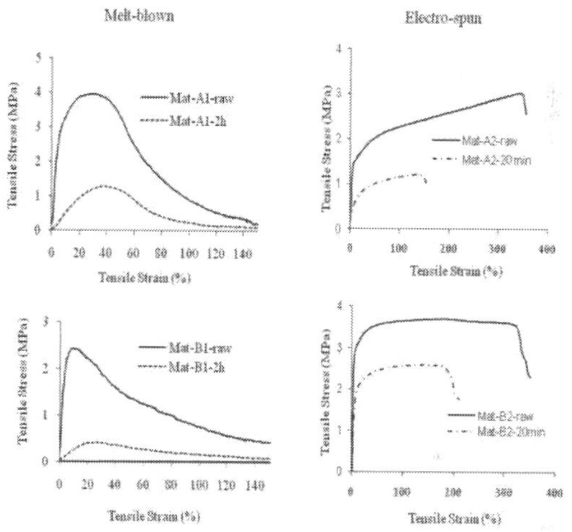

Figure 2: Stress-strain curves of untreated and NaOH treated PET mats

4. Conclusions

These findings suggest that the structure of the matrices and the properties required for its end use, in biomedical applications, should be considered in selecting NaOH treatment condition. In addition, an appropriate structural design and matrices fabrication method in conjunction with a simple NaOH treatment may lead to the production of potential scaffolding materials for tissue engineering applications.

5. Acknowledgments

We are very thankful to NSERC for financial support through the strategic projects program and NRC for access to some laboratories

6. References

Akihiro Suzuki, K. T. "Poly(ethylene terephthalate) nanofibers prepared by CO2 laser supersonic drawing." Polymer **50,** 913–921(2009).

Andrews, K. D., P. Feugier, et al. "Vascular prostheses: performance related to cell-shear responses." J Surg Res **149**(1), 39-46 (2008).

Beatriz Veleirinho, J. A. L.-d.-S. "Application of electrospun poly(ethylene terephthalate) nanofiber mat to apple juice clarification." Process Biochemistry **44,** 353–356 (2009).

Changzhong Chen, L. W. a. Y. H. "A novel shape-stabilized PCM: Electrospun ultrafine fibers based on lauric acid/polyethylene terephthalate composite." Materials Letters **62**(20), 3515-3517 (2008).

Flemming, R. G., C. J. Murphy, et al. "Effects of synthetic micro- and nano-structured surfaces on cell behavior." Biomaterials **20**(6), 573-88 (1999).

Hadjizadeh, A. and C. J. Doillon ("Directional Migration of Endothelial Cells Towards Angiogenesis Using Polymer Fibers In a 3-D Co-culture System." Journal of Tissue Engineering and Regenerative Medicine (2010).

Palmaz, J. C., A. Benson, et al. "Influence of surface topography on endothelialization of intravascular metallic material." J Vasc Interv Radiol **10**(4), 439-44 (1999).

Robin Ng, X. Z., Ning Liu and Shang-Tian Yang "Modifications of nonwoven polyethylene terephthalate fibrous matrices via NaOH hydrolysis: Effects on pore size, fiber diameter, cell seeding and proliferation " Process Biochemistry **44**(9), 992-998(2009).

Ron (Rongguo) Zhao, L. C. W., Christine Sun, Dong Zhang "Properties of PP/PET bicomponent melt blown microfiber nonwovens after heat-treatment." Polymer International **52**(1), 133-137(2003).

SWELLABLE DRESSING MATERIALS FOR WOUND HEALING

N. Saha[1*], A. Saarai[1], T. Kitano[1] and P. Saha[1]

[1]Polymer Centre, Faculty of Technology, Tomas Bata University in Zlin, TGM-275, CZ-76272, Zlin, Czech Republic - nabanita@ft.utb.cz , sa_amarjargal@yahoo.com, t_kitano297@yahoo.com; saha@utb.cz

Abstract - Wound healing is an important biological process involving tissue repair and regeneration. Several topical agents: ointments, creams, lotions are used in wound healings to prevent secondary infection and to aid early healing. Beside application of topical agents on wounds it is required periodic washing, debridement, oxygen permeability and attention on dressing. To make the wound dressing materials more user friendly and comfortable, biopolymer based swellable polymeric dressing materials hava been developed. The hydrogels were prepared with gelatin (G), sodium alginate (SA), polyvinyl alcohol nanofiber (PVA-NF), glycerin, sodium chloride, polyethylene glycol and/or seabuckthron oil (SB) by applying physical stimulation technique under control heat (80°C) and stirring (300 -100 rpm). Finally, the hydrogels achieved in dry form after 48-72 hours and designated as G/SA/PVA-NF and G/SA/SB/PVA-NF. These hydrogels vary from light yellow to orange, show quite a good swelling property. As the base materials of hydrogels are natural in origin, it is predicted that these hydrogels could be useful for the application of specific type of drug release or swellable medicated dressings for wound healing.

Introduction

Wound healing is an important biological process involving tissue repair and regeneration. The healing process is classified into four distinct stages namely: hemostasis, inflammatory, proliferatie and remodelling. The factors which play major roles in wound healing are bacterial infection, nutritional deficiency, drugs, site of wound and health condition of patients. Moist treatment of wounds is considered to be the gold standard of therapy for acute open wounds as well as chronic wounds. One of the disadvantages is the promotion of bacterial growth within the wound on the surface or deep in the tissue. Moist wound treatment improves healing at a possibly increased risk of bacterial infection and many local antiseptics impair healing. Thus, a moist treatment modality with efficient antibacterial activity would be desirable. Several topical agents: ointments, creams, lotions are used in wound healings to prevent secondary infection and to aid early healing. Beside application of topical agents on wounds it is required periodic washing, debridement, oxygen permeability and attention on dressing. The ideal topical wound dressing agents would therefore deliver both moisture and antiseptic effects at the same time. The goal of wound therapy is to keep the wound away from microorganism content as low as possible in order to prevent infection and accordingly to stimulate the repair process [1-2].
To make the wound dressing materials more user friendly and comfortable, efforts have been given to develop natural antibacterial property based dressing materials. Hydrogel dressings are polymeric materials which are gradually becoming more popular than the conventional dressings. Thus, aiming to develop biopolymer based swellable dressing materials without using any chemical crosslinking agent. This paper is reporting about the physico-chemical and morphological appearance of the novel swellable hydrogel wound dressings.

Experimental

The swellable dressing material i.e. hydrogel dressings were prepared basically using natural and synthetic polymeric materials like: gelatin (G), sodium alginate (SA), polyetheleneglycol (PEG), sodium chloride (NaCl), glycerin varying only the amount of G and SA in composition and named as G/SA hydrogel. Finally, the hydrogel with G/SA = 50/50 was modified with polyvinyl alcohol nanofiber (PVA-NF) (diameter 20 µ and average length 3 mm) to improve its mechanical property. Seabuckthron (SB) oil was incorporated to include some nutritive/medicinal value within the hydrogel and to promote moist healing as well as rapid epitheliazation thereby decreasing hypertropic scar formation. All these hydrogels were prepared (without adding any additional chemical crosslinking agents) under controlled physical stimulation i.e. under continous stirring (300 - 100 rpm) and temperature (80 °C). Finally, the polymer mass was casted in a grooved (appoxmately 2 mm thickness and 25 mm diameter) plate to acheive the hydrogel dressing materials in a definite shape [3]. The resultant hydrogels are designated as 'G/SA/PVA-NF' or 'G/SA/SB/PVA-NF' which were dried at room temperature under vacuum drying environment for 3 days. Swelling behaviour of hydrogels with and without PVA-NF was studied at room temperature using distilled water at pH 7.0. The degree of swelling was defined by the following equation ,where W_s and W_d are weights of swollen gel and dried gel, respectively [4].

$$Swelling\ (\%) = (W_s\text{-}W_d) / W_d \times 100.$$

The visual image of G/SA/PVA-NF and G/SA/SB/PVA-NF (before dry, after dry and after 5 h swelling) hydrogels was captured by a normal digital camera. The IR spectra was analysed by using a NICOLET 320 FTIR Spectrophotometer with "Omnic" software package.The sectional and surface view of these hydrogels was investigated by scanning electron microscope (SEM,VEGA\\LMU, TESCAN).

Results and Discussion

The degree of swelling or water uptake capacity of hydrogel dressing materials is one of the important factors to choose them for moist wound therapy.

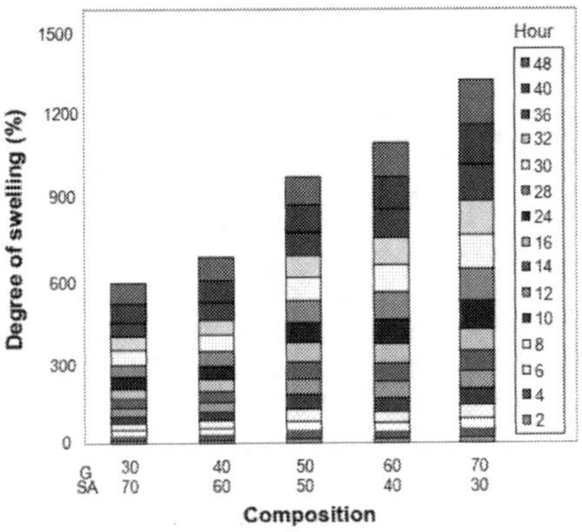

Figure 1.Effect of composition and time on degree of swelling of G/SA hydrogels

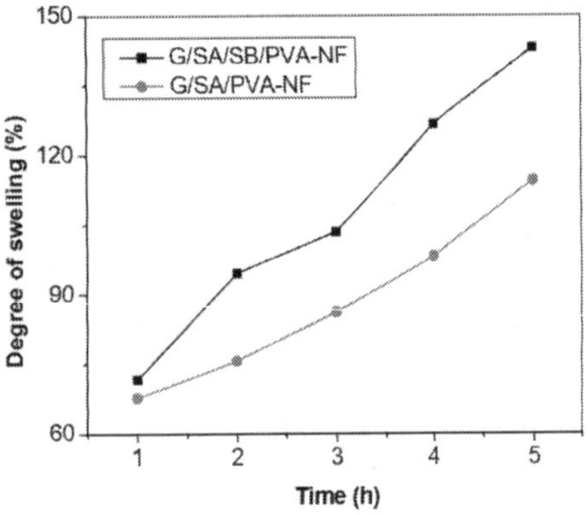

Figure 3.Water uptake capacity of G/SA/PVA-NF & G/SA/SB/PVA-NF hydrogels until 5h swelling

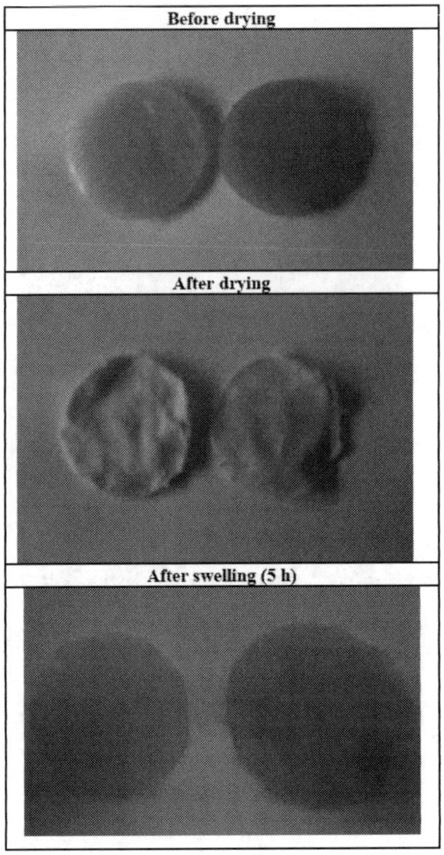

Figure 2.Visual images of G/SA/PVA-NF (left) and G/SA/SB/PVA-NF (right) with 2% PVA-NF at various states.

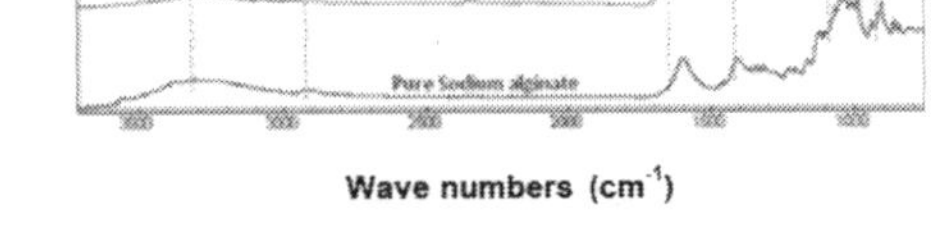

Figure 4.FTIR view of G/SA/PVA-NF hydrogels

It can be seen from the Figure 1 that the quantity of biopolymers (G and SA) in hydrogel and time has great influence on degree of swelling. Hydrogel with higher amount of gelatin (G/SA=70/30) demonstrates maximum degree of swelling within 48 hour compared to the hydrogel with less amount of gelatin (G/SA = 30/70).

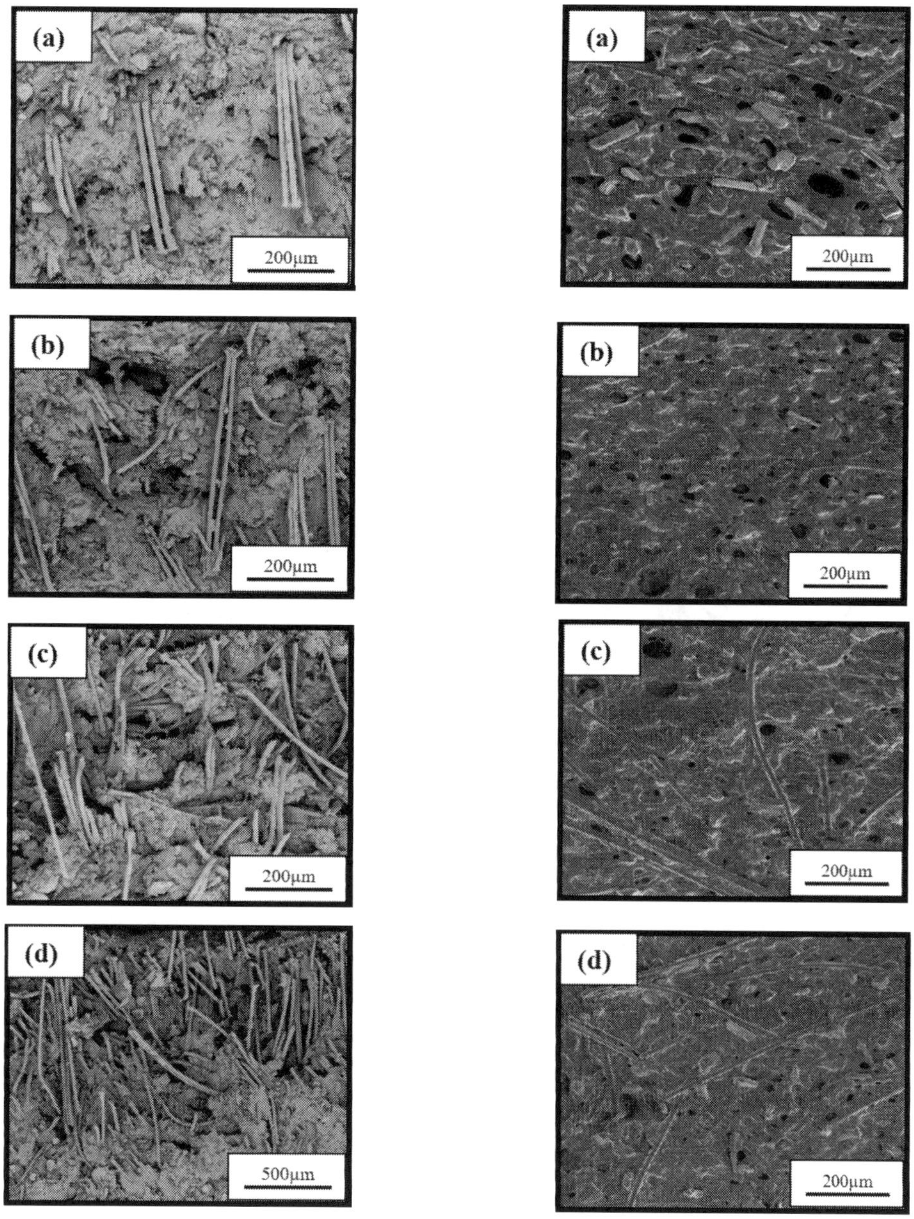

Figure 5.SEM images of sectional (left) and surface (right) views with different contents of PVA-NF in G/SA hydrogels. (a) 1wt.% (b) 2wt.% (c) 3wt.% (d) 4wt.%

It is reported that the natural polymer, sodium alginate (SA) put up with natural antimicrobial properties and water retention capacity. Thus, to maintain water retention and interim antimicrobial properties within the hydrogel G/SA = 50/50 has been selected for its further improvement as hydrogel dressings. PVA-NF has been used to improve the mechanical property of hydrogel and SB oil has been incorporated as scar hydration oilment [5]. The resultant hydrogels are round varies from light yellow (G/SA/PFA-NF) to orange (G/SA/SB/PVA-NF) as shown in Figure 2. Due to the presence of PVA-NF in hydrogels, though they are swelled but not damaged after long time swelling. The water uptake capacities of G/SA/ PVA-NF and G/SA/SB/PVA-NF with 2wt. % of PVA/NF have been shown in Figure 3. It can be seen from the figure that in both cases water uptake trend is similar but G/SA/SB/PVA-NF hydrogel accumulated more water than the G/SA/PVA-NF hydrogel. Figure 4 represents the FTIR view of G/SA/PVA-NF hydrogels having various amount of PVA-NF (1-4 wt. %) and pure G, SA and PVA-NF as well which indicates the physico-chemical nature of hydrogel after modification. The sectional and surface images of SEM micrographs are depicated in Figure 5 which confirm that amount of PVA-NF in hydrogel has commandable influence on external and internal morphology of novel swellable hydrogel.

Conclusions

Both G/SA/PVA-NF and G/SA/SB/PVA-NF hydrogels show quite a good swelling property. Moreover, the said hydrogels are natural in origin thus it can be predicted that G/SA/PVA-NF and G/SA/SB/PVA-NF hydrogels could be useful for the employment of specific type of drug release or swellable medicated dressings for wound healing.

Acknowledgements

The authors are also thankful to the Ministry of Education, Youth and Sports of Czech Republic (MSM 70088352101) for financial support.

References

1. Vogt P.M., et.al., " PVP-iodine in hydrosomes and hydrogel- A novel concept in wound therapy leads to enhanced epithelialization and reduced loss of skin grafts", Burns, 32, 698-705 (2006)

2. Momoh.M.A,et al., "Evaluation of wound healing effect of gelatin and polyethyleneglycol containing Cicatrin Powder", Arch Pharm Sci & Res 1, 1, 54-57 (2009)

3. Saha N., et.al., "Seabuckthron oil incorporated medicated hydrogel based on Gelatin - Sodium alginate", European Conference on Medical Polymers, September 2-4, Belfast, UK, (2008)

4. Kyoung R. P. and Young C. N., "Synthesis of PVA/PVP hydrogels having two-layer by radiation and their physical properties", Radiation Physics and Chemistry ,67 361–365 (2003)

5. Shih R. et.al. "Review of over- the counter topical scar treatment products" Plastic and Reconstructive Surgery,119,3, 1091-1095 (2007)

CHITIN BEADS INCORPORATED WITH CURCUMIN FOR DRUG DELIVERY APPLICATIONS

P. Ratanajiajaroen[1]* and M. Ohshima[1]

[1]*Material Processing Engineering Division, Dept. of Chemical Engineering, Kyoto University, Kyoto 615-8514, Japan*
–fai@cheme.kyoto-u.ac.jp; oshima@cheme.kyoto-u.ac.jp

Abstract - Spherical curcumin-incorporated chitin porous beads, of which porosity could exceed 95%, were successfully fabricated by a coagulation method. Different initial amounts of curcumin (1.6, 2.8 and 4.0 wt% based on the volume of the aqueous solution) were incorporated into the chitin beads. Tween20 was used to enhance the solubility of curcumin in chitin during the incorporation step. By freeze-drying, spider-net liked chitin microfibrils could be created in the internal porous structure and the porous structure could be controlled by changing the amount of incorporated curcumin and freezing conditions. The releasing behavior of curcumin from the chitin beads was investigated by the total immersion method in an acetate buffer solution at 37°C simulating the human skin pH condition of 5.5. The internal pores and the outer non-porous shell structure of the chitin beads could control the releasing behavior: the non-porous outer skin layer dominates the releasing behavior of curcumin and prevents curcumin from immediate degradation at the physiological temperature of 37°C while the internal porous structure could dramatically increase the uptake of the releasing medium, acetate solution, and control the degradation rate of chitin beads.

Introduction

Due to environmental awareness as well as public health concern, biopolymers are interesting choices as alternatives to synthetic polymers, especially in a pharmaceutical field. Chitin is an aminopolysaccharide derived from various fungi, shells of insects as well as crustacean shell wastes. It is composed of β(1→4) linked 2-acetamido-2-deoxy-β-D-glucose units (or *N*-acetylglucosamine) forming a long linear polymer chain. In the presence of the acetamide group which is similar to the amide linkage of protein in living tissues, chitin is biocompatible and safe to be utilized for human (Muzzarelli, 1997). According to many studies, chitin was found to enhance wound healing acceleration (Minagawa et al., 2007). It promoted cell attachment and spreading of normal human keratinocytes as well as fibroblasts (Noh et al., 2006). Moreover, chitin can be degraded by lysozyme presented in the body, and subsequently generates harmless products (Kurita et al., 2000). To enhance the biological properties of chitin, incorporation of another natural substance that possesses distinct biological activities is interesting. Curcumin (1,7-bis(4-hydroxy-3-methoxyphenyl)-1,6-heptadiene-3,5-dione) is a naturally occurring substance that imparts the yellowish pigmentation to turmeric (*C. longa.* L.). It is widely known for the antioxidant (free radical scavenging activity), antimicrobial, and anti-inflammatory activity of curcumin. It was also observed that curcumin is capable of scavenging oxygen free radicals such as superoxide anions and hydroxyl radicals, which are the initiators of lipid peroxidation. However, curcumin has some drawbacks for clinical applications due to its poor solubility in aqueous solutions, as well as a photodegradation under light exposure. Many attempts have been done to increase the stability and safety of curcumin usage. Several biomaterials have been used as matrixes to provide a slow release of curcumin such as cellulose acetate (Suwantong et al., 2007) and collagen (Gopinath et al., 2004). However, there is no publication reporting the release behavior of curcumin from the chitin matrix due to high rigidity, poor solubility, and low processability of chitin. In this study, the curcumin-incorporated porous chitin beads with a non-porous chitin shell were successfully fabricated by a coagulation method. Curcumin was incorporated into the chitin bead by the addition of the surfactant, Tween20. The effects of curcumin incorporation as well as freezing condition on the beads' porous structure were investigated. The releasing behavior of curcumin from the chitin matrix was evaluated by the total immersion method in an acetate buffer solution simulating the human skin condition of 5.5 at the physiological temperature of 37°C.

Experimental

A. Chitin flakes preparation

Firstly, decalcification was conducted by immersing squid pen flakes in 1 M hydrochloric acid (HCl) for 2 days. The squid pen flakes were then separated and treated in a deproteinization process by immersing in 4% (w/w) sodium hydroxide (NaOH) solution for 24 h. The mixture was heated on a hot plate at 80°C for 4 h in order to obtain pure chitin flakes. Finally, the chitin flakes were then washed and dried in an oven at 60°C for 24 h.

B. Fabrication of chitin and freeze-dried chitin beads

A known amount of chitin flakes was suspended in concentrated formic acid (98%). The suspension was then frozen overnight at 0°C. After freeze-thawed at room temperature, chitin-formic acid solution with 0.007 g/ml chitin concentration was obtained. The chitin-formic acid solution underwent a coagulation process (Yilmaz and Bengisu, 2003) where the solution was simply dropped into acetone under magnetic stirring at 250 rpm to obtain homogeneous spherical chitin beads. The chitin solution with 1:10 volume ratio of chitin to acetone was maintained throughout the experiment. After keeping the acetone-suspended chitin beads at room temperature for 24 h so as to allow beads formation and solvent exchange, acetone was then replaced by the same amount of distilled water. The medium was changed everyday until neutral pH. The beads were then frozen by liquid nitrogen. After freezing the beads at a constant freezing rate, the solidified sample was freeze-dried at 263K for 3 days under a vacuum condition to sublimate the ice in the chitin beads.

C. Preparation of curcumin-incorporated chitin beads

Curcumin in various initial amounts (1.6, 2.8 and 4.0 wt% based on the volume of curcumin aqueous solution) was added after the chitin beads formation before freeze-drying. Acetone was replaced by the curcumin aqueous solution, in the presence of the surfactant, Tween20 (2%v/v). The mixture was left at room temperature under a mild stirring for 24 h to incorporate curcumin into the chitin beads. Then, curcumin-incorporated chitin beads were separated from the mixture and re-suspended in distilled water to remove the residual solvent. After neutralization of pH was achieved, some samples were kept in a refrigerator at 4°C before freeze-drying.

D. Determination of curcumin-incorporating efficiency

The curcumin-incorporated chitin bead (~ 2 mm in diameter) was first immersed in 2 ml of ethanol for 3 days. UV-vis spectrophotometer (Shimadzu UV-2500PC) was used to measure absorbance values of ethanol- dissolved curcumin at the wavelength of 424 nm. Then, the actual amount of curcumin in the chitin bead was determined by back-calculating from the obtained data against a predetermined calibration curve of curcumin in ethanol. The incorporating efficiency of curcumin was calculated by Eq. 1:

$$\text{Curcumin-incorporating efficiency (\%)} = \frac{[\text{Actual amounts of curcumin}]}{[\text{Total initial amounts of curcumin}]} \times 100 \quad (1)$$

E. Morphology observation

The surface and cross-sectional morphology of freeze-dried chitin and curcumin-incorporated chitin beads were examined by scanning electron microscopy (Tiny-SEM 1540, Technex Lab Co. Ltd.). The freeze-dried beads were further frozen by liquid nitrogen to prevent the beads from deformation, and cut with a razor blade for SEM observation. The samples were then coated with gold-palladium *in vacuo* prior to SEM observation.

F. Porosity measurement

The apparent volumes of freeze-dried chitin and curcumin-incorporated chitin beads were determined by the solvent displacement method using toluene as a solvent (Yan et al., 2008). A flask was pre-calibrated using distilled water to find the precise volume before testing. The density of toluene determined by weighing the flask filled with toluene on an electronic balance (Shimadzu, AUW220D) was found to be 0.8614 g/cm^3. The apparent volume of the sample was calculated by Eq. 2:

$$V_s = V_f - [(M - M_f - M_s) / \rho_s] \quad (2)$$

where V_s is the apparent volume of the sample; V_f is the volume of the flask; M is the weight of the flask plus the sample and toluene; M_f is the weight of the flask; M_s is the weight of the sample; and ρ_s is the density of toluene.

Porosity (ε) is defined as the volume fraction of air in the sample. In this study, it was calculated by Eq. 3:

$$\varepsilon = 1 - (\rho_a / \rho_t) \quad (3)$$

where ρ_a is the apparent density of the sample calculated from V_s and M_s ; and ρ_t is the true density of chitin, 1.425 g.cm^{-3}.

G. Releasing study

In this study, the release characteristic of curcumin from the curcumin-incorporated chitin beads was investigated by the total immersion assay. Acetate buffer was chosen as a releasing medium simulating the human skin, pH 5.5. Each sample was immersed in 4 ml of the medium at the physiological temperature of 37°C. At a specified immersion period ranging from 0 to 7 days (168 h), 2 ml of a sample solution was dipped out and an equal amount of the fresh medium was then added back to the vial. The amount of curcumin in the sample solutions was determined using the Shimadzu UV-2500PC UV-vis spectrophotometer at the specified wavelength of 426 nm. The obtained absorbance values were back-calculated to determine the amounts of curcumin released from the samples at each immersion time by using the following equation (Eq. 4):

Release of curcumin (%) =

$$\frac{[\text{Released curcumin}]_{t=t}}{[\text{Actual amounts of curcumin}]_{t=0}} \times 100 \qquad (4)$$

where t is a specified immersion time. The experiment was conducted in triplicate and the results were reported as average values.

H. Releasing medium uptake capacity

To investigate effects of the internal porous structure and porosity on the releasing medium uptake capacity, each curcumin-incorporated chitin bead was weighed after immersion in acetate buffer, pH 5.5, at 37°C for 0-3 h. At each time point, the wet samples were taken from the buffer and weighed after removing an excess medium stained on the samples surface with a filter paper. In comparison with the dry weight of the samples prior to the immersion, the percentage of the releasing medium uptake capacity was calculated by Eq. 5:

$$\text{Uptake capacity (\%)} = [(W_1 - W_0)/W_1] \times 100 \qquad (5)$$

where W_1 and W_0 represent the weight of the samples after and prior to immersion. The experiment was repeated 3 times and the average value was taken as the percentage of uptake capacity value.

Results and Discussions

Curcumin-loading

The nonionic surfactant, Tween20, was used to enhance the solubility of curcumin in chitin during the incorporation step. With the same amount of Tween20 (2%v/v) and 24 h of incorporation process, the increase of the initial curcumin contents resulted in the higher amounts of curcumin incorporation in the chitin beads as shown in Table 1. However, the incorporating efficiency was diminished, lower than 1.0 %, and found to be decreased as the initial curcumin contents was increased. This might be due to a short available period of the incorporation time that minimized the incorporating efficiency of curcumin.

Morphology observation

Chitin and curcumin-incorporated chitin beads were successfully fabricated from the chitin-formic acid solution by a coagulation method. After removing residual solvent using distilled water, the samples were frozen by liquid nitrogen with a constant freezing rate. After freezing the sample completely, the solidified sample was freeze-dried at 263K (Fig. 1). Figure 2 shows the surface appearance as well as the cross sectional area of the freeze-dried sample.

Table 1
Curcumin incorporated in chitin beads after 24 h of the incorporation step

Initial curcumin contents (wt%)	Curcumin incorporated in chitin beads*	Incorporating efficiency (%)
1.6	60.04 ± 1.77	0.97 ± 0.03
2.8	88.29 ± 6.37	0.81 ± 0.06
4.0	100.26 ± 2.96	0.65 ± 0.02

*Weight (g) of incorporated curcumin referring to the actual curcumin content divided by weight (g) of chitin
The values represent the mean ± SD, $n = 3$

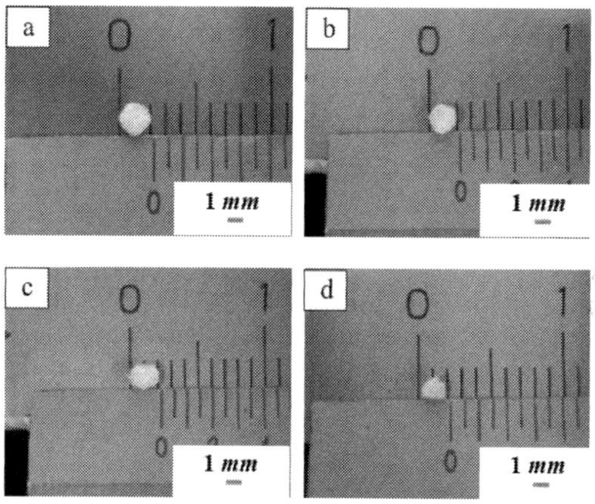

Figure 1. (a) Freeze-dried chitin bead. (b)-(d) curcumin-incorporated chitin beads, (b) 1.6, (c) 2.8, and (d) 4.0 wt% of initial curcumin contents.

Non-porous outer skin layer was observed as shown in Figs 2 a and b. The formation of skin layer and inter porous structure could be explained by a two-step of dehydration during freeze-drying process, in which the sample was first frozen and then the water was removed by sublimation. The ice sublimation created pores inside the beads, the wall of which was shrunk due to surface tension (Krokida et al., 1998). The internal porous structure shown in Figs. 2c and d provided a higher percentage of porosity to the beads, ranging from 98 to 99% (Table 2). Spider-net liked chitin microfibrils were observed within the internal porous structure as illustrated in Fig. 2d. Rinaudo (2006) reported that chitin was crystallized naturally as ordered microfibrils due to tightly binding of intra-hydrogen bonds. A rearrangement of the chitin polymers chain could occur during the freeze-drying process and resulted in a formation of the chitin microfibrils. The porous and microfibril morphology were changed by amount of curcumin incorporated into the chitin bead as shown in Fig. 3. As the initial curcumin content increased (Figs. 3b-d), the microfibrils tended to decrease. It can be considered that Tween20 and curcumin acted as impurities interrupting the intra-hydrogen bonding of the chitin polymer chains. As a result, the rearrangement of the chitin polymer chains was retarded and the numbers of the chitin microfibrils were reduced. However, this could be modified by changing the freezing condition.

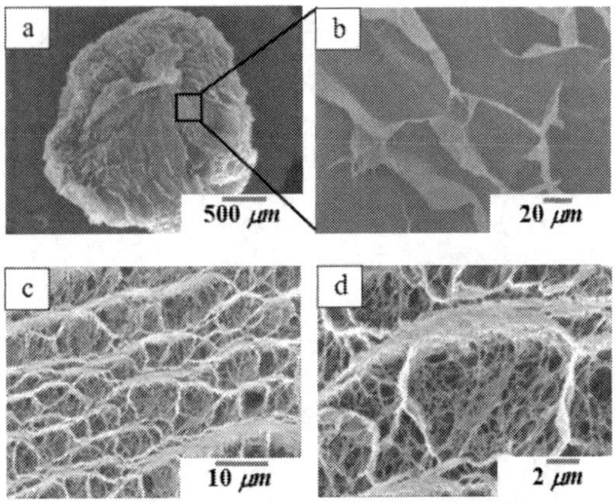

Figure 2. (a) and (b) Surface appearances of outer skin layer, (c) and (d) cross-sectional morphology of freeze-dried chitin beads indicating internal porous structure with spider-network of chitin microfibrils.

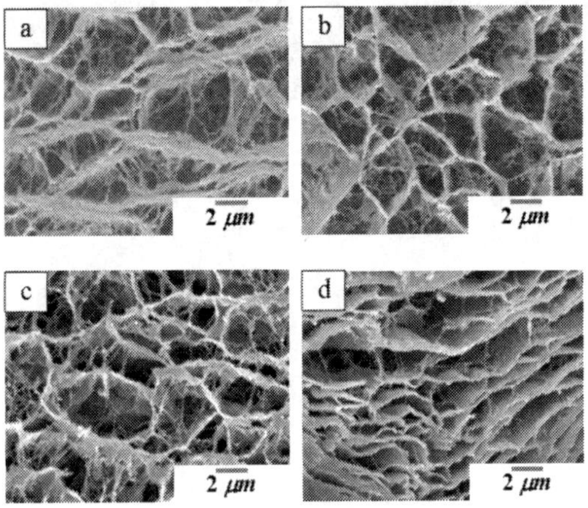

Figure 3. Cross-sectional morphology of (a) freeze-dried chitin beads (b)-(d) curcumin-incorporated chitin beads, (b) 1.6, (c) 2.8, and (d) 4.0 wt% of initial curcumin contents.

Releasing behavior of curcumin

Since the solubility of curcumin in the acetate buffer solution was limited at pH 5.5, the surfactant, Tween20, in the curcumin-incorporated chitin beads was important to control the releasing behavior of curcumin from the beads. In this study, the amount of curcumin released from the beads was evaluated by the percentage of the amounts of curcumin initially contained in the beads. Fig. 4 shows the releasing profiles of curcumin from the curcumin-incorporated chitin beads with three different initial curcumin contents 1.6, 2.8, and 4.0 wt%. During the initial 30 min, a remarkable increase in the amount of releasing curcumin was observed for all samples (Fig. 4a). The higher releasing rate at the beginning of releasing test was caused by the curcumin attached on the surface of chitin beads. The incorporated curcumin was solubilized with the help of Tween 20 incorporated into the beads together with curcumin. A slower releasing rate was observed one hour after releasing started and it was prolonged for more than 7 days (Fig. 4b). The slow releasing behavior was determined by the mass transfer of curcumin in chitin matrix, i.e., diffusivity and concentration of curcumin in the beads. The uptake capacity of releasing medium within the curcumin-incorporated chitin beads was shown in Fig. 5. It was found that the freeze-dried samples could rapidly absorb the releasing medium by the first 10 min due to the high percentage of porosity. After 30 min, the uptake capacity was leveled out, which clarifies a localized time of the releasing medium within the internal porous structure. Hence, during the first 10 minutes the uptake of the releasing medium occurred and then curcumin entrapped with Tween20 was carried and diffused out through the non-porous outer skin layer to the exterior releasing medium. It could be considered that the internal porous structure increased the uptake of the releasing medium while prevented the curcumin from degradation in the human body at the physiological temperature of 37°C. As illustrated in Fig. 4a and b, the amount of releasing as well as the releasing rate decreased with increasing of the initial curcumin contents in the beads. This could be explained by the excess amount of Tween20 added during the incorporating step. At lower initial contents, curcumin could be entrapped with Tween20 in a micelle form and diffused out through the non-porous outer skin layer to the exterior releasing medium as mentioned above. At higher amounts of incorporated curcumin, the amount of Tween20 available to form micelles became lower. This causes the decrease in releasing amounts of curcumin from the chitin beads. The higher porosity of the beads leads to the larger uptake capacity of releasing medium but does not lead to the higher releasing rate. Thus, it could be said that the main factor of controlling the releasing profile of curcumin from the curcumin-incorporated chitin beads was the non-porous outer skin layer as well as the concentration of incorporated Tween20.

Table 2 Percentages of porosity of freeze-dried chitin and curcumin-incorporated chitin beads

Initial curcumin contents (wt% based on the volume of curcumin aqueous solution)	Porosity (%)
0*	98.92 ± 0.71
1.6	98.70 ± 1.09
2.8	98.88 ± 0.29
4.0	99.00 ± 0.85

*Freeze-dried chitin beads
The values represent the mean ± SD, $n = 3$

Conclusion

In this study, chitin and curcumin-incorporated porous chitin bead were successfully fabricated by a coagulation method. The solubility and release profile of curcumin could be controlled by the amount of Tween20 during the incorporation step. The amount of curcumin in chitin beads was controlled by the initial curcumin concentration in the incorporation step as well. At this stage, the incorporating efficiency of curcumin was lowered; however, this could be improved by prolonging the incorporation time or promoting a water treatment after the chitin beads formation step. Higher amounts of water inside the beads might enhance the incorporating efficiency of curcumin entrapped with a micelle form of Tween20 in the aqueous solution. By freeze-drying, the spider-net liked chitin microfibrils could be created within the internal porous structure. Tween20 and curcumin could act as impurity for crystallization of chitin and retard, the formation of the chitin microfibrils. The internal pores and the outer non-porous shell structure could control the curcumin releasing behavior: the non-porous outer skin layer could dominate the releasing behavior of curcumin while the internal porous structure could dramatically increase the uptake of the releasing medium and control the degradation rate of beads.

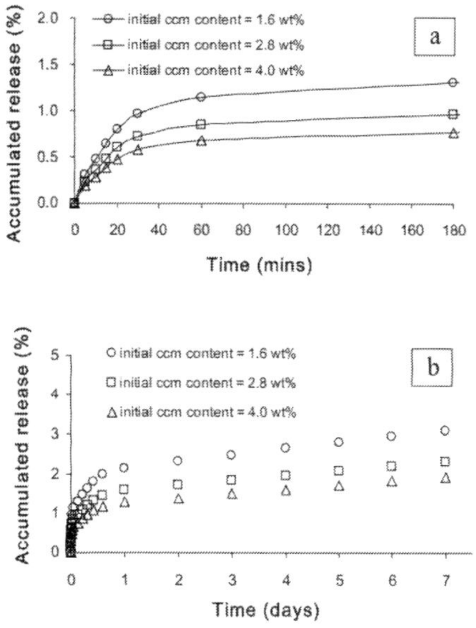

Figure 4. (a) Release profiles of curcumin from freeze-dried curcumin-incorporated chitin beads during the first 3 h and (b) prolonged for 7 days.

References

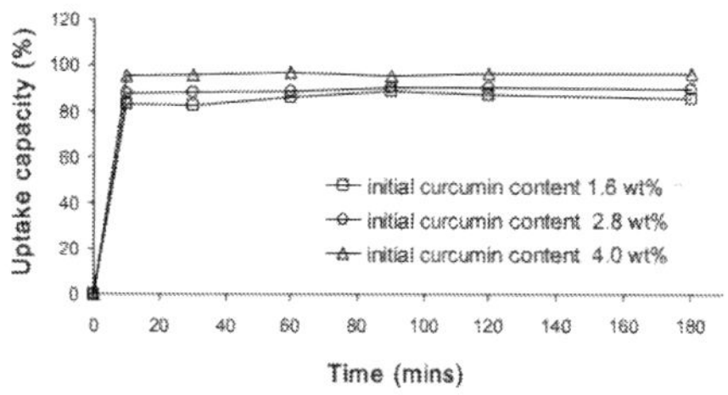

Figure 5. Releasing medium uptake capacity of curcumin-incorporated chitin beads immersed in acetate buffer, pH 5.5, at 37 °C.

Gopinath, D., Rafiuddin Ahmed, M., Gomathi K., Chitra K., Sehgal. P.K., R. Jayakumar. R.,"Dermal wound healing processes with curcumin incorporated collagen films", Biomaterials., **25**, 1911-1917 (2004)

Krokida, M.K., Karathanos, V. T., Maroulis, Z. B.,"Effect of Freeze-drying Conditions on Shrinkage and Porosity of Dehydrated Agricultural Products", Journal of Food Engineering., **35**, 369-380 (1998)

Kurita, K., Kaji, Y., Mori, T., Nishiyama, Y.,"Enzymatic degradation of β-chitin: susceptibility and the influence of deacetylation", Carbohydrate Polymers., **42**, 19–21 (2000)

Minagawa, T., Okamura, Y., Shigemasa Y., Minami, S. Okamoto, Y.,"Effects of molecular weight and deacetylation degree of chitin/chitosan on wound healing", Carbohydrate Polymers., **67**, 640–644 (2007)

Muzzarelli, R. A. A.: Depolymerization of chitins and chitosans with hemicellulase, lysozyme, papain and lipase, R. A. A. Muzzarelli, & M. G. Peter (Eds.), Chitin hand book, Italy (1997)

Noh, H.K., Lee, S.W., Kim, J.M., Oh, J.E., Kim, K.H., Chung, C.P., Choi, S.C., Park, W.H., Min, B.M.,"Electrospinning of chitin nanofibers: Degradation behavior and cellular response to normal human keratinocytes and fibroblasts", Biomaterials., **27**, 3934–3944 (2006)

Rinaudo, M.,"Chitin and chitosan: Properties and applications", Prog. Polym. Sci., **31**, 603–632 (2006)

Suwantong, O., Opanasopit, P., Ruktanonchai, U., Supaphol. P.,"Electrospun cellulose acetate fiber mats containing curcumin and release characteristic of the herbal substance", Polymer., **48**, 7546-7557 (2007)

Yan, Z., Sousa-Gallagher, M., Oliveira, F. A.R., "Shrinkage and porosity of banana, pineapple and mango slices during air-drying", Journal of Food Engineering., **84**, 430–440 (2008)

Yilmaz, E., Bengisu, M.,"Preparation a characterization of physical gels and beads from chitin solutions", Carbohydrate Polymers., **54**, 479–488 (2003)

INFLUENCE OF OPERATIONAL PARAMETERS ON CHARACTERISTICS OF ULTRAFINE CHITOSAN MICROPARTICLES AS A NASAL DRUG DELIVERY SYSTEM

M. Rafienia[1], S. Taranejoo[2], S.A. Monemian[2], M. Janmaleki[1], A. Ramazani Saadatabadi[3]* and A. Kalativahid[2]

[1]*Isfahan University of Medical Science - Isfahan – Iran*

[2]*Nanomedicine and Tissue Engineering Research Center - Shahid Beheshti University - Tehran – Iran*

[3]*Department of Chemical and Petroleum Engineering, Sharif University of Technology, - Tehran – Iran, Email: ramazani@sharif.edu*

Abstract - In this study, chitosan microparticles were prepared with tripolyphosphate (TPP) via spray dry method. For the preparation of the ultrafine chitosan microparticles, as a nasal drug delivery system, sonication was applied in very dilute chitosan solution during crosslinking. The applied sonication power was 15, 45 or 90watt. The microparticles were characterized by scanning electron microscopy (SEM) and optical microscopy. The increasing of sonication power results change in microparticles morphology from spherical shape and smooth surface to irregular pseudo-spherical shape and wrinkled surface. Decrease of crosslinking time from 60 to 30min has a similar effect on the surface morpholgy of the chitosan microparticles. The prepared chitosan microparticles were extremely small and the mean particle sizes varied between 1.19 to 1.32μm.

Introduction

Chitosan, a cationic polymer derived from chitin by deacetylation, is utilized in the medical and pharmaceutical fields such as drug delivery systems and wound dressing because of its notable features i.e. nontoxicity, mucoadhesion, biocompatibility and biodegradability. It has been reported that the chitosan microspheres significantly enhanced the systemic and/or mucosal immune responses after mucosal vaccination. (Jiang et al., 2004; Kuo et al., 2004; Taranejoo et al., 2009).

Spray-drying is the usual method to produce small chitosan microspheres in size from a few microns to several tens of microns with a relatively wide range of variation in diameter and sphericity (Kuo et al., 2009). For intranasal administration of vaccine, particles with an aerodynamic size between ca. 1 and 3μm should be used (Patton et al., 2004).

In this study, we prepared ultra fine chitosan microparticles with the spherical or pseudo-spherical shape via spray dry method and ultrasonication technique and then investigated the influence of operational conditions such as sonication power and crosslinking time on the microparticles characteristics.

Experimental

Materials

Medium molecular weight chitosan was obtained from Sigma-Aldrich Chemie (Steinheim, Germany). The degree of deacetylation of the chitosan was about 80%. TPP and acetic acid was supplied from Merck Co.

Preparation of chitosan microparticles

Four samples were prepared by spry drying methods. Formulations and operational condition of these samples are presented in Table 1.

Medium molecular weight chitosan is dissolved in 2% acetic acid solution. TPP is dissolved in water in order to make 1wt% TPP solution and pH adjusted 4.0. 20cc of TPP solution was added to 200cc of chitosan solution under the sonication.

The suspension of crosslinked chitosan were then spray dried (Buchi® Mini spray drier, type 190, Switzerland) through a 0.2mm nozzle at a feed rate 2.4ml/min. The driving pressure was 8bar. Temperature was maintained at 120°C.

Characterization

The spray-dried microparticles were studied using scanning electron microscopy (SEM). Samples were coated by a thin gold layer observed with microscope (25kV) (Philips XL30).

The particle size was determined by optical microscopy as described by Goud et al. (2005). Briefly, about 1mg of chitosan–TPP microparticles were taken on a glass slide and sizes of about 300 particles were measured by the optical microscope (Nikon TE 2000-J). Zeta-potential of the microparticles prepared was determined by a Zetasizer Nano-ZS-90 (Malvern Instruments) in 0.1mM KCl solution (pH =7.4) at 25°C and automatic mode.

To be sure about acuuracy of experiments, all of the experiments were performed in triplicate (n=3).

Table 1 - Formulation and operational condition of chitosan microparticles preparation

Sample code	Crosslinking time (min)	Sonication power (W)	Zeta potential (mV)	Mean particle size (µm) ± SD
Sample1	60	45	26.5 ± 1.4	1.30 ± 0.47
Sample2	60	90	24.3 ± 1.3	1.19 ± 0.26
Sample3	60	15	25.6 ± 1.7	1.32 ± 0.7
Sample4	30	45	27.8 ± 1.2	1.23 ± 0.4

Statistical analysis

Statistical analysis was performed using the computer based Excel program version 2007. All results were expressed as mean ± standard deviation (SD). Statistical differences between the groups were analyzed with Student's t-test. Differences were considered significant if probability values of $p < 0.05$ were obtained.

Results and Discussion

Figure 1 shows SEM microphotographs of the chitosan microspheres. It can be seen from these images that the microparticles shape and surface morphology vary from spherical shape and smooth surface to irregular pseudo-spherical shape and wrinkled surface with increasing of sonication power.

It seems that sonication as a high energetic process declines microspheres wall stability.

On the other hand, when the crosslinking time decreased from 60 to 30min, sphericity and smoothness of the chitosan microparticles dramatically declined.

This result was similar to observation of Wang et al. (2006). They reported when the crosslinking time of chitosan microparticles by glutaraldehyde was less than 60 min, the reaction of chitosan functional groups and glutaraldehyde was not enough to form a stable gel structure. Therefore, chitosan microparticle could not be formed.

A high decline in crosslinking degree of chitosan due to a decrease in cross-linking time caused chitosan microparticles to lose their ability to remain the spherical structure. It seems that one of the effective parameters on surface morphology is stability of microparticle structure during spray dry process. Spray dry is a kind of solvent evaporation process. The solvent evaporation rate has a significant effect on microspheres surface morphology.

Figure 1 - Scanning electron micrographs of TPP cross-linked chitosan microspheres prepared from: (a) sample 1, (b) sample 2, (c) sample 3 and (d) sample 4.

With decreasing crosslinking time, the degree of chitosan crosslinking dramatically decreased that resulted more compact chitosan microparticles and enhanced the solvent evaporation rate. Depression on the surface may be due to the fast evaporation of water or high shear generated during the spray drying process. The results of Çalış et al. (2000) are in agree with this hypothesis; They found that enhancement of the solvent evaporation rate caused the formation of porous and non-spherical BSA microspheres with wrinkled surface.

The surface charge (zeta potential) and mean particles sizes of the microparticles are tabulated in Table 1. As it can be seen, the ultrafine chitosan microparticles with mean particle size of 1.09 ± 0.34 to 1.46 ±0.64μm were prepared using very low concentration of chitosan and employing ultrasonication technique.

These ranges of particle size are very good for intranasal administration of vaccine. The main factors limiting pulmonary antigen delivery are poor deposition of the antigen at the alveolar area, low absorption from the epithelial barriers in the peripheral airways and the central lungs and the presence of a mucociliary escalator in the central and upper lung, which rapidly removes antigens or particles from the central respiratory tract (Patton et al., 2004).

Conclusions

This study demonstrated the influence of sonication power and crosslinking time as operational parameters of chitosan microparticles preparation on the characteristics of the microparticles. The obtained ultrafine chitosan microparticles which are useful for intranasal administration could be achieved by utilizing sonication technique. Furthermore, it was observed that operational parameters affected the microparticles morphology, significantly.

Acknowledgements

The authors are grateful to Dr. Peirovi, the chief of Nanomedicine and Tissue Engineering Research Center, for his sincere support and guidance.

References

Jiang. H. L, Park. I. K, Shin. N. R, Yoo. H. S., Akaike. T., Cho. C. S.,"Controlled release of Bordetella bronchiseptica dermonecrotoxin (BBD) vaccine from BBD-loaded chitosan microspheres In Vitro", Arch. Pharm. Res, **27**, 346-350 (2004)

Kuo. S. M, Niu. G. C. C. , Chang. S. J. , Kuo. C. H. , Bair. M. S., "A one-step method for fabricating chitosan microspheres ", J. Appl. Polym. Sci. , **94**, 2150-2157 (2004)

Taranejoo, S., et al. "Effect of polymer molecular weight on morphology and particle size of chitosan microspheres prepared via spray drying method", IFMBE Proceedings, Papers, 251-252 (2009)

Patton. S. J, et al.,"The Lungs as a Portal of Entry for Systemic Drug Delivery" Proc Am Thorac Soc. Papers, 338-44 (2004)

Wang. L.Y., Gu Y.H., Zhou. Q.Z., Ma. G.H., Wan. Y.H., Su. Z.G., "Preparation and characterization of uniform-sized chitosan microspheres containing insulin by membrane emulsification and a two-step solidification process", Colloid Surf. B Biointerfaces **50**, 126-135 (2006)

Çalış. S., Bozdağ. S., Kaş. S., Hıncal. A.A., "Formulation and characterization of Albumin microspheres containing Naproxen sodium", J. Control Release **64**, 269-347 (2000)

Polymer Surface Manipulation for reduced thrombogenicity

Subbu Venkatraman (1), Li Buay Koh (1), Isabel Rodriguez (2)*

(1) Nanyang Technological University - Singapore - Singapore, (2) Institute of Materials Research and Engineering - Singapore - Singapore

Strategies for improving the non-thrombogenicity of implanted polymer surfaces are reviewed. Such strategies include chemical modification, physical modification, heparin immobilization and coatings. In our recent work we have focussed on nano-topographical modification of biodegradable polymer surfaces, using templating and lithographic techniques, as well as by incorporation of multi-walled carbon nanotubes. The templating involves the use of anodized alumina modified with colloidal silica; this is followed by e-beam or deep UV lithography, and then by inductively-coupled plasma-deep reactive ion etching (ICP-DRIE). PLGA solution is then cast on the template, dried and peeled off to create physical pillars of specified dimensions (diameter, spacing and height). These modified surfaces are then incubated with platelets to quantify platelt attachment and quantification with respect to unmodified surfaces of the same polymer. Results show that the surface features influence the attachment of plasma proteins, and hence the eventual adhesion of platelets. Both platelet attachemnt density and activation are quantified for various surfaces with pillars of differing aspect ratio. Our work shows that simple topographical modification of a thrombogenic surface, such as that of a biodegradable polymer (PLGA) is sufficient to decrease platelet adhesion and activation (1). We rationalize the behaviour in terms of fibrinogen adsorption on these surfaces. We believe that such surface modification methods may lead to reduced clot formation on blood-contacting devices. References: 1. Li Buay Koh, Isabel Rodriguez, Subbu Venkatraman, Effect of topography of polymer surfaces on platelet adhesion, Biomaterials (2009) http://dx.doi.org/10.1016/j.biomaterials.2009.11.022

COLLAGEN SURFACE MODIFICATION BY MEANS OF PLASMA TREATMENT

Jorge López García[1*], Jiří Pacherník[2], Marián Lehocký[3], Ita Junkar[4], Petr Humpolíček[3], and Petr Sáha[1]
*** Author to whom correspondence should be addressed: Email vextropk@gmail.com**

[1]*Polymer Centre, Faculty of Technology, Tomas Bata University in Zlín, T.G.M Sq 275, 76272, Zlín, Czech Republic - vextropk@gmail.com; saha@utb.cz*

[2]*Institute of Experimental Biology, Faculty of Sciences, Masaryk University Brno, Kotlářska 2, 61137, Brno, Czech Republic - jipa@sci.muni.cz*

[3]*Tomas Bata University in Zlín, T.G.M Sq 5555, 76001, Zlín, Czech Republic - lehocky@post.cz; humpolicek@uni.utb.cz*

[4]*Department of Surface Engineering, Plasma Laboratory, Jožef Stefan Institute, Jamova cesta 39, SI-1000, Ljubljana, Slovenia - ita.junkar@ijs.si*

Abstract – Collagen thin films (Atelocollagen from bovine Achilles tendon) were prepared and their surfaces were modified using argon, nitrogen and air plasma treatment. The untreated and treated samples were characterized by employing surface probe techniques. Moreover, human immortalized keratinocyte (HaCaT) cells were seeded upon the films and their cell proliferation was evaluated by using (MTT) assay test. The surface and biological data confirmed chemical and physical changes on the treated surfaces as well as treated films showed better conditions for cell growth than the untreated ones; thereby this endeavour may be considered as a potentially promising approach for tissue regeneration purposes.

Introduction

Collagen is a fibrillar protein that is present in almost all mammalian tissues, which makes up about 25% of whole-body protein content. Its abundance is particularly centred in connective tissues such as tendons, ligaments and cartilage. Skin also contains this protein. Furthermore, it is involved in important biological functions such as tissue formation and cell attachment (Prockop and Kivirikko, 1995). For these reasons, collagen is widely used in the design of materials with biomedical applications. In this regard plasma treatment is effective and economical tool with a growing interest in biomedical engineering, since this modification is able to enhance surface properties and biocompatibility without affecting bulk characteristics. Likewise, it does not require relatively expensive devices for its operation, it is environmental friendly and depending on the carrier gas can be produced unique surface properties which can be approached for different applications (Chu et al, 2002; Coen et al, 2003 and Desmet et al, 2009). In this study, atelocollagen films were prepared and their surfaces were modified by using argon, nitrogen and air plasma treatment. Both untreated and treated samples were characterized by attenuated total reflection Fourier transform infrared (ATR-FTIR) spectroscopy, scanning electron microscopy imaging (SEM), and X-ray photoelectron spectroscopy (XPS). In addition, HaCaT cells growth was carried out on both samples and the cell proliferation was measured by MTT assay. The data disclosed that the treated samples possess better conditions for cell growth than the untreated ones owing to chemical and physical changes undergone by the treated surfaces; for instance, the incorporation of chemical functionalities such as carbonyl, carboxyl and C-O-C linkages. As for physical changes, SEM images showed that surface roughness was noticeable raised after each plasma treatment. It was observed that by means of these plasma treatments significantly enhanced the cell proliferation extent.

Experimental

Preparations of collagen films
Atelocollagen (Vipo A.S, Slovakia) was solubilized in 0.1M acetic acid to prepare a 0.1 % w/w solution. Then, 2mL of this solution was poured into tissue culture dishes. Thereafter, the solvent was evaporated at ambient conditions for three days.

Plasma treatment
The plasma treatment of these collagen thin films was carried out by using plasmochemical reactor (Femto, Diener electronic, Germany) operated at frequency of 40 kHz, pressure 40 Pa, and power input of 50 W. Argon, nitrogen and air were used as carrier gases. The time of treatment was 5 minutes for each sample.

Spectroscopic and microscopic techniques
Surface chemical composition of both untreated and treated collagen films were evaluated by XPS

(microprobe instrument PHI Versaprobe, Physical Electronics, USA). The samples were excited by X-rays over a 400-µm spot area with a monochromatic Al $K_{\alpha 1,2}$ radiation at 1486.6 eV. The emitted photoelectrons were detected by a hemispherical analyzer positioned at an angle of 45°. Survey-scan spectra were obtained at a pass energy of 187.85 and 0.5 eV step resolution. The elemental concentration analysis was performed by MultiPak v7.3.1 supplied with the spectrometer. ATR-FTIR spectra were collected at a FTIR spectrometer Avatar 320 (Nicolet, USA) equipped with ZnSe Crystal at an incident angle of 45°. Each spectrum was obtained by recording 32 scans having a 2 cm^{-1} resolution. The spectral range was within 4000–650 cm^{-1}. The surface morphology was analyzed on a Vega LMV microscope (Tescan s.r.o, Czech Republic) operated in the high vacuum/secondary electron imaging mode at 5 kV. The samples were coated with a thin layer of Au/Pd alloy and tilted 30° in order to get higher resolution and observation of the surface topography. The images were taken at 10,000x magnification.

Cell culture and proliferation
Human immortalized non-tumorigenic keratinocyte HaCaT cells (Cell Lines Service, Germany) were seeded onto the treated and untreated samples in the culture dishes and incubated at 37 °C for 4 days (Boukamp et al, 1988). DMEM (high glucose) supplemented with 2mM L-glutamine solution and 10% fetal calf serum was used as the culture medium. Cell proliferation was determined after 4 days of culturing by MTT assay (Vybrant® MTT cell proliferation Assay kit, Invitrogen, USA) by measuring formazan concentration at 570nm.

Results and Discussion

Surface spectroscopic analysis
The surface atomic compositions along with the corresponding atomic ratios and the acquired survey spectra of the samples are given in table 1 and figure 1 respectively. The data shows a considerable raise in the oxygen content subsequent to the plasma treatments (mainly with argon and air). This increment is a consequence of surface oxidations which are indeed stimulated by these treatments. Plasma treatment brings reactivity to the samples introducing free radicals which can survive for several days and react with the oxygen and other reactive species from the atmosphere (France and Short, 1997). On the other hand, the N/C ratio was increased after Nitrogen plasma treatment which can be explained by the incorporation of nitro or any nitrogen-containing functionality.

Table 1- Elemental compositions of the untreated and plasma treated films by XPS measurement.

Sample	C1s%	N1s%	O1s%	N/C	O/C
Untreated	66.0	14.1	19.3	0.21	0.29
Ar treatment	58.1	11.3	25.0	0.19	0.43
N_2 treatment	58.2	18.4	22.7	0.32	0.39
Air treatment	58.1	11.3	25.9	0.20	0.46

Figure 1- XPS survey-scan spectra of: (A) Untreated, (B) argon, (C) nitrogen and (D) air plasma treated films.

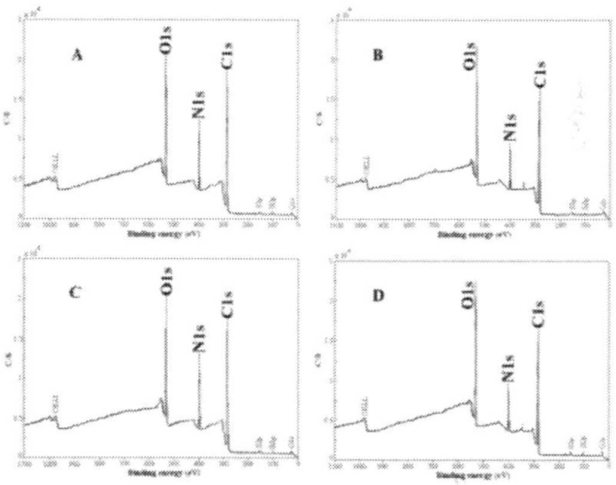

The ATR-FTIR spectra from untreated and treated sample are presented in figure 2. Amide characteristic such as N-H stretching, Amide I, Amide II, Amide III and N-H out of plane can be identified. After plasma treatment, the characteristic amide bands shifts down indicating of an alteration in the surface chemistry. The absorption bands corresponding to aliphatic C-H stretching shift and change in strength. Besides, visible changes at the bands assigned to C-O stretching can be observed. Thus; it was found out that plasma treatment is capable of affecting the surface chemistry via ablation and fictionalization (Shi et al, 2006).

Figure 2 – ATR-FTIR spectra of: (A) Untreated, (B) argon, (C) nitrogen and (D) air plasma treated films.

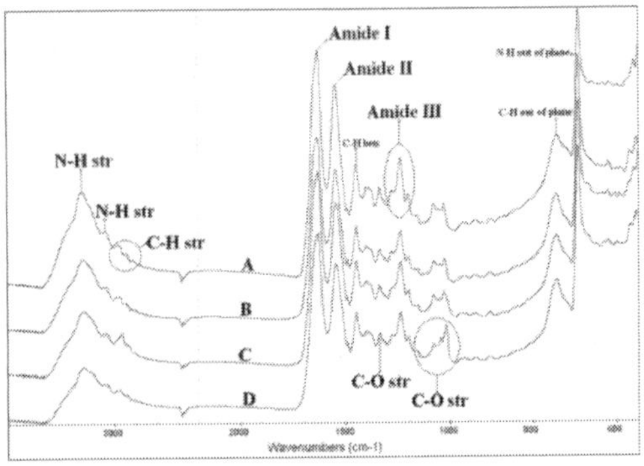

Microscopic techniques

The SEM images (Figure 3) revealed significant changes on surface morphology after each plasma treatment. Treated samples show relatively rougher surface topography compared to untreated specimen which is a result of surface etching and functionalization.

Figure 3- SEM micrographs of: (A) Untreated, (B) argon, (C) nitrogen and (D) air plasma treated films at 10,000 x magnification.

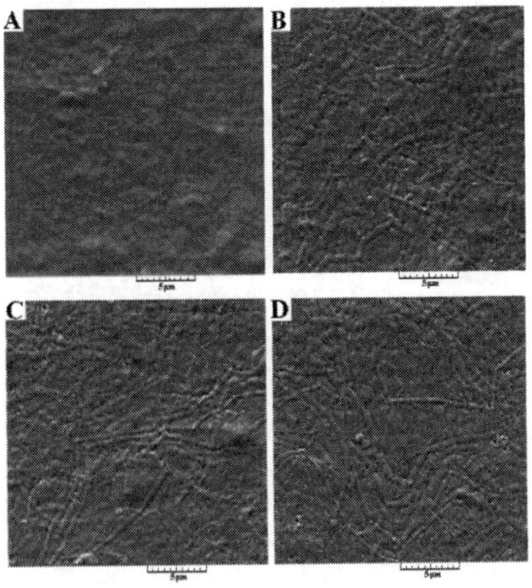

HaCaT cell growth upon collagen films

HaCaT cells behaviour on untreated and treated films evaluated by MTT assay is shown in figure 4. Cell proliferation was approximately 1.6, 1.5 and 1.49 times higher for argon, nitrogen and air plasma treated samples respectively than untreated films. Despite the mechanisms of the HaCaT adhesion and proliferation upon different substrates are still unclear, it is well known that roughness and porosity of films influence cell adhesion and cell proliferation. In addition to surface polar entities content because HaCaT cells are mainly attached by carbonyl and carboxyl groups; therefore, their cell growth tends to be favoured in hydrophilic surfaces (Ishaug-Riley et al, 1999; Lehocký et al, 2009).

Figure 4- Histograms of HaCaT cell proliferation measured by MTT assay (A) Untreated, (B) argon, (C) nitrogen and (D) air plasma treated films (the error bars depict standard deviations).

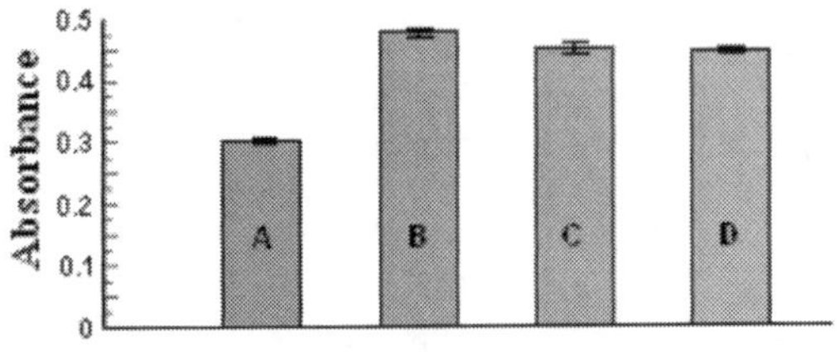

Conclusions

The enhancement of HaCaT cells attachment is most likely due to the increase of oxygen-containing functional groups and roughness. Although argon treated films showed the best cell growth, these results suggest that all of these plasma treated atecollagen films are biocompatible materials and might be used for tissue regeneration applications.

Acknowledgements

The Authors would like to express their gratitude for financing of this research to the Ministry of Education, Youth and Sport of the Czech Republic (grant VZ MSM 7088352101), and the Slovenia Ministry of Higher Education, Science and Technology (Program P2-0082-2, Thin Film and Plasma Surface Engineering).

References

Boukamp, P., et al.,"Normal Keratinization in a Spontaneously Immortalized Aneuploid Keratinocyte Cell Line", J. Cell Biol., **106,** 761-771 (1998)

Chu, P.K., et al.,"Plasma-surface Modification of Biomaterials", Mater. Sci. Eng., **R36,** 143-206 (2002)

Coen, M.C., et al.,"Modification of the Micro and Nanotopography of Several Polymers by Plasma Treatments", Appl. Surf. Sci., **207,** 276-286 (2003)

Desmet, T., et al.,"Nonthermal Plasma Technology as a Versatile Strategy for Polymeric Biomaterials Surface Modification: A Review", Biomacromolecules, **10,** 2351-2378 (2009)

France, R.M., Short, R.D.,"Effects of Energy Transfer from an Argon Plasma on the Surface Chemistry of Poly(styrene), Low Density Poly(ethylene),

Poly(propylene) and Poly(ethylene terephthalate)", J. Chem. Soc., Faraday Trans., **93,** 3173-3178 (1997)

Ishaug-Riley, S.L., et al., "Human Articular Chondrocyte Adhesion and Proliferation on Synthetic Biodegradable Polymer Films", Biomaterials, **20,** 2245-2256 (1999)

Lehocký, M., et al., "Adhesion of Rhodococcus sp S3E2 and Rhodococcus sp S3E3 to Plasma Prepared Teflon-like and Organosilicon Surfaces", J. Mater. Process. Technol., **209,** 2871-2875 (2009)

Prockop, K.L., Kivirikko, K.I., "Collagens: Molecular Biology, Diseases and Potentials for Therapy", Annu. Rev. Biochem., **23,** 403-434 (1995)

Shi, L.-S., et al., "The Investigation of Argon Plasma Surface Modification to Polyethylene: Quantitative ATR-FTIR Spectroscopic Analysis", Eur. Polym. J., **42**, 1625–1633 (2006)

DUAL FUNCTIONAL IONIC LIQUIDS AS PLASTICIZERS AND ANTIMICROBIAL AGENTS FOR MEDICAL POLYMERS

S. Y. Choi[1]*, H. Rodriguez[2], D. Gilpin[3], S, McGrath[3], R. D. Rogers[4], K. Malcolm[3], M.Tunney[3], and T. McNally[1]*

[1] *School of Mechanical and Aerospace Engineering, Queen's Univeristy Belfast, United Kingdom – schoi03@qub.ac.uk; t.mcnally@qub.ac.uk*
[2] *Department of Chemical Engineering, University of Santiago de Compostela, Santiago de Compostela, Spain – hector.rodriguez@usc.es, rdrogers@as.ua.edu*
[3] *School of Pharmacy, Queen's University Belfast, United Kingdom – d.f.gilpin@qub.ac.uk, s.mcgrath@qub.ac.uk, k.malcolm@qub.ac.uk, m.tunney@qub.ac.uk*
[4] *Department of Chemistry, The University of Alabama, Tuscaloosa, Alabama 35487, USA – rdrogers@as.ua.edu*

Composites of poly (vinyl chloride) (PVC) and two ionic liquids, 1-ethylpyridinium docusate (IL1) and tributyl(2-hydroxyethyl) phosphonium docusate (IL2) at loadings of 0, 10, 20 and 30wt.%, were prepared by melt extrusion using a twin screw extruder. Plasticization manifest by a reduction in glass transition temperature (T_g) of the composites was investigated using dynamic mechanical thermal analysis (DMTA), and melt viscosity monitored with changes in torque values from on-line data collection during processing. The T_g of a highly TiO_2 filled PVC was reduced by up to 46°C on addition of 30wt% IL2. The antimicrobial activity of the PVC-IL composites was also determined using disc diffusion testing on 24 strains of microbes, including pathogens and biofilm-forming bacteria, followed by biofilm formation testing on selected samples. Addition of both ILs to PVC inhibited growth of a wide range of bacteria, most notably MRSA and significantly hindered biofilm formation. Thermogravimetric analysis (TGA) was performed to determine the thermal stability of PVC and PVC-IL composite materials. IL addition did not have a significant effect on the thermal stability of PVC.

Introduction

PVC is a commonly used medical polymer which accounts for more than 25% of plastics used in medical devices, such as in lightweight non-breakable containers or bags, catheters and endotracheal tubing, gloves and other protective devices due to its biocompatibility, superior flexibility and clarity (Wisner et al, 2005; Fanelli and Zuccato, 2002). However, phthalates, a family of plasticizers commonly used to soften PVC, have been reported to have potential reproductive and developmental toxicity, which makes preterm infants and dialysis patients particularly vulnerable upon exposure of phthalates (Matsumoto et al, 2008), although this continues to be a matter of great debate as a consequence of conflicting results from a number of different studies into this topic. PVC soft tubing also faces the problem, for example, of catheter-related infections and biofilm formation. Biofilm formation in PVC medical tubing is mainly caused by coagulase-negative Staphylococci, *Staphylococcus epidermidis*, which can result in implant-failures and as such increased healthcare expenditure (Roth et al, 1988). To address these problems, an alternative plasticizer with antimicrobial effect is required.

Ionic liquids (ILs) are a new class of non-molecular organic salts with very low melting points which can be liquids at room temperature. ILs consist of two types of ions, that is, positively-charged cations and negatively charged anions (Kubisa, 2005; Wasserscheid and Keim, 2000; Zhao, 2006). They have been regarded as 'green' due to their non-volatility compared to traditional volatile organic compounds (VOC), therefore they are considered non-contaminating to the environment. ILs are widely used in many applications, such as heat transfer fluids, azeotrope-breaking liquids, lubricants, electrolytes, liquid crystal supported IL membranes. ILs have a range of properties including low melting point, good chemical and thermal stability, ion conductivity, heat capacity, thermal conductivity, non-flammability and a wide electrochemical potential window (Zhao, 2006). Due to their ionic nature, IL's are usually highly polar allowing them to dissolve many organic, inorganic and metallo-organic compounds. The greatest advantage of ILs is versatility, as their properties may be adjusted by tailoring the cation and anion combination (Kubisa, 2005).

In a mixture of PVC and plasticizer, the polarity and moderate solubility parameter of PVC allows miscibility with many polar liquids and plasticization, permitting its stiffness to be tailored to different levels of flexibility (Wisner et al, 2005). The variable polarity of ILs makes them suitable plasticizers for polymers. Early work in developing ILs as plasticizers for polymers was demonstrated by Scott *et al.* using 1-butyl-3-methylimidazolium hexafluorophosphate [C4mim][PF6] as a plasticizer for poly(methyl methacrylate) (PMMA) (Scott, 2002; Scott, 2003a;

Scott, 2003b). Further studies also showed that ILs could act as plasticizers for PMMA (Puyu, 2006), poly(lactic acid) (PLA) (Zhang, et al, 2008; Park and Xanthos, 2009), and PVC (Shoff, et al, 2004; Rahman and Brazel, 2004a; Rahman and Brazel, 2006). A plasticizing effect is usually manifest by a reduced glass transition temperature (T_g), lower melt viscosity, ductility, modulus, tensile strength, hardness but increased flexibility, elongation at break and toughness (Rahman and Brazel, 2004b; Hong, 2000). Methods commonly used to assess plasticizing effect of ILs in polymers are differential scanning calorimetry (DSC) (Scott, et al, 2002; Scott, et al, 2003a; Scott, et al, 2003b; Puyu, 2006; Zhang, et al, 2008), and dynamic mechanical thermal analysis (DMTA) (Rahman and Brazel, 2006).

The published literature on ILs functioning as plasticizers has increased recently in contrast to those reporting the antimicrobial properties of ILs which have been reported for some time. Early work by Pernak *et al.* described the antimicrobial effect of benzimidazolium and pyridinium based ILs with different functional groups (Pernak, et al, 2001a; Pernak, et al, 2001b). Quantitative structure-activity relationship (QSAR) is a method commonly used to determine the relationship between chemical structure of ILs with their anti-microbial activity (Pernak and Chwala, 2003). The antimicrobial effect of choline-derivative-based ILs (Pernak and Chwala, 2003; Pernak, et al, 2007a), imidazolium ILs with different substitution (Pernak, et al., 2003; Pernak, et al., 2004a; Docherty and Kulpa, 2005; Carson, et al., 2009; Kanjilal, et al., 2009) and anions (Pernak, et al., 2004b), pyridinium halide ILs (Docherty and Kulpa, 2005), phosphonium-based ILs (Cieniecka-Roslonkiewicz, et al, 2005), ammonium-based ILs (Pernak and Feder-Kubis, 2005; Saadeh, et al, 2009) and trigeminal tricationic ILs (Pernak, et al, 2007b) was determined using a tube dilution method and minimum inhibition concentration values were used as comparison. However, to the best of our knowledge, there is no literature to date describing ILs when incorporated into a polymer matrix exhibit dual functionality, i.e. antimicrobial activity combined with a plasticizing effect. In this paper we report the results of an investigation of plasticizing and antimicrobial effects of ILs encapsulated in PVC.

Experimental

Materials PVC-IL composite preparation

A PVC (Colorite, Europe Ltd., proprietary formulation) was first mixed manually with two ionic liquids at loadings of 0, 10, 20 and 30wt%, 1-ethylpyridinium docusate (IL1) and tributyl(2-hydroxyethyl) phosphonium docusate (IL2), followed by spin mixing using a Rondol Speedmixer (Rondol Technology Ltd.). The mixtures were then melt mixed using a Thermo Haake MiniLab (L:D ratio of 21.9-7.8, conical screws) intermeshing co-rotating mini twin screw extruder. PVC-IL1 composites were processed at 150°C with a screw speed of 50 rpm, while for the PVC-IL2 composites a processing temperature of 140°C and a screw speed of 40 rpm were used. A slit die of dimension 5mm x 0.5 mm was used to produce tape-shaped extrudate. During processing, on-line torque values were recorded for neat PVC and PVC/IL composites through a data dump port to monitor the processing conditions.

Characterization

Dynamic Mechanical Thermal Analysis

Determination of glass transition temperature (T_g) was carried out using dynamic mechanical testing with a Triton thermoanalyzer, DMA Tritec 200 in tensile mode. A frequency of 1 Hz, clamp distance of 5.32mm, temperature rate of 2°C/min, displacement of 0.01mm and temperature range from room temperature to 120°C, were employed to study PVC and PVC-IL composite samples, with specimens having dimensions of 4.4mm (width) x 0.55mm (thickness).

Thermogravimetric Analysis

Thermal properties of all materials were evaluated using a thermogravimetric analyzer TGA/SDTA 851e/LF/1600/1382 (Mettler Toledo) for a heating range of 25° to 600°C and a heating rate of 10°C/min.

Antimicrobial Testing

Disc diffusion tests were carried out to determine the antimicrobial effect of PVC and PVC-IL composites towards a total of 24 strains of bacteria, which consisted of 6 type strains obtained from National Collection of Type Cultures (NCTC); (*Pseudomonas aeruginosa* NCTC 12934, *Staphylococcus aureus* NCTC 29213, *Enterococcus faecalis* NCTC 12697, *Proteus mirabilis* NCTC 11938, *Klebsiella aerogenes* NCTC 10896, *Escherichia coli* NCTC 12697), 11 clinical isolates (*Staphylococcus epidermidis* RP62A, *Staphylococcus* E-MRSA 16, *Staphylococcus aureus* AH-7, MRSA CFP16, MRSA CFP17, MRSA CFP18, MRSA CFP19 MRSA 4D, *Pseudomonas aeruginosa* B1, *Pseudomonas aeruginosa* C1, *Burkholderia cepacia complex* BB1, *Burkholderia cepacia complex* BB2), and 6 biofilm-forming coagulase-negative *Staphylococcus epidermidis* cultured from continuous ambulatory peritoneal dialysis catheters provided by the Microbiology Department, Belfast City Hospital, Belfast Health and Social Care Trust (B5, B7R, B8R, B10, B11R, B12). PVC and PVC-IL composites were stamped into discs with diameter of 4mm and applied onto the surface of agar inoculated with the microorganism of interest. A zone of inhibition was observed after overnight incubation at 37°C of the agar plates.

Biofilm Formation Test

Samples which gave efficient inhibition at the lowest IL loading in the disc diffusion test were selected for biofilm formation testing. Replicates with similar dimensions were immersed in bacterial suspensions containing biofilm-forming coagulase-negative *Staphylococcus epidermidis* (B5, B7R, B8R, B10, B11R, B12) for overnight incubation at 37°C, followed by ultrasonication at a frequency of 50Hz for 5 minutes after a gentle rinse with sterile phosphate buffered saline (PBS). The total viable count of bacteria was determined following serial 10-fold dilution and overnight incubation, using the Miles & Misra Method (Miles, et al, 1938).

Results and Discussion

On-line measurement of torque values

Torque values are commonly related to changes in viscosity of polymer melts, higher torque values normally indicates higher viscosity of the melt. Figure 1 and 2 shows the torque values collected during processing of PVC with various loadings of IL1 and IL2. All curves in Figure 1 and 2 show a peak before a decrease to constant torque values, which can help explain the 3 stages of PVC processing; fusion or gelation, flow, and solidification. It is believed that the peaks observed resulted from the gelation process, where the particle structure collapsed due to melting under shear stress, high pressure and temperature. The molecular chains diffuse into each other and chain entanglements result, thus giving high torque values. As the torque values reduced to a constant, the flow stage is reached, followed by solidification attained at the die exit when the extrudate is cooled to room temperature (Fujiyama and Kondou, 2003). In Figure 1, addition of 10wt% of IL1 to PVC reduced torque values by more than half compared to neat PVC, from 370 Ncm to 164 Ncm. This is followed by a further reduction in torque, by almost three-fold for two subsequent additions of IL1, 20 and 30wt%, which gave maximum torque values of 60 Ncm and 22 Ncm, respectively.

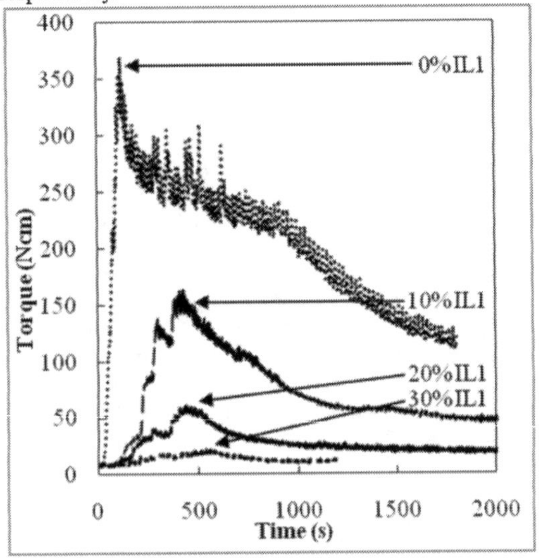

Figure 1 – Variation in extruder torque as a function of time for PVC with 0, 10, 20 and 30wt% of IL1 loadings.

A similar pattern was also observed when IL2 was added to this PVC, see Figure 2. The maximum torque value decreased from 391Ncm to 227Ncm upon addition of 10wt% IL2, and decreased further to 136Ncm and 72Ncm for the composites with 20 and 30 wt% IL2, respectively.

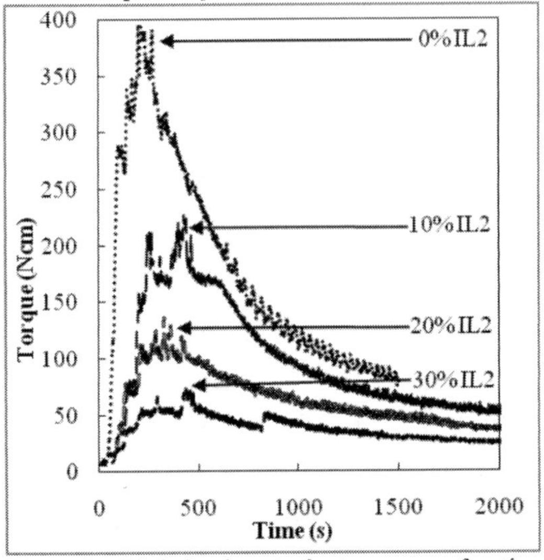

Figure 2 – Variation in extruder torque as a function of time for PVC with 0, 10, 20 and 30wt% of IL2 loadings.

In Figure 1 and 2, there is a slight difference in torque values for neat PVC which can be explained by the slight differences in temperature used in processing PVC around 145°C. It was also observed that that the torque peak positions for PVC appeared at later times when IL1 was added to the PVC, suggesting the addition of IL1 might have delayed the occurrence of gelation. This behavior was not seen when IL2 was added to this PVC.

Dynamic Mechanical Thermal Analysis (DMTA)

The storage modulus at 37.5°C (body temperature) and glass transition temperature (T_g) obtained from the peak maximum in the tan delta peak at 1Hz as a function of IL loading are shown in Figure 3 and 4, respectively. 37.5°C was selected as this is assumed to be working temperature of medical grade PVC used in for example catheters and endotracheal tubing. From Figure 3, it can be seen there was a very sharp decrease in storage modulus of PVC at 37.5°C with increased IL content. The storage modulus of PVC was reduced from 1.8 GPa to 1.1 GPa to 0.3 GPa to 0.1 GPa with each successive addition of IL1. A similar trend was also attained when successive additions of IL2 up to 30wt% was added to PVC, resulting in a reduction of storage modulus from 1.6 GPa to 1.4GPa to 0.5GPa to 0.05GPa for 0, 10, 20 and 30wt% - IL2, respectively.

The difference in storage modulus for neat PVC showed that a slightly higher processing temperature and screw speed resulted in a slightly higher storage modulus, 1.8GPa compared to 1.6GPa. Since storage modulus is an indicator of the amount of recoverable strain energy (Campbell, et al, 2000), the PVC processed at a higher temperature and screw speed was found to have higher elasticity due to more heat from the barrel and shear applied into the PVC melt, which in turn leads to more molten PVC primary crystallites and increased molecular chain diffusion and entanglement (Fujiyama and Kondou, 2003). However, in all instances the storage modulus decreased significantly on IL addition, implying a strong plasticizing effect. This observation was corroborated by the significant changes observed in T_g on IL addition. Variation in T_g can be used as a measure of plasticization which can be accurately determined from DMTA measurments (Campbell, et al, 2000). The change in T_g of PVC with various IL1 and IL2 content is shown in Figure 4. The T_g of PVC decreased from 90.7°C to 85.5°C to 81.8°C to 81.2°C on successive additions of 10wt% IL1 up to 30wt%., However, a much more dramatic reduction in T_g was obtained for IL2, the T_g decreased from 92.3°C to 77.4°C to 63.1°C to 46°C on addition of 0, 10, 20 and 30wt% IL2, respectively. This represents a 46°C reduction in T_g, altogether more remarkable as the PVC comprised 38wt% TiO_2 filler. The plasticizing effect of these ionic liquids can be explained by their ability to occupy PVC free volume, the extent of which is governed by IL molar volume and tailored polarity.

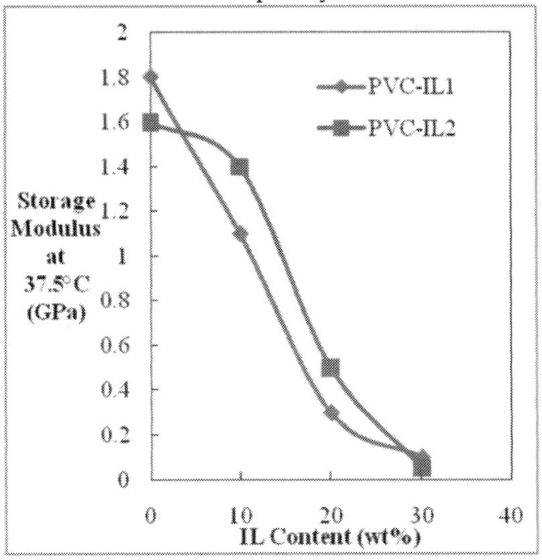

Figure 3 – Storage modulus at 37.5°C as a function of IL content in PVC.

The plasticizing effect was more pronounced for IL2, at least in the composition range studied where a more sharp and constant decrease of T_g was observed. A further more enhanced plasticizing effect may be a possibility for IL1 with a loading above 30wt%.

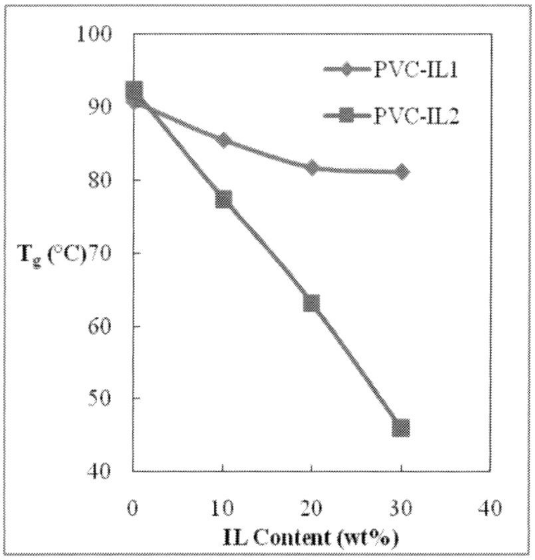

Figure 4 – Variation in T_g as a function of IL content.

Thermogravimetric Analysis(TGA)

The weight loss as a function of temperature for PVC with and without IL was measured from room temperature to 600°C, and plotted in the first derivative form, see Figures 5 and 6. Two characteristic peaks were observed in all PVC and PVC-IL composites, typical of the well described thermal degradation process for PVC: firstly dehydrochlorination, where hydrochloric acid is evolved with formation of a conjugated structure at temperatures between 200°C and 350°C, followed by condensation and fragmentation of the carbon backbone at 400-500°C (Gonzalez, et al, 2006). The mechanism for PVC thermal degradation has been reported previously (Elakesh, 2003; Ahmad, et al, 2008). Both degradation peaks shifted towards lower temperatures with increasing IL content, indicating the addition of IL1 and IL2 to this PVC reduced the thermal stability of PVC by accelerating the process of dehydrochlorination and condensation to lower temperatures, see tables 1 and 2. However, there was one exception in that PVC with 10wt% IL1 added underwent dehydrochlorination at a slightly higher temperature (293.4°C) relative to neat PVC (281.6°C). This indicates that degradation may be delayed, a process which may be related to stoichiometric reaction between IL1 and PVC at certain compositions. Two peaks were also observed for IL1 itself, indicating the possibility of contamination or slight impurities in the IL1 sample used. The presence of this impurity at lower temperatures in IL1 relative to the major degradation of neat PVC, caused the early onset of PVC degradation as the concentration of IL1 in PVC was increased.

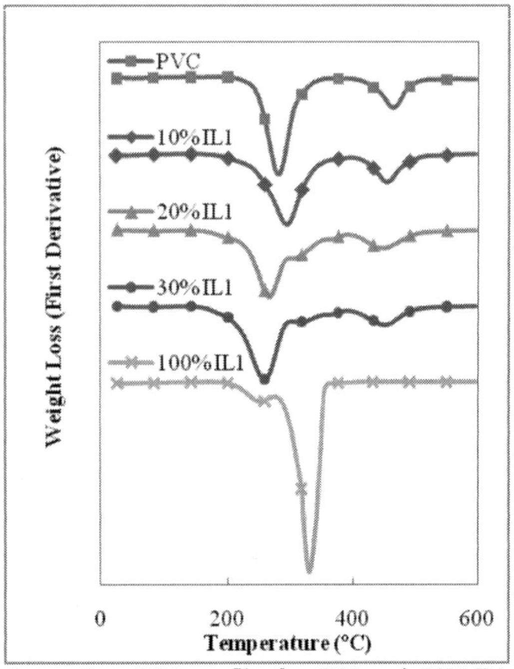

Figure 5 – TGA profile for PVC and PVC-IL1 composites.

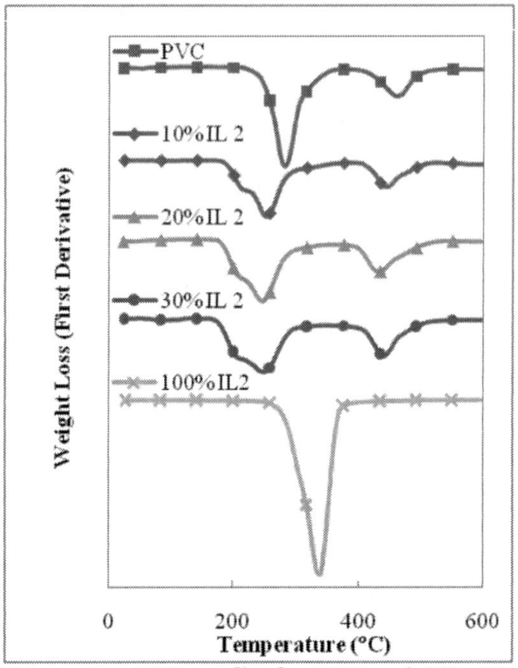

Figure 6 – TGA profile for PVC and PVC-IL2 composites.

A second peak for IL1 occurred at higher temperatures, centered at about 460°C.

Table 1 – Thermal degradation data for PVC-IL1 composites.

	IL 1 in PVC (wt%)				
	0	10	20	30	100
Temp. at 95% wt loss	255.7	236.3	221.5	209.2	235.7
Derivative peak(s)	281.6, 465.2	293.4, 456.4	264.2, 447.6	258.5, 450.6	252.8, 328
Wt. remaining at derivative peak(s)	72.3, 37.4	68, 32.4	76.6, 34.1	69, 28.8	91, 46.9
Onset temp.[1]	-	386.6	293.3	293.5	-

Onset temp.[1] = shoulder peak appeared after first derivative peak

Similar behavior was observed for IL2 and PVC-IL2 composites, although IL2 is more thermally stable than IL1, most probably due to the absence of any impurity.

There is also evidence of a third degradation process for the PVC-IL2 composites manifest by a shoulder to the main dehydrochlorination peak and not observed for either PVC or IL2 alone.

Table 2 – Thermal degradation data for PVC-IL2 composites.

	IL 2 in PVC (wt%)				
	0	10	20	30	100
Temp. at 95% wt loss	263.3	210.3	205.6	205.8	290.3
Derivative peak(s)	281.4, 462.2	253, 445.3	246.6, 433	249.6, 439	339.4
Wt. remaining at derivative peak(s)	80.7, 39.9	76, 45.8	73.3, 38.2	73.4, 40.6	34.9
Onset temp.[2]	-	212.2	211.8	205.9	-

Onset temp.[2] = shoulder peak appeared before first derivative peak

There is some evidence to suggest that the dehydrochlorination process consists of two-stages, which involves dehydrochlorination of labile and more stable chlorine atoms (Gonzalez, et al, 2006; Kim, et al, 2004). IL's are known to be good solvents in facilitating free-radical polymerization (Schmidt-Naake, et al, 2007). The shoulder to this main peak might be due to facilitation of free-radicals by IL1 and IL2, resulting in accentuation and segregation of derivative peaks which were normally observed as single dehydrochlorination peak in conventional TGA, thus observed as overlapped peaks in this study.

Disc Diffusion Test
The antimicrobial activity of PVC-IL1 and PVC-IL2 composites was determined using the standard disc diffusion test. Table 3 shows the disc diffusion test results of PVC-IL1 and PVC-IL2 composites. The test records either a positive or negative effect to different bacterial strains.

Table 3 – Disc diffusion test results of PVC added with IL1 and IL2 on type strains and clinical isolates.

Strains	PVC-IL1	PVC-IL2
Staphylococcus aureus NCTC 29213*	+	+
Staphylococcus epidermidis RP62A**	+	+
Staphylococcus E-MRSA 16**	+	+
Staphylococcus aureus AH-7**	+	+
MRSA CFP16**	+	+
MRSA CFP17**	+	+
MRSA CFP18**	+	+
MRSA CFP19**	+	+
MRSA 4D**	+	+
Enterococcus faecalis NCTC 12697*	-	-
Pseudomonas aeruginosa NCTC 12934*	-	-
Pseudomonas aeruginosa B1**	-	-
Pseudomonas aeruginosa C1**	±	-
Burkholderia cepacia complex BB1**	±	-
Burkholderia cepacia complex BB2**	±	-
Proteus mirabilis NCTC 11938*	±	-
Klebsiella aerogenes NCTC 10896*	±	-
Escherichia coli NCTC 10418*	-	-
Biofilm-forming *Staphylococcus* strains		
B5**	+	±
B7R**	+	±
B8R**	+	±
B10**	+	+
B11R**	+	+
B12**	+	±

*Type strain, **Clinical isolates, + inhibition observed, - No inhibition observed, ± insignificant inhibition

From Table 3, PVC-IL1 composites showed inhibition to all MRSA and biofilm-forming *Staphylococcus* strains, while PVC-IL2 inhibited all MRSA and two biofilm-forming *Staphylococcus* strains. These results showed that both IL1 and IL2 have the tendency to inhibit the growth only of gram-positive bacteria, especially *Staphylococcus* strains. Although another reason for poor inhibition of PVC-IL1 and PVC-IL2 composites on gram-negative bacteria could be due to presence of an additional layer, peptidoglycans which might prevent the penetration of active compounds from PVC-IL composites, but both PVC-IL composites did not inhibit *Enterococcus faecalis* strain, which is also a gram-positive bacterium. Therefore, it is suspected that these 2 ILs may only be species-specific in inhibiting bacteria. The antimicrobial effect of PVC-IL1 and PVC-IL2 may also be caused by the anionic component of IL1 and IL2, the docusate anion. Docusate anion in salt form with a sodium cation is a common anionic surfactant used in oral drugs such as laxatives. Surfactants such as docusate anions are able to enhance membrane permeability (Hough, et al, 2007), and may further induce cell lysis, exerting an antimicrobial effect (Kanjilal, et al, 2009). Comparing both PVC-IL composites, PVC-IL1 has better antimicrobial activity as it inhibited more strains than PVC-IL2. This is in agreement with that reported previously in that phosphonium ILs (IL2) are less inhibitory compared to pyridinium ILs (IL1) (Cieniecka-Roslonkiewicz, et al, 2005; Stock, et al, 2004), while it has been claimed that phosphonium ILs have better biocompatibility with gram-negative bacteria (Baumann, et al, 2005).

Biofilm Formation Test
Based on the results of the disc diffusion tests, PVC with 10wt% IL1 was selected for biofilm formation testing, as this composite gave effective inhibition at the lowest IL content. Using PVC samples without IL addition as control samples, the percentage inhibition was calculated as below:
Percentage of inhibition, % = 100-(total count of bacteria obtained from PVC with 10wt% IL1/total count of bacteria obtained from PVC without IL addition. The biofilm formation test was performed on six strains of biofilm-forming *Staphylococcus* strains using PVC with 10wt% of IL1. Apart from B11R, PVC with addition of 10wt% IL1 resulted in a percentage of inhibition of more than 90% in all cases. Biofilm formation is related to adhesion of the micro-organism to the material surface mediated by Van der Waals forces, acid base or electrostatic interactions (Del Pozo, et al, 2008), hydrophobic, stearic and environmental forces (Macassey and Dawes, 2008), followed by production of a slimy matrix which enables persistent attachment together with protection from antibiotics and phagocytosis of white blood cells (DePaola, et al, 2002). In this experiment, it is possible that the high percentage of inhibition is due to instant killing of tested strains while in its planktonic form before the attachment onto sample surface began. This is a most significant result given the large number of deaths, particularly among the elderly, from hospital derived infections.

Conclusions

Both ILs have shown a plasticizing and antimicrobial effect after addition to PVC. IL2 provided a better plasticizing effect by reducing T_g of a highly filled PVC by 46.3°C compared to a reduction of 9.3°C in T_g by IL1, for a loading of 30wt%. Incorporation of IL1 in PVC gave a better antimicrobial effect by inhibiting 15 out of 24 bacteria strains tested compared to PVC-IL2 composites which inhibited 11 strains tested, but in both cases MRSA growth was significantly inhibited. Incorporation of IL1 into PVC has also exhibited excellent anti-biofilm activity for only 10wt% IL addition. The dual plasticizing and antimicrobial properties of such ILs offers huge potential of tailoring ILs as multi-functional additives for medical polymers.

Acknowledgements

We thank Colorite Europe Ltd. for the provision of PVC grades and SYC acknowledges Queen's University Belfast for funding a studentship.

References

Ahmad, Z., et al, "Thermal degradation studies in poly(vinyl chloride)/poly(methyl methacrylate) blends", Polymer Degradation and Stability, **93** (2), 456-465 (2008).

Baumann, M. D., et al, "Phosphonium ionic liquids for degradation of phenol in a two-phase partitioning bioreactor", Applied Microbiology and Biotechnology, **67**(1), 131-137 (2005).

Campbell, D., Pethrick, R. A., White, J. R.: Polymer Characterization-Physical Techniques, 2nd Edition, Stanley Thornes Ltd., Cheltenham, p. 384-386 (2000).

Carson, L., et al, "Antibiofilm activities of 1-alkyl-3-methylimidazolium chloride ionic liquids", Green Chemistry, **11** (4), 492-497 (2009).

Cieniecka-Roslonkiewicz, A., et al, "Synthesis, antimicrobial activities and anti-electrostatic properties of phosphonium-based ionic liquids", Green Chemistry, **7** (2), 855-862 (2005).

Del Pozo, J. L., et al, "Bioelectric effect and bacterial biofilms. A systematic review", International Journal of Artificial Organs, **31**(9), 786-795 (2008).

DePaola, L. G., et al, "A review of the science regarding dental unit waterlines", Journal of the American Dental Association, **133** (9), 1199-1206 (2002).

Docherty, K.M. and C.F. Kulpa, "Toxicity and antimicrobial activity of imidazolium and pyridinium ionic liquids", Green Chemistry, **7** (4), 185-189 (2005).

Elakesh, E. O., et al, "Thermal decomposition of chlorinated poly(vinyl chloride) (CPVC)", Journal of Vinyl & Additive Technology, **9** (3), 116-126 (2003).

Fanelli, R., Zuccato, E., "Risks and benefits of PVC in medical applications", Boll Chim Farm, **141** (4), 282-289 (2002).

Fujiyama, M., and Kondou, M., "Effect of gelation on the flow processability of poly(vinyl chloride)", Journal of Applied Polymer Science, **90** (7), 1808-1824 (2003).

Gonzalez, N., et al, "Application of high resolution thermogravimetry to the study of thermal stability of poly(vinyl chloride) resins", Polymer Degradation and Stability, **91** (4), 629-633 (2006).

Hong, K. Z., "An overview on polyvinyl chloride (PVC) and alternatives in medical application", Antec 2000:Society of Plastics Engineers Technical Papers, Vol. I-III, 2704-2709 (2000).

Hough, W. L., et al, "The third evolution of ionic liquids: active pharmaceutical ingredients", New Journal of Chemistry, **31**(8), 1429-1436 (2007).

Kanjilal, S., et al, "Synthesis and evaluation of micellar properties and antimicrobial activities of imidazole-based surfactants", European Journal of Lipid Science and Technology, **111** (9), 941-948 (2009).

Kim, S., et al, "Peak separation method for estimating apparent kinetic parameters of dehydrochlorination reactions of PVC: Independent reactions", Journal of Industrial and Engineering Chemistry, **10** (3), 484-491 (2004).

Kubisa, P., "Ionic liquids in the synthesis and modification of polymers", Journal of Polymer Science Part A-Poymer Chemistry, **43**(20), 4675-4683 (2005).

Macassey, E., and P. Dawes, "Biofilms and their role in otorhinolaryngological disease", Journal of Laryngology and Otology, **122** (12), 1273-1278 (2008).

Matsumoto, M., et al, "Potential adverse effects of phthalic acid esters on human health: A review of recent studies on reproduction", Regulatory Toxicology and Pharmacology, **50**, 37-49 (2008).

Miles, A. A., et al, "The estimation of the bactericidal power of the blood", Journal of Hygiene, **38**, 732-749 (1938).

Park, K. I., and M. Xanthos, "A study on the degradation of polylactic acid in the presence of phosphonium-based ionic liquids as PVC plasticizers", Polymer Degradation and Stability, **94** (5), 834-844 (2009).

Pernak, J. and P. Chwala, "Synthesis and antimicrobial activities of choline-like quarternary ammonium chlorides", European Journal of Medicinal Chemistry, **38** (11-12), 1035-1042 (2003).

Pernak, J., and Feder-Kubis, J., "Synthesis and properties of chiral ammonium-based ionic liquids", Chemistry-A European Journal, **11** (15), 4441-4449 (2005).

Pernak, J., et al, "Antimicrobial activities of ionic liquids", Green Chemistry, **5**, 52-56 (2003).

Pernak, J., et al, "Antimicrobial activities of protic ionic liquids with lactate anion", Green Chemistry, **6** (7), 323-329 (2004b).

Pernak, J., et al, "Choline-derivative-based ionic liquids", Chemistry-A European Journal, **13** (12), 6817-6827 (2007a).

Pernak, J., et al, "Ionic liquids with symmetrical dialkoxymethyl-substituted imidazolium cations", Chemistry-A European Journal, **10** (14), 3479-3485 (2004a).

Pernak, J., et al, "Synthesis and antimicrobial activities of new pyridinium and benzimidazolium chlorides", Eur. J. Med. Chem., **36** (4), 313-320 (2001a).

Pernak, J., et al, "Synthesis and antimicrobial activities of some pyridinium salts with alkoxymethyl hydrophobic group", European Journal of Medicinal Chemistry, **36** (11-12), 899-907 (2001b).

Pernak, J., et al, "Synthesis and properties of trigeminal tricationic ionic liquids", Chemistry-A European Journal, **13** (11), 3106-3112 (2007b).

Puyu, L.S.Z., "Synthesis and application of ionic liquid as green plasticizer for PMMA", Journal of Polymer Material, **23** (1), 97-100 (2006).

Rahman, M., and C.S. Brazel, "Effectiveness of phosphonium, ammonium, and imidazolium-based ionic liquids as plasticizers for poly(vinyl chloride): Thermal and ultraviolet stability", Polymer Preprints, **45** (1), 301-302 (2004a).

Rahman, M., and C.S. Brazel, "Ionic liquids: New generation stable plasticizers for poly(vinyl chloride)", Polymer Degradation and Stability, **91** (12), 3371-3382 (2006).

Rahman, M., and C.S. Brazel, "The plasticizer market: an assessment of traditional plasticizers and research trends to meet new challenges", Progress in Polymer Science, **29** (12), 1223-1248 (2004b).

Roth. B, et al, "Di-(2-ethylhexyl)-phthalate as plasticizer in PVC respiratory tubing systems-Indications of hazardous effects on pulmonary-function in mechanically ventilated, preterm infants", European Journal of Pediatrics, **147**(1), 41-46 (1988).

Saadeh, S. M., et al, "New room temperature ionic liquids with interesting ecotoxicological and antimicrobial properties", Ecotoxicology and Environmental Safety, **72** (6), 1805-1809 (2009).

Schmidt-Naake, G., et al, "Ionic liquids-New solvents in the free radical polymerization", Macromolecular Symposia, **259** (1), 226-235 (2007).

Scott, M. P., et al, "Application of ionic liquids as low-volatility plasticizers for PMMA", European Polymer Journal, **39** (10), 1947-1953 (2003b).

Scott, M. P., et al, "Application of ionic liquids as plasticizers for poly(methyl methacrylate)", Chemical Communications, **13**, 1370-1371 (2002).

Scott, M. P., et al, "Plasticizing effects of imidazolium salts in PMMA: High-temperature stable flexible engineering materials", ACS Symposium Series, **856**, 468-477 (2003a).

Shoff, H. W., et al, "Leaching and migration resistance of phosphonium-based ionic liquids as PVC plasticizers: Comparative study of traditional phthalate and citrate plasticizers with ionic liquids", Polymer Preprints, **45** (1), 295-296 (2004).

Stock, F., et al, "Effects of ionic liquids on the acetylcholinesterase - a structure-activity relationship consideration", Green Chemistry, **6**(6), 286-290 (2004).

Wasserscheid, P. and W. Keim, "Ionic liquids-New 'solutions' for transition metal catalysis", Angewandte Chemie-International Edition, **39** (21), 3773-3789 (2000).

Wisner, D., et al, "Chapter 18 PVC Environmental, Health, Safety" in PVC Handbook, Wilkes C. E. et al (Eds.), Hanser Publishers, Munich, p. 664 (2005).

Zhang, P. Y., et al, "Application of ionic liquid [bmim]PF6 as green plasticizer for poly(L-lactide)", e-Polymer, Article number 172 (2008).

Zhao, H., "Innovative applications of ionic liquids as 'green' engineering liquids", Chemical Engineering Communications, **193** (12), 1660-1677 (2006).

IN-VITRO ELUTION OF VANCOMYCIN/AMIKACIN/STEROID FROM SOLVENT-FREE BIODEGRADABLE SCLERAL PLUGS

Shih-Jung Liu[1], Yi-Chuan Kau[1,2], Chau-Wei Liaw[1], Yi-Jie Peng[3]

[1]Department of Mechanical Engineering, Chang Gung University, Taiwan - *shihjung@mail.cgu.edu.tw*
[2]Department of Anesthesiology, Chang Gung Memorial Hospital
[3]Department of Ophthalmology, Buddhist Taipei Tzu Chi General Hospital

Abstract - The purpose of this report was to develop solvent-free biodegradable scleral plugs for vancomycin, amikacin and dexamethasone delivery for endophthalmitis treatment. To fabricate a biodegradable plug, polylactide-polyglycolide copolymers were pre-mixed with the drugs. The mixture was then compression molded and sintered to form a scleral plug of 1.4mm in diameter. An elution method was utilized to characterize the in-vitro release characteristics of the antibiotics and the steroids over a 14-day period. The HPLC analysis and bacterial inhibition test showed that biodegradable scleral plugs released a high concentration resulting in significant activity of vancomycin and amikacin (well above the minimum inhibition concentrations) and dexamethasone in vitro, for the period of time needed to treat intraocular infection. A bacterial inhibition test was carried out to determine the relative activity of the released antibiotics. The activities of the eluted vancomycin and amikacin ranged from 69% to 89% and from 66% to 88% respectively. In addition, the experimental result suggests that one will be able to reduce the drug release rate and prolong the total release period of the plugs by adopting a lower antibiotic/steroid to polymer ratio, increasing the sintering temperature, or increasing the compression pressures.

Introduction

The standard treatment for endophthalmitis is currently done by trans pars plana intravitreal injection of both antibiotics [1-3] and steroids [4, 5]. However, repeated injections may increase the risks of intraocular infection and hemorrhage. This paper was to develop solvent-free biodegradable scleral plugs for the delivery of antibiotics and steroids.

Experimental

The polymers used were polylactide-polyglycolide with a ratio of 50:50 and a molecular weight of approximately 5,000. The antibiotics used were commercial grade vancomycin and amikacin, while the steroid used was dexamethasone.

To fabricate the biodegradable plugs, polylactide-polyglycolide copolymers were pre-mixed with the drugs. Table 1 lists the contents of the mixtures used in the plugs. The mixture was then compression molded [6, 7] to form a scleral plug of 1.4mm in diameter (Figure 1a). The compressed plugs with the mold were then placed in an oven for sintering. The sintering temperature was set at 55°C. The sintering time used was 30 min. in order to attain an isothermal sintering of the materials. Figure 1b shows schematically the mold and the biodegradable plugs.

The relative activity test of vancomycin and amikacin to bacteria were determined using an antibiotic disk diffusion method in Nutrient Broth. The relative activity of the released antibiotics was defined as:

$$\text{Relative activity (\%)} = \frac{\text{diameter of sample inhibition zone}}{\text{diameter of maximum inhibition zone}}$$

An in-vitro elution method was utilized to determine the release characteristics of antibiotics and steroid from the plugs. The antibiotics and steroids concentrations in buffer for the elution studies were determined by a high-performance liquid chromatography (HPLC) assay [6, 8].

Results and Discussion

Biodegradable scleral plugs were successfully manufactured by the compression-sintering method. The thermal stability results with various pharmaceuticals were showed in Figure 2a. The results suggest that the pharmaceuticals used in this study were not degraded at a sintering temperature of 55°C.

The HPLC analysis results in Figure 3 from the in-vitro elution test showed that the plugs released a high concentration of vancomycin and amikacin (well above the minimum inhibition concentration) and dexamethasone for up to 14 days. The result in Figure 3a shows that the release rate of vancomycin decreases with the antibiotic/steroid to polymer ratio. On the other hand, Figure 4 shows the release curves of the plugs subjected to different sintering temperatures. The experimental result suggests that the drug release is higher when the sintering temperature is lower. Figure 5 shows the release curves of vancomycin, amikacin and dexamethasone from the plugs subjected to different compression pressures during processing. The result also suggests that the release rate of the plugs decrease with the compression pressure. The plug that was not compressed before the sintering dissolved within a shorter period of time. Overall the experimental results from the in-vitro elution test showed that the plugs released a high concentration of vancomycin and amikacin (well above the minimum inhibition concentration) and dexamethasone for up to 14 days. The activities of the eluted vancomycin and amikacin ranged from 69% to 89% and from 66% to 88% respectively.

Table 1. Composition of the biodegradable scleral plugs

Drug/ Polymer ratio	Weights of drugs (mg)			Weight of polymer (mg)
	Vancomycin	Amikacin	Dexamethasone	
1:2	1	0.4	0.4	4
1:3	0.8	0.3	0.3	4.5

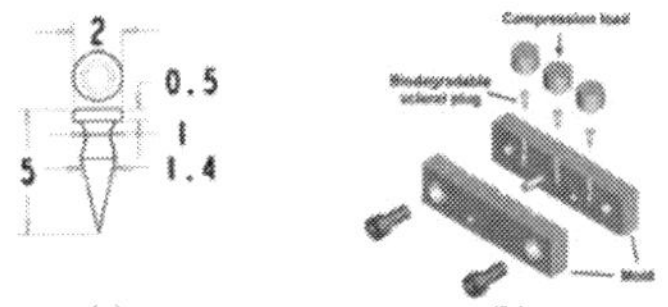

(a) (b)

Figure 1. (a) Dimensions of the scleral plug (unit: mm), and (b) the mold for the manufacture of the plugs

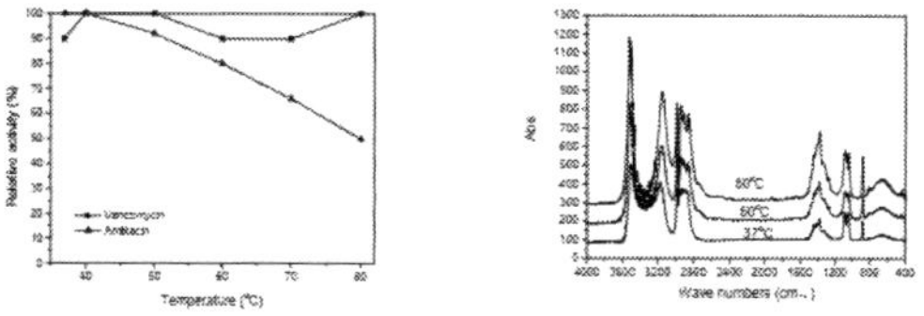
(a) (b)

Figure 2. Thermal stability of (a) the antibiotics, and (b) steroid used in this study

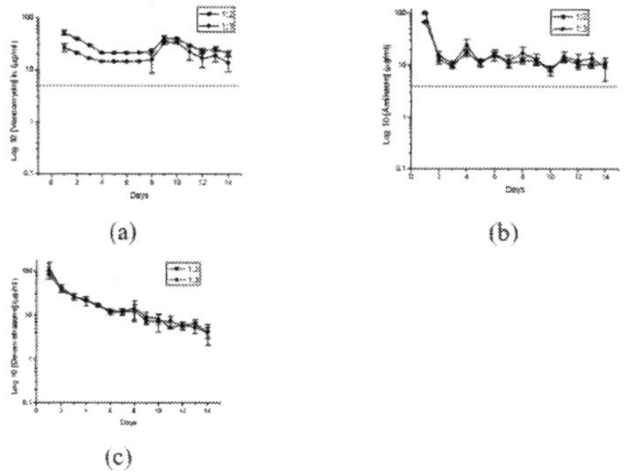
(a) (b) (c)

Figure 3. Release curves of (a) vancomycin, (b) amikacin, and (c) dexamethasone, from biodegradable scleral plugs of different drug mixture to polymer ratios

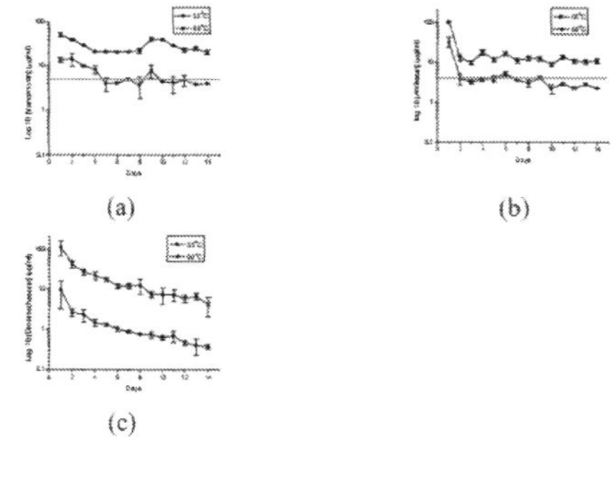
(a) (b) (c)

Figure 4. Release curves of (a) vancomycin, (b) amikacin, and (c) dexamethasone, from biodegradable scleral plugs subjected to different oven temperatures

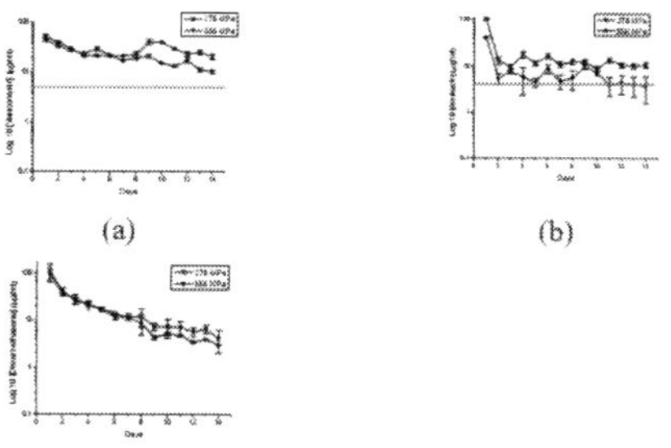
(a) (b) (c)

Figure 5. Release curves of (a) vancomycin, (b) amikacin, and (c) dexamethasone, from biodegradable scleral plugs subjected to different compression pressures

Conclusions

This paper has developed solvent-free biodegradable antibiotic/steroids plugs for a long-term drug release. The experimental result suggests that one will be able to reduce the drug release rate and prolong the total release period of the plugs by adopting a lower antibiotic/steroid to polymer ratio, increasing the sintering temperature, or increasing the compression pressures.

References

1. Han D., et al., "Spectrum and susceptibilities of microbiologic isolates in the endophthalmitis vitrectomy study", Am. J. Ophthalmol. , **122**, 1-17(1996)
2. Benz M.S., et al., "Endophthalmitis isolates and antibiotic sensitivities: A 6-year review of culture-proven cases", Am. J. Ophthalmol., **137**, 38-42(2004)
3. Aguilar H.E., et al., "Vancomycin levels after intravitreal injection", Retina, **15**,428-32(1995)
4. Das T., et al., "Intravitreal dexamethasone in exogenous bacterial endophthalmitis: results of a prospective randomized study", Br. J. Ophthalmol., **83**, 1050-1055(1999)
5. Shah G.K., et al., "Visual outcomes following the use of intravitreal steroids in the treatment of postoperative endophthalmitis", Ophthalmology , **107**, 486-489(2000)
6. Liu S.J., et al., "In vitro elution of antibiotics from biodegradable beads", J. Biomed. Mat. Res. (Applied Biomaterials), **48**, 613-620(1999)
7. Wang G., et al., "The release of Cefazolin and Gentamicin from biodegradable PLA/PGA beads", Int. J. Pharmaceutics, **273**, 203-212(2004)
8. Milojevic Z., et al., "High-performance liquid chromatographic method for the assay of dexamethasone and xylometazoline in nasal drops containing methyl *p*-hydroxybenzoate", J. Chromat ., **949**, 79-82(2002)

MECHANICAL PROPERTIES OF EPOXY HYBRID NANOCOMPOSITE REINFORCED WITH ORGANIC BLOCK-COPOLYMER AND INORGANIC NANOCLAY

M. Bashar and P. Mertiny*

Department of Mechanical Engineering, University of Alberta, Edmonton, AB, Canada – bashar@ualberta.ca; pmertiny@ualberta.ca

Abstract - Mechanical properties and morphology of ternary epoxy nanocomposites modified with acrylic tri-block-copolymer and organophilic layered silicate clay were explored. In cured epoxy, the acrylic block-copolymer phase separated into uniformly dispersed spherical micelle inclusions, and a predominantly intercalated and flocculated structure was observed for layered silicate. The microstructure of the resulting nanocomposites was examined by transmission electron microscopy, scanning electron microscopy, and X-ray diffraction. Mechanical testing indicated that the block-copolymer enhanced toughness and ductility of the epoxy-block-copolymer system by sacrificing stiffness and strength, whereas the nanoclay filled nanocomposite exhibited superior tensile strength and stiffness while compromising ductility. A simultaneous improvement of toughness, stiffness, and strength was achieved by forming ternary nanocomposite containing 1wt% of each of the nanoparticles. This improvement in properties was attributed to the synergistic effect between the intercalated layered silicate and the self-assembled nanostructure of the acrylic block-copolymer. It is expected that implementing the same procedure with higher particle loadings will generate even higher property enhancements for the epoxy hybrid nanocomposite.

Introduction

Nanocomposites are materials comprise of one or more nano-scale reinforcements inside a matrix element. Polymer nanocomposites are emerging as a new breed of high performance materials, outweighing many of the properties of the conventional polymer composites. Polymer nanocomposites often exhibit significantly enhanced mechanical properties, gas barrier properties, thermal stability, fire retardancy, chemical resistance and good dimensional stability. Sometimes a compromise between properties must be tolerated because multifunctionality in all three dimensions and different length scales is difficult to achieve. However, this may be accomplished effectively by adding two or more different types of nanoparticles and incorporating their useful properties in a matrix substrate. Thereby a material possessing various favourable characteristics necessary for certain high-performance applications such as space and cryogenic service can be produced.

General wisdom about rubbery and rigid particulate inclusions is that soft particles increase ductility and toughness, and conversely, hard particles enhance stiffness and strength of the polymer matrix (Kinloch et al., 1985). Epoxy is well known for its high stiffness, creep resistance, chemical inertness and high temperature application but provides limited resistance against crack propagation. Extensive research has been conducted by incorporating a diverse range of dispersed phases into the epoxy matrix, i.e. liquid rubbers (Kinloch et at., 1983; Pearson and Yee, 1983; Sultan and McGarry, 1973), thermoplastic spheres (Höfflin et al., 2000; Kinloch et al., 1994), core-shell particles (Becu-Longuet et al., 1999; Lin and Shieh, 1998), and nanostructured block-copolymers (Dean et al., 2001, 2003; Gerard et al., 2007; Hydro and Pearson, 2007; Wu et al., 2005). In most of these cases, improved toughness was realized whilst lowering the stiffness and strength of epoxy. Inorganic particles such as silica (Moloney et al., 1983), glass beads (Kinloch et al., 1985) and nanoclay (Lan and Pinnavaia, 1994; Messersmith and Giannelis, 1994) are known to enhance stiffness, strength and in some cases toughness (intercalated nanoclay) of the thermoset epoxy while compromising ductility. Thus, blending both rigid and rubbery nanoparticles in a thermosetting polymer would be an alternative to the situations discussed above.

Block-copolymers are a group of chemically dissimilar homopolymers bonded together by covalent linkage to form a macromolecular network. In the molten state or in a selective solvent such as in epoxy, block-copolymers readily self-assemble into nanostructures. Nano-level organization is prompted by chemical interactions (block-block, block-solvent), polymer molecular weight, copolymer/solvent composition and concentration (Dean at el., 2001, 2003). On the other hand, inorganic particulate nanoclay is made of a plate-like structure with aluminum octahedrons sandwiched between silicon tetrahedrons. Layered silicate nanocomposites exhibit three distinct morphologies, i.e. phase separated, intercalated and exfoliated (Lan and Pinnavaia, 1994). Superior mechanical, thermal and barrier properties are obtained for an exfoliated structure when individual silicate lamellae are homogenously distributed inside the polymer matrix.

The focus of this study was to associate organically modified nanoclay I.30E and acrylic tri-block-copolymer M52N into epoxy polymer EPON 826 and thereby synergistically improve mechanical properties of the resulting hybrid nanocomposite. Previous studies have successfully mixed two or more nano-reinforcements in an epoxy (Balakrishnan et al., 2005; Frohlick et al., 2003, 2004; Gam et al., 2003; Liu et al., 2004; Marouf et al., 2009). In the present study, characterization of nanocomposites by morphological examination and mechanical testing was performed.

Experimental

Materials

Epoxy resin EPON 826 (Bisphenol-A) and curing agent EPI-CURE 9551 (non-MDA polyamine) used in this study were from Hexion Specialty Chemicals (USA). Nanoclay I.30E, $CH_3(CH_2)_{17}NH_3$-MMT which is an organically modified Na-montmorillonite (Na-MMT) was supplied by Nanocor Inc. (USA). Acrylic tri-block-copolymer M52N used here as an organic nanoparticle was synthesised by Arkema Inc. (USA).

Preparation of epoxy nanocomposite

M52N copolymer was dissolved directly by gradually adding the block-copolymer powder into the epoxy resin at room temperature by gentle mixing. The temperature was then raised and held stationary at 80°C until a uniform miscible solution and good optical transparency was observed. The dissolution time varied upon block-copolymer concentration.

Prior to dispersion into the epoxy precursor, nanoclay I.30E was dried in an oven at an elevated temperature of 120°C for a period of 24 hours. Afterwards, the nanoclay was allowed to cool down to room temperature. A specific amount of I.30E was first introduced into acetone, and pre-swelling in acetone was performed for six hours at room temperature. The solution was then mixed with the preheated EPON 826 resin at 60°C by gentle stirring. Acetone promoted the migration of epoxy inside the clay interlayer and also reduced the resin viscosity. To obtain the desired level of resin viscosity for efficient ultrasonic mixing, the system temperature was slowly raised to 80°C. Sonication continued for eight hours with a Branson Sonifier operated at its maximum power intensity, followed by mechanical mixing for two hours. Finally, the mixture was degassed under vacuum to completely strip off the excess acetone from the solution.

Identical techniques were followed for epoxy nanocomposite formulations containing 1wt% and 3wt% I.30E or 1wt% and 3wt% M52N. For preparation of the hybrid nanocomposite (combined loading of 1wt% I.30E and 1wt% M52N) an acetone-clay solution was commixed with an initially prepared EPON 826-M52N mixture; remaining processing was similar to epoxy-I.30E.

A stoichiometric amount of curing agent was mixed with the epoxy-nanofiller mixture at 60°C for five minutes by a mechanical stirrer. Formation of bubbles during curing agent addition necessitated a second degassing cycle to be performed for 10~20 minutes. Finally, the resin mixture was poured into a steel mould and cured at 120°C for two hours.

X-ray diffraction

X-ray diffraction (XRD) analysis of the epoxy-I.30E nanocomposites was carried out using a Rigaku Geigerflex 2173 diffractometer with a vertical goniometer. The system is equipped with a Co tube and a graphite monochromator to filter K-beta wavelengths. All the tests were ran at 40 kV and 30 mA, and the samples were scanned at a rate of 0.008 $2\theta sec^{-1}$ up to a range of 2θ = 1 to 30°.

Transmission electron microscopy

Microstructural examination of the cured epoxy nanocomposites was performed in a Morgagni 268 transmission electron microscope (TEM) with an acceleration voltage of 80 kV. 40~60 nm thin sections were prepared by a Reichert Jung ultra microtome equipped with a diamond knife. Ultrathin sections were then laid onto 300 mesh copper grids and placed inside the TEM for scanning.

Scanning electron microscopy

Fractured surfaces of Single-Edged Notch Bend (SENB) specimens were probed using a JEOL 6301F field emission scanning electron microscope (SEM) at an acceleration voltage of 5 kV. Depending on the size of the nano-inclusions, conductive coatings of either gold or chrome were applied on sample surfaces in a Xenosput XE200 sputter coater. The JEOL 6301F can generate high-magnification images of up to 250,000X, thus allowing inspection of the nano-level surface features to a resolution of approximately ~3 nm.

Tensile testing

Uniaxial tensile tests were conducted on dog bone specimen according to the ASTM D 638-03 standard. Using an MTS810 universal testing machine, each sample was tested under stroke control condition at a loading rate of 0.5 mm/min. The modulus was estimated from the slope of the linear section of the stress-strain curve. Tensile strength and elongation at break were also recorded.

Fracture toughness testing

Plane-strain fracture toughness tests by three-point bending were carried out to determine the fracture toughness, K_{IC}, and strain energy release rate, G_{IC}, of nanocomposite samples. The SENB specimen geometry was chosen according to the ASTM D 5045-99 standard, and testing was performed using the MTS810 machine with a crosshead speed of 0.2 mm/min. A sharp pre-crack was generated in each specimen by tapping a fresh razor blade inside a machined notch. The fracture toughness was calculated from the following formulae:

$$K_{IC} = \left(\frac{P}{BW^{1/2}} \right) f(x), \text{ with } x = \frac{a}{W}$$

$$f(x) = 6x^{1/2} \cdots$$

$$\cdots \frac{\left[1.99 - x(1-x)\left(2.15 - 3.93x + 2.7x^2\right)\right]}{(1+2x)(1-x)^{3/2}}$$

where P is the load at failure, B is the specimen thickness, W is the specimen width, and a is the overall crack length. The strain energy release rate, G_{IC} was determined from the area under the load-displacement curve.

Results and Discussion

X-ray diffraction study

Interlayer spacing of the nanoclay was estimated by reading the X-ray diffraction patterns of cured nanocomposite samples. For illustration the d-spacing of the pure I.30E calculated from XRD is approximately $d_{001} = 1.9$ nm. Fig. 1 compares the X-ray diffraction results of the hybrid and epoxy-I.30E nanocomposites for various clay loadings. Disappearance of the diffraction peak for 1wt% I.30E signifies the possibility of either an exfoliated or intercalated configuration. The left shift of the diffraction peak for the hybrid specimen represents a laminated structure with interlayer spacing $d_{001} = 3.9$ nm (Frohlick et al., 2004; Hernandez, et al., 2007). But reappearance of the same peak in case of 3wt% I.30E means some of the nanoclays preserved their initial phase separated morphology.

Microstructure evaluation

TEM allowed for micro and nano-level images of the nanocomposites to be taken as well as a means of measuring the interlayer distance of the layered silicate. As shown in Fig. 2 Epoxy-I.30E nanocomposite exhibits intercalated silicate layers that are 10~15 nm apart from each other, whereas epoxy-M52N shows a uniform distribution of discrete 20~30 nm spherical micelles. Both M52N and I.30E nanocomposites were transparent under the visible light for the entire range of nanoparticle loadings considered. TEM microscopy also revealed aggregation of some of the silicate layers. Messersmith and Giannelis (1994) suggested this phenomenon as the edge-to-edge and edge-to-face interaction, and attributed this to hydrogen bonding between the silicate layer and the N-H group of the organic cation.

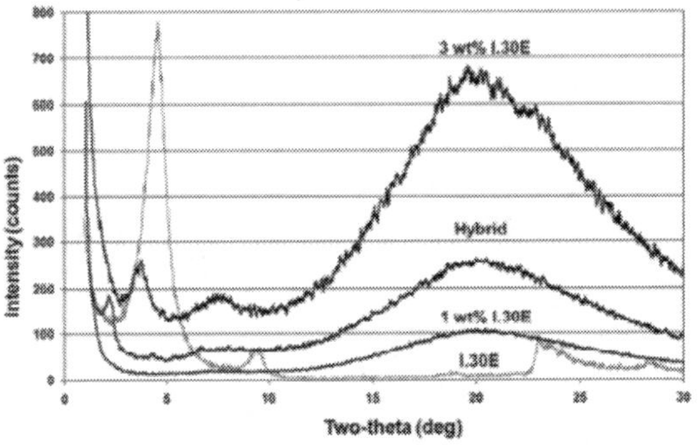

Figure 1: X-ray diffraction patterns of pure I.30E and cured nanocomposites.

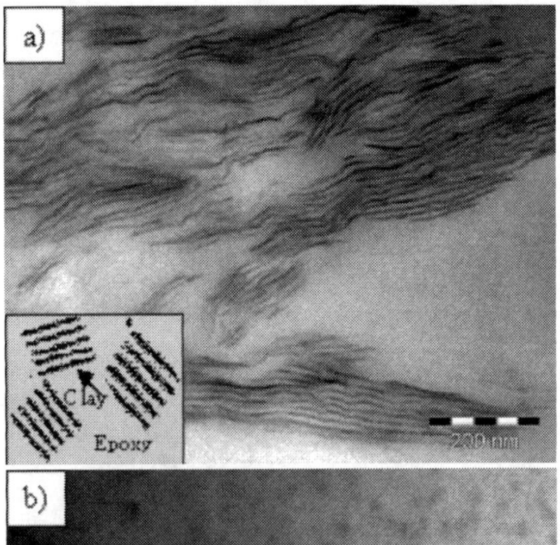

Figure 2: SEM images of the a) I.30E nanoclay and b) M52N block-copolymer modified epoxy.

Tensile results

Uniaxial tensile test results represent average values of at least three test samples. Tensile moduli of the M52N, I.30E and hybrid nanocomposites along with the neat epoxy are shown in Fig. 3. It was observed that stiffness changed significantly with increasing I.30E concentration due to high aspect ratio and rigidity of the clay fillers (Kornmann et al., 2002; Wang and Pinnavaia, 1998). Addition of M52N caused a slight reduction in modulus. Stiffness of the hybrid nanocomposite was in between the values of M52N and I.30E samples with 1wt% filler content. Similar results were observed for tensile strength as illustrated in Fig. 4 (Balakrishnan et al., 2005; Frohlick et al., 2004). However, the tensile elongation for samples made with the nanoclay dropped substantially. Elongation was increased for epoxy-M52N samples due to the formation of spherical micelle which consists of a soft rubbery core surrounded by a rigid shell (Fig. 5). In the case of the hybrid nanocomposite, strain at break was close to the value for neat epoxy. Enhancement of the modulus and strength and retention of ductility is an indication that the combined effect of both the nanoclay and block-copolymer are simultaneous operative for the hybrid nanocomposite.

Fracture toughness results

Fracture toughness of the modified epoxy nanocomposites are shown in Fig. 6 as a function of nanoparticle concentration. All epoxy-M52N nanocomposites greatly enhanced fracture toughness relative to the neat epoxy; an increase of approximately 60% was measured for a modifier loading of 3wt%. Layered silicate did not cause any considerable change to the fracture toughness values. As expected, toughness of the hybrid nanocomposite was higher and close to the epoxy-M52N toughness. It was ascertained from these results that fracture toughness for hybrid epoxy improves significantly. Similar improvements for hybrid nanocomposites have been documented before in the technical literature (Balakrishnan et al., 2005; Frohlick et al., 2003; Kinloch et al., 2009). The strain energy closely resembles the fracture toughness trend as depicted in Fig. 7 for the different nanocomposites.

Fracture surface analysis

Fractured surfaces of the SENB specimens were examined by SEM, and their surface topology images are displayed in Fig. 8 and 9, which represent low and high magnification images respectively. These micrographs exhibit the fast cracking region of a plane strain fracture toughness test. It is apparent from Fig. 8(b) and 9 (b) that the spherical micelles in M52N modified epoxy created matrix rich regions and

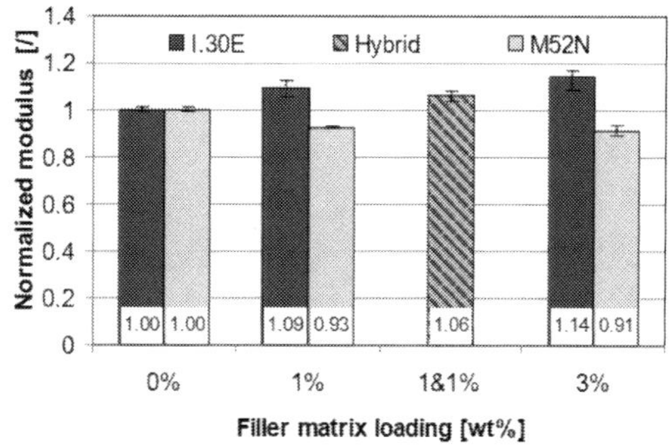

Figure 3: Normalized tensile modulus as a function of nanoparticle concentration.

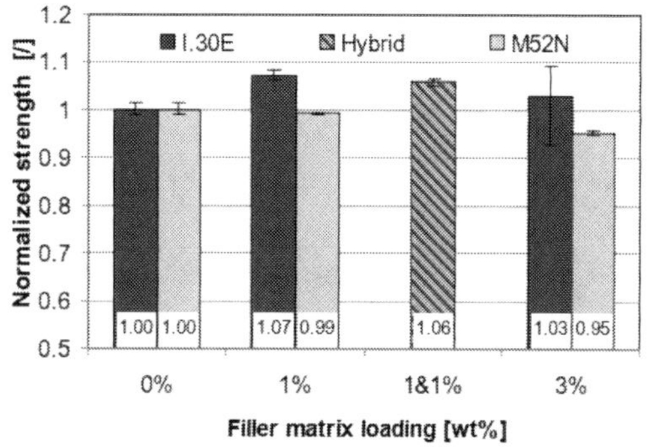

Figure 4: Normalized tensile strength as a function of nanoparticle concentration.

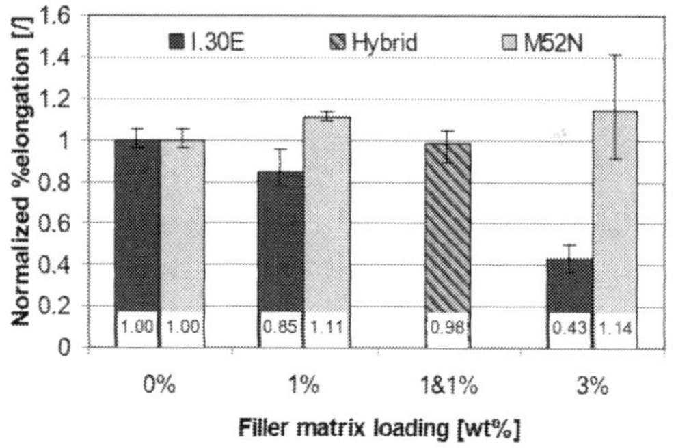

Figure 5: Normalized tensile elongation as a function of nanoparticle concentration.

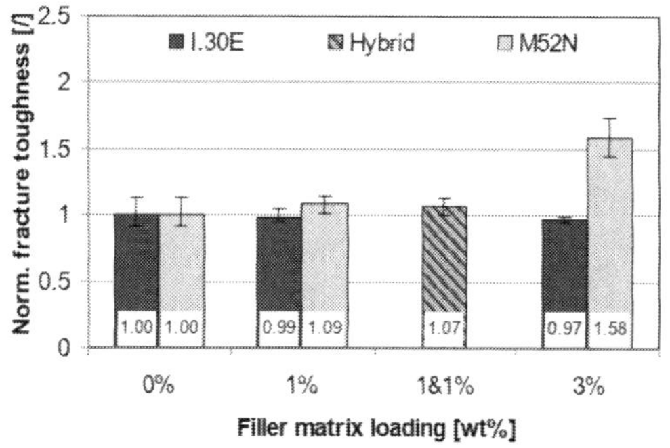

Figure 6: Normalized fracture toughness as a function of nanoparticle concentration.

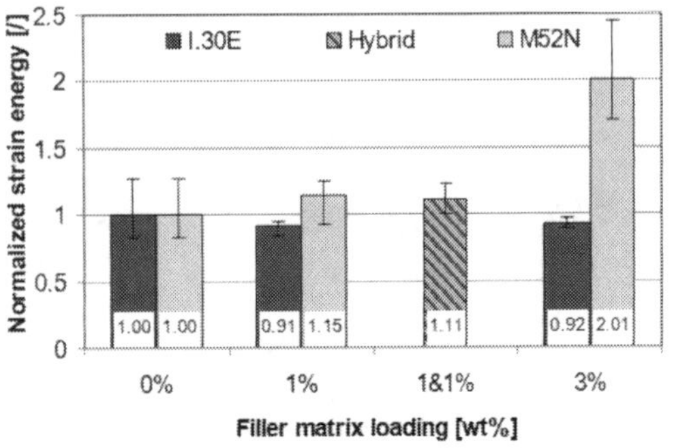

Figure 7: Normalized strain energy as a function of nanoparticle concentration.

matrix damage, which is reminiscent of pure epoxy. In contrast, nanoclay modified epoxy surfaces are extensively textured, more course and rough. The high magnification image in Fig. 9(c) shows the laminated clay layers with epoxy penetration in between adjacent clay sheets. It is surprising that the extensively convoluted morphology of the epoxy-I.30E fracture surface did not generate the expected toughness that was reported by various authors for an intercalated architecture (Zerda and Lesser, 2001; Zilg et al., 1999). It is speculated that flocculated micron size clay clumps are responsible for this exception. Hybrid surface features are similar to the epoxy-I.30E surfaces except that barely noticeable discrete microscopic

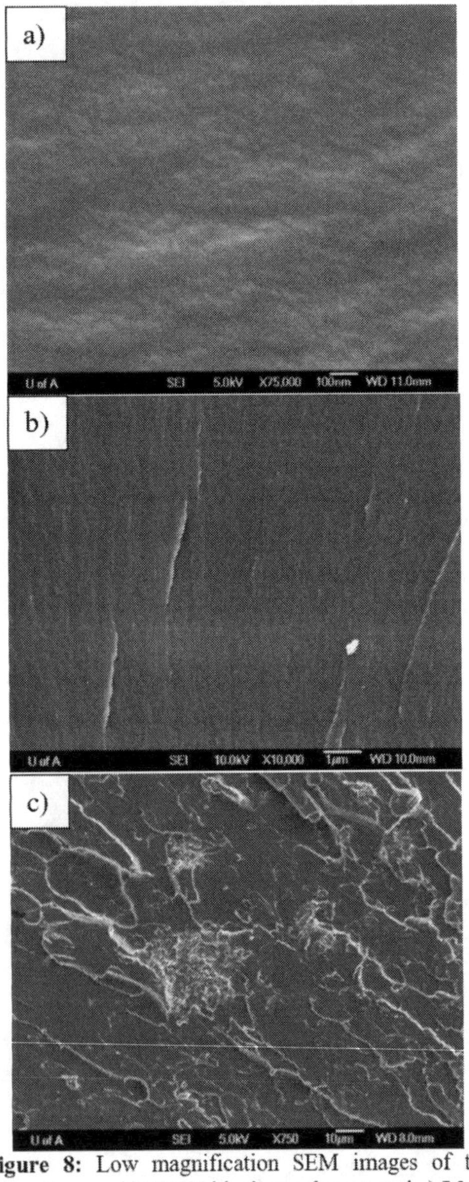

Figure 8: Low magnification SEM images of the a) neat epoxy, b) M52N block-copolymer, and c) I.30E nanoclay fractured surfaces.

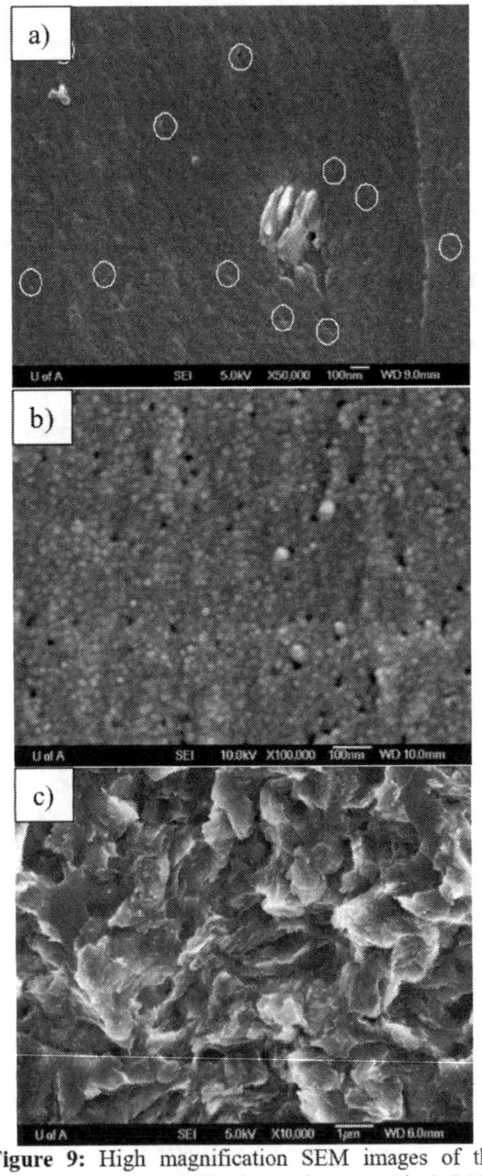

Figure 9: High magnification SEM images of the a) hybrid, b) M52N block-copolymer, and c) I.30E nanoclay fractured surfaces.

surface features are similar to the epoxy-I.30E surfaces except that barely noticeable discrete microscopic holes (created by spherical micelles) can be seen and are marked by white circles in Fig. 9(a).

Conclusions

Different characterization methods evidenced intercalated silicate layers for epoxy-I.30E and uniformly dispersed micelles for epoxy-M52N nanocomposites. Addition of the tri-block-copolymer M52N greatly enhanced toughness of the thermosetting epoxy, which was attributed to a mechanism of particle cavitation (debonding) and subsequent shear deformation of the matrix (Dean et al., 2001, 2003; Wu et al., 2005). Organically modified I.30E significantly improved stiffness of the epoxy nanocomposites while maintaining fracture toughness constant, presumably due to the particle agglomeration which acts as stress concentration points in the glassy epoxy matrix (Marouf et al., 2009). A hybrid nanocomposite formed by combining the nanoclay and block-copolymer was able to harness the favourable properties of both types of nanocomposites, and an improvement was achieved over the neat epoxy properties.

References

Becu-Longuet, L., Bonnet, A., Pichot, C., Sautereau, H., Maazouz, A., "Epoxy networks toughened by core-shell particles: influence of the particle structure and size on the rheological and mechanical properties" J. Appl. Polym. Sci., 72, 849-858 (1999).

Balakrishnan, S., Start, P. R., Raghavan, D., Hudson, S. D., "The influence of clay and Elastomer concentration on the morphology and fracture energy of preformed acrylic rubber dispersed clay filled epoxy nanocomposites" Polymer, 46, 11255-11262 (2005).

Dean, J. M., Lipic, P. M., Grubbs, R. B., Cook, R. F., Bates, F. S., "Micellar structure and mechanical properties of block copolymer-modified epoxies" J. Polym. Sci., 39, 2996-3010 (2001).

Dean, J. M., Grubbs, R. B., Saad, W., Cook, R. F., Bates, F. S., "Mechanical properties of block copolymer vesicle and micelle modified epoxies" J. Polym. Sci., 41, 2444-2456 (2003).

Frohlick, J., Thomann, R., Mulhaupt, R., "Toughened epoxy hybrid nanocomposites containing both an organophilic layered silicate filler and a compatibilized liquid rubber" Macromolecules, 36, 7205-7211 (2003).

Frohlick, J., Thomann, R., Gryshchuk, O., Karger-Kocsis, J., Mulhaupt, R., "High-performance epoxy hybrid nanocomposites containing organophilic layered silicates and compatibilized rubber" J. Appl. Polym. Sci., 92, 3088-3096 (2004).

Gam, K. T., Miyamoto, M., Nishimura, R., Sue, H. J., "Fracture behaviour of core-shell rubber-modified clay-epoxy nanocomposites" Polym. Eng. Sci., 43, 1635-1645 (2003).

Gerard, P., Boupat, N. P., Fine, T., Gervat, L., Pascault, J. P. "Toughness properties of lightly crosslinked epoxies using block copolymers" Macromol. Symp., 256, 55-64 (2007).

Höfflin, F., Könczöl, L., Döll, W., Morawiec, J., Mülhaupt, R., "Toughening of epoxy resins modified with polyetherester block copolymers: the influence of modifier molecular architecture on mechanical properties" J. Appl. Polym. Sci., 76, 623-634 (2000).

Hernandez, M., Sixou, B., Duchet, J., Sautereau, H., "The effect of dispersion on PMMA-epoxy-clay ternary blends: In situ study and final morphologies" Polymer, 48, 4075-4086 (2007).

Hydro, R. M., Pearson, R. A., "Epoxies toughened with triblock copolymers." J. Polym. Sci., 45, 1470-1481 (2007).

Kinloch, A. J., Shaw, S. J., Tod, D.A., Hunston, D. L., "Deformation and fracture of rubber-toughened epoxy" Polymer, 24, 1341-1353 (1983).

Kinloch, A. J., Maxwell, D. L., Young, R. J., "The fracture of hybrid-particulate composites" J. Mater. Sci., 20, 4169-4184 (1985).

Kinloch, A. J., Yuen, M. L., Jenkins, S. D., "Thermoplastic-toughened epoxy polymers" J. Mater. Sci., 29, 3781-3790 (1994).

Kinloch, A. J., Masania, K., Taylor, A. C., "The Fracture of nanosilica and rubber toughened epoxy fibre composites" Composite & Polycon 2009, Tampa, FL (2009).

Kornmann, X., Thomann, R., Mülhaupt, R., Finter, J., Berglund, L., "High performance epoxy-layered silicate nanocomposites" Polym. Eng. Sci., 42, 1815-1826 (2002).

Lan, T., Pinnavaia, T. J., "Clay-reinforced epoxy nanocomposites" Chem. Mater., 6, 2216-2219 (1994).

Lin, K.-F., Shieh, Y.-D., "Core-shell particles designed for toughening the epoxy resins. II. Core-shell-particle-toughened epoxy resins" J. Appl. Polym. Sci., 70, 2313-2322 (1998).

Liu, W., Hoa, S. V., Pugh, M., "Morphology and performance of epoxy nanocomposites modified with organoclay and rubber" Polym. Eng. and Sci., 44, 1178-1186, (2004).

Lee, H. B., Kim, H. G., Yoon, K. B., Lee, D. H., Min, K. E., "Preparation and properties of a carboxyl-terminated butadiene acrylonitrile toughened epoxy/montmorillonite nanocomposite" J. Appl. Polym. Sci., 113, 685-692 (2009).

Moloney, A. C., Kausch, H. H., Stieger, H. R., "The fracture of particulate-filled epoxide resins - Part 1" J. Mater. Sci., 18, 208-216 (1983).

Messersmith, P. B., Giannelis, E. P., "Synthesis and characterization of layered silicate-epoxy nanocomposites" Chem. Mater., 6, 1719-1725 (1994).

Marouf, B.T., Pearson, R.A., Bagheri, R. "Anomalous Fracture Behavior in an Epoxy-Based Hybrid Composite." Mater. Sci. Eng., A, 515, 49-58 (2009).

Pearson, R. A., Yee, A. F., "Toughening mechanisms in Elastomer modified epoxies" J. Mater. Sci., 21, 2462-2488 (1986).

Sultan, J. N., McGarry, F. J., "Effect of particle size on deformation mechanics of glassy epoxy" Polym. Eng. Sci., 13, 29-34 (1973).

Sun, C.T., Subramaniyan, A. K., "Toughening Polymeric Composites Using Nanoclay: Crack Tip Scale Effects on Fracture Toughness." Composites Part A, 38, 34-43 (2007).

Wang, Z., Pinnavaia, T. J., "Hybrid organic-inorganic nanocomposites: Exfoliation of magadiite nanolayers in an elastomeric epoxy polymer" Chem. of Mater., 10, 1820-1826 (1998).

Wu, J., Thio, Y. S., Bates, F. S., "Structure and properties of PBO-PEO diblock copolymer modified epoxy" J. Polym. Sci., 43, 1950-1965 (2005).

Zilg, C., Mülhaupt, R., Finter, J., "Morphology and toughness/stiffness balance of nanocomposites based upon anhydride-cured epoxy resins and layered silicates" Macromol. Chem. Phys., 200, 661-670 (1999).

Zerda, A. S., Lesser, A. J., "Intercalated clay nanocomposites: Morphology, mechanics, and fracture behaviour" J. Polym. Sci., 39, 1137-1146 (2001).

EVALUATION OF RADIATION RESISTANCE OF CHLOROSULFONATED POLYETHYLENE TO Γ-IRRADIATION AS JACKET CABLE

M. Tabaei[1],G. Mir Mohamad Sadeghi[1]* N.Sheikh [2], B. Shamayeli [1]

[1]*Department of Polymer Engineering, Amirkabir University of Technology, Tehran, Iran – tabaei@aut.ac.ir, gsadeghi@aut.ac.ir,b_shamayeli@yahoo.com*
[2] *Nuclear Science and Technology Research Institute, Radiation Applications Research School, Tehran, Iran – nasheikh@aeoi.org.ir*

This study aims to increase the radiation resistance of chlorosulfonated polyethylene (CSP) elastomer, used as a jacket material for electric cables, under γ-radiation environment. Four different formulations of CSP have been prepared to investigate effect of various stabilizers and other parameters. Different cured sheets, CSP named CP(1), CSP stabilized with trimethyl quinoline (TMQ) named CP(2) , CSP stabilized with mixtures of 2,6-Di-tert-butyl-4-(dimethylaminomethyl) phenol, 2-(2H-Benzotriazol-2-yl)-4,6-bis(1-methyl-1-phenylethyl) phenol and TMQ named CP(3) have been prepared. Another formulation based on CP(3) including lead dioxide (PbO_2) named CP(5).
The specimens were irradiated at various doses from 250 up to 1000 kGy. The effect of above mentioned additives and under-curing on performance properties of them has been studied by monitoring the degradation in γ-radiation using electrical & mechanical properties as well as swelling ratio. The radiation resistance of samples according to failure criteria of properties determined according to International Electrotechnical Committee: IEC 544 standard, (guide for determining the effects of ionizing radiation on insulating materials).degradation mechanism in the various formulations has been studied. Additives to stabilize CSP against a total dosage of 500 kGy of γ-radiation have been achieved.

Introduction
γ radiation produces drastic chemical and physical changes in polymeric materials. Many types of basic changes have been observed, such as scission of the main chain, crosslinking and changes in crystallinity.
Various organic materials, including polymeric jackets and insulations of electrical cables are used in radiation environments such as gamma irradiation rooms. Electrical cables used in these environments are required high stability. In some references, the critical dose of 100 kGy is quoted for chlorosulfonated polyethylene (hypalon) (wrigth, 2001).
Loss of cable functionality is usually determined by the changes in mechanical properties, cracking of the insulation preceding electrical failure (Kim, 2005).
Incomplete vulcanization improves radiation resistance that this improvement was due to the initial expenditure of radiation to overcome the deficiency in crosslink density (Anderson, 1960). Lead is resistant to ionizing radiation because its higher atomic number (Z) results in greater photoelectric and pair production absorption (Abdel-Aziz, Gwaily, 1997). Therefore, lead oxide (PbO_2) was used.In the present paper, the effect of some additive as well as under- curing technique on increasing of radiation resistance of CSP was described.

Experimental

Material
chlorosulfonated polyethylene was supplied by DuPont de Nemours under trade name CSM 40.
Antioxidants and stabilizer was used in the study with their chemical Structures and sources are described in the Table 1.

Table 1- Source and Chemical Structure of The Chemicals Used

Name	Source	Chemical structure
2,6-Di-tert-butyl-4-(dimethylaminomethyl)phenol **(phenolic antioxidant)**	Aldrich chemicals	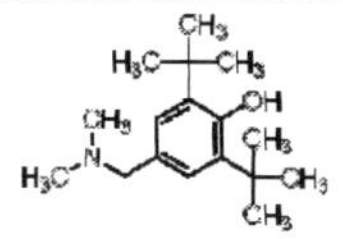
Polymerized 2, 2, 4-trimethyl-1, 2-dihydroquinoline **(TMQ)**	Bayer company	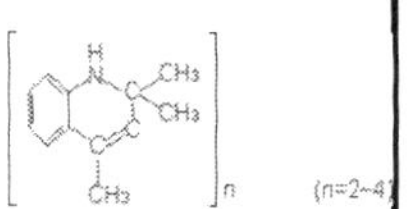
2-(2H-benzotriazol-2-yl)-4, 6-bis (1-methyl-1-phenylethyl) phenol **(UV absorber 234)**	Aldrich chemicals	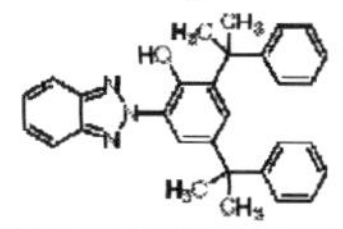

Preparation of samples
Crosslinking can be achieved with metal oxide curing system. Compounding of CSP was carried out with the necessary ingredients as per the compositions given in Table 2, on the open roller mill at 70°C. The sample numbers indicate the different additives such as antioxidants.
The test slabs of 15 cm*10 cm * 0.2 cm were molded using the hot press at 155°C and 7 tons pressure for

optimum cure time determined from the Monsanto Rheometer 100S. As well as incomplete vulcanization for CP(3) sample was carried out that named CP(4).

Table 2-Compoition of samples

CSP Ingredient	CP(1) phr[a]	CP(2) phr	CP(3) phr	CP(5) phr
CSP	100	100	100	100
Paraffin wax	4	4	4	4
Hard clay	20	20	20	20
Aluminum silicate	18	18	18	18
Carbon black	10	10	10	10
Aromatic oil	6	6	6	6
Zinc oxide	5	5	5	5
Magnesium oxide	5	5	5	5
Stearic acid	1.5	1.5	1.5	1.5
ethylene thiourea (ETU)	2	2	2	2
Tetramethylthiuramamidisulfide (TMTD)	0.2	0.2	0.2	0.2
N-cyclohexyl-2-benzothiazyl sulfenamide (CBS)	1	1	1	1
Sulfur	1.8	1.8	1.8	1.8
Vulcasil S	15	15	15	15
diethylene glycol (DEG)	3	3	3	3
TMQ	-	5	5	5
phenolic antioxidant	-	-	4	4
UV absorber 234	-	-	4	4
lead oxide (PbO_2)	-	-	-	4

[a] phr: part per hundred parts of rubber

Irradiation
The samples were irradiated with γ-rays in the presence of air at room temperature, in a ^{60}Co γ-irradiation cell type Gamma Cell-220 .The overall doses were 250, 500, 1000 kGy at a dose rate of 17.46 kGy/h.

Mechanical properties
Dumb-bell shaped specimens were used for the determination of mechanical properties. The tensile properties such as elongation at break were measured using appropriate ASTM method. Elongation at break was determined from stress-strain curves.

Equilibrium swelling measurements
The swelling ratios of CP(1), CP(5) in toluene were measured at 25°C. The samples were immersed in solvent. Then these were removed from the solvent and were frequently weighed after drying with a filter paper to remove excess solvent on the surface. Since there was no observation change in the solvent uptake after 96 hr, the swelling was assumed to have reached its equilibrium state. Equilibrium swelling ratio of the samples was determined gravimetrically through the following equation.

$$Q=1+(w_2/w_1-1)\,\rho_2/\rho_1 \qquad (1)$$

Where Q is the swelling ratio of polymers by volume, w_1 is the weight of the sample before swelling, w_2 is the weight of the sample at equilibrium swelling and ρ_1 and ρ_2 are the densities of the solvent and polymer, respectively.

Electrical properties
Dielectrical behaviour was examined on samples with dielectrical analyser (Tettex, model 2809A). This apparatus give both capacitance (C) and dissipation factor (tanδ). In the measurement of dielectric constant permittivity and dissipation factor a frequency of 50 HZ and constant voltage (1000V) was used.

Results and Discussion
It is very clear from the Figure 1 that the elongation reduces whereas modulus increases with the increased dose of γ-radiation. This trend observed in all samples. Therefore that all the samples deteriorate predominantly via crosslinking mechanism (Ito, 2005).

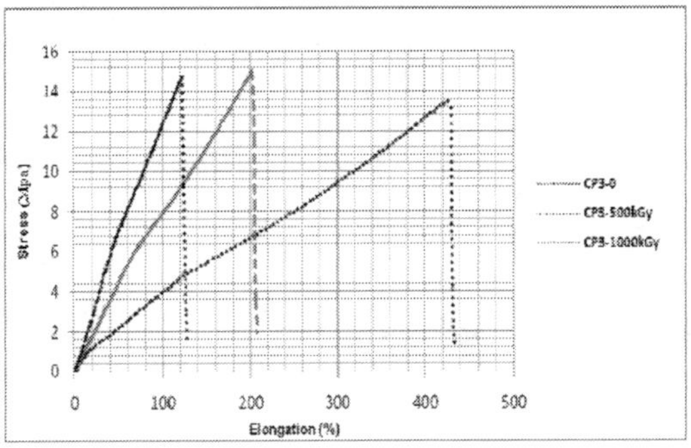

Figure 1-tensile curve for CP(3) sample before and after irradiation

Mechanical properties represent the most important criterion for estimation of radiation resistance of these materials. Elongation at break is the most sensitive property; criterion recommended for useability of any material is a requirement that 50% of the original value must not be exceeded (IEC Standard 544-2, 1991).
Figure 2 shows the retention in ultimate elongation (%) (ratio elongation at break at each dose to elongation at break at before irradiation) all the samples (CP(1), CP(2), CP(3), CP(4) and CP(5)) before and after irradiation at 500, 1000 kGy.
It can be seen that the changes in elongation at break are retarded with addition of TMQ in CP(2) whereas retention in ultimate elongation don't increase with addition of mixtures of stabilizers in CP(3). This phenomena occur due to immiscibility stabilizers together which appeared in the form of leaching on the surface of CP(3) sample. In other words, antagonist effects between stabilizers take placed. As seen from Figure 2, In comparison CP(4) with CP(3), the difference is observed in magnitude of retention in ultimate elongation (%) that CP(4) have shown improved although mixtures of stabilizers were shown antagonistic effect. Among the stabilized samples, the decrease in elongation at break (%) of 'CP (5)' sample is lesser than other samples.

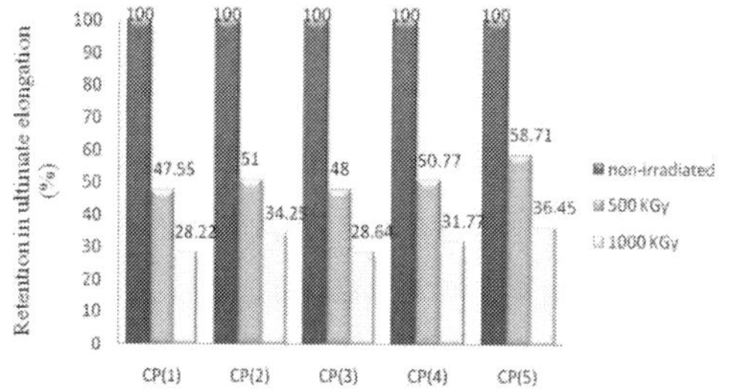

Figure 2- The retention in ultimate elongation (%) before and after exposure to 500, 1000kGy

In other words, this sample shows higher resistant under γ-irradiation which can be attributed to the fact that lead oxide exhibits higher stability against γ-irradiation. But stability was observed only up to 500kGy γ-irradiation that is confirmed by results of equilibrium swelling.

It can be understood from Figure 3 that the increasing in the cross link density of the material. It is clear that lead oxide can protect CSP vulcanizates against gamma radiation up to 500 kGy.

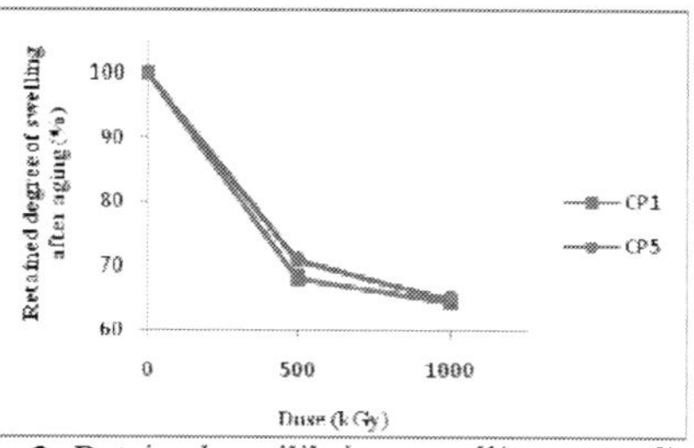

Figure 3- Retained equilibrium swelling vs. radiation dose for CP(1) , CP(5)

The stabilizing efficiency of CP(5) can be confirmed by Dielectric analysis.Figure 4 shows the change in Dielectric Constant of the test specimens. As increasing the irradiation dose, dielectric constant of CP(1) is increased. However, the dielectric constant decrease for CP(5) sample with increasing absorbed dose.

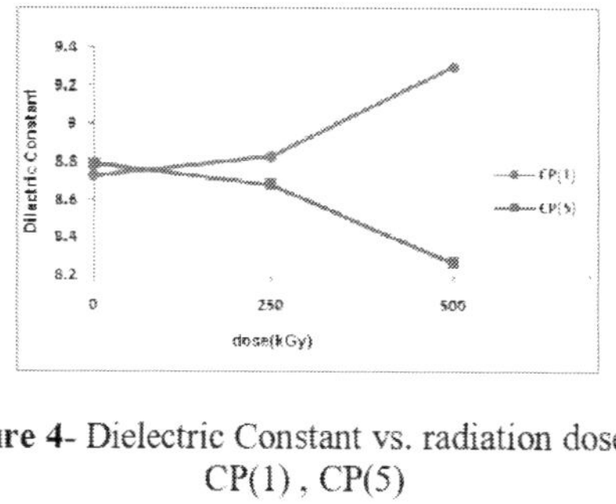

Figure 4- Dielectric Constant vs. radiation dose for CP(1) , CP(5)

Figure 5 shows the dissipation factor CP(1) , CP(5) as a function of irradiation dose. As increasing the irradiation dose, dissipation factor of CP(1) is increased. However, the dissipation factor decrease for CP(5) sample with increasing absorbed dose.

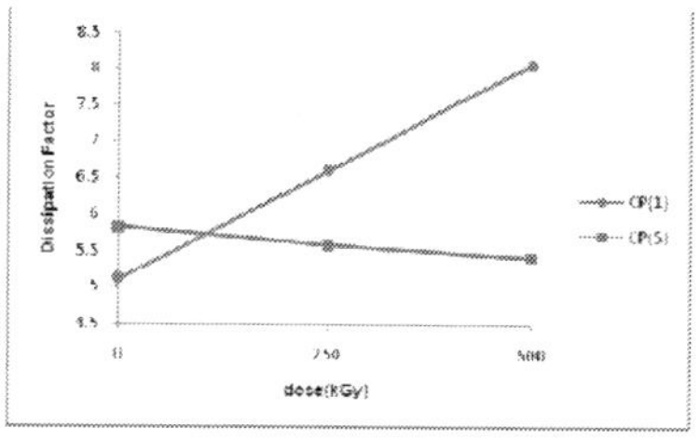

Figure 5- Dissipation Factor vs. radiation dose for CP(1) , CP(5)

By addition lead oxide, conductivity of CP(5) increase. The reduction in dielectric constant and dissipation factor can be attributed to the consumption of lead oxide during γ-irradiation, with higher doses.

Conclusions

In this study, the effects of different additives and under-curing technique on performance properties of them were investigated by monitoring the degradation CSP in γ-radiation using electrical & mechanical properties as well as swelling ratio. The result of tensile Provides evidence of chlorosulfonated polyethylene undergoing degradation predominantly via cross-linking reactions as well as indicate that mixtures of stabilizers were shown antagonistic effect. But addition lead oxide caused to higher resistant under γ-irradiation. Tensile and swelling ratio results show that lead oxide can protect CSP vulcanizates against gamma radiation up to 500 kGy. The increasing γ-rays irradiation dose, the values of dielectric constant and dissipation factor are decreased can be attributed to the consumption of lead oxide during γ-irradiation. Thus, it can be concluded that the stability of chlorosulfonated polyethylene against γ-irradiation can also be improved by suitable additives along with under-curing technique.

References

D.Wright "Failure of Plastics and Rubber Products: Causes, Effects and Case Studies Involving Degradation", 1st ed, Rapra Technology Limited, London (2001)

H.R. Anderson. "Compounding rubber for radiation resistance", J. Appl. Polym. Sci., 3, 9, 316-330 (1960)

International Electrotechnical Committee, Guide for determining the effects of ionizing radiation on insulating materials: procedures for irradiation and test. IEC 544-2(1991)

J.S. Kim. "Evaluation of cable aging degradation based on plant operating condition". J. Nucl. Sci. and Technol., 42, 8, 745–753(2005)

M. M. Abdel-Aziz, S. E. Gwaily. "Thermal and mechanical properties of styrene-butadiene rubber/lead oxide composites as gamma-radiation shields". Polym. Degrad. Stab., 55, 269-274(1997)

M.Ito. "the methodology study of time accelerated irradiation of elastomers", Nucl. Instr. and Meth. in Phys. Res. B 236 ,229–234 (2005)

AMINOACID-ASISTED SYNTHESIS OF TIO_2 NANOCRYSTALS WITH CONTROLLABLE SHAPE AND SIZE: A NOVEL AGENT FOR THE FABRICATION OF POLYMER/TIO_2 PHOTOVOLTAIC MATERIALS

Vu Thi-Thuy-Duong[1], Mighri Frej[1], Do Trong-On[2], Ajji Abdellah[3], Jayesh D. Patel[1]

[1]*CREPEC, Department of Chemical Engineering, Faculty of Science and Engineering, Laval University, Quebec, Canada*

[2]*Department of Chemical Engineering, Faculty of Science and Engineering, Laval University, Quebec, Canada*

[3]*CREPEC, Department of Chemical Engineering, École Polytechnique de Montréal, Quebec, Canada*

Abstract: This paper reports the synthesis of TiO_2 nanoparticles (NPs) with different morphologies using oleic acid (OA) and oleylamine (OM) as surfactants. These TiO_2 nanoparticles were then dispersed in polystyrene (PS) matrix using the drop-coating technique. Two different NPs shapes (nanosphere and nanobar) were achieved by varying the ratio of surfactants. Transmission electron microscopy (TEM) clearly shows these two shapes: The nanosphere shape had an average size of around 5nm and the the nanobar shape had an average size of 10 nm x 20 nm. The dispersion of TiO_2 NPs in PS matrix was characterized by scanning electron microscopy (SEM) technique on TiO_2/PS nanocomposite films. SEM results showed that the best dispersion of TiO_2 was achieved with the nanosphere shape. Optical absorption spectrum of TiO_2/PS nanocomposite films showed a strong absorption in visible region. It also showed that nanobar TiO_2/PS films absorb more in visible region compared to nanosphere TiO_2/PS films. This may be due to a combined effect of NPs size and shape and also to NPs dispersion. Fourier transform infrared spectroscopy of TiO_2/PS films showed a weak interaction between NPs and the host PS matrix.

Introduction

During the past decade, there have been considerable efforts in design and controllable preparation of organic-inorganic nanocomposites with different morphologies due to their potential application in various fields, such as catalysis, microelectronics, optics and photovoltaic devices. PS is considered as a conventional low cost polymer with good physical properties. However, PS does not absorb light in visible range and its optical properties are poor. Zan et al. (2004) found that by adding TiO_2 into PS polymer matrix, a significant enhancement of PS optical properties was achieved because of its proper bandgap (3.2 eV for anatase phase, 3 eV for rutile phase). These materials have promising applications for efficient electron transportation.

In this paper, we report the synthesis and characterization of TiO_2 nanoparticles with different shapes and their use for the preparation of TiO_2/PS composite thin films. The optical properties of PS-TiO_2 composites as well as the dispersion of TiO_2 NPs in the composite have been also evaluated.

Experimental

All chemicals were used as received without further purification. Titanium (IV) butoxide (TB, 97%), oleic acid (OA, 90%), oleylamine (OM, 70%); absolute ethanol and toluene (analytical grade) were purchased from Aldrich. Polystyrene (PS-3900) was purchased from Nova Chemical.

The synthesis of OA capped anatase TiO_2 nanocrystals with various shapes (nanosphere and nanobar) were synthesized by hydrolysis of TB using OA and OM as surfactants at low temperatures followed by the method reported in the literature (Dinh et al., 2009). Typically, 1 mmol of TB was added to a mixture of OA, OM and ethanol. The system was then heated at 150°C for 18 h. The resulting TiO_2 nanocrystals were washed with toluene and ethanol several times then dried at room temperature. After purification, TiO_2 was re-dispersed in toluene for thin film fabrication.

A mixture of PS and TiO_2 was prepared by solution blending process with a mole ratio of PS:TiO_2 is 1:1 in toluene solvent. Ultrasonic agitation was used to break up nanoparticle agglomerates in the solvent and to ensure dispersion homogeneity. The PS/TiO_2 thin film was fabricated by drop-coating method. 5 drops of mixture were dropped on surface of the glass then heated at 50°C to dry the solvent.

TiO_2 size and morphology characterization were done at 120 kV using a JEOL JEM 1230 transmission electron microscope (TEM). Scanning electron microscopy (SEM) analyses were carried out to observe the overall morphology of the composite on the JEOL 6360 instrument with an accelerating voltage of 15 kV. The UV–visible spectrum of the nanocrystals was recorded for the thin films on a Hitachi U-3010 spectrometer. Fourier transform infrared absorption spectroscopy (FTIR) spectra were measured with a FTS 45 infrared spectrophotometer using KBr pellet technique.

Results and Discussion

Figure 1 shows TEM images of TiO_2 nanoparticles capped by OA and OM, which were synthesized using different molar ratios of OA/OM. TiO_2 uniform nanospheres with diameter of 5 nm were obtained using a molar ratio of 6:4 (Fig. 1a) and TiO_2 nanobars (with size of 10 nm in width and 20 nm in length) were achieved using a molar ratio of 8:2 (Fig. 1b).

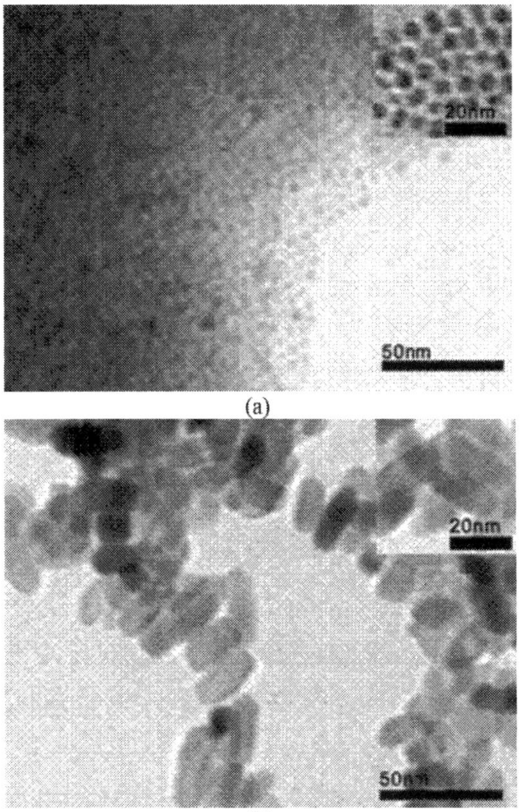

Figure 1 – TEM images of TiO_2 nanoparticles (a) TiO_2 nanospheres with an average size of 5 nm (b) TiO_2 nanobars with an average size of 10 nm x 20 nm.

The dispersion quality of TiO_2 nanoparticles into the PS matrix was evaluated by SEM. It could be obviously seen from figure 2 that TiO_2 nanospheres show better dispersion in PS matrix (Fig. 2b) than TiO_2 nanobars (Fig. 2a). This could be due to the difference in shape and size between nanospheres and nanobars.

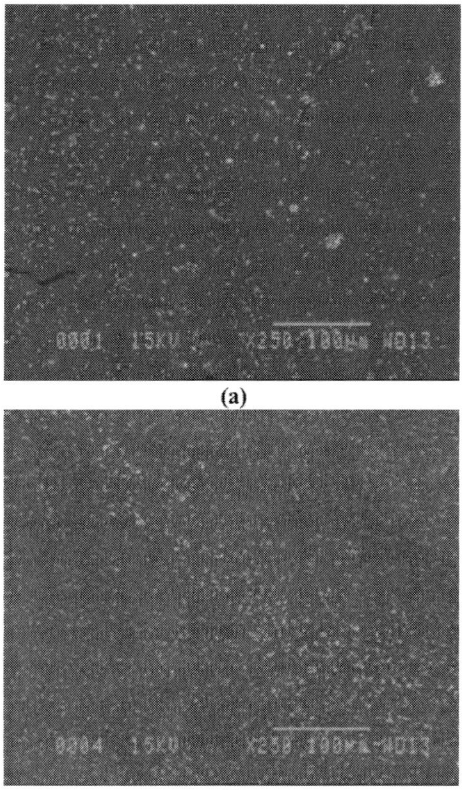

Figure 2 – SEM images of (a) nanobars TiO_2/PS and (b) nanosphere TiO_2/PS films.

Figure 3 shows the UV-vis absorption spectra for the samples of glass, pure PS film, and nanobars TiO_2/PS and nanospheres TiO_2/PS films. With adding TiO_2 NPs into PS matrix, the absorption band of the composite was shifted to visible light range, which is promising for photovoltaic devices. However, the absorption of nanobar TiO_2/PS films was significantly higher than that of nanosphere TiO_2/PS. This could be due to the difference in shape and size of TiO_2 nanoparticles, which can affect their electronic state (energy bandgap) (Li and Wang, 2003).

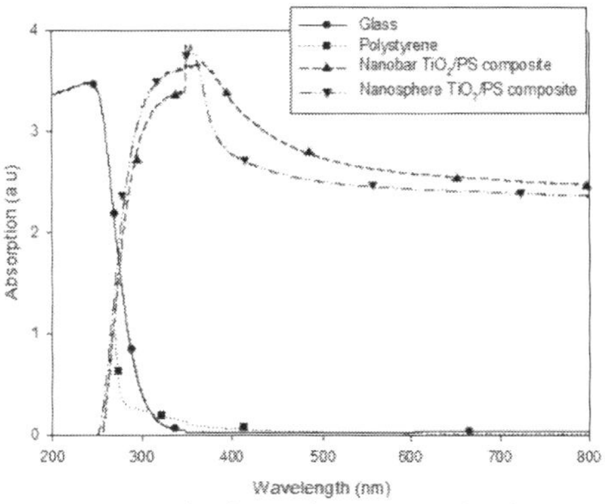

Figure 3 – UV-vis absorption spectra for the samples of glass, pure PS film, TiO_2 nanobars/PS and TiO_2 nanosphere/PS films.

FT-IR spectra corresponding to nanospheres TiO_2/PS, nanobars TiO_2/PS, and pure PS films are shown in figure 4. As seen in the spectrum of TiO_2/PS films, the peaks of phenyl ring are at 1496, 1450, 756 and 701 cm^{-1}, and they are same as in pure PS film. However, there is a significant decrease in intensity at 756 and 701 cm^{-1}. This could be due to the phenyl ring opening reaction in PS. The spectra also show a growth of carbonyl group (C=O) in the band region 1700-1710 cm^{-1} (Zan et al., 2004). The existence of C=O proves the oxidation of some $-CH_2-$ to –CO– on the chain of PS during the heat treatment, even at low temperature. In a previous study done by Su et al. (2007), the authors obtained a C-O-Ti band at 1265.78 cm^{-1}; composite films were prepared using ionic polymerization. However, this band was not clearly observed in our case by using the solution blending technique, which has more advantages in controlling the shape and size of doped-nanoparticles. Unfortunately, the latter technique has disadvantage in creating strong coupling between nanoparticles and polymer matrix.

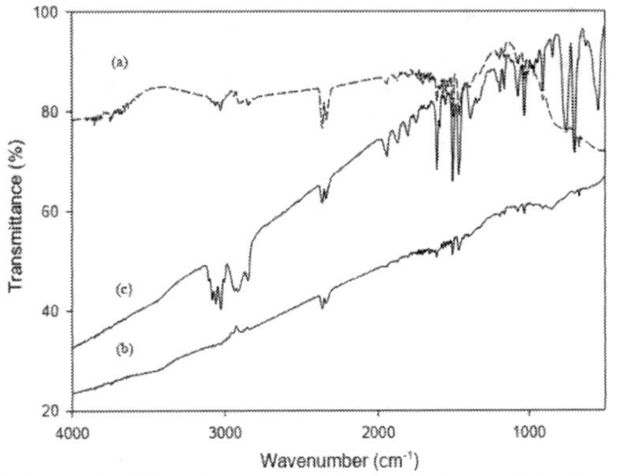

Figure 4 – FT-IR spectra of (a) nanosphere TiO_2/PS film (b) nanobar TiO_2/PS film, and (c) pure PS film.

Conclusions

In summary, various shapes of TiO_2 nanoparticles (nanospheres, nanobars) can be controlled by changing the reaction parameters. TiO_2/PS thin films were successfully prepared by solution blending and drop-coating techniques. The dispersion quality of TiO_2 NPs in PS matrix films as well as optical properties of these films were characterized. Results showed a promising application in solar cell devices. The next step of our study is to replace PS by MEH-PPV (2-methoxy-5-(2'-ethylhexyloxy)-p-phenylene vinylene) conjugated polymer .

References

Dinh, C. et al., "Shape-Controlled Synthesis of Highly Crystalline Titania Nanocrystals", ACS Nano, 3, 3737-3743 (2009).

Li, J., Wang, L., "Shape Effects on Electronic States of Nanocrystals" Nano Letters, 3, 1357-1363 (2003)

Su, B et al., "Preparation of TiO_2/PS complex nanoparticles", Materials Science and Engineering: A, 458, 44-47 (2007).

Zan, L. et al., "A new polystyrene–TiO_2 nanocomposite film and its photocatalytic degradation", Applied Catalysis A: General, 264, 237–242 (2004).

AGEING OF PLASTIC JACKET PIPE FOR DISTRICT HEAT PIPING SYSTEMS - IDENTIFICATION AND SEPARATION OF RELEVANT INFLUENCES

A. Leuteritz[1]*, W. Schuricht[2] and K.-D. Döring[3]

[1]*Processing Department, Leibniz Institute of Polymer Research Dresden – Leuteritz@ipfdd.de - Hohe Str. 6, 01069 Dresden, Germany;*
[2] *GEF Chemnitz mbH, Germany – info@gef-chemnitz.de;*
[3]*IMA Materialforschung und Anwendungstechnik GmbH, Germany - klaus-dieter.doering@ima-dresden.de*

For heat transport buried plastic jacket pipes are used consisting of a steel medium pipe insulated by polyurethane foam and protected by a jacket of polyethylene. In a still running project influencing variables of ageing leading to a loss of shear strength and insulation had to be identified and characterized separately in independent experiments in order to quantify their contribution to a summarized parameter. In an experiment to separate thermic and thermo-oxidative degradation specimen were aged accelerated at a temperature of 150 °C in closed bottles and open bottles allowing unhindered diffusion into the foam, respectively. Apon implementation of density independent reduced relative values, the loss of shear strength could be determined as slope of a straight line, and the magnitude of ageing due to oxidative processes turned out to be much higher then ageing due to thermal stresses. In another series of experiments the cell gass composition was evaluated from end-sealed plastic jacket pipes and compared to a cell gas composition which would be expected from the evaluated values of gas permeation through the casing material. Whereas for nitrogen in both cases similar values were found, the amount of oxygen was much lower than the calculated amount from the permeation rate, indicating a consumption due to a reaction.

Introduction

Europe has obliged itself to drastically cut its CO_2 emissions. The European commission has acknowledged that one of the key actions to be taken to meet this goal is to increase the exploitation of waste incineration, combined heat and power production (CHP), low temperature waste heat and biomass.
A prerequisite to do this is a heavy expansion of the district heating networks across Europe. These ambitious objectives for reduction of CO_2 emissions will also require energy effective cooling solutions.
Similar like in centralization of CHP production in large units can achieve again in cooling significantly higher efficiency through the displacement of small individual units. Hot and cold fluids transported over distance are increasingly being conveyed in buried pre-insulated pipelines. This applies particularly for hot water for space heating and cold water for space cooling.
The majority of piping used for this purpose is insulated with rigid polyurethane foam (PUR) in a PE-HD protective casing. There are even European standards EN 253:2009, EN 13941:2003 and EN 15698-1:2009 covering the product.
The PUR used has an unsurpassed combination of heat resistance, mechanical toughness and low thermal conductivity.
District heating and cooling networks are infrastructure where a service lifetime of 50-100 years may realistically be expected. It therefore has the highest interest and importance that the calculation of long term properties and energy efficiency factors are based on solid evidence. The aim of this investigation is to identify variables contributing to an ageing process of buried pre-insulated pipes, and quantify, as far as possible, their contribution to the process of aging in order to give a meaningful prediction of the lifetime of such plastic jacket pipes.

Experimental

A set of different experimental setups where used to investigate ageing processes in plastic jacket pipes.
In a first set of experiments pipes where aged using the setup described in EN253 for determination of the "Calculated Continuous Operating Temperature (CCOT)" at a medium Temperature of 150 °C was used. Sample pipes of different producers and production technologies with and without diffusion barriers of aluminium films were included in the experiments. In a second setup end sealed plastic jacket pipes were aged at 90 °C in a heating chamber to determine processes due to gas exchange through the casing. At several ageing times pipe samples from both setups where taken out and cut to peaces to determine the thermal conductivity at a mean temperature of 50 °C, the cell gas composition using a method developed by the Danish Technological Institute and the axial shear strength. In a model setup aluminium bottles where filled with polyurethane foams from two different producers in such a way that a foam density

of 75...80 kg/m³ was realized for specimen cut from the middle part of the foam filled bottle. Half of the bottles where closed using a brazen screw plug complemented by a teflon sealing rope in the convolution. The other half was perforated to allow unhindered air access to the foam. The bottles where aged at 150...180 °C and samples were taken after certain time intervals. Each sample bottle was cut into two specimen for determination of the axial shear strength and density. Further the colour of the freshly cut surface was characterized using a spectrophotometer CM-3600d (Fa. Konica Minolta). Gas permeation was determined using a Brugger GDP-C unit.

Results and Discussion

The ageing behaviour of plastic jacket pipes was determined by the setup described in EN 253 with a medium temperature of 150 °C. The shear strength of plastic jacket pipes was characterized directly after production and compared to values up to 10,000 h. In all cases a decrease of shear strength with a mean value of 22 % (5000 h) and 31 % (10.000 h) was found. In case of plastic jacket pipes containing an aluminium film as barrier a lower decrease in shear strength is recognized.

A second effect of ageing is seen in the thermal conductivity of plastic jacket pipes. Therefore thermal conductivity of fresh and aged pipes was compared. Fig. 1 shows the development for thermal conductivity of a sample without and with an aluminium film as barrier within the plastic jacket acting against diffusion of gases.

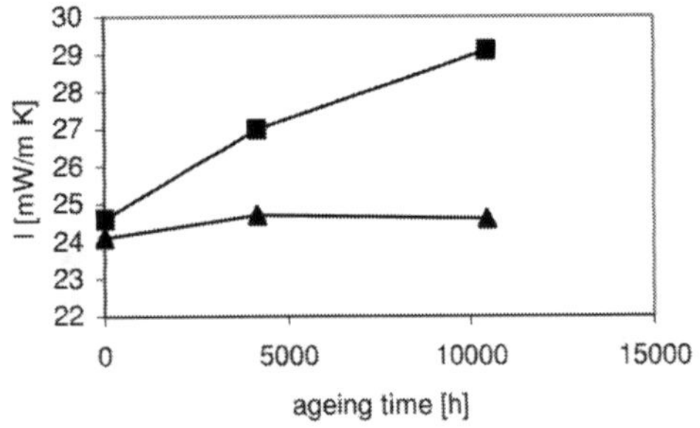

Figure 1 – Development of thermal conductivity from plastic jacket pipes with (triangle) and without (square) aluminium as diffusion barriere in the jacket material.

Thermal conductivity in PU-foams consists of several contributions as generally formulated in eq. (1):

$$\lambda = \lambda_{\text{solid}} + \lambda_{\text{radiation}} + \lambda_{\text{cell gas}} \quad (1)$$

and as a rough estimation 60 % of thermal conductivity arises from cell gases. Therefore it is good to have a control over cell gas composition. Usually PU foams for plastic jacket pipes are propelled from chemical and physical action by CO_2 and cyclopentane, because of their favourable low thermal conductivity. Due to diffusion processes these gases are exchanged with air, which has a higher thermal conductivity. The cell gas composition of fresh and aged pipes therefore was determined with the method developed at Danish Technological Institute. A drilling core was crushed under inert conditions and the evolving gas collected and analyzed using a chromatograph. From the change in gas composition the change in thermal conductivity was calculated. Both values agree quite good, thus proofing that cell gas determination as well as thermal conductivity measurements are trustable. To calculate the theoretical amount of cell gases the gas permeation through the casing material had been determined to 450 ml/m^{2}· 24h· bar O_2, 148 ml/m^{2}· 24h· bar N_2 and 1700 ml/m^{2}· 24h· bar CO_2 on a film of 0,15 mm thickness and at a temperature of 22 °C as well as the temperature dependancy. From these values the theoretical amount of nitrogen and oxygen in the polyurethane foam after ageing at 90 °C in end sealed plastic jacket pipes can be estimated and compared to those values determined by cell gas analysis (Fig. 2), neglecting a rather small diffusion resistance from the foam itself.

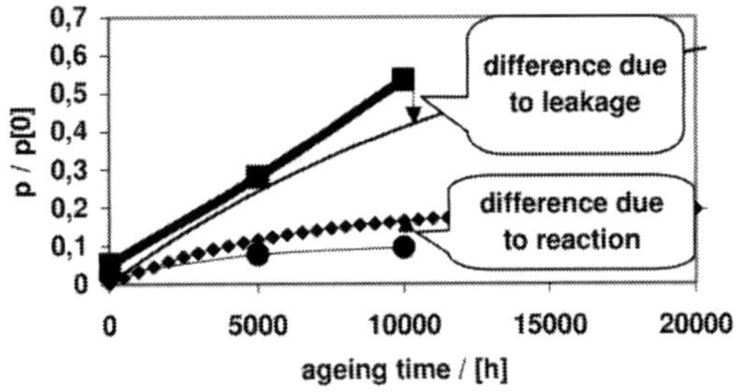

Figure 2 – Comparison of Nitrogen content (square) and oxygen content (dots) in samples after ageing at 90 °C in a heat box.

In Fig. 2 the determined nitrogen value from cell gas measurement always is higher than the expected value from calculation. This is easily explained by some leakage at the end sealants of the plastic jacket pipes. Nevertheless, the values determined for oxygen are always below the values of theoretical values, which is a strong evidence for an oxygen consuming process. This means ageing processes of oxidation even take place at temperatures as low as 90 °C in plastic jacket pipes.

In order to determine the different contributions of ageing due to thermal stress and thermo oxidative stress the bottle model was performed. Samples of freshly prepared foams with a mean density of 79 kg/m³ were characterized by a shear test in flat-wise plane and compared with samples aged in closed bottles as well as open bottles at 150 °C. Fig. 3 shows the determined shear strength in dependence of ageing time for samples under thermo oxidative conditions. It is recognized, that the shear strength is decreasing rapidly, and reaches within 30 weeks of ageing the limiting value of 0.12 N/mm² which is requested in EN253 as lower limit for plastic jacket pipes.

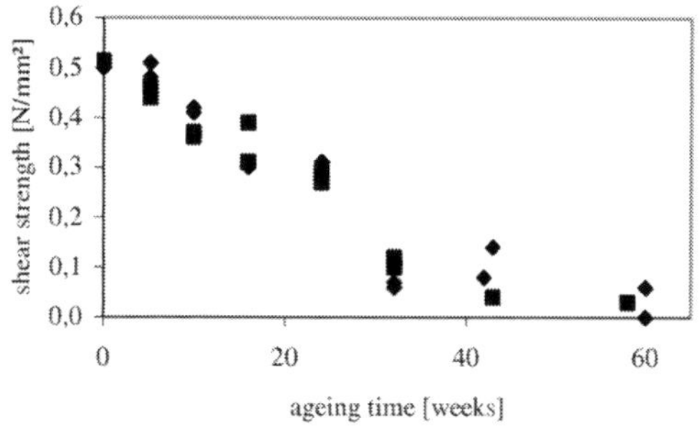

Figure 3 – Development of shear strength of bottle samples aged under thermo-oxidative conditions at 150 °C.

In the same time of ageing the development of mean density in the foam samples was decreasing from 79 kg/m³ to 52 kg/m³. Naturally, a less dense foam has to have different mechanical properties then a dense foam. Therefore, to truly characterize the contribution on thermo oxidative stress and to compare samples with different density, density reduced values have to be compared. It was found out[1], that there exists a power law correlation between the shear strength and the density given in equation (1):

$$\tau = 0.00136 * \rho^{1.35} \qquad (1)$$

with ρ in [kg/m³]. Using this equation reduced relative shear strength is introduced to characterise the loss of mechanical properties of aged foams by dividing the measured shear strength with the calculated value of shear strength from the actual density of the specimen, equation (2):

$$\tau_{\text{red. rel.}} = \tau_{\text{measured}} / \tau_{\text{calculated from density}} \quad (2)$$

Redrawing Fig. 3 with the calculated reduced relative shear strength then gives as slope a density independent gradient for a thermo oxidative parameter of ageing polyurethane with a fairly good slope of s = - 0.02 [a.u.] (see fig. 4).

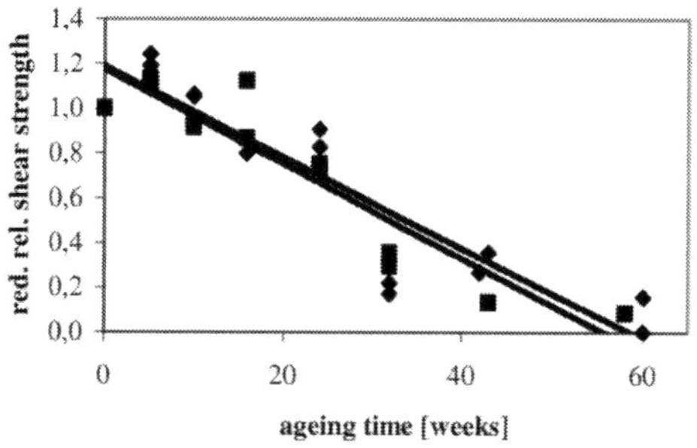

Figure 4 – Development of ageing due to thermo oxidative stress represented as density independed relative shear strength.

Within the limitations of this experiment the different foam systems do not differ. Second, the change in density is identified as additional factor contributing to the loss of shear strength during ageing of polyurethane foams for plastic jacket pipes. Analyzing the results for closed bottles aged at 150 °C a much smaller slope of about s = - 0.0016 results. Therefore the loss in shear strength in PU-foams is mainly caused by oxidative processes.

Conclusions

Ageing processes in plastic jacket pipes lead to a decrease in shear strength and increase in thermal conductivity. In the still running project gas exchange of cell gases with air was identified to cause these ageing processes. After introduction of density independent relative values a separation of thermal and oxidative ageing processes seems possible, thus contributing to a better forecast of product life time compared to the actual method described in EN253. The contribution just due to thermal stress seems rather low in case of shear strength loss. The results of this project already lead to two resolutions of the the European Committee for Standardization (CEN/TC 107/WG 3) with the aim to revise the EN253.

Acknowledgements

Gefördert durch:

Bundesministerium für Wirtschaft und Technologie

aufgrund eines Beschlusses des Deutschen Bundestages

This project is financially suppurted by the Federal Ministry of Economics and Technology.

SEPARATION OF ETHANOL/WATER MIXTURES THROUGH POROUS POLYPROPYLENE MEMBRANES

B. Hararak[1]*, N. Kerddonfag[1], W. Chinsirikul[1] and S. Kulprathipanja[2]

[1] *National Metal and Materials Technology Center (MTEC), Thailand- bongkoth@mtec.or.th ;noppadk@mtec.or.th; wanneec@mtec.or.th*

[2] *UOP LLC Company, USA- Santi.Kulprathipanja@uop.com*

Abstract - Separation of ethanol/water mixtures was performed by a pervaporation (PV) process using developed porous polypropylene membranes (PP-membranes). Several porous PP-membranes were fabricated by melt extrusion process of the PP containing a proprietary nucleating agent, followed by biaxial stretching process. Resulting membranes produced at various stretch ratios in a range of 1.5x1.5 to 4.0x4.0 (machine direction (MD) x transverse direction (TD)) had different pore structures and geometry. Results showed that the structure of the PP-membranes of low stretch ratios of up to 2.0 x 2.0, appeared to be discrete micro-pores while the elongated pores in a layered structure was observed in the membranes produced at higher stretch ratios. The permeation flux and separation factor of ethanol for these developed porous PP-membranes were determined. At a given test condition with 35 wt% ethanol feed concentration, and a test temperature of 50°C, the PP-membranes of a low stretch ratio of 1.5x1.5 (S1.5) could exhibit up to 10 times higher separation factor of ethanol but showed slightly lower permeate flux, as compared to the membranes of higher stretch ratios. The ethanol separation factor of this S1.5 membrane was 3.82. Based on a further study with varying feed concentrations of ethanol (from 10 to 90 wt%), it was found that the S1.5 PP-membranes, with a rather discrete micro-porous structure, possessed the highest ethanol separation factor of 8.5 and ethanol permeation flux of 20 g/m^2hr, at 10 wt% ethanol feed concentration and at 50°C feed temperature. Current findings reveal a potential means in developing an alternative material based on PP-membranes, for a separation of ethanol/water mixtures, in particular for the dilute ethanol solution. Details on the relationships of processing-structure-property of these specially developed porous PP-membranes are also included in this manuscript.

Keywords: Pervaporation; Ethanol/water mixture;

Introduction

The pervaporation (PV) is considered to be a promising alternative to conventional energy intensive technologies such as extraction or distillation in liquid-liquid mixture separation. There has been growing research interest in the application of pervaporation to biotechnology in the area such as recovery of ethanol from fermentation broth [1-3]. Success of pervaporation for ethanol-water mixtures depends largely on the characteristics of membranes, that has high ethahnol permeability and selectivity for development of hydrophobic membrane and high water permeability and selectivity for hydrophilic membrane.

The polymer membrane preparation processes includes solution casting, phase inversion, track-ecthing and extrusion followed by stretching [4-5]. In membranes research field, the solution casting is mainly used to prepare polymer membrane for pervaporation [6-9]. In this study, however, the extrusion followed by stretching were the process of interest. One major advantage of the extrusion and stretching process is no requirement of solvent and thereby eliminating any possible contamination from solvent and reducing processing cost. When preparing membranes based on semi-crystalline polymers via extrusion and stretching, crystals form, % crystallinity developed in the extruded samples (precursor) and the subsequent process parameters of stretching are key factors influencing the pore structure, geometry and the final membrane's properties [10-12].

The purpose of this study was to prepare porous polypropylene membranes (PP-membranes) by extrusion followed by stretching process and use them for separating ethanol from ethanol/water mixture. The special β form of PP crystal was intentionally developed during extrusion of PP sheets utilizing a proprietary nucleating agent. These sheets were precursor materials of the subsequent stretching process. In order to understand the influences of stretch ratios in membrane preparation on the pore structure and geometry of PP-membrane, the morphologies of PP precursor sheets and the porous PP-membranes after stretching were investigated using scanning electron microscope (SEM). Wide Angle X-ray Diffraction (WAXD) measurements were used to characterize the changes in the β-form crystals in the stretched PP membrane samples having different stretch ratios. In addition, the relationships between microstructural changes of the PP membranes and pervaporation performance were discussed.

Experimental

-Materials

Polypropylene homopolymer, HP420J from HMC polymers (Thailand) with MFI of 3 g/10 min was used in this study. A selected β-Nucleator was used.

-Precursor sheet and membrane preparation

A blend of PP with selected β nucleator of hundred ppm content was prepared using a single screw extruder. The β-nucleated PP masterbatch (MB) was diluted with neat PP to was and used to prepare PP precursor sheets by extrusion through a single screw equipped with a sheet die. The melt temperature in the extruder was 230 C. The chill roll temperature was maintained at 120°C by circulating hot oil. Thickness of the PP precursor sheets were about 0.4 mm.

PP precursor sheets were preheated at 110°C for 30 seconds, followed by simultaneous stretching via a lab-scale biaxial film stretching machine (Karo-IV Brueckner). Stretching temperature and speed were kept relatively constant at 110°C and 100 percent per second (pps), respectively. Stretch ratio of the sheet was a key parameter of interest. Stretch ratios in machine direction (MD) and transverse direction (TD) were varied; these were 1.5X1.5 (S1.5), 2.0X2.0 (S2.0), 2.5X2.5 (S2.5), 3.0X3.0 (S3.0), 3.5X3.5 (S3.5), and 4.0X4.0 (S4.0).

- Sheet and membrane characterization

Wide Angle X-ray Diffraction (WAXD) technique was also used to confirm the development of β-form crystalline structure of the PP precursor sheets and the stretched PP-membrane samples. Tests were performed on a JEOL JDX-3530 based on Ni-filtered CuK_α radiation. The diffraction angle (2θ) ranged from 10 to 30 degrees. The relative beta (b) crystal contents, indicated by a so-called K values, of the PP precursor sheet and stretched PP-membrane samples were also analyzed based on the well-known Turner-Jones equation [13]:

$$K_\beta = \frac{H_\beta(300)}{H_\alpha(110) + H_\alpha(040) + H_\alpha(130) + H_\beta(300)}$$

Where K_β is the relative content of the beta-form crystal in the PP precursor sheet and stretched PP-membrane samples and $H_\alpha(110)$, $H_\alpha(040)$, and $H_\alpha(130)$ are diffraction peak intensities of (110), (040), and (130) crystal planes of alpha-form crystals in the PP precursor sheet and stretched PP-membrane samples, respectively.

Pore structure of the PP precursor sheets and stretched PP-membrane samples was examined under scanning electron microscope (SEM), JEOL JSM-5410. SEM micrographs of the cross-sectional specimens were taken. Noted that the specimens were prepared by fracturing PP precursor sheets and stretched PP-membrane samples under liquid nitrogen. In the order to prevent charge accumulation on the specimen surface, all samples were coated with thin layer of gold by sputtering.

-Pervaporation test

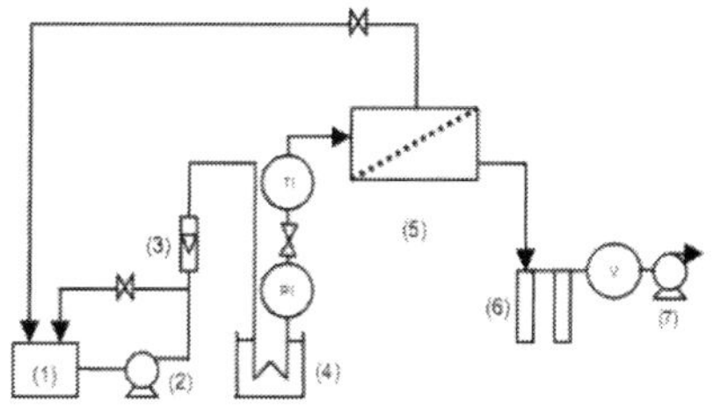

Fig. 1. Schematic of pervaporation test unit. (1) feed tank, (2) feed pump, (3) flow meter, (4) water batch, (5) membrane, (6) cold trap and (7) vacuum pump.

Pervaporation experiments were carried out using a pervaporation set-up, as shown schematically in Fig. 1. The effective area of testing membrane was 0.005413 m^2. The permeate was collected in a cold trap that was immersed in methanol dry ice. To achieve steady-state at a given temperature, the experiment was run for 3 hr prior to collection of the permeate. The ethanol concentrations in the feed and the permeates were determined by gas chromatography (GC). The membrane selectivity is expressed in terms of separation factor, $\alpha_{A/B}$, defined as the concentration ratio of the permeate divided by the concentration ratio of the feed:

$$\alpha_{A/B} = \frac{Y_A/Y_B}{X_A/X_B}$$

Where X and Y are weight fractions of A and B in the feed solution and permeate vapor, respectively.

In this study the subscript A and B refer to Ethanol and water, respectively. Permeation flux is defined by the following equation:

$$J = \frac{Q}{At}$$

Where J, Q, A and t are permeation flux (g/m^{2}hr), total amount of permeate (g), effective area of membrane (m^2) and operating time (hr), respectively.

The feed solution was ethanol aqueous solution with 35 wt% ethanol and kept at 50 °C to study the pervaporation performance of the PP membranes produced at various stretch ratios. The feed temperature of 50 °C was compared with 30° C and ethanol feed concentration was varied from 10 to 90 wt%.

Results and Discussion

-Beta crystal contents in unstretched and stretched samples

In general, based on WAXD measurements, the characteristic diffraction peak of the beta-form PP crystal (300) and (221) can be observed at the diffraction angle (2θ) = 16.2° and 23.1°, respectively , where those for the α-form PP crystal with the crystal planes of (110), (040) and (130) are shown at the diffraction angles 2θ of α_1:14.1°, α_2:16.9 and α_3:18.5, respectively. From Fig. 2, it can be seen that the intensities of β-form crystal peaks tend to decrease with increasing stretch ratios (from S1.0 to S4.0). The relative contents of β-form crystals in these samples are also represented by the 'K-values' (see Fig. 2 and Table 1). Among all sample tested, the K-value or β-crystal content of the PP precursor without stretch (S1.0) is the highest (0.95). In the case of stretched PP-membrane samples, K-values are 0.88, 0.39, 0.04, 0.01, 0.00 and 0.00, for the various stretch ratios of 1.5X1.5, 2.0X2.0, 2.5X2.5, 3.0X3.0, 3.5X3.5 and 4.0X4.0, respectively. Decreases in the β-crystal contents as increasing stretch ratios can be attributed to the transformation of β to α-crystaline structure under stretching at the temperature below the melting of β-form crystal, as also reported in reference [13].

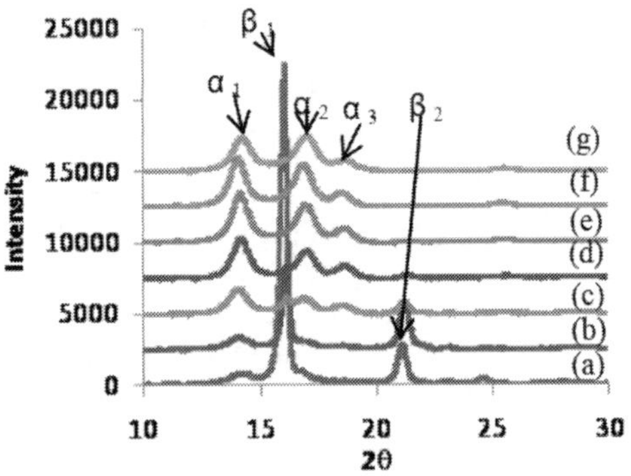

Fig. 2 WAXD patterns of the PP precursor sheet (S1) and stretched PP-membrane samples (S1.5-S4.0) with various stretch ratios: (a) Precursor sheet (S1.0), (b) S1.5, (c) S2.0, (d) S2.5, (e) S3.0, (f) S3.5 and (g) S4.0.

Table 1 Calculated K-Values of the PP precursor sheet and stretched PP-membrane samples with various stretch ratios.

Samples	K-Value*
Precursor sheet (S1.0)	0.95
S1.5	0.88
S2.0	0.39
S2.5	0.04
S3.0	0.01
S3.5	0.00
S4.0	0.00

Remarks: *K-Value is calculated using Turner-Jones equation.

-Pore structure and geometry analysis

Structures of the unstretched PP (PP precursor) and stretched PP-membrane sheets of varying stretch ratios were examined by SEM. As shown in Fig. 3, SEM micrographs of surface structures for all stretched PP-membrane sheets show no significant differences. However, the cross-sectional views illustrate noticeable different structures of stretched samples with varying stretch ratios. For S1.5 (1.5x1.5) and S2.0 (2.0x2.0) (Figs 2(b) and (c)), pores or voids in the micro-scales

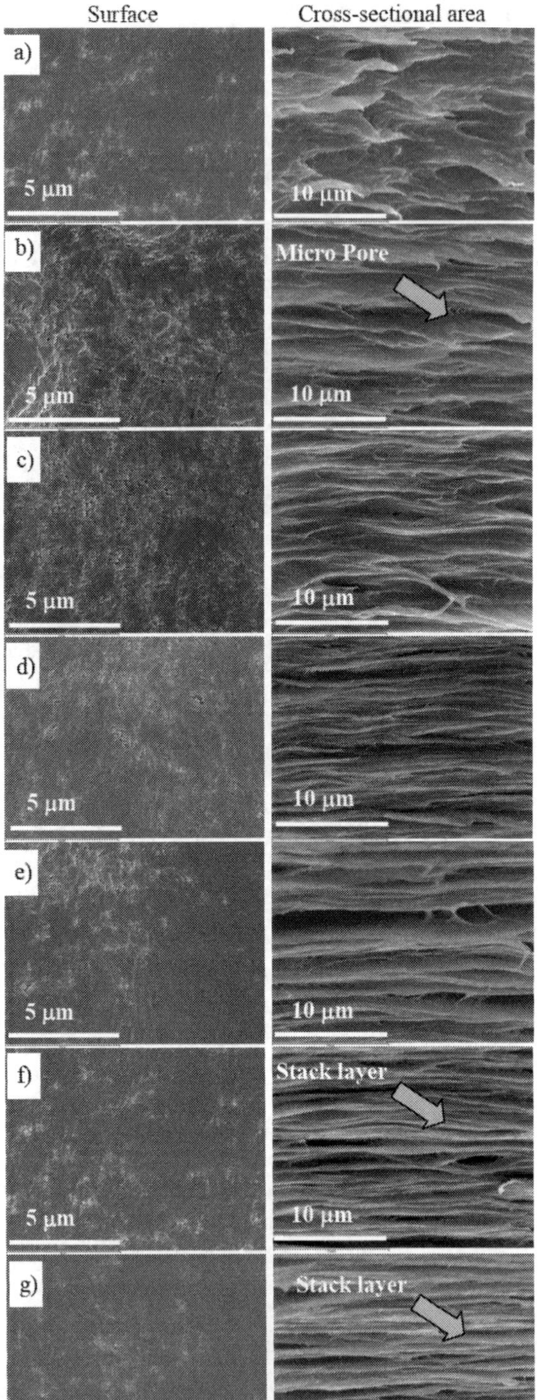

Fig. 3 SEM Micrographs showing surface and cross-sectional structures of PP precursor sheet (S1) and stretched PP-membrane samples with various stretch ratios (S1.5-S4.0): (a) S1, (b) S1.5, (c) S2.0, (d) S2.5, (e) S3.0, (f) S3.5 and (g) S4.0.

can be observed. It is possible that the micro-or nano-pores can be generated as a result of the transformation of β-form to α-form crystals upon stretching to a certain level. Owing to the almost 3% lower packing density of the β-form PP crystal, as compared to the α-form PP [14], one can picture that micro-pores occur to compensate for volume contraction when less dense β-form crystals are converted into a more dense α-form PP crystals. Reduced β-crystal contents (or increased α-crystal proportion) are also apparent for S1.5 and S2.0, as evidenced by the corresponding lower K-values. Further stretching from 2.0x2.0 tends to result in elongated pores, and for even higher stretch ratios of up to 4.0x4.0, pore or cavities of the samples become much thinner and aligned in a more or less stack-layer pattern.

In these stack layers of narrow cavities (Fig.3 (d), (e), (f), (g)) SEM micrographs also reveal the presence of elongated dense PP domains placed between cavities. For high stretched ratio samples (S3.0 to S4.0), their corresponding K-values are close to or equal to zero. K-value of zero is indicative of a complete transformation of β- to α-form crystals. A postulated picture of the structural changes from the micro-pores to stack layers of thin/narrower cavities is illustrated in Fig. 4. Another part of this study also involved measurements of thermal conductivities of the unstretched (S1) and stretched samples (S1.5-S4.0) to explore if any correlation exists between different porous structure and thermal conducting property. However, this series of measurements is not included in this article but was published in reference [15].

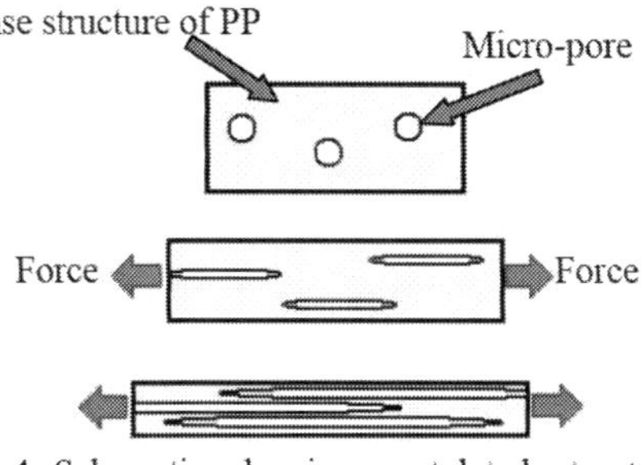

Fig. 4 Schematic showing postulated structures of micro-pores and pores elongated into stack-layers upon higher stretching.

-Effects of stretch ratios on pervaporation performance of PP membranes

A series of pervaporation tests was carried out. Important parameters include PP membranes with various stretch ratios, feed temperatures and feed concentrations of ethanol. This part describes effects of stretch ratios of the PP sheets on pervaporation performance; tests were performed using ethanol feed contration of 35 wt%, and at a temperature of 50 C. Fig.5 (a) and (b) show effects of stretch ratios on separation factor and flux, respectively. It can be seen that the ethanol separation factor decreases with increasing stretch ratios from 1.5x1.5 to 4.0x4.0.

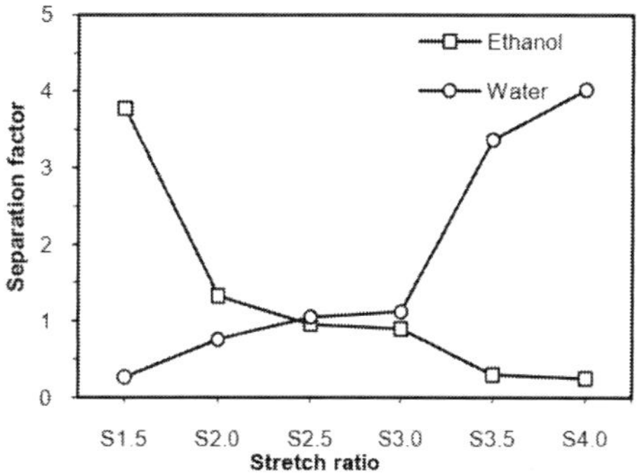

(a)

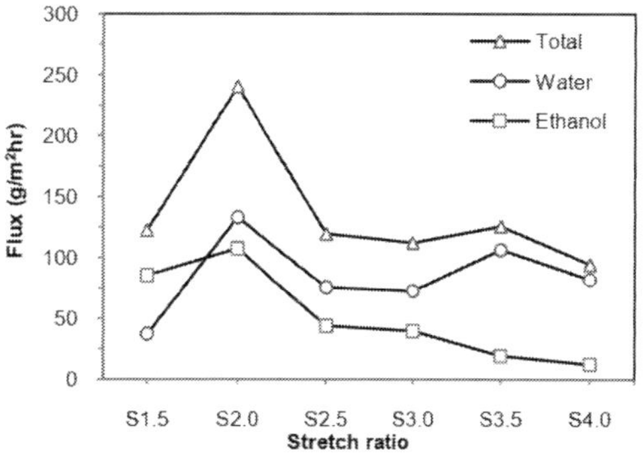

(b)

Fig. 5 Effect of stretch ratios of PP sheets on pervaporation performance: (a) Separation factor vs. Stretch ratio and (b) Flux vs. Stretch ratio (ethanol concentration=35 wt%, temperature= 50°C)

The high ethanol separation factor of 3.82 could be obtained for the stretched PP of 1.5x1.5 (S1.5) while the values tend to drop and become lower than 1 over the higher stretch ratios of 2.5x2.5 (S2.5) to 4.0x4.0 (S4.0), see Fig. 5 (a). Differences in the observed ethanol separation factors of all stretched PP samples could be attributed to different structures of the samples (e.g., pore geometry, porosity and overall structure) developed at various stretching conditions. As earlier described and revealed by SEM, more discrete micro-pores were apparent for the samples with low stretch ratios (S1.5-S2.5), whereas elongated pores and cavities in a stack pattern were present in the samples with higher stretch ratios above 2.5x2.5. In the case of ethanol permeation flux, as shown in Fig. 5 (b), S 1.5 is the only sample having higher ethanol flux than water flux, even though the total flux of most samples are in a similar range of ~100-125 g/m^{2}hr, except for a very high total flux of S2.0 (~250 g/m^{2}hr). It should also be noted that the PP precursor sheet (unstretched sample, S1) had no permeate flux under test. Although a thorough understanding of the influence of the porous membrane structure (sample S1.5) on the high ethanol separation factor and flux is not available at present, overall results tend to suggest that the stretched PP membranes of 1.5x1.5 have a structure favoring ethanol separation from the ethanol/water mixtures at a given test condition. Thus, other two sets of experiments to study effects of feed temperatures and ethanol concentrations were performed using the stretched PP of 1.5x1.5 (S1.5).

-Effects of feed temperature on pervaporation performance

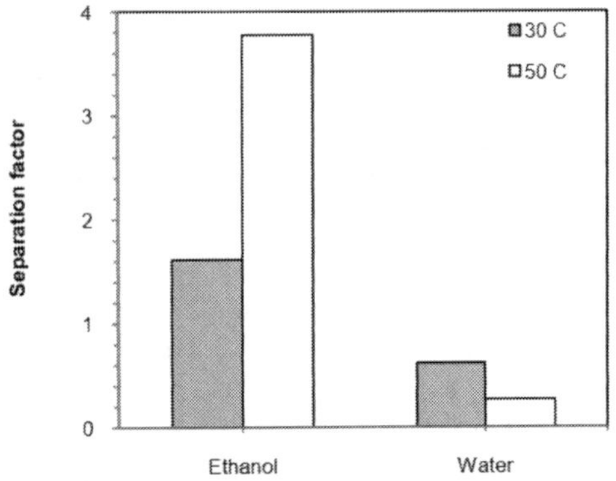

(a)

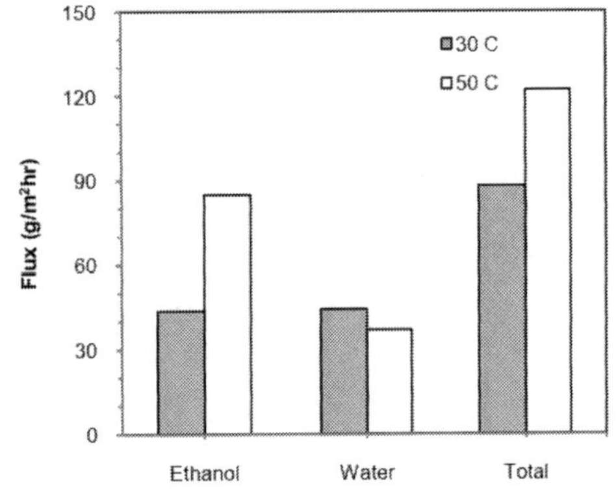

(b)

Fig. 6 Effect of feed temperatures on pervaporation performance of S1.5 membrane: (a) Separation factor vs. Feed temperature and (b) Flux vs. Feed temperature (ethanol concentration=35 wt%)

Fig.6 presents the performance of sample S1.5 as a pervaporation membrane. Results were assessed based on two different feed temperatures: 30°C and 50°C, with 35 wt% ethanol in feed solution. The S1.5 membrane shows a higher ethanol separation factor of 3.82 at 50°C and a slightly lower water separation factor than those obtained at 30°C. Fig. 6 (b) shows the permeation flux of S1.5 membrane for the two different feed temperatures. With increasing feed temperature from 30 °C to 50°C, the total and ethanol permeation flux increases, although the water permeation flux is somehow slightly lower at higher feed temperature of 50°C. As expected, an increase in ethanol and total permeation flux at higher feed temperature can be explained by the general solution-diffusion mechanism, where both solubility of the permeate on the membrane surface and diffusion through the membrane increase with increasing temperature.

-Effects of feed concentrations on pervaporation performance

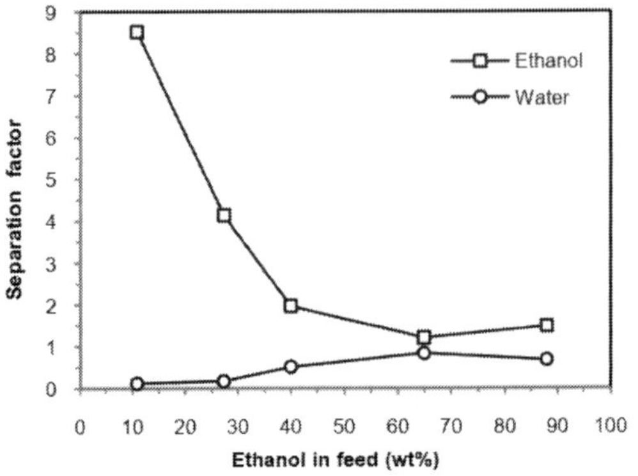

(a)

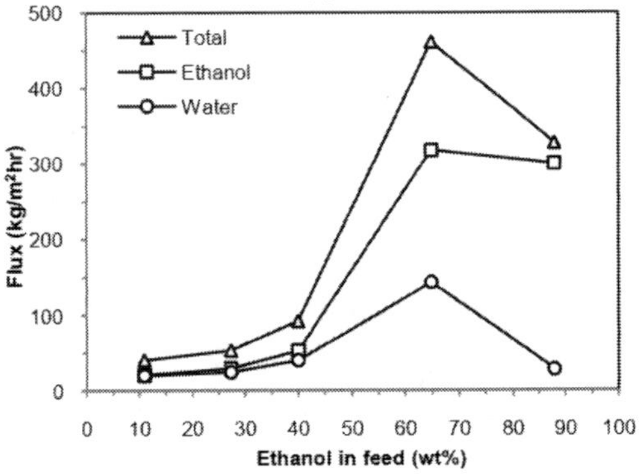

(b)

Fig. 7 Effect of feed concentrations on pervaporation performance of S1.5 membrane(a) Separation factor vs. Ethanol in feed (wt%) (b) Flux vs. Ethanol in feed (wt%) (pervaporation test temperature = 50°C)

Effects of ethanol contents in the feed solutions, on the ethanol separation factor and permeation flux of the S1.5 membrane was examined using the ethanol feed concentrations ranging from 10 to 90 wt% and tests were performed at a feed temperature of 50° C. Fig.7 (a) and (b) illustrate results of ethanol separation factor and permeation flux as a function of ethanol concentration in the feed solution. It is clearly seen in Fig. 7(a) that the highest ethanol separation factor of ~8.5 is observed at 10 wt% ethanol, the values then decrease as increasing ethanol feed concentrations up to 90 wt%. The water separation factors increase slightly with increasing ethanol feed concentration and reach a factor of ~1 at high wt% of ethanol in the feed solution. In terms of the permeation flux (Fig. 7(b)), measured values of total, ethanol and water flux tend to show a general increasing trend with increasing ethanol feed content. Very high flux values are observed at 65 wt% ethanol, followed a decrease at 90 wt% ethanol content. As well realized that an increase in permeation flux with a possible decrease in separation factor can be possibly due to a swollen polymer membrane, especially at a high feed concentration of the permeant (or ethanol in this study). Swelling can cause structural changes of the polymer membrane and in turn its separation performance. Y. Huang et al.[8] found that the ethanol separation factor decreased as ethanaol in feed increased, due to the polydimethylsiloxane (PDMD) membrane became more swollen. Swelling of polymer membranes improved the polymer chain mobility, thus the permeation flux of both ethanol and water increased.

Conclusions

Porous Polypropylene membranes were effectively developed through extrusion and stretching process. Developed β-form PP crystal and its contents in the precursor sheets, together with stretch ratios of the sheets, play an important role in controlling porous structures (pore dimension, geometry and orientation) of the final stretched PP membranes. Based on the pervaporation tests, PP membranes with different porous structures as a result of various stretch ratios showed different ethanol separation factors and permeation flux. The PP-membranes of a low stretch ratio of 1.5x1.5 having a more or less discrete micro-pores appeared to favor ethanol separation from the ethanol/water mixtures at given test conditions (i.e., temperature of 50 C, ethanol feed concentration of 10-90 wt%). At 10 wt% ethanol feed concentration and at 50°C feed temperature, this S1.5-PP porous membrane showed the highest ethanol separation factor of 8.5. The resulting permeate had ~51 wt% ethanol, and total permeation flux was ~40 g/m^2hr. Overall results suggest a potential application of our developed porous PP membranes in separation of ethanol/water mixtures, in particular for the dilute ethanol solution such as in the initial stage of ethanol fermentation broth with 5-12 wt% ethanol.

Acknowledgements

The authors would like to acknowledge a financial support of this research from National Metal and Materials Technology Center (MTEC), NSTDA and Asst.Prof. Paiboon Innachitra, Prince of Songkla University (PSU) for technical supports of pervaporation tests.

References

[1] D.J. O'Brien, L.H. Roth, A.J. McAloon, Ethanol producing by continuous fermentation-pervaporation: a preliminary economic analysis, J. Membr. Sci 166 (2000) 105-111.

[2] M.D. Luccio, C.P. Borges, T.L.M. Alves, Economic analysis of ethanol and fructose production by selective fermemtation couple to pervaporation: effect of membrane cost on process economic, Desalination 147 (2002) 161-166.

[3] L.M. Vane, A review of pervaporation of product recovery from biomass fermentation processes, J. Chem. Technol. Biotechnol. 80 (2005) 603-629.

[4] Encyclopedia Polymer Science and Engineering (1989).

[5] S.P. Nunes, K.V. Peinemann, Membrane Technology in the Chemical Industry 2nd revised (2006).

[6] J.J. Shieh, R.Y.M. Huang, Chitosan/*N*-methylol nylon 6 blend membranes for the pervaporation separation of ethanol-water mixtures, J. Membr. Sci 148 (1998) 243-255.

[7] X. Liu, Y. Sun, X. Deng, Studies in the pervaporation membrane of permeation water from methanol/water mixture, J. Membr. Sci 325 (2008) 192-195.

[8] Y. Huang, P. Zhang, J. Fu, Y. Zhou, X. Huang, X. Tang, Pervaporation of ethanol aqueous solution by polydimethlysiloxane/polyphoshazene nanotube nanocomposites membranes, J. Membr. Sci 339 (2009) 85-92.

[9] S.H. Chen, R.M. Liou, C.S. Hsu, D.J. Chang, K.C. Yu, C.Y. Chang, Pervaporation separation water/ethanol mixture through lithiated polysulfone membrane, J. Membr. Sci 193 (2001) 59-67.

[10] S.H. Tabatabaei, P.J. Carreau, A. Ajji, Structure and properties of MDO stretched polypropylene, Polymer 50 (2009) 3981-3989.

[11] S.H. Tabatabaei, P.J. Carreau, A. Ajji, Effect of processing on the crystalline orientation, morphology, and mechanical properties of polypropylene cast films and microporous membrane formation, Polymer 50 (2009) 4228-4240.

[12] F. Sadeghi, A. Ajji, P.J. Carreau, Analysis of microporous membranes obtained from polypropylene films by stretching, J. Membr. Sci 292 (2007) 62-71.

[13] Turner-Jones, A., Aizlewood, A. M. and Beckett, D. R., *Makromol. Chem.* 75, 1964, p. 134.

[14] P. Jacoby, Beta Nucleating Masterbatch Offers Enhanced Properties in Polypropylene Products, Plastic Additives & Compounding, May/June 2007, 32-35.

[15] B. Hararak, U. Watcharosin, W. Chinsirikul, Thermal Conductivity Measurement of Porous Polypropylene Films, Asian Conference on Thermal Analysis and Application (2009) 167-170.

Concentration of Etyhanolf rom Bio-ethanol through Cellulose Nitrate and Cellulose Acetate membranes during Temperature-difference Controlling Evapomeation

*Tadashi Uragami**

Kansai University - Suita, Osaka - Japan

We have been investigating the permeation and separation characteristics of aqueous ethanol solutions produced from the fermentation of biomass through polymer membranes by pervaporation, evapomeation (EV) and temperature-difference controlled evapomeation (TDEV) techniques. It was revealed that the membrane performance was influenced significantly by the structure of polymer membranes and consequently the control of membrane structure was one of important factors for an improvement of the membrane performance. In this paper, to obtain a fundamental knowledge for the effect on the membrane structure in concentrating bioethanol by TDEV, in which temperatures of the liquid feed solution and the membrane surroundings can be controlled, cellulose nitrate (CN) and cellulose acetate (CA) that have remarkably different affinity for ethanol are selected as membrane material, and various asymmetric membranes which have different physical and chemical structure were prepared by wet method. The permeation and separation characteristics for aqueous bioethanol solutions through these CN and CA membranes in TDEV were studied. When asymmetric CN and CA membranes having similar physical structure were applied to the permeation for an aqueous solution of 10 wt% ethanol in TDEV, in which temperatures of the liquid feed solution and the membrane surroundings were 40 and 25 oC, respectively, the ethanol concentration in the permeate of asymmetric CN membrane was higher than that of asymmetric CA membrane. In cellulose ester membranes that have similar physical structure and different chemical structure, it was found that the ethanol/water selectivity of CN membrane with higher affinity to ethanol was greater. The permeation and separation mechanisms of the ethanol/water selectivity through porous membranes in TDEV are discussed in detail.

ENHANCING THE RECYCLABILITY OF POLY(ETHYLENE TEREPHTHALATE) (PET) FOR HIGH-VALUE APPLICATIONS: NOVEL PROCESSING BY SOLID-STATE SHEAR PULVERIZATION

J. M. Torkelson[1,2]*, C. Pierre[2], and P. J. Brunner[2]

[1]Department of Chemical and Biological Engineering, Northwestern University, Evanston, Illinois 60208 USA – j-torkelson@northwestern.edu
[2]Department of Materials Science and Engineering, Northwestern University, Evanston, Illinois 60208 USA – cynthiapierre2008@u.northwestern.edu, philipbrunner2008@u.northwestern.edu

Abstract - One of the grand challenges associated with the field of "green polymers" is to enhance the effective recyclability for high-value applications of synthetic polymers made from non-renewable resources. Here we apply a novel, industrially scalable process method called solid-state shear pulverization (SSSP) to poly(ethylene terephthalate) (PET) to overcome a stumbling block in such high-value recycling of PET. The SSSP apparatus is a modified twin-screw melt extruder in which the barrel is cooled rather than heated. Process variables are identical to those of twin-screw extrusion (feed rate, temperature, screw design, and screw speed), and the powder or particulate output can be directly melt processed into final products using conventional melt process methods. When SSSP is applied to PET, the product exhibits both a tunable increase in melt viscosity (demonstrated factor of 1.2 to 2.0) and a major enhancement in crystallizability. This increase in melt viscosity originates from mechanochemistry during SSSP leading to lightly branched PET and addresses a long-standing problem regarding the high-value recycling of PET, i.e., melt processing PET at ~550 K into high-value products results in a molecular weight reduction caused by hydrolysis reactions, thus making the viscosity too low for effective use of recycled PET in high-value applications. Additionally, the enhanced crystallizability of the the PET subjected to SSSP indicates that any need for cold crystallization of the blow-molded PET product can be reduced or potentially eliminated, providing some savings in process time and energy.

Introduction

Sustainable polymer chemistry and engineering can involve a number of approaches ranging from the development of monomers and polymers from renewable resources to the effective recycling of non-renewable polymers for high-value applications. At present, only a few types are recycled at substantial levels, including high density polyethylene and poly(ethylene terephthalate).

Because of the relatively low thermal stability of PET above its melt transition temperature, melt processing of virgin PET into relatively high-value products such as bottles for storage of food or beverages can lead to degradation and reduction in molecular weight. This degradation means that attempts to recycle PET for reuse in bottle production often results in failure because the viscosity of the recycled PET is too low to be effectively processed in such applications. As a result, recycled PET is often limited to low-value application.

Various studies have attempted to address this problem via chemical modification of recycled PET to increase its melt viscosity by the production of a branched PET structure [1-7]. While this approach, often called chain extension, can be used to tune the melt viscosity of the recycled PET to levels needed for high-value applications, it nevertheless requires the use of additional synthetic compounds and chemistry, thereby reducing the sustainability benefit of this approach.

Here we demonstrate that a simple processing method called solid-state shear pulverization can be used to increase the melt viscosity of PET by processing neat PET in the pulverizer. Solid-state shear pulverization has been employed in studies focused on producing blends with enhanced properties from recycled, commingled plastic waste [8], compatibilized polymer blends from virgin polymers [9-12], nanostructure polymer blends [13], composites from synthetic and natural, renewable polymers [14], and well-dispersed or well-exfoliated polymer nanocomposites with record property improvements [15-17]. In the study described below, we take advantage of the fact that SSSP can be tuned to yield small levels of chain scission of polymers. This mechanochemistry results in the formation of small levels of polymeric radicals that can participate in *in situ* chain transfer and radical-radical combination reactions, resulting in linear PET being transformed into lightly branched PET with an accompanying increase in melt viscosity.

Experimental

Poly(ethylene terephthalate), PET Eastman 7352, was from Eastman Chemicals. The polymer was used as received without drying.

The SSSP apparatus is a Berstorff ZE-25 pulverizer, a modified intermeshing twin-screw extruder equipped with a cooling system set at 266 K, allowing the polymer to remain in the solid state during processing. The screw diameter is 25 mm with an aspect ratio, L/D, of 26. The screws are modular in nature and comprised of combinations of conveying, mixing, and shearing elements. A detailed description of the SSSP apparatus is found in ref. 18.

The rheological properties of the PET samples before and after SSSP processing were measured using a Rheometrics Scientific ARES strain-controlled rheometer with a 5-cm parallel-plate geometry. The samples were subjected to frequency sweeps at 553 K from 0.01 to 50 Hz and strains of 100% and 30%. Measurements are reported for conditions equivalent to a shear rate of 1 s^{-1}.

The crystallization behavior of PET was characterized via differential scanning calorimetry at cooling rates of 10 K/min. For isothermal crystallization, samples were heated above the melt temperature, T_m, and then quenched to the respective isothermal crystallization temperature.

Dynamic mechanical analysis was done at a frequency of 1 Hz with temperature ramped at 3 K/min from 298 K to 473 K. The DMA samples were made by compression molding SSSP powders into ~0.3 mm thick, rectangular films. Intrinsic viscosity characterization of PET samples in 60/40 phenol/TCE at 303 K was done commercially by Plastic Technologies, Inc.

Results and Discussion

Table 1 demonstrates the significant tenability of the melt viscosity that can be achieved via SSSP of PET. All samples subjected to SSSP were made using the same screw design and feed rate, with screw speed as the only variable. The as-received PET exhibits a shear viscosity of 350 Pa*s at a temperature of 553 K and a shear rate of 1 s^{-1}. At the same temperature and shear rate, the melt viscosity can be increased by a minimum range of 20-125% via simple, single-pass SSSP, with lower screw speeds resulting in greater increases in melt viscosity.

Such increases in melt viscosity can result from either an increase in average linear chain molecular weight or in the production of a branched molecular structure accompanying pulverization. Intrinsic viscosity measurements were done on PET samples before and

Table 1 – Effects of SSSP processing on melt viscosity of PET samples at 553 and a shear rate of 1 s^{-1}.

Sample	SSSP Screw Speed (rpm)	Shear Viscosity (Pa*s)
As-received PET	----	350
PET after SSSP	50	790
PET after SSSP	250	525
PET after SSSP	300	420

after pulverization which revealed a slight decrease in intrinsic viscosity accompanying SSSP. For example, for PET pulverized at a screw speed of 250 rpm, the intrinsic viscosity changed form 0.755 dl/g before SSSP to 0.715 dl/g after SSSP. If SSSP resulted in an increase in linear chain molecular weight, this would lead to an increase in intrinsic viscosity. This picture of small levels of chain scission (with accompanying polymeric radical formation and reactions) leading to a lightly branched PET molecular structure is consistent with both the melt viscosity and intrinsic viscosity measurements.

A hypothetical mechanism for the achievement of lightly branched PET after SSSP processing is as follows. The application of high levels of shear and compressive forces (mechanical action) can lead to low levels of scission of the weakest bonds in PET, those being the C-O backbone bonds. The resulting radicals can participate in hydrogen abstraction reactions, leading to the presence of radicals on the backbone of a PET chain. This radical can recombine with another PET radical (likely with a radical located at a chain end), thereby resulting in branched PET. Thus, *in situ* mechanochemistry accompanying simple SSSP processing of linear PET chains can lead to a lightly branched PET product.

The SSSP processing of PET can yield additional benefits beyond enhancement of melt viscosity via chain branching. Figure 1 provides a comparison of the non-isothermal crystallization curves for PET before and after SSSP processing. The figure demonstrates that the PET crystallization kinetics are strongly enhanced by SSSP. For example, the onset crystallization temperature of PET increases by 11 K after SSSP, and the extent of crystalline structure achieved during the cooling from the melt state increases from 19% to 31% for the PET subjected to SSSP. Isothermal crystallization measurements at 493 K also exhibit a reduction in the crystallization half-time from 17 min for the as-received PET to 4 min for the PET subjected to SSSP. Two previous studies of chemically chain-extended PET have reported small reductions (20%) in isothermal crystallization half-times [4] or slower crystallization kinetics [6] following chain extension and branching of the PET. Here, we find that SSSP leads to major increases in the level of crystallization and crystallization rate achieved

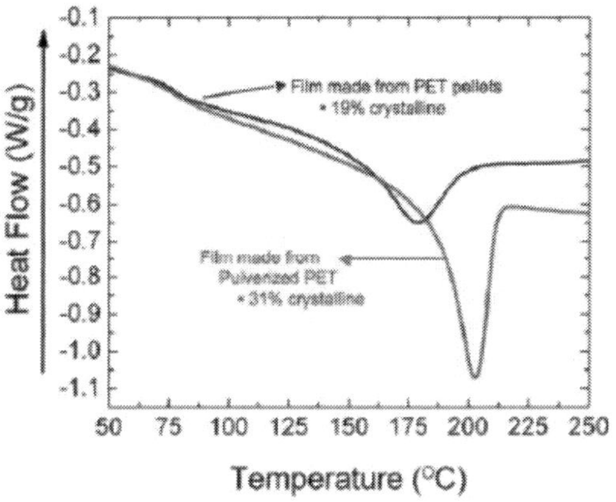

Figure 1 – Non-isothermal crystallization from the melt state of films from pristine PET and pulverized PET.

upon cooling PET rapidly from the melt state (without the use of cold crystallization). We believe that small levels of chain branches may serve a nucleation sites for crystallization and that SSSP can lead to effective dispersion of heterogeneous nucleating agents that may be present in the as-received PET.

Relative to the as-received PET, the tremendous enhancement in crystallization kinetics and crystallinity achieved upon cooling melt-state PET that was previously processed by SSSP also leads to an interesting comparison of the temperature dependence of dynamic mechanical properties. Figure 2 compares DMA data for as-received and pulverized PET samples. The pulverized PET sample has a higher tensile storage modulus near room temperature, reflecting the higher level of crystallinity in the sample. Between 350 K and 375 K, the modulus decreases by a factor of 10 for the pulverized PET sample but by a factor of 200-300 for the as-received PET. With a further increase in temperature, the modulus of the as-received PET increases to a level nearly equal to that of the pulverized PET sample, reflecting the "cold crystallization" that occurs over the temperature range of 385-410 K. Thus, the dynamic mechanical property characterization reveals that pulverization of PET does not lead to overall greater levels of crystallization than can be achieved after cold crystallization of as-received PET. However, the greatly increased crystallization kinetics accompanying SSSP results in a major improvement in the processability of PET that makes SSSP attractive not only for enhancing the recyclability of PET for relatively high-value applications but also for reducing PET processing time.

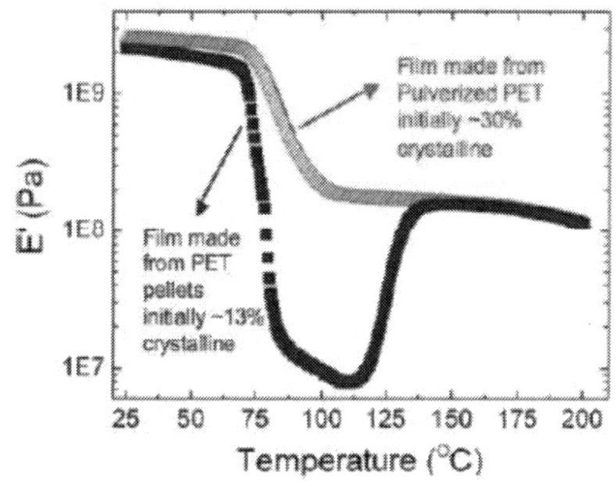

Figure 2 – Non-isothermal crystallization from the melt state of films from pristine PET and pulverized PET.

References

1. Incarnato, L. et al., Polymer **41**, 6825 (2000).
2. Cavalcanti, F.N. et al., Polym. Eng. Sci. **47**, 2155 (2007).
3. Kiliaris, P. et al., J. Appl. Polym. Sci. **104**, 1671 (2007).
4. Hanley, T.L. et al., Polym. Internat. **55**, 1435 (2006).
5. Forsythe, J.S. et al., J. Appl. Polym. Sci. **100**, 3646 (2006).
6. Sorrentino, L. et al., J. Polym. Sci., Part B: Polym. Phys. **43**, 1966 (2005).
7. Awaja, F. et al., Polym. Eng. Sci. **44**, 1579 (2004).
8. Khait, K. and Torkelson, J.M., Polym.-Plast. Technol. Eng. **38**, 445 (1999).
9. Furgiuele, N. et al., Macromolecules **33**, 225 (2000).
10. Lebovitz, A.H. et al, Macromolecules, **35**, 8672 (2002).
11. Lebovitz, A.H. et al., Macromolecules, **35**, 9716 (2002).
12. Tao, Y. et al., Polymer **46**, 4753 (2005).
13. Tao, Y. et al., Polymer **47**, 6773 (2006).
14. Walker, A.M. et al., Polymer **47**, 6773 (2007).
15. Wakabayashi, K. et al., Macromolecules **41**, 1905 (2008).
16. Mu, M. et al., Polymer **49**, 2686 (2008).
17. Masuda, J. and Torkelson, J.M., Macromolecules **41**, 5974 (2008).
18. Furgiuele, N. et al., Polym. Eng. Sci. **40**, 1447 (2000).

RESTABILIZATION OF POLYOLEFIN FILMS CONTAINING COBALT STEARATE AS A PHOTODEGRADANT

W.W. Focke and R.P. Mashele

Institute of Applied Materials, Department of Chemical Engineering, University of Pretoria, Pretoria 0002, South Africa

Littering of polyethylene plastic film packaging represents a major problem because of poor biodegradability in the natural environment. All polymer materials age when exposed to sunlight. Unfortunately, the intrinsic rate is simply too slow. In oxo-biodegradable polymers the rate of mineralization is enhanced by incorporation of a prodegradant additive, e.g. cobalt stearate. However, its presence may cause life expectancy problems when the plastic is recycled into more permanent applications, e.g. polyethylene piping. This study aims to show oxo-biodegradable polyethylene can be restabilized.

Low density polyethylene films containing 0.2 wt.% cobalt stearate and optional stabilizers were blown on a laboratory film blower. Accelerated weathering behaviour of the film samples was studied in a QUV fitted with A340 UV lamps and using a dry cycle at 63 °C and an irradiance of 0.67 W/m^2. The rate of polymer oxidation was followed by IR spectroscopy by measuring the growth of the carbonyl peak at 1710 cm^{-1}. A Carbonyl Index, defined as the ratio of this absorption to that at 720 cm^{-1}, was used to quantify the degradation progress ($CI = A_{1720}/A_{720}$).

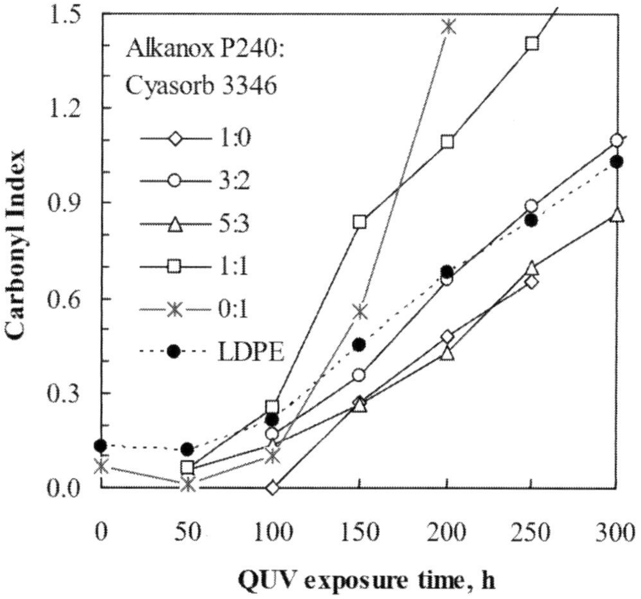

Figure: Effect of the combination of the stabilizers Alkanox P240 and Cyasorb 3346on the artificial QUV ageing of a virgin film and films containing a total 0,2 wt % total stabilizer and/or 0,2 wt % cobalt stearate.

Primary antioxidants, e.g. hindered phenols, significantly improve the oxidative stability at processing temperatures of polyethylene containing metal-based photodegradants. However they did not interfere noticeably with the photodegradation activity the metal catalyst. The Figure shows that the phosphite antioxidant tris(2,4-di-*t*-butyl phenyl)phosphate (Alkanox P240, CAS [31570-04-4]) appear to be able to suppress this kind of photodegradation. The UV degradation efficiency is retained when on partial replacement with the HALS light stabilizer Cyasorb 3346 (CAS [082451-48-7]). These experimental results suggest that it should be possible to recycle low density polyethylene containing cobalt stearate.

The following tentative explanation for the observed effect is offered. The catalytic cycle shown in the Scheme relies on the ability of the transition metal ions to exist in two oxidation states of similar stability and specific interactions with hydroperoxide species. The phosphite antioxidants are hydroperoxide decomposers. They transform hydroperoxides into thermally stable, non-radical products. This means that they can compete with the metal ions for hydroperoxides. It is thus conceivable that they can negate the effect of the metal catalysts by effectively by fast scavenging of the hydroperoxide species.

$$ROOH + M^{2+} \rightarrow RO^{\bullet} + OH^{-} + M^{3+}$$
$$ROOH + M^{3+} \rightarrow ROO^{\bullet} + H^{+} + M^{2+}$$
$$\text{Overall:} \quad 2ROOH \rightarrow RO^{\bullet} + ROO^{\bullet} + H_2O$$

Scheme. Cycle of redox reactions whereby a transition metal ions catalyses the decomposition of hydroperoxides to produce alkoxy and peroxy radicals

Close loop reuse of PET bottle scrap: A new perspective

Yalan Tait, Vineet Jha, Luke Savage, Oana Ghita, Ken Evans*

School of Engineering, Mathematics and Physical Sciences - Devon – UK

5million tonnes of plastics are produced in the UK every year, of which process scrap represents 250,000tonnes. Closed loop recycling operations for in-factory PET bottle scrap are still being avoided by many moulders. This is surprising considering that, unlike post-consumer bottles, in-factory scrap has the advantage of being contaminant free, well separated and from a known source. Whilst modest amounts are recycled (~10% incorporation level) , moulders and customers alike, are often fearful of using higher percentages due to fears over polymer degradation and consequent drop-off in mechanical performance or physical appearance. Consequently the scrap is often ground to flake for reuse further down the value chain. A major concern stems from the fact that PET long chain polymers progressively degrade with each thermal melting/cooling cycle, leading to changes in rheology and therefore, injection moulding parameters, and additionally, affecting the resulting mechanical performance of the final moulding. The present work studies the closed loop recyclability of different PET bottle scrap grades, to establish the relationship between thermal history of the material, mechanical properties and bottle performance (intrinsic viscosity and burst tests). The study compared regrind PET scrap, bottleflakes scrap and particularly, repelletized bottleflakes –that offers advantages to the moulder in terms of processabilty, but at the cost of an extra thermal excursion. Whilst thermal analysis and rheology were used to monitor degradation, a program of injection mould trials and mechanical testing provided performance profiles of each type of scrap, aiming to establish the level of scrap incorporation that can be achieved, without appreciably affecting material properties and injection moulding performance.

Reprocessing of glass fibre reinforced polypropylene through Injection moulding: Investigation of mechanical, physical and rheological properties

S. A. Elsheikhi[1], L.Mulvaney-Johnson[1], I.F. Campean[1], T.Gough[1] and P.D Coates[1]*

[1]*IRC in Polymer Engineering, School of Engineering, Design and Technology, University of Bradford, Bradford, United Kingdom- salahelsheikhi@gmail.com ; l.johnson@Bradford.ac.uk; F.Campean@bradford.ac.uk ; t.gough@Bradford.ac.uk and p.d.coates@Bradford.ac.uk*

Abstract

An investigation is conducted to determine the effect of reprocessing, through injection moulding, short glass fibre filled (20 wt %) polypropylene. Key mechanical and physical properties are monitored throughout the successive reprocessing phases in order to quantify the changes to the material. The fibre length is known to influence the mechanical properties and here the fibre length distribution is measured, again over the successive processing steps. This distribution is observed to show a reduction in average fibre length even within the first processing stage using virgin material. Processing parameters linked with melt plastication (screw rotation rate and back pressure) are observed to influence the average fibre length of the material entering the mould.

Key words: Injection moulding; glass fibre reinforced polypropylene; regrind; reprocessing stages; viscosity; and mechanical properties.

1. Introduction

Thermoplastics constitute 85% of consumed plastics in the world (Meran et al., 2008). Therefore, the reprocessing of the thermoplastics and the usability of the reprocessed materials are gaining significance.

Rising raw material costs, linked to increasing oil prices, increased consumer demand and increased environmental awareness have led to the development of plastic reprocessing programs (La Mantia, 2002).

Polymer reprocessing can be performed with many different technologies: this work concentrates on injection moulding which is one of the most versatile and important technology that widely used to process and reprocess different type of polymers.

Many researchers have conducted studies in the area of polymer reprocessing through the injection moulding technology. These studies concerned the entirely recycled material in form of post consumer materials/ material reclaimed from scrap yards (Meran et al., 2008), entirely virgin material (Lewis and Buser, 1997), and/or entirely mixed materials (Xu et al., 2002). For many manufacturers it is common practice to introduce reground sprues, scrapped parts, and runner systems into the injection moulding process.

*Corresponding author:
Tel: +44(0)1274 234529
Fax: +44(0)1274 235925
Email address: salahelsheikhi@gmail.com

This is advantageous for economic and environmental reasons (La Mantia, 2002) and easy to implement because it doesn't require any sorting, filtering or cleaning operations usually required with material coming from landfill sites or post consumer or rubbish yards. However, it is important to ensure that part quality is not affected; i.e. the properties of the reprocessed material remain the same with virgin material. Therefore, the process set up products in forms of sprues, runners and unused parts were regrind and used in this work.

Common industrial practice is to mix regrind material with virgin material in ratios from 0-50% or higher (Lewis and Buser, 1997) , depending on the customer specifications. Lewis and Buster (1997) have investigated the effect of processing low density polyethylene "LDPE" on tensile properties by adding 25% of regrind to the virgin material with differential melt histories within the same batch. Their result showed that LDPE with 25% of regrind can be added to the process without affecting either the processability or the tensile properties. This allows manufacturers to save money producing parts with regrind without being concerned with the effects it may have on the tensile properties. Another study (Dontula et al., 1994, Bernardo et al., Feb,1996) reported in, based on high density polyethylene "HDPE" with different regrind ratios of 0%, 50% and 100% , found that the melt temperatures didn't change between virgin to 100% regrind, and the Young's modulus was found to be the same between the three regrind ratios(virgin, virgin +50% regrind and 100% regrind).

Reprocessing of recycled materials or reclaimed from post-consumer waste material feedstock is widely used in industry; however the product quality is an important challenge against the reprocessing operations. Therefore, the most published studies of such recycling/reprocessing of such material have dealt with the effect of reprocessing on the physical and mechanical properties of a range of polymers. The possibility of recycling and using recycled refuse LDPE, HDPE and polypropylene "PP" were investigated (Meran et al., 2008), and it was found that the PP is the material which can be most successfully recycled based on the tensile strength of the processed part.

Reprocessing of filled polypropylene is important given that it has many applications especially for injection moulding (Dave et al., 1997) , and hence it is interest to study the behaviour under repetitive reprocessing stages. Studies were reported on glass fibre reinforced PP, paying particular attention to two important aspects of injection process: firstly the heat history to which the material is subjected, and secondly, the number of times the material is subjected to the shear of the screw (Guimond, 1997). The effect of four recycle histories on the properties of 30% glass fibre reinforced polypropylene was investigated (Guimond, 1997).

He found the tensile modulus, percentage of elongation at yield and impact strength does decrease with each reprocessing cycle, but not at an alarming rate. It was found that the glass fibre do seem to shred when exposed to excess shred and grinding, but there was no sign of a major loss in mechanical properties, and hence it can be concluded that the glass fibre reinforced PP can be effectively reprocessed and retain most of its properties (Guimond, 1997). Charbonneau (1997) examined the effect of eight reprocessing cycles using 40% long glass fibre on the mechanical properties. It was found that the tensile strength, tensile modulus and impact strength have a decrease in percentage of 30%, 45% and 25% respectively from virgin to cycle number eight. This implies that the material has been successful in retaining many of its useful properties even after exposing it to repeated heating and shearing cycles (Charbonneau, 1997).

Fibre length plays an important role in determining the mechanical properties like strength, modulus and fracture toughness of short fibre reinforced thermoplastic composites (Fu et al., 2002). Severe fibre breakage occurs in injection moulding process, and resulting composites have short fibre (Gupta and Wolters, 2008). It is well known that fibre reinforced thermoplastics do not achieve their maximum possible stiffness, strength, and toughness properties, because of many factors and one of the main reasons for this is the short fibre length within the composites.

For this reason, in this work, fibre length analysis was carried out, through the reprocessing stages, as good indicator to explain the behaviour of key physical and mechanical properties.

Generally, in the area of polymer reprocessing , in spite of many researchers have conducted studies in the area of polymer reprocessing through injection moulding, however, some essential information still needed especially those that linked to the part quality, physical properties, processing properties and mechanical properties.

Therefore, the lack of comprehensive investigation of these properties over successive reprocessing stages make the examination of these properties is interested and useful, especially, if connected to the industry. On other world, in industry, this could help to know more accurate and varying information for these properties, in order to satisfy the customer who has a lot of requirements, demands and specifications.

In this work a comprehensive study on the current state of reprocessing polypropylene material for many times using 100% regrind material (without considered the reprocessing of post consumer wastes, but well controlled reprocessing of samples) under the same moulding conditions, and a focus on the investigations of varying properties and to find the possible links between these properties. The strategy of this study aims have some advantages such as reducing material consumption/cost, and improving the material productivity.

The key properties under investigations are:

- art quality (mass and in-plane shrinkage).
- hysical properties (density, crystallinity, thermal stability, fibre length, and molecular weight).
- heological property (melt viscosity)
- echanical properties (Young's modulus, percentage of elongation at yield and flexural strength).

The material used through this study was glass fibre reinforced polypropylene "GFRPP" injection moulding grade. GFRPP is widely used in variety of applications for instant; in the automotive it is used to produce battery cases; in medical it is used for disposable syringes , also it is used for appliances and hand tools as well as the rigid packaging containers and housewares [Karian, 2003]. The reprocessing operations (see Figure 1) were done under the same moulding conditions by using 100% of regrind material in form of product set up, runners and sprues without using any type of additive materials.

2. Procedure

2.1 Material and sample preparation

20 wt% glass fibre reinforced polypropylene homopolymer; grade BMW N 601 supplied from A.Schulman Inc. with melt flow rate of 3.5 g/10 min, solid density of 1.08 g/cm^3 based on the technical data sheet of this material, average molecular weight and shear viscosity (at shear rate of 6000 sec^{-1}) based on the laboratory measurements, were 181,000 g/mol and 14.24 Pa.sec, respectively.

The injection moulding machine used in this study was a 100 tonne clamp force FANUC (Roboshot S-2000i-100A) all-electric drive with screw diameter of 32 mm and L/D = 22:1. The injected test part was a ASTM D638 tensile bar specimen with gage length of 90 mm, width of 10 mm and thickness 4mm. The moulding machine was fitted with a special nozzle adaptor that houses a melt pressure with a Kistler melt pressure sensor type 6190A08 and a FOS melt infra-red (IR) temperature sensor type MTS408-HT. These sensors are computer monitored along with the screw position and pressure from a transducer mounted at the back of the screw. The computer monitoring system is a National Instruments (NI) LabVIEW based software application. The signal conditioning and interface system is through a NI SCXI unit connected to a 12-bit PCI 6024E data acquisition card.

Sample collection for subsequent testing on viscosity index results will be discussed in more details in section 2.7.1.

2.2 Injection moulding conditions

The melt processing temperature was set to 240 °C, based on the material technical data sheet. The mould temperature was set to 50 °C which was controled by the mould temperature controllers (Regloplas type 90s). Melt plastication condition was set to 100 rpm for screw speed and 20 bars for back pressure based on the tensile strenght and average fibre length as shown in Figure 2 (a, b and c). Injection velocity was set to 80mm.sec^{-1} based on the tensile strength (see Figure 2-d) and cooling time set to 20 sec. Gate frozen time was determined from a range of holding time against part weight at 300 bar holding pressure, as shown in Figure 2(e). The part at which specimen mass no longer increases with applied pressure is the gate freeze. In this case the hold time is taken to be 30 sec. It should be noted that the tensile sample thickness is 4 mm , but the runner is 8 mm and will solidify after the tensile sample. The processing settings are summerized in Table 1.

2.3 Reprocessing sequence

The reprocessing operations started by introducing 5 Kg of virgin material. The first processing cycle was named virgin as an output from virgin feed stock material; and subsequent reprocessing cycles were labelled stage 1, stage 2 and so on up to five. After each reprocessing stage, a number of specimens were extracted for testing and the remainder (including runners and sprues) granulated in a 'Cumberland granulator' with three tangential knives. A small portion of regrind material (about 500 grams) from every stage was kept for the related tests.

2.4 Part quality

2.4.1 Specimen mass

The mass of a product is commonly used as a "part quality" indicator and changes of specimen mass can often be linked to changes in dimensions, and hence the stability in the specimen mass could indicate to the stability of the part dimension. The specimen mass is obtained by measuring the test sample including the runner and sprue on a standing mass balance for every reprocessing stage then the average specimen mass was calculated.

2.4.2 In-plane shrinkage

Shrinkage behaviour of a moulded plastic part plays a principle role in determining final dimensions of the part. Shrinkage occurs due to the thermal contraction of the molten material during cooling and the relaxation of stretched polymer chains (Fathi and Behravesh, 2007). This phenomenon is affected by many factors including process parameters at filling, packing and cooling phases, cooling system design, and material properties (Fathi and Behravesh, 2007).

In-plane shrinkage was measured in several locations on the specimen. Five samples were selected for each stage to measure the shrinkage; Equation 1 (Mamat et al., 1994) was used to determine the in plane shrinkage in as a percentage of the cavity dimension.

$R = (1 - L/L_0)$..Equation 1

Where L_0 is a dimension measured on the mould cavity and L is the same dimension measured on the specimen. The mould cavity shape and the selected measurement positions are shown in Figure 3.

2.5 Physical properties

2.5.1 Density

Density is one of the important physical features for the polymer as it linked to the volume and crystallinity. It has been measured using AccuPyc^{TM} 1330 Pycnometer. The regrind materials for each stage in form of pallets were used, and then the apparatus calculated the best five readings of the density and provided the average values.

2.5.2 Crystallinity

Crystallinity is indication of amount of crystalline region in polymer with respect to amorphous content. Crystallinity influences many of the polymer properties such like hardness, tensile, stiffness, crease and melting point (Gowariker et al., 1986).

Differential scanning calorimetry "DSC" is a thermo-analytical technique in which the difference in the amount of heat required to increase the temperature of a sample and reference are measured as a function of temperature. It's used to study what happens to polymers when they're heated and give useful information such like actual melt temperature; crystallinity and glass transition temperature. In this work it's used to determine the degree of crystallinity for each reprocessing stage using DSC model Q2000. Universal TA software was used to analyse the data and the experiment was done under standard setting using temperature range of 30 °C - 240°C and liquid nitrogen was used for cooling the system at 100 µL min^{-1}, with heating rate 10°C.min^{-1}. Care was taken to cut and measure the mass of the sample in order to get the same sample weight for TGA test.

The percentage of crystallinity was calculated from the heat of fusion of pure fully crystalline PP, ΔH_f°, *of* 207(J/grams) as shown in Equation 2 (Bureau and Denault, 2004).

$$X = (\Delta H_f / ((1-W_f) * \Delta H_f^{\circ})) * 100 \quad \text{..............Equation 2}$$

Where W_f is the resin weight fraction and ΔH_f is the heat of fusion (in Joule per gram of composite) of the same samples.

2.5.3 fibre length Measurement

Glass fibre is used as a reinforcing material within many polymeric matrix materials. Fiber length plays an important role in determining the mechanical properties like strength, modulus and fracture toughness of short fiber reinforced thermoplastic composites (Fu et al., 2002).

Processing may influence fibre length distribution through the shearing within the machine barrel (screw). The fibres may break of this point.

In this study fibre length analysis was carried out as a preliminary step assist with the choice of process parameters (i.e. back pressure and screw speed) for the subsequent reprocessing stages.

Fibre length was also studied throughout the reprocessing stages. The interaction between the fibre length and the screw speed can be explained as follows(Yilmazer and Cansever, 2002): The shear stress imposed on the polymer melt not only causes chain scission in the polymer molecules, but also leads to a decrease in the length of glass fibres. As the screw speed increases, the shear rate increases and the melt viscosity of the polymer matrix decreases (Yilmazer and Cansever, 2002).

An increase in the screw speed also causes an increase in the melt temperature of the polymer owing to viscous heat generation. Thus, at higher screw speeds, higher temperature and lower viscosity may result in lower shear stress acting between the glass fibre and the polymer matrix at the interface: consequently the fibre length may decrease less than proportionally to the screw speed as observed (Yilmazer and Cansever, 2002).

The first step of analytical process is the extraction of a clean fibre sample. The fibre can be extracted from the matrix by means of burning off the sample in the furnace at specific temperature (Shortall and Pennington, 1982, Patcharaphun and Opaskornkul, 2008) or by dissolving it in chemical solution(Licea-Claverie et al., 2002). In this work fibre was extracted by pyrolyzing it in a vacuum oven. Samples were prepared for back pressure and screw speed investigation by collecting the same amount of extrudate (1.32 grams) for all conditions within plasticating phase (without injection) as shown in Figure 4, then the fibres were extracted from the composite material by pyrolyzing in vacuum oven at 550°C (based on TGA test). The temperature was maintained at 550°C for 5 hours to oxidize completely any carbon deposits. The ash fibre placed on a glass plate. The glass plate was then placed on the observation stage of a calibrated graticule reflectance light microscope. Digital calibrated photos of the images were taken (see Figure 5). The images were loaded into a software package, an image processing analysis program called JMicroVision software. The length of fibres from each sample were measured and recorded.

The fibres were extracted from the moulded and regrind samples in the same way, but the samples were prepared by cutting off the moulded part to extract a 1.32 grams sample from the centre of the part, as shown in Figure 4.

2.5.4 Thermal stability

Thermal degradation of polymeric materials is an importance issue for industrial viewpoint, especially in polymer processing and higher temperature applications.

Thermogravimetric Analysis "TGA" is used to determine the thermal stability of a material by measuring mass loss (material evaporation) over a range of temperature. TGA model Q5000 was used here, and all samples were gently prepared by cutting into round shape (samples weight range is 2.5 – 3 mg) then the samples were subjected under nitrogen atmosphere using the temperature range of 35-600°C and heating rate of 10°C.min^{-1}. Universal TA Instruments software was used to determine fibre content for each stage, and to examine the thermal stability by using isothermal TGA test at setting temperature of 240 °C over 7 hours.

2.5.5 Molecular weight (M_w)

Molecular weight is important because it influences with many properties such as stiffness, strength, viscoelasticitic response, toughness, melt & glass transition temperatures and viscosity. As molecular weight increases, melt and transition temperatures along with the mechanical properties will generally increase. Low M_w materials are typically not useful in structural applications (Gowariker et al., 1986). In this study gel permeation chromatography model "GPC" was used, and two runs for each sample were analysed to determine the average molecular weight for each reprocessing stage.

2.6 *In-line viscosity measurement*

In-line viscosity measurement (Viscosity index "VI") is a process measurement that has been shown to indicate process variation through material viscosity change and/or process control (R.G.Speight, 1993). It has been used here to obtain an indication of material viscosity changes directly from the process and to identify regions of process stability within a reprocessing stage. VI is of absolute metric and well change with injection velocity, temperature...etc. However, where the process settings are constant it can be used to indicate process variations, may be through statistical process control.

VI was obtained from the available nozzle pressure signals recorded from FANUC machine by integrating the area under the nozzle pressure - time curve, where only the portion from 60% to 80% of the velocity controlled phase is considered. Figure 6 shows the selected area under the nozzle pressure curve which is integrated to calculate the viscosity index. A stable value of VI across cycles indicates consistent sample manufacture. Sample here, therefore, selected for stable regions of VI as shown in Figure 7.

2.7 Rheological property (melt viscosity)

Viscosity is the resistance of a fluid to flow, caused by its internal friction, which makes it resist flowing past a solid surface and makes layers of the fluid resist flowing past other layers (Johannaber, 1983). Viscosity is very important to the injection moulding process and it is strongly dependent on the temperature of the material. The effect of shear rate on viscosity is very important as the viscosity can reduce several orders of magnitude when the shear rate is increased (pseudo-plastic shear thinning). This is very important in injection moulding as very high local shear rates can be present within the flow path, in particular at the gate region.

In this work, viscosity has been measured off-line over the six reprocessing stages by using capillary rheometer. Melt viscosity at shear rate range of 20 - 6000 sec^{-1} was studied using a twin bore Rosand RH10 capillary rheometer. Dual pistons of diameter 15 mm were used to drive melt flow, through capillary dies, at speeds up to 20 $mm.s^{-1}$ with a maximum crosshead force of 100 kN. One pair of long and short capillary dies with entry angle of 180° and diameter of 1 mm was utilised. One barrel of the rheometer was fitted with a capillary die with an L: D ratio of 16 (long die) and the other bore was fitted with zero length die. The melt temperature was 240°C for all cases. The material was subjected to precompression pressures of 1.0 and 0.5 MPa for the long and short dies respectively, and subjected to a total preheating time of 540 sec. Rosand Flowmaster control software was used to analyse this data.

2.8 Mechanical properties

Tensile test samples were analysed to determine the tensile properties (Young's modulus and percentage of elongation at yield) along flexural strength. 'Instron testing machine' model 5564 using 'Instron Bluehill software' was utilised for testing and data analysis. The tests were carried out at room temperature (20-22°C) and relative humidity of 50 %. The crosshead speed of 5 $mm.min^{-1}$ was selected according to a ASTM D638 test properties for this material. It should be noted that the specimens were selected for each speed category based on viscosity index as explained in section 2.6.1.

3- Results and Discussion

3.1 Part quality

3.1.1 Specimen mass

In this study, the specimen mass is consistent at any given reprocessing stage, but mass increases slightly over the six reprocessing stages. This is shown in Figure 8, where the mass increases insignificantly from virgin to stage 5 (by 0.08 %). The process parameter settings are consistent for all of the reprocessing stages. An increase in specimen mass may be linked to reductions in melt viscosity, (as described in sections 3.3.1 and 3.3.3) along with increase in crystallinity and density entitled in section 3.2.1.

3.1.2 In-plane shrinkage

It can be noted from Figure 9 that the shrinkage through thickness (D1, D2 and D3) reduces along the flow path. Similarly, the shrinkage through width (W1, W2 and W3) reduces along the flow path. Oppositely, the shrinkage increases along the length of the specimen (L), which is along the flow direction. This is due to the fact that the crystalline materials tend to have high shrinkage and the shrinkage is grater in the direction of flow than across the direction of flow (Bryce, 1996). The shrinkage at "L" position has a relative increase of 8% from virgin to stage 5. Regarding to the thickness, the higher percentage of shrinkage is 0.498, located at D1 (stage 6). Similarly, the width has a percentage of shrinkage of 0.457, located at W1 (stage 5). This is might be as a result of the moulded part has more tendency toward the shrinkage in the posions that more closes to the direction of flow (ie.D1,W1) , because the other positions of the part (ie.D3,W3) are already filled and started to solidify earlier than others. According to ASTM D955, the shrinkage range of polypropylene material related to injection moulding is between 0.254-0.762 mm/mm (Bryce, 1996), this could lead to say the calculated shrinkage in this work is slightly low and hence the moulded parts produced in every reprocessing stages have tendency toward dimensional stability.

3.2 Physical properties

3.2.1 Density & crystallinity

Polypropylene is a semi-crystalline material consisting of amorphous and crystalline regions (Gowariker et al., 1986). The density of the bulk material is dependent upon the relative volume fractions of the amorphous (low density) and crystalline (high density)phases (Gowariker et al., 1986). As shown in Figure 10, the density increases slightly (by 0.34%) with reprocessing stages from virgin to stage 5 reprocessed material. This correlates well with the 4% increases in crystallinity observed over 6 stages as shown in Figure 11.

3.2.2 Fibre length analysis

The approach here is to establish the link between the fibre length and the important physical, rheological, and mechanical properties in every reprocessing stage. It accepted that fibre orientation with also has influence on the mechanical properties. No attempt is made to measure and assess within this work at present. The change in properties through successive reprocessing stages is investigated.

In order to measure the fibre length , the common central tendency parameters (mean, median and standard deviation) were used in the fibre length analysis for feed stock material, reprocessed material, and regrind material as shown in Figure 12(a, b, and c). It can be noted from Figure 12(a) that the mean reduction from feed stock material to stage 5 reprocessed materials is 39%, while this percentage is 38% from feed stock to stage 5 regrind material. Similarly, Figure 12(b, c, and d) show the median, standard deviation, and percentage of fibres longer than 100 μm reductions from feed stock to stage 5 reprocessed materials are 16%, 61%, 60% respectively. While, these percentages are 17%, 63%, and 70% from feed stock to stage 5 regrind material. The analysis of the results marked in Figure 12 shows that through reprocessing stages the fibre length is getting more uniform; this shown by the reduction in standard deviation (see Figure 12.c) and the reduction in the proportion of "long fibres" (see Figure 12.d). A reduction in mean fibre length was also noted in Figure 12.a.

A reduction in median is irrelevant (see Figure 12.b) due to there no clear trend for fibre length data over six stages. It can be noted also from Figure 12 that there is no sharp difference between the reprocessed and regrind material in every stage, this may indicate that the regrinding operation has a little effect on the fibre length.

A "Lognormal" distribution was fitted to the fibre length data illustrated in Figure 13.a, b.

Some authors (Patcharaphun and Opaskornkul, 2008) reported that the possible reasons that cause to reduce the fibre length as follows:

- Fibre–fibre interaction—abrasion of fibre surfaces reduces strength by inducing stress concentrations, leading to direct or subsequent fibre fracture, or bending stresses by fibre overlap may also cause fibre breakage;
- Fibre interaction with polymer matrix—viscous forces imparted by the polymer matrix may cause fibre fracture;
- Fibre contact with equipment surfaces—evident due to the wear of the walls of injection moulding machine.

In this work, fibre attrition due to shear action or regrinding procedure, through reprocessing stages, could be described as a minor negative effect especially the percentage of fibre content has not changed over six reprocessing stages based on TGA test (as illustrated in next section), and there is little change in some properties through the reprocessing stages (i.e. thermal stability, molecular weight , melt viscosity, Young's modulus and flexural strength) as illustrated in the next sections.

3.2.3 Thermal stability

Thermal degradation is an irreversible effect due to the breaking of chemical bonds. Results are reported for both virgin and stage 5, as percentage of mass loss as a function of temperature shown in Figure 14.a. The typical processing temperature based on the material data sheet is located between 230 °C to 270 °C and the experimental set point was 240 °C (based on the actual setting temperature in the Fanuc machine).

It can be noted from Figure 14.a, there is no significant deference in an initial degradation temperatures between the virgin and stage 5, and the fibre content has remained constant for these stages.

Isothermal TGA result (see Figure 14.b) shows insignificant difference in the percentage of weight loss between virgin and stage 5 over 7 hours (the percentage of weight loss difference is only 0.8289%). However, this could be resulted from the consistency of the reprocessing operations, and this result may help to explain the behaviour of mechanical and processing properties.

Based on TGA result, it can be stated that there is no significant change, in practical processing terms of the degradation temperature and percentage of weight loss for these stages. The thermal stability is enhanced by the average molecular weight through the reprocessing stages as illustrated in next section.

3.2.4 Molecular weight "M_w"

The importance to study the M_w behaviour here, due to it has influenced with processing and mechanical properties over 6 reprocessing stages. Figure 15 shows the average molecular weight over six reprocessing stages. As can be seen in this graph, the molecular weight is reduced with increasing number of reprocessing stages due to degradation of the matrix. Some authors (Licea-Claverie et al., 2002) justified the reduction in molecular weight as a result of a thermo-mechanical reaction catalyzed by hydrolysis, where the water is taken from the environment. In the current study, the reduction in molecular weight is very little, where the percentage of reduction is only 2% from virgin to stage 5. A reduction in molecular weight with increasing the number of reprocessing stages may be linked with the reduction of the amount of the material chain branching and the changing in chemical bonds as a result of material degradation. The behaviour of M_w over 6 reprocessing stages has clear correlated with shear viscosity and mechanical properties as mentioned in the sections.

3.3 Rheological property(melt viscosity)

Shear viscosity has been determined using Rosand RH10 two bore capillary rheometer. The data for virgin and all reprocessing stages has been presented on a single axis but the discrimination between the data is difficult (see Figure 16.a). Figure 16.b. shows the rheologized behaviour of the virgin material and that after 6 reprocessing stages over the full range of shear rates studied. At lower shear rate (20 sec^{-1}), the percentage of reducing is 17% from virgin to stage 5, while this percentage is 11% at higher shear rate (6000 sec^{-1}).

A reduction in viscosity, through reprocessing stages, could be as a result of shearing which taken place and causing the material to flow more.

In injection moulding process, as the shear rate increases the number of entanglements of the molecules reduces (Pötsch and Michaeli, 1995). This is because as the polymer melt flows shearing occurs and it is shearing this may cause the random and elongated chains to elongate. Having elongated chains results in chains being more able to slide past each other and hence flow of the plastic melt is easier and viscosity reduces with melt flow rate increases (Pötsch and Michaeli, 1995) . This may cause to reducing the amount of the material chain branching as well as overall molecular weight and hence increasing the crystallinity (La Mantia, 2002).

3.4 Mechanical properties

Mechanical properties have been assessed through tensile (modulus of elasticity, % of EL at yield) and three points bend (flexural strength) tests. In the former case, the strain is determined through crosshead displacement. As can be seen in Figure 17 , Young's modulus is slightly decreased with increasing number of reprocessing stages (by 7%). Percentage of elongation is reduced by 24% from virgin to stage 5(see Figure 18). Some investigators (La Mantia, 2002) justified this by decreasing molecular weight and increasing the crystallinity induce to reducing the elongation –and also a ductile-brittle transition –while having different effect on rigidity. Another investigators presented the same behaviour of the elastic modulus and the elongation of polypropylene material for multiple reprocessing stages (La Mantia, 2002;Dintcheva and La Mantia, 2003; Meran et al., 2008).

Figure 19 shows the flexural strength over six reprocessing stages. As can be seen in this graph, the data has some fluctuations across the stages, but the general trend is for flexural strength result to reduce the flexural strength on 11% from virgin to stage5. This reduction is related with decreasing molecular weight and could be also reducing the fibre reinforced efficiency. According to La Mantia, 2002, the reduction in mechanical properties , through the multiple reprocessing stages, for this material was attributed to the fact that the glass fibre contribute to the material degradation via chain scission due to more extensive viscous heating, and to the fibre length decrease that was found to take place after reprocessing.

Table 1 Moulding settings

Parameter	Setting
Barrel temperature	240 C°
Nozzle temperature	240 C°
Back pressure	20 bar
Holding pressure	300 bar
Cooling time	20 sec
Holding time	30 sec
Shot Size	36 mm
Screw rotation speed	100 rpm
Injection velocity	80 mm/sec

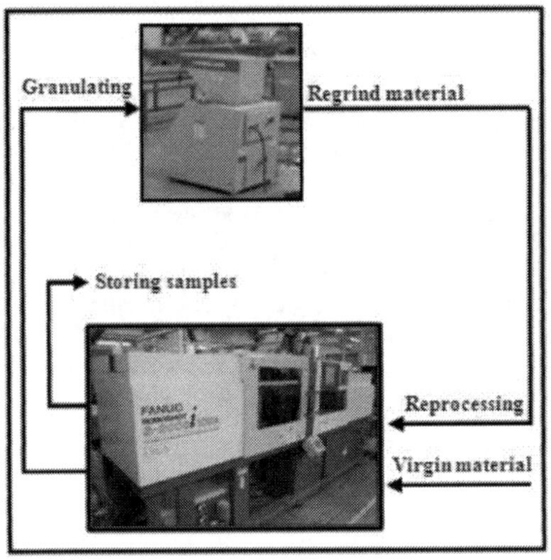

Figure 1 Reprocessing procedure

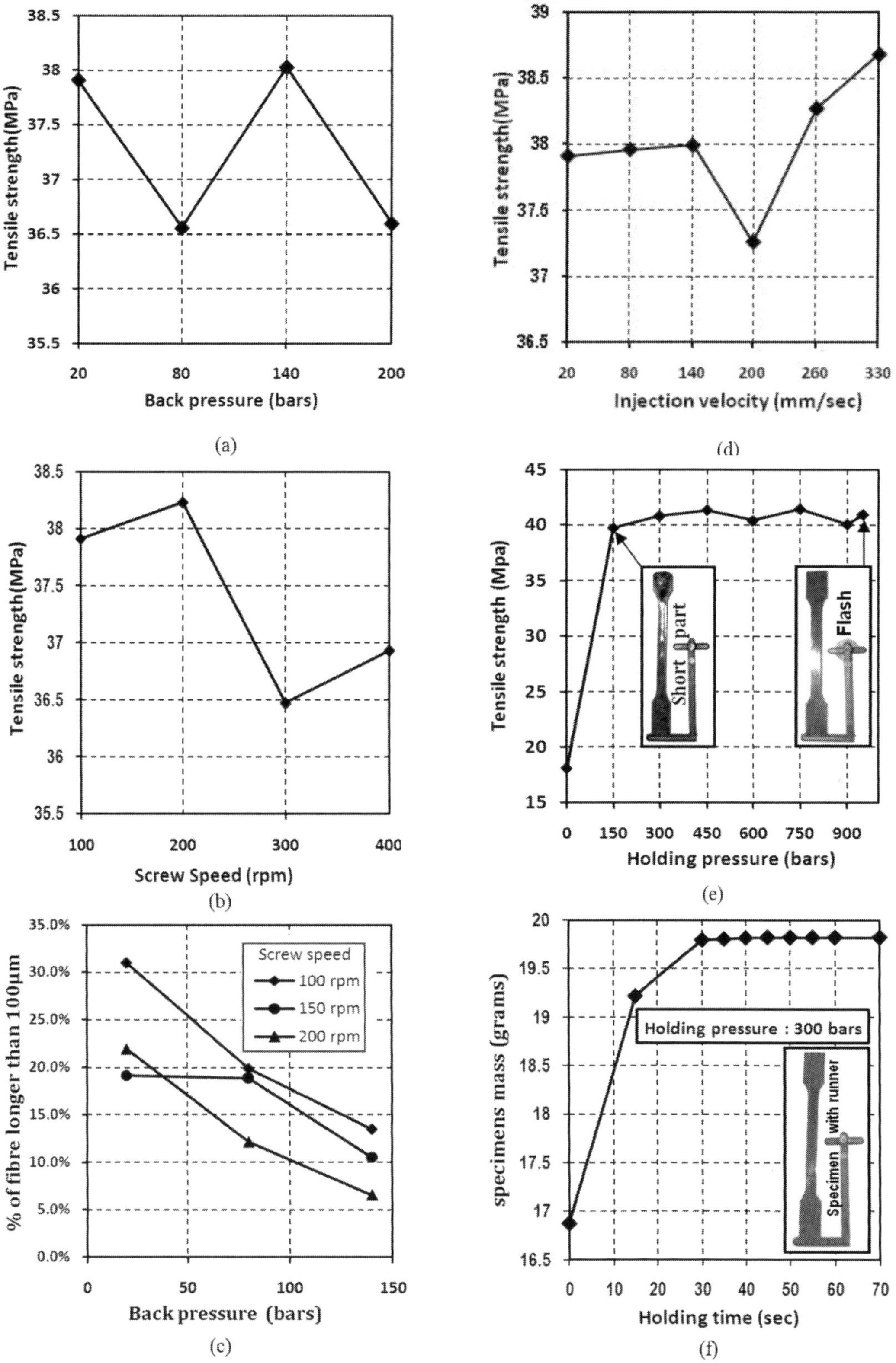

Figure 2 (a, b, c, d, e and f) Moulding conditions selection based on tensile strength & fibre length

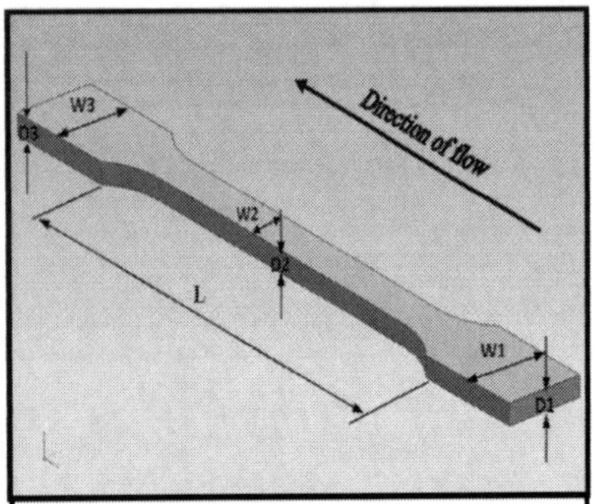

Positions (in mm)		
L =110.49	D1=4.11	W1=20.18
	D2=4.1	W2=10.05
	D3=4.06	W3=20.18

Figure 3 Mould cavity shape and measurement locations

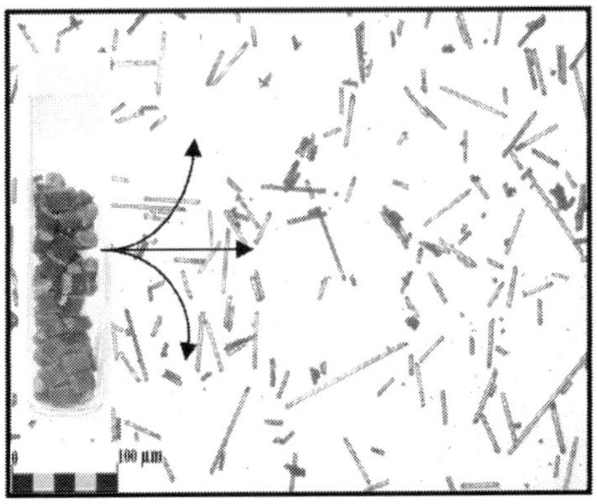

Figure 5 Fibre extracted from feed stock material (magnification 5X)

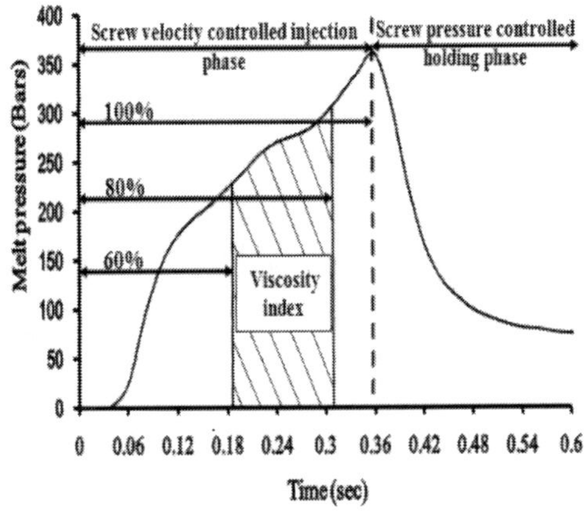

Figure 6 Viscosity index is calculated from a portion of the primary injection pressure-time date

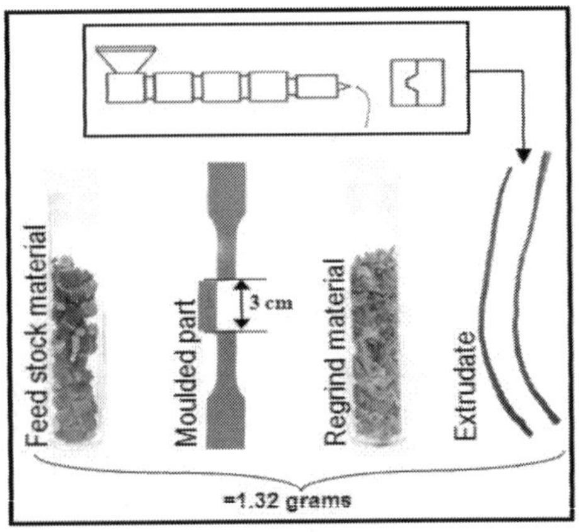

Figure 4 Sample preparations for fibre length measurement

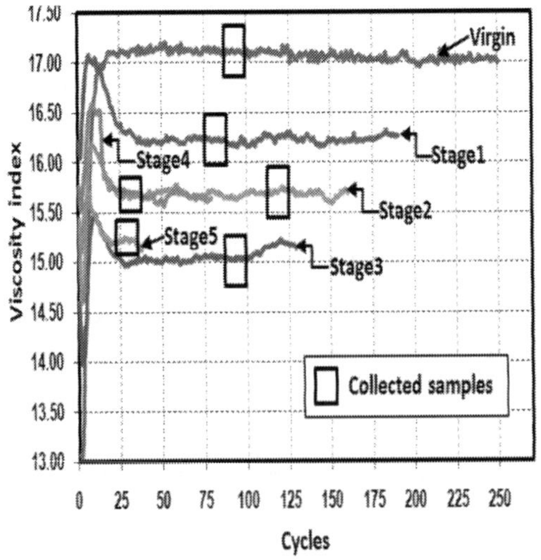

Figure 7 Viscosity index for all stages

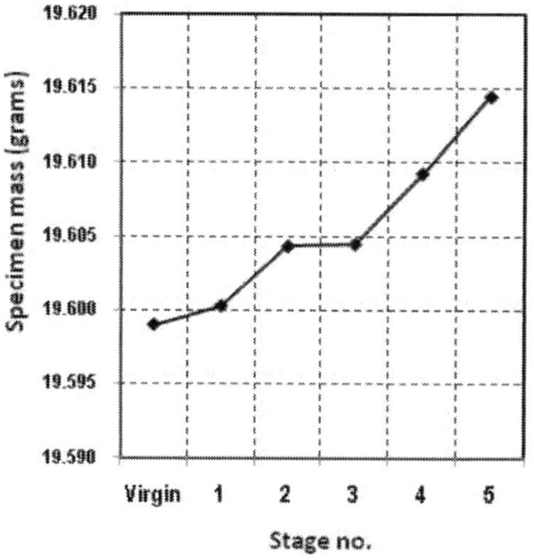

Figure 8 Specimen mass for all stages

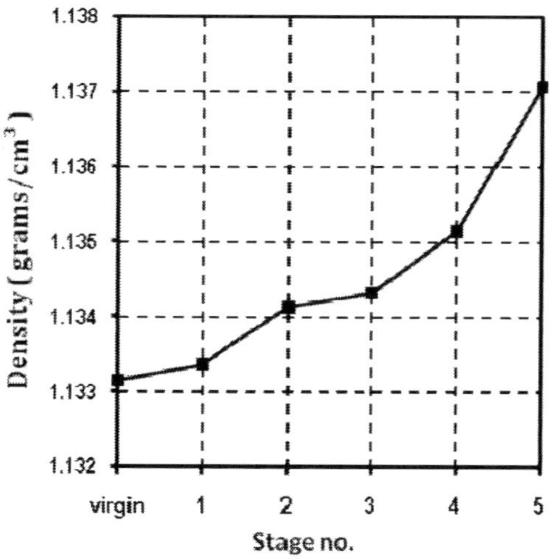

Figure 10 Density over six reprocessing stages

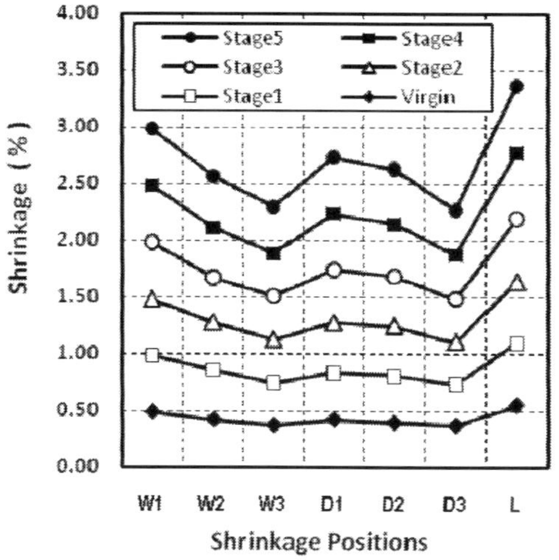

Figure 9 Shrinkage effects over 6 reprocessing stages

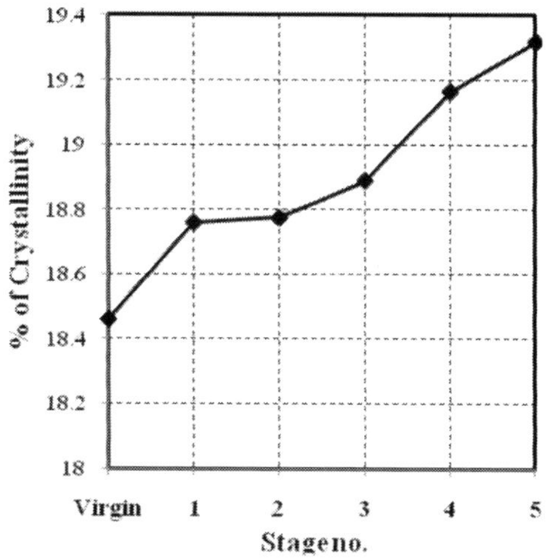

Figure 11 Crystallinity for all stages

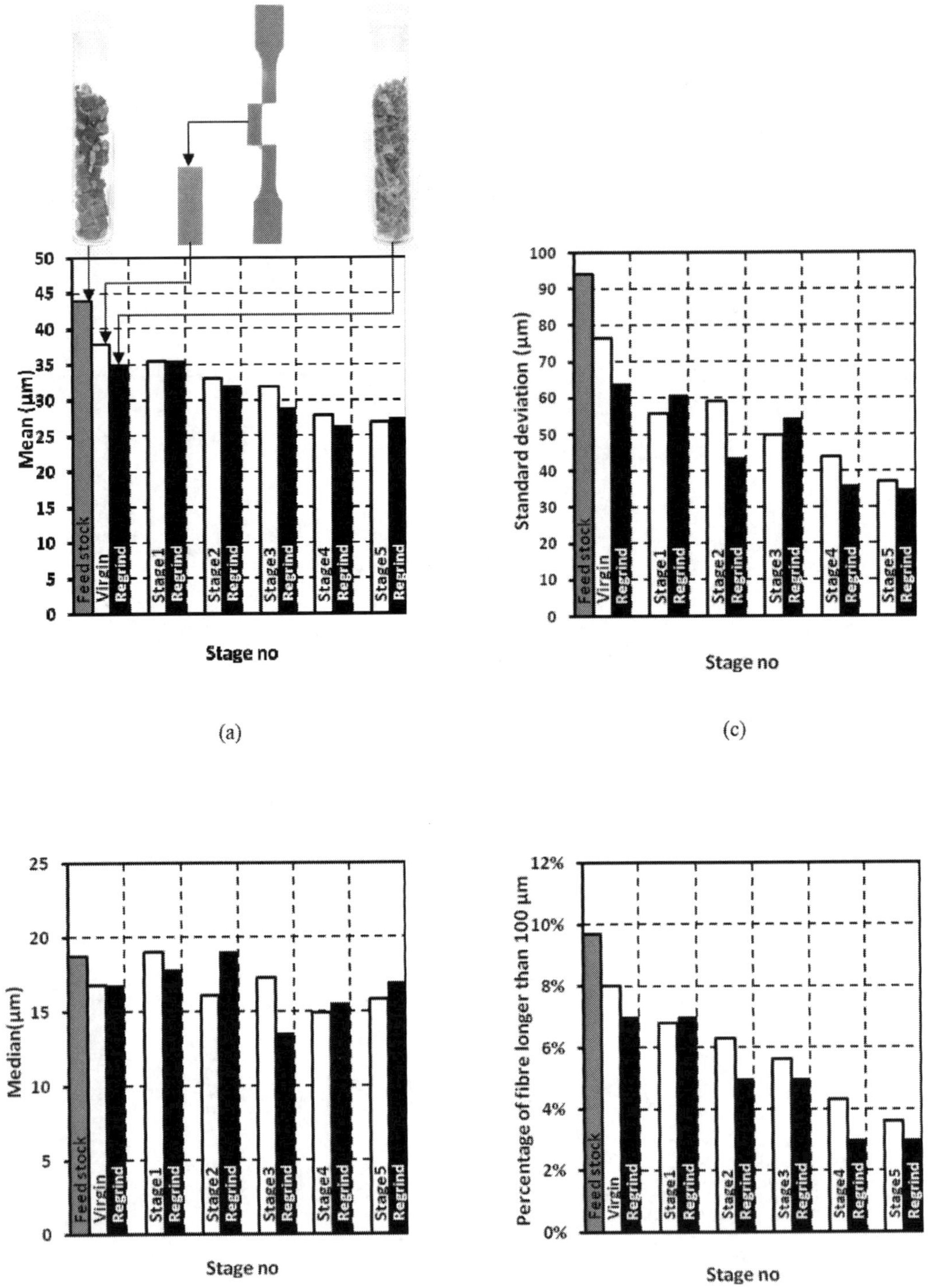

(a) (c)

(b) (d)

Figure 12 Statistical aspects of fibre length behavior for feed stock, reprocessed, and regrind materials (a:Mean; b:Median; c:Stndard deviation; and d: percentage of fibre)

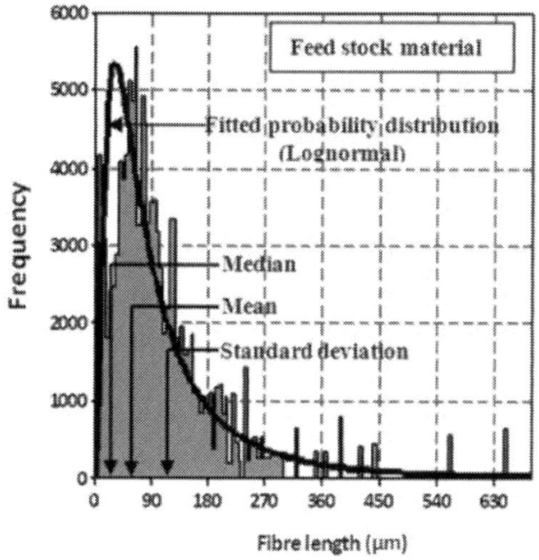

(a)

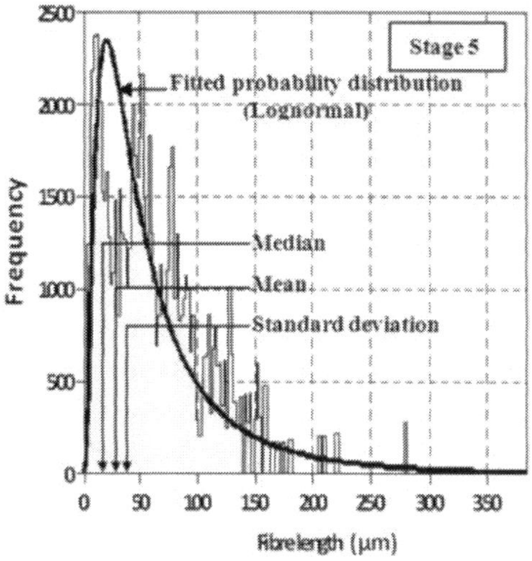

(b)

Figure 13 Fibre length distribution
(a: feed stock material, b: stage 5)

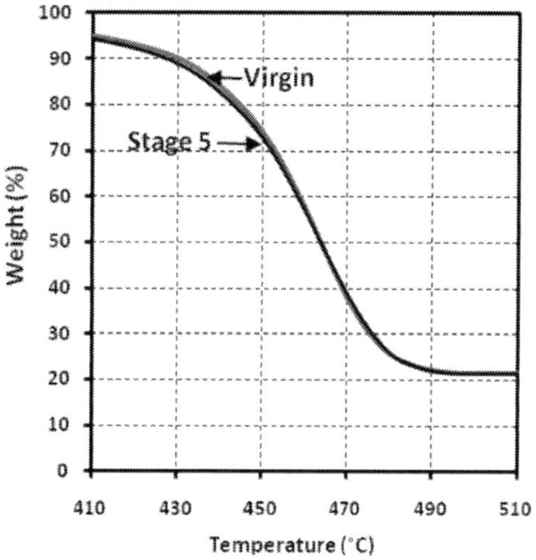

(a)

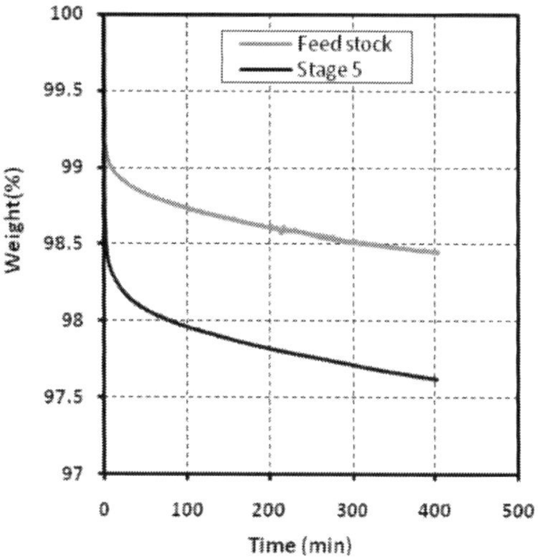

(b)

Figure 14 TGA result (a: mass vs. temperature,
b: mass vs. time

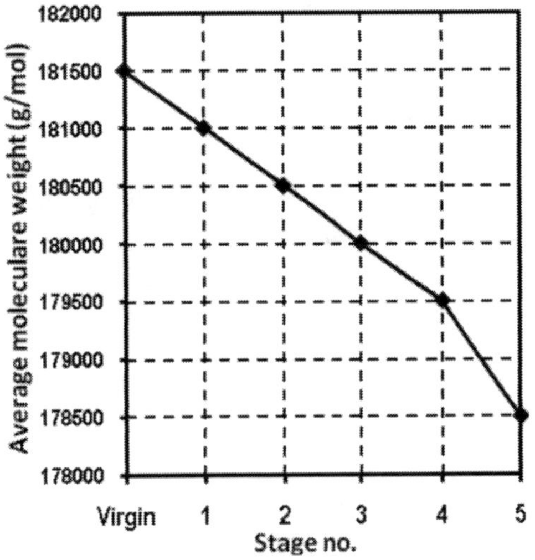

Figure 15 The average Molecular weight

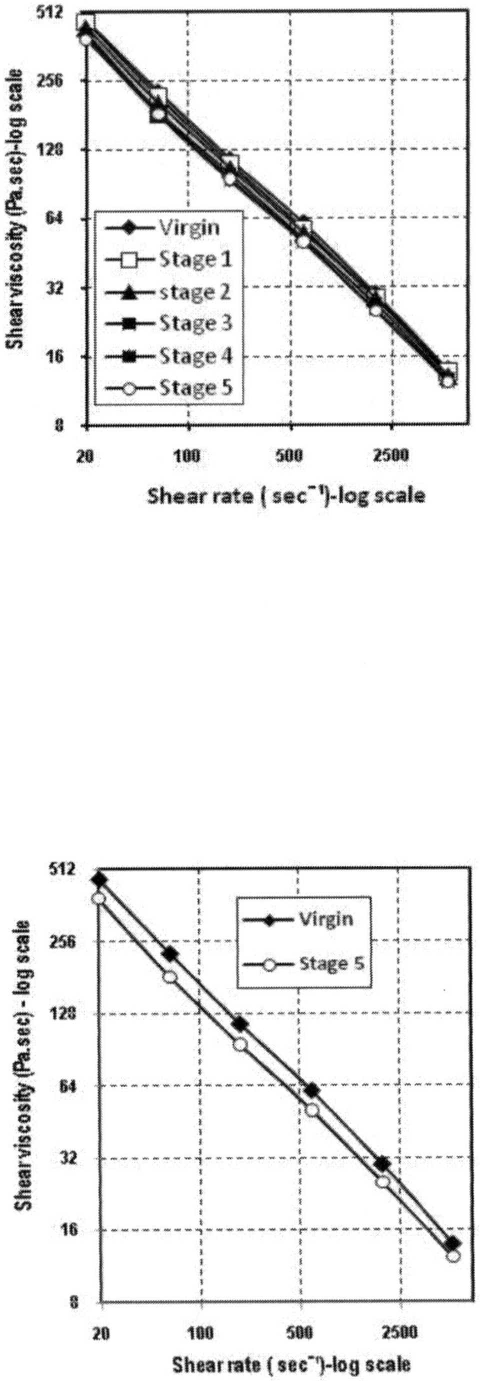

Figure 16 Melt viscosity
(a: over six reprocessing stages, b: for virgin and stage 5)

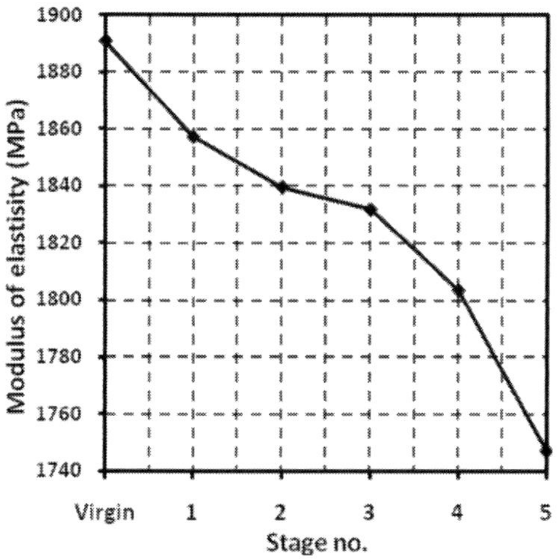

Figure 17 Modulus of elasticity

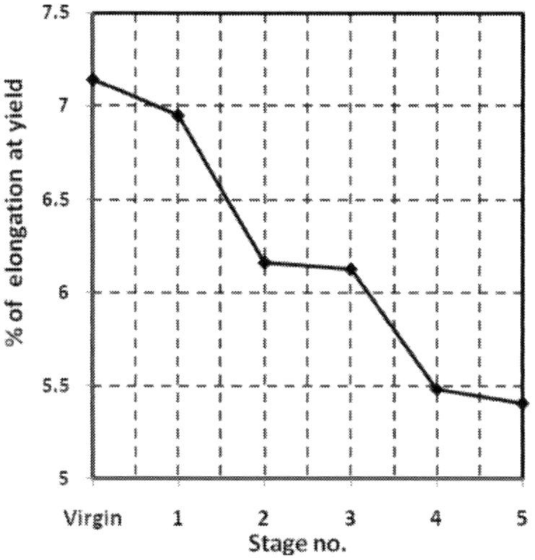

Figure 18 Percentage of elongation at yield

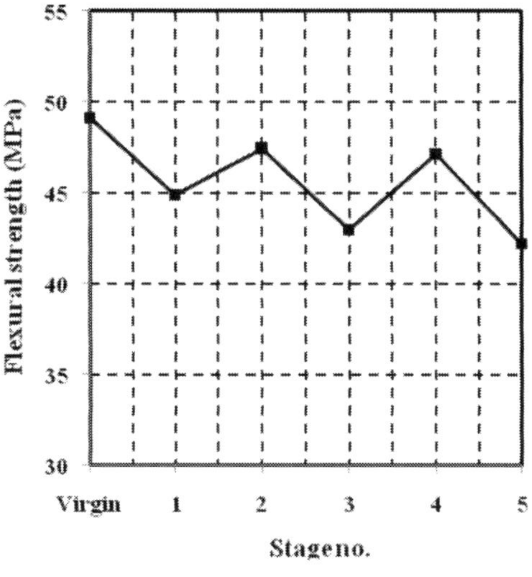

Figure 19 Flexural Strength

4- Conclusions

This study has investigated the effects of reprocessing a glass fibre polypropylene material. Effects have been investigated in terms of:

- Part quality (mass, in-plane shrinkage)
- Physical properties (density, crystallinity, fibre length, thermal stability and molecular weight).
- Rheological property (melt viscosity).
- Mechanical properties (Young's modulus, percentage of elongation at yield and flexural strength).

Tensile samples have been moulded using an injection moulding machine (high precision Fanuc all-electric drive) and the moulding conditions have been maintained consistently for each reprocessing stage. Some samples where set aside for testing whilst others where regranulated using a standard blade regranulating machine. The following points can be concluded:

1- The current work demonstrates the possibility to reprocess 100% of regrind PP material over six successful repetitive stages and full tensile test specimen products can be manufactured under the same moulding conditions.

2- Part quality: no considerable difference in specimen mass and in-plane shrinkage; this could as a result of a consistent moulding conditions (holding pressure, holding time).

3- Physical and rheological properties: a reduction in shear viscosity (-11%) at shear rate of 6000 sec^{-1} over 6 reprocessing stages lead to reduce the overall molecular weight (-2%), this could allowing some of the material chains to align more readily and hence increase crystallinity (4%). Fibre length analysis shows minor negative effect over six reprocessing stages. This is probably, as a result of consistent reprocessing operations (i.e.proper moulding settings, constant amount of fibre content through the reprocessing stages and irrelevant effect of regrinding procedure). Based on TGA result, there is no significant difference between the virgin and stage 5, and hence the degradation behaviour is seemed to be stabilized over 6 reprocessing stages.

4- Mechanical properties: in industry, to accept or refuse the relative decrease related to mechanical properties is absolutely depended on the type of application, and the customer's specifications/requirements. But it could be said that the modulus of elasticity, flexural strength have reasonable change, where Young's modulus and flexural strength have percentage of reduction of 4% for from the virgin to stage 3.

Acknowledgement

The authors would like to thank the technical staff in IRC polymer engineering at the University of Bradford for their help and support.

References

BERNARDO, C. A., et al., "The Recycling of Thermoplastics: Prediction of the Properties of Mixtures of Virgin and Reprocessed Polyolefins", Polymer Engineering and Science, **36**, 511-519(1996).

BRYCE, D. M.: Plastic Injection Molding: Manufacturing Process Fundamentals, Society of Manufacturing Engineers. Michigan(1996).

BUREAU, M. N., DENAULT, J., "Fatigue Resistance of Continuous Glass fiber/Polypropylene Composites: Consolidation Dependence", Composites Science and Technology, **64**, 1785-1794 (2004).

CHARBONNEAU, M. T., "Effects of Recycle History on the Mechanical Characteristics of Long Glass Fiber Reinforced Polypropylene", SPE ANTEC Tech. Papers, **3**, 2920-2923(1997).

DAVE, P., et al., "Injection Molding Applications of Polyolefin-Based Filled & Reinforced Products by the Use of Designed of Experiment Technique", SPE ANTEC Tech. Papers, **1**, 514-520(1997).

DINTCHEVA, N. T., LA MANTIA, F. P., "Thermomechanical Degradation of Filled Polypropylene", Macromolecular Symposia, **194,** 277-286(2003).

DONTULA, N., et al., "Approach Towards Molding Parts with Constant Properties on Addition of Regrind", SPE ANTEC Tech. Papers, 3037-3041(1994).

FATHI, S., BEHRAVESH, A. H., "Visualization of in-Mold Shrinkage in Injection Molding Process", Polymer Engineering and Science, **47**, 750-756(2007).

FU, S.-Y., et al., "Correction of the Measurement of Fiber Length of Short Fiber Reinforced Thermoplastics", Composites Part A: Applied Science and Manufacturing, **33**, 1549-1555(2002).

GOWARIKER, V. R., et al.:Polymer Science, Wiley, New York(1986).

GUIMOND, C. N., "Effect of Recycle History on the Properties of Glass Fiber Reinforced Polypropylene", SPE ANTEC Tech. Papers, **3**, 3166-3169(1997).

GUPTA, V. B., et al., "Characterization of Fiber Length Distribution in Short And long-Glass-Fiber Reinforced Polypropylene During Injection Molding Process", Kasetsart J.(Nat.Sci.), **42**, 392-397(2008).

JOHANNABER, F.:Injection Molding Machines : A User's Guide, Hanser Puplishers, Munich(1983).

KARIAN, H. G.: Hand book of Polypropylene and Polypropylene Composites, 2nd Edition, Marcel Dekker Inc., New York(2003).

LA MANTIA, F. P.:Handbook of Plastics Recycling, Rapra Technology Ltd.. Shrewsbury (2002).

LEWIS, C. L., BUSER, M. A., "Effect of Reprocessing Pre-Comsumer Polyethylene in an Injection Molding Environment", SPE ANTEC Tech. Papers, **1**, 3162-3165(1997).

LICEA-CLAVERIE, A., et al., "Reprocessing Effects on the Properties of a Hybrid Nylon 6,6-Composite Reinforced with Short Glass and Carbon Fibers", Journal of Reinforced Plastics and Composites, **21**, 847-856(2002).

MAMAT, A., et al., "Shrinkage Analysis of Injection Molded Polypropylene Parts", SPE ANTEC Tech. Papers, 513-517(1994).

MERAN, C., et al., "Examination of the Possibility of Recycling and Utilizing Recycled Polyethylene and Polypropylene", Materials & Design, **29**, 701-705(2008).

PATCHARAPHUN, S., OPASKORNKUL, G., "Characterization of Fiber Length Distribution in Short and Long-Glass-Fiber Reinforced Polypropylene During Injection Molding Process", Kasetsart J. (Nat.Sci.), **42**, 392-397(2008).

PÖTSCH, G., MICHAELI, W.: Injection Molding : An Introduction, Hanser Publishers, New York(1995).

SPEIGHT, R.G.: In-Line Process Measurements for Injection Molding Process Control, PhD, University of Bradford, Bradford(1993).

SHORTALL, J. B., PENNINGTON, D., "Characterization of Fibre Length Distribution in Injection Moulded Plastics". Plastics and Rubber Processing and Applications, **2**, 33-40(1982).

XU, G. J., et al., "Characterization of Virgin and Postconsumer Blended High-Impact Polystyrene Resins for Injection Molding", Journal of Applied Polymer Science, **84**, 1-8(2002).

YILMAZER, U., CANSEVER, M., "Effects of Processing Conditions on the Fiber Length Distribution and Mechanical Properties of Glass Fiber Reinforced Nylon-6", Polymer Composites, **23**, 61-71(2002).

ELECTROSTATIC SEPARATION TO SORT AND RECYCLE PLASTIC WASTES
– SEPARATION OF POLYETHYLENE AND POLYPROPYLENE –

V. Albrecht[1], E. Reinsch[2], R. Schünemann[2], U. A. Peuker[2], U. Gohs[1] and F. Simon[1]*

[1]Leibniz Institute of Polymer Research Dresden, Hohe Strasse 6, D-01069 Dresden, Germany – albrecht@ipfdd.de; gohs@ipfdd.de; frsimon@ipfdd.de
[2]Freiberg University of Mining and Technology, Institute of Mechanical Process and Reprocess Engineering, Agricolastrasse 1, D-09596 Freiberg, Germany – edith.reinsch@mvtat.tu-freiberg.de; ralf.schuenemann@mvtat.tu-freiberg.de; urs.peuker@mvtat.tu-freiberg.de;

Abstract – The most polymers are incompatible in their melt or solution. Phase separation processes lead to an unsatisfactory quality of goods made from recycled plastic materials. In order to produce useable consumer goods from plastic waste, the recycling process requires a sorting where the plastic waste is separated into its material fractions.
The electrostatic separation of plastic waste is an effective and reasonably priced technology. It bases on the well known electrostatic charging of polymers: Two polymer grains were electrically charged during their collision. After separation the formed charges remain stable. The known tribo-electric rows show that the charging behaviour of polymers correlates with their chemical structure. In a tribo-charged mixture of two different polymer species one of the two polymers is positively charged and the other is negatively charged. If they drop in an electrostatic field the oppositely charged polymers were separated into sorted polymer fractions.
In actual practice the simple process seems to be more complex. The separation in the electrostatic field requires a high and selective charging of all components in a polymer mixture. However, some of the common polymers, e.g. polyethylene (PE) and polypropylene (PP) do not satisfy this condition. In order to separate mixtures consisting of these polymers, which have a low tendency to be charged, we applied additional treatments to improve the polymers' charging behavior selectively.

Introduction

In the Federal Republic of Germany and the European Union a clear political commitment exists to recycle plastic wastes. Several national laws (BGB, 1994) and supranational directives express this commitment (EU, 2002a; EU, 2002b; EU, 2002c).
Generally, plastic wastes are mixtures of several polymer types. A recycling and reprocessing of such multi-component plastic waste requires a sorting into the waste's individual components. The reason for that is the thermodynamic incompatibility of the most polymers in a mixed melt and solution leading to segregation processes and inhomogeneous distribution of the individual polymer species. Commodities made from mixed plastic wastes show mechanical properties and visual appearances, which cannot be satisfying for their use. Only the separation of the mixed plastic wastes into their polymer components allows to produce commodities with properties comparable with those were made from new production material.
Among other separation techniques, the electrostatic separation seems to be a very effective and efficient process to separate plastic wastes into their individual polymer components. Its main advantage is the separation of shredded plastic wastes in their dry state. The electrostatic separation bases on the well known phenomenon that insolating plastics (or polymers) are charged up in contact with conductors and non-conductors, respectively. The established tribo-electric rows (Coehn, 1898; Brück, 1981) clearly show that the polymer's charging behaviour depends on its chemical constitution (Albrecht et al., 2003), and in many cases it can be considered as material-specific. In former works (Albrecht et al., 2003; Albrecht et al., 2006; Albrecht et al., 2009) we extensively discussed the relation between chemical structures and charging behaviour. According these works, the fundamental step of polymer charging is the dominating ability of a polymer grain surface to donate or accept electrons during the mechanical contact with a second grain made from another polymer material (Albrecht et al., 2003). Electron exchanges in the particles' contact zones induce pattern of oppositely charged domains on the particle surfaces (Albrecht et al., 2006). In competition to the charging up discharging or charge disappearing can take place. The stability of charges on the polymer surfaces is strongly influenced by the environmental conditions. In particular, increased air humidity can force the charge disappearing and control the electrostatic net charge of some polymer species as well as the separation process (Albrecht et al., 2006; Albrecht et al., 2009).
A good quality of the separation process is characterized by a complete separation of all individual

polymer species in a plastic waste mixture. It requires an intensive and material-specific charging of all the polymer species of the mixture. This behaviour is not expected for multi-component mixtures because the sign and the amount of the tribo-electric charging of a polymer sample also depends the other polymer species in the mixture. That implies that it could be necessary to separate common multi-component mixtures stepwise into classes, which can be finally separated into their individual polymer components. Nevertheless, for the successful separation of a two component mixture an intensive and opposite charging of the two individual polymer species is required (Németh et al., 2005). Polymers having polar groups in their chains, such as polyamides, poly(vinyl chloride), polyesters etc. can be easily tribo-electrically charged, and their two-component mixtures can be excellently separate in an electric field (Németh et al., 2005). The tribo-electric charging and electrostatic separation of other polymers, such as polyethylene (PE) and polypropylene (PP), which do not have polar groups in their structure, seem to be a challenge. These polyolefin types are very common polymers with an annual production of several million tons, and therefore one of the commonly components in wastes. Two-component mixtures of PE and PP cannot be tribo-electrically charged and separated in an electric field with good qualities (Németh et al., 2005; Albrecht et al. 2009).

The aim of our work was to develop processes enabling the electrostatic separation of polyolefin waste mixtures, in particular PE and PP. Wastes are mainly characterized by manifold contaminations and impurities. Neither the kind nor amount of the unavoidable impurities should have an influence on the quality of the separation step.

Experimental

Materials

Low-density PE (LDPE), *high-density* PE (HDPE) and PP grains having a mean diameter of 3.6 mm were used as test materials. LDPE (trade name LDPE 150L) was provided by the DOW Olefinverbund GmbH, Schkopau (Germany), HDPE (GD 7255) by the Basell AG, Wesseling (Germany), and PP (HD 120 MO) by Borealis AS (Norway). The polymers did not contain dyes or fillers. In order to remove native charges the grains were carefully washed with de-ionized water and dried in an oven at 50 °C. The samples prepared in this way were completely uncharged.

For all separation experiments two- or three-component mixtures of 300 g were produced by mixing 150 g or 100 g of the polyolefins, respectively.

Electrostatic gravity separator (laboratory test system)

Figure 1 shows the scheme of the electrostatic gravity separator. Details of its build-up and workings are described in Ref. (Németh et al., 2005). The charging unit (*1*) consists of a cylindrical fluidization chamber. It was completely lined with PE. Conditioned air provided by an air-condition system (*4*) flown through mixtures of polyolefin grains (300 g). The adjusted air temperature (25 °C), relative humidity (25 %) and volumetric flow were computer-controlled. At first, the mixtures were conditioned in an air flow of 1 m·s^{-1} for 30 min. For charging up the particles in a fluidized-bed the air flow was set to 5 m·s^{-1} for 4 min. Then, the slit (*1a*) on the bottom was opened and the particles dropped in the electric field of the electrostatic separation zone (*2*).

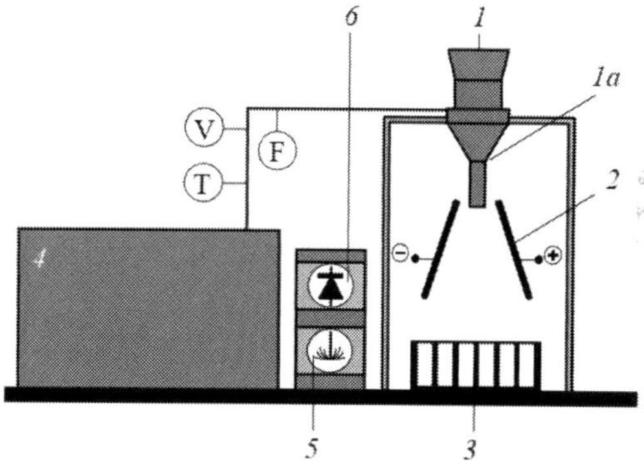

Figure 1 – Schematic build-up of an electrostatic gravity separator (laboratory test system) to charge mixtures of plastic grains and separate them into their individual polymer fractions. Charging unit (*1*) with outflow (*1a*), plate electrodes to generate an electric field (*2*), collecting tanks (Faraday cups) (*3*), air condition unit with controllers for temperature, humidity and volume flow (*4*), electrometer (*5*), and two high-voltage rectifiers (*6*).

The electric filed was generated by two tilted electrodes (*2*). The applied DC voltage was provided by two high-voltage rectifiers (6). The maximum field strength was ca. 4.5 kV·cm^{-1} at 144 kV. The minimum field strength was ca. 2 kV·cm^{-1}. According their net charges the polymer particles were deflected in the electric field. Below the electrodes eight collecting tanks (*3*) collected the grains. In order to measure the charges of the collected grains the tanks worked as Faraday cups. Their charges were measured with an electrometer (type 6517 A, Keithley Instruments, USA) (*5*). In order to determine the mass and charge distributions the products collected in each tank (*3*) were analyzed concerning the polymer species.

Electron-beam treatment

The electron-beam treatment experiments were carried out with an electron accelerator ELV-2 (INP

Novosibirsk, Russia). In order to avoid surface contaminations the samples (mixtures of polymer grains) were wrapped in a PE bag. Parameters of the electron-beam treatment are given below.

Results and Discussion

Separation experiments without pre-treatments

Figure 2 shows the result of the electrostatic separation of a mixture of poly(vinyl chloride)/poly(ethylene terephthalate) (PVC/PET) grains. In the fluidized-bed of the charging unit the PVC and PET grains were oppositely charged. During the grains' passage of the electric field the positively charged PET grains were deflected to be preferably collected in the collecting tanks 1 to 3, while the negatively charged PVC particles were preferably collected in the collecting tanks 6 to 8. Figure 2 demonstrates a successful separation of a two-component mixture because the two individual polymers PET and PVC were collected as homogeneous polymer fractions (Albrecht et al., 2008).

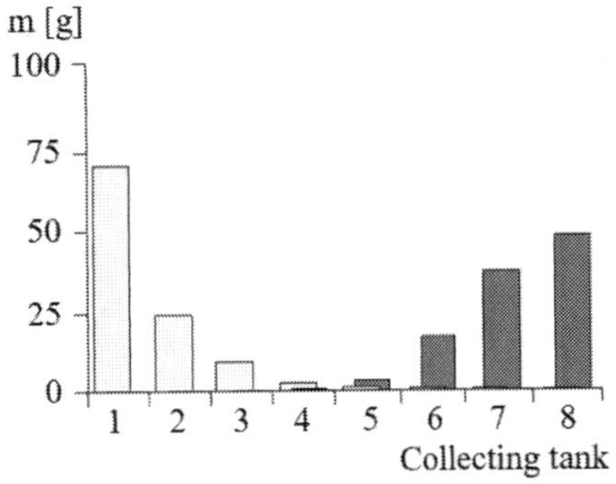

Figure 2 – Mass (m) distribution of PVC (red bars) and PET (yellow bares) fractions from a mixture of PET/PVC grains ($m_{PVC+PET}$ = 300 g; $m_{PVC} : m_{PET}$ = 1:1) after successful sorting in the electrostatic gravity separator.

Typical results of electrostatic separation experiments with mixtures of PE and PP grains are shown in Figure 3. In contrast to PET/PVC mixtures it was impossible to separate PE/PP mixtures electrostatically. The reason for these unsatisfactory results is the insufficient charging of the PE and PP grains in the charging unit. The polyolefin polymers do not have functional groups in their structures, which have the ability to donate or accept electrons. Hence, during the polymer/polymer contact no electrons can be exchanged between the grain surfaces and interacting particles remain uncharged. During their passage of the electric field the non-charged grains were not polymer-specifically deflected, and finally the two polymer species were collected as mixtures in the middle collecting tanks. As mentioned above, the humidity of the process air is able to control the surface net charge via its influence on the discharging or charge disappearing. In other words, the variation of the relative air humidity can be considered as a tool to control the charging of polymers. In the case of the non-polar polyolefins the relative humidity does not have an influence on the particle charging (Schubert et al, 2004) because the primary step the generation of charges via an electron transfer does not take place.

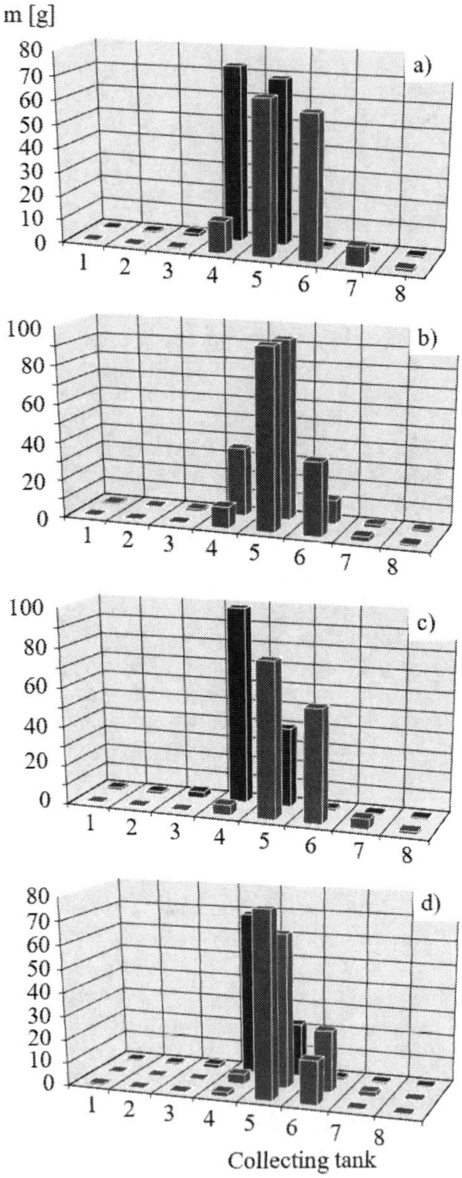

Figure 3 – Results of electrostatic separation experiments with LDPE/PP (a), HDPE/PP (b), LDPE/HDPE (c), and LDPE/ HDPE/PP (d) mixtures (dark-blue bars = LDPE; light-blue bars HDPE, reddish brown = PP). The figures show the mass (m) distribution of the polymers ($m_{LDPE+HDPE+PP}$ = 300 g; $m_x : m_y$ = 1:1; $m_{LDPE} : m_{HDPE} : m_{PP}$ = 1:1:1).

For a successful separation of non-polar olefins, such as LDPE, HDPE, and PP, other charge-controlling substances or process are required.

Separation experiments after an electron beam pre-treatments of the PE/PP mixtures

In our former works we studied the influence of adsorbed ion-carrying substances, such as acids, bases, surfactants, and polyelectrolytes on the charging behaviour of polymers (Schubert et al., 2004). In the case of the non-polar polyolefins the influence of these substance classes on the polymers' charging up is rather small and the slight changes seem to be insignificant. In 1996, Stückrad showed that electron-rich dyes, such as the azine dye *Nicrosin*, have a strong influence on the charging up of non-polar polymers (Stückrad, 1996). Charging experiments with PE and PP mixtures where *Nicrosin* was adsorbed showed that the two polymers were oppositely charged and their particular net charge was sufficient to separate the two polymers successfully in the electric field of the electrostatic gravity separator (Schubert et al., 2004). The main disadvantages of this pre-treatment were the irreversible dying of the polymer grains and the high processing costs.
From polymer chemistry and the paint-spraying in the automotive industries it is well known that plasma treatments are able to modify a few nanometers of polyolefin surfaces. The treatment changes the non-polar character of the polyolefin surface by the implementation of highly polar groups. The method seemed to be suitable to better the charging of PE and PP. Actually, after an oxygen-plasma treatment and a storage time of a few hours it was possible to separate PE/PP mixtures in excellent qualities (Németh et al., 2004). The reason for the polymer-specific charging is the different ability of the two polymers to incorporate oxygen-functional surface groups (Albrecht et al., 2008). The use of nitrogen or air as plasma gas also introduced polar surface groups in both polymers. However, the introduction of nitrogen-containing surface groups counteracts the effect observed after the oxygen-plasma pre-treatment. The application of the low-cost atmosphere plasma technique seems to be unsuitable for the pre-treatment of the polyolefin waste mixtures.
Treatments of polyolefins with electron-beams should also introduce oxygen-containing functional surface groups and turn the surface properties from non-polar to polar. The electron-beam treatment can be carried out under atmospheric conditions. The high-energy beam passing air generates highly reactive species, such as ozone and NO_x, which can react with the sample surface. The main advance of the electron-beam irradiation is that the process can work continuously and guarantees high throughputs. Electron-beam treatments of PE were applied to cross-link the polymer matrix, while PP undergoes degradation reactions (Heger et al., 1994).

Polyolefin mixtures, which were electron-beam irradiated, were stored for 24 h. Then, the samples were tribo-electrically charged and successfully sorted in the electrostatic gravity separator (Figure 4) (Albrecht et al., 2007).

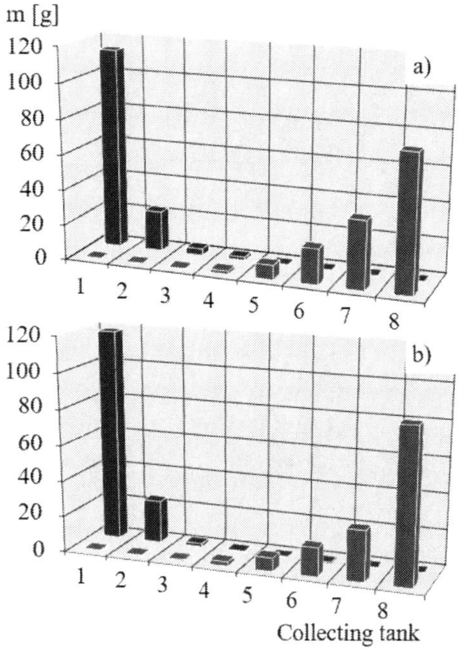

Figure 4 – Results of electrostatic separation experiments with LDPE/PP (a) and HDPE/LDPE (b) mixtures (dark-blue bars = LDPE; light-blue bars HDPE, reddish brown = PP). All sample mixtures were pre-treated by electron-beam irradiation: electron power $E_s = 0.6$ MeV at a current of $I = 4.0$ mA, dose $D = 7.5$ kGy. The figures show the mass (m) distribution of the polymers ($m_{x+y} = 300$ g; $m_x : m_y = 1:1$).

The analysis of the charging and separation experiments showed that PP was negatively charged, while PE was positively charge. These findings are in contrast to the results obtained after an oxygen-plasma pre-treatment where PP was positively charged and PE showed a negative net charge. In addition, the analysis of the surface composition of the electron-beam treated samples showed only a minute modification effect. The unexpected charging behaviour and the low modification degree indicate a charge formation mechanism, which is different to the electron transfers described above.
Time-dependent electron spin resonance (ESR) studies showed that the electron-beam irradiation implements electrons in the bulk of the polymer samples (Figure 5). The stability of these charge carriers in the bulk material is polymer-specific. Electrons implemented in the PE matrix disappear rapidly, while electrons in the PP bulk remain stable over a few days.

The storage of one day guaranteed polyolefin mixtures where the PP fraction is strongly charged, while the PE fraction can be considered as neutral. The additional treatment in the charging unit of the electrostatic gravity separator does not influence the buried charge carriers of the PP grains but charges the PE fraction positively.

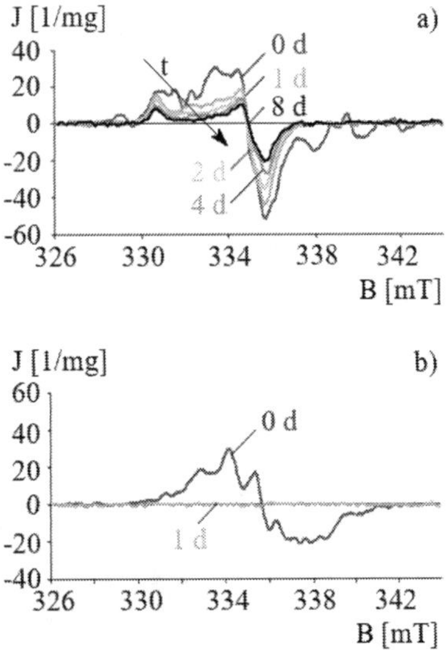

Figure 5 – Time-dependent ESR studies of electron-beam irradiated (a) PP and (b) LDPE (B = magnetic field strength, J = relative intensity, t = storage time in days). The electron-beam irradiation experiments were carried out with an electron power of E_s = 1.5 MeV at a current of I = 4.0 mA, dose D = 7.5 kGy.

As mentioned above, electron irradiation is widely applied to modify the polymer matrix of polyolefins. Cross-linking and polymer degradation reactions cannot be accepted for the polymer processing and the use of the recycled materials. However, the electron-beam irradiation for the pre-treatment of the polyolefins can be carried out under very mild conditions. Optimal separation results required an electron energy of E_s = 0.6 MeV at a current of I = 4.0 mA and a dose of D = 7.5 kGy. Under these conditions we observed neither cross-linkages in the PE matrix nor any degradation of the PP polymer.

Surface contaminations

Waste means products, which are characterized by multifaceted contaminations. Above we briefly described the use of adsorbed dyes to control the charging behaviour of polyolefin samples. This indicates that impurities or contaminations could also have a strong influence on the charging behaviour of polyolefin materials. The technical charging and separation process cannot be tuned or optimized for certain contaminations because the kind and origin of the impurities can be very manifold.

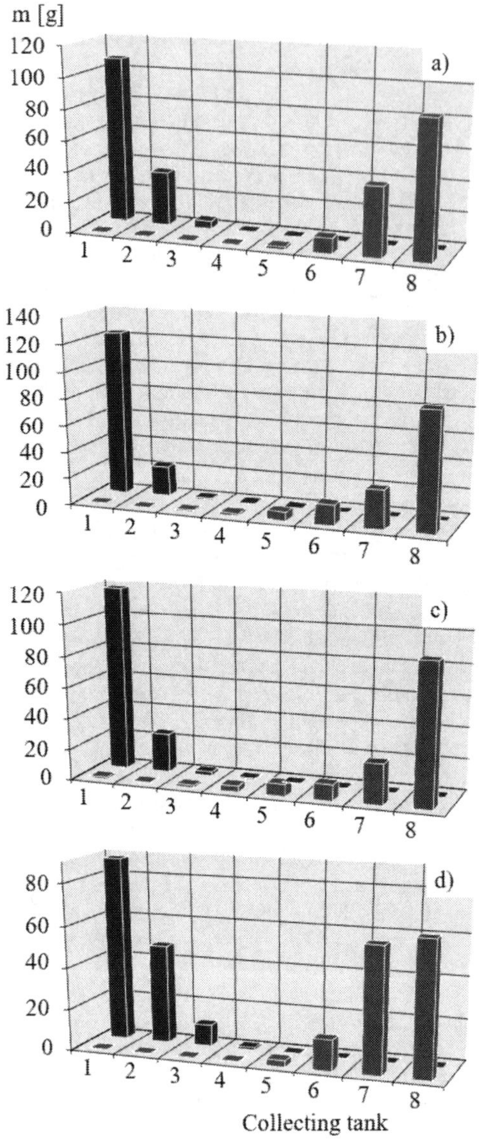

Figure 6 – Results of electrostatic separation experiments of surface-contaminated LDPE/PP mixtures (dark-blue bars = LDPE, reddish brown = PP). Contaminating agents were milk (a), egg-white (b), sugar solution (c), and sunflower oil (d). All contaminated sample mixtures were pre-treated by electron-beam irradiation: electron power E_s = 0.6 MeV at a current of I = 4.0 mA, dose D = 7.5 kGy. The figures show the mass (m) distribution of the polymers ($m_{LDPE+PP}$ = 300 g; $m_{LDPE} : m_{PP}$ = 1:1).

In order to study the influence of impurities on the polyolefin separation we contaminated sample mixtures with substances typical for food packages, such as milk and egg-white (both as typical examples for proteins), sugar solution, and sunflower oil (esters of fatty acids). All these substances are able to carry charges and could influence the charging behaviour of the polyolefins by their charges.

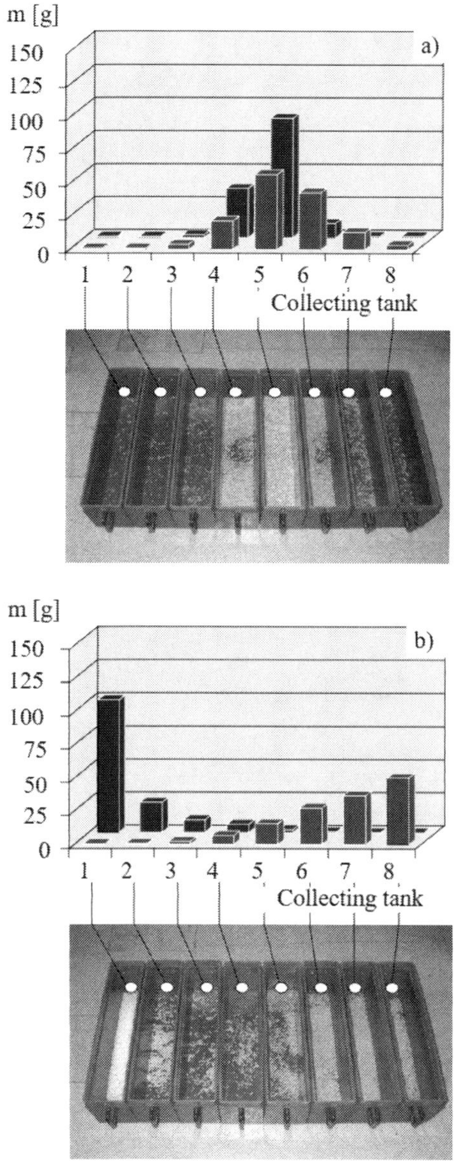

Figure 7 – Results of electrostatic separation experiments of LDPE/PP mixtures (dark-blue bars and white grains = LDPE, reddish brown and yellow grains = PP) where the PP fraction was dyed with 0.5 weight-% of a yellow *diazo*-pigment. The separation experiment in Figure 7a was carried out without an electron-beam pre-treatment, the experiment in Figure 7b shows the successful separation after an electron-beam pre-treatment (E_s = 0.6 MeV at a current of I = 4.0 mA, dose D = 7.5 kGy). The diagrams show the mass (m) distribution of the polymers ($m_{LDPE+PP}$ = 300 g; m_{LDPE} : m_{PP} = 1:1).

The contaminated sample mixtures were pre-treated in the electron accelerator, 24 h stored, then, charged in the charging unit of the electrostatic gravity separator and finally sorted in the separator's electric field. As can be seen in Figure 6 the contaminations do not influence the selectivity of the electrostatic separation process. These findings were expected because the charging of the electron-beam pre-treated polyolefin samples was not preferably controlled by the tribo-electric process but by charges implemented into the polymer bulk phases. Obviously, the inner or buried charges are able to compensate additional charge phenomena on the particle surfaces.

Influences of pigments on the charging

Common plastics contain fillers, pigments, stabilizers etc. in their bulk phase. Inorganic materials, such as metal oxide particles are able to generate (e.g. by the dissociation of the Me–OH surface groups into Me–O$^-$ and H_3^+O ions) and carry charges. Charge effects coming from fillers or polymer additives could be able to over-compensate the charges, which were implemented by the electron-beam irradiation or significantly change the rate of charge disappearing (e.g. conducting fillers, such as antistatic agents, carbon black or carbon nano-tubes).

For our first experiments we used mixtures of LDPE/PP grains dyed with a yellow *diazo*-pigment (0.5 weight-%). As shown above untreated LDPE/PP mixtures cannot be charged in the tribo-charging unit. Possible charging effects observed with the coloured samples must result from the pigment added.

Figure 7a shows a typical result of a separation experiment carried out with a mixture of LDPE/PP grains where the PP polymer was dyed. The result agrees with the experiments shown in Figure 3. A sorting of the mixture into the individual polymer fractions was impossible. Obviously, the small (but for commercial applications typical) amount of the yellow *diazo*-pigment did not influence the charging behaviour of the polyolefins.

The pre-treatment of the polyolefin mixtures with an electron-beam irradiation enabled a electrostatic sorting of the polymer mixture with an excellent selectivity (Figure 7b).

Conclusions

The sorting of plastic wastes is essentially necessary to re-process polymers. The tribo-electric charging and the following electrostatic sorting process is an effective and suitable way to sort plastic wastes into their individual polymer fractions. Nonpolar polyolefin polymers, such as the common polymers PE and PP cannot be tribo-electrically charged and separated in an electrostatic gravity separator. Pre-treatment processes are required to modify the polyolefins to be polymer-specifically charged. We successfully tested the use of electron-rich dyes, oxygen-plasma treatments and electron irradiation. The latter process seems to be very suitable because the employed electron accelerator allows high throughputs and the irradiation does not significantly change the polymer matrices. Surface contaminations and impurities typical for plastic wastes do not influence the charging behaviour of the electron-beam treated polyolefin samples. The influence of fillers and polymer additives on the polymers' charging behaviour is under investigation.

Acknowledgement

The authors acknowledge the financial support of the German research fond *Arbeitsgemeinschaft industrieller Forschungseinrichtungen* (AiF) and the *Federal Ministry of Economy and Technology* (grant #14980 BR).

References

Albrecht, V., Németh, E., Schubert, G., Simon, F., "Polymer Tribo-electric Charging: Dependence on Thermodynamic Surface Properties and Relative Humidity", J. Electrostatics, **58**, 3–16 (2003), DOI: 10.1016/S0304-3886(02)00137-7.

Albrecht, V., Janke, A., Drechsler, A., Schubert, G, Németh, E., Simon, F., "Visualization of Charge Domains on Polymer Surfaces", Prog. Coll. Polym. Sci., **132**, 48–53 (2006), DOI: 10.1007/b104970

Albrecht, V., Reinsch, E., Gohs, U., Schünemann, R., Husemann, K., Simon, F., "Zur elektrostatischen Trennung von Polyolefingemischen – Electrostatic Separation of Polyolefin Mixtures", Aufbereitungstechnik – Mineral Processing, **49**, 26–33 (2008)

Albrecht, V., Janke, A., Németh, E., Spange, S., Schubert, G., Simon, F., "Some Aspects of the Polymers' Electrostatic Charging Effects", J. Electrostatics, **67**, 7–11 (2009), DOI:10.1016/j.elstat.2008.10.002

BGB – Gesetz zur Förderung der Kreislaufwirtschaft und Sicherung der umweltverträglichen Beseitigung von Abfällen KrW-/AbfG (*engl.:* Act for Promoting Closed Substance Cycle Waste Management and Ensuring Environmentally Compatible Waste Disposal) – Kreislaufwirtschafts- und Abfallgesetz vom 27.09.1994, BGBl. I 1994 S. 2705; 1996 S. 1354; 1998 S. 509, 1485, 2455; 2000 S. 632; 27.7. 2001 S. 1950, 2001 S. 2331; 29.10.2001. S. 2785 Art. 57; 21.8.2002 S. 3322 02; 25.1.2004 S. 82 04; Gl.-Nr.: 2129-27-2.

Brück, R., "Chemische Konstitution und elektrostatische Eigenschaften von Polymeren", Kunststoffe, **71**, 234–239 (1981)

Coehn, A., "Über ein Gesetz der Electricitätserregung", Ann. Physik Chem., **64**, 217–232 (1898)

DE Patent 10 2004 024 754 (2004) Németh, E., Simon, F., Albrecht, V., Schubert, G., "Verfahren zur elektrostatischen Trennung von Kunststoffgemischen"

DE Patent 10 2007 055765.7 (2007) Albrecht, V., Gohs, U., Husemann, K., Reinsch, E., Schünemann, R., Simon, F., "Verfahren zur elektrostatischen Trennung von Kunststoffgemischen durch prozeßvorbereitende Elektronenbestrahlung der Gemische"

EU – Directive 2000/53/EC of the European Parliament and of the Council of 18 September 2000 on *End-of-Life Vehicles* – Commission Statements. Official Journal L 269, 21/10/ 2000 S. 34-43.

EU – Directive 2002/95/EC of the European Parliament and of the Council of 27 January 2003 on the *Restriction of the Use of Certain Hazardous Substances in Electrical and Electronic Equipment.* Official Journal L 037, 13/02/2003 S. 19-23.

EU – Directive 2002/96/EC of the European Parliament and of the Council of 27 January 2003 on *Waste Electrical and Electronic Equipment (WEEE)* – Joint declaration of the European Parliament, the Council and the Commission relating to Article 9. Official Journal L 037, 13/02/2003 S. 24-39.

Heger, A., Dorschner, H., Dunsch, L., Ihme, B., Lunkwitz, K., Technologie der Strahlenchemie von Polymeren, Hanser Fachbuchverlag, Munich (1994)

Németh, E., Schubert, G., Albrecht, V., Simon, F., "Triboelektrische Aufladung von Kunststoffgemischen – Triboelectric Charging of Mixed Plastics", Aufbereitungstechnik – Mineral Processing, **45**, 35–46 (2005)

Schubert, G., Németh, E., Simon, F., Albrecht, V., "Beeinflussung der Kontaktaufladung von Kunststoffen durch Modifizierung ihrer Oberfläche", in Partikelwechselwirkung bei Prozessen der Mechanischen Verfahrenstechnik, Husemann, K., Graichen, K. (Eds.), Shaker-Verlag, Aachen, p. 183–212 (2004), ISBN 3-8322-3245-1

Stückrad, O., "Sortieren von Kunststoffgemischen durch Flotation", Doctoral Thesis, Clausthal University of Technology, Clausthal-Zellerfeld (1996)

Effects of novel additives on processability of recycled linear polypropylene

Krishnamurthy Jayaraman, Amit Kumar Chaudhary*

Michigan State University - MI - USA

Linear polypropylene by itself does not have adequate melt strength for shaping by blow molding into bottles, or for producing closed cell foam. The melt strength can be improved sufficiently by compounding the linear polypropylene with 2 to 3 wt% of a suitable organically modified montmorillonite, along with functionalized polypropylene as compatibilizer. The recycled polymer will display effects of both temperature and flow. Hence the object of this paper was to investigate separately the effects of high temperature annealing without flow and of high shear in a melt rheometer on the melt strength of linear PP-clay nanocomposites with and without silane treatment on the organoclay. In this paper, additional silane treatment was carried out with different silane coupling agents on the organoclay and each organoclay was compounded with the same linear polypropylene and compatibilizer and in the same proportions. The results reveal that after annealing for up to five hours at 180C, silane treatment of the organoclay with dialkoxy aminoalkylsilane and with monoalkoxy aminoalkylsilane lead to compounds where the melt strength is preserved to the same extent. But after extrusion through a die at 500 s-1, there is a marked difference between compounds with the two different coupling agents. The melt strength is preserved in the compound with the monoalkoxy silane treated clay. But with the other dialkoxy silane treatment, the melt strength deteriorates significantly after extrusion. These trends are explained by different changes in the nanostructure of the various compounds after annealing and extrusion, as found by X-Ray diffraction and by dynamic shear moduli of the melts.

TECHNOLOGY OF MICRO THERMOFORMING AND MICRO BLOW MOULDING FOR THE INTEGRATION OF SELECTIVE SURFACEMODIFICATIONS IN THREE-DIMENSIONAL MICROSTRUCTURES

M. Heilig[1], M. Schneider[1], T. Ide[1], H. Dinglreiter[1], M. Worgull[1]

[1] *Institute for Microstructure Technology, Karlsruhe Institute of Technology, Hermann-von-Helmholtz-Platz 1, 76344 Eggenstein-Leopoldshafen, Germany – markus.heilig@kit.edu*

Abstract - Thermoforming and Blow Moulding are well established replication technologies. Thermoforming as a replication technology of thin walled parts is mainly used for the fabrication of enclosures, packages and large-scale parts. Process of Thermoforming was transferred to micro-scale at Forschungszentrum Karlsruhe, now part of the Karlsruhe Institute of Technology (KIT). At the Institute for Microstructure Technology (IMT) a high pressure Micro Thermoforming Process is used to fabricate film microchips for capillary electrophoresis (CE) and for three-dimensional cell cultivation. Selective surface modifications on the thermoformed parts are possible by the use of prestructured polymer films. For this purpose, the technologies of nanoimprinting and micro- thermoforming will be combined. In a first step the nanostructures are patterned onto the surface of polymer films by nanoimprinting. In a second step the structured film is formed into the cavities of the mould by microthermoforming.
Thermoforming parts with micro-scale design features require use of thin polymer films. Low heat capacity and fragility of thin (< 100 μm) polymer films requires new technical developments for precise heating and form pressure application. This paper will present results of an improved process based on a new replication machine. Temperature stability and pressure distribution is highly improved, even at reduced cycle times. With an upgraded temperature range microthermoforming of high temperature polymers is possible.
Replication of hollow parts with multiscale features is also possible. Outer dimension in macroscopic length scale and functional design features in microscopic length scale are combined with nanoscale surface structures in an affordable technology. Influence of temperature and pressure application are discussed for amorphous and semi crystalline polymers. Measurements of surface structures before and after Thermoforming will be shown to demonstrate the influence of Thermoforming Process.

Introduction

Replication processes like injection moulding, hot embossing (thermal nanoimprint) and thermoforming are well established and show their performance by a large diversity of moulded parts in different kind of thermoplastic polymers. Especially in hot embossing the replication is typically limited to structures with vertical sidewalls allowing the demoulding of the replicated structures (Worgull 2009). Because of the precise one directional demoulding path the replication of structures, even nanostructures, perpendicular to the demoulding direction will damage the structures. Therefore the replication of nanostructured vertical sidewalls, to influence for example the wetting behaviour of micro channels, is under the use of the common mould insert and replication technology not possible. Anyway there exist other methods like etching or coating processes to modify the surface of replicated structures. Nevertheless these methods produce stochastic nanostructures. To control the wetting behaviour locally or to influence the wetting behaviour over a flow distance, the bottom and the sidewalls of flow channels have to be modified in a controlled way.

Especially in life sciences, a large number of applications will benefit from this concept. In micro fluidics, for example, interfacial interactions between fluids and the channel walls are dominant factors. One major interfacial interaction effect is the wetting behaviour of the channel walls. In many cases a controllable wetting behaviour is of interest to obtain locally defined hydrophilic or hydrophobic surfaces. Super hydrophobic surfaces, for example, are interesting because of their water-repellent properties.
For many bio-inspired surface structures (Abbott and Gaskell 2007) a fabrication technique for the integration of nanostructures on 3-dimensional surfaces is required. Another example is the antireflection characteristic of moth eye structures, which requires the combination of a specific localized nanostructured surface on a complex 3-dimensional micro structured design (Wilson and Hutley 1982).
The integration of selective nanostructured surfaces in freely shaped 3-dimensional cavities is a challenge in replication technology. Unlike to stochastic methods, the patterning by hot embossing or nanoimprint allows the structuring of well designed structures on surfaces but the replication is typically limited to plane surfaces (Worgull at al. 2008a).

Thermoforming is an established fabrication method for structures mainly in macroscopic dimensions and is known to be a cheap and fast process to produce enclosures or packages (e.g. Throne 2008). Process of Thermoforming was transferred to micro-scale at Forschungszentrum Karlsruhe, now part of the Karlsruhe Institute of Technology (KIT). At the Institute for Microstructure Technology (IMT) a high pressure micro thermoforming process is used to fabricate film microchips for capillary electrophoresis (CE) (Truckenmüller et al. 2002) and for three dimensional cell cultivation (Truckenmüller and Giselbrecht 2004). Microthermoforming of thin polymer films, with local surface modifications, was described by Giselbrecht et al. (2006). This approach of microscale thermoforming of thin polymer film substrates overlaid with a split local modification of the films is called 'SMART', which stands for 'substrate modification and replication by thermoforming' (Truckenmüller et al. 2008). Geometrical modifications of nanoscale surface modifications due to uniaxial and biaxial stretching of the polymer film are measured, described, and compared with an analytical model by Heilig et al. (2010).
At the Institute for Microstructure Technology (IMT) processes of hot embossing and microthermoforming are under permanent development. This paper presents first results of an improved microthermoforming process, based on a new thermoforming machine. Temperature stability and pressure distribution is highly improved by the newly designed control unit. Reproducible variations in pressure and temperature distribution can be archived at reduced cycle times. Thermoforming of high temperature polymers up to 350 °C is possible. Extension of process towards the replication of surface structured hollow parts is under current development. Therefore process of twin-sheet-forming and blow moulding will be adapted for the production of hollow microparts with selective surface modifications.

Experimental

The process to fabricate 3-dimensional microstructures with selective surface modifications fuses on a combination of two replication processes: hot embossing and micro thermoforming.

The required process steps can be summarised:

- Replication of the selective surface modifications onto a thin film of polymer by hot embossing (Fig. 1).
- Forming the surface structured polymer film into a 3-dimesional mould by micro thermoforming (Fig. 7).
- Closing of the 3-dimensional microstructures can be done either by thermal bonding of a plane cover or by twin-sheet-forming in parallel to the forming process.

Process and Technology of Hot Embossing
In a first step the local surface modifications will be embossed onto the surface of a thin polymer film by hot embossing. Hot embossing, comparable with thermal nanoimprinting, is a microreplication technology where a melted polymer is embossed into a heated mould insert (Heckele and Schomburg 2004). There are several approaches to fabricate the mould insert. If a plane, periodic pattern of the surface modification is required, interference lithography is a good way to produce consistent nanostructures (Lu and Lipson 2009). Structure geometries may be sinusoidal, parabolic, lamellar or conical. The nanopattern is written into a positive or negative resist by two or three interfering laser beams. After development, the structures can be transferred into nickel by electroplating.
For surface modifications with variable patterns, electron beam lithography is a well-suited method. In this case, the nanopattern is written into the resist, for example polymethylmethacrylat, by a focused and well-guided electron beam. To fabricate the mould insert which is a thin metal plate consisting of the desired structures, the developed resist layer has to be transferred into nickel by electroplating.
The resist layer is coated with a 5–10 nm thick chrome layer by evaporation deposition. The chrome coating is needed as adhesive layer. Then a 20–30 nm thick gold layer is deposited on the chrome layer. The vapour deposited gold layer is needed as electrode in the electroplating process. On the gold layer nickel is deposited during electroplating up to a typical thickness of 500 µm. Finally the structured resist is separated from the electroplated nickel film. The thin nickel metal film is so called shim moulding tool.
The shim moulding tool is mounted on a metal substrate and is called mould insert. The mould insert is mounted on one side of a hot embossing machine as shown in Fig. 1. A metal plate with a rough surface, the so-called substrate plate, is positioned on the opposite side. The polymer film is positioned between the mould insert and the substrate plate.

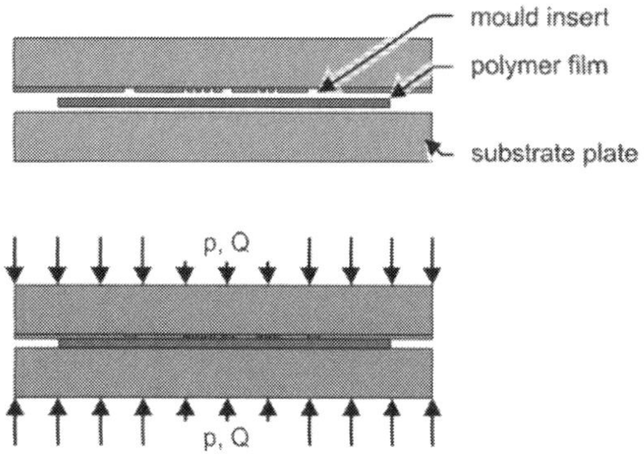

Figure 1 – Schematic view of the hot embossing process. Lateral dimensions of the structures are between 200 nm and 17 µm. The process is similar to thermal nanoimprinting.

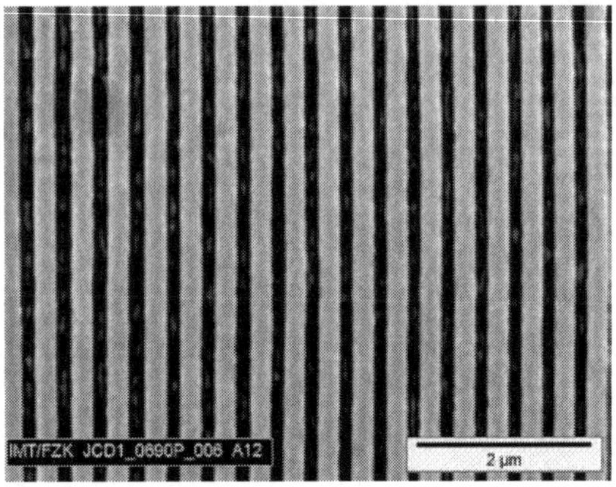

Figure 2 – Nickel mould insert with line and space structures. Lateral dimensions are 200 nm with a pitch around 400 nm. The line and space structure is written into a resist by interference lithography and transferred into the nickel mould insert by electroplating.

After closing the vacuum chamber (not shown in Fig. 1) and evacuation of the enclosed volume, the thin polymeric film is heated above glass transition temperature for amorphous polymers or to the melting temperature for semi crystalline polymers and the embossing force is applied. Moulding force and temperature are the main parameters in this process stage. During force- and velocity controlled moulding, the cavities of the mould insert are filled with the viscous polymer melt. After filling the cavities, mould insert and substrate plate are cooled down. During cooling, the moulding force is maintained to compensate the shrinkage of the polymer.

The replication of periodic nanostructures with lateral dimensions of 200 nm and a pitch around 400 nm was done in polypropylene (NOWOFOL PP-HM 100 µm) at 165 °C and 40 kN moulding force. The size of the structured area is 19 x 19 mm². Fig. 2 shows a SEM image of the mould insert. In Fig. 3 a SEM image of the hot embossed polypropylene film is shown.

This surface modification was also hot embossed into a polymethylmethacrylat film (PMMA Degalan G7E). Because of the large structured surface (65 x 65 mm²) and the difference of shrinkage between polymer and a metal mould insert the demoulding forces increases. In this dimensions of structure sizes already small undercuts in the nanorange can increase the demoulding forces significant which will result in damage of the replicated structures during demoulding. Demoulding fuses on a stronger adhesion of the residual layer on the substrate plate. The average of the measured demoulding stress is typically in a range of 0.2–0.5 N/mm². In this case demoulding of these structures with a periodicity of 400 nm and a surface area of 4225 mm² was the challenge during the replication. Demoulding temperature and the quality of the nanoimprinting tool are the major parameters responsible for successful demoulding (Worgull et al. 2008b and Kabanemi et al. 2009).

Figure 3 – Replication of line and space structures in polypropylene (PP-HM 100 µm). Replication was done with hot embossing at around 165 °C and 40 kN moulding force. Size of the structured area is 19 x 19 mm².

Undercuts and surface roughness are factors that can make demoulding impossible. The visualisation and determination of the lateral dimensions of moulded nanostructures films is challenging if the film was moulded inside a small gap or flow channel. To image the nanostructures itself, a view on a cross section is recommended. Best results have been archived by breaking deep-frozen polymer film. Figure 4 is an in-depth view of the nanostructured surface geometry of a polymethylmethacrylat film (PMMA Degalan G7E 100 µm) after nanoimprint and before thermoforming.

As a second example, cuboids with lateral dimensions of 17 x 17 µm and a height of 21 µm are hot embossed onto PMMA as an example of larger surface-modifications. Figure 5 shows the mould insert and Figure 6 the corresponding surface structured polymer film.

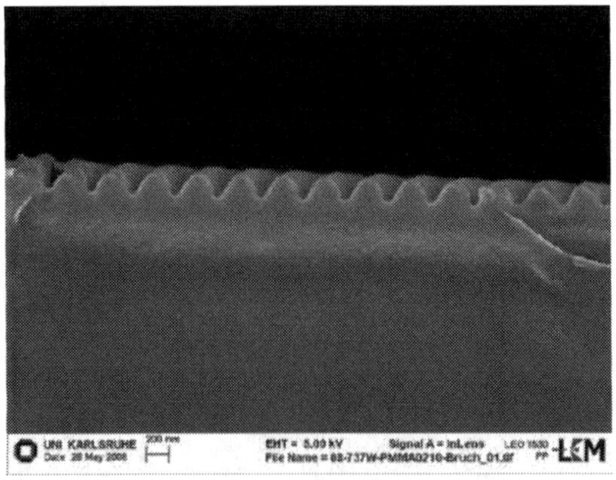

Figure 4 – Cross section of the structured polymer film (PMMA Degalan G7E). The height of the structures is around 200 nm, the thickness of the polymer film is around 80 µm. Deformations of the structures on the left and the right side of the SEM-image is a result of cutting the polymer film.

Figure 5 – Nickel mould insert with cuboid holes. Lateral dimensions of holes are 17 µm x 17 µm, with a depth around 21 µm. Width of bars is 8 µm. The structure is written into a resist (AZ 9260) by UV lithography from a chrome mask and transferred into the nickel mould insert by electroplating.

Figure 6 – Replication of cuboids in PMMA (Degalan G7E 100 µm). Lateral dimensions of pins are 17 µm x 17 µm, with a height around 21 µm. Replication was done with hot embossing at 160 °C and 20 kN moulding force. Size of the structured area is 50 x 50 mm^2.

Process and Technology of Microthermoforming

The major process steps of microthermoforming are shown in Fig. 7. The surface structured polymer film is placed into the thermoforming machine between the microstructured mould and the counter plate. The chamber is evacuated and closed completely to achieve thermal contact between the machine plates and the polymer film for heating. Heating of thin surface structured polymer films (< 100 µm) cannot be done outside of the thermoforming machine (pre heating) because of the low heat capacity and the risk of destroying the surface modifications.

By heating the mould and the counter plate, the polymer film is heated close to glass transition temperature (105–110 °C for PMMA and 155–158 °C for PP) but remains in a rubber-elastic state. In this temperature range the internal polymeric structure of the polymer film and the surface structure are preserved. To mould the surface structured polymer film into the micro cavities of the moulding tool, pressurized gas with a pressure in the range between 0.5 and 5 MPa is applied, while the polymer film is clamped to its surroundings. The polymeric film is formed by the gas pressure into the evacuated micro cavities of the mould. Because of the different temperature ranges of hot embossing and thermoforming the surface modifications will not be damaged while their carrier layer, the polymer film, will be formed into the microcavities of the mould. The cohesion of the polymer film is maintained. Because of the difference between the height of the nanostructures and the thickness of the carrier layer the forming process is determined by forming of the carrier layer. Stretching of the carrier layer corresponds also to the kind of geometry of the mould insert. Here uniaxial or biaxial strain can occur which determine the distribution of the surface modification in the final shape.

The microthermoformed polymer film will be cooled down to demoulding temperature under the load of the gas pressure. After achieving the demoulding temperature, typically below glass transition temperature of the polymer, the gas pressure is set to ambient pressure and the micro-thermoformed polymer film can be demoulded manually.

The mould for microthermoforming is typically made of brass or steel (Fig. 8). Polymeric moulds can be used for prototyping or as carrier layer of the microthermoformend parts (Heilig et al. 2010). Therefore a mould of the same material as the surface structured polymer film is used. It can be structured by hot embossing, but any other structuring method can be used as well. Because of the same material of surface structured polymer film and polymer mould, bonding between the polymer film and the mould can be achieved during microthermoforming.

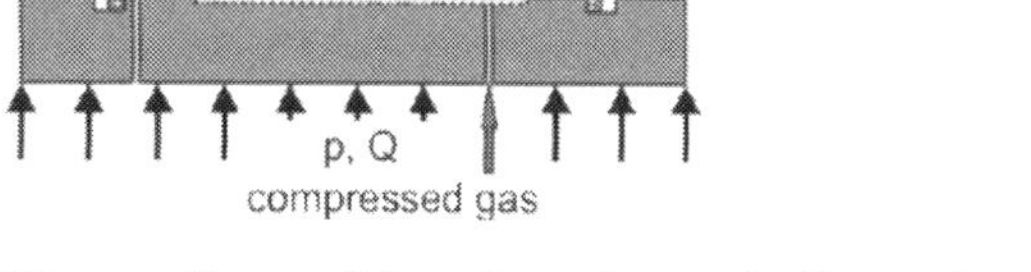

Figure 7 – Schematic view of the micro thermoforming process. A thin surface structured polymer film is clamped and heated up in the range of glass transition temperature and will finally be formed by gas pressure into microcavities.

Figure 8 – Microthermoforming tool with integrated heating and cooling. Brass mould has a diameter of eight inch with a structured area of 85 x 125 mm². The structures consist of 117 structured fields with 144 microcavities of 300 µm in diameter and depth in each field.

Process Parameters of Microthermoforming

The behaviour of polymeric materials during thermoforming is strongly affected by the stress level, the rate of loading and the temperature (Argon 1973). Therefore, a precise control of the process parameters is necessary, especially during microthermoforming with the reduced heat capacity of thin polymer films (film thickness typically below 100 µm) and the small volume of the microcavities.

Microthermoforming has to be done in a small temperature range. The thermoplastic polymer has to be in a softened state, so that large polymer deformations are possible without breaking, nor melting the polymer (Sweeney 2009). For amorphous polymers this temperature range is just above the glass transition temperature. Semi crystalline polymers are typically formed just below melting temperature.

Applied pressure level and rate of pressure rise are the other two technical main parameters. Makradi et. al. (2008) showed that the mechanical characteristics of the final products, which are mainly controlled by the final polymer film thickness and orientation of the polymer molecular chains, depend strongly on the process temperature and the applied pressure profile.

Control of Process Temperatures

For the forming process the polymer film is heated to a softened state. The low heat capacity of thin polymer films requires that the mould and counter plate (Fig. 7) is also heated and hold in the forming temperature range until the full gas pressure is applied. To get better response during heating and reduce cycle times, mass of heated Parts (mould, counter plate and heating plates) is reduced and thermally decoupled from the cooling device. After the full gas pressure is reached, cooling is started by coupling and activating the cooling device. Therefore the heating plates are pressed against water-cooled machine plates.

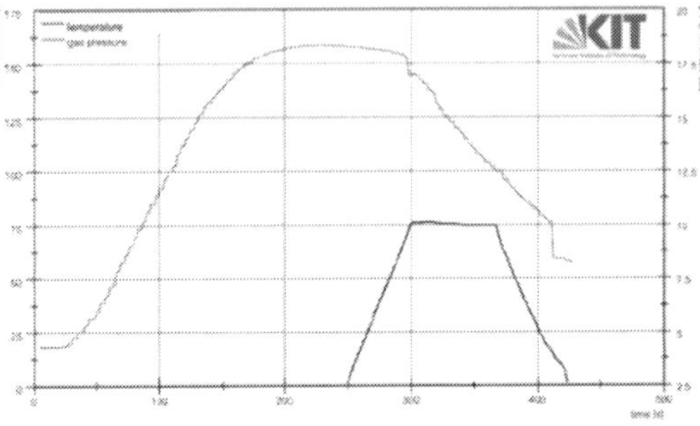

Figure 9 – Temperature and pressure profile during microthermoforming of polypropylene (Nowofol HM). Temperature was set to 156 °C and hold in a temperature range of ± 2 °C. Forming pressure was set to 10 bar (1MPa). A cycle time of 5 min is accomplished (75 °C to 156 °C and back to 75 °C)

In Fig. 9 temperature and pressure profile during microthermoforming of polypropylene (Nowofol HM) is presented. Forming temperature was set to 156 °C and was held in a temperature range of ± 2 °C until forming pressure is applied. Rate of temperature change was 65 K/min during heating and -40 K/min during cooling.

Control of Process Pressure

Microthermoforming requires precise control of the applied pressure profile. An automated forming pressure control unit, for presetting the pressure rise, level and fall was designed. Pressure rise and fall can be set between 10 and 40 kPa/s with a linearized pressure distribution profile. Maximal level of gas pressure can be up to 5 MPa and is hold constant by the controller at the preset value. The challenge was to achieve these parameters at the small (gas-) volumes of the microcavities, typically around 500 mm³.

As shown in Fig. 9, the gas pressure of 1 MPa was applied and hold in a range of ± 30 kPa for microthermoforming of polypropylene (Nowofol HM 100 µm). Pressure rise and fall was set to 20 kPa/s.

Figure 10 – Microthermoformed cell cultivation device in polypropylene (formed with mould of Fig. 8)

Results and Discussion

With the enhanced temperature and forming pressure control unit, it is possible to take advanced control over the microthermoforming process. It will be possible to analyse the process of microthermoforming in detail and compare it with analytical calculations and simulations. The influence of microthermoforming on the surface modifications of the polymer film measured by AFM measurements and illustrated by SEM images.

Measurement of the surface modifications before and after microthermoforming

To verify the distribution and deformation of the surface modifications after the thermoforming process step a combination of scanning electron microscopy (SEM) and atomic force microscopy (AFM) was used to obtain verifiable measurement results.

These experiments fuses on a different design of the microthermoforming mould as described above. The design is characterized by flow channels with a cross section after thermoforming of around 200 x 200 μm^2. Further this design consists of circular reservoirs and a rectangular intersection of two flow channels. Because of this design uniaxial and biaxial stretching during thermoforming can be expected.

Numerical analysis of the geometrical modifications of the surface modifications (Fig. 3 and Fig. 4) due to 3-dimensional forming was accomplished by means of AFM measurements. Four measurement areas of 1.2 μm^2 are defined at characteristically points of the microthermoformed parts (Fig. 11).

Measurements are taken at the same measurement areas (1-4) before and after microthermoforming. Fig. 12 is the visualisation of the measured AFM data of measurement area 4 (shown in Fig. 11) before microthermoforming. Within each measurement area three periods of the periodic nanostructure (line-space) are analysed over the full measurement length of 1.2 µm (Fig. 13). Fig 14 shows the AFM measurement results of all measurement areas before and after microthermoforming.

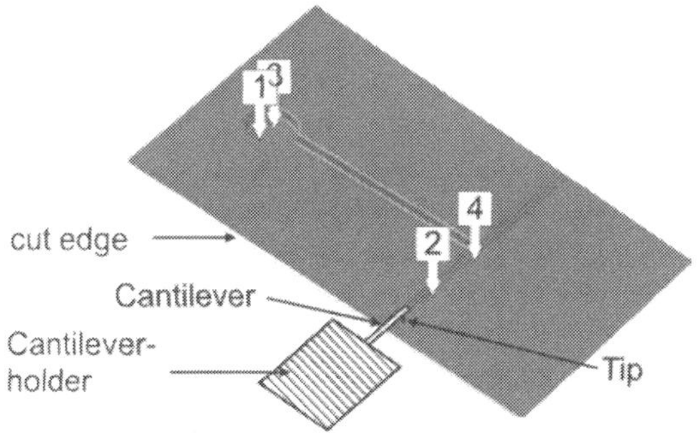

Figure 11 – Illustration of AFM measurements inside a micro channel and position of measurement areas (not to scale). A square of 1.2 μm^2 was measured at each measurement area.

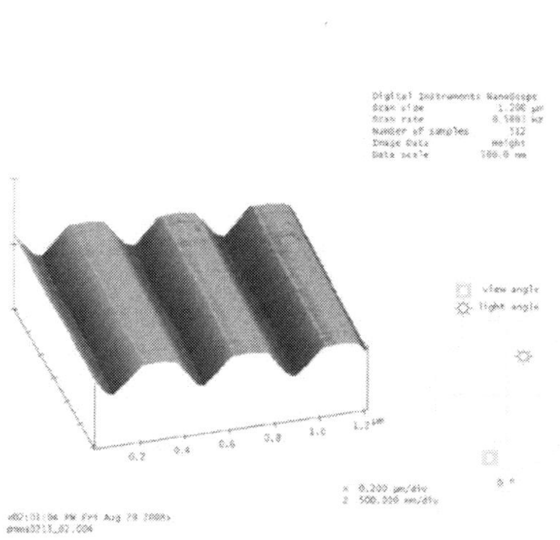

Figure 12 – AFM Characterisation of the replicated surface structured polymer film in the corresponding measurement area of point no. 4 in Fig. 11.

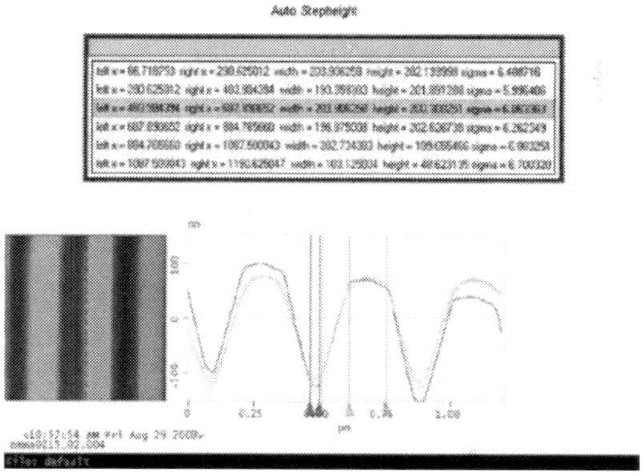

Figure 13 – Analysis of AFM measurement data of point no. 4 in Fig. 11 before microthermoforming. Six slopes of the periodic nanostructure are analysed over the measurement length of 1.2 µm. Average values and standard deviations of width and height of the surface structure are calculated.

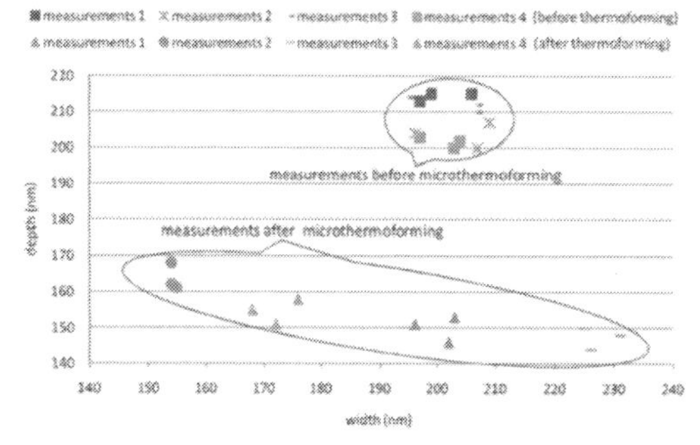

Figure 14 – Depth and width of the line-space nanostructures before and after micro thermoforming. Depending on the measurement area the depth (resp. height) of the structures decreases up to around 25%.

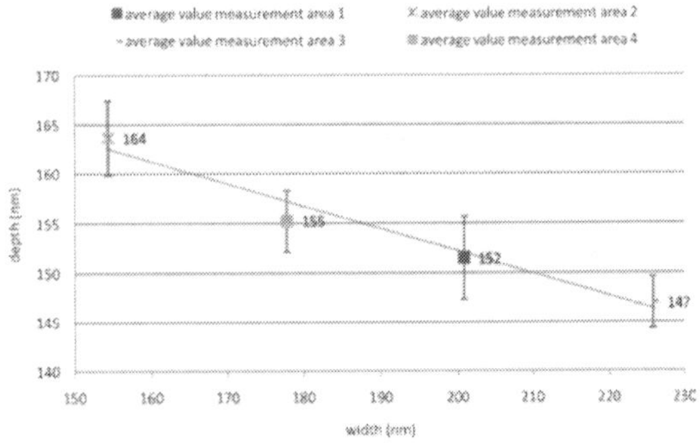

Figure 15 – Mean values of depth and width of the surface modification after microthermoforming in the four characteristic measurement areas. A trendline is plotted as an indication of a linear relationship between the resulting depth and width of surface structures after microthermoforming.

Because of the difference in size between the surface modification (in this example a typical height and width around 200 nm) and the carrier layer (typical thickness around 80 µm) the deformation of the surface structures can be an indication for the direction of deformation of the carrier layer at the measured areas. Characteristic is the decrease of the height of the structures, because of the forming of the carrier layer. Depending on the direction of the strain the width of the structures will also decrease.
The correlation between height (resp. depth) and width of the surface modifications after forming is illustrated in Fig. 15. The average values of depth and width of the nanoscale surface structures fuses on the measurement areas 1–4 illustrated in Fig. 11. It can be shown that the height of the structures decrease if the width increase. A first approximation shows a linear correlation if uniaxial strain occurs.
To illustrate the influence of microthermoforming on the surface modification, the PMMA film with the cuboid surface structure of Fig. 6 is also microthermoformed. In Fig. 16 a SEM image of the microthermoformed part is shown. The deformation is a result of the uniaxial deformation of the carrier layer in the microchannel, which influences the distance of the cuboid surface structure mainly in orthogonal direction to the microchannel.

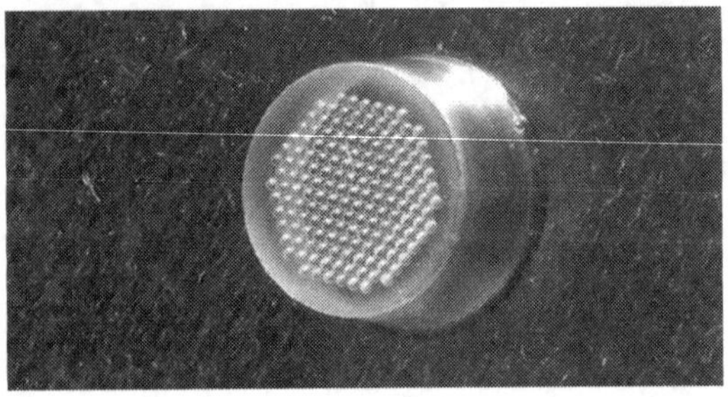

Figure 16 – Surface structure of Fig. 6 on PMMA (Degalan G7E) after microthermoforming. SEM image shows the top view on the microthermoformed part.

Conclusions

The integration of selective surface modifications in freely shaped 3-dimensional cavities is expected to be a very useful technology e.g. for life science applications. It is demonstrated, that a combination of hot embossing and microthermoforming is a suitable approach to fulfil this task. A previous structured thin polymer film can be brought into a multiscale shape in one process step. In Fig. 17 and 18 an example of a multiscale container for cell cultivation is shown.
With the new precise control unit for temperature and gas pressure application, detailed research of the microthermoforming process of prestructured amorphous and semi crystalline polymers is possible and under current development.

Figure 17 and 18 – Backside view of multiscale container for cell cultivation. Outer dimensions of 6 mm in diameter and 5 mm in depth with cups of 300 µm in diameter and depth on its ground (top in picture on the left side). Optical refraction on the cell container due to the nanoscale surface structuring of Fig. 3 and 4 with an intermittency of 400 nm.

References

Abbott SJ, Gaskell PH (2007) Mass production of bio-inspired structured surfaces. In: Proceedings of the institution of mechanical engineers, Part C: J Mech Eng Sci 221:1181–1191. doi:10.1243/09544062JMES540

Argon, AS (1973) A theory for the low temperature plastic deformation of glassy polymers. Philos. Mag. 28:839-865.

Giselbrecht S, Gietzelt T, Gottwald E, Trautmann C, Truckenmüller R, Weibezahn KF, Welle A (2006) 3D tissue culture substrates produced by microthermoforming of pre-processed polymer films. Biomed Microdevices 8:191–199. doi:10.1007/s10544-006-8174-8

Heckele M, Schomburg WK (2004) Review on micro molding of thermoplastic polymers. J Micromech Microeng 14:R1–R14. doi:10.1088/0960-1317/14/3/R01

Heilig M, Giselbrecht S, Guber A, Worgull M (2010) Microthermoforming of nanostructured polymer films: a new bonding method for the integration of nanostructures in 3-dimensional cavities. Microsyst Technol. doi:10.1007/s00542-010-1072-2

Kabanemi KK, Marcotte JP, Hétu JF, Worgull M, Heckele M (2009) Numerical simulation of a thermoviscoelastic frictional problem with application to the hot-embossing process for manufacturing of microcomponents. International Polymer Processing 24:174-84. doi:10.3139/217.2227

Lu C, Lipson RH (2009) Interference lithography: a powerful tool for fabricating periodic structures. Laser Photon Rev 1–13. doi:10.1002/lpor.200810061

Makradi S, Belouettar S, Ruch D, Ahzi S (2008) Thermoforming process of semi-crystalline polymeric sheets: Modeling and finite element simulations. Polym. Sci. Ser. A 50:550-557. doi:10.1134/S0965545X0805009X

Sweeney J, Spares R, Woodhead M (2009) A constitutive model for large multiaxial deformations of solid polypropylene at high temperature. POLYM. ENG. SCI. 49:1902-1908. doi:10.1002/pen.21426

Throne JL (2008) Understanding thermoforming. Hanser, München. ISBN-13:978-3-446-40796-1

Truckenmüller R, Rummler Z, Schaller T and Schomburg WK (2002) Low-cost thermoforming of micro fluidic analysis chips J. Micromech. Microeng. 12 375–9. doi:10.1088/0960-1317/12/4/304

Truckenmüller R, Giselbrecht S (2004) Microthermoforming of flexible, not-buried hollow microstructures for chip-based life sciences applications. IEE Proc Nanobiotechnol 151(2004):163-166. doi:10.1049/ip-nbt:20040823

Truckenmüller R, Giselbrecht S, van Blitterswijk C, Dambrowsky N, Gottwald E, Mappes T, Rolletschek A, Saile V, Trautmann C, Weibezahn KF, Welle A (2008) A flexible fluidic microchips based on thermoformed and locally modified thin polymer films. Lab Chip 8:1570–1579. doi:10.1039/b803619e

Wilson SJ, Hutley MC (1982) The optical properties of 'moth eye' antireflection surfaces. J Mod Opt 29(7): 993–1009. doi:10.1080/713820946

Worgull M, Heckele M, Mappes T, Matthis B, Tosello G, Metz T, Gavillet J, Koltay P, Hansen HN (2008a) Sub-μm structured lotus surfaces manufacturing. Microsyst Technol 15:1327–1333. doi:10.1007/s00542-008-0744-7

Worgull M, Kabanemi KK, Marcotte J-P, Hétu J-F, Heckele M (2008b) Modeling of large area hot embossing. Microsyst Technol 14:1061–1066. doi:10.1007/s00542-007-0493-z

Worgull M (2009) Hot embossing: theory and technology of microreplication. William Andrew, Oxford. ISBN-13:978-0-8155-1579-1

ROTATIONAL MOLDING OF POLYCARBONATE REINFORCED POLYETHYLENE COMPOSITES: PROCESSING PARAMETERS AND PROPERTIES

Shih-Jung Liu and Kang-Ming Peng

Department of Mechanical Engineering, Chang Gung University, Taiwan - *shihjung@mail.cgu.edu.tw*

Abstract - The goal of this report is to study the rotational molding of polycarbonate reinforced polyethylene composites. Polyethylene powder mixed with polycarbonate/polyethylene pellets were mixed by a dry mixer and rotationally molded by a box mold. Characterization of molded part properties was performed after molding. It was found that the impact strength of molded composites increased with the volume fraction of polycarbonate reinforcement, while the tensile strengths decreased with the reinforcement. Higher cooling rates molded composites with higher impact strengths but lower tensile strengths. There exists an optimum internal peak air temperature with which composites of the highest mechanical properties can be obtained.

Introduction

Rotational molding is one of the most important polymer processing methods for producing hollow plastic articles. Despite the advantages associated with the rotational molding method, there are also some unsolved problems that confound the overall success of this technology. One of them is the limited moldable material selection. An important attribute of polymers is the ability to modify their inherent physical properties by the addition of certain fillers, while still retaining their characteristic processing ease. Like other polymer processing methods, reinforcements have been incorporated into the rotationally molded components to increase their mechanical properties [1-4].

The goal of this report is to study the rotational molding of polycarbonate reinforced polyethylene composites. Characterization of molded part properties was performed after molding. The effect of various processing parameters on molded part qualities was examined. Furthermore, scanning electron microscopy (SEM) observation was completed to examine the microscopic structure of the composites and its influence on molded product qualities.

Experimental

Materials

The resin used in this study was high density polyethylene (HDPE) (LH-614, USI Corp., Taiwan). Polycarbonate (S-3000, Mitsubishi Corp., Japan) was used as the reinforcement. Polycarbonate was dry mixed with HDPE with a fixed weight ratio of 85:15. The mixture was then blended by a twin-screw extruder (Desunnano Model SHJ-20B, Taiwan) with a temperature setting of 180, 230, 250, 265, 275, and 280°C from the hopper to die.

Molding machine and mold

A laboratory-scale uniaxially rotating machine was designed and built in our laboratory. A 100x100x165 mm^3 box mold made of a 7 mm thick sheet of aluminum alloy was used for the molding tests. The oven temperature was maintained at 350°C. To mold the composites, various weight ratios of HDPE powder to HDPE/PC pellets were mixed by a dry mixer. The effect of various processing parameters was studied. Table 1 lists the processing variables as well as the values used in this study.

Tensile and impact tests of molded parts

Tensile tests were performed in a LLOYD tensiometer according to the ASTM D638 test. A 2.5 kN load cell was used and the crosshead speed was 50 mm/min. Izod impact tests were performed in a CEAST model Resil 5.5 impact tester according to the ASTM D256 test. Five specimens were tested for each part.

Microscopic observation

Specimens cut from the PC/HDPE blend pellets and rotationally molded parts were observed under a Hitachi Model S-300N scanning electron microscope.

Results and Discussion

Figure 1 shows the morphology of PC in rotationally molded composite parts. Despite the morphology of PC phase not being significantly changed during the molding process. Figure 2 shows photographically, the morphology of tensile fracture surface. It is obvious that despite some PC particles becoming elliptical-shaped, most of the PC particles remain spherical after the tensile test. Figure 3 shows the morphology of PC particles in specimens after the impact test. PC particles exhibit a fibril-shape distribution. These fibrils mainly result from the stretch of the initially spherical particles.

Tensile strengths of rotationally molded HDPE/PC composites were found to decrease with the volume fraction of reinforcement up to 4.5%, as shown in Figure 4. The experimental data in Figure 5a and 5b showed that both the tensile strengths and tensile modulus decreased with the cooling rate. The measured results in Figure 6 suggest that there exists an optimum internal peak air temperature (at 220°C) in which the highest tensile and impact properties could be obtained.

Table 1. The processing variables as well as the values used in the experiments

Volume fraction of reinforcement (%)	Cooling condition	Internal peak air temperature (°C)
1.5	Still air	180
3.0	Blowing fan	**200**
4.5	**Water shower**	220
6.0	--	240
7.5	--	--

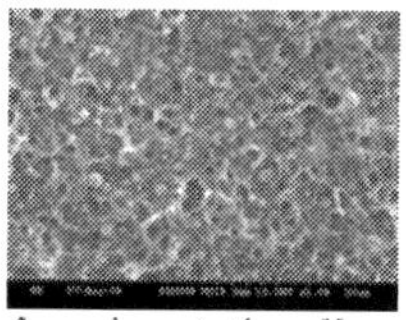

Figure 1. Morphology in rotationally molded HDPE/PC composites.

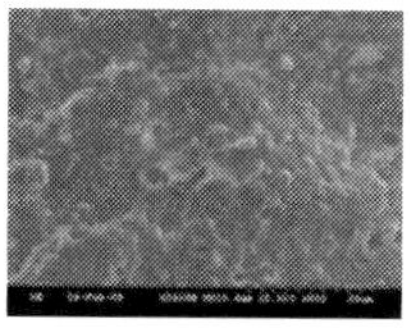

Figure 2. SEM micrograph of the tensile fracture surface.

Figure3. SEM micrograph of the impact fracture surface.

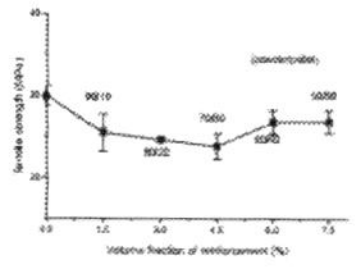

(a)

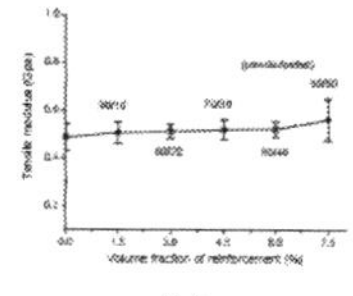

(b)

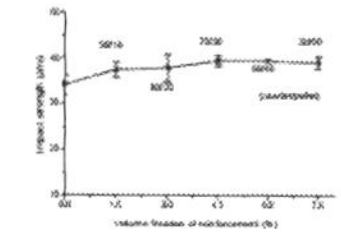

(c)

Figure 4. Effects of volume fraction of reinforcement on (a) tensile strength, (b) tensile modulus, and (c) impact strength of rotationally molded composites.

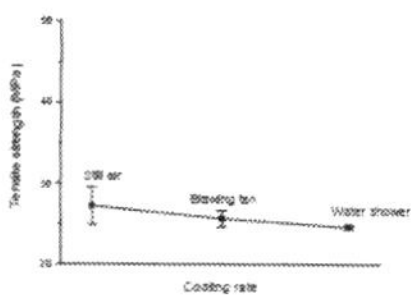

(a)

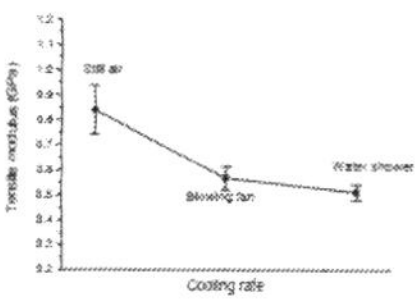

(b)

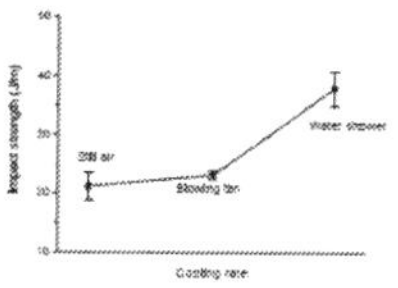

(c)

Figure 5. Effects of cooling rate on (a) tensile strength, (b) tensile modulus, and (c) impact strength of rotationally molded composites.

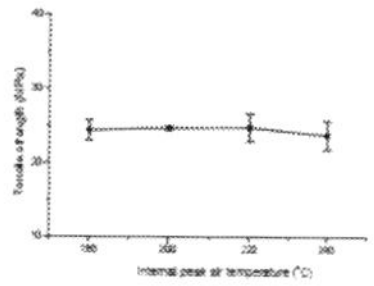

(a)

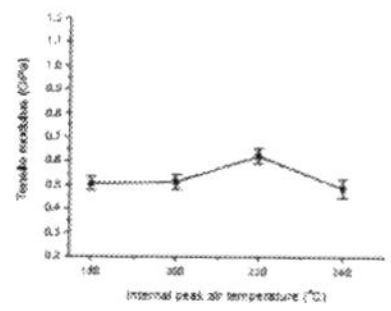

(b)

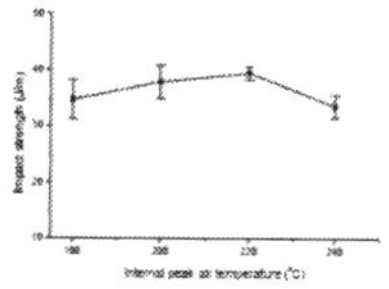

(c)

Figure 6. Influence of internal peak air temperature on (a) tensile strength, (b) tensile modulus, and (c) impact strength of rotationally molded composites.

Conclusions

This report has examined the rotational molding of polycarbonate particulate reinforced composites. The experimental results suggest that the impact strength of molded composites increase with the volume fraction of polycarbonate reinforcement, while the tensile strengths decrease with the reinforcement. Higher cooling rates molded composites with higher impact strengths but lower tensile strengths. There exists an optimum internal peak air temperature with which composites of highest mechanical properties can be obtained.

Acknowledgements

The National Science Council of Taiwan, R.O.C. under the grant NSC95-2221-E-182-032-MY3, has supported this work financially.

References

1. Kontopoulou M., et al., "An. experimental study of rotational molding of polypropylene/Polyethylene copolymers", Int. Polym. Process., **12**, 165-173(1997)
2. Wisley B.G., "The rotational molding of glass fiber reinforced polyethylene", SPE Technical Paper, **44**, 1487-1491(1999)
3. Robert A., Crawford R.J, "The effect of fillers on the properties of rotationally moulded polyethylene", SPE Technical Paper , 44, 1478-1481(1999)
4. Torres F.G., Aragon C.L., "Final product testing of rotational moulded natural fibre-reinforced polyethylene", Polymer Testing, **25**, 568-577(2006)

CRYSTALLIZATION BEHAVIOUR OF SINGLE-SITE ETHYLENE COPOLYMERS IN RELATION TO WARPAGE IN ROTATIONAL MOLDING

C. T. Bellehumeur*, J. Ference, I. Gibbons and W. Lin

NOVA Chemicals Corporation, Calgary, AB Canada – bellehc@novachem.com

Abstract - The authors present a study of the crystalline properties of selected series of single-site ethylene copolymers in relation to their warpage at conditions relevant for rotational molding. The influence of molecular architecture and composition on crystallization kinetics and dimensional stability is of primary interest in this study. Results obtained from experiments carried out on a bench-scale setup were assessed against warpage measurements obtained from molding trials. Our analysis highlights limitation of thermoelastic models and provides insight in the mechanisms controlling warpage in rotational molding.

DEVELOPMENT OF A REALISTIC FINITE ELEMENT THERMOFORMING SIMULATION

P.J. Martin[1]* and C.P.J. O'Connor[1]

[1]*School of Mechanical and Aerospace Engineering, Queen's University Belfast, Belfast, UK*
p.j.martin@qub.ac.uk; ciaran.oconnor@qub.ac.uk

Abstract - Although thermoforming appears to be a relatively straightforward polymer processing technique, efforts to build effective simulations of the process have been greatly hampered by poor process knowledge and difficulties in measuring key process phenomena. Over a number of years many of these impediments to progress have been gradually overcome and realistic process simulations are now starting to emerge. In this paper the development of a more representative finite element thermoforming simulation is reported. The model is sophisticated and now includes a complex viscoelastic material deformation model, derived from biaxial test data, as well as user subroutines to embed relationships for contacting surface effects such as friction and heat transfer. In all cases these relationships have been established from the results of extensive experimental programs. The structure of the finite element software is described and its performance in simulating a number of practical thermoformed parts is presented. In general the predictive capability of this more realistic simulation is shown to be very good and a major step forward from previous process models. Some minor areas for improvement are highlighted and the future development of the simulation is discussed.

Introduction

Thermoforming is amongst the most important and widely used of the polymer processing techniques. It generally involves reheating an extruded polymer sheet and then deforming it further to take the shape of a new product. There are numerous variants of thermoforming that comprise of different combinations of sheet deforming methods such as moving tool surfaces and applied air pressure. Packaging, particularly for foodstuffs, is the most important application of the process, where it is used to produce billions of products annually worldwide. In such applications the most dominant process variant is plug-assisted thermoforming.

The plug-assisted thermoforming process is illustrated in Fig. 1. Thermoforming takes place in two steps. Firstly, the heated polymer sheet is pre-stretched by the movement of a mechanical plug that pushes material partly into the mould (Plugging phase). Secondly, positive air pressure is applied from above stripping the material from the surface of the plug and onto the mould wall where it cools (Pressure phase). The principal objective of the plugging step is to try to distribute material more evenly in the part by using the plug to capture and push material towards the furthest extremities of the mould.

The geometry and quality of output produced by a thermoforming process is dependent on major processing variables such as the tooling design and materials employed, the magnitudes of mechanical displacements and pressures applied, the forming speeds, and the temperatures of the polymer sheet and tool surfaces. Whilst some of these variables can be closely controlled in industry (e.g. sheet temperatures and forming speeds), others such as plug shapes may only be decided prior to manufacturing and then may only be adjusted off-line following on-line testing.

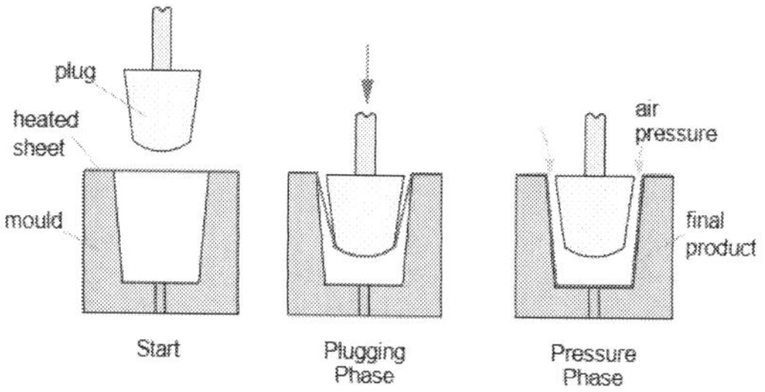

Fig. 1 – Plug-assisted thermoforming

In an ideal thermoforming process the initial sheet thickness would be evenly and gradually thinned through deformation to produce an even wall thickness in the final product. However, such a high level of process control is impossible to achieve in practice and generally thermoformed products contain some degree of unevenness in their wall thicknesses, with the thinnest regions dictating the performance of the final parts.

The effective control of product wall thickness variations is therefore the major manufacturing concern of most thermoforming companies. In recent years this has led to an increasing desire for the development of effective process simulations so that product wall thickness may be evaluated off-line rather than having to resort to expensive trial and error investigations whilst in production.

Research Objective

The major ongoing objective of this work is the development of a realistic, fully representative, reliable and accurate simulation of the plug-assisted thermoforming process.

Process Analysis

Researchers have attempted to model thermoforming processes for more than 20 years, although the body of work in this area is relatively small compared to other processes (McCool et al, 2006). Although the thermoforming process can appear to be quite straightforward in comparison to many other manufacturing processes, this belies a dearth of understanding of some of the key physical phenomena that underlie the process operations. These limitations were highlighted in previous research and were identified as the principal reasons for the relatively slow progress in the development of thermoforming process simulations (Collins et al, 2002). Analyses of this type have demonstrated that thermoforming can be readily simulated using FEA, but that the accuracy of the models is highly dependent on two key input parameters (McCool et al, 2005). These are:

(1) The mathematical representation or model for the deformation properties of the polymer material. This is one of the most challenging research objectives as the model must be able to accurately replicate the response of the polymer under biaxial stretching at very large strains, high strain rates and high temperatures that are representative of the thermoforming process. Work in this area has spawned a substantial body of separate research dedicated solely to the development of mathematical models for the major polymer materials (Makradi et al, 2007). Often the models proposed are mathematically complex and different formulae have been suggested for the different polymer materials (O'Connor et al, 2008). In the past this work suffered from a lack of adequate test methods for accurately measuring the response of a polymer under thermoforming conditions. However, with the development of new biaxial test systems, such as the flexible biaxial stretcher at Queen's University Belfast (shown in Fig. 2), the understanding of the thermoforming behaviour of different polymers and the accuracy of material models is improving rapidly (Martin et al, 2005).

(2) The treatment of contact between the polymer and other interacting surfaces during the process. These include its interaction with the surrounding air and the tooling surfaces (plug and mould). The principal interactions are heat transfer (convection and conduction) and the ability of the polymer to slide across the contacting surface (frictional effects), which is particularly important in the treatment of the plug. These effects are illustrated in Fig. 3, and in much of the reported work they have received little attention and very basic relationships such as the simple Coulomb friction law and a single coefficient of friction have been used for simplicity. However, the improved understanding of the process and particularly the role of the plug (Collins et al, 2002; McCool et al, 2005), are now enabling researchers to more accurately address these issues in process simulations. Researchers have also started to report the development of test methods specifically designed for the measurement of thermal and frictional properties relevant to thermoforming (Hegemann et al, 2003; Choo et al, 2007).

Fig. 2 – Flexible Biaxial Stretcher at QUB

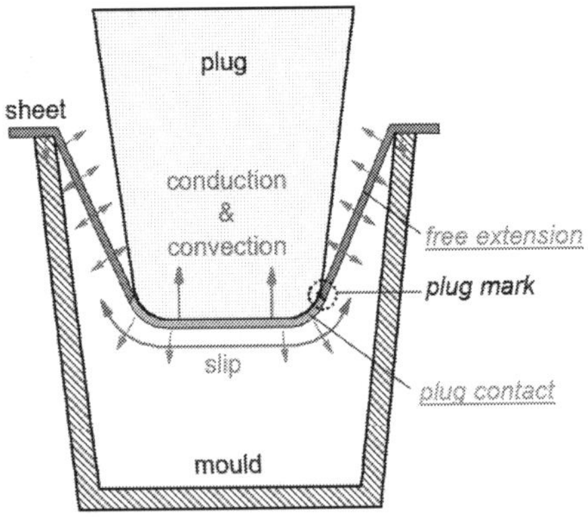

Fig. 3 – Contact Effects in Thermoforming

Model Development

The prototype simulation software under development in this work has been constructed using the commercial finite element (FE) package Abaqus Standard, which has been used for many years for this type of simulation (McCool et al, 2006) This approach significantly eased the programming task and provided great flexibility in the system architecture by permitting user programmed subroutines to be linked to central process simulation. This was of particular importance in providing non-standard experimentally derived models for material deformation behaviour and surface contact effects.

In the development of the process simulation the plug and mould have been modelled as rigid surfaces and the sheet divided into an element mesh, constructed of 4-noded plane stress continuum elements of type CPS4. Contact friction and heat transfer effects are included within the simulation as detailed in the next section.

Material Modelling

Through early work on the measurement and modelling of deformation in thermoforming, it was rapidly established that the accuracy of the material model used in the process simulation was critical in determining the quality of its output (Collins et al, 2000). It was also apparent that traditional tensile test systems were inadequate for providing realistic deformation data that was truly representative of the actual thermoforming process (Martin et al, 2005). Such conclusions have directly provided the impetus for the development of biaxial tensile test machines that are increasingly appearing worldwide. With improvements in the data available, researchers were also able to prove that the simple hyperelastic material laws used by early process modellers were insufficient for accurate representation of thermoforming, particularly for processes employing mechanical plugs. In recent times researchers have capitalized on the improved measurement systems by developing more sophisticated viscoelastic or viscoplastic material models for many of the common thermoforming polymers (O'Connor et al, 2008).

In the current process simulation a dedicated viscoelastic model has been developed to represent the deformation response of polypropylene (O'Connor et al., 2007). After evaluation of a number of alternative constitutive equations a network diagram for the chosen model is illustrated in Fig. 4 (O'Connor et al, 2009). It is based on the work of Sweeney (Sweeney et al, 1998) and represents the behaviour of polypropylene by a double parallel network of an Edwards-Vilgis entropic spring in series with an Eyring dashpot. The Edwards-Vilgis springs mimic the change in entropy of the polymer network through the action of cross-linking and slip-linking behaviours in the material. The Eyring dashpots provide a method to address the viscous time dependent behaviour of the material as it is progressively deformed at elevated temperature.

The constants for the mathematical model were obtained by fitting to experimental data from biaxial material testing. A comparison of the material model to the fitting data from equi-biaxial test results is shown in Fig. 5. In general the fit obtained was quite close, although it was difficult for the model to fully replicate the sharp yield point in polypropylene which is observed at some temperatures. However, the fit was equally as good when compared to data obtained under constant width and sequential testing modes, demonstrating the relative robustness of the model under thermoforming conditions. Temperature dependency of the material response is provided by piecewise fitting of the material model constants from the available biaxial test data.

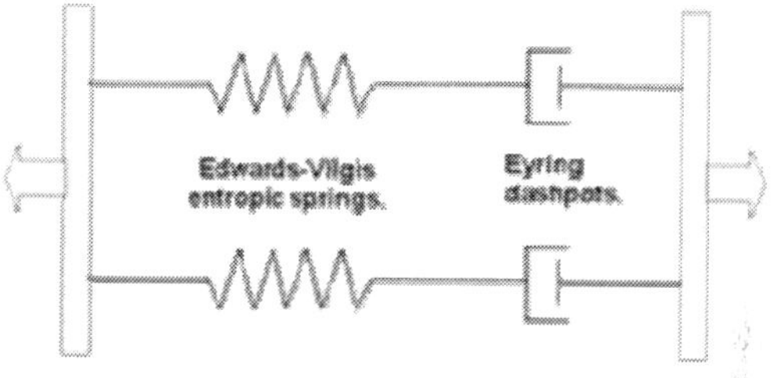

Fig.4 - Viscoelastic Model for Polypropylene

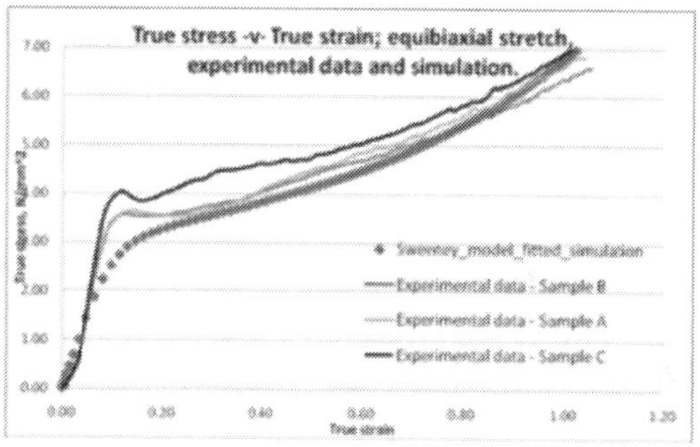

Fig. 5 - Model vs. Biaxial Test Data for PP

Contact Modelling

In simple thermoforming processes such as vacuum forming, contact effects are often ignored entirely and good results may still be obtained However, for plug-assisted thermoforming this is grossly inaccurate as interactions with the plug dominate the process and dictate the final wall thickness distribution. In many reported simulations a single coefficient of friction is applied to contact surfaces to allow for frictional effects. Heat transfer is often ignored altogether and in many cases higher values of friction are applied to offset the changes due to heat transfer. For example, on mould walls the rapid cooling of the sheet on contact is modelled by ignoring heat flow and instead applying a very high coefficient of friction to the surface. This locks the material on contact and prevents further changes in wall thickness, which gives the same effect as rapid heat loss.

In the current simulation user subroutines are used to define the temperature dependent coefficients of friction for the contacting pairs of materials. Again these relationships are obtained from experimental results (Choo et al, 2007) and the program architecture gives provision for the development and inclusion of more complex friction models. Heat transfer is included in the forms of conduction at the contacting surfaces and convection on the free surfaces of the sheet. Realistic heat transfer coefficients derived from experimental results and importantly measured values of gap conductance for the interfaces are included in the simulation (Choo et al, 2008).

Results and Discussion

The performance of the process simulation was evaluated by testing its ability to replicate the wall thickness distribution of a typical thermoformed part. In this case the part chosen was a simple pot for food packaging as shown in Fig. 6 and Fig. 7(a).

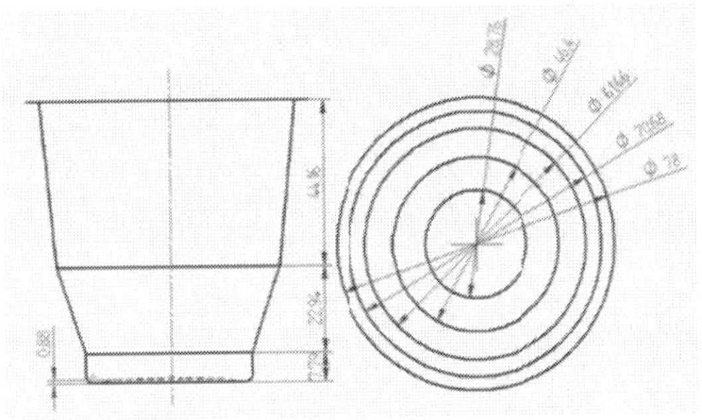

Fig. 6 - Thermoformed Part Design

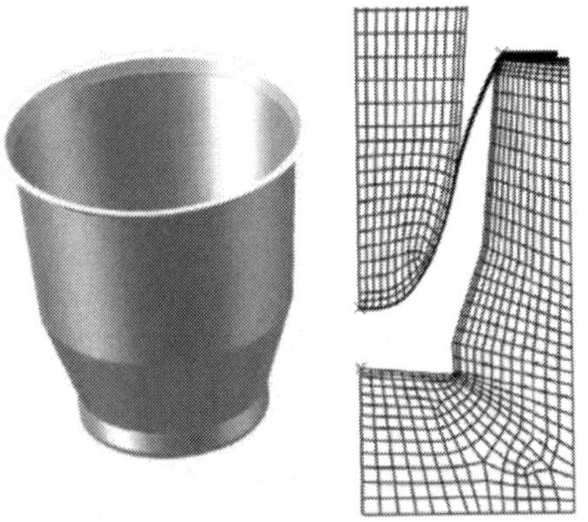

(a) 3D Part Design (b) 2D Axisymetrical Model

Fig. 7 - Part Design & Process Model

The thermoforming of the conical pot was simulated using a 2D axisymetrical geometrical model (as shown in Fig. 7(b)), which included a material model created from biaxial test data for the polypropylene material. All other process conditions were obtained from the pot manufacturers and were replicated in the simulation. An example showing the correlation of the average actual wall thickness measured in the pot compared to that predicted by the simulation is shown in Fig. 8. In both cases the wall thickness distributions are measured from the centre of the base of the pot to its lip (left to right). For the actual pot production settings were those normally used for a product of this type and this data was used to create equivalent parameters for the simulation.

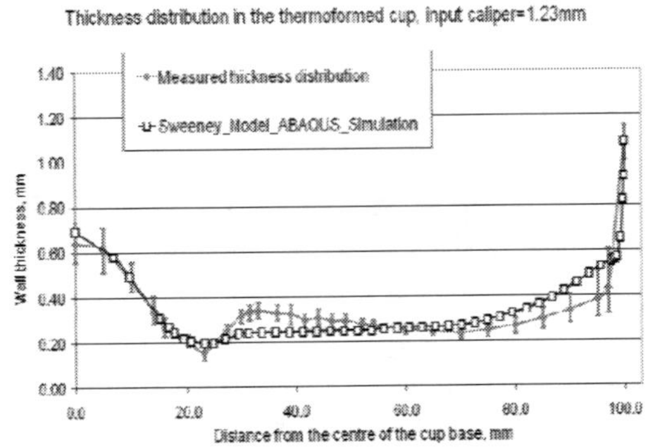

Fig. 8 - Actual vs. Predicted Part Wall Thickness

From the results shown in Fig. 8 and other similar comparisons of performance, it is clear that the simulated results give a good representation of the actual wall thickness distribution in the part. Predictions of the thickness distribution in the base of the pot are very close, as is the average thickness in the sidewall, and importantly all of the major features that are apparent in the wall thickness distribution of the actual part may be seen in the prediction. The major discrepancy is that the simulation tends to overestimate the thickness in the upper sidewall, near the lip, and underestimate the thickness in the lower sidewall near the bottom corner. This pattern of discrepancy is consistently found in comparative tests of the simulation, and it is not thought to arise as a result of any major problem with the material model, which has proven to be extremely stable and consistent across different simulations. Instead it is thought that the problem may be due to the temperature dependent friction model. With greater friction in the early stages of the process, the material forming the sidewall would be forced to deform more during plug displacement and reduce in thickness, but if this high coefficient of friction was to remain then the base of the part would become overly thick. This therefore points towards a more complex friction relationship, where the material would initially stick, or slip very little in the early stages, and then slip more freely in the latter stages of the process. Work to more fully investigate the actual friction behaviour during plug contact is currently ongoing.

Conclusions

A fully working, stable finite element simulation of the plug-assisted thermoforming process has been successfully constructed. For the first time the simulation employs representative models of material deformation response, contact friction and heat transfer that have been provided by dedicated experimental programmes designed to replicate the forming process. The predictive capability has been tested and the results show good agreement with the measured wall thickness distribution of actual parts. The modular nature of the new software architecture provides a readily adaptable platform that can be refined further to increase its predictive capability

References

Choo H. L., Martin P.J., McCool R. & Harkin-Jones E.M. Measurement of the Thermal and Frictional Properties of Contacting Surfaces in Thermoforming. Proc. Polymer Processing Society 23rd Annual Meeting, Salvador, Brazil, May (2007).

Choo H. L., Martin P.J. & Harkin-Jones E.M. Measurement of Heat Transfer for Thermoforming Simulations. Proc. 11th ESAFORM Conference on Material Forming, Lyon, France, April (2008).

Collins P., Lappin J.F., Harkin-Jones E. & Martin P.J. Effects of Material Properties and Contact Conditions in Modelling of Plug Assisted Thermoforming. Plastics, Rubber & Composites, 29(7), 349 (2000).

Collins P., Harkin-Jones E.M.A. & Martin P.J. The Role of Tool/Sheet Contact in Plug-Assisted Thermoforming. International Polymer Processing. Polymer 17 (4), 361 (2002).

Hegemann B., Eyerer P., Tessier N., Bush T. & Kouba K. Polymer-polymer friction at temperatures and rates simulating the thermoforming process. Thermof. Quart. 22(2), 10 (2003)

Makredi A., Belouttar S., Ahzi S. & Puissant S. Thermoforming process of amorphous polymeric sheets: modeling and finite element simulations. Journal of Applied Polymer Science, 106, 1718 (2007).

Martin P.J., Tan C.W., Tshai K.Y., McCool R., Menary G., Armstrong C.G. & Harkin-Jones E.M.A. Biaxial Characterisation of Materials for Thermoforming and Blow Moulding. Plastics, Rubber And Composites: Macromolecular Engineering, 34(5), 276 (2005).

McCool R., Martin P.J. & Harkin-Jones E.M.A. Control of Product Wall Thickness Distribution in Plug-Assisted Thermoforming. Proceedings of PPS-2005 Americas Regional Meeting. Polymer Processing Society, Quebec, Canada, Aug. (2005).

McCool R., Martin P.J., Harkin-Jones E.M. Process Modelling for Control of Product Wall Thickness in Thermoforming. Plastics, Rubber & Composites: Macromolecular Engineering, 35 (8), 340-347 (2006).

O'Connor C.; Martin P.J.; Menary G.H.; McConville E. Modelling the Thermoforming of Polypropylene using ABAQUS. Proc. ABAQUS UK Regional User Meeting, Daventry, England, Nov (2007).

O'Connor C., Martin P.J., Menary G.H., McConville E. Material Models for the Thermoforming of Polypropylene. Proceedings of the Polymer Processing Society 24th Annual Meeting, PPS-24, Salerno, Italy, June (2008).

O'Connor C., Martin P.J., Menary G.H., McConville E. Comparison of Material Models of Polypropylene for Finite Element Analyses of Thermoforming Processes. Proceedings of the Polymer Processing Society Meeting, Larnaca, Cyprus, Nov. (2009).

Sweeney, J., Collins, T.L.D., Coates, P.D., Duckett, R.A. High Temperature Large Strain Viscoelastic Behaviour of Polyporpylene Modeled Using an Inhomogeneously Strained Network. Journal of Applied Polymer Science 72, 13 (1998).

OPTIMIZATION OF DESIGN AND PROCESS CONDITION FOR STRETCH BLOW MOULDED BOTTLES

Fiorotto M.[1]*, Cellere C.[2], Carraro P.[3], Lucchetta G.[4]

[1] *Department of Innovation in Mechanics and Management, University of Padova, Italy – marco.fiorotto@unipd.it; DIMEG, Via Venezia 1, 35131 – Padova – Italy - Phone: +39-049-8276823 – Fax: +39-049-8276816,*

[2] *Dep. of Innovation in Mechanics and Management, University of Padova, Italy - alessandro.cellere@unipd.it*

[3] *Dep. of Innovation in Mechanics and Management, University of Padova, Italy - piertommaso.carraro@unipd.it*

[4] *Dep. of Innovation in Mechanics and Management, University of Padova, Italy - giovanni.lucchetta@unipd.it*

Abstract - Stretch blow molding (SBM) is widely employed for the production of PET containers, in particular in the food and beverage industry as well as in the pharmaceutical sector. SBM is a high volume process with costly tooling for both the preform and the container. This study presents an optimization strategy for the entire stretch blow molding process. The method combines structural and process numerical simulations in an optimization environment. The overall simulation technique provides with a rational means to assist the bottle design and the determination of the optimal process conditions. The first aim of the strategy is to target the best preform thickness distribution which guarantees the highest top load for the bottle. The method is based on coupling an optimization algorithm and finite element (FE) simulations carried out using LS-DYNA. FE simulations were validated using in situ tests and measurements performed on 10-liter PET bottles. The computed container thickness distribution constitutes the target of the optimization of the process conditions. It has been analyzed the stretch and inflation sequence for the preform into a complex-shaped mold cavity. An unsteady large deformation analysis of the preform inflation process was carried out using a three dimensional membrane FE for viscoelastic materials. A non-isothermal flow with free surfaces and viscoelastic properties was described by a K-BKZ integral constitutive equation. The preform geometry was mathematically parameterized. This model was combined to an optimization loop to automatically compute the thickness distribution by manipulating the geometry of the perform and the operating parameters subjected to process and design constraints. The numerical results were verified using experimental data of an industrial case study, in order to assess the validity and the robustness of the proposed strategy.

Introduction

Injection stretch-blow molding (ISBM) is one of the most popular processes for polyethylene terephthalate (PET) bottles due to several benefits that can get associated with relatively low pressure process and easiness in producing parts with complex shapes and a very thin wall. A schematic representation of the stretch blow molding process is shown in Figure 1.

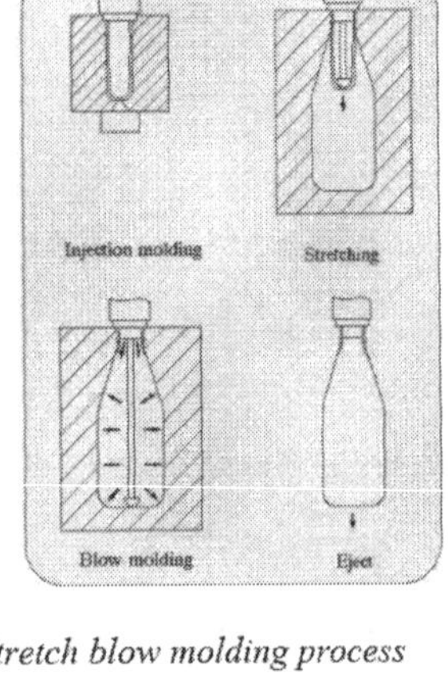

***Figure 1**: Stretch blow molding process*

In this process, a PET is first injected into a mold in order to achieve an amorphous preform. Then the preforms are heated in an infrared oven above the glass transition temperature (typically T_g = 80°C).The hot preforms are stretched using a stretching rod and blown by air at high pressure to create bottles with the desired shape. The bottles are cooled in the mold and finally removed. The whole process of stretch blow molding has a few seconds duration. Performance of PET bottle depends on three main variables:

- initial geometry of preforms,
- initial temperature of preforms,
- balance between speed of stretching rod and blowing.

These parameters control thickness distribution, biaxial orientation and crystallinity (which determines optical properties), mechanical and physical properties of the bottles. Experimental studies and geometric models have been developed to optimize design of preforms and process parameters.

However, numerical studies are far less costly in terms of time consumption and costs. It is also difficult to obtain accurate experimental data from a real process because it takes place in a closed mold. Researchers and industry are therefore concentrating on developing new and accurate numerical models.

The accuracy of numerical simulations depends mainly on material models. There are several models that have been developed in recent years. Hyperelastic and viscoelastic models are most used and show most encouraging results in simulating the stretch-blow molding process. Closely related to the choice of constitutive equations is the choice of the type of test used to characterize the material.

Once a material is characterized, numerical simulations should be used by the designer to quickly and automatically identify the geometry of preforms and operative parameters that yield a bottle with a desired thickness profile. This can be achieved by means of an integrated optimization strategies. The use of optimization algorithms associated with FEM simulations have begun to be common in industrial design.

However, the process consists of several stages highly coupled, and the material rheological behavior during stretching and blowing is highly nonlinear and temperature dependent. In addition to these problems of numerical simulation accuracy, optimization must take into account the process constraints (i.e. design), which further complicate the calculation.

Below a literature review is reported on the issues that have been introduced previously.

Numerical schemes have been applied to several models for the simulation of stretching with large deformations. One approach is based on incorporating a model developed by Rivlin (1948). In this case the deformation energy is assumed to be a function of the deformation invariants. Another model was developed by Odgen (1972), where deformation energy is given by principal stretches, rather than the deformation invariant. Delorenzi (1991) and Nied (1990) developed an extensive membrane model based on mass conservation and mold geometry, developed by Rosenweig (1979), which allows a good thickness approximation.

Christensen (1982) proposed a viscoelastic model in which plastic and viscous components are present. Yang et al. (2004) have proposed a viscous-hyperelastic model changing the Christensen's one. In this model the elastic part is described as in the Mooney-Rivlin model, while the viscous part is extended with an expression of the deflection tensor invariant. This viscous-hyperelastic model is named after Christensen and Yang.

IMI/NRC researchers team (Khayat, 2004; DiRaddo 1997, Martin 1999) has developed a code, BLOWSIM, which simulates the stretch-blow molding process. However, the implemented Mooney-Rivlin and Odgen hyperelastic models do not represent the time-dependent behavior of PET.

Thibault et al. (2007) have implemented in the same code the viscous-hyperelastic Christensen-Yang model. A characterization method based on tensile test using a Bruckner hot biaxial tester was developed. To determine the model parameters a nonlinear curve-fitting program based on the algorithm of Levenberg-Marquardt was used.

In recent years rapid progress has been made in the numerical simulations of real industrial flows, mainly due to the use of realistic type constitutive models, and differential viscoelastic models. Despite the complexity of implementation, in recent years the K-BKZ model, which was proposed by Kaye and Bernstein et al. (Kouba, 1994, 1998, Tanner 2000), is getting a growing consent. It is an integral viscoelastic, non-linear and time dependent model. K-BKZ with multiple relaxation times provides good fits of shear, elongational and normal stress data for polymer melts such as low-density and high-density polyethylene.

Several research teams studied the problem of optimization in extrusion blow molding. Laroche et al. (1999) utilized the finite element method and a gradient-based updating technique to search for the optimal parison thickness distribution that resulted in a uniform part thickness. Gauvin et al. (2003) used a gradient-based numerical optimization scheme to determine the optimal operating conditions to produce a blow molded part with a given performance. The optimization process included two phases, performance optimization and process optimization. Hsu et al. (2004) combined numerical simulation and fuzzy logic to determine the optimal die gap opening for a part with required wall thickness distribution. Lee and Soh (1996) determined the optimal parison thickness distribution for a blow molded part with the required thickness distribution. Thibault et al. (2007) used a numerical approach based on a constrained gradient-based optimization algorithm. The strategy allows for targeting a specified container thickness distribution by manipulating consecutively the preform geometry (thickness and shape) and the operating parameters subject to process and design constraints.

In this work, a new Computer Aided Engineering (CAE) method was developed to predict optimal design and process conditions for stretch blow molded bottles, minimizing the cost associated with the design of a new container. A new method based on FEM and a genetic algorithm manipulates the perform geometry (shape and thickness) and the process conditions to reach the desired container thickness distribution. The tool allows to choose bottle preforms geometry among those available on the market and optimal process parameters that allow the manufacturing of containers that can meet project specifications. A mechanical optimization is performed to predict the container thickness distribution that maximizes the structural properties and satisfies performance specifications, such as top load. A relatively simple material characterization technique was used, based on hot tensile tests, because it is able to provide all information necessary to obtain complex model parameters. For the process simulations, commercial general purpose numerical codes were used. With reference to an industrial case study, the predicted optimization results were compared to the experimental outcomes to assess the validity and the robustness of the proposed approach. The flow chart of the optimization process is shown in Figure 2.

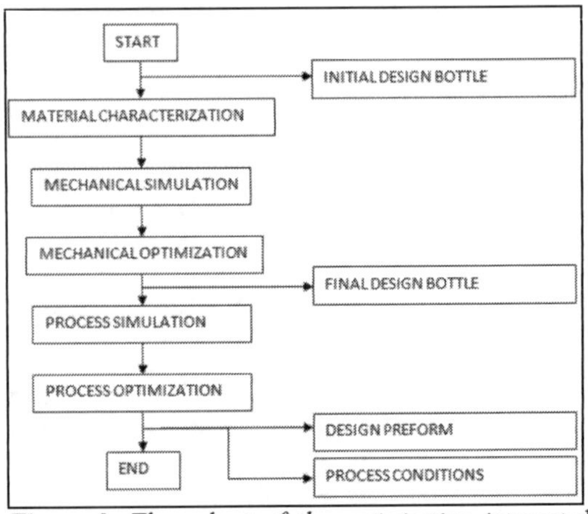

***Figure** 2: Flow chart of the optimization integrated environment*

Experimental

EXPERIMENTAL MATERIAL CHARACTERIZATION

The equipment used for testing is a CETR Micro Universal Material Tester. (UMT) This system, shown in Figure 3, has been modified to carry out uniaxial tension tests at controlled temperature ranging from ambient to 200 °C. The standard equipment was modified by the creation of dedicated grippers and by the design and implementation of a dedicated climatic chamber.

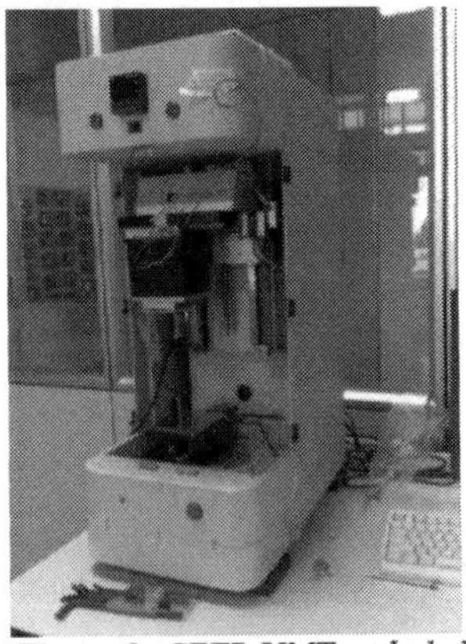

***Figure** 3: CETR UMT with dedicated grippers*

The instrument can perform tensile tests with micro-scale resolution, in particular, it provides a position resolution of 10^{-3} mm and can be equipped with various sensors of force. For this specific case, a 1000 N transducer with a resolution of 10^{-3} N was used.
For material characterization a series of uniaxial tensile tests was carried out at high-temperature. Dog-bone shaped specimens of 10 mm length and 2 mm width were taken from the side of the 10 l bottle. Tests were performed at temperature values of 90, 105 and 120 °C. For the test at 90 °C three different speed values were used: 1, 5 and 10 mm/s. For each test, several replications were performed. Results of the experiments, reported in the next section, led to implementing the numerical model of the material. Stress-strain curves showed that the material elastic behavior is strongly influenced by temperature, as well as by the deformation speed (Figure 5 and 6).

K-BKZ CONSTITUTIVE MODEL

The K-BKZ model is an integral equation containing information about the deformation history of the fluid. The constitutive equation is time dependent. Total extra-stress tensor is decomposed into a viscoelastic component $\mathbf{T_1}$, and a purely viscous component $\mathbf{T_2}$:

$$\mathbf{T} = \mathbf{T_1} + \mathbf{T_2} \tag{1}$$

$\mathbf{T_1}$ is calculated using the full viscoelastic K-BKZ and $\mathbf{T_2}$ is a purely viscous term, which is computed by

$$\mathbf{T_2} = 2\eta_2 \mathbf{D} \tag{2}$$

where $\mathbf{D}$ is the deformation rate tensor and η_2 is the viscosity component of the Newtonian extra-stress tensor.
$\mathbf{T_1}$ is calculated at time t by the following equation:

$$\mathbf{T_1}(t) = \frac{1}{1-\theta}\int_0^\infty \mu(t-s) H(I_1, I_2)\left[\mathbf{C}_t^{-1}(t-s) + \theta \mathbf{C}_t(t-s)\right] ds$$

(3)

where μ is the time-memory function, H is the damping function (with I_1 and I_2 are the first and the second invariants of the Finger strain tensor), $\mathbf{C}_t$ is the Cauchy-Green strain tensor and $\mathbf{C}^{-1}_t$ is the Finger strain tensor. θ is a scalar parameter that controls ratio of the normal-stress differences:

$$\frac{N_2}{N_1} = \frac{\theta}{1-\theta} \tag{4}$$

Explaining the individual terms, the time-memory function is:

$$\mu(t-s) = \sum_{i=1}^{N}\left\{\frac{g_i}{\tau_i^2} \cdot e^{-\frac{s}{\tau_i}}\right\} \tag{5}$$

where g_i are the relaxation moduli, τ_i are relaxation times and N is the number of modulus-time pairs.
The damping function can take the following forms:

$$H(I_1, I_2) = \frac{1}{1+\alpha\sqrt{(I_1-3)(I_2-3)}} \tag{6}$$

$$H(I_1,I_2)=\frac{1}{\exp(\beta\sqrt{\alpha I_1+(1-\alpha)I_2-3})} \quad (7)$$

$$H(I_1,I_2)=\frac{\alpha}{(\alpha-3)+\beta I_1+(1-\beta)I_2} \quad (8)$$

α, β are adjustable parameters, I_1, I_2 are the first and the second invariants of the Cauchy-Green strain tensor given by:

$$I_1=(\mathbf{C}_t^{-1}) \quad (9)$$

$$I_2=(\mathbf{C}_t) \quad (10)$$

In blow molding simulation, damping is usually neglected and the integral model can be simplified as follows:

$$\mathbf{T}_1(t)=\int_0^\infty\sum_{i=1}^N\frac{g_i}{\tau_i^2}\exp\left(\frac{-s}{\tau_i}\right)[\mathbf{C}_t^{-1}(t-s)-\mathbf{I}]ds \quad (11)$$

At each time step, extra viscoelastic stresses are evaluated explicitly by integrating the Equation (10) over the current time step.

The memory integral constitutive equation can be extended to incorporate temperature dependence.

A shift factor and variation of density with the temperature was used. Accordingly, the relaxation times τ_i and relaxation moduli g_i are expressed in terms of temperature by:

$$\tau_i=a_T\tau_i^0 \quad (12)$$

$$g_i=a_Tg_i^0\frac{\rho T}{\rho_0T_0} \quad (13)$$

where the subscript "0" denotes a reference state, T is the absolute temperature, ρ is the density and a_T is the shift factor. In this work, the William–Landel–Ferry (WLF) equation is use to describe the shift factor a_T:

$$\ln[a_T]=-\frac{C_1(T-T_0)}{C_2+T-T_0} \quad (14)$$

where C_1 and C_2 are fitting parameters.

By making a change of variable Finger tensor $\mathbf{C}_t^{-1}$ does not become exclusively a function of the elapsed time t' = s – t, but remains present time t dependent, although not explicitly.

Fluid constitutive equations are reduced as follows:

$$\mathbf{T}_1(t)=\int_{-\infty}^t\sum_{i=1}^N\frac{g_i}{\tau_i^2}\exp\left(-\frac{t-t'}{\tau_i}\right)[\mathbf{C}_t^{-1}(t)-\mathbf{I}]dt'$$

(15)

The integral can be split into two parts using the intervals [-∞,0] and [0,t], since the Finger strain tensor takes distinct values. Thus, the stress tensor is given by:

$$\mathbf{T}_1(t)=\int_{-\infty}^0\sum_{i=1}^N\frac{g_i}{\tau_i^2}\exp\left(-\frac{t-t'}{\tau_i}\right)\mathbf{C}_t^{-1}(t)dt'+$$

$$+\int_0^t\sum_{i=1}^N\frac{g_i}{\tau_i^2}\exp\left(-\frac{t-t'}{\tau_i}\right)\mathbf{C}_t^{-1}(t-t')dt' \quad (16)$$

where the Finger strain tensor is:

$$\mathbf{C}_t^{-1}=\begin{pmatrix}\lambda^2(t,t') & 0 & 0\\ 0 & \lambda^{2m}(t,t') & 0\\ 0 & 0 & \lambda^{-2(1+m)}(t,t')\end{pmatrix} \quad (17)$$

$$\lambda(t,s)=1+\frac{v}{l}(t-t') \quad (18)$$

λ is the extension ratio, v is the deformation velocity, l is the initial simple length and m is a parameter which depends on the deformation kind (m = -1/2 for uniaxial elongation, m = 0 for planar extension, and m = 1 for equibiaxial extension). For negative values of t', t'=0 is considered. The model parameters were fitted by solving the integral using numerical integration through reverse techniques on experimental curves. In order to calculate the first contribution the Laguerre quadrature method provides:

$$\mathbf{T}_{1\text{-}1}=\sum_{k=1}^M\sum_{i=1}^N\exp(t'(k))\frac{g_i}{\tau_i^2}\exp\left(-\frac{t-t'(k)}{\tau_i}\right)\left(1+\frac{v}{l}t\right) \quad (19)$$

where N is the relaxation times number of spectrum and t'(k) are coefficients for Laguerre quadrature of M-points. The second term is calculated by the Cavalieri-Simpson method.

Both integrals were computed numerically by implementing an algorithm in Matlab®. Experimental results were fitted by means of the numerical multi-object optimization program modeFRONTIER®, as shown in Figure 4.

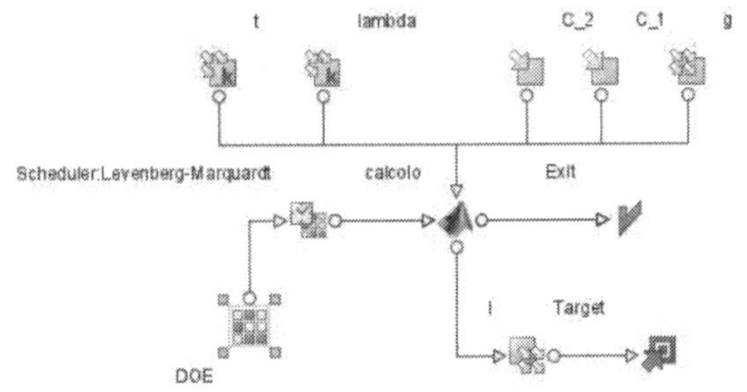

Figure 4: Fitting implementation in the modeFRONTIER® program

Among the input data, extensional ratios, specimen length and the parameter m (related to the type of deformation, in this case uniaxial) were considered constant. Five values of relaxation spectrum times were fixed in order to reduce the computation time. Deformation rate and temperature were modified depending on the test type. The optimization varied the modules of the relaxation spectrum and the WLF constants C_1 and C_2. The code output was a vector containing the values of calculated tensions at the given input strain. The integral modifying variables were calculated within a predetermined interval. The Levenberg-Marquardt optimization algorithm was used for minimizing the following objective function:

$$error = \sum_{i=1}^{U} \left(T_{exp} - T_i\right)^2 \qquad (20)$$

where U is the number of stress measured in uniaxial tensile tests, T_{exp} is the value of tension experimentally measured and T_i is the stress calculated by solving the K-BKZ integral model.
K-BKZ model parameters obtained from the fitting are summarized in Table 1. The stresses obtained from numerical simulations compared to experimental curves are reported in Figures 5 and 6.

K-BKZ model parameters	
g_k(MPa)	τ_k(s)
0.4	0.01
2.5	0.1
2.2	1
5.3	10
510	100
WLF model parameters	
T_{ref}(°C)	105
C_1	6.1
C_2	140.2

Table 1: *K-BKZ and WLF parameters at reference temperature*

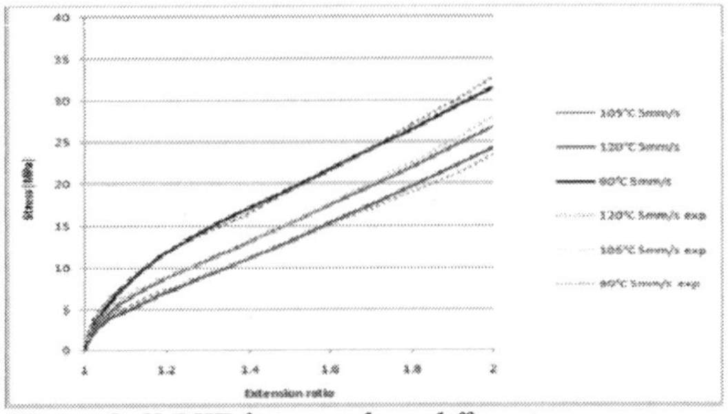

Figure 5: *K-BKZ fitting at three different temperatures*

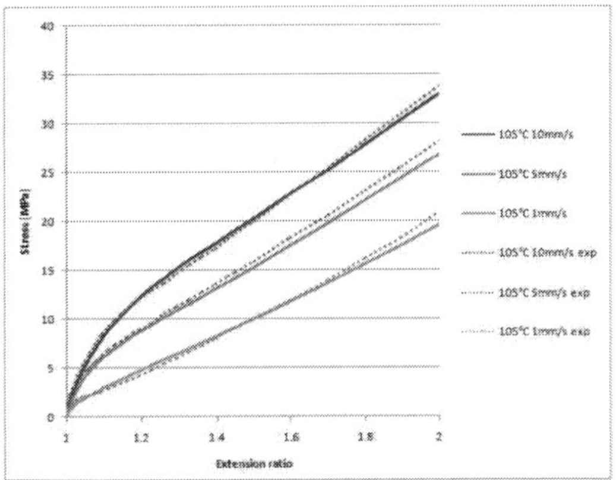

Figure 6: *K-BKZ fitting with three different rates of deformation*

MECHANICAL ANALYSIS

In order to simulate the mechanical behavior of the filled bottle a set of FEM explicit simulations was performed with LS-DYNA. The simulations results were focused on the top load value of a filled and capped bottle. This value is very important in packaging design, because it is strictly related to the maximum stacking load in transportation. Simulations were designed in order to determine the best thickness distribution for a given bottle weight. Five different zones were defined on the bottle, as shown in Figure 7.

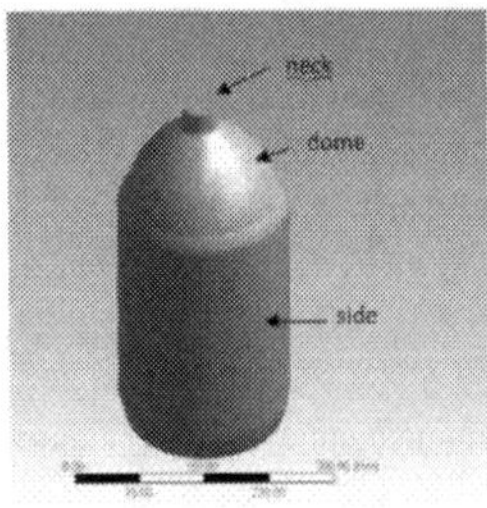

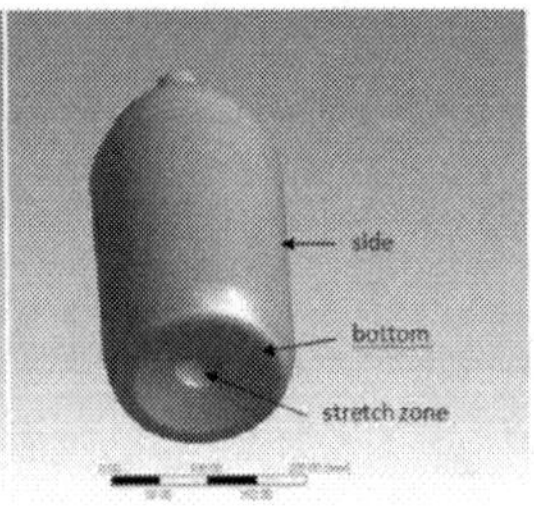

Figure 7: *Bottle zones*

Each zone was modeled with a shell mesh, considering both the plasticity and damage in the material. An elasto-plastic material with a stress versus strain curve and strain rate dependency was defined according to the PET behavior.

The filling was simulated with a 3D mesh, with an elastic fluid material. Air and liquid were considered as the same fluid material, which has the air elastic contribution. Water compressibility was neglected. Pressure in this material was calculated as:

$$p^{n+1} = -K \ln V^{n+1} \qquad (21)$$

where K is the bulk modulus and V is the relative volume.

In order to evaluate the best material thickness distribution an initial set of FEM simulations was run. As reported in Table 2 the thickness distribution was varied in an interval for each zone, considering a constant volume constraint of 136 cm^3.

Zone Name	Thickness Interval [mm]
neck	1.5 – 2.5
dome	0.3 – 0.6
side	0.3 – 0.6
bottom	0.4 – 1.0
stretch zone	1.5 – 2.5

***Table 2** – Definition of design variable limits*

An plan with 200 experiments with random distribution run. The mechanical FEM output was defined as the highest load before the fracture of the external shell and the consequent bottle failure and simulation termination. The best result was chosen as the target thickness distribution in the stretch blow molding process optimization.

NUMERICAL SIMULATION OF STRETCH BLOW MOLDING

The stretch blow molding process was simulated using a full viscoelastic multi-mode model implemented in ANSYS Polyflow 3.12. This method uses a membrane element. Membranes can greatly reduce the required computational time. Viscoelastic numerical models for blow molding are based on an extension of the shell element for fluids. An industrial case was used to validate the simulation strategy. The geometries of this case are shown in the Figures 8 and 9.

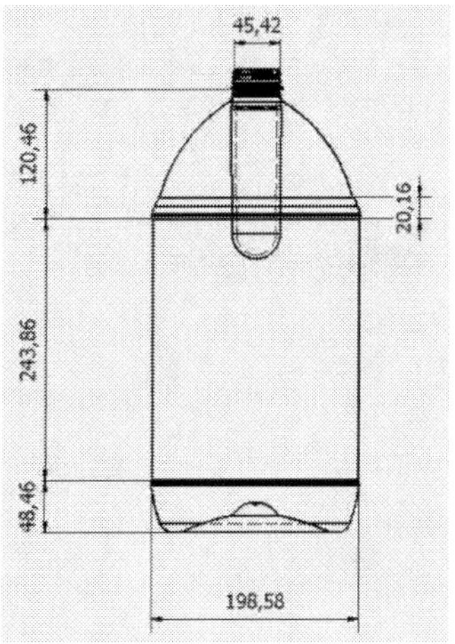

***Figure 8:** Dimensions of analyzed bottle*

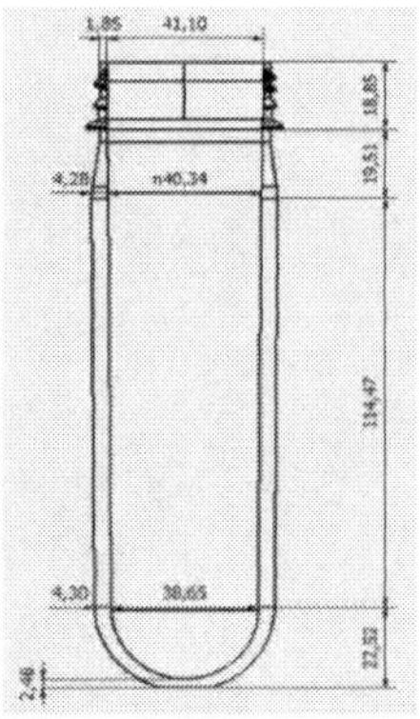

***Figure 9:** Dimensions of analyzed perform*

The problem symmetry was used to further reduce the computation time and only half geometry was considered. The preform surface temperature was obtained from experiments by means of a thermocamera. Then time and modulus of relaxation derived from experimental data fitting was assigned. Thickness was assigned with reference to the normal direction to the shell surface at each node. The initial temperature profile from the top to the bottom of the perform was set as shown in Figures 10.

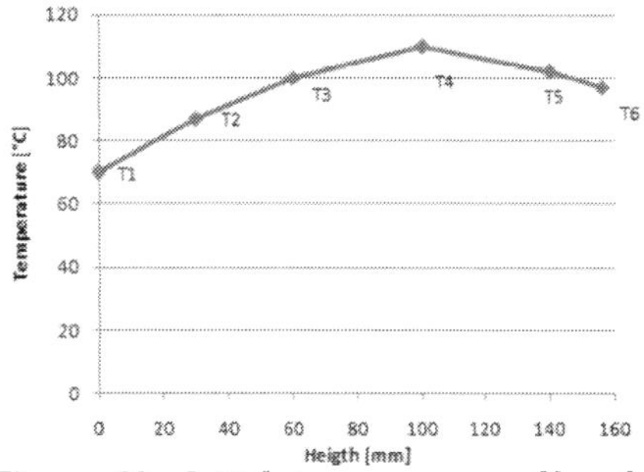

***Figure 10:** Initial temperature profile of preform before stretch blow molding*

The preforms were stretched to the bottom of the mold, then they were blown first at low pressure and later at high pressure, to the desired final shape of the bottle (Figure 11).
High pressure was maintained for 3 s while cooling down the container. Speed profile with a multi ramp was set according to experimental data (Figure 12).

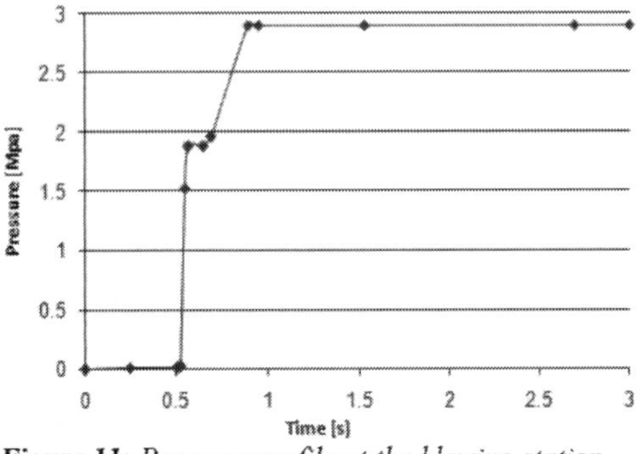

Figure 11: *Pressure profile at the blowing station*

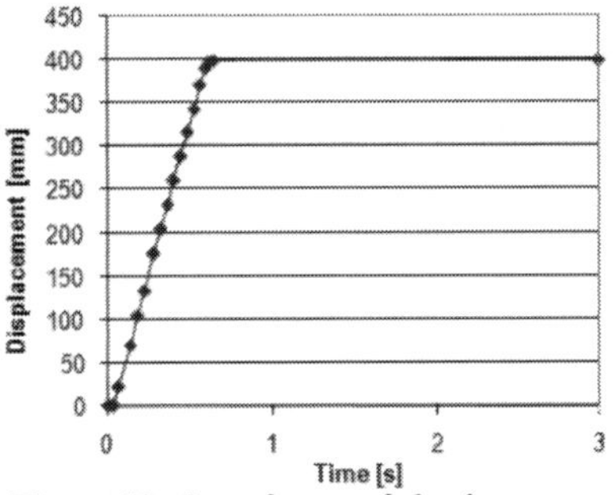

Figure 12: *Stretching rod displacement at the blowing station*

The thickness of the bottle was measured in 8 different locations from the neck to the bottom of the container. The measurement was repeated on different samples to ensure operation repeatability. Thickness values obtained from simulations were very close to those measured experimentally (Figure 13).
Figure 14 shows the preform deformation during the stretch-blowing inside the mold. Five deformation steps are represented.

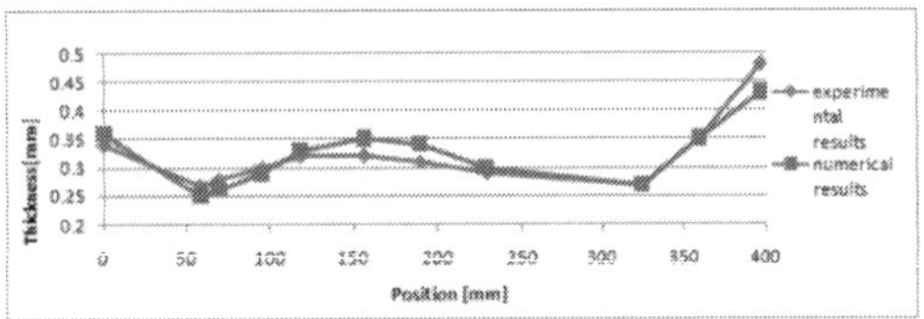

Figure 13: *Final thickness comparison of the bottle*

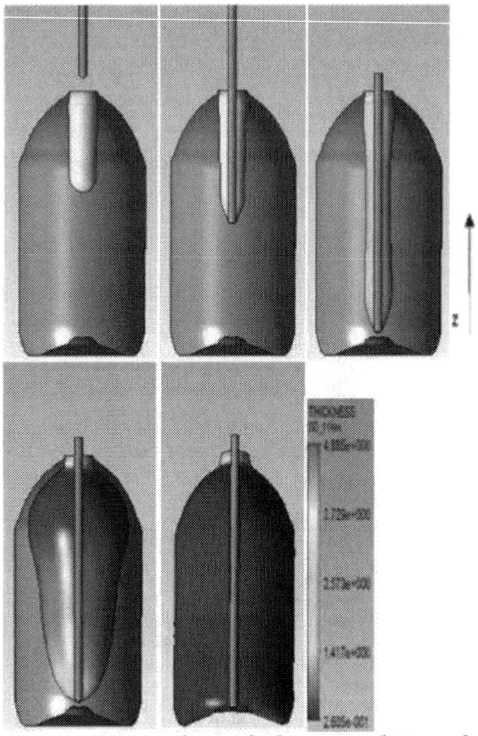

Figure 14: *Preform thickness evolution during stretching and blowing*

PREFORM AND OPERATING CONDITIONS OPTIMIZATION

The second goal of this work was to develop an integrated design environment for the design of preforms and container molds.
The development of an optimization strategy starts with the definition of a target that must be minimized. This target is subjected to a number of boundary constraints limiting the design area through an automatic handling of a number of design variables. Once the objective function is defined, numerical optimization can be performed using an optimization techniques. An optimization algorithm simultaneously manipulates the perform geometry (shape and thickness) and the condition optimization to target the desired container thickness distribution.
The following objective function was developed to evaluate the wall thickness distribution of blow molded parts:

$$\sigma(X) = \sum_{i=1}^{N} (T_{hi} - T_i)^2 \qquad (22)$$

σ(X) represents the container thickness variance around thickness targets (T_i) defined by the designer along the container in z direction. N are the control points in which nodal finite element thickness, T_{hi}, is calculated.

Different types of perform are used in industry (Figure 15). This work focuses on the first three preform types (outside, inside, and dual side taper preforms). Preforms were divided into four parts (finish, taper, body and end-cap) and they were parameterized in such way, varying one parameter at a time and automatically reconstructing their geometry (Figure 16). The finish part was kept constant. To maintain an equilibrium between efficiency of calculation and validation of results, only five preform design variables will be considered to define the preform. The end-cap has been parameterized using a hemi-spherical shape ($x_4 = x_7 + x_{10}$). The end-cap thickness is imposed equal to the body thickness ($x_{10} = x_{11}$).

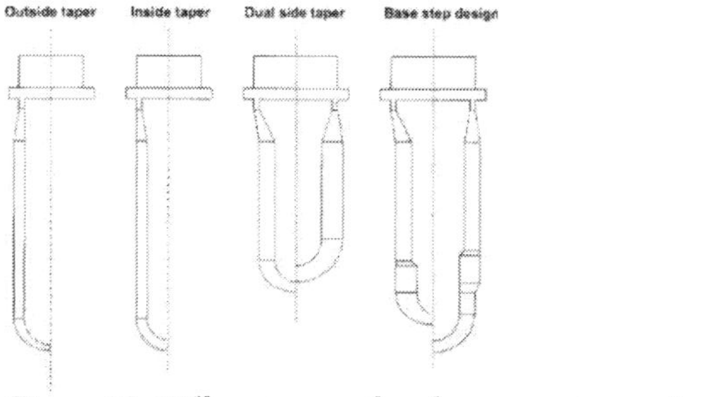

Figure 15: *Different types of preform geometry used in industry*

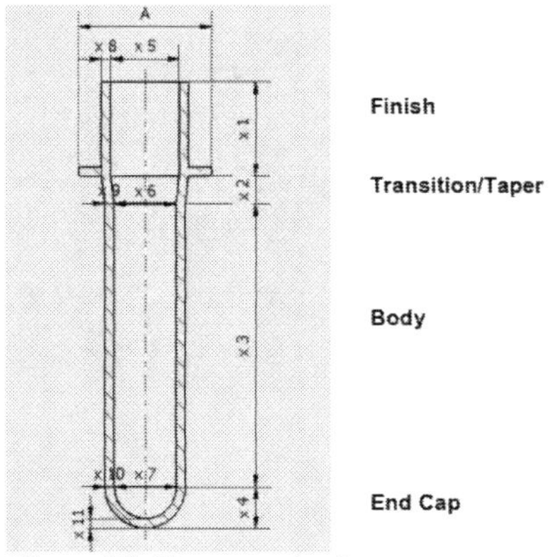

Figure 16: *Parameterization of preform*

As mentioned previously, POLYFLOW adopts the theory of membranes. So it was necessary to calculate a midplane for each configuration of the preforms. A Matlab algorithm was implemented. It automatically rebuilds a new geometry of the preforms and calculates thickness values in a finite number of points along the length of the preform by varying one parameter at a time.

Starting from an existing preform, the application numerically simulates the process and provides the bottle thickness. If the container's thickness distribution is unsatisfactory, then a new preform is designed modifying the previous one, using an appropriate optimization algorithm. Once the optimization is completed, process conditions are properly optimized further improving the final result.

Preforms design must take into account constraints that are imposed by the mold and the stretching rod geometry.

The following operating parameters are simultaneously manipulated:

- Pre-blow pressure (P_1)
- Blow pressure (P_2)
- Stretching rod speed (V_R)
- Temperature profile along the preform ($T_{i,min}$).

The objective function is minimized by manipulating the preform design variables subjected to process and geometric constraints:

$$\begin{aligned} &x_{2,min} < x_2 < x_{2,max} \\ &x_{6,min} < x_6 < x_{6,max} \\ &x_{7,min} < x_7 < x_{7,max} \\ &x_{9,min} < x_9 < x_{9,max} \\ &x_{10,min} < x_{10} < x_{10,max} \end{aligned} \quad (23)$$

and the processing design variables:

$$\begin{aligned} &P_{1,min} < P_1 < P_{1,max} \\ &P_{2,min} < P_2 < P_{2,max} \\ &T_{i,min} < T_i < T_{i,max} \\ &V_{R,min} < V_R < V_{R,max} \end{aligned} \quad (24)$$

T_{min} were chosen to depict strain-hardening effects and T_{max}, the upper bounds to depict minimal blowouts. Temperature profile were imposed in six equispaced points along the perform. Body length (x_3) had to be calculated from the total volume.

The constraints can be formulated as

- stretching rod clearance:

$$\begin{aligned} &x_3 < D_{Rod} + R_{Cl} \\ &x_4 < D_{Rod} + R_{Cl} \end{aligned} \quad (25)$$

- material volume:

$$Vol_{Finish} + Vol_{Taper} + Vol_{Body} + Vol_{End\ cap} = Vol_{TOT} \quad (26)$$

where D_{Rod} and R_{Cl} represents, respectively, the rod diameter and clearance (~1–2 mm). Vol_{Finish}, Vol_{Taper}, Vol_{Body} and $Vol_{End\ cap}$ are volumes of the different parts of the bottle and Vol_{TOT} is the total volume calculated by the mechanical analysis. Eight control point along the bottle were used for the optimization, as shown in Figure 17. The numerical optimization iterates over a simulation tool as it moves towards the optimal condition. The multi-objective optimization algorithm (MOGA-II) used in this work integrates finite element SBM simulations that model the preform material deformation during the consecutive process stages to target a required container thickness profile. A Sobol experimental plan was used to initially identify which input variables mostly affect the experiment results.

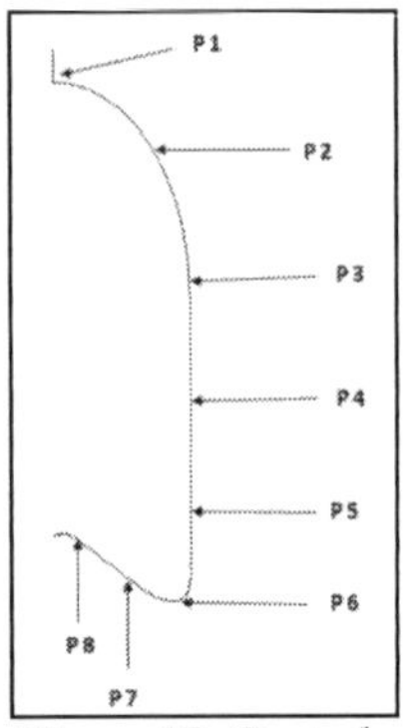

Figure 17: Control point along the bottle

Preforms parameterization, mesh generation, background settings, thickness allocation and the preforms process simulation have been integrated into a single optimization environment, run in modeFRONTIER. Table 3 lists the initial preform design variables and the preform design variable limits for the case studied.

Design variable	Lower bound	Upper bound
Taper length (x_2) [mm]	15	30
End Cap length (x_4) [mm]	40	55
Taper internal diameter (x_6) [mm]	40	55
Body internal diameter (x_7) [mm]	2	6
Body thickness (x_{10}) [mm]	2	6
Pre-blow pressure (P_1) [MPa]	1.5	2
Blow pressure (P_2) [MPa]	2.5	4
Stretching rod speed (V_R) [mm/s]	500	1000
Temperature at position 1 (T_1) [°C]	60	110
Temperature at position 2 (T_2) [°C]	60	120
Temperature at position 3 (T_3) [°C]	60	120
Temperature at position 4 (T_4) [°C]	60	120
Temperature at position 5 (T_5) [°C]	60	120
Temperature at position 6 (T_6) [°C]	60	110

Table 3 – Definition of design variable limits

Results and Discussion

MECHANICAL OPTIMIZATION

Figure 18 shows the graphical output of a top load simulation. During the optimization cycle, a total of 200 process simulations have been run in order to evaluate the optimal thickness profile along the bottle (total CPU time of 48 hours, on a 2.99 GHz PC computer).After a small local deformation near the neck the simulation shows a mayor failure that is usually located at the end of the dome, near the smaller radius rib.

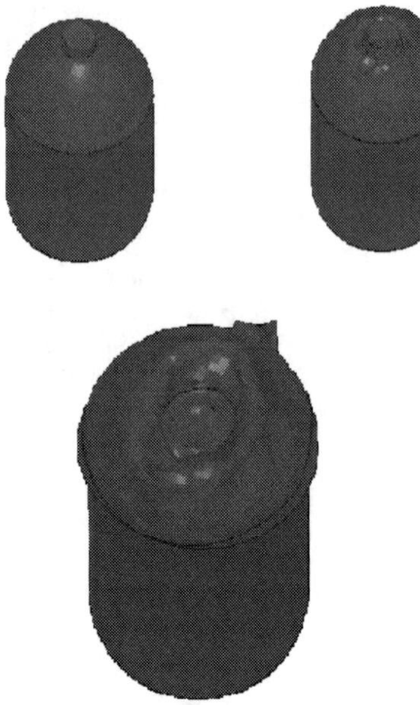

Figure 18: Mechanical simulation evolution

A full set of mechanical analysis was performed and the code was able to automatically choose the best one. Results vary from 502, N for the worst thickness distribution to 915 N for the best case. The best results were achieved with the thickness distribution reported in Table 4.

Zone Name	Best Thickness [mm]
neck	2.25
dome	0.45
side	0.43
bottom	0.55
stretch zone	2.5

Table 4: Comparison between initial and optimal mechanical variables

PREFORM AND OPERATING CONDITIONS OPTIMIZATION

The results are shown in Table 5. Since more design variables are involved in this constrained optimization, the convergence is obtained after 172 process simulations (total CPU time of 48 h). The objective function decreased eventually to 0.011. The optimal preform presented a decreased thickness with reference to the initial design. The comparison between initial and optimal variables reported in Table 5, show that the design variables that have been significantly modified are the preform temperature values and the stretching rod speed. Concerning the comparison between the optimal and experimental blow pressure no significant improvement has been observed. Consequently, the stretching rod speed and temperature can be identified as parameters to manipulate to get a better design.

Design variable	Initial value	Final value
Taper length (x_2) [mm]	19.5	24
Taper internal diameter (x_4) [mm]	40.3	48
Body internal diameter (x_6) [mm]	38.7	47
Taper thickness (x_9) [mm]	4.28	4
Body thickness (x_{10}) [mm]	4.3	4
Pre-blow pressure (P_1) [MPa]	1.8	1.8
Blow pressure (P_2) [MPa]	2.9	3.3
Stretching rod speed (V_R) [mm/s]	700	790
Temperature at position 1 (T_1) [°C]	70	85
Temperature at position 2 (T_2) [°C]	87	100
Temperature at position 3 (T_3) [°C]	100	81
Temperature at position 4 (T_4) [°C]	110	79
Temperature at position 5 (T_5) [°C]	102	96
Temperature at position 6 (T_6) [°C]	97	69

Table 5: Comparison between initial and optimal variables

Conclusions

In this work, an optimization environment was developed to automatically predict the container thickness distribution that maximizes structural properties. In the second step the strategy allows to calculate the preform geometry and process parameters to obtain the specific thickness distribution along the bottle. Optimal process parameters that allow the manufacture of a container with a particular thickness profile could be identified without slow and expensive trials on prototype molds. A computing environment was parameterized in order to choose different preforms among those ones commercially available. Preforms were parameterized using an algorithm implemented in Matlab. This optimization environment provides the designer with a tool to automatically rebuild preforms geometry varying different parameters and assigning the correspondent thickness at each point. Simulations were verified on an industrial case study. Characterization test used to identify constitutive model coefficients are relatively simple to perform. An easy, feasible and economic hot tensile test device was used. The developed approach allows to save time and resources that are generally required to move from preforms choice and process parameters setting to final bottle production.

References

Christensen R. M.,"Theory of viscoelasticity", Academic Press (1982).

DeLorenzi G. and Nied H. F., "Finite Element Simulation of Thermoforming and Blow Molding", Progress in Polymer Processing, Hanser Publishers, New York (1991).

R. DiRaddo, D. Laroche, R. Aubert, D. M. Gao, M. Doordan, and G. Jobin, SPE ANTEC Tech. Papers, 850 (1997).

C. Gauvin, F. Thibault, D. Laroche, "Optimization of blow molded part performance through process simulation", Polymer Engineering and Science, **43**, 1407–1414 (2003).

Y.L. Hsu, T.C. Liu, F. Thibault, B. Lanctot, "Design optimization of the blow moulding process using a fuzzy optimization algorithm", J. Eng. Manuf. **218**, 197–212 (2004).

R. E. Khayat and A. Derdouri, "Stretch and inflation of hyperelastic membranes as applied to blow moldiing", Polymer Engineering and Science, **35**, 1852-1863 (2004).

Kouba K., Ghafur M.O., Vlachopoulos J., Haessly W. P., SPEAntec Tech Papers, 40, 850 (1994).

Kouba K. , Novotny P., "Computer Simulation of Thermoforming Process-The Role of Material Behavior", Thermoforming Quarterly, 13 (1998).

Laroche D., Kabanemi K. K., Pecora L., and Diraddo R. W., "Integrated numerical modeling of the blow molding Process", Polymer Engineering and Science, **39**, 12-23 (1999).

D.K. Lee, S.K. Soh, "Prediction of optimal perform thickness distribution in blow molding", Polymer Engineering and Science, **36**, 1513–1520 (1996).

L. Martin, D. Stracovsky, D. Laroche, A. Bardetti, R. Ben-Yedder, and R. DiRaddo, "Modeling and Experimental Validation of the Stretch Blow Molding of PET", SPE ANTEC Tech. Papers, 238 (1999).

Nied H. F., Taylor C. A., and DeLorenzi H. G., Three dimensional finite element simulation of thermoforming, Polymer Engineering and Science, **30**, 1314-1322 (1990).

Odgen R. W., Proceedings of the Royal Society, Series A **326**, 565 (1972).

Pham X.-T., Thibault F., Lim L-T., "Modeling and simulation of stretch blow molding of polyethylene terephthalate", Polymer Engineering and Science, **44**, 1460-1472 (2004).

R. S. Rivlin, 1948, Large Elastic deformations of Isotropic Materials I – Fundamental Concepts, Philosophical Transactions of the Royal Society of London, A **240**, pp 459-490.

Rosenzweig N., Narkis M., and Tadmor Z., "Wall thickness distribution in thermoforming", Polymer Engineering and Science, **19**, 946-951 (1979).

R. I. Tanner, Engineering Rheology: 2nd Ed., Oxford University Press, New York (2000).
Thibault, A., Malo, B. Lanctot, R. Diraddo, "Preform Shape and Operating Condition Optimization for the Stretch Blow Molding Process", Polymer Engineering and Science, **47**, 289-301 (2007).

Yang Z.J., Harkin-Jones E., Menary G.H., Armstrong C.G., "A Non-Isothermal Finite element Model for Injection Strech-Blow Molding of PET Bottles With Parametric Studies", Polymer Engineering and Science, **44**, 1379-1390 (2004).

ENHANCEMENT OF ANGIOPLASTY BALLOON PERFORMANCE THROUGH ANNEALING

F. Daver[1], P. Girard[2], F. Thibault[2], Christian de Grandpré[2],R. DiRaddo[2]

[1]*RMIT University, School of Aerospace, Mechanical and Manufacturing Engineering, Bundoora, Vic 3083, Australia*

[2]*National Research Council Canada, Industrial Materials Institute, 75 Blvd de Mortagne, Boucherville, Quebec, J4B6Y4*

Angioplasty balloons are used to implement stenting of atherosclerosed blood vessels and put in place heart valve systems. The balloons are required to have thin walls and high burst pressure while having a given compliance (e.g., given balloon diameter at a nominal pressure). In this study 3 by 15 mm and an 8 by 30 mm balloons are formed through a micro-blow moulding process. The balloon forming machine process (BFMP) starts by stretching an extruded tubing of thermoplastic elastomer into a 'parison' by means of a specially instrumented balloon forming machine. The BFMP is identified as a three step stretch micro blow moulding process, where the 'parison' is stretched under pressure and at controlled temperatures to form a 'balloon': Forming of the cylinders, forming of the balloon's cones and annealing of the balloon. In this article we show how the elastic and mechanical properties of the balloon can be fine tuned to the specifications of the medical suppliers. A balloon design method and suggestion for an in-line control method for the balloon forming process are also presented.

Introduction

Angioplasty balloons are commonly used to position stents and heart valve systems. Both these procedures require very controlled dilation parameters from the balloon so that the atherosclerotic lesion is first pushed open at a relatively low pressure and the stent is then deployed and embedded in the vessel wall at a much higher pressure.

The balloon manufacturing process must thus ensure that the balloon will reliably produce a given diameter at a given inflation pressure. These standard diameter vs pressure specifications are generally presented on the balloon packaging as a table, as shown in figure 1.

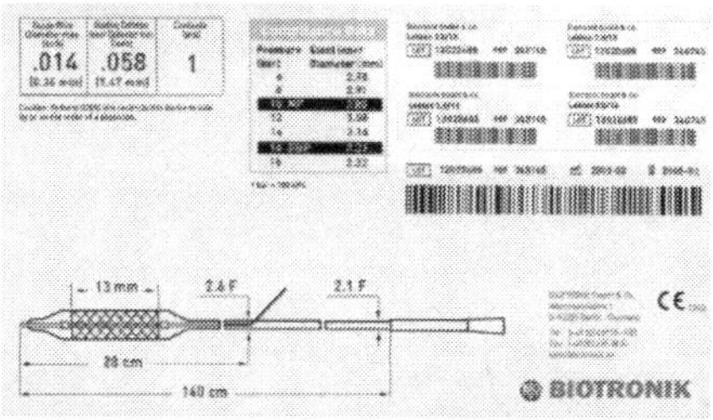

Figure 1: Typical specifications for a balloon

Since we could adjust the residual elongation of the PEBAX (Tradename, see references) balloon material used in this study through heat treatment during manufacturing by a factor of 2 and the manufacturer reports a factor of 3.8 (Despotopoulou and O'Brien), achieving this diameter vs. pressure relationship is not obvious.

The purpose of this article is to present a design method for optimally forming PEBAX balloons with the specified compliance and diameter.

Three processing steps were identified and analyzed for the balloon forming process:

1) Forming of cylinders
2) Forming of cones
3) Annealing phase, which is the most important for adjustment of the balloon's compliance and standard diameter.

Designs of experiments (DOEs) were run in the vicinity of the optimum point for each of these steps. From these DOEs sensitivity analyses were done, from which design procedures to manufacture a balloon with a given compliance and diameter at the specified test pressure were identified.

Experimental

Balloon Manufacturing:

The balloons were produced using a Double-end stretcher (DES), Balloon Forming machine (BFM) and Burst Pressure Testing Machine (BPTM) from Interface Associates (Fig. 2). The BPTM was modified to interface and download files to another computer using the XML language, which can be used for data serialization.

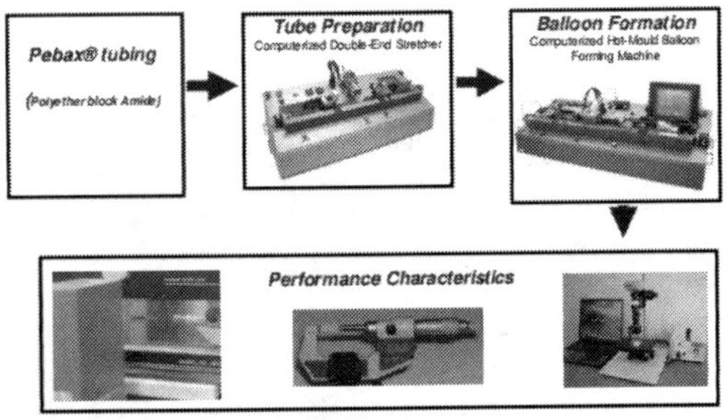

Figure 2: Manufacturing Process: Double-end stretcher (DES), Balloon Forming machine (BFM) and Burst Pressure Tester (BPT)

A balloon preform is first manufactured from the PEBAX tubing on a Double-end stretching machine from "Interface Associates" (Fig. 3).

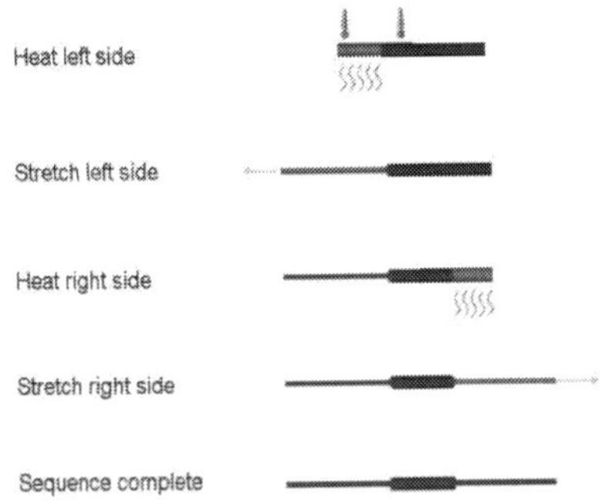

Figure 3: PEBAX tube Double-end stretching process

The preform is then formed into a balloon on a balloon forming machine from the same company (fig. 4).

One can note on figure 5 that:

1) Although the total forming step lasts about 220s the actual forming cycle lasts only about 0.2s or less, making it mostly instantaneous.
2) From the pressure graph the amount of energy used to form the balloon can be measured
3) The force is not symmetrical since the tubing has not the same diameter on both ends
4) The temperature does not vary during forming because of the mould's thermal inertia

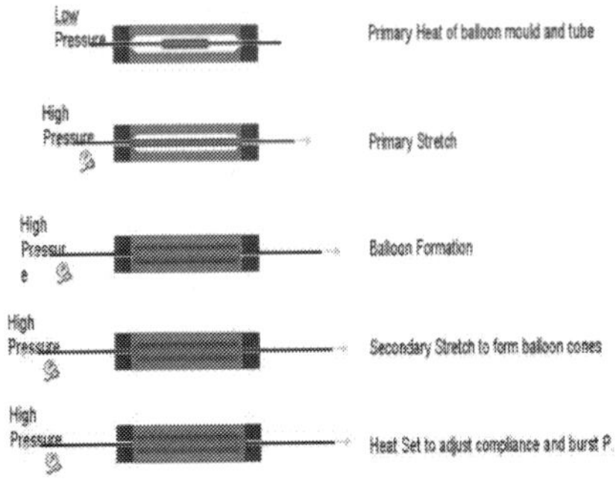

Figure 4: PEBAX balloon forming process

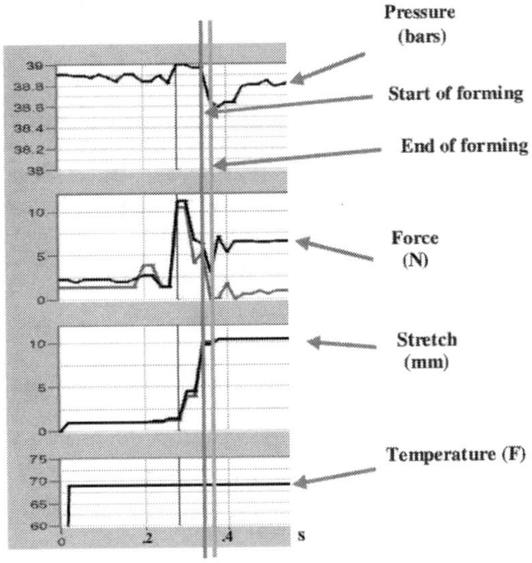

Figure 5: Actual forming step of the balloon on the BFM.

Balloon Testing:

The balloons were then evaluated for quality, thickness and burst pressure (fig. 6).

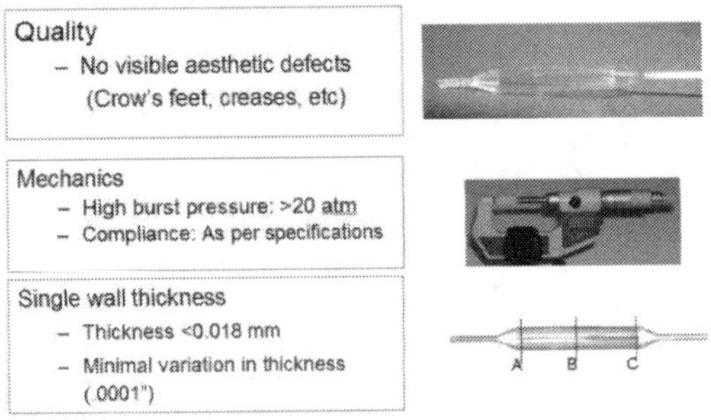

Figure 6: Quality criteria for balloons

Figure 7 presents the burst pressure testing machine (BPTM). The BPTM applies a ramped pressure increase (0-67 bars) to a balloon immersed in water (5 to 40C) and measures the variation of the balloon diameter with a laser beam. For this project it was modified to be able to download files to another computer trough a serial USB connection implemented using the XML language.

Figure 7: Burst Pressure Testing Machine (BPTM)

The interface allows plotting the evolution of the volume and diameter of the balloon against time and pressure (fig. 8).

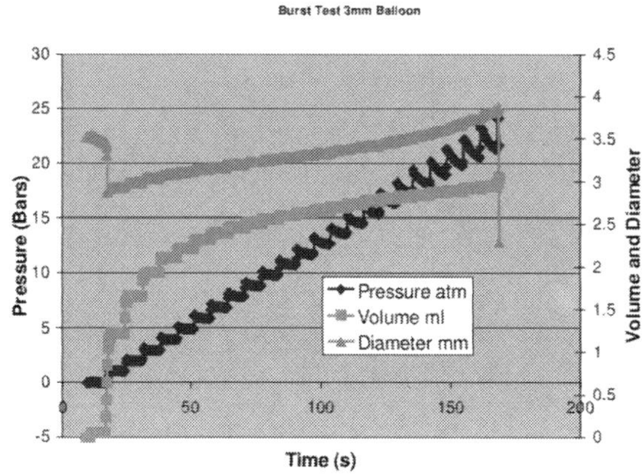

Figure 8: Output of the Burst Pressure Testing Machine (BPTM)

The BPTM can be used as a rheometer by plotting the apparent compliance of the material ΔV/ΔP, (same as a Bruckner rheometer) (fig. 9).

The balloons produced were evaluated according to the following criteria

- No visible aesthetic defects such as Crow's feet, creases, etc.
- Mechanics: Burst pressure >20 atm and compliance as per specifications.
- Wall thickness: Thickness <0.018 mm and minimal variation in thickness.

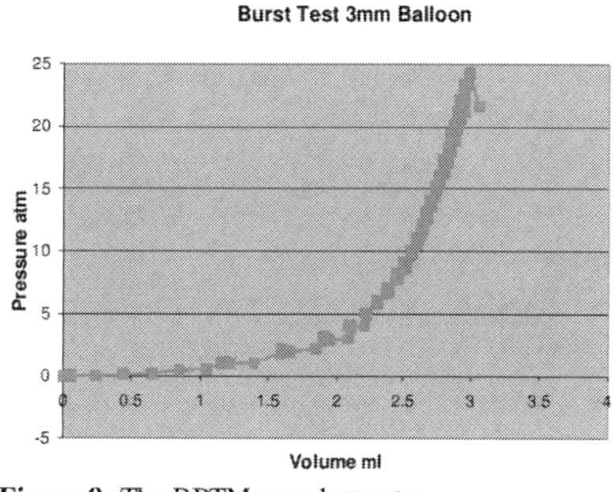

Figure 9: The BPTM as a rheometer

Results and Discussion

Balloon Optimization:

Each of the three steps:

1) Forming of cylinders
2) Forming of cones
3) Annealing phase,

was optimized separately through a design of experiments (DOE) generated around an optimum processing point for a 3mm and 8 mm balloon (table 1). The annealing phase, called heat set phase in the manufacturing process is the most important one since it not only adjusts the compliance and burst pressure of the balloon but also minimizes recoil during sterilisation

Processing Parameters	Units	Initial 8 mm balloon	Optimized 8 mm balloon	Initial 3mm balloon	Optimized 3mm balloon	
Primary θ Cylinder	°C	85	85	70	70	
Primary θ Cones	°C	80	80	70	70	
Heating time	s	140	140	65	55	
Primary Stretch	mm	22	22	14	14	
Prim. St. speed	mm/s	130	130	150	150	
Forming Pressure	atm	30	30	42	42	
Secondary Stretch	mm	25	25	16	16	
Sec. St. speed	mm	110	110	100	100	
Sec. St. θ Cylinder	°C	95	95	>120	>90	*
Sec. St. θ Cones	°C	90	120	>120	>90	*
Heat Set θ Cylinder	°C	90	92	120	90	
Heat Set θ Cones	°C	90	100	120	90	
Heat Set Pressure	atm	30	30	0	42	
Heat set Time	s	60	60	120	60	
Burst Pressure	atm	19.08	19.30	18.07	25	
Diameter at 6 bars	mm	7.96	8.2	3.21	3.01	
Visual Defects		Cones	No defects	Cones	No defects	

Table 1: Initial and optimized 3 mm and 8 mm balloons. (*) Means a given heating rate was realized between set points

The optimized points in table 1 all met the quality criteria defined in figure 5. They also gave the specified diameter at the specified pressure. In general a heat set treatment will substantially increase the burst pressure with regard to no heat set (fig. 10).

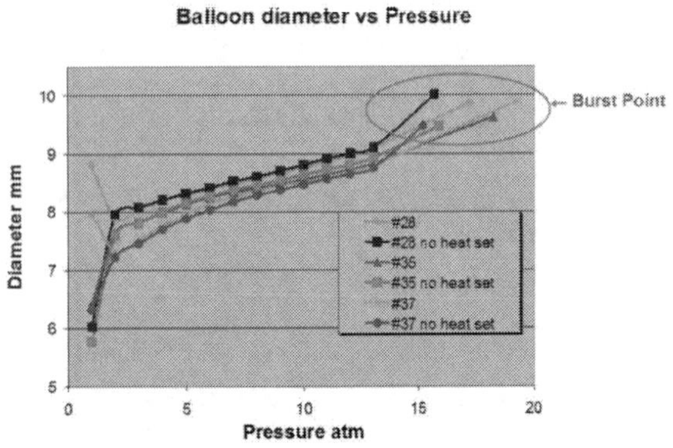

Figure 10: Effect of heat set treatment on diameter and burst pressure on 8 mm balloon (all: 90-95-90 °C heat set, Mould θ: #28 75-85-75 °C, #35 and #37 85-85-85 °C)

Figure 11 presents the variation of the diameter and burst pressure with the heat set temperature. The "No heat set" points show the value of the burst pressure and diameter with no heat set treatment for reference purposes.

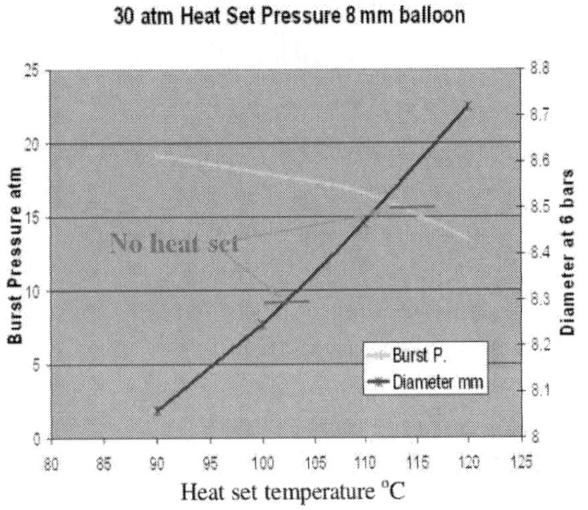

Figure 11: Variation of the diameter and burst pressure with the heat set temperature

Effects of the heat set (annealing) on the balloon:

- As the heat set temperature increases, the crystallinity and mechanical resistance go down.
- No heat set treatment results in a low burst pressure and a high diameter.
- Maximum burst pressure is achieved at low heat set temperatures.
- Diameters less than the mould diameter are achievable because of shrinkage from the mould dimensions.

Balloon Design method:

The balloon design technique is presented in figure 12.

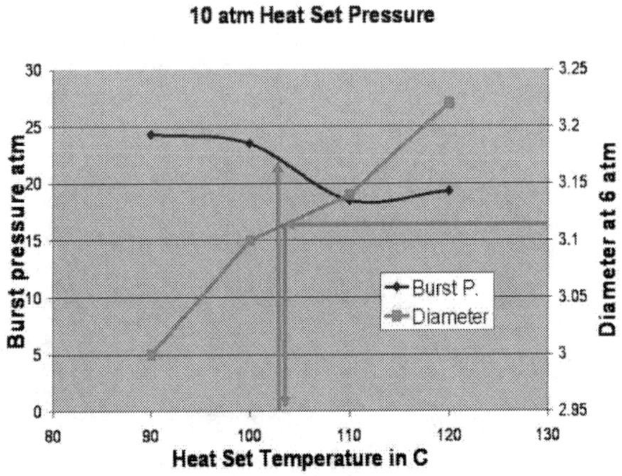

Figure 12: Balloon design method for 3 mm balloon

1. Choose required diameter at 6 bars.
2. Read Heat set Temperature
3. Read burst pressure

Conclusions

General conclusions for Pebax balloons:

- Balloon manufacture can be fine tuned to meet the existing pressure vs. diameter and compliance specifications.
- A lower heat setting temperature results in a higher burst pressure with excellent compliance at the specified pressure.
- Loss of surface definition and defects like crow feet can be avoided by using the right mould temperature.
- A low heat set temperature will also prevent recoil during sterilization.
- For optimum cone aesthetics the cone temperature during the secondary stretch must be higher than in the cylindrical section.
- The mold setup requires three separate zones; the 2 cones and the middle portion. This allows for the defects in the cones to be tuned out independently of the middle section by manipulating the temperature of the cones without affecting the mechanical properties of the cylindrical section.

Specific conclusions for 8 mm Pebax balloons:

- A heat set temperature of 90°C is optimum. Lower than that will produce crow's feet and higher than that will reduce the burst pressure

- The ideal blowing pressure is 30 Atm. Lower will not blow properly and higher will burst the balloon during forming

- A heat set time of 60 seconds is optimum. A longer time will produce bigger crow's feet without increasing the mechanical properties

Future Developments:

- Extension of the method to complex balloon shapes such as heart valve delivery systems.
- Development of on-line rheometer to adjust to batch to batch variations in material.
- Optimization through simulation of the process with the Pressure volume method (Mir et al. 2007).
- Integration of on-line rheometer and simulation for on-line control of balloon forming.

References

Tradename: PEBAX: thermoplastic elastomer, namely Polyether Block Amide (PEBA) supplied by "Arkema Inc." under the trade name of Pebax® 7233

Manufacturer's Pamphlet: Despotopoulou, O'Brien, Arkema (manufacturer), "The superior dynamic properties of Pebax resins".

Proceedings: H. Mir, Z. Benrabah and F. Thibault, 2007. "The Use of Elasto-Plastic Material Model Coupled with Pressure-Volume Thermodynamic Relationship to Simulate the Stretch Blow Molding of Polyethylene Terephtalate", Numiform, 2007, Porto, Portugal.

MECHANICAL PROPERTIES OF KNITTED-FABRIC REINFORCED ELASTOMERIC COMPOSITES

B. Bekisli and H. F. Nied*

Department of Mechanical Engineering and Mechanics, Lehigh University, Bethlehem, PA, USA – hfn2@lehigh.edu

A hierarchical finite element procedure is presented to analyze the large deformation behavior of the knitted fabric reinforcements for use in elastomer-based flexible composites. Manufacture and experimental analysis of these novel composites are also described. It is shown that by fully utilizing the high stretchability of the knitted fabric reinforcements, attractive material properties can be obtained especially for energy/impact absorption and forming applications. The improvement of thermoforming process stability with the use of carefully tailored knitted fabric reinforcements is also presented.

Introduction

Knitted-fabric reinforcements from high performance fibers, have not generally been preferred in composites over their traditional counterparts such as woven reinforcements, mainly due to their lower stiffness/strength performance when embedded in a rigid, thermosetting matrix material (Ruan and Chou, 1996; Khondker, 2001; Gommers, 1996). However, with their unique formability, these fabrics promise great potential in applications where large deformation of the structure is desirable; such as energy/impact absorption and forming applications (Leong, 2000; Hamada, 1999; Ramakrishna, 1997). In order to fully utilize the exceptional stretch properties of the knitted-fabric, the matrix material should be able to deform at least as much as the fabric, and the knitted yarn movements should be restricted by the matrix as little as possible. For this purpose, elastomeric materials such as polyurea and thermoplastic elastomer (TPE) were used to fabricate composites with knitted-fabric. Two different types of arrangements were experimentally studied: knitted fabric embedded in the elastomer and a sandwich of knitted fabric between elastomeric skins. A multi-level finite element procedure was also developed to analyze the deformation characteristics of a plain weft knit and its composites. A database of mechanical properties for various knit geometries was obtained. Using these results, it is shown that a carefully "tailored" knit fabric reinforcement can be used to improve mechanical performance and facilitate polymer forming processes, such as thermoforming.

Deformation Characteristics of Plain Weft Knitted Fabrics

The plain weft knit-fabric architecture, a representative unit cell, and the wale-course coordinate system are shown in Figure 1. Due to the relative simplicity and the availability of mathematical descriptions for this geometry (Ramakrishna, 1997; Hamada, 1999; Ruan and Chou, 1996), it is the most commonly analyzed knit structure from many alternatives. The geometry of the fabric unit cell can be described using three parameters (Ramakrishna, 1997; Hamada, 1999); wale count, W, course count, C, and yarn diameter, d. Wale and course counts (loops/cm) correspond to the reciprocals of the unit cell lengths in the course and wale directions, respectively and as shown in Figure 1. A typical uniaxial load-deformation curve obtained from a test of a knitted fabric in the walewise direction is presented in Figure 2. The fabric is initially in a relaxed state and a large amount of stretch, i.e. elongation per original length, can be obtained by sliding of yarns over each other. After a certain amount of yarn movement, fibers start to bend in order to orient themselves in the load direction. When yarns become as straight as possible, the stiffness of the fabric increases dramatically due to the stretching of individual fibers. The magnitude of the stretch where the mechanical transition of the fabric from a relaxed state to a high stiffness state occurs, will be called the "critical stretch". One way to define this stretch is by obtaining the intersections of two linear fits to the two distinct phases of the curve as shown in Figure 2. For example, for the case shown in Figure 2, the critical stretch of the fabric is estimated to be 55%. Although the curve will look similar for various knit geometries and loading directions, the critical stretch can be very different. Therefore, the determination of critical stretches in relation to geometric parameters, i.e., W, C

and d, and loading conditions is the key for the analysis of these fabrics.

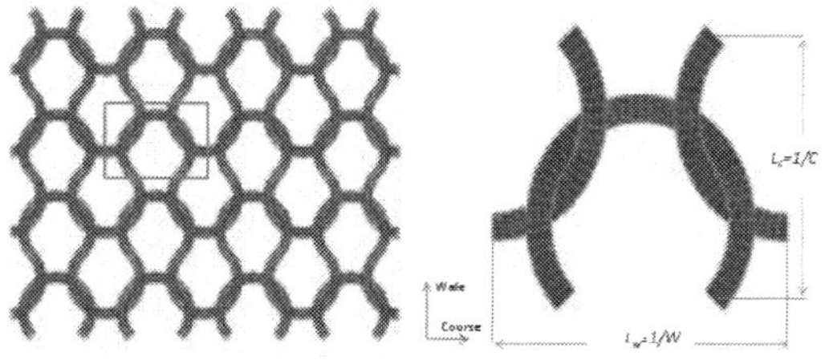

Figure 1. (Left) A Plain weft-knit architecture, (Right) A representative unit cell

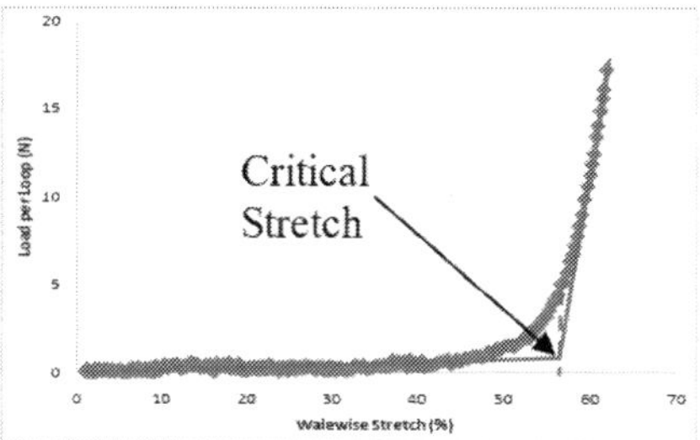

Figure 2. A typical load-deformation curve from a uniaxial test of a plain weft-knitted fabric

Analysis of Plain Weft Knitted-Fabrics

A two-level finite element analysis was performed to analyze the uniaxial and biaxial stretch behavior of plain weft knitted fabrics. The procedure consisted of a 3D unit cell analysis (micro/meso scale) and a 2D global analysis (macro scale).

Micro/Meso Scale

A unit cell analysis is an extremely efficient approach for detailed investigation of periodic patterns under uniform deformation. Using periodic boundary conditions, a nonlinear finite element solution was obtained for four different loading cases; namely uniaxial and pure shear loading in both walewise and coursewise directions (Figure 3). As the geometrical model of the knitted yarns, the mathematical description by Ramakrishna (1997) was used. A single fiber, modeled using beam elements and linear elastic material properties, was placed in the core of the 3D yarn modeled as a soft filler material. This is necessary to determine the accurate contact conditions with other yarns (Figure 4). Contact elements were meshed over the cylindrical yarn surfaces. The meshing and solution were performed in ANSYS 11.0. Critical stretches were determined for various knit parameters and plotted in Figure 5.

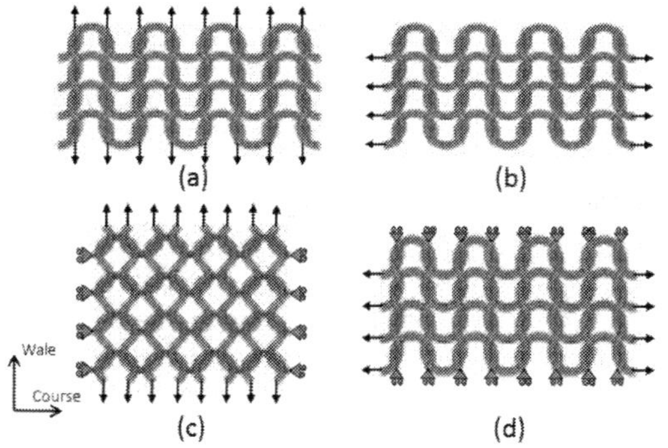

Figure 3. Uniaxial loading in (a) wale and (b) course directions. Pure shear loading in (c) wale and (d) course directions

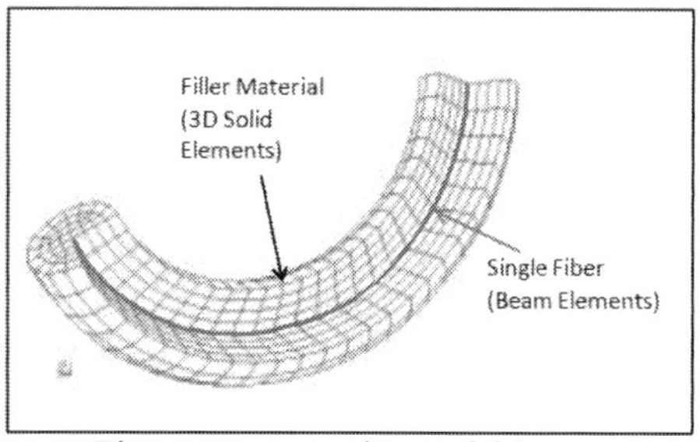

Figure 4. Element types used to model the yarn.

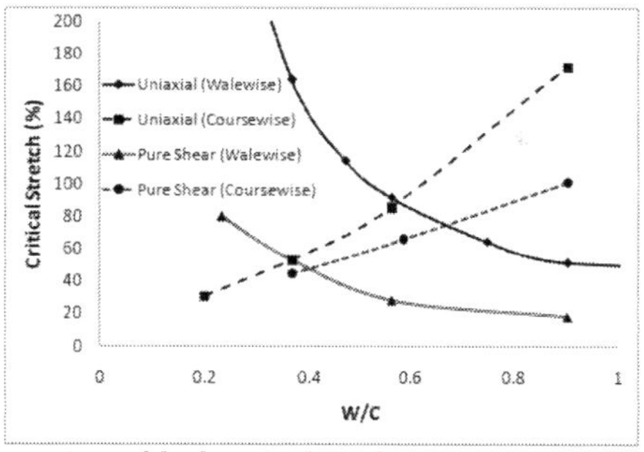

Figure 5. Critical stretch values for varying knit geometries and loading conditions *($C{\cdot}d=0.135$)*

Experimental characterization of raw knitted fabrics was also carried out using glassfiber yarns with a diameter of 0.660 mm. Uniaxial tensile test specimens with a length of 10 cm were knitted with a manual knitting machine (Silver Reed SK840) and tested on an INSTRON tensile testing equipment. Tests were performed for ten specimens of three different knit densities. Average critical stretches obtained from each knit pattern ($C{\cdot}d = 0.23\text{-}0.28$) are plotted in Figure 6 in comparison with the numerical results for the uniaxial walewise loading case. While a good overall numerical prediction is present, the experiments consistently yield smaller critical stretches than numerical predictions. This is mainly due to the presence of friction between yarns, which is neglected in the finite element model. Although an accurate measurement of the inter-yarn friction coefficient is beyond the scope of this study, numerical analysis confirms that an increased friction coefficient slightly reduces the critical stretch.

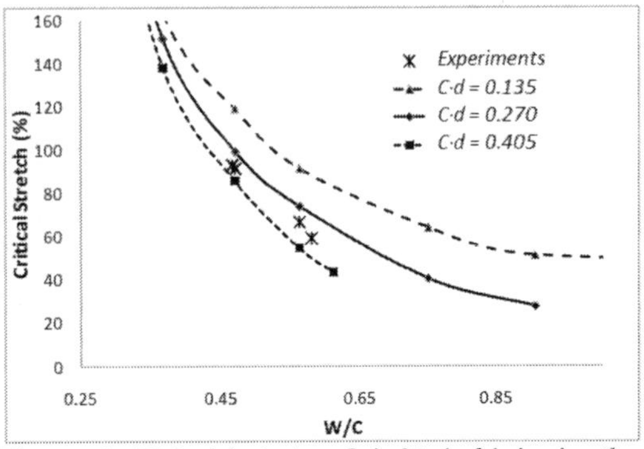

Figure 6. Uniaxial tests of knitted fabric in the walewise direction compared with numerical results

Macro Scale

A typical knitted fabric for structural applications is composed of thousands of loops, which adds significant complexity to the numerical analysis in terms of calculation times and effort. A convenient approach is to benefit from the resemblance of the hexagonal honeycomb structure exhibited by the knitted fabric (Wu, 1994; Araujo, 2004) as illustrated in Figure 7. Nonlinear truss elements, whose mechanical properties can be determined from the micro/meso scale analysis, can be used to construct hexagons. In this manner, a large number of loops can be efficiently represented by hexagons.

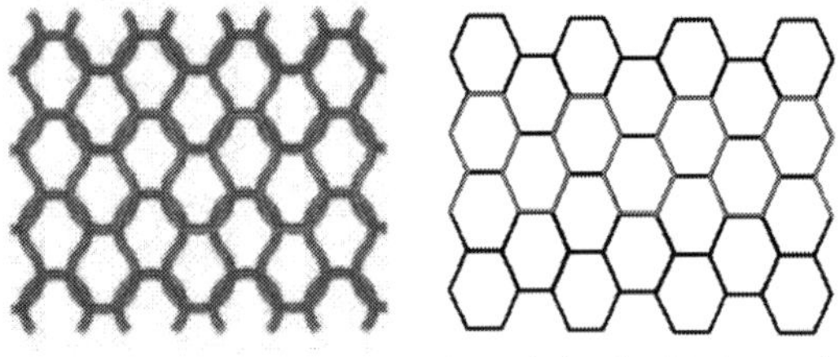
Figure 7. Resemblance of a knitted fabric and hexagonal honeycomb structure.

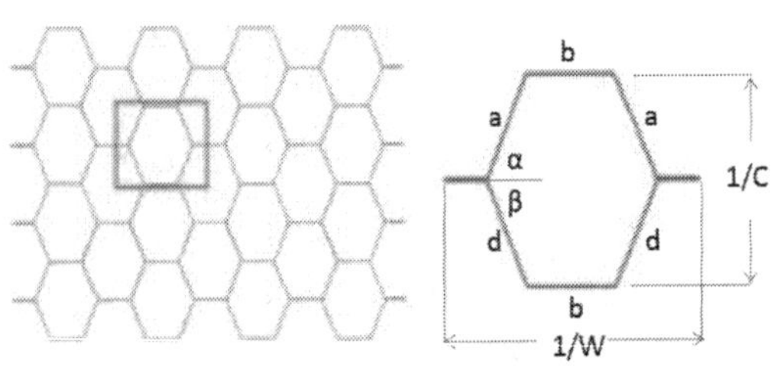

Figure 8. A unit cell in the hexagonal representation

Three different nonlinear springs of initial lengths *a*, *b*, and *d* form the hexagon as shown in Figure 8. Two geometric relations are initially required to model the knitted fabric with wale (*W*) and count numbers (*C*):

$$\frac{1}{W} = 2b + 2a\cos\alpha \quad (1)$$

$$\frac{1}{C} = 2(a\sin\alpha + d\sin\beta) \quad (2)$$

where the relation $a(\cos\alpha)=d(\cos\beta)$ must also hold. When a bilinear load-deformation behaviour of the sort shown in Figure 2 is assumed for each spring, critical stretch variables δa, δb and δd can be introduced as the extensions at which the spring switches to the high stiffness region. Unknown variables a,b,d, α, δa, δb and δd can be found by the use of Equations (1-2) and relations obtained from the micro/meso-scale analysis. For example, when the uniaxial walewise loading case is considered in the hexagonal structure, the critical stretch of the fabric corresponds to a locking behaviour where the springs *a* and *d* are aligned in the load direction and are extended to their specific critical stretches δa and δd. Critical stretch (*CS*) of the unit cell, can then be incorporated into the hexagonal system through a simple relation:

$$CS = \frac{(a + \delta a + d + \delta d)}{a\sin\alpha + d\sin\beta} - 1 \quad (3)$$

Other loading cases similarly create additional relationships between the unknowns. When all four cases designated in Figure 3 are included, a total of 6 equations are available for the 7 unknowns. Using a least squares error minimization procedure, it is possible to find the best fit parameters. These parameters are assigned to the spring elements of the hexagonal model if the error from the best fit procedure is acceptable.

In addition to the critical stretches, the spring constants of the two linear stages, i.e, before and after the critical stretch, are found by investigating the initial and final slopes of the load-deformation curves in the micro/meso-scale analysis. When the spring load-deformation behaviour is completely known for all three types of springs, global models with large numbers of hexagons can be constructed and the resulting nonlinear finite element models can be solved for suitable loading cases. For example, a rectangular steel block was placed on a glassfiber knitted fabric as shown in Figure 9, and a simulation with a controlled displacement loading was performed using the hexagonal representation for the knitted fabric in the global scale. The maximum depth the fabric displaced underneath the steel block was measured to be 29.4 mm in the experiment and 27.8 mm in the finite element model. The slight discrepancy between these two values can be attributed to the manual knitting process, which generally can not produce perfectly uniform knit architectures. In addition, inaccuracies in the micro/meso scale model, e.g. neglected friction between yarns, also contributes to the deviation.

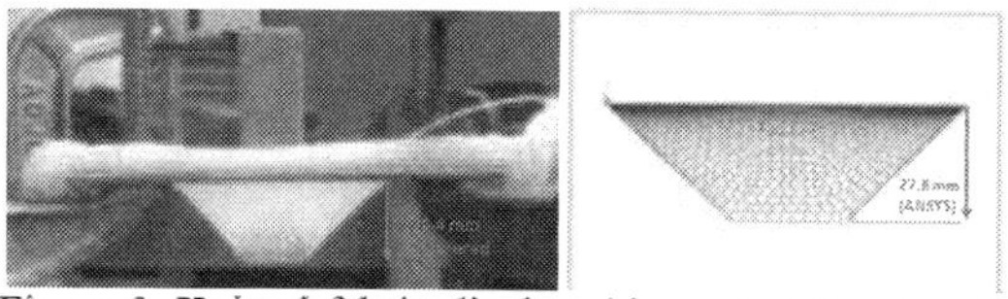

Figure 9. Knitted fabric displaced by a steel block and corresponding finite element simulation.

Experimental Analysis of Flexible Composites

Knitted-fabric reinforced flexible composites were fabricated using two elastomers; polyurea (Airthane PHP-80D combined with Versalink P1000 from Air Products and Chemicals, Inc.) and TPE (PEBAX 2533 from Arkema, Inc.) with glassfiber yarns, in two structural arrangements. First, knitted reinforcements were embedded in the elastomers. Compression molding was used for the TPE composite while polyurea composite was produced by immersing the fabric in a mold filled with polyurea resin and letting the resin cure. Pure elastomers without the fabric were also produced by similar methods. Composite and pure elastomeric sheets of thicknesses between 0.1-0.5 cm were water-jet cut to test specimen dimensions of 18 cm x 1.2 cm.

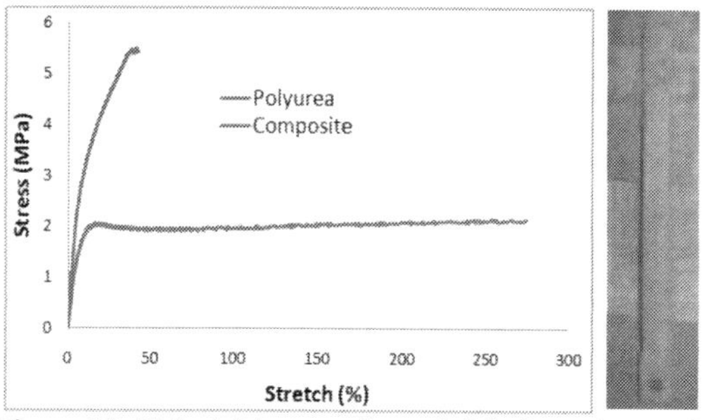

Figure 10. Uniaxial test results and failed specimen of a knitted-fabric embedded polyurea composite in comparison with pure polyurea.

Test results from polyurea and its composite are shown in Figure 10. Although the composite is considerably stiffer than the elastomer, the failure occured at a surprisingly early stretch (at around 45%), and the unique stiffening behavior of the knitted fabric, as seen in Figure 2, was not observed. The curve implied that the unique deformation behavior of the fabric reinforcement was constrained by the surrounding polyurea, and the increase in stiffness was essentially a rule-of-mixtures type of contribution from the reinforcement.

A quite different behavior was obtained from tests of the TPE composites as shown in Figure 11. Although the stiffening behavior expected solely from a knitted fabric reinforcement was still not completely seen, the composite's flexibility was greatly increased. This was mainly due to the fact that TPE matrix did not fail but could carry significant load even when the fabric failed in a localized region (Figure 11). This, in turn, enabled further stretching of the specimen and also of the remaining portions of the knitted fabric, triggering new fabric failures. The sudden load drops in the curve correspond to localized failure of knitted fabric. The different mechanical behavior of TPE composite versus polyurea composite may be explained by the different nature of the bonding between the polymer and the yarn. It was observed from the failed specimens that polyurea resin infused inside the yarns to a much higher degree than TPE infusion during compression molding. Therefore, constraints on the yarn movement were lower in the TPE composites.

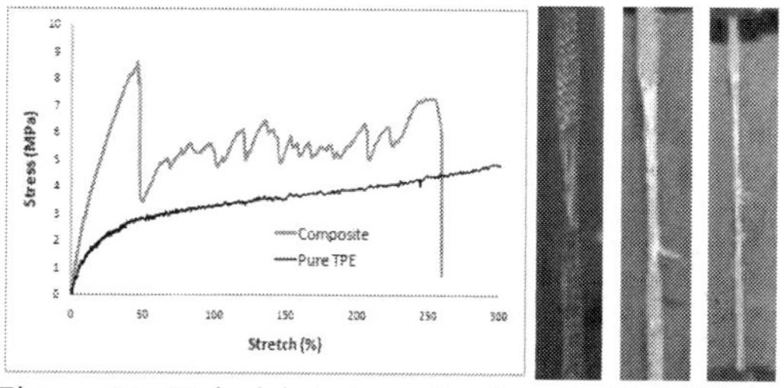

Figure 11. Uniaxial test results from knitted fabric embedded TPE composite

The second type of flexible composite that was analyzed was of a sandwich type. This type of composite can be made using two elastomeric skins and knitted fabric reinforcements in the core (Figure 12). The advantage of this construction is the relative freedom of the yarn movement, resulting in a mechanical behavior similar to that in Figure 2, i.e. behavior expected from the unrestrained fabric. A simple tension test was performed using separate specimens of TPE and raw knitted fabric to analyze the behavior of these composites. For instance, two knitted fabric specimens with expected critical stretches of 86% ($W/C=0.53$, $C{\cdot}d=0.25$) and 58% ($W/C=0.62$, $C{\cdot}d=0.30$) were prepared. These fabrics were clamped to the test grips together with a TPE specimen as shown in Figure 13 (a). Uniaxial test results of this configuration are plotted in Figure 13 (b) as Composite 1 and compared with a pure TPE specimen. As expected, the stiffening behavior of the fabrics appeared in the curve near their corresponding critical stretches. In a second experiment (Composite 2), two knitted fabrics with slightly different critical stretches were combined with the TPE. The first fabric with a critical stretch of 65% ($W/C= 0.60$, $C{\cdot}d=0.30$) and the second fabric with a critical stretch of 72% ($W/C= 0.57$, $C{\cdot}d=0.28$) yielded a test result shown in Figure 13 (b). When compared to the pure TPE test results, improved strain energy dissipation capacity of the composite is very clear. Moreover, the curve can easily be controlled and manipulated in order to obtain the desired energy absorption capacity through a careful selection of knitted fabrics using the database constructed in the micro/meso scale analysis and a graded knitted fabric arrangement depicted in Figure 13 (c).

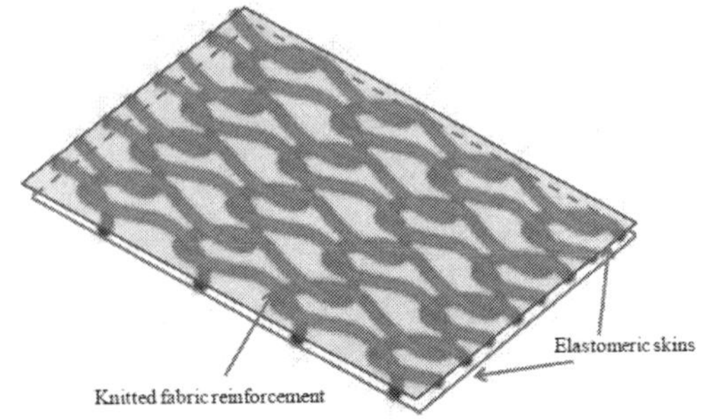

Figure 12. Illustration of a sandwich-type flexible composite

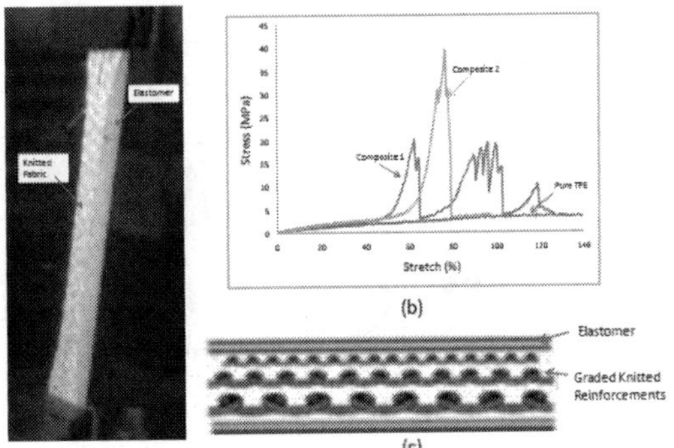

Figure 13. (a) Test including separate specimens of knitted fabric and an elastomer, (b) Test results, (c) A sandwich arrangement of graded knitted fabrics between elastomeric skins

Finite Element Analysis of Flexible Composites

The finite element procedure developed for the knitted fabric can be extended to model the mechanical behavior of the sandwich-type flexible composites. Shell elements were used to model the elastomeric skin layers, while the previously described hexagonal global structure was used to represent the knitted fabric. Contact pairs were required between the truss elements of the hexagons and the shell elements to prevent penetration of two materials. Curve fitting to tensile test data was used to determine appropriate hyperelastic material models for the elastomers. A finite element solution including nonlinearities due to material, contact, and large deformation was obtained for various loading conditions. A sample problem is the inflation of a rectangular composite sheet (uniform surface pressure). Stages of the calculated deformation are shown in Figure 14 (a) and the deformed configurations of shell and hexagonal elements are seen in Figure 14 (b). The out-of-plane displacement of the center point (δ) normalized by the side length of the square L, is plotted in Figure 15 against the applied pressure (P), normalized by first order Ogden constants (α=0.10, μ=2.264 MPa) used to model the polymer. The center point displacement is also shown for the inflation where only the polymer is present. Stiffening behavior of the composite due to the knit fabric reinforcement is clear, although the polymer experiences a softening especially after a displacement δ/L=0.4.

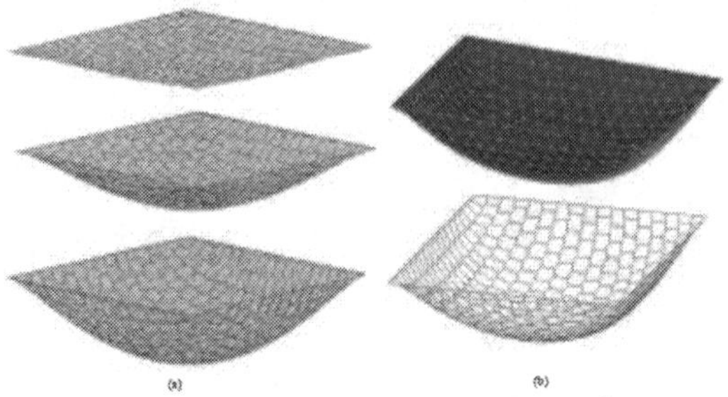

Figure 14. (a) Stages of inflation of an elastomer-knitted fabric sandwich, (b) Deformed shapes shown separately; the hyperelastic shell and hexagonal elements.

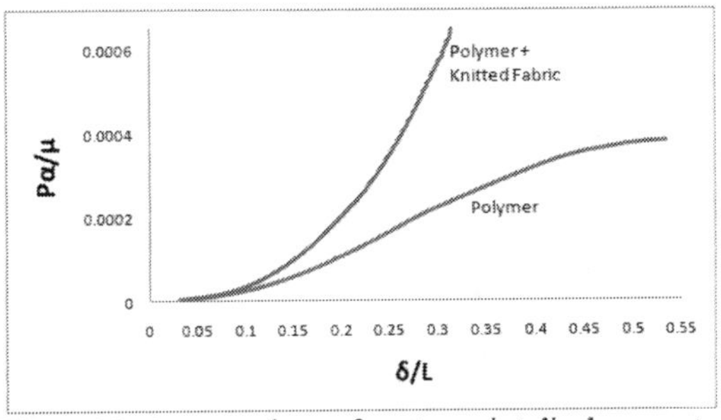

Figure 15. A comparison of center point displacements from inflation of an initially square flexible composite (elastomeric polymer + knitted fabric) and a pure elastomer.

Control of Thermoforming Process

Knitted-fabric reinforcements can also be used to reduce or completely eliminate some of the major problems in thermoforming processes; such as non-uniform thickness distribution on the final part and sensitivity to thermal variations. The thickness variations arise from the fact that different regions of the polymer experiences different amounts of stretch during the forming. For instance, Figure 16 (a) illustrates the thickness distribution of a thermoformed U-shaped long channel. Utilizing symmetry and assuming plane strain along the channel length, only half of a thin strip was modeled. The finite element methodology explained by deLorenzi and Nied (1991) was used and initially homogenous material properties, i.e. isothermal conditions, were assumed. The thickness profile along an arc length passing through the half-width was also plotted. A large variation is clearly present especially between the center and the thin corner which experienced the most stretch until contacting the mold. A knitted fabric can be used to support the polymer and to control the amount of stretch experienced by different regions of the polymer. For instance, especially thin regions can be reinforced with knitted patterns of low critical stretch values, while other regions are reinforced with more relaxed fabric patterns. An example knit pattern can be the one shown in Figure 17. A pattern with a critical stretch of 44% in the coursewise pure shear (*W/C=0.37, C·d=0.135*) is placed under the regions corresponding to the thin corners. The rest of the fabric is formed by a knit pattern with a 70% critical stretch (*W/C=0.63, C·d=0.135*). When the thermoforming process including the tailored fabric was simulated, the improved thickness distribution shown in Figure 18 (a) was obtained. In Figure 18 (b), it can be seen that by distributing the stretch experienced by the corners to a larger region, a more uniform stretch, and thus a more uniform thickness distribution was obtained on the formed strip.

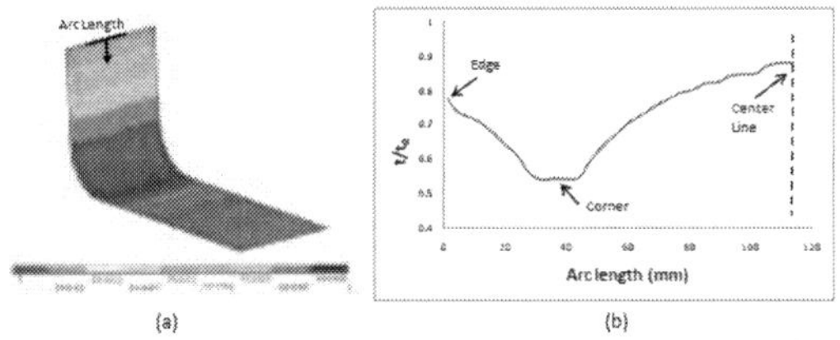

Figure 16. (a) Thickness distribution on a strip of a long thermoformed channel-like part, (b) Thickness profile along the width of the part.

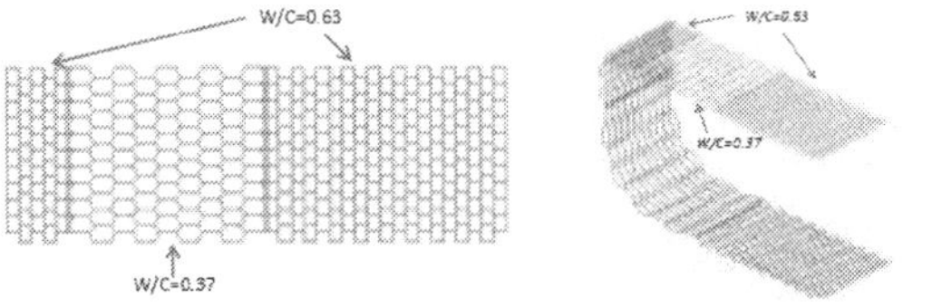

Figure 17. Three-pattern knit fabric used in the sample problem, (b) Deformed configuration of the fabric after forming.

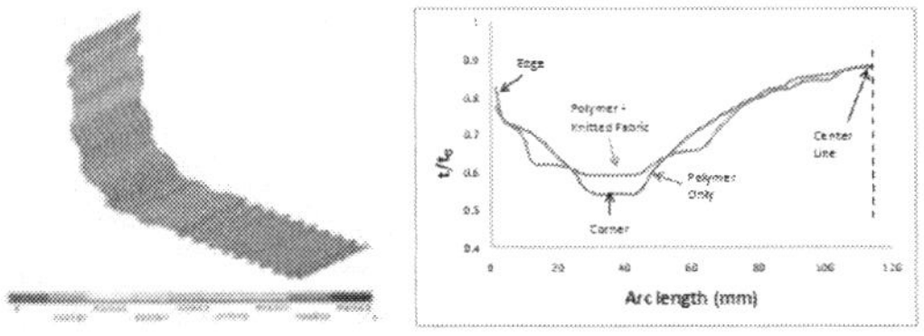

Figure 18. (a) Thickness distribution after thermoforming of knitted-fabric reinforced thermoplastic, (b) Improved thickness profile in comparison to the polymer-only case.

The quality and repeatability of certain types of deep draw thermoforming processes are known to be very sensitive to thermal processing conditions. This is due to dramatic changes in the polymer's mechanical properties near the glass transition temperature with small changes in temperature. The use of knitted fabrics in the thermoforming process would also provide better stability during forming, since the deformation behavior of the fabric does not depend on temperature and since the stiffness of a stretched fabric will dominate the rubbery behavior of the polymer. In other words, the forming process would be dominantly characterized by the thermally insensitive mechanical properties of the fabric.

Conclusions

The deformation behavior of knitted fabric reinforcements and elastomer-based flexible composites with knitted reinforcements were examined, both numerically and experimentally. A multi-level finite element model was shown to be effective for modeling knitted structures with a very large number of loops. Experiments on knitted fabrics showed good agreement with the numerical results.

Two different flexible composite arrangements were fabricated and tested. It was shown that in order to fully utilize the unique mechanical properties of knitted fabrics in flexible composites, arrangements similar to the sandwich-type structure should be used. This is due to the minimal restriction on yarn movements by the surrounding elastomer. Finite element analysis was extended to cover this type of composite structure.

Using the developed numerical procedure, it was also shown that some of the major challenges in thermoforming processes can also be addressed. A simple example confirmed the improvement of thickness distribution on a thermoformed long channel.

Acknowledgements

The support of the US National Science Foundation under grant CMMI-0856611 is gratefully acknowledged. The authors would like to also thank Air Products and Chemicals, Inc. and Arkema, Inc. For providing material samples.

References

Araujo M.D., et al., "Modelling and Simulation of the Mechanical Behaviour of Weft-Knitted Fabrics for Technical Applications- Part III", AUTEX research Journal, **4**, 25-32, (2004).

deLorenzi H.G., Nied H.F., "Finite Element Simulation of Thermoforming and Blow Molding", in Modeling of Polymer Processing, Isayev A.I. (Ed.) Hanser Publ., Munich, p.118-171, (1991).

Gommers B., et al.,"Modeling the Elastic Properties of Knitted-Fabric-Reinforced Composites", Composites Science and Technology, **56**, 685-694, (1996).

Hamada H., et al., "Knitted Fabric Composites", Hamada, Ramakrishna, Huang, in 3D Textile Reinforcements in Composite Materials, Miravete A. (Ed.) Woodhead Publishing, Cambridge, p.180-216, (1999).

Khondker O.A., et al., "An Investigation of the Structure-Property Relationship of Knitted Composites", Journal of Composite Materials, **35**, 489-508, (2001).

Leong K.H., et al., "The Potential of Knitting for Engineering Composites- A Review", Composites: Part A, **31**, 197-220, (2000).

Ramakrishna S., "Characterization and Modeling of the Tensile Properties of Plain Weft-Knit Fabric Reinforced Composites", Composites Science and Technology, **57**, 1-22, (1997).

Ruan X.P., Chou T.W., "Experimental and Theoretical Studies of the Elastic Behavior of Knitted-Fabric Composites", Composites Science and Technology, **56**, 1391-1403, (1996).

Wu W.L., et al., "Computer Simulation of the Deformation of Weft-Knitted Fabrics for Composite Materials", Journal of Textile Institute, **85**(2), 198-214, (1994).

Modeling Fiber Length Attrition in Molded Long-Fiber Composites

Tucker Charles L., Phelps Jay H., abd El Rahman Ahmed I., Kunc Vlastimil, Frame Barbara

Long-fiber thermoplastic (LFT) composites consist of an engineering thermoplastic matrix with glass or carbon reinforcing fibers that are initially 12 to 25 mm long. When an LFT is injection molded, flow during mold filling orients the fibers and degrades the fiber length. Fiber orientation models for injection molding are well developed, and special orientation models for LFTs have been developed. Here we present what we believe is the first quantitative model for fiber length attrition in a flowing fiber suspension. The model tracks a discrete fiber length distribution (FLD) at each spatial node. Key equations are a conservation equation for total fiber length, and a breakage rate equation. The breakage rate is based on buckling of fibers due to hydrodynamic forces, when the fibers are in unfavorable orientations. The FLD model is combined with a mold filling simulation to predict spatial and temporal variations in fiber length distribution in the mold cavity during filling. The predictions compare well to experiments on a glass/PP LFT composite. Fiber length distributions predicted by the model are easily incorporated into micromechanics models to predict the stress-strain behavior of molded LFT materials.

UNIAXIAL DEFORMATION BEHAVIOR AND MECHANICAL PROPERTIES OF POLY(BUTYLENE SUCCINATE)/ORGANO-MONTMORILLONITE NANOCOMPOSITES

Z.A Mohd Ishak[1*], W.S. Chow[1], Y.J. Phua[1]

[1] *School of Materials and Mineral Resources Engineering, Engineering Campus, Universiti Sains Malaysia, Nibong Tebal, 14300 Penang, Malaysia – zarifin.ishak@googlemail.com; chowwenshyang@yahoo.com; yijing84@yahoo.co.uk*

Mechanical properties of polymers can be greatly modified by molecular orientation of polymer chains. This studies addresses the improvement of the properties of poly(butylene succinate) (PBS) and 2% organo-montmorillonite (OMMT) filled PBS nanocomposites by altering its molecular chain orientation via uniaxial cold rolling. PBS nanocomposite was prepared in an internal mixer and compression molded into sheet form before subjected to the cold rolling process. The cold rolling process was carried out at various thickness reductions (20-60%). Distinct improvement in mechanical properties, in terms of strength and elongation at break were observed for rolled samples when tested in machine direction (MD). This great improvement was attributed to the better molecular orientation of polymer chains after rolling which can be confirmed by wide angle X-ray diffraction (WAXD), typically at high thickness reduction. As observed from transmission electron microscope (TEM), degree of clay alignment increased after rolling. This may be one of the reason that causing the increment in mechanical properties for PBS nanocomposites. However, mechanical properties degraded along transverse direction (TD). WAXD and differential scanning calorimetry (DSC) reported on the decrement in crystallinity after cold rolling. This indicates that any enhancement in mechanical properties may not be the result of crystallinity increment. Besides that, No notable changes in thermal properties after cold rolling as shown in DSC thermograms.
***Keywords**: Poly(butylene succinate), Nanocomposites, Molecular Orientation, Cold Rolling*

Introduction

Poly(butylenes succinate) (PBS) is a biodegradable semi-crystalline polymer which is classified under the polyester family. Fillers such as organo-montmorillonite (OMMT) are often incorporated into PBS in order to make it useful in a wide range of applications (Okamoto et al., 2003; Okada et al., 2003). However, in order to further enhance the mechanical properties of PBS/OMMT nanocomposites, numerous post-compounding techniques such as cold drawing, stretching and cold rolling can be applied. The structure and properties of the materials can be modified after the post-compounding process (Nielsen & Robert, 1994).

Orientation structure is an important factor in determining the properties of polymeric materials. The strength and ductility of polymers can be greatly modified by molecular orientation of the polymer chains (Lee et al., 1996; Wu & Lee, 2002). In this study, uniaxial cold rolling has been carried out on PBS and its nanocomposites at different thickness reduction (20-60%) in order to introduce molecular orientation to the polymer chains. The relationship between structure and properties of PBS and its nanocomposite is investigated through various characterizations such as tensile properties, wide angle X-ray diffraction (WAXD), birefringence, differential scanning calorimetry (DSC) and transmission electron microscope (TEM).

Experimental

Materials

Poly(butylene succinate) (PBS) (Bionolle #1020) with MFI value 25g/10min (190°C,2.16kg), melting temperature 115°C, density 1.26g/cm^3, was obtained from Showa Highpolymer Co., Ltd., Tokyo. Organo-montmorillonite (OMMT) Nanomer® I.30TC (CEC= 110 mequiv/100g) was supplied from Nanocor.

Sample Preparation

PBS and OMMT were dried in vacuum oven at 60°C for 24 hours. It is followed by a melt mixing process by using internal mixer (Haake PolyDrive R600, Germany) for 5 minutes at 135°C and rotary speed of 50rpm. PBS nanocomposite was prepared by incorporated 2wt% of OMMT into PBS matrix. The compounds were then molded into 1mm sheets at 135°C for 3 minutes on a compression molding machine (GoTech GT7014-A30C, Taiwan). The molded 1mm sheets of PBS and nanocomposites were rolled in one direction by using two-roll mill machine XK-160 (Shanghai rubber-machine work) without heating the rollers. The sheets with 1mm original thickness were rolled to various reductions in thickness (i.e. 20%, 40% and 60%).

Characterizations

Tensile test was performed at ambient temperature through Instron 3366 machine according to ASTM D638-03 (Type IV). Gauge length of 50mm and cross-head speed of 5mm/min were used. Tensile test was carried out at both machine direction (MD) and transverse direction (TD) of rolling.

The molecular orientation of rolled samples was studied by using CuKα radiation in RINT2100FSL instrument (Rigakudenki Ltd). WAXD patterns and diffractograms were recorded.

Birefringence is generated via Microwave Molecular Orientation Analyzer (MOA-6015, Oji Scientific Instrument, Japan). The sheet is rotated at a speed of 6.0s per turn and the transmitted wave intensity is measured at every 1° of rotation angles. Birefringence (Δn) was determined as the difference between the refractive indices in the MD and TD.

DSC analysis was carried out using Perkin-Elmer DSC-6 in nitrogen atmosphere from 30°C to 150°C at a heating rate of 10°C/min.

The morphology of nanocomposite sample was observed by Philips TEM microscope operated at 200kV. The ultrathin sections (80-90nm) of sample were prepared via ultramicrotomy technique using Reichert Ultramicrotome Supernova.

Results and Discussion

Tensile Properties

Tensile characteristics of PBS and PBS/2%OMMT before and after uniaxial cold rolling are investigated in both MD and TD, which presented in Table 1. A great enhancement in the tensile strength and elongation at break are observed at the MD. This is attributed to the increased molecular orientation and enhanced clay exfoliation after rolling (Abu-Zurayk et al., 2009). Degree of clay alignment will become more pronounced at higher thickness reduction, which is verifiable by TEM. Besides that, elongation at break is increased significantly when tested at MD, mainly due to the brittle-to-ductile transition in the sample after rolling. Better ductility is observed at MD, yet leads to a higher elongation (Nielsen & Robert, 1994). The enhancement in strength and elongation at break become more pronounce at a higher thickness reduction (60%). Brittle-to-ductile transition is a reason of the slight decrement in modulus after rolling.

Theoretically, anisotropy weakens a polymer sample perpendicular to the direction of orientation (TD). Cracks or other imperfections in the samples tend to orient parallel to the orientation direction. They could act as strong stress concentrators, typically perpendicular to the orientation direction (Abu-Zurayk et al., 2009). Thus, strength reduces when tested in the TD. During rolling process, the previously oriented molecular chains in TD will first deorient and then reorient in the direction of the applied force. The deorientation and reorientation process bring about plastic deformation and leads to the unusual high elongation at break at TD. Modulus increased with the thickness reduction after rolling, attributed to the lowering of chain mobility during applying the tensile force at TD and caused the whole material become non-flexible or stiff (Nielsen & Robert, 1994; Miller, 1996).

Table 1 - Tensile properties of PBS and PBS/2%OMMT nanocomposites after cold rolling

	Thickness Reduction (%)	Compound	Tensile Strength (MPa)	Modulus (MPa)	E_b (%)
Unrolled	-	PBS	32.62	589.1	10.9
		PBS/2%OMMT	33.56	631.8	12.9
MD	20	PBS	38.32	526.0	38.3
		PBS/2%OMMT	38.51	577.7	22.3
	40	PBS	57.41	563.4	119.0
		PBS/2%OMMT	58.81	598.9	113.0
	60	PBS	96.73	526.3	123.2
		PBS/2%OMMT	98.12	567.7	115.8
TD	20	PBS	32.30	606.4	27.5
		PBS/2%OMMT	30.84	640.6	20.5
	40	PBS	31.04	688.1	17.8
		PBS/2%OMMT	30.43	747.8	17.0
	60	PBS	28.83	757.3	17.4
		PBS/2%OMMT	28.00	837.3	10.0

Wide Angle X-Ray Diffraction (WAXD)

X-ray images for PBS and PBS/2%OMMT at different extent of rolling are presented in Figure 1 and Figure 2. Before uniaxial cold rolling, the azimuthal intensity distribution for each reflection is uniform, verifying the isotropic nature of PBS. Upon uniaxial cold rolling, a clear difference in the intensities between meridian and equator can be observed, indicating molecular orientation had occurred (Dean et al., 1998). The difference of intensities becomes more significant at high thickness reduction.

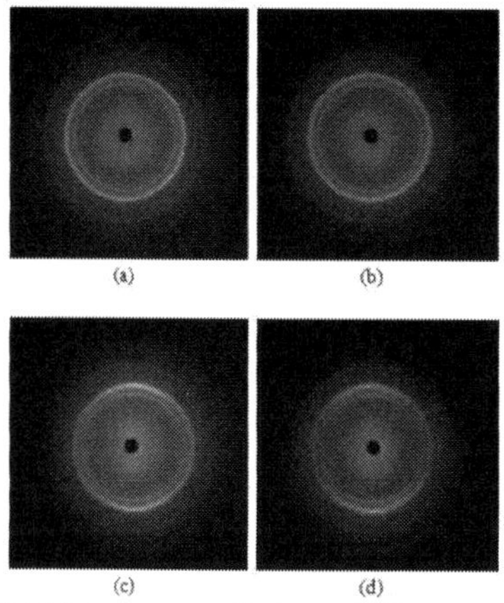

Figure 1 - WAXD patterns of PBS at different thickness reductions: (a) 0%, (b) 20%, (c) 40%, (d) 60%

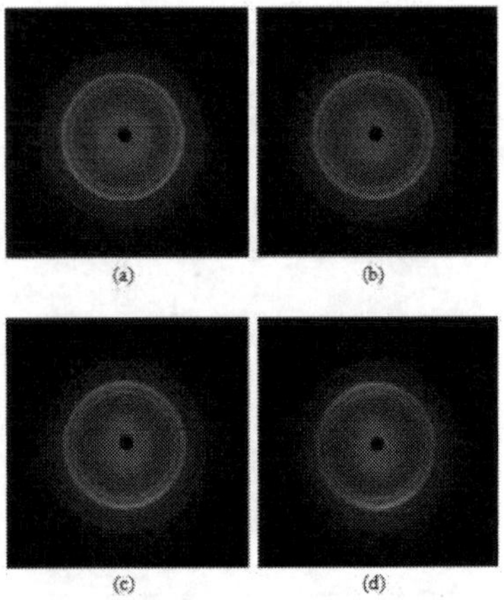

Figure 2 - WAXD patterns of PBS at different thickness reductions: (a) 0%, (b) 20%, (c) 40%, (d) 60%

Furthermore, it could be seen that rolled PBS is showing a better molecular orientation as compared to PBS/2%OMMT, with a more significant changes in the intensities. This result can be correlated with the mechanical properties, where rolled PBS was able to provide greater enhancement in strength.

Birefringence

Birefringence is a measurement of the total molecular orientation of both crystalline and amorphous region in a semicrystalline polymer (Golik et al., 1967). Birefringence of PBS and PBS/2%OMMT as a function of thickness reduction is shown in Figure 3. Δn increases with the thickness reduction, indicates that higher thickness reduction during uniaxial cold rolling is able to induce higher molecular orientation to the polymeric chains. The Δn of PBS is higher than that of PBS/2%OMMT after rolled, which is conforms to the WAXD results.

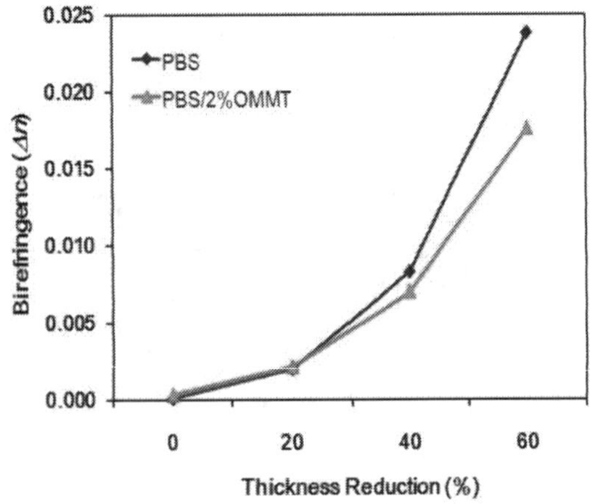

Figure 3 - Birefringence as a function of thickness reduction

Transmission Electron Microscope (TEM)

Figure 4 shows the TEM images of PBS/2%OMMT before and after cold rolling to 60% thickness reduction. From the images, the unrolled nanocomposite shows a random clay orientation. Upon uniaxial rolling, the clay platelets aligned in the direction of rolling.

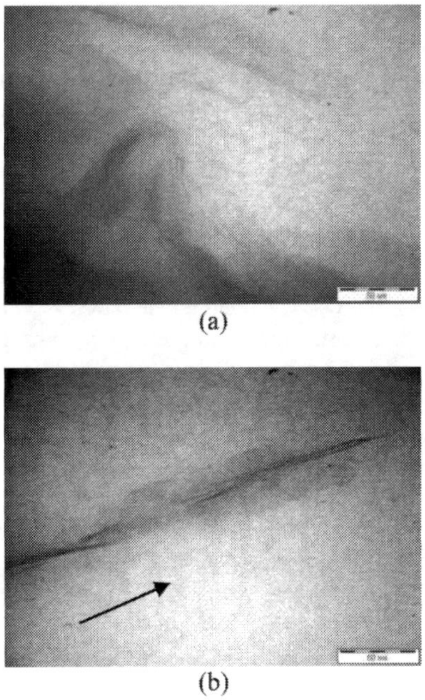

Figure 4 - TEM image of (a) unrolled and (b) rolled PBS/2%OMMT nanocomposites at 60% thickness reduction (arrowed: rolling direction)

Differential Scanning Calorimetry (DSC)

Thermal behavior of PBS and PBS/2%OMMT were studied at both unrolled and rolled (60%) conditions. DSC results are presented in Figure 5 and Table 2. It shows that there are no distinct differences of DSC patterns, melting and crystallization peaks for rolled and unrolled samples. Two melting peaks has been observed in PBS, corresponds to the two different types of crystalline lamella presented in PBS (Sinha Ray, 2005). No significant change has been observed in melting temperatures, $T_{m,x}$ and $T_{m,}$ for PBS after cold rolling process. A slight reduction of 0.81°C in $T_{m,x}$ was observed in PBS/2%OMMT samples after rolling. Since melting temperature is dependent on the crystallite sizes, the decrement may suggests the reduction of lamellae size for PBS/2%OMMT after rolling action.

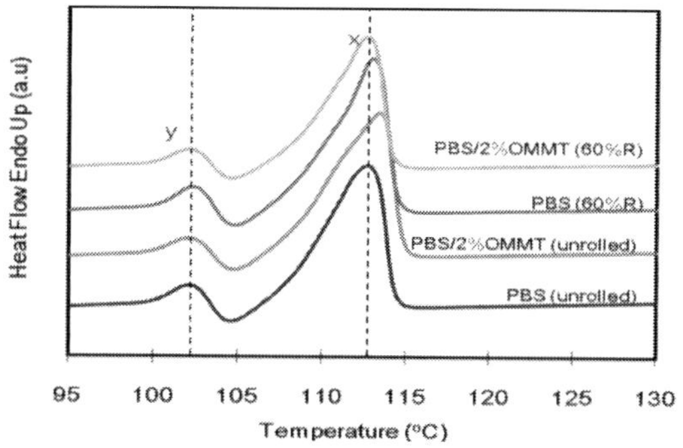

(a)

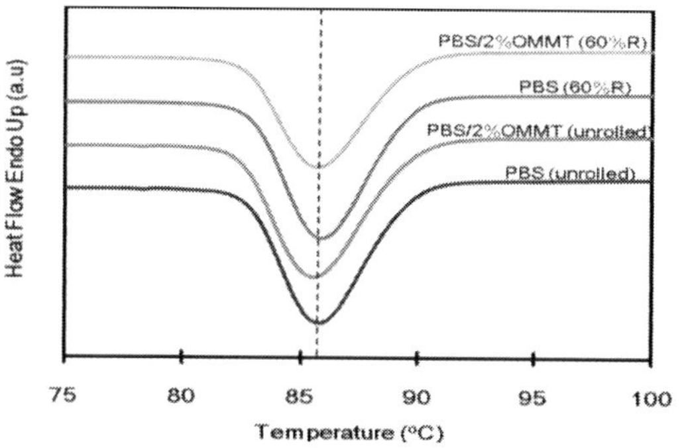

(b)

Figure 5 - DSC thermograms of PBS and PBS/2%OMMT nanocomposite

Table 2 - DSC results of rolled and unrolled PBS and PBS/2%OMMT nanocomposite

Compound	$T_{m,x}$ (°C)	$T_{m,y}$ (°C)	ΔH_m (J/g)	T_c (°C)	χ_c (%)
PBS (unrolled)	112.44	102.22	63.58	85.88	57.64
PBS (60%)	112.92	102.40	59.78	85.91	54.20
PBS/2%OMMT (unrolled)	113.40	102.05	60.25	85.57	55.74
PBS/2%OMMT (60%)	112.59	102.23	50.90	85.74	47.09

The crystallization temperature, T_c remained unchanged. However, the degree of crystallinity, χ_c decreased after cold rolling. This is due to the destruction of crystalline region as the folded chain structures were subjected to rolling to form an extended chain structures, which leads to a thinner lamella and lower crystallinity (Tiaw et al., 2009).

Conclusions

As a conclusion, tensile strength and elongation at break is greatly enhanced after subjected to uniaxial cold rolling, which is attributed to the better molecular orientation of polymer chains. Brittle-to-ductile transition was observed on the rolled sample at MD, explained by the great increment of elongation at break and reduction of modulus. Samples tested at TD shows a poor tensile strength due to poor molecular orientation. However, modulus and elongation at break are increased. WAXD and birefringence provides a useful indication in the molecular orientation of the rolled samples. Molecular orientation of polymer chains becomes more significant at high thickness reduction. Moreover, uniaxial cold rolling of PBS/2%OMMT nanocomposite leads to a better clay exfoliation and intercalation, which is verifiable by TEM. DSC results shown that no notable change of melting and crystallization temperatures after cold rolling. However, degree of crystallinity had been reduced after cold rolling.

Acknowledgements

The financial support of USM Research University Grant and USM Fellowship, and the technical support from Kyoto Institute of Technology are gratefully acknowledged.

References

Abu-Zurayk, R., Harkin-Jones, E., McNally, T., Menary, G., Martin, P., Armstrong, C., "Biaxial Deformation Behavior and Mechanical Properties of a Polypropylene/Clay Nanocomposite", Compos. Sci. Technol., **69**, 1644-1652 (2009)

Dean, D.M., Rebenfeld, L., Register, R.A., "Matrix Molecular Orientation in Fiber-Reinforced Polypropylene Composites", J. Mater. Sci, **33**, 4797-4812 (1998)

Golik, A.Z., Kuchinka, M.Y., Fridman, A.Y., "Relationship Between Molecular Orientation Characteristics Determined from Birefringence and Acoustic Data and the Effect of Orientation on the Strength of Polycaprolactam Fibers", Polym. Mech., **3**, 3-7 (1967)

Lee, W.B., Wu, S.Z., Song, M.S., "Characterization of the Orientation Structure and Distribution in Rolled Polypropylene", J. Mater. Eng. Perform., **5**, 637-645 (1996)

Miller, E.: Introduction to Plastics and Composites: Mechanical Properties and Engineering Applications, Marcel Dekker, USA (1996)

Nielsen, L.E., Robert, F.: Mechanical Properties of Polymers and Composites, Marcel Dekker, New York (1994)

Okada, K., Mitsunaga, T., Nagase, Y., "Properties and Particles Dispersion of Biodegradable Resin/Clay Nanocomposites", Korea-Aus. Rheo. J., **15**, 43-50 (2003)

Okamoto, K., Sinha Ray, S., Okamoto, M., "New Poly(Butylene Succinate)/Layered Silicate Nanocomposites.

II. Effect of Organically Modified Layered Silicates on Structure, Properties, Melt Rheology, and Biodegradability", J. Polym. Sci., **41**, 3160-3172 (2003)

Sinha Ray, S., Mosto Bousmina, Okamoto, K., "Structure and Properties of Nanocomposites Based on Poly(Butylene Succinate-co-Adipate) and Organically Modified Montmorillonite", Macromol. Mater. Eng., **290**, 759-768 (2005).

Tiaw, K.S., Teoh, S.H., Chen, R., Hong, M.H., "Processing methods of ultrathin poly(ε-caprolactone) films for tissue engineering applications", Macromolecules, **8**, 807-816 (2009).

Wu, S.Z, Lee, W.B., "Modeling of the Elastic Properties of Ccold-Rolled Polypropylene Sheets", Scr. Mater., **46**, 587-592 (2002)

STRUCUTRE-PROPERTY RELATIONSHIP OF POLY(LACTIC ACID)/POLY(ε-CAPRO LACTONE)/TALC COMPOSITE FILMS FOR PACKAGING APPLICATIONS

Shikha Jain[1, 3], Murali M. Reddy[1], Manjusri Misra[1, 2], Anup K. Ghosh[3], and Amar K. Mohanty[1, 2]

[1]*Bioproducts Discovery and Development Centre, Department of Plant Agriculture, Crop Science Building, University of Guelph, Guelph, N1G 2W1, ON, Canada-shikhaiitdjain@gmail.com; mmreddy@uoguelph.ca; mmisra@uoguelph.ca; mohanty@uoguelph.ca*

[2]*School of Engineering, Thornbrough Building, University of Guelph, Guelph, N1G 2W1, ON, Canada*

[3]*Centre for Polymer Science and Engineering, Indian Institute of Technology Delhi, New Delhi, 110016, India-anupkghosh@gmail.com*

Abstract - In this study, biodegradable and green polymer composites from poly (lactic acid) (PLA)/poly (caprolactone) (PCL) blend reinforced with micro-talc were prepared for high barrier packaging applications. Morphology of PLA/PCL/talc composites were studied by scanning electron microscopy (SEM) and correlated with performance properties to understand the effect of talc loading on structure-properties relationship. It was observed from SEM analysis that PLA/PCL blend exhibited two phase morphology where more viscous PLA was appeared as domain phase and less viscous PCL was remained in continuous bulk phase. Addition of talc into PLA/PCL blend has improved phase morphology and compatibility of two polymer phases resulting further improvement in barrier properties of the composites. Talc was mainly dispersed in less viscous PCL phase and at interface of PLA and PCL phase of PLA/PCL/talc composites. Domain size of PLA phase was reduced in presence of talc in PLA/PCL blends. Talc provide tortuous path to permeant gas molecule and resulted significant improvement in barrier properties of the PLA/PCL blend. Addition of 3 wt.-% of talc into PLA/PCL blend has significantly improved oxygen and water vapor barrier properties of composite by 33 and 25%, respectively. DSC analysis revealed that talc acted as nucleating agent and increased the crystallinity of the PLA/PCL/talc composites compared to its base blend. Furthermore, talc has improved the tensile properties and thermal stability of PLA-PCL blend. Tensile and storage modulus was linearly increased with increased talc content. Oxygen permeability and tensile modulus was also predicted theoretically by various equations and compared with experimental values.

Introduction

Limited resources of petroleum based polymers and increased environmental concern has attracted a great interest towards biodegradable and biobased plastics for packaging applications. Among wide range of biobased and biodegradable polymers available, poly(lactic acid) (PLA) with high tensile strength, gloss and clarity, biodegradability, and recyclability (Gralotta, 2001; Bhardwaj and Mohanty, 2007; Ray et al., 2003; Lim et al., 2008). The major challenges associated with PLA are its poor elongation at break, inferior barrier properties, narrow processing window and low melt strength (Gralotta, 2001; Bhardwaj and Mohanty, 2007; Ray et al., 2003; Lim et al., 2008). Blending of other flexible polymers such as poly (caprolactone) with PLA can improve its elongation at break (Yeh et al., 2009; Simoes et al., 2009). Addition of platelets filler such as talc can improve barrier, mechanical and thermal properties of the polymer matrix (Whaling et al., 2006; Huda et al., 2007; Sekelik et al., 1999). Talc is a high performance reinforcing filler and provides better dispersion and orientation in the polymer matrix than other fillers. Also, talc platelets provide tortuous path to permeant gas molecule and improves the barrier properties of the polymer matrix. Addition of inorganic filler to binary blends improves the compatibility between two polymer phases by reducing the viscosity ratio of the constituent polymers and thereby reduces the domain size of high viscous polymer phase (Wu et al., 2009). Addition of talc reduces the elongation of the PLA while blending of PCL reduces the barrier properties of PLA hence optimization is desired. The aim of this work was to develop biodegradable packaging sheets from PLA/PCL/talc composites with improved mechanical, barrier and thermal properties. Also, to study the effect of talc content on the structure-property relationship of PLA/PCL/talc composites.

Experimental

Materials-Poly (lactic acid) (PLA) (Biomer L9000) was obtained from Biomer, Inc., Germany. PLA resin was dried in a convection oven at 80 °C for 4 hr before processing. Poly (caprolactone) (PCL) was obtained from Perstorp UK Limited, UK. Talc Jetfill 700C was obtained from Luzenac America Inc., USA.

Processing of PLA/PL/talc Composites-PLA/PCL/talc composites were processed by melt mixing technique in twin screw extruder, LabTech Engineering Company Ltd., Thailand (L/D = 32:1, D = 26 mm) with co-rotating screw type at 100 rpm and 180 °C temperature. In the first step, master batch of PCL and talc at 70/30 weight fraction was prepared. PCL/talc masterbatch was diluted with PLA and PCL where talc loading was 1, 3 and 5 wt.-% and PLA/PCL was 40/60 ratio in the PLA/PCL/talc composites. The composites was named as PLA/PCL/talc-(1, 3 and 5) for 1, 3, 5 wt.-% of talc. PLA/PCL/talc composites pellets were processed into the compression molded sheet by hydraulic press model 4128, Carver Inc., USA at 180 °C and 20,000 psi where melting time was 5 min and cooling time was 6 min.

Scanning Electron Microscopy (SEM)-Morphology of the cryogenic fractured surface of the PLA/PCL blend and its talc reinforced composites was studied by using Hitachi instrument model S-570 SEM (Hitachi High Technologies, Japan).

Tensile Testing-Tensile testing was carried out in Universal Test System (UTS) load frame Instron (model 3382) according to ASTM D-638 with a crosshead speed of 5 $mm.min^{-1}$ for PLA, 500 $mm.min^{-1}$ for PCL, 50 $mm.min^{-1}$ for PLA/PCL blend and 5 $mm.min^{-1}$ for PLA/PCL/talc composites.

Dynamic Mechanical Analysis (DMA)-DMA was performed by using a DMA Q-800 (TA Instruments, USA) on tension film clamp. Samples were analyzed from -70 °C to 40 °C at a heating rate of 3 $°C.min^{-1}$, 1 Hz frequency, preload of 0.01 N, in multi frequency strain mode and 15 μm of amplitude.

Differential Scanning Calorimetry (DSC)-Thermal behavior was determined by using DSC Q-200 system (TA Instruments, USA). The characterization was done as heat/cool/heat where sample was first heated from -70 °C to 200 °C at a heating rate of 10 $°C.min^{-1}$ and then cooled to 200 °C to -70 °C with a cooling rate of 5 $°C.min^{-1}$ and again heated from -70 °C to 200 °C at a heating rate of 10 $°C.min^{-1}$.

Thermo-gravimetric Analysis (TGA)-Thermal stability was determined by using TGA Q-500 (TA Instruments, USA) at the rate of 20 $°C.min^{-1}$ from room temperature (~25 °C) to 600 °C in a nitrogen environment.

Oxygen Transmission Rate (OTR)- OTR was measured according to ASTM D-3985 standard, by using Illinois series 8000 (Illinois Instruments, Inc., USA) instrument at 23°C and 0% RH.

Water Vapor Permeability (WVP)-WVP was measured according to ASTM E-96 at 30 °C and 90% relative humidity.

UV Spectroscopic Analysis-The absorbance of light was examined by using Varian Cary 300 Bio UV Visible spectrophotometer (LabCommerce, Inc., USA) in the range of 800 to 200 nm.

Results and Discussion

Scanning Electron Microscopy (SEM)-Morphology of PLA/PCL/talc composites was studied by the SEM analysis and shown in Figure 1. It was observed that PLA/PCL blend exhibited phase separation behavior where less viscous PCL was appeared as continuous bulk phase and more viscous PLA was appeared as discrete domain phase. Addition of talc into PLA/PCL blend has reduced the domain size of PLA phase hence improved the compatibility between PLA and PCL phase. It was observed that in PLA/PCL/talc composites, talc was mainly dispersed in the PCL bulk phase and at the interface of PLA and PCL phases. Presence of talc into PLA/PCL blend has resulted reduction in extent of voiding at the interface of PLA and PCL contributing significant improvement in barrier properties of the PLA/PCL/talc composites.

Tensile testing-Tensile properties of the PLA/PCL/talc composites are given in the Table 1. Tensile modulus of PLA/PCL/talc composites was increased by 5, 34 and 50% with 1, 3 and 5 wt.-% of talc content, respectively because of stress transfer ability of talc platelets (Whaling et al., 2006; Huda et al., 2007). Various equations were used to predict the tensile modulus of PLA/PCL/talc composites and compared with experimental values to understand the morphology and dispersion of the talc particles in the PLA/PCL matrix. Blending of ductile PCL with brittle PLA improves elongation at break of PLA and it was observed 5 and 375% for PLA and PLA/PCL blend, respectively. Elongation at break was reduced with increased talc content into PLA/PCL blend because of immobilization of macromolecular chains. Addition of talc into PLA/PCL blend didn't provide any further improvement in the tensile strength.

Dynamic Mechanical Analysis (DMA)- Storage modulus of the PLA/PCL/talc composites was measured by DMA analysis and given in Table 1 and Figure 2. It was observed that blending of PCL into PLA has reduced the storage modulus of PLA. Addition of talc into PLA/PCL blend has increased the storage modulus by 14, 27 and 58% with 1, 3 and 5 wt.-% of talc loading, respectively. Also, addition of talc into the PLA/PCL blend has increased the glass transition temperature by 4 °C with 5 wt.-% because of reduction in molecular mobility of polymer chain.

Differential Scanning Calorimetry (DSC)-Degree of crystallinity was calculated from melting enthalpy values and given in Figure 3. Addition of talc into PLA/PCL blend has increased the crystallinity of PCL phase compare to its base blend. It was observed that PLA/PCL blend has 34 % degree of crystallinity and PLA/PCL/talc composites have 57, 52 and 47% degree of crystallinity with 1, 3 and 5 wt.-% of talc loading, respectively. PLA/PCL blend was immiscible blend and exhibited two separate melting peaks for PLA and PCL phase at 170 °C and 57 °C, respectively (Figure 4). In presence of talc, PLA melting peak was changed to bimodal melting peak due to the formation of two types of crystal structure. Also, in presence of talc, PLA melting peak shifted from 170 °C to 166 °C and this could be due to heterogeneous nucleation effect of talc resulting low melting crystal structure.

Thermo-gravimetric Analysis (TGA)- Addition of talc improves the thermal stability of the polymer matrix, hence PLA/PCL/talc composites have higher thermal degradation onset temperature compared to PLA/PCL blend.

Oxygen Permeability- Effect of talc content on barrier properties of PLA/PCL/talc composites was studied and given in Figure 5. It was studied that addition of talc into PLA/PCL blend has resulted significant improvement in oxygen barrier property of PLA/PCL/talc composites. Oxygen permeability for PLA/PCL blend was observed 35 × 10^{-3} cc.m^{-1}.day^{-1}.atm^{-1}. Oxygen barrier property were improved by 8, 33 and 25% with 1, 3 and 5 wt.-% of talc loading into PLA/PCL blend, respectively. Oxygen permeability of PLA/PCL/talc composites was predicted theoretically by Nielsen model. Improvement in barrier properties with experimental values was higher compared to theoretically predicted values.

Water Vapor Permeability (WVP)-Water vapor barrier properties of PLA/PCL/talc composites was studied and given in Figure 5. It was observed that addition of talc into PLA/PCL blend resulted significant improvement in water barrier property of the polymer matrix. Water vapor permeability of PLA/PCL blend was observed as 4.05×10^{-14} kg.m^{-1}.s^{-1}.Pa^{-1} it was improved by 15, 25 and 24% with addition of 1, 3 and 5 wt.-% talc in PLA/PCL blends, respectively. Water vapor permeability depends on various factors i.e. crystallinity of the polymer matrix, temperature and relative humidity, nature of the penetrate molecule and size of penetrate molecule (Auras et al., 2003; Tsuji et al., 2006).

UV Spectroscopic Analysis-It was observed that addition of talc into PLA/PCL blend gives higher absorbance in UV spectroscopic analysis.

Table 1

Table 1-Mechanical properties of PLA, PCL, PLA/PCL blend and PLA/PCL/talc composites

Sample	Tensile Strength at Yield (MPa)	Tensile Modulus (MPa)	Elongation at Break (%)	E" (MPa)
PLA	62.2 ±1.2	2438.8±41.2	5.07±0.6	3819
PCL	14.8 ±0.5	276.6±10.5	> 800.0±32.9	359
PLA/PCL 40/60	20.6±1.7	675.2±35.2	375.7±127.9	873.3
PLA/PCL/ talc1	19.4±0.9	709.9±23.7	25.5±14.3	999.5
PLA/PCL/ talc3	22.1±1.6	906.4±33.8	22.8±10.7	1105
PLA/PCL/ talc5	21.6±2.9	1015.6±62.2	10.7±2.9	1377

E" – Storage modulus at 25 °C from DMA analysis

Figure

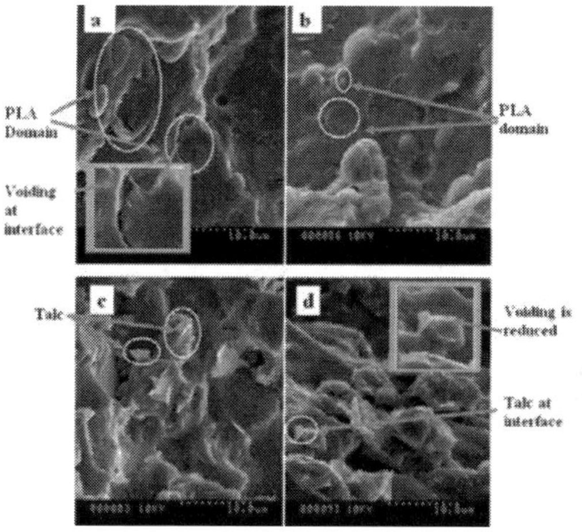

Figure 1-SEM micrographs of the cryogenic fractured (a) PLA/PCL blend (b) PLA/PCL/talc1 (c) PLA/PCL/talc3 (d) PLA/PCL/talc5 composites (magnification: X3000)

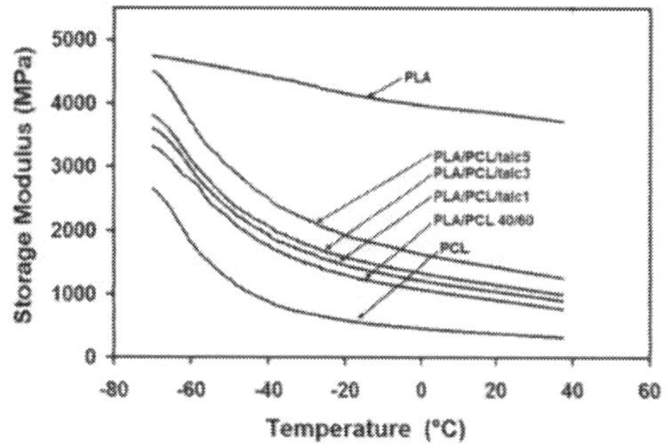

Figure 2-Temperature dependence of storage modulus of PLA, PCL, PLA/PCL blend and PLA/PCL/talc composites

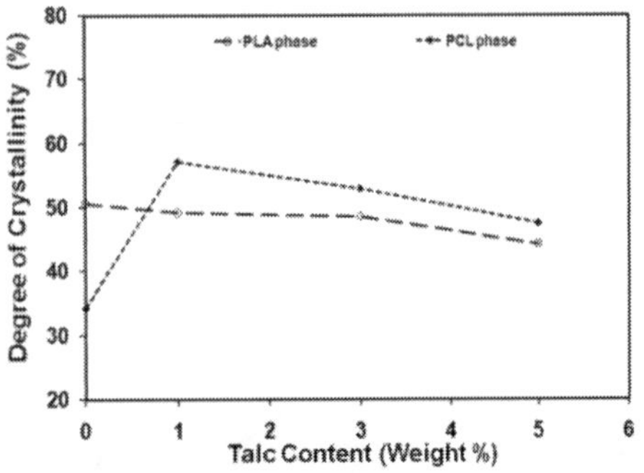

Figure 3-Degree of crystallinity of PLA and PCL phase in PLA/PCL blend and PLA/PCL/talc composites

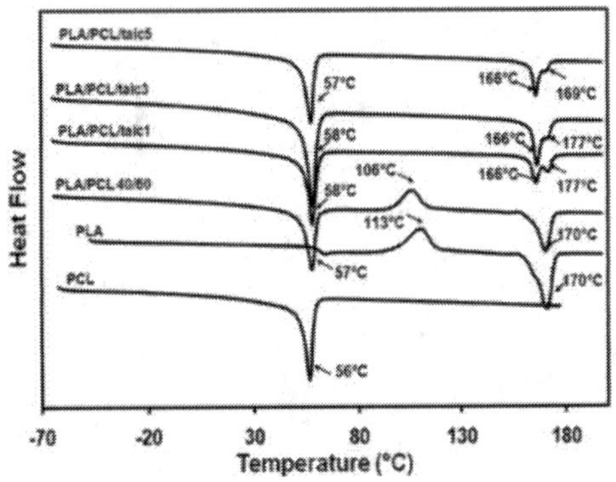

Figure 4-DSC thermograms of PLA, PCL, PLA/PCL blend and PLA/PCL/talc composites (representing endothermic melting peak from 2nd heating cycles)

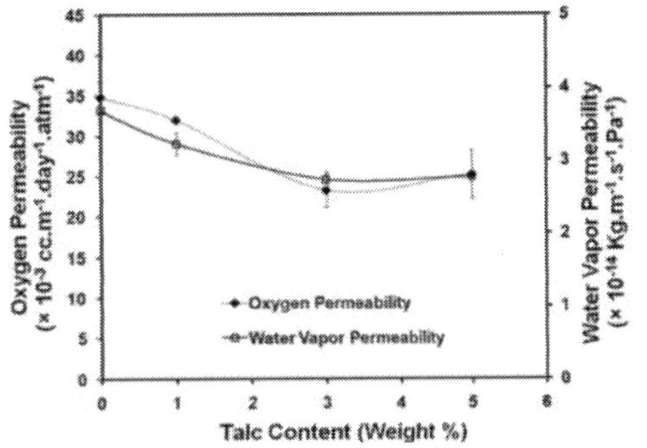

Figure 5-Oxygen and water vapor permeability of PLA/PCL blend and PLA/PCL/talc composites

Conclusions

Effect of talc on structure-properties relationship of PLA/PCL/talc composites was studied. Morphology of PLA/PCL/talc composites was studied to understand the effect of talc on phase behavior and barrier and mechanical properties. It was observed that talc improves the compatibility between PLA and PCL phase by reducing domain size of PLA phase of PLA/PCL/talc composites. Also, it was found that addition of talc into PLA/PCL blend has reduced extent of voiding at the interface of PLA and PCL phase resulting significant improvement in oxygen and water barrier properties. Talc was mainly dispersed in the PCL phase compared to PLA phase in the PLA/PCL/talc composites. Tensile and storage modulus of PLA/PCL/talc composites was increased with increased talc content. There were 33 and 25% improvement in oxygen and water vapor barrier properties with 1, 3 and 5 wt.-% of talc loading into PLA/PCL/talc composites, respectively. It was observed from DSC analysis that talc acted as nucleating agent for PCL phase and degree of crystallinity was increased by 54% with addition of 3 wt.-% of talc loading into the PLA/PCL composites. Improvement in crystallinity by addition of talc has also contributed to barrier properties improvement of PLA/PCL/talc composites. This study could be very useful to develop a compostable material with improved barrier performances for food packaging applications.

Acknowledgements

Authors thank to Department of Foreign Affairs and International Trade, Canada (DFAIT) and Canadian Bureau for International Education (CBIE) for providing Graduate Student Exchange Program (GSEP) 2009-2010 fellowship to S. Jain. Council of Scientific and Industrial Research (CSIR), New Delhi, India is acknowledged for providing senior research fellowship (SRF) to S. Jain. Authors are also thankful to the Ontario Ministry of Agriculture, Food and Rural Affairs (OMAFRA)–2009 New Directions & Alternative Renewable Fuels 'Plus' Research Program project number SR9223 for the partial financial support to carry out this research.

References

1. Auras, R. A., Harte, B., S. Selke, S., Hernandez, R., "Mechanical, Physical and Barrier Properties of Poly(lactic acid)", Jour. of Plast. Fil. and Sheet., **19**, 123-135 (2003)
2. Bhardwaj, R., Mohanty. A. K., "Advances in the Properties of Polylactides Based Materials: A Review", Jour. of Bio. Mater. and Bioen., **1**, 191-209 (2007)
3. Garlotta, D., "A Literature Review of Poly(lactic acid)", Jour. of Polym. and Env., **9**, 63-83 (2001)
4. Huda, M. S., Drzal, L. T., Mohanty, A. K., Misra, M., "The Effect of Silane Treated and Untreated-Talc on the Mechanical and Physico-mechanical Properties of Poly(lactic acid)/Newspaper Fibers/Talc Hybrid Composites", Composites: Part B, **38**, 367-379 (2007)
5. Lim, L. T., Auras, R., Rubino, M., "Processing Technologies for Poly(lactic acid)", Prog. Polym. Sci., **33**, 820-852 (2008)
6. Ray, S. S., Yamada, K., Okamoto, M., Ueda, K., "New Polylactide Layered Silicate Nanocomposites. 2. Concurrent Improvements of

Material Properties, Biodegradability and Melt Rheology", Polymer, **44**, 857–866 (2003)
7. Sekelik, D. J., Stepanov, E. V., Nazarenko, S., Schiraldi, D., Hiltner, A., Baer, E., "Oxygen Barrier Properties of Crystallized and Talc-Filled Poly(ethylene terephthalate)", Jour. of Polym. Sci. Part B: Polym. Phys., **37**, 847-857 (1999)
8. Simoes, C. L., Viana, J. C., Cunha, A. M., "Mechanical Properties of Poly(ε-caprolactone) and Poly(lactic acid) Blends", Jour. of Appl. Polym. Sci., **112**, 345-352 (2009)
9. Tsuji, H., Okino, R., Daimon, H., Fujie, K., "Water Vapor Permeability of Poly(lactide)s: Effects of Molecular Characteristics and Crystallinity", Jour. of Appl. Polym. Sci., **99**, 2245–2252 (2006)
10. Whaling, A., Bhardwaj, R, Mohanty, A. K., "Novel Talc-Filled Biodegradable Bacterial Polyester Composites", Indus. Eng. Chem. Res., **45**, 7497-7503 (2006)
11. Wu, D., Zhang, Y., Zhang, M., Yu, W., "Selective Localization of Multiwalled Carbon Nanotubes in Poly(ε-caprolactone)/Polylactide Blend", Biomacromolecules, **10**, 417-424 (2009)
12. Yeh, J. T., Wu, C. J., Tsou, C. H., Chai, W. L., Chow, J. D., Huang, C. Y., Chen, K. N., Wu, C. S., "Study on the Crystallization, Miscibility, Morphology, Properties of Poly(lactic acid)/Poly(ε-caprolactone) Blends", Polym. Plast. Tech. and Eng., **48**, 571-578 (2009)

PROCESSING OF MULTICOMPONENT COMPOSITES OF EPOXY/SILSESQUIOXANE REINFORCED WITH CARBON FIBERS AND CARBON NANOTUBES BY RESIN TRANSFER MOLDING

M. A. Farias[1], R. D. S. Silva[2], S. C. Amico[2], L. A. F. Coelho[1] and S. H. Pezzin[1]*

[1]Center of Technological Sciences, Santa Catarina State University, Santa Catarina, Brazil – pezzin@joinville.udesc.br; coelholaf@gmail.com; marcelo_adf@hotmail.com
[2]Materials Engineering Department, Federal University of Rio Grande do Sul, Rio Grande do Sul, Brazil – rafasonaglio@hotmail.com; amico@ufrgs.br

The good mechanical properties of composite materials combined with low weight has lead to a growing demand for these materials and the suitability for mass production at an affordable cost has become crucial. The Resin Transfer Molding (RTM) is a process for manufacturing composite materials which is based on the use of fibrous reinforcements in a closed mold, followed by resin injection, curing of the composite, and demolding of the rigid part. The composite formed by the combination of carbon nanotubes, carbon fibers (CF) and epoxy resin may present promising mechanical and electrical properties for practical applications. In this work, epoxy resin was modified by silsesquioxane oligomers containing amino and phenyl groups and used to produce, by RTM, nanocomposites reinforced with CF and different contents of multi-walled carbon nanotubes (MWCNT). Inorganic precursors 3-aminopropyltriethoxysilane and phenyltriethoxysilane were condensed by sol-gel process under acid conditions and incorporated into an epoxy resin based on bisphenol A diglycidyl ether (BADGE) at a molar ratio of 19:1 epoxy:silsesquioxane. MWCNT (0.25 and 0.50 wt %) were then dispersed in the modified epoxy by sonication. The composites were further characterized by mechanical testing (tensile, Izod impact and flexural), scanning electron microscopy (SEM) and dynamical mechanical analysis (DMA). The results showed that the incorporation of silsesquioxane and the application of sonication increased the tensile strength (from 44 ± 4 to 62 ± 1 MPa) and impact resistance (from 103 ± 83 to 290 ± 131 J/m) of the BADGE resin. The incorporation of CFs (20 wt%) and MWNT (0.25 wt%) to the matrix by RTM increased the modulus to 34.7 ± 7.6 GPa, the tensile strength to 303 ± 41 MPa and the impact strength to ca. 1 kJ/m, demonstrating the great potential of this multicomponent composite for coatings and structural applications.

Introduction

Resin Transfer Molding (RTM) is a process widely used in the automotive and aerospace fields because large parts with complex shapes can be efficiently manufactured by this process compared to traditional ones (Rudd, 1997). RTM can be used with many kinds of low-viscosity thermoset polymers and it is based on the use of a fibrous reinforcement in a closed mold, followed by resin injection, curing of the composite, and demolding of the rigid part (Amorim Júnior et al., 2006; Cheng et al., 2010; Schrank et al., 2005; Zhao et al., 2004).

Epoxy resin is one of the most used thermoset matrices. The main advantages of this resin are its chemical resistance, high tensile and compression strength, low shrinkage during curing, easy processing, good adhesive properties and structural stability (Chen et al., 2007; Jain et al., 2003; Ramírez et al., 2008). Carbon fibers (CF) are a typical fiber reinforcement for composites molded by RTM. Carbon fibers combine exceptional mechanical properties and low weight, making them ideal reinforcements for composite materials ranging from aerospace to sport applications (Jin et al., 2006; Montes-Morán et al., 2004; Zhang et al., 2009; Zhao et al., 2004).

The chemical modification of epoxy resins has been widely used in an attempt to improve its properties, for instance, the incorporation of inorganic structures (i.e. silsesquioxanes) to produce functional hybrids (Lee et al., 2005; Li et al., 2001; Ni and Zheng, 2006). Studies comprising the chemical modification of epoxy resins by silsesquioxanes, reported an increase in glass transition temperature, corrosion resistance and Young's modulus (Sun et al., 2008).

The combination of epoxy resin, carbon nanotubes (CNT) and carbon fiber shows a great potential for a number of applications. When properly designed, carbon nanotubes may work together with carbon fibers to produce multifunctional, lightweight structural composites with superior strength (Zhang et al., 2009). CNT have exceptional aspect ratio (over 1000) combined with low density, high strength and toughness, making the CNT an excellent reinforcement for polymeric materials (Fiedler et al., 2006; Gojny et al., 2004). In order to achieve these excellent properties, an efficient dispersion is needed. For polymer composites, sonication and mechanical mixing at high shear rates have shown to be the most effective ones to disperse the carbon nanotubes in a polymer matrix (Suave et al., 2009; Xie et al., 2005).

In this study, static and dynamic mechanical properties of epoxy/silsesquioxane/CF/CNT multi-component nanocomposites produced by RTM were investigated. Dynamical mechanical analysis (DMA) was also used to determine the glass transition temperature (T_g) of the nanocomposites, and scanning electron microscopy (SEM) was carried out to characterize the morphology of fracture surfaces.

Experimental

Materials

3-Aminopropyltriethoxysilane (APES, 99.9%) and phenyltriethoxysilane (PTES, 99.5%) were purchased from Sigma-Aldrich. Diglycidyl ether of bisphenol A monomer (BADGE) LY 1316 and an aliphatic polyamine hardener, Aradur HY 1208, were purchased from Huntsman. Multi-walled carbon nanotubes (MWCNT) Baytubes® C 150 HP (Lot-No: E0009AAD07; Drum-No: 034) were supplied by BAYER Material Science. Carbon-fiber plain weave fabrics (RC200P) with 12.5 yarn/inch, weight of 197 g/m^2 and thickness of 0.25 mm, were supplied by Barracuda Advanced Materials. Other analytical grade reactants like hydrochloric acid and absolute ethanol were supplied by Cinética (Brazil).

Synthesis of silsesquioxane olygomers by sol-gel process

Absolute ethanol (19.5 g), APES (23.7 g) and PTES (25.7 g) were added to a beaker and stirred until complete homogenization. Then, 0.1 mol.L^{-1} HCl (0.35 g) was added to the system (Pezzin et al., 2008), which remained at 30°C under magnetic stirring for 18 h. To remove the solvent, the system was then heated to 70°C until constant mass.

The detailed synthesis of silsesquioxane oligomers containing amino and phenyl groups and the subsequent preparation of hybrid epoxy/silsesquioxane are described elsewhere (de Farias et al., 2010).

Preparation of epoxy/silsesquioxane hybrids

BADGE hybrids with 7.5 wt% (in relation to the total hybrid mass) of APES/PTES silsesquioxane were prepared. Epoxy/silsesquioxanes mixtures were stirred for 10 min at 500 rpm and later placed in an ultrasonic bath for 5 min at 40 °C to minimize microvoids.

Preparation of epoxy/silsesquioxane/MWCNT nanocomposites

First of all, acetone (20 wt%) and MWCNT (0.25 or 0.50 wt%) were sonicated (Sonics Vibra Cell, model VCX 750 W) at 165 W during 20 min (acetone and MWCNT contents refer to the hybrid mass). After addition of the hybrid (epoxy/silsesquioxane) to the system (acetone + MWCNT) another sonication was conducted during 30 min at 187.5 W. Magnetic stirring was used simultaneously with sonication to achieve better MWCNT dispersion (Suave et al., 2009, Xie et al., 2005). Full solvent removal was verified by applying a mass balance to the system after degasification.

Production of multicomponent composites by RTM

The stainless steel RTM mold (300 × 300 × 2 mm) was covered with a thin layer of polyvinyl acetate. After that, five layers of the plain-weave carbon fiber fabric, approximately 30 wt% of the total composite mass, were placed on the mold, which was then closed with a 25 mm thick tempered glass plate.

The resin was mixed with 12 phr of hardener for 5 min at 250 rpm and finally injected from the center of the inner cavity of the mold following a radial pattern. The pressure required to infiltrate the different systems are shown in Table 1. The curing of the systems was performed at room temperature during 12 h. The plates were then post-cured at 50 °C for 45 min.

Table 1: Pressure used to infiltrate the carbon fiber for the various systems.

Systems	Pressure (bar)
Without carbon fiber	0.1
With carbon fiber	0.5
With carbon fiber and 0.25% (w/w) MWCNT	0.6
With carbon fiber and 0.50% (w/w) MWCNT	0.6 – 2.0

Measurements

Morphological characterization of the fracture surface of the samples submitted to tensile tests was conducted in a scanning electron microscope (SEM) Zeiss DSM 940 A at 20 kV. The samples were coated with a thin layer of gold on a Baltec SCD 050 sputter coater.

Viscoelastic properties of the various samples were characterized using an oscillatory rheometer (Anton Paar Physica MCR 101) and 30 x 10 x 2 mm specimens. The samples were heated from room temperature to 200 °C, with a heating rate of 5 °C/min, deformation amplitude of 0.1% and frequency of 1 Hz.

Mechanical properties under tensile and flexural loadings were evaluated in an EMIC DL 30000 universal testing equipment, according to ASTM D638 and D790, respectively. Impact properties were analyzed in a CEAST Resil 25 impactor, according to ASTM D256. In all mechanical tests, 10 specimens were analyzed for each system.

Results and Discussion

RTM processing

The critical stage in RTM is the infiltration and impregnation of the fibers, which should be as fast as possible to increase productivity, yet minimizing undesirable factors such as the presence of microvoids or non-uniform wetting of the reinforcement, in this case, carbon fibers (Schrank et al., 2005). Roughly, it can be said that there are two flow fronts in the mold, the macroscopic convective front determined by the pressure gradient, and the microscopic diffusive front (which occurs within the fiber bundles) determined by capillary pressure and dependent on the surface tension. If the two flow fronts do not coincide, voids may result due to the difficulty in expelling the entrapped air once the macroscopic front has advanced (Nam et al., 2002; Rudd et al., 1997).

When no CF fabric is present, the system finds nearly no obstacle to flow and, with a pressure of just 0.1 bar, the mold is filled in about 120 s. When the CF fabric was in the mold, the flow was hindered (Figure 1) and the pressure had to be increased to 0.5 bar for an adequate filling up of the mold before the resin reached its gel time.

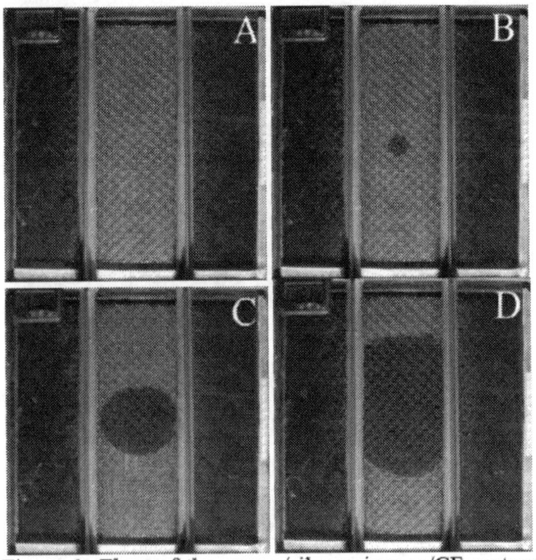

Figure 1: Flow of the epoxy/silsesquioxane/CF system for distinct filling times: A = 0 s; B = 1 s; C = 20 s; D = 60 s.

The MWCNT addition in the resin caused an even greater barrier to the flow through the carbon fibers, due to an increase in viscosity and, because of that, for 0.25 wt% and 0.50 wt% MWCNT the pressure had to be increased up to 0.6 and 2.0 bar, respectively (Figure 2).

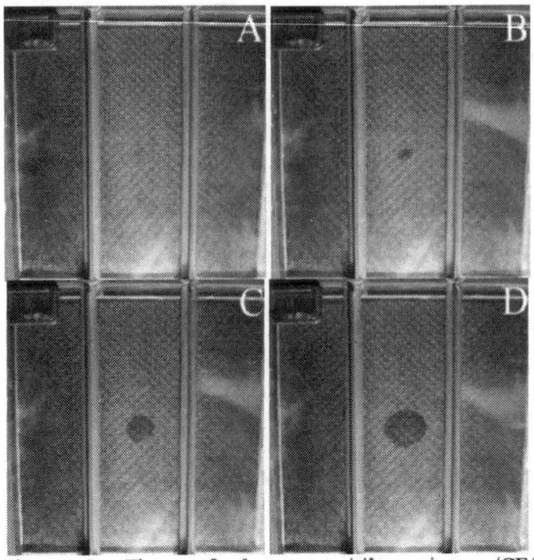

Figure 2: Flow of the epoxy/silsesquioxane/CF/050CNT system for distinct filling times: A = 0 s; B = 1 s; C = 20 s; D = 60 s.

Scanning electron microscopy (SEM)

The morphology of fracture surfaces after tensile tests were characterized by SEM. The addition of silsesquioxanes and/or CNT to the BADGE resin did not alter the typical brittle nature of epoxy resins.

SEM images were also used to evaluate the wettability of CF by the hybrid matrix. Figure 3 shows the fracture surface of epoxy/silsesquioxane/CF, where high wettability of the carbon fiber by the matrix was achieved, with a low amount of voids.

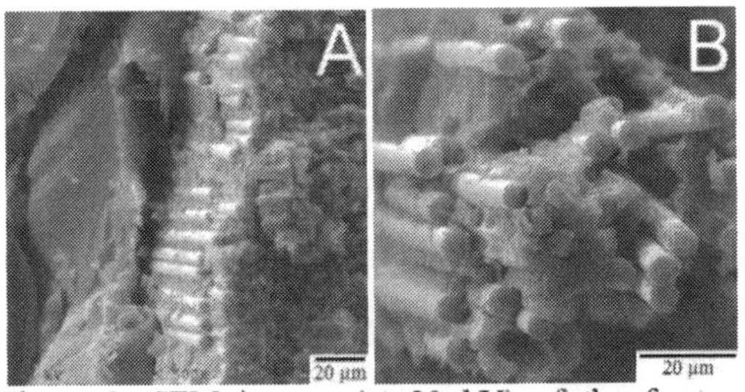

Figure 3: SEM images (at 20 kV) of the fracture surface of a epoxy/silsesquioxane/CF composite. (A) fracture surface after tensile tests and (B) zooming showing adequate resin infiltration between CF.

Figure 3(A) shows some delamination occurring between the infiltrated CF fabric and the neat hybrid resin-rich region at the surface. This was a consequence of the difficult impregnation of the fabric when the nanotubes were used. Moreover, the morphology of the fracture surface suggests an adequate resin infiltration, but poorer interfacial adhesion (Figure 3(B)).

Viscoelastic properties

Dynamical mechanical analysis (DMA) curves are shown in Figure 4. Figure 4(A) shows the storage modulus (E') versus temperature of the various samples. It can be observed a large increase in storage modulus due the incorporation of both CF and CNT. However, the glass transition temperature (T_g), determined by the maximum of the loss modulus (E") as a function of the temperature (Figure 4(B)), decreased in comparison with the T_g of the neat epoxy.

In addition, an increase in the E" peak intensity, in comparison with neat epoxy, was noted for all samples and the incorporation of CF increased the base width and height of the E" peak. This enlargement can be attributed to a restriction on the relaxation process in certain segments of the polymer chains due to the presence of rigid chain segments (Hameed et al., 2007), in this case, possibly APES/PTES silsesquioxane domains.

Figure 4(C) illustrates tan delta (E'/E") curves. The maximum of tan delta as a function of the temperature was also used to estimate T_g of the samples, which are summarized in Table 2. It can be noted that all composites showed a lower glass transition temperature than the neat epoxy resin. Although the incorporation of CF could be responsible for the decrease in T_g, the same effect was observed for sonicated systems without CF.

Table 2: T_g values obtained by DMA (tan delta) ("s" designates samples sonicated before injection).

Samples	T_g (°C)
neat epoxy	99
s-epoxy	64
s-epoxy/silsesquioxane	67
epoxy/silsesquioxane/CF	66
s-epoxy/silsesquioxane/CF	63
epoxy/silsesquioxane/CF/0.25CNT	63
epoxy/silsesquioxane/CF/0.50CNT	62

Tensile and flexural mechanical properties

Table 3 presents the Young's modulus (E) and tensile strength (σ_{max}) values of the samples molded by RTM. Sonification of the neat epoxy resin was also performed to evaluate this effect on the matrix. No significant change was observed in E and σ_{max} for the sonicated and the non-sonicated neat epoxy. For s-epoxy/silsesquioxane systems, the Young's modulus remained the same, but the tensile strength increased 40% compared to neat epoxy.

In contrast, the use of carbon fiber in the hybrid matrix (epoxy/silsesquioxane/CF) has provided a significant increases in Young's modulus (ca. 950%) and tensile strength (ca. 430%), as expected, although no extra enhancement was achieved with sonication.

The addition of 0.25 wt% MWCNT provided significant enhancement in the properties of epoxy or the hybrid matrices reinforced with CF. An increase of about 14 and 20% in Young's modulus was observed for epoxy/silsesquioxane/CF and s-epoxy/silsesquioxane/CF, respectively, while tensile strength increased ca. 30 and 27%, respectively.

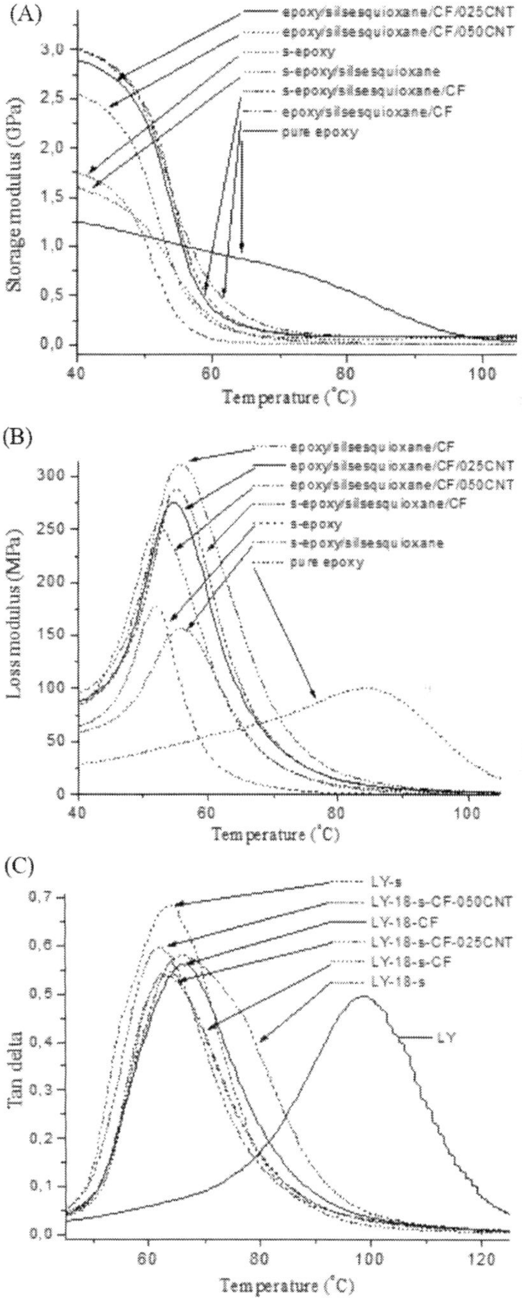

Figure 4: Storage modulus (A), loss modulus (B), and tan delta (C) curves. "s" designates samples sonicated before injection in RTM.

The addition of 0.50 wt% MWCNT caused a decline of 32% in Young's modulus and tensile strength compared to the nanocomposite containing only 0.25% (w/w) MWCNT. This effect was probably a consequence of the difference between diffusive and convective flows, the latter being much faster (due to the higher required pressure) than the microscopic flow. The non-coincident flow rates could generate microvoids, as there is no time for expelling the trapped air from CF woven layers (Nam et al., 2002; Rudd et al., 1997; Schrank et al., 2005). Furthermore, the dispersion of a greater amount of MWCNT may have been more troublesome, leading to agglomerates which would lower the mechanical performance of the composites.

Table 3: Elastic modulus (E) and tensile strength (σ_{max}) results obtained from tensile tests.

Samples	E (GPa)	σ_{max} (MPa)
neat epoxy	2.9 (± 0.5)	44 (± 4)
s-epoxy	3.2 (± 0.3)	48 (± 10)
s-epoxy/silsesquioxane	2.7 (± 0.2)	62 (± 1)
epoxy/silsesquioxane/CF	30.4 (± 6.5)	233 (± 57)
s-epoxy/silsesquioxane/CF	28.8 (± 1.8)	239 (± 32)
epoxy/silsesquioxane/CF/0.25CNT	34.7 (± 7.5)	303 (± 41)
epoxy/silsesquioxane/CF/0.50CNT	23.4 (± 2.6)	207 (± 43)

The elastic modulus and tensile strength results obtained from flexural tests are presented in Table 4. The sonication process and the incorporation of silsesquioxane yielded similar results to non-sonicated neat epoxy. Whereas, as expected, the incorporation of carbon fiber fabric promoted a significant increase in these values. For epoxy/silsesquioxane/CF, the increase was above 720 and 280% for flexural modulus and strength, respectively. Sonication has not significantly changed these properties in epoxy/silsesquioxane/CF samples.

A distinct behavior was noted in the flexural response of the nanocomposites containing MWCNT. Unlike what had been observed for the tensile and impact results, the addition of 0.25 wt% MWCNT caused a decrease in flexural modulus and strength compared to epoxy/silsesquioxane/CF, which was recovered when 0.50 wt% MWCNT was incorporated.

Impact resistance

Impact strength of samples produced by RTM was analyzed in Izod impact tests and the results are shown in Table 5. The incorporation of silsesquioxane oligomer with amino and phenyl groups promoted over 100% increase when compared to neat epoxy, suggesting that the incorporation of inorganic domains promotes an increase in bond strength between polymer chains, agreeing with what was observed for tensile tests (Table 3).

Table 4: Elastic modulus (E) and tensile strength (σ_{max}) values obtained from flexural tests.

Samples	Flexural modulus (GPa)	Flexural strength (MPa)
neat epoxy	2.1 (± 0.1)	91 (± 5)
s-epoxy	2.2 (± 0.1)	93 (± 5)
s-epoxy/silsesquioxane	2.1 (± 0.1)	86 (± 3)
epoxy/silsesquioxane/CF	17.2 (± 4.1)	346 (± 48)
s-epoxy/silsesquioxane/CF	20.5 (± 2.4)	396 (± 67)
epoxy/silsesquioxane/CF/025CNT	7.0 (± 0.9)	190 (± 50)
epoxy/silsesquioxane/CF/050CNT	26.7 (± 1.9)	400 (± 95)

With the incorporation of the carbon fiber fabric, over 450% increase in impact strength was found in comparison with neat epoxy. This increase was even more pronounced for sonicated epoxy/silsesquioxane/CF composites, reaching over 850% increase.

The incorporation of 0.25 wt% MWCNT also promoted a significant increase (70%) in impact strength compared to epoxy/silsesquioxane/CF composites. As observed in tensile tests, the addition of 0.50 wt% MWCNT decreased impact strength in comparison to nanocomposites containing 0.25 wt% MWCNT. However, in this case the deviations were too large and the difference was not considered significant.

Table 5: Values of Izod impact strength.

Samples	Impact strength (kJ/m)
neat epoxy	0.1 (± 0.1)
s-epoxy	0.2 (± 0.1)
s-epoxy/silsesquioxane	0.3 (± 0.1)
epoxy/silsesquioxane/CF	0.6 (± 0.3)
s-epoxy/silsesquioxane/CF	1.0 (± 0.4)
epoxy/silsesquioxane/CF/025CNT	1.0 (± 0.3)
epoxy/silsesquioxane/CF/050CNT	0.8 (± 0.5)

Conclusions

Composites and nanocomposites based on hybrid epoxy/silsesquioxane matrix reinforced with carbon fibers and carbon fibers + MWCNT were successfully produced by RTM. These composites and nanocomposites showed significant improvement in tensile, flexural and impact properties. Further increase may be achieved with sonication due to the minimization of entrapped air inside the system. SEM images showed good wettability of the carbon fibers by the hybrid epoxy resin, and the expected brittle behavior of the thermoset material. Glass transition temperature, obtained by dynamical mechanical analysis, decreased for all composites produced by RTM in comparison with neat BADGE resin. The overall results show that these composites and nanocomposites can be produced on large scale and used as multifunctional materials for practical applications, such as, coatings, structural components and electronic devices.

Acknowledgements

The authors are grateful to the Brazilian Space Agency (AEB Uniespaço)for the financial support, to, CAPES-DS for the scholarship to Mr. Marcelo Alexandre de Farias and to Prof. Ademir José Zattera for the support in the mechanical testing.

References

Amorim Júnior, W. F., Chaves, G. Q., Calado, V. M. A., Bastian, F. L. "Processamento de Placa Espessa de Compósito através de Moldagem por Transferência de Resina". Revista Matéria, **11**, 316-323 (2006).

Chen, H., Jacobs, O., Wu, W., Rudiger, G., Schadel, B. "Effect of dispersion method on tribological properties of carbon nanotube reinforced epoxy resin composites". Polymer Test, **26**, 351-360 (2007).

Cheng, Q. F., Wang, J. P., Wen, J. J., Liu, C. H., Jiang, K. L., Li, Q. Q., Fan, S. S. "Carbon nanotube/epoxy composites fabricated by resin transfer molding". Carbon, **48**, 260–266 (2010).

De Farias, M.A., Coelho, L.A.F., Pezzin, S.H. "Epoxy/silsesquioxane organic-inorganic hybrids: Sol-gel synthesis of inorganic precursors containing amino and phenyl groups". Polym. Eng. Sci., submitted (2010).

Fiedler, B., Schulte, K., Gojny, F. H., Wichmann, M. H. G., Nolte, M. C. M. "Fundamental aspects of nano-reinforced composites". Compos. Sci. Tech., **66**, 3115-3125 (2006).

Gojny, F. H., Wichmann, M. H. G., Kopke, U., Fiedler, B., Schulte, K. "Carbon nanotube-reinforced epoxy-composites: enhanced stiffness and fracture toughness at low nanotube content". Compos. Sci. Tech., **64**, 2363-2371 (2004).

Hameed, N., Sreekumar, P. A., Francis, B., Yang, W., Thomas, S. "Morphology, dynamic mechanical and thermal studies on poly(styrene-co-acrylonitrile) modified epoxy resin/glass fibre composites". Compos. Appl. Sci. Manuf., **38**, 2422-2432 (2007).

Jain, P., Choudhary, V., Varma, I. K. "Effect of structure on thermal behaviour of epoxy resins". Eur. Polymer J., **39**, 181-187 (2003).

Jin, Z., Zhang, Z., Meng, L. "Effects of ozone method treating carbon fibers on mechanical properties of carbon/carbon composites". Mater. Chem. Phys., **97**, 167–172 (2006).

Lee, T.-M., Ma, C.-C. M., Hsu, C.-W., Wu, H.-L. "Effect of molecular structures and mobility on the thermal and dynamical mechanical properties of thermally cured epoxy-bridged polyorganosiloxanes". Polymer, **46**, 8286–8296 (2005).

Li, G. Z., Wang, L., Toghiani, H., Tyrone, L., Daulton, T. L., Koyama, K., Pittman Jr, C. U. "Viscoelastic and Mechanical Properties of Epoxy/Multifunctional Polyhedral Oligomeric Silsesquioxane Nanocomposites and Epoxy/Ladderlike Polyphenylsilsesquioxane Blends". Macromolecules, **34**, 8686–8693 (2001).

Montes-Morán, M. A., Gauthier, W., Martínez-Alonso, A., Tascón, J. M. D.. "Mechanical properties of high-strength carbon fibres. Validation of an end-effect model for describing experimental data". Carbon, **42**, 1275–1278 (2004).

Nam, J., Lee, K. J., Jang J., Lee, N. J., Lee, G. W. W. "Effects of surface modification on the resin-transfer moulding (RTM) of glass-fibre/unsaturated-polyester composites". Compos. Sci. Tech., **62**, 9-16 (2002).

Ni, Y., Zheng, S. "Epoxy Resin Containing Polyphenylsilsesquioxane: Preparation, Morphology, and Thermomechanical Properties". J. Polymer Sci. Polymer Chem., **44**, 1093-1105 (2006).

Pezzin, S. H., Stock, N., Shishatskiy, S., Nunes, S. P. "Modification of proton conductive polymer membranes with phosphonated polysilsesquioxanes". J. Membr. Sci., **325**, 559–569 (2008).

Ramírez, C., Rico, M., Torres, A., Barral, L., López, J., Montero, B. "Epoxy/POSS organic–inorganic hybrids: ATR-FTIR and DSC studies". Eur. Polymer J., **44**, 3035–3045 (2008).

Rudd, C. D., Long, A. C., Kendall, K. N., Mangin, C. G. E: Liquid moulding technologies. Resin transfer moulding, structural reaction injection moulding and related processing techniques. Woodhead Publishing Limited and the Society of Automotive Engineers, INC (1997). England.

Schrank, C., Schwarz, B., Mayerhofer, K., Neubauer, E. "Influence of thermal treatment on the adhesion of copper coatings on carbon substrates". Vacuum, **80**, 122-127 (2005).

Shen, J., Huang, W., Wu, L., Hu, Y., Ye, M. "Thermo-physical properties of epoxy nanocomposites reinforced with amino-functionalized multi-walled carbon nanotubes". Compos. Appl. Sci. Manuf., **38**, 1331-1336 (2007).

Suave, J., Coelho, L. A. F., Amico, S. C., Pezzin, S. H. "Effect of sonication on thermo-mechanical properties of epoxy nanocomposites with carboxylated-SWNT". Mater. Sci. Eng., **509,** 57-62 (2009).

Sun, D., Hu, L., Zhang, X., Lu, Z. "Nanostructures and properties of hybrid films based on DGEBA and modified with APES/PGE polyhedral oligomeric silsesquioxanes". Colloid Surface Physicochem. Eng. Aspect, **313/314**, 278–281 (2008).

Xie, X. L., Mai, Y. W., Zhou, X. P. "Dispersion and alignment of carbon nanotubes in polymer matrix: A review". Mater. Sci. Eng. R, **49**, 89–112 (2005).

Yaping, Z., Aibo, Z., Qinghua, C., Jiaoxia, Z., Rongchang, N. "Functionalized effect on carbon nanotube/epoxy nano-composites". Mater. Sci. Eng. **435/436**, 145-149 (2006).

Zhao, J., Park, H., Han, J., Lu, J. P. "Electronic Properties of Carbon Nanotubes with Covalent Sidewall Functionalization". J. Phys. Chem. B, **108**, 4227-4230 (2004).

Zhang, Q., Liu, J., Sager, R., Dai, L., Baur, J. "Hierarchical composites of carbon nanotubes on carbon fiber: Influence of growth condition on fiber tensile properties". Compos. Sci. Tech., **69**, 594–601 (2009).

INVESTIGATION OF THE CROSSLINK DENSITY OF NOVEL BINDERS FOR STRUCTURAL COMPOSITES

Edmonds Neil [1*], Zhang Xinyue [2]

[1] *Department of Chemistry, The University of Auckland, Auckland, New Zealand* - nr.edmonds@auckland.ac.nz

[2] *Department of Chemistry, The University of Auckland, Auckland, New Zealand* - zhangxinyue85@hotmail.com

Current technologies for the manufacture of structural composites using unsaturated polyester and epoxy liquid resins produce harmful by-products and often have long cure times. Unsaturated polyester resins emit styrene monomer vapour during composite manufacture, while amine hardeners in liquid epoxy resins can induce skin sensitivity. Urea-formaldehyde resins used in wood fibre composites may have problems with formaldehyde emissions both during composite manufacture and in product end use. This research aimed to develop novel binders for structural composites based on polymers traditionally used for powder coatings. Previous studies have shown that powder coating resins can be used to replace amino resins in wood fibre composites. Resins systems investigated in the current project include epoxy and epoxy hybrid systems, plus polyester/HAA and polyester/TGIC systems.

When used for powder coatings, the above materials form crosslinked networks on heating to 180°C in the presence of suitable crosslinkers. Formulations had to be adapted to provide increased crosslink density, this was determined by DMTA. Different grades of epoxy powders and different crosslinking compounds were examined, and fibreglass reinforced laminates successfully produced and tested. The results confirm that the technology can be used to produce structural laminates. A potential advantage of using powdered resins is that thermoplastic laminates can be made by heating composites to ~100°C which could be formed/shaped as required and converted to thermosetting products by heating to 180°C. This would allow added flexibility to manufacturing processes, something that cannot be achieved using traditional liquid resins.

Introduction

The main objectives of this research were the development and evaluation of thermosetting powder coating resins with increased reactivity to give crosslink densities similar to liquid unsaturated polyester resins, and evaluation of different resin systems to determine suitability for composite manufacture. The powder used for the process is a mixture of finely ground particles of reactive resin and crosslinkers, other additives may include fillers and pigment. Powder coatings are typically electrostatically applied to metal substrates and cured by heating to 180 °C. The powder melts at ~95 °C, flowing and fusing into a strong, continuous and strongly adhered coating film at the final cure temperature. This provides a high quality coating with corrosion protection, improved durability and attractive finish. While powder coatings have obvious environmental advantages over conventional solvent-borne coatings, the manufacturing process results in a significant waste stream of powder that is outside the required particle size range necessary for electrostatic application. Combined with overspray and waste powder from other sources, this material is now used as a matrix for the manufacture of composite boards. The current study utilises virgin resin to make structural composites using fibreglass reinforcement. For this purpose, optimisation of crosslink density was a critical consideration, this was measured using DMTA. For optimum physical properties, the most important characteristics that can be controlled are glass transition temperature and crosslink density. While Tg is regularly determined using differential scanning calorimetry (DSC) and Dynamic Mechanical Thermal Analysis (DMTA), crosslink density is seldom determined.

Crosslink density is defined as Ve = the number of moles of elastically effective network chains per cm^3 of sample. The unit of ve is moles/cm^3. It is easier to visualize a network based on the value of Mc which is defined as
Mc = weight of sample in grams that contains one mole of elastically effective chains.

The relationship between *ve* and *Mc* is
$Mc = \rho/ ve$ where ρ is film density in g/cm^3 [1].

DMTA is used for calculation of crosslink density, the value of E' in the rubbery plateau region above Tg being observed. The length of the rubbery plateau as well as its modulus is dependent on the molecular weight between crosslinks. This is normally calculated during a stress-relaxation experiment.

With DMTA, modulus measurements can be made in a short time over a wide temperature range without destroying samples. Most importantly, DMTA provides the required modulus values with very small deformation. For short-chain networks, it is important to keep the sample deformation small during modulus measurements. In this way, the sample can respond to deformation by changes in chain conformation. When deformation is accommodated by conformational changes, without bending or breaking of bonds, the modulus in the plateau region is proportional to the number of crosslinks.

The relationship is $ve = G'/RT = E'/3RT$

Storage modulus is obtained in the rubbery plateau. T is the temperature in °K corresponding to the storage modulus value, and R is the gas constant. A wide range of variation of E' has been observed from approximately 4×10^6 Pascal for lightly crosslinked films to approximately 2×10^8 Pascal for very highly crosslinked films [1].

Experimental

Only pure resin and crosslinking agent without any additives were used to make samples for testing. Five systems were studied, these were polyester hybrid, polyester with Primid crosslinker and polyester with TGIC crosslinker. The other two products were urethane and epoxy systems. Four different materials were used to make three polyester based powders: a commercial polyester (K4057), a commercial solid epoxy (CHS-EPOXY 130), a cross linker (Primid®XL-552) and a commercial grade of TGIC (Araldite PT180). Four types of materials were used with urethane and epoxy systems: a commercial polyester (Crylcoat TM 2818-0), a commercial solid epoxy (D.E.R. TW 642U), a curing agent (D.E.H. TW 81) and a crosslinking agent (Vestagon®B 1530).

Each type of resin was ground in a Retsch SM100 mill. All powders were dried in a vacuum oven at 50°C for more than 48 hours. Powders were then weighted and thoroughly mixed to ensure uniform dispersion. Samples were placed in Teflon moulds supported on an aluminium plate and cured in a heated press at 190°C and at a pressure of 8.28MPa for 10 minutes.

Unreacted powders and cured samples were characterised by both FTIR and DSC before determination of crosslink density of cured samples by DMTA. The change in Tg indicates the degree of crosslinking and hence the final performance of the system when used as a matrix in a composite.

Theoretically, increased crosslinking increases the Tg. From the results of the three polyester based powders coating, it was deduced that the samples cured by heating to 250°C were well crosslinked, but the curing of polyester powder in a press at 190°C for 10 minutes could not ensure complete crosslinking. Similarly for urethane systems, curing the powder at 190°C for 10 minutes could not complete crosslinking, whereas epoxy systems could be considered to be completely crosslinked under these conditions.

Dynamic mechanical thermal analysis was carried out in a nitrogen atmosphere with a Rheometric scientific DMTA IV. Measurement of the tensile storage modulus (E') and loss tangent (tan δ) of fully cured samples was carried out at a frequency of ω = 1Hz and working in single cantilever with rectangular samples. The temperature range was selected to span the glassy region, transition region, and rubbery plateau region [13]. For these experiments the initial temperature was 20-30°C and the final temperature varied from 110-125°C. Dynamic properties were determined at 1min intervals with scans run at 1°C/min or 2°C/min [2].

The strain was set to 0.01% as all the samples were very brittle. Tg was taken as the maximum in the damping peak associated with α transition (maximum of the tan δ curve) [3].

Results and Discussion

The dynamic properties of polyester hybrid powders and the calculated crosslink densities are listed in Table 1 and polyester TGIC in Table 2, these are typical of results obtained. Corresponding DMTA scans are shown in Figures 1 and 2 and comparison of crosslink densities in Figures 3 and 4. A highly crosslinked thermosetting material would typically have a *ve* value of approximately 1×10^{-3} moles/cm^3 [2]. The highest value was achieved by the polyester Primid system but results were not always consistent, this was due to the difficulty both in making uniform films and mounting the films in the DMTA. Considering the results of both DSC and DMTA, urethane and epoxy systems were superior to the three polyester based systems for use in composites.

In the next phase, glass fibre was introduced to increase sample strength. Dry powder was coated directly onto a chopped strand mat (CSM). Four types of materials were used to achieve two types of composites: a commercial polyester (Crylcoat TM 2818-0), a commercial solid epoxy (D.E.R. TW 642U), a curing agent (D.E.H. TW 81) and a crosslinking agent (Vestagon®B 1530). Chopped strand mat (CSM, 450 g/m^2) was used as reinforcement at a 70:30 ratio of glass:resin, samples were hot pressed in a cavity aluminium mould. The chopped strand mat was first placed into the mould and one third of the measured powder mixture was distributed as uniformly as possible onto the mat. Another Teflon sheet was then placed on the top to avoid the sample sticking to the press during curing. The mould was then loaded into a preheated press, cured at a temperature of 190°C and at a pressure of 10.34MPa (1500 psi) for 10 min. For comparative purposes, laminates were also prepared using unsaturated polyester resin at a 2:1 ratio of resin:glass. As expected, this sample had a high crosslink density level, averaging 9.295×10^{-2} moles/cm^3, expressed as the laminate crosslink density. However some other laminate samples based on powders gave comparable values.

Trial	Tg (°C)	E'(Pa)	Temp of E'(°C)	Crosslink Density (mole/cm^3)
1	70.52	3.26E+06	91.26	3.58E-04
2	74.21	6.87E+05	90.27	7.59E-05
3	73.36	1.37E+06	90.06	1.51E-04
Ave	72.69	3.83E+06	90.53	1.95E-04

Table 1. Data obtained from DMTA measurements of Hybrid system

Trial	Tg (°C)	E'(Pa)	Temp of E'(°C)	Crosslink Density (mole/cm^3)
1	72.5	7.29E+06	90.32	8.05E-04
2	74.3	2.86E+06	100.41	3.07E-04
3	77.9	2.45E+06	94.99	2.67E-04
Ave	74.9	4.20E+06	95.24	4.60E-04

Table 2. Data obtained from DMTA measurements of Polyester & TGIC system

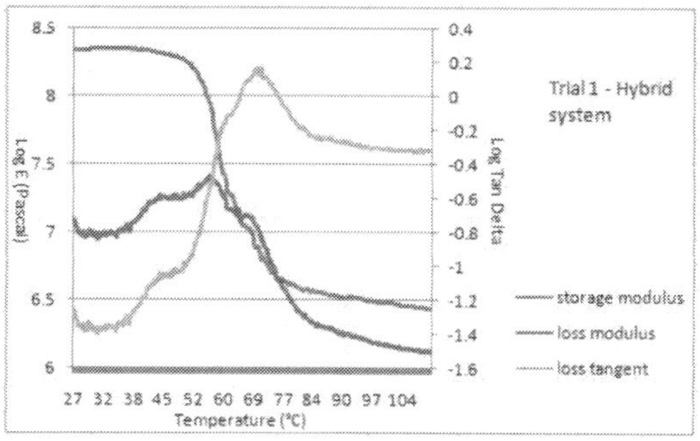

Figure 1. DMTA of Polyester Hybrid sample

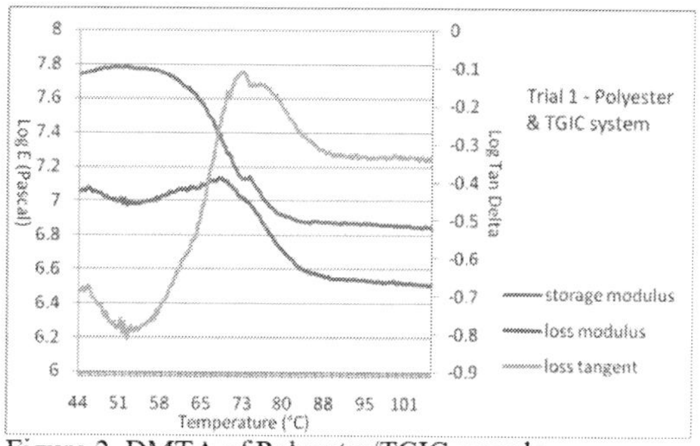

Figure 2. DMTA of Polyester/TGIC sample

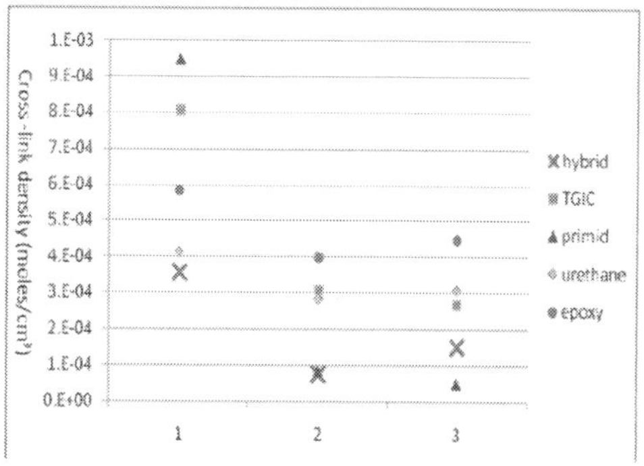

Figure 3. Comparison of Crosslink Densities

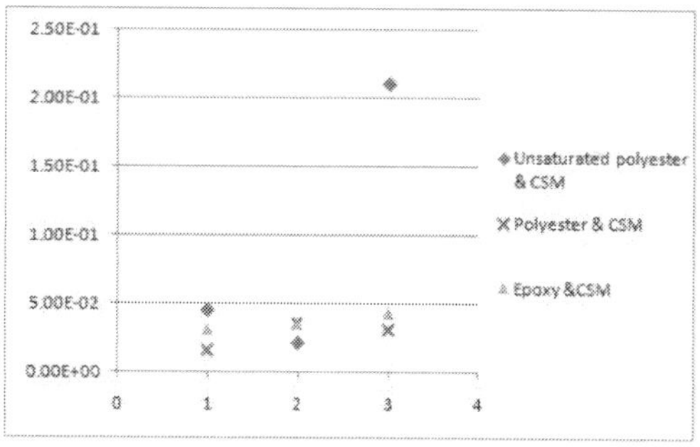

Figure 4. Comparison of Crosslink Densities

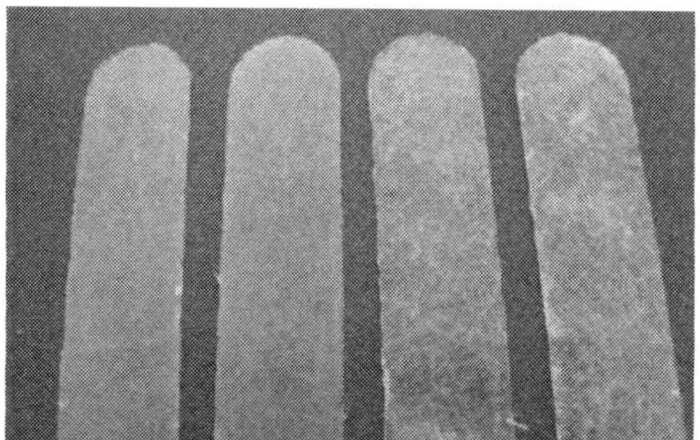

Figure 5. Composite samples

Figure 6. Composite sample

Conclusions

It can be concluded that while epoxy and urethane powders cannot achieve the same crosslink density as unsaturated polyester, the results also showed that powders used for powder coatings have considerable potential for use as the matrices in composites, plus they are not harmful to the environment.

By using chopped strand mat as a reinforcement, the manufacturing of test samples was straight forward and the glass fibre provided good mechanical properties to the network. The test results were more accurate and more consistent compared to samples without the reinforcement.

However part of the modulus measured can be attributed to the modulus of the glass fibre, not the result of crosslinking between resins and curing agent. For this reason the formula cannot accurately predict the crosslink density of powder coating and therefore the calculated laminate crosslink density does not reveal the crosslinking level between resin and curing agent. However comparison of unsaturated polyester and powder composites made with the same chopped strand mat and same resin to CSM ratio, the results are still valuable.

In further work, some modifications need to be made to the powders to give more highly crosslinked samples and a more accurate method is required to determine the level of crosslinking of powder systems in laminate samples.

References

1. Loren W.Hill, "Dynamic Mechanical and Tensile Properties", *Paint and coating testing manual,* p 541

2. Loren W. Hill, "Structure/property relationships of thermoset coatings", *The Journal of Coatings Technology* 64.n808: pp29

3. J.S. Martin, J.M. Laza, M.L. Morras, M. Rodriguez, L.M. Leon, Polymer 41 (2000) 4203.

GAP-IMPREGNATION: A HIGH VOLUME CAPABLE PROCESS FOR MANUFACTURING STRUCTURAL COMPOSITE PARTS

W. Michaeli, J. Strohhaecker, K. Fischer and R. Bastian*

Institute of Plastics Processing at RWTH Aachen University (IKV), Pontstr. 49, D-52062 Aachen, Germany
zentrale@ikv.rwth-aachen.de; bastian@ikv.rwth-aachen.de; fischer@ikv.rwth-aachen.de;
strohhaecker@ikv.rwth-aachen.de
http://www.ikv-aachen.de

Compression Liquid Composite Molding processes (CLCM) offer a high potential for a high volume production of structural composite parts. The idea of CLCM processes is the impregnation of a non-compressed endless fiber reinforcement - which is equivalent to a high permeability of the reinforcement - and a following compression step leading to high fiber volume content of the composite part. One CLCM-variant is the so called gap-impregnation-process which has been developed at IKV. The gap-impregnation-process enables fast impregnation of the reinforcing fibers at a high level of automation of the manufacturing process.
First, the fiber reinforcement (preform) is placed into a rigid two-sided mould. Then a flow-gap is generated over the preform by moving one side of the mould. In that gap the thermoset resin can be distributed quickly and nearly without a significant pressure drop. Then the flow-gap is closed in a wedge-shaped manner to promote the impregnation and subsequently to compress the preform. The wedge-shaped closing procedure facilitates a very fast impregnation and a compression within 19 s. Hence, very fast resin systems can be used which crosslink very quickly. So, the whole production cycle (from preform positioning to part demolding) for endless fiber reinforced plates at 50 % fiber volume content takes actually 280 s. Extensive results of the analysis of this process will be presented.

Introduction

Parts made of fiber reinforced plastics (FRP) with a high content of oriented continuous fibers can contribute in many fields of machine and vehicle design to increase efficiency. Reasons for that are the excellent lightweight capabilities through high weight-specific mechanical properties, the freedom of design, potentials to produce highly integrated structures and the good resistance against environmental influences.
Until today FRP parts with continuous fiber reinforcement are only produced in small to middle series. It is causal for that, that for the production of such parts no high volume capable manufacturing processes are available (Eberth et al., 2006). This is due to the long cycle times and the only partial automatable manufacturing processes. The manufacturing processes that are mainly used today are the prepreg/autoclave process, and so-called Liquid Composite Molding (LCM) processes like the Resin Transfer Molding (RTM) process and Resin Infusion (RI) process. LCM processes offer a high potential for the economic manufacturing of composite parts (Eberth et al., 2006; Ehrenstein, 1992; Flemming et al., 1999; Michaeli et al. 1989; Mitschang et al., 2004).
In order to enter new markets for composite parts, it is therefore necessary to develop fully automated manufacturing processes which allow a clear reduction of the cycle time and thereby facilitate a more economic part manufacturing (Mitschang, 2005; Herrmann, 2006).
This can be achieved by so called Compression Liquid Composite Molding (CLCM) processes (Young, 1995). When using these processes, the preform is impregnated at low fiber volume contents. Thereby the resin can be distributed very quickly and a fast impregnation is facilitated. After impregnation, the preform is compressed to high fiber volume contents, which are essential for high weight specific mechanical properties.
The use of such CLCM processes for the manufacturing of components with a fiber content over 50 % and continuous fiber reinforcement in large numbers (n > 20,000 pcs./a, cycle time < 10 min.), which is the demand of the automotive industry, is still a great challenge. It is necessary for this kind of production to develop fully automated process chains with robust manufacturing processes that enable short cycle times (Herrmann, 2006; Mitschang, 2005).
One approach to this is the process chain "In-mold impregnation" of the research unit 860 "New process chains for production of continuous fiber reinforced plastic parts: Integration of preforming, impregnation, molding and curing" that ist funded by the Deutsche Forschungsgemeinschaft (German Research Foundation, DFG). The chain links of the process chain are shown in Figure 1.

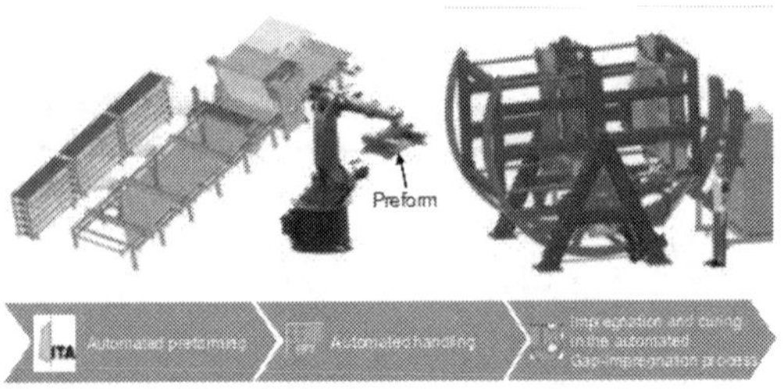

Figure 1 – Process chain "In-mold impregnation"

Three institutes of the RWTH Aachen University take part in this study. First the Institut fuer Textiltechnik (ITA), which develops new preforming technologies, second the Fraunhofer Institute for Production Technology (IPT), which develops handling and tooling technologies and last Institute of Plastics Processing (IKV), which develops technologies for the processing of composites. For the impregnation, molding and curing a novel CLCM process with the potential for manufacturing large series of high performance composite parts has been developed at the IKV. The so-called Gap-Impregnation process leads to a clear reduction of the cycle time in comparison to the RTM process. This is achieved by a specific compression technique.

Introduction of the Gap-Impregnation process

For the realization of the process control the Gap-Impregnation process requires a specific tool technology (Meyer-Noack, 2006). Figure 2 shows the components of an experimental tool for the production of monolithic parts available at the IKV and the schematic procedure.

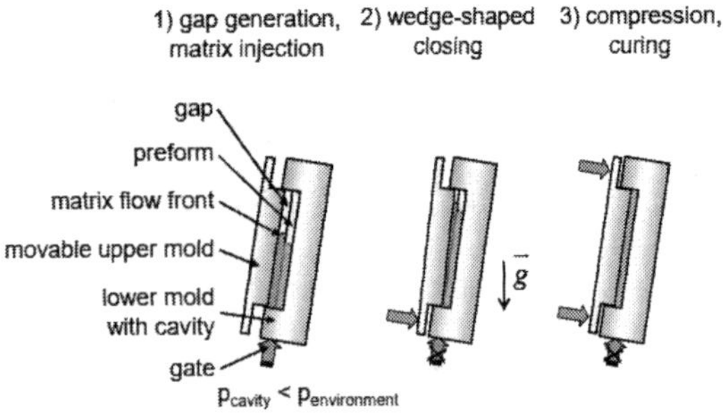

Figure 2 – Schematic procedure of the Gap-Impregnation process

The tool consists of a lower mold with the cavity and a movable upper mold. For the part production a net-shape preform is inserted into the cavity. Then the cavity is shut airtightly with the movable upper mold by using a silicone o-seal and is evacuated by applying vacuum. In the next step a flow gap is generated on one side of the preform by lifting the upper mold.

Then the gate is opened and the resin is fed into the cavity. In order to guarantee an optimal removal of air bubbles, the tool is turned so that the flow direction of the resin is oriented against the force of gravity. The resin can be distributed very quickly over the preform through the flow gap. The impregnation of the preform occurs in thickness direction so that the resin has to cover only a short flow path within the preform. Since the preform is decompressed during the impregnation, the impregnation is significantly improved. After a sufficient volume of resin material is fed into the cavity, the gate is shut and the flow gap is closed in a wedge-shape manner. The necessary clamping forces during the closing procedure of the flow gap are generated by the pressure difference between the evacuated cavity and the environment. The closing of the flow gap leads to the further distribution of the resin over the preform and impregnation of the last dry areas of the preform. In a last step the flow gap is completely shut and the laminate is compressed to the final fiber volume content. After the resin is cured, the part can be demolded from the tool. The necessary finishing is restricted to the deburring of the part. The wedge-shaped closing of the flow gap in combination with the injection of the resin against the force of gravity facilitates a fast compression and a high laminate quality.
The Gap-Impregnation process is also suitable for the manufacturing of integrated sandwich structures (Fischer et al., 2008). For this a net-shaped foam core with the reinforcement on both sides is fixed in the mold and flow gaps are generated on both sides of the preform by using two movable molds.

Experimental

To support the investigation of the processes during the manufacturing of composite parts in the Gap-Impregnation process different lab tools for the manufacturing of monolithic and sandwich parts are available at the IKV.

Experimental setup

The subsequently presented investigations were carried out with a Gap-Impregnation tool which allows the production of flat monolithic parts with a flow path length of 800 mm and a width of 250 mm in variable thicknesses up to 8 mm. The tool can be equipped with different upper mold configurations, so that the following two types of investigations could be carried out:

- Investigation of the flow front advancement during the mold filling
- Investigation of process and preform parameters on laminate quality

For the investigation of the flow front advancement during the mold filling an upper mold with a transparent polymethylmethacrylate (PMMA) mold insert is available. The flow front advancement is recorded and documented with a video camera. As the upper mold with the transparent mold insert is not suitable for processing of hot curing resins, a heatable aluminium upper mold is available for the part manufacturing. Figure 3 shows the experimental setup for manufacturing parts with hot curing resins.

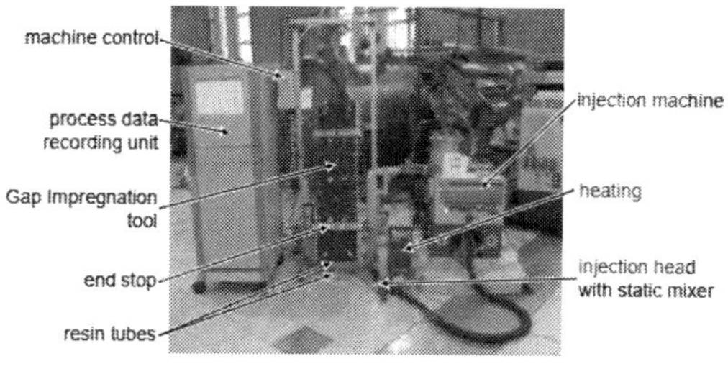

Figure 3 – Semi-automated Gap-Impregnation machine

This semi-automated Gap-Impregnation machine fundamentally consists of the Gap-Impregnation tool and an injection system for mixing and handling the resin. A vacuum pump LVS 610 Tp EcoFlex (Ilmvak GmbH, Ilmenau, Germany) is used for evacuation of the cavity at a defined absolute pressure of 200 mbar. The vacuum is applied through two punctual outlets. For the injection of the resin, an injection machine Eldo-Mix 101 (Hilger & Kern GmbH, Mannheim, Germany) with a static mixer is used. The resin is injected through two point gates into the cavity. The complete impregnation and compression process is controlled by a numerical machine control unit.
The geometry of the cavity, the position of the gates and the integrated sensors are shown in Figure 4.

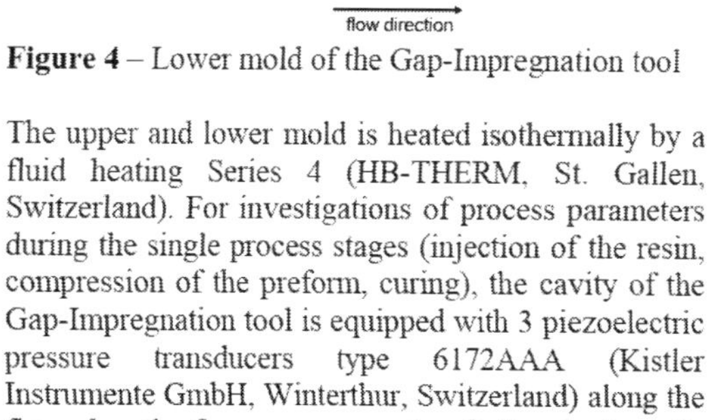

Figure 4 – Lower mold of the Gap-Impregnation tool

The upper and lower mold is heated isothermally by a fluid heating Series 4 (HB-THERM, St. Gallen, Switzerland). For investigations of process parameters during the single process stages (injection of the resin, compression of the preform, curing), the cavity of the Gap-Impregnation tool is equipped with 3 piezoelectric pressure transducers type 6172AAA (Kistler Instrumente GmbH, Winterthur, Switzerland) along the flow length for measurement of the hydrostatic pressure distribution and inductive displacement transducer to monitor the height of the flow gap. Additionally 3 dielectric 1" tool-mount-sensors (Netzsch Geraetebau GmbH, Selb, Germany) are integrated in the cavity, which are used for online monitoring of the cure kinetics of the resin during impregnation, compression and curing phase.

Materials used for mold filling studies

For the investigation of the mold filling studies preforms which consist of 1 layer non-crimp fabric (NCF) type 64005 (Ahlstrom Corporation, Helsinki, Finland) with an areal weight of 1170 g/m^2 have been used. Type 64005 consists of 4 layers with orientations +45° (160 tex), 90° (600 tex), -45° (160 tex) und 0° (600 tex) (0° = production direction of textile). The areal weight of the single layers are respectively 195 g/m^2, 390 g/m^2, 195 g/m^2 und 390 g/m^2. The total thickness amounts to 1.06 mm for the decompressed preform, respectively 0.92 mm for the compressed preform at a fiber volume content of 50 %. As fluid a non-reactive model fluid Mesamoll, Lanxess AG, Leverkusen, Germany has been used to prevent damages of the transparent PMMA mold insert. Mesamoll has a viscosity of 125 mPas at 20 °C, which is comparable to a heated resin-system.

Materials used for part manufacturing

For the investigation of process and preform parameters on the laminate quality, preforms with a total areal weight of 2400 g/m^2 and a symmetric, quasi-isotropic laminate setup (0°, 90°, +45°, -45°)$_S$ have been used. The preforms were manufactured by using 4 layers of two custom designed non-crimp fabrics type A and type B (Owens Corning Vetrotex, Brussels, Belgium) with an areal weight of 600 g/m^2 each. Type A consists of 2 layers with orientations 0° (1200 tex) and 90° (1200 tex) (0° = production direction of textile). Type B consists of 2 layers with orientations +45° (1200 tex) and -45° (1200 tex) (0° = production direction of textile). As the preform is not compressed during the mold filling, two different kinds of preform fixing strategies have been investigated. On the one hand the NCF layers have been stitched together. On the other hand a thermoplastic adhesive web Spunfab PA1300 (Keuchel Associates Inc., Cuyahoga Falls, USA) has been used as binder to fix the single NCF layers in order to prevent fiber washing during the compression phase. To achieve short cycle times, a highly reactive 2-component epoxy system LY 564/Aradur HY 3458 (Huntsman Advanced Materials, Basel, Switzerland) has been used. The curing duration of this system can be adjusted between 10 min at 80 °C and 3 min at 100 °C. As the duration for impregnation and compression of the preform is limited due to the maximum flow rate of the injection machine, the mold temperature has been set to 80 °C. This reduces the total time from start of the process until demolding to 10 min.

Mechanical testing

The mechanical properties of the produced laminates have been characterized by measuring the interlaminar shear strength by short beam bending tests according to DIN ISO 14130. This test has been chosen, in order to characterize the impregnation quality and the influence of the content of the thermoplastic adhesive web, which is located between the interfaces of the single layers of the reinforcement textiles, on the mechanical properties. The interlaminar shear strength is determined according to Equation 1.

$$\tau_{ILSS} = \frac{3}{4} \cdot \frac{F_{failure}}{b \cdot t} \quad [1]$$

τ_{ILSS}: Interlaminar shear strength
$F_{failure}$: Force which leads to a shear failure of the sample
b: Sample width
t: Sample thickness (laminate thickness)

The interlaminar shear strength has been measured at 3 positions of each laminate: near the gate, in the middle and near the suction. 6 samples with the dimensions 20 x 10 x 1.9 mm^3 have been tested lengthwise and crosswise at each position to calculate the mean values and the standard deviation. The testing speed was set to 1 mm/min. The testing was carried out at room temperature and without a special moisture conditioning of the samples. The test setup is shown in Figure 5.

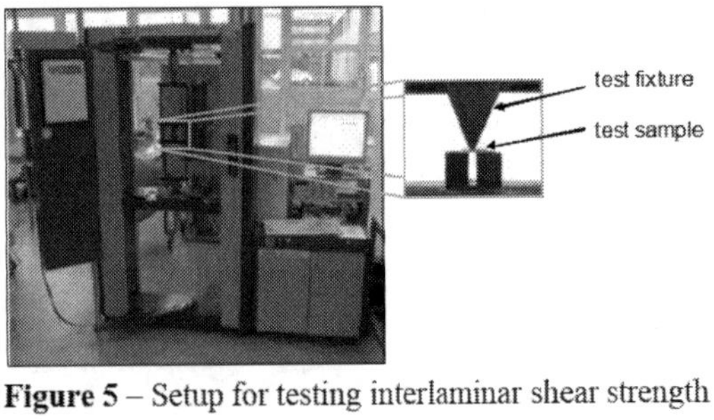

Figure 5 – Setup for testing interlaminar shear strength

Results and Discussion

Mold filling studies

Following, results of the mold filling studies, which have been carried out with the transparent upper mold, are presented. All investigations have been carried out at room temperature of 23 °C with Mesamoll as model fluid. The mold filling and the impregnation of a preform with glass fiber reinforcement is shown in Figure 6.

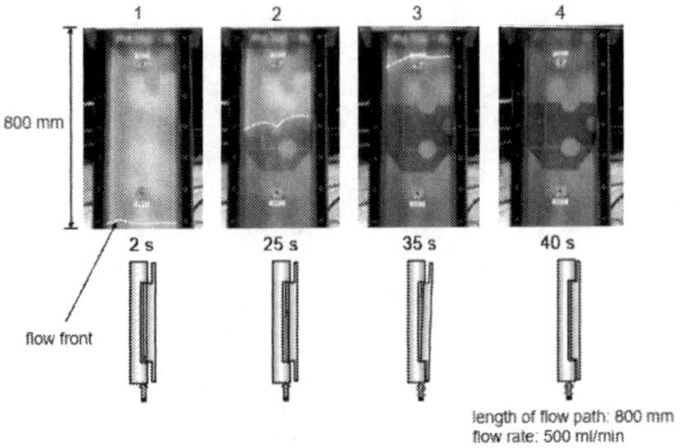

Figure 6 – Impregnation and consolidation of a glass fiber preform (areal weight = 1170 g/m^2)

The first and second image in Figure 6 show the matrix distribution in the flow gap and the impregnation of the preform. The impregnation can be clearly observed, as the transparency of the glass fiber preform increases when the fibers are wetted by the model fluid. After a partial filling of the cavity the gate is shut and the flow gap is closed in the area of the gate. By closing the flow gap the resin is further distributed over the preform (Figure 6, image 3). In the last step the flow gap is completely closed and the preform is compressed to the final fiber volume content (Figure 6, image 4). As the sequence of images in Figure 6 shows, the preform with a flow path length of 800 mm can be impregnated and consolidated with a fiber volume content of 50 % in less than 60 seconds.
Compared to the RTM process the Gap-Impregnation process offers a further important advantage next to the faster impregnation. During the part manufacturing in the RTM process it can occur that the resin advances beside the preform (Bickerton et al., 1999). This effect referred to as "race-tracking" is caused for example by inaccurately tailored preforms. "Race-tracking" can lead to an incomplete impregnation of the preform. For the investigation of the influence of inaccurately tailored preforms on the Gap-Impregnation process, preforms were placed into the cavity, so that a 5 mm wide channel forms between the lateral wall of the cavity and the preform. The investigations were carried out under variation of the flow gap size and the flow velocity. The comparison of the flow front distribution in dependence of the flow gap size is shown in Figure 7.

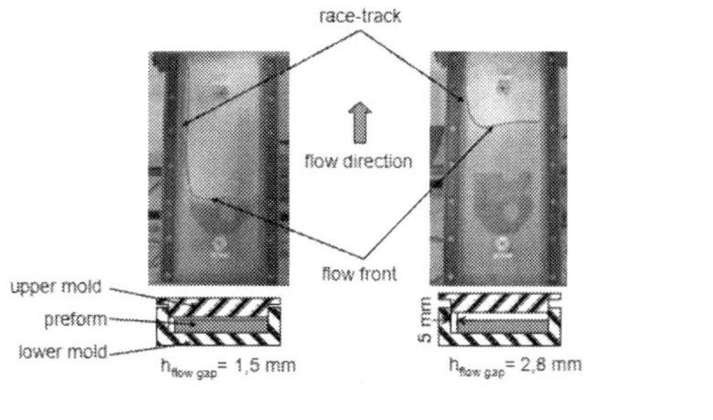

Figure 7 – Development of "race-tracks" in dependence of flow gap height

The left image in Figure 7 shows the flow front advancement when using a flow gap with a height ($h_{flow\ gap}$) of 1.5 mm. One recognizes clearly, that the resin advances significantly at the left side of the preform. The comparison with the resin distribution in the right image in Figure 7 shows, that the enlargement of the flow gap on 2.8 mm leads to a more even distribution of the advancing resin. The robustness of the impregnation process can be increased clearly by use of a ratio of about 3:1 between the height of the flow gap and the height of the "race-track". A higher process robustness helps to reduce the risk of the production of poor parts and thereby contributes to a higher economic efficiency.

Concerning the mold design, the influence of the gate design and the position of the resin injection in the cavity on the mold filling has been investigated. For the injection of the resin and the connection of the cavity with the vacuum pump, 2 point gates and 2 punctual outlets which are symmetrically arranged at both end of the cavity, are integrated in the tool (Figure 8).

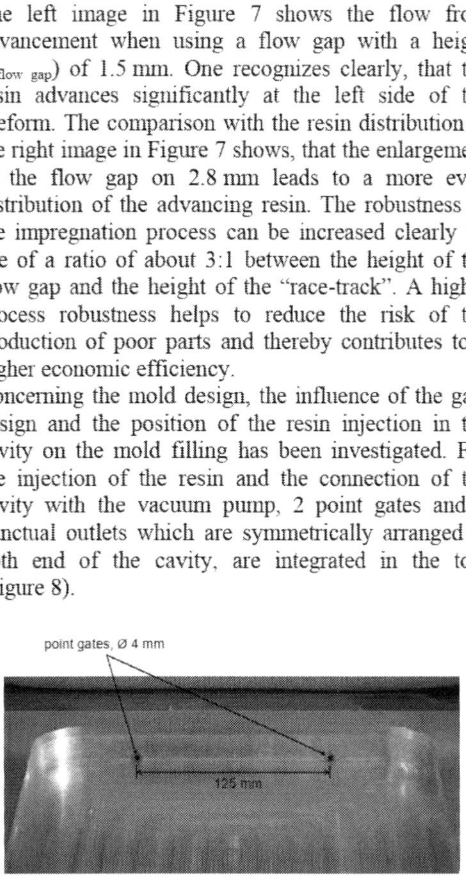

Figure 8 – Arrangement and geometry of the gates and outlets

Investigations have been carried out, where the resin has been injected into the cavity asymmetrically through one gate only. Furthermore, the influence of the design of the upper mold near the gates and the height of the flow gap on the mold filling behavior has been studied.

When injecting the resin asymmetrically through the left point gate at a flow gap height of 1.5 mm, the flow front advances as shown in Figure 9.

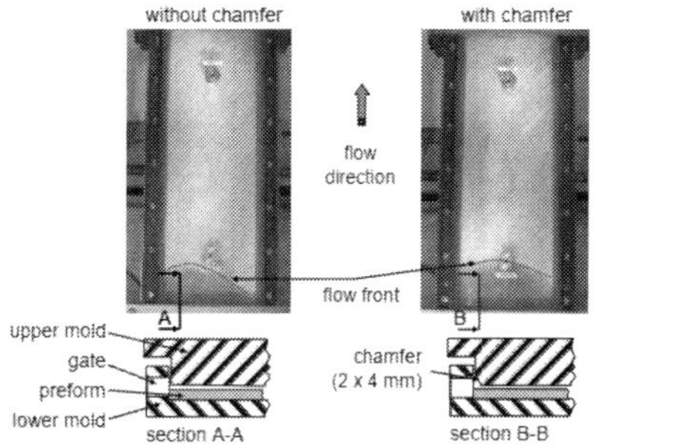

Figure 9 – Influence of mold design on the flow front advancement ($h_{flow\ gap}$ = 1.5 mm)

The left image in Figure 9 shows the flow front, when using an upper mold with a sharp-edged gating area. It can be observed, that the flow front advances asymmetrical at the beginning of the injection. It advances on the side of the cavity, where the resin is injected. The flow front can be more evenly distributed, when using an upper mold with a chamfer of 2 x 4 mm at the edge of the gating area (Figure 9, right image). When injecting the resin, it distributes in the channel, which is generated between the lateral wall of the cavity and the upper mold by the chamfer and then starts to flow into the cavity. So the even distribution of the flow front can be achieved.

These investigations have been carried out under variation of the flow gap height. They have shown, that the flow front is distributed more evenly, when the flow gap height is increased.

Investigations of preform and process parameters on the laminate quality

Based on the results of the mold filling studies, investigations of preform and process parameters on the laminate quality have been carried out. Therefore, preforms with a quasi-isotropic laminate setup and an areal weight of 2400 g/m^2 have been manufactured. 4 layers non-crimp fabric have been mounted with different contents of the binder Spunfab PA1300 for 180 min at a temperature of 140 °C by using a vacuum bag to compress the setup. The binder content between each reinforcement layer has been varied in 3 grammages (6 g/m^2, 12 g/m^2, 18 g/m^2). Afterwards, the preforms have been cut out of the fixed NCF layers by using a CNC controlled cutter Zuend M-1200 (Zuend Systemtechnik AG, Altstaetten, Switzerland). An inspection of the preform quality showed, that higher binder contents lead to a higher bending stiffness of the dry preform and a higher quality of the outer contour. This contributes to a better handling and a faster positioning of the preforms in the cavity.

Laminates with a thickness of 1.9 mm (respectively 50 % fiber volume content) have been manufactured under variation of binder content and injection temperature of the resin. The preform and process parameters are listed in Table 1.

Table 1 – Variations of preform and process parameters

cavity pressure (mbar)	resin flow rate (ml/min)	flow gap height (mm)	mold temperature (°C)	injection temperature (°C)	binder content (g/m²/layer)
100	760	3	75	40	6
100	760	3	75	40	12
100	760	3	75	40	18
100	760	3	75	60	6
100	760	3	75	60	12
100	760	3	75	60	18

The time necessary for injecting the resin limits the mold temperature when using high-reactive epoxy resins. A higher mold temperature speeds up the curing process of the epoxy resin, and thereby leads to a faster viscosity buildup of the resin. A higher viscosity of the resin complicates the impregnation of the fibers and makes higher forces to compress the preform necessary. With respect to a high laminate quality and short cycle times, an isothermal mold temperature of 75 °C has been evaluated as a good compromise within preliminary tests. 3 laminates have been manufactured and tested per combination of parameters. The laminates have been demolded 10 min after starting the process. A visual inspection of the laminates showed, that all laminates could be impregnated completely with a high impregnation quality. The visual inspection did not show significant differences of the impregnation quality in dependence of the variations. All laminates have been post-cured for 30 min at 80 °C to guarantee a complete cure of the epoxy resin.
The samples for evaluation of the interlaminar shear strength have been prepared out of the laminates and tested.

Visual examination

In order to predict the processing characteristics of preforms, which were bindered and stitched, the laminates were put to a visual examination. A selection of the sample laminates is shown in Figure 10.

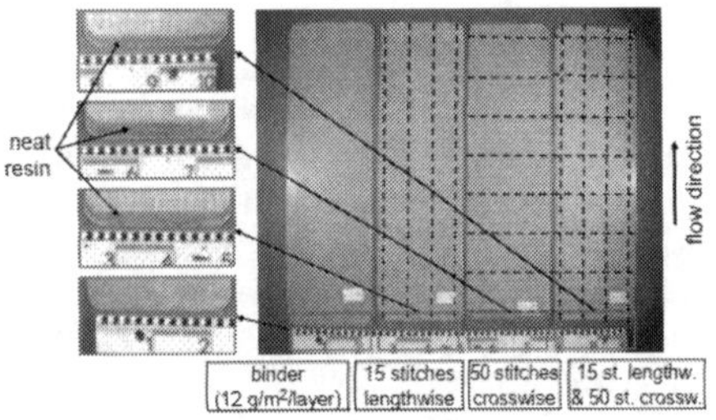

Figure 10 – Comparison of the processability between bindered and stitched preforms

All produced laminates have a very high quality regarding impregnation of the preforms and pore content. In the gate area of the stitched preforms with different seam patterns are clearly areas of neat resin visible. These areas of resin are caused by the displacement of the entire preform in the direction of the outlets. These shifts are caused by the flow forces that occur during the compression of the preform by closing the gap. The preforms are not fixed during the impregnation and the beginning of compression. Just because of their inherent stability the preforms are not displaced.
By the impregnation of the preforms, which are stitched lengthwise or crosswise to the resin flow direction the preform are shifted approximately 25 mm. The preforms, which are stitched lengthwise and crosswise, are shifted approximately 15 mm. The bindered preforms show no shift behaviour.
The results show that a two-dimensional fixation by using the thermoplastic adhesive web is best suited for processing in the Gap-Impregnation process. The influence of the stitched preforms in one direction can not be ascertained. In contrast, a seam pattern in two directions leads to a higher inherent stability and thus to a lower displacement of the preforms.

Results mechanical testing

The mean values and the standard deviations of the interlaminar shear strength of laminates with an injection temperature of 40 °C are shown in Figure 11, for 60 °C in Figure 12.

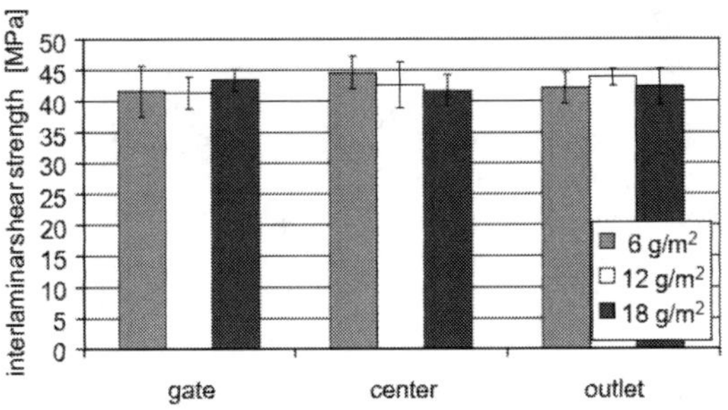

Figure 11 – Interlaminar shear strength, resin injection temperature 40 °C

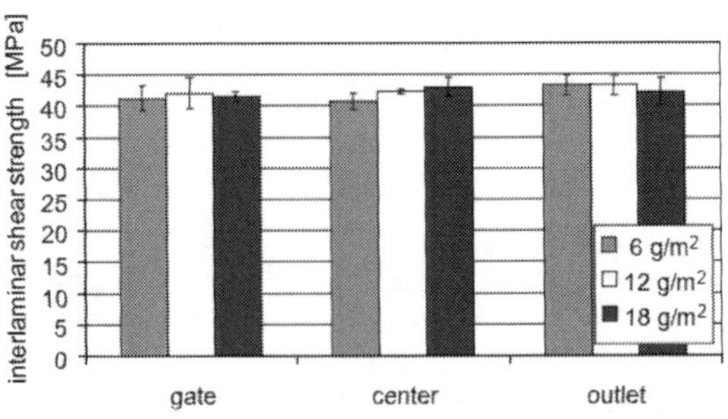

Figure 12 – Interlaminar shear strength, resin injection temperature 60 °C

The diagrams point out that there are no significant effects of the binder content and the injection temperature detectable. The measured values fluctuate within the range of the standard deviation. The same results have been achieved for the position along the flow path. No significant difference of the averaged interlaminar shear strengths of the laminates has been measured along the flow path.

As a result of the mechanical testing, it can be summarized, that neither the investigated injection temperatures nor the binder content and the flow path have significant influence on the mechanical properties.

Conclusions and Outlook

Liquid composite molding processes become more and more important, as they often facilitate production of high-performance composite parts at lower manufacturing costs compared to the expensive prepreg/autoclave process. An approach to reduce cycle times by a faster impregnation and to increase the robustness of the impregnation process is the Gap-Impregnation process. The mold filling studies have shown that the generation of a temporary flow gap facilitates a fast distribution of the resin and a high robustness of the impregnation process. The used thermoplastic binder Spunfab PA 1300 showed no negative influence on the impregnation of the preforms compared to the stitched preforms. The thermoplastic binder also had no influence on the interlaminar shear strength under the investigated conditions. The handling of the preform and the quality of the cut edges are improved by increased binder content. The investigated laminates with a fiber volume content of 50 % have been impregnated, compressed and cured within 10 minutes.

The Gap-Impregnation tool used for the presented investigations facilitates a semi-automated impregnation, compression and curing process. The preparation of the Gap-Impregnation tool and the demolding still needs manual effort. Furthermore, the forces available for compression of the preform and the flow rate for injecting the resin are limited. In order to reduce the time for the complete manufacturing cycle, including preparation and demolding, a completely automated manufacturing system has been developed and built up in cooperation with Hille Engineering, Aachen, Germany (Figure 13).

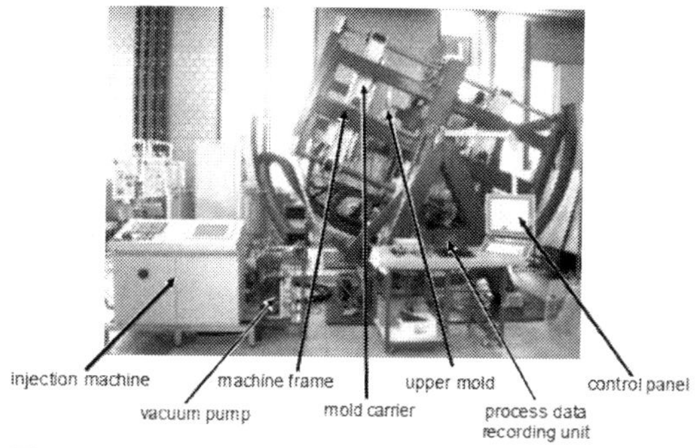

Figure 13 – Automated Gap-Impregnation machine

The manufacturing system consists of an injection machine which can realize an injection of the necessary resin within a few seconds. The defined movement of the upper mold is controlled by a clamping unit with special kinematics. The clamping unit facilitates clamping forces of 250 kN and closing speeds of the flow gap up to 30 mm/s. The fast mold filling in combination with a fast compression due to the high clamping forces facilitate a higher isothermal mold temperature and thereby contribute to shorter curing times.

This newly developed Gap-Impregnation machine facilitates an impregnation and compression to the final fiber volume content of 50 % in only 19 seconds. By using an isothermal mold temperature of 100 °C the curing of the resin can be reduced to 180 seconds. Through this time reduction a cycle time of 5 minutes for monolithic composite parts (500 mm x 500 mm) can be realized (Fischer et al., 2009).

Future research needs to be done regarding the influence of the preform parameter and the process variables on the laminate quality. Furthermore, it is necessary to transfer the so far developed knowledge to the production of components with higher complexity. These investigations should be carried out for the production of uniaxially curved parts with thickness differences, parts with openings, parts with load transition elements (inserts) and parts with integrated stiffening profiles.

Acknowledgements

The depicted research has been funded by the German Forschungsgemeinschaft (DFG) as part of the research unit 860 "New process chains for production of continuous fiber reinforced plastic parts: Integration of preforming, impregnation, molding and curing", to whom we extend our thanks.

References

Bickerton, S., Advani, S.G.: "Characterization and modeling of racetracking in liquid composite molding processes", Composites Science and Technology, **59**, 2215-2229 (1999)

Eberth, U., Herrmann, A.S.: Technische Textilien für Luftfahrt- und Fahrzeuganwendungen, Technische Textilien, chapter 6: Mobiltextilien, Deutscher Fachverlag GmbH, Frankfurt am Main (2006)

Ehrenstein, G. W.: Faserverbundkunststoffe: Werkstoffe - Verarbeitung – Eigenschaften, Carl Hanser Publishers, Munich, Wien (1992)

Flemming, M., et al.: Faserverbundbauweisen - Fertigungsverfahren mit duroplastischer Matrix, Springer Publishers, Berlin, Heidelberg (1999)

Fischer, K., Michaeli, W.: Untersuchungen zur Fertigung von flächigen Bauteilen aus faserverstärkten Kunststoffen mithilfe des Spaltimprägnierverfahrens. Zeitschrift Kunststofftechnik **4** (2008)

Fischer, K., Michaeli, W.: Automated Manufacturing of High-Performance Composite Parts by the Gap-Impregnation Process. Proceedings 2nd European Conference on Materials and Structures in Aerospace (EUCOMAS), Augsburg (2009)

Herrmann, A.S .: "CFK – Prozessbeherrschung – Wege und Methoden", 12. Nationales Symposium SAMPE Deutschland e.V., Braunschweig (2006)

Michaeli, W., Wegener, M.: Einführung in die Technologie der Faserverbundwerkstoffe, Carl Hanser Publishers, Munich, Wien (1989)

Mitschang, P., Neitzel, M.: Handbuch Verbundwerkstoffe – Werkstoffe, Verarbeitung, Anwendung, Carl Hanser Publishers, Munich, Wien (2004)

Mitschang, P.: "Schlüsseltechnologien für die Preform-LCM-Prozesskette", 8. Internationale AVK-TV Tagung für verstärkte Kunststoffe und duroplastische Formmassen, Baden-Baden (2005)

Meyer-Noack, S.: Konzeption und Erprobung des Spaltimprägnierformens, Dissertation RWTH Aachen (2006)

Young W.B., Chiu C.W.: "Study on compression transfer molding", Journal of Composite Materials, **29**, 2180–2091 (1995)

INFLUENCE OF BLENDING CONDITIONS ON PROPERTIES OF COMPOSITES BASED ON POLYANILINE

M.T. Cortés[1*], M. Cote[2], P. Ortiz[2], D. Beltrán[2]

[1*] *Department of Chemistry, Universidad de los Andes, Bogotá, Colombia –marcorte@uniandes.edu.co*
[2] *Department of Chemical Engineering, Universidad de los Andes, Bogotá, Colombia*

Abstract
A composite based on polyaniline (PANI) and low density polyethylene (LDPE) with electrical conductivity was developed. PANI-LDPE composites were prepared via melt blending, and was studied the influence of three variables (temperature, [PANI], rotor speed) on conductivity, microstructure and mechanical properties of the composite. It was found that the PANI concentration is the most influential variable in the conductivity and the elongation at break of the composite.

Introduction

Polyaniline (PANI) has become one of the most promising intrinsically conductive polymer (ICP) due to its chemical stability and physicochemical properties (Laakso, Osterholm et al. 1989; Han, Motobe et al. 1991; Bhattacharya and De 1996). Additionally, the relative inexpensive monomer let to use this polymer in potential technological applications such as the development of semiconducting polymeric materials for electronic applications (Wang and Jing 2005). However, a disadventage of PANI is his poor processability and poor physical and mechanical properties (Mitzakoff and Paoli 1999; Zhang, Wang et al. 2004). One alternative for the obtention of processable conducting materials from PANI is the preparation of composite materials using a conventional resin and a filler. Composites based on polyethylene (PE) and PANI had been prepared by several methods such as electrochemical deposition, powder dispersion, melt and solution blending. The miscibility between PANI and PE had proved to be a critical factor in the resulting features of the composite. Some studies evidence that the tensile strength is particularly sensitive to the adhesion between the filler and the matrix and the elongation break is severely affected by the poor mechanical features of PANI. This is observable when the PANI concentration increases. In general, the addition of conducting polymer has a negative influence on mechanical properties of the composite, especially, at a low temperature (Zhang, Gao et al. 2003; Zhang, Wang et al. 2003). This study aims at a better understanding of the interaction between some blending variables and of the influence of these variables on electrical and mechanical characteristics of the composite. This goal has been achieved from a factorial design, which allows to establish meaningful statistically relationships between the variables.

Experimental

PANI was synthesized by chemical oxidation of the monomer with ammonium persulfate and NaCl in aqueous acid media and doped with DBSA (Dodecylbenzensulfonic Acid) (Cortés and Sierra 2005). After polymerization PANI was dried at 45°C during 48 h. LDPE/PANI blends were prepared using a Brabender mixer with Banbury blades. As is shown in Table 1, the concentration of PANI, the processing temperature and the rotor speed were analyzed following a two level factorial design (2^3)(Cote, Cortés et al. 2009)

Table 1 – 2^3 Factorial design related to variables evaluated

	Levels		
Variables	-		+
Rotor speed (rpm)	40		80
Temperature (°C)	170		190
% PANI (% wt)	3		8
Test number	V*	T	%P
1	-	-	-
2	+	+	-
3	-	+	-
4	+	-	-
5	-	-	+
6	+	-	+
7	-	+	+
8	+	+	+

* Rotor speed

Initially, the matrix was added to the chamber until melting was completed, then the conducting polymer was added. The blending time was 8 min followed by the torque curve stability. The morphology of PANI powders was evaluated by Scanning Electron Microscopy (Quanta 200–FEI Instruments scanning electron microscope) at 30 kV. The microstructure of the composites was studied through conventional microscope with connected camera (Mitutoyo). Compression film molding was made following a protocol for PE matrix film molding in a hydraulic press (Drake) at 190°C and 70,000 psi. The conductivity of the films was measured using a 12.4 cm diameter aluminum disc, as electrodes, connected to a high voltage DC source (Daedalon®) and a digital multimeter (Fluke® 80 Serie III) at room temperature. Tensile properties were measured in a SW-501 equipment following NTC 942. The statistical studies were carried out with Design Expert software 7.0. The presented results represent the average over all four identical samples with standard deviation below 12% for tensile strength, 17% for elongation, and 8% for the conductivity.

Results and Discussion

The results of overall DC electrical conductivity of polymerized PANI (3.4×10^{-2} S cm^{-1}) lies inside the recommended values exposed elsewhere (Kuramoto and Tomita 1997; Haba, Segal et al. 1999)and suggest a reproducible method for PANI synthesis. The micrograph in Fig. 1 allows observing that the PANI particles not showing interconnected network, this morphology is usually found for polymerization by chemical oxidation(Jia, Li et al. 2007)

Figure 1 – SEM micrographs x1600 magnification.of PANI powders used in the blending.

The results of tensile properties of PANI/LDPE composites are shown in Table 2. Break stress and yield stress do not show large variations in all tests, while the elongation at break and Young's modulus expose considerable changes. The most relevant change observed is the elongation at break with %PANI, so the higher the PANI concentration the lower the elongation at break. The result is related to the poor mechanical properties of the PANI, which is very inflexible due to the aromatic rings and double bonds in its structure. The poor adhesion between PE and PANI could increase the stress exerted on PANI particles due to the slipping between PE chains, this would be the beginning of the fracture of the composite (Chipara, Hui et al. 2003).

The nonsignificant effects and interactions were eliminated from ANOVA model for the elongation at break. The significant interactions obtained from the refined model are shown in Fig. 2. According to this figure, there is no interaction between the effects since there is a slight improvement in the elongation at 190°C for the two PANI concentrations.

Table 2 – Mechanical properties of PANI(DBSA)/LDPE composites

Test number	Elongation at break (%)	Break stress (Mpa)	Yield stress (Mpa)	Young modulus (Mpa)
1	57	11	10	143
2	34	10	10	139
3	69	11	11	130
4	65	10	11	117
5	36	12	11	158
6	28	11	10	123
7	35	10	11	130
8	37	11	11	138

On the other hand, was observed that the contribution percentage of PANI concentration on conductivity is quite higher than the other contributions. In summary PANI% is the most influential variable and is independent of the others therefore the interactions with temperature and speed are negligible.

Figure 2 – Interaction diagram between %PANI and blending temperature for the elongation at break at 40 rpm (dotted line: 3%PANI w/w, dashed line: 8%PANI w/w) of PANI(DBSA)/LDPE composites.

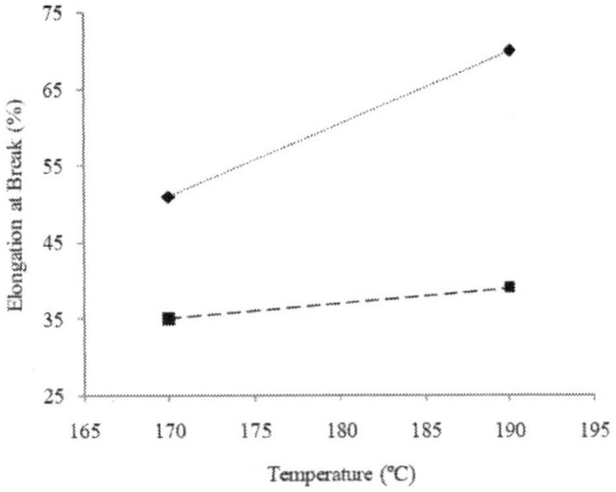

The composites obtained showed similar microstructures for each PANI concentration (tests 1-4 and 5-8). In Figure 3 are shown the typical microstructures found for each concentration. The analysis of the binary images showed the same type of particle size distributions for these composites with an average diameter lover than 10µm. This agreed with the observations for the SEM micrograph of the PANI (Figure 1) and demonstrates that under the experimental conditions used, the processing of the blends don't modify the particle size of the PANI. On the other hand, the dispersion of PANI particles in the PE matrix is random with clusters in all the obtained composites. The distance between particles was about 40μm for the composite 2 and about 28μm for the composite 8. The last is directly related to the observed conductivities changes, the composites with highest PANI concentration and shortest particle size showed the highest conductivity.

Figure 3 – Microstructures of the composites obtained in test 8 (a) and 2 (b)

a.

b.

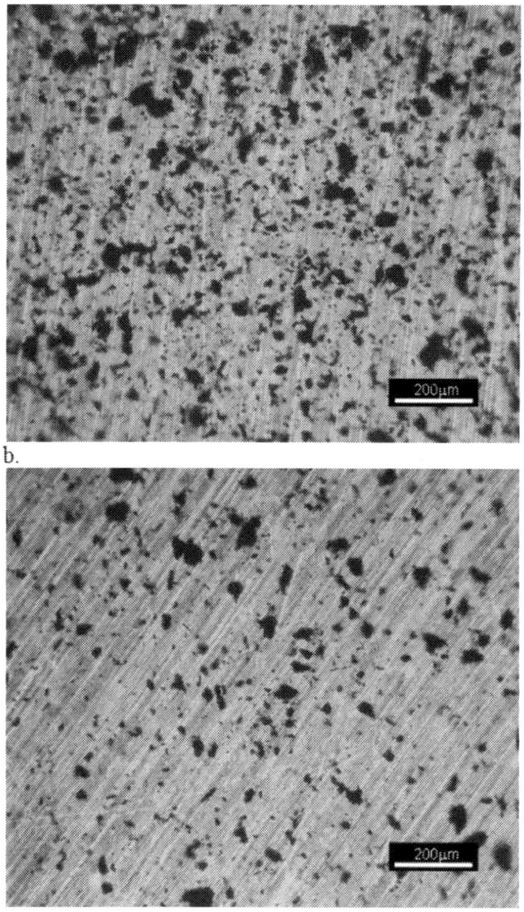

Finally, the results show that the processing temperature and speed have no significant effect on microstructures of the composites, which is also related to the observed mechanical properties.

Conclusions

PANI/LDPE composites were elaborated by melt blending and molded by compression films. Of the tensile properties studied only the elongation at break suffered significant changes. The most important factor that contributes to the change of the tensile and electrical properties was the concentration of PANI added in the matrix. The microstructures of the composites showed a PANI with a wide particle size distribution and the morphology was similar although the processing conditions were changed. Finally, the interparticle distance decreased for composites containing the highest PANI concentration, which is related to a conductivity increase.

Acknowledgements

We thanks to Wiley Publishers by the partial reproduction of these results from the paper: "PANI-LDE composites: effect of blending conditions", M. Cote, M.T. Cortés, D. Beltrán, P. Ortiz, Polymer Composites, 30(1) 2009. The authors are also thankful to Science Faculty and CIPP, Universidad de los Andes, Colombia for the support in this work, and to J. A. Delgado for the edition assistance.

References

Bhattacharya, A. and A. De, "Conducting composites of polypyrrole and polyaniline a review", Prog. Solid State Chem., **24**(3), 141-181 (1996)

Cortés, M. T. and E. V. Sierra, Polym.Bull, **56**, 37-45 (2005)

Cote, M., M. T. Cortés, et al., "PANI-LDPE composites: Effect of blending conditions", Polymer Composites, **30**(1), 22-28 (2009)

Chipara, M., D. Hui, et al., "On polyethylene-polyaniline composites", Composites:Part B, **34**(7), 637-645 (2003)

Haba, Y., E. Segal, et al., "Polymerization of aniline in the presence of DBSA in an aqueous dispersion", Synth.Met., **106**(1), 59-66 (1999)

Han, J. H., T. Motobe, et al., "Highly electrically conducting polymer blends", Synth.Met., **45**(2), 261-264 (1991)

Jia, Q. M., J. B. Li, et al., "Electrically conductive epoxy resin composites containing polyaniline with different morphologies", Mater.Sci.Eng.,A, **448**(1-2), 356-360 (2007)

Kuramoto, N. and A. Tomita, "Aqueous polyaniline suspensions: Chemical oxidative polymerization of dodecylbenzene-sulfonic acid aniline salt", Polymer, **38**(12), 3055-3058 (1997)

Laakso, J., J. E. Osterholm, et al., "Conducting polymer blends", Synth.Met., **28**(1-2), 467-471 (1989)

Mitzakoff, S. and M.-A. D. Paoli, "Blends of polyaniline and engineering plastics", Eur. Polym. J., **35**, 1791-1798 (1999)

Wang, Y. and X. Jing, "Intrinsically conducting polymers for electromagnetic interference shielding", Polym.Adv.Technol., **16**(4), 344-351 (2005)

Zhang, Q.-H., X.-H. Wang, et al., "Electrically conductive, melt-processed ternary blends of polyaniline/dodecylbenzene sulfonic acid,

ethylene/vinyl acetate, and low-density polyethylene", J. Polym. Sci., Part B: Polym Lett., **42**(20), 3750-3758 (2004)
Zhang, Q. H., J. Gao, et al., "Phase separation of polyaniline blends with thermoplastic polymers", Synth.Met., **135-136**, 479-480 (2003)
Zhang, Q. H., X. H. Wang, et al., "Dynamic mechanical properties of melt processable PANI-DBSA / LDPE blends", Synth.Met., **135-136**, 481-482 (2003)

Properties of polyamide 6.6 composites containing recycled glass particles

Samuel M. Toffoli, Irina M. Factori*

University of São Paulo - São Paulo - SP - Brazil

Polyamide 6.6 is one of the most important members of the polyamide family, mainly for its excellent engineering properties such as good mechanical and thermal performances. The possible application areas for such material are enlarged by the addition of inorganic fillers, among which, glass fibers, talc (natural), wollastonite (natural), and glass microspheres are the most frequently used in industrial scale. On the other hand, glass powder from glass cullet has never been studied as a polyamide reinforcement. Glass is infinitely recyclable. However, the smaller glass particles are rejected by the glass industry because of the carry-over phenomenon inside the glass furnace, what increases the cost of the smoke cleaning, as well as the possibility of increasing refractory corrosion and the consequent reduction in the useful life of the furnaces. In this paper, PA-6.6 composites, reinforced with different percentages (up to 20 wt.%) of recycled, irregularly shaped, glass powder, and the other common industrial fillers, were processed, in laboratory scale, in a double–screw extruder. Specimens for testing were obtained by injection molding, and the following composite properties were measured: tensile strength, elastic modulus, elongation at rupture, notchless Charpy impact strength, and dimensional stability. The results were compared and discussed and indicated that it is possible to use particles of recycled glass in a PA-6.6 matrix, once the final composite properties are compatible to the ones of composites containing usual commercial fillers, particularly glass microspheres.

STOKESIAN DYNAMICS SIMULATIONS OF THE RHEOLOGICAL PROPERTIES OF FIBER SUSPENSIONS IN UNIAXIAL EXTENSIONAL FLOWS

M. Yamanoi[1*] and J. M. Maia[2]

[1] *I3N - Institute for Nanostructures, Nanomodelling and Nanofabrication, Department of Polymer Engineering, University of Minho, Guimarães 4800-058, Portugal – yamanoi@dep.uminho.pt*
[2] *Department of Macromolecular Science and Engineering, Case Western Reserve University, 2100 Adelbert Rd. Cleveland, OH 44106-7202, USA – joao.maia@case.edu*

Abstract – Fiber/polymer composites are widely used industrially. Orientation of fibers is important to improve their properties and its characteristics under extensional flows are key because fibers don't show tumbling, which is observed under shear. Stokesian Dynamics simulation has been applied for analyzing the relations between structure and rheological properties in uniaxial extensional flows. The predictions were compared with published experimental data and the orientation tensor was discussed as a function of flexibility, aspect ratio and volume fraction. The results show that orientation depends on strain only and that rheological properties may be described by a constitutive equation.

Introduction

For improving performance of polymer products, fillers like silica particles, glass fibers, carbon fibers, and so on are commonly mixed to the polymeric matrix [1]. For fiber composites, orientation of fillers becomes important, which is not a factor in sphere particle composites, and for improving mechanical properties, methods to align orientation of fibers would be key. Simple shear induces tumbling motions [1], which makes orientation disordered. Extensional flow does not have rotation effects and thus yield better aligned structures. A Stokesian Dynamics Simulation Software for fiber composites has been recently developed by our group and in this paper we use this code to probe the relation between extensional rheology and structures. First, the performance of the software is validated and then the relation between rheological properties and structure is discussed.

Modeling

Modeling of Fiber
To discuss the effects on uniaxial extensional rheological behavior of fiber flexibility, fibers are modeled as a string of beads. Between adjacent beads, a stretch force and bending and torsion torques are introduced.

Stokesian Dynamics
Hydrodynamic interaction (HI) is calculated by Stokesian dynamics [2] and the expression for HI between spherical particles, which is known [3], was applied in the research. HI was then calculated with all the segments in the system by calculating the inverse of the resistance matrix. Such calculation is costly, so HI was calculated for each fiber [4].

Results and discussion

The validation of the software was performed with experimental data (HDPE/glass fiber; volume fraction ϕ =0.1, aspect ratio Ar=13) [5]. The results are shown in Fig. 1. When the strains are small, the prediction fits to experimental data; at large strains the predictions differ from the experiments because the latter shows rupture of the samples. From this, it can be said that software developed in the research has performance to discuss rheological properties.

For the next step, the second order orientation tensor, defined as

$$\mathbf{a}_2 = \int_{\mathbf{p}} p_i p_j \Psi(\mathbf{p}) d\mathbf{p}$$

is discussed as a function of (a) aspect ratio; (b) rigidity; (c) volume fraction in Fig. 2, where the component of flowing direction $a_{2,11}$ is shown.

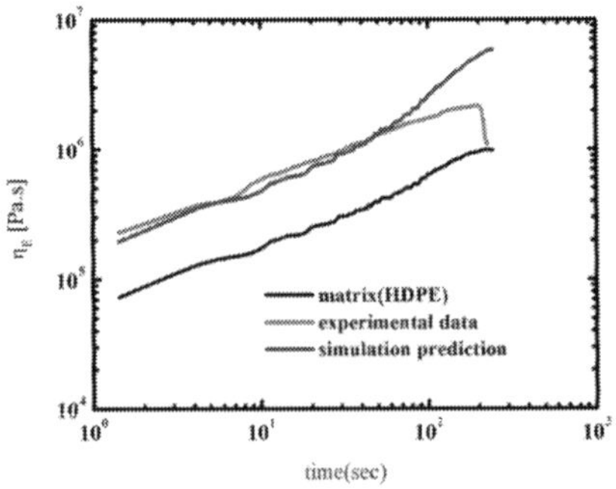

Figure 1 – Comparison of extensional viscosity η_E between simulation predictions and experimental data [5].

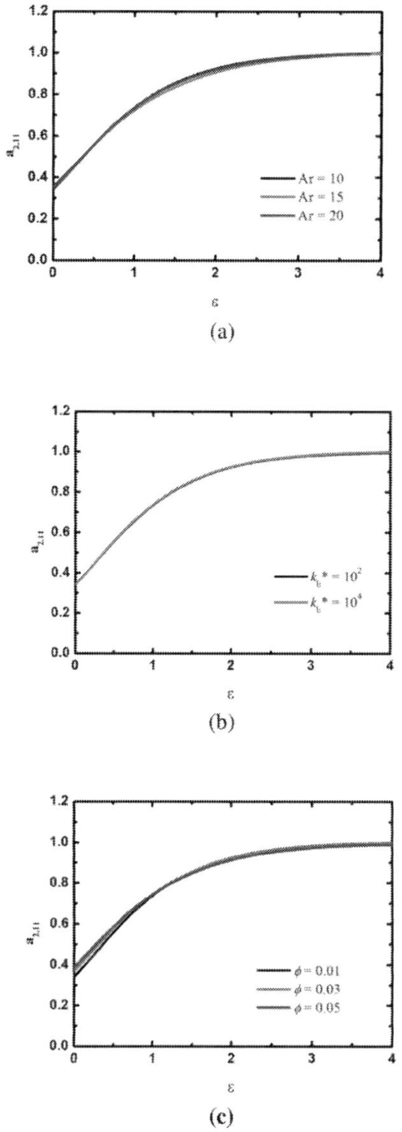

Figure 2 – Prediction of orientation tensor as a function of (a) aspect ratio Ar; (b) rigidity (non-dimensional bending constant $k_b^* = k_b / \eta_m a^3 \dot{\varepsilon}$, where η_m is viscosity of matrix, a radius of glass fiber, $\dot{\varepsilon}$ Henkey strain rate); (c) volume fraction ϕ.

From this, it is obvious that the growth of orientation is independent of aspect ratio, rigidity, and volume fraction. Fig. 3 shows η_E / η_m at ε=0.14 as a function of ϕ, just before sample rupture. The coefficient of ϕ is around 70 for both (a) and (b) which suggests that a proper constitutive equation can

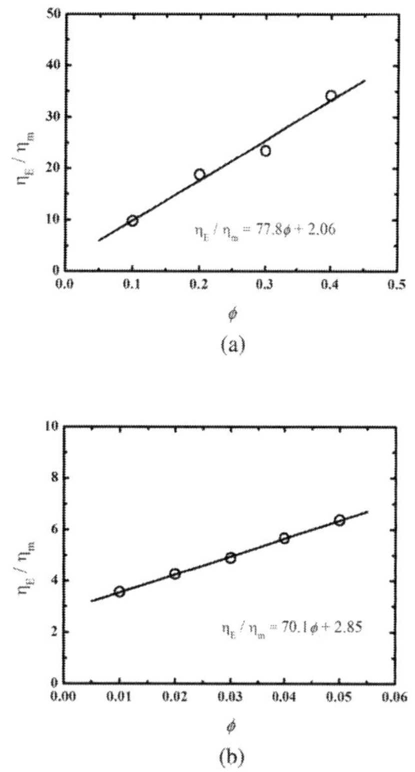

Figure 3 – Relative uni-axial viscosities normalized by the viscosity of matrix: (a) experimental data [6], (b) simulation prediction at strain 0.14. The approximation line is shown in the graphs.

describe the time-dependent extensional viscosity of fiber composites.

References

[1] R. Pal, Rheology of Particulate Dispersions and Composites, CRC, Boca Raton, 2007.

[2] G.B. Jeffery, The motion of ellipsoidal particles immersed in a viscous fluid, Proc. Roy. Soc. **A102** (1922) 161.

[3] J. F. Brady, and G. Bossis, Stokesian Dynamics, Ann. Rev. Fluid Mech, **20** (1988) 111.

[4] S. Kim, S. J. Karrila, Microhydrodynamics: Principles and Selected Applications, Dover Publications, 1991.

[5] S. Yamamoto, T. Matsuoka, Dynamic simulation of fiber suspensions in shear flow, J. Chem. Phys. **102** (1995) 2254.

[6] T.J. Fiske, A.N. Wagner, R. Yazici, D.M. Kalyon, Uniaxial extensional flow behavior of fiber filled suspensions as affected by fiber aspect ratio and concentration, SPE/ANTEC 1996 Proceedings (1996) 1134.

Evaluation of Mechanical Properties of Poly (propylene) and poly(lactic acid) reinforced by raw or corona treated miscanthus fibres

Mohamed Ragoubi, Stephane Molina, Beatrice George*

LERMAB - Nancy - France

ABSTRACT Composite materials based on cellulosic fibres reinforcements show very important environmental and economical advantages, they offer as well benefits in terms of mechanical properties. These composite materials can be considered as an alternative to conventional ones made from technical fibres (glass fibres mainly) in different areas for technical applications (automotive, building, packaging...). Nevertheless, the hydrophilic character of fibres and hydrophobic nature of polymer remains a major problem because of poor and weak interfacial compatibility. Several chemical (maleic anhydride, acetylation) or physical treatments are suitable (cold plasma, corona, UV...) to enhance the fibres wettability and increase the adhesion « reinforcements / matrix » [1, 2]. In the first part of our study, the effects of Corona Discharge Treatment (CDT) onto the surface of miscanthus fibres were investigated. Then, the mechanical behaviour of polypropylene (PP) and polylactic acid (PLA) reinforced with miscanthus fibres was evaluated. The surface modifications resulted under corona discharge treatment were characterized by different technical analyses (X-ray Photoelectron Spectroscopy XPS, Scanning Electon Microscopy SEM,...). XPS findings revealed that corona treatment causes an increase in the O- containing species and a decrease of C carbon content on the fibres surface. This confirms that corona discharge treatment induces a surface oxidation by introduction of polar groups (such as carboxyl, carbonyl and hydroxyl groups) on the fibre surface. From analyses by SEM, it appears that roughness increases with corona treatment time inducing a scale effect generated by etching as a result of electron irradiation [3,4].

THE INFLUENCE OF PEROXIDE INITIATED CROSSLINKING ON THE MORPHOLOGY AND PHYSICAL PROPERTIES OF PE/SISAL FIBRE COMPOSITES

EEM Ahmad and AS Luyt*

Department of Chemistry, University of the Free State (Qwaqwa Campus), South Africa

(E-mails: AhmadM@qwa.ufs.ac.za, LuytAS@qwa.ufs.ac.za)

Abstract

The present study has dealt with the effectiveness of organic peroxide initiated crosslinking as compatibilization method in sisal fibre/polyethylene (LDPE, LLDPE, and HDPE) composites. Dicumyl peroxide was used to initiate both grafting and crosslinking that may improve the interfacial adhesion in the composites. Both micro- and nano-sisal fibres were used to prepare the composites. The microcomposites (10, 20, and 30% fibre) were prepared by melt mixing, followed by hot-melt pressing. The nanocomposites were prepared with the aid of Pluronic F98 (PEO-PPO-PEO triblock copolymer) as a surfactant to enhance the fibre dispersion in the polymers. The aqueous surfactant solution was added to an aqueous nanofibres suspension, followed by sonication, mechanical stirring, and freeze-drying. The interaction between the surfactant and the nanofibres was established by ATR-FTIR spectroscopy. Different fibre loadings (3, 6, and 10%) were used to prepare the nanocomposites, following the same method used for microcomposites. The thermal, mechanical, dynamic mechanical and morphological properties of the micro- and nanocomposites were investigated, but only the results obtained with the microcomposites will be discussed in this paper. In general, the addition of the peroxide was found to enhance the tensile strength and modulus of the resulting composites, with a few exceptions. The interfacial adhesion was improved significantly, as substantiated through polarized optical microscopy (POM), scanning electron microscopy (SEM), and FTIR microscopy. The TGA results showed that the presence of the sisal fibres led to thermally less stable composites compared to the neat matrices, whereas marginal differences in stability were observed between the untreated and peroxide treated composites. The gel content and DSC results show interesting trends in terms of the influence of fibre size and content, as well as dicumyl peroxide treatment, on the morphology and crystallization behaviour of the polymers and composites. The DMA results show differences in the viscoelastic properties of the samples, depending on whether the samples were prepared in the presence or absence of dicumyl peroxide. The influence of fibre content, as well as dicumyl peroxide initiated compatibilization, on the tensile and impact properties of the samples will also be presented.

Introduction

The utilization of lignocellulosic fibres as reinforcing materials in polymer matrix composites has been an accepted route to achieve enhancement in material properties, due to their unique characteristics. Additionally, the growing environmental awareness as well as the lower cost from an economic point of view is also of great interest (Bledzki and Gassan, 1999; Mohanty et al. 2000; Reddy and Yang, 2005a; Wambua et al. 2003; Kalia et al. 2009; Bogoeva-Gaceva et al. 2007; John and Thomas, 2008; Saheb and Jog, 1999; Li et al. 2000). The advantages of natural fibres, compared to traditional reinforcing materials, are their low density, high specific strength and modulus, wide availability, recyclability, biodegradability, low cost, and renewability (Bledzki and Gassan, 1999; Wambua et al. 2003; Mohanty et al. 2000; Huda et al., 2005; Nishino et al., 2004; Puglia, 2003).
A significant amount of research work has been done on the applications of natural fibre composites, particularly in the field of the automotive industry (Alves et al., 2010; Pervaiz and Sain, 2003; Bledzki et al., 2006; Zah et al., 2007), and construction (Riedel and Nickel, 1999; Dweib et al., 2004, 2006; Burgueño et al., 2004; Mathur, 2006). Despite the fact that natural fibre-reinforced composites offer many benefits compared to traditional reinforcing materials, there are many shortcomings of these materials. They are incompatible with hydrophobic polymers, have high moisture absorption, low thermal resistance, quality variations, and poor wettability.

Sisal fibre is a hard fibre extracted from the leaves of the sisal plant (Agava sisalana). Many researchers (Li et al., 2000; Mukherjee and Satyanarayana, 1984; Silva et al., 2008; Munawar et al, 2007; Mishra, 2004) have reported the physical properties and chemical composition of sisal fibre. Due to their higher mechanical and physical properties sisal fibre has been used as a reinforcing material in many thermoplastic (Mokoena et al, 2004; Joseph, 1996; Malunka, 2006; Djidjelli et al., 2006; Zhou et al., 2002) and thermosetting (Pavia and Forollini, 2006; Megiatto et al., 2007; Singh et al, 1996; Rong et al., 2002;

Zhong, 2007) matrices. Owing to the incompatibility between natural fibres and hydrophobic polymers, as mentioned above, methods of promoting adhesion are frequently needed. Various chemical (Li et al., 2000; Jackson, 2007; Botaro et al., 2009; Mishra, 2004) and physical (Martin et al., 2003) methods were applied to achieve this goal in sisal fibre-polymer composites, either by modifying the surface of the sisal fibres, the polymer matrix, or the mix during compounding. The use of organic peroxides as compatibilizers in natural fibre (sisal) reinforced composites has been reported by only a few researchers (Joseph, 1996; Mokoena et al, 2004; Malunka et al., 2006; Luyt and Malunka, 2005). Joseph et al. (1996) have studied the effect of chemical treatment in the tensile properties of short sisal fibre reinforced LDPE. The enhancement in tensile properties after addition of small amounts of peroxides was attributed to the peroxide induced grafting of PE onto sisal fibre surfaces. SEM micrographs of the fracture surfaces were used as evidence. The influence of fibre content and different dicumyl peroxide concentrations on the thermal, mechanical, and viscoelastic properties of short sisal fibre-LLDPE composites, were investigated by Mokoena et al., (2004). The study revealed a significant improvement in the tensile strength after peroxide induced grafting between thefibres and the PE matrix. The stress relaxation measurements of the samples prepared by addition of 1% DCP indicated better stability upon prolonged loading. However, detrimental effects of higher DCP concentrations on the PE matrix, especially at lower fibre content, was also reported. In another study, the physical-, morphological as well as the chemical interactions between EVA and sisal fibres, in short sisal reinforced EVA composites prepared in the presence of dicumyl peroxide was examined by Malunka et al., (2006). The results suggested both crosslinking and grafting of PE. The grafting of EVA onto sisal fibres was confirmed by FTIR analyses, as well as porosity and surface energy measurements.

Kalaprasad et al., (2004) examined the effect of fibre length and chemical modifications on the tensile properties of sisal/glass hybrid fibres reinforced LDPE composites. Various treatments, such as akali, acetic anhydride, stearic acid, $KMnO_4$, maleic anhydride, silane, DCP and benzoyl peroxide (BPO) were applied. The results showed maximum increase in tensile strength and modulus for the samples treated with BPO and DCP compared to others treatments. BPO and DCP induced grafting of LDPE onto sisal surfaces were claimed to be the reason behind this improvement as evident from SEM micrographs. According to these studies it is clear that organic peroxides have an excellent effect as compatibilizers on the polymer-fibre interface. In addition, the number of scientific publications in using organic peroxides as compatibilizers is very small and most of the studies were concerned about the effect of these materials in the composites properties, without giving evidence that are more detailed about the interface itself.

The aim of this study was the preparation and characterization of polyethylene (LDPE, LLDPE, and HDPE) based composites reinforced with sisal fibres (micro- and nano) with particular attention to the optimization of the filler-polymer interactions through dicumyl peroxide treatment during the process of melt mixing. The effect of this compatibilization on the filler-polymer interface was examined thoroughly to determine the type of the interaction between these components. Furthermore, the compatiblized samples were compared with uncompatibilized ones.

Experimental

Materials

Sisal fibre was obtained from the National Sisal Marketing Committee in Pietermaritzburg, South Africa. Polyethylene (LDPE, LLDPE, and HDPE) were supplied in pellet form by Sasol Polymers, Johannesburg, South Africa. They have the following specifications (LDPE: MFI = 2 g/10 min at 190 °C / 2.16 kg, density = 0.922 g cm^{-3}; LLDPE: MFI = 1 g / 10 min at 190 °C / 2.16 kg, density = 0.924 g cm^{-3}; HDPE: MFI = 2 g / 10 min at 190 °C / 5 kg, density = 0.956 g cm^{-3}). Dicumyl peroxide (Bis(α,α-dimethylbenzyl)peroxide) was supplied by Merck, South Africa. It has a minimum assay of ≥ 98%. Petroleum ether was also supplied by Merck.

Treatments of the sisal fibres

Sisal fibres were cut into an average length between 5 and 10 mm, and then the fibres were soaked in distilled water for 24h to remove any surface impurities, filtered out and washed thoroughly with distilled water before allowed to dry in an oven at 80 °C for 48h. The dried fibres were washed with petroleum ether for 5h at 50 °C to remove waxy materials and natural oils, followed by oven drying at 80 °C.

Composites preparation

The polyethylene-sisal fibre composites were prepared in the 55 ml mixing chamber of a Plastograph-W50EHT (Brabender®OHG, Germany) at 140 °C, 30 rpm and 10 min for LLDPE and LDPE composites, and at 150 °C, 30 rpm and 10 min for HDPE composites. Dicumyl peroxide treated samples were prepared by addition of 1 phr DCP to the mixture one minute before the end of the mixing time. The melt-pressing of the prepared samples was performed at 175 °C, 50 bar for 10 min.

Characterization techniques
Differential scanning calorimetry (DSC)
DSC analysis was performed in a Perkin Elmer-DSC7 differential scanning calorimeter. Sample weights in the range of 5 to 10 mg sealed in aluminum pans were heated from 30 to 170 °C at a heating rate of 10 °C min^{-1} under nitrogen flow (20 mL min^{-1}), cooled and re-heated under the same conditions. The melting and crystallization temperatures as well as melting and crystallization enthalpies of the samples were determined from the cooling and re-heating runs. The crystallinities of the samples were calculated according to Equation 1.

$$\% \text{ Crystallinity} = (\Delta H_m/\Delta H_m^0) \times w \times 100 \quad (1)$$

where ΔH_m^0 = 288.8 J g^{-1} is the heat of fusion for 100% crystalline polyethylene (Peacock, 2000) and w is the weight fraction of the polymer in the composite.

Thermogravimetric analysis (TGA)
Thermogravimetric analysis was used to study the thermal stabilities of the resultant composites in a Perkin Elmer TGA7 thermogravimetric analyzer. Samples with weights in the range 5-10 mg were heated from 30 °C to 700 °C at a heating rate of 10 °C min^{-1} in nitrogen atmosphere.

Tensile analysis
The tensile properties of the composites were determined on a Hounsfield H5KS tensile tester. Eight samples were tested for each composition with a crosshead speed of 50 mm min^{-1} at ambient temperature.

Dynamic mechanical analysis (DMA)
The dynamic mechanical properties of the samples were determined in a Perkin Elmer Diamond dynamic mechanical analyzer to determine the viscoelastic properties in bending mode. Samples were heated from -100 °C to 100 °C at a heating rate of 5 °C min^{-1} at a frequency of 1 Hz under nitrogen flow (30 ml min^{-1}).

Scanning electron microscopy (SEM)
The morphology of the composites was examined by using a Shimadzu SSX-550 Super-Scan scanning electron microscope. The surfaces of cryofractured samples were sputter-coated with gold before viewing.

Results and discussion

Differential scanning calorimetry (DSC)
DSC measurements were performed to characterize the thermal behaviour of neat and DCP treated polyethylenes, and their DCP modified and unmodified sisal composites. Tables 1 to 3 show the melting peak temperatures (T_m) and the melting enthalpies, which were determined from the DSC curves, as well as the calculated melting enthalpies and degrees of crystallinity, whereas the DSC heating curves are shown in Figures 1 to 3. Table 1 and Figure 1 summarize the thermal behaviour of LDPE and its sisal composites. It can be seen that the T_m values of LDPE in all the unmodified samples are the same (about 110 °C) regardless of the fibre content. This is an indication that the size of the crystalline domains is retained in the matrix, because T_m is directly linked to the size of the crystalline domains of the polymer matrix (Pasquini. et al., 2008). The melting enthalpies, however, decreased with increasing fibre load, as expected, due to the decrease in polymer matrix quantity. The degrees of crystallinity, calculated from the respective melting enthalpies, remained fairly constant with increasing fibre content in the absence of DCP. This indicates that the fibre had little influence on the crystallization behaviour of LDPE. In the case of the DCP treated LDPE the DSC melting peak is less intense but broader than that of pure LDPE, although the melting temperature and melting enthalpy, as well as the degree of crystallinity, are very similar to those of pure LDPE. This could be due to a reduction in the structural regularity caused by crosslinking, which caused a broader distribution of crystallite sizes. On the other hand, the DCP treated composites show observably lower melting temperatures than the untreated samples and the DCP treated LDPE, while the degrees of crystallinity did not change appreciably. This is probably the result of grafting between the fibre and the LDPE chains, and the accompanying restriction of lamellar growth giving rise to thinner lamellae. These results are consistent with the results of previous studies (Joseph et al., 1996; Malunka et al., 2006; Mokoena et al., 2004; Luyt and Malunka, 2005).

The DSC melting behaviour of LLDPE and its composites are given in Table 2 and Figure 2. The melting temperature of LLDPE slightly increases with increasing fibre content for the unmodified composites, but the temperature increase is not significant enough to relate it to specific morphological changes in the samples. The degree of crystallinity also did not change appreciably with increasing fibre content. For the DCP treated samples both the melting temperature and the degree of crystallinity show obvious decreases with increasing fibre content. In this case there was obviously a higher extent of crosslinking and grafting compared to the LDPE samples, which inhibited the crystallization of the polymer chains. Table 3 and Figure 3 represent the DSC melting behaviour of HDPE and its composites. The melting temperatures and degrees of crystallinity did not change significantly with increasing filler content for the untreated samples, which implies that the presence of the fibre did not significantly influence the crystallization behaviour of HDPE. The degrees of crystallinity of the treated samples were observably lower than those of the untreated samples, which indicates a stronger influence of crosslinking and grafting on the crystallization behaviour of samples with high degrees of crystallinity.

Table 1 Melting characteristics of LDPE composites: melting temperature (T_m), observed melting enthalpy (ΔH_m^{obs}), calculated melting enthalpy (ΔH_m^{calc}) and degree of crystallinity (χ_c)

LDPE/Sisal (w/w)	T_m (°C)	ΔH_m^{obs} (J g^{-1})	ΔH_m^{calc} (J g^{-1})	χ_c (%)
Untreated LDPE composites				
100/0	110.6	55.6	55.6	19.3
90/10	110.6	46.1	50.0	17.7
80/20	110.9	42.9	44.5	18.6
70/30	110.7	36.3	38.9	18.0
DCP-treated LDPE composites				
100/0	108.6	56.0	56.0	19.4
90/10	107.0	44.7	50.0	17.4
80/20	106.9	43.3	44.8	18.7
70/30	106.3	35.4	39.0	17.5

Table 2 Melting characteristics of LLDPE composites: melting temperature (T_m), observed melting enthalpy (ΔH_m^{obs}), calculated melting enthalpy (ΔH_m^{calc}) and degree of crystallinity (χ_c)

LLDPE/ Sisal (w/w)	T_m (°C)	ΔH_m^{obs} (J g^{-1})	ΔH_m^{calc} (J g^{-1})	χ_c (%)
Untreated LLDPE composites				
100/0	124.7	58.5	58.5	20.3
90/10	125.6	52.1	52.7	20.0
80/20	125.0	46.6	46.8	20.2
70/30	126.6	43.2	41.0	21.4
DCP-treated LLDPE composites				
100/0	119.0	60.9	60.9	21.0
90/10	121.0	46.2	54.8	17.8
80/20	116.8	39.2	48.7	17.0
70/30	114.4	32.2	42.6	15.9

Table 3 Melting characteristics of HDPE composites: melting temperature (T_m), observed melting enthalpy (ΔH_m^{obs}), calculated melting enthalpy (ΔH_m^{calc}) and degree of crystallinity (χ_c)

HDPE/Sisal (w/w)	T_m (°C)	ΔH_m^{obs} (J g^{-1})	ΔH_m^{calc} (J g^{-1})	χ_c (%)
Untreated HDPE composites				
100/0	132.7	176.7	176.7	61.2
90/10	133.5	160.0	159.0	61.7
80/20	132.9	146.0	141.4	63.2
70/30	132.7	124.8	123.7	61.7
DCP-treated HDPE composites				
100/0	132.0	148.9	148.9	51.6
90/10	132.2	136.7	134.0	52.6
80/20	130.9	118.6	119.0	51.0
70/30	130.7	112.8	104.0	55.8

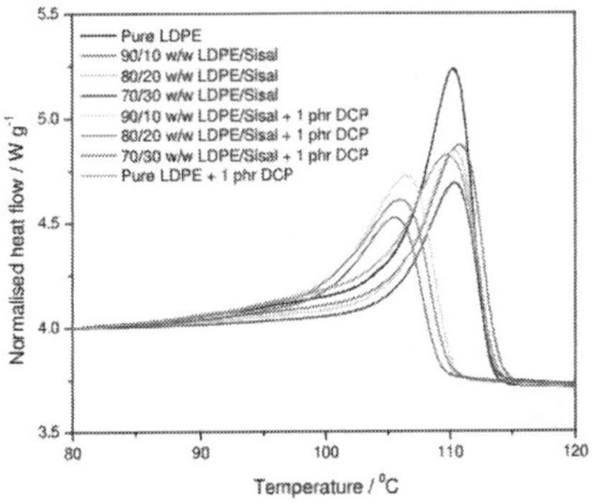

Figure 1 DSC heating curves of LDPE and LDPE-Sisal composites

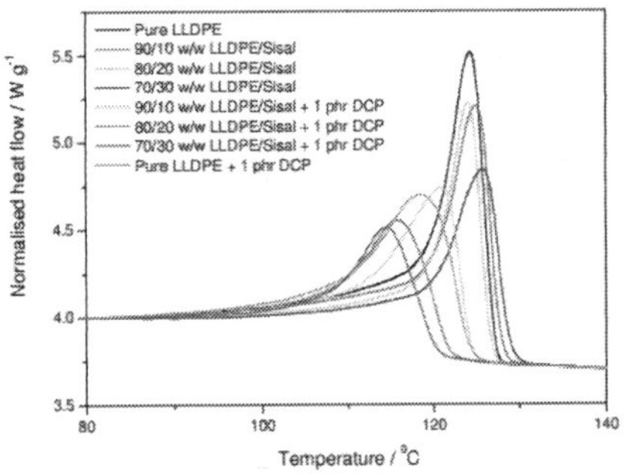

Figure 2 DSC heating curves of LLDPE and LLDPE-Sisal composites

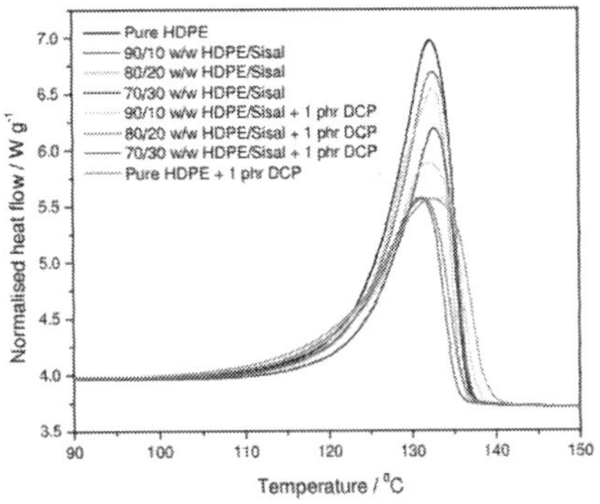

Figure 3 DSC heating curves of LDPE and LDPE-Sisal composites

Thermogravimetric analysis (TGA)

Figures 4 to 6 show the weight loss curves of the polyethylenes, DCP treated polyethylenes, and their modified and unmodified sisal composites. Figure 4 represents the thermal degradation behaviour of sisal fibre, LDPE, treated LDPE and its sisal composites. In the case of sisal fibre, the TGA curve shows several decomposition steps. The first step around 80 °C is due to dehydration (the release of absorbed moisture) from the fibres, the step around 277 °C corresponds to the thermal decomposition of hemicellulose and the glycosidic links of cellulose, while the step around 344 °C is the result of the thermal decomposition of α-cellulose.

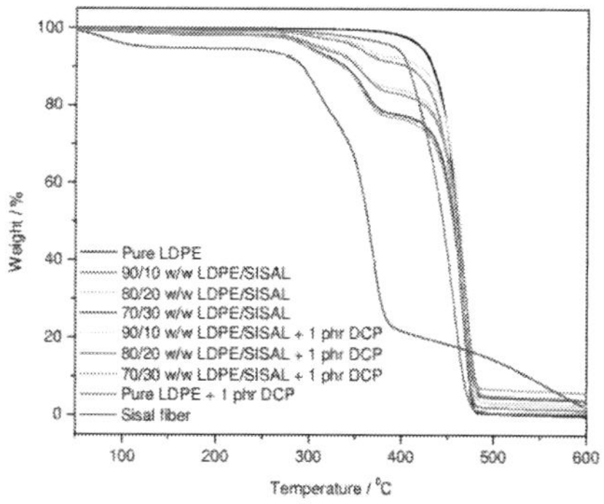

Figure 4 TGA curves of sisal fibre, virgin LDPE, DCP treated LDPE, untreated composites and DCP treated composites

Pure LDPE shows only one decomposition step around 413 °C, whereas DCP treated LDPE seems to have lower thermal stability with the decomposition temperature shifted to a lower temperature of 396 °C. This could be due to some thermal degradation of the LDPE chains induced by DCP. Another reason may be that crosslinking introduces tertiary carbons in the system, which are more prone to thermal decomposition. For the unmodified and DCP modified composites no significant differences in the thermal stability can be seen, and the percentage sample left after the first decomposition step corresponds well with the amounts of fibre initially mixed into the samples.

Figure 5 shows the TGA curves of LLDPE, LLDPE treated with DCP, sisal fibre, as well as untreated and DCP treated composites. For both pure LLDPE and DCP treated LLDPE the thermal degradation occurs in a single step, but in this case the DCP treated LLDPE also shows lower thermal stability. The reasons for this are probably the same as in the case of LDPE. As in the case of LDPE, the DCP treatment did not seem to influence the degradation behaviour of the composites at all the investigated fibre contents, and there is a very good correlation between the mass loss after the first degradation stel and the amount of fibre initially mixed into the composite. This seems to indicate that the DCP initiated grafting between the polymer and the fibre rather than degradation of the polymer. The good correlation between the % mass loss after the first step and the amount of fibre initially mixed into the polymer also indicates a fairly good distribution of fibre in the polymer.

Figure 6 shows the TGA curves of HDPE, HDPE treated with DCP, sisal fibre, as well as untreated and DCP treated HDPE/sisal composites. A considerable difference in thermal stability is observed between neat HDPE and DCP treated HDPE. This is attributed to thermal degradation initiated by DCP. The thermal stabilities of the polymers in the modified composites are lower than those of the unmodified composites. The reason for this is probably that the DCP in this case did not initiate grafting between the polymer and the fibre, but rather degradation of the polymer chains.

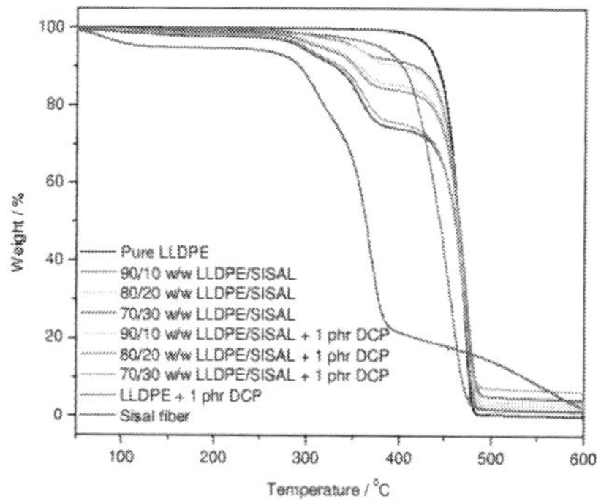

Figure 5 TGA curves of sisal fibre, virgin LLDPE, DCP treated LLDPE, untreated composites and DCP treated composites

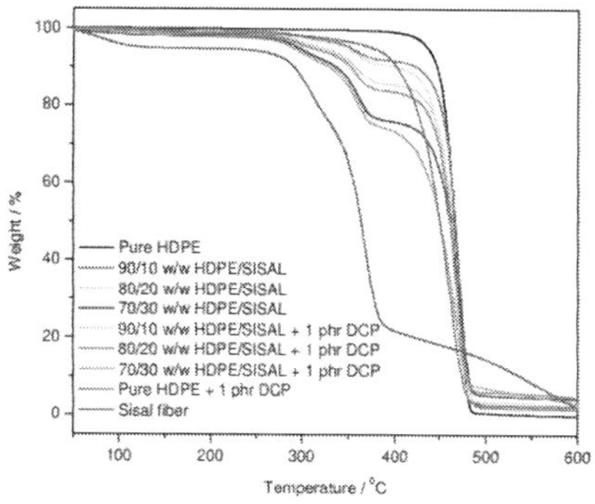

Figure 6 TGA curves of sisal fibre, virgin HDPE, DCP treated HDPE, untreated composites and DCP treated composites

Tensile properties
The tensile results for the LDPE based samples are summarized in Table 4. The tensile strength and elongation at break for the unmodified composites decrease with increasing fibre. In contrast the Young's modulus increases. The decrease in tensile strength can be attributed to poor fibre-matrix adhesion which gives rise to inefficient stress

transfer from the matrix to the fibres. On the other hand, the presence of sisal fibres imparts rigidity and brittleness to the LDPE matrix that leads to a decrease in elongation at break as well as to an increase of Young's modulus. The composites treated with DCP show an increase in tensile strength with increasing filler content, with the stress at break for the 30% containing fibre higher than that of pure LDPE. Young's modulus increases more significantly for the DCP treated samples. This improvement in mechanical properties of the treated samples is probably due to better interfacial adhesion caused by DCP initiated grafting.

Table 4 Tensile properties of neat LDPE, unmodified and DCP modified sisal composites

LDPE/ sisal (w/w)	Elongation at break (%)	Tensile strength (MPa)	E-modulus (MPa)
100/0	510 ± 34	10.0 ± 0.9	82 ± 6
90/10	45.0 ± 8.9	7.6 ± 0.4	102 ± 7
80/20	21.0 ± 4.0	6.0 ± 0.3	111 ± 7
70/30	11.9 ± 1.9	6.5 ± 0.7	147 ± 15
90/10	54.0 ± 8.0	8.0 ± 0.5	108 ± 8
80/20	20.8 ± 3.7	9.0 ± 1.0	129 ± 10
70/30	13.0 ± 2.0	12.0 ± 1.0	189 ± 18

Table 5 summarizes the tensile behaviour of LLDPE as well as the treated and untreated composites. The presence of fibre significantly reduces the tensile strengths, although DCP initiated grafting slightly improved these values. As in the case of LDPE, there was a significant increase in Young's modulus with increasing fibre content, and this increase is even more significant for the DCP treated composites.

Table 5 Tensile properties of neat LLDPE, unmodified and DCP modified sisal composites

LLDPE/sisal (w/w)	Elongation at break (%)	Tensile strength (MPa)	E-modulus (MPa)
100/0	880 ± 133	20.0 ± 3.6	114 ± 7
90/10	61.0 ± 4.8	7.0 ± 0.6	114 ± 10
80/20	35.8 ± 9.8	7.0 ± 0.5	153 ± 12
70/30	19.7 ± 3.0	7.0 ± 0.4	171 ± 13
90/10	99.0 ± 27.7	10.0 ± 0.8	110 ± 5
80/20	25.9 ± 7.8	10.5 ± 0.4	167 ± 11
70/30	15.0 ± 3.0	12.6 ± 2.0	213 ± 18

Typical results showing the tensile behaviour of HDPE and its composites are summarized in Table 6. In this case neither the tensile strength at break nor the modulus was significantly influenced by the presence or content of fibre, or by DCP treatment. It seems as if the polymer crystallinity has a bigger influence on these properties than the presence of sisal fibre, and because there probably was little grafting between the fibre and the polymer, the DCP treatment did not have a significant influence. As was expected, and also observed in the case of LDPE and LLDPE, the elongation at break decreased significantly with increasing fibre content.

Table 6 Tensile properties of neat HDPE, unmodified and DCP modified sisal composites

HDPE/ sisal (w/w)	Elongation at break (%)	Tensile strength (MPa)	E-modulus (MPa)
100/0	471 ±190	16.0 ± 1.9	304 ± 19
90/10	13.0 ±1.7	20.9 ± 1.5	353 ± 19
80/20	9.6 ±1.0	17.0 ± 0.8	346 ± 13
70/30	7.9 ±1.6	15.0 ± 1.9	313 ± 30
90/10	24.6 ±7.0	20.5 ±1.0	291 ± 31
80/20	13.0 ± 5.0	18.0 ± 3.0	312 ± 36
70/30	7.0 ±1.6	12.7 ±2.0	287 ± 21

It may be concluded that the effect of the DCP treatment on the mechanical properties of the PE-sisal composites depends on the type of PE, which probably determines the extent of grafting between the polymer and fibre. The effect of crystallinity, particularly at lower fibre loading, also seems to overshadow any possible influence of fibre or DCP treatment on the tensile strength and stiffness of the polymer.

Dynamic mechanical analysis (DMA)

Dynamic mechanical measurements were performed on neat polyethylene, DCP treated polyethylene, as well as treated and untreated polyethylene/sisal fibre composites. The curves of storage modulus versus temperature at 1 Hz for LDPE and LLDPE, as well as their composites, are shown in Figures 7 and 8. The storage modulus increases with fibre content, and almost all the composites exhibited higher values than those of the neat matrices, especially at temperatures above the glass transition. The treated samples show higher storage modulus values than the untreated samples, except for the DCP treated polymers (in the absence of fibres). These observations are in line with the Young's modulus values, and can be explained in the same way. A clear transition is seen at temperatures around -20 °C for both polymers. This is the β-relaxation, which is generally accepted as the glass transition in polyethylenes. For the treated composites there is a slight shift to higher temperatures of this transition, which may be the result of the immobilization of the amorphous polymer chains.

The curves of storage modulus versus temperature at 1 Hz for HDPE and its composites are shown in Figure 9. An increase in storage modulus for the untreated samples with increasing fibre load can be seen, while the storage modulus for the treated samples decreases compared to untreated samples. This can be attributed to the decrease in the crystallinity (as was observed from DSC) of the matrix after DCP treatment.

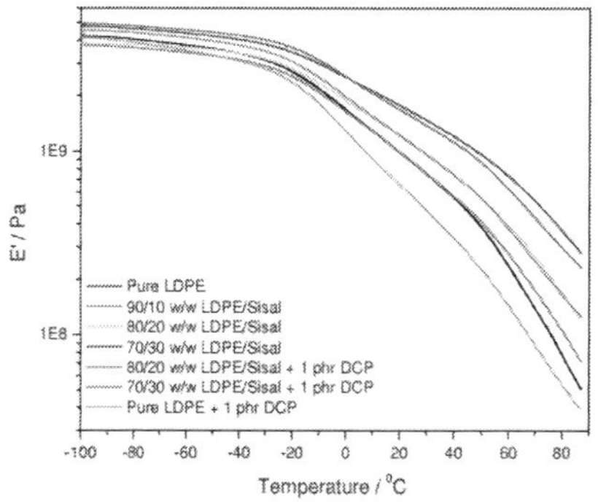

Figure 7 Storage modulus versus temperature LDPE, treated LDPE, untreated and DCP treated composites

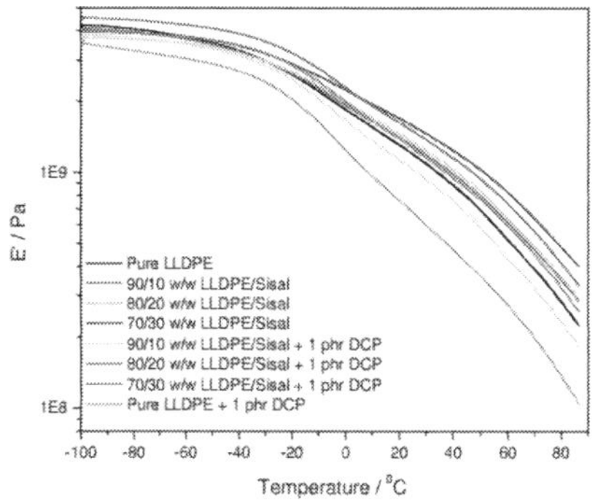

Figure 8 Storage modulus versus temperature for LLDPE, treated LLDPE, untreated and DCP treated composites

Morphological characterization

SEM was used to characterize the morphology of polyethylene/sisal fibre reinforced composites, treated and untreated. The scanning electron micrographs of untreated and DCP treated 30% sisal reinforced polyethylene composites are shown in Figures 10 to 12. For the composites treated with DCP a strong adhesion between the sisal fibre and the polymer matrix is clearly observed. No fibre pullout can be seen, and the fibres are completely impeded within the matrix, which indicates better wetting by the polymer. In addition, fibre defibrillation and tearing, as well as fibre breakage, are visible on the fractured surfaces of the composites.

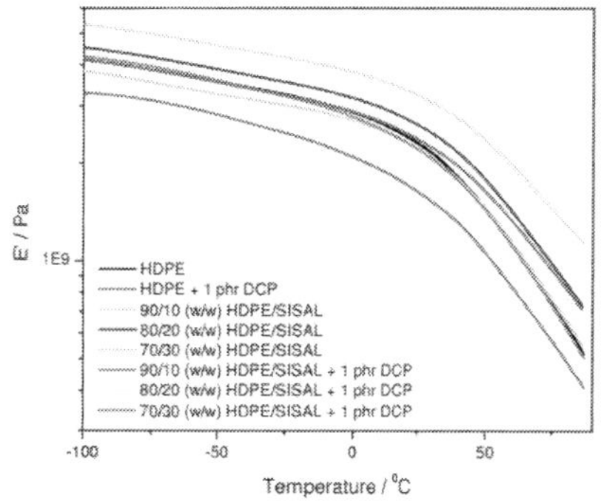

Figure 9 Storage modulus versus temperature for HDPE, treated HDPE, untreated and DCP treated composites

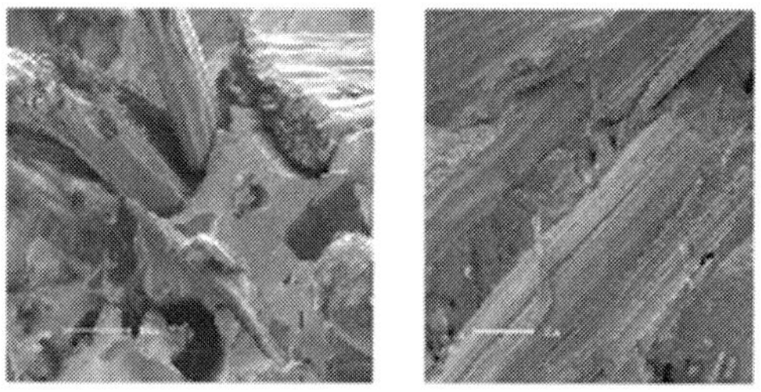

Figure 10 Scanning electron micrographs of 30% sisal fibre reinforced LDPE untreated (left) and DCP treated (right)

The effect of treatment seems to be less effective in the case of the HDPE composites compared to the other composites, which can be seen in the presence of holes, fibre pullout and poor wetting of the fibres by the polymer matrix. On the other hand, the fibres in the untreated composites appear to be free of any matrix interactions, and cracks, voids, and fibre pullout are very obvious. These observations are clear evidence of poor adhesion between the fibres and the polyethylene matrices.

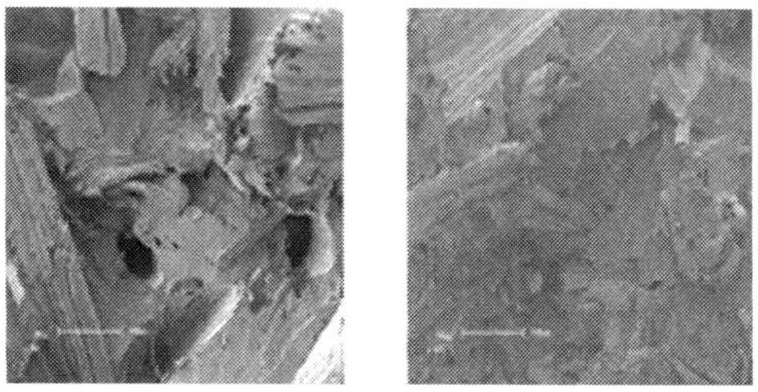

Figure 11 Scanning electron micrographs of 30% sisal fibre reinforced LLDPE untreated (left) and DCP treated (right)

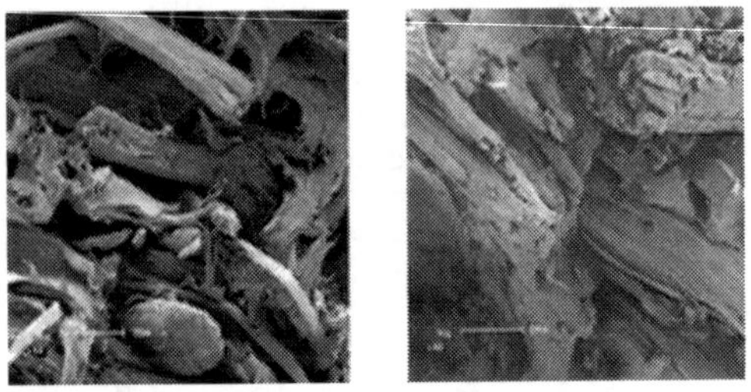

Figure 12 Scanning electron micrographs of 30% sisal fibre reinforced HDPE untreated (left) and DCP treated (right)

Conclusions

Polyethylene based sisal fibre composites, unmodified and DCP modified, were prepared by melt mixing and hot melt pressing. DSC measurements were performed to examine the thermal behaviour of the neat matrices and the composites. These results indicate a very good possibility of polymer-fibre grafting for samples prepared in the presence of DCP. They also show that the presence of DCP during sample treatment more significantly influenced the crystallization behaviour of polyethylenes with higher degrees of crystallinity. TGA was used to investigate the thermal stability of the neat matrices, as well as the unmodified and DCP modified composites. All the composite samples were thermally less stable compared to the neat matrices, and the TGA curves of the DCP treated composite samples were almost exactly the same as those of the untreated composite samples for LDPE and LLDPE, but the polymer in the DCP treated HDPE composite samples were thermally less stable than the untreated samples. The tensile properties of the composites showed different trends following DCP treatment, depending on the type of the polyethylene matrix. A significant increase in tensile strength and tensile modulus of the treated LDPE composites was observed compared to neat matrix and the unmodified composites. DCP treatment only slightly improved the tensile strengths of the LLDPE composites, while it had virtually no influence on the tensile strengths of the HDPE composites. DCP treatment significantly improved the tensile modulus of the LLDPE composites, but had almost no influence on the tensile modulus of the HDPE composites. The increase in the storage moduli of the composites compared to those of the neat matrices was attributed to a reinforcing effect of the natural fibres and to the better adhesion between the fibres and the polymer brought about by DCP treatment. SEM was used to characterize the morphology of the composites. Better adhesions for all the treated samples were observed in comparison to the untreated ones.

Acknowledgements

The authors would like to thank the South African National Research Foundation and the University of the Free State for their financial support.

References

Aves, C., Ferrão, P.M.C., Silva, A.J., Reis, L.G., Freitas, M., Rodrigues. L.B., Alives, D.E., "Ecodesign of automotive components making use of natural fibre composites", Journal of Cleaner Production., 18, 313-327 (2010)

Bledzki, A.K., Faruk, O., Sperber, V.E., "Cars from Biofibres", Macromol. Mater. Eng., 291, 449-457 (2006)

Bledzki, A.K., Gassan, J., "Composites reinforced with cellulose based fibres", Prog. Polym. Sci. 24, 221-274 (1999)

Bogoeva-Gaceva, G., Avella, M., Malinconico, M., Buzarovska, A., Grozdanov, A., Gentile, G., Errico, M.E., "Natural fibre eco-composites", Polym. Comp., 28, 98-107 (2007)

Botaro, V.R., Siqueira, G., Megiatto, J.D. Jr., Frollini, E., "Sisal fibres treated with NaOH and benzophenonetetracarboxylic dianhydride as reinforcement of phenolic matrix", J. Appl. Polym. Sci., 115, 269-276 (2010)

Burgueño, R., Quagliata, M.J., Mohanty, A.K., Metha, G., Drazl, L.T., Misra, M., "Load-bearing natural fibre composite cellular beams and panel", Composites: Part A., 35, 645-656 (2004)

Djidjelli, H., Boukerrou, A., Founas, R., Rabouhi, A., Kaci, M., Farenc, J., Martinez-Vega, J-J., Benachour. D., "Preparation and characterization of poly (vinyl chloride)/virgin and treated sisal fibre composites", J. Appl. Polym. Sci., 103, 3630-3636 (2007)

Dweib, M.A., Hu, B., O'Donnell, A., Shenton. W.H., Wool, P.R., "All natural composites sandwich beams for structural applications", Composites Structures., 63, 147-157 (2004)

Dweib, M.A., Hu, B., O'Donnell, A., Shenton. W.H., Wool. P.R., "Bio-based composites roof structure: Manufacturing and processing issues", Composites Structures., 74, 379-388 (2006)

Garcia, M., Garmendia, I., Garcia, J., "Influence of natural fibre type in eco-composites", Journal of Applied Polymer Science. 107, 2994-3004 (2008)

Huda, S.M., Drazl, T.L., Misra, M., Mohanty, K.A., Williams, K., Mielewski, F.D., "Study on biocomposites from recycled newspaper fibre and poly (lactic acid)", Ind. Eng. Chem. Res. 44, 5593-5601 (2005)

John, J.M., Thomas, S., "Biofibres and biocomposites", Carbohydrate Polymers., 71, 343-364 (2008)

Kalia, S., Kaith, B.S., Kaur, I., "Pretreatment of natural fibres and their application as reinforcing material in polymer composites – a review", Polym. Eng. Sci., 49, 1253-1272 (2009)

Li, Y., Mai, Y-M., Ye, L., "Sisal fibre and its composites: a review of recent development", Compos. Sci. Technol., 60, 2037-2055 (2000)

Luyt, A.S., Malunka, M.E., "Composites of low-density polyethylene and short sisal fibres: the effect of wax addition and peroxide treatment on thermal properties", Thermochimica Acta, 426, 101-107 (2005)

Malunka, M.E., Luyt, A.S., Krump, H., "Preparation and characterization of EVA-Sisal fibre composites", J. Appl. Polym. Sci., 100, 1607-1617 (2006)

Martin, A.R., Denes, S.D., Rowell, R.M., Mattoso, L.H.C., "Mechanical behaviour of cold plasma-treated sisal and high density polyethylene composites", Polym. Compos., 24, 464-474 (2003)

Mathur, V.K., "Composite material from local resources", Construction and Building Material., 20, 470-477 (2006)

Megiatto, J.D. Jr., Hoareau, W., Gardrat, C., Frollini, E., Castellan, A., "Sisal fibres: surface chemical modification using reagent obtained from a renewable source; characterization of hemicelluloses and lignin as model study", J. Agric. Food. Chem., 55, 8576-8584 (2007)

Megiatto, J.D. Jr., Oliveira, F.B., Rosa, D.S., Gardrat, C., Castellan, A., Frollini, E., "Renewable resources as reinforcement of polymeric matrices: composites based on phenolic thermosets and chemically modified sisal fibres", Macromol. Biosci., 7, 1121-1131 (2007)

Mishra, S., Mohanty, A.K., Drazl, L.T., Misra, M., Hinrichsen, G., "A review on pineapple leaf fibres, sisal fibres and their biocomposites", Macromol. Mater. Eng., 289, 955-974 (2004)

Mohanty, A.K., Misra, M., Hinrichsen, G., "Biofibres, biodegradable polymers and biocomposites: An overview", Macromol. Mater. Eng. 276/277, 1-24 (2000)

Mokoena, M.A., Djoković, V., Luyt, A.S., "Composites of linear low density polyethylene and short sisal fibres: the effects of peroxide treatment", Journal of Materials Science, 39, 3403-3412 (2004)

Mukherjee, P.S., Satyanarayana, K.G., "Structure and properties of some vegetable fibres-part1: sisal fibre", J. Mater.Sci., 19, 3925-3934 (1984)

Munawar, S.S., Umemura, K., Kawai, S., "Characterization of the morphological, physical, and mechanical properties of seven nonwood plant fibre bundles", J. Wood Sci., 53, 108-113 (2007)

Nishino, T., Matsuda, I., Hiro, K., "All-cellulose composites", Macromolecules, 37, 7683-7687 (2004)

Pasquini, D., Teixeira, E.M., Curvelo, A.A.S., Belgacem, M.N., Dufresne, A., "Surface esterification of cellulose fibres: processing and characterization of low density polyethylene/cellulose fibres composites", Compos. Sci. Technol., 68, 193-201 (2008)

Pavia, J.M.F., Frollini, E., "Unmodified and modified surface sisal fibres as reinforcement of phenolic and lignophenolic matrices composites: thermal analyses of fibres and composites", Macromol. Mater. Eng., 291, 405-417 (2006)

Peacock, A.J.: Handbook of polyethylene: structures, properties, and applications, Marcel Dekker, Inc, New York (2000)

Pervaiz, M., Sain, M.M., "Sheet- molded polyolefin natural fibre composites for automotive applications", Macromol. Mater. Eng., 288, 553-557 (2003)

Premalal, B.G.H., Ismail, H., Baharim, A., "Comparison of the mechanical properties of rice husk powder filled polypropylene composites with talc filled polypropylene composites", Polymer Testing, 21, 833-839 (2002)

Puglia, D., Tomassucci, A., Kenny, M.J., "Processing, properties and stability of biodegradable composites based on Mater-Bi® and cellulose fibres", Polym. Adv. Technol., 14, 749-756 (2003)

Reddy, N., Yang, Y., "Biofibres from agricultural byproducts for industrial applications", Trends in Biotechnology. 23, 1 (2005)

Riedel, U., Nickel, J., "Natural fibre-reinforced biopolymers as construction materials-new discoveries", Die Angewandte Makromolekulare Chemie, 272, 34-40 (1999)

Rong, M.Z., Zhang, M.Q., Liu, Y., Yan, H.M., Yang, G.C., Zeng, H.M., "Interfacial interaction in sisal/epoxy and its influence on impact performance", Polymer Composites, 23 182-192 (2002)

Saheb, N.D., Jog, P.J., "Natural fibre polymer composites: a review", Adv. Polym. Technol. 18, 351-363 (1999)

Silva, F.A., Chawla, N., Toledo Filho, R.D., "Tensile behaviour of high performance natural (sisal) fibres", Compos. Sci. Technol., 68, 3438-3443 (2008)

Singh, B., Gupta, M., Verma, A., "Influence of fibre surface treatment on the properties of sisal-polyester composites", Polymer Composites, 17, 910-918 (1996)

Wambua, P., Ivens, J., Verpoest, I., "Natural fibres: can they replace glass in fibre reinforced plastics?", Composites Science and Technology, 63, 1259-1264 (2003)

Zah, R., Hischier, R., Leão, A.L., Braun, I., "Curauá fibres in automobile industry a sustainability assessment", Journal of Cleaner Production, 15, 1032-1040 (2007)

Zhong, J.B., Lu, J., Wei, C., "Mechanical properties of sisal fibre reinforced urea-formaldehyde resin composites", Express Polymer Letters, 1, 681-687 (2007)

Zhou, X.P., Li, R.K.Y., Xie, X.L., Tjong, S.C., "Reinforcement of polypropylene using sisal fibres grafted with poly (methyl methacrylate)", J. Appl. Polym. Sci., 88, 1055-1064 (2003)

Effect of Phenyl-TPOSS and Epoxy-POSS on Mechanical and Surface Properties of TEOS and TEOS/APTES-based Aerogel

*Yannan Duan, Jason Randall, Sadhan C. Jana**

University of Akron - OH - USA

We incorporated trisilanolphenyl polyhedral oligomeric silsesquioxanes (Phenyl-TPOSS) and glycidyl POSS® Cage Mixture (Epoxy-POSS) to prepare high-performance tetraethoxysilane (TEOS) and tetraethoxysilane/ amino propyltriethoxysilane (TEOS/APTES)-based aerogels. Two series of aerogels with different contents of phenyl-POSS and epoxy-POSS, respectively, were produced using the two-step sol-gel process followed by supercritical drying. The effects of phenyl-TPOSS and epoxy-POSS on the structure, and mechanical and surface properties of the resultant aerogels were studied respectively using scanning electron microscopy (SEM), Instron tensile tester and contact angle measurement. It was observed that a small amount of POSS significantly enhanced the hydrophobicity and compressive strength and modulus of the aerogels.

HYGROTHERMAL AGEING OF POLYPROPYLENE/GLASS FIBER COMPOSITES: INFLUENCE OF MONTMORILONITE AND MALEATED PP COMPATIBILIZER CONTENT

J. E. Motezuki[1], M. H. M. Dias[1], M. C. Branciforti[1] and J. A. Sousa[1]*

[1] *Materials Engineering Dept., Universidade Federal de São Carlos, Brazil – jasousa@ufscar.br*

Abstract – Improved mechanical performance of short glass fiber (GF) reinforced polypropylene (PP) composites can be achieved by the combined use of adequate GF coupling agents (aminosilanes) with a maleated-PP (PP-g-MAH) interfacial compatibilizer, added to the chemically inert and non-polar polymer matrix in order to increase the fiber-polymer interfacial interactions of the composite. However, due to the introduction of polarity in both the matrix and the fiber-polymer interface through the addition of maleic anhydride molecules, it is important to monitor the integrity of the fiber-polymer interfacial adhesion against the erosive action of a hydrothermally active environment, in order to assure the long-term mechanical performance of these PP composites. Also, the use of nanoclay (O-MMT) in PP / PP-g-MAH / FV composites is interesting not only due to the expected mechanical properties improvement as a reinforcement and but also a barrier to moisture permeating through the composite, when adequately intercalated / exfoliated in the ternary composite matrix. Therefore, injection molded tensile test-specimen of 40 wt.% GF-reinforced PP composites with varying content of PP-g-MAH (from 0 up to 10 wt.%) and with organophilic montmorillonite (O-MMT - 2wt.%) were subjected to hygrothermal ageing in water at 80 °C for 48/120/240/480/720 hours, in order to investigate the influence of the polar compatibilizer content and also the effect of barrier property of O-MMT lamellar particles on the fiber polymer interface integrity upon ageing. The tensile strength (TS) data of aged PP/PP-g-MAH/40%GF composites, clearly indicated a critical compatibilizer content which minimizes the deterioration of this property with increasing ageing time. On the other hand, the addition of O-MMT had negative influence on both aged and non-aged PP/PP-g-MAH/40%GF composites, due to the abrasive action of the O-MMT particles on the GF's sizing treatment.

Introduction

Glass fiber (GF) reinforced polypropylene (PP) is still object of substantial technological and scientific investigation due to its attractive low cost alternative composite material option for engineering applications. Improved mechanical performance of these composites has been achieved by the combined use of adequate GF coupling agents (aminosilanes) with a maleated-PP (PP-g-MAH) interfacial compatibilizer, in order to increase the fiber-polymer interfacial adhesion of this composite with a chemically inert polymer matrix [Karian, H.G. (2003)].

In order to assure the long-term mechanical performance of these PP composites, it is important to assure the integrity of the fiber-polymer interfacial adhesion against the erosive action of a hydrothermally active environment induced by combined action of water and heat. Hygrothermal ageing of composite materials has two main sources for material deterioration; one which influences through chemical activity and the other which is essentially of physical nature. The physical source is reflected by water molecules plasticization mechanism of highly hygroscopic matrices like polyamides. This plasticization effect leads to polymer Tg depression and, consequently, mechanical deterioration of the composite through reduction of the polymer-fiber interfacial shear strength [Karian, (2003); Kim and Mai (1998); Maron, G. (1994)]. However, in non-polar and chemically inactive polymer matrices like PP, the hygrothermal deterioration of the composites is essentially induced through chemical interactions, mainly due to the moisture attack at the fiber-matrix interface in the presence of polar maleic anhydride (MAH) molecules of the compatibilizer. Therefore, it is important to monitor the integrity of the fiber-polymer interfacial adhesion against the erosive action of hydrothermally active environment, as the concentration of the PP-g-MAH is increased in GF-reinforced PP composites, so as to determine the optimum compatibilizer content when improved interfacial bond strength is offset by posterior water attack under service conditions [Karian, (2003), Cruz, M.C.A. et al (2008)].

Another possible solution for the expected hygrothermal deterioration could be achieved by the inclusion of nanoclay particles in GF-reinforced PP composites. Silicate platelets of nanometer dimensions, such as organophylic Montmorillonite nanoclays (O-MMT), have been largely used in polymer nanocomposites not only for their mechanical and thermal properties improvements, but also for their high barrier efficiency to gases and liquids [Akkapeddi, M.K.(2000)].

Conceptually, even at very low loading levels, such ultra-fine silicate mineral dispersions with their high surface area and aspect ratio can introduce a high tortuousity factor for the permeating water molecules and, consequently, lead to significantly improvements in the composite's permeability against the humidity action. [Akkapeddi, M.K. (2000)].

In an attempt to reduce the hygrothermal ageing effects on the mechanical performance of short GF-reinforced PP composites, the influence of the PP-g-MAH compatibilizer content, as well as the incorporation of O-MMT in the composite, was analyzed through tensile tests carried out as a function of ageing time.

Experimental

The PP/GF composites were compounded in a co-rotating twin-screw extruder (ZSK-30 from Coperion) with a screw profile and operating conditions adequate for compounding GF-reinforced PP composites, as detailed in previous publications [Lopes, P.E (2002)].

Two grades of isotactic-PP homopolymer (grade HP648S with ρ = 0.905 g/cc and MFI = 40g/10min and grade HS 501K with ρ = 0.905 g/cc and MFI = 3.5 g/10min.), both supplied by Basell-Lyondell Brasil, were mixed in a mass ratio of 3:1 (HP648S:HP550K) so as to compose the polymer matrix. The formulated PP matrix was stabilized with an adequate antioxidant-thermal stabilizer system (0.2% Irganox 1010 and 0.4% Irgafos 168) during the extrusion compounding process.

The short GF used was supplied by Vetrotex/Owens South America (E type P968 chopped strands with ρ = 2.54 g/cc; D = 13 μm and L= 4.5 mm.), pre-treated with PP compatible sizing system with aminosilane coupling agent, as reported in literature [Karian, H.G.(2003); Lopes, P.E.(2002)]. The short GF content was fixed at 40 weight percent.

The choice of the PP-g-MAH compatibilizer used in this work (Polybond 3200, with 1% nominal MAH content and high MFI = 250 g/10 min., as supplied by Crompton/Chemtura Brazil) for all PP/40%GF binary composites, was based on an earlier work reported by the authors [Lopes, P.E. et al (2002)]. For the binary PP/40%GF composites, the PP-g-MAH compatibilizer was incorporated in nominal concentrations of 0; 0.25; 0.5; 1; 2; 5 and 10 wt% in the composite.

The organoclay (O-MMT) used was Cloisite 20A (d001 = 24.2 Å) from Southern Clay Company. The O-MMT content in the ternary PP/PP-g-MAH/40%GF/O-MMT composites was fixed at 2 wt%. Since the O-MMT content in the ternary composite is low and considering the known difficulties for intercalation/exfoliation of O-MMT particles in polyolefin matrices during the extrusion compounding process, a masterbatch concentrate containing 20 wt% of O-MMT was first prepared in a previous extrusion run. This masterbatch concentrate contained PP-g-MAH compatibilizer in a mass proportion of 3:1 of O-MMT, as recommended in literature for PP/O-MMT nanocomposites [Kim,D.H. (2007)]. It was then diluted with the composed matrix PP in a second extrusion run with short GF being incorporated downstream into the twin-screw extruder. A different batch of ternary PP/PP-g-MAH/40%GF/2%O-MMT composites were extrusion compounded with the same grade PP matrix used in this work.

The test specimen for tensile tests were injection molded in an automatic Arburg Allrounder machine (270V/300-120), with nearly uniform processing conditions for all composite formulations and a barrel temperature profile of 195 to 240 °C, injection pressure of 1,200 bar and mold temperature of 50 °C. The test specimen of all composites, as well as those of the matrix polymer (PP control), were submitted to the hygrothermal ageing process in a water bath with demineralized distilled at 80 °C for periods of 48, 120, 240, 480 and 720 hours, being subsequently removed, dried superficially and conditioned for 48h at room temperature (23 ± 2°C) for mechanical testing.

Results and Discussion

The tensile strength (TS) test data of the hydrothermally aged PP/40%GF composites, with varying PP-g-MAH content, is presented in Fig. 1 as a function of ageing time. Firstly, in the case of the unaged composites (zero ageing time), the introduction of PP-g-MAH improves substantially the TS from around 75 MPa for the composite without compatibilizer to a maximum TS value of 107 MPa at approximately 1.8 wt% and then tends to decrease slightly at higher PP-g-MAH content. This indicates clearly the existence of an optimum compatibilizer content of around 2 wt%, when all the fiber-polymer interface is saturated with the formation of an interphase of copolymer of PP-co-siloxane, generated from the chemical coupling imidization reactions between the amine functionalities of the silane treated GF surface with the MAH co-reactive groups present in the maleated PP, as detailed elsewhere in the literature [Cruz, M.C.A. et al (2008); Karian, H.G.(2003); Lopes, P.E. et al(2002)].

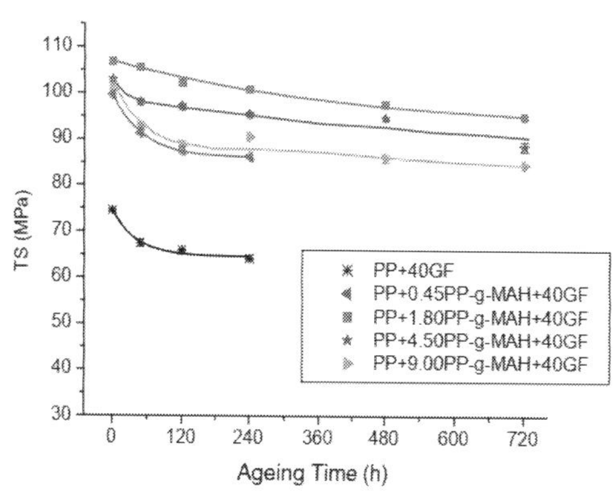

Figure 1 – Influence of hygrothermal ageing time on the tensile strength of PP/40%GF composites, as a function of PP-g-MAH compatibilizer content.

With increasing hygrothermal ageing time, the composite with 2% of compatibilizer showed the lowest drop in TS values of up to 11% after 720 hours with respect to its unaged property. At compatibilizer contents below the optimum level, the drop in TS tends to be much more pronounced then that verified at PP-g-MAH contents above the optimum level. Therefore, this sequence of curves clearly indicates that at the optimum compatibilizer content, the fiber-polymer interface is most impervious to the hygrothermal activity, when a minimum thickness interphase of PP-co-siloxane copolymer is chemically bound at the fiber surface and completely saturates the fiber-polymer interface. At compatibilizer contents below this saturation level, the interface is permeable to water absorption by capillary action and, thereby, leading to substantial fiber-polymer debonding with ageing time. At compatibilizer contents above optimum level, the presence of chemically unreacted PP-g-MAH molecules at the interface contributes to a reduction in the fiber-polymer interfacial shear strength and, consequently, reducing the TS values of unaged composites. With increasing ageing time of these composites, a slight drop in TS values is verified, probably due to possible solvation of the unreacted PP-g-MAH at the fiber-polymer interphase.

Hence, the PP40%GF composite with 2 wt% of PP-g-MAH optimum compatibilizer content was chosen for the ternary composite preparation, by introducing 2wt% of O-MMT in its matrix. The choice of 2 wt% of O-MMT content was defined from previous moisture (water vapour) permeability tests results carried out for some PP/PP-g-MAH/O-MMT nanocomposites with varying O-MMT content. The addition of 2wt% of O-MMT showed the best performance in the permeation, XRD and TEM (results not presented here).

In Fig. 2, the TS test data of the PP composite's matrix (PP/2%PP-g-MAH) and that of PP/PP-g-MAH/40%GF composites are shown, both with and without O-MMT, as a function of ageing time.

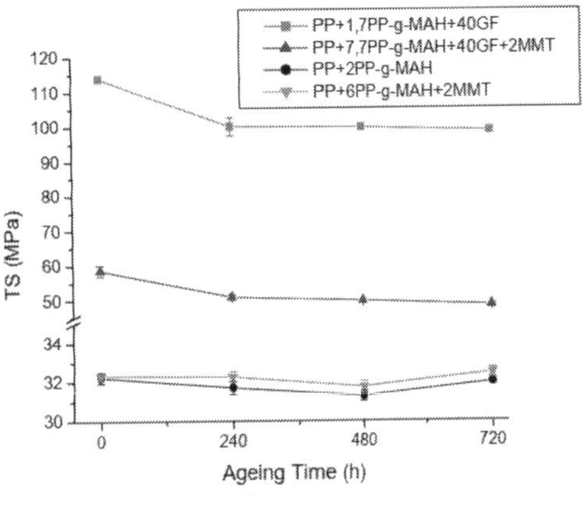

Figure 2 – Tensile strength of PP/PP-g-MAH/GF composites, with and without O-MMT, as a function of hygrothermal ageing time.

Initially, comparing the TS values of the unaged (zero ageing time) binary PP/1.7%PP-g-MAH/40%GF and ternary PP/7.7%PP-g-MAH/40%GF/2%O-MMT composites, it is evident that the presence of nanoclay particles contributes to a substantial reduction in the TS value of the ternary composite. This unexpected negative result is credited to a probable erosion action of the nanoclay particles at the GF surface, affecting the fiber-polymer interfacial adhesion. Thus, the presence of O-MMT in the GF-reinforced PP composites does not result in a synergic effect in the system.

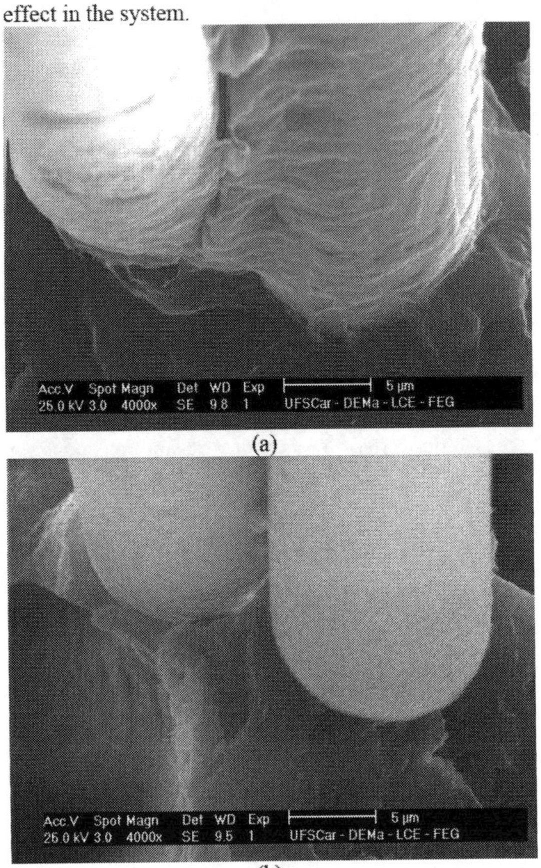

Figure 3 – SEM microphotographs of cryo-fractured surfaces of (a) PP/1.7%PP-g-MAH/40%GF and (b) PP/7.7%PP-g-MAH/40%GF/2%O-MMT composites.

Some authors [Ravindran & Cho (2006); Vlasveld (2005)] also reported this negative effect of nanoclay addition on the TS property of short GF-reinforced polar polymer matrices, such as in the case of epoxy and PA 6. However, no definitive conclusions were given for this behavior. In Fig. 3 are shown the SEM microphotographs of cryo-fractured surfaces from cross sections of PP/GF tensile specimen of PP/PP-g-MAH/40%GF composites, without (Fig. 3-a) and with (Fig. 3-b) 2% of O-MMT. The existence of strong interfacial adhesion and fiber surface wetting due to chemical coupling can be seen in the Fig. 3(a) for the PP/1.7%PP-g-MAH/40%GF composite. However, in

Fig. 3(b), we can see that very little polymer matrix is still adhered to the fiber surfaces and it presents a rough surface appearance due to the probable abrasive action of the nanoclay particles present in the polymer matrix.

Finally, Fig. 2 indicates that the hygrothermal ageing effect does not significantly affect the matrix polymer and neither the PP/O-MMT nanocomposite, both showing nearly the same initial TS value of the unaged samples. So, from this observation, it can be inferred that the presence O-MMT nanofiller apparently leads to no deterioration of the polymer matrix with hygrothermal ageing time. Therefore, with the increasing ageing time, the TS property of the ternary composite appears to be affected in nearly the same proportion as the binary composites up to 240h, and then both tend to level off at higher ageing times.

Conclusions

The hygrothermal ageing test contributed to the loss in the tensile strength (TS) properties of the PP/PP-g-MAH/40%GF composites, as a function of the compatibilizer content. The composite with optimum 2 wt% of PP-g-MAH compatibilizer indicated the lowest drop in TS properties, even at long ageing time of up to 720 hours. At compatibilizer concentrations below the optimum level, a steep decrease in TS values can be attributed to the drastic deterioration of the fiber-polymer interfacial adhesion induced by the action of the water diffused by capillarity effect. The addition of O-MMT in the ternary PP composite substantially deteriorates the mechanical properties due to the abrasive action of the nanoclay particles on the sizing treatment on the fiber surface.

Acknowledgements

The authors wish to thank FAPESP for the research funding and also the companies Basell-Lyondell Brasil and Vetrotex/Owens South America for kind supply of the materials used. Some of the authors gratefully acknowledge their graduate research grants from CNPq and CAPES.

References

-Akkapeddi, M.K., "Glass fiber reinforced polyamide-6 nanocomposites", Polymer Composites, vol. 21, No. 4, p. 576 - 585 (2000).

-Cruz, M.C.A., et al., "Influence of Maleated Polypropylene on Hygrothermal Ageing of Polypropylene/PP-g-MAH/Glass Fiber Composites", Proceedings in (CD-ROM) of 24th Annual Meeting of Polymer Processing Society, Salerno, Italy, p. 15-19 (2008).

-Karian, H.G., Handbook of Polypropylene and Polypropylene Composites, 2nd edition, Marcel Dekker, New York (2003).

-Kim, D.H., et al., "Structure and properties of polypropylene-based nanocomposites: Effect of PP-g-MA to Organoclay Ratio", Polymer, 48, p. 5308-5323 (2007).

-Kim, J-K., Mai, Y-W., "Surface Treatments of Fibers and Effects on Composite Properties", Chapter 5 in Engineered Interfaces in Fiber Reinforced Composites, Elsevier Science Ltd, Oxford, p. 72-183 (1998).

-Lopes, P.E., Sousa, J.A., "Influence of PP-g-MAH Compatibilizer Characteristics on Interphase and Mechanical Properties of Glass Fiber Reinforced Polypropylene Composites", Proceedings in (CD-ROM) of 18th Annual Meeting of Polymer Processing Society, Guimarães, Portugal, p. 9. (2002).

-Marom, G., "The Role of Water Transport in Composite Materials", Chapter 9 in Polymer Permeability, Chapman &Hall, London, p. 341-374 (1994).

-Ravindran, N., E.H. Cho, "Durability of Glass-Fiber-Reinforced Polymer Nanocomposites in an Alkaline Environment", J. Vinyl and Additive Tech., DOI 10.1002/vnl.20063, p. 25-32 (2006).

-Vlasveld, D.P.N., et al., "Fibre–matrix adhesion in glass-fibre reinforced polyamide-6 silicate nanocomposites", Composites: Part A 36, p. 1-11 (2005).

Nucleating Agent Assisted Dispersion of POSS in PP: Properties of Nanocomposite Fibers and Films

Sayantan Roy, Byoung J. Lee, Sadhan C. Jana*
Department of Polymer Engineering, University of Akron, Akron, OH 44325-0301

Abstract: The study developed fundamental understanding of nanocomposite formation by "bottom-up" self-assembly of inorganic-organic hybrid molecules (POSS) in polypropylene (PP) matrix. A nucleating agent compatible with POSS was used to promote dispersion of POSS molecules and to provide templates for self-assembly of POSS molecules into spherical nanoparticles of 25-200 nm diameter. Typical formulations containing 0.3-0.7 wt% nucleating agent and 0-30 wt% POSS were spun into fibers and converted into films and their mechanical properties studied. It was found from the mechanical properties of spun fibers that optimum mechanical properties resulted with 10 wt% POSS. Microscopic POSS particles formed at POSS concentration higher than 10 wt%. The optimum formulation resulted in a 50% reduction in fiber diameter, 70% increase in tensile modulus, and 30% increase in tensile strength compared to unfilled PP. The interactions between POSS and nucleating agents were studied using infra-red spectroscopy, differential scanning calorimetry, and oscillatory shear rheology. It was found that fibrillation of nucleating agent was subdued due to hydrogen bonding interactions with POSS.

1. Introduction

The research interest on polymer nanocomposites has grown dramatically in last one decade with occasional news of commercial success and reporting of substantial research findings in peer-reviewed journals from universities and laboratories all across the globe. Research on polymer nanocomposites can be broadly classified into two categories based on how nanoparticulate fillers are dispersed in the polymers – a top-down or a bottom-up approach. Our survey revealed that a large majority of research literature is focused on top-down approach where bulk nanofillers are blended with polymers in extruders and dispersion is achieved by shear. Examples of top-down nanocomposites are exfoliated clay composites of nylon [1-2] and polyurethanes [3], and carbon nanofibers in polymethylmethacrylate [4]. A major issue of top-down approach is the lack of control of the degree of dispersion of nanoparticles. In addition, the nanoparticles cannot be consistently exfoliated in the polymer matrix or in many cases severe particle attrition occurs. In this work, an alternative scheme was used for production of nanocomposites by *"bottom up"* approach whereby small, isotropic particles of molecular dimension, such as POSS, are allowed to assemble into nano-scale structures. These nanoscale structures then reinforce the polymers. The results presented in this paper pertain to reinforcement of PP fibers by POSS nanoparticles created by self-assembly of POSS molecules in the presence of a nucleating agent. The structure of the nucleating agent was such that POSS molecules were more compatible with it than with PP. Consequently, the nucleating agent acted as a template for dispersion of POSS molecules in PP. As will be seen, the interactions between POSS and nucleating agents were ramified in terms of excellent dispersion of POSS in PP and interference of POSS with fibrillar network formation by the nucleating agent.

2. Materials and methods

Isotactic PP (grade PP31S07A, Equistar, MFI 0.9, crystalline melting temperature 168°C), 1,3:2,4-dibenzylidenesorbitol (DBS, m.p. 225°C Milliken grade Millad 3905), trisilanol phenyl-POSS (T-POSS, grade SO1458, m.p.>400°C, Hybrid Plastics) and octaisobutyl POSS (M-POSS, MS0825, m.p. ~270°C) were used to produce the nanocomposite materials. The materials were mixed in a batch mixer at 185°C for 10 minutes to obtain compounds. In some compounds the nucleating agent DBS was not used at all to determine the potential of nanoparticle formation by POSS molecules alone. Approximately 0.3 and 0.7 wt% DBS were used and the amount of POSS was restricted to 10wt%. The polymer compounds were subsequently spun into fibers using a capillary die (1.6 mm diameter) and a draw down ratio (DDR) of 50-1000. In another case, the compound with 10 wt% POSS and 0.3 wt% DBS were subjected to film blowing at 230°C and 70 rpm with blow-up ratio (BUR) of ~2.0 and draw down ratio (DDR) between 5-15. The fibers were microtomed to obtain transmission electron microscopy (TEM) images and to reveal the size of POSS nanoparticles. The values of tensile modulus, tensile elongation, and tensile strength of fibers and films were determined. The hydrogen bonding between POSS and DBS molecules was studied using infra-red spectroscopy.

3. Results and discussion

It was found that the maximum drawdown ratio and hence the minimum fiber diameter varied with POSS concentration for a given amount of DBS, as shown in Figure 1(a). An optimum occurs at approximately 10 wt% POSS. Accordingly, all the tensile properties also showed optimum values at 10 wt% POSS. Figure 1(b) presents the values of tensile strength. The morphology of POSS particles at various concentrations of POSS was investigated using transmission electron microscope. An example is shown in Figure 2. At a given concentration of DBS, the POSS molecules formed nanoparticles up to approximately 10 wt% of POSS. At higher concentrations, microparticles are formed in addition to the nanoparticles as seen in Figure 2(b), which prevented drawing of fibers to smaller diameter.

The presence of POSS created a stark contrast in the morphology in films as shown in Figure 3. The spherulites appeared oriented in the machine direction in the presence of POSS. Consequently the tensile modulus increased from 890 MPa for PP to 1270 MPa for the composite.

The infra-red spectroscopy of mixtures of POSS and DBS was taken at room temperature and at 190°C. It was found that considerable hydrogen bonding between POSS and DBS molecules existed even at 190°C.

4. Conclusions

The following conclusions can be drawn from the results obtained in this study.

- For a given concentration of DBS, there is an optimum concentration of POSS, below which only the nanometer size self-assembled particles are formed and above which the microscopic POSS particles are generated. The microscopic POSS particles prevent formation of small diameter fibers and as a result the tensile properties are reduced.

- The films obtained by film-blowing process showed oriented spherulite morphology in presence of POSS. Much higher tensile modulus and tensile elongation at break were obtained when POSS was used.

- The investigation using infra-red spectroscopy revealed existence of strong hydrogen bonding between POSS and DBS which supports the the idea that nanoscale dispersion of POSS was promoted by DBS.

5. Acknowledgements: Financial support from National Science Foundation in the form of award CMMI-0727231 is gratefully acknowledged.

6. References:

1. Pinnavaia, T.J., Beall, G. W., Eds., *Polymer-Clay Nanocomposites*, Wiley, Chichester and New York (2000).
2. Yoon, K. H., Polk, M. B., Min, B. G., Schiraldi, D. A., Kumar, S., Structure and Property Study of Nylon-6/Clay Nanocomposite Fiber, *Polym. International, 53, 2072-8* (2004).
3. Pattanayak, A., Jana, S.C., Synthesis of thermoplastic polyurethane nanocomposites of reactive clay by bulk polymerization methods. *Polymer*, 46(10), 3275-3288 (2005).
4. Jimenez, G., Jana, S.C., Electrically conductive polymer nanocomposites of polymethylmethacrylate and carbon nanofibers prepared by chaotic mixing. *Composites: Part A*, *38*, 983-993 (2007).

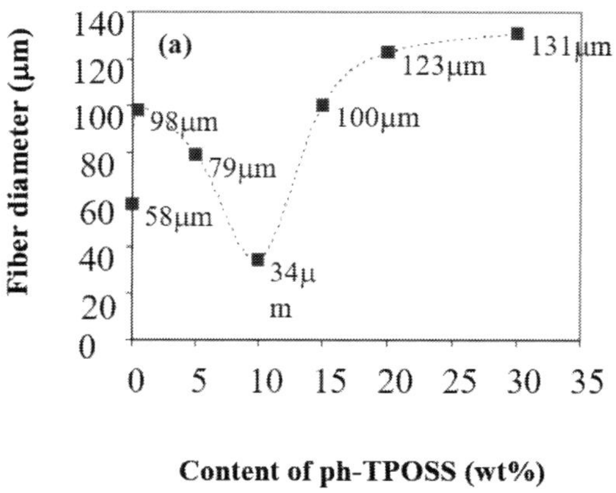

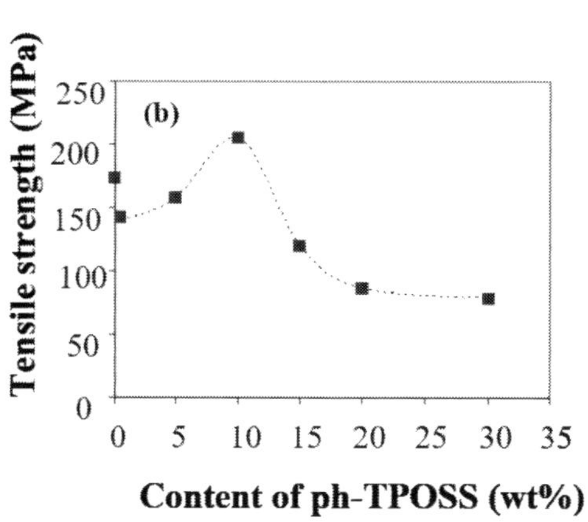

Figure 1. Effect of POSS concentration on (a) fiber diameter and (b) tensile strength. DBS content was 0.3 wt%. Dashed lines are used only to guide the eye.

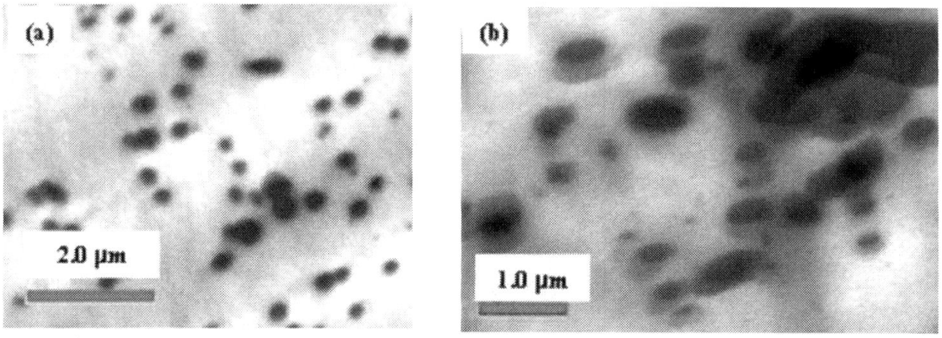

Figure 2. Transmission electron microscope image showing the diameter of POSS particles in spun fibers with (a) 10 wt% and (b) 20 wt% POSS. In both cases, DBS was 0.3wt%. Note the existence of larger particles in (b).

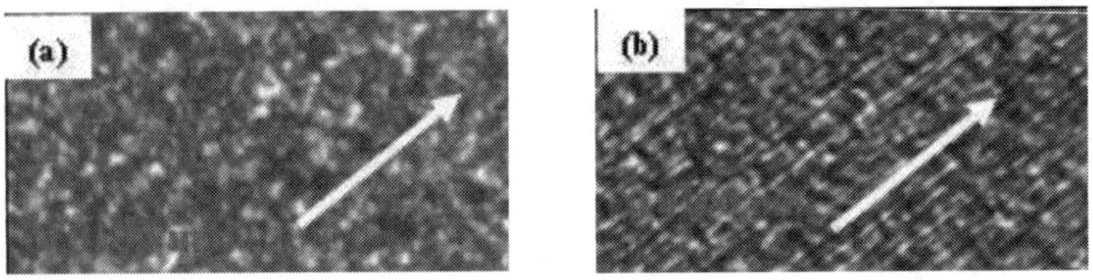

Figure 3. Optical microscope image showing the spherulitic structures of PP in blown film (a) unoriented; PP without POSS, BUR 2.0; DDR 8.0 (b) oriented; PP with 0.3 wt% DBS and 10 wt% POSS; BUR 2.0; DDR 6.0. Machine direction is shown by the arrow.

NANOTUBE CONDUCTIVE NETWORKS IN MICROINJECTED POLYMERS AND POLYMER BLENDS

S. Abbasi[1*], A. Derdouri[2] and P.J. Carreau[1]

[1] *CREPEC, Department of Chemical Engineering, Ecole Polytechnique Montreal, Montreal, Canada – samaneh.abbasi@polymtl.ca; pierre.carreau@polymtl.ca*
[2] *CREPEC, Industrial Materials Institute, National Research Council Canada, Boucherville, Canada – Salim.Derdouri@imi.cnrc-nrc.gc.ca*

Abstract - In this study the electrical conductivity and percolation behavior of composites using multiwalled carbon nanotubes (MWCNT) in a polycarbonate (PC), a polypropylene (PP) and a blend of polypropylene and cyclic butylene terephthalate (CBT) were investigated and compared. The effects of microinjection conditions on the morphology, microstructure and properties of composites were also studied. Samples of various MWCNT loadings were prepared by diluting commercial masterbatches of nanotubes using optimized melt-mixing conditions. SEM and TEM techniques were then used to examine the quality of dispersion and formation of nanotube networks within the polymer matrix. In PP and PC composites the results showed that the nanotubes are well aligned in microinjected samples although the level of alignment is less in the crystalline polymer. It was also found that the orientation of crystalline phase in sheared samples decreased with incorporation of nanotubes into the polypropylene. The conductivity properties were also significantly altered by nanotube alignment in both PP and PC nanocomposites; however, the presence of the crystalline phase improved the conductivity even for high shear conditions through the concept of double percolation threshold. The electrical conductivity of PP/CBT blend nanocompositess was also improved through a double percolation that is the basic requirement for the conductivity of the blends. It was also found that the presence of nanotubes affected the morphology of the blends resulting in a smaller and more elongated dispersed phase, which was more favorable for the formation of a co-continuous structure in the nanocomposites.

Introduction

For the last two decades, carbon nanotubes (CNTs) have attracted a strong interest due to their unique electronic structure and extraordinary properties (Meyyappan 2005). Their intrinsic structure, size scale and aspect ratio suggest a variety of applications in nanoelectronics, sensors and field emission as well as high performance nanocomposites. Particularly, carbon nanotube based polymer composites are suitable candidates for the fabrication of microscale components where the nano size of the added particles is an advantage and only grams of materials are necessary to manufacture high-valued products. A number of processing technologies have been explored to produce polymer-based microdevices amongst which microinjection molding is one of the most suitable processes for producing microparts cheaply and with high precision.

The intense interest in carbon nanotubes incorporated into polymeric materials stems from their potential to reach thermal, electrical and rheological percolations at relatively small concentrations (Abbasi, Carreau et al. 2009; 2010). At the percolation point, the formation of a filler network or interconnecting structure creates additional and large contributions to the nanocomposites properties. For sphere and ellipsoids fillers the percolation behavior is well understood (Fangming, Fischer et al. 2005), while the onset of percolation in the case of high aspect ratio fillers such as nanotubes strongly depends on their alignment. When the nanotubes are aligned in the polymer matrix the probability of tube-tube contacts decreases and consequently the percolation threshold raises (Wu, Liang et al. 2007; Abbasi, Carreau et al. 2010). Accordingly, the full exploitation of the nanotube properties in polymer composite applications will require an exceptional control of the nanotube alignment in macroscopic parts. Incorporation of carbon particles into incompatible polymer blends can provide them with electrical conductivity at very low filler content due to a double percolation phenomenon (Sumita, Sakata et al. 1991; Sumita, Sakata et al. 1992). Double percolation refers to the structure in which there are two types of percolation in the same composite material. The first one, which is typically observed in the reinforced composite materials, is associated with the electrical continuity of the conductive particles in the filler-rich phase and the second one is associated with the continuity of this phase in the blend. For the attainment of high electrical conductivity, both types of percolation are required and the conductivity improvement will depend on the conductive particle concentration in the filler-rich phase and on the volume fraction of the filler-rich phase in the blend (Sumita, Sakata et al. 1992).

The idea of the current study is to investigate the effect of the crystalline structure on the nanotube alignment and consequently on the microstructure and electrical conductivity of the nanocomposites in high shear conditions. The attention is focused on two kinds of polymer: a polycarbonate (PC) as a polar amorphous polymer and an isotactic polypropylene (iPP) as a non-polar semi-crystalline polymer. The effect of the selective location of carbon nanotubes and the double percolation phenomenon on the electrical conductivity of the nanocomposites was also investigated on blends of a cyclic butylene terephthalate (CBT) with a PP containing MWNCTs with different nanotube loadings.

Experimental

Masterbatches of 20 wt% MWCNT in PP and of 15 wt% MWCNT in PC were purchased from Hyperion Catalysis International, Cambridge, MA. The nanotubes are about 15-50 nm in diameter and 1-10 μm in length. The masterbatches were then diluted with an isotactic polypropylene (PP HD120) supplied by Borealis Co. and with a polycarbonate (Calibre 1080) supplied by Dow Chemical to prepare nanocomposite samples of various loadings using a lab-twin screw extruder, Leistritz ZSE 18HP operated at optimized mixing conditions. A polypropylene grafted maleic anhydride (PP-g-MA Polybond PB3150 with 0.5 wt% MA) was used as a compatibilizer to try improving the dispersion in PP/MWCNT nanocomposites. A cyclic butylene terephthalate (CBT) supplied by Cyclics Co. was used to prepare the PP/CBT blends at different CBT concentrations. The prepared nanocomposites were then microinjection molded using a Battenfeld Microsystem 50 molding machine and a mold having a dog-bone shaped cavity.

The morphology of the nanocomposites was studied through scanning electron microscopy (SEM) and transmission electron microscopy (TEM). For SEM we used a high resolution Hitachi S-4700 microscope for observing ultramicrotomed surfaces of PC samples and etched surfaces of the PP and PP/CBT nanocomposites. Prior to SEM examination, all samples were coated with a vapor deposit of Pt for 25s. TEM was done on ultrathin nanocomposites cross sections using a Hitachi HD-2000 microscope.

Raman spectroscopy was used for nanotube quantitative alignment measurements using a Renishaw spectrometer equipped with an inVia Raman microscope fitted with a 20x objective. The samples were excited using a NIR laser (785 nm) with a grating of 1200 g/mm in a regular mode. Measurements were carried out at two different orientations of the polarized laser beam with respect to the flow direction in the samples (Fig. 1).

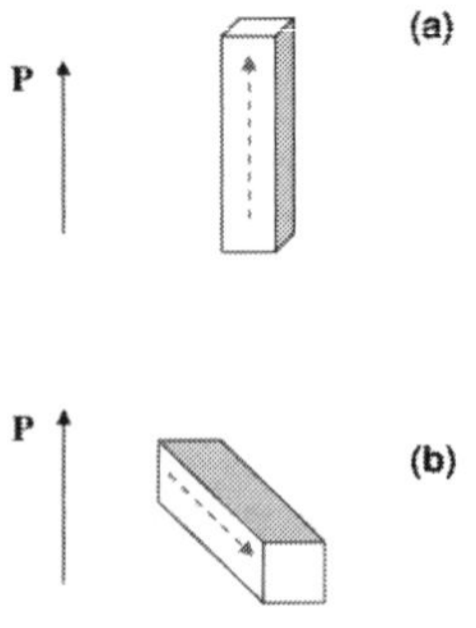

Figure 1. Schematic for Raman spectroscopy. Characteristic scattering configurations for the Raman spectroscopic measurements: (a) Parallel position of specimen, corresponding to 0° angle between the polarization direction (P) and the flow direction (dashed arrow); (b) Perpendicular position of specimen, corresponding to 90° angle between the polarization direction (P) and the flow direction (dashed arrow).

Wide angle X-ray diffraction (WAXD) measurement was carried out using a Bruker AXS X-Ray goniometer equipped with a Hi-STAR two-dimensional area detector. The orientation factors from WAXD are mainly due to the crystalline part.

The volume resistivity of the tensile bars prepared by both micromolding and compression molding was determined using a Keithley electrometer model 6517. For each sample the *I-V* curve was obtained across the thickness of the tensile bar and the sample resistance was determined from the slope of the curve. The resistance was then converted to volume resistivity, ρ_v, using the formula

$$\rho_v = WDR_v / L \quad (1)$$

where W is the width, D the thickness, L the length of the sample, and R_v is the measured resistance.

Results and Discussion

SEM micrographs of the surfaces of compression molded nanocomposites of 5wt% MWCNT/PC and 3wt% MWCNT/PP are shown in Figure 2 (top micrographs).

The nanotubes are dispersed almost individually in the polycarbonate and the distribution at the micro level is quite uniform (Fig. 2a, SEM top micrograph). For polypropylene, however, nanotube aggregates are formed and the distribution is not uniform (Fig. 2b, SEM top micrograph).

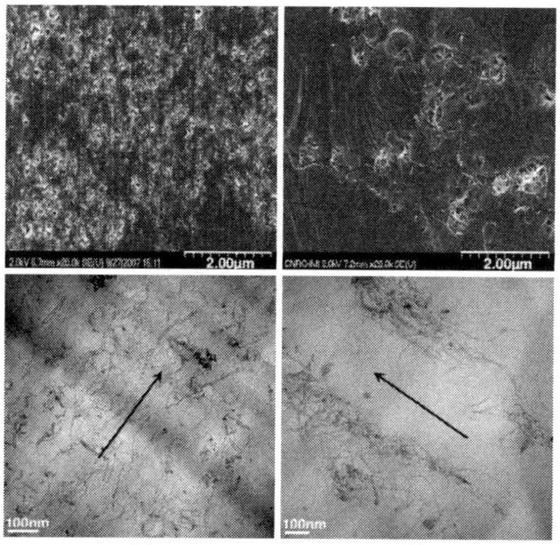

(a) (b)

Figure 2. Micrographs on the top: SEM micrographs of (a) crayon surfaces of 5wt% MWCNT/PC and (b) etched surface of 3wt% MWCNT/PP nanocomposites. Micrographs on the bottom: TEM micrographs of thin sections of (a) 5wt% MWCNT/PC and (b) 3wt% MWCNT/ PP nanocomposites

On the other hand, a well dispersed system may not be necessarily required for the formation of an electrically conductive network. As it is observed (Fig.2 , top) for both nanocomposites, PP and PC/MWCNT, an interconnected network of entangled MWCNTs exist forming a percolated network. This network seems to be even stronger and more efficient in the case of PP nanocomposites in spite of worse dispersion quality. The presence of a crystalline phase in the PP matrix creates a structure similar to a two-phase blend with one phase crystalline and the other amorphous (Nielsen and Landel 1993). Nanotubes are expected to be found only in the amorphous phase since from the thermodynamic point of view fillers and other impurities can never exist in a crystal (Khunová, Smatko et al. 1988). Consequently, according to the concept of double percolation threshold strong nanotube networks are formed even at lower content compared to what is observed in PC/MWCNT nanocomposites where there is no crystalline phase. The overall percolation threshold in PP is equal to the threshold of its amorphous phase, which is only a fraction of the polymeric matrix and fewer nanotubes are therefore needed to create a conductive network in this case.

It has been shown that the nanotubes are well aligned when they are subjected to high shear rates (Abbasi et al. 2010). The nanotubes alignment in microinjected samples is shown in the TEM micrographs of Figure 2 (bottom micrographs). The images were taken from thin sections cut along the flow direction. In the case of polycarbonate (Fig. 2a, bottom) the nanotubes are well aligned and most of them are perfectly stretched along their length axis. In PP nanocomposites (Fig. 2b, bottom), however, we mostly observe nanotube aggregates with an overall orientation in the flow direction (as illustrated by dashed arrows in Fig. 1). Individual nanotubes are not well aligned and in some area there are bent nanotubes, which connect the oriented aggregates together.

Raman spectroscopy was used to further investigate MWCNT alignment within the polymeric matrices (Abbasi, Carreau et al. 2010). The intensity ratios $D_{\Pi}/D_{\perp}$ and $G_{\Pi}/G_{\perp}$ parallel/perpendicular to the flow direction are used to assess the degree of nanotube alignment. As indicated in Table 1, both ratios are much larger for the PC nanocomposites, demonstrating that the nanotubes are highly aligned in PC while the lower ratios obtained for the PP nanocomposites indicate that the alignment is less in PP. These results are completely in agreement with the morphological analysis and clearly show the different level of alignment in the nanocomposites. The presence of the crystalline phase in PP significantly reduces the nanotube alignment.

Table 1. Intensity ratios parallel/perpendicular to the flow direction

	Raman intensity ratio parallel/ perpendicular	
	$D_{\Pi}/D_{\perp}$	$G_{\Pi}/G_{\perp}$
PC/5 wt% MWCNT	1.95	1.75
PP/5 wt% MWCNT	1.35	1.24

WAXD was employed to study the orientation of the crystalline and amorphous phases in the presence of nanotubes using microinjected PP parts. Diffraction patterns obtained from a two-dimensional WAXD analysis are shown in Figure 3.

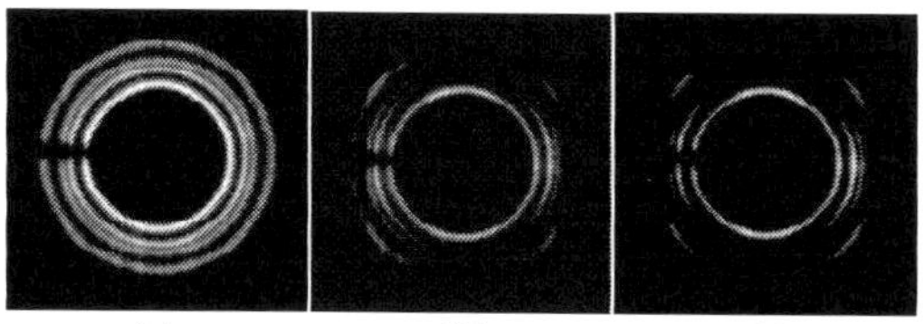

(a) (b) (c)

Figure 3. WAXD pattern of (a) compression molded PP (b) microinjected PP (c) microinjected 5 wt% MWCNT / PP nanocomposite.

Within the compression molded or non-injected neat polypropylene samples (Fig. 3a) there is no preferred orientation and whole rings are observed indicating random orientation; microinjected PP samples (Fig. 3b), however, exhibit distinct and bright asymmetric arcs, a clear indication of a relatively high level of orientation. Figure 3c shows the effect of nanotubes on the orientation of PP phase in the nanocomposites. Although the bright arcs are still observed as an indication of crystal orientation, compared to what is presented in Fig. 3b, the arcs are less sharp indicating that adding nanotubes to the PP matrix can change the crystal orientation to some extent.

Figure 4 presents the effect of nanotube loading on the electrical conductivity of the PP and PC nanocomposites. The electrical percolation threshold can be found by applying a power-law expression to the electrical conductivity data (Abbasi, Carreau et al. 2009; 2010).

$$\sigma = \beta_{c,\sigma}\left(\frac{m - m_{c,\sigma}}{m_{c,\sigma}}\right)^{q} \qquad \text{for } m \geq m_{c,\sigma} \qquad (2)$$

where $\beta_{c,\sigma}$ and q are power-law constants, and $m_{c,\sigma}$ is the electrical percolation threshold (wt%). This power-law expression is found to describe very well the data of Figure 4 where the left limit of the dashed line corresponds to the percolation threshold value in each case.

For compression molded samples the percolation threshold is found to be around 3 wt% of nanotubes for PC nanocomposites and around 1 wt% for PP nanocomposites. These values result from the effect of the crystalline structure on the electrical properties of the nanocomposites. In the semi-crystalline polypropylene a nanotubes network is formed at a lower nanotube loading according to the concept of double percolation threshold as discussed in the morphology results section.

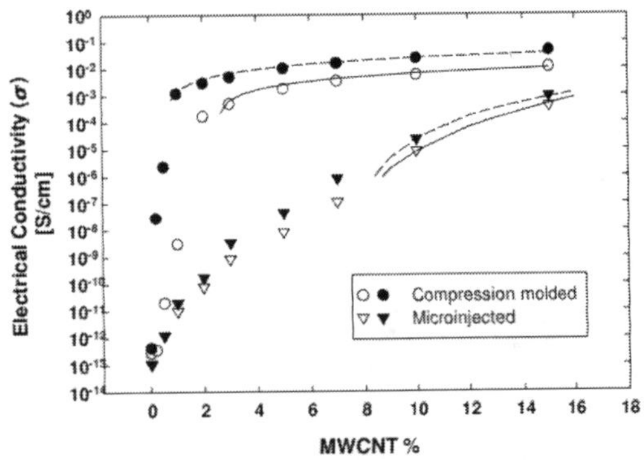

Figure 4. Electrical conductivity of compression molded and microinjected nanocomposites: PP/MWCNT (filled symbols) and PC/MWCNT nanocomposites (blank symbols)

In the microinjected samples the nanotubes are well aligned in the flow direction and are less likely to be connected (Abbasi, Carreau et al. 2009; 2010). To form a connected or percolated network in such a system, more nanotubes are required and consequently the percolation threshold rises. Even in this case, a lower percolation threshold and a higher electrical conductivity level can be observed in the case of PP nanocomposites relative to PC nanocomposites. As discussed previously, the presence of a crystalline phase decreases the nanotube alignment resulting in a more isotropic arrangement, which is more favorable for the creation of nanotube networks.

The morphology and electrical properties of MWCNT-filled PP/CBT blends were also investigated in this study.

Figure 5 shows the SEM micrographs of an iPP/CBT (80/20) blend without nanotubes (Fig. 5a) and filled with 1 wt% MWCNTs (Fig. 5b). The CBT phase forms the minor or dispersed phase,as nearly spherical domains in the absence of nanotubes, indicating that at this composition the blend is still far from having a co-continuous structure. By incorporating nanotubes into the blend, the minor CBT phase becomes completely elongated (Fig.5b), changing from droplets into strips, which are more favorable to form a co-continuous structure.

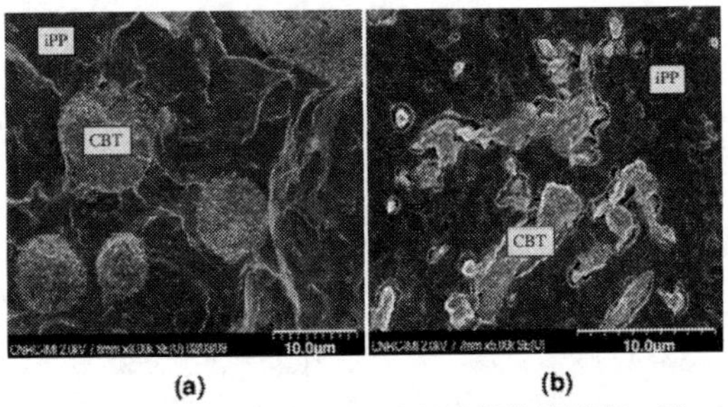

Figure 5. SEM micrograph of iPP/CBT (80/20) without nanotube (a) and filled with 1 wt% MWCNTs (b)

It is well known that the viscosity ratio of the two components of a blend is an important factor affecting the morphology of the blend. When this ratio is closer to 1, the size of the minor phase domains becomes smaller and more irregular(Gubbels, Jerome et al. 2002). The viscosity of PP is much larger than that of CBT and as a consequence nanotubes are preferably located in the less viscous phase, CBT. However, the nanotubes more likely increase the CBT viscosity and the kinetic of the phase coalescence is perturbed during the molding process.

Figure 6 shows the effect of nanotube loading and the blend composition on the electrical conductivity of MWCNT-filled PP/CBT blends.

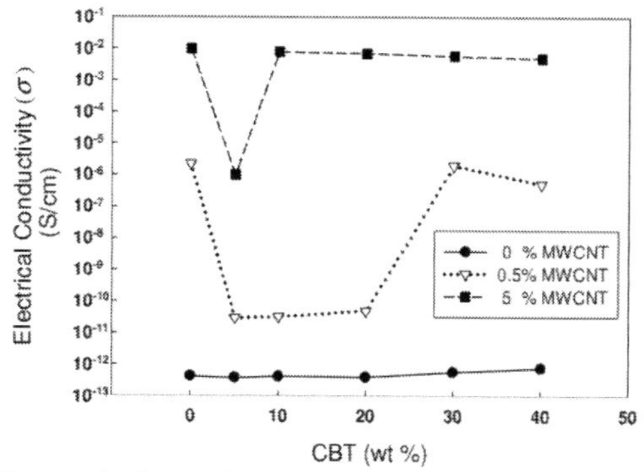

Figure 6. Dependence of electrical conductivity of MWCNT-filled PP/CBT blends on CBT content at various nanotube loadings

As this figure shows clearly, incorporation of 5 wt% CBT into PP decreases sharply the electrical conductivity and therefore one can expect that the addition of CBT at a low level of concentration would increase the nanotube percolation threshold. But the trend is not the same for all compositions of the blends. When the CBT content reaches 30 wt% the blend becomes conductive with only 0.5 wt% nanotube loading. In fact, based on the concept of double percolation threshold, in order to get conductive blends, the filled phase or CBT phase should exhibit a continuous phase in a co-continuous blend. Accordingly, in the 30 wt% CBT blend filled with 0.5 wt% nanotubes the second percolation or the continuity of the CBT phase creates a conductive blend at a significantly low nanotube loading. However, with increasing CBT content at this nanotube loading the electrical conductivity decreases because the local concentration of nanotubes in the CBT phase decreases and 0.5wt% loading would not be sufficient to create conductive blends. However, for nanotube loading of 5wt%, a lower CBT percolation threshold is observed, beyond which the electrical conductivity increases and reaches a constant value irrespective of the CBT concentration.

Conclusions

The effect of the crystalline structure and nanotube alignment on the electrical conductivity of PP/MWCNT and PC/MWCNT nanocomposites prepared at high shear rates were examined in this study and it was found that the nanotubes are less aligned in the presence of a crystalline phase in the nanocomposites. The results also showed that addition of carbon nanotubes reduced the orientation of crystalline phase in the crystalline polymer (PP). The electrical percolation thresholds rose significantly with increasing orientation level; however, the presence of a crystalline structure improved the electrical conductivity and decreased the percolation threshold according to the concept of double percolation threshold. The effect of nanotube loading on the electrical conductivity of PP/CBT blends was also analyzed and it was found that a double percolation is the basic requirement for the conductivity of these nanocomposites. Percolation of nanotubes in a filler-riched phase and the continuity of this phase in the blend both affected the electrical conductivity of the nanocomposites. The nanotubes also affected the morphology of the blends by increasing the viscosity of the filler-rich phase. As a consequence, smaller and more elongated dispersed phase was observed in the presence of nanotubes.

Acknowledgements

Financial support from NSERC (Natural Science and Engineering Research Council of Canada) is gratefully acknowledged. We are also thankful to Ms. Weawkamol Leelapornpisit for her great help in morphological studies.

References

Abbasi, S., Carreau, P., Derdouri, A., Moan, M., "Rheological properties and percolation in suspensions of multiwalled carbon nanotubes in polycarbonate." Rheologica Acta, **48**(9): 943-959 (2009)

Abbasi, S., Carreau, P., Derdouri, A., "Flow induced orientation of multiwalled carbon nanotubes in polycarbonate nanocomposites: Rheology, conductivity and mechanical properties." Polymer, **51**(4): 922-935 (2010)

Fangming, D., Fischer, J.E., Winey, K.I., "Effect of nanotube alignment on percolation conductivity in carbon nanotube/polymer composites." Phys. Rev. B, **72**(12): 121404-1 (2005)

Gubbels, F., Jerome, R., Teyssie, P., Vanlathem, E., Deltour, R., Calderone, A., Parente, V., Bredas, J.L., "Selective Localization of Carbon Black in Immiscible Polymer Blends: A Useful Tool To Design Electrical Conductive Composites." Macromolecules, **27**(7): 1972-1974 (2002)

Khunová, V., Smatko, V., Hudec, I., Beniska, J.: Relationships of Polymeric Structure and Properties, Springer, Berlin / Heidelberg (1988)

Meyyappan, M.: Carbon nanotubes : science and applications, CRC Press, Boca Raton (2005)

Nielsen, L.E., Landel, R.F.: Mechanical properties of polymers and composites , Marcel Dekker, New York (1993)

Sumita, M., Sakata, K., Asai, S., Miyasaka, K., Nakagawa, H., "Dispersion of fillers and the electrical conductivity of polymer blends filled with carbon black." Polymer Bulletin, **25**(2): 265-271(1991)

Sumita, M., Sakata, K., Hayakawa, Y., Asai, S., Miyasaka, K., Tanemura, M., "Double percolation effect on the electrical conductivity of conductive particles filled polymer blends." Colloid Polym. Sci., **270**(2): 134-139 (1992)

Wu, D., Liang, W., Yurong, S., Ming, Z., "Rheological properties and crystallization behavior of multi-walled carbon nanotube/poly(ε-caprolactone) composites." J. Polym. Sci. B, **45**(23): 3137-3147 (2007)

PLA-based organic-inorganic hybrid materials

A. Prébé[1,2], P. Cassagnau[1]*, P. Alcouffe[1,2], JF Gérard[2]

[1] *Université Lyon 1, Ingénierie des Matériaux Polymères, 69622 Villeurbanne, France, Philippe.cassagnau@univ-lyon1.fr*
[2] *INSA de Lyon, Ingénierie des Matériaux Polymères, 69621 Villeurbanne, France*

Abstract - The objective of the present study is to prepare poly(L-lactide) based organic-inorganic hybrid materials. For that propose two strategies have been developed:
1- In situ bulk polymerization of L-Lactide in the presence of 5%wt. of modified fumed silica. The polymerization kinetics was followed from the changes of the complex shear modulus versus reaction time. Adding fumed silica particles into the monomer leads to the formation of a physical (percolated) network from particle-particle interactions, i.e. silica, in the L-Lactide probably due to hydrophilic interactions. It was found that basic hydrophilic silica leaded to a microcomposite with highly dense agglomerates in the polymer matrix whereas with a less hydrophilic silica it was possible to decrease the size of the agglomerates increasing the dispersion. The best dispersion state was obtained with the "initiating" functionalized silica leading to a "grafting from" polymerization of the L-Lactide.
2-In situ synthesis (sol-gel method) of silica particles in PLA. The generation of PLA-based O/I hybrids has been study from the in situ synthesis of silica particle in molten PLA. The water physically absorbed into the PLA at saturation combined with the linear polydiethoxysiloxane precondensate was used to synthesis the inorganic rich phase through typical hydrolysis-condensation reaction. The presence of a majority of Q_3 species showed by solid state ^{29}Si NMR assures that condensation occurred. Interfacial agents, i.e. 3-trimethoxysilylpropylmethacrylate and γ-Aminopropyltriethoxysilane, at different ratios were used to bring physical and/or chemical interactions between the inorganic rich phase and the PLA matrix. TEM images show that even if 3-trimethoxysilylpropylmethacrylate concentrates at the interface, the size of the inorganic rich phase is not lowered and has a diameter of 200-300 nm.

Introduction

Polymers issued from the biomass have become of great interest as they are an interesting alternative to polymers coming from fossil resources. The unique physical characteristics that PLA possesses make it suitable for many different applications. PLA has good crease-retention and crimp properties, excellent grease and oil resistance, easy low-temperature heat sealability, and good barrier to flavours and aromas. Unfortunately, to be able to compete with those commodity polymers, its mechanical properties need to be at least even. It is, now, well known that it is possible to enhance them by adding nanofillers in order to nanostructure the polymer and/or to enhance its crystallization rate. On the other hand, different fillers have already been studied in PLA matrices but only a few report on a common filler used in formulations such as fumed silica to improve mechanical properties. Presently, fumed silica are versatile nanofillers which can offer different specific surface (50-400 $m^2.g^{-1}$) and a well known surface chemistry to manage interactions, surface reactions as well as grafting initiating groups for ring opening polymerization. In this work, we present two strategies for the development of PLA nanocomposites: i) the *in-situ* polymerization of L-Lactide in the presence of modified fumed silica and ii) the generation of silica from a linear precondensed alkoxysilane with or without interfacial agent in the molten PLA by reactive extrusion. These two methods are compared and discussed in term of morphology development.

Experimental

The PLA Natureworks grade 2002D was selected. The linear precondensed polydiethoxysiloxane (PDEOS) was purchased at ABCR and used as received. γ-Aminopropyltriethoxysilane and 3-trimethoxysilylpropylmethacrylate, referred as APTES and TMSPM respectively, were purchased from ABCR and used as received. Anti-oxidant B225 was kindly provided by Ciba and was used as received

L-Lactide was purchased from Boehringer Ingelheim Co. and was used without any further purification. The water content was titrated to be close to 200 ppm. Tin (II) octanoate, $Sn(Oct)_2$, purchased from Aldrich was also used as received as well as the triphenylphosphine. The fumed silica of different specific surfaces and organically modified was kindly provided by Wacker Chemie Co.

Results and Discussion

PLA polymerisation

For this purpose, the rheological changes are studied during polymerisation in bulk with and without fumed silica. This remains of a large importance in order to analyze the effect of the nanofiller and its characteristics. As a matter of fact, a rheometer can be used as a reactor for *in situ* polymerisation of homogeneous media. The advantage of this method is the possibility to anticipate the effect of the polymerization reaction on the rheology and forecast what will happen in an extruder [Cassagnau 2009].

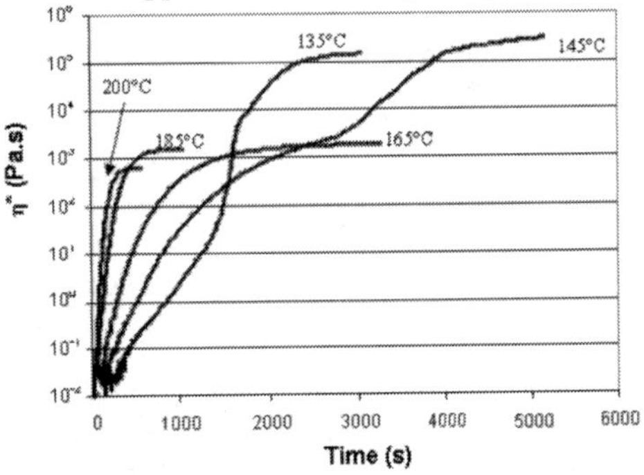

Fig 1: Complex viscosity, η*, during isothermal polymerization of L-Lactide using Tin (II) octanoate (R=1000) as catalyst at different reaction temperatures (Frequency: 10 rad.s^{-1}; strain amplitude: 0.5%)

As shown in Fig 1, the complex shear viscosity increases during polymerization, i.e. from the behaviour of a low viscosity liquid (liquid monomer) to that of a viscoelastic liquid at the end of the polymerization reaction. Obviously, it can be also seen that the higher the temperature, the faster the polymerization is completed. The average weight molar mass of Poly(L-Lactide) polymerized at different temperatures are close to 90,000 mol.g^{-1}. For temperatures below 150°C, i.e. temperature of polyL-Lactide crystallization, the polymer crystallizes while the polymerization occurs leading to the appearance of a second inflexion point on the viscosity curves.

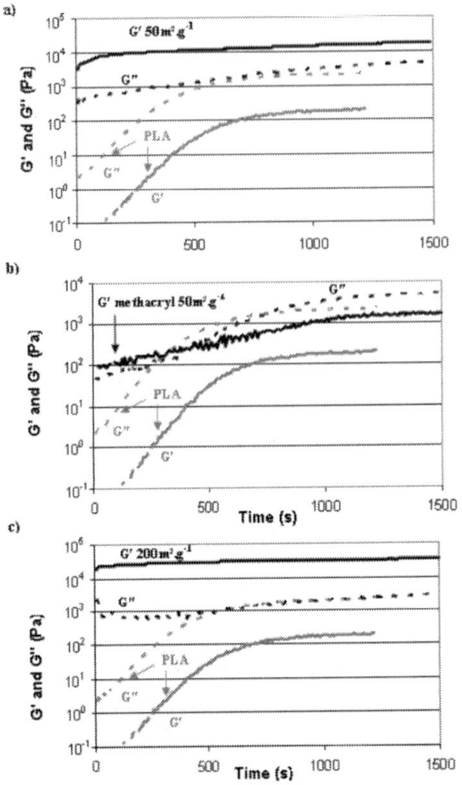

Figure 2 – Rheological behaviour during polymerization of L-Lactide using Tin (II) octanoate (M/I=1,000) as catalysts compared to the polymerization of L-Lactide in the presence of 5%wt. of fumed silica. a) 50m^2.g^{-1}; b) methacrylate-functionalized 50m^2.g^{-1} ; c) 200m^2.g^{-1}. (185°C, strain = 0.5%, ω = 10 rad.s^{-1})

The influence of fumed silica particles on the L-Lactide polymerization is studied at 185°C, i.e. above crystallization temperature. It can be seen in Fig 2a that the addition of 5 wt% of fumed silica drastically changes the reactive media rheology. It can be seen at the early stage of polymerization that the storage modulus is much higher that the loss modulus, i.e. the reactive medium displays a solid-like behaviour. Fumed silica fillers are known to form nanostructures from particle-particle interaction leading to the formation of a three-dimensional network in reactive media. As a consequence, percolated structures are formed which displays rheological behaviour of conventional gels (G'>G''). Furthermore, this gel-like behaviour is kept while polymerization proceeds, i.e. the storage modulus remains larger than the loss one all along the polymerization. It has been reported that a percolation threshold needs to be reached, in terms of volume fraction of silica, to form a network in the polymer. This percolation threshold, which depends on the balance between particle-particle and particle-organic medium interactions has been determined to be very low (2-3%wt.). Even if the L-lactide and the PLA

can be considered as hydrophilic, the particle-particle interactions are favoured compared to the particle-monomer or particle-polymer interactions.
Another interesting feature of the fumed silica is their specific surface area. Indeed, it is possible to consider silica from $50m^2.g^{-1}$ to $400m^2.g^{-1}$. The polymerization in bulk of neat L-Lactide at 185°C or containing 5%wt. of fumed silica of $200m^2.g^{-1}$ between the plates of the rheometer can be compared (Fig 2c). Again, it is possible to evidence that the storage modulus is above the loss shear modulus from the beginning of the polymerization to the end meaning that a physical network of fumed silica is generated from particle-particle interactions in the monomer and kept as percolated particles all along the reaction. A higher storage modulus is obtained with the $200m^2.g^{-1}$ compared to the one obtained with the $50m^2.g^{-1}$ contributing to the idea of having of a stronger silica network bonded by stronger particle interactions, i.e. Van der Waals and H-bonding, with the growing number of Si-OH groups on the silica surface.
These interactions are driven by the concentration of silanol groups onto the silica surface which can form strong hydrogen bonds between silica particles. One way to avoid these particle-particle interactions is to lower the silanol group concentration on the surface as well as to render the silica surface reactive with the monomer. Consequently, the rheological behavior of several other organically-modified fumed silica was considered in order to avoid the aggregation. For example, with the methacrylate functionalized fumed silica it is possible to see that, during the polymerization, the loss modulus crosses the storage one (≈400s) meaning that the silica network formed by methacrylate-functionalized fumed silica breaks down during the polymerization of L-Lactide (Fig 2b). This result tends to imply that this type of function could allow a good dispersion of the silica in the PLA matrix due to a better matching of the polarity of methacrylate groups with the growing species at a given extent of reaction, i.e. monomer content.

Silica by in situ sol-gel

The addition of polydiethoxysiloxane to PLA was carried out in mini-extruder and samples at four different times (10, 20, 30 and 40min) of mixing were collected. Morphologies observed by TEM for different magnification of PLA-hybrid materials at different times of mixing are shown in Fig 3. After 20 minutes of extrusion, the inorganic-rich phase is composed of two populations: 200-300nm large phase and very small particles of only a few nm. After processing for one hour, the morphology has dramatically changed with only one population of particles having 100-200 nm diameter. This particular change in morphology can be explained by the extent of the condensation reaction combined with the shear energy provided by the extruder which leads to the coalescence of Si-rich particles in larger ones. Indeed, after 20 minutes the condensation reaction has not progressed enough to prevent the dispersion of the silicone-rich phase in the PLA matrix under shear due to the fact that the viscosity of the dispersed phase remains low, i.e. the original viscosity of PDEOS is equal to 4.10^{-3} Pa.s. Oppositely, after one hour, the extent of condensation is high enough to lead to a high value of the viscosity of the dispersed phase. This viscosity tends forward to be infinite as a gel is rapidly reached due to the very high functionality of the polyalkoxysilane avoiding its disaggregation into small particles under shear.

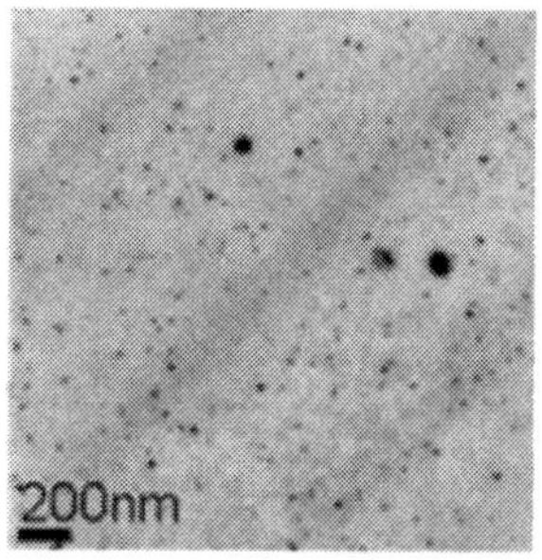

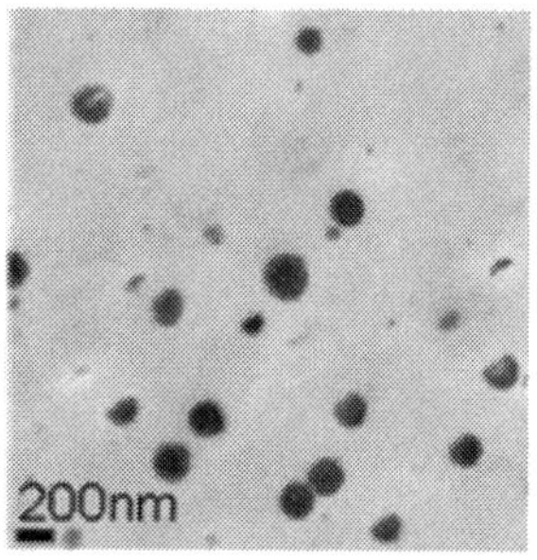

Fig 3: TEM pictures of PLA- based O/I hybrid films.
Upper picture: after 20 minutes in the extruder
Lower picture: 1 hour at 185°C in the extruder
(Composition: 11%wt. of PDEOS in PLA).

Interfacial agents having both organic functionalities to interact and/or react with the PLA matrix and hydrolysable functions, i.e. able to be hydrolyzed and co-condensed with the silanol counterparts of the preconsensate, were considered in the next part. The two interfacial agents used are 3-trimethoxysilylpropylmethacrylate (TMSPM) and Aminopropyltriethoxysilane (APTES) respectively. TMSPM was considered for its polarity approaching the one of PLA and its ability to be marked by Osmium or Ruthenium through its double bond for TEM images. APTES was considered for its ablility to react with the PLA matrix through amidification creating a covalent bond between the matrix and the inorganic-rich phase. The two interfacial agents were added simultaneously with polydiethoxysiloxane (PDEOS) to the melted PLA using the same method described before.

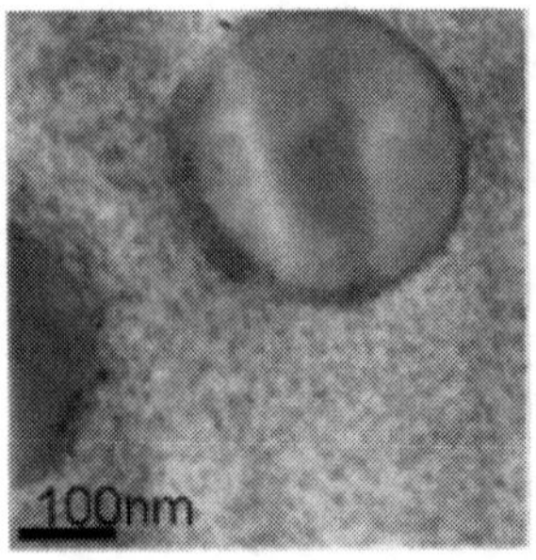

Fig 4: TEM images of PLA-hybrid films containing TMSPM (marked with Ruthenium tetroxide)

The sample marked by Ruthenium tetroxide (Fig 4) shows a black corona around the inorganic rich phase. This feature means that the TMSPM is concentrated at the interface between the PLA matrix and the inorganic-rich phase. Unfortunately, even if the interfacial agent is at the interface it seems that it does not have a strong impact on the sizes of the inorganic rich phase as their diameter is around 200 to 300 nm as for the PDEOS alone.

Conclusions

The study of the bulk *in situ* polymerization of L-Lactide in the presence of organically-modified fumed silica showed that it was possible to obtain a nanocomposite morphology in a one step reaction. It was found that the morphology obtained is directly linked to the surface modification of the silica surface. Indeed, lowering the hydrophilicity of the surface leaded to a better dispersion of the filler.

The "grafting from" polymerization of L-Lactide on the initiating group of the silica surface was achieved leading to highly dispersed silica on a nanoscale. The main drawback lies in the low molar mass obtained.

The generation of PLA-based O/I hybrids in the melt via sol-gel chemistry in extruder was achieved. Indeed it was possible to generate inorganic-rich phase into the PLA via reactive extrusion as proven by the solid state ^{29}Si NMR. Moreover, the TMSPM interfacial agent was evidenced to concentrate, as supposed, at the interface between the PLA matrix and the inorganic-rich phase.

Acknowledgements

The authors would like to thank the Nanofun-poly network of Excellence for their financial support as well as Wacker Chemie Co. for kindly providing the fumed silica and the microscopic centre of Lyon1 for the TEM microscopy

References

Cassagnau P, Fenouillot F, Bounor-Legaré V, Reactive processing of thermoplastics polymers: A review of fundamental aspects, Int. Polym Processing, 2007, 22(3):218-258

NOVEL SEGREGATED NETWORK ELECTRICALLY CONDUCTIVE POLYMERIC NANOCOMPOSITES USING METAL NANOWIRES

Genaro A. Gelves, Mohammed Al-Saleh, Uttandaraman Sundararaj*

Department of Chemical and Petroleum Engineering
Schulich School of Engineering
University of Calgary, Calgary, Alberta, Canada
u.sundararaj@ucalgary.ca

Abstract - Highly electrically conductive metal nanowire/polymer nanocomposites have been produced using a novel solution mixing method that enables the formation of unique nanowire segregated networks in single polymer matrix without the use of surfactants or surface modification of nanofillers and/or polymers. Segregated networks of nanowires provide excellent electrical properties at low nanowire concentrations. The nanocomposites exhibit electrical conductivities of up to 10^4 S·m^{-1} and high Electromagnetic Interference Shielding effectiveness.

Introduction

Polymer nanocomposites containing metal nanowires are emerging and promising materials for applications like electrostatic dissipation (ESD), electromagnetic interference shielding (EMI), conductive adhesives, sensors, etc. The use of electrically conductive and high aspect-ratio nanowires and nanotubes enables significant changes in the electrical properties of nanocomposites at much lower concentrations than those required with conventional fillers (i.e. carbon black, carbon fibers, metal fibers and stainless steel fibers). We have previously reported low electrical percolation thresholds (< 1.0 vol. %) of copper and silver nanowires in polymer nanocomposites,(Gelves et al., 2006) which are comparable to those of carbon nanotube-based polymer nanocomposites.(Chang et al., 2006; Curran et al., 2006; Ramasubramaniam et al., 2003; Safadi et al., 2002; Zhang et al., 2005) However, the formation of metal nanowire networks can result in nanocomposites with high electrical conductivities that can outperform nanotube-based nanocomposites for conductive applications.

In previous works, we have observed strong agglomeration of nanowires resulting in agglomerates with dimensions in the order of tens of micrometers, which prevents electrical percolation at lower concentrations.(Gelves et al., 2006; Gelves et al., 2005) In addition, electrical conductivities of the composites close to that of the metals were not attained. In this work, we present highly conductive polymer nanocomposites containing segregated electrically conductive nanowire networks with electrical conductivities approaching the conductivity of bulk copper. Copper nanowire/polymer nanocomposites showed high performance for shielding of electromagnetic interference, an application that typically requires electrically conductive materials.

Electromagnetic interference (EMI) is the generation of undesired signals in an electrical or electronic device due to the interaction between external electromagnetic radiation and internal electrical signals. Thus, electromagnetic interference (EMI) shielding refers to the ability of materials to prevent electromagnetic radiation to penetrate or be emitted from an electronic device and prevent malfunction of electronic equipment. The miniaturization of electronics continously introduces higher requirements for EMI shielding and materials with improved performance are required.(Chung, 2001) Metal nanowires with high aspect ratios can form electrically conductive networks at low concentrations, which result in nanocomposites with high EMI shielding effectiveness, improved processability and light- weight.

Experimental

Synthesis of nanowires. Copper nanowires were prepared using porous alumina templates prepared in-house. The alumina templates are prepared by two-step anodization process of Al foil in 0.3 M sulfuric acid (H_2SO_4) solutions at 25 V. Al electrodes of 25×10×0.1 cm in size were utilized. Copper was electrodeposited in the pores of the alumina templates using ac electrodeposition and then released using 1.0 M Sodium hydroxide (NaOH) solutions. In the final step, CuNWs are transferred to methanol and dispersed using ultrasound (135 W average power, 38.5-40.5 KHz). Copper nanowires in methanolic suspensions were used to prepare polymer nanocomposites by a solution technique as explained below.

Preparation of polymer nanocomposites. Polymer solutions were prepared in methylene chloride (CH_2Cl_2). Polystyrene (Styron 666D, M_w 200,000 g/mol, MFI 7.5, Tg 100 °C) was kindly provided by Dow Chemical. CuNW-filled PS nanocomposites were prepared using a solution method. CuNW/PS composites were formulated by solution processing. Composites with different concentrations of CuNW were produced by mixing different volumes of ~3.3 mg/ml CuNW/methanol solution with 28.5 mg/ml or 20 mg/ml PS/methylene chloride solution. Each mixture was treated in ultrasound. The mixture was then placed in an evaporation dish for 16 hours in a fume hood. After that, the evaporation dish was transferred to a vacuum oven for 2 hr at 40 °C to remove the residual solvents from the composite powder. For EMI SE, electrical resistivity and composite morphology characterization, the dry composite powder was annealed in a Carver compression molder at 250 °C, 31 MPa for 30 min to produce films 210 μm in thickness, 2.5 cm in width and 4.2 cm in length.

Characterization of polymer nanocomposites. Thermogravimetric analysis (TGA) of CuNW/polymer nanocomposites were carried out using a Perkin Elmer Thermogravimetric Analyzer (Pyris1TGA) to determine the concentration of CuNWs in the composites produced by solution processing. The analysis was performed by heating 10 mg of the CuNW/PS composite from 25 °C to 650 °C in a nitrogen environment at a rate of 20 °C/min. The sample was maintained at 650 °C for 10 min before cooling down to room temperature. The sample was held at 650 °C for 10 min before cooling. TEM micrographs of CuNW/PS composites were taken using Philips Morgagni 268 microscope at an acceleration voltage of 70 kV. Thin sheets (210 μm in thickness) of nanocomposite specimens used for EMI SE characterization were fractured and cross sections of the samples were ultramicrotomed to sections of ~70 nm in thickness at room temperature using a Leica EM UC6 Ultramicrotome.

Electrical properties of polymer nanocomposites: Electrical resistivity measurements were carried out according to ASTM 257-75 using a Loresta GP resistivity meter and ESP four point probe. EMI SE in the X-band frequency was measured using an HP 8757D scalar network analyzer, HP 8375A sweep oscillator, three HP 11664A detectors, coaxial to waveguide adapters and two HP X752C wave directional couplers. The directional couple opening is 22.5 mm × 10 mm. The dynamic range of the setup (maximum EMI SE) is 50 dB.

Results and Discussion

Gram amounts of metal nanowires have been produced by template-directed synthesis using porous aluminum oxide templates. Copper nanowires are electrodeposited into the pores using AC electrodeposition, then released from the template using basic solutions, and finally dispersed in methanol using ultrasound techniques. Copper nanowires are 25 nm in diameter and show an average length of 1.29 μm.(Gelves et al., 2006) Figure 1 shows SEM images of representative samples of Cu nanowires used for nanocomposite preparation. Individual and bundles of several nanowires are observed. Results of surface composition (XPS), X-ray diffraction, thermogravimetric analysis and aspect ratio characterization of these nanowires have been published previously.(Gelves et al., 2008; Gelves et al., 2006) Other metal nanowires (eg. Ag, Fe, Ni, Au, Pt) can be produced in gram amounts produce using similar methods.(Gelves et al., 2006)

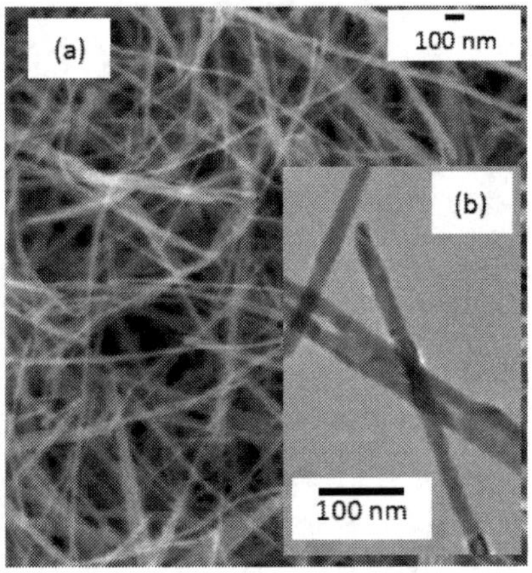

Figure 1. (a) SEM and (b) TEM images of copper nanowires produced in gram amounts and used to prepare electrically conductive polymer nanocomposites.

In this work we have used a new method of preparation of polymer nanocomposites that prevents post-synthesis agglomeration of nanowires and enables the formation of segregated nanowire networks. The method consists in mixing of nanowire suspensions ($CuNW/CH_3OH$) and polymer solutions (PS/CH_2Cl_2) in a miscible solvent mixture (CH_3OH/CH_2Cl_2); the solvent ratio is then modified (CH_3OH concentration is increased) to decrease the solubility of the polymer in the three- component mixture leading to the precipitation of nanowire/polymer nanocomposite particles. The nanocomposite particles are easily separated from the solvent mixture (i.e filtration, rotovaporation, solvent drying). We refer to the method as Miscible Solvent Mixing and Precipitation (MSMP) method. The nanocomposites can be further processed by traditional polymer processing and manufacturing techniques (eg. Masterbatching).

Particles of nanowire/polymer nanocomposites that result from the MSMP method were further processed by compression molding in order to produce specimens for characterization of electrical properties. It is worth mentioning that surface modification of copper nanowires was not used in these experiments. We have demonstrated that the presence of organic layers on the copper nanowire surface introduces electron barriers that affect the electrical conductivity of the composites. Therefore, mixing bare copper nanowires and polymers should enable the highest electrical conductivity of nanowire networks in the nanocomposites.

The morphology of copper nanowire/polystyrene nanocomposites prepared by MSMP method is presented in Figure 2. Thin sections of about 70 nm were obtained by Ultramicrotoming of specimens obtained after compression molding for TEM analysis. The TEM image shows a typical morphology of the cross section of the nanocomposites. Copper nanowires are dispersed throughout the polystyrene matrix forming segregated networks. Unlike other works of segregated nanostructures in polymers, the segregated nanowire networks have formed in a single polymer phase. The mechanism of the formation of segregated copper nanowire networks in a single polymer matrix will be published elsewhere. These segregated networks increase the likelihood of interconnections between nanowires dispersed in the polymer matrix and the reduction of electrical percolation thresholds.

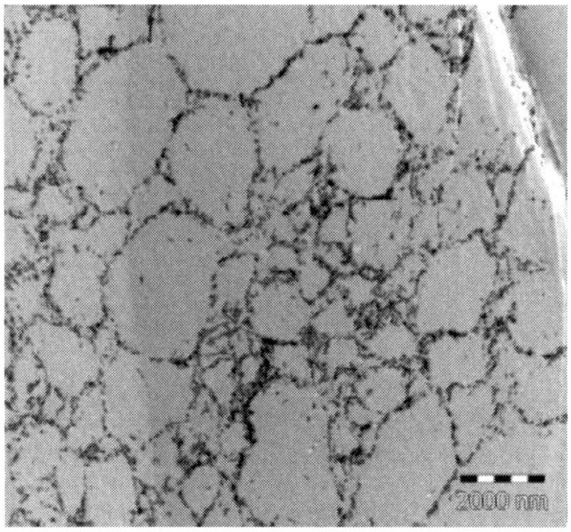

Figure 2. Morphology of copper nanowire-polystyrene nanocomposites. The formation of segregated and interconnected networks of copper nanowires have been facilitated by the precipitation of polymer during the MSMP method.

CuNW/PS nanocomposites have been prepared with different concentrations of nanowires using MSMP method. Figure 2 shows the electrical conductivity of polymer nanocomposites as a function of concentration of copper. Two different setups were used for the characterization of the electrical properties of the nanocomposites given the wide range of electrical conductivities that spans from 10^{-15} for pristine polystyrene to close to 10^{5} S/m for electrically percolated nanocomposites as indicated in Figure 3. The electrical conductivity of the nanocomposites shows a significant change between 0.5 vol. % and 0.75 vol. % copper nanowires. Above, 0.75 vol. %, the electrical conductivity of the nanocomposites gradually changes (from 10^{1}-10^{4} S/m) due to a larger number of electrical paths created by the incorporation of more nanowires in the network. The electrical conductivity of nanocomposites containing 2.8 vol. % copper nanowires was in the order of 10^{4} S/m. The electrical conductivity of CuNW/PS nanocomposites is higher than that reported in previous works. We believe that two factors contributed to the high value of electrical conductivity: (1) The formation of segregated networks that contributes to decrease the effective distance between the nanowires and (2) the use of higher pressures during compression molding of the nanocomposites. It is worth mentioning that the electrical conductivity of copper nanowires used in this work might be lower than that of bulk copper (~5.8×10^{7} S/m at 20°C).(Toimil-Molares et al., 2003) Previous works of characterization of electrical properties of copper nanowires have reported lower electrical conductivity than bulk copper. Therefore, lower electrical conductivities than bulk copper can be expected for CuNW/PS nanocomposites.

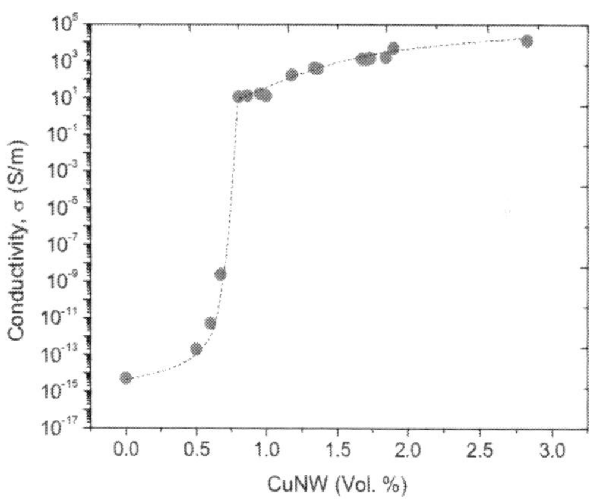

Figure 3. Electrical conductivity of CuNW/PS nanocomposites as a function of volume concentration of copper nanowires.

The electrical percolation threshold (ϕ_c) can be calculated using the power law equation $\sigma=\sigma_o(\phi-\phi_c)^t$.(Lux, 1993; Weber Kamal, 1997) Here, σ is the composite electrical conductivity, σ_o is scaling factor, ϕ is the volume fraction of filler, and t is a critical exponent revealing the lattice dimensionality. The electrical percolation threshold, ϕ_c, for CuNW/PS nanocomposites calculated from fitting the results to the power law is 0.67 vol. % CuNW. The critical exponent is 2.9 and indicates that the lattice is the result of nanowire-nanowire contacts.

Highly electrically conductive nanowire networks in polymers such as those produce by the MSMP method can be used for applications in which high conductivity are typically required like EMI shielding. The EMI shielding of electrically conductive CuNW/PS nanocomposites was measured in the X-band range of frequency. The performance of materials for EMI shielding is calculated as the ratio from the power of incident radiation to the power of the radiation transmitted by the sample. Thus, the EMI shielding effectiveness of a material is given by

$$\text{EMI } SE = 10Log10(P_i/P_t)$$

where P_i and P_t are the power of the incident and transmitted radiation, respectively. EMI SE is expressed in dB. Thus, if a material shows EMI SE of 20 dB it means that the material is capable of shielding 99 % of the power of the incident radiation, and similarly, if the material shows 30 dB, it means that it is capable of shielding 99.9 % of the power of the incident radiation.

The EMI SE of CuNW/PS nanocomposites prepared by MSMP method and compression molding show outstanding performance. For applications like laptops and portable electronics, 20 dB EMI SE is typically required. We found that CuNW/PS nanocomposites with a concentration of 1.3 vol. % (10 wt. %) CuNW showed EMI SE of 26 dB. In addition, polymer nanocomposites containing concentrations higher than 2.0 vol. % showed EMI SE beyond the dynamic range of the equipment (50 dB). CuNW/polymer nanocomposites containing low concentration of nanowires can attain even higher levels of shielding such as those required for military applications (100 dB).

Table 1 shows a comparison of filler concentration, thickness and average EMI SE in the X-band of polymer nanocomposites reported in literature and this work. The EMI SE of CuNW/PS nanocomposites is outstanding compared to those previously reported with other emerging filled-polymers containing vapor-grown carbon nanofibers and carbon nanotubes. For instance, in order to achieve an EMI SE of 26 dB, 1mm thick specimens made of ~7 wt. % MWCNT/PS nanocomposites would be required. However the same level of shielding is obtained using a 0.2 mm film (5 times thinner) made of 10 wt. % CuNW/PS nanocomposites.

It is worth mentioning that the level of shielding of nanocomposite specimens prepared with thickness of 1.0 mm and with concentrations of 10 wt. % CuNW was beyond the range of measurement of the equipment (>50 dB). Thus, it is clear that CuNW/PS nanocomposites can be light-weight and thinner materials for EMI SE applications and can be further developed for applications in automotive, consumer, military, and medical electronics.

Table 1. Average EMI shielding effectiveness (SE) in the range of frequency 8.2-12.4 GHz for polymer composites containing different fillers reported in literature and this work.

Filler / Polymer	Filler Conc. (wt.%)	Thickness (mm)	Average EMI SE (dB)
SWCNT / PU (Liu et al., 2007)	20	2.0	17
SWCNT / Epoxy (Huang et al., 2007)	15	2.0	16
SWCNT / EVA (Das Maiti, 2008)	15	2.5	22
MWCNT / PS Foam (Yang Gupta, 2005)	7	1.0	19
MWCNT / PS (Yang et al., 2005)	7	1.0	26
VGCNF / PS (Yang et al., 2005)	20	1.0	20
CuNW /PS (this work)	**10**	**0.2**	**26**

MWCNT: Multi-walled carbon nanotube
SWCNT: Single-walled carbon nanotube
VGCNF: Vapor grown C-nanofiber
CuNW: Cooper nanowires
EVA: Ethyl vinyl acetate,
PU: Polyurethane,
PS: Polystyrene.

Conclusions

Highly conductive copper nanowire/polystyrene nanocomposites have been prepared by a novel method termed Miscible Solvent Mixing and Precipitation. The method results in polymer nanocomposite particles that can be further processed by conventional polymer processing techniques. In this work, compression molding was used to produce nanocomposites specimens for electrical and EMI shielding characterization. TEM characterization of the morphology of the composites showed that copper nanowires form segregated networks in the polystyrene matrix without the use of surfactants or surface modification of nanowires and/or polymer. Segregated networks contribute to the low electrical percolation threshold and high electrical conductivity of the nanocomposites. These novel conductive materials showed outstanding properties for EMI shielding applications compared to other emerging materials.

Acknowledgements

The authors acknowledge financial support from NSERC strategic grants program and Dr. Bin Lin for preparation of specimens by ultramicrotoming for TEM characterization.

References

Chang, T. E., et al. Conductivity and mechanical properties of well-dispersed single-wall carbon nanotube/polystyrene composite. Polymer, **47**(22), 7740-7746. (2006).

Chung, D. D. L. Electromagnetic interference shielding effectiveness of carbon materials. Carbon, **39**(2), 279-285. (2001).

Curran, S. A., et al. Dynamic electrical properties of polymer-carbon nanotube composites: Enhancement through covalent bonding. Journal of Materials Research, **21**(4), 1071-1077. (2006).

Das, N. C. Maiti, S. Electromagnetic interference shielding of carbon nanotube/ethylene vinyl acetate composites. Journal of Materials Science, **43**(6), 1920-1925. (2008).

Gelves, G. A., et al. Electrical and rheological percolation of polymer nanocomposites prepared with functionalized copper nanowires. Nanotechnology, **19**(21). (2008).

Gelves, G. A., et al. Low electrical percolation threshold of silver and copper nanowires in polystyrene composites. Advanced Functional Materials, **16**(18), 2423-2430. (2006).

Gelves, G. A., et al. Multigram synthesis of copper nanowires using ac electrodeposition into porous aluminum oxide templates. Journal of Materials Chemistry, **16**(30), 3075-3083. (2006).

Gelves, G. A., et al. Electrostatically dissipative polystyrene nanocomposites containing copper nanowires. Macromolecular Rapid Communications, **26**(21), 1677-1681. (2005).

Huang, Y., et al. The influence of single-walled carbon nanotube structure on the electromagnetic interference shielding efficiency of its epoxy composites. Carbon, **45**(8), 1614-1621. (2007).

Liu, Z. F., et al. Reflection and absorption contributions to the electromagnetic interference shielding of single-walled carbon nanotube/polyurethane composites. Carbon, **45**(4), 821-827. (2007).

Lux, F. Models proposed to explain the electrical conductivity of mixtures made of conductive and insulating materials Journal of Materials Science, **28**(2), 285-301. (1993).

Ramasubramaniam, R., et al. Homogeneous carbon nanotube/polymer composites for electrical applications. Applied Physics Letters, **83**(14), 2928-2930. (2003).

Safadi, B., et al. Multiwalled carbon nanotube polymer composites: Synthesis and characterization of thin films. Journal of Applied Polymer Science, **84**(14), 2660-2669. (2002).

Toimil-Molares, M. E., et al. Electrical characterization of electrochemically grown single copper nanowires. Applied Physics Letters, **82**(13), 2139-2141. (2003).

Weber, M. Kamal, M. R. Estimation of the volume resistivity of electrically conductive composites. Polymer Composites, **18**(6), 711-725. (1997).

Yang, Y. L. Gupta, M. C. Novel carbon nanotube-polystyrene foam composites for electromagnetic interference shielding. Nano Letters, **5**(11), 2131-2134. (2005).

Yang, Y. L., et al. A comparative study of EMI shielding properties of carbon nanofiber and multi-walled carbon nanotube filled polymer composites. Journal of Nanoscience and Nanotechnology, **5**(6), 927-931. (2005).

Zhang, B., et al. Preparation and characterization of gas-sensitive composites from multi-walled carbon nanotubes/polystyrene. Sensors and Actuators B-Chemical, **109**(2), 323-328. (2005).

POLYPROPYLENE –MMT CLAY NANOCOMPOSITES: PROCESS AND POST-PROCESS STRUCTURING

R.Patel[1], H.Benkreira[1], A. Khan[1] and P.D.Coates[1]*,
Y. Shen[2], S Xie[2], R Abu-Zurayk[2], E. Harkin-Jones[2], T. McNally[2], and P.Hornsby[2]

[1]*Polymer IRC, School of Engineering, University of Bradford, Bradford BD7 1DP, UK - p.d.coates@bradford.ac.uk*
[2]*Mechanical & Aeronautical Engineering, Queens' University Belfast, UK*

Abstract

A unique mini-mixer, able to reproduce the intense mixing created in larger scale twin extruders but utilizing only small batches of materials, typically less than 100g, has been used to explore structuring of selected PP-clay nanocomposites in a design of experiments approach, assessing exfoliation number and degree of dispersion, and a wide range of rheological and physical properties in a study across two universities and six companies. The further effect of post-processing on the nanocomposite structures has been explored by biaxial stretching, to simulate the extensional deformations which will be experienced in stretch blow moulding and thermoforming. We aim for a fundamental understanding of the interactions between material formulation, process parameters and the microstructure and properties of polymer nanocomposites, providing generic processing information to guide nanocomposite production, and apply the understanding to the development of proof of concept applications. Results will be presented from the DOE experiments and subsequent property assessments (including morphological, dynamic rheology, mechanical and barrier properties), together with evidence of the significant effect of post-processing. We will also briefly refer to our findings from similar research into PET-clay nanocomposites, which are presented in detail in a separate paper, and initial scale-up trials.

1. Introduction

Polymer nanocomposites form an important research area of our laboratory, including a major programme on modeling of nanocomposites across the length scales (Bradford, Queens Belfast and Oxford Universities, plus an industrial consortium); various projects on nanoclays in PETs and polyolefins (including Sabic- sponsored projects); carbon nanotubes in various polymer matrices (including a collaborative project with Sichuan University), and the process structuring of nanocomposites (Bradford and Queens Belfast, plus a consortium of six companies including Danone) – this latter project is the main focus here.

Polymer nanocomposites are formed by seeding a polymer melt with large aspect ratios particles of nanometer dimensions so that with the addition of only a few percent of such particles, a huge interfacial area is created between the polymer and the particles. It is this area platform and the associated interface chemistry which has the potential to transform a commodity polymer into a high added value, highly performing nanocomposite. Since their discovery, much has been made of the potential growth of polymer nanocomposites [1] across a wide material and product range but the industrial reality is that only a handful of nanocomposites polymers are currently available, mainly Nylon and PP, from a limited number of main suppliers worldwide. The shortfall in the predicted availability of polymer nanocomposites and applications is mainly due to the fact that the issues of processing and scale up have largely been neglected in research efforts [2, 3]. Clearly, mixing conditions play a key role- in the optimum formulation balance of intercalation/exfoliation; and without a comprehensive dispersing/distributing compounding device with the required balance of shear and extensional mixing, the creation of the large interfacial areas required is not possible. This is precisely the subject of this paper which uses a newly developed laboratory minimixer as a bench top processor to formulate nanocomposites of polypropylene.

The uniqueness of this mini-mixer is that it is able to reproduce the intense mixing created in large twin extrusion machines but utilizing only small batches of materials, typically less than 100grammes. Previous research using a range of masterbatches tested in this minimixer and in range of larger twin screw extruders supports this claim [4]. The principal aim here is to thus achieve, with a common but commercially significant polymer, polypropylene, a fundamental understanding of the interactions between material formulation, the mixer parameters and the properties and microstructure of polymer nanocomposites and to apply this to the development of proof of concept applications which provide generic processing

information to guide nanocomposites production. To validate further the work, compounding with a larger scale extruder (1kg output in relation to the 100 g mini-mixer) was also performed using an APV 19mm twin screw co-rotating twin screw extruder.
Finally and importantly, an assessment of the performance of the nanocomposite during further processing was carried out. Because of its importance to PP processing, biaxial stretching was performed on the nanocomposite using the biaxially stretcher to evaluate what gain the nanoclays provided to the mechanical properties and permeability.

2. Experimental set-up

2.1 The Mini-mixer

In order to achieve the required structural transformation of the clay particles during melt compounding high levels of both shear and elongation are required to break up the organoclay particles first into dispersed stacks of silicate tactoids and then into thin stacks of layered-silicates which then peel apart through a combination of shear and diffusion in the organoclay galleries [5 [Kim 2002a]. The compounding device of choice in industry is thus the co-rotating twin screw extruder which combines intermeshing screws and mixing cams to obtain intense distributive and dispersive mixing of polymer melts.

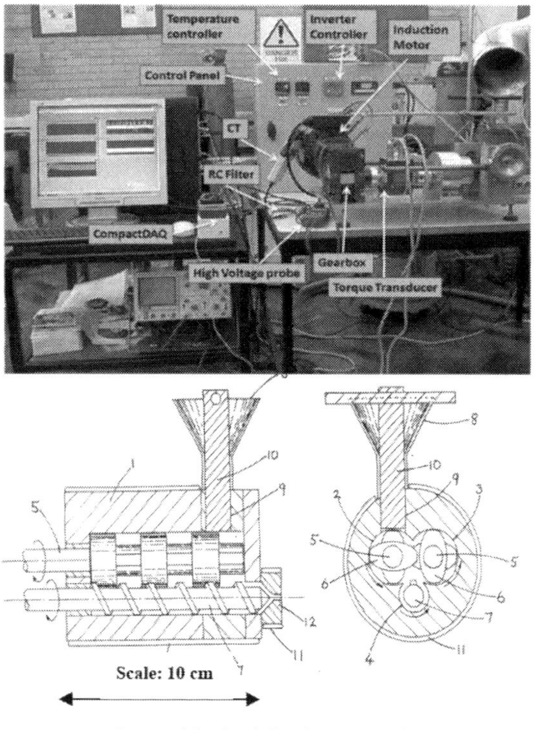

Figure 1 The Bradford Mini-Mixer Experimental Set-up

The design rationale of the mini-mixer is based on established principles [5-6] underpinning the operation of co-rotating twin screw compounding extruders Miniature versions of these machines are unsuitable because of residence time limitations. Re-circulating modes of operation of miniature versions [7-8] such as the ThermoHaake PolyLab, the Rheomix batch mixer and the Brabender mixer are also limited in their use because in the re-circulating channel, the molten polymer will experience very low velocity at the channel surface hence experience very little exposure to the screw mixing an may even stagnate and thermally degrade. Also, the material remaining in the channel when mixing has been completed will require manual removal.

In our mini-mixer design, we circumvent the need for a recirculation channel by incorporating a third screw positioned axially and tangentially with the mixing elements of the co-rotating intermeshing twin screw. Axial movement of the material required for uniformity of mixing is thus achieved. This simple but effective mean of re-circulating the material is at the heart of this new design. It then enables us to reduce the size of the whole apparatus down to a very small device holding typically 10 to 100g of material whilst retaining the ability to discharge the mixed material as one or more strands or as a strip for testing.

Figure 1 shows this novel design with three screws housed in a barrel heated to a controlled temperature. The third screw- the novel aspect of the design-conveys the material away from the entry point (hopper and feed pipe assisted by a piston) and as it is fed, the material is continuously transferred between the two intermeshing chambers and this conveying screw. Eventually with assistance from the piston, the mixer will be completely filled and the material being mixed contained within the barrel by the piston which closes the feed during mixing.

Another novel aspect of the design is the provision of variable speed of the conveying screw and the mixing element rotors via a splitter gearbox which gives rotor speeds of 3.5 times the screw speed and being nearly double the diameter of the screw the rotors impose much higher shear and elongation rates than the screw. When mixing is completed the screw and mixing elements are reversed and the screw conveys the mixed material to the outlet and extrudes it out through the die. During this emptying period, the rotating cams transfer material from the intermeshing chambers to the channel of the screw.

This mixer design allows the mixing elements to be in the form of lobe cams, cylinders with eccentric centres or any combination, to provide a controlled amount of shear and elongation deformations. .Also, the mixer, which is driven with a variable speed control, has on its drive a torque-measuring device. Torque measurement during mixing will provide a measure of work input which can be related to degree of dispersion. One additional important feature of this minimixer is that the die through which the extrudate exits is designed as a precision made rheological slit (45mm x 8.4mm x 0.84mm) fitted with three miniature (3mm diameter head) pressure transducers (Kulite Sensors). Thus from this novel minimixer, we can assess independently the effect of changes in shear rate, elongation rate, residence time and temperature as well as measure the rheology of the nanocomposite as it exits the die.

2.2 The APV Co-Rotating 19mm Twin Screw Extruder (TSE)

This table mounted compounder of L/D=25 which operates under starved conditions is in comparison with the mini-mixer a bigger machine (x 10 and larger output) and served as a scale tester. Essentially, this extruder, as shown in Figure 2 comprises a metering section of intermeshing screws (14 turns x 10mm pitch), followed by the main section of mixing lobes (18 elements x 4mm pitch). This is then followed by another metering section of intermeshing screws (8 turns x 10mm pitch) which transfer material into the next section of mixing lobes (8 elements x 4mm pitch). From here, material is then pumped towards the die by another section of intermeshing screws (12 turns x 10mm pitch + 5 turns x 5mm pitch). The key dimensions for the purpose of scaling are screw barrel/cam gap= 125 microns, cam diameter 30mm and maximum cam speed of 185 rpm. Using these dimensions, conditions achieved in the mini-mixer were duplicated in this device (which was suitably calibrated to establish its residence time per pass for each given screw speed) and these include of course temperature but also shear rate and residence time. The shear rate was duplicated by scaling the corresponding screw speed /gap in the TSE. The residence time was duplicated by re-extruding in the TSE to match the batch residence time in the mini-mixer.

Figure 2 The APV 19mm co-rotating twin screw extruder

2.3 The Nanocomposite Model System

Though in our programme we have also studied a range of polymers-clay-compatibilisers system, we report here work on a single model system, PP with an organomodified montmorillonite (MMT) clay grades 20A (Southern Clay Products Inc., USA) otherwise referred to as Cloisite 20A and a maleic anhydride compatibiliser in a masterbatch form, Polybond 3200 (Crompton, USA) otherwise referred to as PP-g-MA. The clay content of the nanocomposite was varied from 0 to 6%. In order to assess, the effect of the polymer grade, two grades of PP were used: PP 575P (Sabic) with a MFI of 10.5 g/10min and Moplen HP420M (LyondellBasell Industries) with a MFI of 8g/10min. The rheological properties of the PP used were measured, and are shown in Figure 3. Table 1 shows the physical properties of the clay and the compatibiliser used.

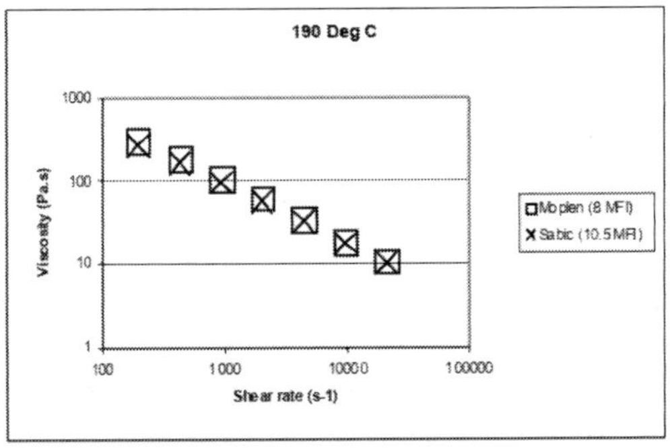

Figure 3 Rheological Properties of PP used

As for typical compounding conditions, previous work with PP-Clay Cloisite 20A nanocomposites indicates melt temperature in the range 160-200 ^{0}C, screw speed of 300rpm in the large twin screw machines and 50rpm with a residence time of up to 10mins in laboratory mixers such as the ThermoHaake PolyLab, the Rheomix batch mixer and the Brabender mixer.

Table 1 Physical Properties of the nano-clay and compatibiliser used

Cloisite 20A		Polybond 3200	
Colour	Off White	Melt Flow Rate (190/2.16)	115g/10min
Density	1.77 g/cc	Density @23 °C	0.91 g/cc
Dry Particle Size	10% < 2μ; 90% <13μ	Melting Point	160 - 170 °C
		Maleic Anhydride Level	1.0 weight %

2.4 The Compounding Experiments

The bulk of the compounding experiments were done on the minimixer with a limited number performed on the twin screw extruder to assess if the performance could be scaled at a larger scale. Clearly, many parameters are at play during the compounding of the nanocomposite and an experimental design protocol was necessary to optimise the experimental effort in the mini-mixer experiment. Also, careful preparation of the feed stock was necessary to prevent contamination during processing and prior to characterisation of the finished product which for some testing required moulding transformation of specimen according to measurement standards. These steps, details of the instruments and the procedure used are described below.

2.4.1 Design of Experiments (DOE)

In order to minimise experimentation, a 1/4fractional factorial experiments that provide sufficient degrees of freedom to resolve all main parameter effects as well as all two-factor interaction were utilised. The experimental design of experiments (DOEs) was generated using Design-Expert 7.0 (Stat-Ease Inc., Minneapolis, USA). Table 2 describes the factor and factor levels and Table 3 the DOE run order which shows: four 'centrepoint' runs (11, 17, 19 and 20), two at low MFI and two at high, included to estimate curvature and pure error and three 'control' runs (21-23), containing no nanoclay with the factors set at low, medium and high levels.

Table 2 The factor and factor levels of the experiments

Factor	Name	Units	Type	Low	High
A	Speed	rpm	Numeric	20	60
B	Residence Time	mins	Numeric	2	8
C	Temperature	°C	Numeric	190	230
D	Nanoclay Loading	%	Numeric	2	6
E	Compatibiliser Loading	%	Numeric	2	6
F	Polymer MFI	g/10min	Categorical	8	10.5

Once the experimental responses, i.e. the nanocomposite properties and structures have been quantified, an analysis of variance was carried out in order to identify which factors/interactions are most significant in determining a particular response. Such analysis also enables models to be generated to maximize/minimize a particular response or to optimise the response for example to maximize dispersion and minimize degradation. This is essential to test the results intelligently in the larger scale trials with fewer experimental variables. Note also that the objectives of this research are twofold, first to obtain nanocomposites with the greatest degree of exfoliation and secondly identify the parameters and interactions that will allow to creation of nanocomposites with varying degrees of intercalation/exfoliation in a controlled manner. This is necessary in order to determine the influence of the initial structural state of compounded nanocomposites on subsequent processing routes such as biaxially blown film extrusion.

Table 3 The DOE run order of the experiments

Run No.	Speed (rpm)	Resid. Time (min)	Temp (C)	Cloisite 20A (%wt)	PP-g-MA (%wt)	MFI (g/10min)
1	60	8	190	6	2	8
2	20	2	190	6	2	10.5
3	60	8	230	2	6	8
4	60	2	190	6	6	10.5
5	20	8	190	2	6	10.5
6	20	8	190	6	6	8
7	60	8	230	6	6	10.5
8	20	2	230	2	6	10.5
9	20	2	230	6	6	8
10	20	8	230	2	2	8
11	40	5	210	4	4	8
12	20	8	230	6	2	10.5
13	60	8	190	2	2	10.5
14	60	2	230	6	2	8
15	20	2	190	2	2	8
16	60	2	190	2	6	8
17	40	5	210	4	4	10.5
18	60	2	230	2	2	10.5
19	40	5	210	4	4	10.5
20	40	5	210	4	4	8
21	20	2	190	0	2	8
22	40	5	210	0	4	8
23	60	8	230	0	6	10.5

2.4.2 Preparation of Feed Stock

The nanocomposite feed system was prepared using the following procedure. First the PP resin in the form of pellets as supplied by the manufacturer was ground into a powder using an industrial grinder (Wedco SE-12-SP). A small amount (15g) of the powdered PP was manually blended with the clay and compatibiliser in the required proportion and this 'masterbatch' fed to the mini-mixer. For the TSE experiments, a feed stock of 1 kg was prepared in a similar manner.

2.4.3 Nanocomposite Production

The minimixer was set for the required conditions and run to produce the nanocomposite which exited the mixer die in the form of a cylindrical (3mm diameter) strand which was hauled off at constant speed of 0.85 m/min through an ambient temperature air ring and then pellitised into 3mm long pellets samples which were then stored in sealed plastic bags prior to analysis or further preparation (moulding) for analysis. Moulding which was required for preparing tensile testing, DMTA, oxygen barrier testing, rheology, GPC, DSC and XRD test specimens in accordance with ASTM standards were produced using a Dr Collin P200P Platen press.

2.5 The Nanocomposite Characterisation Protocol

Having prepared the various nanocomposites in the form required for analysis, the full characterization then proceeded using a number of methods to assess performance and identify the critical control variables of the compounding process described above.

2.5.1 Microscopy, POM and TEM

Clearly, the first characterization must be the assessment of the extent of mixing of the nano-particles into the polymer melts, principally the extent of agglomeration and tactoids (undesirable as these may be) and that of intercalation and exfoliation. This structural analysis is achieved by using first light or optical microscopy and then electron microscopy working in transmission mode to zoom into the nanocomposite structure. For the light microscopy, film samples with thicknesses of about 30 μm were prepared by hot-pressing samples of the extrudate obtained from the minimixer between glass slips at 200 °C then cooled at 5°C/min to ambient. These were then viewed under a Nikon polarising optical microscope (POM) system fitted with a Nomarski prism that could detect agglomerates of size 5μm or larger. For the transmission electron microscopy (TEM), nanotomes of 50-70nm thickness were cut and observed using a Philips CM100 TEM operated at an accelerating voltage of 100 kV. The cutting of the nanotomes was critical and performed on a strip 0.5 mm x 0.5 mm x 5 mm of the nanocomposite embedded in a cured epoxy resin using a DiATOME diamond knife.

2.5.2 Structural Analysis, SAXS/WAXDand GPC

Following the microscopic observation, a Philips X'Pert type PW3040 was used to determine the d-spacing between clay platelets in the materials. The X-ray beam was Cu Kα1 (λ = 1.5418A). Data were collected from 1 to 40°. GPC analysis was carried out on a Polymer Laboratories GPC220 with refractive index and Viscotek differential pressure detectors using a solvent of 1, 2, 4-trichlorobenzene. The GPC tests were carried out at a flow rate 1.0 ml/min and temperature of 160°C.

2.5.3 Thermal Analysis, DCS, TGA, DMTA

Essentially, these methods measure structural changes with temperature changed in a controlled manner, either to investigate the melt crystallization behaviour (DSC), thermal degradation (TGA) and phase transition molecular relaxation (DMTA). The DSC experiments were performed with a DSC6 Modular Thermal Analyzer (Perkin Elmer Inc, USA) at heating/cooling rates of 10 C/min, between 3°C and 230°C, under a nitrogen atmosphere. The samples were held at 230°C for 2 min to remove previous thermal history; they were then cooled to 30°C, and reheated again to 230°C. Dynamic mechanical thermal analysis was performed on sample strips (20×10×0.9 mm) using a TA Q800 DMA operated in single cantilever bending mode from -60 to 130°C at a heating rate of 3 °C/min with a displacement of 0.05 mm and a frequency of 1 Hz. Two tests were completed for each run. Thermal degradation was analysed with a with a TGA/SDTA 851e/LF/1600 (Mettler Toledo) that uses samples of 15 mg heated from 30 to 800 °C at a heating rate of 30 °C /min. in a nitrogen purged furnace to eliminate degradation due to thermal oxidation.

2.5.4 Rheological Analysis

Here the objective is to measure_the shear and extensional flow behaviour of the various compounds in their melt form to assess the effect of the dispersed nano-particles on flow. The analysis was performed either on line using the slit die which is integrated in the mixer or off-line in a Rosand RH7 dual capillary rheometer operating in a shear rate range of 10 to 5000 s^{-1}. Also used was the Gemini™ 200 Advanced Rheometer (Malvern Instruments Ltd., UK) operated in the parallel plate geometry at a strain of 0.01 in a range of frequency sweeps, 0.01Hz to 100Hz. In addition to this, the torque transducer in the minimixer drive also traced the power requirements during compounding. Die swell of the extrudate compounds was also monitored and measured.

2.5.5 Mechanical Properties

Here the objective is to measure the gain in the mechanical properties resulting from the addition of the nano-particles, ultimately the key desirable outcome of the compounding process. Tensile testing was performed according to BS EN ISO 527-1: 1996 using an Instron 5564 Universal Tester (Instron, UK) with a clip-on extensometer at a cross-head speed of 50mm/min. A minimum of five tests were completed for each run.

2.6 Nanocomposite Performance during Biaxial Stretching

An important aspect of the research was to measure the performance of the nanocomposite produced during further processing to produce finished products. To that effect biaxial stretching, key in the processing of PP was investigated. Biaxial stretching was conducted using the Queen's Biaxial Stretcher, which was developed in order to duplicate the deformation behavior of polymeric materials in thermoforming and blow molding under controllable conditions [see Martin et. al 16, for design and operation details]. The parameters investigated here were temperature, strain rate and stretch ratio in equal biaxial stretching mode on compression moulded sample sheets of dimensions 76 mm x 76 mm x 1 mm. Each test was repeated twice to give the true stress (σ_T) and the normal strain (ε_n) which enabled the evaluation of the mechanical properties of the stretch films with and without nano-clays.

A structural analysis of the stretched specimen was also carried out in order to image and measure agglomerates, tactoids and the d-spacing between the clay platelets. To assess agglomeration, deemed to occur when particles greater than 5μm were observe, POM was carried out at 200°C on hot-pressed prepared 30μm film samples. Tactoids were viewed using a Philips TEM microscope at a magnification of 28500 of the microtomed samples (60-90nm thickness). For ease of cutting and repeatability, the stretch film sample were subdivided into 6 sub-samples and sandwiched between two thick supporting plastic plates to produce a large number of microtomes for viewing. A Philips X'Pert type PW3040 was used to determine the clay platelets d-spacing using an X-ray beam of Cu Kα1 (λ = 1.5418 A°) operated in the range 1-40° on samples 2 cm diameter, rotated during the test so that an average plot is produced for each sheet.

The addition of nanoparticles is particularly used to affect permeability of polymers to gases and vapours which was measured on the biaxially stretched specimen using an Oxygen Transmission System, OX-TRAN 2/21 (MOCON®, USA) at 23°C and 0% relative humidity on a sample test area of 1cm^2, using an aluminium mask. The oxygen transmission rate was recorded in the steady state when samples equilibrated. The gas permeability coefficient was calculated after the oxygen transmission rate was normalized with respect to thickness of the samples.

3. Results & discussion

The results are presented first in term of changes in observed structure to assess if the minimixer has achieved adequate dispersion and distribution of nanoclay. The POM images will thus be the first proof and they will enable to measure the extent of the undesirable agglomerate and tactoids. The second stage is in zooming further into the structure to evaluate the extent of intercalation and exfoliation and this will be provided by the TEM images and corresponding quantifiers. Once we have elucidated the structure, we present properties, those supporting changes in structures (thermal data) but particularly mechanical and degradation temperature which are generally the strong indicators of gain in adding nano-clays. Rheological properties are then discussed as they provide the link between structure and flow behaviour. Finally, as biaxial stretching is key in the processing of PP, we examine performance using the same steps as those highlighted above.

3.1 Microscopy Assessment of Structure

In addition to gross agglomeration which can be viewed under the POM, Figure 4 describes structures viewed with TEM under increased magnification showing the formation of agglomerates and tactoids as well as exfoliation in a random run indicating the difficulties in achieving a complete uniform dispersive and distributive mixing.

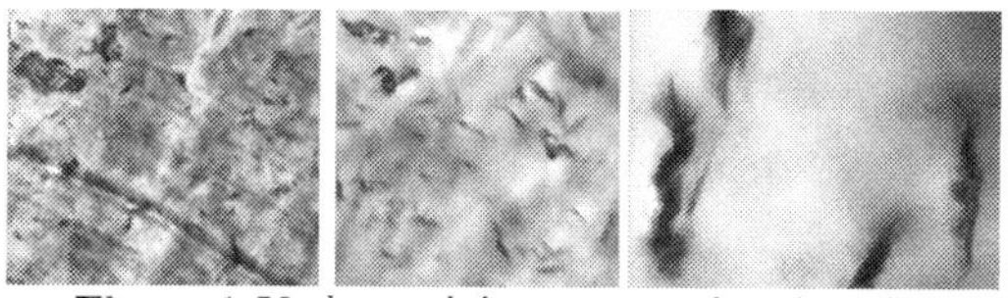

Figure 4 Various mixing states and scales of PP nanocomposite, PP(SABIC 575P)/Clay(Cloisite 20A)/PP-g-MA(Polybond 3200)=92g/6g/2g, at 190°C, 20 rpm, 2 min mixing time in minimixer. The scale bars from left to right correspond to 5 μm, 1 μm and 200 nm respectively.

In regions, where very good mixing is achieved, we can clearly see exfoliation of the nanocomposite as shown in Figure 5.

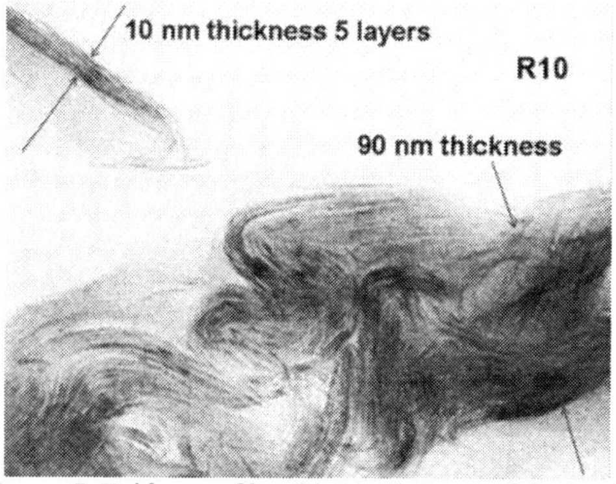

Figure 5 Evidence of local intercalation /exfoliation in Run 10.

Figure 6 indicates how mixing changes depending on the mixing conditions. This figure immediately reveals that larger stresses are desirable as manifested by the better dispersion achieved at the lower temperature of 190°C in comparison to 230°C even when the mixing speed is reduced from 60 rpm to 20 rpm. However, keeping all other conditions the same, the PP with the higher MFI appears to perform better. These observations and the almost identical shear behaviour of the two polymers (see Figure 3) refer us to the need to assess carefully how the rheology of the nanocomposite, not the polymer, is changing to optimize the mixing speed and temperature, i.e. the prevailing viscosity, for any given MFI.

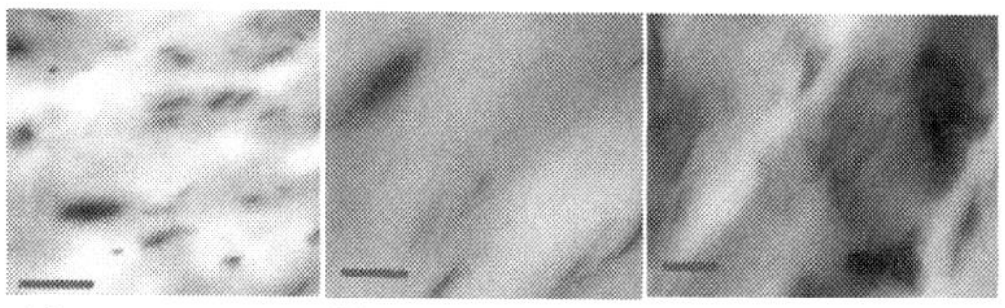

(a) (b) (c)

Figure 6 Mixing dependence of mixing condition for the same nano-composition. Scale bars are from left to right: 1 μm, 200 nm and 200 nm, respectively;

(a) PP(Moplen HP420M) /Clay (C20A)/PP-gMA = 92g/2g/6g, 230°C, 60 rpm, 8min
(b) PP(SABIC 575P) /Clay (C20A)/PP-g-MA = 92g/2g/6g, 190°C, 20 rpm, 8min
(c) PP(Moplen HP420M) /Clay (C20A)/PP-g-MA = 92g/2g/6g, 190°C, 20 rpm, 8min

Figure 7 describes the important effect of the concentration of the stabilizer on the intercalation of the nanocomposite. It shows that in spite of decreased mixer speed, R06 achieves a better intercalation than R01 when a larger amount of compatibiliser is added.

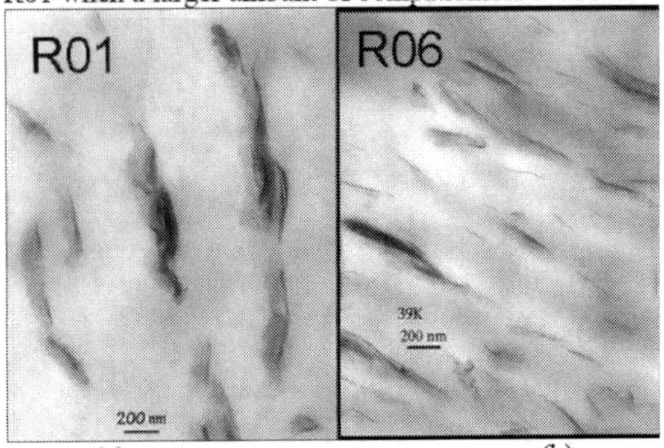

(a) (b)

Figure 7 Mixing dependence of compatibiliser concentration

(a) RO1: PP(Moplen HP420M) /Clay (C20A)/PP-gMA = 92g/6g/2g, 190°C, 60rpm, 8min

(b) RO6: PP(Moplen HP420M) /Clay (C20A)/PP-gMA = 92g/6g/6g, 190°C, 20rpm, 8min

Figures 4 to 7 clearly indicate that the quantification of mixing depends on the scale of scrutiny and a number of mixing indices can be defined, depending on the scale of assessment as viewed under POM or TEM. These indices are defined in Table 4 and can be obtained by image analysis (essentially highlighting the dark areas or "particles" in the images) and using computer based measurements of numbers, lengths, gaps, perimeters and areas of these particles to obtain indices that reflect on the extent of mixing.

Table 4 Qualitative and Quantitative Definition of Mixing Indices depending on scale of scrutiny

% Agglomeration	$Agglo\% = \dfrac{Vol_{agglo}\%}{Vol_{theor}\%}$	
Polymer-clay interfacial area per unit volume	$S_V = \dfrac{4L_A}{\pi}$	L_A: perimeter of particles per unit area
Volume fraction	$V_V = A_A$	A_A: Volume or Area particles fraction
Total clay interfacial area	$(S_V)_{total} = \dfrac{V_V}{0.94}$	0.94: theoretical clay particle thickness
Degree of Dispersion	D=EN×(1-Agglo%)	$(0 \le DD$
Interparticle Distance per volume % clay	IPD/(V_V×100)	

Using the indices in Table 4, a systematic assessment of the various runs carried out in the minimixer was obtained from the microscopic images. The results are displayed in Figure 8a-8d with regard to the important index, the exfoliation number EN, a dimensionless quantity that measures the total clay interfacial area exposed to the polymer matrix. EN ranges from 0-100 with 0 indicating no exfoliation and 100 indicating complete exfoliation.

The data indicates that Run 5 produces the largest EN=30, followed by Run 4 then 6, 13 and 17 which performed similarly. This is confirmed by the data in Figure 8c, which displays the dispersion degree DD. All the data point to better performance at the lower temperature of 190°C. Further examination of the data interestingly reveals with respect to Run 5 that a good exfoliation of small amount of nanoclay can be achieved if the mixing is undertaken at lower speed but over a longer time and with a larger amount of compatibiliser than clay. Interestingly also the PP with the larger MFI performs better. Figure 8d, which displays the inter clay particle distance also confirms that low MFI and high temperatures (Runs 10 and 3) are not conducive to good nano-compounding.

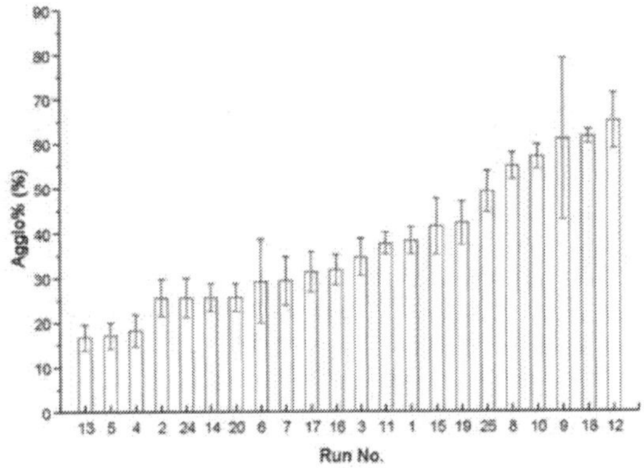

Figure 8a Agglomeration % in each of the runs tested

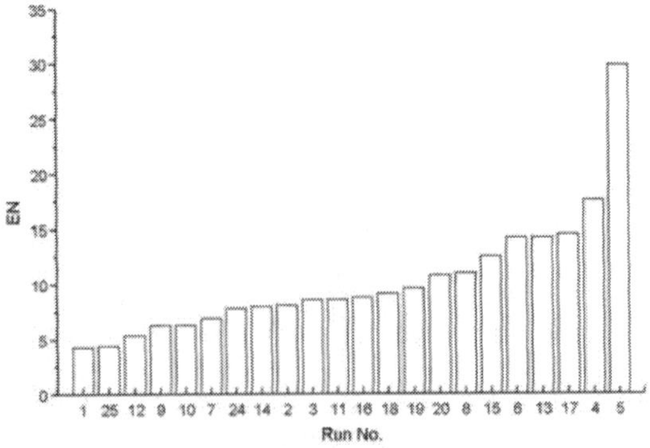

Figure 8b Exfoliation number in each of the runs tested

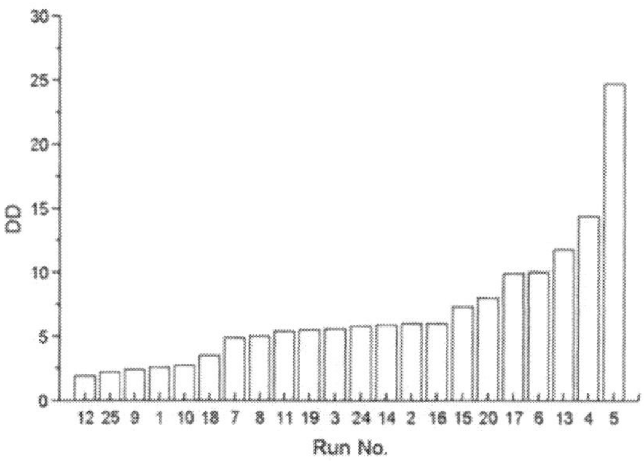

Figure 8c Dispersion Degree in each of the runs tested

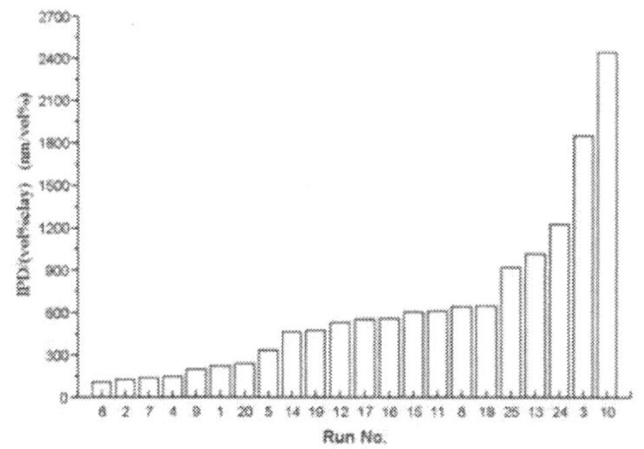

Figure 8d Interparticle distance per volume % clay in each of the runs tested

3.2 Properties (Thermal, Mechanical, Degradation and Rheological)

Here we test the relation between structure and properties to establish if the indicators of low melt temperature, low speed mixing but over a longer time with a high MFI PP and an amount of compatibiliser larger than clay content produce nanocomposites with enhanced properties. All runs show an increase in the tensile modulus (see Figure 9), in one case with Run 4 increasing from about 923 (reference sample) to nearly 1200 MPa, equivalent to 32% increase. Run 4 coming superior in term of showing the largest storage modulus may be attributed, in relation to Run 5, to the larger content of clay (6% as compared to 2%).

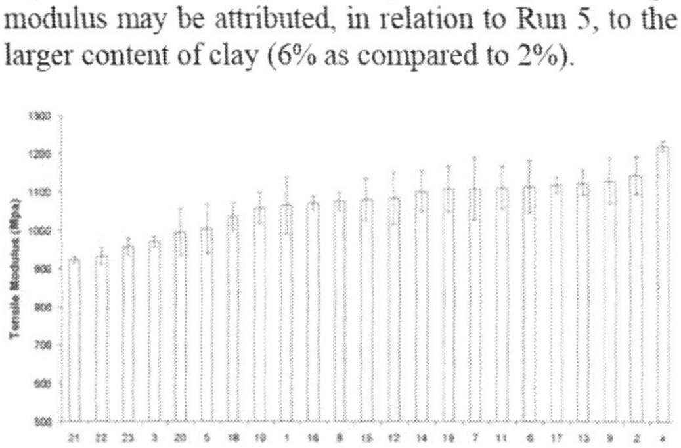

Figure 9 Enhancement in the Tensile Modulus in the various test runs

Similar behaviour is observed when we examine the elongational data which are reduced with the addition of clay as shown in Figure 10, from about 13 % with the virgin polymer to about 4% with the best performing runs. The interesting observation is again the better performance at lower temperature.

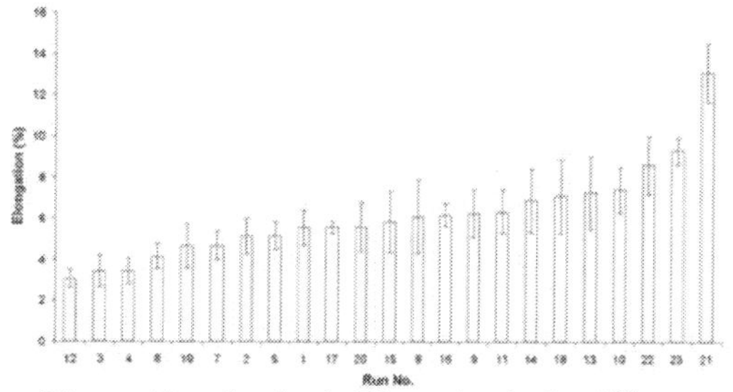

Figure 10 reduction in Elongation in the different tests.

3.3 Performance of Biaxially Stretched Samples

Here we assess the change in structure and in properties upon biaxial orientation. Figure 11 demonstrates the merit of biaxial orientation in reducing agglomeration, increasing exfoliation and decreasing inter-particle distance. This suggests that if further elongational mixing was possible in the minimixer, a better performance could be achieved. However, biaxial orientation is achieved at much lower temperature (145°C) and this allows very large stresses to be applied in both directions to affect structure. The end effect is best seen in the gain of permeability (Figure 12), which is reduced upon orientation (permeability coefficient increases from 8% in the unstretched film to 30% with the biaxially stretched film). This is best understood in terms of tortuosity as pulling the particles apart creates a longer path for the gas to diffuse through.

Figure 11: Biaxial Stretching of Run 4 showing in the upper set of images an enhancement in dispersion as defined by agglo%, EN and IPD and in the lower set of images evidence of decreased interparticle distance

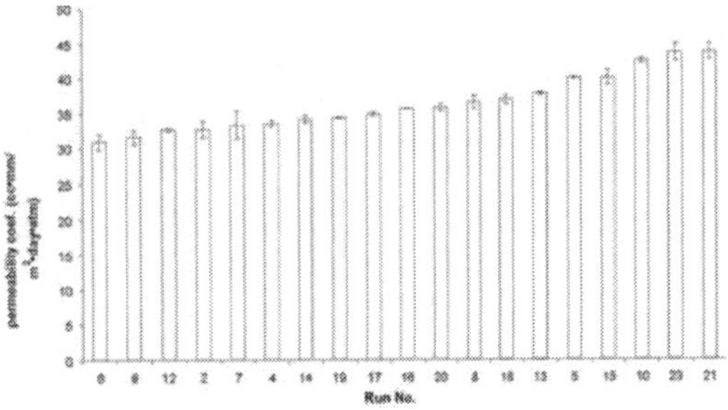

Figure 12: Permeability coefficient in the different runs.

4. Conclusions

A new mini-mixer has been developed and proven to be effective for the study of nano-compounding. The mixer has proved easy, convenient and safe to use as a research tool to study dispersive mixing and as a formulation tool to develop new nanocomposites. A number of data have presented to support this claim. The data indicates that a gain in properties has been achieved and this is manifested in the change of the structure as observed and measured using various mixing indices. It is however necessary to test these finding in larger machines to ascertain if indeed the miminmixer is the tool of choice for this type of research.

References

1. http://www.omnexus.com/resources/articles/article.aspx?id=5957: Nanocomposites-from research to reality
2. http://www.omnexus.com/resources/articles/article.aspx?id=5957#5957: Report on Nanocomposites 2006
3. http://www.nanoclay.com/faq/faq.html
4. Benkreira, H., ' Step-Change in Enhancing Extrusion as a Unit Operation', World Congress of Chemical Engineering, Glasgow, Scotland, July 10-14, 2005
5. Kim S. W., Jo W. H., Lee M. S., Ko M. B. and Jho J. Y. [2002a] Polym. J. 34 (3), 103-111. "Effects of Shear on Melt Exfoliation of Clay in Preparation of Nylon 6/Organoclay Nanocomposites"
6. Manas-Zloczower, I. and Tadmor, Z ., Editors, 'Mixing and compounding of polymers; Theory and practice', Hanser, Cincinnati, OH (1994)
7. Wang, Y., 'Compounding in co-rotating twin screw extruders', RAPRA Review Report 116, 10:8 (2000)
8. DSM ResearchBV, Geleen, The Netherlands 'Twin screw mini-extruder'
9. http://www.rheologysolutions.com/rheo_pdfs/haakel.pdf 'Minilab microrheology compounder'

Acknowledgements

We acknowledge the support of EPSRC, who funded this programme under grant EPSRC EP/EO40667 in collaboration with industrial partners, Boran Mopack, Danone, Innovia Films, Perrite and Smith & Nephew.

ANOMALOUS ELASTIC PROPERTIES OF POLYMER NANOCOMPOSITES

Sati N. Bhattacharya*, Rahul K Gupta

Rheology and Materials Processing Centre,
School of Civil, Environmental and Chemical Engineering,
RMIT University, Melbourne, Vic 3001, Australia
Sati.bhattacharya@rmit.edu.au; rahul.gupta@rmit.edu.au;

ABSTRACT

Elastic properties of polymeric micro composites have been widely studied. Studies reveal that the contribution of filler particles to elasticity is a strong function of the inter and intra particle attractions. As the particle size decreases this interaction becomes larger, and affects the elastic properties more significantly. For polymer nanocomposites, particle size and surface functionality have a strong bearing on yield stress, normal stress and extensional viscosity. Normal stress function becomes dependent on the morphology of the system and cannot just be explained in term of the applied stresses.

In this paper the normal stress behaviour of clay based polymer nanocomposites will be discussed and compared with the published result for silica based macro composites. Presence of clay doesn't affect the normal stress for intercalated systems, but normal stress is found to decrease with increase in clay for exfoliated systems. Reduction of normal stresses for exfoliated clay-polymer nanocomposites will be explained in terms of particle-polymer chain contacts, interaction and mobility of polymer chains within the matrix.

Keywords: Nanocomposites, Elastic properties, Normal stresses, Extensional viscosity

Introduction

The addition of fillers and reinforcements has played a major role in the plastics industry (Giannelis, 2001). Many different types of fillers have been introduced in plastics to provide a synergistic improvement to processability and final product properties like tensile strength, heat distortion temperatures, thermal and electrical conductivities, and enhanced gas barrier properties (Giannelis, 2001; Lim and Park, 2000). Over the last two decades, the addition of nano-sized layered silicates in plastics has been found to offer improvements to above-mentioned properties with just a small quantity, typically in the range of 5 wt%. These materials, normally called polymer-layered silicate nanocomposites (PLSN) have gained tremendous interest in both the academic and research fields due to their unique structure as well as properties (Dennis et al, 2001).

Layered silicates have been found to be useful in the design of nanocomposites due to their lamellar elements that have high in-plane strength and stiffness and a high aspect ratio (> 50). The clay material has a very high surface area of about 750 m^2/g (e.g. montmorillonite). Almost all groups of lamellar solids have these defining characteristics, but smectite clays (e.g. montmorillonite), which belong to the structural family called 2:1 phyllosilicates, is the main choice for nanocomposite material for two reasons (Solomon et al, 2001; Krishnamoorti and Silva, 2000):

- Their rich intercalation chemistry allows them to be chemically modified and made compatible with organic polymers for dispersal on a nanometer scale.
- They can be easily acquired at low costs.

Melt intercalation is one of the major techniques used in the preparation of polymer nanocomposites. Depending on the polymer-silicate interactions and processing conditions (Giannelis et al, 1999; LeBaron et al, 1999; Dennis et al, 2001), three polymer-clay morphologies of are possible. They are phase-separated, intercalated and exfoliated (Solomon et al, 2001). 'Phase-separated' refers to composites that maintain the immiscibility between the polymer and their inorganic filler. 'Intercalated' structures are obtained when polymer chains are intercalated between the layers of silicates resulting in ordered, alternating layers of polymer/clay hybrids with a repeat distance of a few nanometers. 'Exfoliated' morphologies result when individual layers are well dispersed throughout the polymer matrix and the distance between them become dependent of the filler concentration (Solomon et al, 2001; Krishnamoorti and Silva, 2000).

To gain a thorough knowledge of the processing and applications of this relatively new class of material, it is important to understand its rheological and mechanical properties. Melt rheological properties are dictated by a combination of the mesoscopic structure and the strength of the interaction between the polymer and the layered silicate. Further, the mesoscopic structure would be crucially dependent not only on the strength of the polymer/layered silicate interaction, but also on the inherent viscoelastic properties of the matrix in which the layers or collection of layers are dispersed (Krishnamoorti and Silva, 2000, Bhattacharya et al, 2007). Many of the rheological studies conducted centered on shear rheology where the addition of fillers had been reported to enhance viscoelastic properties with a mere 5 wt% loading.

In this paper, we critically analyze and present first normal stress difference (N1) arising from four different ethylene-vinyl acetate copolymers (EVA)/layered silicate nanocomposites. These four types of EVAs present intercalated to almost exfoliated structure as vinyl acetate content increases from 9 wt% to 40 wt%. These N_1 values will be related to the microstructure of these nanocomposites.

2. Experimental Materials and Techniques

2.1 Materials:

The EVAs used in this work differed in their vinyl acetate (VA) concentrations, giving rise to dissimilar properties. The VA concentrations in these EVAs were 9 wt%, 18 wt%, 28 wt% and 40 wt% respectively. In this text, these differing EVAs will be referred as EVA9, EVA18, EVA28 and EVA40 respectively. All these EVAs were obtained from DuPont Packaging and Industrial Polymers (Australia). The organically modified montmorillonite (OMMT) clay used in this work was Cloisite® 15A (15A) for EVA9 and EVA28 while Cloisite® 30B (C30B) was used with EVA18 and EVA40 respectively. Both of these clays were obtained from Southern Clay Products. These groups of modified MMTs are suitable for the less hydrophobic polymers like EVA9- EVA40.

2.2 Preparation of EVA Nanocomposites:

Compounding of different EVA pellets and clay were carried out using a Brabender twin-screw extruder. The pellets and C30B were initially premixed before introducing into the twin-screw extruder. The compounding temperature was set at 100°C with an extrusion speed of 70 rpm. EVA9, EVA18, EVA28 and EVA40 nanocomposites with clay loadings of 2.0 wt% and 5 wt% were produced. The extruded materials were then pelletized and compression molded at 120°C to yield samples of about 2 mm thickness.

WAXS and TEM techniques were used to characterise all the nanocomposites prepared.

2.2 Shear Rheology

The 2 mm thick compression molded samples were cut into 25 mm diameter discs. Shear rheological properties were studied using the Advanced Rheometrics Expansion System (ARES) in both dynamic and steady shear modes with parallel plate geometry. The tests were conducted using 25 mm diameter plates at 110°C. All measurements were performed with a force transducer with a range of 0.2 to 200 g-cm torque.

3. Results and Discussion

3.1. WAXS and TEM

WAXS and TEM images for EVA9, EVA18, EVA28 and EVA40 nanocomposites are shown in figures 1 and 2 respectively. Generally, all EVA nanocomposites show shifts in WAXS peaks to lower angles, indicating penetration of on EVA chains into the silicate layers. Unlike EVA9 and EVA18 nanocomposites, this did not show any Bragg peaks (corresponding to exfoliated morphologies), EVA28 and EVA40 nanocomposites displayed weak shoulders at lower angles relating to possible disordered intercalated structures.

Unlike the WAXS results, TEM images showed that all sets of EVA nanocomposites possessed mixed intercalated/exfoliated morphologies. These findings contradict that of WAXS, which suggested exfoliation for EVA28 and EVA40 and disordered intercalation for EVA9 and EVA18. The images also reveal that the melt mixing process had not only segregated the silicate agglomerates, but also distributed them quite well in the EVA28 and EVA40 matrices.

It must be noted that although WAXS techniques are very useful in determining basal spacing (or interlayer *d*-spacing) of ordered layered silicate structures, it may fall short when dealing with disordered intercalated or exfoliated structures where no scattered intensity peaks will be observed. The lack of peaks may be construed as exfoliation (Morgan and Gilman, 2003). For this reason, TEM was found to be very useful as it gave a pictorial view of what structure exists in the sample.

VA is a polar functional group that is attached to the ethylene backbone. The higher the VA content, the higher the polarity. The advantage here is that besides the ethylene backbone providing a point of contact between the EVA chain and clay layers via the surface modifier, the VA group can help by attaching itself to unmodified region of the clay layers that is virtually hydrophilic. Therefore, EVAs with higher VA (EVA18, EVA28 EVA40 compared to EVA 9) will be able to have a greater degree of interaction with layered silicates, thus producing a much greater enhancement of melt properties.

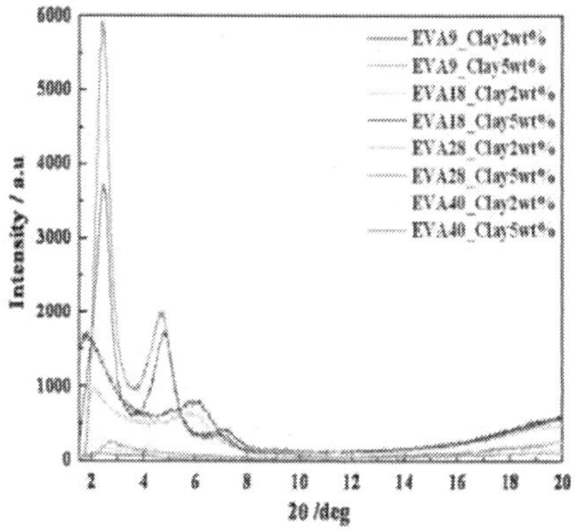

Figure 1: WAXS of EVA9, EVA18, EVA28 and EVA28 nanocomposites at loadings of 2.0.wt% and 5.0wt% ..

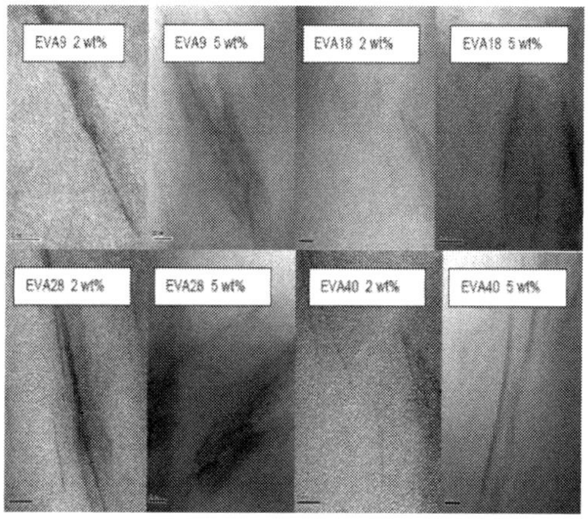

Figure 2: TEM images of different EVA nanocomposites. The bar in the figure represents20 nm.

3.2. Shear Rheology

Steady shear viscosity profiles of EVA9, 18, 28 and 40 nanocomposites (not shown), show that at high shear rates, the viscosity of EVA nanocomposites was found to be independent of silicate loadings. The convergence of viscosity at high shear rates for unfilled and filled EVAs tends to suggest that the contribution of silicates to the viscosity is negligible (Bhattacharya et al, 2009. This could be attributed to the preferential orientation and alignment of silicate layers and polymer chains in the flow direction. All EVA nanocomposites exhibited rapid shear thinning at high shear rates, but the shear viscosities did not converge. This is an interesting observation considering that all materials were characterized as having intercalated/exfoliated morphologies. The difference in viscosity response could be due to the higher degree of polar interactions between carboxyl groups (VA) of the EVA28 and EVA40 chains and hydroxyl groups of the silicate layers compared to that of EVA9 and EVA18 (Li and Ha, 2003).

At low shear rates, the shear viscosity of EVA nanocomposites increased monotonically with silicate loadings. This increase in viscosity is a result of increased interactions between the silicate layers and polymer chains (Choi et al, 2001). The unfilled EVAs quite clearly had Newtonian plateau. The shear thinning nature of these polymeric nanocomposites means that they can be melt processed.

3.3 Effect of Silicates on Elasticity

Han and Jhon (1986) noted the importance of plots of N_1 vs. τ_{12} and how these correlations become virtually independent of temperature and that they can be used to explain the elastic behavior of polymers. Figure 3 shows the first normal stress difference (N_1) plotted against shear stress (τ_{12}) for EVA28 and EVA28 nanocomposites. The first normal stress difference is found to reduce with increase in clay concentration for exfoliated system. Once interaction is attained the normal stress monotonically increases with increase in clay loading remaining almost independent of clay concentration (Figure3) with further increase in stress there is a significant decrease in N1 with increase in clay content.. In Figure 4, on the other hand, N1 seems to be independent of filler content due to the ability of 2D silicate layers to orientate in shear flow. It is to be noted that from morphological studies, EVA9 nanocomposites are dominantly intercalated. These results are similar to reported by Krishnamoorti et al (2001) for intercalated polystyrene-polyisoprene block copolymer nanocomposites.

It has been widely acknowledged that adding fillers to polymers generally leads to reduction in their elastic nature (Tanaka and White (1980), Ren and Krishnamoorti (2003), Pasanovic-Zujo et al (2004) and Krishnamoorti et al (2001)). This is equivalent to observing reduced die swell, which can be advantageous when processing. White and Huang (1981) attributed the reduced die swell in filled polymers to the existence of yield value. They explained that as the material exits from the die, the stress drops below the yield value leading to zero elastic recovery. Generally in the case of all EVA and EVA nanocomposites having different VA contents, increasing filler loading seems to decrease N_1 for a given value of τ_{12}. The highest decrease in elasticity is quite evidently seen for EVAs having higher VA content. In the limit of experimental error, this deviation, together with the convergence of N_1 at high shear stresses could be due to the ability of the highly anisotropic silicate layers to align themselves in the flow field. This alignment, according to Krishnamoorti et al (2001), reduces the contribution of the fillers to the elasticity, as the filler-filler interaction becomes two-dimensional rather than three-dimensional that exists in the quiescent state. Prasad et al (2004) have previously shown that intercalated EVA nanocomposites displayed concentration–independent elasticity, attributing this sort of behaviour to the strong alignment of silicate layers in the direction of shear (Figure 4).

Normally melt strength or elasticity of the samples increases with increase in clay concentration as indicated by the increase in zero shear viscosity. The observed peculiarity in the behavior of first normal stress difference is in the domain of exfoliated morphology. The degree of interaction between the clay particles and the polymer chains is higher in the exfoliated regime as compared to the intercalated regime which is characterized by ordered clay particles with reduced surface area. The delamination of the clay particles improves the degree of interaction between the polymer chains and the clay surface enhancing the overall force of attraction between the clay particles in the exfoliated state due to the long range Vander Waals forces. With increased particle to polymer chain contact and interaction the ability of the composite to develop a resisting force to an applied shear increases. As the clay concentration is now further increased in the exfoliated region the particle to particle interaction force increases due to reduction in distance between the particles leading to a tendency to aggregate. This behaviour improves the mobility of the polymer chains and hence the resistance to an applied shear decreases causing a subsequent reduction in the normal stress difference.

Once an intercalated morphology is established an equilibrium energy state is attained for the clay particles as indicated by constant gallery spacing at intercalation. Hence further increase in clay concentration doesn't cause a change in the structure of the clay particles. Hence now a three dimensional network of ordered particles and polymer chains is established making the first normal stress difference independent of clay concentration. The formation of a network is due to the ordering of the particles. Interestingly all this behaviour is in the linear viscoselatic region since in the linear region there is no permanent change in structure, the changes are only temporary and the original structure is attained once the applied stress is released.

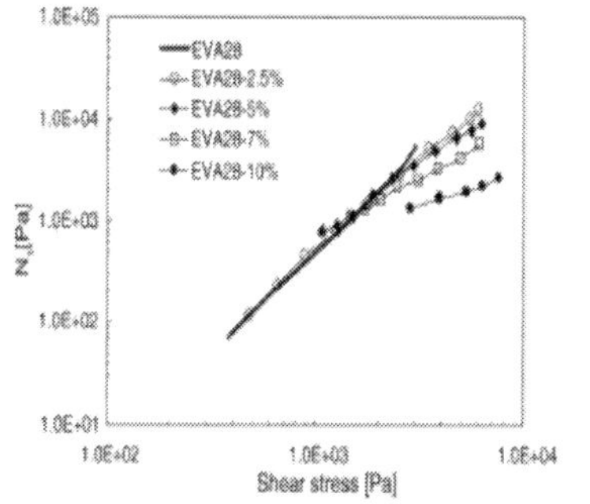

Figure 3: Plot of First Normal Stress Difference (N1) of EVA28 and EVA28 nanocomposites.

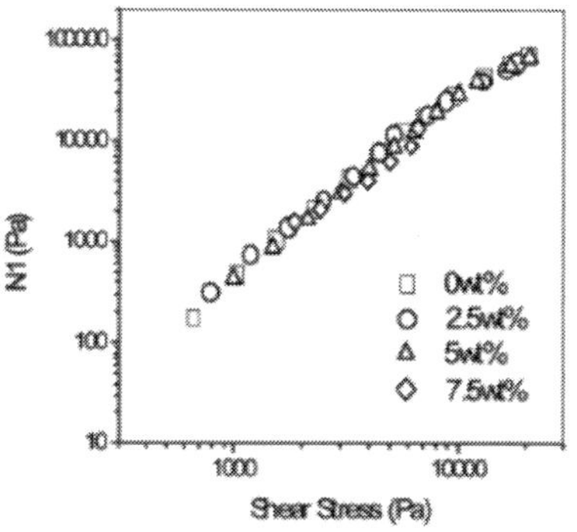

Figure 4: Plot of First Normal Stress Difference (N1) of EVA9 and EVA9 nanocomposites.

3.4 Deviations from general behavior of filled systems and Moddelling Studies

Modeling studies conducted by Pryamitsyn and Ganeshan (2006) reveals an interesting fact .In case of frictional interaction between polymer chains and the particlesthe nanoparticles act as nodes which immobilize polymer chains diffusing in the overall matrix. Hence addition of more particles would lead to the formation of transient interparticle network which would reduce the overall diffusion of the polymer chains into the matrix and hence increase melt elasticity. Now in case of normal force as observed in the previous studies it is the addition of polymer chain relaxation a (elasticity positive) and the particle mobility (negative) within the nanocomposite. Hence if the influence of the particles is higher then the overall normal force would be reduced.. It is interesting to note that the influence of the particle is governed by the overall morphology of the system. If the particles are independently dispersed with very less short range interaction and more of a long range interaction, then the applied hydrodynamic force overcomes the long range force and is able to orient particles accordingly. But t since the chain relaxation or recoil is still high due to the presence of long range forces and more contact between particle and polymer chains and hence the overall normal stress reduces or goes down. In case the particles are aggregated or close to each other the particle interaction potential is higher. In that case the hydrodynamic force is not strong enough to overcome the particle to particle interaction. Therefore the influence of the particle on the polymer chain mobility is lesser, which would actually cause an increase in the normal force.

4. Conclusions

The EVA nanocomposites having different concentrations of VA contents were produced and then analyzed using WAXS and TEM. The nanocomposites were deemed to have mixed intercalated/exfoliated morphologies. The polymer nanocomposites were found to have higher shear and elastic properties with increase in silicate content. However elasticity of the nanocomposites expressed in terms of first normal stress difference-shear stress plots showed clearly that for intercalated morphology, N_1 is independent of silicate loading. This observation is similar to the one reported by other researchers. However, in the case of exfoliated systems of EVA28 nanocomposites, N1 is found to decrease with increase in clay concentration.

5. References

Bhattacharya, S.N., Gupta, Rahul K. and Kamal, M.R., Polymeric Nanocomposites-Theory and Practice, Carl Hanser Verlag, Munich (2007).

Bhattacharya, S., Gupta, Rahul K. and Bhattacharya, S.N., in Polymer Nanocomposites Handbook, Materials Science and Chemical Engineering, Taylor & Francis/CRC Press (July 2009).

Choi HJ, Kim SG, Hyun YH, Jhon MS, Macromol Rapid Commun **22,** 320-325 (2001).

Dennis HR, Hunter DL, Kim S, White JL, Cho JW, Paul DR, Polymer **42,** 9513-9522 (2001).

Giannelis EP, Appl Organometal Chem **12**, 675-680 (2001).
Giannelis EP, Krishnamoorti R, Manias E, Adv Polym Sci **138**, 107-147 (1999).
Han CD, Jhon MS, J Appl Polym Sci **32**,3809-3840 (1986).
Krishnamoorti R, Ren J, Silva AS, J Chem Phys **114**:4968-4973 (2001).
Krishnamoorti R, Silva: in: Pinnavaia, TJ, Beall GW (Eds) Polymer-clay nanocomposites, John Wiley & Sons, 315-343 (2000).
LeBaron PC, Wang Z, Pinnavaia TJ, Appl Clay Sci **15,** 11-29 (1999).
Li X, Ha CS, J Appl Polym Sci **87,** 1901-1909 (2003).
Lim YT, Park OO, Rheol Acta **40,** 220-229 (2001).
Morgan AB, Gilman JW, J Appl Polym Sci **87,** 1329-1338 (2003).
Pasanovic-Zujo V, Gupta RK, Bhattacharya SN, Rheol Acta **43,** 99-108 (2004)
Prasad R, Pasanovic-Zujo V, Gupta RK, Cser F, Bhattacharya SN, Polym Eng Sci, **44**, 1220-1330 (2004).
Pryamitsyn, V. and Ganesan, V., J Rheol., **50 (5),** 655-683 (2006).
Ren J, Krishnamoorti R, Macromolecules **36,** 4443-4451(2003).
Solomon MJ, Almusallam AS, Seefeldt KF, Somwangthanaroj A, Varadan P, Macromolecules **34**, 864-1872 (2001).
Tanaka H, White JL, Polym Eng Sci **20,** 949-956 (1980).

Carbon and organic nanotubes and some of their Bio-applications

*Mosto Bousmina**

INANOTECH - Rabat - Morocco

We will present some new results about the fabrication of carbon and organic-based nanotubes. The subject is vast and hence it is a daunting task to summarise this remarkably rich and multifaceted are. Therefore, the presentation will make the focus on few applications in the area of biomaterials, bio-medicine and in the area of sensors. Both results from the literature and those obtained in our laboratory will be presented.

Polymer composites based on nanocrystalline cellulose

Karen Stoeffler (1), Minh-Tan Ton-That (1), Johanne Denault (1), Khaled Mahmoud (2), Alfred Leung (2), John Luong (2)*

(1) Industrial Materials Institute – National Research Council - Quebec - Canada
(2) Biotechnology Research Institute – National Research Council - Quebec - Canada

Green composites can be obtained by combining matrix and fillers derived from natural resources. Cellulose, which is a semi-crystalline polysaccharide contained in plant cells, is widely available. Nanocrystalline cellulose (NCC) can be isolated from cellulose fibers though acid hydrolysis. NCC ranges from 100-1000 nm in length and from 5-60 nm in diameter. NCC exhibits a longitudinal elastic modulus of about 140 GPa which makes it suitable for mechanical reinforcement in polymers. However, NCC is highly hydrophilic, rendering its incorporation in polymers problematic. By now, the use of NCC has mainly been restricted to hydrophilic matrices: composites were either prepared via a solution process in water or via a latex. Some authors [1, 2] prepared polylactide (PLA)/NCC composites by extrusion, feeding the NCC in a suspension form. The composites exhibited poor performance which was attributed to the presence of residual solvent. Polyethylene (PE)/NCC composites were prepared by extrusion, feeding the NCC in a powder form [3]. A significant improvement of elongation at break was observed when NCC was modified with a proper organic treatment. This result was related to a better dispersion of the filler. In this work, we incorporated pristine NCC and organophilic NCC in PLA and PE matrices by micro-extrusion. The dispersion state was analyzed by electron microscopy. The thermal and mechanical properties were analyzed by DSC and DMA. The effect of the addition of a coupling agent (PLA-g-MA or PE-g-MA) was also investigated. [1] K. Oksman, A.P. Mathew, D. Bondeson, I. Kvien, Composites Science and Technology, 66, 2776–2784 (2006). [2] D. Bondeson and K. Oksman, Composites: Part A, 38, 2486–2492 (2007). [3] A.J. de Menezes, G. Siqueira, A. A.S. Curvelo, A. Dufresne, Polymer, 50, 4552–4563 (2009).

TIME DEPENDENT RHEOLOGICAL PROPERTIES OF POLYPROPYLENE NANOCOMPOSITES

S. Laske[1], A. Witschnigg[1], M. Kracalik[1], C. Holzer[1]*

[1] *Institute of Plastics Processing, University of Leoben, Franz-Josef Strasse 18, 8700 Leoben, Austria – stephan.laske@unileoben.ac.at; andreas.witschnigg@unileoben.ac.at; milan.kracalik@unileoben.ac.at; clemens.holzer@unileoben.ac.at*

Layered silicates as nanoscale fillers have a great potential in improving polymer material properties. Depending on the composite structure (agglomerated, intercalated or exfoliated) a significantly higher level of reinforcement can be achieved with a very small amount of filler. Rheological properties of polypropylene nanocomposites have been intensively studied over the years using for example rotational, capillary or extensional rheometry. Meanwhile the principle of reinforcement is known as the formation of a 3D network of single layers in the polymer matrix.
But there is still a lack of knowledge about the ageing of the exfoliation and stability of the physical network respectively. There are no studies on the storage-time dependent behavior of the layered structure in polymer matrix. As most of the nanocomposite applications have more or less a long-term shelf life, this investigation is of major interest.
In this study, the rheological properties of different polypropylene nanocomposites were measured immediately after processing. To evaluate the structural stability, the same samples were measured after defined periods under constant storage conditions to see the changes in material properties.

Introduction

Nanocomposites are modified polymers with particles, where at least one dimension is in the order of nanometers. Polymers reinforced by nanoscale particles exhibit significantly higher performance, for example higher elastic modulus, tensile strength, thermal resistance, lower gas and liquid permeability, reduced flammability and enhanced rheological properties (e.g. increased melt stiffness) already for small amounts of the filler [1, 2]. For preparation of polymer nanocomposites, layered silicates (especially montmorillonite, MMT) are the most commonly used nanofillers due to the possibility to achieve aspect ratios up to 1000.
Thereby, melt mixing (compounding) is the industrially most attractive method due to its technological simplicity. According to the dispersion of MMT platelets in the polymer matrix, three composite structures (conventional composites, intercalated and exfoliated nanocomposites) can be formed. During the process the structures, which are responsible for the reinforcement level, are formed by physical bonding between the hydrophilic clay, the hydrophobic polymer matrix and the compatibilizer.
Rheological properties of polymer nanocomposites have been intensively studied over the years identifying the 3D network of single layers in the polymer matrix as the principle of reinforcement [3, 4]. But there is still a lack of knowledge about the stability of this network respectively the influence of the layers on the material ageing.
As already published [5, 6], common fillers like calcium carbonate have a major influence on the rheological properties of polymer composites. Thereby the interactions between filler and polymer matrix are influenced by [7]

- particle surface area,
- particle surface functionality,
- hydrophilicity,
- thermal- and photosensitization properties,
- transition metal ion content,
- particle aspect ratio.

In the case of nanofillers, the (photo-)oxidative degradation of polymers is accelerated, because of [7]

- larger surface for interaction,
- more or less perfect distribution in exfoliated state,
- transition metals as impurities,
- degradation of the organic clay modification (Hofmann elimination) and the compatibilizer

Therefore, the stability of the 3D network and the possible faster material degradation is of major interest, as most of the nanocomposite applications have a more or less a long-term shelf life.

Experimental

Preparation of polymer nanocomposites
The used polymer matrix was an isotactic polypropylene homopolymer (HC600TF; MFR 2.8 g/10 min, 230°C/2.16 kg; Borealis/A). As nanofiller a layered silicate intercalated with dimethyl distearyl ammonium chloride (Nanofil15, Rockwood /D) and as a compatibilizer a PP grafted with maleic acid anhydride (PP-MA, Scona TPPP 2112 Fa. Kometra/D) were used. For the compounding process,

a co-rotating twin screw extruder Theysohn/A TSK30/40D with a string die has been employed. The feed rate was set at 10 kg/h with varying screw speed. The extruder temperature profile has been set to 160 – 200 °C from the hopper up to the die. The different formulations, process conditions and ageing times are listed in Table 1 respectively Table 2.

Table 1 - Process conditions of the nanocomposites

Nr	Screw geometry	Screw speed (rpm)	Formulation clay / Compatibilizer (wt%)
1	1	100	20/20
2			5/5
3	1	200	5/2.5
4			5/12.5
5	3	100	5/5
6	3	300	

Table 2 - Materials ageing time and nomenclature

Nr.	Difference	Ageing Time (months)	Nomenclature
1	Amount MMT	36	MMT20
2			MMT5
3	Amount Compatibilizer	36	Comp25
4			Comp125
5	Screw Speed	18	Screw100
6			Screw300

Rotational-Rheometry

For the rheological properties, the primary granulate obtained from the compounding process was molded to plates with a thickness of 2 mm using a hydraulic vacuum press machine 200 PV (Dr. Collin/D). The rheological properties (zero shear viscosity) were determined using a Physica MCR501 (Anton Paar/A) rheometer.

Test procedure

The different materials were examined right after production as well as after 18 respectively 36 month. During shelf life, all the materials were stored under the same conditions (20 °C, 60 rel% humidity).

Results and Discussion

Figure 1 shows the results of the materials with different MMT content evaluated right after respectively 36 months after production.

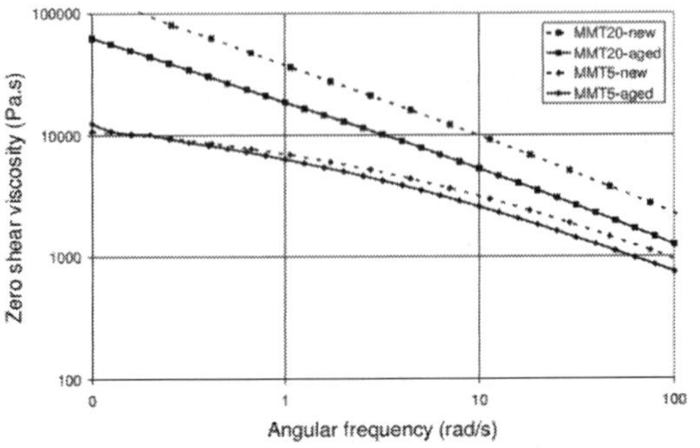

Figure 1 - Results of the materials with different MMT content

First of all, Figure 1 shows clearly the effect of different clay content on the rheological behavior of polypropylene nanocomposites (e.g. the Newtonian plateau). Furthermore the content of MMT has a major influence on the ageing effect. The decrease of the viscosity of the 20wt% clay nanocomposite (MMT20) is significantly higher then that with 5wt% (MMT5). Additionally the MMT20 shows a difference over the whole angular frequency range (possible polymer chain splitting and destroyed 3D network) whereas the MMT5 only differ in the higher range which can be interpreted as intact polymer chains but abolished 3D network (reduced virtual chain enhancement, caused by the layers). As polymers are generally sensitive to material degradation (e.g. oxidation), especially when they are filled with minerals, the enhanced surface of the nanoscale particles of the MMT are enforcing degradation. As the deterioration is acceptable with MMT5, nanocomposites containing higher amounts of MMT (e.g. masterbatches containing >20wt% clay) are not suitable of being stored for longer time.

Figure 2 shows the results of the materials with different compatibilizer content evaluated right after respectively 36 months after production.

Legendenreihenfolge wie oben

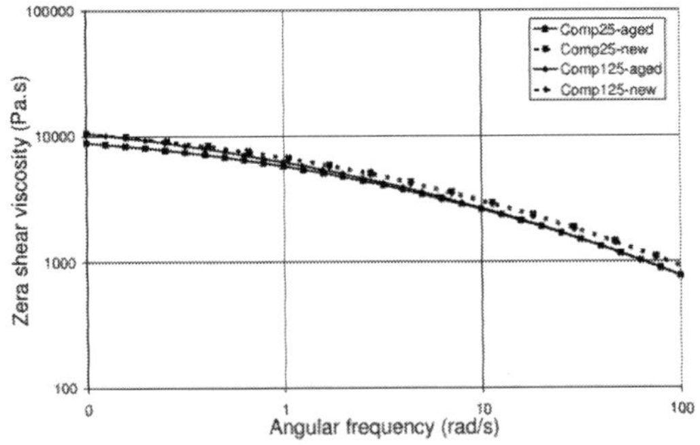

Figure 2 - Results of the materials (5 % MMT) with different compatibilizer content (2,5 % and 12.5 %)

Figure 2 shows, that the amount of compatiblizer has no influence on the viscosity of the aged nanocomposites. As the compatibilizer is a maleic acid anhydride (MAH) grafted polypropylene with a graft-grade of only 0.001 wt% it can be expected, that this small amount of MAH has no effect. Generally spoken, there is a straight coherence between material properties and clay/compatibilizer ratio; however this has no influence on the storing capability. This confirms, that nanocomposites can be long-term stored or utilized below a certain clay content (in this case 5 wt%).

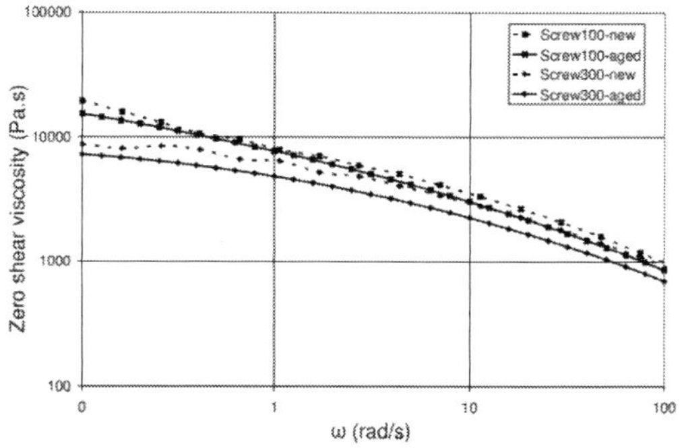

Figure 3 - Results of the materials processed at different screw speeds

Figure 3 shows the results of the aged nanocomposites, processed at different screw speeds. As the induced shear energy differs with varying screw speeds, the rate of exfoliation is diverse. As already said, a higher exfoliation rate ensures higher surface area, aspect ratio and therefore more interaction prospects.

Following this, the better the clay is exfoliated the more sensitive the nanocomposite is to degradation. The results of Figure 3 confirm clearly, that a higher degree of exfoliation (Screw300, screw speed at 300 rpm) indicates a higher decrease of viscosity over the whole range of angular frequency while Screw100 exhibits almost no difference.

Another interesting fact is that the degradation runs very fast, as there is almost no change between 18 month and 36 month of ageing (Comp25-FIG vs. Screw100-FIG).

Conclusions

In this work the rheological properties of different nanocomposites (formulation, process conditions) where measured right after respectively 18 and 36 months after production.

The results show clearly, that the long-term stability is influenced by the clay content and the degree of exfoliation. The grafted compatibilizer with very low graftgrades has no influence itself but is a major factor to achieve a high exfoliated nanocomposite.

Generally spoken, the higher the clay content is and/or the higher the degree of exfoliation is, the higher is the risk of degradation during long-term storage or utilization. This can be explained through

- larger surface area for interaction,
- higher aspect ratio and therefore
- faster degradation of the organic clay modification.

Due to the investigations special attention must be paid to the stabilization of the polymer by means of higher amounts or more effective additives.

Additionally further investigation on other properties (mechanical or barrier) must be made to ensure if the loss in chain length and 3D structure over shell-life also leads to deterioration in for example semi-finished parts like films.

Acknowledgements

This research has been supported by the PlaCompl Project, which is part of the Nanocomp cluster founded by Austrian Nanoinitiative.

References

[1] Gilman, J. W.; Kashiwagi, T.; Lichtenhan, J. D. Sampe J., 33, 40 (1997)

[2] Ray, S. S.; Okamoto, M. Progr. Polym. Sci., 28, 1539 (2003)

[3] Witschnigg, A., Laske, S., Kracalik, M., Feuchter, M., Pinter, G., Maier, G., Märzinger, W., Haberkorn, M., Langecker, G. R., Holzer, C., In-line characterization of polypropylene nanocomposites using FT-NIR, J. Appl. Polym. Sci., in press (2010)

[4] Laske, S., Kracalik, M., Feuchter, M., Pinter, G., Maier, G., Märzinger, W., Haberkorn, M., Langecker, G. R., Journ. App. Polym. Sci., 114, 2488-2496 (2009)

[5] Cabedo, L., Plackett, D., Gimenez, E., Lagaron, J. M., Studying the Degradation of Polyhydroxybutyrate-co-valerate during Processing with Clay-Based Nanofillers, J. Appl. Polym. Sci., 112, 3669-3676 (2009)

[6] Allen, N. S., Edge, M., Corrales, T., Childs, A., Liauw, C. M., Catalina, F., Peinado, C., Minihan, A., Aldcroft D., Ageing and stabilisation of filled polymers: an overview, Polym. Degr. Stab., 61, 183-199 (1998)

[7] Pfaendner, R., Additives to Enhance Manufacturing and Long-term Performance of Nanocomposites, NanoEurope 2009 Proceedings (2009)

STABILIZATION OF IMMISCIBLE POLYMER BLENDS BY INTERFACIAL CONFINEMENT OF CARBON NANOTUBES

A.-C. Baudouin*, J. Devaux and C. Bailly

Bio- and Soft- Matter, Institute of Condensed Matter and Nanosciences, Université catholique de Louvain, Croix du Sud 1, 1348 Louvain-la-Neuve, Belgium – anne-christine.baudouin@uclouvain.be, jacques.devaux@uclouvain.be, christian.bailly@uclouvain.be

Multiphase polymer blends have attracted tremendous attention to achieve high properties materials. However, the stabilization of immiscible blends is required. In this context, the morphology of blends of ethylene methyl acrylate copolymer (EA), polyamide (PA) and multiwall carbon nanotubes (MWNTs) has been studied. Two mixing strategies are studied: the predispersion of MWNTs in PA and the simultaneous mixing of the three components. The experimental results show in both cases a partial interface confinement of the MWNTs, in agreement with themodynamical predictions. As a consequence of the interfacial confinement, we first demonstrate that unpurified and unfunctionalized MWNTs enhance phase dispersion. MWNTs exhibit a length comparable or higher than the diameter of the dispersed phase. Thus, when they are simultaneously mixed with both polymers, they enhance phase dispersion during processing by splitting the polymer droplets into smaller particles, leading to finer dispersion. However, when MWNTs are premixed in PA, the sea-island morphology is slightly coarser as pure blend

Second, MWNTs can be, at least partially, confined at the interface between two immiscible polymer, taking advantage of thermodynamics, irreversible polymer adsorption on MWNTs and their subsequent migration. Due to the interfacial localization, the blends behave like so called "Pickering" emulsions. The droplets of the dispersed phase are stabilized against coalescence by the MWNTs adsorbed at the interface with the matrix. Moreover, this stabilization is effective for very long mixing times (at least 60 minutes). This effect is observed for MWNTs content as low as 0.5wt.-% in simultaneously blended samples and is observed for a total content of 0.2wt.-%MWNTs when premixed in PA.

When located at the interface, MWNTs can thus enhance both phase dispersion and stability of the dispersed phase.

Introduction

Multiphase polymer blends have attracted tremendous attention to achieve valuable properties materials by taking advantage of the attractive features of each component. However, binary blends of immiscible polymers generally exhibit poor properties due to their coarse and unstable morphology (Tucker and Moldenaers, 2002). To generate a material exhibiting the desired properties, it is therefore essential to produce a blend with fine and stable dispersed particles. Coalescence can be reduced by using block copolymers, acting as surfactants in stabilization of emulsions, at the interface of the blend (Van Puyvelde et al., 2001). The compatibilization can also be achieved by using nanoparticles, as it has been reported for a few decades (Callan et al.; 1971; Binks, 2002; Vermant et al., 2004, 2008; Elias et al., 2008). The particles can be trapped at the interface between polymers to minimize the total interfacial energy of the system. These stabilized emulsions are called Pickering emulsions. The effectiveness of emulsion stabilization depends on various parameters such as the size, the wettability and the level of interparticle interactions of the particles (Tambe and Sharma 1993). The mechanisms of stabilization are, however, not fully understood yet.

The stabilization by nanoparticles can nevertheless present some advantages, compared to other compatibilization methods, by adding interesting properties to the blend such as thermal or electrical properties.

Among the nanofillers, carbon nanotubes present exceptional properties such as electrical and thermal conductivity. Moreover, Madivala et al. (2009) demonstrated that sufficiently long particles can be efficient emulsifiers of immiscible fluid phases, even when spherical or lower aspect ratio particles presenting the same wetting properties are inefficient.

Therefore, the use of carbon nanotubes as colloidal particles for the stabilization of polymer blends could be original and innovative.

In this context, the morphology of blends of polyamide, methyl acrylate copolymer and carbon nanotubes (MWNTs) has been studied. The influence of mixing strategy and nanotubes content on morphology and polymer droplets size distribution has been evaluated using both TEM and SEM.

Experimental

Materials

The blends under study are composed of two immiscible polymers: a polyamide and a ethylene acrylate copolymer. The polyamide (PA6/12) is a copolyamide 6/12 (Grilon CF6S, EMS Grivory) with a low melting point. The ethylene-acrylate copolymer (EA) used is a random copolymer of ethylene and 26-30 wt% methyl acrylate (Lotryl 28MA07, Arkema).
The MWNTs are Nanocyl®-7000 (Nanocyl S.A., Belgium) produced via a catalytic carbon vapour deposition (CCVD) process. They present 7 graphene shells, their average diameter is of 9.5 nm and their average length is of 1.5 μm. The carbon purity is 90% and the surface area is around 250-300 m^2g^{-1}.

Preparation

The samples were mixed in a co-rotating twin-screw minicompounder (DSM Microcompounder 15 cm^3) with two stainless steel screws and a bypass allowing continuous recycling of the material at the head of the mixing chamber. Rotation speed and mixing time can be independently modified. The polymers are not dried prior to compounding.
Two methods are used to prepare the samples. First, the components are simultaneously mixed at 180°C during 10 minutes. Second, the MWNTs are predispersed in PA6/12 at 180°C during 10 minutes and this composite was the mixed with the EA copolymer.
The torque was measured during the compounding of neat polymers and composites to follow their degradation with extrusion time. The relative torque measured in the compounder versus the compounding time is reported on Figure 1. The degradation of polymers is limited with time, indicating that the molecular weights and thus the viscosity of both phases remain approximately constant with increasing compounding time. Moreover, this degradation is not promoted in presence of carbon nanotubes (Figure 1 - insert).

Characterization

Scanning electron microscopy was performed on cryo-fractured surfaces using a LEO 982 (Zeiss) operating at 1kV.

Transmission electron microscopy was performed using a LEO 922 Transmission Electron Microscope operating at 200 kV. The specimens were cut perpendicularly to the axes of the extruded samples using a Reichert Microtome. Ultrathin sections of approximately 95 nm in thickness were cut using a cryo-diamond knife with a cut angle of 35° (Diatome, Switzerland) and collected in a blend of 60 vol% dimethylsulfoxide (DMSO)/40 vol% demineralised water on copper grids (Diatome).
The nodules size was evaluated using the iTEM software. The nodules diameter presented here is thus not the real diameter of the nodules but the diameter measured on a cut of 95 nm thickness of the sample.

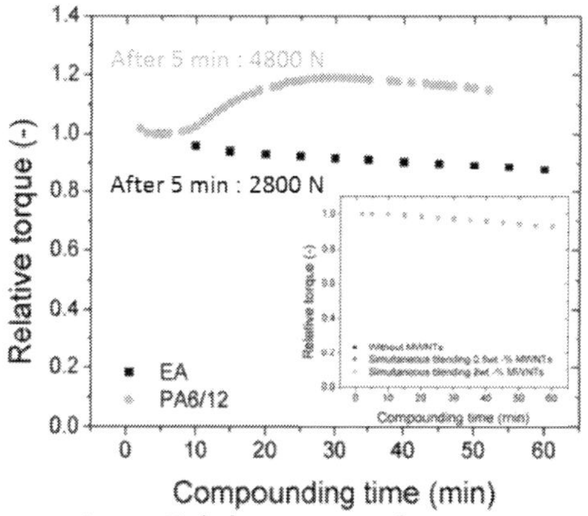

Figure 1 – Relative measured torque versus compounding time for pure components; insert: unfilled and filled blends

Results and Discussion

Blend morphology

Two different blending strategies are analyzed to produce 10wt.-%PA / 90wt.-%EA blends: simultaneous mixing of the three components and premix in PA.

The morphology of blends of simultaneously mixed EA, PA and MWNTs has been analyzed by TEM and SEM. MWNTs are mainly located at the interface after 10 minutes mixing (Figure 2, Figure 4) or 60 minutes mixing (Figure 3).

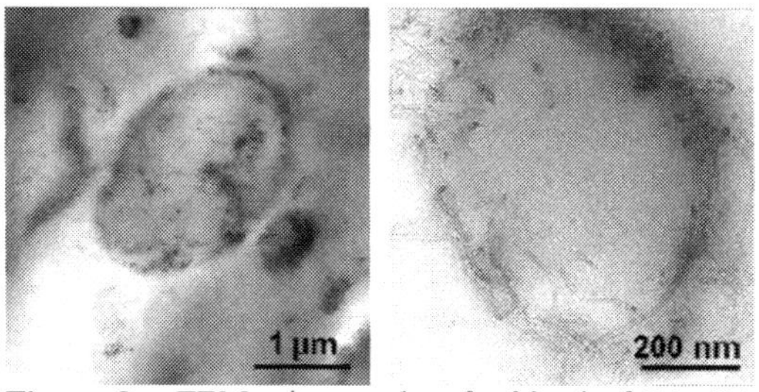

Figure 2 – TEM micrographs of a blend of EA, PA and MWNTs simultaneously mixed during 10 minutes

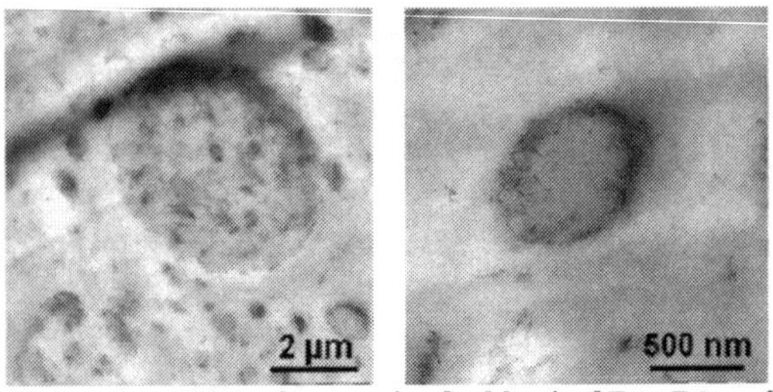

Figure 3 – TEM micrograph of a blend of EA, PA and MWNTs simultaneously mixed during 60 minutes

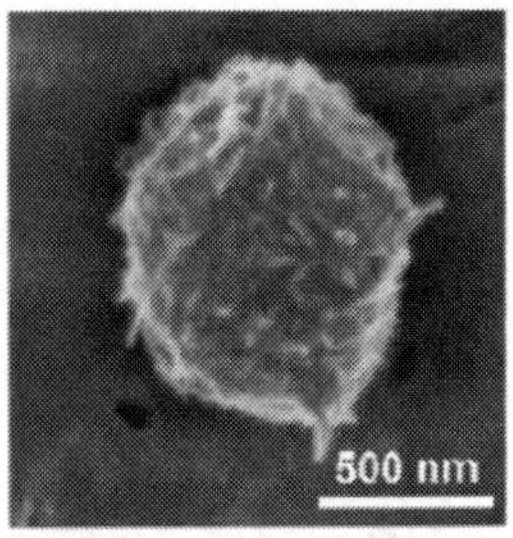

Figure 4 – SEM micrograph of a blend of EA, PA and MWNTs simultaneously mixed during 10 minutes

When MWNTs are first dispersed in the PA phase, TEM (Figure 5) and SEM (Figure 6) micrographs show that part of them migrate torwards the interface. After 60 minutes the MWNTs are still present at the interface indicating that this morphology is stable (Figure 5).

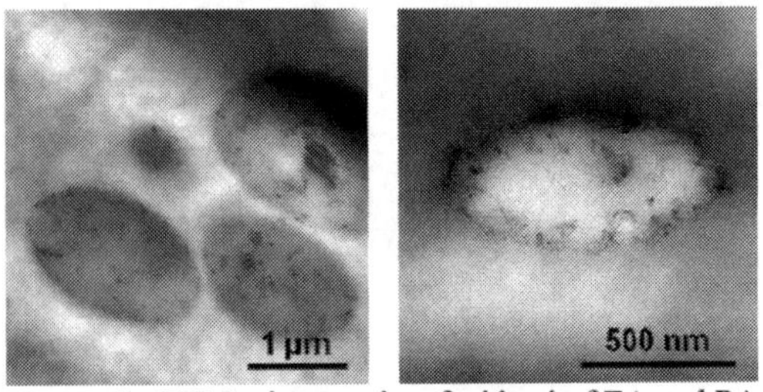

Figure 5 - TEM micrographs of a blend of EA and PA containing 2wt.-% MWNTs mixed during 60 minutes

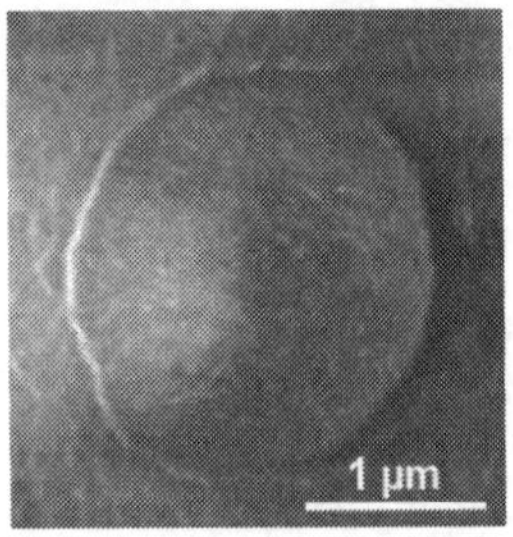

Figure 6 – SEM micrograph of a blend of EA and PA containing 2wt.-% MWNTs mixed during 60 minutes

The localization of a filler in a blend of two immiscible polymers can be predicted using the wetting coefficient, ω_a, analysis. This coefficient is directly linked to the minimization of the interfacial energy of the system (Wu, 1982):

$$\omega_a = \frac{\gamma_{MWNTs-PA} - \gamma_{MWNTs-EA}}{\gamma_{EA-PA}}$$

Where $\gamma_{MWNTs-PA}$, $\gamma_{MWNTs-EA}$, γ_{EA-PA} are the interfacial energies respectively between MWNTs and PA, between MWNTs and EA and between PA and EA. If the wetting coefficient is greater than 1, the fillers will preferentially be located in PA; if the wetting coefficient is lower than -1, the fillers will preferentially be located in EA, and if the wetting coefficient is between -1 and 1, the fillers will be located at the interface between both polymers.
The interfacial energy is evaluated using the surface energies of the components. Two main approaches can be used depending on the type of surfaces: the harmonic-mean equation, which is valid between low-energy materials, and the geometric mean-equation, valid between a low energy material and a high energy material.
Harmonic-mean equation:

$$\gamma_{12} = \gamma_1 + \gamma_2 - 4\left(\frac{\gamma_1^d \gamma_2^d}{\gamma_1^d + \gamma_2^d} + \frac{\gamma_1^p \gamma_2^p}{\gamma_1^p + \gamma_2^p}\right) \quad (2)$$

Geometric-mean equation:

$$\gamma_{12} = \gamma_1 + \gamma_2 - 2\left(\sqrt{\gamma_1^d \gamma_2^d} + \sqrt{\gamma_1^p \gamma_2^p}\right) \quad (3)$$

Where γ_i is the surface tension of component i; γ_i^d is the dispersive part of the surface tension of component i; and γ_i^p is the polar part of the surface tension of component i.
In the concerned blend, the thermodynamics predict an interface localization of the carbon nanotubes (Table 1) for both harmonic-mean and geometric-mean equation. These evaluations are in agreement with the observed morphologies.

Table 1 – Interfacial energies as calculated using harmonic and geometric mean equations at 180°C

Materials	Wetting coefficient according to harmonic-mean equation (mJm-²)	Wetting coefficient according to geometric-mean equation (mJm-²)
EA/PA/MWNTs	0.03	-0.02

Nodules size analysis

The morphology of the blends in absence or in presence of 2wt.-% MWNTs and for both mixing strategies has been investigated using TEM and SEM. In absence of MWNTs, the morphology coarsens when increasing mixing time (Figure 7a,b , Figure 8a,b).
On the contrary, in presence of MWNTs, the morphology is stabilized against coalescence.

Figure 7c, d and Figure 8c, d show that when EA and PA are simultaneously mixed with 2wt.-% MWNTs, the nodules are stable. No increase of their diameter is observed and small nodules with MWNTs on the interface are still present even after 60 minutes compounding (Figure 8c' and d').

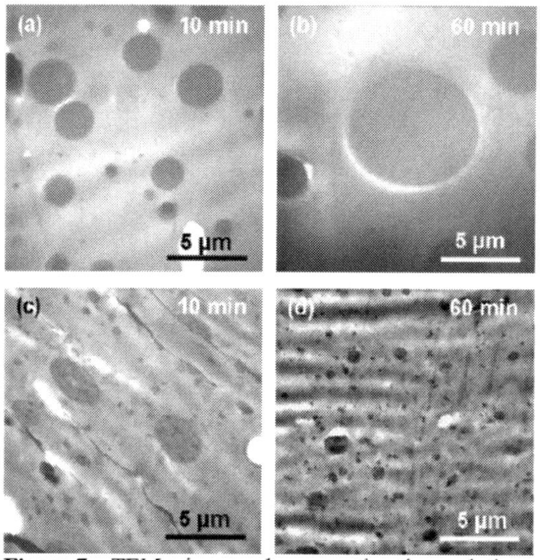

Figure 7 – TEM micrographs presenting the evolution of the morphology with compounding time: (a) and (b) in absence of MWNTs; (c) and (d) in presence of 2wt.-% MWNTs, simultaneous blending.

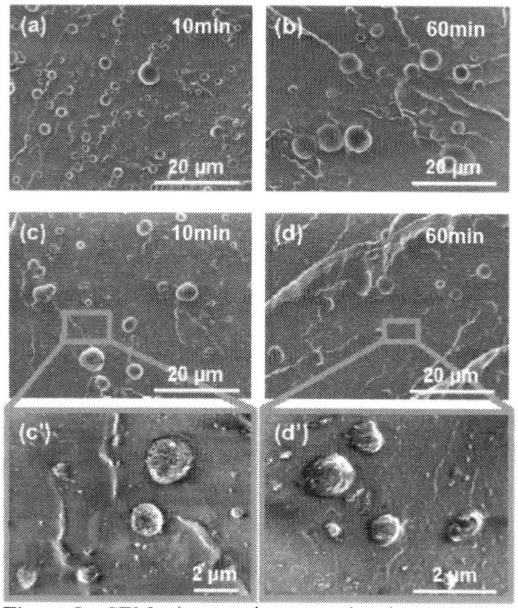

Figure 8 – SEM micrographs presenting the evolution of the morphology with compounding time: (a) and (b) in absence of MWNTs; (c) and (d) in presence of 2wt.-% MWNTs, simulatneous blending; (c') and (d') enlargement of (c) and (d)

The size of the PA nodules were measured on the TEM micrographs. The cumulative frequency versus the nodules diameter measured are presented on Figure 9.

Two effects are clearly highlighted on this graph. First, for an equivalent time of compounding, the nodules are smaller when MWNTs are simultaneously mixed with the immiscible polymers. Unpurified and unfunctionalized MWNTs can thus enhance phase dispersion. Indeed, the as produced MWNTs exhibit a length comparable or higher than the diameter of the dispersed phase (average length 1.5µm). They are therefore able to enhance phase dispersion during processing by splitting the polymer droplets into smaller particles, leading to finer dispersion as it has already been observed by Li et al. (2007).

Second, MWNTs are, at least partially, confined at the interface between two immiscible polymer, taking advantage of thermodynamics, irreversible polymer adsorption on MWNTs and their subsequent migration (Baudouin, 2010a, 2010b). Due to the interfacial localization, the blends behave like Pickering emulsions. The droplets of the dispersed phase are stabilized against coalescence by the MWNTs adsorbed at the interface with the matrix forming a rigid layer. Moreover, this stabilization is effective for very long mixing times (at least 60 minutes) and low CNTs contents (0.5wt% CNTs).

This stabilization is also observed when MWNTs are first dispersed in EA. However, the stabilization occurs at higher nodules size. This effect could be explained by the change in rheological parameters of the blend. Adding MWNTs in the PA phase increases the viscosity of this phase, which is already higher than the viscosity of the PA phase. Indeed, after 5 minutes compounding the measured torque is around 4800N for pure PA and around 2800N for pure EA. The addition of MWNTs is therefore not in favour of a fine dispersion of the PA nodules.

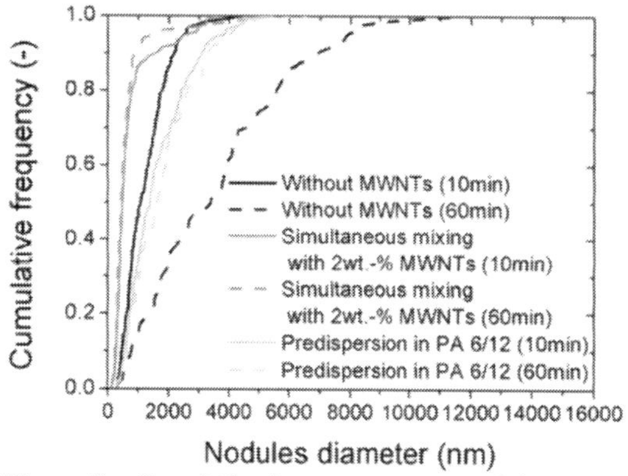

Figure 9 – Cumulative frequency versus nodules diameter for unfilled and filled blends (2wt.%MWNTs) for 10 and 60 minutes blending (measured on TEM micrographs)

Conclusions

Using nanoparticles to affect the morphology of polymer blends is a topic of great scientific and technological interest.

We studied here the effect of mixing strategy and particle concentration on the blend morphology.

When MWNTs are simultaneously mixed with EA and PA, they are lacated at the interface. They enhance both phase dispersion and stability of the dispersed phase, even for long mixing time (60min) and low filler content (0.5wt.%MWNTs).

When MWNTs are premixed with PA, the nodules are coarser but are also stabilised against coalescence for long mixing time and at very low content of MWNTs (0.2 wt.-% MWNTs)

The presence of MWNTs at the interface of the blend acts thus as a rigid layer providing a mechanical barrier against coalescence of the nodules. Blends stabilized by solid particles, like MWNTs, could therefore offer an interesting altenative to blends compatibilized by block-copolymers.

Acknowledgements

The authors thank Nanocyl S.A. for providing the Nanocyl®-7000 and Arkema for providing LY28MA07.

References

Aveyard R, Binks BP, Clint JH, "Emulsions stabilised solely by colloidal particles, Adv. Colloid. Interface Sci., 100, 503-546 (2003)

Baudouin A.-C., Bailly C., Devaux J., "Interface localization of carbon nanotubes in blends of two copolymers, Polym. Degrad. and Stab. 95, 389-398 (2010)

Baudouin A.-C., Devaux J., Bailly C., "Localization of carbon nanotubes at the interface in blends of polyamide and ethylene–acrylate copolymer", Polymer, 51, 1341–1354(2010)

Binks BP, Lumsdon SO, "Pickering emulsions stabilized by monodisperse latex particles: effects of particle size, Langmuir, 17, 4540-4547 (2001)

Binks BP, "Particles as surfactants - similarities and differences", Curr. Opin. Colloid Interface Sci., **7**, 21-41 (2002)

Callan, J.E., Hess, W.M., and Scott, C.E., "Elastomer Blends: Compatibility and Relative Response to Fillers", Rubber Chem. Technol. 44, 814 – 837 (1971)

Elias L., Fenouillot F., Majesté JC., Alcouffe P., Cassagnau P., "Immiscible polymer blends stabilized with nano-silica particles : Rheology and effective interfacial tension", Polymer, **49**, 4378-4385 (2008)

Li Z.-M., Li S.-N., Xu X.-B., "Carbon Nanotubes can Enhance Phase Dispersion in Polymer Blends", Polym Plast Technol Eng, 46, 129-134 (2007)

Madivala B., Vandebril S., Fransaer J., Vermant J., "Exploiting particle shape in solid stabilized emulsions", Soft Matter, 5, 1717-1727 (2009)

Tambe DE, Sharma MS, "Factors controlling the stability of colloid stabilized emulsions, Langmuir 157, 244-253 (1993)

Tucker CL. and Moldenaers P., "Microstructural evolution in polymer blends", Annu. Rev. Fluid. Mech., 34, 177-210 (2002)

Van Puyvelde P, Velankar S., Moldenaers P., "Rheology and morphology of compatibilized polymer blends", Curr. Opin. Colloid Interface Sci., 6, 457-463 (2001)

Vermant J., Cioccolo G., Golapan Nair K., Moldenaers P., "Coalescence suppression in model immiscible polymer blends by nanosized colloidal particles", Rheol. Acta, **43**, 529-538 (2004)

Vermant J., Vandebril S., Dewitte C., Moldenaers P., "Particle-stabilized polymer blends", Rheol. Acta, **47**, 835-839 (2008)

Wu S.: Polymer Interface and Ahesion, 2nd edition, Marcel Dekker, New York (1982)

*** Formatting Issue - Best Available Paper/Graphic ***

TAILORING OF NOVEL RUBBER NANOCOMPOSITES FOR ADVANCED APPLICATIONS

Saritha.A[1], Kuruvilla Joseph*[2]

1. Amrita Vishwavidyapeetham University, Amritapuri, Kollam, Kerala, India- sarithatvla@gmail.com

2. Indian Institute of Space Science and Technology (IIST), ISRO, Thiruvananthapuram
*kjoseph.iist@gmail.com, Fax: +91-471-2564806

Abstract

Polymer nanocomposites are recognized as one of the most promising research areas in polymer science and technology in the 21st century. Most of the polymer-clay nanocomposites display best physico-mechanical properties at a significantly low loading of nanoclay unlike conventional filler, where higher loading not only increases weight but also affect some of the properties of pristine polymer. The interest in rubber nanocomposites is mainly due to the reinforcement anisotropy and barrier properties that can be achieved by exfoliated silicate layers having nanoscale thickness. Further, novel organophilic silicates are envisaged to be developed that would be specifically ''tailored'' for elastomers. Overall, a lot of scope exists to create novel rubber products using nanoclay as the reinforcing material in rubber matrices. In the present work we aim to show that understanding the influence of surafce characteristics is the basis for selecting the ideal oraganoclay for the given rubber matrix. The organoclays used in the present study were selected to explore the effects of the amine surfactant structure on the dispersion of clay particles in different rubber matrices. Understanding the interaction between organically modified clay platelets and organic solvent molecules as well as the corresponding structure of organoclays in a suspension is a critical step toward tailoring and characterizing rubber nanocomposites. Moreover, the use of other nano fillers like titanium dioxide, calcium phosphate etc. to obtain multifunctional materials with enhanced mechanical, thermal, and electrical properties are also envisaged. The enhanced mechanical properties and the excellent barrier as well as transport properties pave way for the effective utilization of these nanocomposite materials in sophisticated industrial, automotive, defence and space related applications.

Introduction

At present nanocomposites based on polymer matrices and organo-layered silicates represent an interesting opportunity for the design of high characteristic materials mainly from the mechanical properties and thermal stability points of view (Ray et al.2003).While clay nanocomposites have been investigated in combination with many thermoplastics and thermosetting polymers, much less attention has been paid to the elastomers as matrices (Mohammed, Sadhu et al. 2005, Kader et al. 2006, Cataldo 2007). Properties of rubber/organoclay nanocomposites were investigated for a number of rubber matrices, e.g. natural rubber (NR) (Lopez et al. 2003, Varghese et al. 2003), EPDM (Sadhu et al. 2004, Gatos et al. 2004, Li et al. 2004), SBR (Ganter et al. 2001), HNBR (Hermann et al. 2006), silicon rubber (Wang et al. 1998) or polybutadiene rubber (Wang et al. 2005). In addition to nano-scale silica the incorporation of layered silicate into the polymer matrix is a field of intensive research. In the literature an extensive variety of reports can be found describing the improvement effects concerning for instance mechanical and tribological properties (Tsuruk et al. 2001) as well as thermal and chemical stability. Due to the intercalation and exfoliation of organolayered silicate and consequently a large increase in the active surface of the filler the low concentration of clay is one of the most important aspects of rubber layered silicate nanocomposites.

Experimental

Layered silicate was swollen in a suitable solvent and kept aside for 24 hours. CIIR is also swollen separately in the same solvent .The two solutions are mixed with vigorous strirring for 12 hours. The nanocomposites so obtained are cast on a petri dish at room temperature to get a thin film. The solvent was allowed to evaporate at room temperature and dried in a vacuum oven at 60^{0}C till there was no weight variation. The film so casted is mixed with the compounding ingredients on a two roll mill of 170 mm diameter, working distance 300 mm, speed of slow roll 18 rpm, and gear ratio 1 : 4 by careful control of temperature, nip gap, and time of mixing. After complete mixing the stock was passed six times through tight nip and finally sheeted out at a fixed nip gap. The CIIR composites were left for a day before vulcanization in the normal air.

Results and Discussion

Chlorobuty rubber nanocomposites containing three different types of clays (cloisite 10 A, cloisite 15 A,

Results and Discussion

Keeping the clay same but on changing the concentration of modifier different properties have been observed for the rubber clay nanocomposite system. When cloisite 15 A and 20 A are compared both have the same organic modifier but the modifier concentration is higher in the case of 15 A. As modifier concentration increases more will be the incorporation of amine and more will be the organophilicity of clay. So the polymer chain incorporation will be higher or most facilitated when modifier concentration is high. More the organophilicity of clay better will be the polymer filler interaction, which leads to better strength of the nanocomposites. Of the three clays studied two (15 A and 20 A) are having aliphatic groups in the modifier whereas the clay 10 A is having an aromatic group.. The organoclay with a broader d spacing can have much more possibility to be exfoliated in polymer/ organoclay nanocomposites because polymer molecules penetrate into the gallery more easily. Lower degree of intercalation is obtained with aromatic salts as modifiers due to steric effect of the aromatic rings. Layered silicates act as better reinforcing agent at high filler concentrations due to the high aspect ratio when compared with the other fillers like titania (TO) and calcium pl\hosphate (TO).

Figures

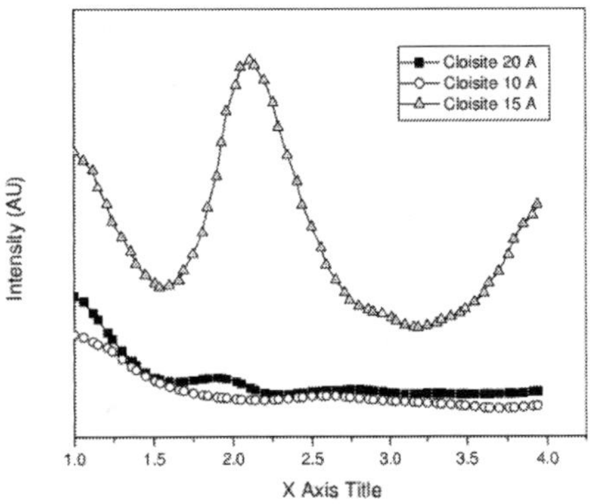

Fig1.XRD plots of the chlorobutyl rubber nanocomposites containing 5 phr of different clays

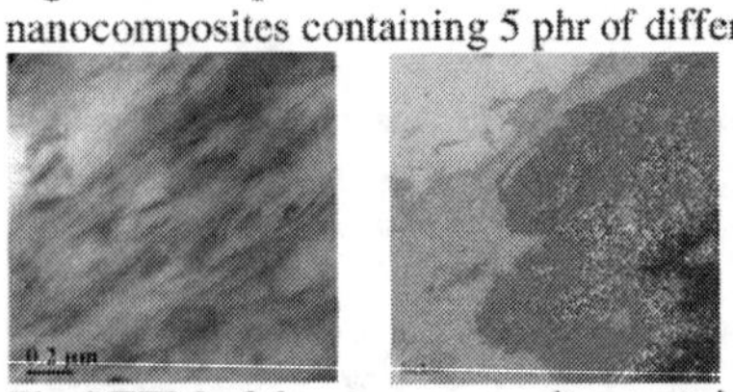

Fig 2.TEM of the nanocomposites containing 5 phr clays

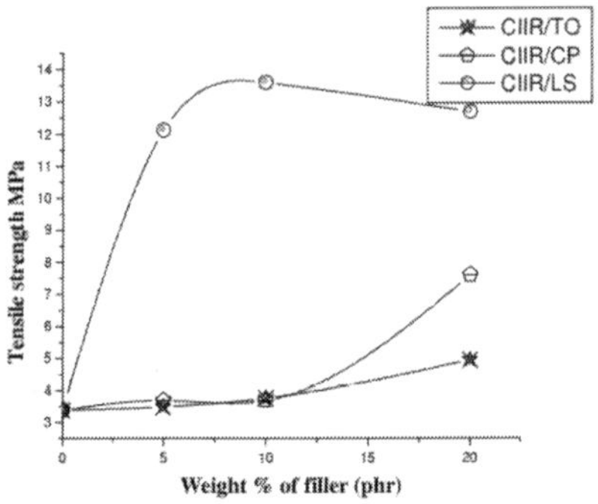

Fig 3 Comparison of various fillerslike titanium dioxide (TO), calcium phosphate (CP) with layered silicates

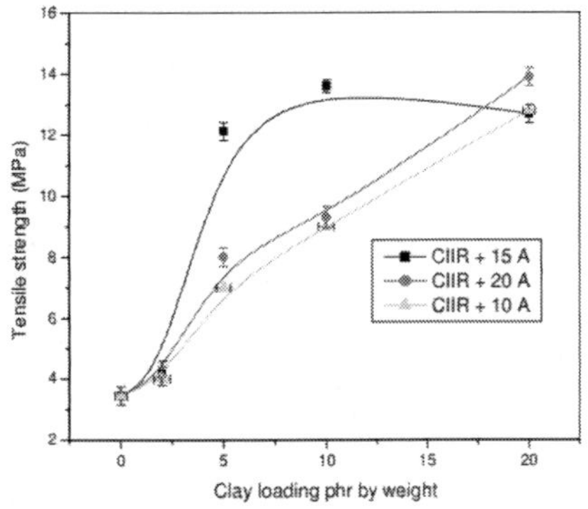

Fig:3 Effect of clay type on the tensile strength of chlorobutyl rubber nano

Conclusions

Chlorobuty rubber nanocomposites containing three different types of clays (cloisite 10 A, cloisite 15 A, cloisite 20 A) were prepared and the mechanical, thermal, kinetic, dielectric as well as nitrogen permeability studies were done. The mechanical properties of nanocomposites containing cloisite 15 A were superior when compared to the other varieties of nanoclay. This can be explained due to the increased d spacing in 15 A which facilitates easier intercalation. The mechanical and barrier properties exhibited by the nanocomposites are promising. Moreover it can be shown that the rubber nanocomposites containing organically modified clay varieties can be tailored so as to make them suitable for automotive and space related applications.

*** Formatting Issue - Best Available Paper/Graphic ***

Acknowledgements

The authors are thankful for the financial support extended by the Indian Space Research Organisation (ISRO/RES/3/587/2007-08).

References

1. Cataldo F, Preparation and properties of nanostructured rubber composites with montmorillonite, Macromol Symp 247:67, 2007.
2. Ganter M., Gronski W., Reichert P., Mülhaupt R ,Rubber nanocomposites: morphology and mechanical properties of BR and SBR, Rubber Chemistry and Technology, 74, 221–235 (2001).
3. Gatos KG, Thomann R, Karger-Kocsis J, Characteristics of ethylene propylene diene monomer rubber/organoclay nanocomposites resulting from different processing conditions and formulations ,Polym Int, 53:1191, 2004.
4. Herrmann W., Uhl C., Heinrich G., Jehnichen D, Analysis of HNBR-Montmorillonite Nanocomposites, Morphology, Orientation and Macroscopic Properties, Polymer Bulletin, 57, 395–405 (2006).
5. Kader MA, Kim K, Lee Y-S, Nah C, Preparation and properties of nitrile rubber/montmorillonite nanocomposites via latex blending, Journal of Materials Science, 7341-7352, 2006
6. Li W., Huang Y. D., Ahmadi S. J, Preparation and properties of ethylene-propylene-diene rubber/organomontmorillonite nanocomposites , Journal of Applied Polymer Science, 94, 440–445 (2004)
7. Lo´pez-Manchado MA, Herrero B, Arroyo M, Organo-montmorillonite as substitute of carbon black in natural rubber compounds, Polym Int, 52:1070, 2003.
8. Mohammad A, Simon GP, chap 12, Polymer nanocomposites,Mai Y-W, Yu ZZ Ed, p 297, CRC Press, Boca Raton
9. Ray SS, Okamoto, layered silicate nanocomposites: a review from preparation to processing, Prog Polym Sci 2003;28(11):1539–641.
10. Sadhu S, Bhowmick AK, Morphology study of rubber based nanocomposites by transmission electron microscopy and atomic force microscopy, J Mater Sci 40:1633, 2005 J Mater Sci 41: 7341, 2006
11. Sadhu S, Bhowmick AK, Characteristics of ethylene propylene diene monomer organoclay nanocomposites resulting from different processing conditions and formulations, J Appl Polym Sci 92:698, 2004.
12. Tsuruk V. V Tribology Letters, Nanocomposite polymer layer for molecular tribology. Tribology Letters, 10, 127–132, (2001).
13. Varghese S, Karger-Kocsis J, Natural rubber based nanocomposites by latex compounding with layered silicates, Polymer 44:4921, 2003
14. Wang S, Long C, Wang X, Li Q, Qi Z , Synthesis and properties of silicone rubber/organomontmorillonite hybrid nanocomposites, J Appl Polym Sci, 69:1557, 1998
15. Wang S., Zhang Y., Peng Z., Zhang Y, A new method for preparing polybutadiene rubber/ clay composites, Journal of Applied Polymer Science, 98, 227–237, (2005).

INCORPORATION OF MICROFIBRILLATED CELLULOSE IN POLYMERS VIA TWIN-SCREW EXTRUSION

N. Quiévy[1*], N. Jacquet[2], M. Sclavons[1], J. Soulestin[3], M. Paquot[2], J. Devaux[1]

[1]*Institute of Condensed Matter and Nanosciences, Université Catholique de Louvain, Louvain-la-Neuve, Belgium – nicolas.quievy@uclouvain.be, michel.sclavons@uclouvain.be, jacques.devaux@uclouvain.be*
[2]*Department of Industrial Biological Chemistry, Gembloux Agricultural University, Belgium – nicolas.jacquet@ulg.ac.be, mpaquot@ulg.ac.be*
[3]*Polymers & Composites Technology & Mechanical Engineering Department, Ecole des Mines de Douai, France - jeremie.soulestin@mines-douai.fr*

Mechanical shearing (homogenization) was used to defibrillate cellulose fibres into microfibrillated cellulose (MFC) which is composed of microfibrils and microfibril bundles. Coating the obtained MFC gel with either low molecular weight polypropylene grafted with maleic anhydride (PP-g-MA) suspensions or either polyethylene glycol (PEG) was performed before freeze-drying in order to prevent hydrogen bonding between microfibrils.
Thermal stability, critical in melt processing application, of freeze-dried coated MFC was found to be different compared to cellulose fibres and freeze-dried MFC.
Lab and semi-pilot scale twin-screw extruder were used to disperse MFC in polyolefins or poly(lactic acid) (PLA). Injection of pressurized water during extrusion allowed to limit the thermal degradation of MFC during processing. Mild conditions, without water injection, were used for producing MFC-PLA composites. Morphological characterization of the elaborated composites exhibited a defibrillated structure. Rheological properties of coated MFC composites showed a percolating effect which was not observed with equivalent contents of cellulose fibres and untreated MFC. This phenomenon was due to the interconnected network of microfibrils in the polymer.

Introduction

Natural fibres are increasingly used to reinforce polymers due to their renewable origin, good specific properties, availability and low price, etc. Wood and plants contain cellulose molecules which arrange into microfibrils (diameter of 4-35 nm) with properties such as high aspect ratio, Young's modulus around 140 GPa (Sakurada et al. (1963)) and gel formation in water. Cellulose microfibrils are difficult to individualize because of hydrogen bonding between them. Mechanical treatment like homogenization, using high shearing forces, is used to isolate microfibrils in water suspension (Turbak et al. (1981), Sain and Bhatnagar (2005)). The obtained product, called microfibrillated cellulose (MFC) is in majority composed of entangled microfibrils and microfibril bundles forming a physical entangled network. Another treatment using strong acid hydrolysis (Nickerson and Harble (1947)) leads to the formation of structures called cellulose (nano)whiskers or nanocrystals or crystallites having lower aspect ratios (length on diameter).

Surface modification or coating is generally applied to MFC, allowing microfibril preservation. Otherwise, formation of agglomerates due to interfibrillar hydrogen bonding takes place (Heux et al. (2000)). Scientific literature abounds of treatments applied on nanowhiskers or microfibrils like coating with polymers (Anglès and Dufresne (2000), Dufresne et al. (2000), Azizi Samir et al. (2004a, 2004b, 2004c) or surfactants (Heux et al. (2000)) or surface modification using poly(ε-caprolactone) (Habibi et al. (2008), isocyanate (Siqueira et al. (2009)), silane (Goussé et al (2004)), TEMPO-mediated oxidation (Saito et al. (2006)),etc.

Nowadays, nanocomposite elaboration draws a specific attention. Generally, twin-screw extrusion is required to achieve a good dispersion of fillers in the polymer melt. When a polar filler such as cellulose and an apolar matrix such as polyolefins are used, compatibilization agent like PP-g-MA are used to improve interfacial properties (Palaniyandi et al. (2007), Gaultier et al. (1998)). However, dispersion of cellulose nano-fillers in polymers is most of the time realized by solution casting. A very few has been made using melt processing (Wang and Sain (2007a, 2007b, 2007c), Nakagaito and Yano (2008), Özgür Seydibeyoglu and Oksman (2008)). According to a recent review about cellulose nanocomposites (Hubbe et al. (2008)), nothing was reported concerning the use of twin-screw extrusion for the MFC composite elaboration.

The first objective of this work was to disperse coated MFC in polyolefins using twin-screw extrusion. Coated MFC was produced by homogenization and by applying embodiments with aqueous suspensions of low molecular weight PP-g-MA or PEG.

Morphological characterization and thermal stability of coated MFC were first evaluated. Then morphological characterization of extruded MFC-polyolefin or MFC-PLA composites was performed.

The second objective was to characterize composite properties. MFC dispersion was evaluated using rheology. Indeed, physical gels like MFC, can be considered as percolating structures (Kelarakis et al. (2007). The addition interconnected arrangements influences the viscoelastic behaviour of molten polymers (Durmus et al. (2005), Pötschke et al. (2004)). Above the gel point, if MFC is properly dispersed in the polymer, a rheological percolation effect is expected.

Experimental

Materials

LDPE powder (1965Z900) was provided Sabic. PP powder (Moplen HF500N) was provided by Basell. PLA 4032D was provided from NatureWorks®. Purified cellulose fibers were purchased from Mikro-technik GmbH under the trade name Albafibre C-200. The average fibre particle size is around 250 µm length with a diameter of 20-50 µm.

Aqueous suspensions of PP-g-MA were kindly provided by Dick Peters N.V. The suspensions are Permanol DP6 (PPgMA1), Permanol KP5 (PPgMA2), Permanol 603 (PPgMA3) and Permanol 601 (PPgMA4). Suspensions with different molecular weights were chosen. Surfactants with different nature are present to stabilize the PP-g-MA. Cationic and anionic surfactants should be comparable to alkylammoniums and fatty acid salts respectively.

Polyethylene glycol 2000g/mol purchased from Sigma-Aldrich was used as coating agent.

Methods

Homogenization technique was used to produce (MFC). It was realized on a LAB Model Homogenizer APV "Rannie". Water suspensions of 10 g/l of cellulose Albafibre C-200 were passed 50 times through the homogenizer at a pressure of 500 bars.

Suspensions of PP-g-MA or PEG were then mixed with the MFC gel and stirred. The selected weight ratio between the solid content of the suspension and the MFC was 2. The coated filler was then freeze-dried and a mat-shaped structure was obtained. The mat was grinded in order to be dry-blended with the polymer before melt processing.

Lab scale processing of MFC composites was performed on a DSM Micro15 (Research B.V., SR Mini HG System) at a screw speed of 200 rpm, a processing temperature of 150°C for LDPE, 180°C for PP and 187°C for PLA with a mixing time of 3 minutes.

MFC-PP composites were also elaborated using a semi-pilot scale co-rotating extruder WP ZSK25 WLE from Coperion Krupp having a high shear capacity (up to 1200 rpm screw rotation speed) and a pressurized solution injection system (Korbee and Van Geenen (1999)). The screw length is 1000 mm and its L/D ratio is 40. Processing temperature was set at 180°C, screw speed at 600 rpm and a throughput at 10 kg/h.

MFC-PLA composites were also elaborated using a semi-pilot scale co-rotating twin screw extruder (Clextral BC45). Temperature of the barrels ranged from 170°C to 215°C and screw speed was set at 80 rpm. The PLA was dried wit a low pressure air dryer from Sotralem industry before processing.

FTIR spectroscopy coupled to thermogravimetric analysis measurements were performed using a Nicolet Nexus 870 FT-IR (ThermoElectron Corporation) spectrometer with a gas cell and a Mettler Toledo TGA/SDTA 851e. The temperature profile started from 30°C until 600°C with a heating rate of 20°C/minute under nitrogen flow. The gases released in the TGA were carried immediately to the cell for detection.

The carrying tube and gas cell were heated to 230°C, to avoid condensation or adsorption of volatile products. Weight of samples was set at 5mg for cellulose fibers and freeze-dried MFC, 15mg for coated MFC and 10mg for the dried suspensions enabling comparisons in terms of intensity between the spectra.

Morphological characterization of the cellulose fibers, MFC and composites were examined by Optical (Olympus AX70) or Scanning Electron (LEO 982 from Zeiss) Microscopy.

Rheological characterization was performed on a strain controlled rheometer (ARES - TA Instruments) for MFC-polyolefin composite and on a Malvern Gemini stress-controlled rheometer for MFC-PLA composites. The measurements were performed with 25mm parallel plates in Small Amplitude Oscillatory Shearing (SAOS) mode under nitrogen atmosphere. The test temperature was set at 190°C for MFC-polyolefin composites and at 170°C for MFC-PLA composites. The gap was set around 1 mm.

Results and Discussion

Cellulose microfibrils can link together by hydrogen bonding to form cellulose fibres. Mechanical shearing by homogenization was used to isolate microfibrils in water suspension. The obtained MFC, in majority composed of entangled microfibrils and microfibril bundles, behaved like a gel.

The most efficient way to incorporate MFC in a polymer using melt processing is by first drying it and making a dry-blend with the polymer. Unfortunately whatever the drying process, formation of microfibril agglomerates, due to interfibrillar hydrogen bonding, takes place. Coating MFC with low molecular weight PP-g-MA (MFC-PPgMA1, MFC-PPgMA2, MFC-PPgMA3, MFC-PPgMA4) or PEG (MFC-PEG) was thus performed before freeze-drying. Microstructure of the freeze-dried coated MFC is presented in Fig. 1.

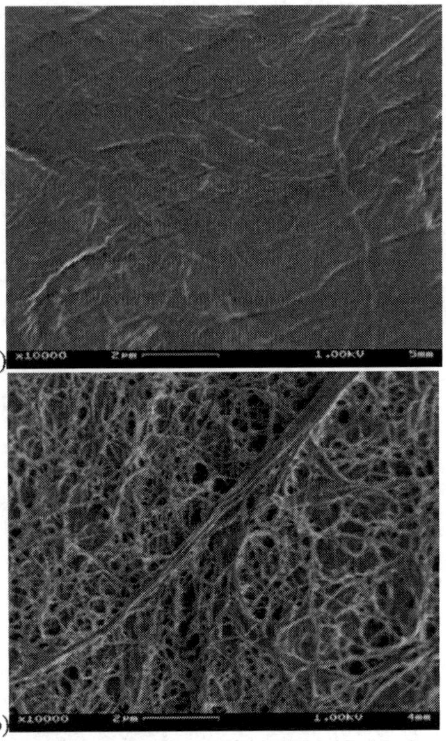

(a) (b)

Figure 1: SEM pictures of (a) MFC-PPgMA2 and (b) MFC-PEG. Magnification 10000x.

MFC-PPgMA2 was composed of large flakes where individual microfibrils were detected on the edges. Nevertheless, increasing magnification on the flake surface (Fig. 1a) showed that microfibrils were well present and entrapped in the solid PP-g-MA matrix.
MFC-PEG formed also large flakes. By looking at the surface, microfibrils were well present and dispersed in the waxy PEG (Fig. 1b). In a general way, the applied coating preserved cellulose microfibrils aspect after the drying step.

As extrusion temperatures of polymers are close to the thermal degradation temperature of natural fillers, the thermal stability of the coated MFC was checked by TGA (Fig. 2) and compared to cellulose fibres and freeze-dried MFC.

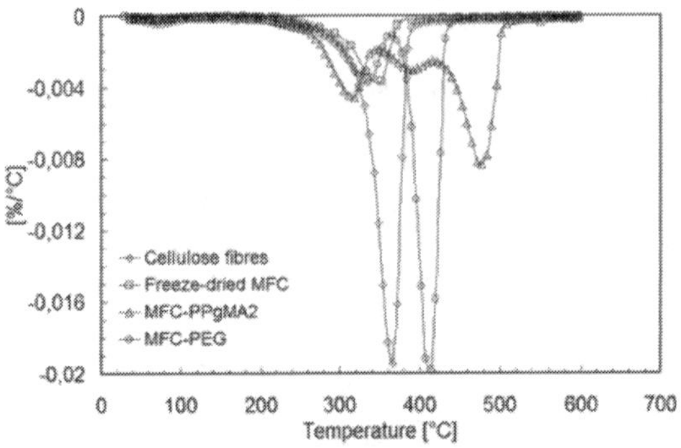

Figure 2: Derivative of mass loss of cellulose fibres freeze-dried MFC, MFC-PPgMA2 and MFC-PEG analysed by TGA at 20°C/min.

The derivative of the mass loss shows that one peak is encountered for cellulose fibres (370°C) and freeze-dried MFC (350°C) while two peaks are observed for MFC-PEG (335°C and 410°C) and three peaks for MFC-PPgMA2 (320°C, 390°C and 480°C).
FTIR analysis of volatile thermal degradation products emitted during the temperature increase showed that, for each sample, the first peak corresponded to cellulose thermal degradation. The second peak observed for MFC-PEG corresponded to PEG thermal degradation. The second peak observed for MFC-PPgMA2 was due to the combination of cellulose, PP-g-MA and surfactant degradation. The third peak corresponded to the PP-g-MA degradation.
Homogenization and freeze-drying of cellulose were found to decrease its thermal stability (Quiévy et al. (2010)). The preservation of the microfibrils using coating agents does not improve the thermal properties compared to freeze-dried MFC. The high specific surface of the preserved microfibrils could make them more reactive at high temperature especially when the coating agent is melted.
Choosing a polymer matrix with adapted processing temperature seems thus essential for the MFC stability.

MFC-PPgMA2 was incorporated in powder PP using high shear extrusion with and without pumping water into the extruder. The MFC content in the polymer was set at 5wt% (meaning 10wt% of coating agent). Fig. 3 shows the visual aspect of MFC composites and references after injection moulding of tensile dog bones.

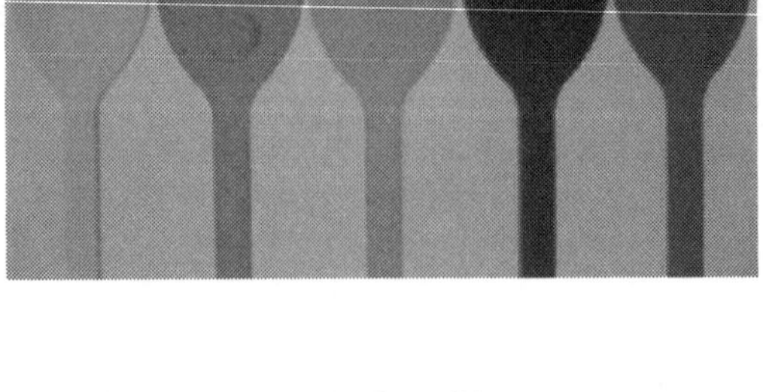

Figure 3: Picture of tensile dog bones of (from left to right): PP+H_2O, PP/cellulose fibres 5wt%, PP/cellulose fibres 5wt% +H_2O, PP/MFC-PPgMA2 5wt% and PP/MFC-PPgMA2 5wt% +H_2O.

The addition of cellulose fibres or MFC gives coloration to the composite. This yellowing is ascribed to cellulose thermal degradation during melt compounding. Due to its lower thermal stability, the coloration of composites containing MFC is markedly increased compared to cellulose fibre composites. However the coloration of the coating agent (PPgMA2) is also involved in the visual aspect of the composites. Water injection limited the yellowing of extruded composites (Soulestin et al. (2007)) by: 1) decreasing the melt temperature with as consequence the preservation of cellulose from thermal degradation, and 2) acting as stripping agent by releasing the chromophore groups produced by cellulose degradation. By looking to the MFC composites presented in Fig. 3, the use of water injection seems necessary to preserve the filler integrity.

Morphological characterization of the elaborated composites was performed with SEM on compressed films. Fig. 4a and b show microstructure of PP/MFC-PPgMA 5wt% and PLA/MFC-PEG 5wt% respectively.

(a)

(b)

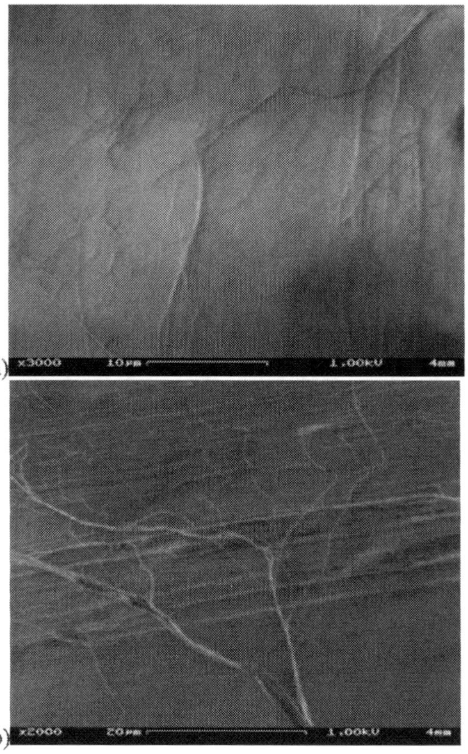

Figure 4: SEM pictures of (a) PP/MFC-PPgMA2 5 wt% and (b) PLA/MFC-PEG 5wt% composites.

Dispersion of microfibrillated cellulose was achieved in a polyolefin and PLA matrix as shown in Fig. 5. Moreover microfibrils and microfibril bundles seem to be entangled forming an interconnected network. This kind of microstructure was not observed for cellulose fibres and freeze-dried MFC dispersed in the polymer matrices.

The addition of interconnected arrangements has a marked influence on the viscoelastic properties of molten polymers. At the gel point, changes in frequency dependence of storage (G') and loss (G") moduli, complex viscosity (η^*) and/or loss tangent tan(δ) are observed due to the transition from a liquid-like to a pseudo solid-like behaviour. Fig. 5a gathers measurement of G' for increasing content of MFC in LDPE/MFC-PPgMA1 system during a frequency sweep at 190°C. Fig. 5b shows a similar test with the PLA/MFC-PEG system at 170°C.

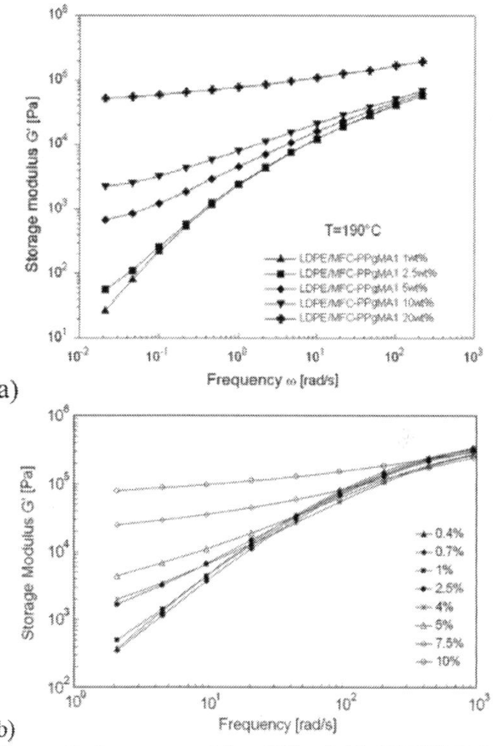

(a)

(b)

Figure 5: Storage modulus (G') of (a) LDPE/MFC-PPgMA1 and (b) PLA/MFC-PEG composites for increasing MFC contents.

From Fig. 5a and b, it can be observed that above a critical concentration of MFC, the dynamic modulus flattens at low frequencies reaching a plateau.
Similar observations can be made in terms of viscosity. The values of η^*, above the gel point, does not reach the Newtonian plateau at low frequencies. This phenomenon corresponds to the presence of a percolating network of MFC in the polymer. Such behaviour was not observed for equivalent contents of cellulose fibres or freeze-dried MFC.

Conclusions

The dispersion of coated microfibrillated cellulose (MFC) in polymers (polyolefins and poly(lactic acid)) was achieved using lab and semi-pilot scale twin-screw extrusion.

Coating MFC low molecular weight PP-g-MA or PEG did not improve the thermal stability compared to freeze-dried MFC but allowed the preservation of microfibrillated structure.

Thanks to the applied coating agents, the dispersion of MFC in polyolefins and poly(lactic acid) was improved. Rheological properties showed a percolating effect due to the dispersion of MFC while nothing was observed for uncoated MFC or cellulose fibre composites with equivalent amounts of filler.

Acknowledgements

The authors are grateful for the financial support of the Walloon Region in the frame of the NA-CELL program (2005-2009, convention n°031/5575, 041/5699 and 041/5700).

The authors are grateful for the financial support of the Frond National de la Recherche Scientifique (FNRS).

References

Anglès M.N. and Dufresne A., "Plasticized Starch/Tunicin Whiskers Nanocomposites. 1. Structural Analysis", Macromolecules, **33**, 8344-8353 (2000).

Azizi Samir M.A.S., Alloin F., Sanchez J. and Dufresne A., "Cellulose nanocrystals reinforced poly(oxyethylene)", Polymer, **45**, 4149–4157 (2004).

Azizi Samir M.A.S, Alloin F., Gorecki W., Sanchez J-Y. and Dufresne A., "Nanocomposite Polymer Electrolytes based on Poly(oxyethylene) and cellulose nanocrystals", J. Phys. Chem. B, **108**, 10845-18852 (2004).

Azizi Samir M.A.S, Alloin F., Gorecki W., Sanchez J-Y., El Kissi N. and Dufresne A., "Preparation of Cellulose Whiskers Reinforced Nanocomposites from an Organic Medium Suspension", Macromolecules, **37**, 1386-1393 (2004).

Dufresne A., Dupeyre D. and Vignon M.R., "Cellulose microfibrils from potato tuber cells: Processing and characterization of starch-cellulose microfibril composites", J. Appl. Polym. Sci,, **76**, 2080-2092 (2000).

Durmus A., Kasgoz A. and Macosko C.W., "Linear low density polyethylene (LLDPE)/clay nanocomposites. Part I: Structural characterization and quantifying clay dispersion by melt rheology", Polymer, **48**, 4492-4502 (2007).

Gauthier R, Joly A., Coupas A.C., Gauthier H. and Escoubes M., "Interfaces in polyolefin/Cellulosic fiber composites: chemical coupling, morphology, correlation with adhesion and aging moisture", Polymer composites, 19, 287-300 (1998).

Goussé C., Chanzy H., Cerrada M.L., Fleury E., "Surface silylation of cellulose microfibrils: preparation and rheological properties", Polymer, **45**, 1569–1575 (2004).

Habibi Y., Goffin A-L., Schiltz N., Duquesne E., Dubois P. and Dufresne A., "Bionanocomposites based on poly(ε-caprolactone)-grafted cellulose nanocrystals by ring-opening polymerization", J. Mater. Chem., **18**, 5002-5010 (2008).

Heux L., Chauve G. and Bonini C., "Nonflocculating and Chiral-Nematic Self-ordering of Cellulose Microcrystals Suspensions in Nonpolar Solvents", Langmuir, **16**, 8210-8212 (2000).

Hubbe M.A., Rojas O., Lucia L.A., Sain M., "Cellulosic nanocomposites: a review", BioResources, **3**, 929-980 (2008).

Kelarakis A., Yoon K., Somani R.H., Chen X., Hsiao B.S. and Chu B., "Rheological study of carbon nanofiber induced physical gelation in polyolefin nanocomposite melt", Polymer, **46**, 11591-11599 (2005).

WO9929767 (A1) (1999), Korbee R.A., Van Geenen A.A.

Nakagaito A.N., and Yano H., "The effect of fiber content on the mechanical and thermal expansion properties of biocomposites based on microfibrillated cellulose", Cellulose, **15**, 555–559 (2008).

Nickerson F.R. and J.A. Habrle J.A., "Cellulose intercrystalline structures-study by hydrolytic methods", Journal of Industrial and Engineering Chemistry, **39**, 1507-1512 (1947).

Özgür Seydibeyoglu M. and Oksman K., "Novel nanocomposites based on polyurethane and microfibrillated cellulose", Composites Science and Technology, **8**, 908–914 (2008).

Palaniyandi V. and Simonsen J., "Effect of compatibilizers on the crystallization kinetics of cellulose filled high density polyethylene", Composite Interfaces, **14**, 73–83 (2007).

Pötschke P., Abdel-Goad M., Alig I., Dudkin S. and Lellinger D., "Rheological and dielectrical characterization of melt mixed polycarbonate-multiwalled carbon nanotube composites", Polymer, **45**, 8863–8870 (2004).

Quiévy N., Jacquet N., Sclavons M., Deroanne C., Paquot M. and Devaux J.. "Influence of the homogenization and drying processes on the thermal stability of the microfibrillated cellulose", Polymer Degradation and Stability, **95**, 306-314 (2010).

Canadian Patent, 2437616 (2005), Sain M. and Bhatnagar A.

Saito T., Nishiyama Y., Putaux J-L, Vignon M. and Isogai A., "Homogeneous Suspensions of Individualized Microfibrils from TEMPO-Catalyzed Oxidation of Native Cellulose", Biomacromolecules, **7**, 1687-1691 (2006).

Sakurada I., Nukushima Y., Ito T., "Experimental determination of the elastic modulus of crystalline regions in oriented polymers", J. Polym. Sci., **57**, 651-660 (1963).

Siqueira G., Bras J. and Dufresne A., "Cellulose Whiskers versus Microfibrils: Influence of the nature of the nanoparticle and its surface functionalization on the thermal and mechanical properties of nanocomposites", Biomacromolecules, **10**, 425-432 (2009).

Soulestin J., Quiévy N., Sclavons M., Devaux J., "Polyolefins-Biofibre : A new way for an industrial production", Polymer Engineering and Science, 47(4), 467-476 (2007).

FR Patent, 2472628 (A1) (1981), Turbak A.B., Snyder F.W. and Sandberg K.R.

Wang B. and Sain M., "The effect of chemically coated nanofiber reinforcement on biopolymer based Nanocomposites", BioResources, **2**, 371-381 (2007).

Wang B. and Sain M, "Isolation of nanofibers from soybean source and their reinforcing capability on synthetic polymers", Composites Science and Technology, **67**, 2521–2527 (2007).

Wang B. and Sain M., "Dispersion of soybean stock-based nanofiber in a plastic matrix", Polymer International, **56**, 538-546 (2007).

IMPACT BEHAVIOR OF INJECTED PP/NANOCLAY PARTS

V. Pettarin[1], A. Pontes[2], G. Viau[1], J. C. Viana[2], P. M. Frontini[1], A. S. Pouzada[2]*

[1] *Grupo de Ciencia e Ingeniería de Polímeros, Instituto de Ciencia y Tecnología de Materiales(INTEMA),CONICET-Universidad de Mar del Plata, Argentina* – pettarin@fi.mdp.edu.ar ; pmfronti@fi.mdp.edu.ar
[2] *IPC-Institute for Polymers and Composites, University of Minho, Guimarães, Portugal* – pontes@dep.uminho.pt ; jcv@dep.uminho.pt ; asp@dep.uminho.pt

This work attempts to contribute to bridge the gap between scientific challenges and industrial stakes regarding PP/nanoclay composites. Pieces of nanocomposites were obtained by direct injection of commercial PP mixed with a commercial MB of PP with 50% of organoclay, with a double-gated hot runner mould, which produced mouldings with a weld line. The moulding microstructure was assessed by POM and XRD, while the distribution and exfoliation grade of clay was evaluated by TEM and XRD. The typical skin-core structure was found, with a skin thickness wider in bulk than in weld line zones. Regarding clay platelets mostly intercalated structures were seen. The impact properties at room temperature were assessed by means of tensile and biaxial tests. Properties were monitored at different sites of the mouldings. At the weld line zone less energy was consumed under tensile conditions and exhibited higher apparent impact toughness under biaxial conditions than the bulk zone. Visual inspection of biaxially impacted samples showed that the orientation of polymer molecules and clay platelets induced by melt flow prevailed, and the weld line was not the determinant of the toughness of the mouldings. An optimum in impact performance was found for moulding with 3% of clay, since at larger clay contents platelets agglomerated and acted as stress raisers.

Introduction

Polypropylene (PP) belongs to the group of thermoplastics produced in large quantities. Great efforts are undertaken to transform PP into a thermoplastic suitable for engineering use. Research work has therefore been focused on upgrading the mechanical performance of PP by toughening (to overcome the moderate cold and impact resistance) by filling and reinforcing (to increase stiffness, strength and temperature resistance and to decrease shrinkage) or both (Karger-Kocsis, 1993).

Polymeric nanocomposites have emerged as new materials showing potential for industrial application and quite exceptional properties at low filler contents (Ray and Okamoto, 2003; Alexandre and Dubois, 2000). The interest in polymer/layered silicate nanocomposites is still large and intensive research is going on in many research groups in this field in spite of the fact that a real breakthrough has not been achieved yet.

An important commodity polymer such as PP, when filled with a low incorporation level of nanoclay (typically less than 5%), can be used in applications with requirements typical of engineering plastics with additional advantages. However, only well-dispersed and well-exfoliated nanoparticles can lead to the expected improvement of properties (Zhang et al, 2004; Ding et al, 2005; Chen et al, 2000). Raw material producers, converters and end-users have therefore to tackle both compounding and processing issues. Surface modification of nanofillers with organic surfactant and adaptation of compounding conditions (high shear, high residence time, special screw profile design in case of melt compounding for example) may help to get rid of most of compounding issues. Research groups have made significant progress in that field. With the development of masterbatches (MB) the final injection – or extrusion – moulded part should be easily obtained by mixing/diluting the MB in the appropriate polymer matrix. The nanoparticle dispersion and exfoliation is usually assumed to be achieved during the MB compounding. Experience unfortunately often shows that the industrial reality is quite different (Krawczak, 2007).

Injection moulding is the most important process to manufacture plastic parts, since complex geometries become available in one automated production step. Nonetheless, the influence of processing parameters is critical in the performance of these products (*e.g.* Cunha and Pouzada, 1995). Moreover, if weldlines are likely to occur in injection moulded products, things become more complicated. Weld lines are an unavoidable reality in the injection moulding of complex parts. Multiple gating, splitting of the melt flow due to inserts in the cavity or through holes, as well as changes of thickness give rise to regions in the impression where the melt flow fronts recombine and weld. Weld lines are formed along with the mould filling process: particularly in the moulding of very complex components a multiplicity of weld lines is generated. In the region were weld lines occur the properties are different from the bulk (Ersoya and Nugay, 2004; Morelli et al, 2007). Whereas the complete elimination of weld lines is not

always successful, their adverse effects on part appearance and mechanical performance can be minimized by changing mould design and process conditions (Wu and Liang, 2005).
Impact strength is often one of the fundamental material requirements used to measure the performance of plastics products. Falling-dart impact tests on disc or plate specimens are widely used by the plastics industry to characterize the "practical" resistance of rigid thermoplastics, because they are intended to reproduce the real behaviour of moulded parts under impact loading. They are product tests on representative components, rather than measurements of basic material properties. One additional advantage of the falling-dart impact test is that it provides a convenient method for studying changes induced by flow modification in part performance (Cunha et al, 1997; Bucknall, 2000).
This work attempts to contribute to bridge the gap between scientific challenges and industrial stakes regarding PP/nanoclay composites. Impact behaviour of PP/nanoclay injection moulded parts obtained by direct compounding of commercial PP and MB was studied. Influence of processing parameters and weld lines on in-plane and out-of-plane impact behaviour was explored.

Experimental

Materials

This study was carried out on a propylene homopolymer, F-045,D2 (from SUNOCO Chemicals) and a commercial MB of PP with 50% of organoclay, Nanomax-PP P-802 (from Nanomax Polyolefin Masterbatch Products).

Injection moulding

Nanocomposites were obtained by direct injection of mixtures of PP and MB. Different amounts of incorporation of nanoclays were used by diluting the MB in the PP matrix. Rectangular boxes of 1.4-mm thickness (Fig.1), were injection moulded in a double-gated hot runner injection mould using a Klöckner Ferromatic FM20 injection machine of 200 kN clamping force. The injection pressure was monitored during injection. The processing conditions are listed in Table 1.
The Taguchi method (Rosse, 1996) was applied to determine the effect of moulding parameters on the impact behaviour of pieces. Three moulding parameters were selected: masterbatch content (MBp), melt temperature (Tm), and injection speed (Qj). Because the influence of these factors may vary non-linearly, each factor was assigned with three levels. According to the levels of each parameter, an L9-orthogonal table was used to inject the pieces (Table 2). The nine conditions were repeated with pure PP mouldings for comparison. Additional pieces were injected according to experiment #2 with various MBp contents (2, 6, 10 and 14%).

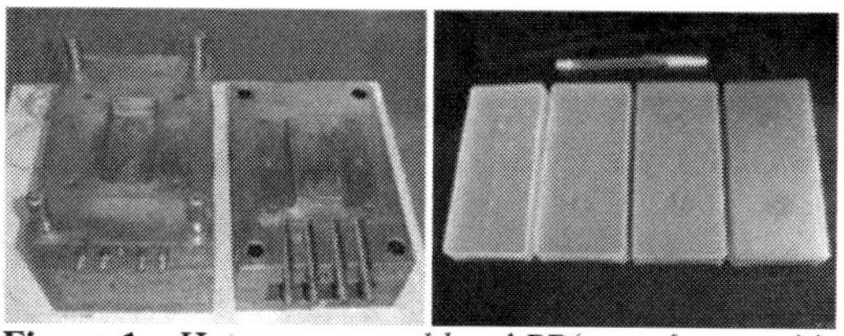

Figure 1 – Hot runner mould and PP/nanoclay mouldings.

Table 1 – Processing settings

Processing parameter	Value	
Barrel temperature	Zone 1	160°C
	Zone 2	185°C
	Zone 3	215°C
	Zone 4	235°C
Screw rotation velocity	250 rpm	
Injection pressure	5 MPa (50 bar)	
Injection time	1 s	
Packing pressure / time	3 MPa, 10 s	
Cooling time	15 s	

Table 2 – Design of experiment and factors and levels for the injection moulding experiments

Experiment #	MBp (%)	Tm (°C)	Qj (mm/s)
1	6	205	100
2	6	220	350
3	6	235	400
4	10	220	400
5	10	235	100
6	10	205	350
7	14	235	350
8	14	205	400
9	14	220	100

Microstructure

Polarized light microscopy was used to observe the morphology of mouldings. 15-μm thick specimens were microtomed with a Leitz 1401 microtome and observed with an Olympus BH2 polarized light microscope.
Distribution and exfoliation level of clay was evaluated by TEM and XRD. XRD analysis was performed on pieces surface using a Phillips X'PERT MPD diffractometer (CuKα radiation λ=1.5418 Å, generator voltage=40 kV, current=40 mA). Measurements were recorded every 0.02 θ for 1 s each varying 2 h from 2 θ to 40 θ. The interlayer distance of clay was calculated from the (001) peak using the Bragg equation. TEM microphotographs were obtained with a TEM Jeol 100 CX microscope using an acceleration voltage of 200 kV. Samples were ultramicrotomed at room temperature with a diamond knife to a 70 nm-thick section.

Impact tests

The impact properties at room temperature of the mouldings were assessed by means of tensile (in-plane) and biaxial (out-of-plane) tests.
Tensile-impact tests were carried out at room temperature using a pendulum Ceast 6545 of 7.5 J. For these tests, rectangular samples were cut with a G-WEIKE LG900 laser engraver at different locations in the moulding: along the region of weld line formation at different melt collision angle (WL1 180°, WL2 140°, and WL3 110°); and away from the weld line, in a radial position along the flow direction (WWL), as shown in Fig. 2a. A 2 mm hole was drilled at the centre of the samples. The toughness, or tensile impact resistance, of the samples in the tensile test (J/m^2) was measured by the ratio between the fracture energy (J) and the broken surface area (m^2).

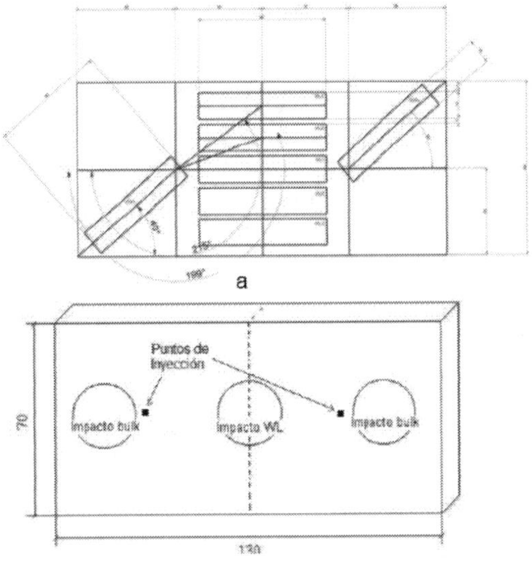

Figure 2 – Layout of tensile (a) and disc (b) test specimens.

Biaxial impact tests (with loading perpendicular to the plane containing one of the injection points) were carried out according to ASTM D 3763-93 in two locations of the pieces (on (WL) and out (bulk) of weld line) (Fig. 2b) using a Ceast Fractovis 6787 falling weight equipment. This equipment allows the load on the specimen to be continuously recorded as a function of time prior to fracture. This gives a more complete description of an impact than a single calculated value. The specimens were clamped between two steel plates with a circular opening of 40 mm in diameter. A velocity at impact of 1 $m.s^{-1}$ was used. This technique is widely accepted for the out-of-plane fracture response assessment of polymeric composites because it gives a realistic view of in-service impact situations, hence being closer to real life conditions (Cunha et al, 1992; Bucknall, 2000). Moreover, it was found that this type of test was sensible to variation in skin-core structure induced by variations in processing settings (Cunha et al, 1992; Karger-Kocsis et al, 1999).

Results and Discussion

Microstructure of mouldings

The typical skin-core structure was found in mouldings (Fig. 3). Near the cold mould surface a frozen layer is formed adjacent to a shish-kebab or row structure developed, probably, in the region of highest shear rate.

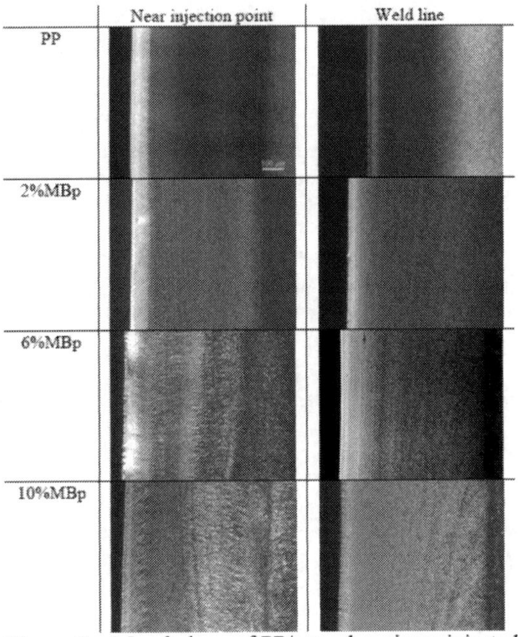

Figure 3 – Morphology of PP/nanoclay pieces injected according to experiment 2 (Table 2) observed by polarized light microscopy.

The shish-kebab structure is common in PP and other polyolefins and is a consequence of the fast crystallization of the polymer chains under high stresses near the skin region, as described by Fujiyama et al. (1988). In the vicinity of this sheared layer, β-spherulites can be found as a result of the crystallization under shear. In the core α-spherulites are formed under quiescent crystallization conditions. It is possible to observe the decrease of the skin thickness with the increase in MBp content (Fig. 4). This may be due to the nucleating effect of the nanoclay on PP. The distance from the injection points also influence the morphology. Thus, a reduction in skin thickness of samples on weld lines, away from the injection gates, when compared with samples near injection point could be verified (Fig. 4).

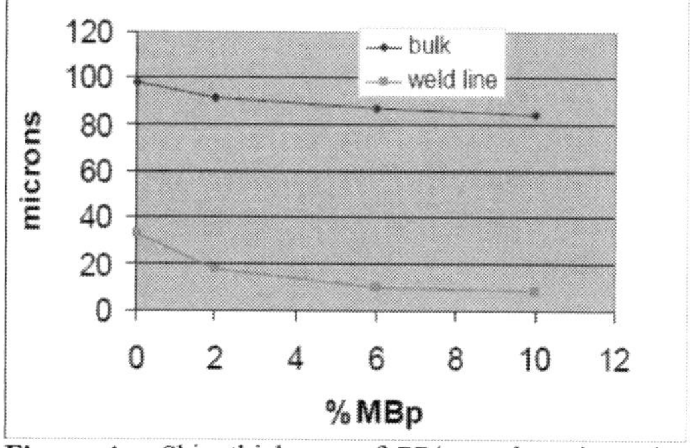

Figure 4 – Skin thickness of PP/nanoclay pieces injected according to experiment #2 (Table 2).

XRD confirmed the α nucleated effect of clay (Fig 5). Besides the main α–form crystallites (peaks at 2θ = 14.2, 17, 18.8 y 25.5°, corresponding to (110), (040), (130) and (131) planes), a small amount of β-form (peaks at 2θ = 16.2 y 21.2°, associated with (300) and (301) planes) was found in the skin layer of PP mouldings. This result is in agreement with other results reported in the literature (Kalay and Bevis, 1997). When clay was added to mouldings, intensities of peaks at 16.2° and 21.2° (β polymorph) diminished, while intensities of peaks at 17° and 25.5° (α polymorph) increased. It has been suggested that clay platelets have a similar effect as quenching, that is, they limit chain mobility (Fornes and Paul, 2003). This may be the cause of the decrease of β-form content in skin mouldings with increasing clay content.

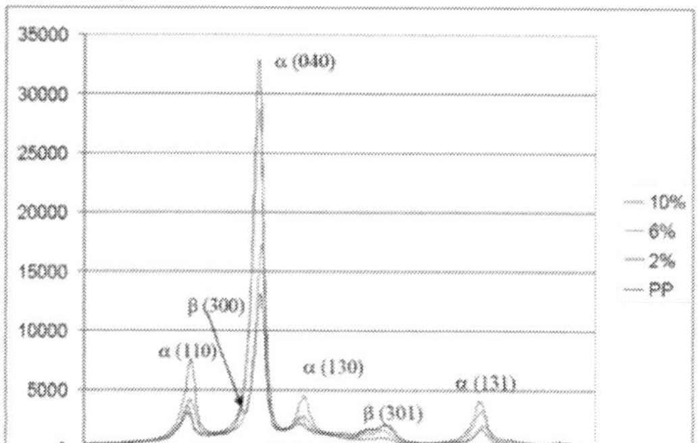

Figure 5 – XRD patterns (high 2θ) of PP/nanoclay pieces injected according to experiment 2 (Table 2).

Regarding clay exfoliation, XRD patterns indicated that clay platelets are only intercalated (Fig. 6), while in TEM pictures clusters of stacked sheets are clearly seen (Fig. 7). This is because the degree of clay dispersion not only depends on the affinity and compatibility of the organoclay with the matrix (which is an intrinsic factor dependent on the materials), but also on the shear stress (which is an extrinsic factor dependent on processing conditions and clay loading) (Dennis et al, 2001). Hence, the fact that, in our case, organoclay exfoliation is not complete is in part because the imposed processing conditions do not generate a shear force in the mix strong enough to delaminate completely the clay agglomerates.

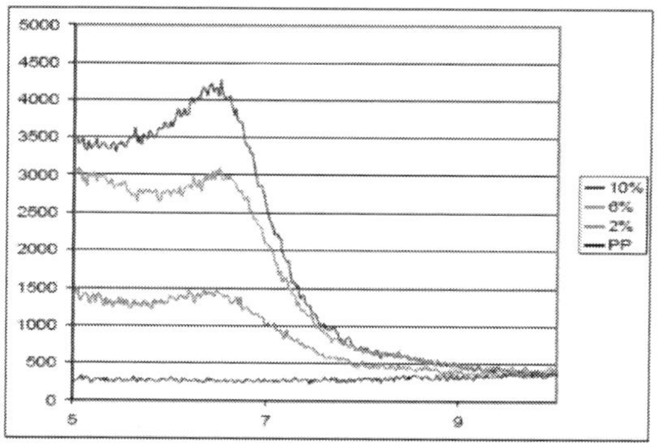

Figure 6 – XRD patterns (low 2θ) of PP/nanoclay pieces injected according to experiment 2 (Table 2).

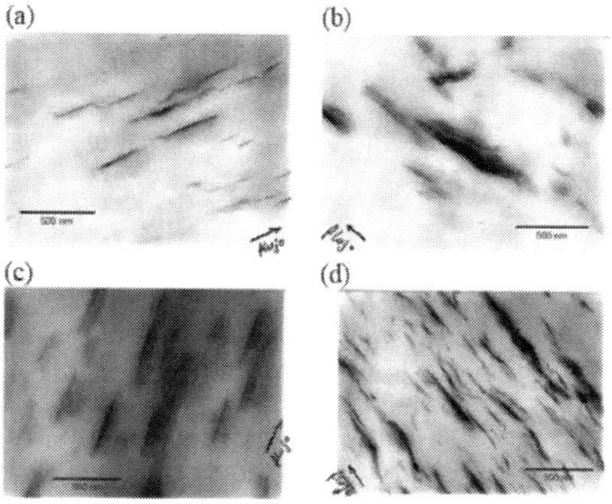

Figure 7 – TEM pictures of some PP/nanoclay pieces (a) 2% MBp (b) 6% MBp (c) 10% MBp (d) 14% MBp.

Influence of processing conditions on impact toughness

Figure 8 illustrates the effect of processing factors on the tensile and biaxial impact toughness following an ANOVA analysis (Keppel and Zedeck, 1989). The results show that MBp is the most influencing parameter, and impact toughness is practically not affected by the other processing conditions as reported before for other composites by Koster (1999) and Martinez Gamba et al (2009).

Pressure measurements inside the mould cavity go with this result (Fig. 9). Melt viscosity increased with clay content, inducing smaller pressures inside mould cavity. However, differences in pressure curves due to different clay content were not significant, indicating that clay did not affect processing conditions of boxes and changes in their mechanical properties are mainly due to MBp content.

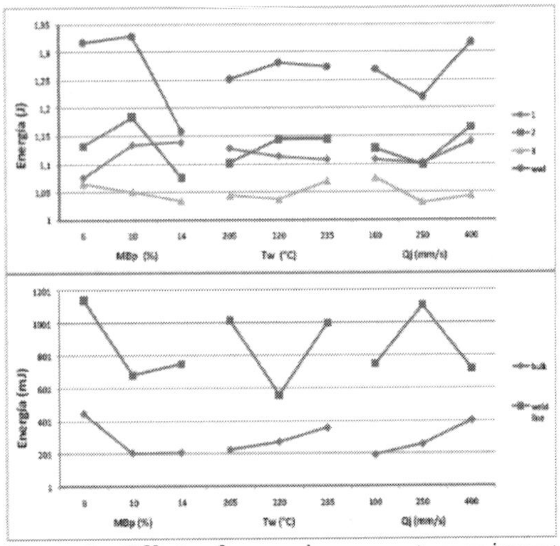

Figure 8 – Effects of processing parameters on impact toughness under (a) tensile impact and (b) biaxial impact.

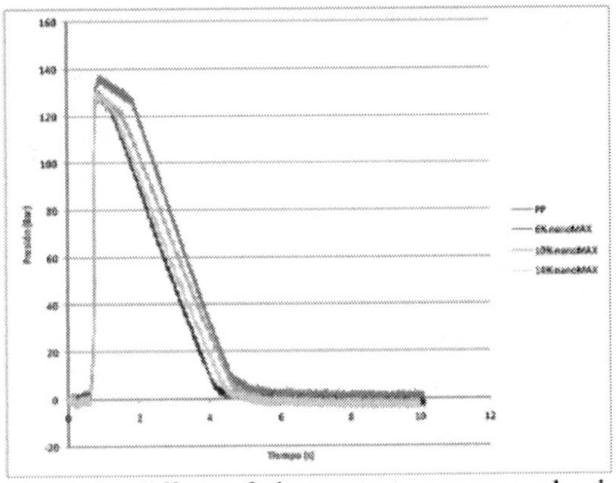

Figure 9 – Effects of clay content on pressure developed inside mold cavity.

Influence of the weld line on impact toughness

As expected, at the weld line the amount of consumed energy is lower than at the bulk under tensile conditions (Fig. 10).

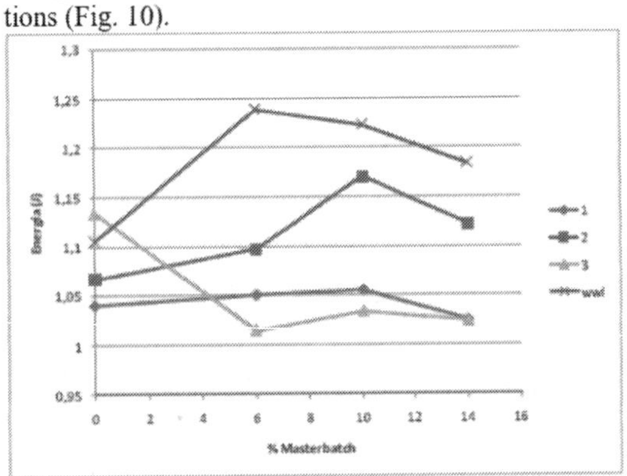

Figure 10 – Tensile impact toughness as a function of position in box and clay content.

It is possible to verify that the impact resistance tends to increase from a minimum associated to the sample WL1, where there is a frontal meeting of the flow fronts, to WWL, the reference sample in a location well away from the weld line in the flux direction. This trend may result both from geometrical aspects of the part and from the filling process of the moulding. It is useful to recall here the orientation of polymer molecules and clay platelets in the mouldings (Fig. 11).

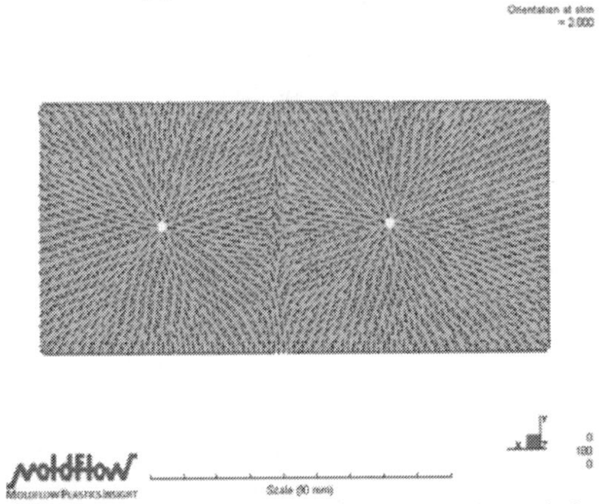

Figure 11 – Orientation of polymer molecules and clay particles as assessed by Moldflow.

It can be observed that the inclination of the weld line plane with respect to the specimen axis varies from a maximum of 90° (sample WL1) to a minimum around 45° (sample WL3). Therefore, at the region WL1, the polymer chains and the clay platelets are oriented in a plane transversal to the applied force, thus contributing to the weakness of this sample. In the other regions (WL2 and WL3), the weld line formed is not exactly perpendicular to the imposed tensile stress, which attenuates the depreciative effect of the weld line. The samples at location WWL show the best performance because of the absence of the weld line and the solicitation made in the flux direction.

A surprising result was that in PP mouldings weld line zone exhibited a more ductile behaviour (Fig. 12) and higher apparent impact toughness than the bulk under biaxial conditions (Fig. 13). The same result was previously reported by others (Bucknall, 1986). Visual inspection of biaxially impacted samples confirmed a larger fracture surface involved in the fracture of WL specimens (Fig. 14). Forming a larger fracture surface requires higher energy absorption. The macroscopic analysis of the fractured surfaces shown in Fig. 14 provides additional insight of the differences in the mechanical behaviour. This test does not impose a preferential direction of failure specimen. Consistently, isotropic or orthotropic materials are expected to display a symmetric deformation pattern to the impact site. The failures originate at the weakest point in the sample and propagate from there due to the high radial and circumferential tensile stresses (Alcock et al, 2006). The images show that the fracture patterns are

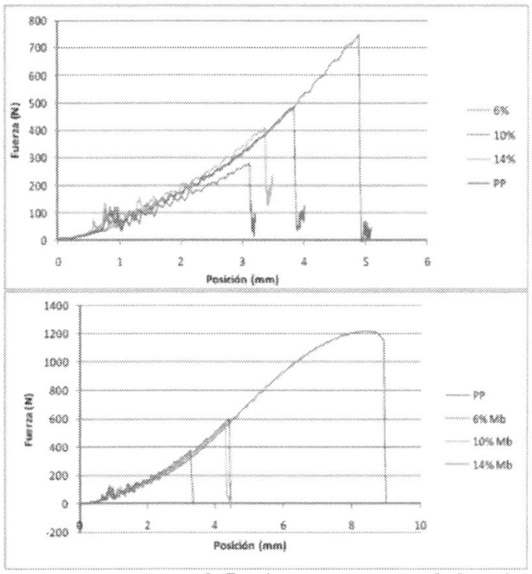

Figure 12 – Force-deflection curves recorded under biaxial impact in (a) bulk and (b) weld line zones.

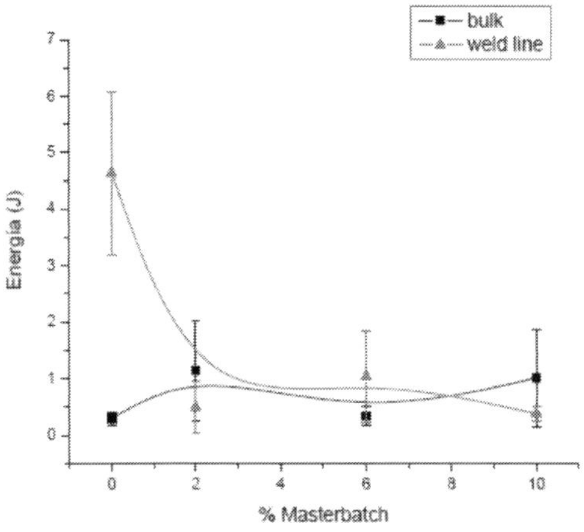

Figure 13 – Impact toughness of PP mouldings under biaxial loading.

significantly different. At the bulk of PP mouldings bending along the clamping ring becomes obvious. The plate failed by single splitting along the melt flow direction. Conversely, at the weld lines it can be seen that some cracks run radially from the central point of impact, while others follow a circular path around the same point (circular cracks occur where the depression caused by the striker produces a large bending moment and a high level of out-of-plane curvature), and there are evidence of plastic deformation (whitened zones). Therefore it is clear that the molecular orientation induced by melt flow prevailed, and the weld line were not the determinant of the toughness of the mouldings. On the other hand, weld lines of PP/nanoclay mouldings did not exhibit higher impact toughness than respective bulks. It is clearly seen that there is not only a main crack path along the melt flow direction in bulk samples, but also a main crack path along the weld line in weld line samples (Fig. 14).

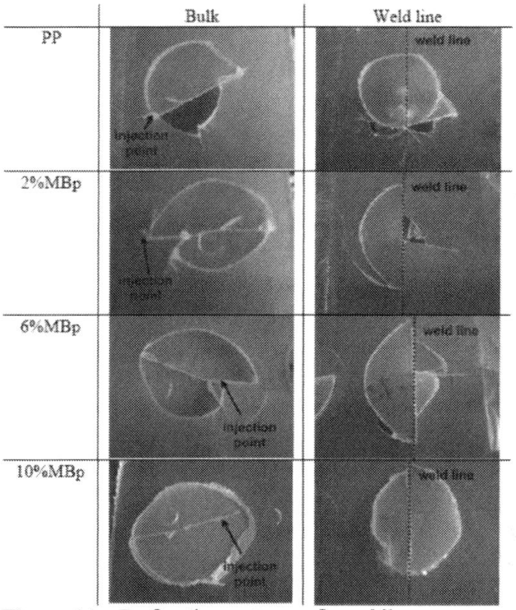

Figure 14 – Perforation patterns of mouldings.

This result is a consequence of lower molecular mobility and chain entanglement in the weld line, because of the viscosity increment resulting from the clay addition, and unfavourable orientation of the clay particles in the weld line plane. However, there is not a depreciation of impact toughness due to weld line formation, indicating that in these mouldings the molecular and particles orientation is the determinant of the toughness of the mouldings under biaxial loading, and not the weld line.

Influence of clay content on impact toughness

In the case of tensile tests, the lower values of the relationship between toughness at weld line and toughness at bulk were found with the sample WL1 (transversal plane of weld line), thus indicating the worse detrimental effect of the weld line in the tensile-impact resistance when it is perpendicular to the impact direction. This finding is a consequence of the orientation of the clay particles in the plane of the weld line, unfavourable to the direction of mechanical loading, and the higher melt viscosity of the composites with clay that decreases the molecular mobility, the chain entanglement, and the healing of the weld line. In biaxial tests, differences in perforation energy with clay content are negligible.

Analyzing all impact results, an optimum in impact performance was observed for mouldings with 6% of MBp (corresponding to 3% of nanoclay). It seems that at higher clay contents, clay platelets agglomerated and acted as defects, concentrating stresses.

Conclusions

In this work the uniaxial and biaxial impact response of injected PP/nanoclay mouldings was assessed.

The typical skin-core structure developed in the mouldings. A reduction of the skin with the increase in MBp content was observed probably due to the nucleating effect of clay on PP, also confirmed by XRD patterns. Besides a reduction in skin thickness of samples at the weld lines, away from the injection gates, when compared with samples near injection point was verified.

XRD and TEM analysis of mouldings showed that clay particles are at the most intercalated, indicating that the imposed processing conditions do not generate a shear force in the mix strong enough to delaminate completely the clay agglomerates.

ANOVA analysis revealed that the impact performance of the mouldings was practically not affected by the processing conditions (Tm and Qj), being MBp the most influencing parameter, which suggests that the matrix morphology may be not the leading factor in this type of materials.

Weld line zone of mouldings consumed less energy to fracture under tensile conditions than bulk zone due to molecular and clay orientation: in weld line the polymer chains and the clay platelets are oriented in a plane transversal to the applied force, while in the bulk molecules and clay platelets are oriented in the direction of the applied force.

Weld line zone of PP mouldings exhibited a more ductile behaviour and higher apparent impact toughness than bulk zone under biaxial conditions. Visual inspection of impacted samples showed that the molecular orientation induced by melt flow prevailed, and the weld line was not the determinant of the toughness of the mouldings. Differences between perforation energies of PP/nanoclay mouldings on bulk zones and weld line zones were negligible. Both main crack paths along the melt flow direction and along the weld line were seen, due to the lower molecular mobility and chain entanglement in the weld line, (because of the viscosity increment resulting from the clay addition) and unfavourable orientation of the clay particles in the weld line plane. In summary, in these mouldings the molecular and particles orientation was the determinant of the perforation energy, and not the weld line.

An optimum in impact performance was found for mouldings with 6% of MBp (corresponding to 3% of nanoclay). It seems that at higher clay contents, clay platelets agglomerated and acted as defects, concentrating stresses.

Acknowledgements

The authors thank CONICET and ANPCyT from Argentina for financial support, and also the covenant MINCyT (Argentina) - FCT (Portugal) between the Universities Nacional de Mar del Plata and Minho.

References

- Alcock, B., Cabrera, N.O., Barkoula, N.M., Peijs T., "Low velocity impact performance of recyclable all-polypropylene composites", Comp. Sci. Technol., **66**, 1724-37 (2006).
- Alexandre M., Dubois P., "Polymer-layered silicate nanocomposites: preparation, properties and uses of a new class of materials", Mat. Sci. Eng., **28**, 1-63 (2000).
- Bucknall, C.B., "Application of fracture mechanics to the prediction of the ductile-brittle transition", Pure & Appl. Chem, **58,** 999-1014 (1986).
- Bucknall, C.B., "Characterizing toughness using Standard Empirical Tests", in Polymer Blends, Volume 2: Performance, Paul, D.R., Bucknall, C.B. (Eds.), Wiley, New York, p. 25–27 (2000).
- Chen L., Wong S.-C., Liu T.X., Lu X.H., He C.B., "Deformation mechanisms of nanoclay-reinforced maleic anhydride-modified polypropylene", J. Polym. Sci. Part B: Polym. Phys., **42**, 2759-68 (2000).
- Cunha, A.M., Pouzada, A.S., Crawford, R.J., "A study of the impact behaviour of injection moulded PP using different modes of testing", Plast. Rubb. and Comp. Proc. Appl., **18,** 79-90 (1992).
- Cunha, A.M., Pouzada, A.S., "The impact behaviour of polypropylene and the injection moulding conditions", in Impact and Dynamic Fracture of Polymers and Composites". ESIS 19, Williams, J.G., Pavan, A. (Eds.), Mech.Eng.Publ., London, p. 315-325 (1995).
- Dennis, H.R., Hunter, D.L., Chang, D., et al., "Effect of melt processing conditions on the extent of exfoliation in organoclay-based nanocomposites", Polymer, **42**, 9513–22 (2001).
- Ding, C, Jia, M., He, et al., "How organo-m.ontmorillonite truly affects the structure and properties of polypropylene", Polymer Testing, **24**, 94-100 (2005).
- Ersoya, O.G., Nugay, N., "A new approach to increase weld line strength of incompatible polymer blend composites: selective filler addition", Polymer, **45,** 1243-1252 (2004).
- Fornes, T.D., Paul, D.R., "Crystallization behavior of nylon 6 nanocomposites", Polymer, **44**, 3945–61 (2003).
- Fujiyama, M., Wakino, T., Kawasaki, Y., "Structure of skin layer in injection molded polypropylene", J. Appl. Polym. Sci., **35**, 29-49 (1988).
- Kalay, G., Bevis, M.J., "Processing and physical property relationships in injection-molded isotactic polypropylene. 2. morphology and crystallinity", J. Polym. Sci. Part B: Polym. Phys., **35**, 265-291 (1997)
- Karger-Kocsis, J., "Instrumented impact fracture and related failure behavior in short- and long-

glass-fiber-reinforced polypropylene", Compos. Sci. Technol., **48**, 273-283 (1993).

- Karger-Kocsis, J., Mouzakis, D.E., Ehrenstein, G.W., Varga J, "Instrumented tensile and falling weight impact response of injection-molded alpha- and beta-phase polypropylene homopolymers with various melt flow indices", J. Appl. Polym. Sci., **73,** 1205-1214 (1999).
- Keppel, G., Zedeck, S., "Data Analysis for Research Designs: Analysis of Variance and Multiple Research/Correlation Approaches", W.H. Freeman, New York (1989).
- Koster R.P., "Importance of injection molding parameters for mechanical performance of cold flow weld lines", J. Inj. Mold. Technol., **3**, 154-158 (1999)
- Krawczak, P., "Editorial corner – a personal view, Compounding and processing of polymer nanocomposites: from scientific challenges to industrial stakes", eXPRESS Polymer Letters, **1**, 188 (2007).
- Martinez Gamba, M., Pouzada, A.S., Frontini P.M., "Impact properties and microhardness of double-gated glass-reinforced polypropylene injection mouldings", Polym. Eng. Sci., **49**, 1688-1695 (2009).
- Morelli, C.L., Sousa, J.A., Pouzada, A.S., "Assessment of weld line performance of PP/talc moldings produced in hot runner injection moulds", J. Vinyl Appl. Techn., **13**, 159-165 (2007).
- Ray, S.S., Okamoto, M., "Polymer/layered silicate nanocomposites: a review from preparation to processing", Prog. Polym. Sci., **28**, 1539-1641 (2003)
- Rosse, P., "Taguchi Techniques for Quality Engineering, Loss Function, Orthogonal Experiments, Parameter and Tolerance Design", McGraw-Hill, Toronto (1996)
- Wu, C.-H., Liang, W.-J., "Effects of geometry and injection-moulding parameters on weld line strength", Polym. Eng. Sci., **45,** 1021-30 (2005).
- Zhang, Y.-Q., Lee, J.-H., Rhee, J.M., Rhee K. Y., "Polypropylene–clay nanocomposites prepared by in situ grafting-intercalating in melt", Compos. Sci. Technol., **64,** 1383-89 (2004).

INFLUENCE OF PROCESSING TECHNIQUE ON MECHANICAL AND PHYSICAL PROPERTIES OF POLYMER NANOCOMPOSITES

A. Witschnigg[1]*, S. Laske[1], M. Kracalik[1], C. Holzer[1]

[1] *Institute of Plastics Processing, University of Leoben, Leoben, Austria - Andreas.Witschnigg@unileoben.ac.at, Stephan.Laske@unileoben.ac.at, Milan.Kracalik@mu-leoben.at, Clemens.Holzer@unileoben.ac.at, Franz Josef Strasse 18, 8700 Leoben, Austria*

The use of nanofillers (especially layered silicates) is steadily growing in polymer processing due to their great potential of enhancing mechanical and physical properties. In order to obtain this improvement, a homogenous dispersion of the nanofiller must be achieved. Depending on the particle structure (agglomerated, intercalated or exfoliated) a significantly higher level of improvement of the mechanical and physical properties can be achieved with a very small amount of filler. The degree of dispersion is exceedingly linked with the energy and time used for processing nanocomposites. The energy and time themselves are depending on the type of processing technique and processing conditions.

The influence of the residence time on the exfoliation and the amount of energy needed to create a stable 3D physical network is still not known precisely, but would be of major interest according to an optimization of mechanical and physical properties of the nanocomposite.

In this study, the effect of different processing techniques (compounding, extrusion and injection molding) on the dispersion of the nanofiller is investigated. The differences in the degree of dispersion according to dissimilarity of the chosen processing techniques are compared via various methods such as barrier test, scratch test and tensile test.

Introduction

Nanoclays (especially layered silicates) are the most common nanofiller used in polymer processing due to their great potential of enhancing material and the fact that they are not expensive compared to other nanofillers. The right use of layered silicate can enhances barrier properties, flammability, Young's modulus and other mechanical and physical properties. Beside this, the size of the interface is also of great importance as it is increasing with increasing exfoliation of the silicate layer in the polymer melt and therefore a good degree of dispersion. From the processing point of view there are two main influencing factors: the shear energy and the residence time. Both influences have to be cumulated for all processing steps starting from the polymer pellets to the test specimen to make statements for the whole polymer layered silicate system.

The effect of screw speed, which is linked with the shear energy, on the Young's modulus as well as oxygen permeation rate of different nanocomposites has been already investigated [1, 2].

The effect of duration and the speed of mixing on a nanocomposites solute in glycerol on mechanical properties (Young's modulus) have also been published [3, 4].

Experimental

Materials

A polyethylene with a melt flow rate (MFR) of 0.22 g/10 min; 190 °C/ 5 kg has been used. For homogenous dispersion a modificated nanofiller (montmorillonite intercalated with dimethyl distearyl ammonium chloride) was used in combination with a compatibilizer.

Preparation of the compounds

For the preparation of the compounds an intermeshing, co-rotating twin screw extruder Theysohn TSK30/40D (Theysohn Holding, Vienna, Austria) was used. The feed rate was set at 6 kg/h for the masterbatch and 5 kg/h for the final polyethylene nanocomposite with a screw speed of 100 rpm. The masterbatch formulation was 50 wt% compatibilizer and 50 wt% layered silicate. For the finally produced composite (referred to as PE-C and PE-M) the formulation was 90 wt% polyethylene, 5 wt% compatibilizer and 5wt % layered silicate. A melt pump (Δp ~-100 bar) was used to increase the residence time and shear rate and therefore improve the dispersion of the nanofiller. A melt temperature of 200 °C was chosen. The residence time for the masterbatch was 365 seconds compared to 330 seconds for the diluted composite. The difference between the two compounds is a second compounding step (for the PE-C) to achieve the final composite and only one compounding step (masterbatch production) with dilution directly in the form giving process (pipe extrusion, film extrusion and injection molding of the test specimen) for the PE-M. The temperature profile can be seen in table 1.

Table 1 – Temperature profile for the compounding process

Cylinderzones	Temperature (°C)
1	cooling
2	160
3	180
4	190
5	200
6	200
7	200
8	200
9	200
10	200

The two different compounds have then been used to produce pipes, films as well as tensile and scratch test specimens.

Production of the pipes

For the production of pipes, a single screw extruder Proton A 45-28G (Cincinnati Extrusion, Vienna, Austria) with a melt temperature of 225 °C at the screw die and a screw speed of 40 rpm was used. The pipes where produced with 1 m in length, a diameter of 50 mm and a wall thickness of 2.5 mm. The residence time for the pipe production was measured with 200 s. The temperature profile for pipe extrusion can be seen in table 2.

Table 2 – Temperature profile for pipe extrusion

Cylinderzones	Temperature (°C)
1	cooling
2	175
3	200
4	205
5	220
6	225
7	225

Production of the polymer films

The films were produced by using a single screw extruder E 30 P (L=30 D) (Dr. Collin GmbH, Ebersberg, Germany) with a 250 mm flat film die. The thickness of the films was adjusted at 200 µm. The melt temperature was set at 240 °C with a screw speed of 10 rpm. The residence time was 1013 s. The temperature profile for the extruder including the temperature for the die can be seen in table 3.

Table 3 – Temperature profile for film extrusion

Cylinderzones	Temperature (°C)
1	cooling
2	160
3	180
4	190
5	210
6	220
7	240
8	240
die	230

The chill roll temperature has been chosen with 40 °C.

Production of tensile and scratch test specimen

The test specimens for tensile and scratch tests have been produced with an ES330/80H, 800 kN closing force, injection molding machine (Engel, Schwertberg, Austria). The melt temperature was set to 230 °C with a tool temperature of 50 °C. The temperature profile for the injection molding machine can be seen in table 4.

Table 4 – Temperature profile for injection molding

Cylinderzones	Temperature (°C)
1	215
2	220
3	225
4	230

Mechanical properties – tensile test

A universal tensile testing machine Z010 (Zwick and Co. KG, Ulm, Germany) was used to carry out tensile tests according to ISO 527-1. All specimens were tested under standardized conditions (23 °C/ 50 %r.H.). The data were evaluated using the testXpert II software (ZWICK, Ulm, Germany).

Mechanical properties – scratch test

The scratch tests were carried out with a microtribometer UMT-2 (CETR, Campbell, USA) and a microscope for measuring the scratches on the surface of the specimen.

Mechanical properties – pipe test

The long term stability tests were carried out with a pipe facility of IPT (Todtenweis, Germany). The test conditions have been set with 80 °C and 5.3 MPa test tension according to ÖNORM EN ISO 1167.

Physical properties – barrier properties

For carrying out the O_2-barrier tests a $PermeO_2$ double cells s.n. 32 (ExtraSolution S.r.I., Gescher, Germany) was used. All specimen have been tested under standardized conditions (23 °C/ 50 % r.H.) according to DIN 53380.

Calculation of the shearing force for extrusion and compounding

The energy balance for an extruder is given with:

$$P \pm H = \frac{dm}{dt} \cdot \Delta h + p \cdot \frac{dV}{dt} \qquad \text{(eq.1)}$$

P	driving power
H	heating/cooling power
$\frac{dm}{dt} \cdot \Delta h$	enthalpy change
$p \cdot \frac{dV}{dt}$	compression work

With the negligence of the heating power the energy introduced into the system must be equal to the driving power minus the compression work and has to be transferred into shear energy (=S).

$$\frac{dm}{dt} \cdot \Delta h = P - p \cdot \frac{dV}{dt} = S \qquad \text{(eq.2)}$$

With this equation the shear energy introduced into the system for the twin screw extruder and the other two single screw extruders can be calculated.
Although the comparison between single and twin screw extruders is very difficult an attempt was taken to consider the influence of throughput and residence time. The idea behind it was to normalize the measured shear energy making the different extruder types comparable.

$$S_{norm} = \frac{S}{\dot{m} \cdot t} \qquad \text{(eq.3)}$$

Calculation of the shearing force for injection molding

The major part of shear exposure is composed of yielding in the injection die and in the tool. The shear energy evolving while plastification has also been calculated, but compared to the shearing while injection it can be neglected. The injection work which comprises the shear exposure in the injection tool can be calculated by integrating the injection pressure along the injection distance.

$$W_{inj} = A_S \cdot \int_{s_0}^{s_1} p_S(s) \cdot ds \qquad \text{(eq.4)}$$

A_S	screw area
p_S	injection pressure

The injection energy can then be calculated with:

$$S_{inj} = \frac{W_{inj}}{t_{inj}} \qquad \text{(eq.5)}$$

t_{inj} injection time

The injection energy comprises shear energy, compression energy and strain energy. The compression energy is calculated with equation 6:

$$S_{compress} = T \cdot \frac{1}{V_{spez}} \cdot \frac{\Delta V_{spez}}{\Delta T} \cdot \frac{\Delta p}{\Delta t} \cdot V \qquad \text{(eq.6)}$$

The combined shear and strain energy can than be calculated with:

$$S_{comb} = S_{inj} - S_{compress} \qquad \text{(eq.7)}$$

This combined shear and strain energy cannot be compared with the shear energy of extruding processes, due to the fact that the strain energy part is very high and not neglectable as far as injection molding is considered. So the calculated values appear higher than those for extrusion and compounding. Nevertheless an attempt was started by normalizing the combined energy by considering the mass of the specimen.

$$S_{norm} = \frac{S_{comb}}{m} \qquad \text{(eq.8)}$$

Results and Discussion

Influence on Young's modulus

The shear energy generated through compounding and the combined shear and strain energy caused by injection molding in the production of a tensile test specimen as well as the Young's modulus measured through tensile test for both materials is shown in table 5.

Table 5 – shear energy through compounding, shear and strain energy for injection molding and Young's modulus for two compounds

	PE-C	PE-M
norm. shear energy **compounding step 1** (W/kg)	3438	3438
norm. shear energy **compounding step 2** (W/kg)	+ 5218	/
norm. shear energy **compounding** (W/kg)	**= 8656**	**= 3438**
norm. comb. shear/strain energy **injection molding** (W/kg)	**64550**	**64354**
Young's modulus (MPa)	837	854

It can be seen that the Young's modulus for PE-C is slightly lower although the shear energy introduced during compounding is much higher. The reason for that is possibly the very high combined shear and strain energy generated during injection molding which potentially flattens the effect of the second compounding step. It might also be possible that a slight material degradation is beginning to occur during injection molding and that this degradation is slightly lower for a polymer system, which is less exposed.

Influence on O_2-barrier properties

The shear energy generated through compounding and film extrusion as well as O_2-barrier properties are shown in table 6.

Table 6 – shear energy through compounding and shear energy introduced through film extrusion as well as O_2-barrier properties for two compounds

	PE-C	PE-M
norm. shear energy **compounding step 1** (W/kg)	3438	3438
norm. shear energy **compounding step 2** (W/kg)	+ 5218	/
norm. shear energy **film extrusion** (W/kg)	+ 1642	+ 1642
norm. shear energy (W/kg)	**= 10298**	**= 5080**
O_2-permeability (ml/m².100µm bar)	371	404

The energy introduced into the polymer system according to film extrusion is, as expected, by far lower than those of compounding.
As can be seen the permeability for the PE-C is lower than those for the PE-M. The reason for that is that the shear energy is nearly twice as high for PE-C than for the second compound. Therefore the dispersion of the layered silicate is by far lower resulting in an improved O_2- permeability. Material degradation could not be observed.

Influence on scratch behaviour

The specimens used for scratch tests had the same previous processing history then those of the tensile tests so the energy introduced into the polymer system is equal. Table 8 shows the measured scratch hardness as well as the energy introduced through compounding and injection molding.

Table 8 – shear energy through compounding, shear and strain energy for injection molding and scratch hardness for two compounds

	PE-C	PE-M
norm. shear energy **compounding** (W/kg)	**8656**	**3438**
norm. comb. shear/strain energy **injection molding** (W/kg)	**64550**	**64354**
Scratch hardness (MPa)	136	141

It can be seen that the scratch hardness is slightly lower for PE-C although the shear energy during compounding is higher. The reason for that is the same like it was for the Young's modulus and is explained earlier.

Influence on stability of pipes

Table 9 – shear energy through compounding and shear energy through pipe extrusion for two compounds

	PE-C	PE-M
norm. shear energy **compounding step 1** (W/kg)	3438	3438
norm. shear energy **compounding step 2** (W/kg)	+ 5218	/
norm. shear energy **pipe extrusion** (W/kg)	+ 4630	+ 4936
norm. shear energy (W/kg)	**= 13286**	**= 8374**

Table 9 shows the shear energy generated through compounding and pipe extrusion. Stability tests executed on the pipes are currently under progress and will be of major interest. The values measured will on the one hand show if a higher shear energy of 13286 W/kg compared to 8374 W/kg will improve the properties and on the other hand give more information about the existence of an energy plateau on which the properties cannot be raised and are even slightly decreasing because of shear induced degradation of polymer chains.

Conclusions

The investigations show first indications that the amount of shear energy generated during processing is influencing the physical and mechanical properties. An attempt was taken to calculate the shear energy for different processing techniques and make them comparable. It can also be seen that the shear energy generated during injection molding cannot be separated from the strain energy in the die and in the tool and therefore is present in the calculated values in big extent. This fact is responsible for the incomparableness of the injection molding energy values with the normalized shear energy values of all extrusion processing techniques. Nevertheless the shear energy comparison between the extrusion processes shows good compliance with the measured properties.
Further investigations should lay emphasis on a step by step rising of the shear energy generated during production to find an energy plateau on which the property of interest is no longer improving. This would lead to a characteristic ratio for a polymer-compatibilizer-nanofiller system to achieve good homogeneity.
As far as injection molding is concerned a new way of determining only the shear energy during injection should be found to sum this shearing part directly to those of extrusion.

References

[1] Hasook A., et al., "Preparation of nanocomposites by melt compounding polylactic acid/ polyamide 12/ organoclay at different screw rotating speed using a twin screw extruder", Polymer Composites 29 (1), pp. 1-8 (2008)

[2] Thellen C., et al., "Influence of Montmorillonite layered silicate on plasticized poly(L-lactide) (PLA) blown films", Journal of Membrane Science 340 (1-2), pp. 45-51 (2009)

[3] Xuechen W., et al., "Impact of pre-processing of montmorillonite on the properties of melt-extruded thermoplastic starch/montmorillonite nanocomposites", Starch/Staerke 61 (9), pp. 489-494 (2009)

[4] Wang N., et al., "Effect of citric acid and processing on the performance of thermoplastic starch/montmorillonite", Carbohydrate Polymers 76 (1), pp. 68-73 (2009)

MELT COMPOUNDING OF ZIRCONIUM PHOSPHATE (ZrP) LAMELLAR NANOCOMPOSITES USING AQUEOUS SLURRY

Y. Hung*, C. Carrot, Y. Chalamet

Université de Lyon, F-42023, Saint-Etienne, France; CNRS, UMR5223, Ingénierie des Matériaux Polymères, F-42023, Saint-Etienne, France ; Université de Saint-Etienne, Jean Monnet, F-42023, Saint-Etienne, France - yvong.hung@univ-st-etienne.fr; carrot@univ-st-etienne.fr; chalamet@univ-st-etienne.fr

Abstract - This work focuses on the processing of materials obtained from a synthetic lamellar filler, α-zirconium phosphate (α-ZrP), dispersed within a synthetic rubber of the SBR type.
The main topic is the melt compounding using aqueous slurry of α-ZrP. However, this requires to overcome three major issues: at first, strong physical and chemical incompatibility (hydrophilic vs hydrophobic), secondly the high viscosity ratio between the slurry and the polymer matrix (low viscosity vs very high viscosity) and at third, elimination of the aqueous phase.
Processing of the filler intercalated with alkylamine surfactants using a batch mixer was tested. X-ray diffraction and transmission electron microscopy were used to quantify the state of dispersion of the filler within the nanocomposite. Melt or solid dynamic rheological analysis was used to connect different classes of morphologies observed and the viscoelastic behavior of nanocomposites.

Introduction

Polymer nanocomposites have been extensively studied since 1990's. This class of materials exhibits various significant improvements in physical, mechanical, thermal and gas barrier properties. In particular, only a small amount of inorganic nanoplatelets filler, dispersed in the polymer matrix, allow to produce reinforced polymer. However, the efficiency of layered nanofiller on enhanced properties of polymer nanocomposites depends on the combination of parameters like size, aspect ratio, degree of exfoliation, interfacial adhesion filler/matrix and processing. Unfortunately, except some polymers such as polyamide, which present favorable affinity, most of polymers exhibit poor surface affinity with inorganic layered fillers. This lack of affinity leads to poor dispersion of platelets and conventional micro-composites. To overcome this issue, the layered fillers such as the montmorillonite (MMT) are interesting because of their cation exchange capacity (CEC) which allows to modify the surface affinity of platelets by intercalating small organic cations like alkylamines. By the way, the d-spacing between platelets increases and the friendly environment facilitates the access of polymer chains to exfoliate fillers into several individual platelets.
Among the different kind of techniques for nanocomposite preparation, melt processing is the preferred method because of it uses widely spread conventional tools of the plastic industry and it has high producing output.

Hasegawa et al. (2003) reported a novel process to prepare PA6 nanocomposites, using aqueous slurry of unmodified MMT. According to the authors, the exfoliation of platelets is enhanced by the favorable affinity between MMT, PA6 matrix and water but also by an adapted extrusion process.

Recently, some works focused on synthetic α-zirconium phosphate (α-ZrP), $Zr(HPO_4)_2,H_20$ instead of traditional MMT. Compared with MMT, ZrP exhibit a number of additional advantages, including a higher cation exchange capacity, a better controllable aspect ratio and a higher surface functionality. Sue et al. (2003) reported a method to obtain ZrP nanocomposites based on epoxy by exfoliation of platelets with tetrabutylammonium hydroxide in aqueous phase.

Different characterization methods are coupled to analyze the efficiency of platelets dispersion. Among them, X-ray diffraction (XRD) and transmission electron microscopy (TEM) have been widely considered. Nevertheless, these methods have shown several drawbacks due to the local analysis. The association with macroscopic characterization as dynamic mechanical analysis (DMA) allows a structure-property relationship understanding of the nanocomposite.

The present paper describes the preparation of ZrP nanocomposite based SBR by melt compounding using ZrP aqueous slurry. Attention will be placed on studying the processing conditions. The work focuses on the way to overcome three major issues: at first, strong physical and chemical incompatibility (hydrophilic vs hydrophobic), secondly the high viscosity ratio between the slurry and the polymer matrix (low viscosity vs very high viscosity) and at third, elimination of the aqueous phase. Processing of the filler intercalated with alkylamine surfactants using a batch mixer was tested. XRD and TEM were used to quantify the state of dispersion of the filler within the nanocomposite. DMA was used to connect different classes of morphologies observed and the viscoelastic behavior of nanocomposites.

Experimental

Materials

The polymer matrix used in this study is a synthetic copolymer of styrene and butadiene (SBR) from MFP Michelin. This SBR has the following microstructure: 25 wt-% of styrene, 24-% of butadiene 1-2 and 50wt-% of butadiene trans 1-4. The main molecular characteristics are summarized in the table 1. The rheological behavior in the melt can be well described by a power-law with the parameters of table 1.

Table 1 - Main characteristics of SBR matrix

M_n(g/mol)	115200
M_w(g/mol)	225000
Ip	1.96
Density	0.938
*K; n (80°C)	114600; 0.35
*K; n (100°C)	81150; 0.39

*K: consistency; n: shear thinning coefficient

The α-ZrP used is a ZrP aqueous slurry from Rhodia. The slurry content of ZrP is 20 wt%. α-ZrP has layered structure with a basal space composed by zirconium atoms bridged by phosphate groups. The phosphate groups are placed alternatively on each side of the basal space. Globally, the platelets present a negative charge. A hydrogen atom placed between the layers counterbalances the negative charge of the platelets. Moreover, a water molecule is caged in the center of each cavity. The α-ZrP crystal has an interlayer distance of 0.756 nm. α-ZrP exhibits a high cation exchange capacity thanks to the hydrogen mobility. Its CEC is 8 meq/$g_{\alpha\text{-}ZrP}$. The hydrogen mobility allows direct substitution of a variety of intercalants such as alkylamines.

Table 2 - Main characteristics of ZrP

Basal spacing	20 nm x 150 - 200 nm
Platelet thickness	0.63 nm
D-spacing	0.756 nm
Density	2.72 g/cm

In this study, only the aqueous slurry ZrP is used. Consequently, further ZrP term is corresponding to the aqueous slurry α-ZrP (without others precisions). Two alkylamine intercalants are used to modify the surface and the interlayer distances of ZrP. Propylamine is used as a short intercalant and equals the CEC of ZrP in aqueous phase (P100-ZrP). Some water is added to adjust the consistency of the solution. The intercalated P100-ZrP exhibits a whitish gel-like appearance and has the following contents: 15 wt% of ZrP, 78 wt% of water and 7 wt% of propylamine. The D-spacing becomes 1.74 nm corresponding to a trans-trans position of propylamine into the interlayer space perpendicular to ZrP sheet as suggested MacLachlan et al. (1992). Octadecylamine is used as a long alkyl chain and is added to equal 56% of the CEC of ZrP in aqueous phase (O56-ZrP). Some water is added to adjust the consistency of the solution. The intercalated O56-ZrP exhibits a whitish toothpaste-like appearance and has the following contents: 11 wt%, of ZrP, 75 wt% of water and 14 wt% of octadecylamine. The D-spacing measured is 4.56 nm. Dal Pont et al. (Personal Communication, 2009) have shown that 56% of the CEC is sufficient to obtain the highest interlayer space with a trans-trans conformation of octadecylamine chains.

Figure 1 - XRD pattern of pristine ZrP, ZrP modified propylamine P100-ZrP and ZrP modified octadecylamine O56-ZrP.

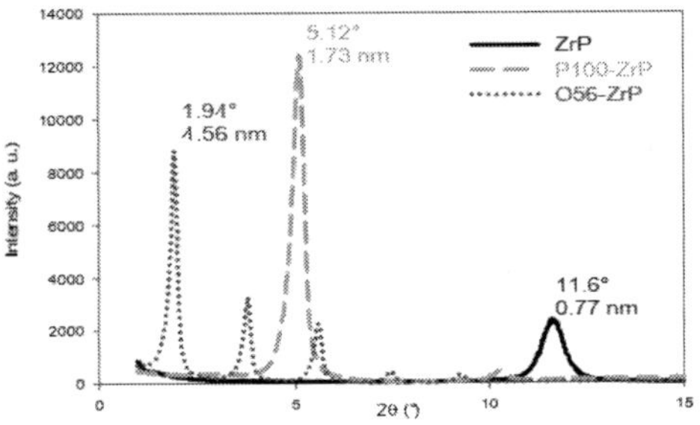

Melt processing and preparation of ZrP nanocomposites based SBR

Melt blending was carried out in a Haake internal batch mixer Rheomix 600 with roller rotor type to achieve ZrP nanocomposites based SBR with 3 vol% (8 wt%). The slurry ZrP and the SBR matrix are incorporated in one step into the mixing chamber. The rotor speed was 80 rpm corresponding to a shear rate of 45 s^{-1} as shown Brouillet-Fourmann et al. (2002). The filling factor was 0.7 and the ratio SBR/slurry was around 55/45. The mixing is first realized at 80°C with water phase during 10 min then the temperature is raised to 110°C to eliminate water by evaporating trough the feeding orifice during 10 min. The same method of preparation was applied for the P-ZrP and the O-ZrP with SBR.

Table 3 – Samples nomenclature

Reference	Nature of filler	Composite obtained
ZrP	Un-modified	SBR_ZrP
P100-ZrP	Modified par propylamine	SBR_P100-ZrP
O56-ZrP	Modified par octadecylamine	SBR_O56-ZrP

Characterization

Before analysis, the samples were molded by compression at 100°C under 150 bars during 45 min.
X-ray diffraction pattern of slurry ZrP, slurry ZrP modified by propylamine and octadecylamine and the nanocomposite samples were obtained for 2θ in the range from 1° to 30°, with a step interval of 0.0217°. Bruker D8 Advance diffractometer using a Cu-Ka radiation source is investigated.
TEM analysis was carried out by a HITACHI H800-3 using an accelerating voltage of 200 kV. Before examination in the microscopy, all specimens were prepared by using ultra cryo-microtome at -110°C. Frozen sections are cut on a MTXL RMC microtome using a diamond knife to obtain section with thickness from 50 to 80 nm.
Dynamic mechanical analysis was performed using a controlled strain dynamic spectrometer ARES from Rheometrics at a fixed frequency of 10 $rad.s^{-1}$ and a temperature sweep of 2°C/min ranging from -80°C to 100°C was applied. Samples were strained from 0.3 to 2.3% within the linear viscoelastic domain. The dynamic storage modulus (G') and loss factor tan δ were plotted against temperature.

Results and Discussion

Compounding

Fig. 2 presents the variation of torque and temperature during mixing of blend systems SBR_ZrP, SBR_P100-ZrP and SBR_O56-ZrP. The torque signal is similar for SBR_ZrP and SBR_P100-ZrP systems. At the feeding of materials, no significant torque signal is observed despite of the matrix presence because of the important volume of aqueous phase ZrP. Because of the high viscosity ratio between matrix and the slurry, the mixing is not efficient and the final blend is composed of a "damp powder" of SBR and slurry ZrP. The same phenomenon is observed with the system SBR_P100-ZrP. If the temperature is raised to 110°C to eliminate water during mixing, the torque increases which corresponds to an efficient mixing and a composite material is obtained with systems SBR_ZrP and SBR_P100-ZrP. The SBR_ZrP presents a dark color whereas the SBR_P100-ZrP presents a similar transparency to that of pure SBR matrix.
The SBR_O56-ZrP system displays a different mixing behavior. As soon as the materials are fed in the mixing chamber, the torque raises and quickly reaches a constant value. When the temperature is increased to eliminate water, a drop of the torque value followed by an increase is observed. At the end of mixing, the SBR_O56-ZrP nanocomposite material is molded like previous samples. The different samples were summarized in the table 3.

Figure 2 - mixing curves for SBR_ZrP, SBR_P100-ZrP and SBR_O56-ZrP. Torque progress and temperature progress against mixing time.

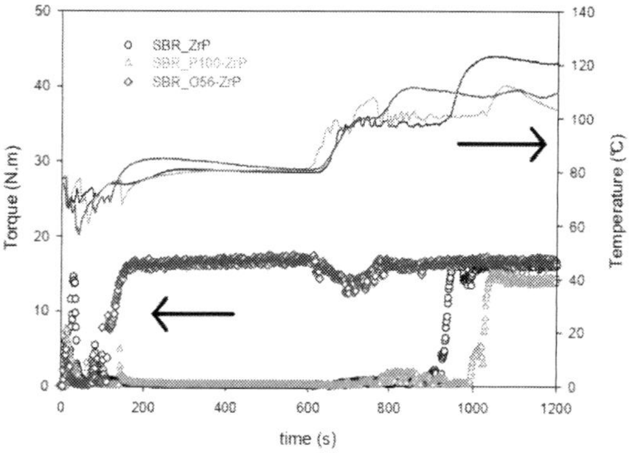

Characterization

Fig. 3a presents XRD pattern of SBR_ZrP system. The pattern does not show any change between pristine ZrP and the studied systems. The un-modified ZrP inside the composites remain with their initial interlayer distance. After water removing and even with a significant torque applied, the un-changed interlayer distance suggests that despite of shear stress, SBR chains could not intercalate the interlayer gap because of unfavorable surface affinity and small gap between platelets.
Fig. 3b displays XRD pattern SBR_P100-ZrP nanocomposite. The pattern reveals a high intensive peak (5.2°) for SBR_P100-ZrP which corresponds to the interlayer distance of ZrP intercalated by propylamine. It suggests that the ZrP intercalated by propylamine predominates. Nevertheless, it is noted that in the pattern of SBR_P100-ZrP, a small broad peak (2.9°) appears. This is not observed on the pattern of P100-ZrP. The broad peak matches with an interlayer distance of 3 nm. It suggests that in the case of SBR_P100-ZrP mixing, after water removal, a few SBR chains intercalate the ZrP interlayer space under shear stress. But the ZrP intercalated by SBR is a minor phenomenon because of the strong interaction between platelets and an un-sufficient affinity between the propylamine intercalants and the matrix.
Fig. 3c shows the pattern of the system SBR_O56ZrP. The pattern of O56-ZrP is also plotted. A very intensive peak is observed at 1.94° which corresponds to the initial interlayer distance of 4.56 nm observed in ZrP intercalated by octadecylamine. No other peak can be observed. Consequently, it can be concluded that the ZrP platelets are mostly modified by octadecylamine. No shear stress effect is observed.

Figure 3 - XRD pattern of a) SBR_ZrP, b) SBR_P100-ZrP, c) SBR_O56-ZrP.

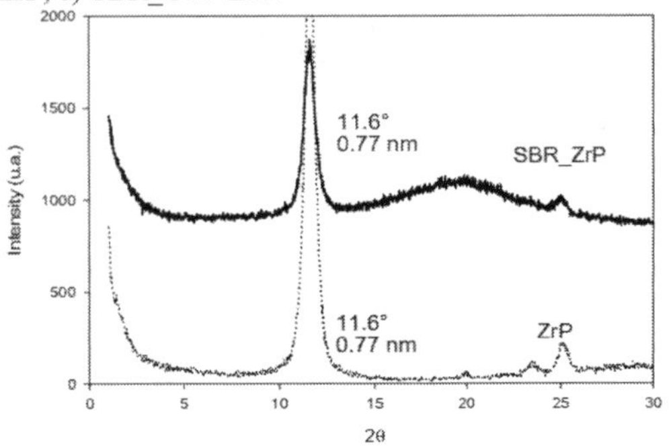

a)

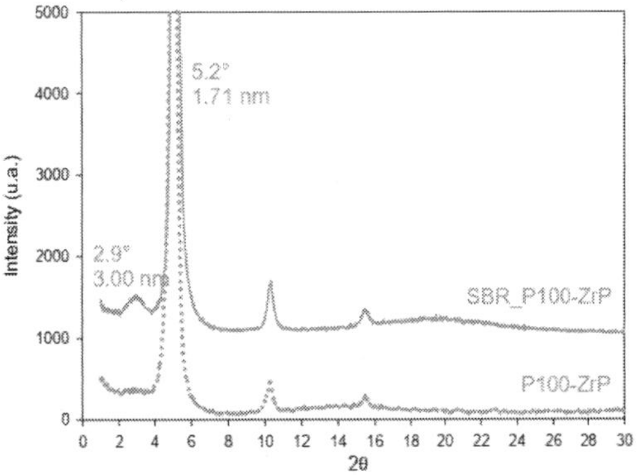

b)

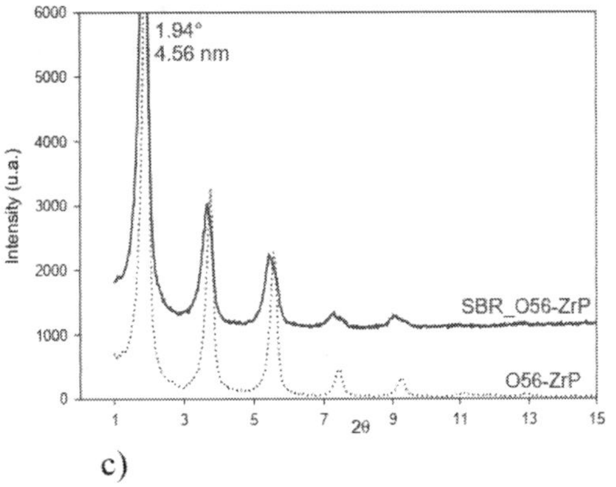

c)

To confirm the state of dispersion of each system studied, TEM investigation was performed. For both systems SBR_ZrP and SBR_P100-ZrP, the ZrP particles appear as more or less well dispersed as shown fig 4a and fig 5a. The dispersion is the best and the most homogeneous with the ZrP treatment by propylamine. Fig 4b presents the morphology of SBR_ZrP with a heterogeneous dispersion and a particle size ranging between 200 nm and 1 μm with an average aspect ratio of 3. For SBR_P100-ZrP, a very good dispersion of ZrP particles is observed with a particle size under 300 nm and an average aspect ratio of 6 as shown fig 5b. Comparing with the SBR_ZrP system, in the case of ZrP treatment by propylamine, the intercalants reduce the interaction between platelet and weaken the global structure. Thus, during mixing, the shear forces break the ZrP particles more easily by peeling the stacks of platelets. At very high magnification (x 200,000, x 600,000), interlayer spaces are observed and the measured distance is 13 nm (fig 5c) and 2-4 nm (fig 5d). The distance of 13 nm was not observed with XRD because it reaches angular limits of the equipment whereas the distance of 2-4 nm matches with the small broad peak observed with XRD.

Figure 4 - TEM picture of SBR_ZrP nanocomposite showing (a) low magnification (x 5,000) and (b) high magnification (x 30,000) of dispersion of ZrP layer particles

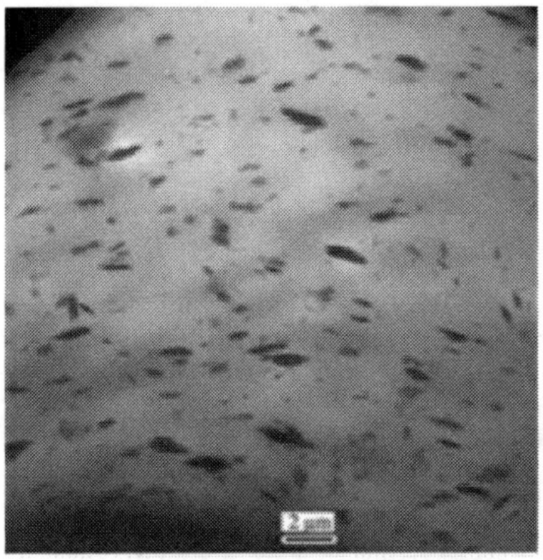

a)

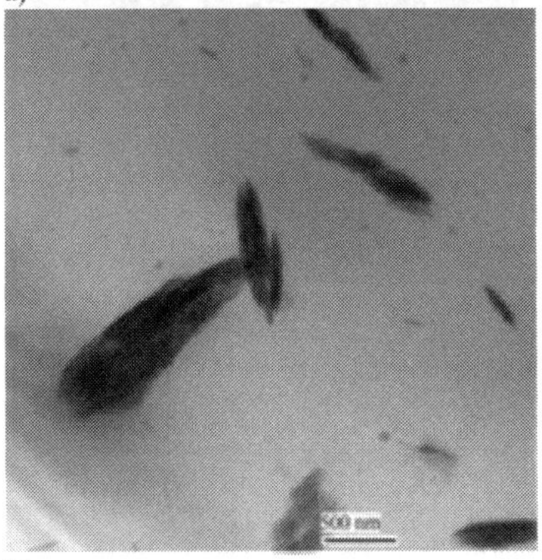

b)

Figure 5 - TEM picture of SBR_P100-ZrP nanocomposite showing (a) low magnification (x 5,000), (b) high magnification (x 30,000) and (c, d) very high magnification (x 200,000 and x 600,000) of dispersion of ZrP layer particles.

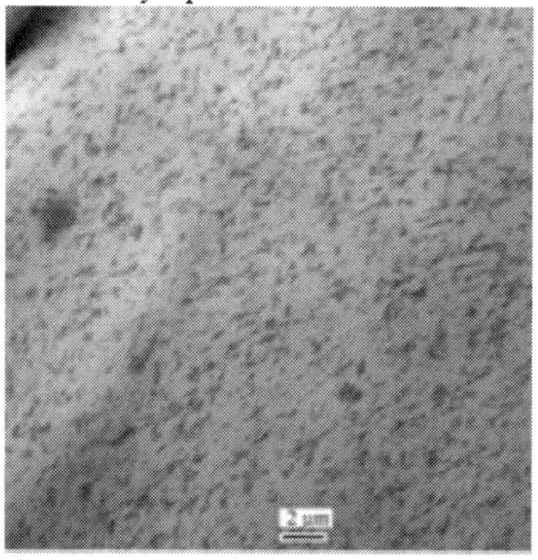

a)

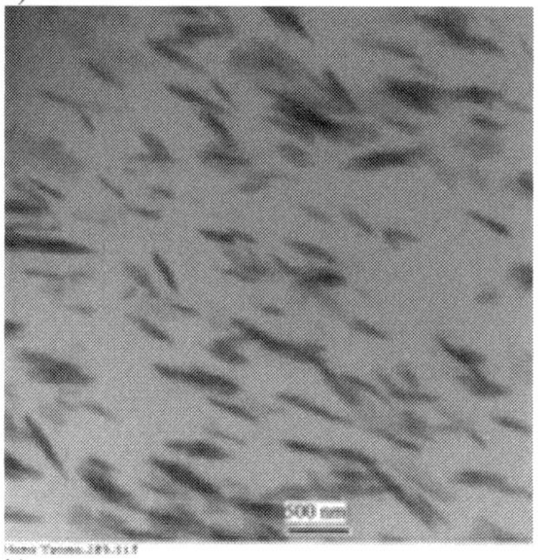

b)

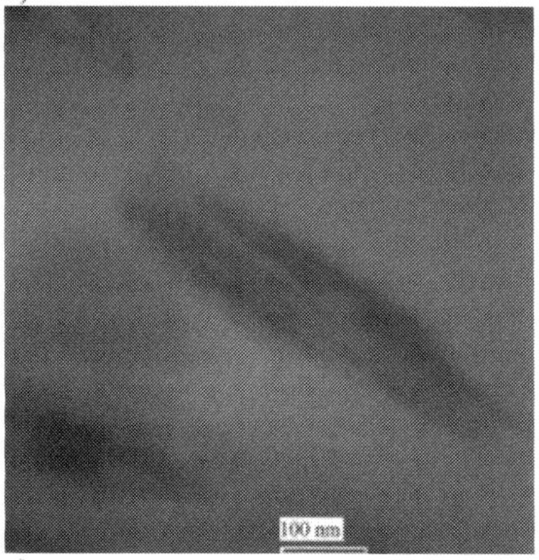

c)

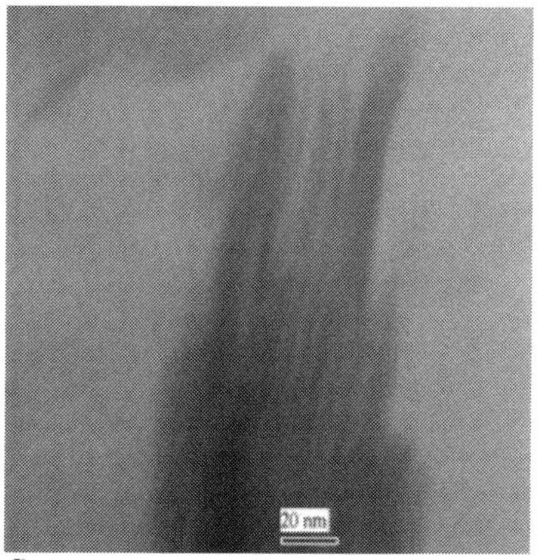

d)

Fig 6 presents TEM pictures of system SBR_O56-ZrP. At low magnification (x 5,000), the dispersion state seems homogeneous without large agglomerates. At higher magnification (x 30,000), it appears the dispersion is nice with a particle size under 500 nm and an average aspect ratio of 6. Some stacks of ZrP platelets still remain. At very high magnification (x 200,000 and x 600,000), the interlayer space is visible and the distance measured is 3 nm which is closed to the distance measured with XRD (4.56 nm). This corresponds to the interlayer distance of ZrP modified by octadecylamine. No larger interlayer space is observed which suggests that the ZrP platelets are not intercalated by SBR matrix chains. This is due to the weak miscibility between octadecylamine and the matrix.

Figure 6 - TEM picture of SBR_O56-ZrP nanocomposite showing (a) low magnification (x 5,000), (b) high magnification (x 30,000) and (c, d) very high magnification (x 200,000 and x 600,000) of dispersion of ZrP layer particles.

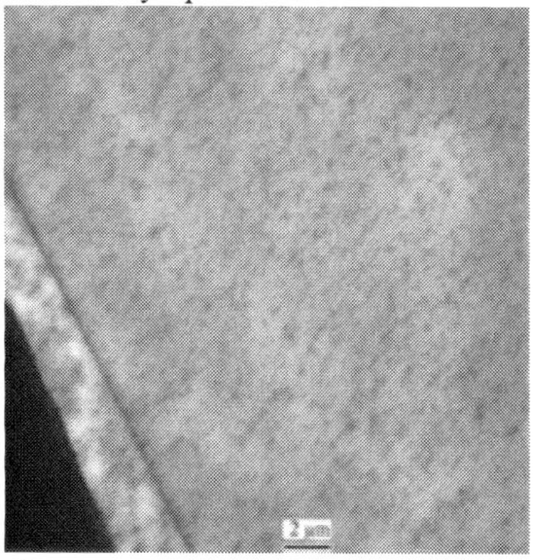

a)

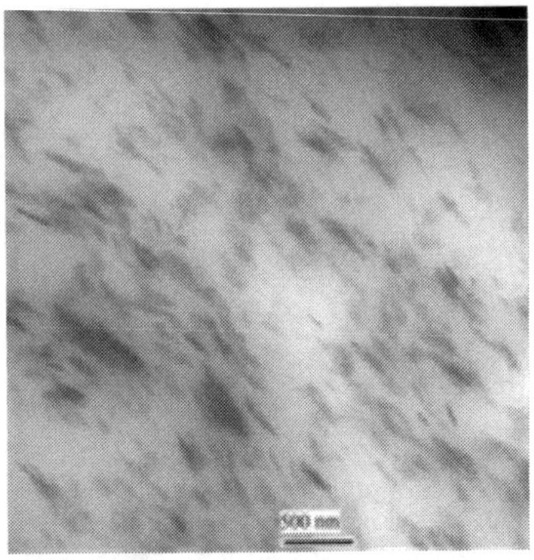

b)

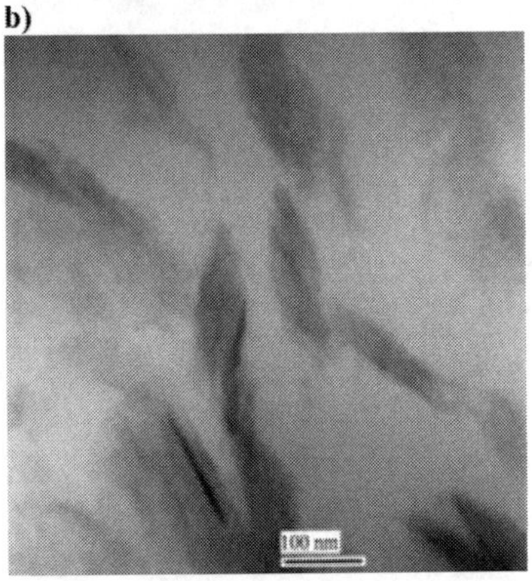

c)

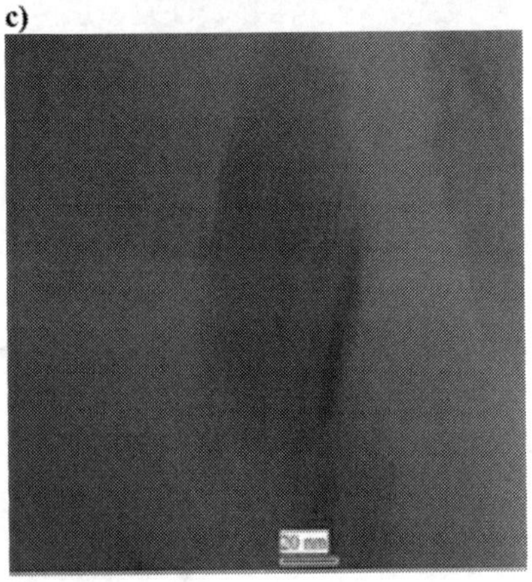

d)

Fig 7 shows the storage modulus G' against temperature for the SBR matrix, SBR_ZrP, SBR_P100-ZrP and SBR_O56-ZrP. A blend of SBR_octadecylamine with the same content of octadecylamine (10 wt%) in the system SBR_O56-ZrP is obtained and its modulus is also plotted to assess the octadecylamine influence on the nanocomposite behavior. SBR_ZrP and SBR_P100-ZrP have similar behavior. In particular, an increase of G' is observed in comparison to the neat matrix. This increase is more significant around -20°C. It is the same with unmodified ZrP and ZrP modified by propylamine for the same 8 wt-% (3 vol-%) of ZrP. Therefore, the treatment of ZrP does not impact the behavior of the nanocomposite. The G' enhancement is only related to the ZrP volume ratio. The system SBR_O56-ZrP presents a larger increase of G' despite of the same ZrP ratio. For the SBR_octadecylamine blend, an improvement of G' is also observed. This difference is nevertheless lower than with the system SBR_O56-ZrP. In fact, the increase G' of SBR_O56ZrP is due to a combined effect of octadecylamine and ZrP on the volume fraction. Fig 8 presents tan δ against temperature of SBR neat matrix, SBR_ZrP, SBR_P100-ZrP, SBR_O56-ZrP and the blend of SBR_octadecylamine. Except for the SBR neat matrix, a drop of maximum of tan δ is observed for all systems. The drop is smaller with SBR_ZrP and SBR_P100-ZrP. However, for SBR_O56-ZrP, the important drop is induced by the presence of octadecylamine. Finally, the drop of maximum of tan δ can be attributed to hindrance of the mobility of ZrP particles by closed SBR chains.

Figure 7 - Dynamic storage modulus G' of SBR neat matrix, SBR_ZrP, SBR_P100-ZrP, SBR_O56-ZrP and SBR_ctadecylamine blend.

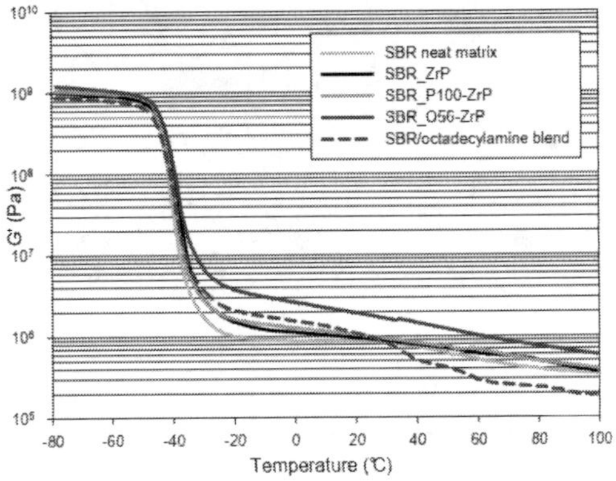

Figure 8 - Loss factor tan δ of SBR neat matrix, SBR_ZrP, SBR_P100-ZrP, SBR_O56-ZrP and SBRàctadecylamine blend.

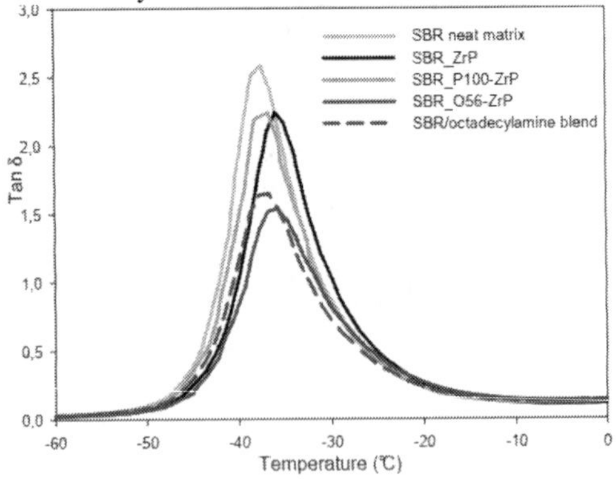

Conclusions

Preparation of ZrP nanocomposites based SBR using aqueous slurry ZrP was carried out in an internal batch mixer. ZrP treatment by propylamine and octadecylamine within the slurry were also investigated. For both systems studied, an increase of the processing temperature over 110°C is required to eliminate water and to obtain the nanocomposite materials. Depending on the treatment of ZrP, materials having different states of dispersion are obtained. The nanocomposite based on un-modified ZrP shows a heterogeneous dispersion of ZrP particles without any intercalation because of the strong interactions between platelets but also because of the un-favorable affinity between the ZrP surface and the SBR matrix. The SBR_P100-ZrP presents a very homogeneous fine dispersion of the ZrP aggregates without reaching exfoliation. The ZrP treatment by propylamine reduces the pristine ZrP strong platelet interaction and weakens the global cohesion of ZrP stacks. Thus, after removing water during processing, the shear stress breaks more easily the ZrP particles by peeling stacks of platelets. Moreover, a few amount of ZrP layer are intercalated by SBR matrix under shear stress. The third nanocomposite SBR_O56-ZrP also shows homogeneously dispersed ZrP aggregates without obtaining ZrP exfoliation despite of the intercalation of octadecylamine. Moreover, no intercalation by SBR matrix can be observed. This might be caused by the weak miscibility between octadecylamine and the matrix. For both three systems concerned, the increase of dynamic storage modulus on the rubber plate and the drop of the maximum of tan δ are only connected to the volume ratio of ZrP in the nanocomposite.

Acknowledgements

This work was sponsored by the Pôle de compétitivité AXELERA, the CNRS, MFP MICHELIN and RHODIA for which grateful acknowledgement is made.
The TEM acquisition was supported by the Centre de Microscopie Electronique Stéphanois (St-Etienne)
The XRD acquisition was supported by the Centre de Diffractométrie Henri Longchambon (Lyon)

References

Journal:

Boo, W. J. et al., "Effect of nanoplatelets dispersion on mechanical behavior of polymer nanocomposites", Journal of Polymer Science: Part B: Polymer Physics, **45**, 1459-1469 (2007)

Boo, W. J. et al., "Morphology and mechanical behavior of exfoliated epoxy/α-zirconium phosphate nanocomposites", Composites Science and technology, **67**, 262-269 (2007)

Brandão, L. S. et al., "Thermal and mechanical properties of poly(ethylene terephtalate)/lamellar zirconium phosphate nanocomposites", Journal of applied polymer science, **102**, 3868-3876 (2006)

Brouillet-Fourmann, S. et al., "On the use of an internal mixer for the rheological characterization of maize starch", Applied Rheology, **12**, 192-199 (2002)

Casciola, M. et al., "Gels of zirconium phosphate in organic solvents and their use for the preparation of polymeric nanocomposites", J. Mater. Chem., **15**, 4262-4267 (2005)

Cho, J. W. and Paul, D. R., "Nylon 6 nanocomposites by melt compounding", Polymer, **42**, 1083-1094 (2001)

Dennis, H. R. et al., "Effect of melt processing conditions on the extent of exfoliation in organoclay-based nanocomposites", Polymer, **42**, 9513-9522 (2001)

Fornes, T. D. et al., "Effect of organoclay structure on nylon 6 nanocomposite morphology and properties", Polymer, **43**, 5915-5933 (2002)

Giannelis, E. P., "Polymer layered silicate nanocomposites", Advanced materials, 8, 29-35 (1996)

Hasegawa, N. et al., "Nylon 6/Na-montmorillonite nanocomposites prepared by compounding Nylon 6 with Na-montmorillonite slurry", Polymer, **44**, 2933-2937 (2003)

Kim, H.-N. et al., "Characterization of zirconium phosphate/polycation thin films grown by sequential adsorption reactions", Chem. Mater., **9**, 1414-1421 (1997)

MacLachlan, D. J. and Morgan, K. R., "Phosphorus-31 solid-state NMR studies of the structure of amine-intercalated α-zirconium phosphate. Titration of α-zirconium phosphate with n-propylamine and n-butylamine", J. Phys. Chem., **96**, 3458-3464 (1992)

Russo, G. M. et al., "Rheological and mechanical properties of nylon 6 nanocomposites submitted to reprocessing with single and twin screw extruders", Polymer Degradation and Stability, **92**, 1925-1933 (2007)

Schön, F. et al., "Shear controlled morphology of rubber/organoclay", Macromol. Symp., **189**, 105-110 (2002)

Sue, H.-J. et al., "Epoxy nanocomposites based on the synthetic α-zirconium phosphate layer structure", Chem. Mater., **16**, 242-249 (2004)

Sue, H.-J. Et al., "Fracture behavior of α-zirconium phosphate-based epoxy nanocomposites", Acta Materiala, **52**, 2239-2250 (2004)

Sun, L. et al., "Barrier properties of model epoxy nanocomposites", Journal of membrane science, **318**, 129-136 (2008)

Sun, L. et al., "Effect of cristallinity on the intercalation of monoamine in α-zirconium phosphate layer structure", Chem. Mater., **17**, 5606-5609 (2005)

Sun, L. et al., "Preparation of exfoliated epoxy/α-zirconium phosphate nanocomposites containing high aspect ration nanoplatelets", Chem. Mater., **19**, 1749-1754 (2007)

Sun, L. et al., "Preparation of intercalating agent-free epoxy/clay Nanocomposites", Polym. Eng. Sci., **47**, 1708-1714 (2007)
Vaia, R. A. and Giannelis, E. P., "Lattice model of polymer melt intercalation in organically-modified layered silicates", Macromolecules, **30**, 7990-7999 (1997)

Supercritical Carbon Dioxide as an Exfoliating Agent in the Preparation of a Layered Silicate Polymer Nanocomposite

J. Liu, W. R. Rodgers, P. D. Fasulo*, and M. R. Thompson*

Department of Chemical Engineering, McMaster University, Hamilton, Ontario, Canada

** General Motors R&D and Planning Facilities, Warren, Michigan, USA*

ABSTRACT

This work examined the influence of supercritical CO_2 in the production of a highly exfoliated thermoplastic polyolefin (TPO) nanocomposite using a high pressure batch mixing system. The polymer matrix, organoclay and compatibilizer were melt compounded under extensional flow in the presence of either CO_2 or nitrogen at various conditions. Morphology of the resulting nanocomposites was analyzed using X-ray diffraction (XRD) and rheological analysis. Evidence pointed to the need to control the plasticizing aspects of the supercritical fluid to allow for greater intercalation and exfoliation.

INTRODUCTION

Supercritical CO_2 ($scCO_2$) has been studied recently as an exfoliating aid for the production of polymer layered silicate (PLS) nanocomposites [1-8]. Technologies used in these studies can generally be divided into four categories. The first and most extensively pursued is direct gas injection into a mixing device where all components are melt compounded, with the expectation of improved intercalation and exfoliation through the introduction of the supercritical fluid [1-3]. Another approach focuses on pre-exfoliating the layered silicate filler using the supercritical fluid. Gulari et al. used a technique known as 'particles from gas-saturated solutions/suspensions (PGSS)' in the hopes of fracturing the layered structure of an organoclay before it was added to the compounding process [4, 5]. Baird et al. modified the second technology from a batch to a continuous process [6]. The third method is recently proposed by Rodgers et al. [7] who added the supercritical fluid and organoclay together into a highly pressurized zone of the compounding process which maintains the CO_2 gas in its supercritical state during mixing, at least for the initial stages. The last approach, proposed by the authors, is batch processing a mixture of clay and compatibilizer with $scCO_2$ and compounding the mixture into a polymer matrix [8]. All of the methods refer to improved exfoliation. Each of these methods varies the way the different components of the nanocomposite are exposed to the supercritical fluid; not all components are exposed simultaneously except by the method of direct gas injection. This work examines whether there is significance to how the different components are exposed to CO_2 in order for it to be an effective exfoliating aid.

EXPERIMENTAL

A medium-to-high flow thermoplastic olefin (TPO) (13 MFR) supplied by LyondellBasell, an organoclay, Cloisite 20A (C20A) from Southern Clay Products, and a maleated polypropylene, Polybond® 3200 (PB3200) from Chemtura Corporation were selected as the components for the nanocomposite in this study. The three materials were melt compounded in an environment of either $scCO_2$ or N_2 using extensional flow in a custom-built batch mixer. The mixer has two chambers which allows for the components to be separated as desired before compounding. The different combinations for separation of the three components are shown in Table 1. Materials listed in a pair of brackets means that they were put together in one chamber and the un-bracketed component was placed in the other chamber. In all cases, the composition of the final nanocomposite was 90:5:5 (%w/w) for the matrix, compatibilizer, and clay, respectively. The three components were heated for approximately 50 minutes under an environment of either $scCO_2$ or N_2 until the apparatus reached 200°C. Mixing was conducted over a 10 minute period by moving the heated components back and forth between the two chambers through an orifice plate. The morphology of the prepared nanocomposites is discussed as determined by X-ray diffraction (XRD) and by rheology measured using a parallel plate rheometer.

Table 1 Design of experiments using the mini-mixer

Mixture state during heating	Gas during heating (8.3MPa, 200°C)
(TPO + PB3200 + C20A)	CO_2 or N_2
TPO + (PB3200 + C20A)	CO_2 or N_2
(TPO + PB3200) + C20A	CO_2 or N_2

RESULTS AND DISCUSSION

XRD analysis

Figure 1 displays the XRD patterns of the three pairs of nanocomposites prepared under either $scCO_2$ or N_2 conditions. Compared to their corresponding N_2 case, the (001) diffraction peaks for both (TPO+PB3200+C20A) and TPO+(PB3200+C20A) formed under $scCO_2$ showed a shift to lower 2θ angle by 0.3° and 0.5°, respectively indicating larger basal spacing. Differences for (TPO+PB3200)+C20A between the CO_2 and N_2 cases were not readily apparent (Fig. 1c). These results showed that the supercritical gas could aid the destructuring of clay, but that the manner of combining the three components in its presence was a significant concern. Based on the detected changes in basal spacing, the favored order seemed to be TPO+(PB3200+C20A) > (TPO+PB3200+C20A) > (TPO+PB3200)+C20A.

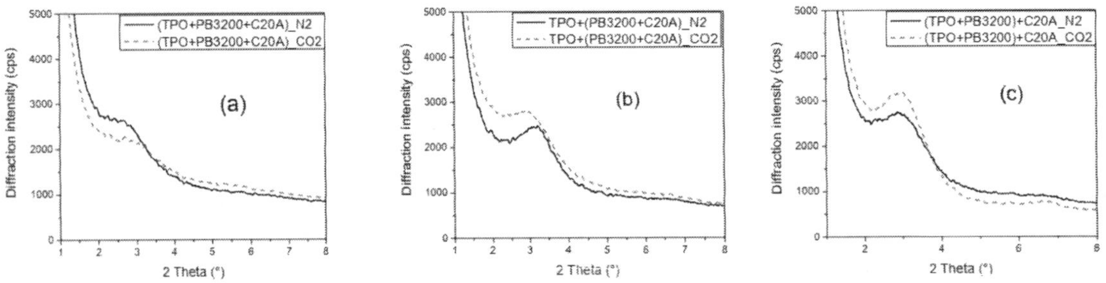

Figure 1. XRD patterns of the TPO nanocomposites prepared under $scCO_2$ and N_2 with formulation

Rheological analysis

The storage moduli of these composite melts are shown in Figure 2. In the low frequency region, all three $scCO_2$-based nanocomposites exhibited higher storage moduli than their N_2-counterparts, indicating better development of the percolating network of clay within the TPO, confirming a significant effect of the CO_2. These results help clarify the XRD observations. For the $scCO_2$ cases, the storage modulus of TPO+(PB3200+C20A) in the low shear rate region was significantly higher (75% greater at 0.3 rad/s) than the other two combinations which were similar. This result implies that greater clay dispersion occurred when the compatibilizer and organoclay were first heated together under CO_2; that is the $scCO_2$ was localized at the PB3200/C20A interface during melting. For the N_2 cases, there was no significant difference in modulus curves between the three combinations, indicating that the manner by which the three components were combined had no effect on intercalation and exfoliation of the nanocomposite prepared without the presence of $scCO_2$.

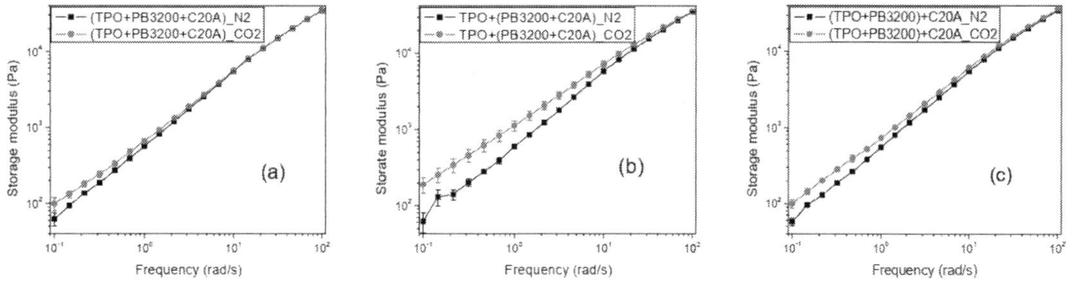

Figure 2. Comparisons in storage moduli of the three pairs of TPO nanocomposites

CONCLUSIONS

Supercritical CO_2 was found to significantly improve the level of exfoliation for the organoclay, Cloisite 20A, when dispersed in a TPO nanocomposite. When $scCO_2$ was present, the manner in which the three components making up our nanocomposite were combined had an effect on the modulus of the final material. Concentrating CO_2 at the compatibilizer/clay interface was more effective in this regards than when these components were annealed separately before mixing or when the TPO matrix was also present with the clay and compatibilizer to dilute the effectiveness of the CO_2.

ACKNOWLEDGEMENTS

This work was made possible by funding from the Natural Science and Engineering Research Council (NSERC) of Canada and General Motors of Canada Limited (GMCL). The authors wish to especially thank Dr. Doug Hunter from Southern Clay Products for his assistance in the preparation of the organoclays as well as for his technical advice, Ingenia Polymers for the cryogenic grinding of the compatibilizer, and LyondellBasell for providing TPO materials.

REFERENCES

1. Y. Zhao, H.-X. Huang. Polymer Testing 2008, 27, 129-134
2. M.A. Treece, J.P. Oberhauser, J Appl. Polym. Sci. 2007, 103, 884-892
3. J. Ma, E. Bilotti, T. Peijs, J. A. Darr. European Polym. J. 2007, 43, 4931-4939
4. S. Horsch, G. Serhatkulu, E. Gulari, R.M. Kannan, Polymer 2006, 47, 7485-7496
5. C. W. Manke, E. Gulari, D. F. Mielewski, E.C. Lee. US Patent 6,469,073 (2002)
6. Q. T. Nguyen, D. G. Baird. Polymer 2007, 48, 6923-6933
7. R. A. Ottaviani, W. R. Rodgers, P. D. Fasulo, D. A. Okonski. US Patent US 7,462,666 B2, Dec. 9, 2008
8. M.R. Thompson, W. R. Rodgers, P.D. Fasulo, J. Liu, Z. Zhuang, Method of Producing Nanocomposite Polymer Using Compatibilizer and Supercritical Fluid, US Patent, Applied, Dec. 2009

PREPARATION AND CHARACTERIZATION OF CONDUCTING POLYMER/CARBON NANOTUBE COMPOSITES

Y. W. Lin, S. H. Lin and T. M. Wu*,

Department of Materials Science and Engineering, National Chung Hsing University, Taichung, Taiwan--
tmwu@dragon.nchu.edu.tw; *d9366104@mail.nchu.edu.tw*

This presentation reports the synthesis and characterization of conducting polymer with multi-walled carbon nanotubes (MWCNT) by *in situ* chemical oxidation polymerization. Raman, FTIR, FESEM and HRTEM were used to characterize their structure and morphology. Structural analysis by FESEM and HRTEM showed that the conducting polymer/MWCNT composites are core-shell structures with various diameters of nanometers, depending on the content of conducting polymer. The electrical conductivities of these composites are higher than those of conducting polymer without MWCNTs.

Introduction

Since the discovery of carbon nanotube (CNTs) in 1991 by Iijima [1], CNTs have been extensively studied by researchers in various fields such as chemical, physical, and materials science. CNTs have unique nanostructure with remarkably mechanical, thermal and electrical properties [2], which made them highly attractive for the use as reinforcement in nanotube-based composite materials. Recently, extensive efforts have been made to prepare functional conducting polymer/CNT composites, which are expected to exhibit excellent electrical conductivity and optical properties compared to the original conducting polymer matrix [3, 4].
Among virous conducting polymers, polyaniline (PANI) and polypyrrole (Ppy) have potential uses in synthesizing conducting polymer/CNT composites owing to its environmental stability and good processability. This presentation reports the synthesis of PANI and PPy with multi-walled carbon nanotubes (MWCNT) by in situ chemical oxidation polymerization. To improve the dispersion and homogeneity of the MWCNTs within the conducting polymer matrix, the as-prepared MWNCTs, which contained carboxylic acid groups (designated as c-MWCNTs) at the defect sites, were treated using a 3:1 mixture of concentrated H_2SO_4:HNO_3. The structure, morphology and conductivity of PANI/c-MWCNT and PPy/c-MWCNT composites were characterized.

Experimental

Synthesis of PANI/c-MWCNT composites

The MWCNTs used in this study were synthesuzed by ethylene CVD using Al_2O_3 supported Fe_2O_3 catalysts. The purity of MWCNTs was highrt than 90% after purification. The as-prepared MWNCTs were ultrasonically treated using a 3:1 mixture of concentrated H_2SO_4 and HNO_3 at 50℃ for 24 h, producing carboxylic acid groups at the defect sites and thus improving the solubility of the c-MWCNTs in HCl solution.
The composite of PANI with c-MWCNT was synthesized via in situ chemical oxidation polymerization. In a typical synthesis experiment, the c-MWCNTs were dissolved in 1.0 M HCl solution and ultrasonicated over 3 h, then transferred into a 500 ml four-neck flask with an ice-bath. Aniline monomer also dissolved in 1.0 M HCl solution was added to the above c-MWCNTs suspension. A 1.0 M HCl solution containing 0.125 M ammonium persulfate was slowly added dropwise into the suspension with constant mechanical stirring at a reaction temperature of 0-5℃ for 1 h. The reaction mixture was stirred for a further 2 h at 0-5℃, and then the resulting green suspension, indicating the formation of insoluble PANI in its emeraldine salt form, was filtered and rinsed several times with distilled water and methanol. The powder of PANI/c-MWCNT composites thus obtained was vacuum dired at 60℃ for 24 h.

Characterization of PANI/c-MWCNT composites

The ATR-FTIR spectroscopy, UV-Vis spectroscopy and Raman spectroscopy were used to characterize the structure of PANI/c-MWCNT composite. The field-emission scanning electron microscopy (FESEM) and high-resolution transmission electron microscopy (HRTEM) were used to characterize the morphology of PANI/c-MWCNT composite.

Results and Discussion

Fig. 1 illustrates the FTIR spectra of MWCNTs and c-MWCNTs. As compared with the spectrum of

MWCNTs (Fig. 1a), the new peaks at around 1718 and 1186 cm^{-1} were from the stretching vibration of C=O and C-O groups in c-MWCNTs (Fig. 1b), respectively. This result indicated that the oxygen-containing functional groups, such as hydroxyl (-OH), carboxyl (-COOH), and carbonyl (-C=O) have been introduced on the surface defects and more hydrophilic surfaces have been formed after treatment of chemical oxidation [5].

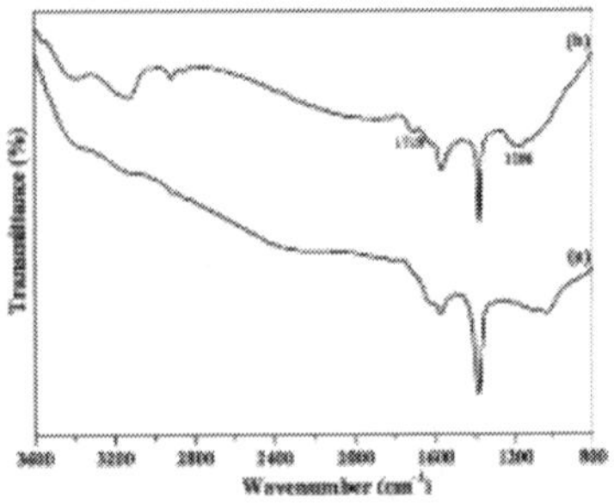

Figure 1 FTIR spectea of (a) MWCNTs and (b) c-MWCNTs.

To verify the structure of the synthesized composites, Fig. 2 presents typical field-emission scanning electron microscopy (FESEM) and high-resolution transmission electron microscopy (HRTEM) images of 3 wt% MWCNT-containing PANI/c-MWCNT composites. Clearly, a tubular layer of a highly uniformly coated PANI film is present on the c-MWCNT surface and the diameters of the PANI/c-MWCNT composites range from several tens to hundreds, depending on the PANI content. Therefore, this composite has a typical core-shell structure, and the c-MWCNT serves as the core and is dispersed individually into the PANI matrices. The aniline monomer is uniformly polymerized on the surface of the c-MWCNT and forms a tubular shell of the composites.

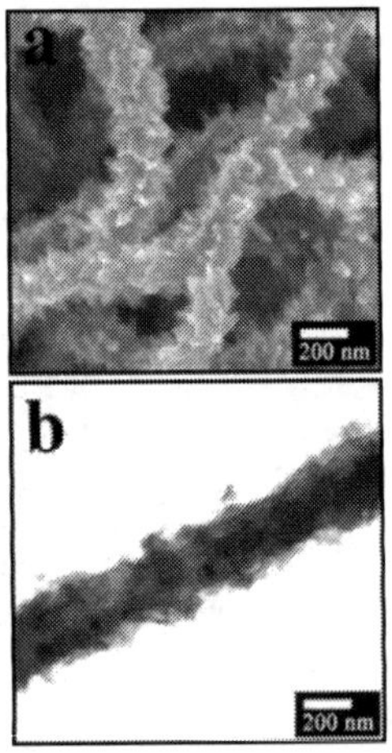

Figure 2 (a) FESEM and (b) HRTEM images of 3 wt% MWCNT-containing PANI/ c-MWCNT composites

Fourier transform infrared (FTIR) and Raman spectra were used to characterize the molecular structure of the resulting PANI/c-MWCNT composites. Obviously, these FTIR and Raman spectra of the composites are approximately the same to those of conducting polymer, supporting the idea that MWCNTs served as the core in the formation of a coaxial nanostructure for the composites. In addition, the electrical conductivities of conducting polymer and composites were measured with the standard Van Der Pauw dc four-probe method [6]. The conductivities of c-MWCNT and protonic acid-doped PANI in its conductive emeraldine salt form at room temperature were 200 and 29.6 S/cm, respectively. Meanwhile, the conductivities of 0.5, 1 and 3 wt% MWCNT-containing PANI/c-MWCNT composites were 48.0, 51.8, and 58.4 S/cm, respectively. The electrical conductivities of all composites with low c-MWCNT content at room temperature are higher than those of PANI without MWCNTs, perhaps because MWCNTs have a large aspect ratio and surface area, and so could serve as a "conducting bridge" between the PANI conducting domains, increasing the effective percolation.

Figures 3a-3c present TEM images of external doping of sulfonated PANI (ED-SPANI)/c-MWCNT composites with various c-MWCNT to explicate the nanostructure of these composites. Clearly, numerous tubular structures were randomly and individually distributed in the ED-SPANI matrix, implying that c-MWCNTs do not aggregate during the solution mixing procedure. Additionally, these tubular composites formed a network in the ED-SPANI matrix. This network might act as a new conductive bridge to enhance the conductivity of composites. To further confirm the nanostructure of composites, Figures 3d-3f show HRTEM images of ED-SPANI/c-MWCNT composites. The resulting ED-SPANI/c-MWCNT composites clearly have coaxial tubular structures. These figures reveal that the surface of c-MWCNTs was coated by a thin layer of ED-SPANI. The thickness of the ED-SPANI layer on the c-MWCNT surface was about 10-18 nm estimated from the HRTEM images. These results demonstrate that ED-SPANI molecules were initially absorbed on the surface of c-MWCNTs by the electrostatic interaction, and the polymer chains were then encapsulated to form a thin layer of ED-SPANI. Therefore, these ED-SPANI/c-MWCNT composites had the typical core-shell structures, and the oxidized MWCNTs served as the core, and were dispersed separately into the ED-SPANI matrix.

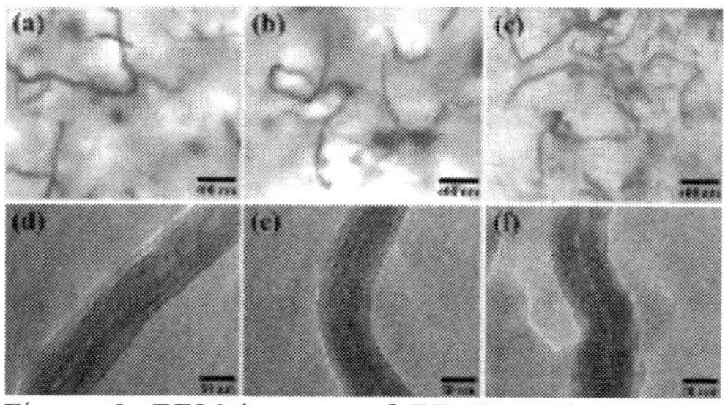

Figure 3: TEM images of ED-SPANI/c-MWCNT composites containing (a) 1 wt% c-MWCNTs, (b) 3 wt% c-MWCNTs and (c) 5 wt% c-MWCNTs. HRTEM images of ED-SPANI/c-MWCNT composites containing (d) 1 wt% c-MWCNTs, (e) 3 wt% c-MWCNTs and (f) 5 wt% c-MWCNTs.

Conclusions

PANI/c-MWCNT composites were prepared via the in situ polymerization. MWCNTs that contained carboxylic acid groups were used as core in the formation of tubular shells of PANI/c-MWCNT composites.Moreover, the conductivities at room temperature of 0.5 wt% MWCNT-containing PANI/ c-MWCNT composites are about 50% higher than that of PANI without MWCNTs. That is because the addition of as little as 0.5 wt% MWCNT could be randomly dispersed in PANI matrix and serve as a "conducting bridge" between the PANI conducting domains, thus increasing the effective percolation.

Acknowledgements

The financial support provided by National Science Council through the project NSC96-2212-E-005-049-MY3 is greatly appreciated.

References (Times New Roman 10, bold)

1. Iijima, S. Helical microtubules of graphitic carbon. *Nature*, **354** (1991) 56-58.
2. Baughman, R. H. et al. Carbon nanotubes-the route toward applications. *Science*, **297** (2002) 787-792.
3. Ago, H. et al. Composites of carbon nanotubes and conjugated polymers for photovoltaic devices. *Adv. Mater.*, **11** (1999) 1281-1285.
4. Kymakis, E. and Amaratunga, G. A. Single-wall carbon nanotube/conjugated polymer photovoltaic devices. *Appli. Phys. Lett.*, **80** (2002) 112-114.
5. Woo, H. S. et al. Organic light emitting diodes fabricated with single wall carbon nanotubes dispersed in a hole conducting buffer: the role of carbon nanotubes in a hole conducting polymer. *Synth. Metal.*, **116** (2001) 369-372.
6. Zhang, J. et al. Effect of chemical oxidation on the structure of single-walled carbon nanotubes. *J. Phys. Chem. B*, **107** (2003) 3712-3718.
7. Van Der Pauw, L. J. A method of measuring specific resistivity and Hall effect of disc or arbitrary shape. *Philips Research Report*, **13** (1958) 1-9.

SEMI-SOLID STATE DEFORMATION OF POLYETHYLENE TEREPHTHALATE/ MULTIWALL CARBON NANOTUBE NANOCOMPOSITES

B. Mayoral[1]*, P.R. Hornsby[1] and T. McNally[1]*

[1]*School of Mechanical and Aerospace Engineering, Queen's University Belfast, Belfast BT9 5AH, United Kingdom – bmayorallahuerta01@qub.ac.uk ; t.mcnally@qub.ac.uk; peter.hornsby@qub.ac.uk .*

The effect of semi-solid state deformation on the dispersion of multiwall carbon nanotubes (MWCNT) in a polyethylene terephthalate (PET) matrix and on composite mechanical, thermal and electrical properties was investigated. PET/MWCNT composites were prepared via melt compounding in a twin-screw extruder. The MWCNTs formed a well-dispersed and highly interconnected structure in the PET matrix as observed from SEM, HRTEM and WAXD studies. As the weight percent addition of MWCNTs to the PET matrix increased, efficient MWCNT dispersion became more difficult to achieve and the MWCNTs tended to form agglomerates. Uniaxial and biaxial deformation of the PET nanocomposites induced orientation of the CNTs in the stretching direction and improved CNT distribution and dispersion throughout the PET matrix. A considerable improvement in mechanical, thermal and electrical properties of PET was achieved with addition of low loadings of MWCNTs. A further increase in mechanical properties was obtained due to the enhanced dispersion of MWCNTs in the PET/MWCNT composites after uniaxial and biaxial stretching. An electrical percolation threshold was attained for a MWCNT loading of 0.8 wt%. As the stretch ratio increased, electrical conductivity decreased because the distance between CNTs was bigger than the minimum distance required for electron hopping, a consequence of the destruction of the CNT network and electrically conducting pathways.

Introduction

Since Ijima (1991) first identified carbon nanotubes (CNT), they have attracted great interest in the academic and industrial world. CNTs exhibit extraordinary electrical, mechanical and thermal properties: elastic modulus of ~1TPa, thermal conductivity twice as high as diamond, and electric conductivity a thousand times higher than that of copper wire (Bin et al.,2003). The interest in the area of polymer/CNT composites increased dramatically since 1994 when Ajayan et al. (1994) first described them. Over the last two decades, extensive research has been undertaken to develop high performance polymer nanocomposites including the incorporation of nanoreinforcements into polymer matrices. The addition of CNTs to polymers can increase stiffness, strength, electrical and thermal conductivity of the composite materials relative to the neat polymer. However, there are some limitations due to the propensity of CNTs to form aggregates which make it more difficult to disperse and distribute CNTs in polymer matrices. These aggregates are formed because CNTs are strongly affected by Van der Waal's attraction due to their small size and large surface area. Such aggregation presents a major problem irrespective of the method of composite preparation, therefore control of de-agglomeration and dispersion of nanoparticles throughout the polymer matrix has been recognized as a major challenge (Du, J.H. et al., 2007; McClory et al., 2009). The overall objective of this research is to understand how the dispersion and distribution of carbon nanotubes in polymers can be improved during both melt-mixing (primary) and secondary processing. This is essential if such composites are to be processed using conventional polymer processing equipment, permitting scale-up for industrial applications.

Poly (ethylene terephthalate) (PET) has excellent properties and is widely used in many applications such as synthetic fibers, films, packaging containers, for thermoforming applications and as an engineering resin in electronic applications. PET/MWCNT composites will potentially improve PET thermal, mechanical, electrical and other properties if good levels of dispersion can be obtained. While PET/ MWCNT composites produced by melt compounding have already been investigated by other researchers, most of this work has been carried out on the effect of the extrusion process on chain comformation, crystalline content and crystallization behaviour of PET. Tzavalas et al. (2008a) prepared PET/MWCNT composites by melt-mixing in a twin screw and reported an increase of the *trans* conformations and crystallinity at low MWCNT content followed by a sharp decrease for 1wt% of CNTs due to tube aggregation. In other research, Tzavalas (2008b) uniaxially stretched PET and PET/MWCNT composites to investigate the effect of MWCNTs on chain conformation of PET via IR spectroscopy. Kim et al. (2007) prepared PET/MWCNT composites via melt compounding in a twin extruder and reported that the incorporation of MWCNTs altered the crystallization behaviour through heterogeneous nucleation, improved thermal stability and mechanical properties of the PET. Each report highlighted the importance of CNT dispersion and its influence on composite properties. However, little attention has been focused into methods for improving the dispersion of MWCNTs in PET. Some authors have reported the dispersion of PET/MWCNTs composites prepared by other techniques: Hu et al. (2008) produced homogeneous PET/MWCNTs composites via a coagulation method with few agglomerates of MWCNTs obvious in the coagulated PET/MWCNTs composites. In contrast, numerous agglomerates of MWCNTs in the melt compounded PET/MWCNTs were observed. The agglomerates were found for a 0.5wt% MWCNT loading in melt compounded samples prepared in a micro-scale ($2cm^3$) mixer. Wang et al. (2007) prepared PET/MWCNT composites by *in situ* polymerization with very low MWCNT content (0.1wt%) and achieved good dispersion and a significant enhancement in composite properties. Jin et al. (2007) also prepared PET/MWCNT composites by *in situ* polymerization, but prior to composite formation, the surface of the MWCNTs were modified with carboxylic acid and diamine functional groups and the rheological and mechanical properties of the composites studied.

Many polymers, including PET, undergo secondary processing such as blow moulding and thermoforming where the polymer/composite undergoes semi-solid state deformation. The effect of such deformation on CNT dispersion and composite properties has not been investigated in detail. In this paper, we report the effect of semi-solid state deformation of PET/MWCNT composites on composite properties and CNT dispersion and distribution. The effect of uniaxial and biaxial stretching on mechanical, thermal and electrical properties was investigated. The extent of dispersion of PET/MWCNT composites was investigated using a combination of Scanning Electron Microscopy (SEM), High Resolution Transmission Electron Microscopy (HR-TEM) and Wide Angle X-Ray Diffraction (WAXD).

Experimental

Materials

The PET used in the study was Polyclear F019 (Intrinsic viscosity 0.895 dl/g) in pellet form and was provided by Invista Resins & Fibers Gmbh. The MWCNTs used were Nanocyl® 7000 with a purity of 90% and produced via catalytic chemical carbon vapour deposition (CCVD). A 10wt% MWCNT masterbatch in PET (Polyclear F019) was obtained in pellet form from Nanocyl S.A. Belgium. The MWCNTs have a typical diameter of 9.5 nm and average length of 1.5 μm.

Nanocomposite Preparation

The compounding process of masterbatch dilution was performed using a co-rotating intermeshing twin-screw extruder, Collin GmbH, with a screw diameter of 25mm and a barrel length of 750mm (L/D=30). Both PET and the 10wt% MWCNTs Masterbatch were dried at 120°C for 12 hours prior to mixing. The melt blending process was carried out using a screw speed of 160 rpm, a residence time of 1min and temperature settings from 230 to 260°C, resulting in a melt temperature of 280°C. By this melt dilution procedure, concentrations of 1, 2 and 4wt% MWCNTs in PET were produced.

Characterization

The morphology and the degree of dispersion of MWCNTs in the PET matrix were investigated using a combination of Field Emission Scanning Electron Microscopy (FESEM), High Resolution Transmission Electron Microscopy (HRTEM) and wide-angle X-ray diffraction (WAXD). Samples for SEM were prepared by cryogenic fracture after immersion in liquid nitrogen or by mechanical tensile failure and examined using a JEOL 6500F thermal field emission SEM with an operating voltage of 5 kV. Specimens for HRTEM examination were microtomed to 90nm thickness using an Ultracut-E Reichert Jung ultramicrotome and imaged using an FEI Tecnai F20 field emission high-resolution transmission electron microscope operating with an accelerating voltage of 200 kV. WAXD spectra of extruded and stretched samples of PET and PET nanocomposites were recorded using a PANalytical X'Pert PRO diffractometer with Cu-K_α radiation (λ=1.5406 Å) at a scanning rate of 0.02°/min over the range 2–60°(2θ). Biaxial Stretching experiments were carried out using QBS, a specially developed biaxial stretching machine constructed by Queen's University (Martin et al., (2002). The samples were heated from top and bottom by convection heaters up to 100°C for 2 min to ensure an even temperature distribution before stretching. Differential scanning calorimetry (DSC) was performed using a Perkin–Elmer Diamond DSC model 6 under an inert nitrogen atmosphere using a heating and cooling rate of 10°C/min between 30 and 300°C. In all cases the samples were held at 300°C for 5 min. prior to cooling to remove previous thermal history. The samples were then cooled to 30°C at 10°C/min. and reheated again to 300°C at 10°C/min. The mechanical properties were measured using an Instron 4411 Universal Tester and testing was carried out on specimens cut parallel and transverse to the extrusion direction for unstretched, uniaxial and biaxial stretched samples. Volume resistivity measurements were performed on extruded samples of all nanocomposites of thicknesses 0.5mm and 1 mm. For high resistivity samples, a Keithley electrometer (Model 6517A) equipped with an 8009 test fixture with circular samples of diameter 60 mm was used (ASTM D-257). For more conductive samples (< 10^7 Ωcm) measurements were taken with a Keithley electrometer

(Model DMM 2000) using a two-point probe test fixture.

Results and Discussion

Scanning Electron Microscopy (SEM)

The morphology and degree of dispersion of MWCNTs in the PET matrix were investigated using FESEM. By way of example, Figure1 shows that MWCNTs form a well-dispersed and distributed network in the PET matrix.

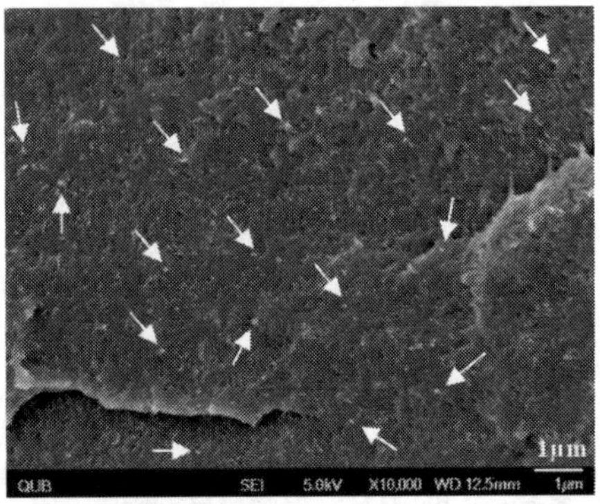

Fig.1 – SEM micrograph x10K of PET/2wt%MWCNT

Wide-Angle X-ray Diffraction Analysis (WAXD)

The WAXD traces for the virgin PET, PET/1wt%, 2wt% and 4wt%MWCNTs and MWCNTs alone are shown in Fig.2a. The characteristic sharp diffraction peak for MWCNTs which occurs at a scattering angle of 2θ=26.1° is derived from the ordered arrangement of the concentric cylinders of graphitic carbon (Zhou et al., 1994). This peak is absent from the XRD patterns of PET/MWCNT composites due to the fact that the shear forces achieved during mixing have destroyed the short range order of the concentric geometry of the MWCNTs. These results are further evidence for efficient mixing of the MWCNTs in PET.
The PET crystal faces have been assigned previously to the following diffraction peaks at 2θ angles: $(0\bar{1}1)$= 16.3°, (010)= 17.5°, $(\bar{1}11)$= 21.5°, $(\bar{1}10)$= 22.7°, (100)= 26.1°, $(1\bar{1}1)$= 27.65° (Fakirov et al., 1975). They are labelled in Fig. 2a as 1=$(0\bar{1}1)$, 2=(010), 3=$(\bar{1}11)$, 4=$(\bar{1}10)$, 5=(100), 6=$(1\bar{1}1)$. As expected, the neat PET diffractogram doesn't show any major crystaaline peaks. However, as the content of MWCNT increased, the diffraction peaks start to appear at the same 2θ angles relative to the assigned crystalline PET crystal faces, and become narrower and more distinct when 4wt% MWCNT was added. This behaviour, previously reported by Tongying, may be partially due to the gradual perfection of the crystallite. The rapid crystallization due to the nucleating effect of CNTs did not allow the polymer to reach a highly regular semicrystalline framework (Tongying et al., 1983).

The spectra of PET/1wt% MWCNTs unstretched and biaxial stretched at stretch ratios of 1.5% and 2% are shown in Fig.2b. It can be seen that as the stretch ratio increased the crystallisation peaks remain in the same position, no effect on the crystal cell parameters.
However, peaks 2=(010), 4=$(\bar{1}10)$, 6=$(1\bar{1}1)$ exhibit higher intensity due to strain induced crystallisation, now peak 6=$(1\bar{1}1)$ is the dominant peak.

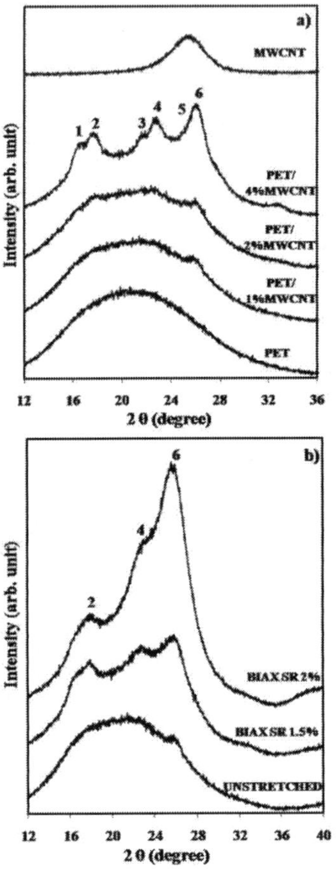

Fig.2 – XRD traces for **a)** virgin PET, PET/1wt%, 2wt% and 4wt% MWCNT and pristine MWCNTs and **b)** PET/1wt% MWCNT unstretched and biaxial stretched at a stretch ratio of 1.5% and 2%.

High Resolution Transmission Electron Microscopy (HRTEM)

In order to assess the effect of the biaxial stretching on the degree of dispersion of MWCNTs in the PET matrix, samples were microtomed and examined using HRTEM. The TEM micrographs at low magnification for samples of PET/2wt% MWCNT unstretched and biaxial stretched are shown in Fig. 3. It can be clearly seen in the TEM micrographs that biaxial stretching orients and aids dispersion of the CNTs in the PET matrix.

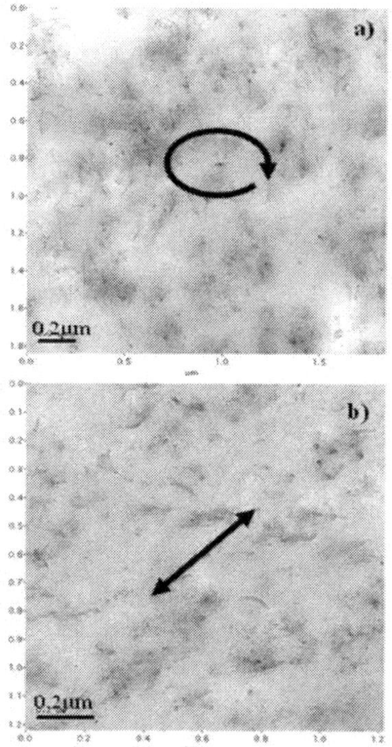

Fig.3 – TEM micrographs of PET/2wt%MWCNTs x 9.9K a) unstretched and b) biaxial stretched.

Mechanical Properties

Figure 4 shows the stress-strain behaviour for virgin PET, PET/1wt% MWCNT and PET/2wt% MWCNT for a strain rate of $2s^{-1}$ and a stretch ratio of 1.5% for biaxial stretching.

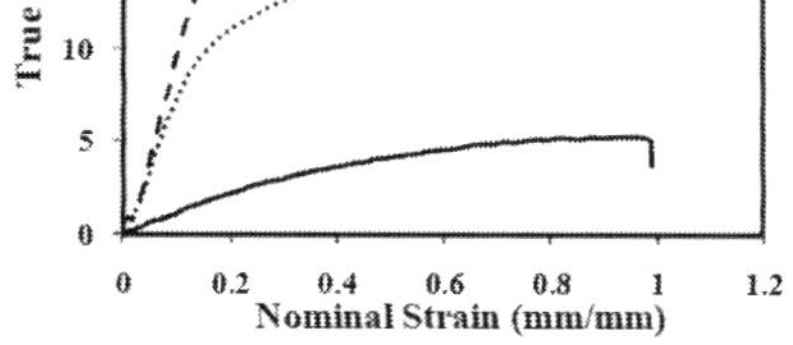

Fig.4 – Effect of MWCNT loading on the behavior of PET during biaxial stretching for a stretch ratio of 2%.

Similar behaviour was obtained for both uniaxial and biaxial stretching, for a stretch ratio of 1.5%: the yield stress for PET/1wt% MWCNT increased by ~400% and for PET/2wt% MWCNT by ~700% relative to the unfilled biaxial stretched PET. Similar trends were found by Abu-Zurayk et al. (2009) with PP/clay composites. This increase in yield stress in the composites is equivalent to the force required to deform the material and may be caused by the greater restrictions in molecular mobility or entanglement of polymer chains due to the presence of carbon nanotubes. The effect of addition of 1%, 2% and 4wt%MWCNT and uniaxial/biaxial stretching on the mechanical properties of PET, tested in the direction parallel to the extrusion direction are shown in Table 1.

Table 1- Effect of MWCNT addition (wt%) on the mechanical properties of PET and PET/MWCNT composites, unstretched and after uniaxial and biaxial stretching. Tested in the direction parallel to the extrusion direction.

	Y. Mod (MPa) (+/-change) (%)*	Y. Stress (MPa) (+/-change) (%)*	Strain (%) (+/-change) (%)*
PET virgin unstretched	1170	57.4	6.8
PET/1%MWCNT unstretched	1380 (+18%)	66.6 (+16%)	5.9 (-13%)
PET/1%MWCNT uniaxial stretched	2096 (+79%)	93.9 (+64%)	6.7 (-2%)
PET/1%MWCNT biaxial stretched	1879 (+61%)	90.1 (+57%)	7.6 (+13%)
PET/2%MWCNT unstretched	1447 (+24%)	65.4 (+14%)	6.2 (-9%)
PET/4%MWCNT unstretched	1483 (+27%)	68.2 (+19%)	6.1 (-9%)

*with respect to virgin PET unstretched

The Young Modulus, yield stress and % strain results were very similar for samples cut in parallel and transverse to extrusion direction, being slightly lower for transverse direction. It can be observed that the Young modulus and yield stress exhibit an 18% and 16% improvement respectively with addition of 1wt% MWCNT.This enhancement could be attributed to the reinforcement effect of the MWCNTs having a high aspect ratio and their uniform dispersion in the PET matrix. There are other factors that may affect this, such as the enhanced crystallinity due to the addition of MWCNTs and secondary interactions, such as Van der Waals bonding between PET chains and MWCNTs (Ahn et al., 2008). The improvement in mechanical properties was found to be less significant at higher MWCNT content, in agreement with the results obtained reported by Kim et al. (2007). These results may be related to the fact that Van der Waals attractions between the individual carbon nanotubes create the agglomerates preventing efficient load transfer to the polymer matrix and leading to lower mechanical properties than expected. The results show that the Young modulus and yield stress increased when the samples were uniaxial and biaxial stretched. In the case of 1wt% MWCNT addition, uniaxial stretching resulted in an enhancement in Young's modulus of 79% and for biaxial stretching 61%, when compared with the virgin PET unstretched sample. The yield stress follows a similar trend. For the PET/2wt% MWCNT composite the improvement is lower due to the fact that a higher content of MWCNTs leads to higher amounts of agglomerates and therefore less efficient load transfer to the polymer matrix.

Differential Scanning Calorimetry (DSC)

The glass transition temperature (T_g), crystallization temperature (T_c), melting temperature (T_m) and percentage crystalline content for PET and PET/MWCNT composites are shown in table 2.

Table 2- Effect of MWCNT loading (wt%) on the thermal properties of PET and PET/MWCNT composites, unstretched and after uniaxial and biaxial stretching.

	First Heating			Cooling
	T_g (°C)	T_m (°C)	X_c (%)	T_c (°C)
PET virgin unstretched	84.5	254	32	194.1
PET/1%MWCNT unstretched	85.1	254.7	38	217.1
PET/1%MWCNT uniaxial stretched	85.2	254.5	53	221.2
PET/1%MWCNT biaxial stretched	85.4	254.2	55	227.9
PET/2%MWCNT unstretched	84.9	252	37.1	220.4
PET/4%MWCNT unstretched	85	255.8	37.3	227.9

Addition of MWCNTs to PET had little effect on the glass transition (T_g) and melting (T_m) temperatures. This behaviour was also reported by Kim et al. for similar nanocomposites (2007). The crystallization temperature (T_c) shows a substantial increase with the addition of MWCNT, increasing by 23°C for 1wt% MWCNT and by up to 34°C for the PET/4wt% MWCNT composite. As the MWCNT content increases, the crystalline content of PET/MWCNT composite increased, showing the maximum increase with 1wt% MWCNT addition. This increase in crystallisation temperature and of crystalline content suggests that the MWCNTs are acting as nucleating agents for PET crystallization. By acting as a nucleating agent, MWCNTs promote the formation of heterogeneous nuclei, with lower energy consumption required to reach critical stability for crystal growth. This means that the incorporation of a very small quantity of MWCNT could enhance the crystallisation of the PET matrix through heterogeneous nucleation (Tzavalas et al., 2008b), an observation corroborated by the WAXD data shown in figure 2. Uniaxial and biaxial stretching of PET/1wt% MWCNT composites did not significantly affect the glass transition temperature (T_g) and melting temperature (T_m), see Table 2. However, the crystallization temperature (T_c) and crystallinity showed a slight increase due to uniaxial stretching and an additional increase when biaxial stretched. This increase may be due to strain induced crystallization.

Electrical conductivity

Electrical conductivity of filled polymer systems can be attributed to the formation of conductive pathways when the filler content exceeds a critical volume fraction. According to Gojny et al. (2006), MWCNTs can exhibit a low percolation threshold due to the relatively low surface area and high aspect ratio. The volume resistivity (Ω.cm) of the nanocomposites as a function of the MWCNT content is shown in Fig. 5. A percolation threshold was attained for a MWCNT loading of 0.8 wt%, above which the composites can be regarded as electrically conductive. A slightly higher value of percolation threshold (0.9wt %) was found by Hu et al. (2006) for PET/MWCNT composites prepared by a coagulation method. Other lower values have been recently reported by Logakis et al. (2010) for PET/MWCNT composites prepared by *in situ* polymerisation (p_c=0.06 wt.%), by direct mixing (p_c=0.10-0.20 wt.%) and by dilution of a masterbatch (synthesized via *in situ* polymerization) using melt mixing (p_c=0.05-0.1 wt.%).

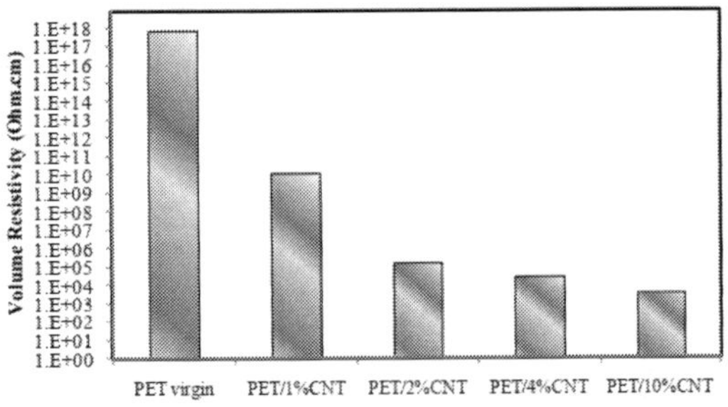

Fig.5 – Volume resistivity of the PET/MWCNT nanocomposites as function of MWCNT content.

The behaviour obtained is due to the fact that when stretching the sample, the distance between CNTs is increased, and the network pathways are interrupted (See sketch in Fig.6). It has been reported by Hu et al. (2008) that when nanotubes reach the electrical percolation threshold in a polymeric matrix, they do not need to be always physically touching each other, they can be just close enough to allow electron hopping or tunneling processes. The electron hopping/tunneling mechanism requires a critical minimum distance between neighbouring nanotubes and it has been estimated to be ~1.8nm from Monte Carlo simulations (Maultzsch et al., 2002).

The resistivity of polymer/MWCNT composites under strain have also been investigated by Bokobza (2007) and Du, F. et al. (2003) and both ascribed the increase in resistivity with strain to the orientation of nanotubes.

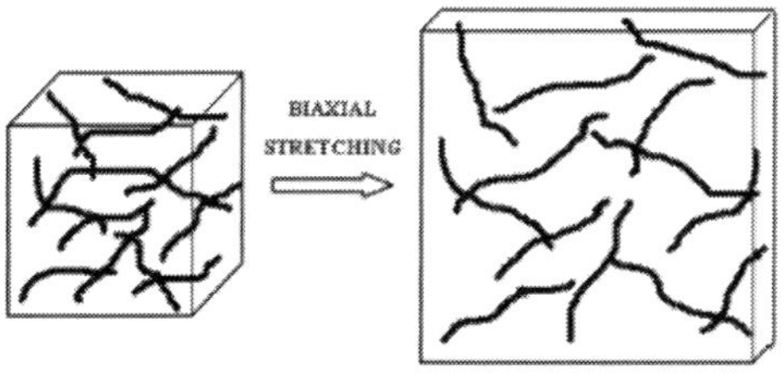

Fig.6– Illustration of MWCNT network unstretched and after biaxial stretching.

Further experiments are being performed in order to investigate the recovery of the conductive network of stretched samples during annealing. Some researchers have investigated the formation of the conductive network in polymer nanocomposites systems after extrusion, as Alig et al. (2007) when PP/MWCNT samples were annealed above T_m in a slit die. The same group studied other polymer systems, such as polycarbonate and reported the destruction and formation of CNT network in polymer melts (Alig et al., 2008).

Conclusions

PET/MWCNT composites were prepared using twin screw melt mixing. The MWCNTs formed a well-dispersed and highly interconnected structure in the PET matrix as indicated by SEM, HRTEM and WAXD results. However, as the percentage of MWCNTs added to the PET matrix increased, efficient dispersion became more difficult to achieve because the MWCNTs tended to form agglomerates. Uniaxial and biaxial stretching induced orientation of the carbon nanotubes in the stretching direction and therefore improved their distribution and dispersion throughout the polymer matrix. The addition of MWCNTs to the PET matrix increased X_c and T_c due to a CNT nucleation effect. The Young's modulus and yield stress exhibited a maximum increase of 18% and 16% respectively with the addition of 1wt% MWCNT. Uniaxial and biaxial stretching yielded an increase in both the degree of CNT dispersion and strain induced crystallization and as a consequence, the mechanical properties of the composites were greatly improved. For uniaxial and biaxial stretching at a 1.5% stretch ratio, the yield stress for PET+1wt% MWCNT increased by ~400% relative to the unfilled PET and for PET+2wt% MWCNT by ~700% relative to the unfilled PET. An electrical percolation threshold was attained for a MWCNT loading of 0.8 wt%. However, when the samples were uniaxial and biaxial stretched, electrical conductivity decreased because the network pathways were interrupted.

Annealing at $T>T_g$ and $T<T_m$ of the stretched PET/MWCNT composites is being investigated in order to examine the effect of deformation on network formation before and after stretching.

Acknowledgements

The authors would like to thank Nanocyl S.A. for the provision of the materials used in this study.

References

Abu-Zurayk, R., Harkin-Jones, E., McNally, T., Menary, G., Martin, P., Armstrong, C., "Biaxial deformation behavior and mechanical properties of a polypropylene/clay nanocomposite", Composites and Technology, **69**, 1644-1652 (2009)

Ahn, B.W., Chi, Y.S., Tae, J.K., "Preparation and characterization of multi-walled carbon nanotube /poly(ethylene terephthalate) nanoweb", J. Appl. Polym. Sci., **110**, 4055-4063 (2008)

Ajayan, P.M., Stephan, O., Colliex, C., Trauth, D., "Aligned carbon nanotube arrays formed by cutting a polymer resin-nanotube composite", Science, **265**, 1212-1214 (1994)

Alig, I., Lellinger, D., Dudkin, S., Pötschke, P., "Conductivity spectroscopy on melt processed polypropylene–multiwalled carbon nanotube composites: Recovery after shear and crystallization" Polymer, **48**, 1020-9 (2007)

Alig, I., Skipa, T., Lellinger, D., Pötschke, P., "Destruction and formation of a carbon nanotube network in polymer melts: Rheology and conductivity spectroscopy", Polymer, **49**, 3524-3532 (2008)

Bin, Y., Kitanaka, M., Zhu, D., Matsuo, M., "Development of Highly Oriented Polyethylene Filled with Aligned Carbon Nanotubes by Gelation /Crystallization from Solutions", Macromolecules, **36**(16), 6213-6219 (2003)

Bokobza, L., "Multiwall carbon nanotube elastomeric composites: A review", Polymer, **48**, 4907-4920 (2007)

Du, F., Fischer, J.E., Winey, K.I., "Coagulation method for preparing single-walled carbon nanotube /poly(methyl methacrylate) composites and their modulus, electrical conductivity, and thermal stability", Journal of Polymer Science, Part B: Polymer Physics, **41**, 3333-3338 (2003)

Du, J.H., Bai, J., Cheng, H.M., "The present status and key problems of carbon nanotube based polymer composites", eXPRESS Polymer Letters **1**(5), 253-273 (2007)

Fakirov, S., Fischer, E.W., Schmidt, G.F., "Unit cell dimensions of poly(ethylene terephthalate)", Die Makromolekulare Chemie", **176**(8), 2459-2465 (1975)

Gojny, F.G., Wichmann, M.H.G., Fiedler, B., Kinloch, I.A., Bauhofer, W., Windle, A.H., Schulte, K., "Evaluation and identification of electrical and thermal conduction mechanisms in carbon nanotube/epoxy composites", Polymer, **47**, 2036-2045 (2006)

Hu, G., Feng, X., Zhang, S., Yang, M., "Crystallization behavior of poly (ethylene terephthalate) /multiwalled carbon nanotubes composites", J. Appl. Polym. Sci., **108**, 4080-4089 (2008)

Hu, G., Zhao, C., Zhang, S., Yang, M., Wang, Z., "Low percolation thresholds of electrical conductivity and rheology in poly (ethylene terephthalate) through the networks of multi-walled carbon nanotubes", Polymer, **47**, 480-488 (2006)

Iijima, S., "Helical microtubules of graphitic carbon", Nature (London), **354**, 56- 58 (1991)

Jin, S.H., Yoon, K.H., Park, Y-B., Bang, D.S., "Properties of surface-modified multiwalled carbon nanotube filled poly (ethylene terephthalate) composite films", Journal of Applied Polymer Science, **107**, 1163-1168 (2008)

Kim, J.Y., Park, H.S., Kim, S.H., "Multiwall carbon nanotube-reinforced poly (ethylene terephthalate) nanocomposites by melt compounding", Journal of Applied Polymer Science, **103**(3), 1450-1457 (2007)

Logakis, E., Pissis, P., Pospiech, D., Korwitz, A., Krause, B., Reuter, U., Pötschke, P., "Low electrical percolation threshold in poly (ethylene terephthalate) /multi-walled carbon nanotube nanocomposites", Eur. Polym J., Article in press (2010)

Martin, N., Harkin-Jones, E.M.A., Martin, P.J., "Characterizing the biaxial properties of materials used in thermoforming and blow molding" ANTEC 2002, San Francisco, **1**, 940-945 (2002)

Maultzsch, J., Reich, S., Thomsen, C., "Raman scattering in carbon nanotubes revisited", Phys.Rev.B, **65**, 233402 (2002)

McClory, C., Chin, S.J., McNally, T., "Polymer/ Carbon Nanotubes Composites", Aust. J. Chem., **62**, 762-785 (2009)

Tongyin, Y.U., Haishan, B.U., Jiacong, H.U., Wei, Z., Qiaoying, G., "The double melting peaks of poly(ethylene terephthalate)", Chinese Journal of Polymer Science, **1**(1), 83-91 (1983)

Tzavalas, S., Gregoriou, V.G., "Uniaxially stretched PET/PET-MWNT nanocomposites: Effect of the MWNTs on the chain conformations of PET", Vibrational Spectroscopy, **46**, pp.135-140 (2008b)

Tzavalas, S., Mouzakis, D.E., Drakonakis, V., Gregoriou, V.G., "Poly(ethylene terephthalate)-multiwall nanotube nanocomposites: Effect of nanotubes on the conformations, crystallinity and crystallisation behaviour of PET", Journal Polymer Science: Pt. B: Polymer Physics, **46**, 668-676 (2008a)

Wang, Y., Deng, J., Wang, K., Zhang, Q., Fu, Q., "Morphology, crystallisation and mechanical properties of poly(ethylene terephthalate)/multiwall carbon nanotube nanocomposites via *in situ* polymerisation with very low content of multiwall carbon nanotubes", Journal of Applied Polymer Science, **104**, 3695-3701 (2007)

Zhou, O., Fleming, R.M., Murphy, D.W., Chen, C.H., Haddon, R.C., Ramirez, A.P., Glarum, S.H., "Defects in Carbon Nanostructures", Science, **263**, 1744-1747 (1994)

CARBOXYLATE CLAYS AS A NEW ROUTE FOR POLYPROPYLENE NANOCOMPOSITES FROM SODIUM-MONTMORILLONITE

D.D.J. Rousseaux*, M. Sclavons, P. Godard, J. Marchand-Brynaert**

Institute of Condensed Matter and Nanosciences, Université catholique de Louvain, Louvain-la-Neuve, Belgium
** dimitri.rousseaux@uclouvain.be ; ** jacqueline.marchand@uclouvain.be*

Carboxylate clays were prepared by intercalating sodium acetate inside untreated clay interlayers. Those materials revealed to be interesting for the production of PP/clay nanocomposites. The carboxylate salts within the clay layers, are used to partially neutralize the PP-g-MA anhydride groups, in situ, during the melt compounding. The ionic groups of the partially neutralized PP-g-MA offer favourable interactions with the clay, hence reinforcing the interfacial junction between polymer and clay and improving the composite properties. Finally, model PP-g-MA/clay nanocomposites were synthesized from carboxylate clays and compared with a reference nanocomposite from untreated clay to elucidate the mechanism. Nanocomposites were also produced from untreated clay by in situ synthesizing the carboxylate clays. The use of carboxylate clays clearly improves the clay dispersion into the polymer matrix and rises the thermal and rheological properties of the nanocomposites.

Introduction

Polypropylene (PP)/clay nanocomposites are mainly prepared by melt-blending from alkylammonium-montmorillonite (organoclay) and PP, with compatibilizers like PP-graft-maleic anhydride (PP-g-MA) [1-2]. The main drawback of using organoclays is the weak thermal stability of these surfactants. Indeed, alkylammoniums often degrade at extrusion temperatures leading to a collapse of the silicate layers which limits the clay dispersion and consequently the material properties [3-6].

This work reports the synthesis and characterization of carboxylate clays and examines the feasibility of using these carboxylate-clays to synthesize polyolefin/clay nanocomposites. Carboxylate clays are interesting for the production of PP/clay nanocomposites. The carboxylate salts within the clay layers, are used to partially neutralize the PP-g-MA anhydride groups, in situ, during the melt compounding. The ionic groups of the partially neutralized PP-g-MA offer favourable interactions with the clay.

Model PP-g-MA/carboxylate clay nanocomposites are prepared to elucidate the mechanism and compared to PP-g-MA/untreated clay nanocomposites.

Experimental

Materials

Untreated clay sodium-montmorillonite (Cloisite Na+, labeled CNa) was purchased from Southern Clay Product. For the preparation of carboxylate clays, Cloisite Na+ was dried (dCNa) at 100°C under vacuum. Anhydrous sodium acetate (NaAc) and trihydrate sodium acetate ($NaAc.3H_2O$) both of analysis grade, were purchased from Sigma-Aldrich. Methanol (MeOH) of analysis grade, was purchased from Fluka. Commercial PP-g-MA Polybond 3200 (1 wt.% of maleic anhydride, labeled Plb) was purchased from Chemtura. Commercial organoclay Nanofil 948 (labeled NF948) used for comparison in thermogravimetric analysis, was purchased from Süd Chemie.

Carboxylate clay synthesis

dCNa and the carboxylate salt (added in equivalent [equiv.] quantities in respect to the cation exchange capacity of CNa = 0.92 meq./g) are dispersed and mixed into methanol at room temperature until complete evaporation of the solvent. The residue is dried at 100°C under vacuum and ground.

PP-g-MA/clay nanocomposites

References (samples with no clay) and nanocomposites samples (5 wt.% of clay in inorganic content) were prepared in a DSM Xplore 15 ml microcompounder with Polybond 3200 (Plb) as matrix. The samples were extruded at a screw speed of 200 RPM at 200°C during 2 minutes.

Characterizations

The interlayer spacing of clays and nanocomposites was measured by X-ray diffraction (XRD) on a Bruker D8 Advance diffractometer using CuKα radiation at 40 kV and 30 mA. The thermal stability of the different clays and nanocomposites was studied by thermogravimetric analysis (TGA) on a Mettler Toledo TGA/SDTA 851e using a heating rate of 20°C/min under nitrogen or air atmosphere. Fourier Transform Infrared spectroscopy (FTIR) spectra were obtained on a Nicolet Nexus 870 FTIR spectrometer equipped with a Continuum microscope in attenuated total reflectance (ATR) mode using a slide-on accessory with a Si crystal. The spectra for clay samples were recorded in ATR mode and the spectra for nanocomposite samples in transmission mode. The nanocomposite spectra were normalized by setting the peak height of the 2723 cm^{-1} absorption band (characteristic C-H stretching of PP) to an absorbance of 1. The clay dispersion of nanocomposites was examined on ultra-microtomed samples by transmission electron microscopy (TEM) with a LEO 922 microscope operating at 200 kV. Rheological measurements were carried out in an oscillatory mode on a Bohlin Gemini II rheometer equipped with a parallel plate geometry using 25 mm diameter plates. Frequency sweep tests were performed under nitrogen atmosphere, at 180°C with strain of 10%.

Results and Discussion

Figure 1 shows XRD patterns of the dried untreated clay (dCNa) and of carboxylate clay prepared with 5 equiv. of NaAc (dCNa-5NaAc). The intercalation of NaAc into dCNa doubles its basal spacing (001) to 2.0 nm. Some reflexions belonging to NaAc are visible at 8.8, 11.3, 16.9 and 19.0 °2θ, suggesting that all NaAc is not completely intercalated. Attempts to wash carboxylate clays with MeOH in order to remove excess NaAc failed; both intercalated and excess NaAc were eliminated and untreated clay was recovered.

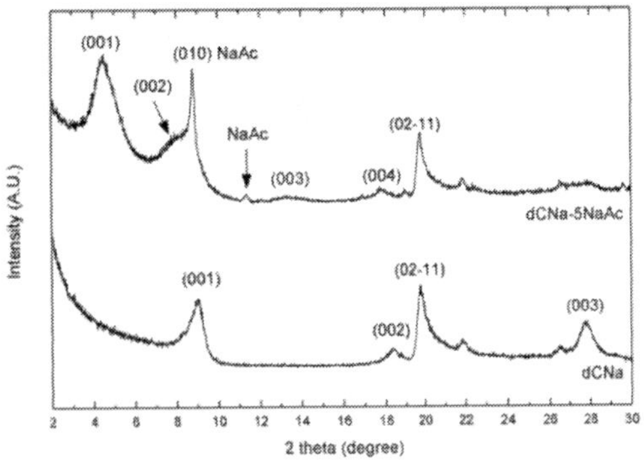

Figure 1 – Carboxylate clay (CNa-5NaAc) and dCNa XRD patterns (the patterns are vertically shifted).

The thermal stability of the carboxylate clay was investigated by TGA under nitrogen atmosphere and compared to an usual alkylammonium clay NF948. TGA results reveal that dCNa-5NaAc carboxylate clay is more thermally stable than NF948 alkylammonium clay and can be used in the extrusion temperature range, between 200 to 300°C, without thermal degradation drawback.

The interactions between the clay and NaAc were investigated by FTIR. It appears that the separation between the carbonyl symmetric and asymmetric stretchings, called the carboxylate shift, is 164 cm^{-1} for NaAc and 137 cm^{-1} for CNa-5NaAc. This decrease of the carboxylate shift suggests, according to Deacon and Philips [7], that the carboxylate coordination which is ionic in pure NaAc, becomes of bridging-type in CNa-5NaAc carboxylate clay (Figure 2).

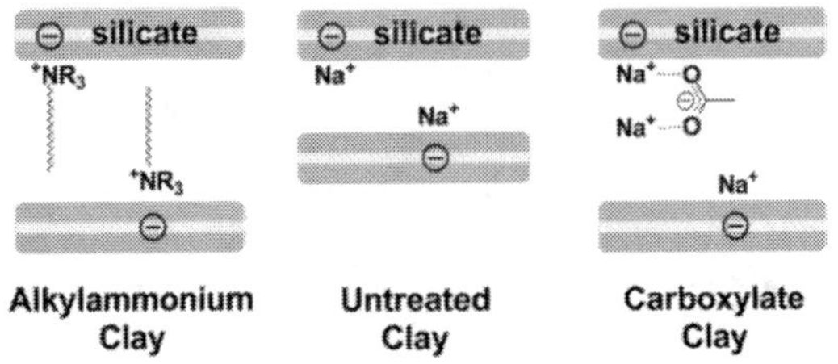

Figure 2 – Proposed structure of carboxylate clay, comparatively to untreated clay and alkylammonium clay.

PP-g-MA/carboxylate clay nanocomposites were prepared by two different methods: (1) the sequential process in which preformed carboxylate clay is melt-blended within PP-g-MA matrix; (2) the one-pot process in which carboxylate clay is synthesized in situ during melt-blending of the nanocomposite. Only the results for references and nanocomposites prepared by one-pot process (Table 1) are shown hereafter.

Table 1 – Composition of references and nanocomposites prepared by the one-pot process.

Samples	Matrix	Clay	Clay %	Carboxylate salt	Salt equiv.
R0	Plb	/	/	/	/
R1	Plb	/	/	NaAc	5
R2	Plb	/	/	$NaAc.3H_2O$	5
NC0	Plb	CNa	5	/	/
NC1	Plb	CNa	5	NaAc	5
NC2	Plb	CNa	5	$NaAc.3H_2O$	5

The clay dispersion in nanocomposites was investigated by TEM (Figure 3). NC0, the reference nanocomposite from untreated clay, shows, as expected, only large clay aggregates of several microns. Using NaAc, in NC1, clay aggregates are broken into smaller ones and the overall clay dispersion is slightly better. A best clay dispersion is obtained using $NaAc.3H_2O$ where a mixed morphology of small clay stacks with individual clay platelets, homogeneously distributed into the Plb matrix, is observed.

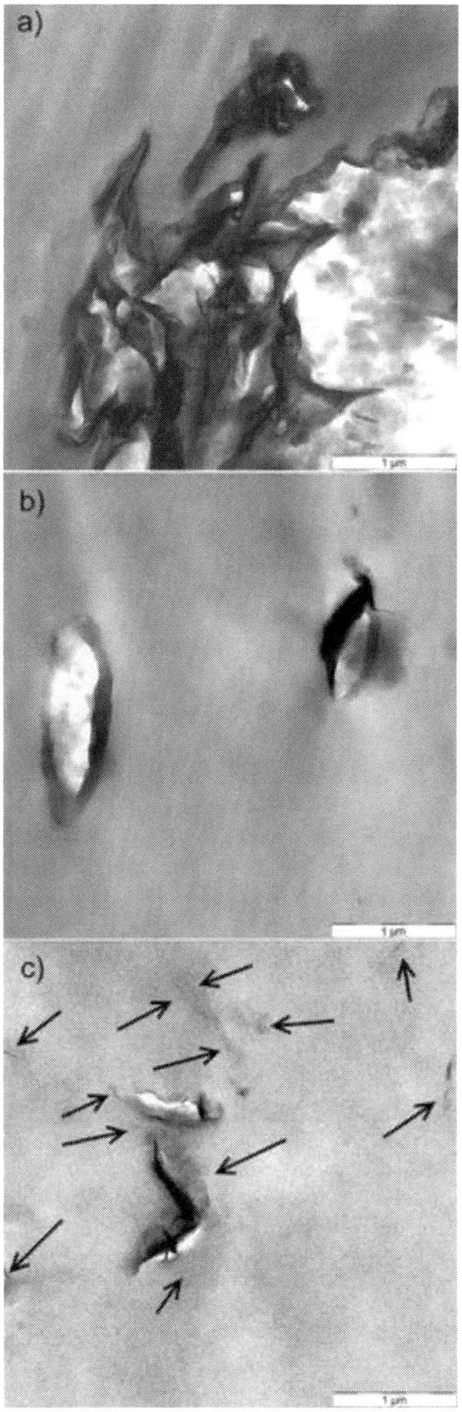

Figure 3 – TEM pictures of a) NC0, b) NC1 and c) NC2, all at a magnification of 5,000.

FTIR investigations on references indicate a greater efficiency of $NaAc.3H_2O$ to neutralize the Plb matrix than NaAc. Indeed, the neutralization reaction requires water to ionize the base [8]. For nanocomposites prepared with $NaAc.3H_2O$, the clay amplifies the Plb neutralization while it is not the case when using NaAc. In NC1 and NC2, both hydrogen-bond and ionic interactions are available to link Pbl to the clay. As ionic interactions are stronger than hydrogen-bonds, the interfacial junction between Plb and clay is reinforced. A better clay dispersion is observed in NC1, and an even better with NC2, than in NC0 where only hydrogen-bonds interactions link the clay to the Plb matrix

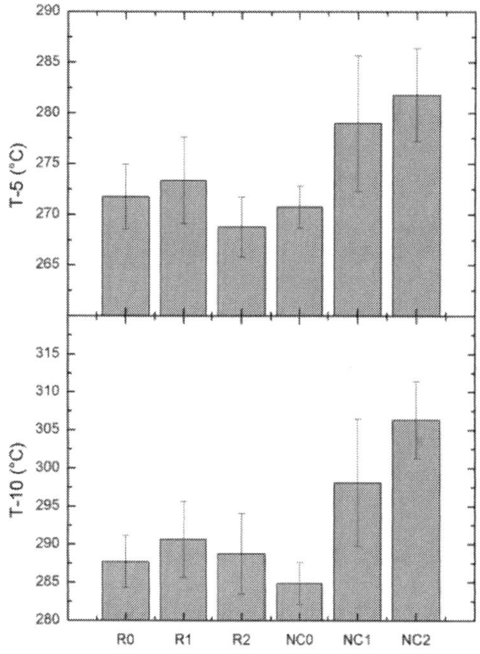

Figure 4 – Temperatures, from TGA curves, at 5% (T-5) and 10% (T-10) weight loss for references and nanocomposites.

The thermal stability of references and nanocomposites was investigated by TGA (Figure 4). The addition of NaAc, $NaAc.3H_2O$ or CNa to Plb matrix does not change significantly the T-5 and T-10 temperatures of R1, R2 and NC0 in comparison to R0. Conversely, the simultaneous addition of CNa and NaAc or $NaAc.3H_2O$ to Plb matrix clearly raises the T-5 and T-10 temperatures of NC1 and NC2.

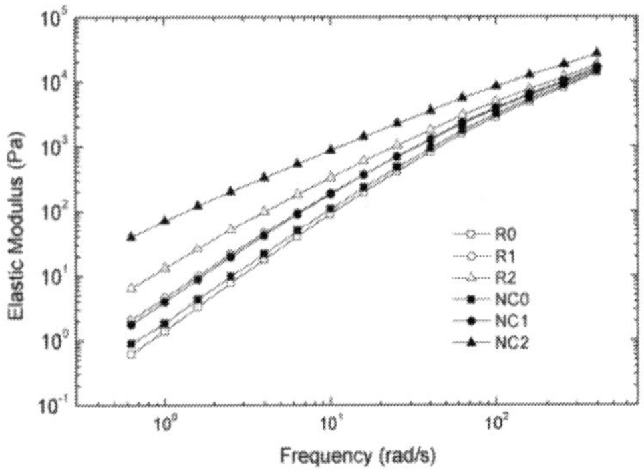

Figure 5 – Elastic modulus at 180°C as a function of the frequency for references and nanocomposites.

Rheological characterization of references and nanocomposites is presented in Figure 5. An increase of the elastic modulus is observed for R1 and R2 in comparison to R0. The addition of CNa in NC0 and NC1 does not change significantly their elastic modulus in comparison to R0 and R1. Conversely, the simultaneous addition of CNa and $NaAc.3H_2O$ increases NC2 elastic modulus which is markedly superior to the sum of the increases of elastic modulus for NC0 and R2.

Conclusions

Carboxylate-clays were successfully synthesized and characterized. The clay intercalated with sodium acetate is more thermally stable than a classical alkylammonium clay and can be used in the extrusion temperature range (between 200 to 300°C) without thermal degradation drawback.

Model PP-g-MA/carboxylate clay nanocomposites were successfully prepared and compared with a reference nanocomposite from untreated clay. The use of carboxylate clays clearly improves the clay dispersion into the polymer matrix and rises the thermal and rheological properties of the nanocomposites.

This work paves the route towards the use of untreated montmorillonite for the preparation of PP/clay nanocomposites in a one-pot process including $NaAc.3H_2O$ as additive. Works are in progress considering home-made PP-g-MA as compatibilizer and PP as matrix.

Acknowledgements

The authors thank La Région Wallonne and FRIA (Fonds pour la formation à la Recherche dans l'Industrie et dans l'Agriculture) for financial support. J. Marchand-Brynaert is Senior Research Associate of the FRS-FNRS (Fonds National de la Recherche Scientifique, Belgium). Prof. Ph. Dubois (Université de Mons, UMH), Prof. J. Devaux (Université catholique de Louvain, UCL) and Prof C. Bailly (UCL) are acknowledged for useful discussions and advices and P. Lipnik for her contribution to the experimental work.

References

1. Hasegawa, N., et al., "Preparation and mechanical properties of polypropylene-clay hybrids using a maleic anhydride-modified polypropylene oligomer", J. Appl. Polym. Sci. , **67**, 87-92 (1998).
2. Kawasumi, M., et al., "Preparation and Mechanical Properties of Polypropylene-Clay Hybrids", Macromolecules, **30**, 6333-6338 (1997).
3. Cervantes-Uc, J.M., et al., "Thermal degradation of commercially available organoclays studied by TGA-FTIR", Thermochim. Acta **457**, 92-102 (2007).
4. Hedley, C.B., et al., "Thermal analysis of montmorillonites modified with quaternary phosphonium and ammonium surfactants", Appl. Clay Sci. , **35**, 180-188 (2007).
5. Shah, R.K. and Paul, D.R., "Organoclay degradation in melt processed polyethylene nanocomposites", Polymer, **47**, 4075-4084 (2006).
6. Xie, W., et al., "Thermal Degradation Chemistry of Alkyl Quaternary Ammonium Montmorillonite", Chem. Mater. , **13**, 2979-2990 (2001).
7. Deacon, G.B. and Phillips, R.J., "Relationships between the carbon-oxygen stretching frequencies of carboxylato complexes and the type of carboxylate coordination", Coord. Chem. Rev., **33**, 227-250 (1980).
8. Rousseaux, D.D.J., et al., "Polypropylene Ionic Thermoplastic Elastomers: Synthesis and Properties", Polym. Degrad. Stab. , **95**, 363-368 (2010).

POLYPROPYLENE-BASED NANO-COMPOSITE FORMATION: DELAMINATION OF ORGANICALLY MODIFIED LAYERED FILLER VIA SOLID-STATE PROCESSING

T. Saito[1]* and M. Okamoto[2]

[1]*Research Laboratories, Toyota Boshoku Corporation, 1-1, Toyoda-cho, Kariya, Aichi, Japan – tomotaka_saitou@toyota-boshoku.co.jp*
[2]*Advanced Polymeric Nanostructured Materials Engineering, Graduate School of Engineering, Toyota Technological Institute, 2-12-1, Hisakata Tempaku, Nagoya, Aichi, Japan – okamoto@toyota-ti.ac.jp*

Abstract - Solid-state processing for the preparation of polypropylene (PP)-based nano-composites having finely dispersed layered fillers was conducted. The mixture of PP and organically modified layered filler (OMLF) (95:5 wt./wt.) was subjected to the processing using alumina mortar heated 65 °C, below T_m of PP (i.e., PP is still at the solid-state), and ground for 8 h before melt compounding. On X-ray diffraction, the $d_{(001)}$ peak of OMLF was broaden and peak position shifted slightly. The mixture prepared by solid-state processing exhibited disorder and delaminated layer structure with the thickness of 3-7 nm into PP matrix through TEM observations. On the contrary, nano-composite prepared by melt compounding at 180 °C for 3 min (without solid-state processing) showed the large stacked silicate layers in the PP matrix. Furthermore, instead of using alumina mortar, we carried out solid-state processing using internal mixer. X-ray diffraction pattern and TEM observation exhibited similar results. The solid-state processing led to delaminate of the silicate layers and attained the discrete dispersion.

Introduction

Polymeric nano-composite is one of the fast-growing categories on a worldwide basis. A significant amount of work has already done on various aspects of polymeric nano-composite containing organically modified layered fillers (OMLFs) (Ray and Okamoto 2003). However, complete exfoliation of OMLFs in continuous polymer matrix is still challenging issue because it could not be satisfactorily attained.

In the preparation of nano-composites based on commercial thermoplastic polymers, it is fair to state that intercalation of the aggregates of OMLFs has been observed more often than exfoliation. There are several factors that play important roles in producing exfoliated nano-composites; (1) chemical affinity (or compatibility) between the surface of OMLF and polymer matrix, (2) melt viscosity of the polymer matrix, and (3) capillary pressure drop within the nano-galleries (Saito et al., 2007). The capillary pressure drop plays a very important role in controlling the effective mixing between OMLFs and polymer to attain the exfoliation.

For this reason, we have conducted solid-state shear processing of poly(*p*-phenylenesulfide)/OMLF system (Saito et al., 2007), in which the delamination of silicate layers was achieved in the nano-composite as revealed by transmission electron microscope (TEM) observation.

The solid-state shear processing may be an innovative technique to delaminate the layered fillers in overcoming the pressure drop within the nano-galleries. Therefore, successful delamination of OMLFs could broaden the scope of application of this procedure. The objective of this work is to gain insight into the solid-state processing for the preparation of nano-composite based on polypropylene (PP) having finely dispersed layered fillers.

Experimental

Montmorillonite (MMT) modified with *n*-hexadecyltri-*n*-butyl phosphonium ($C_{16}TBP$, T_m=61.1°C) (MMT-$C_{16}TBP$) of 5 wt% was mixed with polypropylene powder (SEISHIN ENTERPRISE Co., Ltd; T_m=151°C, average particle size ~5μm). The mixture was subjected to the processing by using alumina mortar heated at 65 °C, below T_m of PP (i.e., PP is still at the solid-state), and ground for 8 h. After solid-state processing, the mixture of PP and OMLF (PP/OMLF) was melt compounded using miniature mixer of gram scale and to convert into sheet for the characterization. The sheet samples were characterized by wide-angle X-ray diffraction (WAXD), polarized optical microscope (POM) and TEM. For analyzing the features of the micrographs, we carried out fast Fourier transform (FFT) analysis on digitally saved images of TEM micrographs using the commercial image analysis software (Ultimage®, Graftek, France), which allowed to provide us information equivalent to the scattering analyses.

For comparison, the PP-based nano-composite preparation was conducted via conventional melt compounding operated at 180 °C for 3 min.

Results and Discussion

WAXD patterns for the mixture powder of PP and OMLF (95:5 wt./wt.) are presented in Fig. 1(a). The mean interlayer spacing of the (001) plane ($d_{(001)}$) for the MMT that was modified by C_{16}TBP [MMT-C_{16}TBP] obtained by WAXD measurements is 2.26 nm (diffraction angle, 2Θ = 3.90 °). After solid-state processing for 2 h, a small remnant shoulder is observed around $2\Theta \cong 4.00$ ° (Fig. 1(c)). In the following processing from 3 to 8 h, the peak was almost featureless diffraction, only exhibiting an monoclinic (α-form)/triclinic unit cell (γ-form) of the crystallized PP (Meille et al., (1990)).

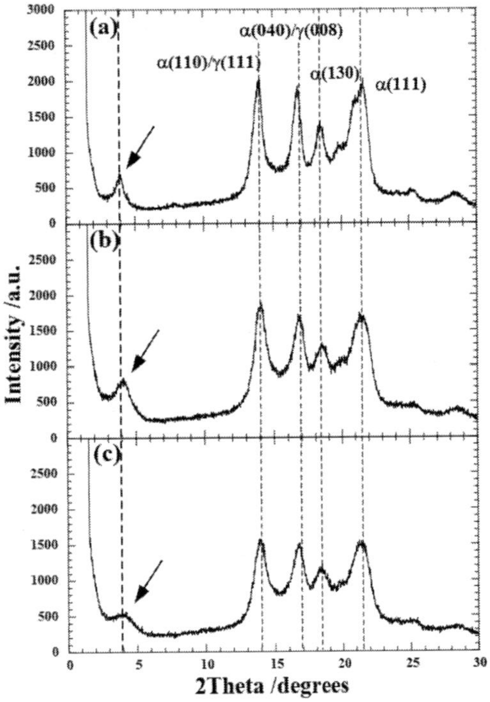

Fig.1 WAXD patterns of (a) mixture of PP and OMLF (95:5 wt./wt.) before solid-state processing, (b) and (c) processed mixture of PP/OMLF at 65 °C for 1 and 2h, respectively. Arrows in each panel indicate $d_{(001)}$ peak of OMLF.

Fig. 2 shows the time variation of $d_{(001)}$ and corresponding intensity during processing. With increasing processing time the intensity from $d_{(001)}$ gradually decreases with peak shift to lower angle (layer shrinkage). Beyond 4 h, the layer spacing finally reached a constant value (~ 2.05 nm). To elucidate the morphologies before and after solid-state processing, we conducted POM observation at 180 °C.

Fig. 3(a) shows the POM photograph of the mixtures prepared by annealing at 180 °C for 30 s (without shear processing). It is clear from the POM photographs that stacked-and-agglomerated structure of layers is evident in the un-processed mixture, whereas a good dispersion appears in the processed sample for 8 h (Fig. 3(b)). The dispersion state in the nanometer scale was directly observed *via* TEM analysis.

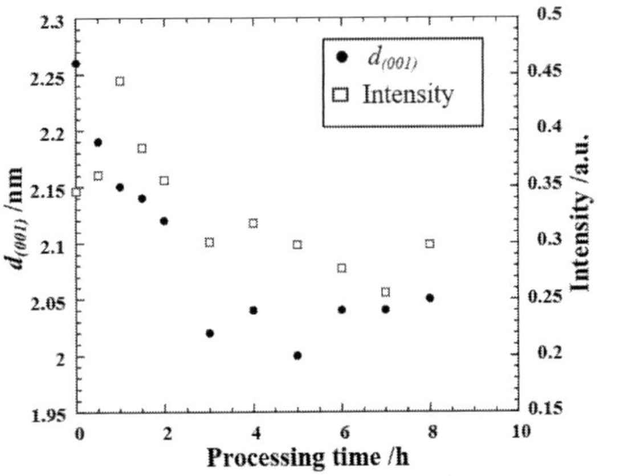

Fig.2 Time variation of $d_{(001)}$ and corresponding intensity during processing. Intensity is normalized by α(110)/γ(111) peak.

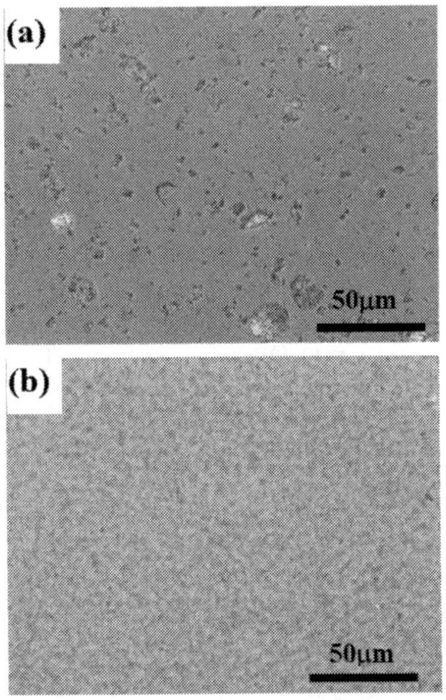

Fig.3 POM photographs of the mixture (a) un-processed sample and (b) solid-state processing for 8 h.

Fig. 4 shows the results of TEM bright field images with their FFT patterns of the mixtures corresponding to the POM experiments, in which dark entities are the cross section of layered nano-fillers. The disorder and delaminated silicate layer structure with the thickness of 3-7 nm are observed in TEM image compared with unprocessed sample.

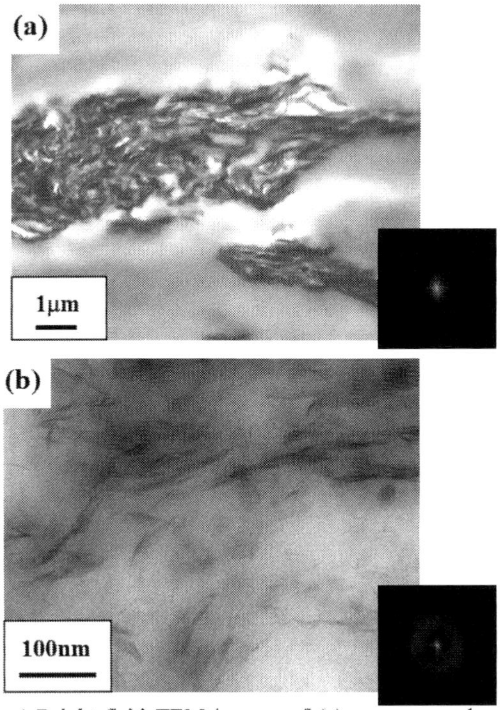

Fig.4 Bright field TEM images of (a) un-processed sample and (b) solid-state processing for 8 h. The insert in each image is a computed FFT spectrum of the micrograph.

Fig. 5 shows the one-dimensional scattering pattern obtained from FFT analysis. Here, the scattering intensity ($I(q)$) is shown as a function of the magnitude of the scattering vector (q). We found a broad shoulder around $q \sim 0.02$ nm^{-1} and small remnant peaks in the q range from 0.04 to 0.09 nm^{-1}. The broadness is due to the fact that the dispersed OMLF particles are rather disordered and short-range ordered. The decay of intensity in the q range from 0.02 to 0.1 nm^{-1}can be expected by a power law in the form $I(q) \cong q^{-1}$, suggesting the form factor is the randomly oriented rod. Apparently, this is consistent with the shape of the dispersed OMLF particles in the TEM image (Fig. 4(b)). We estimated the form factors obtained from TEM images and FFT analysis combined with Debye-Bueche (DB) theory, i.e. average value of the particle length (L_{MMT}), thickness (t_{MMT}), of the dispersed particles and the correlation length (ξ_{MMT}) between the particles. The details of the evaluation were described in our previous paper (Ray et al., 2003; Okamoto and Inoue 1993). The results are presented in Table 1.

The sample prepared by annealing without solid-state processing, L_{MMT} and t_{MMT} are in the range of 6-10 μm and 1-3 μm, respectively. On the other hand, PP/OMLF prepared by solid-state processing exhibits a small value of L_{MMT} (67±6 nm) with almost double of stacking of the silicate layers (t_{MMT} =3-7 nm). ξ_{MMT} value of PP/OMLF with processing (45±2 nm) suggests that a more uniform dispersion of the silicate layers seems to be attained due to the solid-state processing. This is a unique observation of the discrete silicate layers. The intercalants into nano-galleries act as a lubricant at processing temperature.

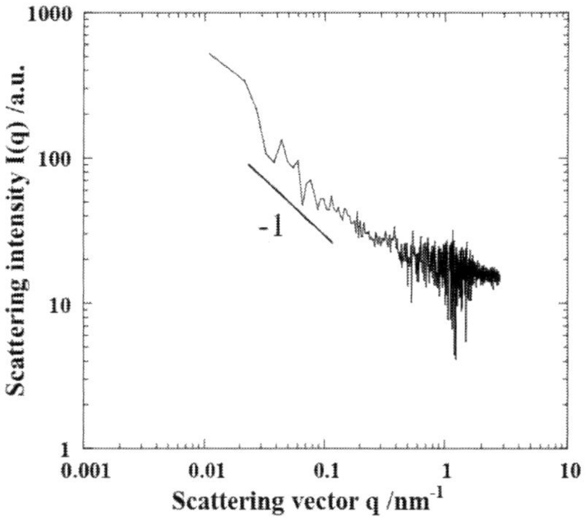

Fig.5 One-dimensional scattering pattern of processed PP/OMLF for 8 h obtained from FFT analysis.

Furthermore, we conducted solid-state processing by using internal mixer instead of alimina mortar. For processing for 5 h at 50 °C the dispersed morphology of the silicate layers exhibits same trend with applying pressure (see Fig. 6).

From these facts, the solid-state processing is extremely effective method to collapse the stacked structure. This processing leads to the layer delamination accompanying the discrete dispersion.

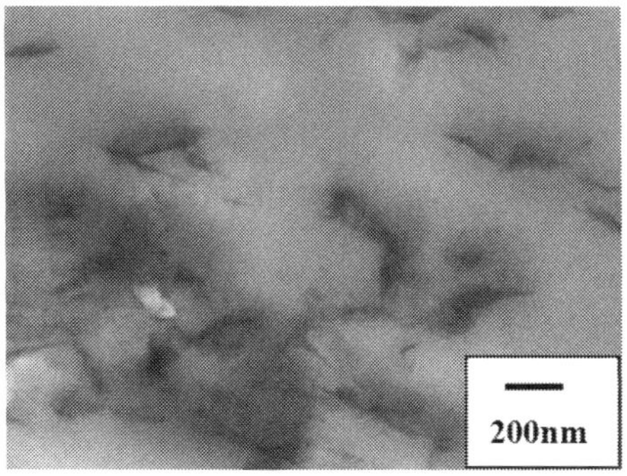

Fig.6 Bright field TEM image PP/OMLF prepared by internal mixer for 5h at 50 °C.

Table 1. Form factors of two nano-composites obtained from TEM and FFT analyses

Parameters	TEM	FFT with DB theory
L_{MMT} /nm	67±6	---
t_{MMT} / nm	5.8±0.5	3.7
ξ_{MMT} / nm	45±2	157

Conclusions

We have reported a novel and economic method for the nanoscale control of the dispersed layered fillers via solid-state processing. This processing led to delaminate of the silicate layers and attained the discrete dispersion. This approach can be extended to prepare polymeric nano-composites with delamination of the nano-fillers in overcoming the pressure drop within the nano-galleries.

References

Meille S, Bruckner S, Porzio W "γ-Isotactic polypropylene. A structure with nonparallel chain axes" Macromolecules, 23: 4114- 4121(1990).

Okamoto M, Inoue T "Reactive Processing of Polymer Blends: Analysis of Change in Morphological and Interfacial Parameter with Processing" Polym. Eng. Sci., 33, 175-182 (1993)

Saito T, Okamoto M, Hiroi R, Yamamoto M, Shiroi T "Poly(p-phenylenesulfide)-based nano-composite formation: Delamination of organically modified layered filler via solid-state processing" Polymer, 48, 4143- 4151(2007)

Sinha Ray S, Okamoto M "Polymer/Layered Silicate Nanocomposites: A Review from Preparation to Processing" Prog. Polym. Sci., 28, 1539-1641 (2003)

Sinha Ray S, Yamada K, Okamoto M, Ogami A, Ueda K "New Polylactide/Layered Silicate Nanocomposites.3. High-Performance Biodegradable Materials" Chem. Mater., 15, 1456-1465 (2003)

COMPOSITES OF POLY(ε-CAPROLACTONE) WITH $Mo_6S_3I_6$ NANOWIRES

S. J. Chin[1*], T. McNally[1]*, P. Hornsby[1], D. Vengust[2] and D. Mihailovic[2]

[1]*School of Mechanical and Aerospace Engineering, Queen's University Belfast, Belfast BT9 5AH, UK- schin02@qub.ac.uk; t.mcnally@qub.ac.uk; peter.hornsby@qub.ac.uk*
[2]*Complex Matter Department, Jozef Stefan Institute, Jamova 39, Ljubljana, SI-1000 Ljubljana, Slovenia, and Mo6 d.o.o., Teslova 30, SI-1000 Ljubljana, Slovenia-damjan.vengust@ijs.si; dragan.mihailovic@ijs.si*

Molybdenum sulphur iodine (MoSI) nanowires are electrically conducting nanofillers which have monodisperse properties. Composites of poly(ε-caprolactone) with $Mo_6S_3I_6$ were prepared using twin screw compounding with $Mo_6S_3I_6$ loadings of 0.05, 0.1, 0.3, 0.5, 1, 3, 5 wt%. Tensile strength and elongation at break decreased by up to 30%, particularly with higher $Mo_6S_3I_6$ loading, due to the presence of agglomerations, confirmed from field emission scanning electron microscopy (FESEM) studies. However, One Way Analysis of Variance (ANOVA) analysis showed statistically that the Young's modulus of PCL was unaffected by $Mo_6S_3I_6$ addition. Both differential scanning calorimetry (DSC) and hot stage polarised optical microscopy (HSPOM) analysis confirmed the nucleation effect of $Mo_6S_3I_6$ to PCL. The incorporation of $Mo_6S_3I_6$ to PCL resulted in a slight increase in spherulite size and increased crystallisation temperature, by up to 5°C. The electrical conductivity of PCL-$Mo_6S_3I_6$ composites were significantly enhanced by 7 orders of magnitude and a very low percolation threshold of 0.025wt% achieved for the PCL-$Mo_6S_3I_6$ system, superior than that reported for any polymer CNT composites prepared using melt mixing.

Introduction

There has been intense interest in the development of electrically conducting polymers for a range of applications including electromagnetic interference (EMI) shielding, electrostatic discharge (ESD) and as substrates for flexible electronics. By adding electrically conducting nanofillers, such as CNTs, electrons can hop between nanofillers dispersed throughout the polymer (insulator) matrix, allowing current flow. Researchers recognize that good dispersion of nanofillers is essential in attaining high performance polymer nanocomposites i.e. with much improved electrical, mechanical, thermal and other properties. Attempts to achieve complete dispersion and distribution of CNTs in polymer matrices typically involve two distinct approaches: a mechanical approach and/or surface modification of CNTs to promote polymer/CNT interaction. The former can damage the nanotube walls (McClory et al. 2009) by fragmenting CNTs if the process is excessively aggressive such as during ultrasonication and high shear compounding; the latter might introduce structural defects and deteriorate the perfect structure of tubes as a consequence of covalent attachment (Spitalsky et al. 2010). Non-covalent attachment can be employed to overcome this problem however the exact mechanism of interaction between CNTs and different surfactants/polymers is still uncertain (Hu et al. 2009). Therefore, from a melt mixing perspective there will always be a trade-off between the efficiency of the mixing technique and functionalisation of the nanotube. More recently, the synthesis of MoSI nanowires may offer a route to achieving agglomerate-free nanocomposites due to the very weak van der Waals attraction between such nanowires. MoSI offer one alternative to CNTs because of their monodisperse properties, the relative ease at which they can be functionalised, ease of fabrication with one step synthesis and comparable mechanical and electronic properties to CNTs (Mihailovic, 2009). MoSI grows in bundles with weak van der Waals bonding between wires, which exhibit low inter-wire shear strength and thus are relatively easy to disperse into single nanowires (Nicolosi et al. 2006) in many organic solvents without functionalization (McCarthy et al. 2007; Nicolosi et al. 2005a; Nicolosi et al. 2005b; Nicolosi et al. 2007). Sulfur based MoSI can be functionalized more readily than CNTs, by a single step dispersion in water, for example with gold nanoparticles or biomolecules (Ploscaru et al. 2005; Ploscaru et al. 2006; Ploscaru et al. 2007; Strle et al. 2009). The Young's modulus of MoSI was reported to be approximately 430GPa, although it is not as high as that proposed for CNTs, at 1000GPa (Kis et al. 2007). It was reported that the Young's modulus increased by 15% upon addition of 0.5wt% $Mo_6S_3I_6$ into a Nylon 6 matrix by melt mixing (Huskic, 2008). Unlike CNTs where electronic properties depend on tube chirality, all MoSI nanowires are metallic in nature regardless of stoichiometry (Yang et al. 2008), with electrical conductivity values up to 8.5 $\times 10^5$S/m being recorded (Strle, 2007). Murphy at al. (Murphy et al. 2006) studied a solution mixed PMMA-$Mo_6S_{4.5}I_{4.5}$ composite, achieving an electrical conductivity of 10^{-3} S/m for a MoSI volume fraction of 10^{-3} and the lowest electrical percolation threshold to date for any one dimensional conductor/insulator composite, 1.3 x 10^5 vol.%.

However, the two short studies cited with regard polymer-MoSI composites represent the total literature published on this topic to date. In this paper, we report the preparation of composites of PCL-$Mo_6S_3I_6$ by melt compounding and the effect of $Mo_6S_3I_6$ addition on the mechanical, electrical and thermal properties, and crystalline behaviour of PCL. To the best of our knowledge, the PCL-$Mo_6S_3I_6$system has not been reported previously.

Experimental

Preparation of PCL-$Mo_6S_3I_6$ Composites

As received pristine $Mo_6S_3I_6$ nanowires (Mo6 Ltd., Slovenia) were incorporated into CAPA® 6500 PCL (Solvay, Belgium) at loadings of 0.05, 0.1, 0.3, 0.5, 1, 3 and 5wt% by melt blending in a Dr Collin GmbH co-rotating intermeshing twin screw extruder at 60 to 70°C using a screw speed of 60rpm.

Characterisation Methods

Injection moulded samples of composites were freeze-fractured in liquid nitrogen and gold coated to facilitate imaging using a JEOL JSM-6500F field emission scanning electron microscopy (FESEM) at an accelerating voltage of 5kV. An Instron 5564 twin column tensile tester with 5kN load cell and 2603-08 long travel extensometer was used to measure the static mechanical properties of PCL-$Mo_6S_3I_6$ composites. At least five dog-bone specimens for each loading were prepared by adhering to ISO 527-2:1996. Two different test speeds of 5mm/min and 500mm/min were employed to test the specimen until 10mm maximum strain and failure respectively for a gauge length of 25mm. One Way Analysis of Variance (ANOVA) was applied to ascertain if addition of $Mo_6S_3I_6$ had a statistically significant effect on tensile properties across any of the composites under study. A value of $P < 0.05$ and F ratio>F critical was accepted as statistically significant. Calorimetric measurements was performed using a sample mass of 5±0.1mg using a PerkinElmer Diamond differential scanning calorimetry (DSC) coupled with Pyris™ software for data collection and quantitative analysis, with a ramp rate of 250°C/min over a thermal and cooling cycle from -110°C to 120°C. Crystallisation behaviour of composites were recorded in real time video with Nikon Eclipse ME600 microscope equipped with LINKAM hot stage and Nikon Digital Sight DS-5M charge-coupled devide (CCD) camera, for subsequent analyis using Lucia G software. Electrical resistivity was measured at ambient temperature, to ASTM D257-07 where a direct voltage of 1V was applied to two electrodes of a Keithley Model 8009 resistivity test fixture in contact with thin films and a read-off time of 1min. Current was distributed through the volume of the specimen and a volume resistivity reading displayed on a Keithley Model 6517A electrometer.

Results and Discussion

Dispersion and Distribution of $Mo_6S_3I_6$ in the PCL matrix

Different levels of $Mo_6S_3I_6$ dispersion in the PCL matrix were distinguishable from the FESEM micrographs. In Figure 1(a) the $Mo_6S_3I_6$ nanowires, (indicated by arrows), are well aligned with preferential direction, which may be caused by the extensional forces achieved during the extrusion process (Wong et al. 2003). As the $Mo_6S_3I_6$ content increased, as shown in Figure 1(b) the particle-particle distance are thought to be closer until intermolecular van der Waals forces become strong enough to let $Mo_6S_3I_6$ form agglomerations in the composites. Such an observation has been reported in other polymer nancomposite systems (Huang et al. 2008; Lee et al. 2008). The nanowires in Figure 1(c) show the nanowires/ropes to be reasonably distributed in the PCL with a small amount of aggregrates (as indicated by circles) observed.

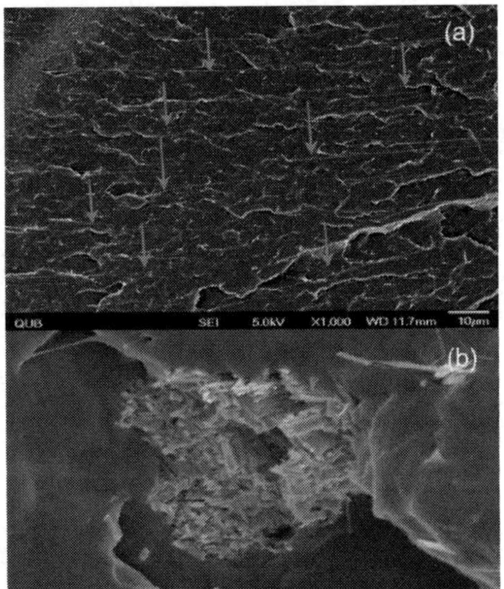

Figure 1 Cryofractured surface of PCL composites with $Mo_6S_3I_6$ content of (a) 0.5wt% (b, c) 5wt%.

Mechanical Properties

The ultimate tensile strength and englogation at break decreased by up to 20% and 30%, respectively, with higher $Mo_6S_3I_6$ content, as shown in Figure 2. The former may be attributed to the formation of aggregates that serve as stress concentrators and accelerate stress build-up, hence initiate premature failure in the composites compared to unfilled PCL. This was verified by FESEM image analysis, as described above. The latter may be associated with poor interfacial interaction between $Mo_6S_3I_6$ and PCL thus reducing any reinforcement effect. Figure 1(c) suggests poor bond strength between $Mo_6S_3I_6$ and PCL since many nanowires protrude (as depicted by arrows) above the polymer surface (Lee et al. 2008). The embrittlement of other polymers with the addition of other nanofillers such as CNTs or carbon black has also been reported elsewhere (McNally et al. 2005; Zhou et al. 2006). The Young's moduli of the PCL-$Mo_6S_3I_6$ composites, see Figure 3, did not change statistically with the addition of $Mo_6S_3I_6$ when compared with unfilled PCL ($P>0.05$ and F ratio>Fcritical, ANOVA). This finding is very important because the aforementioned defects did not significantly reduce the final stiffness perfomance of the composites.

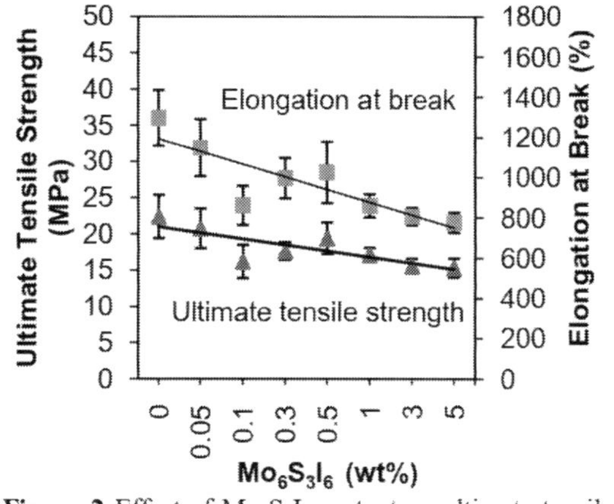

Figure 2 Effect of $Mo_6S_3I_6$ content on ultimate tensile strength and elongation at break of PCL composites.

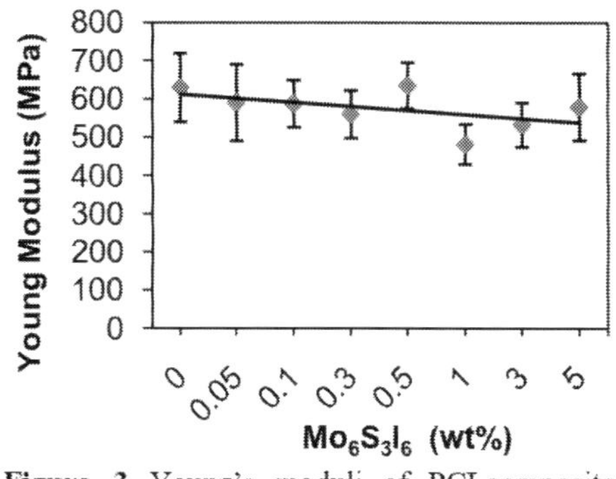

Figure 3 Young's moduli of PCLcomposites with various loading of $Mo_6S_3I_6$.

Electrical Properties

In Figure 4, the unfilled PCL shows a volume resistivity of around 3 x $10^{14}\Omega cm$, however, the incorporation of $Mo_6S_3I_6$ in PCL reduces resistivity by 7 orders of magnitude to about $10^7\Omega cm$. An abrupt decrease is observed between 0 to 0.05wt% $Mo_6S_3I_6$ addition and thus a percolation threshold was attained at approximately 0.025wt%.. Although this percolation (0.025wt%) is above that reported for the solution prepared PMMA-$Mo_6S_{4.5}I_{4.5}$ composite in (Murphy et al. 2006), it is still significantly smaller than that reported for other polymer-CNT composites (0.05wt%) (McClory et al. 2009). It is proposed that a percolated network of conductive MoSI nanowires is created throughout the PCL matrix, as reported for CNTs composites (Du et al. 2004). A previous study on the electrical conductivity of $Mo_6S_3I_6$ (Bercic et al. 2006) indicated that the presence of second phases, for example interstitial iodine, water or impurities present in the nanowires, typically, $Mo_6S_6I_2$ crystals, can contribute to the conductivity improvement due to their metallic nature by forming good electrical contacts between $Mo_6S_3I_6$ nanowires (Dvorsek et al. 2007). The $Mo_6S_6I_2$ crystal is a cubic structure and has been observed in these composites, see Figure 5.

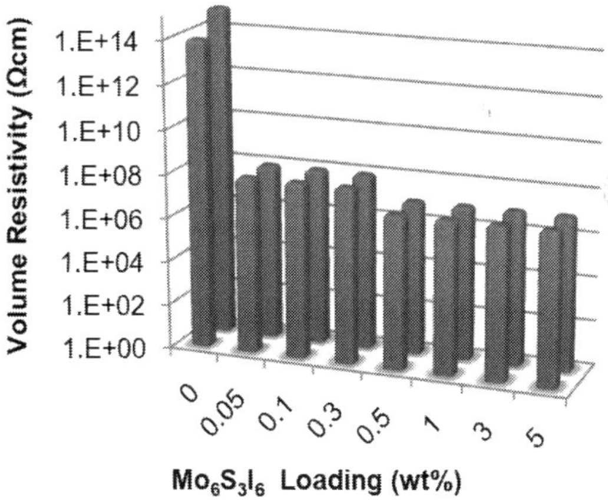

Figure 4 The influence of $Mo_6S_3I_6$ content on volume resistivity of PCL.

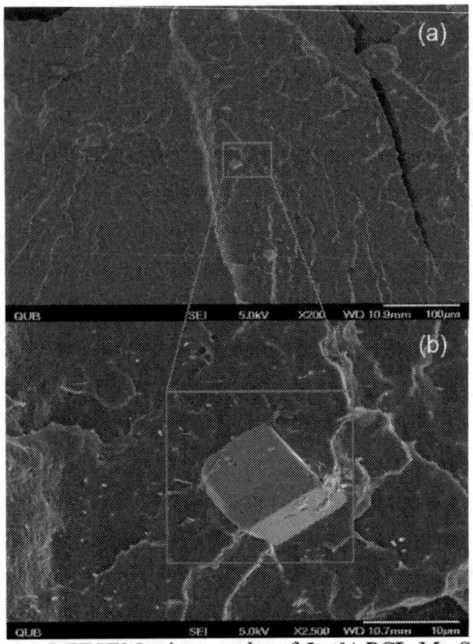

Figure 5 FESEM micrographs of 5wt% PCL-$Mo_6S_3I_6$, (a) $Mo_6S_6I_2$ crystals dispersed in the PCL matrix, highlighted by boxes and circles (b) $Mo_6S_6I_2$ crystals at higher magnification .

Thermal Properties

Table 1 lists the temperature of melting (T_m), glass transition temperature (T_g), enthalpy of melting (ΔH_m), enthalpy of crystallization (ΔH_c) and full width at half height ($FWHH_m$) for PCL and the PCL-$Mo_6S_3I_6$ composites. Interestingly, T_m, T_g and X_c of PCL are largely unaffected by $Mo_6S_3I_6$ addition. Despite little or no change in X_c, there is an indication of a nucleation effect as T_c increased with increasing $Mo_6S_3I_6$ by up to 5°C for 5wt% $Mo_6S_3I_6$ addition. This behaviour was confirmed from observations made from hot-stage polarised optical microscopy (HSPOM) experiments, see Figure 6. The average spheruite size for unfilled PCL was 6.7μm (Figure 6 (a)), however, even incorporation of 0.1wt% $Mo_6S_3I_6$ to PCL yielded some larger spherulites up to 9.9μm, although significant spherulite impingment was also evident and an increase in nucleation density compared to unfilled PCL in Figure 6(a). A large number of uneven small crystals can be seen almost covering all the interspherulitic boundaries which become indiscernible. The change in spherulite size may explain the slight reduction in yield stress and elongation at break of PCL on nanowire addition. The time for complete crystallisation (t_c) was also recorded. By way of example, for 0.1wt% $Mo_6S_3I_6$, t_c was 920s, shorter than that obtained for the unfilled PCL, with t_c =960s. This clearly indicated that the crystallisation rate increased for the composites and is similar to that reported for other PCL composite systems (Pan et al. 2008; Qiao et al. 2009; Wu and Chen. 2006). As previously reported extensively for a wide range of PCL nanocomposites (Chen and Wu. 2007; Saeed and Park. 2007; Wu and Chen. 2006), large spherullites with the well described Maltese-cross are visible. This correlates well with the broader crystallisation distribution revealed by the minor change in the $FWHH_c$, especially for the 1, 3 and 5wt% composites. This effect may be associated with two factors: firstly, there are more irregular spherullites in the composites which are not capable of penetrating each other in the confined structure; secondly, the broader transition may be attributed to the effect the presence of impurities i.e. $Mo_6S_6I_2$ from $Mo_6S_3I_6$, have on crystallisation behaviour (Lepoittevin et al. 2002).

Table 1 DSC results of PCL and PCL-$Mo_6S_3I_6$ composites.

$Mo_6S_3I_6$ (wt%)	T_m [a] (°C)	T_c [a] (°C)	T_g [a] (°C)	ΔH_m [a] (J/g)	ΔH_c [a] (J/g)	X_c [a, b] (%)	$FWHH_m$ [a] (°C)	$FWHH_c$ [a] (°C)
0	62	9.6	-52	25.7	-40	18.9	18.5	11.6
0.05	64	12.2	-52	26.5	-41	19.5	19.1	11.6
0.1	65	12.1	-51	23.8	-40	17.5	19.4	13.6
0.3	61	14.4	-52	23.1	-40	17.0	15.6	11.2
0.5	62	13.7	-53	24.0	-40	17.7	15.8	9.6
1	65	13.6	-52	24.7	-39	18.3	18.8	13.2
3	65	13.1	-52	22.7	-39	17.2	20.5	14.0
5	64	14.8	-52	23.6	-37	18.3	18.7	15.2

[a]The reported values correspond to melting point (T_m), crystallisation temperature (T_c), glass transition temperature (T_g), enthaply of melting (ΔH_m), enthaply of crystallisation (ΔH_c), degree of crystallinity (X_c), full width at half height of endothermic peak ($FWHH_m$) and full width at half height of exothermic peak ($FWHH_c$). [b] X_c is calculated from ΔH_m based on 100% crystalline PCL=136J/g (Khambatta et al. 1976).

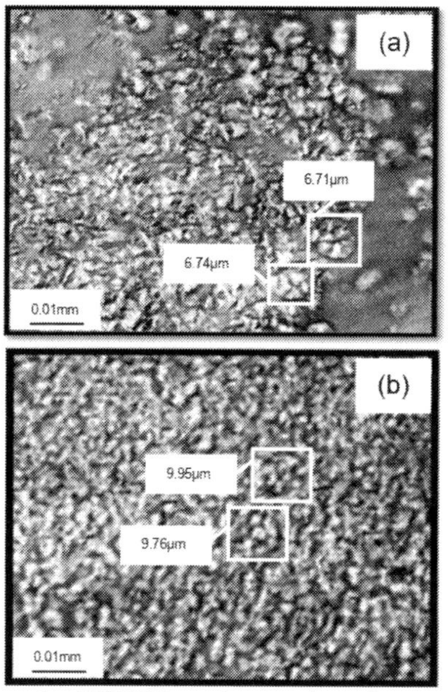

Figure 6 HSPOM micrographs demonstrating spherulitic morphology (a) PCL (b) 0.1wt% PCL-$Mo_6S_3I_6$ at ca. 30°C during cooling from melts.

Conclusions

1. FESEM micrographs show clearly the aggregation of $Mo_6S_3I_6$ in PCL matrix and poor interfacial bond strength in PCL-$Mo_6S_3I_6$ composites. This explains why both the ultimate tensile strength and elongation at break of the PCL composite are not improved by the addition of $Mo_6S_3I_6$. Nevertheless, the Young's moduli of the PCL-$Mo_6S_3I_6$ composites are statistically unaffected.
2. $Mo_6S_3I_6$ nanowires/bundles act as nucleating sites for PCL crystallisation, manifest by higher crystallsation temperatures, faster crystallisation rates and on average larger spherulite sizes in PCL-$Mo_6S_3I_6$ composites. T_m, T_g and X_c were unaffetced by $Mo_6S_3I_6$ addition.
3. The PCL-$Mo_6S_3I_6$ composites exhibited a sevenfold increase in electrical conductivity compared to the unfilled PCL with a low percolation threshold of 0.025wt%, much lower than that reported for polymer-CNT composites prepared via melt mixing methods.
4. It can concluded that only very low $Mo_6S_3I_6$ content is required to achieve the electrical conductivity levels required for electrostatic dissipation applications, without impairing the elastic modulus, despite a change of spherulite size.

Acknowledgements

The authors acknowledge the technical assistance of Dr B. Millar, Dr P. Douglas, Mr G. Garrett, Mr S. McFarland in composite fabrication and characterization. S. J. Chin is much indebted to QUB for funding a studentship.

References

Bercic B., et al, "Transport Properties of Mo6S3I6 Nanowire Networks", Appl. Phys. Lett., **88,** 173103 (2006).

Chen E., Wu, T., "Isothermal Crystallization Kinetics and Thermal Behavior of Poly(ε-caprolactone)/Multi-walled Carbon Nanotube composites", Polym. Degrad. Stab., **92,** 1009-1015 (2007).

Du F., Scogna, R. C., Zhou, W., Brand, S., Fischer, J. E., Winey, K. I., "Nanotube Networks in Polymer Nanocomposites: Rheology and Electrical Conductivity", Macromolecules, **37,** 9048-9055 (2004).

Dvorsek D., Zumer, M., Nemanic, V., Mihailovic, D., Vengust, D., "Growth and Field Emission Properties of Vertically Aligned Molybdenum-sulfur-iodine Nanowires on Molybdenum and Quartz Substrates", J. Appl. Phys., **102,** 114308 (2007).

Hu C., Xu, Y., Duo, S., Zhang, R., Li, M., "Non-Covalent Functionalization of Carbon Nanotubes with Surfactants and Polymers", J. Chin. Chem. Soc., **56,** 234-239 (2009).

Huang X., Jiang, P., Kim, C., Ke, Q., Wang, G., "Preparation, Microstructure and Properties of Polyethylene Aluminum Nanocomposite Dielectrics", Composites Sci. Technol., **68,** 2134-2140 (2008).

Huskic, M., et al. MoSI-PA6 Composites Shows Enhanced Mechanical Properties, http://www.mo6.com/News/ (accessed October, 2009).

Khambatta F. B., Warner, F., Russell, T., Stein, R. S., "Small-Angle X-ray and Light Scattering Studies of the Morphology of Blends of Poly(ε-caprolactone) with Poly(vinyl Chloride)", J. Polym. Sci. Polym. Phys. Ed., **14,** 1391-1424 (1976).

Kis A., et al, "Nanomechanical Investigation of Mo6S9-xIx Nanowire Bundles", Small, **3,** 1544-1548 (2007).

Lee K. Y., Kim, K. Y., Han, W. Y., Park, D. H., "Thermal, Electrical Characteristics and Morphology of Poly(Ethylene-Co-Ethyl Acrylate)/CNT Nanocomposites", IEEE Trans. Dielectr. Electr. Insul., **15,** 205-213 (2008).

Lepoittevin B., et al, "Poly(ε-caprolactone)/Clay Nanocomposites by in-Situ Intercalative Polymerization Catalyzed by Dibutyltin Dimethoxide", Macromolecules, **35,** 8385-8390 (2002).

McCarthy D. N., et al, "Dispersion and Purification of Mo6S3I6 Nanowires in Organic Solvents", J. Appl. Phys., **101,** 014317 (2007).

McClory C., Chin, S. J., McNally, T., "Polymer Carbon Nanotube Composites", Aust. J. Chem., **62,** 1-24 (2009).

McNally T., et al, "Polyethylene Multiwalled Carbon Nanotube Composites", Polymer, **46,** 8222-8232 (2005).

Mihailovic D., "Inorganic Molecular Wires: Physical and Functional Properties of Transition Metal Chalco-halide Polymers", Prog. Mater. Sci., **54,** 309-350 (2009).

Murphy R., et al, "Observation of Extremely Low Percolation Threshold in Mo6S4.5I4.5 nanowire/polymer Composites", Scr. Mater., **54,** 417-420 (2006).

Nicolosi V., McCarthy, D. N., Vengust, D., Mihailovic, D., Blau, W. J., Coleman, J. N., "Exfoliation of Mo6SxI9-x Nanowires in Common Solvents", Eur. Phys. J. Appl. Phys., **37,** 149-159 (2007).

Nicolosi V., Vengust, D., Mihailovic, D., Blau, W. J., Coleman, J. N., "Debundling by Dilution: Observation of Significant Populations of Individual MoSI Nanowires in High Concentration Dispersions", Chem. Phys. Lett., **425,** 89-93 (2006).

Nicolosi V., et al, "Solubility of Mo6S4.5I4.5 Nanowires in Common Solvents: A Sedimentation Study", J. Phys. Chem. B, **109,** 7124-7133 (2005a).

Nicolosi V., et al, "Solubility of Mo6S4.5I4.5 Nanowires", Chem. Phys. Lett., **401,** 13-18 (2005b).

Pan P., Zhu, B., Dong, T., Serizawa, S., Iji, M., Inoue, Y., "Kenaf Fiber/Poly(ε-caprolactone) Biocomposite with Enhanced Crystallization Rate and Mechanical Properties", J. Appl. Polym. Sci., **107,** 3512-3519 (2008).

Ploscaru, M., Kokalj, S. J., Uplaznik, M., Vengust, D., Turk, D., Mihailovic, D., "Force Spectroscopy Measurements of MoSIx Nanowires with Functionalized Molecules", 20th International Winterschool/Euroconference on Electronic Properties of Novel Materials, 3325-3328 (2006).

Ploscaru M., et al, "Self-assembly of Gold Particles to MoSxIy Nanowires Ends", Electronic Properties of Novel Nanostructures, **786,** 374-377 (2005).
Ploscaru M. I., et al, "Mo6S9-xIx Nanowire Recognitive Molecular-Scale Connectivity", Nano Lett., **7,** 1445-1448 (2007).

Qiao X., Li, W., Sun, K., Xu, S., Chen, X., "Nonisothermal Crystallization Behaviors of Silk-Fibroin-Fiber-Reinforced Poly(ε-caprolactone) Biocomposites", J. Appl. Polym. Sci., **111,** 2908-2916 (2009).

Saeed K., Park, S., "Preparation and Properties of Multiwalled Carbon Nanotube/Polycaprolactone Nanocomposites", J. Appl. Polym. Sci., **104,** 1957-1963 (2007).

Spitalsky Z., Tasis, D., Papagelis, K., Galiotis, C., "Carbon Nanotube–Polymer Composites: Chemistry, Processing, Mechanical and Electrical Properties", Progress in Polymer Science, **35,** 357-401 (2010).

Strle, J., Self-Assembled Mo6S9-xIx Networks, http://videolectures.net/slonano07_strle_smn/video/1/slides/ (accessed November, 2009).

Strle J., Vengust, D., Mihailovic, D., "Inorganic Molecular-Scale MoSI Nanowire-Gold Nanoparticle Networks Exhibit Self-Organized Critical Self-Assembly", Nano Lett., **9,** 1091-1095 (2009).

Wong M., Paramsothy, M., Xu, X. J., Ren, Y., Li, S., Liao, K., "Physical Interactions at Carbon Nanotube-Polymer Interface", Polymer, **44,** 7757-7764 (2003).

Wu T. M., Chen, E. C., "Crystallization Behavior of Poly(ε-caprolactone)/Multiwalled Carbon Nanotube Composites", J. Polym. Sci. Part B, **44,** 598-606 (2006).

Yang T., Berber, S., Tomanek, D., "Compositional Ordering and Quantum Transport in Mo6S9-xIx Nanowires: Ab Initio Calculations", Phys. Rev. B, **77,** 165426 (2008).

Zhou Z., Wang, S., Zhang, Y., Zhang, Y., "Effect of Different Carbon Fillers on the Properties of PP Composites: Comparison of Carbon Black with Multiwalled Carbon Nanotubes", J. Appl. Polym. Sci., **102,** 4823-4830 (2006).

Polymer Processing Society (PPS)
26th Annual Meeting, July 4-8, 2010, Banff, Canada

EFFECTS OF FEEDING ROUTE AND STATE OF CURE UPON PERCOLATION THRESHOLD AND RHEOLOGICAL BEHAVIOR OF ELECTRICALLY CONDUCTIVE DYNAMICALLY VULCANIZED PP/EPDM/EXPANDED GRAPHITE NANOCOMPOSITE

Hossein Mirzazadeh[1], Ali Asghar Katbab[1, 2,*], Andrew N. Hrymak[3]

[1] Amirkabir University of Technology, Polymer Engineering Department
[2] Amirkabir Nanotechnology Research Institute
[3] University of West Ontario, Faculty of Engineering, Canada

Abstract
Electrically conductive thermoplastic vulcanizates (TPV) nanocomposites based on polypropylene (PP), ethylene-propylene-diene rubber (EPDM), maleic anhydride grafted polypropylene (PP-g-MA) and maleic anhydride grafted EPDM (EPDM-g-MA) as compatibilizers, and expanded graphite (EG) have been successfully prepared via dynamically vulcanization of the EPDM phase during melt mixing process in an internal mixer. The influence of the feeding route upon partitioning of the conductive expanded graphite nanolayers between PP matrix and dispersed EPDM phase, and hence conductivity percolation threshold (σc) has been investigated. Correlation between the TPV morphology developed during various times of dynamic vulcanization process (SOC) and conductivity threshold has also been studied. Results showed that incorporating EG in the form of masterbatch with compatibilizers leads to the longer conductivity threshold (> 5 vol %) compared with direct melt blending (3.5 vol %). This was found to be attributed to the encapsulation of the EG by EPDM phase as well as nanolayers breakdown during long time of mixing process, and hence retardation of the formation of conductive networks throughout the PP matrix. This was consistent with the lower dynamic melt elastic modulus (G') with less non-terminal behavior measured by RMS for the TPV nanocomposites prepared by pre-blending method. Higher residing of EG in EPDM phase when it was masterbatched with EPDM-g-MA was evidenced by micro-morphological development during dynamic vulcanization of the blend. Conductivity showed to be dependent on the state of cure (SOC) once conductive system was incorporated.

Keywords: conductivity, feeding route, expanded graphite, nanocomposite.

Introduction

Polymer composites with electrical conductivity have attracted great attention over the past few decades, as they can be used for numerous purposes such as antistatic materials, pressure and gas sensors, electromagnetic shielding materials, etc [1-3]. Electrically conductive nanocomposites based on conductive nanofillers have shown distinct features including low conductivity threshold and hence low weight with better processability and mechanical properties [2-4].

Natural graphite flake (NGF) is a layered material consisting of graphene layers formed by sp2 hybridized carbons, resulting in delocalization of π bonds and high electrical conductivity of graphite layers (1.567×103 S.cm-1). Single graphite sheets are stacked parallel to each other with gallery spacing of 0.335nm [1-5].

Natural graphite flake intercalated with acid (graphite intercalated compound or GIC), can be expanded up to hundreds of times of its initial volume, resulting in the separation of graphene nano sheets along the c-axis of graphene layers, and this material is called expanded graphite (EG) [5].

Thermoplastic vulcanizates (TPVs) with a matrix-disperse type of morphology are blends of a crystalline polymer, e.g. polypropylene (PP), as the matrix and a rubbery phase in the form of aggregates of crosslinked droplets, and prepared via melt mixing process [6]. This process yields a fine dispersion of micron-sized crosslinked rubber in the thermoplastic matrix with intensified interface. TPVs offer a wide range of interesting properties such as rubbery behavior at low temperatures, and processability like ordinary thermoplastics [7]. The expanded honey combed structure of EG with –OH and –COOH

*: Corresponding Author: Prof. A.A. Katbab, katbab@aut.ac.ir

functional groups on the edges of nanolayers gives good miscibility with polymer materials. Nevertheless, the dispersion and size distribution of EG in the polymer matrixes govern conductivity behavior and insulator-conductor transition threshold. These are controlled by the route of mixing, processing conditions, and degree of compatibilization.

The main objective of the present work is to prepare dynamically vulcanized thermoplastic elastomer nanocomposites based on PP/ethylene-propylene-diene rubber (EPDM) and EG as electrically conductive filler via in-situ dynamic vulcanization of EPDM phase in the presence of EG during melt mixing process and evaluating the effect of feeding route and state of cure (SOC) on the electrical percolation threshold and rheological behavior of nanocomposites. Therefore, the electrical conductivity behaviors of the different nanocomposite TPVs have been evaluated and compared and the relationship between melt rheological characteristics and insulator to conductor threshold (σ_c) has been investigated.

Experimental

Materials

PP with the trade name of Poliran PI800, Bandar Imam Petrochemical Co. (Iran), and EPDM with the trade name of Keltan 2340A, DSM (Netherlands), were used for preparing in-situ dynamically vulcanized TPV samples. Maleic anhydride grafted PP (PP-g-MAH), from Honam Petrochemicals Co., Korea, with the trade name of Popelen CM 1120 and maleic anhydride grafted EPDM (EPDM-g-MAH), from Pluss Polymers Pvt. Ltd., India, with the trade name of OPTIM P-635 were used as compatibilizers. GIC was supplied from Beijing Invention Biology Engineering & New Material Co. (China), with the trade name of EX 9550-200N.

Preparation of EG

Worm-like expanded graphite composed of nanolayers with the thickness of 100-400nm was prepared by rapid heating of GIC at 1000 °C for 30 seconds in a furnace.

Preparation of nanocomposite samples

A brabender internal mixer operating at 180 °C and at a rotor speed of 80 rpm and filling factor of 75% was used for melt mixing process. The as-prepared mixes were ground into powder and then formed into sheets with the dimensions of 100×100×1 mm^3 using a hydraulic hot press at 180 °C.

TPV nanocomposite samples were prepared by direct melt mixing of PP and EPDM (with the ratio of PP/EPDM = 40/60) in the presence of the curing ingredients (ZnO: 5 phr, stearic acid: 1.5 phr, TMTD: 1 phr, MBTS: 0.5 phr, sulfur: 2 phr). The composition and mixing information of prepared samples are presented in Table 1. In the case of samples which were prepared via masterbatch feeding route, the expanded graphite and compatibilizer were mixed with their corresponding matrix prior to the dynamic vulcanization process. In the case of samples which were prepared via direct melt mixing process, expanded graphite and compatibilizer were fed into the mix before the incorporation of the curing ingredients. The ratio of compatibilizer to EG was 3 for all of the prepared samples. Mixing process was continued till the mixing torque reached to a constant level. The time of addition of curing system for the samples which were prepared via masterbatch and direct feeding route was 6th and 9th minutes, respectively. The schematic graph of internal mixer's torque-time for preparation of a sample of nanocomposite TPV is presented in Figure 1.

Table 1- Coding of the prepared samples

Sample Code	EG Content (Vol %)	Mixing Time (min)	EG Feeding Method
TPV-Ref	0	12	-
TPV-PP-3%	1.2	12	Masterbatch with PP phase
TPV-PP-4%	1.6	12	
TPV-PP-5%	2	12	
TPV-EPDM-3%	1.8	12	Masterbatch with EPDM phase
TPV-EPDM-4%	2.4	12	
TPV-EPDM-5%	3	12	
TPV-Both-4%	4	12	Masterbatch with both PP and EPDM phases
TPV-Both-5%	5	12	
TPV-D-5%-PPg-EPg	5	15	Direct Feeding of EG to the blend of PP and EPDM at minute 4, and before addition of curing system
TPV-D-5%-PPg	5	15	
TPV-D-5%-No	5	15	
TPV-D-3%	3	15	
TPV-D-1%	1	15	
TPV-D-5%-1 min	5	10	
TPV-D-5%-30 sec	5	9.5	
TPV-D-5%-0sec	5	9	

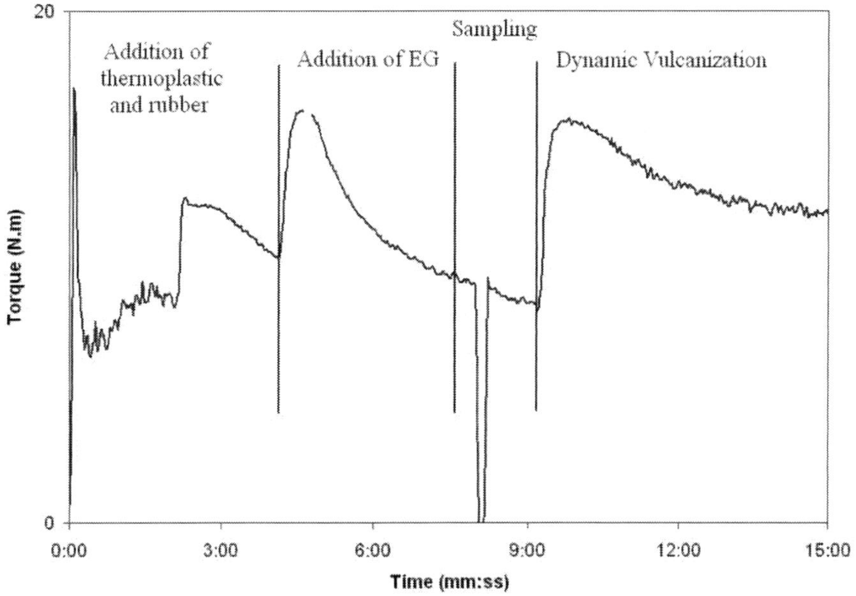

Figure 1 – schematic graph of internal mixer's torque-time for preparation of a sample of nanocomposite TPV

Characterization

The volume conductivity (σ) of the plate like samples was measured at room temperature according to the ASTM D991 standard test method with a conductometer. Scanning electron microscopy (SEM) (Seron, AIS-2100, Korea) was performed on cryofractured surfaces of samples to observe the dispersion of graphite nanolayers within the nanocomposite matrix. Melt rheological characterization was also performed on the samples using a parallel plate rheometer (Paar Physica USD 200, Austria) and disk shaped samples, with 2.5cm in diameter and 1 mm thickness. Strain sweep test was first done in the range of 0.1-100% at 220°C, in order to verify that measurements were within the linear viscoelastic regime, and then frequency sweep test was carried out between 0.1 and 1000 s-1.

Results and Discussion

Morphology Studies

Figure 2 presents a comparison between SEM micrographs of different nanocomposite TPV samples. The droplet size of EPDM disperse phases were measured using image analysis software (ImageJ 1.42q) and the results are presented in Table 2. It was observed that the presence of EG nanolayers has increased the size of EPDM droplets in TPV nanocomposite samples. This is related to the effect of EG nanolayers on the interfacial interaction between two phases in the TPV which was increased by addition of EG to PP and EPDM phases.

In addition, measuring the lateral size of EG clusters by image analysis software, showed that the lateral size of EG clusters is smaller the samples which are made by masterbatch mixing method in compare of the samples which are made by direct feeding method. It can be concluded that longer mixing time and higher amount of applied shear rate in masterbatch mixing method is the reason for EG nanolayers' break down and smaller lateral size of EG clusters.

Table 2- measured droplet size of EPDM in different nanocomposite TPV samples

Sample Code	EG Content (Vol %)	Droplet Size (μm)
TPV-Ref	0	0.34
TPV-EPDM-5%	3	6.18
TPV-Both-5%	5	1.28
TPV-D-5%-PPg-EPg	5	2.73
TPV-D-5%-PPg	5	0.87
TPV-D-5%-No	5	0.56

Table 3- measured lateral size of EG clusters in different nanocomposite TPV samples

Sample Code	EG Content (Vol %)	Lateral Size of EG Clusters (μm)
TPV-EPDM-5%	3	60.1
TPV-Both-5%	5	58.0
TPV-D-5%-PPg-EPg	5	134.6
TPV-D-5%-PPg	5	117.1
TPV-D-5%-No	5	109.8

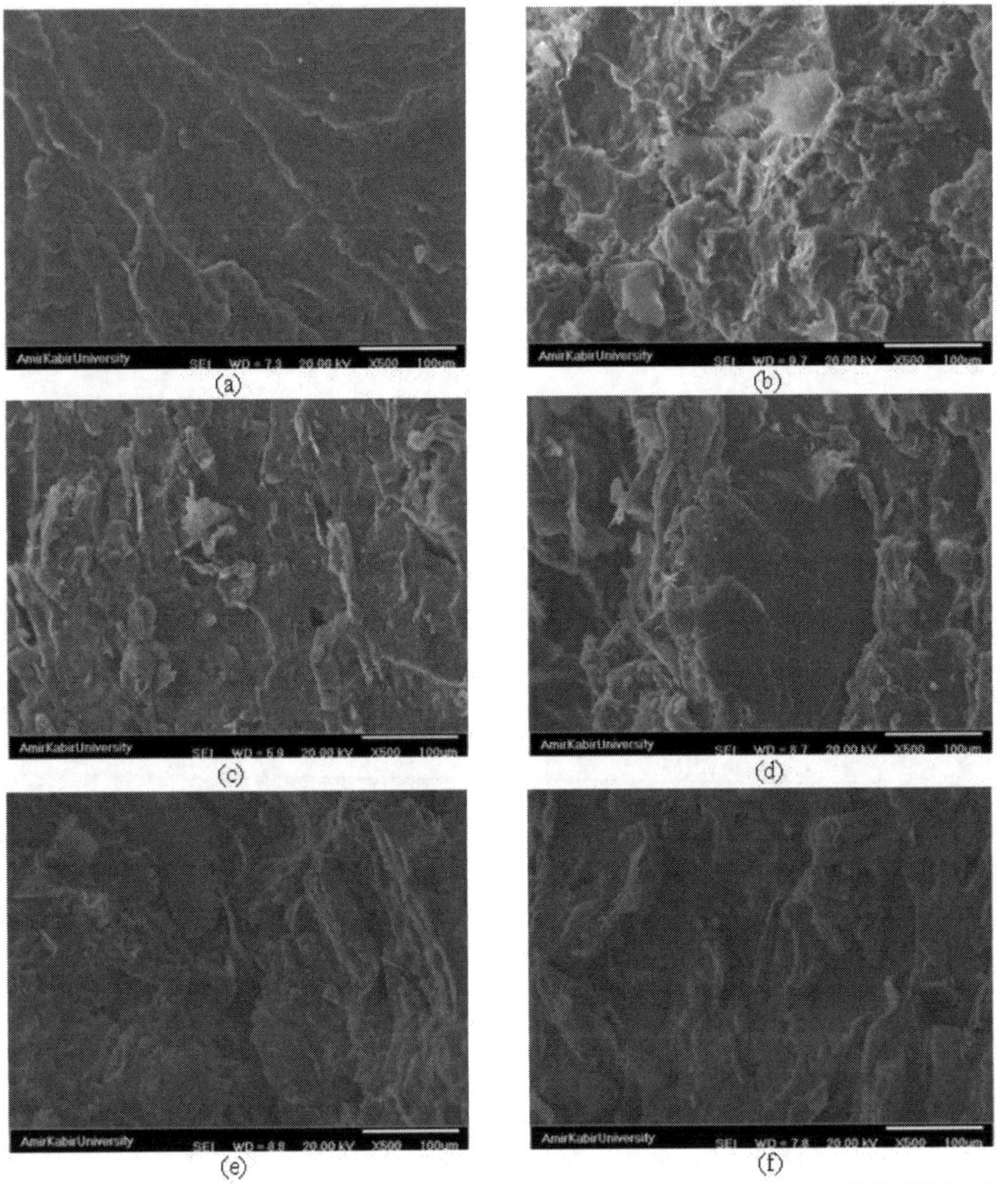

Figure 2 - SEM micrographs of different nanocomposite TPV samples: a) TPV-Ref, b) TPV-EPDM-5%, c) TPV-Both-5%, d) TPV-D-5%-PP-EPg, e) TPV-D-5%-PPg and f) TPV-D-5%-No

Linear Viscoelastic Properties

Melt viscoelastic behavior of polymer/clay nanocomposites can provide useful information to understand melt rheology and microstructure relationship for these materials. In rheological and melt viscoelastic analysis, dynamic strain sweep is first performed on the nanocomposite samples in order to characterize strain dependency of the viscoelastic behavior and determine linear viscoelastic region. Since the melt viscosity (η) and storage modulus (G') are mainly sensitive to the microstructural changes which might occur during dynamic melt rheometry, therefore these two parameters have been followed in this study. It was observed that all nanocomposite TPV samples exhibited a linear melt viscoelastic behavior bellow the strain amplitude of 10%. Therefore, the frequency sweep tests were carried out in the strain level of 5%. The dependency of storage modulus (G') to frequency for different nanocomposite TPV samples is presented and compared in Figures 3.

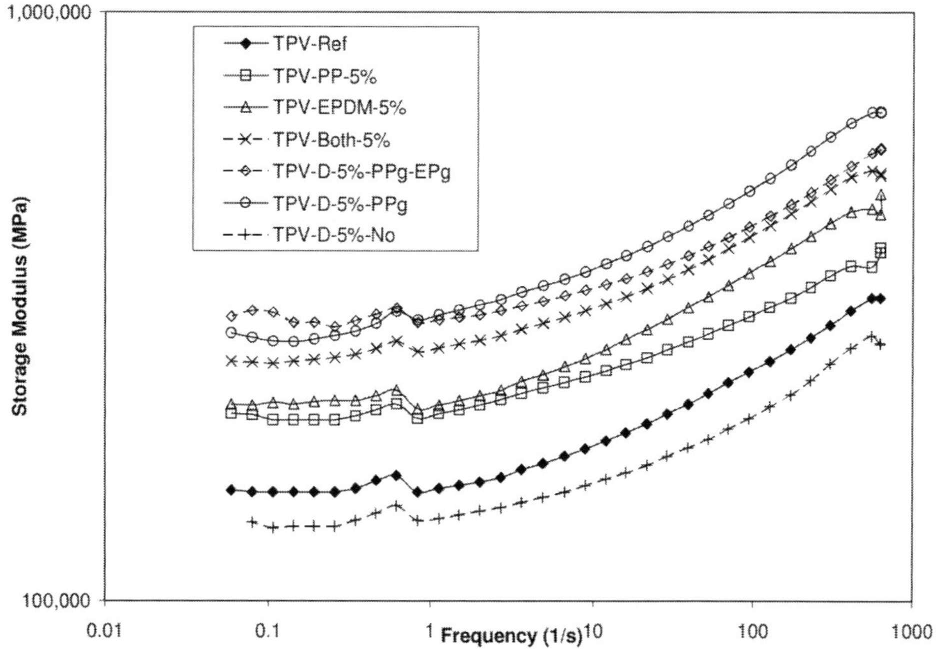

(a)

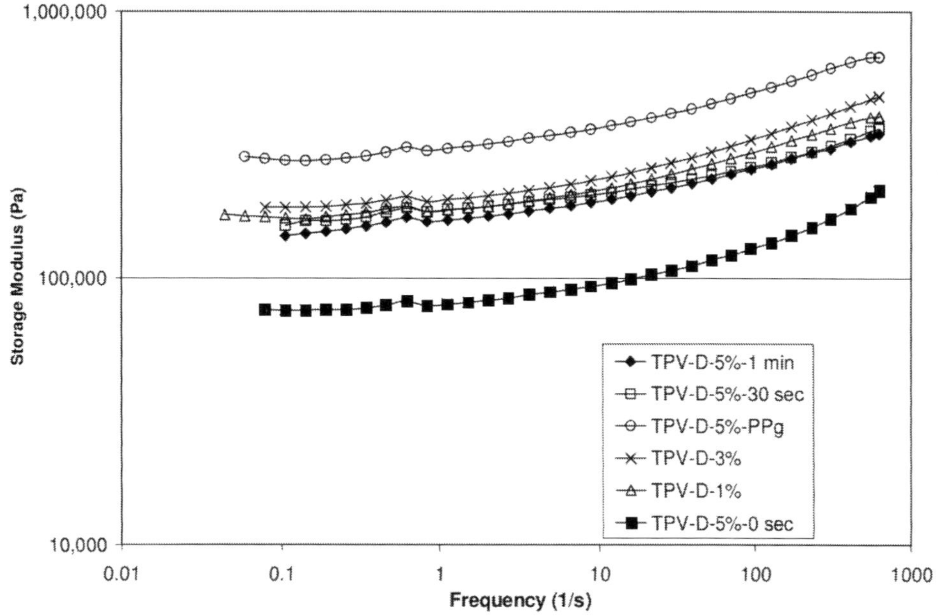

(b)

Figure 3 – comparison of storage modulus (G') of different nanocomposite TPV samples

By comparing the storage modulus of TPV-Ref sample with other nanocomposite samples in Figure 3 (a), it can be concluded that incorporation of EG nanolayers to the TPV matrix has made an increase in the extent of storage modulus of nanocomposite samples. It is related to the interaction between polymer chains and graphite nanolayers surface which retards the elastic response of polymer chain and increase the storage modulus. In addition, comparing the storage modulus of samples TPV-Both-5% and TPV-D-5%-PPg-EPg, which had the same composition, results that direct mixing method makes higher level of storage modulus in compare of the samples which are produced by masterbatch mixing method. It is related to the larger lateral size of EG clusters which makes stronger three-dimensional network between EG nanolayers.

Figure 3 (b) shows the comparison between storage modulus of the samples which has different level of EG and also different states of cure (SOC). It can be seen that higher content of EG in the nanocomposite sample increases the storage modulus, because of the higher level of interaction between the polymer chains and EG nanolayers surface. In addition, state of cure showed a great important role in the viscoelastic behavior of TPV nanocomposite samples. It can be seen that the sample which have no dynamic vulcanization has very low storage modulus, which shows the effect of dynamic vulcanization in the existence of interlocked polymer chains in the interface between PP and EPDM phases.

Electrical Conductivity

The results of electrical resistivity measurements of different nanocomposite samples are shown in Table 4. It can be observed that the nanocomposite TPV samples which were produces by masterbatch mixing method has higher percolation threshold (>5 vol %) than the samples which are produced by direct mixing method (< 5 vol %). This is related to the smaller lateral size of the EG clusters in the masterbatch feeding route which prevents from connection of EG nanolayers and existence of conductive path through the polymer matrix.

In addition, the nanocomposites which have shorter dynamic vulcanization time, shows higher electrical resistivity which is related to the higher percolation threshold of these samples. It shows that the SOC of TPV samples has great impact on the percolation threshold of nanocomposite TPV samples.

Conclusions

It the present work, in-situ dynamically vulcanized electrically conductive TPV nanocomposites have been prepared and evaluated. For this purpose, PP and EPDM have been used as the polymeric phases and expanded graphite (EG) as conductive filler. Effects of feeding route and state of EPDM vulcanization upon the developed microstructure and conductivity behavior have been investigated. The nanocomposite TPV samples generated by direct mixing of EG with polymeric phases exhibited lower percolation threshold, whereas the samples fabricated via masterbatch melt mixing showed better dispersion of EG nanolayers but higher percolation threshold. In addition, it was observed that state of cure (SOC) of TPV samples has important effect on the rheological behavior and electrical percolation threshold of nanocomposite samples.

Table 4- electrical resistivity of different TPV samples

Sample Code	EG Content (Vol %)	Mixing Time (min)	Electrical Resistivity (KΩ.cm)
TPV-Ref	0	12	>400,000
TPV-PP-3%	1.2	12	>400,000
TPV-PP-4%	1.6	12	>400,000
TPV-PP-5%	2	12	>400,000
TPV-EPDM-3%	1.8	12	>400,000
TPV-EPDM-4%	2.4	12	>400,000
TPV-EPDM-5%	3	12	>400,000
TPV-Both-4%	4	12	>400,000
TPV-Both-5%	5	12	>400,000
TPV-D-5%-PPg-EPg	5	15	61728.4
TPV-D-5%-PPg	5	15	4.4
TPV-D-5%-No	5	15	246913.5
TPV-D-3%	3	15	>400,000
TPV-D-1%	1	15	>400,000
TPV-D-5%-1 min	5	10	>400,000
TPV-D-5%-30 sec	5	9.5	>400,000
TPV-D-5%-0sec	5	9	>400,000

References

1. H. Kim and C. W. Macosco, Macromolecules, 41, 3317-3327, 2008

2. K. Kalaitzidou, H. Fukushima, L. T. Drzal, Carbon, 45, 1446-1452, 2007

3. L. Chen, G. Chen, L. Lu, Adv. Func. Mater., 17, 898-904, 2007

4. A.A. Katbab, A. N. Hrymak, K. Kasmadjian, J Appl Polym Sci, 107, 3425–3433, 2008

5. A. Celzard, J.F. Mareche, G. Furdin, Prog. Mater. Sci., 50,93-179,2005

6. J.L. Leblanc, Rheol Acta, 46,1013–1027,2007

7. A.Y. Coran, R. Patel, Rub. Chem. Technol., 53,141-150, 1980

PROCESSIN-MICROSTRUCTURE-PROPERTY RELATIONSHIP IN CONDUCTIVE POLYMER NANOCOMPOSITES

Mohammed H. Al-Saleh and Uttandaraman Sundararaj*

Department of Chemical and Petroleum Engineering, University of Calgary, Calgary, Canada – mhsaleh@just.edu.jo; u.sundararaj@ucalgary.ca

Polymer composites based on high aspect ratio nanofillers such as carbon nanotubes and nanofibers are receiving considerable attention because of their unique multifunctional properties at very low filler loading [1-4]. The filler aspect ratio is crucial for many properties such as electrical properties. Electrical properties are the emphasis of this work. The electrical percolation threshold, which is the critical filler concentration that is required to initiate a conductive network within an insulating matrix, is known to decrease with increase in the filler aspect ratio. Percolation thresholds in the range of 0.0021-0.005 wt% were experimentally reported for several multiwall-carbon nanotube (MWCNT)/epoxy nanocomposites [5-8]. Nevertheless, for CNT/thermoplastic nanocomposites, where melt mixing is an industrially preferred processing method [1, 9], much higher percolation thresholds have been reported [10-12]. This high threshold is due to the poor dispersion of nanofiller under low shear conditions or to significant degradation of the nanofiller aspect ratio under high shearing conditions.

In addition to the filler aspect ratio, several other factors affect the properties of polymer composites including the polymer properties; the filler intrinsic properties; dispersion, distribution, and orientation of the filler; and adhesion between filler and polymer matrix. Understanding the influence of the these parameters, especially those related to the material's microstructure, on the composite's final properties is of great scientific importance. For conductive polymer composites, there is no direct experimental evidence on the influence of filler dispersion on its final properties. This work focuses on the influence of processing conditions on the filler dispersion and subsequently on the electrical and electromagnetic interference (EMI) shielding properties of conductive polymer nanocomposites, i.e. we study the processing-microstructure-property relationship. To achieve this target we studied the influence of processing temperature on the microstructure, i.e. basically the dispersion and aspect ratio, of high density polyethylene (HDPE)/vapor grown carbon nanofiber (VGCNF) nanocomposite. The dispersion and aspect ratio of carbon nanofiber (CNF) were characterized. CNF dispersion was directly characterized by microscopic techniques and indirectly utilizing the dynamic rheological analysis technique. Final electrical and EMI shielding properties were measured and linked to the nanocomposites' microstructure and processing conditions.

The processing-microstructure-property relationship in conductive polymer nanocomposites was investigated. Nanocomposites of vapor grown carbon nanofiber (VGCNF)/high density polyethylene (HDPE) with different levels of nanofiber dispersion were formulated by changing the nanocomposites' compounding temperature. Direct (SEM and optical microscopy) and indirect methods (linear viscoelastic properties) were used to characterize the dispersion of nanofiller. VGCNF aspect ratio before and after mixing was measured. Increasing processing temperature was found to increase the nanofiller agglomeration and reduce the breakage of nanofiller because of the decrease in the mixing shear stress and energy. The electrical and electromagnetic interference (EMI) shielding properties of the VGCNF/HDPE nanocomposites decreased with increase in processing temperature from 180 °C to 220 °C because the increase in the agglomeration of VGCNF was more significant than the preservation of the VGCNF aspect ratio. This finding does not mean that the increase in processing temperature will always lead to decrease in the electrical conductivity and EMI shielding properties for all polymer composites. For some composites, it is possible to preserve the filler aspect ratio enough so that the increase in agglomeration is less of a factor.

References

1. Al-Saleh MH, Sundararaj U. A review of vapor grown carbon nanofiber/polymer conductive composites. Carbon 2009:47(1):2-22.

2. Breuer O, Sundararaj U. Big returns from small fibers: A review of polymer/carbon nanotube composites. Polym Compos 2004:25(6):630-645.

3. Moniruzzaman M, Winey KI. Polymer nanocomposites containing carbon nanotubes. Macromolecules 2006:39(16):5194-5205.
4. Winey KI, Kashiwagi T, Mu M. Improving electrical conductivity and thermal properties of polymers by the addition of carbon nanotubes as fillers. MRS Bulletin 2007:32(4):348-353.
5. Sandler JKW, Kirk JE, Kinloch IA, Shaffer MSP, Windle AH. Ultra-low electrical percolation threshold in carbon-nanotube-epoxy composites. Polymer 2003:44(19):5893-5899.
6. Bauhofer W, Kovacs JZ. A review and analysis of electrical percolation in carbon nanotube polymer composites. Compos Sci Technol 2009:69(10):1486-1498.
7. Bryning MB, Islam MF, Kikkawa JM, Yodh AG. Very low conductivity threshold in bulk isotropic single-walled carbon nanotube-epoxy composites. Adv Mater 2005:17(9):1186-1191.
8. Martin CA, Sandler JKW, Shaffer MSP, Schwarz MK, Bauhofer W, Schulte K, Windle AH. Formation of percolating networks in multi-wall carbon-nanotube-epoxy composites. Compos Sci Technol 2004:64(15):2309-2316.
9. Krause B, Pötschke P, Häußler L. Influence of small scale melt mixing conditions on electrical resistivity of carbon nanotube-polyamide composites. Compos Sci Technol 2009:69(10):1505-1515.
10. Meincke O, Kaempfer D, Weickmann H, Friedrich C, Vathauer M, Warth H. Mechanical properties and electrical conductivity of carbon-nanotube filled polyamide-6 and its blends with acrylonitrile/butadiene/styrene. Polymer 2004:45(3):739-748.
11. McNally T, Pötschke P, Halley P, Murphy M, Martin D, Bell SEJ, Brennan GP, Bein D, Lemoine P, Quinn JP. Polyethylene multiwalled carbon nanotube composites. Polymer 2005:46(19):8222-8232.
12. Zhou Z, Wang SF, Zhang Y, Zhang YX. Effect of different carbon fillers on the properties of PP composites: Comparison of carbon black with multiwalled carbon nanotubes. J Appl Polym Sci 2006:102(5):4823-4830.

Effect of organic and inorganic particles on conductivity of water soluble conductive polymer-polyvinylalcohol composites

S. Konagaya[1*], M. Terada[2], M. Mutoh[1], K. Tada[2], T. Yamada[2]

[1]*Department of Applied Chemistry, Graduate School of Engineering, Nagoya University, Furo-cho, Chikusa-ku, Nagoya, 464-8603 Japan - konagaya@apchem.nagoya-u.ac.jp; mutoh.misako@apchem.nagoya-u.ac.jp*
[2]*: Graduate School of Natural Science and Technology, Kanazawa University, Japan- m-yumeneko@hotmail.co.jp; tyamada@t.kanazawa-u.ac.jp; ktada@t.kanazawa-u.ac.jp*

Abstract - The effect of non-conductive organic and inorganic fillers with the primary particle size of around 15 nm such as water-dispersible polyester particles (PEs), alumina particles (Al_2O_3), and two kinds of titanium dioxide particles (TiO_2-A and –B) on the conductivity of the composites of water-soluble conductive polymer, polyaniline sulfonic acid (PAS), and water soluble polyvinyl alcohol (PVA). It was found that the species of fillers have a large effect on the conductivity of the PAS/PVA composite. PEs and TiO_2-A increased the conductivity of the PAS/PVA composites while Al_2O_3 and TiO_2-B decreased it. The increase of the conductivity was possibly caused by the increase of PAS molecule density and efficient formation of conductive PAS networks around nano-particles.

Introduction

Konagaya et al have been studying preparations, properties and nano-structure of thin conductive film composites prepared from polyaniline sulfonic acid (PAS) as shown in **Figure 1**, and organic polymer binders in order to develop a new conductive coating agent, which gives plastic films with high transparency and excellent conductivity[1,2,3].

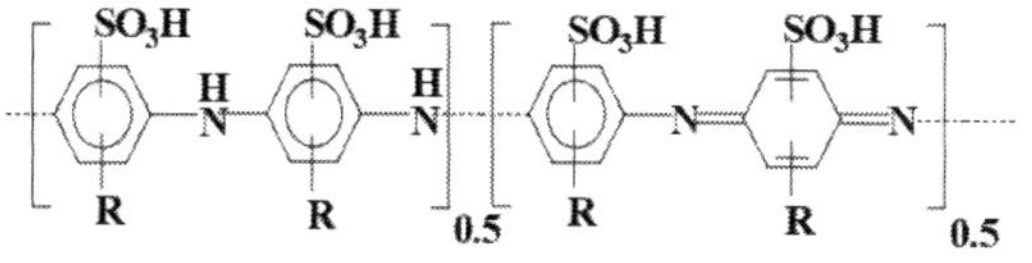

Figure 1 Chemical structure of PAS

The composite of PAS and a water-dispersible polyester binder (PEs, nano-particle (less than 100 nm)/water slurry) shows good transparency and an excellent conductivity (low Rs value (10^7 Ω/□) in the wide range humidity).

Figure 2 shows a TEM photo of the cross section of the PAS/PEs composite layer coated on a PET film. Dark lines (networks) show PAS molecules oxidized by OsO_4, and white parts show PEs polymer solid. **Figure 2** suggests that nano-particles such as PEs are helpful to make conductive networks of PAS molecules efficiently, since PAS is insoluble in PEs.

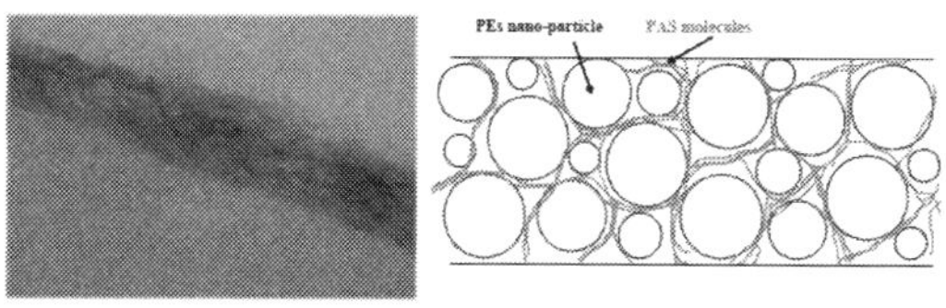

Figure 2 TEM Photo of PAS/PEs cross section and its probable structure

Experimental

Materials

The water-soluble conductive polymer, PAS was given by Mitsubishi Rayon Co., Ltd. Water soluble poly (vinyl alcohol) was supplied by Kuraray Co. Ltd. Water dispersible organic nano-particle, PEs, was given by TOYOBO Co. Ltd. Titanium dioxide A (TiO_2-A) was given by Ishihara Sangyo Co. Ltd. Titanium dioxide B (TiO_2-B) and alumina (Al_2O_3) by Nippon aerosol Co., Ltd. Pure water was used for the coating solvent. PET film with the thickness of 38 μm as a substrate was supplied by TOYOBO Co., Ltd. The inorganic nano-particles are non-conductive as shown in **Table 1** and their primary particle sizes are around 15 nm.

Table 1 Conductivity of particles [4]

Particle	Conductivity (S/cm)
TiO_2-A	10^{-14}～10^{-13}
TiO_2-B	10^{-14}～10^{-13}
Al_2O_3	10^{-15}

Preparation of PET films laminated with PAS/PVA/nano-particle composites.

The aqueous mixtures of PAS, PVA and nano-particles were prepared with a wet miller, "Star Burst Mini (Sugino Machine Limited)", and coated on the PET film. The coated layer was dried at 80℃ for 16 hours. The coated PAS/PVA/nano-particle layer had a thickness of 3 μm or more.

Evaluation of surface resistivity (Rs) of PAS/PVA/nano-particle composites.

Rs (Ω/□) values were measured with Model 1152 (Treck Japan Co. Ltd) under the following conditions; temperature: 20-25℃, humidity:45-70 %, voltage: 100 V. Lower Rs means higher conductivity.

Results and Discussion

Effect of organic PEs particles on Rs values of the PAS/PVA composites

The Rs values of PAS/PVA composite was decreased with the increase of PEs nano-particles at the PAS concentration of 5, 9 and 13 wt% as shown in **Figure 3**. The decrease of Rs value is ascribed to the increase of PAS molecule density, since PAS is not dissolved in PEs.

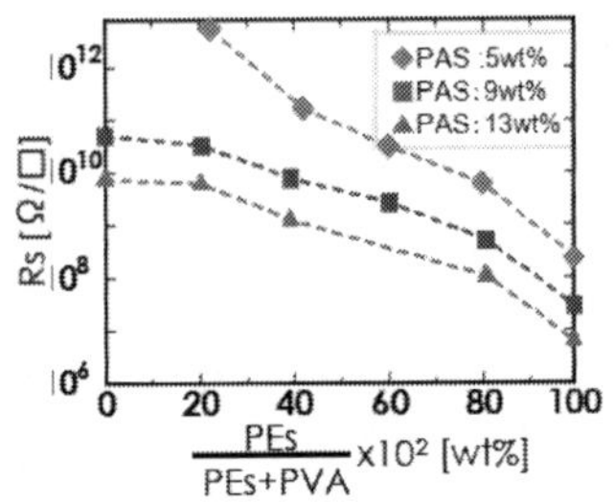

Figure 3 Effect of PEs on the Rs value of PAS/PVA/PEs composites

Effect of TiO_2-A, TiO_2-B and Al_2O_3 nano-particles on the Rs value of PAS/PVA/nano-particles composites

Figure 4 shows the Rs values of the PAS/PVA/nano-particles (PAS/PVA=15 wt %). There is a large difference in Rs value among the species of nano-particles added in the composites. The Rs values decreased with the increase of TiO_2-A up to its concentration of 12 vol%, but increased at the TiO_2-A concentration of 17 vol%. However, the addition of TiO_2-B showed little change of Rs values of the PAS/PVA composites. On the other hand the addition of Al_2O_3 brought the increase of Rs values of the PAS/PVA composites.

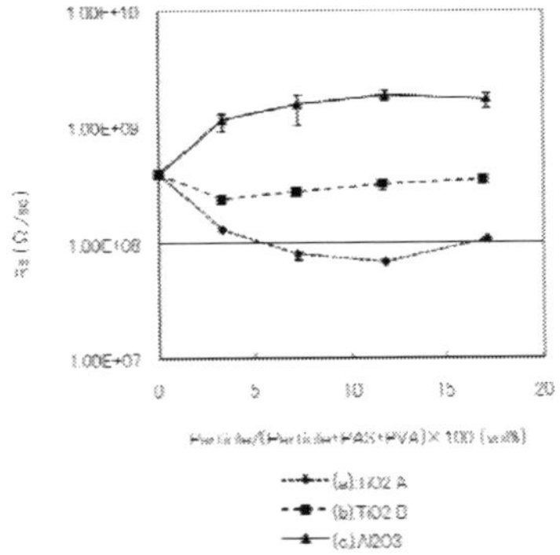

Figure 4 Effect of inorganic particles on Rs values of PAS/PVA composites

Effect of PAS and TiO_2-A concentration on the Rs values of PAS/PVA/TiO_2-A composites

The decrease of Rs values of PAS/PVA/TiO_2-A composites by the addition of TiO_2-A was observed at a wide range of PAS/PVA volume ratio (12.3/87.7～35.6/64.4) as shown in **Figure 5**.

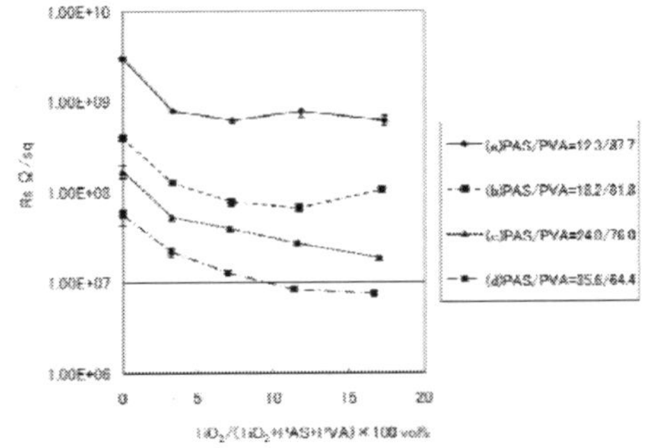

Figure 5 Effect of TiO_2-A concentration on the Rs values of PAS/PVA/TiO_2-A composites

Probable mechanism of nano-particle effect

The decrease of Rs value is possibly ascribed to the increase of PAS molecule density and efficient formation of conductive PAS networks around nano-particles as shown in **Figure 6**. The mechanism is now under study.

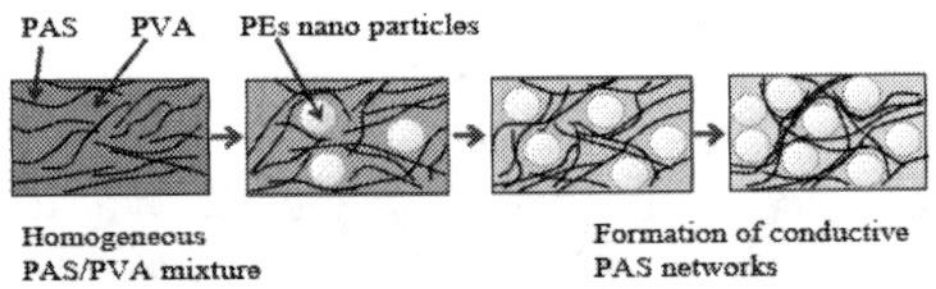

Figure 6 Probable mechanism of additon effect of PEs and TiO_2-A particles

Conclusions

1) The addition of PEs and TiO_2-A particles was effective for the decrease of Rs values of PAS/PVA composites.
2) The addition of Al_2O_3 particles did not improve the conductivity of PAS/PVA composites.
3) The decrease of Rs values by PEs and TiO_2-A particles was possibly caused by the increase of the PAS molecule density and efficient formation of conductive PAS networks around nano-particles.

Acknowledgements

The authors would like to thank Mitsubishi Rayon, Ishihara sangyo, Kuraray, Nippon aerosil and Toyobo companies for the offer of PAS, PVA, PEs, TiO_2, Al_2O_3 and PET films, Sugino Machine Limited for the help of the wet milling and JSPS (Japan Society for the Promotion of Science) for the financial support.

References

1. S. Konagaya et al, Plastics, Rubber and Composite, **31**, No5, p201-204 (2002)
2. S. Konagaya et al, Polymer Preprints, **46**(1), 503-504 (2005)
3. S. Konagaya et al, Seikei-Kakou, **17**, No8, 543-547 (2005)
4. Filler Society of Japan: Practical Filler Dictionary, Taisei-sya publisher, Tokyo (1994)

STIMULUS-RESPONSIVE POLYMER/NANOPARTICLE COMPOSITES

B. Nearingburg, L. Nguyen, and A.L. Elias[1]*

[1]*Department of Chemical and Materials Engineering, University of Alberta, Edmonton, Canada – benn@ualberta.ca, ltnguyen@ualberta.ca, aelias@ualberta.ca*

Abstract

Stimulus-responsive or "smart" materials are able to sense a change in their environment and react by altering their physical or chemical properties. Polymer/nanoparticle composites consisting of magnetic nanoparticles embedded in a polymer matrix are an emerging example of a stimulus-responsive material. In these materials, embedded nanoparticles couple energy from an electromagnetic field into heat, causing the polymer matrix to deform in a reversible and repeatable manner. A variety of interesting devices based on these systems including sensors, actuators, and microvalves have been demonstrated in the literature. The purpose of our work is to investigate the physical properties of composite films which contain iron oxide nanoparticles, and optimize the responsivity of these materials. Paraffin wax, polyethylene glycol (with a range of molecular weights), and PDMS have been employed as host materials in these systems. Composite samples containing between 0-5.0 wt% nanoparticles were prepared by dispersing nanoparticles in a liquid matrix using a combination of mechanical and ultrasonic mixing. In the case of polyethylene glycol, the resulting mixtures were photopolymerized in the presence of an appropriate initiator. Transmission electron microscopy was used to characterize the nanoparticles. Heating experiments were undertaking using an Ameritherm Novastar 10kW inductive heater operating at a frequency of 151 kHz. Heating rates between 0.4°C/s and 13.5°C/s were observed, depending on the particle concentrations and matrix materials. Large temperature differences (> 40°C) were achieved, with sufficient heat flow to surmount thermal transitions (glass transition, or melting) and alter the mechanical properties of the composites. The temperature profile of the system has been correlated to the heat capacity of the polymer matrix, suggesting that copolymers or other mixed composite systems may be combined to yield functional devices.

Introduction

Stimulus Responsive Polymers

Stimulus responsive polymers are a class of "smart" materials which are able to respond to their surroundings through factors such as pH (Ryan et al., 2005), temperature (Mohr et al. 2006), and exposure to visible light (Suzuki et al. 1996). In recent years a number of stimulus-responsive polymer/nanoparticle composites have been developed which can be actuated via non-contact electrical induction. In these systems nanoparticles absorb the electromagnetic energy provided by an inductive heater and transform this energy into heat, which leads to a thermal expansion of the polymer matrix; the deformation of the matrix can be harnessed to achieve controlled motions which can be remotely initiated. As an example, a polymer/nanoparticle composite has been used to create medical devices which could be inserted into the body in a collapsed state and then remotely deformed into a desired shape *in vivo* by exposure to a magnetic field (Bukley et al. 2006). Another area of application in which magnetically responsive polymer/nanoparticle composites could be of use is in microfluidic or lab-on-a-chip devices, in which miniaturized components such as switches, valves and heaters must be hard wired to macroscopic off-chip controllers such as power sources. Conventional devices of this nature require complex patterning and connections to be introduced into the device, whereas remotely-actuated materials could be used to simplify some of these components. For example, wireless heaters for cell lysis have been created using solid nickel heaters which could be integrated into the chip and heated inductively (Baek et al. 2009). Stimulus-responsive polymer/nanoparticle composites could potentially be used in this type of application if suitable temperatures gradients could be achieved. An added advantage of employing nano-particle/composites in microfluidic systems is that these materials could be simply patterned on the microscale using techniques such as microcontact printing and micromoulding, which require limited equipment and are easily adapted into large scale fabrication.

The main purpose of our work is to investigate how the composition and processing of magnetic nanoparticle/polymer composites affects their ability to transduce heat from a commercial inductive heater system. It is known that the size and surface properties of magnetic nanoparticles influence both their magnetic properties and specific absorption (Zeng et al. 2007; Baker et al. 2006). In this work we therefore test a number of particles with different characteristics in solution and in solidified polymer composites for heating using a standard inductive furnace, to identify materials that undergo the largest temperature increase. Different host polymers are also screened to compare heating profiles in certain polymer matrices. To investigate the importance of dispersion, we characterize inductive heating of composites of the same materials which have been prepared using different methods. The effects of particle concentration on heating profile are also investigated. Finally, thermal analysis is undertaken to observe how addition of nanoparticles to our systems affects the properties of the host materials themselves.

Inductive Heating

Inductive heating is a term given to a collection of processes through which energy can be transferred from an external source through oscillating electro-magnetic fields into a system. In conventional industrial applications inductive heaters are used for metallurgical and materials processing wherein metallic objects must be heated for mixing, annealing, and/or bonding where conventional crucible heating is not viable. In systems of nanoparticles, energy is transferred from the alternating field primarily via magnetic loss processes that occur as the magnetization of the nanoparticles oscillates with the external field (Hergt et al., 2006). The energy transferred by these mechanisms depends strongly on the nanoparticle size, microstructure, and geometry (Hergt et al., 1998). In addition, the frequency of the field which can be used to couple energy into a system of magnetic nanoparticles has been shown to vary as a function of the size of the particles, which may be effectively larger if the particles become agglomerated (Eggeman et al. 2007; Baker et al. 2006). The frequency of the inductive heater must therefore be carefully matched to the nanoparticles employed.

Experimental

Materials

We utilized several types of iron-oxide (Fe_2O_3/Fe_3O_4) nanoparticles obtained from commercial sources. Particles were obtained in suspensions of water (fluidMAG-PEG/P, Chemicell) and toluene (magnetic nanoparticle solution #700312, Sigma-Aldrich), and as a dry nanopowder (iron (II,III) oxide #637106, Sigma-Aldrich). A variety of polymers were utilized, including: polydimethylsiloxane (PDMS, Sylgard 184, Dow Corning), poly(ethylene-glycol) (PEGDMA Mn~575 and PEGDMA Mn~258), and paraffin wax (#327204, Sigma Aldrich). Toluene (Sigma-Aldrich) and oleic acid (Sigma-Aldrich) were used to suspend and then stabilize nanoparticle solution derived from nanopowders as described below. All substances were used as received unless otherwise specified.

Sample Preparation

Nanoparticles prepared from an iron-oxide nanopowder were processed by first mixing the purchased iron oxide nanoparticles into a uniform fluidic suspension.

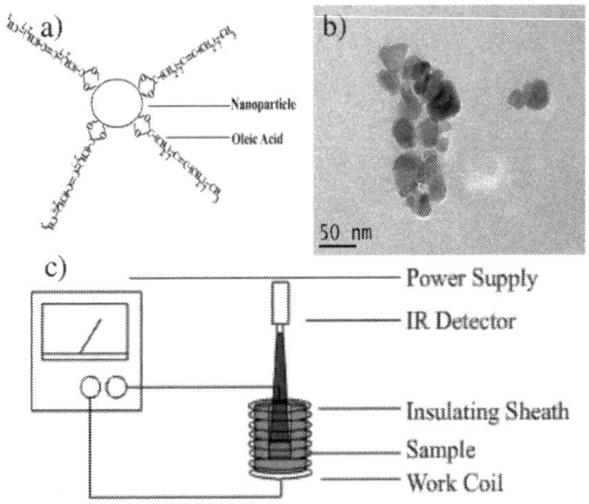

Figure 1 – a) Schematic diagram of iron-oxide nanoparticles with oleic acid coating. b) TEM image of particles prepared from iron oxide nanopowder that have been stabilized with oleic acid. c) Schematic diagram of inductive heating apparatus.

To create well-dispersed solutions some particles were functionalized with oleic as noted. To do so, nanopowder samples were suspended in toluene and dispersed by immersing the samples in a heated (65°C) ultrasonic bath for 30 minutes after which 0.1 wt% oleic acid was added to stabilize the free nanoparticles (to yield particles as shown in Figures 1 a and b). The particles were then magnetically isolated from the toluene solution by placing the mixture over an external magnet and extracting the supernatant with a pipette.

Isolated, stabilized nanoparticles or as-received particles in powder form were dispersed in the polymer mixtures through several mixing procedures. PEG composites were formed by mechanically mixing particles with liquid PEG, vortex mixing for 5 minutes, followed by 10 minutes in an ultrasonic bath. Crosslinking was accomplished by adding 0.01 wt% 2,2-dimethoxy-2-phenylacetophenone photocrosslinker and exposing the samples to UV light at 365 nm from an 8W source for 15 minutes. Uncrosslinked PDMS is a viscous fluid and as such forming composites with high levels of dispersion required additional mixing. Isolated particles were added to uncrosslinked PDMS and mixed for 2 minutes by vortex mixing followed by immersion in a heated (65°C) ultrasonic bath for 120 minutes; polymerization was initiated by adding 9.1 wt % crosslinking agent and heating the samples at 60°C overnight. Composites with a paraffin matrix were formed by melting paraffin on a hot plate and holding at 60°C while introducing nanoparticles through mechanically stirring. Composites were formed by pouring the paraffin/particle solution into moulds and solidifying.

Inductive Heating

Inductive heating was carried out using an Ameritherm Novastar 10 kW inductive heater operating at a frequency of 151 kHz. Samples were exposed to oscillating EM fields in a custom built thermally insulating sheath inside a copper coil consisting of 10 turns, with a radius of 2 cm, and a height of 10 cm; a diagram of the apparatus is given in Figure 1c. The temperature of the samples was continuously measured using a calibrated Omega OS53x-CF IR sensor throughout the heating experiments. Samples were heated at 23 % power for 200 s durations.

Transmission Electron Microscopy

Transmission electron microscopy (TEM) was undertaken to investigate the size, shape, and dispersion of the Fe_3O_4 nanoparticles. Particles were imaged by casting a dilute nanoparticle solution directly onto carbon-coated copper grids.

Thermomechanical Analysis

The thermomechanical properties of the composite materials were examined using dynamic mechanical analysis (DMA) on a Perkin-Elmer DMA 8000 and differential scanning calorimetry (DSC) on a TA Q1000 DSC. Each sample was tested a minimum of three times to ensure accurate measurements. Thin films of the composite materials were examined by DMA under tension mode to test for variations in elastic modulus at 25°C at a frequency of 1 Hz and a displacement of 0.01 mm. Modulated DSC with a heating rate of 2°C/min and a temperature modulation of 0.4°C every 60 seconds was used to determine composite heat capacity as a function of particle concentration.

Results and Discussion

Inductive Heating

Inductive heating experiments conclusively demonstrated energy transfer into heat in a medley of composite materials to a high degree of repeatability. A sample heating curve for differing nanoparticle concentrations in PDMS is given in Figure 2b. All of the measured heating rates followed an exponentially decaying increase in temperature with time; heating rate curves were fit to functions given by

$$T(t) = T_o + T_\Delta(1-e^{-t/\tau}) \qquad (1)$$

Where T_o is the initial temperature of the sample, T_Δ is the difference between the equilibrium and initial temperatures, and τ represents a time constant characteristic of the heating profile. To compare heating effects in different matrix materials, composites were prepared from four polymers commonly used in microfabrication. 0.1-1.0 wt% Sigma-Aldrich nanopowder was dispersed in each polymer and the resulting composites were examined for their normalized heating rate constants, the results of which are given in Table 1. Of the materials

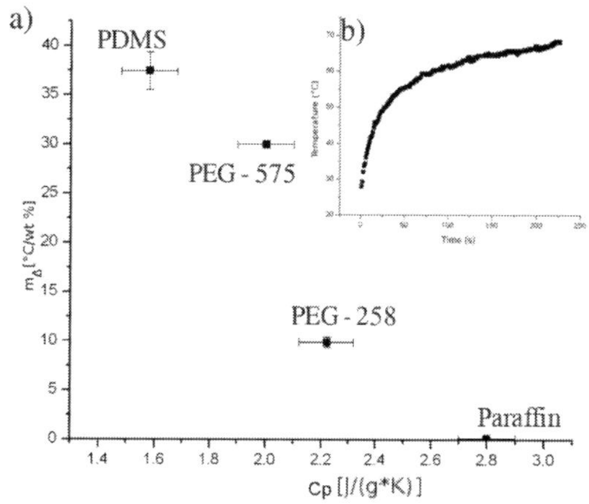

Figure 2 – a) Relationship between rate heating rate constant and composite heat capacity as measured by DSC. b) Typical heating curve for a 0.8 wt% PDMS composite.

Table 1 – Normalized equilibrium temperature constants as a function of polymer matrix material.

Polymer Matrix Material	τ_{ave} [s]	m_Δ [°C/wt %]
Paraffin	14±6	0.11±0.01
PEGDMA Mw~575	90±50	29.9±4
PEGDMA Mw~258	58±17	9.9±.0.5
PDMS	40±6	37±2

Table 2 – Heating rate constants as a function of nanoparticle material for particles in solution.

Nanoparticle Material	τ [s]	T_Δ [°C]
Chemicell FluidMag (in water)	356.±14	29.2±1
Sigma-Aldrich Fe_3O_4 Fluid (in toluene)	No heating observed	No heating observed
Sigma- Aldrich Nanopowder (in toluene)	24.5±.5	41.4±.4

examined it was found that PDMS demonstrated the highest heating rates without undergoing a phase change (as was found for the paraffin sample), and achieved the largest temperature change. PDMS was therefore used in subsequent experiments. The equilibrium temperature constant T_Δ was measured to vary linearly with particle concentration for each material tested as

$$T_\Delta = m_\Delta * x + b \qquad (2)$$

where m_Δ is the concentration normalized equilibrium temperature constant, x is concentration in wt % and b is an offset term. Variations in m_Δ were compared to composite heat capacity as shown in Figure 2a. The results indicate that increasing heat capacity leads to a decreased m_Δ constant indicative of less overall temperature change; this relationship is reasonable as heat capacity is a metric of the quantity of energy required to manifest a temperature change.

Heating experiments performed with different types of nanoparticles in solution (of 1 wt% concentration) are given in Table 2. Differences in heating rates can be attributed to variations in particle size, and surface functionalization (as in Hergt et al. 1998). The Chemicell particles had an average diameter of 50 nm, with a PEG/P surface coating. The nanofluid obtained from Sigma-Aldrich consisted of particles with an average diameter of 10 nm stabilized with oleic acid. Each of these types of particles were smaller than the particles in the nanopowder, which was shown to have the most favourable heating rates. This suggests that field oscillations of 151 kHz inherent to our inductive heater couple best with particles with a diameter greater than 50 nm. As the nanopowder derived nanoparticles demonstrated the best heating profile, they were used for subsequent future experiments.

Achieving high levels of dispersion can be challenging when preparing polymer/nanoparticle composites. As a part of our study we explored how different methods of composite preparation lead to different levels of dispersion, and how the degree of particle aggregation affected the overall heating of the composite. Composites were prepared using two different methods. In the first method, the as-received particles were manually stirred into the PDMS elastomer, which was then combined with the crosslinker (10 wt%) and polymerized for four hours at 60°C. The resulting composites contained large agglomerated flakes of material which were visible to the naked eye. To create well-dispersed samples, a different preparation method was used wherein the particles of the nanopowder were functionalized with oleic acid and dispersed using a two hour sonication cycle, as described in the Experimental section. The resulting samples had homogeneous appearance, and were brown in color. Agglomerations were not visible to the naked eye or using a 50x microscope objective, suggesting that the particles were much better dispersed than the previous samples. For comparison, agglomerated samples and well-dispersed samples were prepared at concentrations ranging from 0.2 - 1.0 wt%. Differences in the heating behaviour of these composites are shown in Figure 3, where the heating rate (τ) is smaller and equilibrium temperature (T_Δ) are larger for the well-dispersed materials. Further work is required to achieve a quantitative method of measuring dispersion, and to verify if even faster rates and larger temperature changes can be achieved by further improving the dispersion of the nanoparticles within the samples.

The heating profiles for PDMS composites of varying concentrations are favourable as they suggest that

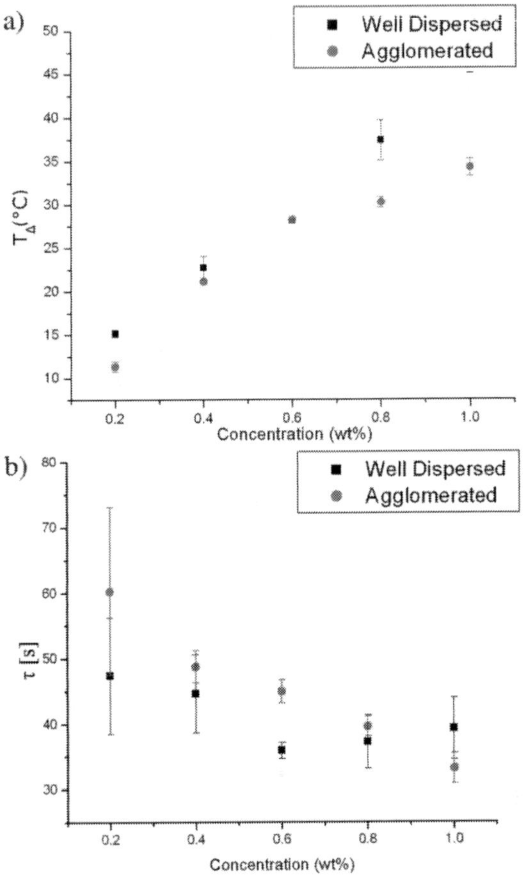

Figure 3 – a)T_Δ and b)τ as functions of particle concentration for well dispersed ultrasonic bath processed samples (black), and agglomerated 'as received' samples (red).

Table 3 – Thermomechanical data for PDMS/nanoparticle composite systems

Nanoparticle Concentration [wt%]	Heat Capacity at 25°C [J/gK]	Elastic Modulus at 25°C [10^6 Pa]
0.0	1.6±1	1.9±0.1
0.2	4.6±0.2	Not measured
0.3	5.55±0.07	1.7±0.1
0.7	6.4±0.1	1.3±0.1
0.8	6.90±0.06	1.2±0.1

energy coupling mechanisms within the composites behave the same in high and low concentrations and changes in heating rate are caused only by an increasing density of absorbing particles. These results provide indications that through precise control of particle concentration throughout a material heater devices could be created yielding complex, and controllable heating profiles.

Thermomechanical Analysis

Thermomechanical analysis demonstrated tunable properties as a functions of particle concentration, and structure. Through variation of nanoparticle concentration in PDMS from 0 wt% to 1.0 wt% composites were found to exhibit decreases in elastic modulus measured by DMA as well as increasing heat capacity measured by DSC as shown in Table 2.

The value of the elastic modulus is representative of the quality of crosslinking within a PDMS elastomer. The measured results suggest that incorporation of nanoparticles may alter the crosslinking process which converts the PDMS resin into an elastomeric composite, leading to a less crosslinked and less rigid material.

DSC analysis of the composite samples demonstrated increasing heat capacity with increasing particle concentration. Further investigation is required to fully determine the effect of nanoparticle incorporation on composite thermal behaviour including thermal conductivity and other effects such as glass transitions.

Conclusions

Our results show that iron-oxide/nanoparticle composites are able to achieve transduction of electrical energy through inductive heating. Commercial nanoparticles which can generate controllable changes in temperature throughout a material when subjected to a 151 kHz field in a standard inductive heater were identified. It was shown that the heat capacity of the matrix material is critical for determining the final temperature which can be achieved. It was demonstrated by varying the concentration of the particles between 0.1% and 1%, the change in temperature (T_Δ) achieved could be tuned. We found that the method used to disperse the particles in the matrix had a strong effect on the heating of the composites. Overall higher heating rates were achieved in samples in which particles were well dispersed rather than in samples in which the nanoparticles were visibly agglomerated.

It was also demonstrated that both the elastic modulus and heat capacity of PDMS composite materials could be varied by changing the particle concentration. These results are useful as they contain the basic thermal and structural properties that are required to design and implement advanced functional devices from these materials.

Acknowledgements

The authors would like to thank Brian Worfolk for his assistance with TEM characterization. Funding from the MicroSystems Technology Research Initiative (MSTRI) and the Natural Sciences and Engineering Research Council of Canada (NSERC) is gratefully acknowledged.

References

Ahn. S. et al., "Stimuli-responsive polymer gels", Soft Materials, **4** , 1151-1157 (2008)

Baek. S. et al, "Wireless induction heating in a microfluidic device for cell lysis", Lab of a Chip, **10**, 909-917 (2010)

Baker. L et al.,"Heat deposition in iron oxide and iron nanoparticles for localized hyperthermia", Journal of Applied Physics, **99**, 08H106 (2006)

Buckley. P. R. et al., "Inductively Heated Shape Memory Polymer for the Magnetic Actuation of Medical Devices", IEEE Transactions on Biomedical Engineering, **53**, 2075-2083 (2006)

Eggeman A. S. et al., "Size and Concentration Effects on High Frequency Hysteresis of Iron Oxide Nanoparticles", IEEE Transactions On Magnetics, **43**, 2451-2453 (2007)

Hergt, R. et al., "Physical limits of hyperthermia using magnetite fine particles," IEEE Transactions on Magnetics, **34**, 3745-3754 (1998)

Hergt. R. et al., "Magnetic particle hyperthermia : nanoparticle magnetism and materials development for cancer therapy ", Journal of Physics Condensed Matter, **18**, S2919-S2934 (2006)

Jordan. L. et al, "Inductive heating of ferrimagnetic particles and magnetic fluids: Physical evaluation of their potential for hyperthermia", **25**, 499-511 (2009)

Lin. Y. and Haynes. C., "Synthesis and characterization of biocompatiable and size-tunable multifunctional porous silica nanoparticles", Chem. Mater., **21**, 3979-3986 (2009)

Lu. A. et al., "Magnetic nanoparticles: synthesis, protection, functionalization, and application", Angew. Chem. Int. Ed., **46**, 1222-1244 (2007)

Mohr. R et al., "Initiation of shape-memory effect by inductive heating of magnetic nanoparticles in thermoplastic polymers", PNAS, **103**, 3540-3545 (2006)

Ryan. A. J. et al., "Mechanical Actuation by Responsive Polyelectrolyte Brushes and Triblock Gels", Journal of Macromolecular Science Part B Physics, **44**, 1103-1121 (2005)

Suzuki. A. et al., "Optical switching in polymer gels", Journal of Applied Physics, **80**, 131-136 (1996)

Zeng. Q. et al., "Fe/Fe oxide nanocomposite particles with large specific absorption rate for hyperthermia", Applied Physics Letters, **90**, 233112 (2007)

PREPARATION AND CHARACTERIZATION OF POLYETHYLENE NANOCOMPOSITES USING PRISTINE CLAY VIA NOVEL WATER INJECTION METHOD IN A TWIN-SCREW EXTRUDER

K. Sarikhani1 and H. Garmabi1*

[1]*Polymer Engineering and Color Technology Department, Amirkabir University of Technology, Tehran, Iran*
garmabi@aut.ac.ir; kaveh.sarikhani@gmail.com

Abstract – In order to modify clay surface to obtain polyethylene nanocomposites in one step melt state process, we used saline water as a pre-intercalating agent. Saline water contains Dodecyl Trimethyl Ammonium Chloride salt in distilled water and was injected into the twin-screw extruder using a high pressure dosing pump. Polyethylene nanocomposite was prepared via injection of saline water into the mixture of polymer melt and pristine clay within a lab scale twin-screw extruder.

The resultant compounds were characterized using X-Ray Diffraction (XRD) and dynamic rheological measurements by means of RMS. Tensile and notched-Izod impact properties of the samples were measured, as well. The XRD results showed no peak for sample containing 2wt% pristine clay and prepared via injection of salt solution representing exfoliated microstruture, and in other sample the XRD results showed intercalated morphology. In addition, the slop of the storage modulus has been decreased and cross over frequency of loss and storage moduli shifted to lower frequencies as a result of increasing the interaction and change in dynamics of polymer chains due to intercalation/exfoliation of clay. Tensile results also showed an increase of tensile modulus in the range of about 22% for sample containing 2 Wt. % of clay prepared via injection of salt solution and fracture energy of these samples which obtained through notched-Izod impact was comparable to neat polyethylene.

Keywords: Polyethylene (PE), Nanocomposites, Pristine Clay, Melt Compounding, Water Injection

Introduction

Nanocomposites are a new class of composites, for which at least one dimension of the dispersed particles is in the nanometer range. Clay contain polymeric nanocomposites currently resulted in a tremendous interest from both academics and industries [1,2].

Beside conventional method of nanocomposites preparation, there are new innovative methods which lead to reducing the cost and steps of preparation [3-6]. In these techniques, one of the characteristic of clay minerals i.e their swelling in water is used in order to disperse and exfoliate clay platelets. Hasegawa et al [3], introduced a novel melt compounding process using the Na–montmorillonite water slurry for preparing the nanocomposites consisted of nylon 6 and Na–montmorillonite without any surfactant.

Yu et al [4] fabricated Nylon nanocomposites without pretreatment with water by small amount of water. In this method, water is injected into the extruder barrel to aid swelling of MMT. TEM micrograph showed uniformly dispersion of clay platelets in polymeric matrices. Fedullo et al [5] prepared Nylon 6 nanocomposites by extrusion with a water injection system.

Nylon 6 was not the only polymer which is used in water injection melt compounding process, Kato et al [6] used similar method for preparation of Polypropylene clay nanocomposites. The main difference of their work is usage of two compatibilizers to prevent aggregation of clay platelets in matrices; it was because of much more hydrophobic nature of PP in compare to Nylon 6.

The aim of this study is to introduce novel method of producing of polyethylene matrix in order to find a commercialized approach of preparation. Saline water was used as a pre-intercalating agent for clay surface to assist clay dispersion in one step melt state process. Polyethylene nanocomposite was prepared via injection of saline water into the mixture of polymer melt and pristine clay. The solution contains Dodecyl Trimethyl Ammonium Chloride salt in distilled water which at the first step to the clay surface. Afterwards, with evaporation of water, polyethylene grafted anhydride maleic acts to stabilize clay surface in the polyethylene matrices.

Experimental

Material

High density polyethylene provided by Sabic® with trade name of HDPE B4660 (MFI= 0.7 g/10 min, 190 °C and 2.16 kg) based on ASTM D1238, with a density of 0.961 g/cm3 is used as the matrix. Maleic anhydride grafted polyethylene, PE-g-MA, (Fusabond® E100) was purchased from DuPont™ to enhance matrices compatibility with nanoclay, its melt flow rate is 2 g/10 min with a nominal density of 0.954 g/cm^3. Sodium-montmorillonite (Na-MMT) under the trade name of Cloisite Na+ with a cation exchange capacity of 0.92 meq/g., was provided by Southern Clay Products (Gonzalez TX). Dodecyl Trimethyl Ammonium Chloride is also used for in-situ treatment of clay to enhance its compatibility with matrix. It was obtained from Xiamen Pioneer Technology Inc. with purity >99%. The water used for injection is a distilled one.

Preparation of Nanocomposites using saline water

Compounding of Nanocomposites has been done using a laboratory co-rotating intermeshing twin screw extruder from Brabender with length to diameter ratio of 40 and diameter of 20 mm. There are three mixing zones composed of kneading blocks to enhance the clay dispersion. The temperatures of five heating zones were set at 180, 190, 200, 205, and 205°C, respectively. The screw speed was 200 rpm and the feeding rate in all the samples was 1.3 kg/hr. The Saline solution contains Dodecyl Trimethyl Ammonium Chloride salt in distilled water was injected into to the barrel through DLS 2/20 dozing pump made by ETATRON, Italy which can produce 20 bar pressure. This pressure is high enough to compensate saturated pressure of water at 205 ° C (about 16 bars). The extruded clay/polyethylene compounds were cooled in a water bath and then pelletized. The residence time in twin screw extruder was approximately 2 min. The screw design contains blocks to prevent back flow of water vapor to the hopper. At the final step of compounding, the water was removed from the vent by vacuum.

Nanocomposite of polyethylene was prepared with 2 w.t % of clay and salt amount of 0.25 equivalent of clay is abbreviated to PE-Nano1. For the sake of comparison, another sample is prepared with the same composition prepared with different method of incorporation of alkyl ammonium salt; in this sample, distilled water injected and the salt fed to the extruder with other materials, abbreviated as PE-Nano2. Besides, neat polyethylene sample was produced under the same condition with injection of pure water for comparison to nanocomposites, this sample is named PE.

Characterization

Dispersion of the silicate layers in the composites was evaluated by an X-ray diffractometer. X-ray diffraction (XRD) patterns were obtained through a Philips X'Pert equipped with Cu KR generator (λ= 1.5404 Å) at a scan rate of 0.02°/s ranging from 1.5° to 30°. Generator tension is 40 kV and generator current is 40 mA.

AFM results have been obtained by Dualscope/ Rasterscope C26, DME, Denmark operating in ambient temperature. Topography and phase images were obtained under tapping mode using a Mikromasch CSC17 Silicon tip with force constant of 0.05 -0.30 N/m and Resonance frequency of 150-190 kHz.

The storage modulus (G') and complex viscosity (η) were measured as a function of angular frequency (ω) using a Paar Physica, UDS200 rheometer. The rheometer was operated in the dynamic oscillatory mode with parallel plate geometry of 25 mm diameter and 1 mm gap. Dynamic frequency sweep tests were executed in a frequency range of 0.05 – 1000 sec-1 using strain amplitude of 1%, proven to be in the linear viscoelasticity range by means of strain sweep measurements at temperatures of 205 °C.

Tensile tests were conducted on Galdabini instrument based on ASTM D638 at a cross speed of 50 mm/min and 2 N pre-load. Impact test are done by U-F impact tester from UESHIMA SEISAKUSHO LTD Japan under conditions of ASTM D256 and 30 kgf.cm load at room temperature.

Results and Discussion

The XRD results for PE-Nano1, PE-Nano2 and neat PE samples are shown in Fig 1. No diffraction peak is observed for PE-Nano1 sample at low 2θ values but a small peak at about 6° is visible for PE-Nano2 samples which corresponds to basal spacing of 1.5nm, it can be related to either alkyl ammonium salt or absorbed water molecules layers

In case of PE-Nano 1, due to better dispersion of the alkyl ammonium chains in water and good interaction with clay surface ,consequently by donating hydrophobicity nature, the basal spacing of silicate layers increase which results into easier diffusion of polymer and compatibilizer chains between layers. Although in PE-Nano 2 sample diffraction peak is appears in basal spacing of 1.5 nm, the intensity of peak is low and it shows that other clay stacks dispersed by interaction of salt modifierand clay surface.

Tapping Mode AFM height and phase images of PE-Nano1 sample in 3 μm field of view is presented in Fig 2. Clay particles in light streaks are visible in phase image which shows uniform distribution of them in polyethylene matrix. The clay particles have dimension of 100-200 nm length with 20-30 nm width in PE-Nano1 sample. These results showed that solution of ammonium salt is more efficient in distribution than incorporation with other materials. This distribution is in agreement with complex viscosity results which discuss later.

As Fig.3 shows, both samples showed deviation from neat polyethylene in terminal zone but those of PE-Nano1 indicates higher deviation and its slope has been changed dramatically in compare to PE-Nano2 which is evidence of solid-like behavior due to percolated network of clay platelets as reported before

[7-9]. The more dispersion and distribution of platelets, the more possibility of network formation. These results are in agreement with XRD and AFM results.

The exponent of n in power law equation in low frequencies is used as a measure of clay dispersion [10] it is based on the fact that at low angular frequencies, rheological response is characteristic of the unperturbed filler structure in system. Complex viscosity against angular frequency of samples is shown in Fig.4. Both samples showed deviation from neat polyethylene behavior at low frequency region; the slop and n exponent has changed and viscosity is higher than those of polyethylene. In high frequency region the house of cards structure cannot follow the shear-induced disturbance, and the dynamic of nanocomposite system is mainly controlled by the liquid polymer melt. For PE-Nano1, the change in slope and n exponent is more prominent and a viscosity upturn can be seen at low frequencies. The change in slope comes back to ease of shear thinning behavior in polyethylene nanocomposites based on both anisotropic filler and structure formation by high aspect ratio silicate layers [11].

Table 1 shows the tensile properties of PE-Nano1 and PE-Nano2 samples compared to the neat PE. As can be observed, both of the composites showed an enhancement of the tensile modulus of polymer as compared to the pure PE. The enhancement in tensile modulus of PE-Nano1 is about 22% but the PE-Nano2 showed 14% increment in tensile modulus. This different comes back to variation of dispersion and distribution of samples. Based on the storage and complex viscosity results, distribution of silicate layers were in proper condition in PE-Nano1, due to better dispersion and distribution of the clay particles in PE-Nano1 where interfacial interaction between polymer and silicate layer has been increased in it. It seems as a result of better distribution of alkyl ammonium salt in saline water, the interaction of silicate layers with alkyl ammonium salt has been improved.

Fracture energy acquired through impact test also showed in Fig 5. As it is shown, PE-Nano1 has comparable impact strength with PE but those of PE-Nano2 showed about 35% lower than impact strength than PE sample. The brittle behavior of nanocomposites probably originates from the formation of microvoids due to deboning of clay platelets from the polymer matrix upon failure [12]. These microvoids will merge into larger cracks causing embrittlement, ultimately resulting in reduced toughness [13]. It is probable that in PE-Nano2, formation of these microvoids have led to reduction in toughness but in PE-Nano1, because of its proper dispersion, the size of microvoids are comparable with well dispersed silicate layers, therefore, their size will be below the critical size of the crack, eventually their fracture energy is comparable to polyethylene. Other researcher also reported reduction in impact strength in PE-based nanocomposites, indicating that the incorporation of clay into semicrystalline thermoplastics usually results in toughness reduction [12].

Table 1. Tensile properties of PE composites in compare to neat PE.

sample Name	Tensile Modulus (N/mm^2)	Tensile Strength (N/mm^2)	Stress at Break (N/mm^2)	Elongation at Break (%)
PE	988±12	25.2	12.8	205
PE-Nano1	1213±17	28.8	13.9	154
PE-Nano2	1131±20	27.7	15.8	183

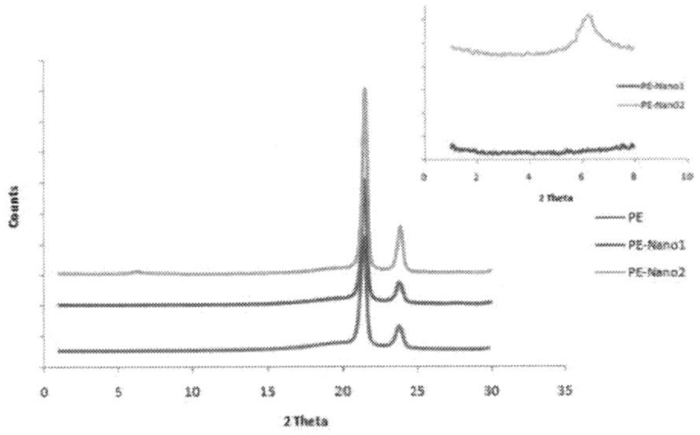

Fig 1. XRD patterns for PE, PE-Nano 1 and PE-Nano 2 samples.

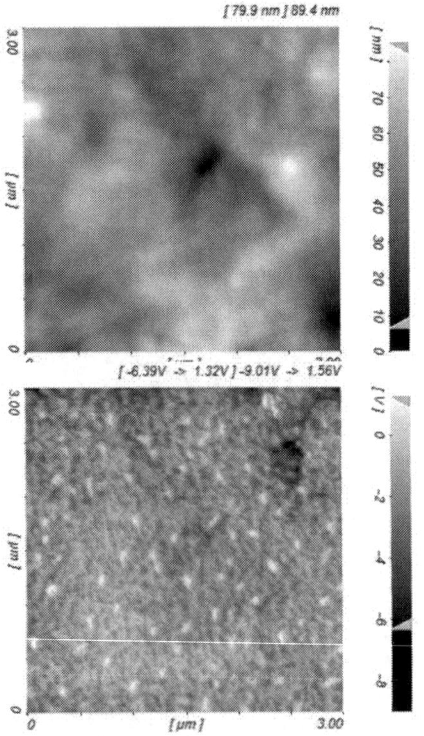

Fig 2. Tapping Mode AFM images of PE-Nano2 sample, height (left) and phase (right) are shown in 3*3 µm field of view

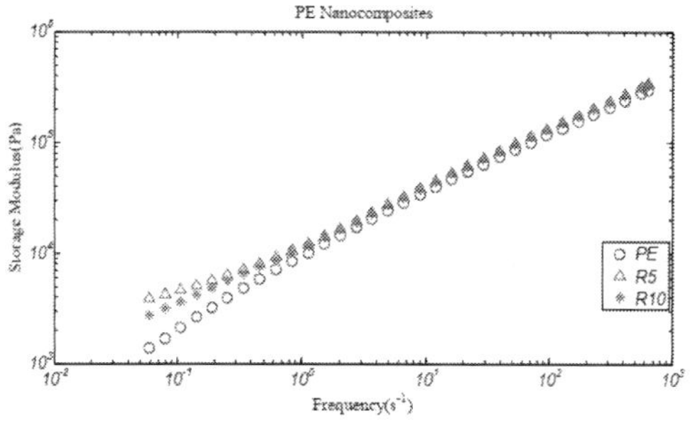

Fig 3. The storage modulus against angular frequency of PE-Nano1 and PE-Nano2 samples

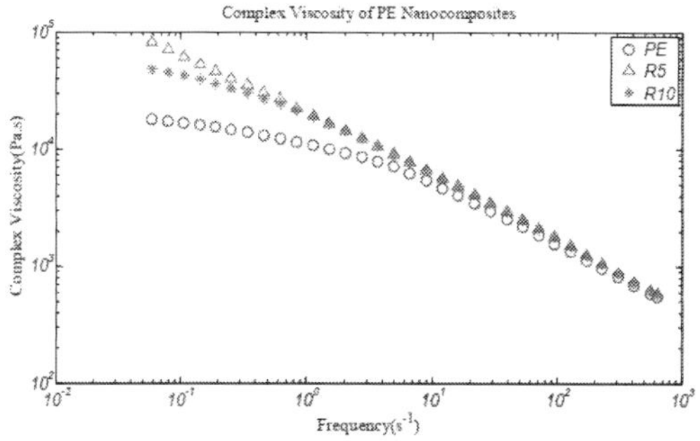

Fig 4. The complex viscosity against angular frequency of PE-Nano1 and PE-Nano2 samples

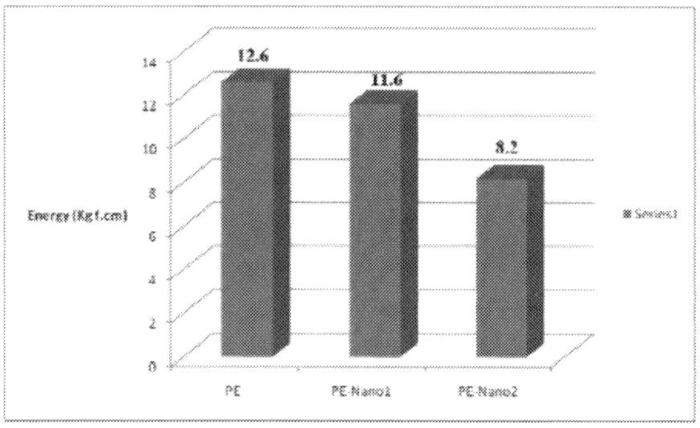

Fig 5. Fracture energy of PE composites in compare to neat PE.

Conclusions

Polyethylene nanocomposite prepared with the aid of saline water as a pre-intercalating agent for clay surface to assist clay dispersion in one step melt state process. In this process saline water was injected into the mixture of polymer melt and pristine clay within a lab scale twin-screw extruder. The solution contains Dodecyl Trimethyl Ammonium Chloride salt in distilled water and was injected into the twin-screw extruder using a high pressure dosing pump.
The XRD results showed no peak for sample containing 2wt% pristine clay which prepared via injection of salt solution representing exfoliated microstruture. AFM in tapping mode images also showed uniform distribution of clay particles in polyethylene matrix. Furthermore, for the saline water injected sample, RMS results showed that the slop of the storage modulus has been declined and has been reached to a near plateau which is evidence of network structure for this sample. In Complex viscosity curves, deviation from neat polyethylene behavior in low frequency region was observed; the slope and n exponent has changed and viscosity is higher than those of polyethylene. Change in slope and viscosity upturn in saline water solution injected sample, confirms better dispersion and formation of network in this nanocomposite. Tensile results also showed an increase of tensile modulus in the range of about 22% for sample prepared via injection of salt solution. Furthermore, Izod impact strength in sample prepared via injection of salt solution had energy of fracture comparable to neat polyethylene, while other sample showed about 35% reduction in impact strength. The crystallization data in terms of degree of crystallinity, melting and crystallization temperatures of samples had not noticeable variation comparing with the neat polyethylene.

References (Times New Roman 10, bold)

1. S. Pavlidou, C.D. Papaspyrides, Progress in Polymer Science 2008, 33, 1119–1198.
2. M. Alexandre,G. Beyer, C.Henrist, R. Cloots, A.Rulmont,R.Je´ro^me, Ph.Dubois, Chem. Mater. 2001, 13, 3830-3832,
3. N.Hasegawa, H.Okamoto, M.Kato, A.Usuki, N.Sato. Polymer 2003, 44, 2933–2937.
4. Zh.Zh. Yu, G.H. Hu, J.Varlet, A. Dasari, Y.W. Mai. J Polym Sci Part B: Polym Phys 2005,43,1100–1112.
5. N.Fedullo, M.Sclavons, Ch.Bailly, J.M. Lefebvre, J. Devaux. Macromol. Symp; 2006, 233, 235–245.
6. M.Kato, M.Matsushita, K. Fukumori. Polym. Eng. Sci. 2004, 44,1205–1211.
7. G. Galgali, C. Ramesh, A. Lele, Macromolecules, 2001, 34, 852.
8. J. Ren, B.F. Casanueva, C.A. Mitchell, and R. Krishnamoorti, Macromolecules 2003, 36, 4188.
9. A. Hedayati, A. Arefazar. Polymr Composite., 30:1717–1731, 2009.
10. D.P.N. Vlasvelda, M. de Jonga, H.E.N. Berseec, A.D. Gotsisa, and S.J. Picken, Polymer 2005, 46, 10279.
11. Fornes TD, Yoon PJ, Keskkula H, Paul DR. Polymer 2001,42,9929.
12. Morawiec J, Pawlak A, Slouf M, Galeski A, Piorkowska E, Kransnikowa N. Eur Polym J 2005,41,1115–22.
13. Phang IY, Liu T, Mohamed A, Pramoda KP, Chen L, Shen L, et al. Polym Int 2005,54,456–64.

SYNTHESIS OF NANOCOMPOSITE OF POLYURETHANE-EPOXY INTERPENETRATING POLYMER NETWORKS AND INVESTIGATION OF MECHANICAL, MORPHOLOGICAL AND THERMAL PROPERTIES

A. Sayadi [1], G. Mir Mohamad Sadeghi [1*], M. Ghafghazi [2], J. Morshedian [2]

[1] *Polymer Engineering Department, Amirkabir University of Technology, Tehran, Iran – Afsoon.sayadi@aut.ac.ir,* gsadeghi@aut.ac.ir

[2] *Iran Polymer & Petrochemical Institute, Tehran, Iran – m_ghafghazi_ppt@yahoo.com, j.morshedian@ippi.ac.ir*

Abstract - Interpenetrating polymer networks (IPNs) are special polymer blends consisting of at least two polymers in a network, which is held together by permanent entanglement and shows excellent thermal stability and mechanical properties because of a synergetic effect induced by the forced compatibility of the individual components. Polyurethane-Epoxy interpenetrating polymer networks (IPN) was prepared by simultaneous method as explained and after the formation of IPN was proved by Fourier Transform Infrared Spectroscopy (FT–IR) and Differential Scanning Calorimetric (DSC), different weight ratios of Polyurethane-Epoxy (PU-EP) were investigated by mechanical, morphological and thermal experiments. Results of mechanical experiments showed that PU-EP IPN in 25/75 weight ratio has the most tensile strength and elongation at break in comparing with other weight ratios. It is an evidence for better penetration of two phases and more compatibility between them. In other hand Scanning Electron Microscopy (SEM) that used to study morphological properties of samples of different weight ratio IPNs indicated a finely dispersion of Polyurethane phase in Epoxy matrix, also it's fracture surface consisted of fine crazes that was observed beside shear yielding mechanism. Thermal Gravimetric Analysis (TGA) was used to observe thermal resistance of prepared IPNs. The most degradation temperature was related to 25/75 PU-EP IPN. Synthesis of nano composite of PU-EP were done in a similar manner and the effect of adding 1,3,5 wt% of Closite 30B were studied by mechanical, morphological , thermal and X-Ray Diffraction (XRD) analysis. Exfoliated structure were found by addition of 1 wt% Closite 30B while intercalated and agglomerated structure were obtained by using 3 wt% and 5 wt%. More toughness of IPN 1(1 wt% of Closite 30B in IPN) was the result of better dispersion and miscibility of final system that arises from the formation of hydrogen banding between polymer chain and Closite 30B and higher cross linking density. These observations were accepted by mechanical analysis by getting more energy absorption and in addition with morphological and thermal analysis.

Introduction

Epoxy resins are commonly used as thermosetting matrices for glass, carbon, and aramid fiber-reinforced composites used in structural and isolation applications (Bucknall, 1977). Cured epoxy resins (ER) are extensively employed in several important applications such as structural adhesives, surface coatings, advanced composites for the aerospace and electronic industries, etc, because of their good engineering properties which include high stiffness and strength, creep resistance, chemical resistance and good adhesion to many substrates (May and Tanka, 1973). However, some unsolved problems encountered with some types of epoxy resins during application, such as relatively poor toughness, still remain. Several attempts have been made to alleviate this problem. The use of rubber particles as a second phase to the epoxy resins to increase the toughness of the brittle matrix has been considered extensively (Bucknall, 1977). One method that is employed in using rubber in epoxy matrices is formation of IPNs. The interpenetrating polymer networks (IPNs) are a unique type of polyblend which define as a mixture of two or more crosslinked polymer networks that have partial or total physical interlocking between them (Ma et al., 1994; Kim et al., 1976; Jeevananda and Siddaramaiah, 2003). In fact, formation of IPNs is the only way of intimately combining crosslinked polymers (Frisch et al., 1982). If a high degree of intermixing is intended, the Simultaneous IPN (SIN), in situ mixing and polymerization of monomers or prepolymer, is preferred. This is due to the compatibility of the monomeric mixture, which is much higher than that of a polymeric mixture (Kim and Kim, 1987). The resulting mixture exhibiting (at worst) only limited phase separation while normal blending or mixing of polymers results in a multiphase morphology due to the well-known thermodynamic incompatibility of polymers. However, if mixing is accomplished simultaneously with crosslinking, phase separation may be kinetically controlled by permanent interlocking of entangled chains (Frisch et al., 1982; Kim and Kim, 1987; Ghafghazi et al. 2009). The entanglements in IPNs must be of a permanent nature and are made so by homo-crosslinking of the two or more polymers (Frisch et al., 1982). These entanglements of multiple crosslinked polymers lead to forced miscibility compared to simple blends and the resulting IPN materials exhibit excellent size stability and a possibly synergistic combination of the properties of their components (Ghafghazi et al. 2009; Zhu et al., 2007; Chen and Chen, 2004). These make IPNs interesting in several applications (Zhu et al., 2007; Chen and Chen, 2004). The object of this research is to prepare EP/PU interpenetrating polymer networks and to investigate the effect of Epoxy/Urethane weight ratio and use of chain extender modified nanoclay (Closite 30B) on the mechanical, morphological and thermal and swelling properties of the IPN systems.

Experimental

Material

Epoxy resin used was diglycidylether of bisphenol A(DGEBA,Epikote828,HuntsmanEpoxyCo,Mw=380), Hydroxylterminatedpolybutadiene(HTPB,Hydroxyl equivalent 0.075-0.08eq/100gr), TDI (Toluene Di isocyanate, an 80/20 mixture of 2,4/2,6 isomers),Dibutyltindilaurate (urethane catalyst), Tributylamine (epoxy catalyst), Aradur 917, was obtained from Huntsman Advanced Materials (Switzerland).

Characterization

Fourier transform infrared spectroscopy (FTIR), Tensile and Izod impact properties and thermal gravimetric analysis (TGA) were obtained to study the properties of samples. Other thermal analyses were performed by Dynamic mechanical thermal analysis (DMTA) and Differential Scanning Calorimetry (DSC) finally SEM experiment was done to study fracture surface of samples.

Synthesis of Epoxy/Urethane IPN

Simultaneous method was used for synthesis of EP/PU IPN Predetermined amount of HTPB (NCO: OH ratio was 1:1) and epoxy was mixed at room temperature and under dry nitrogen atmosphere by high shear mixer. It. A reactor which was equipped with a mechanical stirrer and dry nitrogen inlet was obtained to perform synthesis. Specified Weight of epoxy, Aradur hardener was added and agitated for some minutes then Tributylamine as the epoxy catalyst, predetermined TDI and DBTDL as urethane catalyst were added respectively. Mixture was cast in to a silicon mold and was cured. In next stage, for determining the effect of chain extender presence in IPN properties, butanediol chain extender was added at first step and other steps were performed according to previous parts as explained.

Synthesis of Epoxy/Urethane Nanocomposite

Determined weight ratio of Closite 30B was added to epoxy resin and mixed using high shear mixer then it was put in ultrasonic bath at last HTPB and Butanediol were added and other steps were past as before.

Results and Discussion

Investigation of IPN formation

FTIR and DSC were used to analyze the formation of epoxy/polyurethane IPN. Curing peak of EP/PU IPN 50/50 was observed broader than cured epoxy. That was an evidence of simultaneous polymerization of polyurethane and epoxy. In addition FTIR spectra showed the presence of NCO and epoxid groups at 2273 cm^{-1} and 915 cm^{-1} respectively in spectra before curing while they disappeared after curing. Also surface area related to NH band in urethane linkage which occurs at 3340-3420 cm^{-1} increased in in cured sample as are shown in Figure 1.

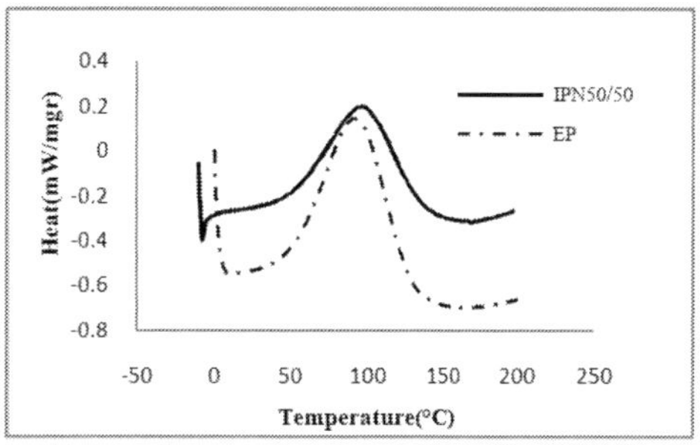

Figure 1 – Curing peaks of Epoxy and IPN50/50

Mechanical properties of EP/PU IPNs

The results of mechanical properties for IPN systems at various EP/PU weight ratios (HTPB M_w=2800 g/gmol) are shown in Table 1. As it is shown, there is an ascending order for the tensile modulus and strength with increasing EP/PU weight ratio but the trend for the tensile strength is reversed at epoxy weight ratio above 75 (wt%). Tensile strength experiments were done to observe mechanical properties of IPN.

Table 1 – mechanical properties of samples

Epoxy (wt%)	Modulus (MPa)	Tensile (MPa)	Elongation at break (%)	Toughness (J)	Izod Impact (J/m)
0	8.61	4.43	126.26	33	-
50	401	8.92	3.8	12	68/78
75	542	19.2	6.81	52	79.85
100	1319	32.25	3.45	9	64.91

Modulus and tensile strength increased with increment of epoxy content while toughness and elongation at break decreased with an increment of epoxy content. This is due to characteristic properties of epoxy that has lower toughness and elongation at break. best mechanical properties was observed in 75/25 EP/PU IPN. The reason was related to better diffusion of two phases and more hydrogen bonding between them. improvement of Izod impact results were an emphasis on having toughening mechanism on polyurethane addition to epoxy matrix.

Morphological properties of EP/PU IPNs

SEM results showed that epoxy and polyurethane have smooth fracture surface but the fracture surface of

EP/PU IPN 50/50 and EP/PU IPN 75/25 contained matrix-disperse morphology. Finer particle size distribution and shear yielding mechanism were observed for EP/PU IPN 75/25 that confirmed developed mechanical results about EP/PU IPN SEM image of EP/PU IPN 75/25 is indicated in Figure 2 in comparing with epoxy.

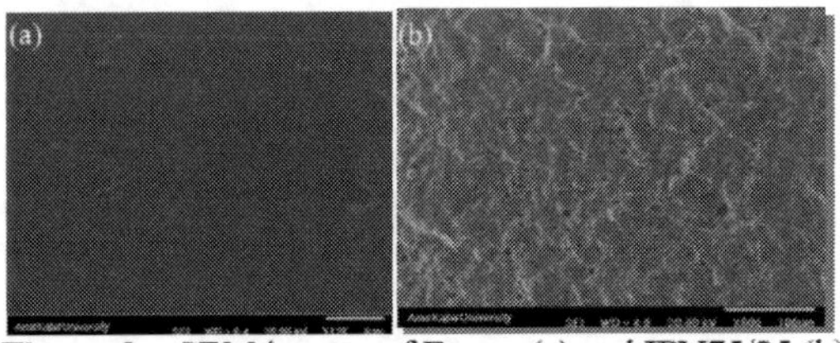

Figure 2 – SEM images of Epoxy (a) and IPN75/25 (b) (500X).

Effect of chain extender addition on mechanical properties

As it was pointed in synthesis section, the aim of this study is also, determining the effect of butanediol as chain extender in IPN. In this way EP/PU 75/25 IPN was chosen to study the behavior of IPN with use of chain extender. Butanediol was added in three weight ratios(8%,16% and 24% hard segment) and was named according to table 2. Results indicated that butanediol did not have its effect as a chain extender of polyurethane in EP/PU 75/25-BD1 with lowest hard segment (8%), but acted like a plasticizer and reduced plastic resistance of IPN (Argon et al., 1994). In higher contents of butanediol EP/PU IPN 75/25-BD2 (16% hard segment) a sudden increment was observed in modulus and tensile strength although toughness and elongation at break became smaller than EP/PU IPN 75/25-BD1. In EP/PU IPN 75/25-BD3 due to shorter gelation time for polyurethane and higher content of diisocyanate in system phase separation happened intensively and in conclusion weaker mechanical properties was reported by tensile strength test. So EP/PU IPN 75/25-BD2 was chosen as the best sample with suitable mechanical properties.

Table 2 – Mechanical properties of EP/PU IPN 75/25 with BD wt%

Samples	Hard Segment Content (%)	Tensile Modulus (MPa)	Tensile Strength (MPa)	Elongation at Break (%)	Toughness (J)	Izod Impact Strength (J/m)
IPN75/25	0	542	19.2	6.81	52	79.85
IPN BD1	8	293	9.7	43.7	207	53.33
IPN BD2	16	481	24.74	10.08	97	82.89
IPN BD3	24	627	21.56	7.6	63	64.41

Morphological properties of EP/PU IPN 75/25 with and with out chain extender

As it was mentioned in previous section, EP/PU IPN 75/25-BD2 was selected as the chosen sample among 3 different hard segments. Comparison of SEM images of EP/PU IPN 75/25 and EP/PU IPN 75/25-BD2 with the same magnification in figure 3 showed that particle size distribution of EP/PU IPN 75/25-BD2 was finer than EP/PU IPN 75/25 and also shear yielding mechanism was seemed to be more dominate in EP/PU IPN 75/25-BD2 in contrast with EP/PU IPN 75/25. The reason related to better compatibility of two phases resulting from more hydrogen bonds between urethane bands formed from chain extender and oxygen and hydroxyl of epoxy group.

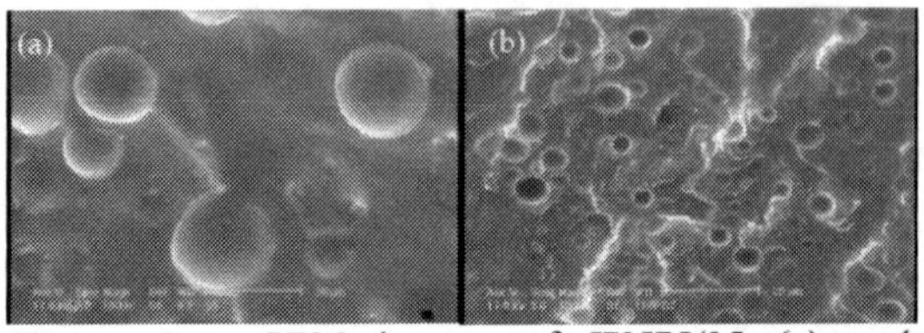

Figure 3 – SEM images of IPN75/25 (a) and IPN75/25-BD2 (1000X).

Mechanical properties of Nano IPNs

Nano IPNs with1 wt%,3 wt% and 5wt% of nanoclay that called NIPN1, NIPN3 and NIPN5 were examined by extensometer in tensile strength and Izod impact experiment, results are shown in table 3 In low content of nanoclay in NIPN1 lower tensile strength, modulus and Izod impact strength were reported, but higher toughness and higher elongation at break was also observed. It was defined that in NIPN1 due to good dispersion of nanoclay that will verified by XRD result, lower modulus was accounted. the reason was related to lower modulus of nanoclay layers also.

Table 3 – Mechanical properties of NanoIPNs

Samples	Nanoclay content (%)	Tensile Modul. (MPa)	Tensile Strength (MPa)	Elongation at Break (%)	Toughness (J)	Izod Impact Strength (J/m)
IPN 75/25	0	542	19.2	6.81	52	79.85
IPNBD2	0	481	24.74	10.08	97	74.89
NIPN1	1	422	13.41	20.43	68	51.85
NIPN3	3	596	16.22	5.47	24	74.07
NIPN5	5	583	13.64	4.91	17	51.95

XRD analysis

The XRD patterns are shown in table 4. oMMT had characteristic diffraction peak at $2\theta=4.61$ on the basis of Bragg equation, corresponding to a basal spacing of about 18.5 an. NIPN1 showed no diffraction peak, that can be interpreted to formation of exfoliated structure, some peaks were observed in NIPN3 and NIPN5. reported peaks related to intercalate structure (Alexandre and Dubois, 2000), but weaker mechanical properties caused that partly flocculated structure could be intended.

Table 4 – XRD analysis of samples

Sample code	2θ(°)	d-spacing(Å)	Height(cts)
NIPN3	2.70	62/32	31.91
	34/5	16.52	19.57
NIPN5	2.55	34.49	46.44
	5.19	16.99	30.21

SEM analysis

SEM micrographs of nanocomposites are provided in Figure.4 . Surface fracture of NIPN1 contained bigger polyurethane holes. More phase separation was guessed as the reason of this happening that was verified by DSC and DMTA results. Observation of higher toughness in low nanoclay content (NIPN1) has also been accounted by Gupta and Maharsia (2005). Foundation of voids amplifies toughening mechanism that was found in development of toughness in mechanical test. Dimension of polyurethane cavities decreased considerably in NIPN3 and NIPN5. Two reasons existed for this observation, increase of hydrogen bonding between hydroxyl group on nanoclay layers with urethane linkage and oxygen of epoxy main chain cased to develop of physical cross linking and smaller polyurethane holes.

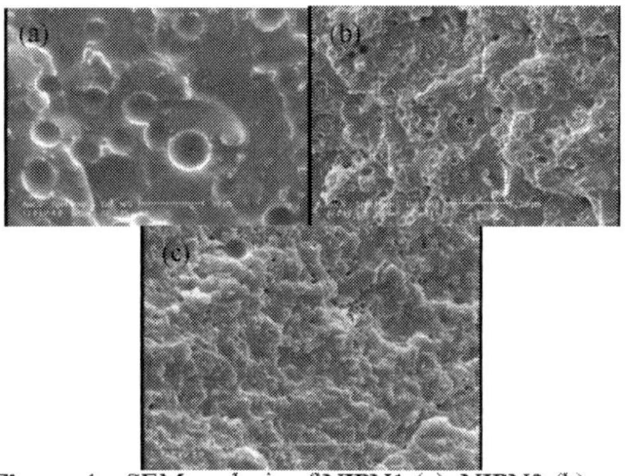

Figure 4 – SEM analysis of NIPN1 (a), NIPN3 (b), NIPN5 (c).

Thermal analysis

DSC test were applied to study thermal behavior of synthesized IPN and nanocomposites and their glass transition temperature (T_g).

According to figure 5 two distinct T_g was observed for EP/PU IPN 75/25 and NIPN1 while just one peak was assesed for EP/PU IPN 75/25-BD2, NIPN3 and NIPN5.

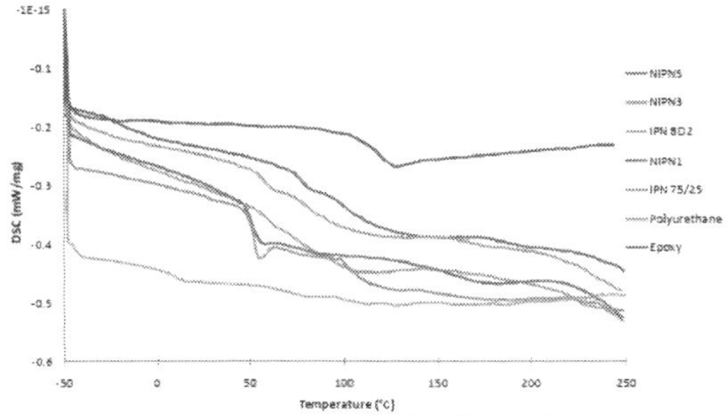

Figure 5 – DSC analysis of samples

These observations were rationed to development of phase separation of connected polyurethane chains to epoxy phase and a broad area was an explanation for such happening.Following to DSC results, DMTA acheivements verified last observations.

Conclusions

In this study EP/PU IPNs were synthesis by simultaneous method. Mechanical and morphological results disclosed that EP/PU IPN 75/25 contained suitable properties in mentioned experiments. Checking the effect of butadiene on EP/PU IPN 75/25 indicated that higher extent of physical cross linking in EP/PU IPN 75/25-BD2 caused to have more miscibility between two phases. Finally utilization of 1wt% nanoclay increased toughness in tensile strength test. Suitable dispersion of nanoclay in NIPN1 was verified by XRD and intercalated structure was also assigned for NIPN3 and NIPN5 by XRD method. Thermal properties of EP/PU IPN developed by addition of butadiene in EP/PU IPN-BD2 by formation of more hydrogen bonds and more miscibility. at last addition of nanoclay in higher extents (NIPN3 and NIPN5) increased T_gs of samples.

References

Alexandre, M., Dubois, P., "Polymer-layered silicate nanocomposites: preparation, properties and uses of a new class of materials", Mater. Sci. Eng. R., **28**, 1–63, (2000)

Argon, A. S., et al., "Mechanisms of Toughening Brittle Polymers", Mater. Sci. Eng. A., **176**, 79–90 (1994)

Bucknall C. B.: Toughened Plastics, Applied Science Publishers, London (1977)

Chen, L.; Chen, S., "Latex interpenetrating networks based on polyurethane, polyacrylate and epoxy resin", Prog. Org. Coat., **49**, 252–258 (2004)

Frisch, K. C., et al., "Recent advances in interpenetrating polymer networks", Polym. Eng. Sci., **22**, 1143–1152 (1982)

Ghafghazi, M., et al., "Preparation of epoxy/urethane graft interpenetrating polymer networks and study of mechanical, thermal and morphological properties", e-Polymers, **69**, 1–13 (2009)

Gupta, N., Maharsia, R., "Enhancement of Energy Absorption in Syntactic Foams by Nanoclay Incorporation for Sandwich Core Applications", Appl. Compos. Mater., **12**, 247–261, (2005)

Jeevananda, T., Siddaramaiah, "Synthesis and characterization of polyaniline filled PU/PMMA interpenetrating polymer networks", Eur. Polym. J., **39**, 569–578 (2003)

Kim, J. H., Kim, S. C., "Analysis of reaction injection molding process of polyurethane-unsaturated polyester blends. Part II: Mechanical properties", Polym. Eng. Sci., **27**, 1252–1257 (1987)

Kim, S. C., et al., "Polyurethane Interpenetrating Polymer Networks. II. Density and Glass Transition Behavior of Polyurethane-Poly(methyl methacrylate) and Polyurethane-Polystyrene IPN's", Macromolecules., **9**, 263–266 (1976)

Ma, C. C. M., et al.,"Interpenetrating polymer networks of polyurethane and furfuryl alcohol. I. Morphology and mechanical properties", Die. Angew. Makromol. Chem., **214**, 39–56 (1994)

May C. A., Tanka G. Y.: Epoxy resin chemistry and technology, Marcel Decker, New York (1973)

Zhu, Y., et al., "Influence of MMT content on the microstructure and miscibility for PU/EP IPN nanocomposites by positron", Phys. Stat. Sol. C., **4**, 3780–3784 (2007)

PREPARATION AND CHARACTERIZATION OF POLYMER NANOCOMPOSITES WITH TITANIUM DIOXIDE NANOPARTICLES

P.R.A. Bernardo[1], A.J.F. Carvalho[2] and L.A. Pessan[3]*

[1] *Graduate Program in Materials Science and Engineering, Federal University of São Carlos - (PPG-CEM/UFSCar), São Carlos, Brazil - professorpaulorodrigo@yahoo.com.br*
[2]*Department of Materials Engineering, University of São Paulo - (EESC/USP), São Carlos, Brazil - toni@sc.usp.br*
[3]*Department of Materials Engineering, Federal University of São Carlos - (DEMa/UFSCar), São Carlos, Brazil - *pessan@ufscar.br*

Abstract - Titanium dioxide nanoparticles (TiO_2) have attracted a special interest due to its ability in absorbing ultraviolet radiation (UV) and also to the possibility of its use either to reduce or to accelerate the degradation process in the polymer matrix depending on each for is used, rutile covered or anatase not covered form, respectively. The objective of this work was to develop polymer nanocomposites with TiO_2 nanoparticles and correlate its structure to its photodegradation properties. A new technique based on the use of spray drying process to obtain high content TiO_2 polymer nanocomposites was developed and tested. TiO_2 nanoparticles were prepared through the sol-gel process and used in the preparation of a stable mixture with hydrosoluble polymer PVOH. The aqueous suspension of TiO_2/PVOH was then spray dried to obtain dry powder that was used to incorporate in another polymer matrix. The TiO_2/PVOH concentrated powder obtained through the spray drying process was incorporated in polyethylene terephthalate (PET), ethylene vinyl acetate (EVA) and low density polyethylene (LDPE) by melt mixing their films compared with films of LDPE nanocomposites, with TiO_2 rutile and anatase nanoparticles, prepared through a conventional melt mixing technique. The results showed that the spray drying process is an innovative and powerful way to obtain a nanocomposite with high TiO_2 nanoparticles content (50%), isolated and encapsulated. The mechanical properties of the films after the photodegradation for UV radiation indicate that the films with TiO_2/PVOH concentrated powder show potential to accelerate the effects of the photodegradation in the LDPE matrix.

Introduction

The dispersion of nanofillers in a polymeric matrix to produce polymer nanocomposites represents one of the main challenges in the nanocomposite technology and a great effort is dedicated to this research area. The dispersion of nanoparticles is a drawback in nanocomposite production processes, because nanoparticle shows a strong tendency of agglomeration during processing, reducing the performance of the composite. Several inorganic materials are used as additive in polymers such as: mica, silica, carbon, clay, calcium carbonate and titanium dioxide.

Titanium dioxide (TiO_2) has attracted a special interest due to its functional properties. TiO_2 can be produced in three polymorphic phases: rutile (tetragonal), anatase (tetragonal) and the unstable bruquite (orthorhombic). Rutile is formed at high temperatures (>1000 °C), while anatase is formed starting from low temperatures (up to about 450 °C). Anatase TiO_2 is non toxic and absorb ultraviolet radiation (UV) to produce free radicals which confers to these materials antimicrobial properties. Due to these properties, TiO_2 is used in several industrial applications. The properties of these materials are highly dependent on their size, and consequently are directly related to its surface area. Other relevant factors are the dispersion, homogeneity and chemical decomposition temperature (Costa et al., 2006).

TiO_2 nanocomposites are used for UV protection due to its UV-absorption properties and due to its transparency in the visible light (Chen, 2002). However, if TiO_2 particles are not protected with a protective layer, usually SiO_2, it has the potential to cause the degradation of the matrix (DuPont, 2007) allowing its use as a photodegradant material especially in the anatase phase (ThreeBond, 2004). The use of anatase nanoparticles for the production of photodegradable polymers is of great interest for several applications, such as packaging and agricultural mulch films.

Several methods of chemical synthesis might be used for the preparation of TiO_2 nanometric powders. The Pechini method is based on the polyesterification of titanium citrate and ethylene glycol to produce a polymer which can be pyrolysed rendering TiO_2 nanoparticles. The particle size can be controlled by the pyrolyzes parameters. Anatase phase nanoparticles with medium size of 21 nm have been prepared by this method (Costa et al., 2006). TiO_2 nanoparticles can also be prepared in aqueous suspension by the sol-gel method. For this purpose alquil titanates (Huguenin et al., 2005) or titanium tetrachloride ($TiCl_4$) are hydrolyzed under controlled conditions (Chen, 2002).

The preparation of nanoparticles in suspensions shows several advantages, such as simple preparation procedure and good control of particle size and particle size distribution in nanometric scale. However, its incorporation in hydrophobic matrixes is not easy and particle agglomeration is a sever drawback.

In this work polymer nanocomposites with TiO_2 nanoparticles were obtained and studied regarding its mechanical and photodegradation properties. A new technique to obtain high content TiO_2 polymer nanocomposites, based on the use of the spray drying process, was developed and tested.

Experimental

Materials

PET resin was obtained from Rhodia Ster S.A. (RHOPET S80, IV 0.78 dL/g). EVA resin was obtained from Petroquimica Triunfo (TN2020, 8-9% vinyl acetate, ρ 0.931 g/cm^3). LDPE resin was obtained from Braskem S.A. (BF0323HC, IF 0.32 g/10min, ρ 0.923 g/cm^3). The stabilization agent used to avoid TiO_2 agglomeration was polyvinyl alcohol (PVOH - M_w ~10,000, 80% hydrolyzed, from SigmaAldrich). The colloidal dispersion of anatase TiO_2 was prepared by hydrolysis of tetra-n-butyl titanate from Du-Pont (Tyzor). TiO_2 rutile (300 nm, SiO_2 layer) and anatase (25 nm) nanoparticles were purchased from DuPont (Ti-Pure) and SigmaAldrich respectively.

Preparation of TiO_2 nanoparticles through the sol-gel process

A mixture of 30 mL of the organic titanate and 30 mL of 2-propanol was slowly added (15 min) to 300 mL of distilled water under vigorous stirring. To this mixture was added 2 mL of a 70% nitric acid (under stirring). The resulting solution was stirred for 2 h at room temperature. The mixture was then heated to 80 °C under stirring for 4 h, giving rise to a stable and transparent (slightly cloudy) TiO_2 suspension (pH = 2). Even after cooling no change in the suspension's transparency was observed. The suspension remained stable for at least 12 months. The anatase phase in the films prepared by this colloidal dispersion was detected by X-ray diffraction (XRD). According to previous work (Bernardo et al., 2009) the effective diameter of the colloidal TiO_2 particles was around 24 nm, as determined by dynamic light scattering (DLS); the XRD of thin films, obtained after casting the TiO_2 dispersions deposited on a glass substrate, was recorded on a Siemens D5005 diffractometer using monochromatic Cu Kα radiation; the diffractogram displayed two diffraction peaks at 2θ values of 25° and 48° which are characteristic of (101) and (200) of anatase TiO_2 respectively.

Preparation of mixture of hydrosoluble polymer PVOH and TiO_2 and spray drying of TiO_2/PVOH aqueous suspension

PVOH were dispersed in distilled water and mixture with the colloidal dispersion of anatase TiO_2 at the ratio of 1:1. The aqueous suspension of TiO_2/PVOH was atomized through the use of the spray drying process (Mini Spray Dryer B-190 Büchi) (W.I.P.O. Patent WO/2009/132406). During the process, the water was removed by evaporation generating particles of the TiO_2/PVOH nanocomposite with high nanoparticles content.

Preparation of the nanocomposites

The TiO_2/PVOH concentrated powder obtained by spray drying were incorporated in PET, EVA and LDPE. The PET nanocomposite was prepared in an internal mixer HAAKE Rheomix 600p via melt blending (at 230 °C, rotor speed 100 rpm and 10 min) with TiO_2/PVOH powder system to a final TiO_2 concentration of 3wt%. The EVA nanocomposite was prepared in the same internal mixer via melt blending (at 170 °C, rotor speed 100 rpm and 10 min) with TiO_2/PVOH powder system to a final TiO_2 concentration of 3wt%. The LDPE nanocomposites were prepared in an twin-screw extrusion machine via melt mixing (B&P; diameter 19mm; L/D 25, 100 rpm, temperature profile: T1 180 °C, T2 200 °C, T3 200 °C, T4 200°C, T5 200 °C) with LDPE resin and TiO_2/PVOH powder system to a final TiO_2 concentration of 1, 3 and 5wt%. Films of LDPE nanocomposites were prepared in a single-screw extrusion machine (AX Plastics, diameter 16mm, L/D 26, 70 rpm, temperature profile: T1 180 °C, T2 200 °C, T3 210 °C). Conventional LDPE nanocomposites with TiO_2 rutile and anatase nanoparticles, obtained from DuPont and Sigma-Aldrich respectively, were prepared at the same conditions.

Characterization of the nanocomposites

Scanning electron microscopy (SEM) surface images of TiO_2/PVOH powder system were obtained with a Philips XL-30 FEG microscope. X-ray analyses were performed with the EDX Oxford Beryl Detector.

TGA analyses were performed in a thermogravimetric analyzer from Shimatzu (TGA-50). Samples were heated from 25^{0}C to 700^{0}C at a heating rate of 10^{0}C/min under nitrogen atmosphere.

Specimens for the transmission electronic microscopy (TEM) were cut using a Reichert Ultracut S50 ultramicrotome and then analyzed in a Philips CM120 microscope at an acceleration voltage of 120kV.

Samples of the LDPE nanocomposites films (ribbons with dimensions of 20 by 100 mm) were exposed to the ultraviolet radiation UV artificial and natural (14 days of solar light, Sao Carlos city, Brazil). UV artificial was performed in an accelerated method for exposure to the UV light in laboratory, through a high intensity discharge mercury lamp (250 W). These samples were submitted to mechanical tests, according to a procedure based on the standard ASTM D882-00. The tests were performed in an equipment INSTRON model 5569, load cell 500 N, claws distance 60 mm, 5 samples for composition, with deformation speed 5 mm/min up to 0.2% of deformation and 50 mm/min for the remaining of the test.

Results and Discussion

Figure 1 shows the SEM micrographs for TiO_2/PVOH powder system after spray drying process. It can be observed in that the powder obtained is quite fine and it shows spherical form.

Figure 1 – SEM micrographs for TiO_2/PVOH powder system after evaporation (1000x - scale = 20 µm).

It is also observed in Figure 1 the formation of a morphology characteristic of the dry titanium dioxide observed when the colloidal solution is evaporated without PVOH.

The x-ray analyses performed with the EDX Oxford Beryl Detector during the scanning electron microscopy (SEM) for TiO_2/PVOH powder system indicate the presence of titanium and also of organic material in the powder analyzed.

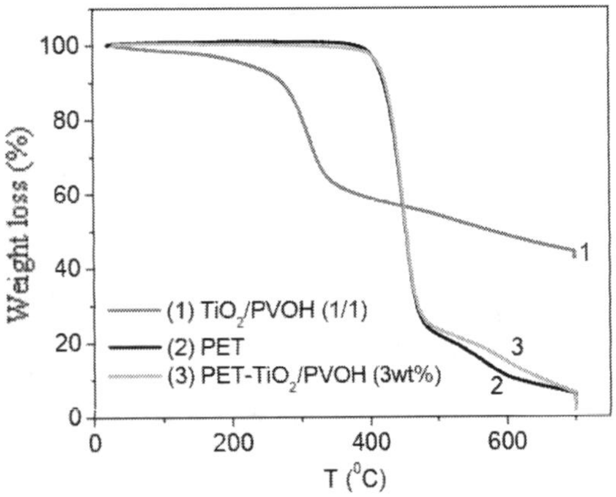

Figure 2 – TGA curves of TiO_2/PVOH powder system, PET resin and PET-TiO_2/PVOH nanocomposite (3wt% of TiO_2).

Figure 2 shows the TGA curves for the TiO_2/PVOH powder system, PET resin and PET-TiO_2/PVOH nanocomposite at TiO_2 concentration of 3wt%. As observed in Figure 2, the weight loss at 600°C for TiO_2/PVOH powder system was about 50%. The maximum of weight loss occurred at ~310°C to the TiO_2/PVOH system and ~450°C to the PET resin and PET-TiO_2/PVOH nanocomposite which correlates to degradation of the polymer chains. These results indicate that the powder of the TiO_2/PVOH system is composed approximately by 50% of TiO_2 nanoparticles and 50% of PVOH.

Figure 3 shows a TEM micrograph of PET-TiO_2/PVOH nanocomposite (3wt% of TiO_2). It can be observed in Figure 3 the formation of a cluster of TiO_2/PVOH at the central region of the PET matrix.

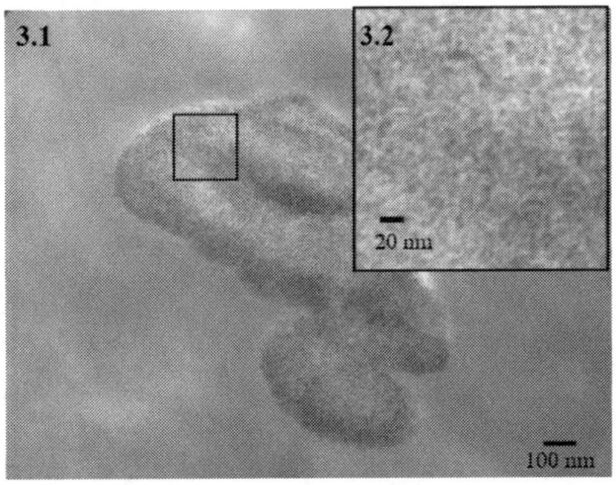

Figure 3 – TEM micrographs of PET-TiO_2/PVOH nanocomposite (3wt% of TiO_2): Figure 3.1 - matrix of PET and central region with TiO_2/PVOH (scale = 100 nm); Figure 3.2 - enlargement of central region of TiO_2/PVOH (scale = 20 nm).

However, it can also be observed in enlarged Figure (3.2) that this region show dense particles totally dispersed and encapsulated in a material with different density. According to previous work (Bernardo et al., 2008), these observations indicate that the shear applied during the preparation of the nanocomposite was not high enough to lead to a complete dispersion of the TiO_2 nanoparticles in the polymer matrix. However, the PVOH was a good stabilization agent to avoid TiO_2 agglomeration, like demonstrate the TEM results.

Figures 4 and 5 shows TEM micrographs of LDPE-TiO_2/PVOH nanocomposite (3wt% of TiO_2) and EVA-TiO_2/PVOH nanocomposite (3wt% of TiO_2) respectively. It can be observed in Figure 4 that similar result was obtained in the PET nanocomposite, that is, the formation of an agglomerate of TiO_2/PVOH at the central region of the LDPE matrix. It can be observed in Figure 5 the same result was obtained with the EVA nanocomposite. These observations indicate that the shear applied during the preparation of the LDPE and EVA nanocomposites was not also high enough to lead to a complete dispersion of the TiO_2 nanoparticles in the polymer matrixes.

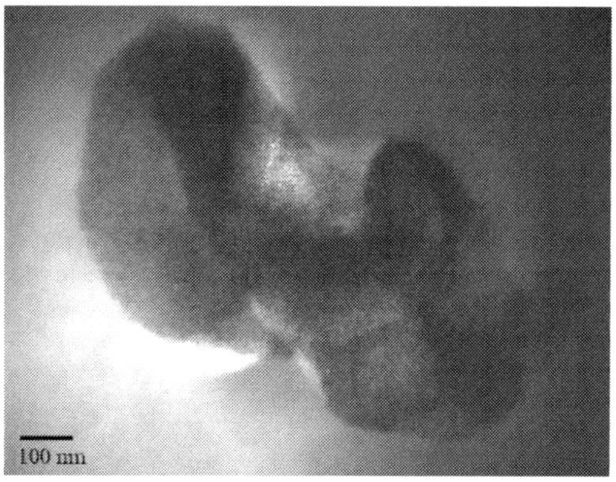

Figure 4 – TEM micrographs of LDPE-TiO_2/PVOH nanocomposite (3wt% of TiO_2) (scale = 100 nm).

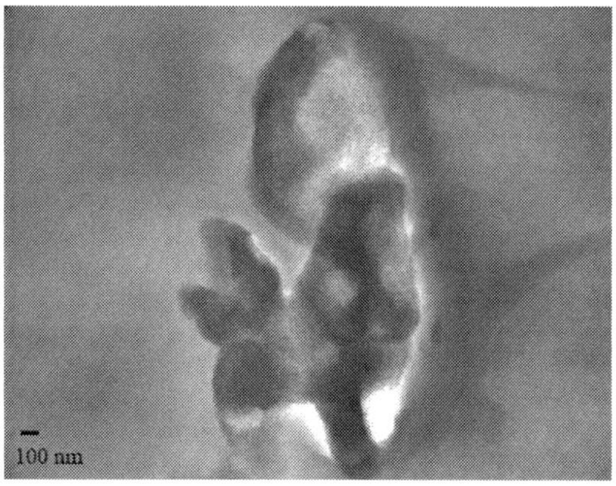

Figure 5 – TEM micrographs of EVA-TiO_2/PVOH nanocomposite (3wt% of TiO_2) (scale = 100 nm).

The LDPE films with TiO_2/PVOH concentrate powder system at TiO_2 final concentration of 1, 3 and 5wt%, showed in the mechanical tests, rupture stress and yield strength below the values obtained for the films of neat LDPE. The rupture stress and yield strength presents decrease tendency with the increase of the time of expose to the artificial ultraviolet radiation, however the most significant results were detected in the analysis of the elasticity modulus.

The Figure 6 presents the behavior of the modulus of elasticity as a function of photodegradation time. The LDPE films modulus at TiO_2 concentration of 1, 3 and 5wt% are relatively close, but lower than the values measured for the films of neat LDPE. It can be observed the trend of a decrease in modulus for the LDPE films with TiO_2/PVOH concentrated powder system with the increase of the time of expose to UV radiation. The average standard deviation of the values of modulus was of 20 MPa.

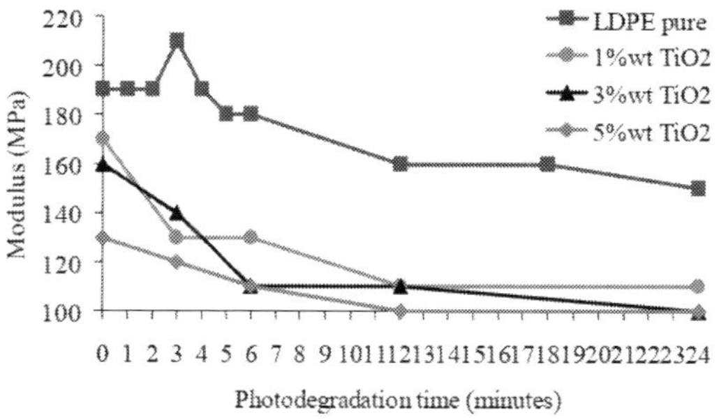

Figure 6 – Modulus of elasticity in function of photodegradation time for LDPE-TiO_2/PVOH nanocomposites (time in minutes of expose to the artificial ultraviolet radiation UV).

The Figure 7 shows the behavior of the modulus of elasticity as a function of photodegradation time to the LDPE films with TiO_2/PVOH concentrated powder system, TiO_2 rutile and TiO_2 anatase purchased from Sigma-Aldrich, at the TiO_2 concentration of 3wt%, in comparison with LDPE pure and LDPE/PVOH pure. It can be observed the different behavior of the films with TiO_2 rutile and TiO_2 anatase.

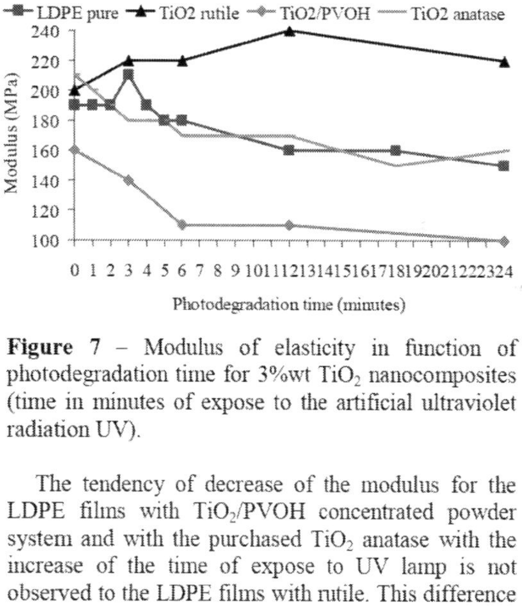

Figure 7 – Modulus of elasticity in function of photodegradation time for 3%wt TiO_2 nanocomposites (time in minutes of expose to the artificial ultraviolet radiation UV).

The tendency of decrease of the modulus for the LDPE films with TiO_2/PVOH concentrated powder system and with the purchased TiO_2 anatase with the increase of the time of expose to UV lamp is not observed to the LDPE films with rutile. This difference indicates that LDPE films with rutile are protected against photodegradation.

Besides, the tendency of decrease of the modulus for the LDPE films with TiO_2/PVOH concentrated powder system indicate that the nanocomposite show potential to accelerate the effects of the photodegradation in the LDPE matrix, which confirm that the technique developed is quite promising for applications where the photodegradation is a requested.

This tendency of decrease of modulus was observed in the samples of LDPE-TiO_2/PVOH nanocomposites exposed to the natural ultraviolet radiation UV (14 days of solar light). LDPE-TiO_2/PVOH films at TiO_2 final concentration of 1, 3 and 5wt% presented decrease of 30, 40 and 30 MPa in the modulus, equivalent to a reduction of 17%, 25% and 23% respectively. Those reductions were the largest among all tested films in natural ultraviolet radiation UV, indicate that the nanocomposite show potential to accelerate the effects of the photodegradation in the LDPE matrix.

Conclusions

A new technique based on the use of the spray drying process was proposed and tested to obtain high content TiO_2 polymer nanocomposites. Nanocomposites of TiO_2 nanoparticles dispersed in PVOH were prepared and dispersed in PET, EVA and LDPE matrixes. The preparation of TiO_2 nanoparticles through the sol-gel method was efficient and allowed the preparation of nanoparticles with particle size controlled and in the anatase phase, as required for applications where the photodegradation is needed. PVOH was used successfully to stabilize and isolate TiO_2 nanoparticles.

The results showed that the spray drying process is an innovative way to obtain a nanocomposite with high TiO_2 nanoparticles content, isolated and encapsulated. The TiO_2/PVOH powder system obtained was composed approximately by 50% of TiO_2 nanoparticles and 50% of PVOH. This high concentration of TiO_2 nanoparticles is very interesting when it is used in the preparation of polymer nanocomposites because high concentration of TiO_2 in the final nanocomposite can be obtained without changing significantly the matrix composition.

TEM micrographs indicated that the shear applied during the preparation of the nanocomposite was not high enough to lead to a completely dispersion of the TiO_2 nanoparticles in the polymer matrix. However, it was observed that the regions with formation of an agglomerate of TiO_2/PVOH show dense particles of TiO_2 totally dispersed and encapsulated in a material with different density (polymer). This is an indication that the PVOH was a good stabilization agent and was effective in avoid TiO_2 agglomeration.

The mechanical properties of the films after the photodegradation for UV radiation indicate that the films with TiO_2/PVOH concentrated powder show potential to accelerate the effects of the photodegradation in the LDPE matrix, which makes the technique developed in this work quite promising for applications where the photodegradation is a requested.

References

Bernardo, P.R.A.; Vidotti, S.E.; Cavalho, A.J.F.; Pessan, L.A. "Nanocomposite of poly(ethylene terephthalate with titanium nanoparticles encapsulated with PVOH." Polymer Processing Society PPS-24, Italy, papers... (2008).

Bernardo, P.R.A.; Cavalho, A.J.F.; Pessan, L.A. "Preparation of low-density polyethylene nanocomposites with anatase titanium dioxide from colloidal suspensions." Brazilian Polymer Congress 10°CBPol, Brazil, papers... (2009).

Chen, X. "Preparation and property of TiO_2 nanoparticle dispersed polyvinyl alcohol composite materials." Journal of Materials Science Letters, 21, 1637-1639 (2002).

Costa, A.C.F.M.; Vilar, M.A.; Lira, H.L.; Kiminami, R.H.G.A.; Gama, L. "Synthesis and characterization of TiO_2 nanoparticles." Ceramic., 52, 255-259 (2006).

DuPont. "Ti-pure DuPont." Coatings brochure (2007).

Huguenin, F.; Zucolotto, V.; Carvalho, A.J.F.; Gonzalez, E.R.; Oliveira Jr, O.N. "Layer-by-layer hybrid films incorporating WO_3, TiO_2, and Chitosan." Chem Mater., 17, 6739-6745 (2005).

ThreeBond. "TiO_2 Photocatalyst." ThreeBond technical brochure (2004).

W.I.P.O. Patent WO/2009/1322406 (2009) Pessan, L.A.; Cavalho, A.J.F.; Bernardo, P.R.A.; Vidotti, S.E. "Method of producing nanocomposites, thus produced nanocomposites and compositions of said nanocomposites with polymer matrixes." PCT/BR2009/000116, PI0801349-7.

EXTENSIONAL AND SHEAR RHEOLOGY OF ORIENTED BLOCK COPOLYMERS / CLAY NANOCOMPOSITES

D. J. Carastan[1,2], L. G. Amurin[1], A. F. Craievich[3] and N. R. Demarquette[1*]

[1] *Metallurgical and Materials Department, Polytechnic School, University of São Paulo, Av. Prof. Mello Moraes, 2463, Cidade Universitária, São Paulo – SP – Brazil, ZIP: 05508-030, danilo.carastan@gmail.com; leicita.amurin@yahoo.com.br; *nick@usp.br*

[2] *present address: Center for Engineering, Modeling and Applied Social Sciences (CECS), Federal University of ABC (UFABC), Santo André – SP, Brazil*

[3] *Applied Physics Department, Physics Institute, University of São Paulo, São Paulo – SP, Brazil, craievich@if.usp.br*

Block copolymers and nanoparticles can be combined to form materials with interesting properties and structures. The phase-separated structures usually formed in the solid state and even in the molten state within a certain temperature range may have different shapes and dictate the rheological properties of the copolymers in the ordered state. When nanoparticles are added, the interactions between each phase and the particles will result in more complex structures. It is possible to align their microstructure using special processing procedures, and the resulting materials will have very anisotropic properties. Extensional rheometry is a suitable technique to study the properties of anisotropic samples. It is sensitive to the morphology of the materials, and the extensional flow aligns further the anisotropic phases and particles. It is also well established that shear flows are able to align block copolymer domains, but there are not many studies concerning the alignment of nanoparticles within the domains and its stability with time. In this study shear and extensional rheological properties were carried out in different pure styrenic block copolymers and their clay nanocomposites. The effects of their microstructure on rheological properties in randomly oriented and aligned samples were evaluated. The materials were then studied by small angle x-ray scattering to analyze the microstructural changes caused by processing and the rheological tests. It was possible to observe that the extensional flow readily realigns samples tested in a direction perpendicular to the original flow direction, and the clay particles affect the rheological behavior when they are also realigned by the extension.

Introduction

The study of the rheological properties of block copolymers is a complex subject due to their phase-separated nature at the nanoscale. Two of their most common morphologies – the cylindrical and lamellar structures – are highly anisotropic. Polymer nanocomposites are also multiphase materials with complex rheological properties. Some of the most common nanoparticles, such as clays, which are also very anisotropic, tend to aggregate within the polymer matrix forming networks, strongly affecting the rheological behavior of these materials (Lim and Park, 2001). The combination of block copolymers and nanoparticles results in materials with even more complicated structures, and their rheological properties depend on the interactions between each phase, the degree of dispersion of the nanoparticles and the relative orientation between the block copolymer domains and the nanoparticles (Silva et al., 2001; Lee and Ha, 2003).

For most practical applications it is desirable to have materials with highly controlled morphologies, and sometimes it is important to align the polymer phases and fillers in order to maximize some property at a desired direction (Vaia, 2007). Some procedures are capable of aligning some or all of these anisotropic structures, such as shear (Lee et al., 2003) or extensional (Daniel et al., 2000) melt processing, mechanical deformation (Park et al., 2006) or applying external electric fields, such as electric (Koerner et al., 2004) or magnetic (Koerner et al., 2005). In particular, there are few studies regarding the effects of melt extension on block copolymer nanocomposites.

In this work, nanocomposites of styrenic block copolymers and organoclays were prepared by different techniques, such as melt mixing and extrusion, and the effects of shear and extension orientation during rheological experiments were evaluated using small angle x-ray scattering (SAXS). Some samples were aligned using tape extrusion, and extensional rheological tests were carried out in two perpendicular directions, in order to evaluate the morphology evolution during the tests.

Experimental

The copolymers used in this study were different styrene-butylene/ethylene-styrene triblock copolymers (SEBS), one having 30 wt% of PS blocks with cylindrical structure (SEBS-30 – Kraton G1652), another one with 13 wt% of PS and 29 wt% of diblocks, forming cylindrical and spherical structures (SEBS-13/29 – Kraton G1657) and a 30 wt% PS-containing SEBS modified with maleic anhydride (SEBS-MA – Kraton FG1901). All copolymers were provided by Kraton Polymers. The organoclays used were Cloisite 15A and 20A, from Southern Clay. Their structures and composition can be found elsewhere (Carastan and Demarquette, 2007).

The SEBS-30 and SEBS-MA copolymers and their nanocomposites were melt processed in a Haake Rheomix PTW-16 double screw extruder attached to a ThermoHaake PolyLab 900 torque rheometer, using a 25 mm wide and 1 mm thick ribbon die. The processing conditions were 220 °C and screw speed of 100 rpm. The amount of clay in the composites was 5 wt%. Other nanocomposites were prepared by melt mixing in a chamber attached to the torque rheometer.

The shear rheological tests were carried out in an Anton Paar MCR 500 rheometer. The samples were 25 mm wide 1 mm thick disks prepared by melt pressing at 200 °C. Steady extensional tests were carried out using the SER (Sentmanat Extensional Rheometer) fixture attached to the MCR 500 rheometer. The samples were prepared either by pressing or cutting films less than 1 mm thick into rectangular samples of the materials studied. Figure 1 shows schematically how the samples used in the extensional experiments were obtained from the tape extruded materials. Rectangular samples (20x6 mm) were cut from the tape in the longitudinal (L) and transversal (T) directions with respect to the extrusion direction (x). The extensional test direction and its perpendicular at the tape plane were assigned as **u** and **v**, respectively, as indicated in Figure 1.

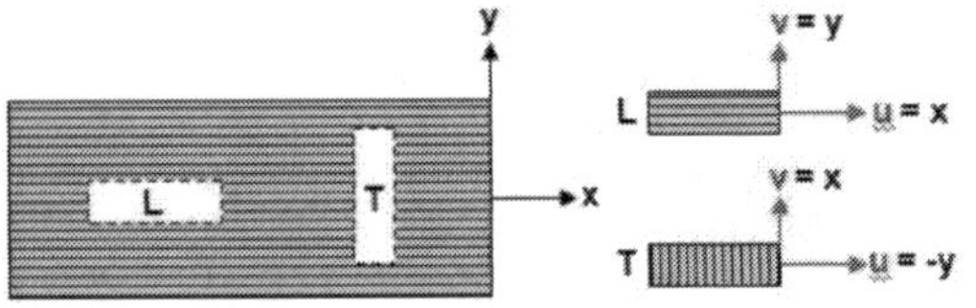

Figure 1 – Scheme of how the extensional test samples were obtained from the tape samples.

Bidimensional small angle x-ray scattering (SAXS) tests were carried out in Brucker's NanoSTAR, using 1.5 kV CuKα radiation and a sample-to-detector distance of 650 mm. Some SAXS experiments were also carried out in the National Laboratory of Synchrotron Light (LNLS) in Campinas, Brazil, using a wavelength of 1.488 Å and a sample-to-detector distance of 950 mm.

Results and Discussion

Shear Orientation

Block copolymers in the ordered state can be easily aligned by shear, especially the cylindrical and lamellar structures. The copolymers SEBS-30 and SEBS-MA, for instance, have a cylindrical morphology even at high temperatures, and their domains can be aligned if large amplitude oscillatory shear tests (LAOS) are carried out in isotropic samples. Figures 2 and 3 show the modulus of complex shear viscosity curves as a function of time for these two pure copolymers and their nanocomposites.

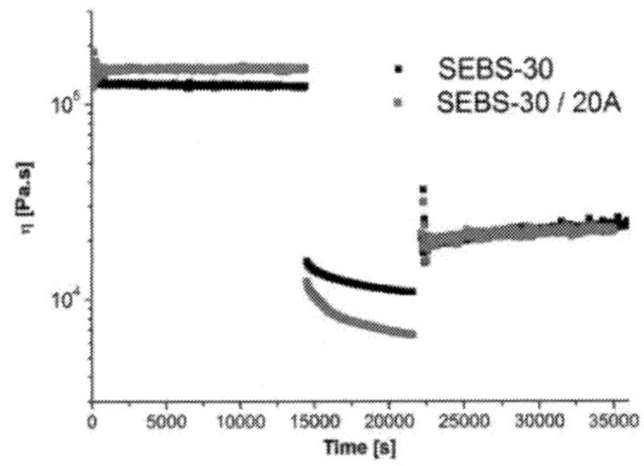

Figure 2 – Time sweep-LAOS-time sweep tests carried out at 200 °C for the SEBS-30 samples.

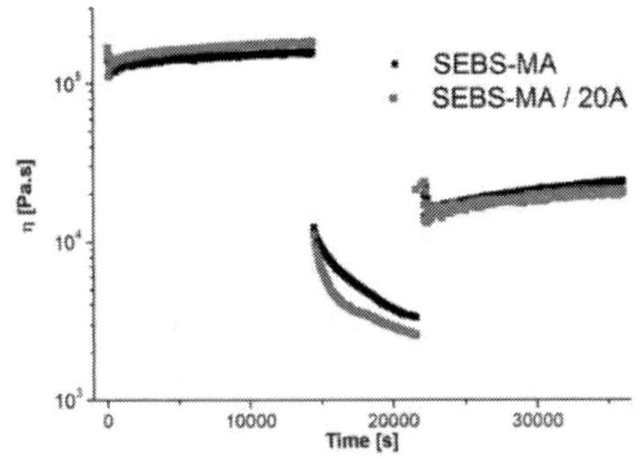

Figure 3 – Time sweep-LAOS-time sweep tests carried out at 200 °C for the SEBS-MA samples.

Initially, time sweep tests were carried out for 4 hours at 0.01 rad/s and strain amplitudes ranging from 0.2% to 0.5%, in order to assure that the measurements are within the linear viscoelastic regime for each sample. It can be seen that the behavior of SEBS-30 and its nanocomposites is more stable with time than the SEBS-MA samples, probably due to interactions caused by the presence of maleic anhydride in SEBS-MA. The middle curves correspond to the LAOS tests carried out at a strain amplitude of 100%. The viscosity drops during this test, because the samples are being aligned. After LAOS new time sweep tests within the linear viscoelastic regime were carried out in order to monitor the behavior of the newly formed structures. It can be seen that the viscosity is much lower than before the LAOS tests, indicating that the copolymer domains and clay particles were aligned in the flow direction, in order to reduce the resistance to flow caused by randomly oriented PS cylinders and clay particles. It can also be observed that the aligned structures remain stable for long times, although there is a tendency for the viscosity to slightly increase with time, an effect again more pronounced in the case for SEBS-MA samples.

Extensional Rheometry - Isotropic samples

Rheological tests in extension were first carried out in non-aligned block copolymer and nanocomposite samples in order to evaluate the average effect of the block domains and clay particles in the rheological properties of systems which are more isotropic.

Figure 4 shows the extensional viscosity of the SEBS-13/29 copolymer and its nanocomposites containing 15A clay measured at 120 °C and at a strain rate of 0.01 s^{-1}. This copolymer may present different morphologies in the ordered state, depending on the temperature and method of preparation. The cylindrical structure is the most stable structure at 120 °C, commonly obtained when the sample is prepared from the melt at the disordered state (T > 200 °C). The spherical structure was prepared by annealing the samples at 150 °C for more than 5 hours, time long enough for the order-order transition to occur at this temperature, according to other studies (Carastan et al., 2008). The lamellar structure was obtained by casting the copolymer from a toluene solution. The formation of all these structures was confirmed by SAXS.

The extensional viscosity curves of the pure copolymer having cylindrical and lamellar structures are very similar. As the PS domains in these structures are continuous in one (cylinders) or two (lamellae) directions, they become aligned in the flow direction during the test, decreasing the melt resistance to flow due to immiscibility between the blocks. The spherical structure, on the other hand, has a stronger resistance to flow, because the PS spheres are arranged in a body-centered-cubic lattice, and as they are not interconnected, they increase the solidlike behavior of the melt (Kossuth et al., 1999). Therefore, the extensional viscosity is higher than for the other structures, and it increases up to a certain strain, and then drops, showing strain softening behavior. This occurs probably due to a disruption of the BCC spherical domains, forming the more stable cylindrical structure, induced by the strong extension, as confirmed by SAXS in Figure 5.

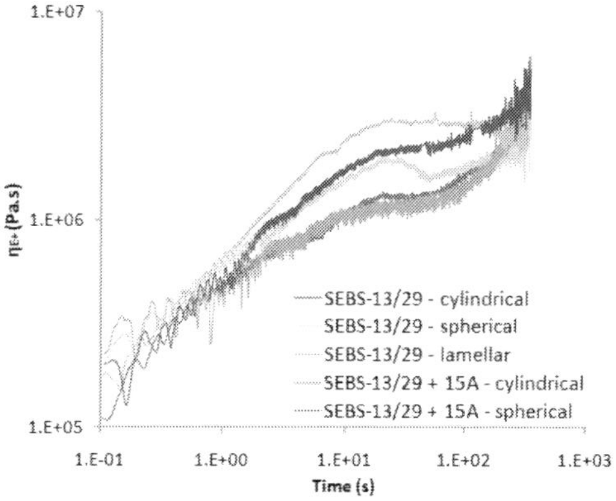

Figure 4 – Extensional viscosity curves of different SEBS-13/29 samples at 0.01 s^{-1} and 120 °C.

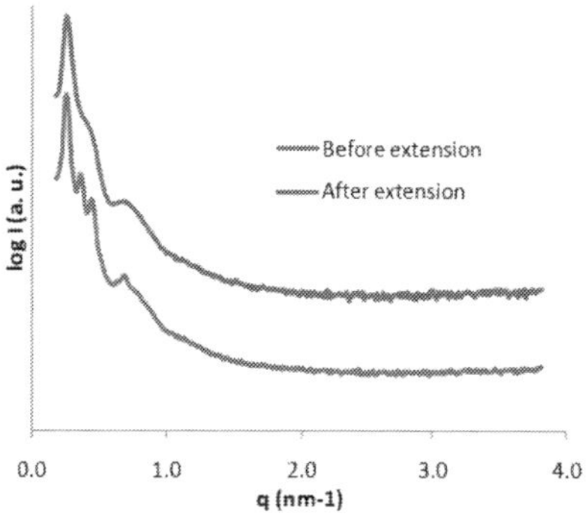

Figure 5 – SAXS curves for pure SEBS-13/29 having spherical structure before extension. These results were obtained in LNLS.

The addition of clay changes the extensional rheological behavior, as seen in Figure 4. The viscosity significantly increases for the cylindrical structure, as the clay particles increase the solidlike behavior of the material. The rheological behavior for the spherical structure, on the other hand, is not strongly affected by the presence of clay, as the spherical domains already cause a strong solidlike behavior, an effect also observed in shear studies (Carastan et al., 2008). The main difference between the spherical structure curves is that the addition of clay hinders the strain softening behavior, probably due to the alignment of clay particles during the extension, masking the sensitivity for changes in the copolymer structure.

Extensional Rheometry -Oriented samples

The SEBS-30 and SEBS-MA copolymers, which are far more viscous than SEBS-13/29, are more easily aligned during melt processing, and this alignment is very stable with time, as seen in the shear section. Therefore, it is a good material to be used in anisotropy studies. Figures 6 and 7 present the extensional viscosity curves as a function of time for the tape samples of SEBS-30 and its nanocomposite evaluated in the L and T directions, respectively. The steady shear curve obtained for the isotropic pure copolymer is also included.

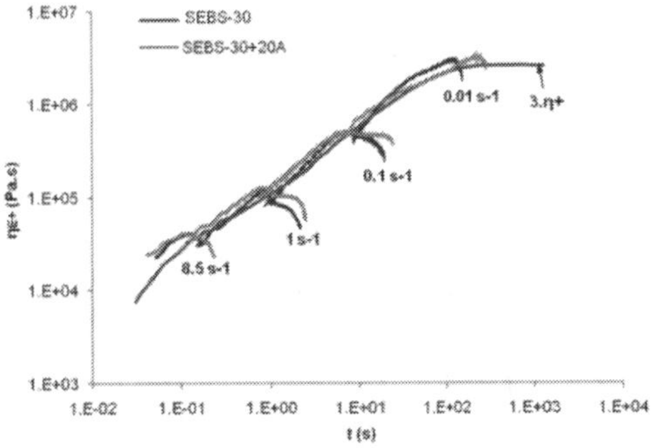

Figure 6 – Extensional viscosity curves of SEBS-30 and its nanocomposite in the L direction at 200 °C.

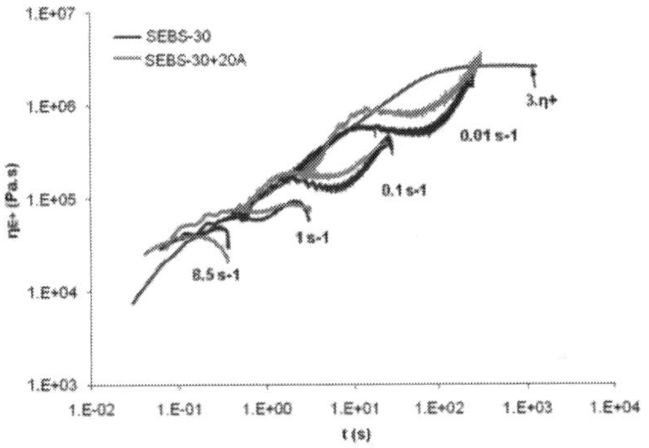

Figure 7 – Extensional viscosity curves of SEBS-30 and its nanocomposites in the T direction at 200 °C.

It can be seen that the extensional curves in the L direction are similar to the triple of the shear curve, showing that the Trouton rule holds for this case. No sensible effect caused by the addition of clay can be observed. The curves for the T direction show a completely different rheological behavior. There is a strong deviation from the Trouton rule, and the materials show strain softening behavior, usually observed in block copolymers (Takahashi et al, 1998; Kotaka et al., 2001). It can be seen that the 20A clay increases the viscosity of the strain softening part of the curves, an effect more clearly observed at low strain rates. The SEBS-MA copolymer and its nanocomposite were also tested in the same conditions. The rheological behavior of the aligned samples is different from the isotropic samples studied in the previous section. The results can be interpreted with the aid of bidimensional SAXS images of the samples after extension, as they keep the history of deformation in their structures along the strain direction. Figures 8 and 9 show the examples for the SEBS-30 + 20A samples elongated at a strain rate of 0.001 s^{-1} in the L and T directions, respectively.

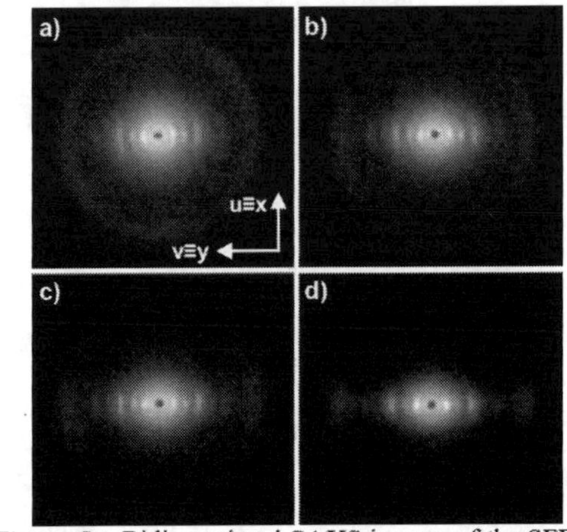

Figure 8 – Bidimensional SAXS images of the SEBS-30 + 20A sample tested in extension at 0.01 s^{-1} and 200 °C in the L direction at four different positions, corresponding to strain values of: a) 0.28, b) 0.82, c) 1.24 and d) 2.43.

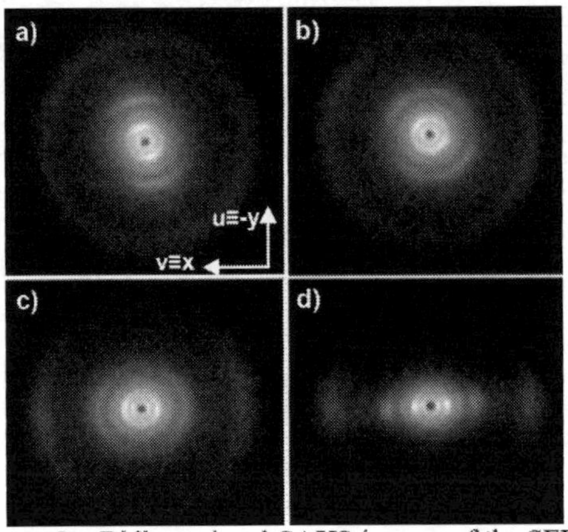

Figure 9 – Bidimensional SAXS images of the SEBS-30 + 20A sample tested in extension at 0.01 s^{-1} and 200 °C in the T direction at four different positions, corresponding to strain values of: a) 0.12, b) 0.25, c) 0.82 and d) 2.35.

Figure 8a shows the SAXS pattern of the tape in the beginning of extension in the L direction. It can be seen that the cylindrical domains are aligned in the extension direction, and, therefore, they maintain the same direction during the test. There is a halo in Figure 8a corresponding to the diffraction peak for the 20A clay. This halo is weak because most of the clay is aligned parallel to the x-y plane, but it shows that there is some clay randomly dispersed perpendicularly to the plane of the tape. As the sample is elongated, it can be seen that the halo becomes two peaks that show that the clay particles become aligned in the same direction of the cylinders. Interestingly, this rotation of clay particles is not strong enough to affect the extensional viscosity of the copolymer, as seen in Figure 6.

Figure 9 shows the SAXS patterns for the same sample elongated in the T direction. In this case, the cylinders are initially oriented perpendicular to the extension direction, and as the deformation proceeds, a rotation of the domains can be observed, so that they become aligned to the extension direction. This is the reason for the strain softening behavior observed in the viscosity curves of Figure 7. The clay halo is also present, and the clay particles are also aligned during extension. In this case, the rotation of clay particles affects the extensional viscosity, probably because they act as barriers for the rotation of the cylindrical domains.

Conclusions

In this study it was possible to evaluate the alignment of block copolymers and their clay-containing nanocomposites in the ordered state under shear and extension experiments. The anisotropic cylindrical domains and clay particles are easily aligned with flow, and their rheological properties are strongly affected by this effect.

Acknowledgements

To Fapesp and CNPq for financial support, to Kraton for materials supply, to the Crystallography Laboratory of the Physics Institute at the University of São Paulo and to LNLS for the SAXS experiments.

References

Carastan, D. J.; Demarquette N. R.; Vermogen A.; Masenelli-Varlot, K., "Linear viscoelasticity of styrenic block copolymers–clay nanocomposites" Rheol. Acta, **47**, 521-536 (2008)

Carastan, D.J.; Demarquette, N.R., "Polystyrene/clay nanocomposites". Int. Mater. Rev. **52**, 345-380 (2007)

Daniel, C.; Hamley, I.W.; Mortensen, K., "Effect of planar extension on the structure and mechanical properties of polystyrene–poly(ethylene-*co*-butylene)–polystyrene triblock copolymers", Polymer, **41**, 9239-9247 (2000)

Koerner, H.; Jacobs, J. D.; Tomlin, D. W.; Busbee, J.; Vaia, R., "Tuning polymer nanocomposite morphology: AC electric field manipulation of epoxy-montmorillonite (clay) suspensions" Adv. Mater.,**16**, 297-302 (2004)

Koerner, H.; Hampton, E.; Dean, D.; Turgut, Z.; Drummy, L.; Mirau, P.; Vaia, R., "Generating triaxial reinforced epoxy/montmorillonite nanocomposites with uniaxial magnetic fields" Chem. Mater., **17**, 1990-1996 (2005)

Kossuth, M. B.; Morse, D. C.; Bates, F. S., "Viscoelastic behavior of cubic phases in block copolymer melts", J. Rheol., **43**, 167-196 (1999)

Kotaka, T.; Okamoto, M.; Kojima, A.; Kwon, Y. K.; Nojima, S.; "Elongational flow-induced morphology change of block copolymers. 2. A polystyrene-*block*-poly(ethylene butylene)-*block*-polystyrene triblock copolymer with cylindrical microdomains", Polymer **42**, 3223-3231 (2001)

Lee, J. Y.; Park, M. S.; Yang, H. C.; Cho, K.; Kim, J. K., "Alignment and orientational proliferation of HEX cylinders in a polystyrene-block-polyisoprene-block-polystyrene copolymer in the presence of clay", Polymer, **44**, 1705-1710 (2003)

Lee, K. M.; Ha, C. D., "Linear dynamic viscoelasticity properties of funcionalized block copolymer/ organoclays nanocomposites", Macromolecules, **36**, 804-815 (2003).

Lim, Y. T.; Park, O. O.,"Phase morphology and rheological behavior of polymer/layered silicate nanocomposite", *Rheol. Acta,* 40:220-229, 2001.

Park, S.-Y.; Cho, Y.-H.; Vaia, R. A., "Three-dimensional structure of the zone-drawn film of the nylon-6/layered silicate nanocomposites" *Macromolecules,* **38**, 1729-1735 (2005)

Silva, A. S.; Mitchell, C. A.; Tse, M. F.; Wang, H.-C.; Krishnamoorti, R., "Templating of cylindrical and spherical block copolymer microdomains by layered silicates", J. Chem. Phys., **115**, 7166-7174 (2001)

Takahashi, T.; Takimoto, J.-I.; Koyama, K., "Elongational viscosities of random and block copolymer melts", J. App. Polym. Sci., **69**, 1765-1774 (1998)

Vaia, R. A.; Maguire, J. F., "Polymer nanocomposites with prescribed morphology: Going beyond nanoparticle-filled polymers", Chem. Mater., **19**, 2736-2751 (2007)

REACTIVE INJECTION MOULDING OF POLYAMIDE 6 – AN INNOVATIVE APPROACH FOR THE PRODUCTION OF HIGH PERFORMANCE COMPOSITE PARTS

L. F. Berg[1]*, P. Elsner[12], F. Henning[13] , B. Thoma[3]

[1] *Fraunhofer Institute for Chemical Technology, Pfinztal, Germany – lars-fredrik.berg@ict.fraunhofer.de; peter.elsner@ict.fraunhofer.de;*

[2] *Karlsruhe Institute of Technology – Institute of Material Science and Engineering I*

[3] *Karlsruhe Institute of Technology - Institute of Vehicle System Technology, Karlsruhe, Germany – frank.henning@kit.edu; bernd.thoma@ict-extern.fraunhofer.de*

In order to respond to the marked needs regarding energy efficient and large volume capable processing technologies for the production of high performance composites, Fraunhofer ICT has developed an innovative reactive injection moulding process.

The process has been developed to enable the production of continuous fibre reinforced high performance composite parts with the engineering thermoplastic matrix Polyamide 6. The developed processing technology is based on the well established and worldwide available injection moulding technology. During the process development, various relevant machine components such as the configuration of the screw and the non return valve have been redesigned to enable the processing of low viscous thermoplastic polymer intermediates such as Caprolactam. Furthermore a special mixing nozzle and suitable mould technologies have been developed to enable an anionic in-situ polymerisation of Caprolactam to high molecular Polyamide 6 in the presence of continuous fibre reinforcements made of glass and carbon fibres to high performance composites.

The developed processing technology enables the production of continuous fibre reinforced thermoplastic composite structures with a high fibre volume content in a cycle time of less than 5 min. The process development is aimed to enable an upgrade of suitable injection moulding machines to enable an efficient production of high performance thermoplastic composite parts with continuous fibre reinforcements.

This paper presents a brief overview of the processing concept and shows the feasibility and potential of the process by presenting results from the characterization of both non-reinforced and continuous fibre reinforced composite parts with a Polyamide 6 matrix.

Introduction

The answer to many of tomorrow's challenges in the development of eco efficient and sustainable transportation solutions can be found by utilising innovative high performance light-weight materials. In particular within road transportation it has been demonstrated how the use of high performance light weight materials, such as high performance continuous fibre reinforced polymer composites, can help enhancing vehicle performance and passenger safety while reducing fuel consumption and emissions [Stewart, 2009]. However the broad use of these materials in the commodity market requires the availability of suitable material formulations and highly effective production technologies in order to cut the part cost to an acceptable level. For this reason the use of high performance continuous fibre reinforced composites has so far been limited to niche products like super sport cars or smaller vehicle series where the high material costs, and the limitations regarding a large volume production, only plays a secondary role [Marsh, 2006]. In order to bring high performance composites into the commodity market, innovative low cycle time capable material formulations must be combined with new large volume capable, highly automated production processes.

Within the product development for the road transportation market, the requirements for environmentally friendly material solutions and recyclability after end of life is affecting the choice of the material formulations in composites and has recently led to a strong focus on thermoplastic resin systems. On the other hand the thermoplastic resin must be of such a kind that it can interact with continuous fibre reinforcements to form a highly reinforced composite. This issue leads to complications which cannot be fulfilled by utilising established thermoplastic melt processing technologies. Therefore the development of a new process technology is required before the step of going into a large series production of high performance continuous fibre reinforced thermoplastic composite structures can be taken.

By exploiting the benefits of the in-situ polymerisation of thermoplastic intermediates such as Caprolactam and in the combination with continuous fibre reinforcement products, the challenge to produce environmentally friendly high performance thermoplastic composite can be met. Together with the development of an automated and efficient processing technology which can be designed for a short cycle time and large volume production, the future market need for composites can be fulfilled.

Processing concept for reactive injection moulding

The concept for an automated reactive injection moulding process for Caprolactam is based on the standard injection moulding process technology utilising the highly automated features of the injection moulding machine to enable an efficient process. Compared to the standard injection moulding process where thermoplastic polymer granules are dosed and molten in the injection unit to a viscous melt which is then injected into the mould cavity where it is cooled to solidification, the processing of the polymer intermediate Caprolactam requires a completely different approach. Due to the difficult material feeding, the very low melt viscosity, the necessity to mix in reactive components such as catalyst and activator and the in-situ polymerisation in the mould at elevated temperatures, the reactive injection moulding process of Caprolactam requires several fundamental developments and changes to the injection moulding equipment in order to establish a stable process. Even though the injection moulding equipment must be adapted to fulfil the requirements for reactive processing of Caprolactam, the goal of the development was to leave the basic machine setup and functions unchanged. Utilising many of the available and well automated functions, the motivation for not changing key machine elements is based on the thought that it should be possible to upgrade existing injection moulding machines which are available to also be able to produce continuous fibre reinforced thermoplastic composites without investing in completely new equipment. Such an upgrade would enable moulders to upgrade their production portfolio to also include continuous reinforced high performance thermoplastic composites in the future.

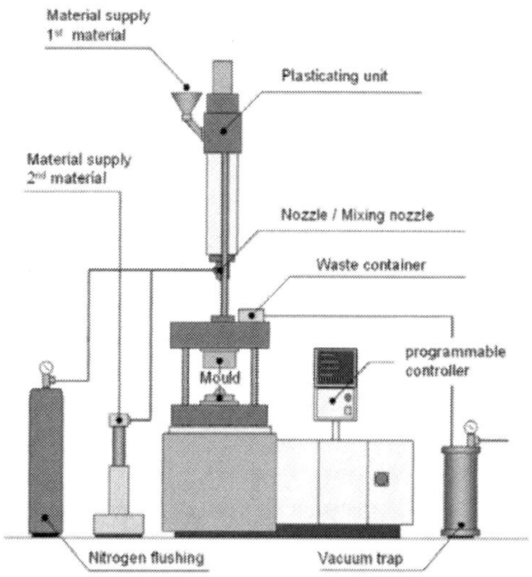

Figure 1 – Vertical injection moulding machine for reactive injection moulding of Caprolactam.

Figure 1 shows the set-up of a vertical injection moulding machine fitted for the reactive processing of Caprolactam. Here the plasticising unit is fed with a premix of Caprolactam pellets and catalyst which are molten and mixed to a low viscosity melt within the unit using the screw. An external high precision pump unit is used to supply the activator to the mixing nozzle which is a key element in the processing equipment. The mixing of the reactive components takes place directly before the injection of the melt into the mould and a cleaning system helps to ensure that Polymerisation cannot take place in the injection unit itself.

Due to the very low viscosity of the Caprolactam, one of the main challenges is to adapt the processing hardware in such a way that it enables an exact volume controlled flow of the materials during injection. This is necessary for the exact dosing of the low viscosity components and to enable in-situ polymerisation of the Caprolactam in the mould.

Figure 2 shows the dynamic viscosity plot for each of the three components used for the in-situ polymerisation within the temperature range of 70°C to 170°C. The activator and catalyst product, in this case both supplied by Brüggemann Chemical has a slightly higher viscosity than the Caprolactam itself. During processing the components are injected into the mould at a temperature of 120°C at which the reactive blend has a viscosity of 5 mPas.

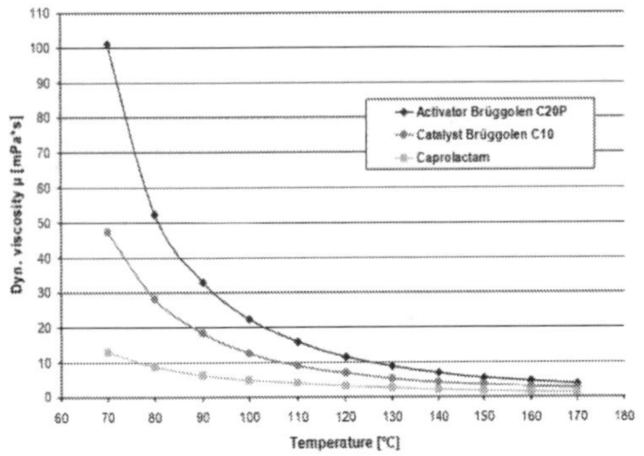

Figure 2 – Dynamic viscosity plot for the reactive components processed during the reactive injection moulding process.

To enable a controlled injection the design of the screw and especially the non-return valve is of special importance. While normal non-return valve systems are not able to seal properly at such low melt viscosities, a special non-return valve featuring an alternative sealing concept was developed and utilized.

The main benefit of the low melt viscosity compared to the classically melt processed Polyamide 6 (> 500 Pas) is the possibility to impregnate dense continuous fibre reinforced structures or very high contents of fillers. During the process development the focus was put on the impregnation of continuous fibre reinforcements to produce high performance composites with high fibre volume fractions.

Unlike the classic injection moulding process where the mould is tempered at temperatures around 80°C to cool the Polyamide 6 resin, the reactive injection moulding process requires significantly higher mould temperatures in the range of 150-165°C to enable the anionic polymerisation. Together with the material recipe (concentration of activator and catalyst) and melt temperature at injection, the mould temperature also significantly influences the kinetics of the anionic Polymerisation and can be used to speed up or slow down the polymerisation process. Due to the fact that the anionic in-situ polymerisation of Caprolactam takes place directly in the mould and is carried out at temperatures below the melting point of the final Polyamide 6 product. (216 - 220°C) the demoulding can take place without cooling the mould. This enables a quick and energy efficient process which could also be able to produce larger volumes of composite parts.

Experimental

During the production of specimen various recipe were tested. The additive concentrations were varied between 2.2wt% - 3.7wt% catalyst and 1.6wt% - 2.9wt% activator while the mould temperature was held constant at 155°C.

During the process development phase a mould closing time varying from 3-10 minutes was realized depending on the material recipe and reinforcement fibre type.

The continuous fibre reinforcement performs used to produce composite parts were manufactured from direct glass fibre rovings (2400 tex) with a selected sizing suitable for the anionic Polymerisation. For the manufacturing of performs the radial braiding technology was used. Different perform variations with fibre volume fractions in the range of 15 to over 40 vol% all with a braiding angle of 60° were produced. The reinforcement structures were pre-dried before injection to prevent issues where moisture could disturb the polymerisation of the Caprolactam.

In order to evaluate the quality and the reproducibility of the developed process both non-reinforced and fibre reinforced parts were produced for characterisation. In order to produce comparable samples for the characterisation the processing parameters were defined as shown in table 1.

Table 1 – Processing parameters used during the reactive injection moulding to produce samples for process characterisation.

Recipe	Parameter	Value
Low concentration	Catalyst cons. [wt%.]	2.2
	Activator cons. [wt%.]	1.6
High concentration	Catalyst cons. [wt%.]	3.7
	Activator cons. [wt%.]	2.9
	Melt temp. at injection	117°C
	Mould temperature	155°C
	Mould closing time	8 min.

Results and Discussion

The characterisation of the non-reinforced Polyamide 6 specimen was carried out using DSC and DMA analysis. In order to have a relevant reference low concentration samples were also produced by classic RIM processing using the same processing parameters listed in table 1. All characterisations were conducted dry as moulded.

Figure 3 shows the melt temperature of the non-reinforced Polyamid 6 specimen produced with the reactive injection moulding process and with a high and a low concentration recipe. (see Table 1) As a reference low concentration RIM samples were produced as well.

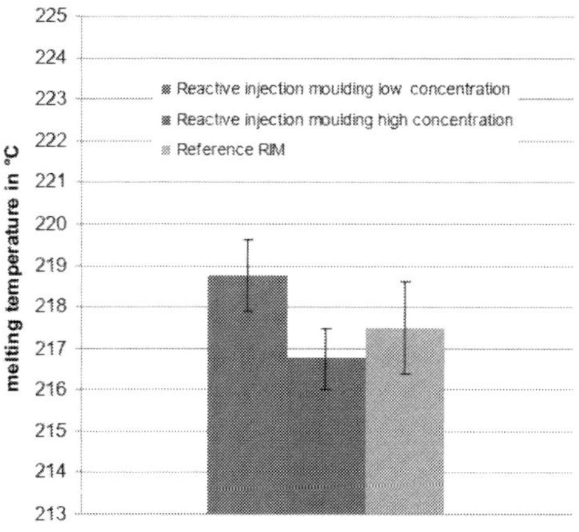

Figure 3 – Melting temperature for Polyamide 6 specimen produced with reactive injection moulding and RIM

It was observed that the low concentration recipe processed in the reactive injection moulding process produces a Polyamide 6 with a higher melting temperature (figure 3) and crystallinity (figure 4) than the high concentration recipe. Further it was observed that the specimen produced with the reactive injection moulding process has a higher melt temperature and crystallinity than the reference samples produced with the state of the art RIM processing technology under similar conditions.

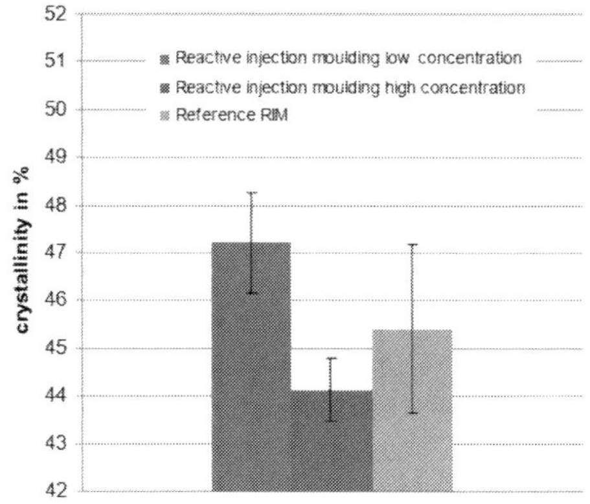

Figure 4 – Degree of crystallinity for Polyamide 6 specimen produced with reactive injection moulding and RIM

For the mechanical characterisation a DMA analysis was conducted. Figure 5 shows the DMA plot for a non-reinforced sample produced with a low concentration recipe in the reactive injection moulding process.

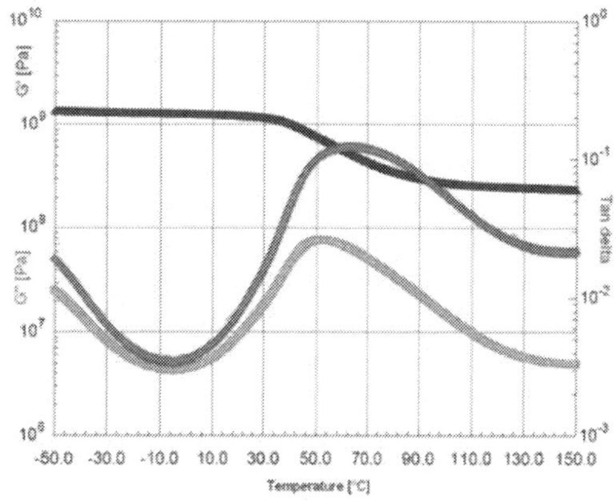

Figure 5 – DMA plot for a non-reinforced Polyamide 6 specimen produced with the reactive injection moulding process (low concentration recipe).

From figure 5 it can be observed that the glass transition temperature for the produced specimen is 62°C which is in the region for dry Polyamide 6 (60°C).

In order to see how the implementation of a continuous fibre reinforcement structure effects the mechanical performance, DMA characterisations were conducted on fibre reinforced samples as well. Figure 6 shows the DMA plot of a specimen containing 40 vol%. braided continuous glass fibres.

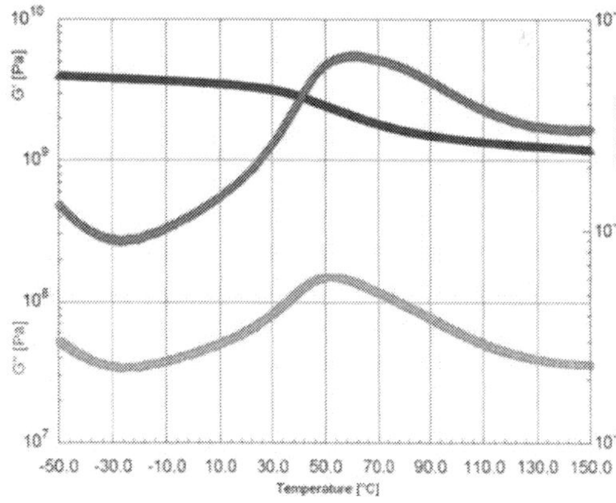

Figure 6 – DMA plot for a continuous fibre reinforced Polyamide 6 specimen (FVF 40 vol.%) produced with the reactive injection moulding process (low concentration recipe).

Based on the DMA results the modulus was calculated for 23°C using the formula. [Domininghaus, 2007]

$$E = G' * 2(1 + \nu)$$

The Poissons ratio ν was defined to 0,33 according to literature [Bottenbruch, 1998]

Table 2 lists the calculated modules for both a non-reinforced specimen as well as a reinforced specimen. As a reference the modules for a basic Polyamide 6 grade is listed as a reference.

Table 2 – Calculated modulus values for produced non-reinforced and reinforced specimen compared to a reference non-reinforced Polyamide 6. [LanXess, 2010]

Material	Modulus [GPa]
Non-reinforced PA6 sample	2.55
Reinforced PA6 sample (40 vol %)	8.75
Reference PA6 sample LanXess B30S	3.3

Conclusions

The developed reactive injection moulding process enables the production of continuous fibre reinforced composites with an engineering thermoplastic matrix such as in situ polymerized Polyamide 6 on an injection moulding machine. In order to be able to process and inject the low viscosity Caprolactam melt a sealing non-return valve system was developed and applied. Further a mixing nozzle system has been developed to mix Caprolactam, catalyst and activator prior to the injection into the mould. Continuous fibre reinforced composites could be produced by placing a dry fibre perform in the mould which is impregnated during the injection of the low viscosity melt.

The used continuous fibre performs were made from direct roving with a suitable sizing system for Caprolactam. Composites with fibre fraction up to 40 vol.% could be produced successfully.

The developed process was characterised by DSC and DMA, where it could be demonstrated that the developed process is able to produce non-reinforced and reinforced samples with a performance that is comparable with samples produced using the state of the art RIM processing technology. Further it was demonstrated that the developed reactive injection moulding can be used to produce continuous fibre reinforced composites with a thermoplastic matrix ion an injection moulding machine.

Future investigations will explore the full potential of the process with the goal to exploit the high degree of process automation the technology offers to shorten cycle time to < 5 min. and large production volume capability.

References

R. Stewart. "Light weighting the automotive marked" Reinforced plastics p.14-21 (2009)

G.Marsh. "Composites conquer with carbon supercars" Reinforced plastics p.20-24 (2006)

Domininghaus. H.: Die Kunststoffe und ihre Eigenschaften, 7th edition, Springer Berlin (2007)

Bottenbruch. L , Binsack. R,: Polyamide, 4th edition, Hanser München (1998)

LanXess Durethan B30S http://techcenter.lanxess.com (2010)

ENHANCING THE PROCESSABILITY OF HIGH DENSITY POLYETHELYNE USING DIFFERENT NANOCLAYS

A. A. Adesina[1], A. Al-Juhani[1], I. A. Hussein*

[1] *Department of Chemical Engineering, King Fahd University of Petroleum and Minerals, Saudi Arabia - ayuadesi@kfupm.edu.sa*

* *Department of Chemical Engineering, King Fahd University of Petroleum and Minerals, Saudi Arabia - ihussein@kfupm.edu.sa*

Abstract

This research involves investigating the effect of different modifiers and aspect ratio of the organoclay on the processability of high density polyethylene (HDPE). Improvement in the processing was assessed on the basis of shifting the melt fracture behavior of HDPE resins to higher shear rates; hence increasing the production. Also, the impact of these organoclays on the shear thinning and extensional rheology of HDPE was assessed. Two different compounding strategies were examined. The dispersion and the orientation of these organoclays was investigated by High Resolution Scanning Electron Microscopy (HR-SEM) and X-Ray diffraction (XRD) techniques. The effect of mixing on the HDPE instability and its rheology was critically analyzed. Thereafter, the comprehensive shear and extensional rheological characterization will be performed based on very small amount of this organoclay (500 to 5000ppm) in HDPE matrix to relate rheology to processing.

Introduction

It is the wish of any processing engineer to continuously increase the output of the extruded polyolefins by increasing its production rate. At high shear rate, the output rate is limited by the onset of surface and/or gross instabilities which made the glossy polymer surface rough. Many studies had been carried out in the past to study these instabilities, causes, effects and ways to postpone or eliminate them (Achilleos E. et al. 2002, Carreras E. S., et al. 2006, Hatzkiriakos and Migler 2005). Among methods used to combat these instabilities depending on the type of the instabilities are modification of the extruder especially the die head (Howells and Benbow 1962, Fujiyama and Inata 2002a, 2002b), conditioning the die surface (Anastasiadis and Hatzikiriakos 1998, Kolnaar and Keller 1997, Pudjijanto and Denn 1994, Perez-Gonzalez and Denn 2001), modification of the polymer (Kazatchkov et al. 1999, Filipe et al. 2008) and addition of processing additives (Hatzikiriakos et al. 1997). Addition of fluoropolymers has proved successful in the elimination or postponement of surface melt fracture in the polyolefins (Rudin A et al. 1986, Priester D. E. et al. 1993). Boron nitride, a platey structure material, added at a very small amount (less than 5000ppm) has been used as well (Seth et al. 2000, Vogel R. et al. 2003). Recently, Hatzikiriakos et al. (2005) showed that organically modified clay is likely a very good candidate for the elimination or postponement of these polyolefin fracture especially gross melt fracture. A lot of work has been conducted to prove beyond doubt the applicability of some of these processing additives but little works have been done with respect to application of organoclay. Hatzikiriakos et al.(2005) in their work mentioned that dispersion of organoclay in the polyolefin matrix is one of the key factors in effective elimination or postponement of the instabilities. In this work, the effect of dispersion on the polyolefin processing using two different methods of mixing will be examined. It has been observed for more effective dispersion, a compatibilizer- maleated polyolefin can be used (Chong et al. 2002, Chrissopoulou et al. 2005, Durmus A. et al. 2007, Hotta and Paul 2004). The effect of this will be explored using both HR-SEM and XRD. Also, the effect of various compositions (less than 5000ppm) of organoclay in the polymer matrix on processing and rheology will be discussed so that relationship between processing and rheology can be better understood. There is an issue of organoclay degradation at high temperature. Even though a lot of work had been done in this respect, many issues are yet to be resolved. A quick thermal analysis will be carried out to know the extent of degradation in our own work and way to move round it since this is not the major aim of this work.

Experimental

Materials

Commercial grade high density polyethylene (HDPE) (density: 0.952g/cm3, melt flow index: 0.05g/10mins) was supplied by Saudi Basic Industries Corporation (SABIC). Polyethylene grafted maleic anhydride (PE-g-MA) was used as compatibilizer, purchased from

Aldrich and was used as received. The maleic anhydride is approximately 3wt% as the co-monomer while the viscosity of the compatibilizer was 4,500cP at 140^{0}C. Cloisite$^{(R)}$ 15A- dimethyl, dehydrogenated tallow, quaternary ammonium was from Southern Clay. It has a d_{001} spacing of 31.5Angstrom.

Melt Blending and Morphology Characterization

The Brabender 50 EHT mixer supplied with the single screw extruder was used in the preparation of the nanocomposite. The organoclays were first heated in a vacuum oven at 108^{0}C for more than 24 hours to remove physico adsorbed water. The grounded polymer sample was pre mixed with organoclay before introduced into the mixer using manual feeder. Then, a masterbatch of 2wt% clay with ground resin was prepared with and without PE-g-MA as compatibilizer. A desired final concentration of a particular blend was obtained by mixing virgin HDPE to the masterbatch using the same mixer. The blending was done for 10minutes at 50rpm and 200^{0}C. Addition of 5000ppm of antioxidant (AO) was used to avoid degradation of the organoclay during the melt blending. The AO is a 50/50 blend of Irganox 1010 and Irgafos 168 from Ciba- Geigy Speciality.

The structure of the PE/organoclay composites was characterized by HR-SEM and XRD. The XRD was performed using a XRD-6000 Shimadzu diffractometer, operating at 40kV and 30mA, with 2θ scan range of 1-10^0 at room temperature, at a scanning speed of 10/min, using CuKα radiation (λ=0.154nm). HR-SEM was carried out on a Philips. Samples were cryogenically fractured by dipping the sample in liquid nitrogen.

Thermal Analysis

Thermal stability analysis was performed with a TA Instruments Q1000 MDSC. Sample (2–5mg) was compressed into hermetic aluminum pans with a hole on the lid for testing. The cloisite 15A was heated from 20 to 300^{0}C at 10^{0}C/min in a nitrogen environment flowing at 50ml/min.

Set-up for Melt Flow Instabilities

Extrusion was carried out in a measuring single screw extruder 19/25D from Brabender equipped with a specially developed slit die with 0.3mm height, 3mm width and 30mm length. The slit die has piezoelectric transducers located at three different positions. The time dependent pressure oscillations were collected by the use of these fast acquisition piezoelectric transducers combined with oversampling techniques. The extrusion was done at temperature 180^{0}C applying a range of speed up to the limit of the extruder. The measurement were done to specifically identify and analyse the time dependent pressure fluctuations associated with smooth surface and melt flow instabilities. The pressure was continuously monitored at the end of the extruder, and at three different positions along the die length. The final extrudates were quenched immediately after leaving the die, in order to allow the establishment of correlations between the instabilities developed at the extrudate surface and the pressure fluctuation occurring at the die.

The collected pressure was further analyzed using Fourier transformation to obtain a spectrum with the peak of maximum intensity at frequency of 0Hz. This intensity corresponds directly to the mean value of the pressure. The pressure oscillation during melt flow instabilities can be observed as additional peaks, located at higher frequencies. Details of this method can be seen in the work of Filipe S. et al. (2008).

Rheological Measurement

The shear experiments were conducted under both steady and extensional rheology. For extensional flow, experiment on start up of steady extension was performed.

The samples of the shear and extensional experiments were prepared at a temperature of 200^{0}C and a pressure up to 7tons on a Carver press. The samples for extensional flow were prepared with dimensions 20x10x1 mm^3, whereas for shear, discs with 25mm diameter and 2mm thickness were prepared.

ARES rheometer was used for all the rheological measurements. It is a constant strain rheometer equipped with heavy transducer (range 2-2000g for normal force; 2-2000g-cm for torque). For steady shear measurement, parallel plate (diameter 25mm) was used at temperature of 200^{0}C and shear rate from 0.01 to 10s^{-1}.

In case of extensional rheology, the extensional Viscosity Fixture-EVF was used. The sample was pre stretched with a strain rate 0.04s^{-1} to remove sagging. The sample was left for 5mins to relax any accumulated stress before the start of the experiment. Hencky strain rates of 0.05 to 20s^{-1} were used for these experiments.

Results and Discussion

The thermal analysis of Cloisite 15A using DSC is as shown in Figure 1. The thermal degradation of Cloisite 15A occurred at temperature of about 204°C (209°C with hole on the lid). Gases were evolved during the heating of the organoclay as shown with arc-shaped baseline. The gases, as seen from the onset of degradation, slightly affected the degradation of the organoclay. Some of the gases evolved are alkenes (Edwards G. et al. 2005) and aliphatic compounds (Jose M. et al. 2007). The effects of the evolution of gases on polyolefin processing are yet to be fully researched but suggestions from the literature showed that they may actually aid the processing of polyolefin. This will be explored in the future work. Frankowski D. J. et al. (2007) suggested that degradation of organoclay in polymer processing is delayed to higher temperature. This, according to them, is still open for discussion. To avert this problem, anti oxidant is added to the polymer samples during the nanocomposite preparation.

Considering the Fourier analysis performed on the raw pressure data from the slit die, it shows that for a smooth surface no peak is detected at any frequency. For other instabilities, the pressure variations lead to peaks at different frequencies depending on the extent and types of instabilities (Figures 2 and 3). The effect of organoclay on these instabilities will be measured and be presented later. It is expected that these peaks will disappear or reduce depending on the concentration of the organoclay in the HDPE matrix and the type of the instabilities.

It is recognized that extensional property of a sample plays a major role in influencing the development of melt flow instabilities (Migler K. B., et al., 2002; Hatzikiriakos S. G. et al. 2005). The characterization of the pure HDPE and its organoclay-based nanocomposite in uniaxial extensional flow is shown in Figures 4 and 5 at low and high Hencky strain rates. There is a reduction in elongational stress due to inclusion of organoclay at both Hencky strain rates as depicted in both Figures. At low Hencky strain rate, the compatibilizer has effect on the dispersion of the organoclay to the extent that there is reduction in elongational strain at break compared to the nanocomposite without compatibilizer. At high Hencky strain rate, the reduction in elongational stress in the presence of compatibilizer is approximately 26% compared to 18% when there is no compatibilizer though both occur at relatively the same Hencky strain at break. These results further strengthen the importance of dispersion of organoclay in the polymer matrix for the effective control of instabilities during polyolefin processing as reported by Hatzikiriakos et al. (2005). The concentration effect of cloisite 15A on the extensional rheology of the HDPE will be presented later.

The influence of Cloisite 15A on normal stress difference (N_1-N_2) is shown in Figure 6. There is a slip at a very low rate for all the samples and this occurs at approximately the same shear rate. N_1–N_2 for both nanocomposites follows the same trend at low shear rates; however with increasing shear the organoclays (Cloisite 15A) suppress the normal stresses of the pure resin. This observation supports the results of extensional rheology and explains the reported ease of processing in the presence of organoclays (Hatzikiriakos et al. 2005)

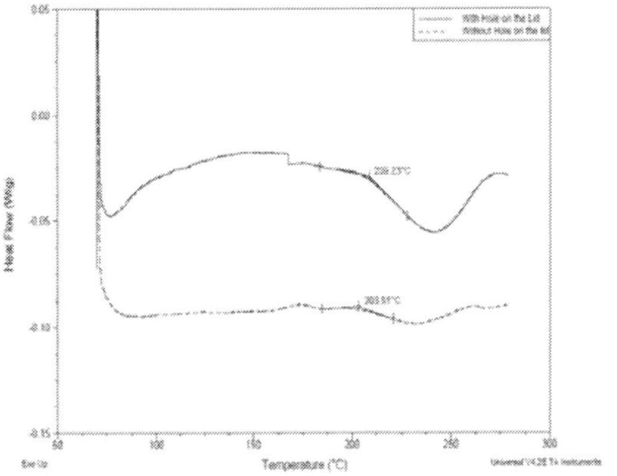

Figure 1 – DSC Thermogram for Cloisite 15A with and without hole on the lid

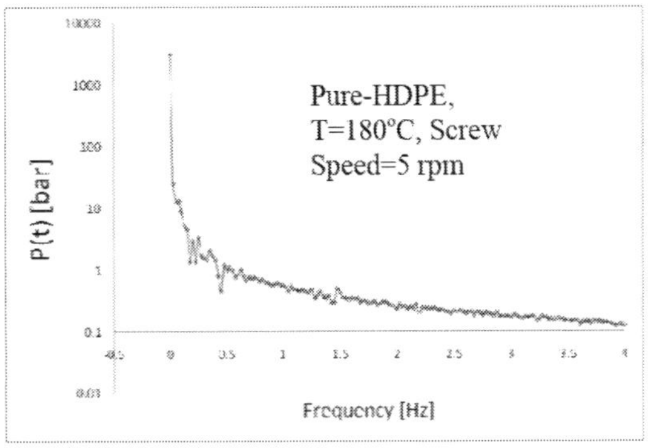

Figure 2 – Fourier Spectra of the pressure profile at position close to the die entrance. No instabilities are detected via Fourier analysis

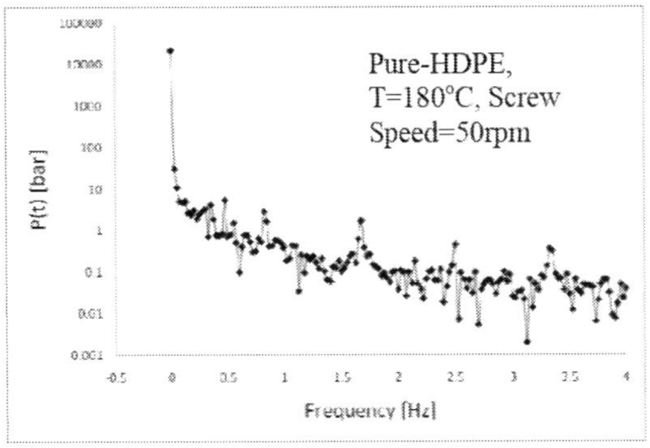

Figure 3 – Fourier Spectra of the pressure profile at position close to the die entrance. Instabilities are detected via Fourier spectra by a peak located at 0.475Hz and multiples of this frequency

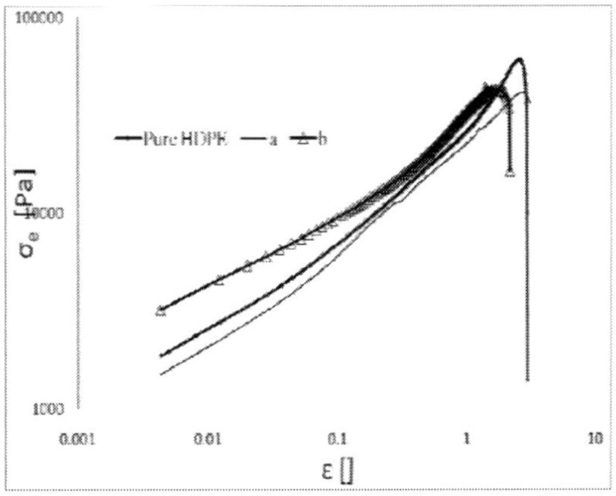

Figure 4 – Effect of organoclay Cloisite 15A on the extensional rheology of HDPE with and without compatibilizer at 200^{0}C and Hencky Strain Rate of 0.05s^{-1}. a: HDPE+Cloisite 15A without compatibilizer b: HDPE+Cloisite 15A with compatibilizer

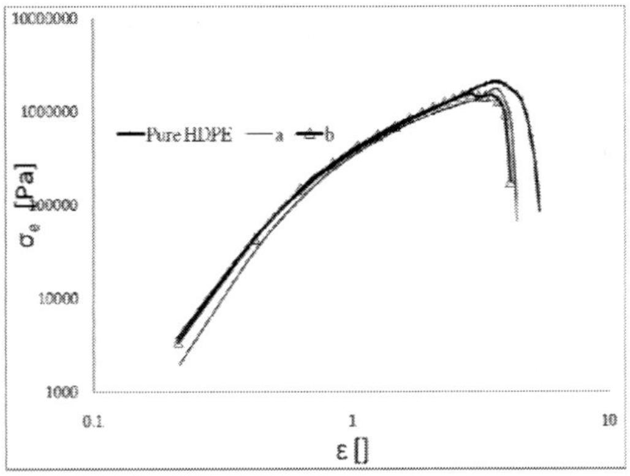

Figure 5 – Effect of organoclay Cloisite 15A on the extensional rheology of HDPE with and without compatibilizer at 200^{0}C and Hencky Strain Rate of 20s^{-1}. a: HDPE+Cloisite 15A without compatibilizer b: HDPE+Cloisite 15A with compatibilizer

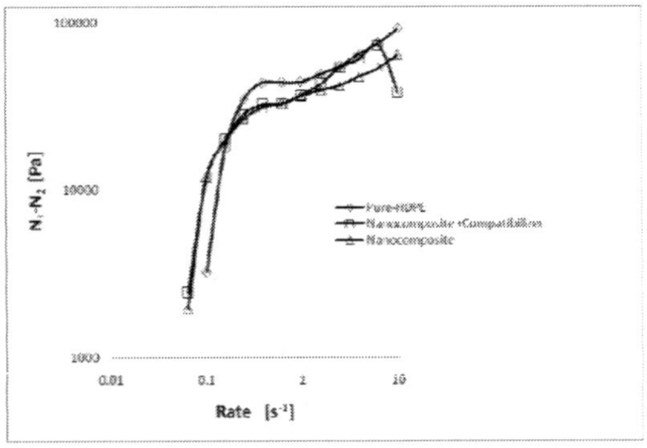

Figure 6 – Effect of organoclay Cloisite 15A on the Normal Stress Difference (N_1- N_2) of HDPE with and without compatibilizer at 200^{0}C

Conclusions

The effect of organoclay on the processing and rheology of HDPE was studied. The dispersion of the organoclay in HDPE was enhanced by the introduction of PE-g-MA as a compatibilizer. The XRD and HR-SEM characterization of the dispersion will be presented later. The effect of the compatibilizer was pronounced in some cases especially in the extensional rheology and normal stress difference. This improvement in the dispersion will be explored in the shifting and elimination of melt fractures during the extrusion of the HDPE. At present, we have shown that the new slit die with three piezometric transducer and the post fast Fourier transform rheological analysis of the time-dependent pressure can be effectively used to characterize these melt instabilities. Study on other types of organoclay will be discussed later.

Acknowledgements

The authors are grateful to King Abdul Aziz City for Science and Technology (KACST) for providing financial support for this research under research Grant # AT-27-107. We also acknowledge the support of King Fahd University of Petroleum and Minerals (KFUPM). Acknowledgement is also extended to Saudi Arabia Basic Industries Corporation (SABIC) for providing the polymer samples in our laboratory.

References

Achilleos E,et al. "On numerical simulation of polymer extrusion instabilities", Appl. Rheol. 12, 88-104 (2002)

Anastasiadis S. H., Hatzikiriakos S. G., "The Work of Adhesion of Polymer/Wall Interfaces and its Association with the Onset of Wall Slip", J. Rheol., 42, 795-812 (1998)

Carreras E. S., et al.," Pressure effects on the viscosity and flow stability of polyethylene melts during extrusion", Rheol. Acta 45,209-222 (2006)

Chrissopoulou K. et al., "Controlling the miscibility of polyethylene/layered silicate nanocomposites by altering the polymer/surface interactions", Polymer 46, 12440–12451 (2005)

Durmus A. et al.,"Linear low density polyethylene (LLDPE)/clay nanocomposites. Part I: Structural characterization and quantifying clay dispersion by melt rheology", Polymer, 48, 4492-4502 (2007)

Edwards G., et al., "Thermal stability analysis of organo-silicates, using solid phase microextraction techniques", Thermochimica Acta, 429, 13-18 (2005)

Filipe S., et al.,"Evaluation of Melt Flow Instabilities of High Density Polyethylenes Via an Optimised Method for Detection and Analysis of the Pressure Fluctuations in Capillary Rheometry", Applied Rheo. 19, 23345-1 to 23345-14 (2009)

Francowski, D.J., et al.,"Stability of Organically Modified Montmorillonites and their Polystyrene Nanocomposites after Prolonged Thermal Treatment", Chem. Mater., 19 (11), 2757–2767 (2007)

Fujiyama M., Inata H., "Melt Fracture Behavior of Polypropylene- type Resins with Narrow Molecular Weight Distribution. 1. Temperature Dependence", J. Appl. Polym. Sci., 84, 2111-2119 (2002a)

Fujiyama M., Inata H., "Melt Fracture Behavior of Polypropylene- type Resins with Narrow Molecular Weight Distribution. II. Suppression of Sharkskin by Addition of Adhesives Resins ", J. Appl. Polym. Sci., 84, 2120-2127 (2002b)

Hatzikiriakos S. G. et al., "Interfacial Phenomena in the Capillary Extrusion of metallocene Polyethylenes", J. Rheol., 41, 1299-1316 (1997)

Hatzkiriakos S.G., Migler K.B."Polymer Processing Instabilities: Control and Understanding",Marcel Dekker (2005)

Hatzikiriakos S. G., "The Effect of Nanoclays on the Processibility of Polyolefins", Polym. Eng. Sc., 1098-1107 (2005)

Hotta S. and Paul D. R., "Nanocomposites formed from Linear Low Density Polyethylene and Organoclays", Polymer, 45, 7639–7654 (2004)

Howells E. R. and Benbow J., "Flow Defects in Polymer melts", Trans. Plastics inst., 30, 240-253 (1962)

Jose M., et al., "Thermal Degradation of Commercially Available Organoclays Studied by TGA–FTIR", Thermochimica Acta, 457(1–2), 92–102 (2007)

Kazatchkov I. B. et al., "Influence of Molecular Structure on the Rheological and Processing Behavior of Polyethylene Resins", Polym. Eng. Sci., 39, 804-815 (1999)

Kolnaar J. W. H., Keller, A," A Singularity in the Melt Flow of polyethylene with Wider Implications for Polymer Melt Flow Rheology", J. Non-Newton. Fluid Mech., 69, 71-98 (1997)

Migler K. B., et al., "Extensional deformation, Cohesive failure, and Boundary Conditions during Sharkskin Melt Fracture, J. Rheol., 46, 383-400 (2002)

Perez-Gonzalez J., Denn M. M., "Flow Enhancement in the Continuous Extrusion of Linear Low-Density Polyethylene", Ind. Eng. Chem. Res., 40, 4309-4316 (2001)

Pudjijanto S., Denn m. M., " A Stable Island in the Slip-stick Region of Linear Low Density Polyethylene", J. Rheol., 38, 1735-1744 (1994)

Rudin A. et al., "Improvements n Polyolefins Processing with a Fluorocarbon Elastomer Processing Aid", ANTEC Tech. Pap., 32, 1154-1158 (1986)

Seth M. et al., "Combining Boron Nitride with a Fluoroelastomer: An Enhanced Polymer Processing Additive", Proc. ANTEC Tech. Pap., 47, 2649-2653 (2000)

Vogel R. et al., "Improved Spinnability of Metallocene Polyethylenes by using Processing Aids", Int. polym. Process, 18, 67-74 (2003)

EFFECT OF NANOSILICA ON THE MORPHOLOGY AND MECHANICAL PROPERTIES OF TPOs CONTAINING VARIOUS AMOUNTS OF A POLYOLEFIN ELASTOMER

S.-H. Lee[1], M. Kontopoulou[2*]

[1] *Production Technology & Research Center, Chemical R&D Center, Samsung Cheil Industries Inc., Korea*
[2] *Department of Chemical Engineering, Queen's University, Kingston, Ontario, Canada – kontopm@queensu.ca*

Thermoplastic Olefin (TPO) blends containing polypropylene (PP) and ethylene-α-olefin copolymer-based polyolefin elastomers (POE) were prepared at various blend ratios using a co-rotating twin-screw extruder. The effect of nanosilica addition on the morphology and mechanical properties of the blends was examined in the presence of a maleated PP compatibilizer. Addition of nanosilica resulted in a finer dispersed domain size, and finer co-continuous morphology at the phase inversion composition. Nanosilica also affected the coarsening dynamics upon annealing of the blends. The nanocomposites exhibited improvements in flexural and Young's moduli, whereas the impact strength improved slightly. These improvements were associated to a morphology wherein the filler partitions preferentially in the rigid PP phase.

Introduction

The polymer industry strives for impact modified thermoplastic polyolefin blends (TPOs) exhibiting a good balance of toughness and stiffness, suitable for demanding engineering applications. Ternary nanocomposites containing a soft elastomeric phase, such EPR, EPDM and elastomeric polyolefins (POEs) dispersed within a PP matrix, and containing a rigid inorganic filler provide a route to accomplish this.

During the recent decade nanofillers, such as clays, nano-calcium carbonate ($CaCO_3$) and nanosilica (SiO_2) have found increased use in TPO formulations as replacements to traditional micron-size fillers (Mishra et al. 2005; Mehta et al 2004; Ma et al., 2007). As reviewed by Ray and Okamoto (2003) and Zou et al. (2008) for composites containing layered silicates and nanosilica respectively, addition of nanofillers in polymeric matrices typically offers great enhancement in strength and modulus, but at the expense of ductility and toughness. On the other hand, impact modified polymers generally gain in toughness, but loose in strength (Liang and Li, 2000).

Nanoparticles having dimensions comparable to the size of the dispersed polymeric phase can readily localize preferentially in one of the phases of a polymer blend, depending on thermodynamic and kinetic considerations. As a consequence, the incorporation of nanoparticles provides flexibility in tailoring the properties of each individual phase. Liu and Kontopoulou (2006) and Bailly and Kontopoulou (2009) used nanosilica to preferentially reinforce a PP matrix, while the toughening capability of the POE dispersed phase was maintained. They showed that when silica resides in the PP matrix, the improvements in Young's and flexural moduli are accompanied by unchanged or even slightly improved impact strength.

Furthermore a transformation in PP/ POE blends from co-continuous to droplet-matrix morphology has been reported by Lee et al. (2010) at compositions close to the phase inversion. This was accompanied by the appearance of a yield stress, and a higher modulus than the unfilled blend, while the ductility was maintained.

The purpose of this study is to complete a detailed examination on the morphology and mechanical properties of TPO blends consisting of PP and POE, and containing a maleated PP compatibilizer to achieve improved dispersion of the filler. The phase morphology and mechanical properties of these blends have been analyzed quantitatively over the entire composition range.

Experimental

Materials

Polypropylene homopolymer, Escorene PP 1042, MFR 1.9 g/10 min at 230 °C, was supplied by ExxonMobil Chemical. Maleated PP (PP-g-MA), Fusabond® PM 613, MFR 49 g/10 min at 230°C containing 0.55 wt.% of maleic anhydride, was supplied by E.I. DuPont Canada. The polyolefin elastomer (POE) was a commercial grade poly (ethylene-co-octene)

(Engage™ 8100), containing 24 wt.% of comonomer with density of 0.87 g/cm^3 and melt flow index of 1.0 g/10min (at 190°C) and was provided by Dow Chemical. Nanosilica (SiO_2), Aerosil® R805, with a specific surface area of 150±25 m^2/g and mean particle size of 12 nm, surface modified by octylsilane was supplied by Evonik Industries (formerly Degussa Corp.). Irganox 225 antioxidant was obtained from Ciba-Geigy.

Compounding and characterization

All blends were prepared in a Werner & Pfleiderer ZSK-30 (L/D=40, D=30.7mm) co-rotating twin screw extruder. Blends containing POE ranging from 30% to 70% by weight were prepared at a feed rate of 2.5 kg/hr and a screw speed of 200 rpm. Composites containing 5 phr of SiO_2 and 10 phr of PP-g-MA were prepared. Tensile properties were measured using an Instron 3369 universal tester, at crosshead speeds of 50 mm/min. Dumbbell-shaped specimens were cut with a Type V die according to ASTM D638 from 1.5 mm thick sheets, which were prepared by compression molding of the compounded samples at approximately 200°C using a Carver press. Flexural tests were performed according to ASTM D790, procedure B, at a speed of 13.65 mm/min. Rectangular bars of dimensions 127x12.7x3.2 mm were produced by compression molding at 200 °C. Notched Izod impact tests were carried out on an Instron BLI impact tester at room temperature according to ASTM D 256. Fracture surfaces were observed using a JEOL JSM-840 scanning electron microscope. The POE phase was extracted using heptane. TEM imaging was carried out on samples stained with RuO_4 using an FEI Tecnai 20 instrument.

Results and Discussion

According to the SEM photomicrographs shown in Figure 1 a droplet-matrix structure was obtained at compositions up to 40% POE. The morphology became co-continuous at about 50/50 PP/POE composition by weight.

Addition of nanosilica to the blends led to a refinement in morphology, which was especially pronounced in the co-continuous region, as discussed previously (Lee et al., 2010). Changes in the morphology of immiscible blends in the presence of nanosilica have been reported extensively in the literature when the nanofiller partitions selectively in the matrix. Since the rheological properties of the blends were only marginally affected by the presence of nanosilica at the loadings investigated in this work, the changes in morphology must be due to differences in the mechanism of droplet breakup and coalescence during compounding. Indeed, annealing experiments performed using a hot stage at 180°C revealed that the coalescence rate was much lower in the presence of the nanosilica (Figure 2).

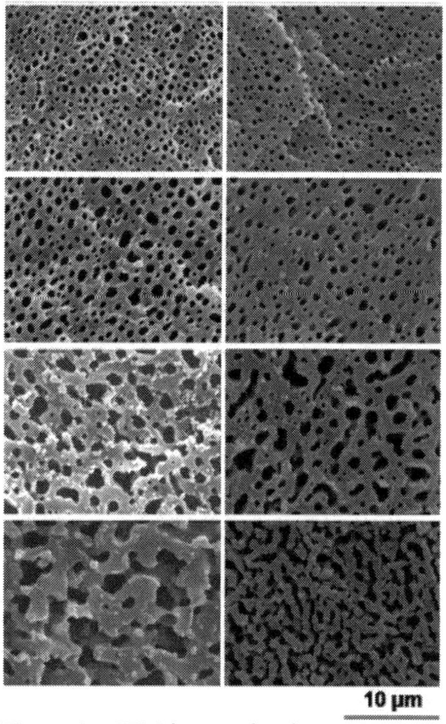

Figure 1 – SEM images showing composite structure as a function of POE content for composites containing 5 phr SiO_2. The PP/POE ratio is: (a) 70/30; (b) 60/40; (c) 50/50; (d) 30/70.

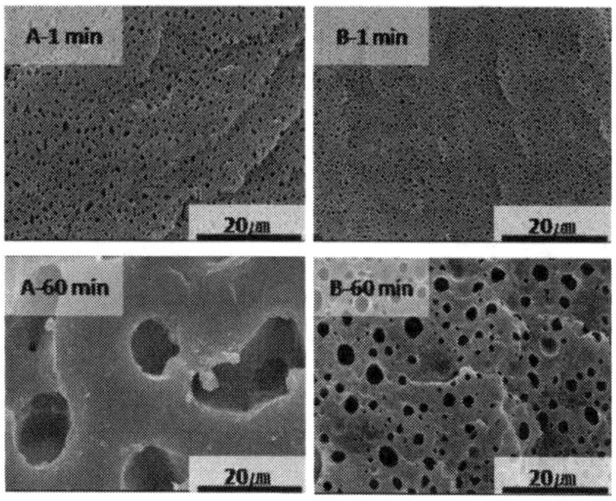

Figure 2. SEM images of PP/POE 60/40 blends with and without silica at 1 min and 60 min of annealing time. (a) unfilled blend; (b) blend containing 5phr SiO_2.

TEM micrographs of PP/POE/nanosilica composites (Fig. 3) confirmed that the finely dispersed nanosilica is located exclusively in the PP/PP-g-MA matrix phase and forms a separated microstructure. As explained below, this selective partitioning affects significantly the physical properties of the composites.

Generally addition of SiO_2 results in a significant reinforcement effect, as reflected by the flexural test results in Figure 4. Depending on the composition, up to a 50% enhancement in flexural modulus can be obtained. This may be attributed to the fine dispersion of the filler that serves to reinforce the matrix without causing weak spots. The significant increase in flexural modulus seen at the 50/50 and 30/70 composition could also be due to the different morphologies arising when nanosilica is added to the blends. Similar trends were noted on the tensile tests.

While it is well known that addition of POE to PP results in an increase in impact properties, most literature reports show decreased in toughness upon addition of nanofillers. However in the present blends the impact properties were not compromised, but in fact exhibited a slight increase (see Figure 5). We attribute this good balance between flexural and impact properties to the fact that the matrix is reinforced selectively, while leaving the toughening ability of the elastomeric phase intact. The slight reduction in the particle size of the dispersed POE phase in the presence of the silica, may also act favourably toward improving toughness.

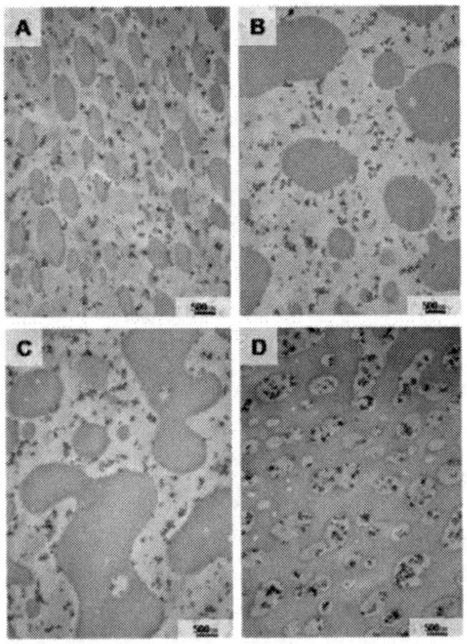

Figure 3. TEM micrographs of composites as a function of POE content. (a) 70/30; (b) 60/40; (c) 50/50; (d) 30/70. The stained elastomer phase appears darker. Scale bars represent 500 nm.

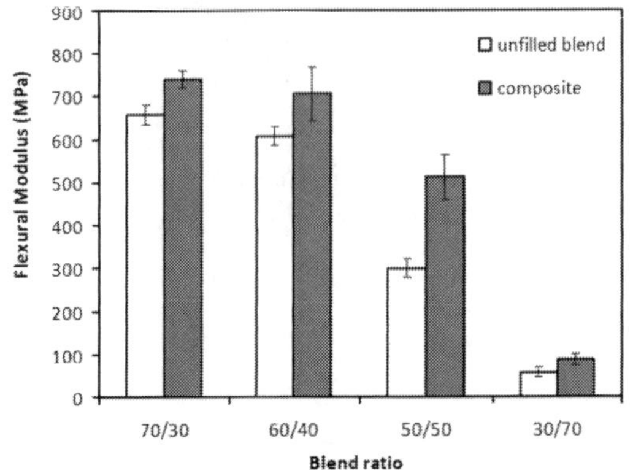

Figure 4. Effect of SiO_2 on the flexural modulus as a function of PP/POE ratio.

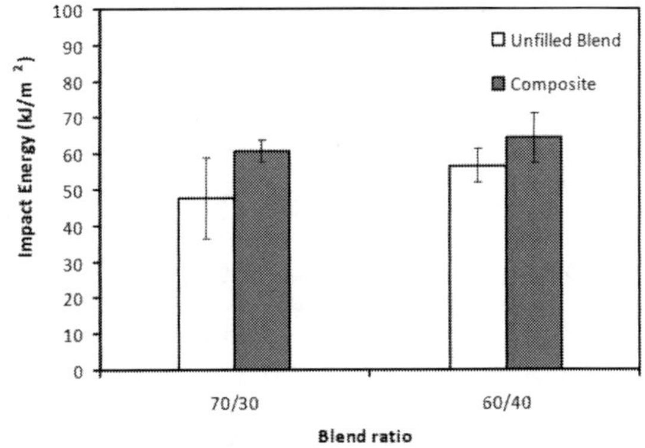

Figure 5. Effect of SiO_2 on the impact energy as a function of PP/POE ratio. Blends having a ratio of 50/50 and 30/70 did not fracture.

Conclusions

Presence of nanosilica in TPOs containing above 30 wt% of a polyolefin elastomer resulted in a finer blend morphology, with the filler residing preferentially in the PP phase, irrespective of the composition. Annealing experiments revealed that the droplet coalescence rate was significantly lower in the filled blend. Filled TPOs exhibited improvements in flexural and impact properties over the entire composition range.

Acknowledgements

The authors gratefully acknowledge funding from the Ontario Research Fund (ORF) – Research Excellence (RE) program, by the Ministry of Research and Innovation of Ontario. We would also like to thank Evonik Industries, ExxonMobil, Dow Chemical and E.I. DuPont Canada for providing the materials used in this work.

References

Bailly, M., Kontopoulou, M., "Preparation and characterization of thermoplastic olefin/nanosilica composites using a silane-grafted polypropylene matrix", Polymer, **50**, 2472-2480 (2009)

Liang, J. Z., Li, R. K. Y., "Rubber toughening in polypropylene: A review", J. Appl. Polym. Sci., **77**, 409-417 (2000)

Liu, Y., Kontopoulou, M., "The structure and physical properties of polypropylene and thermoplastic olefin nanocomposites containing nanosilica", Polymer, **47**, 7731-7739 (2006)

Lee, S. H., Kontopoulou, M. and Park, C. B, "Effect of nanosilica on the co-continuous morphology of polypropylene/ polyolefin elastomer blends", Polymer, **51**, 1147-1155 (2010)

Ma, C. G., Mai, Y. L., Rong, M. Z., Ruan, W. H. and Zhang, M. Q. "Phase structure and mechanical properties of ternary polypropylene/elastomer/nano-$CaCO_3$ composites", Compos. Sci. and Technol., **67**, 2997-3005 (2007)

Mehta, S., Mirabella, F. M., Rufener, K. and Bafna, A., "Thermoplastic olefin/clay nanocomposites: Morphology and mechanical properties", J. Appl. Polym. Sci., **92**, 928-936 (2004).

Mishra, J. K., Hwang, K. and Ha, C., "Preparation, mechanical and rheological properties of a thermoplastic polyolefin (TPO)/ organoclay nanocomposite with reference to the effect of maleic anhydride modified polypropylene as a compatibilizer", Polymer, **46**, 1995-2002 (2005)

Ray, S. S., Okamoto, M., "Polymer/layered silicate nanocomposites: a review from preparation to processing", Progr. in Polym. Sci, **28**, 1539-1641 (2003)

Zou, H., Wu, S. and Shen, J., "Polymer/ silica nanocomposites: Preparation, characterization, properties, and applications", Chem. Rev., **108**, 3893-3957 (2008)

EFFECT OF NANO-CACO$_3$ ON THERMAL AND MECHANICAL PROPERTIES OF TOUGHENED POLY(LACTIC ACID)

Y. W. Leong[1*], S. Thumsorn[1], W. Klinklai[2], H. Hamada[1]

[1]*Department of Advanced Fibro-Science, Kyoto Institute of Technology, Kyoto, Japan – leong@kit.ac.jp; nooh17@yahoo.com; hhamada1@aol.com*
[2] *Faculty of Engineering, Rajamangala University of Technology Thanyaburi, Pathumthani, Thailand – warunee.a@en.rmutt.ac.th*

Abstract – Poly(lactic acid) (PLA) has been known to be extremely brittle even without notching, which greatly limits its application. As such, a novel toughening agent was introduced in order to significantly enhance the toughness as well as deformability of PLA. In order to further enhance its mechanical performance and thermal stability without sacrificing biodegradability, naturally derived fillers were incorporated into PLA. In this study, nano-sized $CaCO_3$ obtained from precipitation and deposition techniques was used as the filler for the toughened PLA. The effects of $CaCO_3$ on the structure, crystallization, thermal stability and mechanical performance of PLA were evaluated.

Introduction

Poly(lactic acid) (PLA) is a bio-based polymer derived by using renewable resources. PLA has been popular due to its good recyclability and superior strength and clarity. However, PLA is known to exhibit brittleness and would tend to deform significantly above its glass transition temperature (Urayama et al. 2003). Various methods have been presented for improving the toughness of PLA such as fiber reinforcement (Oksman et al.2003), blend with other polymers (Aslan et al., 2000; Ramiro et al., 2001) and incorporation of nano-filler (Urayama et al., 2003; Jian et al., 2007).

Calcium carbonate ($CaCO_3$) is a type of mineral filler that is commonly incorporated into polymers for improving their stiffness and toughness. The effectiveness of the filler, however, would largely depend on the filler size. Nano-sized $CaCO_3$-filled composites have been known to exhibit good tensile and thermal properties (Gu et al., 2009; Gong et al., 2009; Li et al., 2006). Urayama et al. (2003) have studied the effect of incorporating of various inorganic filler on properties of poly(L-lactic) (PLLA) and they concluded that the crystallization of PLLA matrix as a result of the incorporation of nano-$CaCO_3$ can prevent the thermal deformation of the PLLA. Polystyrene-nano-$CaCO_3$ composite has been studied by Li et al. (2006) and found that the filler was able to plasticize the polystyrene matrix, which led to greatly reduced melt shear rates during processing.

The objective of this research is study the effect of nano-$CaCO_3$ on the rheological, mechanical, and thermal properties of toughened PLA.

Experimental

Materials and Compounding of Nano-Composites

The toughened poly(lactic acid) used in this study was Lapol™ 33 concentrate supplied by Lapol, LLC, USA, which contains 33 wt% of Lapol polyester resin as the plasticizer and will be referred as Lapol_PLA. The calcium carbonate nano-particles were prepared by precipitation technique in our laboratory, with an average particle size of about 200 nm. Compounding was performed by using a Labo Plastomill (4C150, ToyoSeiki, Japan) internal mixer set at 160 °C with rotor speed of 10 rpm for 10 minutes. After mixing, the compound was cut into grinded into powder and compression molded at 160 °C to form 1 mm-thick sheets. The compound was pre-heated for 3 minutes prior to applying a pressure of 10 MPa for another 3 minutes. The mold was subsequently rapid cooled by flowing water while the specimens were kept under constant pressure.

Characterization

Rheological and viscoelastic properties of PLA nanocomposites were measured by using a parallel plate rheometer (AR1000, TA Instruments) set at 160 °C. A frequency sweep mode was chosen with the angular frequency varied from 0.0 to 100 rad/s at a constant strain of 0.5% to maintain linear viscoelasticity in the specimens.

Morphology of the blends was characterized by observing the cryogenic fractured surface of the specimens. A field emission scanning electron microscope (Hitachi, S-4200) was used for surface observation. Gold was sputtered onto the specimens for electron conductivity.

Dynamic mechanical properties of the nanocomposites were characterized by using a dynamic mechanical analyzer (TA instruments, DMA2980) equipped with a single cantilever setting. The temperature range was set from -30 to 80 °C at a heating rate of 2 °C /min and frequency of 1 Hz.

Thermal analysis was performed by using a differential scanning calorimeter (TA instruments, MDSC 2920). The scanning was performed at a temperature range of between 20 to 200 °C under nitrogen atmosphere. The heating and cooling rates were set at 10 and 2 °C/min, respectively. All specimens were first heated to 200 °C and held isothermal for 5 minutes to erase thermal history. After cooling to 30 °C, the specimens were then subjected to a second heating cycle. Crystallization process was analyzed during the cooling cycle.

Thermal stability of the composites were investigated by using a thermogravimetric analyzer (TGA2950, TA Instruments) set at a heating rate of 20 °C/min from room temperature to 600 °C in air atmosphere.

Results and Discussion

Figure 1 shows the SEM micrograph of cryogenic fractured surface of Lapol_PLA and 10 wt% nano-$CaCO_3$ filled composites. Generally, Lapol has been used as a bioplasticizer that was reported to be compatible and miscible with PLA. The micrographs presents the phase morphology of the toughened Lapol_PLA. Large amounts of cavities could be observed from the fracture surface, which should not be confused as voids or being caused by interfacial debonding. These cavities could have been generated as a result of extensive crazing prior to failure of the material. Similar instances could be found in toughened poly(ethylene terephthalate). The presence of these cavities highlights the effectiveness of Lapol resin in enhancing the toughness of PLA. The number and depth of the cavities, however, reduced with the incorporation of nano-$CaCO_3$, which suggests that the filler could have reduced molecular motion of the matrix and thereby enhancing its stiffness.

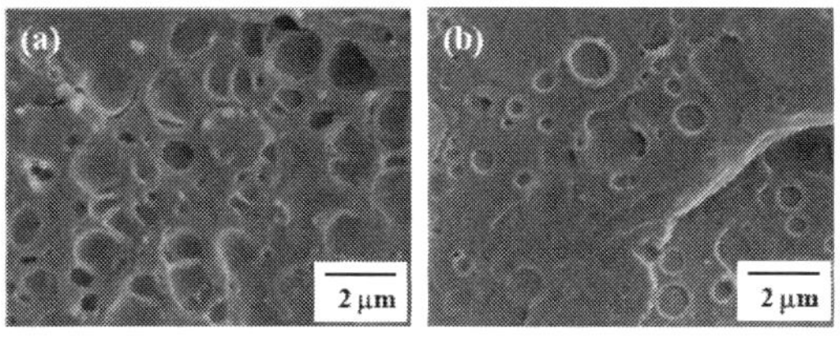

Figure 1 SEM micrographs of (a) Lapol_PLA and (b) 10 wt% nano-$CaCO_3$ filled Lapol_PLA composite.

Dynamic rheological measurement is useful for understanding the influence of filler concentration on the mobility of macromolecular chains (Gong et al., 2009). Figure 2 (a)-(c) present the complex viscosity and dynamic rheological properties as a function of nano-$CaCO_3$ content in Lapol_PLA. In Figure 2 (a), the complex viscosity (η^*) of the nanocomposites were significantly lower than that of unfilled Lapol__PLA. This could indicate that the nano-$CaCO_3$ could act as a lubricant when the composite is in the molten state (Gu et al., 2009; Li et al., 2006).

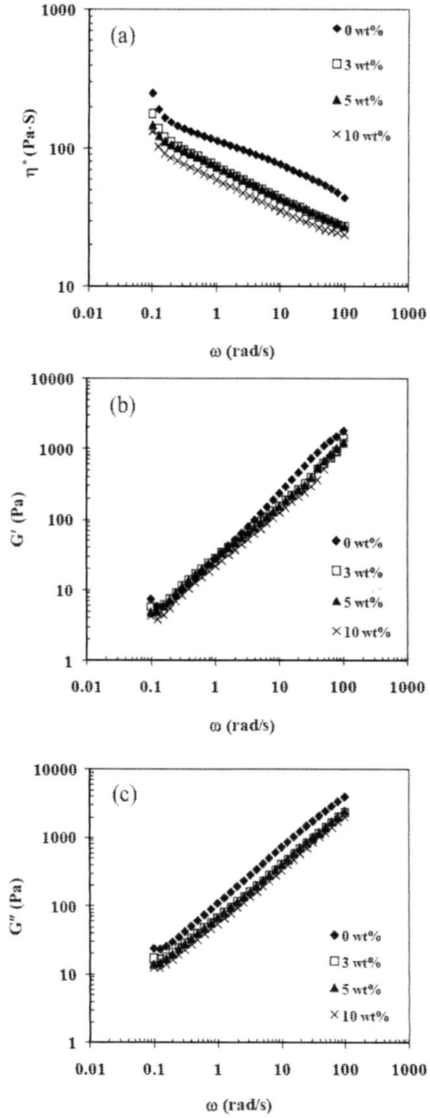

Figure 2 Complex viscosity and dynamic rheological properties of nano-$CaCO_3$ filled Lapol_PLA composite (a) complex viscosity (b) Storage modulus G′ (c) loss modulus G″.

The dynamic rheological properties of the composites are represented by the storage modulus G′ and loss modulus G″ as a function of angular frequency as shown in Figure 2 (b) and (c) respectively. The results show a decrease in storage modulus G′ and loss modulus G″. The storage modulus G′ is proportional to the elastic deformation, while the loss modulus G″, represents the viscous component, and is proportional to the dissipated energy in the material (Kually et al., 2007). The reduction of both moduli represented less elastic and viscous properties of nanocomposites during molten stage.

Dynamic mechanical properties of nano-$CaCO_3$ filled PLA composites are display in Figures 3 and 4. The results indicated phase separation between Lapol and PLA with the incorporation of nano-$CaCO_3$. Figure 3 shows the storage modulus as a function of temperature of nano-$CaCO_3$ filled Lapol_PLA composites. At low temperature, the storage modulus of toughened Lapol_PLA increased to 4.38 GPa when incorporated with 10wt% of nano-$CaCO_3$. This increment in bulk stiffness is due to the higher stiffness of the nanoparticles.

The tan δ peaks from DMA measurements corresponding to the glass transition temperatures (T_g) of Lapol and PLA phases in the nanocomposites are shown in Figures 4 (a) and (b), respectively. The tan δ peak of Lapol phase has evidently shifted to higher temperatures with the incorporation of nano-$CaCO_3$, which suggest inhibition of chain mobility by the filler. Meanwhile, the tan δ of the PLA phase shifted towards a lower temperature with a prominent emergence of a shoulder next to the maximum peak, as shown in Figure 4 (b). This could indicate that the restraint in Lapol chain mobility by the fillers provided some free volume for the movement of PLA molecules.

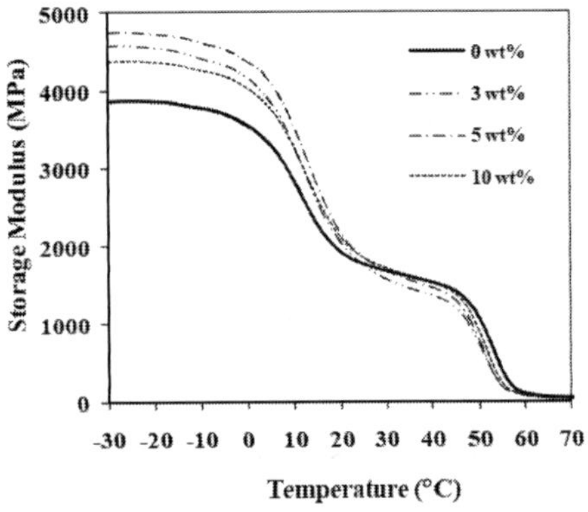

Figure 3 Storage modulus of various content of nano-$CaCO_3$ filled Lapol_PLA composite.

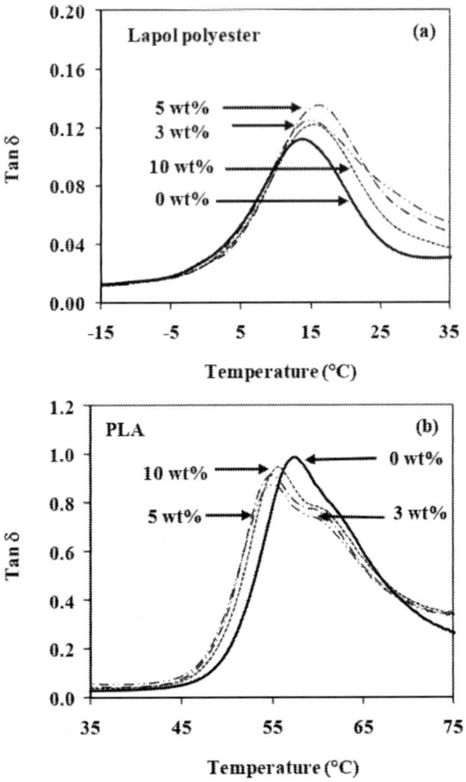

Figure 4 Tan δ of various content of nano-$CaCO_3$ filled Lapol_PLA composite (a) Lapol phase and (b) PLA phase.

The non-isothermal crystallization of nano-$CaCO_3$ filled Lapol_PLA is shown in Figure 5. The PLA phase seemed to crystallize very slowly, which could be due to inhibition by the Lapol resin. The incorporation of nano-$CaCO_3$ yielded exothermic peaks during cooling, which suggests that the filler has aided the crystallization of PLA. It should be noted that nano-$CaCO_3$ promoted crystallization of the PLA phase by heterogeneous nucleation. The lower tan δ peak registered by the PLA phase in the nanocomposites would confirm that more free volume was available for the PLA molecules to crystallize. Table 1 presents the thermal properties of the nanocomposites. The nucleating effect of the nano-$CaCO_3$ has prompted an increase in crystallinity of the filled nanocomposites without interference of the PLA crystalline structure, hence the melting temperature was unchanged. The higher in crystallinity in the nanocomposites could have also contributed to the increased storage modulus and stiffness of the material.

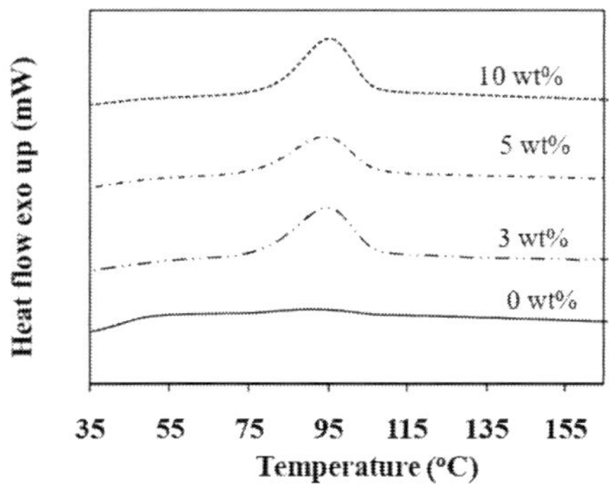

Figure 5 Crystallization temperature of Lapol_PLA composites filled with nano-$CaCO_3$ at various contents.

Table 1 Thermal properties of nano-$CaCO_3$ filled Lapol_PLA composites.

$CaCO_3$ (wt%)	T_m (°C)	T_c (°C)	%X_c	T_d (°C)
0	148.8	90.6	4.80	271
3	148.5	94.7	15.71	272
5	148.5	94.2	14.01	276
10	148.5	95.2	16.95	279

T_m = melting temperature
T_c = crystallization temperature
X_c = crystallinity
T_d = degradation temperature

Effect of nano-$CaCO_3$ on thermal stability of nanocomposites is presented Table 1. The onset thermal degradation temperature (T_d) corresponds to the temperature at 5% weight loss. The T_d of Lapol_PLA slightly increased with increasing nano-$CaCO_3$ content. This could be attributed to the filler particles acting as a barrier to prevent the resin from extensive thermo-oxidative decomposition (Kim et al., 2009).

Conclusions

The nano-$CaCO_3$ was found to be a good nucleating agent for PLA, judging from the enhanced crystallization peak with increasing filler content. The nano-filler could also act as a barrier to improve the thermal stability of the matrix resin. Moreover, the nano-fillers were able to act as a lubricant for the matrix in the molten state, which is expected to enhance the processability of the material. In the solid state, however, the nano-$CaCO_3$ was able to improve the stiffness of the composites while maintaining its toughness.

Acknowledgement

The authors are grateful to Department of Materials and Metallurgy Engineering, Faculty of Engineering, Rajamangala University of Technology Thanyaburi, Thailand for preparing precipitated nano-$CaCO_3$.

References

Aslan S., et al., "Poly (D,L-lactic acid)/Poly (ε-caprolactone) Blend Membranes: Preparation and Morphological Characterisation", J. Mater. Sci, **35**, 1615-1622 (2000)

Gong G., et al., "Dynamic Rheological Behavior of Isotactic Polypropylene Filled With Nano-Calcium Carbonate Modified by Stearic Acid Coating", J. Macromol. Sci. Phys., **48**, 329-343 (2009)

Gu J., et al., "Effect of Unsaturated Hydroxyl-Fatty Acid Modified Nano-$CaCO_3$ on the Morphological and Rheological Behavior of PP", Front. Chem. China, **4**, 75-82 (2009)

Jiang L., et al., "Comparison of Polylactide/Nano-sized Calciumcarbonate and Polylactide/Monmorillonite Composites: Reinforcing Effects and Toughening Mechanisms", Polymer, **48**, 7632-7644 (2007)

Kaully T., et al., "Rheology of Highly Filled Natural CaCO3 Composites. II. Effect of Solid Loading and Particle Size Distribution on Rotational Rheometry", Polm. /comps., **28**, 524-533 (2007)

Kim J. Y., et al., "Thermal Decomposition Behavior of Poly(ethylene 2,6-naphthlate)/Silica Nanocomposites", Polm. Compos., **30**, 1779-1787 (2009)

Li G., et al., "Plasticization of Nano-$CaCO_3$ in Polystyrene/Nano-$CaCO_3$ Composites", J. Appl. Polm. Sci., **99**, 2138-2143 (2006)

Oksman K., et al., "Natural Fibers as Reinforcement in PLA composites", Comp. Sci. Technol., **63**, 1317-1324 (2003)

Ramiro D. E., et al., "Immiscible Polymer Blends of Semicrystalline Biocompatible Components: Thermal Properties and Phase Morphology Analysis of PLLA/PCL Blends", Polymer, **42**, 7831-7840 (2001)

Urayama H., et al., "Mechanical and Thermal Properties of Poly(L-lactide) Incorporating Various Inorganic Fillers with Particle and Wisker Shapes", Macromol. Mater. Eng., **288**, 562-568 (2003)

MORPHOLOGY AND PROPERTIES OF MELT INTERCALATED POLYSTYRENE/HALLOYSITE NANOTUBES NANOCOMPOSITES

D. Acierno[1], P. Russo[1], P. Iannelli[2], P. Scarfato[3,4]*

[1]*Department of Materials and Production Engineering, University of Naples Federico II, Naples, Italy – acierno@unina.it; pietro.russo@unina.it*

[2] *Department of Pharmaceutical Science, University of Salerno, Fisciano (Sa), Italy – piannelli@unisa.it*

[3] *Department of Chemical and Food Engineering, University of Salerno, Fisciano (Sa), Italy – pscarfato@unisa.it*

[4] *NANO_MATES – Researh Centre of NANOMAterials and nanoTEchnologies, University of Salerno, Fisciano (Sa), Italy – pscarfato@unisa.it*

Abstract – Polystyrene based composites containing up to 5 wt% of tubular structured clays, best known as halloysite, have been prepared by melt compounding and characterized always taking the pure PS, extruded under the same condition, as the reference material. The results collected in this abstract confirmed the achievement of an adequate level of distribution of the rigid filler within the hosting matrix. Moreover, a slight influence of the reinforcement on the degradation behavior of the matrix was observed especially for halloysite nanotube loadings higher than 3% by weight.

Introduction

In the last two decades, the development of nano-filled composites through the incorporation of nano-sized additives in polymer matrices has drawn much attention due to the peculiar properties as well as the successfully commercial applications of these systems [1,2]. Polymer-based nanocomposite formulations exhibit specific features even by the addition of low weight percentages of nanofillers essentially because of their very high interfacial area. Anyway, it is well established that valuable synergies between filler and matrix can occur only if an optimal dispersion of the included additives within the hosting matrix and an appropriate interaction at the interface to ensure an adequate level of stress-transfer are achieved.

Recently, melt compounded organic–inorganic nanohybrid materials, such as polymer-layered silicate nanocomposites with inclusions of natural and organic montmorillonites (MMTs), having at least one characteristic dimension in the order of the nanometer, have been intensively investigated for a lot of polymer matrices [2-4]. However, only few papers have considered the possibility to use as nano-filler some type of tubular nanoclays (NTs), having two nanometric dimensions. Preparation of such nanocomposites was achieved either by melt intercalation or by the intercalation of the monomer followed by polymerization. Examples of the first approach were reported in the cases of polyethylene glycol, polyethylene oxide and polyhydroxybutyrate and the second in the case of polyacrylonitrile, nylon-6 and polystyrene [5-9].

In the present work we have prepared, by melt intercalation, polystyrene-based nanocomposites incorporating small amounts (≤ 5 wt%) of halloysite nanotubes (HNTs) [10]. The produced samples have been analyzed by several methods (SEM, rheological tests, TGA, DMTA, FT-IR, DSC) to investigate their morphology and thermo-mechanical properties.

Experimental

Materials

The polymeric matrix used in this study was a polystyrene (PS N2982, Mw/Mn=1.98, supplied by Polimeri Europa), whereas the nanometric filler was the halloysite ($Al_2Si_2O_5(OH)_4 \cdot 2H_2O$, supplied by Aldrich), a natural two-layered aluminosilicate nanotube. Halloysite has a predominantly hollow tubular structure, with sizes within 1-15 microns of length and 10-150 nm of inner diameter (Fig. 1). Typical specific surface area of this halloysite is 65 m^2/g.

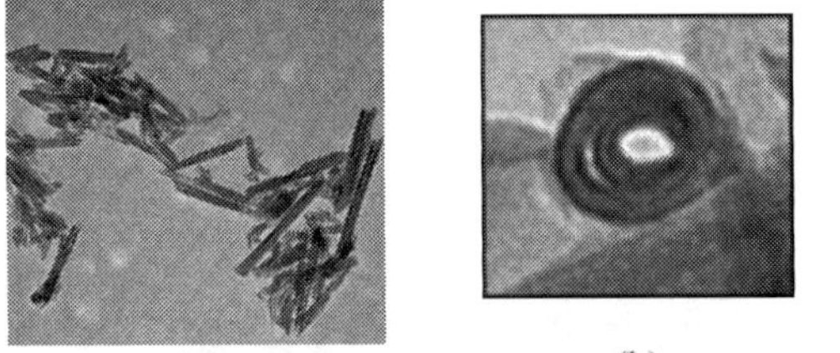

(a) (b)

Figure 1: TEM image of 50 nm diameter halloysite nanotubes (a) and tube cross-section (b) [10].

Processing

Mechanical polymer-silicate blends at three different weight percentages of halloysite (1%, 3% and 5%) were prepared and dried in a vacuum oven at 90°C for 18 hrs to avoid moisture induced degradation reactions. The melt compounding of the blends was performed

using a Haake twin screw extruder having counter-rotating intermeshing conical screws (L/D=40); this screw design allows to have varying degrees of mixing along the length of the screws. A temperature profile of 200-210-220-230-235-235-230-230-230°C from hopper to die was imposed and a cylindrical die (ϕ=1 mm) was used. Nanocomposite extrudates at different organoclay contents were extruded at screw speed of 70 rpm.
In the following the nanocomposite samples were coded as PS/xHNT, where x is the weight percentage of the filler, whereas the extruded pure matrix was reported as extruded PS.

Analyses

Scanning electron micrographs were obtained using a LEO 420 apparatus (LEO Electron Microscopy Ltd.). The samples were analyzed on surface after etching by immersion in a solution of sulfuric and orthophosphoric acid (10:4 by volume) at 1% w/v of $KMnO_4$, at room temperature for 4 hours. All samples were coated with a AuPd alloy using a high vacuum sputter coater before analysis.
Thermogravimetric analyses (TGA) were carried out on neat PS and all its blends extruded with different clay content and extrusion rate. TGA tests were performed using a TA Q500 apparatus in the temperature range of 25÷900°C with a heating rate of 20°C/min under nitrogen atmosphere.
Calorimetric measurements (DSC) were performed with a Mettler Differential Scanning Calorimeter mod. DSC-1. The samples were heated at a rate of 20°C/min from 25°C to 200°C under nitrogen gas flow.
The flow properties of the molten materials were measured with an ARES rheometer AR-G2 (TA Instruments), using a parallel plates geometry (plate radius=25 mm, gap=0.8 mm). Dynamic frequency sweep tests were performed in the frequency range ω=0.01÷100 Hz, with a strain amplitude of 1% (proven to be in the linear viscoelasticity range). All samples were tested at 235°C after drying at 90°C in a vacuum oven for 18 hrs, under a nitrogen gas purge in order to minimize thermo-oxidative degradation phenomena.
Dynamic-mechanical (DMA) properties were analyzed using a Triton 2000 Analyzer. Dynamic mechanical spectra were recorded in single cantiler bending mode, at a frequency of 1Hz and a heating rate of 4°C/min in the range of 0÷160°C.

Results and Discussion

Hybrid systems such as nanostructured materials offer interesting properties and, consequently, promising applications if included nanometric fillers show an adequate level of distribution within the hosting matrix. In this context all investigated materials were analysed by electron microscopy in order to evaluate the actual dispersion of HNTs into the PS matrix. Preliminary observations showed, on one side, that clay nanotubes are quite homogeneously distributed within the hosting matrix. On the other side, no relevant failures of the dispersed filler seemed to be verified.
Regarding thermal analyses, main aims of authors were both to verify the actual content of HNTs in the nanocomposites and to investigate the effect of the filler on the thermal response of the samples. Thermal parameters collected from thermogravimetric and calorimetric measurements are summarized in Table 1.

Table 1 – Thermal parameters of neat and nanocomposite PS samples.

Samples	*T_d (*) [°C]*	*DTG [°C]*	*residue weight at 750°C [%]*	*T_g [°C]*
extruded PS	397	435	0	106
PS/1HNT	399	435	0.89	104
PS/3HNT	405	440	2.38	105
PS/5HNT	406	441	4.26	106

(*) measured at 5wt% of weight loss

TGA data demonstrated that a good control on the HNTs loading is obtained in all cases. Moreover, it is clear that the inclusion of nanometric reinforcement does not worsen the thermal behavior of nanocomposite systems compared with the extruded matrix, taken as the reference material, and involves a slight increase in temperature of incipient degradation. The latter effect is in accordance with the appearance of a shoulder in the profile of the derivative weight loss curves of PS/3HNT and PS/5HNT samples (see Figure 2), suggesting a change in their degradation mechanisms. On the other hand, regarding the glass transition temperature T_g, no detectable effects have been observed.

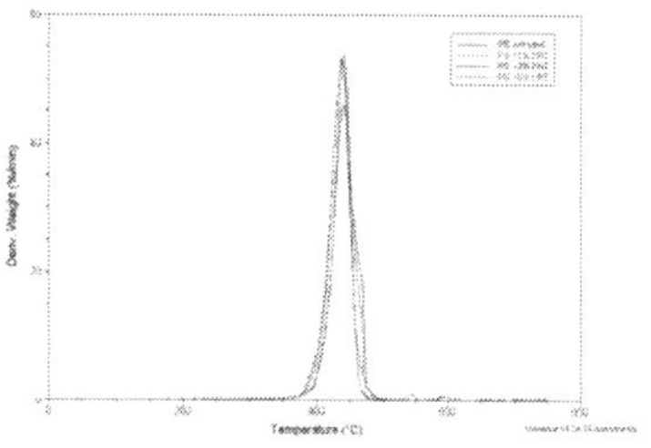

Figure 2: TGA derivative weight loss curves of neat and nanocomposite PS samples.

In order to evaluate the influence of HNT fillers on the flow behavior of the hybrids, rheological tests in the low shear rate region were performed on nanocomposite samples at different loadings. At this regard, main results in terms of storage modulus (G') and complex viscosity (η*) curves are shown in Figures 3 and 4, respectively.
In both cases, at least on the examined oscillation frequency range, inclusions of the rigid nanosized tubular clay always induce an upward shift of the storage modulus curve and an increase of the complex viscosity, with respect to the reference material. Moreover, the absence of significant viscoelastic

effects at low frequencies, usually accompanied by apparent dynamic viscosity deviations from Newtonian behavior, demonstrates that, with our preparation conditions and within the range of compositions considered, no special structures of the dispersed phase occurs in the hosting matrix.

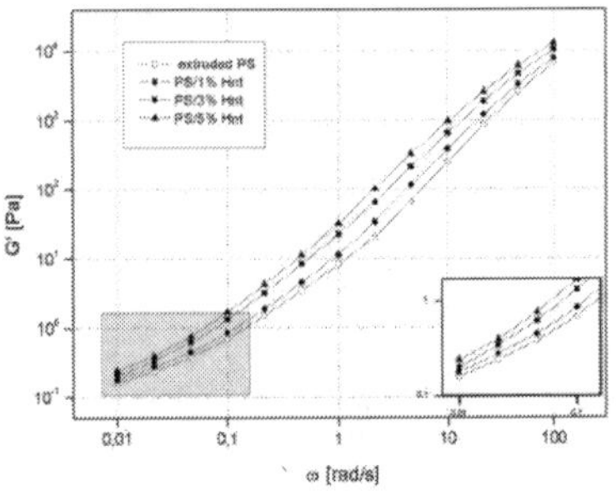

Figure 3: Dynamic storage modulus curves of neat and nanocomposite PS samples obtained in dynamic frequency sweep.

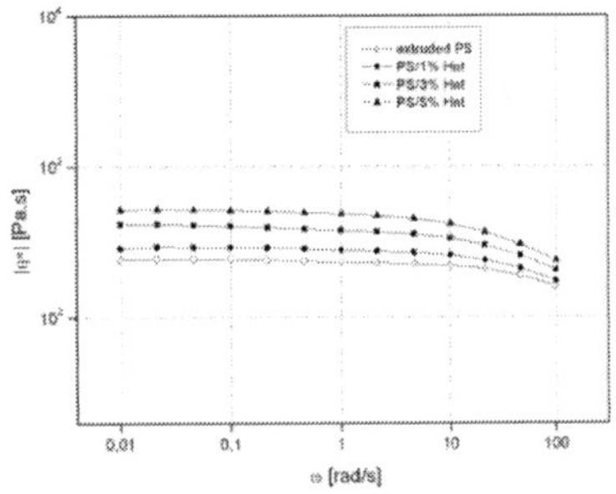

Figure 4: Complex viscosity curves of neat and nanocomposite PS samples obtained in dynamic frequency sweep.

Finally, preliminary investigations in terms of dynamical-mechanical results emphasize a certain degree of interaction at polymer-filler interphase, as witnessed by the lowering of the tanδ peak signal.

Conclusions

Polystyrene based composites containing relatively low contents of halloysite nanotubes were prepared by melt blending and analysed in terms of morphological features, thermal properties, rheological behaviour and dynamic-mechanical performances.

Main results achieved by now demonstrated that the inclusion of the nanometric filler does not significantly modify the thermal behavior of the matrix, but changes the viscoelastic behavior of hybrid materials in the molten state with respect to the extruded matrix, selected as the reference material, and reduces the mechanical damping capacity of the same.

References

1. Jordan, J., Jacob, K.I., Tannenbaum, R., Sharaf, M.A., Jasiuk, I., "Experimental trends in polymer nanocomposites - a review", Materials Science and Engineering **A 393**, 1-11 (2005).
2. Giannelis, E.P., "Polymer-Layered Silicate Nanocomposites: Synthesis, Properties and Applications", Appl. Organometal. Chem., **12**, 675 (1998).
3. Ray, S.S., Okamoto, M., "Polymer/layered silicate nanocomposites: a review from preparation to processing", Prog. Polym. Sci. **28**, 1539-1641 (2003).
4. Pinnavaia, T.J., Beall, G.W., Editors. Polymer-Layered Silicate Nanocomposites, N.Y.: John Wiley & Sons Ltd. (2001).
5. Li, C., Liu, J., Qu, X., Yang, Z., "A General Synthesis Approach toward Halloysite-Based Composite Nanotube", J. of Appl. Polym. Sci., **112**, 2647–2655 (2009).
6. Zhao M., Liu, P. "Halloysite nanotubes/polystyrene (HNTs/PS) nanocomposites via in situ bulk polymerization", J. of Thermal Analysis and Calorimetry, **94**, 103-107 (2008).
7. Ning, N.-Y., Yin, Q.-J., Luo, F., Zhang, Q., Du, R., Fu, Q., "Crystallization behavior and mechanical properties of polypropylene/halloysite composites", Polymer **48,** 7374-7384 (2007).
8. Hedicke-Höchstötter, K., Lim, G.T., Altstädt, V., "Novel polyamide nanocomposites based on silicate nanotubes of the mineral halloysite", Composites Science and Technology **69**, 330-334 (2009).
9. Guo, B., Zou, Q., Lei, Y., Du, M., Liu, M., Jia, D., "Crystallization behavior of polyamide 6/halloysite nanotubes nanocomposites", Thermochimica Acta **484**, 48-56 (2009).
10. http://www.sigmaaldrich.com/materials-science/nanomaterials/nanoclay-building/halloysite-nanotubes.html

POLYETHYLENE INONOMER BASED NANO-COMPOSITE FOAMS PREPARED BY A BATCH PROCESS AND MUCELL® INJECTION MOLDING

Hidetomo Hayashi [1], Tomoko Mori [1], Baiju John [1], Masami Okamoto [1]*, Satoshi Yamasaki [2], Hiroshi Hayami [2]

[1] *Advanced Polymeric Nanostructured Materials Engineering, Graduate School of Engineering, Toyota Technological Institute, Hisakate 2-12-1, Tempaku, Nagoya, 468-8511, Japan- okamoto@toyota-ti.ac.jp*
[2] *Polymer Materials Technology, Department of Electronics & Materials R&D Laboratories, Sumitomo Electric Industries, Ltd., Shimaya, Konohana-ku, 1-1-3, Osaka, 554-0024, Japan.*

Abstract

To understand the correlation between formability and melt rheology of polyethylene-based ionomers having different degrees of the neutralization and corresponding nano-composites, foam processing via a batch process in an autoclave and microcellular foam injection molding (FIM) process using the Mucell® technology. The obtainable morphological properties in both foaming processes were discussed. All the cellular structures were investigated by using field emission scanning electron microscopy. The competitive phenomenon between the cell nucleation and the cell growth including the coalescence of cell was discussed in light of the interfacial energy and the relaxation rates revealed by the modified classical nucleation energy and rheological measurement, respectively. The FIM process led to the opposite behavior in the cell growth and coalescence of cell as compared with that of the batch process, where the ionic crosslinked structure has significant contribution to retard the cell growth and coalescence of cell. The mechanical properties of structural foams obtained by FIM process were discussed.

Introduction

The effect of organo-clay on the crystal structure [1] and foam processing of the matrix semi-crystalline polymer has been investigated in nano-composites. The grown cells having diameter of ~ 200 nm were localized along the dispersed nano-clay particles in the cell wall. The dispersed nano-clay particles acted as nucleating sites for cell formation and the cell growth occured on the surfaces of the clays [2]. On the other hand, microcellular foam injection molding (FIM) have been largely applied to produce structural foams that consist of two compact polymer skin layers enclosing a foamed core, it can be described as an inhomogeneous material with varying properties across the part thickness.

The correlation between the foamability and melt rheology of polyethylene-based ionomers having different degrees of the neutralization and corresponding nano-composites via a batch process in an autoclave was investigated. The ionic crosslinked structure in the ionomer contributed to the retardation of the cell growth and thereby the coalescence of cell. The use of nano-clay in the nano-composite foaming will lead to an increase in the cell density. The batch process and the FIM process using the MuCell®technology are adopted to generate structured nano-composite foams. Such a comparison between batch and FIM will be worthwhile not only to assess the foam morphology but also for studying the foaming performance of the nano-composite.

Experimental

An ionomer resin (Himilan 1706: MA content = 15 wt%, Zinc oxide (ZnO) content = 3.70 wt%, neutralization = ~ 55 % and melt flow index (MI) = 4.1g/10 min) was used as the polymer matrix. The organo-clay used in this study was synthesized by replacing Na+ ions in montmorillonite (MMT) (cation exchange capacity (CEC) of 90 mequiv/100g, thickness ≈ 1nm, average length of 150-200 nm) with di-octadecyl di-methylammonium (DC18DM) cation using the ion exchange method.

The nano-composites were prepared by dry-mixing the organo-clay and ionomer and subsequent melt extrusion using a twin-screw extruder (PCM-30, Ikegai Corp.) operated at 200 °C to yield fine nano-composite strands. The strands were pelletized and dried under vacuum at 60 °C for 48h to remove water. The nano-composite prepared with 20 wt% of organo-clay was denoted as ionomer/clay 20/0/58. The abbreviations of various ionomers and the corresponding nano-composites prepared by using zinc oxide (ZnO - as a neutralizating agent) are presented in Table 1. The degree of neutralization of the nano-composites including ZnO exhibited large increment due to the presence of organo-MMT. Presumably, the dispersed organo-clay acted as catalytic sites for the neutralization. For ionomer 3/89, the degree of neutralization is controlled by using ZnO (3 phr) and dodecyl tri-methyl ammonium chloride (as a neutralizating catalyst).

Results and Discussion

Table 1 – Composition and characteristic parameters of ionomers and nano-composites

Samples	ZnO / phr	Neutralization[b] / %	Organo-clay / wt%
Ionomer 0/55	0	55	0
Ionomer 3/89[a]	3	89	0
Ionomer/clay 20/0/58	0	58	20
Ionomer/clay 20/3/78	3	78	20
Ionomer/clay 20/6/97	6	97	20

a) prepared by ZnO and dodecyl tri-methyl ammonium chloride
b) value measured by titrimetric analysis

Figure 1 shows the results of TEM bright field images of ionomer/clay 20/0/58, in which dark entities are the cross section of layered MMT nano-fillers. It is clear that there are some intercalated-and-stacked silicate layers co-existing with the disordered or exfoliated MMT layers in the nano-composite structure. The dark areas represent the nano-sized ionic aggregates similar to what is observed for other ionomers. After a higher magnification, shown in the enlarged views, the spheres with a domain size of approximately 2-3 nm (indicated with arrows) are seen. The ionic aggregates are randomly arranged in the ionomer matrix. The calculated value of the number of the aggregates was 0.08-0.12 nm^{-2} for ionomer/clay 20/0/58 and 0.98-0.15 nm^{-2} for ionomer/clay 20/6/97. This result suggests that the ionic cross-linked density in the system of 20/6/97 is higher than that of matrix in the absence of ZnO (6 phr).

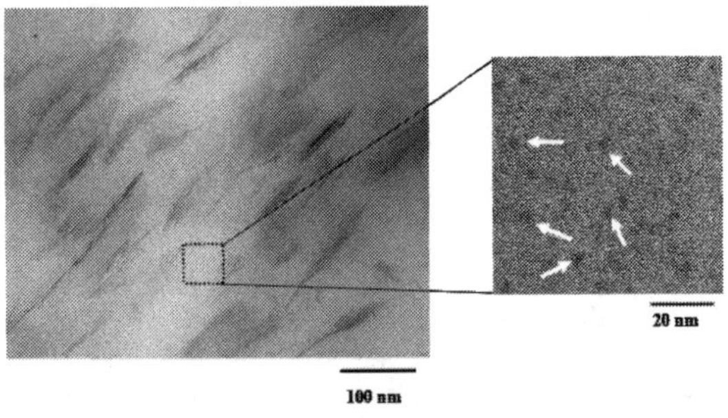

Figure 1 – Bright field TEM images of ionomer/clay 20/0/58.

Figure 2 shows the FE-SEM images of the fracture surfaces of the materials foamed at 90 °C under the isobaric saturation conditions (30MPa) prepared by batch process. All foams exhibited the closed-cell structure, have smaller cell size (2d) and larger cell density (N_c) with increasing saturation pressure condition. The incorporation of nano-clay shows smaller cell size and larger cell density

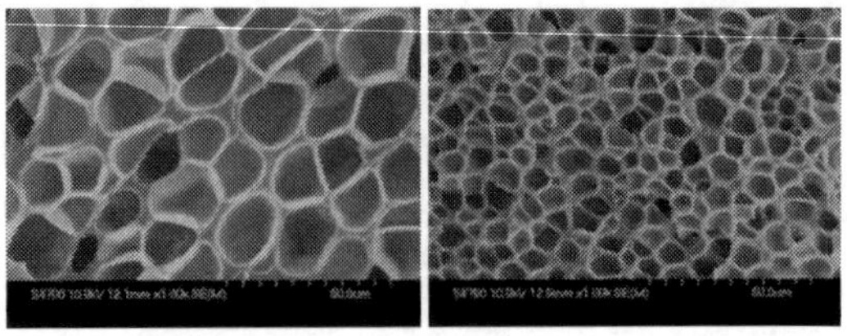

Figure 2 – Typical results of SEM images of (a) ionomer 0/55 and (b) ionomer/clay 20/0/58 foamed at 90 °C under 30 MPa (batch process).

compared with neat ionomer 0/55 foams. Experimentally, nano-clay particles lead to an increase in N_c. The controlled structure is from 2d = 1-20 μm with $N_c = 10^8$-10^{11} cells cm^{-3}. The cell nucleation in the heterogeneous nucleation system such as nano-composite foams may take place in the boundary between the matrix and the dispersed nanoparticles.

The influence of internal surfaces of the dispersed nano-clay was characterized by the

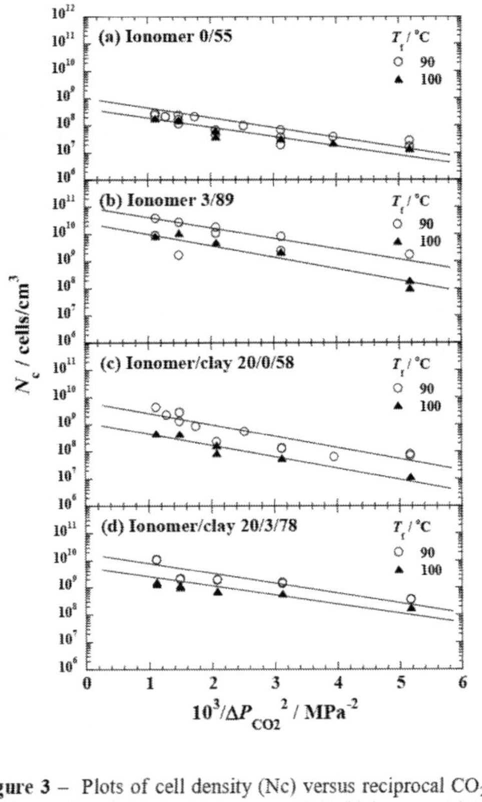

Figure 3 – Plots of cell density (Nc) versus reciprocal CO_2 pressure ($1/(PCO_2)_2$) for (a) ionomer 0/55, (b) ionomer 3/89, (c) ionomer/clay 20/0/58 and (d) ionomer/clay 20/3/78 foamed at two different T_f (batch process).

interfacial tension between the bubble and matrix by using modified classical nucleation theory [3] proposed by Suh and Colton. Assuming no effect of the coalescence of cell on the value of N_c, the interfacial energy ($\gamma S(\theta)1/3$) of the systems, i.e., the slope of the

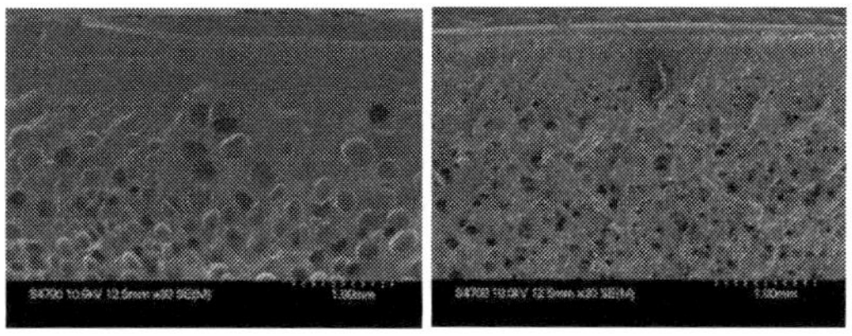

Figure 4 – Typical results of FE-SEM images of ionomer 0/55 processed at two different conditions (a) MC-1 and (b) MC-2 (FIM process).

plots (N_c versus $1/(\Delta P_{CO2})^2$) was estimated. For all cases examined the plots conformed to the straight lines (Figure 3). The results showed almost same values (~ 6 mJ/m^2). However, the nano-composite (ionomer /clay 20/0/58) foams exhibited smaller cell size and larger cell density without a significant difference in $\gamma S(\theta)1/3$ as compared with neat ionomer 0/55 foams, suggesting that the dispersed silicate particles do not act as nucleating sites for cell formation. This trend reflects the relative importance of the cell growth and the coalescence of cell. That is, the viscosity effect (relaxation time) on the cell growth and coalescence of cell has to be taken into consideration. The viscosity effect of the cell coalescence was investigated using the flow behavior of the ionomers and the corresponding nano-composites. For ionomer 3/89, the estimated relaxation rates were about 0.082-12.5 s^{-1} in the temperature range of 100 and 180 °C. These values are much smaller than those of ionomer 0/55 (0.24-84.5 s^{-1}), suggesting the ionic cross-linked density (the degree of the neutralization) is dominant factor for the relaxation time scale.

The ionic cross-linked structure has significant contribution to retard the cell growth and coalescence of cell. For this reason, ionomer 3/89 foams lead to a large value of N_c accompanied with the small value of 2d as compared with that of ionomer 0/55 foams. That is, in the ionomer foaming the cell growth and coalescence of cell are dominant factors rather than the heterogeneous nucleation. Meanwhile, in ionomer/clay 20/0/58 foaming, a similar feature is observed as that of ionomer 3/89, implying that the nucleating effect of the dispersed nano-clay particles is overwhelmed (i.e., no enhancement of the nucleation).

Figure 4 shows the typical results of FE-SEM images of the fracture surfaces of the structural foams processed at two different conditions under FIM-process. The processing conditions of MC-1 and MC-2 are set at the density reduction values of 10 and 5 %, respectively. All the structural foams thus prepared consisted of two compact polymer skin layers enclosing a foamed core. However, in the case of MC-2 (ionomer/clay 20/0/58 and ionomer/clay 20/3/78) the second layer is observed at the end of skin layer to the central area of the foamed core. The cell size in this area exhibited much smaller one (~10-30 μm) than that of the foamed core. The characteristic features of the obtained foams are presented in the Figure 5.

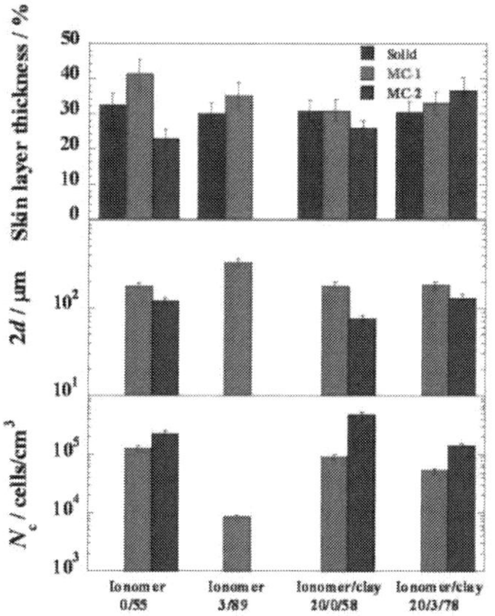

Figure 5 – Skin layer thickness, cell size (2d) and the cell density (N_c) of the structural foams prepared by FIM process. The values of the skin layer thickness in solid samples are shown for comparison.

It should be specially noted that in the case of MC-1, all foams exhibited larger cell size and smaller cell density as compared with those of MC-2 condition except ionomer 3/89. The coalescence effect in the cellular core is dominant factor rather than the nucleation due to the low viscosity of the materials with a SCF N_2. For this reason, the FIM-process in this study leads to bigger cell size and smaller cell density than that of batch-process. The development of solidified skin layer can be explained easily by heat transfer. However, the skin layer thickness has no significant difference between these processing sets including injection molding of the solid. Interestingly, the FIM-process leads to the opposite behavior in the cell growth and coalescence of cell as compared with that of the batch -process, where the ionic cross-linked structure has significant contributed to the retardation of the growth and coalescence of cells. The mechanical properties of solids and foamed materials obtained by FIM-process were measured by the specific dynamic storage modulus (G') at -50 °C and the specific thermal expansion coefficient (Ω) in the temperature range of -150 °C to 0 °C.

In the case of solid, there is a significant increase in G' and a decrease of α for all nano-composites as compared with that of ionomers. After structural foaming prepared by MC-1 condition, all nano-composite foams exhibited much higher enhancement in G' (200 % for ionomer/clay 20/0/58, 210 % for ionomer/clay 20/3/78) than those of the structural foams of the corresponding ionomers. This is due to the mechanical reinforcement by the dispersed nano-clay particles in the cell wall. The dispersed nano-clay particles seem to act as a secondary cloth layer to protect cells from being deformed by the external forces [4]. At the same time, the reduction of the thermal expansion coefficient in the nano-composite foams seems to stem from the mechanical constraint by the dispersed nano-clay particles. For this reason, all nano-composite foams showed significant reduction of α value and the values decreased up to 66 % for ionomer/clay 20/0/58 and 53 % for ionomer/clay 20/3/78 with MC-1 condition when compared to the structural foams prepared from the ionomers.

Conclusions

The correlation between foamability and melt rheology of polyethylene-based ionomers having different degrees of the neutralization and corresponding nano-composites via a batch process in an autoclave and the FIM process using the MuCell® technology were discussed. In the batch-process, the ionic cross-linked structure exhibited significant contribution to retard the cell growth and coalescence of cell. The degree of the neutralization is mainly responsible for the cell formation as compared to the incorporation of nano-clay particles. The ionomer/clay 20/6/97 foams enhanced the value of N_c accompanied with the small value of 2d than with the other nano-composite foams. The controlled structure is from 2d = 1-20 μm with $N_c = 10^8$-10^{11} cells cm^{-3}. The structural foams using FIM process was processed at two different conditions. The coalescence effect in the cellular core is dominant factor rather than the promoting nucleation due to the low viscosity of the materials with a SCF N_2.

The FIM-process led to bigger cell size and smaller cell density when compared to that of batch process. After structural foaming prepared by MC-1 condition, all nano-composite foam exhibited much higher enhancement in G' than those of the structural foam of the corresponding ionomers. At the same time, all nano-composite foams showed significant reduction of α value and decreased up to 66 % as compared to their structural foam from ionomers.

Acknowledgement

This work was supported by the STAS Project (2005-2009) and the MEXT "Collaboration with Local Communities" Project (2005-2009).

References

1. Asai, K., Okamoto, M., Kohji, T.,"Crystallisation Behavior of Nano-composite based on Poly(vinyldene fluoride) and Organically Modified Layered Titanate", Polymer, **49**, 4298-4306 (2008)
2. Ema Y., Ikeya, M., Okamoto M., "Foam Processing and Cellular Structure of Polylactide based Nanocomposites", Polymer, **47**, 5360-5363 (2006)
3. Colton, J.S., Suh, N.P., "The Nucleation of Microcellular Thermoplastic Foams with Additives: PartI: Theoretical Considerations", Polym. Eng. Sci., **27**, 485-492 (1987)
4. Okamoto, M., Nam P.H., Maiti, M., Kotaka, T., Nakayama, T., Takada, M., Ohshima, M., Usuki, A., Hasegawa, N., Okamoto, H., "Biaxial Flow-Induced Alignment of Silicate Layers in Polypropylene/Clay Nanocomposite Foam", Nano Lett., **1**, 503-505 (2000)

RHEOLOGICAL, THERMAL AND MECHANICAL PROPERTIES OF PET NANOCOMPOSITES

A. Alfouzan*, P.D. Coates and T.D. Gough

IRC in Polymer Engineering, School of Engineering, Design and Technology, University of Bradford, Bradford BD7 1DP, UK – a.m.i.al-fouzan@brad.ac.uk, t.gough@brad.ac.uk, p.d.coates@brad.ac.uk

Polyethylene Terephthalate nanocomposites (PET/nanoclay) were prepared by melt mixing using a twin-screw extruder incorporating dual motor feeders. The rheological behaviour of PET with various nanoclays (Somasif MAE, Somasif MTE and Cloisite 25A) at different concentrations (1, 3, 5 and 20%) was studied in the molten state under a wide range of shear rates (0.1 to 5000 s^{-1}). The thermal properties were also investigated for these nanocomposites. Differential Scanning Calorimetry (DSC) measurements were performed to study the crystallinity changes due to the presence of the nanoclays. It has been observed that the clays enhanced the crystallisation temperature and accelerate the crystallisation rate of PET nanocomposites with increasing nanoclay content. Oscillatory tests were conducted on the PET nanocomposites in order to study the effect of adding the nanoclay on the viscoelastic properties of the PET at low shear rates while capillary rheometry was used to study the viscosity behaviour at high shear rate. At low shear rates, an increase in the shear viscosity, storage modulus and loss modulus was observed in the PET nanocomposites with increasing nanoclay content while the viscosity decreased with increasing nanoclay content at high shear rates. It has been shown that the PET nanocomposites do not obey the Cox-Merz rule. Films of each nanocomposite were manufactured using cast film extrusion and tested using uniaxial and biaxial stretching.

Introduction:

Poly (ethylene terephthalate) (PET) is a semicrystalline thermoplastic polyester with good mechanical properties, thermal stability and low permeability to gases. PET has many applications such as beverage and food packaging. The low rate of crystallisation and low modulus lead to limitations in the application of PET as an engineering plastic. Polymer nanocomposites can show improvements in the physical, optical and thermal properties when compared to the pure polymer. The crystallization of PET first involves the formation of nuclei and their subsequent growth whether through heating of amorphous PET from the glassy state or cooling from the melt state. PET has molecular chains connected with aromatic ethyl and ester groups. The twist or torque motion between the PET rigid aromatic segments and the ethyl connector greatly slow down its crystallization rate and thus prolong its processing cycle (Ke et al. (2007)). Pegoretti et al. (2004) manufactured and characterized recycled PET/clay nanocomposites using recycled PET and two montmorillonite clays (Cloisite Na^+ and Cloisite 25A). PET nanocomposites with 1,3 and 5 wt.% were melt compounded with co-rotating screw extruders. Scanning transmission electron microscopy (STEM) was used to evaluate the dispersity of the nanoclays following extrusion.

It was found that the Cloisite 25A gave much better dispersion in the nanocomposite compared to Cloisite Na^+. Wide Angle x-ray Scattering (WAXS) measurements indicated increased lamellar periodicity of Cloisite 25A in the composition which means intercalation occurred. Cloisite 25A also led to a modulus increase in the recycled PET composites. McConnell et al. (2006) studied the effect of various organoclays (MAE, MTE and others) on the mechanical properties of the PET nanocomposites. Processing methods included compounding via single screw extrusion and injection moulding to produce test specimens. It was found that the organoclay enhanced the mechanical performance and it was suggested that the organoclays are acting as nucleating agents to form the crystalline phases in the PET during fibre spinning. Through capillary rheometry their results showed that the nanofiller negatively affected the intrinsic viscosity (IV) of the PET due to thermal degradation. The researchers also used single screw extrusion which may decrease the mixing efficiency compared to twin screw extrusion and lead to increased residence time which in turn could cause thermal degradation. Gurmendi et al. (2006) studied the effect of three types of organoclays to produce PET nanocomposites. Cloisite 15A and 20A have the same surfactant but at different levels thus allowing the effects of surfactant level to be studied.

Cloisite 30B was also studied since it has same level of surfactant as 20A but has a more polar chemical nature. The researchers concluded that a higher surfactant content (Cloisite 15A) does not have a positive effect on the ability of the polymer to intercalate into the clay galleries. This was attributed to the initial interlayer distance of Cloisite 15A being larger which leads to a decrease in the strength necessary to overcome the attractive interlayer force when the polymer is intercalated inside the clay galleries. In addition, it was found, after comparing Cloisite 30B and 20A, that the polarity was favored for intercalation. It was observed that the modulus increased and was independent of the level and nature of the organic modification of the clay which means that the parameters that lead to a large increase in the modulus of elasticity differ from those that lead to a large degree of intercalation. The increasing modulus of elasticity correlates with wide clay dispersion. Anand et al. (2006) used single walled carbon nanotubes (SWNTs) to study their influence on PET crystallization. It was found that SWNTs at low concentration (0.03%) enhance the crystallization temperature during melt cooling by increased it from 200 °C to 210 °C. Calcagno et al. (2006) studied the effect of the organic modifier of clays Cloisite 10A, 15A and 30B on the crystallization and morphology properties of PET nanocomposites. The crystallization temperature (T_c) of the pure PET was lower than that of the PET nanocomposites which possibly means that the nanoclay has a nucleation effect on the PET. It was also found that the PET 30B and PET 10A nanocomposites showed exfoliated and intercalated morphologies owing partially to the polarity of the surfactant and its chemical structure. In additional, they reported that the nanocomposites showed smaller mean spherulite size and more nuclei than for pure PET.

Experimental:

Poly (ethylene terephthalate) (PET) grade T74F9IV080 was supplied by Tergal Industries and has an IV of 0.8 dl/g and density of 1400 kg/m^3. Three nanoclays were used in this study; these being Somasif MAE, Somasif MTE (both from CO-OP Chemical Co., Japan (now CBC Co. Ltd)) and Cloisite 25A from Southern Clay Products.

All the organoclays are chemically modified by the producer to improve dispersibility and miscibility.

An APV MP19TC twin screw co-rotating extruder (L/D = 25, D=19 mm) was used in the compounding process. Dual motor feeders were used to load the clay and PET simultaneously at constant feed rates direct to the screws. Prior to the compounding processes, the organoclay powder was vacuumed dried (R.S. Wolverhampton Ltd) at 80 °C for 20 hours. Meanwhile, the PET pellets were dried by dehumidifier dryer (Motan Luxor CA) at 160 °C for 20 hours. The screws have two mixing zones to allow good mixing and interaction between the polymer and the clay. The temperature profile along the barrel was 240, 260, 265, 265, 260 °C from the feed zone to the die. The extrusion speed was held constant at 100 rpm. The PET nanocomposite pellets were re-crystallized in the vacuum dryer at 160 °C for 20 hours prior casting the PET nanocomposite films. This was achieved using a single screw extruder (Betol 2525, D= 25 mm) with set temperature of 260 °C and quenched between counter-rotating chill rolls at 16 °C. Differential Scanning Calorimetry (TA instrument DSC, Q2000) was used to study the thermal properties of the compounded nanocomposite pellets through heating under a nitrogen atmosphere to 280 °C at a ramp rate of 10 °C/min. The percentage of crystallinity (Xc) were calculated from the following equation:

$$Xc\ (wt\%) = \frac{\frac{(\Delta H_m - \Delta H_{gc})}{(1-x)}}{\Delta H_m^{\circ}}$$

Where $\Delta \mathbf{H_m}$ is the enthalpy of melting, $\Delta \mathbf{H_{gc}}$ is the enthalpy of crystallization from the glassy state, **x** is the organoclay weight fraction and $\Delta \mathbf{H_m}^{\circ}$ of 140 J/g is the enthalpy of melting for a 100% crystalline PET. [Minkova (2003), Kok (2009)]

Rheological properties, at 260 °C, of virgin PET (vPET), extruded virgin PET (ext. vPET) and PET/clay nanocomposites were measured using a capillary rheometry (Rosand RH10) for shear rates from 50 to 5000 s^{-1} and oscillatory rheometry (Anton Paar MCR 301 with 25mm diameter parallel plates). All samples were dried in a vacuum oven for 20 hours at 150 °C. Measurements of tensile mechanical properties of the PET nanocomposites films (thickness 110 ± 10 µm) were performed on a Instron 5564 using a cross head speed of 5 mm/min.

Some abbreviations were used in DSC results explanations are described in table 1.

Table 1: DSC abbreviation keys

Key	Description
T_g	Glass transition temperature
T_{gc}	Crystallisation temperature from the solid
T_{gc1}	Onset crystallisation temperature from solid
T_{mc}	Crystallisation temperature from the melt
T_{mc1}	Onset crystallisation temperature from melt
T_m	Melting temperature.
$t^{1/2}$	Crystallisation half-time

Results and discussion:

Thermal properties:

Differential Scanning Calorimetry (DSC) was used to study the effect of the nanoclays on the crystallisation temperatures, glass transition temperature, crystallisation half-time and the crystallinity of the PET nanocomposite (NC). It was also used to check the quality of the process mixing through analysis of many samples from the same run to check for consistency.

Figure 1 shows the relation between the PET NC at various nanoclay contents (0, 1, 3, 5 and 20%) with the thermal analysis temperatures. It is clear that the crystallisation temperature, T_{gc}, decreases with nanoclay content while T_{mc} increases with the addition of clay. A decrease in the glass transition temperature, T_g, is also apparent and can be expected since the clay increases the flexibility between the PET molecules which leads to easier relative motion. It is very important to avoid a big effect on the T_g especially if the PET product is used in hot countries where the temperature variation can be of the order of 60 °C which may lead to deformation of the product.

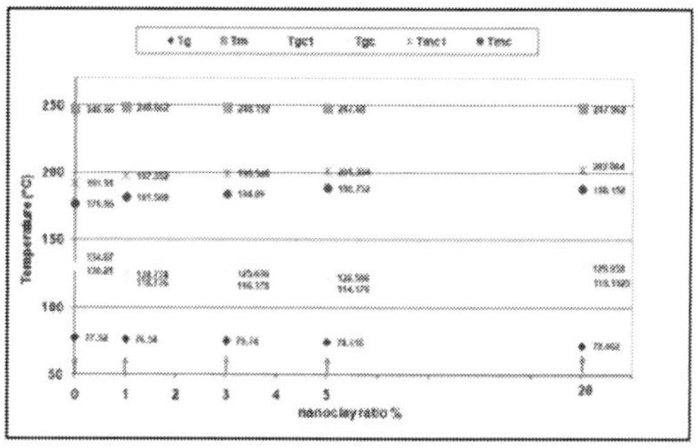

Figure 1: DSC average results for PET/MTE nanocomposites for varying MTE content

The small increase in the melt temperature, T_m, could be a result of the good dispersion of the nanoclay as this can lead to the PET segments forming 'more perfect' crystals during the crystallisation as observed by Liu et al. (2003) in their study of PET/silica nanocomposites. Mucha et al. (2000) attribute the increase in T_m in PP/carbon black to some hindrance at the beginning of the melting process. This increase can be neglected because the melt processing temperature of the PET is 15 °C or more over T_m.

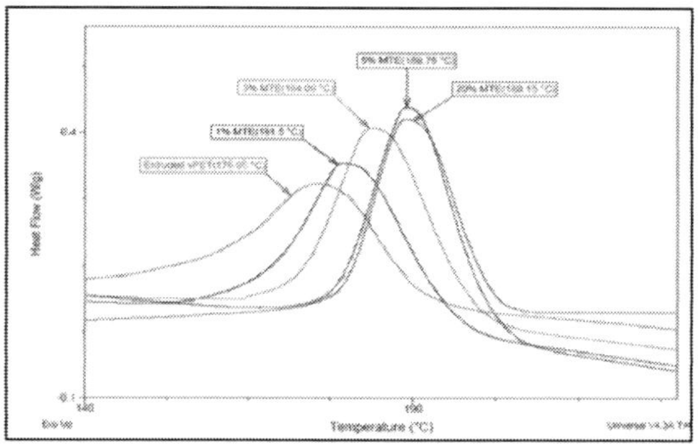

Figure 2: T_{mc} curves obtained from DSC for PET/MTE nanocomposites at various contents.

No significant change is evident in the degree of crystallinity (as table 2 shows) though the crystallisation exothermal peak (for T_{mc} at figure 2) is gradually shifted to higher temperatures with increasing clay content. This is indicative of the enhancement of the crystallization rate of the PET with addition of clay. In addition the peaks get sharper (decreasing $t^{1/2}$) which indicates the crystallization rate is increased (as figure 3 shows). There are many reasons for this enhancement of the crystallisation rate in nanocomposites. It is possible that the structure of the nanoclay can help the molecules of PET to stack on each other to grow into crystallites or that the clay may act as an effective heterogeneous nucleating agent.

Table 2: Shows the degree of crystallinity of vPET and PET nanocomposites

Sample name	Xc (%)
Ext. vPET	10.06
PET/MTE (1%)	11.61
PET/MTE (3%)	11.4
PET/MTE (5%)	11.12
PET/MTE (20%)	8.41
PET/MAE (1%)	8.65
PET/MAE (3%)	10.29
PET/MAE (5%)	10.1
PET/MAE (20%)	8.96
PET/Cloisite25A (1%)	9.42
PET/Cloisite25A (3%)	11.36
PET/Cloisite25A (5%)	9.76
PET/Cloisite25A (20%)	10.73

The results of thermal properties for PET nanocomposites for all nanoclay types at 1(wt.%) are shown in figure 4. Negligible changes in the Tg and Tm are apparent when comparing with non-filled PET while all nanoclays seem to act as a nucleation enhancer thus leading to the results for the crystallisation temperatures either from solid or melt.

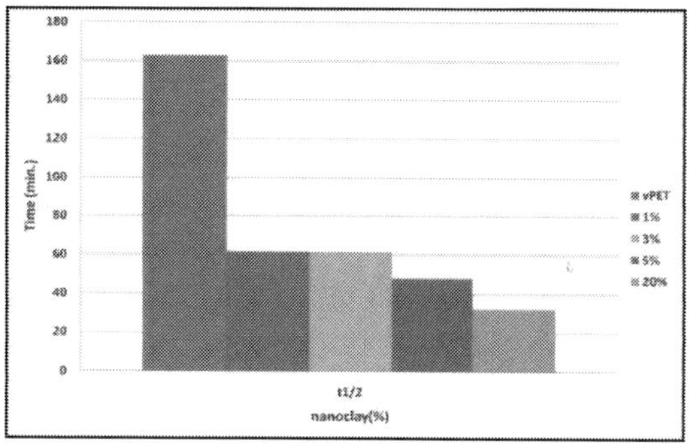

Figure 3: Crystallisation half-time for PET/MTE NC for crystallisation from melt states (T_{mc}).

All clays enhanced the T_{gc} to a similar extent reducing the value by around 10 °C compared to the non-filled PET. Between the clays, MTE shows lower enhancement for the T_{mc} while PET/Cloisite25A and PET/MAE show good results (6 °C and 11 °C higher than PET/MTE and non-filled PET respectively).

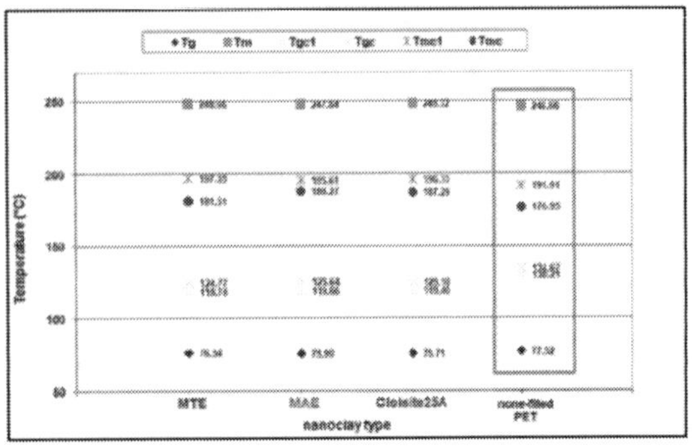

Figure 4: DSC results of PET nanocomposites at 1% for all nanoclay types

Rheological properties:

Capillary and oscillatory rheometers were used to study the behaviour of the material viscosity with increasing the shear rate to aid understanding of material behaviour during melt processing.

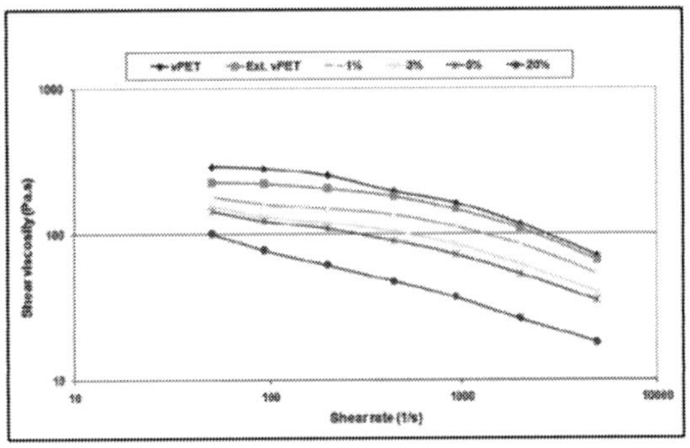

Figure 5: Shear viscosity variation of PET/MAE nanocomposites with shear rate

Figure 5 reveals that the shear viscosity decreases systematically with addition of the nanoclays. For example, 20% MAE loading gives a three-fold decrease in shear viscosity compare to 1% MAE loading at a shear rate of $5000s^{-1}$. The shear viscosity also decreases with increasing shear rate. This is observed for all three nanoclays studied.

Figure 6 shows the shear rate dependence of complex viscosity for PET/MTE nanocomposites at different concentrations. Large increases in complex viscosity are observed between 1% and 3% loading depending on the particular nanoclay utilised.

The base PET (vPET), the extruded base PET (ext. vPET) as well as the PET nanocomposites (1 and 3%) exhibit almost Newtonian behaviour for low shear rate ranges.

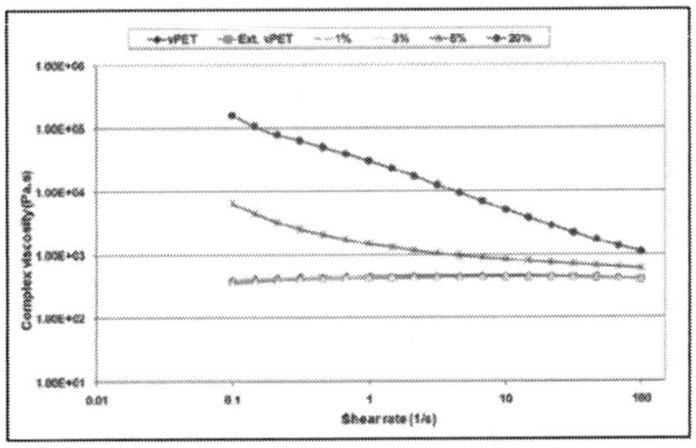

Figure 6: Complex viscosity variation of PET/MTE nanocomposites with low shear rate

The PET nanocomposites exhibit shear thinning behavior resulting from the random orientation of entangled molecular chains in the polymer nanocomposites during the applied shear deformation. The formation of network-like or interconnected structures in the PET polymer-nanoclay and nanoclay-nanoclay interphases causes the viscosity of the PET nanocomposites to increase with nanoclay content, this being clearer at the lower shear rates.

The relationship between the shear viscosity and the clay content for different PET nanocomposites at a shear rate 100 s^{-1} (typical of extrusion) is shown in figure 7 clearly showing a decrease in shear viscosity with increasing addition of nanoclay.

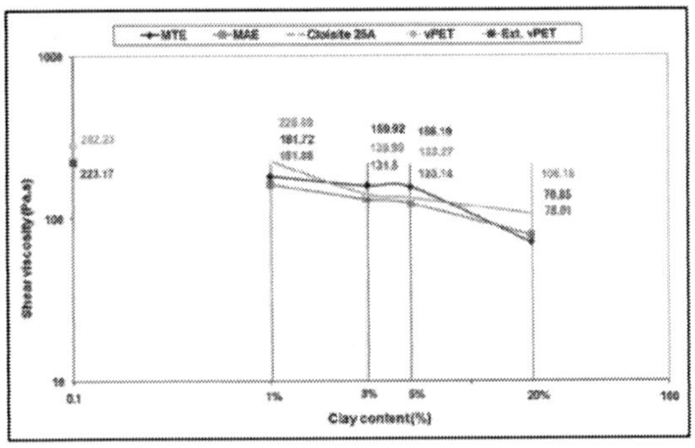

Figure 7: Variation in shear viscosity at 100 s^{-1} for different PET nanocomposites.

For many polymer melts, it has been shown that there is a close relationship between the shear viscosity $\eta(\dot{\gamma})$ and complex viscosity $\eta^*(\omega)$. This is known as the Cox-Merz rule [Mezger (2006)]:

$$\eta(\dot{\gamma}) = \left|\eta^*(\omega)\right|$$

Kelly et al. (2009) observed that polypropylene exhibits this correspondence and Han et al. (1995) observed a good agreement with Cox-Merz rule in polystyrene homopolymer. The Cox-Merz rule is not useful for complex structured fluids, such as liquid crystalline polymers, crosslinked or gels.

Many researchers have shown that the Cox-Merz rule does not apply for polymer nanocomposites and blends [Fornes el at. (2001)]. Han et al. (1995) studied the effect of flow geometry on the rheological properties of PS/PMMA (poly(methacrylate)) blends. It was found that the logarithmic plots of shear rate versus shear viscosity obtained by a capillary rheometer do not overlap with those obtained by a cone-plate rheometer – this being due to the difference in morphological states of PS/PMMA blends between the non-uniform shear flow in a capillary rheometer (due to the effect of exit and entrance) and the uniform shear flow in cone-plate rheometer. Krishnamoorhi and Yurekli (2001) observed non-adherence to this rule in polystyrene-polyisoprene block copolymer nanocomposites. The relation also breaks down for glass fibre reinforced nylon 6 and polypropylene as Bailey et al. (1992) reported. Nakajima et al. (1977) reported deviations from the Cox-Merz rule in carbon black/rubber composites. It was observed that the complex viscosity obtained by dynamic shear measurements was higher that the shear viscosity measured by capillary rheometer.

Figure 8 illustrates the shear viscosity variation with shear rate over a wide shear rate range for the PET nanocomposites. The results match the results obtained by others [Nakajima et al. (1977), Krishnamoorti et al. (2001), Bailey et al. (1992)] showing that the Cox-Merz rule does not hold for PET nanocomposites.

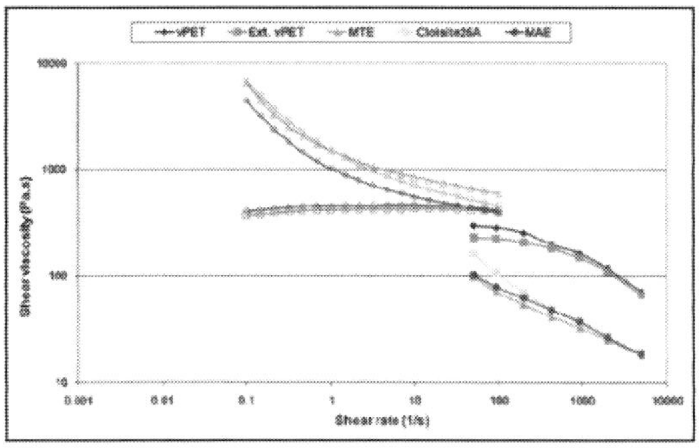

Figure 8: Plots of log shear rate vs. log of shear viscosity for PET nanocomposites at 5% nanofiller concentration, where filled symbols denote capillary data and open symbols denote parallel plate data.

Figures 9 and 10 show increases in the storage modulus and loss modulus respectively for the PET nanocomposites when compared to vPET over the whole shear rate range which suggests that the interaction between the nanoclay-polymer and nanoclay-nanoclay interfaces can lead to the formation of crosslinking structures in the polymer nanocomposites. Similar results have also been reported by Potschke et al. (2002) and Kim et al. (2007, 2009).

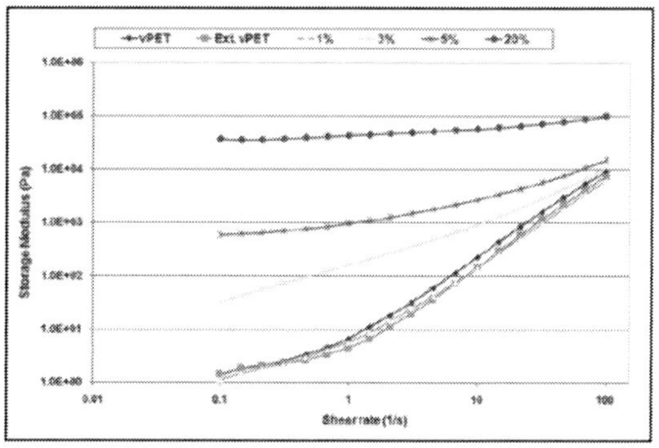

Figure 9: Shear rate (s^{-1}) vs. Storage Modulus (Pa) for PET/Cloisite 25A nanocomposites

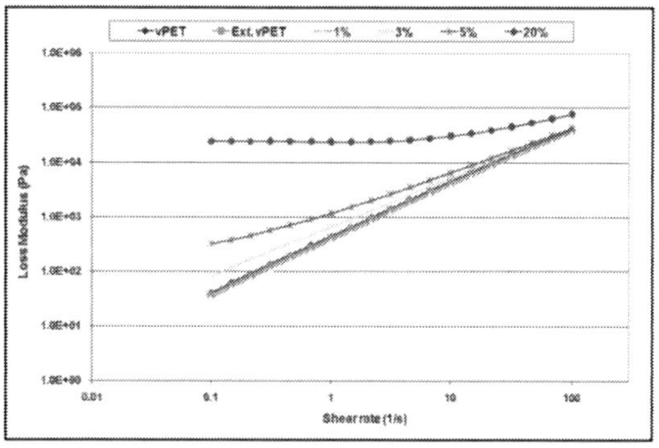

Figure 10: Shear rate (s^{-1}) vs. Loss Modulus(Pa) for PET/Cloisite 25A nanocomposites

Mechanical properties:

The variation in mechanical properties of the PET nanocomposites with addition of the nanoclays are shown in figure 11.

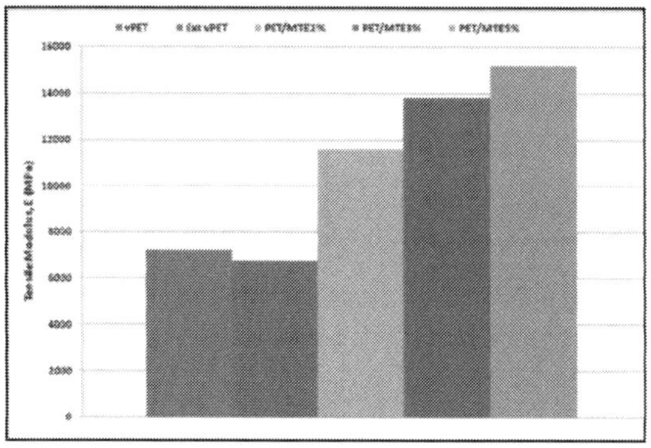

Figure 11: Tensile modulus for vPET, ext.vPET and PET/MTE nanocomposites (5mm/min)

The tensile modulus of the base PET slightly decreases after extrusion for the vPET. Adding 1% of nanoclay increases the tensile modulus by 60% and by 100% when adding 5%. This improvement in tensile modulus is indicative of good dispersion and thus strong interfacial interaction between the PET and the nanoclay. Meng et al. (2008) studied the mechanical properties of modified CNT/PA nanocomposites and found higher tensile modulus than those of unmodified CNT/PA nanocomposites due to the good interfacial adhesion and uniform dispersion in the modified CNT/PA nanocomposites.

Conclusion:

Rheological, thermal and mechanical properties of three nanoclays compounded into PET at a range of concentrations have been well-characterised.

It has been shown that the crystallization rate of PET can be significantly enhanced through the employment of these clays while mechanical (tensile) properties are also significantly enhanced.

Rheological testing over a wide range of shear rates shows the breakdown of the Cox-Merz rule for PET nanocomposites.

Preliminary work also indicates a significant improvement in gas barrier properties thus implying that nanoclays provide a viable route to improved packaging when employed as nanofillers within a PET matrix. These results will be presented at the conference.

References:

Anoop Anand, K., Agarwal, U.S. and Rani, J., "Carbon nanotubes-reinforced PET nanocomposite by melt-mompounding", Polymer v. 47 pp. 3976-3980 (2006).

Bailey R.S., Bellamy A.M., Groves D.J., Stocks D.M. and Young R.C., Theor. Appl. Rheol. v. 2 p. 838 (1992).

Calcagno, C.I.W., Mariani, C.M., Teixeira, S.R. and Mauler, R.S., " The effect of organic modifier of the clay on morphology and crystallization properties of PET nanocomposites" Polymer 48 pp. 966-974 (2007).

Forness T.D., Yoon O.J, Keskkula H. and Paul D.R, "Nylon 6 nanocomposites: the effect of matrix molecular weight", Polymer, v. 42 9929-9940 (2001).

Gurmendi, U., Eguiazabal, J.I. and Nazabal, J., "Structure and properties of nanocomposites with a poly(ethylene terephthalate) matrix" Macromolecular Mater. Eng. Volume 292, Issue 2 , Pages 169 – 175 (2007).

Han, J.H., Chin Choi-Feng, De-Jie Li and Chang Dae Han, "Effect of flow geometry on the rheology of dispersed two-phase blends of polystyrene and poly(methyl methacrylate)", Polymer v. 36 (12), pp. 2451-2462 (1995).

Ke, Y.C., Wu, T.B. and Xia, Y.F., "The nucleation, crystallization and dispersion behavior of PETe monodisperse SiO2 composites", Polymer 2007, 48, 3324.

Kelly, A.L., Gough, T., Whiteside, B.R. and Coates, P.D., "High shear strain rate rheometry of polymer melts", J. App. Poly. Sci., v. 114, pp. 864-873 (2009).

Kim, Jun Young, "The effect of carbon nanotube on the physical properties of poly(butylene terephthalate) nanocomposite by simple melt blending", J. App. Poly. Sci., v. 112, pp. 2589–2600 (2009).

Kim, Ju Young, Park, Hawe Soo and Seong, Hun Kim, "Multiwall-carbon-nanotube-reinforced poly(ethylene terephthalate) nanocomposites by melt compounding", J. App. Poly. Sci., v. 103, pp. 1450–1457 (2007).

Krishnamoorti R. and Yurekli K. "Rheology of polymer layered silicate nanocomposites", J. Chem. Phys., v. 114 (11), pp. 4968-4973 (2001).

Liu, Z., Chen, K. and Deyue, Y, "Crystallization, morphology, and dynamic mechanical properties of poly(trimethylene terephthalate)/clay nanocomposites" European Polymer Journal, v.39, 12, p. 2359 (2003).

McConnell, D.C., Hornsby, P. R., Lew, C.Y. and Qua, E.H., "Structure-property relationships in PET nanocomposite fibres" ", SPE ANTEC Tech. Papers pp. 387-392 (2006).

Meng, H., Sui, G.X., Fang, P.F. and Yang, R., "Effects of acid- and diamine-modified MWNTs on the mechanical properties and crystallization behavior of polyamide 6", Polymer, v. 49, p. 610 (2008).

Mezger T.G., "The rheology handbook (2nd edition)", Vincentz Network GmbH & Co. (2006).

Minkova, L. and Peneva, Y., "Microhardness of PET-based liquid crystalline copolyesters: influence of the microstructure", Polymer, v. 44, pp. 6483–6488 (2003).

Mucha, M.; Marszalek, J.; Fidrych, A., "Crystallization of isotactic polypropylene containing carbon black as a filler", Polymer, v. 41, p. 4137 (2000).

Nakajima N., Bowerman H.H., Collins W.A., "Nonlinear viscoelastic behavior of butadiene-acrylonitrile copolymers filled with carbon black", J. Appl. Polym. Sci., v. 21, p. 3063 (1977).

Pegoretti, A., Kolarik, J., Peroni, C. and Migliaresi, C., "Recycled poly(ethylene terephthalate)/layered silicate nanocomposites: morphology and tensile mechanical properties", Polymer, v. 45, p. 2751 (2004).

Potschke, P., Fornes, T.D. and Paul, D.R., "Rheological behavior of multiwalled carbon nanotube/polycarbonate composites", Polymer, v. 43, p. 3247 (2002).

Soon, K.H., Harkin-Jones, E., Rajeev, R.S., Menary, G., McNally, T., Martin, P.J. and Armstrong, C., "Characterisation of melt-processed poly(ethylene terephthalate)/synthetic mica nanocomposite sheet and its biaxial deformation behavior", Polym. Int., v. 58, pp. 1134-1141 (2009).

ON-LINE OPTICAL MONITORING OF THE CRYSTALLIZATION OF PA6/MMT NANOCOMPOSITE DURING INJECTION MOLDING

C. A. G. Beatrice[1], A. B. Bonel[2], M. M. Favaro[3] and R. E. S. Bretas[1]*

[1]*Programa de Pós-Graduação em Ciência e Engenharia de Materiais, Universidade Federal de São Carlos, São Carlos, SP, Brazil – cagbeatrice@gmail.com – *bretas@ufscar.br*
[2] *Department of Materials Engineering, Universidade Federal de São Carlos, São Carlos, SP, Brazil – alan_bonel@yahoo.com.br*
[3]*Department of Chemistry, Universidade Estadual de Campinas, Campinas, SP, Brazil – mmfavaro@gmail.com*

Abstract – This work had as main objectives to study the flow induced crystallization of nanocomposites of polyamide 6 (PA6) with a lamellar nanoclay (MMT) and to optically monitor "on-line" the crystallization during the injection molding. The nanocomposite had an exfoliated and percolated network, with the nanoclay's lamellas homogeneously distributed thru the polymeric matrix. From the flow-induced crystallization analyses, it was observed that at low shear rates (0.01 s^{-1}), independent of the cooling rate, the times for crystal growth were smaller in the nanocomposite than in the pure PA6; however, when the shear rate was high (0.1 s^{-1}), also independent of cooling rate, the nanocomposite needed a higher time for crystal growth than the pure PA6. Two laser sources were used: one of λ=405 nm and the other of λ=632 nm; the use of a laser with a smaller wavelength did not bring any advantages, because a superimposition of scattering and light absorption by the polymer occurred.

Introduction

Polymeric nanocomposites usually have better mechanical, optical and transport properties than the pure matrixes. The amount of nanofiller that is normally used is between 1 and 5 wt%, extremely low when compared to the traditional composites which use concentrations between 20 and 30wt%. However, the optimization of the nanocomposite's properties requires a complete dispersion and homogeneous distribution of the nanofiller thru the polymeric matrix, allied to a high interfacial adhesion between the macromolecular chains and the nanoparticles (Alexandre et al., 2000; Ray and Okamoto, 2003; Zeng et al., 2005).

It is known that the end properties of a nanocomposite is directly correlated to its processing (Dennis et al., 2001; Bousmina, 2006; Rao and Pochan, 2007); in injection molding, for example, the high deformation and temperature gradients can induce orientation of the nanofillers and change the crystallization kinetics of the materials. This orientation can increase the elastic module, while the change in crystallinity and crystal morphology can change the impact and transport properties of the end product. Highly exfoliated structures of nanoclay's lamellas and thick crystalline lamellas can increase the transport properties of the nanocomposite, besides increasing its thermal stability.

Therefore, numerous works have been made to understand the influence of nanofillers on the crystallization kinetics and polymeric morphology under quiescent conditions (Maio et al., 2004; Razavi-Nouri et al., 2009). Usually the nanofillers act as nucleating agents increasing the global crystallization rate and changing the size and shape of the crystals. However, under flow, the polymeric crystallization is modified; for example, in some cases (Moretti et al., 2010; Favaro et al., 2010) a retardation effect on the polymeric crystallization promoted by the nanoclay is observed. Thus, the accelerating or retarding effect seems to be influenced by the flow gradient.

This work had as main objectives to study the crystallization kinetics of a nanocomposite under flow and to compare the results with the ones obtained from an optically monitoring of this crystallization during injection molding.

Experimental

Materials

A PA6, donated by Rhodia Engineering Plastics from Brazil (trade name Technyl C 402 M) was used as polymer matrix. This resin had residual monomer (0.60 wt %), density of 1.14 g/cm^3, melting temperature of 222°C, melt flow index of 2.5 g/10 min (235°C/2.16 kg), a concentration of carboxyl terminal groups of 35 Meq/kg, and a concentration of amine terminal groups also of 35 Meq/kg.

The nanoclay was an organically modified montmorillonite (MMT) Cloisite® 30B (C30B) from Southern Clay Products Inc with density of 0.36 g/cm3. The original gallery's distance was 1.85 nm. The modification was done with a surfactant, a quaternary ammonium salt, which has unsaturated aliphatic chains with two hydroxyl groups, named diethanol-methyl-tallow quaternary ammonium salt. The surfactant concentration was 900 Meq/kg.

Preparation and characterization of the Nanocomposite

The PA6/MMT nanocomposite was prepared in a corotating twin screw extruder, as described in Beatrice et al (2010). The final composition of the PA6/MMT nanocomposite was 95/5 wt %. The extent of intercalation of the nanocomposite and the formation of a percolated network were evaluated by WAXS, transmission electron microscopy (TEM) and dynamic rheological properties, as also described in Beatrice et al. (2010). It was concluded that the nanoclay's lamellas were exfoliated and well distributed thru the PA6 matrix.

Isothermal Flow-Induced Crystallization Studies by Rheometry

Studies of the flow-induced crystallization were done using a strain-controlled rheometer (ARES) from Rheometrics Scientific, with parallel-plates geometry (25 mm diameter) and gap of 1.0 mm. The procedure as described by Farah and Bretas (2004) was used. The material was melted at 250ºC and after 3 min at this last temperature, the sample was quenched down to the crystallization temperature T_C = 190ºC. After temperature stabilization, shear rates of 0.01 and $0.1 s^{-1}$ were imposed to the sample and the shear tension (τ) was monitored as a function of time. From these measurements, a t_{growth_ARES} time was calculated; this time was defined as the time at which the shear tension increased abruptly.

Optical Monitoring During Injection Molding

The extruded nanocomposite was injection molded using an Arburg 270V injection molding machine, with clamping force of 300 kN, screw diameter of 25 mm and a disk cavity where the optical sensor was installed as already described in (Moretti et al., 2010; Favaro et al., 2008) and shown in Figure 1a. Two different laser sources, with wavelength (λ) of 405 and 632 nm were used; the light of the laser source passed thru the sapphire window and the mold cavity and it was collected in a collector fiber, as shown in Figure 1b. The transmitted (or collected) light flux $\Phi_{transmitted}$ is given by Eq. 1:

$$\Phi_{transmitted} = \Phi_{initial} - \Phi_{absorbed} - \Phi_{scattered} \qquad \text{(eq. 1)}$$

where Φ = IA = light flux, I = light intensity vector normal to the area A, $\Phi_{initial}$ = initial light flux, $\Phi_{absorbed}$ = absorbed light flux and $\Phi_{scattered}$ = scattered light flux.

The mold temperature (T_w) was changed (30, 63 and 96ºC) while the other injection-molding parameters (injection temperature, T_{inj} = 260ºC, packing pressure, P_{emp} = 400 bar, and flow rate, Q = 35 cm^3/s) were kept constant; the packing time (t_{emp}) was also changed depending on the condition.

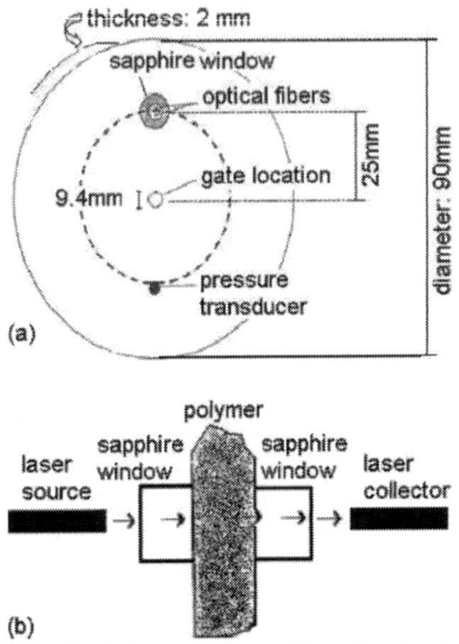

Figure 1 – (a) Scheme of the mold cavity showing the position of the optical fiber and cavity dimensions. (b) Scheme of the light path inside the cavity mold.

Morphological Characterization of the Injection Molded Samples

The morphological characterization of the injection molded samples was done in the region around the sapphire window. A block of 1.5 cm of length, 0.6 cm of width and 2 mm of thickness was cut. This block was microtomed along the thickness with a MICROM microtome, model HM360, at room temperature, producing slices 2 mm wide and 10 µm thick. To observe the morphology of the slices a polarized light optical microscope (PLOM) DMRX, from Leica coupled to a video camera and a PC was used.

Results and Discussion

Isothermal Flow-Induced Crystallization by Rheometry

Early work (Beatrice et al., 2010) showed that the post-polymerization of the residual monomer in the pure PA6 produced more branched chains than linear ones, while in the nanocomposite, part of this monomer reacted first with the nanoclay's surfactant avoiding the complete formation of branching. Figure 2 shows the shear tension τ as a function of time after the application of a shear rate for the pure PA6 and the PA6/MMT nanocomposite. A standard behavior is observed: during the first seconds of the test, the shear tension is constant or slightly increases (Region A); however, after certain period, the tension begins to increase abruptly (Region B). This behavior was slightly different from the one found by Farah and Bretas (2004) in their studies with pure PP.

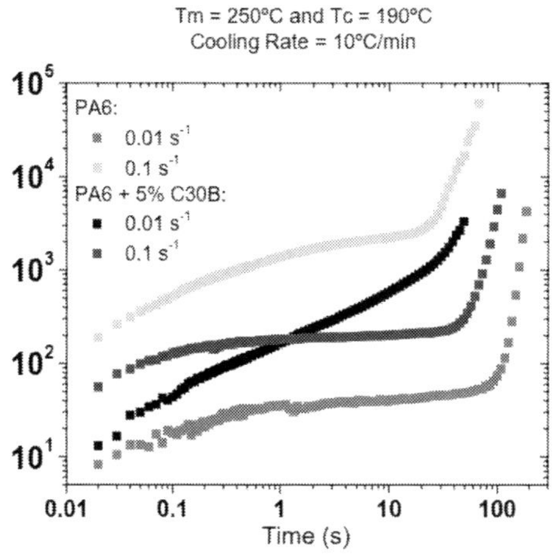

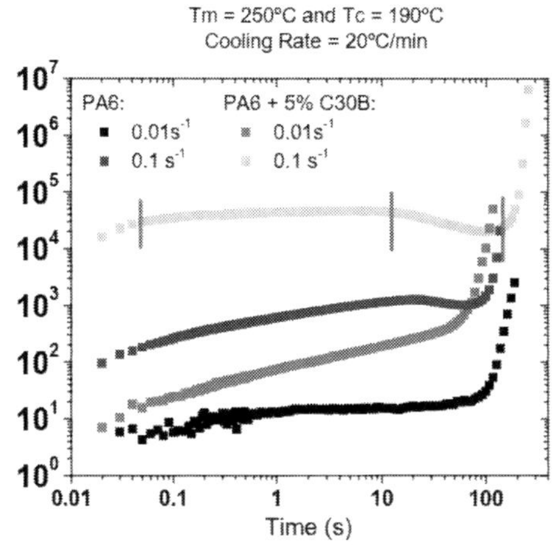

Figure 2 – Shear tension as a function of time for PA6 and PA6 + 5%C30B: (a) 10°C/min and (b) 20°C/min.

The authors showed that the flow-induced crystallization of PP had four distinct regions in which τ behaved differently with respect to time: (i) Region 1, at low shear times, where τ was constant or slightly decreased with time (the slight decrease was attributed to melt disentanglement); (ii) Region 2, at intermediate times, where τ increased with time due to the beginning of crystallization (nucleation); (iii) Region 3, at high times, where τ again decreased with time (the decrease was attributed to the heat release during crystallization which increased the local temperature, decreasing the viscosity and the shear stress; (iv) Region 4 (or B), at very high times, where τ increased again with time, abruptly, due to crystal growth. It was assumed that the induction crystallization time t_i was the time where region 2 begun. However, PA6 has a lower viscosity (η_0 = 1474 Pa.s^{-1}) than PP (η_0 = 2500 Pa.s^{-1}); therefore, regions 2 and 3 were not observed. Thus, only the time when there was an abrupt increase of tension (crystal growth) was calculated and named t_{growth_ARES}.

Table 1 shows the measured t_{growth_ARES}. When the shear rate was low (0.01s^{-1}), independent of the cooling rate, the times for crystal growth were smaller in the nanocomposite than in the pure PA6; however, when the shear rate was high (0.1s^{-1}), also independent of cooling rate, the nanocomposite needed a higher time for crystal growth than the pure PA6.

Table 1 – Kinetics parameters of the flow-induced crystallization.

Tm (°C) / Tc (°C) / shear rate (s^{-1}) / cooling rate (°C/min)	t_{growth_ARES} (s)	
	PA6	PA6 + 5% C30B
250 / 190 / 0.01 / 10	84.6	27.8
250 / 190 / 0.01 / 20	91.7	61.6
250 / 190 / 0.1 / 10	20.2	41.4
250 / 190 / 0.1 / 20	84.7	141.5

The flow-induced crystallization of a polymer occurs when fibril nucleus are formed by chain orientation; the higher the deformation rate, the higher the chain orientation and the lower the induction time for crystal growth. Analyzing the behavior of the pure PA6 it can be observed that, independent of cooling rate, the time for crystal growth at high shear rates was lower than the time at low shear rates. It is also observed that the low shear rates curves have an initial constant viscosity region while the high shear rates curves have an initial increasing viscosity region; that is, the high shear rates accelerated the post-polymerization process in the PA6, resulting from the residual monomer that did no react during the extrusion. When nanoclay lamellas are present, however, chain orientation will take more time than in a pure polymer because the percolated network will difficult the chains alignment due to the strong linking between the chains and the nanoclay's surfactant. Then it is expected that the time for crystal growth will be higher than in a pure polymer as found in other works (Favaro et al., 2010). This behavior was observed at high shear rates, independent of cooling rate; however at low shear rates, t_{growth_ARES} of the nanocomposite was lower than of the pure PA6. The curves of the nanocomposite at low shear rates had an initial increasing viscosity region; that is, reactions between the residual monomer and the nanoclay' surfactant were taken place and accelerating the nucleation process. Probably, the residual monomer did not have time to react with the nanoclay′ surfactant at high shear rates, reacting instead with the PA6.

Optical Monitoring During Injection Molding

Three standard curves of light intensity versus time can be obtained during the optical monitoring of materials when a laser source of 632 nm is used and are shown in Figure 3. Changes between I_a (light intensity when the mold was open) and I_o (light intensity when the mold was closed) are observed for the three kind of materials.

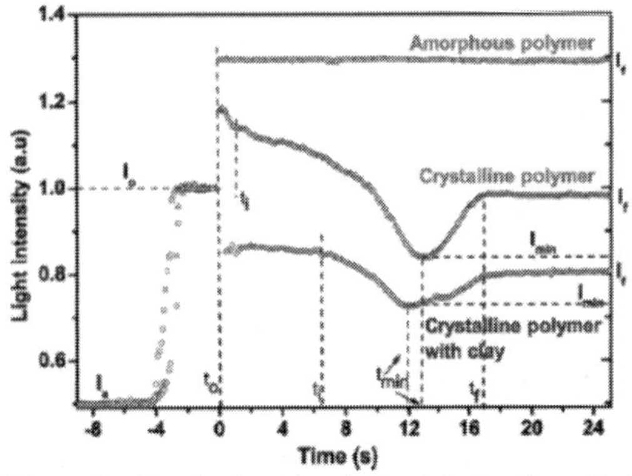

Figure 3 – Standard curves obtained during the optical monitoring: I_a = light intensity obtained with the open mold, I_o = light intensity obtained with the closed mold, t_o = initial monitoring time, t_i = induction time for the detection of scattering structures, I_{min} = minimum light intensity during the monitoring, t_{min} = time to reach I_{min}, I_f = final light intensity (final plateau) and t_f = final crystallization time (Moretti et al., 2010).

The standard curve of an amorphous polymer is a straight line because no scattering and absorbing structures are present. On the other hand, the standard curve of a crystalline polymer shows a continuous decrease after I_o due to crystallization. The growing crystals created new interfaces (increasing the interfacial area) which scattered (and absorbed) light, decreasing the signal from I_o to a minimum value (I_{min}); after attaining a given size, however, the spherulites begun to impinge one with another diminishing the interfacial area (and the light scattering) and increasing the light intensity signal. When the crystallization ended, a plateau (I_f) is attained.

The standard curve of a crystalline polymer with nanoclay is slightly different; it has the same continuous decrease of the light signal from I_o to I_{min}, but the valley of the curve was shallower than of the pure polymer.

To be able to compare all the collected signals, all the curves were normalized with respect to I_o. The time for the beginning of crystallization, or induction time t_i, was defined as the time when the first scattering (and absorbing) structures were detected by the optical sensor, that is, when the light intensity begun to decrease after t_o while the time for the end of crystallization, t_f, was assumed to be the beginning of the I_f plateau. Figures 4 and 5 show the standard curves obtained with the different lasers, while Table 2 shows the induction and end times.

It can be observed that, independent of the mold temperature and laser wavelength, the induction times were similar for the pure PA6 and the nanocomposite. Also, the laser with 405 nm detected similar end crystallization times for both materials; however, the laser with 632 nm detected a crystallization ending time higher for the nanocomposite than for the pure PA6.

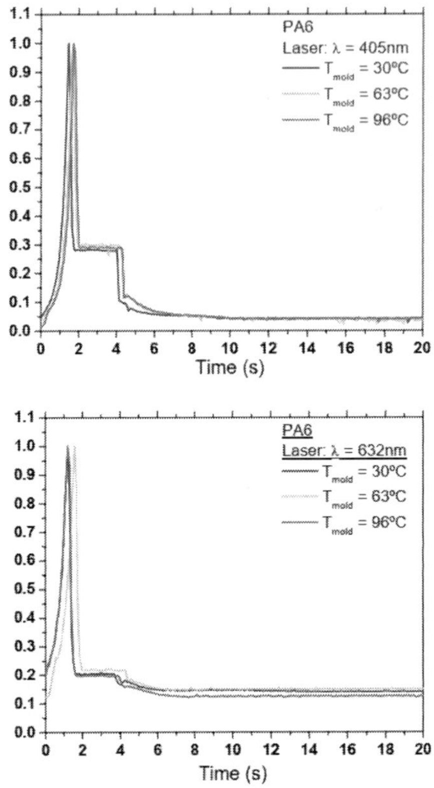

Figure 4 – Light intensity curves as a function of the mold temperature, Tw, for the pure PA6, using two different lasers.

Table 2 – Values of t_i and t_f.

λ_{Laser} (nm)	T_w (°C)	PA6		PA6 + 5% C30B	
		t_i(s)	t_f(s)	t_i(s)	t_f(s)
405	30	1.7	11.1	2.0	12.7
	63	2.0	12.0	1.9	14.3
	96	2.1	14.1	1.9	14.9
632	30	1.6	6.7	2.0	10.0
	63	2.0	7.6	2.1	10.6
	96	1.6	8.8	2.1	12.3
T_{inj} = 260°C, P_{emp} = 400 bar and Q = 35 cm^3/s.					

The laser with λ = 405nm was used because it was expected that a lower wavelength would allow to detect smaller scattering structures than a laser with λ = 632nm; therefore, it was expected that the induction times of the laser with λ = 405nm would be smaller than of the laser with λ = 632nm. However, the measurement of the time at which the scattering and absorbing structures are detected by the optical sensor depends also on the polymer light absorption; as the laser wavelength decreases, smaller scattering particles can be detected, but the polymer can also absorb light. Thus, both phenomena will superimpose. Therefore, the results at 405 nm are a combination of scattering and absorption of light by the polymer and the nanoclay's tactoids.

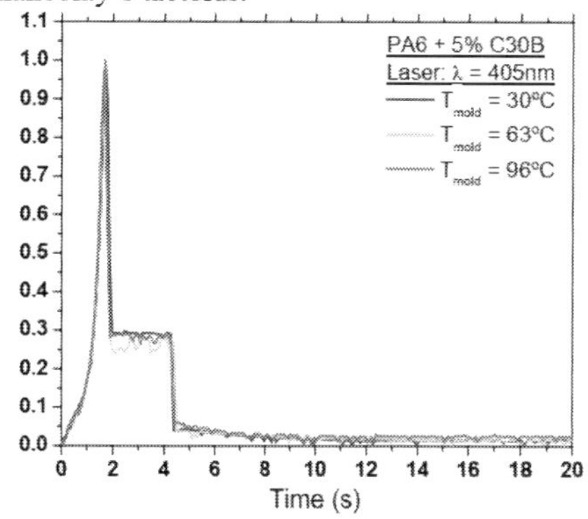

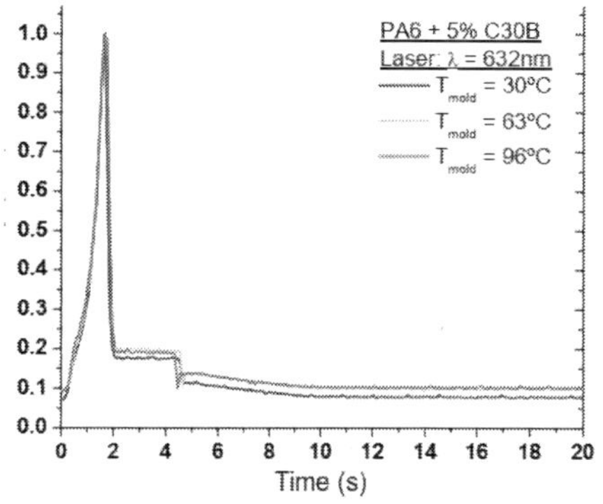

Figure 5 – Light intensity curves as a function of the mold temperature, Tw, for the PA6 + 5% C30B, using two different lasers.

Morphological Characterization of the Injection Molded Samples

The morphologies of the pure PA6 and the nanocomposite are shown in Figure 6.

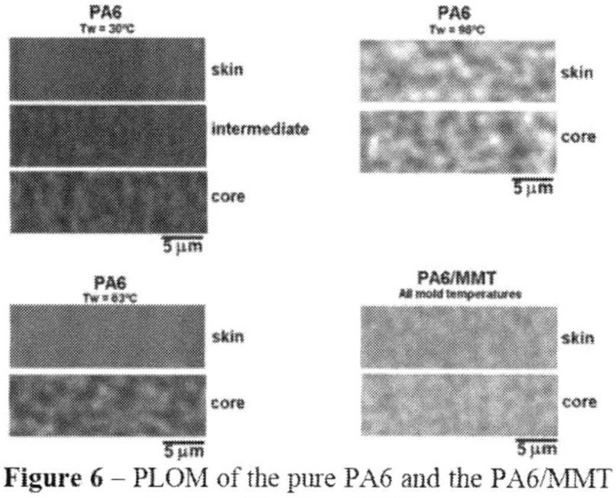

Figure 6 – PLOM of the pure PA6 and the PA6/MMT nanocomposite. Magnification: 500X.

Conclusions

The flow induced crystallization of nanocomposites of PA6/MMT showed that reactions between the residual monomer and the surfactant can accelerated the beginning of the crystallization under flow.

The use of a laser with a smaller wavelength did not bring any advantages, because a superimposition of scattering and light absorption by the polymer occurred.

Acknowledgements

The authors thank FAPESP for the financial support and Rhodia Engineering Plastics from Brazil for the PA6 donation.

References

ALEXANDRE, M.; DUBOIS, P. "Polymer-layered silicate nanocomposites: preparation, properties and uses of a new class of materials". Materials Science and Engineering, v. 28, p. 1 – 63, 2000.

BEATRICE, C. A. G.; BRANCIFORTI, M. C.; ALVES, R. M. V. and BRETAS, R. E. S. "Rheological, mechanical, optical, and transport properties of blown films of polyamide 6 / residual monomer / montmorillonite nanocomposites". Journal of Applied Polymer and Science, v. 116, p. 3581 – 3592, 2010.

BOUSMINA, M. "Study of intercalation and exfoliation processes in polymer nanocomposites". Macromolecules, v. 39, p. 4259 – 4263, 2006.

RAO, Y.; POCHAN, J. M. "Mechanics of polymer-clay nanocomposites". Macromolecules, v. 40, p. 290 – 296, 2007.

DENNIS, H. R.; HUNTER, D. L.; CHANG, D.; KIM, S.; WHITE, J. L.; CHO, J. W.; PAUL, D. R. "Effect of melt processing conditions on the extent of exfoliation in organoclay-based nanocomposites". Polymer, v. 42, p. 9513 – 9522, 2001.

FARAH, M. and BRETAS, R. E. S. "Characterization of i-PP shear induced crystalline layers developed in a slit die". Journal of Applied Polymer and Science, v. 91, p. 3528 – 3541, 2004.

FAVARO, M. M.; MARINELLI, A. L.; FARAH, M. and BRETAS, R. E. S. "Optical monitoring of polypropylene crystallization during injection molding". Polymer Engineering and Science, v. 48, p. 257 – 266, 2008.

FAVARO, M. M.; REGO, B. T.; BRANCIFORTI, M. C. and BRETAS, R. E. S. "Study of the quiescent and shear-induced crystallization kinetics of intercalated PTT/MMT nanocomposites". Journal of Polymer Science: Part B: Polymer Physics, v. 48, p. 113 – 127, 2010.

MAIO, E.; IANNACE, S.; SORRENTINO, L.; NICOLAIS, L. "Isothermal crystallization in PCL/clay nanocomposites investigated with thermal and rheometric methods". Polymer, v. 45, p. 8893 – 8900, 2004.

MORETTI, F.; FAVARO, M. M.; BRANCIFORTI, M. C. and BRETAS, R. E. S. "Optical monitoring of the injection molding of intercalated polypropylene nanocomposites". Polymer Engineering and Science, published on-line in Wiley Interscience. DOI 10.1002/pen.21662.

RAY, S. S.; OKAMOTO, M. "Polymer/layered silicate nanocomposites: a review from preparation to processing". Progress in Polymer Science, v. 28, p. 1539 – 1641, 2003.

RAZAVI-NOURI, M.; GHORBANZADEH-AHANGARI, M.; FEREIDOON, A.; JAHANSHAHI, M. "Effect of carbon nanotubes content on crystallization kinetics and morphology of polypropylene". Polymer Testing, v, 28, p. 46 – 52, 2009.

ZENG, Q.H.; YU, A. B.; LU, G. Q.; PAUL, D. R. "Clay-based polymer nanocomposites: research and commercial development". Journal of Nanoscience Nanotechnology, v. 5, p. 1574 – 1592, 2005.

DISPERSION OF MULTI-WALLED C NANOTUBES INTO CROSS-LINKABLE POLYETHYLENE GLYCOL MONOACRYLATE: EFFECTS ON DEVELOPMENT OF CROSSLINK DENSITY, VISCOELASTICITY AND SWELLING

Seda Vural, Kimberly Dikovics, Dilhan M. Kalyon*

Stevens Institute of Technology, Castle Point Station, Hoboken, NJ 07030
svural@stevens.edu, kdilovic@stevens.edu, * dkalyon@stevens.edu

Multi-walled carbon nanotubes, CNT, were compounded into a crosslinkable Newtonian fluid (photo initiator mixed with polyethylene glycol monoacrylate (PEGMA)) in the volume percent, φ, range of 0.01 to 0.5%. The linear viscoelastic material functions of the uncured PEGMA/CNT suspensions depended on the state of the dispersion of the nanotubes as affected by the sonication conditions. The dynamic properties, especially the storage modulus, increased with the better dispersion of the CNTs into an interconnected network, with the behavior of the PEGMA/CNT suspensions approaching gel-like behavior as the concentration of CNT reached φ=0.5%. Upon uv-curing of PEGMA/CNT the dynamic properties decreased significantly with the increasing concentration of CNTs indicating that the formation of the CNT network prior to curing decreased the mobility of PEGMA matrix to give rise to a reduction in the crosslink density of the PEGMA/CNT. The decrease of the crosslink density of the hydrogels gave rise to the increases in their rates of swelling and degradation in water, suggesting that the ultimate properties of CNT incorporated hydrogels depend on the microstructure that is developed during the compounding of the CNT, which in turn affects the crosslink density and hence the ultimate properties of the hydrogels.

Introduction

Water-swollen cross-linked hydrophilic polymers, hydrogels, are widely used, especially in the biomedical field [1], including controlled drug release via the encapsulation of bioactive agents into hydrogels and their in vivo release as a function of the physiological conditions [2]. For example, polyethylene glycol acrylate hydrogels are used in encapsulation and sustained release of human growth hormones [3]. Surface-functionalized multi-walled carbon nanotubes are considered for drug delivery to targeted sites, including chemotherapy drugs into solid tumors [4]. Thus, the combinations of carbon nanotubes, CNT, with hydrogels potentially provide additional routes to drug delivery, provided that the mechanical properties, swelling and biodegradation rates of the hydrogels can be predictably controlled. The properties of the resulting nanocomposites should be functions of the processing history [5].

One of the major factors that govern the ultimate properties of cross-linked hydrophilic polymers is their degree of crosslinking. However, previous studies furnish conflicting conclusions on the role played by carbon nanofibers and nanotubes on the rate and degree of crosslinking of thermosetting polymers [6-9]. For example, differential scanning calorimetry studies focusing on the crosslinking of CNT-filled epoxies report that CNTs and/or nanofibers increase, have no effect or decrease the degree of crosslinking of the epoxy matrix into which they are incorporated [6- 9]. Sources of this discrepancy can arise from the state of the dispersion of the nanotubes and nanofibers in the polymeric matrix and the type of surface functionality or the coating applied to them [6]. In general, achieving an adequate dispersion of the rigid particles [10-12] and especially the nanotubes [13- 22] into a polymeric matrix is a challenge, prompting the application of specialized methods including in situ polymerization [13- 15], surface modification of CNTs [16- 19], surfactants [20, 21], and spraying a CNT solution onto a polymeric powder [22].

The relatively high aspect ratios and the surface to volume ratios of nanotubes and the formation of an interconnected network structure by the CNTs can generate electrical percolation and rheological percolation at surprisingly low concentrations of nanotubes [23, 24]. The rheological properties, i.e., especially the linear viscoelastic material functions, of suspensions of carbon nanotubes are very sensitive indicators of the state of aggregation and dispersion of the nanotubes [25, 26]. Formation of interconnected CNT networks in the polymer matrix and the elasticity associated with the bonding between the nanotubes principally give rise to this sensitivity. This will be demonstrated here by focusing on the behavior of suspensions of a uv-crosslinkable polymer, i.e., polyethylene glycol monoacrylate, PEGMA, incorporated with CNTs and comparisons with PEGMA suspensions incorporated with traditional fillers with larger particles sizes including graphite and carbon black. Furthermore, we vary the dispersion conditions to generate a series of suspensions of PEGMA with different concentrations of carbon nanotubes and show that the linear viscoelastic material functions are very sensitive to the state of dispersion of the nanotubes. Finally, we show that the state of the dispersion of the nanotubes in PEGMA affects the degree of crosslinking of PEGMA, as revealed by significant changes in the linear viscoelastic properties, which in turn affects the degree of swelling of the PEGMA/CNT hydrogels.

Experimental

PEGMA/CNT Suspensions

Polyethylene glycol monoacrylate with an average molecular weight of 336 g mol-1 was obtained from Cognis Inc. and has a density of 1120 kg/m3. The photo initiator, 2-hydroxy-1-[4-(2-hydroxyethoxy) phenyl]-2-methyl-1-propanone, with an average molecular weight of 224.3 g mol-1, was from Ciba Specialty (Irgacure 2959). Weight ratio of photo initiator to polyethylene glycol monoacrylate is 0.02. Polyethylene glycol monoacrylate, prior to curing, exhibits magnitude of complex viscosity, η^*, values of 0.08-0.1 Pa-s over the frequency, ω, range of 0.1 to 100 rps (at 25 °C). Storage modulus, G', values of polyethylene glycol monoacrylate are $O(10^{-3})$ Pa and the loss modulus, G'', values are in the 0.01 to 7 Pa range for $0.01 \leq \omega \leq 100$ rps. The relatively small sensitivity of the magnitude of complex viscosity, η^*, values to deformation rate and $G'' >> G'$ suggest that the behavior of the polyethylene glycol monoacrylate approaches the behavior of a Newtonian fluid. The linear viscoelastic properties of the polyethylene glycol monoacrylate mixed with 2% by weight photo initiator are similar to those of pure polyethylene glycol monoacrylate. The mixture of polyethylene glycol monoacrylate with 2% by weight photo initiator is referred to as PEGMA.

Carbon black was obtained from Asbury Graphite Mills (# 5346). Carbon black nanoparticles are in the 30±10 nm diameter range and are clustered in the size range of 10-20 μm (nitrogen surface area of 91 m2 g-1). Non-colloidal graphite A60 is manufactured by Asbury Carbons and exhibits an average diameter of 44 ± 10 μm. Multi-walled carbon nanotubes (CNTs) with a reported density of 2100 kg/m3 were obtained from Cheap Tubes Inc. (SKU #030104) and exhibit outside diameter, d, of 25 ± 5 nm, inside diameter of 5-10 nm, length, L, of 10-30 μm, aspect ratios of 330 to 1500. They consist of 15-20 concentric rolled-up graphene layers. CNTs were not surface-functionalized so as not to further complicate the crosslinking reaction of PEGMA with the photo initiator in the presence of the CNTs and not to deteriorate the aspect ratio of the CNTs. Significant reductions of the aspect ratio of CNTs during surface functionalization are commonly encountered, especially when strong acids are used [41]. For the dispersion of CNTs in PEGMA a Misonix XL2020 ultrasonic processor was used at 150 W to generate liquid jet streams resulting from ultrasonic cavitation to overcome the bonding forces between the nanotubes, and separate the tubes [42, 43] in the volume fraction, φ, range of 10^{-4} to $5x10^{-3}$.

During sonication the samples were cooled upon circulation of water at 20 °C in a cooling jacket. A thermal imaging camera (Inframetrics PM290) was employed to assure that the bulk temperature of the suspension during sonication does not surpass 25°C. Nanosuspension samples were collected at different sonication times, i.e., 5 to 90 minutes for optical microscopy (Nikon, Optiphot2-POL polarizing microscope) and for the characterization of their linear viscoelastic properties using an ARES rheometer (TA Instruments). Within the typical time spans of the rheological characterization and optical microscopy (about 1 hour) the PEGMA/CNT samples were sufficiently stable.

UV Curing and Swelling

The sonicated PEGMA/CNT suspensions were transferred into a two-piece Teflon mold (0.12 g per cavity). The suspension samples were cured under uv-light (100 mW cm^{-2}, 320-420 nm), at a distance of 2 cm from the uv source. The ARES rheometer was used for the characterization of the linear viscoelastic properties of the cured samples. For each φ, series of uv-curing experiments, in which the curing time was varied in the 15 to 240 min range, were carried out. The swelling and degradation behavior of the hydrogel samples prior to and upon uv-curing are characterized by immersing the PEGMA/CNT samples into either de-ionized water at 25°C or into phosphate buffered saline (PBS) solution held at 37°C.

Results and Discussion

The linear viscoelastic behavior of PEGMA/CNT nanosuspension is compared in Figure 1 with those of PEGMA compounded with graphite nanoparticles (30±10 nm diameter range and are clustered in the size range of 10-20 μm) and carbon black particles (34-54 μm diameter) at $\varphi=3x10^{-4}$.

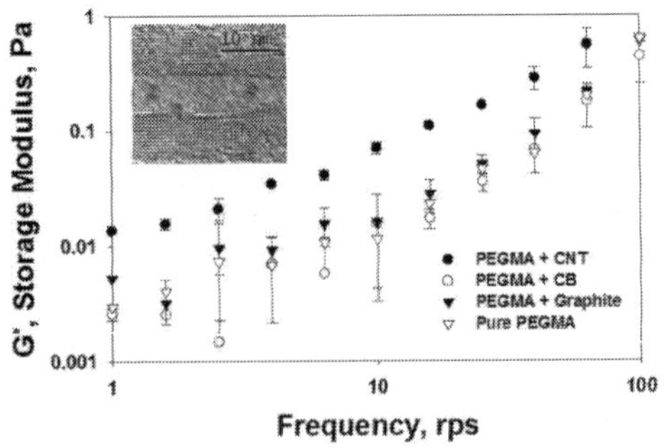

Figure 1: Storage modulus, $G'(\omega)$, of PEGMA suspensions incorporated with CNT, carbon black and graphite at $\varphi=0.003$. Inset TEM image of MWCNT layers.

The CNTs, with their significantly higher aspect ratios in comparison to graphite and carbon black, give rise to an appreciable increase of the storage modulus, i.e., the elasticity of the suspension, whereas the suspensions of PEGMA with carbon black and graphite exhibit similar elasticity to pure PEGMA. At the same particle concentration of $\varphi=3x10^{-4}$ the loss modulus and the magnitude of the complex viscosity values of the CNT incorporated PEGMA suspensions are also significantly higher than the PEGMA compounded with carbon black and graphite- again emphasizing the important role played by the very high aspect ratios of the CNTs. Such high aspect ratios allow the formation of a CNT network on the basis of the van der Waals interactions between nanotubes.

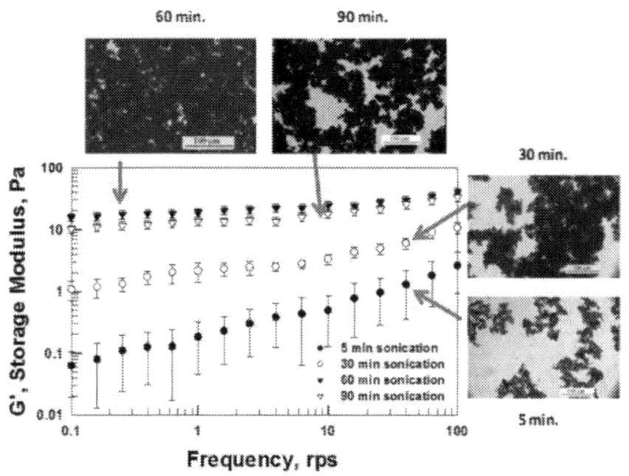

Figure 2: Storage modulus, G'(ω), as affected by the dispersion state of the carbon nanotubes at φ=0.003

Effect of Dispersion on Dynamic Properties
Figure 2 shows the typical development of the storage modulus, G′(ω), of PEGMA/CNT at the CNT volume fraction, $\varphi=3x10^{-3}$ as a function of sonication time used in the mixing of the CNTs into the PEGMA. The insets in the figure show the characteristic micrographs of the nanosuspensions obtained upon different durations of sonication using a light microscope. Relatively large clusters of CNTs are observed at sonication times of 5 and 30 minutes. For these two sonication times part of the sample is devoid of any clusters of CNTs, whereas other parts of the PEGMA/CNT samples exhibit relatively large CNT clusters. At a sonication time of 60 minutes the CNTs appear to be well distributed in the PEGMA.

The G′ (ω) values of these samples sonicated for different durations follow the states of the dispersion suggested by the micrographs (Figure 2). Segregated CNT-rich zones are prevalent at 5 minutes of sonication and the resulting G′ (ω) values are small, i.e., are in the range of 0.1-2 Pa. The increase of the sonication time to 30 minutes again gives rise to segregated PEGMA regions but the CNTs now form a percolated network that spans the entire structure with encapsulated PEGMA islands. The formation of the volume-spanning CNT network gives rise to G′(ω) values which are about one order of magnitude greater than those observed at 5 minutes of sonication. The increase of the sonication time to 60 minutes leads to better dispersion as manifested by the CNT network spanning the entire volume. Correspondingly the G′(ω) values increase to 15-40 Pa (two orders of magnitude increase with better dispersion going from 5 min of sonication to 60 at ω=0.1 rps).

When the sonication time is further increased to 90 minutes the degree of dispersion of the CNTs deteriorates in comparison to that achieved at 60 min. The deterioration of the CNT dispersion reduces the G′(ω) values by about 30% in comparison to those observed at a sonication time of 60 minutes. The total volume of the PEGMA islands observed at the sonication time of 90 minutes is smaller than observed at a sonication time of 30 minutes giving rise to the G′(ω) values of samples sonicated for 90 minutes to be greater than those observed for the PEGMA/CNT samples sonicated for 30 minutes. Such a deterioration of the degree of dispersion with increased sonication time (as observed going from 60 minutes of sonication to 90 minutes) is not observed for $\varphi < 3x10^{-3}$. At these lower concentrations the homogeneity of the CNT distribution increases asymptotically to a maximum with increasing sonication time, which coincides with maxima in G′(ω) values. Therefore, for $\varphi < 3x10^{-3}$ the structure and linear viscoelastic properties become insensitive to the sonication time when a critical sonication time is reached.

At $\varphi = 3x10^{-3}$ (Figure 2) the G′(ω) values for all sonication times show little sensitivity to frequency in the 0.1 to 5 rps range. Furthermore, for all sonication times, G′>> G′′. The lack of sensitivity to the frequency and G′>> G′′ suggest that gel-like behavior is approached [28, 29], with the strength of the gel depending on the bonding between the CNTs [25] and hence on the state of the dispersion of the CNTs. G′>> G′′ suggests that relatively long relaxation times prevail, i.e., a precursor of solid like behavior. The significant differences in G′(ω) values observed for different sonication times indicate that the networking of the CNTs occurs very differently as a function of the sonication conditions with clusters of CNTs tied with individual strands of CNTs at low sonication times versus well connected clusters of CNTs at longer sonication durations, with the G′(ω) values, i.e., elasticity, increasing with the formation of a volume spanning network of CNTs. Overall, these results suggest that the dispersion state of the CNTs depends on the sonication conditions and in turn the dynamic properties of the resulting nanosuspensions are very sensitive to the state of the dispersion. The dynamic properties were thus used as an indicator of the state of dispersion.

Dynamic Properties of Well-dispersed CNTs in PEGMA

The optimum sonication times necessary for achieving well-dispersed CNTs and corresponding maxima in the moduli (duration of sonication at which the linear viscoelastic materials functions G′ and G′′ exhibit maxima) were determined for each φ. With decreasing φ the optimum sonication time also decreases, indicating that the dispersion of the nanotubes is easier with decreasing φ. The optimum sonication times are 5 minutes for $\varphi = 1x10^{-4}$, 30 minutes for $\varphi = 3x10^{-4}$ and $8x10^{-4}$ and 60 minutes for $\varphi = 10^{-3}$, $3x10^{-3}$ and $5x10^{-3}$. The strain amplitude, γ_0, at which the transition from linear to nonlinear viscoelastic behavior is observed decreases with increasing concentration of carbon nanotubes for well-dispersed samples (γ_0 of 10% at $\varphi = 0$ and γ_0 of 0.01-0.2% at $\varphi = 5x10^{-3}$), indicative of the development of solid-like behavior as φ increases.

With increasing φ the values of G′(ω) also increase. Furthermore, the slope of the G′(ω) versus ω decreases. Inspection of the other dynamic properties, i.e., the loss modulus, G′′(ω), and tan δ= G′′(ω)/G′(ω) suggests that G′′(ω) > G′(ω) at relatively low φ (for example, tan δ values are in the 7-14 range for $\varphi = 3x10^{-4}$) and that G′(ω)> G′′(ω) for higher φ (for example, tan δ values are in the 0.2-1 range for $\varphi = 3x10^{-3}$). The increase of G′(ω) in comparison to G′′(ω) and the decrease of the dependence of the dynamic properties on ω (reduced slope of G′(ω) versus ω) reflect the relative ease of the development of a volume-spanning CNT network with increasing φ. This is especially evident for $\varphi \geq 10^{-3}$, leading to the observed gel-like behavior of the PEGMA/CNT suspensions with increasing φ.

The storage modulus at small frequencies over which the storage modulus becomes independent of frequency is the plateau modulus, G_0'. The values of G_0' scale as $\varphi^{3.1}$. Considering G_0' versus φ as $G_0' \alpha (\varphi - \varphi^*)^v$, where φ^* is the rigidity percolation volume fraction of CNTs [25], φ^* can be taken as $6x10^{-4}$, around which a two orders of magnitude increase of $G_0'$with φ is observed. The best fit of the G_0'versus (φ- φ^*) data provides a percolation exponent, v, of 2.35±0.5 (R^2=0.95) which is similar to the percolation exponent, 2.35±0.1 determined by Hough et al. [25] for single walled nanotubes (surfactant stabilized aqueous suspensions) and close to the theoretical percolation exponent of 2.1 ±0.2 that was calculated on the basis of the existence of percolating bonds between nanotubes that are free to rotate but resist stretching under stress [30].

The critical concentration at which rheological material functions significantly increase for carbon nanotube systems may be assumed to have similar values to those of polymer solutions [31, 32]. Carbon nanotubes occupy very high hydrodynamic volumes and generate critical concentrations in the order of 0.1 wt % for overlap of nanotube aggregates and hence significant step changes in rheological properties [32]. For the CNTs and their concentrations used in this study, i.e., $10^{-4} \leq \varphi < 10^{-3}$, $cL^3 >> 1$ and $cL^2d \leq 1$ (where c is the number density, d is the outside diameter and L is the length of the nanotubes). These indicate that the nanotube concentrations in this range are in the semidilute regime, in which the nanotubes cannot rotate freely without being impeded by their neighbors [33]. The transition to the concentrated isotropic concentration regime, at which the tubes should begin to have difficulty packing isotropically, occurs at $\varphi \approx \pi d/4L$ i.e., $O(10^{-3})$ [34]. The transition in rheological behavior, i.e., from a liquid like behavior to a solid like behavior is reported to occur between loadings of 0.25 and 0.5 wt% multi-walled CNTs in polycarbonate [35], consistent with our findings. As shown in Figure 2, however, besides the concentration of the CNT, the state of the dispersion of the CNTs also plays a significant role in this transition, with better dispersion of the nanotubes giving rise to greater elasticity, i.e., orders of magnitude increases in G′.

UV-Cured PEGMA/CNT Samples

Dynamic Properties and Dispersion

It is striking that the ultimate storage modulus values of the cured PEGMA/CNT samples decrease with increasing CNT concentration, φ. The theory of rubber elasticity suggests that particulate fillers consisting of finely divided solids, for example carbon black, increase the modulus of elasticity, E, akin to the increase of the shear viscosity of a suspension with volume fraction of the spherical solid particles, φ [36] i.e., $E(\varphi) = E_0(1 + 2.5\varphi + 14.1\varphi^2)$, where E_0 is the modulus of elasticity of the elastomeric binder. Furthermore, under certain cross-linking reactions, for example when rubber containing carbon black is vulcanized using S, the cross-linking of the polymer units to the filler surface can occur on the basis of the carbon black surfaces being unsaturated to allow additional cross-linkages to form on the carbon black surfaces [36, 37]. The higher crosslink density developed next to the surface of the carbon black gives rise to further increases of the modulus of elasticity of the rubber/carbon black composite, beyond what is predicted by the above $E(\varphi)$ relationship. The stiffening of the matrix as a whole due to the formation of bonds between the polymer and the CNT surface is not evident in our data with $G'(\varphi)$ decreasing with increasing φ.

Additional evidence on the mechanism of the hindering effect of CNTs on the crosslinking of PEGMA is obtained by investigating the effects of the state of dispersion of the CNTs on the elasticity of the crosslinked PEGMA/CNT samples. With better dispersion of the CNTs in the PEGMA the rate of photo-polymerization of PEGMA and the resulting crosslink density, (v_e/V) are reduced significantly, as suggested by the one order of magnitude smaller storage modulus, $G'(\omega)$, values observed at 30 minutes of sonication versus those at 1 and 3 minutes of sonication. It appears that the presence of well-dispersed CNTs and the associated initial gel like behavior prior to curing disrupts the progress of the PEGMA polymerization under uv-curing and the associated intra molecular crosslinking process as φ increases.

The disruption of the crosslinking process due to the presence of CNTs should be affected by the increase of the shear viscosity of the reaction medium upon CNT network formation and gelation. The very high surface to volume ratio of the nanotubes, i.e., $O(10^8 \text{ m}^{-1})$ leads to a vast interfacial area between the nanotubes and the polymer, to reduce the mobility of the polymer in the vicinity of the nanotubes for several radii of gyration [38]. This extensive inter-phase region of reduced macromolecule chain mobility surrounding the nanotubes permeates through a greater volume of the nanocomposite as the state of dispersion of the CNTs improves.

With better dispersion, possibly little of the bulk polymer would remain unaffected giving rise to the ultimate crosslink density of the PEGMA/CNT decreasing significantly with increasing φ, as revealed by the appreciable decrease of the storage modulus values with increasing φ. The neglect of the consideration of the state of the dispersion of the nanotubes can therefore obscure the intrinsic effects of the nanotubes on the crosslinking behavior of the thermosetting polymers into which they are incorporated.

Swelling Behavior

The change in the crosslink density due to the presence of the CNT should have a significant effect on the degree of swelling of the cured PEGMA/CNT hydrogel samples. Three concentrations of CNTs (φ = 0, 10^{-4}, $3x10^{-4}$) and multiple uv-exposure times were used as the parameters of the swelling studies. The swelling ratio, q, is defined as $q= V/V_0$, where V is the volume of the swollen hydrogel and V_0 is the volume of the unswollen hydrogel (with or without CNT). The equilibrium swelling ratio, q_m, is achieved when the swelling ratio, q, no longer changes with increasing time in water or in phosphate buffered saline, PBS.

The equilibrium degree of swelling, q_m, is related to the crosslink density, (v_e/V), as $1/q_m^{5/3}$ α (v_e/V) [36]. Thus, q_m should increase with decreasing extent of crosslinking (v_e/V). This is indeed the behavior observed from the swelling experiments and is consistent with earlier studies which have also shown that the degree of swelling of polymers increased upon incorporation of CNTs [37, 38]. For example when cured PEGMA gels were immersed into water, the equilibrium degree of swelling, q_m, values increased from q_m =3.2 at φ=0 to q_m =4.4 at $\varphi=3x10^{-4}$, and to q_m=4.7 at $\varphi=5x10^{-4}$, indicating that the crosslink density, (v_e/V), decreases with increasing φ. This is consistent with the observed viscoelastic behavior of the cured samples. Similar swelling behavior is also obtained in PBS. Furthermore, it was determined that the hydrogels with greater φ decompose faster in water or PBS as a consequence of their reduced degree of crosslinking, consistent with the swelling and viscoelastic behavior.

Conclusions

Multi-walled carbon nanotubes, CNTs, are compounded into a cross-linkable mixture of polyethylene glycol monoacrylate, PEGMA, via ultrasonication. Microscopy and linear viscoelastic material functions of the PEGMA/CNT suspensions indicate that with increasing sonication time the CNTs can be dispersed better in the PEGMA leading to the formation of a CNT network that spans the entire volume and gives rise to increased elasticity, i.e., orders of magnitude increase of the storage modulus, $G'(\omega)$, with the behavior reaching that of a percolated gel with increased CNT volume fraction, φ. This significant increase in elasticity does not occur with carbon black or graphite particles with greater diameters and smaller aspect ratios in comparison to CNTs at similar φ. The presence of the CNTs reduces the elasticity of the cured PEGMA/CNT with the extent of crosslinking decreasing with increasing concentration of CNTs and with better dispersion of the CNTs. The degree of swelling of the PEGMA/CNT in water and PBS, i.e., PEGMA/CNT hydrogels, increase with increasing CNT concentration, as affected by the reduced crosslink density, i.e., an effect which becomes more pronounced as the CNTs are better dispersed. The observed significant changes in the viscoelasticity and swelling behavior of PEGMA/CNT hydrogels suggest that the degree of dispersion and the concentrations of CNTs can be used as additional parameters to tailor the ultimate properties of hydrogels especially in the tissue engineering and controlled drug release applications.

Acknowledgements

We thank Dr. H. Gevgilili, Prof. F. Fisher and Dr. G. Mago of Stevens Institute of Technology for their valuable comments and input.

References

1. P. Krsko, M. Libera, Mater. Today, 8, 12, 36-44 (2005).
2. D. Schmaljohann, Adv. Drug Deliv. Rev., 58, 1655-1670 (2006).
3. P. van de Wetering, A.T Metters, R.G. Schoenmakers, J.A. Hubbell, J. Controlled Release, 102, 619-627 (2005).

4. Y. Liu, and H. Wang, Nature Nanotech., 2, 20-21 (2007).
5. S. Bhattacharya, R. Gupta and M. Kamal, "Polymeric Nanocomposites: Theory and Practice, Hanser, NY (2008).
6. J. Wu, D. D. L. Chung, Carbon, 42, 3003 (2004).
7. D. Puglia, L. Valentini, I. Armentano, J. M. Kenny, Diamond Relat. Mat., 12, 827 (2003).
8. J. Bae, J. Jang, S. H. Yoon, Macromol. Chem. Phys., 203, 2196 (2002).
9. K. Tao, S. Yang, J. C. Grunlan, Y. Kim, B. Dang, Y. Deng, R. L. Thomas, B. L. Wilson, X. Wei, J. App. Polym. Sci., 102, 5248–5254 (2006).
10. R. Yazici and D. M. Kalyon, Rubber Chem. and Tech. 1993. 66, 527-537 (1993).
11. D. M. Kalyon, D. Dalwadi, M. Erol, E. Birinci and C. Tsenoglu, Rheologica Acta, 45, 641-658 (2006).
12. M. Erol and D. Kalyon, Int. Polym. Proc., 20, 228-237 (2005).
13. C. A. Mitchell, J. L. Bahr, S. Arepalli, J. M. Tour, and R. Krishnamoorti, Macromol., 35, 8825–8830 (2002).
14. T. Fukushima, A. Kosaka , Y. Yamamoto, T. Aimiya , S. Notazawa, T. Takigawa, T. Inabe, T. Aida, Small, 2, 554-560 (2006).
15. B. Zhu, S. Xie, Z. Xu, Y. Xu, Comp. Sci. Tech., 66, 548-554 (2006).
16. C. A. Mitchell and R. Krishnamoorti, Macromol., 40, 1538–1545 (2007).
17. E. K. Hobbie, D. J. Fry, J. Chem. Phys., 126, 124907 (2007).
18. M. Abdalla, D. Dean, D. Adibempe, E. Nyairo, P. Robinson, G. Thompson, Polymer, 48, 5662-5670 (2007).
19. A.M. Shanmugharaj, J.H. Bae, K. Y. Lee, W. H. Noh, S. H. Lee, S. H. Ryu Comp. Sci. and Tech., 67, 1813-1822 (2007).
20. X. Gong, J. Liu, S. Baskaran, R. D. Voise, J. S. Young, Chem. Mater., 12, 1049-1052 (2000).
21. F. H. Gojny, K. Schulte, Comp. Sci. Tech. 2004, 64, 2303-2308 (2004).
22. Q. Zhang, D. R. Lippits, S. Rastogi Macromol., 39, 658–666 (2006).
23. G. Hu, C. Zhao, S. Zhang, M. Yang, Z. Wang, Polymer, 47, 480-488 (2006).
24. E. T. Thostenson, T. W. Chou, Polymer, 44, 3022-3029 (2006).
25. L. A. Hough, M. F. Islam, P. A. Janmey, and A. G. Yodh, Phys. Rev. Lett., 93, 168102 (2004).
26. W. K. Ma, F. Chinesta, A. Ammar, and M. R. Mackley, J. Rheol., 52, 1311-1330 (2008).
27. M. Doi and S. F. Edwards, Theory of Polymer Dynamics, Oxford University Press, New York, (1986).
28. F. Chambon, H. H. Winter, J. Rheol., 31, 8, 683-697 (1987).
29. M. E. DeRosa, H. H. Winter, Rheol. Acta, 33, 220-237 (1994).
30. M. Sahimi, S. Arbabi, Phys. Rev. B 1993, 47, 703
31. W. W. Graessley, Polymer, 21, 258–262 (1980).
32. S. Rahatekar, K. Koziol, S. Butler, J. Elliott, M. Shaffer, M. Mackley and A. Windle, J. Rheol., 50, 599-610 (2006).
33. D. Fry, B. Langhorst, H. Wang, M. Becker, B. Bauer, E. Grulke and E. Hobbie, J. Chem. Phys., 124, 54703 (2006).
34. R. Larson, Structure and Rheology of Complex Fluids, Oxford University Press, New York (1999)
35. M. Abdel-Goad, P. Pötschke, J. Non-Newtonian Fluid Mech., 128 2–6 (2005).
36. P. Flory, Principles in Polymer Chemistry, Cornell Univ. Press, Ithaca, NY (1953)
37. R. Stearns, B. and Johnson, Ind. Eng. Chem., 43, 146 (1951).
38. T. Ramanathan, H. Liu and C. Brinson, J. Polym. Sci. Part B: Polym. Phys., 43, 2269-2279 (2005).
39. X. Tong, J. Zhen, Y. Lu, Z. Zhang, H. Cheng, Mat. Lett., 61, 1704-1706 (2007).
40. M. Xu, T. Zhang, B. Gu, J. Wu, Q. Chen, Macromol., 39 3540-3545 (2006).
41. Z. Fan and S. Advani, J. Rheol., 51, 585-604 (2007).
42. J. Hilding, E.A. Grulke, Z.G. Zhang, and F. Lockwood, J. Dispersion Sci. Tech., 24, 1–41 (2003).
43. A. Gedanken, Curr. Sci., 85, 1720–1722 (2003).

INFLUENCE OF SHEAR AND RESIDENCE TIME IN TWIN-SCREW EXTRUSION OF PCL – CNT COMPOSITES

T. Villmow, B. Kretzschmar, and P. Pötschke*

Leibniz Institute of Polymer Research Dresden, Hohe Strasse 6, 01069 Dresden, Germany
villmow@ipfdd.de, kretzsch@ipfdd.de, poe@ipfdd.de

Melt processing of thermoplastic based nanocomposites is the favoured route to produce electrically conductive or electrostatic dissipative polymer composites based on conductive fillers, like carbon nanotubes (CNT). As these properties are desired at low fractions of the filler a high state of dispersion is required in order to benefit from the intrinsic properties of the individual CNT. This study discusses the influence of extrusion parameters like screw configuration, rotation speed, and throughput on the minimum residence time and resulting macroscopic CNT dispersion within polycaprolactone (PCL) based masterbatches containing 7.5 wt.% multiwalled carbon nanotubes (MWNT). A Berstorff ZE 25, which is a co-rotating twin-screw extruder, was used as melt mixing device.
The processing conditions were found to have a strong influence on the CNT dispersion. Whereas an increase of rotation speed, resulting in high specific energy inputs, leads to a significant increase of the degree of CNT dispersion, an increase of throughput showed the opposite effect. Beside from these machine parameters, the design of screw configuration can promote the CNT dispersion, when using mixing elements in combination with back-conveying elements.
The lowest area fraction of undispersed primary CNT agglomerates of all investigates masterbatches was determined when using an extended screw containing mixing elements. These remaining agglomerates after the masterbatch step could be nearly entirely dispersed in an additional dilution process. The high degree of CNT dispersion within the diluted composites resulted in a very low percolation threshold of 0.35 wt.% within the PCL, which corresponds to 0.23 vol.%.

Introduction

A favoured strategy for the development of new functional polymer materials is the mixing of a conventional matrix polymer with further polymer components (polymer blends) and/or solid fillers (composites). The big advantage of this route and such composite materials is based on the unlimited number of possible property combinations, much better conditions for the market launch, and higher costumer acceptance in comparison to new homopolymers.
Carbon Nanotubes (CNT), whose growth was firstly described in 1991 [1], are one of these functional fillers, which got into the focus of research due to their outstanding electrical [2-4], thermal [5-7], and mechanical properties [8-10]. These excellent characteristics predestine them as filler for polymer-based composites, although a transfer of properties to the polymer matrix is challenging, as CNT tend to form strong agglomerates resulting from physical interactions like van-der-Waals forces. Possible physical entanglements of high-aspect ratio CNT resulting from the synthesis process often further enhance the strength of primary agglomerates. These primary agglomerates have to be dispersed and distributed within the polymer in order to benefit from the properties of individual CNT. Recently, a correlation between dispersability of different CNT in aqueous solutions and resulting composite properties after incorporation of these CNT into a polymer matrix via melt mixing was published by Krause et al. [11]. Depending on the polymers nature, different processing techniques are discussed in literature for CNT incorporation like ultra-sonification, high-shear mixing, solution mixing, in-situ polymerisation and melt mixing. The latter process is the preferential method for the processing of thermoplastic-based composites, as melt processing has the highest relevance for industrial applications. It is based on conventional industrial processing technologies like twin-screw extrusion or injection moulding. So far, a limited number of publications deal with industrial relevant melt processing techniques and the influence of processing parameters on composite properties. In order to produce injection-moulded specimen, the dilution process of commercial available, solution mixed, or self-made masterbatches is described in literature [12-14]. From investigations regarding the processing conditions during the shaping processes injection moulding [12] and compression moulding [15, 16] it is well known, that secondary agglomeration of dispersed primary agglomerates during shaping is important in order to create electrically percolated composites. But so far, there is no systematic study

about the influence of extrusion conditions on the dispersion of such primary CNT agglomerates available.

For this reason, this contribution will discuss the influence of extrusion conditions like screw configuration, throughput, and rotation speed on the macroscopic CNT dispersion within a polycaprolactone matrix. In order to interpret the observed phenomena, the minimum residence time and specific mechanical energy input were determined for all experiments. The basis for this study are thereby recently published results on the influence of twin-screw extrusion conditions on the dispersion of CNT in a polylactic acid matrix, where a set of two screw profiles and a variation of rotation speed was used to produce masterbatches containing 7.5 and 15.0 wt.% CNT [17].

A few publications report about the processing and properties of nanocomposites based on polycaprolactone and CNT, as this filler can increase the thermomechanical properties like storage and loss modulus, which enables a broader range of application for this thermoplastic [18]. Two different approaches of nanocomposite processing like melt compounding [19] and in-situ polymerisation [20] are described to result in percolated network structures within the PCL matrix at MWNT contents of around 2.0 wt.%, whereas the percolation concentration was determined by means of rheological experiments. A significantly lower percolation threshold of 0.08 wt.% CNT was discussed for PCL/SWNT composites, which were prepared by solution mixing with the aid of zwitterionic surfactant [21].

Experimental

Materials

PCL CAPA 6800 (Perstorp, U.K.) is a biodegradable polyester with a low melting point of around 60°C and a glass transition temperature of about −60°C. The grade CAPA 6800 is a homopolymer with a molecular weight of 80.000 g/mol and a MFI of 3 (2.16 kg at 160°C). The PCL having a density of 1.14 g/cm^3 was melt mixed with NC 7000 (Nanocyl S.A., Belgium), which are thin multi-walled carbon nanotubes (MWNT) with a purity of 90 % produced via the catalytic carbon vapour deposition process. The nanotubes have a typical diameter of 9.5 nm and an average length of 1.5 μm. The density of MWNT after incorporation into a polymer is approximately 1.75 g/cm^3 [22].

Composite preparation via twin-screw extrusion

In order to investigate the influence of twin-screw extrusion conditions, several PCL/CNT masterbatches containing 7.5 wt% MWNT were produced under variation of screw configuration, rotation speed, and throughput. The temperature profile was kept constant on a relative high level with a mean barrel temperature of 200°C to ensure the processability of PCL for all parameter combinations. The compounding was performed using a co-rotating twin-screw extruder ZE25 (Berstorff, Germany) having a screw diameter of 25 mm and barrel lengths of 900 mm (L/D = 36) or extended towards 1200 mm (L/D = 48). PCL pellets and the powdery MWNT material were fed simultaneously into the hopper by gravimetric dosing.

The design of five screw configurations was based on two screws, which were published by our group recently [17]. This set of screws was extended to get more information about the influence of screw configuration on the mean residence time and CNT dispersion within the PCL matrix. Again, two classes of screw configurations were used for the compounding processes to enable different types of mixing like dispersive and distributive mixing. The former results in the break-up of cohesive agglomerates, while the latter increases the homogeneity of the mixture by generating new interfacial areas. The two dispersive screw configurations that are used in the presented study contain kneading blocks and conveying elements. The number of kneading zones was kept constant, whereas the number of right-handed conveying elements, enabling an increased back pressure, was varied to control the minimum residence time during extrusion (SC1: distributive and less back conveying elements, SC2: distributive and added back conveying elements). Both screws were designed to have a length to diameter ratio of 36.

Another three distributive screw configurations were designed on the basis of the dispersive ones regarding the number and position of mixing elements. The number and location of conveying elements of screw profile SC1 were maintained and the kneading blocks were substituted by mixing elements to create SC3. Screw configuration SC4 was than again designed to have additional right-handed conveying elements. The fifth screw configuration SC5 was designed based on SC4 as a distributive mixing screw, but having an extended length to diameter ratio of 48. Furthermore, this screw profile contains two additional mixing zones to have five of them in sum.

Processing of PCL/CNT composites by masterbatch dilution and electrical testing

One masterbatch was diluted under optimised conditions (SC5, 500 min-1, and 5 kg/h) to produce composites with 0.125, 0.25, 0.375, 0.5, 0.75, 1.0, 1.5, 2.0, 3.0, and 4.0 wt.% CNT in order to determine the percolation behaviour of CNT within the PCL matrix. Masterbatch pellets and pure PCL were premixed prior feeding into the hopper for this purpose.

The volume resistivity of the composites was determined according to the standards ASTM D 4496 and ASTM D 257 on compression-moulded plates. Plates having a diameter of 65 mm and a thickness of 0.5 mm were prepared from the pelletised materials using a PW 40 EH hot press (Paul-Otto Weber GmbH, Germany). The pellets were compression-moulded at 220°C for 3 min with a pressing velocity of 6 mm/min at a pressing force of 20 kN. The pressed plates were cooled down at room temperature.

The resistivity measurements were performed at room temperature. A Keithley 8009 Resistivity Test Fixture combined with a Keithley 6517A Electrometer was used to determine the electrical resistance of highly resistive samples, whereas the measurements were performed on the cylindrical plates. In case of low resistance samples, small strips (30 x 3 mm²) were cut from these plates and electrical resistance was measured using a self-made four-point Test Fixture in combination with a Keithley multimeter Model 2000. All measured volume resistances R_v were converted to volume resistivity ρ_v.

Determination of minimum residence time

The minimum residence time $_{tR,min}$ of PCL during twin-screw extrusion was evaluated by means of an optical method, whereas a PCL masterbatch with 7.5 wt% MWNT was used as colourant. After reaching steady state extrusion conditions of the neat PCL, five granules of the masterbatch were introduced into the hopper in a pulsed manner. The time $t_{R,min}$ was determined by taking into account the time when the first blackened material left the extrusion die. Residence time distribution (RTD) curves could not be recorded, as the extruded unfilled PCL could not be collected as a continuous strand, although the CNT filled masterbatches could be collected properly. As classical RTD curves show a certain tail depending on the processing conditions, the minimum residence times were recorded, as they are directly comparable and reliable for all experiments.

Specific mechanical energy

The specific mechanical energy (SME) is the amount of mechanical energy, which is dissipated as heat inside the material during the extrusion process and can be calculated using **Equation 1**. Specifically, it is the work input from the drive motor into the material being extruded and thus provides a good characterization of the extrusion process. Usually SME increases with increasing viscosity, increasing screw speed, and decreasing throughput. Both the extrusion operation and the screw performance can be well characterized by the SME value. It is also an important parameter in extrusion processes, as it provides a measurable parameter, which can be correlated with the quality characteristics of the final product. In case of the presented study, SME values will be correlated with the macroscopic CNT dispersions.

$$SME = \frac{\tau * N}{\dot{m}}\left[\frac{kJ}{kg}\right]$$

Equation 1: Specific mechanical energy

Determination of macroscopic CNT dispersion

The state of macroscopic CNT dispersion in the form of remaining primary MWNT agglomerates within the PCL matrix was investigated by light transmission microscopy on thin sections of extruded pellets, which were prepared following the standard ISO 18553. The thin sections cut from masterbatch pellets had a thickness of 2.5 µm, whereas the thickness of thin sections from composite pellets was in the range between 10 and 20 µm. Sections were cut from at least six different pellets within one experiment using a JUNG RM 2055 microtome (Leica, Germany) and a histo-diamond knife (Diatome, Switzerland) at cryo conditions (–40°C). The micrographs were imaged with a BH2 microscope (Olympus, Germany) using an objective with a 10x magnification. In order to quantify the quality of dispersion a particle analysis was performed using the software ANALYSIS (Olympus, Japan) to evaluate the area ratio A_r between the area occupied by primary CNT agglomerates A_{CNT} and the overall area A_0 (**Equation 2**). According to the ISO 18553 standard, agglomerates with diameters smaller than 5 µm were neglected. Furthermore, the primary agglomerate diameter distribution was analysed and will be discussed in context with processing conditions.

$$A_r = \frac{A_{CNT}}{A_0} * 100\%[\%]$$

Equation 2: Area ratio as a characteristic value for CNT macro dispersion in PCL based composites

As the ZE25 extruder enables the removal of the extrusion screw during the extrusion process, the development of macroscopic CNT dispersion along the extrusion length was investigated for two experiments, where SC2 and SC5 were used, while rotation speed (500 min^{-1}) and throughput (5 kg/h) were kept constant. Small samples of approximately 1 g were taken after every mixing/kneading zone for light microscopic investigations.

Results and Discussion

Morphology of CNT material

A SEM study of the NC 7000 was carried out to investigate the initial primary agglomerate size and structure. Both parameters are important to judge the state of dispersion after incorporation into the PCL matrix and to estimate their principle dispersibility. Although CNT are a nano-filler, NC 7000 appears in the shape of macroscopic primary agglomerates with diameters up to several millimetres. Besides the strongly pronounced agglomeration, the SEM study revealed that NC 7000 form rope-like superlattices, whereas these ropes are built by a large number of individual CNT, which are packed in a relatively loose way.

Influence of processing conditions on residence time

When dispersing any kind of filler in a polymeric matrix, mixing time is a crucial factor to control the filler dispersion. Especially the "erosion mechanism" described for carbon black in a Newtonian fluid is time depending and follows a first order rate [23], as erosion takes place from the agglomerate surface and is relatively slow compared to the "rupture mechanism", where big agglomerates are fragmented into a small number of relatively large agglomerates. Many publications dealing with melt mixing of polymer/CNT composites discussed the influence of mixing time on the CNT dispersion when using microcompounders [24, 25]. These microcompounders are discontinuous mixers, where rotation speed and mixing time can be set up independently, whereas the screw configuration is generally unchangeable. In case of twin-screw extrusion the residence time is a complex factor, which is a function of rotation speed, throughput, and screw configuration. As sufficient residence time is needed to disperse CNT in polymer matrices, it was the aim of this study to evaluate the influence of each of these factors on the residence time of PCL in the extrusion process.

The choice of mixing elements like kneading or mixing has a significant influence on the minimum residence time of PCL during extrusion. The use of kneading blocks (SC1) instead of mixing elements (SC3) results in 10 % longer minimum residence times, as these elements exhibit an active conveying effect, if they are designed to have negative staggering angles, whereas mixing elements act neutral regarding their conveying behaviour (**Figure 1**). This effect is independent from rotation speed. The further addition of back conveying elements for screw profile SC2 in comparison to SC1 leads to a significant increase of the minimum residence time of 9 (100 min^{-1}) and 33 % (500 min^{-1}). A similar trend was found for the screw configurations containing mainly mixing elements, whereas SC4 led to 10 (100 min^{-1}) and 25 % (500 min^{-1}) longer minimum residence times. The biggest effect on the minimum residence time was achieved with increasing the extruders processing length to 1200 mm (L/D 48). An increase of minimum residence time of 50 % was found for 100 and 500 min^{-1}. Temperature profile and throughput (5 kg/h) were kept constant for all experiments.

Figure 1 illustrates $t_{R,min}$ of PCL during extrusion in dependency on screw configuration and rotation speed, whereas the last-mentioned parameter was varied on two levels at 100 and 500 min^{-1}. The $t_{R,min}$ decreases in the range of 25 to 40 % with increasing rotation speed for all investigated screw configurations. The presented effect is moderate, as the rotation speed was increased by the factor of five.

The influence of throughput on the minimum residence time was investigated in the range between 5 and 15 kg/h, whereas the rotation speed was kept constant at 500 min^{-1}. **Figure 2** shows that $t_{R,min}$ decreases exponentially with increasing throughput independently for all screw configuration. The reduction of residence is in the range of 50 to 60 %, when increasing the throughput from 5 to 15 kg/h.

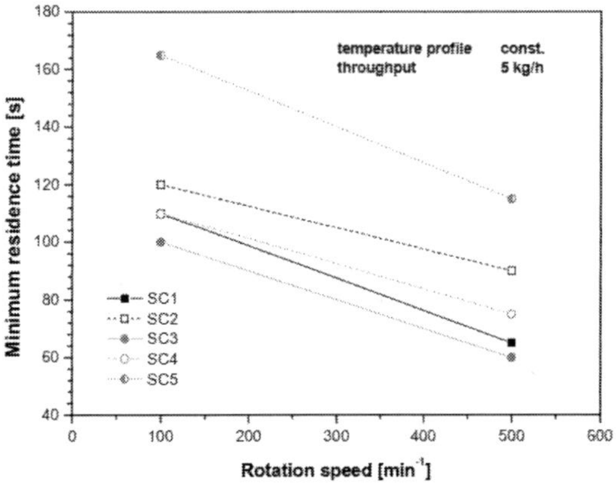

Figure 1 - Minimum residence time of PCL depending on screw configuration and rotation speed

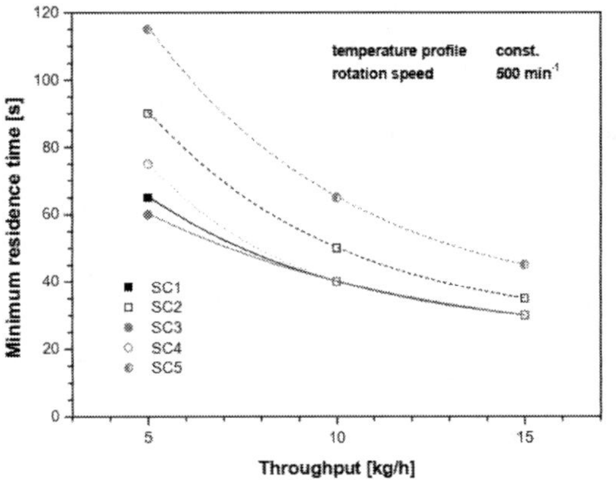

Figure 2 - Minimum residence time of PCL depending on screw configuration and throughput

Influence of processing conditions on CNT dispersion in PCL based masterbatches

As CNT agglomerates have high cohesive strengths due to van-der-Waals forces and physical entanglements, critical shear stresses have to be exceeded during melt mixing to disperse them. Rotation speed and matrix viscosity are the major parameters, to control the shear forces acting on the CNT agglomerates. The increase of rotation speed in twin-screw extrusion of PCL masterbatches containing 7.5 wt% CNT, while keeping throughput and temperature profile constant, was found to have a significant influence on the macroscopic CNT dispersion. The investigated macroscopic CNT dispersion within PCL based masterbatches increased significantly for all investigated screw profiles. In all cases, aside from SC5, where no masterbatch could be produced with 100 rpm due to a torque overload, the area ratio decreased by about 70 % with increasing rotation speed from 100 to 500 min^{-1} (**Figure 3**).

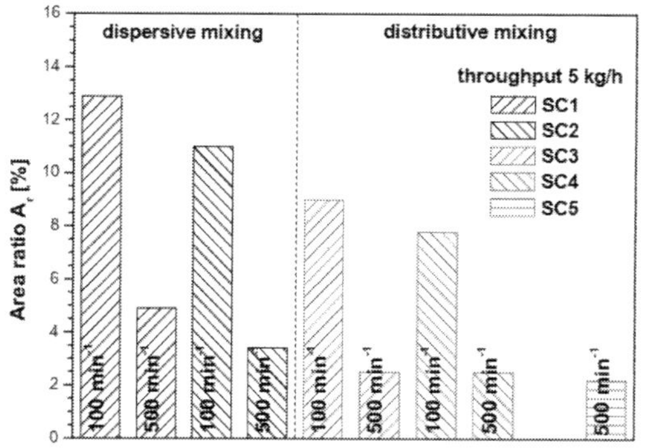

Figure 3 - A_r of PCL masterbatches depending on rotation speed and screw configuration

This finding can be explained with the increasing SME input, which is introduced into the polymer melt during the melt mixing process, with increasing rotation speed according to **Equation 1**. Due to a torque reduction at the higher rotation speed, caused by shear thinning and a certain thermal reduction of PCL, the SME is tripled at increased rotation speed instead of quintupled. A power law relationship between observed area ratio A_r and SME input was found for both, dispersive and distributive mixing screws (**Figure 4**). It can be concluded, that the use of distributive mixing screws leads to a higher degree of CNT dispersion at lower SME inputs, which can be explained with a significant higher mixing efficiency of the mixing elements. Furthermore, it was shown, that there is a certain limitation in dispersing the CNT within one extrusion run, as a further increase of SME would not lead to a significantly better state of dispersion.

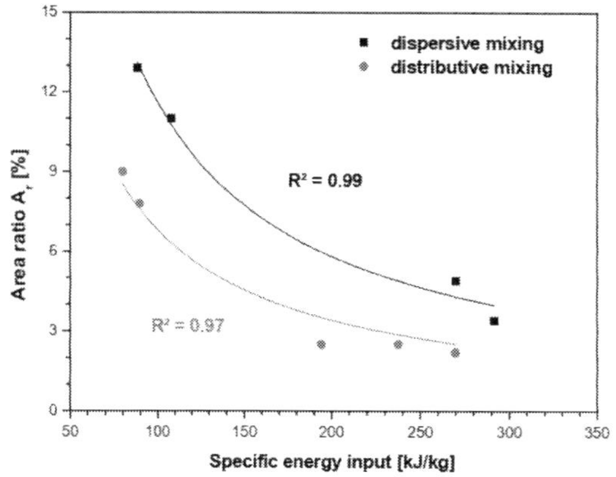

Figure 4 - A_r of PCL masterbatches versus SME input at constant throughput of 5 kg/h

A closer look on the analysis of agglomerate size distribution reveals detailed information about the change in CNT dispersion, when increasing the rotation speed. An agglomerate size distribution diagram for SC1 and SC2 is shown in **Figure 5**. The first important information concerns the width of the curve, which depends on the rotation speed. An increase of this parameter leads to a significant width reduction. In case of the dispersive screw configurations, an increase in rotation speed leads to a reduction of the maximum primary agglomerate size from 120 to 100 µm (SC1) and 150 to 120 µm (SC2). Despite from this, the relative reduction of occupied are by large primary agglomerates is small in comparison to the reduction of area resulting from smaller agglomerates, which explains the finding of a certain plateau that hinders the further increase of CNT dispersion with increasing SME input.

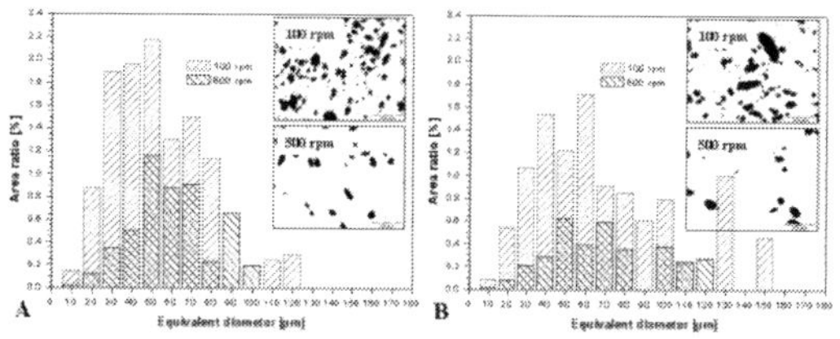

Figure 5 - Primary agglomerate distribution in PCL masterbatches depending on rotation speed and screw configuration, A: SC1, B: SC2

A significant reduction of A_r in PCL masterbatches was also found for the distributive mixing screws, although there is a principle difference in their agglomerate size distribution curves. In comparison to the masterbatches, which were produced with the dispersive screws, the masterbatches processed with distributive screws contain CNT agglomerates, whose maximum size is significantly reduced to 100 µm, independent from the rotation speed used for extrusion, resulting in a significant reduction of A_r. A further increase of processing length and therefore increased minimum residence time from 75 to 115 seconds, did not lead to a further reduction of A_r, although a minor variation in the distribution curve was observed.
Besides the rotation speed, which is useful to control the shear forces acting on the CNT agglomerates leading to their dispersion, the throughput is an important parameter to control the minimum residence time and overall SME input during extrusion. The investigation of minimum residence time has shown that it decreases significantly with increasing throughput. Whereas the increase in rotation speed leads to increasing macroscopic CNT dispersions, the increase of throughout shows the opposite effect. The increase of throughput in twin-screw extrusion of PCL masterbatches containing 7.5 wt.% CNT, while keeping rotation speed and temperature profile constant, was found to decrease CNT dispersion for all investigated screw profiles. Additionally, the increase of throughput lowers the SME input during mixing, which is a further explanation for this finding. The

SME depending A_r for masterbatches produced with a constant rotation speed of 500 min^{-1} and varied throughputs is presented in **Figure 6**.

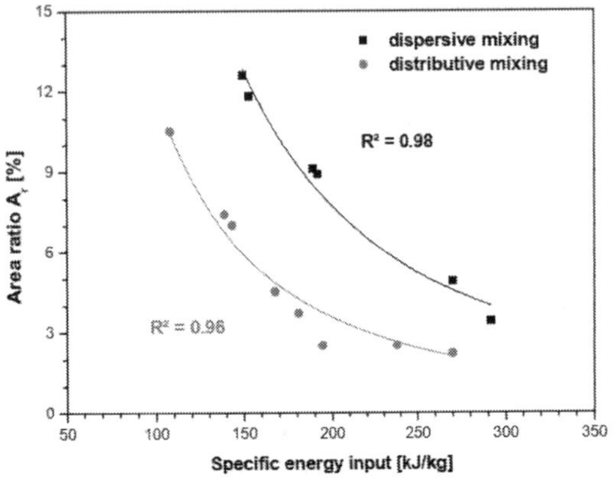

Figure 6 – Area ratio A_r of PCL masterbatches and SME input at constant rotation speed of 500 min^{-1}

Figure 7 shows the throughput dependent macroscopic CNT dispersion of the PCL masterbatch processed with screw configuration SC1, where a significant increase of A_r can be observed with increasing throughput. The area ratio increases from 4.9 (5 kg/h) to 9.1 (10 kg/h), and 11.8 % (15 kg/h). Interestingly, the distribution width is not affected by the throughput depending increasing residence time. The maximum CNT agglomerate size within the PCL masterbatches for all three processing conditions is 100 µm. Furthermore, the number of CNT agglomerates in the rage between 70 and 100 µm remains nearly constant. The differences in area ratio result from significant reduced numbers of agglomerates in the range of 10 to 60 µm, where the particle number is strongly reduced with increasing residence time. This finding leads to the conclusion that big agglomerates above 100 µm in diameter are dispersed very fast, independently from the residence time. A residence time dependent dispersion of small primary CNT agglomerates within the PCL could be an indicator for ongoing erosion, where CNT are eroded continuously from the agglomerate surfaces. This process was found to be strongly time dependent in case of carbon black [26].

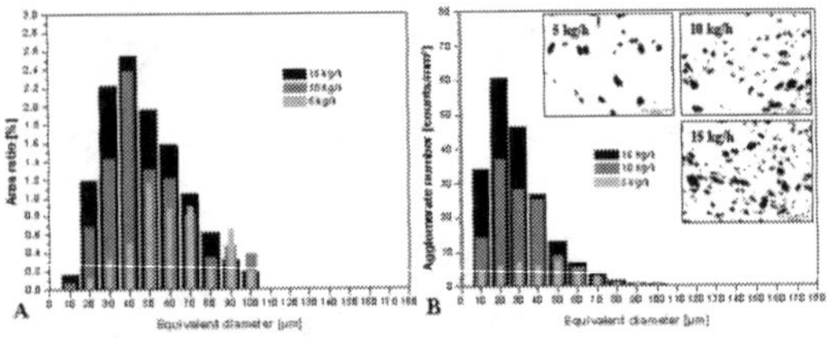

Figure 7 - Influence of throughput on the primary CNT agglomerate size distribution for screw configuration SC1, A: area ratio A_r, B: agglomerate number

In comparison to the results obtained for the processing of PCL masterbatches with screw configuration SC1, the findings were confirmed for SC2 and SC4, too. Again a significant reduction of agglomerate numbers was determined for agglomerate sizes between 10 and 60 µm, whereas the number of lager agglomerates above 60 µm remained constant with variation of throughput.

In order to get an impression how the different mixing and kneading elements act on the CNT dispersion and where the CNT dispersion takes place, the screw was pulled out during extrusion of PCL masterbatches. Two different processes with the use of screw configurations SC2 and SC5 were investigated. Up to ten samples of around one gram were collected along the extrusion length starting directly behind the mixing zone and after every mixing or kneading section. As shown in **Figure 8**, the area ratio of CNT within PCL masterbatches develops exponentially along the extruder. Whereas the highest dispersion capacity was found for the first half of the extruder length, the dispersion rate is smaller afterwards and levels on a plateau. Interestingly, the agglomerate size distribution width is not influenced along the extruder.

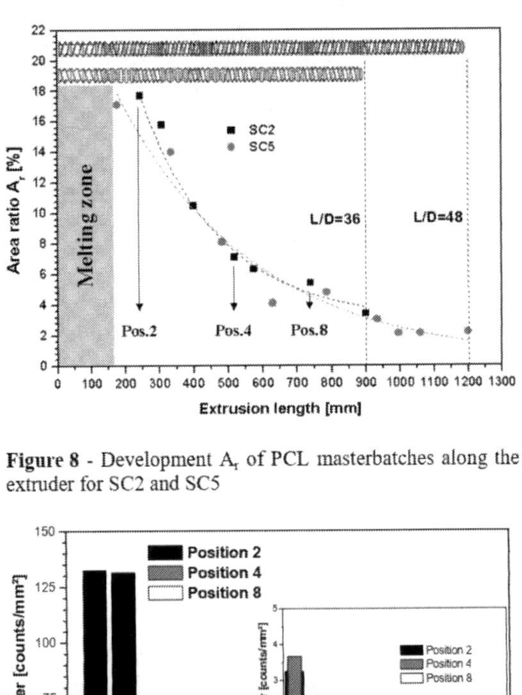

Figure 8 - Development A_r of PCL masterbatches along the extruder for SC2 and SC5

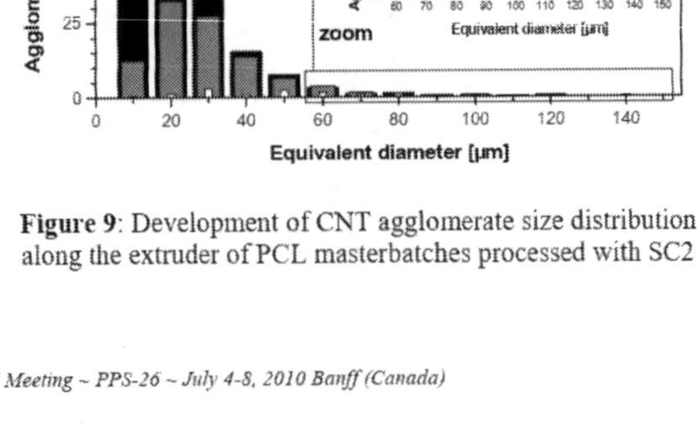

Figure 9: Development of CNT agglomerate size distribution along the extruder of PCL masterbatches processed with SC2

The maximum agglomerate size of 120 µm observed directly behind the melting zone was also found in the material close to the extrusion die (**Figure 9**). This finding indicates again, that the particle size is not influenced by the extrusion length or residence time. Moreover, it is the particle size distribution, which changes significantly for agglomerates in the range between 10 and 60 µm. It has to be assumed, that the rupture of very big agglomerates to fragments of 100 µm or smaller, happens in the melting zone of the twin-screw extruder in the very beginning of the mixing process.

Masterbatch dilution

The masterbatch, which was processed with screw configuration SC5 using a rotation speed of 500 min^{-1} and a throughput of 5 kg/h exhibited the highest degree of CNT macro dispersion and was therefore diluted with neat PCL CAPA 6800 in order to produce composites with desired CNT contents in context with future applications. Masterbatch pellets and pure PCL were premixed prior feeding into the hopper for this purpose. The processing was performed under the same conditions as the masterbatch production. PCL based composites had CNT contents of 0.125, 0.25, 0.375, 0.5, 0.75, 1.0, 1.5, 2.0, 3.0, and 4.0 wt.%.
Light microscopic analysis PCL composites reveals a nearly perfect macroscopic CNT dispersion within the matrix, as just a few very small remaining agglomerates in the micron scale were observed. The measurement of electrical resistivity of PCL based composites filled with NC N7000 showed a very low percolation threshold of 0.35 wt.%, which corresponds to 0.23 vol.% of filler (**Figure 10**).

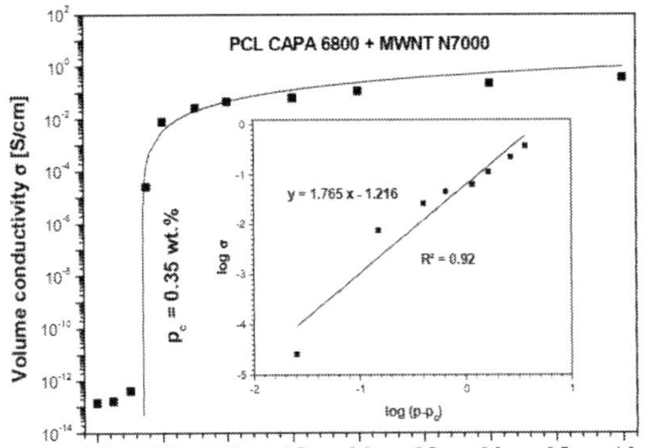

Figure 10 - Electrical percolation behaviour of MWNT N7000 in PCL based composites

Summary and conclusions

The presented article discussed the extrusion parameter dependent CNT macro dispersion within a PCL matrix, whereas the focus was on the processing of masterbatches. The reason for that is the assumption that high quality masterbatches with highly dispersed CNT should increase the quality of further diluted composites. Therefore, these studies are essential as they can help to increase the state of CNT dispersion within nano-composites of further investigations in the CNT research society.
The processing parameters in twin-screw extrusion highly influence the CNT macro dispersion in PCL based masterbatches. The presented study revealed the fundamental relationship between screw profile, rotation speed, and throughput in extrusion on the minimum residence time of the extrudates, specific mechanical energy input, and resulting CNT macro dispersion, which was quantified in terms of the area ratio A_r between area occupied by primary agglomerates and investigated area by means of light microscopy on thin sections.
The rotation speed, which mainly controls the shear conditions during extrusion, was found to have a high influence on the CNT macro dispersions in PCL. The determined area ratio A_r decreased significantly about 70 % when increasing the rotation speed from 100 to 500 min^{-1}, although a decrease of minimum residence time in the range of 25 to 40 % was found. The much better macro dispersion has therefore to be attributed to the three times higher input of specific mechanical energy. As the macro dispersion was found to follow a power law dependency on SME input, it has to be assumed, that a further increase of rotation speed does not result in the entire dispersion of CNT in the PCL matrix. The reason for this was discussed on the basis of primary CNT agglomerate size distribution curves. Whereas an increase of rotation speed leads to a certain decrease of maximum CNT agglomerate size in the range of 20 to 30 microns and a significant reduction of CNT agglomerates of smaller size, undispersed CNT agglomerates in the range of 100 µm were observed in the masterbatch processed with 500 min^{-1}.
The minimum residence time during extrusion of PCL was found to depend strongly on the throughput, whereas an increase of throughput from 5 to 15 kg/h was found to result in a decrease of $t_{R,min}$ in the order of 50 %. Although the variation of throughput had no influence on the CNT primary agglomerate size distribution width, a significant reduction of the agglomerate number of small agglomerates with sizes between 10 and 60 µm was observed. This finding leads to the conclusion that the dispersion of small agglomerates is strongly time dependent, whereas the dispersion of larger agglomerates is mainly shear depending.
The third important processing parameter in twin-screw extrusion is the design of the screw configuration, where two principle design possibilities exist. An extrusion screw can contain either mixing or kneading elements. The investigations of the influence of screw design on $t_{R,min}$ have shown, that the choice of screw elements like kneading or mixing has a significant influence. As kneading blocks exhibit an active conveying effect, the use of these elements

results in longer minimum residence times in comparison to the use of mixing elements. This effect is independent from rotation speed. The further addition of back conveying elements leads to a significant increase of the minimum residence time of about 10 to 30 %, depending on the applied rotation speed. This trend was found for dispersive and distributive screw configurations. The biggest effect on the minimum residence time was achieved with increasing the extruders processing length to 1200 mm (L/D 48). An increase of minimum residence time of 50 % for both investigated rotation speeds was found.
Although all distributive screw configurations exhibit shorter minimum residence times in comparison to the dispersive pendants, the use of these screws led to significantly better CNT dispersions. This finding can be related to the much higher mixing efficiency of such mixing elements. Furthermore, the distributive screw configurations mixing elements resulted in significantly lower SME inputs, which can be advantageous in order to minimise the CNT shortening, which is expected to occur during extrusion process.
As a major result of the presented study, the masterbatch exhibiting best macroscopic CNT dispersion was processed with an extended distributive screw having a length to diameter ratio of 48. The area ratio A_r of remaining CNT primary agglomerates was 2.2 %, whereas the agglomerates where further dispersed in the subsequent dilution process. The diluted composites containing highly dispersed CNT showed furthermore a very low percolation threshold of 0.35 wt.%, which corresponds to an overall filling of 0.23 vol.%.

Acknowledgements

We gratefully acknowledge financial support from the project INTELTEX (Intelligent multi-reactive textiles integrating nanofiller based CPC-fibres) - a European Integrated Project supported through the Sixth Framework Programme for Research and Technological development. In addition, we thank Nanocyl S.A. for providing the nanotubes within the INTELTEX project.

References

1. Iijima S, Helical Microtubules of Graphitic Carbon, Nature, 1991;354(6348):56-58.
2. Saito R, et al., Electronic-Structure of Chiral Graphene Tubules, Applied Physics Letters, 1992;60(18):2204-2206.
3. Wildoer JWG, et al., Electronic structure of atomically resolved carbon nanotubes, Nature, 1998;391(6662):59-62.
4. Odom TW, et al., Atomic structure and electronic properties of single-walled carbon nanotubes, Nature, 1998;391(6662):62-64.
5. Berber S, et al., Unusually high thermal conductivity of carbon nanotubes, Physical Review Letters, 2000;84(20):4613-4616.
6. Kim P, et al., Thermal transport measurements of individual multiwalled nanotubes, Physical Review Letters, 2001;8721(21):
7. Che JW, et al., Thermal conductivity of carbon nanotubes, Nanotechnology, 2000;11(2):65-69.
8. Wong EW, et al., Nanobeam mechanics: Elasticity, strength, and toughness of nanorods and nanotubes, Science, 1997;277(5334):1971-1975.
9. Treacy MMJ, et al., Exceptionally high Young's modulus observed for individual carbon nanotubes, Nature, 1996;381(6584):678-680.
10. Yakobson BI, et al., Nanomechanics of carbon tubes: Instabilities beyond linear response, Physical Review Letters, 1996;76(14):2511-2514.
11. Krause B, et al., Correlation of carbon nanotube dispersability in aqueous surfactant solutions and polymers, Carbon, 2009;47(3):602-612.
12. Villmow T, et al., Influence of injection molding parameters on the electrical resistivity of polycarbonate filled with multi-walled carbon nanotubes, Composites Science and Technology, 2008;68(3-4):777-789.
13. Yuan JM, et al., Preparation of polystyrene-multiwalled carbon nanotube composites with individual-dispersed nanotubes and strong interfacial adhesion, Polymer, 2009;50(14):3285-3291.
14. Chen L, et al., Study on polycarbonate/multi-walled carbon nanotubes composite produced by melt processing, Materials Science and Engineering a-Structural Materials Properties Microstructure and Processing, 2007;457(1-2):287-291.
15. Pegel S, et al., Dispersion, agglomeration, and network formation of multiwalled carbon nanotubes in polycarbonate melts, Polymer, 2008;49(4):974-984.
16. Kasaliwal G, et al., Influence of Processing Conditions in Small-Scale Melt Mixing and Compression Molding on the Resistivity and Morphology of Polycarbonate-MWNT Composites, Journal of Applied Polymer Science, 2009;112(6):3494-3509.
17. Villmow T, et al., Influence of twin-screw extrusion conditions on the dispersion of multi-walled carbon nanotubes in a poly(lactic acid) matrix, Polymer, 2008;49(16):3500-3509.
18. Xu GY, et al., Nonisothermal crystallization kinetics and thermomechanical properties of multiwalled carbon nanotube-reinforced poly(epsilon-caprolactone) composites, Polymer International, 2008;57(9):1052-1066.
19. Wu D, et al., Rheological properties and crystallization behavior of multi-walled carbon Nanotube/Poly(epsilon-caprolactone) composites, Journal of Polymer Science Part B-Polymer Physics, 2007;45(23):3137-3147.

20. Saeed K and Park SY, Preparation and properties of multiwalled carbon nanotube/polycaprolactone nanocomposites, Journal of Applied Polymer Science, 2007;104(3):1957-1963.
21. Mitchell CA and Krishnamoorti R, Dispersion of single-walled carbon nanotubes in poly(epsilon-caprolactone), Macromolecules, 2007;40(5):1538-1545.
22. Shaffer MSP and Windle AH, Fabrication and characterization of carbon nanotube/poly(vinyl alcohol) composites, Advanced Materials, 1999;11(11):937-941.
23. Rwei SP, et al., Characterization of Agglomerate Dispersion by Erosion in Simple Shear Flows, Polymer Engineering and Science, 1991;31(8):558-562.
24. Krause B, et al., Influence of small scale melt mixing conditions on electrical resistivity of carbon nanotube-polyamide composites, Composites Science and Technology, 2009;69(10):1505-1515.
25. Pötschke P, et al., Dielectric spectroscopy on melt processed polycarbonate--multiwalled carbon nanotube composites, Polymer, 2003;44(17):5023-5030.
26. Rwei SP, et al., Observation of Carbon-Black Agglomerate Dispersion in Simple Shear Flows, Polymer Engineering and Science, 1990;30(12):701-706.

MODELLING POLYMER/CLAY NANOCOMPOSITES USING STATISTICAL AND CONTINUUM APPROACHES WITH EXPERIMENTAL VERIFICATION.

J Sweeney*, A Alshehri, P E Spencer and P D Coates

School of Engineering, Design and Technology/IRC in Polymer Science and Technology, University of Bradford, Bradford, UK – j.sweeney@bradford.ac.uk

Abstract – Finite element modeling is used to simulate the reinforcement effect of clay platelets in a polypropylene matrix. A Monte Carlo approach is used to insert model platelets at a random distribution of orientation angles into a two-dimensional model. Small strain constant width stretches are used to predict the macroscopic modulus of the nanocomposite. A simple two-phase model gives low modulus predictions in comparison with experimental measurement. The introduction of an interphase adjacent to the platelets, of higher modulus than the matrix, is shown to give more realistic predictions. The extent and modulus of the interphase are consistent with basic polymer physics.

Introduction

We have developed a model of the mechanical behavior of platelet-reinforced composites which we have applied to polypropylene/clay nanocomposites. This enables us to explore the effects of platelet concentration, orientaton distribution, level of exfoliation and curvature on the macroscopic properties of the composite (Spencer and Sweeney 2009). Here we compare its predictions of tensile modulus with experiments conducted in our own laboratory. We investigate the 'nano-effect', the observation that when reinforcement is on the nanoscale, its effects are greater than that observed for similar properties and concentration at the microscale. Clearly simple two-phase (platelet/matrix) modeling will not reproduce this effects, as the results will be scale-independent. However, the introduction of an interphase around the nanoparticles, with physical dimensions and properties dependent on the local physics and, opens up the possibility that macroscopic properties will depend on the absolute filler size.

Experimental and modeling

The essentials of the modeling technique have been outlined elsewhere (Spencer and Sweeney 2009). Figure 1 shows a model consisting of a fully exfoliated system of nanoparticles, of aspect ratio 100, randomly distributed and embedded in a relatively soft elastic matrix. With appropriate modulus values (~2GPa for a polypropylene modulus and 160 GPa for clay platelets), the stretching of the model predicts an increase in macroscopic modulus of around 15-20% for filler loadings of 5 – 7 wt%. This is much lower than observed for effective nanocomposites with this system, which can produce reinforcement amounting to around 50% in modulus (Lui and Wu 2001).

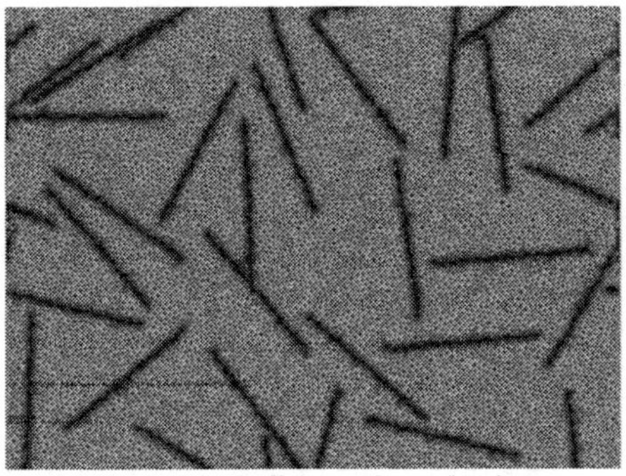

Figure 1. Model including random platelets

We have introduced interphases around the nanoparticles which are stiffer than the matrix material. Such zones are believed to exist as a result of the confinement effect of the presence of a nanoparticle surface; the motion of polymer chains is restricted, producing a stiffening effect of the same order as a glass transition. The extent of the zone is assumed to be determined by the radius of gyration of a polymer chain. In our case, the molecular weight of the polypropylene is given by $\bar{M}_n = 80,000$, which gives from standard calculations a radius of gyration of 3.9nm. This is related to the zone width around the platelet in our model. The modulus of the polypropylene is 1.3 GPa, and this is assumed to be enhanced by a factor of 3 in the interphase. This is in line with molecular dynamics calculations, and of the same order as is associated with a glass transition. Two types of zones were introduced: 'step zones', of extent equal to the radius of gyration r, and the modulus falling immediately to the matrix value at the edge of the zone; and 'linear zones', where the modulus falls as a linear function of distance from the point defined by the radius of gyration, to reach the matrix value at 3r from the platelet. For a realistic model, we have assumed that the exfoliation of the model is not perfect, with the plates occurring in stacks of two, separated by a single chain radius of gyration. This is appropriate for an intercalated system. A typical mesh for the model using step zones, with contours associated with modulus, is shown in Figure 2.

Nanocomposites were made at various clay loadings and loadings of the compatibiliser PPMA. The modulus results are shown in Figure 3.

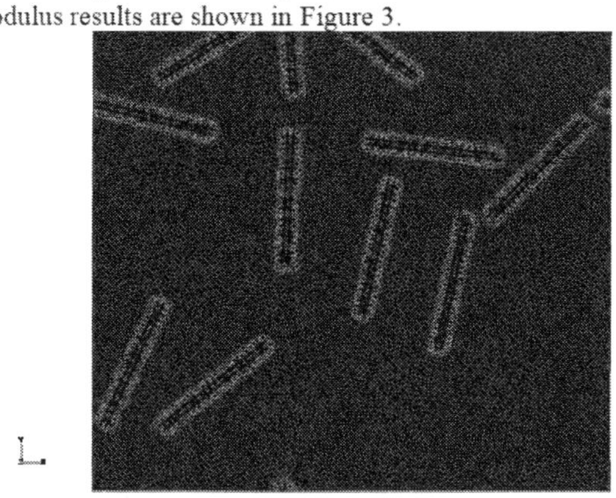

Figure 2. Mesh for model including interzones.

Results and Discussion

Results, at 5 wt% clay loading, are compared with experiment in Figure 3. The 'step zone' predictions are

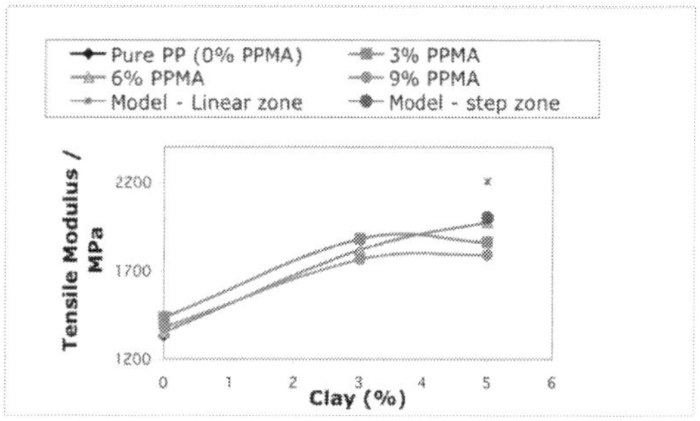

Figure 3 Experimental results and model predictions.

close to the observations, and best for the optimum modulus results where the particle bonding may be assumed to be most effective.

Conclusions

We have made reasonable estimates of the extent and modulus of the interphase zone, based on relevant polymer physics. The predictions are close to our observations, which are of around 50% increase in modulus, consistent with the results of other workers (Liu and Wu 2001). This supports the contention that the 'nano-effect' is indeed the result of the interphase, with properties in line with conventional theory.

References

Liu, X, Wu,Q "PP/clay nanocomposites prepared by grafting-melt intercalation", Polymer **42**, 10013-10019 (2001).

Spencer, P.E. and Sweeney, J "Chapter 15 Modeling of polymers clay nanocomposites for a multiscale approach", in Nano- and Micro-Mechanics of Polymer Blends and Composites, Karger_Kocsis, J. and Fakirov, S(Eds.), Hanser Publishers, Munich, p. 545–578 (2009).

A relationship between microstructure and surface paintability of PP/EPDM/organoclay nanocomposites interfacially compatibilized with ethylene vinyl acetate

Ali Asghar Katbab, Ramin Yousefzadeh Tabasi*

Amirkabir University of Technology - Tehran - Iran

Thermoplastic olefins(TPO)/ Clay nanocomposites with improved surface paintability and pull off strength using ethylene vinyl acetate(EVA) as interfacial compatibilizer have been prepared via melt mixing process. XRD analysis exhibited that incorporation of EVA with both polar functional group and rubber-like behavior into the pre-blended PP/EPDm prior to the feeding of nanoclay facilitates the intercalation level and dispersion of the clay nanolayers. EVA showed to be effective in controlling the polar component of the surface tension and γp hence tailoring the surface paintability of the TPO nanocomposite. The smoothness of the painted surface and pull off strength found to be influenced by the clay microstructure and the micromorphology of the dispersed rubber droplets. Qualitative analysis by AFM and SEM revealed reduction in the size of EPDM droplets by inclusion of the clay particles in the presence of EVA, leading to the increased impact resistance of the TPO nanocomposites.

POLY (ETHYLENE TEREPHTHALATE) NANOCOMPOSITES BASED ON TURKISH ORGANO-BENTONITES TREATED WITH THERMALLY STABLE PHOSPHONIUM AND IMIDAZOLIUM SALTS

W. Abdallah and U. Yilmazer*

Department of Chemical Engineering, Middle East Technical University, Ankara, Turkey – w.k.abdallah@hotmail.com; yilmazer@metu.edu.tr

Purification of bentonite clays mined from Resadiye (Turkey) and their modification with two phosphonium and two imidazolium organic salts were investigated. The organo-bentonites were subsequently melt compounded with amorphous Poly (ethylene terephthalate), with and without the use of an elastomeric compatibilizer. The thermal stability, morphology and mechanical properties of the binary and ternary nanocomposites were studied.

The onset decomposition temperature (5 %wt. loss) of the unpurified clay and purified clay were determined as 666°C and 769°C respectively. Inorganic minerals present in the raw clays were removed upon purification, and the clay cation exchange capacity was increased. The phosphonium and imidazolium treated organo-bentonites exhibited maximum onset decomposition temperature of 380°C and 454°C, respectively. The onset decomposition temperature of pure PET was measured as 400°C. Upon compounding PET with phosphonium and imidazolium salts, the onset decomposition temperature increased to a maximum value of 408°C and 411°C, respectively. The basal spacing of the purified bentonite is 1.1 nm, and those of the two phosphonium organo-bentonites used were measured as 2.6 nm and 1.8 nm. The interlayer spacing of the imidazolium-bentonites was between 1.34 nm and 1.43 nm indicating a more compact structure. These interlayer spacings did not change significantly upon compounding with PET. The tensile strength and Young's modulus of pure PET were 47 MPa and 1750 MPa respectively. The tensile strength and Young's modulus reached a maximum of 57.8 MPa and 2064 MPa respectively for binary imidazolium-nanocomposites with organo-bentonites only, and 53 MPa and 1810 MPa for ternary nanocomposites with organo-bentonites and elastomer, whereas phosphonium organo-bentonite nanocomposites exhibited lower mechanical properties. The impact strength of PET increased from 2.9 kJ/m^2 to a maximum of 4.7 kJ/m^2 and 5.1 kJ/m^2 for phosphonium and imidazolium based nanocomposites, respectively.

Introduction

Organoclays are abundant, inexpensive, environment friendly and above all essential to develop polymer nanocomposites. All these properties attracted researchers to surface modification of clays which creates new materials to be used in a wide spectrum of new applications (Bergaya et al., 2001). Smectites are widely used clays by virtue of their high cation exchange capacity, excellent ability to swell, high platelet aspect ratio and ease with which their surface can be modified (Tiwari et al., 2008).

Polymer-clay nanocomposites have received great interest in the past decade, because these materials can exhibit enhanced mechanical properties, improved gas barrier properties, superior flame retardancy, higher thermal stability, and higher heat distortion temperature at low clay loadings compared to conventional polymer composites containing traditional fillers (Mert and Yilmazer, 2008). Ion-exchange reactions with cationic surfactants, such as ammonium ions, render the normally hydrophilic silicate surface organophilic, which makes intercalation of many engineering polymers possible and improve the wetting characteristics with the polymer (Giannelis, 1998) by decreasing the surface energy of montmorillonite and expanding the basal spacing (Favre and Lagaly, 1991).

The first approaches to montmorillonite-based organoclays adopted ammonium surfactants which usually contain long aliphatic chain (C12–C18) to cause expansion of the distance between the layers (Calderon et al., 2008). The thermal stability and flammability performance of polymer nanocomposites are strongly influenced by the processing stability of both the polymer and the organic-treated layered silicate. The low thermal stability of ammonium surfactants (onset of thermal decomposition at about 200°C or below) presents a problem for melt compounding and processing (injection molding) of polymer nanocomposites into final molded products, where high processing temperatures exceeding 200°C are commonly encountered such as: Polyamides (PA6, PA66) and poly (ethylene terephthalate) (Awad et al., 2004). Alternative clay surfactants, like phosphonium or imidazolium salts, have been used in order to enhance the thermal stability of organoclays and circumvent the problem presented by ammonium surfactants. The onset of the thermal decomposition of the organic modifier sets the ceiling temperature for polymer processing (Xie et al., 2002).

Impurities like non-clay minerals and other clay minerals exist in bentonites. Some of these minerals that are porous materials are montmorillonite, beidellite, nontronite, saponite and hectorite. Some common impurities in bentonite clay are quartz, calcite, feldspar, mica and organic matter (Grim, 1953). These impurities negatively affect the cation exchange capacity and thermal stability of bentonites. Therefore, isolation of some smectite group minerals from bentonites by purification is of great importance before the surface modification step (Onal et al., 2003).

Poly (ethylene terephthalate) (PET) is a thermoplastic polyester having poor impact resistance and high notch sensitivity. In PET/clay nanocomposites, the addition of clay sometimes imparts drawbacks to the resulting material such as brittleness. Impact modification of PET can be achieved by dispersing elastomeric polymers in the polymer matrix. In addition, the presence of a thermally stable organoclay will also overcome the problem of melt compounding and processing at high temperatures.

The main objective of the present investigation is aimed at exploring the effects of purification and organic modification of Turkish layered silicates on the thermal stability, morphology, and mechanical properties of PET/organo-bentonite nanocomposites.

Experimental

Materials

The montmorillonite rich sodium bentonite clay used in this study was mined from Resadiye/Tokat region in Turkey. The raw bentonite clay (B) contained 90% Montmorillonite. Two phosphonium (Ph_1 and Ph_2) and two imidazolium (Im_1 and Im_2) surfactants were used as clay modifiers (Figure 1). Amorphous Poly (ethylene terephthalate) pellets were used as the polymer matrix. Lotader® AX8900 resin (a random terpolymer of ethylene, methyl acrylate and glycidyl methacrylate (E-MA-GMA)) in pellet form was used as an impact modifier. The epoxy functionality of GMA can react with the carboxyl and hydroxyl end groups of PET in the melt phase to form a graft copolymer.

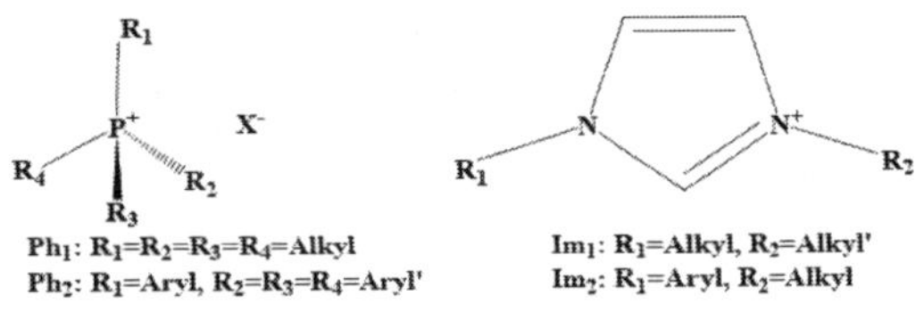

Figure 1 Chemical structure of the surfactants used.

Preparation of Organoclays

Purification of the bentonite clay before surface modification is necessary in order to obtain highly pure clays by removing non-clay minerals. This step increases the intercalation of the polymer and clay, and also increases the thermal stability of the clay. The purification procedure was accomplished via the separation of the clay minerals from other minerals by sedimentation. The IR spectrum of natural montmorillonite indicated a moderate Fe^{3+} content (875 cm^{-1}) with the presence of detectable quartz (700 cm^{-1}) and silica phase (800 cm^{-1}) impurities. The cation exchange capacity of the bentonite clays as received and after purification was determined as 67.5 meq/100g clay and 95 meq/100g clay respectively in accordance with the procedure in ASTM Test Method C 324. To prepare phosphonium and imidazolium organoclays; purified bentonites (PB) were initially dried in vacuum at 120°C for 12 hrs. The amount of the surfactants added to all clays was 1.1 * the CEC. After the ion exchange process, washing was repeated at least three times, and no halide traces were detected with silver nitrate. After washing, the organoclay was dried overnight at room temperature then, at 120°C for 24 hrs under vacuum. The phosphonium organoclays ($PB.Ph_1$ & $PB.Ph_2$) and imidazolium ones ($PB.Im_1$ & $PB.Im_2$) produced were ground in a mortar (<106μm) and dried again under same conditions and then stored in a dessicator.

Preparation of PET/organoclay Nanocomposites

PET based nanocomposites were prepared in air atmosphere by melt blending in a co-rotating, intermeshing Thermoprism TSE 16 TC twin screw extruder. During the experiments, the feed rate, screw speed and the temperature profiles were kept constant at 25 g/min, 350 rpm and 200-275-275-275-280°C from the main hopper to the die respectively. Organo-bentonite and elastomer were fed for all composites at 2% and 5%, respectively. After the first extrusion step, the extrudate was cooled in a cooling bath and pelletized using a chopper. The pellets of PET containing compounds obtained at the end of the process were stored in polyethylene bags and were all dried at 110°C under vacuum for 15 hours before the next extrusion step or injection molding. Drying was performed under vacuum for 24 hours at 120°C for the organoclays and for 12 hours at 40°C for the elastomers. Drying step was repeated before each processing step since the presence of even small traces of moisture can cause significant hydrolytic degradation of the materials. The specimens were injection molded by DSM Xplore laboratory scale micro injection molding equipment. During the molding process, the barrel temperature was adjusted to 275°C and the mold temperature was set to 21°C. The samples were injected at a pressure of 8 bars after waiting 4 minutes in the melt state.

Characterization

TGA analysis was performed using a Shimadzu DTG-60H thermal analyzer under nitrogen atmosphere at a scanning rate of 15°C/min, from room temperature up to 800°C for the clays, and up to 600°C for the nanocomposites. RIGAKU D/MAX 2200/PC X-Ray diffractometer that generates a voltage of 40kV and current 40 mA from monochromatic Cu Kα radiation source (λ = 1.5418) was used to analyze the morphological properties of the purified clays and the produced nanocomposites and the dispersion effect between the clays and the polymer matrix. TEM samples were examined by a Tecnai™ G2 F30 Transmission Electron Microscope at an acceleration rate of 120 kV. Tensile tests were performed by using Shimadzu Universal Testing Machine AG-IS (100kN) according to ISO 527. Charpy impact strength of one sided notched specimens with the dimensions of 80x10x4 mm was measured by pendulum Ceast Resil Impactor according to ISO 179.

Results and Discussion

Thermal Properties

The onset decomposition temperatures (5 % wt. loss) of the unpurified and purified clay were determined as 666°C and 769°C, respectively. Non-clay minerals causing thermal instability in the structure of the clays were removed upon purification, and the clay cation exchange capacity was increased from 67.5 meq/100g up to 95 meq/100g. The organoclays PB.Ph$_1$, PB.Ph$_2$, PB.Im$_1$ and PB.Im$_2$ exhibited onset decomposition temperatures of 338°C, 380°C, 454°C and 438°C, respectively (Figure 2). The onset decomposition temperature of pure PET was measured as 400°C. Upon compounding PET with PB.Ph$_1$, PB.Ph$_2$, PB.Im$_1$ and PB.Im$_2$ organo-bentonites, the onset decomposition temperature of the nanocomposites became 395°C, 408°C, 411°C and 410°C, respectively (Figure 3). The addition of the elastomer had no significant effect on the decomposition of the nanocomposites.

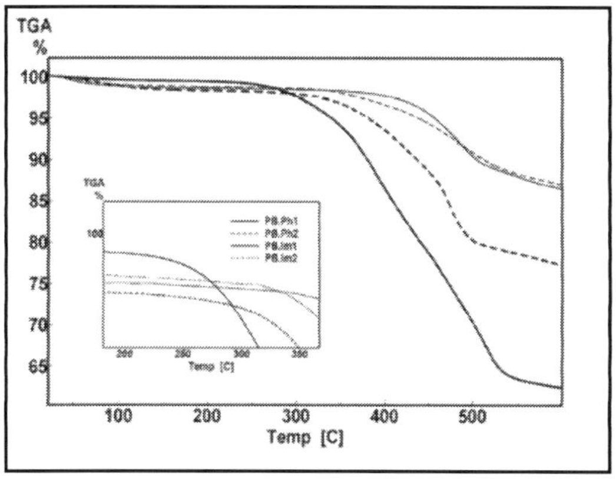

Figure 2 TGA patterns of the organo-bentonites

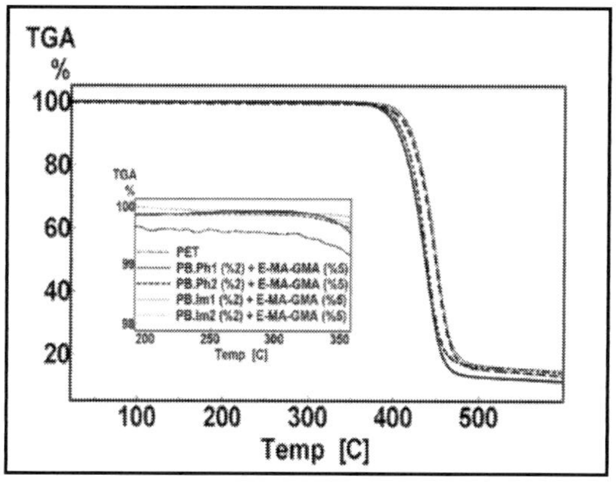

Figure 3 TGA patterns of PET Nanocomposites

Morphology Properties

X-Ray diffraction analyses of the unmodified natural bentonite (B) and the purified bentonite (PB) are presented in Figure 4. The analysis of the natural bentonite (B) by XRD confirmed the major presence of montmorillonite. The unpurified bentonite clay contained clinoptilolite (Cln), analcime (An), feldspar (F), calcite (C), dolomite (D), quartz (Q) and opal-C (OC) as non-clay minerals (Onal et al., 2003); (Ramos Filho et al., 2005). After purification, these minerals did not exist, except for negligible amounts of dolomite (D), quartz (Q) and opal-C (OC). The interlayer distance of bentonite (B) is 1.2 nm (2θ = 7.42). It did not change significantly after purification (PB), and it became 1.1 nm (2θ = 8.08). The basal spacing of the organo-bentonites PB.Ph$_1$, PB.Ph$_2$, PB.Im$_1$ and PB.Im$_2$ used was measured as 2.6 nm, 1.8 nm, 1.34 nm and 1.43 nm, respectively. These interlayer spacings did not change significantly upon compounding with PET (Figure 5) but intercalation was confirmed with TEM results (Figure 6).

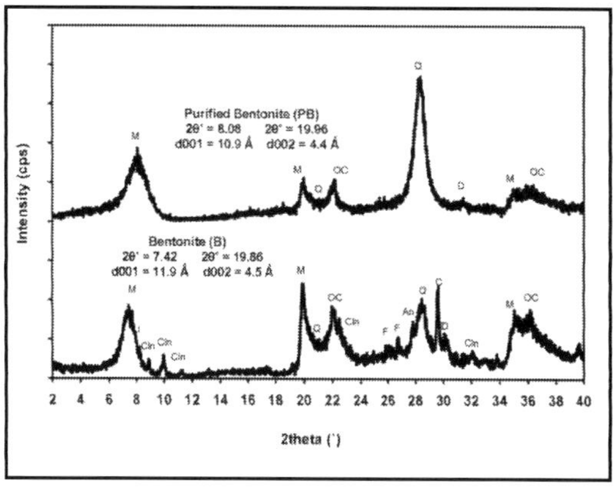

Figure 4 XRD patterns of bentonite (B) and purified bentonite (PB)

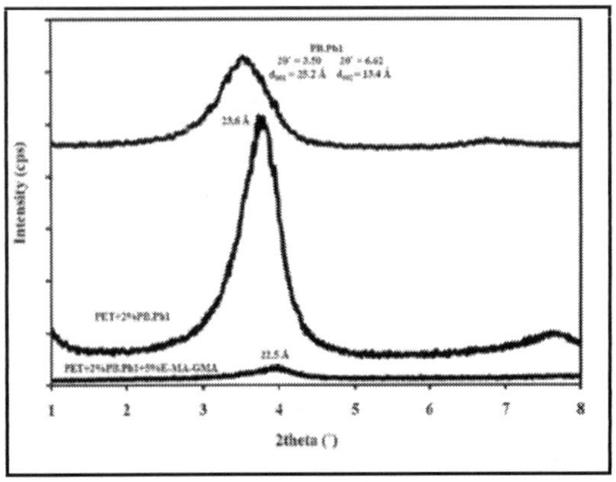

Figure 5 XRD patterns of PET/Ph_1/E-MA-GMA nanocomposite

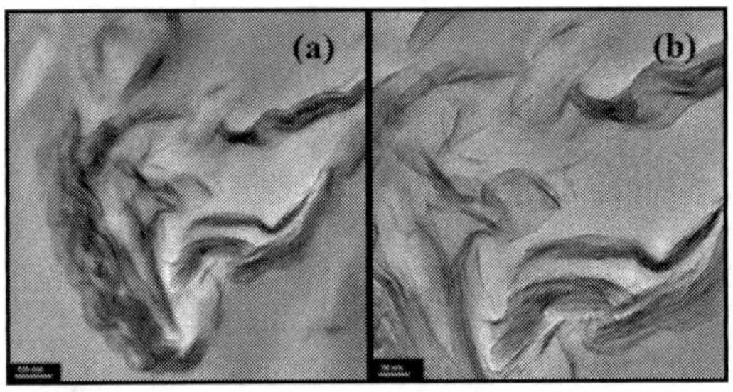

Figure 6 TEM micrographs of the PET/PB.Ph_1/E-MA-GMA nanocomposite (a) 100 nm (b) 50 nm

Mechanical Properties

The tensile strength and Young's modulus of pure PET were 47 MPa and 1750 MPa respectively. The tensile strength (Figure 7) and Young's modulus (Figure 8) of PB.Ph_1 and PB.Ph_2 based binary-nanocomposites were 56.9 MPa, 56.1 MPa, and 1959 MPa, 2145 MPa, respectively. The impact strength of PET and PB.Ph_1 and PB.Ph_2 based binary-nanocomposites (Figure 9) was 2.9 kJ/m^2, 3.1 kJ/m^2 and 3.2 kJ/m^2, respectively. Thus, the use of alkyl (Ph_1) and aryl (Ph_2) based organoclay increased the tensile strength, Young's modulus and the decomposition temperature of PET without changing the impact strength significantly. Imidazolium binary nanocomposites exhibited similar results in which, the tensile strength (Figure 7) and Young's modulus (Figure 8) of PB.Im_1 and PB.Im_2 based binary-nanocomposites were 57.8 MPa, 48.2 MPa, and 2064MPa, 1705 MPa, respectively. The impact strength of PET and PB.Im_1 and PB.Im_2 based binary-nanocomposites (Figure 9) was 2.9 kJ/m^2, 3.1 kJ/m^2 and 3.1 kJ/m^2, respectively.

The tensile strength and Young's moduli for phosphonium based ternary nanocomposites were 49.6 MPa, 46.1 MPa, and 1640 MPa, 1641 MPa. The impact strength of PET increased from 2.9 kJ/m^2 to a value of 4.7 kJ/m^2 and 3.4 kJ/m^2 for PB.Ph_1 and PB.Ph_2 based ternary nanocomposites, respectively. The tensile strength and Young's moduli for imidazolium based ternary nanocomposites were 49.6 MPa, 53.0 MPa and 1810 MPa, 1635 MPa. The impact strength of PET increased from 2.9 kJ/m^2 to a value of 5.1 kJ/m^2 and 3.9 kJ/m^2 for PB.Im_1 and PB.Im_2 based ternary nanocomposites, respectively. The use of alkyl (Ph_1 and Im_1) based ternary nanocomposites results in increase in the impact strength with little variation in the tensile strength and modulus.

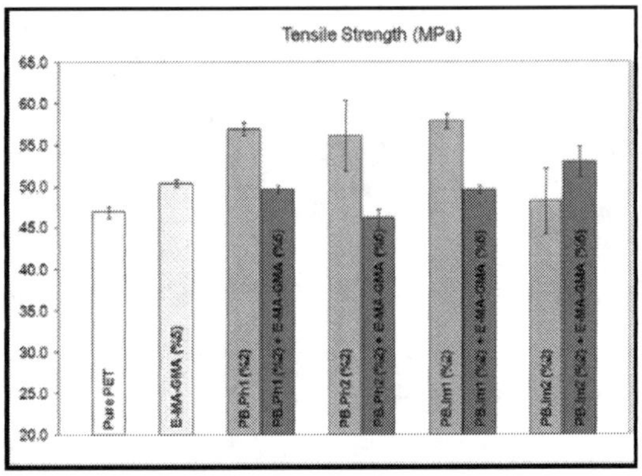

Figure 7 Tensile strength of PET nanocomposites

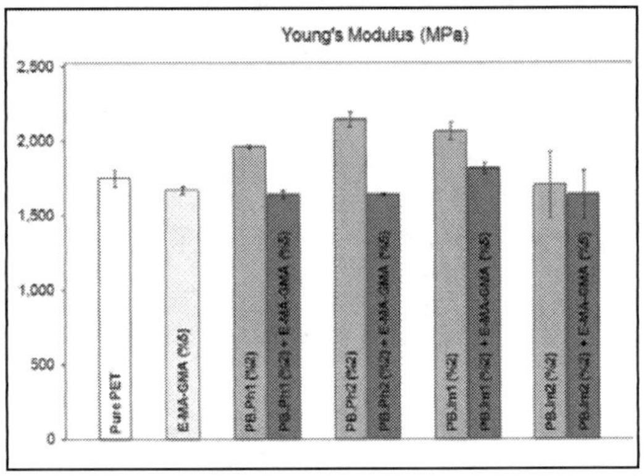

Figure 8 Young's Modulus of PET nanocomposites

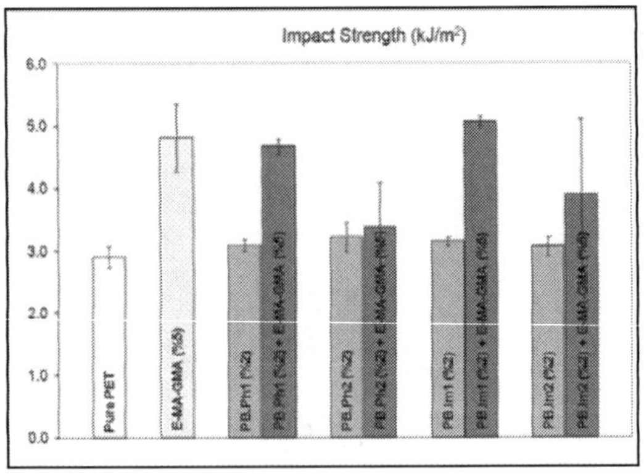

Figure 9 Impact strength of PET nanocomposites

Conclusions

Purification of bentonites by sedimentation isolates the smectite portion and ends up with highly pure MMT with enhanced properties such as cation exchange capacity, chemical structure and thermal stability. The surface modification of the pure MMT with suitable phosphonium and imidazolium surfactants results in thermally stable organo-bentonites which overcome the thermal degradation problem of conventional organo-bentonites produced by alkyl ammonium surfactants when used with high melting-point polymers during compounding and processing.

Acknowledgements

Special thanks to the Scientific and Technological Research Council of Turkey (TUBİTAK) for their financial support during my study.

References

Awad, W. H., Gilman, J. W., Nyden, M., Harris Jr., R. H., Sutto, T. E., Callahan, J., Truvole, P. C., DeLong, H. C., Fox, D. M., "Thermal Degradation Sudies of Alkyl-Imidazolium Salts and Their Application in Nanocomposites", Thermochim. Acta, **409**, 3–11 (2004)

Bergaya, F., Lagaly, G., "Surface Modifications of Clay Minerals", Appl. Cl. Sci., **19**, 1–3 (2001)

Calderon, J. U., Lennox, B., Kamal, M. R., "Thermally Stable Phosphonium-Montmorillonite Organoclays", J. Appl. Cl. Sci., **40**, 90–98 (2008)

Favre, H. and Lagaly, G., "Organo-Bentonites with Quaternary Ammonium Ions'', J. Clay Miner., **26**, 19–32 (1991)

Giannelis, E. P., "Polymer-Layered Silicate Nanocomposites: Synthesis, Properties and Applications'', J. Appl. Organometal. Chem., **12**, 675–680 (1998)

Grim, R. E.: Clay Mineralogy, 1st Edition, McGraw-Hill, New York (1953)

Mert, M. and Yilmazer, U., ''Processing and Properties of Modified Polyamide 66-Organoclay Nanocomposites'', J. Appl. Polym. Sci., **108**, 3890–3900 (2008)

Onal, M., Sarikaya, Y., Alemdaroglu, T., Bozdogan, I., "Isolation and Characterization of a Smectite as a Micro-Mesoporous Material from a Bentonite", Turk. J. Chem., **27**, 683–693 (2003)

Ramos Filho, G. R., Melo, T. J. A., Rabello, M. S., Silva, S. M. L., "Thermal Stability of Nanocomposites Based on Polypropylene and Bentonite", J. Polym. Deg. and Stab. **89**, 383–392 (2005)

Tiwari R. R., Khilar K. C., Natarajan U., "Synthesis and Characterization of Novel Organo-Montmorillonites", Appl. Cl. Sci. **38**, 203–208 (2008)

Xie, W., Xie, R., Pan, W., Hunter, D., Koene, B., Tan, L., Vaia, R., "Thermal Stability of Quaternary Phosphonium Modified Montmorillonites", J. Chem. Mater., **14**, 4837–4845 (2002)

EFFECT OF NANO-CACO3 ON THERMAL AND MECHANICAL PROPERTIES OF TOUGHENED POLY(LACTIC ACID)

Y. W. Leong[1*], S. Thumsorn[1], W. Klinklai[2], H. Hamada[1]

[1]*Department of Advanced Fibro-Science, Kyoto Institute of Technology, Kyoto, Japan – leong@kit.ac.jp; nooh17@yahoo.com; hhamada1@aol.com*
[2]*Faculty of Engineering, Rajamangala University of Technology Thanyaburi, Pathumthani, Thailand – warunee.a@en.rmutt.ac.th*

Abstract – Poly(lactic acid) (PLA) has been known to be extremely brittle even without notching, which greatly limits its application. As such, a novel toughening agent was introduced in order to significantly enhance the toughness as well as deformability of PLA. In order to further enhance its mechanical performance and thermal stability without sacrificing biodegradability, naturally derived fillers were incorporated into PLA. In this study, nano-sized $CaCO_3$ obtained from precipitation and deposition techniques was used as the filler for the toughened PLA. The effects of $CaCO_3$ on the structure, crystallization, thermal stability and mechanical performance of PLA were evaluated.

Introduction

Poly(lactic acid) (PLA) is a bio-based polymer derived by using renewable resources. PLA has been popular due to its good recyclability and superior strength and clarity. However, PLA is known to exhibit brittleness and would tend to deform significantly above its glass transition temperature (Urayama et al. 2003). Various methods have been presented for improving the toughness of PLA such as fiber reinforcement (Oksman et al.2003), blend with other polymers (Aslan et al., 2000; Ramiro et al., 2001) and incorporation of nano-filler (Urayama et al., 2003; Jian et al., 2007).

Calcium carbonate ($CaCO_3$) is a type of mineral filler that is commonly incorporated into polymers for improving their stiffness and toughness. The effectiveness of the filler, however, would largely depend on the filler size. Nano-sized $CaCO_3$-filled composites have been known to exhibit good tensile and thermal properties (Gu et al., 2009; Gong et al., 2009; Li et al., 2006). Urayama et al. (2003) have studied the effect of incorporating of various inorganic filler on properties of poly(L-lactic) (PLLA) and they concluded that the crystallization of PLLA matrix as a result of the incorporation of nano-$CaCO_3$ can prevent the thermal deformation of the PLLA. Polystyrene-nano-$CaCO_3$ composite has been studied by Li et al. (2006) and found that the filler was able to plasticize the polystyrene matrix, which led to greatly reduced melt shear rates during processing.

The objective of this research is study the effect of nano-$CaCO_3$ on the rheological, mechanical, and thermal properties of toughened PLA.

Experimental

Materials and Compounding of Nano-Composites

The toughened poly(lactic acid) used in this study was Lapol™ 33 concentrate supplied by Lapol, LLC, USA, which contains 33 wt% of Lapol polyester resin as the plasticizer and will be referred as Lapol_PLA. The calcium carbonate nano-particles were prepared by precipitation technique in our laboratory, with an average particle size of about 200 nm. Compounding was performed by using a Labo Plastomill (4C150, ToyoSeiki, Japan) internal mixer set at 160 °C with rotor speed of 10 rpm for 10 minutes. After mixing, the compound was cut into grinded into powder and compression molded at 160 °C to form 1 mm-thick sheets. The compound was pre-heated for 3 minutes prior to applying a pressure of 10 MPa for another 3 minutes. The mold was subsequently rapid cooled by flowing water while the specimens were kept under constant pressure.

Characterization

Rheological and viscoelastic properties of PLA nanocomposites were measured by using a parallel plate rheometer (AR1000, TA Instruments) set at 160 °C. A frequency sweep mode was chosen with the angular frequency varied from 0.0 to 100 rad/s at a constant strain of 0.5% to maintain linear viscoelasticity in the specimens.

Morphology of the blends was characterized by observing the cryogenic fractured surface of the specimens. A field emission scanning electron microscope (Hitachi, S-4200) was used for surface observation. Gold was sputtered onto the specimens for electron conductivity.

Dynamic mechanical properties of the nanocomposites were characterized by using a dynamic mechanical analyzer (TA instruments, DMA2980) equipped with a single cantilever setting. The temperature range was set from -30 to 80 °C at a heating rate of 2 °C /min and frequency of 1 Hz.

Thermal analysis was performed by using a differential scanning calorimeter (TA instruments, MDSC 2920). The scanning was performed at a temperature range of between 20 to 200 °C under nitrogen atmosphere. The heating and cooling rates were set at 10 and 2 °C/min, respectively. All specimens were first heated to 200 °C and held isothermal for 5 minutes to erase thermal history. After cooling to 30 °C, the specimens were then subjected to a second heating cycle. Crystallization process was analyzed during the cooling cycle.

Thermal stability of the composites were investigated by using a thermogravimetric analyzer (TGA2950, TA Instruments) set at a heating rate of 20 °C/min from room temperature to 600 °C in air atmosphere.

Results and Discussion

Figure 1 shows the SEM micrograph of cryogenic fractured surface of Lapol_PLA and 10 wt% nano-$CaCO_3$ filled composites. Generally, Lapol has been used as a bioplasticizer that was reported to be compatible and miscible with PLA. The micrographs presents the phase morphology of the toughened Lapol_PLA. Large amounts of cavities could be observed from the fracture surface, which should not be confused as voids or being caused by interfacial debonding. These cavities could have been generated as a result of extensive crazing prior to failure of the material. Similar instances could be found in toughened poly(ethylene terephthalate). The presence of these cavities highlights the effectiveness of Lapol resin in enhancing the toughness of PLA. The number and depth of the cavities, however, reduced with the incorporation of nano-$CaCO_3$, which suggests that the filler could have reduced molecular motion of the matrix and thereby enhancing its stiffness.

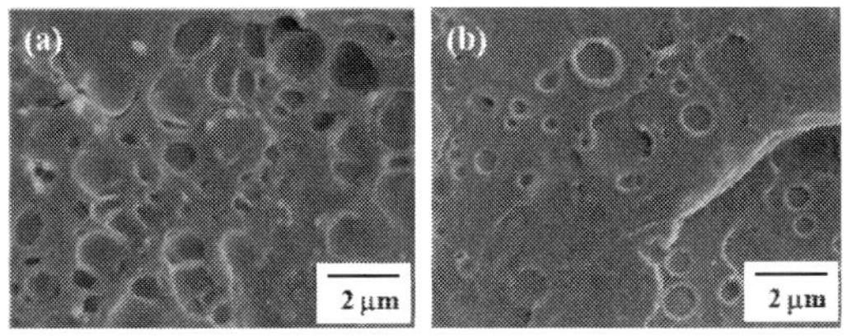

Figure 1 SEM micrographs of (a) Lapol_PLA and (b) 10 wt% nano-$CaCO_3$ filled Lapol_PLA composite.

Dynamic rheological measurement is useful for understanding the influence of filler concentration on the mobility of macromolecular chains (Gong et al., 2009). Figure 2 (a)-(c) present the complex viscosity and dynamic rheological properties as a function of nano-$CaCO_3$ content in Lapol_PLA. In Figure 2 (a), the complex viscosity (η^*) of the nanocomposites were significantly lower than that of unfilled Lapol__PLA. This could indicate that the nano-$CaCO_3$ could act as a lubricant when the composite is in the molten state (Gu et al., 2009; Li et al., 2006).

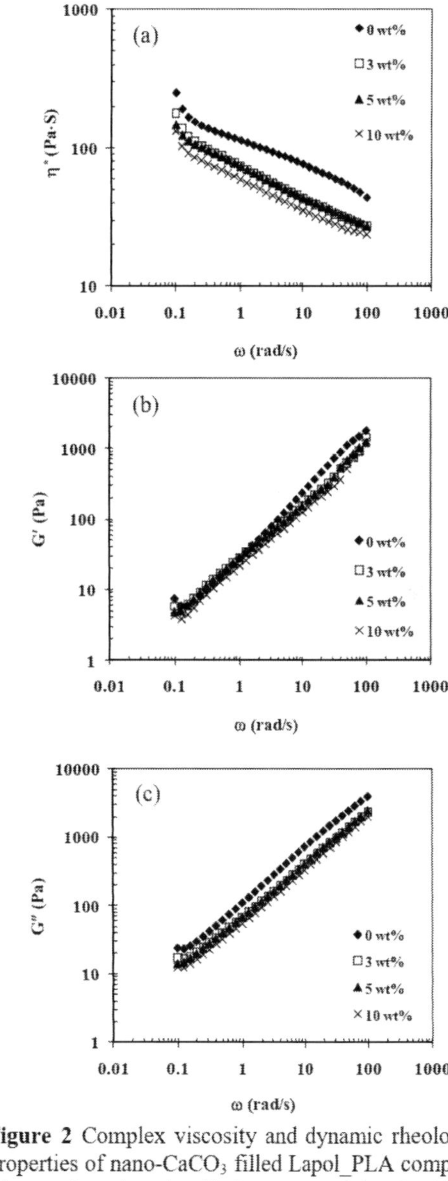

Figure 2 Complex viscosity and dynamic rheological properties of nano-$CaCO_3$ filled Lapol_PLA composite (a) complex viscosity (b) Storage modulus G′ (c) loss modulus G″.

The dynamic rheological properties of the composites are represented by the storage modulus G′ and loss modulus G″ as a function of angular frequency as shown in Figure 2 (b) and (c) respectively. The results show a decrease in storage modulus G′ and loss modulus G″. The storage modulus G′ is proportional to the elastic deformation, while the loss modulus G″, represents the viscous component, and is proportional to the dissipated energy in the material (Kually et al., 2007). The reduction of both moduli represented less elastic and viscous properties of nanocomposites during molten stage.

Dynamic mechanical properties of nano-$CaCO_3$ filled PLA composites are display in Figures 3 and 4. The results indicated phase separation between Lapol and PLA with the incorporation of nano-$CaCO_3$. Figure 3 shows the storage modulus as a function of temperature of nano-$CaCO_3$ filled Lapol_PLA composites. At low temperature, the storage modulus of toughened Lapol_PLA increased to 4.38 GPa when incorporated with 10wt% of nano-$CaCO_3$. This increment in bulk stiffness is due to the higher stiffness of the nanoparticles.

The tan δ peaks from DMA measurements corresponding to the glass transition temperatures (T_g) of Lapol and PLA phases in the nanocomposites are shown in Figures 4 (a) and (b), respectively. The tan δ peak of Lapol phase has evidently shifted to higher temperatures with the incorporation of nano-$CaCO_3$, which suggest inhibition of chain mobility by the filler. Meanwhile, the tan δ of the PLA phase shifted towards a lower temperature with a prominent emergence of a shoulder next to the maximum peak, as shown in Figure 4 (b). This could indicate that the restraint in Lapol chain mobility by the fillers provided some free volume for the movement of PLA molecules.

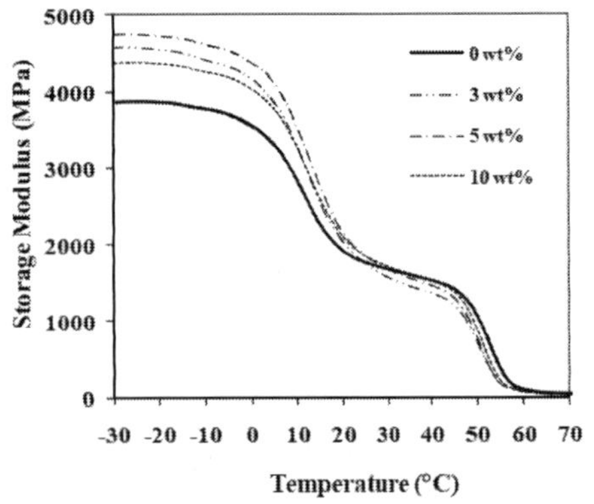

Figure 3 Storage modulus of various content of nano-$CaCO_3$ filled Lapol_PLA composite.

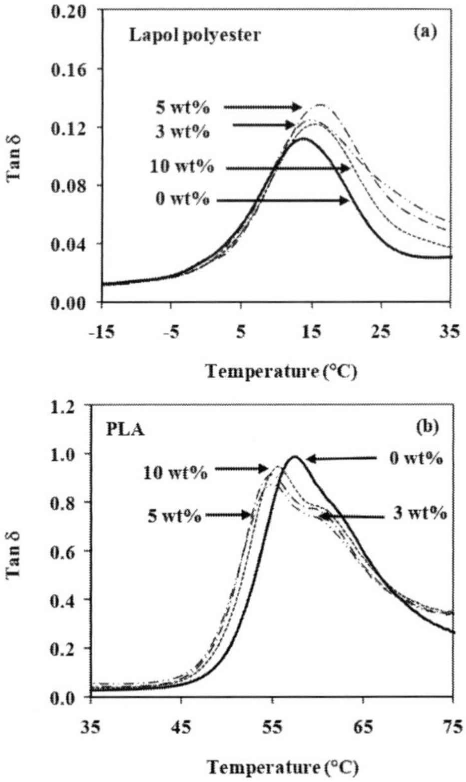

Figure 4 Tan δ of various content of nano-$CaCO_3$ filled Lapol_PLA composite (a) Lapol phase and (b) PLA phase.

The non-isothermal crystallization of nano-$CaCO_3$ filled Lapol_PLA is shown in Figure 5. The PLA phase seemed to crystallize very slowly, which could be due to inhibition by the Lapol resin. The incorporation of nano-$CaCO_3$ yielded exothermic peaks during cooling, which suggests that the filler has aided the crystallization of PLA. It should be noted that nano-$CaCO_3$ promoted crystallization of the PLA phase by heterogeneous nucleation. The lower tan δ peak registered by the PLA phase in the nanocomposites would confirm that more free volume was available for the PLA molecules to crystallize. Table 1 presents the thermal properties of the nanocomposites. The nucleating effect of the nano-$CaCO_3$ has prompted an increase in crystallinity of the filled nanocomposites without interference of the PLA crystalline structure, hence the melting temperature was unchanged. The higher in crystallinity in the nanocomposites could have also contributed to the increased storage modulus and stiffness of the material.

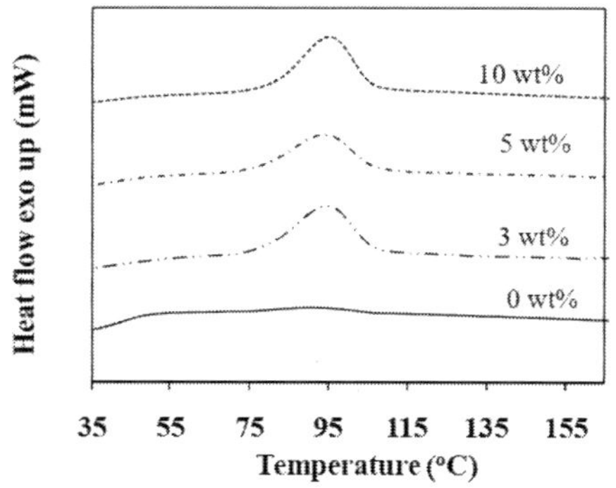

Figure 5 Crystallization temperature of Lapol_PLA composites filled with nano-$CaCO_3$ at various contents.

Table 1 Thermal properties of nano-$CaCO_3$ filled Lapol_PLA composites.

$CaCO_3$ (wt%)	T_m (°C)	T_c (°C)	%X_c	T_d (°C)
0	148.8	90.6	4.80	271
3	148.5	94.7	15.71	272
5	148.5	94.2	14.01	276
10	148.5	95.2	16.95	279

T_m = melting temperature
T_c = crystallization temperature
X_c = crystallinity
T_d = degradation temperature

Effect of nano-$CaCO_3$ on thermal stability of nanocomposites is presented Table 1. The onset thermal degradation temperature (T_d) corresponds to the temperature at 5% weight loss. The T_d of Lapol_PLA slightly increased with increasing nano-$CaCO_3$ content. This could be attributed to the filler particles acting as a barrier to prevent the resin from extensive thermo-oxidative decomposition (Kim et al., 2009).

Conclusions

The nano-$CaCO_3$ was found to be a good nucleating agent for PLA, judging from the enhanced crystallization peak with increasing filler content. The nano-filler could also act as a barrier to improve the thermal stability of the matrix resin. Moreover, the nano-fillers were able to act as a lubricant for the matrix in the molten state, which is expected to enhance the processability of the material. In the solid state, however, the nano-$CaCO_3$ was able to improve the stiffness of the composites while maintaining its toughness.

Acknowledgement

The authors are grateful to Department of Materials and Metallurgy Engineering, Faculty of Engineering, Rajamangala University of Technology Thanyaburi, Thailand for preparing precipitated nano-$CaCO_3$.

References

Aslan S., et al., "Poly (D,L-lactic acid)/Poly (ε-caprolactone) Blend Membranes: Preparation and Morphological Characterisation", J. Mater. Sci, **35**, 1615-1622 (2000)

Gong G., et al., "Dynamic Rheological Behavior of Isotactic Polypropylene Filled With Nano-Calcium Carbonate Modified by Stearic Acid Coating", J. Macromol. Sci. Phys., **48**, 329-343 (2009)

Gu J., et al., "Effect of Unsaturated Hydroxyl-Fatty Acid Modified Nano-$CaCO_3$ on the Morphological and Rheological Behavior of PP", Front. Chem. China, **4**, 75-82 (2009)

Jiang L., et al., "Comparison of Polylactide/Nano-sized Calciumcarbonate and Polylactide/Monmorillonite Composites: Reinforcing Effects and Toughening Mechanisms", Polymer, **48**, 7632-7644 (2007)

Kaully T., et al., "Rheology of Highly Filled Natural CaCO3 Composites. II. Effect of Solid Loading and Particle Size Distribution on Rotational Rheometry", Polm. /comps., **28**, 524-533 (2007)

Kim J. Y., et al., "Thermal Decomposition Behavior of Poly(ethylene 2,6-naphthlate)/Silica Nanocomposites", Polm. Compos., **30**, 1779-1787 (2009)

Li G., et al., "Plasticization of Nano-$CaCO_3$ in Polystyrene/Nano-$CaCO_3$ Composites", J. Appl. Polm. Sci., **99**, 2138-2143 (2006)

Oksman K., et al., "Natural Fibers as Reinforcement in PLA composites", Comp. Sci. Technol., **63**, 1317-1324 (2003)

Ramiro D. E., et al., "Immiscible Polymer Blends of Semicrystalline Biocompatible Components: Thermal Properties and Phase Morphology Analysis of PLLA/PCL Blends", Polymer, **42**, 7831-7840 (2001)

Urayama H., et al., "Mechanical and Thermal Properties of Poly(L-lactide) Incorporating Various Inorganic Fillers with Particle and Wisker Shapes", Macromol. Mater. Eng., **288**, 562-568 (2003)

TAILORED STRUCTURING OF POLYETHYLENE TEREPHTHALATE / CLAY NANOCOMPOSITES

Y. Shen[1]*, E. Harkin-Jones[1], S. Xie[1], R. Abu-Zurayk[1], T. McNally[1] and P. Hornsby[1]
R. Patel[2], H. Benkreira[2], A. Khan[2] and P.D. Coates[2]

[1] *School of Mechanical & Aerospace Engineering, Queen's University Belfast,Belfast, UK*
yshen02@qub.ac.uk; e.harkinjones@qub.ac.uk; s.xie@qub.ac.uk; rabuzurayk01@qub.ac.uk; t.mcnally@qub.ac.uk; peter.hornsby@qub.ac.uk
[2] *Polymer IRC, School of Engineering, University of Bradford, Bradford, UK*
r.patel@bradford.ac.uk; h.benkreira@bradford.ac.uk; ahkhan5@bradford.ac.uk; p.d.coates@bradford.ac.uk

Abstract: Polyethylene terephthalate (PET) and nanoclay were mixed and prepared by melt compounding in a mini-mixer. A statistically based design of experiments (DoE) approach was used to investigate the effect of screw speed, residence time, extrusion temperature and clay loading on the structural parameters and mechanical properties of the nanocomposites. Quantitative characterization of clay dispersion was achieved by the combination of transmission electron microscopy and optical microscopy. DMTA, Tensile, Flexural and DSC tests were used to characterize the mechanical properties and crystallization of the materials. Results show that clay loading has the most significant effect on the overall improvement of mechanical properties. Various combinations of processing factors have also been identified that involve the enhancement of properties. Most importantly, there is a good correlation between structural parameters and mechanical properties; hence tailored structuring of the PET nanocomposites can be achieved.

Introduction

Nanofillers offer many potential property enhancements to polymers. Improvements in properties such modulus, barrier and flame retardance have been reported by many authors (Paul and Robeson, 2008). However, despite the large volume of literature on the subject very few polymer nanocomposites materials are commercially available and relatively few materials have achieved the predicted improvements in properties.

It is true that for a successful nanocomposite the chemistry of the polymer-filler system must be favorable but the subsequent effect of the deformation experienced in processing is also likely to have an important effect that needs to be investigated. Although much work has been published on the effect of processing and / or components on the structure and properties of polymer/clay nanocomposites (Abu-Zurayk et al., 2009; Soon et al., 2009; Dennis et al., 2001; Vermogen et al., 2005; Chavarria et al., 2007; Kim et al., 2007), the relationship between processing conditions, structure and mechanical properties has not yet been fully understood. A design of experiments (DoE) approach is a statistically sound and highly organized method that has proved a good means to determine the significant factors affecting studied responses. Porfyrakis et al. (2002) adopted DoE to investigate the interrelationship between processing conditions, microstructure and mechanical properties for a series of injection moulded rubber-toughened PMMA samples. The back pressure was identified as the single most important factor affecting the sample properties. Dong and Bhattacharyya (2008) used a DoE method to investigate the effects of clay type, clay/compatibiliser content and matrix viscosity on the mechanical properties of PP/organoclay nanocomposites and showed that the type of PP used had the most significant effect on the overall improvements of mechanical properties. In addition, clay content was determined as the second significant factor to enhance the tensile properties. However, only material effects were investigated and possible interactions between various processing factors were not considered in their studies.

Quantitative characterization of clay dispersion in polymer/clay nanocomposites has been developed in our previous studies (Xie et al., 2009). Based on image analyses of transmission electron microscopy (TEM) and optical microscopy micrographs. Average interparticle distance per unit volume of clay and agglomerates percentage will be used, as well as average tactoid thickness and average tactoid length, to measure the level of clay dispersion in this study. The aim of this work is to identify the effect of processing on the mechanical properties of PET-clay nanocomposites by determining the links between processing, composite structure and properties.

Experimental

Materials

The polyethylene terephthalate (PET) was a food grade

ARTENIUS PURA PET, with an intrinsic viscosity of 0.74 dl/g. Synthetic micas (fluorohectorites) Somasif MTE was purchased from Uni-Coop Japan (now CBC Co. Ltd). The Somasif MTE contains methyl trioctyl ammonium chloride as surfactant.

DoE method

The chosen factors to vary and their operating values were based on initial trials. In order to minimize experimentation, a fractional factorial experiment was utilized which provided sufficient degrees of freedom to resolve all main factor effects as well as all two-factor interactions. For this work, a 2-level, 4 factor fractional factorial design of experiments was constructed using Design-Expert 7.1.3 (Stat-Ease Inc., Minneapolis, USA). This required 13 runs, which include 1 run of centre point (run 3) and 4 runs of controls (run 10 - 13) as shown in Table 1.

Table 1. The DOE run order of the experiments

Run Order	Speed (rpm)	Temperature (° C)	Time (min)	Clay Loading (%)
1	20	270	1	2
2	60	285	1	2
3	40	277.5	2	4
4	60	270	1	6
5	20	285	3	2
6	60	270	3	2
7	20	285	1	6
8	20	270	3	6
9	60	285	3	6
10	20	270	1	0
11	60	270	3	0
12	20	285	1	0
13	60	285	3	0

Preparation

The ground PET powder and clay were pre-mixed and dried under vacuum for 12 h at 100 °C prior to melting processing. They were then compounded in a mini-mixer at different processing parameters and clay loadings. The extruded filament was air cooled and pelletized. The pre-dried pellets were compression molded with heating at 260 °C for 2.5 minutes and crash cooling using a cold water cassette for 1 minute. Full holding pressure of 150 bars was maintained.

X-ray diffraction

XRD studies were carried out with a Philips PW3040 Console X-ray diffractometer using Cu Kα radiation with a wavelength of 1.54 Å. The step size was 0.01 ° and scan rate was 0.3 °/min. The data were collected from 1 ° to 12 °. Samples of 2 cm diameter were tested for each sheet, which was rotated during the test so that an average plot is produced for each sheet. The d-spacing was calculated according to Bragg's law.

Transmission electron microscopy

TEM micrographs of the samples were obtained using a FEI Tecnai F20 instrument recorded at 200 kV. 50 nm thin slices were prepared from compression molded samples using a Leica ultra microtome. Image analysis was conducted using Photoshop (Adobe) and Fovea Pro (Reindeer Graphics) for binary conversion. The resulting binary images were analyzed by Scion Image (Scion) to automatically measure the perimeter and area of each particle.

Optical microscopy

OM observations were conducted at 260 °C to eliminate any influence of PET spherulites using a Nikon microscope and a magnification of ×20.

Tensile and Flexural test

Tensile tests were carried out using an Instron 5564 Universal Tester with a clip-on extensometer at a crosshead speed of 1mm/min for modulus test and 50 mm/min for strength and elongation test at room temperature. Samples were prepared by compression moulding sheets 1 mm thick, from which dumbbell-shaped samples (type 1BA) were cut using a stamping press. A minimum of five tests was performed for each reported value. Flexural tests were performed in 3 point method with crosshead speed of 1.07 mm/min. Span distance was 25.4 mm, and 3-5 samples were used for each run.

DMTA

Dynamic mechanical measurements were performed on a Tritec 2000 DMA instrument in dual cantilever bending mode from 30°C to 180°C at a ramp rate of 3°C /min, a frequency of 1Hz and a displacement of 0.05mm. Two tests were used for each run.

Thermal analysis

DSC was performed on a Perkin Elmer DSC 6 using a heating and cooling rate of 10 °C/min between 30 °C and 285 °C under a nitrogen atmosphere. The samples were held at 285 °C for 2 min to remove previous thermal history; they were then cooled to 30 °C, and reheated again to 285 °C.

Statistical Analysis of Correlation

Correlations between the response variables were analyzed using a 'bootstrap' technique. Bootstrapping involves generating further samples of data by random re-sampling with replacement from the original data set. A correlation analysis is then carried out on each sample, yielding a range of estimates for the correlation coefficient. A 95% confidence interval for the correlation coefficient can thus be established to determine whether or not the correlation is statistically significant. If the upper and lower levels of the correlation coefficient confidence interval are both positive or both negative then the correlation is deemed significant, if the interval contains zero then the correlation can be deemed not significant. The bootstrap analysis was executed with 5000 resamples applying the Bias Corrected and accelerated (BCa) interval using Matlab.

Results and Discussion

Table 2 shows the values of d-spacing for MTE nanoclay in powder form and PET/clay nanocomposites in compression moulded sheets form from XRD. The d-spacing decreases in the DOE runs compared to the nanoclay, and there is only slight difference in the d-spacing between different runs. This may indicate that the nanoclays are not well intercalated, however, XRD should not be used as a stand-alone characterisation technique for polymer nanocomposites. First, if there is a large amount of clay, a distribution of interlayer distances, uneven distribution of surfactant or random orientation of the clay, it may give a misleading interpretation. Second, the melt processing temperature was above the degradation temperature of the nanoclay surfactant, and the reduction of the d-spacing may be due to the degradation of the surfactant during processing, resulting in a collapse of the nanoclay layers. Also, Gurmendi et al. (2007) have also pointed out that the intercalation may still occur without an increase of d-spacing if the PET could intercalate and substitute the surfactant in the clay interlayer region.

Table 2. Values of d-spacing for MTE nanoclay and PET/clay nanocomposites

Sample	d-spacing (A)
MTE powder	23.28
Run1	22.30
Run2	22.14
Run3	22.09
Run4	22.12
Run5	22.17
Run6	22.09
Run7	22.20
Run8	22.20
Run9	22.02

The microscopical dispersion of tactoids smaller than 5 μm, including exfoliated platelets and intercalated tactoids is characterized by TEM as shown in Figure 1. There is a generally homogeneous dispersion of nanoclays with tactoids containing multiple layers. In contrast, the macroscopical dispersion of clay in PET was characterized by OM at low magnification (×20), where micron-size agglomerates larger than 5 μm were observed. A typical OM micrograph from Run 3 is shown in Figure 2. No large scale agglomerates are observed. Further quantitative characterization of nanoclay dispersion in PET is

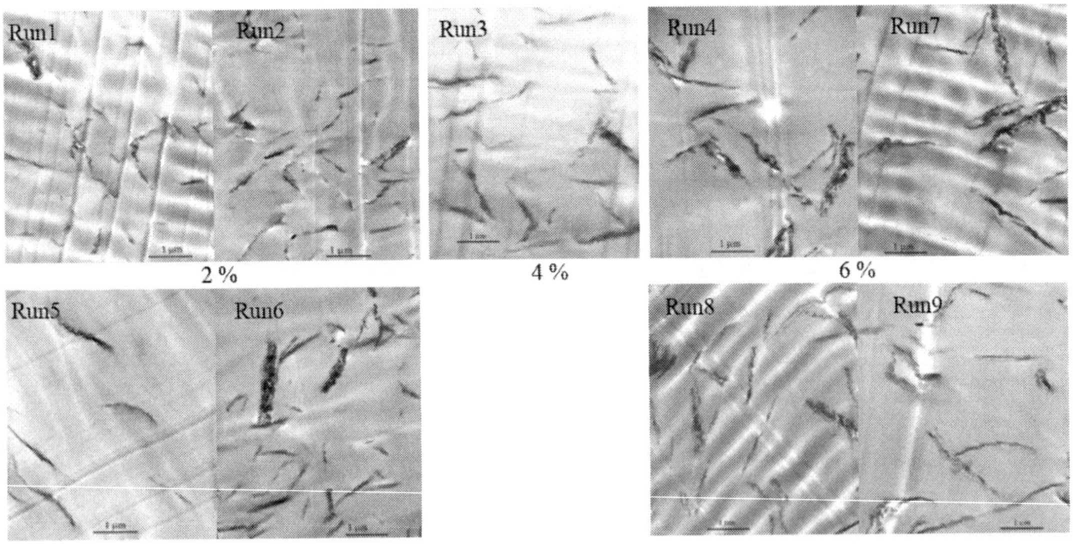

Figure 1. TEM micrographs of runs with nanoclay loading of 2% (Run1, Run2, Run5, Run6), 4% (Run3) and 6% (Run4, Run7, Run8, Run9).

shown in Figure 3. IPD/vol% clay is a measure of spatial separation between particles relative to clay loading, and agglomerates percentage is the proportion of micron size agglomerates. Average tactoid thickness and average tactoid length are the number average values measured from TEM micrographs.

Table 3 reports the mechanical properties for each run. Data for run11 and run13 were not obtained, because they were too brittle to obtain tensile bars using the stamping press. From the data reported and Figure 4, a significant increase in tensile modulus with clay loading is observed, with a maximum modulus enhancement of 38% for run4 with 6% clay loading compared to unfilled run10. The tensile strength and elongation decrease at increased clay loading. Run 3 (4wt %) had the highest modulus (31.5% enhancement over run 12) and yield strength (12% enhancement over run 12), this may be because low clay content (2wt %) is not enough to provide enhancement in flexural properties, while at higher content (6wt %)

Figure 2. Optical micrograph of Run 3.

agglomerates may form that affect the flexural properties negatively. Runs with 2 wt% nanoclay show higher flexural yield compared to ones with 6wt% nanoclay, which could be again due to the formation of agglomerates at high clay content. Runs with 4 wt% and 6wt% clay fractured during flexural testing. In order to properly allocate any change of mechanical properties in the nanocomposites to the correct source, the crystallinity of all DoE runs was measured by DSC. As shown in the Figure 5, the maximum difference of crystallinity among DoE runs is about 5%, and for most of runs, it is even within 2%. Therefore, it is considered that the changes in crystallinity will not be the source of any enhancement.

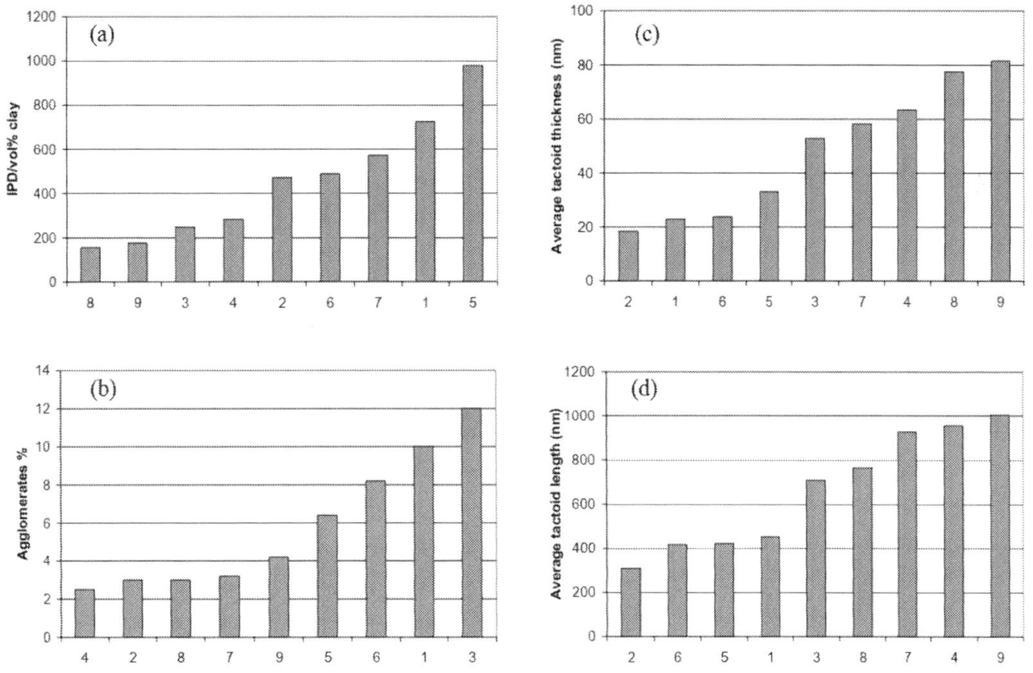

Figure 3. Quantitative characterization of nanoclay dispersion (a) IPD/vol% clay, (b) Agglomerates %, (c) Average tactoid thickness, (d) Average tactoid length as a function of run numbers.

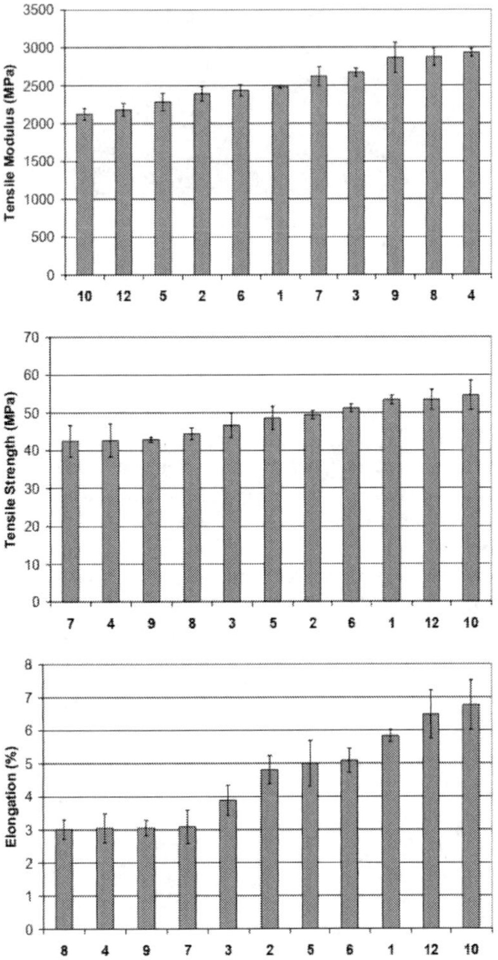

Figure 4. Tensile properties of compression moulded DoE Runs.

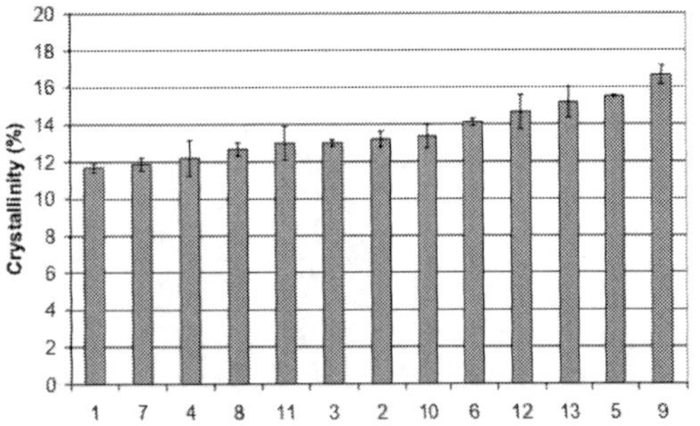

Figure 5. Crystallinity of compression moulded DoE Runs.

For the purpose of analysis the storage modulus results for the DMTA are taken at 40 °C. The result of dynamic mechanical properties of DoE runs are shown in Figure 6. At 40°C, there is a good enhancement in storage modulus for PET/clay nanocomposites

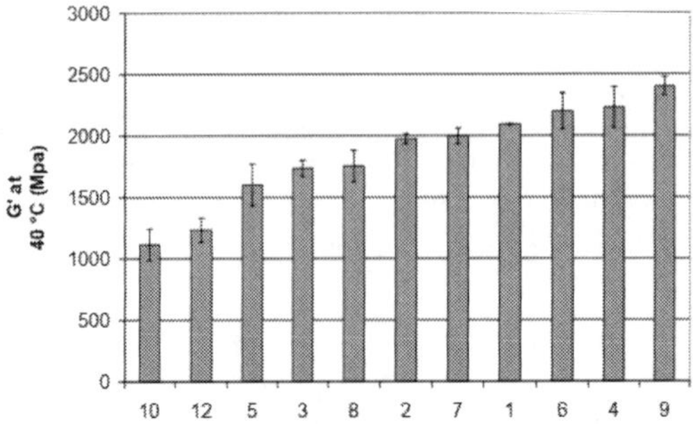

Figure 6. Storage modulus G' at 40 °C of compression moulded DoE Runs from DMTA.

compared to the unfilled PET. The best enhancement was found in run9 (94%), with 60rpm screw speed, 285°C processing temperature, 3 min residence time and 6wt% nanoclay loading.

Figure 7 and Figure 8 show the correlation between dispersion of nanoclay and tensile modulus with 95% confidence interval. There is a good correlation between tensile modulus and IPD/vol% clay, average tactoid thickness, average tactoid length. As expected, the higher clay loading leads to thicker and longer clay tactoids. The correlation between tensile modulus and agglomerates percentage was also analyzed, but it is under 95% confidence interval, hence considered not significant.

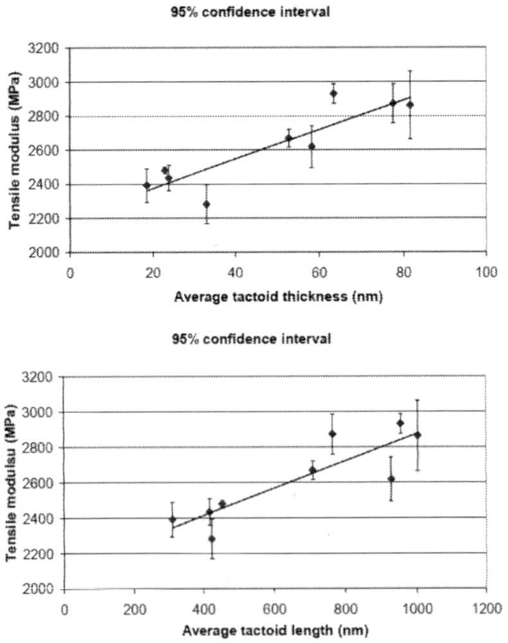

Figure 7. Tensile modulus as a function of average tactoid thickness and average tactoid length.

Table 3. Mechanical properties of PET/clay systems

Run No.	Tensile						Flexural			
	Modulus (MPa)	Standard deviation	Strength (MPa)	Standard deviation	Elongation (%)	Standard deviation	Modulus (MPa)	Standard deviation	Yield (MPa)	Standard deviation
1	2480	19	53.41	1.19	5.82	0.18	2193	255	53.19	6.98
2	2391	97	49.43	1.08	4.8	0.42	1984	170	50.34	6.15
3	2668	52	46.71	3.27	3.88	0.45	2480	78	60.53	2.64
4	2930	56	42.7	4.41	3.05	0.44	2046	257	28.75	6.79
5	2282	114	48.6	3.08	4.99	0.69	1997	154	54.22	5.11
6	2435	75	51.22	1.01	5.08	0.36	2068	57	57.92	1.63
7	2618	124	42.52	4.20	3.09	0.50	2237	143	46.62	2.59
8	2871	114	44.48	1.56	3.01	0.29	1984	268	35.94	11.97
9	2863	199	42.95	0.64	3.05	0.23	1618	226	16.19	6.15
10	2122	75	54.62	3.84	6.76	0.75	1758	113	51.30	1.21
12	2178	85	53.43	2.69	6.48	0.73	1885	104	54.08	3.55

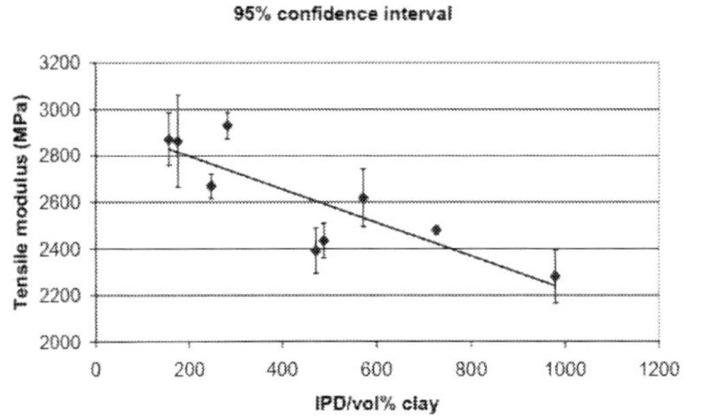

Figure 8. Tensile modulus as a function of IPD/vol% clay.

Half-normal plots and Pareto analysis were performed in DoE software to identify the significant factors. Figure 9 shows clay loading is the most significant factor for tensile modulus. The processing parameters are not considered as significant factors influencing tensile modulus.

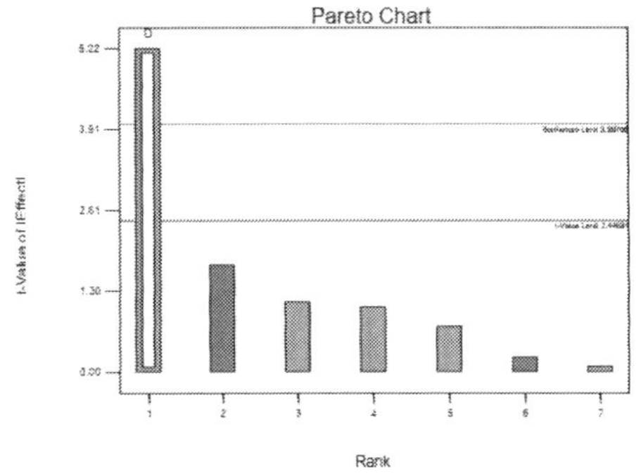

Figure 9. Pareto chart of tensile modulus from DoE analysis.

Figure 10 shows the DoE analysis of storage modulus at 40°C. A two-factor interaction is identified between speed and time. At short residence time (1 min), speed has a limited effect on the storage modulus. At long residence time (3min), speed has a significant effect on the storage modulus.

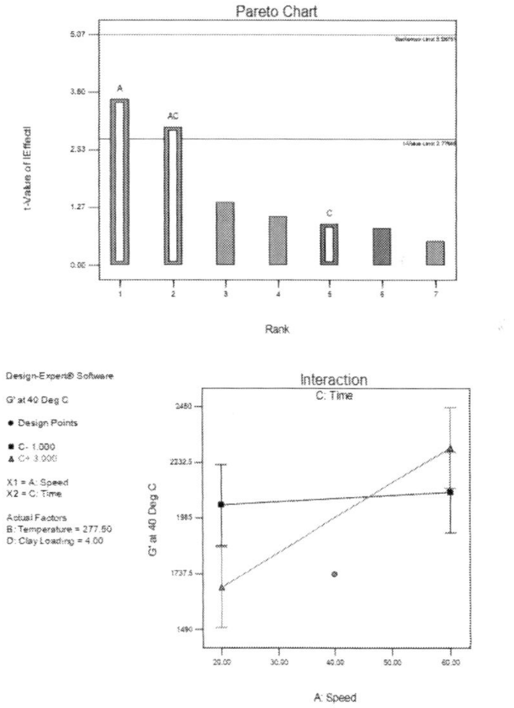

Figure 10. Pareto chart and model graphs of storage modulus at 40°C from DoE analysis.

The effect of clay loading and processing parameters was also studied by DoE analysis as shown in Figure 11. Only the clay loading was considered as the significant factor for IPD/vol% clay.

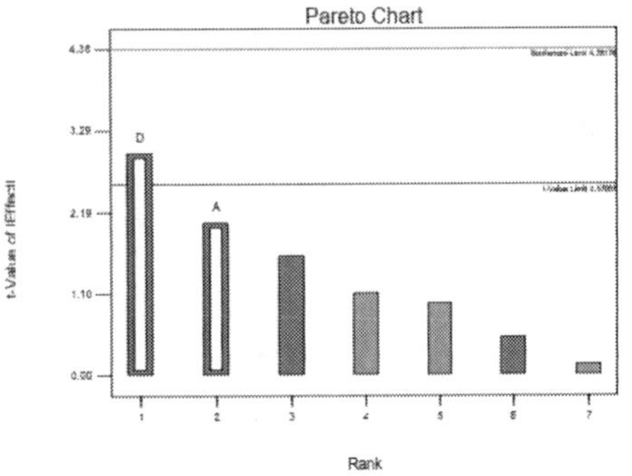

Figure 11. Pareto chart of IPD/vol% clay from DoE analysis.

Conclusions

The effect of processing on the mechanical properties and clay dispersion in PET/clay nanocomposites was studied using a design of experiments methodology. Results show that there is a good correlation between structural parameters and mechanical properties. Speed and time were involved as a two-factor interaction on the storage modulus at 40°C. Overall clay loading has the most significant effect on the improvement of mechanical properties.

Acknowledgements

The authors wish to acknowledge Engineering and Physical Science Research Council (EPSRC), UK for funding this work.

References

Abu-Zurayk, R., Harkin-Jones, E., McNally, T., Menary, G., Martin, P.J., Armstrong, C.G., "Biaxial deformation behavior and mechanical properties of a polypropylene/clay nanocomposite", Composites Science and Technology, **69**, 1644-1652 (2009)
DOI:10.1016/j.compscitech.2009.03.014

Chavarria, F., Shah, R.K., Hunter, D.L., Paul, D.R., "Effect of Melt Processing Conditions on the Morphology and Properties of Nylon 6 Nanocomposites", Polymer Engineering and Science, **47**, 1847-1864 (2007)
DOI: 10.1002/pen.20894

Dennis, H.R., Hunter, D.L., Chang, D., Kim, D., White, J.L., Cho, J.W., Paul, D.R., "Effect of melt processing conditions on the extent of exfoliation in organoclay-based nanocomposites", Polymer, **42**, 9513-9522 (2001)
DOI: 10.1016/S0032-3861(01)00473-6

Gurmendi, U., Eguiazabal, J.I., Nazabal, J., "Structure and Properties of Nanocomposites with a Poly(ethylene terephthalate) Matrix", Macromolecular Materials and Engineering, **292**, 169-175 (2007)
DOI: 10.1002/mame.200600376

Kim, D.H., Fasulo, P.D., Rodgers, W.R., Paul, D.R., "Effect of the ratio of maleated polypropylene to organoclay on the structure and properties of TPO-based nanocomposites. Part I: Morphology and mechanical properties", Polymer, **48**, 5960-5978 (2007)
DOI: 10.1016/j.polymer.2007.08.010

Paul, D.R., Robeson, L.M., "Polymer nanotechnology: Nanocomposites", Polymer, **49**, 3187-3204 (2008)
DOI: 10.1016/j.polymer.2008.04.017

Porfyrakis, K., Assender, H.E., Robinson, I.M., "The interrelationship between processing conditions, microstructure and mechanical properties for injection moulded rubber-toughened poly(methyl methacrylate) (RTPMMA) samples", Polymer, **43**, 4769-4781 (2002)
DOI:10.1016/S0032-3861(02)00286-0

Soon, K., Harkin-Jones, E., Rajeev, R.S., Menary, G., McNally, T., Martin, P.J., Armstrong, C.G., "Characterisation of melt-processed poly(ethylene terephthalate)/syntheticmica nanocomposite sheet and its biaxial deformation behaviour", Polymer International, **58,** 1134-1141 (2009)
DOI: 10.1002/pi.2641

Xie, S., Harkin-Jones, E., Shen, Y., Hornsby, P.R., McAfee, M., McNally, T., Patel, R., Benkreira, H., Coates, P.D., "Quantitative characterization of clay dispersion in polypropylene-clay nanocomposites by combined transmission electron microscopy and optical microscopy", Materials Letters, **64**, 185-188 (2010)
DOI: 10.1016/j.matlet.2009.10.042

Vermogen, A., Masenelli-Varlot, K., Seguela, R., Duchet-Rumeau, J., Boucard, S., Prele, P., "Evaluation of the Structure and Dispersion in Polymer-Layered Silicate Nanocomposites", Macromolecules, **38**, 9661-9669 (2005)
DOI: 10.1021/ma051249

Halloysite Nanoclays, a Safe Alternative to CNT's

Edmonds Neil [1*], Mok LingHon [1], Ray Sudip [1]

[1] *Department of Chemistry, The University of Auckland, Auckland, New Zealand*

nr.edmonds@auckland.ac.nz; enoch743@gmail.com;

Halloysite clays are naturally occurring silicate nanotubules formed by geological processes. Halloysite nanotubes are recognised as a valuable raw material for use with plastics because of their unique structural and chemical properties.

Halloysite can be used in nanocomposites to provide enhanced mechanical and fire retardant properties and potentially for the self-healing of polymers, this functionality being provided through the loading of healing agents into the tubules.

This presentation covers the use of halloysite in nylon 6 and PET, including

- Loading mechanism into halloysite lumens
- Dispersion behavior of halloysite in nanocomposites
- Chemical and thermal analysis of nanocomposites

DMTA was used to evaluate the effect of different halloysite loadings on the mechanical properties of specific polymers. The storage modulus (in tensile test mode) was found to vary at different mixing ratios. DSC results showed halloysite acted as a nucleating agent in both PA6 and PET, the presence of halloysite significantly increasing the overall crystallization rate and the level of crystal formation. SEM/EDX confirmed the excellent distribution of nanotubules in the polymer matrix, this was attributed to the unique structure/charge properties of halloysite. When the tubules were loaded with sodium ascorbate, FTIR results suggested that the active compound was present in lumens rather than on the surface of tubules.

In summary, a successful loading system was developed for active compounds, and halloysite based PET and PA6 nanocomposites were produced. The clear changes in mechanical/crystallization properties were demonstrated and halloysite dispersion in polymers verified. Evaluation of the potential of halloysite as a carrier for self-healing agents and other active compounds is continuing, as is research into novel nanocomposites with superior mechanical properties

Introduction

Halloysite is a naturally occurring inorganic microtubular material belonging to the kaolinite group of clay minerals (Figure 1,2). It is formed by the curling of platelets sheared from kaolin clays during geological processes, the asymmetric structure of the platelets causing the curling behaviour. Halloysite clay deposits found in New Zealand are derived from weathering of rhyolite domes, clay is mined from these domes and processed to produce a fine product with a purity of over 90% halloysite and low concentrations of iron and titania. Tubular forms of halloysite known as halloysite nanotubes (HNT) can be separated from other forms by filtration and centrifugation [1].

An ultrafine grade of halloysite (UH) was provided by Imerys Tableware Asia Limited, this was used for all experiments. UH is a hydrated aluminum silicate containing 90-93% of tubular halloysite with the following elemental composition: SiO_2 49.5; Al_2O_3 35.5; Fe_2O_3 0.29; TiO_2 0.09 (%w/w). The specific gravity of UH is 2.55g/cm^3 and surface area approximately 20m^2/g. The particles of halloysite can adopt a variety of morphologies, the most common of which are elongated tubules. Tubule lengths range from 100 nm to approximately 750 nm, with internal diameters of ~5-10 nm. In both hydrated and dehydrated forms, halloysite comprises layers of silica tetrahedral and alumina octahedral units.

Halloysite has the chemical formula $Al_2O_3.2SiO_2.nH_2O$. Silica is mainly positioned on the outer surface of the tubule, whereas the alumina is present mainly on the inner surface and edges of tubules [2]. Since the outside of tubules is mainly silica, halloysite tends to have an anionic surface except at very low pH, and readily binds cationic compounds [2,3]. Halloysite partially dehydrates at about 110°C, all structural water is lost at about 575°C. The water interlayer can be replaced by an organic liquid such as ethylene glycol, diethylene glycol or glycerine [4].

Experimental

The aim of the research program was to

1. Investigate the loading of actives into halloysite lumens.
2. Evaluate the potential for using halloysite as a carrier for self-healing agents.
3. Investigate the dispersion behavior of halloysite in nanocomposites.
4. Investigate the use of silane coupling agents.

1. Halloysite can be used for delivery of active agents such as pesticides, drugs, antifoulings, dyes, enzymes, peptides and hormones. These compounds can be adsorbed onto the inner surface or outer surface of halloysite tubules, or on both surfaces. Coating on the interior of tubules allows for controlled release of the active agent. In this study, sodium ascorbate was used as a model compound for investigation of loading conditions. 5g of UH were immersed in 10 %w/w aqueous solutions of sodium ascorbate under vacuum and stirred for 24h. The treated UH was washed to remove surface sodium ascorbate on the outside surfaces of tubules, filtered and dried.

2. Unlike nanoclays, HNT's do not require exfoliation and can therefore be easily dispersed in a polymer matrix without degrading the polymer chains, even at high loading rates. They are therefore a suitable carrier for self-healing agents, which can be loaded into the tubules. When the filled halloysite contacts a micro-crack, the self-healing reagent will be released and repair the break. Subsequent polymerization within the crack may be initiated by heat, radiation or oxygen [5].

3. As a result of the surface charge, tubules do not agglomerate when dispersed in polymer matrices, this is a major advantage over other reinforcing materials used for nanocomposites. Nanocomposites based on Nylon 6 (PA6, Zytel 7335F NC010 from Du Pont™) and PET with HNT were prepared by melt blending on a Brabendar Plastograph followed by injection moulding of test specimens.

The storage modulus (G') of PA6 and PET halloysite nanocomposites was investigated using DMTA. Thermal properties were evaluated by DSC.

4. Nan-ying Ning [6] and Du [7] showed that halloysite nanotubes treated with a silane or cationic surfactant improved the dispersion of HNT in polypropylene (PP). The modification of halloysite using 3-glycidoxypropyl-trimethoxysilane was investigated.

The silane group can bond to silica groups on the HNT outer wall, leaving the glycidyl (epoxy) groups available for reaction with functionality on the polymer matrix.

Results and Discussion

Ascorbic acid was successfully loaded into nanoclay lumens. Treated samples were examined by FTIR and SEM/EDX.

DMTA testing of nanocomposites based on PA6 and PET with HNT showed a significant increase in stiffness at low halloysite content. However, while the storage modulus (G') of PA6 significantly increased at a 2% HNT loading it decreased at a 5% loading, suggesting that the nanotubules had not been fully incorporated into the PA6 matrix (Figure 3). In contrast, 2% and 5% loadings in PET both showed significantly increased stiffness with added halloysite (Figure 4).

SEM/EDX images of silicon and aluminium distribution on the surface of a PA6 nanocomposite (Figures 5,6) show that HNT is fully dispersed within the polymer matrix.

When the epoxy modified HNT was blended with PET, the modified HNT dispersed evenly into the polymer matrix compared to the unmodified HNT control and increased the tensile modulus.

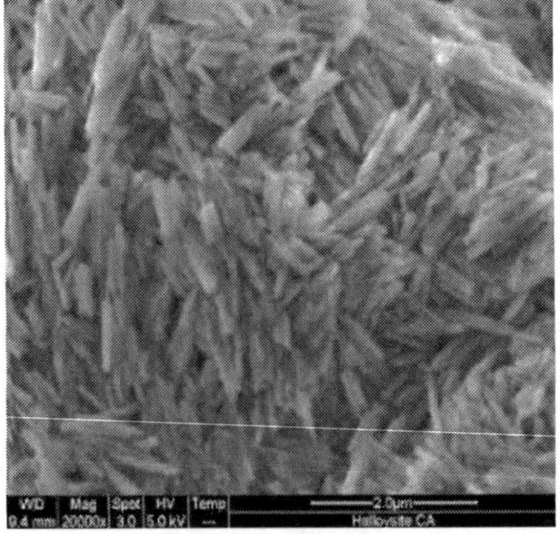

Figure 1. SEM Image of Halloycite Nanoclay

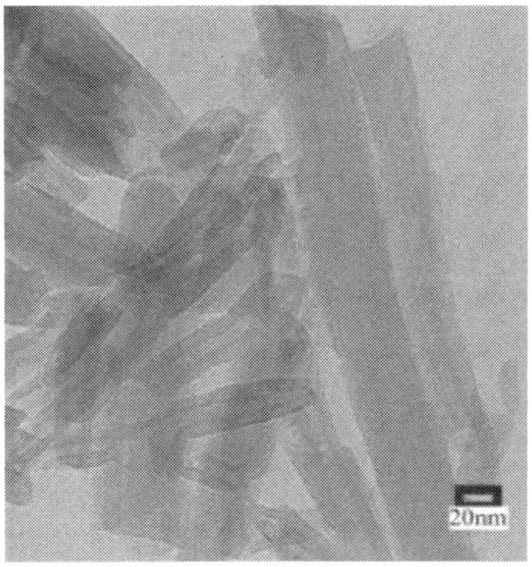

Figure 2. TEM Image of Halloycite Nanoclay

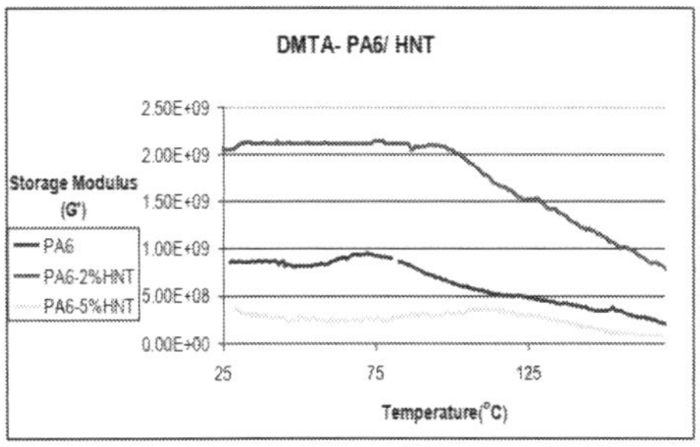

Figure 3. Storage modulus of 2% and 5% HNT in a PA6 nanocomposite.

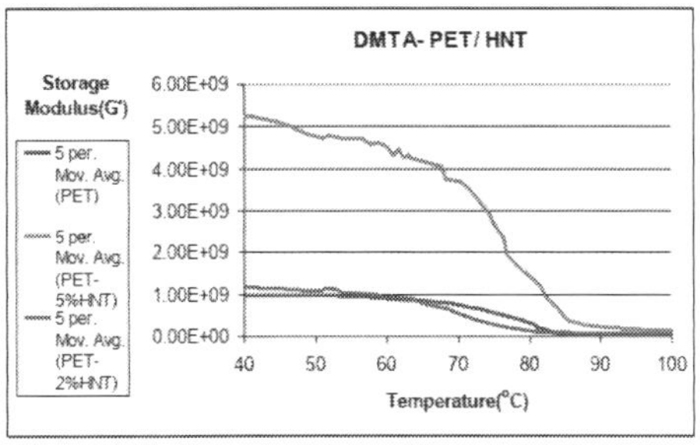

Figure 4. Storage modulus of 2% and 5% HNT in a PET nanocomposite.

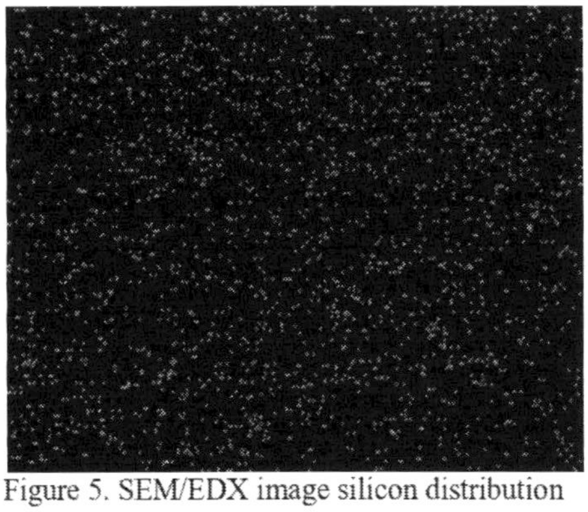

Figure 5. SEM/EDX image silicon distribution on the surface of a PA6 nanocomposite.

Figure 6. SEM/EDX image silicon distribution on the surface of a PA6 nanocomposite.

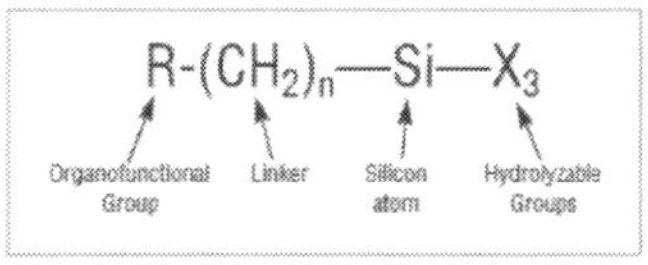

Figure 5. Silane coupling agents

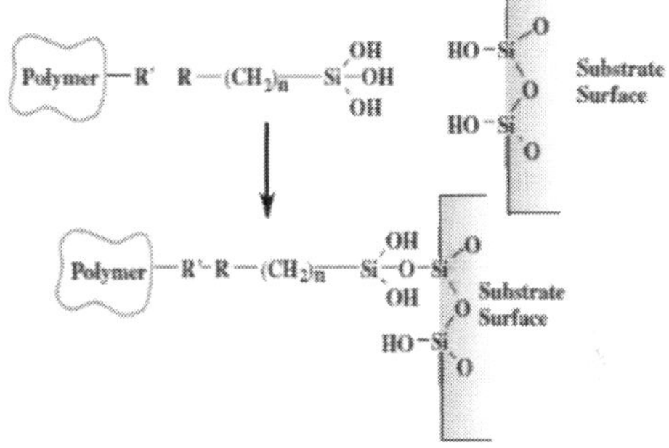

Conclusions

The ease of dispersion of halloycite in polymer matrices allows for stiffening of nanocomposites without risk of polymer degradation. If the nanotubule lumens are first loaded with actives, this provides a route for self-healing action. Modification of the outer walls of HNT prior to composite formation can enhance bonding between matrix and nanotubules.

References

1. M G Townsend, K A Luke and R B Evans, *Recent development in the exploration and uses of halloysite clay deposits in Northland.* Geology and Exploration of New Zealand Mineral Deposits
2. Levis, S.R. and P.B. Deasy, *Characterisation of halloysite for use as a microtubular drug delivery system.* International Journal of Pharmaceutics, 2002. 243(1-2): p. 125-134.

3. Zhao, M. and P. Liu, *Adsorption behavior of methylene blue on halloysite nanotubes.* Microporous and Mesoporous Materials, 2008. 112(1-3): p. 419-424.
4. Price, R.R.S., MD), Gaber, Bruce P. (Bethesda, MD), *Controlled release of active agents using inorganic tubules.* 1997, The United States of America as represented by the Secretary of the Navy (Washington, DC): United States.
5. Yin, T., et al., *Self-healing epoxy composites - Preparation and effect of the healant consisting of microencapsulated epoxy and latent curing agent.* Composites Science and Technology, 2007. 67(2): p. 201-212.
6. Ning, N.-y., et al., *Crystallization behavior and mechanical properties of polypropylene/halloysite composites.* Polymer, 2007. 48(25): p. 7374-7384.
7. Du, M., B. Guo, and D. Jia, *Thermal stability and flame retardant effects of halloysite nanotubes on poly(propylene).* European Polymer Journal, 2006. 42(6): p. 1362-1369.

MONITORING OF NANOSTRUCTURE RESPONSE ON MECHANICAL TESTS IN ORIENTED NANOCOMPOSITES

N. Stribeck*, **A. Zeinolebadi**

Department of Chemistry, University of Hamburg, Hamburg, Germany — norbert.stribeck@desy.de; zeinoleb@chemie.uni-hamburg.de

Nanocomposites with uniaxial orientation are monitored during mechanical fatigue tests by two-dimensional small-angle X-ray scattering. High-resolution, low-noise images are recorded at repeat rates of 30 s. In the near future low-noise data will be recordable at repeat rates of several Hertz, and automated methods must be designed for the quantitative evaluation of the data stream. We present novel adapted methods and demonstrate their application. The focus of this extended abstract is on a review of the methods.

Introduction

Technical progress in X-ray scattering

Presently we are experiencing that intense, reliable X-ray sources and fast detectors become generally available. At DESY in Hamburg PETRA III will provide one of the most brilliant X-ray sources worldwide. There less than a second will be sufficient to expose a low-noise two-dimensional (2D) scattering pattern of a polymer sample. Moreover, in detector development a similar break-through has been achieved. Cycle times of less than a second can easily be realized with the novel PILATUS (Broennimann et al. (2006); Hülsen-Bollier (2005)) detectors.

It may be objected that for a long time low-noise scattering curves can be recorded with repeat frequencies of several Hertz. Nevertheless, the value of those older data is limited. Either the material is exhibiting isotropic scattering – in this case a comprehensive analysis requires assumptions on the structure (e.g. the assumption of an ideal lamellar stack), or the material is anisotropic. In this case one measured curve does not describe the complete scattering pattern, or the 2D pattern of the past is noisy. On the other hand, by application of the new technology it will become possible for many polymer scientists to monitor the structure evolution in *oriented* polymer materials in real time. The results of such experiments offer the potential to strengthen the understanding of structure evolution. Moreover, the new instruments provide very narrow X-ray beams with diameters of less than 1 μm ("microbeams") that permit to study the variation of structure inside the material with respective spatial resolution. A complete sweep of a polymer fiber by a microbeam will take ca. 30 s. Thus, spatial resolution and time resolution may be combined to study the response of different annular zones in a polymer fiber to, e.g. mechanical or thermal load.

Uniaxially oriented parts monitored by X-ray scattering

Parts with fiber symmetry – like fibers, rods or tubes – are not only of practical interest because they are ubiquitous in everyday life. Moreover, they appear particularly suited for investigation by means of scattering methods, because the recorded information is complete: As the part is rotated about its axis, the 2D scattering pattern does not change. Thus, there is no need to take patterns at different rotation angles, as it would be necessary for materials with less symmetry. The recorded 2D scattering pattern of a fiber is oriented and contains all the accessible information on the structure of the sample.

Adapted data evaluation methods

Nevertheless, it will not be sufficient to simply engage a novel instrument – the anticipated data must be adequately evaluated in due time, as well. In this connection two big problems are arising. On the one hand, the data flood is increasing to such an extent that automated evaluation methods must be developed. On the other hand, the data structure has changed, in principal. Instead of scattering curves scattering images must be evaluated. The goal can be achieved by combining digital image processing and scattering theory (Stribeck (2007)). In this paper several methods are presented that have recently been developed in our group in order to quantitatively evaluate extensive sets of 2D scattering patterns from polymer materials with fiber symmetry. Two previously published reviews deal with other aspects of the scattering experiments from polymer materials, namely a presentation of scattering theory (Stribeck (2009c)) from the point of view of the materials scientists and a presentation of instrumental development (Stribeck (2009a)).

Monitoring the WAXD of Fibers Under Load

Motivation and Method Design

In wide-angle X-ray diffraction (WAXD) experiments the diffraction patterns must be mapped into reciprocal space before they can be analyzed quantitatively. For this purpose interactive computer programs (Rajkumar et al. (2005); Bian et al. (2006)) are utilized that rest upon unnecessary (Stribeck (2009b)) approximations. Such a design is no disadvantage in crystallography, because sophisticated interactive refinement methods are required anyway for the exact determination of crystal structure parameters in manageable series.

In contrast, in materials science frequently time-resolved experiments are carried out, and voluminous series of diffraction patterns must be processed. The materials scientist already knows the unit cell parameters. Thus, minor inaccuracy of the mapping can be tolerated, if in the experiment variation of peak intensity or shape shall be monitored. Here it is important to carry out the mapping fast and automatically. Because the fiber tilt may change during the experiment, the algorithm must be able to track and to compensate such variation. By revisiting the theoretical treatment of the fiber mapping it has been demonstrated (Stribeck (2009b)) that there is no principal reason to refine an approximate center of the fiber pattern iteratively. Moreover, instead of an approximation (Franklin & Gosling (1953); Fraser et al. (1976)) of the tilt angle β of the fiber an exact equation (Stribeck (2009b)) can be employed. In the methodical paper (Stribeck & Nöchel (2009)) an algorithm is presented by which the mapping can be performed automatically. Its design rests on the application of the mentioned findings. Intricate parametrization is simplified, and slow trigonometric functions are avoided to a large extent. The method is unsuitable for diffuse scattering patterns. If inaccuracies of 2 pixels can be tolerated, a pattern is automatically mapped into reciprocal space in real time.

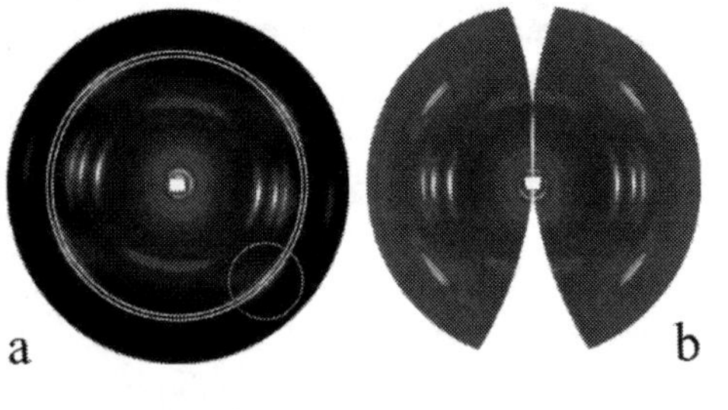

Figure 1: (a) Interactive mode of wf_premap: Draw the reflection circle through the centers of the reflection spots of an internal standard reflection (here: polypropylene (131) reflection), and widen the circle into a belt that contains the maxima of the reflection spots. Finally input 4 circles (one indicated in red). (b) The procedure `wf_map` has mapped the fiber diffraction pattern into reciprocal space

Actions Required By The User

For each series of diffraction patterns from a time-resolved experiment, some mapping parameters must once be determined interactively. Our procedure `wf_premap` assumes that the studied material exhibits a sharp reflection that is located neither on the equator nor on the meridian. In the example we investigate polypropylene and select the (131) reflection as the internal standard. With the crystallographic c-axis parallel to the meridian, the reflection is characterized by the parameters (Mencik (1972)) $d_{hk\ell} = d_{131} = 0.406\,\mathrm{nm}$ and by $c/\ell = 0.6504\,\mathrm{nm}$ that defines the position of the reflection ring on the Polanyi sphere (Stribeck (2009b); Polanyi (1921); Polanyi & Weissenberg (1923)). The wavelength of X-radiation (here: $\lambda = 0.15\,\mathrm{nm}$) must be known, too. In this case the first pre-mapping run of the series is started by

```
wave>ab = wf_premap(ss,a,0.15,0.406,0.6504)
```

The procedure requires a diffraction pattern as input (`a`). It generates output both in a "save-set" (`ss`), and in a background-corrected diffraction pattern (`ab`). If the provided save-set has never been used before, the procedure enters interactive mode and the user is presented the pattern as shown in Fig. 1.

In Fig. 1a the meridian is not vertical and the fiber is tilted. By means of the pointing device the user draws the reflection circle through the maxima of the reflection spots of the (131)-reflection. Thereafter he transforms the circle into a belt that is wide enough to contain the maxima of the spots (Fig. 1a). Now the procedure cuts out this belt and presents it to the user, who finally specifies 4 disjoint clips (regions of interest on the reflection circle). This is done by punching out circular regions from the belt (the bottom right one is shown in Fig. 1a). The clips are the intersections of the belt (white double-band) and the smaller circles. The program determines the positions of the maxima in the clips. From these 4 maxima-positions the best reflection circle is determined by regression. After that the orientation of the meridian is computed both from the upper and from the lower pair of spots. Based on this information the diffraction pattern is centered and aligned. Finally, the program computes the tilt angle β of the fiber from the orientation angles (Stribeck (2009b)) δ and δ' of the spots using the exact equation (Stribeck (2009b))

$$\tan\beta = \frac{\sqrt{4-\lambda^2 s_r^2}}{2\lambda s_r}\left(\cos\delta' - \cos\delta\right).$$

Here $s_r = 1/d_{131}$ is the radius of the Polanyi sphere of the reference (131)-reflection. Finally, the save-set `ss` is filled with several data: the tilt angle β; the true radius p_r of the reference-reflection circle on the detector; the crystallographic data of the reference reflection; the wavelength λ; and the geometrical data of the 4 clips. This information is sufficient for automated processing of the complete series of diffraction patterns recorded in a time-resolved X-ray diffraction experiment of materials with fiber symmetry. A more detailed description is in the original paper (Stribeck & Nöchel (2009)). Figure 1b shows a typical result obtained by automatic direct mapping.

Application

In a study (Stribeck et al. (2009)) that applies the method, uniaxially oriented polypropylene (PP) is molten and crystallized isothermally from the oriented, quiescent melt. The results show that nucleation and growth of differently oriented sets of crystallites (c-set and a^*-set) are decoupled. After shallow quench crystallization is preceded by (spinodal) decomposition. Peak integrals (→crystallinity) and minimum crystallite size are tracked. In the commercial starting material a^*-set crystallites melt at 158°C. The c-set melts at 170°C furnace temperature. After recrystallization both sets melt at 170°C. Isothermal crystallization is divided in two distinct phases. During nucleation the crystallinity stays low. The second phase is dominated by crystallinity growth. At 150°C the c-set is seeded first. At 145°C and 140°C a^*-oriented crystallites are the first. The first-seeded set starts to grow first, as well. c-set crystallinity is always growing faster than a^*-set crystallinity. The evolution of the corresponding SAXS (Stribeck et al. (2007)) cross-diagram in the growth phase can both be explained by lamellae growing at right angles, and by block merging.

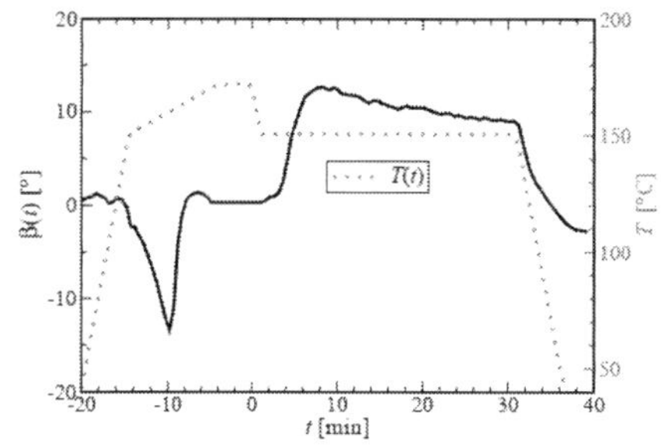

Figure 2: Tilt-angle tracking curve $\beta(t,T)$ from the automatic fiber-mapping procedure in an experiment in which β is changing considerably (hard-elastic polypropylene; melt-annealing at 171°C and recrystallization at 150°C)

Figure 2 shows $\beta(t,T)$ of one of the experiments. The tracking curve appears smooth and demonstrates the reliability of the tilt-angle determination. Tilt-angle variation is an issue, because the oriented PP film is heated until it becomes a viscous melt. Therefore the material shrinks and bends in the synchrotron X-ray beam.

After the mapping the intensity distribution is known almost everywhere in reciprocal space except for a wedge region near the meridian (cf. Fig. 1b). Thus, reflection intensities can readily be integrated in reciprocal space. In reciprocal space the relation between scattering and structure is clear from scattering theory (Stribeck et al. (2009); Stribeck (2007)), and structural data can be computed (Stribeck et al. (2009)), e.g., from total intensities (crystallinity) and integral breadths (crystallite sizes).

Figure 3 shows the evolution of the weight crystallinities of the two sets of crystallites at 3 different crystallization temperatures. As shown by Ruland (Ruland (1961)), such reflection integrals that are complete in reciprocal space are proportional to the weight crystallinity of the perfect crystallites that produce the reflections. Latency periods between the quench and the start of the crystallization, as well as crystallization velocities of the two kind of crystallites can be extracted from these data. Finally, conclusions concerning the crystallization mechanisms can be drawn (Stribeck et al. (2009)).

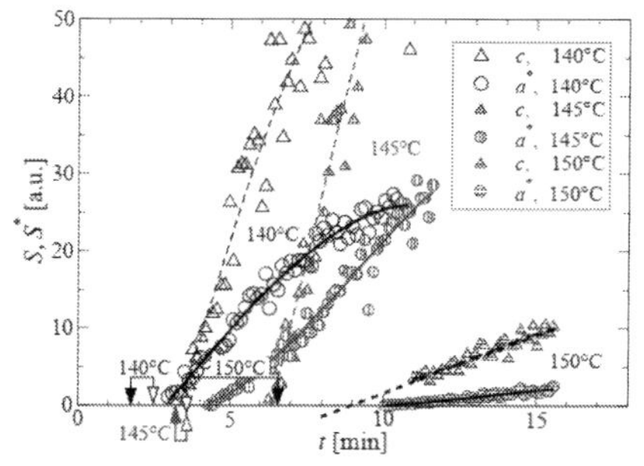

Figure 3: Evolution of relative weight crystallinities S, and S^* of c-oriented crystallites (dashed regression lines) and a^*-oriented crystals (solid regression lines), resp. during isothermal, oriented crystallization of hard-elastic PP from a quiescent melt as a function of crystallization temperature. Double-head arrows pointing at the t-axis indicate the first sighting of the a^*-set (full arrow head), and of the c-set (open arrow head), respectively

X-Ray Scattering Fiber Tomography

Motivation

The structure inside a polymer part is not necessarily homogeneous. With fibers, extrudates, and latex particles (Dingenouts et al. (1999)) frequently a core-shell structure is reported. Moreover, engineers deliberately generate structure gradients in polymer parts for special functionality (Kieback et al. (2003); Pompe et al. (2003)). If such a part is irradiated by an X-ray beam, the recorded pattern is an integral superposition of all the SAXS patterns emerging from the sequence of volume elements (voxels) along the beam path. From the mathematical point of view such a superposition is a projection, and a single projection is of little use for the study of graded materials.

A first step towards a study of structure gradients has been the development of the X-ray microbeam technique (Riekel & Engström (1995); Waigh et al. (1999); Kolb et al. (2001)). Here only the diameter of the beam is limiting the lateral spatial resolution. Nevertheless, the longitudinal spatial resolution is simply the thickness of the sample. Microbeam scanning experiments in particular of fibers have been performed for many years and the raw data have been discussed, although the corresponding shortcoming has been known (Paris et al. (2007): "The long-term goal is to proceed from microbeam scanning experiments to a real imaging technique"). The solution of the problem is tomographic reconstruction. Problems arise from the fact that scattering patterns are multidimensional but not simply a number (like the absorption in classical tomography). Thus, approximate tomographic reconstruction of scattering data with a manageable amount of artifacts is difficult. An exception is the case of part with uniaxial symmetry. In this case a reconstruction of the scattering patterns is possible that would have emerged from individual voxels in a plane perpendicular to the "fiber" axis (Stribeck et al. (2006)). Nevertheless, this method is only of academic value. The exposure time for the recording of scattering data of one cross-section of the part is in the order of days. The computing time for the tomographic reconstruction of the scattering patterns is typically a week (Stribeck et al. (2006)).

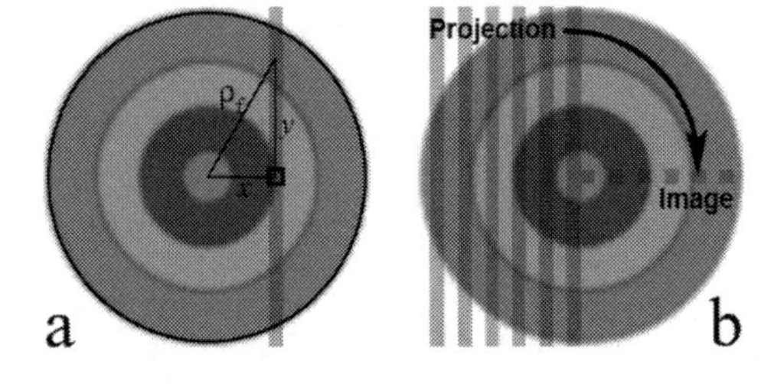

Figure 4: (a) Fiber cross-section irradiated by an X-ray beam at an offset x from its center. The structure $\rho\ (\rho_f)$ shows fiber symmetry. From all structures along the beam path a superposition is probed. y is the variable of the integration. (b) One-dimensional tomographic reconstruction turns the measured series of *projected* scattering patterns that carry the accumulated the structure information passed by the beam (vertical bars) into the *image* patterns from voxels (quadratic boxes) residing on the fiber radius

Introduction of the Method

A more practical tomographic method can be applied, if the part to be studied both exhibits macroscopic fiber symmetry, and the structure only varies as a function of the distance from its central axis. By means of this method fibers, pipes, and extruded strands can be investigated. Thus, we call it "X-ray scattering fiber computer tomography" (XSF-CT). A complete set of projected scattering patterns is collected in a single microbeam scan across the fiber, because the set of projections does not change as the sample is rotated about its axis. Such an experiment is completed in about 30 min. Moreover, compared to the general tomography the mathematics of image reconstruction is simplified considerably and the computational effort decreases by 5 orders of magnitude. A set of 40 measured scattering patterns is reconstructed in typically less than 10 min. From medicine and other fields of science the considerable potential of information increase after tomographic imaging is well-known.

In a general tomographic X-ray experiment (Bonse & Busch (1996)), a voluminous sample is scanned by a thin X-ray beam. As a function both of the position x of the scanning beam on the sample, and of the sample rotation angle ϕ, projections (notation: $\{\}$) of the absorption $\{A\}(x,\phi)$ or even of complete scattering patterns $\{I\}(\mathbf{s},x,\phi)$ are measured, in order to analyze the structure variation in the plane of the sample that is scanned by the X-ray beam (Schroer et al. (2006b); Stribeck et al. (2006); Schroer et al. (2006a)). Here $\mathbf{s}$ is the scattering vector with $|\mathbf{s}| = s = (2/\lambda)\sin\theta$, the X-ray wavelength λ and the scattering angle 2θ. In the examples mentioned, a tomographic image reconstruction (Kak & Slaney (1999); Bracewell (1999); Bonse & Busch (1996)) returns either the spatial variation of the absorption in the plane, or of the scattering emanating from the resolved voxels in the plane. The smearing caused from projection is eliminated by application of the Fourier transform theory, and a clear image of the inner structure is obtained.

If the studied material shows cylindrical symmetry, the results of the measurement are no function of ϕ any more, and the complete image information is in a single microbeam scan. The fundamental geometry is sketched in Fig. 4a. The information in the measured signal $\{A\}(x)$ or $\{I\}(\mathbf{s},x)$, respectively, does not represent the sought information $A(x)$ or $I(\mathbf{s},x)$ originating from the small square (voxel) around the position x. Instead, to a first approximation it is represented by the projection integral

$$\{I\}(\mathbf{s},x) = 2\int_x^\infty I\left(\mathbf{s},\sqrt{x^2+y^2}\right)dy \qquad (1)$$

$$= 2\int_x^\infty \frac{I(\mathbf{s},\rho_f)\,\rho_f d\rho_f}{\sqrt{\rho_f^2-x^2}}. \qquad (2)$$

The sought information in image space $(I(\rho_f))$ along the radius ρ_f of the fiber has to be reconstructed from the information in projection space $(\{I\}(x))$.

Eq. (1) is the definition of the Abel transform (Bracewell (1999)). In X-ray scattering Eq. (1) is established textbook knowledge (Guinier & Fournet (1955); Hosemann & Bagchi (1962); Alexander (1979); Glatter & Kratky (1982); Feigin & Svergun (1987); Baltá Calleja & Vonk (1989); Stribeck (2007)). There it describes the slit smearing. Even the inverse Abel transform

$$I(\mathbf{s},x) = -\frac{1}{\pi}\int_0^\infty \frac{d\{I\}(\mathbf{s},\rho_f)}{d\rho_f}\frac{dy}{\rho_f} \qquad (3)$$

$$= -\frac{1}{\pi}\int_x^\infty \frac{d\{I\}(\mathbf{s},\rho_f)}{d\rho_f}\frac{d\rho_f}{\sqrt{\rho_f^2-x^2}}, \qquad (4)$$

which has first been derived by Niels AbelAbel (1826) is in scattering textbooks since Guinier (Guinier & Fournet (1947)) and DuMond (DuMond (1947)). Similar to the filtered backprojection algorithm of the general tomography, low-noise reconstruction algorithms (Bitter et al. (2001); Dasch (1992); Dribinski et al. (2002)) for the tomography of materials with cylindrical symmetry are readily available in the field of "one-dimensional tomography". The principle of one-dimensional tomographic reconstruction is sketched in Fig. 4b. It must be mentioned that here we implicitly assume that the scattering from every irradiated voxel in the fiber shows fiber symmetry itself (local fiber symmetry). Otherwise characteristic reconstruction aberrations are expected (Stribeck et al. (2006, 2008a)). The meaning of local fiber symmetry is sketched in Fig. 5. Deviations from local fiber symmetry (i.e. tangential or radial grain, resp. (Stribeck et al. (2008a))) cause restricted or shifted visibility of scattering features along the fiber radius. These aberrations can be detected and result in additional information on the structure inside the fiber (Stribeck et al. (2008a)).

Figure 5: Sketch of local fiber symmetry inside the cross-section of the fiber. Each voxel has its own fiber axis. Inside the voxel there is only axial grain

Application

In an application-oriented feasibility study (Stribeck et al. (2008a)) precursors of polymer microfibrillar-reinforced composites (MFC) containing poly(ether)-*block*-amide (PEBA) and poly(ethylene terephthalate) (PET) with varying cold-draw ratio are studied. The studied strands are relatively thick, because presently the achievable "microbeam" at HASYLAB in Hamburg is relatively wide. The results from a direct analysis of the smeared measured patterns are compared to results obtained after tomographic reconstruction. Ideas for advanced practical applications of the XSF-CT method are discussed.

Data are presented from a cold-drawn (draw ratio $\lambda_d \approx 3$) co-extrudate of 70 wt.-% PEBA and 30 wt.-% PET (abbreviated: MFC73). In the scanning-microbeam experiment the strand shows an isotropic long period and an equatorial streak at almost every beam position (Fig. 6, left column). Only the tomographically reconstructed patterns (right column) exhibit that the long-period ring-reflection is not present in the core of the fiber. As the feature becomes visible, it first shows up at the equator. With increasing distance from the fiber axis, reflection arcs are growing towards the meridian. Above $\rho_f = 300\,\mu$m the arcs join into a closed circle. Because such behavior has been observed with neat PEBA as well, this orientation phenomenon is not indicating some interaction between the PEBA and the PET microfibrils. The reconstructed central voxel ($\rho_f = 0$) is dominated by one of the reconstruction aberration effects (from voxels with tangential grain) (Stribeck et al. (2008a)). Examination of the equatorial streak exhibits only in the reconstruction that it grows broader towards the center of the fiber. The streak is allocated to needle-shaped domains, and is only observed with strands from co-extruded blends. If these needles are thin PET microfibrils, the tomography shows that in the center of the fiber these microfibrils are shorter than in an intermediate region. The structure gradient in the outer region of the strand is discussed in the original paper (Stribeck et al. (2008a)).

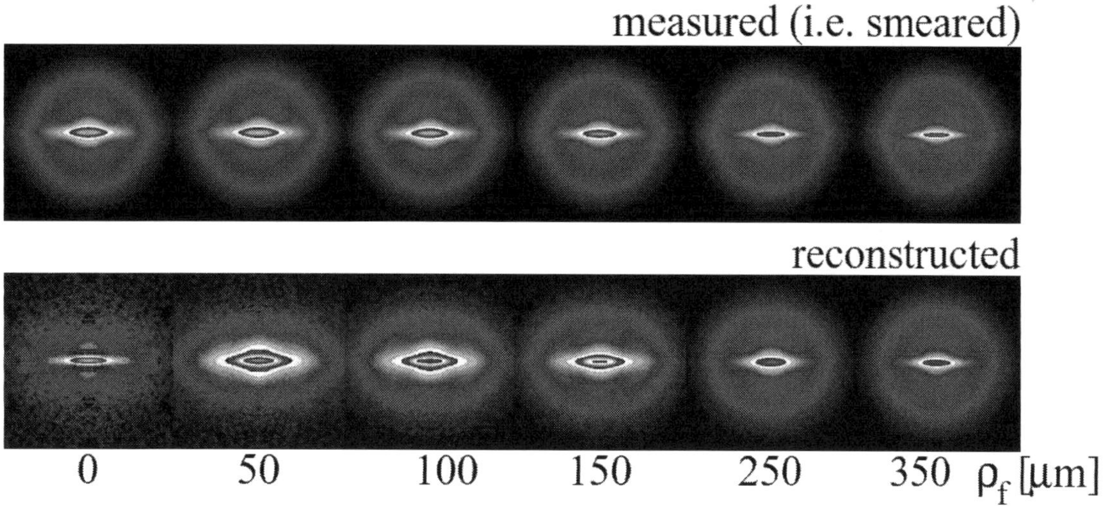

Figure 6: Cold-drawn ($\lambda_d \approx 3$) MFC73 in a scanning microbeam experiment. Measured scattering intensity $\{I\}(s_{12}, s_3, \rho_f)$ (top row) and reconstructed scattering $I(s_{12}, s_3, \rho_f)$ (bottom row) for short distances ρ_f from the fiber axis. The patterns display the range -0.1 nm$^{-1} \le s_{12}, s_3 \le 0.1$ nm^{-1} in uniform logarithmic scale ($s = (2/\lambda)\sin\theta$)

SAXS Monitoring of Mechanical Tests

Motivation and Method Development

Advanced polymer materials are urgently sought after e.g. in automotive industry in order to accomplish the goals of climate protection by reduction of weight. Such newly engineered materials have to prove their serviceability in mechanical tests. Classical tensile tests are performed to determine the modulus and ultimate properties. Load cycling experiments are carried out to determine the fatigue behavior. In order to reveal the mechanisms of, e.g., failure or fatigue, it is desirable to monitor mechanical tests by SAXS. In this way the response of the nanostructure to mechanical load is revealed. Since about 2005 the advance of instrumentation at synchrotron beamlines facilitates considerable reduction of the exposure required for the recording of low-noise SAXS patterns. Such quality is required for a quantitative analysis by, e.g., the CDF method (Stribeck (2001); Stribeck et al. (2008c)). Concerning the monitoring of tensile tests by SAXS this advance has meant that change-over from the stretch-hold technique (Wu et al. (2000)) to the more practical continuous straining has become possible (Stribeck et al. (2008b)), although the achievable strain rate is frequently still by a factor of 100 lower than that relevant in industry and service. With the advent of very high-brilliance synchrotron sources even this limitation is presently being abolished.

In order to gain full control on tensile and load-cycling tests a tensile tester has been built for operation at synchrotron beamlines (Stribeck et al. (2008b)). The control program written in LabVIEW® is continuously adapted to more complex experiments. Presently it can handle different kinds of continuous load-cycling programs that now are strictly synchronized with both the SAXS detector and a video grabber module (recording sequences of pictures of the sample with fiducial marks).

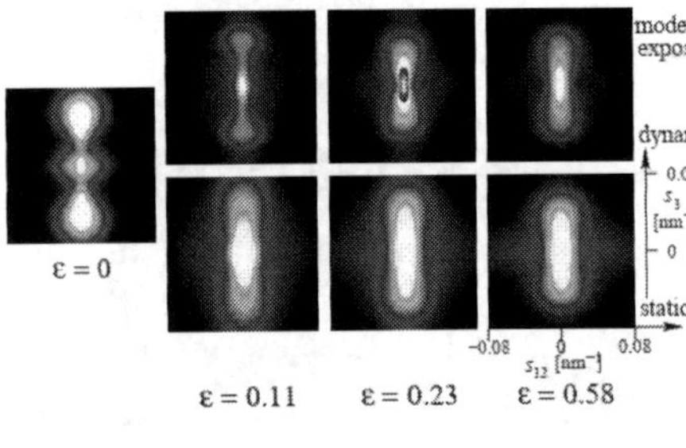

Figure 7: Oriented PP films in tensile tests. Comparison between SAXS patterns $I(s_{12}, s_3)$ recorded during continuous straining (top row, $\dot{\varepsilon} \approx 10^{-3} s^{-1}$) and patterns from an experiment in stretch-hold technique (bottom row). Straining direction and meridian is vertical (s_3). Equator (s_{12}) is horizontal

The latest methodical development (Denchev et al. (2010)) is an automatic evaluation method of the video frames, by which the true macroscopic elongation of the sample can be determined with an accuracy of 3 decimals. Such high accuracy is required for thermoplastic materials that fail at low elongations (ca. 0.1). The method extracts the grating of the fiducial marks on the sample from the actual video image, computes the 1D correlation function (Vonk & Kortleve (1967); Strobl & Schneider (1980)), and evaluates its "long period".

Results

In a study (Stribeck et al. (2008b)) of hard-elastic polypropylene (PP), scattering patterns recorded during continuous straining differ considerably from those recorded in the step-hold technique. Even though during exposure the elongation is no longer constant when applying the dynamic technique, the images collected in stretch-hold technique appear much more blurred (Fig. 7). This result indicates relaxation of nanostructure while the extensometer stands still. Quantitative analysis shows that during relaxation the extension of crystalline lamellae is increasing. Lamellae thicknesses are becoming non-uniform. The range of order is shortening. Cross-hatched lamellae are formed.

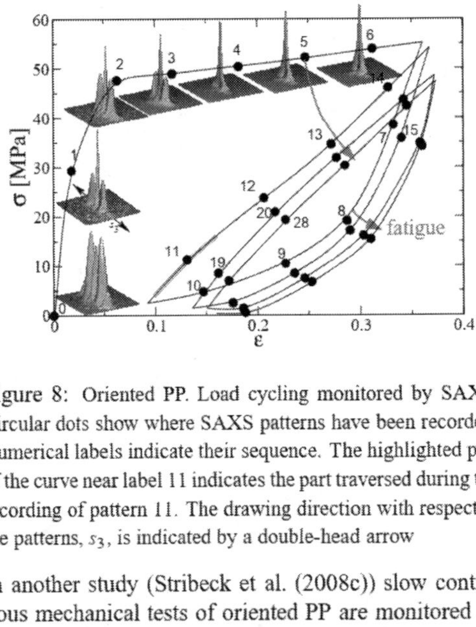

Figure 8: Oriented PP. Load cycling monitored by SAXS. Circular dots show where SAXS patterns have been recorded. Numerical labels indicate their sequence. The highlighted part of the curve near label 11 indicates the part traversed during the recording of pattern 11. The drawing direction with respect to the patterns, s_3, is indicated by a double-head arrow

In another study (Stribeck et al. (2008c)) slow continuous mechanical tests of oriented PP are monitored by SAXS and quantitatively evaluated. A continuous-strain test exhibits fracture and release of weak lamellae (2 - 10% strain). Beyond that conversion of lamellae into needles is observed. As all layers are consumed, the material breaks. Fatigue is studied in a load-reversal experiment (between 10% and 35% strain, Fig. 8). In each cycle crystallization, layer break, and relaxation melting are observed. Figure 9 presents the result. The top chart shows the true elongation, $\varepsilon(t)$, imposed to the material and measured at the point of X-ray irradiation. The four load-reversal cycles are easily identified. The chart in the middle reports the extracted nanostructure parameters, $L(t)$, $e_{cac}(t)$, and $S(t)$. The bottom diagram presents the macroscopic resistance, $\sigma(t)$, which the material is opposing to strain. The figure shows that crystallization, rupture of lamellae and melting of fragments are continuously reshaping the domains of the PP material in the cycles of the fatigue test. The long-period cycle exhibits a phase shift with respect to the imposed strain cycling. Moreover, there is an indication of amplitude attenuation. Fatigue is demonstrated in the curve $\sigma(t)$ by decreasing stress peaks. An in-depth discussion of the zones indicated in Fig. 9 is in the original paperStribeck et al. (2008c). In particular, the combination of SAXS and fatigue test shows the transition from stress-induced crystallization to crystallite rupture $\sigma(t) \approx 20$ MPa.

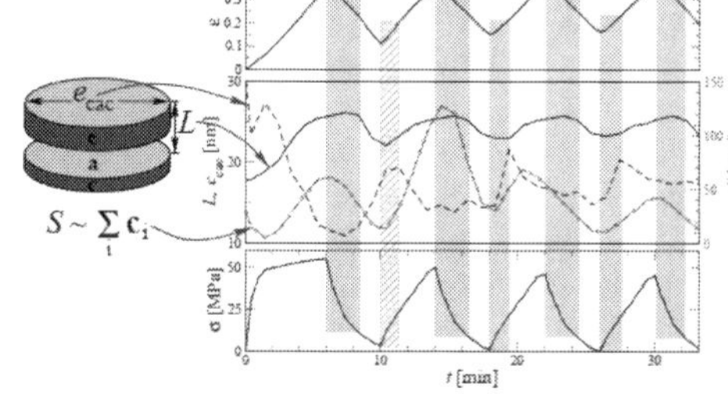

Figure 9: Dynamic load-reversal mechanical test of hard-elastic PP film. As a function of the elapsed time t the macroscopic parameters elongation, ε (top graph), and tensile stress, σ (bottom graph) are displayed together with topological nanostructure parameters (middle). In the middle diagram the solid line shows the long period, L. The broken line displays the lateral extension, e_{cac}, of a sandwich made from two crystalline lamellae. The line with circular dots exhibits the variation of the strength, S, of the CDF. Vertical bars indicate zones of strain-induced crystallization (dark green) and relaxation-induced melting (light gray), respectively

In a recent studyDenchev et al. (2010) of microfibrillar reinforced blends based on polyethylene we have switched from hard-elastic materials to materials that fail at the low elongations typical for plastic polymers. Because of the much lower elongation at break ($\varepsilon_b \approx 0.1$), now the macroscopic elongation ε_m and nanoscopic elongation ε_n must be determined with high precision. Respective methods are presented. The results show that the hardest materials exhibit a very inhomogeneous nanodomain structure. During straining their domains appear to be wedged together and inhibit transverse contraction on the nanometer scale.

Further components are polyamides (PA6, PA12) (20–30%) and as compatibilizer Yparex® 8102 (YP) (0–10%). Some HDPE/PA6 blends are additionally loaded with nanoclays (Nanomer® or Cloisite®). Blending of HDPE with PA12 causes no synergistic effect. In the absence of nanoclay, PA6 and HDPE form a heterogeneous nanostructure with high Young's modulus. After addition of YP a more homogeneous scaffold structure is observed in which some of the PA6 microfibrils and HDPE crystallites appear to be rigidly connected, but the modulus has decreased. Both kinds of nanoclay induce a transition from a structure without transverse correlation among the microfibrils into a macrolattice with 3D correlations among HDPE domains from neighboring microfibrils. For extensions between 0.7% and 3.5% the scattering entities with 3D correlation exhibit transverse elongation instead of transverse contraction. The process is interpreted as overcoming a correlation barrier executed by the crystallites in an evasion-upon-approaching mechanism. During continued straining the 3D correlation is reduced or removed.

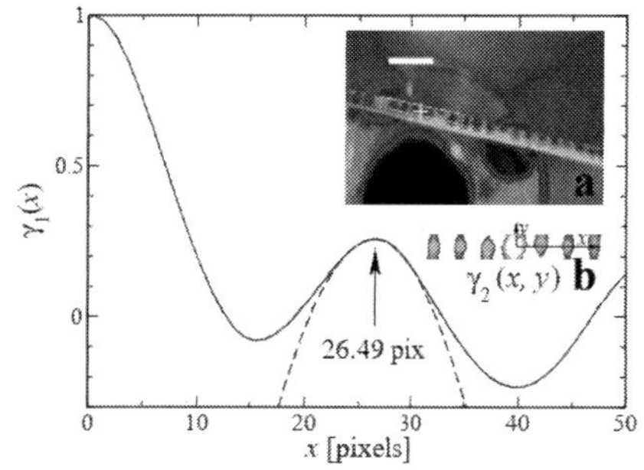

Figure 10: Elongation from recorded video frames. Inset **a**: In the first video frame a region of interest (ROI) with fiducial marks is defined. Inset **b**: From the ROI the 2D correlation function $\gamma_2(x,y)$ is computed. Main drawing: The center of the long-period peak in $\gamma_1(x) = \gamma_2(x,0)$ is fitted by a parabola (dashed line) to compute the distance between the fiducial marks

The true macroscopic elongation ε_m is determined from video frames taken of the sample that carries fiducial marks (see Fig. 10a). Once for an experiment the user has to provide some input. It is based on the first image (Fig. 10a) of the series. The center of the X-ray beam on the sample is marked by a cross in the image. Close to this center the user defines a rectangular region of interest (ROI), $\rho'_m(x,y)$. In Fig. 10a this region is bordered by a dashed line. x and y are pixel coordinates in the direction of strain and perpendicular to it, respectively. The same ROI is applied to all video frames of the experiment. The ROI is structured by the fiducial marks running perpendicular to the straining direction. As is known from scattering theoryStribeck (2007), from such a structure function $\rho_m(x,y)$ the 2D correlation function

$$\gamma_2(x,y) = \frac{\rho_m^{*2}(x,y)}{\rho_m^{*2}(0,0)}$$

can be computed, with

$$\rho_m(x,y) = \rho'_m(x,y) - \bar{\rho}_m$$

representing the fluctuation of $\rho'_m(x,y)$ about its average $\bar{\rho}_m$, and the autocorrelation being defined by the integral

$$f^{*2}(x,y) = \iint_{-\infty}^{\infty} f(u,v)\, f(u+x, v+y)\, du dv.$$

In Fig. 10b the colored caps demonstrate, where $\gamma_2(x,y)$ is positive. The macroscopic elongation ε_m in straining direction is the section $\gamma_1(x) = \lceil \gamma_2(x,y) \rceil_1(x)$ of γ_2 in straining direction. Figure 10 presents this curve and its analysis. Its first positive peak is the long-period peak that is related to the actual average distance of the fiducial marks, ℓ. A parabola (dashed line) is fitted to the long-period maximum, and the position of its vertex is determined (arrow). Thus, ℓ can be determined with an accuracy of 0.01 pixels. Let ℓ_0 the initial distance between the marks, then the macroscopic elongation is $\varepsilon_m = \ell/\ell_0 - 1$.

Concerning the scattering patterns, similar analysis is possible. Nevertheless, the peaks observed in the scattering pattern or in the CDFsStribeck (2001) computed from the pattern are no longer one-dimensional but two-dimensional – positioned in the plane (s_{12}, s_3) or (r_{12}, r_3), respectively. Again, the user defines a ROI in which the analysis procedure searches for the peak maximum, extracts the peak cap, and fits now a bivariateHall (1980) polynomial of 2nd order to it. From the coefficients of the polynomial several parameters are readily determined. Most important are the peak position and the widths of the peak in meridional and equatorial directions, respectively. Thus, the evolution of these parameters during the experiment can be tracked automatically.

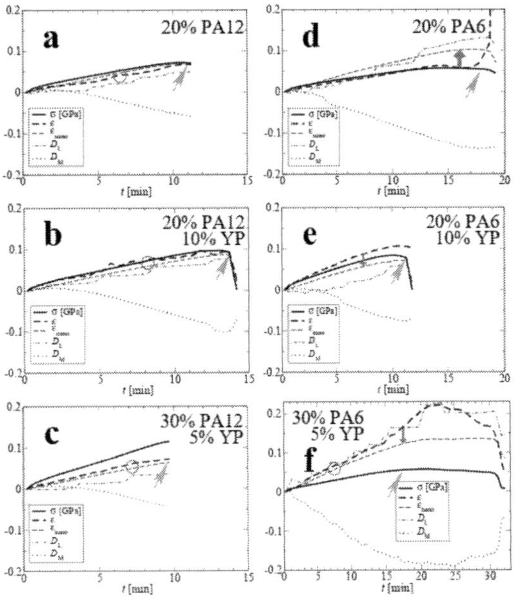

Figure 11: MFC precursor blends from HDPE, two different polyamides (PA12, PA6) and a compatibilizer (YP) in tensile tests. Evolution of macroscopic stress and strain (σ, ε) as well as of nanostructure parameters. ε_{nano} is the nanoscopic elongation computed from the HDPE long period. D_L is the relative change of the width of the long period distribution. D_M is the relative change of the extension of the microfibrils in transverse direction

Figure 11 presents the results of a quantitative nanostructure analysis for the blends which do not contain nanoclays. Elongations are illustrated by dashed lines. Bold lines show the macroscopic elongation, ε. Thin lines report the nanoscopic elongation ε_{nano} of the HDPE matrix. Circular marks indicate regions in which $\varepsilon \approx \varepsilon_{nano}$. All materials reinforced by PA12 exhibit this similarity of macroscopic and nanoscopic deformation. Dashed-dotted lines show D_L, the relative variation of the breadth of the long-period distribution. The increase of all curves demonstrates increasing inhomogeneity of the long periods with increasing elongation. Dotted lines show D_M, the relative variation of the microfibrillar diameter during the straining process. In all materials the elongational deformation causes the microfibrils to thin. Variation of the material composition does not cause considerable changes. This is different with the samples reinforced by PA6 (Fig. 11d-f). Here an increase of the PA6 content to 30% causes strong thinning of the microfibrils. In the PA6 blends the strong transverse decrease is going along with only moderate nanoscopic elongation ε_{nano} of the HDPE. An explanation for this finding could be microfibrillation by fracture of crystalline domains of the polyethylene. Moreover, the diagrams in Fig. 11d-f demonstrate a considerable difference (vertical arrows) between the two dashed curves. In Fig. 11e-f ($\varepsilon_{nano} < \varepsilon$) the nanoscopic elongation of the HDPE phase is considerably lower than the macroscopic elongation. Similarity is only observed during the initial deformation in Fig. 11f (circular mark). In the 80/20 HDPE/PA6 blend (Fig. 11d) the nanoscopic elongation of the HDPE microfibrils is considerably longer than the macroscopic elongation ($\varepsilon_{nano} > \varepsilon$). Although this finding appears to be unreasonable, an indication for a possible mechanism is in the strong increase of D_L (Fig. 11d). This is discussed in the original paper-Denchev et al. (2010).

Conclusions

Considering the present instrumental development at synchrotron radiation facilities the development of advanced data evaluation methods appears to be both promising and necessary in order to master the future data flood. The 3 presented methods demonstrate the potential of such work. In order to discharge the user, a part of the data evaluation may be carried out at the synchrotron facility. Such added service would require not only considerably computing power, but also additional manpower. In addition to the beamline scientist an evaluation specialist would become necessary. It would be his job to detect if raw data must be smoothed. He would have to eliminate the machine background, would generate detector masks, would center and align each scattering pattern and would fill blind areas from consideration of symmetry. As an added service, the community of the evaluation specialists could select some standard experiments for which complete user-friendly programming environments could be built. Nevertheless, for the predominant fraction of individually designed setups it will remain necessary for the polymer scientist himself to familiarize with adapted programming techniques. This will be of particular importance, if methods shall be developed that grow with the growth of instrumental capacity. Ultimately, it is expected that an increase of the quality of results returned from scattering experiments will closely be correlated to the manpower dedicated to the programming of data evaluation modules.

Acknowledgment. The author thanks the Hamburg Synchrotron Radiation Laboratory (HASYLAB) for beam time granted in the frame of project II-20080015. Development of the reported methods has been supported by the 7th framework program of the European Union (Project NANOTOUGH FP7-NMP-2007-LARGE-2.1.1).

References

Abel, N. H., "Auflösung einer mechanischen Aufgabe", in J. Reine Angew. Math., **1**, 153–157 (1826)

Alexander, L. E., X-Ray Diffraction Methods in Polymer Science, Wiley, New York (1979)

Baltá Calleja, F. J., Vonk, C. G., X-Ray Scattering of Synthetic Polymers, Elsevier, Amsterdam (1989)

Bian, W., Wang, H., McCullough, I., Stubbs, G., "WCEN: a computer program for initial processing of fiber diffraction patterns", in J. Appl. Cryst., **39**, 752–756 (2006)

Bitter, I., Kaufman, A. E., Sato, M., "Penalized-Distance Volumetric Skeleton Algorithm", in IEEE Trans. Visualization and Computer Graphics, **7**, 195–206 (2001)

Bonse, U., Busch, F., "X-ray Computed Microtomography (μCT) Using Synchrotron Radiation", in Prog. Biophys. Molec. Biol., **65**, 133–169 (1996)

Bracewell, R., The Fourier Transform and Its Applications, Mc Graw-Hill, New York, 3rd edn. (1999)

Broennimann, C., Eikenberry, E. F., Henrich, B., Horrisberger, R., Hülsen, G., Pohl, E., Schmitt, B., Schulze-Briese, C., Suzuki, M., Tomizaki, T., Toyokawa, A., Wagner, A., "The PILATUS 1M detector", in J. Synchrotron Rad., **13**, 120–130 (2006)

Dasch, C. J., "One-dimensional tomography: a comparison of Abel, onion-peeling, and filtered backprojection methods", in Applied Optics,, **31**, 1146–1153 (1992)

Denchev, Z., Dencheva, N., Funari, S. S., Motovilin, M., Schubert, T., Stribeck, N., "Nanostructure and mechanical properties studied during dynamical straining of microfibrillar reinforced HDPE/PA blends", in J. Polym. Sci. Part B: Polym. Phys., **48**, 237–250 (2010)

Dingenouts, N., Bolze, J., Potschke, D., Ballauff, M., "Analysis of polymer latexes by small-angle X-ray scattering", in Adv. Polym. Sci., **144**, Epoxide Resins, Polyampholytes), 1–Epoxide Resins, Polyampholytes),47 (1999)

Dribinski, V., Ossadtchi, A., Mandelshtam, V. A., Reisler, H., "Reconstruction of Abel-transformable images: The Gaussian basis-set expansion Abel transform method", in Rev. Sci. Instr., **73**, 2634–2642 (2002)

DuMond, J. W. M., "Method of Correcting Low Angle X-ray Diffraction Curves for the Study of Small Particle Sizes", in Phys. Rev., **72**, 83–84 (1947)

Feigin, L. A., Svergun, D. I., Structure Analysis by Small-Angle X-Ray and Neutron Scattering, Plenum Press, New York (1987)

Franklin, R. E., Gosling, R. G., "The Structure of Sodium Thymonucleate Fibres. II. Cylindrically Symmmetrical Patterson Function", in Acta Cryst., **6**, 678–685 (1953)

Fraser, R. D., Macrae, T. P., Miller, A., Rowlands, R. J., "Digital Processing of Fibre Diffraction Patterns", in J. Appl. Cryst., **9**, 81–94 (1976)

Glatter, O., Kratky, O. (Eds.), Small Angle X-ray Scattering, Academic Press, London (1982)

Guinier, A., Fournet, G., "Correction of Measurments of Low-Angle X-Ray Scattering", in Nature, **160**, 501 (1947)

Guinier, A., Fournet, G., Small-Angle Scattering of X-Rays, Chapman and Hall, London (1955)

Hall, E. L., Computer Image Processing and Recognition, Academic Press, London (1980)

Hosemann, R., Bagchi, S. N., Direct Analysis of Diffraction by Matter, North-Holland, Amsterdam (1962)

Hülsen-Bollier, G., The PILATUS 1M Detector. A Novel Large Area Pixel Detector, Ph.D. thesis, Dept. Physics, University of Erlangen-Nürnberg, Germany (2005)

Kak, A. C., Slaney, M., Principles of Computerized Tomographic Imaging, IEEE Press, New York (1999)

Kieback, B., Neubrand, A., Riedel, H., "Processing techniques for functionally graded materials", in Mater. sci. eng. A Struct. mater. prop. microstruct. proces., **362**, 81–106 (2003)

Kolb, R., Wutz, C., Stribeck, N., v. Krosigk, G., Riekel, C., "Investigation of secondary crystallization of polymers by means of microbeam X-ray scattering", in Polymer, **42**, 5257–5266 (2001)

Mencik, Z., "Crystal Structure of Isotactic Polypropylene", in J. Macromol. Sci. Phys., **B6**, 101–115 (1972)

Paris, O., Li, C., Siegel, S., Weseloh, G., Emmerling, F., Riesemeier, H., Erko, A., Fratzl, P., "A new experimental station for simultaneous X-ray microbeam scanning for small- and wide-angle scattering and fluorescence at BESSY II", in J. Appl. Cryst., **40**, s466–s470 (2007)

Polanyi, M., "Das Röntgen-Faserdiagramm I. (The X-Ray Fiber-diagram I.)", in Z. Phys., **7**, 149–180 (1921)

Polanyi, M., Weissenberg, K., "Das Röntgen-Faserdiagramm II. (The X-Ray fiber-diagram II.)", in Z. Physik, **9**, 123–130 (1923)

Pompe, W., Worch, H., Epple, M., Friess, W., Gelinsky, M., Greil, P., Hempel, U., Scharnweber, D., Schulte, K., "Functionally graded materials for biomedical applications", in Mater. sci. eng. A Struct. mater. prop. microstruct. proces., **362**, 40–60 (2003)

Rajkumar, G., AL-Khayat, H., Eakins, F., He, A.,

Knupp, C., Squire, J., "FibreFix - A New Integrated CCP13 Software Package", in Fibre Diffraction Rev., **13**, 11–18 (2005)
Riekel, C., Engström, P., "Diffraction and diffuse scattering from materials with microfocussed X-rays", in Nuclear Instr. Meth. Phys. Res., **B97**, 224–230 (1995)
Ruland, W., "X-ray Determination of Crystallinity and Diffuse Disorder Scattering", in Acta Cryst., **14**, 1180–1185 (1961)
Schroer, C. G., Kuhlmann, M., Günzler, T. F., Benner, B., Kurapova, O., Patommel, J., Lengeler, B., Roth, S. V., Gehrke, R., Snigirev, A., Snigireva, I., Stribeck, N., Almendárez Camarillo, A., Beckmann, F., "Full-field and scanning microtomography based on parabolic refractive x-ray lenses", in Proc. SPIE, **6318**, 6318H (2006a)
Schroer, C. G., Kuhlmann, M., Roth, S. V., Gehrke, R., Stribeck, N., Almendarez Camarillo, A., Lengeler, B., "Mapping the local nanostructure inside a specimen by tomographic small-angle x-ray scattering", in Appl. Phys. Lett., **88**, 164102 (2006b)
Stribeck, N., "Extraction of domain structure information from small-angle X-ray patterns of bulk materials", in J. Appl. Cryst., **34**, 496–503 (2001)
Stribeck, N., X-Ray Scattering of Soft Matter, Springer, Heidelberg, New York (2007)
Stribeck, N., "Deformation behavior of nanocomposites studied by X-ray scattering: Instrumentation and methodology", in Karger-Kocsis, J., Fakirov, S. (Eds.), "Nano- and Micromechanics of Polymer Blends and Composites", Hanser Publisher, München, vol. 1, chap. 8, 269–300 (2009a)
Stribeck, N., "On the determination of fiber tilt angles in fiber diffraction", in Acta Cryst., **A65**, 46–47 (2009b)
Stribeck, N., "Scattering of Soft Condensed Matter. From Fundaments to Application", in Ezquerra, T. A., Garciá Gutiérrez, M., Nogales, A., Gómez, M. (Eds.), "Applications of Synchrotron Light to Scattering and Diffraction in Materials and Life Sciences", Springer, Berlin Heidelberg, vol. 776 of *Lect. Notes Phys.*, chap. 2, 25–62 (2009c)
Stribeck, N., Almendarez Camarillo, A., Nöchel, U., Schroer, C., Kuhlmann, M., Roth, S. V., Gehrke, R., Bayer, R. K., "Volume-Resolved Nanostructure Survey of a Polymer Part by Means of SAXS Microtomography", in Macromol. Chem. Phys., **207**, 1239–1249 (2006)
Stribeck, N., Nöchel, U., "Direct mapping of fiber diffraction patterns into reciprocal space", in J. Appl. Cryst., **42**, 295–301 (2009)
Stribeck, N., Nöchel, U., Almendárez Camarillo, A., Roth, S. V., Dommach, M., Bösecke, P., "SAXS Study of Oriented Crystallization of Polypropylene from a Quiescent Melt", in Macromolecules, **40**, 4535–4545 (2007)
Stribeck, N., Nöchel, U., Fakirov, S., Feldkamp, J., Schroer, C., Timmann, A., Kuhlmann, M., "SAXS-Fiber Computer-Tomography. Method Enhancement and Analysis of Microfibrillar-Reinforced Composite Precursors from PEBAX and PET", in Macromolecules, **41**, 7637–7647 (2008a)
Stribeck, N., Nöchel, U., Funari, S. S., "Melting and Crystallization of Differently Oriented Sets of Crystallites in Hard-Elastic Polypropylene", in Macromolecules, **42**, 2093–2101 (2009)
Stribeck, N., Nöchel, U., Funari, S. S., Schubert, T., "Tensile Tests of Polypropylene Monitored by SAXS. Comparing the Stretch-Hold Technique to the Dynamic Technique", in J. Polym. Sci. Polym. Phys., **46**, 721–726 (2008b)
Stribeck, N., Nöchel, U., Funari, S. S., Schubert, T., Timmann, A., "Nanostructure Evolution in Polypropylene During Mechanical Testing", in Macromol. Chem. Phys., **209**, 1992–2002 (2008c)
Strobl, G. R., Schneider, M., "Direct Evaluation of the Electron Density Correlation Function of Partially Crystalline Polymers", in J. Polym. Sci., Part B: Polym. Phys., **B18**, 1343–1359 (1980)
Vonk, C. G., Kortleve, G., "X-Ray Small-Angle Scattering of Bulk Polyethylene", in Colloid Polym. Sci., **220**, 19–24 (1967)
Waigh, T. A., Donald, A. M., Heidelbach, F., Riekel, C., Gidley, M. J., "Analysis of the native structure of starch granules with small angle x-ray microfocus scattering", in Biopolymers, **49**, 91–105 (1999)
Wu, J., Schultz, J. M., Yeh, F., Hsiao, B. S., Chu, B., "In-Situ Simultaneous Synchrotron Small- and Wide-Angle X-ray Scattering Measurement of Poly(vinylidene fluoride) Fibers under Deformation", in Macromolecules, **33**, 1765–1777 (2000)

STRUCTURE-PROPERTIES RELATIONSHIP IN MICROFIBRILAR COMPOSITES BY MEANS OF SIMULTANEOUS X-RAY/STRETCHING EXPERIMENTS

Z. Denchev[1]*, M. Motovilin [1], N. Stribeck[2], N. Dencheva[1]

[1] *i3N – Institute for Polymers and Composites, University of Minho, Guimarães, Portugal – denchev@dep.uminho.pt; mmotov@dep.uminho.pt; nadiad@dep.uminho.pt*
[2] *Institute of Technical and Macromolecular Chemistry, University of Hamburg, Germany – norbert.stribeck@desy.de*

This study exemplifies the use of X-rays from synchrotron for studying in real time the nanostructure development and the mechanical properties of some new composite hybrids. They comprise an isotropic high-density polyethylene (HDPE) matrix reinforced by polyamide fibrils. For further reinforcement, organically treated montmorillonite (MMT) was dispersed within the fibrils. Synchrotron small-angle X-ray scattering (SAXS) patterns were collected while subjecting the samples to static or cyclic load. The resulting two-dimensional SAXS patterns were reconstructed, calibrated and the structural information extracted from them by means of new image treatment and data evaluation procedures involving the computation of the Chord Distribution Function (CDF). True stress-strain curves were obtained in real time for each SAXS data frame. Thus, it was possible to visualize how the processing conditions and nanoclay content could change the structure at nanometer level and to assess the structure-mechanical properties relationship.

Introduction

Monitoring of nanostructure evolution during mechanical failure is of both fundamental and practical relevance. The mechanisms revealed in such studies should help increase the significance of modeling in predicting of the mechanical properties of polymer materials. Understanding the relationship between the structure and mechanical properties can direct the research for new composite materials and the optimization of the already existing ones. To this purpose, generation and analysis of time-resolved structural data obtained during the mechanical tests are needed.

X-ray scattering methods are being used for non-destructive structural investigations in polymers in different forms and levels of sophistication for 50-60 years. During the last decade there has been a substantial progress in the areas of synchrotron facilities, construction and development of new X-ray detectors, which brought to pass the extensive use of synchrotron radiation in materials science. Nevertheless, performing X-ray scattering and mechanical tests simultaneously continues to be a challenging task, especially as far as the automatic acquisition and processing of large data sets is concerned. The scattering patterns are often processed and interpreted reflection by reflection, so as to extract indirectly the structural information. The latter can be often distorted or even damaged due to various reasons, related either to the data collection or to the data treatment.

In this study various polymer composites were stretched in a testing machine mounted in a small-angle X-ray scattering (SAXS) beamline. It was possible to acquire simultaneously good quality SAXS patterns together with the respective real stress - real strain values. New SAXS data handling methodologies allowing for a direct transformation of the X-ray patterns into real-space images of the nanostructure were applied. The SAXS and mechanical data sets were used to construct adequate structural and failure models.

Experimental

Sample Description

The samples represent microfibrilar reinforced composites (MFC). In them, the matrix and reinforcements are obtained *in situ* from polymer blends as described previously (Dencheva et al, 2008): high-density polyethylene (HDPE, 65-90 wt.%), polyamide 6 (PA6) or polyamide 12 (PA12) (10-35 wt. %) and a commercial compatibilizer (0-10 wt. %). The PA6 fibrils of some blends were loaded with 1.0-7.5 wt.% of commercially available montmorillonite (MMT) nanoclays as described by Denchev et al (2010)). The samples investigated were the oriented precursors - blends in which both HDPE and PA phases are fibrillated by cold drawing, and the respective MFC comprising isotropic HDPE matrix reinforced by fibrils of PA – with or without additional reinforcement by MMT.

Data Acquisition and Processing

Both SAXS and mechanical data were collected in the A2 beamline at HASYLAB, Hamburg, Germany. The SAXS patterns were recorded during dynamical stretching. The data analysis includes the calculation of the multidimensional chord distribution function (CDF), $z(\mathbf{r})$ (Stribeck, 2001), from the fiber-symmetric SAXS patterns. High-precision determination of the macroscopic elongation is accomplished by application of a correlation method. The reduction of the sample dimensions is considered assuming constant volume.

Results and Discussion

HDPE/PA oriented blends

Fig. 1 (a-c) shows the time evolution of the nano-structure as revealed by the corrected and reconstructed SAXS patterns (upper part of the images) and by the respective CDFs (lower part) in three different oriented blends subjected to stretching while being irradiated. After automated processing of each CDF data set, the time dependence of three nanostructural parameters can be obtained, namely or ε_{nano} representing the nanoscopic elongation, that was computed from the HDPE long period; D_L — the relative change of the width of the long period distribution and D_M — the relative change of the microfibril extension in transverse direction [Fig. 1 (d-e)]. The same graphs contain also the respective stress-strain curves as obtained from the stretching machine. The critical evaluation of the structural and mechanical data allows the construction of models of the sample structure and its failure [Fig. 1(f)].

With the materials containing no MMT, reduction of Young's modulus is clearly related to the decrease of the HDPE microfibril diameter, D_M, whereby the stronger the decrease the lower the modulus. Reverting this reasoning, suppressing the transverse compression appears to be an effective way to harden the material. Adding of compatibilizer has different effect in PA6 and PA12 containing blends. In the first case the compatibilizer has a homogenizing effect on the longitudinal lattice inside the HDPE microfibrils. In the HDPE/PA12 blends, it promotes more regular arrangement of the HDPE crystallites, thus lowering the modulus.

All four HDPE/PA6/MMT blends in which $\varepsilon_{nano} < \varepsilon_{macro}$ show identical mechanical behavior. The highest-modulus material shows $\varepsilon_{nano} > \varepsilon_{macro}$ and the decrease of the mircofibril diameter D_M is the highest. Therefore, the hardness in this case seems to be primarily controlled by the inhomogeneities of the nanostructure. They cause wedging of the domains in transverse direction that, on its turn, suppresses the transverse contraction.

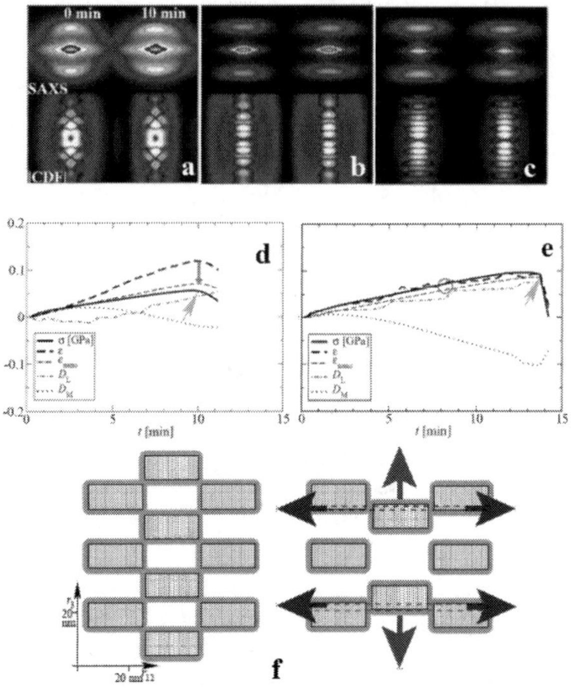

Figure 1 – Example of typical output data in the case of HDPE/PA oriented blends: SAXS patterns (upper row) and the respective CDFs (lower row) of (a) HDPE/PA12/YP, (b) HDPE/PA6/YP and (c) HDPE/PA6/MMT oriented blends; (d) – time dependence of the macroscopic and nanoscopic parameters for HDPE/PA6/YP = 70/20/10 + 5% Cliosite and (e) – HDPE/PA612/YP = 70/20/10; (f) structural and failure models for the latter sample.

HDPE/PA microfibrilar composites

As the MFCs are obtained by compression molding of oriented blends selectively isotropizing the HDPE phase, the interpretation of the initial SAXS patterns is limited by the presence of large fractions of isotropic material hiding the diffraction of the oriented fibrils. Therefore, before calculating the CDF it is necessary to separate the SAXS intensity of the oriented and non-oriented scatterers. Figure 2 visualizes the data processing for MFCs. The CDF functions (d)-(e) reflect the nanosctructire of the oriented fraction of the MFC, *i.e*, of the reinforcing fibril. These CDFs are unique and quite complex. At this point their processing reveals that the reinforcing fibril comprises an oriented core made of polyamide and an oriented shell of transcrystalline HDPE. The macro- and nanoscopic parameters can be extracted as for the oriented blends (Figure 2g). It can be concluded that that the compatibilizer has a reduced influence on the PA and transcrystalline HDPE long spacings, while the amount and type of the MMT affect the CDF stronger.

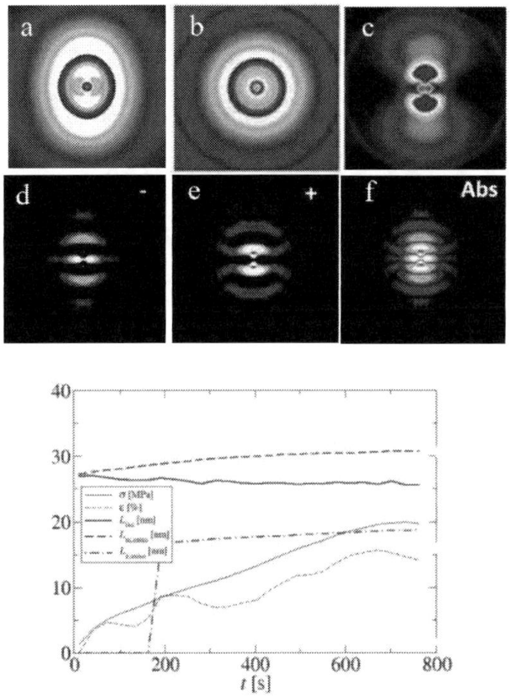

Figure 2 – Example of typical output data in the case of HDPE/PA/YP MFC materials: (a) – initial SAXS pattern; (b) – amorphous SAXS intensity; (c) – oriented SAXS intensity; (d) – negative face of the CDF; (e) – positive face of the CDF; (f) – superposition of the (d) and (e).The graph below is obtained analogously as in Fig. 1, (d), (e).

Conclusions

In this study the information increase by transformation of SAXS patterns into CDFs is clearly demonstrated. By means of the two proposed peak-fitting methods, small variations of macroscopic elongation and nanostructure parameters are clearly resolved. It was indicated how relations between properties and nanostructure in oriented HDPE/PA blends and the MFCs obtained thereof may be setup.

Acknowledgements

The authors thank the Hamburg Synchrotron Radiation Laboratory (HASYLAB) for beam time granted in the frame of projects II-04-039 and II-07-011. Development of the reported methods has been supported by the 7th framework program of the European Union (Project NANOTOUGH FP7-NMP-2007- LARGE). Z. D. is grateful for the tenure of a sabbatical grant awarded by the Fundação para a Ciência e Tecnologia, Portugal (grant SFRH/BSAB/ 812/2008) and for the hospitality of the Institute of Technical and Macromolecular Chemistry, Hamburg

References

Dencheva, N., et al., "Preparation and properties of in-situ composotes based on HDPE/PA oriented blends", Mat. Sci. Forum **587-588,** 515–519 (2008)

Denchev, Z., et al., "Structure and mechanical properties during straining of HDPE/PA blends", J. Polym. Sci: Part B – Polym. Phys, **48**, 237–250 (2010)

Stribeck, N., "Extraction of domain structure from SAXS patterns of bulk materials" J. Appl. Crystallography, **34**, 496–503 (2001).

INFLUENCE OF THE CALENDERING STEP ON THE ADHESION PROPERTIES OF COEXTRUDED STRUCTURES

J.F.Agassant*, M.C.Serrat*,, J.Bikard*, S.Devisme****

** Mines-Paristech, CEMEF, UMR CNRS 7635 Sophia-Antipolis (France)*
*** Arkema, Cerdato research centre, 27470, Serquigny – France)*

Abstract - Coextruded sheets of Polyamide and Polypropylene are often calendered in order to adjust the final thickness of the sheet, but also to improve adhesion between the different layers. In fact, the influence of the calendering parameters remains controversial. Thermomechanical models will help in understanding the relative influence of these processing parameters. Successive models of increasing complexity have been developed :
An isothermal monolayer mechanical model, based on the lubrication approximations, allows to capture the order of magnitude of pressure, shear and elongationnal strains in the calender nip. A multilayer temperature computation, accounting for the crystallisation kinetics of both polymers, allows to capture the temperature profile especially at the interface between the two layers. Finally, a temperature dependent multilayer thermomechanical model has been developed using the finite element software Forge 2007 with 2D plane deformation assumptions. Even using remeshing techniques it requires huge computation times. However it captures the main features of the process: temperature, strain and stress fields in the nip and especially at the interface between the coextruded layers.

Introduction

Multilayer sheets of polypropylene and polyamide are widely used for packaging applications. There are produced by coextrusion and then calendered in order to adjust the final thickness, but also to improve adhesion between the different layers (figure 1).

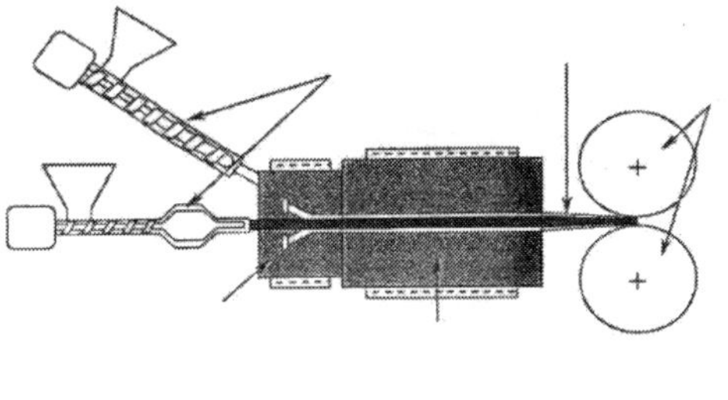

Figure 1 - General scheme of the coextrusion / calendering production line

For that purpose, a thin layer of polypropylene grafted with maleic anhydride is introduced between the incompatible polymers (figure 2).

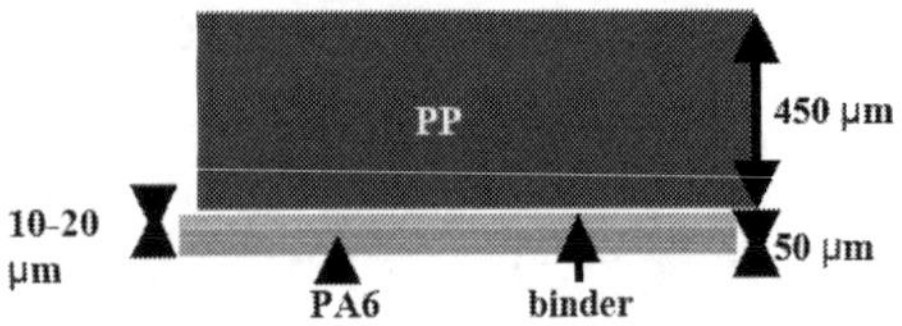

Figure 2 - Multilayer coextruded sheet

Depending on the thermomechanical history all along the process (i.e. in the black box in which molten polypropylene and polyamide are assembled, in the coextrusion die, during the stretching step between die and calender , and finally in the calender nip itself) a copolymerisation mechanism between modified polypropylene and polyamide will develop and promote adhesion. Calendering parameters (rotation velocity and temperature of the calender rolls, importance of the bank- the polymer reservoir upstream the nip, see figure 3) will influence adhesion.

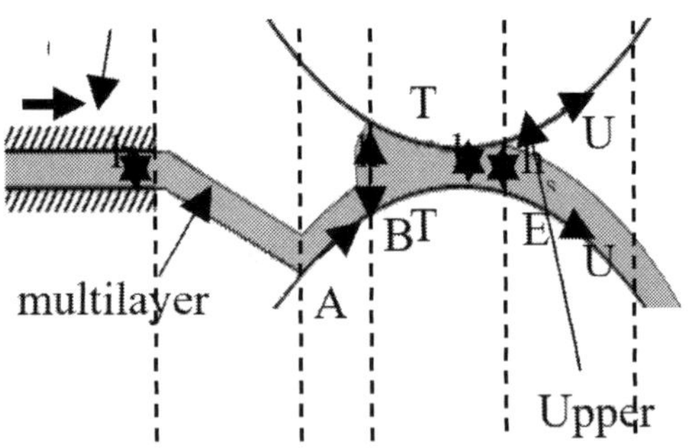

The influence of these parameters is controversial: opening the nip does not influence adhesion at low roll rotation velocity, but decreases adhesion at high roll rotation velocity; in both situations, increasing roll velocity decreases adhesion (figure 4)

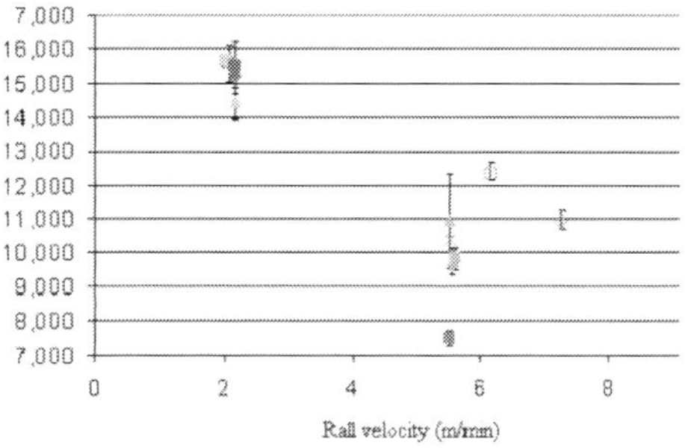

Figure 4 - influence of the roll velocity on the peeling force (□, the calender is open; △ , the calender is closed)

One may believe that the copolymerisation mechanism will be governed by the temperature history along the interface; on the other hand, diffusion mechanisms towards the interface will be governed by the history of shear and elongation around this interface, but kinematic and temperature histories are strongly coupled which necessitates to build a global thermomechanical model of the calendering process. Successive models of increasing complexity will be proposed.

Isothermal monolayer thermomechanical model

Early calendering models have been developed for PVC or rubber sheet production (Ardichvili ,1938; McKelvey,1962). The bank upstream the nip is very important (between 20 and 50 times the calender nip) and the problem is complex, 3D and involving free surface (figure 5)

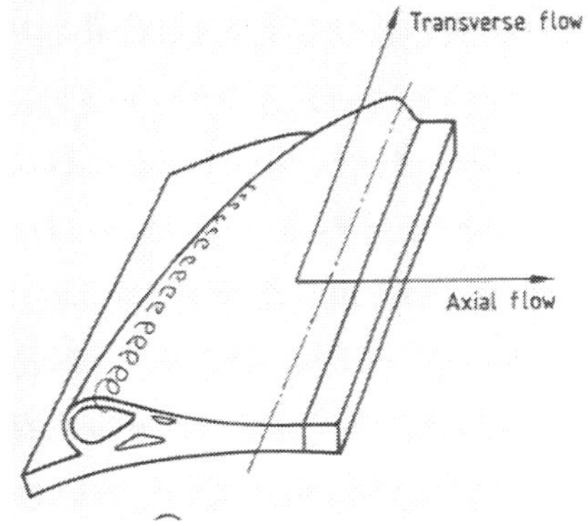

Figure 5 - Scheme of the flow in the calender bank (Unkruer,1972)

Most of the models assumed the so called "lubrication approximations": the flow in the calender is symmetric which means that the bank is not feeded on one of the two rolls (figure 6).

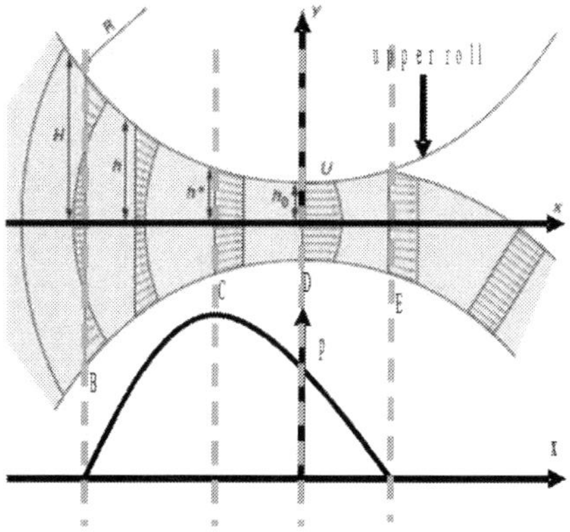

Figure 6 - Kinematic in the calender nip

These models throw light on very important features (Agassant and Avenas, 1977):
- polymer does not leave the rolls at the nip; there is a spread height which origin does not lie on viscoelasticity (like die swell in extrusion) but on pure kinematic considerations,
- there is a pressure peak between the two rolls which maximum value is upstream the nip,
- both spread height and pressure peak are dependant on bank size and reach an asymptotic value when it becomes very important (figure 7).

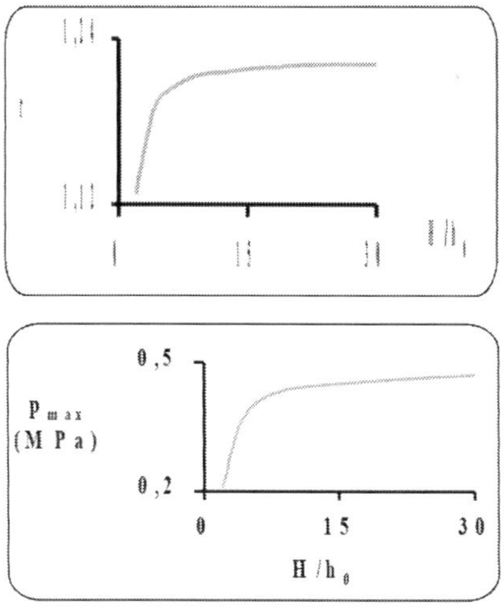

Figure 7: Spread height r and maximum pressure as a function of the bank size

2D finite element models (Agassant and Espy, 1985) allow to compute the complex, non symmetric kinematics within the bank (figure 8) and it is obvious that such kinematics would destroy the coextruded structure.

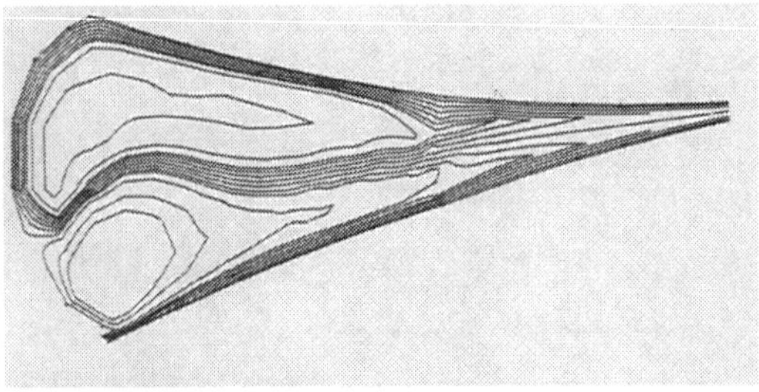

Figure 8: 2D finite element computation of the flow lines within the calender bank (5)

For that reason, calendering of coextruded sheets will use very small banks, but this makes the process difficult to master:

- The flow rate between the calender rolls needs to be equivalent to the flow rate delivered by the different extruders. If they are not balanced, the bank could grow up progressively without limit or, on the contrary disappear and, in that case, there is no calendering step anymore.
- In fact it is however possible to obtain coupled coextrusion / calendering conditions within a substantial processing window because increasing the extrusion rate increases the calendering bank, which in turn increases the pressure between the rolls. This induces a deflexion of the rolls which increases the calendering rate and so stabilizes the process.

Figure 9 shows the 2D finite element pressure peak in a typical coextrusion/calendering processing case. Surprisingly the pressure distribution is very near the one obtained with the above mentioned lubrication approximations. As a consequence, a simple model allows to compute the order of magnitude of shear rates and elongationnal rates within the calender gap, but in the real life, the process is non isothermal and the rheology of both polymers is different.

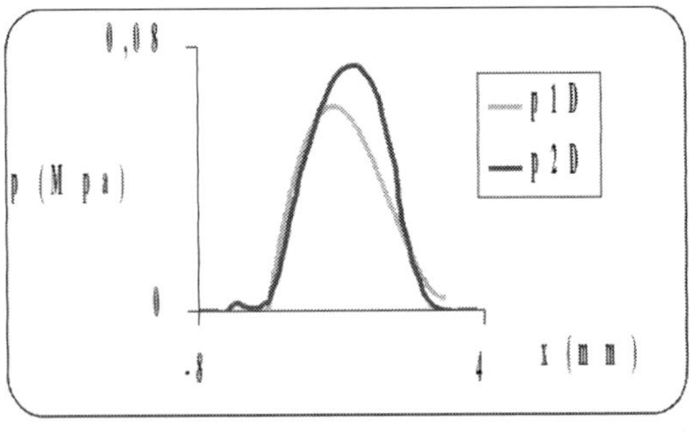

Figure 9 – Pressure peak in the small bank calendering process , comparison between 2D finite element results and lubrication approximation models (Serrat et al., 2009)

Purely thermal model

One neglects now the polymer flow between the two rolls which means that we consider a constant thickness of the coextruded sheet with a perfect thermal contact with the calender roll (the contact distance on the lower cylinder is much more important than on the upper one (figure 10). The crystallisation kinetics of both polymers is accounted for. A finite difference method (Therm Software, Devisme et al., 2007) allows to compute the temperature field and the crystallisation rate in the whole domain.

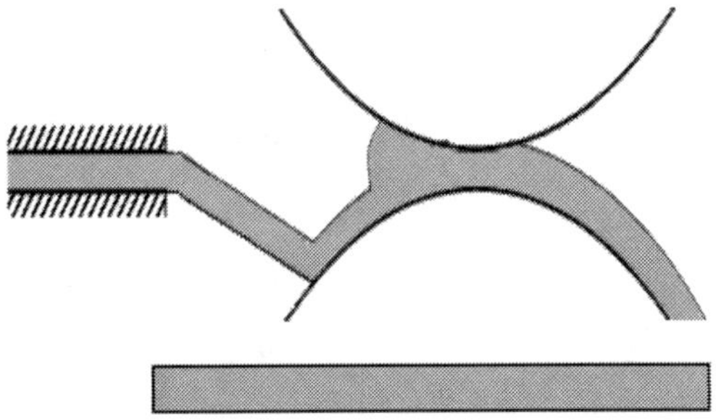

Figure 10 - 2D equivalent domain for the purely thermal computation

On figure 11, one observes that polyamide will crystallize on a very short distance, as soon as the coextruded sheet is in contact with the lower roll. On the contrary, polypropylene will crystallize within the calender nip at a distance from the contact point which depends on the calendering parameters, and especially on the roll velocity. One may believe that the grafting reaction between modified polypropylene and polyamide may occur as long as one (or both) polymer(s) are in the molten state. As a consequence the reaction will take place on a distance which depends on the roll rotation velocity, but this does not mean that the time open to the reaction increases too.

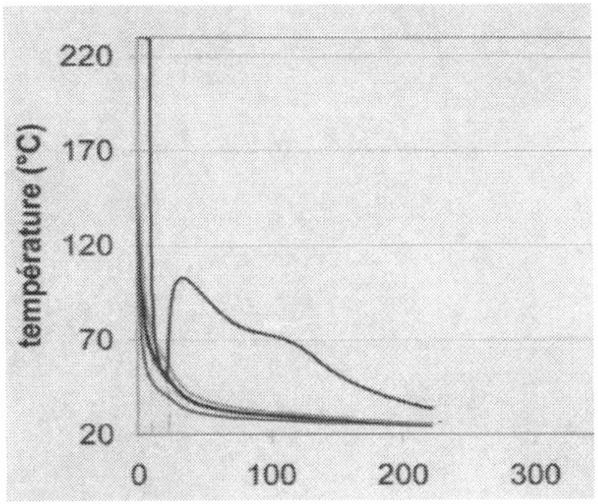

Figure 11 - Temperature field along the interface between the two polymers

In fact one observes on figure 4 a global decrease of the adhesion force between polyamide and polypropylene when the calendering velocity increases.

It is so necessary to develop a global thermo-mechanical model for the calendering process of the coextruded structure.

Thermomechanical model

A bilayer 2D temperature dependent finite element has been developed using Forge 2 software (Piqué et al., 2006). It requires important computation resources, as the mesh needs to be refined especially around the interface between both polymers. The enthalpy of crystallisation has not been accounted for.
Figure 12 presents the temperature field in the whole calendering domain for relevant processing conditions. As pointed out previously, polyamide crystallizes quickly, polypropylene crystallizes at the interface with polyamide within the nip, but an important fraction of the polypropylene layer remains in the molten state at the nip exit which means that an important pressure may be exerted through the remaining molten polymer on the interface in the whole contact.

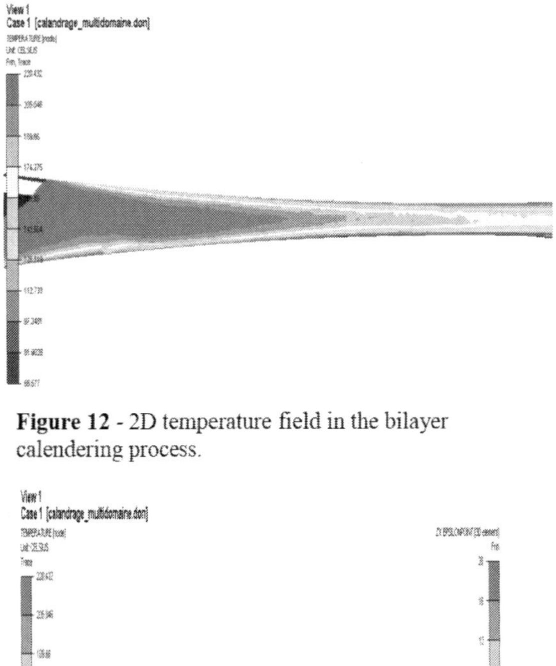

Figure 12 - 2D temperature field in the bilayer calendering process.

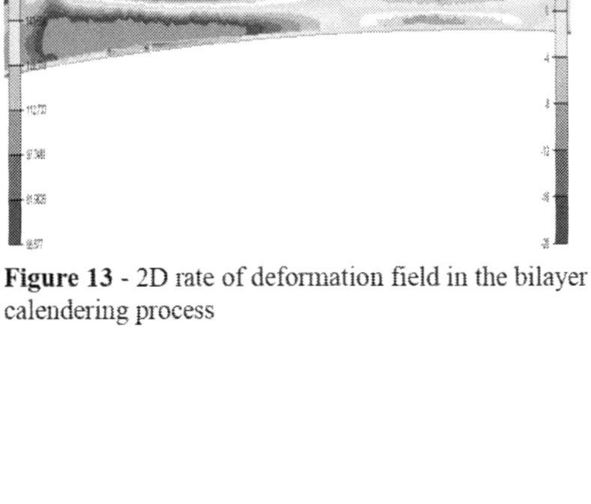

Figure 13 - 2D rate of deformation field in the bilayer calendering process

Figure 13 presents the shear rate distribution within the contact in the same conditions.

Discussion

We have now at our disposal a global picture of temperature, rate of deformation, pressure which allows to propose a tentative explanation of the surprising results presented on figure 4:
- When the calender is open, the crystallisation kinetics on the lower roll in independent on the roll velocity but, as the extrusion rate increases, the grafting reaction in the coextrusion die is not completed and the adhesion force decreases.
- When the calender is in action (which means that there is a calender bank and the coextruded sheet is laminated between the two rolls) there is no influence on the adhesion force at low roll velocity, but a significative increase of the adhesion force at high roll velocity, which means that pressure, shear and elongationnal rates will influence adhesion when they overcome a step value. Identifying the relevant parameters (rate of deformation, stress at the interface, pressure…) as well as these step values require additional experiments.

Bibliography

Agassant J.F., Avenas P., "Calendering of PVC, prediction of stress and torque", J.Macromol. Sci. Phys.,B14, 345 (1977)
Agassant J.F.,Espy M., "Theoretical and experimental study of the molten polymer flow in the calender bank", Polym. Eng. Sci., 25, 118 (1985)
Ardichvilli G.,"An Attempt at a rational determination of the Cambering of Calendar Rolls" Kautschuk,14,23 (1938)
Devisme S., Haudin J.M., Agassant J.F.; Rauline D., Chopinez F., « Numerical simulation of extrusion coating, » Int. Polym. Proc., 22, 90 (2007)
McKelvey J.M.,"Polymer Processing", Wiley, New York (1962)
Picqué B.,Bouchard P.O., Montmitonnet P., Picard M.,"Mechanical behaviour of iron oxide scale" Wear, 260-3,231 (2006)
Serrat M.C., Devisme S., Laffargue J., Agassant J.F., Bikard J., "Calendering of coextruded polymer structures", 25th annual meeting of the Polymer Processing Society , Goa (India) (2009)
Unkruer W., Kunststoffe, 62, 7 (1972)

DEVELOPMENT OF FLOW INSTABILITIES DURING EXTRUSION OF POLYMER MELTS AND POLYMERIC SUSPENSIONS: IMPORTANCE OF THE USE OF PRESSURE-DEPENDENT WALL SLIP CONDITION

Moinuddin Malik[1], Hansong Tang[2], Dilhan M. Kalyon[1]*

(1) Stevens Institute of Technology, Castle Point Station, Hoboken, NJ 07030
mmalik@stevens.edu, * dkalyon@stevens.edu
(2) City University of New York, NY

A number of studies have indicated that various types of flow instabilities exhibited by polymeric melts and concentrated suspensions of polymeric suspensions are affected by the total normal stress at the wall (which for concentrated suspensions can be assumed to be equivalent to pressure) and the pressure dependence of the wall slip velocity [1-3]. For polymeric liquids and polymeric suspensions the pressure dependence of the wall slip velocity can be significant with the slip velocity decreasing with increasing pressure [4-7]. Since the flow curves characterized over relatively high apparent shear rates are subject to wall slip, and since such slip is a function of wall shear stress, as well as pressure, new methodologies, including the use of inverse problem solutions are necessary for the characterization of the parameters of pressure-dependent wall slip and shear viscosity material. Pressure dependence of wall slip needs to be incorporated into the mathematical models of extrusion processes and die flows. In this presentation an overview of the importance of the pressure dependence of wall slip behavior will be made and comparisons of experimental and numerical analysis results will be made for rectangular slit and capillary dies. The incorporation of the pressure dependent wall slip behavior to the 3-D FEM based mathematical modeling of the twin screw extrusion and die flows will be described and typical results will be used to highlight the ramifications on processability of polymeric melts and suspensions.

References

1. Hatzikiriakos, S. and J. Dealy, J. Rheol. 36, 703 (1992).
2. Den Doelder, C., Koopmans, R. and J. Molenaar, J. Non-Newtonian Fluid Mech. 79, 503 (1998).
3. Georgiou, G, J. Non-Newt. Fluid Mech. 109, 93 (2003).
4. Vinogradov, G. and L. Ivanova, Rheol. Acta 6, 209 (1967)
5. Person, T. and M. M. Denn, J. Rheol. 41, 249 (1997)
6. H. Tang and D. Kalyon, J. Rheol, 52, 507 (2008 a)
7. H. Tang and D. Kalyon, J Rheol, 52, 1069 (2008 b)

INTRODUCTION OF THE CONICAL CO-ROTATING TWIN SCREW EXTRUDER

S. Schuschnigg[1]*, G Klammer[2], C Holzer[1]

[1] *IKV-Institut of Plastics Processing, University of Leoben, Austria, stephan.schuschnigg@unileoben.ac.at*
[2] *M-A-S, Maschinen und Anlagen Schulz GmbH, Pucking, Austria, klammer@mas-austria.com*

The new conical technology (NCT) developed by M-A-S is a tightly intermeshing co-rotating twin screw extruder with conical screws. This new development in the extrusion sector is very suitable for extrusion of materials with low bulk density, such as film flakes and other recycling materials. It is also very successfully used for the extrusion of WPC (wood plastics compounds) and PPC (polymer paper compounds). Contrary to conventional parallel co-rotating twin screw extruders, no melt pumps are needed for pressure generation using the conical design. The pressure build-up in the metering zone allows direct extrusion of profiles or pelletizing of material.
The NCT usually has six different zones, which can be equipped with different screw elements for the varying purposes of the system. After the hopper there is the intake zone with a large channel volume. This volume reduces, so that the filling grade rises and the material starts to melt. Next there is a flow restriction zone followed by mixing and shearing elements. The venting section, especially for moisture sensitive materials like PET or heavily printed polymer films, is followed by the final metering zone. In this zone, the needed pressure for the die is generated. A comparison of different extrusion systems in respect to their specific properties is carried out.

Introduction to the NCT

The classification of extrusion systems differ between single-, twin- and multiple screw extruders. The twin screw extruders can further be divided in co- and counter-rotating and into parallel and conical screw axes.

The patent of the first co-rotating, self cleaning twin screw extruder for the polymer extrusion was given in the year 1944 and in 1953; Werner & Pfleiderer obtained the worldwide license. (Kohlgrüber, Bierdel 2007). The development of the counter-rotating twin screw extruders started in the year 1939 (Schenkel 1963), while the concept of conical counter-rotating screws was developed in the 1960 (Wortberg, Kaczmarek 2004).

The conical co-rotating self wiping concept for production was first presented at the international Trade Fair for Plastics and Rubber 2007 in Düsseldorf, Germany. Up to date the NCT (new conical technology) is available in two dimensions. The first was developed for throughputs up to 500 kg/h and the second one can process up to 1000 kg/h.

The extruder (figure 1) has two tightly intermeshing conical co-rotating screws, which are positioned in the segmented barrel. This cylinder can be swivelled out to change the screws. The material comes from either a horizontal or a crammer feeder, depending on the bulk density of the material and the desired throughput. With the feeder system the filling grade of the screw and therefore also the pressure built up and temperature development in the intake zone can be controlled.

For many polymers processed on the extruder the evaporation zone is crucial for the product quality. The vacuum unit is positioned beneath the barrel. For a good temperature control the barrel has an electrical heating and also an oil cooler.

The needed power is provided by an asynchronous motor and transferred via the gearbox to the screw shafts.

The NCT can be used for direct extrusion, but also as a compounder for pelletizing. The materials can vary from PET, polyolefin to WPC (Wood Plastic Compounds) and PPC (Polymer Plastic Compounds) (Bittermann, Sykacek, 2007). Due to the high free volume at the intake the extruder is also capable to process materials with a very low bulk density as flakes and regrinds.

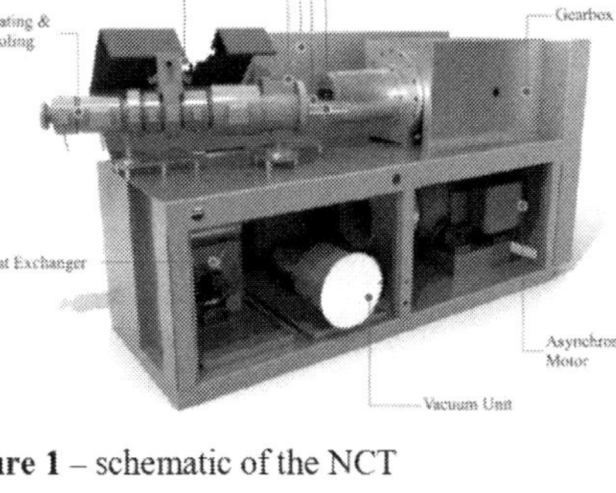

Figure 1 – schematic of the NCT

In the screw the material occurs in four different states. During the initial stage, the material is in loose pellet state, which is compacted to a solid bed in the second stage. This will be partly molten in the third stage and at the final stage the material is totally molten. For every state different zones with different tasks are needed.

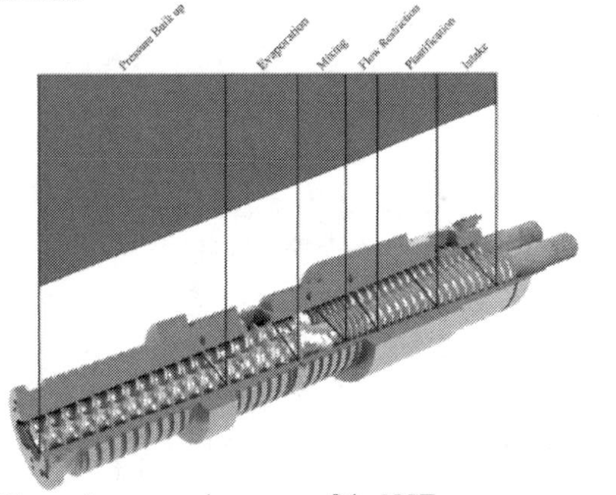

Figure 2 – processing zones of the NCT

After the hopper is the intake, where the polymer enters the channels and is conveyed forward (figure 2). The channels are mostly not fully filled except a crammer feeder is used. The material will be pressed together and the entrapped air will be pushed back to the hopper. Until the filling grade, the ratio of the volume of the material and the channel geometry, reaches one there is only the flow induced through the pushing flights flank of the screw. After that there will be a pressure build up in the channel. The use of conical screws leads to an automatic compression, due to the reduction of the volume of the channels. In parallel systems this is done through changing the pitch (Kohlgrüber, Bierdel 2007). Because of the high pressures the friction between the cylinder and the material will generate heat, which will lead to the first melting of the polymer.

The mathematical model for parallel twin screw extruders for melting of the polymer is described by Potente 1994. A similar approach will be done in our further work on the conical screws. The problem with the varying channel geometry will be the biggest challenge to calculate the needed length of the melting zone. The approach to divide the channel into small segments with constant geometries and temperatures should simplify it.

For melting, the channels must be fully filled. Therefore a restricting zone with backward-conveying elements or with kneading-blocks must be mounted. These elements also help to finish the melting process, so that in the mixing zone only molten polymer will be processed. The mixing zone can have kneading discs, screw mixing-, neutral mixing- or backward pumping mixing elements (Kohlgrüber, Bierdel, 2007).

In the devolatilization zone the surface of the molten polymer should often be replaced, so that all the volatile substances can migrate to the open surface and can be taken away by the vacuum. The degree of filling must therefore be low, so that a big free surface is available. After this comes the pressure build up zone, where the screws must first raise the DOF (degree of filling) to the value of one, so that the pressure can be build up. Here the conical design is very helpful, as the free channel space reduces towards the die. In this part the flow in the channels is similar to the flow in the single screw extruder with a drag flow towards the die and a pressure flow backwards. If the compression of the screw is high enough the pressure flow can reverse. With the good performance of the pressure build up zone this extruder can operate without a melting pump. In an experiment with the NCT a pressure of 600 bar at a mass flow rate of 400 kg/h was reached.

The experience of M-A-S has lead to the design of screw elements, which can be selected and mounted on the screw shaft (figure 3). The screw elements cannot be exchanged, as in parallel systems, because the elements can only be placed at the correct axial position. Different screw elements with different geometries like numbers of flights (one or two), different pitches, forward or backward conveying and various lengths are available.

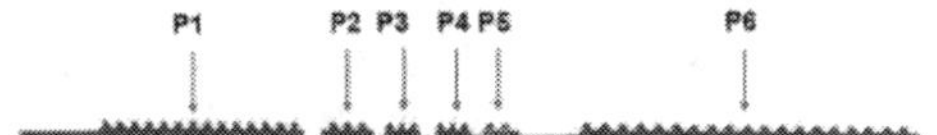

Figure 3 – different screw elements

The screw elements in tightly intermeshing twin screw extruders have a special design. The design most often used is based on the screw speeds and the number of flights. The Erdmenger profile (standard profile) is very good described by Booy (1978) and the patent (DE 813154) was given in the year 1949 to Rudolf Erdmenger. Further developments of the profile resulted in a variety of shapes for excentric profiles, different rotational speeds etc.

As modification to the Erdmenger profile, shear edge or box profiles (double shear edge) where designed (figure 4).

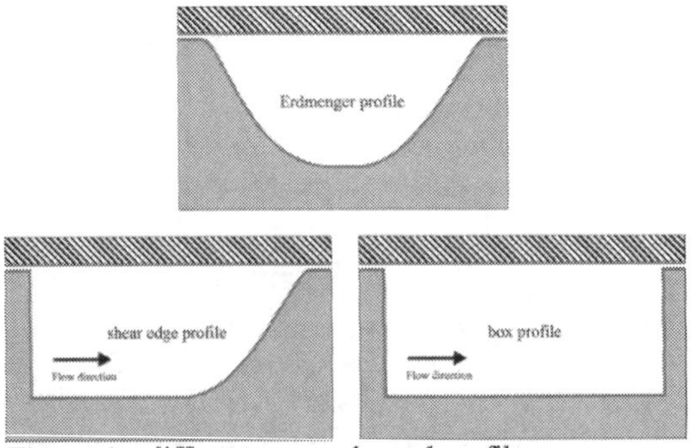

Figure 4 – different screw channel profiles

The Erdmenger profile has the smallest free volume; the box profile can convey far more material. The disadvantage of the box profile is that the self wiping of the screw is no more valid. The comparison of the Erdmenger profile with the box profile for a parallel twin screw system shows, that the channel area ratio

(box to Erdmenger) has about the value of two and so is the ratio of the calculated volume rate for a given pressure gradient.
In the last time a great variety of different screw elements where developed, with various advantages and disadvantages.

Comparison of different extrusion systems

Before in this paper other screw systems where mentioned. In this section a short overview over these systems are given (figure 5).
The single screw extruder has one or more open channels from the beginning to the end. The melt is forwarded in a double helical path around the screw. The pressure build up depends either on the melt conveying or, with grooved barrels, on the feed section. Through the melt flow path the mixing of the melt is good, but the residence time is much longer than in other systems. The biggest advantage is the low cost of the system (Rauwendaal 2001).

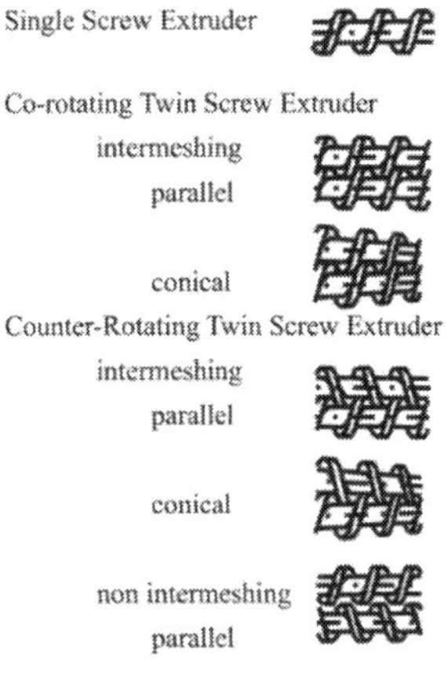

Figure 5 – Extruder classification

Twin screw extruders are classified in intermeshing and non intermeshing systems. The first is important for polymer processing and will be discussed here; the latter is mostly used in food processing and will not be discussed.
The co-rotating intermeshing parallel twin screw extruder has two separate screws, which mostly will have the same rotational speed. The geometry of a self whipping system is fixed with the diameter, the centreline distance, the helix angle and the number (n) of flights. The flow pattern is eight shaped and divided in 2n-1 channels. Very often the system is not fully filled so the degree of filling is less than one. This system has like the single screw open channels from the beginning to the screw tip. In normal processing it is used as a compounder to blend different materials together. The mixing quality can further be improved with kneading and mixing elements. The pressure build up is not so good, because the material can flow backwards. The residence time distribution is comparatively better / shorter than in the single screw extruder.
The counter rotating twin screw extruder form C-shaped chambers, which convey the material according to the geometry and proportional to the screw speed. As there is nearly no interchanging between these chambers the mixing quality is poor. The pressure build up can, with the right design of the screw, be very good. The residence time spectra is very narrow, as the particles that enter at a time will all be forwarded with the same velocity.
The conical twin screw extruder systems are considerably complex in the design, as with the geometric interaction also the variation of the diameter comes into account. The conical counter-rotating twin screw is mostly used in the processing of PVC and was developed at the beginning of the 60s.
The conical design has some advantages over the parallel systems. The first one is the shorter length with the same channel length, the second is the compression in the channels, as the diameter and so the channel width decreases. There are some conical designs, which diminish the compression as the channel depth increases toward the screw tips (Jäcker 2000).
The conical co-rotating twin screw extruder, on industrial scale, was developed by M-A-S and has some big advantages . The pressure build up is very good, so that there is no need for a melting pump for the following processing steps. On the other side with its high volume at the feeding zone it can process materials with very low bulk densities. The mixing quality of the conical co-rotating twin screw extruder can be compared with the parallel one.

Measurement and Calculation

Measurements were made on the NCT with a die, of which the gap could easily be changed. With the derived model the dependence of pressure, at the screw tip, against the volume rate can be calculated.
The model of the conveying zone is based on the isothermal, Newtonian, steady flow, where there is no slip at the wall. The viscosity dependence is calculated from the shear rate at each point. The derivation begins with the equation of momentum (Bird et.al. 2002) and the boundary conditions of the Erdmenger profile. After several calculation steps the following equation was obtained, which shows the relationship of the pressure gradient to the volume rate.

$$\dot{V} = \left\{ \begin{array}{l} 2 \cdot v_{bz} \cdot \left[\dfrac{\alpha \cdot t \cdot \cos(\phi_0) \cdot h_0}{4 \cdot \pi} + \displaystyle\int_{\frac{\alpha \cdot t \cdot \cos(\phi_0)}{4 \cdot \pi}}^{\arccos\left(\frac{\rho_c^2}{2} - 1\right) \frac{t}{2 \cdot \pi \cdot \cos(\phi_0)}} h(x)\, dx \right] \\ -2 \cdot \dfrac{1}{12 \cdot \eta} \cdot \dfrac{\partial p}{\partial z} \cdot \left[\dfrac{\alpha \cdot t \cdot \cos(\phi_0) \cdot h_0^3}{4 \cdot \pi} + \displaystyle\int_{\frac{\alpha \cdot t \cdot \cos(\phi_0)}{4 \cdot \pi}}^{\arccos\left(\frac{\rho_c^2}{2} - 1\right) \frac{t}{2 \cdot \pi \cdot \cos(\phi_0)}} h(x)^3\, dx \right] \\ + \dfrac{1}{2} \cdot k_{zw} \cdot \left(\dfrac{t}{D_s} \right) \cdot R_s^3 \cdot n \end{array} \right\} \cdot (2 \cdot i - 1)$$

With a computer program it is easy to solve this equation stepwise and also to integrate the changing

conditions at each step forward. These are the geometrical conditions, the viscosity and the velocities. They all change from one step to another. The geometric constants are derived by Booy (1978). With this equation it is also possible to implement changes in the channel profile h(x).
The flow has three major parts, the one due to the drag flow (upper terms), the pressure flow which points backward (middle term) and the flow in the nip zone, where the screws wipe each other.
The pressure characteristic of measured values (figure 6) is a dimensionless plot, where the dimensionless pressure is plotted against the dimensionless volume rate. The measured points should lie on one line, but the problem is the changing temperature with the different die gap.

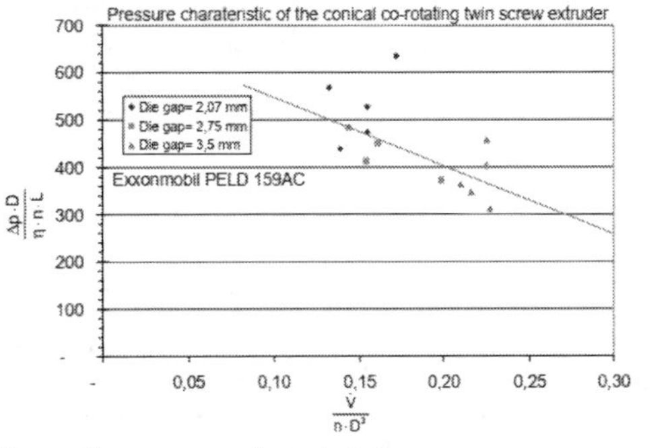

Figure 6 – pressure characteristic

The gained results where also calculated with the above equation. In the moment the comparison is not satisfying, but with further measurements we hope to gain better results.

Conclusions

The conical, co-rotating, self wiping twin screw extruder, developed by M-A-S, is presented in this paper and the advantages are outlined. The concept is, from the processing point of view, extremely suitable for processing materials with low bulk densities, which need a high compression to process. The materials mostly used with these systems are WPC (Wood Polymer Compounds), PET flakes from the recycling but also polyolefin granules. The processing begins at the feed zone where the channels are not completely filled. After the compression the DOF will be one and the material can start to melt. After the melting zone the flow restriction guarantees, that the material in melting zone is properly melted. The mixing zone ensures that a homogeneous melt is delivered to the venting zone. After that the pressure build up takes place. The extruder concept can build a pressure at the screw tip up to 600 bar, which is suitable for mostly all extrusion dies. Not only granules can be produced but also profiles, pipes and films. In the moment there are two models available ranging from 500 to 1000 kg/h.
The calculation and the measurements must be improved to optimize the prediction in the pressure build up zone. The description for the other zones is far more challenging, as we do not know the exact location of the point, where the DOF reaches one and the melting begins. The melting profile will be derived based on the work of Potente (1994).
The goal of this project is the scaling of the NCT, so that the development time could be reduced and better data on the individual screw configuration could be gained.

Acknowledgements

M-A-S Maschinen und Anlagen Schulz GmbH for providing the possibility to perform this project, which is financed by the FFG (Österreichische Forschungsförderungsgesellschaft mbH).

References

Bird R. B., Stewart, W. E., Lightfoot E. N. Transport phenomena, 2nd EditionJohn Wiley & Sons, (2002)

Bittermann R., Sykacek E., "From Tetra Pak to Raw Materials", Kunststoffe International, 10, 190-191, 2007

Booy M.L., "Geometry of Fully Wiped Twin Screw Equipment", Polymer Engineering and Science, Vol. 18, No.12, (1978)

Booy M.L., "Isothermal Flow of Viscous Liquids in Corotating Twin Screw Devices", Polymer Engineering and Science, Vol. 20, No.18, (1980)

Jäcker M., "The Conical Concept of the Future", Kunststoffe Plast Europe, 90 (12), (2000)

Klammer G., "BraInDay 2009 NCT New Conical Technology" Presentation, 16.11.09 Linz, Austria

Kohlgrüber K., Bierdel M.," Der gleichläufige Doppelschneckenextruder, Grundlagen, Technologie, Anwendungen" ; 1st. Edition, Hanser Fachbuch, Munich, (2007)

Meijer, H.E.M., Elemans, P.H.M., "The Modeling of Continuous Mixers. Part 1: The Corotating Twin Screw Extruder", Polymer Engineering and Science, 28 (5)

Potente H., Ansahl J., Klarholz B., "Design of Tightly Intermeshing Co-Rotating Twin Screw Extruders", Int. Polym. Process., 9, 11-25 (1994)

Rauwendaal C., "Polymer Extrusion", 4th Edition, Carl Hanser Verlag, Munich, (2001)

Schenkel G., "Kunststoff-Extrudertechnik", 2nd Edition Carl Hanser Verlag, München, (1963)

Wortberg J., Kaczmarek D., "Kunststoff-Maschinenführer, Chapter Extruder und Extrusionsanlagen", Carl Hanser Verlag, München, (2004)

AN ALTERNATIVE ANALYSIS OF SINGLE-SCREW MELTING

Gregory A. Campbell[1] *and Mark A. Spalding*[2]

[1] *Castle Research Associates, 1145 Mason Bay Road, Jonesport, ME 04649 swcgac@maineline.net*
[2] *The Dow Chemical Company, Midland, MI 48667 maspalding@dow.com*

Abstract

The focus of this investigation was to develop a new and alternative model for the melting dynamics of the solid bed in single-screw extruders. Re-analysis of literature single-screw melting data has determined the rate of loss of material in the cross channel (x) direction and the solid bed thickness direction (y) of several melting data sets. The new screw rotation based analysis demonstrates that the polymer solid bed goes to zero dimension in the thickness direction well before the solid bed width is consumed consistant with the analysis of the literature single screw melting data. This observation is contrary to the historical assumption that the cross-channel width of the solid bed dominates the melting rate. In this paper we present a new physical concept for melting in single-screw extruders that is consistent with the new data analysis. We developed a dynamic melting model and then simulated this first-order theory. The simulation qualitatively fits the literature observations for melting.

Introduction

Experiments for studying the single-screw extrusion melting mechanism were first carried out by Maddock (1959) and then by Street (1961). The melting mechanism was observed by performing Maddock solidification experiments. For these experiments, the extruder was fed with a mixture of natural and colored (1-3%) polymer. When steady state was achieved, screw rotation was stopped, and full cooling was applied to the barrel, solidifying the polymer in the channels. Usually, the original conditions existing in the extruder were retained. By examining the cross sections of the solidified polymer after taking it out of the screw channel, a clear picture of the melting behavior was obtained, as shown in Figure 1. Their famous solidification experiments revealed that melting in many situations takes place in a specific order. The experiments showed that after melting began, there was a continuous solid bed and a melt film over the inner barrel surface. Later the solid bed was completely surrounded by melt. Farther downstream, a melt pool developed between the pushing flight and the solid bed. The solid bed disappeared gradually all the way down the channel (Maddock, 1959). Tadmor's research group (1966, 1967, 1978) observed similar melting phenomena. They studied many different materials which included low density polyethylene (LDPE), high density polyethylene (HDPE), acrylonitrile butadiene styrene (ABS), rigid polyvinyl chloride (PVC), and polypropylene (PP) resins using different operating conditions and extruder geometries. The results were interpreted to demonstrate that the Maddock melting mechanism prevails in quite a large range of melting processes with most resins. Other investigators' experiments led to other melting phenomena, such as the acceleration of the solid bed, break up of the solid bed, and a melt pool existing by the trailing flight (Agassant et al., 1991). Maddock first proposed the physical description of the melting process in 1959. In his description, the energy for the melting process came from two sources: energy conduction from the barrel, and viscous dissipation from the film at the inner barrel surface. Beginning from 1966 when Tadmor (1966) set up the first mathematical model based on the Maddock melting mechanism, he proposed a reorganizing solid bed that led to a focus on the x-dimension as rate controlling in the melting process. That is, the rate of melting was defined as the ratio of the solid bed width to the width of the channel.

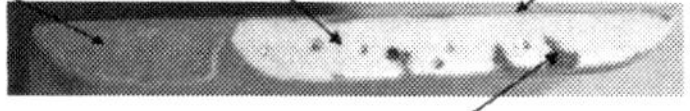

Figure 1. Photograph of resin solidified in the transition section after a Maddock solidification experiment.

New Data Analysis

Many other investigators continued working on improving the melting model in later years. Numerous papers have since been written which refine the Tadmor model; all are based on the hypothesis that the solid bed reorganizes as it melts leading to the bed being consumed in the cross-channel direction until it is completely melted. We will in this paper re-examine the assumptions in the literature models in order to address the classical melting data. The classical melting data originally presented by Tadmor and Klein (1978) for PE resin was examined using digital dimensional analysis (Campbell and Tang, 2004) and the melting results for three specific runs are shown in Figure 2.

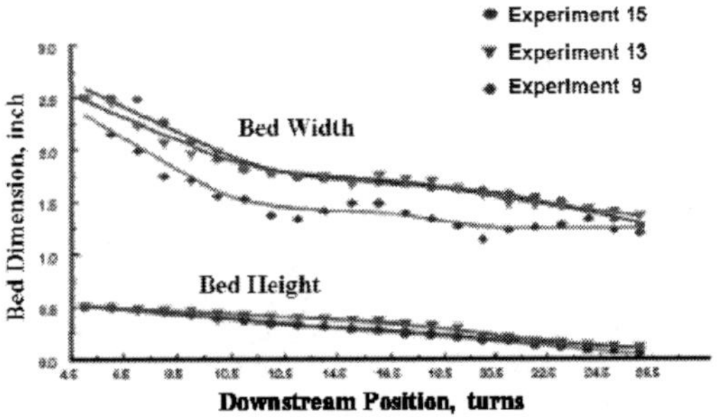

Figure 2. Melting data from Tadmor for PE resins (Tadmor and Klein, 1978; Campbell and Tang, 2004). At a downstream position of 25 turns, the solids were completely melted and thus the bed width and bed height were zero.

For essentially all of the data sets found in Tadmor and Klein (1978) the melting results resemble the data shown in Figure 2. The data analysis is of course most difficult near the end of melting, and in essentially all cases the fraction of the bed remaining in the cross-channel direction is much greater than in the thickness direction. Also, in general the last third of the x direction material appears to melt in the last turn of the channel, creating nearly a step function in the melting rate with respect to the axial distance, as shown in Figure 2. Step functions, however, do not in general occur in processes that are heat conduction dominated. To reiterate, the digital analysis conclusively demonstrated that the y direction thickness always goes to zero before the x direction width (Campbell et al., 2003, 2004). This observation led us to the development of a new physical model that is based on these new observations. It is interesting to note that Lindt recognized this melting behavior experimentally for polypropylene (PP) resin in 1976 (Lindt, 1976).

Theory Development

The new single-screw melting concept presented here was developed based on recognizing that the rate limiting melting dynamics for single-screw solid bed consumption was in the channel height and not the channel width direction. Thus, we used a diagram that focused on the four melt films surrounding the solid bed to set the foundation for the development of the physics of melting, as shown in Figure 3.

Consistent with the results presented in Figure 2, separate melting dynamics were defined for the four melt films as originally defined previously (Pearson et al., 1976), and as depicted in Figure 3; C is the film between the solid bed and the barrel wall, D is the film at the screw root (or core) polymer interface, E is the film at the trailing flight of the screw, and B is the film (or melt pool) on the pushing side of the screw flight.

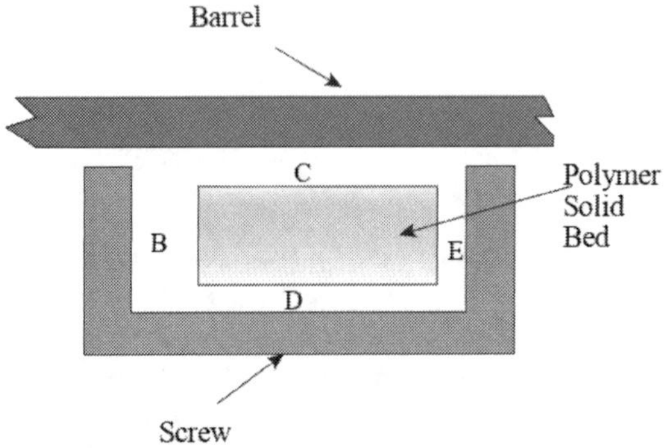

Figure 3. Schematic for the zones of the new melting concept.

Previous analysis of dissipation in extrusion (Campbell et al., 1995) over predicted the polymer melt temperature when barrel rotation dissipation was used. We therefore use the screw rotation velocities in this analysis because the process is dissipation dominated (Campbell et al., 2008, 1992a, 1992b, 1996, 2001a, 2001b) and dissipation is not frame indifferent. We start with classical analysis (Tadmor and Klein, 1978). One major difference is that the barrel is not moving so the velocities that are used in the simulation are adjusted to account for screw rotation dissipation. The diagram of this adaptation of the Tadmor analysis is shown in Figure 4. We start by unwrapping the screw into the x, y, and z coordinates. The barrel is hot as depicted by the red color and its velocity is zero. The velocity gradient in the z direction goes from V_{sz}, the solid bed velocity in the downstream directions (z direction), to 0 at the barrel interface. A similar diagram was developed for the other three fluid interfaces around the solid bed, remembering that for these liquid interfaces the metal velocity is the screw velocity, as shown in Figure 5.

In the literature analysis the system equations were manipulated to eliminate V_{sy}, the velocity of the solid bed consumption in the thickness direction (y direction), from the analysis by using the assumption that the solid bed reorganizes. This allowed a straightforward differential analysis and a closed form solution in the cross channel x direction for solids melting.

We retain V_{sy} as a variable because this allows us to calculate the change in bed thickness in our re-analysis of the literature data, as shown in Figure 2.

The nomenclature used here follows that used by the literature (Tadmor and Klein, 1978) as much as possible. Where deviations occur, a full explanation will be presented. A nomenclature list is presented at the end of the paper.

The velocity of the solid bed in the z direction, V_{sz}, is as defined by Tadmor and Klein (1978). The velocity of the solid bed in the x direction, V_{sx}, due to screw rotation is as follows:

$$V_{sx} = \frac{N\pi D}{60}\sin\theta \tag{1}$$

$$\tan\theta = \frac{\pi L}{D} \tag{2}$$

where N is the screw rotation speed in rpm, D is the diameter at the surface of the solid bed, θ is the helix angle at diameter D, and L is the lead length of the flight.

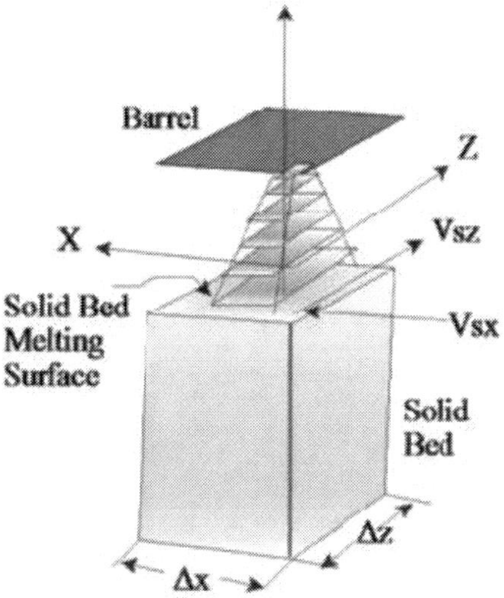

Figure 4. Melting differential element for a rotating screw for the melt film in zone C.

We calculated the thickness of the melt film, δ_c, at film C using similar assumptions as the literature (Tadmor and Klein, 1978) and thus solve their equation (5-23) for $V_{sy,C}$, the velocity of the loss of the solid bed due to melting at δ_c:

$$V_{sy,C} = \frac{\left[\frac{k_m}{\delta_c}(T_b - T_m) + \frac{\mu V_j^2}{2\delta_c}\right]}{\rho_s\lambda + \left[\rho_s c_s(T_m - T_s)\right]} \tag{3}$$

where k_m is the thermal conductivity of the melt, T_b is the barrel temperature, T_m is the melting temperature of the resin, μ is the shear viscosity, V_j is the vectorial velocity in the film (discussed later), ρ_s is the density of the compacted solid bed, λ is the heat of fusion, c_s is the heat capacity, and T_s is the temperature of the solid bed. The term $V_{sy,C}$ here is equivalent to Tadmor and Klein's (1978) V_{sy}. Equation (3) is simply an energy balance for the melting process. Melting velocities at the other three interfaces $V_{sy,D}$, $V_{sx,B}$, and $V_{sx,E}$ are essentially identical to that for Equation (3) except that the vectorial velocity used corresponds to the specific film. These equations were omitted here.

The melting velocities of the solid bed interfaces are the foundation of this new approach. The mathematical equality to calculate the rate of loss of the solid at the interface next to the barrel is found in Equation (4).

$$\left.\frac{dY}{dt}\right|_{\delta c} = V_{sy,C} \tag{4}$$

where Y is the local thickness of the solid bed and t is the independent time variable.

The melt film thickness for zone C, δ_c, is determined using Equation (5). Note that V_{sx} is essentially numerically equal to the traditional V_{bx} as specified by Tadmor and Klein (1978) because of the small value of the film thickness.

$$\delta_c = \left\{\frac{\left[\left[2k_m(T_b - T_m) + \mu V_j^2\right]X/\alpha\right]}{V_{sx}\rho_m\left[c_s(T_m - T_s) + \lambda\right]}\right\}^{\frac{1}{2}} \tag{5}$$

where X is the local width of the solid bed and α is a fitting factor. Equation (5) is a mass balance for the zone C film and is essentially identical to (Tadmor and Klein, 1978) the literature equation (5-31). The flow equation is a first-order approximation since it neglects flows due to pressure gradients. Ideally, the value of α is set to 2 so as to provide an average thickness for the film in the cross-channel direction. The value of α was set to 0.5 for the simulation here to force the melting lengths to be equivalent to those in the literature.

The film thickness, δ_e, for zone E was determined using a similar mass balance equation as follows:

$$\delta_e = \left\{\frac{\left[\left[2k_m(T_b - T_m) + \mu V_j^2\right]Z/\alpha\right]}{V_{sz}\rho_m\left[c_s(T_m - T_s) + \lambda\right]}\right\}^{\frac{1}{2}} \tag{6}$$

where Z is downstream position of the solid bed. The melt film thicknesses, δ_b and δ_d for zones B and D were found by difference:

$$\delta_b = W - X - \delta_e \tag{7}$$

$$\delta_d = H - Y - \delta_c \tag{8}$$

where W is the width of screw channel and H is the local depth of channel.

The boundary conditions for the velocity at the screw root interfaces are shown in Figure 5 below.

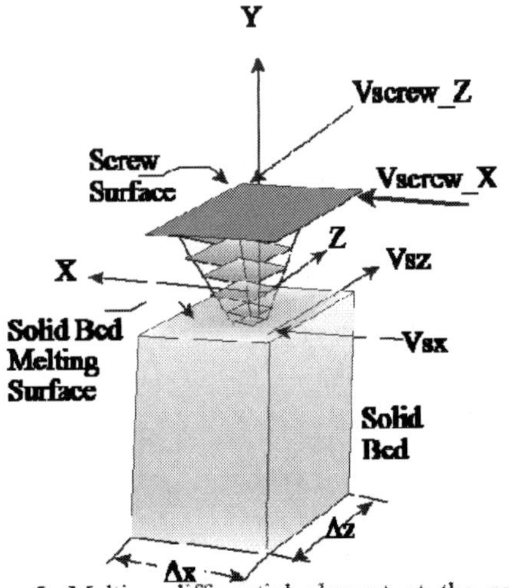

Figure 5. Melting differential element at the screw surfaces for a rotating screw reference frame.

The x and z component velocities of the screw are calculated as follows and they are used as the boundary conditions at the screw surface:

$$Vscrew\_x = \frac{N\pi D_s}{60}\sin\theta_s \tag{9}$$

$$Vscrew\_z = -\frac{N\pi D_s}{60}\cos\theta_s \tag{10}$$

$$\tan\theta_s = \frac{\pi L}{D_s} \tag{11}$$

where $Vscrew\_x$ is the velocity of the screw root in the x direction, $Vscrew\_z$ is velocity of the screw root in the z direction, D_s is the diameter at the screw root, θ_s is the helix angle at the screw root. The direction of $Vscrew\_z$ is in the opposite direction of V_{sz}.

The vectorial velocities for the four polymer films that account for the heat transfer and dissipative melting are as follows:

$$V_j\big|_C = \left(V_{sx}^2 + V_{sz}^2\right)^{1/2} \tag{12}$$

$$V_j\big|_D = \left|V_{screw\_z}\right| + V_{sz} \tag{13}$$

$$V_j\big|_E = \left|V_{screw\_z}\right| + V_{sz} \tag{14}$$

$$V_j\big|_B = \left|V_{screw\_z}\right| + V_{sz} \tag{15}$$

The new concepts presented here remove the literature assumption that the solid bed reorganizes, and it allows melting at all solid bed interfaces. The Tadmor model allows melting at only a single interface; i.e., as specified by the melting velocity V_{sy}. The model presented here allows realistic melting at all four interfaces: two in the y-direction and two in the x-direction.

We calculated the δ_d in film D by subtracting from the local channel depth, H, the δ_c at C and the remaining bed thickness Y_{remain}. Y_{remain} is the original bed thickness minus the loss predicted by the solid bed melting velocities, the two $V_{sy,C}$ and $V_{sy,D}$ equations at both the zone C and D fluid and solid bed surfaces. An artistic representation of the solid bed in the channel is shown in Figure 6. We depict the four films as the solid bed is consumed in the melting section of the screw.

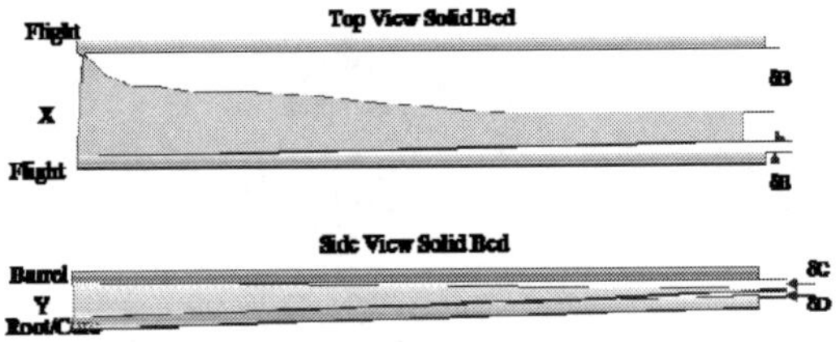

Figure 6. Qualitative shape of x and y bed dimension and melt film thickness for compression.

The calculations for the x-direction melting are done in a parallel manor. First evaluate the melting at the δ_e solid interface, the surface adjacent to the trailing flight, and then calculate the size of δ_b at zone B, normally described as the melt pool, in a similar manner as was done for film D. Essentially all of the experimental data show an increase in the thickness of δ_e as the polymer solid bed progresses from the initial melting film to the end of melting at or near the end of the compression zone. We have incorporated this observation in our theory. We at this time do not address solid bed breakup.

Results and Discussion

We are now in a position to determine if the proposed theory will produce results that are consistent with the observations of the literature data shown in Figure 2.

This new theory requires the solution of five coupled differential equations where x, y, and z are the local coordinates of the solid bed:

$$\left.\frac{dY}{dt}\right|_{\delta c} = V_{sy,C} \quad (16)$$

$$\left.\frac{dY}{dt}\right|_{\delta d} = V_{sy,D} \quad (17)$$

$$\left.\frac{dX}{dt}\right|_{\delta e} = V_{sx,E} \quad (18)$$

$$\left.\frac{dX}{dt}\right|_{\delta b} = V_{sx,B} \quad (19)$$

$$\frac{dZ}{dt} = V_{sz} \quad (20)$$

The first four of these equations were developed using physics and the analysis that led to Equations (3) through (6). We used a differential equation solver to obtain the results of the initial value problem.

Typical simulation results are presented in Figure 7. As shown by this figure, the Y thickness goes to zero at a down channel position of 250 cm into the melting section. We see that the change in the thickness, x direction, leaves about 60% of the original bed width remaining just at the end of melting. The end of melting is defined as when the y direction thickness of the solid bed goes to zero. This simulation was performed using a PE resin with a shear viscosity of 880 Pa-s.

We now examine the melting energy at the four solid interfaces for this new alternative model and for the historical model (or literature model) (Tadmor and Klein, 1978). The data from the simulations are summarized in Table 1 for the resin with a viscosity of 880 Pa-s. We first observe that the vectorial velocities (V_j) for zones C and D are very different for the assumption of barrel and screw rotation, as presented previously and as shown in Table 1. For the historic model, all energy is dissipated in the zone C melt film, and the cumulative energy for melting was calculated at 1980 J/s. The dissipation energy is assumed to be zero for the other films; i.e., the vectorial velocity (V_j) for zones B, D, and E were ignored and thus assumed to be zero. For the new alternative analysis here with screw rotation boundary conditions, the vectorial velocity in the zone C film is about 30% of that for the historical analysis (barrel rotation boundary conditions). This reduced vectorial velocity leads to the energy dissipation in zone C to be about 60% of that using the historical analysis. The remaining 40% of the energy dissipation occurred in the remaining films, as shown in Table 1. Almost 20% of the total energy dissipation occurred in film D.

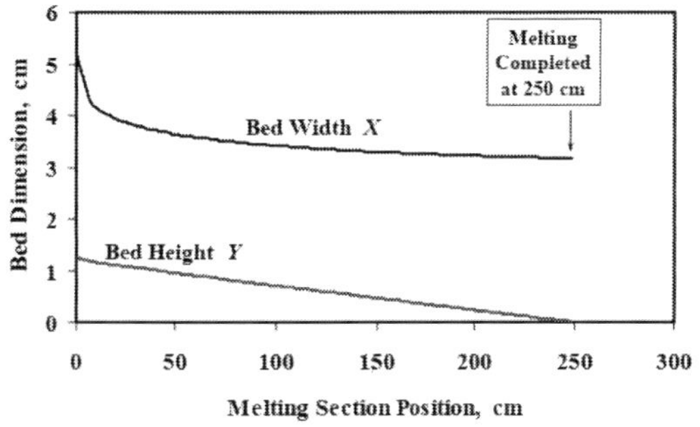

Figure 7. Simulation results using Example 5.1 (Tadmor and Klein, 1978) data. Melting was completed at 250 cm downstream from the start of melting for a resin with a shear viscosity of 880 Pa-s.

Table 1. Calculated vectorial velocities for zones C and D and the melting energies developed in the four melt films. The simulations were for a PE resin with a shear viscosity of 880 Pa-s.

Model Parameter	Alternative Analysis	Historical Analysis [4]
Vectorial Velocity (V_j) in zone C, cm/s	8.7	29.5
Vectorial velocity (V_j) in zone D, cm/s	20.3	2.4
Melting energy developed in zone C, J/s	1140	1980
Melting energy developed in zone E, J/s	480	0
Melting energy developed in zone D, J/s	340	0
Melting energy developed in zone B, J/s	20	0
Total melting energy, J/s	1980	1980

The model is also consistent with observations for flow surging due to solids blockage in a channel (Klein, 1972). That is, as a blockage occurs the local pressure upstream of the blockage increases causing the local melting rate to increase. As the melting rate increases, enough of the blockage will melt such that solid polymer fragments will flow downstream. With the blockage removed the pressure is relatively low in the region and melting rates are low, causing more solids to accumulate and increasing the size of the next blockage. Now the local pressure starts to increase and the cycle is repeated. Although pressure terms were not included in the new model, the additional pressure would likely cause the melt film thicknesses to decrease and thus cause the melting rate to increase.

The results presented here are encouraging but only qualitative and have been produced using this first-order model. We will be continuing to refine this model and intend to evaluate its quantitative predictions in the future. Current limitations of the model are the use of a constant viscosity function independent of temperature and shear-rate. Also, the dynamic local temperature of the barrel and screw (Pham et al., 2008) must be incorporated into the model; they are currently set as constants. An enhanced model for the film thickness at both the barrel and screw surfaces will also be added to the current model along with flows induced by pressure gradients.

Conclusions

A new alternative model is proposed for the melting of the solid polymer bed in a single-screw extruder. The model alters the heat transfer and melting velocities originally proposed by Tadmor and Klein (1978). The model releases three of the original literature assumptions: a) elimination of the V_{sy} melting velocity in the y direction of the bed, b) the concept that the bed rearranges, and c) barrel rotation dissipation dynamics. This required recasting the heat conduction and energy dissipation analysis using screw velocities and extending the material balances to all four bed solid interfaces at the fluid films. The model predicts the dynamics of melting previously reported (Campbell and Tang, 2004) as determined by digital dimensional analysis of the solids remaining in the bed, as shown in Figure 2. The model also predicts qualitatively the experimental observation that the melting length decreases as the polymer viscosity is decreased. It would appear that further effort to make this model quantitative is indicated.

Nomenclature

c_s heat capacity of the solid.
D diameter at the surface of the solid bed.
D_s diameter at the screw root.
H local channel depth.
k_m thermal conductivity of the melt.
L lead length of the flight.
N screw rotation speed in rpm.
T independent time variable.
T_b barrel temperature.
T_m melting temperature of the resin.
T_s temperature of the solid bed.
V_j vectorial velocity in the films.
V_{bx} velocity of the barrel in the x direction for a barrel rotation reference frame (not used).
V_{sx} velocity of the solid bed in the x direction.
$V_{sx,B}$ velocity of the solid bed consumption in the thickness direction at the zone B interface (x direction).
$V_{sx,E}$ velocity of the solid bed consumption in the thickness direction at the zone E interface (x direction).
V_{sy} velocity of the solid bed consumption in the thickness direction at the zone C film (Tadmor).
$V_{sy,C}$ velocity of the solid bed consumption in the thickness direction at the zone C interface (y direction).
$V_{sy,D}$ velocity of the solid bed consumption in the thickness direction at the zone D interface (y direction).
V_{sz} velocity of the solid bed in the z direction.
V_{screw_x} velocity of the screw root in the x direction.
V_{screw_z} velocity of the screw root in the z direction.
W width of the channel perpendicular to the flight.
X width of the solid bed.
x independent variable in the cross-channel direction.
Y thickness of the solid bed in the y direction.
Y_{remain} original bed thickness minus the loss predicted by solid bed melting.
y independent variable in the channel-depth direction.
Z downstream position of the solid bed.
z independent variable in the down-channel direction.
α fitting factor.
θ helix angle at diameter D.
θ_s helix angle at the screw root.
σ_m density of the molten polymer.
σ_s density of the compacted solid bed.
λ heat of fusion.
δ_b thickness of melt film B.
δ_c thickness of melt film C.
δ_d thickness of melt film D.
δ_e thickness of melt film E.
μ shear viscosity.

Acknowledgments

We would like to thank The Dow Chemical Company, Becton Dickinson, Eastman Chemical, XALOY, American Leistritz, 3M, and Pechiney. Much of this work was initiated in the Polymer Extrusion Consortium at Clarkson University.

References

Agassant, F., Avenas, P., Carreau, P.J., and Sergent, J.Ph., "Polymer Processing: Principles and Modeling," Oxford University Press, New York 1991.

Campbell, G.A., Sweeney, P.A., and Felton, J.N., "Experimental Investigation of the Drag Flow Assumption in Extruder Analysis," *Polym. Eng. Sci.*, **32**, 1765 (1992a).

Campbell, G.A., Sweeney, P.A., and Felton, J.N., "Analysis of an Alternative Extruder Screw Pump Design," *Intern. Polym. Proc.*, **7**, 240 (1992b).

Campbell G.A. and Dontula, N., "Solids Transport in Extruders," *Intern. Polym. Process.*, **10**, 30 (1995).

Campbell, G.A., Sweeney, P.A., Dontula, N., and Wang, C., "Frame Indifference: Fluid Flow in Single Screw Pumps and Extruders," *Intern. Polym. Proc.*, **11**, 3 (1996).

Campbell, G.A., Cheng, H., Wang, C., Bullwinkel, M., and te-Riele, M.A., "Temperature Rise in Single Screw Pump-Extruders," *SPE-ANTEC Tech, Papers,* **47**,152 (2001a).

Campbell, G.A., Cheng, H., Wang, C., Bullwinkel, M., and te-Riele, M.A., "An Investigation of Flow Rate and Viscous Dissipation in the Single Screw Pump-Extruder," *Intern. Polym. Proc.*, **16**, 323 (2001b).

Campbell, G.A., Tang, Z., Wang, C., and Bullwinkel, M., "Some New Observations Relative to Melting in Single Screw Extruders," *SPE-ANTEC Tech. Papers*, **49**, 213 (2003).

Campbell G.A. and Tang, Z., "Solid Bed Melting In Single Screw Extruders - An Alternative First Order Mechanism," *SPE-ANTEC Tech. Papers*, **50**, 162 (2004).

Campbell, G.A., Spalding, M.A., and Carlson, F., "Prediction of Melt Temperature Rise in Single-Screw Pumps," *SPE-ANTEC Tech. Papers*, **54**, 267 (2008).

Klein, I., "Melting in Plasticating Extruders and its Effect on Axial Pressure Profile," *SPE J.*, **28**, 47 (1972).

Lindt, J.T., "A Dynamic Melting Model for a Single-Screw Extruder," *Polym. Eng. Sci.*, **16**, 284 (1976).

Maddock, B.H., "A Visual Analysis of Flow and Mixing in Extruder Screws," *SPE J.*, **15**, 383 (1959).

Pearson, J.R.A., Halmos, A.L., and Shapiro, J., "Melting in Single Screw Extruders," *Polymer*, **17**, 905 (1976).

Pham, H.T., Spalding, M.A., Crabtree, S.L., Pavlicek, C.L., and Altınkaynak, A., "Experimental Measurement of the Surface Temperature of a Single-Screw Extruder Screw," *SPE-ANTEC Tech. Papers*, **54**, 289 (2008).

Street, L.F., "Extrusion Plastication," *Int. Plast. Eng.*, **1**, 289 (1961).

Tadmor, Z. "Fundamentals of Plasticating Extrusion. I. A Theoretical Model for Melting," *Polym. Eng. Sci.*, **6**, 185 (1966).

Tadmor, Z., Duvdevani, I.J., and Klein, I., "Melting in Plasticating Extuder Theory and Experiments," *Polym. Eng. Sci.*, **7**, 198 (1967).

Tadmor, Z. and Klein, I., "Engineering Principles of Plasticating Extrusion," Robert E. Krieger publishing company, New York, 1978.

CO-EXTRUSION OF CONDUCTIVE AND DIELECTRIC FILLED POLYMERS FOR CAPACITOR APPLICATIONS

Lisa A. Mondy*, Rekha R. Rao, Lothar Bieg, John L. Schroeder, Mark Stavig, Scott Spangler, Michael Winter[1] Randy A. Mrozek and Joseph L. Lenhart[2]

[1]*Sandia National Laboratories, PO Box 5800, Albuquerque, NM 87185-0346, USA– lamondy@sandia.gov; rrrao@sandia.gov; lfbieg@sandia.gov*
[2]*Army Research Laboratory, Adelphi, MD 20783-1197, USA – randy.mrozek@arl.army.mil; joseph.lenhart1@arl.army.mil*

We are developing structured composites composed of alternating layers of conductive and dielectric polymers through a process called multi-layered co-extrusion. Here, the two polymers are extruded through a feed block, and then the extruded material is split vertically and recombined by stacking the segments horizontally to multiply the number of layers. One advantage of the co-extrusion process is that large amounts of materials can be made very quickly. Intrinsically conductive polymers are typically difficult to extrude due to their conjugated backbones. Our approach is to melt process a polymer composite composed of a polystyrene matrix, conductive nickel particulates, and a conductive eutectic alloy. The eutectic alloy melts during processing, leading to a combination of emulsified liquid metal and solid metal particles, with a lower composite viscosity than using solid particles alone for the same fraction of metal and composite conductivity. Extrudable dielectric materials can also be formed using dielectric particle fillers in a polymer matrix. Rheological characterization of the filled polymers proved to be critical to producing stable flows and uniform layers. Finite element modeling is used to examine the practical limits of density or viscosity mismatches of the two materials in the co-extruder. Examination of microstructure and measurements of electrical properties, combined with the modeling, is used to improve the formulations and process parameters.

Introduction

Multi-layered co-extrusion combines multiple polymers into a layered structure to produce composite properties that are not found in a single polymer. Specifically, multi-layered structures with alternating mechanical, electrical, or optical properties can be produced to enable a variety of functional devices [Khomenko et al. 2005, Gerard et al. 2002, Chung 2001]. There are several emerging technologies that could potentially benefit from multi-layered structures including energy storage devices, sensors, optical devices, barrier materials, membranes, micro-composites, or armor applications.

The equipment described here is based on a co-extruder developed at the University of Minnesota [Cole & Macosko 2000, Zhao & Macosko 2002] and combines the polymer flows from two extruders into an initial layered structure that then passes through a series of dies to multiply the number of layers (Figure 1). The first die splits the flow vertically; the two components are stacked, and recombined to effectively double the number of layers. As the number of layers is increased in the same die cross-sectional area, the individual layer thickness is decreased. This method has been implemented successfully by companies like 3M and Dow to produce multilayered structures of ~ 20 μm and 110 nm, respectively [3M 2005, Dow 2007]. In our equipment, the initial layered structure consists of 8 layers, rather than Minnesota's four, to decrease the number of multiplications needed. As a result, the number of multiplication sections, n, will result in 2^{n+3} layers in a 1 mm tape. In theory, these layers should not intermix provided there is a uniform laminar flow. However, instabilities, especially as the layers get thinner, can lead to non-uniformity of the layers or even layer break up.

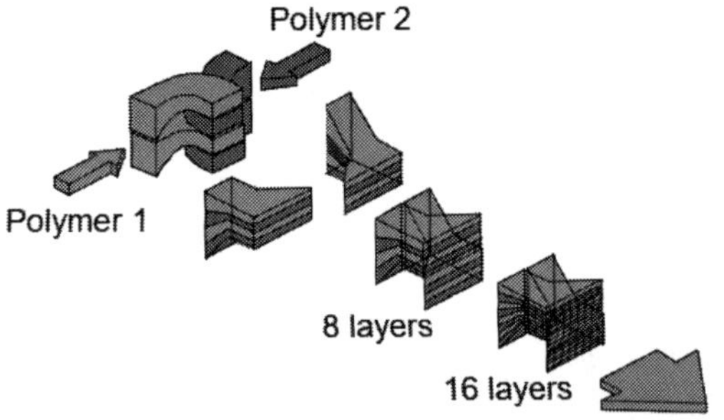

Figure 1. Multi-layer die system that multiplies the number of layers in a given cross-section. Although our system begins with eight layers, only four are shown for clarity.

The incorporation of fillers into the two polymers can enhance the material properties of the matrix including the thermal stability, flame retardancy, chemical resistance, permeability, conductivity, and stiffness. The targeted application here is to extrude a capacitor structure consisting of alternating conductive and non-conductive layers. To produce a capacitor structure the biggest technical challenge is to produce a conductive polymer composite that is melt processable. Inherently conductive polymers, however, typically consist of a conjugated backbone that is relatively rigid and will produce very high melt viscosities. In addition, the conjugated backbone will raise the melting point often above the degradation temperature of the polymer [Wessling 1998]. Conversely, conductivity obtained through incorporation of a filler typically requires very high loadings. At these high loadings it can be difficult to disperse particles resulting in aggregation and inconsistent properties in the material. More importantly, there is a dramatic increase in the viscosity at high loadings of solids, which can damage the processing equipment.

To minimize the impact of filler loading, we explored the addition of a low-melting-point eutectic to take advantage of the fact that emulsions in general have a lower viscosity than suspensions of solid particles at the same volume fraction of the dispersed phase. The viscosity of the molten metal is typically on the order of 0.001 Pa-s, negligible compared to the viscosity of the polymer of 100-1000 Pa-s, giving a vanishing viscosity ratio. Figure 2 demonstrates the exponential increase in the viscosity as the volume fraction of spherical particles is increased, as well as the less dramatic effect of an emulsion where the viscosity of the dispersed phase is much lower than that of the continuous phase. The points in this figure representing suspension viscosity are taken from the semi-empirical Krieger-Dougherty correlation for the relative viscosity of a suspension of solid particles [Krieger & Dougherty 1959]:

$$\frac{\eta_s}{\eta_0} = \left(1 - \frac{\phi}{\phi_m}\right)^{-[\eta]\phi_m} , \qquad 1$$

where η_s is the apparent viscosity of the suspension, η_0 is the viscosity of the continuous phase (the suspending medium), ϕ is the volume fraction of solids in suspension, ϕ_m is the maximum packing of solids, and $[\eta]$ is the intrinsic viscosity of the suspension, which is defined as $\lim_{\phi \to 0} \frac{\eta_s - \eta_0}{\eta_0 \phi}$.

Einstein showed that for hard spheres, $[\eta]$ is 2.5 [Einstein 1906, 1911]. Here we take ϕ_m to be 0.65 and $[\eta]$ to be 2.5, values that give a curve through Thomas' survey of measurements on suspensions of hard spheres [Thomas 1965].

Also plotted in Figure 2 are two formulas to describe the relative viscosity (η_e/η_0) of emulsions, where the droplet viscosity is much lower than that of the suspending phase. The Taylor-Mooney formulation [Taylor 1934, Mooney 1951, Prud'homme and Khan 1996] is:

$$\frac{\eta_e}{\eta_0} = \exp\left(\frac{K\phi_d}{1-\phi_d}\right) , \qquad 2$$

where, for emulsions in which the viscosity ratio of the droplet viscosity to the continuous phase viscosity is zero, K=1, and ϕ_d is the volume fraction of droplets in the emulsion. Frankel and Acrivos [1970] proposed a more complex constitutive relationship that takes into account deformation of the droplets and gives rise to memory effects if surface tension is taken into effect. For finite concentrations and vanishing shear rates and viscosity ratios, the formula reduces to [Grizzuti et al. 2000]:

$$\frac{\eta_e}{\eta_0} = 1 + \phi_d + 2.5\phi_d^2 . \qquad 3$$

For high shear rates the viscosity is somewhat lower (again for vanishing viscosity ratio):

$$\frac{\eta_e}{\eta_0} = (1 + \phi_d + 2.5\phi_d^2)\left(\frac{3+12\phi_d}{3+20\phi_d}\right) . \qquad 4$$

From Figure 2, one can see that above a dispersed phase volume concentration of about 20%, the emulsion viscosity can be significantly lower than the suspension viscosity.

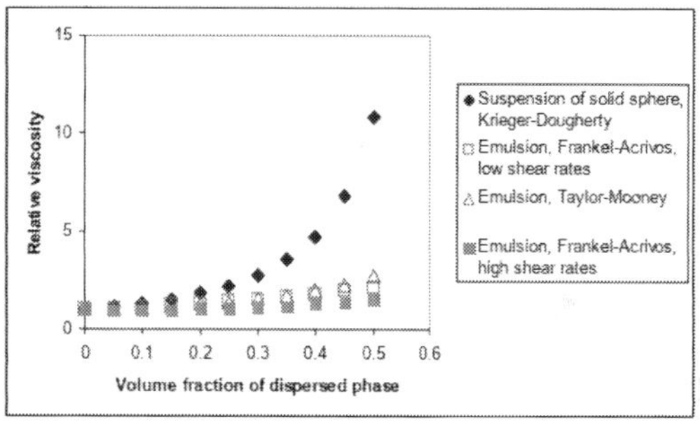

Figure 2. Relative viscosity of suspensions of solids compared to emulsions of inviscid drops.

Experimental

Materials and Mixing Methods

To produce a conductive melt processable material, we explored the incorporation of a low melting eutectic metal which would be molten during processing to reduce the impact on the viscosity but would be solid at the anticipated operational temperature of the device. The eutectic material used consisted of 58 % bismuth and 42 % tin, a commercial material called Cerrotru (McMaster-Carr). The polymer was Styron 666 polystyrene (Dow), chosen as a matrix simply because of its availability. The eutectic melted at 138 °C well below the polystyrene extrusion temperature of approximately 260 °C. Also, the eutectic material had a very low net expansion of 0.05 % when changing from a solid to a liquid, which would reduce any internal strain after the extrudate cools.

A Randcastle RCP-0750 single-screw extruder, equipped with a special mixing screw with reflow sections was used in the initial tests to combine the polymer and eutectic. It was found that the eutectic could be mixed into the polymer only in the presence of solid particles. In the absence of the solid filler, the eutectic was initially dispersed but during processing dispersed eutectic droplets coalesced and phase-separated out of the material leaving only ~ 2 volume % eutectic metal well dispersed. A nickel particulate was chosen because of its conductivity, relative stability to oxidation, and good availability. For our proof-of-principle test, we chose a filler with a fairly large particle diameter (2-4 µm nickel particulate from Aldrich), primarily because of its low price, but also because smaller particulates on the market were oxidized and not conductive.

There was a strong interaction of the particulate and the eutectic, which had a pronounced impact on the microstructure. The constituent elements were mapped using Scanning Electron Microscopy-Energy Dispersive Spectroscopy (SEM-EDS) for a sample containing 10 volume % nickel and 50 volume % eutectic. From these images, it appeared that there was a high concentration of nickel in the areas where there was eutectic but outside of those areas there was very little nickel. Inside the metal rich regions, the nickel and tin signals overlapped and were surrounded by the bismuth. This indicated that during phase separation, typical of eutectic materials, the tin was nucleating on the nickel. Whether this indicated that the tin had some affinity for the nickel or whether the nickel just provided a solid surface to facilitate nucleation was unclear.

The composites were also mixed using two different twin-screw extruders. The first, a Haake Minilab (Thermo Scientific), included recirculation for higher residence time and better mixing. The main disadvantage of this equipment was the small volumes of material that could be mixed. The second was a large extruder, a Haake Rheomex OS PTW16 twin screw mixer driven with a PolyLab OS Rheodrive 7 drive unit (Thermo Scientific), to make batches of a reasonable size for testing in the co-extruder. This instrument has multiple heating zones and can process much more material than the smaller Minilab in the same amount of time. However, the material does not recirculate and must be taken from the extruder exit, pelletized, and returned to the inlet in order to better mix the final product. Mixing the most highly concentrated suspensions, especially with high amounts of eutectic, was as difficult in these two machines as in the single-screw extruder.

Conductivity and capacitance measurements

The composite resistance of the extruded materials was determined by simply measuring the resistance (R) between two alligator clips a known length apart, L, on cylindrical (diameter $2r$) extrudate. The results can be interpreted as a volume conductivity (σ) which is equivalent to the reciprocal of the bulk resistivity (ρ_r); both material properties are independent of the geometry. These properties are related to the measured R in Ohm/cm by:

$$\sigma = 1/\rho_r = L/RA = L/R(\pi r^2) \quad \textbf{5}$$

Conductivity σ is in units of S/cm. This is a way of comparing data taken on samples of different diameter. However, this assumes that the material is homogeneous throughout the volume, which in a multiphase system is not necessarily true.

Samples of several materials were tested as the conductive material in a simple capacitor. Here each composite material was pressed into a disc 0.157 cm thick, then placed in a stack with a copper plate on the bottom, followed by a Kapton sheet in the middle, then the composite on the top. Capacitance C was measured using a four-wire bridge. Measured values were compared to theoretical (equation 6).

$$C\ (pF) = K^*Eo^*A/D, \quad \textbf{6}$$

where:

K is the dielectric constant of the Kapton,
A is the overlapping surface area of the plates (cm^2),
D is the distance between the plates (cm), and
Eo= 2.2489 x 10^{-11} (pF/cm)

Modeling

A numerical model of a multilayer co-extrusion process has been developed based on a finite element discretization and two different free surface algorithms: an arbitrary-Lagrangian-Eulerian (ALE) moving mesh implementation and an Eulerian level set method. The free surface algorithms help us to understand the moving boundary problem associated with the polymer-polymer interface. For this study, elastic instabilities are ignored and we focus on Newtonian and shear-thinning fluids only.

Numerical models of varying degrees of complexity were investigated including stability analysis and direct finite element free surface approaches. Linear stability analysis was used in simplified two-dimensional geometries to help understand the limit on density mismatch between layers. Finite element free surface flows were used in three different three-dimensional (3D) geometries to understand stable operating conditions such as pressure or flow rate mismatch or viscosity mismatch.

Details of the modeling approach and equations, the numerical methods used and the finite element implementation, and the complete results can be found in a report by Rao and coworkers [2010].

Results and Discussion

Modeling

3D free surface flows can give us a great deal of information about the shape of the extrudate and how materials properties and processing parameters affect final film thickness and layering. Because of the complexity of the numerical approaches and because boundary conditions at the wall were not known, an important goal was to validate a model with experimental data. We undertook simulations on a simplified model of a two-layer structure split and multiplied to be a four-layer one (Figure 3).

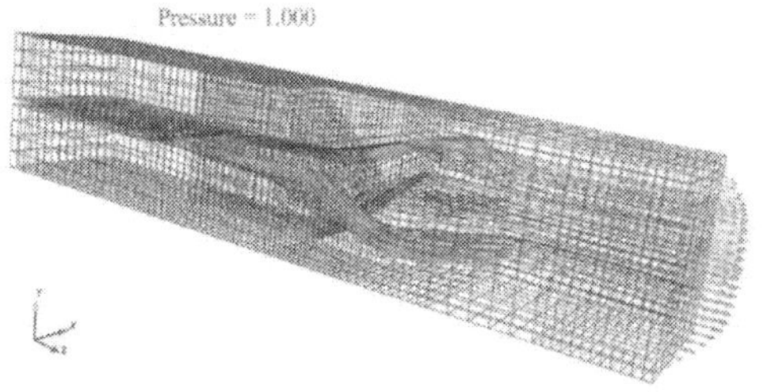

Figure 3. Co-extruder mesh for initial multiplication creating a four-layered structure by splitting and restacking a two-layered one. Velocity vectors on mesh show acceleration and redirection of fluid through the splitter.

We wanted to determine the correct boundary conditions at the wall. To this end, we performed a validation experiment in the co-extruder to look at materials with the same density and viscosity and run at the same flow rate. To differentiate the two polymers, we added 3 volume % carbon black to create a black material and 3 volume % titanium dioxide to the other layer, to create a white material. This was an amount suitable to change the color of the polymer without significantly affecting the properties. Results from the validation experiment (Figure 4) compared to the numerical model showed that slip boundary conditions are the most appropriate.

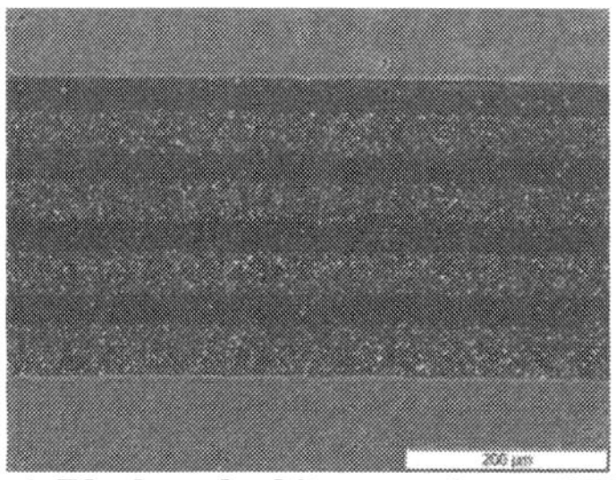

Figure 4. Black and white experiment SEM.

When we increased the injection rate of one material over that of the other with a pressure mismatch in the model, the results predicted thinner layers of the slower flowing material (Figure 5). Here the red and blue materials once again had equal properties. SEMs of the experiment in which the ratio of the extrusion speed of the carbon black to that of the titanium dioxide was mismatched is also shown in Figure 5, where one can see that the layers start to lose integrity.

Using this model we also determined that the layers would be dangerously close to mixing if the ratio of viscosities of the two phases was greater than about 5.5 (keeping the densities the same) or the density ratio greater than about 4 (keeping the viscosities the same). We wanted to understand how the stability of the flow would be affected by both dissimilar densities and viscosities. As an example, in which we used properties similar to what was expected for our filled polymers, we used a red phase viscosity of 25,000 Poise and density of 4.09g/cc to represent the conductive polymer and a blue phase viscosity of 15,000 Poise and density of 2.06g/cc to represent the dielectric. We assumed that the polymers were Newtonian for this study, with no interfacial tension between the phases. The inflow velocity was set to 5 cm/s. The results from this test case are given in Figure 6, showing that the combination of mismatched viscosity (ratio = 1.667) and mismatched density (ratio = 1.99) resulted in problems in the splitter where the conductive red fluid can become encapsulated by the lower viscosity blue dielectric fluid.

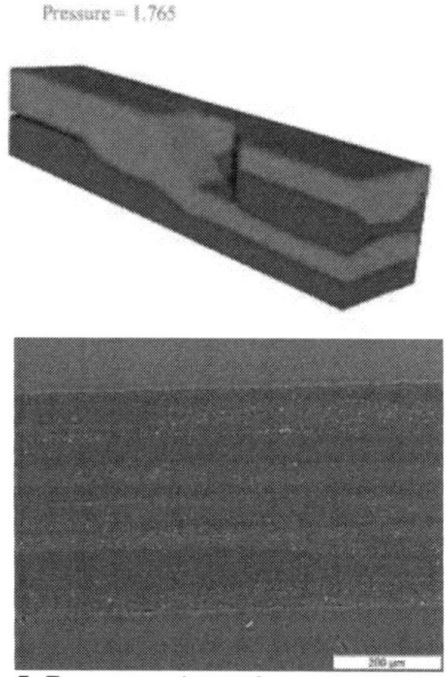

Figure 5. Pressure mismatch creates thinner layers of the slower flowing material (blue) in the model (top). Ratio of red to blue pressure =1.76. SEM of co-extruded material (bottom) shows distortion of layers at a ratio of flowrate = 1.75, and layers lose integrity. Here light is equivalent to blue in the model.

Figure 6. Level set simulation results showing an early time when the bottom red sheet breaks. It later reforms but remains very thin.

Conductive and Dielectric Properties of Materials and Measured Capacitance

The conductivity of the nickel/eutectic materials are shown as a function of composition in Figure 7. Unfortunately, the material at the highest loading of metal was also difficult to extrude consistently. Furthermore, we found that as we mixed the material more thoroughly, the conductivity often went down. From SEMs of the materials, we noted that the best conductivity was often given by materials with long droplets stretched in the direction of extrusion (and conductivity measurement), indicative of poorly mixed materials. However, for our application, we actually need conductivity in the direction perpendicular to that of extrusion, and we would prefer the repeatability and consistency of a well mixed state.

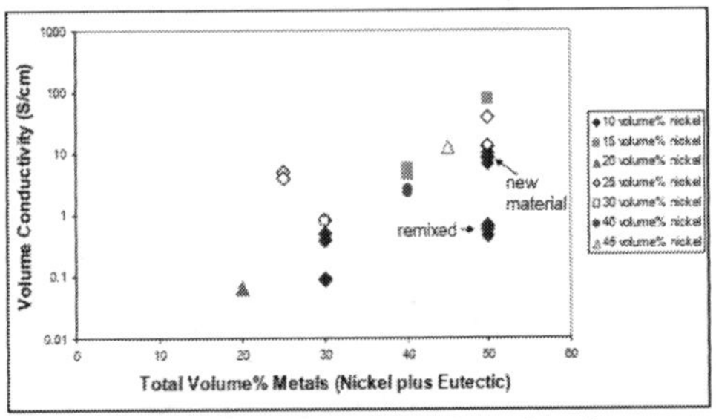

Figure 7. Conductivity of 15-cm lengths of extrudate from the Haake minilab. Various compositions are shown, where the volume % of the eutectic is the total volume % of metal (the abscissa) less the amount of nickel shown in the legend.

The ratio of the measured capacitance to the calculated theoretical values (equation 6) correlated with the amount of conductive material distributed within the polymer. For example, for a sample with a total metals loading of 50 volume %, the calculated theoretical value was 61.9 pF, but the measured value was 31 pF. Another sample with 40 volume % total metals loading had a theoretical capacitance of 49 pF, but a measured value of 20 pF. Another sample with 10 volume % nickel (and no eutectic), when joined with a layer of Mylar (instead of the Kapton used in the other tests), had a theoretical capacitance of 980 pF, but a measured value of 98 pF. We believe that the reduced capacitance is the result of the actual A not being that of the total surface, but reduced to only that of the conductive material in that surface.

In order to both improve the dielectric properties of the polymer and better match the viscosity and density of the conductive layer, we decided to fill the polystyrene with barium titanate ($BaTiO_3$) powder (diameter >2 μm), a material with a very high dielectric constant. Another idea to create a better dielectric material that better matched the density of the conductive material was to use nonconductive (oxidized) submicron nickel particles as a filler. As shown in Figure 8, this proved to be a viable option at higher particle loading based on the measured dielectric constant. However, for our proof of principle we chose to use the $BaTiO_3$.

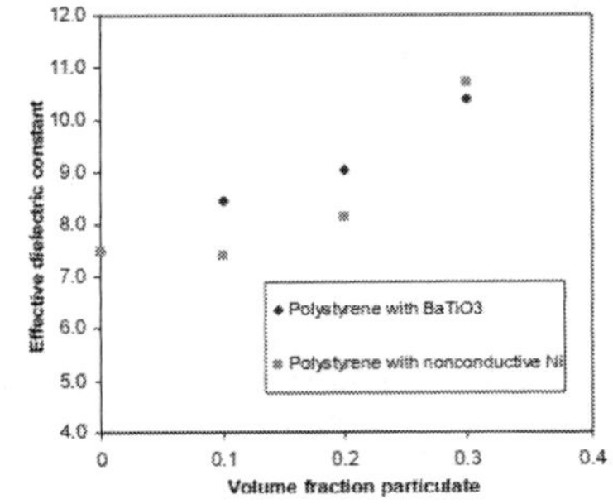

Figure 8. Effect of filler addition on the dielectric properties of polystyrene

Rheology of Filled Polymers

The rheology of the candidate conductive material, based on a compromise between processability and conductivity (12 volume % nickel, 28 volume % eutectic, and 60 volume % polystyrene), was measured with both a shear sweep using parallel plates on an TA ARES controlled strain-rate rheometer and on a Dynisco LCR 7000 capillary rheometer. Polystyrene samples filled with 20 volume % of $BaTiO_3$ were also tested with both rheometers (Figure 9). The results were compared to approximations based on equations 1 through 4 and a mixture relationship for the viscosity of the conductive emulsion/suspension material. The mixture relationship was based on work by Farris for suspensions of mixtures of two particle sizes [Farris 1968]. Like Farris, we assumed that the relative viscosity of the mixture could be considered the same as the relative viscosity of the solid particles in a continuous phase that was represented by the viscosity of the emulsion without the solid particles. In other words, the relative viscosity of the nickel-eutectic-polystyrene mixture, $\eta_{Ni\text{-}eu}/\eta_0$, would be the relative viscosity of the suspension of nickel particles alone times the relative viscosity of the eutectic droplets alone, or, for low shear rates

$$\frac{\eta_{Ni-eu}}{\eta_0} = \frac{\eta_{Ni-eu}}{\eta_{eu}}\frac{\eta_{eu}}{\eta_0} = \left\{(1-\frac{\phi_{Ni}}{\phi_m})^{-[\eta]\phi_m}\right\}\left\{1+\phi_{eu}+2.5\phi_{eu}^{\ 2}\right\},$$ **7**

and for high shear rates

$$\frac{\eta_{Ni-eu}}{\eta_0} = \frac{\eta_{Ni-eu}}{\eta_{eu}}\frac{\eta_{eu}}{\eta_0} = \left\{(1-\frac{\phi_{Ni}}{\phi_m})^{-[\eta]\phi_m}\right\}\left\{(1+\phi_{eu}+2.5\phi_{eu}^{\ 2})(\frac{3+12\phi_{eu}}{3+20\phi_{eu}})\right\},$$ **8**

where $\phi_{Ni} = 0.12$ and $\phi_{eu} = 0.28$. Figure 9 shows that the conductive material appears much more shear thinning than the neat polystyrene, contrary to the theory outlined in equations 1-4 and 7-8. Although the theory predicted that these compositions would have similar viscosities, this only occurred at the higher shear rates.

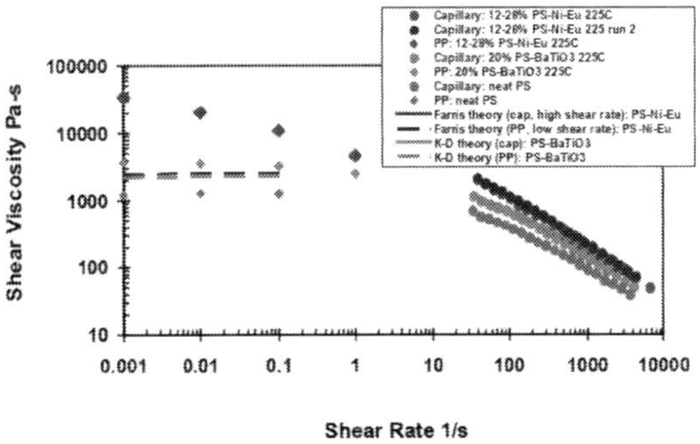

Figure 9. The rheology of the conductive material (labeled PS-Ni-Eu) and the dielectric (labeled PS-BaTiO3) and the theoretical predictions (equations 1-4, 7-8).

Co-extrusion of Filled Polymers

The final conductive material (12 volume % of nickel, 28 volume % of eutectic, and 60 volume % of polystyrene) and a dielectric material made of 25 volume % $BaTiO_3$ in polystyrene were co-extruded to form an 8-layered structure. As shown by the model, the density and viscosity mismatches between these materials (at the low shear rates we eventually used, see below) were far from optimal from the standpoint of flow stability.

The co-extruder can be run at various flow rates (screw rpm), with each material independently controlled. Other processing parameters include temperature and the take up rates of the two chilled rollers at the exit of the die. A complete design-of-experiment would include a matrix of values for the parameters. The goal of this proof-of-principle experiment was merely to get experience running filled polymers and to see if the final layered structure could be useful as a capacitor.

We found that, contrary to our estimate from our experience processing in the twin-screw and the preliminary rheology measurements, the co-extruded material looked better at lower flow rates. This was primarily due to bubble formation at the drawdown, a phenomena not seen in our co-extrusion with neat polymers. The bubbles could have been formed from cavitation, but we think it more likely that they formed from air entrained during incorporation of the particles into the polystyrene or from de-adhesion of the polystyrene from the particles. Flowing at a significantly lower rate minimized the bubble formation and prevented stretching of the remaining bubbles.

When the exit temperature was cooled to 200°C and the flow of the two materials were regulated so that the higher viscosity conductive material was pushed at twice the rate of the lower viscosity dielectric material (20 and 10 rpm respectively) a fairly nice layered ribbon was formed, albeit one still containing some flaws. An SEM of a random representative sample of this ribbon is shown in Figure 10.

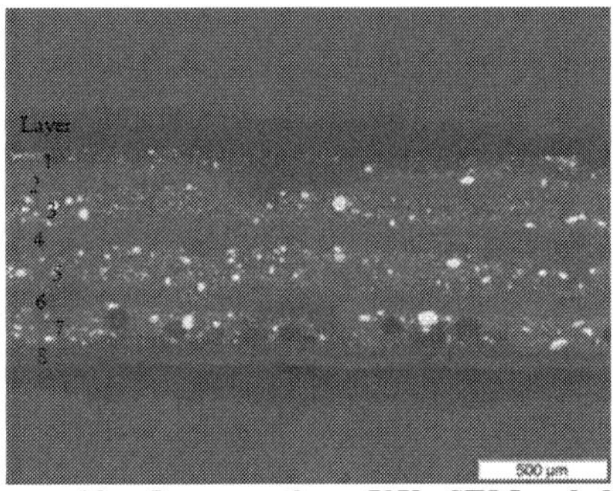

Figure 10. Cross-section 50X SEM of layered structure formed at an exit temperature of 200 °C, where the conductive layer is extruded at 10 rpm and the dielectric at 20 rpm.

The resistance of the exposed surface conductive layer was measured to be between 1000 and 2000 Ohm/cm and that of the exposed dielectric layer had no measurable conductance. A capacitor measurement was performed by treating the structure as a 2-plate capacitor using a 4 wire bridge type capacitance measurement. By connecting the 4 wire bridge capacitance meter to the aforementioned conductive surfaces and piercing through the opposing non-conductive surface until conductive material was touched, a capacitance measurement of 275 pF and a dissipation factor (DF) of 0.5 was recorded. The large DF number is due to the high internal resistance of the structure, whereas a DF of 0.005 would be a more realistic number for a good capacitor.

Conclusions

The great advantage of the co-extrusion process is that large amounts of materials can be made very quickly. Our rate of production was greater than 1 cm/s of ribbon. The ribbon is malleable until cooled, so can be bent, twisted, or rolled into a variety of final geometries.

Numerical modeling has been valuable for elucidating optimal processing regimes from viscosity and density mismatch to flow rates. Rheology measurements have proved to be critical to determining the final quality of the co-extruded part. Once we had measured the shear-dependent rheology of the conductive and dielectric material, it became clear from the modeling that the viscosity mismatch of 10 between the two phases would lead to flow instabilities. These problems were seen when we co-extruded the material.

Although the early studies on using a eutectic to form an emulsion rather than solid particles were promising, much work remains in order to optimize the formulation and improve the incorporation of the eutectic. Fundamental studies on process control of droplet size formation would be advised to aid in reducing droplet size for use in fine layers. Using only solid materials could be a viable alternative if the size distribution could be improved to lower the viscosity. For example, bimodal, trimodal, and even higher modal size distributions have been used for decades in propellant formulations to achieve very high particle packing while maintaining processability [Farris 1968]. Looking at a range of other conductive particles including carbon nanotubes is strongly advised since these materials percolate at much lower volume fractions and have the possibility of remaining flexible in the final part. Using smaller particles would allow for much thinner layers, greatly enhancing the possible range of capacitance, for example. From a material development standpoint, we should also investigate different polymeric suspending fluids. The polystyrene used here had a high viscosity and was extremely shear-thinning, resulting in a composite material that followed these properties. Using a low molecular weight polyethylene would have resulted in a superior rheology for co-extrusion and less likelihood of cavitation. In addition, using different polymers for the conductive and dielectric phase polymer matrix would stabilize the layers by introducing an interfacial tension between layers and reduce mixing between the layers. Finally, another important idea to explore is the addition of processing aids such as surfactants used as dispersive agents. [Huang and Cheng 2009]. It has been found that such dispersive agents can allow much higher concentrations of particles to flow smoothly by dispersing the particles, adhering them more strongly to the polymer matrix, and reducing the viscosity of the composite system.

Acknowledgements

Phil Cole and his graduate work at University of Minnesota were the inspiration for this project. The authors would like to acknowledge useful discussions with Randy Schunk, David Noble, Anne Grillet, and Chris Diantonio, as well as Alice Kilgo's help with imaging, and Adam Lester's help in capacitance measurements. This work was supported by the Laboratory Directed Research and Development program at Sandia National Laboratories. Sandia is a multiprogram laboratory operated by Sandia Corporation, a Lockheed Martin Company, for the United States Department of Energy's National Nuclear Security Administration under contract DE-AC04-94AL85000.

References

3M, U.S. Patent 6837698 (2005), "Multilayer coextrusion die and method."

Chung, D. D. L., "Electromagnetic interference shielding effectiveness of carbon materials," *Carbon*, **39**, 279-285 (2001).

Cole P.J. and Macosko C.W., "Polymer-polymer adhesion in melt-processed layered structures," J. Plastic Film & Sheeting, **16**, 213-222 (2000).

Dow Global Technology, (2007) "Multilayer polymer sheets," U.S. Patent Application 20070184259.

Einstein, A., "A new determination of the molecular dimensions," *Annalen der Physik*, **19**, 289-306 (1906).

Einstein, A., "A new determination of the molecular dimensions (vol 19, pg 289, 1906)," *Annalen der Physik*, **34**, 591-592 (1911).

Farris, R. J., "Prediction of the Viscosity of Multimodal Suspensions from Unimodal Viscosity Data,"*J. Rheol.*, **12**, 281-301 (1968).

Frankel, N. A. and Acrivos, A., "The constitutive equation for a dilute emulsion," *J. Fluid Mech.* **44**, 65-78 (1970).

Gerard, M.; Chaubey, A.; Malhorta, B. D., "Application of conducting polymers to biosensors," *Biosens. Bioelectron.*, **17**, 345-359 (2002).

Grizzuti, N.; Buonocore, G.; Iorio, G., 2000, "Viscous behavior and mixing rules for an immiscible model polymer blend," *J. Rheol.* **44**, 149-164 (2000).

Huang, C. K. and Cheng, F. S., "Study on Kneading and Molding of PP/TiO_2 Nanocomposite," *Intern. Polymer Processing*, **2009/03**, 267-271 (2009).

Khomenko, V.; Frackowiak, E.; Beguin, F., "Determination of the specific capacitance of conducting polymer/nanotubes composite electrodes using different cell configurations," *Electrochim. Acta* , **50**, 2499-2506 (2005).

Krieger, I. M. and Dougherty, T. J., "A mechanism for non-newtonian flow in suspensions of rigid spheres," *Trans. Soc. Rheol.*, **3**, 137-152 (1959).

Mooney, M., "The viscosity of a concentrated suspension of spherical particles," *J. Colloid Sci.*, **6**, 162-170 (1951).

Prud'homme, R. K. and Khan, S. A., "Experimental Results on Foam Rheology," in *Foams, Theory, Measurements and Applications,* Prud'homme and Khan (Eds.), Marcel Dekker, Inc. New York, p. 234 (1996).

Rao, R. R; Mondy, L; Collins, R.; Baer, T; Mrozek, R; Schunk, P. R.; Hopkins, M.; and Lenhart, J., *Finite Element Analysis of Multilayer Co-Extrusion*, SAND report, Sandia National Laboratories, Albuquerque (2010).

Taylor, G. I., " The formation of emulsions in definable fields of flow," *Proc. Roy. Soc. Ser. A.* **146**, 0501-0523 (1934).

Thomas, D. G., "Transport characteristics of suspensions .8. A note on viscosity of Newtonian suspensions of uniform spherical particles," *J. Colloid Science*, **20**, 267-277 (1965).

Wessling, B., "Dispersion as the link between basic research and commercial applications of conductive polymers (polyaniline)," *Synth. Met.*, **93**, 143-154 (1998).

Zhao, R; Macosko, C. W., "Slip at polymer-polymer interfaces: Rheological measurements on coextruded multilayers," *J. Rheol.* **46**, 145-167 (2002).

REAL TIME INSPECTION OF DIFFERENT PARTICLES IN FLOWING POLYMER MELTS DURING EXTRUSION PROCESSING

M.Stephan[1*], A.Rudolph[2], S.Grosse[2], M.Stintz[3]

[1] *Leibniz Institute of Polymer Research Dresden, Germany – stephan@ipfdd.de*
[2] *Topas GmbH, Dresden, Germany – office@topas-gmbh.de*
[3] *TU Dresden, Germany – michael.stintz@tu-dresden.de*

This paper presents the application of a light extinction method to perform a continuous real time particle analysis within flowing polymer melts. The optical particle sensor which has been developed can be connected to various types of polymer processing machines (extruders or injection moulding machines) via a ½" 20 UNF thread. Its design, setup and adaptation have been described in recent papers [1-5]. The sensor has proven its reliability and robustness at typical processing conditions during long-term operation under plastics industry environmental conditions. By applying different kinds of signal evaluation a broad range of particle sizes and particle concentrations can be covered. This gives the opportunity to use that sensor system for a versatile number of applications in polymer processing.
Detection and size classification of particle impurities within flowing transparent/opaque polymer melts has reached a final stage in terms of a continuous documentation of melt quality at extrusion processes at industrial scale. These disturbing particles which are detected by means of a *single particle counting* algorithm can be for instance gel particles, black spots, gas bubbles or any other kind of particle material with a minimum size of approximately 20 microns. Therefore extrusion startup's and batch changes can be controlled in real time as well as steady state processing. At the moment research is focused on real time evaluation of nanofiller dispersion during extrusion processing by means of an integral signal analysis in terms of a classical *turbidity measurement*. There are first results showing sensors sensitivity to detect changes in nanofiller concentration and average nanofiller particles size. The effect of different nanofiller/polymer systems and modified extrusion processing conditions on nanofiller dispersion can be monitored in real time without any sampling.
The results which will be presented are demonstrating a remarkable potential of this method to be used as a new and powerful Process Analytical Technology (PAT).

Introduction

Gel particles, black spots, gas bubbles and other disturbing particles are a well known problem at industrial extrusion processing. Especially at extrusion start-up periods and at batch changes a significant number of such impurity particles usually appear. But there are also remarkable quality deviations observable in many cases from batch to batch of the raw material. To avoid such impurity particles melt filters are widely used even their efficiency is limited as these particles often show an elastic behaviour. By applying standard offline methods (visual inspection, microscopy, film inspection) including discontinuous melt sampling from the running process production of large amounts of low quality batches can not be avoided due to the lag of time between the sampling and the inspection results. By using the new optical particle sensor system melt quality can be monitored without any lag of time continuously directly at extrusion processing.
In contrast to these disturbing single particles polymer melts are filled with nanoparticles in order to have an improved property profile. By such a particle modification it is possible to modify standard polymers with a tailor-made property profile comparable or even better than a high performance polymer. As for all heterogeneous materials there is a strong correlation between the nanocomposite morphology and the resulting product properties. For that reason it is very useful to receive real time information on nanofiller dispersion directly during nanocomposite processing by applying the new sensor system.

Experimental

Materials
Experiments on detection and classification of impurity particles have been performed at different grades of a number of amorphous (PS, PC, PMMA) and semicristalline (LDPE, HDPE, PP, PA6) thermoplastic polymers as well as for thermoplastic elastomers (TPU). For nanocomposite investigations different montmorillonites (Nanofil, Dellite, Cloisite) and needle silicates (Sepiolite) have been added to a PP and a PA6 matrix polymer. Furthermore a reactive extrusion process resulting in a nanocomposite with titanium dioxide particles in a PP matrix has been realised.

Sensor Setup
The used sensor system is applying a light extinction method which is a known method for particle analysis in many other applications such as aerosol analysis and purity control of aqueous liquids. This method has now been transformed also to polymer processing

applications. The sensor system uses a 633nm laser source in combination with a diode detector which is placed at the opposite site for acquiring light extinction signals coming from any type of particle within the flowing polymer melt interacting with that crossing laser beam. There are different optical setups possible in order to adjust the sensor system to a certain application. The sensor is connected to a data acquisition unit which handles further signal processing. As a special patented [6] feature a classical turbidity measurement is performed simultaneously with a single particle counting algorithm by acquiring extinction signals at different time resolutions from 0,1 Hz to 10 kHz. Furthermore the sensor control box also contains an industrial PC unit for result presentation and documentation. For further information refer to former publications [1-5].

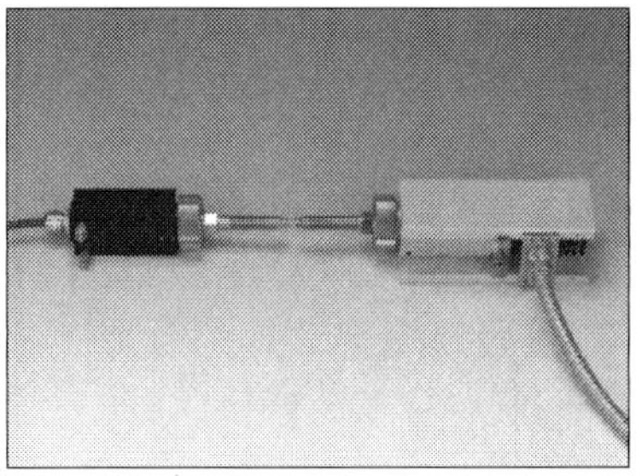

***Fig.1**: Polymer Melt Particle Sensor PMP694 (Topas GmbH)*

Extruder Setup

Most of the experiments have been done at the Leibniz Institute of Polymer Research (IPF). The sensor system has been adapted to different types of single and twin screw extruders by using a ½" 20UNF connector. This makes the sensor system compatible to standard extruder ports. This light extinction measurement gives reliable results for flow channels with a thickness in a range from 2 ... 20 mm. From this limitation it has to be decided if an inline adaptation of the sensor system is possible or if a bypass has to be used for online measurements.

***Fig.2**: Inline adaptation of the PMP 694 sensor to a single screw extruder*

The following table gives an overview on the range of the main extrusion processing parameters of the different single and twin screw extruders which have been used for the described experiments.

length / diameter	(/)	20 ... 44
mass flow	(kg/h)	5 ... 20
screw speed	(min^{-1})	50 ... 400
barrel temperature	(°C)	170 ... 260
melt temperature	(°C)	180 ... 300
melt pressure	(bar)	0 ... 250

***Table 1**: extrusion processing conditions*

Results and Discussion

The first part of presented results is regarding to the detection and classification of disturbing particles in transparent matrix melts by means of a single particle counting signal algorithm. For closer illustration the following figure shows exemplary measured counting rates of impurity particles larger than 100 microns. At the start-up of the extruder a large number of black spots have been detected within the melt. From the real time measurements it was shown that the extruder self cleaning takes approximately one hour of processing in this special case. But also for steady state extrusion processing continuous monitoring of melt quality is strongly recommended. At Fig.3 short term contaminations are observable from the optical particle sensor measurements which will not be recognized reliably with standard offline methods. For that reason larger low quality batches can be avoided a priori. The quality of each batch is documented continuously and retraceably in that way.

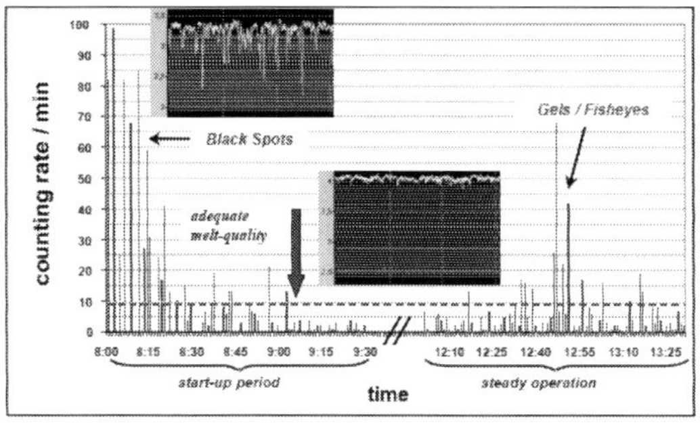

***Fig.3**: Continuous counting of melt impurities during extrusion processing using a PMP694 sensor*

The second important sensor application is focussing on the characterisation of the morphology of nanocomposites. For that reason a simultaneous evaluation of two different signal evaluation algorithms is applied. This is necessary as a wide range of particles sizes (100nm ... 100µm) is present at nanocomposite systems. A classical turbidity measurement gives integral information on the concentration and the size of the primary nanoparticles. On the other hand the single particle counting signal algorithm indicates larger nanofiller agglomerates larger than 20 microns.

In the following figure sensitivity of the method against changes of nanofiller concentration is demonstrated. With an increasing nanofiller concentration an increasing light extinction has been measured by the sensor systems which relates to an increasing melt turbidity. Non-linearity of the curves at larger concentrations is caused by multiple scattering effects.

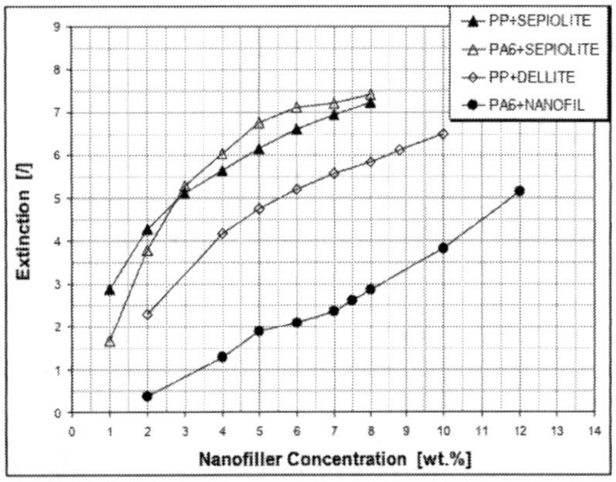

***Fig.4**: Measured light extinction vs. nanofiller concentration for 4 nanocomposites*

After sensor calibration with a set of known nanofiller concentrations it is possible to implement a continuous quantitative concentration monitoring. This is a very useful tool to control stability and reliability of the nanofiller feeding at nanocomposite extrusion processing.

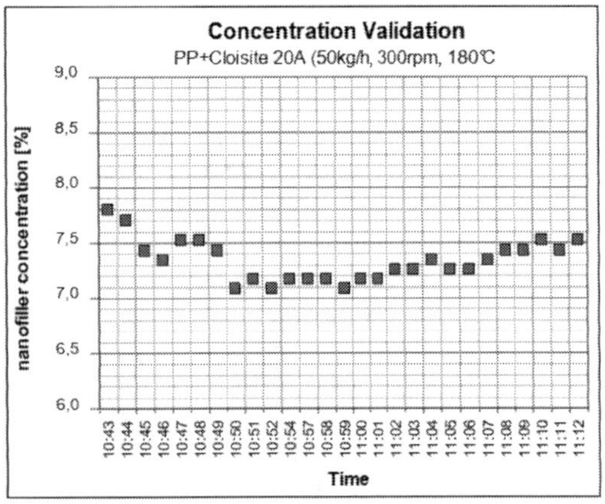

***Fig.5**: Real time monitoring of nanofiller concentration during nanocomposite extrusion processing*

Besides changes in nanofiller concentration the used sensor method is also sensitive to detect differences in nanocomposite morphology for different nanofiller types. Figure 6 shows the result for three PP based nanocomposites. There are two signal curves included in this diagram. The average detector signal (*left axis*) gives information on melt turbidity. The agglomerate counts (*right axis*) provide additional information on the nanofiller amount which has not been successfully dispersed to primary particles. Because of their small concentration these large agglomerates do not contribute to melt turbidity. From the results in Figure 6 it is shown that PP/Cloisite20A and PP/Dellite67G show a comparable morphology. But there is a significant difference for PP/Cloisite15A. In this case a much smaller melt turbidity has been observed in terms of a larger detector signal which is due to a highly increased number of nanofiller agglomerates which have been counted. That means nanofiller dispersion seems to be much different for this nanocomposite compared to the two systems mentioned before.

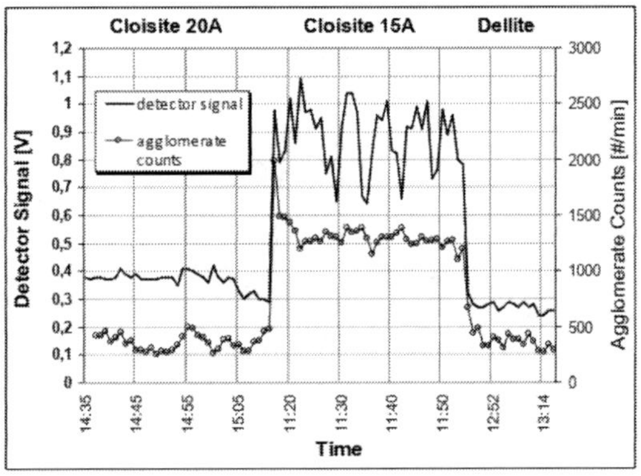

***Fig.6:** Simultaneous turbidity measurement and single particle counting for characterization of nanofiller dispersion*

Another sensor application is regarding to monitoring of reactive extrusion processing of nanocomposite materials. Experiments have been done on in-situ formation of titanium dioxide particles in a PP matrix melt. For that reason two liquid feeding systems (titanium-n-butoxide, water) have been adapted to a twin screw extruder (Leistritz Micro 27). During these experiments the amount of the added water has been increased slowly. Figure 7 shows the light transmission values measured inline with the optical particle sensor. At first there is no signal change when the titanium-n-butoxide is added to the PP matrix. With 1mol water the reaction starts as it is indicated by the reduced light transmission. This results from the generated TiO_2 particles which act as light scatterers. With an increasing amount of water there is a further reduction of the measured light transmission. This demonstrates that the reactive conversion can be followed by means of this inline monitoring method.

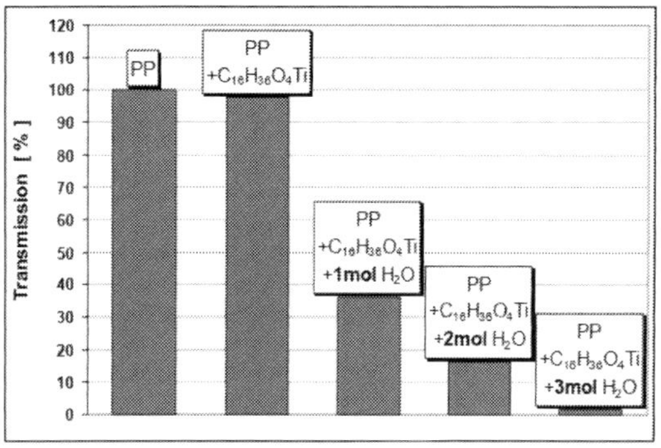

Fig.7: *Inline monitoring of reactive extrusion processing (PP/TiO2)*

Conclusions

The optical particle monitoring technique described in this paper shows a strong potential to be implemented as a process analytical technology (PAT) to existing polymer processing environments. As different algorithms of signal evaluation are applied simultaneously there is a wide range of particles which can be analysed. First major application is regarding to the detection of impurity particles (gel particles, black spots, gas bubbles) within flowing transparent melts. This gives the opportunity to have a continuous and retraceable documentation of melt quality. Another important application is the characterisation of nanocomposite morphologies. It was shown that the method is sensitive to detect changes in nanofiller concentration. It is also possible to distinguish the morphology of different nanofiller types mixed into the same matrix polymer. Furthermore reactive extrusion processes can be followed in real time. From a series of experiments and the feedback from industrial users it was proven that the used sensor system is robust at typical extrusion processing conditions.

References

1. M. Stephan, S.Große et al.: Microphotometric Detection of Particles/Inhomogeneities in Flowing Polymer Melts, Macromolecular Symposia, Volume 217, Issue 1, p. 343-356, Wiley 2004
2. S. Djoumaliisky, M. Stephan, S. Große: In-line/real time monitoring of extrusion foaming processes using microphotometric measuring methods, Proceedings of the Scientific and Technical Union of Machine Building (Bulgaria), Vol. XI, Issue 1, pp. 277-281, 2004
3. M. Stephan, S.Große et al.: Production Monitoring: Particle Monitoring on the Extruder, Plast Europe, issue 2005/09
4. S. Große, M. Stephan et al.: Real-Time Detection of particulate Heterogeneities in Polymer Extrusion Processes using Microphotometric Measuring Method, Plastics, Rubber and Composites: Macromolecular Engineering, Vol.35, No.10, 432-438, 2006
5. S. Große, M. Stephan et al.: Microphotometric Inline Determination of Polymer Blend Morphologies during Extrusion Processing, Journal of Applied Polymer Science, Vol. 103, 258-262 (2007)
6. Patent EP 1 252 499 A2

MELT-MIXING BY PITCHED-TIP KNEADING DISKS IN A TWIN-SCREW EXTRUDER

Yasuya Nakayama[1]*, Eiji Takeda[2], Kajiwara Toshihisa[2], Takashi Shigeishi[3] and Hideki Tomiyama[3]

[1] *Dept. of Chemical Engineering, Kyushu University, Fukuoka 819-0395, Japan – nakayama@chem-eng.kyushu-u.ac.jp*
[2] *Dept. of Chemical Engineering, Kyushu University, Fukuoka, Japan*
[3] *The Japan Steel Works, Hiroshima, Japan*

Melt-mixing in twin-screw extruders is a key process for development of polymer composites, and quantification of mixing performance of kneading elements based on the physical process in them is a challenging problem. We discuss melt-mixing by novel kneading elements, called pitched-tip kneading disks. Disk-stagger angle and tip angle are the main geometrical parameters of the pitched-tip kneading disks. We studied four typical arrangements of the pitched-tip kneading disks, which are forwarding and backwarding disk-staggers combined with forwarding and backwarding tips, respectively. Numerical simulations under a certain feed and screw revolution speed were performed and mixing process was investigated by using Lagrangian statistics. It is found that the four types had different mixing characteristics, and their mixing processes were explained by a coupling effect of drag flow by disk staggering and pitched-tip and pressure flow, which is controlled by operational conditions.

Introduction

A twin-screw extruder (TSE) is widely used in various kind of industries. In the plastic industry, a TSE is used for development of products of some desired properties by compounding and blending different kinds of polymers, fillers, and other additives. In this process, controling the mixing process in TSE is a key factor to optimize and improve productivity and functionality of products.

The control of the mixing process is based on the design of the kneading screw element. In the past, among different kneading screws, kneading disks (KDs) are the commonly used for rather general purpose of compounding due to its mixing efficiency. One main geometric parameter of a KD is the disk-stagger angle by which both pumping capability and inter-disk leakage flow are simultaneously controlled. However, these two factors have different effects on the mixing process.

For controling the these two factors separately, we have proposed a novel design of a kneading screw element, called "pitched-tip kneading disk (ptKD)" [Shigeishi *et al.*(2006)]. In the ptKDs, disk tips are pitched to the screw axes in contrast to the tips parallel to the screw axes in the conventional KDs. By arranging the pitched-tip angle along with the disk-stagger angle, the pumping capability and the inter-disk leakage flow are independently controlled. This flexibility can make possible to provide better control of distributive mixing and dispersive mixing than the conventional KD.

We investigate the flow characteristics and mixing process in melt-mixing by several types of ptKD. For this purpose, we conduct a course of three-dimensional numerical simulation of melt-mixing zone in a TSE. Material kinetics is solved by Lagrangian manner, and the characteristics of distributive and dispersive mixing in melt-mixing by the several types of ptKD are investigated.

Pitched-tip kneading disk

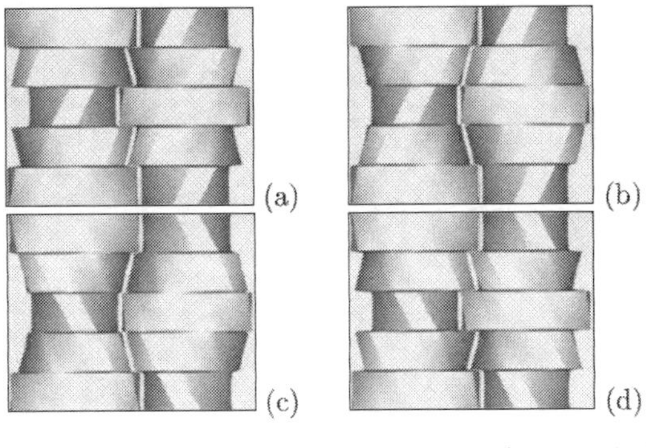

Figure 1: Top view of typical four types of the pitched-tip kneading disks: (a) backwarding stagger and forwarding tip (BS-FT), (b) forwarding stagger and forwarding tip (FS-FT), (c) backwarding stagger and backwarding tip (BS-BT), (d) forwarding stagger and backwarding tip (FS-BT).

One element of a ptKD is composed of several kneading block, as is the conventional KD, but each kneading block of ptKD has non-parallel tips to the screw axes in contrast to the conventional KDs. When a tip is pitched so as to pump a fluid forwardly (backwardly) along the screw rotation, it is called "forwarding (backwarding) tip". When the disk-stagger angle is arranged so as to pump a fluid forwardly (backwardly) along the screw rotation, it is called "forwarding (backwarding) stagger". According to these two main geometrical characteristics, ptKDs are typically classified to four types, namely backwarding-stagger-forwarding-tip (BS-FT), forwarding-stagger-forwarding-tip (FS-FT), backwarding-stagger-backwarding-tip (BS-BT), and forwarding-stagger-backwarding-tip (FS-BT). Figure 1 shows the top views of the four typical types of ptKDs. These four types of ptKDs in Fig. 1 composed of five blocks in one element are chosen for our investigation.

Numerical Simulations

We assumed that the melt flow is pseudo-steady to the screw rotation. Incompressible flow of a non-Newtonian fluid of the Cross model is solved. Temperature dependence of the viscosity is modeled by exponential law. Stick boundary conditions on the moving screws and on the barrel walls are imposed. Temperatures on the barrel and on the inlet are set to be 473 K and 453 K, respecively. Based on the flow field, Lagrangian trajectories of tracers are computed.

Evaluation of Mixing

Tracers, which are initially distributed at a certain section, are advected by the flow field untill they reach the outlet. Along each trajectories, several quantities including residence time, expansion rate, flow type, and stress magnitude, are computed.

Results and Discussions

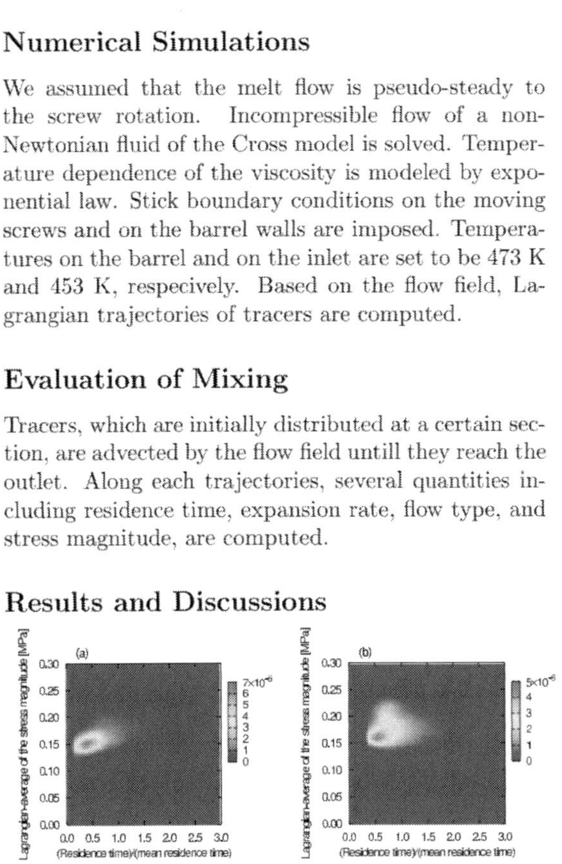

Figure 2: Joint probability density of the residence time and Lagrangian-history average of the stress magnitude under the volumetric flow rate of 60 cm^3/s and the screw rotation speed of 200 rpm: (left) FS-BT ptKD, and (right) BS-FT ptKD.

In counter-pump combinations of pitched-tip angle and disk-stagger angle, namely forwarding-stagger-backwarding-tip (FS-BT) and backwarding-stagger-forwarding-tip (BS-FT), inter-disk axial transport is mainly occured at the pitched-tip. Enhanced axial mixing by pitched-tip in addition to circumferrential mixing along screw rotation promotes the gross distributive mixing during extrusion in FS-BT and BS-FT. Moreover, rather homogeneous stress histories in FS-BT and BS-FT ptKDs were observed (Fig.2). As a result, main advantages of FS-BT and BS-FT are the gross distributive mixing and controlled dispersive mixing. It is expected that, by adjusting the combinations of the pitched-tip angle and the disk-stagger angle in FS-BT and BS-FT, the optimal balance of distributive mixing and dispersive mixing are found for various extrusion process.

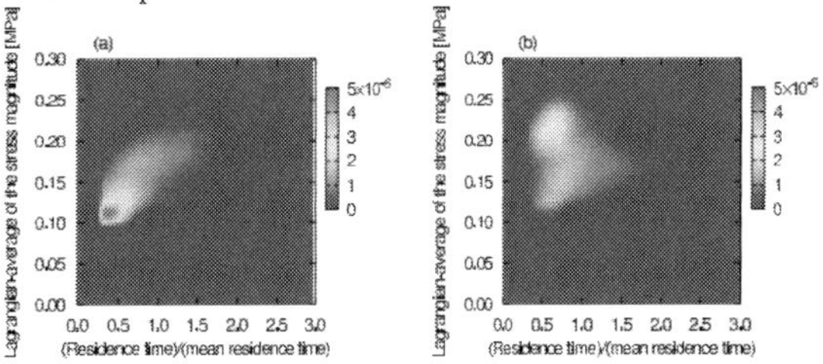

Figure 3: The same plot as Fig.2 but for (left) FS-FT ptKD, and (right) BS-BT ptKD.

In co-pump combinations of pitched-tip angle and disk-stagger angle, namely forwarding-stagger-forwarding-tip (FS-FT) and backwarding-stagger-backwarding-tip (BS-BT), flow pattern and axial transport is inhomogeneous. Tracer trajectories are splitted into the pitched-tip regions and the other. In FS-FT ptKD, negative pressure drop is generated due to the forward pump ability of the forwarding-stagger and forwarding-tip. Faster axial transport occured at the forwarding-tip regions, which results in large fluctuation of the residence time (Fig.3(left)). In BS-BT ptKD, the trajectories are to split as in FS-FT ptKD, but their group velocities are comparable, thus the fluctuation of the residence time is rather small. As a result of the splittd path, the history of the stress becomes inhomogeneous (Fig.3(right)).

The results reveals that the different flow pattern by four typical ptKDs. The geometrical degress of freedom of ptKDs provides a good control of the distributive and dispersive mixing. It is expected that the results above gives a basic guideline to select a suitable type of ptKDs for various processing and to the optimization of the operation along with the geometrical parameters of ptKDs.

Acknowledgments

The computation was mainly carried out using the computer facilities at Research Institute for Information Technology in Kyushu University, and Supercomputing Division of Information Technology Center in the University of Tokyo.

References

[Shigeishi *et al.*(2006)] Shigeishi, T., *et al.* "Development of special kneading screw in twin screw extruder", *Proceedings of 22nd Annual Meeting of the Polymer Processing Society*, G04–12 (2006)

ALTERNATIVE EXTRUSION CONCEPTS – ONE ANSWER TO THE DEMAND OF HIGH EFFICIENT SINGLE SCREW EXTRUSION

Prof. Dr.-Ing. J. Wortberg[1*], Dipl.-Ing. M. Bussmann[1], Dipl.-Ing. M. Großmann[1]

[1]Universtity of Duisburg-Essen, Institute for Product Engineering, Engineering Design and Plastics Machinery, Duisburg, Germany – marcel.grossmann@uni-due.de

The focus on increasing plasticating capacity of single screw extruders led to the development of several alternative prototype extrusion concepts. All concepts were designed to meet the following demands: Optimal (low) melt temperature; high plasticating capacity; lower process length; low losses by convection; flexible in processing a wide range of materials.
The separation of solid material and melt within a short transition zone by a melt permeable barrel leads to high plasticating capacities and lower melt temperatures due to a significant reduction of viscous heating. An almost exclusive and constant solid bed in the screw channel intensifies and accelerates the plasticating process significantly.
Using a separate driven coaxial feeding screw the decrease of specific throughput especially for materials with varying physical properties and higher screw speeds can be reduced in a wide range. To be independent of material properties the capacity of vented extruders can always be on maximum, particularly with regard to the varying bulk density of regrind material.
In contrast to almost all plasticating concepts which use electrical mechanical power for the conveying and transition process another prototype was developed using natural gas as primary energy source. According to a calculation the operator can almost save half of the energy costs if he is using natural gas than electricity. By the development of a heat exchanger up to 90 % of the total energy amount which is needed for conveying and melting is generated by natural gas. The high-energy exhaust can be used to preheat the solid material in the hopper. Thereby energy efficiency as well as the environmental balance can be improved significantly.
Practical results of all alternative extrusion concepts document a realistic chance to make the extrusion process much more efficient in the future than estimated so far.

Introduction

What does improving perfomance of single screw extrusion mean in detail? In the current age of emission regulations and increasing energy costs improved performance can be described by the following demands:

- Increasing throughput
- Improving quality (consistency)
- Flexibility
- Minimizing losses
- Sustainability

For example the increase of throughput of the pipe and sheet extrusion industry has reached more than 400 % over the last 40 years. Enlarged process lengths from 24D up to 40D have improved plasticating capacity for constant srew diameters [Stieglitz and Schneider 2006]. However flexibilty gets lost with a view to process different materials with one and the same screw design. Varying specific enthalpies as well as unequal physical properties of bulk materials cause problems relating to the stability of specific throughput and melt temperature.

Increasing throughput is more often reached by higher screw speeds especially in thermoforming and flat/blown film extrusion. A long time resistance to high screw speeds was widespread. The potential incomplete melting of partial crystalline materials and the concern to high melt temperatures caused by the enlarged induced dissipation energy due to high screw speeds have been two essential arguments. Several research activities concerning the influence of high screw speeds to material properties did show no significant degradation of molecular structures [Pohl 2003]. Processing Polystyrene with a 75 mm extruder and a screw speed of 1100 1/min confirm this assumption by unchanged good product properties [Roth 2006].
Nearly all research activities are focussed on improvement of single procedural elements like screw and feeding zone design, barrel heaters, mixers and so on. Just a few ideas for complete new and alternative extrusion concepts have been developed so far [Jones 1987, Grünschloß 2001, Pohl and Pape 2001, Barr 1998]. Research activities of the Institute of Product Engineering are traditionally focussed on the development of new extrusion technologies. In addition to an increase of throughput, aspects of flexibility, energy balance and sustainability are

important decision factors for the design of new prototype concepts.

Prototype KoAx-S-Truder

The need of a flexible high efficient plasticating concept with an influenceable specific throughput and low rate of heat dissipation results in the development of the KoAx-S-Truder which are shown in figure 1.

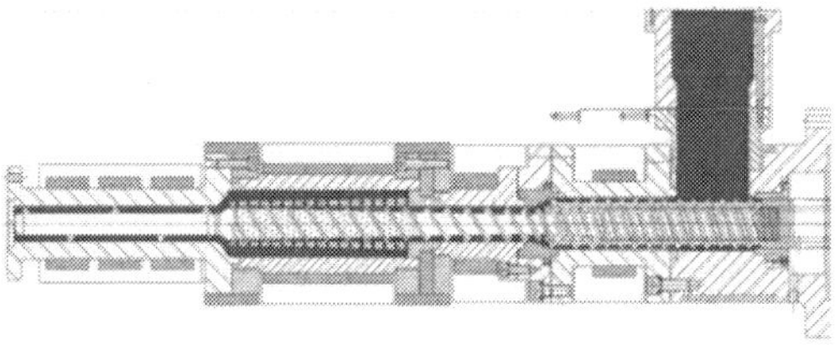

Figure 1 - Prototype of KoAx-S-Truder (Ø 50 mm)

This patent pending prototype has two main features which offer high plasticating capacities despite of its short transition zone and low L/D ratio of 20D. The specific throughput can be influenced by a separately driven coaxial feeding screw (figure 2).

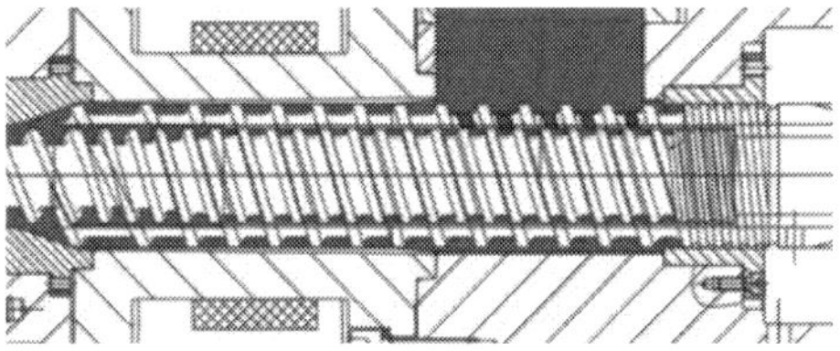

Figure 2 - Coaxial feeding screw (Ø 70 mm)

This 70 mm feeding screw allows diversifying the spec. throughput by a separate control of rotation speed. Elongated holes in the screw channel provide a second material flow between both coaxial screws. The inner surface of the feeding screw can be designed smooth or with grooves, both types have been researched. At the end of the grooved feeding section the material flow between feeding screw and cylinder is brought together with the flow between both screws. Keeping the rotation speed of the plasticating screw constant the spec. throughput can be controlled by varying the feeding screw speed.
The high plasticating capacity is the result of a solid/melt separation in the transition zone by a melt permeable cylinder. By separating solid material from melt two fundamental advantages are achieved. First of all plasticating capacity increases through the lower proportion of melt in the screw channel.

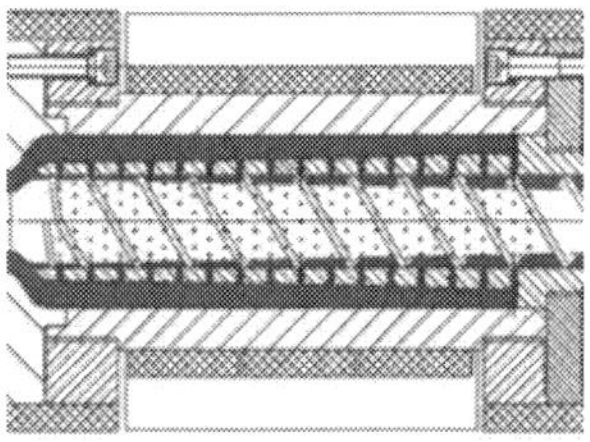

Figure 3 - Solid/melt separation

Directly after the formation of a thin melt film at the cylinder surface the melt is forced to flow out of the screw channel at the next bore line. In this way no melt pool is generated and a maximum surface of solid material improves the melting process. In addition already molten material does not get further heated by the rotating screw particularly for high screw speeds. Thereby overheat of melt temperature can be avoided for a wide range of screw speed.

Experimental results

Until now different materials have been tested for maximum throughput. In every experiment the spec. throughput is tried to be kept on the highest level and constant over the whole speed range by varying the feeding screw speed. The following results of plasticating tests were reached using a double flight plasticating screw and a grooved feeding screws with a grooved inner surface.

A challenge for the short plasticating concept is processing a high molecular material like a PE 100 which is often used for extrusion of pipes. Results of this material (Eltex Tub 121 N20259) are shown in figure 4. The just 10 % decrease of specific throughput permits a throughput of 270 kg/h for a screw speed of 330 1/min. Although a dynamic mixing device was used in this test, the melt temperature was almost constant between 180 and 190 °C. Such high plasticating capacities of a PE 100 are hardly to achieve by a standard extruder with same screw diameter. The temperature was rather to low and for screw speeds more than 300 1/min these results seem not to be reproducible with a standard 50 mm extruder.

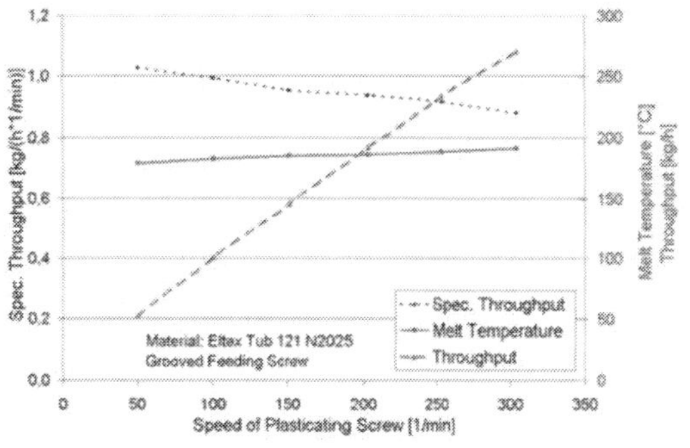

Figure 4 - Plasticating capacity processing PE 100

Stablization of the spec. throughput for higher screw speeds in combination with an increased plastication capacity of a solid/melt separation makes manageable melt temperatures development possible, which is the major problem in today's high speed extrusion processes. According to the experiments with PE100 in figure 4 results of processing a PE-LLD (Dowlex NG5056G) with a MFR of 1,1 g (190 °C/2,16 kg) compared to a standard 50 mm extruder (barrier screw; 28D) are shown in figure 5.

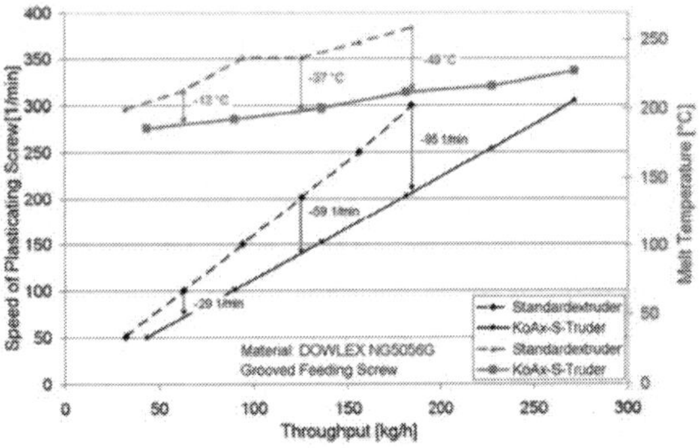

Figure 5 - Reducing screw speed and melt temperature (PE-LD)

The spec. throughput of 0.85 can be kept constant within the whole speed range (max. 330 1/min). Keeping in mind that the main task of this prototype is the reduction of screw speed and melt temperature by a simultaneous increase of throughput the following results validate this operating behavior. Processing a throughput of 180 kg/h the screw speed of the KoAx-S-Truder needs to be 95 1/min (-30 %) less compared to a standard extruder. Because of lower speed and higher spec. throughput melt temperature is decreased about 50 °C.

Alternative plasticating concept based on natural gas

All known extrusion concepts have one procedural aspect in common. The melting energy is always generated by friction power respectively dissipation. Considering today's increasing energy costs and efforts to conserve natural resources alternative plasticating concepts may help to reduce costs and emissions at the same time. The goals of the consecutive study are therefore verbalized as follows: Development of an alternative plasticating concept which reduces energy costs and carbon dioxide emissions at the same time.

In cooperation with e.on Ruhrgas (Essen), WEMA (Luedenscheid), Motan (Buende) and Kreyenborg (Muenster) several concepts based on natural gas have been developed and designed. Natural gas driven barrel heaters and IR-rotary dryers for examples are used in several plastic processing factories. These research projects have shown the potential of natural gas as an alternative energy source for the plastic processing industry.

In conventional extrusion processes the energy costs represent round about 5-10 % of the total product costs. Using the energy of natural gas in terms of heat without conversion into electrical power, conversion and transport losses can be reduced at the same time. Figure 6 shows the specific amount of primary energy to create process heat of 1 kWh by electrical energy compared to natural gas.

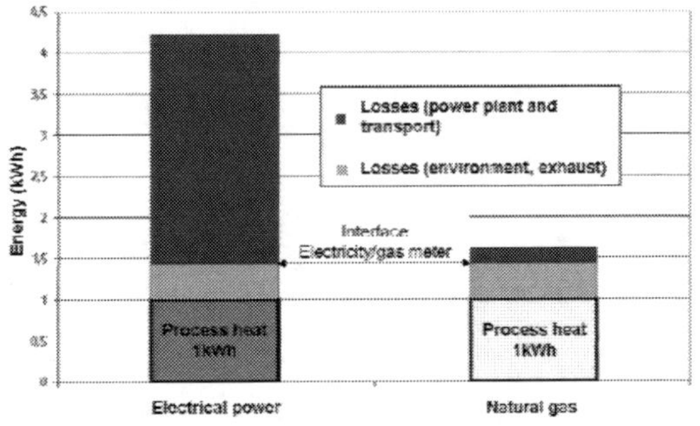

Figure 6 – Plasticating concept [Adelt 2008]

The conversion and transport losses to provide 1 kWh electrical power are more than 10 times higher compared to the primary energy source natural gas. Concidering the overall energy costs once more, up to 70 % can be reduced by the use of natural gas compared to electrical power. In addition natural gas is a relatively environment-friendly energy source which causes low carbon dioxide emissions during combustion.

To clarify the potential reduction of energy costs an extrusion line with a throughput of 250 kg/h is assumed with an annual production capacity of 2.000 tonnes. Considering plasticating costs of 0,025 €/kg (0,033 $/kg) the annual overall costs only for plasticating represent 50.000 Euro (66.700 $) whereof 70 % could be reduced by the use of natural gas.

In common extruders the procedural demands of heating, shearing and compression of a primary solid material into melt is realized almost totally by electrical drive power. To use natural gas for the plasticization the internal energy must be released. The combustion of the gas provides an opportunity to generate heat energy which can be used to melt the plastic. According to this process step a plasticating concept should be developed which prohibits any dissipative heating caused by electrical mechanical power.

Concept

These demands led to the development of a prototype plasticating concept which is based on insertion of heat energy by convection as well as by thermal conduction. The developed concept consists of two coupled heat exchanger, a solid conveying unit and a melt pump, shown in figure 7.

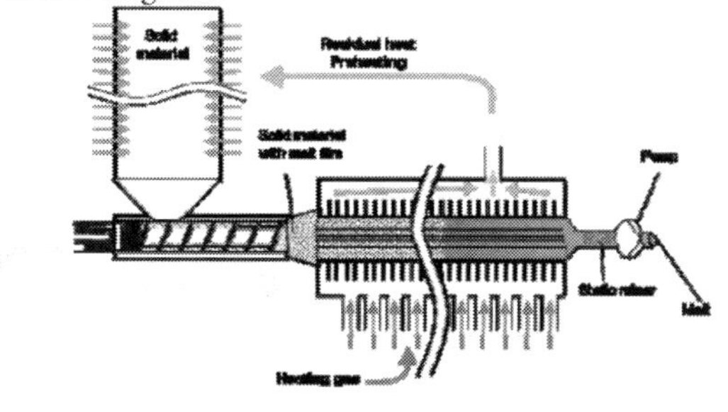

Figure 7 – Plasticating concept

The solid material is stored in the first heat exchanger, where it gets pre-heated. By pre-heating the solid material a significant increase of enthalpy can be realized which shortened the following plasticating process again.
Another much more important argument for pre-heating consists in the exhaust gas flow of the second gas driven heat exchanger. The high temperature of this exhaust gas flow makes a direct pre-heating of pellets possible and minimizes energy losses of the total concept significantly. Pre-heated the materials gets into the conveying unit. It consists on a grooved barrel to ensure efficient conveying of solid material. The directly driven screw is designed to generate a certain build-up pressure which is necessary to force the material passing the following second heat exchanger. The deep-flighted and short screw design makes a minimal energy insertion due to dissipation possible. The small proportion of melt which is generated so far helps the solid material flow into several seperated flow channels of the second heat exchanger. The second heat exchanger gets its thermal energy from a gas-driven barrel heater. To plastisize the material within this process unit completely the cylinder diameter is expanded to get a large surface for heat transfer. To improve the heat transfer the total mass flow is seperated by several singular flow channels within the heat exchanger (figure 8).

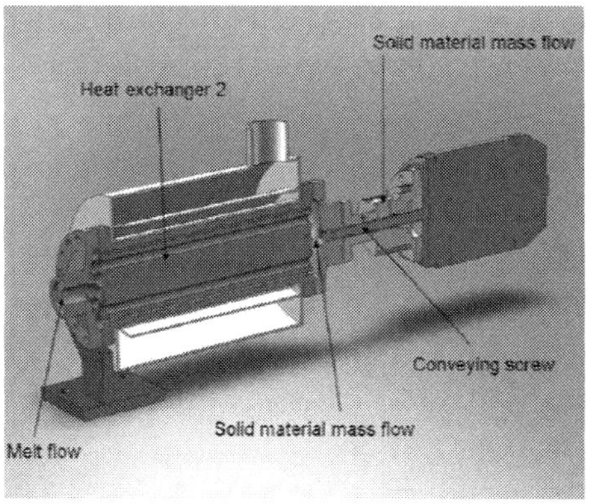

Figure 8 – 3-D Model of gas-driven heat exchanger

After melting is completed the melt flows are combined again and get homogenized by a static mixer. A melt pump at the end of the plasticating concept is necessary to overcome the pressure loss of a following die.

Energy balance

To calculate the proportion of energy from electrical mechanical power and natural gas used for this plasticating concept an energy balance seems to be helpful. The concept is developed for a proportion of the energie sources natural gas 70 % to electrical power of 30 %. For the total concept (figure 9) the energy balance of the defined calculatory area can be described as follows:

$$P_S + P_{\Delta p} + \dot{Q} = \dot{m} \cdot \Delta h + \dot{Q}_V + \dot{V} \cdot \Delta p \qquad \text{Eq.: 1}$$

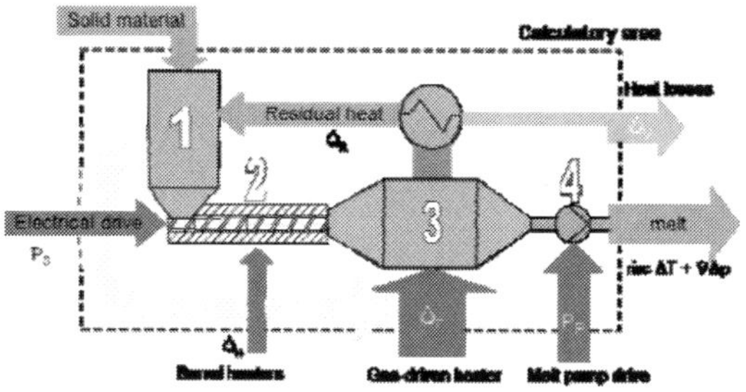

Figure 9 – Energy balance of calculatory area

P_s and $P_{\Delta p}$ are induced by electrical power and necessary predominant to build up pressure.
Heat energy Q is almost exclusive induced by the heat exchangers (Q_E) working with energy of combusted natural gas.
For a given melt temperature the increase of enthalpy can be calculated. It combines the temperature increase due to friction of conveying with the increase from both heat exchangers.

$$\Delta h = \Delta h_{fri} + \Delta h_{WT1} + \Delta h_{WT2} \qquad \text{Eq.: 2}$$

Only Δh_{fri} is caused by electrical power P_s. P_s was calculated to overcome the pressure loss of the gas-driven heat exchanger, the pressure loss of a static mixer as well as the primary pressure of the melt pump.

$$P_S = P_{fri} + P_{\Delta p_P} = \dot{m} \cdot \Delta h_{fri} + \dot{V} \cdot \Delta p_P \qquad \text{Eq.: 3}$$

The pressure loss of the heat exchanger was calculated for a simplified laminar melt flow to merly 50 bar. P_{fri} seems to be proportionally small in spite of high screw speeds, considering the deep-fligthed screw with a diameter of 25 mm and length of 8D. The prototype was designed to plasticate polyolefins with a throughput of 100 kg/h. Theoretical calculations promise to deliver succesful results and the first prototype (figure 10) is already designed for practical tests.

Figure 10 – Prototype Extrusion concept

Conclusions

Experimental results of the KoAx-S-Truder document that an alternative way of thinking can be one answer to the question for more efficient extrusion concepts. Throughputs of 300 kg/h for PE-LD, PE-HD, PE-LLD, PE100 and PP (Homopolymer) within the melt permeable cylinder with a length of only 350 mm are currently limited only by the installed drive. A further increase of screw speed seems to be successful to reach a throughput of more than 500 kg/h (50 mm screw 24D) without overheating the melt. The coaxial feeding screw implicates flexibilty to process any kind of solid plastic material with a high and constant spec. throughput. Particularly for degassing applications or processing inhouse recycling material with varying proportions and physical properties a coaxial separate driven feeding screw can be of use. At present the Institute of Product Engineering is working on a High-Speed 35mm directly driven extrusion concept with a solid/melt separation which combines high throughputs, compact design and energetical efficiency with improved homogeneity reached by an innovative new static mixer.

Until today natural gas as primary energy source is not widespread the the plastic processing industry but the advantages regarding to lower emissions and energy costs is going to make it more and more interesting. Theoretical investigations in combination with the first practical tests have shown the potential of an alternative extrusion concept based on natural gas and heat exchangers. The present development to stricter emmision regulations and increasing energy costs will help to improve accaptance of natural gas also for new applications within the plastic processing industry.

Acknowledgements

This project is supported by the
German National Science Foundation (Deutsche Forschungsgemeinschaft, DFG)
Project-No. WO-302/33-1

References

Proceedings: Stieglitz, H., Schneider F.: „Das 36D Verfahrenskonzept eröffnet neue Potenziale in der Rohrextrusion",VDI Tagung - Der Einschnecken-extruder von Morgen, Baden-Baden (2006)

Dissertation: Pohl, T.: „Entwicklung schnelldrehender Einschneckensysteme für die Kunststoffverarbeitung auf Basis theoretischer Grundlagenuntersuchungen", Universität-GH Paderborn, Paderborn (2003)

Proceedings: Roth, M.: „Entwicklungsstand der Hochgeschwindigkeitsextrusion – erfolgreiche Anwendungen", VDI Tagung - Der Einschneckenextruder von Morgen, Baden-Baden (2006)

Proceedings: Jones, D.: "New concept of extrusion: The rotary extruder", RETEC Polyolefins V, Society of Plastics Engineers, Houston USA, 151 – 167 (1987)

Proceedings: Grünschloß, E.: „Helibar® - Ein neuartiger Einschnecken-Extruder mit verbesserter Plastifizier- und Förderleistung", 17. Stuttgarter Kunststoff-Kolloquium, Stuttgart (2001)

Journal: Pohl, T., Pape, J.: „Prozessinnovationen stärken die Extrusion", Maschinenmarkt J. 47, 46-47 (2001)

Proceedings: Barr, R.: "New Concept Maximizes Melting", Polymers Laminations and Coatings Conference, Vol. 1, 469 (1998)

Proceedings: Adelt, M.: "Aktuelle Entwicklungsaktivitäten für die Kunststoffindustrie", Fachseminar zur Fakuma 2008, Erdgas in der Kunststoffvereinbarung, Lindau (2008)

CONCEPT FOR DIE MODULARISATION APPLIED ON PROFILE EXTRUSION

W. Michaeli[1], C. Windeck[1] and L. Ederleh[1]*
Y. Queudeville[2], T. Ivanov[2]
M. Probst[3] and S. Elgeti[3]

[1]Institute of Plastics Processing at RWTH Aachen University, Aachen, Germany - zentrale@ikv.rwth-aachen.de; windeck@ikv.rwth-aachen.de; ederleh@ikv.rwth-aachen.de; phone: +49 (0) 241 80-93806; fax: +49 (0) 241 80-92262
[2] Foundry institute, RWTH Aachen University, Aachen, Germany – y.queudeville@gi.rwth-aachen.de; t.ivanov@gi.rwth-aachen.de; phone: +49 (0) 241 80-95880; fax: +49 (0) 241 80-92276
[3]Chair for Computational Analysis of Technical Systems, RWTH Aachen University, Aachen, Germany - probst@cats.rwth-aachen.de; elgeti@cats.rwth-aachen.de; phone: +49 (0) 241 80-99900; fax: +49 (0) 241 80-99910

In general, the level of modularisation in profile extrusion dies is quite low. Usually, only parts like flanges, heater bands or sensors are standardised. All other parts are designed individually. An increasing level of modularisation could accelerate the designing process and lower the costs for dies that consist of standardised components. Hence, it is the aim to develop a concept that helps a die manufacturer to decide which and how modules can be standardised. A group of institutes at the RWTH Aachen University therefore work together in order to develop a general approach for die modularisation. This concept consists of three phases: In the first phase, the range of products and production processes are examined. The results of this phase are the most promising die types of a company that are best suited for modularisation. In the second phase, one of those die types is examined more closely with emphasis on the functional dependencies of the different die components and an evaluation of the possible components which can be standardised. The third phase is about the actual design of standardised designs of a specific component and an evaluation whether these designs represent a wide area of profile variations. By testing this approach for different production processes, the general validation is shown. One of the chosen processes is the plastics profile extrusion. For this process, the die part that pre-distributes the melt in the die is standardised so that one standard component can be used for different variations of an extruded profile. The part of the flow channel close to the die land is still individually designed based on an automated geometry optimisation algorithm.

Introduction

The processing technologies for mass productions have the advantages of high throughput and therefore a low price per unit as well as a highly automated production. The costs for the die design and manufacturing are relatively low compared to other costs like material price and machinery. On the other hand, these technologies for mass production lack the ability to react fast on individual customer requests. For small lot sizes, it is challenging for companies using these manufacturing technologies to produce economically. But the ability to react fast on individual orders is certainly one of the best arguments to stay competitive in the market.
Therefore, it is the aim to improve the flexibility of forming technologies for mass production. The percentage of costs for dies increases with smaller lot sizes. This effect can be counteracted by developing modular dies.
In this project, a systematic modularisation of forming dies enables to produce individual products faster and cheaper in a high quality is examined. Basically, the focus is to enhance the amount of components in the dies that can be used for more than one product. This concept for the layout of modular dies will enable companies to bridge the existing gap between individual manufacturing and mass production (Queudeville et al., 2009).
The concept for designing modular dies has to consider technological and economical conditions to enable an individualised mass production.
Furthermore, the general concept has to be transferable on dies of different production technologies. In our research, we developed the concept and adopted it to two different production technologies for mass production: die-casting of light metals and profile extrusion of plastics. Although the processes are quite different, they have an increased demand for smaller lot sizes in common. A successful practice of modularising the die parts of both processes shall support the generality of the concept.
In this context, we focus the concept of the application of a profile extrusion die.

State of the art

The design of profile extrusion dies is an iterative process (Figure 1).

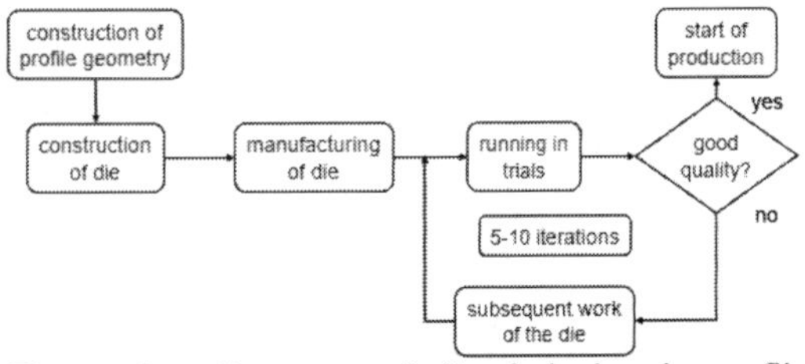

Figure 1 – Sequence of die designing in profile extrusion

Based on the desired profile geometry, the designer constructs the channel geometry according to his experience as there do not exist analytical formulae to calculate the flow through the profile die. After the construction of the die, so called running-in trials are carried out: The die is connected to an extruder and the quality of the extrudate is examined. As long as the quality does not match the required quality standards, the die maker changes the flow channel geometry in an iterative trial and error approach. This is expensive and time-consuming, especially as it mainly depends on the expert knowledge of the die maker. After 5 or more iterations of changing the flow channel geometry, the die geometry is usually good enough to be used in the profile production.
The design of a profile extrusion die is highly individual. Each profile is formed in its own, individual designed die. Often, however, a company produces quite a lot of similar profiles which differ only slightly from one another. This interferes with the increasing demand of the market for small lot sizes in low developing time and acceptable prizes.
Most of the components in profile extrusion are individually-designed components. Besides the die, this applies for the heating of the die, the adapter connecting die and extruder, core-inlays and so on. But there are also some standardised components like the extruder itself and temperature or pressure sensors.
The components of dies can be categorised in certain groups. This is shown in Figure 2. These groups differ according to their degree of modularisation. Standard modules are components with the highest degree of modularisation. These modules are defined in all characteristics without being restricted to a certain area of application. The group of standard modules can be subdivided in standard parts that are standardised by internationally accepted organisations (DIN, ANSI, BSI etc.) and in company-specific standard parts.
Another group consists of modules with individual subsequent work. These components have a high modularisation potential in particular characteristics (e.g. outer dimensions), but other characteristics have to remain individually-designed.
The last group contains all those components with no or minor modularisation potential. There is no benefit in modularising these components, so those have to be designed individually.
By following the die modularisation concept the number of company standard parts and modules with individual subsequent work shall be increased.

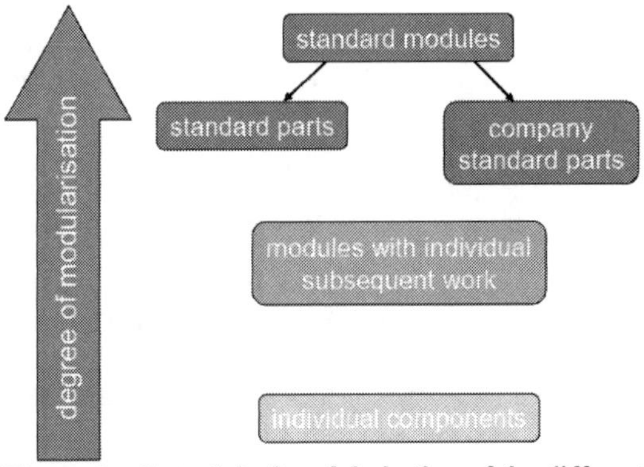

Figure 2 – Potential of modularisation of the different die components

One general approach for die modularisation and validation is described by Boos, 2008. The work attempts to transfer modularisation tools on the die making industry. His procedure, however, is taken as basis for our investigations. We expand the concept as it lacks of precise application instructions.

Results and Discussion

In order to match the demand of the market for individual products, we developed a concept that enables the die making company to evaluate which dies and, to be more precise, which die component has the highest potential for standardisation. Furthermore, guidelines are shown how components should be standardised and how components can be designed in an optimal way if those have to remain designed individually.
The concept consists of three phases:
In phase 1 (initial phase), the range of products and production processes is examined. The result at the end of this phase is a conclusion whether modularisation is reasonable as well as a listing of die types with the highest potential.
In phase 2 (analysis phase), one of those die types is chosen and the interrelation of the different components of this die type and the environment is investigated more closely. At the end of this phase, there results a diagram of all components and their degree of modularisation potential.
In phase 3 (design phase), one chosen component is designed in standardised sizes or the proceeding of a more systematic individual design is applied. At the end, a validation is carried out whether the new module can be implemented into the die or has to be redesigned.

In the following, we focus on the application of phase 3 on profile extrusion dies:

At the end of phase 2, there is an overview about the components of a die type arranged in a system grid similar to Lindemann, 2005. The system grid is a description of all components depending on their active and passive value. The values represent a qualitative extent of how sensible a component reacts to other changes in the system (passive) and how a changed component influences the whole system (active). The system grid is a description of all components depending on their active and passive value.

For the example of a regular profile extrusion die, a system grid is shown in Figure 3. Some of the components are already standardised; mainly those which have a low active and a low passive value. This means they can be modified without causing a big impact on other components. For the modularisation of further components, it is the next step to design components with a high active value and a low passive value (active elements). The interaction of such a component is quite low so it seems to have good conditions for standardisation. In our case, we select the inlet of the flow cannel for further investigations. Components that have high passive values due to the strong influence from other components are hard to standardise (passive elements). The design progress of those components, however, can be executed in a more standardised approach. Exemplary, this is shown in this project by enabling an automated flow channel optimisation to improve the die design. The borderline of high modularisation potential to low potential is soft, so also components with high passive values and few dependencies can be standardised in certain characteristics, whereas other characteristics which result in the high passive value can be kept individual.

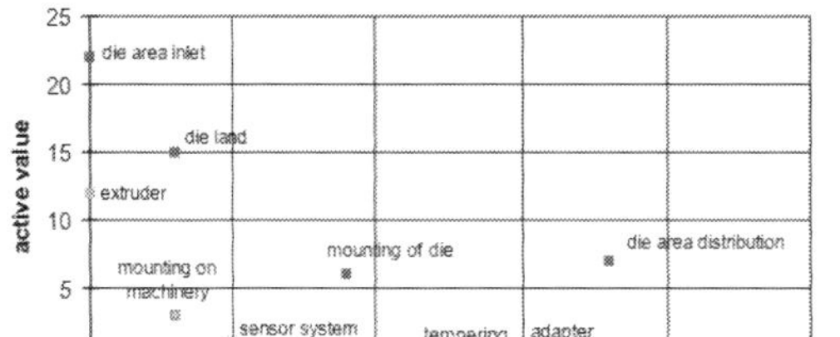

Figure 3 – System grid of the components in a profile extrusion die

In the design phase, the die is observed from the components view. Therefore, this phase can be passed through several times for different components.

In order to standardise the inlet, it is necessary to determine a shape of the flow channel and a shape of the interface to the next component that is valid for different profile geometries. Alike, the shape of the interface has to be determined out of the geometry of the profile so that a general approach can be established.

In the first step, we take a L-shaped profile geometry. The length of the module inlet is about one third of the total die length. This length is usually sufficient to pre-distribute the melt without stressing the material too much caused by big changes in section. The remaining two thirds of the die length consist of the component die area distribution and die land.

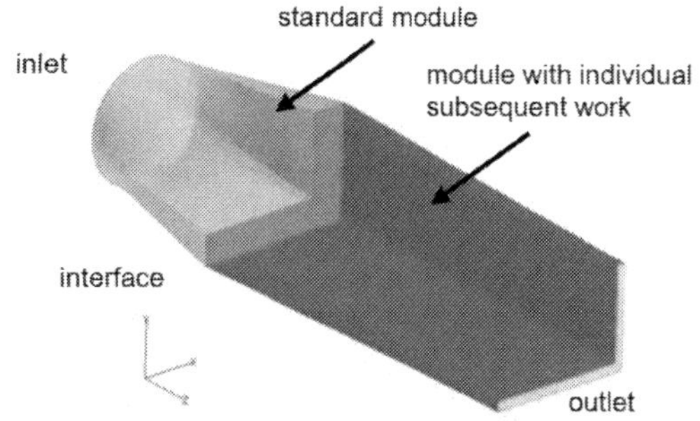

Figure 4 – Die modules for a L-shaped profile

From the outline of the L-profile, the outline of the interface of the inlet is calculated based on a formula that takes width and height of the arms of the L into account. Furthermore, the center point of shear is regarded. By using this formula the outline is scaled up differently for width and height. Critical, however, are different heights in the arms: When the height of the arms differ more than 10 percent, the formula is not valid anymore as the pre-distribution is so unequal that the flow cannot be corrected anymore in the following flow channel sections. In that case; a homogeneous velocity distribution is only achievable with separated flow channels for each arm until the die land. By separating the melt flow for each arm it is easier to control and improve the flow in each arm without having influence on the other arm.

With the application of the formula, we get a shape of the interface of the inlet for each L-profile. The flow channel geometry between outlet and interface is designed using morphing algorithms. The next step is to reduce the number of shapes and set discrete steps. In order to detect whether a velocity distribution through different outlets is homogeneous simulations are made. In the computational fluid dynamics (CFD) simulations, the shape of the inlet and its interface are kept constant and the outlet geometry is changed by increasing or decreasing the height and width of an arm of the L-profile. As a result, the width of an arm at the outlet can be changed in a wide range. This is shown in Figure 5. The quality is defined as percentage of the averaged velocities in the two arms. That means a quality of 1 is best. The interface is designed based on an outlet shape of 30 mm width each arm and 1,5 mm height.

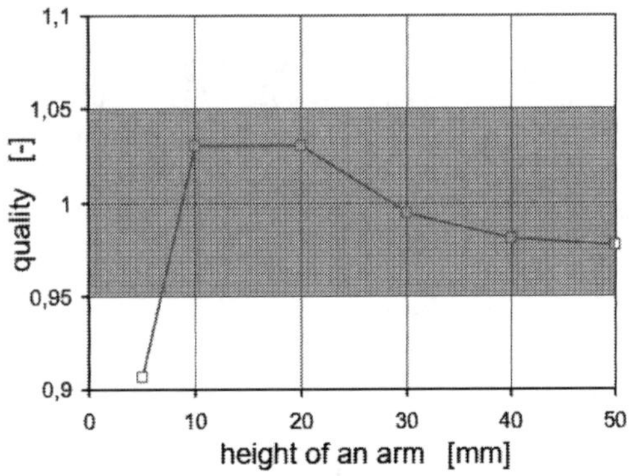

Figure 5 – Quality of the velocity distribution with changing height of an arm and constant interface

On the other side, the width can only be changed in a narrow range, otherwise the velocities differ massively (Figure 6).

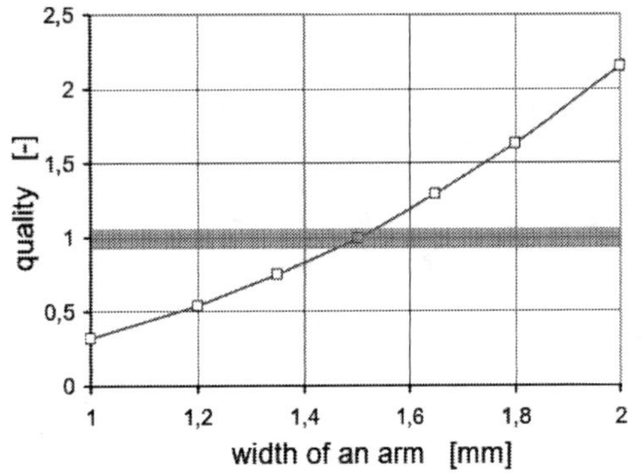

Figure 6 – Quality of the velocity distribution with changing width of an arm and constant interface

This means that for L-profiles with different heights it is possible to design one inlet module with a standardised interface for those profiles. Different widths, however, are hard to produce with only one inlet module. As it is easier to apply a die that is designed for a big size of a certain profile shape on a smaller size than the other way round, the first discrete design of an interface is taken from the biggest version of the examined profile type. Then the discrete steps are chosen in a wide range for different widths and narrow for different heights. Then next step will be to examine how this can be transferred on other geometries.

For individually designed parts, it is recommended to use a standardised approach. One option is to use automated optimisation algorithms in order to find the best solution. This can be applied on the die area distribution and the die land to get a flow channel geometry with an homogeneous velocity distribution at the outlet.

It is important to remember the hierarchy of different components. The newly designed module forces other components to adopt its interface design on common interfaces with other components.

The die area distribution and die land have to be designed individually. Therefore, an automated geometry optimisation algorithm is developed and applied on the flow channel geometry. This component is responsible for the precise distribution of the melt, so the design has to be made individually. In this case, the interfaces are already set by the profile geometry at the outlet and the design of the interface of the inlet part. The rest of the flow channel geometry is created by morphing the path between outlet and interface and then calculated by an automated optimisation algorithm that analyses the flow through the die of computational fluid dynamics (CFD-) simulations. This optimisation algorithm changes the shape of the flow channel geometry until a certain quality function is minimised. Here, the quality function is a mathematical description of a homogeneous velocity distribution at the outlet. First investigations on optimising simple profile channels (e.g. L-shaped profile) show already reasonable improvement of the velocity distribution at the outlet (Figure 7). In order to improve the distribution in the edge of the profile, the geometry description needs a finer mesh.

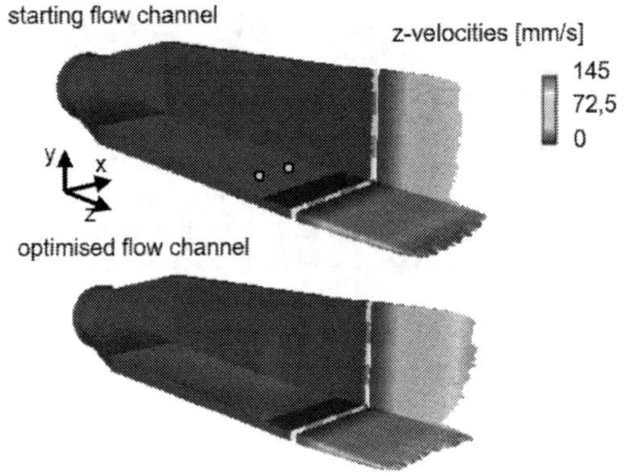

Figure 6 – Velocity distribution for a starting and an optimized flow channel geometry

Conclusions

The modularisation of dies in profile extrusion and other processes is just getting started. The reason is the individual design of each die.

With the described concept, it is possible to detect areas of the design process and die components that have a high potential for modularisation. Components with small interaction with forming parts of the die are best suited for standardisation, some of these components are already standard parts. Forming parts of the die have to be examined closely. The die area inlet is quite similar for a variety of profiles, so here it might be useful not to use an individually designed component. This is exemplarily shown on a die of a L-shaped profile. Even though components like the die area distribution and die land can have new standard values for outer dimensions etc. some individually-designed areas remain as these areas are responsible for the precise melt distribution. An optimisation algorithm helps to design the flow channel geometry of these parts so that some of the cost intensive running in trials can be avoided.

Acknowledgements

The depicted research has been funded by the German Forschungsgemeinschaft (DFG) as part of the program Cluster of Excellence "Integrative Production Technology for High-wage Countries" at RWTH Aachen University.

References

Boos, W.: Methodik zur Gestaltung und Bewertung von modularen Werkzeugen, Apprimus Wissenschaftsverlag, (2008)

Lindemann U.: Methodische Entwicklung technischer Produkte, Springer (2005)

Queudeville Y., et al.: Decision and design methodologies for the lay-out of modular dies for high-pressure-die-cast-processes, Materials Science Forum Vols. 618-619, p. 345-348 (2009)

CLOSED-LOOP QUALITY CONTROL OF THE PROCESSING OF UNDRIED POLYCONDENSATES USING MELT DEGASSING SINGLE-SCREW EXTRUSION

Prof. Dr.-Ing. Dr.-Ing. E.h. W. Michaeli[1], Dipl.-Ing. L. Ederleh[1]*, Dipl.-Ing. H. Seidel[1]

[1]*Institute of Plastics Processing at RWTH Aachen University (IKV), Germany – zentrale@ikv.rwth-aachen.de*

Processing undried polycondensates as polyethylene terephthalate (PET) or polyamide (PA) using a melt degassing extruder is a real alternative to the conventional energy- and time-consuming pre-drying process. Due to the industrial demand of processing recycled and low priced materials the moisture content and/or the resin mixture can substantially vary from batch to batch. This leads to a deviant molecular weight degradation and an unsteady final product quality. The paper presents an approach to compensate these variations in the feedstock in a closed-loop control. It is based on the direct influence of the degassing pressure on the moisture evaporation of the melt. The product quality is estimated online in the extrusion process (via melt viscosity or pressure drop in the extrusion die) and is controlled by the melt degassing pressure in the extruder.
The installation of the closed-loop quality control with a 60 mm-single-screw extruder is realised stepwise. In reference experiments the influence of the polymer and the moisture content in combination with the degassing pressure on the product quality parameters are analysed. On the basis of size exclusion chromatography (SEC) and Intrinsic Viscosity (IV) measurements, the processing range of the control is identified.
Due to the complex control process with its high reaction time and inertia, the adaption of the parameters of the PID controller are done by analysing the system reaction on an instantly changed degassing pressure. The analysing is done by established graphical as well as by numerical methods. The following simulation of the automatic controller action highly correspond with the extrusion trials.
The validation is done by the occurring of disturbance variables (e. g. moisture content) at the extrusion process. The closed-loop control can provide a constant product quality in the PET- and PA 6-extrusion in spite of varying moisture content of the resin.

Introduction

When processing polyethylene terephthalate (PET) and polyamide (PA), the challenge of coping with its various possibilities of degradation has to be met. Mechanical, thermal, thermo-oxidative and especially hydrolytic degradation exert negative influence on the resulting properties of the product and have to be kept within certain limits (Michaeli et al, 2008, Schmitz, 2005).

Michaeli et al (2009) analysed the influence of the moisture on the PET and PA 6 molecular weight degradation as well as build up by using a high pressure capillary rheometer (HPCR). Although, the resin is melted by heat and discharged through a capillary using a constant shear rate. The measured pressure drop of the polymer flow can be converted to the melt viscosity which has a direct connection to the molecular weight (see also Menges et al, 2002). As shown in Figure 1, the pressure drop of the dry PET melt in the HPCR decreases only marginally due to the thermal molecular weight degradation. Because of the presence of water and high reaction rates, the moistly PET melt degrades already significantly during the heating time.

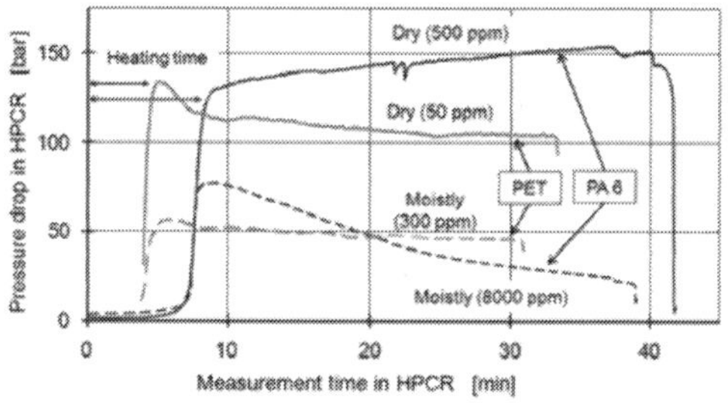

Figure 1 – Measurement of the pressure drop in a high pressure capillary rheometer (HPCR) with constant shear rate.

The reaction rate of the hydrolysis of PA 6 is lower and the degradation effects can be measured subsequently. On the other hand the dry PA 6 melt shows during the HPCR analysis an increasing pressure drop (with it the molecular weight) by reason of molecular weight build up.

Due to their hydrolytic sensitivity, polycondensates have to be processed with minimal moisture contents, usually less than 50 ppm (parts per million) for PET and less than 1000 ppm for PA 6. The presence of water in combination with high temperatures and sufficient residence time in the extrusion line otherwise result in rapid degradation of the polymer and unsatisfactory product quality.

Conventionally, the low moisture content is realised by an energy- and time-consuming pre-drying process. An alternative is the evaporation of the moisture directly out of the polymer melt by using a melt degassing extruder. Compared to the PET melt degassing process, Schmitz (2005) identified an additional energy consumption of the extrusion with pre-drying of up to 30 %.

As shown in Figure 2 for a single-screw extrusion process (comparable for other extrusion concepts), the undried resin has to be molten prior the evaporation in the following functional unit "degassing zone". A lower degassing pressure emphasises the flash vaporisation as well as the water diffusion in the melt degassing zone and results in a more effective dehydration.

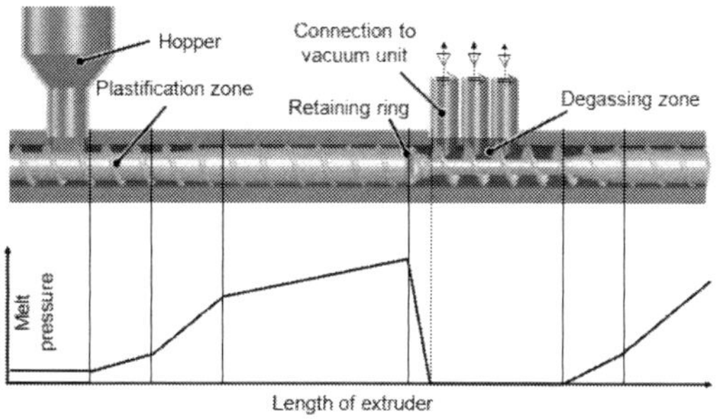

Figure 2 – Principle of a single-screw melt degassing extruder.

The unavoidable presence of water in the plastification zone results in hydrolytic degradation which depends – beside the temperature and the residence time – on the moisture content of the resin. The following melt degassing cannot compensate the hydrolysis completely. The bulk of the molecular weight degradation proceeds in the melting zone of the extruder. Keeping melt temperature and residence time in certain limits causes insignificant further degradation effects behind the degassing zone.

In production lines, the moisture content and/or the resin mixture can substantially vary from batch to batch. This leads for an otherwise constant extrusion process to a deviant molecular weight degradation and an unsteady final product quality. The closed-loop quality control in this paper balances these variations and provides a constant product quality in spite of varying moisture content of the resin.

Experimental

Set-up

Realising an automatic quality control in the extrusion process implies an online data logging of the final product property as a controlled parameter. The established quality criteria like average molecular weight and its distribution as well as the Intrinsic Viscosity (IV) for PET or the Relative Viscosity (RV) for PA 6 are not suitable. The measurements via Size Exclusion Chromatography (SEC) and solution viscosimetry are time consuming and offline.

Due to the higher flow ability of a polymer melt with reduced molecular weight, the degradation can additionally be evaluated online by a rheometer (inline or bypass) or by the pressure drop in the extrusion die. This enables a direct intervention in the extrusion process when detecting an unsatisfying product quality.

As shown in Figure 3, the undried polycondensate resin is molten in a 60 mm-single-screw extruder with a melt degassing zone (length/diameter-ratio 38).

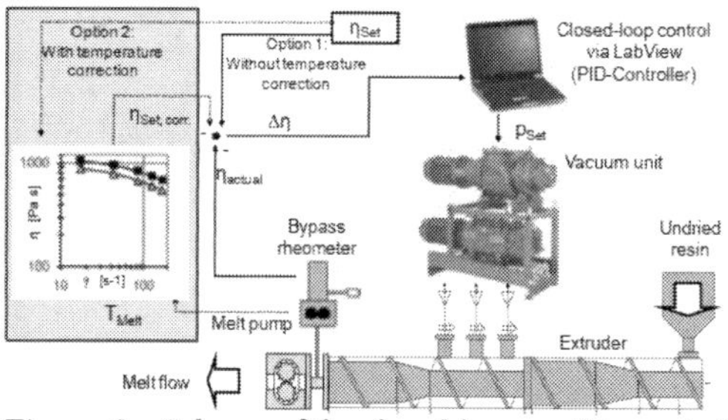

Figure 3 – Scheme of the closed-loop quality control.

The melt viscosity is measured continuously online in a bypass rheometer at a constant shear rate. The viscosity varies in reference to the resin, the moisture content and the degassing efficiency. The PID controller (neuronal networks are also possible) compares the actual with the defined set value and adapts the pressure in the degassing zone via the vacuum unit (Option 1).

An increasing melt temperature also increases the flow ability of the polymer melt and reduces the melt viscosity. This effect is comparable to a lower molecular weight. Therefore, a material dependent temperature correction of the set melt viscosity is additionally implemented in the closed-loop control (Option 2).

Operation of the control

As shown in Figure 4 in principle, for an otherwise constant extrusion process with an unchanged moisture content the degassing pressure influences the melt viscosity directly. A higher (absolute) degassing pressure reduces the efficiency of the water evaporation and leads to a higher hydrolytic degradation and therefore to a lower melt viscosity.

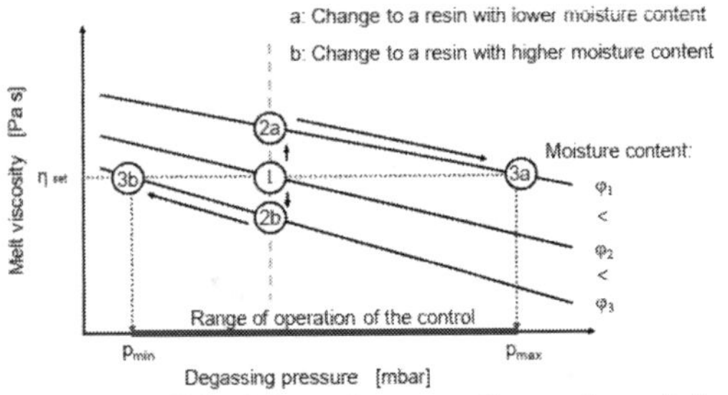

Figure 4 – Principle and range of operation of the closed-loop quality control.

Processing a resin with varying moisture content results in varying degradation effects in the plastification zone and yields to an inconstant product quality, detected online by the melt viscosity. The intervention of the control adapts the degassing pressure to the actual material and readjusts the product quality.

The range of the operation is limited. The minimum degassing pressure is defined by the vacuum unit. The maximum degassing pressure depends on the realisation of the sufficient product quality and can physically not be higher than 1000 mbar.

Results and Discussion

Reference tests

The static process behaviour is analysed in reference tests, where the influence of the polymer (PA 6, PET) and the moisture content in combination with the degassing pressure and melt temperature on the product quality is evaluated. As shown in Figure 5 for PET, the degassing pressure has a direct influence on the melt viscosity as well as on the IV. Processing a resin with higher moisture content results in increasing hydrolytic degradation and a reduced melt viscosity and IV.

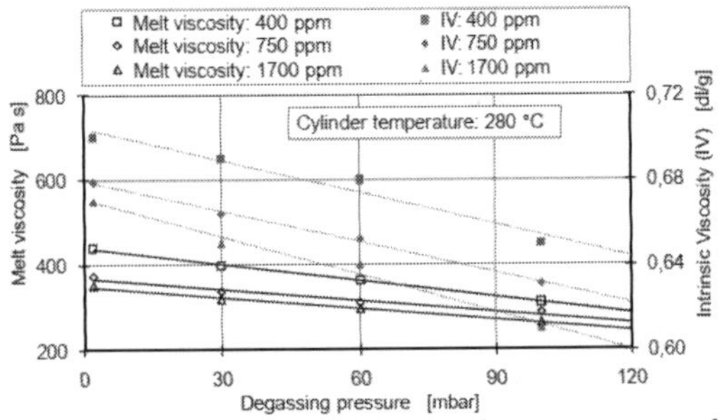

Figure 5 – Influence of degassing pressure and moisture content on the melt viscosity of PET.

On the basis of SEC or IV/RV measurements, the processing range of the control is identified. Especially the maximum moisture content has to be fixed to ensure the specificly needed product quality by using an optimal degassing process with a minimal degassing pressure. The correlation of the online (e.g. pressure drop) and offline (e.g. RV) quality criteria for a constant melt temperature is good (Figure 6).

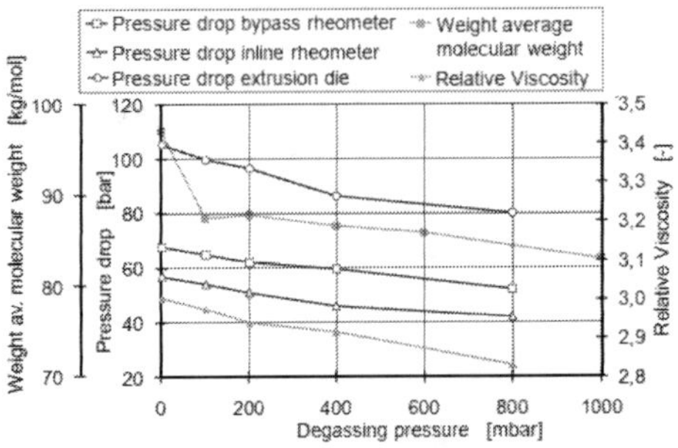

Figure 6 – Correlation of online and offline quality criteria for a constant melt temperature.

Avoiding a false interpretation of the closed-loop control for a changing melt temperature, the temperature correction of Option 2 in Figure 3 is implemented by using the established approach of Williams, Landel and Ferry (WLF), see Menges et al. (2002).

Because of the reactivity of the polycondensate melt in the extrusion line, as shown in Figure 1, the parameters of the WLF approach has to be adapted to the extrusion equipment and polymer material. This is done by the processing of the material with a constant moisture content, constant degassing pressure and varying melt temperatures. The resulting WLF parameters are listed in Table 1.

Parameter	Literature values	Adapted values
C_1 [-]	-8,86	-8,86
C_2 [°C]	101,6	101,6
T_S [°C]	90	-20

Table 1 – Literature and fitted parameters of the WLF temperature correction.

As shown in Figure 7, the correlation of online and offline quality criteria for a varying melt temperature is insufficient. The online measured melt viscosity is influenced by the for polymers typical temperature shift.

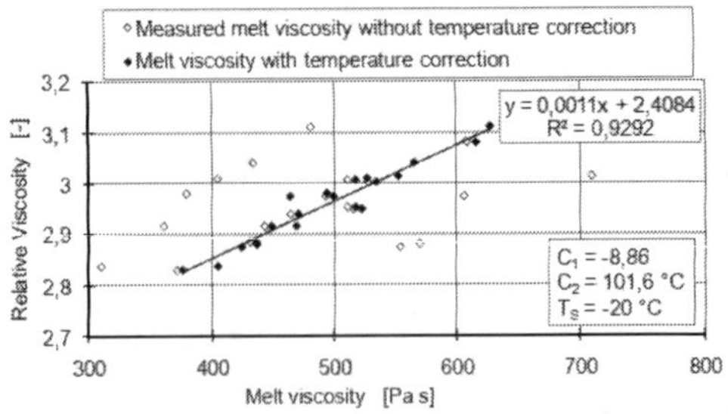

Figure 7 – Effect of temperature correction on the correlation of online and offline quality criteria.

Using the WLF approach as a temperature correction allows to compensate these effect and avoids a false interpretation of the closed-loop control.

Controller calibration

Due to the complex control process with its high reaction time and inertia, the adaption of the parameters of the PID controller requires the knowledge of the dynamic process behaviour.

The extrusion process possess several dead times. The polymer has to be transported from the hopper into the extruder, to the degassing zone, to the die as well as to the online measurement equipment. This is shown in Figure 8 and the analysis of the residence time using coloured tracer particles temporarily in the running extrusion process.

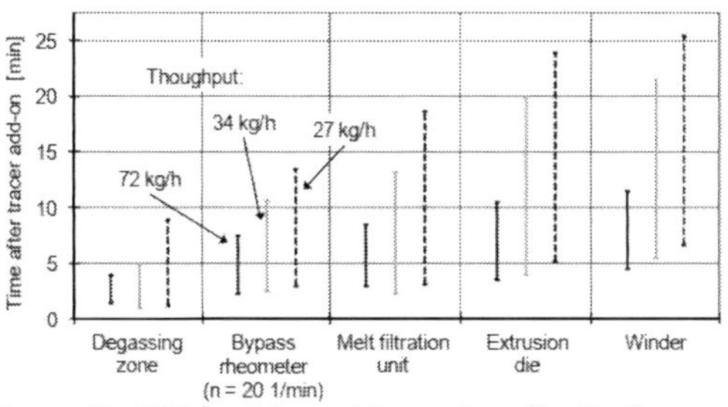

Figure 8 - Width of the residence time distribution.

In the closed-loop control, a changed moisture content can firstly be detected in the bypass rheometer after a delay of 2 to 14 min. The extrusion die is passed even after a time delay of 4 to more than 20 min.

Additionally, an abruptly changed degassing pressure results in a delayed reaction of the polymer melt due to the balance achievement of the evaporation.

The systematic calibration of the PID parameters is done by the analysis of the influence of an instantly changed degassing pressure on the extrusion system and established control adjustment methods. More control-oriented details of the PID calibration procedure and the adjustment methods can be found in Alstroem and Haegglund (1995).

Covering the operation range of the closed-loop control requires extensive extrusion tests with varying material, moisture content, melt temperature and level and direction of the change of the degassing pressure. Sufficient PID parameters has to be selected for each polymer PET and PA 6.

The system reaction for an instantly decreased degassing pressure and with it an advanced degassing effect is shown in Figure 9. After a delay of 2 min the online measured melt viscosity for a constant shear rate is increasing and reachs the final value asymptotically.

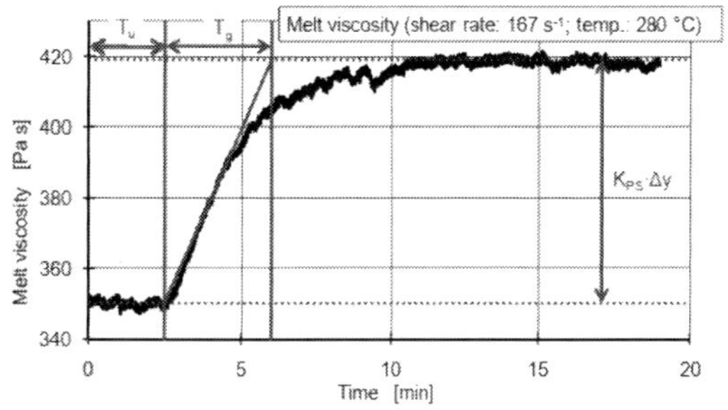

Figure 9 - Calibration of PID parameters using instantly changed degassing pressure (PET).

The interpretation of each trial can be done graphically as well as numerically. Compared to the numerical procedure, the graphical by e. g. the inflectional tangent method in Figure 9 is imprecise and user dependent. The resulting characteristic values significantly depend on the location and gradient of the inflectional tangent. The values are the input for the control adjustment methods.

The numerical interpretation enables a fast and precise analysis and an additional simulation of the final control behaviour. This procedure supports the evaluation of average PID parameters for the operation range, the comparison of different control adjustment methods, increases the control comprehension and detects insufficient control parameters prior the extrusion trials. This reduces the number of extensive and cost-intensive tests.

The influence of the proportional parameter K_{PR} on the control process is exemplary shown in Figure 10. A high K_{PR} value advances the responding characteristic after a disturbance appliance, but results in an inappropriate oscillation with a time-consuming balancing.

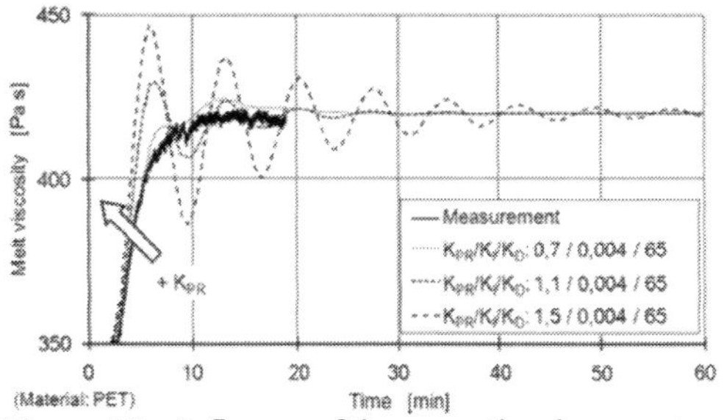

Figure 10 – Influence of the proportional parameter on the control behaviour.

The reference trials enable the user to define the operation range of the control and select material dependent PID parameters. The parameters are shown in Table 2 and the implemented closed-loop control is validated as follows.

PID Para-meters	PET		PA 6
	Control via bypass rheometer	Control via die pressure drop	Control via bypass rheometer
K_{PR}	1,1	2,5	1,1
K_I (T_n)	0,004 (275 s)	0,01 (250 s)	0,0058 (190 s)
K_D (T_v)	66 (60 s)	120 (300 s)	52,2 (47,5 s)

Table 2 – Material dependent PID parameters for the closed-loop quality control.

Validation

As shown in Figure 11, the set value of the melt viscosity is reduced from 350 to 300 Pa s in the running PET extrusion process at the time of 10 min.

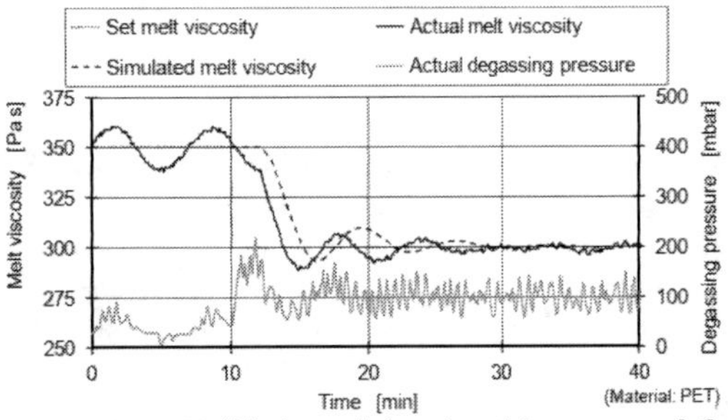

Figure 11 – Validation of the closed-loop control for the processing of PET.

After the reduction of the set point, the actual melt viscosity is higher than the set value. This results in a corrective action and a rising degassing pressure from approximately 50 to 100 mbar. It reduces the degassing efficiency, allows more hydrolytic degradation and reduces the actual melt viscosity until it reaches the new set point. The simulated control behaviour possesses a high correlation to the extrusion results. Especially the frequency and the amplitude of the oscillation correspond to the measured progress.

The aim of the closed-loop quality control is to balance the effect of changing moisture content in the feedstock. The validation is realised by the use of a two stage hopper system. The extruder is fed with the material with the moisture content 1 out of the lower hopper. In the running extrusion process, the material is discharged via a bypass above the extruder access. At the time of material change, material 1 is completely drained. The connection with the second hopper is opened and the material with moisture content 2 is delivered to the running extruder. During the material change, the extrusion process is uninterrupted.

As shown in Figure 12 for a change of the PA 6 moisture content from 600 to 5000 ppm, the higher water amount results in a rising hydrolytic degradation. This leads - after a dead time of 2 min - to a decreasing melt viscosity. Without the closed-loop control and the use of a constant degassing pressure, the melt viscosity would reach the value of approximately 450 Pa s. A constant product quality cannot be ensured.

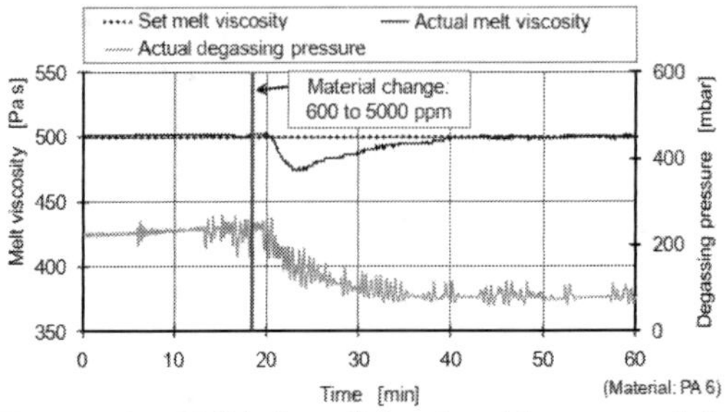

Figure 12 – Validation of the closed-loop control for the processing of PA 6 and a changing moisture content.

The automatic control detects the difference between set and actual melt viscosity and decreases the degassing pressure. This rises the degassing efficiency and the melt viscosity reaches the set point after a period of 5 min after the material change. Without an overshooting of the set point, the control intervention is finished after a period of approximately 20 min. The disturbance of a changed moisture on the product quality content is balanced successfully.

Controlled variable: Pressure drop of the extrusion die

Up to now, the presented closed-loop control requires an additional, cost-intensive and high-maintenance bypass rheometer. The functionality of the rheometer is based on the measurement of the pressure drop of the polymer flow with constant volume flow in a defined and constant capillary.

In principle, every already measured pressure drop of the polymer flow in the extrusion line can be used as a qualitative control parameter. The passed geometry as well as temperature and volume flow of the polymer melt have to be held constant. Additionally, the PID parameters have to be calibrated as presented prior and possible extra disturbance variables have to be respected. The approach of the use of a pressure drop as the controlled variable avoids the implementation of the bypass rheometer.

The validation of the closed-loop control via the pressure drop of the extrusion die is shown in Figure 13. At the time of 15 min, the material is changed from a moisture content of 700 to 2300 ppm. This leads to a rising degradation and decreasing pressure drop in the extrusion die due to the higher amount of water. The closed-loop control balances the effect of the changed moisture content successfully by reducing the degassing pressure. After a control intervention time of approximately 15 min, the product quality is comparable to the process prior the material change.

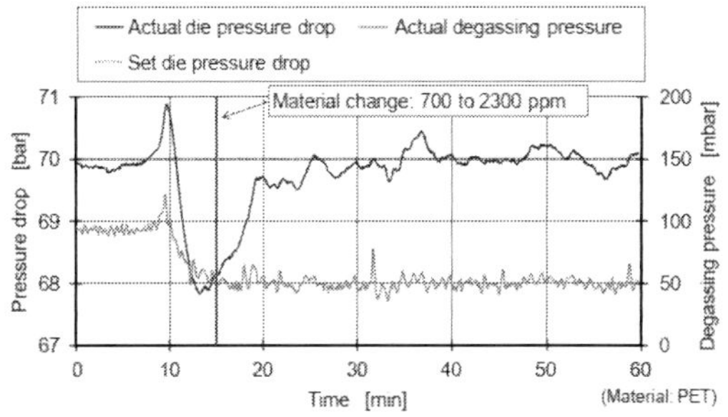

Figure 13 – Validation of the closed-loop control via the pressure drop of the extrusion die.

Conclusions

Processing undried polycondensates (as PA 6 or PET) using melt degassing single-screw extrusion can reduce the energy consumption in comparison to the pre-drying process significantly. Due to the process-related presence of water during the plastification and the resulting hydrolytic degradation, the existing processes cannot guarantee constant product qualities.

The presented closed-loop quality control can compensate the varying moisture content and allows a constant product quality. The systematic selection of sufficient PID parameters enables a material- and cost-saving adaption on industrial applications.

The basis of the closed-loop control is independent from the extrusion equipment and resin. In future research works, it can be adapted to other extrusion concepts (e. g. twin-screw extruder) as well as other polycondensates (e. g. polylactid acid, PLA).

Accelerating the intervention of the automatic control, additional information about the material prior the processing is needed. Therefore, a contactless online measurement of the moisture content of the bulk material has to be implemented in the extrusion line as well as in the closed-loop control.

Acknowledgements

The research project No. 15285 N of the Forschungsvereinigung Kunststoffverarbeitung was sponsored as part of the "Industrielle Gemeinschaftsforschung" (IGF) by the German Bundesministerium für Wirtschaft und Technologie (BMWi) through the AiF, to whom we extend our thanks.

We additionally like to thank BASF SE, Ludwigshafen/Germany, Artenius PET Packaging GmbH, Mendig/Germany, as well as Asentec GmbH, Heilbronn/Germany, for the allocation of polymer material and melt pressure measurement equipment.

References

K. Alstroem, T. Haegglund: *PID-controllers: Theory, Design and Tuning*, Research Triangle Park: The International Society of Measurement and Control, 1995, ISBN: 3-86130-890-8

G. Menges, E. Haberstroh, W. Michaeli, E. Schmachtenberg: *Werkstoffkunde Kunststoffe*, 5th Edition, Hanser Publishers, Munich (2002)

W. Michaeli, L. Ederleh, H. Seidel: *Degassing of Polycondensates*, Final report, Federal Ministry of Economics and Technology by the AiF 14629N (2008).

W. Michaeli, L. Ederleh, H. Seidel: *Closed-loop molecular weight control*, Final report, Federal Ministry of Economics and Technology by the AiF 15285N (2009).

T. Schmitz: *Processing Polyethylene Terephthalate on a Single Screw Extruder with Hopper- and Melt Degassing*, Doctoral thesis (2005), RWTH Aachen University, Germany, ISBN: 3-86130-507-0

A MODEL-BASED SCALING METHOD FOR COMPOUNDING PROCESSES ON CO-ROTATING TWIN SCREW EXTRUDERS

A. Thümen

BASF SE, Polymer Research GK, Carl-Bosch-Str. 38, 67056 Ludwigshafen, Germany, anne.thuemen@basf.com

Due to the incomplete geometrical similarity of different sized twin screw extruders, scaling of compounding processes continues to be a challenging task. In particular for industrial processes experience based or semi-quantitative approaches are still the rule. In order to close this gap, a model-based approach, developed by Potente and Preuß, was applied and tested for realistic compounding processes. This approach is based on the principle of energetic similarity and was developed for the melt conveying zone of twin screw extruders.

In our study, systematic experiments on extruders of different diameters were carried out and measured process data were compared with predictions from the scaling model. In general, consistency of experimental and calculated data is promising. Moreover, we found that the melting zone does not significantly influence the scaling results. As a consequence, there is no need to further develop specific model laws for the melting zone.

Introduction

Scaling extrusion equipment is considerably challenging, if compared to scaling other polymer engineering apparatus or chemical engineering apparatus. It becomes even more difficult for fully intermeshing co-rotating twin screw extruders (hereinafter abbreviated as TSE). The underlying reason for this is the complex geometry of this machine (Figure 1), which cannot be scaled between different screw diameters with full similarity. The maximum channel depth

$$h_{max} = D_S - a \qquad (1)$$

is fixed for a certain TSE.

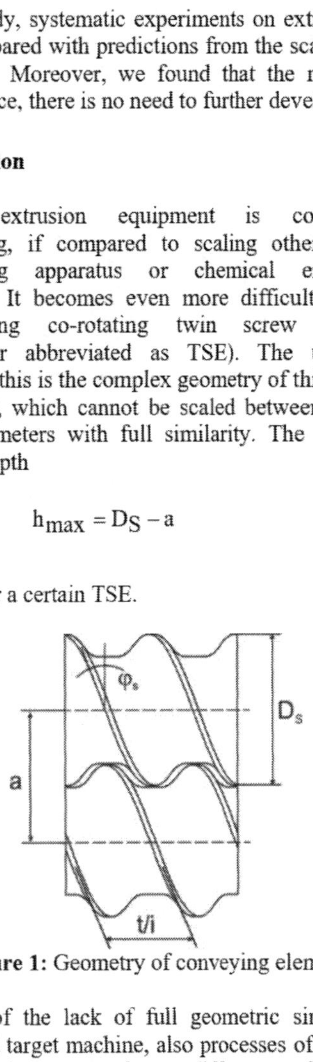

Figure 1: Geometry of conveying elements

Because of the lack of full geometric similarity of model and target machine, also processes of flow, heat transfer and mass transfer are different on both TSEs. Hence, aim of a scale-up/ scale-down can only be to achieve as similar as possible processes on both machines and with that as similar as possible product qualities.

Although there are some publications (Agur 1986, Christiano 1994, Potente 1981, Preuß 2009, White 1991), which deal with the topic of how to scale TSE processes appropriately, the published approaches are still not utilized in industry. Main reason for this is the complexity of the theoretical methods. The approach used within this study – a combination of model theory and one-dimensional simulation – offers a possibility to bring scaling theory into practice.

Common Scaling Approach

The most common method to scale TSE processes is to calculate mass throughput $\dot{m}_1$ of the target extruder by equating the ratio of the mass throughputs with the ratio of the outer screw diameters to the power of the scale-up exponent X.

$$\frac{\dot{m}_1}{\dot{m}_0} = \left(\frac{D_{S,1}}{D_{S,0}}\right)^X \qquad (2)$$

Herein, $\dot{m}_0$ indicates the mass throughput of the model machine, D_0 the outer screw diameter of the model extruder and D_1 the outer screw diameter of the target extruder. In most cases the scale-up exponent X takes values between 2 and 3 and there are some rules of thumb how to set X for a certain case. However, these rules are very rough so that only rough estimation of the appropriate target process results.

Moreover, shear rates in both extruders should approximately be the same.

$$\bar{\dot{\gamma}}_0 \equiv \bar{\dot{\gamma}}_1 \qquad (3)$$

Commonly, the average shear rates $\bar{\dot{\gamma}}_0$ and $\bar{\dot{\gamma}}_1$ for both TSEs are also calculated in a very rough way: the shear rates are seen as viscosity independent and assessed by means of the peripheral velocity v_0 only. From this the screw speed of the target machine $n_{0,1}$ results:

$$n_{0,1} = \frac{\bar{\dot{\gamma}}_{0,1} \cdot \bar{h}_1}{\pi \cdot D_{S,1}} = \frac{v_{0,1}}{\pi \cdot D_{S,1}} \qquad (4)$$

Model-based Scaling Approach

Compared to the number of publications dealing with scaling single screw extruders there are only a small number of publications on scaling up and down TSE processes (Agur 1986, Christiano 1994, Potente 1981, Preuß 2009, White 1991). Although these approaches exist, most scalings are still done by means of an experimental course of action. Reason for this is that these rules are complex and with that not easy to handle. One option to cope with the complexity of such an approach is to use a combination of process simulation and model theory that is offered by the simulation software SIGMA. SIGMA is a so-called one-dimensional process simulation software for co-rotating TSE developed at the University of Paderborn (http://ktpweb.de/en/software/sigma). Since 2004 this simulation tool includes the model theory for scaling TSEs developed by Preuß (2009). However, this model-based approach was developed for the zone of the extruder of pure melt conveying. Therefore, the aim of this study was to evaluate, if the implementation of additional theory for the melting zone would result in an increased quality of the scalings.

The following shows the underlying approach of Potente and Preuß (Potente 1981, Preuß 2009) in brief.

Aim of this model theory was to develop two types of model laws: model laws for the transfer of the geometry and model laws for the transfer of the operating point. The first had to be product and operating point independent while the latter are product and operating point dependent.

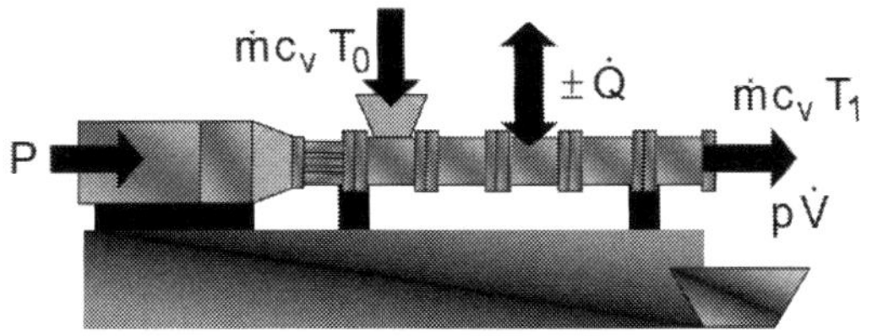

Figure 2: Energy balance

Starting point for the scale-up or scale-down is an energy balance at the extruder (see Figure 2). The sum of power input P and heat transfer via the barrel $\dot{Q}$ equals the sum of enthalpy increase between hopper and die $\dot{m}\, c_v \Delta T$ and the volumetric strain energy $\Delta p \dot{V}$.

$$\begin{aligned} & P \pm \dot{Q} = \dot{m} \cdot c_v \cdot \Delta T + \Delta p \cdot \dot{V} \\ \Leftrightarrow\ & \frac{P}{\dot{m} \cdot c_v \cdot \Delta T} \pm \frac{\dot{Q}}{\dot{m} \cdot c_v \cdot \Delta T} = 1 + \frac{\Delta p \cdot \dot{V}}{\dot{m} \cdot c_v \cdot \Delta T} \\ \Leftrightarrow\ & \Pi_P \pm \Pi_{\dot{Q}} = \Pi_p \end{aligned} \qquad (5)$$

In order to achieve full similarity of model and target process it is mandatory to keep the dimensionless power Π_P, the dimensionless heat flow $\Pi_{\dot{Q}}$ and the dimensionless pressure gradient Π_p constant for both processes. For the reason of incomplete geometric similarity of both machines this demand cannot be fulfilled. This means that the user of this method has to set a boundary condition to solve the set of model equations for the present scaling case. We will come to that later.

The flow profile is decisive for the pressure build-up and power consumption. Therefore, the flow conditions have to be the same in both machines. According to this the normalized velocity profiles in the channel have to be the same in both machines. And from this results the demand for an constant dimensionless pressure gradient Π_p, which is the ratio of pressure volume flow $\dot{V}_p$ and drag volume flow $\dot{V}_S$. For the Newtonian case the following applies:

$$\begin{aligned} & \dot{V} = \dot{V}_s - \dot{V}_p \\ \Leftrightarrow\ & \frac{\dot{V}}{\dot{V}_s} = 1 - \frac{\dot{V}_p}{\dot{V}_s} \\ \Leftrightarrow\ & \Pi_{\dot{V}} = 1 - \Pi_p \end{aligned} \qquad (6)$$

The dimensionless pressure gradient can only be constant, if the lead angle is the same for both extruders (in the considered section). This demand can be met by maintaining the ratio of lead to screw diameter t/D_S for both TSEs.

Due to the fact that the total volume flow $\dot{V}$ is in proportion to the drag volume flow $\dot{V}_S$ the following is implied

$$\dot{V} \sim \dot{V}_s \sim b\, h\, D_S\, n_0 \qquad (7)$$

Herein b indicates the channel width and, h the channel depth.

In the next step formal relations are introduced

$$v \sim D_S\, n_0 \tag{8}$$

$$b \sim D_S \tag{9}$$

$$h \sim D_S^{\ \Psi} \tag{10}$$

$$n_0 \sim D_S^{\ -\chi} \tag{11}$$

With

$$\frac{b_1}{b_0} = \left(\frac{D_{S,1}}{D_{S,0}}\right) \tag{12}$$

$$\frac{h_1}{h_0} = \left(\frac{D_{S,1}}{D_{S,0}}\right)^{\Psi} \tag{13}$$

$$\frac{n_{0,1}}{n_{0,0}} = \left(\frac{D_{S,1}}{D_{S,0}}\right)^{-\chi} \tag{14}$$

and if constant material properties and a constant lead angle are assumed, the following results

$$\begin{aligned} &\frac{\dot{m}}{\dot{m}_0} = \frac{\dot{V}}{\dot{V}_0} \\ \Leftrightarrow\ &\frac{\dot{m}}{\dot{m}_0} = \frac{b_1}{b_0}\frac{h_1}{h_0}\frac{D_{S,1}}{D_{S,0}}\frac{n_{0,1}}{n_{0,0}} \\ \Leftrightarrow\ &\frac{\dot{m}}{\dot{m}_0} = \left(\frac{D_{S,1}}{D_{0,0}}\right)^{2+\psi-\chi} \end{aligned} \tag{15}$$

Ψ is called the channel depth exponent and χ the screw speed exponent. In order to solve the equation it is necessary to couple the exponents. More detailed information regarding the solution of the system of equations and the therefore required assumptions can be found in Preuß 2009 and Potente 2006.

Important is that for the reason of the incomplete geometric similarity of both TSEs a boundary condition has to be chosen by the user for the respective scaling case. The following boundary conditions are available:

- constant specific energy input (CSE)
- constant specific heat flow (CSH)
- constant residence time (CRT)
- constant wall temperature difference (CWTD)
- constant average shear rate (CSR)

The resulting process is considerably dependent on the chosen boundary condition and has to be set in view of the "needs" of the respective polymer/ polymer system.

Practical Implementation of the Approach

As it was mentioned previously, the above described model theory for scaling TSE processes was implemented into the software tool SIGMA. The resulting approach is divided into three steps (see Figure 3). Step 1 comprises the simulation of the model machine at the operating point to be scaled. For that purpose material properties of the used polymer/ polymer system (shear viscosity h, thermal properties such as heat capacity c_p and heat conductivity λ of the melt), geometry data of the used TSE (e.g. screw diameter Ds, lead t, ...) and parameters of the respective process (screw speed n_0, mass throughput $\dot{m}$ and barrel temperatures T_{Zyl}) are required as input data. The simulation should show a good correlation of the resulting process values (mass temperatures, residence times, screw torque, ...) with the process values determined in the lab, because of that these values are needed as input for the second step of the approach – the scaling of the process.

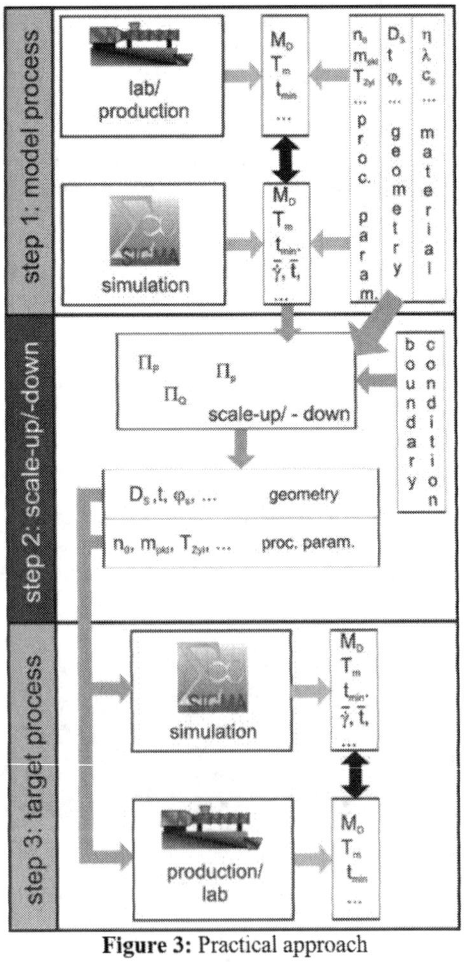

Figure 3: Practical approach

The above described model theory is matter of the second step. By means of the process values of the model process simulation, geometry data, material data and choosing one of the available boundary conditions the user gets the geometry and the process settings for the target process.

Moreover, the user has to set some more boundary conditions for the scale-up/ scale-down, for instance the user is able to set the length of the target extruder and define whether the radial clearance should be considered or not. However, in most cases this results in a deviation from the model theory. In the majority of cases this means that the processes of model and target machine are not as similar as possible.

In the third step the resulting target process is simulated with SIGMA and can be modified where appropriate.

Experimental

Two different types of setups were used for the experimental investigation. For the analysis of pure melt flow processes the main extruder was fed with melt coming from a feeding extruder (see Figure 4). As main extruder a ZSK 58 and a ZSK 26 MC were used, which have the similar geometries. The feeding extruder was for both setups the same and was operated under the same operating conditions for model process (ZSK 58) and target process (ZSK 26 MC).

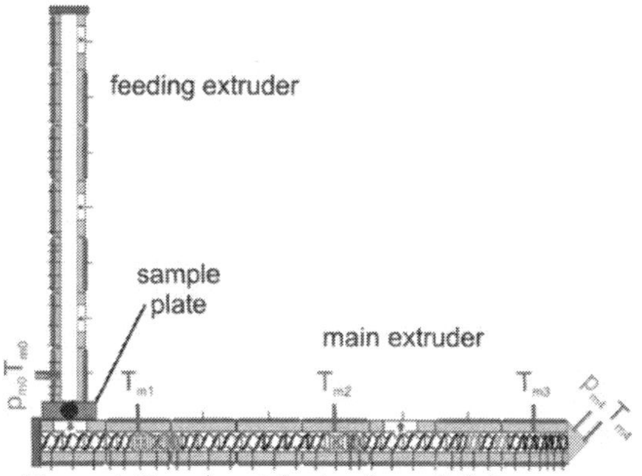

Figure 4: Experimental setup for melt fed processes

The in Figure 5 shown setup was used for the analysis of solid fed processes. Again, as main extruder a ZSK 58 and as target extruder a ZSK 26 MC were used. In addition a ZSK 30, whose geometry is not similar to the one of the ZSK 58, was used.

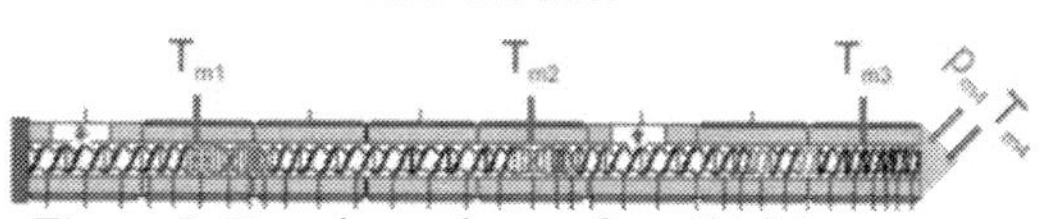

Figure 5: Experimental setup for solid fed processes

The measurement locations for mass temperature and pressure are indicated in Figure 4 and Figure 5. Mass temperature values were evaluated by a hand measurement at the die exit. The power values were not determined by an exact measurement; in fact the evaluated values are the values given by the machine control – knowing that these values are usually not very exact. While one of the possible boundary conditions for the scale-up/ scale-down is to achieve equal average residence times on both TSEs, minimum residence times were measured by means of a simple method, which only gives rough results. Coloured masterbatch was added to the feeder and the time was measured until the first colouration was visible at the outlet.

Two different polymers were used for the experimental investigation: a polystyrene PS 143E and a polyamide PA6 B3K. The shear viscosities of both materials are shown in comparison to usual processing temperatures (PS 143E @ 230°C and PA6 B3K @ 280°C) in Figure 6.

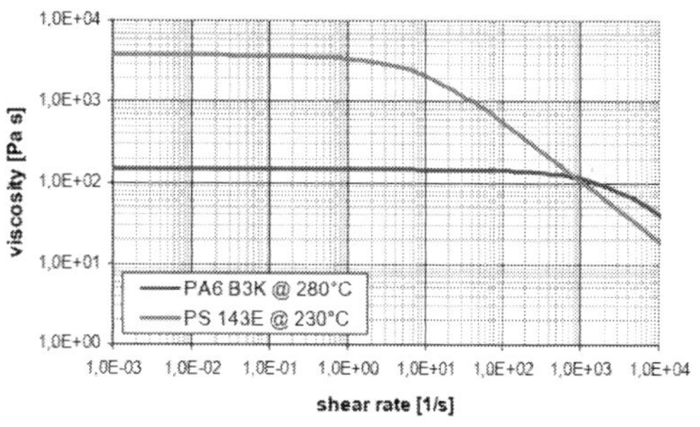

Figure 6: Viscosity of the used polymers

Results and Discussion

One aim of this study was to evaluate, if additional model laws could increase quality or quantity of the scalings. This question can be answered with help of Figure 7, Figure 8 and Figure 9. Figure 7 shows the mass temperatures at the outlet for the melt and solid fed processes and the scale-downs from the ZSK 58 to the ZSK 26MC under assumption of different boundary conditions. Although the mass temperature are higher on the ZSK 58 than on the ZSK 26 MC for melt fed temperatures and lower on the ZSK 58 than on the ZSK 26 MC for solid fed processes, there is not a significant qualitative or quantitative difference between processes with or without melting.

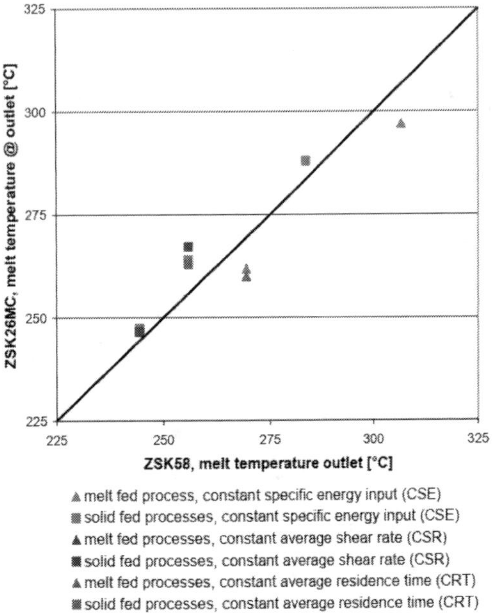

Figure 7: Experimental melt temperatures at the outlet

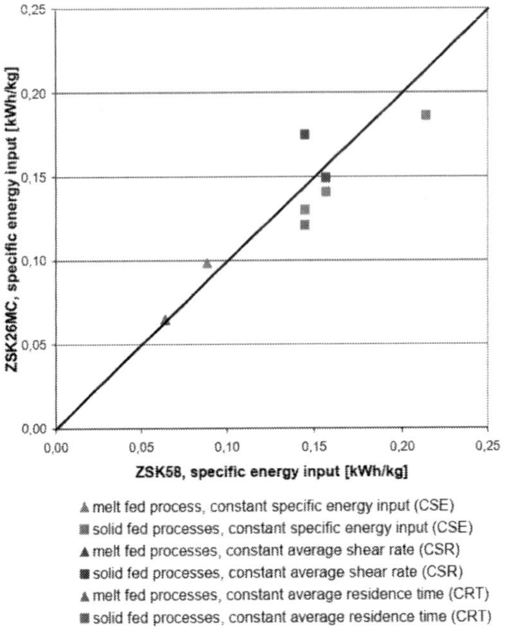

Figure 8: Experimental specific energy inputs

As well, the specific energy inputs shown in Figure 8 do not show a distinct quality or quantity difference between the resulting scalings for melt and solid fed processes. In most cases the specific energy input is higher for the processes on the extruder with larger diameter than for the processes performed on the smaller extruder. It can be assumed that the differences are caused by the different volume to surface ratios of the machines and the with that resulting different heat flows between barrel and melt. On the other hand different approaches for the calculation of the power consumption from the electrical signals of the motor can be assumed. However, the differences of the specific energy inputs for the comparable processes on both machines are within an acceptable range.

Figure 9 depicts the minimum residence times measured in the trials with ZSK 58 and the ZSK 26 MC for solid fed processes. For the melt fed processes on the larger machine the minimum residence times could not be measured.

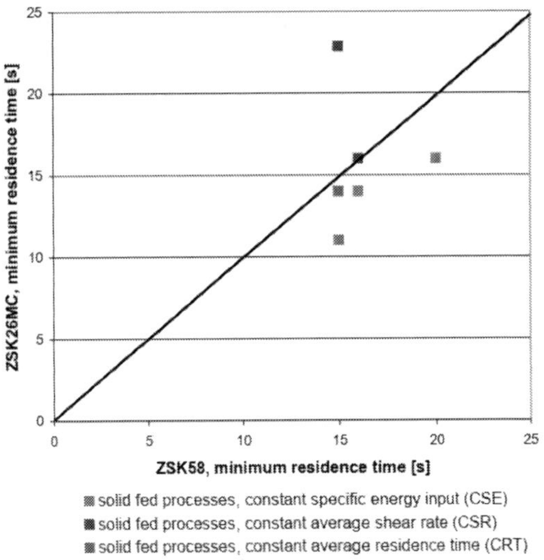

Figure 9: Experimental minimum residence times

In most cases the minimum residence times are higher on the bigger machine. Looking at these results it should be kept in mind that the average residence time is considered to be equal, if the "constant average residence time" boundary condition is considered.

On the whole there are differences between melt and solid fed processes noticeable. Nevertheless, they are not significant and do not indicate that additional model laws could improve the scaling of TSE processes.

Figure 10, Figure 11 and Figure 12 show exemplary experimental results for mass temperature, specific energy input and minimum residence time for solid fed processes scaled down under the "constant specific energy input" boundary condition from the ZSK 58 to the ZSK 26 MC and the ZSK 30. In view of the geometry the ZSK 58 and the ZSK 26 MC are as similar as possible, while the ZSK 58 and the ZSK 30 are less geometrically similar. Moreover, these figures give results from the model-based approach in comparison to the "traditional" approach. Aim for the traditional approach was to achieve scale-up exponents of 3, where possible.

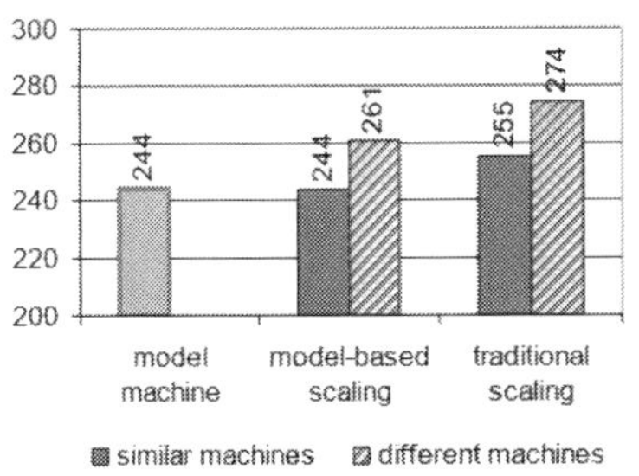

Figure 10: Comparison of experimental melt temperatures for different series of TSEs

Melt temperatures (Figure 10) are the same for model process and target process on the similar machine, if the model-based approach is applied. If the traditional approach is applied, the deviation of the melt temperatures at the outlet is larger. If the same comparison is done for model process, and target process on geometrically different machines is done, the mass temperatures difference is significantly increased. Reason for this is that the differences of the flow conditions within the TSEs are more distinct.

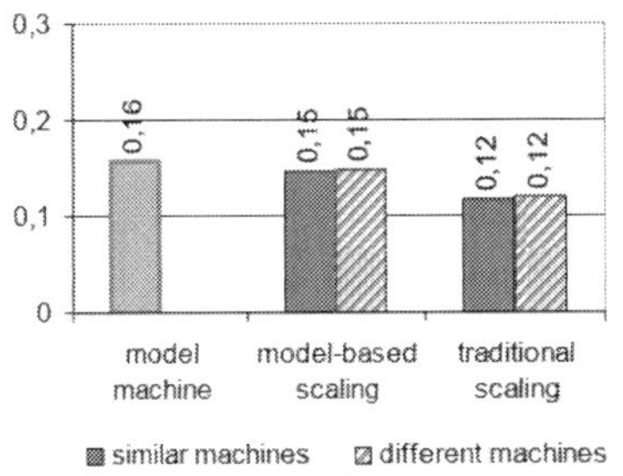

Figure 11: Comparison of experimental specific energy inputs for different series of TSEs

The specific energy inputs are nearly the same for processes scaled down on the geometrically similar and the geometrically different extruder. Nevertheless, it makes a difference whether the model-based or the traditional approach is applied.

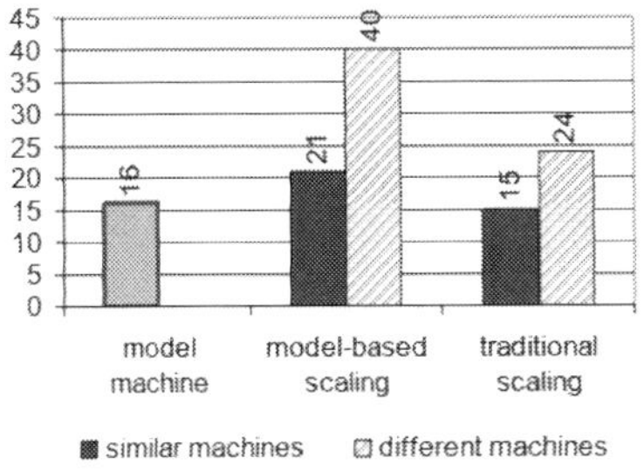

Figure 12: Comparison of experimental minimum residence times for different series of TSEs

A comparison of the minimum residence times depicts a significant difference between processes on geometrically similar and geometrically different TSEs. While the minimum residence times of processes on similar machines show a good agreement, this is not true for processes on different machines. In order to achieve as similar as possible average residence times on geometrically different machines of different diameters, a bigger difference of all other process parameters has to be accepted. The mass temperatures differ not as much as the average residence times, due to fact that they largely correspond to the specific energy inputs.

Conclusions

The method according to Potente and Preuß provides reasonable scalings of compounding processes on TSEs.

Neither a qualitative nor a quantitative difference between scaling melt fed and solid fed extrusion processes can be observed, which indicates that additional model laws and their implementation into the method would not result in an improvement of the scaling results.

Differences between processes on different series of machines cannot be avoided, but can be minimized and driven in a target-oriented way.

Acknowledgements

My special thanks go to Prof. Dr.-Ing. V. Schöppner of the University of Paderborn and his SIGMA team for their support.

References

Agur, E. E.: "Extruder scale-up in a corotating twin-screw extrusion compounding process", Advances in Polymer Technology, Vol. 6, Issue 2, p. 225-231 (1986)

Christiano, J. P.: "Scale-up study of co-rotating fully intermeshing twin screw extruders using 47mm, 69mm and 96.5mm diameters", SPE ANTEC Tech. Papers, p. 239-247 (1994)

Potente, H.: Kunststoff-Fortschrittsberichte, Band 6, "Auslegen von Schneckenmaschinen-Baureihen – Modellgesetze und Ihre Anwendung", Carl Hanser Verlag, Munich Vienna (1981)

Potente H: „Modellgesetze für gleichläufige, dichtkämmende Doppelschneckenextruder",final report for the AiF (German Federation of Industrial Research Associations) project 13020 N (2006)

Preuß, Th.: Polymerforschung in Paderborn, Band 23, "Übertragung von Betriebspunkten gleichläufiger dichtkämmender Doppelschneckenextruder in Theorie und Praxis", Shaker Verlag, Aachen (2009)

White, J. L., Chen, Z.: "Dimensionless Non-Newtonian Isothermal Simulation and Scale-up Considerations for Modular Intermeshing Corotating Twin Screw Extruders", Intern. Polymer Proc., VI, 4, p. 304-310 (1991)

List of Symbols

a	centreline distance
b	channel width
b_{max}	maximum channel width
b_0	channel width model extruder
b_1	channel width target extruder
c	heat capacity
D_S	outer screw diameter
$D_{S,0}$	outer screw diameter model extruder
$D_{S,1}$	outer screw diameter target extruder
h_{max}	maximum channel depth
h_0	channel depth model extruder
h_1	channel depth target extruder
$\bar{h}_1$	average channel depth target extruder
i	number of flights
$\dot{m}_0$	mass throughput model extruder
$\dot{m}_1$	mass throughput target extruder
n_0	screw speed
$n_{0,0}$	screw speed model machine
$n_{0,1}$	screw speed target machine
p	pressure
P	Power
Δp	pressure difference
p_{m0}, p_{m4}	melt pressures
$\dot{Q}$	heat flow
t	lead
$T_{m0}, T_{m1}, \ldots$	melt temperatures
T_0	mass temperature at the inlet
T_1	mass temperature at the outlet
TSE	twin screw extruder
ΔT	temperature difference
$v_{0,1}$	peripheral velocity target extruder
$\dot{V}$	total volume flow
$\dot{V}_p$	pressure volume flow
$\dot{V}_S$	drag volume flow
X	scale-up exponent
Π_p	dimensionless pressure gradient
Π_P	dimensionless power
$\Pi_{\dot{Q}}$	dimensionless heat flow
$\Pi_{\dot{V}}$	dimensionless volume flow
$\bar{\dot{\gamma}}_0$	average shear rate model extruder
$\bar{\dot{\gamma}}_1$	average shear rate target extruder
φ_S	lead angle
χ	screw speed exponent
Ψ	channel depth exponent

THREE-DIMENSIONAL EXTRUDATE SWELL SIMULATION OF POLYMER MELTS

S. Tanoue[1]*, T. Sunao[1], Y. Iemoto[1], S. Nojima[2] and S. Nomura[2]

[1]*Fiber Amenity Engineering, Graduate School of Engineering, University of Fukui, Fukui, Japan – tanoue@matse.u-fukui.ac.jp*
[2]*Fukuvi Chemical Industry Co., Ltd., Fukui, Japan*

Abstract: Three-dimensional extrudate swell simulation is a basic calculation technique for the consideration of die shape on a profile extrusion. In this study, we simulated the three-dimensional extrudate swell simulation of polymer melt by the finite element method. The Phan-Thien Tanner model was employed as the constitutive equation. First, we carried out the three-dimensional extrudate swell simulation through the straight die having the square cross-section. The extruded fluid swelled at once near the die exit because of elastic recovery. The degree of swelling phenomena near the die exit depends on the elasticity of polymer fluid. The changing rate of profile of the extruded fluid in the draw region from the profile of die exit increases a little with fluid elasticity. This profile would depend on the swelling phenomena near the die exit. Next, we carried out the three-dimensional extrudate swell simulation through the straight die having other cross-section shapes. The profile of the extruded fluid at the draw point is almost independent of the fluid elasticity in case using the straight die having the rectangular cross-section. On the contrary, in case using the straight die having the squared tubular cross-section, the profile of the extruded fluid depends on the flow rate and the fluid elasticity. As mentioned above, the profile of the extruded fluid on the profile extrusion would depend on the profile after swelled by elastic recovery near the die exit.

Introduction

Extrudate swell phenomenon was not important on profile extrusion which is known as the molding method for long goods with a complex cross-section, e.g. the windows frame, because the cross-section shape of the extrudated product is made up in the sizing just after extruded from the extruder. Recently, the profile extrusion products require the complex cross-section shape, using various material, and high accuracy of the dimension. Therefore, the prediction of the extrusion shape by the three-dimensional extrudate swell simulation becomes necessary on the profile extrusion. In this study, we try to do the three-dimensional extrudate swell simulation with viscoelasticity of polymer melt as the basic study for the prediction of the profile extrusion shape, and we discuss the influence of the viscoelasticity of the test fluid on the shape of the extrudates.

Calculation

The isothermal incompressible viscoelastic fluid was assumed. The high density polyethylene (HDPE) melt was focused as the test fluid. The inertia term in the momentum equation was negligible as the test fluid has high viscosity. In addition, the gravity term in the momentum equation was negligible, and the fixed velocity to the main flow direction is imposed at the exit boundary conditions in the flow problem. The Phan-Thien Tanner (PTT) model (Phan-Thien and Tanner, 1977) was employed as a constitutive equation for the viscoelastic fluid. The constitutive equations are as follows.

$$\boldsymbol{\tau} = \sum_{i=1}^{N} \boldsymbol{E}_i \tag{1}$$

$$\left\{\exp\left(\frac{\varepsilon_i}{G_i}\mathrm{tr}\boldsymbol{E}_i\right)\right\}\boldsymbol{E}_i + \lambda_i\left\{\left(1-\frac{\xi_i}{2}\right)\overset{\nabla}{\boldsymbol{E}}_i + \frac{\xi_i}{2}\overset{\Delta}{\boldsymbol{E}}_i\right\} = 2\lambda_i G_i \boldsymbol{D} \tag{2}$$

Here, λ_i and G_i are the relaxation time, relaxation modulus for the i-th mode, ε_i and ξ_i are the material constants for the i-th mode, and $\boldsymbol{D}$ is the rate of deformation tensor. In addition, the signs Δ and ∇ shown in the upside of $\boldsymbol{E}_i$ are the upper-convected and lower convected time derivative for $\boldsymbol{E}_i$. Figure 1 shows the experimental data of the shear viscosity and first normal stress difference as a function of shear rate at steady-state (Orbey and Dealy, 1984), and their fitted curves by the PTT model. In this study, we focused the two kinds of test fluid; Fluid A (Material constant of ε_i is 0.4 in each i-mode) and Fluid B (Material constant of ε_i is 0.02 in each i-mode). The different point of these two fluids is the elongational flow characteristic. Fluid B has the strain-thickening elongational flow characteristic, i.e. strain-hardening. This characteristic of Fluid A is weaker than that of Fluid B although the shear viscosity curve of Fluid A is almost the same as that of Fluid B. In addition, the first normal stress difference of Fluid B is a little larger than that of Fluid

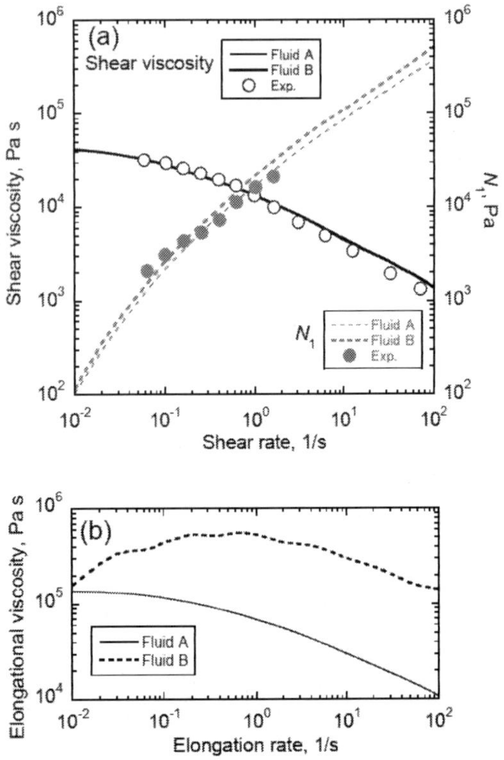

Fig.1 Steady-state flow characteristics as a function of shear rate for HDPE melt (Orbey and Dealy, 1984) and its fitting curves for the Phan-Thien Tanner (PTT) model.
(a) Shear flow characteristics for the PTT model and their experimental data (Orbey and Dealy, 1984)
(b) Elongational viscosity for the PTT model.

A at same shear rate. Then, the elasticity of Fluid B is larger than that of Fluid A.

A Finite element method was used in this study. In order to solve stably, the streamline-upwinding finite element method (Marchal and Crochet, 1987) was employed for the solution of the constitutive equation. In addition, the calculation results for the Carreau-Yasuda model (Yasuda et al., 1981), which is the purely-viscous non-Newtonian fluid, were employed as the initial values for the iteration scheme for the purpose of the improvement of convergent validity.

Results and Discussions

Extrudate swell passed through the straight die having a square cross-section

First, we discuss the extrudate swell passed through the straight die having a square cross-section. Here we discuss the extrudate swell simulation problem as follows; 1) The polymer melt was extruded from the straight die having square cross-section which length on a side is 1 cm. 2) The extrude pull on the point 25 cm from the die exit at the mean velocity on the die exit $\langle u \rangle$. In addition, no-slip condition imposed on the wall in the die, and the drawing velocity of the extruded fluid was imposed on all section of the exit boundary.

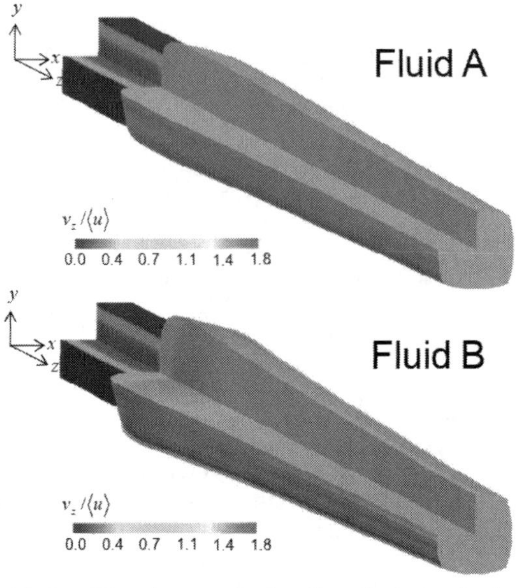

Fig.2 The shape and the dimensionless velocity distribuion for z-direction (extrusion direction) of the extruded fluid at $\langle u \rangle/H = 40$ s^{-1}.

Figure 2 shows the shape of extruded fluid and dimensionless velocity distributions for z-direction (extrusion direction) at $\langle u \rangle/H = 40$ s^{-1}. Here, H is the reference length defined by the half length on a side of the square cross-section at die exit, and we set up H = 0.5 cm. In these figures, the extruded fluids were cut 1/4 size for the easy view of the velocity distributions in the extruded fluid and the shape of that along the flow direction. The straight region in the upper stream side is the flow region in the die having a square cross-section. The swelling phenomena can be seen near the die exit because of elastic recovery. The degree of swelling near the die exit for Fluid B is larger than that for Fluid A. The fluid with strain-hardening, i.e. large elasticity swells largely near the die exit. As might be expected, the cross-section area of the extruded fluid becomes reduced to the downstream because of draw operation.

Figure 3 shows the cross-section shape of extruded fluid at the draw location, i.e. exit boundary, z = 25 cm (z/H = 50). The cross-section shape of an extruded fluid at the point of the most swelled (z/H = 5.5 for Fluid A, and z/H = 4.0 for Fluid B) is shown in addition as the reference. The cross-section shape of the extruded fluid is different from that at the die exit for both test fluids. The large swelling phenomenon is occurred at point of the large shear rate (x/H, y/H) = (0, 1) or (x/H, y/H) = (1, 0). Therefore, the thickness of the extruded fluid at this point is about 1.05 times larger than the die size at the die exit. In addition, the corner of the extruded fluid at the draw point moved to the center axis, i.e. the direction of (x/H, y/H) = (0, 0) because of the momentum balance with imposed

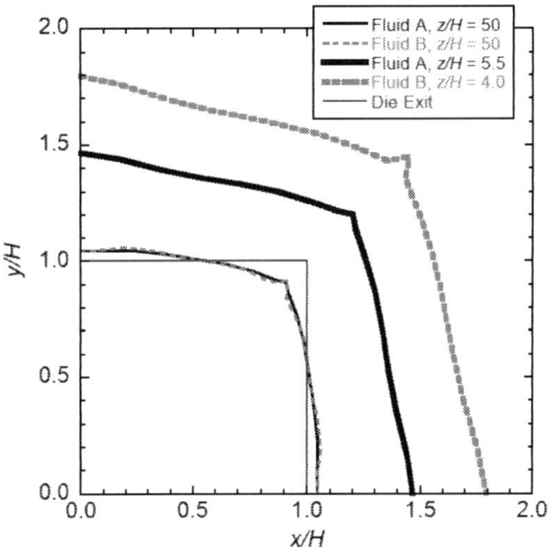

Fig.3 The Cross-section shape of extruded fluid at the draw point (z/H = 50 for both fluids) and maximum swelling point (z/H = 5.5 for Fluid A, z/H = 4.0 for Fluid B) at $\langle u \rangle /H = 40\ s^{-1}$.

uniform velocity on the exit boundary. The cross-section shape at the draw point, i.e. the cross-section shape of final extrusion product, would depend on the swelling phenomenon near the die exit as the cross-section shape at the draw point would be analogously the same as that at the most swelling point. By the way, the final cross-section shape of the extruded fluid is almost independent of the degree of fluid elasticity as long as this case.

The extrudate swell through the die with other cross-sections

Next we discuss the shape of extruded fluid passed through the straight die with other cross-section shapes. In each case, the calculation region size is set that the straight die length is 5 cm and the length from the die exit to the draw point, i.e. the extruded fluid length is 25 cm. In addition, an axial velocity at the draw point was 1.1 times larger than the mean velocity at the die exit $\langle u \rangle$, defined as the reference velocity.

First, we discuss the case using the straight die having the rectangular cross-section. The sizes of cross-section shape at die exit are 40 mm for the width and 2 mm for the gap (The aspect ratio defined by the ratio of die width length to gap one is 20). And a half size of the gap is defined as the reference length H, i.e. H = 1 mm. Figure 4 shows the shape and the dimensionless axial (z-component) velocity distribution of the extruded fluid at $\langle u \rangle /H = 20\ s^{-1}$. The swelling phenomenon in the y-direction (width direction) can be seen near the die exit. The degree of the swelling phenomenon in case of Fluid B with strain-hardening characteristic is larger than that in case of Fluid A with typical strain-softening one. In case of the die having a rectangular cross-section, the effect of fluid elasticity on cross-section shape of an extruded fluid at the draw point was almost same even if the aspect ratio of the cross-section shape of the straight die is changed.

Second, we discuss the shape of extruded fluid passed through the die having a squared tubular cross-section. The size of cross-section shape at die exit is that the length of a side is 20 mm; the length between the walls is 2 mm. And a half size of the length between the walls is defined as the reference length H, i.e. H = 1 mm in this case. Figure 5 shows the shape and the dimensionless axial (z-component) velocity distribution of the extruded fluid at $\langle u \rangle /H = 25\ s^{-1}$. The swelling phenomenon outside of the extruded fluid can be seen near the die exit just like the case of extruded fluid shape through the die with square cross-section shown in Fig.2. And the degree of the swelling phenomenon in case of Fluid B is larger than that in case of Fluid A. In comparison of a cross-section shape of the extruded fluid at the draw region between the

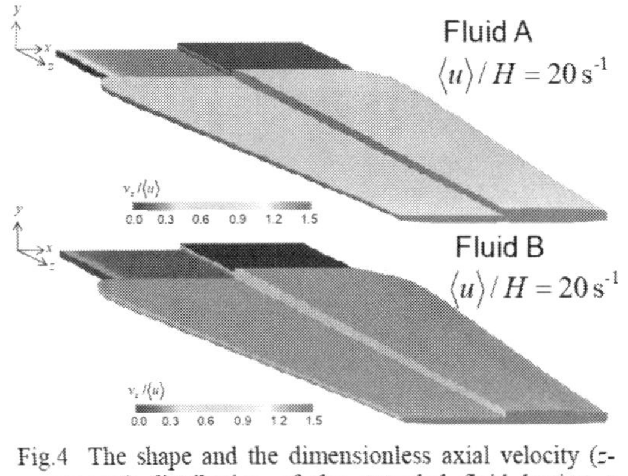

Fig.4 The shape and the dimensionless axial velocity (z-component) distribution of the extruded fluid having a rectangular cross-section at $\langle u \rangle /H = 20\ s^{-1}$.

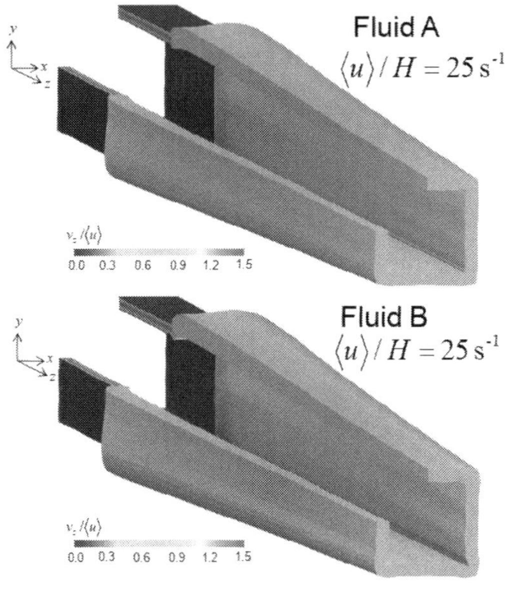

Fig.5 The shape and the dimensionless axial velocity (z-component) distribution of the extruded fluid having a squared tubular cross-section at $\langle u \rangle /H = 25\ s^{-1}$.

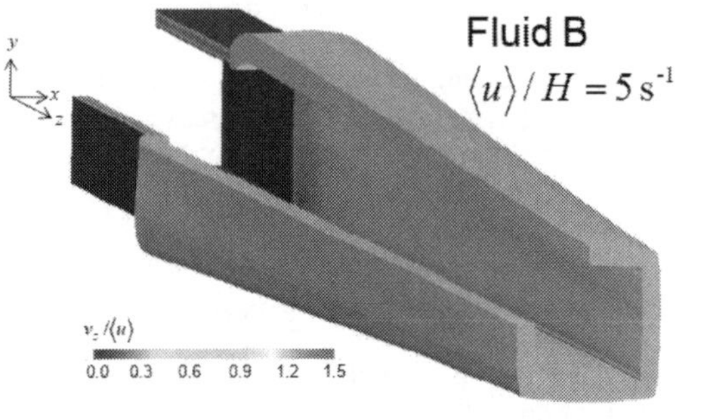

Fig.6 The shape and the dimensionless axial velocity (z-component) distribution of the extruded fluid having a squared tubular cross-section at $\langle u\rangle/H = 5\ s^{-1}$.

case of Fluid A and Fluid B, the thickness of the extruded fluid at center on the side in case of Fluid B is larger than that in case of Fluid A. In parallel, the outer length at center on the side in case of Fluid B is smaller than that in case of Fluid A. In addition, the cross-section shape of the extruded fluid at draw point is closely similar to that at maximum swelling point. This implies that the cross-section shape of an extruded fluid at draw point depends on that at maximum swelling point. In order to consider the effect of flow rate on the cross-section shape of the extruded fluid at draw point, the shape and the dimensionless axial (z-component) velocity distribution of the extruded fluid for Fluid B at $\langle u\rangle/H = 5\ s^{-1}$ is shown in Fig. 6. The cross-section shape at draw point seems to be different from that at $\langle u\rangle/H = 25\ s^{-1}$ shown in Fig. 5, and the swelling phenomena near the die exit is smaller than that at $\langle u\rangle/H = 25\ s^{-1}$ as a matter of course.

As mentioned above, the profile of the extruded material on the profile extrusion would depend on the profile after swelled by elastic recovery near the die exit. In addition, it is necessary to consider the viscoelasticity of a test fluid in order to understand exactly the tendencies of the cross-section shape of an extruded fluid with a profile cross-section using the extrudate swell simulation.

Conclusions

In this study, we simulated the three-dimensional extrudate swell simulation of polymer melt by the finite element method.

The extruded fluid through the straight die having a square cross-section swelled at once near the die exit because of elastic recovery. The degree of swelling phenomena near the die exit depends on the elasticity of polymer fluid. The changing rate of the extruded fluid near the draw region from the profile of die exit increases a little with fluid elasticity.

The profile of the extruded fluid at the draw point is almost independent of the fluid elasticity in case using the straight die having a rectangular cross-section. On the contrary, in case using the straight die having a squared tubular cross-section, the profile of the extruded fluid depends on the flow rate and fluid elasticity.

As mentioned above, the profile of the extruded material on the profile extrusion would depend on the profile after swelled by elastic recovery near the die exit.

References

Marchal, J. –M., Crochet, M. J., "A new mixed finite element for calculating viscoelastic flow", J. Non-Newton. Fluid Mech., **26**, 77-114 (1987)

Orbey, N., Dealy, J. M., "Isothermal swell of extrudate from annular dies; effects of die geometry, flow rate, and resin characteristics", Polym. Eng. Sci., **24**, 511-518 (1984)

Phan-Thien, N., Tanner, R. I., "A new constitutive equation derived from network theory", J. Non-Newton. Fluid Mech., **2**, 353-365 (1977)

Yasuda, K., et al., "Shear flow properties of concentrated solutions of linear and star branched polystyrenes", Rheol. Acta, **20**, 163-178 (1981)

Blown Film Processing of Multilayer Polyethylene and Nylon for Military Food Packaging Applications

Jo Ann Ratto (1), Sarah Schirmer (1), Jay Swamy (2), Gail Becke (2)*

(1) U.S. Army Natick Soldier Research Development and Eng. Center - MA - USA
(2) Berry Plastics - MA - USA

Low density polyethylene (LDPE) and nylon multilayer films are being investigated as high oxygen barrier alternatives to current linear low density polyethylene (LLDPE) monolayer films for meal bags that hold the Meal, Ready-to-Eat™ food components. An 80:20 blend of MXD6 and Nylon 6, respectively, was processed into blown film with maleic anhydride modified LLDPE tie layers and LDPE/LLDPE blended skin layers. Three different samples with MXD6/Nylon 6 core layer thicknesses of 10, 15 and 20%, of the total multilayer film thickness, were produced and analyzed for barrier and mechanical properties. It was found that the 80:20 MXD6/Nylon 6 blend produced a brittle multilayer film, with low tear strength, therefore additional MXD6/Nylon 6 blends were processed in order to decrease the brittleness of the samples and increase their tear strength. A 15% core sample using 40:60 and 60:40 MXD6/Nylon 6 blends were investigated and two additional multilayer films were produced. These films were significantly less brittle and exhibited improved barrier and mechanical properties. In order to improve the oxygen barrier properties of the films, ethylene co-vinyl alcohol (EVOH) was added as a layer in the multilayer structure. Two additional multilayer films incorporating an EVOH core, sandwiched by 40:60 and 60:40 MXD6/Nylon 6 blends, with maleic anhydride modified LLDPE tie layers and LDPE/LLDPE blended skin layers, were processed and as expected, the oxygen barrier performance improved considerably. The 40:60 MXD6/Nylon 6 blend, which was expected to have improved high temperature performance over the 60:40 MXD6/Nylon 6 blend, was processed into both 7-mil and 8-mil thick blown films for further analysis. Additional barrier, mechanical and thermal analysis and insect infestation tests will be completed to determine if either the 7-mil or 8-mil multilayer structures will be suitable for inclusion in the military rations.

NANOFIBER TECHNOLOGY FROM THE ELECTROSPINNINTG

Y.Yamashita[1], T.Kanzawa[2] and K. Tokumitsu[1]

[1]Department of Materials Science,, The University of ShigaPrefecture, Hikone, Japan – yamashit@usp.ac.jp
[2] Notheastern Industrial Research Center of Shiga Prefecture, Nagahama, Japan

Abstract - The creation of the nanofiber that used the electro-spinning process and the examination to the mass production were done. In addition, we examined the possibility of the melt-spinning method of the nanofiber for PP and PE polymer. It has been understood that productivity is low though the nozzle method of the electro-spinning can obtain a stable nanofiber nonwoven.

1. Introduction

The nanofiber is produced by the electro spinning process, and it is widely utilized in medical filters, electronic materials, etc.. Moreover, the production of the manufacturing electro spinning machines of the solution-spun type started also in Japan [1]. In this paper, the challenge of Japan for application and mass production of the nanofiber noticed at present is explained.

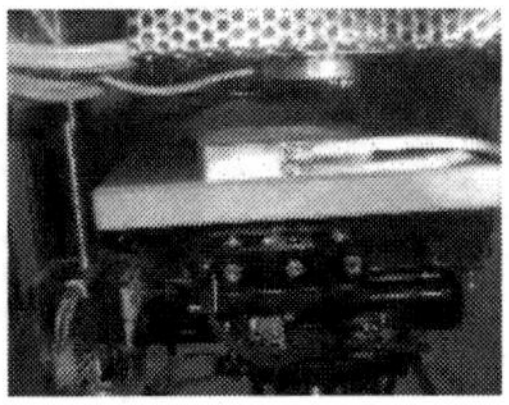

(a) (b)

Fig.1(a) Labo melt electrospinning device (b) Nozzle and target

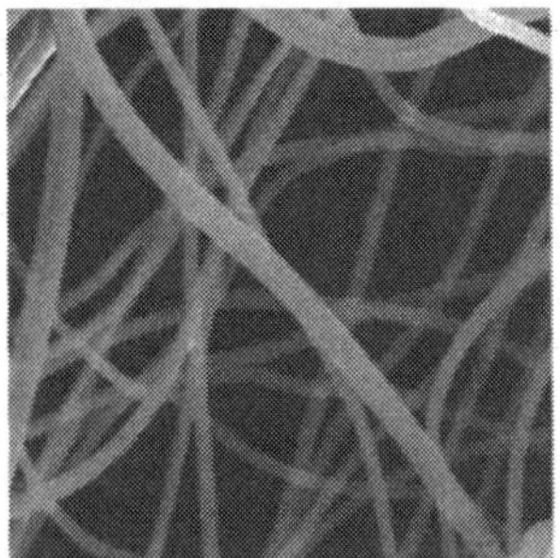

Fig.2 PP nanofiber from the melt electrospinning

2. Melt electro spinning

Electro spinning equipment of the melting type which does not use the solvent is very much noticed, and the research using the laser is detailed for the report of Ogata [2-4] of Fukui Univ. and Ohkoshi [5-7] of Shinshu Univ.. We tried the farther experiment using the general-purpose melt spinning equipment. In the melt electrospinning from low molecular weight PP, the fiber diameter of 1-5 micrometers was obtained (Fig.1,2).

3. Manufacturing of nanofiber

3.1 The solution sending method.

Solution is straightened to the tank in the drum and the bubble type. Only it is to supply solution to the tank, it is very easy in this method. The solution is supplied at syringe or air pressure in case of the nozzle system. Theoretically, the air pressure is easier. However, since concentration is 5 - 25wt% (the viscosity 50-1000mPa・s), viscosity passes over the solution used for electro spinning low to control by air pressure. And, though there is some a method using gear pump and tube pump, it is necessary to make the part of those pumps and tubes even in the case in the insulating material. There is danger of transmitting the charge in solution and causing the spark in the pump because a high voltage is applied to solution. The advantage of the syringe system is that the solution supply is stable. However, a syringe pump is a batch type and needs to stop equipment for supply of a solution. The production cost increases because the mechanism to extrude solution enlarges moreover, too, when there is a lot of capacity of the syringe pump. The cleaning in the syringe pump after the use is also troublesome.

3.2 Approach of Shiga Prefecture on manufacturing device development

As for the nozzle method and the drum method, the number of possible spinning is limited at a time. If a lot of nanofiber is spun at once, evaporation of a solvent will become insufficient and adhesion of fibers and also the dissolution of nanofiber on a target will start. This is still difficult though the irregular thickness of nano fiber nonwoven control is also important. It is necessary to disturb the nano fiber by the methods in which the nozzle minute vibrates in order to equalize thickness nonuniformity of the nano fiber from the nozzle, and methods for blowing the air out from the nozzle, etc.. In Shiga Prefecture, the manufacturing equipment was developed on the basis of the support from Ministry of Economy, Trade and Industry. The features are described in the following.(Fig.6)

3.3 Resin nozzle

Though metal nozzles are necessary for the manufacturing equipment, the resin nozzle of the disposable type is suitable, when the cleaning of the nozzle by solution exchange and nozzle blockage is considered in test machine of the continuous production. We produced multi- nozzle with 10 nozzles in the 1cm interval experimentally. To make the quantity of delivery from each nozzle constant, the taper is applied to an internal resin pool part, the diameter in the nozzle is smaller in the center part, and the outside is enlarged. The nozzle tip can attach a commercial pipette chip, it is cheap, and this resin nozzle is the sale schedule. In addition, the resin nozzle can prevent the nozzle dirt with the re-attachment of the Nano fiber that drifts in air. We are planning 20cm in length and 20 disposable nozzles for the base of 150cm in width(Fig.3,4).

3.4 High speed camera

The confirmation of whether they are nano fiber or nanoparticle can not be judged the observation by the high magnification microscope. However, it is not easy in the manufacturing process. It is also difficult that the sample is taken out from the part of the product. Then, the method for watching the condition of the spinning from the nozzle by the high speed camera was used. The high speed camera observes continuously the image in 1/1000 seconds. Fig.5(a) are images in producing the nano fiber well. The length of the spray from the nozzle is long, and it is understood a thin nano fiber is made afterwards by electrostatic repulsion. On the other hand, Fig. 5(b) is a nanoparticle. Spray length from the nozzle tip to the electrostatic repulsion is short. The segmentation by the electrostatic repulsion is immediately generated, because the entanglement of the molecular chain does not happen for low concentration and low molecular weight. Moreover, the particle is being sprayed afterwards.

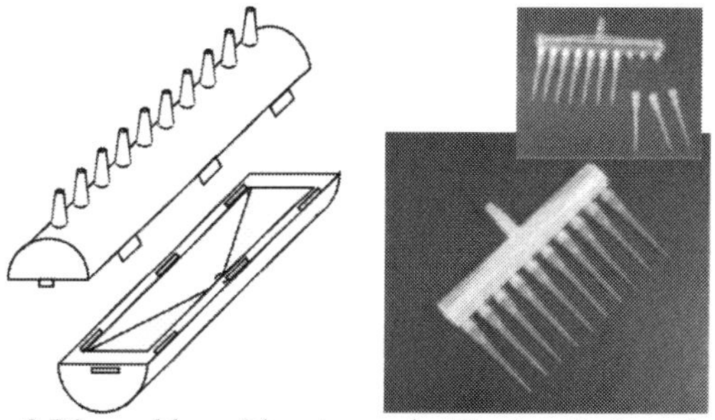

Fig.3 Disposable multi resin nozzle (The installation of a disposable pipet is more possible in the nozzle tip).

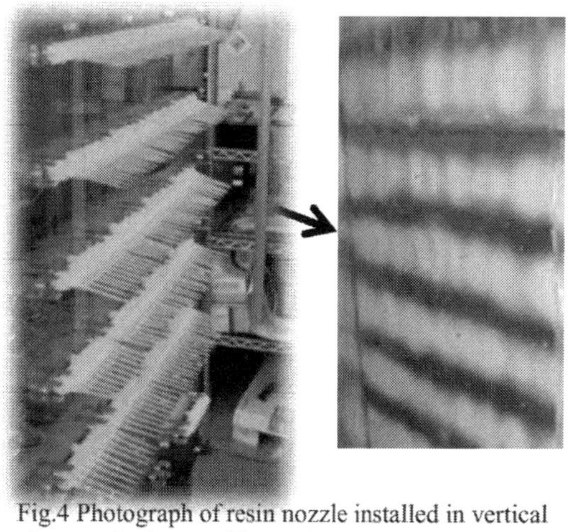

Fig.4 Photograph of resin nozzle installed in vertical electro spinning machine

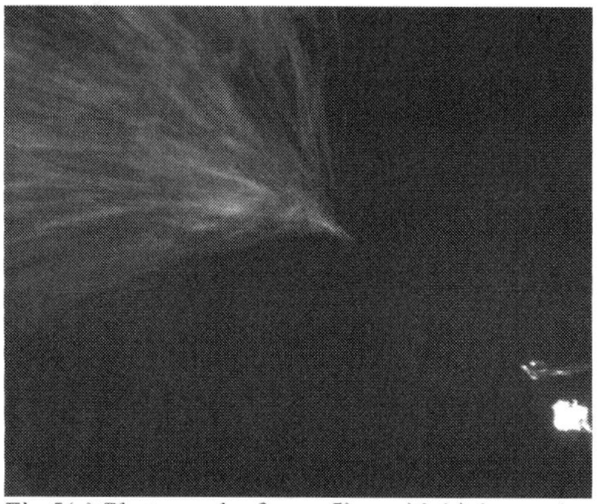

Fig.5(a) Photograph of nanofiber with high speed camera (1/1000s)

Fig.5(b) Photograph of nanospray (nanoparticle) with high speed camera (1/1000s)

Fig.5 Photographs of the electrospinning & electrosplay. Whether the nanofiber is made 6(a) or a nanoparticle 6(b) can be judged with the high speed camera.

Fig. 6(a)

Fig 6(b)

Fig.6 Electrospinning machine for industry use. Width 150cm, Nozzle number 400, roll speed 0.1~10m/min (manufactured by KATO-TECH CO. LTD in Japan)

Reference

[1] Yamashita Y, "Electorospinning –The latest in nanotechnology", Sen-Isha, ISBN978-4-9902580-1-6 (2007)

[2] OGATA Nobuo, SHIMADA Naoki, YAMAGUCHI Shinji, NAKANE Koji, OGIHARA Takashi, J Appl Polym Sci , Vol.105 No.3 Page.1127-1132 (2007)

[3] OGIHARA Takashi, J Appl Polym Sci , Vol.104 No.3 Page.1640-1645 (2007)

[4] OGATA Nobuo, IWATA Toshiharu, YAMAGUCHI Shinji, NAKANE Koji, OGIHARA Takashi, LU Gang, J Appl Polym Sci , Vol.104 No.2 Page.1368-1375 (2007)

[5] TAKASAKI Midori, FU Hao, NAKATA Kazuhiro, OHKOSHI Yutaka, HIRAI Toshihiro, Senigakaishi, Vol.64 No.1 Page.29-31 (2008)

[6] NAKATA Kazuhiro, OHKOSHI Yutaka, GOTOH Yasuo, NAGURA Masanobu, KIM Seong Hun,Senigakaishi, Vol.63 No.12 Page.307-312 (2007)

[7] UDDIN Ahmed Jalal, MASHIMA Yusuke, OHKOSHI Yutaka, GOTOH Yasuo, NAGURA Masanobu , SAKAMOTO Akinobu, KURODA Ryuma, J Polym Sci Part B, Vol.44 No.2 Page.398-408 (2006)

Microporous membranes obtained from polypropylene/high density polyethylene multilayer films by stretching

Seyed Hesamoddin Tabatabaei, Abdellah Ajji, Pierre Carreau*

École Polytechnique of Montreal - Quebec - Canada

Polypropylene/high density polyethylene (PP/HDPE) multilayer films were prepared to develop microporous membranes using cast film extrusion followed by stretching. The effects of draw ratio (DR), cooling air flow rate (AFR), and annealing on the crystalline structure and orientation of the monolayer and components in the multilayer films were investigated using wide angle X-ray diffraction (WAXD) and Fourier transform infrared (FTIR). Scanning electron microscopy images of the membrane surface and cross-section obtained for the cold and hot stretch ratios of 55% and 75%, respectively, showed larger pores and higher porosity for the HDPE layer compared to the PP. As the level of the applied extension during cold stretching increased, the water vapor transmission rate (WVTR) of the HDPE monolayer improved while the effect was inversed for the PP monolayer. In addition, compared to the monolayer membranes, the multilayer ones showed smaller WVTR. Tensile properties of the precursor films and microporous membranes in the machine and transverse directions (MD and TD, respectively) as well as puncture resistance in the normal direction (ND) were evaluated.

UTILIZATION OF HIGH-SPEED MELT SPUN FIBERS FOR THE PREPARATION OF REFLECTIVE POLARIZING FILM

Tomonori Furuyama, Wataru Takarada, Takeshi Kikutani*

Department of Organic and Polymeric Materials, Tokyo Institute of Technology, 2-12-1, O-okayama, Meguro-ku, Tokyo 152-8552, Japan –Furuyama.t.aa@m.titech.ac.jp ;Takarada.w.aa@m.titech.ac.jp; kikutani.t.aa@m.titech.ac.jp

Abstract - Reflective polarizing films (RPF) consisting of multi-layered structure with controlled optical anisotropy attracts much attention these days since such films can be utilized for the improvement of the brightness of liquid crystalline displays. In this research, we tried to utilize the high-speed melt spinning process for the preparation of fibers applicable for the production of RPFs. When sheath-core bicomponent fibers were prepared using Poly(ethylene 2-6-naphthalene dicarboxylate) (PEN) as the sheath and polystyrene (PS) as the core, refractive index of both components in the direction perpendicular to the fiber axis, n-perp, were similar whereas that in the direction parallel to the fiber axis, n-para, showed significant difference at the take-up velocity of around 3.5 km/min. On the other hand, when sheath-core bicomponent fibers were prepared using poly(1-4- cyclohexylene dimethylene terephthalate) (PCDT) as the sheath and PS as the core, n-para of both components were similar where as n-perp were different at the take-up velocity of around 4.5 km/min. The islands-in-the-sea type bicomponent fibers of PEN/PS and PCDT/PS were also prepared under similar spinning conditions and wavelength-dependent transmittance of those fibers was evaluated. When the incident light was polarized parallel to the fiber axis, PEN/PS fiber showed transmittance of around 50% whereas PCDT/PS fiber showed that of close to 100%. On the other hand, when the incident light was polarized perpendicular to the fiber axis, transmittance of PEN/PS fiber was slightly lower than 100 % whereas that of PCDT/PS fiber was around 80%. High reflectivity of PEN/PS is due to the significant difference of n-perp between the two components whereas high transmittance of PCDT/ PS is due to the perfect matching of n-perp between the two components. The RPFs were prepared embedding the aligned PEN/PS or PCDT/PS fibers in UV-curable resins of controlled refractive indices.

Introduction

Reflective polarizing films (RPF) consisting of multi-layered structure with controlled optical anisotropy attracts much attention these days since such films can be utilized for the improvement of the brightness of liquid crystalline displays. On the other hand, it has been proposed[1)] that the RPF can be prepared by embedding aligned bicomponent fibers into film base, in that cross-sectional configuration and optical anisotropy of each component in the bicomponent fibers are highly controlled. However, technologies for the production of bicomponent fibers applicable for such device have not been established yet. In this research, we tried to utilize the high-speed melt spinning process for the preparation of fibers with high optical functionality. In the high-speed melt spinning of bicomponent fibers, birefringence of each component are known to be affected significantly depending on the differences in the composition, temperature dependence of viscosity, solidification temperature etc. between the two components.

Experimental

High-speed spinning process was utilized for the preparation of sheath-core type and islands-in-the-sea type bicomponent fibers. Poly(ethylene 2-6-naphthalene dicarboxylate) (PEN) and poly(1-4-cyclohexylene dimethy-lene terephthalate) (PCDT) were used as the sheath or sea component whereas polystyrene (PS) was used as the core or islands component. Spinning conditions are summarized in Table 1. Refractive indices of each component in the directions parallel and perpendicular to the fiber axis ($n_{//}$ and $n_{\perp}$) were analyzed using an interference microscope. Dependence of anisotropic transmittance of obtained fibers on wave length was analyzed using a spectrophotometer.

Table 1 Conditions for bicomponent fiber spinning.

sheath/core (S/C) or Sea-islands (S-I)	Extrusion Temp. (°C)	Through-put (g/min)	Take-up Velocity (km/min)
S/C=PEN/PS	320	4.5 / 1.5	0.5 ~ 7.0
S/C=PCDT/PS	280	4.5 / 1.5	0.5 ~ 6.0
S-I=PEN-PS	320	3.0 - 3.0	0.5 ~ 7.0
S-I=PCDT-PS	300	3.0 - 3.0	0.5 ~ 6.0

Results and Discussion

Variation of refractive indices of sheath and core components with take-up velocity

Relation between take-up velocity and refractive indices of sheath and core component in PEN/PS and PCDT/PS bicomponent fibers are shown in Figs.1 (a) and (b). With an increase in the take-up velocity, $n_{//}$

and $n_{\perp}$ of PEN increased and decreased, respectively, indicating that there was a significant development of molecular orientation. On the other hand $n_{\perp}$ of PS remained constant. These orientation behaviors can be explained considering the difference in glass transition temperature (Tg) of both components, in that PEN has higher Tg than PS. It was confirmed that the $n_{\perp}$ of PEN and PS matched at the take-up velocity of around 3.5 km/min. In the case of PCDT/PS fibers, $n_{//}$ and $n_{\perp}$ of PCDT increased and decreased, respectively, whereas $n_{//}$ of PS decreased with an increase in the take-up velocity. Since PS has negative intrinsic birefringence, it can be understood that the molecular orientation of both components increased with the take-up velocity. It was also confirmed that $n_{//}$ of PCDT and PS matched at around 4.5 km/min.

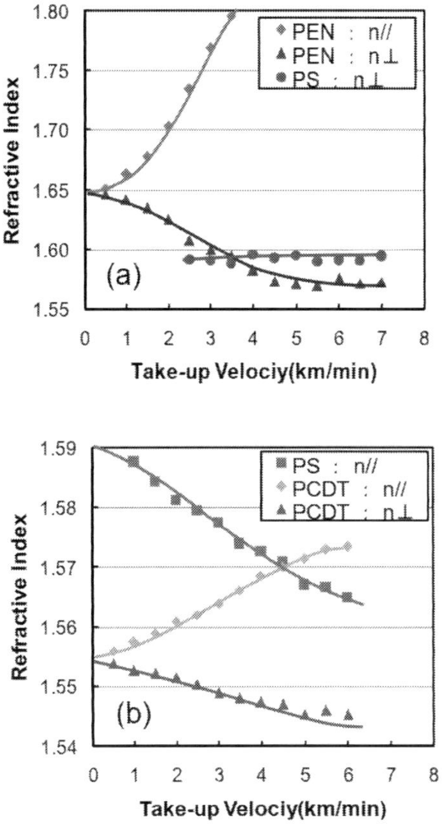

Figure 1 Refractive index versus take-up velocity for high-speed spun sheath-core bicomponent fibers of (a)PEN/PS and (b)PCDT/PS.

Interference fringe patterns of PEN/PS fiber prepared at 3.5 km/min and PCDT/PS fiber prepared at 4.5 km/min observed under an interference microscope are shown in Figs.2 (a) and (b). When the PEN/PS fiber was observed with the light polarized perpendicular to the fiber axis, there was a continuity of the interference fringe indicating that the refractive indices of both components were similar. Oppositely, when the PCDT/PS fiber was observed with the light polarized parallel to the fiber axis, there was a continuity of the interference fringe. Slight distortion of fringe pattern in PEN/PS fiber is probably due to the development of orientation distribution in the cross section of PEN component.

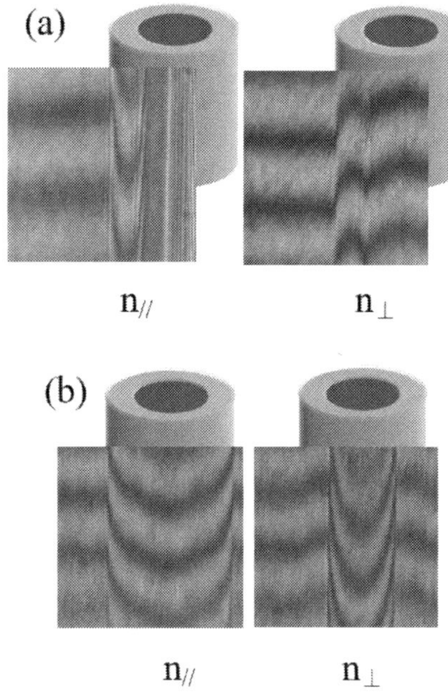

Figure 2 Interference fringe patterns for (a)PEN/PS sheath-core fiber obtained at 3.5 km/min and (b)PCDT/PS sheath-core fiber obtained at 4.5 km/min.

Transmittance of prepared fibers for the lights polarized parallel or perpendicular to the fiber axis

Wavelength-dependent transmittance of PEN/PS and PCDT/PS fibers with the sheath-core type cross-sectional configuration obtained at appropriate spinning velocities are compared in Fig.3. When the incident light was polarized parallel to the fiber axis, PEN/PS sheath-core fiber showed transmittance of around 60% whereas PCDT/PS sheath-core fiber showed that close to 100%. On the other hand, when the incident light was polarized perpendicular to the fiber axis, transmittance of PEN/PS sheath-core fiber was slightly lower than 100 % whereas that of PCDT/PS fiber was around 90%.

In other words, if the RPFs are prepared using the PEN/PS and PCDT/PS sheath-core fibers, those films should have the characteristics of high reflectivity and high transmittance, respectively. High reflectivity of PEN/PS is due to the significant difference of $n_{//}$ between the two com-ponents whereas high transmittance of PCDT/ PS is due to the perfect matching of $n_{//}$ between the two components.

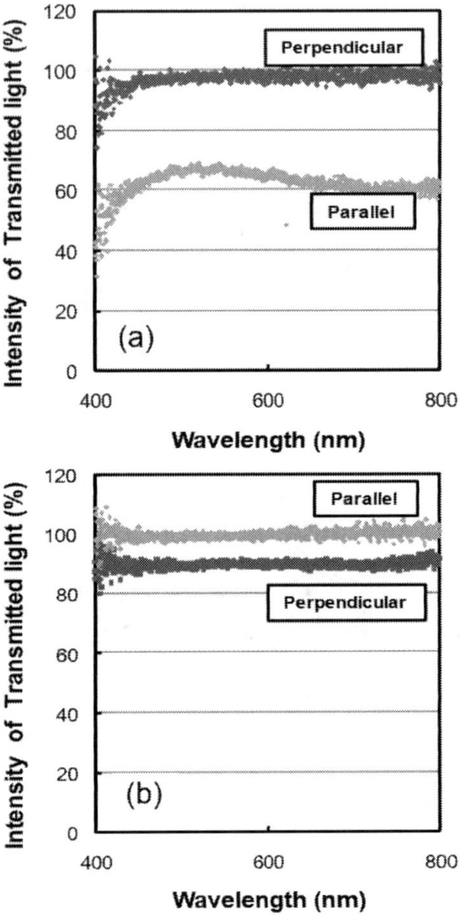

Figure 3 Intesity of transmitted light for parallel and perpendicular polarizations to the fiber axis for sheath-core fibers of (a)PEN/PS and (b)PCDT/PS .

If change of cross-sectional configuration from the sheath-core type to the islands-in-the-sea type is considered, reduction of transmittance can be expected because of the significant increase of interfacial area between the two components. Transmittance of the islands-in-the-sea type fibers with 261 islands are shown in Fig.4. In case of PEN/PS fiber, transmittance for the perpendicular polarization decreased slightly whereas that for the parallel polarization decreased more significantly to the level of around 50%. On the other hand, in case of PCDT/PS fiber, transmittance for the parallel polarization remained at around 100%, whereas there was a decrease of transmittance for the perpendicular polarization from 90% to 80%. These results indicate that the change of the cross-sectional configuration is useful for the improvement of the optical functionality of the fibers when those are applied for the preparation of reflective polarizing films.

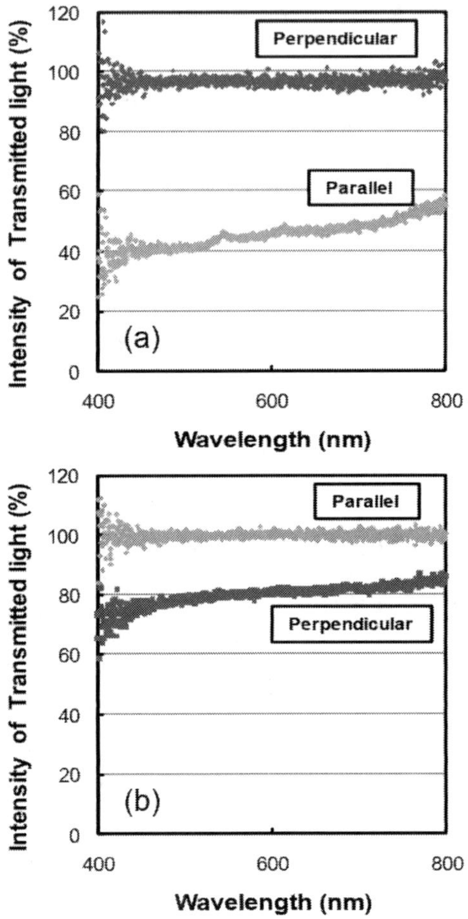

Figure 4 Intesity of transmitted light for parallel and perpendicular polarizations to the fiber axis for islands-in-the-sea fibers of (a)PEN/PS and (b)PCDT/PS . Number of islands in each fiber is 261.

For the further improvement of the optical properties, the islands-in-the-sea type PEN/PS fibers with 1519 islands were prepared. Cross-sections of the sheath-core and islands-in-the-sea fibers were observed under a scanning electron microscope after the extraction of PS component as shown in Fig.5.

Result of the transmittance measurement is shown in Fig.6. From this result, it was confirmed that the prepared PEN/PS fibers with the diameter of around 40 μm showed the transmittance of less than 40% (reflectance of higher than 60%) for the parallel polarization keeping the transmittance for the perpendicular polarization at a level close to 100%.

(a)

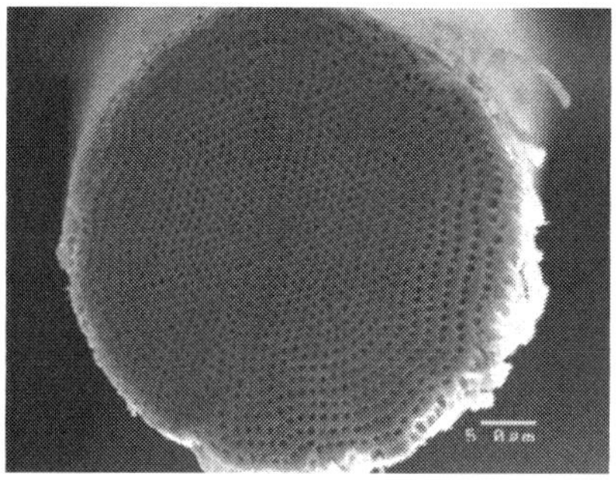

(b)

Figure 5 Scanning electron micrographs of (a)PEN/PS sheath-core and (b)PEN/PS islands-in-the-sea fiber with 1519 islands

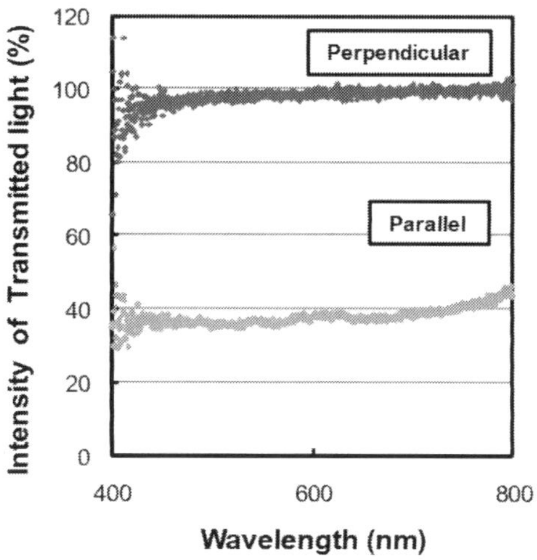

Figure 6 Intesity of transmitted light for parallel and perpendicular polarizations to the fiber axis for islands-in-the-sea fibers of PEN/Pswith 1516 islands.

Preparation of reflective polarizing film

Prepared PEN/PS fibers were aligned in parallel and embedded in UV-curable resin. The refractive index of the resin was controlled close to that of the fiber in the direction perpendicular to the fiber axis. The prepared film was observed using a polarizing film. As shown in Fig. 7, the film was almost translucent when direction of polarization was perpendicular to the fiber axis, whereas there was a significant reflection of light when direction of polarization was parallel to the fiber axis.

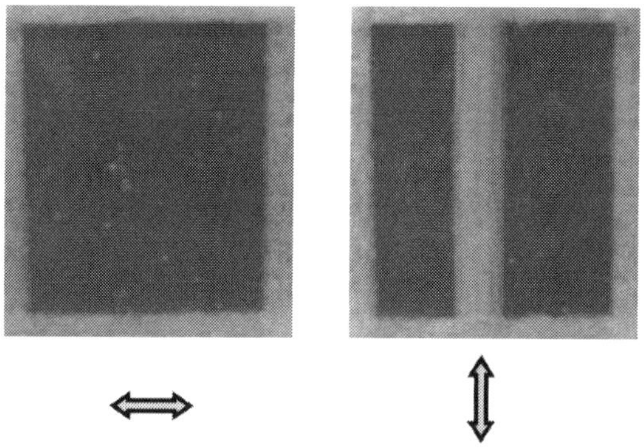

Figure 7 Observation of PEN/PS islands-in-the-sea fiber bundle embedded in UV resin. Direction of polarization is indicated in the figure.

Conclusions

High controllability of molecular orientation and resultant optical anisotropy of each component in high-speed melt spinning of bicomponent fibers was utilized for the preparation of fibers applicable for the construction of reflective polarizing film.

References

1. Toshiyuki Watanabe, Kenrou Totani, Hideki Hayashi, Masaaki Kubota, Masamitsu Ishitobi, Yoshiki Matsuoka,WO2005-008302

Functional fibres prepared by melt spinning of conducting polymer composites

*Martin Karl Straat, Bengt Hagstrom**

Swerea IVF - Sweden

Melt spinning is the preferred technology for high volume production of fibres. The process induces high anisotropy and sometimes dramatic changes are seen in the physical properties of the material, both wanted and unwanted. Melt spun fibres prepared from conducting polymer composites (CPC) offer great possibilities in tailoring the properties of fibres for production of yarns with a high degree of functionality. Conducting fibers prepared from CPC however are limited to antistatic and ESD purposes due to limitations in processing such as reduction of ultimate draw down ratio with increasing filler content and loss of conductivity during drawing. New fillers such as carbon nanotubes (CNT) have been suggested as promising solutions. In the presented work we mix polyethylene (PE) or polypropylene (PP) with relatively high filler concentrations from two CNT-masterbatches, where one was prepared by in-situ polymerisation and one by melt compounding. A highly conductive grade of CB was also tested as a reference. Electrical percolation curves between 2 and 14 wt% were made as well as rheological measurements for the blends. The influence of drawing on conductivity was studied and the ultimate draw down ratio at 4 wt% filler content for different die geometries and processing rates was determined. A relationship was found between shear viscosity and conductivity of the materials. The ultimate draw down ratio decreased significantly with increasing filler content above the percolation threshold of each mixture, materials with high conductivity generally had low ultimate draw down ratio. Capillary geometry and extrusion rate showed significant influence on the ultimate draw down ratio as expected. Conductivity was lost gradually with increasing draw ratio when CB was used. For CNTs however, no relationship could be established since conductivity was reduced below the level of detection for all draw ratios and concentrations studied. Using bicomponent technology on a labscale equipment, a fiber with both high conductivity and high strength was produced.

RHEOLOGICAL AND MORPHOLOGICAL STUDIES OF PLA/CLAY NANOCOMPOSITES BLOWN FILM

Satpal Singh[1], Anup K. Ghosh[1], S. N. Maiti[1], Rahul Gupta[2] and S.N. Bhattacharya[2]
[1]*Centre for Polymer Science & Engineering, Indian Institute of Technology, Delhi, New Delhi, India*
E-mail: anupkghosh@gmail.com
[2]*Department of Chemical and Metallurgical Engineering, Rheology and Materials Processing Centre, RMIT University, Melbourne, Vic., Australia E-mail: rahul.gupta@rmit.edu.au*

Abstract- Polylactic acid (PLA) is a highly versatile biodegradable polymer derived from 100% renewable resources. PLA nanocomposites containing 1, 3 and 5 phr of nanoclay were compounded in a co-rotating twin screw extruder, to study the effect of nanoclay content on film processibility of composites as well as properties of blown films. Film blowing processing conditions such as temperature, rpm, feed rate, air pressure and nip roll speed were optimized. Differential scanning calorimetry (DSC) data showed that the cold crystallization and melting temperatures were influenced by the presence of nanoclay. The thermogravimetric analysis (TGA) showed increase in the decomposition temperature for all the nanocomposites. The measurement of rheological properties of Polylactide/nanoclay helped in understand the processability as well structure-property relationship in nanocomposites. Temperature dependent (DMTA) and Frequency dependent (Oscillatory shear Rheology) viscoelastic spectra showed an increase in the storage and loss moduli with the increase in the organoclay content. Wide angle X-ray diffraction (WAXD) and Transmission electron microscopy (TEM) were used to determine the degree of intercalation and exfoliation as a function of nanoclay content of nanocomposites. XRD data showed complete exfoliation in the film at lower nanoclay content. On increasing the nanoclay content both the exfoliated and partially intercalated structures were obtained. Nanocomposites blown film specimen showed the improvement in tensile properties.
Keywords: Poly (Lactic Acid), Nanoclay, Rheology, Morphology, DMTA, Thermal..etc

Introduction: In an attempt to alleviate the global solid waste disposal problem and to reduce the dependence on petroleum-based plastics for packaging materials, renewable resources based biodegradable and compostable thermoplastic polymers are required. These renewable resources based systems are expected to play an important role as materials for humanity through their exploitation in a progressively more elaborated fashion. Poly (lactic acid), PLA is at present one of the most promising renewable resources based biodegradable polymer[1-4]. PLA is commercially and largely available in a wide range of grades. Hence, fundamental understanding of its rheological and morphological properties becomes critical for effective processing of this material. In a recent study have been done based on Poly (L-lactic acid)/layered silicate nanocomposite blown film for packaging application [5].

Experimental:

Compounding of PLA/Nanoclay: Poly (L-lactic acid), PLA (Nature works ® 4032D), was procured from Cargill/Dow LLC (Minnetonka, MN, USA). Organically treated montmorillonite layred silicate, Cloisite® 30B containing methyl-bis(2-hydroxy-ethyl) tallow alkyl ammonium cations was obtained from Southern Clay Ltd (Texas, USA). Different composition of Polylactic acid and Nanoclay (1%, 3% and 5%) were compounded in 16mm thermoprism co-rorating twin – screw extruder (TSE-16) (L/D 40:1). The processing temperature 180-200 °C at screw speed 100 was maintained.

Blown film processing: The compounded PLA/Nanoclays were extruded in a laboratory Haake Lab scale blown film system (Figure 1).

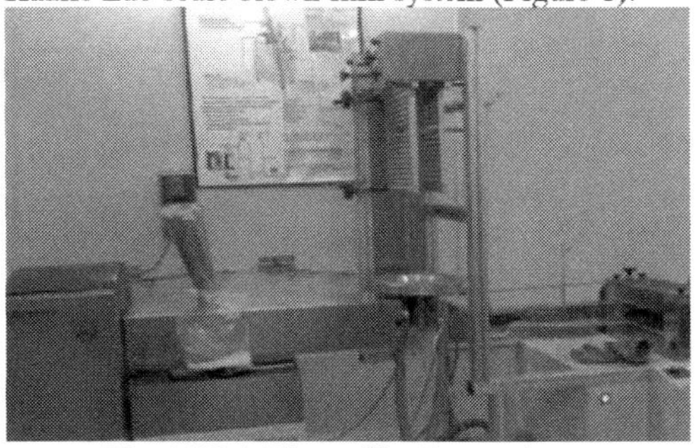

Fig. 1 PLA Blown film Processing.

The processing temperature 200 °C, screw speed 60 rpm and Nip Roll Speed 700 cm/min. were used and resulting film thickn where range of 38-62μm.

Characterization:

Thermal characterization: The thermal properties of the developed PLA nanocomposite were investigated with the PERKIN-ELMER DSC 7 differential scanning calorimeter (DSC). Samples were heated from 0 to 200°C at 10°C/min. Thermal stability was assessed using PERKIN-ELMER Pyris 6 TGA with a heating rate of 20°C/min up to 750°C under flowing nitrogen.

Dynamic mechanical properties were measured with an TA instrument DMA Q 800 in a single cantilever bending mode. The dynamic storage and loss moduli were determined at a constant frequency (ω) of 1 HZ or as a function of temperature from -90°C to 150°C at a heating rate of 5°C/min.

Morphological characterization: Freshly cut glass knives with cutting edge of 45° were used to get the cryosections of 50nm thickness by using a leica ultracut UCT microtome. JEOL-2100 electron microscope (Tokyo, Japan) having LaB6 filament and operating at an accelerating voltage of 200 kv was utilized to obtain the bright field images of the cryomicrotomed samples. Blown film samples were investigated in the transmission mode (coupled θ/2θ) to check the effect of the nano-particle orientation. A wide angle goniometer was coupled to a sealed-tube source of filtered Cu Kα radiation operating at 40 kv and 30 ma (PW 3040/60 X' Pert PRO).

Rheological characterization:

Melt rheological measurements were carried out with an advanced research grade rheometer (ARES; Rheometric scientific). The measurements were performed with a parallel plate geometry using 25 mm diameter plates and the sample thickness was ~ 1.75 mm. The compression molded samples were investigated at temperature 180°C.

Mechanical properties: The mechanical properties of the developed blown film samples were studied in tensile loading conditions using a ZWICK/Z010 tensile tester according to ASTM D 882 method. The cross head speed was kept at 5mm/min. The gauge length of the test samples was 70mm and the width of the sample was 15mm.

Result and Discussion:

DSC thermogram data showed that the cold crystallization temperature (T_{cc}) and melting point (T_m) slightly decreased with increasing clay content possible due to nucleating effect of the nanoclay. TGA thermogram data showed that the degradation temperatures increased by (6-7°C) the addition of nanoclay content.

Tabe 1: Thermal and Mechanical Properties of PLA/clay nanocomposite and their fims.

Sample	PLA/ nanoclay	T_{cc} (°C)	T_m (°C)	Degradation Temp. (°C)	Tensile Modulus (MPa)	Tensile Strength (MPa)	Elongation (%)
PNC0	100/0	110.8	170.1	374	1310 ± 100	28.65 ± 0.5	11.66 ± 1.0
PNC1	100/1	107.9	169.5	378	1468 ± 80	28.09 ± 1.0	16.30 ± 0.5
PNC3	100/3	106.8	169.5	381	1061 ± 40	20.37 ± 1.1	17.98 ± 0.3
PNC5	100/5	101.1	168.5	381	884 ± 60	19.52 ± 1.0	10.98 ± 1.0

The rheological data (figure 2 & 3) showed the G' and G'' of PNCs were higher than those of pure PLA at all frequencies and increased with increasing clay content especially at low frequencies (< 10 rad/s). In the case of homopolymer samples, it is expected that they would exhibit the characteristic low-frequency terminal behavior expressed by the power laws $G' \alpha \omega^2$ and $G'' \alpha \omega$, as was seen for pure PLA. The PLA nanocomposite samples, however, showed a gradual deviation from the terminal liquid-like behaviour to a solid-like behaviour with increasing clay concentration, as evidenced by the decrease in the slope of low frequency G' and G'' with respect to frequency, plotted on log-log scale. For PNC 5 the solid-like behavior was most pronounced. PNC5 showed largest increase in both G' and G'' at low frequencies. The complex viscosity (η*) graph of PNC0 showed a Newtonian plateau at low and mid frequency ranges followed by a small shear-thinning behaviour at the highest frequencies (Figure 4). The Newtonian plateau viscosity was also observed for the nanocomposite with the lowest loading, PNC1. However, at higher clay loadings i.e. for PNC3 and PNC5, no Newtonian plateau could be observed within the frequency range and the viscosity continued to increase with decreasing frequency indicating a pseudo-yield-stress behaviour at low frequencies. Such pseudo-yield-stress behaviour was most prominent for the nanocomposite (PNC5) having the highest clay loading. The multi wave ramp test showed that for PNC5 at low frequency (1 rad/s) G' was slightly improved with time due to formation of percolating network structure (Figure 5). At low shear rate, steady shear viscosities were increased with increasing clay content due to flow restriction. But at high shear rate, shear viscosity values decreased with increasing clay content, which may be due to the orientation of silicate layer along the shear direction (Figure 6 & 7).

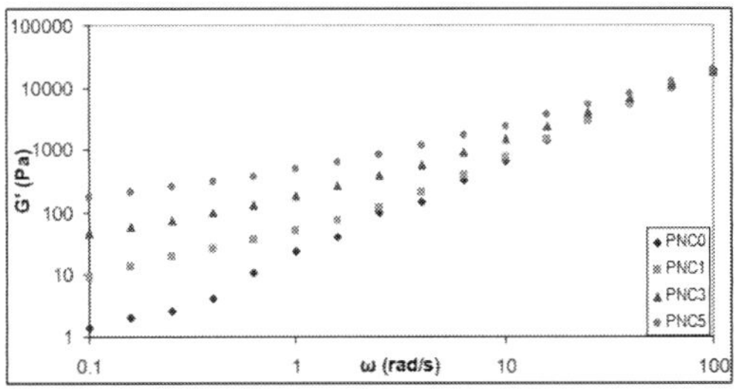

Fig.2: Storage Modulus (G') of PLA nanocomposites

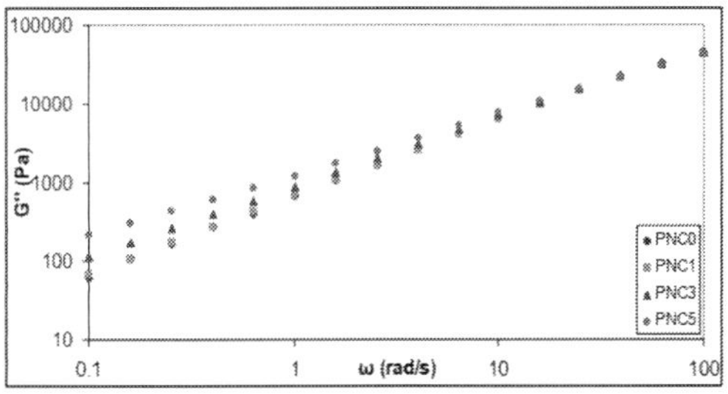

Fig.3: Loss Modulus (G') of PLA nanocomposites

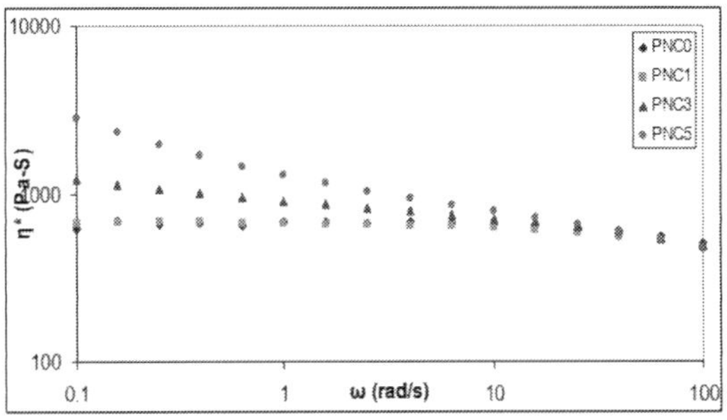

Fig.4:Complex Viscosity of PLA nanocomposites

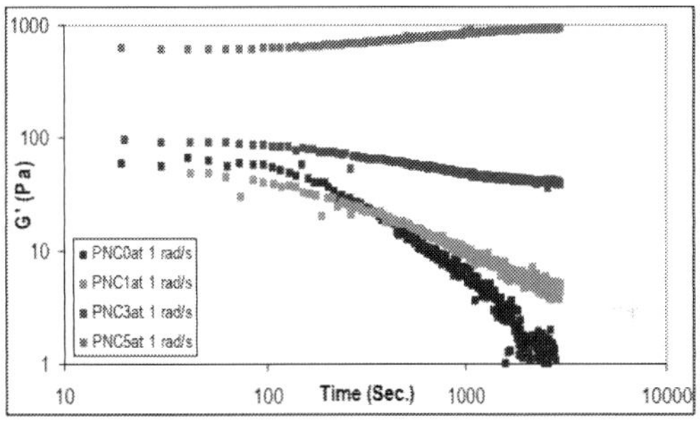

Fig.5:Multi Wave Temperature Ramp Test at 1 rad/s

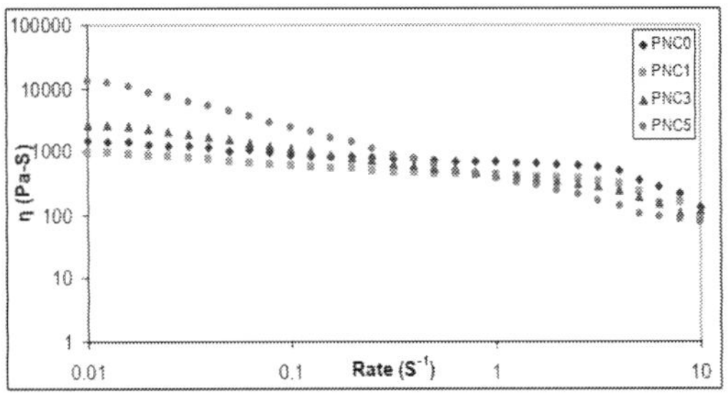

Fig. 6: Steady Shear Viscosity at low shear rate

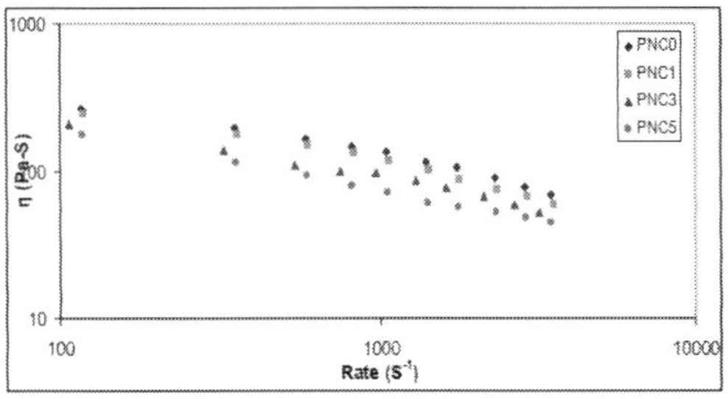

Fig. 7: Steady Shear Viscosity at high shear rate

DMTA studies were performed on solid nanocomposite samples to examine the effect of temperature on their storage modulus, loss modulus and tan δ. With addition of nanoclay in PLA matrix, increase in the storage and loss modulus values were observed. The degree of PLA/Nanoclay intercalation was assessed using both X-ray and TEM methodologies. Figure 8 illustrated the difference in WAXD patterns of the nanocomposite PNC1, PNC3 and PNC5. A shift to lower 2θ values was clearly observed in all nanocomposite samples. The corresponding d-spacings of the processed films indicated intercalated dispersion state with d-spacing corresponding to the (001) reflection of the nanoclay doubling from 18.7 to 35 A° for PNC5.

The Transmission electron microscopy (TEM) images corresponding to the samples corroborates the XRD measurements. PNC1 showed the exfoliated structure, PNC3 showed exfoliated and partially intercalated structure and PNC5 showed exfoliated and intercalated structure. Tensile properties (Table 1) showed the improvement of % elongation, tensile strength and tensile modulus up to PNC1. PNC1 showed the improvement in elongation % due to the orientation of nanoclay plates along the machine and transverse direction.

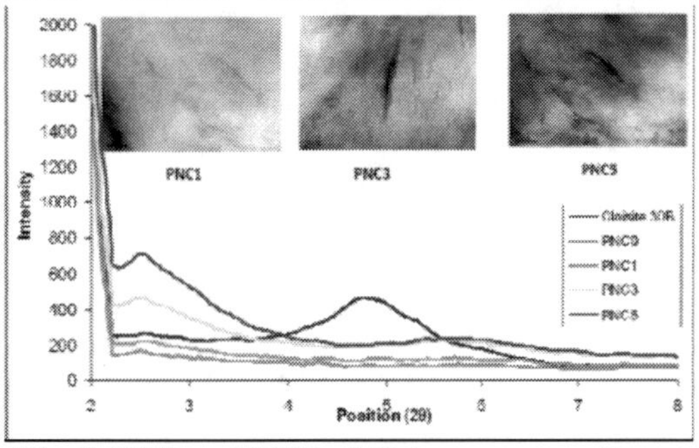

Fig. 8: XRD and TEM of PLA nanocomposites

Conclusions:

This paper presents a comprehensive melt rheological and morphological study of PLA nanocomposite. The frequency sweep data showed a monotonic increasing in storage moduli, loss moduli and dynamic viscosities with nanoclay content. In the case of PNC5, a pseudo-solid-like behavior was observed. Wide angle X-ray diffraction (WAXD) and Transmission electron microscopy (TEM) results indicated that PNC1 showed the exfoliated structure, PNC3 showed exfoliated and partially interclated structure and PNC5 showed exfoliated and interclated structure.

Acknowledgements:

1. Department of Science & Technology (DST) New Delhi , India
2. RMIT University, Melbourne, Australia

Reference:

1. Anderson KS, Lim SH, Hillmyer MA. *Journal of Applied Polymer Science,* **89**: 3757;2003.
2. Defeng W, Liang W, Lanfeng W, Ming Z. *Polymer Degradation and Stability,* **91**:3149;2006.
3. Thellen C, Orroth C, Froio D. *Polymer,* **46**: 11716; 2005.
4. Lee SK, Seong DG, Youn JR *Fibers and Polymers*, **6**: 289; 2005.
5. Satpal Singh[1], Anup K. Ghosh[1], S. N. Maiti[1], Rahul Gupta[2] and S.N. Bhattacharya[2] *Special issue of journal of Polymer Engineering*, **accepted (In Press).**

ANALYSIS OF THE DEFORMATION BEHAVIOR OF MULTILAYER FILM IN A TRANSVERSE STRETCHING PROCESS WITH A TENTER

T. Tosa[1]*, T. Yamada[2] , H. Toya[1] and T. Sawasato[1]

[1] *Polymer Technology Institute, Denki Kagaku Kogyo K. K., Chiba, Japan – takahiro-tosa@denka.co.jp*
[2] *Faculty of Engineering ,Kanazawa University, Ishikawa, Japan – tyamada@t.kanazawa-u.ac.jp*

In recent years, plastic films have become widely in a variety of for wide fields, such as packaging materials and optical applications. Since the need for high-performance films continues to increase, it is necessary to impart a high degree of functionality to films. A multilayer film composed of two or more polymers has been developed in order to achieve a higher functioning material. Though a uniform film thickness is a very important property, attempts to predict it in the film manufacturing process have been limited to monolayer films and have are not known for multilayer films. Generally, flat films are made by a successive stretching method composed of a longitudinal stretching process, employing the difference in speed between rolls, and a transverse stretching process with a tenter.
This work attempts to simulate the deformation behavior of multilayer film in a tenter with a finite element method. The differing polymer characteristics of each layer composing a multilayer film makes it difficult to simulate the deformation behavior, such as the thickness distribution of film in a tenter. In this work, we present a new simulation model to predict the thickness distribution of a multilayer film in a tenter. We have confirmed that the proposed model can make such predictions judging from the fact that the predicted thickness distribution by computer simulation is in good agreement with experimental ones obtained from a pilot plant.

Introduction

Plastics film products are used in many fields, including food packaging materials and industrial products, components of optical devices such as flat panel displays, and electronic devices such as flexible printed circuits. The performance requirements for these films are becoming more severe every year. In order to impart a high degree of functionality to film products, multi layered films with resins of different characteristics have been developed.

The use of a tenter is a common aspect of the film manufacturing process. The transverse direction stretching behavior of films in the tenter causes significant impacts on the process stability. Also, the transverse direction stretching process is a critical process in determining the thickness distribution that is the most important characteristic of films. Resin characteristics such as elongational viscosity and elastic modulus and stretching conditions such as stretching speed and process temperature all affect the thickness distribution of the resulting films. Their influences on the thickness of films in the stretching process are too complicated to easily forecast. Therefore, many studies have examined analysis methods to predict film stretching behavior. However, the only analytical models that exist are for monolayer films.

Multilayered films are laminated with resins having different processing characteristics. Therefore, the deformation behavior of these films is much more complicated than those of monolayer films. As a result, better methods of precision analysis have not yet been proposed. In this study, the behavior analysis of film deformation while stretching multilayered films was attempted with the finite element method. In conventional approaches, the rigid plastic model was often used for the calculation of film deformation from the viewpoint of convergence. To better model multilayer films, we attempted calculations with an elastoplastic model in this study.

The results of the calculations of the thickness distribution in the width direction were in good agreement with those of stretching experiments in the pilot plant. Hence, we feel that our analysis method has been proved to be appropriate. The calculations and experimental results were also in good agreement with regard to the changes in the thickness distribution when the stretch ratio was changed. In addition, we examined how differences in the thickness composition in each layer of sheets before the stretch affected the thickness distribution of post-stretching films.

Experimental

Resin

Amorphous styrene-butadiene-styrene block copolymer (SBS resin) was used for this study. The properties of this resin change depending on the ratio between stylene and butadiene. Two types of SBS resin, named SBS-A and SBS-B were used to make multi layered films. The stress-strain characteristic of each resin was measured with the uni-axial elongational viscometer (Toyo Seiki Seisaku-sho, Ltd.) shown in Figure 1. An example of such a measurement is shown in Figure 2. The SBS-A resin shows elastoplasticity with a strong strain-hardening property. On the other hand, the SBS-B resin shows elastic behavior and its elastic modulus is low compared with that of SBS-A.

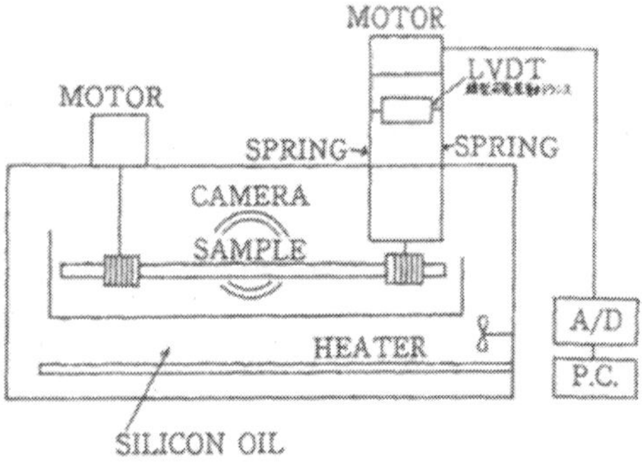

Figure 1 – Schematic diagram of the elongation viscometer.

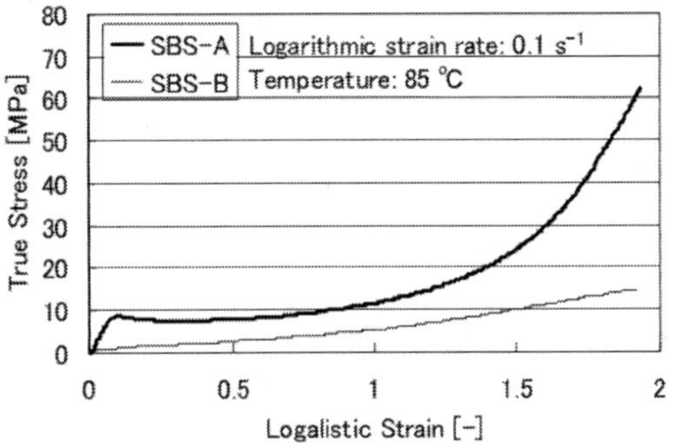

Figure 2 – Comparison of stress-strain curves for SBS-A and SBS-B at a temperature of 85° Celsius and a strain rate of 0.1 s^{-1}.

Making multi layered film

Using a successive biaxial stretcher, we made two types of three-layer films whose surfaces are SBS-A and central layer is SBS-B. A schematic diagram of the successive biaxial stretcher is shown in Figure 3. In this study, machine direction stretching was not performed (the machine direction stretch ratio is 1.0 times) but three types of multi layered films were obtained using transverse direction stretch ratios of 2.5, 3.5 and 4.5 times. The line speed was 10m/min and the temperature for the transverse direction stretching process inside the tenter was 85° Celsius.

Thickness measurement samples were cut from the sheet before stretching in front of the tenter inlet. Cross sections were examined with the optical microscope and the thickness of each layer of the sample determined. The thickness distribution of each layer in the width direction is shown in Figure 4. The thickness of the central layers varies depending on the width direction, being thicker in the center and thinner at the ends. Central layers have convex distribution patterns.

The thickness of each layer of the film stretched in the transverse direction in the tenter was measured by observing the cross-section of the film with a scanning electron microscope. The thickness distribution over the entire width direction of transverse direction stretched films for the different stretch ratios is shown in Figure 5. Both ends of the film are thicker because they were pinched with chucks. The thickness in the width direction becomes uniform as the stretch ratio becomes larger.

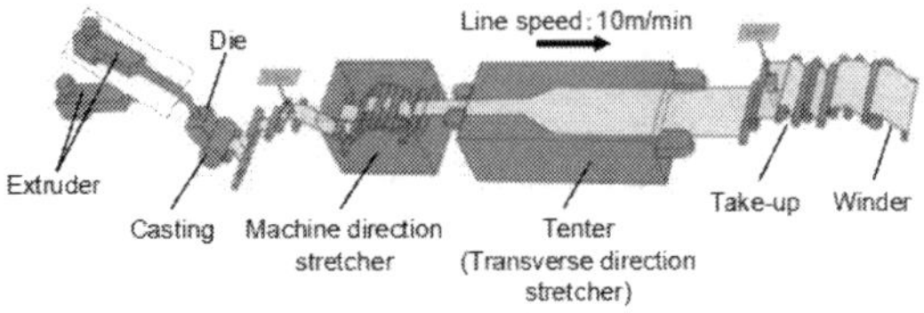

Figure 3 – Schematic diagram of the biaxially oriented stretcher.

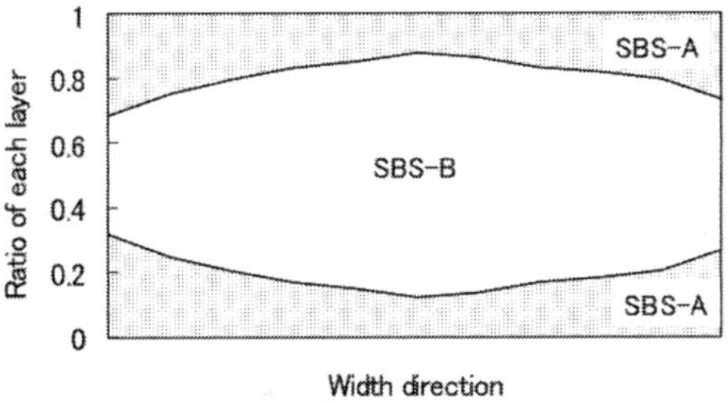

Figure 4 – Schematic diagram of the ratio of each layer in width direction for an unstretched sheet.

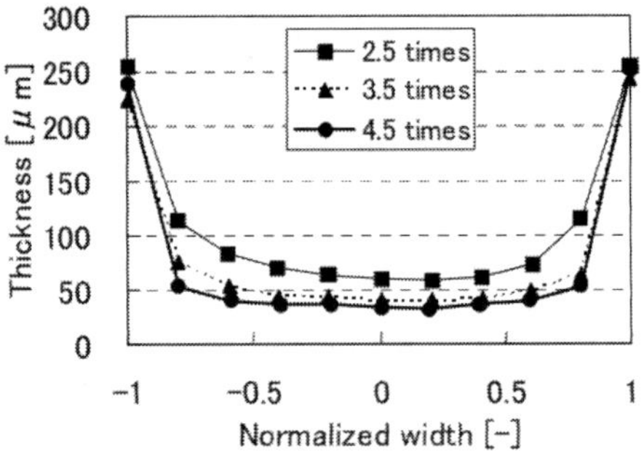

Figure 5 – Thickness distributions after stretching.

Analysis

Analytical conditions

The target of the analysis is the transverse direction stretching process inside the tenter. The stretching behavior was calculated with the same stretch ratios as the experiment, 2.5, 3.5 and 4.5 times. The line speed was 10m/min, the stretching temperature 85° Celsius and the initial sheet thickness was 300μm. The thickness of each layer in the width direction of the sheet before stretching sheet is shown in Figure 6. The program Marc (MSC.Software Corporation), general purpose nonlinear structural analysis software, was used for the calculations.

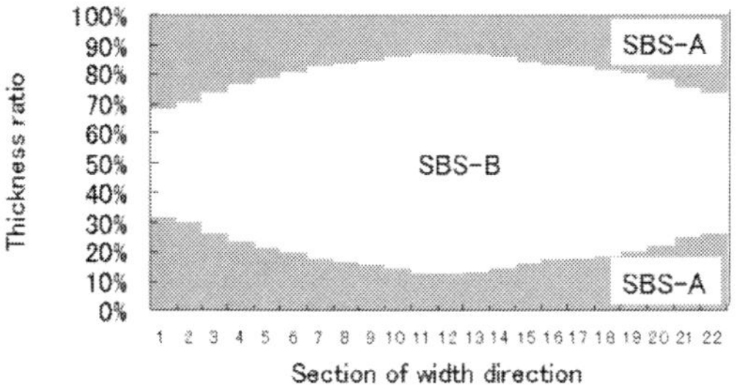

Figure 6 – Schematic diagram of the ratio of each layer in width direction used for analysis..

Material properties

A composite option that enables definition of the material properties for each layer was used. Since the surface SBS-A exhibited noticeable strain-hardening property, a elastoplastic model was employed, and stress was defined as the strain-speed dependent functions in the equations (1) to (6).

$$\sigma = a + b\varepsilon + c\left(\frac{\varepsilon}{2}\right)^{5.5} - \frac{d}{(1+\varepsilon)^{e}} \quad (1)$$

$$a = 26.59\dot{\varepsilon} + 3.817 \quad (2)$$

$$b = 13.82\dot{\varepsilon} + 2.160 \quad (3)$$

$$c = 4.495\dot{\varepsilon} + 0.880 \quad (4)$$

$$d = 26.59\dot{\varepsilon} + 3.817 \quad (5)$$

$$d = 26.59\dot{\varepsilon} + 3.817 \quad (6)$$

wher, σ is stress, ε is strain, $\dot{\varepsilon}$ is strain-speed.

The stress-strain relations shown in equations (1) to (6) for SBS-A at a temperature of 85° Celsius and strain-speeds of 0.05 and 0.1 s^{-1} are compared with the measurement results in Figures 7 and 8. In both cases, the predictions and measurements agree well with each other. Since SBS-B in the central layer showed little strain-hardening property, an elastic model was used. The calculation of the elastoplastic model was stabilized by an elemental decomposition method and used suitably small time steps.

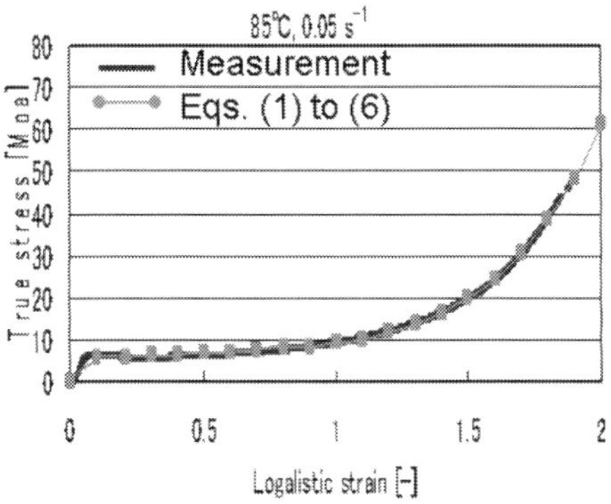

Figure 7 – Comparison of stress-strain curves between experiment and Eqs.(1) to (6) for SBS-A at a temperature of 210° Celsius at an elongational strain rate of 0.05 s^{-1}.

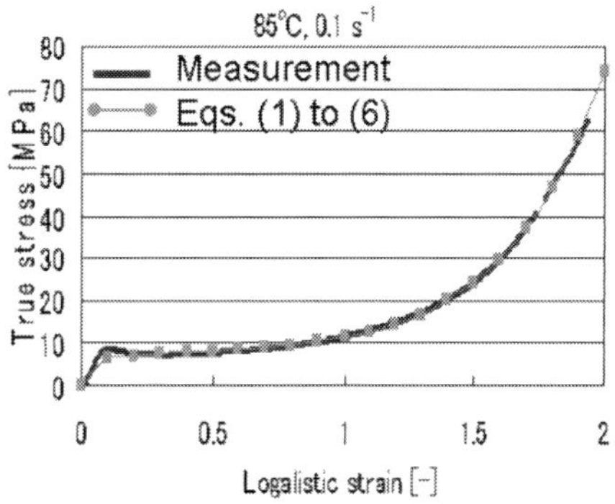

Figure 8 – Comparison of stress-strain curves between experiment and Eqs. (1) to (6) for SBS-A at a temperature of 210° Celsius at an elongational strain rate of 0.1 s^{-1}.

Results and Discussion

Validity of analysis

A comparison between the experimental thickness distribution in the width direction and the analysis result at a stretch ratio of 2.5 times is shown in Figure 9. The experimental and analysis results agree well with each other, indicating that our analysis method is appropriate.

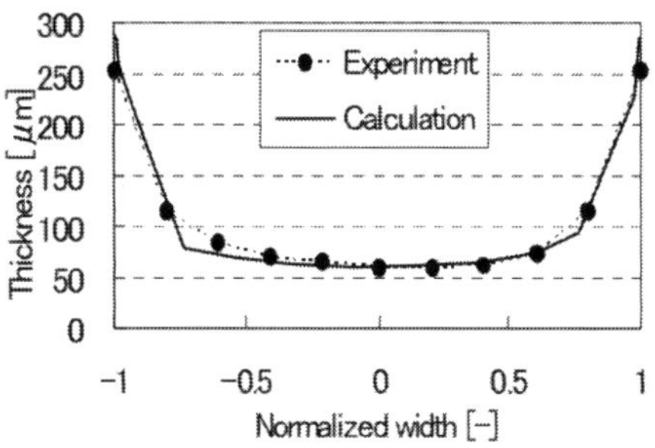

Figure 9 – Comparison of thickness distribution found by experiment and the results of calculations.

Effect of stretch ratio

The analysis results for the thickness distribution in the width direction at stretch ratios of 2.5, 3.5 and 4.5 times is shown in Figure 10. The thickness distribution in the width direction showed a tendency to become even as the stretch ratio became higher. This agrees well with the experimental trend seen in Figure 5.

Effect of initial thickness distribution

The thickness of the central layer of the unstretched sheets used for the experiment had a convex distribution - thicker in the center and thinner at both ends. Where thickness distribution of the central layer varies, that of the stretched film is also expected to vary. Hence, we also calculated the overall thickness distribution in the case where the central thickness was uniform in the width direction. The thickness distribution of stretched films calculated for the uniform case was then compared with that where the central layer was not uniform (Figure 4). The results are shown in Figure 11. The results of both cases varied greatly, showing that the effect on the overall thickness caused by the thickness distribution of the central layer was large.

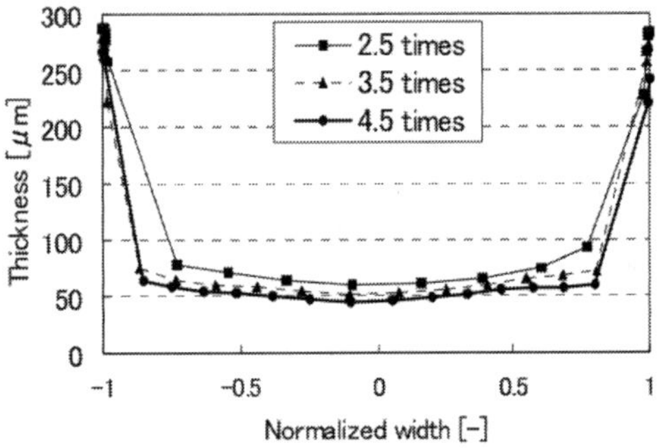

Figure 10 – Results of calculations of the thickness distribution.

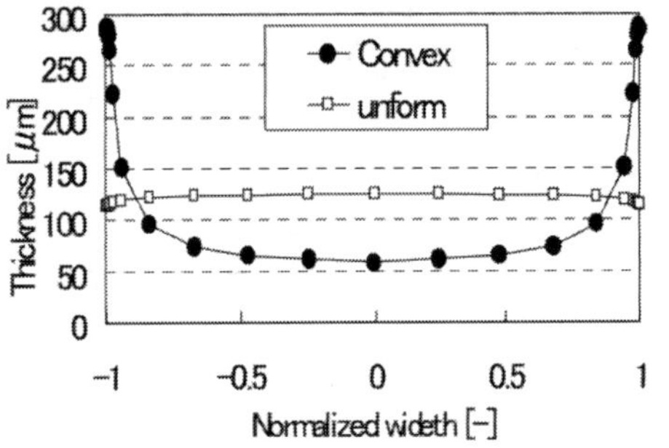

Figure 11 – Comparison of the thickness distribution for convex and uniform central layers.

Conclusions

The film stretching process has been conventionally analyzed by targeting monolayer films. Examples of the analysis of multi layered films have not been reported, because the handling of their material properties in the laminated structure and stretching behavior is very complicated. In this work, we have studied the transverse direction stretching process of multilayered films in the tenter.

An elastoplastic model that correctly represents the stress-strain characteristic of the material was adopted in this study. Calculations with the elastoplastic model are unstable at the inflection point from an elastic to a plastic region. However, the calculation was stabilized by an elemental decomposition method and using suitably small time steps. Moreover, the laminated structure of the films was taken into consideration by assigning different material properties to the layers.

In this way, multilayered films were accurately modeled and the transverse direction stretching behavior was analyzed with the finite element method to yield the following results:

1) The calculated and experimental thickness distributions of the transverse direction stretched film in the width direction agreed well with each other, indicating that this analysis method was appropriate.

2) The thickness distribution in the width direction of the stretched film depended on the stretch ratio; the larger the stretch ratio, the more even the thickness distribution becomes, resulting in a film of better quality.

3) Variations in the thickness distribution of the central layers of the unstretched produces changes in the overall thickness distribution of the stretched film. This suggests that the thickness distribution of the entire film could be controlled by defining the layer composition of the unstretched sheet.

References

1. Yamada T. and Nonomura C., "An Attempt to Simulate the Bowing Phenomenon in Tenter with Simple Models", J. Appl. Polym. Sci., Vol.48, No.8, 1399-1408 (1993)
2. Yamada T. and Nonomura C., "Attempt to Reduce Bowing Distortion in Tentering of Film", J. Appl. Polym. Sci., Vol.52, No.10, 1393-1403 (1994)
3. Yamada T., et al., "Influence of Temperature Distribution on Film Thickness in Tentering Process", Proceedings of the 2004 Annual Meeting and 20th Anniversary Celebration of the Polymer Processing Society (PPS-20), Paper Filed in CD-ROM, 1-8 (2004).
4. Senda, et al., "Influence of Process Conditions on Distribution of Film Thickness in Tentering Process", Proceedings of the 22nd Annual Meeting of the Polymer Processing Society (PPS-22), Paper Filed in CD-ROM, 1-2 (2006)
5. Yamada T., et al., "Prediction of Deformation Behavior in the Film Stretching Process", Proceedings of Asian Workshop on Polymer Processing 2006 (AWPP2006), 177-180 (2006).
6. Yamada T., et al., "Sequential analysis of roll drawing and tentering processes in successive biaxial stretching by FEM", Proceedings of the Europe/Africa Regional Meeting 2007 Polymer Processing Society (PPS-2007), Paper Filed in CD-ROM, pp.1-6 (2007)

Polypropylene-based long-life insecticide-treated mosquito netting

Walter W. Focke[1] and Walter van Pareen[2]

[1]Institute of Applied Materials, Department of Chemical Engineering, University of Pretoria, Lynwood Road, Pretoria 0001, South Africa
[2]Sasol ChemCity, 1 Klasie Havenga St, Sasolburg 1947, South Africa

Abstract

The World Health Organization (WHO) is promoting the use insecticide treated nets (ITN's) in order to "Roll Back Malaria". Members of the class of pyrethroids are suitable insecticides for this application. Long-life netting is preferred and can be achieved by incorporating the active in the fibre polymer as demonstrated by Sumitomo's Olyset^{TM} high density polyethylene mono filament based netting. This study considered the use multifilament polypropylene yarn. The advantage of the PP multifilament yarn is that it provides for a fabric with a softer feel.

Processing temperatures in polypropylene fibres production often exceed 200°C. This disqualifies some WHO approved insecticides owing to insufficient thermal stability. Nevertheless, multifilament polypropylene yarns containing WHO approved insecticides were successfully produced on a production spinning line. The fume generation during extrusion with pyrethroids necessitates superior ventilation, extraction and scrubbing systems.

The tenacity of the polypropylene yarn prepared at a draw ration of 2,37 was about 3,6 N/dtex and the elongation about 60 – 90 %. The incorporation of actives via LDPE masterbatches lowered the tenacity of the yarns by up to 35 % and increased the elongation to break by up to a factor of two. These dramatic property changes did not have a detrimental effect on knitting behaviour.

Efficacy testing results for unwashed socks showed encouraging results. Most unwashed samples containing more than 0,19 % active complied with the WHO requirements (> 95 % knockdown after 60 minutes and > 80 % mortality after 24 h). Analysis reveals that blooming from the thin fibres in the multifilament yarn will be rapid with a time constant a low as one single day. Washing is known to remove in excess of 30 % active on the fibre surface per wash. This implies that it is unlikely that fibres containing amorphous additives such as cyfluthrin will achieve the minimum of 20 washes stipulated by WHO. Better results were obtained with the crystalline insecticides such as β-cyfluthrin as shown in the Figure.

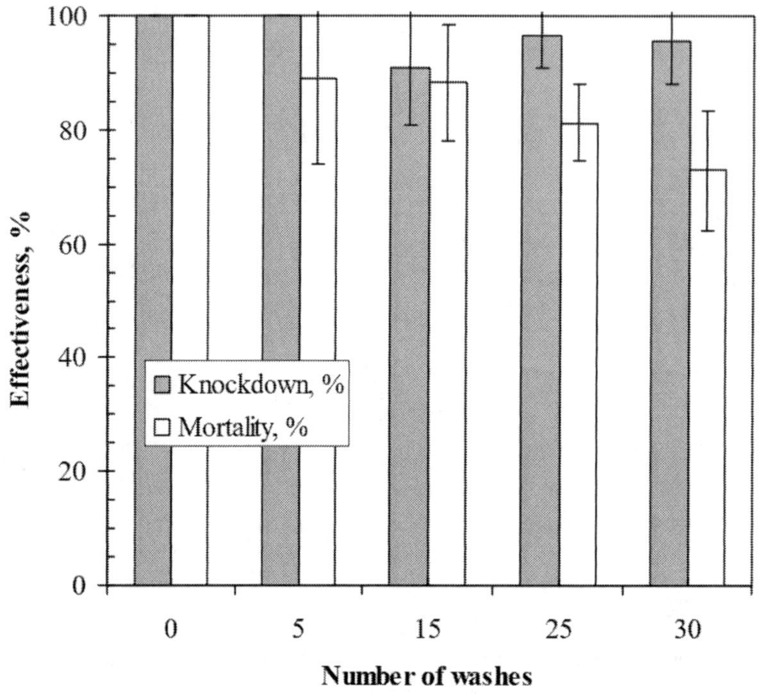

Figure. The effect of washing on the bioactivity of a net containing 0,38 % β-cyfluthrin. The nets were repeatedly washed on consecutive days and performance evaluated using the WHO cone test several days after the last wash.

Prediction of Deformation Behavior of Film during Transverse Stretching with Consideration of Heated Air Flow in a Tenter

*Masayoshi Tokihisa [1], Yoshiyuki Kushizaki [1], Hideki Tomiyama [2], Yousuke Yamamoto [3], Toshiro Yamada [3]
[1] Machinery Research Laboratory, The Japan Steel Works, LTD.
[2] Plastics Processing Machinery Department, The Japan Steel Works, LTD.
[3] Graduate School of Natural Science & Technology, Kanazawa University
E-mail: masayoshi_tokihisa@hiro.jsw.co.jp

The deformation behavior of polypropylene (PP) film during transverse direction (TD) stretching with a tenter was newly predicted under consideration of heated air flow in a tenter in order to obtain the more realistic heat transfer coefficient than that in the previous works based on an impinging jet model or a wall jet model. The heated air flow in a tenter was analyzed with a finite volume method in order to calculate the distribution of heat transfer coefficient on film. The distribution obtained has been confirmed to be more realistic than that in the previous works. The film thickness and the bowing distortion after stretching were analyzed by applying the distribution of heat transfer coefficient based on air flow analysis under the assumption that the deformation of film obeys an elastic-plastic rule. The film deformation behavior such as film thickness distribution, the bowing distortion and the successive distortion patterns of the drawn line on film both machine direction (MD) and TD was experimentally measured with a pilot plant. It has been shown that the experimental results with a pilot plant agree better with the calculated results in this work based on the distribution of heat transfer coefficient obtained from air flow analysis than the previous works based on that obtained from an impinging jet model or a wall jet model. It is revealed that the distribution of heat transfer coefficient has a strong influence on the difference of film temperature and the difference of longitudinal stress, and results in the difference of uniformity of film thickness and the bowing distortion.

Keywords: Transverse Stretching / Heat Transfer Coefficient / Thickness / Bowing Distortion

Introduction

Film and sheet products are widely used for many applications such as packaging and optical film. And the demands for improving qualities such as thickness uniformity, surface flatness and for improving productivity are increasing. In order to produce film with satisfactory qualities, the manufacturing machinery for these products must have high performance and accuracy. In particular, transverse direction (TD) stretching process with a tenter is a key process because it greatly influences the physical properties of final products. In order to control the film qualities precisely with the tenter, optimum deformation must be given to film by applying carefully chosen operation conditions such as temperature, stretching speed and stretching ratio. In these conditions, film temperature, which is determined by the heat transfer coefficient between blowing air and film, is difficult to control. The main reason for it is the complexity of air flow in the tenter. In previous studies, heat transfer coefficient calculated using an impinging jet model has been employed in order to predict the deformation behavior of film in the tenter [1), 2)]. In this study, the air flow in the tenter is analyzed and the distribution of heat transfer coefficient is calculated in order to obtain the more realistic distribution of heat transfer coefficient than that in the previous works. And the deformation behavior of film during TD stretching is analyzed by applying the distribution obtained from air flow analysis.

Air Flow Analysis

A tenter is generally constructed as a series of units for heated air blowing in the machine direction (MD). Air flow analysis for one unit was conducted under the assumption that the performance is the same in each unit. Four nozzles with a slit as air outlet were installed to both the upper and the lower sides of each unit. Temperature and velocity of air from the slit were 418K and 15m/s, respectively. The line speed was 0.833m/s. Air flow analysis with a finite volume method (FVM) was conducted using commercial heat fluid analysis software, Solidworks Flow Simulation. Air temperature adjacent to the film surface (T_∞), the surface temperature of film (T_W) and the heat flux which pass through the film surface (q) were extracted from the results of air flow analysis. And the heat transfer coefficient on the film surface (h) was calculated from Equation (1)[3].

$$q = h\,(T_\infty - T_W) \qquad (1)$$

Constitutive Equation

Equation (2) was employed as a constitutive equation[4)-6)] for polypropylene (PP, F-300SP, PRIME POLYMER CO., LTD.). In order to determine material constants in Equation (2), tension tests with clasping the edges of MD were conducted by using small scale stretching machine. The size of specimen is 90mm in length × 90mm in width. A stretching speed is 0.134m/s. The temperatures are 418K, 423K, 428K and 433K. The constants were obtained from the results of tension tests by using the method of least square. The experimental true stress-logarithmic strain curves and approximated ones are shown in Figure 1.

$$\sigma_t = a\varepsilon_t + b\left(\frac{\varepsilon_t}{2}\right)^{5.4} + c\left\{1 - \frac{1}{(1+\varepsilon_t)^d}\right\} \qquad (2)$$

[σ_t: True stress ε_t: Logarithmic strain
a, b, c, d: Constant]

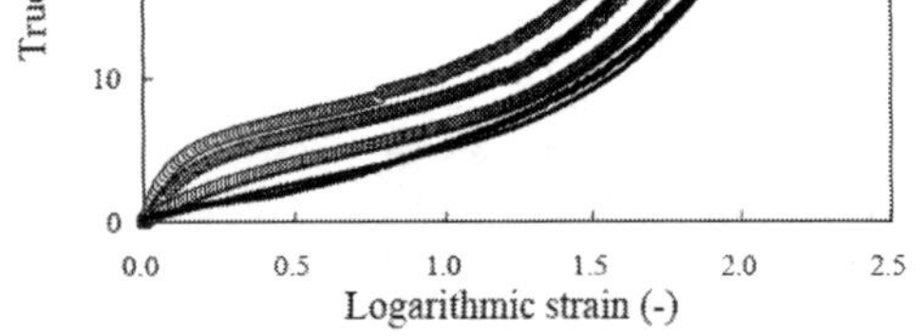

Figure 1 True stress – logarithmic strain curve

Film Stretching Analysis

The film stretching analysis was carried out under the assumption that the deformation of film obeys an elastic-plastic rule. The stretching condition was the same as the experimental condition. A commercialized nonlinear solver, MSC Marc 2008, was used.

Experimental

A casted film was set in the entrance of pilot tenter and successively stretched in TD only. The lines on film both MD and TD were drawn at the entrance of tenter in order to measure the bowing distortion and the successive distortion patterns. The number of unit was 3 for the preheating zone and the stretching zone, 4 for the thermosetting zone, respectively. The temperature was 483K in the first and second preheating zone, 418K in the third preheating zone, 418K in the stretching zone and 433K in the thermosetting zone. The line speed was 0.833m/s. The stretching ratio was 8.

Bowing Distortion

The bowing distortion B is defined as the percentage of bowing distance b by the film width W as shown in Figure 2, that is, Equation (3) as defined in the previous studies[7)-11)]. When the film center lags behind the film edge, the bowing distance b is positive.

$$B = b/W \times 100\ (\%) \qquad (3)$$

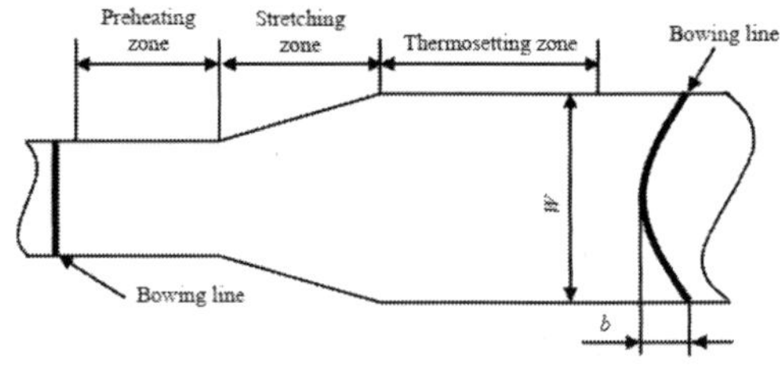

Figure 2 Schematic diagram of tenter

Results and Discussion

Figure 3 shows the distribution of heat transfer coefficients on the film calculated by using air flow analysis as well as an impinging jet model or a wall jet model (previous study's distribution). The heat transfer coefficients right under air outlets had the maximum values like previous study's distribution. And the heat transfer coefficients on the other parts of film were higher than those based on previous study's distribution, especially so near the air outlets. This is because an expanded air flow from the air outlet and the reproduction of impinging air toward the film occurred. From these results, the distribution of heat transfer coefficient based on air flow analysis takes the complicated air flow into consideration. And the distribution obtained is more realistic than the previous study's distribution.

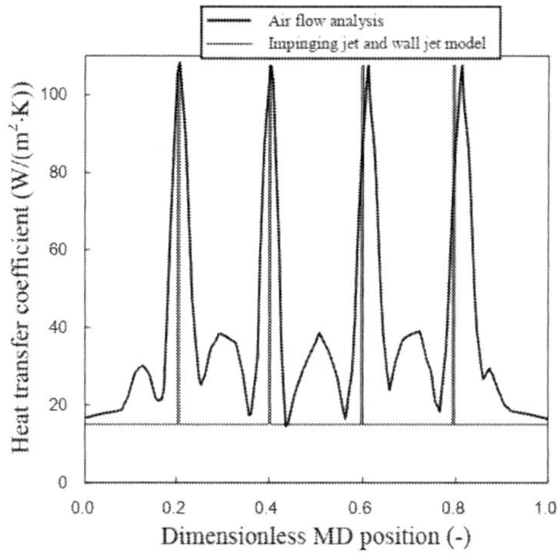

Figure 3 Distribution of heat transfer coefficient

Figure 4 shows the calculated distribution and the experimental result of film thickness after stretching. The result calculated by using the distribution of heat transfer coefficient based on air flow analysis (Case1) was almost the same as the experimental result. However, the result calculated by using previous study's distribution (Case2) predicted that there existed a considerably thick part, which was not the case in the experimental result.

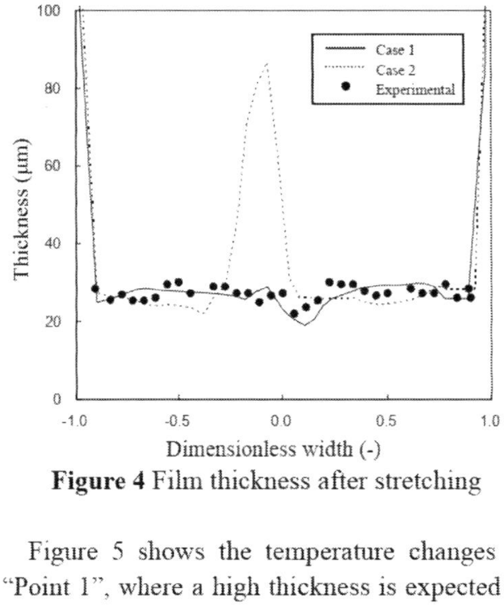

Figure 4 Film thickness after stretching

Figure 5 shows the temperature changes in "Point 1", where a high thickness is expected in Case2 analysis, and "Point 2", which locates at a distance of a quarter of film width from the transverse edge, respectively. In Case2, the temperature at the inlet of stretching zone did not reach the setting value, and the temperature at Point 1 rose slower than the temperature at Point 2. This indicates that non-uniformity of temperature during stretching in Case2 analysis leads to the thickness unevenness after stretching.

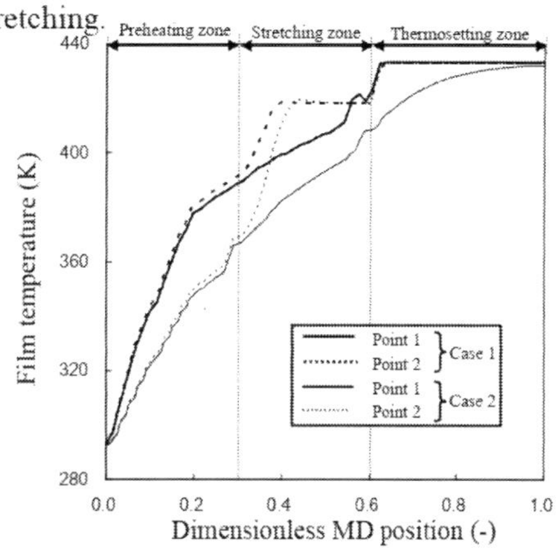

Figure 5 Change of film temperature along MD

Figure 6 shows the calculated bowing lines after stretching in comparison with the experimental result. The result of Case1 was close to the experimental result than the result of Case2 is.

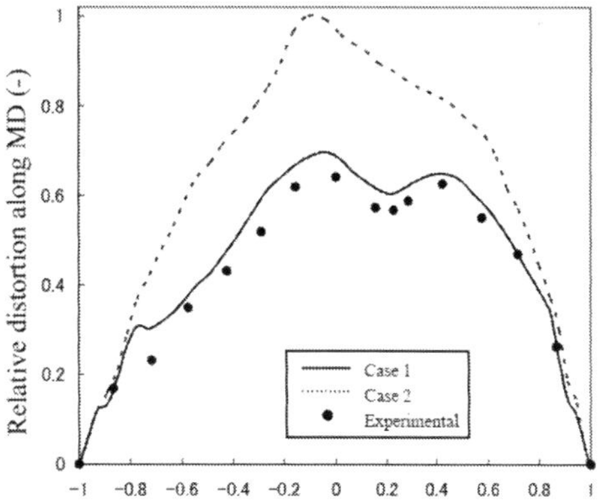

Figure 6 Bowing line after stretching

In order to make the difference between Case1 and Case2 result clear, the change of bowing distortion was compared. Figure 7 shows the calculated bowing distortion profiles along MD. When the film enters the stretching zone, the bowing distortion takes negative value, followed by a quick surge to positive value, and then it develops the maximum value in the thermosetting zone in both Case1 and Case2 analysis in a similar way described in previous studies[7)-11)]. However, the absolute values of the bowing distortion of Case2 were larger than that of Case1 in the upstream of stretching zone and thermosetting zone.

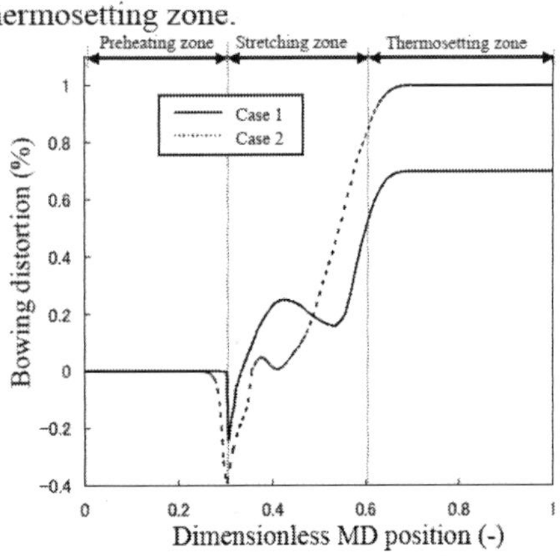

Figure 7 Bowing distortion profile along MD

Figure 8 shows the longitudinal stress profile along TD at the point, where the bowing distance (*b*) reached the maximum value. The stress near the center of film in Case2, where *b* becomes larger than the other parts due to lower the constraining force of clip, is lower than that in Case1. This part corresponded to a considerably thick part (cf. Figure 4).

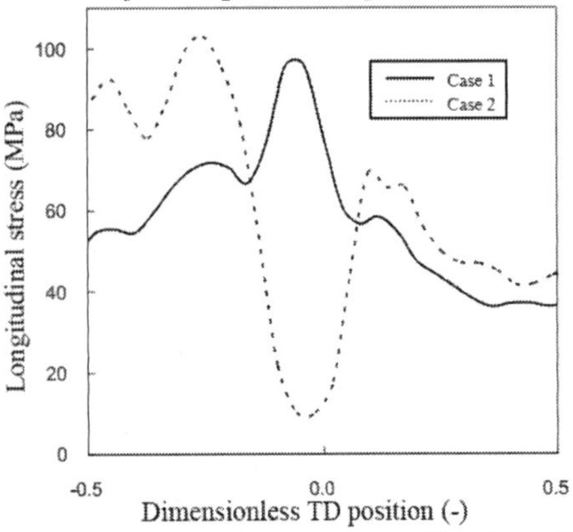

Figure 8 Longitudinal stress profile along TD

From these examinations, the discrepancy between the experimental results and Case2 analysis is that the bowing distance of high thick part in Case2 became large due to lower the stiffness rigidity than other parts. In Case1, the film was stretched uniformly than in Case2 and the stiffness rigidity become high. Therefore, applying the distribution of heat transfer coefficient based on air flow analysis makes the prediction of film deformation behavior more realistic.

Conclusions

The deformation behavior of film during TD stretching calculated by using the distribution of heat transfer coefficient based on air flow analysis was more realistic than that calculated by using the distribution based on an impinging jet model or a wall jet model.

References

1) Iwasaki T, Sugihara G, Yamada T, Asanuma M, Toya H, Yokoyama A, Fukunaga M (2003) JSPP'03 Sympo Papers: 303-306
2) Senda H, Yamada T, Tada K, Asanuma M, Toya H, Yokoyama A (2005) JSPP'05 Sympo Papers: 301-304
3) Bird R B, Stewart W E, Lightfoot E N (2002) Transport Phenomena Second Edition, John Wiley & Sons, Inc.
4) Yamada T, Tsukamoto A, Takahashi T, Tada K, Fukunaga M, Yokoyama A (2004) The 16th International Congress of Chemical and Process Engineering (CHISA 2004) Paper Filed in CD-ROM: 1-15
5) Yamada T, Takahashi T, Kunishima K, Yokoyama A (2005)The Polymer Processing Society, America's Regional Meeting-2005 (PPS-2005) Paper Filed in CD-ROM: 1-2
6) Yamada T, Takagi K, Asanuma M, Toya H, Yokoyama A (2007) The Europe/ Africa Regional Meeting 2007 of the Polymer Processing Society(PPS-2007) Paper Filed in CD-ROM: 1-6
7) Nonomura C, Yamada T, Matsuo T (1992) Seikei-Kakou Vol.4 (5): 312-317
8) Nonomura C, Yamada T (1993) Seikei-Kakou Vol.5(10): 703-708
9) Yamada T, Nonomura C (1993) J Appl Polym Sci vol.48 (8): 1399-1408
10) Yamada T, Nonomura C (1994) J Appl Polym Sci vol.52 (10): 1393-1403
11) Yamada T, Nonomura C, Matsuo T (1995) Intern Polym Process vol.X (4): 334-340

STUDY OF SOLIDIFICATION PROCESS IN MELT-SPINNING OF PET/CNT FIBERS

S. Mazinani[1]*, F. Sharif[2], C. Dubois[3], A. Ajji[4]

[1]*Department of Plymer Engineering, Amir Kabir University of Technology, 424 Hafez Ave., Tehran, Iran – saeedeh.mazinani@polymtl.ca*
[2]*Department of Plymer Engineering, Amir Kabir University of Technology, Tehran, Iran – sharif@aut.ac.ir*
[3]*CREPEC, Department of Chemical Engineering, Ecole Polytechnique de Montreal, Montreal, QC, Canada – charles.dubois@polymtl.ca*
[4]*CREPEC, Department of Chemical Engineering, Ecole Polytechnique de Montreal, Montreal, QC, Canada – Abdellah.ajji@polymtl.ca*

Abstract - Polymer melts containing CNT have been processed to produce fibers. An important aspect of fiber production from melt-spinning is the solidification stage which determines the structure and properties of the produced fibers. In this paper, solidification of polyethylene terephthalate (PET)/CNT monofilament is experimentally examined and compared with pure melt-spun PET. Alignment of the nanotubes and draw ratios (DRs) are reported as a measure of intensity of the elongational flow on the solidification process. We used differential scanning calorimetry (DSC) to study the role of CNTs on crystallization and the results show that increasing CNT content results into fibers formation containing more amorphous phase. Different methods such as transmission electron microscopy (TEM) and Raman spectroscopy were employed to assess the predictions obtained from DSC analysis and the most aligned structure was obtained at highes CNT concentration and DR. In addition, the final mechanical properties showed that it was possible to produce fibers with more amorphous structure and high draw-ability using CNT during the fiber processing.

Introduction

Polyethylene terephthalate (PET) has specific properties such as high drawability which has made this polymer one of the most important polymers in textile industry, as demonstrated in previous works on this subject (Chan and Isayev, 1994; Rwei et al., 2004). PET fibers properties have been modified by different methods including electrical conductivity using carbon nanotubes (CNTs) (Ahir et al., 2009). The maximum orientation of CNTs in CNT nanocomposites is of considerable interest because orientation considerably enhances mechanical and electrical properties even for low concentration of CNTs. Therefore, there is few works showing the improvement of drawability and conductivity of PET fibers by using nanofillers (Li et al., 2006). We believe solidification is an importantly controlling factor of orientation. Thereafter, the role of CNTs on solidification of single fiber is studied in this work. The extensive study of PET/CNT properties has been done elsewhere (Mazinani et al., 2010); here, we only focus the role of CNT on solidification and its effect on controlling the CNT orientation under spinning condition. Moreover, the final crystallinity of the melt-spun fibers at various CNT content is studied employing DSC method and it is correlated to final CNT orientation obtained by TEM and Raman experiments. Moreover, the final mechanical properties are studied considering the role of CNTs on changing the crystallinity.

Experimental

Melt-Spinning Process

We used PET (Selar 7086) obtained from DuPont Co. and a commercial PET/CNT masterbatch (MB-6815-00), composed of PET/15% multiwall carbon nanotube (MWCNT) produced by Hyperion Catalysis Co. A melt mixing method was employed for masterbatch dilution and CNT dispersion. A lab-scale twin-screw extruder (Leistritz Group Co.; with 1.78 mm diameter and length of 40D) was used for mixing at controlled temperature and speed. The mixing procedure was performed to optimize CNT dispersion in the matrix and to obtain optimal dispersion at different CNT concentrations. Followed by optimizing the mixing condition to obtain the best CNT dispersion, a capillary rheometer (Rosand) along with a take-up device was used to produce single fibers at different DRs. Diluted PET/CNT compounds were spun to obtain fibers at different CNT concentrations (containing up to 3 wt% MWCNT) and DR. The fibers produced in this work at different processing conditions (different DR) and MWCNT concentrations are given in Table 1.

Characterization Methods

Differential scanning calorimetry (DSC, Q1000; TA Instruments) was used to study the crystalline behavior of PET/CNT nanocomposites. TEM (JEOL, JEM-2100

F) and Raman spectroscopy (Renishaw spectrometer equipped with an inVia Raman Microscope) were conducted on the resulted fibers to evaluate the amount of orientation. In addition, the mechanical properties of the samples produced at different processing conditions were obtained employing a micro tester 5548 (Instron) under the tensile test with a load cell of 5N and a speed of 50 mm/min.

Table 1: Average diameter (µm) of fibers

	Pure PET	1% MWCNT	3% MWCNT
DR= 112 *(20% of DR_{max})*	97 ± 14	97 ± 14	97 ± 14
DR=202 *(40% of DR_{max})*	66 ± 14	66 ± 14	66 ± 14
DR=291 *(60% of DR_{max})*	57 ± 7	57 ± 7	57 ± 7
DR=380 *(80% of DR_{max})*	*Impossible to produce*	50 ± 9	50 ± 9
DR=470 *(DR_{max})*	*Impossible to produce*	48 ± 9	48 ± 9

Results and Discussion

CNTs orientation along the fiber axis is affected by solidification as described previously. Since PET is a semi-crystalline polymer; the solidification is controlled by cooling rate (processing parameter) and the rate of crystallization (material parameter). Therefore, we used DSC method as the first step to evaluate the role of CNTs on crystallinity in this system.

DSC Results

Rate and kinetics of crystallization were compared in different samples through the measurement of the time required to reach 50% relative crystallinity ($t_{1/2}$) to compare the rates of crystallization of different samples as shown in Fig. 1.

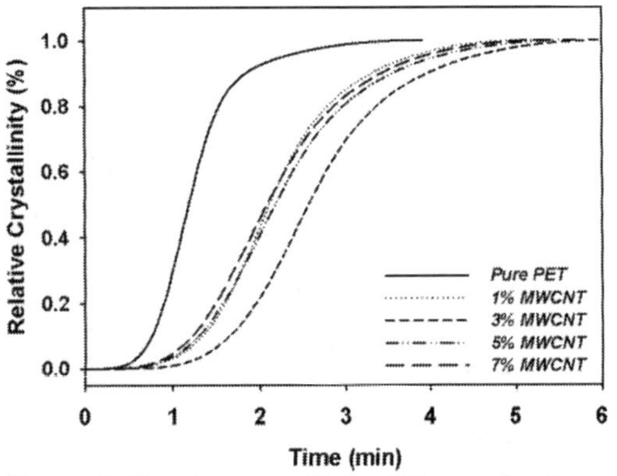

Figure 1: Relative crystallinity (%) as a function of time at different carbon nanotube concentrations (Mazinani et al., 2010)

As it is shown in Fig. 1, the samples containing CNTs have higher values of $t_{1/2}$ (half-time of crystallization). Therefore, the crystalline content is reduced by increasing CNTs. Similar result has been reported previously in crystalline behavior of other polymer nanocomposite system (Wurm et al. 2010). The crystalline content of as-spun fibers was also measured and the results are given in Table 2. As depicted, the overall crystallinity is reduced when CNT content is increased.

Table 2: Crystallinity (%) as a function MWCNT concentration at different DRs (Mazinani et al., 2010)

	20% DR_{max} (112)	**60% DR_{max} (291)**
Pure PET	17.2	23.5
1% CNT	16.2	15.8
3% CNT	16.3	15.4

CNTs could accelerate the solidification by increasing the rate of heat conduction and heat transfer. Therefore, they might decrease the chance of nanotubes orientation because of rapid solidification. However, increasing CNT concentration decreases the crystalline content from the material point of view. Retardation the crystallization causes delay in solidification and more aligned nanotube formation along fiber axis contrary to expectation. Both accelerating the solidification and retardation of crystallization result into fibers containing less crystalline content as shown in Table 2. Nevertheless, retardation of crystallization by adding CNTs is in favor of producing fibers containing more aligned CNTs.

CNTs Alignment Characterization

We used Raman spectroscopy and TEM method to study the CNT orientation at different processing condition.

CNTs include two strong peaks located at 1580 cm^{-1} (G) and 1350 cm^{-1} (D) (Mazinani et al., 2010) which are sensitive to oriented structure formation. The results obtained from Raman spectroscopy of the fibers containing 1wt% and 3wt% MWCNT as different DRs are shown in Fig. 2. The intensity of the peaks due to the presence of oriented MWCNT is quite high and as it is shown the orientation is quite high at the maximum DR (DR_{max}) as expected. Especially, the fibers containing 3 wt% MWCNT at DR_{max} show the most intensified peaks and maximum amount of orientation. As described in the DSC results, the fibers containing 3 wt% MWCNT at DR_{max} expresses the least crystalline content (14.7 %) and the highest CNT orientation as a result (Fig. 2).

The consequences of morphological analysis (TEM method) show the same behavior for the existence of aligned CNT along fiber axis (Fig. 3).

Therefore, controlling the solidification (Crystallization and cooling rate) by CNT content could be an important determining factor on controlling the amount of CNT alignment and final electrical conductivity as a result. However, it might also be a controlling parameter of final mechanical properties.

Mechanical Properties

The mechanical results obtained at different DRs for as-spun fibers at different MWCNT concentrations were studied for 1 wt% and 3 wt% as given in Fig. 4.

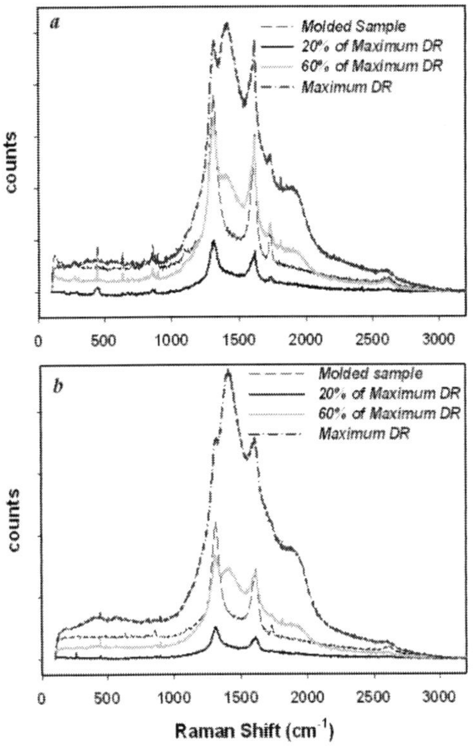

Figure 2: Raman spectra of as-spun fibers at different DRs: a) PET/1%MWCNT b) PET/3%MWCNT

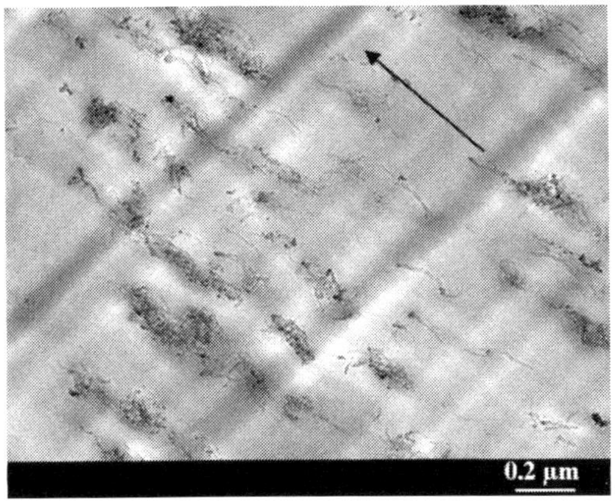

Figure 3: TEM images of PET/3% MWCNT and CNT orientation at 20% of DR_{max} along fiber axis

As it is shown, changing the MWCNT concentration and DR does not affect the modulus (Fig. 4a). However, increasing the CNT content causes a considerable enhancement in drawability of as-spun fibers (Fig. 4b). As depicted, addition of MWCNT to 1 wt% and 3 wt% concentrations increase drawability three times more compared to pure PET fibers. Since the fibers containing MWCNT are less crystalline as discussed previously, they show more tensile elongation and drawability which is of considerable practical interest in fiber industry.

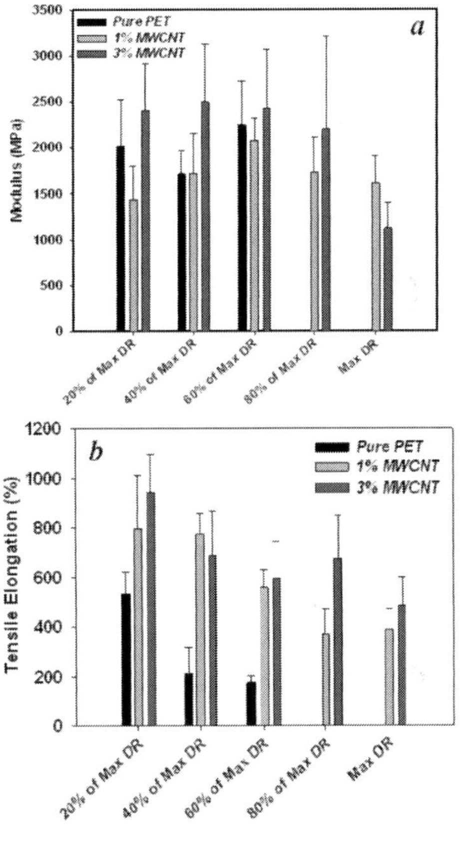

Figure 4: Mechanical properties of fibers: a) Modulus, b) Tensile elongation.

Conclusions

Orientation of CNT in PET/CNT melt-spun nanocomposite fibers at different MWCNT concentrations and DR were studied in this work. Microscopy observations and Raman spectroscopy measurements prove a high degree of MWCNT orientation after fiber spinning. It was also shown that the most oriented CNT structure was obtained at higher CNT content. It was depicted through DSC analysis that the addition of MWCNT decreases the crystalline content of final nanocomposite fibers. Since DSC results showed that MWCNTs increase the half-time of crystallization, and reduce the crystalline phase formation. The effect of MWCNT in the reduction of crystalline formation could considerably increase the drawability of as-spun fibers and the elongation at break in mechanical test. These results show the important role of CNTs on modifying the final properties of melt-spun fiber such as draw-ability in addition to elongation at break during fiber formation besides enhancement of electrical conductivity.

References

Ahir, S.V., et al., "Polymers with Aligned Carbon Nanotubes: Active Composite Materials", Polymer, **49**, 3841-3854 (2008).

Chan, T.W., Isayev, A.I., "Quiescent Polymer Crystallization: Modelling and Measurements", Polym. Eng. Sci., **34**, 461-471 (1994).

Li, Z., et al., "Microstructure of Carbon Nanotubes/PET Conductive Composites Fibers and Their Properties", Compos. Sci. Technol., **66**, 1022 (2006).

Mazinani, S., et al., "Structure and Properties of Melt-Spun PET/MWCNT Nanocomposite Fibers", Polym. Eng. Sci., **In Press**, DOI: 10.1002/pen.21727 (2010).

Rwei, S.P., et al., "PBT/PET Conjugated Fibers: Melt-Spinning, Fiber Properties, and Thermal Bonding", Polym. Eng. Sci., **44**, 331-344 (2004).

Wurm, A., et al., "Retarded Crystallization in Polyamide/Layered Silicates Nanocomposites Caused by an Immobilized Interphase", Macromolecules, **In Press**, DOI: 10.1021/ma902175r (2010).

Processing conditions to control structure-property of nucleated/non-nucleated polypropylene nanocomposite films

Manash Jyoti Kashyap, Anup Kumar Ghosh*

IIT Delhi - Delhi - India

Processabilty of nanoclay reinforced polypropylene films and properties imparted thereby have attracted applications in packaging industries. However, control of film thickness uniformity as also the effect of nanoclay on the blow-up ratio is intimately related to its (nanoclay) effect on the crystallinity and film morphology. In the present work film blowing of PP (polypropylene) as also of PPNC (polypropylene nanocomposites) with nanoclay in increasing concentrations was carried out. The subject of investigation was correlation of PP film blowability with film morphology. PP films with blow-up ratio of 2.6 were prepared. Although nanoclay reinforced films of 1 and 2 and 3 phr were prepared, it was observed that the bubble stability decreased as the content of nanoclay was increased above 3 phr. However quite interestingly, there was a marked decrease in film thickness variation as compared to films of extruded polypropylene. Films with increase in take-off speeds at fixed extruder outputs were also prepared. The films were analysed for crystallinity and film morphology as also the mechanical properties. Rheology as a tool to predict changes in microstructure of nucleated/non-nucleated PPNC films was carried out. Film morphology /crystallinity were investigated with polarizing-optical microscopy. From the various studies the structure properties were endeavored to be correlated with the processing conditions for preparation of nucleated/nucleated polypropylene nanocomposite films.

Keywords: polypropylene, nanoclay, processability, crystallinity, morphology

EFFECT OF NANOFIBER SPECIFIC SURFACE AREA ON IONIC CONDUCTIVITY OF ELECTROSPUN POLY(ACRYLONITRILE-*CO*-METHYL ACRYLATE) – ELECTROLYTE MEMBRANES

P. Zamani*, M. Rafizadeh

Deportment of Polymer Engineering & Color Technology, Amirkabir University of Technology, Tehran, Iran, pouyan.zamani@aut.ac.ir; mehdi@aut.ac.ir

Abstract – Polymer electrolyte membranes (PEM) with pores of nanometer to micrometer size are employed to uptake and retain liquid electrolyte. In the present research, electrospun poly(acrylonitrile-*co*-methyl acrylate) (P(AN – MA)) nanofiber membrane is produced using 10 %wt of P(AN – MA) in dimethyl formamide (DMF) solution. Consequently, nanofibers are soaked in a mixture of diethyl carbonate (DEC)/dimethyl carbonate (DMC) and lithium perchlorate ($LiClO_4$) with various compositions. To do so, response surface methodology is applied to study relationship between ionic conductivity of membranes and three factors: specific surface area, solvent composition (DMC ratio) and soaking time based on central composite design. Ionic conductivities of P(AN – MA) electrolyte membranes are in order of 10^{-3} S cm^{-1}. It is observed that specific surface area has the most effect on ionic conductivities. No interaction is observed among factors. Finally, a linear relationship is fitted on data with a very good precision.

Introduction

During this decade, polymer electrolyte membranes (PEM) have absorbed considerable attention for application in rechargeable batteries because they have advantages in terms of less weight, more specific energy, flexibility in design and better safety compared with their liquid counterparts [1, 2]. In porous PEMs, pores of nanometer to micrometer size are soaked in a liquid electrolyte [2, 3].

Some polymers, such as poly(vinylidenefluoride) (PVdF) [4 – 6], poly(vinylidenefluoride-*co*-hexafluoropropylene) (P(VdF-HFP)) [1, 2, 7 – 11], poly(acrylonitrile) (PAN) [3, 9, 12 – 14], Poly(vinyl chloride) (PVC) [14], poly(methyl methacrylate) (PMMA) [6], poly(ethylene oxide((PEO) [15 – 17] and Nafion [18, 19] have been used in power generation and PEMs owing to their electrochemical activity and electrolyte uptake.

Electrospinning is the best technique for fabrication of continuous polymer nanofibers with porous morphologies. Geometry of nanofibers like pore shapes and size, pore size distribution, free volume and specific surface area (SSA e.g. surface area to volume ratio) are important. When the average fiber diameters (AFD) are in order of nanometer (e.g. 100 to 1000 nm), several features appear such as extremely high SSA (for a nanofiber it can be 1000 times more than for a microfiber) and more flexibility in surface treatment [20]. Since the membranes have high porosity, they present high electrolyte uptake. However, there are not many studies that reported the usage of PEMs based on electrospun nanofibers for lithium batteries [1].

One of the most important properties of PEM batteries is proton or ionic conductivity. The proton conductivity of membranes is measured using AC impedance [17]. The conductivity (σ) was calculated by the following equation:

$$\sigma = \frac{L}{R\,S} \qquad (1)$$

where R is the bulk resistance of membrane measured by AC impedance spectroscopy, L is the reference electrodes distance and S is the cross area or surface of the membrane [17,19].

Soaked PEMs consist a large amount of liquid electrolytes in the pores of the polymer membrane and display a good ionic conductivity more than 1 mS cm-1 at room temperature [5].

In this research, PEMs were produced from electrospinnig of 10 %wt solutions of poly(acrylonitrile-*co*-methyl acrylate) (P(AN – MA)) in dimethyl formamide (DMF). Consequently, PEMs soaked in a mixture of diethyl carbonate (DEC)/dimethyl carbonate (DMC) and lithium perchlorate ($LiClO_4$) with various compositions. Specific surface area (SSA), solvent composition (DMC ratio) and soaking time (based on central composite design) were selected as main factors. Relationship between ionic conductivity of membranes and three factors were derived by response surface methodology based on the CCD. Polymer electrolyte membranes displayed a good ionic conductivity in order of 10^{-3} S cm^{-1} at room temperature.

Experimental

Materials

poly(acrylonitrile-*co*-methyl acrylate) (P(AN – MA)) (% 94 acrylonitrile, % 6 methyl acrylate as comonomer) (M_V = 70000 g mol^{-1}) was obtained from Iran Polyacryl Company. N, N- dimethylformamide (DMF), diethyl carbonate (DEC), dimethyl carbonate (DMC) and lithium perchlorate ($LiClO_4$ battery grade) were supplied by Merck.

Electrospinnig process

A Gamma High Voltage Research RS232C power supply was used to produce voltages in the range of 20 – 24 kV. A circular piece of aluminum foil was used as collector. Distance between needle tip and the collector was fixed at 15 cm. Both the needle and collector were enclosed in a chamber to reduce effect of air currents on trajectory of electrospun jet. Solutions of 10 %wt P(AN – MA) in DMF was electropsun applying electric voltages of 20, 23, 24 kV and flow rate of 200 $\mu l\ hr^{-1}$ at room temperature. This condition was selected based on the previous studies in our group.

Preparation of membrane electrode assembly

P(AN – MA) membrane electrolytes were prepared by soaking the membranes with 5 cm diameter in electrolyte solutions. Three solutions were applied; LiClO4 was solved in 1) DMC, 2) DEC, 3) a mixture of DMC/DEC (1:1 by weight) with concentration of 1M. Solution preparation was conducted under nitrogen atmosphere, in order to avoid from undesirable reactions with Oxygen or moisture. Soaking times were chosen 1.0, 1.25 and 1.5 hr based on the experimental design. Ionic conductivity cells were assembled by sandwiching PEMs between two circular stainless steel electrodes with 2.5 cm diameter.

Proton conductivity

The ionic conductivities were measured at room temperature using AC impedance spectroscopy of stainless steel/PEM/stainless steel applying Autolab/PGSTAT30 (Eco Chemie B.V. Co.) over the frequency range of 0.01 Hz – 1 MHz at amplitude of 0.01 V.

Experimental design

Response surface methodology and central composite design with three factors in two levels were applied. Table 1 shows the levels of the factors. Under this circumstances there would have been 17 samples and AC impedance runs. The low and high levels of each variable were designated as −1 and +1, respectively.

Table 1. The three – factors and two – levels used in central composite design

Actual factors	Coded factors	Level 1 (code = -1)	Level 2 (code = +1)
SSA ($m^2\ gr^{-1}$)	A	6.37	8.67
Soaking Time (s)	B	3600	5400
DMC Ratio	C	0.0	1.0

Results and Discussion

Morphology of electrospun nanofiber membrane, derived from 10 %wt P(AN – MA) in DMF, is presented in Fig. 1a and 1b. The membranes consist of network of interlayer and fully bead fibers with AFDs in range of 100 – 160 nm. Membranes showed high initial electrolyte uptake and high retain (final uptake) due to their unique SSA. Excellent SSA is because of nanofibers unrivaled geometry resulting in an extremely high $L/_D$ in which L is nanofiber length and D is fiber diameter.

The SSA values were calculated by SSA-diameter in literature [21]. Hence, 153 and 112 nm AFDs are consistent with SSA values of 6.37 and 8.67 $m^2\ gr^{-1}$, respectively. To do so, 10 %wt solution and applied voltages of 24 and 20 kV were used.

In Fig. 1 the SEM images of dry (1a and 1b) and soaked (1c and 1d) membranes can be seen. Due to geometrical equations, fibers with less AFD (1a) have more SSA and they have fewer beads and absorb more electrolyte solution on their surface and display more ionic conductivity [20]. Fig. 1c and 1d are respectively electrolyte – soaked membranes of 1a and 1b in a larger scale than the counterpart images. They both had DMC weight ratio of 0.5 and soaked for 75 min due to the design. It is observed that sample with less AFD or more SSA (1a) has more initial and final electrolyte uptake (1b).

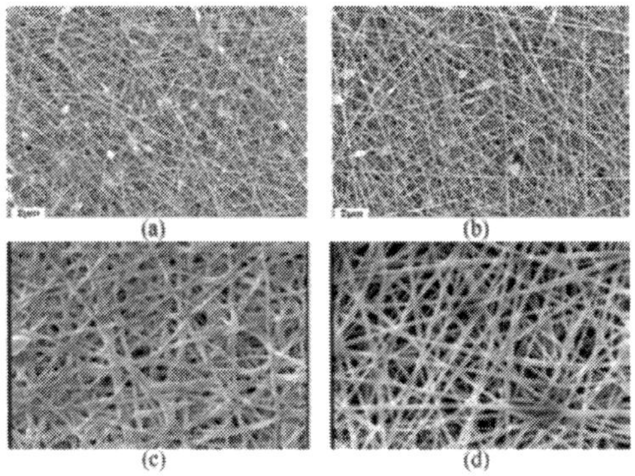

Figure 1. SEM images of electrospun nanofibers derived from 10 %wt P(AN – MA) in DMF with applied voltage of (a) 20 kV, (b) 23 kV before wetting, and their P(AN – MA) – based polymer electrolytes derived from (c) a, (d) b

Table 2 shows results of the experiments. The design consisted of 3 centre points (No. 2, 5 and 15 with factor A = 7.52 $m^2\ gr^{-1}$, factor B = 75 min, factor C = 0.5) for better precision of the experiment.

The most value of ionic conductivity as in this table (run 1) was related to the most SSA, soaking time and the most DMC ratio and the least one (run 9) was related to the least SSA, soaking time and the most DEC ratio. The most differences in ionic conductivity results were related to vary in SSA values which represent the importance of SSA comparing with other factors. A linear model was fitted to the yield response (Eq.2), based on small P-value for linear terms.

$$\sigma = -1.89 + 1.15\,S + 6.44 \times 10^{-5}\,t + 0.71\left(\frac{W_{DMC}}{W_{DMC} + W_{DEC}}\right) \quad (2)$$

where σ (mS cm^{-1}) is the anticipated ionic conductivity response. Characters S (m^2 gr^{-1}) and t (s) are specific surface area and soaking time, W_{DMC} and W_{DEC} denote solvent weights. The term including t is negligible and can be omitted to simplify the equation. So equation (2) will be:

$$\sigma = -1.89 + 1.15\,S + 0.71\left(\frac{W_{DMC}}{W_{DMC} + W_{DEC}}\right) \quad (3)$$

Table 2. Results of the experimental design used in this research

Run	SSA (m^2 gr^{-1})	Soaking Time (s)	DMC ratio	Ionic Conductivity (mS cm^{-1})
1	8.67	5400	1.0	9.03
2	7.52	4500	0.5	7.63
3	6.37	5400	0.0	5.85
4	7.52	4500	0.0	6.98
5	7.52	4500	0.5	7.55
6	8.67	5400	0.0	8.38
7	7.52	5400	0.5	7.52
8	7.52	3600	0.5	7.56
9	6.37	3600	0.0	5.47
10	8.67	4500	0.5	8.71
11	6.37	5400	1.0	6.37
12	6.37	4500	0.5	6.09
13	8.67	3600	0.0	8.34
14	6.37	3600	1.0	6.30
15	7.52	4500	0.5	7.77
16	7.52	4500	1.0	8.01
17	8.67	3600	1.0	8.90

The P-value in Analysis of Variance (ANOVA) was less than 5% and so far, the linear model was best and R-squared was R^2 = 0.97. Fig 2 shows three-dimensional response surface for the yield response in terms of SSA and DMC ratio. There was no optimum point in the regions being detected and from examination the contour plots, it was noted that ionic conductivity was dramatically more sensitive to changes in SSA.

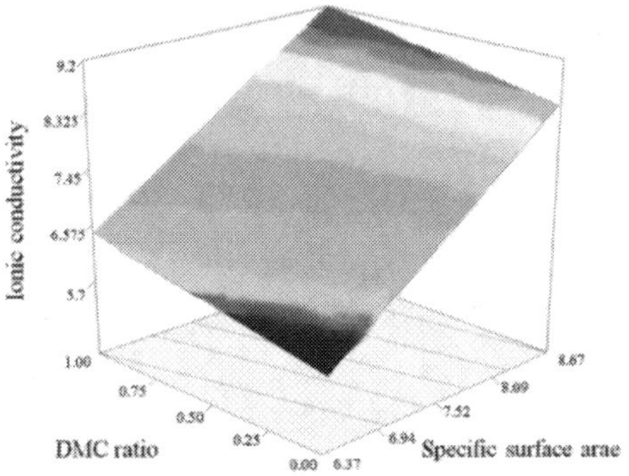

Figure 2. Response and contour plots of ionic conductivity versus SSA and DMC ratio

In Fig 3 the relationship between ionic conductivity results of the membranes and SSA values is shown. The results proved that with increasing in SSA (less AFD), ionic conductivity also improved dramatically and a linear graph was fitted for them.

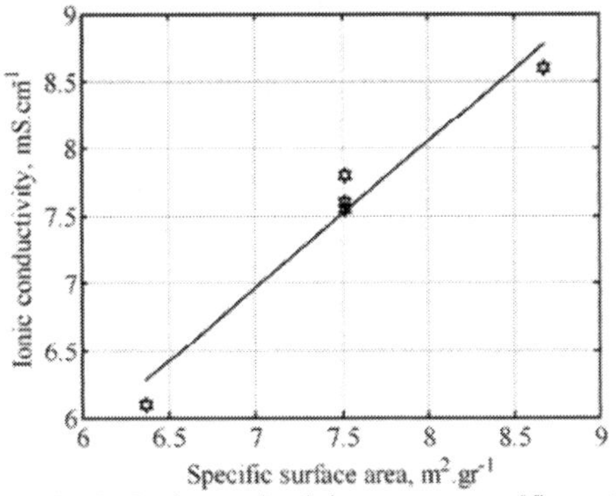

Figure 3. Ionic conductivity versus specific surface area of P(AN – MA) membranes

Based on the results, soaking time had negligible effect but DMC ratio was important. It was because of difference in dielectric constant, molecular surface energy and viscosity of solvents that effected on uptake, retain and final ionic conductivity of membranes. Although dielectric constant of DMC (ε = 3.13) is more than DEC (ε = 2.83) and seems to be more suitable for electrical and electrochemical purposes, at high DMC amount, samples were aggregated and caused unreal errors. Therefore it is better to use a mixture of both solvents.

Conclusions

Solutions of 10 %wt poly(acrylonitrile-*co*-methyl acrylate) in dimethylformamide were electrospun applying voltages from 19 to 24 kV and average fiber diameters were in range of 100 – 160 nm. Polymer electrolyte membranes were prepared by soaking membranes in electrolyte solutions of 1M $LiClO_4$ in solvents of diethyl carbonate and dimethyl carbonate. Central composite design and response surface methodology were applied to study relationship between ionic conductivity and nanofibers specific surface area, solvent ratio and soaking time. Polymer electrolyte membranes displayed good ionic conductivity in order of 10^{-3} S cm^{-1} at room temperature. Results proved that with increasing in specific surface area (decreasing in average fiber diameter) ionic conductivity improved dramatically and a linear graph was fitted for the results with respect to the design. Based on the results, soaking time had negligible effect but DMC ratio was important.

References

1. Cheruvally et al., "Electrospun polymer membrane activated with room temperature ionic liquid: Novel polymer electrolytes for lithium batteries" J. Power Sources 172 (2007) 863–869
2. Li et al., "Polymer electrolytes based on an electrospun poly(vinylidene fluoride-co-hexafluoropropylene) membrane for lithium batteries" J. Power Sources 167 (2007) 491–498
3. Wang et al., "Electrospun carbon–cobalt composite nanofiber as an anode material for lithium ion batteries" Scripta Materialia 58 (2008) 405–408
4. Choi et al., "Electrospun PVDF nanofiber web as polymer electrolyte or separator" Electrochimica Acta 50 (2004) 339–343
5. Kim et al., "Electrospun PVdF-based fibrous polymer electrolytes for lithium ion polymer batteries" Electrochimica Acta 50 (2004) 69–75
6. Cui et al., "Preparation of PVDF/PMMA blend microporous membranes for lithium ion batteries via thermally induced phase separation process" Materials Letters 62 (2008) 3809–3811
7. Choi et al., "An Electrospun Poly(vinylidene fluoride) Nanofibrous Membranes and its Battery applications" Adv. Mater. 23 (2003) 2027 – 2032
8. Oh et al., "Preparation and characterization of gel polymer electrolytes for solid state magnesium batteries" Electrochimica Acta 50 (2004) 903–906
9. Bansal et al., "Gelled membranes for Li and Li-ion batteries prepared by electrospinning" J. Power Sources 178 (2008) 848–851
10. Raghavan et al., "Novel electrospun poly(vinylidene fluoride-co-hexafluoropropylene)–in situ SiO2 composite membrane-based polymer electrolyte for lithium batteries" J. Power Sources 184 (2008) 437–443
11. Kim et al., "Preparation and electrochemical characterization of electrospun, microporous membrane-based composite polymer electrolytes for lithium batteries" J. Power Sources 178 (2008) 815–820
12. Nicotera et al., "Some physicochemical properties of PAN-based electrolytes: solution and gel microstructures" Solid State Ionics 167 (2004) 213–220
13. Akashi et al., "Practical performances of Li-ion polymer batteries with LiNi0.8Co0.2O2, MCMB, and PAN-based gel electrolyte" J. Power Sources 112 (2002) 577–582
14. Subramania, "Preparation of a Microporous Gel Polymer Electrolyte with a Novel Preferential Polymer Dissolution Process for Li-Ion Batteries" J. Applied Polymer Science 98 (2005) 1891–1896
15. Seki, "Fabrication of High-Voltage, High-Capacity All-Solid-State Lithium Polymer Secondary Batteries by Application of the Polymer Electrolyte/Inorganic Electrolyte Composite Concept" Chem. Mater. 17 (2005) 2041-2045
16. Parcher, "Solid-state Voltammetry and Polymer Electrolyte Plasticization as a Basis for an Electrochemical Gas Chromatographic Detector" Anal. Chem. 61 (1989) 584-589
17. Lee, "Importance of Proton Conductivity Measurement in Polymer Electrolyte Membrane for Fuel Cell Application" Ind. Eng. Chem. Res. 44 (2005) 7617-7626
18. Hasani-Sadrabadi et al., "Nafion®/bio-functionalized montmorillonite nanohybrids as novel polyelectrolyte membranes for direct methanol fuel cells" J. Power Sources 190 (2009) 318–321
19. Hasani-Sadrabadi et al., "Preparation and characterization of nanocomposite membranes made of poly(2,6-dimethyl-1,4-phenylene oxide) and montmorillonite for direct methanol fuel cells" J. Power Sources 183 (2008) 551–556
20. Huang et al., "A review on polymer nanofibers by electrospinning and their applications in nanocomposites" Composites Science and Technology 63 (2003) 2223–2253
21. Frank K. Ko et al, Nanofiber Technology: Bridging the gap between nano and macro world. Drexel University, Philadelphia, USA, (2004)

ON-LINE MEASUREMENT OF BIREFRINGENCE DEVELOPMENT DURING STRETCHING OF CYCLO OLEFIN COPOLYMER FILMS

Ikeda K., Menjyu H., Takarada W. and Kikutani T.*

Department of Organic and Polymeric Materials, Graduate School of Science and Engineering, Tokyo Institute of Technology, Tokyo, Japan-
ikeda.k.aa@m.titech.ac.jp; menju.h.aa@m.titech.ac.jp; takarada.w.aa@m.titech.ac.jp; kikutani.t.aa@m.titech.ac.jp

Cyclo olefin copolymer (COC) is copolymerized from norbornene and ethylene using a metallocene catalyst. COC possesses excellent properties like high refractive index, low moisture absorption, and high rigidity, which are important properties when it is used in optical parts. In addition to that, flow properties and heat resistance of COC can be controlled by changing the copolymer content. The purpose of this study is to clarify the stress and birefringence behaviors of different types of COC films during uni- and biaxial drawing processes. In the uniaxial drawing, stress-optical behavior follows the "modified stress-optical rule (MSOR)". However, when the drawing temperature is very close to the glass transition temperature birefringence becomes lower than expected by the MSOR because of its chemical structure. In the sequential biaxial drawing, stress-optical behavior during the first drawing and the subsequent relaxation between the first and second drawing is nearly same as in the uniaxial drawing. On the other hand, the stress-optical behavior during the second drawing and the subsequent relaxation is different from the uniaxial drawing. It was found out that the structure development behavior at the second drawing is much more influenced by the structure which is formed during the first drawing.

Introduction

The market of flat panel displays is expanding these years. Polymeric materials for optical devices, like cyclo olefin polymers (COPs) and cyclo olefin copolymers (COCs) attract much attention because of their low intrinsic birefringence, high transparency, low moisture absorption and high glass transition temperatures. In addition to that, flow properties and heat resistance of COCs can be controlled by changing the copolymer content. The molecular orientation is very important for the optical properties of these polymers. The development of molecular orientation is influenced both by processing conditions and copolymer content. In order to clarify the mechanism of structure development of COC films in the uniaxial drawing and the sequential biaxial drawing processes, in-situ stress and birefringence measurements were performed during the stretching and relaxation processes with different types of COC films.

Experimental

Material
The chemical structure of COC is shown in Fig. 1. Two types of COC films, supplied by Polyplastics Co. Ltd. were used. The initial thickness of those films was 350 μm. Those films were made of TOPAS 6013 and TOPAS 6015. The glass transition temperature (Tg) of those films are 138 °C and 158 °C. TOPAS 6015 has a higher Tg due to its higher norbornene content.

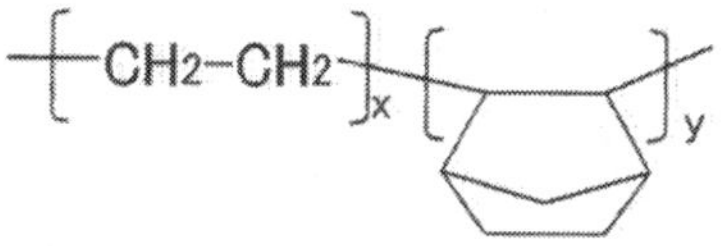

TOPAS 6013 x=0.48 y=0.52
TOPAS 6015 x=0.43 y=0.57

Figure 1 - Chemical structure of COC.

Uniaxial and biaxial drawing
The uniaxial drawing was performed by a small tensile testing machine with a small heat chamber. The biaxial drawing was performed by a tabletop tentering machine which was built in a hot air oven. The drawing was done sequentially. First, the film was drawn in one direction and then drawn in transverse direction. Both drawings were performed with a constant width (planer drawing). In this paper, first drawing direction is referred to as MD and second drawing direction as TD. The drawing conditions of uni- and biaxial drawing are shown in Table 1.

Table 1 - Drawing conditions.

	Uni-axial	Bi-axial
Temperature	Tg + 2 ~ 22 °C	Tg + 7 °C
Draw ratio	x 1.4	x 1.5 x 1.5
Strain rate	100, 500 %/min	100 %/min
Holding time after MD drawing		5 ~ 600 s

In-situ measurements
Tensile force and optical retardation were measured during the uni- and biaxial drawing. The tensile force was measured with a load cell built in the drawing machine. The tensile stress was calculated from tensile force and cross-sectional area. The cross-sectional area at each point was calculated from the drawing ratio under the assumption that the deformation is an affine deformation with constant volume.

Optical retardation was measured by an instrument with laser source, rotating polarizer, quarter wavelength plate and two detectors with a polarizer. Each detector measured the change of laser intensity before and after passing the drawing film. Basically, the measured intensities were sinusoidal waves and the phase shifting between the two waves corresponded to an optical retardation of the drawn film. Birefringence was calculated from optical retardation and film thickness. The film thickness at each point was calculated with the same method as the cross-sectional area. The specifications of this instrument can be obtained from a previous paper [1].

Results and Discussion

Uniaxial drawing
Stress and birefringence during the uniaxial drawing is shown in Fig. 2. At a high temperature (Tg + 17 ~ 22 °C) (Fig. 2 (a)), the stress and birefringence curves overlapped each other in drawing and relaxation for a fitted scale of the diagram axes. This means that birefringence was proportional to stress under this drawing conditions. Therefore, the plot in the birefringence versus stress diagram, which is shown in Fig.3 became a straight line. This effect is well known as the "Stress-optical rule" on polymer deformation in rubbery state. TOPAS deformed rubbery at a high temperature.

At a lower temperature (Tg + 12 ~ 15 °C) (Fig. 2 (b)), the stress and birefringence curves overlapped only in relaxation. At the beginning of the drawing, stress increased steeply, and then the slope of stress became lower. It showed a knee point on the stress curve like yielding. On the other hand, birefringence increased smoothly similar to the curve at a high temperature. After the knee point, stress and birefringence increased parallel to each other. At the end of drawing, stress dropped rapidly and merged into the birefringence curve. This phenomenon indicates the existence of a structure which is hard to deform like plastics. Therefore, the birefringence - stress diagram in Fig.3 shows a hysteresis. This offset of stress is also found on the deformation of polystyrene [2] and it shows that TOPAS deformed as mixture of plastic and rubber at a lower temperature. Osaki et al. propounded a modified stress-optical rule (MSOR) to describe this behavior [3].

At a temperature very close to Tg (Tg + 2 °C) (Fig. 2 (c)), the stress and birefringence curves overlapped only at a late stage of relaxation. At the beginning of the drawing, stress increased more steeply and showed

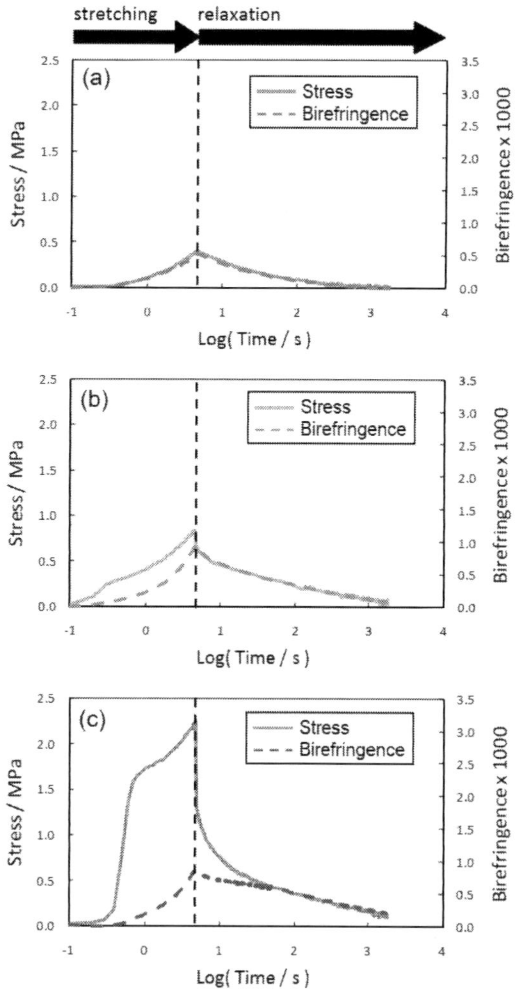

Fig. 2 - Changes of stress (solid line) and birefringence (dotted line) during uniaxial drawing and relaxation processes of TOPAS® 6013 at Tg+22 °C (a), Tg+12 °C (b) and Tg+7 °C (c). Strain rate $\dot{\epsilon} = 500\%/min$

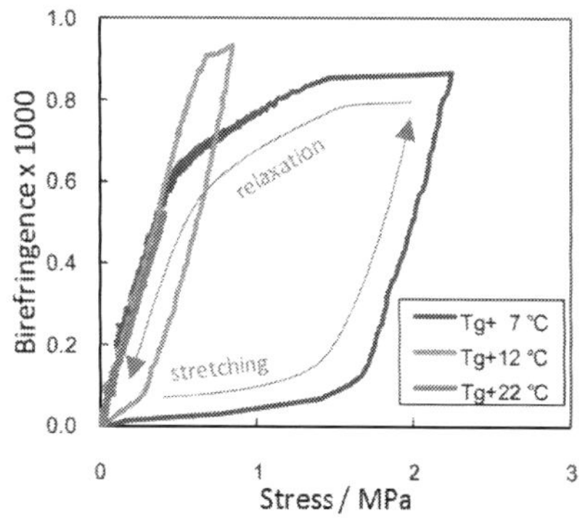

Fig. 3 - Birefringence vs stress relation during uniaxial stretching and relaxation processes of TOPAS®6013 at various temperatures. Strain rate $\dot{\epsilon} = 500\%/min$.

a sharper yielding than under other drawing conditions. On the other hand, birefringence increased smoothly, similar to other drawing conditions. At the end of drawing, the stress dropped rapidly, but still showed a higher level than the birefringence. During the relaxation, stress decreased smoothly. But the decrease of birefringence was unnaturally slow in the early stage of relaxation. In the late stage of relaxation, birefringence decreased smoothly and merged into the stress curve. This indicates the existence of a structure which has negative birefringence and a mid-level (10 ~ 100 s) relaxation time. This phenomenon is unexplained by the MSOR. This Phenomenon could be originated by the unique chemical structure of TOPAS.

Biaxial drawing

The changes of stress and birefringence during the biaxial drawing and relaxation processes of TOPAS 6015 are shown in Figs. 4 and 5. At MD stretching, the stress increased steeply and the birefringence increased smoothly. This drawing behavior is basically the same as in the uniaxial drawing. However, in the late stage of MD drawing, the stress increased more steeply whereas birefringence increase became slower. Probably strain hardening occurred because of the higher draw ratio. (The film thickness after 1.5 times planer drawing is 21% thinner than that after 1.4 times uniaxial drawing.)

During the holding time after MD drawing, stress decreased steeply whereas birefringence decreased slowly. This behavior was nearly the same as in relaxation after uniaxial drawing and indicated the existence of a negative birefringence structure.

During the TD stretching, TD stress increased steeply to the same level as the MD stress at the end of TD drawing. However, the birefringence decreased to a negative value in the late stage of the second stretching whereas the stress difference was still positive. This behavior indicated that the molecular orientation behavior toward TD was enhanced by the structure which was formed during the first drawing.

During the holding time after the TD drawing, the TD stress showed a faster relaxation than MD stress. This was caused by the structure with a short relaxation time which was developed along the TD direction during the TD drawing. Consequently, the stress difference became positive and the birefringence asymptotically showed MD orientation.

Conclusions

Structure development behavior of TOPAS during uni- and biaxial drawing was investigated. In the uniaxial drawing, stress-optical behavior followed the MSOR. However, when the drawing temperature was very close to the glass transition temperature, birefringence became lower than expected by MSOR probably because of the chemical structure of TOPAS. In the sequential biaxial drawing, the stress-optical behavior during the MD drawing and the subsequent relaxation between MD and TD drawing is nearly the same as in the uniaxial drawing. On the other hand, the stress-optical behavior during the TD drawing and the subsequent relaxation is different from the uniaxial drawing. It was found that the structure development behavior at the TD drawing is highly influenced by the structure formed during the MD drawing.

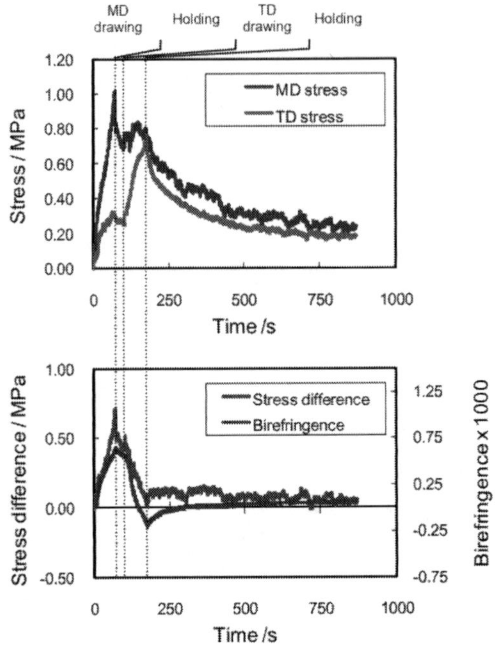

Fig. 4 - Changes of stress and birefringence during biaxial drawing and relaxation processes of TOPAS 6015 at Tg + 7 °C, holding time after MD stretching 30 s, strain rate $\dot{\epsilon} = 100\%/min$.

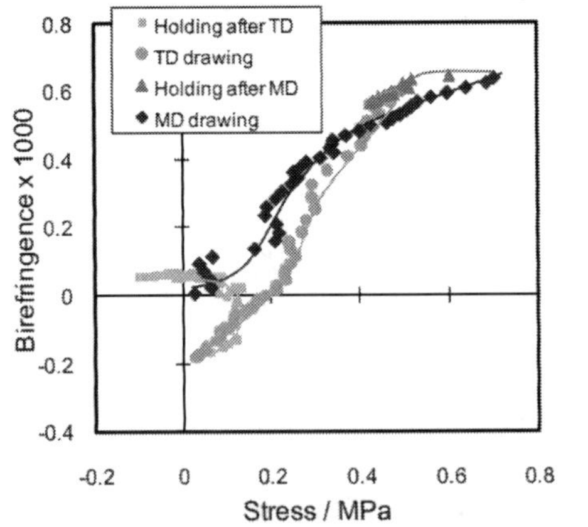

Fig. 5 - Birefringence vs. stress during sequential biaxial drawing and relaxation processes of TOPAS 6015 at Tg + 7 °C. Holding time after MD stretching 30 s, strain rate $\dot{\epsilon} = 100\%/min$.

References

1. Kikutani T. et al., *Polym. Eng. Sci.*, **39**, p.2349 (1999)
2. M. Kröger et al., *Macromolecules*, **30**, p. 526 (1997)
3. Osaki K. et al., *Macromolecules*, **28**, p. 3625 (1995)

STUDY ON PARAMETERS AFFECTING MICROFIBRILLAR MORPHOLOGY DEVELOPMENT IN PP/PBT/NANOCLAY BLEND NANOCOMPOSITE FIBERS

A. Bigdeli [1], H. Nazockdast *[2], A.S. Rashidi [1], M.E. Yazdanshenas [3]

[1]*Department of Textile Engineering, Science and Research Branch, Islamic Azad University, Iran.*
[2]*Department of Polymer Engineering, Amirkabir University of Technology, Iran. Nazdast@aut.ac.ir,*
[3]*Department of Textile Engineering, Yazd Branch, Islamic Azad University, Iran.*

Abstract
The main objective of this work was to study the effect of presence of organically modified montmorillonite (organoclay) and the compatibilizer (iPPgMA) on microfibrillar morphology development in PP/PBT/Nanoclay nanocomposite fibers. The samples with the same blend ratio (80/20) but varying in organoclay content were prepared with and without the compatibilizer by using a melt intercalation process in a co-rotating twin screw extruder. The microfibrillar morphology evolution of blend nanocomposite fibers was achieved by transferring stress from matrix to disperse phase by using organoclay. This enhanced fibrillation of PBT droplets was attributed to the role of nanoclay particles as a compatibilizer to improve the interfacial adhesion and suppress the interfacial slip between PBT and PP phases, so that the shear stress was effectively transferred to the disperse phase.

Introduction

Polymer blending process has been extensively used to produce new materials with modified properties. This method has also been successfully employed to prepare blend fibers with microfibrillar morphology with enhanced properties. Compared to other processes, fiber spinning is the most desirable method to produce fibers with fibrillar morphology due to existing of the elongational flow field. Recently blend nanocomposite fiber has attracted many scientific and industrial groups. Although a few research work has been devoted to parameters affecting microstructure development of the blend nanocomposite fibers, the number of research emphasizing on the role of nanoclay ,in particular, nanoclay partitioning on the extent of microfibrillar morphology are rare.

Experimental

Materials

Commercial fiber grade iso tactic PP, known as V30S, was obtained from Arak petrochemical Co., Iran. PP had a melt flow index (MFI) of 16 g/10min, density of 0.92 g/cm^3 and T_m of $166^{\circ c}$. PBT was produced by Bayer chemical company with trade name of Pocan. PPgMA (trade name PB3150) was received from Uniroyal chemical Co., USA. This polypropylene had MFI of 50 g/10min, MAH index of 1.5% and density of 0.9 g/cm3. The organically modified Montmorillonite nanoclay (Cloisite30B) supplied by Southern clay products Inc. were used .

Sample Preparation

The PP/PBT blend nanocomposite samples with the same blend ratio (80/20) but varying in organically content (1,3and5) with and without compatibilizer (iPPgMA) were considered. All the samples were prepared by melt intercalation process in a twin screw extruder and then melt spun into fiber using single screw extruder. A drawing apparatus was also employed to perform hot drawing of the blend nanocomposite fibers. For morphological structure observation of PBT microfibrils in the compatibilized and uncompatibilized fibers, the matrix PP was etched away by decahydronaphtaline(decalin).

Characterization
The extent of the melt intercalation was studied by the wide angle X-ray diffraction(WAXD). Morphology of the blend samples as well as the microfibril formation of blend nanocomposite fibers were examined by SEM. The melt linear viscoelatic properties of the samples were studied by means of Rheometeric Mechanical Spectrometer(RMS).

Results and Discussion
The X-ray diffraction patterns and melt viscoelastic results suggested much greater extent of melt intercalation and consequently better dispersion of organoclay in PBT compared with PP due to higher affinity between organoclay and PBT. This could be related to partitioning of organoclay in PBT droplet and the interface. This was evidenced by the melt rheological behavior of blend nanocomposite samples which, in contrary to PBT containing nanoclay sample that showed a low frequency nonterminal storage modulus, exhibited a behavior similar to PP matrix. By comparing the SEM micrographs shown in figure 1, one can notice that incorporation of PP-g-MA into the blend samples reduced the PBT droplet size. These results also show that presence of organoclay can have a reducing effect on PBT droplet size but not as effective as PPgMA. Figure2 compares the SEM micrographs of the blend fibers with and without compatibilizer after being etched in decalin. As it can be seen fine microfibril morphology can be developed in simple blend sample. Despite reducing the PBT droplet size by using PPgMA, it could enhance the microfibril formation via increasing interfacial interaction and therefore, stress transformation between two phases. These results also show that the presence

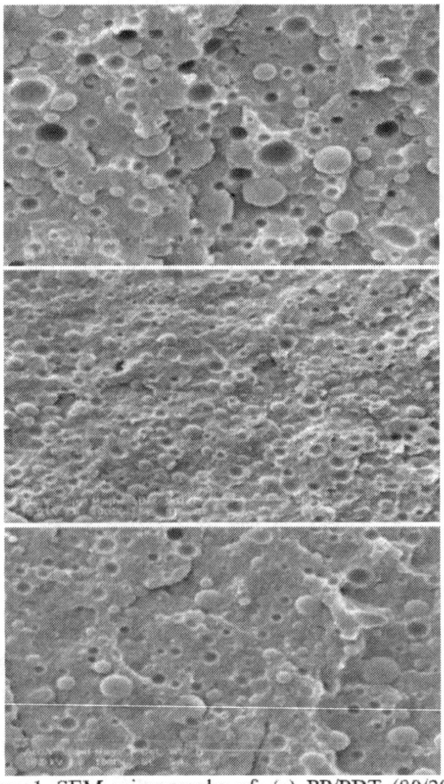

Figure1–SEM micrographs of: (a) PP/PBT (80/20by wt%) uncompatibilized blend , (b) PP/PBT/PPgMA (77/20/3 by wt%), (c) PP/PBT/Organoclay (77/20/3 by wt%)

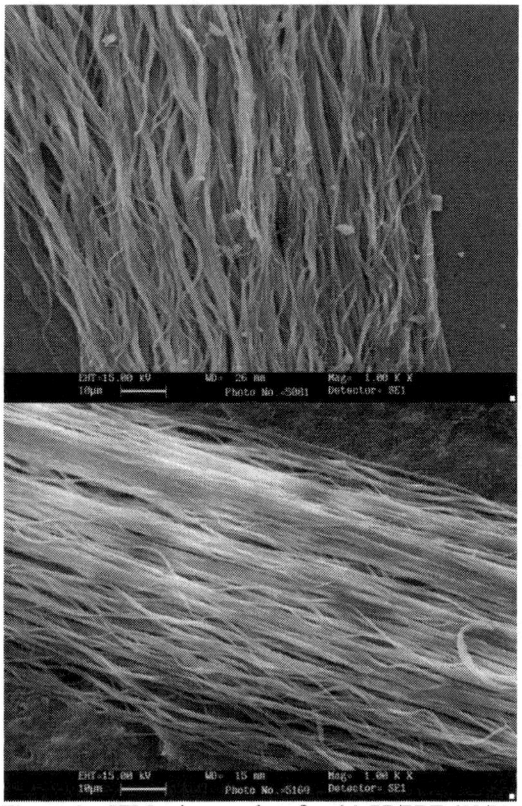

Figure 2 –SEM micrographs of : (a) PP/PBT (80/20 by wt%) uncompatibilized blend , (b) PP/PBT/ PPgMA (77/20/3 by wt%) compatibilized blend,

of PPgMA in spite of reducing the PBT droplet size ,which is expected to reduce the microfibril formation, has not limited the microfibrillar development process. The presence of organoclay was found to have two opposite effects on the microfibrillar morphology development. Depending on organocly concentration, at low organoclay loading, the tactoids and/or platelets may mostly located in the interface which can assist the microfibril formation. while at higher nanoclay loading a large portion of organoclay can selectively be partitioned in the PBT droplet which can reduce the fibril formation(Fig3). However, presence of compatibilizer(PP-g-MA) in blend nanocomposite samples can assist part of tactoids or platlets to be transferred from PBT droplet into PP matrix; a process

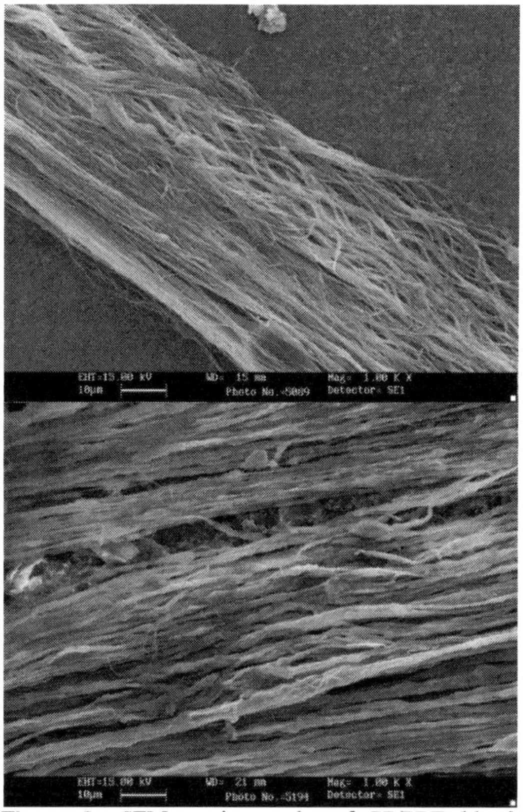

Figure 3- SEM micrographs of PP/PBT blend nanocomposite fibers : (a) containing 1% nanoclay, (b) containing 3% nanoclay

Figure 4- SEM micrograph of PP/PBT blend nanocomposite containing 3% PPgMA and 3% nanoclay

which can promote microfibril formation in blend nanocomposite fibers.

Conclusions

It was shown that presence of organoclay in PP/PBT blend fiber systems can have different effects on the extent of microfibril formation depending on organoclay concentration. While in the blend sample containing 1% nanoclay, the organoclay enhanced microfibril formation process, the higher concentration of organoclay limited the microfibrillar development due to partitioning of organoclay inside the PBT droplet. These results could be explained in terms of organoclay partitioning in the interface and PBT droplet. It was also demonstrated that incorporation of compatibilizer (PPgMA) into PP/PBT blend samples, although reduce droplet size; it could improve microfibrillation as result of interfacial enhancement and therefore assisting to transform stress between two phases. The compatibilizer was also found to play significant role in improving the microfibrillation in the blend nanocomposite fiber systems by increasing compatibility between two phases and therefore, assisting the organoclay to be transferred from droplet into interface and PP matrix. This was supported by melt viscoelastic results.

References

[1]Afshari.M., et al., "Effect of blend ratio on bulk properties and matrix-fibril morphology of polypropylene/nylone6 polyblend fibers", Polymer 43, 1331-1341(2002) .

[2]Calcagno,C.I.W. , et al., "The role of the MMT on the morphology and mechanical properties of the PP/PET blends", Composite Science and Technology 68 , 2193-2200(2008).

[3]Wang.D., et al., "Formation and morphology of cellulos acetate butyrate (CAB)/polyolefin and CAB/polyester in situ microfibrillar and lamellar hybrid blends".European Polymer Journal, 43,3587-3596(2007).

[4]Yi,X., et al., "Morphology and properties of isotactic polypropylene/poly(ethylene terephthalate) in situ microfibrillar reinforced blends:Influence of viscosity ratio", European Polymer Journal .46 ,719-730(2010).

SIMULTANEOUS AND SEQUENTIAL BIAXIAL STRETCHING OF POLYPROPYLENE FILMS CONTAINING β-FORM CRYSTALLINE STRUCTURE

N. Kerddonfag[1]*, W. Chinsirikul[1], O. Saravari[2], W. Takarada[3] and T. Kikutani[3]

[1] *National Metal and Materials Technology Center (MTEC), National Science and Technology Development Agency (NSTDA), Ministry of Science, Pathumthani, Thailand - noppadk@mtec.or.th, wanneec@mtec.or.th*
[2] *Department of Materials Science, Faculty of Science, Chulalongkorn University, Bangkok, Thailand - sonusa@chula.ac.th*
[3] *Department of Organic and Polymeric Materials, Graduate School of Science and Engineering, Tokyo Institute of Technology, Tokyo, Japan -takarada.w.aa@m.titech.ac.jp, kikutani.t.aa@m.titech.ac.jp*

Abstract - This study investigated the biaxial stretching behavior of PP containing various β-crystal contents, by comparing between simultaneous and sequential processes. Extruded precursor sheets consisted of a very small amount in a ppm level of a proprietary nucleating agent. Precursor sheets were fabricated using different chill roll temperatures (60, 70, 80, 90, 100, 110 and 120°C) and resulted in high β-crystal contents, indicated by K-values of 0.77, 0.81, 0.84, 0.87, 0.87, 0.85 and 0.85, respectively. Biaxial film stretching of β-nucleated sheets was performed on a lab-scale stretching machine (Bruckner KARO IV) at the temperature of 145°C with a stretching speed of 400%/s. For stretching behavior comparison, all β–nucleated PP sheets were stretched to 5.5 x 5.5. Based on the simultaneous stretching results, biaxial yield stress of normal PP (α-PP) was 3.3 MPa, where much lower biaxial yield stress values were apparent for all β-nucleated PP, in a range of 1.5-2.4 MPa. Interestingly low yield stress and low maximum stress during biaxial stretching of the β-PP films were dependent upon the β-crystals (both content and morphology), initially developed in the sheet extrusion process. Chill roll temperature or cooling was a critical parameter controlling morphology, the stretch behaviors and the resulting film properties. β-nucleated PP fabricated under optimum cooling condition clearly required lower energy than the normal α-PP, to deform under both simultaneous and sequential film stretching processes. The stretched β-PP films also showed good optical property, evidenced by low haze values of <1% which were close to that for the normal BOPP films. The β-PP stretched films were characterized by thermal analysis to correlate between morphology, stretch behavior and final film optical properties.

Introduction

Isotactic polypropylene (PP) is one of the important polyolefins having versatile properties and low cost. It is a semi-crystalline polymer consisting of 3 different crystal structures such as α-monoclinic crystal, β-hexagonal crystal and γ-triclinic crystal. The common α-crystal structure could be formed in general processing technique or induced by using a variety of nucleating agent e.g. mineral or organic compounds. In contrast, a formation of β-form crystalline structure requires specific condition and nucleating agent. There are many chemicals that have been reported as β-nucleating agents such as organic pigment compounds, metal salts or their mixtures with dicarboxylic acids, aromatic amide compounds and rare earth substances. Polypropylene containing high proportion of β-form crystal exhibit interesting properties differed from α-PP such as low modulus, high impact strength and tensile toughness. Since β-form crystal is less stable structure, it could be rearanged or recrystalized to be α-form crystal under heat or stress (Lesak et al., 2006). A number of literatures and patents have already reported on using enriched β-PP sheet as a precursor to create a porous structure based on its crystal transformation under biaxial stretching process (Chu and Kimura, 1996). This porous film is opaque white due to high content of microvoids. In the case of common biaxially oriented PP (BOPP) films for flexible packaging application, it possesses high light transmission or high clarity and gas barrier properties. Biaxial stretching process results in the high quality product in terms of transparency and thickness tolerances of films. Stretched film properties are dependent on structural development during orienting process and such properties are well-related to initial morphology of the unoriented sheet from extrusion cast process (precursor). There have been extensive work reporting about the relationship between processing parameter of cast sheet process of normal α-crystal PP (cooling conditions, stretching ratio) as well as the influence on final film properties (Capt et al., 2003; Lupke et al., 2004; Uejo and Hoshino, 1970). Apart from processing parameters, tacticity also affects crystallinity and stretching behavior of polypropylene. The metallocene catalyst has been demonstrated as a tool to control the molecular structure of polypropylene. The final physical properties of metallocene PP (m-PP) could also be designed to differentiate it from normal Zeigler-Natta PP (ZN-PP) such as lower melting temperature. Even with a low melt temperature, a rather narrow melting characteristic of m-PP seems to limit its uses as homopolymer in BOPP application. This limitation leads to the route of blending between ZN-PP and m-PP (Phillips, R.A., Nguyen, T., 2001). However, another material candidate offering an advantage of low melt temperature and in a board range is PP having a special β-form crystalline structure (Chinsirikul et al., 2008). It is interesting to carry out a systematic investigation to understand the characteristics or behaviors of the β-form crystal structure of PP under biaxial orientation process. Thus, the objective of this work is to study the biaxial stretching behavior of PP containing various β-crystal contents, by comparing between simultaneous and sequential processes. It is also hoped that the obtained information can be useful for the production of biaxially-oriented PP films.

Experimental

Polypropylene homopolymer, HP420J from HMC polymers (Thailand) with MFI of 3 g/10min was used in this study. The proprietary type and content of "fine particle" β-Nucleator masterbatch was prepared at MTEC. The PP sample sheets were melt extruded using a flat film unit CR-136 (Collin, Germany). The extrusion melt temperature was 230°C, take up speed was 0.8 m/min. Chill roll temperatures were controlled and varied in a range of 60 to 120°C during sheet forming process to investigate effects of cooling on the crystallization of β-crystal nucleated PP samples (as described in Table 1). While the temperature of chill roll for preparing neat PP sample was kept at 60°C, thicknesses of all cast sheets were precisely controlled to be about 0.6 mm.

Table 1 Sample identification of neat PP and β-nucleated sheet and chill roll temperature during cast sheet process

Samples	Materials	Chill roll temperature (°C)
n-PP60C	Neat PP	60
b-PP60C	β-nucleated PP	60
b-PP70C	β-nucleated PP	70
b-PP80C	β-nucleated PP	80
b-PP90C	β-nucleated PP	90
b-PP100C	β-nucleated PP	100
b-PP110C	β-nucleated PP	110
b-PP120C	β-nucleated PP	120

Specimens for film stretching were cut from cast sheet with a dimension of 96 mm x 96 mm. Simultaneous biaxial film stretching was performed on a laboratory film stretcher KARO IV (Bruckner, Germany) with a drawing area (difined by tentering clips) of about 75 mm x 75 mm. Test specimens or raw sheets (precursor) were rapidly preheated, using hot air blowing system at a stretching temperature of ~145°C for 30 seconds before stretching to 5.5 x 5.5 ratio with an engineering strain rate of 400%/sec. The stretched films were also prepared at different draw ratios ranging from 1.0-5.5 in order to follow the morphology development and orientation. During stretching process, the stress vs. extension ratio curve was recorded via a measurement by 2 force transducers which were mounted on tentering clips in machine and transverse directions. In the case of simultaneous stretching process, sheets were stretched biaxially with equal speed and ratio, the stretching forces in machine and transverse directions were almost the same. Thus, MD curve was selected to represent the biaxial draw behavior of each sample. It should also be noted that the cast raw sheets (precursors) prepared in this experiment had no preferred orientation.

The β-nucleating efficiency in PP cast sheets and films was analyzed using a Perkin Elmer DSC-7 with a scanning rate of 50°C/min from 50 to 230°C. WAXD technique was used to confirm the β-crystalline structure of PP using Rigaku R-Axis (Rigaku, Japan) with 2D-CCD detector (flat plate). X-ray diffraction experiment was performed at Department of Organic and Polymeric Materials, Graduate School of Science and Engineering, Tokyo Institute of Technology. The X-ray generator was operating at 45kV and 60 mA, based on Ni-filtered CuKα radiation. The diffraction experiments were performed on specimens aligned perpendicular to the X-ray beam. An optical property, in terms of haze value, was investigated using haze gard plus (BYK Gardner, USA) according to ASTM D1003. The silicone oil was applied on film surface in order to minimize the haze from surface roughness contribution.

Results and Discussion

Crystalline structure of cast PP sheet

The neat PP and β-nucleated PP sheets were analyzed by DSC technique to determine the differences of crystal structure, using high heating rate of about 50°C/min to minimize the recrystallization of β- to α-crystal structure during heating. From figure 1, it was found that neat PP sample exhibits a single melting peak of α crystal at a temperature of about 162.3°C. β-nucleated samples, however, reveal additional lower melting peak(s) and a relatively broader melting peak of the α crystal than that of the neat PP sample. β-nucleated PP sheets fabricated using chill roll temperature of 60°C exhibit two melting peaks of β-crystals at the temperatures of 145.4°C and 151.0°C, indicating the existence of different β-crystalline phases, possibly in terms of crystal perfection and/or dimension. When the chill roll temperatures increase from 70 to 90°C, these two peaks tend to be less visually identifiable with larger areas deviating towards the lower melting peaks of ~145.4°C. Such two melting peaks become even close to a single peak at higher temperature for the PP sheets cast at higher chill roll temperatures from 100-120°C. Overall DSC results clearly state the importance of cooling in controlling the structures of the β-nucletaed PP phase. How these developed different structures in the PP samples affecting the stretching behaviors, will be later discussed.

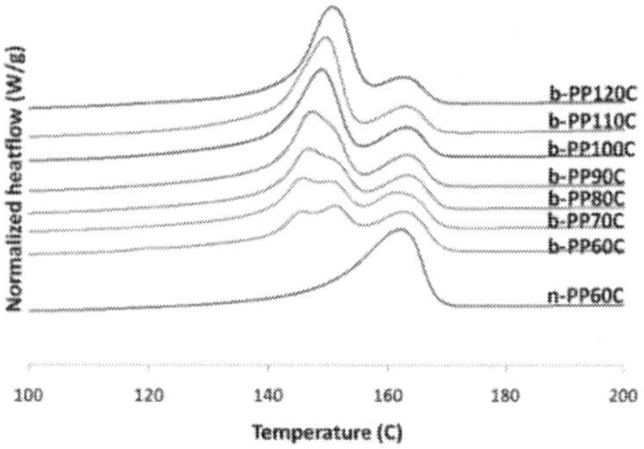

Figure 1 DSC thermograms of neat PP (n-PP60C) and β-nucleated PP sheets (as labeled).

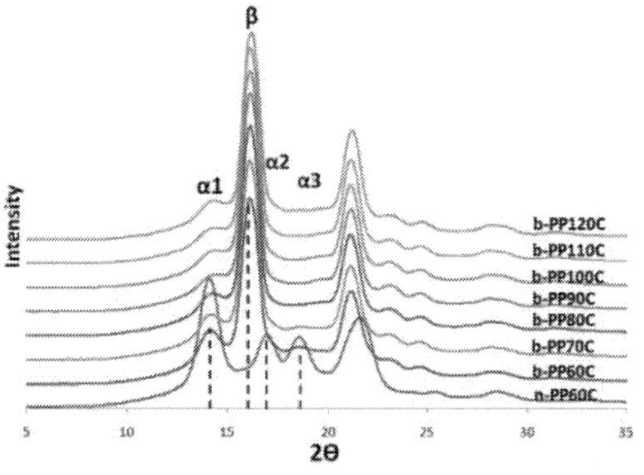

Figure 2 WAXD results of neat PP (n-PP60C) and β-nucleated PP sheets (as labeled).

WAXD measurements of cast sheets, neat PP and β-crystal nucleated PP, prepared under different chill roll temperatures were shown in Figure 2. The β-crystal characteristic peaks at 2θ ~ 16.2° were found in samples prepared at all cast temperatures, even at 60°C, except for the neat PP sheet. The observed peak intensities were influenced by cast temperatures. The amount of β-phase crystal relatively to the total crystal content was calculated using the *Turner-Jones equation:* K value = H_β / [H_β + $H_{\alpha 1}$ + $H_{\alpha 2}$ + $H_{\alpha 3}$]. A comparison on calculated K values of the β-nucleated PP samples as a function of chill roll temperatures was shown in Figure 3. It was found that K values increase (from 0.77 to 0.87) with increasing chill roll temperatures of the cast process from 60 to 90°C, and then tend to be plateau in a small variation range of 0.85-0.87 at the chill roll temperature of 100, 110 and 120°C. These findings tend to indicate that the chill roll temperature of ~90°C seems to be suitable for developing high β-form crystal contents in the PP cast sheets. Based on DSC results (Figure 1), it is also apparent that the melting endotherms of the β-form crystals (or the lower melting peaks of 145-151 °C) become more pronounced with large areas, for the β-nucleated PP samples cast on the chill roll temperatures of 90, 100, 110 and 120 °C. Such large melting areas also reflect high crystallinity of the β-form crystals in these samples.

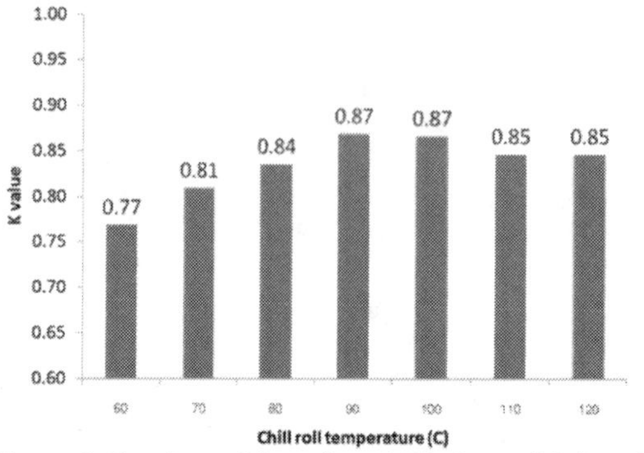

Figure 3 K values of β-nucleated PP sheets fabricated using different chill roll temperatures.

Simultaneous biaxial stretching behavior

With regard to the biaxial stretching temperature of 145°C, this stretch temperature is higher than the onset melting zone and closes to the melting temperature of the β-crystal endothermic peak, revealed by DSC (145.4°C). By this reason, a partial melting of β-form crystals can occur under stretching; this results in a reduction of energy requirement to deform β-nucleated PP samples during the biaxial stretching process. The stretching behavior illustrated by 'biaxial stress vs. stretching ratio' curves were shown in Figure 4. Neat PP sample exhibits the highest biaxial yield stress of about 3.58 MPa, stress then drops with a strain hardening (starting from a high stretch ratio of ~3.0), as seen in a typical tensile stress-strain curve of a semi-crystalline polymer. However, it can be seen that all β-nucleated PP samples show lower biaxial yield stress values than that of the neat PP (values are depicted in Table 2). Reduced load requirements observed in all β-nucleated PP could be owing to a proportion of the lower melting characteristics based on β-form crystalline structure (melting within a range of ~140-150 °C) as compared to that of α-form crystal PP (melting of ~160 °C). Upon stretching at 145 °C, the PP sheets containing β-form crystals tend to require lower energy to initiate disruption of crystalline morphology in the precursor sheets. Of all stretched samples, biaxial yield stress of the b-PP80C appears to be the lowest with a value of 1.69 MPa (or merely 53 % of the stretched neat PP (n-PP60C)).

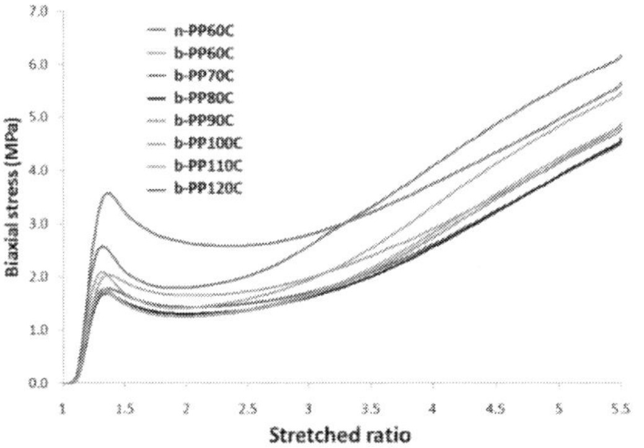

Figure 4 Biaxial stretching curve of neat PP and β-crystal nucleated samples, prepared at different chill roll temperatures.

In the case of β-PP cast on the chill roll temperatures of higher than 80 °C, biaxial yield stress values of those samples increase with increasing chill roll temperatures (see Table 2). These results could be attributed by further crystal developments when β-PP sheets were subjected to high chill roll temperatures of 90, 100, 110 and 120 °C, as evidenced by the larger DSC melting peaks of β-crystals (Figure 1) along with the higher β-crystal contents (K-values) by WAXD (Figure 2) as compared to the b-PP80C sample. Nevertheless, if comparing the b-PP120C and b-PP80C, these two samples possess similar K values but the b-PP120C shows a much higher biaxial yield stress (~2.58 MPa) than that of b-PP80C (~1.69 MPa). As evidenced by DSC, the β-crystal melting of the b-PP120C sample is a rather sharp peak, with a peak temperature shifted towards higher temperature. This may suggest an existence of large and/or perfect crystals. Therefore, the critical factors influencing different biaxial behaviors among all PP samples (among β-nucleted PPs or between β-nucleted PP versus neat PP) could be not only the crystal content but also the crystal size as well as crystal perfection developed under different thermal history conditions, in the PP precursor preparation process.

Note that strain hardening is still evident in all β-nucleted PP samples, as found in the neat PP sample. It has been suggested that the fibrillar network development in the stretched or drawn PP films, can be an origin of strain hardening (as reported by Lupke et al., 2004).

The area under stretching curve or biaxial stretching energy can also indicate energy consumed during film forming process. It is also found that b-PP80C requires the lowest energy among all samples (neat PP and all β-nucleated PP). The optical property, in terms of haze value, of the b-PP80C sample is about 0.73. This value is higher than that of n-PP60C sample (haze value = 0.29). Nevertheless, the haze value of b-PP80C with silicone oil coated on the film surface reveals a much lower value of ~0.19 which is comparable to that of neat PP (~0.15). Haze results imply that a major contribution to the haze of the b-PP80C samples comes from the surface property such as surface roughness or any possible inhomogeneity. However, the total haze of lower than 1% is considered a practical material for general flexible packaging application.

Table 2 Biaxial stretching stress results and haze of neat PP and β-nucleated samples.

Sample	Biaxial yield stress (MPa)	Final stress (MPa)	Area under curve* (%)	Haze value**
n-PP60C	3.58	5.65	100.0	0.29/0.15
b-PP60C	2.03	4.79	74.2	0.43/0.16
b-PP70C	1.78	4.55	67.0	0.48/0.19
b-PP80C	1.69	4.60	65.1	0.73/0.19
b-PP90C	1.7	4.87	67.9	1.19/0.25
b-PP100C	1.77	4.88	69.3	1.77/0.29
b-PP110C	2.08	5.48	79.9	2.94/0.59
b-PP120C	2.58	6.17	97.2	5.99/3.88

*as compared to area under stretching curve of neat PP sample.
**as-stretched sample/ with silicone oil coated on sample surface.

Study of changes in PP containing β-form crystals upon stretching

In this series of study, the b-PP80C sample was selected as it exhibits the lowest biaxial yield stress and energy requirement during biaxial stretching. It is of an interest to perform a systematic investigation on the crystal changes or developments of the β-form PP, compared to the neat-PP, during stretching from 1.0x1.0 to 5.5x5.5, at 145 C. Samples with different stretch ratios were prepared for crystal characterization utilizing WAXD and DSC techniques. As shown in Figure 5, the K values from WAXD measurement decreases as a function of stretching ratio. These could imply a possible crystal transformation, under stress and high temperature, from less stable β-crystal to more stable α-crystal structure. It can be seen that K values become close to zero when stretching ratios are higher than 4.0. This suggests that the β-crystals tend to disappear in highly stretched b-PP80C samples.

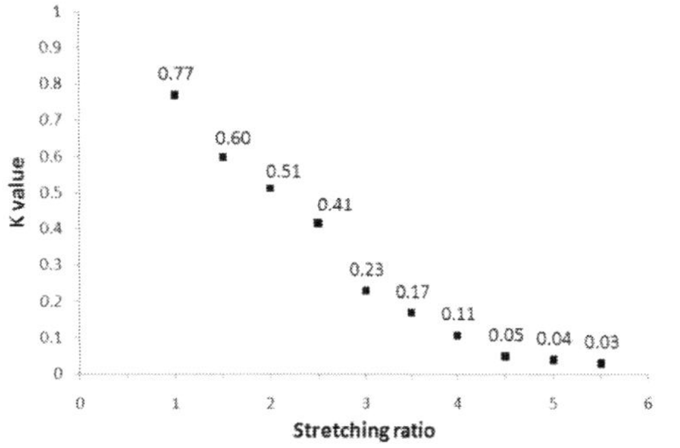

Figure 5 K values of b-PP80C sample with different stretching ratios.

From figures 6 and 7, DSC results show crystal morphology changes/development during biaxial stretching process of the n-PP60C and b-PP80C sheets. As described earlier, the melting peak temperature of α-crystal PP, in the n-PP60C sample is about 162.3°C and the melting ends at about 172.0°C (Figure 1). Obviously seen in both figures 6 and 7, the melting characteristics of stretched films change with increasing stretch ratios. The n-PP60C with 1.5x1.5 stretch ratio, beyond its yield point, has a relatively sharp melting peak but at a lower temperature than the unstretched sample. This could be explained by a destruction of lamellae from the highly ordered-spherulitic morphology of the starting precursor of n-PP60C. By further stretching above 1.5x1.5, the melt temperatures of stretched films increase as stretching ratios increase. The stretched film of n-PP60C at 5.5x5.5 ratio has melting peak at a temperature of 169.2°C and melting ends at about 180.5°C. The higher meltings of highly stretched n-PP60C is possibly due to the perfection of the developed crystals and higher degree of orientation. These result in a required higher energy of melting. As previously reported by Elias et al. (2000) and Tabatabaei et al. (2009), spherulitic structure of polypropylene could be destroyed and converted into an oriented fibrillar structure which attributed to a higher melting temperature in oriented film.

In the case of b-PP80C sample, by further stretching, the crystal transformation from β- to α-phase occurs and results in reducing area of β-crystal melting peak (Figure 7). Melting peaks finally disappear at high stretching ratios, as previously explained based on K value reduction. The oriented b-PP80C sample, with a stretch ratio of 2.0x2.0, exhibits broaden α-melting peak that tends to shift to a higher temperature than that of unstretched specimen. At a stretch ratio of 2.5x2.5, where strain hardening begins (Figure 4), the b-PP80C sample stretched to this stage shows a small melting peak with higher temperature of presumably 'newly formed α-crystal'. This is also associated with an abrupt decrease of β-melting peak. For the samples with stretch ratio of higher than 2.5x2.5 to final stretch ratio of 5.5x5.5, the areas and peaks of the α-crystal melting endotherms increase owing to changes of the crystalline morphology. At a similar stretch ratio of 5.5x5.5, the melting enthalpy of stretched b-PP80C is about 80.0 J/g which is higher than that of stretched n-PP60C (~73.4 J/g).

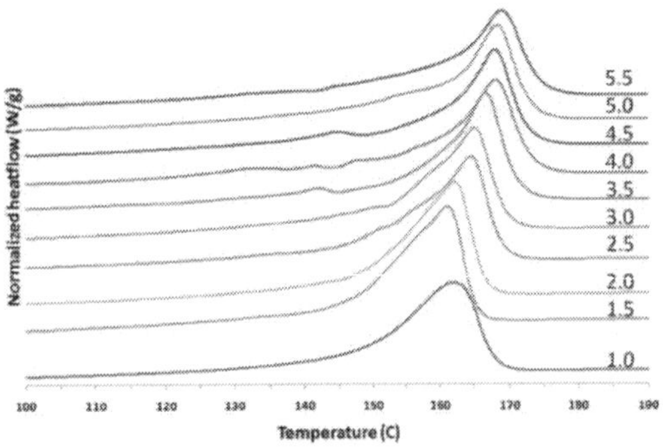

Figure 6 DSC thermograms of n-PP60C with different stretching ratios.

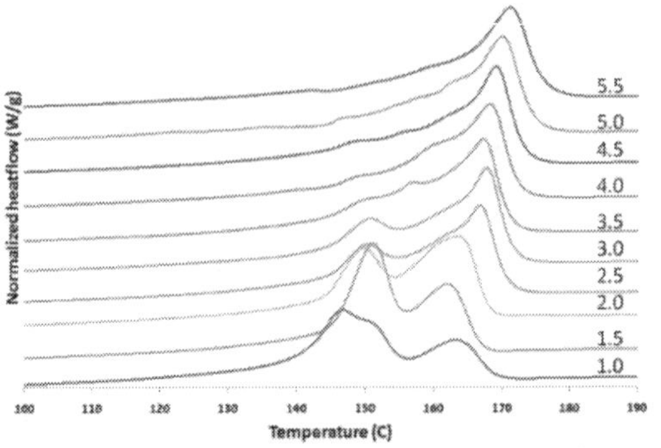

Figure 7 DSC thermograms of b-PP80C with different stretching ratios.

Sequential biaxial stretching behavior

Sequential biaxial stretching behaviors of neat-PP and β-nucleated PP samples were demonstrated in Figure 8 (a) and (b). The sheet deformation under sequential stretching process was performed in two steps: the first stretch was in machine direction (MD) to a final stretch ratio of 5.5, while keeping the film width in transverse direction (TD) constant, followed by the second stretch in TD with the same final stretch ratio of 5.5. The stretching forces in both machine and transverse directions were continuously recorded through the whole process. The temperature in stretching oven was kept constant at about 145°C. As seen in Figure 8 (a), the neat PP sample exhibits typical yield stress characteristics in both MD or the first stretch and TD or the second stretch with recorded biaxial yield stress values in the two steps of ~3.15 MPa and ~5.09 MPa, respectively. For the β-nucleated PP sample (Figure 8 (b) results reveal different biaxially stretched behaviors from the neat-PP sample. Differences are the lower MD-biaxial yield stress during the first stretching step (~2.11 MPa) and no noticeable TD-yield point in the second stretching step. Moreover, a clear strain hardening is also present in the second stretching (TD curve) of β-nucleated PP sample. These could suggest that any structural development during the first stretch (in MD) of the β-nucleated PP sheets may influence and give rise to observed stretching behavior with strain hardening on the second step of sequential stretching process for the b-PP80C. The total area under stretching curve of the b-PP80C sample was ~ 26 % lower than that of n-PP60C. Like the stress-deformation results obtained under simultaneous stretching of PP (Figure 4 and Table 2), lower energy is also required for sequential stretching of the PP precursor containing β-form crystal, as compared with the neat PP. With regard to clarity, resulting haze values of both neat PP and β-nucleated PP films (with similar stretch ratio of 5.0x5.0) are relatively the same, regardless of stretching sequences. Low haze values of both simultaneously and sequentially stretched neat PP and β-PP films are ~0.37+0.10 and 0.74+0.05, respectively.

Further investigation is also underway to better understand the effects of β-form crystal in the precursor sheets on the interesting sequential biaxial stretching behaviors of polypropylene films.

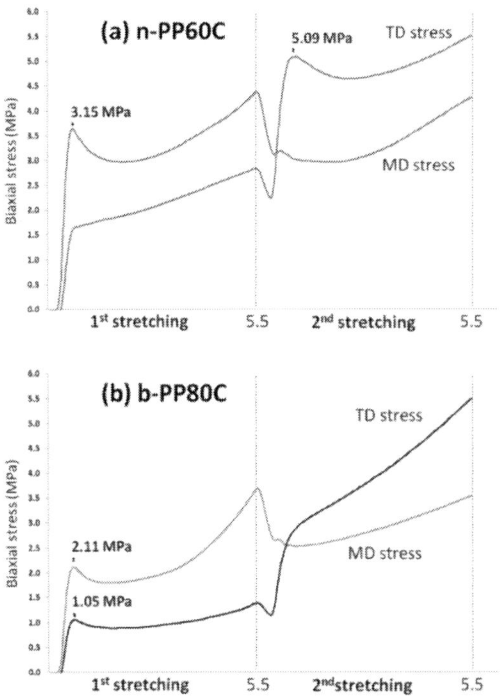

Figure 8 Sequential biaxial stretching curves of n-PP60C (a) and b-PP80C samples (b), with stretching ratio of 5.5x5.5.

Conclusions

High transparent biaxially oriented films from β-nucleated PP were successfully developed by simultaneous and sequential stretching technique. Haze values of the highly stretched β-nucleated PP films (5.5x5.5) appeared to be lower than 1%. The biaxial stretching behavior of β-nucleated PP exhibited a lower yield stress than that of α-form PP. Total energy requirement for stretching β-nucleated PP films to high ratio of 5.5x5.5 was much less than that for the neat PP having α-form crystal. The processing conditions during cast sheet process or thermal history, in particular, cooling of sheets with different chill roll temperatures had a crucial influence on biaxial stretching behavior and final film clarity. The optimum chill roll temperature for β-nucleated PP was found to be 80°C, sheets produced at this condition required the lowest stretching energy among all β-PP samples and the resulting stretched films possessed high clarity. The stretched films of both neat and β-nucleated PP exhibited sharp and higher melting temperatures than unstretched sheets, indicating more uniformity and perfection of highly ordered crystal morphology.

Acknowledgements

The authors would like to acknowledge financial support of this research from National Metal and Materials Technology Center (MTEC) and staffs of the Thermal Analysis Laboratory at MTEC for technical support and valuable discussions. The first author is also grateful to Tokyo Institute of Technology for an invitation for foreign young researchers program.

References

Capt, L., et al.,"Simultaneous Biaxial Deformation Behavior of Isotactic Polypropylene Films", Polymer Engineering and Science, **43**, 1428-1441 (2003)

Chinsirikul, W., et al., "Draw Behavior of PP and β-nucleated PP Extruded Sheets during Biaxial Stretching: Processing-Property Aspects", Asian Workshop on Polymer Processing 2008 Papers, Tokyo, Japan (2008)

Chu, F., Kimura Y., "Structure and Gas Permeability of Microporous Films Prepared by Biaxial Drawing of β-Form Polypropylene" Polymer, **37**, 573-579 (1996)

Elias, M.B., et al.,"Thermal and Dynamic-Mechanical Characterization of Uni- and Biaxially Oriented Polypropylene Films", Journal of Thermal Analysis and Calorimetry, **59**, 143-155 (2000)

Lesak, E., et al.,"Plastic Deformation Behavior of β-Phase Isotactic Polypropylene in Plane-Strain Compression at Room Temperature", Polymer, **47**, 8562-8574 (2006)

Lupke, Th., et al.,"Sequential Biaxial Drawing of Polypropylene Films", Polymer, **45**, 6861-6872 (2004)

Phillips, R.A., Nguyen, T.,"Structure, Processing, Morphology, and Property Relationships of Biaxially Drawn Ziegler-Natta/Metallocence Isotactic Polypropylene Film", Journal of Applied Polymer Science, **80**, 2400-2415 (2001)

Tabatabaei, S.H., et al., "Structure and Properties of MDO Stretched Polypropylene", Polymer, **50**, 3981-3989 (2009)

Uejo, H., Hoshino, S.,"Structure of Biaxially Oriented Polypropylene Film", Journal of Applied Polymer Science, **14**, 317-328 (1970)

APPLICATION OF LOW TACTICITY POLYPROPYLENE TO NON-WOVEN FABRICS AND FILM

Toshitaka.Kanai[1*], Tomoaki.Takebe[1], Yohei Kohri[1], Yutaka Minami[1]

[1]Performance Materials Laboratories, Idemitsu Kosan Co., Ltd. 1-1, Anesaki-Kaigan, Ichihara, Chiba, 299-0193, JAPAN, toshitaka.kanai@si.idemitsu.co.jp;
tomoaki.takebe@si.idemitsu.co.jp;yohei.kohri@si.idemitsu.co.jp;yutaka.minami@si.idemitsu.co.jp

Abstract
Novel polypropylene having very low isotacticity, low modulus and narrow molecular weight distribution (LMPP) was obtained using the specific C_2 symmetric metallocene complexes.
Spunbond non-woven fabrics using this novel polypropylene had high elastic recovery which reached 85% using bi-component (BICO) or multi- layered structure under high line speed.
A small amount of very low tacticity polypropylene added to standard PP improved the spinnability and could produce very fine denier and soft touched non-woven fabrics. It has a potential of reducing non-woven fabrics weight.
It was also added to the biaxially oriented polypropylene grade (BOPP). Blending of very low tacticity PP to BOPP influenced spherulite size and lamella thickness and decreased the yield stress of stretching force and the stress at the high stretching strain without reducing melt temperature which improves film thickness uniformity and prevents film break during the stretching process.

1. Introduction

Low isotacticity and low modulus polypropylene (LMPP) was studied the applications for high elastic recovery non-woven fabrics and for improving the spinnability and stretchability by adding a small amount of LMPP to conventional PP. LMPP has potentials to be used a disposal diaper, a sanitary napkin, hygiene products, bandages and biaxially oriented films. This paper introduces textile and non-woven fabrics and film application and their physical properties.

2. Experimental

LMPP(M_w=110000, M_w/M_n=2.0, T_m=70C) was polymerized by using the specific C_2 symmetric metallocene complexes. In terms of processability and physical properties, Primepolymer iso-PP Y6005GM (MFR=60, T_m=165℃) as a spinning grade and F300SP (MFR=3、Tm=161℃) and F300SV (MFR=3,Tm = 165℃) as biaxially oriented film grades were blended to LMPP. Reicofil 4 (#4 generation machine of Reicofil Co., Ltd.) was used to produce the spunbond non-woven fabrics. The Opt-Rheometer was used to evaluate the stretchability.

3. Results and Discussion

3-1 Elastic Recovery Non-Woven Fabrics

1) Spinnability of high elastic recovery spunbond non-woven fabrics

Addition of a small amount of high tacticity polypropylene to LMPP as crystallinity control agent gave high recoverable and non-sticky spunbond non-woven fabrics[1)].

In case of LMPP 100%, as the web was easy to stick to an emboss roll, it was hard to produce non-woven fabrics. But bi-component structure of LMPP 100% as core layer and LMPP blended with a small amount of iso-PP as the sheath layer gave stable continuous production.

The multi-layered non-woven fabrics composed of blend of iso-PP to LMPP as outer layers and 100% LMPP as an inner layer was produced at high speed 400m/min.

2) Elastic Recovery Data[1),2)]

Fig.1 shows the load-strain hysteresis in MD and TD for multilayer non-woven fabrics. 1st load-strain shows not only elastic deformation, but also plastic deformation, so upcurved line was shown. The second load-strain curve shows downcurved line which means elastic deformation was mainly dominated. The stress level in MD was much larger than that in TD and shows anisotropy. This result comes from high line speed. As each fiber was oriented to belt conveyer line direction, the fiber orientation increases with increasing the line speed.

Fig.2 shows the dependence of elastic recovery on the contents of polypropylene for monolayer, multilayer and BICO non-woven fabrics produced using Kasen Nozzle and Reicofil4 spunbond machines. The elastic recovery increases with decreasing conventional iso-polypropylene. It is considered that the crystal phase which causes plastic deformation decreases. As the bicomponent structure composed of only a small amount of high tacticity polypropylene included in the sheath layer and 100% LMPP in the core layer could reduce the total high tacticity polypropylene, high elastic recovery was obtained.

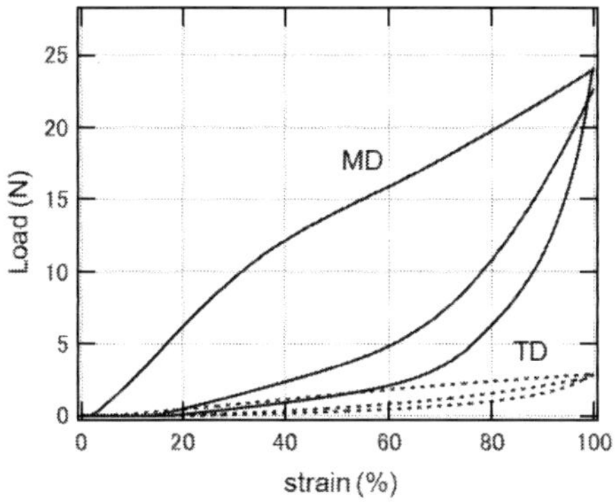

Figure 1. Load-strain hysteresis in MD (solid line) and TD (broken line) for multilayer nonwoven fabric (SMS).

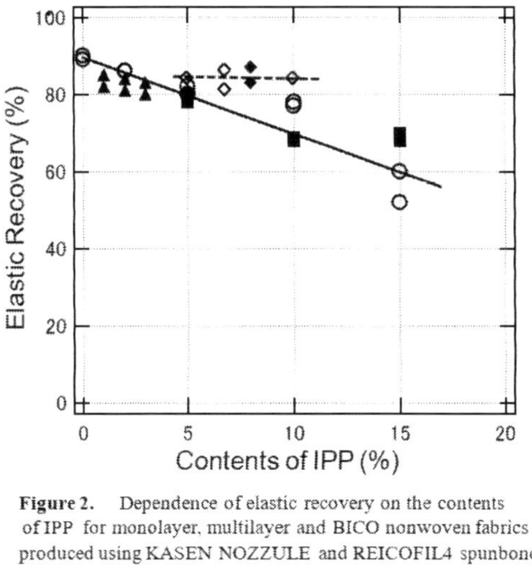

Figure 2. Dependence of elastic recovery on the contents of IPP for monolayer, multilayer and BICO nonwoven fabrics produced using KASEN NOZZULE and REICOFIL4 spunbond

3-2 Improvement of spunbond spinnability

Addition of 10% LMPP to high tacticity PP improved maximum spinning speed from 5km/min to 9km/min[3)]. In application to spunbond process, the soft touch spunbond non-woven fabrics composed of fiber denier 0.85 and low web weight 10g/m2 with good web distribution could be produced without any break at high line speed. The fiber denier was predicted by using the theoretical analysis which could set the process conditions of spunbond process.

3-3 Improvement of PP Film Stretchability

1) Reducing yield value by adding LMPP to high tacticity PP

PP grade for biaxially oriented PP film A (F300SP M_w=37.7×10^4, M_w/M_n=4.3, [mmmm]=90.0mol%) and high tacticity polypropylene B (F300SV; M_w=31.8×10^4, M_w/M_n=3.9, [mmmm]=97.7mol%) and 10wt% added low tacticity polypropylene to B (L4-10 were extruded by T-Die and formed 300μm thickness sheet and stretched uniaxially at 140℃.

The stress and strain curves for the blend of LMPP in the conventional PP are shown in Fig.3. The blending of very low tacticity PP decreases the yield value of stress and a higher blending ratio causes a reduction in the yield stress by more than 30%. The blending of low tacticity PP in F-300SP may reduce the crystallization speed, which decreases crystallinity.

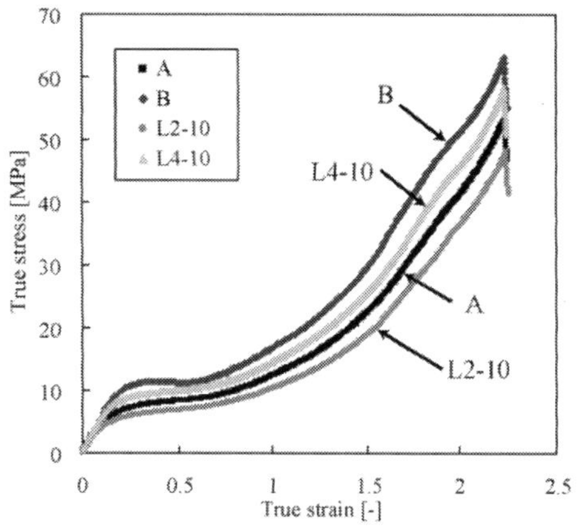

Figure 3.Stress-Strain curve of PP and PP/LMPP Blends

Adding the small amount of very low tacticity polypropylene (LMPP) to high tacticity polypropylene can improve stretchability and thickness uniformity, because the crystallization speed can be controlled by using LMPP which has slow crystallization speed.

As a result, blending of very low tacticity PP to BOPP influenced spherulite size and lamella thickness and decreased the yield stress of stretching force and the stress at the high stretching strain without reducing melt temperature which improves film thickness uniformity and prevents film break during the stretching process (Fig.4).

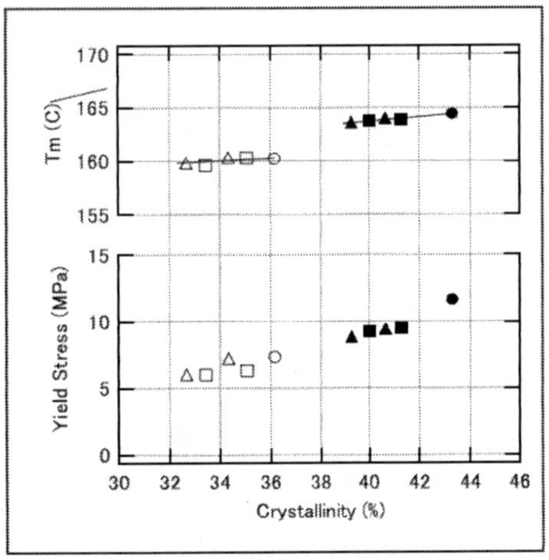

Figure 4. (a)Yield Stresses and Crystallinities of 5% or 10% Blends(■,▲), (□,△) of Various Different Molecular Weight LMPP (L1:4.3×10^5, L2:4.5×10^4, L3:7.0×10^4, L4:1.1×10^5) to BOPP A (●) and High Tacticity PP(○). (b)Melting Temperature and Crystallinity of Blends

Fig.4(a) shows yield Stresses and crystallinities of 5% or 10% blendsb of various different molecular weight LMPP to BOPP A and high tacticity PP. It was found that yield stress is proportional to crystallinity. 4(b) shows melting temperature is almost constant and does not depend on crystallinity of blends.

4. Reference

1) T.Takebe,Y.Minami,T.Kanai; Seikeikakou, 21(4)202(2009)

2) T.Takebe, T.Kanai,Y.Minami; JSPP '09 Annual Meeting proceedings, 21(4)202(2009)

3) Y.Kohri, W.Takarada, H.Itoh, T.Kanai, T.Takebe, Y.Minami, T.Kikutani; JSPP '09 Annual Meeting proceedings, 21(4)202(2009)

4) T.Kanai; Seikeikakou , 18(1),53 (2006)

5) T.Yamaguchi, T.Kanai, T.Takebe,T.Yamada; JSPP '08 Symposia proceedings, 85(2008)

SIGNIFICANCE OF BLEND RATIOS OF H-LLDPE AND LDPE ON THE TEAR PROPERTIES OF BLOWN FILMS

Khaled Mezghani[1]*, Sarfaraz Furquan[1], and Rached Ben-Mansour[1]

* [1]*Mechanical Engineering Department, King Fahd University of Petroleum and Minerals, Dhahran, Saudi Arabia* - mezghani@kfupm.edu.sa; sfurquan@kfupm.edu.sa; rmansour@kfupm.edu.sa.

Abstract - Polyethylene is widely used for packaging applications. From the different grades of polyethylene, LLDPE and LDPE play a major role in the film blowing industry. In our study, blown films of h-LLDPE, LDPE, and their blends were produced using a twin screw extruder. The film blowing process parameters which significantly affect the properties of the film were all optimized and kept constant. All films were processed at a constant extruder temperature profile, blow ratio of 1.6, draw ratio of 21, mass flow rate of 8g/min, and die temperature of 230°C. The tear properties of all films were determined in the machine direction (MD) and the transverse direction (TD) using an Elmendorf tear tester. On one hand, similar values of the MD tear-resistance for the two homopolymers, h-LLDPE and LDPE, are measured to be 100 and 120 KN/m, respectively. On the other hand, the TD tear-resistance value of h-LLDPE is 180 KN/m, four times higher than that of LDPE, 45 KN/m. Even though LDPE film has low value of TD tear-resistance, small additions of this polymer to h-LLDPE produced blends with better TD tear-resistance films. It was observed that small additions of 5%, 10%, 15%, and 20% of LDPE to h-LLDPE increased the TD tear-resistance value up to 400 KN/m (220 KN/m increase).

Introduction

Polymers play a significant role in the modern day engineering world. They are fast replacing the traditionally used materials in almost all major applications [1-3]. The commonly used polymers are polyethylene teraphthalate (PET), Polycarbonate (PC), Polyethylene (PE), Polystyrene (PS) and nylon. Among these, the most commonly used for film making is the polyethylene. PE comes in different grades. The major classification of PE consists of Low Density Polyethylene (LDPE), Linear Low Density Polyethylene (LLDPE) and High Density Polyethylene (HDPE) [4]. PE is widely used in film blowing industry for producing films for various applications. Within the family of PE resins LDPE and LLDPE resins have found an extensive application in the packaging area using film blowing process. The film blowing is a complex process, and the industry is exposed to a lot of difficulties due to processing variability in producing good quality films. The different process parameters which affect the quality of the blown film are the draw ratio (DR), blow ratio (BR), extrusion temperatures, mass flow rates and the die temperatures. Each of these variables contributes to the quality of the film. While comparing LDPE and LLDPE, it has been observed that the LDPE offers a very good processability with lower mechanical properties; on the other hand LLDPE exhibits better mechanical properties with lower processability [5]. Thus, the material variability is also a big concern while producing the blown films. The morphology and property differences among HDPE, LDPE, and LLDPE films have also been demonstrated and explained in several recent publications [6-13]. Processability of the material has been defined in various ways such as, energy consumption [5], bubble stability [8, 14, 15], torque required to turn the extruder screws, etc. Among of the most desired properties of film is its toughness and tear strength. The diverse applications of the PE films demand a vast range of mechanical properties that can be achieved by blending different types of resins. However, due to blending, the processability of the polymer and the product quality of the film are significantly affected. The aim of this research is to study the effect of blend ratio of hexene-LLDPE (h-LLDPE) and LDPE on the film processability and its tear properties.

Experimental

Material

Commercial grade of SABIC, Saudi Arabia made h-LLDPE and LDPE polymer pellets were used to produce the blown films. The melt index for h-LLDPE was 0.8 g/10 min as provided by the supplier. The density value of h-LLDPE was

0.921 g/cm^3. The melt index of LDPE was 2.0 g/10 min with a density value of 0.922 g/cm^3.

Film Blowing Setup

The film blowing extrusion was carried out on a Thermo Haake twin screw extruder. The L/D ratio for the extruder used was 40. A constant temperature profile of 120/150/180/200/200/200/200 °C was maintained in the extruder. A constant feeding and extruder screw speed was maintained to obtain a pressure of 14 to 16 bar at the extruder exit. A constant melt pump speed was set to pump the material to the film blowing die. The maximum mass flow rate at these feed rates was around 8 g/min. The film blowing die temperature was 230 °C. The blow ratio (BR) of 1.6 and the draw ratio (DR) of 21 were used while producing the blown films. The different blend ratios that were considered are 5%, 10%, 15%, 20% and 50% weight fractions of LDPE to h-LLDPE.

DSC

Crystallinity in the films was determined using Differential Scanning calorimetry. Mettler DSC 882 was used in the present study for thermal analysis. Samples were prepared by cutting circular discs from the film to fit into the aluminum pans. These discs were stacked until the weight was in the range of 3 mg to 5 mg. Samples were scanned on Mettler DSC 882 unit from 20 °C to 180 °C for all the prepared samples at a rate of 10 °C/min. The enthalpy value used was 293.6 J/g [[16, 17]]. Three samples at each condition were tested and its average value has been reported.

Birefringence

The orientation induced in the films was measured as a birefringence index on an optical microscope along with a compensator. The degree of orientation in the films was determined in both machine direction (MD) and transverse direction (TD). The orientation was measured as the retardation wavelength with respect to the dial increment of the compensator. The thickness of the film played an important role in deriving the degree of orientation in the film. The difference in retardation wavelength with the film and without the film was divided upon its thickness to obtain the birefringence in the plane. The birefringence along the MD direction was calculated.

Elmendorf Tear Test

Tear tests were conducted using the Elmendorf tear tester provided by Thwing Albert Company. Tear tests were carried out in accordance with ASTM D 1922 [[19]]. Tear tests were conducted in both machine and transverse directions of the film.

Results and Discussion

DSC

The h-LLDPE/LDPE film samples processed at five different blend ratios and the homo-polymers h-LLDPE and LDPE films were analyzed and their first heating curves are shown in Figure 1. The melting range of h-LLDPE was observed to be in the range of 115°C to 124°C. The melting range of LDPE was observed to be in the range of 104°C to 114°C. The DSC profiles clearly show the presence of one or more sizes of crystallites being developed in the film. The blend ratios exhibit double peak indicating the co-existence of separate crystals with different lamellar thickness. The heating curves obtained from DSC results were used to determine the crystallinity content that was induced in the film at different blend ratios. The area under the curve was calculated and the crystallinity content is shown in Figure 2. When comparing the individual materials, h-LLDPE has higher crystallinity then that of LDPE homo-polymer. The crystallinity contents for all blends above 10wt% of LDPE are observed to be higher than that of individual materials. This indicates that the addition of LDPE has enhanced the molecular mobility of LLDPE and promoted its crystallinity. In addition, the melting peaks of the blends, especially 15 and 20 wt%, are higher than that of the homopolymers. It is well established that the melting of polymers are directly related to the thickness of the lamellae crystal. The higher the melting peak of a crystalline polymer, the thicker is its lamellar structure. Therefore, apparently, the 15 and the 20wt% blends have thicker lamellar structure than that of the homopolymers. Currently, the morphological development of these crystals are under investigation with SEM techniques.

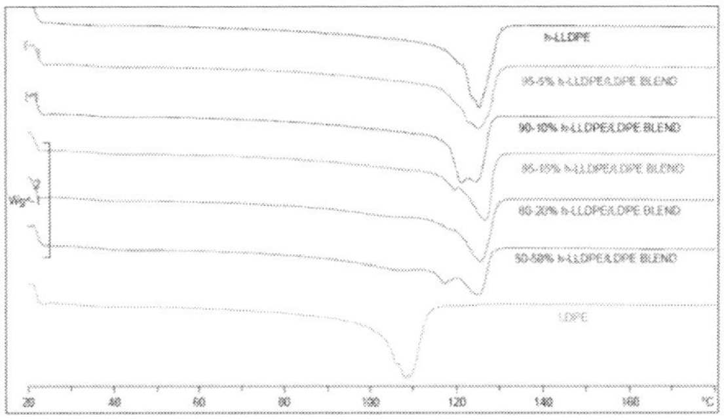

Figure 1. Heating cycles of h-LLDPE, LDPE, and their blended films.

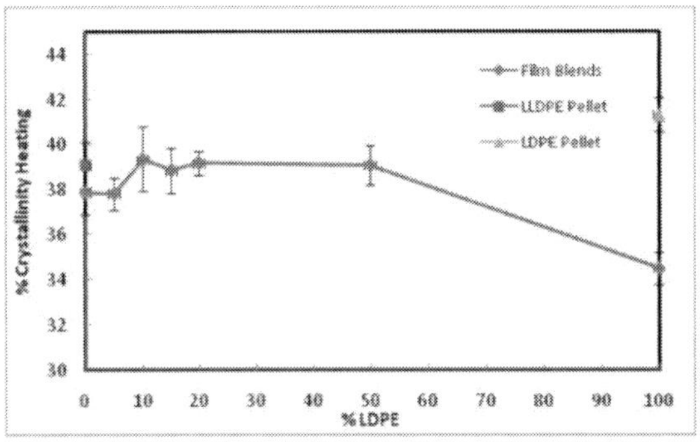

Figure 2. Crystallinity of h-LLDPE, LDPE, and their blends measured from Figure 1.

Birefringence

The birefringence (Δn) when viewed from MD direction in the plane of the film is shown in Figure 3 for the homopolymers and their blends. The birefringence values increased with addition of LDPE to h-LLDPE. The birefringence sign gradually changes from a negative to a positive sign when LDPE is added to h-LLDPE. The homopolymer h-LLDPE shows a negative birefringence sign (-3) which indicates that the preferred lamellar orientation is in the TD direction. However, LDPE polymer shows a positive birefringence (+3), indicating that the lamellar orientation is in the MD direction. According to Figure 3, the addition of LDPE to h-LLDPE promotes the lamellar orientation in the MD direction. The birefringence value of 50%-50% blend ratio was higher when compared to both the h-LLDPE and LDPE homo-polymers studied.

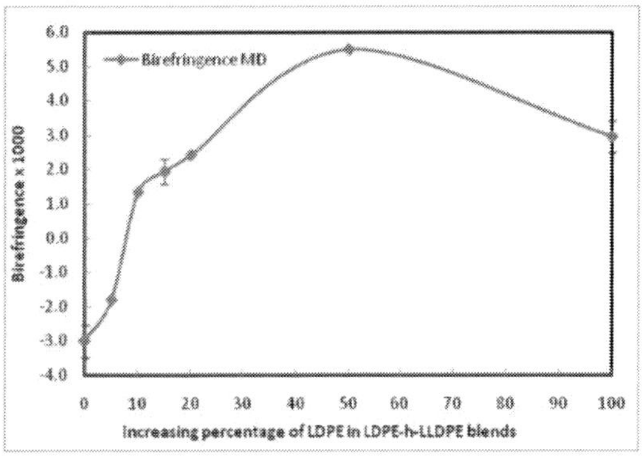

Figure 3. Birefringence of h-LLDPE, LDPE, and their Blends.

Elmendorf Tear Test

The Elmendorf tear test results measured along the TD direction are shown in Figure 4. The MD tear resistance for the two homo-polymers was almost in the same range with a value 100 kN/m for h-LLDPE and a value 120 kN/m for LDPE. On the other hand, the TD tear resistance of h-LLDPE is 180 kN/m when compared to that of 45 kN/m for LDPE. The TD tear resistance of h-LLDPE is about four times higher than that of LDPE. Thus, if the rule of mixture is applicable for the blend of these two polymers, then adding LDPE to h-LLDPE is expected to decrease the tear resistance of the blends. However, the only blend that has obeyed the rule of mixture is the 50-50wt% blend. All remaining blends have shown better tear resistance in the TD direction than their homopolymers. Thus, with small addition of LDPE to h-LLLDPE up to 20wt% the TD tear resistance increased from 180 kN/m to 400 kN/m. This dramatic change of tear resistance with small blend ratios could not be directly correlated the small increase of crystallinity as indicated in Figure 2. The birefringence measurements, on the other hand, show an increase of lamellar orientation towards the MD direction with the blend ratio. This can explain the relative improvement of low blend ratios. However, the birefringence alone cannot explain the tear behaviors of all blends. For example, the 50wt% has the highest MD orientation but shows lower tear resistance than that of the 20wt%. Therefore, more details of the internal structure developed in the blends are needed. Currently, investigation of the microstructure of the blends using SEM techniques is in progress.

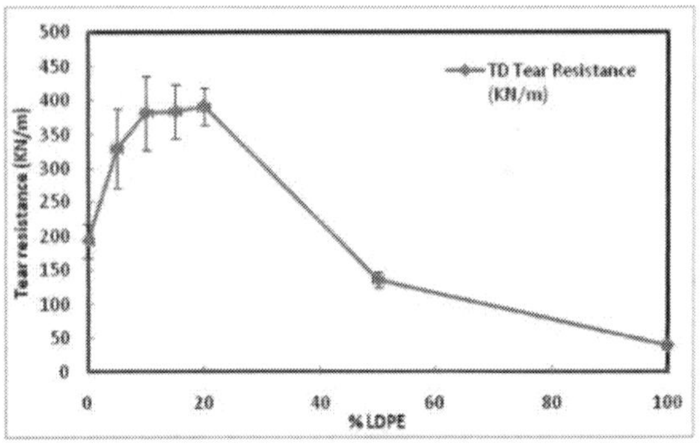

Figure 4. TD tear resistance of h-LLDPE /LDPE blends.

The processability of these polymers in terms of the torque required to turn the screw can be observed in Figure 5. The torque required to process LDPE is much lower when compared to processing h-LLDPE. The addition of LDPE to h-LLDPE has improved the processability. There was a slight increase in torque for processing the blend with 5% LDPE. For the other blends, the processability almost followed the additive rule.

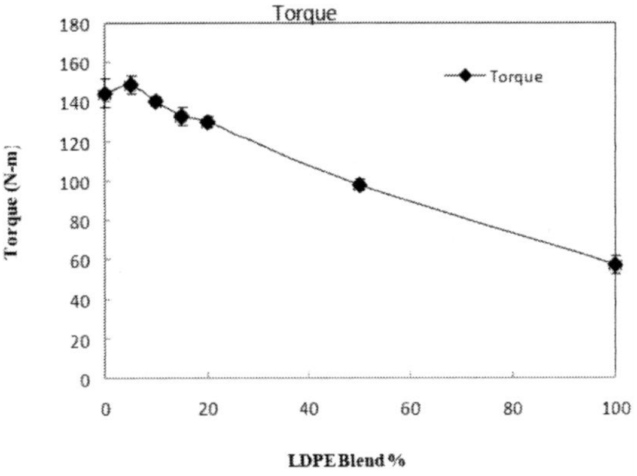

Figure 5. Torque required in turning the extruder screw against increasing blend percentage.

Conclusion

The blends of h-LLDPE and LDPE significantly influenced the tear properties of these blown films. The TD tear resistance values of these blended films were much higher when compared to the individual tear resistance of h-LLDPE and LDPE. The tear resistance has reached a value of 400 kN/m for blend ratios in the range of 10 to 20 wt% compared to 200 kN/m for pure h-LLDPE. The DSC results showed an improvement of crystallinity with the addition of LDPE to h-LLDPE; however, this slight increase of crystallinity cannot explain the large increase in tear resistance. The birefringence results can partially explain the tear resistance for the low blend ratios but it is not sufficient to cover all blends. Therefore, studies pertaining to structural– morphological changes are suggested for further investigation. On the other hand, the processability followed the mixture rule with a decrease of torque requirement when LDPE content is increased in the blend.

Acknowledgement

The authors would like to acknowledge King Abdul Aziz City for Science and Technology (KACST) for funding this research project under grant # AT 19-26. Acknowledgment is also due to King Fahd University of Petroleum & Minerals (KFUPM) for providing research facilities to carrying out this study.

References

1. R. J. Crawford, *"Plastics Engineering"*, Butterworth Heinemann, (2005)
2. I.M Ward and J. Sweeney, *"An Introduction to The Mechanical Properties of Solid Polymers"*, Wiley Interscience, (2004)
3. James E. Mark, *"Physical properties of Polymers"*, American Institute of Physics (AIP Press), (1996)
4. MacAdams JL Ross JF, *"The Polymeric Materials Encyclopedia"*, New York: CRC Press, (1996 - Polyethylene 'Commercial', pg. 5953 - 65)
5. C. M. Wong, Shih H. H. and Huang C. J., *"Effect of various polyethylene structures on film extrusion"*, Journal of Reinforced Plastics and Composites, Vol.17, No.10, (1998), P.945-954
6. Darin Huizenga, Chornoby Kurt and Engelmann Paul V., *"Optimized Lldpe/Ldpe Film Properties Through Correlation of Blend Analyses"*, Journal of Plastic Film and Sheeting, Vol.6, No.4, (1990), P.318-326
7. A. S. Bawa S. Nadanasabapathi J. H. Jagannath, *"Effect of starch on thermal mechanical, and barrier properties of low density polyethylene film"*, Journal of Applied Polymer Science, Vol.99, No.6, (2006), P.3355-3364

8. S. N. Bhattacharya P. Micic, G. Field,, *"Transient elongational viscosity of LLDPE/LDPE blends and its relevance to bubble stability in the film blowing process"*, Polymer Engineering & Science, Vol.38, No.10, (1998), P.1685-1693
9. P. Micic and Bhattacharya S. N., *"Rheology of LLDPE, LDPE and LLDPE/LDPE blends and its relevance to the film blowing process"*, Polymer International, Vol.49, No.12, (2000), P.1580-1589
10. P. Miller, Sbarski I., Kosior E., Masood S. and Iovenitti P., *"Correlation of rheological and mechanical properties for blends of recycled HDPE and virgin polyolefins"*, Journal of Applied Polymer Science, Vol.82, No.14, (2001), P.3505-3512
11. Y. L. Fang, Carreau P. J., Lafleur P. G. and Ymmel S., *"Properties of mLLDPE/LDPE blends in film blowing"*, Polymer Engineering and Science, Vol.45, No.3, (2005), P.343-353
12. C. Silvestre, Cimmino S., Raimo M., Duraccio D., Fernandez B. D., Lafuente P. and Sanz V. L., *"Structure and morphology development in films of mLLDPE/LDPE blends during blowing"*, Macromolecular Materials and Engineering, Vol.291, No.12, (2006), P.1477-1485
13. J. Lu, Sue H. J. and Rieker T. P., *"Dual crystalline texture in HDPE blown films and its implication on mechanical properties"*, Polymer, Vol.42, No.10, (2001), P.4635-4646
14. S. Kim, Fang Y. L., Lafleur P. G. and Carreau P. J., *"Dynamics and criteria for bubble instabilities in a single layer film blowing extrusion"*, Polymer Engineering and Science, Vol.44, No.2, (2004), P.283-302
15. Y. L. Fang, Carreau P. J. and Lafleur P. G., *"Rheological effects of polyethylenes in film blowing"*, Polymer Engineering and Science, Vol.43, No.7, (2003), P.1391-1406
16. Q. Fu, Men Y. and Strobl G., *"Understanding of the tensile deformation in HDPE/LDPE blends based on their crystal structure and phase morphology"*, Polymer, Vol.44, No.6, (2003), P.1927-1933
17. J. Morawiec, Pawlak A., Slouf M., Galeski A., Piorkowska E. and Krasnikowa N., *"Preparation and properties of compatibilized LDPE/organo-modified montmorillonite nanocomposites"*, European Polymer Journal, Vol.41, No.5, (2005), P.1115-1122
18. ASTM-D882, *"Annual Book of ASTM Standards"*, ASTM, (1995),
19. ASTM-D1922, *"Annual Book of ASTM Standards"*, ASTM, (1995),

THE EFFECT OF TEMPERATURE ON THE ELECTROSPINNING OF CHITOSAN/PEO SOLUTIONS

M. Pakravan, M.C Heuzey* and A. Ajji

CREPEC, Department of Chemical Engineering, Ecole Polytechnique de Montreal, Montreal, Canada – m.pakravan@polymtl.ca; marie-claude.heuzey@polymtl.ca; abdellah.ajji@polymtl.ca

Abstract

Chitosan is a modified natural polymer derived from chitin, one of the most abundant renewable natural polymers in the world. Recently chitosan has been used in a broad range of applications due to its valuable properties such as biocompatibility, biodegradability and antibacterial activity. Our focus in this study is to produce chitosan in the form of films and membranes for filtration and antimicrobial packaging purposes using the electrospinning technique. The electrospinning process is a simple and efficient technique to produce polymeric nanofibers, resulting in non-woven porous mats. Chitosan and polyethylene oxide (PEO) were dissolved in an aqueous solution containing 50 wt% acetic acid , and various blends of these solutions with different chitosan/PEO ratios were prepared (4 wt% total polymer content). A modified electrospinning set up have been designed to control the solution temperature in the spinnning process. The impact of temperature and blend composition on morphology and fiber diameters of electrospun materials have been investigated. Electrospinning at moderate temperatures (40 to 80 °C) improves the quality of the produced nanofibers by lowering the number of beads and droplets formed. Using this method enabled defect free nanofibers preparation with fiber diameters of less than 160 nm and chitosan content of up to 90 wt% at spinning temperatures varying from 40 to 80 °C.

Introduction

Electrospinning or electrostatic spinning is a recently developed technology to produce nanofibers. The average diameter of these nanofibers is typically in the range of 50-500 nanometers which is one or two orders of magnitude less than fibers obtained by conventional fiber spinning methods. The produced nanofibers are collected as three dimensional nanoporous non-woven mats that provide unique properties of surface area, porosity and uniformity, characterizations that are highly interesting for a number of applications including filters, membranes, composites and biomedical materials [Shin et al. 2001; Li and Xia 2004]. In this process a high electric field is applied on the polymer solution that is pumped out by a precise syringe pump through a needle. Electrostatic charging of the fluid forms the well-known Taylor cone at the tip of the nozzle and after deformation; a jet is ejected out from its apex. The jet accelerates and stretches during its whipping and splaying motion to the grounded target. The solvent evaporates immediately and finally the nanofibers are deposited on the collector as a thin fabric [Greiner and Wendorff 2007; Reneker and Yarin 2008]. A number of polymers have been specifically used in the electrospinning process with different degrees of success due to the required critical properties needed for electrospinnability. Among them only few polymers have shown appropriate properties for use in critical applications such as antibacterial packaging. Chitosan is one of those polymers that can reveal such characteristics. Chitosan is a modified natural polymer produced by deacetylation of chitin, one of the most abundant natural polymers in the world with similar structure to cellulose. Recently chitosan has attracted a great deal of attention and has been used in various applications due to its valuable properties such as biocompatibility, biodegradability and antimicrobial activity. Ease of formation to films, is also another characteristics that makes chitosan suitable for filtration and packaging purposes. The nanoporous morphology of films and membranes obtained from the electrospinning process enhances their efficiency because of their large specific area. Therefore, nanoporous chitosan mats can not only present the specific physicochemical properties of chitosan but also can benefit from the physical characteristics of nanoporous films. However, chitosan is a challenging polymer to electrospin, mainly because of its polycationic nature and rigid chemical structure. Furthermore, the formation of strong hydrogen bonds prevents the free movement of polymeric chains exposed to the electrical field [Geng et al. 2005; Li and

Hsieh 2006]. Therefore finding the best technique to produce chitosan nanofibers is still under investigation. The focus of this work is to electrospin chitosan solutions. Polyethylene oxide (PEO) is a water soluble, biodegradable synthetic polymer that has ether (- O -) groups in its backbone and thus can form hydrogen bonds with chitosan. It is added as a co-spinning agent to the chitosan solution and works as a carrier to make it electrospinnable. In addition, a modified electrospinning set up has been used to control the temperature of the charged solution being pumped through the syringe and needle and to allow spinning at moderate temperatures. Temperatures higher than ambient conditions reduce the solution viscosity and enable to increase the chitosan content in the final produced nanofibers. The influence of temperature on the electrospinnability of chitosan solutions and their rheological properties is investigated

Experimental

Materials

A chitosan with a degree of deacetylation (DDA) of 75% and weight-average molecular weight of ~ 750 kDa was purchased from Marinard Biotech (Rivière-au-Renard, QC, Canada). Polyethylene oxide with a molecular weight of 600 kDa was obtained from Scientific Polymers Inc. (USA). Acetic acid (99.7 %, Aldrich, WI, USA) was used to prepare the aqueous solutions.

Methods

Solution preparation. 50 wt% acetic acid in water was used as the solvent for all solutions; chitosan and PEO were dissolved in it separetly at 4 wt% concentration. The mixing was performed at room temperature using a laboratory magnetic stirrer (Corning Inc., MA, USA). Chitosan-PEO blend solutions were then prepared by mixing the two solutions at different chitosan/PEO ratios with total polymer concentration of 4 wt%. The prepared solutions were left to rest for 4 hours for degassing and kept in a sealed container at room temperature. Fresh samples were used in order to avoid aging effects due to polymer degradation or complexation.

Electrospinning procedure. The electrospinning of the solutions was performed using a set up that contains a variable high voltage power supply (Gamma High Voltage Research, USA) and a programmable syringe pump (Harvard Apparatus, PHD 2000, USA). The solutions were poured into a 10 ml stainless steel syringe connected to a needle and mounted on the syringe pump. Based on our previous experimental results, a flow rate of 0.25 ml/hr, distance of needle tip-to-collector plate of 15 cm and applied voltage of 30 kV are optimized process parameters for these solutions. A cartridge heater connected to a controler was placed around the needle and syringe to keep the solution temperature at the desirable value during the process. Nanofibers were collected on an aluminium foil attached to a rotating drum that could be easily removed for subsequent characterization.

Characterization

The morphologies of the electrospun nanofibers were observed by scanning electron microscopy (SEM, Hitachi, Japan) at accelerating voltages of 5 or 7.5 kV. All samples were sputtered with gold before the microscopic observation.

Results and Discussion

Preliminary results at room temperature showed various difficulties in the electrospinning of chitosan solutions. At optimized processing conditions, only an unstable jet was obtained after an explosion-like phenomena and a small quantity of the splayed solution was collected on the plate, with only a few beaded fibers. Therefore, as mentioned in the introduction PEO was used as a co-spinning agent with the goal of maximizing the chitosan content and reaching stable conditions for producing chitosan-based nanofibers in a continuous manner. In addition, the modified heated set-up was used to increase the solution temperature during electrospinning. Since the cartridge heater is installed all around the needle and syringe, thermal equilibrium should be achieved rapidly and it can be safely assumed that the output temperature is the same as the temperature set on the heater controller. Increasing temperature leads to a large reduction in solution viscosity of chitosan solution, as shown in Figure 1.

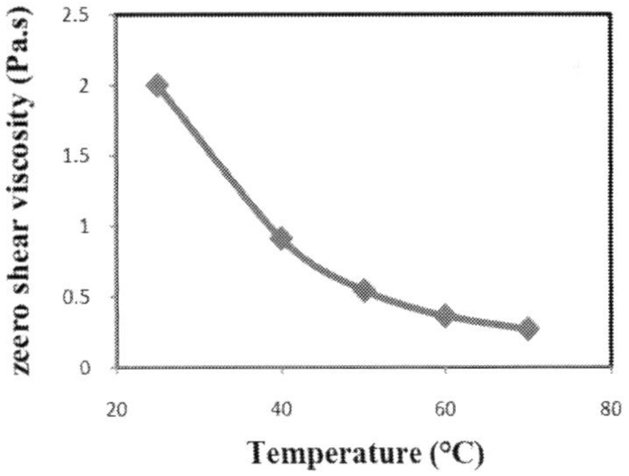

Figure 1 – Effect of temperature on zero shear viscosity of chitosan (4 wt% in 50 wt% acetic acid).

The effect of moderate temperature on the electrospinnability of chitosan solutions can be seen in Figure 2. Chitosan shows very poor fiber formation at room temperature and only some nanobeads and droplets have formed at 25 °C. This can be related to the low amount of chain entanglements in this chitosan solution, and also to its polyelectrolyte nature that causes repulsive force between the macromolecules in the solution state [Desai et al. 2008]. As temperature increases (between 40 and 60 °C) fiber formation slightly improves and the morphology changes to bead and fibers.

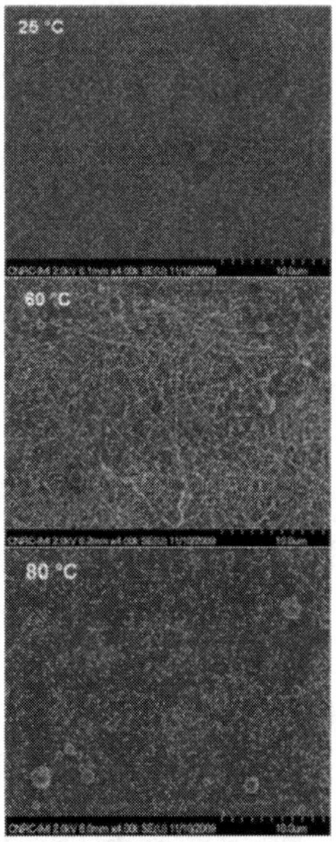

Figure 2 – SEM images of electrospun chitosan at various temperatures, 4 wt% polymer in 50 wt% acetic acid.

However, at the highest temperatures (70 to 80 °C) the number of beads rises again so that at 80 °C the result is nearly the same as at room temperature, with only beads instead of fibers. This behavior can be explained by three phenomena competing at elevated temperatures: an increasing rate of solvent evaporation, a decreasing surface tension and viscosity. The reduction in viscosity and surface tension may stabilize the jet in the spinning process, while faster solvent evaporation rate can cause enhanced drying of the whipping jet and increase chain entanglements that improves spinability [Desai and Kit 2008]. However at higher temperatures, the jet may dry too fast without having enough time to be stretched by the electrical field and it may result in the disappearance of fibers and return to the bead morphology.

Effect of blend ratios and spinning temperature

A PEO solution of 4 wt% was added to the chitosan solution of the same concentration, both in 50 wt% acetic acid at different blend ratios (4 wt% total polymer concentration). At room temperature beadless nanofibers could be obtained from mixtures of 50/50 to 70/30 of chitosan/PEO. Higher chitosan content resulted in a large presence of beads in the final microstructure (micrographs not shown). These results show that the addition of PEO can facilitate the electrospinning process of chitosan up to 70 wt% chitosan in the mixture. Moreover, PEO may break down the interactions between the chitosan macromolecules by making new hydrogen bonds between ether groups on its chain and amino groups of chitosan. This may increase chain entanglements in solution and make chitosan electrospinnable [Li and Hsieh, 2006].

On the other hand, at higher temperatures beadless morphologies and more stable jets during the spinning process were obtained, as shown in Figure 3 for a 90/10 chitosan/PEO blend. This can be attributed to a reduction in viscosity and surface tension, and also to a faster solvent evaporation rate that helps the charged jet to be further stretched and stabilized in its whipping motion toward the collector [Desai and Kit 2008]. In blend solutions higher temperatures (70 to 80 °C) did not have the same effect as for chitosan alone. That may be due to the presence of the PEO chains which increase chain entanglements, so that faster evaporation rate cannot change the morphology from fibers to beads.

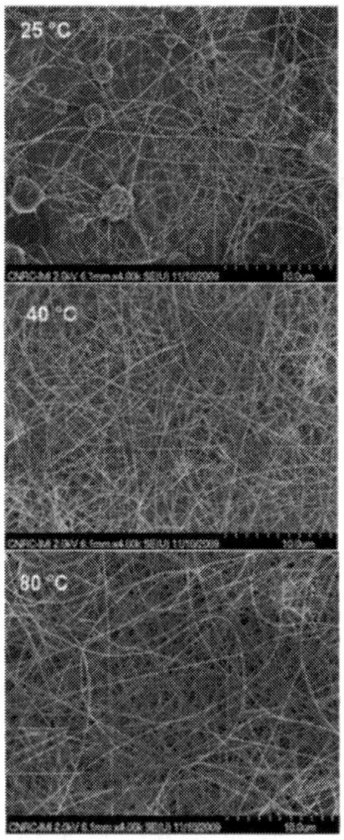

Figure 3 – SEM images of electrospun 90/10 chitosan/PEO blend at various temperatures, total polymer concentration 4 wt% in 50 wt% acetic acid.

At higher chitosan content (95 wt %), the number of beads increased even at high temperature due to the large amount of chitosan in solution. The effect of chitosan content and spinning temperature on the distribution of fiber diameters is shown in Figure 4. It reveals that fiber diameter decreases with increasing chitosan content due to higher repulsion forces between the chitosan chains. No uniform trend in fiber diameter with temperature is observed, but in most blends, as temperature increases, slightly larger values are observed. This could be correlated to a faster evaporation rate, allowing smaller stretching on the jet.

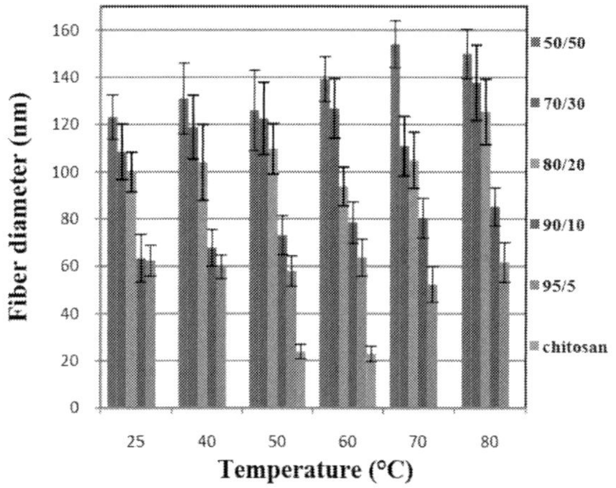

Figure 4 – Effect of blend ratio (chitosan/PEO) and spinning temperature on fiber diameter, total polymer concentration 4 wt% in 50 wt% acetic acid.

Conclusions

Defect free chitosan-based nanofibers could be obtained using chitosan/PEO blends and a modified electrospinning set up that can control the solution temperature in the spinning process. Electrospinning at moderate temperatures (40-80 °C) helped to stabilize the jet and improved the spinnability of the chitosan blend solutions, so that higher chitosan content could be reached. Non-woven mats containing 90 wt% chitosan were prepared from 90/10 blend solutions of chitosan/PEO at 40-80 °C. Increasing solution temperature caused different competing phenomena to occur during the spinning process: decreasing surface tension and viscosity and faster solvent evaporation rate. These have different effects on the final produced nanofibers depending on the electrospun solution. Finally, increasing chitosan content in the blend solutions lead to a significant reduction in nanofiber diameters, mainly due to repulsive forces between chains in solution.

References

Desai, K. and Kit, K., "Effect of spinning temperature and blend ratios on electrospun chitosan/polya(crylamide) blend fibers", Polymer, **49**, 4046-4050, (2008)

Desai, K., Kit, K., Li, J. and Zivanovic, S. "Morphological and surface properties of electrospun chitosan nanofibers", Biomacromol., **9**, 1000-1006 (2008)

Geng, X, Kwon. O.H. and Jang. J., "Electrospinning of chitosan dissolved in concentrated acetic acid solution", Biomat., **26**, 5427-5432, (2005)

Greiner A. and Wendorff J.H., "Electrospinning: A fascinating method for the preparation of ultrathin nanofibers", Angew. Chem. Int. Ed., **46**, 5670-5703 (2007)

Li, D. and Xia, Y. "Electrospinning of nanofibers, reinventing the wheels?", Adv. Mat, **16**, 1151-1170, (2004)

Li, L. and Hsieh, Y.L. "Chitosan bicomponent nanofibers and nanoporous fibers", Carbohydrate Research, **341**, 374-38 (2006)

Reneker, D. and Yarin, A.. "Electrospinning of jets and polymer nanofibers", Polymer, **49**, 2387-2425 (2008)

Rinaudo M., "Chitin and chitosan: Properties and applications", Prog. Polym. Sci., **31**, 603-632 (2006)

Shin, Y.M., Hohman, H., Brenner, M.P. and Rutledge, G.C. "Electrospinning: A whippig fluid jet generates submicron polymer fibers", Appl. Phys. Let., **78**, 1149-1151 (2001)

Structure-Properties Relationship of Polypropylene Cast Films

Seyed Hesamoddin Tabatabaei, Abdellah Ajji*

École Polytechnique of Montreal - Quebec - Canada

Three polypropylene cast films of different initial morphologies (only spherulitic structure, coexisting rows of lamellae and spherulites, and lamellar structure) were prepared by extrusion followed by stretching using a chill roll. The effects of the original morphology on the orientation, mechanical respo-nses, tear resistance, and oxygen permeability were investigated. The precursor film with the lamellar morphology showed much larger crystalline alignment, Young modulus, tensile strength, and tensile toughness along the machine direction (MD) and greater oxygen permeability compared to the precursors with the only spherulitic structure and coexisting rows of lamellae and spherulites.

Gas Counter Pressure Control for Microcellular Injection Moulding and Its Application to Rheological Characterization for Polymer Melts Dissolved with Supercritical Fluid

S. C. Chen[1,2,3 *], P. S. Hsu[1,2,3] and Y. W. Lin[1]

[1]Mechanical Engineering Department of Chung Yuan Christian University, Chung Li, Taiwan 32023, ROC – shiachun@cycu.edu.tw; soon1231@gmail.com; rd5@kingley.com
[2]R &D Center for Mold and Molding Technology, Chung Li, Taiwan 32023, ROC – shiachun@cycu.edu.tw; soon1231@gmail.com
[3]R &D Center for Membrane Technology, Chung Li, Taiwan 32023, ROC – shiachun@cycu.edu.tw; soon1231@gmail.com

Abstract - The purpose of this study is to investigate the rheological behavior of neat PS melt and PS melt dissolved or partially dissolved with SCF of nitrogen. Gas counter pressure technology was developed and applied during the melt front advancement in the filling stage. The application of gas into the mold cavity prior to melt filling provides a counter force against the melt front advancement, restricting the foaming process during the melt filling stage. Under various gas counter pressures, different percentage of foaming can be achieved during the melt filling process that allows one to investigate the rheological characteristics of polymer melt partially dissolved with SCF. A slit cavity is designed to measure the pressure drop of polystyrene mixed with 0.4wt% supercritical nitrogen fluid under different melt/mold temperatures, injection speeds as well as counter pressures. It was found that the glass-transition temperature, Tg, can be reduced by about 48% when applied GCP is 30 MPa. In addition, compared to conventional injection moulding, melt viscosity is reduced by 30% when GCP is increased from 5 MPa to 25 MPa. The non-nucleation mixture melt obtained by using a GCP of 30 MPa has viscosity value lowered down by about 50% compared to that in neat PS melt.

Introduction

The MuCell process was initially investigated by Wang et al. [1] followed by Shimbo [2]. In 2001, Trexel Inc. successfully developed a microcellular injection molding process [3]. The supercritical fluids (SCF) used in the MuCell process offer many advantages such as shrinkage reduction, light weight, low material cost, and the avoidance of sink marks. In addition, MuCell products also allow production within closer part tolerances and, in many cases, reduced cooling times. In recent years, MuCell technology viscosity reduction of polymers has been of great interest in the realm of polymer processing, and investigators have presented rheological behavior for microcellular injection molding. The viscosity of polymer/ SCF cannot be measured by standard methods. Initially, gas counter pressure technology was used in industry in the 1980s. Recently, an injection molding process called AMOTEC (Asahi Molding Technology of CO_2)[4] was invented to control surface roughness and prevent surface defects in solidified plastic products by exploiting the plasticization effects of dissolved CO_2.

In this study, a gas counter pressure system was established to maintain the single phase during MuCell injection molding. We established a rheology measuring system for MuCell injection molding with gas counter pressure under different injection shear rates. The purpose is to investigate the rheological behavior of neat PS melt and PS melt dissolved or partially dissolved with SCF of nitrogen.

Experimental

Polystyrene (POLYREX PG-33, Chemicals) resin was used in this rheology study. The glass-transition temperature (T_g) was confirmed to be 96°C. Nitrogen (N_2) was used as the gas source and the dissolved N_2 level was set at 0.4%. The volume expansion ratio was 5%. A high temperature mold for measuring MuCell with a Gas Counter Pressure (GCP) system was established to measure the rheology behavior of PS/ScN_2 in the MuCell injection moulding process as shown as Figure 1. A stable high mold temperature was provided by a high mold temperature controller and an isolated design in the mold. The system was required to measure the pressure and temperature within the slit cavity above the bubble pressure ensuring that a one–phase mixture is maintained during the injection process. The pressure drop and melt temperature across the slit cavity are measured via three pressure sensors in the cavity and two temperature sensors at the core. A high temperature seal was used to prevent gas escaping from the cavity, and a relief control valve device connected to the mold was used to adjust and monitor the real time gas pressure in the mold cavity.

Results and Discussion

The pressure drop was measured by three pressure sensors located along the length of the slit cavity, while the shear rate was calculated from the velocity of the screw as measured by a displacement transducer.

The modified Cross-WLF model is used to calculate the depressions of T_g for PS rein contain dissolved N_2. It was found that Tg is not regular when GCP is between 5~30 MPa at different injection speeds. In addition, when GCP reaches 30 MPa, the T_g falls from 96 °C to about 50 °C. It means when GCP is 30 MPa, the PS/ScN_2 maintains a single phase during the injection molding process.

The viscosity of MuCell injection moulding at various GCP levels is shown in Figure 2. Different GCP levels can influence the viscosity value. The higher the GCP levels, the more reduction in melt real viscosity. The results showed that the real viscosity obviously decreased when the GCP is 30 MPa. The viscosity curves were at different shear rates when temperature is at 185~205 °C. When the mold temperature is rising, the viscosity is decreasing at an accelerating rate.

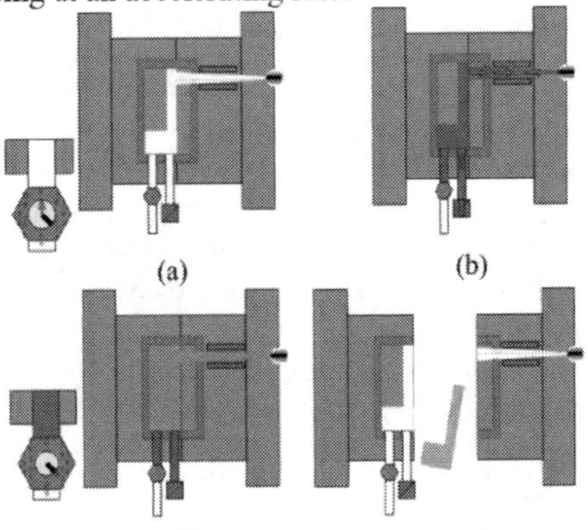

Figure 1 Process for MuCell with Gas Counter Pressure (a) Mold close and setting of gas pressure; (b) gas injection into cavity; (c) melt injection and the cavity pressure controlled by the gas controller valve; (d) product injection.

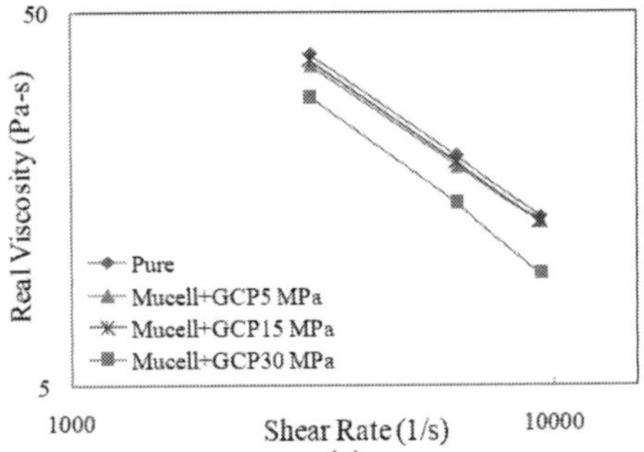

(a)

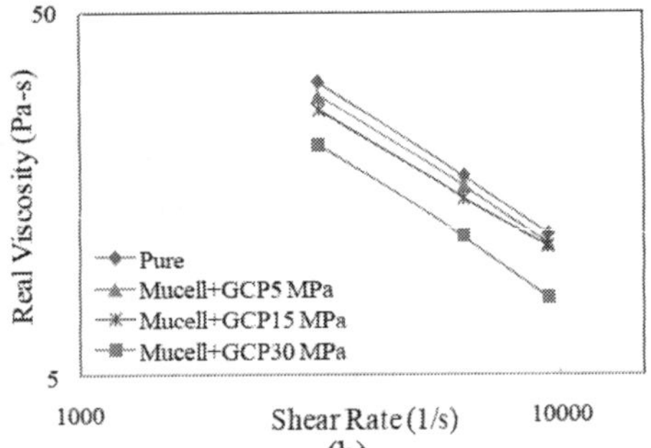

(b)

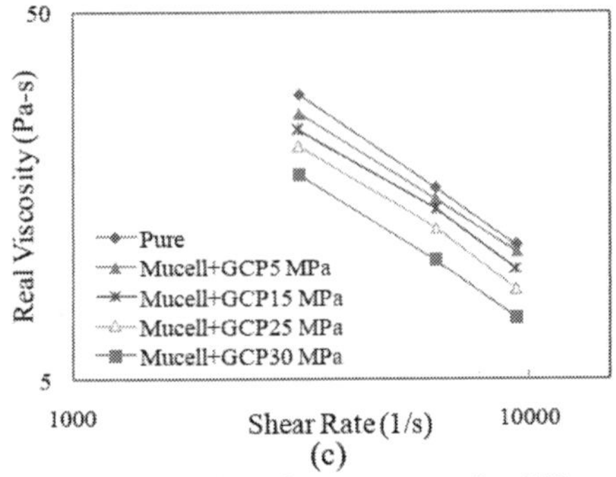

(c)

Figure 2 Real viscosity curve of different mold temperatures and GCP (a)185°C; (b)195 °C; (c)205 °C

Conclusions

In this study, a high pressure Gas Counter Pressure (GCP) system and a slit die rheometer were established to measure a wide range of viscosity for PS/ScN_2. Applying gas into the mold cavity prior to the melt filling provides a counter force against melt front advancement, restricting the foaming process. We observed that during the MuCell injection molding process, the injection pressure and viscosity obviously decreases. The results of this study are as follows:

(1) It was found that the T_g can be reduced by about 48% (fell from 96 °C to 50 °C) when applied GCP is 30 MPa.

(2) MuCell injection molding combined with a GCP system can control the cell size uniformly, compared to conventional injection molding, melt viscosity is reduced by 30% when GCP is increased from 5 MPa to 25 MPa.

(3) When the GCP is at 30 MPa, the viscosity of the one-phase injection melt was decreased by as much as 50% compared to that in neat PS melt, depending on the injection conditions.

Acknowledgements

Current research is supported by The Center-of-Excellence Program on Membrane Technology from the Ministry of Education and the project of the specific research fields in the Chung Yuan Christian University, Taiwan, under grant CYCU-98-CR-ME.

References

1. M. Lee, C. B. Park, and C. Tzoganakis, “Measurements and Modeling of PS/Supercritical CO Solution Viscosities”, Polym. Eng. Sci., **39**, 1, 99-109 (1999)
2. S. M. Lee, J. H. Han, K. Y. Kim, Y. J. Ahn, and J. W. Lee, “High-Pressure Rheology of Polymer Melts Containing Supercritical Carbon Dioxide”, Korea-Aust. Rheol. J., **18**, 2, 83-90 (2006)
3. Xu, J., Pierick, D., “Microcellular Foam Processing in Reciprocating-Screw Injection Molding Machine” J. Inject. Mold. Technol., **5**, 3, 152-159 (2001)
4. J. R. Royer, Y. J. Gay, J. M. Desimone, and S. A. Khan, “High-Pressure Rheology of Polystyrene Melts Plasticized with CO_2: Experimental Measurement and Predictive Scaling Relationships”, J. Polym. Sci. Pt. B-Polym. Phys., **38**, 3168-3180 (2000)

THE EFFECTS OF VARIOUS VARIOTHERM PROCESSES AND THEIR MECHANISMS ON INJECTION MOLDING

Chao-Tsai Huang*, I-Sheng Hsieh, and Cheng-Han Tsai
CoreTech System (Moldex3D) Co., Ltd., ChuPei City, Hsinchu, Taiwan
cthuang@moldex3d.com; cliffhsieh@moldex3d.com; olivertsai@moldex3d.com

Abstract - To eliminate surface defects and improve the quality of molded parts, increasing the mold temperature is one of the solutions. Using high mold temperature can eliminate weld lines, reduce molding pressure, residual stress, clamping force and improve part surface quality. However, with the increasing of mold temperature, the cycle time will also be increased. Hence, people have paid the attention to mold temperature control technologies. Among them, the variotherm molding technologies, including Rapid Heat Cycle Molding (RHCM), Induction Heating Molding (IHM), and Electricity Heating Mold (E-mold), are some effective methods. Although those variotherm technologies have been proposed, how does the external or internal heating source affect the injection molding process and the final product? The true function and the efficiency study of each technology still remain vague. Hence, in this paper, we have systematically conducted various technologies, including Conventional Injection Molding (CIM), RHCM, IHM, and E-mold by using true 3D transient cool technology. Through the inside mechanism investigation from time to time, the functions and the heating-cooling efficiency for each technology can be visualized. Finally, experimental study and verification of IHM is also performed.

Introduction

In recent years, people applied high mold temperature to eliminate weld lines, reduce molding pressure, residual stress, clamping force and improve part surface quality. However, with the increasing of mold temperature, the cycle time will also be increased. Hence, people have paid the attention to mold temperature control technologies. Among them, the variotherm molding technologies, including Rapid Heat Cycle Molding (RHCM), Induction Heating Molding (IHM), and Electricity Heating Mold (E-mold), are some effective methods described briefly as following (Chen 2004, Chen 2006, Chiou 2007, and Chiou 2009).

RHCM process generally is applied hot steam to raise the mold temperature during filling phase, and rapidly cool the mold at the beginning of packing. In each cycle, the mold temperature is changed from high to low back and forth. The disadvantages of RHCM process are full moldbase heating, long cycle time, special heating/cooling channel design, complex switchover control and vapor heating equipment. Moreover, IHM process is electromagnetic field inducing eddy current which generates heat within worked piece. IHM designs induction coil for cavity shape, and the coil's current and the distance to mold surface determine the heating rate. The advantages of IHM process are localization heating, i.e. it is not necessary to redesign mold, but the disadvantage is non-homogeneous heating in large areas may occur possibly. Furthermore, the application of E-mold process is to heat the moldbase electrically up to extremely high temperature. Its advantage is uniform high temperature distribution, but the disadvantages are high energy wasted, lower heating rate, and mold design collected. Although various variotherm technologies have been proposed, how does the external or internal heating source affect the injection molding process and the final product? The true function and the efficiency study of each technology still remain vague.

Hence, in this paper, we have systematically conducted various technologies, including Conventional Injection Molding (CIM), RHCM, IHM, and E-mold by using Moldex3D 3D Transient Cool technology. Through the inside mechanism investigation from time to time, the functions and the heating-cooling efficiency for each technology can be visualized. Finally, experimental study and verification of IHM is also performed.

Investigation Procedures

The geometrical model and runner system for CIM are shown in Figure 1. The dimension of this part is 225 mm (length) x 164 mm (width), and the main thickness is 1.5 mm. Moreover, the operation conditions are listed in Table 1. The polycarbonate (PC) produced by Teijin was used. To remove surface defects of injection molded part and improve its quality, the temperature increase on the surface of cavity side up to 150°C at the beginning of filling for RHCM, IHM, and E-mold processes is employed. In order to reach 150°C on the surface of cavity side, 180°C steam (see Figure 2) is set to heat cavity side for 25 seconds in RHCM process. Similarly, apply the induction heater to heat cavity side for 4 seconds in IHM process. Also, use the electricity heater to heat cavity side for 8 seconds in E-mold process.

The time sequence of each cycle for different processes is in Figure 3. In CIM process, the water temperature on both cavity side and core side is 80°C, and the cooling time and cycle time are 10.7 and 19.2 seconds respectively. In RHCM process, the water temperature on the core side is 80°C. The 180°C steam is used to heat cavity side from open mold time to filling time. During packing and cooling phases, water with 55°C is used to cool the cavity side. The cooling time and cycle time are 25 and 58.5 seconds respectively. In IMH process, the water temperatures on the core side and cavity side are 80°C and 60°C. The cooling time and cycle time are 12 and 24.5 seconds respectively. In E-mold process, the water temperatures on the core side and cavity side are 80°C and 60°C. The cooling time and cycle time are 50 and 62.6 seconds respectively.

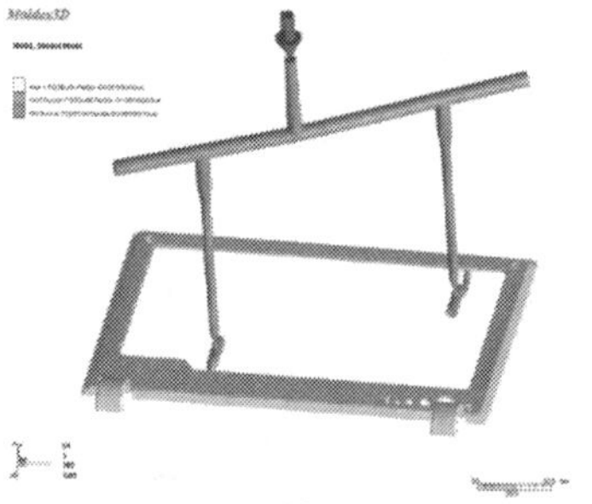

(a)

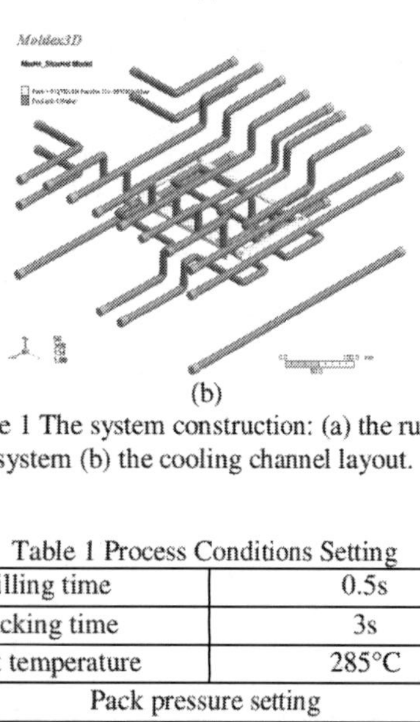

(b)

Figure 1 The system construction: (a) the runner system (b) the cooling channel layout.

Table 1 Process Conditions Setting

Filling time	0.5s
Packing time	3s
Melt temperature	285°C
Pack pressure setting	
0~0.5s	160MPa
0.5s~1s	140MPa
1s~1.5s	120MPa
1.5s~3s	80MPa

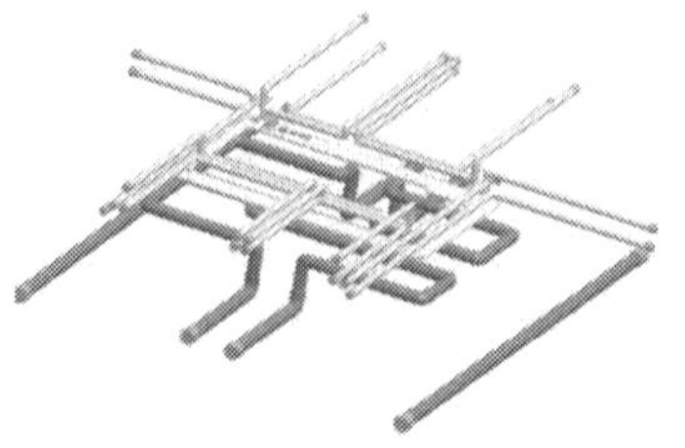

Figure 2 the coolant channel of RHCM

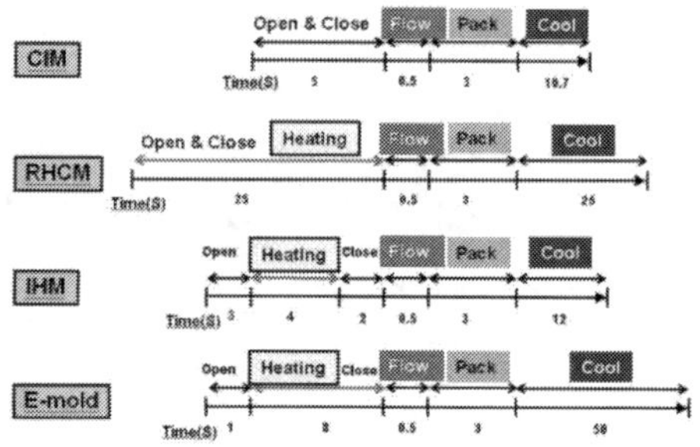

Figure 3 The time sequence of each cycle for different processes.

Result and Discussion

Figure 4 shows the temperature distribution of moldbase at the end of filling in CIM process. The high temperature region is located in the center of moldbase with highest temperature of 106°C. Figure 5 shows the temperature distribution of moldbase during filling phase in RHCM. Obviously, the high temperature region is located around steam channel during filling phase. Figure 6 shows the temperature distribution of moldbase in filling phase in IHM process. Since the heating method is via the mold surface using electrical power, the high temperature region is located on the surface of cavity side at the beginning of filling. Also, at the end of filling, the temperature is cooled down. Finally, Figure 7 shows the temperature distribution of moldbase during filling phase in E-mold process. A high temperature region occurs around the heater when using the electrical heating.

Furthermore, the usage benefits of high mold temperature for the quality improvement of injection parts can be verified via the temperature distribution around the weldline location as shown in Figure 8 and Table 2. Apparently, the application of variotherm technologies, including RHCM, IHM, and E-mold, makes the surface temperature raised up at the end of filling. The strength of the weldline is therefore enhanced.

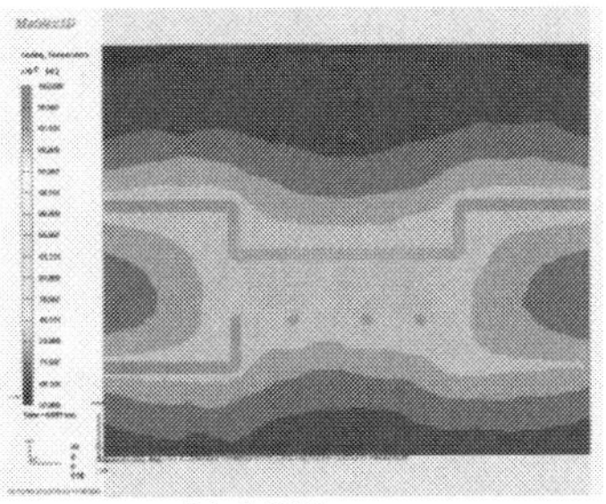

(a) 0.001s

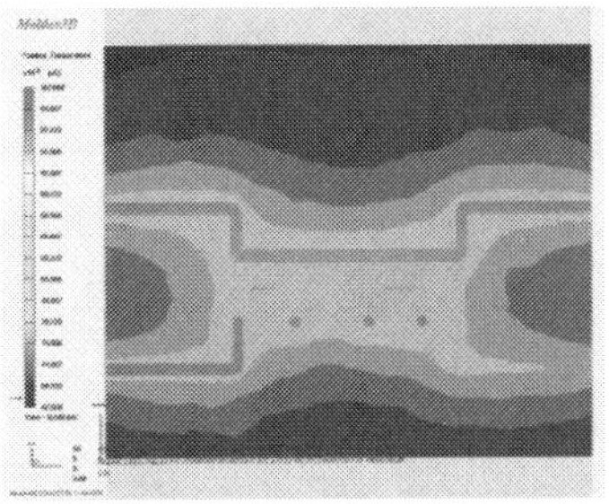

(b) 0.5s (end of filling)

Figure 4 Temperature distribution of moldbase during filling phase in CIM process.

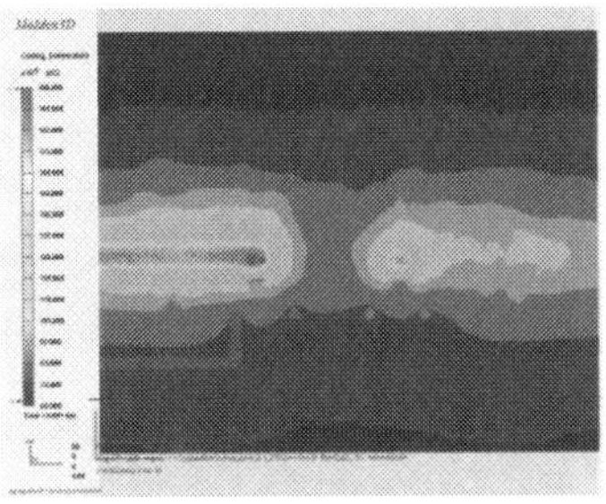

(a) 0.001s

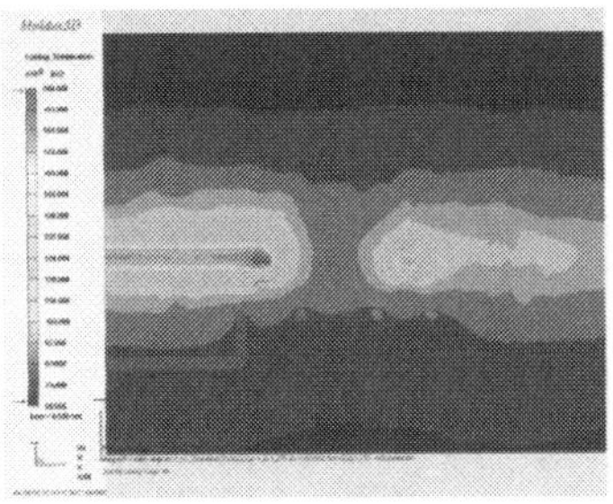

(b) 0.5s (end of filling)

Figure 5. Temperature distribution of moldbase during filling phase in RHCM process

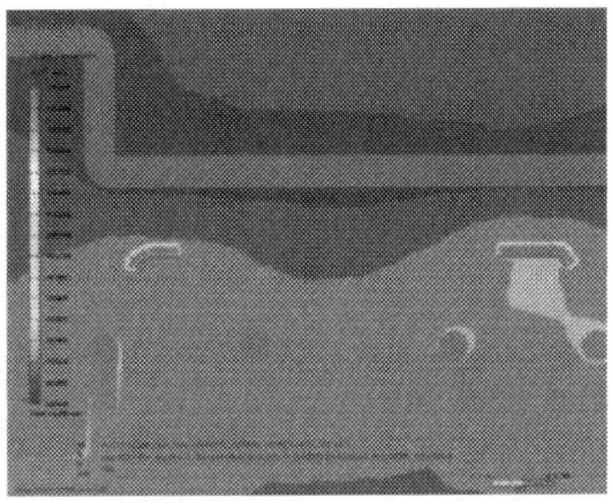

(a) 0.001s

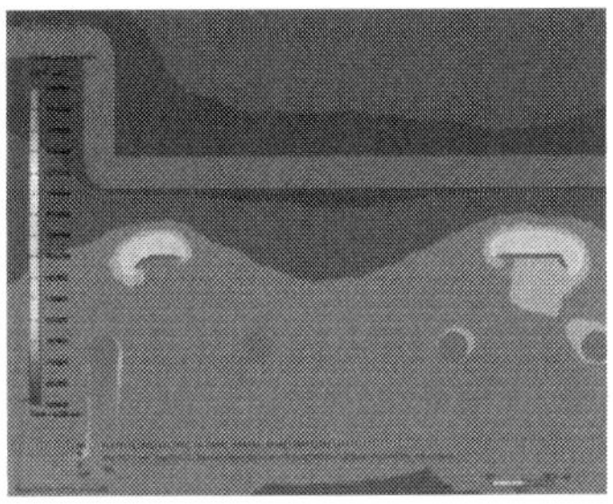

(b) 0.5s (end of filling)

Figure 6 Temperature distribution of moldbase during filling phase in IHM process

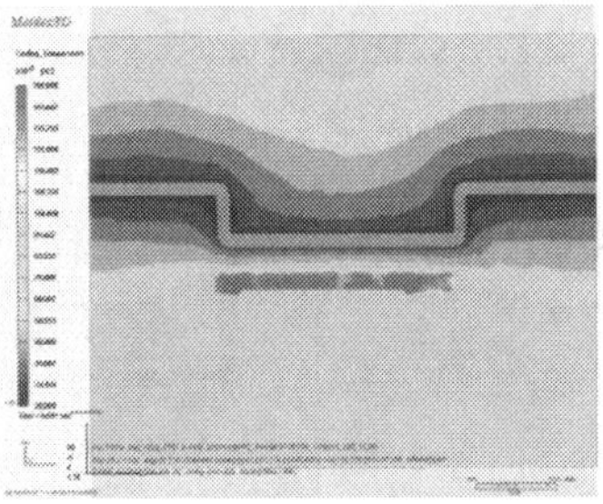

(a) 0.001s

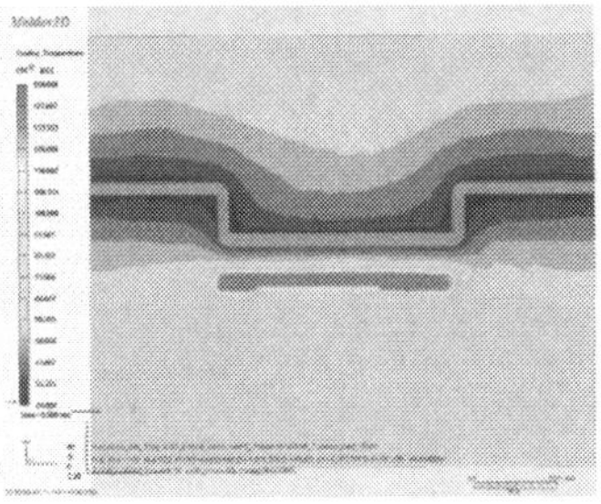

(b) 0.5s (end of filling)

Figure 7 the temperature distribution of moldbase during filling phase in E-mold process

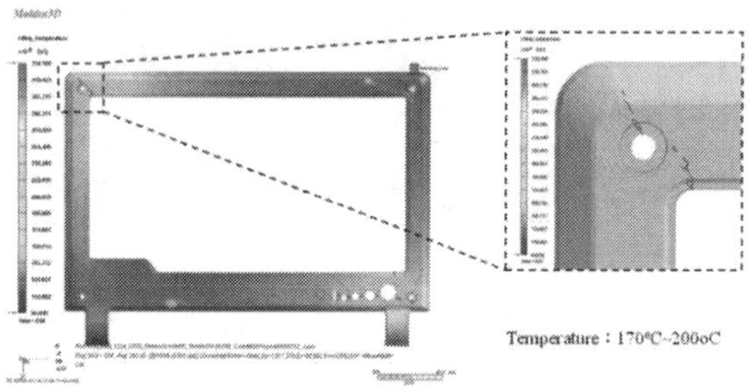

Figure 8 Temperature distribution around the weldline location at the end of filling for CIM.

Table 2 Temperature distribution around the weldline location for various technologies

	Weldline temperature
CIM	170°C ~200°C
RHCM	190°C ~220°C
IHM	205°C ~235°C
E-mold	210°C ~240°C

Moreover, we also considered the temperature history after cooling phase. Figure 9 shows the temperature distribution inside the mold at the end of cooling for various processes. In CIM and IHM processes, it needs only 12 seconds to cool down to 90°C. However, in RHCM and E-mold processes, it needs more than 25 seconds to cool down to 90 oC. In IHM process, it could reach 90°C just similar to that of CIM. Of course, the cooling period depends on how much heating source we have introduced.

Regarding the quality issue, warpage behavior is of vital importance. Figure 10 shows the Z-displacement. It is clear that when the thermal system is changed, the warpage behavior is changed accordingly. However, since the introduced heating amount and source is not optimized, the warpage trend doesn't seem to improve much either Z-displacement or total displacement.

Furthermore, to verify the simulation, one of variotherm technologies based on IHM is performed. To catch the temperature variation along the injection history from cycle to cycle, the sensors are set and located as shown in Figure 11. The basic construction and operation conditions are described in Table 1 and Figure 3. Figure 12 shows the temperature history curve sensed from three sensors numerically. The temperature is heating up to 230°C and cooling down back and forth from cycle to cycle. Also, Figure 13 displays the comparison between the numerical simulation and experimental results for temperature history curve at sensor S1. Clearly, the trend is in a reasonable agreement. Finally, the warpage behavior is also evaluated both numerically and experimentally. Figure 14 demonstrates the locations and numbers of each measure node for displacement measurement. Figure 15 shows the Z-displacement behavior from the front view for both experimental and simulation results. The trend is matched well. Table 3 shows the amount of Z-displacement. Obviously, both numerical simulation and experimental result are in a good agreement.

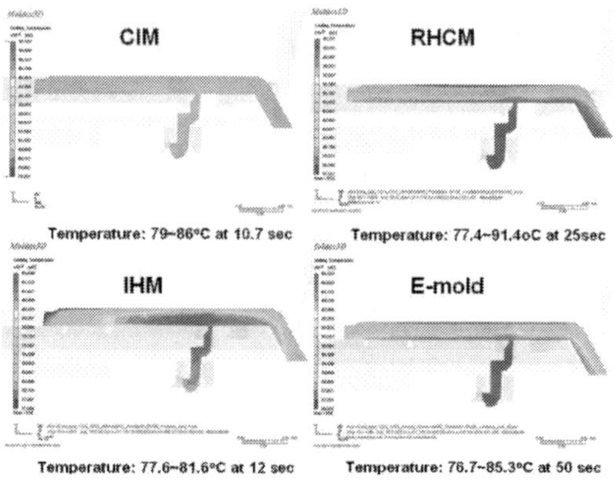

Figure 9 the temperature distribution inside the mold at the end of cooling

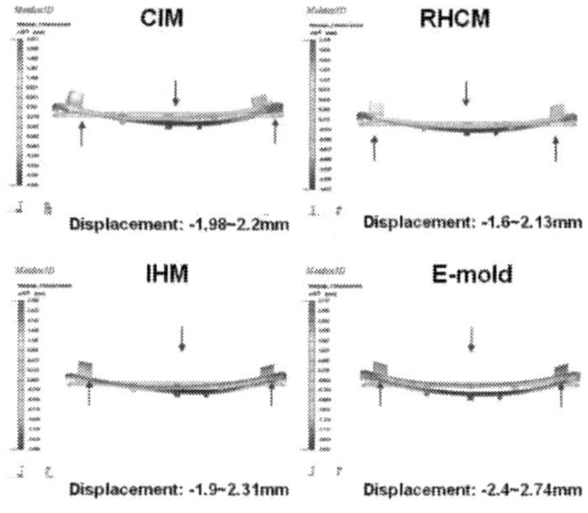

Figure 10 Z-displacement results

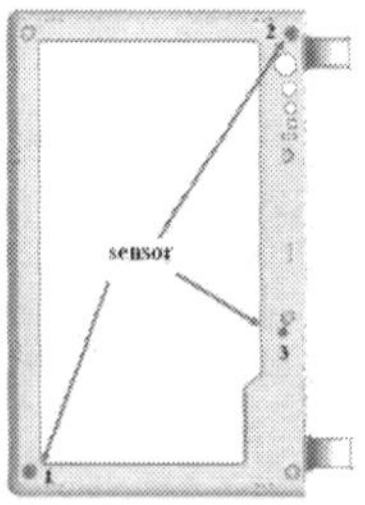

Figure 11. The location of sensors to measure the temperature variation in IHM experiments.

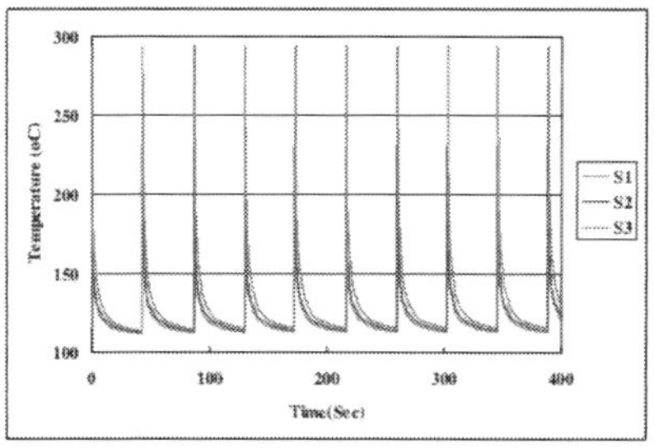

Figure 12 Temperature history curve for three pre-installed sensors.

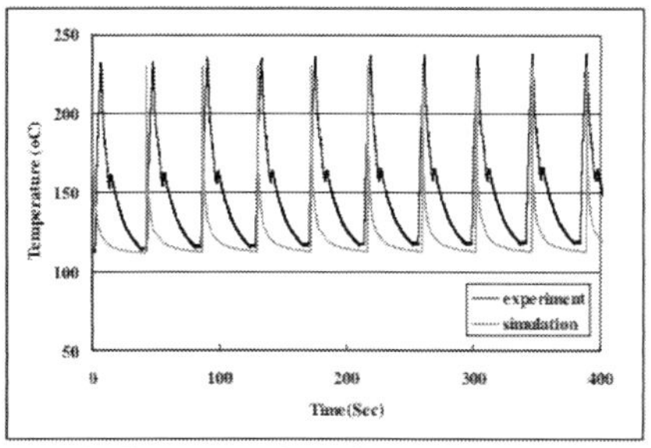

Figure 13. The comparison between the numerical simulation and experimental results for temperature history curve at sensor S1.

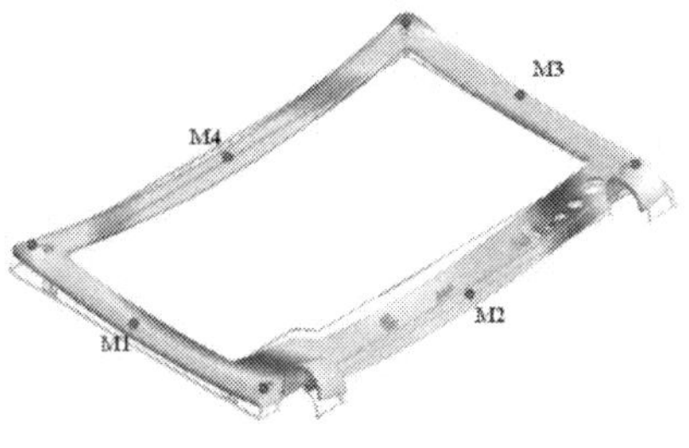

Figure 14. The locations and no. of measure nodes for displacement measurement.

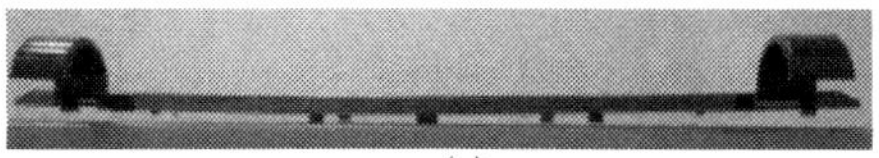
(a)

(b)

Figure 15 The displacement measurement based on IHM: (a) experimental result, (b) simulation result.

Table 3 the compare of Z-displacement between experiment and simulation.

Displacement (mm)	M1	M2	M3	M4
Experiment	0.2	1.4	0.2	0.8
Simulation	0.2	1.29	0.17	0.66

Conclusions

In this study, we have systematically studied CIM and various variotherm technologies, including RHCM, IHM, and E-mold, numerically. Through 3D transient behavior investigation inside the moldbase, the dynamic feature of the temperature can be obtained. This technique can help us to realize the thermal history from location to location and also from time to time for various technologies. To verify this technique, the experimental study based on IHM is also performed. Through the temperature history curve and Z-displacement measurement, both numerical and experimental results are in a good agreement.

References

Chen, S.C. et al., "Three-dimensional simulation of transient temperature distribution for lens mold embedded with heaters", *SPE ANTEC Tech. Papers*, 631–635 (2004).

Chen, Y. et al., "True 3D and Fully Transient Mold Temperature Simulation for RHCM Process", *SPE ANTEC Tech. Papers*, 1295–1298 (2006).

Chiou, Y. et al., "Integrated True 3D Simulation of Rapid Heat Cycle Molding Process", *SPE ANTEC Tech. Papers*, 2577–2580 (2007).

Chiou, Y. et al., "Thermal Feature of Variotherm Mold in Injection Molding Processes", *SPE ANTEC Tech. Papers*, 2491–2495 (2009).

IMPROVEMENTS IN THE SIMULATION OF ORIENTATION IN INJECTION MOLDING OF SHORT FIBER THERMOPLASTIC COMPOSITES

G. M. Vélez-García[1], S. M. Mazahir[2], P. Wapperom[3] and D. G. Baird[2]*

[1]*Macromolecular Science and Engineering Department, Virginia Tech, Blacksburg, VA 24061, USA – gvelez@vt.edu*
[2] *Chemical Engineering Department, Virginia Tech, Blacksburg, VA 24061, USA – syedm@vt.edu; dbaird@vt.edu*
[3] *Mathematics Department, Virginia Tech, Blacksburg, VA 24061, USA – pwappero@math.vt.edu*

The mechanical properties of injection molded short-fiber reinforced thermoplastic composite parts are highly dependent on the orientation distribution of the fibers. A simulation tool capable of predicting fiber orientation accurately as a function of mold design and processing conditions is required as the predicted fiber orientation capabilities in commercial software show large discrepancies when compared with experimentally measured orientation. In this work a two dimensional coupled Hele-Shaw approximation for predicting the flow-induced orientation of glass fibers in injection molded composite parts is presented. In addition to coupling the stresses to fiber orientation for a highly concentrated short glass fiber PBT suspension, the model considers the slowdown of the evolution of orientation due to fiber interaction. Material parameters in the model are determined from basic rheometry rather than using data from injection molding experiments. The equation of motion coupled with stress equations are discretized using the discontinuous Galerkin Finite Element Method. Flow simulations are performed using a measured orientation profile at the gate instead of random orientation assumed in previous studies. Finally, the evolution of fiber orientation in the cavity is determined experimentally using a modified version of the method of ellipses and results are compared against the predicted values of orientation. The fiber orientation predicted in the entry region and the core layer structure at the end of fill region are now in closer agreement with the experimental values, but there are still some discrepancies.

Introduction

Fiber reinforced thermoplastic made by injection molding is an attractive technology to develop lightweight, high-performance materials. The lightweight molded composites consist of a polymeric matrix reinforced with fibers because of the excellent mechanical properties obtained in the final product, the high throughput, and the cost reduction. The desired properties are only obtained when the orientation of the fibers in the direction of mechanical interest is met. However, the fiber orientation varies through the part as a consequence of flow within the mold during the forming; a concept called flow induced orientation. The fibers considered here are short glass fibers defined as fibers with high aspect ratio ($a_r > 30$) and absolute length below 1 mm.

Optimization of the technology requires a prediction tool using a computer model capable of designing the correct molding machinery, molding and processing conditions, and consistently controlling fiber orientation. The actual capabilities of simulations are unable to make a quantitative prediction of the orientation due to several limitations in the modeling and numerical techniques used to solve the system of equations. Some of these limitations are the use of models which ignore the fiber interactions (Jeffery model), or account for them in a concentration below the commercial interest (Folgar- Tucker model (Folgar 1984)), or ignore the viscoelastic effects of the polymer, and use a decoupled approach to solve the system of equations.

This paper develops a 2D solution for a coupled flow and fiber orientation using the Hele-Shaw (HS) approximation, which is the typical method of flow description in commercial simulators. The material behavior is modeled using a modified version of Folgar-Tucker model (Eberle 2010) and a Newtonian model for the polymer matrix, with parameters determined from simple flow rheology. The modification in the Folgar-Tucker accounts for the slow-down in the fiber motion due to inter-particle interaction occurring in highly concentrated fiber suspension. The objective of this work is to develop an accurate numerical tool capable of predicting the flow-induced orientation of glass fiber in a commercial range of concentrations using a coupled approach.

Governing Equations

The HS flow approximation for an isothermal flow in a center gated disk is described by

$$\frac{1}{r}\frac{\partial}{\partial r}(hr\bar{v}_r)=0 \qquad (1)$$

$$-\frac{\partial p}{\partial r}+2\eta\frac{\partial^2 v_r}{\partial z^2}+\frac{1}{r}\frac{\partial}{\partial r}(rT_{rr})+\frac{\partial T_{zr}}{\partial z}-\frac{T_{\theta\theta}}{r}=0 \quad (2)$$

where r represents the radial or flow direction, z the gapwise direction, h the half gap width and $\overline{v}_r$ the average radial velocity along the gapwise coordinate, p the pressure, T_{ij} the fiber-contribution to the extra-stress tensor components. These equations are supplemented by the typical boundary conditions for pressure and velocity.

The hydrodynamic extra-stress model is used to represent the fiber-contribution and is defined as

$$\mathbf{T}=\nu\zeta_{str}\left((\nabla\mathbf{v})^T+\nabla\mathbf{v}\right):\mathbf{R}_4 \quad (3)$$

where ν represents the fiber concentration, , ζ_{str} the viscous drag, $\mathbf{A}$ and $\mathbf{R}_4$ the second and fourth order orientation tensor, respectively. R_4 can be expressed in terms of a quadratic closure:

$$\mathbf{R}_4=\mathbf{AA} \quad (4)$$

The evolution of the second order orientation tensor is governed by:

$$\frac{\partial\mathbf{A}}{\partial t}+\mathbf{v}\bullet\nabla\mathbf{A}=\alpha\Big[(\nabla\mathbf{v})^T\bullet\mathbf{A}+\mathbf{A}\bullet(\nabla\mathbf{v})-2\left((\nabla\mathbf{v})^T+\nabla\mathbf{v}\right):\mathbf{R}_4$$
$$-6C_I II\left(\mathbf{A}-\frac{1}{3}\delta\right)\Big] \quad (5)$$

where α represents the slip parameter, C_I the interaction coefficient and II the magnitude of the rate of deformation tensor. The inter-fiber interactions in a semi-dilute and concentrated regime are represented by the last term and the slip parameter in Eq. (5), respectively. With α=1 and C_I =0, Eq. (5) reduces to the Jeffery model, which only accounts for the kinematic effects typical in dilute suspensions.

The second order orientation tensor represents an average orientation of all fibers in a particular region. Experimentally it is computed by

$$A_{ij}=p_i p_j \quad (6)$$

where p_i and p_j are components of the vector of orientation. They are defined by

$$p_1=\sin\theta\cos\phi \quad p_2=\sin\theta\sin\phi \quad p_3=\cos\theta \quad (7)$$

where ϕ and θ are the in-plane and out-of-plane angles.

Problem Description and Numerical Methods

Two 75% short shot center-gated disk of 30wt% short glass fiber PBT with average mass of 18.06, average internal diameter (d_i) of 5.96 mm, average outer diameter (D) of 48.55, and average thickness ($2h$) of 1.38 mm were molded. The filling time of the part was approximately 1 s and the injection pressure was estimated to be 20 MPa. The material parameters were determined from steady shear and start-up of shear experiments (Eberle 2009). Over a wide range of shear rates the PBT matrix behaves like a Newtonian fluid with viscosity η = 350 Pa·s. The coefficients were determined as α = 0.40, $\nu\zeta_{str}$ = 7417 Pa and $C_I = 0.02^2$. The weighted average length of the fiber was 364μm.

The initial orientation of the fibers was experimentally determined at the gate using method of ellipses (Clarke 2002) with several modifications to obtain unambiguous orientation measured over small sampling region. The specimen was metallographically polished in r,z-planes, plasma etched , and gold sputtered. A single-column of images were taken using reflective optical microscopy (20X) with a motorized stage. The images were stitched and the ϕ and θ were obtained using an in-house program written in Matlab. Then, the **A** was computed using Eq (6). The samples were cut along a line of constant θ. For validation purposes the orientation was determined in a similar manner along the gapwise direction at 10%, 40% and 90% of (R-r_i). The filling of a disk having the characteristic values described above was simulated. The standard Galerkin method (GFEM) is used in the discretization of the HS flow approximation of the balance equations. The discontinuous Galerkin method (DGFEM) with a standard explicit Euler scheme is used to discretize the evolution equations. In the simulations only the top half of the domain was considered with no slip velocity at the wall and symmetry boundary conditions at z=0.

Results and Discussion

Experimental orientation

There are a limited number of studies which include both experimental and numerical work, therefore the initial conditions of orientation have typically been assumed as random orientation, especially for a center-gated disk (Bay 1992). Figure 1 and 2 respectively illustrate the diagonal and off-diagonal components of orientation determined experimentally from a 75% short shot center-gated disk. The asymmetric profile seen in A_{rr} along the cavity thickness at the gate region corroborates predictions indicted in the literature but never validated (Chung 2002; VerWeyst 2002). Two plateaus with different values of A_{rr} can be seen in Fig. 1 indicating different levels of orientation with respect to the main flow direction for the upper half cavity and the lower half cavity. The higher values of A_{rr} in the upper half cavity have been attributed to higher velocity in the flow direction in the upper half(Chung 2002; VerWeyst 2002). However, these experimental results add two observations not indicated before in the literature. Firstly, the orientation at the gate region becomes almost immediately planar orientation as seen from the values of $A_{zz} \approx 0$. Secondly, there is a cavity-wise asymmetry in $A_{r\theta}$, while the rest of the off diagonal components remain close to 0, as seen in Fig. 2. Values of off-diagonal components close to 0, indicate tilt of fibers in 2 opposite directions that when approximately equally averaged make $A_{ij} \approx 0$.

However, the $A_{r\theta}$ seen in Fig. 2 shows several regions along the thickness with a local tilt.

Orientation along the thickness measured at 40% and 90% (R-r_i) for PBT filled with short glass fibers

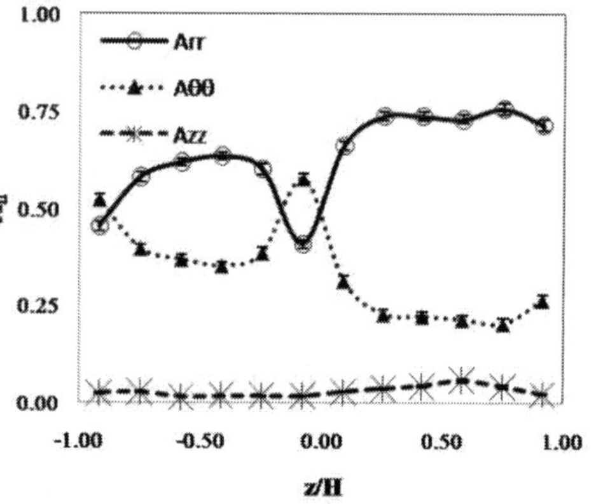

Figure 1. Experimental values of the diagonal components of **A** at the gate region. The symbols ○, ▲, and * represent the A_{rr}, $A_{\theta\theta}$, and A_{zz}, respectively.

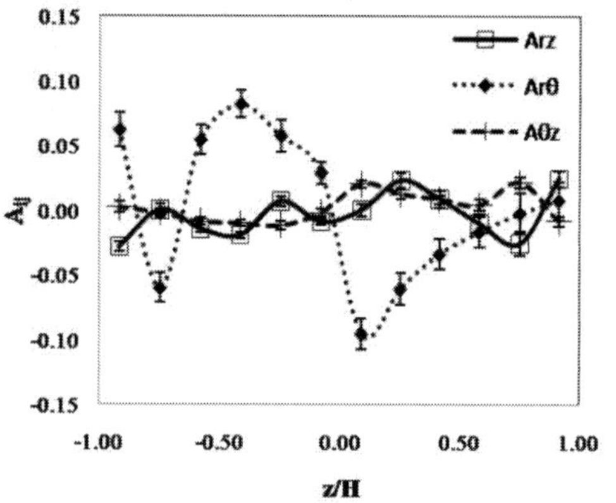

Figure 2. Experimental values of the off-diagonal components of **A** at the gate region. The symbols □, ♦, and + represent the A_{rz}, $A_{r\theta}$, and $A_{\theta z}$, respectively.

indicate changes in the magnitude and distribution of the orientation components compared to the initial orientation at the gate. The initial asymmetry fades as a function of the radial location. However, experimental results suggest development of a symmetric orientation or stable structure of orientation for disks at r ≥ 40%R as postulated by Altan and Rao (Altan 1995). Therefore, this observation suggests the formation of a secondary structure of orientation different from the accepted layered structure. This additional structure which evolves from the gate to about 40%R along the flow direction has not been previously reported. The impact of this structure on the mechanical properties can be critical in determining areas of weakness around the gate for large parts, especially when they have multiple gates.

Prediction of orientation

The experimental orientation in the 3 characteristic regions of flow of the center gated disk, i.e. entry, lubrication, and front were compared with results from standard commercial simulations and the proposed approach. The standard method of simulation is described by Folgar-Tucker model (α = 1), decoupled simulations, and random orientation at the gate while, the proposed approached is described by delayed Folgar Tucker (α = 0.4), coupled simulation, and experimentally measured orientation at the gate. Fig. 3 illustrates contrast in the cavity-wise A_{rr} orientation profile at a radial location in the entry region (10% (R-r_i)) between experimental results, standard and proposed approach. The generalized reduction in A_{rr} around the midplane (z/H=0) at entry region causes an asymmetric expansion of the core probably caused by a strong extensional flow in this area (Ranganathan 1993). In this region, the standard and proposed approach fails to reproduce the experimental behavior. However, the asymmetry in the profile is still seen in the proposed approach, but with the minimum A_{rr} located in the lower half cavity. The orientation close to the wall can not be reproduced by both simulation approaches.

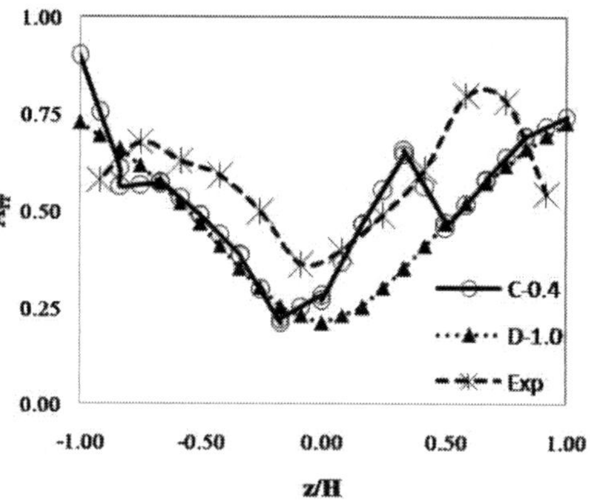

Figure 3. Experimental values of A_{rr} orientation component at the entry region (10% of radial flow length) (*) with the predicted values of A_{rr} orientation component using coupled simulation with measured orientation at the gate (○) and decoupled simulation with measured random orientation at the gate (▲).

Figure 4 shows the A_{rr} at a radial location in the lubrication region. The experimental orientation shows almost a symmetrical profile of orientation with a maximum A_{rr} = 0.88 for the upper and lower half of the cavity. Now the standard approach is capable of reproducing the orientation in the large portion of the core and transition layer, but fails in the shell layer. The proposed approach under-estimates the complete orientation profile and some asymmetry. The maximum A_{rr} for both simulations is found in plateau next to the wall, having a value of 0.83.

The A_{rr} orientation profile at a region close to the front is shown in Fig. 6. The experimental orientation profile shows a symmetrical behavior for the core and transition layer. However a large difference is seen in the upper and lower shell layers. The upper shell shows a generalized reduction in orientation towards $z/H = 0.75$, not seen in other radial locations while the lower shell shows a constant value. At this location the maximum A_{rr} is found close to the walls, but in general the magnitude of A_{rr} is lower than the magnitudes found in the lubrication region. When the simulations are compared with the experimental values, the standard approach over predicts A_{rr} in the complete profile. The proposed approach becomes completely symmetric, and overlaps the results of the standard approach for the transition and shell layers. However, this is the only approach capable of reproducing a portion of the core region at a radial location close to the end-of-fill.

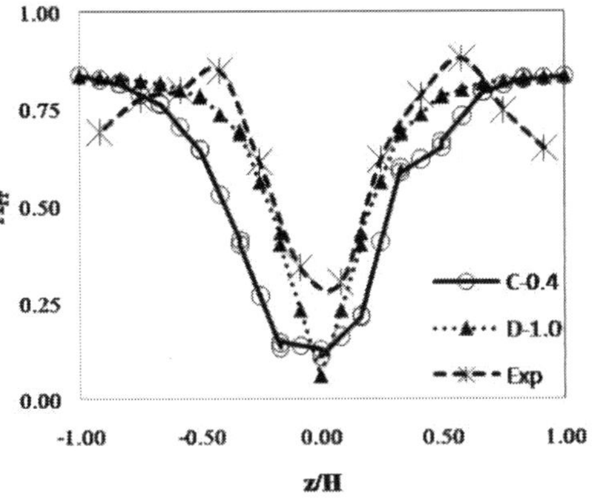

Figure 4. Experimental values of A_{rr} orientation component at the lubrication region (40% of radial flow length) (*) with the predicted values of A_{rr} orientation component using coupled simulation with measured orientation at the gate (○) and decoupled simulation with measured random orientation at the gate (▲).

The complete frontal region was qualitatively inspected to find reasons for the irregular behavior seen in the shell layer. The images of the advancing front (Fig 6) showed an irregular surface which is almost parabolic in shape. This region is severely affected by void spaces which are mainly found close to the free surface. The upper portion of the advancing front showed higher concentration of void spaces and a different structure of orientation when compared with the lower portion. Most of the fibers at the center of the advancing front seem to be oriented in θ-direction. However there is a tendency to have large amount of fiber aligning parallel to the flow in the upper but it rapidly goes down towards $z/H=0$. Therefore, the difference in dispersion of void spaces and fiber orientation at the front can be responsible for the irregular profile seen in the shell layer at a radial location close to the front.

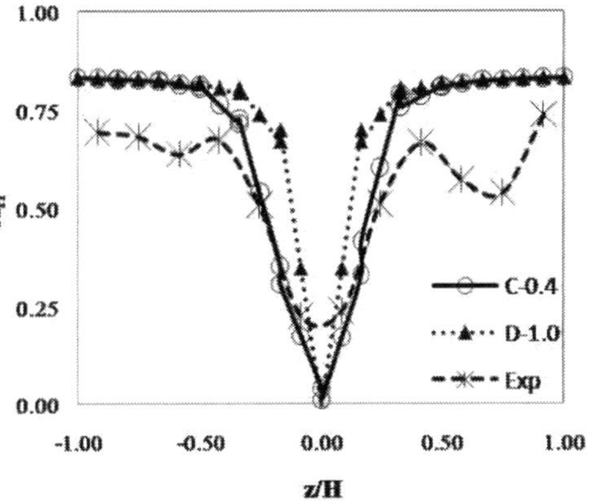

Figure 5. Experimental values of A_{rr} orientation component at a region close to the end-of-fill (90% of radial flow length) (*) with the predicted values of A_{rr} orientation component using coupled simulation with measured orientation at the gate (○) and decoupled simulation with measured random orientation at the gate (▲).

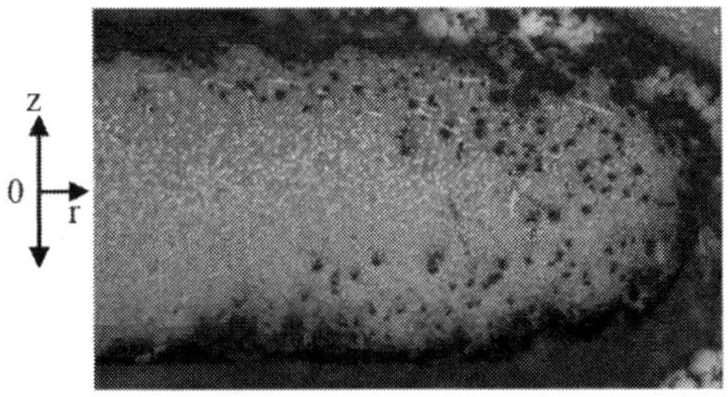

Fig 6. Reflective optical microscopy image (5X) of advancing front of PBT matrix with 30wt% short glass fibers.

Conclusions

The experimental results indicate that the profile of orientation at the gate was asymmetric and evolved into a steady layered structure of orientation at a radial location larger than 40% of the flow length. This suggests that the typical assumption of initial orientation as a symmetric random profile is not very accurate for a center-gated disk. An asymmetric profile was also found in the $A_{r\theta}$ component at the gate. The advancing front causes a reduction in A_{rr} at radial location of the end-of fill and distortions in the profile at the shell layers.

The use of model parameters, determined independently from the geometry, and a coupled fiber and orientation scheme help delay the evolution of A_{rr} component of fiber orientation compared to the decoupled scheme typically assumed in existing literature. The proposed method delays too much the evolution of A_{rr} causing under estimation of A_{rr} at the entry and lubrication. However, the proposed method was able to reproduce the washout of asymmetric orientation profile observed experimentally. The standard approach was able to reproduce values of A_{rr} at the core and transition layers in the lubrication region. Experimental values of A_{rr} in core layer in a region close to the front were captured well by the proposed method. Many of the experimental observations were satisfactorily captured by the proposed model. However some additional modifications must be included in the proposed method to locally increase or reduce the predicted A_{rr}.

Acknowledgements

The financial support of NSF/DOE: DMI-052918 is gratefully acknowledged. Gregorio M. Vélez-García also acknowledges support from the NSF-MS&IE-IGERT and from the University of Puerto Rico-Mayagüez.

References

Altan, M. C., Rao, B. N., "Closed-form solution for the orientation field in a center-gated disk" *J Rheol* **39**, 581-599 (1995).

Chung, D. H., Kwon, T. H., "Numerical studies of fiber suspensions in an axisymmetric radial diverging flow: the effects of modeling and numerical assumptions" *J Non-Newtonian Fluid Mech* **107**, 67-96 (2002).

Clarke A. R., Eberhadt C. N. Microscopy techniques for materials science. CRC (2002).

Bay R. S., Tucker, C. L., "Fiber orientation in simple injection moldings. 2. Experimental results" *Polym Comp* **13**, 332-341 (1992).

Eberle, A., et al, "Fiber orientation kinetics of a concentrated short glass fiber suspension in startup of simple shear flow" *J Non-Newtonian Fluid Mech* **165**, 110-119 (2002)

Eberle, A., et al, "Obtaining reliable transient rheological data on concentrated short fiber suspensions using a rotational rheometer", *J Rheol* **53**, 1049-1068 (2009).

Folgar F. P., Tucker, C. L. "Orientation behavior of fibers in concentrated suspensions" *J Reinf Plast Comp* **3**, 98-119 (1984).

Ranganathan S., Advani, S. G., "A simultaneous solution for flow and fiber orientation in axisymmetric diverging radial flow" *J Non-Newtonian Fluid Mech* **47**, 107-136 (1993).

VerWeyst B. E., Tucker, C. L., "Fiber suspensions in complex geometries flow/orientation coupling" *Can J Chem Eng* **80**, 1093-1106 (2002).

Polymer rheology at high shear rate for microinjection moulding

Michel Vincent, Cheima Mnekbi, Jean-Francois Agassant*

MINES ParisTech - CEMEF - Alpes Maritimes - France

In microinjection molding, high velocity combined to small flow gaps lead to specific flow behavior compared to classical injection molding. One of the questions is the viscosity data at high shear rate, as numerical simulation of mould filling requires precise rheological data. The aim of the work is to obtain such data and to test them by comparing the results of numerical simulation to measurements in an instrumented mould. First, we present the rheology of a high-density polyethylene. We used a capillary rheometer. With conventional diameter (1 mm) and flow rate, we obtained viscosity data up to an apparent shear rate of 103 s-1. The curves were fitted with a Carreau-Yasuda law. When using smaller diameters (0.3 and 0.5 mm) at high flow rate, the apparent shear rate reached 106 s-1. Pressure became high enough so that viscous heating and pressure-dependence of the viscosity cannot be neglected. Moreover, spurt and chaotic defects related to wall slip were encountered. Nevertheless, we used again classical data processing. We obtained viscosity at high shear rate lower than the viscosity obtained with the Carreau-Yasuda extrapolation of the moderate shear rate data. In order to discuss the origin of this difference, we developed a flow numerical simulation in the capillary where we take into account pressure and temperature dependent viscosity, compressibility and wall slip. The relative importance of these phenomena and their coupling is presented. Second, injection molding experiments were carried out in a plaque mould 0.5 and 0.3 mm thick, instrumented with three pressure sensors, three temperature sensors and a light-absorption sensor to detect polymer crystallization. Different molding parameters were tested. The average shear rate reached 104 s-1. A three dimensional injection molding numerical simulation, using Rem3D software, was carried out. The quality of the different sets of rheological data is assessed by comparing computed and measured pressure.

INJECTION MOLDING OF LONG FIBRE REINFORCED THERMOPLASTIC SYSTEMS: FIBRE LENGTH ANALYSIS AND IMPACT PROPERTIES

M. Rohde[1]*, A. Ebel[1], V. Altstädt[1]

[1]Department of Polymer Engineering, Bayreuth University, Bayreuth, Germany – melanie.rohde@uni-bayreuth.de

Injection molding of long fibre reinforced thermoplastics is a well established method in automotive industry to produce high quality structural parts in mass production without finishing. However, the processing parameters as injection velocity, holding pressure, revolution speed and back pressure directly influence the final fibre length. This decrease in fibre length can have a negative effect on the resulting mechanical properties of the part. For instance, especially in terms of impact behavior, the fibre length is seen as a key factor.
The aim of the present work is to quantify the effect of the injection molding parameters on the impact behaviour of long fibre reinforced compounds. In a first step the resulting fibre length of injection molded glass fibre reinforced polypropylene is analyzed as a function of the injection velocity, holding pressure, revolution speed and back pressure. For this reason the single steps of fibre length analysis were certified in terms of feasibility and potential for optimization. In general, these steps can be divided in fibre-matrix-separation by pyrolysis, dispersion and separation of the fibres in an adequate solvent, digitalization and finally measurement of the fibre length. Subsequently, the impact behaviour of the PP-GF compounds is quantified by falling dart experiments, and the results are then correlated to the fibre length and the processing parameters to establish processing-structure-properties relationships.
The certification of the fibre length analysis shows that a high pyrolysis temperature leads to embrittlement of the glass fibres and a decrease of tensile strength of the single fibres. This falsifies the fibre length analysis due to fibre breakage. First results of the statistical investigation of the processing parameters indicate a significant influence of the back pressure on impact energy. In consequence it is possible to tailor the composite properties by process control.

Introduction

The combination of fibrous materials with polymeric matrix systems is established for several decades. Due to advantages like outstanding performance and high potential for integration of long fibre reinforced thermoplastic compounds the development of this material class was visibly promoted within the last years. This led to a constant annual growth rate about 8 % (Schemme 2003). Thereby automotive applications like frontends and underbody structures definitely dominate this market segment (Martin 2005). By incorporation of reinforcing fibres into a polymer it is possible to generate new material classes with systematically influenced and application friendly properties. By addition of glass fibres to polypropylene or polyamide it is possible to gain significant advancement in terms of stiffness, strength, heat resistance, warpage and also of mechanical behaviour under impact load. This way lightweight designed parts with excellent mechanical properties allow the substitution of conventional metallic materials.
The most relevant material properties of fibre reinforced thermoplastic compounds are impact strength, Young's modulus and tensile strength (Schijve 2002). On one hand these characteristics are strongly dependent on fibre content and orientation, fibre-matrix-bonding and the morphology of the matrix system (Menges, Geisbuesch 1982; Thomason, Adzima 2001; Thomason 2001; Thomason 2007). Thomason has already shown that in a fibre reinforced polyamide 66 the modulus increases linearly and the tensile strength and impact strength increase follow polynomial correlation with increasing fibre content (Thomason 2006). But it is also well known that fibre length or the fibre length to diameter ratio respectively is also strongly correlated to Youngs modulus and tensile strength and can be seen as a key factor for impact strength (Thomason 2002; Thomason 2007).
Injection molding provides an economical method to process fibre reinforced thermoplastic materials. In automotive industry it is already a well-established method to produce high quality structural parts in mass production without finishing. However, during injection molding the processing parameters as injection velocity, holding pressure, revolution speed and back pressure directly influence the final fibre length in the molded part. Consequently a decrease in fibre length can have a negative effect on the resulting mechanical properties of the part.
The objective of this present study is to determine the most significant injection molding parameters on both impact properties and fibre length, and therefore establishing a correlation between these parameters. The first step regarding this work is consequently to develop a suitable method for fibre length analysis.

Experimental

Fibre Length Analysis

In order to measure comparable fibre length values, it is of major importance to establish a method for fibre length analysis. In doing so the single steps of fibre length analysis were certified in terms of feasibility and potential for optimization. According to Figure 1 these steps can be divided into fibre-matrix-separation either by chemical digestion or pyrolysis, dispersion and separation of the fibres in an adequate solvent, digitalization and finally measurement of the fibre length. For the evaluation of fibre length analysis, unprocessed fibre reinforced granules of PP, PA 6 and PA 66 with fibre contents of 30 wt.-% were provided by FACT Future Advanced Composites Technology GmbH. For single fibre tensile tests Tufrov 4510 rovings from PPG Industries were used.

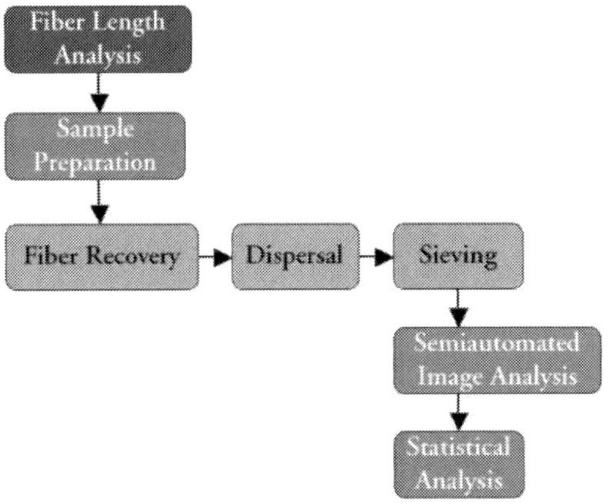

Figure 1 – Schematic illustration of the single steps of Fibre Length Analysis

Injection Molding

Long-fibre reinforced polypropylene granules were used in order to prepare impact plates by injection molding. The granules were provided by FACT Future Advanced Composites Technology GmbH and contained 30 wt.-% of 11 mm long glass fibres. The injection molding was done in an Arburg Allrounder 320 S 500 – 150 with a screw diameter of 30 mm. The test plates had dimensions of 60 mm x 60 mm and a thickness of 4 mm.
Preliminary tests were conducted in order to identify a process window within the parts have an acceptable quality and the injection molding runs steadily. The preliminary tests demonstrated that injection velocity, holding pressure, revolution speed of the screw and back pressure during metering are suitable parameters to change during the process. The limitations are founded by technical limitations of the injection-molding machine.
Both the temperature of the barrel and the temperature of the mold were kept steady, Table 1.

Table 1 – Constant process parameters

Parameter	Values
Barell temperatures [°C]	80 230 250 260 270 280
Mold temperature [°C]	50
Metering stroke [m]	85
Switchover point [mm]	15
Holding pressure time [s]	20
Cooling time [s]	15

Table 2 gives the processing parameters for the design of experiment. The variables were changed during the tests from a minimum [-1], to a medium [0] and finally to a maximum value [+1].

Table 2 – Independent Variables

	Min [-1]	C [0]	Max [+1]
Injection velocity [mm/s]	10	80	150
Holding pressure [bar]	90	160	230
Revolution speed [m/min]	5	27,5	50
Back pressure [bar]	20	50	80

Two full factorial plans were conducted for the evaluation of both target values impact energy and fibre length. In order to provide a sufficient statistical confidence the single parameter combinations for impact energy were repeated 12 times according to DIN EN ISO 6603 and for fibre length three times.

Characterization of the Injection Molded Samples

For the mechanical testing a falling impact tester Ceast Fractovis Plus was used. With the falling dart test according to DIN EN ISO 6603 – 2 it is possible to determine the multiaxial impact behavior by piercing the samples vertical to the sample surface with a constant striker velocity of 4,4 mm/s. Meanwhile the force-deformation diagram is stored. From this diagram the maximum impact force and the impact energy can be analyzed. Thereby the impact energy is the energy value until the impact force decreases to the level of the half of the maximum force.
Since falling dart reveals multiaxial material data and thus orientation independent values it is a very suitable method to characterize fibre reinforced materials.
The tests were conducted in a temperature chamber with 80° C. To guarantee a stable sample temperature the samples were stored in the chamber for 45 minutes before starting up the testing.
The evaluated method for fibre length analysis was applied in order to measure the fibre length of the

samples after injection molding and to correlate impact properties and fibre length.

Results and Discussion

Fibre Length Analysis

First of all the complete separation of the fibres from the matrix system is required. In principle this is possible on two ways – chemically and thermally.
The soxhlet apparatus for chemical digestion consists of an oil bath, a flask, an extraction support, an extraction thimble and a reflux condenser. Figure 2 illustrates the reaction setup in overview and the extraction chamber in detail.

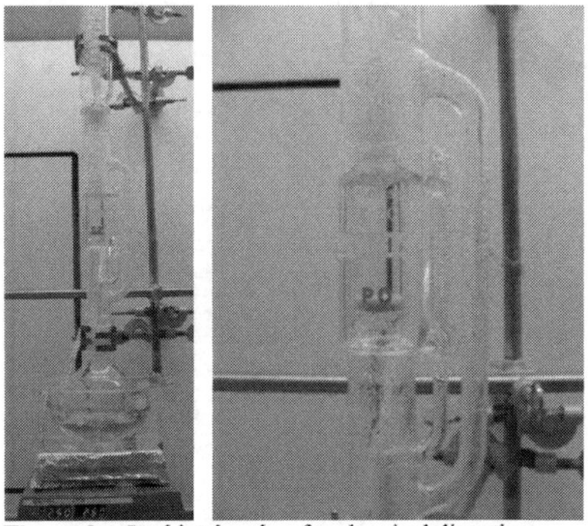

Figure 2 – Soxhletchamber for chemical digestion

Before starting the process the sample and the extraction thimble were dried to estimate afterwards the efficiency of chemical digestion by weight loss. The flask was filled with the respective solvent. As solvents formic acid is suitable for PA 6 and PA 66 and xylene for PP. 2 g of the sample was filled into the thimble before positioning the thimble in the extraction chamber. To start the reaction the oil bath heated up the solvent to the boiling point, the solvent evaporated and condensed over the reflux condenser and trickles continuously into the extraction chamber, where the solvent dissolved the matrix polymer. The mixture of solvent and polymer drains back through an overflow tube back into the flask and the cycle restarted. This procedure runs autonomously and needs approximately 5 to 10 hours until the fibre recovery is completed.

For pyrolysis a microwave furnace from the company CEM, type phoenix was used. The samples were put into the furnace chamber with an area about 100 cm^2. For compete separation of the fibres a burning time of approximately 1 to 2 hours is required dependent on temperature and matrix system. Thereby temperatures between 400° and 500° C are necessary.

Consequently in comparison to pyrolysis, the chemical digestion is characterized by a very gentle thermal treatment of the fibres. Hence the embrittlement of the fibres can be avoided when using this method. On the other hand, pyrolysis appears as an easy and matrix independent method for fibre separation. Also concerning the sample size the pyrolysis allows to process parts with an area approx. up to 10 cm^2, while with the soxhlet apparatus only sample weights up to 2 g are processable. Lastly also the recovery time is of interest, for which reason the pyrolysis is more time efficient compared to the soxhlet process.

In terms of completeness of the fibre recovery both methods are comparable when choosing the right temperature, a proper solvent or a sufficient digestion time respectively. The surfaces of fibres after different methods of fibre recovery are shown in figure 3 in order to compare the recovery completeness from compounds of PP, PA 6 and PA 66. The soxleth extraction was conducted with formic acid for 5 hours for both types of polyamide and with xylene for 20 hours for polypropylene. The thermal treatment for the three compounds was conducted at 500° C for one hour. The figure reveals satisfactory recovery completeness for all processes since very less remainders of the matrix systems are visible on the fibre surfaces. Furthermore it is apparent that a thermal treatment at 500° C for 1 hour is absolutely adequate to guarantee a complete fibre recovery.

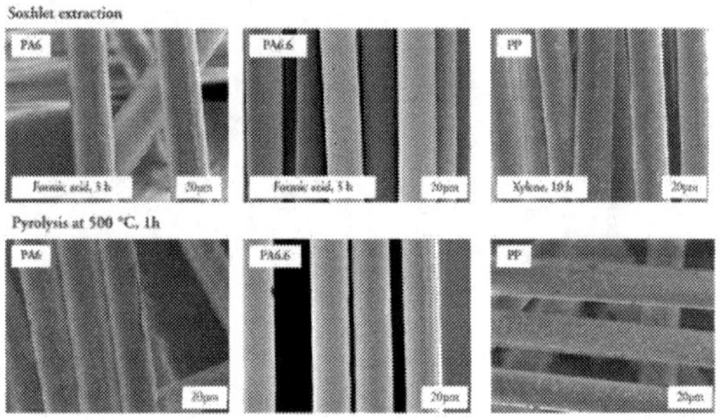

Figure 3 – Comparison of the completeness of fibre recovery

However, due to the high temperatures during pyrolysis an embrittlement is expected which leads to fibre length degradation in the following handling steps. In order to quantify the embrittlement of the fibres during pyrolysis glass fibre rovings with a diameter of 17 µm were treated by different temperature profiles for 1 hour, e. g. 400° C, 500° C and 600° C. A single fibre tensile test was performed and the results were compared to the modulus and strength of virgin fibres. A tensile test machine Bose Electroforce 3200 was employed. For each experiment 5 valid tests were conducted. The single fibres were taken out of the thermal treated roving, sticked on a frame of polyimide and clamped mechanically into grips, shown in figure 4. The tests were conducted at a test speed of 0.01 mm/s.

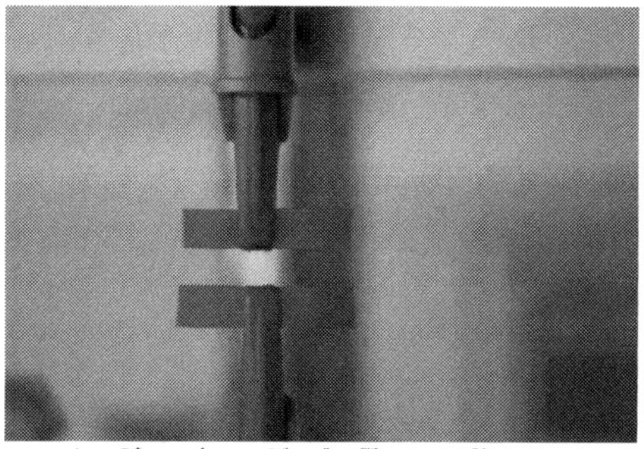

Figure 4 – Clamping - Single fibre tensile test

The stress-strain-curves for the different treated fibres are shown in figure 5. Until breakage the fibres behave linear-elastic. The elongation decreases from 6 % at the virgin fibre to 1.5 % for a treatment of 600° C. Furthermore from 500° C to 600° C the modulus increases from approximately 60 MPa to 80 MPa which indicates an explicit embrittlement of the fibres with higher temperature treatment.

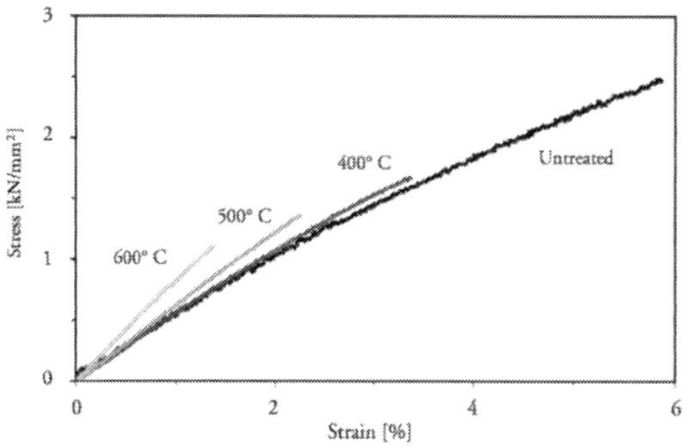

Figure 5 – Stress-strain-diagram – single fibre tensile test

Figure 6 shows the significant loss of the tensile strength under consideration of the scattering bars. After treatment with 400° C the fibres have still 85 % of the original strength. After treating the fibres with 600° C the strength already decreased to 42 % of the original value.

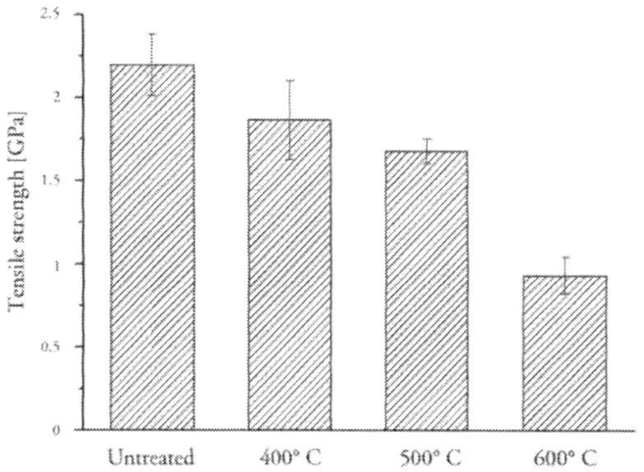

Figure 6 – Tensile strength – single fibre tensile test

As a result fibre recovery by pyrolysis is proposed for all common fibre reinforced thermoplastic composites like PA-GF or PP-GF.

The advantages are:

- Time-efficiency
- Bigger samples are possible to handle, possibly whole injection molded parts
- Uncomplicated handling
- No chemical waste

As temperature profile 500° C at 1 h is suggested since the fibre recovery completeness is sufficient and the loss of mechanical properties is not as pronounced as at 600° C. A temperature of 400° C has shown an insufficient recovery completeness for which reason it is proven as not suitable.

For dispersion the fibres were transferred carefully to a beaker glass with a mixture of distilled water, 0.25 % silane, 0.25 % surfactant and acetic acid up to pH 4 recommended by Sawyer (Sawyer 1979). In doing so the monofunctional silane reacts with the silane on the sized fibre surface and builds methylsiloxane in order to keep the fibres separated. The surfactant overcomes the surface forces of the water and the acid catalyzes the glass-silane reaction. In this dispersion medium a dilution series has to be done until very few fibres are left in the last beaker glass. Therefore the mixture was stirred carefully that on the one hand no fibre breakage occurs and on the other hand all fibres are in motion, since the shorter fibres tend to sink on the bottom of the glass. Afterwards the glass fibres in the medium were passed to a Petri dish. The fibre dispersion is sufficient when very few up to zero crossing fibres are on the Petri dish. The Petri dish was placed on a flat bed scanner, which scans the sample on the dish. The image was analyzed with Fasep, a fibre length analysis software by IDM, Darmstadt and Coesfeld respectively, which is based on the image processing software Image Pro Plus. In a semiautomated mode the well-dispersed single fibres on the dish are gauged automatically while fibre clusters i. e. crossing fibres are counted semiautomatically. Therefore the program gives a suggestion for each cluster, the operator either accepts the proposal or refuses it and determines the length of the single fibres himself.

After measurement of each single fibre of the sample it is possible to calculate an average fibre length or to display the fibre length within a histogram. First of all it is important for both cases to divide the range of measured fibre length into different classes and to classify each single fibre into these classes. From the number of single fibres within each class and with the average length of the length class the fibre length of the whole sample can be calculated in two established ways – number weighted and volume weighted.

Number weighted average fibre length:

$$L_N = \frac{\sum_{i=0}^{n} n_i \cdot L_i}{n}$$

Volume weighted average fibre length:

$$L_V = \frac{\sum_{i=0}^{n} n_i \cdot L_i^2}{\sum_{i=0}^{n} n_i \cdot L_i}$$

where:
L: Average fibre length per class,
n: number of fibres within the class.

Since the contribution of longer fibres is higher when calculating the volume weighted average fibre length this value was chosen for comparison in the following investigation. The depiction as a histogram of each fibre length is abandoned, due to better comparison of single values.

Impact Properties and Fibre Length

The impact properties were determined by impact falling dart test according to DIN EN ISO 6603 – 2. The results of the statistical investigation of the process parameters on the impact energy are illustrated in figure 7 as a pareto plot. This kind of depiction shows all independent and thus variable process parameters serialized in order of their effect. The confidence range is a threshold that divides significant from non-significant parameters. Effects that exceed the confidence range are significant with a probability of 99 %.

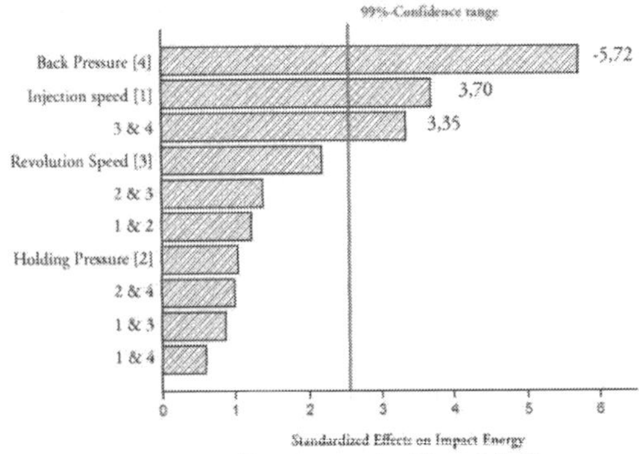

Figure 7 – Pareto diagram – significant influences on impact energy

In this case the pareto plot shows evidently that the back pressure and the injection speed and a mixture of back pressure and screw revolution speed exhibit a certain significance, which is guaranteed with a probability above 99 %. That means the impact energy decreases with increasing back pressure. This fact shows that the higher the pressure charge during plasticizing the lower the impact strength of the molded plates. In comparison the impact strength increases with increasing injection velocity.

An explanation for this material behaviour can be given by the examination of the statistical analysis of the process parameters on the resulting fibre length. The fibre length was measured by the method introduced before.
The pareto plot in figure 8 indicates a significance of back pressure as well as a mixture of the effects of holding pressure and back pressure on the fibre length. Consequently the impact properties correlate highly with the fibre length in the molded parts, since back pressure during plasticizing affects strongly both properties. Thus the effect of back pressure on impact strength correlates also with its effect on fibre length. Since only back pressure and not the revolution speed is significant in this case, apparently only the pressure during plasticizing and not the shear stress is of importance during injection molding.

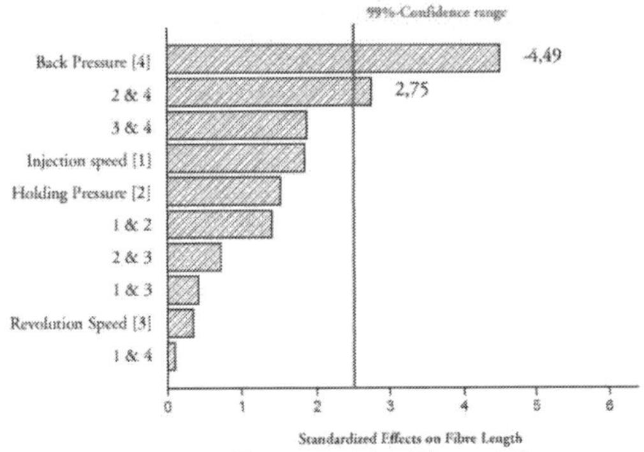

Figure 8 – Pareto diagram – significant influences on fibre length

In comparison the significance of injection velocity on impact strength is not explained by correlation with fibre length. But in processing of fibre reinforced thermoplastic materials by injection molding it is well known that higher injection velocities lead to a smoother surface of the molded parts. Due to phase segregation and freezing at the melt front at lower injection speed the surfaces appear rough. In this case as well the more structured surfaces were observed when molding with lower injection velocities. The standard DIN EN ISO 6603 gives the hint that parts that have a structured surface on one side should be always tested in the same manner i. e. from the same side (structured or unstructured). It may be possible that the worse surface quality of the plates molded with lower injection speed lead to lower impact forces and thus energy values, since the surface is finely notched.

Table 3 gives the settings for the serial runs. The average values of impact energy and fibre length and their standard deviation for each production run in comparison to the central setting run (right column) are demonstrated in figure 9 and 10.

Table 3 – Serial production runs - Settings

Run	Back pr.	Inj. Veloc.	Hold. pr.	Rev. Sp.
1	-1	-1	-1	-1
2	-1	-1	-1	+1
3	-1	-1	+1	-1
4	-1	-1	+1	+1
5	-1	+1	-1	-1
6	-1	+1	-1	+1
7	-1	+1	+1	-1
8	-1	+1	+1	+1
C	0	0	0	0
9	+1	-1	-1	-1
10	+1	-1	-1	+1
11	+1	-1	+1	-1
12	+1	-1	+1	+1
13	+1	+1	-1	-1
14	+1	+1	-1	+1
15	+1	+1	+1	-1
16	+1	+1	+1	+1

The plot above shows in both cases all serial runs with the minimum setting of backpressure. According to DIN EN ISO 6603 each setting was tested by falling dart 10 times plus reliability of two: 12 times.

The diagrams indicate in detail what the pareto plots have already shown: The higher the back pressure the lower the impact strength and the lower the fibre length. It is evident that in the case of impact energy the standard deviation is nearly comparable. The average value of impact energy for all test runs is about 14.09 J while the average value for all settings with lower backpressure is about 15.00 J and with higher backpressure about 13.09 J. This shows clearly a tendency for increase of the impact energy with decreasing backpressure.
With injection velocity the effect is not that pronounced. With increasing injection speed the impact energy raises from 13.73 J to 14.36 J.
Picture 10 shows the results of fibre length analysis for each serial run. Also in this case a decrease of fibre length is evident with increasing backpressure. The volume weighted fibre length is reduced from 1.3622 mm to 1.7196 mm, which gives again the correlation between fibre length and impact strength of the composites. The overall average volume weighted fibre length is 1.5945 mm.

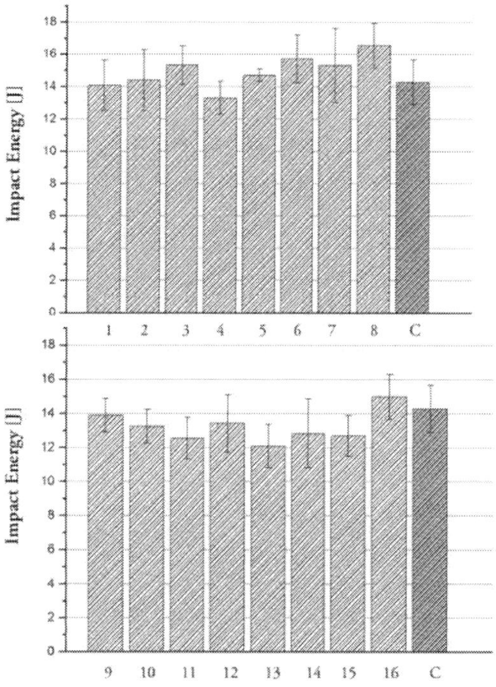

Figure 9 – Comparison of the impact energy for every production run

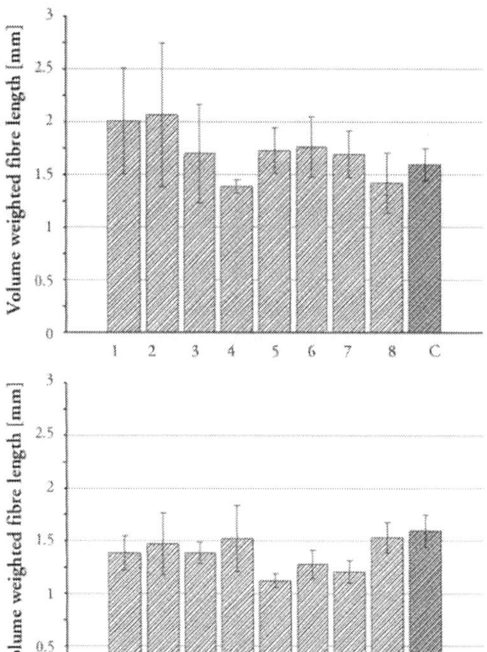

Figure 10 – Comparison of the volume weighted fibre length for every production run

Conclusions

The single steps of fibre length analysis were evaluated in terms of feasibility and potential for optimization. Therefore the steps were divided into fibre-matrix-separation by pyrolysis, dispersion and separation of the fibres in an adequate solvent, digitalization and finally measurement of the fibre length. It could be shown that high pyrolysis temperature led to embrittlement of the glass fibres and a decrease of tensile strength of the single fibres. This falsifies the fibre length analysis due to fibre breakage. For this reason the preparation of the samples and the handling of the fibres is very important. A reliable method was found in order to prepare a fibre reinforced sample properly and to measure the fibre length comparably for each sample.
Furthermore injection molding of long fibre reinforced polypropylene was conducted to investigate the effect of the processing parameters and their possible interaction on impact properties and fibre length. Subsequently, the impact behaviour of the PP-GF compounds was quantified by falling dart experiments, and the results are then correlated to the fibre length and the processing parameters to establish a processing-structure-property relationship. The results clearly show a strong negative influence of the back pressure during injection on both impact energy and fibre length. This allows to conclude that fibre length correlates strongly with the impact energy. Also the injection speed has a significant effect on the impact properties, which may be attributed to the smoother surface of the samples when molding with high injection velocities.
In consequence it is possible to tailor the composite properties by process control and to give a correlation between structure and the properties of fibre reinforced compounds.

Acknowledgements

The authors acknowledge the support of the company FACT for providing the material and the Friedrich-Baur-Institute for biomaterials for the use of the tensile testing machine for single fibre tensile tests. Furthermore we like to thank Mr. Mainz for the support in all terms of mechanical testing.

References

Martin, G.A., "Neue Wege in der Kunststoffverarbeitung", Kunststoffe, 8, 26-33 (2005)

Menges, G, Geisbuesch, P., "Die Glasfaserorientierung und ihr Einfluss auf die mechanischen Eigenschaften thermoplastischer Spritzgießteile – Eine Abschätzmethode", Colloid & Polymer Science, **260**, 73-81 (1982)

Sawyer, L.C., " Determination of Fieberglass Length: Sample preparation and automatic image analysis", Polymer Engineering and Science, **19** (1979)

Schemme, M., "Long fibre reinforced thermoplastics", Kunststoffe, **8**, 106-109 (2003)

Schijve, W., "Common errors and honest material comparison for long glass PP materials", AVK-Conference, 2002

Thomason, J.L., "Micromechanical parameters from macromechanical measurements on glass reinforced polyamide 66", Composites Science and Technology, **61**, 2007-2016 (2001)

Thomason, J.L., "The influence of fibre length and concentration on the properties of glass fibre reinforced polypropylene: 5. Injection molded long and short fibre PP", Composites: Part A, **33**, 1641-1652 (2002)

Thomason, J.L., "Structure-Property Relationships in Glass-Reinforced Polyamide, Part 1: The effects of fibre content", Polymer Composites, 552-532 (2006)

Thomason, J.L., "Structure-Property Relationships in Glass-Reinforced Polyamide, Part 2: The effects of Average Fibre Diameter and Diameter Distribution", Polymer Composites, 552-562 (2007)

Thomason, J.L., Adzima, L.J., "Sizing up the interface: An insiders guide to the science of sizing", Composites Part A, **32**, 313-321 (2001)

MORPHOLOGICAL DEVELOPMENT IN WATER ASSISTED INJECTION MOLDED POLYETHYLENE/POLYAMIDE-6 BLENDS

Shih-Jung Liu, Wei-Ru Lin, Kun-Yeh Lin

Department of Mechanical Engineering, Chang Gung University, Taiwan - *shihjung@mail.cgu.edu.tw*

Abstract - This study examined the morphological development in water assisted injection molded high density polyethylene (HDPE)/polyamide-6 (PA-6) blends. Samples for microscopic observation were prepared by an 80-ton injection-molding machine equipped with a tube cavity and with a water injection unit. A distinct skin layer, core region, and channel layer were observed across the thickness. High water pressures were found to mold parts with smaller polyamide particle distributions. Additionally, the morphology of water assisted injection molded parts was compared to that of gas assisted injection molded products. It was found that water molded parts with a smaller polyamide particle distribution than gas.

Introduction

Water assisted injection molding (WAIM) [1, 2] has gradually become one of the most important techniques for manufacturing hollow plastic parts. The physical properties of a water assisted injection molded part can be strongly influenced by its microstructure such as molecular orientation and morphology. To present, while there is literature available for phase morphology in conventional injection molding [3-5] and gas assisted injection molding [6, 7], no research has ever studied the morphology in water assisted injection molded parts.

This report is devoted to studying the morphological development in water assisted injection molded immiscible blend. The influence of water pressure on the morphology of molded blends was also studied. Furthermore, a comparison was made between the morphology of water assisted injection molded parts and that of gas assisted injection molded products.

Experimental

Materials

The resins used in this study were high density polyethylene (HDPE) (LH-614, USI Corp., Taiwan) and polyamide-6 (PA-6) (Mapex-NT0110GN, Ginar Tech., Taiwan).

Preparation of specimens

The weight ratios of HDPE/PA-6 mixture was 80:20. The mixture was blended by a twin-screw extruder (Desunnano Model SHJ-20B, Taiwan) with a temperature setting of 180, 220, 235, 240, 245, and 230°C from the hopper to die.

All specimens were molded by an 80-ton conventional injection-molding machine. Figure 1 shows the structure and dimensions of the parts. A lab scale water injection unit developed earlier in our lab was used to water assisted mold the HDPE/PA blends. The gas assisted injection molded samples were manufactured by using a commercial available high-pressure nitrogen gas injection unit (Gas Injection model PPC-1000, U.K.). Table 1 lists the processing variables as well as the values used in the gas and water assisted injection molding experiments.

Microscopic observation

After molding, specimens cut from the front, middle and rear sections (see Figure 1) in parts molded by gas and water assisted injection moldings were observed under a Hitachi Model S-300N scanning electron microscope.

Results and Discussion

Morphology in water assisted injection moldings

A distinct skin layer, core region, and channel layer could be observed across the thickness (Figures 2 and 3). The PA phase was of the greater degree deformation in the skin layer (Figures 2a and 3a), followed by the channel layer (Figures 2c and 3c), and it exhibited the least deformation in the core region (Figures 2b and 3b). Higher water pressure molded skin layer (Figures 2a) exhibited greatly deformed PA phase than lower pressure molded skin layer (Figure 2a).

A comparison between water and gas assisted injection moldings

In the skin layer, when compared to that of water molded parts, gas molded parts exhibited less stretched PA phase. The observed result implies that water has a better capability of shearing the materials than gas.

In the core region, while the PA particles in molded parts do not show significant difference at various sections of the core region, the PA phase at the rear section of the core layer showed much greater size particles (Figures 4b-R and 5b-R) than those at the front and middle sections(Figures 4b-F and 4b-M, and Figures 5b-F and 5b-M).

In the channel layer, contrary to that in the skin layer, gas assisted molded PA phase was more stretched than water molded PA phase.

Table 1. The processing variables as well as the values used in the experiments.

Parameter	Gas	Water
Melt filling pressure (MPa)	83.3	83.3
Melt temperature (°C)	230	230
Mold temperature (°C)	40	40
Short-shot size (%)	57	57
Fluid temperature (°C)	25	45

Fluid injection delay time (s)	1	1
Fluid injection pressure (MPa)	9.5 or 16.2	11.0 or 16.5
Fluid packing time (s)	5	5
Cool time (s)	15	15

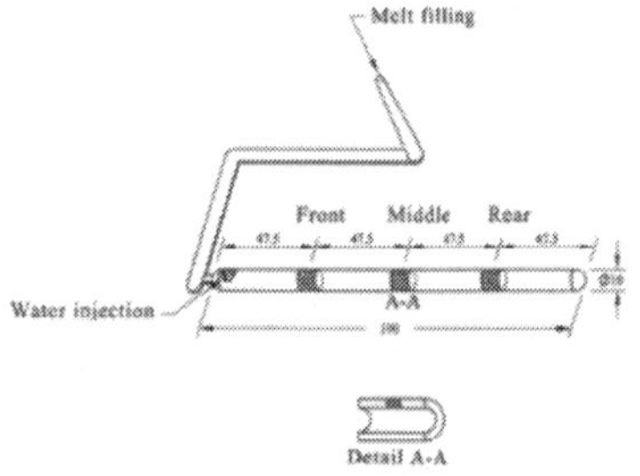

Figure 1. Layout and dimensions of the tube parts (unit: mm)

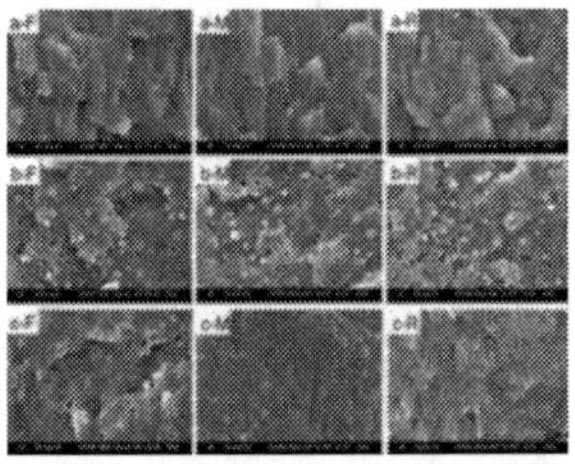

Figure 2. SEM photos at the front (F), middle (M) and rear sections (R) of low pressure water assisted injection molded parts: (a) skin layer, (b) core layer, and (c) water channel layer

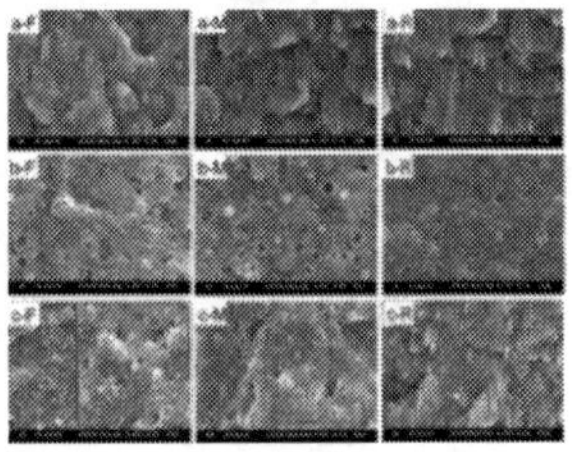

Figure 3. SEM photos at the front (F), middle (M) and rear sections (R) of high pressure water assisted injection molded parts: (a) skin layer, (b) core layer, and (c) water channel layer

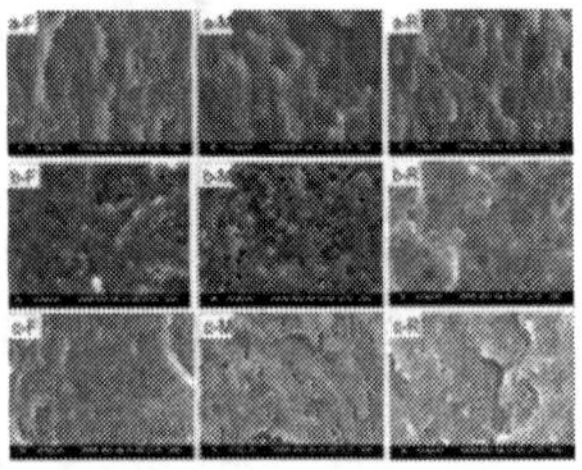

Figure 4. SEM photos at the front (F), middle (M) and rear sections (R) of low pressure gas assisted injection molded parts: (a) skin layer, (b) core layer, and (c) water channel layer

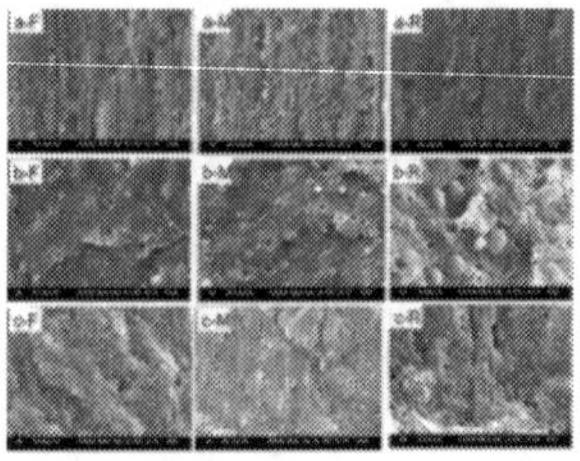

Figure 5. SEM photos at the front (F), middle (M) and rear sections (R) of high pressure gas assisted injection molded parts: (a) skin layer, (b) core layer, and (c) water channel layer

Conclusions

This study has examined the morphological development in water assisted injection molded high density polyethylene (HDPE)/polyamide-6 (PA-6) blends. Distinct skin layer, core region, and channel layer were observed across the thickness in both gas and water assisted injection molded tubes. The shape and size of the dispersed phase depended on the position both across the part thickness and along the flow direction. Water molded parts with a smaller polyamide particle distribution than gas. Additionally, high water pressures were found to mold parts with smaller polyamide phase domain.

Acknowledgements

The National Science Council of Taiwan, R.O.C. under the grant NSC95-2221-E-182-032-MY3, has supported this work financially.

References

1. Knights M., "Water Injection-it's all coming together", Plast. Technol., 54-61(2005)
2. Michaeli W., "Chapter 6 Water Injection Techniques, in Injection Molding: Technology and Fundamental", Kamal M., Isayev, A.I., Liu S.J. (Eds.), Hanser, Munich, p. 329-373(2009)
3. Fellahi S., et al., "Morphological stability in injection-moulded high-density polyethylene/ polyamide-6 blends", Polymer, **13**, 2615-2626(1996)
4. Son Y., et al., "Morphology of injection molded modified poly(phenylene oxide)/polyamide-6 blend", Polym. Eng. Sci., **40**, 1376-1384(2000)
5. Li Z., Narh K.A., "Morphological development during injection molding of self-reinforcing composites. II: numerical simulation and analytical predictions", Polym. Compos., **21**, 978-987(2000)
6. Zheng G., et al., "Morphology of gas-assisted and conventional injection molded polycarbonate /polyethylene blend", J. Appl. Polym. Sci., **102**, 3069-3077(2006)
7. Zheng G.Q., et al., "The role of gas penetration on morphology formation of polycarbonate /polyethylene blend molded by gas-assisted injection molding", J. Mater. Sci., **42**, 7275-7285(2007)

CT DATA EVALUATION OF FIBRE REINFORCED POLYMERS TO DETERMINE FIBRE LENGTH DISTRIBUTION

D. Salaberger[1*], K.A. Kannappan[1], J. Kastner[1], J. Reussner[2] and T. Auinger[3]

[1] *University of Applied Sciences Upper Austria, Stelzhamerstr. 23, 4600 Wels, Austria,* d.salaberger@fh-wels.at, k.arunachalam@fh-wels.at, j.kastner@fh-wels.at

[2] *Borealis Polyolefine GmbH, St. Peterstr. 25, 4021 Linz, Austria,* jens.reussner@borealisgroup.com

[3] *Transfercenter für Kunststofftechnik GmbH, Franz Fritschstr. 11, 4600 Wels, Austria,* thomas.auinger@tckt.at

For this study Sub-μm-CT was used to determine fiber orientation and fiber length distribution in injection molded Polypropylene, filled with short glass fibers. For data evaluation a previously developed method was used that was developed for μ-CT data. Since data quality and resolution is much better with Sub-μm-CT the capability of the data analysis concept to extract orientation and length was investigated. For this reason the algorithms were applied on specimens with four different fiber contents: 1, 5, 10 and 30 % wt. To quantify the error of fiber extraction a minimum and average probability to extract the fibers correctly were estimated.

The results show a strong dependence of error on fiber content. While the orientation is not affected much, the fiber length is affected for fiber contents above 5%. For fiber length distribution also the size of the analyzed dataset has to be taken into account. It could be shown that for the given specimens a cubic sample volume with an edge length of ca. three times the mean fiber length is sufficient to represent the fiber length distribution properly.

Introduction

To be able to predict mechanical properties of a structural component made of heterogeneous polymers the knowledge of the geometrical details of the different phases is necessary. For short glass fiber filled polymers the parameters fiber orientation distribution (FOD), fiber length distribution (FLD) and fiber content have been identified as very important for the mechanical behavior of this material. [von Bradsky et al., 1997] In recent years methods have been established to measure FOD and FLD. The best established methods are destructive methods that can usually gather either to information of orientation or length distribution. [Zak et al, 2000] For orientation thin cross sections can be analyzed either with light optical microscopy (LOM) or x-ray methods. [Phelps et al., 2008] The fiber length can be determined by LOM- methods after burning the matrix or estimating the length from cross sections. A correction of the fiber lengths derived from cross sections might be necessary due to statistical effects that have to be taken into account when the number of analyzed cross sections is low. [Shao-Yun Fu et al., 2002, Clarke et al. (1995)]
Recently also three dimensional methods based on x-ray tomography (xCT) have been developed. Depending on the tomographic method that has been used, the results can be only an estimation or more accurate. Since data quality is determined by the resolution as well as the appearance of artefacts the best xCT data quality can be achieved from Synchrotron measurements at the moment. The industrial CT devices are most commonly μ-CT devices that allow for resolutions of a few μm.
The disadvantage of the destructive methods is usually the time consumption and the fact that for fiber length and fiber orientation two different methods have to be applied. Synchrotron CT is cost intensive and the availability is a limiting factor.
xCT devices that deliver CT data with quality near SCT are available. These Sub-μm-CT devices usually work according to the cone beam CT principle. [Kastner, a, 2008] The specimen is rotated in an x-ray beam that is generated by an x-ray tube. According to density and atomic number the x-rays are attenuated. The resulting projection is grabbed by a digital detector that usually works according to the scintillation principle. Usually more than thousand projections are generated from different angles. These projections are reconstructed using mathematical algorithms resulting in a 3D data set which consists of volumetric pixels (Voxels) that have different gray values. After the specimen was rotated once by 360° the full representation of the specimen is available.
Sub-μm-CT devices are equipped with x-ray tubes with very small focal spots to be able to minimize the effect of blurring. X-ray detectors are well established with more than 2000 pixels in one direction. Best achievable resolutions are in the range of a few 100 nm. These devices are built in a desktop- style and can be bought from different manufactures.
The achievable resolution depends on one hand on the CT equipment and on the other hand on the size of the specimen. To be able to determine length and orientation very accurately a compromise has to be

found between small sample size and representative volume.
The quality of CT data also determines possible ways of analysis. Many different algorithms have been developed to determine FOD on many different three dimensional data. [Requena et al, 2009, Vincent et al., 2005, Park et al., 2001] It could be seen that it was mainly possible to segment single fibers in SCT data whereas global or averaged orientations were determined on µ-CT data.
The in- house developed algorithms for CT data analysis were designed to determine fiber orientation using µ-CT data. [Kastner et al., b, 2008] The resolution was in the range of the diameter of the fibers. The orientation could be determined accurately whereas the fiber length could not be determined since data quality was not sufficient.
This publication deals with the application of the previously developed algorithms on Sub-µm-CT data to determine the fiber length distribution most accurately. It will be discussed how the accuracy could be estimated and how the algorithms could be improved further. Also the influence of fiber content on the quality of the analysis will be shown.

Experimental

The material that was used for this study is injection molded Polypropylene filled with short glass fibers. Fiber contents of 1, 5, 10 and 30 % wt. were analyzed. The mold was a standard tensile bar geometry. The average diameter of the fibers was 13.5 µm.

For the generation of CT data the Sub-µm-CT device Nanotom (Phoenix-xray, Wunstorf, Germany) was used. The Nanofocus® tube was operated at 50 kV and a focal spot size of about 2 µm. A Hamamatsu detector with 2300 x 2300 pixels was used.
For comparison, µ-CT data was generated using the RayScan 250 E (RayScan Technologies, Meersburg, Germany) Microfocus tube.
To be able to visualize the whole cross section of a tensile bar test specimen with 4 by 10 mm edge length, the resolution could be set to 6 µm per Voxel. To improve image quality and reduce artefacts measurements with a resolution of 2 µm per Voxel were performed on a smaller part of the specimen. For the dataset of 1 % fiber content two sample volumes were cut out, a smaller and a bigger one. It should be documented how the FLD changes with sample volume.

For data analysis a sequence of algorithms was applied on the 3D datasets according to Kastner et al. (c, 2008). The fiber extraction tool uses different algorithms that are applied subsequently. The first step is anisotropic diffusion to reduce noise. Then the fibers are separated from the matrix by binarization using the principle proposed by Otsu (1979). In the next step binary thinning is applied to extract only a single line of voxels for each fiber. The last step treats regions where fibers are touching, so called clusters in a way that the fibers are separated into single objects.
Following parameters were set for the algorithms:
Anisotropic diffusion: Iteration (10), timestep (0.0625), conductance (5);
Binarization: Otsu- value * 1.25.
Cluster analysis: kink angle: 160°, cluster distance:
14 µm (ca. fiber diameter)

The software that was used to determine the accurate number of fibers was the standard 3D data analysis tool 'VG Studio MAX 2.0' (Volume Graphics, Heidelberg, Germany). The toolbox 'Geometry analysis' was used therefore to fit cylinders by manually selecting every individual fiber.

Specimen	1% a	1% b	5%	10%	30%
Data size (MB)	285.5	2080	79.7	33.0	3.8
Voxel X	533	1064	388	210	125
Voxel Y	497	1001	217	204	124
Voxel Z	552	1000	485	388	126
No. of fibres (manually counted)	82	--	106	175	112

Table 1: Overview of analyzed data, Voxelsize: 2 µm

Synchrotron-CT-measurements were done at the ESRF (European Synchrotron Radiation Facilities) in Grenoble using monochromatic x-rays with 21.2 kV and a resolution of 0.7 µm.

Results and Discussion

The result of a CT scan with a Sub-µm-CT device shows much better quality for polymeric samples than a more common µ-CT device. (Figure 1) The reason is on one hand the better quality of the x-ray beam since the focal spot is very small and the energy spectrum has its maximum at lower energies which leads to a better contrast for less absorbing materials like polymers. On the other hand the used detectors have 1024 (µ-CT) and 2300 (Sub-µm-CT) Pixels in a row which leads to half the voxelsize for the Sub-µm-CT device at the same size of the specimen.

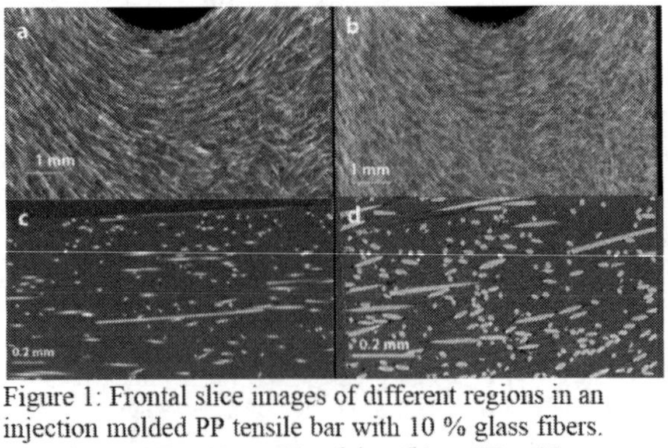

Figure 1: Frontal slice images of different regions in an injection molded PP tensile bar with 10 % glass fibers. (a) Sub-µm-CT at 12 µm Voxelsize; (b) µ-CT at 10 µm

Voxelsize; (c) Sub-µm-CT at 2 µm Voxelsize; (d) Synchrotron-CT at 0.7 µm Voxelsize

Figure 1 shows slice images of data sets that were generated using different CT devices. It is clearly visible that the Synchrotron CT measurement represents each and every fiber in a very good way. For Sub-µm-CT and high resolution (c, 2 µm) the data quality is also good enough to separate single fibers. When resolution decreases (a, 12 µm) single fibers cannot be resolved properly any more. A similar result is gained with µ-CT with the disadvantage of only half the size of measurement volume compared to Sub-µm-CT.
Even at limited data quality or limited resolution the orientation of the fibers is visible and can be quantified using proper software algorithms.

Evaluation of data analysis pipeline

To be able to point out how many fibers within the analyzed volume are represented correctly the total number of fibers was determined by a manual data analysis. For each fiber a cylinder was fit to guarantee to count each fiber exactly once. In Figure 2 the fibers are displayed semi-transparent to see the fitted cylinders as well as the fibers. For the evaluation of the algorithms all CT data were scanned at 2 µm Voxelsize.

Figure 2: 3D image of transparent fibers and fitted cylinders for each fiber (Fiber content 5 %)

In a first step the result of the automated analysis was evaluated in terms of correct number of fibers.

Fiber content (wt.)	Manually Counted fibers	Total fibers	**Error in No. of fibers (%)**
1 % a	85	99	**16.5**
5 %	107	112	**4.7**
10 %	175	179	**2.3**
30 %	112	130	**16.1**

Table 2: Comparison of number of fibers counted manually and total fibers extracted automatically

For the investigation of the reason for the errors in number of fibers, images of the binarized, the binary thinned as well as the extracted fibers were analyzed. (Figure 3) These showed that problems only occur in points where fibers are touching. Free fibers are represented correctly.

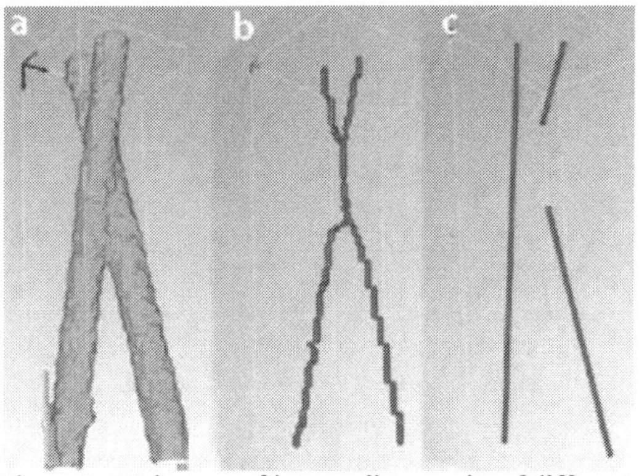

Figure 3: 3D images of intermediate results of different stages of the data analysis pipeline in a cluster region: binarization (a), binary thinning (b), extracted fibers (c)

Two reasons were identified for the wrong extraction: binarization and Binary Thinning. Due to artefacts that influence the gray values for the fibers in different regions of the data set, global thresholding will lead to the loss of small fibers, if the threshold is too high. On the other hand the touching areas will increase if the threshold is too low.
In any case areas of touching fibers, the clusters, affect the result of binary thinning in a way that the number of fibers cannot be reconstructed in this area.
These two effects lead to the conclusion, that taking only the error in number of fibers into account is not sufficient to estimate the overall error of the whole fiber extraction algorithm.

From the matter of fact that free fibers are always extracted in a correct way, we can define a minimum probability PM to extract the fibers correctly:

$$P_M = \frac{Free\ fibers}{Total\ fibers} * 100\%$$

Fiber content (wt.)	Total fibers	Free fibers	**Minimum probability P_M (%)**
1 % a	99	66	**67**
5 %	112	77	**69**
10 %	179	87	**49**
30 %	130	19	**15**

Table 3: Estimation of Minimum Probability to extract the fiber correctly from the number of free fibers

For the cluster fibers it turned out that, depending on the fiber content, the error varies. Figure 3 shows the main disadvantage of global thresholding and binary thinning. Depending on the gray value threshold in regions where fibers are close together or touching, the area of touching is varying. In the subsequent step of binary thinning this area is transformed into a line of voxels. The available cluster analysis was not capable to extract and combine the fiber fragments properly.

Since the complexity of the analysis and also artefacts increase with increasing fiber content a trust factor t was defined. The factor t was chosen by experience and was used to estimate an average probability to extract fibers correctly.

$$P_A = \frac{Cluster\ fibers}{Total\ fibers} * t * 100\% + P_M$$

Fiber content (wt.)	Cluster fibers	t	**Average probability P_A(%)**
1 % a	33	0.5	**84**
5 %	35	0.5	**85**
10 %	92	0.25	**62**
30 %	111	0.25	**36**

Table 4: Estimation of Average Probability to extract fibers correctly

Influence of Volume size on FLD and FOD

Since every fiber is represented by its start and endpoint after extraction it is easy to calculate the orientation tensor for every fiber. In Table 5 the average values for a11 (x-direction), a22 (y-direction) and a33 (z-direction) for the whole specimens are shown.

Fiber content (wt.)	Number of fibers	a 11	a 22	a 33	Mean fiber length [µm]	Max fiber length [mm]
1% a	99	0.12	0.18	0.7	300	1.12
1% b	738	0.24	0.08	0.68	300	1.65

Table 5: Number of extracted fibers, elements of the main diagonal of the orientation tensor and mean and maximum fiber length

Figure 4 shows a comparison of original CT data and the result of fiber extraction. The top image shows the manually fitted cylinders and the fibers whereas the image at the bottom shows the extracted fibers using glyphs that connect every start and endpoint. Most of the extracted fibers have corresponding fibers in the original data. For bent fibers a straight line is drawn since the algorithm was not designed for bent fibers and the represented glyphs are straight lines.
The color of the glyphs gives information about the orientation. One can see that many fibers are colored red, which means that there is a preferred orientation.
In this case it is the z-direction which is the direction of injection. Also the quantitative analysis shows the same result as a33 has the highest value in both sample volumes.

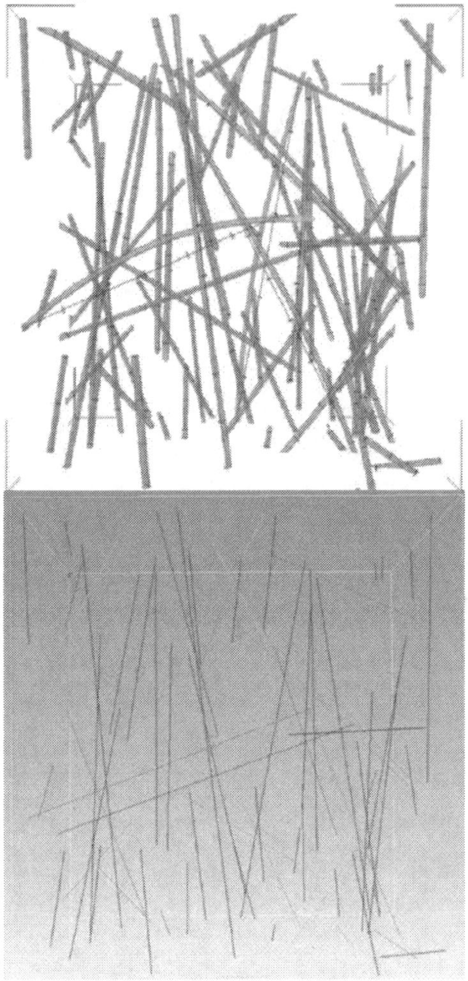

Figure 4: 3D images of original CT data (top) and extracted fibers (bottom) for fiber content 1% a

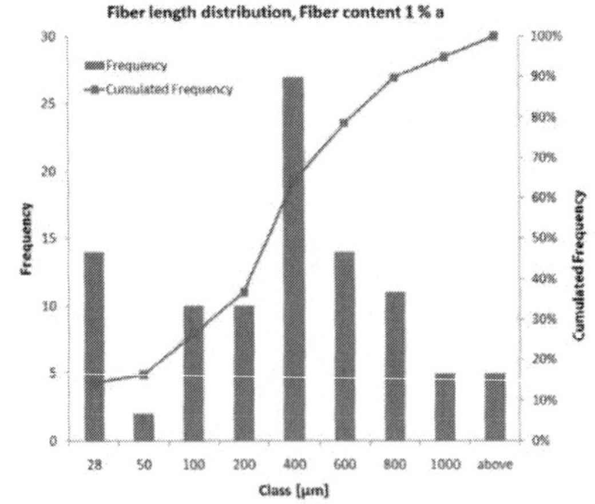

Figure 5: FLD for fiber content 1% and small sample volume a (edge length 1.1 mm)

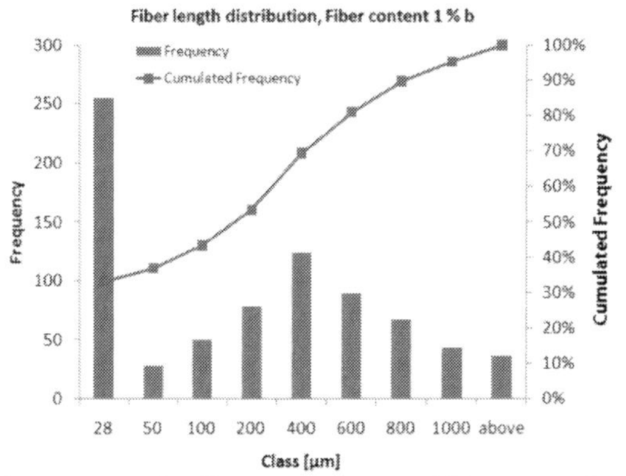

Figure 6: FLD for fiber content 1 % and bigger sample volume b (edge length 2 mm)

Small sample volume is needed to achieve high resolutions. Long fibers may not be completely contained in the sample volume. To investigate this effect, two different data sets were generated for one measurement of the 1% specimen. The edge length of the sample volume a) was ca. 1.1 mm and of sample volume b) it was ca. 2 mm. Figure 5 and 6 show the results of the automated fiber extraction in terms of fiber length. The first length class was set to twice the fiber diameter (28 µm). Elements in this class are more or less fiber fragments that don't have the ideal cylindrical shape of a fiber any more.

FLD for both sample volumes show a similar result. The average fiber length without fiber fragments is ca. 300 µm (in the class between 200 to 400 µm). In the bigger sample volume the frequency of these fragments is with 33 % higher than in the small volume with 15 %. An error will be underlying the results of FLD according to the previous results of data analysis evaluation. Nevertheless the results are in good accordance to the visual inspection since for example many fiber fragments are visible in the slice images as well. The fraction of long fibers is in both sample volumes very similar. The longest fiber in the smaller sample volume a) is 1.12 mm and in the bigger sample volume b) 1.65 mm long.

Conclusions

Investigations were presented on the application of x-ray computed tomography to determine fiber length and fiber orientation distribution.

In pre- investigations data quality of different x-ray tomography methods were investigated. Synchrotron delivers the best possible CT data but has the disadvantage that there are only a few facilities available and that it's very cost intensive. The size of the specimen has to be very small to reach the best resolution and therefore data quality. µ-CT data can be generated with commercially available devices. For glass fiber filled polymers data quality is sufficient to determine fiber orientation. Sub-µm-CT in principle is capable of generating data with quality that is sufficient to determine both, fiber orientation and length distribution. Data acquisition using Sub-µm-CT takes reasonable amount of time and money.

A CT data analysis pipeline was applied on specimens with different fiber contents to discuss the accuracy of this method.

For lower fiber contents up to 5 % wt. the fiber extraction works well and the error is around 15 % in terms of number of correct represented fibers. The reason of errors was found only in the separation of touching fibers. Since there is usually a preferred fiber orientation in injection molded polymers, the touching fibers will have almost the same orientation. Therefore the result for fiber orientation for a whole specimen will not be affected much by the error in data analysis.

By increasing the fiber content the amount of fibers that are touching is increasing and respectively the error increases too.

For 1 % fiber content the sample volume was increased fourfold to investigate the influence of small sample volumes and the problem that not all (especially long) fibers are included fully in the sample volume.

The result for the original and the increased sample volume show a very similar FLD and the same mean fiber length without fiber fragments. The maximal detected fiber length is a bit higher in the bigger volume. The amount of long fibers is quite low which implies that FLD is calculated quite accurately and that only a few fibers are not represented in the analyzed volume completely.

A cubic sample volume with an edge length of ca. 3 times the mean fiber length seems sufficient to lead to representative results for FOD for the analyzed samples without increasing the error for long fibers.

The main advantages of 3D tomographic methods are, that both FOD and FLD can be determined using the same data set without destroying the specimen.

Future work

The focus of the current work lies on the minimization of error by replacing global thresholding for binarization and binary thinning by other, better suited algorithms. Especially for higher fiber content the error has to be reduced since fiber contents of 30 % wt. are commonly used in industry.

Also bent fibers should be represented correctly since this method should also be applicable for cellulose fibers.

Acknowledgements

The project is supported by the Austrian Research Promotion Agency (FFG).

References

[1] G. J. von Bradsky et al., “Characterisation of finite length composites – Part IV- Structural studies on injection moulded composites”, Pure & Appl. Chem., **Vol 69, No. 12**, 2523-2539 (1997)

[2] H. Shen et al., “Direct observation and measurement of fiber architecture in short fiber-polymer composite foam through micro-CT imaging”, Comp. Sci. & Tech., **64,** 2113-2120, (2004)

[3] A. R. Clarke et al., “A novel technique for determining the 3 D spatial distribution of glass fibers in polymer composites“, Comp. Sci. & Tech., **55,** 75-91, (1995)

[4] J. Kastner (Editor) (a). Proceedings Industrielle Computertomografietagung,Wels, Austria, 26.+27. February 2008, Shaker Verlag, Aachen, 2008.

[5] J. Kastner et al. (b), “Determination of diameter, length and three-dimensional distribution of fibres in short glass-fibre reinforced injection moulded parts by μ-X-ray computed tomography”, Proceedings PPS24, (2008)

[6] J. Kastner et al. (c), “Method for three-dimensional evaluation and visualization of the distribution of fibres in glass-fibre reinforced injection molded parts by μ-X-ray computed tomography”, WCNDT Proceedings CD, Shanghai (2008)

[7] G. Zak et al., “Estimation of average fiber length in short-fibre composites by a two-section method”, Comp. Sci. & Tech., **60,** 1763-1772, (2000)

[8] J. H. Phelps et al., “New models for fiber orientation in injection-molded composites”, Proceedings PPS24, (2008)

[9] G. Requena et al., “3D-quantification of the distribution of continuous fibres in unidirectionally reinforced composites”, Comp. PartA, **40**, 152-163, (2009)

[10] M. Vincent et al., “Description and modeling of fiber orientation in injection molding of fiber reinforced thermoplastics”, Polymer **46**, 6719-6725, (2005)

[11] CH. Park et al., “A Study on fibre orientation in the compression molding of fiber reinforced polymer composite materials”, Mat. Proc. Tech., **11,** 233-239, (2001)

[12] N. Otsu, “A threshold selection method from grey level histograms”, IEEE Transactions on Systems, Man, and Cybernetics, **9, (**1979)

COMPARISON OF RECENT FIBER ORIENTATION MODELS IN AUTODESK MOLDFLOW INSIGHT SIMULATIONS WITH MEASURED FIBER ORIENTATION DATA

Jin Wang* and Xiaoshi Jin

Moldflow R&D Center, Autodesk, Inc., 2353 N. Triphammer Rd., Ithaca, NY 14850
jin.wang@autodesk.com; xiaoshi.jin@autodesk.com

Abstract - The *reduced strain closure* (RSC) and *anisotropic rotary diffusion* (ARD) models have been implemented in a research version of Autodesk Moldflow Insight software to improve the accuracy of fiber orientation predictions in simulations of injection-molded parts. The RSC model captures the slow fiber orientation kinetics, which are observed in experiments but over-predicted by the widely used Folgar-Tucker model. The ARD model accounts for fiber-fiber interactions using anisotropic diffusion to rectify the problem that for long-fiber materials, the Folgar-Tucker model does not match all aspects of measured fiber orientation data. Simulating the injection molding process in Autodesk Moldflow Insight software, the fiber orientation is predicted for a number of end-gated ISO plaques and center-gated disks of different thicknesses (from 1.5 to 6 mm) filled at different injection rates, with Midplane meshes as well as 3D meshes, using the RSC model for short-fiber materials and the ARD-RSC model for long-fiber materials. In Midplane models, with an appropriate inlet condition for fiber orientation applied at the gate, the RSC and ARD models show good agreement with measured fiber orientation data. In 3D models with random inlet orientation applied at the gate, the RSC model gives good agreement with one orientation component of the experimental data for short-fiber materials.

Introduction

Plastic materials are widely used as metal replacement in industrial applications owing to the advantages of higher strength-to-weight ratio and high corrosion resistance. Glass and carbon fibers are commonly added into plastic materials to reinforce the mechanical and thermal properties with negligible change of weight. The addition of fibers introduces anisotropy of the properties in injection-molded parts, mainly due to the preferential orientation of fibers induced by the flow during processing. For example, shearing flow, which is usually dominant during injection-molding processing of a thin part, tends to align fibers in the shearing direction. The part is much stronger and stiffer in the direction along which most fibers are aligned than in the direction crossing the major fiber alignment. In order to determine the properties in injection-molded parts and aid the design of mold and part and the selection of processing conditions, the fiber orientation must be accurately predicted.

Currently, most programs and commercial software, including Autodesk Moldflow Insight, use the Folgar-Tucker model (Folgar and Tucker, 1984) to predict the fiber orientation. The Folgar-Tucker model was developed based on Jeffery's equation (1922) for the motion of a single rigid fiber in an infinite Newtonian fluid by adding an isotropic rotary diffusion to represent interactions between fibers, and then the equation was recast in the term of the second-order orientation tensor for affordable computations.

However, recent experiments, including measurements of fiber orientation in injection-molded parts (O'Gara, J. F., personal communication, 2003) as well as measurements of rheology of fiber suspension (Sepehr et al., 2004), indicate that the Folgar-Tucker model predicts much quicker fiber orientation change with respect to strain than has been observed, and thus, usually over-predicts the overall fiber alignment. Wang et al. (2008) modified the Folgar-Tucker model by reducing the growth rates of the eigenvalues of the fiber orientation tensor by a constant scalar factor but keeping the rotation rates of the eigenvectors unchanged and developed the *reduced strain closure* (RSC) model. The RSC model exhibits slower orientation kinetics than the Folgar-Tucker model but exhibits a similar steady-state orientation and also provides excellent agreement with experimental data (Wang, 2007). United States patent No. 7,266,469 (Tucker et al., 2007) is held on the RSC model by Delphi Automotive LLP, and the model is used under license in this work.

Fibers longer than 1 mm are generally considered as long fibers. Usually, the fiber alignment in the flow direction is smaller in long-fiber materials than in short-fiber materials in injection-molded parts (Phelps and Tucker, 2009). The isotropic diffusion used in the Folgar-Tucker and RSC models is unable to capture the behavior of fiber-fiber interactions in long-fiber materials and cannot accurately predict all fiber orientation components simultaneously. The isotropic diffusion was replaced with the *anisotropic rotary diffusion* (ARD), which is defined on the surface of the unit sphere traced by all orientations of the unit vector, developed by Phelps et al. (2009). The RSC model can also be combined with the ARD model.

Both the RSC and ARD models were implemented in a research version of Autodesk Moldflow Insight software and were used to predict fiber orientation in injection-molded parts. The fiber orientation results were compared with the measured data.

Fiber Orientation Models

A concise description of fiber orientation is the second-order orientation tensor (Advani and Tucker, 1987), which is defined as

$$\mathbf{A} = \langle \mathbf{pp} \rangle , \qquad (1)$$

where the unit vector **p** directs along the fiber length, and the bracket "⟨ ⟩" is the average over a volume domain. The trace of **A**, tr**A**, is unity due to the unit length of **p** and the normalization condition.

The Folgar-Tucker model in the term of the orientation tensor **A** is

$$\frac{D\mathbf{A}}{Dt} = (\mathbf{W}\cdot\mathbf{A} - \mathbf{A}\cdot\mathbf{W}) + \xi(\mathbf{D}\cdot\mathbf{A} + \mathbf{A}\cdot\mathbf{D} - 2\mathbb{A}{:}\mathbf{D}) + 2C_I\dot{\gamma}(\mathbf{I} - 3\mathbf{A}). \qquad (2)$$

Here, $D\mathbf{A}/Dt$ is the material derivative of **A**; **W** and **D** are the vorticity and the rate-of-deformation tensors, respectively; ξ is the particle shape parameter and is very close to unity due to the large length-to-radius ratio of fibers; C_I is the interaction coefficient, and a larger C_I implies more fiber-fiber interactions; and $\dot{\gamma}$ is the scalar magnitude of **D**. $\mathbb{A}$ is the fourth-order orientation tensor, defined as $\mathbb{A} = \langle \mathbf{pppp} \rangle$, and is approximated as a closure function of components of **A**.

In Autodesk Moldflow Insight software, for Midplane and Dual Domain meshes, the Folgar-Tucker model is further revised by a thickness moment of interaction coefficient D_z ($0 \leq D_z \leq 1$) (Autodesk Moldflow Insight help topic "Fiber orientation models"), and therefore Moldflow's orientation model is

$$\frac{D\mathbf{A}}{Dt} = (\mathbf{W}\cdot\mathbf{A} - \mathbf{A}\cdot\mathbf{W}) + \xi(\mathbf{D}\cdot\mathbf{A} + \mathbf{A}\cdot\mathbf{D} - 2\mathbb{A}{:}\mathbf{D}) + 2C_I\dot{\gamma}(\mathbf{I} - (2 + D_z)\mathbf{A}). \qquad (3)$$

Here, a smaller D_z implies less out-of-plane orientation. However, if $D_z < 1$, $D(\mathrm{tr}\mathbf{A})/Dt \neq 0$, given by Eq. (3). Thus, among the diagonal components of **A**, two in-plane components are calculated by Eq. (3), and the out-of-plane component is calculated by enforcing tr**A** = 1.

The RSC model introduces a scalar factor $\kappa < 1$ to slow the fiber orientation kinetics, and the equation is

$$\frac{D\mathbf{A}}{Dt} = (\mathbf{W}\cdot\mathbf{A} - \mathbf{A}\cdot\mathbf{W}) + \xi(\mathbf{D}\cdot\mathbf{A} + \mathbf{A}\cdot\mathbf{D} - 2[\mathbb{A} + (1-\kappa)(\mathbb{L} - \mathbb{M}{:}\mathbb{A})]{:}\mathbf{D}) + 2\kappa C_I\dot{\gamma}(\mathbf{I} - 3\mathbf{A}). \qquad (4)$$

Here, the fourth-order tensors $\mathbb{L}$ and $\mathbb{M}$ are functions of the eigenvalues λ_i (i = 1,2,3) and the eigenvectors $\mathbf{e}_i$ (i = 1,2,3) of the orientation tensor **A**, defined as $\mathbb{L} = \sum_{i=1}^{3} \lambda_i \mathbf{e}_i \mathbf{e}_i \mathbf{e}_i \mathbf{e}_i$ and $\mathbb{M} = \sum_{i=1}^{3} \mathbf{e}_i \mathbf{e}_i \mathbf{e}_i \mathbf{e}_i$. The scalar factor κ is a phenomenological parameter, and its value is determined by fitting the fiber orientation or rheology prediction to experimental data. Setting $\kappa = 1$ reduces the RSC model of Eq. (4) to the Folgar-Tucker model of Eq. (2).

The ARD model replaces the isotropic diffusion term with an anisotropic diffusion and the RSC version of the ARD model (ARD-RSC) is

$$\begin{aligned}\frac{D\mathbf{A}}{Dt} = {} & (\mathbf{W}\cdot\mathbf{A} - \mathbf{A}\cdot\mathbf{W}) + \xi(\mathbf{D}\cdot\mathbf{A} + \mathbf{A}\cdot\mathbf{D} \\ & - 2[\mathbb{A} + (1-\kappa)(\mathbb{L} - \mathbb{M}{:}\mathbb{A})]{:}\mathbf{D}) \\ & + \dot{\gamma}(2[\mathbf{C} - (1-\kappa)\mathbb{M}{:}\mathbf{C}] - 2\kappa(\mathrm{tr}\mathbf{C})\mathbf{A} \\ & - 5(\mathbf{C}\cdot\mathbf{A} + \mathbf{A}\cdot\mathbf{C}) \\ & + 10[\mathbb{A} + (1-\kappa)(\mathbb{L} - \mathbb{M}{:}\mathbb{A})]{:}\mathbf{C}).\end{aligned} \qquad (5)$$

Here, the rotary diffusion tensor **C** is constructed from **A** and **D** as (Phelps and Tucker, 2009)

$$\mathbf{C} = b_1\mathbf{I} + b_2\mathbf{A} + b_3\mathbf{A}^2 + b_4\frac{\mathbf{D}}{\dot{\gamma}} + b_5\frac{\mathbf{D}^2}{\dot{\gamma}^2}, \qquad (6)$$

where b_i (i = 1, …, 5) are scalar constants and selected by matching experimental steady-state orientation and requiring stable orientation. Setting $b_1 = C_I$ and $b_i = 0$ (i = 2, …, 5) reduces the ARD-RSC model of Eq. (5) to the RSC model of Eq. (4).

Experiments

A series of end-gated ISO plaques and center-gated disks were molded by Delphi Automotive LLP and the Pacific Northwest National Laboratory (PNNL) and the Oak Ridge National Laboratory (ORNL), to obtain the measured fiber orientation data which were compared with simulation results. The ISO plaques are 90 mm long and 80 mm wide, and the disks have a diameter of 177.8 mm. Delphi used a polybutylene terephthalate reinforced by 30% wt short glass fibers (General Electric Plastics Valox 420) and molded parts in thicknesses of 1.5, 2, 3, and 6 mm, filled at three different injection rates (slow, medium, and fast) for each thickness. PNNL-ORNL used a polypropylene with 40% wt glass fibers to mold plaques and disks of 3 mm thick, filled at two different injection rates (slow and fast) for each geometry, and the fibers have a nominal length of 13 mm and are considered as long fibers.

Fiber orientation through the thickness was measured by Delphi and PNNL-ORNL for their respective moldings using the optical system developed by Hine et al. (1996). Measurements were made at selected locations on the polished cross-section cut along the centerline of plaques and the radial direction of disks. Further details of the sample preparation and the orientation measurements can be found in O'Gara et al. (2003) and Nguyen et al. (2008).

Simulations

The research version of Autodesk Moldflow Insight software with the implementation of the RSC and ARD models was then used to simulate the injection molding process and predict the fiber orientation in these plaques and disks, with Midplane meshes as well as 3D meshes. The models and meshes for selected parts are illustrated in Figs. 1–4.

For Delphi parts molded with a short-fiber material, we applied the RSC model and used the scalar factor κ = 0.05, the same value which produces good agreement with experimental data in Wang (2007). For PNNL-ORNL parts molded with a long-fiber material, we applied the ARD-RSC model and used the same scalar factor κ = 1/30 and the same b_i parameters as suggested by Phelps and Tucker (2009).

As an essential nature of the RSC model, the development of fiber orientation is slowed down compared to the Folgar-Tucker model, and hence the predicted fiber orientation undergoes less change along the flow length of the part and is more strongly influenced by the inlet condition for fiber orientation at the gate. An accurate inlet orientation is therefore crucial to the accurate prediction of fiber orientation in the part. In the research version of the Autodesk Moldflow Insight Midplane solver, we added an option to specify the inlet orientation across the thickness at the gate. For the Midplane simulations presented in this paper, the fiber orientation measured near the gate is used as the inlet orientation for Delphi parts, and the fiber orientation similar to the measured orientation is used as the inlet orientation for PNNL-ORNL parts. However, in a 3D mesh, there is no generic definition of the thickness direction and no option for the inlet orientation, hence a random orientation is used at the gate in the 3D simulations.

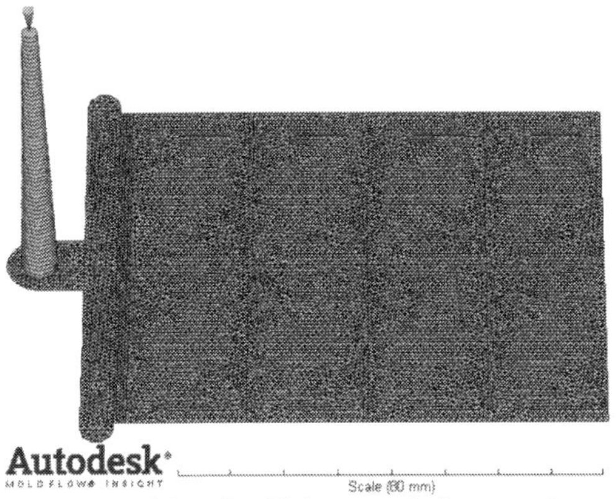

Figure 1. Model and Midplane mesh for an end-gated ISO plaque. The mesh size is about 1 mm.

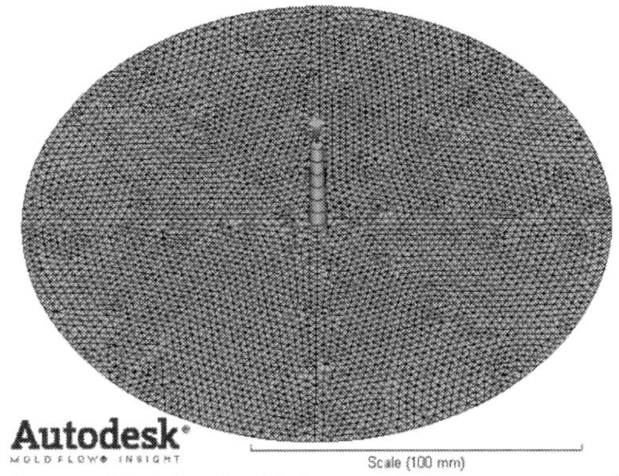

Figure 2. Model and Midplane mesh for a center-gated disk. The mesh size is about 2.47 mm.

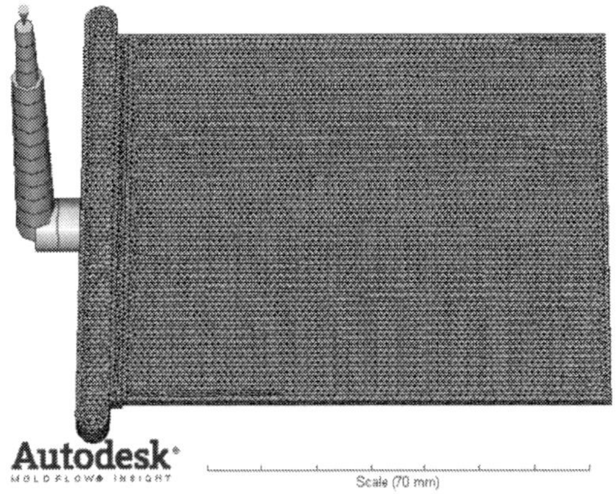

Figure 3. Model and 3D mesh for a 1.5 mm thick, end-gated ISO plaque. The surface mesh size is about 1.15 mm, and there are 12 mesh layers through thickness.

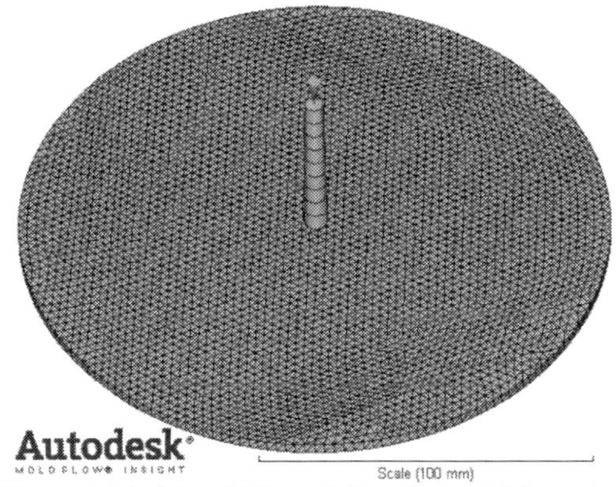

Figure 4. Model and 3D mesh for a 2 mm thick, center-gated disk. The surface mesh size is about 2.8 mm, and there are 12 mesh layers through thickness.

Results and Discussion

The orientation predictions by Autodesk Moldflow Insight are compared with measured fiber orientations in Figs. 5–10. Here, A_{11} denotes the component of the fiber orientation tensor in the flow direction, and A_{22} denotes the component in the cross-flow direction.

The fiber orientation data in plaques and disks exhibits the typical shell-core-shell structure through the thickness. The A_{11} component is small near the center, which is referred to as the core. At the core, A_{11} for plaques is similar to the inlet orientation due to low shearing in the core, while for disks it is transversely aligned due to strong stretching. The A_{11} component is large on both sides of the core, which are referred to as the shell layers. In the shell layers, fibers usually have strong alignment due to high shearing near the mold surface.

The Folgar-Tucker model over-predicts the fiber orientation change and hence gives a much narrower core than observed in experiments. Moldflow's D_z model attempts to broaden the core by decreasing the out-of-plane orientation, but the improvement is limited. Neither Moldflow's D_z model nor the Folgar-Tucker model is able to predict adequately the correct width of the core, as demonstrated in Figs. 5 and 6 and Figs. 7 and 8 for Midplane and 3D meshes, respectively, for Delphi plaques.

Using the RSC model with Midplane meshes for the plaques and disks models accurately captures the core orientation and width, and also predicts the orientation in the shell layers reasonably well. The comparisons of the predicted fiber orientation with data are plotted in Figs. 5 and 6.

For 3D meshes of Delphi plaque and disk moldings, the RSC model gives an excellent fit to the A_{11} component through the thickness, but gives a poor fit to the A_{22} component, particularly in the core, as shown in Figs. 7 and 8. This poor prediction for the A_{22} component is probably due to the random orientation given at the gate. The A_{22} component at the center is always close to 1/3, which is almost the same as the inlet orientation. A more appropriate inlet orientation is therefore required in order to produce a good prediction.

The predictions by the RSC and the ARD-RSC models are compared with experimental data in Figs. 9 and 10 for a PNNL-ORNL plaque and a disk, respectively, molded with a long-fiber material. The RSC model is able to match the A_{11} component in the shell layers, but under-predicts the A_{22} component. Furthermore, adjusting the C_I value can produce a good match to either one component or the other but not to both simultaneously. The ARD-RSC model shows good agreement with the measured A_{11} and A_{22} components for both plaque and disk.

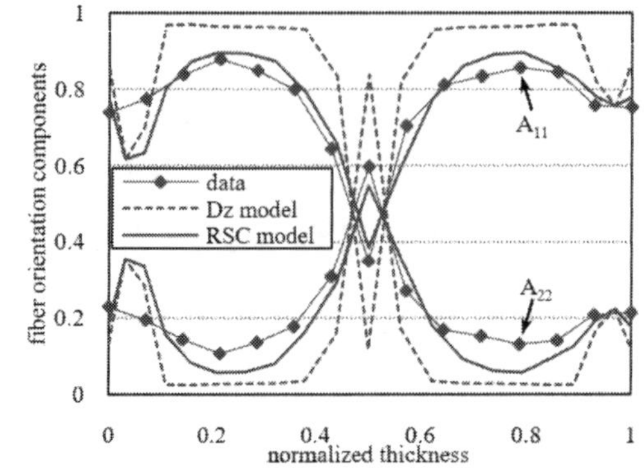

Figure 5. Comparison of fiber orientation data with Autodesk Moldflow Insight Midplane predictions in the Delphi 2 mm thick plaque filled at a slow injection rate. Moldflow's D_z model and the RSC model are used. The C_I and D_z values are automatically calculated by Autodesk Moldflow Insight, and $\kappa = 0.05$.

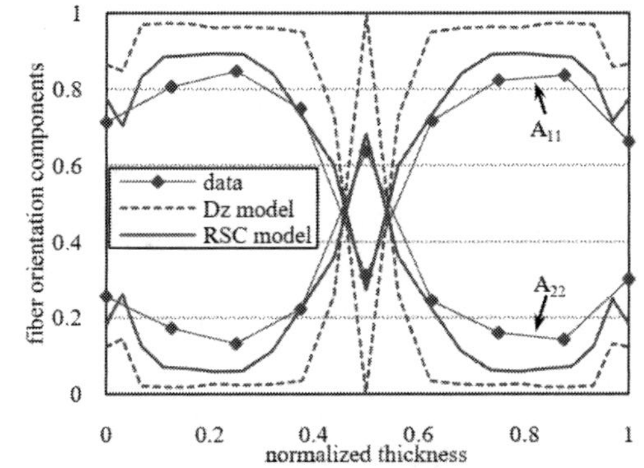

Figure 6. Comparison of fiber orientation data with Autodesk Moldflow Insight Midplane predictions in the Delphi 1.5 mm thick disk filled at a slow injection rate. Moldflow's D_z model and the RSC model are used. The C_I and D_z values are automatically calculated by Autodesk Moldflow Insight, and $\kappa = 0.05$.

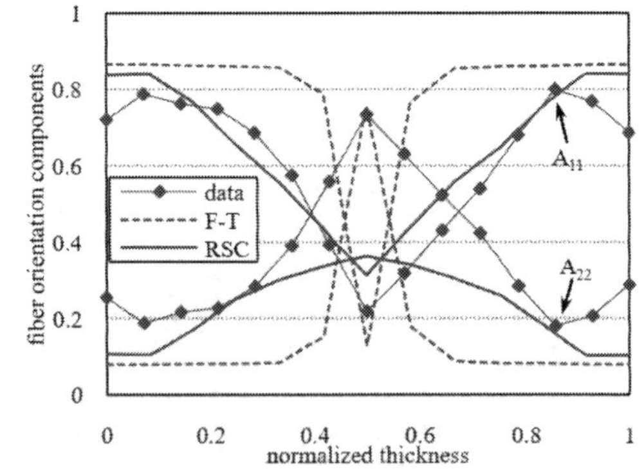

Figure 7. Comparison of fiber orientation data with Autodesk Moldflow Insight 3D predictions in the Delphi 3 mm plaque filled at a fast injection rate. The Folgar-Tucker and RSC models are used. The C_I value is automatically calculated by Autodesk Moldflow Insight, and $\kappa = 0.05$.

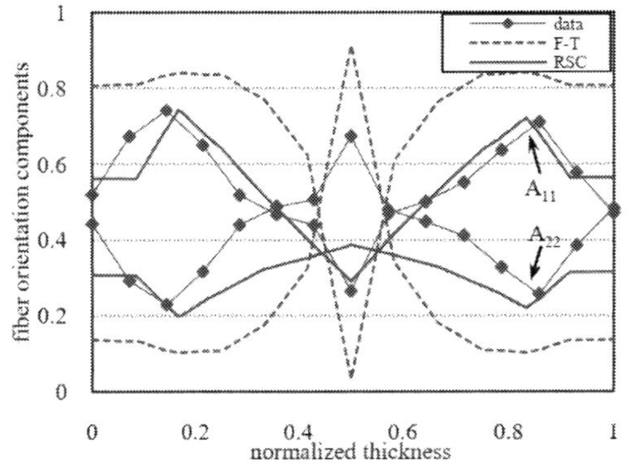

Figure 8. Comparison of fiber orientation data with Autodesk Moldflow Insight 3D predictions in the Delphi 6 mm disk filled at a slow injection rate. The Folgar-Tucker and RSC models are used. The C_I value is automatically calculated by Autodesk Moldflow Insight, and $\kappa = 0.05$.

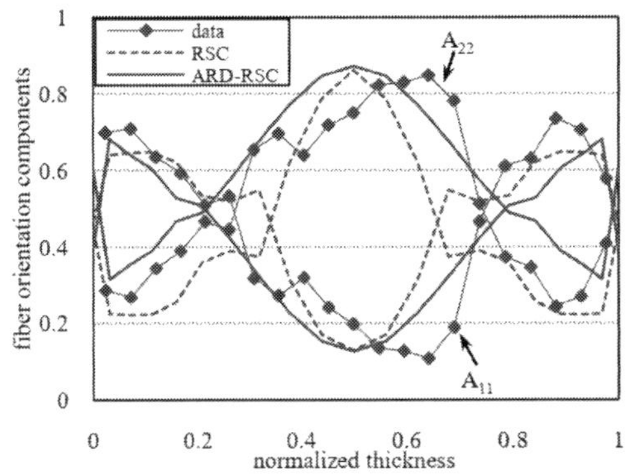

Figure 9. Comparison of fiber orientation data with Autodesk Moldflow Insight Midplane predictions in the PNNL-ORNL 3 mm thick plaque filled at a fast injection rate. The RSC and the ARD-RSC models are used. $C_I = 0.03$ for RSC, and $\kappa = 1/30$ for RSC and ARD-RSC.

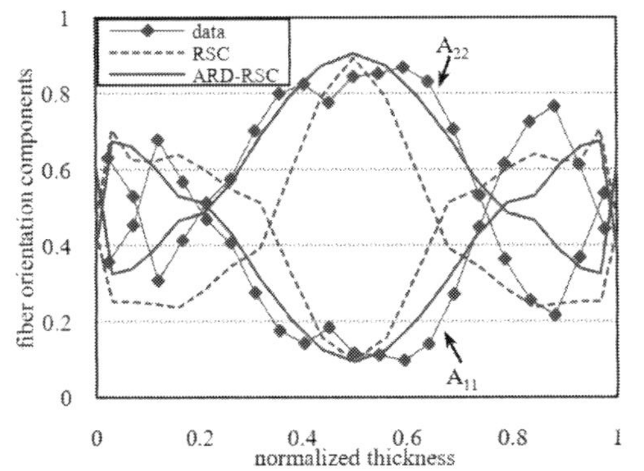

Figure 10. Comparison of fiber orientation data with Autodesk Moldflow Insight Midplane predictions in the PNNL-ORNL 3 mm thick disk filled at a slow injection rate. The RSC and the ARD-RSC models are used. $C_I = 0.03$ for RSC, and $\kappa = 1/30$ for RSC and ARD-RSC.

Conclusions

Two recent orientation models, the RSC and ARD models, were successfully implemented in a research version of Autodesk Moldflow Insight software used for fiber orientation analysis. The program reasonably predicts the fiber orientation, and the prediction agrees well with the measured orientation data for a short-fiber material using the RSC model and for a long-fiber material using the ARD-RSC model. Overall, the RSC and ARD models are superior to the commonly used Folgar-Tucker model and the default Moldflow's D_z model implemented in Autodesk Moldflow Insight software available commercially.

The RSC model slows down the orientation kinetics, and the fiber orientation in a part strongly depends on the inlet orientation at the gate. A careful selection of the inlet orientation condition is required for an accurate orientation prediction in a part. How to specify the inlet orientation across thickness in a 3D mesh and how to automatically give a reasonable inlet orientation need to be addressed to make the RSC model more useful for general cases.

Acknowledgements

The authors would like to thank Delphi Automotive LLP, and the Pacific Northwest National Laboratory and the Oak Ridge National Laboratory for providing experimental measurements and CAD models of their molded parts. The work on long-fiber materials was conducted under the Cooperative Research and Development Agreement (CRADA) (Nr. 260) between Battelle as Operator of the Pacific Northwest National Laboratory and Autodesk, Inc.

References

Advani, S. G. and C. L. Tucker III. "The Use of Tensors to Describe and Predict Fiber Orientation in Short Fiber Composites." Journal of Rheology **31**: 751–784 (1987).

Folgar, F. and C. L. Tucker III. "Orientation Behavior of Fibers in Concentrated Suspensions." Journal of Reinforced Plastics and Composites **3**: 98–119 (1984).

Hine, P. J., N. Davidson, R. A. Duckett, A. R. Clarke, and I. M. Ward. "Hydrostatically Extruded Glass-Fiber Reinforced Polyoxymethylene. I. The Development of Fiber and Matrix Orientation." Polymer Composites **17**: 720–729 (1996).

Jeffery, G. B.. "The Motion of Ellipsoidal Particles Immersed in a Viscous Fluid." Proceedings of the Royal Society **A102**: 161–179 (1922).

Nguyen, B. N., V. Kunc, B. Frame, J. H. Phelps, C. L. Tucker III, S. K. Bapanapalli, J. D. Holbery, and M. T. Smith. "Fiber Length and Orientation in Long-Fiber Injection-Molded Thermoplastics. Part I. Modeling of Microstructure and Elactic Properties."

Journal of Composite Materials **42**: 1003–1029 (2008).
O'Gara, J. F., M. G. Wyzgoski, and G. Novak. "Development of an ISO Standard for Determining Anisotropic Properties of Glass-Filled Thermoplastics." SPE ANTEC Conference Proceedings **3**: 2883–2887 (2003).
Phelps, J. and C. L. Tucker III. "An Anisotropic Rotary Diffusion Model for Fiber Orientation in Short- and Long-Fiber Thermoplastics." Journal of Non-Newtonian Fluid Mechanics **156**(3): 165–176 (2009).
Sepehr, M., G. Ausias, and P. J. Carreau. "Rheological Properties of Short Fiber Filled Polypropylene in Transient Shear Flow." Journal of Non-Newtonian Fluid Mechanics **123**: 19–32 (2004).
Tucker III, C. L., J. Wang, and J. F. O'Gara. "Method and Article of Manufacture for Determining a Rate of Change of Orientation of a Plurality of Fibers Disposed in a Fluid." U.S. Patent No. 7,266,469 (2007).
Wang, J.. "Improved Fiber Orientation Predictions For Injection-molded Composites." Ph.D Thesis, University of Illinois at Urbana-Champaign, Urbana, IL (2007).
Wang, J., J. F. O'Gara, and C. L. Tucker III. "An Objective Model for Slow Orientation Kinetics in Concentrated Fiber Suspensions: Theory and Rheological Evidence." Journal of Rheology **52**(5): 1179–1200 (2008).

THE WARPAGE AND BIREFRINGENCE AFFECTED BY MULTI-COMPONENT INJECTION MOLDING OF PURE/SHORT FIBER REINFORCED COMPOSITES

Yi-Cheng Chen[1], Chao-Tsai Huang[2], Yuan-Rong Chang[2] and Shi-Chang Tseng[1],

1:Dept. of Mechanical Engineering, National Yunlin University of Science & Technology-tsengsc@yuntech.edu.tw

2:CoreTech System (Moldex3D) Co., Ltd., ChuPei City, Hsinchu, Taiwan-cthuang@moldex3d.com

Abstract

Sequential multi-component molding is one of great methods to fabricate the modern injection products. Previous literatures have shown that various parameters including the product geometries and materials selection, will affect the warpage of final products significantly. However, how the interaction between the first shot and the second shot is not clear. In this study, various process-induced properties of the first shot are taken into consideration for the further processing. Results show that the process-induced properties can affect the warpage and optical property significantly. Results can further help us to discover the inside mechanism more comprehensively.

Introduction

In modern plastic molded product fabrication, multi-component molding (MCM) provides great methods to the injection molding development. In fact, the fundamental idea of MCM was first proposed and patented in 1962 regarding the development of multiple material tailing light by G. Carozzo [1]. During the past decades, many new technologies, methods, and related material combination have been proposed [2-4]. In the recent years, people have tried to realize the physical mechanism and to handle this complicated process to be more understanding. In our previous studies, we have focused on the geometrical effect, material combination, and the physical mechanism. However, some physical properties, such as fiber orientation effect, birefringence, and so on, are still not clear comprehensively [7-10]. Due to its complicated nature for MCM processes, a conventional trial-and-error method cannot catch the crucial factors effectively. Hence, in this study, firstly, we have applied CAE tool to simulate MCM process to get better understanding. Specifically, we have conducted the Process-Induced Property Variation (PIPV) on the warpage and optical property for a U-type injection parts in MCM. Results show PIPV (for example, fiber orientation effect) in the 1st shot will affect the final product quality significantly. Furthermore, the PIPV can also influence the optical property of parts.

Theory Approach and Assumption

The major analysis procedures for injection molding processes include filling, package, cooling, and warpage. In filling/packing stages, the polymer melt is assumed to behave as Generalized Newtonian Fluid (GNF). Hence the non-isothermal 3D flow motion can be mathematically described. The FVM (finite volume method), due to its robustness and efficiency, is employed in this study to solve the transient flow field in complex three-dimensional geometries. During the mold cooling stage, a three-dimensional, cyclic, transient heat conduction problem with convective boundary conditions on the cooling channel and mold base surfaces is involved. The basic mathematical models are described in detailed in Ref. [5-7].

To handle the viscoelastic behavior of polymeric melts, τ is obtained by the constitutive equation of White Metzner Model:

$$\tau+\lambda\left(\frac{\partial\tau}{\partial t}+V\cdot\nabla\tau-\nabla V^{T}\cdot\tau-\tau\cdot\nabla V\right)=\eta\left(\nabla V+\nabla V^{T}\right) \quad (1)$$

$$\eta\left(T,\dot{\gamma}\right)=\eta_0\left(T_f\right)\cdot a\left(T,\dot{\gamma}\right) \quad (2)$$

$$\lambda\left(T,\dot{\gamma}\right)=\lambda_0\left(T_f\right)\cdot a\left(T,\dot{\gamma}\right) \quad (3)$$

$$a\left(T,\dot{\gamma}\right)=\frac{\eta\left(T,\dot{\gamma}\right)}{\eta_0} \quad (4)$$

where λ is the relaxation time, λ_o is the relaxation time at zero shear rate, η is the shear viscosity, η_o is the shear viscosity at zero shear rate, and a is shift factor.

The fiber orientation state at each point in the part is represented by a 2nd-order orientation vector A,

$$A_{ij}=\int(p_ip_j)\varphi(p)dp \quad (5)$$

The equation of orientation change for the orientation tensor proposed by Advani and Tucker is employed for the analysis,

$$\frac{\partial A_{ij}}{\partial t}+u_k\frac{\partial A_{ij}}{\partial x_k}=A_{ik}\Omega_{kj}-\Omega_{ik}A_{kj}+$$

$$\lambda(A_{ik}E_{kj} + E_{ik}A_{kj} - 2A_{ijkl}E_{kl}) + 2C_{I}\dot{\gamma}(\delta_{ij} - 3A_{ij}) \quad (6)$$

Where C_I is the interaction coefficient with the value ranged from 10^{-2} to 10^{-3}. For the fourth-order tensor A_{ijkl}, a closure approximation is needed.

In addition, to handle the MCM process, the molded part (the previous shot) will be setup into the cavity to become the boundary for the next shot. The geometry of 1st shot is assumed as a perfect U-type shape. The overall heat transfer phenomenon is governed by a three-dimensional Poisson equation. We assume there is a cycle-averaged mold temperature that is invariant with time. The warpage analysis assumes the mechanical properties are linear elasticity. The stress-strain equilibrium equations enable us to solve the problems. Also, since the materials used in this study are all PC, the interfacial adhesion between sequential shots is assumed perfect.

Furthermore, to predict the optical property in plastics, the mathematical models and equations, including Stress-Optics (or Brewster's Law) for birefringence, retardation definition, and fringe order prediction, are described in detailed in our previous paper (please see Ref. [7]).

Investigation Procedures

A. Numerical Study

To get better understanding of PIPV effects on the property of parts in MCM processes, we have conducted using Moldex3D® R9.1 software numerically. The geometrical model which includes runner system and dimensions is as shown in Fig. 1 and table 1. The dimension of 1st shot, or insert, is 19mm X 9.5mm X 2mm. The dimension of 2nd shot is (19+2**x**) mm X (9.5+2**x**) mm X 2mm, where **x** is 0.5, 1, and 2 mm. The materials of 1st shot or 2nd shot are PC (means pure PC without glass fiber filled) or PC+GF (means PC with glass fiber filled). Specifically, PC is TEIJIN_Panlite L-1250Y, and PC+GF is TEIJIN_Panlite G-3430R. Under different material combination of 1st shot and 2nd shots, when the width of 2nd shot varies, the injection molding processing for 2nd shot will be effected from filling to warpage performance.

B. Experimental Investigation

To catch PIPV on the warpage behavior and optical property, we also performed the experimental investigation. The testing model and its dimensions are described in Fig. 1 and Table 1. The main injection molding procedures are exactly same as those of numerical study described earlier. The only difference is that the sample of 1st shot has slightly warped before installed into mold for the 2nd shot. Also, the verification of numerical results is performed based on those experimental data.

Results and Discussions

Firstly, we investigate the fiber orientation effect on the 1st shot as shown in Fig. 2 and 3. In Fig. 2, the fiber is aligned along the flow direction. This feature will affect the warpage behavior in Fig. 3. The displacement could be lower down from 0.075 mm to 0.062 mm for the 1st shot part. Furthermore, we also conducted how the 1st shot PC+GF influences the final products. In Fig. 4 (a), the 1st and 2nd shots are all PC. The warpage amount is about 0.1405 mm. Fig. 4 (b) shows that the warpage amount is lower down to 0.139 mm when the 1st shot is PC+GF. Fig. 4 (c) shows that the warpage amount is further lower down to 0.085 mm when the 1st shot and 2nd shots are all PC+GF. Therefore, both fiber filled in either 1st shot or/and 2nd shots can enhance part strength to overcome warpage defect.

To analyze warpage behavior quantitatively, we defined the **inward** and **outward** phenomena as shown in Fig. 5. It defines the length of one side as A and of the other side as B. The warpage is regarded as **outward** when A is greater B and contrarily regarded as **inward** when B is greater than A. For example, in the presence of runner system, after completed injection molding, the U-type parts can be examined. The single shot shows inward warpage in Fig. 6, while the two-shot overmolding is also completed and measured as shown in Fig. 7. Obviously, when the width of 2nd shot is 2 mm, the warpage of the whole part displays outward. Moreover, to predict the PIPV with PC+GF, we have systematically conducted for various geometrical models from 0.5 mm to 2.0 mm in width. In Fig. 8, when we increase the width from 0.5 mm to 2.0 mm for PC/PC system, the warpage tendency will from inward to outward. The transition region is around 0.75 mm in width. It is the same as we obtained in our previous study [7]. Similarly, when we consider the 1st shot with PC+GF and 2nd shot with PC, the trend is almost same as PC/PC system, except the warpage amount is smaller. However, when we consider both shots with fiber filled, there us no inward region happened, although the warpage tendency is still the same.

Moreover, to verify the numerical results, the experimental study is executed. Fig. 9 shows the comparison of warpage behavior between experimental and numerical analyses where 1st shot material is PC/GF, and 2nd shot is PC. It is clear that both numerical and experimental results are matched with the outward feature qualitatively. Fig. 10 shows the warpage tendency from numerical analyses and experimental studies for (1st shot PC +2nd shot PC) materials. Clearly, when the width of 2nd shot keeps increased, the outward warpage tendency is increased. Both of numerical and experimental studies are in a good agreement. Furthermore, when we consider the flow-induced fiber orientation effects, the warpage tendency for (1st shot PC+GF and 2nd shot PC) and (1st shot PC+GF and 2nd shot PC+GF) materials are also in a reasonable agreement for both numerical and experimental studies as shown in Fig. 11 and Fig. 12.

In addition, we also focused on how the PIPV affect the optical characteristics of injection parts. Fig. 13 shows the fringed pattern for the single shot. Obviously, the fiber orientation effect will influence the optical property (shown as fringed pattern) significantly. Fig. 14 displays the fringed pattern for the 2nd shot when the 1st shot is either PC or PC+GF. From Fig. 14 (a) and (b), it shows that when the 2nd shot is PC, the fringed pattern looks similar for both 1st shot with and without fiber filled systems. The difference is only around region (3) to (5). However, this fringed pattern becomes very different when the 2nd shot is PC+GF as shown in Fig. 14 (c).

To realize the PIPV on the optical property along the geometrical structure, we have conducted various material combinations and focused on the melt flows from upstream (line A) to downstream (line D) as shown in Fig. 15. In Fig. 16 (a), it is clear when the 2nd shot is PC+GF (i.e., PC+GF/PC+GF system) flowing through Line A, the tendency and amount of the birefringence is very different with others. The maximum value of birefringence is 2 to 3 times to other systems. Also, when the 2nd shot is PC+GF (i.e., PC+GF/PC+GF system) flowing down to Line B, the difference is significantly.

Conclusion

In this study, we have investigated the process-induced property variations (PIPV) effects on the warpage and optical features in MCM products. In the presence of PIPV (for example, fiber orientation effect), either in 1st shot or 2nd shot, the warpage and optical property of the final product can be influenced significantly. To verify this investigation, numerical and experimental studies have been performed. Results showed that both numerical and experimental studies are in a very good agreement in various material combination systems.

References

1. US patent #3,051,994.
2. Multi-material Technology, Battenfeld.
3. Web source http://www.engelmachinery.com
4. Chao-Tsai Huang et. al, SPE ANTEC, 1888 (2006).
5. R. Y. Chang and W. H. Yang, Intern. J. Num. Meth. in Fluids, 37: 125 (2001).
6. Rong-Yeu Chang et. al, SPE ANTEC, 496 (2004).
7. Yuan-Rong Chang, et. al, SPE ANTEC, 2490 (2007).
8. Chao-Tsai Huang et. al, SPE ANTEC, 761 (2007).
9. A. I. Isayev and T.H. Lin, SPE ANTEC, 2485 (2007).
10. G. D. Shyu, A. I. Isayev and T.H. Lin, J. Polym. Sci., Polym. Phy., Vol. 39, 2252 (2001).

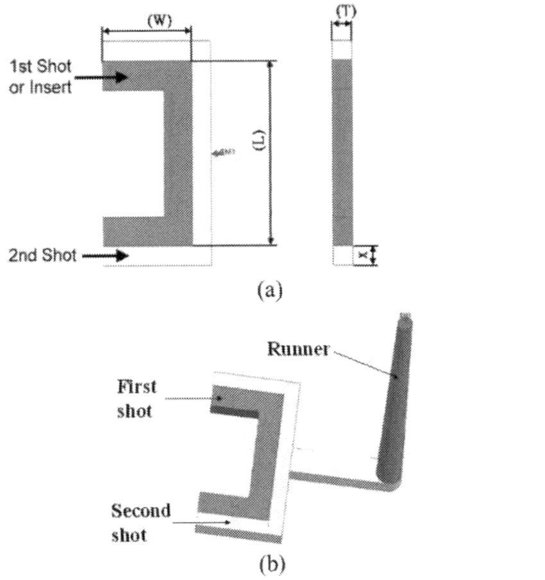

Fig. 1 System construction: (a) the model structure and dimension, (b) the runner structure.

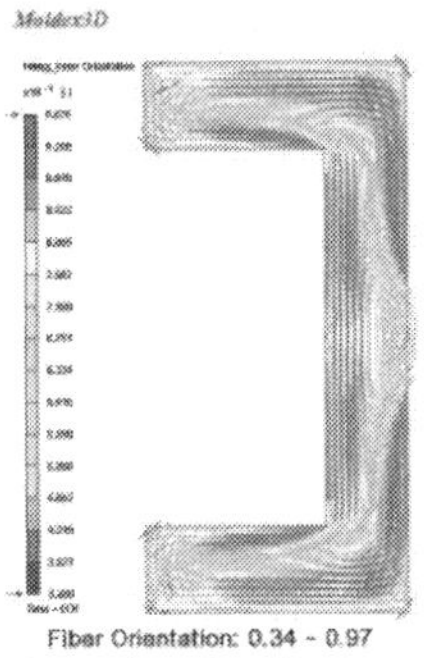

Fig. 2 The fiber orientation distribution of the 1st shot.

Table. 1 Geometrical structure and material selection

Geometrical Structure	Dimension in shot$^{\alpha}$		Material in shot		Injection process focused$^{\beta}$
	1st	2nd	1st	2nd	1st & 2nd shot
U type	L=19 W=9.5 T=2 (mm)	L=19+2a W=9.5+a T=2 a*=0.5~2 (mm)	PC	PC	F/P/C/W
			PC+GF	PC	

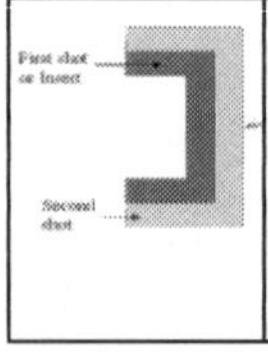			PC+GF	PC+GF	

α: L is length, W is width, and T is thickness.
β: F is filling; P is packing; C is cooling; W is warpage.
PC: TEIJIN_Panlite L-1250Y (PC is pure material without GF filled).
PC+GF: TEIJIN_Panlite G-3430R (with 30% GF); (PC+GF means PC with GF filled).
GF: means glass fiber.

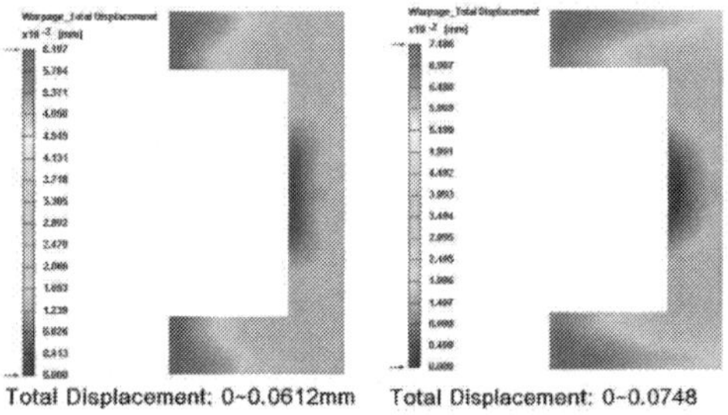

(a) PC+GF (b) PC

Fig. 3 The warpage behavior for the 1st shot with or without fiber.

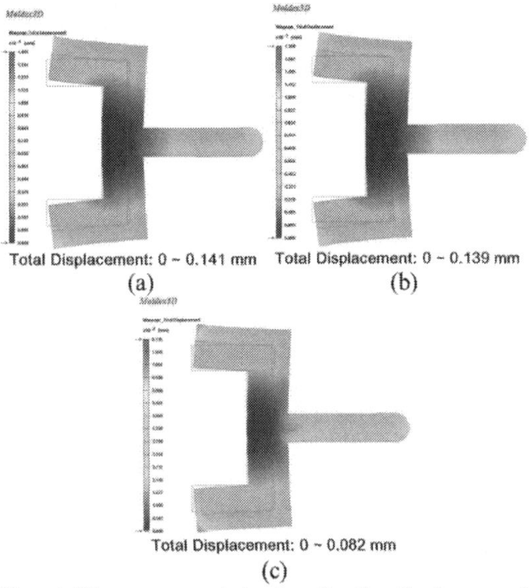

(c)

Fig. 4 The warpage behavior for the final parts: (a) PC/PC, (b) PC+GF/PC, (c) PC+GF/PC+GF; where PC+GF means PC with glass fiber filled, and PC means pure PC material without glass fiber filled.

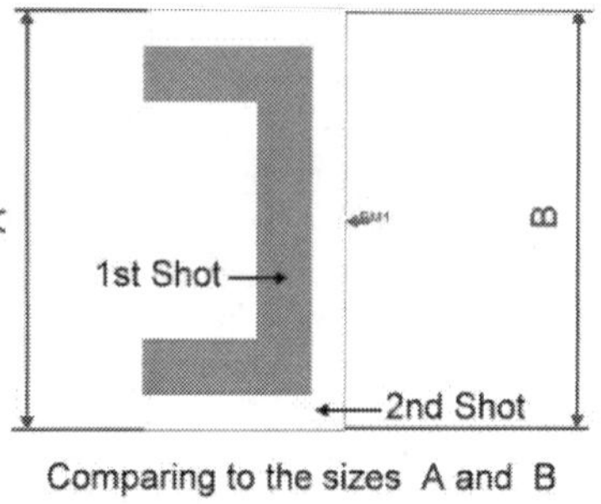

Fig. 5 Warpage behaviors definition for Inward or Outward from experiment study: (a) A>B is Outward, (b) A<B is Inward.

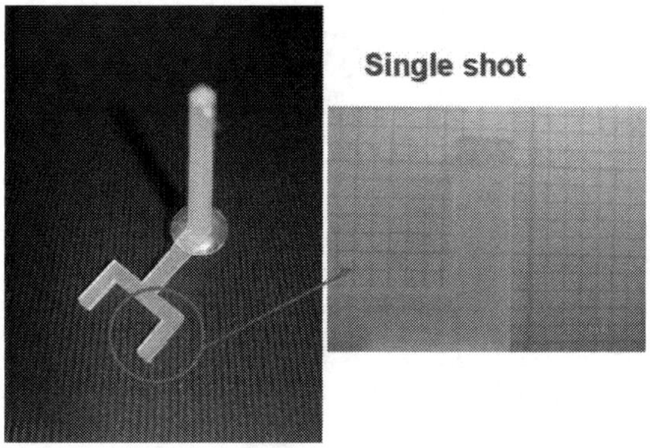

Fig. 6 Warpage behavior for a single shot: experimental result shows that it is inward.

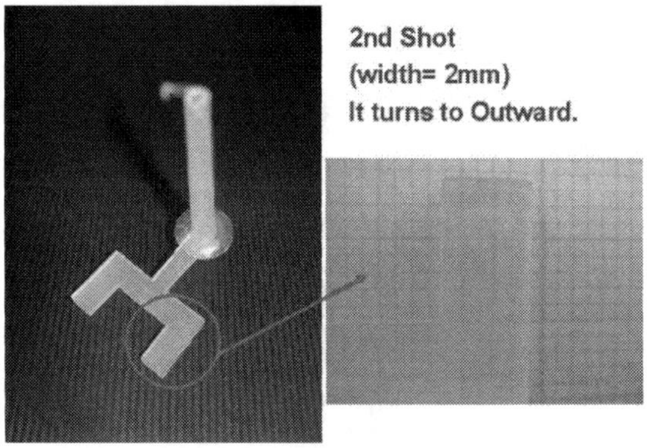

Fig. 7 Warpage behavior for a MCM with two shots: the experimental result shows that it is outward.

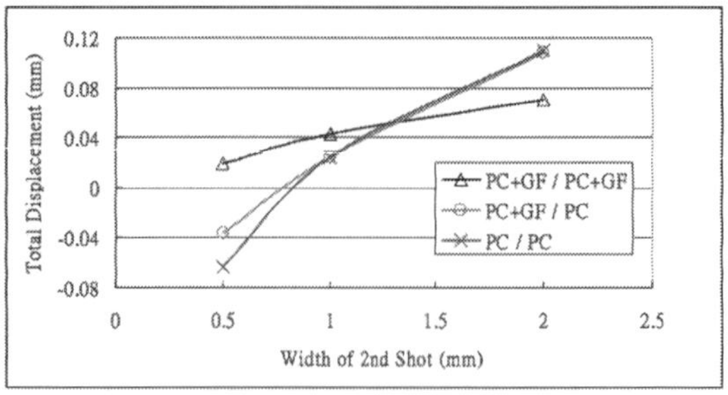

Fig. 8 Warpage tendency results from numerical

simulation for three material combinations.

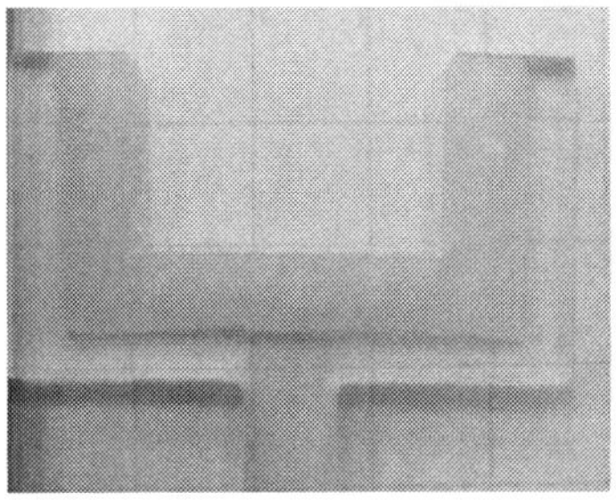

(a)

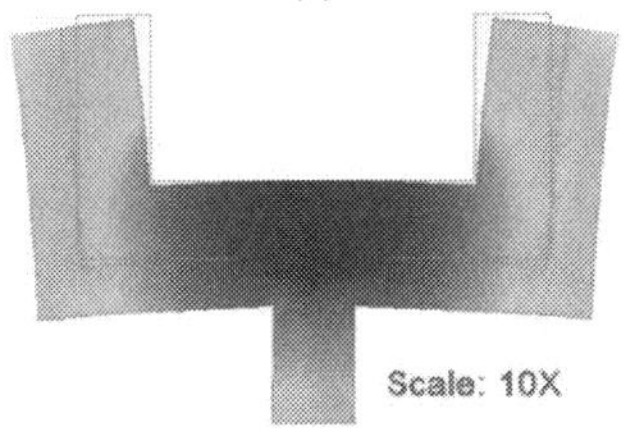

(b)

Fig. 9 The comparison of warpage behavior between experimental and numerical analyses; 1st shot material is PC/GF, and 2nd shot is PC.

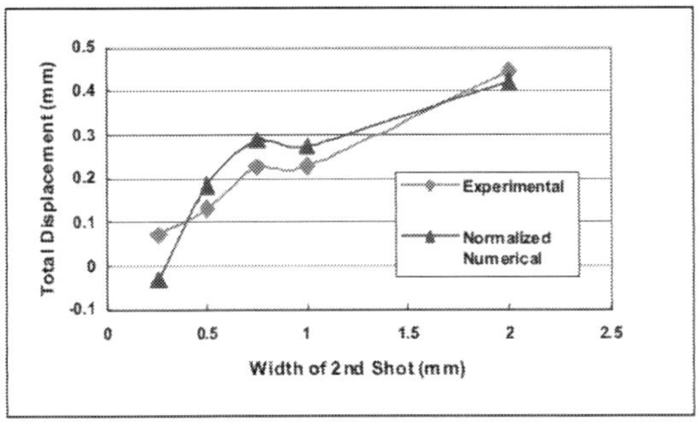

Fig. 10 Warpage tendency investigation using numerical analysis and experimental study in different 2nd shot width in the 1st shot material: 1st shot material is PC, and 2nd shot is PC.

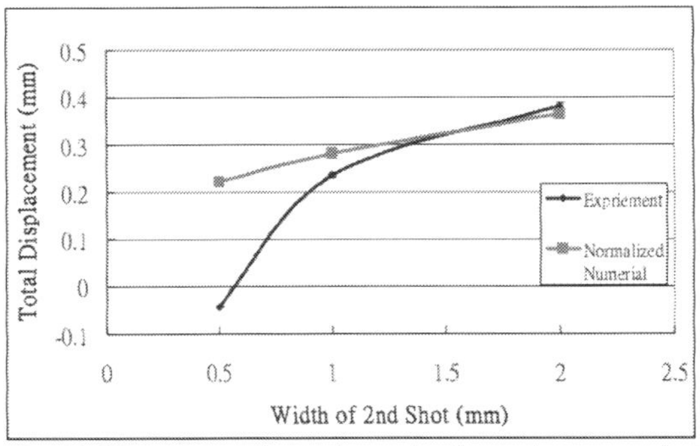

Fig. 11 Warpage tendency investigation using numerical analysis and experiment study in different 2nd shot width in the 1st shot material: 1st shot material is PC+GF, and 2nd shot is PC.

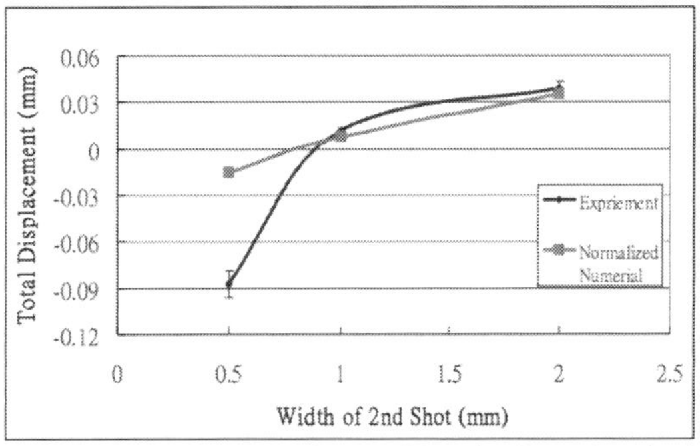

Fig. 12 Warpage tendency investigation for both numerical analysis and experiment study in different 2nd shot width in the 1st shot material: 1st shot material is PC+GF, and 2nd shot is PC+GF.

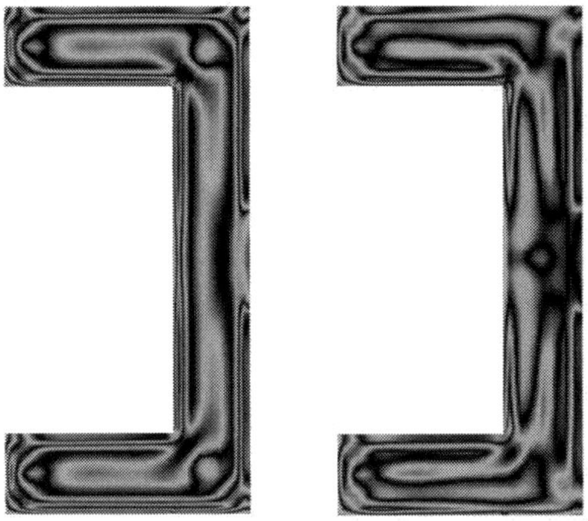

(a) PC (b) PC+GF

Fig. 13 Fringed pattern for the single shot (1st shot): (a) material is PC, (b) material is PC+GF.

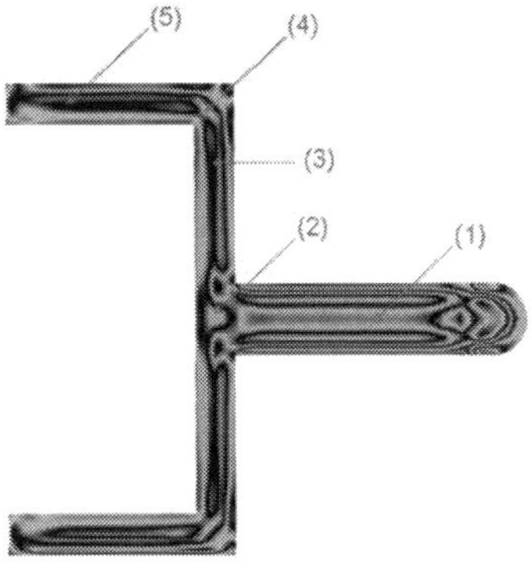

(a)

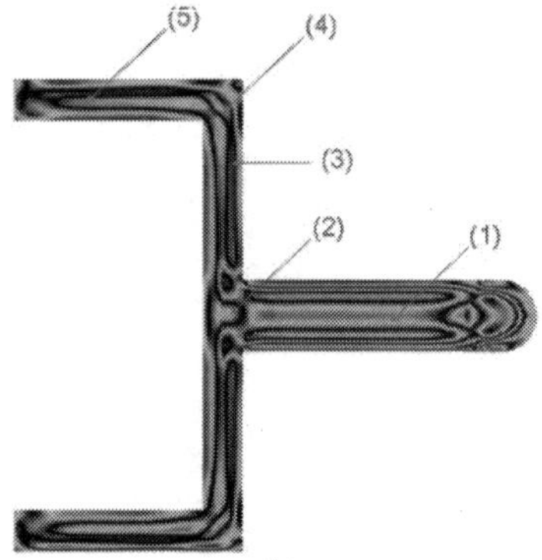

(b)

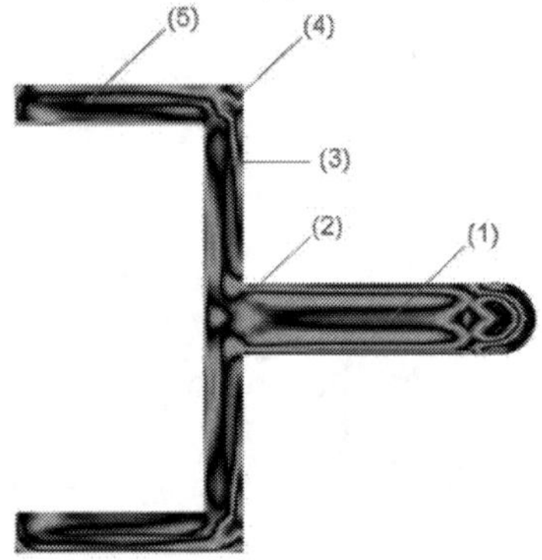

(c)

Fig. 14 Fringed pattern for the 2nd shot: (a) system is PC/PC, (b) system is PC+GF/PC, (c) system is PC+GF/PC+GF.

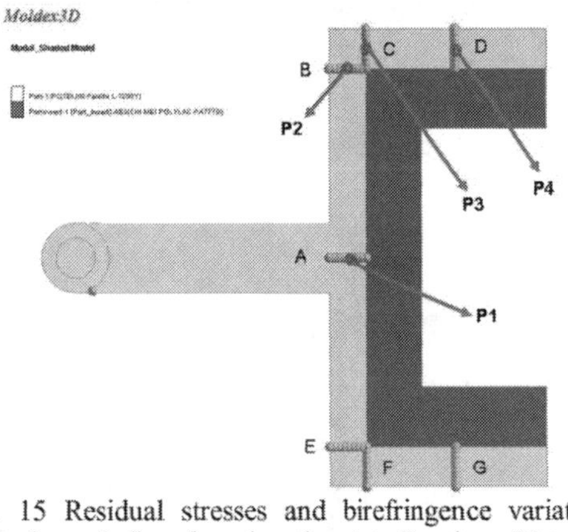

Fig. 15 Residual stresses and birefringence variation study for melt flow from location A to D, where P1 to P4 are at the center points for each line.

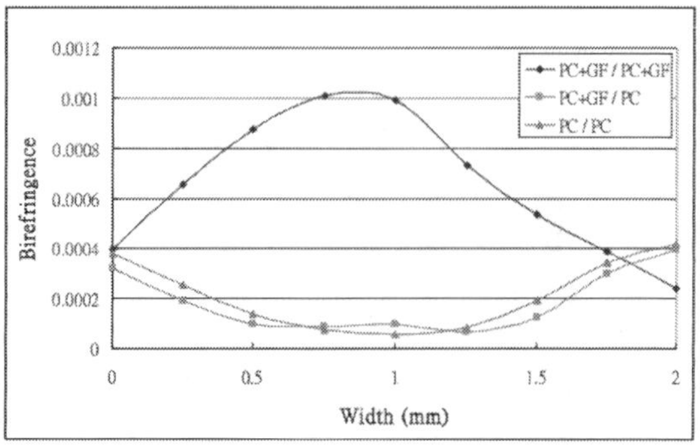

(a)

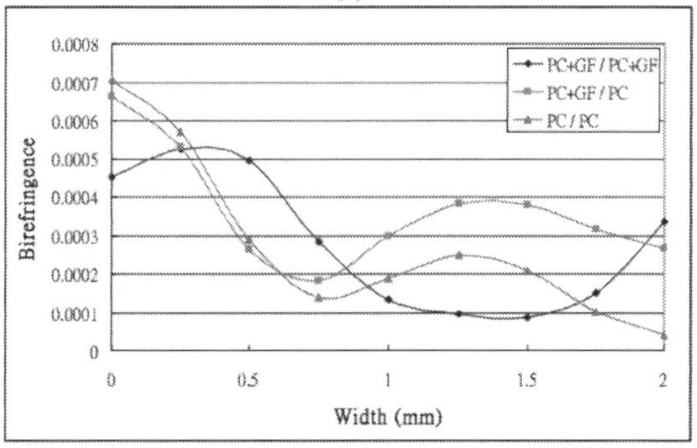

(b)

Fig. 16 Birefringence variation along flow direction with λ=0.1 and width=2.0 mm: (a) Line A, (b) Line B.

Key Words

multi-component molding (MCM), Process-induced Property Variation (PIPV), birefringence, warpage.

TEMPERATURE CYCLE INJECTION MOULDING FOR PRODUCTION OF WELDLINE-FREE, HIGH-GLOSS PARTS

P.G. Wlodarski[1], J. F. T. Pittman[1]*, J. Sienz[1], K. Crow[2], R. Foad[2]
[1]*School of Engineering, Swansea University, SA2 8PP, UK*
p.wlodarski@swansea.ac.uk; j.f.t.pittman@swansea.ac.uk; j.sienz@swansea.ac.uk
[2] *Gas Injection Worldwide Ltd, 31 Bostock Hall, Bostock Rd., Middlewich, CW10 9JN, UK*
ken.crow@gasinjection-ww.com; raymond.foad@gasinjection-ww.com

Abstract

In temperature cycle injection moulding (TCIM) of thermoplastics, the mould cavity surface is heated rapidly to a temperature close to the glass transition or crystalline melting point of the resin before melt injection, and then cooled after injection is complete. A number of important benefits of the process are briefly listed, among which is the facility to produce parts with high surface quality. Results are reported for ABS/PMMA mouldings made conventionally and with TCIM, in which the tool surface is pre-heated to 120°C. Weld line dimensions and surface roughness are determined using white light inteferometry. Weld lines, which on conventional parts are up to 17 µm deep and 70 µm wide, are not detectable on TCIM parts using a scale of inspection below 5nm. Surface roughness, *Ra*, measured by interferometry, is found to be 37nm for conventional parts, reducing to 20nm using TCIM.

Introduction

Temperature cycle injection moulding (TCIM) is a thermoplastics moulding process in which the mould tool, or at least the surface of the cavity, is heated to a temperature close to the glass transition or crystalline melting point before melt injection. After filling, the tool is cooled in the usual way. Several proprietary names are in use for the technology and a variety of means are used to bring about the tool pre-heating. The process has come to prominence recently, mainly as a result of a commercial incentive to produce high-gloss mouldings. Elimination of the frozen skin formation during injection has, however, several other important potential benefits, as indicated in a number of reviews (Smith, 2004; Anon, 2004; Michaeli and Cramer, 2006; Cronau and Schmidt, 2006; Mapleston, 2008; Anon, 2009a; Anon, 2009b). These include:

- Reduction of injection pressure, making possible lower clamp tonnage, longer flow paths and thin wall moulding.
- Reduction in part thickness without excessive injection pressures, resulting in savings in material and part weight. Thinner parts also cool more rapidly.
- Reduction of melt injection temperature without the risk of incomplete filling, leading to lower shrinkage, and allowing the use of lower packing pressures.
- Improvement of surface finish. This includes gloss on unfilled resins, and an improved resin-rich surface on filled materials, including long glass fibre filled mouldings. Importantly, foamed mouldings can also be produced with good surface. Replication of small surface features (micron scale) is made substantially improved.
- Elimination of visible weldline defects, providing greater freedom in gating arrangements, and possible elimination of painting or metalizing finishing processes.
- Improvement of optical properties in clear mouldings. Birefringence and moulded-in stresses are reduced.

These features are elaborated upon in an earlier article (Crow, et al. 2008), where we also reviewed the requirements and technologies used to implement the process. Simulation results were presented showing how injection pressure could be lowered by tool pre-heating, and how cycle times could be reduced as a result of intensified cooling and part thickness reduction. Additional energy input for tool pre-heating can be offset by the use of lower tonnage machines, increased productivity, material economies and elimination of finishing processes. Preliminary results were shown demonstrating improved surface gloss and elimination of visible weldlines on ABS mouldings. In the present article this work is extended by detailed quantitative examination of weldlines and surface finish using white light interferometry. A companion paper (Wlodarski et al. 2010) shows how major surface improvements are also obtained for foamed mouldings and for those filled with long glass fibre.

Experimental

Mouldings and materials
The test part is a 2mm thick, 200mm diameter, centre-gated, dish-shaped moulding with several cut outs to produce weld lines. The A-surface is smooth and polished, see Fig.1. Mouldings were made in an ABS/PMMA blend.

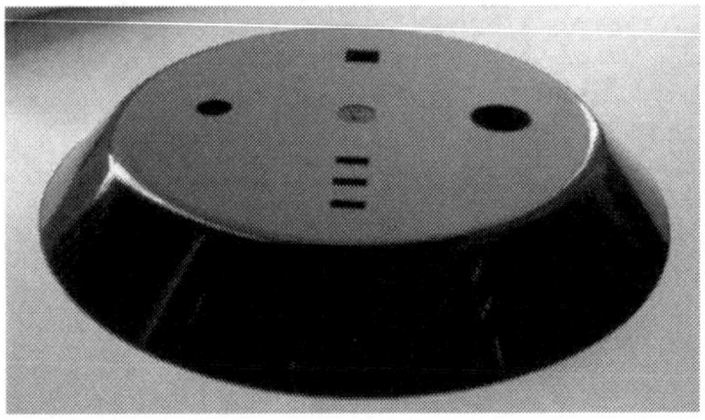

Figure 1. The test part.

The tool and temperature cycling

For efficient temperature cycle moulding it is essential that the tool has low thermal mass; that is, the energy input for tool heating must be minimized by appropriate tool design. Heating and cooling must be confined to a region close to the cavity surface, both to reduce energy use and to provide rapid heat transfer to and from the cavity surface. For the dish mouldings the fixed half of the tool is formed as an assembly of front and back plates. As seen in Fig. 2. the back plate incorporates channels and integral manifolds for intensive circulation of heat transfer media. The assembly is supported with minimum contact to the bolster, to reduce heat transfer into the back of the tool.

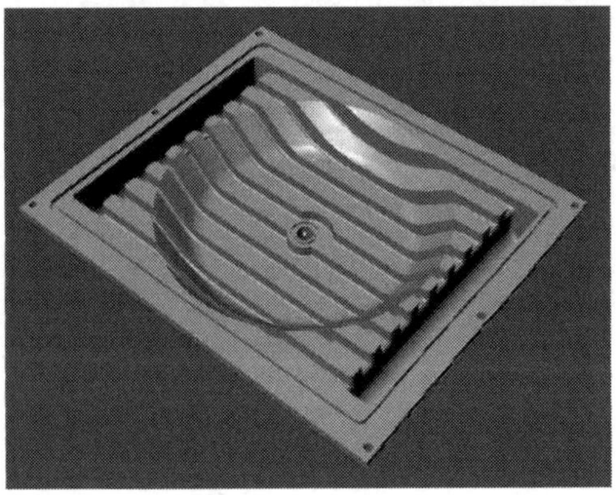

Figure 2. View of the back plate of the fixed half of the tool showing integral manifolds and channels for circulation of heat transfer media.

Temperature cycling

Temperature cycling of the tool was achieved using steam heating and water cooling. The media are supplied via a controller interfaced with the moulding machine. The cycle is initiated by a mould close signal, which triggers steam supply to the tool from a portable electric boiler. When the target tool temperature is reached, as sensed by a thermocouple in the fixed half, melt is injected. A compressed air blast clears condensate from the channels and cooling water is supplied. When the tool temperature falls to a required value, or after a fixed time, the tool opens and the part is ejected. Compressed air clears water from the channels, and the cycle is repeated. The controller is from Gas Injection Worldwide Ltd (GIWW), and a schematic of the system is shown in Fig. 3.

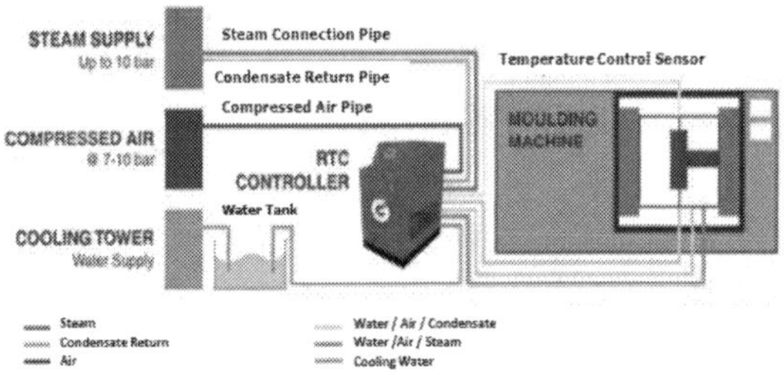

Figure 3. Schematic of the temperature cycling system. RTC™ (Rapid Temperature Cycle moulding) is a trade mark of GIWW.

Processing conditions

Processing conditions were as follows: Melt injection temperature 240°C; filling time 3s; packing time 2s; cooling time 15s; time to pre-heat tool 12s; tool target temperature 120°C; steam temperature 150°C; coolant temperature 35°C.

White light interferometry

Weldlines and surface finish on the mouldings were examined using white light interferometry (Veeko WYKO NT2000). A light beam passes though a microscope objective and a beam splitter. Beams reflected from a reference surface and from the sample are combined to produce interference fringes. The microscope scans vertically through focus capturing the interference data at regular closely spaced intervals. The fringe contrast is maximized as each part of the sample surface comes into focus, and, by recording the vertical positions at which this happens, the system software is able to generate a 3D topographic image of the surface with a height resolution of better than 5nm. This can be processed in various ways, to produce surface profiles along selected lines and to provide a number of statistical surface roughness measures.

Results

Weld lines with conventional processing

To provide a basis for comparison, weldlines formed in conventional processing are first examined (the tool is cooled by continuous water flow). Fig. 4 identifies the positions where weld lines are formed.

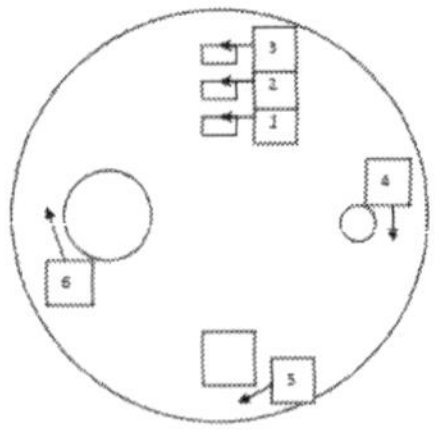

Figure 4. Identification of weldline positions on the test part in conventional processing.

Fig. 5 is a photograph showing a weldline clearly visible at position 6.

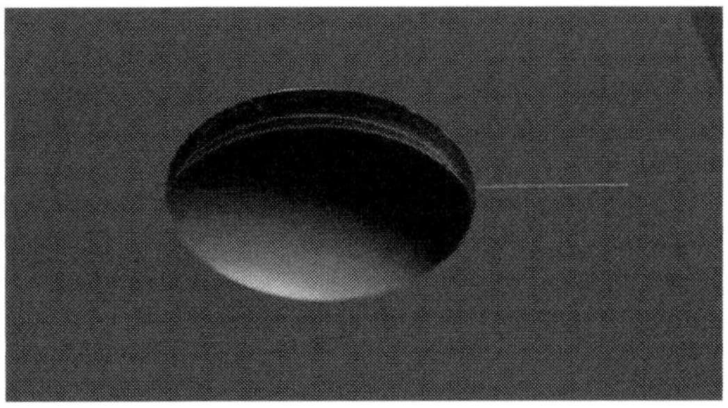

Figure 5. Weldline formed at position 6.

A plot of the surface height data provided by the interferometer at position 1 is given in Fig. 6. The height datum is defined from a flat area to the sides of the weldline. The interferometer has some difficulty capturing data from the steeply sloping sides of the weldline, which therefore appear as black bands. Reflection from the base is, however, captured well and provides the blue line giving the weld line depth.

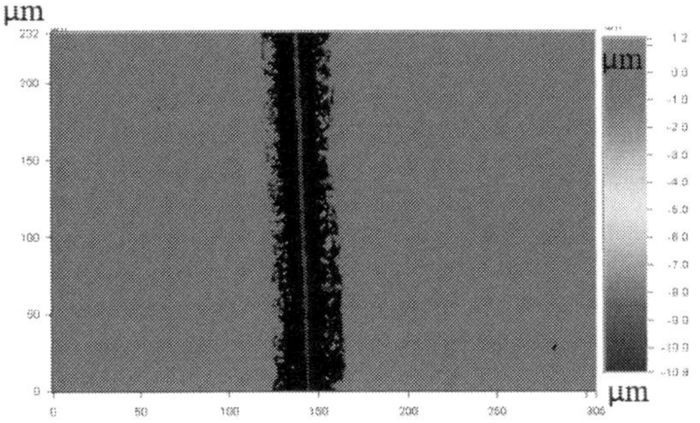

Figure 6. Surface height plot showing weldline at position 1.

The profile of sections through the weldline is in Figs. 7, showing a sharp V-shaped groove, varying from about 10µm deep and 40µm wide near the obstruction to 5 µm by 20 µm towards its further end. Definition of the profile has been improved by pooling data from a band 20 µm wide.

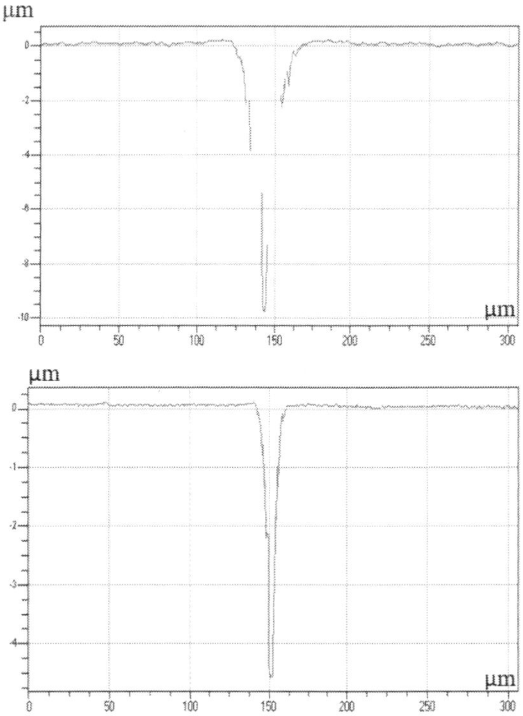

Figure 7. Profiles of sections through the weldline at position 1. Upper figure – close to the obstruction. Lower figure – further along the weldline.

The depth of the weldline along its length is plotted in Fig. 8, showing variations of a couple of µm between samples, which may be due to small variations in processing conditions.

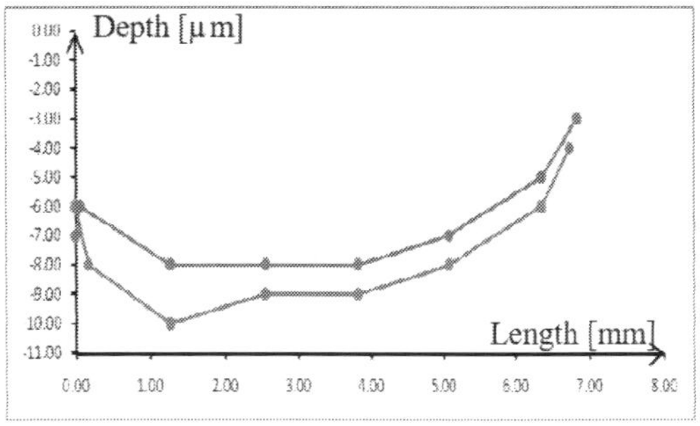

Figure 8. Depth along its length of the weldline at position 1, for two moulding samples.

Figs. 9 show depth and width (defined at 0.5 µm below the datum surface) of the weldline at position 6. Similar results are found for the other positions. Weldline dimensions found here are consistent with data in the literature (e.g. Akkerman et al. 1995, Fellahi et al., 1995Yamada et al. 2005). Further results and details of the experimental techniques are contained in Wlodarski (2009).

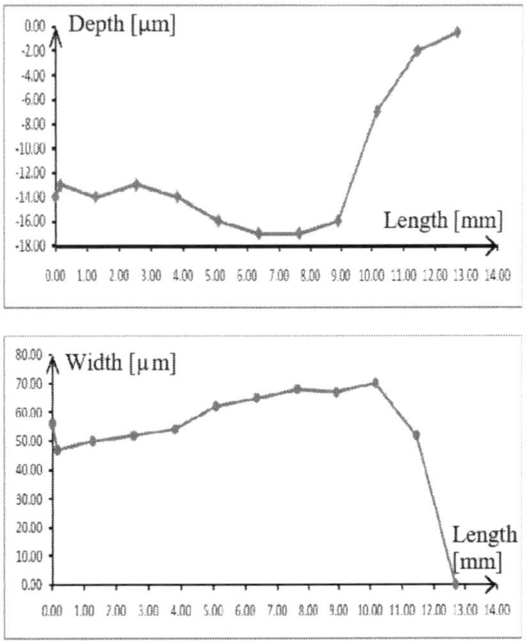

Figure 9. Depth (upper figure) and width (lower figure) along its length of weld line at position 6.

Weldline elimination using TCIM

The photograph in Fig. 9 is to be compared with Fig. 5 and shows how the visible weld line is eliminated using TCIM.

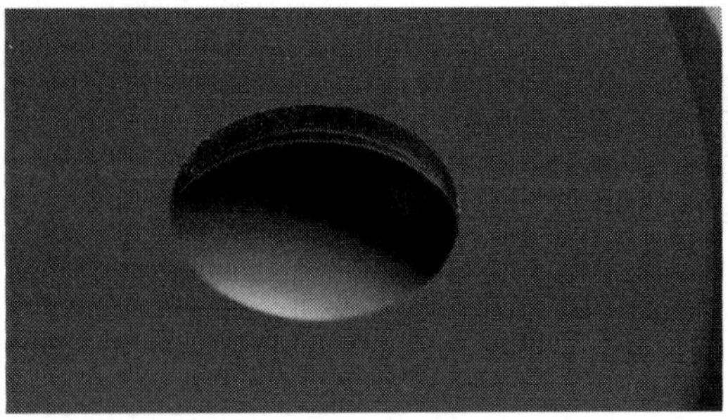

Figure 10. Elimination of weldline at position 6 using TCIM.

Fig. 11 compares surface height plots at position 6 for conventional and TCIM processing. No trace of the weldline is to be seen.

Surface finish, conventional processing vs. TCIM

A number of surface roughness measures are provided by the interferometer software.

Ra is the arithmetic mean deviation of the surface z(x) from the best fit line over the sampling length *L*.

$$Ra = \frac{1}{L}\int_0^L |z(x)|dx$$

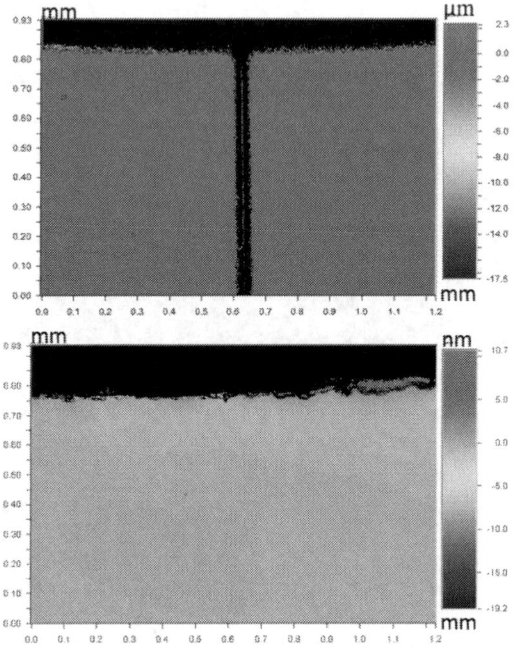

Figure 11. Surface height plots at position 6 for conventional moulding (upper figure) and TCIM (lower figure), showing elimination of the weld line. The black upper region in the figures is the cut-out formed by the obstruction.

Rq is the analogous root mean square value, which gives more weight to larger departures.

$$Rq = \sqrt{\frac{1}{L}\int_0^L [z(x)]^2 dx}$$

Among further measures of interest is *Rt*, the absolute distance between the highest and lowest points within the sampling length.

Surface roughness was determined for each moulding, averaging results from 10 scan areas each 305x232 µm. Variation is seen between scans and the standard deviation of the mean is tabulated. Table 1 summarizes the averaged results.

Table 1. Comparison of surface roughness for conventionally moulded and TCIM parts. Values are averages from 10 scans, area 233µm x 306µm. Standard deviations of the means are shown in parentheses.

	***Ra* nm**	***Rq* nm**	***Rt* nm**
Conventional	37 (10)	65 (15)	2540 (700)
TCIM	20 (5)	34 (10)	1750 (1480)
Reduction	47%	49%	31%

Discussion and Conclusion

Temperature cycle moulding in its various implementations has been widely reported to eliminate visible weldlines; however, to our knowledge no results of a quantitative surface examination have previously been available to confirm this. The present work shows that the typical, V-groove of the weld line is entirely removed. The effect of TCIM on mechanical properties in this region is also important and will be the subject of further work.

The mould tool used here was polished, and the ABS/PMMA blend is a naturally glossy material. Mouldings produced conventionally had a glossy surface, as evident in Fig. 5. The visible surface improvement provided by TCIM (see Fig. 10) is therefore not very obvious in the photographs. Previous results moulding a matt ABS showed a dramatic gloss development with TCIM (Crow et al. 2008). The present measurements with the interferometer, however, show a significant reduction in the statistical surface roughness measures – by approximately 50% for *Ra* and *Rq*. *Rt*, which is sensitive to a small number of high and low points, reduced rather less – by approximately 30%.

Gloss on a surface is produced when a substantial part of the reflection is specular – that is when the angles of incidence and reflection are equal as measured relative to a plane fitted through the surface. For this to occur, the surface needs to be substantially flat over areas a few times larger than the wave length of visible light (390 – 750nm). The *Ra* and *Rq* values seem consistent with this, though *Rt* indicates the presence of some larger asperities.

Injection into a cavity with a heated surface is known to improve replication of tool surface detail on mouldings, at least on the scale of microns (see e.g. Crow et al. 2008 for citations). Here, we are seeing surface roughness on a much finer scale, which we believe to be governed by the polymer morphology rather that by replication of the tool surface. The origins and nature of the surface morphology, which may limit the smoothness that can be obtained, and its relationship to TCIM conditions, will be the subject of further work.

Acknowledgement

This work is supported by the Welsh Assembly Government, UK, through its A4B (Academic Expertise for Business) programme as Cooperative Industrial Research Project 'Rapid temperature Cycle Injection Moulding: development and applications'.

References

Akkerman R., Rekers G., Huetink J., 'Finite Element Simulation of Meltlines', Simulation of Materials Processing: Theory, Methods and Applications – Numiform, 95, 1007-1012 (1995)

Anon, 'The Smooth Way to Mold Microcellular Foam Parts', Plastics technology, 50(12) (2004)

Anon, 'Cyclical Developments', European Plastics, February, 16-17 (2009a)

Anon, 'Flowing Hot and Cold Brings product Improvement' British Plastics and Rubber, April, 4-6 (2009b)

Cronau C., Schmidt H., 'A Mirror Finish with no Visisble Weld Seams', Kunststoffe, **11**, 49-51 (2006)

Crow K., Foad R., Pearson T. C., Polynkin A., Pittman J. F. T., 'Multiple Benefits from Rapid Temperature Cycle Injection Moulding', PPS24 (24th International Meeting of the Polymer Processing Society), Salerno, Italy, Paper SO3-1310, 9pp (2008).

Fellahi S., Meddad A., Favis B. D., 'Weldlines in Injection-Molded Parts – A Review', Advances in Polymer Technology, 14(3), 169-195 (1995)

Mapleston P., 'Turning up the Heat: Mould Temperature Control', Plastics Engineering, October, 10-16 (2008)

Michaeli W., Cramer A., 'Improved Surfaces in Foam Injection Moulding', Kunststoffe International, **12**, 21-27 (2006)

Smith C., 'Expanding the Bubble', European Plastics News, November, 16-17 (2004)

Yamada K., Tomari K., Ishiaku U. S., Hamada H., 'Evaluation of Mechanical Properties of Adjacent Flow Weldline', Polymer Engineering and Science, 45, 1180-1186 (2005)

Wlodarski P. G., 'Temperature Cycle Injection Moulding: Benefits, Tests and Development', Bachelorarbeit, Fachhochschule Aachen, Fachbereich Maschinenbau und Mechatronik / Swansea University, School of Engineering (2009).

Wlodarski P. G., Pittman J. F. T., Sienz J., Crow K., Foad R., 'Quantified Surface Improvement for Foamed and Long-fibre Filled Parts from Injection into a Hot Cavity', PPS26 (26th International Meeting of the Polymer Processing Society), Banff, Canada (2010).

PRESSURE AND TEMPERATURE DEPENDENCE OF THERMAL CONDUCTIVITY FOR INJECTION MOLDING SIMULATION

R. G. Speight[1,2], F. S. Costa[1*] and C. Hadinata[1]

[1] *Autodesk Australia Pty Ltd, 259-261 Colchester Rd, Kilsyth, VIC 3137, Australia - franco.costa@autodesk.com*
[2] *Honorary Visiting Research Professor, Advanced Materials Engineering Centre, University of Bradford, UK*

Thermal conductivity is an important material property as it influences polymer temperature distribution during the whole injection molding cycle. This property is typically measured in the laboratory at pressures which are low (5-10 MPa) compared to those which occur in the injection molding process (30-200MPa). Measurement of thermal conductivity may be for a single temperature or multiple temperatures over the processing range of a material, including the solid state. Implementation of the transient line source technique (ASTM D5930) with bench top twin bore plunger capillary rheometry, allows pressure control of the polymer sample, with two significant enhancements: (i) the ability to improve the contact of the probe with the sample and thus eliminate the effects of contact resistance. This can be significant for semi-crystalline materials, and (ii) allowing the pressure dependence of thermal conductivity to be investigated. Experimental data is presented, outlining the typical effect of pressure on thermal conductivity for both semi-crystalline and amorphous materials. The ultimate aim is to improve the accuracy of injection molding simulations by analytical modelling of thermal conductivity for both temperature and pressure behaviours, providing a method by which experimental error can be reduced, while ensuring the measured behaviour follows theoretical assumptions.

Introduction

Injection molding simulation by the finite-element method began in the 1970's, and has continually developed in sophistication to the present day. The accuracy of injection molding simulation is influenced by many factors such as modelling of part geometry, runner and nozzle, mesh type and density, mathematical finite element solution, process settings and the focus of this paper, material data. Material testing technology is continually evolving to meet the increased demands of simulation as outlined by Speight and Costa (2009). The importance of thermal conductivity in achieving high accuracy simulation is outlined in this paper.

Thermal conductivity (k, W/m.K) defines the heat flux that flows through the material if a temperature gradient exists over the sample. It is defined as the quantity of heat Q, transmitted in time t through a thickness L, in a direction normal to a surface of area A, due to a temperature difference ΔT, under steady state conditions and when the heat transfer is dependent only on the temperature gradient:

$$k = \frac{Heat\ Flow \cdot Distance}{Area \cdot Temperature\ Difference} = \frac{Q}{t} \cdot \frac{L}{A \cdot \Delta T} \quad (1)$$

Considering the case of a cylinder of infinite length, with an infinitesimally thin, infinitely long heating element located along its axis, from Fourier's conduction equation, it is possible to derive an expression for the transient temperature rise at any point in the cylinder. Starting at $t = 0$, a constant finite quantity of heat Q is produced per unit length of the heating element. Lobo and Cohen (1990) explain that at a fixed point of radius r, the change in temperature $(T_2 - T_1)$ in the time interval $(t_2 - t_1)$ is then given by:

$$k = \frac{QCln(t_2/t_1)}{4\pi k(T_2 - T_1)} \quad (2)$$

The transient line source technique uses a probe constant, C, to compensate for theoretical deviations in the experimental set up of the device. The constant is determined by calibrating the probe against a material of known thermal conductivity, such as Polydimethylsiloxane (PDMS) fluid, (60,000 centistokes), with referenced thermal conductivity of 0.16 W/m.K, Chakravorty and Brown (1995). Further investigations are required to determine the stability of thermal conductivity with respect to temperature for Polydimethylsiloxane (PDMS).

To demonstrate the importance of thermal conductivity as outlined by Sridhar and Narh (1999), a sensitivity study is implemented on typical semi-crystalline and amorphous materials. These studies are carried out using Autodesk Moldflow Insight 2010, on a model of a highly instrumented experimental test plaque, dimensions 200x40x2mm. The influence of perturbing thermal conductivity on the prediction of maximum injection pressure and the material freeze time are investigated.

Experimental

The experimentation described uses two laboratory thermal conductivity devices that are present in both our material testing laboratories in Melbourne,

Australia and Ithaca, New York. The first device is an AC-Technology, K-system II. This has been used by our laboratories since the early 1990's, as described by Lobo and Newman (1990), and allows thermal conductivity to be measured in the melt and solid states. The second device is a CEAST SmartRheo 2000 (maximum force 20kN) bench top twin bore plunger capillary rheometer, with transient line-source thermal conductivity option. The CEAST SmartRheo 2000 rheometer has been used since 2008 and allows pressure control of the polymer sample during the measurement of thermal conductivity (up to 200MPa). Experimentation is also presented in this paper from INSTRON - CEAST Division, specifically pressure effect on thermal conductivity data. Transient line source probes for this device were developed together jointly between CEAST and Hukseflux Thermal Sensors B.V., with the capability to measure thermal conductivity in the melt state, Bronzoni and Lanteri (2007).

The transient line-source method can produce very precise results when good contact between the sample and the probe is achieved, eliminating effects of thermal contact resistance. Measurements are typically for a period of 45 seconds, sampled at 10Hz. Radial losses can be seen: (i) in the initial region of the transient through a non-linearity of the curve, and are due to the initial heat wave propagating through the walls of the probe, and (ii) for sample periods that are too long.

The probe casing (hypodermic needle) is 50 mm in length, 1.3 mm diameter; as shown in figure 1. Constantan wire is used for the heater element because of its low temperature coefficient of resistivity. A single thermocouple is used for the temperature measurement (either Type J or K), where the thermocouple bead is halfway down the length of the probe and is in physical contact with the walls. The wires are fixed in place using a high-temperature ceramic cement to maintain electrical continuity.

For injection molding simulation, thermal conductivity may be measured using the following test methods: (i) single point at recommended melt temperature; (ii) multi-point, over a temperatre scan from melt to solid, where the average thermal conductivty is determined, or (iii) where the multi-point thermal conductivty data series, with respect to temperature, and (iv) where the multi-point thermal conductivty data series, with respect to both temperature and pressure. Thermal conductivity is typically reported as a temperature scan of up to 12 points, where the test procedure involves multiple scans to ensure accuracy. Intra-laboratory testing on a Polystyrene (PS) sample shows a repeatability of 3%. However, this result is highly dependent on ensuring good sample contact resistance.

Results and Discussion

The mean transient temperature delta during 6 separate runs is shown in figure 2, using Polydimethylsiloxane (PDMS) at 30C, plotted against natural logarithm (time). The output of the transient line source technique is theoretically a straight line on a semi-log plot of temperature against time. The initial portion of the transient displays some curvature, as the heat moves through the metal wall of the probe, between -2.0 and 1.50 ln(seconds). Since the mass of the probe is finite the temperature difference across the steel wall becomes constant quickly and the termperature sensor will thereafter reflect the temperature rise of the polymer. The temperature rise at the line-source varies linearly with the logarithm of time in the region where the process is dominated by conduction and other effects can be neglected, between 1.70 and 3.70 ln(seconds). This linear zone is characteristic of the sample properties. The different regions of the temperature versus time diagram of the probe are illustrated in figure 2 by review of the coefficient of variation. During the linear region the measurements on PDMS are within 0.1% variation, demonstrating the potential accuracy of the device.

Polymers will shrink during cooling, therefore it is necessary to maintain a constant pressure on the probe to prevent voids forming and gaps at the surface interface when in the solid phase. This is especially important for semi-crystalline materials. Figure 3 highlights that the experimental variation in the solid stage (<110C) is higher then the melt stage (> 150C). These result are measured using the K-System II. Future experimention will investigate the performance of the CEAST SmartRheo 2000 during polymer transitions from melt to solid state under pressure control.

Figures 4 and 5 show the effect of perturbing thermal conductivity, as the input to an injection molding simulation, for a typical unfilled Polypropylene and Acrylonitrile-butadiene-styrene. For this research, thermal conductivity was variated by $\pm$50%, so that the trends are clearly observable. For materials such as the Polypropylene as shown in figure 3, a 20% change in thermal conductivity between melt and solid states will account for a difference in injection pressure and freeze time (cycle time) calculations.

Bronzoni and Lanteri (2007) showed the effect of pressure on thermal conductivity for Low Density Polyethylene (LDPE), Polystyrene (PS) and Polycarbonate (PC), refer Figure 6. At the same test temperature, LDPE and PS show a different behaviour, in particular, LDPE exhibits a near linear pressure dependence on conductivity, compared to the response of the PS. The thermal conductivity data obtained for PC at 300°C are similar in response to the LDPE. The percentage increase in thermal conductivity between 60 to 200MPa for LDPE, PS and PC are 29%, 20% and 14% respectively. These results are presently being validated in our own laboratories.

Extending the sensitivity study as outlined in figures 4 and 5 to a range of materials, typical grades are selected and the relative importance of changes in thermal conductivity can be observed. Figures 7 and 8 show comparative data for injection pressure and material freeze time, for a range of materials. These results are for the specific material grades selected from the Autodesk Moldflow Insight 2010 database and may not be representative of all grades. Thermal conductivity variations will account for a difference in injection pressure and freeze time (cycle time) calculations.

Conclusions

Accuracy of injection molding simulation is influenced by thermal conductivity. The transient line source method has the potential to provide accurate thermal conductivity data, when good contact resistance between the probe and material is maintained. Experimental practice for thermal conductivity measurement will improve by pressure control of the plunger ensuring: (i) the ability to improve the contact of the probe with the sample and thus eliminate the effects of contact resistance, and (ii) allowing the pressure dependence of thermal conductivity to be investigated. Together with these experimental improvements, it is proposed that an analytical model for thermal conductivity be introduced, thereby ensuring experimental data complies with the expected material behaviours.

Acknowledgements

The authors would like to acknowledge the technical assistance regarding the CEAST SmartRheo 2000 rheometer and for the permission to present experimental data received from Marco Bronzoni, Research & Development Product Application Manager, INSTRON - CEAST Division, ITW Test and Measurement Italia S.r.l Instron CEAST Division, Via Airauda 12, 10044 Pianezza, Torino, Italy.

References

Standard: ASTM D5930 - 01 Standard Test Method for Thermal Conductivity of Plastics by Means of a Transient Line-Source Technique, 2001 (by ASTM Committee D20 on Plastics).

Proceedings: Bronzoni and Lanteri (2007), Determination of the thermal conductivity of polymer melts as a function of temperature and pressure, AERC 2007 - 4th Annual European Rheology Conference April 12-14 (2007), Napoli, Italy.

Chakravorty and Brown (1995), Significance of thermal material parameters in polymer processing, NPL report DMM (A) 167, National Physical Laboratory, Teddington, Middlesex, TW11 0LW, UK.

Journal: Lobo and Cohen (1990), Measurement of Thermal Conductivity of Polymer Melts by Line-Source Method, Polymer Engineering and Science, January 1990, Vol. 30, No.2, p65-70.

Proceedings: Lobo and Newman (1990), Thermal Conductivity of Polymers at High Temeperatures and Pressures, ANTEC 1990, Dallas, May 7-11, p862-865.

Proceedings: Speight and Costa (2009), Material Testing for Injection Molding Simulation, Seventh International Conference for Polymer Process Engineering, University of Bradford, UK, 27-28 October, Institute of Materials, p47-83.

Proceedings: Sridhar and Narh (1999), Effect of Temperature Dependent Thermal Properties on the Accuracy of Injection Molding, ANTEC 1999, New York, May 2-6, p496-499.

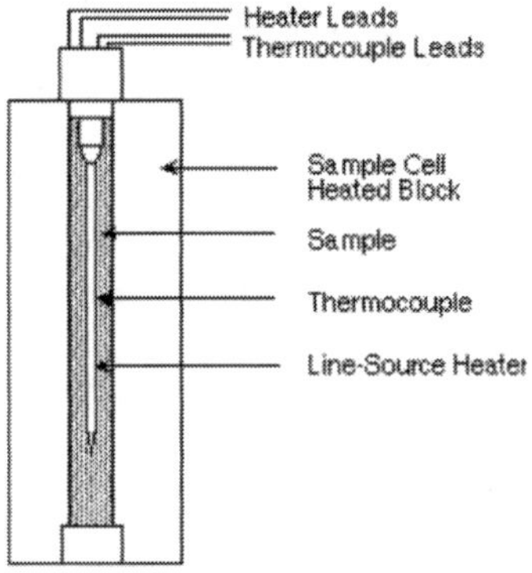

Figure 1, Schematic diagram of AC Technology K System II, transient line source thermal conductivity test cell, bore diameter 12.8mm, probe length 50mm.

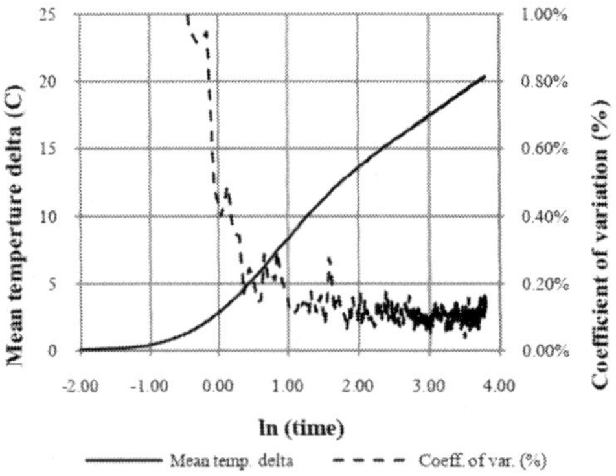

Figure 2, Mean output tranisent temperature delta and coefficient of varation, for of 6 samples, of Polydimethylsiloxane (PDMS) fluid at 30C.

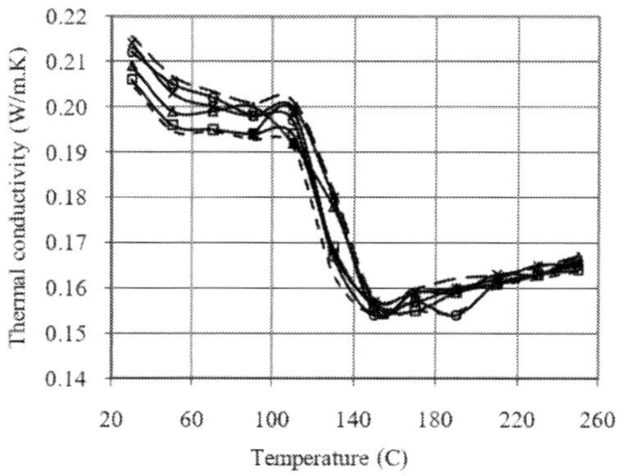

Figure 3, Thermal conductivity against temperature, showing typcial experimental variation for a typical unfilled Polypropylene (PP).

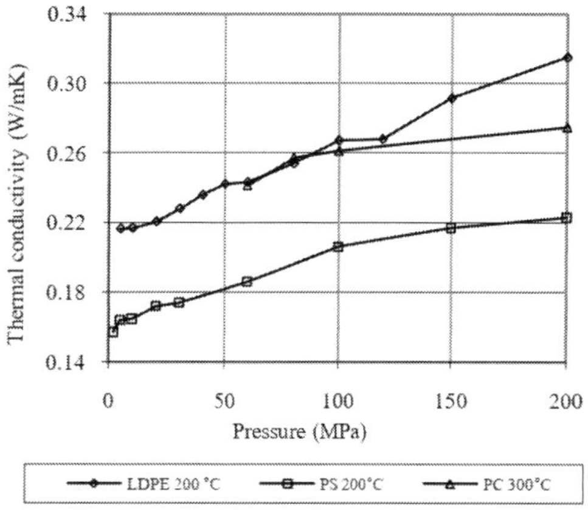

Figure 6, Influence of pressure on the thermal conductivity of LDPE, PS and PC.

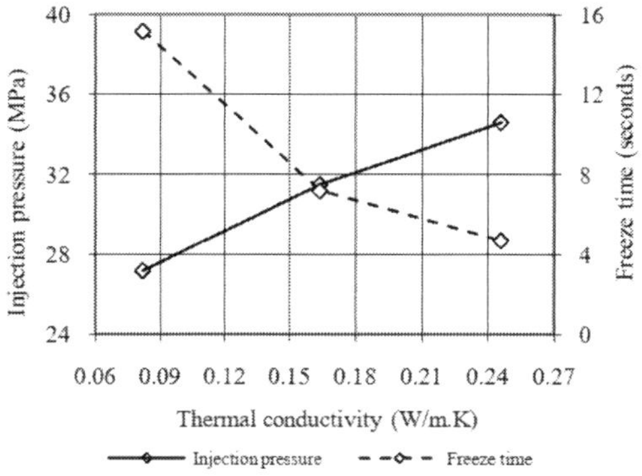

Figure 4, Influence of thermal conductivity on injection pressure and freeze time for a typical unfilled Polypropylene (PP).

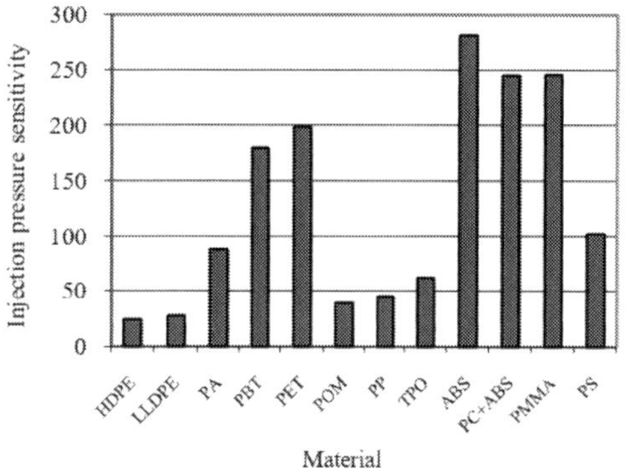

Figure 7, Injection pressure sensitiviy to thermal conductivity (MPa / W/m.K) for a range of materials.

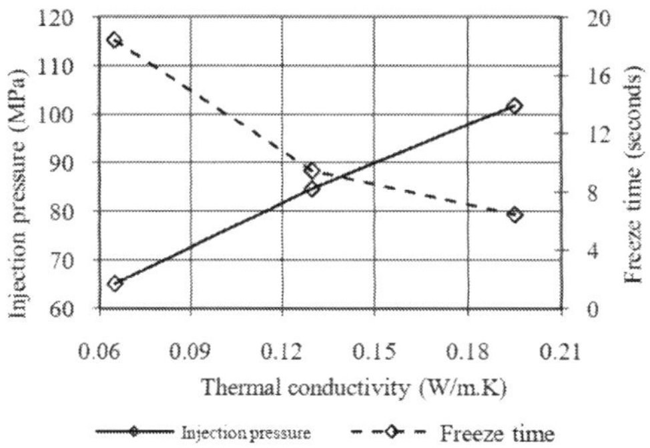

Figure 5, Influence of thermal conductivity on injection pressure and freeze time for a typical Acrylonitrile-butadiene-styrene (ABS).

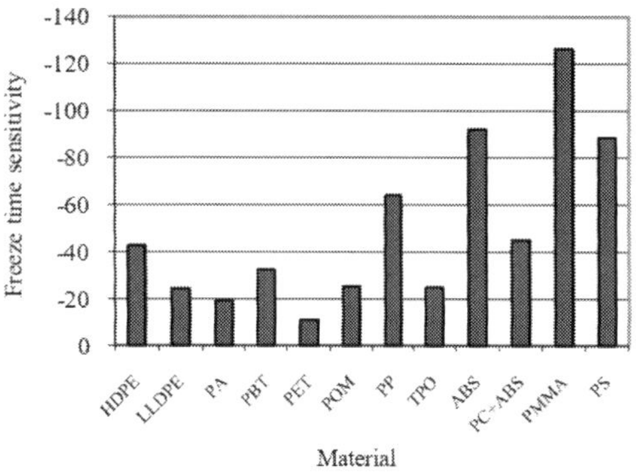

Figure 8, Freeze time sensitiviy to thermal conductivity (Seconds / W/m.K) for a range of materials.

INFLUENCE OF PRESSURE ON VOLUME, TEMPERATURE AND CRYSTALLIZATION OF THERMOPLASTICS DURING POLYMER PROCESSING

Dr.-Ing. N. Rudolph[1]*, Prof. Dr.-Ing. Dr. h. c. G. W. Ehrenstein[2], Prof. Prof. hon. Dr. T. A. Osswald[1]

[1] *University of Wisconsin-Madison, Polymer Engineering Center, WI – nrudolph2@wisc.edu; osswald@engr.wisc.edu Department of Mechanical Engineering, 1065 Mechanical Engineering Building, 1513 University Avenue, Madison, WI, 53706*
[2] *University of Erlangen-Nuremberg, Institute of Polymer Technology, Germany – ehrenstein@lkt.uni-erlangen.de*

The main shrinkage and dimensional stability impact factors are processing temperature and pressure. For that reason, the pvT-behaviour is often used to estimate plastic part shrinkage. As there are still differences between predicted and actual part dimensions, a closer look was taken at these influencing factors. During processing, temperature changes occur due to heat conduction to the mold walls, and the pressure varies in the filling stage and holding time along the flow path. However, these pressure gradients are typically not taken into account in plastics processing. Hence, the influence of pressure was examined in detail and further investigations on pvT-behavior of polymers were performed.
For that purpose, fundamental examinations of the behavior of amorphous thermoplastics during cooling and compression were made. These include analysis in the different phases and at different compression speeds as well as variation of the pressure-temperature-cycles and their succession. It was found that the specific volume is not defined by one value of p and T, but is dependent on the pathway of the process. That applies for both thermal expansion and compressibility. The adiabatic compression heating was also examined and a numeric solution was found to easily adapt the pvT-results on the base of compressibility measurements. These findings are discussed by means of the free volume theory. Additionally, a choice of these investigations was performed with semi-crystalline thermoplastics. It was found, that the compression heating is even more important for these polymers because it can superpose crystallization heating at appropriate parameters. Furthermore, it can be seen that the density, morphology and lamellae thickness is influenced by pressure.
In the outlook of this paper, the impact of these results on the injection molding process is discussed in detail.

Introduction

The main reason for shrinkage of plastic parts is the thermodynamic and thermo mechanic behavior of the plastic. In the so called pressure-volume-temperature-behavior (pvT) differences exist between compressibility and thermal expansion of the material. Figure 1 shows the change of specific volume due to changes in temperature at different pressures (solid black lines). The glass transition T_g is shifted to higher temperatures with increasing pressure. The cooling rate also affects the specific volume, as slower cooling rates shift the glass transition $T_{g(slow)}$ to lower temperatures. As a result, a lower specific volume can be reached (dashed black lines). Therefore, T_g as well as the specific volume in the solid state are time-dependent properties.
The course of the injection molding process can be traced in such a pvT-diagram (bold dot and dash line in Figure 1). After filling (1) the packing pressure is applied (2) as long as the gate remains open. After freezing (3), the plastic cools isochorically until atmospheric pressure is reached (4). From that point the plastic part shrinks without contact to the mold until it reaches room temperature (5). The absolute shrinkage (v_s) is defined as the difference between point 1 and 5. The part shrinkage (p_s) starts as soon as atmospheric pressure is reached at 4 until room temperature is approached at 5 (Menges et al., 1977).

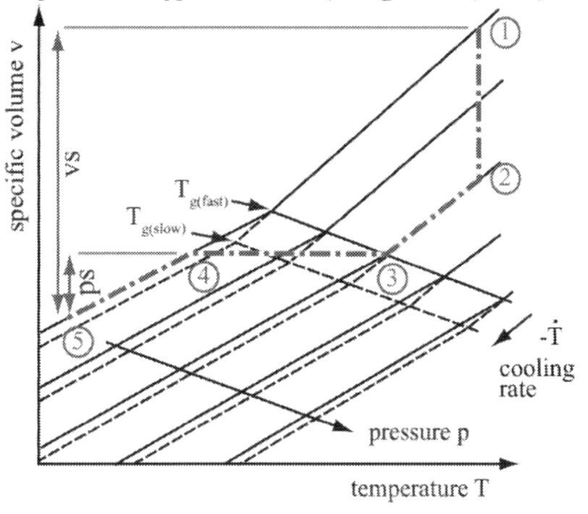

Figure 1 Influence of pressure, temperature and cooling rate on the specific volume of an amorphous thermoplastic (pvT-diagram); bold dot and dash line: volume course during cooling, vs = volume shrinkage, ps = part shrinkage
1: filling, 2: packing pressure, 3: freezing of the gate, 4: reaching atmospheric pressure, 5: reaching room temperature.

Although a lot of knowledge about shrinkage and pvT-behavior has been gained during the last decades, there are still uncertainties between the predicted and resulting part dimensions. By taking a closer look it becomes obvious, that neither the influence of cooling rate nor pressure or temperature deviations in the part is regarded in process control or simulation. However, during the injection molding process the temperature and pressure of the melt change locally. Temperature changes due to heat conduction to the cold mold wall, and pressure deviations occur in the filling stage and holding time over the flow path. That results in local deviations in specific volume, meaning the process path in Figure 1 reflects the behavior of the melt at only one specific location in the part. However, this spacial gradient is typically not taken into account in plastics processing. Hence, further investigations on the pvT-behavior were performed. Hereby, the influence of pressure on amorphous thermoplastics was examined in detail, a factor often left out from analysis. Additionally, the influence of pressure in the crystallization process was studied, since crystallinity and specific volume are affected by cooling rate and pressure, too. Results of these tests are presented in this paper. In conjunction with the knowledge of the time-dependence and local deviations of temperature and pressure the impact on injection molding is discussed.

Experimental

Material

In this study, an amorphous polycarbonate PC Makrolon 2805 from Bayer Material Science AG is used. This resin has a solid density of 1.20 g/cm^3 and a glass transition temperature of 150 °C (pvT-measurements). The melt volume rate (MVR) for the resin is 9.5 (300 °C, 1.2 kg).
The semi-crystalline thermoplastic used is an isotactic PP Moplen HP 500N from Basell that is suitable for general purpose injection molding applications. It has a solid density of 0.90 g/cm^3 and a crystallization temperature of 124 °C (pvT-measurements). The MVR for the resin is 16.0 (230 °C, 2.16 kg).

Devices

A piston driven pvT device (SWO pvT 100, now Thermo Fischer Scientific) was used to analyze the isobaric pvT behavior as well as the compression experiments on the semi-crystalline polymer (Menges et al., 1977; Schmidt et al., 2000; Engelsing, 2000). The temperature ranges from 50 °C to 400 °C and the pressure ranges from 150 to 2500 bar. The compression speed is limited to a slow compression of 750 bar/min.
For the measurement with fast compression speeds between 100 and 100,000 bar/min on the amorphous thermoplastic a new press tool was constructed in the style of the piston driven pvT instrument, Figure 2. The cavity designed to take up the plastic material is determined by a positive connection between upper and lower piston as well as an element in the center of the mold. The upper and lower pistons are mounted on their corresponding side in affixed position, while the center portion is allowed to move, so that the plastic specimen can easily be demolded after the test is completed. A press stroke moves the upper and lower piston into a vertical inter-connection. A distance plate

between lower and central element defines the central element's position within the tool, according to the specimen's height, resulting in a symmetrical temperature control. By means of electric cartridge heaters inside the individual heating elements (drill holes in Figure 2), i.e. the so-called heating zone, the plastic can be heated to a desired temperature. Air is fed by an external blower into circumferential aluminum cooling fins, which chill the mold, and consequently the test specimen inside, at cooling rates of up to 40 K/min. The temperature ranges from 50 °C to 300 °C and the pressure ranges from 10 to 3500 bar.

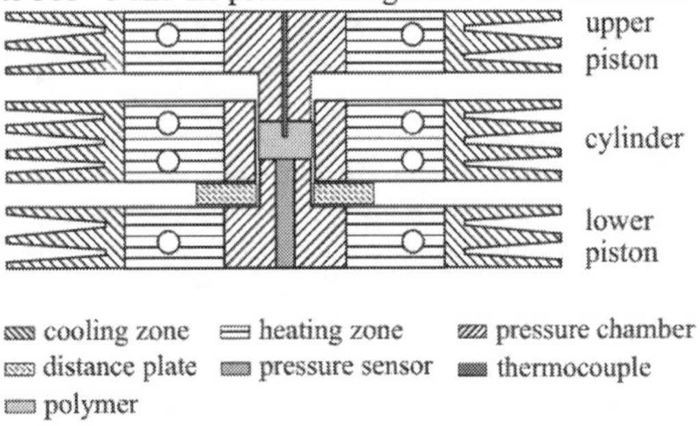

Figure 2 Schematic presentation of the press tool.

Results and Discussion

To develop a profound understanding of the impact of pressure on polymer properties and the process, the influence on volume, temperature and crystallization must be determined. Therefore, fundamental investigations on compressibility, compression heating and properties like e. g. density were made. The findings obtained during research will be presented in the following.

Compressibility

Submitting amorphous thermoplastics to compression, the individual volume contents' values of compressibility are of crucial importance. The volume consists of the occupied (molecular volume and oscillation volume) and the free volume. The molecular volume is regarded as incompressible. Inside the melt, the volume of oscillation expansion may decrease, as can the free volume. In a solid body, only the volume of oscillation expansion can be compressed (Rudolph et al., 2009a). This is why compressibility in the melt constantly decreases with temperature, suddenly gets lower at the point of glass transition, and remains at a constantly low level in the solid body (Breuer and Rehage, 1967; Hellwege et al., 1962). It was therefore required to measure compression volumes, in order to investigate processes occurring during melt compression and glass transition. Additionally, the effect of compression speed was to be determined.

Thus, for the purpose of compressibility measurement, pressure was applied at various, but constant compression speeds at various melt temperatures, and the course of volume determined. Figure 3 presents the course of volume for various melt temperatures and under pressurization from 200 bar to 3300 bar at 10,000 bar/min.

This figure shows that volumes are high at the beginning of the trials. This is due to thermal expansion leading to high specific volumes mostly resulting from the actual free volumes (Dlubek et al., 2007). The higher the volume, the better compressible is the plastic. Curves do not show a parallel, but intersecting course. For high melt temperatures, the intersection can be found at high pressures.

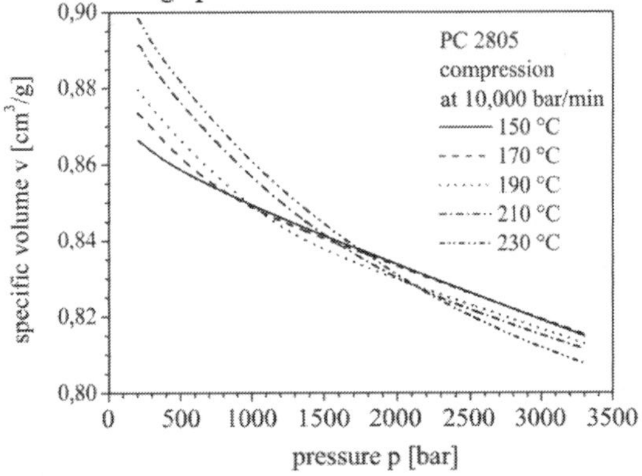

Figure 3 Courses of volumes under compression at 10,000 bar/min, from 200 bar to 3300 bar, shown for melt temperatures of 150, 170, 190, 210 und 230 °C.

What is more, it can be seen that volume changes over the measured pressure range are most significant with the highest melt temperature, and is least pronounced with the lowest melt temperature. This must be contributed to the fact that glass transition is only achieved at high pressures, if temperatures are high. Following solidification, which takes place here, further compression causes less pronounced volume changes in the solid body, due to its low compressibility. Under the same pressure, but at higher melt temperature, the melt still remains molten, meaning it can more easily be compressed. The volume can be further compressed, until, as pressure is increased, glass transition is reached here too, and compressibility altered. These findings comply well with Breuer and Rehage (1967).

Comparing the individual compression speeds at constant melt temperature (Figure 4), it reveals that under slow compression a lower specific volume can be achieved, than is possible under fast compression. From Dlubek et al. (2007), it may be deducted, that the occupied volume changes abruptly. Time-dependant relaxation must be attributed to free volume or to the macro-Brownian oscillations, respectively. Under fast compression, the material has less time to get close to the new state of equilibrium. The volume is unable to follow the change in pressure fast enough, resulting in higher volume.

Comparing compression volumes at a specific pressure to the volumes resulting from isobaric cooling, significant deviations are revealed in the specific volume. To allow a comparison, cooling as well as compression were performed at slow rates. The results are presented in a pvT diagram in Figure 5.

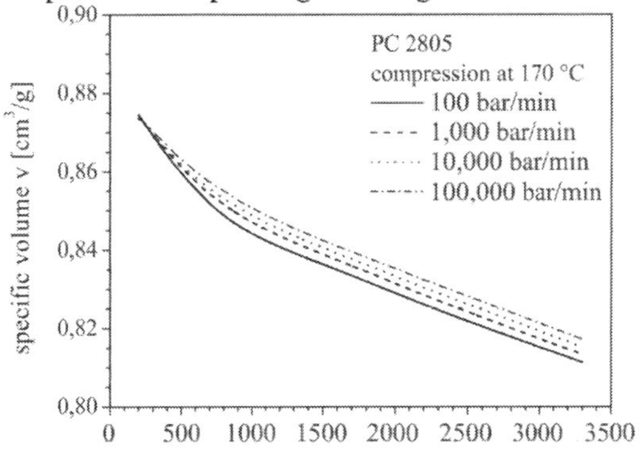

Figure 4 Courses of volumes during compression from 20 MPa to 330 MPa, at various compression speeds from 100 to 100,000 bar/min and 170 °C melt temperature.

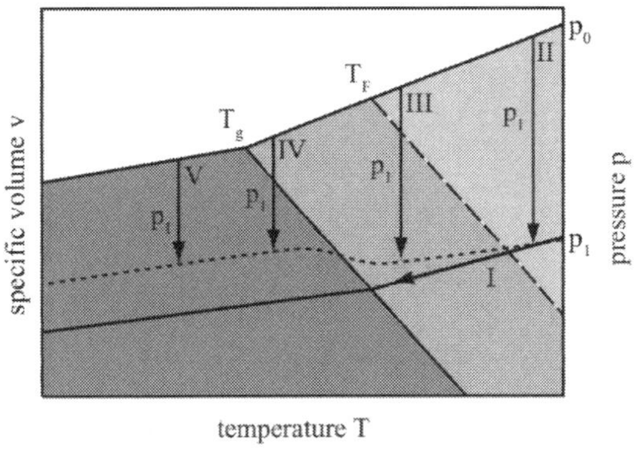

Figure 5 Volumes resulting from compression under constant pressure p_1 in relation to the plastic's original state; I: isobaric cooling, II: melt compression, III: melt and rubber compression, IV: rubber and solid body compression, V: solid body compression.

The figure displays glass transition T_g, as well as the softening temperature T_s, representing the end of the entropy-elastic range. It is assumed that the dependence of softening temperature on pressure (dashed line) is analogous to glass transition. The isobaric course of cooling at pressure p_1 can be followed along path I. The material is compressed at high melt temperatures and than cooled at constant pressure. If the melt is cooled, followed by compression, while still a melt (II), the compression volume corresponds to the volume at isobaric cooling. In case the compression ends in the entropy-elastic range (III), the achieved compression volume is higher than the volume shown in isobaric measurements. This deviation is even more significant, if compression leads to solidification (IV), or when a solidified body is compressed (V). In contrast to shrinkage during isobaric cooling, the entropy-elastic range only affects the compressibility of the polymer. Figure 5 also presents the isobaric path obtained from the compression measurements (dotted line). Here, deviations are even more pronounced.

As compression speed is increased (not shown), deviation between the individual paths increase, as the dotted line shifts towards higher volumes.

Compression heating

The specific volume given in the figures shown above are all related to temperature. During compression the material experiences a temperature increase, that might have to be taken into account. To determine this compression heating, melts were submitted to pressure at various temperatures. Several tests were performed, each time stepping up pressure by the same factor.

Figure 6 shows the compression heating measured for PC 2805 at various melt temperatures, in relation to the pressure applied. Under fast compression, the temperature increase is close to adiabatic, so heat conduction to the mold wall can be neglected. As expected, the course of temperature increase is stepped up with higher melt temperature, and even more with higher pressure. Expansion cooling reveals the same rate as compression heating (not shown). The process is thus reversible.

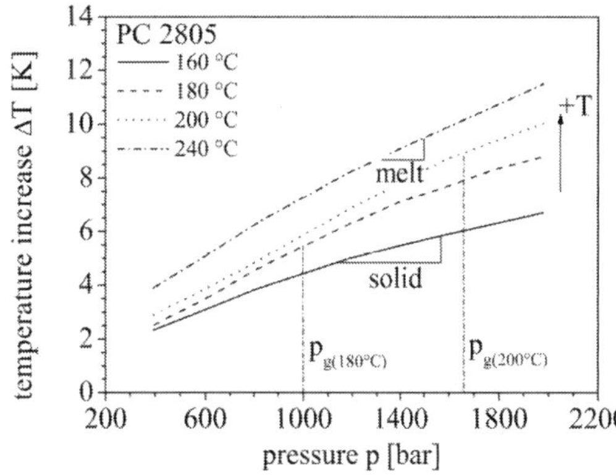

Figure 6 Compression heating of PC at various temperatures in relation to amount of pressure, $T_{g(1\ bar)}$ = 150°C.

The choice of appropriate parameters can determine the type of process, i.e. either melt compression at 240 °C, compression with gradual solidification (180 °C and 200 °C) or mere solid body solidification at 160 °C. It can be seen that the temperature increase of the solid is smaller than that of the melt (slope of the 160 °C and 240 °C-isotherms in Figure 6). The phase change along the isotherms of 180 °C and 200 °C can be seen as a change in the slope of the isotherms. This change is identical to the glass transition pressure p_g for the respective temperature (p_g(180°C) = 100 MPa, p_g(200°C) = 166 MPa/ dot and dash line in Figure 6).

To calculate compression heating, the well-known Joule-Thomson approach (Breuer and Rehage, 1967; Gilmour et al., 1978) was used, which applies best to incompressible media:

$$\Delta T_{Joule-Thomson} = \frac{\alpha \cdot T}{\rho \cdot c_p} \cdot \Delta p \qquad (1)$$

The density was taken directly from the compression tests. The volume expansion coefficients of the melt and the solid were derived by differentiation of the isobars of a pvT-diagramm and taken as constant ($\alpha_{vmelt} = 6{,}71*10^{-4}\,K^{-1}$; $\alpha_{vsolid} = 2{,}24*10^{-4}\,K^{-1}$). The linear thermal expansion coefficient is a third of the volume coefficient. The temperature dependence of the specific heat capacity was used as derived from DSC runs.

The best compliance between measurement and the numeric solution was achieved (Figure 7)

- for the melt by taking the volume expansion coefficient of the oscillation volume $\alpha = \alpha_{v,osc}$,
- for the solid (changing at the at the glass transition pressure p_g) by using the linear thermal expansion coefficient α_l

The first finding indicates that the compression heating of the melt is more dependent on the oscillation of the molecular chains than of free volume between the chains. No energy is needed to squeeze out the free volume. The second indicates that a change from hydrostatic pressure to a linear force occurs during compression. This is another proof for the occurrence and the importance of the solidification and the glass transition pressure p_g (Rudolph et al., 2009b).

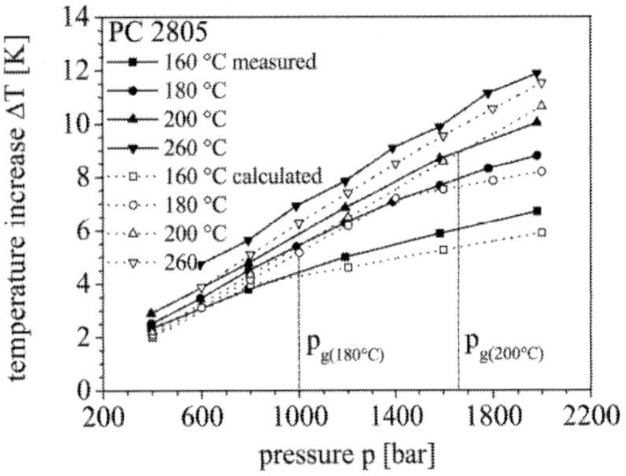

Figure 7 Measured temperature increase compared to the calculated increase obtained from equation 1.

In a nutshell, the temperature increase ranges between 3-5 % of the initial temperature. For processes with more than 1000 bar pressure the temperature decrease should be regarded and the temperature values should be used for the pvT-diagram.

With semi-crystalline polymers the influence of pressure on the temperature is even higher because in addition to compression heating they generate a heat of crystallization. Figure 8 shows the temperature change upon compression with 1000 and 2000 bar in the pvT-device at slow compression. As the melt temperature is the same in both runs and the compression with 2000 bar takes twice the time as with 1000 bar, the temperature increase should be equal. However, the measured temperature increase is almost twice as high at higher pressure. This is due to the additional heat of crystallization which is released during the 2000 bar compression. This can be seen even better in Figure 9. Here the material was compressed with 2000 bar at a higher melt temperature than in the runs in Figure 8. The crystallization occurs not instantly but under isothermal conditions with a short delay.

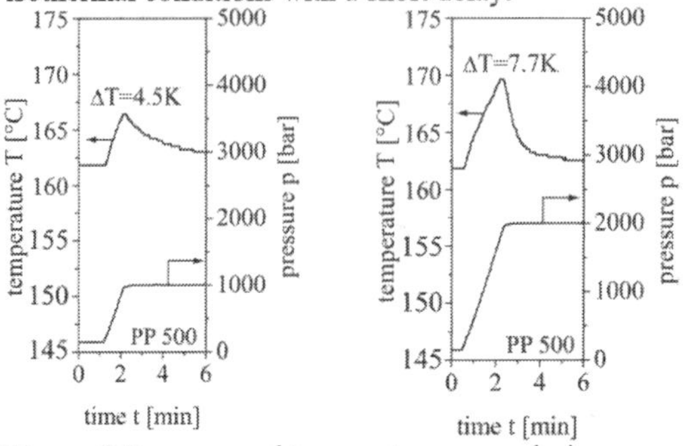

Figure 8 Pressure and temperature curve during compression of PP at 162 °C; left: 1000 bar, right: 2000 bar.

The first peak indicates the compression heating and has a similar height as the 1000 bar-peak in Figure 8. The second peak occurs due to crystallization. It starts just before the compression heat has declined completely. The crystallization heat is even higher than the compression heat. Within the shown time frame the temperature has not reached the base line again. That means at the given temperature and pressure crystallization is not finished yet.

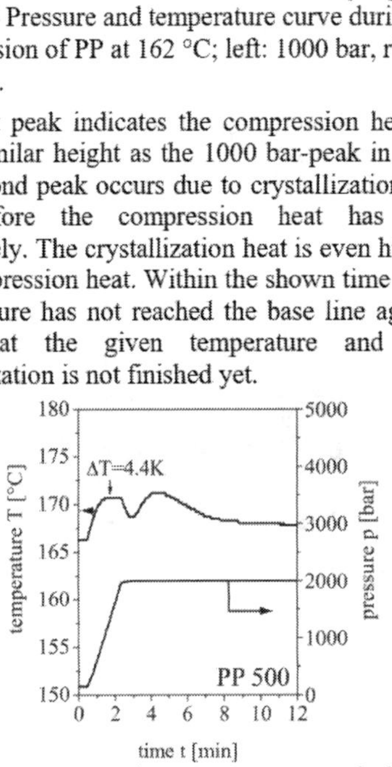

Figure 9 Pressure and temperature curve during compression of PP at 167 °C and 2000 bar.

Compression Induced Crystallization

It is well known that the melt temperature is increased with pressure. That means on the one hand nucleation starts at higher temperatures, and on the other hand crystallization growth can be slowed down or hindered (van Krevelen, 1978; Moneke, 2001). Kuenkel (2005) showed that the spherulites are growing bigger with increasing pressure. Additionally, the increased melt temperature influences the lamellae thickness, which is related to the mechanical properties of the material. As the fold length of crystal lamellae is increasing with temperature, it is increased by pressure as well. The limiting factor is the total chain length (van Krevelen, 1978; Kazmierczak et al., 2005).

To investigate the impact of pressure on density, morphology and lamellae thickness specimens were molded in the pvT device. During the runs the specific volume, cavity temperature and the applied pressure was recorded. Figure 10 shows an example of such a run.

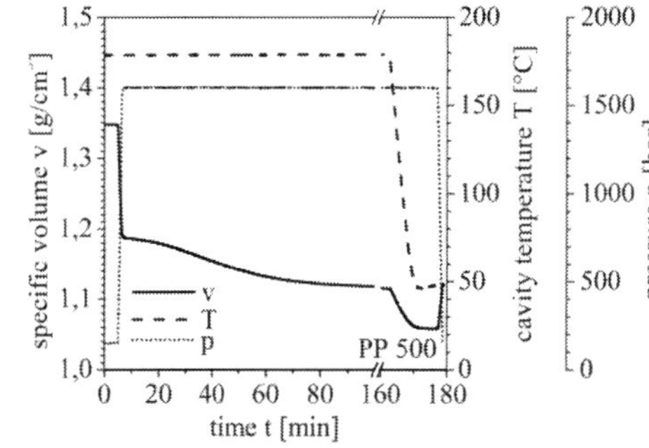

Figure 10 Course of volume under pressure and temperature changes; cavity (=melt) temperature): 180 °C, applied pressure: 1600 bar, holding time: 155 min.

Due to the increase in pressure (dotted line) at the melt temperature of 180 °C (dashed line), the specific volume (solid line) is decreasing. As soon as the pressure has reached 1600 bar, the specific volume is constant for some minutes. But as soon as crystallization starts, the material keeps shrinking for 155 minutes. This shrinkage time is defined as the isothermal crystallization time t_c. Subsequently, the temperature is decreased to 50 °C, which results in a thermal shrinkage of the specimen. As a last step the pressure is released, which causes an expansion of the specimen, and the specimen is demolded. All runs were performed at 180 °C and for a) p = 1600 bar, t_c = 155°C, b) p = 1800 bar, t_c = 98°C and c) p = 2000 bar, t_c = 30°C.

Compared to the isobaric compression volume at the specific temperature and pressure, the specific compression volume is higher (not shown). The difference is increasing with pressure.

After molding, the density of the specimens is measured and a DSC run is performed to determine the melting enthalpy of the specimens as well as to compare the peak width and length. It was found that the isothermal crystallization time t_c as well as the density and the melting enthalpy are decreasing with increasing pressure, Figure 11.

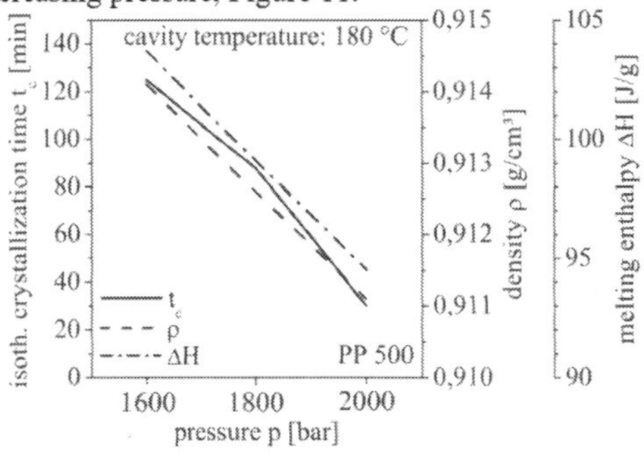

Figure 11 Decrease of isothermal crystallization time, density and melting enthalpy with increasing pressure at a temperature of 180 °C.

A comparison of the peak´s width and length reveals that the lowest pressure leads to the highest and widest peak, Figure 12. The peak temperature is also getting higher with decreasing pressure. By taking the theory of lamellae thickness into account (Ehrenstein et al., 2004) this leads to the conclusion that the widest peak (1600 bar) has the thickest lamellae, because the most energy is consumed to melt those.

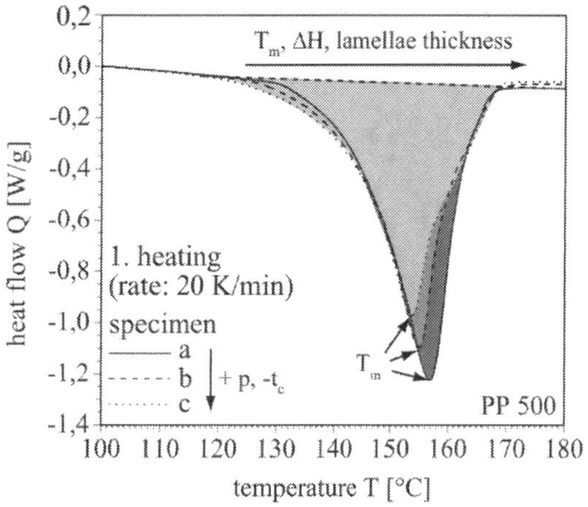

Figure 12 DSC-runs (1. heating) for the 3 different pressures in relation to the lamellae thickness (Ehrenstein et al., 2004)

In Figure 13 the morphology of the specimens is compared. The biggest and most regular spherulites are gained by the lowest pressure (1600 bar). Crystallization at the highest pressure leads to an almost lamellar and inhomogeneous structure.

This is contrary to the expected results from the work of Kuenkel (2005) but can be explained with the interaction of nucleation rate and crystal growth speed. With increasing pressure the chains are pushed closer together. Viscosity as well as the nucleation rate is increasing.

When there is still enough free volume for movement between the chains, the crystals grow to spherulites of a certain size, depending on the surrounding free volume. The spherulites become larger and the lamellae thicker. In the case shown here, the benefits of pressure are on a turning point. When the pressure is increased, the crystal growth is hindered by the reduction in molecular free volume. However, for the same reason, crystallization occurs faster, reflected by the crystallization time t_c. The resulting crystallinity is smaller. As the spherulites are the densest possible package, the specimens exposed to highest pressure have the lowest density in the studied pressure range.

Conclusions

The results of this work show that the specific volume of amorphous thermoplastics is dependent on the state of the material by the time the pressure is applied. As main impact factors the pressure, temperature and also the compression speed were detected. It was found that the viscoelasticity is not negligible as it is during cooling. From these results as well as the state of the art, we can conclude, that temperature, cooling rate, pressure and compression speed determine the specific volume and therefore shrinkage of a part (Rudolph et al., 2009b).

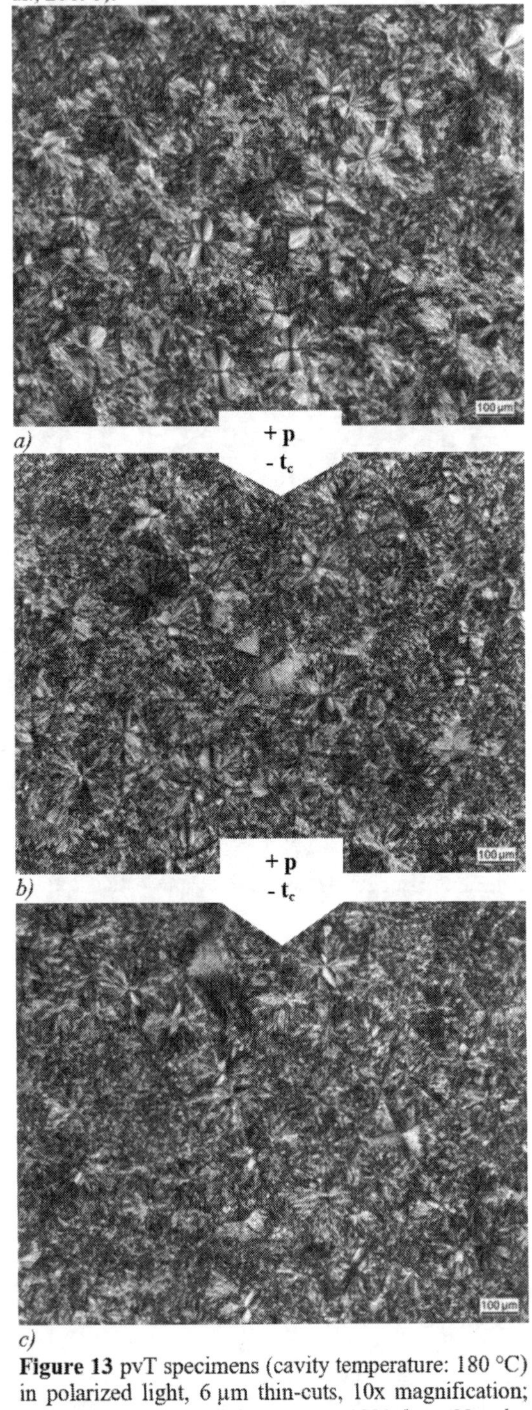

c)

Figure 13 pvT specimens (cavity temperature: 180 °C) in polarized light, 6 µm thin-cuts, 10x magnification; top: 1600 bar, 155 min; center: 1800 bar, 98 min; bottom: 2000 bar, 30 min;
p = pressure, t_c = isothermal crystallization time.

What is more, the volume is not defined by one value of p and T, but is dependent on the pathway determined by processing and the succession of cooling and compression. Therefore, the speed of compression, or injection, affects the material behaviour of the polymer. In addition, the specific volume of the semi-crystalline PP is also pathway dependent, as the volume in the melt reached by isobaric cooling differs from the isothermal compression volume.

In detail, these findings affect *thermal expansion* and *compressibility*. The two techniques used to measure these properties, lead to different values. Both properties are defined as the volume change in relation to the initial volume due to changes in temperature or pressure, respectively, Figure 14. Compressibility should be calculated from a compression test and thermal expansion from an isobaric cooling process.

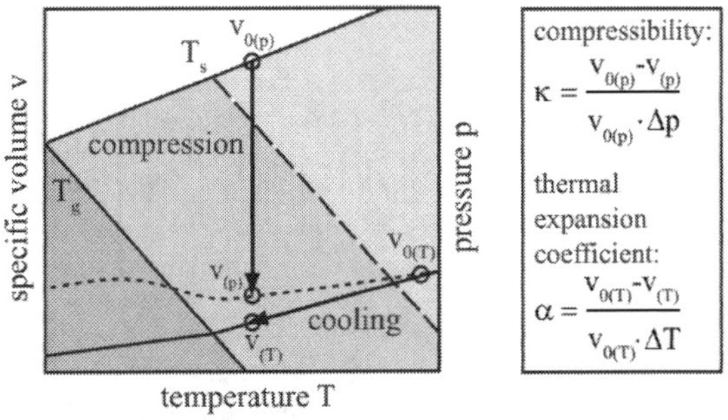

Figure 14 Difference in compressibility and thermal expansion coefficient due to different procedures; detail from Figure 5.

As both values are normally calculated from one test, compressibility can be calculated as $v_{(T)}$ and thermal expansion as $v_{(p)}$, leading to different results. For example, the volume change during isothermal compression is smaller – consequently smaller thermal expansion coefficient and compressibility – than the values that result from an isobaric cooling process. Additionally, the change in compressibility during compression and the differences between temperatures, are not observable in the isobaric cooling procedure.

For injection molding all these results lead to the conclusion that during filling, packing as well as cooling the temperature and pressure field, as well as their rate and sequence, has to be taken into account. In addition to the thermal expansion and compressibility, also crystallization, viscosity and heat conductivity are dependent on these parameters (Rudolph et al., 2009c). Filling or cooling simulation can be improved by using the appropriate values in the calculation. From that knowledge, there is a demand for more flexible simulation tools that allow the implementation of more complex models.

Regarding *crystallization*, pressure is affecting the nucleation rate and the crystal growth speed. On the one hand the nucleation rate is accelerated as the chains are moved towards one another. On the other hand the crystal growth can be hindered as the free volume is reduced. In Figure 15 the temperature dependence of crystallization is shown. According to the presented results, the crystallization maximum is not only affected by temperature but also by pressure. That is why the pressure effect is introduced in this figure. The maximum is shifted to higher temperatures, if pressure is applied. At every temperature between T_g and T_c one specific pressure level results in a crystallization maximum. Even so the experiments where performed isothermally, some conclusions can be drawn for the injection molding process. As described before, there are changing temperature and pressure fields within the part during processing. As can be seen in Figure 11, the density of semi-crystalline thermoplastics is also affected by pressure. That means that the crystallinity gradient in plastic parts is not only due to different cooling rates and shear at the mold wall, but also due to the pressure gradient within the part.

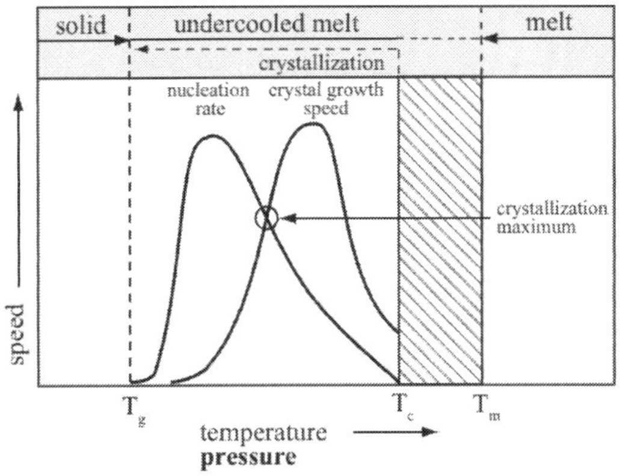

Figure 15 Temperature dependence of the nucleation rate and crystal growth speed by cooling from the melt; T_g = glass transition temperature, T_c = crystallization temperature, T_m = melting temperature (Ehrenstein et al., 2004).

This leads to the conclusion that there is an ideal pressure and temperature set for crystallization that leads to parts with high density, homogeneous structure and, due to the bigger lamellae thickness, improved mechanical properties.

The occurring *solidification by pressure* is also an important topic, normally not regarded. In other words, in addition to the melt solidification at the mold surfaces due to heat conduction, the melt will also solidify as a result of pressure increases during filling and/or packing. The typically unsolidified melt core can be hardened as it reaches p_g, (1) in Figure 16. In that case, the gate already freezes during filling as the injection pressure reaches p_g.

The lower the melt temperature, the lower are the pressures at which the material solidifies. Depending on the melt and mold temperature a proportionate zone solidifies at the border of the part.

The solidified, rigid border is pulling on the warmer and formable melt, (2) in Figure 16. As the shrinkage of the solid is smaller than that of the melt, the contraction at the mold wall is smaller than in the center. That leads to a density gradient over the part volume. In the worst case sink holes are found in the center of the part.

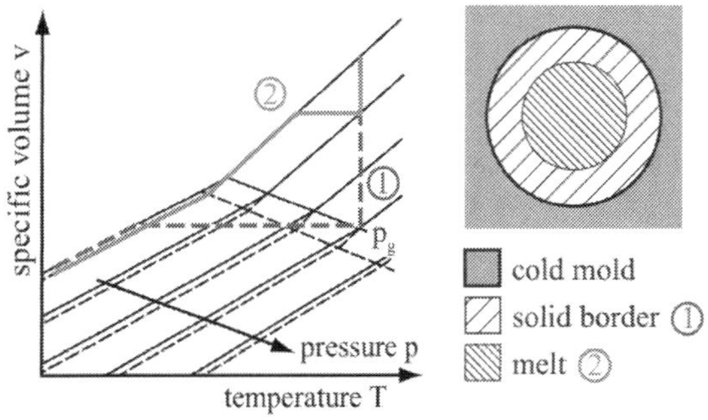

Figure 16 Example for a cooling and solidification scenario during injection molding; left: volume course of border and core in a pvT-diagram; right: corresponding mold and cavity.

This is only one example of the influence of the temperature and pressure field on shrinkage and part quality. By analysing the impact and resulting properties with the extended knowledge on pvT-behavior, new ways of optimizing and dimensioning plastic parts with higher precision can be derived. The next step in this research is to focus on the analysis of the interactions between these properties as well as the implementation of the pressure dependence of properties into injection molding simulation.

Acknowledgements

We thank the DFG for the financial support of this work and Bayer Material Science AG for providing the material. We would also like to thank Mrs. Dr. Kühnert for her permanent support during these studies.

References

Menges et al., 1977: Menges, G.; Thienel, P.; Kemper, W., "Das physikalische Verhalten von Thermoplasten bei der Aufnahme von pvT–Diagrammen unter verschiedenen Messbedingungen", Plastverarbeiter, **28**, 632-637 (1977).

Schmidt et al., 2000: Schmidt, M.; Olsson, M.; Maurer, F.H.J., "Macroscopic pressure–volume–temperature properties versus free-volume characteristics of isotropic pressure-densified amorphous polymer glasses", Journal of Chemical Physics, **112**, 11095-11106 (2000).

Engelsing, 2000: Engelsing, K.; "Einfluß des freien Volumens auf das verarbeitungsabhängige Deformationsverhalten spritzgegossener amorpher Thermoplaste", PhD Thesis, Chemnitz, University of Technology (2000).

Rudolph et al. 2009a: Rudolph, N.; Kühnert, I.; Schmachtenberg, E.; Ehrenstein, G. W., "Pressure solidification of amorphous thermoplastics", Polymer Engineering & Science, **49**, 154 (2009).

Rudolph et al. 2009b: Rudolph, N,; "Druckverfestigung amorpher Thermoplaste", PhD Thesis, Erlangen, University of Erlangen-Nuremberg (2009); www.kunststofftech.com.

Rudolph et al. 2009c: Rudolph, N.; Kuehnert, I.; "High Precision Parts by Compression Induced Solidification (CIS)", Proceedings of the Polymer Processing Society 25th Annual Meeting, Goa, India (2009)

Breuer and Rehage, 1967: Breuer, H.; Rehage, G., "Zur Thermodynamik der glasigen Erstarrung", Kolloid-Zeitschrift & Zeitschrift für Polymere, **216–217**, 159-179 (1967).

Hellwege et al., 1962: Hellwege, K. H.; Knappe, W.; Lehmann, P., "Die isotherme Kompressibilität einiger amorpher und teilkristalliner Hochpolymerer im Temperaturbereich von 20-250°C und bei Drücken bis zu 2000 kp/cm²", Kolloid-Zeitschrift & Zeitschrift für Polymere, **183**, 110-120 (1962).

Dlubek et al., 2007: Dlubek, G.; Pionteck, J.; Shaikh, M. S.; Häußler, L.; Thränert, S.; Hassan, E. M.; Krause-Rehberg, R., "The free volume in two untreated, pressure-densified, and CO2 gas exposed polymers from positron lifetime and pressure-volume-temperature experiments", e-Polymers, **108** (2007).

Gilmour et al., 1978: Gilmour, I. W.; Trainor, A.; Haward, R. N., "The thermoelastic effect in glassy polymers", Journal of Polymer Science, **16**, 1277-1290 (1978)

van Krevelen, 1978: van Krevelen, D.W., "Cristallinity of polymers and the means to influence the crystallization process", Chimia, **32**, 279-294 (1978).

Moneke, 2001: Moneke, M., "Die Kristallisation von verstärkten Thermoplasten während der schnellen Abkühlung und unter Druck", PhD Thesis, University of Darmstadt (2001).

Kuenkel, 2005: Kuenkel, R.., "Auswahl und Optimierung von Kunststoffen für tribologisch beanspruchte Systeme", PhD Thesis, Erlangen, University of Erlangen-Nuremberg (2005).

Kazmierczak , 2005: Kazmierczak, T.; Galeski, A.; Argon, A. S., "Plastic deformation of polyethylene crystals as a function of crystal thickness and compression rate", Polymer, **46**, 8926-8936 (2005).

Ehrenstein et al., 2004: Ehrenstein, G. W., Riedel, G., Trawiel, P.: Thermal Analysis of Plastics, Hanser Gardner Publications, Cincinnatti (2004)

INJECTION MOULDING PROCESS CONTROL THROUGH AT-PROCESS PRODUCT MASS ESTIMATION

M. Umar [1], R. Speight [2], P. D. Coates [1], L. Mulvaney-Johnson [1]*

[1] *IRC, University of Bradford, West Yorkshire, UK* – m.umar1@bradford.ac.uk; p.d.coates@bradford.ac.uk; l.johnson@bradford.ac.uk

[2] *Autodesk Moldflow – Melbourne, Australia* - russell.speight@autodesk.com

It is accepted that accurate control of the injection moulding process is important in order to maintain satisfactory product quality, however that is measured. The dimensions of a product and other quality features have often been linked to the mass of the product and is a popular process quality measurement within a production setting. A strategy has been developed that utilises key injection moulding machine process signals along with material pressure specific volume temperature (pvT) data to calculate the mass of a moulded product. Key issues such as the leakage of melt past the check-ring are considered and monitored using screw torque measurements. In addition, modelling of possible flow through the check-ring is considered along with the effect of melt decompression and pack pressure settings. The model is shown to be accurate within 3% under certain conditions and has potential applications for the conventional injection moulding process. It can be shown that accurate determination of a short shot is possible that could have possible applications with the gas assisted injection moulding process.

Introduction

The accurate control of the injection moulding process has long been an issue, where various methods have been researched in order to maintain product stability. A particular method is to control the cavity pressure in order to maintain high quality injection moulded products (Michaeli and Gruber, 2004). The quality of injection moulding products has increasingly become important within the industry, where product mass is regarded as one of the most important product quality measurements (Fung et al., 2005). A method has been developed to determine the product mass, where the swept volume by the injection screw coupled with thermodynamics of the material is used. Accurately determining the product mass will provide accurate control of the injection moulded products. A key variable in accurately determining product mass is accounting for the loss of melt back through the check-ring. The check-ring is a key feature of the injection moulding screw. During melt plastication the check-ring pushed forward into its open position to allow polymer melt to flow to the front of the injection screw. During melt injection into the mould cavity as the injection screw moves forward the check-ring is pushed back and closes to. The initial model to estimate product mass does not account for this loss of melt back through the check-ring. The work here presents a case study using the model to determine product mass, to demonstrate the errors which occur between the actual product mass and the predicted mass. A typical check-ring has been modelled to determine the loss of melt back through the check ring.

Experimental

Estimation of product mass

A model has been developed based on the theory that the amount of polymer melt delivered to the mould cavity is equal to the quantity of material leaving the injection moulding machine.

$$Weight = \frac{\pi D_b^2}{4}\left[\left(\frac{X_{start\_inj}}{V_{start\_inj}} - \frac{X_{end\_inj}}{V_{end\_inj}}\right) + \left(\frac{X_{start\_hold}}{V_{start\_hold}} - \frac{X_{end\_hold}}{V_{end\_hold}}\right)\right]$$

The estimation of product mass is determined through splitting the melt delivery into the mould into 2 stages, the velocity controlled melt injection stage and the pressure controlled pack/hold stage. Figure 1a) illustrates the injection moulding process through using a process signal map. With reference to Figure 1, A-B represents the velocity controlled melt injection stage, B-C represents the pressure controlled pack/hold stage, and beyond C is cooling and melt plastication. Point A is the start of melt injection (Xstart_inj) and is determined through extrapolating the nozzle pressure profile down to 0 bar pressure (Figure 1b). The corresponding screw position is then selected as the start of melt injection (Xstart_inj). Point B represents the end of velocity controlled melt injection (Xend_inj), and is selected as the screw position at which the moulding process transfers from velocity control to pressure control. This point is also taken as the screw position at the start of pressure controlled pack/hold pressure stage (Xstart_hold). Point C is the end of the hold pressure stage and is selected as the screw position at the corresponding elevated hold pressure. Vstart_inj, Vend_inj and Vend_hold are the specific volume at the start of melt injection, end of melt injection and end of hold pressure stage respectively. The specific volume, a function of pressure and temperature, is often referred to as the pressure-volume-Temperature (pvT) relationship and is modelled using the modified 2-domain pvT Tait equation (Chang et al., 1997). The model coefficients are determined through testing carried put at Moldflow Plastics Lab (MPL).

Hold pressure investigation

As mentioned previously the injection moulding process consists of the 2 stages of melt delivery into the mould cavity: - the velocity controlled melt injection stage and the pressure controlled pack/hold stage. Thus one of the key process parameters used within the injection moulding process is the pack/hold pressure. Upon completion of the velocity controlled stage (95-99% filling of mould cavity) the moulding process transfers to pressure controlled stage. The hold pressure is applied to deliver additional polymer melt into the mould cavity in order to compensate for shrinkage of the polymer which occurs during cooling. Increasing the hold pressure effectively increases the amount of polymer melt delivered into the mould during this pressure controlled stage. Therefore varying the hold pressure provides a good indicator to the accuracy of the model. Hence an experiment is carried out where a range hold pressures, shown in Table 2, are applied to a constant velocity controlled stage (i.e. the process parameters in relation to the velocity controlled stage are kept constant throughout the hold pressure range). Furthermore, a second experiment is carried out where a dwell before extrusion (prior to melt plastication commencing) of 1 bar pressure for a duration of 5 seconds is used. This allows a comparison to be made between the predicted product mass for both cases. The experiments are carried out using a 100 tonne all-electric drive FANUC A-series moulding machine (Figure 2a)). The mould is a tensile bar (Figure 2b)) and the material used for this experiment is Rigidex HDPE 5050EA.

Effect of decompression on backflow

During the melt plastication the check-ring is in its open position so the polymer melt can convey forward through the check-ring and to the front of the injection screw. However, as melt is injected into the mould cavity the check-ring slides forward and closes to prevent polymer melt flowing back through the check-ring and into the screw channel. While the check-ring closes, during melt injection, melt flows back through the check-ring and into the screw channel. It is thought that applying a larger decompression than normal will allow sufficient screw travel, prior to contact being made between the injection screw and the polymer melt, in order for the check ring to close before the melt is injected to the mould cavity. Decompression is regarded as the amount screw movement which is applied through linear retraction of the injection screw upon completion of melt plastication (through screw rotation). This linear retraction of the injection screw is applied in order to decompress the prepared melt volume in front of the injection screw. Therefore a range of decompression values (Table 2) are used, where all other parameters in relation to the velocity controlled stage are kept constant. The investigation is carried out on the velocity control stage only, and the pressure control stage is eliminated by applying a 0 bar hold pressure for 40 seconds.

Modelling the check-ring

The closing of the check-ring can be determined through injection screw torque. A theory has been implemented to the FANUC B-series moulding machines, where the point at which the check-ring closes is determined through screw torque. FANUC state that the maximum screw torque which occurs during the velocity controlled melt injection stage is the point at which the check-ring closes. As melt flows back through the check-ring this causes the injection screw to rotate due to the screw channels (i.e. melt can only flow through the check ring through rotation of the injection screw). Therefore a torque is registered due to the rotation of the injection screw (FANUC). Using this principal the current FANUC A-series moulding machine at the

University of Bradford has been modified such that the check-ring closing point can be determined, through injection screw torque and is represented as extruder current. A typical profile showing the extruder current is shown in Figure 3, where the peak (point A) represents the point at which the check-ring closes and point B is V/P position. In order to determine the amount of polymer melt flowing back through the check-ring during the velocity controlled melt injection stage the check-ring has been modelled using Autodesk Moldflow Insight (AMI) software. Figure 4 shows the check-ring from the injection screw from the FANUC A-series moulding machine. The path the polymer melt takes as it flows back through the check-ring is modelled within Solid Edge and analysed using AMI (shown in Figure 5). The extruder current profile is used in conjunction with the nozzle pressure profile (Figure 3b)), where a linear relationship is assumed between the check-ring fully opened (time = 0seconds) and the check-ring fully closed (maximum extruder current- peak).

Incorporating backflow within the model to determine product mass

The initial model to predict product mass does not account for the flow of melt back through the check-ring throughout the early stage of velocity controlled melt injection. However, the analysis from modelling the check-ring is used to improve the accuracy of the model to predict product mass. The finalised model which incorporates the flow of melt back through the check-ring is tested on the experiment which uses a range of hold pressures described earlier in the hold pressure investigation.

Results and Discussion

Hold pressure investigation

Figure 6 shows the effect of hold pressure on the actual product mass and the predicted mass, where it is observed that as the hold pressure increases both the actual product mass and predicted product mass increases. As the hold pressure increases up to 200 bar hold pressure the actual and predicted product mass increase significantly and this is due to 'short shot' of the product. Beyond 200 bar hold pressure, with increasing hold pressure a slight increase in the actual product mass is observed, and this represents complete filling of the product. However the predicted product mass continues to increase significantly and the error between the actual and predicted product mass increases with increasing hold pressure. Figure 7 shows the effect of hold pressure on the actual product mass and predicted mass when using dwell of 1 bar for 5 seconds before extrusion. It is observed from Figure 7, between 0 bar hold pressure to 200 bar hold pressure the actual and predicted product mass follows the same trend as that observed in Figure 6. However the significant effect in the predicted product mass is seen from 200-700 bar hold pressure, where as a small increase in actual product is observed and a similar effect is seen in the predicted product mass. The error between the actual product mass and predicted product mass remains constant with increasing hold pressure. Therefore, the when applying dwell of 1 bar for 5 seconds prior to extrusion (melt plastication) to the injection moulding process, the model generates a more accurate estimation of product mass. The percentage error observed between the actual product mass and the predicted product mass is +8%. It is believed that the over prediction is purely due to the model not accounting for the loss of melt from the front of the injection screw back through the check ring and into the screw channel.

Effect of decompression on backflow

The effect of decompression on the actual product mass and the predicted product mass is shown in Figure 8. With increasing the decompression value (linear retraction of the injection screw - post melt plastication) the actual product mass increases. However the predicted product mass remains stable throughout the decompression range. The increase in product mass implies, the amount of melt lost from the front of the injection screw back through the check-ring and into the screw channel reduces at higher decompression values. However, this would then lead to the predicted product mass to decrease with increasing decompression. It is observed from Figure 8 that the predicted product mass does not decrease with increasing decompression. Thus, the increase in actual product mass with increasing decompression is due to the increase in the screw position at the start of melt injection and this is shown in Figure 9. The increase in the screw position at the start of melt injection (Xstart_inj) indicates that the shot size (melt volume in front of the injection screw prior to start of melt injection) increases as decompression increases. The increase in shot size is due to melt being drawn in front of the injection screw during the linear retraction of the injection screw. The larger the decompression applied the greater the amount of melt is drawn to the front of the screw, hence the actual product mass increases.

Modelling the check ring

Figure 11 shows the pressure at selected points throughout the check-ring. It can be seen the pressure at the injection location is constant throughout the velocity controlled melt injection stage. Similarly the pressure of the melt at locations close to the gap, at which the polymer melt flows through and into the screw channel, is also constant throughout the filling stage. This indicates that during filling a constant flow rate is present throughout the check-ring. Therefore the range of pressures observed at the injection moulding nozzle (Figure 3b)) nozzle are matched with the corresponding flow rates shown in Figure 12. As the pressure increases the injection time increases and the check-ring gap reduces to a point at which it closes and no further melt flows through the check-ring.. The flow rates can then be used to determine the amount flow back through the check ring given different injection rates.

Incorporating backflow within the model to determine product mass

Research is currently been undertaken to further improve the estimation of the amount of flow back through the check-ring through analysis using AMI. The model to estimates product mass is then to be modified accordingly to account for the loss of melt back through the check-ring.

Table 1 Range of hold pressures

Hold Pressure (bar)
0
50
100
150
200
300
400
500
600
700

Table 2 Range of decompression

Decompression (mm)
1
2
3
4
5
6
7
8
9
10
15
20
25
30
35

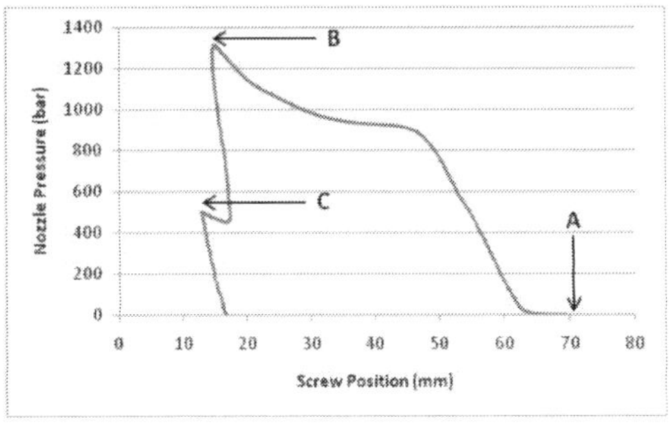

a) IM process described through screw position and nozzle pressure

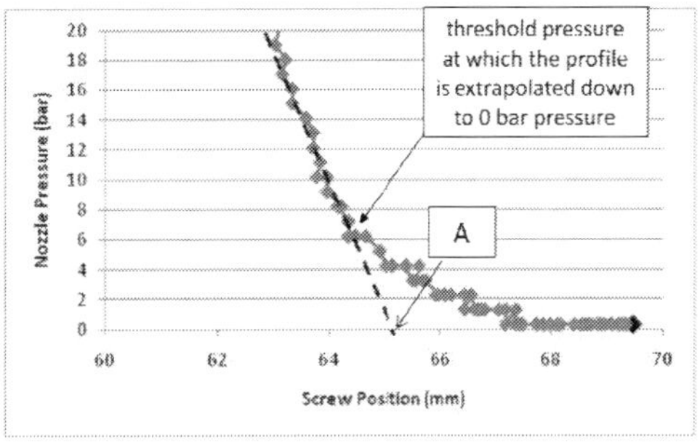

b) The point at which melt contacts the tip of the advancing injection screw

Figure 1 Injection moulding process maps (Umar et al., 2009)

a) Fanuc A-series injection moulding machine

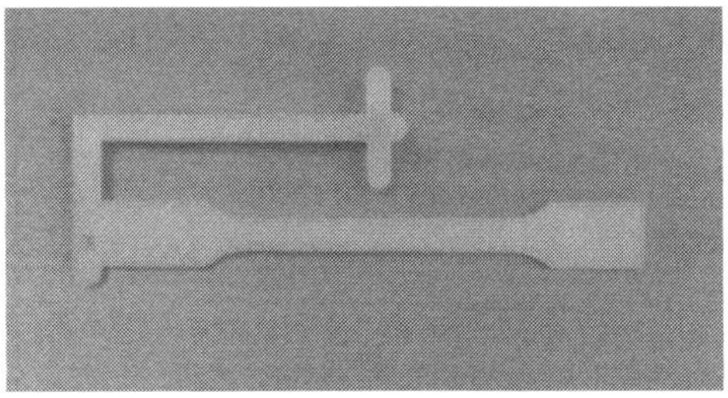

b) Injection moulded sample (tensile bar)

Figure 2 Injection moulding machine and moulded sample

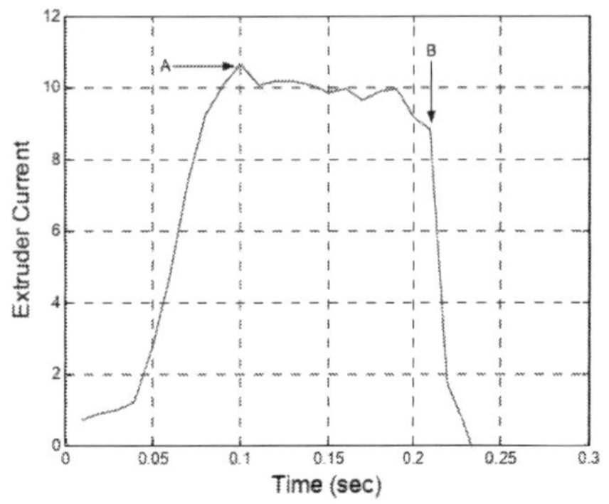

a) Extruder current against time

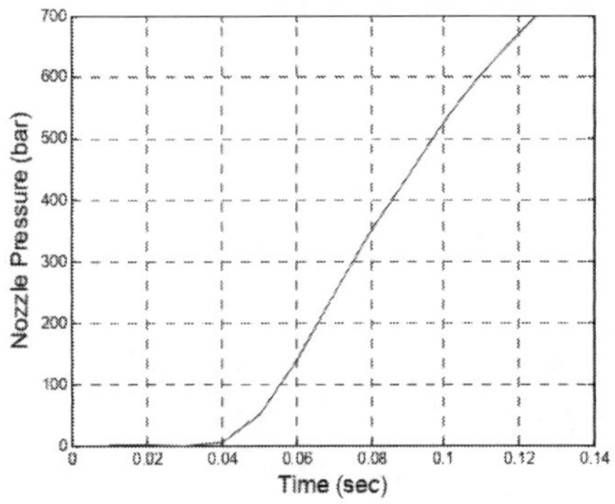

b) Nozzle pressure against of time

Figure 3 Extruder current and nozzle pressure profiles during the open - closing of the check-ring

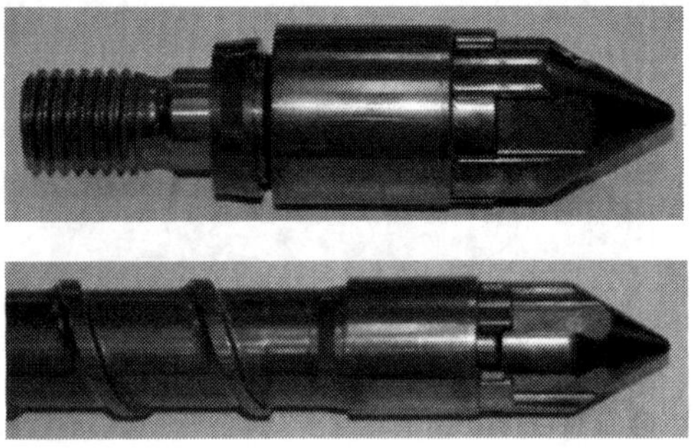

Figure 4 Check-ring from the FANUC injection screw

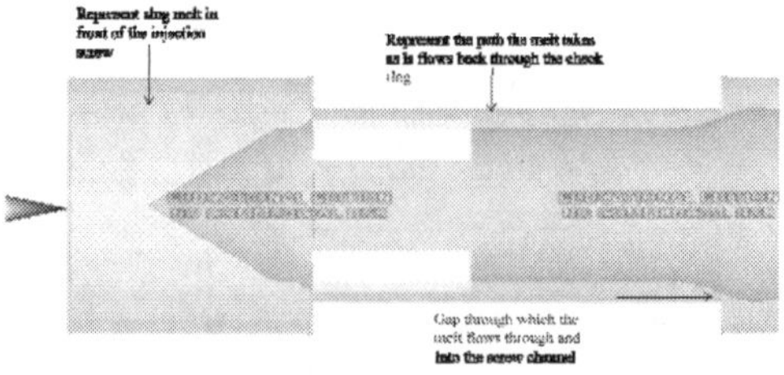

a) 3D solid representation of the flow of melt through the check-ring

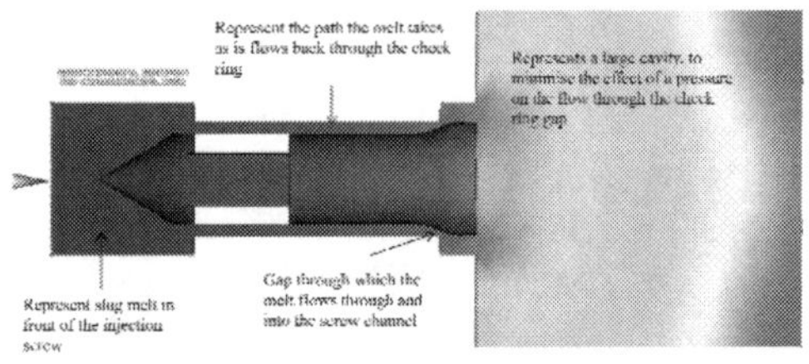

b) Cross-section of the flow of melt back through the check-ring

Figure 5 3D model of the path of the polymer melt as it flows back through the check-ring

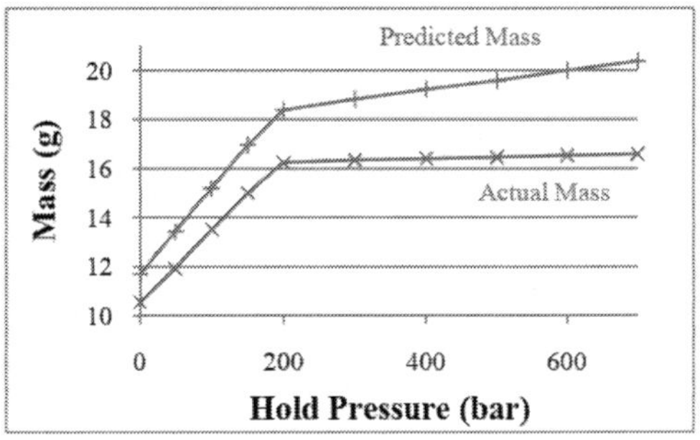

Figure 6 Effect of hold pressure on actual product mass and predicted product mass

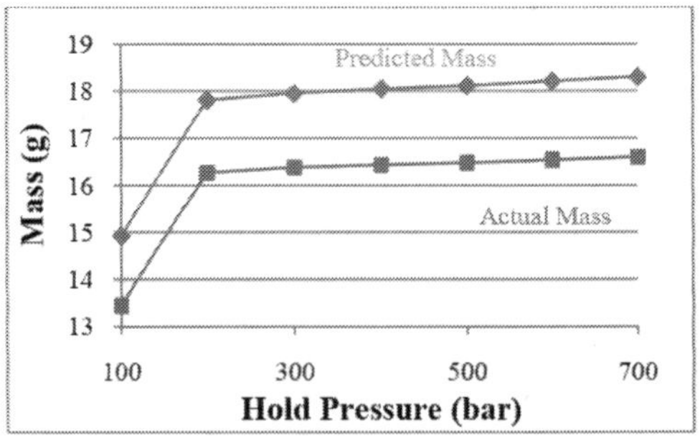

Figure 7 Effect of hold pressure with dwell (1 bar pressure, 5 seconds) before extrusion on actual product mass and predicted product mass

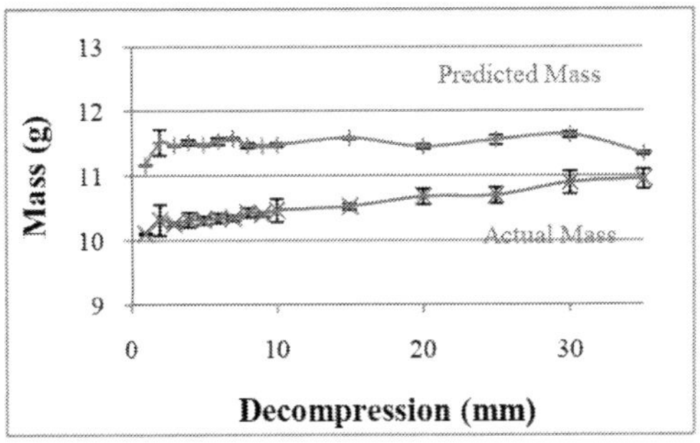

Figure 8 Effect of decompression on actual and predicted product mass

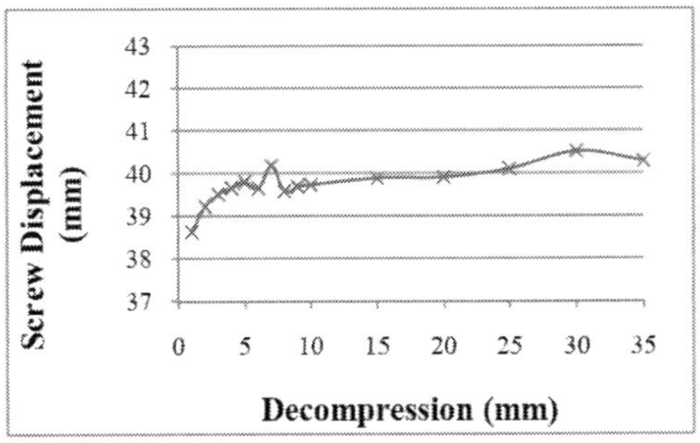

Figure 9 Effect of decompression on the screw position at the start of melt injection (Xstart_inj)

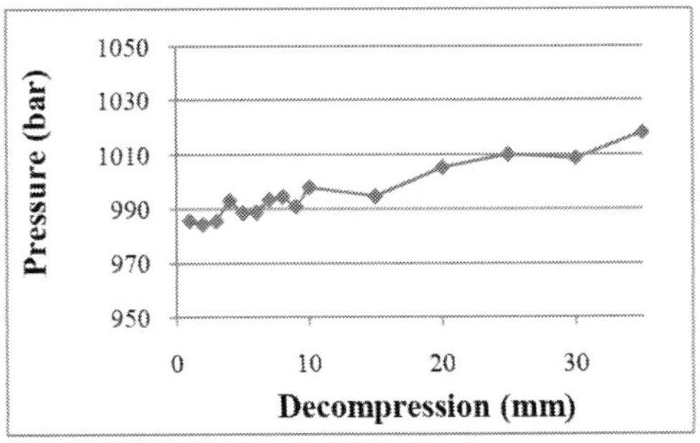

Figure 10 Effect of decompression on the peak pressure registered at V/P

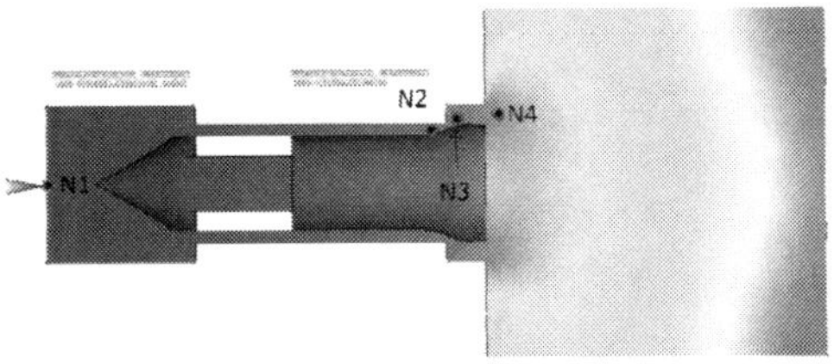

a) Representation of the selected nodes

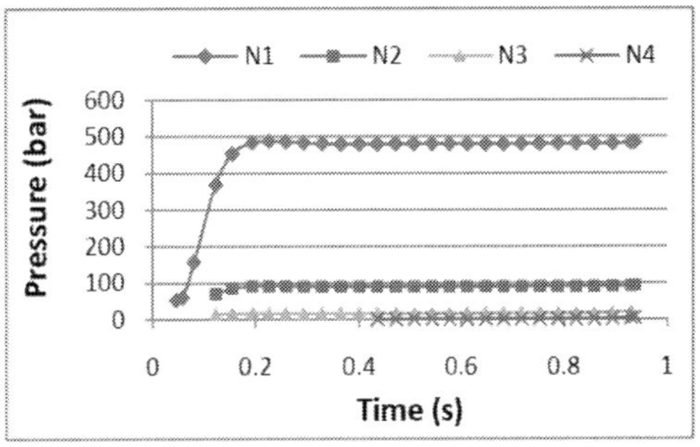

b) Pressure at the selected nodes

Figure 11 Pressure profile at the selected nodes as melt flows back through the check-ring

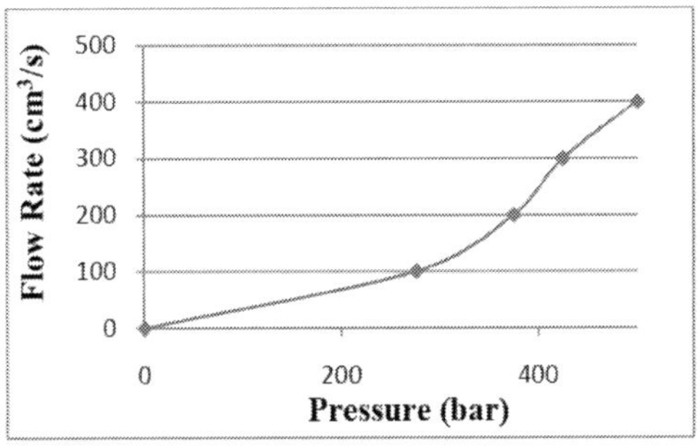

Figure 12 Flow Rate against pressure registered at the injection location (N1) for a fully open check ring

Conclusion

The injection moulding process has 2 stages of melt delivery into the mould cavity, the velocity controlled melt injection stage and the pressure controlled pack/hold stage. The actual product mass has been compared with the predicted product mass through a range of hold pressures. The results showed both the actual and predicted product mass increase with increasing hold pressure. However, at hold pressures greater than 200 bar the predicted product mass continued to increase significantly. During the same range of hold pressures a small change in actual product mass is observed. However when applying a dwell of 1 bar for a duration of 5 seconds before extrusion to the injection moulding process the predicted product mass follows the same trend as the actual product mass. Here the percentage error (+8%) remains constant throughout the hold pressure range. The over prediction is due to the loss of mass back through the check ring during the velocity controlled melt injection stage. The effect of decompression on the flow of melt back through the check-ring is investigated. A range of decompression values are used, where it is though that applying a larger decompression than normal may have a significant effect on the check-ring closing. It is concluded that the amount of decompression applied to the process has no effect on the flow of melt back through the check-ring. Rather decompression has a significant effect on the shot size (melt volume in front of the injection screw). Applying to little decompression can result in 'stringing' effect on the moulded sample and applying too much decompression can result in an increase in shot size. Therefore the correct amount of decompression required to be applied to the moulding process is a key issue. To further improve the accuracy of the estimation of product mass the check-ring from on the injection screw from the FANUC A series 100 tonne clamp injection moulding machine is modelled using Autodesk Moldflow Insight (AMI). The principle of monitoring the screw torque in order to determine the point at which the check ring closes (FANUC) is used combined with the nozzle pressure profile. Assuming linear behaviour between fully opened check ring and fully closed check-ring flow rates can be determined as the check-ring closes. The pressures at the corresponding flow rates are associated to a particular position of the check-ring, which can then be used to determine the total loss of melt back through the check-ring. Further study is to be carried out to in order to improve the accuracy of the estimation on product mass, using this method.

Acknowledgments

The authors would like to thank Autodesk EPSRC for the funding of this project. In addition the authors would like to thank for Autodesk Moldflow for both the funding and support of Autodesk Moldflow Insight (AMI) simulation package.

References

CHANG, R.-Y., HSIEN, Y.-C., LIN, F.-H. & HSU, C.-H. (1997) Modifying the Tait equation with cooling-rate effect to predict the pressure-volume-temperature behavior of semi-crystalline polymers: Model and experiments. Toronto, Can, Soc of Plastics Engineers, Brookfield, CT, USA.

FANUC B-SERIES INJECTION MOULDING MACHINE MANUAL

FUNG, K. T., GAO, F. & CHEN, X. (2005) Part weight prediction using capacitive transducer in injection molding. Boston, MA, United States, Society of Plastics Engineers, Brookfield, CT 06804-0403, United States.

MICHAELI, W. & GRUBER, J. (2004) Increasing quality by online control of the cavity pressure. Chicago, IL., United States, Society of Plastics Engineers.

MOLDFLOW Moldflow Testing Overview. *Moldflow Plastics Lab*.

UMAR, M., MULVANEY-JOHNSON, L., SPEIGHT, R., BRINCAT, P., BAKHAREV, A. & COATES, P. D. (2009) Real-time monitoring of injection moulding for part mass determination. Chicago, IL, United states, Society of Plastics Engineers.

OPTIMUM DESIGN TECHNIQUE FOR INJECTION MOLDING DIE BY DISCRETE OPTIMIZATION METHOD

Atsushi YOKOYAMA

Division of Advanced Fibro Science, Kyoto Institute of Technology, Kyoto, Japan–yokoyama@kit.ac.jp

Abstract

The injection molds consume costs and time for manufacture. Therefore, it is important to predict the molding method. So, The CAE technology that injection molds design using the resin flow analysis. This method is possible for checking the validity of a design. So, this technology cannot make the indication or the automatic correction of a point that improve design variables, such as the runner diameter, and the amount of resin, when fault arises in initial setting. Therefore, a design change is fixed by the trial and error of the engineer.

In this research, improvement of a genetic algorithm (GA) is advanced so that the multi-cavity molding of an injection molding can be automatically determined using the GA, which is one of the discrete optimization method.

Introduction

The injection molding is one of the plastic molding processing technologies, and is used in the manufacturing process of a lot of plastics products. The die configuration especially has a big influence on the product though the factor (resin temperature and Tamotsu pressure time, etc.) that influences exists in the product shape in the injection molding. It is produced by using CAE technology to need a lot of time and the cost for the die making now.

In the present study, the metal mold design was optimized by paying attention to the warp that was one of the defective products, and using genetic algorithm (GA) that it was resin flow analysis and was the optimization technique. acceptable. Personal communications should be cited in the text.

The Optimization Multi Design Variables

The analytical object used for Figure 1 in the present study is shown. The amount of the warp was assumed to be a displacement magnitude in A point by using the amount of the warp as an index of the analytical result in the present study. An analytical model was made by using the Shell element. These two gates were selected from 15 gate candidates who showed with gray circle of Figure 1 though the resin was injected from two gates into this analytical object. Moreover, the rib is installed in an analytical object, and the number has been changed between 2 and 13. In the present study, the combination of the positions of the gate and the number of ribs were optimized as a design variable. The genetic algorithm is one of the discrete optimization techniques, and the one thought about referring to genetically modified. In the genetic algorithm, all the design variables are converted into the binary number character as shown in Figure 2, and it processes it as a chromosome. The chromosome is first done the evaluation by the target function, is selected only the one with high evaluation value, is done the gene manipulation (selection, intersection, and mutation), and generates a new chromosome as shown in Figure 3. The evaluation value of the chromosome is improved by repeating this gene manipulation, and it optimizes it. It paid attention generality of the genetic algorithm height and easily of the application, and it applied to the optimization of the metal mold design in the present study.

Figure 4 is the design area of the multi design problem. The design area that the design variable individually had interferes when there are two or more design variables as shown in Figure 4 each other, and the design variable becomes vaster. Being need a lot of time by the time the best design is requested because the searched area broadens when the design area is such a state is predictable.

The flow chart of the optimization technique used for Figure 5 in the present study is shown. When intersecting and mutating, it proposed the technique for shortening the required time to settling the solution by reducing the design area on one chromosome by individually doing the gene manipulation to gate 1, gate 2, and the chromosome corresponding to the design variable of the rib, and reducing the range of the search though it optimized to the selection that was the gene manipulation as shown so far.

Results and Discussion

The result of optimizing the result of optimization by using the genetic algorithm and the resin flow analysis of the past by using the genetic algorithm and the resin flow analysis considering Mix operation and the multi design problem is assumed to be separate operation, an analytical frequency of the resin flow analysis that requires it is shown to settling the solution, and compared as shown in Figure 6. It succeeded by dividing the chromosome and processing it settling the solution doing a great reduction in the resin flow analysis that required it. Here, when the design condition that the area of the design for one is wide is treated, a better result can be expected to be obtained because it is forecast that becoming duller is not generated though becoming duller was seen in the reaction between the chromosome and the design variable as for this research object because the area of the design per design variable was small.

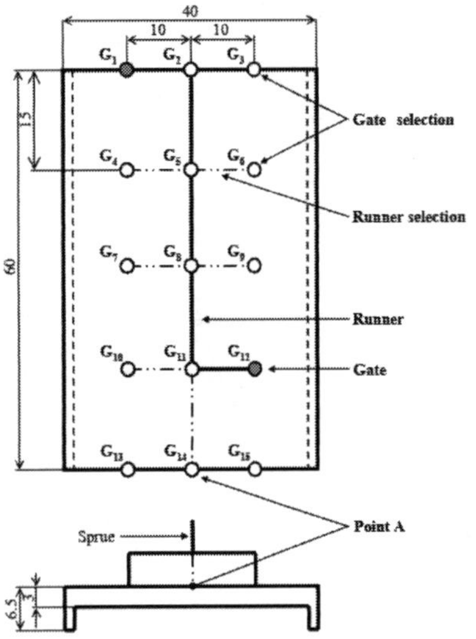

Fig.1 Analytical model

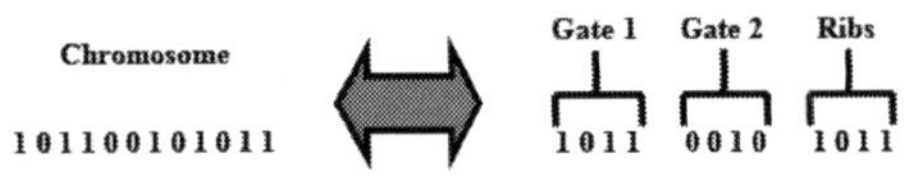

Fig.2 Correlation of a chromosome and a design variable

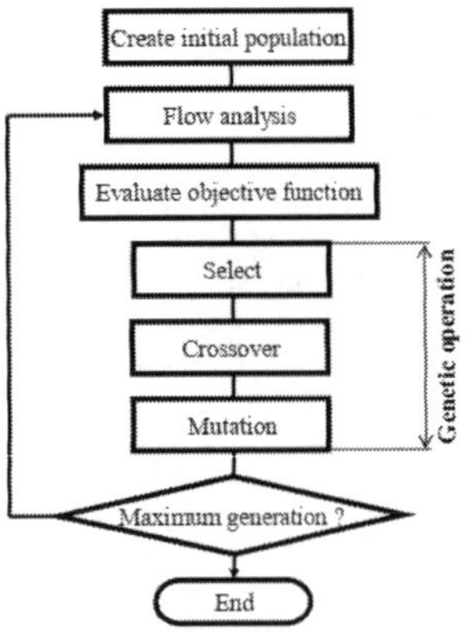

Fig.3 Flowchart of Optimization

Conclusions

In the traditional GA optimization procedure having many design variables such as the multi-cavity injection mold, there are problems as follows.

・The technique for the combination of the genetic algorithm and the neural net work and forecasting the numerical result is effective as the shortening technique of the best solution search.

In the compound optimization design problem where a lot of target functions exist, it needs it for the optimization technique for an injection molding practicable the reduction and efficiency improvement in the search area.

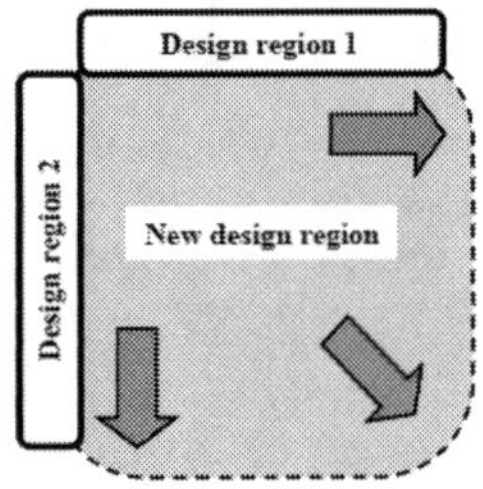

Fig.4 Design region under multi-design condition

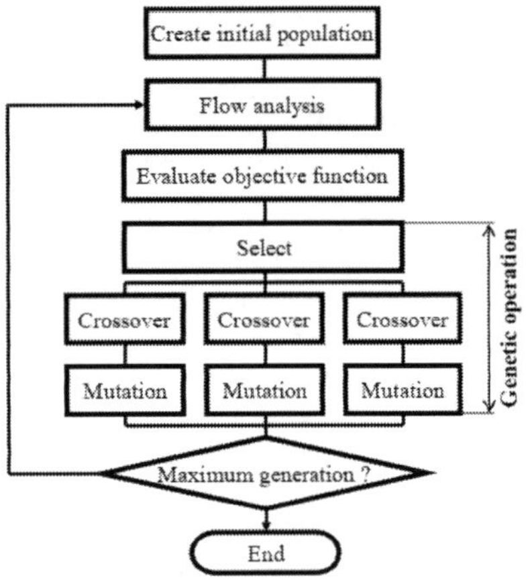

Fig.5 Parallel operation

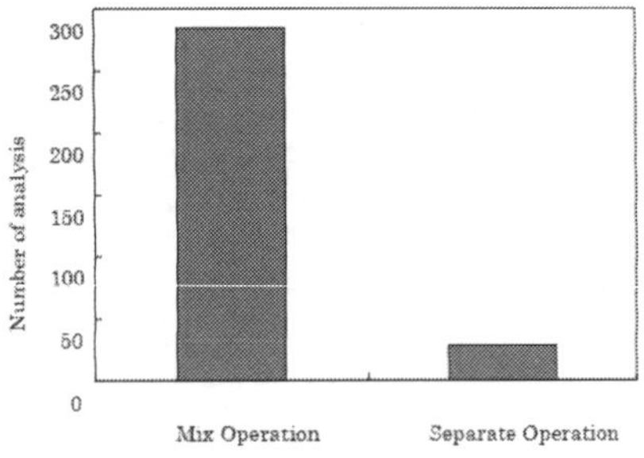

Fig.6 Comparison of number of analysis

Flow visualisation and mechanical analysis of weld line formation in micromoulding

Ben Richard Whiteside, Rob Spares, Phil Coates*

University of Bradford - West Yorkshire - United Kingdom

The high surface area to volume ratios and correspondingly high cooling rates encountered in the micromoulding process can cause significant weaknesses in products where there are weld lines present. This study investigates weld line formation and resulting mechanical properties for both amorphous and semi-crystalline polymers in double gated micromoulding geometries analogous to small scale ISO standard tensile test specimens having wall thickness of 250um. Samples have been moulded according to a design of experiments approach with process variables of melt/mould temperatures, injection velocities, hold parameters and gate design. During moulding, the process data is captured using two techniques. 1) Extensive instrumentation of a Battenfeld Microsystem micromoulding machine with data acquisition and 2) High speed flow visualisation of the flow front evolution, cooling and shrinkage of the product using a bespoke windowed mould and high speed camera system. These systems are linked to a common trigger which allows for direct comparison of data and provides a detailed account of weld line formation and evolution directly within the cavity. Mechanical property measurements to determine weld line strength have been performed on the specimens using a Bose Electroforce tensile test machine, and the weld line has been imaged using sectioning and polarised light microscopy techniques. This paper presents the results of the work and statistically significant micromoulding process conditions for achieving adequate weld line strength.

HETEROGENEOUS TENSILE BEHAVIOR OF IMPACT-MODIFIED POLYPROPYLENE PLATES PROCESSED BY INJECTION MOULDING

O. De Almeida[1,2], F. Touchard[3], L. Chocinski-Arnault[3]

1- Université de Toulouse ; INSA, UPS, Mines Albi, ISAE ; ICA (Institut Clément Ader); Campus Jarlard, F-81013 Albi cedex 09, France

2- Ecole des Mines Albi, Campus Jarlard, F-81013 Albi, France – olivier.dealmeida@mines-albi.fr

3 – Institut P', UPR 3346, Département Physique et Mécanique des Matériaux ; ENSMA, 1, avenue Clément Ader, 86961 Futuroscope Chasseneuil – fabienne.touchard@lmpm.ensma.fr; laurence.chocinski@lmpm.ensma.fr

Abstract – Due to the fast cooling of polymer on the mould surface, the injection moulding process is known to be responsible of property gradients through thickness. Nevertheless, the mechanical behaviour is usually characterized by considering an homogeneous polymer, even when tensile test specimens processed by injection moulding are used. In order to characterize the local three-dimensional behaviour, a new set up based on digital image correlation has been developed and applied to specimens machined from injection moulded plates. This experimental set up allows to measure the strain fields on both front and lateral sides of specimens with a good accuracy. In this study, two impact modified polypropylenes have been analyzed under tensile loading. These polymers are made of ethylene-propylene rubber nodules and lamellar talc particles embedded in a polypropylene matrix. They differ from the volumetric amount of talc (4% in material A, 0,5% in material B). Unlike neat polypropylene, these materials do not exhibit a necking phenomenon in tension. The axial strain and transverse strain through specimen width remain homogeneous along the gauge length during the tensile test. On the other hand, the transverse strains through thickness differ for both polymers. Polymer B exhibits large gradients of ε_{zz} strain between the centre and the plate surfaces while polymer A presents an homogeneous transverse deformation. As confirmed by microstructure investigation, these results reveal a large volumetric dilatation due to void growth for both polymers, but for polymer B, the large transverse gradient measured is due to a high volumetric dilatation gradient through plate thickness. The good thermal conductivity of talc could explain the different behaviour between the two polymers and the origin of volumetric dilatation gradient. In polymer B, the different cooling rates between plate core and surfaces induce large residual stresses which could create damage and initiate voids in plate core.

Introduction

Injection moulding is one of the most widely used plastic part processing and the mechanical properties of thermoplastics are usually identified on moulded samples in considering that the material is homogeneous within the entire gauge volume.
Many studies have though revealed the existence of a skin/core structure in injection moulding process (Viana, 2004). In particular, Pennington et al (1999) and Stegge and Urban (2001) have shown that the skin structure of impact-modified polypropylenes is characterized by a stratification composed of different microstructure layers under the surface of injected plates.
The aim of this study has been to analyse the mechanical behaviour and the deformation mechanisms upon tensile loading of two impact-modified polypropylenes processed by injection moulding.
In order to characterize the deformation mechanisms upon stretching, the strain fields on the surface of the specimens have been measured by digital image correlation. In that way, an experimental set-up has been developed and has allowed to measure strains with a good accuracy on the front and lateral surfaces of the specimens.

Experimental

Materials
The materials investigated in this study are two commercial impact-modified polypropylenes, used for automotive applications as car-bumpers. They were manufactured by Sabic under the references PP Compound 7510 (A) and 108MF97 (B) and provided in granule form. Composite A is a blend of polypropylene, ethylene-propylene-diene terpolymer (EPDM) and talc particles, while composite B is a copolymer of polypropylene and ethylene-propylene rubber (EPR) blended with talc.
The elastomeric and polypropylene phases are immiscible which induces a similar microstructure for both materials A and B. It is characterized by micrometric talc and elastomeric particles (0,5µm diameter) embedded in a polypropylene matrix.
Thermogravimetric analysis (TGA) and weight measurements have been performed on materials A

and B in order to quantified the phase weight proportions and the materials density (Delaval et al., 2009). The results are summarized in Tab. 1. They show that the main difference between these materials is the talc content : composite A contains 10%wt of talc, whereas composite B contains only 0,5%wt. It corresponds for polymers A and B to a volumetric ratio of 4% and 0,5% respectively, in considering a talc density of 2,7g.cm^{-3}.

		A	B
Density (g.cm^{-3})		0,955	0,895
Weight ratio	Polypropylene	0,695	0,775
	Elastomer	0,200	0,220
	Talc	0,105	0,005

Tab. 1 – Composition of impact-modified polypropylenes

As confirmed by differential scanning calorimetry (DSC), the organic phases of the materials have similar physical properties. In polymers A and B, the polypropylene melting temperature is 167°C and the degree of crystallinity is about 49-50%wt. Moreover, in both materials, a crystalline structure has also been identified whose melting temperature has been recorded by DSC between 113 and 115°C. This melting temperature corresponds to polyethylene polymerized during the elastomer synthesis.

Sample preparation
Both polymers were injection moulded into 300×100×3mm^3 plates at the LRMP of St Etienne[1] and tensile specimens with gauge volume dimensions of 10×10×3mm^3 were machined from the moulded plates. A minimum distance of 20mm between the specimen gauge volume and the plate edges has been respected for all the specimens.

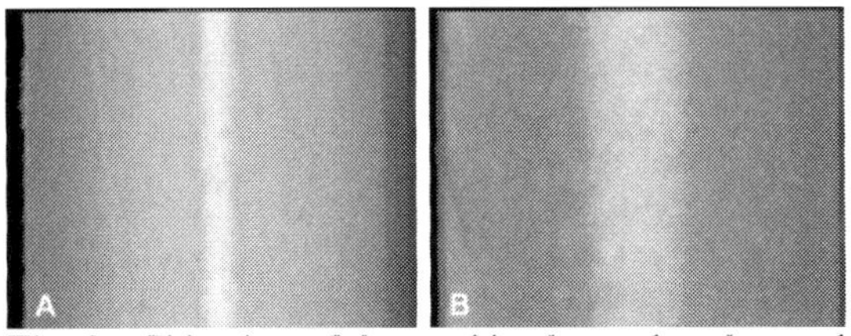

Fig. 1 – Side view of the machined samples observed by optical microscopy (thickness = 3mm)

As observed by optical microscopy after machining, the polymer plates exhibit a different aspect between the core and the surface of the plates (Fig. 1). This optical heterogeneity supposes a different microstructure through plates thickness and is observable in both polymers. Nevertheless, these heterogeneities could not be confirmed by SEM investigation.

[1] Laboratoire de Rhéologie des Matières Plastiques, CNRS, UMR 5156 - 23, rue du Docteur Paul Michelon - 42023 St Etienne Cedex 02 - France. Tel: 0477481554 - Fax: 0477485126

Experimental set-up
Deformation mechanisms of the impact-modified polypropylene have been studied upon stretching and characterized by digital image correlation. Tensile tests have been performed with an electromechanical test machine Instron 4505, at room temperature and with a constant cross head speed of 10mm.min^{-1}.
The front and lateral sides of the tensile specimens have been previously covered with a random speckle pattern whose average white and black grain diameter is 30μm ± 5μm (Brillaud et Lagattu-Touchard, 2002). Due to the grain fineness, a special experimental set-up has been developed so as to take the pictures of the speckle patterns during the tensile tests. It is composed of a mechanical device supporting a digital camera and a mirror (Fig. 2). This device allows to position the camera close to the specimen and in front of a constant centred volume. The mirror is also fastened on the device in such a way to reflect the lateral pattern of the specimen to the camera sensor.
In this configuration, the CCD matrix of 2272×1704 pixels² is separated from the specimen by about 10cm.
Thus, an area of 18×24 mm² is in shot of the digital camera and one pattern grain is defined by about 3×3 pixels² on the CCD matrix.

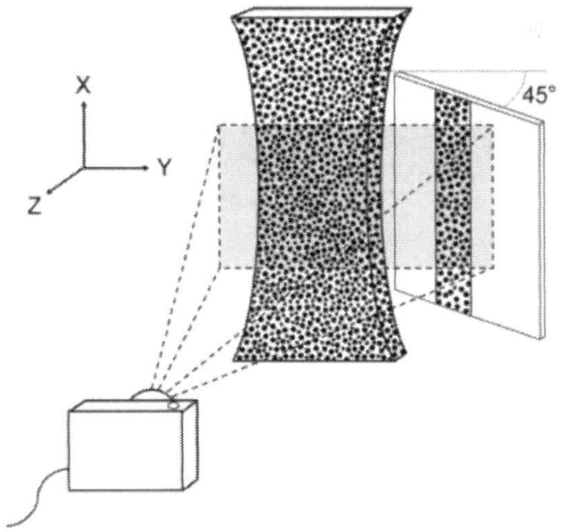

Fig. 2 - Schematic representation of the experimental set-up.

By using this set-up, pictures of the front and the lateral patterns have been taken regularly during the tensile tests. The use of a camera has required short stops during the tests to record the images of front and lateral patterns. Nevertheless, it has been confirmed that the stops did not have influence on the tensile behaviour of the polymers. For further information about the testing procedure, the reader can refer to De Almeida et al. (2008).

Digital images exploitation

The digital image correlation (DIC) of the speckle patterns has been performed with a home developed software Granu. This software is based on the direct correlation of the discrete matrix of the pixel grey levels of a subwindow (Brillaud and Lagattu-Touchard, 2002). The displacement of each subwindow is ascribed to the centre point of the subwindow and the spatial resolution is defined as the real size of the subwindows.

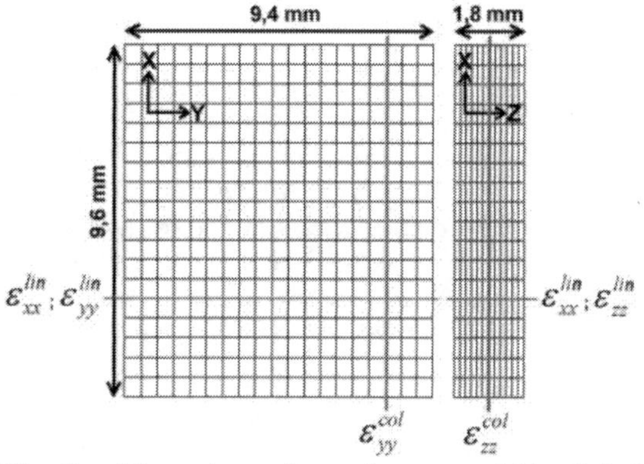

Fig. 3 – Dimensions of sample zones within digital image correlation has been performed to measure displacements (each mesh node corresponds to a measurement point on the sample surface).

For all experiments, the in-plane displacement fields are incrementally obtained, starting with the undeformed image, by correlating subwindows of the speckle pattern from one image to the next with an adaptive mesh. The configuration chosen in this study allows in-plane displacement measurements with good accuracy, on both front and lateral sides of the specimens. With a magnification of 11.5 µm/pixel for front images and 11.1 µm/pixel for lateral images, the local displacements have been measured with respectively a spatial resolution, i.e. a correlation subwindow area, of 460×460 µm² and 440×440 µm².
Digital image correlation has been performed in about 400 measurement points on the front side and about 250 on the lateral side on areas corresponding respectively to 9.4×9.6 mm² (Width×Length) and 1.8×9.6 mm² (Thickness×Length) (Fig. 3). The Granu software processing time is about five seconds for 400 measurement points with a current personal computer.

The Abaqus software has then been used to derive experimental data using large-strain formalisms (Hencky strain), and using a small-strain formalism, afterwards called nominal strain (Cauchy strain). This can be performed with an Abaqus file within the subwindow centre points are written as nodes of a mesh and displacements are treated as boundary conditions. Local true axial strains ε_{xx} are thus calculated from both front and lateral vertical displacements (x-axis), while local front transverse strains ε_{yy} (y-axis) and local lateral transverse strains ε_{zz} (z-axis) are respectively determined from horizontal front and lateral displacements.
Because of the mesh regularity, the average of the local axial strain ε_{xx}^{av} and of the transverse strains ε_{yy}^{av} and ε_{zz}^{av} can been calculated from the results of digital image correlation. ε_{xx}^{av} has been used to characterize the tensile test progress i.e. the global strain applied to the specimen.
The average strains ε_{xx}^{lin}, ε_{yy}^{lin}, ε_{zz}^{lin}, ε_{yy}^{col}, ε_{zz}^{col} have also been calculated from the local strain measurements. As shown in Fig. 3, they correspond to the average of the local strains per line (lin) or per column (col) in all three dimensions.

Results and Discussion

Homogeneous axial deformation

For both materials, the measurements have been performed regularly during the tensile tests until grain pattern degradation. The last step, for which digital image correlation was possible and for which the results were conclusive, corresponds to an average axial strain ε_{xx}^{av} of 49% for both polymers. The axial strain fields of front and lateral surfaces for ε_{xx}^{av} = 49% are presented in Fig. 4 for polymer B. A good agreement can be noticed between the strain fields measured on both surfaces and this has been observed for each measurement step.
Fig. 5 shows the comparison of the average axial strain per line ε_{xx}^{lin} measured for a neat polypropylene and for the two impact-modified polypropylenes. The necking phenomenon observed in polypropylene is characterized by a large axial gradient. The axial strain reaches about 80% in the neck while it is only 27% close to the analysed surface. On the other hand, for both impact-modified polypropylenes, the axial strain is ranged between 45 and 55% along the specimen for an average axial strain ε_{xx}^{av} of 49%. The deformation of this materials upon stretching can thus be considered as homogeneous along the gauge

length, and the average strains ε_{yy}^{col} and ε_{zz}^{col} can be used to analyse the deformation behaviour of polymers A and B through specimen width and thickness. This homogeneous axial deformation is induced by the elastomer phase which has already be noticed in high-impact PS by G'sell et al. (2002).

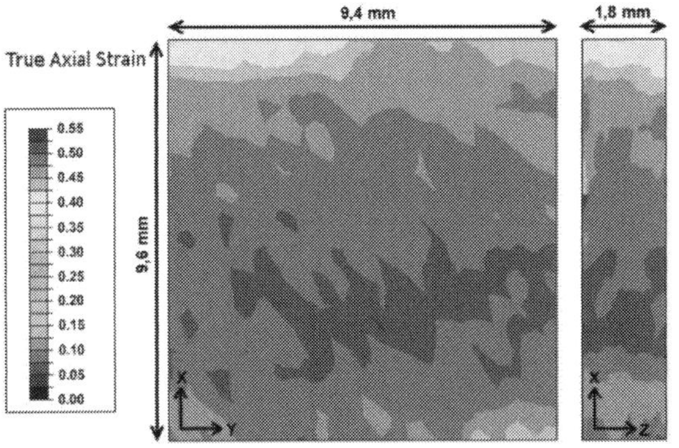

Fig. 4 – Axial strain fields of polymer B measured on front and lateral sides of the specimen for an average axial strain ε_{xx}^{av} of 49%.

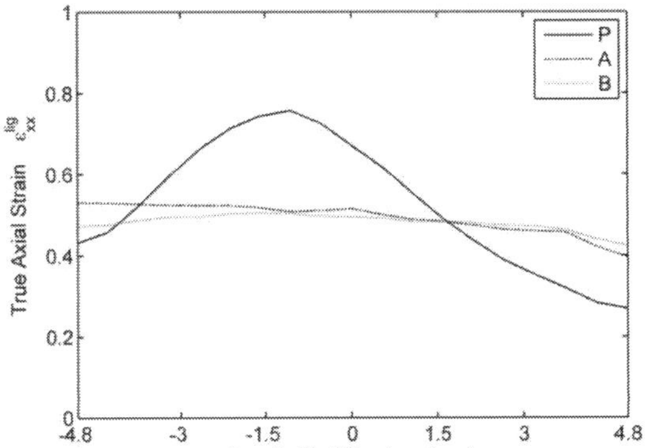

Fig. 5 – Axial strain profile along the strain gauge for polymers A and B. Comparison with the strain profile measured for a neat polypropylene P.

Transverse strains

The transverse strain fields of material B for ε_{xx}^{av} = 49% are shown in Fig. 6 and the transverse strain profiles through specimen width (W) and thickness (T) are presented in Fig. 7 and Fig. 8. As the deformation of the impact-modified polypropylenes are axially homogeneous, the profiles are plotted with the average transverse strain per column ε_{yy}^{col} and ε_{zz}^{col}. On one hand, these average strains are representative of the whole specimen length and on the other hand, they provide a smoother strain profile.

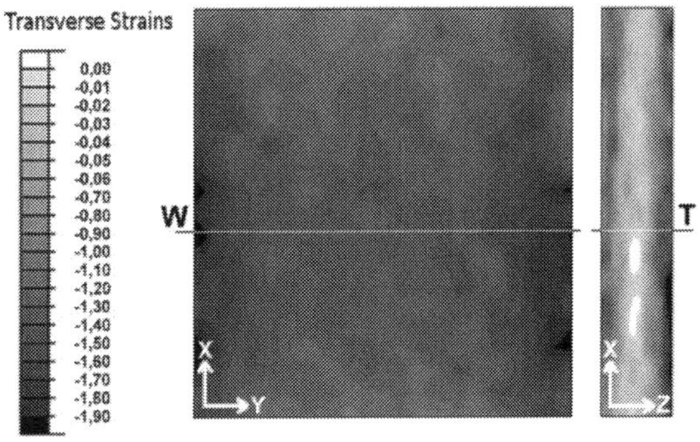

Fig. 6 – Transverse strains fields of polymer B measured on front and lateral sides of the specimen for an average axial strain of 49%.

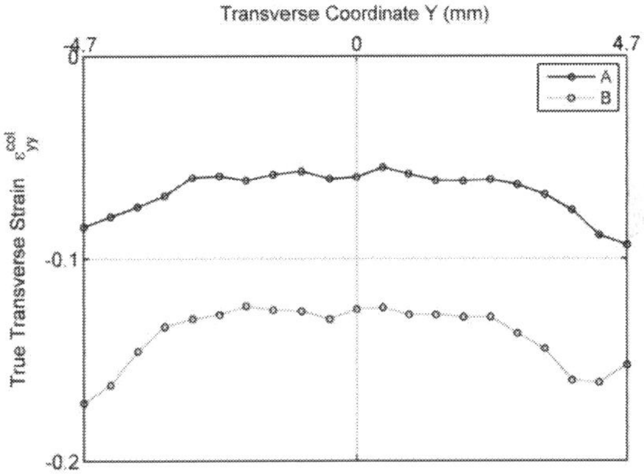

Fig. 7 – Transverse strain profiles measured in polymers A and B through specimen thickness (ε_{zz}^{col}) for ε_{xx}^{av} = 49%.

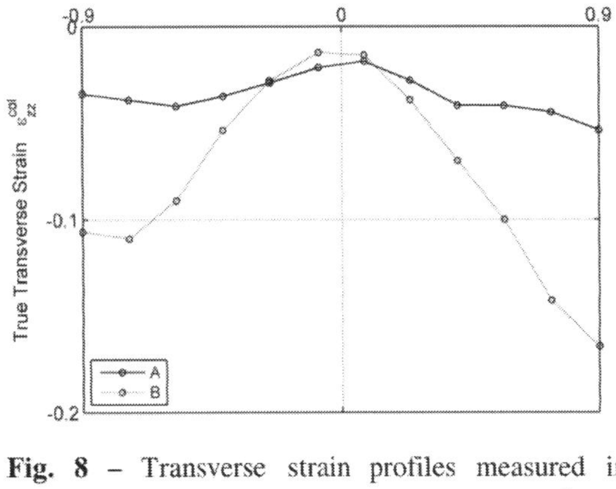

Fig. 8 – Transverse strain profiles measured in polymers A and B through specimen width (ε_{yy}^{col}) for ε_{xx}^{av} = 49%.

As shown in Fig. 7, the transverse strain through specimen width is around -0.15 in material B when it

is only –0.07 in material A for an axial strain of 49%. The width reduction of the specimen is then larger in polymer B than in polymer A. Moreover, the strain profiles are not constant through specimen width. On both sides of the centre 6mm of the specimen gauge, the transverse strain decreases.
Through specimen thickness, shown in Fig. 8, the two impact-modified polypropylenes exhibit different strain profiles. In specimen core, the transverse strain ε_{zz}^{col} is almost nil (around –0.01) for polymers A and B. But whereas polymer A exhibits a constant transverse strain ε_{zz}^{col} through specimen thickness, transverse strains of polymer B largely decrease near the edges of the analysed surface and reach around –0.16.

Strain anisotropy

Due to the large transverse strain gradient measured for polymer B, the analysed surfaces are no longer enough to characterize the global deformation behaviour of the samples. Indeed, the displacement field measurements are performed on 1.8mm of the 3mm thick specimens.
The global transverse strains have thus been calculated with Eq. 1 and Eq. 2 from the entire specimen width and thickness measured in pixels on the recorded pictures. The width and the thickness of the specimens have been measured at three different positions along the gauge length on the pictures in order to get an average strain of the entire specimen.

$$\varepsilon_{yy}^{av}, images = \ln\left(\frac{w}{w_0}\right) \qquad \text{(Eq. 1)}$$

$$\varepsilon_{zz}^{av}, images = \ln\left(\frac{t}{t_0}\right) \qquad \text{(Eq. 2)}$$

In Eq. 1 and Eq. 2, w and t are respectively the width and the thickness of the specimen from edge to edge and w_0 and t_0 are respectively the width and the thickness measured on the reference image, i.e. the image of the undeformed sample.
The Fig. 9 and Fig. 10 show for both polymers the evolution of the transverse strains ε_{yy}^{av} and ε_{zz}^{av} measured with both methods versus the axial strain ε_{xx}^{av}. For both polymers, the transverse strain ε_{yy}^{av} decreases almost linearly and a good agreement can be noticed between the two measurement methods. Indeed, in this direction, the analysis is realized on 94% of the specimen width and the error on the global strain is then limited.

On the other hand, through thickness, Fig 9 and Fig 10 show that the measurement method has induced a large error on the transverse strain ε_{zz}^{av}. While DIC suggests a stabilization of the average transverse strain, i.e. the stabilization of thickness reduction, the values measured on the images show that the thickness contraction is continuous and that ε_{zz}^{av} is always larger than ε_{yy}^{av} until ε_{xx}^{av}=49%.

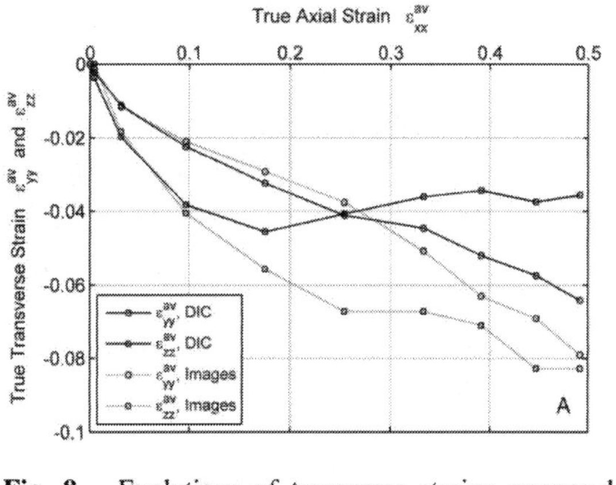

Fig. 9 – Evolutions of transverse strains measured with both methods versus axial strain for polymer A.

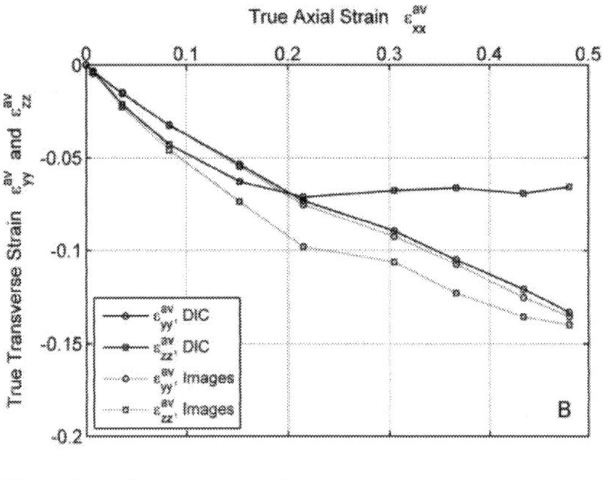

Fig. 10 – Evolutions of transverse strains measured with both methods versus axial strain for polymer B.

The difference between the measurement results obtained with both methods can be explained by the difference of the analysed thickness on the sample surface. Digital image correlation gives a local information of the deformation behaviour but, as shown in the Fig. 9 and Fig. 10, the studied zone is too small to be representative of the global transverse strain behaviour of the tensile specimen. For all that, both results are complementary and not contradictory.

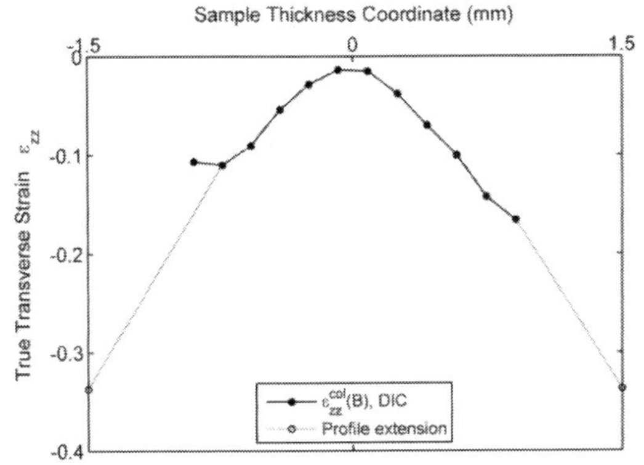

Fig. 11 – Transverse strain gradient measurements by DIC and direct method for polymer B measured for $\varepsilon_{xx}^{av} = 49\%$.

The concordance of both methods can be checked by estimating the transverse strain profile over the DIC studied zone. As shown in Fig. 11 for polymer B, this estimation consists in adding two points to the transverse strain gradient ε_{zz}^{av} on the specimen edges in such a way to find back the thickness variation measured directly on the images. The results show that these two points are in the continuity of the DIC transverse strain profile and so, they confirm the truthfulness of the results. Above all they suggest that the measured strain gradient occurs over the entire specimen thickness.

Macroscopic stress-strain behaviour

The macroscopic true axial stress – true axial strain behaviour of the polymers A and B has been determined by calculating the true stress with the Eq. 3 and Eq. 4. In these equations, $\varepsilon_{xx^N}^{av}$ is the average axial nominal strain, and $\varepsilon_{yy^N}^{av}$ and $\varepsilon_{zz^N}^{av}$ are the average transverse nominal strains. The results are plotted in Fig. 12 for polymer A and in Fig. 13 for polymer B.

The true stress has first been calculated with Eq. 3, by using the transverse strains measured by digital image correlation (Fig. 12 and Fig. 13, DIC). It has also been calculated by using the strains determined from the specimen width and thickness on the recorded pictures (Fig. 12 and Fig. 13, images).

$$\sigma_{xx} = \frac{F}{w_0 \times t_0} \times \frac{1}{\left(1+\varepsilon_{yy^N}^{av}\right)\left(1+\varepsilon_{zz^N}^{av}\right)} \quad \text{(Eq. 3)}$$

$$\sigma_{xx} = \frac{F}{w \times t} \approx \frac{F}{w_0 \times t_0} \times \left(1+\varepsilon_{xx^N}^{av}\right) \quad \text{(Eq. 4)}$$

As shown in Fig. 12 and Fig. 13, the measurement width and thickness has poor influence on the true stress – true strain behaviour. Indeed, even if the large transverse strain gradients induce a large error on the average transverse strain evolutions, the strain intensity is small and the influence on the stress level is then poor.

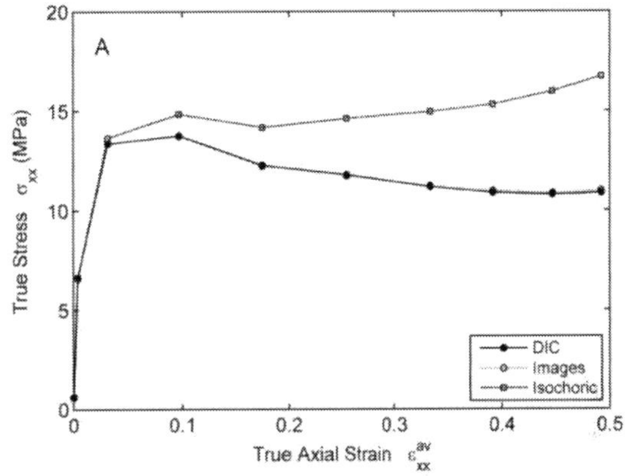

Fig. 12 – Stress-strain behaviour of polymer A.

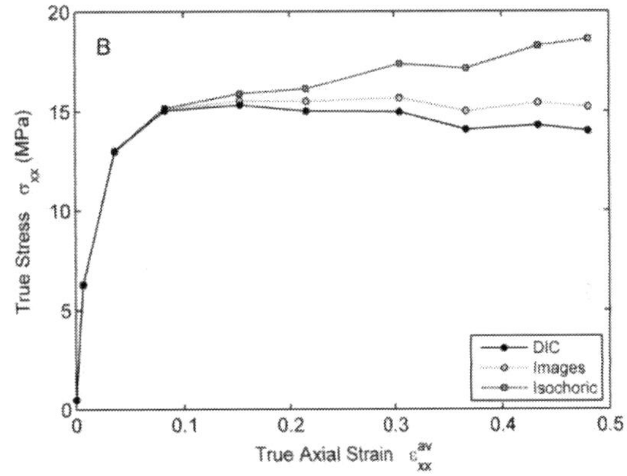

Fig. 13 – Stress-strain behaviour of polymer B.

When the axial stress is calculated with Eq. 4, the deformation is assumed to be isochoric. As shown in Fig. 12 and Fig. 13, this assumption induces a different mechanical response which is observable for both impact-modified polypropylenes.

Although this assumption is frequently used when the strain is only measured with an axial extensometer, it induces a large overestimation of the plastic behaviour in these impact-modified polypropylenes. For a true axial strain of 49%, the error on the true stress is about 5MPa, i.e. about 30%.

Moreover, whereas the stress calculated with Eq. 3 results in a true stress – true strain curve characterized by a maximum stress at the yield point, the isochoric

assumption induces a maximum strength during the strain hardening.
Until 10% of axial strain, the true stress – true strain curve is however the same whatever the stress calculation method. From this point, which can be considered as the limit of the elastic stage, the mechanical behaviour is governed by a non isochoric deformation process.
Furthermore, the comparison of both polymers indicates that polymer A has a higher elastic modulus than polymer B, but a lower yield stress. This result is in agreement with the literature and is the consequence of the talc content in the impact-modified polypropylene A (Pukanszky et al., 1994).

Volumetric strain evolution

The volumetric dilatation of the tensile specimen has been calculated with Eq. 5 within $\varepsilon^{av}_{xx^N}$, $\varepsilon^{av}_{yy^N}$ and $\varepsilon^{av}_{zz^N}$ are the average nominal strains determined by digital image correlation in the three directions.

$$\frac{\Delta V}{V_0} = \left(1+\varepsilon^{av}_{xx^N}\right)\left(1+\varepsilon^{av}_{yy^N}\right)\left(1+\varepsilon^{av}_{zz^N}\right)-1 \qquad \text{(Eq. 5)}$$

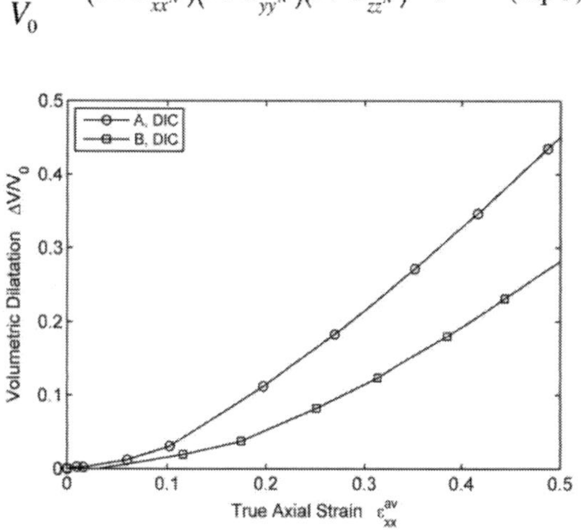

Fig. 14 – Evolution of the volumetric dilatation of the impact-modified polypropylenes with the average axial strain.

As displayed in Fig. 14, the mechanical behaviour of the impact-modified polypropylenes is characterized by a large volume growth during stretching. After the elastic stage, where a conventional Poisson's effect is observed, an significant dilatation process occurs, leading to a large volumetric dilatation of 45% for polymer A and 28% for polymer B for an average axial strain of 49%.
These volumetric strain results prove that non-cohesive deformation mechanisms and void growth support a part of the plastic strain and that the plastic deformation is nearly not supported by plastic shearing (G'Sell et al., 2002). Moreover, the faster volumetric dilatation in polymer A indicates that talc has a large influence on the void development, which tallies with the volumetric strain measurements on calcium carbonate high-density polyethylene mentioned by Parsons et al. (2004, 2005).

Volumetric dilatation gradients

The macroscopic tensile behaviour has been analysed until now by using the average strains and the spatial resolution of the measurement method has not yet been used to characterize the local deformation of the material. The global approach has however revealed that a too small studied surface can be unsuitable to describe the global deformation behaviour.
Thus, in this study, because of the large transverse strain gradients measured through specimen thickness, a local volumetric analysis has been performed on the materials. As described in Fig. 15, the front and lateral correlation meshes (a), for which strain measurements have been performed for each node, have been projected into the specimen gauge volume respectively along z- and y-axes (b). The intersection of the projected surfaces results in a three-dimensional mesh which correspond to the core part of the specimen gauge volume (c). The dimensions of the obtained fictive specimen volume are 9.4×9.6×1.8mm^3.
As a result of the projection, each node of the 3D mesh includes the information of local strains in all three directions. The local volumetric dilatation can thus be calculated for each three-dimensional mesh node within the fictive volume. Eq. 5 has been used to calculate the local volumetric dilatation at each node by replacing the average nominal strains with the nominal strains.
However, the projection of experimental data is done assuming that the in-plane strain fields measured on the specimen surfaces are representative of the specimen core behaviour and such an assumption can distort the result interpretations in presence of inside strain gradients. Nevertheless, observations of the deformed cross-sections at different strain levels for both composites show that the specimen cross-section remains rectangular during the deformation. Therefore, such strain gradients within the polymers are unlikely and this result reinforce the validity of the previous hypothesis.

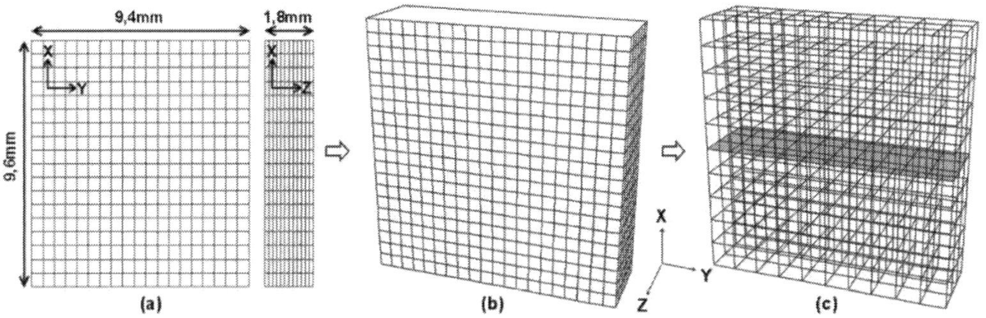

Fig. 15 – Schematic representation of the experimental data extrapolation to reconstruct the specimen volume.

The local experimental volumetric dilatation values in the central cross-section of the specimen are plotted in Fig. 16 and Fig. 17 for polymer A and B at different loading levels.

One can see in Figs. 10 and 11 that the volumetric dilatation exhibits large gradients within the cross-section which increase like the transverse strain gradient ε_{zz}.

In composite B, filled with only 0,5% of talc, the large transverse strain gradients observed on both surfaces lead to a volumetric dilatation mainly concentrated along a thickness middle line. For an average axial strain of 52%, the specimen centre shows a volumetric dilatation of about 43%. Near specimen edges, the transverse strains compensate part of the axial strain and, as a result, the local volumetric dilatation is reduced to only 14%.

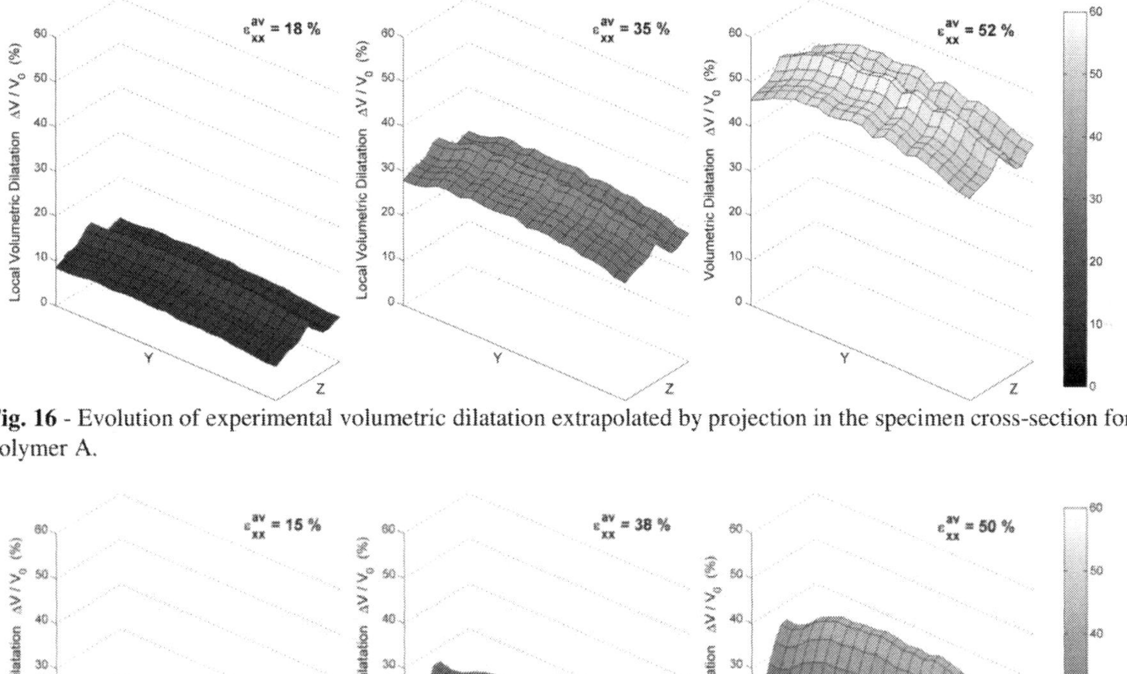

Fig. 16 - Evolution of experimental volumetric dilatation extrapolated by projection in the specimen cross-section for polymer A.

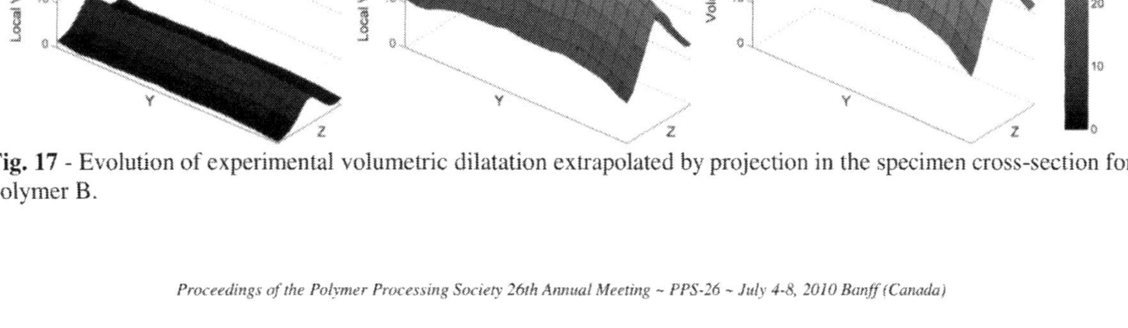

Fig. 17 - Evolution of experimental volumetric dilatation extrapolated by projection in the specimen cross-section for polymer B.

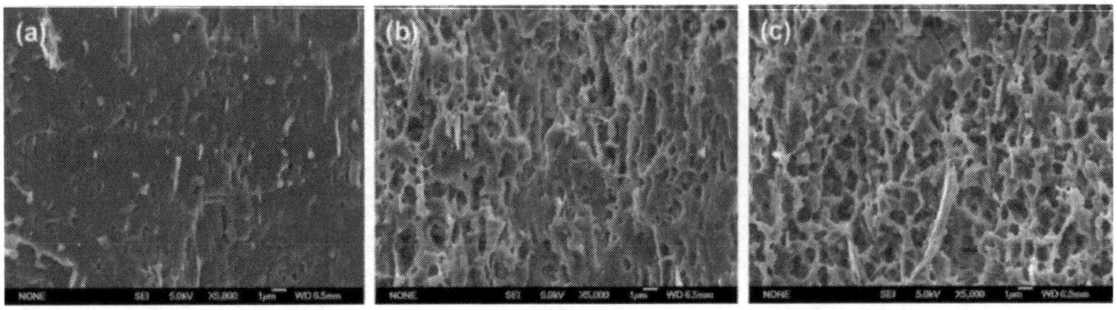

Fig. 18 – Observation of the deformed microstructure in polymer A for ε_{xx}^{av}=23% ; (a) under the front surface, (c) in the centre of the specimen thickness, (b) at mid-distance of both other pictures.

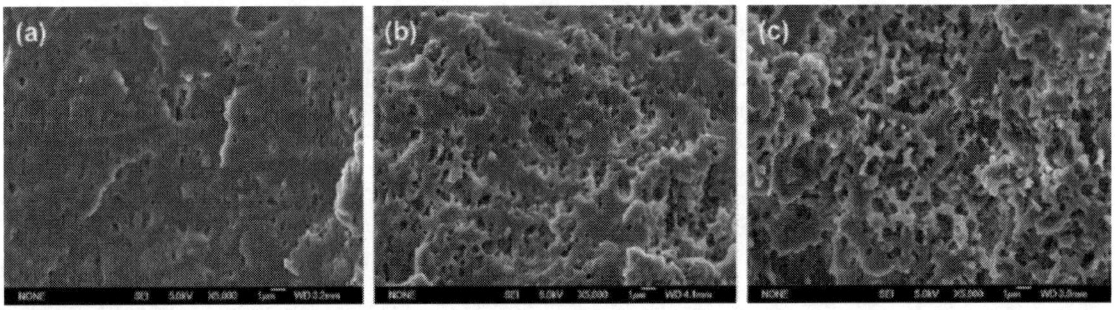

Fig. 19 - Observation of the deformed microstructure in polymer B for ε_{xx}^{av}=23% ; (a) under the front surface, (c) in the centre of the specimen thickness, (b) at mid-distance of both other pictures.

The local volumetric dilatation gradients in composite A, which contains 4% of talc, are lower than in composite B. The thickness middle line still presents a higher volumetric dilatation level but volumetric dilatation is roughly constant in the entire specimen cross-section. For an average axial strain of 52%, the volumetric dilatation values vary from 55% to 38% from the specimen centre to the edges.
If the evidence of void growth was revealed with the measure of the average volumetric dilatation, the analysis of the local volumetric dilatation through specimen cross-section gives additional information about the distribution of this phenomenon.

Comparison with the deformed microstructure
As a complement of volumetric dilatation measurements, the deformed microstructures have been investigated by FEG microscopy. The observed surfaces have been obtained by cryogenic fracture performed on specimens previously deformed to an average axial strain of 23% (Fig. 18 and Fig. 19). For both polymers, pictures of the deformed microstructure have been taken under the front surface (a), in the centre of the specimen thickness (c), and at mid-distance of both pictures (b).
The pictures (b) and (c) illustrate the deformed microstructure at mesh nodes for which local volumetric dilatation has been determined after strain field projection.
The deformed microstructures through specimen thickness, displayed in Fig. 18 for polymer A and in Fig. 19 for polymer B, confirm the existence of volumetric dilatation gradients through thickness. In polymer A (Fig. 18), the pictures (b) and (c) show an advanced void expansion process within specimen cross-section leading to a foam microstructure. In polymer B, this process is delayed in comparison with polymer A, and the dilatation gradient through thickness is obvious. However, in both polymers, the amount of cavities is limited close to specimen front surface (a), on a depth lower than 100μm. According to Stegge and Urban (2001) and Pennington et al. (1999), this depth is the place of surface stratification processes in thermoplastic olefins, which could explain such particular deformed microstructure.

Discussion
The development of volumetric dilatation in impact-modified-polypropylenes was expected since many authors previously observed the initiation of different kinds of voids in such materials at microscopic scale. In particular, the micromechanical deformation processes of toughened and particle-filled semicrystalline polymers have been well described by Kim and Michler (1998). They have shown that void

formation is caused by either cavitation in the stretched rubbery particles or debonding at the interface between particles and matrix.

Cavitation occurs when the phase adhesion between the modifier and the matrix is good. The plastic deformation then occurs via a cavitation process inside the particles. As shown by Fond et al. (2002), cavitation is a damage mechanism energetically favorable when the applied tension induces large hydrostatic stress to a particle.

Under the effect of hydrostatic stress, debonding can be an important damage mechanism when the adhesion between the different phases is poor. It is consistently observed around rigid particles and it can lead to a considerable damage rate (G'sell et al., 2004).

In the present studied impact-modified polypropylenes, the void initiation could not be identified, but the different nature of the particles suggest that both damage mechanisms are active.

Apart from the nature of the fillers, the higher particle content in polymer A can explain a part of the high damage rate in this material. Nevertheless, talc debonding has been observed in both materials on cryogenic fracture surface of the deformed samples and this mechanism is probably partly responsible for the high damage rate of polymer A upon stretching.

But if the composition of the polymers can explain the global volumetric behaviour of both impact-modified polypropylenes, they are not sufficient to clarify the existence of the cavity gradients through specimen cross section as revealed by FEG investigation and volumetric dilatation measurements. Indeed, these gradients may be induced either by scale/edge effects, i.e. geometrical consideration, or by the moulding process conditions and in particular by the cooling rate gradients through specimen thickness.

The specimen were machined from 3mm thick injected plates and so the microstructure can be considered as homogeneous along the specimen width. Moreover, strain gradients are small and are only visible for a few millimetres on both sides of the centre part of the specimen. As a result, the volumetric gradients in this direction may be the consequence of the specimen geometry.

Through thickness, the larger volumetric dilatation has been measured along a centre line in both materials which remind the optical heterogeneity already mentioned when presenting the materials. In material B, the larger dilatation gradient matches with the wider optical heterogeneity and on the contrary, the polymer A exhibits a narrow heterogeneity within specimen thickness.

The results of the present investigation then suggest that this white centre line is actually the sign of damage inside the unloaded material, as a consequence of the moulding process. This damage would be induced by the accumulation of residual stress in the specimen core during cooling.

Indeed, elastomer has a noticeably lower thermal conductivity than the matrix and as a result, the fillers reduce the cooling rates in specimen core. This effect is however only perceptible in specimen core because the difference of thermal conductibility does not affect the polymer that is in contact with the mould which is quenched immediately.

Talc particles exhibit a higher thermal conductivity than the matrix (Droval et al., 2005). Thus, in polymer A, the cooling rate of the specimen core could be enhanced and the white centre line width be reduced.

Thus, when the polymer is deformed, the void extension process would be accelerated by this inside pre-damage and would lead to cavity gradients through specimen thickness.

Conclusions

The plastic deformation of two impact-modified polypropylenes have been studied experimentally by using a novel measurement technique based on digital image correlation. The set-up and the measurement procedures have been developed in order to measure the local strains in the three directions on the surface of tensile test specimens with good accuracy. By using the strain fields, the mechanical response of the materials can be determined without using any isotropic assumption.

Both impact-modified polypropylenes mainly differ in their talc content (4% in A while it is only 0.5% in B). The influence of talc can be noticed on the mechanical response (modulus, yield strength) and on the volumetric dilatation. The true stress – true strain behaviour, as well as the volumetric dilatation could be identified and have confirmed the non isochoric plastic deformation of both materials due to void extension during the tension tests.

But above all, the local strains have revealed the existence of transvers strain gradients that have an incidence on the mechanical behaviour of the materials. In polymer B, the transverse strain gradients induce large dilatation gradients through specimen cross-section while in polymer A, the deformation is roughly homogeneous. The presence of such gradients can be explained by the difference of thermal conductivity of the different phases inside the material and their influence on the local cooling rates.

Acknowledgements

The authors would like to thank the ADEME (Agence pour le Développement de l'Environnement et de la Maîtrise des Energies), the Région Poitou-Charentes and Research Ministry - through ACI Recypro B - for sponsoring this work.

References

Brillaud J, Lagattu-Touchard F., "Limits and possibilities of laser speckle and white light image correlation methods: theory and experiments", Applied Optics, 41, 6603-6613 (2002)

De Almeida O., Lagattu F., Brillaud J., "Analysis by a 3D DIC of volumetric deformation gradients: Application to polypropylene/EPR/talc composites", Comp. Part A : Applied Sci. and Manufacturing, 39, 1210-1217 (2008)

Delaval D., Casetta M., Delobel R., Traisnel M., Guillet J., Raveyre C., Bourbigot S., ''Thermal and fire degradation of recycled and polluted polypropylene-based materials", J. Applied Pol. Sci., 112, 2270-2279 (2009)

Droval G., Glouannec P., Feller J.F., Salagnac P., "Simulation of electrical and thermal behavior of diphasic conductive polymer composites heating elements", J Thermophys Heat Transf, 19, 375–381 (2005)

Fond C., Géhant S., Schirrer R., "Effects of mechanical interactions on the hydrostatic stress in randomly distributed rubber particles in an amorphous polymer matrix", Polymer, 43, 909-919 (2002)

G'Sell C., Hiver J.M., Dahoun A., "Experimental characterization of deformation damage in solid polymers under tension, and its interrelation with necking", Int. J. of Sol. and Struct., 39, 3857-3872 (2002)

G'Sell C., Bai S.-L., Hiver J.-M., "Polypropylene/polyamide 6/polyethylene–octene elastomer blends. Part 2: volume dilatation during plastic deformation under uniaxial tension", Polymer, 45, 5785-5792 (2004)

Kim G.-M. and Michler G. H., "Micromechanical deformation processes in toughened and particle-filled semicrystalline polymers: Part 1. Characterization of deformation processes in dependence on phase morphology", Polymer, 39, 5689-5697 (1998)

Parsons E., Boyce M.C., Parks D.M., "An experimental investigation of the large-strain tensile behavior of neat and rubber-toughened polycarbonate", Polymer, 45, 2665-2684, (2004)

Parsons E.M., Boyce M.C., Parks D.M., Weinberg M., "Three-dimensional large-strain tensile deformation of neat and calcium carbonate-filled high-density polyethylene", Polymer, 46, 2257-2265 (2005)

Pennington B.D., Ryntz R.A., Urban M.W., ''Stratification in thermoplastic olefins (TPO); photoacoustic FT-IR depth profiling studies'', Polymer, 40, 4795-4803 (1999)

Pukánszky B, Belina K., Rockenbauer A., Maurer F.H.J., "Effect of nucleation, filler anisotropy and orientation on the properties of PP composites", Composites, 25, 205-214 (1994)

Stegge J. M., Urban M. W., "Stratification processes in thermoplastic olefins monitored by step-scan photoacoustic FT-IR spectroscopy", 42, 5479-5484 (2001)

Viana J.C., "Development of the skin layer in injection moulding: phenomenological model", Polymer, 45, 993-1005 (2004)

HYBRID MOULDS WITH EPOXY-STEEL FIBRE CORES – EFFECTS OF MATERIALS AND PROCESSING ON MOULDING SHRINKAGE

P. G. Martinho[1], A. C. Sabino Netto[2], C. H. Ahrens[3], G. V. Salmoria[3] and A. S. Pouzada[4]

[1]*Polytechnic Institute of Leiria, Leiria, Portugal – pmgm@estg.ipleiria.pt*
[2] *Centro Federal de Educação Tecnológica de Pelotas (CEFET-RS), Sapucaia do Sul/RS, Brasil – asabino@sapucaia.ifsul.edu.br*
[3] *CIMJECT-EMC, Universidade Federal de Santa Catarina, Florianópolis, SC, Brasil – chahrens@emc.ufsc.br; gsalmoria@cimject.ufsc.br*
[4]*IPC-Institute for Polymers and Composites, Universidade do Minho, Guimarães, Portugal – asp@dep.uminho.pt*

Hybrid moulds are a cost efficient solution to produce small series of plastics parts, and incorporate moulding blocks produced by Rapid Prototyping and Tooling routes. Epoxy based composites are used for those blocks, the composites with metallic fillers being increasingly considered for industrial applications. They are being considered in the context of hybrid moulds. Issues were identified in the mixing of the composite components, the curing agent and the relatively poorer thermal and mechanical properties in comparison with conventional tooling materials. The thermosetting matrices used in the composite have a T_g around 150°C, and lose part of their rigidity and strength above 100°C. This characteristic influences the cycle time, the stabilized mould temperature, and the shrinkage of the mouldings.
This paper contributes with the analysis of the case of a complex tri-dimensional shape that was moulded in a pressure and temperature instrumented standard steel mould and an alternative hybrid mould with steel cavity and epoxy/short steel fibre composite core. The mouldings were produced in polypropylene homopolymer. Pressure and temperature were recorded at two locations in the cavity.
The processing conditions were simulated using the software Moldex3D. The effect of the processing was observed on some dimensions of the part that were used for assessment of the shrinkage. The results suggest the importance of predicting the regime mould temperature when low thermal conductivity materials are used in moulding blocks. This temperature depends on the desired cycle time. Results also evidenced that in hybrid moulds with resin blocks the shrinkage of the mouldings is a consequence of the specific polymer shrinkage and the deformation of the non-metallic core deformed by the processing pressure, this being not predicted by the simulators.

Introduction

Rapid tooling (RT) is being increasingly used for industrial applications to reduce the time-to-market and improve the quality in plastic parts. The casting of filled resins is a RT process widely used in the early phases of the plastic part design (Hilton and Jacobs 2000). The main advantage of this process over the conventional machining is the lower cost to obtain freeform moulding geometries (Martinho *et al.* 2005). Hybrid moulds with moulding blocks produced with epoxy composites have mechanical and thermal properties lower than conventional steel tooling. This means that for the same moulded part the mould has a lower margin of safety in service and a shorter life. Understanding the thermo-mechanical behaviour during the injection moulding cycle is essential to optimize the performance and to extend the life of this type of moulds.
It is not possible to use the same processing parameters for conventional steel tooling and have the same plastic part with a mould obtained by casting process. It is known that the shrinkage of injection mouldings is governed mainly by the thermal and pressure history during the moulding cycle (Martinho *et al.* 2009). Mould temperatures and holding pressure mainly affect the final dimensions of the mouldings. Generally, as observed by Pontes, increasing the mould temperature or decreasing the holding pressure leads to higher shrinkage parts (Pontes *et al.* 2004; Pontes and Pouzada 2004).
I this paper some moulding characteristics dependants on the injection conditions are assessed. A complex three-dimensional moulding was studied and specific dimensions evaluated. The plastics parts were produced in a research hybrid mould. Tool steel and an epoxy-based composite were used in the core moulding blocks. Thermal and mechanical characterization tests were made in the epoxy composite.

Hybrid Mould

Design

The hybrid mould used in this work is shown in Figure 1. Its design and specific performance details were described elsewhere (Martinho *et al.* 2007).

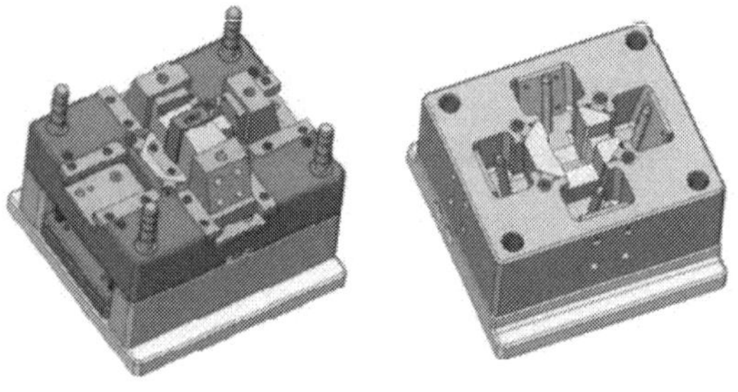

a) Ejection side b) Injection side

Figure 1 - Hybrid mould: a) Ejection side b) Injection side

The ejection is made with a stripper ring and a central ejector pin to remove the gating sys-tem. The moulding blocks are shown in Figure 2.

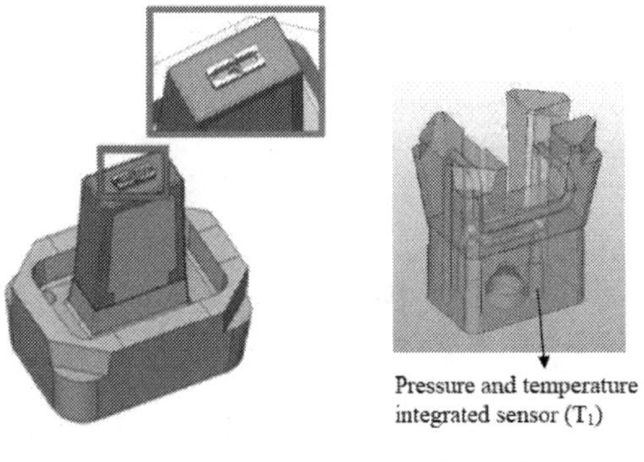

a) Core b) Cavity

Figure 2 - Moulding blocks. a) core; b) cavity

The parts are injected through a spider gating system (Figure 2a). The part and its main dimensions are shown in Figure 3.

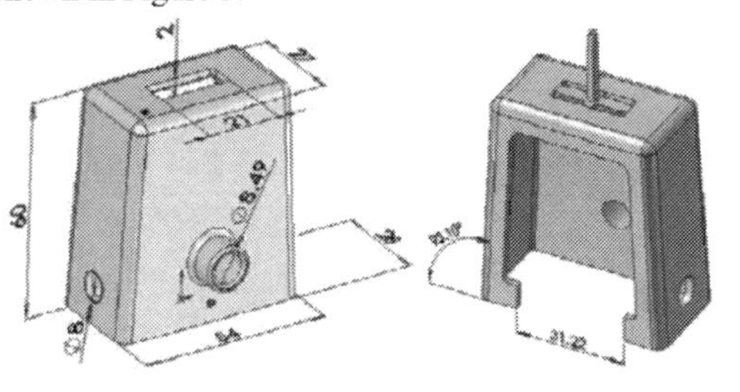

Figure 3 – The moulding

Instrumentation

The hybrid mould was fitted with two integrated temperature and pressure Kistler sensors (Types 6190BA and 6190CA). The integrated pressure and temperature sensors were placed in the cavity and into the backward slide group to collect data values close to the gate location (1) and the welding line (2) of the part (Figure 3). These sensors were connected to a Kistler SCP-Compact data acquisition system (Kistler Instruments, Winterthur).

Experimental

Tooling and materials

The mouldings were produced using two combinations of materials in the moulding blocks: (1) Steel in core and in cavity (identified as Steel/Steel or SS combination) and (2) Epoxy/15% (in volume) short steel fibres (Biresin® L74/15% SSF) in core and steel in cavity (identified as Resin/Steel or RS combination).
The standard steel and the epoxy/SSF composite core moulding blocks are shown in Figure 4.

a) Steel core b) Epoxy/SSF core

Figure 4 - Core moulding blocks a) tool steel 1.2311 b) Epoxy/SSF

The standard moulding blocks were produced in DIN WNr.1.2311 steel by conventional mouldmaking techniques. The epoxy composite cores were produced by vacuum casting in silicone moulds. These cores were made using the actual steel core as a master pattern. The vacuum casting route included a preliminary degassing of the resin composite during 15 min, the degassing of the cast resin during 20 min, the cure of the cast resin at room temperature during 24 h, and the post cure at 60ºC during 3 h and at 140ºC during 3 h. The heating and cooling rates were set at 10ºC/h.

The main thermal and mechanical properties of the materials of the moulding blocks are shown in Table 1.

Table 1 - Properties of the moulding blocks materials

Material	Tool steel	Epoxy/SSF
Properties	(DIN W Nr. 1.2311)	(Biresin® L74+15% SSF)
Specific gravity [$Mg.m^{-3}$]	7.80	2.10
Specific heat [$J.kg^{-1}.K^{-1}$]	460	-
Thermal conductivity [$W.m^{-1}.K^{-1}$]	29	0.392
Coefficient of thermal expansion [K^{-1}]	1.20×10^{-5}	-
Flexural Modulus @20ºC [GPa]	200	3.0-3.5

The short steel carbon fibres have an average length of 453 μm and average thickness of 45 μm (Sabino-Netto *et al.* 2008) and specific gravity of 7.87 $Mg.m^{-3}$. The epoxy Biresin® L74 used as composite matrix has a heat deflection temperature of 160°C and a specific gravity of 1.10 Mg.m^{-3} (Table 2).

Table 2 - Epoxy resin properties (Biresin® L74)

Properties	**Material** Biresin® L74
Hardness* [shore D]	85
Impact resistance* [kJ.m^{-2}]	29
Heat deflection* temperature [°C]	160
Specific gravity [Mg.m^{-3}]	1.1

* Values after post curing: 3h / 60°C + 3h / 140°C

Characterization techniques

The epoxy/short steel fibre composite was determined with conductivity equipment Alambeta (Sensora, Liberec) for the thermal properties, dynamic mechanical analyzer Tritec 2000 (Triton, Nottingham) and universal testing machine Zwick Z100 (Ulm) for the mechanical properties. This alternative material was proposed by Sabino Netto (Sabino-Netto *et al.* 2008) and its density estimated with the law of the mixtures.

Injection moulding

The mouldings were produced in polypropylene homopolymer, PP Homo Domolen 1100N (DOME Polypropylene, Rozenburg) of MFR 12 g/10min (230°C/2.16 kg) using a Ferromatik Milacron K85 injection moulding machine of 850 kN clamping force. The processing conditions were set according to each combination of core and cavity materials, as shown in Table 3.

When using the epoxy composite core the processing conditions (pressure and cycle phases) were adjusted to cope with the poorer mechanical properties of this material compared to steel. The injection and the holding pressures were reduced for epoxy core from that used for steel by 20%. The cycle time was increased up to 60 s in the resin core combination.

Pressure monitoring

Typical pressure data for the moulding blocks combinations and processing conditions are shown in Figure 5 where data variation over the injection cycle is recorded.

Temperature monitoring

The evolution of temperature during moulding was monitored as shown in Figure 6 for a complete moulding cycle. It is shown how the temperature varies during each cycle at the moulding surface of the cavity (sensor T1) and in the core (sensor T2).

Table 3 – Processing conditions

Parameter	**SS**	**RS**
Core/Cavity materials	**Steel/Steel**	**Epoxy-SSF/Steel**
Injection temp [°C]	230	230
Cooling temp [°C]	40	40
Injection time [s]	1.3	1.3
Injection pressure [MPa]	91	70
Holding time [s]	5	10
Opening, closing and ejection time [s]	12	12
Holding pressure [MPa]	70	56
Cooling time [s]	25	35

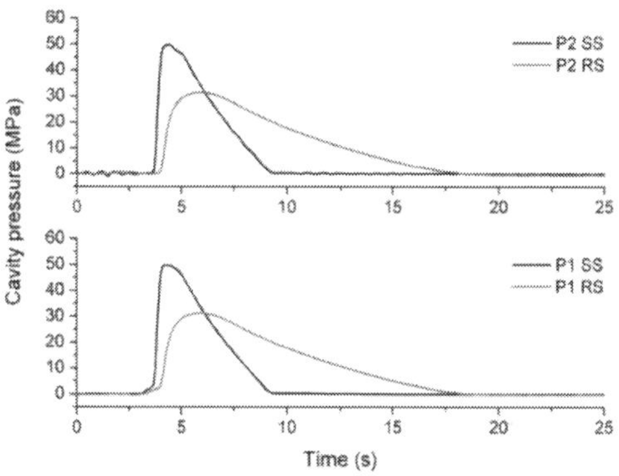

Figure 5 – Pressure data for the SS and RS core/cavity combinations. P1 - cavity; P2 - core

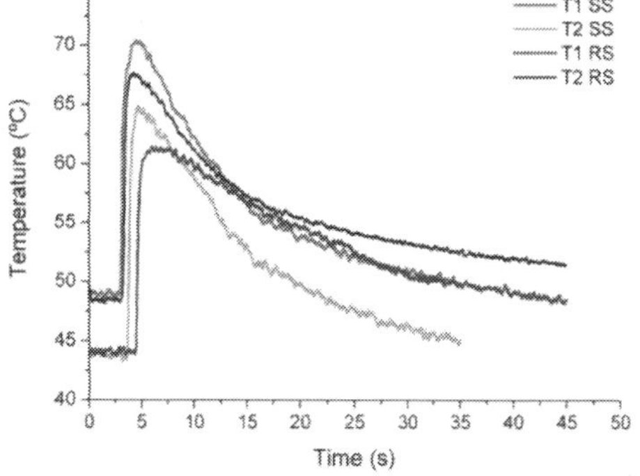

Figure 6 – Temperature data for the SS and RS core/cavity combinations. T1 - cavity; T2 – core

Shrinkage

The shrinkage was analyzed at the specified dimensions of the mouldings. Measurements in the testing part and moulding blocks were made with a 150 mm digital calliper with 0.02 mm resolution.

Simulation

The software Moldex3D® release 9.0 was used for the injection moulding simulations. The total number of elements used in the simulation was of 1 144 902 tetrahedrical elements. 177 792 elements were used in the part. The architecture of the cooling system (Figure 7) was considered in the simulation.

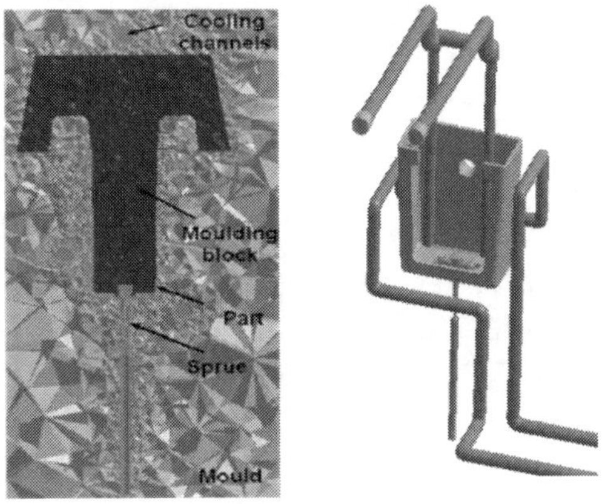

Figure 7 – Model and cooling architecture used in the Moldex3D simulation.

The shrinkage prediction of the variation was made in the part region dimensions shown in Figure 8.

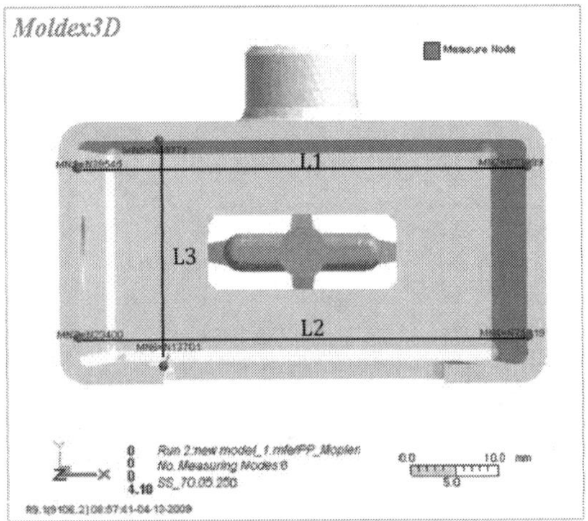

Top view of the part

Figure 8 - Nodes and dimensions L1, L2 and L3 considered for the shrinkage analysis in Moldex3D

Results and discussion

Tooling

The use of epoxy/SSF composites with high density fillers as the short steel fibres require curing methods that prevent sedimentation until the gel time is achieved. The dynamic process at 5 rpm using the apparatus recommended by Sabino-Netto (Sabino-Netto *et al.* 2008) was adequate for this purpose.

Epoxy composites are only adequate for short production series. The combination used to produce the plastics parts, epoxy/SSF composite in the core and tool steel in the cavity, starts to deteriorate close to the gate region, as shown in Figure 9. However it was noticed especially in sliding zones, that small steel particles start to detach from the core causing some seizing in the steel mould surfaces in contact. To avoid this problem it is recommendable to use higher gap tolerances in sliding zones.

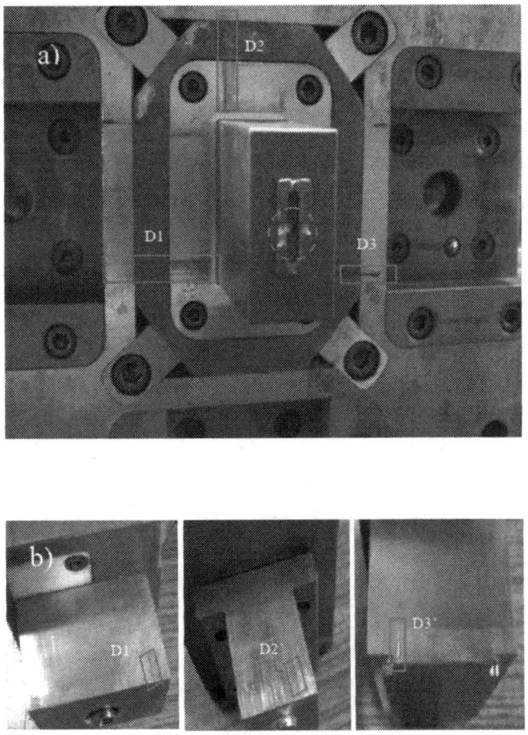

Figure 9 - Seizing in steel components when epoxy/SSF composites are used in the core. a) Damaged gate and scratching zones; b) Seized components

Material characterization

The temperature variation description of the modulus of elasticity is required as the moulding blocks work over a range of temperatures where this property varies, typically from 30°C to 100°C. DMA was used for the characterization in the 3-point bending mode. A typical curve is shown in Figure 10.

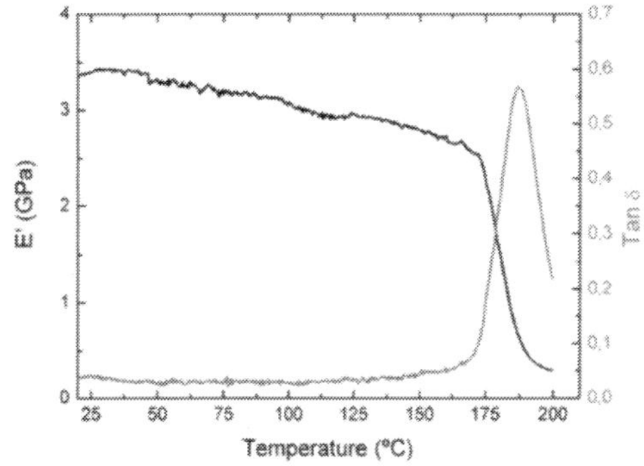

Figure 10 - Elasticity modulus *versus* temperature of the epoxy/SSF composite in three point bending.

The sharp decrease of the modulus above 120°C is related to the approximation of the glass transition temperature of the material. Within the expected operating range from 20°C to 100°C the experimental data can be described by the linear equation:

$$E(T) = -5.24\ T + 3.58\times10^{3}$$

where T is the temperature (°C) and E is the modulus (MPa).

When these composites are used in moulding blocks they are essentially subjected to a compression state of stress. Therefore, the DMA test data were complemented with compression data according to the ISO 604:1993 and tensile data according to the ISO 527-1 standards. The properties of the epoxy composite at 20°C are listed in Table 4 alongside the data obtained in flexure.

Table 4 - Mechanical properties of the Biresin L74+15%SSF composite at 20°C

Property	Tensile	Compression	DMA flexural
Maximum strength [MPa]	38.67	60.32	-
Modulus [GPa]	2.43	4.45	3.40

The temperature data obtained with sensors T1 and T2 (Figure 6) highlight the influence of the higher thermal resistance of the moulding blocks with lower thermal conductivity (RS combination).

Simulation

The evolution of the temperature during moulding was monitored as shown in Figure 11 for a complete moulding cycle. The figure shows how the temperature varies at the moulding surface of the cavity (sensor T1) and in the core (sensor T2).

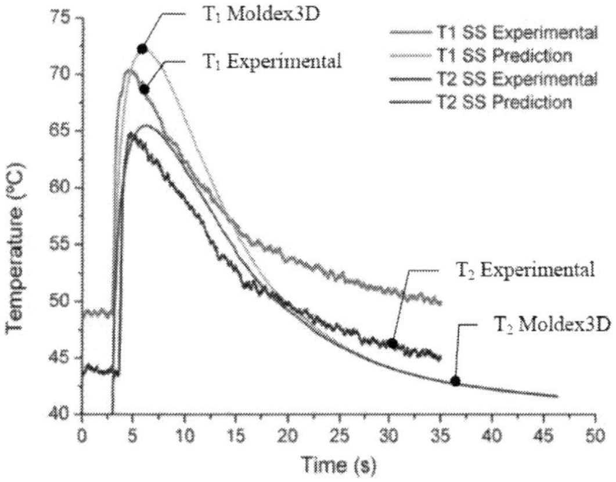

Figure 11 - Temperature data for the steel/steel (SS) cavity/core combination, T_1 - cavity; T_2 – core, and Moldex3D prediction.

The temperature evolution in the initial injection cycles is important when low thermal conductivity materials are used and it is necessary to know the stabilized mould temperature. Figure 12 shows the temperature evolution in the initial injection cycles of the SS combination and that predicted using Moldex3D. The temperature difference between the experimental and the predicted results are less than 5°C.

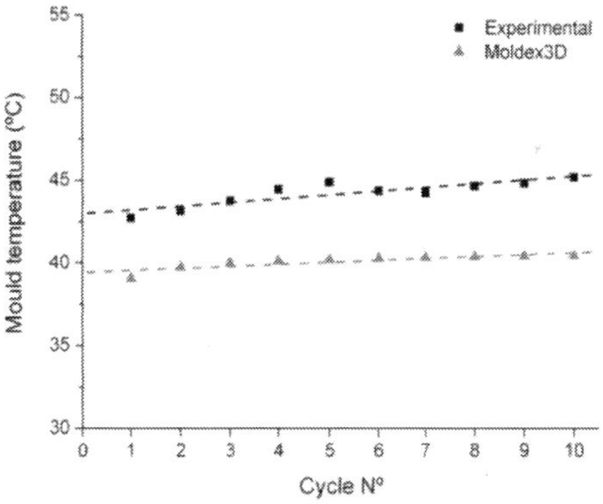

Figure 12 - Temperature evolution during initial injection cycles of the SS combination, and temperature data for the steel/steel (SS)

As shown in Figure 13, the mould temperature stabilizes sooner in the case of SS combination than in the case of RS. In this figure it is also evident the cycle variation repeatability after the stabilization.

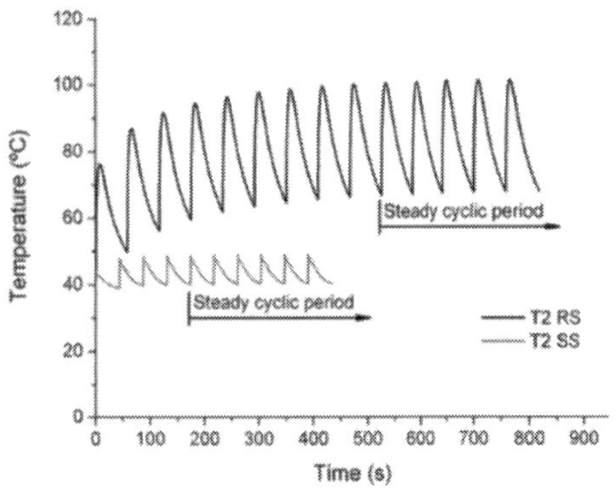

Figure 13 - Temperature evolution in first injection cycles predicted by Moldex3D (SS cycle time = 43.3 s; RS cycle time = 58.3 s)

The definition of the regime temperature depends on the imposed cycle time. Rapid cycling leads to higher service temperatures (as in Figure 14). Moldex3D simulations were made for the case of RS combination to determine an optimal solution.

The analysis shown above (Figure 14) is very important when materials with lower thermal properties, as thermal conductivity and glass transition temperature, are used in moulding blocks.

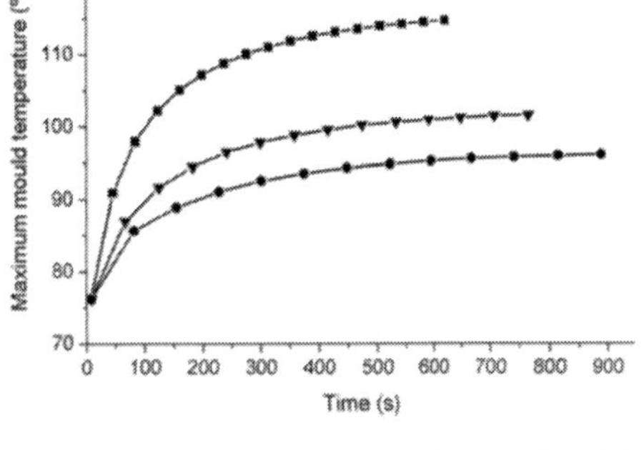

Figure 14 - Temperature evolution in first injection cycles of the RS combination predicted by Moldex3D for different injection cycle times.

Shrinkage

The experimental shrinkage was calculated based on the dimensions, L1, L2 and L3, shown in Figure 9 of the mouldings and corresponding in the moulding blocks. In the case of the shrinkage predicted using Moldex3D the dimensions between the nodes shown in Figure 13 were considered.

In semicrystalline materials the expected shrinkage percentage varies from 1% to 2.5%. The shrinkage measured in this study, as shown in Table 5, is within this range for the SS combination.

Table 5 - Experimental shrinkage and Moldex3D prediction

Shrinkage (%)	Experimental*		Moldex3D	
	SS	RS	SS	RS
L1	0.98	3.65	1.03	1.32
L2	0.84	12.58	0.86	6.57
L3	1.26	11.63	1.33	5.48

* Shrinkage related to core dimensions

For the RS combination larger shrinkage was observed at the specified locations with respect to the SS counterparts (Table 4). It is known that the shrinkage of injection mouldings is governed mainly by the thermal and pressure history during the moulding cycle. In these circumstances these results appear somewhat strange, as the variables involved are not that different apart. Therefore, one could expect that this situation must not be related only to the actual plastics shrinkage but also to the actual deformation of the core during the injection and holding phases of the moulding cycle (Martinho *et al.* 2009). Therefore the differences obtained between the experimental and predicted values of shrinkage, for the case of RS combination, can be associated to the fact that Moldex3D not taking into account the moulding block deformation when soft materials are used, due to injection pressures.

Conclusions

Hybrid moulds are a suitable solution for producing mouldings of relatively complex shapes. When less stiff materials are used as an alternative to steel, it is important for the integrity of the moulding blocks to establish appropriate processing conditions.

The results suggest the importance of predicting the regime mould temperature when low thermal conductivity materials are used in moulding blocks. This temperature depends on the desired cycle time.

Results also evidence that when resin moulding blocks are used the final shrinkage of the mouldings is a consequence of the specific polymer shrinkage and the deformation of the non-metallic core deformed by the processing pressure. Typical manufacturer data are not adequate to predict the moulding part dimensions in this case, and even Moldex3D simulation that is accurate for all steel mouldings is not applicable.

Acknowledgements

This research work was supported by the Portuguese Foundation for Science and Technology under the grant (SFRH/BD/28113/ 2006) and by the Calouste Gulbenkian Foundation. The authors also thank CoreTech Systems (Taiwan) for their technical support on the numerical simulations.

References

Hilton, P. D. and Jacobs, P. F., "Rapid tooling: technologies and industrial applications", Marcel Dekker, New York (2000).

Martinho, P.G., Bártolo, P.J., Queirós, M.P., Pontes, A. J. and Pouzada, A.S., "Hybrid Moulds: the Use of Combined Techniques for the Rapid Manufacturing of Injection Moulds." Virtual Modelling and Rapid Manufacturing - Advanced Research in Virtual and Rapid Prototyping, P. J. Bártolo, ed., Taylor and Francis, London, 421 (2005).

Martinho, P.G., Bártolo, P. J. and Pouzada, A.S., "Hybrid moulds: effect of the moulding blocks on the morphology and dimensional properties." Rapid Prototyping Journal, 15(1), 71-82 (2009).

Martinho, P.G., Bártolo, P. J. and Pouzada, A.S., "Efficient design solutions for hybrid moulds and the widening of the lifecycle of injection moulds." PMI 2007 - Int. Conf. on Polymers & Moulds Innovations, Ghent/Belgium (2007).

Pontes, A.J., Oliveira, M.J. and Pouzada, A.S., "The effect of holding pressure on the shrinkage and birefringence of injection moulded polypropylene plates. "Materials Science Forum, **455-456**, 814-817 (2004).

Pontes, A.J. and Pouzada, A.S., "Ejection force in tubular injection mouldings. Part I: Effect of processing conditions." Polymer Engineering and Science, **44** 891-898 (2004).

Sabino-Netto, A.C., Salmoria, G.V., Ahrens, C.H. and Pouzada, A.S., "On the effect of the cure process on the mechanical properties of steel fibre reinforced epoxy composites." PPS 24th Annual Meeting, Salerno/Italy (2008).

Influence of the injection moulding conditions on the in-line measured electrical conductivity of polymer-carbon nanotube composites

Ingo Alig, Dirk Lellinger, Donghua Xu, Alexander Ohneiser*

Deutsches Kunststoff-Institut - Darmstadt - Germany

We will report on in-line measurements of the electrical conductivity during injection moulding using a recently develop setup [1]. The experimental setup will be described and examples for in line measurements depending on the processing conditions will be given. Time resolved electrical conductivities of polycarbonate- and polyamide-multiwalled carbon nanotube (MWNT) composites and an acrylonitrile-butadiene-styrene/ polycarbonate blend with MWNT were measured inline in an injection moulding machine. During the first few seconds of an injection moulding cycle, the electrical conductivity varies with melt temperature and pressure, but then stays nearly constant for some seconds, until the part vitrifies and breaks off of the conductivity sensor. The conductivity values extracted from the time interval of constant conductivity correlate well to off-line measured conductivities at the same position of the plastic parts. The conductivities are recorded for different injection moulding conditions: injection velocity, melt temperature and mould temperature. It turns out that the conductivity varied by seven orders of magnitude by only small variations of the processing parameters. Injection velocity and melt temperature showed the largest influence, whereas the mould temperature has only minor importance. The inline measured values were compared to offline measured conductivity values, and good correlation was found. The observed behaviour of the electrical conductivity is explained in the framework of destruction/rebuilding of a conductive carbon nanotube network, where the actual state of destruction is frustrated by vitrification or crystallization. [1] D. Lellinger, D. Xu, A. Ohneiser, T. Skipa, I. Alig, Phys. Stat. Sol. (b) 245 (2008) 2268.

REPLICATION OF STOCHASTIC AND GEOMETRIC MICRO STRUCTURES. EFFECTS ON FUNCTIONALITY AND VISUAL APPEARANCE.

G.R. Berger[1]*, W. Friesenbichler[2], and Chr. Teichert[3]

[1]*Polymer Competence Center Leoben (PCCL), Leoben, Austria – gerald.berger@pccl.at*
[2]*Institute of Plastics Processing, University of Leoben, Leoben, Austria – walter.friesenbichler@unileoben.ac.at*
[3]*Institute for Physics, University of Leoben, Leoben, Austria –Teichert@unileoben.ac.at*

The surface quality of injection molded parts depends on the processing conditions, the cavity surface structure, the runner system, the used polymer, and the part geometry. Differences in the replication of the cavity surface structure to the molded part are influencing the surface function and its visual appearance.
A descriptive model of the replication process has been developed, taking into account the thickness of the frozen layer during injection. The expected dependencies were proven by systematic injection molding tests, where geometric surface microstructures, having depths of 45 and 100 microns, and widths from 400 to 2000 microns, were replicated into high-crystalline polypropylene (HCPP) and polycarbonate (PC). The volumetric structure replication ratio increased with rising mold temperature, melt temperature, and cavity pressure. As expected, the smaller the micro feature was, the more important the mold temperature became.
The RMS roughness R_q was found to be a suitable replication parameter for mirror-finished as well as draw-polished stochastic cavity surface structures. A change in wall thickness increased the surface gloss, since the roughness deteriorated as well as a several microns high "step" remained, due to different shrinkage, ascending from the thicker to the thinner part area.

Introduction

Replication of stochastic surface structures is a key issue in manufacturing decorative injection molding parts, e.g. for automotive industry. Common structuring technologies are spark erosion, etching, laser ablation, grinding, and polishing. Although their replication may be described by statistical parameters like the RMS roughness Rq, a detailed analysis of the injection process influences on the replication is hindered due to poor significance.
In the present work the replication of etched surfaces is studied. Microfeatures with geometric defined footprints were used, since their replication could be determined by volume measurements. Typical feature sizes as well as footprints with more or less sharp angles were manufactured.
The replication of a draw-polished and a mirror-finished mold surface into a high-crystalline polypropylene grade is examined by topography analysis. Influences on the visual appearance are discussed.

Experimental

Machines and devices

The injection molding tests were carried out on a fully-electric Battenfeld 1000/200 CDK-SE injection molding machine. The test mold is a single cavity mold with a series of interchangeable mold inserts. Two small pin-shaped mold inserts, having engraved geometric microstructures, were integrated into the structured cavity. Therefore, parts with stochastic as well as geometric surface structures were manufactured simultaneously.
A Kistler 6190A combined pressure-temperature sensor was installed close to the runner to measure the cavity pressure. Data logging and processing were carried out using a HBM Spider 8 system and the AMS BEAM™ software package. Topography and roughness were determined by using an FRT Micro Prof® CWL-sensor and a standard Atomic Force Microscope (AFM).

Geometric microstructure

The geometric microstructure consists of several micro cavities etched into the surface of the mirror-finished pin inserts. Each micro cavity has a geometric footprint, e.g. a circle (Fig. 1). The replicated microcavity is defined as microfeature.

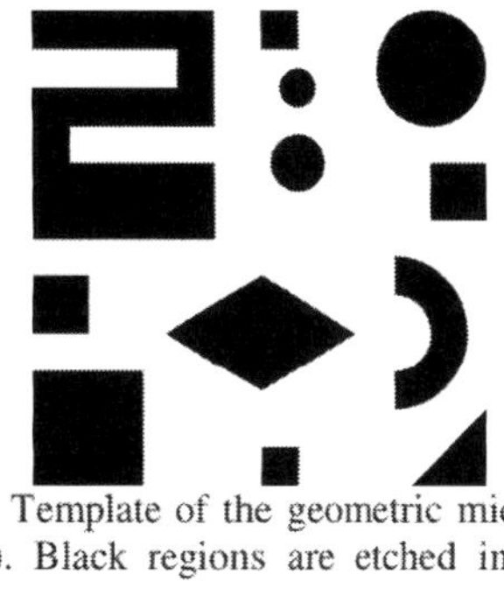

Figure 1 – Template of the geometric micro structure (8x scaled). Black regions are etched into the steel surface.

Design of Experiments

Two polymers with different morphologies and different viscosities in the processing range were selected for the injection molding tests: a semi-crystalline HCPP (high-crystalline polypropylene copolymer), Hostacom PPU X9067HS (BASELL Polyolefine); and an amorphous polycarbonate (PC), Makrolon 2407 (BAYER AG), both colored in dark grey. The levels for the full factorial design of experiments can be seen in Table 1. The maximum values of the cavity pressure were fixed at 3 levels; the required values of holding pressure were determined in pre-trials for each test. The used microstructure was 45 µm deep. Replication results of the square (L = 1200 µm) and the rhomboidal (D_{max} = 2040 µm, D_{min} = 1200 µm) microcavities will be analyzed.

Table 1 – Full factorial experimental design for PC and HCPP.

		HCPP	PC
Maximum cavity pressure, bar	low	180	330
	medium	300	460
	high	420	590
Melt temperature, °C	low	+200	+290
	high	+240	+315
Mold temperature, °C	low	+30	+80
	high	+60	+120

Additional tests were done using HCPP to determine the influence of the micro feature size on the replication. These were carried out at mold and melt temperatures of 30 and 180°C and maximum cavity pressures of 150 to 650 bar. The used microstructure was 100 µm deep. Replication results of the circular elements (D = 400, 600, and 1200 µm) will be shown.

Determination of replication ratios

The geometric replication ratio AG_V is determined by the volumes of the microcavity V_{mc} and the microfeature V_{mf}. Both were measured by the FRT device. The calculation procedure was published earlier (Berger, 2005).

$$AG_V = \frac{V_{mf}}{V_{mc}} * 100\% . \quad (1)$$

Since it is impossible to determine microfeature volumes of smooth stochastic surfaces, statistical parameters will be used to quantify their replication into polymers, e.g. the RMS roughness R_q, which is the root mean square of the magnitude of the deviation of the roughness profile z(x) from the mean line. L is the sampling length (Whitehouse, 2004). The roughness of the draw-polished surfaces was measured by the FRT device, the mirror-finished one by AFM.

$$R_q = \sqrt{\frac{1}{L} * \int_0^L z^2(x)\, dx} \quad (2)$$

Results and Discussion

A summary of the influences on the replication of the geometric microstructure can be seen in Table 2. The cavity pressure and the melt temperature were most important for structure replication into HCPP, whereas for PC cavity temperature and cavity pressure were dominating. There is an influence of the microfeature's geometry since areas with sharp angles (rhombic cross-section) were more difficult to fill.
The dependencies for a PVD TiN coated mold surface were found to be the same; however, the maximum replication ratio of about 94 % was already reached at low mold temperature. Due to the lower thermal effusivity of TiN (4780 W $s^{1/2}$ m^{-2} K^{-1}) compared to steel (9440 W $s^{1/2}$ m^{-2} K^{-1}), the initial contact temperature is about 9°C higher with TiN coating; this leads to a thinner cooled skin which improves the filling of the microcavities (see model of replication process).

Table 2 – Influences on replication.

	Polycarbonate		**High-crystalline Polypropylene**	
increasing	**R**	**S**	**R**	**S**
Mold temperature	++	++	0(+)	+
Cavity pressure	++	++	++	++
Melt temperature	0(+)	0	++	++

++ strongly increases replication; + increases replication
0 no significant dependence; (+) increasing trend
R rhombic cross section; S square cross section

Descriptive model of replication process

Based on the experimental results (Berger 2006) a descriptive 4-phase model of the replication process is defined.

a) Filling of base plate

In conventional injection molding there is a fountain flow region close to the flow front. In this region, melt will be transported from the midplane of the flow field and out towards the cavity wall. In this fountain flow, the polymer will be stretched biaxially causing high molecular orientation. Both, the stretching as well as the contact to the cold air increase the flow front skin's viscosity. This skin is first rapidly freezed as soon as it contacts the cold cavity wall and it then grows to the frozen layer during further injection (Kukla, 1994).

b) Initial filling of the microcavity

As the melt passes the microcavity, the flow front is divided and the microfeature is initially filled. Since the flow front is pressure-less, this filling is determined by polymer viscosity, apparent air pressure, wall adhesion (capillary forces), and inertial forces to some extent. The replication at this state is low and there is no contact to the cavity wall at the top of the microfeature.

c) Pressure-driven filling of the microcavity

As the flow front advances the local polymer pressure increases; the pressure deforms the microfeature's stretched skin. However, the skin is already cooled by lateral heat conduction to the microfeature regions which are already in contact with the cold microcavity wall. Due to local cooling rates of some 1000 K/s the high-viscous skin grows rapidly. The filling process is now similar to thermoforming. Thus, the replication is determined by the thickness and the mechanical properties of the cooled skin as well as of the local polymer pressure. A higher cavity pressure and a thinner cooled skin promote the replication. Fast pressure increase, due to high injection rates, will increase the replication too. In case of a small microfeature the complete microfeature will be part of the frozen layer by the end of the packing phase. Microfeatures with lateral dimensions of more than 400 µm will still have a liquid polymer core region.
Fig. 2 shows a short shot of a HCPP part where intensive shrinkage of the base plate and even of the broader microfeatures is visible. The sidewalls of the microfeatures, which soon had contact to the microcavity wall persisted as a solid frame.

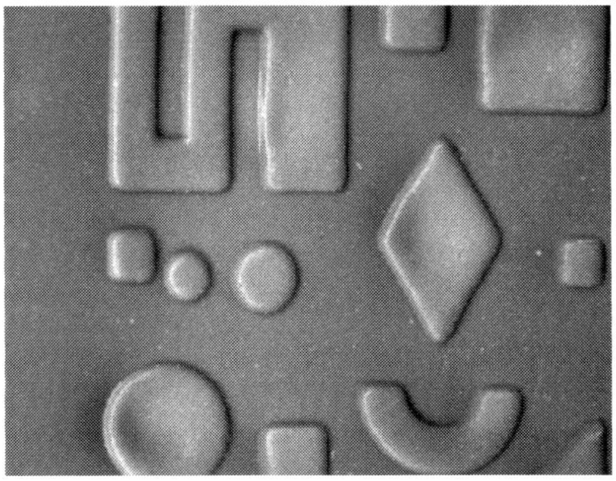

Figure 2 – Short shot of HCPP specimen with partially replicated microfeatures (feature height in between 70 and 90 µm).

d) Shrinkage compensation and thermoforming of the frozen layer

In the holding phase, hot polymer is pressed into the part to compensate the cooling-driven volume drop. The hot melt reduces the frozen layer thickness in the microfeatures. The frozen layer is deformed depending on the feature size and the local pressure, again similar to thermoforming. Although the filling of the microfeatures could be 100% now, post-shrinkage and stress relaxation reduce the final feature size again.

Influence of microfeature size

If maximum cavity pressures less than 300 bar were applied, the small microfeature (D 400 µm) was worse replicated than the others (D 600 µm or 1200 µm), see Fig. 3. The cooling of the small microfeature is faster; thickness and strength of the skin are increased. Due to the higher surface to cross-section ratio, the degree of stretching of the viscous skin is very high. The resulting elastic forces hinder the further deformation and the low pressure in the holding phase is not able to compensate the shrinkage. Thus, a poorer replication of the small microfeature results.

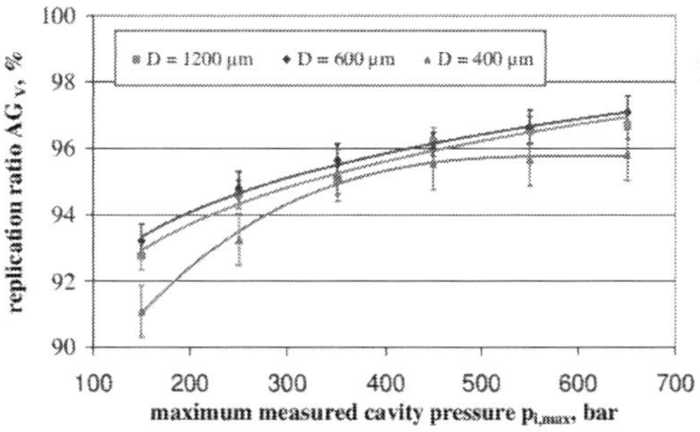

Figure 3 – Replication of HCPP depending on the microfeature size (circular cross-section).

If cavity pressures from 300 to 470 bar were applied, the replication was nearly the same for all microfeatures. On the one hand, higher cavity pressures cause further deformation of the small microfeature's skin, which increases the replication. One the other, in the bigger microfeatures, increased deformation is compensated by higher shrinkage.
If cavity pressures more than 470 bar were applied, shrinkage compensation is further increased. The stretchability of the small microfeature's high viscous skin reaches its maximum at 450 bar, the replication stops, despite increased pressures.

Replication of mirror-finished mold surface into HCPP

Two different mirror-finished steel surfaces, with R_q values of 4.4 and 11.9 nm, were replicated into PC where values of 4.3 and 8.9 nm were reached (Berger, 2010).
However, the replicated HCPP surfaces were significantly rougher (Fig. 4) and diffuse reflecting, since the smooth regions were interrupted by many very deep caverns (Fig. 5). The replicated coarse polishing tracks even in the caverns, as illustrated in Fig. 6, prove the full contact of the polymer surface

and the cavity surface during injection. This shows that the local retraction of the polymer surface occurred afterwards. Morphological dominated "micro shrinkage effects" are more important than replication itself.

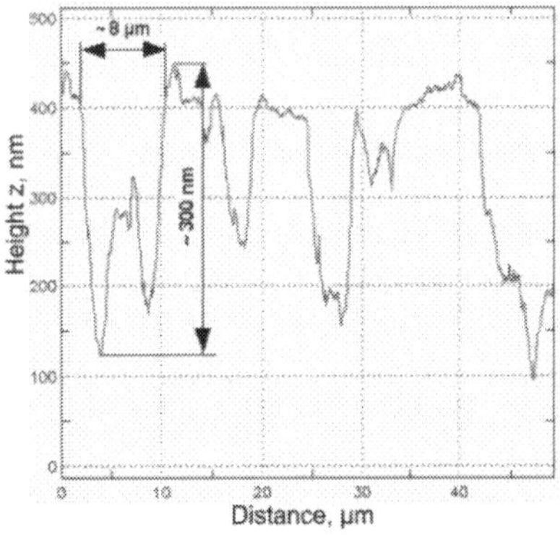

Figure 4 – Roughness profile on HCPP surface.

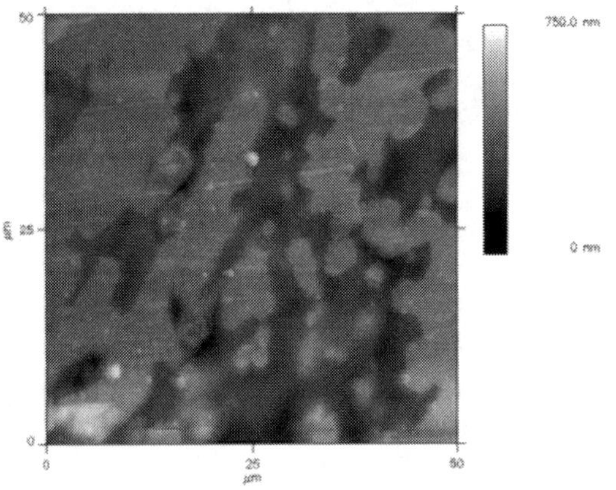

Figure 5 – Color coded topography image of HCPP surface.

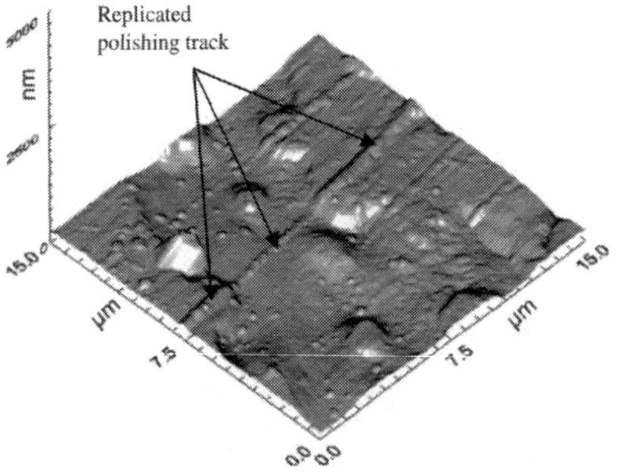

Figure 6 – Topography of HCPP surface with replicated polishing tracks.

Descriptive model for the "micro sink marks"

As soon as the hot, liquid polymer melt contacts the cold cavity surface during injection, a frozen layer is formed, where, due to the local high cooling rates, even semi-crystalline polymers behave "quasi-amorphous", which leads to a good structure replication. While cooling proceeds, surface-near crystalline regions in the polymer shrink more than amorphous ones. These stronger contractions generate higher tensile forces which locally pull down the frozen layer for some hundred nanometers, the caverns are formed.

Replication of draw-polished surfaces

The RMS roughness R_q was also used to determine the replication of a draw-polished steel surface (R_q = 360 nm) into HCPP (Fig. 7). The reduction of part thickness results in a significant gloss difference. A decreasing replication from R_q = 310 to 190 nm was determined. Moreover, a several microns high "step" was found, caused by thickness-dependent different shrinkage. This step increases from the thicker to the thinner part area (Fig. 8). Both, the step as well as the different roughness influence the visual appearance significantly.

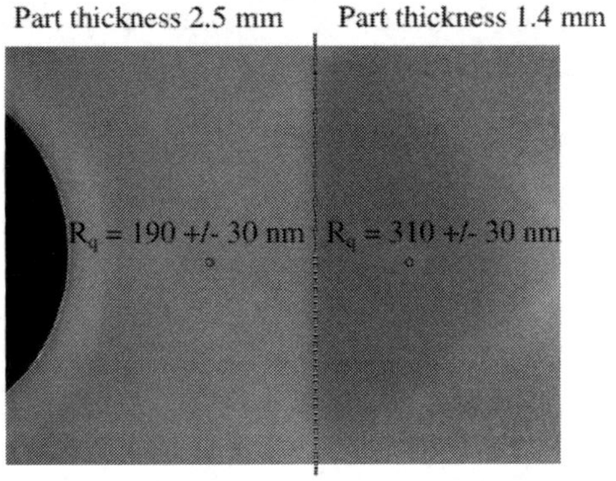

Figure 7 – Photograph of HCPP surface. Decreasing replication of draw-polished cavity surface, due to a reduction in part thickness, leads to major gloss differences.

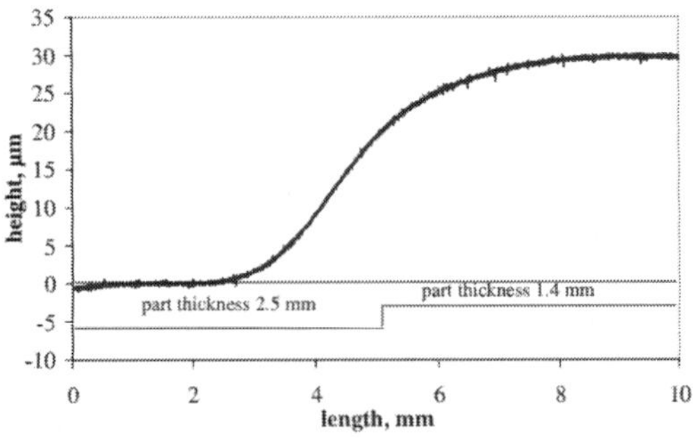

Figure 8 – Shape of HCPP surface opposite to a part thickness decrease.

Conclusions

Replication results of geometric as well as stochastic surface microstructures have been shown.
As expected, the structure replication increased with rising mold and melt temperatures and cavity pressure.
The smaller the micro feature was, the more important the mold temperature became.
Geometric microfeatures with lateral dimensions of 400 µm and smaller were not perfectly replicated if conventional mold temperature of 40 °C was used.
A descriptive model of the replication process has been developed, taking into account the thickness of the frozen layer during injection.
Morphological dominated "micro shrinkage effects" roughened the high crystalline polypropylene surface much more than different replication of the high-gloss polished mold surface. A descriptive model for this so-called "micro sink marks" has been stated.
A change in wall thickness increased the surface gloss, since the roughness deteriorated as well as, due to different shrinkage, a several microns high step remained, ascending from the thicker to the thinner part area.

Acknowledgements

The research work of this paper was performed at the Polymer Competence Center Leoben GmbH (PCCL, Austria) within the framework of the K_{plus}-program of the Austrian Federal Ministry of Transport, Innovation and Technology with contributions by University of Leoben (Institute of Plastics Processing) and Oerlikon Balzers Coating Austria GmbH, Wittmann Battenfeld GmbH, Bohler Edelstahl GmbH, HTP Automotive GmbH, IB STEINER Ingenieurbüro für Kunststofftechnik and JUTZ GmbH. The PCCL is funded by the Austrian Government and the State Governments of Styria and Upper Austria.
The authors thank Dr. Christian Hofer and Michael Reitter for their contributions.

References

Proceedings: Berger, G.R., Friesenbichler, W. et al., "Characterisation of surface reproduction in the injection moulding process", Annual Meeting of the Polymer Processing Society (2005).

PhD Thesis: Berger, G.R.: Replication of geometric and stochastic surface microstructures by injection molding, University of Leoben (2006).

Journal: Berger, G.R., Friesenbichler, W. et al "High-gloss injection molded part surfaces", MP Materials Testing, **5** (2010).

PhD Thesis: Kukla, C.: Rand- und Grenzschicht-bildung beim Spritzgießen, University of Leoben (1994).

Book: Whitehouse, D.: Surfaces and their Measurement, 1st Edition, Kogan Page Science, London (2004), p. 52f.

SURFACE REPLICATION AND STRUCTURE IN MICRO-AND NANO-MOLDIMG WITH HEAT INSULATOR MOLD

H. Ito [1*], S. Takahagi [1], K. Zouda [2], and Y. Kayano [2]

[1] *Department of Polymer Science and Engineering, Graduate School of Science and Engineering, Yamagata University, Yamagata, Japan – ihiroshi@yz.yamagata-u.ac.jp*

[2] *Mitsubishi Gas Chemical Co. Inc., Tokyo, Japan*

Abstract - Injection molding with micro- and nano-surface features and thin-wall plates was performed to fabricate zirconia with heat-insulated molds. Effects of process parameters on processability and surface replication of the molded optical parts such as a optical diffraction grating and anti-reflective plates were evaluated. The replication ratio and higher-order structure of molded parts were analyzed using SEM and a polarizing optical microscope. Using an insulated mold, the flow length and surface replication ratio of molded parts increased. The molecular orientation of molded parts decreased using this mold, as it did in molding conditions of higher mold temperatures and injection temperatures.

Introduction

In micromolding and nanomolding processes, the major factors influencing the molding quality are not only mold and injection temperatures, but also injection speed [1–4]. Especially, the mold temperature is a crucial factor controlling the filling of micro-feature and nano-feature products. For nanoscale features or those with a high aspect ratio, it is difficult to force the polymer to fill the geometry completely. Consequently, the mold must be close to the glass transition temperature of the material and must maintain a sufficiently high temperature to ensure good replication of the nano-features. Maintaining high mold temperatures during filling and lower mold temperatures below solidification temperature during cooling, without increasing the cycle time was not easy. In conventional molding, direct heating and cooling of the entire mold is not an economic option because of the large heat mass of the mold. Various methods to achieve dynamic mold surface-temperature control have been reported [5–7]. Although flame heating efficiency is quite good, safety issues and repeatable and precise temperature control persist as problems. Saito et al. used direct infrared heating of the polymer to control the polymer temperature [8].

An example of a technology that could be used to replace these processes and improve the surface quality of products is rapid heat cycle molding. This is a commercialized technology to heat and cool molds rapidly by flowing hot steam or cool water through a circuit used for mold cooling. However, circulation of two fluids with different temperatures necessitates a boiler and a hot water running system to provide steam.

Yao et al. proposed rapid mold surface variation by coating heat conducting (pyrolytic graphite) and insulating layers (silicon carbide) on the mold surface [9–10]. The heat-conducting layer conducts electrical current to raise the temperature rapidly and provide direct heating to the polymer, whereas the insulating layer prevents the heat from entering the mold base.

This study investigated the processability and higher-order structure of thin-wall parts with various nano-surface features by insertion into a zirconia insulated mold for heating efficiency. The replication of molded micro-features and nano-features can be improved without a marked increase in the cycle time if the mold surface temperature can be controlled efficiently during micromolding and nanomolding processes. Effects of process parameters on processability, surface replication, physical properties, and the internal morphology of parts were discussed.

Experimental

Material

For the experimental work, two commercial polycarbonate (PC) resins (Mitsubishi Engineering-Plastics Corp.) were used. The resin had a solid density of 1.20 g/cm^3. The melt index (MI) for one resin was 63 dg/min (300°C, 1.2 kg) (optical grade); that for the other resin was higher than 63 dg/min as high flow (HF grade). The glass transition temperature of both resins was 145°C.

Mold and molding processes

We used an intelligent analysis system that consisted of a precision injection molding machine, a multi-pass rheometer, and a pressure–volume–temperature (PVT) properties measurement apparatus for micromolding processes [1–4]. A small electric injection-molding machine (modified ELJECT AU3E; Nissei Plastic Industrial Co. Ltd.) was used in this system for injection molding. Process monitoring was also done to elucidate the process behaviors. The cylinder temperature, nozzle temperature, plunger position, and its pressure were measured using this system.

In this study, we produced the mold, made of STARVAX, with surface roughness of Ra 2 nm. The mold block used metal and a heat insulator. Figure 1 portrays a picture of the mold and insert parts of zirconia. We used the dimensions of a square flat mold of 20 × 20 mm. The cavity thickness is 0.5 and 1.0 mm. We prepared three different pattern inserts: 0.3-mm-thick silicon plates. The surface patterns of plates are the diffraction grating and anti-refractive structure. We also used a mirror polishing plate for evaluation of the molecular orientation in nanomolding with a heat insulator mold. One piece of the mold part was made from the zirconia with the heat insulation effect; the same part also was prepared using a steel alloy that is generally used for the metal mold. Surface structures with nano-scale dimensions were fabricated using electron beam lithography. Nickel alloy was electroplated up to a thickness of ca. 100 □m on mother alloy steel. The positive electron beam resist was spin-coated onto the nickel alloy surface. The surface topography of the nano-structure was transferred to nickel alloy on the steel alloy. Regarding diffraction grating, these lines were trapezoid-grooved with 1 μm pitch and 600 nm width, with 530 nm depth. The bottom width was 400 nm.

For injection molding, the injection speed, holding pressure, and maximum injection pressure were, respectively, 200 mm/s, 200 MPa, and 250 MPa. The injection unit temperature was 250–280°C; the mold temperature also was changed: 80°C to 120°C. We measured the process behavior using a data-logger system inside the molding machine.

Characterization for molded parts

These depths and heights were analyzed using a scanning electron microscope (SEM, JSM-6330F; JEOL) and a scanning probe microscope with thermal analysis (SPM, micro-TA, 2990; TA Instruments) using a long probe for nano-pattern measurement.

We analyzed the molecular orientation of the parts using a polariscope and a polarizing optical microscope (POM, BX51-P; Olympus Corp.) with a Berek compensator. Furthermore, a skin-shear-core structure of molded parts was analyzed using polarizing microscopy after a thin section was cut parallel to the MD from the molded parts using a microtome (every 30μm thickness, 2055 Autocut; Leica Microsystems). Thermal stability of nano-surface structures also was analyzed after annealing molded parts. These depths and heights of the surface structure were analyzed using SPM at various annealing temperatures and times.

Results and Discussion

Figure 2 presents a photograph of the thin-wall product of 0.5 mm thickness at different mold inserts. The product length increased the with-heat insulated mold. No difference of flow length at different surface patterns was observed. For mirror polishing of the mold using the heat insulator mold, the molded part flow length showed an increase of 23% at a mold temperature of 120°C and injection temperature of 260°C. Increases in the flow length with the heat-insulator to the metal mold at different mold temperatures are shown in Fig. 3. However, the flow length efficiency decreased with higher injection temperature. Consequently, the mold insulator effect depends on the molding conditions. Results show that the development of a skin-shear region inside molded parts influenced the flow length.

Figure 3 portrays polarized photographs of molded parts produced under different molding conditions. The moldings were performed using a mirror polished plate with metal and zirconia block. The intrinsic birefringence of PC is large, thereby confirming the interference color. The interference fringe correlates closely with optical retardation and residual strain. The interference color was observed clearly near the gate; it faded toward the flow end, implying that the residual strain near the gate was greater than at any other position. The interference color became unclear at the higher injection temperature of 280°C.

In each molding condition, a short shot occurred; a different flow length is shown at each line in Fig. 3. The interference color was different along the flow direction. These tendencies revealed that the cooling rate of resin influenced the polymer flow in the cavity. Contacting the mold surface accelerated polymer cooling. At a higher cooling rate, the molten resin viscosity increased and solidification of the molten resin occurred rapidly. The polymer flow was lengthened using a heat-insulated mold. Because of moderate cooling, the polymer flowed extremely well at the heat insulator.

In this case of 1.0 mm cavity, the interference color was not observed clearly, except near the gate. The color usually was observed clearly at a long optical path, i.e. the thick molded part. However, the interference color was white (less than ca. 300 nm). In this molding condition, the skin-shear region became small because of the high flow rate of 200 mm/s. Using the zirconia mold, optical retardation appeared to be slight at the gate.

Regarding high flow grade PC (HF grade), its flow-ability was effective. For higher mold temperatures and injection temperatures, the parts were packed fully inside the cavity at 0.5 and 1.0 mm thickness. On the other hand, at the lower mold temperatures of 250°C, the flow length changed at different mold temperatures.

Figure 4 shows SEM images of surface patterns of the diffraction grating for HF grade PC. Using the heat insulator, we observed high replication of surface nano-features (Fig. 6, right side). Figure 5 also shows SEM images of surface patterns of the anti-reflection for HF PC. This portion is located near the gate. However, the nano-surface pattern was not observed clearly at a certain condition in the case of molding with a metal block. The replication property was influenced markedly by the kind of mold material. The heat insulator cooled and reduced the surface region of the flowing polymer, thereby improving replication accuracy. However, the replication ratio is almost identical at the higher mold temperature and injection temperature. Consequently, the greatest advantage of using the insulated mold is its possible production at lower mold temperatures and injection temperatures.

Conclusions

We investigated the processability and higher-order structure of thin-wall parts with various nano-surface features using a zirconia-insulated mold to increase heating efficiency. Effects of process parameters on the processability, surface replication, and optical retardation of parts were examined. The flow length and surface replication ratio of molded parts increased using the heat-insulated mold of zirconia. Especially, the replication accuracy was improved using a heat insulator and a high flow grade polymer. The molecular orientation of molded parts near the gate was decreased using this mold, similarly to the molding conditions of higher mold temperatures and injection temperatures. Results show that molding using a heat-insulated mold presents advantages of low energy costs through lower mold temperatures and injection temperatures.

Acknowledgements

This research was partially supported by the Ministry of Education, Science, Sports and Culture, Grant-in-Aid for Scientific Research (C), 2009, #21560710.

References

1. H. Ito, et al., *SPE-ANTEC Tech. Papers,* 2005, **LXIV**, 693 (2005)
2. H. Ito, et al., *Theoretical Appl. Mech. Jpn.*, **54**, 263 (2005)
3. H. Ito, et al., *Macromolecular Symposia, Advanced Polymers for Emerging Technologies*, **249–250**, 628 (2007)
4. H. Ito, et al., *Proc. of the 2007 International Manufacturing Science and Engineering Conference*, ASME-MSEC2007-31035 (2007)
5. D. Yao and B. Kim, *J. Injection Molding Technol.* **6**, 11 (2002)
6. S.C. Chen, et al., *Int. Commun. Heat Mass Transfer*, **31**, 971 (2003)
7. S.C. Chen, et al., *SPE-ANTEC Tech. Papers,* **LXIV**, 556 (2005)
8. T. Saito and I. Satoh, *Polym. Eng. Sci.*, **42**, (12) 2418 (2002)
9. D. Yao, et al., *SPE-ANTEC Tech. Papers*, **LXVI**, 704 (2001)
10. D. Yao and B. Kim, *Polym. Eng. Sci.*, **42**, 2471 (2002)

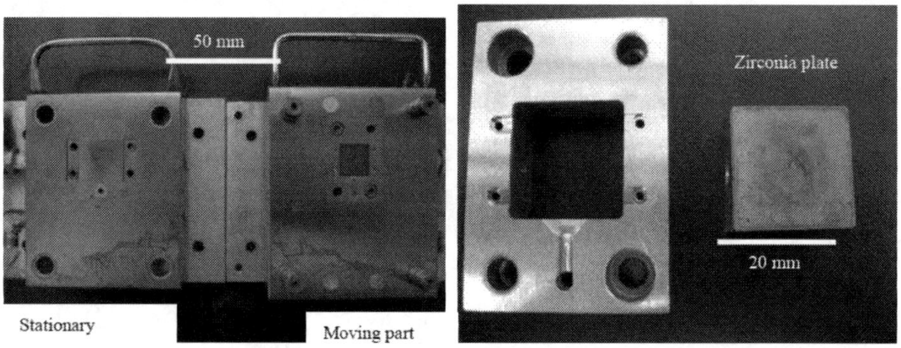

Figure 1 Mold and insert parts of zirconia.

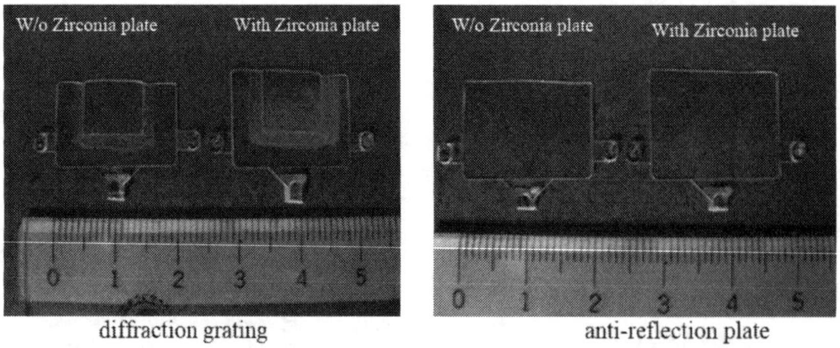

Figure 2 Photograph of the thin-wall product of 5 mm thickness for different mold inserts. The PC is optical grade. Injection and mold temperatures were 260°C and 80°C, respectively.

Without Zirconia plate | With Zirconia plate

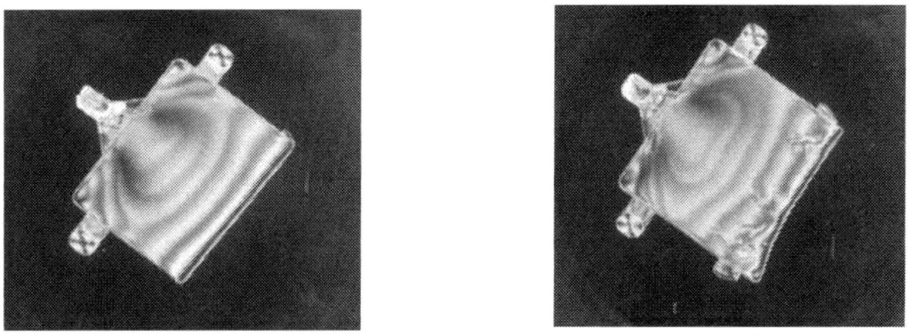

Figure 3 Polarized photographs of molded parts under different molding conditions. Cavity thickness is

Without Zirconia plate | With Zirconia plate

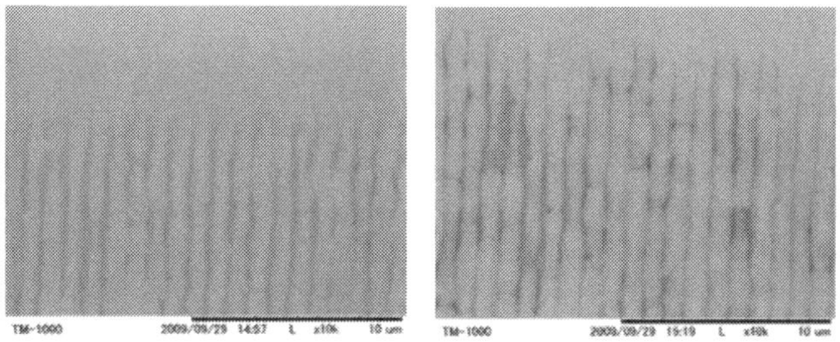

Figure 4 SEM images of surface patterns of the diffraction grating. The PC is HF grade.

With Zirconia plate

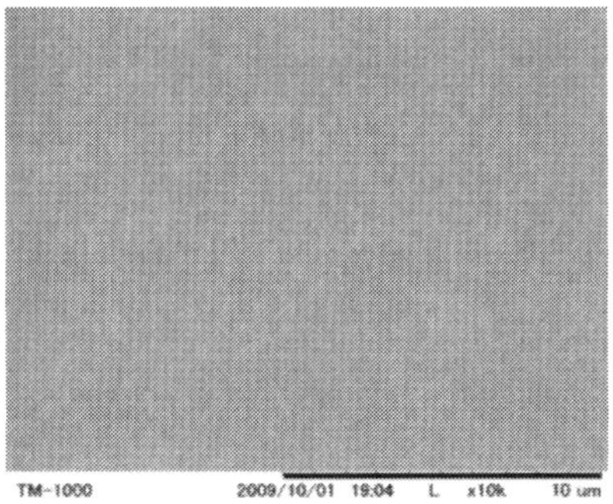

Figure 5 SEM image of surface pattern of the anti-reflection. The PC is HF grade.

OPTICAL MONITORING OF NANOCOMPOSITES´ CRYSTALLIZATION DURING INJECTION MOLDING

M. M. Favaro and R. E. S. Bretas*

Department of Materials Engineering, Universidade Federal de São Carlos – Rod. Washington Luiz, km 235 – São Carlos, Brazil, zip code 13565-905
*mmfavaro@gmail.com; * bretas@power.ufscar.br*

This work had as main objective to study the crystallization of intercalated polyesters/nanoclay nanocomposites during injection molding using an optical monitor inside a mold cavity. The light of a laser source was passed thru a sapphire window and the mold cavity and it was collected in a collector fiber at the opposite wall. Two He-Ne laser sources were used, one with a wavelength (λ) of 632 nm and another one with 405 nm. The optical system was able to detect the beginning and the end of the polyesters' crystallization. The time for the beginning of crystallization was defined as the time when the first scattering (and absorbing) structures were detected by the optical sensor. The growing crystals created new interfaces, decreasing the optical signal. It is worthwhile to point out those particles with sizes between λ/10 and 10λ scattered light. When the crystallization ended, a plateau was attained. From the results, it was concluded that the nanoclay's particles accelerated the overall crystallization of the polyesters, acting as nucleating agents. The nucleating effect of the nanoclay was predominant over the crystallization process of the polyester with slow crystallization kinetics. It was also noted that the nanoclay's particles scattered almost zero during the injection molding when the laser source of 632 nm was used. However, when the laser source with 405 nm was used both, pure polymers and nanocomposites samples, absorb different light quantity, showing that, with the lowest laser source wavelength, the nanoclays' tactoids were detected; therefore this optical system probably can be used to detect systems with different particles' sizes.

Introduction

It is known that the final properties and structures of polymeric products are directly linked to their processing. In the case of engineering thermoplastics nanocomposites (like poly(butylene terephthalate), PBT, and poly(trimethylene terephthalate), PTT, with nanoclays like montmorillonite, MMT), the increase in mechanical properties, for example, is dependent on the dispersion and distribution levels of the nanoclay and on the morphology and crystallinity of the polymeric matrix (Drown et al., 2007; Gurmendi et al., 2008; Narkhede and Shertukde, 2008).

Thus, the study of the influence of nanofillers on the crystallization and morphology of polymers, under static conditions, is intense (Favaro et al., 2009; Hu and Lesser, 2003; Jog, 2006; Tjong and Bao, 2004; Wu et al., 2006). These results have shown that under quiescent conditions the nanoclays' platelets can act as nucleating agents, enhancing the overall crystallization rate and also influencing the size and shape of the developed crystals. However, under non-isothermal and non-quiescent flows, as the ones that occurred in injection molding, where the polymer crystallizes under high temperature, pressure and deformation gradients, the polymer crystallization is different.

To study crystallization under those gradients requires non-conventional techniques capable of simulating the injection molding environment. Computational techniques can be used; however, the modeling requires the calculation of the amount of energy dissipated during the crystallization, which is not a trivial task.

Therefore, our main research objective is the development of an optical tool capable of monitoring the crystallization during the injection molding process and which allows analyzing the surging of the different crystalline structures along the thickness of the cavity mold.

Experimental

Materials

-Poly(trimethylene terephthalate) (PTT): Shell Chemicals (Corterra 200);
- Poly(butylene terephthalate) (PBT): GE Plastics (Valox 315);
-Polycarbonate (PC): Unigel (Durolon IR-2200);
-Nanocomposites PTT/MMT, PBT/MMT and PC/MMT: These materials were produced by twin-screw extrusion using a co-rotational extruder from Werner Pfleider, model ZSK 30, with diameter of 30 mm and length of 1070 mm. The nanoclay was an organophilic montmorillonite (MMT) from Southern Clay Products Inc. (Cloisite 30B) modified with a methyl, tallow, bis-2-hydroxy-ethyl, quaternary ammonium salt. The final nanoclay composition in all the nanocomposites was 5%wt.

Injection Molding and Optical Monitoring

The materials were injection molded using an Arburg 270V injection molding machine, with clamping force of 300 kN and screw diameter of 25 mm; the cavity has a disk geometry where the optical sensor was installed as already described in our previous work (Favaro et al., 2008). The light of the laser fiber (source) passed thru a sapphire window and the mold cavity and it was collected in a collector fiber, as shown in Figure 1. Two He-Ne laser sources were used, one with a wavelength λ = 632 nm and another one with λ = 405 nm.

Three injection-molding parameters were changed, as shown in Table 1: mold temperature (Tw), packing pressure (Pp) and flow rate (Q). The packing time was maintained at 20s and the applied cooling time was 30s, except for condition G, where the cooling time was 50s.

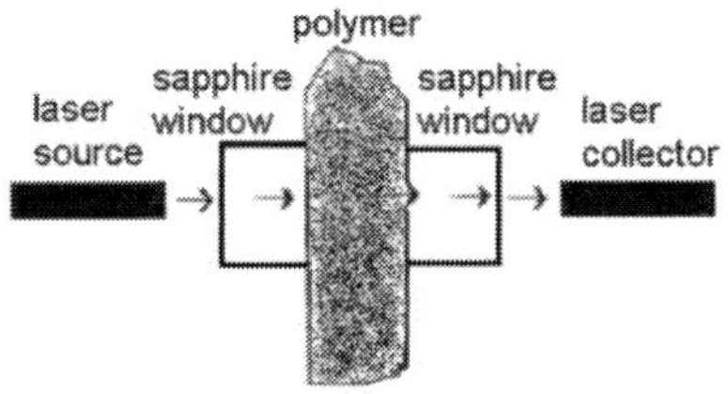

Figure 1: Scheme of the light path inside the mold disk cavity.

Table 1: Injection molding conditions for PTT, PBT, PTT/MMT and PBT/MMT samples.

Condition	Tw (°C)	Q (cm^3/s)	Pp (bar)
A	30	35	400
B	63	5	400
C	63	65	400
D	63	35	400
E	63	35	250
F	63	35	550
G	96	35	400

Results and Discussion

Graphs of the transmitted light intensity thru the samples as a function of time were built for each injection molding cycle. Some of these graphs, using the laser of 632 nm are shown in Figure 2. At the beginning of the molding cycle, when the mold was open, the transmitted light intensity was I_a. This light intensity was constant until the closing of the mold. Afterwards, the intensity signal increased abruptly up to a constant value I_0. When a crystalline polymer like PBT or PTT was injection molded, a gradual decrease of the transmitted light was observed, from a time defined as t_i , which was the crystallization induction time or the time at which the first scattering and absorbing structures were detected by the optical sensor (Favaro et al., 2008; Moretti et al., 2010; Thomas and Bur, 1996). After attaining a minimum value, the transmitted light intensity increased again up to I_F, when the crystallization ended. On the other hand, when an amorphous polymer like PC was injected, the transmitted light signal was constant, indicating that no scattering or absorbing structures were formed.

The transmitted light intensity of the nanocomposites was different; from a determined time, defined as t_C, the light intensity was constant and the scattering valley was not observed. This constant value (t_C) was associated to the end of the crystallization.

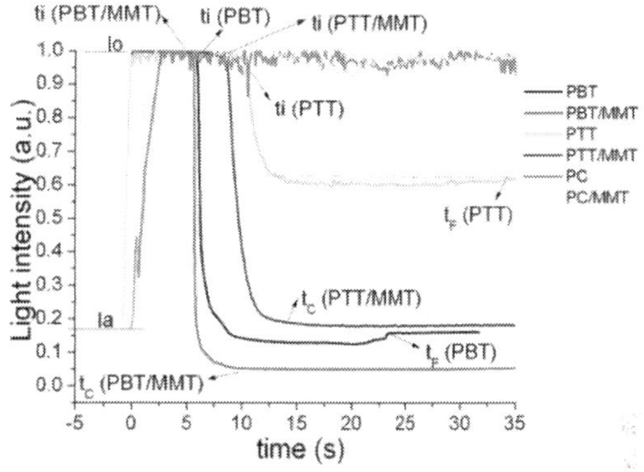

Figure 2: Scheme of the optical signal patterns with the laser of λ = 632nm.

Tables 2 and 3 show the values of the crystallization times (t_i, t_C and t_F) of the pure polyesters and their nanocomposites obtained from the optical signals using the laser of λ = 632nm. It is observed that the PTT had a larger crystallization time than the PBT, which was expected because it is known that PBT has faster crystallization kinetics than PTT due to its more flexible macromolecules. It was also observed that the induction time for the detecting of scattering or absorbing structures was lower in the nanocomposites than in the pure polymers; that is, the nanoclay accelerated the bulk crystallization of both polymers during the injection molding. This difference was stronger between the PTT and the PTT/MMT nanocomposite (a difference of approximately 19%) than between the PBT and the PBT/MMT nanocomposite (a difference of approximately 5%); in other words, the accelerating or nucleating effect was stronger in polymers with slow crystallization kinetics.

Table 2: Values of t_i and t_F using the laser of λ = 632nm, for PBT and PBT/MMT nanocomposite.

	PBT		PBT/MMT	
condition	t_i (s)	t_F (s)	t_i (s)	t_C (s)
A	2.5	12.7	2.4	5.8
B	5.6	13.7	5.0	9.3
C	2.2	9.8	2.2	5.5
D	2.6	12.3	2.4	5.9
E	2.6	15.2	2.5	6.2
F	3.2	20.1	2.4	6.0
G	2.3	28.8	2.3	6.1

Table 3: Values of t_i and t_F using the laser of λ = 632nm for PTT and PTT/MMT nanocomposite.

	PTT		PTT/MMT	
condition	t_i (s)	t_F (s)	t_i (s)	t_C (s)
A	6.4	25.5	3.7	10.4
B	7.4	39.3	6.9	13.3
C	5.3	26.9	4.1	11.1
D	--	ND	3.9	12.4
E	4.7	29.6	4.0	11.1
F	4.6	29.8	4.3	10.1
G	4.8	ND	3.8	13.5

Figure 3 shows the standard light intensity collected with the laser of λ = 405nm. Some differences between both lasers can be observed. The transmitted light intensity of the amorphous polymer PC with the laser of 405nm was not coincident with the signal of the closed and empty mold, as occurred with the laser of 632nm. After certain period t_{abs}, an abrupt decrease of the light intensity occurred. This decrease was related to the absorbed light by the PC. After this decrease, the signal was constant and it was called I_{F^*}.

For the PBT and PTT, an abrupt decrease of the light intensity was also observed; however, after this abrupt decrease the signal was not constant, but continued to decrease until a constant value I_{F^*} was attained. This last light intensity decrease was the result of the developing crystallization. Also, with the laser of 405nm the scattering valley was not observed during the PBT and PTT crystallization.

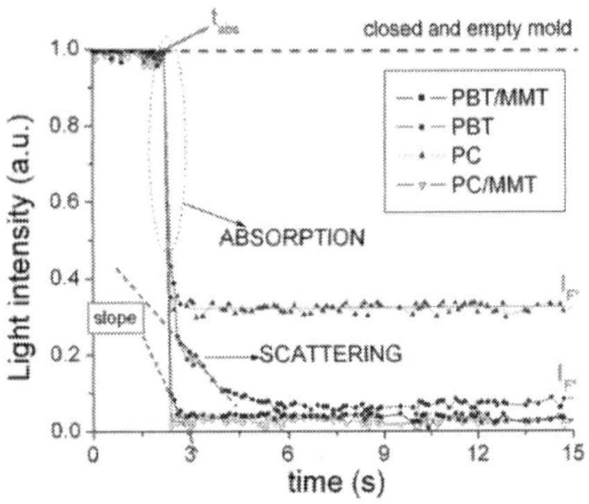

Figure 3: Scheme of the optical signal patterns with the laser of λ = 405nm.

The light intensity signals of the nanocomposites were also different; the signal of the PC/MMT did not superimpose to the signal of the PC as observed with the laser of 632 nm. In this case, the signal decrease was stronger when the PC/MMT passed along the sapphire window, showing that at a wavelength of 405nm, the nanoclay's particles also absorbed or scattered the light. After the strong decrease, the PC/MMT signal was again stable and constant during the remaining of the injection molding cycle.

With the PBT/MMT and PTT/MMT nanocomposites an initial signal decrease was also observed; however, due to the presence of the nanoclay particles, this decrease was stronger than in the pure PTT and PBT. In both nancomposites, the transmitted signal continuously decreased up to a constant value. This last decrease was due to the crystallization process.

The values of t_i and t_F using the laser of 405nm are shown in Tables 4 and 5.

Table 4: Values of t_i and t_F using the laser of λ = 405nm, for PBT and PBT/MMT nanocomposite.

	PBT		PBT/MMT	
Condition	t_i (s)	t_F (s)	t_i (s)	t_C (s)
A	2.9	6.5	2.4	2.9
B	5.0	9.5	5.1	6.1
C	2.7	5.7	2.3	3.5
D	3.2	6.5	2.4	2.9
E	3.0	7.7	2.4	3.4
F	3.0	6.1	2.4	2.9
G	3.0	10.0	2.4	3.3

Table 5: Values of t_i and t_F using the laser of λ = 405nm for PTT and PTT/MMT nanocomposite.

	PTT		PTT/MMT	
Condition	t_i (s)	t_F (s)	t_i (s)	t_C (s)
A	3.5	9.5	2.6	7.8
B	6.8	10.2	7.8	13.3
C	3.2	5.7	2.7	8.0
D	3.1	10.0	3.5	9.8
E	3.3	5.4	3.5	10.2
F	3.4	6.6	2.8	9.5
G	3.5	6.0	3.4	7.9

From these results it is observed that with the laser of 405nm, again, in almost all the injection molding conditions, the t_i and t_F of the PTT were higher than of the PBT, which was expected because PTT has slower crystallization kinetics than the PBT. For the nanocomposites, it was observed that the t_i's were slightly lower than of the pure polymers; that is, the surging of scattering particles was accelerated by the presence of the nanoclay's particles. On the other hand, the t_C's of the PBT/MMT nanocomposite were lower than of the pure PBT, while the t_C's of the PTT/MMT nanocomposite were higher than of the pure PTT. These last results do not agree with the results of the 632nm laser and it can be related to the strong light absorption by the nanoparticles. However, this explanation needs further research.

The laser with λ = 405nm was used because it was expected that a lower wavelength would allow to detect smaller scattering structures than a laser with λ = 632nm; therefore, the induction times of the laser with λ = 405nm would be smaller than of the laser with λ =

632nm. However, after comparing the induction times obtained from both lasers, no change was detected for the PBT, PBT/MMT and PTT/MMT. The only exception were the t_i's of the PTT which were actually smaller than the ones obtained with the 632 nm laser; that is, because PTT has a slow crystallization kinetics it was possible to detect the surging of very small scattering structures.

Thus, for slow crystallizing polymers it is worthwhile to use a laser with a smaller wavelength. For polymers with fast crystallization kinetics, the use of lasers with smaller wavelength does not seem to bring any advantages, because a superimposition of the detection of scattering structures and light absorption by the polymer occurs.

That is, the measurement of the time at which the first scattering and absorbing structures are detected by the optical sensor depends also on the polymer light absorption; as the laser wavelength decreases, smaller scattering particles can be detected, but the polymer can also absorb light. Thus, both phenomena will superimpose.

Also it was observed that using the laser with the smaller wavelength allowed the detection of the nanoclay's tactoids. This optical sensor probably can be used to differentiate between intercalated and exfoliated structures.

Conclusions

Two laser sources of different wavelength, one with 632 nm and the other with 405 nm, in an optical sensor inserted in a mold disk cavity were used to measure the crystallization kinetics of nanocomposites of PTT and PBT with MMT. It was concluded that for PTT (a slow crystallizing polymer), it was worthwhile to use the laser with the smaller wavelength. For PBT (a fast crystallization polymer), the use of the laser with the smaller wavelength did not bring any advantages, because a superimposition of scattering and light absorption by the polymer occurred.

Also it was observed that using the laser with the smaller wavelength allowed the detection of the nanoclay's tactoids. This optical sensor probably can be used to differentiate between intercalated and exfoliated structures.

Acknowledgements

The authors thank FAPESP for financial support and Shell Chemicals, GE Plastics and Unigel for materials donation.

References

Drown, E.K., Mohanty, A.K., Parulekar, Y., Hasija, D., Harte, B.R., Misra, M., Kurian, J.V., "The Surface Characteristics of Organoclays and their Effect on the Properties of Poly(trimethylene terephthalate) Nanocomposites", Comp. Sci. Tech., **67**, 3168-3175 (2007). DOI: 10.1016/j.compscitech.2007.04.011

Favaro, M.M., Marinelli, A.L., Farah, M., Bretas, R.E.S., "Optical Monitoring of Polypropylene Crystallization during Injection Molding", Polym. Eng. Sci., **48**, 257-266 (2008). DOI: 10.1002/pen.20858

Favaro, M.M., Branciforti, M.C., Bretas, R.E.S., "Study of the Quiescent and Shear-Induced Crystallization Kinetics of Intercalated PTT/MMT Nanocomposites" J. Polym. Sci. Part B – Polym. Phys, **48**, 113-127 (2009). DOI: 10.1002/polb.21852

Gurmendi, U., Eguiazabal, J.I., Nazabal, J., "Structure and Properties of Nanocomposites with a Poly(trimethylene terephthalate) Matrix", Eur. Polym. Sci., **44**, 1686-1695 (2008). DOI: 10.1016/j.eurpolymj.2008.04.001

Hu, X., Lesser, A.J., "Effect of a Silicate Filler on the Crystal Morphology of Poly(trimethylene terephthalate) / Clay Nanocomposites", J. Polym. Sci. Part B – Polym. Phys, **41**, 2275-2289 (2003). DOI: 10.1002/polb.10592

Jog, J. P., "Crystallization in Polymer Nanocomposites" Mater. Sci. Tech., **22**, 797-806 (2006). DOI: 10.1179/174328406X101300

Moretti, F., Favaro, M.M., Branciforti, M.C., Bretas, R.E.S., "Optical Monitoring of the Injection Molding of Intercalated Polypropylene Nanocomposites", Polym. Eng. Sci., published online (2010). DOI: 10.1002/pen.21662

Narkhede, J.S., Shertukde, V.V., "Study on Thermal and Mechanical Properties of PBT / Clay Nanocomposites", J. Polym. Mater., **25**, 1-14 (2008).

Thomas, C.L., Bur, A.J., "Optical Monitoring of Polypropylene Injection Molding", Polym. Eng. Sci., **39**, 1291-1302 (1996).

Tjong, S.C., Bao, S.P., "Preparation and Nonisothermal Crystallization Behavior of Polyamide 6 / Montmorillonite Nanocomposites", J. Polym. Sci. Part B – Polym. Phys, **42**, 2878-2891 (2004). DOI: 10.1002/polb.20161

Wu, D.F., Zhou, C.X., Fan, X., Mao, D.L., Zhang, B., "Nonisothermal Crystallization Kinetics of Poly(butylene terephthalate) Montmorillonite Nanocomposites", J. Appl. Polym. Sci., **99**, 3257-3265 (2006). DOI: 10.1002/app.22782

MICROINJECTION MOULDING – WALL-SLIP EVALUATION OF POM FLOW IN MICROCHANNELS

J. C. Vasco[1,2], J. M. Maia[2,3] and A. S. Pouzada[2]*

[1] *Department of Mechanical Engineering, Polytechnic Institute of Leiria, Leiria, Portugal – jvasco@estg.ipleiria.pt*
[2] *IPC - Institute for Polymers and Composites, University of Minho, Guimarães, Portugal – asp@dep.uminho.pt*
[3] *Department of Macromolecular Science and Engineering, Case Western Reserve University, Cleveland, USA – jmaia@dep.uminho.pt*

Abstract - Microinjection is recognized as one of the most efficient replication methods for microcomponents and microsystems in several domains of application. Features such as ultrathin walls or microchannels may be present in the geometry of the plastic parts causing microscale flow phenomena to occur. These phenomena can not be predicted accurately by simulation software packages so far. A micromould with commercial microinjection equipment was used to evaluate the polymer flow in microchannels and the results are seen as useful contributes to enhance the knowledge of the melt flow at the microscale. These findings may have interest in industrial applications. In this paper, rheological phenomena such as wall-slip, melt pressure drop and polymer flow length were studied and compared with standard commercial injection moulding simulator data.

Introduction

Microinjection is one of the most efficient replication methods for microcomponents and microsystems in domains of microengineering, such as optical grating elements, micro pumps, micro fluidic devices and micro gears (Griffiths *et al.*, 2006). The market for microcomponents obtained by microinjection is growing, showing high business potential and is surely becoming one of the key technologies for micromanufacturing (Sha *et al.*, 2005). The need for quickly and accurately getting the moulding zone of moulds for microinjection moulding originated the development of the Variotherm concept. These systems are important for the accurate replication of high aspect-ratio microstructures, and precise process control as is the case optical components (Piotter *et al.*, 2002; Giboz *et al.*, 2007). Despite the success on the replication of microdetails, as in CDs or DVDs, the polymer filling behaviour of mould cavities still requires further study since the simulation tools currently available cannot fully predict flow effects at the microscale. Aspects such as wall slip, surface tension and heat transfer phenomena, require further study to avoid or minimize re-engineering costs or even to prevent wrong options on microcomponents or microsystems design (Kim *et al.*, 2002; Piotter *et al.*, 2002; Su *et al.*, 2004).

This paper is based on the knowledge gathered on previously performed microrheometry studies. Such knowledge is here applied to a test part that replicates situations that may occur in an industrial product.

Experimental

Microrheometry tool

Conventional capillary rheometers are difficult to operate, especially with channels at the microscale. Alternative setups guaranteeing the flow through the microchannel must generate pressure high enough and support the high stresses developed during flow (Chien *et al.*, 2005). A concept tool for microrheometry evaluations at the microscale consisting of a set of interchangeable moulding blocks with calibrated microchannels, assembled on an injection mould structure was used. Its operation requires accurate temperature control during the moulding cycle, as well as pressure control at the extremities of the microchannel (Vasco *et al.*, 2009). The use of miniature pressure sensors and a data acquisition system makes it possible to gather the information required for rheological characterization of the flow in the microchannel. The tool is sketched in Figure 1.

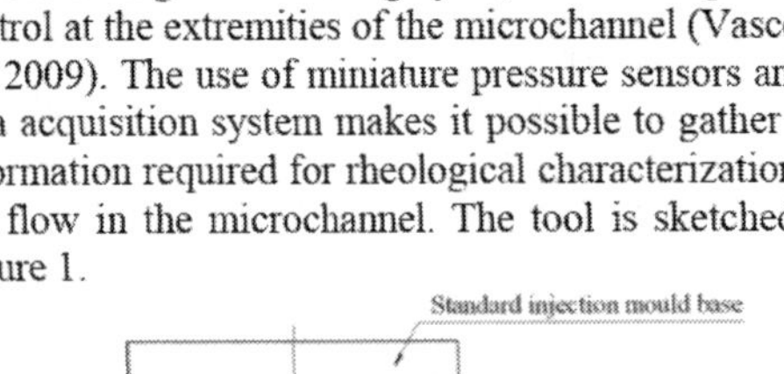

Figure 1 – Microrheometry/microinjection tool

For this study, two microchannel cross sections were tested: a slit with rectangular section, 400 μm wide by 100 μm deep, and a square section of 200×200 μm. The length of both microchannels is identical, 5 mm. The pressure sensor location is depicted in Figure 2.

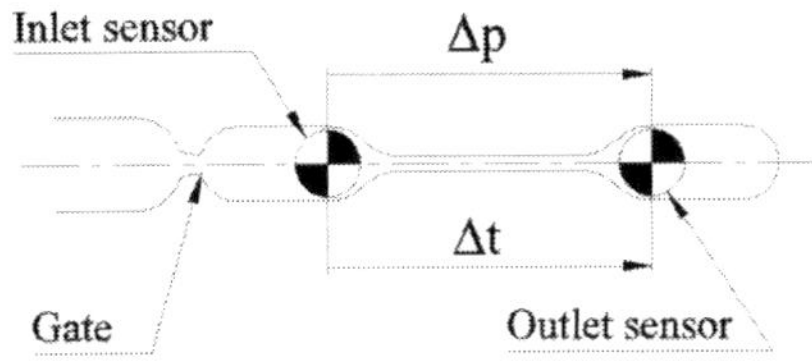

Figure 2 – Variables for microrheometry: Δp – pressure drop; Δt – time interval

The physical dimensions of the sensors used are from a different scale of the microchannel dimensions. This makes necessary the sensors being located outside the microchannel which is a source of error. Therefore, additional zero length microchannels moulding blocks were manufactured to correct for entry effects.

The injection moulding machine was a Boy 12A (Dr Boy, Germany) that combines technical characteristics for microinjection and affordability (Piotter *et al.*, 1997; Song *et al.*, 2007). It is adequate for microinjection, for being able to meter with high precision an injection volume as small as 0,1 cm^3 at flow rates up to 15,6 $cm^3.s^{-1}$. The maximum injection pressure is 240 MPa. The microinjection cell consists of the Boy 12A injection moulding machine with Ø14 mm injection screw, a mould temperature regulator and an external control unit for the cartridge heaters used in the temperature control system of the mould (Figure 3).

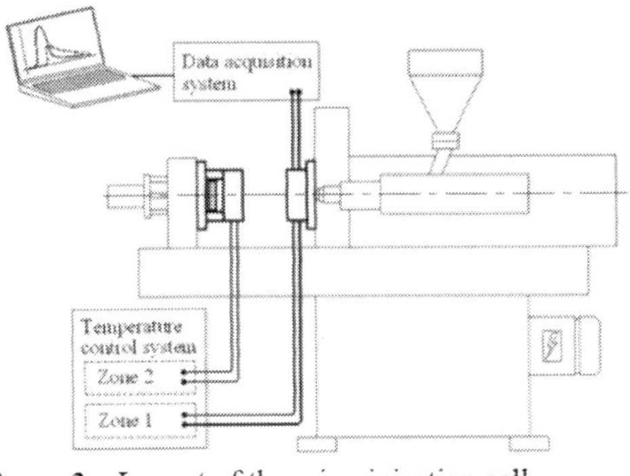

Figure 3 – Layout of the microinjection cell.

Data for microrheometry

Conventional capillary rheometry relies on the variation of the piston speed to impose different shear rates, establishing the characteristic flow curves. In this case, such variation can only be made cycle-by-cycle and not, within the injection cycle, which requires several injection cycles with different melt velocities. Each point of the flow curve must be obtained directly from a single standardized injection cycle. Unlike the microinjection process, no packing pressure was used since the goal was only to obtain a fully established flow in the microchannel. The other process parameters, namely the mould and the melt temperatures had to be raised gradually until complete mouldings were obtained. The 14 mm-screw injection speed was varied within the capability of the equipment from 20 to 200 $mm.s^{-1}$. The required mould temperature was of 145°C and the melt temperature was set at 235°C. For each microchannel the data for analysis were those corresponding to full-shot mouldings, when pressure data were gathered by both pressure sensors.

In which concerns microrheometry evaluation, the pressure drop, Δp, calculated from the pressure sensor data is required for working out the shear stress. The pressure drop between the two locations and the average melt flow velocity are used to relate the shear viscosity to the shear rate (Chien *et al.*, 2005; Chen *et al.*, 2008). The flow time between the two pressure sensors, Δt, is directly obtained from the data acquisition system, enabling the immediate calculation of an average flow rate.

Results and Discussion

The wall-slip phenomenon

The concept of wall slip was first addressed in 1931 by Mooney, who found that certain flow curves depended on the radius of the capillary once the shear stress exceeded a critical value (Lee and Mackley, 2000). The effect takes place when the shear stress at the microchannel wall exceeds a critical value, typically, above 0,1 MPa (Hatzikiriakos and Dealy, 1992; Yao and Kim, 2002; Mitsoulis *et al.*, 2005). As the magnitude of the shear stresses varies, polymer melts flow in a way that is a superposition of slipping at the wall and shearing within the melt (Gleissle and Windhab, 1985). The data gathered initially from the experimental setup enabled important conclusions concerning flow behaviour within both microchannels. The wall-slip effect described above was clearly observed in this study, since a clear shift on the flow regime occurs when critical values of shear rate are achieved (Figure 4).

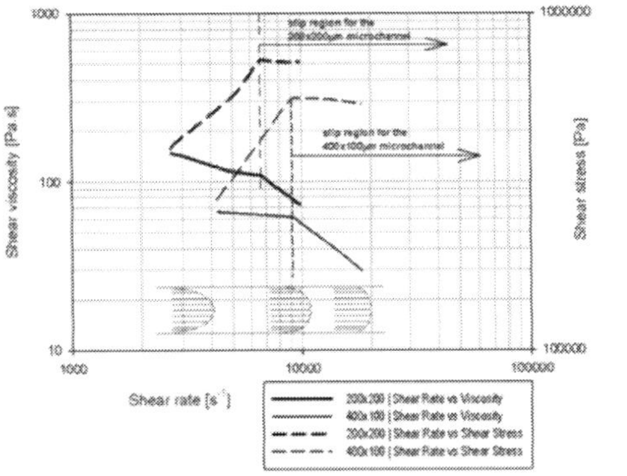

Figure 4 – Flow curves from rheometry calculations based on the 200×200μm and 400×100μm microchannels pressure data.

Both curves show a transition from a normal flow to a new type of flow where the index of non-Newtonian behaviour, n, is close to 1, indicating the onset of wall-slip. The horizontal component of both curves at constant shear stress clearly states that a change in behaviour from shear to plug flow due to the wall-slip effect as occurred. The non-horizontal component of the curves corresponds to shear flow, immediately before the critical shear rate is achieved on both microchannels. It is also possible to observe in this figure that the critical shear rate for this particular material and the 200×200 μm microchannel is around 5000 s^{-1}. The critical shear rate for the 400×100 μm microchannel, for which successful shots were difficult to obtain at lower melt speeds, is estimated to be approximately 6000 s^{-1}.

Comparing these flow curves with conventional rheometry ones, another important difference arises. Despite of the processing temperature used for flow in the microchannels being 235°C, the corresponding flow curve is located above the 180°C flow curve, as shown in Figure 5. This means that the actual temperature on the microchannel is much lower than the imposed processing temperature.

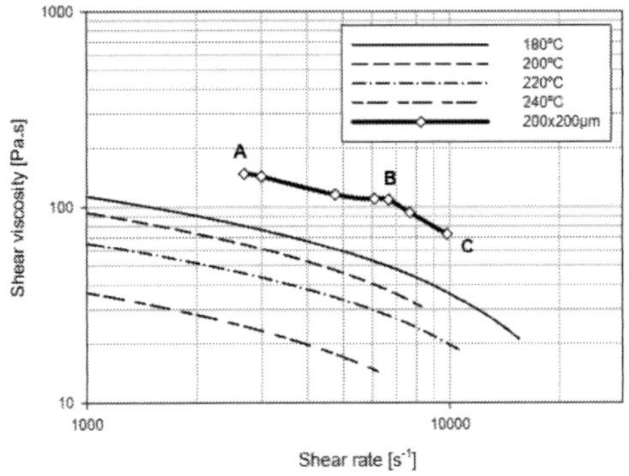

Figure 5 – Experimental data from conventional capillary rheometry and calculations based on 200×200 μm microchannel pressure data.

Such fact is certainly due to the high surface area to volume ratio, which makes conventional heat transfer equations insufficient to describe what happens at this scale. Analyzing the flow curve for the 200×200 μm microchannel, two segments can be observed. From A to B, shear flow is found although the end of this segment represents the transition to the next segment, B to C, in which plug flow is fully developed. The cooling of the molten material in the microchannel means that the temperature of the flowing melt is expected to be well below the initial 235°C. However, the amplitude of the decrease can not be predicted from the TTS principle. In fact, an application of TTS to the microchannel flow data to superimpose it with the other data yields an apparent flow temperature of less than 80°C. This is impossible, not only because it is more than 60°C lower than the mould wall temperature, but is also lower than the no-flow temperature of POM which is quoted as 160°C, according to manufacturer data.

The evaluation of the local wall-slip velocity was made using a simplified power-law slip model introduced for the slip velocity without considering any pressure or temperature dependence (Rosenbaum and Hatzikiriakos, 1997). The melt front profile was estimated as a gradual shifting from shear flow to plug flow as shear stress increases. The data obtained for the 400×100 μm microchannel may be inconclusive since there is not enough data in shear flow conditions to establish a reliable trend. However, the few calculated local wall-slip velocity values for this microchannel seem to corroborate the conclusions already obtained from the shear stress *vs.* shear rate plot (Figure 6).

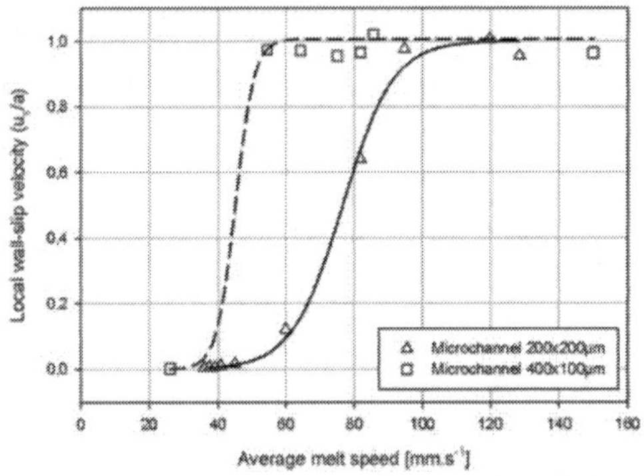

Figure 6 – Average melt speed *vs.* local wall-slip velocity

Nevertheless, both curves in Figure 6 show the same trend: when shear stress is less than the critical shear stress, the local wall-slip velocity is nearly zero. As predicted, the local wall-slip velocity becomes close to 1 if the shear stress keeps increasing (Rosenbaum and Hatzikiriakos, 1997).

Industrial applications

An immediate conclusion of this study is that the flow inside the mould is non-isothermal, this meaning that the data to be obtained eventually are not to be regarded as quantitatively accurate rheometrical data. At this stage it is important to let clear that the equipment setup used in this study is not a microrheometer, but rather a viscometer that replicates the flow conditions in real flow situations, *i.e.*, it is a process analyser and simulator.

This setup, besides its use on the microrheometry studies, also enables a knowledge enhancement on the flow behaviour in microinjection tests of sample test parts. Thin walls and microdetails are typical features in microsystem components such as gyroscopes, accelerometers, microswitches or transducers.

a) Case study

A sample test part was designed for this case study, depicting institutional logos. In the moulding there is thickness variation, with thin walls down to 100 µm and microdetails with low aspect-ratio (Figure 7).

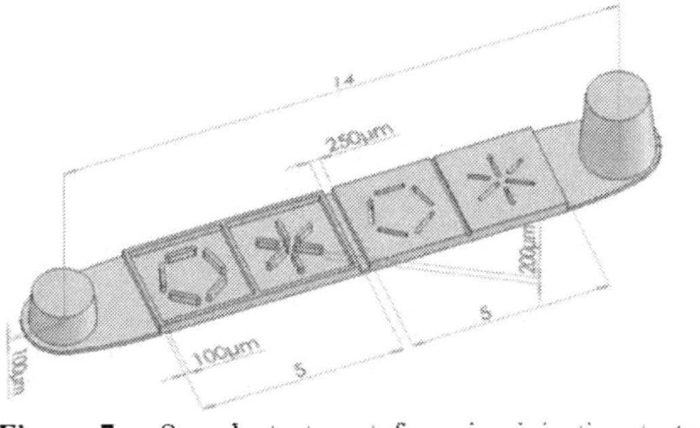

Figure 7 – Sample test part for microinjection tests (dimensions in mm)

This test part presents several challenges since the gate is located on the right tip of the part. The melt must flow through all the obstacles in the first section and then, fill the microfeatures in the second section. The first section contains low-relief logos with local thickness of 300 µm. The second section contains the 200 µm high-relief microfeatures. Throughout the part, except for the protrusions located at the extremities and the first section, the overall thickness is 100 µm. Such geometry enables the evaluation of polymer flow length on a thin wall depending on processing conditions.
The melt pressure drop is monitored by the same acquisition system used previously on the microchannels, enabling the determination of a possible wall-slip flow despite the fact that thickness decreases from the first section to the second section of the part.

b) Processing conditions
The same POM grade, Hostaform® C27021 used on the microrheometry study was also used here. Considering that the injection unit has a plasticizing screw with L/D ratio of 18, recommended processing temperatures for this grade are, respectively, 190 to 230°C for melt temperature, and 60 to 120°C for mould temperature. Pressure of 120 MPa is recommended, both for filling and packing. Preliminary injection tests were performed to establish processing conditions for full-shots of the sample test part. The set of values that achieved that purpose are detailed on Table 1.

Table 1 – Processing conditions

	Pressure	Time	Temperature	
Injection	60 MPa	0,2	Melt	215°C
Packing	20 MPa	2	Mold	100°C

c) Experimental data
From the pressure sensors installed on the cavity, it is possible to obtain i) the pressure at the sensors, enabling the calculation of pressure drop between them, and ii) the time span for the melt touching each sensor, enabling the determination of an average melt speed, since the thickness varies along the part. The evaluation of an average polymer flow length can be established by measurements carried out on the moulded parts. Figure 8 illustrates the average data measured by the pressure sensors.

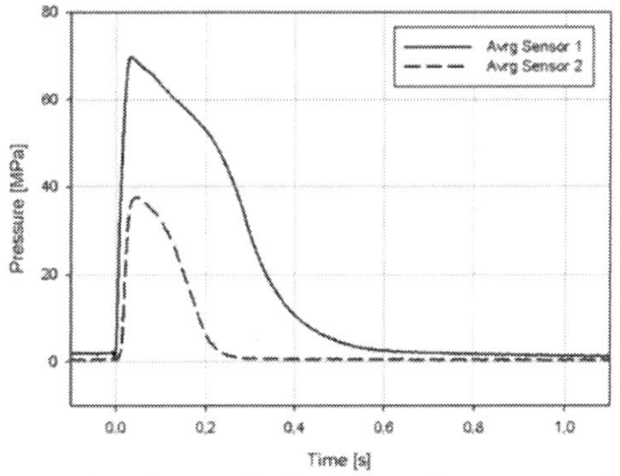

Figure 8 – Curves obtained from the average values obtained by the pressure sensors

A preliminary conclusion arises from the analysis on the previous figure: the melt speed is very high since the time delay between the signals of both pressure sensors is very short. Zooming at the beginning of both curves, it is possible to see that the time delay is around 15 ms, as shown in Figure 9.

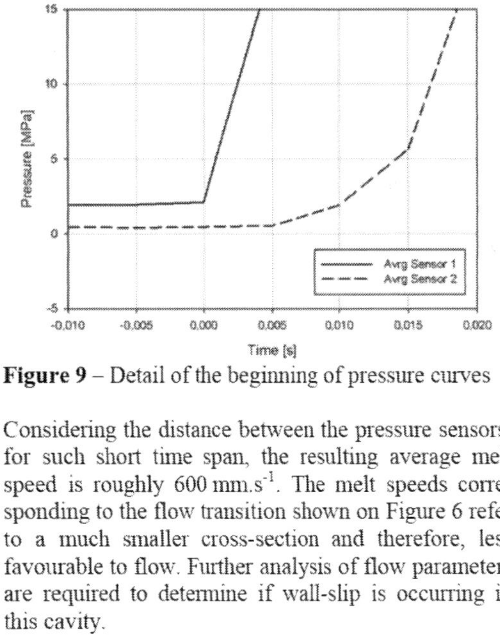

Figure 9 – Detail of the beginning of pressure curves

Considering the distance between the pressure sensors, for such short time span, the resulting average melt speed is roughly 600 mm.s^{-1}. The melt speeds corresponding to the flow transition shown on Figure 6 refer to a much smaller cross-section and therefore, less favourable to flow. Further analysis of flow parameters are required to determine if wall-slip is occurring in this cavity.

d) Data analysis
Preliminary analysis of the pressure drop obtained from the various microinjection tests enabled the establishment of a trend on flow curves. Following a data gathering procedure similar to the one used previously on the microchannels, small variations on the melt speed were imposed, allowing the plotting of shear rate *vs.* viscosity on Figure 10.

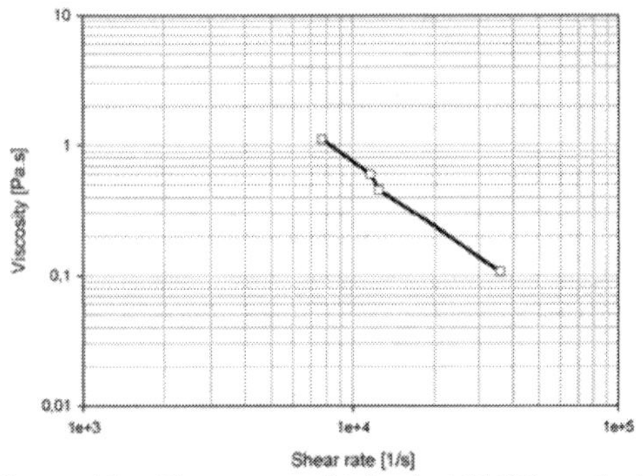

Figure 10 – Flow curve obtained at 215°C on the test part cavity

The curve plotted in Figure 10 shows a particular trend since the calculated index of non-Newtonian flow is always negative, allowing the suspicion that wall-slip is occurring.

Numerical simulations

Filling simulations were performed with the software Moldex 3D R9.1 (CoreTech, Taiwan). This software release incorporates the latest refinements concerning heat transfer, which is the most important factor, determining the process behaviour at the microscale. Filling simulations were performed with the same processing parameters used on microinjection. Furthermore two sensor nodes were located at the same place of the pressure sensors located in the cavity. A hybrid solid mesh was used, optimizing the number of elements and calculation time. At least, four elements are present on the part thickness improving simulation reliability. On the bottom face of the simulation model is located the sensor node as shown on Figure 11.

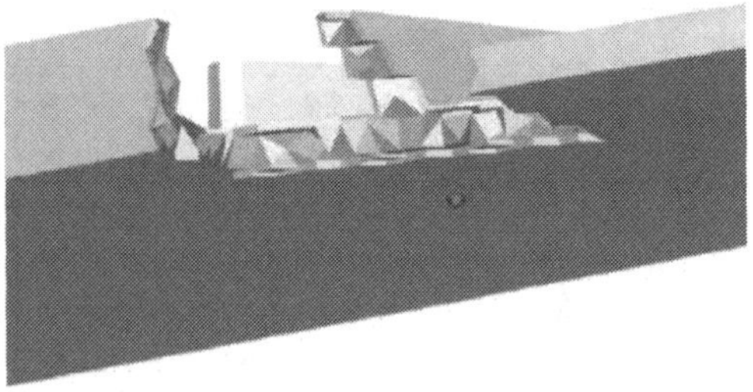

Figure 11 – Mesh section showing the solid hybrid elements across the part thickness

The filing simulations resulted on full-shots, as in the microinjection tests. The same processing parameters were used and the heat transfer coefficient was adjusted for this kind of numerical simulation. The filling results are shown in Figures 12 and 13. The sprue was removed.

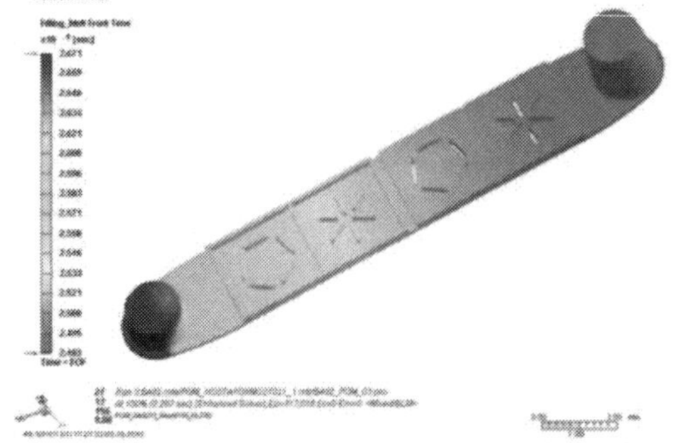

Figure 12 – Melt front time from numerical simulation

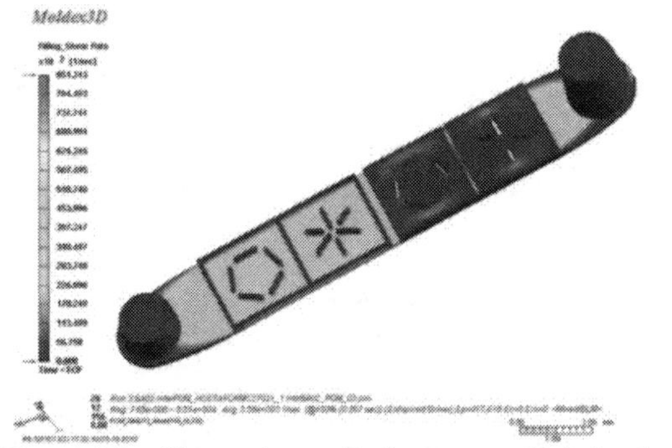

Figure 13 – Shear rate results from numerical simulation

From Figure 12 it is possible to conclude that the filling of the part took slightly less than 0,2 s, which is coherent with the values used on microinjection. The shear rate plot on Figure 13 shows the areas of the part with higher shear rates, coherent with flow in thin walls. The highest shear rate, as expected, occurred at the gate of the part. However, significant shear rates (*ca.* 60.000 s^{-1}) are also obtained at the second section where thickness is almost constant.

The sensor nodes located on the simulation model enable the comparison of the simulation pressure values on the cavity with the average pressure values obtained on microinjection. Figure 14 shows the overlapping of numerical data on experimental data. The pressure values are different concerning their magnitude, but the melt speed determined from numerical simulation is similar, since the time spans where pressure peaks occur are coincident in the time scale.

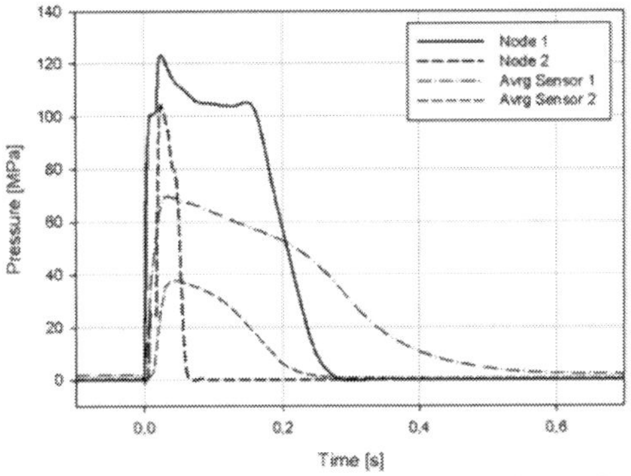

Figure 14 – Overlapping of numerical and experimental values of pressure

Conclusions

The data gathered in this study provided useful information for wall-slip evaluation. From the preliminary conclusions, it is possible to foresee the existence of a mixed-flow, a combination of classic shear flow and plug-flow that occurs when the wall-slip is perfectly established.

The measurement of the pressure drop on a test part shows that wall-slip can also be present on microinjection moulded parts.

The use of numerical simulation at this scale is currently much more reliable. However, the difference between the pressure data is still substantial. This fact may enable the conclusion that when wall-slip occurs and the material detaches from wall cavity, the cavity pressure decreases.

Acknowledgements

The authors would like to thank to Core-Tech Systems (Taiwan) and Simulflow (Portugal) for their technical support on the numerical simulations. This research work was supported by the Portuguese Foundation for the Science and Technology under the grant SFRH/BD/36982/2007 and by the Calouste-Gulbenkian Foundation.

References

Chen, C.-S., Chen, S.-C., Liaw, W.-L., *et al.*, "Rheological behavior of POM polymer melt flowing through micro-channels", European Polymer Journal, **44**, 1891-1898 (2008)

Chien, R.-D., Jong, W.-R., Chen, S.-C., "Study on rheological behavior of polymer melt flowing through micro-channels considering the wall-slip effect", Journal of Micromechanics and Microengineering, **15**, 1389-1396 (2005)

Giboz, J., Copponex, T., Mélé, P., "Microinjection molding of thermoplastic polymers: a review", Journal of Micromechanics and Microengineering, **17**, R96-R109 (2007)

Gleissle, W., Windhab, E., "The "Twin capillary" a simple device to separate shear- and slip-flow of fluids", Experiments in Fluids, **3**, 177-180 (1985)

Griffiths, C. A., Dimov, S. S., Pham, D. T. Micro injection moulding: the effects of tool surface finish on melt flow behaviour. Multi-Material Micro manufacture. W. Menz, S. S. Dimov and B. Fillon. Oxford, Elsevier Ltd. 373-376 (2006)

Hatzikiriakos, S. G., Dealy, J. M., "Wall slip of molten high density polyethilenes II. Capillary rheometer studies", Journal of Rheology, **36**, 703-741 (1992)

Kim, D. S., Lee, K.-C., Kwon, T. H., *et al.*, "Micro-channel filling flow considering surface tension effect", Journal of Micromechanics and Microengineering, **12**, 236-246 (2002)

Lee, K., Mackley, M. R., "The significance of slip in matching polyethylene processing data with numerical simulation", Journal of Non-Newtonian Fluid Mechanics, **94**, 159-177 (2000)

Mitsoulis, E., Kazatchkov, I., Hatzikiriakos, S., "The effect of slip in the flow of a branched PP melt: experiments and simulations", Rheological Acta, **44**, 418-426 (2005)

Piotter, V., Hanemann, T., Ruprecht, R., *et al.*, "Injection molding and related techniques for fabrication of microstructures", Microsystem Technologies, 129-133 (1997)

Piotter, V., Mueller, K., Plewa, K., *et al.*, "Performance and simulation of thermoplastic micro injection molding", Microsystem Technologies, **8**, 387-390 (2002)

Rosenbaum, E., Hatzikiriakos, S., "Wall slip in the capillary flow of molten polymers subject to viscous heating", AIChE Journal, **43**, 598-608 (1997)

Sha, B., Dimov, S. S., Pham, D. T., *et al.*, "Study of Factors Affecting Aspect Ratios Achievable in Micro-injection Moulding", International Conference 4M2005, 107-110 (2005)

Song, M. C., Liu, Z., Wang, M. J., *et al.*, "Research on effects of injection process parameters on the moulding process for ultra-thin wall plastic parts", Journal of Materials Processing Technology, **187**, 668-671 (2007)

Su, Y.-C., Shah, J., Lin, L., "Implementation and analysis of polymeric microstructure replication by micro injection molding", Journal of Micromechanics and Microengineering, **14**, 415-422 (2004)

Vasco, J. C., Maia, J. M., Pouzada, A. S., "Thermo-rheological behaviour of polymer melts in microinjection moulding", Journal of Micromechanics and Microengineering, **19**, 1-8 (2009)

Yao, D., Kim, B., "Simulation of the filling process in micro-channels for polymeric materials", Journal of Micromechanics and Microengineering, **12**, 604-610 (2002)

Prediction of optical properties for a transparent injection molded part and measurement of residual stress using birefringence pattern

Jinsu Hong[1], Min-Young Lyu[2]* and Jae-In Suh[3]

[1] *Seoul National Univ. of Technology, Graduate School of NID Fusion Technology, Korea- fixngbox@snut.ac.kr*
[2*] *Seoul National Univ. of Technology, Depart. of Product Design and Manufacturing Eng, Korea- mylyu@snut.ac.kr*
[3] *Samsung Electronics Co., Ltd, Korea – jaein.suh @samsung.com*

Abstract Residual stress is being developed in the part during injection molding process. Residual stress varies along the thickness direction and location in the part. Residual stress in a transparent part causes birefringence. Optical properties such as birefringence, retardation, and fringed order for a transparent injection molded part have been predicted using commercial software, Modex3D. Measurement method of residual stress for a transparent injection molded part through birefringence pattern has been investigated. Birefringent material is located between two polarized films those are perpendicular to each other. Birefringence is developed according to the level of residual stress. Residual stress can be estimated quantitatively using the fringe pattern.

Introduction

Plastic parts are distorted after molding by residual stresses (Jung Hoon Jung and Jae Ryoun Youn, 1997; Sang Sik Yang and Tai Hun Kwon, 2002). Residual stresses are classified into flow stress and thermal stress. Flow stresses are formed by shear stress during mold filling and called flow induced residual stress. Orientation occurs during flow by shear stress thus the flow stress involves orientation. Thermal stresses are originated from temperature differences along the thickness direction of the part. Measurement of residual stress using photoelasticity has been investigated for injection molded transparent part. Photoelasticity equipment has been used to observe and improve mechanical and optical property of precision products (Aben, et al., 2000; P. Postawa and D. Kwiatkowski, 2006). In this method, birefringence is proportional to the residual stress in the material according to the rheo-optical law. Birefringence and fringe pattern have been simulated using commercial program, Moldex3D. Experimental observation and predicted results were compared.

Experimental and Simulation

Specimen and photoelastic equipment

Flat square plastic plates were prepared by injection molding. The size of specimen is 100mm x 100mm x 3.2mm (thickness). Fig. 1 shows specimen and gate is indicated. Injection molding machine was Woojin Selex ELF25-B and material was LG-DOW CALIBRE200 (polycarbonate). Residual stresses were measured by photoelasticity using PS-100 of Strainoptics Co..

Computer simulation

Moldex3D was used for the simulation of injection molding including birefringence and fringe pattern. Fig. 2 shows mesh for three dimensional simulation.

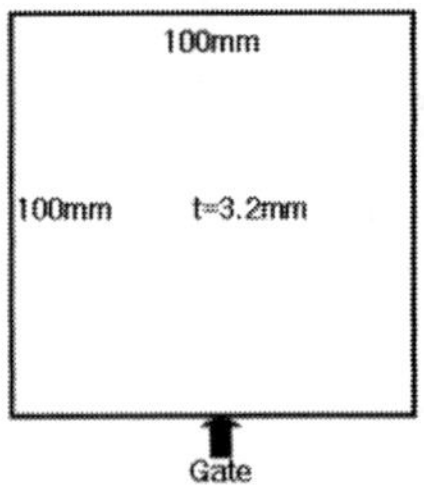

Figure 1. Injection molded specimen.

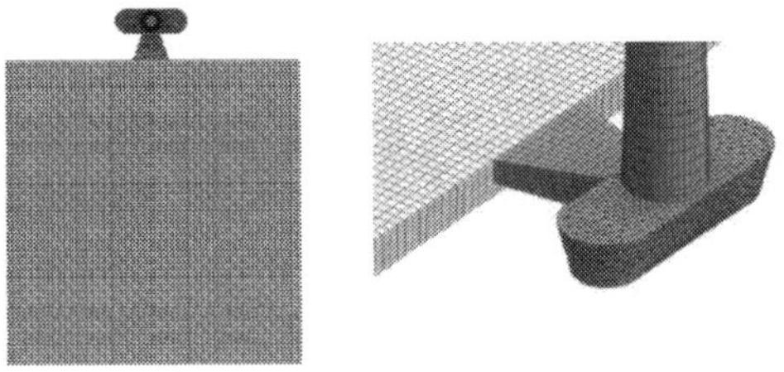

Figure 2. Mesh for computer simulation using Moldex3D.

Results and Discussion

Fig. 3 shows measurement points of residual stresses represented by numbers 1 to 5. Fig. 4 shows residual stresses (differences of principal stresses, $\Delta\sigma=\sigma_1-\sigma_2$) at each point. There exist some deviations according to measurers. Roughly 0.8 MPa of residual stress was developed in injection molded flat plate. Slitting method was adopted to separate the principal stresses as shown in Fig. 5. If one slits the specimen in the 1st principal direction at point, 2nd principal stress would be zero at that point. Fig. 6 shows a measurement of 1st and 2nd principal stresses using slitting method. Fig. 7 shows principal stresses at each point. Fig. 8 shows computational results, birefringence, retardation, and fringed pattern. If we compare the birefringence of specimen shown in Fig. 3 tendency is coincidence. Difference between experimental and computational results comes from material data used in simulation, stress-optical constant.

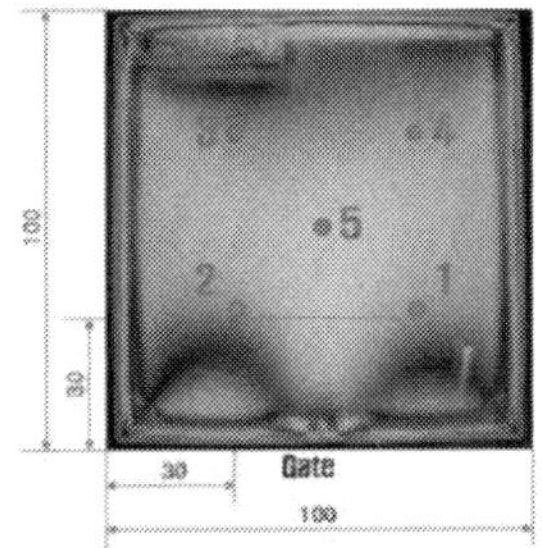

Figure 3. Measurement points of residual stresses.

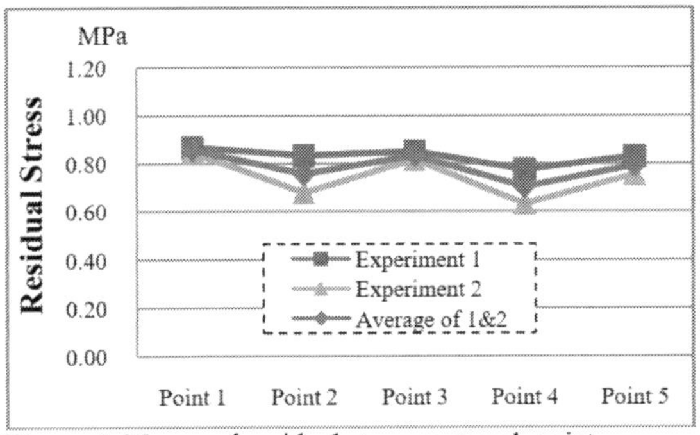

Figure 4. Measured residual stresses at each point.

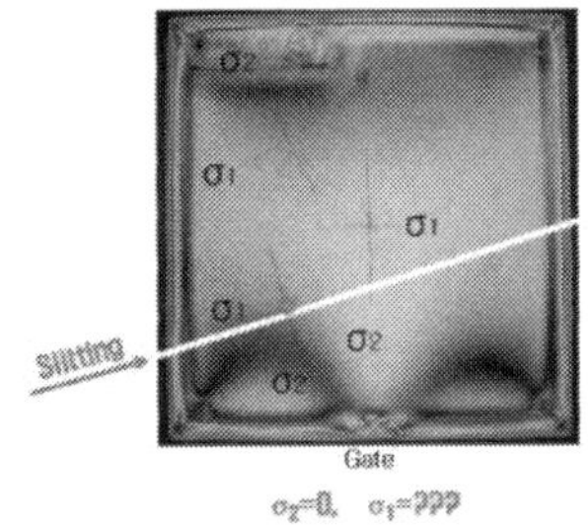

Figure 5. Slitting to eliminate one of principal stresses.

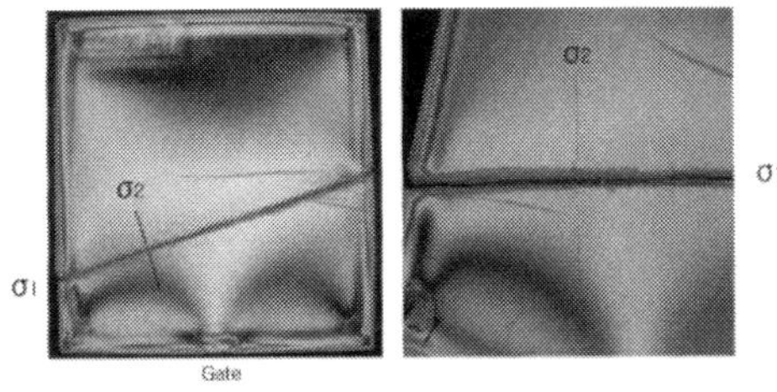

A value of residual stress before slitting	$\sigma_1-\sigma_2$=0.76MPa	σ_1=1.09MPa σ_2=0.33MPa
A value of residual stress after slitting	σ_1=1.09MPa	

Figure 6. Measurement of principal stresses at point 2.

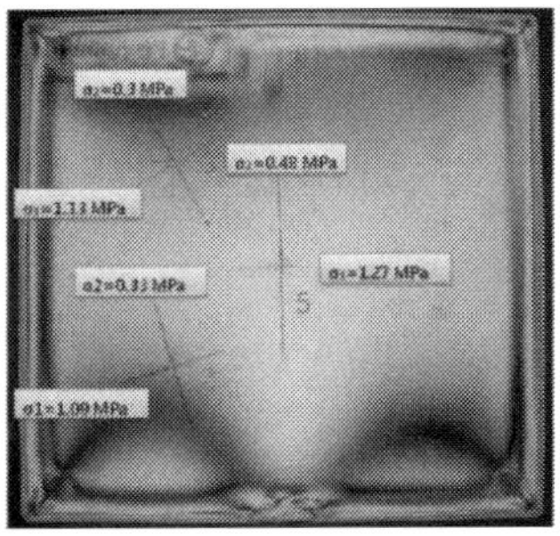

Figure 7. Principal stresses at each point.

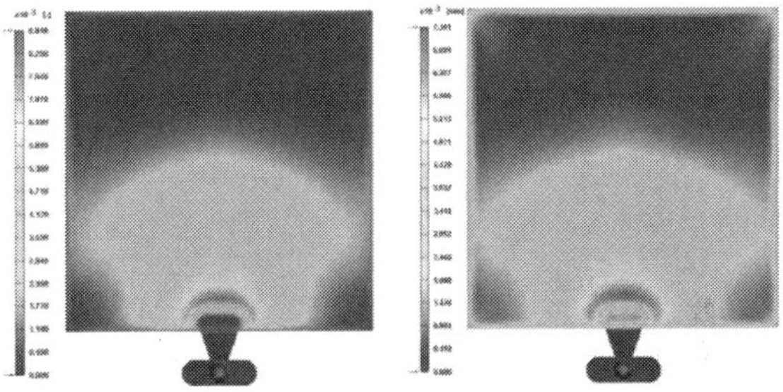

(a) Total birefringenece **(b)** Total retardation

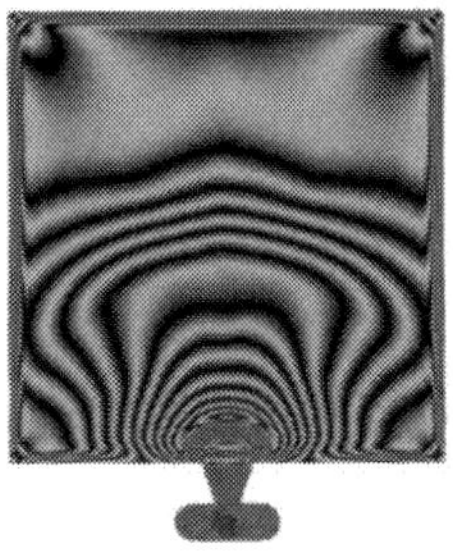

(c) Total fringed pattern

Figure 8. Predicted optical characteristics by computer simulation.

Conclusions

Residual stresses in an injection molded part were measured quantitatively using photoelasticity. Residual stresses for principal directions were measured using slitting method. About 0.8MPa of residual stresses were distributed in a flat plate. Birefringence and fringed pattern were predicted by computer simulation of injection molding. Computational results and experiments are good agreement in their tendency.

Acknowledgements

This work(research) is financially supported by the Ministry of Knowledge Economy(MKE) and Korea Institute for Advancement in Technology (KIAT) through the Workforce Development Program in Strategic Technology.

References

1. Sang Sik Yang and Tai Hun Kwon, "Deformation Analysis of Injection Molded Articles due to In-mold Residual Stress and Subsequent Cooling after Ejection", Transactions of The Korean Society of Mechanical Engineers A, **26(2)**, 340-348 (2002).
2. Jung Hoon Jung and Jae Ryoun Youn, "Prediction of Residual Stresses in Injection Molded Parts Considering Packing and Cooling Stages", The Korean journal of rheology, **9(1)**, 16-26 (1997).
3. H. Aben, et al., "Intergrated photoelasticity for nondestructive residual stress measurement in glass", Optics and lasers in engineering, **33**, 49-64 (2000).
4. P. Postawa and D. Kwiatkowski, "Residual stress distribution in injection molded parts", Journal of Achievements in Materials and Manufacturing Engineering, **18**, 171-174 (2006)

VISUALIZATION ANALYSIS OF MATERIAL FLOW INSIDE SIMPLE RIB CAVITY IN PULP INJECTION MOLDING

K. Matsuzaka[1]*, H. Miyashita[2], N.Masuda[3] , and H.Yokoi[4]

[1]*Daiho Industrial Co., Ltd., Japan, matsuzaka@daiho-hq.com*
[2]*Nissei Plastic Industrial Co., Ltd., Japan, haru_miya@nisseijushi.co.jp*
[3]*Institute of Industrial Science, the University of Tokyo, Japan, nmasuda@iis.u-tokyo.ac.jp*
[4]*Institute of Industrial Science, the University of Tokyo, Japan, hiyokoi@iis.u-tokyo.ac.jp*

Pulp injection molding (hereafter called PIM) is an injection molding technique using molding material made up mainly of pulp and starch, which aims to replace resin moldings with low environmental-load materials. In normal injection molding processes using thermoplastic resins, when the mold has flow channels consisting of partial widening areas such as ribs and areas with enlarged thickness, the resins tend to flow first to thick-walled areas of the mold due to the low flow resistance of these widening areas. Then from these partial thick-walled areas, the resins diverge, as if overflowing, to return to the original flat cavity in the downstream area behind the thick-walled areas. On the contrary, with PIM, the molding material dries as it flows and sometimes the hard tip of the flowing material gets caught by the step-change areas located at the end of thick-walled area while returning to the thin-walled one. Thus in such cases, these step-change areas act like obstacles blocking flow. Sometimes, weld-lines or areas with reduced strength are also formed along the downstream of these thick-walled areas. Therefore, the clarification of material flow behavior at thick-walled areas is a very important theme for preventing such defects in PIM products.

In this study, we applied pigment line-markers containing aluminum powder on both sides of the cavity surface before injection, and after molding, visualized the flow traces of the markers inside the molded samples using X-ray computed tomography (hereafter called X-ray CT) technology. As a result, we succeeded in clarifying material flow behavior near the rib cavities with different thickness and height combinations.

1. Introduction

Pulp injection molding (PIM) is an injection molding technique using molding material made up mainly of pulp and starch. To date, the authors have conducted evaluation on the various characteristics of PIM molded products[1], analysis of correlation between formation of voids in molded products and warp deformation[2], and investigations on high-cycle molding methods [3]. We have also been carrying out numerous observations of flow-front behavior inside the mold using a glass-inserted mold and high speed video camera[4]. Through such basic research processes, PIM has been found to be similar to resin molding in terms of the molding machine, molds, and forming processes used, but differ in many ways, for instance, it fills into the cavity in a sliding manner, etc. Given that PIM is nearly ready for practical application on a full-scale is approaching, detailed clarification of related molding phenomenon is anticipated.

In normal injection molding processes using thermoplastic resins, when the mold has flow channels consisting of partial widening areas such as ribs and areas with enlarged thickness, the resins tend to flow first to thick-walled areas of the mold due to the low flow resistance of these widening areas. Then from these partial thick-walled areas, the resins diverge, as if overflowing, to return to the original flat cavity in the downstream area behind the thick-walled areas. On the contrary, with PIM, the molding material dries as it flows and sometimes the hard tip of the flowing material gets caught by the step-change areas located at the end of thick-walled area while returning to the thin-walled one. Thus in

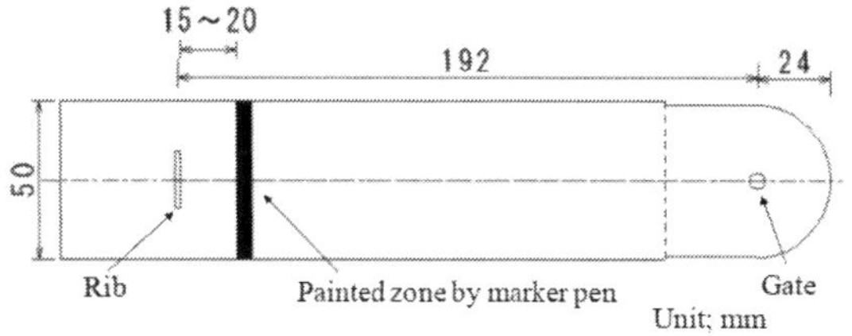

Fig.1 Cavity shape

Table 1 Combination of rib height and thickness

Sample No.	Rib	
	Height(mm)	Thickness(mm)
(a)	0.5	1.0
(b)	1.0	1.0
(c)	2.0	1.0
(d)	1.0	2.0
(e)	1.0	3.0
(f)	1.0	4.0

Table 2 Molding conditions

Cylinder temp.(℃)	(Nozzle)90/80/65(Hopper)
Injection speed	10mm/s
Metering stroke	16.0mm
Rotational speed	50rpm
Back pres.	3.5MPa
Mold temp.	180℃
Injection pres.	157MPa

such cases, these step-change areas act like obstacles blocking flow. Sometimes, weld-lines or areas with reduced strength are also formed along the downstream of these thick-walled areas. For this reason, the clarification of material flow behavior at thick-walled areas is a very important theme in PIM.

In this study, we built a flow process model to observe the flow of PIM material, particularly focusing on how PIM material fills into a simple rib located at vertical to the flow. The results are reported below.

2. Static Visualization in Pulp Injection Molding

As a result of observing flow of PIM material inside a visualization mold using a high speed video camera, it was found that the material flows as if sliding, and the shape of the tip of the flowing material does not change. However, the three-dimensional flow behavior of the material inside the cavity has not been clarified yet.

n this study, we employed the material marking method to visualize internal flow, where pigment line-markers containing aluminum powder are applied onto both sides of the cavity surface, so that these markers adhere to the material when it flows near the walls, thus allowing observation of the distribution of the markers on the surface and inside the rib cavities located at the downstream of the marking area. Aluminum powder was mixed into the markers (specific weight of 2.7) to vary the density of the molding material (specific weight of about 1.0), and enable non-destructive observation of the flow traces of the markers inside the molded product by X-ray CT.

3. Experiment Method

The injection molding machine used in the experiments was the NEX220-50EPI pulp injection molding machine with cylinder diameter of φ56 mm (Nissei Plastic Industrial Co., Ltd.). Standard PIM material with water content adjusted to 35wt% was used (pulp: 60wt%, starch: 30wt%, and PVA: 10wt%). The cavity shape was rectangular with a size of 50×200×1[mm]. This mold with a direct gate was also used in the previous report [2)] for visualization (Fig.1). At the moveable side of the cavity, a rib measuring 18 mm in length is installed at vertical to the flow direction, at a distance of 192 mm from the gate. Table 1 shows the combination of different rib heights and thicknesses used. At conditions (a) to (c), the rib thickness was fixed at 1 mm, and the rib height was varied. At conditions (b), and (d) to (f), the rib height was fixed at 1mm and the rib thickness was varied. Table 2 shows other molding conditions. The marker used was the Mitsubishi paint marker PX-3 made of oil type non-transparent pigment-based ink. The marker was positioned about 15 to 20 mm from the front of the rib. The X-ray CT used was the micro-focus X-ray CT unit SMX-130T (Shimadzu Corp.) scanning the $4mm^2$ area at the center of the rib. The radiation conditions were 35kV, 160μA, number of views of 1200, and average number of shoots of six times. Based on the CT data, distribution of markings was extracted by volume rendering (VR).

4. Analysis Results and Discussion

Fig.2 (1) shows the exterior of a marked molded product with a rib height of 0.5mm and thickness of 1mm. (2) shows the image of the marker extracted using VR after X-ray CT scanning. In (1), inconsistent colored patterns where white dots are scattered over a wide area at the downstream of the rib can be seen. In (2), uncolored white material is seen to distribute and cover the rib from the tip to the downstream, while directly inside, colored material can be seen to flow and distribute as if detouring to the tip of the rib.

Next, Fig.3 shows a comparison of the marker traces (white) inside the rib. In Fig.3 (a) to (c) where

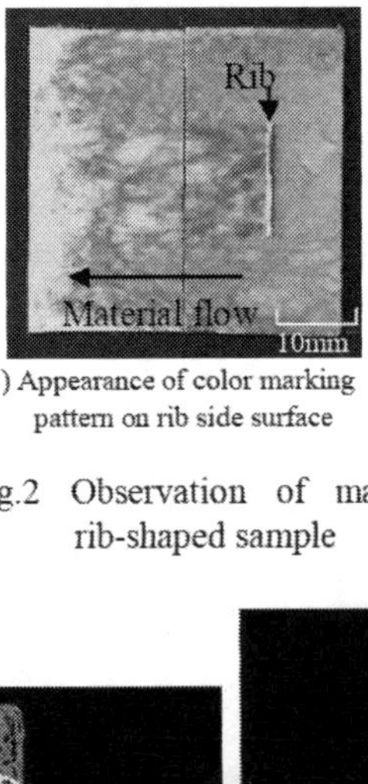

(1) Appearance of color marking pattern on rib side surface

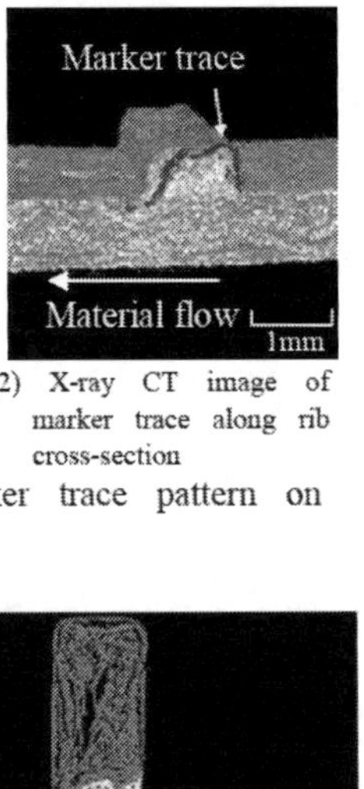

(2) X-ray CT image of marker trace along rib cross-section

Fig.2 Observation of marker trace pattern on rib-shaped sample

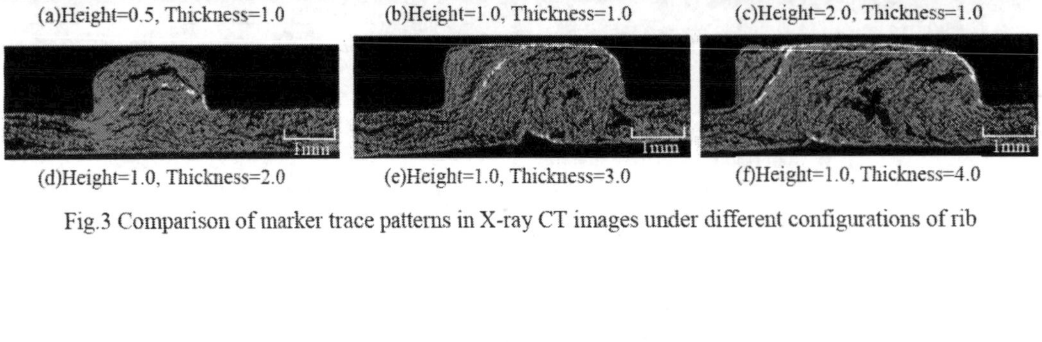

Fig.3 Comparison of marker trace patterns in X-ray CT images under different configurations of rib

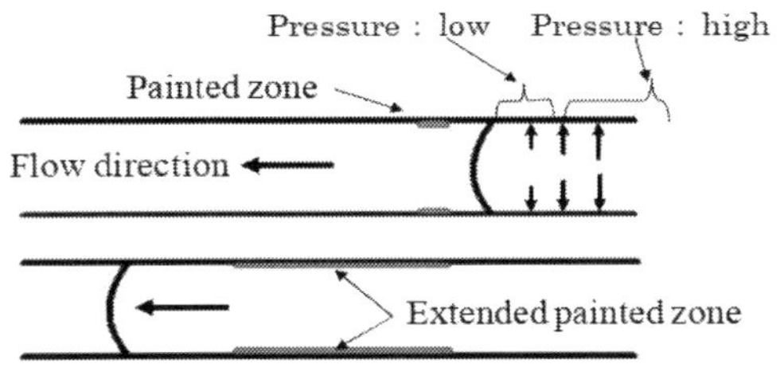

Fig.4 Model of marker adhesion

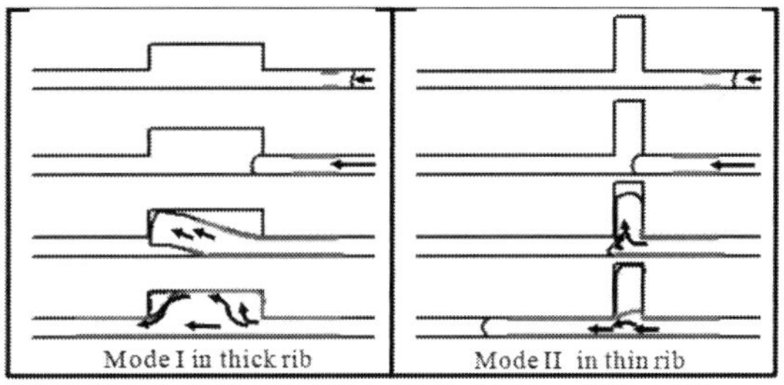

Fig.5 Flow model inside rib cavity

the rib height is varied, the marker trace can be seen to slightly protrude towards the rib regardless of the rib height, bypass the rib, and return to the flat cavity area. In Fig.3 (b) and (d) to (f), (b) shows that the marker has not reached the rib tip, but as the rib thickness increases, the marker trace rises deep inside the rib, and when the rib thickness is 3mm (e), the marker trace protrudes to the tip of the rib. Furthermore, when the rib thickness is 4mm (f), the protruded area at the top of the rib tends to increase. In (e) and (f), a dent that looks like a sink mark is formed at the opposite side of the rib. These are all thought to be due to the trail left behind by the flow of the material as it tilts and enters the rib area, and also because the internal pressure could not rise easily during expansion due to the large rib volume.

Based on these results, Fig.4 and Fig.5 show the marker adhesion process and a model of the generation of marker traces inside the rib. When the PIM material bypasses the marker applied on the cavity surface, the tip of the flow front has already started drying, and the marker does not adhere due to the small contact pressure. After bypassing the tip, the internal pressure increases, the contact pressure becomes large, and for the first time, the marker starts to adhere (Fig.4). When this material flows into the rib, the flow pattern can be broadly categorized into the following three types; (I) Enlarged flow; When the rib is thick and short, the tip of the flow front enters the rib in a slanting manner, gets caught by the step-change areas located at the end of thick-walled area, the rib is filled with material, and until the internal pressure of the rib rises again, the material does not flow to the downstream side. When the material starts flowing again, uncolored material flowing first to the corners of the rib tip are left behind, and material adhered with marker arriving later passes the inside layer of the uncolored material. (II) Bypass flow; When the rib is thin and tall, material not adhered by marker enters the rib first to fill the rib, and it is cut into two by the material adhered with marker following as it traverses the rib, and bypasses it without entering the rib. (III) Detour flow; This is the extreme version of the enlarged flow as described above. As it takes time for the material to fill into the rib and for the internal pressure to rise, the material detouring from both sides of the rib overtakes the material in the rib and precedes it. This may be the reason why weldlines are formed behind the rib.

5. Conclusions

(1) By applying markers containing aluminum powder on the cavity wall and observing molded products using X-ray CT, it was demonstrated that the flow process of PIM material can be visualized and analyzed nondestructively.

(2) Results of observing filling behavior at ribs of various heights and thicknesses show that the PIM material flows in a slipping manner at the tip of the rib for thick ribs, but flows crossing the base of the rib for thin ribs.

In future studies, we hope to correlatively analyze the effects of the rib shape on strength and flow behavior inside the rib.

Acknowledgements

The authors thank Mr. Mitsuyoshi Maruno of Daiho Industrial and Mr. Kazuo Nitta of Olympus for their cooperation in this research.

Reference

1) H. Yokoi, K. Matsuzaka, M. Maruno, H.Miyashita, N. Masuda; Abstracts for the PPS-22, 182, 2006, CD
2) K. Matsuzaka, H. Miyashita, N. Masuda, H. Yokoi; Abstracts for PPS 2009, 114, 2009, CD
3) M. Maruno, H. Yokoi, H. Miyashita, N. Masuda; Abstracts for PPS 2009,119, 2009, CD
4) H. Yokoi, Y. Yamawaki, N. Masuda; Abstracts for PPS 2009, 121, 2009, CD

INFLUENCE OF REINFORCEMENT VOLUME FRACTION ON FIBERS LENGTH REDUCTION IN INJECTION MOLDING OF LGFRT

M. Berton[1], A. Cellere[2] and G. Lucchetta[1]

[1] *Dep.of Innovation in Mechanics and Management, University of Padova, Italy – matteo.berton@unipd.it;*
[1] *Dep.of Innovation in Mechanics and Management, University of Padova, Italy – alessandro.cellere@unipd.it;*
[1] *Dep.of Innovation in Mechanics and Management, University of Padova, Italy – giovanni.lucchetta@unipd.it;*

Abstract - Long glass fiber-filled polypropylene composites are widely used in industry because of their low material and production cost combined with excellent mechanical performances and lightness. Mechanical properties of the parts, in particular elastic modulus, strength and impact resistance, strongly depend on the volume fraction of the reinforcement, on the average fiber length and on its distribution and orientation. During the processing of fiber-reinforced thermoplastics, such as in extrusion and injection molding, fibers are subjected to breakage, reducing the average fiber length. The phenomena governing the fiber degradation are very complex to analyze since both materials properties and process parameters play a key role in the fiber fracture mechanisms, which can be divided into three categories of interaction: fiber-fiber interactions, fiber – polymer matrix interactions, and fiber collision with the manufacturing system. The aim of this study is to analyze the influence of the reinforcement volume fraction on the residual fibers length during injection molding. Experimental tests in different process condition were conducted on an injection molding machine, in order to determine the lower degradation induced on the fibers with the increasing of the reinforcement volume fraction. The traditional measurement procedure, that is limited by the field of view of the optical CMM used, was improved in order to measure even the longer fibers. The application of image mosaic techniques allowed to merge different images, bypassing the limit imposed by the instrument.

Keyword: injection molding, reinforced thermoplastics, fiber degradation, image mosaic

Introduction

In recent years polymer composites are gaining an extraordinary success. The ability to provide high mechanical properties and low density at the same time makes them suitable for a wide range of applications. The employment of these materials at a cost that is becoming more competitive has made them attractive not only for specific sectors, such as aeronautics or automotive for which they were originally developed, but also for mass production of consumer goods. In this context, thermoplastic composite materials are becoming particularly interesting, because they combine the well known physical-mechanical characteristics with low environmental impact, and with all the benefits derived from being suitable for processing by injection molding. Indeed, this technology allows the production of large size components, with very complex geometries in low cycle times. It is therefore possible to mold parts with high performance at low cost. The polymer matrix composites most commonly used are the long glass fiber reinforced thermoplastics. This type of reinforcement aims at improving the mechanical properties, especially the tensile strength and elastic modulus, provided by the presence of the dispersed phase within the matrix. As the load is supported by the reinforcement, the longer the fibers, the greater the increase in the property itself. In previous studies it was highlighted that manufacturing processes led to the breakage of the glass fibers, reducing the benefits given by the presence of the reinforcement. The phenomena governing the fibers degradation are very complex to analyze since they depend on various damage mechanisms involving both materials properties and process parameters. These phenomena can be divided into three main categories, as done by Turkovich R.E. and Erwin R. L., (1983): fiber - fiber; fiber - matrix and fiber - manufacturing system interactions.

In this work the fibers degradation in composite materials with different long glass fibers weight fractions is analyzed. In particular the aim is to investigate the role of the reinforcement volume fraction on the residual fiber length during the plastication phase of the injection molding process. In order to do this it was first analyzed and improved an experimental procedure commonly used for measuring the length of reinforcement fibers, overcoming some of its critical drawbacks by means of image analysis techniques. Experimental tests on a polypropylene reinforced with LGRF at two different values of volume fraction, (30% and 40%), were carried out. Fiber degradation occurred at different process conditions was investigated thorough the use of the Design of Experiments (DOE) method.

Experimental

As already mentioned, fibers length reduction is caused by many factors which can be divided into three types of interactions:

1. Fiber – fiber interactions: bending stresses can be introduced in the fibers due to the entanglements, and the consequent abrasion between fibers can lead to stress concentration higher than the fibers strength.
2. Fiber – manufacturing system interactions: the contacts between fibers and the system surfaces (e.g. screw, barrel, nozzle and mold cavity) can lead to fibers attrition, and consequently to fibers breakage.
3. Fiber – polymer matrix interactions: the filling flow is characterized by high shear rates which induce fibers rotation, and significant shear stresses that can lead to fibers breakage.

As the geometry of the plastication system is kept constant for the whole campaign, the fiber-injection system interactions can be considered independent from the other two breakage mechanisms. The first two mechanisms instead, although distinct, cannot be considered independent: if the variation of some process parameters affects fiber-fiber interactions, then it is likely that it has also some effect on the fiber-matrix interactions. A simple example would be to use two types of reinforcement fibers with different aspect ratio. It is evident that in both cases the influence of fiber-fiber interactions will be different, because shorter fibers can move and rotate (and consequently interact) differently than long fibers. Simultaneously, also the phenomena of transmission of elongational and flexural forces between matrix and fibers will be different in the two cases, and therefore also the influence of the fiber-matrix interactions will not be the same. In the same way, a change in the percentage of reinforcement will have an effect on the phenomena of both fiber-fiber and fiber-matrix interactions. A higher fibers density, in fact, affects both the mobility degree of the fibers and the ability of the matrix to impregnate the reinforcement, inevitably altering the mechanisms of transmission of the stresses between the two components.
To analyze the relationships between fiber-fiber and fiber-matrix interactions, and to better understand the effect of these interactions on the degradation suffered by the fibers, it was necessary to consider the variation of other factors. Since the analysis was carried out considering the material plastication phase, two of the main process parameters of this phase were taken into account: backpressure and melt temperature. The melt temperature level values were selected considering the range suggested by the material producer. The polymers used for the experimental test are two types of polypropylene produced by Ticona Gmbh under the Celstran trade name, reinforced respectively with 30% and 40% of long glass fibers. The optimal ranges of process temperature for the two materials are not the same, even if they are partially overlapped. To be able to test the two materials in the same process conditions, three suitable level values of melt temperature were selected (240°C, 260°C and 280°C). Three level values of backpressure were set to 50, 100 and 200 bar. The experimental tests were conducted using an ENGEL E-Motion 440/100 injection molding machine, equipped with a general purpose screw with a diameter of 40 mm. The injection parameters were set and maintained constant for all the tests, in order to keep constant their effects on fibers degradation:

- Injection speed: 20 mm/s
- Screw speed: 36 rpm
- Dosing stroke: 150 mm

Melt temperature and backpressure were set according to a factorial design showed in Figure 1.

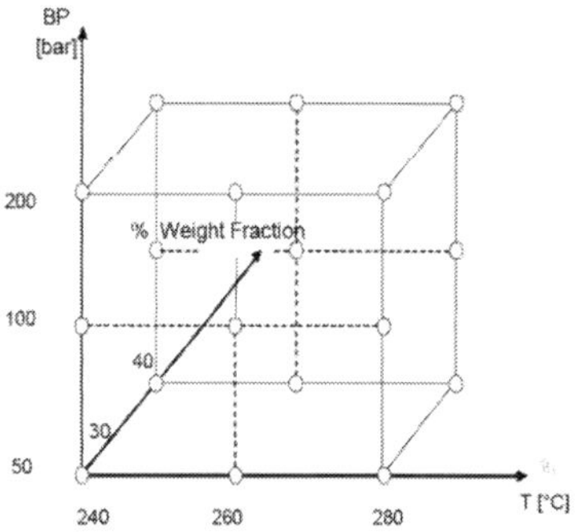

Figure 1. DOE experimental plan

Fiber Length Measurements

In order to evaluate fiber degradation at different process conditions, accurate measurements of the average fiber length and its distribution are required. There are several methods available for determining a fiber length distribution, as described by Fu S.Y., et al. (2002). The most widely used technique consists in performing an image analysis of micrographs of the fibers which are obtained through a long and laborious process. First of all the polymer matrix is burned off by means of an high temperature oven. A small portion of the fibers, obtained by the burned-out matrix, is then put in an ultrasonic bath in order to facilitate their dispersion within the specimen. Finally the specimen is positioned on a optical microscope and pictures are taken at an opportune magnification.
If the fibers are long, as in this case, it is important to take into account the problem related to the size of the images with respect to the length of the fibers: only those fibers that are entirely contained into the picture can be measured, while those intersecting the edges of

the picture cannot be counted. The probability to intersect the edges of the image is higher for longer fibers than for shorter ones, so the next count will return a value of average length underestimated than the real, as discussed by Fu S.Y., et al (2002).

Since the ratio between the length of the fibers to be measured and the field of view of the available vision tool was too high, the measure of the longer fibers was impossible. By applying the image mosaic technique, which transforms a sequence of images partly overlapping each other into one large composition, these limitations of the traditional measurement procedure have been overcome. Thanks to this, it was possible to consider even the contribution of the fibers whose length is longer than the available field of view. The calculated mean fiber length and its distribution could therefore be computed more accurately and the correct correlation between process parameters and the fiber damage was obtained.

For each point of the experimental plan represented in Figure 1, the polymer was melted and injected without a mold (i.e. by purging), to neglect the degradation due to the filling phase. The cooled samples were cataloged and analyzed. In particular, for each sample two slides were taken and analyzed, and for each slide about 30 images were captured. Using the image mosaic technique it was possible to merge all the images to get the complete slide photograph, and to measure all the fibers by means of an image analysis software. Figure 2 shows an example of image mosaic reconstruction, obtained by the merging of all the images of a slide.

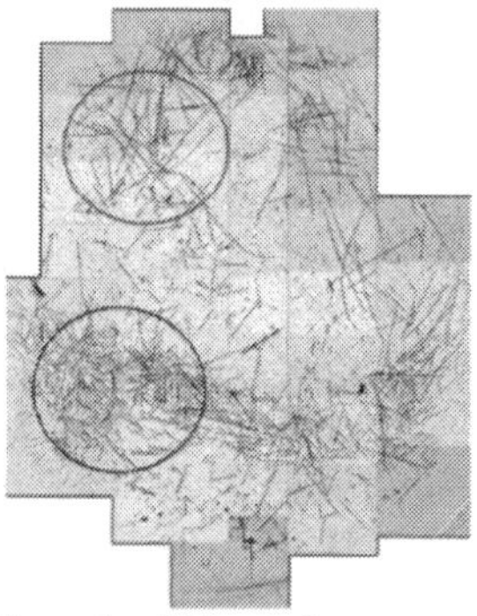

Figure 2. Example of Mosaic Image reconstruction

From this figure, obtained merging 39 images, it is possible to note that inside the same slide there are areas characterized by very different length distributions. Also, a simple qualitative analysis can easily point out that the average fibers length of the blue region is lower than the average of the red one. If the measure is taken only in one of this area, the final average fiber length will not describe the real fibers distribution in the slide. The ability to measure all the fibers within the specimen therefore allows also to neglect the error due to the exclusion of areas with unique length distributions.

Results and Discussion

Once measured all of the fibers in the slides, the average fiber length and the length distribution for each of them was estimated. A statistical analysis of variance (ANOVA) was conducted on the experimental data, to identify the significant main factors and interactions.

The results of the ANOVA analysis on the design is shown in Figure 3. All the main factors are significant, because the p-value is lower than the confidence level of 0.05. The interaction factor of backpressure and temperature is not significant. On the contrary all the other interaction factors of second and third order have p-value lower than 0.05.

Factor	Type	Levels	Values
T	fixed	3	240; 260; 280
BP	fixed	3	50; 100; 200
% Carica	fixed	2	30; 40

Analysis of Variance for L, using Adjusted SS for Tests

Source	DF	Seq SS	Adj SS	Adj MS	F	P
T	2	6,2163	6,2163	3,1081	323,08	0,000
BP	2	5,6923	5,6923	2,8462	295,84	0,000
% Carica	1	25,1728	25,1728	25,1728	2616,60	0,000
T*BP	4	0,0540	0,0540	0,0135	1,40	0,273
T*% Carica	2	1,1166	1,1166	0,5583	58,03	0,000
BP*% Carica	2	0,7982	0,7982	0,3991	41,49	0,000
T*BP*% Carica	4	0,1303	0,1303	0,0326	3,38	0,031
Error	18	0,1732	0,1732	0,0096		
Total	35	39,3537				

S = 0,0980838   R-Sq = 99,56%   R-Sq(adj) = 99,14%

Unusual Observations for L

Obs	L	Fit	SE Fit	Residual	St Resid
8	6,14000	5,93583	0,06936	0,20417	2,94 R
26	5,73167	5,93583	0,06936	-0,20417	-2,94 R

R denotes an observation with a large standardized residual.

Figure 3. ANOVA results of the factorial design

First of all, the effects of process parameters (melt temperature and backpressure) on the residual fibers length with respect to the single materials, PP-GF30 and PP-GF40 will be analyzed.

From Figure 4 and 5 it is possible to note as, in both cases, an increase of the melt temperature induces an increase in the residual fibers length; on the contrary, an increase in the backpressure induces more damage on the fibers, reducing the final length. With the increase of the melt temperature, the melt viscosity decreases, so the stresses induced on the fibers with higher melt temperature are lower than for lower temperature, preserving the fibers length. It can be deduced that increasing the melt temperature reduces the effects of fiber-matrix interactions. On the other hand, incrementing the backpressure the attrition phenomena inside the melt are stronger, as well as the interaction with the injection system, inducing more stresses on the fibers.

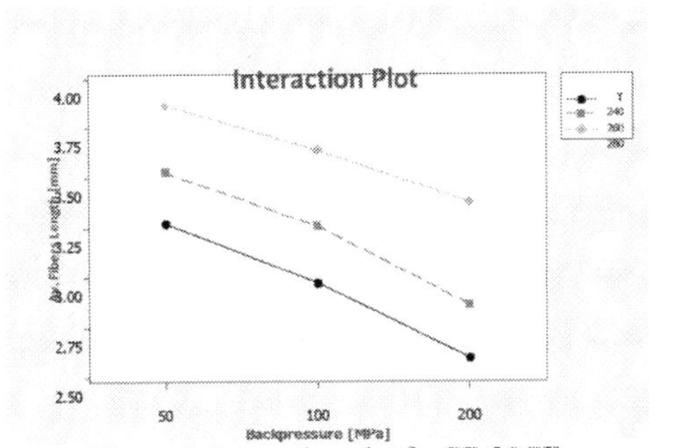

Figure 4. Interaction plot for PP-30GF

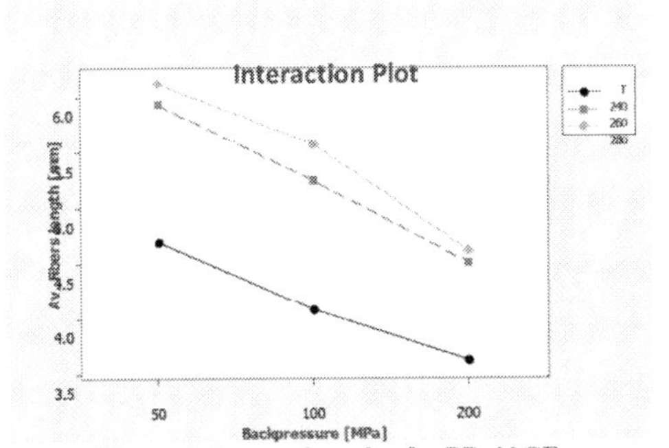

Figure 5. Interaction plot for PP-40GF

From Figure 5 it can be noted that the influence of backpressure for PP-40GF is the same as for the PP-30GF. The influence of melt temperature instead, is similar only in the transition from low to intermediate values. For higher fibers weight fraction in fact, the effects of the transition to high temperatures is much less significant when compared with the result obtained in the case of 30% reinforcement.

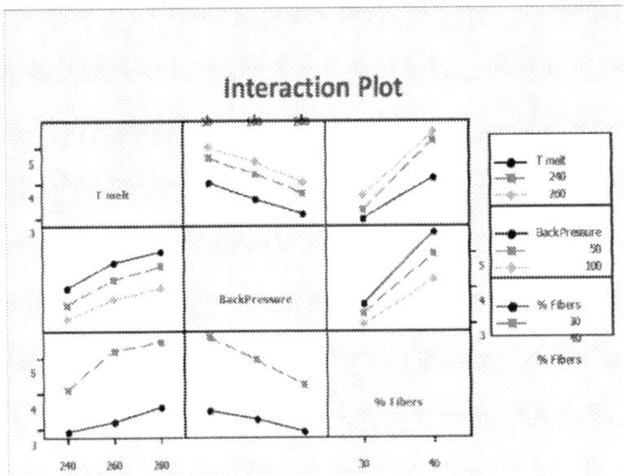

Figure 6. Interaction effects on residual fibers length for complete design

From Figure 6, which summarizes all the results obtained for the two materials, it is evident the effect of the variation of fiber weight fraction: higher percentage of reinforcement induces less damage on the fibers themselves. This can be attributed to the cluster of fibers that is formed with high weight fractions, that are more resistant to the external stress. The high level of entanglement allow to the fiber to act as a single block, so that the interaction with the polymer matrix has a lower influence because the fibers inside the clusters are blocked and cannot rotate or bend. Nevertheless, the reduction in length is very high: starting from an average length of 13 mm, it reaches a value that ranges from 6 mm to 3 mm at the end of plastication. The average length is reduced from 50% to about 75%.

High fiber weight fraction

A specific investigation on a 50% long glass fiber polypropylene was carried out, to analyze if the degradation mechanism were the same also at higher fiber weight fraction, and if the considerations on the process deduced by previous analyses could be generalized for any volume fraction of reinforcement in thermoplastic polymer composites.

The experimental campaign replicated the study carried out with 30% and 40% reinforcement. The same full factorial design represented in Figure 1, was considered. All experimental tests were conducted using a Celstran-GF50, a reinforced polypropylene produced by Ticona Gmbh. The polymer matrix was the same as the material tested in the previous sections; the only change was in the percentage of reinforcement. The results obtained from the ANOVA were significantly different from those obtained with 30% and 40% reinforcement.

General Linear Model: L versus T-melt; BackPressure

Factor	Type	Levels	Values
T-melt	fixed	3	240; 260; 280
BackPressure	fixed	3	50; 100; 200

Analysis of Variance for L, using Adjusted SS for Tests

Source	DF	Seq SS	Adj SS	Adj MS	F	P
T-melt	2	0.17000	0.17000	0.08500	4.53	0.044
BackPressure	2	0.86839	0.86839	0.43419	18.03	0.001
T-melt*BackPressure	4	0.11646	0.11646	0.02911	1.21	0.371
Error	9	0.21673	0.21673	0.02408		
Total	17	1.37157				

S = 0.155180   R-Sq = 84.20%   R-Sq(adj) = 70.15%

Figure 7. ANOVA results for PP-50GF

As reported in Figure 7, considering a confidence level of 5%, the main factors are both significant, while the interaction between them has no significant effect. These results are in accordance to the ones obtained with the other materials, however the effect of the two factors on the response is very different. In polypropylene reinforced with 30% and 40% long glass fibers, the increase of backpressure led to a reduction of the average fibers length. On the contrary, with 50% reinforcement, the average fibers length suddenly increased at higher value of backpressure (Figure 8).

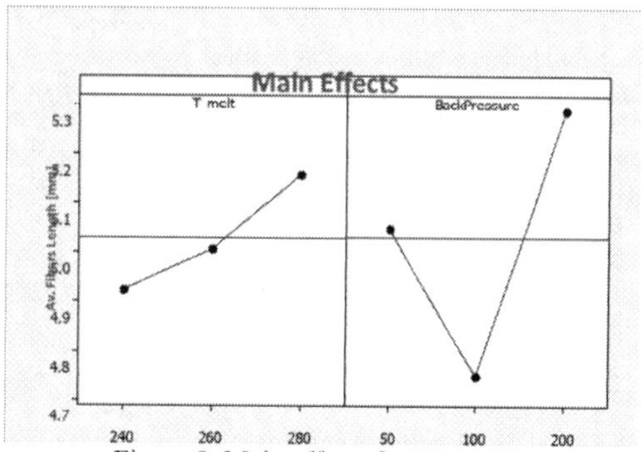

Figure 8. Mein effects for PP-50GF

The effect of the melt temperature on the response is the same for different weight fractions, but a different effects was measured for the backpressure: for low values the increase of backpressure increase the damage on the fibers, while for higher values the damage decreased. This result can be explained considering the melting mechanism acting during the plastication.

According to the most diffuse model used to describe the polymer melting process during the plastication phase (Maddock,B. H., 1959 and Tadmor Z. et al., 1967), polymer pellets are compacted in a solid bed in the first stages of the screw. In the melting zone, a thin film of melt forms between the solid bed and the barrel wall. The melt film is dragged away toward the leading flight flank. The molten plastic accumulates in a melt pool at the leading flight flank, pushing the solid bed against the trailing flight flank. Continuing the process, the fusion mechanism continues incrementing the melt pool, until the whole solid bed is melted. This fusion mechanism, confirmed by several experimental investigations is called Continuous Solid Melting model (CSM). Another important fusion mechanism is the Dispersive Solids Melting (DSM) model, proposed by Rauwendall (2001). In this model the solid particles are not concentrated in a continuous solid bed, but are dispersed in the melt pool. The DSM model assume that the solid particles are uniform, spherical and dispersed in a melt matrix, and that their melting is uniform, and not dependent on the location of the particles themselves.

It was observed that in certain process conditions the pressure on the solid bed is high enough to break the solid bed: with high pressure the melt polymer can fill in the cracks of the solid bed, breaking it. This phenomenon generates some solid particles immersed in the melt pool, that continue the melting process according the DSM model. Jin Z and Gao F. (2002) verified that the backpressure has two opposite effects on the solid bed break-up. On one hand, a higher backpressure increases the solid bed strength, compacting the pellets. On the other hand, higher backpressure values help the melt to fill the solid-bed cracks; as the bed moves along the screw channels, it becomes weaker until it breaks.

The stronger the solid bed, the more blocked the fibers inside it when the interactions with the melt and the barrel force them to brake. This can explain why the increasing of the backpressure decreases the average fibers length. If the backpressure is high enough to break the solid bed, the DSM mechanism is induced. With this melting mechanism the fibers are not blocked inside the solid bed, so the effect of the interactions with the polymer matrix is reduced, obtaining an higher average fiber length. As with the 50% of weight fraction of reinforcement the cracks inside the solid bed are larger, because it is more difficult to compact the pellets, the solid bed is weaker than with lower percentage of reinforcement. This is the reason why with 50% weight fraction high values of backpressure can break the solid bed, while with 30% and 40% it is not high enough to break it. So in composites with high fibers weight fractions a different melting mechanism, induced by high values of backpressure, led to a reduction of the fibers damage, increasing the material properties.

Conclusions

An innovative procedure for the fiber length measurement, based on the image mosaic thecnique, has been used for the analysis of the influence of the weight fraction on the residual fibers length in the injection molding process. Thanks to the new procedure, it was possible to consider even the contribution of the fibers whose length is longer than the field of view of the available vision tool, as often happens in the case of long fiber reinforcement.

Experimental tests were carried out on two different materials with the same polypropylene matrix and reinforced respectively with 30% and 40% long glass fibers. In order to analyze the influence of the factors governing fibers degradation, in particular the fiber-fiber and the fiber-matrix interactions, the variation of the main plastication parameters. The backpressure and the melt temperature were varied on three levels: 240°C, 260°C and 280°C, and 50, 100 and 200 bar respectively. The results obtained show how for both the materials considered, an increase of the melt temperature induces an increase in the residual fibers length; on the contrary, an increase in the backpressure induces more damage on the fibers, reducing the final length.

The most interesting results concern the effect of the variation of fiber weight fraction: an higher percentage of fibers induces less damage on the fibers themselves. This was attributed to the cluster of fibers that forms with high fibers fractions, that are more resistant to the external stresses. The high level of entanglement allows the fibers to act as a single block: so the interactions with the polymer matrix have a lower influence because the fibers inside the clusters are blocked and cannot rotate or bend.

These results confirm that the usage of high weight fraction reinforced materials can lead to higher

mechanical properties in the component, because there are more fibers in the part, but also because they are longer.
Furthermore a specific investigation on polypropylene reinforced with 50% long glass fibers was carried out following the same experimental plan used for the 30% and 40% weight fractions. In this case the effect of the melt temperature on the response was the same as the one found in the previous analysis, but a different behavior was registered for the backpressure. In particular it was found that high value of this parameter led to higher residual fiber length. An explanation for this result was proposed by analyzing the different melting mechanism acting during the plasticating phase.

References

Fu S.Y., et al., "Correction of the measurement of fiber length of short fiber reinforced thermoplastics." Composites: Part A, **33**, 1549–1555 (2002)

Jin Z, Gao F., Solid-bed break-up during plastication in injection molding Society of Plastics Engineers, SPE ANTEC Tech. Papers, 572-575 (2002)

Maddock,B. H., "A Visual Analysis of Flow and Mixing in Extruder Screws", SPE Journal, 383-389, (1959)

Rauwendaal C., Polymer Extrusion 4th Edition, Cal Hanser Verlag Publishers, Munich (2001)

Tadmor Z., Duvdevani I., Klein I., "Melting in Plasticating Extruders Theory and Experiments", Polymer Engineering and Science, **7**, 198 (1967).

Turkovich R.E., Erwin R. L., "Fiber fracture in reinforced thermoplastic processing." Polymer Engineering and Science, **23**, 743- 749 (1983)

Vu-Khanh T, Denault J., "The effects of injection moulding on the mechanical behavior of long fibrereinforced PBT/PET blends." Compos Sci Technol., **40**, 423–35 1991

MELT FLOW BEHAVIOR AND THE EFFECTS OF GAS IN GAS-ASSISTED INJECTION MOLDING

Chih-Chung Hsu[2]*, Hsien-Sen Chiu[1], Wen-Hsien Yang[1], Rong-Yeu Chang[2]

[1] *Department of Chemical Engineering, National Tsing-Hua University, Hsinchu, Taiwan 30043, R.O.C. – s9832822@m98.nthu.edu.tw; rychang@mx.nthu.edu.tw*
[2] *CoreTech System (Moldex3D) Co., Ltd., ChuPei City, Hsinchu, Taiwan –ethanchiu@moldex3d.com; anthonyyang@moldex3d.com*

Although Gas-Assisted Injection Molding (GAIM) has gained acceptance in eliminating sink marks and preventing warpage in the plastic industry, many problems are still encountered for the complicated gas-melt flow interactions. As for the controlling the processing conditions, such as delay time, gas holding time, injected gas pressure, and the amount of injected melt, are critical to ensure consistent gas penetration in the cavity. In this study, a full 3D numerical approach is proposed to model the polymer melt filling and gas penetration in GAIM. The melt-air and melt-gas interfaces are tracked through the algebraic volume-tracking method. Simulation results are in the form of fully featured 3-D animations, giving highly realistic images of flow fields. The effects of the processing conditions and model geometry on the behavior of gas penetration are investigated numerically and systematically. The predicted melt front location and gas penetration behavior show a good agreement with the experimental results. This proposed 3D approach proves to be a highly effective tool for process simulation and mold design for GAIM.

Introduction

Gas-assisted injection molding process provides an aid in the molding of thick walled parts, producing significant weight saving, avoiding part warpage and sink marks due to reduced filling pressure and residual stress, and resulting in shorter cooling time by coring out thick part of the part[1] [2]. The process utilizes compressed gas as the "packing medium"; hence a lower injection/packing pressure and clamp tonnage are required than that in the conventional injection molding process.

The 2.5D generalized Hele-Shaw (GHS) approximation has been extensively adopted to model the conventional injection molding as well as the GAIM [3]. However, the 3D flow nature inherent in the gas penetration phase invalidates the 2.5D GHS assumptions. Moreover, the irregular cross-section of the gas channel cannot be accurately modeled through the use of 1D rod element in the 2.5D approach. An unreliable estimation of geometry-caused flow resistance and heat transfer, and furthermore, the prediction errors for the melt front location and gas penetration length may occur[4-6].

However, controlling the processing conditions, such as delay time, gas holding time, injected gas pressure, and the amount of injected melt, are critical to ensure consistent gas penetration in the cavity. In this study, the collocated cell-centered FVM-based 3D numerical approach developed in our previous work is further extended to describe the melt flow and gas penetration in GAIM [7]. The numerical method is basically a SIMPLE-like FVM with improved numerical stability. The melt-gas interfaces are tracked through the algebraic volume-tracking method. Besides, simulation results of the applications including plastic filling and gas penetration during GAIM are shown.

Governing Equations and Numerical Method

In this study, the fluids are considered to be incompressible, Newtonian (for air and injected gas phase) or generalized Newtonian (for the polymer melt phase). Surface tension at the fluid front is neglected. The governing equations for 3D transient non-isothermal motion are:

$$\frac{\partial \rho}{\partial t} + \nabla \cdot \rho \mathbf{u} = 0 \tag{1}$$

$$\frac{\partial}{\partial t}(\rho \mathbf{u}) + \nabla \cdot (\rho \mathbf{u}\mathbf{u} + \tau) = -\nabla p + \rho \mathbf{g} \tag{2}$$

$$\rho C_P \left(\frac{\partial T}{\partial t} + \mathbf{u} \cdot \nabla T \right) = \nabla(\mathbf{k}\nabla T) + \eta \dot{\gamma}^2 \tag{3}$$

where $\mathbf{u}$ is the velocity vector, T the temperature, t the time, p the pressure, τ the stress tensor, ρ the density, η the viscosity, k the thermal conductivity, C_p the specific heat and $\dot{\gamma}$ the shear rate. For the polymer melt, the stress tensor can be expressed as:

$$\tau = -\eta\left(\nabla \mathbf{u} + \nabla \mathbf{u}^T\right) \tag{4}$$

The modified-Cross model with Arrhenius temperature dependence is employed to describe the viscosity of polymer melt:

$$\eta(T,\dot{\gamma}) = \frac{\eta_o(T)}{1+\left(\eta_o\dot{\gamma}/\tau^*\right)^{1-n}} \tag{5}$$

with

$$\eta_o(T) = BExp\left(\frac{T_b}{T}\right) \tag{6}$$

where n is the power law index, η_o the zero shear viscosity, τ^* is the parameter that describes the transition region between zero shear rate and the power law region of the viscosity curve. A volume fraction function f is introduced to track the evolution of the melt front and gas penetration. Here, f=0 is defined as the air or gas phase, f=1 as the polymer melt phase, and then the melt front is located within cells with $0<f<1$. In this study, the injected gas and the air are assumed to be the same fluids with identical properties. The advancement of f over time is governed by the following transport equation:

$$\frac{\partial f}{\partial t} + \nabla \cdot (\mathbf{u}f) = 0 \tag{7}$$

During the polymer melt filling phase, the velocity and temperature are specified at the mold inlet. While the gas is injected, the gas pressure is specified at the gas inlet. On the mold wall, the no-slip boundary condition is applied, and fixed mold wall temperature is assumed for energy equation. For the hyperbolic volume fraction advection equation, only inlet boundary condition is needed, i.e., f=1 for the polymer injection and f=0 for the gas penetration.

Furthermore, the volume-tracking method based on a fixed framework is incorporated in the flow solver to track the evolutions of melt-gas and melt-air interfaces during injection. The whole solution procedure is stated below. After solving the thermal flow governing equations by the FVM, the advancement of interface at each time step is determined by solving the fractional volume advection equation according to the velocities obtained. Calculate the material properties from the updated volume fraction function, and then initiate the next computation of flow field. Repeat this procedure until the cavity is completely filled.

Moreover, second-order accuracy is carefully maintained during discretization. Pressure, velocity and temperature are solved in a segregated manner. It is this feature that makes the present approach efficient and robust to solve the thermal flow field in complex three-dimensional geometries.

Results and Discussion

The first case was performed by Chen's work [8] of experimental investigations and Hele-Shaw numerical simulation. The geometric details of the cavity are shown in Figure 1. The filling of a gas-assisted injection molded thin plate with a semicircular cross section of gas channel is analyzed. Here this case is analyzed to assess the accuracy of the proposed three-dimensional numerical model. The process parameters and the material properties are the same with reference in Table 1 & 2.

The predicted 3D filling pattern of polymer injection in Figure 2 compares the predicted and measured polymer melts positions at 73.7% and 94% of the full melt-injection stroke. Almost identical results are obtained. The penetrations of the gas-core along the z-axis are compared with experiments. The predicted primary gas penetration length is in good agreement with the experimental data in Figure 3 & 4.

For the following case study, the handler and gear simulation results of two complex geometries are demonstrated. The geometric details are shown in Figure 5 and Figure 7. The process parameters and the material properties are the same in Table 3 & 4, where the same setting is applied.

In GAIM, insufficient amount of the pre-filled polymer melt will result in the "blow through", i.e., the gas depletes all the surrounding polymer melt and bursts at the melt front. The complex interface topology associated with the blow through process is simulated to demonstrate the capabilities of the method proposed in this study in Figure 6.

The uneven gas penetration is another common problem encountered in GAIM. Because of the tendency of seeking minimum flow resistance toward the melt front, the gas bubble will travel along the path closer to the inner corner of the bended channel as show in Figure 8. Besides, a "finger" like penetration is observed while the bubble tip approaches to the melt front. Accordingly, the thickness of polymer skin around the corner with sharp angle will be very thin, which may introduce structural weakness.

Melt temperature ($^\circ C$)	230
Mold temperature ($^\circ C$)	40
Filling Time (sec)	1.4
Prefill (%)	94
Delay Time (sec)	0
Gas Injection time (sec)	5.0
Gas pressure (bar)	35

Table 1 Molding conditions

n	0.2838
τ * (Pa)	1.791×10^4
B (Pa-s)	2.591×10^{-7}
Tb (K)	11680
Density (kg/m^3)	940
Specific Heat Cp (J/kg K)	2100
Thermal Conductivity k (W/m K)	0.18

Table 2 The parameters of PS(ChiMei/PG33) for modified cross model

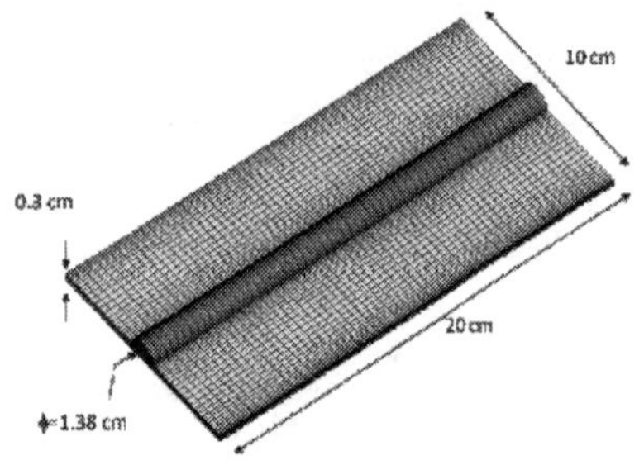

Figure 1: Geometrical parameters of a thin plate with a gas channel of semi-circular cross-section[8]

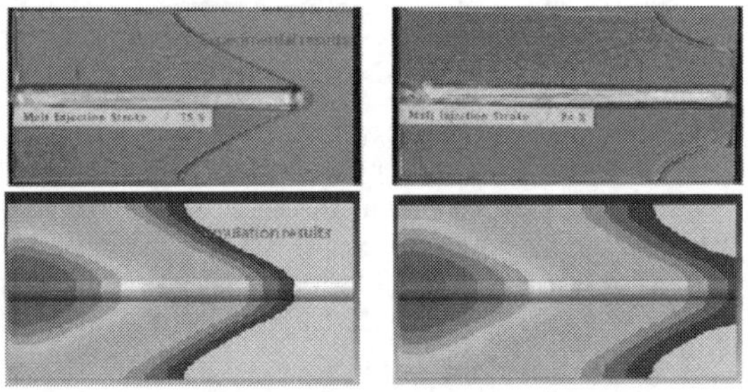

Figure 2: Comparison between 3D analysis(below) and experimental (above) results of melt flow at 73.7 and 94% filled.

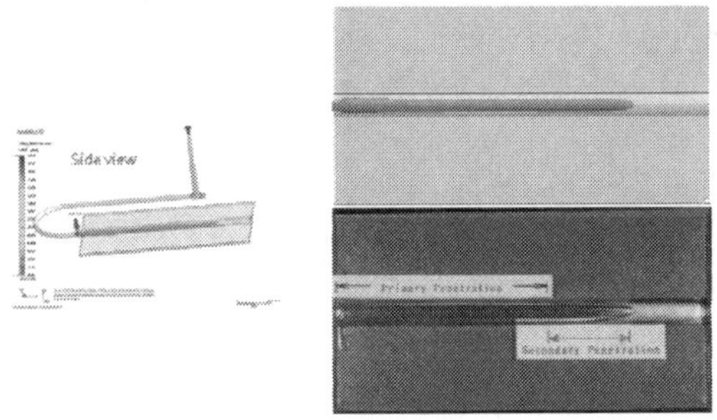

Figure 3: Gas-core area ratio distribution along the gas channel[8]

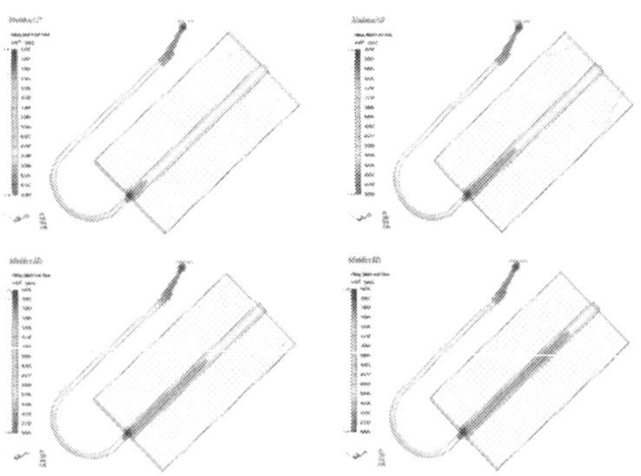

Figure 4: Melt-front locations and gas penetration along the gas channel

Melt temperature ($^{\circ}C$)	205
Mold temperature ($^{\circ}C$)	50
Filling Time (sec)	0.36
Prefill (%)	70
Delay Time (sec)	0
Gas Injection time (sec)	2.0
Gas pressure (bar)	50

Table 3 Molding conditions for car handle

n	0.2169
τ * (Pa)	$1.152\text{x}10^{6}$
B (Pa-s)	$1.005\text{x}10^{-7}$
Tb (K)	13294
Density (kg/m^3)	1034
Specific Heat Cp (J/kg K)	1900
Thermal Conductivity k (W/m K)	0.18

Table 4 The parameters of ABS(ChiMei/PA707) for modified cross model

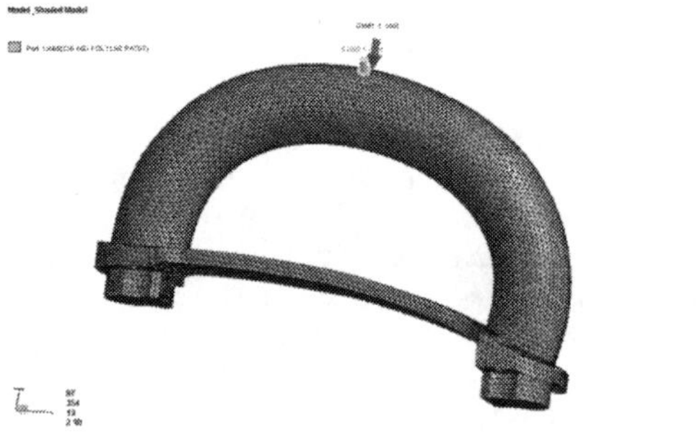

Figure 5: Geometry of a handle

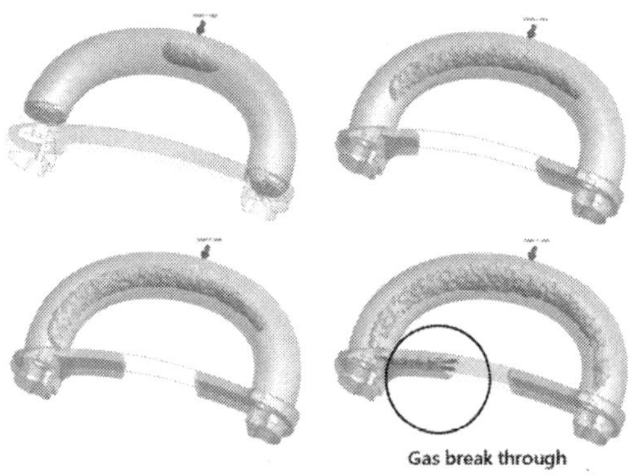

Figure 6: 3D show of the melt and gas filling sequence in GAIM process

Figure 7: Geometry of a gear product

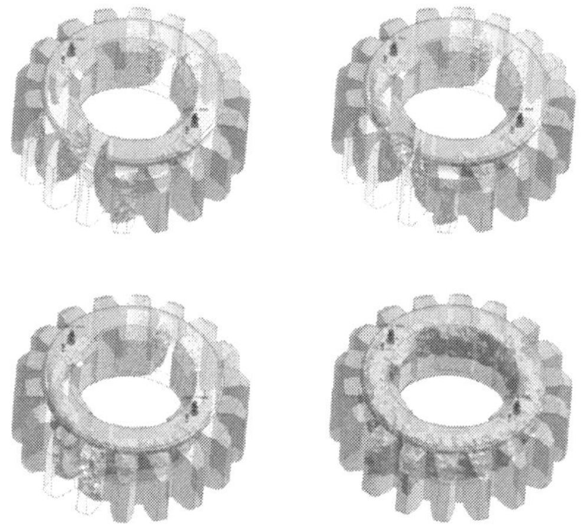

Figure 8: 3D show of the melt and gas filling sequence in GAIM process

Conclusions

This study proposes a full 3D numerical approach to simulate the melt flow behavior and the effects of gas in GAIM. The melt-air and melt-gas interfaces are tracked through the algebraic volume-tracking method. The finite volume method (FVM) is utilized to solve the 3D governing equations of flow and heat transfer. The predicted melt front location and gas penetration behavior show good agreement with the experimental results, and several complex applications are shown as well. The results illustrate the potential practice of an effective 3D approach tool for further process simulation and mold design for GAIM.

References

1. J.F.Stevenson (Ed.), Innovation in Polymer Processing Molding, Hanser (1996)
2. S.Shah and D.Hlavety, SPE Tech. Paper, 1479, ANTEC (1991)
3. C.A.Hieber and S.F.Shen, J. Non-Newtonian Fluid Mech. 7, 1 (1980)
4. L.S.Turng, Adv. Polym. Tech. 14, 1 (1995)
5. R.Y.Chang, M.H.Tsai and C.H.Hsu, SPE Tech. Paper, 176, 598, ANTEC (1996)
6. D.M.Gao, K.T.Nguyen, A.Garcia-Rejon and G.Salluom, J. Mater. Proc. Tech., 69, 282 (1997)
7. R.Y.Chang and W.H.Yang, submitted to Int. J. Numer. Methods Fluids (2000).
8. S.C.Chen, N.T.Cheng and K.S.Hsu, Int. J. Mech. Sci, 38, 335 (1996)

ANALYSIS AND OPTIMIZATION OF MIXING IN TWIN SCREW EXTRUDERS USING THE MAPPING METHOD

A. Sarhangi Fard[1], H.E.H Meijer[1], N.M.H. Famili[2], P.D. Anderson Patrick[1*]

[1]*Materials Technology, Eindhoven University of Technology, Eindhoven, Netherlands– a.sarhangi.fard@tue.nl, h.e.h.meijer@tue.nl, p.d.anderson@tue.nl*

[2]*Polymer engineering group, Tarbiat Modares University, Tehran, Iran– nfamili@modares.ac.ir*

Abstract - Recent advances in finite element modeling have made it possible to accurately determine the velocity field in twin-screw extruders operating under realistic conditions. Analysis and even optimization of mixing based on particle tracking is far from trivial since no volumetric data is available. Here we continue on the application of the mapping method, which does provide volumetric quantities, to quantitatively compare different screw layouts and find optimal designs. The mapping method has proven its merits for static mixers and, in previous work, all popular static mixers are evaluated and guidelines resulted to design new mixers based on either compactness of minimum pressure drop. Now we use the same approach to analyze mixing in dynamic mixers and we compare several screw configurations and different types of screws, changing the pitch and gap width. For kneading elements a variety of staggering angles have been in a sequence. The flux-weighted intensity of segregation is used as a mixing measure to evaluate different screw designs.

Introduction

The most commonly used mixing device in polymer processing is the closely intermeshing, co-rotating twin screw extruder. Although in essence simply a polymer melting and transporting device, its screws are constructed in a modular way in order to create different screw geometries to increase mixing performance. Some common elements are shown in Figure 1.

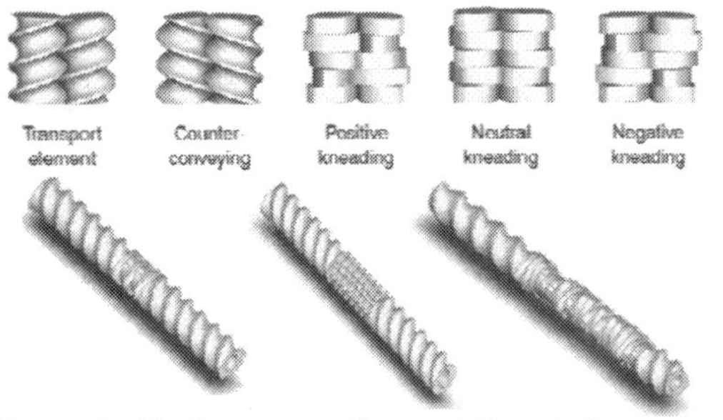

Figure 1. Basic types of co-rotating twin screw extruder elements combined to give different screw configurations.

The main goal of this work is creating a simulation tool, based on the mapping method, for characterization and optimization of mixing in twin screw extruders (TSE) for different screw elements.

Numerical methods

A distribution matrix Φ is formed to store information about the distribution of fluid from one discretized space to the next due to a specified flow (Singh et al. 2008). To obtain the coefficients of Φ the initial volume of the flow domain is subdivided into a large number of discrete cells (N) . During flow, the material from a donor cell is transferred to different recipient cells. The fraction of material that is transferred from the donor cell j to a recipient cell i gives the distribution coefficient Φ_{ij} of the donor cell with respect to the recipient cell (Figure2). Thus, in total N cells form a distribution matrix of the order $N \times N$:

$$\Phi_{ij} = \frac{\int_{\Omega_j|_{t1+\Delta t} \cap \Omega_i|_{t1}} \mathrm{d}A}{\int_{\Omega_j|_{t1}} \mathrm{d}A}$$

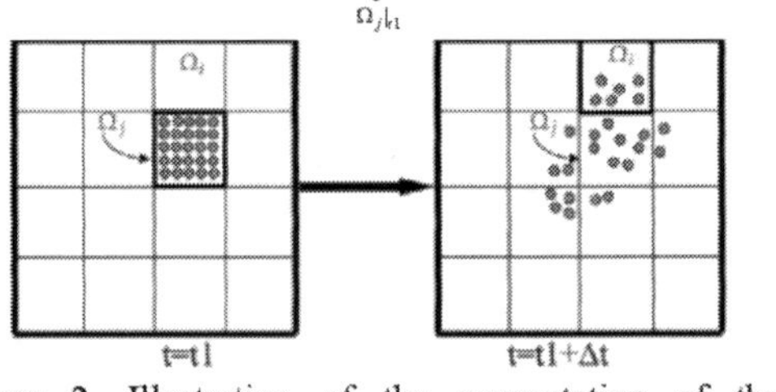

Figure 2. Illustration of the computation of the coefficients Φ_{ij} of the mapping matrix Φ, the geometry is repeated after Δt.

The extruder is time-periodic in the sense that after each revolution, the screw is back in its initial position (Kruijt et al. 2001; Sarhangi Fard et al. 2008). For example for a screw element with two lobes, an identical orientation is also reached after a half rotation, so the mapping matrix is calculated for a half screw complete revolution. The twin screw extruder geometry is discretized with tetrahedron cells (Figure 3) . These cells are filled with particles (500 particles per each cell). Particles are tracked (Figure 4) and finally the mapping matrix is calculated.

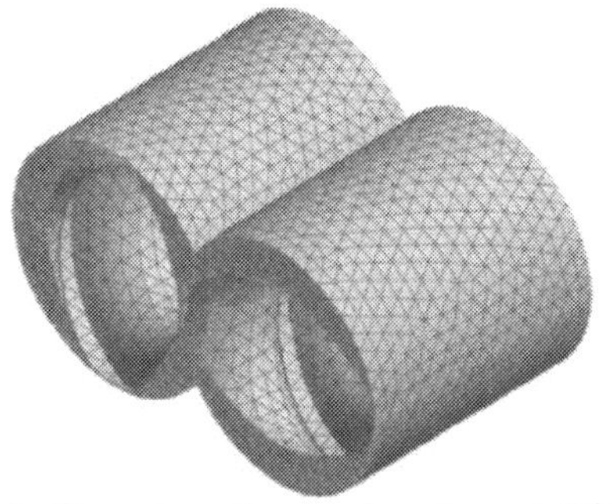

Figure 3. Geometry of conveying element (Pitch=30 mm) with mapping gird.

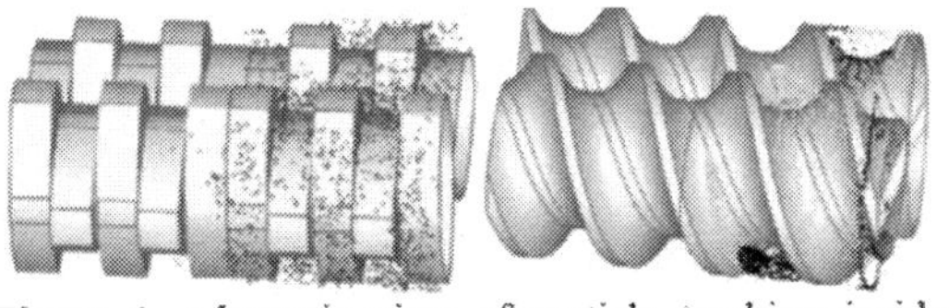

Figure 4. schematic view of particle tracking inside TSE for kneading and conveying elements

If the initial concentration distribution of material 1 for each cell is shown by column vector C_0; the concentration of material 1 after Δt (C) is obtain:

$$C=\Phi\times C_0$$

The screw is basically an assembly of different elements; the mapping matrix is calculated for each screw element. Identical elements have the same mapping matrix. For example if we suppose the screw consists of $n+1$ conveying elements, with mapping matrix Φ_s and Φ_n and the initial concentration of material 1 shown with C_{in}; the method to calculate the final concentration distribution (C^*) is shown in Figure 5.

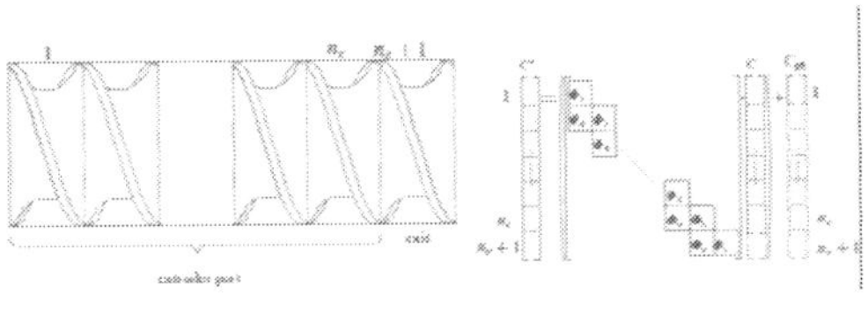

Figure 5. Final assembly of mapping matrix for a screw with n+1 conveying elements.

The accuracy of the mapping method depends on the accuracy of velocity field. We used the 3D XFEM method to simulate Stokes flow inside the TSE for a Newtonian, isothermal polymeric melt. The Intensity of Segregation (I_d) is defined to measure the mixing efficiency:

$$I_d=\frac{1}{\bar{C}(1-\bar{C})}\frac{1}{N}\sum_{i=1}^{N}(C_i-\bar{C})^2$$

Results and Discussion

As test case study, we select two screw configurations with only conveying elements differing in pitch: 15.3 mm and 30.0 mm. It is assumed that the screw is filled with white material and at time t=0 black material is fed into the first element of the left screw, see Figure 6. The materials have the same rheological parameters. The screw speed is 80 RPM.

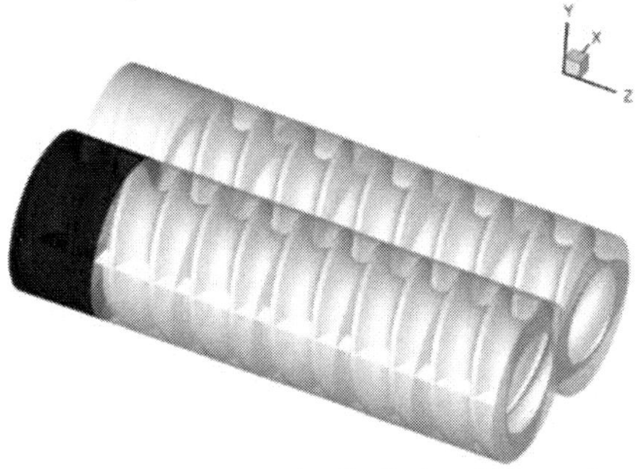

Figure 6. Initial concentration distribution of the material

The concentration distribution of materials across the screw channel is shown in Figure 7. Increasing the pitch increases the positive transport but decreases mixing performance. The intensity of segregation curve is shown in Figure 8.

Conclusions

It is shown that the mapping method is a suitable method for the analysis of dynamic, periodic devices. It is capable of determining concentration distributions and corresponding mixing measures (like the intensity of segregation) as well as average residence time.

Acknowledgements

ASF thanks FP6 program PEPTFLOW for financial support.

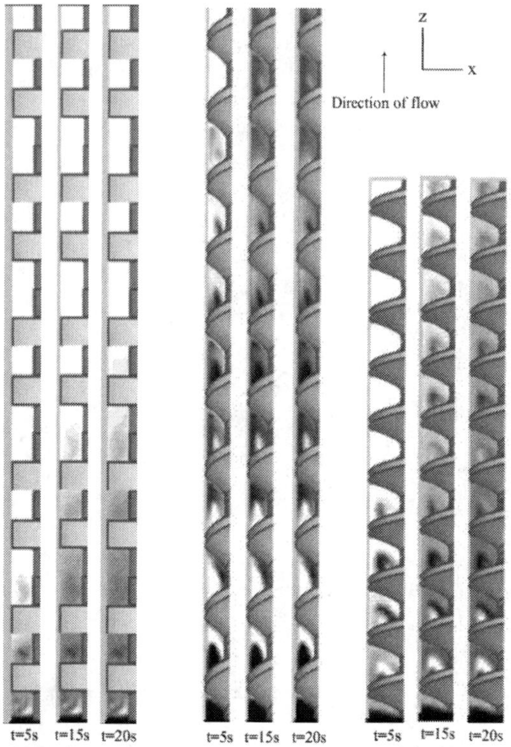

Figure 7. Concentration distribution in the transport section (conveying elements with two pitch size; 15.3 mm and 30.0 mm) and mixing section (kneading element with 90^0 staggering angle and pitch=30 mm) of a co-rotating twin screw extruder. The images illustrate the 'filling' with black material in time of screws.

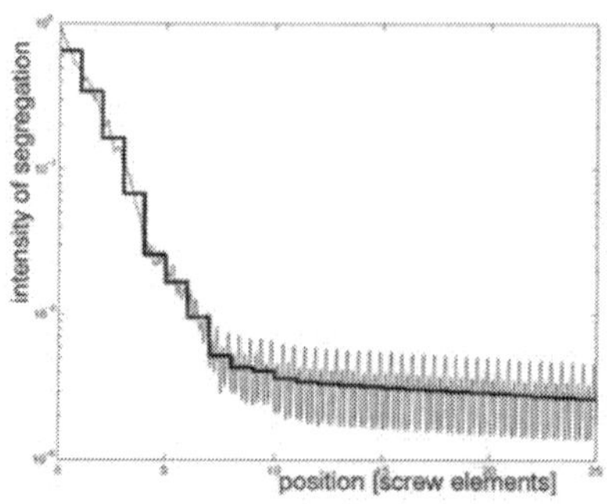

Figure 8. Intensity of segregation along the screws; the thin line indicates the intensity of segregation per layer of sub-domains; the thick line indicates the intensity of segregation per screw element.

References

Kruijt, P.G.M., et al., "The Mapping Method for Mixing Optimization. Part II: Transport in a co rotating Twin Screw Extruder", Int. Polym. Process., **XVI**(2), 161-171 (2001)

Sarhangi Fard, A., et al., "A new adaptation of mapping method to study mixing of multiphase flows in mixers with complex geometries", Computers & Chemical Eng., **32**, 1471–1481 (2008)

Singh, M.K., et al., "The mapping method as a toolbox to analyze, design, and optimize Micromixers", Microfluid Nanofluid, **5**, 313–325 (2008)

TWIN SCREW EXTRUDERS USED FOR REACTIVE COMPOUNDING

Tadamoto Sakai

Tokyo Office, Shizuoka University, 3-3-6 Shibaura, Minato, Tokyo 108-0023, Japan, mtsakai@ipc.shizuoka.ac.jp

Abstract - The use of polymer processing equipment to conduct reactive processing, particularly polymerization and reactive blending, can create newly functionalized polymers with very sophisticated morphology. Recent morphology control is requested at the level of nano-scale. Screw type extruders, in particular twin screw extruders, are the preferred devices for these purposes, but must be solved the major issues of required residence time, temperature and morphology development. This study deals with extruder design to provide the necessary reaction temperature and residence time, intensity of mixing, and devolatilization to remove by-products and residual reactants/solvent.

1. Introduction

A screw type extruder, in particular, twin screw extruder, has many advantages when other types of conventional reactors are comopared with;

(1) Continuous conveying process for solid, powder and liquid

(2) Excellent distributive and dispersive mixing work for highly viscous materials

(3) Easy control of polymer temperature, pressure and relatively short residence time with shearing forces

(4) No solvent requirement due to a bulk and melt phase process

(5) Removal of residual monomer and by-products from molten polymers

(6) Self wiping operation when a closely-intermeshed co-rotating twin screw extruder is used

(7) Consecutive process from chemical reaction, mixing, devolatilization to pelletizing and/or shaping.

(8) Many kinds of diversified polymers, composites and alloys can be produced with a single process line

These advantages have been strongly utilized to creat new polymers and alloys/composites using reactive processing technology.

2. Twin screw extruder classification

Screw extruders are practically classified into single and twin screw types. A single screw extruder is mainly used for film/sheet extrusion, blow molding and injection molding. On the other hand, a twin screw extruder is widelly used for blending/controlling the microstructure of a polymer and chemical mofication such as grafting, as well as pipe/profile extrusion. Twin screw extruders are usually defined by intermeshing type and rotational direction. In general, intermeshed co-rotational twin screw extruders are widely used for diversified usage, in particular for reactive processing, mainly because of its flexible mixing capability with wide ranges, self-wiping effect and a great deal of accumulated operational knowledge in the industry. For laboratory-scale, small screw diameter twin screw extruders such as 20-30mm are commonly utilized. And 30-65mm twin screw extruders for bench-scale production, 65-150mm for industry-scale are widely used in reactive blending applications. For large scales more than 150mm twin screw extruders are applied.

During several decades a twin screw extruder has been progressed and widely diversified, depending on the purposes. These are shown in the followings;

(1) Deeper screw channel depth: D_0/D_i=1.20~1.80

(2) Higher screw rotational speed: 50~2500rpm

(3) Longer screw/barrel length: L/D=10~120

(4) Various sophisticated screw mixing elements

Table 1 shows an example of twin screw specification changes during a decade.

Table 1. Typical specification changes of a twin screw extruder for industrial use.

JSW TEX (Co-rotation)

	Screw Diameter(mm)	Permissible Torque (Nm)	Driving Motor (kW)	Screw Speed (Standard)	Screw Speed (High Speed)
TEX 48V	48	1480	90 (45)	580 (rpm) (340)	1750 (rpm) (565)
TEX 59V	59	2864	160 (75)	530 (rpm) (300)	1450 (rpm) (530)
TEX 72V	72	5174	315 (135)	580 (rpm) (300)	1150 (rpm) (520)

( ) : Twin screw data in 1990's

3. Mixing/kneading screw element design

Fig. 1 shows typical screw elements used for mixing and kneading with an intermeshed co-rotation twin screw extruder. The operational characteristics of the combined screw elements can be theoretically analyzed using commercialized software packages.

In addition to these screw elements, the new screw design has been developed using the progress in twin screw simulation technology. Fig. 2 shows the results on the kneading element combination designed on the concept of the theoretical calculation.

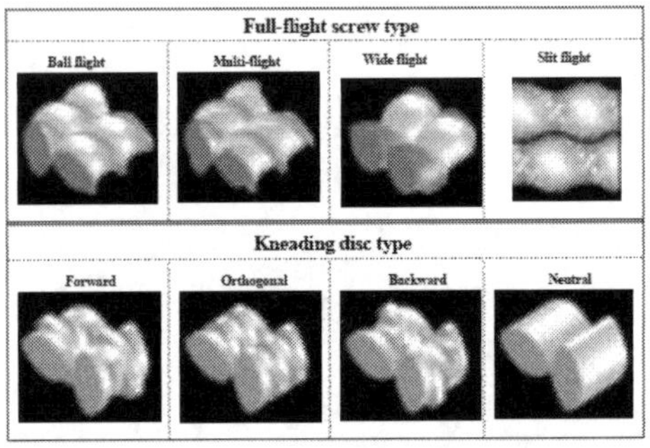

Fig.1 Conventional mixing screw elements

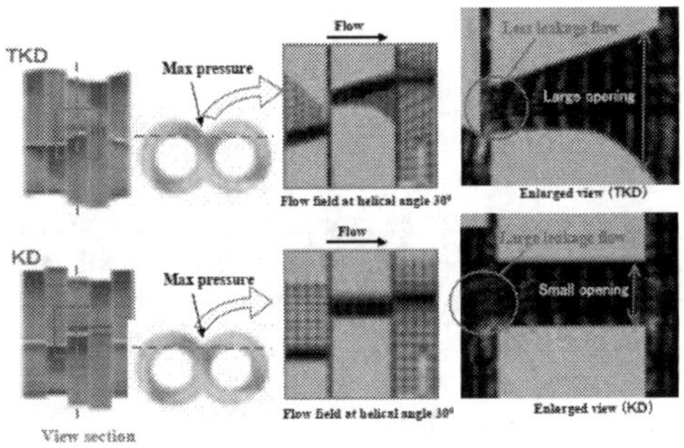

Fig.2 Velocity profile comparison between typical mixing/kneading elements and newly designed ones[1] (leakage flow comparison)

Newly developed screw design, that is, the twisted kneading screw combination from the simulation shows better mixing effects through sufficient flow of high shear between each kneading element tip. The key point of this screw design is how to prevent the leakage flow between each kneading element so as to work more efficient dispersive and distributive mixing[1].

4. Reactive processing using a twin screw extruder

Typical chemical reactions applied to a twin screw extruder are described below;

(1) Bulk polymerization: the production of high MW polymers from monomer or prepolymer with low MW. (PMMA, PA, PLA, thermoplastic PU, etc.)

(2) Graft copolymerization: reaction between polymer and ionic monomer to form graft copolymers. (maleanated PP, modified elastomer, etc.)

(3) Coupling reaction: reactions with multi-functional branching agents to extend the polymer chain or to increase elongational viscosity.
(polyolefin, PET, Thermoplastic elastomer, etc.)

(4) Mechano-chemical degradation or pyrolysis: the production of low MW components or monomer from a high MW polymer.
(PP, PET, PMMA, PS, PLA, etc.)

Table 2 shows typical polymer blends which are commercially used. Today, however, the morphology of the polymer alloy has been gradually changing from micron/sub-micron scale to nano scale.

Table 2. Various polymer blends commercially used

Product	Supplier	Components (matrix/disperse)	Morphology	Property
Dynamar	Dyneon	PE/PTFE	dilute drops	processability (adds wall slip)
MB	Dow Corn.	PP/PDMS	dilute drops	lubricity
Noryl GTX	GE	PA6/PPO/SB	drops (60%)	dimension (H_2O uptake)
Zytel ST	Dupont	PA66/EP	double emuls.	toughness
Selar	Dupont	PE/PA66	lamellar	diffusion barrier
Vectra	Hoechst	PET/LCP	fibers	thermal expans.
TSOP	Mitsubishi/ Toyota	PP/PE/EP	co-continuous	toughness
Statrite	Goodrich	PP/PU	co-continuous	electrical conductivity

Fig.3 shows the comparison between conventional morphology in the micron/sub-micron scale and the one in nano-size. If the morphology in the nano size is obtained, various physical/chemical properties are considerably enhanced as shown in Fig.3.

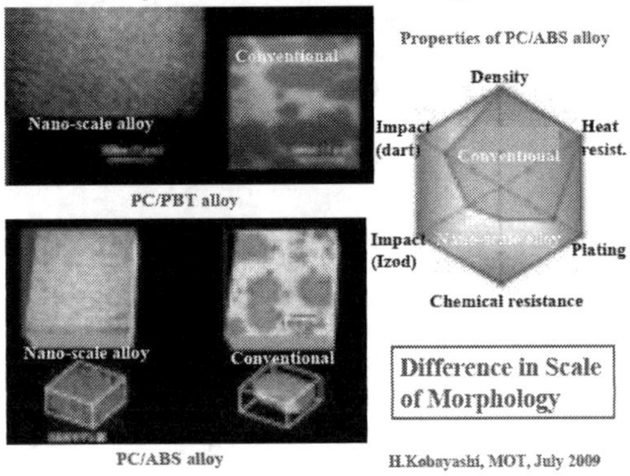

Fig.3. Comparison of the morphology scale[2]

5. Applications

Some applications for reactive processing are shown.

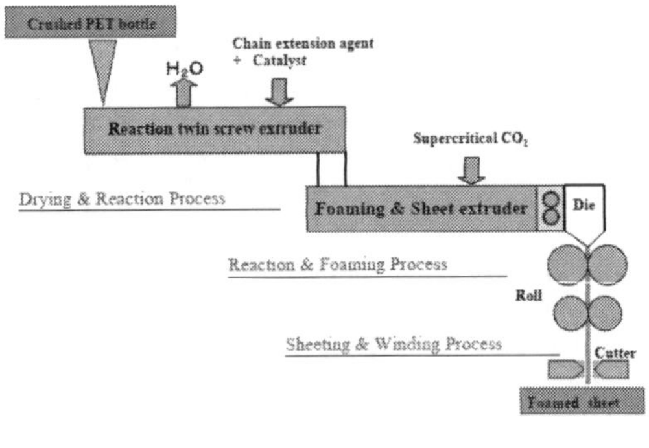

Fig.4. Hybridized reactive processing system[3]

(1) Waste PET recycle/foaming extrusion

Fig.4 shows the combination of a twin screw extruder for chemical modification with a single screw extruder for foaming and sheet making. In the first step, PET is chemically modified with a branching agent to improve elongational melt viscosity for smooth foaming and in the next step SCF(CO_2 or N_2) is injected to molten PET to foam at the high expansion ratio in the single screw extruder.

(2) Waste rubber recycling to obtain higher values

Fig.5 shows a direct reactive process changing from waste rubber to dynamically vulcanized thermoplastic elastomer (TPE).

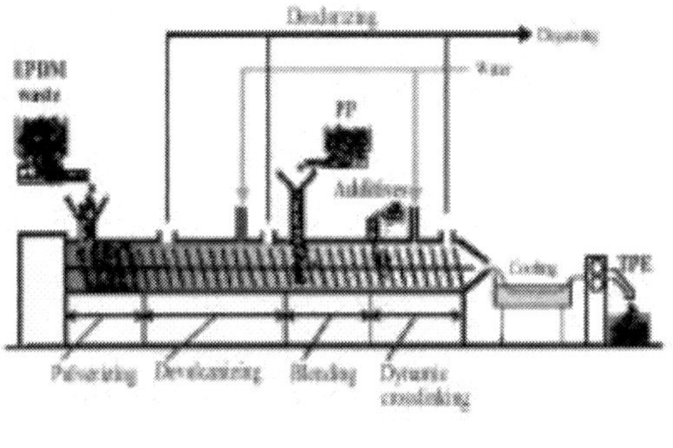

Fig.5. Direct TPE production from waste EPDM[4)]

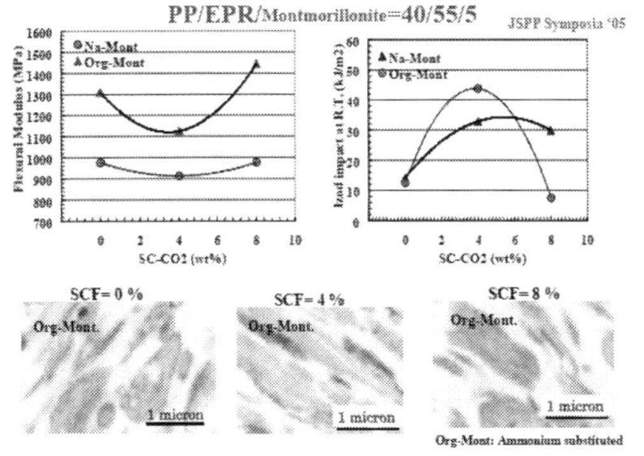

Fig.6. Morphology control using SCF-CO_2[5)]

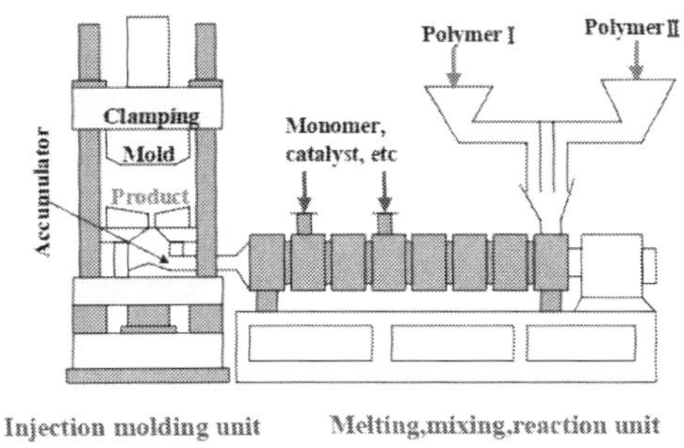

Fig.7 Hybridized reactive processing system[6)]

(3) Morphology control of blends used with SCF

When SCF is utilized as a special solvent, the morphology of polymer blends can be controlled because of the miscibility change between each polymer component. Fig.6 shows the adjustment of the polymer blend morphology by the addition of SCF.

The hybridized system, for example, the combination of twin screw compounding machine with injection molding machine, has the following advantages:

(1) Minimization of thermal degradation through direct molding or shaping.

(2) Easy fixation and control of the morphology obtained from twin screw reactive blending, which is easily changeable during molding and/or shaping.

(3) Consecutive process from polymerization to shaping, even for a polymeric material with poor processability such as low fluidity or high heat-sensitivity.

6. Conclusion

Using recently progressing theory for calculating twin screw extrusion behavior, we can estimate the effects on mixing/kneading, polymer temperature and filling pressure distributions in a twin screw extruder to fit with the usage and product properties, as well as the operational and design parameters.

Furthermore, the design on a twin screw extruder has been rapidly progressingby by means of advanced theoretical simulation, experimental techniques and on-line monitoring methods. Accompanying with these progresses, the screw channel depth, screw/barrel length (*L/D*), screw rotational speed and screw transmittable torque are dramatically improving toward higher productivity and functionality.

Under these conditions the applications of reactive processing technology have been widely expanding from polymerization/depolymerizaiton to various kinds of chemical/physical modification, in connection with more sophisticated polymerization methods and material characterization techniques on the morphology, structure and properties. In addition, the moreimportant requirements in reactive processing are both eco-friendliness and sustainability in recent years.

7. References

1) T.Sakai, AWPP (Penang), December, (2009)
2) H.Kobayashi, MOT-Seminar (Chigasaki), July (2009)
3) T.Sakai, et al, JSPP Annual Meeting (Tokyo), June, (2000)
4) K.Fukumori, et al, Toyota R&D Report, **38**(1) p39 (2003)
5) T.Fukushima, et al, JSPP Symposia (Sendai), Nov. (2005)
6) T.Sakai, Int. Polymer Processing, **16**(1), p3 (2001)

STUDY ON THE OPTIMIZATION OF THE SCREW SEGMENT COMPOSITION OF THE NANO-CLAY REINFORCED ELASTOMER COMPOSITE BY TWIN SCREW EXTRUDER

Kazuya Maeda[1]*, Atsushi Chikaraishi[1] , Yutaka Imaida[2], Tatsuya Tanaka[2]

[1]*Department of Engineering, University of Doshisha, Kyoto, Japan*
[2]*Faculty of Science and Engineering, University of Doshisha, Kyoto, Japan —tatanaka@mail.doshisha.ac.jp*

Abstract - A great deal of attention has been given to the preparation of nanocomposites with the aid of supercritical carbon dioxide (Sc-CO_2) to improve the dispersion of the clay. But, when the clay is compounded to elastomer, composites become high temperature by shear stress. The purpose of this study is the optimization of seal ring segment composition when the nano-clay compounds to elastomer by using twin screw extruder. First, the pressure distribution in the seal ring portion was analyzed by using the finite-element-based CFD software. Then, the pressure in the seal ring portion was measured by using a laboratory-scale extruder. The optimal seal composition was decided by comparing the experimental pressure value with the analytical pressure value. When the melt-blending is performed by using the optimal seal composition with the aid of Sc-CO_2, a mechanical characteristic of the material has improved.

Introduction

Generally, it is well known that a mechanical property and the gas barrier characteristic will improve by adding the clay of the nano-particle level in small quantities to polymer. When compounding the clay and polymer, melt-blending using twin screw extruder is industrially most advantageous. But, when the clay is compound to elastomer, composites become high temperature by shear stress. Therefore, because segment composition uses low shear stress type, it is difficult to improve dispersion only by optimizing operating conditions.

On the other hand, there is the report to be able to improve the dispersion of the clay by using supercritical carbon dioxide (Sc-CO_2). And in order to use Sc-CO_2, we have to make the zone which maintains CO_2 at a supercritical state within the twin screw extruder. There is a seal ring segment as an element used in order to make the zone. And polymer is dammed up near the seal ring portion, and high pressure occurs by the filled polymer. As a result, a supercritical state is maintainable. However, in arrangement of the seal ring used in a supercritical process, elastomer will cause its thermal degradation by high shear stress near the seal ring portion. As a result, even if the dispersion of clay improves by the supercritical process, the mechanical properties of materials decrease.

The purpose of this study is the optimization of seal ring segment composition when the clay compounds to elastomer by using twin screw extruder. First, the pressure distribution was analyzed by using the finite-element-based CFD software. Then, the pressure measurement in the seal ring part was performed by using a laboratory-scale extruder. The optimal seal composition was decided by comparing the experimental pressure value with the analytical pressure value. When the melt-blending is performed by using the optimal seal composition with the aid of Sc-CO_2, a mechanical characteristic of the material has improved. Moreover, when the optimal seal composition is used, the operating range of the extruder is extensible.

Experimental

Simulation

In order to use Sc-CO_2, we have to make the zone which maintains CO_2 at a supercritical state within the twin screw extruder. Therefore, the optimal operating conditions of co-rotating twin screw extruder (screw diameter 30 mm, length/diameter ratio 60) and seal ring compositions that can generate critical pressure were decided by the simulation analysis and experiment. The operating conditions used in simulation are listed in Table 1. The seal ring compositions used in simulation are shown in Figure 1. We used the viscosity data of olefinic elastomer measured by capillary rheometer. The simulation was performed under isothermal condition.

Table 1 Operating condition of twin screw extruder (D=30mm)

screw speed (rpm)	100			100	125	150
feed rate (kg/h)	3.0	3.5	4.0	3.5		

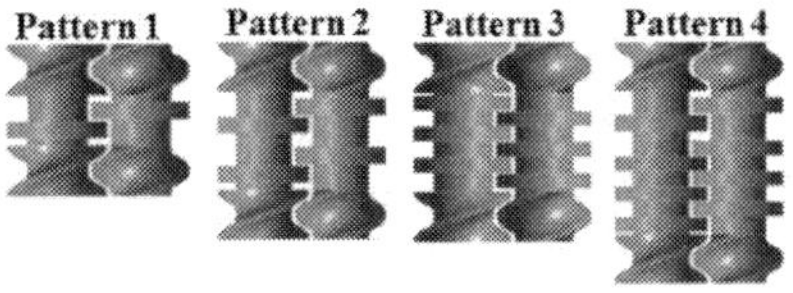

Figure.1 Seal ring configuration

Scale-down

To verify the simulation result, the operating condition and the seal ring configuration of twin screw extruder (screw diameter 30 mm) that was decided by the simulation were scale-downed to those of laboratory-scale twin screw extruder (ZSK18, Coperion, Germany) with screw diameter 18 mm, length/diameter ratio 40. Then, we analyzed pressure distribution on those conditions. When boundary condition was decided, shear-rate was assumed to be constant, and feed rate was determined as follows:

$$\frac{Q_2}{Q_1} = \left(\frac{D_2}{D_1}\right)^3 \quad (1)$$

where Q is the feed rate and D the screw diameter. Subscript 1 refers to the big machine (D=30 mm), 2 to the small one (ZSK18). The dimension of the seal ring used for the pressure measurement was decided by the result of the scale-down. It is shown in Table 2.

Table 2 Geometry of seal ring after scale-down

Geometry	Sign	(mm)
Disc gap	Dg	**0.34**
Clearance	Cl	**0.25**
Calender gap	Cg	**0.94**
Disc radius (large)	Dr-l	**8.90**
Disc radius (small)	Dr-s	**5.16**
Disc width	D-w	**3.00**
Barrel radius		**9.15**
Seal ring wide	Sw	**9.00**
Center distance	Cd	**15.0**

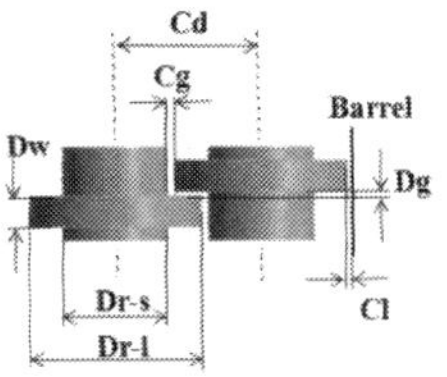

Pressure measurement

Twin screw extruder ZSK18 was used in pressure measurement. The screw was arranged with the scale-downed seal composition. Barrel temperature was set at 210℃. Pressure was measured under some operating conditions, and the experimental pressure value and the analytical pressure value were compared. Operating conditions are shown in Table 3. The optimal seal composition in which critical pressure was able to be kept was decided by the relation between an experimental value and an analytical value.

Table 3 Operating condition of ZSK18

screw speed (rpm)	120			120	200	250
feed rate (mm^3/s)	240	280	320	280		

Melt-blending with the aid of Sc-CO_2

Olefinic elastomer was used as the matrix polymer, and montmorillonite (MMT) was used as clay. Melt-blending processes were performed in a twin screw extruder (screw diameter 30 mm). The melt and Sc-CO_2 were mixed using a screw configuration as shown in Figure.2. Between the CO_2 injection port and the CO_2 vent port, kneading and mixing elements were added to improve the mixing efficiency. Samples of neat elastomer, elastomer/MMt (90/10 w/w), elastomer/MMt (90/10 w/w) with Sc-CO_2 were prepared. The tensile properties of the prepared samples were studied with a crosshead speed of 50 mm/min with a tensile tester machine. To prepare the test specimens, sheets 2 mm thick were cut into dumbbell-shaped specimens.

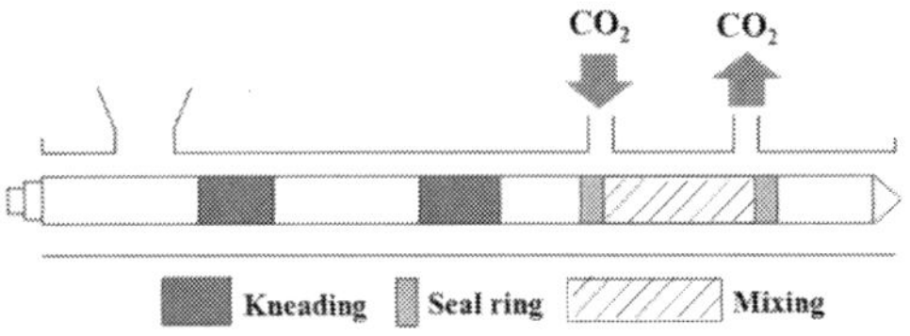

Figure.2 Screw configuration

Results and Discussion

Simulation

Figure.3 gives the results of simulation. Figure.3 (a) shows the effect of screw speed on pressure, (b) shows the effect of feed rate on pressure. In Figure.3 (a), pressure decreases as the screw speed increases, and the maximum pressure generates at 100 rpm. In Figure.3 (b), pressure increases as the feed rate increases. These indicate that 100 rpm and 3.5 kg/h are the best conditions to generate high pressure. Figure.4 shows the simulation results of each seal ring configuration under these operating conditions. It can be confirmed to reach critical pressure with pattern 1.

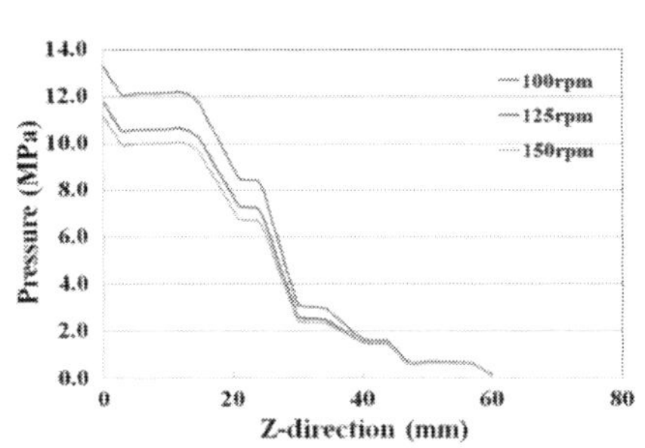

(a) Effect of screw speed on pressure
(Feed rate 3.0 kg/h)

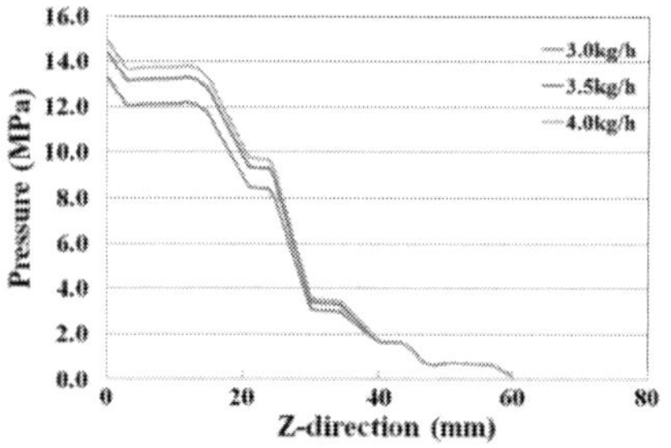

(b) Effect of feed rate on pressure
(Screw speed 100 rpm)

Figure.3 Effect of operating condition on pressure

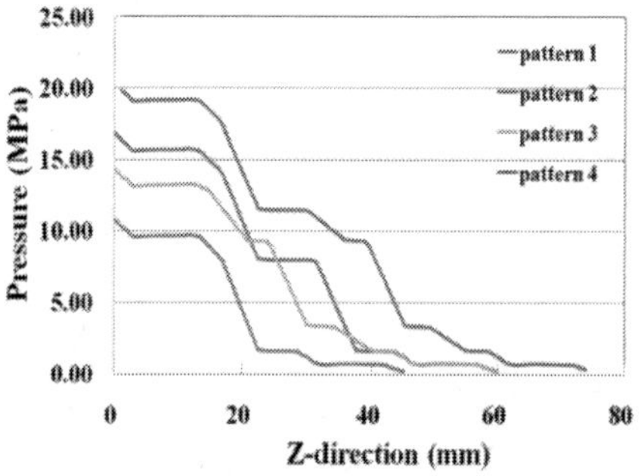

Figure.4 Pressure distribution of each seal ring configuration

Pressure measurement

Pressure was measured in the scale-downed seal rings portion while changing the operating condition and the number of seal rings. Figure.5 (a),(b),(c) give the results of pressure measurement. It is clear that the experimental pressure value is lower than the analytical pressure value though they indicate the same tendency. This may mean that we assumed material was completely filled in the full-flight part in the simulation, but it was not filled in the experiment. There is also a possibility that the viscosity decreased for generation of heat because of the shear stress.

Figure.6 shows the relation between the analytical pressure value and the experimental pressure value. In Figure.6, it is necessary to generate over 15.4 MPa in the simulation to generate over 8.0 MPa (critical pressure of CO_2) in the experiment. In Figure.4, it is understood that if pattern 4 is used, 15.4 MPa can be generated in the simulation. Therefore, this may indicate that critical pressure can be generated by pattern 4.

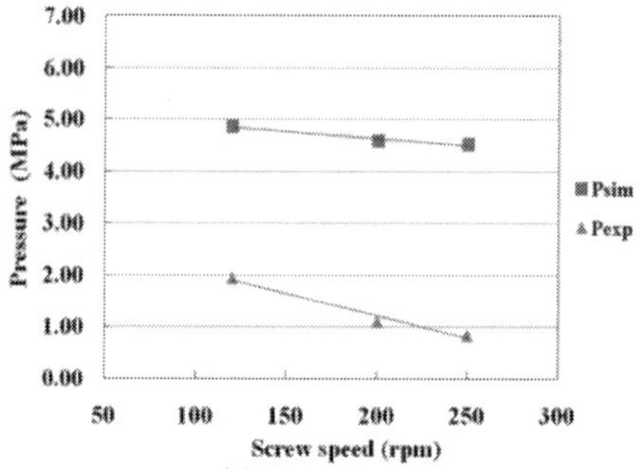

(a) Screw speed

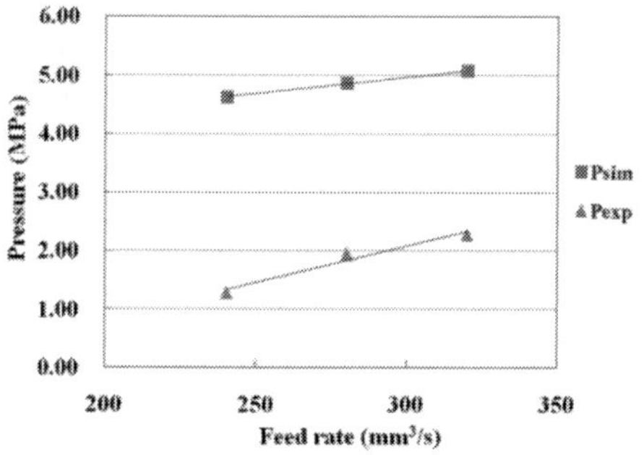

(b) Feed rate

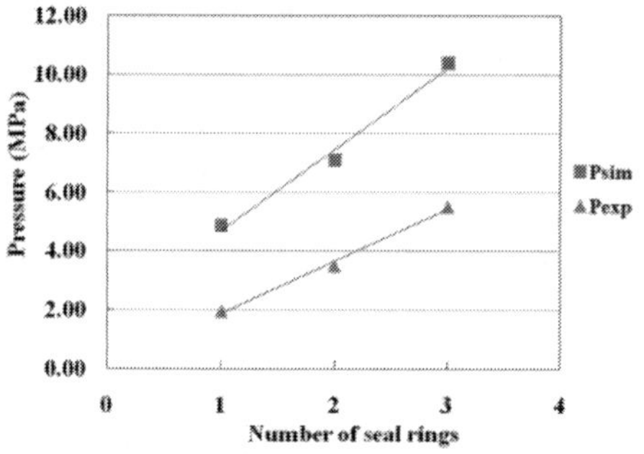

(c) Number of seal rings

Figure.5 Experimental and analytical value of each condition

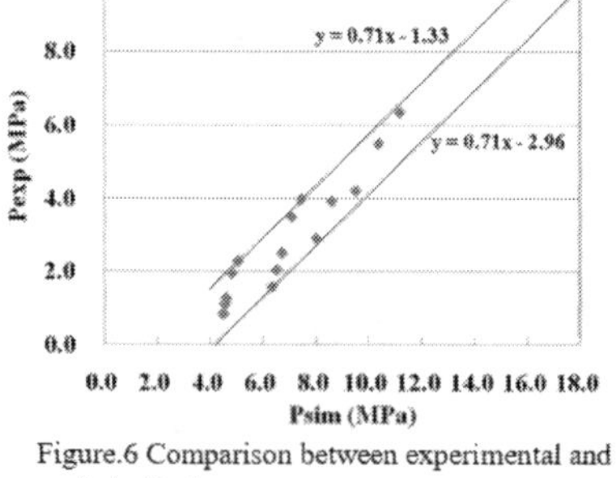

Figure.6 Comparison between experimental and analytical value

Melt-blending with the aid of Sc-CO_2

Melt-blending was performed with the aid of Sc-CO_2 by using twin screw extruder (screw diameter 30 mm) which has arranged the seal composition pattern 4. The result of the tensile test is shown in Figure.7 (a) and (b).

Figure.7 (a) shows tensile strength and 100%modulus of each samples, (b) shows elongation at break of each samples. In Figure.7, it can be confirmed that a mechanical characteristic is improved.

The scale-downed seal ring composition reduced the material temperature and load electric by 11.3 % and 23.3 % respectively from previous experiment. This indicates that the operating range of the extruder can be extensible by the use of the optimal seal composition.

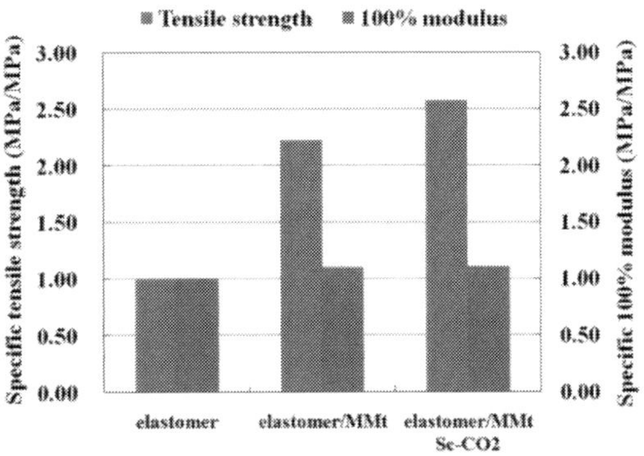

(a) Tensile strength and 100%modulus

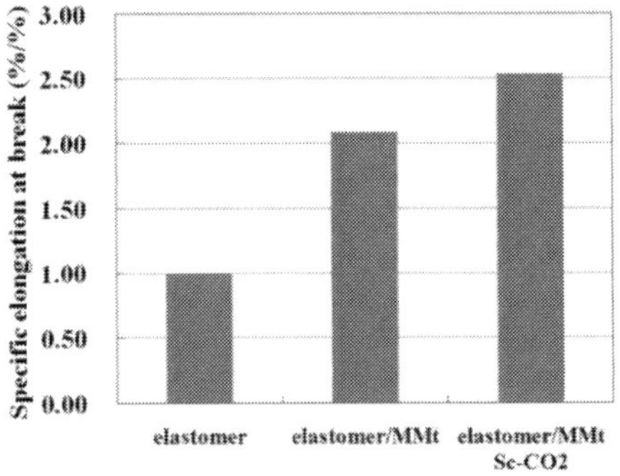

(b) Elongation at break

Figure.7 Results of tensile test

Conclusions

In the present work, to optimize the seal ring composition when the clay is compounded to elastomer with the aid of Sc-CO_2 by using the twin screw extruder, pressure distribution in the seal ring portion was analyzed. And, to verify the analytical results, scale-down of the analytical results and pressure measurement were performed. Melt-blending was performed with the aid of Sc-CO_2 by using the seal composition decided by the analysis and the experiment.

Results are as follows:

1. The pressure change of the simulation and that of the experiment showed a similar tendency.
2. The operating condition and the seal composition that was able to generate critical pressure of carbon dioxide were specified.
3. When the optimal seal composition was used, the supercritical carbon dioxide was effective to improve the mechanical characteristic, and the temperature of the material and the load electric power were able to be lowered.

References

1.Yang Zhao, Han-Xiong Huang, "Dynamic rheology and microstructure of polypropylene/clay nanocomposites prepared under Sc-CO2 by melt compounding", Polymer Testing 27, 129-134(2008)

2. Quang T. Nguyen, Donald G. Baird, "An improved technique for exfoliating and dispersing nanoclay particles into polymer matrices using supercritical carbon dioxide", Polymer 48, 6923-6933(2007)

In-situ characterisation of elementary dispersion mechanisms of silica granules in a polymer matrix under the action of shear

*Innocent Boudimbou, Edith Peuvrel-Disdier**

MINES ParisTech - Sophia-Antipolis - France

The paper reports on the characterisation of elementary mechanisms of dispersion observed on silica granules embedded in a styrene-butadiene rubber (SBR) matrix under the action of shear. Dispersion experiments were performed using a rheo-optical device. The behaviour of two grades of precipitated amorphous silica (different by their specific area) in a common polymer matrix was investigated. Dispersion mechanisms were studied both in diluted and model concentrated systems. Two dispersion mechanisms were observed in the diluted regime: the successive disintegration (breakage where the detached fragments are immediately dispersed down to a few micron size) or the rupture (sudden breakage resulting into a few large fragments) of granules. Both dispersion mechanisms were found to occur above a critical hydrodynamic shear stress, which increases as the granule size decreases. The dispersion mechanism and the level of hydrodynamic stress for dispersion were found to depend on the intrinsic parameters of the silica granule. The concentrated model medium was constituted by a suspension of glass beads in the SBR matrix in which the behaviour of silica granules was observed. Similar dispersion mechanisms were observed in the diluted and concentrated regimes. Erosion by friction was observed in the concentrated regime and not evidenced in the diluted one. The size of the glass beads was found to affect the dispersion kinetics. This is explained by the stress generated by the passing by of two particles which depends on the particle size.

Keywords: silica, granule, shear, dispersion, desintegration, rupture.

RESIDENCE TIME DISTRIBUTION: A POWERFUL CONCEPT FOR BUILDING UP EMULSIFICATION CURVES FOR POLYMER BLENDS

Guo-Hua HU[1*, 2], Cai-Liang ZhANG[3], Lian-Fang FENG[3*], Sandrine HOPPE[1]

1 *Laboratory of Reactions and Process Engineering, CNRS, Nancy Université-INPL-ENSIC, 1 rue Grandville BP 20451, 54001 Nancy, France – hu@ensic.inpl-nancy.fr;sandrine.hoppe@ensic.inpl-nancy.fr*

2 *Institut Universitaire de France, Maison des Universités, 103 Boulevard Saint-Michel, 75005 Paris, France*

3 *State Key Laboratory of Chemical Engineering, Department of chemical and biological engineering, Zhejiang University, 310027 Hangzhou, China – zhangclzju@hotmail.com; fenglf@zju.edu.cn*

Abstract: An original method for constructing the emulsification curve of a polymer blend in a continuous mixer such as a twin-screw extruder was developed based on the tracer-emulsifier concentration distribution (i.e. the residence time distribution) and the corresponding dispersed phase domain size distribution of the blend system. The resulting emulsification curves exhibited two unique features not yet reported in the literature.

Introduction

The efficiency of a copolymer as a compatibilizer is often evaluated by a so-called emulsification curve. The latter essentially follows the evolution of the dispersed phase domain size of the polymer blend as a function of the copolymer concentration. Studies reported in the literature concerning emulsification curves often use batch mixers and not screw extruders. The main reason is the amount of the copolymer required for building up an emulsification curve in a batch mixer is much smaller than in a twin-screw extruder.

This work aims at extending the residence time distribution (RTD), an old concept in chemical engineering, to polymer blending in order that a very small amount of copolymer still allows building up an emulsification curve in a pilot or industrial-scale screw extruder. The idea is based on RTD transient experiments (Zhang et al. 2009). Unlike a steady-state experiment in which given compositions of the polymer components of the blend and the copolymer are charged to the hopper of the extruder altogether, in a transient experiment the polymer components are first charged to the hopper of the extruder. When the process reaches its steady state, a given amount of the copolymer is introduced to the hopper of the extruder as a pulse. Samples are taken at the die exit as a function of time. Both the evolution of the copolymer concentration and that of the morphology as a function of time can be obtained. The former provides the RTD and the dispersed phase domain size distribution (DSD). From both distributions, the emulsification curve can be easily deduced.

Experimental

The emulsifier, a graft copolymer with PS as the backbone and PA6 as the grafts, was synthesized by our lab. The graft copolymer also contained a very small amount of anthracene moieties along the PS backbone. The presence of the anthracene moieties in the graft copolymer allowed in-line measurement of the copolymer concentration during the blending process. For that reason, it will be designated as tracer-emulsifier. Transient experiments were carried out on a standard twin-screw extruder of which the diameter and length-to-diameter ratio were 35 mm and 48, respectively. An in-line RTD measuring device with a bifurcated optical probe capable of exciting and emitting fluorescent light was installed in the die. The barrel temperature was set at 230 °C. When the process reached its steady state, the tracer-emulsifier was injected to the hopper as a pulse. At the same time, the in-line RTD measuring device started measuring the concentration of the tracer-emulsifier and the corresponding extrudate was collected and then quenched immediately in liquid nitrogen to freeze-in its morphology. A scanning electron microscope of type JEOL JSM-T330A was used to measure the morphology of the extrudate.

Results and Discussion

Figure 1 shows RTD and DSD measured for the PS/PA6 (80/20) and (90/10) blend systems. As expected, the two RTD curves for the PS/PA6 (80/20) and (90/10) blend systems almost superimpose. More interestingly, the DSD follows the overall trend of the RTD. A lower value of the DSD corresponds more or less to a higher valued of the RTD. These results show that the use of a very small amount of a tracer-emulsifier allows measuring the RTD and the DSD of a polymer blend at the same time.

Figure 2 shows the emulsification curves corresponding to the results in Figure 1. The emulsification curve from batch mixer (Zhang et al., 2007) is also shown for comparison in Figure

2.

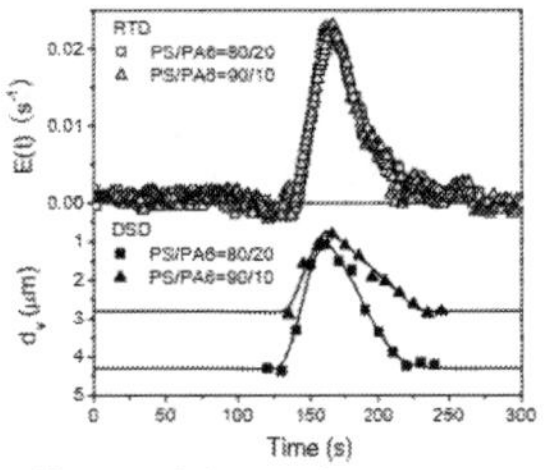

Figure 1 Effects of the PS/PA6 mass ratio on the RTD and DSD curves of the PS/PA6 blend.

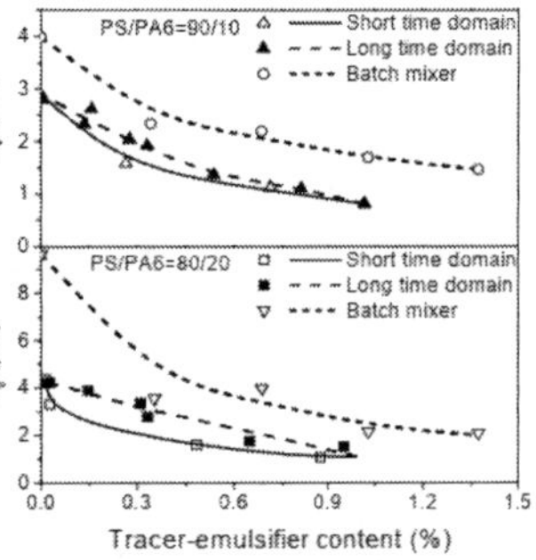

Figure 2 The effect of the PS/PA6 mass ratio and the type of mixer on the emulsification curve of the PS/PA6 blend. Data for twin-screw extruder deduced from Figure 1. Data for batch mixer is obtained in haake torque rheometry.

The striking feature of those emulsification curves from twin-screw extruder is different with that from batch mixer in that they are not single curves but loops, regardless of the PS/PA6 mass ratio used. These loops result from the fact that a RTD curve can be divided into short and long time domains which are demarcated by the maximum of the residence time distribution density function. In the short time domain, the tracer-emulsifier concentration increases with increasing residence time and in the long time domain it decreases with increasing residence time. It is intriguing to note that the curves corresponding to the short time domain are below those corresponding to the long time domain. However, mechanisms responsible for this phenomenon remain poorly understood. Nevertheless, coalescence of the dispersed phase domains of the polymer blend could be one. For a given tracer-emulsifier concentration, coalescence is supposed to be more important at longer time or higher PS/PA6 ratio. Another reason may be related to the fluorescent detection. Figure 3 shows the relationship between the detected fluorescent signal and the depth of polymer. It can be seen that the maximum detected depth by fluorescent instrument is about 1.2mm, which is much lower than the width of extrusion die (5mm). In the extrusion process, the detected concentration of tracer-emulsifier at the die corresponding to the short time domain is lower than the real concentration, while the one corresponding to the long time domain is higher than the real concentration, as shown in Figure 4. The second argument is more likely valid for the system under investigation.

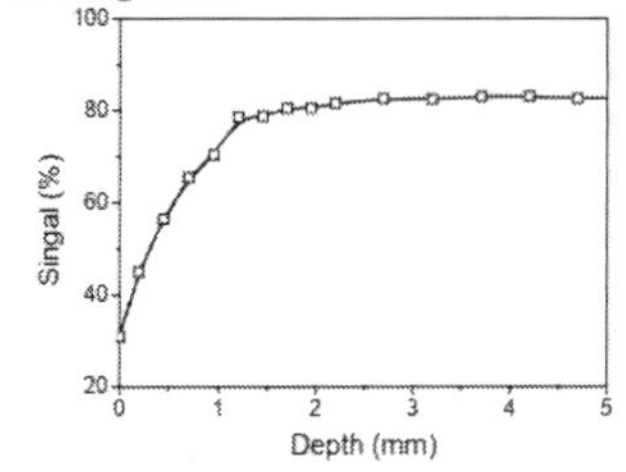

Figure 3 Relationship between detected signal and the depth of polymer

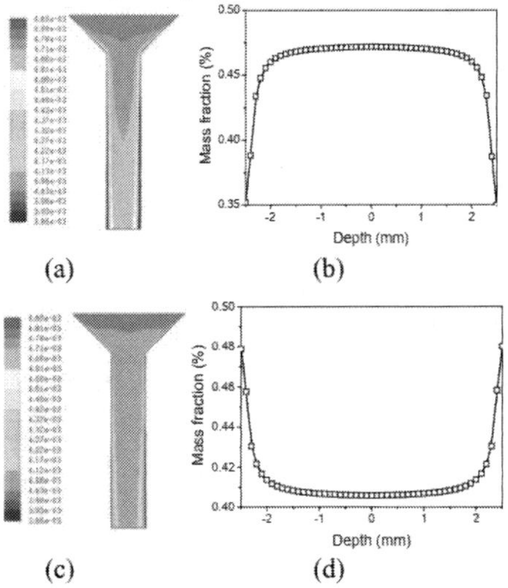

Figure 4 Concentration distribution of tracer-emulsifier in the die for the short time domains (a and b) and the long time domain (c and d)

Conclusions

In this study, a so-called concept of tracer-emulsifier has been developed to build up an emulsification curve in screw extruder. In this manner, for the first time the use of a very small amount of a tracer-emulsifier allows obtaining at the same time the tracer-emulsifier concentration distribution (i.e. the residence time distribution) and the corresponding dispersed phase domain size distribution. From these two distributions, an emulsification curve can be easily obtained.

References

Zhang, C. L., et al., "Residence time distribution: an old concept in chemical engineering and a new application in polymer processing", AIChE J., **55**, 279-283 (2009).

Zhang, C. L., et al., "Efficiency of graft copolymers as compatibilizers for immiscible polymer blends", Polymer, **48**, 5940~5949 (2007).

ANALYSIS AND CHARACTERIZATION OF THE DIRECT-SMC (D-SMC) PROCESS FOR COMPRESSION MOULDED THERMOSET COMPOSITE PARTS

T. Potyra[1]*, F. Henning[1], P. Elsner[1]

[1] *Fraunhofer Institute for Chemical Technology, Joseph-von-Fraunhofer-Str. 7, 76327 Pfinztal (Berghausen), Germany; tobias.poytra@ict.fraunhofer.de*

Sheet Moulding Compound (SMC) is currently being manufactured as a semi-finished product which is separate of the part production process. This separation leads to disadvantages in the manufacturing process, the corresponding logistics and also in the quality of the SMC parts.

These disadvantages were the motivation for developing a one-step direct process in which the production step involving the semi-finished material is eliminated. By this means costs can be reduced and an integrated process chain from the raw material to the part is possible. The properties of the raw material, the moulding compound and the part can be determined at any time. Furthermore changes in the formulation can be made online.

The content of the paper describes a process development for direct SMC technology. The process itself is characterized, with special regard to the maturation behavior, the resulting fibre length and the machine parameter settings. By this means the achievable limits of the process will be demonstrated. Additionally the different parameters will be related to the achievable mechanical properties. Finally the integrated process chain of the Direct SMC process will be evaluated as a whole with respect to the chances and opportunities it creates.

Introduction

Today, exterior panels for the automotive sector which meet the requirements for class A surfaces are often manufactured from sheet molding compound (SMC) material. The manufacturing process sequence starts by mixing resin, filler and additives into a paste in a discontinuous batch mixer. In the next process step, reinforcement fibers are introduced into the resin-filler paste, resulting in a semi-finished SMC sheet material. This sheet is stored for a certain time to allow the material to undergo a dynamic viscosity change called maturation before being formed into a component using compression molding with heated molds (Kia, 1993; Neitzel, 2004).

This process which is based on semi-finished materials results in deviations in semi-finished part quality and hence component quality. Therefore quality can only be determined after a certain period of time. Another disadvantage concerns the limited flexibility of the process. Formulation adaptations or changes will need several days (Kia, 1993).

The direct process allows certain restrictions on the processing of SMC to get rid of. Because D-SMC is a continuous process in which the raw materials are processed to a component within only a few minutes, it is possible to establish a control loop in order to guarantee a high and constant level of quality. The resulting reduced scrap has a positive effect on component costs. Component costs can also be reduced by the elimination of the maturation step. Another advantage of the process is that it offers high flexibility in terms of raw material selection and close-to-real-time formulation changes. Styrene emissions can be reduced for using encapsulated manufacturing equipment (Bräuning, 2008).

Processing technology

The main idea of direct technology for thermoset compression moulded composite parts is to establish a continuous integrated process chain from raw material to the ready moulded parts. From the long fibre reinforced thermoplastics (LFT) the shift from a semi finished product technology to a direct processing technology is well known. Parts made of LFT used to be compression moulded based on long or continuous fibre reinforced semi finished products. In order to get rid off the semi finished product step a direct in-line compounding technology was successfully established. As SMC can be considered as the thermoset type of LFT, the idea was to follow a similar process (Tröster, 2000). For direct processing of thermoset compression moulded composite parts a two step process has been developed as it is schematically shown in Figure 1 (Bräuning, 2008).

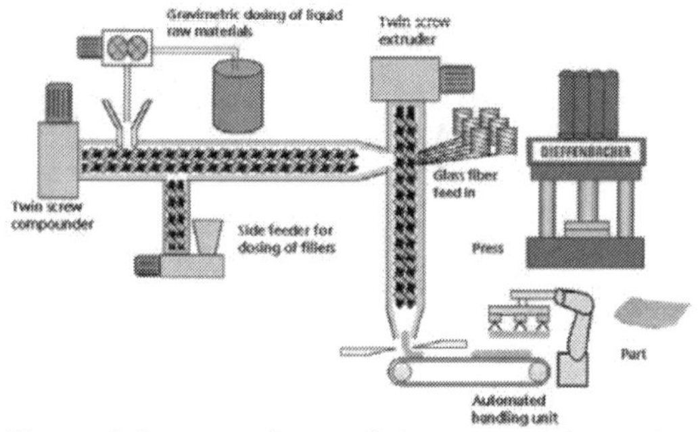

Figure 1 Process scheme of the new and innovative direct processing technology for thermoset compression moulded composite parts

As SMC consists of up to a dozen raw materials (cf Table 1) a major point in the process is dosing and compounding them accurately. The shown processing scheme represents the pilot plant state, where the mixing of the liquid raw materials is done manually. This mixture of liquid raw materials is then being dosed, gravimetrically controlled, in the compounding extruder. This mixture consists of unsaturated polyester resin, low profile additives, processing and mould release additives, peroxides and inhibitor. The filler, which mostly is $CaCO_3$, is then gravimetrically dosed in a side feeder and fed into the twin screw extruder. A Dieffenbacher ZSE40 has been used, In the following a homogeneous resin filler additive compound is made. The achieved compound quality is higher than in a conventional dissolver process. In Figure 2 the mixing quality of dissolver unit is compared with the mixing quality of the twin screw extruder process. In the direct process, the resin filler mixture is prepared using a continuously mixing extruder. The liquid and solid raw materials are forced through this machine, thereby generating an improved mixing effect. The dispersing effect of the extrusion step can be increased by selecting various screw elements such as kneaders and spiked mixers. The amount of air inclusions are minimized because this is a closed mixing system. In addition, it is possible to establish a vacuum zone. It is possible to see that the fillers are finely dispersed and very homogeneously distributed in the twin screw extrusion step. Furthermore, the number of air inclusions is very small (Bräuning, 2008).

Table 1 Typical SMC formulation for Class A parts

Raw material	**Content [phr]**
UP resin	58 phr
Low Profile Additive	42 phr
Dispersing Additive	1,5 phr
Air Release Agent	0,7 phr
Demoulding Agent	4,5 phr
Peroxide	1,0 phr
Inhibitor A	0,4 phr
Inhibitor B	0,1 phr
Raw material	**Content [phr]**
$CaCO_3$	200 phr
Thickening Agent (MgO)	2,5 phr
Glass fiber	28 %wt

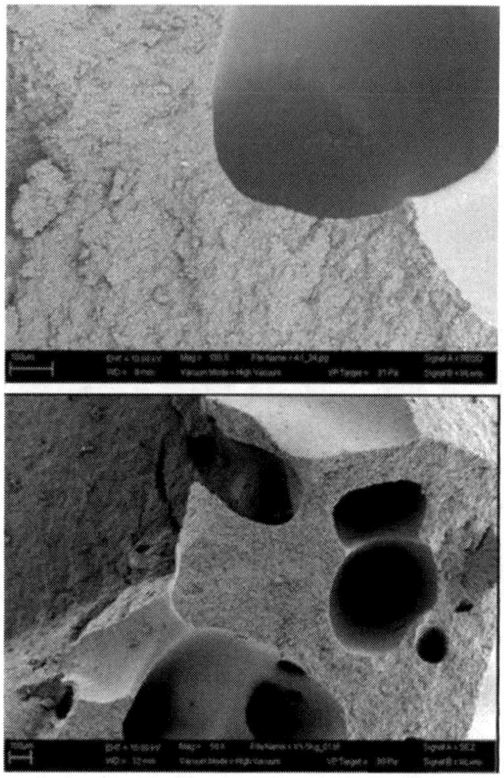

Figure 2 SEM images of a resin filler mixture processed using a blade dissolver (lower) or a twin screw extruder (upper); less air entrapments in the twin screw extruder process;

The resin additive filler mixture is subsequently being transferred into a second twins screw extruder, where the reinforcing fibers are added. A Dieffenbacher ZSG40 has been used. In Figure 3 it is shown how glass fiber roving are fed in directly. The roving is first passed through a system of rollers. The rovings are then fed directly into the extruder and are broken at a defined position using an edge in the cylinder of the extruder. This method of working in is very good for glass fibers because they break at the breaking edge. An advantage of this system is that the fibers are force-wetted. The forced transport of the fibers also means that very high glass concentrations can be achieved. However, not all fibers break under all conditions, so that it is not always possible to achieve a defined maximum fiber length in the material (Bräuning, 2008).

Figure 3 Direct working in of glass fibers in to a twin screw extruder

As shown in Figure 4 a compact extrudate is leaving the die. For adjusting the part thickness it is cut into certain length with a certain weight. This extrudate is being placed on a conveyor belt. A robot picks up the extrudate and places it in the mould where the part being made. After compression molding the part is de molded. For compression molding a Dieffenbacher press DYL630 has been used with testing plaque mould (460 x 460mm). In this new process lies only a period of a couple of minutes between raw material and ready molded part (Bräuning, 2008).

Figure 4 Cutting the extrudate to length and compression molding of sample plaques in the direct process

Experimental Work

For characterizing this new process a Design of Experiment study has been carried out. This new process has bears lot of parameters that all affect the mechanical properties of the final compression molded part. The parameters can be distinguished between process related like throughput, speed of the screws, temperature and material related like viscosity, fiber content, filler content etc. As these parameters interact with each other the major target of this study was to identify the important parameters of process and material. Therefore a two level screening in D-optimal design was chosen (Eriksson, 2000). All parameters as well as the formulation (cf. Table 1) have been set to a defined valued for constant boundary conditions. The variable parameters are listed in Table 2.

Table 2 Variable Parameters in the DoE study for D-SMC process evaluation

Parameter	Unit	Lower limit	Upper limit
Filler content	parts [phr]	170	200
Fiber content	[%wt]	22	38
Number of rovings	number	12	23
Amount of maturation agent	parts [phr]	2	10
Maturation time	hours [h]	0,1	2

The filler content was varied in the range of 170 phr to 200 phr, which is a typical range for Class A SMC formulations. The glass fiber content of 22wt% to 38wt% covers the usual range of fiber reinforcement in the SMC applications (Kia, 1993). The number of rovings determines the rotations of the screws at a given fiber content and was chosen on experience. The amount of maturation agent is higher than in conventional SMC and was chosen on experience. As SMC needs a certain time of maturation this parameter was chosen to demonstrate its influence in the Direct SMC process (Potyra, 2007).
For evaluating these parameters mechanical properties have been chosen. From the resulting sample plaques specimen for tensile testing and unnotched Charpy flexural impact strength have been taken out by water jet cutting. The properties used as responses in the DoE model are tensile E-modulus, tensile strength and Charpy impact strength.
Based on these range of parameters and responses a DoE design was generated with Modde Umetrics software (Eriksson, 2000). 16 Trials have been carried out and evaluated with the Modde Umetrics DoE software.

Results and Discussion

The generated model showed R2 values well above 0.9 for tensile modulus, tensile strength and Charpy impact strength so that the fit of the model is very good. The

Q2 value as expression for the prediction of the variation is above 0.5 for the Charpy impact strength. For tensile strength and tensile modulus Q2 value is above 0.25 which is acceptable. The model validity is well above 0.5 and the reproducibility is above 0.9 for all mechanical properties. The model can be used for further evaluation in this study (Eriksson, 2000).

The influence of both material and process parameters on the mechanical properties is evaluated and shown in Figure 5. The parameters with a VIP value above 1.0 show have a major influence and parameters with a VIP value beneath 1.0 have minor importance. The material formulation parameters glass fiber content and chalk content have are the most important ones, respectively. The maturation time, the revolutions of the screw and the squared chalk content as an indicator for the compound viscosity are of minor importance. For fitting the model the squaring of the parameter chalk content is necessary. As the parameters interact with each other this fact is expressed by multiplication of some factors in the model.

It can be concluded that material parameters have a stronger effect on mechanical properties than process parameters. For further evaluation of the model it is distinguished between process parameters, which are maturation time, revolution of the screw and squared chalk content, and material parameters, which are glass fiber content and chalk content.

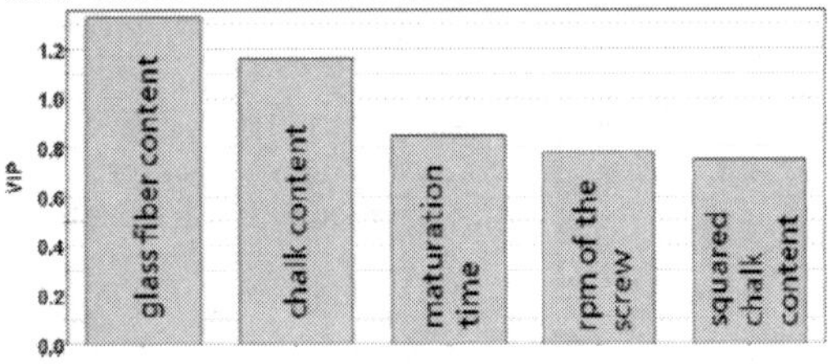

Figure 5 Influence of the parameters on the mechanical properties

The process parameters show a medium effect on the tensile modulus as shown in Figure 6. With increasing maturation time only a small increase of the stiffness can be observed. This behavior is as expected. There is a slight drop of stiffness with increasing speed of screw rotation. That could be explained by higher shear forces at higher rotation speed. This would lead to a higher degree of fiber length degradation. That could result in a decreased tensile modulus. On the other hand the tensile strength increases with higher speed of the screws (cf. Figure 7). This is a sign that fiber wetting is improved and the fiber length is still sufficient enough for not decreasing tensile strength. The speed of the screws does not affect the Charpy impact strength (cf. Figure 8). From this fact one can conclude that fibers a breaking which would lead to lower tensile strength and Charpy impact strength, as these properties are strongly dependent on fiber length (Potyra, 2007). A compensation of this effect could be achieved by a better fiber wetting quality at higher speeds of the screw.

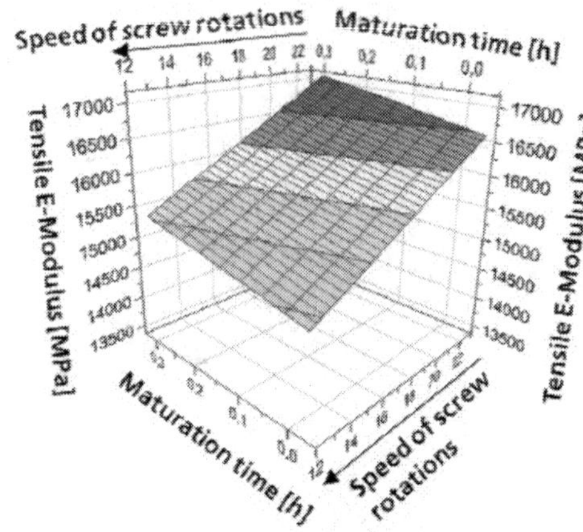

Figure 6 Effect of speed of screw rotation and maturation time on tensile E-modulus

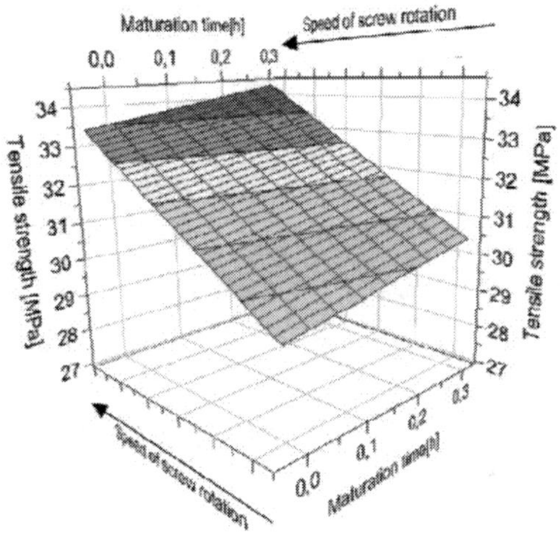

Figure 7 Effect of speed of screw rotation and maturation time on tensile strength

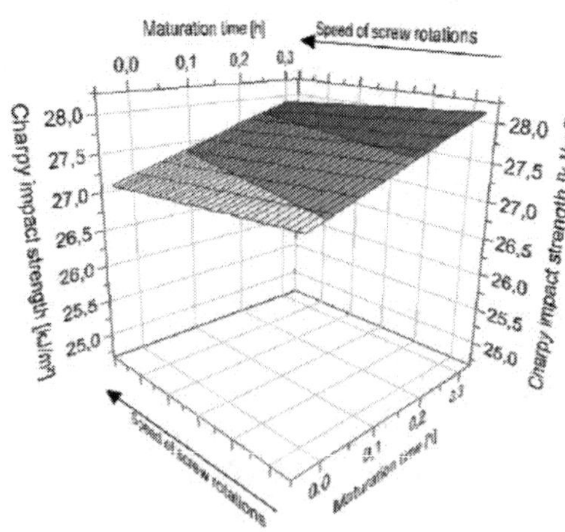

Figure 8 Effect of speed of screw rotation and maturation time on Charpy impact strength

The material parameters show a stronger effect on the material properties than the process parameters. In Figure 9 the effect of chalk content and glass fiber content on tensile E-modulus is illustrated. With increasing glass fiber content the stiffness increases linearly as expected. With increasing filler content the tensile E-modulus reaches a minimum and increases again. Two effects could explain this behavior: on the one hand the viscosity of the compound increases with higher filler content. That would lead to higher fiber length degradation and explain the decrease of the modulus as in Figure 6. On the other hand the increase of filler content leads to higher stiffness in general. These two effects could take place in this behavior: in the first phase the drop of stiffness dominates and in the second phase the increase of stiffness dominates.

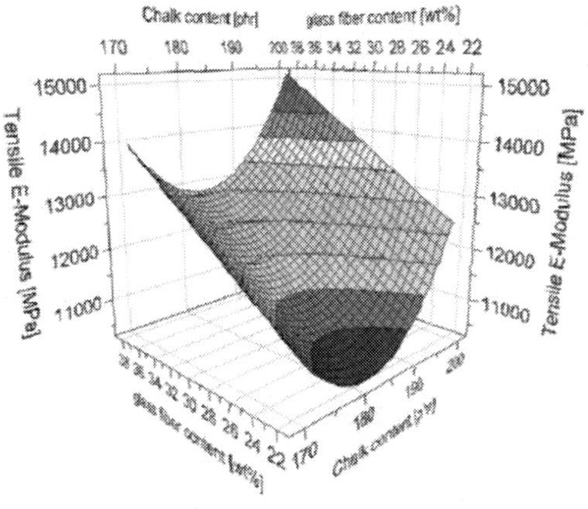

Figure 9 Effect of chalk content and glass fiber content on tensile E-modulus

Different behavior can be observed in Figure 10, where the effect of chalk content and glass fiber content on tensile strength is illustrated. With increasing glass fiber content the tensile strength decreases. This can explained by the long fibers contained in the compound. With fiber length around 20 mm the fiber are not only affected by the viscosity of the compound but also by each other. Certain entanglements take place. If the fiber content is increased the number of entanglements is increase, respectively. As an effect this leads higher fiber length degradation and hence to decreased tensile strength. The first increasing and secondly decreasing of the tensile strength with increase of filler content cannot be explained. It might be a weakness of the model.

Similar behavior of Charpy impact strength compared to tensile strength can be observed in Figure 11. The Charpy impact strength is in the same way affected by the increasing glass fiber content and can be explained in the same way. The increasing filler content leads to higher Charpy impact strength. In the model this coherence is based on squaring assumptions which cannot be explained. But even the absolute values of the Charpy impact strength increase with increasing filler content. A possible explanation is that the chalk content effects the viscosity which leads to different fiber degradation and fiber wetting. At this point further investigation is required in future studies.

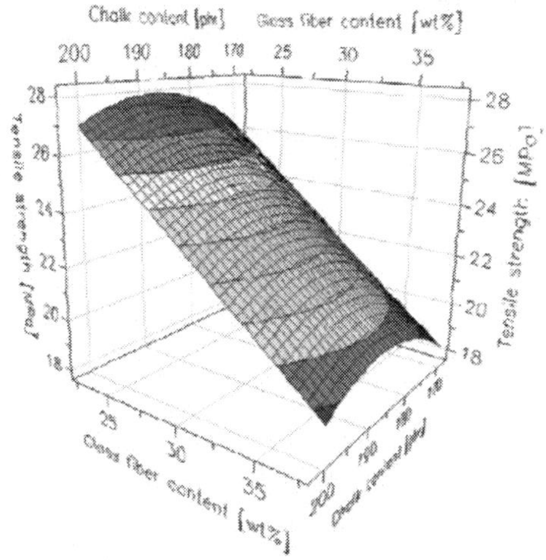

Figure 10 Effect of chalk content and glass fiber content on tensile strength

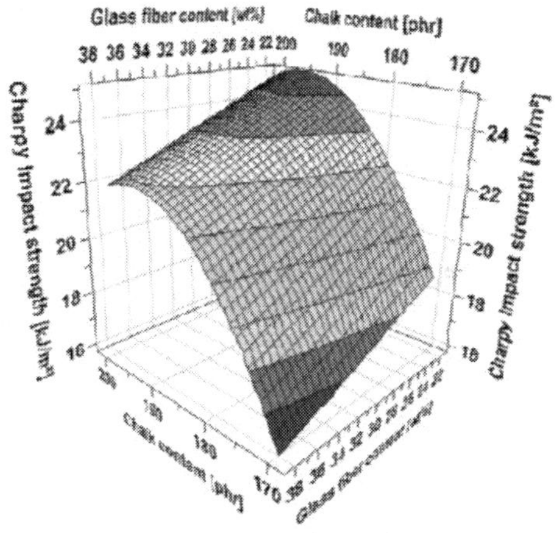

Figure 11 Effect of chalk content and glass fiber content on Charpy impact strength

Conclusions

A Design of Experiment study for screening the new Direct SMC process has been carried out for identifying the importance of the process and material parameters on the mechanical properties tensile E-modulus, tensile strength and Charpy impact strength. The model shows that the material parameters have major importance and the process parameters have minor importance, respectively. The glass fiber content has the strongest effect. While increasing the fiber content the tensile strength and Charpy impact strength decrease and the tensile E-modulus increases. Increasing the speed of the screws leads to decreased modulus, increased tensile strength and does not affect the Charpy impact strength. The maturation time shows only low effect on the mechanical properties in the investigated limits. The filler content shows controversy effects on the mechanical properties. For distinguishing whether it is a weakness of the model or real effects, further investigations need to be carried out.

Acknowledgements

The authors are particularly grateful to Projektträger Jülich (PTJ) as well as the German Federal Ministry for Education and Research who provided public funding (funding code: 03X3000) to enable this joint project to be carried out. In addition, the authors are grateful to the project partners BYK Chemie, Johns Manville, M-Base, Polytec Composites Germany, Reichhold, Respecta and Volkswagen.

References

Bräuning, R.: Material- und Verfahrensentwicklung zur Herstellung von glasfaserverstärkten duromeren Bauteilen im Direktverfahren (Direkt-SMC), Fraunhofer Verlag, Munich (2008)

Eriksson, L. et. al.: Design of Experiments - Principles and Application, Umetrics, Stockholm (2000)

Kia, H. G.: Sheet Moulding Compounds, 1st edition, Hanser Publishers, Munich (1993).

Neitzel, M.; Mitschang, P.: Handbuch Verbundwerkstoffe, Hanser Publishers, Munich (2004)

Potyra, T. et. al.: "Direct SMC: Influence of the Maturation and Fiber length on the Material Properties", 7th ACCE, (2007)

Tröster, S.; et. al.: "LFT-D – A New Process Technology for Cost Reduction of Large Scale Production of Fiber Reinforced Thermoplastic Components - Affordable Composites Manufacturing", 15th Annual Technical Conference of the American Society for Composites (ASC), 689-698 (2000).

INFLUENCE OF EXTRUSION CONDITIONS ON THE RHEOLOGICAL BEHAVIOR OF NUCLEAR BITUMINIZED WASTE PRODUCTS (BWP)

M. Mouazen[1, 2], A. Poulesquen[1*] and B. Vergnes[2]

[1] *CEA Marcoule, DEN, DTCD/SPDE/L2ED, BP17171, 30207 Bagnols sur Cèze (France)*
mouhamad.mouazen@cea.fr, arnaud.poulesquen@cea.fr
[2] *MINES ParisTech, CEMEF, UMR 7635, BP 207, 06904 Sophia Antipolis Cedex (France)*
bruno.vergnes@mines-paristech.fr

Abstract - In France, bitumen has been used since four decades as a container material for low/intermediate activity and long lifetime radionuclides. Industrial Bituminized Waste Products (BWP) are produced by continuous extrusion of 60 wt.% of bitumen and 40 wt.% of salts, in which a very small fraction is radioactive. These incorporated salts are, in majority, of two types: soluble or insoluble. The organic matrix subjected to irradiation produces gases, which can give rise to swelling of BWP in its container. Then, the extent of gas production is directly linked to the degree of dispersion of salts into the bitumen matrix. In order to quantify this dispersion, the effect of extrusion operating conditions, such as feed rate and screw speed, were studied using a laboratory co-rotating twin screw extruder. The rheological characterization of BWP shows the appearance of a pseudo-solid plateau at low frequency and a decrease of the linear viscoelastic domain when screw speed and feed rate increase, indicating an improvement of the dispersion state. Samples characterization with environmental scanning electron microscopy (ESEM) confirms in the case of insoluble salts a decrease of the number of aggregates, accompanied by an improvement of the dispersion at high screw speed but it is less remarkable with the feed rate.

Introduction

Bitumen matrix has been used in the world, especially in France, to embed the radioactive wastes from low to medium activity and long life, coming from chemical treatment of liquid effluents.

The industrial process consists in mixing the bitumen with inorganic soluble or insoluble salts containing the radioactive elements (40 wt .%) at high temperature by a twin screw extrusion process. The final material can be considered as a non Brownian suspension dispersed in an emulsion. Indeed, bitumen is constituted of asphaltenes dispersed in maltenes, a complex mixture of molecules ranging from non polar saturated hydrocarbons to polar polynuclear aromatics. Its composition is often conveniently reported in weight percents of Saturates (S), Aromatics (A), Resins (R), and Asphaltenes (A), collectively called SARA fractions (Lesueur, 2009). The composition of the bitumen used in this study is given in Table 1

From a rheological point of view, we have shown that the bitumen behaves as a yield stress fluid with a yield stress value at 22°C between 0.1 and 1 Pa. The Newtonian viscosity varies with temperature following an Arrhenius law with activation energy around 153 kJ/mol (Mouazen et al., 2009).

Bitumen matrix is very sensitive to irradiation which leads to gas generation (mainly hydrogen) due to radiolytic process. Consequently, the dispersion of radioactive salts plays a major role in the gas production. Therefore, the extrusion conditions have a direct impact on the degree of gas production.

Maltenes (wt. %)	74.9
Saturates (wt. %)	29.9
Aromatics (wt. %)	31.9
Resins (wt. %)	13.1
Asphaltenes (wt. %)	23.6
Penetration grade (1/10 mm)	70/100
Density at 25°C (g/cm^3)	1.02
Softening	47

Table 1 - Some physico-chemical characteristics of the AZALT 70/100 bitumen

For security and safety reasons, inactive samples have been prepared on a laboratory co-rotating twin screw extruder to be the most representative of the industrial samples in term of final composition. The procedure of sample preparation was optimized to recover 40% in weight of salts in the final material.

Compared to the complexity of the industrial material (bitumen + salts like $BaSO_4$, $NaNO_3$, PPFNi, CoS…), the study will be limited to a single salt sample (bitumen + insoluble barium sulfate $BaSO_4$). The influence of the operating conditions (screw speed N and feed rate Q) on the dispersion state will be charaterized in this paper by rheological measurements.

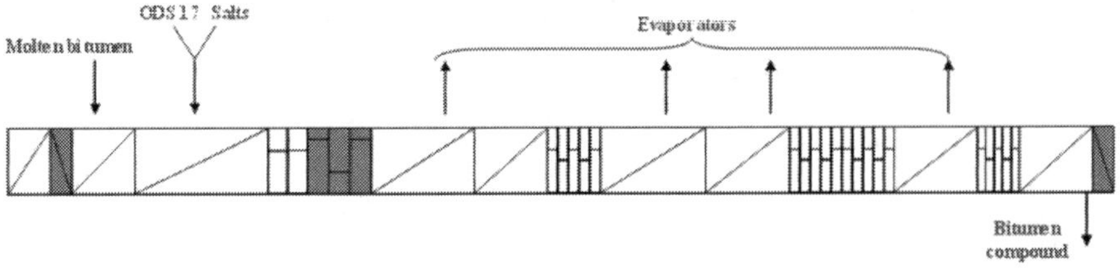

Figure 1 - Screw profile used in the experiments.

Experimental

Preparation of the bitumen compound

Pure bitumen is introduced at a temperature of 140°C in the feeding zone of the co-rotating twin screw extruder (Werner ZSK25WLE: L = 1000 mm, L/D = 40, with L the length and D the screw diameter). The $BaSO_4$ suspension ($BaSO_4$ + water) is inserted 5 cm downstream at room temperature. Toluene and surfactant at 1.2 wt.% are added at the same location to fluidize the mixture. Four evaporators take place along the screws to evaporate the water. The screw profile is represented in Figure 1.

On each sample collected at extruder exit, we proceed to a separation salts/bitumen (in toluene) in order to check the final salts mass. The quantity of residual water is also measured by Karl Fischer method.

Five compounds have been prepared using three different screw speeds N and three feed rates Q. The operating conditions and the experimental final mass of salts are given in Table 2.

Compound	N (rpm)	Q (g/h)	%wt salts
W1	70	606	38.9
W2	150	552	39
W3	300	603	39.7
W4	70	798	37.5
W5	70	1140	40.4

Table 2 - Operating conditions and sample compositions

Samples characterizations

The rheological properties are analyzed using a controlled stress rheometer (DSR 301, Anton Paar) equipped with 25 mm diameter parallel plate geometry, with 1.5 mm gap. The rheometer has been used in strain mode.

For dynamic tests, the strain sweeps are conducted with a frequency of 100 rad/s by varying strain from 10^{-6} to 100 to determine the linear viscoelastic region of the samples. The frequency sweeps are performed from 100 to 0.01 rad/s at a strain value within the linear viscoelastic region.

Environmental scanning electron microscopy ESEM (FEI Quanta 200 ESEM FEG) is used to observe the salts dispersion in the bitumen matrix, and the size evolution with extrusion conditions.

Results

Effect of screw speed

Figures 2 and 3 present the evolution of the dynamic storage modulus G' and its normalized value with strain amplitude measured at 50°C for the three compounds W1, W2 and W3.

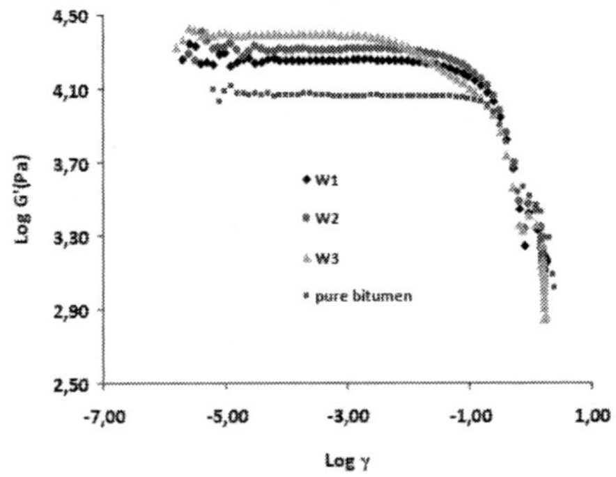

Figure 2 - Storage modulus G' vs. strain for bitumen compounds prepared with different screw speeds

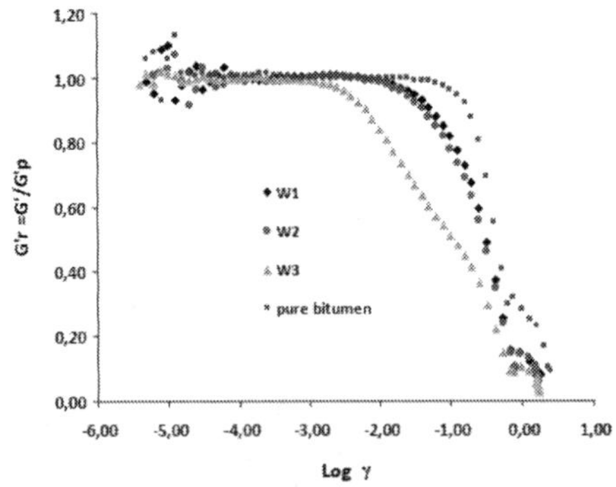

Figure 3 - Normalized storage modulus vs. strain for bitumen compounds prepared with different screw speeds

It is observed that the plateau modulus slightly increases and the critical strain decreases with the screw speed. These results show that the quality of the dispersion is improved when the screw speed increases.

To further investigate the dispersion state and to quantify the interactions between the salts, dynamic frequency sweep tests have been performed. Figure 4 presents the plot of G' versus frequency in the linear region ($\gamma = 0.001$) for the three compounds measured at 50°C. At low frequency, a plateau appears which is characteristic of a solid-like behavior (Durmus et al., 2007). This effect is well pronounced for the highest screw speed with a large increase of G'. This result indicates strong interactions between the $BaSO_4$ particles. At higher frequency, the bitumen matrix contribution dominates over the aggregates contribution (Vermant et al., 2007).

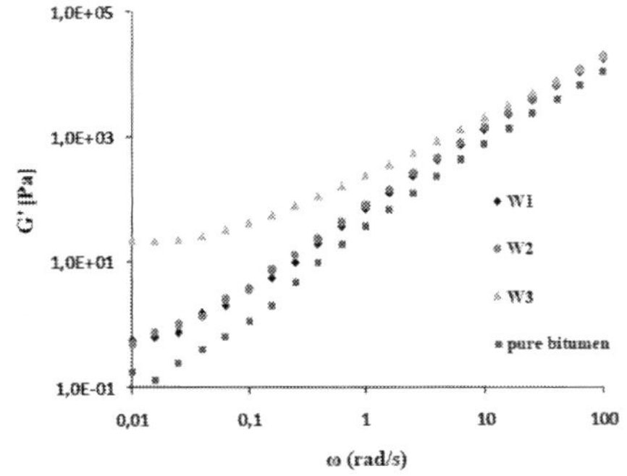

Figure 4 - Storage modulus G' vs. frequency for bitumen compounds prepared with different screw speeds.

As a consequence, the dynamic viscosity η^* shows an increase at low frequency (Figure 5). These results clearly indicate that the dispersion of salts is improved when the screw speed is increased.

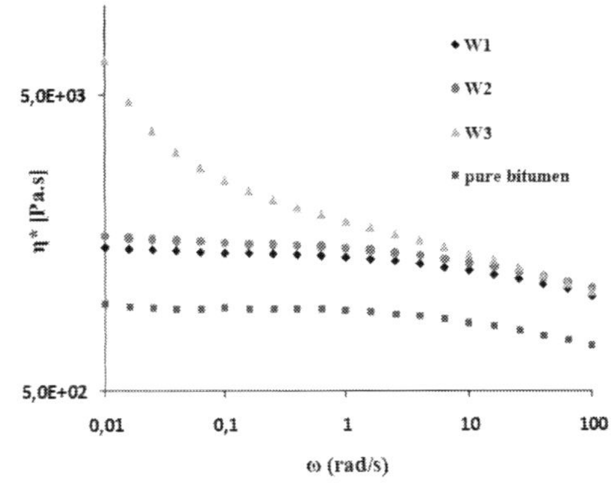

Figure 5 - Dynamic viscosity vs. frequency for bitumen compounds prepared with different screw speeds

Effect of feed rate

As for the last section, dynamic strain sweeps and frequency sweeps have been performed on the samples extruded at different feed rates. We observe that the storage modulus increases (Figure 5) and the linear viscoelasticity region decreases (Figure 6) with the feed rate. These results seem to show an improvement of salts dispersion, as for the influence of the screw speed.

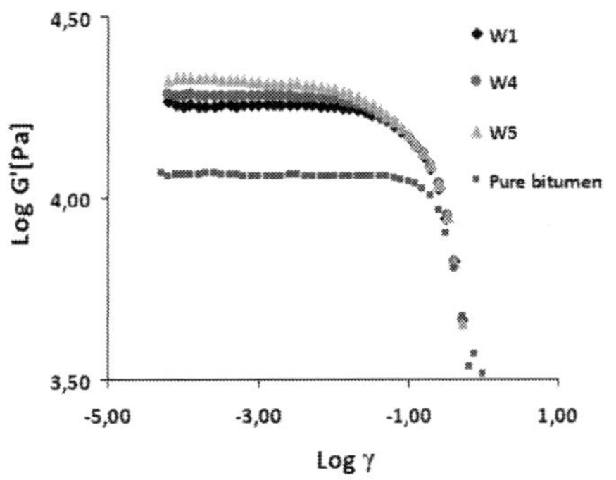

Figure 6 - Storage modulus G' vs. strain for compounds prepared with different feed rates.

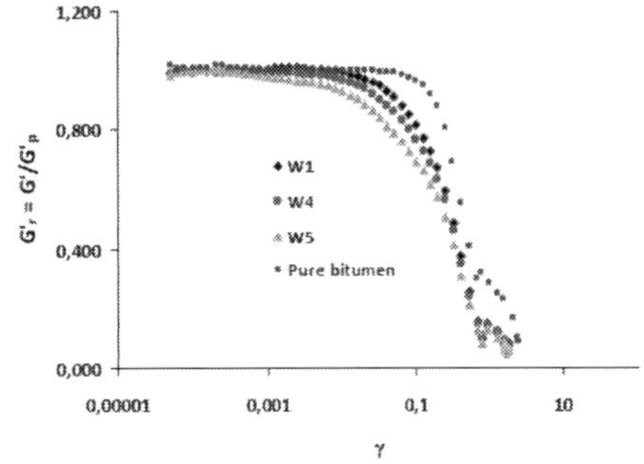

Figure 7 - Normalized storage modulus vs. strain for compounds prepared with different feed rates.

On the frequency sweep experiments, the same trends are observed: at high frequency, G' values are close since the bitumen matrix is dominating over the salts aggregates. At low frequency, a pseudo-solid plateau is observed for the highest feed rate. However, the level of this plateau is less pronounced than for the highest screw speed. For the intermediate feed rates, no difference is noticed.

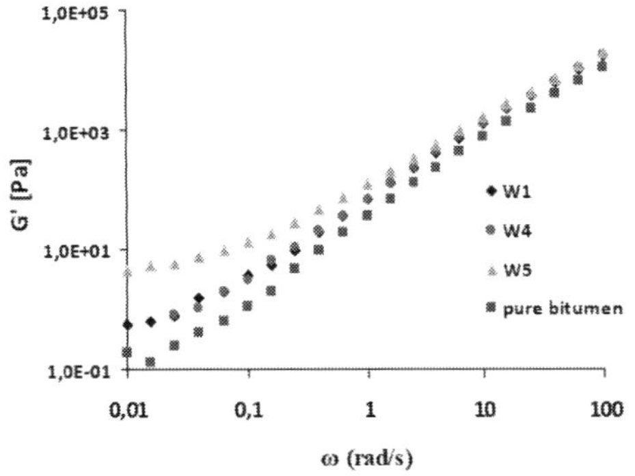

Figure 8 - Storage modulus G' vs. frequency for compounds prepared with different feed rates.

Discussion

As previously mentioned, the increase of screw speed or feed rate has the same effect on the rheological behavior. However, in the literature, the dispersion state is generally improved at low feed rate and high screw speed, conditions resulting in high shear rates and long residence times (Lertwimolnun and Vergnes, 2005). In our case, the increase of the screw speed improves the quality of dispersion, if we consider the linear decrease of critical strain with the screw speed (Figure 9).

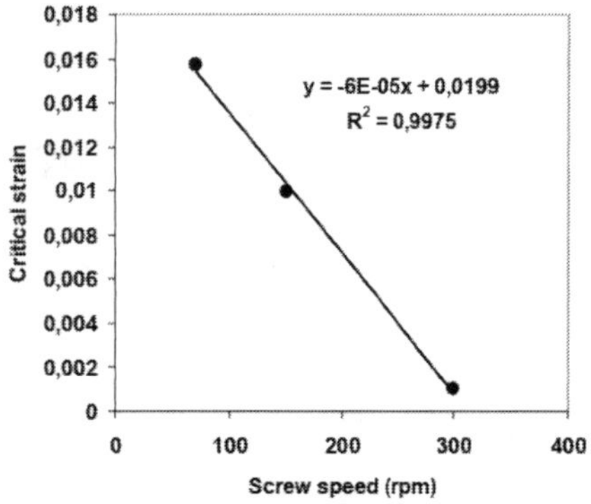

Figure 9 - Dependence of critical strain on the screw speed rate.

ESEM observations (Figure 10) confirm that the quality of dispersion is improved. Moreover, it was observed that the number of large aggregates decreased when the screw speed increased.

Surprisingly, the same results were obtained with the increase of the feed rate: the critical strain decreases linearly with the feed rate (Figure 11). Nevertheless, the extent of the linear viscoelasticity region is larger for the samples extruded at different feed rate than for the other samples. This indicates that the quality of dispersion is probably less pronounced when the feed rate increases. On the other hand, on ESEM observations for the dispersion state and the number of aggregates, the results are unclear and do not permit to conclude.

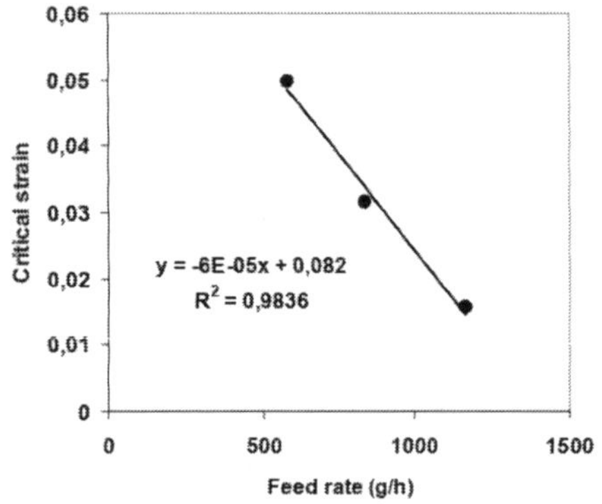

Figure 10 - Dependence of critical strain on the feed rate.

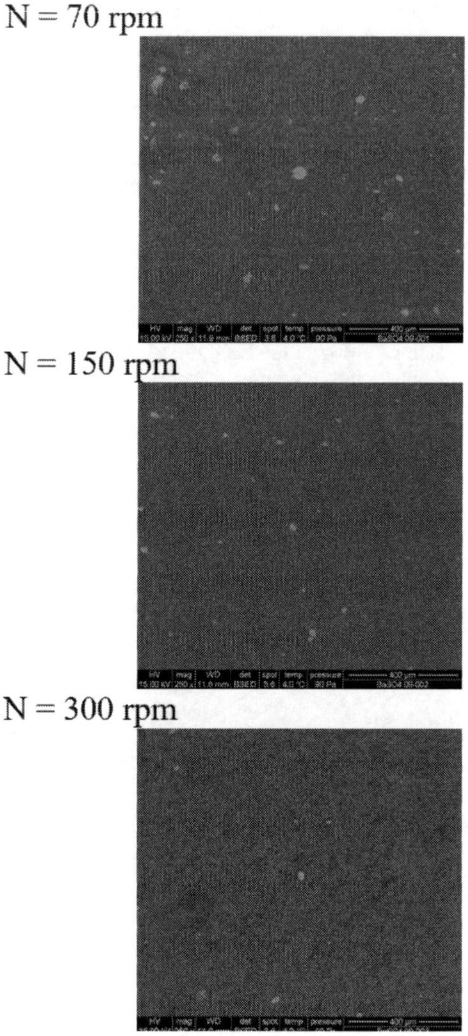

Figure 10 - ESEM (×250) images of bitumen compound prepared with different screw speeds

The screw profile may have an influence on the results. Contrarily to classical extrusion process, no die is used and the samples are collected through a hole into the lower part of the barrel, before the last left-handed element (Figure 1). Consequently, the increase of feed rate entails an increase of the filling ratio in this section, leading to a longer residence time which can explain a better dispersion of salts at high feed rate.

Unlike to polymer composites for which the dispersion quality of fillers has to be improved, it is preferable for BWP to have a worse dispersion of salts. Indeed, because the radioactive elements are mostly embedded in these salts, a better dispersion entails an increase of the radiolytic process which generates gas and could promote container alteration.

Conclusion

The influence of extrusion operating conditions on the dispersion state of salts in bitumen matrix has been described in this paper. The increase of the screw speed leads to an improvement of the dispersion and a decrease of the number of aggregates (ESEM observations). The influence of feed rate is similar, but less obvious to explain. From a rheological point of view, a better dispersion takes place at high feed rate but it is difficult to confirm this result by ESEM observations. The feed rate influence may be explained by the particular screw profile, leading to an increase of residence time in the last section of the screws, inducing therefore a better dispersion.

Upon a screw speed of 150 rpm and a feed rate of 833 g/h, the dynamic storage modulus *G'* increases and exhibits a pseudo solid-like behavior at low frequency.

In order to quantify and to approach the industrial process, we will address in a next future the effects of processing conditions on the dispersion states of other types of salts, such as $NaNO_3$ (soluble), and mixtures (1/3 $NaNO_3$ + 2/3 $BaSO_4$).

References

A. Durmus, A. Kasgoz and C. W. Macosko (2007). "Linear low density polyethylene (LLDPE)/clay nanocomposites. Part I: Structural characterization and quantifying clay dispersion by melt rheology." Polymer 48: 4492-4502.

W. Lertwimolnun and B.Vergnes (2005). "Influence of compatibilizer and processing conditions on the dispersion of nanoclay in a polypropylene matrix." Polymer 46: 3462-3471.

D. Lesueur (2009). "The colloidal structure of bitumen: Consequences on the rheology and on the mechanisms of bitumen modification." Adv. Colloid Interf. Sci. 145: 42-82.

M. Mouazen, A. Poulesquen and B. Vergnes (2009). "Caractérisation rhéologique de bitume 70/100 utilisé comme matrice de confinement de déchets radioactifs". Proceedings Congrès Annuel du Groupe Français de Rhéologie, Strasbourg (France).

J. Vermant, S. Ceccia, M. K. Dolgovskij, P. L. Maffettone and C. W. Macosko (2007). "Quantifying dispersion of layered nanocomposites via melt rheology " J. Rheol 51(3): 429-450.

MODELLING THE FIBRE-DEGRADATION DURING COMPOUNDING OF SHORT FIBRE REINFORCED COMPOSITES BASED ON GLASS FIBRES AND POLYPROPYLENE

P. Kloke[1]*, H. Potente[1], V. Schoeppner[1], J. Rudloff[2], M. Bastian[2], K. Kretschmer[2]

[1] *University of Paderborn, Kunststofftechnik Paderborn KTP, Paderborn, Germany – Philipp.Kloke@ktp.upb.de*
[2] *Sueddeutsches Kunststoff-Zentrum SKZ - KFE GmbH, Würzburg, Germany - J.Rudloff@skz.de*

Abstract - The glass fibre length resulting and the distribution of the glass fibre lengths are influenced by different process conditions during the compounding on a co-rotating twin-screw extruder. Among others these are the screw speed, the throughputs, the processing temperatures, the machine configuration and the properties of the glass fibre chosen.
In a joint research project funded by the DFG (Deutsche Forschungsgemeinschaft) between the SKZ (Sueddeutsches Kunststoff-Zentrum) and the KTP (Kunststofftechnik Paderborn) a physical-mathematical model was developed which describes the glass fibre length resulting during the compounding step. This model is dependent on the current residence times, the shear rates and the viscosities in the screw zones examined. The breaking of the fibre is determined by means of the critical compressive force and compared to the forces resulting from the flow process of non-newtonian materials. In this way a statement can be made about the average fibre lengths in the mixing zone. In order to verify this approach the model was implemented into the twin-screw simulation software SIGMA developed by the KTP. Because of the fact that SIGMA calculates the process behaviour like the residence time, the shear rates and the viscosity this model can be connected with these already existing results to estimate the average fibre length. In first experiments compounds with a constant glass fibre amount are used. Therefore an experimental design was generated that varies the different process conditions like the pressure at the screw tip, the screw speed, the throughput, the barrel temperatures and the screw configuration in the mixing zone for two polypropylenes. The final fibre lengths modelled from the simulations are compared to the average glass fibre lengths from these experimentally produced compounds mentioned. In future the stress appearing in the flow can be derived from numerical simulations and can be compared to those results from SIGMA.

Introduction

Glass fibre reinforced polymers have received considerable attention in industrial use. The attractiveness of those modified polymers is characterised by the improved physical and mechanical properties in comparison to unmodified polymers. The properties of the mentioned polymers are defined by the choice of the glass fibre content, the fibre diameter, the existing fibre length distribution, the orientation of the fibres within the spare part and the adhesion between polymer matrix and fibre. Especially in the field of designing components made of reinforced polymers, this procedure is assisted by software tools that simulate the mechanical behaviour of, in most cases, injection molded parts (Tucker et al., 2008). This forecast of failure implicates that the fibre distribution of the compounds used is known.
The fibre distribution within the compound depends on the way it was produced. In most cases the glass fibres are incorporated using a modular intermeshing co-rotating twin-screw extruder. Generally, the split feed technique is used, i.e. in the first step the polymer is plasticized and in the second step the fibres are inserted into the machine. In this continuous compounding step the resulting average fibre length mainly depends on the process conditions, the choice of the matrix material and bonding agents (Wall, 1989; Shon, White, 1999).
Hence, in a joint research project funded by the DFG (Deutsche Forschungsgemeinschaft) between the SKZ (Sueddeutsche Kunststoffzentrum) and the KTP (Kunststofftechnik Paderborn) the degradation of glass fibres during compounding into a polypropylen matrix has been analysed. The first step of these investigations has been preliminary tests to indicate the main effects concerning the degradation. Afterwards a model to describe the fibre damage was developed. The results of this model have been compared to measurements of the average fibre lengths of produced compounds.

Preliminary Tests

In the first step preliminary tests have been conducted on a ZSK 26 compounder from Coperion to show the effects of the process conditions concerning the degradation of the glass fibres during the incorporation. Within these tests a common screw configuration was used (see Figure 1).

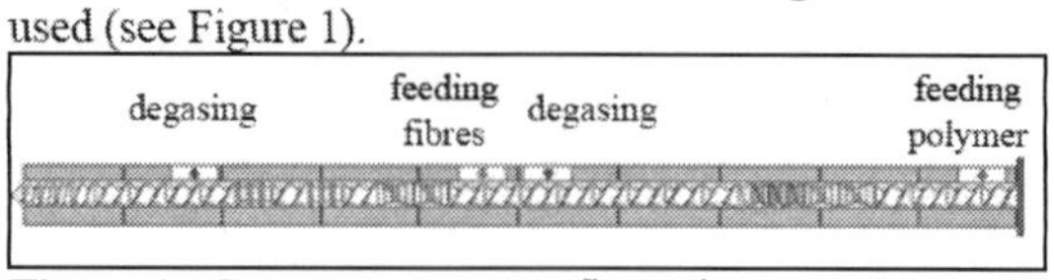

Figure 1 - Common screw configuration

This screw consisted of a kneading block system for the first dispersion of the fibres after inserting them into the system. The mixing section of the screw was a mix of conveying and mixing elements. The Polypropylene HP501N from Basell and short glass fibres with an average fibre length of about 4000 µm from Johns Manville has been used. Screw speed and throughput with a constant mass content of glass fibres have been varied and in these cases Dead-Stopp-Experiments, at which the machine was stopped abruptly and the screws were taken out, have been realised. Along the mixing section, samples could be prepared and analysed with a view to the average glass fibre length at these positions.

These preliminary investigations showed that the fibres are strongly reduced immediately after the incorporation and that the fibre length converts to a final value. Furthermore, the average fibre length increases with a rising mass throughput but they decrease with higher screw speeds. A rising throughput results in shorter residence times. These relations have also been observed in (Shon, White, 1999, 2005). Concluding from this, the damage of the fibres depends on the shear stress that is acting on the fibres and the residence time of the fibres in the system.

Calculation Model

Modelling the damage of the fibres requires a closer look at the melt flow inside a twin-screw extruder. With regard to (Ansahl, 1993; Melisch, 1998; Kretschmer, 2004; White, Potente, 2003), the melt flow in starved channels and in the gap regions can be regarded as a simple shear flow. In fully-filled screw sections there is a superposition of the shear flow and a pressure flow. In the gap regions there will always be a shear flow. This leads to the assumption that the fibres will orientate in flow direction but the fibres will not be damaged by the range of the shear stress in the channel region. The shear stress in the intermeshing region and in the gap region is much higher than in the channel section, so that the damage of the fibre will be expected in those sections.

Depending on the residence time (see Figure 2), the average glass fibre length shows an exponential decrease during the compounding step with regards to the investigations mentioned above.

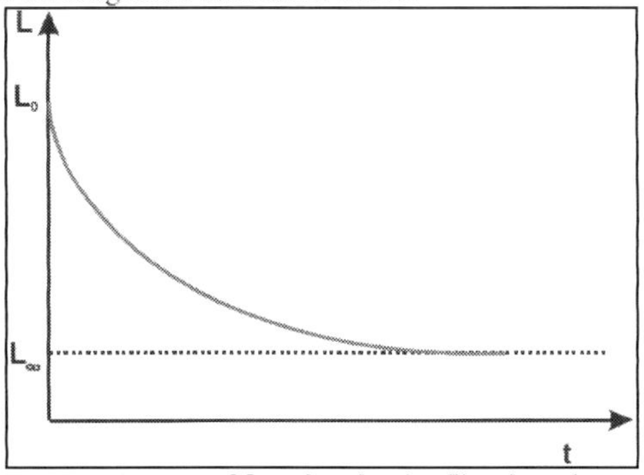

Figure 2 - Course of function for the fibre length

This effect can be described by the following differential equation:

$$\frac{dl}{dt} = c \cdot l^{\alpha} \text{ ; with } l = \frac{L - L_\infty}{L_0 - L_\infty} \qquad \text{Eqn. 1}$$

The origin of this function represents the initial average fibre length L_0 and l is a dimensionless fibre length that represents the ratio of the current and initial fibre length. The curse of function converges asymptotically to a process-limited fibre length l_∞. The additional parameter c indicates how fast and α how easily the fibre will break.

The first approach to describe the process-limited fibre length is that the fibre is going to break in the centre and the fibre-fibre interaction will be neglected. A single fibre in the shear flow is forced by the shear stress that acts at its generated surface. The resulting force can be written as:

$$\tau = \frac{F_{flow}}{A_{gen}} \Rightarrow F_{flow} = \tau \cdot A_{gen} \qquad \text{Eqn. 2}$$

In case that the fibre is stopped by a barrier, like somewhere in the gap or in the intermeshing region, the ratio of the critical compressive force, coming from Euler buckling, to the resulting force from the flow decides if the fibre is going to break or not. The critical compressive force is defined as follows

$$F_{Euler} = \frac{\pi^2 \cdot E \cdot I}{L^2} \qquad \text{Eqn. 3}$$

If the force from the flow dominates the critical compressive force, the fibre will break in the centre. In the next step half of the fibre length is regarded. If the ratio mentioned is still bigger than one, the fibre will again break in the centre. This procedure will go on until the ratio is smaller than one, so that the critical compressive force dominates the force from the flow and the process-limited fibre length is found.

The parameter α can be described by the ratio of the load of tension to the tensile strength of the fibre. The load of tension is proportional to the fibre length l squared:

$$\sigma = \frac{M}{I} \cdot \frac{D}{2} \sim L^2 \qquad \text{Eqn. 4}$$

The tensile strength is a material parameter and independent of the fibre length itself:

$$R_m \sim L^0 \qquad \text{Eqn. 5}$$

For the parameter α follows:

$$\text{sensitivity} = \frac{\text{load of tension}}{\text{tensile strength}} \sim L^2 \text{ ; } \alpha = 2 \qquad \text{Eqn. 6}$$

The solution for the differential equation is:

$$l(t) = -\frac{1}{c \cdot t + k} \qquad \text{Eqn. 7}$$

At the time step t=0, the constant k will be:

$$l(t=0) = -\frac{1}{c \cdot 0 + k} = 1 \text{ ; } \Rightarrow k = -1 \qquad \text{Eqn. 8}$$

The parameter c can be calculated with the help of the time step t_0. At this time the fibre is cut into two halves:

$$l(t=t_0) = -\frac{1}{c \cdot t_0 - 1} = \frac{1}{2}; \Rightarrow c = -\frac{1}{t_0} \qquad \text{Eqn. 9}$$

For the estimation of this time step, the dissipated energy and the energy for breaking a fibre are taken into account. The dissipated energy is the product of the shear stress, the surrounding area, the velocity of the melt and the residence time.

$$E_{diss} = \tau \cdot A \cdot v \cdot t_{RT} \qquad \text{Eqn. 10}$$

The consideration of the energy for breaking can be expressed by:

$$E_{break} = n \cdot \int F dS \qquad \text{Eqn. 11}$$

The force F per length S related to the number of n fibres in the system represents the energy for breaking.

Expressing the force by the tensile stress at break and the surrounding area and solving the integral, the equation can be written as:

$$E_{break} = V \cdot \Phi \cdot \frac{E \cdot \varepsilon_b}{2} \qquad \text{Eqn. 12}$$

The volume V of n fibres in the system can be described by the channel volume of the screw section related to the fibre volume content. The product of the Young's Modulus and the fibre elongation characterise the tensile stress at break.

If the dissipated energy prevails the energy of breaking, the fibre will break. Balancing these forms of energy at the point of break

$$\frac{E_{diss}}{E_{break}} = \frac{2 \cdot \eta \cdot \gamma^2}{\Phi \cdot E \cdot \varepsilon_b} \cdot t = 1 \qquad \text{Eqn. 13}$$

the time step t_0 becomes:

$$t_0 = \frac{\Phi \cdot E \cdot \varepsilon_b^2}{2 \cdot \eta \cdot \gamma} \qquad \text{Eqn. 14}$$

In this case the time t is equal to the time step t_0, after breaking of the fibre into two halves.

By linking Eqn. 14 to Eqn. 7:

$$l(t_{RT}) = \frac{\Phi \cdot E \cdot \varepsilon_b^2}{t_{RT} \cdot 2 \cdot \eta \cdot \gamma + \Phi \cdot E \cdot \varepsilon_b} \qquad \text{Eqn. 15}$$

the dimensionless fibre length can be expressed by a function depending on the residence time, the viscosity, the shear rates and the material data of the fibres.

Experimental

Alongside the theoretical aspects to describe the degradation of the glass fibres during compounding, experimental investigations have been conducted. Therefore a d-optimal design of experiment was chosen to vary the screw geometry, the screw speed, the througputs, the pressure drop at the screw tip, the barrel temperatures and the material. The materials chosen have exclusively been polypropylenes. The PP HP500V and the HP501N from Basell with different ranges of MFR have been analysed. During these experiments the mass content of 25% glass fibres has been kept constant. The glass fibres used have been the same as those in the preliminary tests.

The focus of work was the investigation of a constant screw geometry in the mixing section. The experiments have been performed on a ZSK 26 Mcc Compounder from Coperion. On this machine three screw geometries consisting of conveying elements with two different pitches and one with screw mixing elements were examined. An incorporation zone consisting of kneading blocks has been analysed on a ZSE 27 MAXX compounder from Leistritz. Here the staggering angles (90° and 30°) of the kneading blocks have been varied. Figure 3 shows exemplarily the assembly of the mentioned screw configurations.

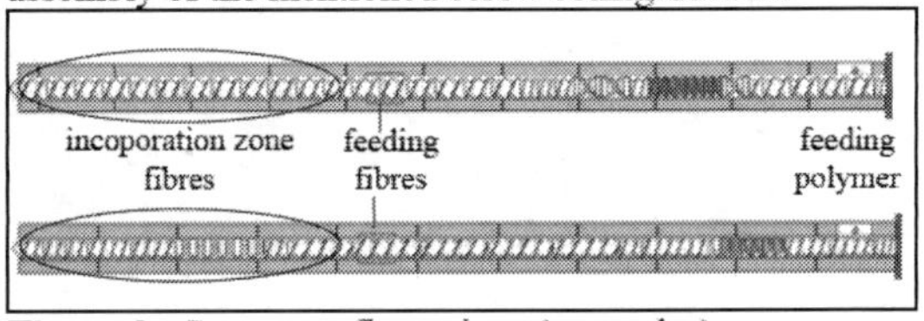

Figure 3 - Screw configurations (examples)

The identification of the average glass fibre length and the glass fibre length distribution in the resulting compounds requires the separation of the polymer from the filler. Ashing those samples seemed to be the best method to do so. After the ashing, the fibres have been separated on specimen slides, so that a scan system could digitalise these samples. A particle measurement software (Fibershape) was used to identify the distribution and to calculate the average glass fibre length.

Results and Discussion

For the comparison of the different settings, the average glass fibre length was taken into account.

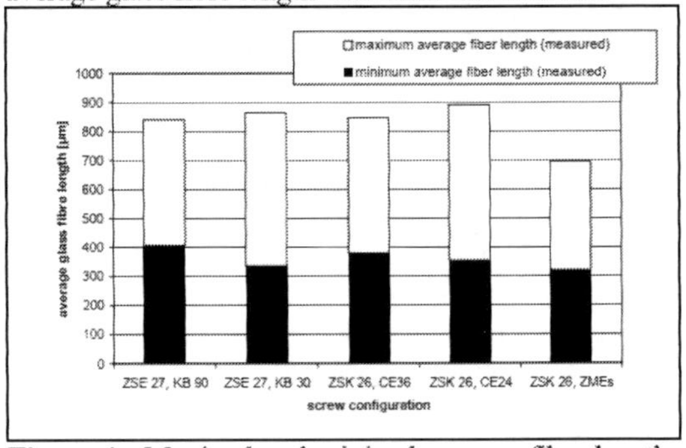

Figure 4 - Maximal and minimal average fibre lengths per screw configuration

Figure 4 shows the range of the average fibre lengths between 300 µm and 900 µm. In this figure the range of the results of the different screw geometries is presented and it shows the maximal and minimal fibre lengths resulting from each process condition per screw geometry. The screws consisting of the conveying elements (CE) and the kneading blocks (KB) are on the same level while the mixing elements (ZMEs) differ from this level. These results point out that the process conditions have a greater influence than the screw geometry. The results of the single screw speeds and throughputs present an obvious influence on the fibre length as it has been detected during the preliminary tests. A rising pressure at the screw tip displays an increase of the fibre length as well. The comparison of the matrix material reveals that the average fibre lengths increase with a decrease of the viscosity. The effect of the different barrel temperatures causes the same. Higher barrel temperatures end up in longer fibre lengths but this influence can be explained by the fact that the extruders used are very small. Thus the barrel temperatures have a great influence on the mass temperature in the system. Hence, this circumstance leads to a change of the viscosity and this influence has been detected by the use of materials with different viscosity levels.

Furthermore, the results of the experimental investigations concerning the glass fibre lengths within the compounds have been used to verify the results that have been calculated by the mentioned process model. In respect to this, the model was integrated into the existing simulation software SIGMA. SIGMA is a simulation tool which evaluates the process behavior of tightly intermeshing, co-rotating twin-screw extruders in advance. SIGMA is based on describing differential equations which are simplified by appropriate assumptions and boundary conditions (Potente, Thümen, 2005).

Besides other calculation methods, SIGMA offers the values for the residence time, the shear rates and the viscosities along the screw length that are needed for the model mentioned. These calculation methods are presented in (Ansahl, 1993; Melisch, 1998; Kretschmer, 2004).

A comparison between the experiments and the theory are shown in Figure 5 and Figure 6.

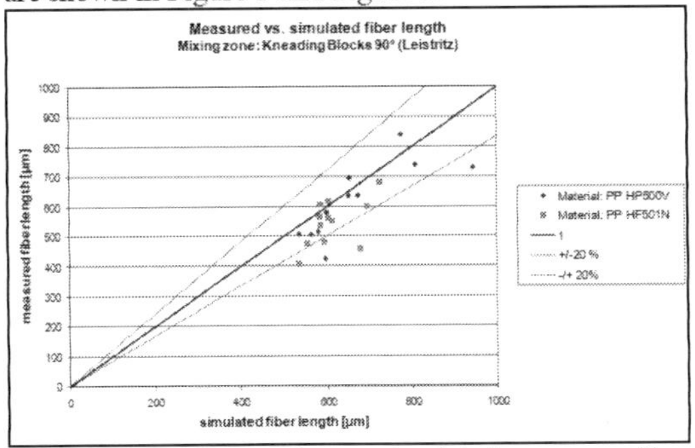

Figure 5 - Measured vs. simulated average fibre lengths, Kneading blocks (90° staggering angle)

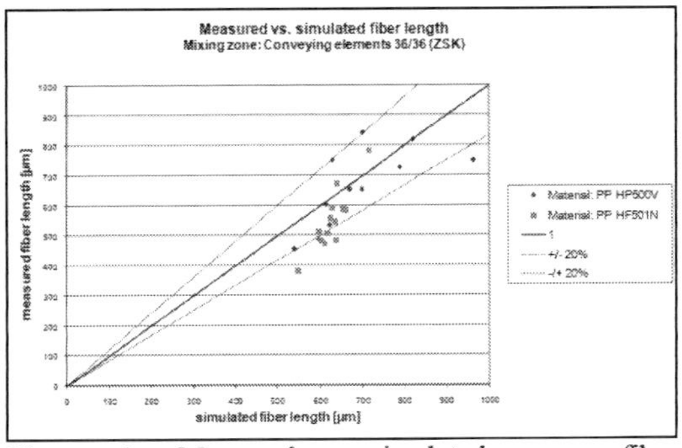

Figure 6 - Measured vs. simulated average fibre lengths, Conveying elements (36 length/36 pitch)

Conclusions

Notwithstanding the assumptions that have been made during this step of modelling, the calculated results are very close to the experimental measurements of the average fibre lengths. Summing up the aim of this work, the results of the experimental investigations can very well be predicted by the presented physical-mathematical model. The next step of modelling will be improving the simulaton of the distribution of the fibre length that results during compounding. Additionally, the influence of the mass content of the fibre, the position of the fibre feeding position along the screw, the initial fibre lengths and the use of bonding agents will be analysed. In future steps the flow inside the screw elements will be calculated in a numerical way to complete this work.

Acknowledgements

Our special thanks goes to the DFG (Deutsche Forschungsgemeinschaft) for funding this research project. Additional thanks goes to the Basell Polyolefine GmbH for supplying the polymer materials and to Johns Manville Inc. for supplying the glass fibre material. Without these supporters the results shown could not have been reached.

List of Symbols

A_{surr}	surrogated surface of the fibre
c	parameter indicating how fast the fibre breaks
D	fibre diameter
E	Young's Modulus
E_{break}	energy of breaking
E_{diss}	dissipated energy of the flow
F_{Euler}	critical compressive force (Euler buckling)
F_{Flow}	force from melt flow
I	moment of inertia (of the fibre)
k	constant
l	dimensionless fibre length
l_{∞}	process-limited fibre length
L	current fibre length
L_0	initial fibre length

M	bending moment
n	number of fibres
R_m	tensile strength of fibre
t	time
t_0	time at which the fibre breaks into two halves
t_{RT}	residence time
v	velocity of the melt
V	volume
α	parameter indicating how easily the fibre breaks
η	viscosity
τ	shear stress
ε_b	elongation at break
$\dot{\gamma}$	shear rate
Φ	fibre volume content

References

Ansahl, J.: PhD Thesis, University of Paderborn, Germany, 1998

Kretschmer, K.: PhD Thesis, University of Paderborn, KTP, 2004.

Melisch, U.: PhD Thesis, University of Paderborn, Germany, 1998

Phelps, J. H., Wang, J., Tucker III, C. L.:"New Models For Fiber Orientation in Injection-Molded Composites", PPS-24, Salerno, Italy, 2008

Potente, H., Thümen, A.,"SIGMA- A Simulation Tool For Tightly Intermeshing Co-Rotating Twin Screw Extruders" in WMSCI 2005, Orlando, Florida, USA.

Shon, K., White, J. L.,"A Comparative Study of Fiber Breakage in Compounding Glass Fiber-Reinforced Thermoplastics in a Buss Kneader, Modular Co-Rotating and Counter Rotating Twin Screw Extruder", Polymer Engineering and Science, Vol. **39**, No. 9, 1757–1768 (1999)

Shon, K., Liu, J., White, J. L.,"Experimental Studies and Modeling of Development of Dispersion and Fiber Damage in Continuous Compounding", Intern. Polymer Processing, Vol. **20**, No. 3, 322–331 (2005)

Wall, D.,"The Processing of Fiber Reinforced Thermoplastics Using Co-Rotating Twin Screw Extruders", Polymer Composites, Vol. **10**, No. 2, 98–102 (1989)

White, J. L., Potente, H.: Screw Extrusion, Hanser Publishers, Munich (2003)

MECHANICAL PROPERTIES OF POLY(CARBONATE) /VAPOR-GROWN CARBON FIBER COMPOSITE PREPARED BY MELT COMPOUNDING

J. Nithikarnjanatharn[1], H. Ueda[1], S. Tanoue[1]., Y. Iemoto[1]

[1]*Graduate School of Engineering, University of Fukui, Bunkyo, Fukui 910-8507, Japan*

Abstract: We discuss the effect of Polycarbonate (PC) type and Vapor-Grown Carbon Fiber (VGCF) size on the mechanical properties of PC/VGCF composites prepared by melt compounding using a twin-screw extruder. Four kinds of PC which have different melt volume flow rate, and two kinds of VGCF which have different aspect ratio (VGCF-S which is the aspect ratio of 100, and VGCF-H which is the aspect ratio of 40.) were used in this study. In addition, the several VGCF contents were set up in each PC/VGCF composites. The Young's modulus of PC/VGCF composite increased with VGCF content in each case, and these tendencies would be almost independent of VGCF size and melt viscosity of PC matrix except for the case using PC with low melt viscosity at high VGCF content. In case using PC with low melt viscosity, the relationships between Young's modulus and VGCF content would depend a little on the VGCF size. Tensile strength, which is maximum stress during tensile testing, of PC/VGCF composite increases a little with VGCF content at low melt viscosity of PC matrix. On the contrary, tensile strength of PC/VGCF composite decreaseds with an increase of VGCF content at high melt viscosity of PC matrix. According to the SEM photographs, the state of fracture cross section of the specimens after tensile testing would depend on the VGCF size.

Introduction

Due to good property balance of polycarbonate (PC) has been widely used in many technical applications where properties like toughness, rigidity or thermal stability are required (Sánchez-Soto et al., 2009). Polycarbonate has been modified and made for a particular purpose in many different ways, particularly by mixing with other polymers for use in demanding applications particularly when its outstanding impact strength is important (Yoon et al., 2003).

Then, Carbon fiber is widely used in polymer matrix composites because of its high tensile strength and modulus in the axial direction, as well as its excellent electrical conductivity of properties. Vapor grown carbon fiber (VGCF) is a new class of carbon fiber that is distinct from other types of carbon fibers in its method of production (Choi et al., 2005). Nowadays VGCF's size is smaller than traditional carbon fibers (diameters typically below 100-200 nm, and lengths below 100 μm).

Polymer composites containing Vapor Grown Carbon Fiber (VGCF) have attracted much attention because of high quality compact. Polymer/VGCF composites possess good thermal conductivity and excellent mechanical properties. In addition, twin screw extrusion was widely used in melt compounding process (Fang et al., 2006). It produced uniform particle dispersion by the large shear forces.

Twin screw extrusion will make case study on properties of polymer/VGCF composites prepared by melt compounding. In this study, We discuss the following items, 1) the effect of VGCF content on the properties of PC/VGCF composites, 2) the effect of PC grade on the properties of PC/VGCF composites prepared by melt componding, 3) to investigate PC/VGCF composites by using different size of VGCF.

Experimental

Material

We used four kinds of commercialized PC which were kindly supplied by Teijin Co., LTD., Japan. In addition, we also used two kinds of VGCF, VGCF-S and VGCF-H, which have the different size. The characteristics of PC and VGCF used in this study are listed in Tables 1 and 2

Table 1 Characteristics of PC used in this study

Name	Materials (Commercial code)	Specification		
		Mn	Mw	MVR $cm^3/10min$
PC-A	K-1300	22,093	57,821	2.5
PC-B	L-1250	16,276	44,756	8
PC-C	L-1225L	16,243	36,412	18
PC-D	AD-5503	12,807	27,654	54

Table 2 Characteristics of VGCF used in this study

Name	Diameter (nm)	Length (μm)	Aspect ratio	Thermal Conductivity (W/(m·K))	Electrical Conductivity (Ω·m)
VGCF-S	100	10	100	1200	10^{-4}
VGCF-H	150	6	40	1200	10^{-4}

Melt compounding

In this study, PC/VGCF composites were prepared in a co-rotating twin screw extruder (TSE) (S1 KRC kneader by Kurimoto Tekkosho, screw diameter of 25

mm, screw length/diameter ratio of 10.2). Nine cases of PC/VGCF composites with the compounding ratios listed in Table 3 were prepared in each type of PC.
Compounding was carried out at a uniform barrel temperature of 260 ℃, a screw rotation speed of 150 rpm. Before melt compounding, PC and VGCF were dried in an oven at 80 ℃ for a minimum of 24 h. The feed rate is 1,000 g/h for all extrusion cases.

Table 3 Compounding ratios for various kinds of PC/VGCF composites

Case number of materials	PC	VGCF-S	VGCF-H
1	100	0	0
2	100	1	0
3	100	2	0
4	100	4	0
5	100	7	0
6	100	0	1
7	100	0	2
8	100	0	4
9	100	0	7

Testing and Measurement

After extrusion, the samples for the measurement of the mechanical properties were manufactured by injection moulding.

Before injection molding, each PC/VGCF composite was dried in an oven at 80 ℃ for a minimum of 24 hrs. When we did the injection molding, we set up a barrel temperature of 300-320 ℃. The tensile properties were measured according to JIS K7162 using the Shimazu Autograph AGS-J at room temperature. The cross head speeds were 1 mm/min for measurement of Young's modulus, and 10 mm/min for measurement until break point. An extension meter with a gauge length was used for measurement of Young's modulus.

Results and Discussion

Tensile testing result

Figure 2 shows the Young's modulus and tensile strength as a function of VGCF content for each composite. The Young's modulus of each composite increases with VGCF content. The degree of an increase rate of the Young's modulus to increase in the VGCF content has become small in the composites based on PC-A with high melt viscosity and become large in the composites based on PC-B and PC-C with middle-value melt viscosity. In this study, the composite based on PC-C has the largest degree of an increase rate of the Young's modulus to increase in the VGCF content. The tensile strength of the composite based on PC-A with high melt viscosity decreases with an increase of VGCF content. However, the tensile strength of the composites based on PC-C and PC-D slightly increases with VGCF content, and the tensile strength of the composite based on PC-B is almost independent of VGCF content on the whole. Moreover, the tendency that the Young's modulus reaches a constant value only as for the case to the composites based on PC-D with the lower melt viscosity when VGCF content is large. When the melt viscosity of the matrix polymer is large, the shear stress imposed on VGCF becomes large on the mixing process. As a result, VGCF may break during the mixing process. This may cause the dependency of VGCF content on tensile properties of PC-A based composite. On the contrary, when the melt viscosity of the matrix polymer is small, the shear stress imposed on VGCF becomes small on the mixing process. As a result, it may be not easy to cause the dispersive mixing of VGCF. This may be one reason that the Young's modulus of PC-D based composite is almost independent of VGCF content at high VGCF content.
Next we discuss the effect of VGCF size on the tensile properties of PC/VGCF composite. Strictly speaking, we obtained the different relativity between the magnitude correlation of Young's modulus and tensile strength and that of VGCF size for each PC/VGCF composite. However, the tensile properties of the composite using VGCF-S with high aspect ratio seem to be slightly larger than that of the composite using VGCF-H with low aspect ratio on the whole. The volume per unit fiber of VGCF-S is ¾ of VGCF-H. The number of VGCF-S is larger than that of VGCF-H at same VGCF content. The surface area per unit fiber of VGCF-S is about 10% larger than that of VGCF-H. The total contact area between VGCF-S and PC matrix is larger than that between VGCF-H and PC one. This may cause the difference of tensile properties. However, the difference between the tensile properties of PC/VGCF composites adding VGCF-S and those adding VGCF-H seems to be smaller than about 10% whose value is the difference of surface area per unit fiber between VGCF-S and VGCF-H

SEM observations

Figure 3 shows the SEM photographs of the fracture surface for PC/VGCF composite with the compounding ratio PC/VGCF = 100/7 (actual content of 6.54 wt%) after tensile testing. The dark regions and white bands represent the mixture polymer and VGCF, respectively. On the whole, VGCF is dispersed in the polymer matrix in each case, although bundles with a small quantity of aggregates (broken in circle) can be found. The VGCF dispersibility seems to be independent of PC melt viscosity from SEM observation results. In case of PC/VGCF composite using VGCF-S with large aspect ratio, VGCF fibers appear clearly on the fracture surface. On the contrary, VGCF fibers are included in the matrix in case of PC/VGCF composite using VGCF-H with small aspect ratio. These results imply that VGCF-S may be not enough to support the tensile properties of the composite because of low miscibility between PC and VGCF-S.

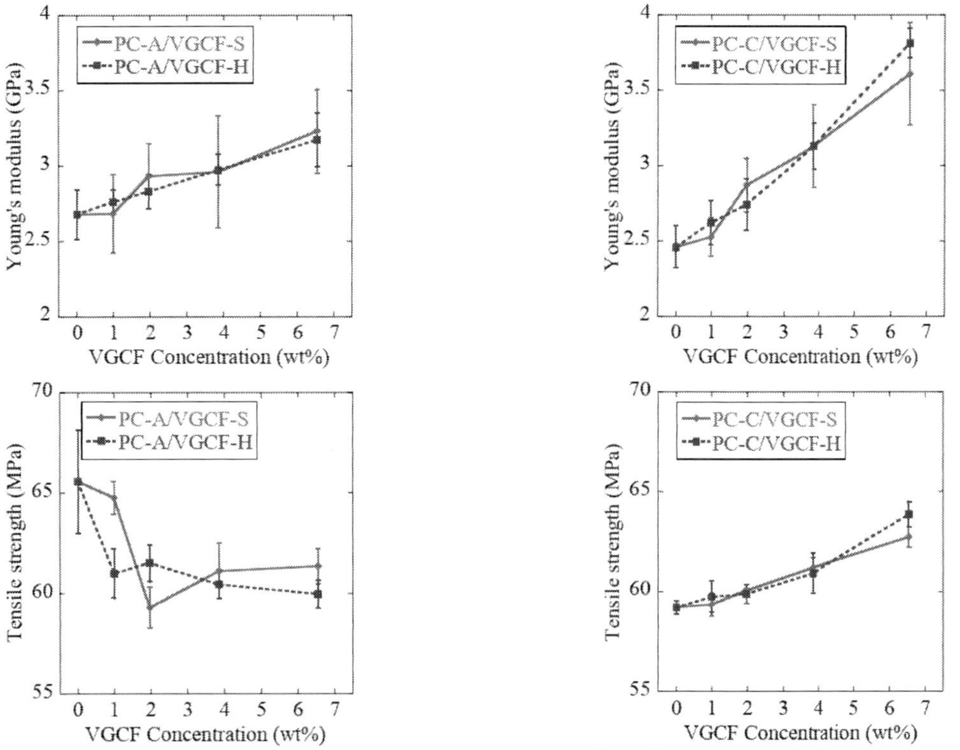

(a) PC/VGCF composite based on PC-A

(c) PC/VGCF composite based on PC-C

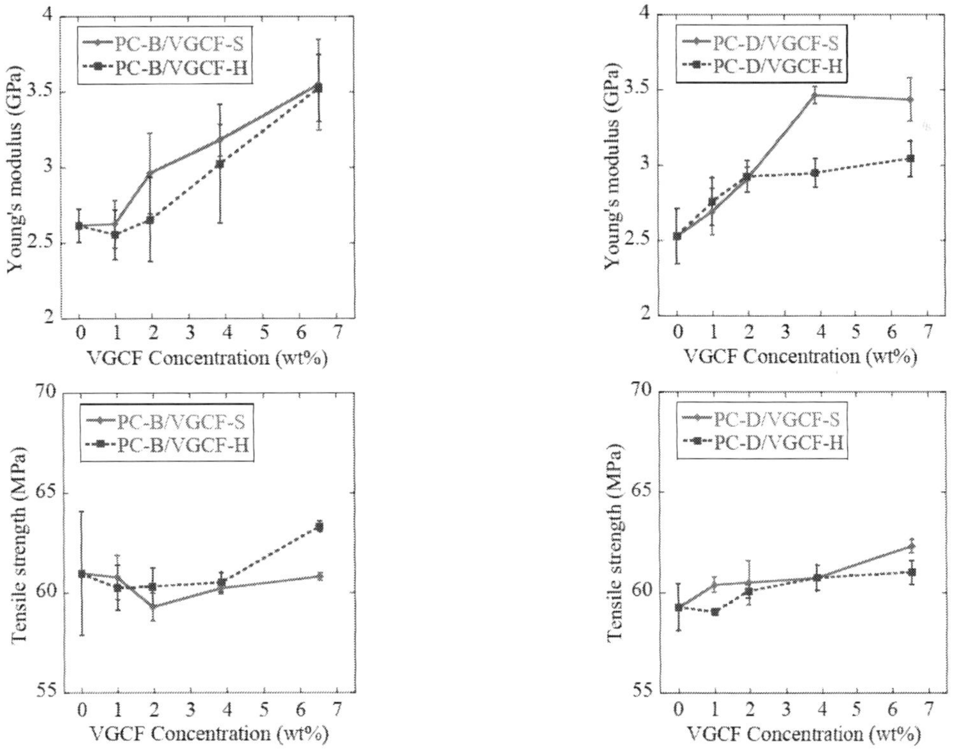

(b) PC/VGCF composite based on PC-B

(d) PC/VGCF composite based on PC-D

Fig.2 Young's modulus and tensile strength as a function of VGCF content for various PC/VGCF composites prepared by melt compounding.

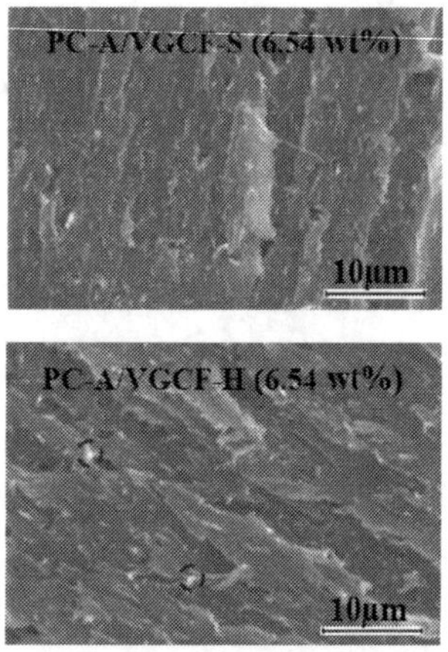

(a) PC/VGCF SEM photographs of PC-A

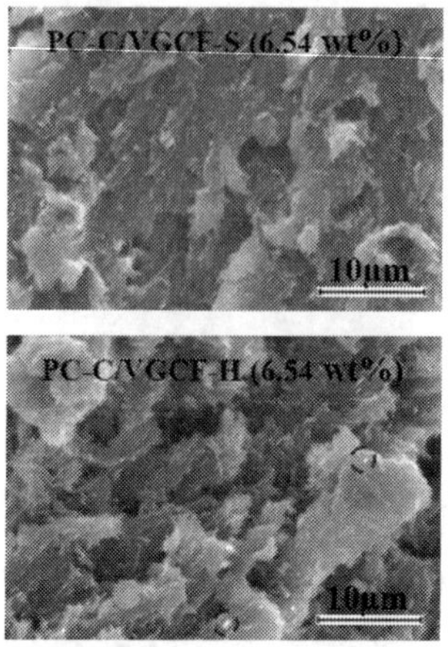

(b) PC/VGCF SEM photographs of PC-C

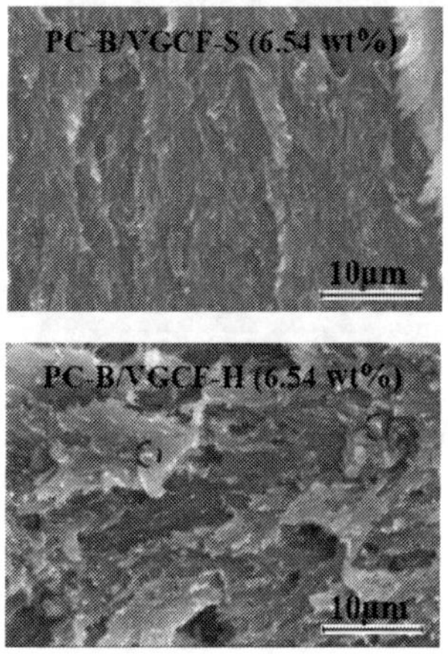

(c) PC/VGCF SEM photographs of PC-B

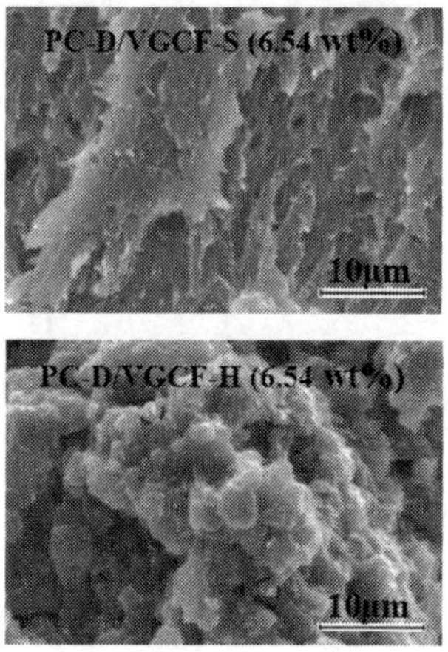

(d) PC/VGCF composite based of PC-D

Fig. 3 SEM photographs of the fracture surface for PC/VGCF composite

Conclusions

1) The Young's modulus of PC/VGCF composite increased with VGCF content in each case, and these tendencies would be almost independent of VGCF size and melt viscosity of PC matrix, except for the case using PC with low melt viscosity at high VGCF content. In case using PC with low melt viscosity, the relationships between Young's modulus and VGCF content would depend on the VGCF size.

2) Tensile strength of PC/VGCF composite increased with VGCF content at low melt viscosity of PC matrix. On the contrary, tensile strength of PC/VGCF composite decreased with an increase of VGCF content at high melt viscosity of PC matrix.

3) According to the SEM photographs, the state of facture cross section of the specimens after tensile testing would depend on the VGCF size.

Acknowledgements

The authors express their thanks to Mr. Tsunemune Unryu, Industrial Technology Center of Fukui Prefecture, Japan, for technical support of the injection molding

References

Choi, Y. K., et al., "Mechanical and thermal properties of vapor-grown carbon nanofiber and polycarbonate composite sheets", Mater. Lett., **59**, 3514-3520 (2005).

Fang, L., et al., "Processing and mechanical properties of HA/UHMWPE nanocomposites", Biomaterials, **27**, 3701-3707 (2006).

Sánchez-Soto, M., et al., "Study of the morphology and properties of melt-mixed polycarbonate-POSS nanocomposites", European Polym. J., **45**, 341-352 (2009).

Yoon, P.J., et al., "Polycarbonate nanocomposites. Part 1. Effect of organoclay structure on morphology and properties", Polymer, **44**, 5323-5339 (2003).

A STUDY ON FABRICATING METAL COMPOSITE FOAM PANELS WITH THE USE OF CLAY FILLERS FOR THE CONTROL OF CELL STRUCTURE

SAEKI Akiya [1*], KINNAN Takashi [1], TANAKA Tatsuya [1], IMAIDA Yutaka [1], MIURA Hodaka [2] and FUJIURA Takayasu [2]

[1]*Faculty of Science and Engineering, Doshisha University, Kyoto, Japan,*
[2]*Kobe Steel, Ltd., Kobe, Japan* tatanaka@mail.doshisha.ac.jp

Sandwich-structure composites, consisting of cellular polymeric core laminated with thin metal facings, have been attracting interests in wide range of industries where high strength and stiffness as well as lightweight are essential design requirements. This study was performed on aluminum/foamed polypropylene (PP) panels to investigate the effect of addition of clay fillers in PP on cellular size and mechanical properties of the composites.

PP samples with and without clay fillers were prepared into sheet form using a twin screw extruder in various kneading conditions, followed by adhering to aluminum plates and heat-foaming of PP at several temperatures.

Composites with clay-filled PP showed quicker expansion in heat-foaming and larger final expansion ratio than composites without fillers at each set temperature, even though the apparent cell sizes and cell densities measured from the X-ray CT images were almost same in those samples. Addition of fillers increased the viscosity of resin and was supposed to restrain coalescence and rupture of cells, which actually occurred in samples without fillers, and helped to make uniform closed cell structure. Better dispersion of fillers raised the viscosity of resin and was found to be effective for achieving higher expansion ratio.

Introduction
Sandwich-structure composites, consisting of cellular polymeric core laminated with thin metal facings, have been attracting interests in wide range of applications where high mechanical properties, vibration damping and sound insulation as well as lightweight are neccesary. In order to meet those requirements, cell structures and expansion ratio are supposed to be the most essential features. In previous studies, the addition of organic-modified clay fillers into thermoplastic/bloating agent formulation was proven to be effective for increasing expansion ratio
This study, using organo-bentonite as the filler for PP foam core of sandwich panels, explored the formation and growth of the cell structure in regard to foaming conditions and resin viscosity.

Experimental

Materials
Polypropylene F-744NP and maleic anhydride denature modified polypropylene 1001 and VINYFOR AC#R were used as the matrix PP and bloating agent, recpectively. Organo-bentonite S-BEN was used as clay fillers.

Preparation of coumpounds
Composition of samples were shown in Table 1. Materials were mixed and compounded with a twin strew extruder into sheet forms. These sheets were sandwiched by aluminum plates using adhesive. Concerning samples with clay fillers, the numbers of kneading passses were varied for 1,2,3,5,7 and 10 times, to investigate the effect of filler dispersion in PP matrix. Additive amounts of clay fillers were also varied in 3, 5 and 7 mass% of PP matrix.

Table.1 Composition of Materials

	sample.A1	sample.B1
MAPP[wt%]	10	10
S-BEN[wt%]	0	5
RPP[wt%]	89	84
Bloating agent[wt%]	1	1

Foaming and characterization
Foaming experiments were done using an electric oven at temperatures from 210℃ to 300℃. Heating periods were varied from 1 to 15 min. In the course of heating, resin temperatures were monitored with thermocouples. Expansion ratios at every one minite were obtained as the change in thickness of samples. Cell structures were observed with an X-ray CT scanner and the average cell diameter and the cell density were calculated.

Results and Discussion

Effect of addition of fillers on overall thermal foaming behaviors
The expansion patterns of sample A1 and B1 at 210℃ is shown in Figure 1. In both samples, the expansion rate increased after 2 min. and reached the maximum expansion ratio at 8min., then gradual shrinkage were observed. Sample B1, containing clay fillers, exhibited earlier initiation of expansion and higher final expansion ratio than sample A1.
Figure 2 shows the change of resin temperatures of sample A1 and B1 in the course of heat-foaming.

After 2 min., temperature of sample B1 kept higher than that of sample B2. It is supposed that addition of clay fillers raised the thermal conductivity of resin, resulting in prompt initiation of foaming in sample B1, as shown in Figure 1.

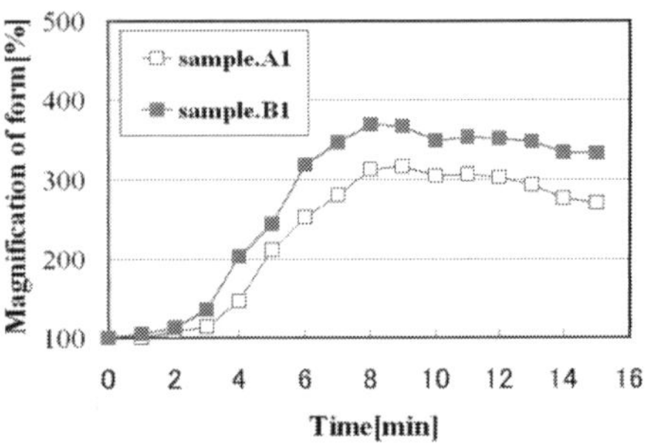

Figure 1 Expansion behaviors of compounded PP sheet samples heated at 210℃

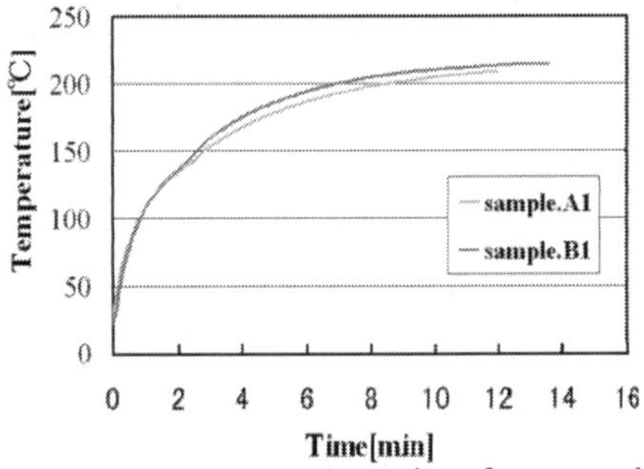

Figure 2 Temperature excursion of compounded PP sheet samples heated at 210℃

Analysis on cell growth and structue

Figure 3 shows examples of X-ray CT image of heat-foamed sample A1 and B1 . The average cell diameter and the cell densities estimated by image processing are summarized in Table 2. Figure 4 and Figure 5 exhibit changes of the average cell diameter and the cell densities in tems of heating period. From those results, it is supposed that the foaming behavior in course of heating comprises 3steps; cell generation, multiplilcation and expansion of cells and termination of cell growth. Although sample B1 showed earlier cell multiplilcation and growth, cell diameters and densities after completion of foaming were almost identical between two samples. The drop in cell density of sample A1 after 7min. implies the coalescense and rupture of cells.

Table 2 Changes of average cell diameters and cell densities

	Form Diameter[mm]		Density of Form[cell/cm^3]	
Time[min]	sample.A1	sample.B1	sample.A1	sample.B1
1	0.00	0.00	0.00	0.00
2	0.00	0.00	0.00	0.00
3	0.11	0.31	25.10	172.65
4	0.26	0.50	280.83	555.96
5	0.43	0.61	820.77	673.42
6	0.58	0.75	871.45	611.04
7	0.67	0.83	905.94	615.14
8	0.83	1.05	680.09	525.38
9	0.95	0.97	679.86	682.30
10	0.89	0.91	737.24	619.91
11	0.92	0.96		
12	1.05	1.05		
13	1.07	0.97		
14	1.12	1.01		
15	1.19	1.05		

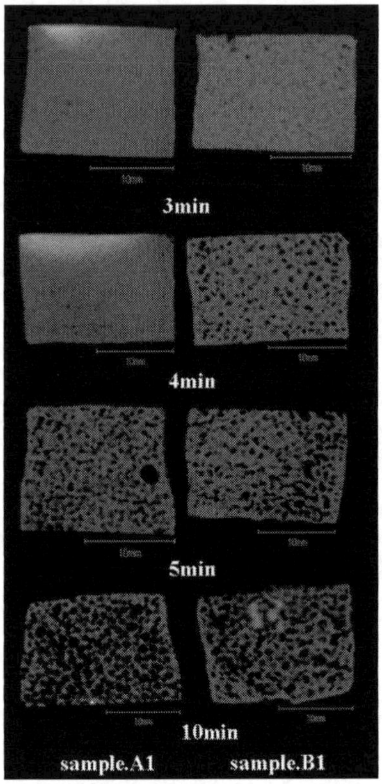

Figure 3 Examples of X-ray CT image of heat-foamed samples

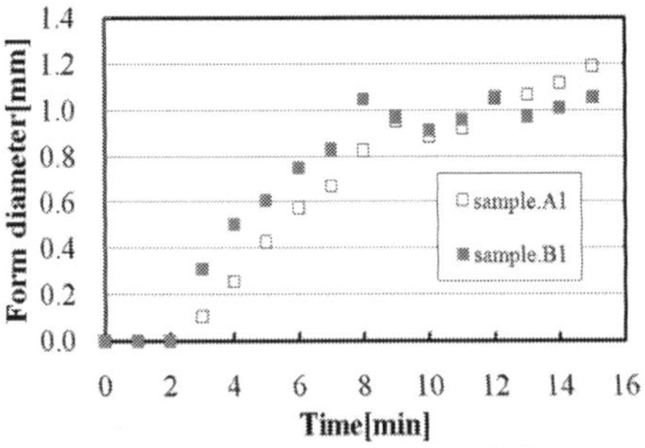

Figure 4 Changes of average cell diameters

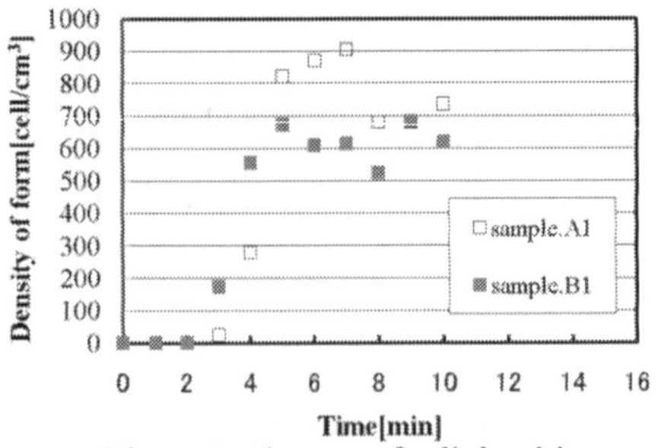

Figure 5 Changes of cell densities

Effect of resin viscosity on foaming behaviors

Melt viscosities of compounds used for sample A1 and B1 are shown in Figure 6. Sample B1, clay-filled compound, had much higher viscosity than sample A1. In sample A1, the resin viscosity may be insufficient to prevent the cells from coalescing and rupturing, which lead the collapse of cell structure. Appearances of heat-foamed samples at 300% expansion are shown in Figure 7. The foam expanded out to the lateral wall in sample A1, while no deformation of lateral wall was observed in sample B1. Those resluts concludes that additon of fillers to the resin matrix can help the foam to preserve the stable closed cell structures and attain higher expansion ratios at the termination of heat-foaming.

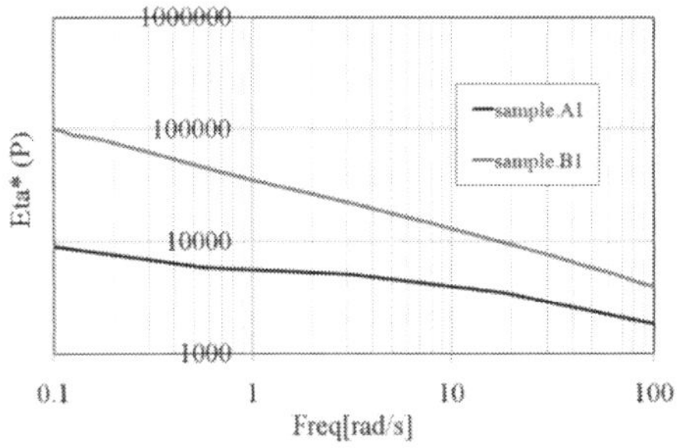

Figure 6 Melt viscosities of compounds

Sample A1 Sample B1

Figure 7 Appearance of heat-foamed samples

Effect of compounding conditions and additive amount of fillers on foaming behaviors

In the preparation of compounds, the number of times for kneading passage through the twin screw extruder were varied to investigate the effect of dispersion state of fillers on foaming behaviors. Figure 8 and Figure 9 show the results of heat-foaming tests and melt viscosities of those compounded samples, respectively. As seen in Figure 8, expansion ratios were enlarged in the case of 2 and 3 kneading passages, though

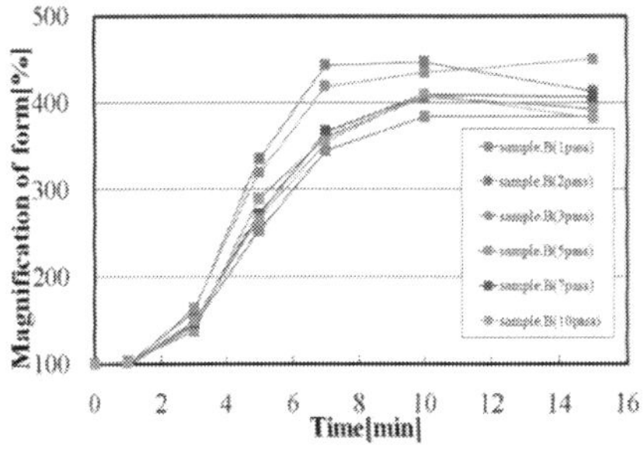

Figure 8 Result of heat-foaming tests on compounds prepared in various conditions

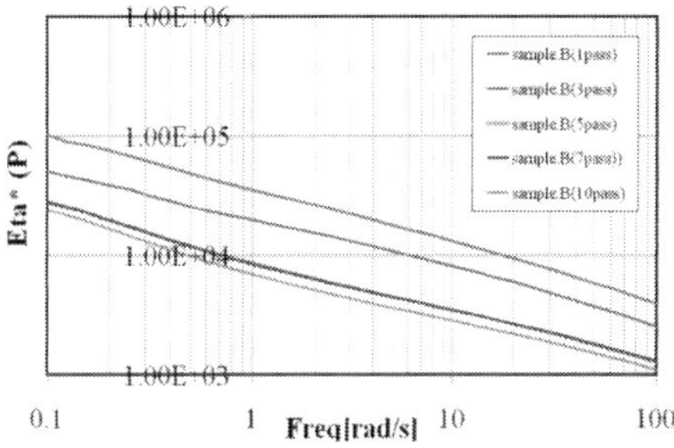

Figure 9 Melt viscosities of compounds prepared in various consditions

the melt viscosity decreased with the number of times of kneading. Increasing the passages of kneading up to may be effective for better dispersion of fillers, however, it is supposed that deterioration of PP matrix by heat progressed with the times of kneading, resulting in the reduction of melt viscosity. Figure 10 describes the expansion ratios of samples containing 3, 5,and 7 mass% of clay fillers in PP matrix. Increasing the additive mount of fillers accelerated the initiation of foaming, while it was not necessarily correlative to the terminated expansion ratios.

Figure 11 and Figure 12 show the expansion behaviors of samples heat-foamed at various temperatures,and actual temperature excursion of resin in the course of heating, respectively. The maximum expansion ratio was almost constant irrespectively of heating temperatures. On the other hand, the terminated expansion ratio drastically decreased at higher heating temperatures. Higher heating temperature is effective for earlier initiation of foaming, however, may cause collapse of cells by the rapid elevation of resin temperature, which hinders the proper growth of closed cell structure.

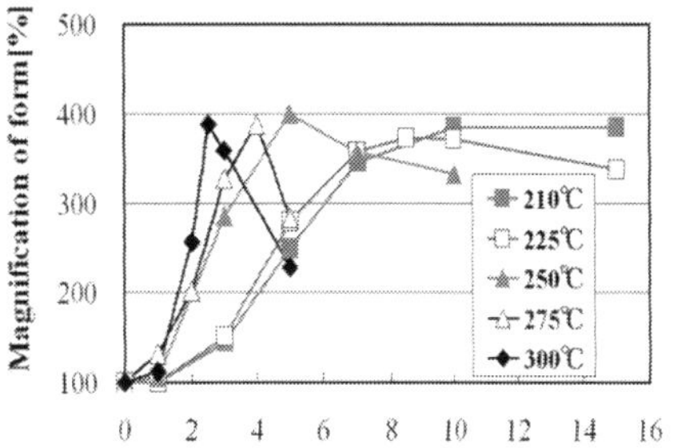

Figure 11 Expansion behaviors at various heat-foaming tenperatures

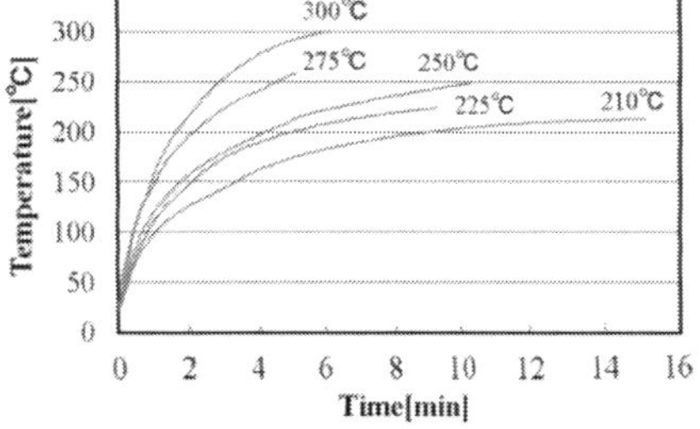

Figure 12 Temperature elevation of compounds heated at various set temperatures

Conclusions

(1) Addition of clay fillers into PP matrix was effective for earlier initiation of foaming by heat and attaining higher terminated expansion ratio.

(2) Increase in viscosity by the addition of fillers was supposed to restrain coalescence and rupture of cells, attaining higher expansion ratio.

(3) Better dispersion of fillers raised the viscosity of resin and was found to be effective for achieving higher expansion ratio.

(4) Higher heating temperature is effective for earlier initiation of foaming, however, it may cause collapse of cells which hinders the proper growth of closed cell structure.

References

(1) Fukazawa Y., Plastic age, 48, 3, 74-78 (2002)

(2) Okamoto M, Polymer nano-composite new technology and intended end-usage expansion, , 9-12 (2004)

(3) Mayama N, Plastics, 45, 11, 49-51 (1994)

A STUDY OF FIBER BREAKAGE DURING COMPOUNDING OF GLASS FIBER REINFORCED COMPOSITES

F. Inceoglu[2], J. Ville[3], N. Ghamri[2], A. Durin[1], R. Valette[1], B. Vergnes[1]

[1] MINES ParisTech, CEMEF, UMR CNRS 7635, BP 207, 06904 Sophia Antipolis (France)
[2] Arkema, Cerdato, 27470 Serquigny (France)
[3] Polytechs, BP 14, 76450 Cany Barville (France)

Abstract – The main problem in the preparation of glass fiber reinforced composites by twin screw extrusion is the reduction of the fiber length by mechanical breakage. In this study, we have characterized the influence of processing conditions (screw speed, feed rate, screw profile…) on the breakage of glass fibers in a polyamide matrix, in order to optimize these conditions to limit fiber degradation. Experiments were carried out in different equipments (internal mixer, twin screw extruders) and fiber length distribution was measured by image analysis. A relationship between fiber length (average length and length distribution) and specific energy is proposed, based on previous model of Shon, Liu and White. Comparisons with experiments validate nicely this model.

Introduction

Fiber glass reinforced composites are widely used in polymer industry when enhanced mechanical properties are required. These properties strongly depend on the fiber properties (modulus), the adhesion between fiber and matrix, the fiber orientation and finally the aspect ratio, i.e. the length to diameter ratio. A high aspect ratio (long fibers) is necessary to achieve good mechanical properties. However, during the compounding process used for dispersing the fibers into the polymer matrix, fibers are broken and their length largely reduced, impacting the final properties. The objective of the present work is thus to investigate the mechanisms of glass fiber breakage during compounding, in order to derive theoretical models able to allow the optimization of the process by limiting fiber length reduction.

Experimental

Materials

The chosen matrix is a polyamide 12 Rilsan© provided by Arkema. It has a melting point of 183°C, a weight average molecular weight of 37 000 g/mol and a polydispersity index of 1.85.

Glass fibers have an average length of 3 380 μm and a diameter of 10 μm (aspect ratio = 338). Their distribution is given in Fig. 1. It is rather monodisperse, with a minimum length around 1350 μm and a maximum around 6000 μm.

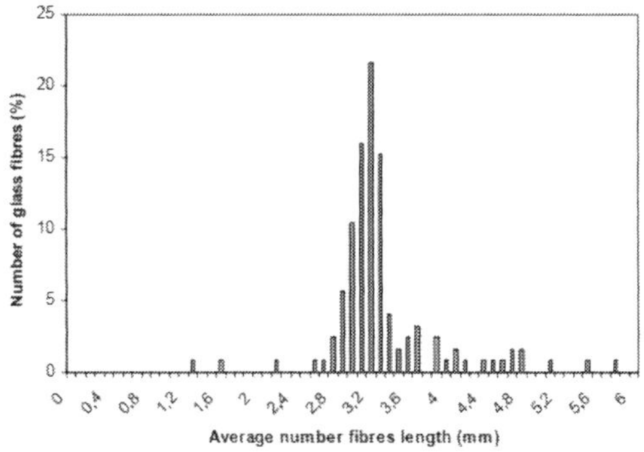

Figure 1 – Initial fiber length distribution

Composites preparation

Three different systems have been used for the compounding of fiber glass reinforced composites:

- an internal mixer Haake Rheocord 600P;

- a laboratory scale co-rotating twin screw extruder ThermoFisher PTW 24 (D = 24 mm, L/D = 40);

- an industrial scale co-rotating twin screw extruder Werner Pfleiderer WP40 (D = 40 mm, L/D = 41).

In all systems, the fibers (30 wt%) were introduced after the melting of the PA matrix. Experiments were carried out at different rotor speeds and different mixing times for the internal mixer (barrel temperature = 240°C), and at different screw speeds and different feed rates for the extruders (barrel temperature = 260°C). Samples were collected after mixing and at the die exit and, in some cases, along the screw profile.

The fiber length distribution was measured after burning of the collected samples, optical microscopy and image analysis on about 300-400 fibers. Average length in number and volume were deduced from these measurements. In the following, we will focus on the number average length L_n.

Results and Discussion

Internal mixer

Figure 2 shows the change in fiber length with time at a fixed rotor speed of 80 rpm. We have plotted not only the average length L_n, but also the maximum and the minimum length measured. We observe a regular decrease with time of the different lengths, followed by a plateau above 60 s. Fibers are broken by buckling [1] in the mixer, until reaching a minimal length under which breakage becomes impossible. In these conditions, the minimal average length is around 725 µm, but it is important to point out that long fibers of more than 3 mm are still present at the end of the mixing process.

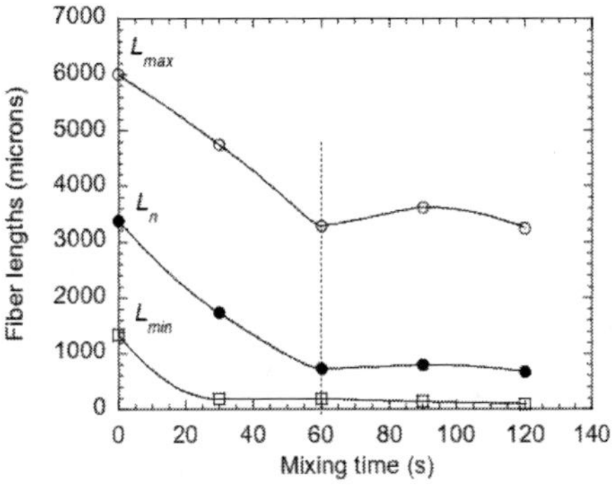

Figure 2 – Change in fiber length with mixing time (80 rpm, 240°C)

Figure 3 shows that the kinetic of fiber breakage is strongly dependant on the rotor speed. When rotor speed increases, breakage is faster and the limit fiber length is shorter: 937 µm at 40 rpm, 665 µm at 80 rpm and 576 µm at 120 rpm. These results are similar to the ones obtained by Burns and Pennington [2] on a Z-blade mixer.

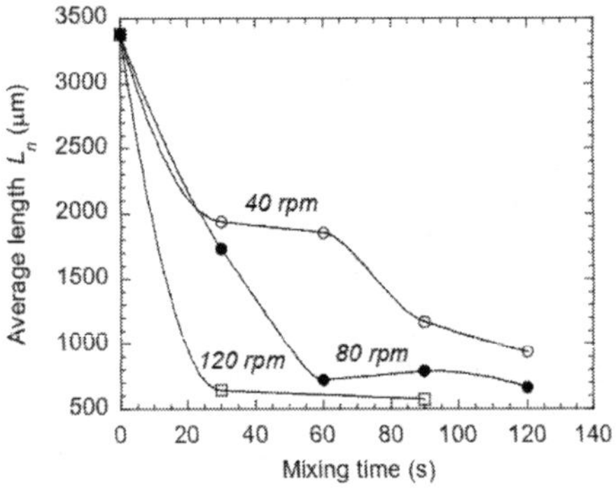

Figure 3 – Change in average fiber length with mixing time and rotor speed

Previous results can be gathered on a single mastercurve if we consider the number of turns, i.e. the product of rotor speed by mixing time (Fig. 4). Same results, with the same orders of magnitude, were obtained by Shimizu et al. [3] on a polypropylene reinforced by 10% of glass fibers.

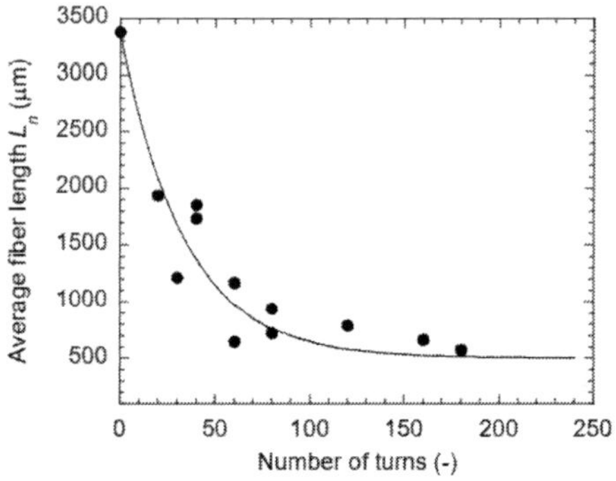

Figure 4 – Change in average fiber length with number of turns

Laboratory twin screw extruder

Different processing conditions have been tested on the extruder: at constant feed rate (Q = 5 kg/h), we have varied the screw speed N between 200 and 600 rpm. Similarly, at constant screw speed (N = 200 rpm), feed rate has been changed between 5 and 9 kg/h. We can see in Fig. 5 that fiber lengths decrease with screw speed. On the contrary, fiber breakage is reduced when increasing feed rate (Figure 6). If the effect of screw speed is expected and was previously reported [4], the effect of feed rate is, to our knowledge, original.

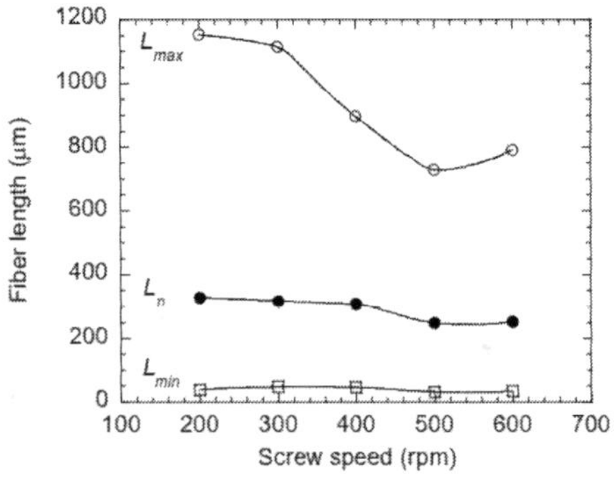

Figure 5 – Change in fiber length with screw speed (5 kg/h, 260°C)

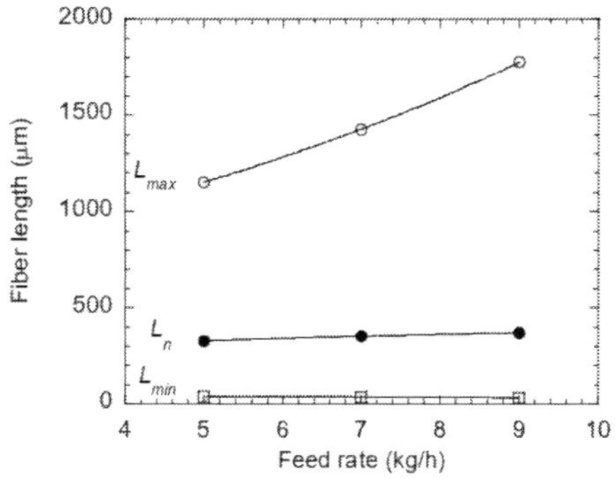

Figure 6 – Change in fiber length with feed rate (200 rpm, 260°C)

Industrial twin screw extruder

Same types of experiments have been performed on the WP40 extruder, using a similar screw profile, with screw speeds in the range 10-300 rpm and feed rates in the range 10-100 kg/h. Results are shown in Figure 7. They confirm that the fiber breakage increases with the screw speed and is less important when feed rate is increased. The values of average fiber length (280 – 380 µm) are in the same range as the ones obtained with the laboratory scale extruder (240 -370 µm).

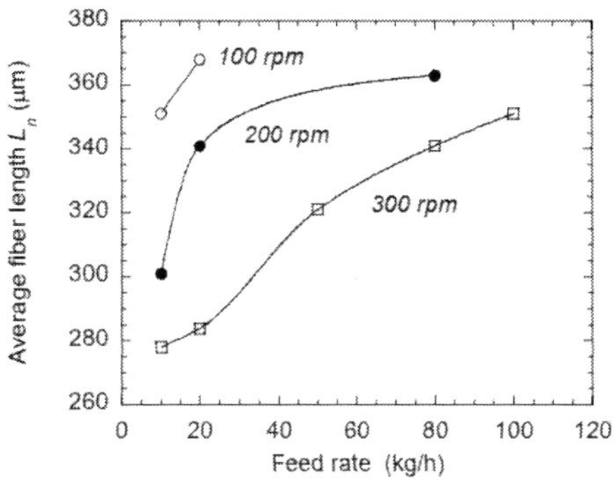

Figure 7 – Change in fiber length with feed rate and screw speed

A screw pulling experiment has been carried out at 80 kg/h and 300 rpm. After stopping the screws and cooling the barrel, the screws have been extracted and samples collected at different locations after the feeding port: in the first and the second blocks of kneading discs and in the left-handed element. Average fiber length is shown in Figure 8 as function of axial position. We observe a fast decrease of fiber length (from 3380 to 720 µm) in the screw conveying elements before the first block of kneading discs, followed by a regular decrease along the different restrictive elements, until reaching a final value of 412 µm. Same rapid change in fiber length along a screw profile has already been reported by Shon and White [5] and Shon et al. [6].

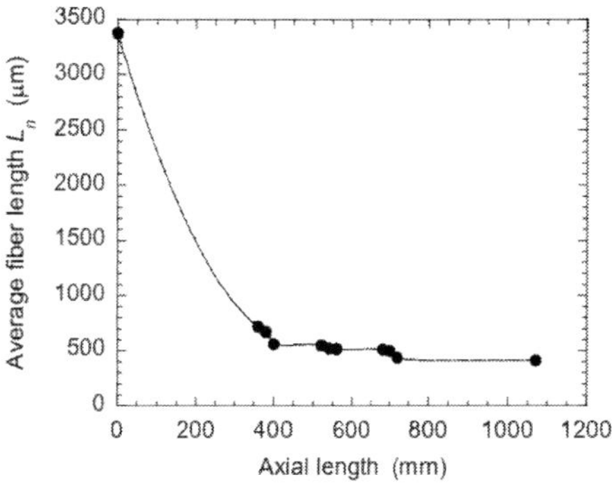

Figure 8 – Change in average fiber length along the screws (80 kg/h, 300 rpm)

Discussion and interpretation

Previous experiments show that important fiber breakage (initial length is divided by a factor 10) occurs during compounding, whatever the machine used for that purpose. In all cases, screw or rotor speed is detrimental, leading to enhanced breakage. On the contrary, it seems that a reduced mixing time is favorable, as it is clearly seen with the internal mixer. We can imagine that it is also the case in twin screw extrusion where it could explain the influence of feed rate. Indeed, at constant screw speed, an increase in feed rate induces a drastic reduction of the residence time [7].

In a recent study, Shon et al. [6] have proposed an evolution law for the fiber length of the form:

$$L_n(t) = [L_0 - L_\infty] \exp(-k\,t) + L_\infty \qquad (1)$$

where L_0 is the initial fiber length, L_∞ the limit one and k is a parameter depending on geometry, applied stress and fiber properties.

In order to interpret the experimental results, we will apply this type of relationship. For example, the curve in Figure 4 has been drawn using Eq. (1) by replacing t by $N\,t$ (number of turns). We can see that the length evolution is nicely described. However, we do not believe that the strain is an adequate parameter to calculate fiber length. As shown by Shimizu et al. [3], the applied stress plays also a key role. Consequently, we propose to replace the term in $k\,t$ which controls the fiber evolution by an energetic parameter, for example the specific mechanical energy (*SME*). We thus propose the following law:

$$L_n(t) = [L_0 - L_\infty] \exp(-\,k'\,SME) + L_\infty \qquad (2)$$

where k' is a parameter depending on fiber properties.

SME can be easily deduced from torque measurements on the internal mixer or on the twin screw extruders. As an example, Figure 9 shows for the laboratory scale extruder that, whatever the values of screw speed and feed rate, the global evolution of average length L_n, but also of minimum and maximum values, are nicely described by the *SME*.

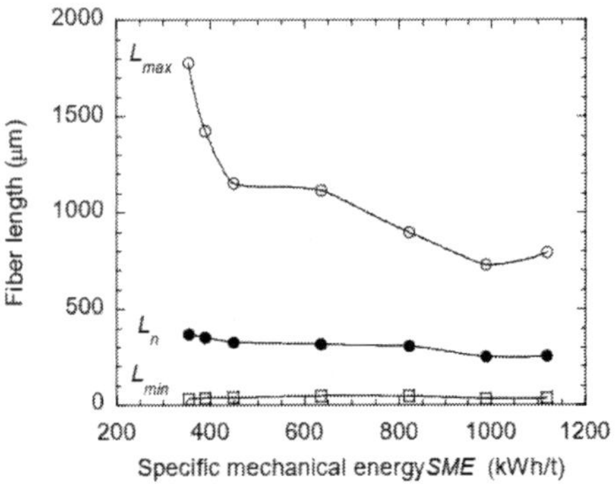

Figure 9 - Change in fiber length with *SME* (laboratory extruder)

Moreover, Eq. (2) allows to gather all the data obtained with the three different systems: it is shown in Figure 10 that the different machines provide a regular evolution of the fiber length according to *SME*.

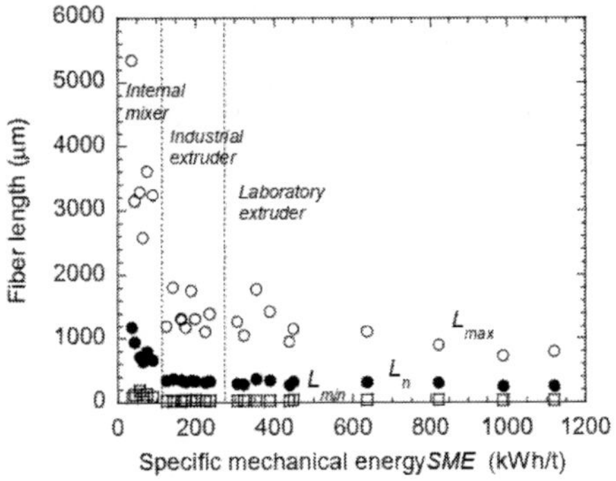

Figure 10 - General comparison of changes in fiber length with *SME*

If we focus on the average fiber length, it can be seen in Figure 11 that Eq. (2) allows a very good fit of the experimental data with the following parameters:

L_0 = 3380 μm, L_∞ = 338 μm, k' = 0.035

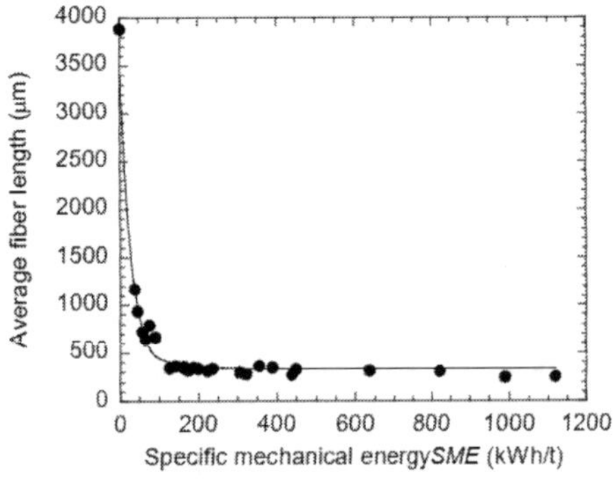

Figure 11 - Comparison between experimental fiber length (symbols) and predictions of Eq. (2) (line)

L_{min} and L_{max} can also be fitted with the same equation and the following parameters:

For L_{min}: L_0 = 1350 μm, L_∞ = 35 μm, k' = 0.035

For L_{max}: L_0 = 6000 μm, L_∞ = 700 μm, k' = 0.0075

Conclusion

A systematic study of fiber breakage during compounding of glass fiber reinforced polyamide has been carried out, using different types of machines. Fiber length is systematically reduced during compounding, by a factor close to 10. It has been shown that fiber breakage increases with screw or rotor speed and with mixing time. We have proposed to describe fiber length evolution using a model based on specific energy, derived from the original expression of Shon et al. This model allows a correct prediction of fiber length (average, minimal, maximal) evolution and shows a nice coherence of data obtained with internal mixer, laboratory scale extruder and industrial scale extruder.

Acknowledgements

This work was carried out during Fonlimics project, supported by the FUI in the frame of "pôle de compétitivité Plastipolis".

References

[1] Forgacs, O.L., Mason, S.G., "Particle motions in sheared suspensions: X. Orbits of flexible threadlike particles", J. Colloid Sci., 14, 457 (1959).

[2] Burns, R., Pennington, D., SPE ANTEC Tech. Papers, **25**, 1 (1979).

[3] Shimizu, Y., Arai, S., Itoyama, T., Kawamoto, H., "Experimental analysis of the kneading disk region in a co-rotating twin screw extruder: Part 2. Glass-fiber degradation during compounding", Adv. Polym. Tech., **16**, 25-32 (1997).

[4] Wall, D., "The processing of fiber reinforced thermoplastics using co-rotating twin screw extruders", Polym. Comp., 10, 98-102 (2004).

[5] Shon, K., White, J.L., "A comparative study of fiber breakage in compounding glass fiber-reinforced thermoplastics in a Buss kneader, modular co-rotating and counter-rotating twin screw extruders", Polym. Eng. Sci., 39, 1757-1768 (1999)

[6] Shon, K., Liu, D., White, J.L., "Experimental studies and modelling of development of dispersion and fiber damage in continuous compounding", Intern. Polym. Proc., **20**, 322-331 (2005).

[7] Poulesquen, A., Vergnes, B., Cassagnau, P., Michel, A., Carneiro, O.S., Covas, J.A., "A study of residence time distribution in co-rotating twin screw extruders; Part II: experimental validation", Polym. Eng. Sci., **43**, 1849-1862 (2003).

MULTICOMPONENT POLYMERIC MATERIALS BY HIGH-SHEAR-PROCESSING: FROM MACRO TO NANO

Yongjin LI*, Hiroshi Shimizu

Nano-system Research Institute, National Institute of Advanced Industrial Science and Technology (AIST), Tsukuba Central 5, 1-1-1 Higashi, Tsukuba, Ibaraki 305-8565 Japan
E-mail: yongjin-li@aist.go.jp

Several multicomponent polymeric materials have been fabricated using newly developed high-shear-processing technique. The dependence of the morphology and properties of the materials on the exerted shear rates during the melt processing have been investigated. It was found that both high shear processing leads to the decreased phase size and improved physical properties.

Introduction

A growing fraction of plastic resins produced today are multi-components polymer systems, namely, polymer blends of two or more polymers and polymer composites with inorganic fillers dispersed in a polymer matrix. The compounding of different polymers and/or inorganic fillers in a polymer presents the possibility of enhancing the overall properties of a material through the synergistic combination of the desirable properties of each component in one system. However, most commercial blends are immiscible because nearly all polymer pairs cannot be soluble in each other due to the high molecular weigh and weak interactions. Therefore, a blend material with big phase size (in the micrometers scale) and weak interface is usually obtained. This type of materials cannot provide expected properties for real application. The same problem occurs for the preparing of polymer composites based on inorganic fillers and a polymer due to the different polarity between the fillers and polymer. Therefore, significant researches on multi-components polymer systems have overwhelmingly focused on decreasing the phase size of polymer blends and improving the dispersion of fillers in a polymeric matrix. In this work, we will show that the high-shear-processing method is an effective way to fabricate both the nanostructured polymer blends and the polymer nanocomposites. The shear effects on the morphology and properties of multicomponent polymeric materials were emphasized.

High-shear processing technique [1]

High-shear processing was performed using an NHSS2-28 high-shear extruder (Niigata Machine Techno Co., Ltd., Japan). The extruder contains a continuous plasticizing injector to ensure the constant sample weight during each run. A specially designed screw with a flow hole inside is used in this extruder to exert a simultaneous shear and an extensional flow during melt blending. The schematic diagram of the polymer melt flow route in the screw during mixing is shown in Figure 1. The L/D ratio of the screw was 1.78. The sample temperature during processing is highly controlled with the variation of less than 10 °C and is independent on the exerted shear rate using a strong water-cooling system. A good balance between the cooling system and the viscous heating during the melt mixing was obtained by carefully adjusting the flow rate of cooling water. The compounded samples were then extruded from a T-die.

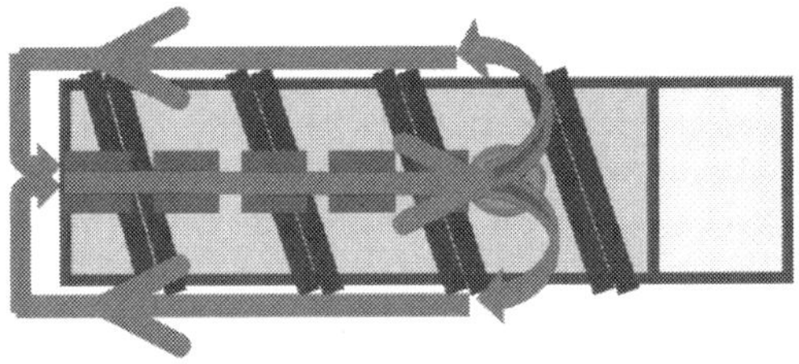

Figure 1. The schematic diagram for screw used in the high-shear extruder and the flow route (red line) of polymer melt during mixing.

PC/PMMA blends by high-shear processing [2]

Figures 2 (a) and 2 (b) shows SEM images of polycarbonate (PC)/polymethyl methacrylate (PMMA)=80/20 samples processed at the 300rpm and 2250rpm, respectively. The low-shear-processed sample has a coarse phase morphology with the PMMA domain size ranging from 200 nm to 2 μm. In addition, a weak interface can also be discerned for this sample. On the other hand, all the PMMA domains are homogeneously dispersed in the PC matrix for the high-shear-processed sample. The domain size of PMMA is less than 50 nm, which is significantly smaller than that in the low-shear-processed sample.

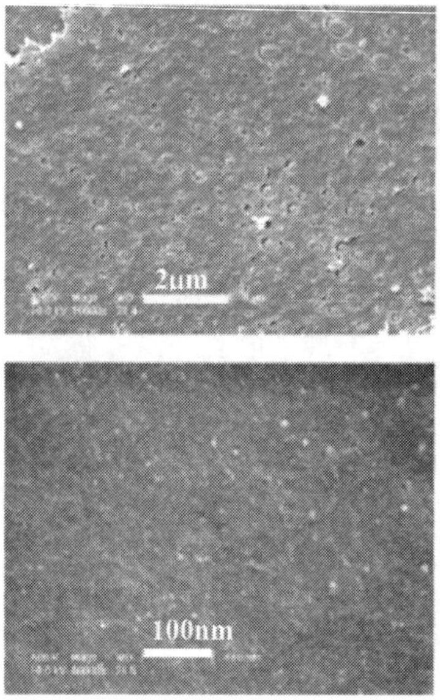

Figure 2. SEM images for (a) PC/PMMA (=80/20) blend processed under 300rpm, and (b) high-shear-processed PC/PMMA (=80/20) blend processed under 2250rpm.

Figure 3 shows the quantitative UV–vis transmittance spectra of pure PC, pure PMMA, and PC/PMMA (=80/20) blends prepared at the high and low shear rates. All samples have a thickness of about 500 μm. PMMA exhibits a very high transmittance of about 93% above a wavelength of 450 nm, while pure PC has a transmittance of about 82% in the same wavelength region. The high-shear-processed blend exhibits a UV-vis spectrum between those of PC and PMMA. The transmittance in the visible region is 91%, indicating the high clearance of the high-shear-processed blend. In contrast, the sample processed under the low screw rotation speed, i.e. 300 rpm, has no transmittance in the entire wavelength region. In addition, the high-shear-processed blend exhibits excellent optical transparency, while the low-shear-processed sample is opaque by the apparence.

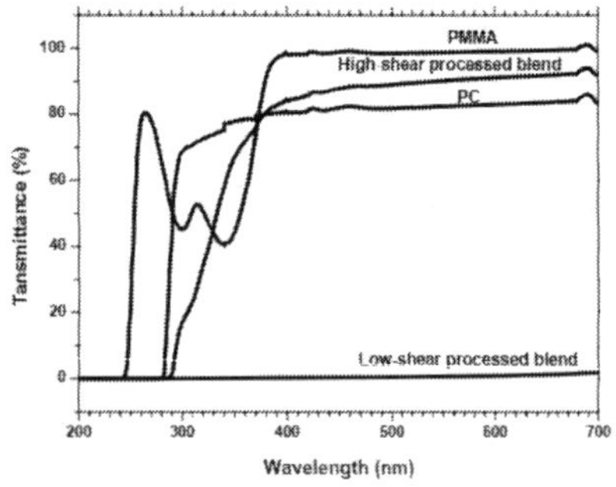

Figure 3. UV-vis transmittance spectra of pure PMMA, pure PC, low-shear-processed PC/PMMA (=80/20) blend, and high-shear-processed PC/PMMA (=80/20) blend

SBBS/CNTs composites by high-shear processing [3]

The dispersion of fillers in a polymer matrix has been reported to be dependent upon the shear stresses exerted on the fillers during melt mixing. The shear stress can overwhelm the electrostatic and van der waals interactions in the fillers and lead to the breakup of the filler agglomerates. This principle should also be applicable to CNT/polymer systems.

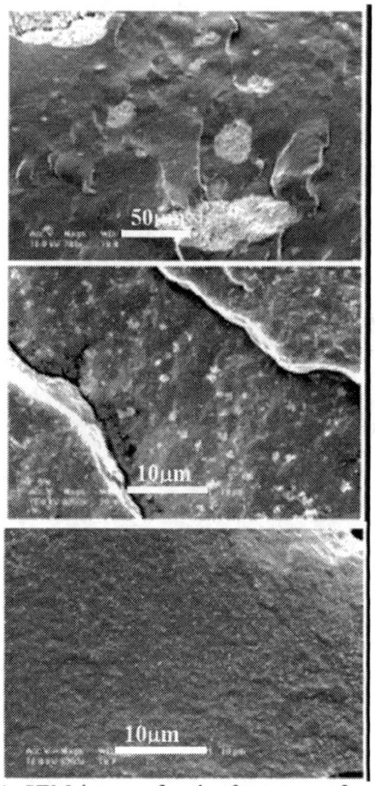

Figure 4. SEM images for the fracture surface of the MWNT/SBBS composites with the shear rates of (a) 300rpm, (b) 1000rpm, and (c) 2000 rpm

Figure 4 shows SEM images of the fracture surfaces of the poly(styrene-b-butadiene-co-butylene-b-styrene) (SBBS) / unmodified multiwalled carbon nanotubes (UMWNTs) composites prepared under various screw rotation speeds. The bright parts in the images are the MWNTs as a result of their high conductivity. Many MWNT aggregates with sizes ranging from 5 to 50 μm are clearly observed in Figure 4 (a). This means that the MWNTs are not well dispersed in the SBBS matrix and most of the MWNTs are stacked together when processed at the screw rotation speed of 300 rpm. Note that the screw rotation speed of 300 rpm is even higher than that of conventional extruders and injection molding machines. The size of the CNT aggregates was significantly decreased to sub-μm scale upon increasing the screw rotation speed to 1000 rpm, as seen in Figure 4 (b). The MWNTs are homogeneously distributed in the SBBS matrix without any aggregates for the composites under 2000 rpm. The typical diameter of the bright regions in the SEM image in Figure 4 (c) ranges from 20 to 50 nm, which is consistent with the diameter of a single carbon nanotube, indicating that most of the MWNTs are dispersed in the matrix as individual tubes. It is apparent that the dispersion of pristine MWNTs is greatly improved upon increasing the exerted shear rate and a homogeneous dispersion is successfully achieved by using the screw rotation speed of 2000 rpm. It is considered that high screw rotation speed imparts high shear stresses on the MWNTs agglomerates, which may skew the stacks of carbon nanotubes and finally separate them into individual tubes.

Figure 5 shows the stain-stress curves of the neat SBBS and the composites prepared at different shear rates. The modulus of the all composites is higher than that of the neat SBBS because of the strengthening effects of the inorganic filler. However, the strain-stress behaviors are different for the composites prepared under different shear rates even with the same MWNTs loading contents. The modulus of the nanocomposites is higher than that of macrocomposites under low shear. The better reinforcement effects of MWNTs in the nanocomposites than in the macrocomposites are obviously due to the fine carbon nanotube dispersion in it. Moreover, the elongation at break of the nanocomposites is as high as 917%, which is also much larger than the value (628%) for the nanocomposites. The lower elongation at break for the low shear processed sample can again be attributed to the poor dispersion of MWNTs in the polymer matrix. The existence of MWNT aggregates may decrease the elongation of nanocomposites.

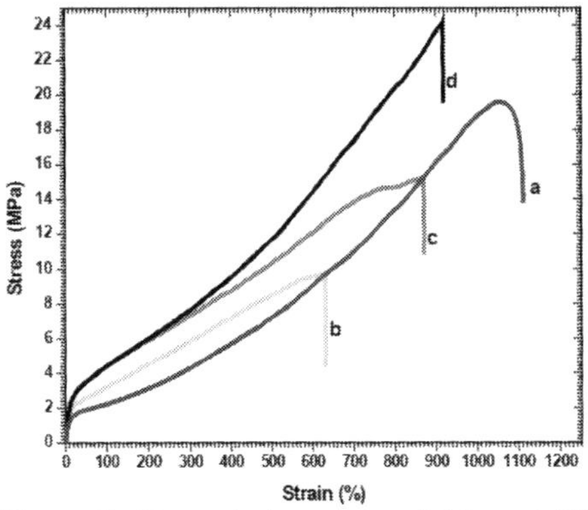

Figure 5. Stress-strain curves of (a) neat SBBS, (b) composites with the shear rate of 300rpm, (c) composites with the shear rate of 1000rpm, and (d) composites with the shear rate of 2000rpm.

It should be mensioned that the high-shear processing technique can also be used for preparing other polymer blends and polymer composites.[4-9]

Conclusion

The investigation indicated that the high-shear processing technique is an effective method for fabricating the high performance multicomponent polymeric materials. The miscibility between the components (both for resin/resin and for resin/filler systems) is improved by the shear stress during melt mixing. The prepared materials exhibit improved physical properties.

References

[1] Shimizu, H.; Li, Y.; Kaito, A.; Sano, H. *Macromolecules* **2005**, *38*, 7880.
[2] Li, Y; Shimizu H. Submitted for publication.
[3] Li, Y.; Shimizu, H. *Polymer* **2007**, *48*, 2203.
[4] Shimizu, H.; Li, Y.; Kaito, A.; Sano, H. *J. Nanosci. Nanotech.* **2006**, *6*, 3923.
[5] Li, Y.; Shimizu H. *Eur. Polym. J.* **2006**, *42*, 3202.
[6] Li, Y.; Shimizu H. *J. Polym. Sci. Part b: Polym. Phys.*, **2007**, 45, 2707.
[7] Li, Y.; Shimizu, H. *Macromolecules* **2009**, *42*, 2587.
[8] Chen, G.X.; Li, Y.; Shimizu, H. *Carbon* **2007**, *45*, 2334.
[9] Chen, G. X.; Shimizu, H. *Polymer* **2008**, *49*, 943.

EFFECT OF THE PROCESSING CONDITIONS ON THE RESIDENCE TIME OF A RUBBER MELT IN A TWIN-SCREW MICRO-COMPOUNDER

Y. Hung*, C. Carrot, Y. Chalamet

Université de Lyon, F-42023, Saint-Etienne, France; CNRS, UMR5223, Ingénierie des Matériaux Polymères, F-42023, Saint-Etienne, France ; Université de Saint-Etienne, Jean Monnet, F-42023, Saint-Etienne, France - yvong.hung@univ-st-etienne.fr; carrot@univ-st-etienne.fr; chalamet@univ-st-etienne.fr

Abstract - Twin-screw extrusion is a common processing technique to obtain composite materials. The final properties are related to the processing conditions of the melt. In particular, the residence time in the twin-screw extruder is a reliable parameter to characterize the thermal and mechanical mechanisms that drive the dispersion of filler within the matrix of a composite material. At the laboratory scale, this parameter remains still relevant with micro-compounder tools. This work focuses on the thermal and mechanical history through residence time measurements during melt compounding of a rubber matrix SBR by with a twin-screw micro-extruder of the Haake Minilab II® type. The system is based on two conical intermeshing screws ending in a by-pass that drives the polymer into an integrated backflow channel or alternatively into the die. Two pressure transducers are integrated in the backflow channel. The system allows compounding in circulation mode (closed route) or extrusion mode (opened route).
The residence times were measured by pressure control and by using color tracers. The influence of the compounding mode, the effect of the temperature and the rotation speed are discussed. Associated flow rates were compared with two models whose one depends on the shear thinning behavior of SBR.

Introduction

Processing of polymer nanocomposite is one of the leading topics of polymer material research. In this area, twin screw extrusion is a common processing technique to obtain composite materials. The final properties depend on the processing conditions of the melt as shown by Dennis et al. (2001) and Cho and Paul (2001). Among the processing conditions, feed rate, screw design, screw speed and temperature are essential parameters. Other parameters like residence time distribution is also necessary to understand the thermal and mechanical mechanisms that drive the dispersion of filler within the composite material. Unfortunately, in most of the twin-screw extruders, residence time data are not directly provided because the measurement equipment is not available. In addition, the residence time distribution depends of the screw design that can be complex. Also, in some cases the composite experiences multiple runs into the extruder to improve the quality of filler dispersion, as discussed by Russo (2007). In any case, the residence time becomes a critical parameter of the compounding process
At laboratory scale, the problem is still relevant within small scale twin-screw micro-compounder tools like Haake Minilab II. This device is based on a conical twin screw compounder with an integrated backflow channel. Two pressure transducers are integrated in the backflow channel. The system allows compounding in circulation mode (closed route) or extrusion mode (opened route). Thanks to recirculation mode and the size of the device, processing polymer with the micro-compounder allows multiple successive passages into the twin screw area where the highest shear rate values are encountered. In this case, the analysis of the processing conditions and their relation to the final properties of materials requires the knowledge of the residence time distribution or at least a mean residence time. .
This work focuses on the description of a method that enables to measure a mean residence time in the twin-screw region. The influences of the compounding mode, temperature and screw speed are investigated. The associated rate of flow is studied and compared with two models.

Experimental

Materials

The polymer matrix used in this study is a synthetic copolymer of styrene and butadiene (SBR) from MFP Michelin. This SBR has the following microstructure: 25 wt-% of styrene, 24 wt-% of butadiene 1-2 and 50 wt-% of butadiene trans 1-4. The main molecular characteristics are summarized in the table 1. The rheological behavior in the melt can be well described by a power-law with the parameters of table 1.

Table 1 - Main characteristics of SBR matrix

M_n(g/mol)	115200
M_w(g/mol)	225000
Ip	1.96
Density	0.938
*K; n (80°C)	114600; 0.35
*K; n (100°C)	81150; 0.39

*K : consistency; n : shear thinning coefficient

Processing

The micro-compounder used is a Haake Minilab II from Thermo Scientific. The device is based on a conical twin screw to transport and mix the polymer, with and integrated backflow channel which allows the matter to move back to the feeding zone. The system allows compounding in circulation mode (closed route) or extrusion mode (opened route). The raw material is inserted in the feeding hopper upstream from the twin screw zone which works in intermeshing co-rotating mode. Two pressure transducers (P1, P3) are integrated in the backflow channel to allow the measurement of pressure drop in circulation mode (figure 1). The torque, the temperature, the screw rotation speed and the pressure data are measured during the compounding. In this investigation, the SBR is fed through the hopper and the experiments are carried out in circulation or extrusion mode at two temperatures (80°C and 100°C) with four screw rotation speeds (10, 40, 80, 120 rpm).

Figure 1 – Picture (above) and scheme (below) of the Haake Minilab II device

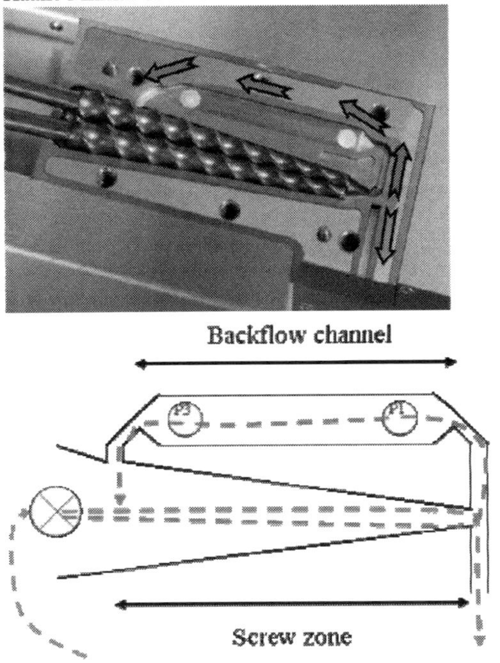

Residence time measurement method

Residence time measurement is made with the help of a color tracer. A color pellet tracer is inserted in the flow during the compounding and its progress at different times is followed. Several cases need to be considered depending on the running mode and the transient or steady flow rate:

- In circulation mode, after feeding the SBR, a transient state settles during the first lap. In addition to measurements by color pellet tracer, the flow front can also be followed by the pressure transducers P1 and P3. The end of the cycle matches with the stabilization of pressure on P1 and P3 (figure 2). Thus, in addition to the transient cycle time $t_{transient\ cycle}$, the residence time $t_{transient\ screw}$ in the screw zone is obtained when the flow front reaches the pressure transducer on P1.

Figure 2 - Descriptive scheme of the pressure measurement in recirculation mode and transient state

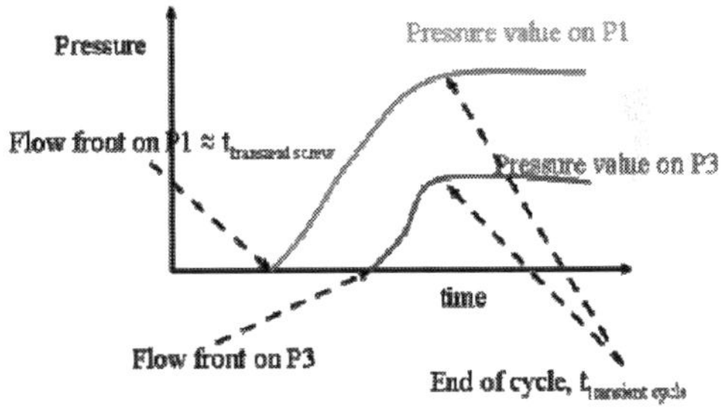

- In circulation mode, from the second lap, the steady state settles and the residence times are measured by color pellet tracer. The method allows only the measurement of the steady cycle time $t_{steady\ cycle}$. It is not possible to obtain directly the residence time in the screw zone in steady state. To value the $t_{steady\ screw}$, we need first to calculate the residence time $t_{backflow\ channel}$ in the backflow channel. For this purpose, we consider a parallel plate flow in the backflow channel between pressure transducer on P1 and P3. In these conditions, the relationship between the flow rate Q and the pressure drop ΔP between P1 and P3 is described by (1). Besides, the relationship between the flow rate Q and the residence time $t_{backflow\ channel}$ in the backflow channel is described by (2).

$$Q = \frac{n}{2n+1}.W.H.(\frac{\Delta P}{K.L})^{\frac{1}{n}}.(\frac{H}{2})^{\frac{n+1}{n}} \quad (1)$$

$$t_{backflow\ channel} = \frac{Q}{W.H.L} \quad (2)$$

W, H and L are respectively the width, the height and the length of the backflow channel. From (3), we deduce the residence time in the screw zone $t_{steady\ screw}$:

$$t_{steady\ screw} = t_{steady\ cycle} - t_{backflow\ channel} \quad (3)$$

Figure 3 - Pressure drop function of the time in recirculation mode

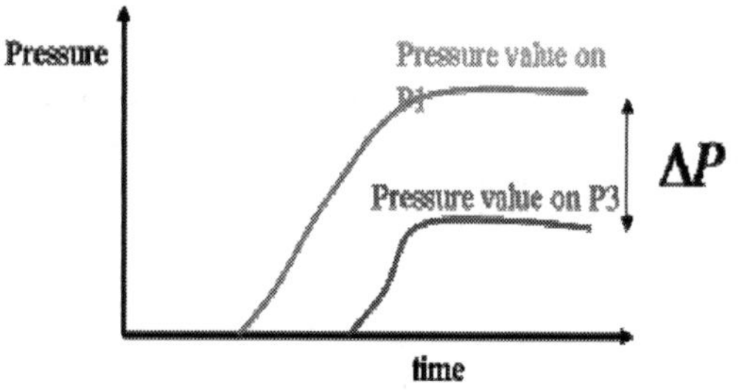

- In extrusion mode with multiple runs, the polymer which ways out the die is eventually fed again in the hopper. After torque stabilization indicating that a steady state is settled, the color tracer is introduced and the residence time $t_{extrusion\ screw}$ measurement starts (figure 4).

Figure 4 – Descriptive scheme of the residence time in extrusion mode

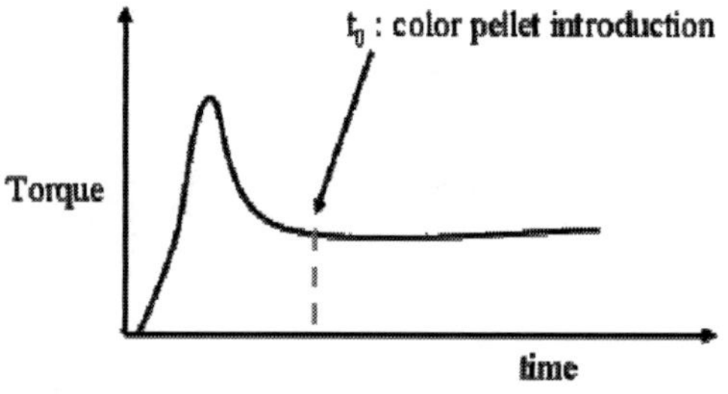

Results and Discussion

Residence times

Figures 5 and 7 present the influence of the screw rotation speed on the residence times in the screw zone for various running modes and natures of flow (transient or steady) at 80°C and 100°C. Whatever the mode and the nature of flow (transient or steady), a dependence of the residence time with the temperature is clear. A higher processing temperature induces a shorter residence time because of the temperature dependence of the matrix viscosity. Similarly, an increasing screw rotation speed causes a drop of the residence time. This result is related to the shear thinning behavior of the matrix. Moreover, whatever the temperature and the screw rotation speed, the residence times can be ranked in this manner:

$$t_{transient\ screw} < t_{screw\ extrusion} << t_{steady\ screw}$$

This can be explained because during the first lap in the recirculation mode and in transient flow, the feeding is forced by the operator. The same reason explains the values obtained in extrusion mode, since the polymer that ways out the die is re-fed into the hopper by the operator. In the recirculation mode and in steady state conditions, the system is self-operating, the feeding is not forced and the residence time is longer.

Figure 5 - SBR matrix residence times in the screw zone function of the screw rotation speed at 80°C depend on the different modes: recirculation in transient state (blue), recirculation in steady state (red), extrusion in steady state (green). Comparison with the model $t=\alpha^{-1}.N^{-1}$ (---) and the model $t=\alpha^{-1}.N^{-(1-n)}$ (—)

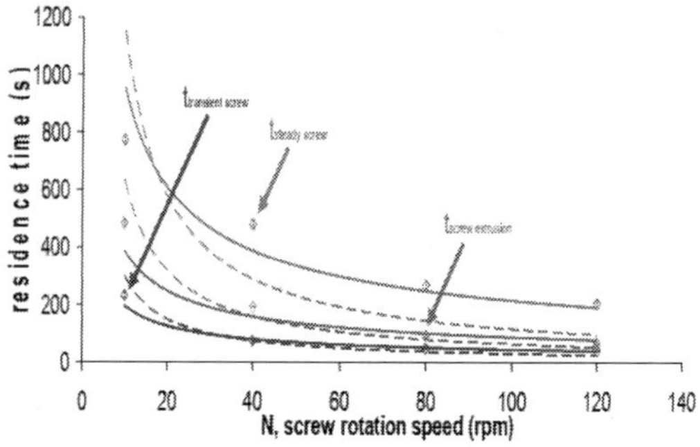

Figure 6 – Volume flow rate function of the screw rotation speed at 80°C depend on the different modes: recirculation in transient state (blue), recirculation in steady state (red), extrusion in steady state (green). Comparison with the model $Q_{vol.} = \alpha.N$ (---) and the model $Q_{vol.} = \alpha.N^{(1-n)}$ (—)

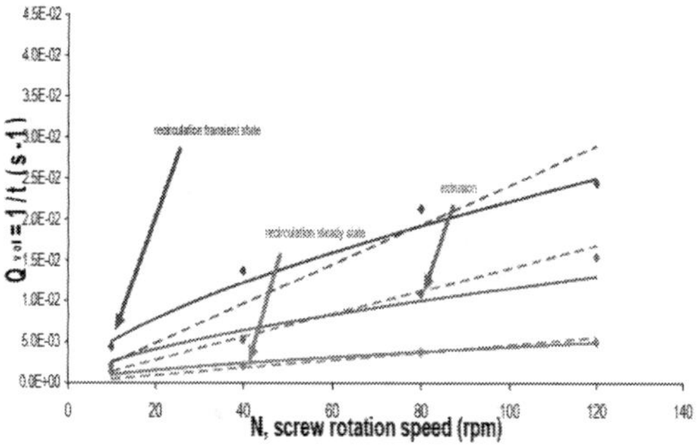

Associated flow rates

In extrusion processing, for polymer melts with Newtonian behavior, it is usually convenient to fit the flow rate by a Maddock's law:

$$Q = \alpha.N - \frac{\beta}{\eta(N)}.\frac{\partial P}{\partial z} \quad (4)$$

Where α and β are coefficients that only depend on the screw design and η is the Newtonian viscosity. If the material displays a shear thinning behavior as for SBR, it is necessary to take into account the shear rate (screw rotation speed N) influence on the viscosity. Besides, it is more convenient to fit flow rate than the residence time. Thus, from the residence time values, the volume flow rate Q_{vol} ($1/t_{residence\ time}$) were compared with two simple theoretical models:

$$Q_{vol} = \alpha.N \quad (5)$$

$$Q_{vol} = \alpha'.N^{(1-n)} \quad (6)$$

The model (5) shows a linear dependence between the flow rate and the screw rotation speed whereas the model (6) takes into account the shear thinning behavior of SBR through the shear thinning coefficient n. Figures 6 and 8 shows that the shear thinning model (6) describes consistently the experimental values.

Figure 7 - SBR matrix residence times in the screw zone function of the screw rotation speed at 100°C depend on the different modes: recirculation in transient state (blue), recirculation in steady state (red), extrusion in steady state (green). Comparison with the model $t=\alpha^{-1}.N^{-1}$ (---) and the model $t=\alpha^{-1}.N^{-(1-n)}$ (—)

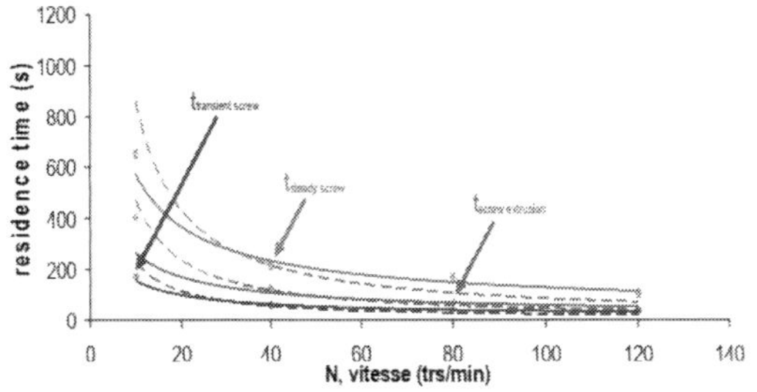

Figure 8 - Volume flow rate function of the screw rotation speed at 100°C depend on the different modes: recirculation in transient state (blue), recirculation in steady state (red), extrusion in steady state (green). Comparison with the model $Q_{vol.} = \alpha.N$ (---) and the model $Q_{vol.} = \alpha.N^{(1-n)}$ (—)

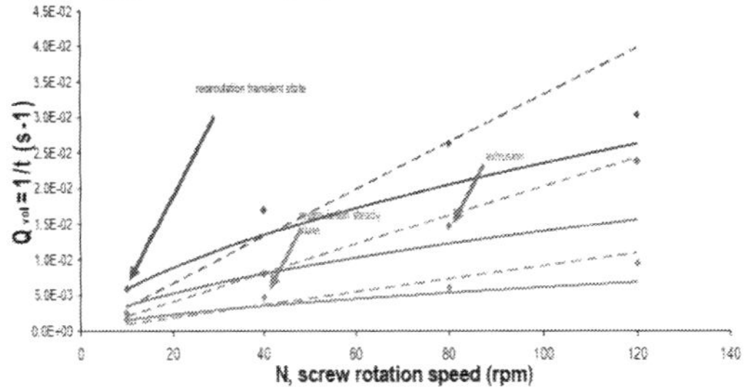

Conclusion

The residence time is a key parameter to understand and control the thermal and mechanical mechanisms that drive the processing of composite materials in an extruder. This work presents a method to estimate residence time in the screw zone of a laboratory scale Haake Minilab II extruder device. This was performed by use of a color pellet tracer and by pressure measurements. The results show that the residence time depends on the running mode. Thus, the comparison between the different mode reveals the following ranking: $t_{transient\ screw} < t_{extrusion\ screw} << t_{steady\ screw}$. The associated flow rate was fitted by two simple models. The results show that the shear thinning fitting model describes efficiently the experimental values. The results presented in this work will be used to improve the compounding of lamellar zirconium phosphate nanocomposites based SBR by melt processing.

Acknowledgements

This work was sponsored by the Pôle de compétitivité AXELERA, the CNRS, MFP MICHELIN and RHODIA for which grateful acknowledgement is made.
The TEM acquisition was supported by the Centre de Microscopie Electronique Stéphanois (St-Etienne)
The XRD acquisition was supported by the Centre de Diffractométrie Henri Longchambon (Lyon)

References

Journal: Cho, J. W., Paul, D. R., "Nylon 6 nanocomposites by melt compounding", Polymer, **42**, 1083-1094 (2001)
Journal: Dennis, H. R. et al., "Effect of melt processing conditions on the extent of exfoliation in organoclay-based nanocomposites", Polymer, **42**, 9513-9522 (2001)
Journal: Hasegawa, N. et al., "Nylon 6/Na-montmorillonite nanocomposites prepared by compounding Nylon 6 with Na-montmorillonite slurry", Polymer, **44**, 2933-2937 (2003)
Journal: Russo, G. M. et al., "Rheological and mechanical properties of nylon 6 nanocomposites submitted to reprocessing with single and twin screw extruders", Polymer Degradation and Stability, **92**, 1925-1933 (2007)

AUTHOR INDEX

AUTHOR INDEX

AUTHOR INDEX

AUTHOR INDEX

AUTHOR INDEX

AUTHOR INDEX

AUTHOR INDEX

AUTHOR INDEX

AUTHOR INDEX

26th Polymer Processing Society Annual Meeting 2010 (PPS-26)

Banff, Canada
4 - 8 July 2010

Volume 2 of 2

ISBN: 978-1-5108-3973-1

Printed from e-media with permission by:

Curran Associates, Inc.
57 Morehouse Lane
Red Hook, NY 12571

Printed by Curran Associates, Inc. (2017)

For permission requests, please contact Polymer Processing Society
at the address below.

Polymer Processing Society
810 E. 10th St.
Lawrence, KS 66044
USA

Phone: +1-800-627-0326
+1-785-865-9403
Fax: +1-785-843-6153

pps@allenpress.com

Additional copies of this publication are available from:

Curran Associates, Inc.
57 Morehouse Lane
Red Hook, NY 12571 USA
Phone: 845-758-0400
Fax: 845-758-2633
Email: curran@proceedings.com
Web: www.proceedings.com

TABLE OF CONTENTS

VOLUME 1

PLENARY SESSIONS

S01 – BIOMEDICAL MATERIALS

S02 – POLYMERS FOR ENERGY: OIL & GAS, FUEL CELLS AND SOLAR CELLS

S03 - RECYCLING AND REUSE OF POLYMERS

G01 – BLOW MOLDING, THERMOFORMING AND ROTOMOLDING

G02 - COMPOSITES

G03 – POLYMER NANOCOMPOSITES

G04 – EXTRUSION AND EXTRUSION PROCESSES

G05 – FIBER, FILMS AND MEMBRANES

G06 – INJECTION MOLDING AND MOLDS

G07 – MIXING AND COMPOUNDING

VOLUME 2

G08 – MORPHOLOGY AND STRUCTURE DEVELOPMENT

G09 – POLYMER BLENDS AND ALLOYS

G10 – POLYMER FOAMS

G11 – POLYMERS FROM RENEWABLE RESOURCES

G12 – PROCESS MODELING AND SIMULATION

G13 – PROCESS MONITORING, CONTROL AND SENSORS

G16 – RUBBER PROCESSING

I01 – CHRIS MACOSKO SESSION (ONLY BY INVITATION)

I02 – JOHN VLACHOPOULOS SESSION (ONLY BY INVITATION

I03 – JAMES L. WHITE SESSION (ONLY BY INVITATION)

OXYGEN ABSORPTION KINETICS OF SODIUM ERYTHORBATE BLENDS

Ángela García[1], Jorge Medina[1]*, Ronald Joven[1] and Oscar Álvarez[1]
[1]*Grupo Materiales y Manufactura CIPP-CIPEM. Universidad de los Andes. Bogotá, Colombia.*
jmedina@uniandes.edu.co

ABSTRACT
The oxygen scavenging kinetics of sodium erythorbate was studied for its implementation in food packaging as active agent. The oxygen scavenging efficiency of the sodium erythorbate was enhanced when this substance was blended with heptahydrated ferrous sulphate. To make possible the dispersion of these oxygen scavengers in a polymeric matrix, the active substances were encapsulated in SiO_2. This process had strong influence in both the kinetic constant and the order of reaction, since the oxidation of sodium erytorbate was reduced. A rheological characterization of a masterbatch composed of low density polyethylene (LDPE) and erythorbate blended with ferrous sulphate will be presented in this article to determine the changes in viscosity and rigidity modulus for possible conventional transformation processes. Strong potential for possible applications in packaging structures was found.

Introduction
The current consumer trend for food and beverage products is toward health, convenience, and enjoyment. However, the pre-processing needed to meet market requirements often means that natural protections are replaced with synthetic alternatives, as with packaging. Initially, barrier or passive packaging was proposed; this type of technology should be accompanied by techniques of controlled atmosphere or vacuum packaging.
Recently, the packaging industry has sought to go further using active packaging (Day., 2008) (M. Ozdemir, 2004) (P. Suppakul, 2003). Active packages are made of materials that serve not only as a barrier but also interact with compounds that reduce shelf life, to keep them within allowed limits for food consumption.
The oxygen present in the headspace of the package degrades the food since it can affect the flavor compounds by the olefins oxidation, the loss of nutritional value (Kurata & Sakurai, 1967) and the growth of aerobic pathogen microorganisms. Currently, active packaging with oxygen scavenging properties is mainly based on iron and oxidizable organic compounds (Frydrych, Foltynowicz, Kowalak, & Janiszewska, 2007) (Rogers, 2002), that have processing restrictions because they have low compatibility with polymeric materials, migration of oxidable agent, presence of color, limited capacity, and high cost (Brody, 2002).
This study proposes sodium erythorbate as a component of active packaging. Sodium erythorbate is a salt of the stereoisomer of vitamin C, erythorbic acid, and is widely used in the U.S.A and Asia as a food additive (E-316) with antioxidant and color preservation properties. Since 1999, erythorbate has been approved in Europe for conservation of meat and fish (Giorgia Spigno, 1999).
The advantage of using sodium erythorbate instead of common oxygen scavengers as ascorbic acid or vitamin C derives from the fact that it does not have vitamin power and can be obtained by fermentation rather than a bioprocess as with the vitamin.
Figure 1 shows the oxidation reaction of erythorbate, since it is an electrochemical reaction, adding to the system an electron acceptor like iron (II), it could accelerate the rate of reaction (1).

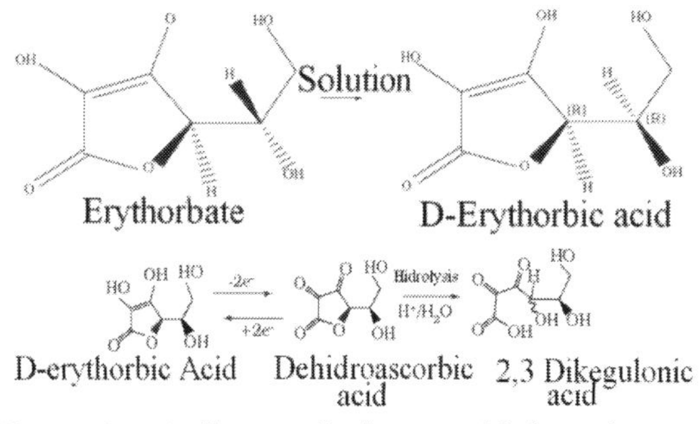

Figure 1 – Sodium erythorbate's oxidation (Jeane Santos de Rosa, 2007).

$$Fe^{+2} + 2e^{-} \rightarrow Fe^{0} \qquad (1)$$

Chemical kinetics
The relationship between the concentration either of reactants or products, and reaction time is expressed by kinetic models. The rate at which the oxygen diminishes as a function of its own concentration could be expressed by (2).

$$\frac{dC}{dt} = kC^{n} \qquad (2)$$

$C(t)$ is the oxygen concentration on time (t) in days, and k is the kinetic constant. The order of the reaction n can be chosen amongst the real numbers including zero.

Experimental
USP grade sodium erythorbate (E-316) with concentration 98% (Tecnas S.A.) and industrial grade iron sulphate was used as main materials.

In the sol-gel reaction hydrochloric acid to 37% w/w, analytical grade (Merck), and industrial grade of sodium silicate (Jadesi Ltd) with a ratio Na_2O:SiO_2 of 1:2,10 to 1:2,40 were used. To analyze the dispersion of the active agents, this particles were mixed in a matrix of LDPE (Dow 722) using a Brabender Plasticorder double screw internal mixer, at different angular velocity until torque stabilization at 160 °C.
The results were analyzed in software Minitab® 14.1 which a 95% of uncertainty.

Measurement of oxygen scavenger capacity
Oxygen scavenger capacity was assessed by packing a specific amount of the substance or mixture to be analyzed in cardboard boxes of constant volume, wrapped in a hermetic envelope of Aluminum foil sealed by heat with a Packaging Industries Model 12AS pneumatic hot sealer (temperature: 232.2 ° C and air pressure 344.7 kPa). Each package was punctured a given day with a Quantek 901 needle, in order to take an air sample and record the percentage of oxygen inside the specimen.

Activation
The conditions necessary for the effective reaction of sodium erythorbate with oxygen are called "activation". Three factors were studied: the ratio of sodium erythorbate to ferrous sulphate (w/w), moisture, and humidity exposure time, using two levels for each parameter. Table 1 presents the factors and levels and Table 2 specifies the conditions of each treatment.

Table 1 - Experimental design for sodium erythorbate activation.

	Level	
Factor	-1	+1
Weight relation Erythorbate/Ferrous sulphate (w/w)	0.39	2.00
Humidity (% HR)	50	90
Humidity treatment time (h)	0.5	2

The response variable of this experimental design is the percentage decrease of oxygen within the sample in each of the treatments after 5 days.

Encapsulation reaction sol-gel
To protect the active particles against the excessive heat typical of polymer processing, an encapsulation method is proposed. Sol-gel encapsulation is a neutralization reaction between sodium silicate (concentration 560g/l) and 1.55M hydrochloric acid. The oxygen absorbing compounds are added during the reaction at pH between 2 and 3, before obtaining a gel (Brinker, Ashley, Bhatia, & Singh, 2002).

The gel obtained is aged, washed with distilled water and filtered under vacuum (-15 mmHg) twice. Afterwards, it is dried until constant weight (less than 1% variation) in a furnace, with atmosphere of air at a constant temperature of 60°C.

Table 2- Treatments for studying the activation of sodium erythorbate

Standard order	Weight ratio	Humidity relative	Humidity exposure time
OE 1	-1	-1	-1
OE 2	1	-1	-1
OE 3	-1	1	-1
OE 4	1	1	-1
OE 5	-1	-1	1
OE 6	1	-1	1
OE 7	-1	1	1
OE 8	1	1	1

To determine the viability of dispersion of these oxygen scavengers in a real polymeric package, a rheological analysis is presented.

Viscoelastic properties
The behavior of the LDPE in the presence of sodium erythorbate was analyzed by a L4 orthogonal array, where the control parameters are (Table 3):
• Particle size.
• Weight percent sodium erythorbate in LDPE.
• Mixture rate

Table 3- Parameter of orthogonal array for experimental study of erythorbate's effect on viscoelastic properties.

	1	2
Particle	< 25μm	90<x<180 μm
%wt.	5%	10%
ω	50rpm	30rpm

Designed experiments in L4 orthogonal array shown in Table 4.

Table 4 - Experiments of the orthogonal array for experimental study of the effect of sodium erythorbate on viscoelastic properties.

Experiment number	A Particle size	B (g)	ω
1	< 25μm	2,57 g	50rpm
2	< 25μm	5.14 g	30rpm
3	90<x<180 μm	2,57 g	30rpm
4	90<x<180 μm	5.14 g	50rpm

Behavior of viscosity under shear loads
Assessment of the influence of particles of sodium erythorbate and SiO_2 over LDPE conducting a robust design of experiments by the method of G. Taguchi (Taguchi, 2005), with the parameters described in Table 5.

Table 5 - Control parameters for robust design of viscosity under shear loads.

	1	2
Temperature (°C)	140	150
Particle	SiO_2	Sodium erythorbate
%wt.	2,5%	7,5%
ω	50rpm	30rpm

Results and Discussion

Activation

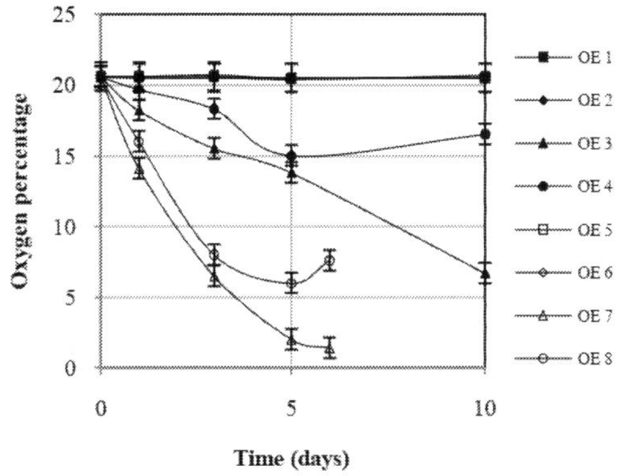

Figure 2 – Evolution of oxygen concentration with different activation treatments.

Figure 2 presents the results of the activation study. The Pareto chart (Figure 3) presents the absolute effect of each factor. With 95% confidence, the humidity level has a significant effect because erythorbate oxidation is promoted by the presence of compounds in solution.

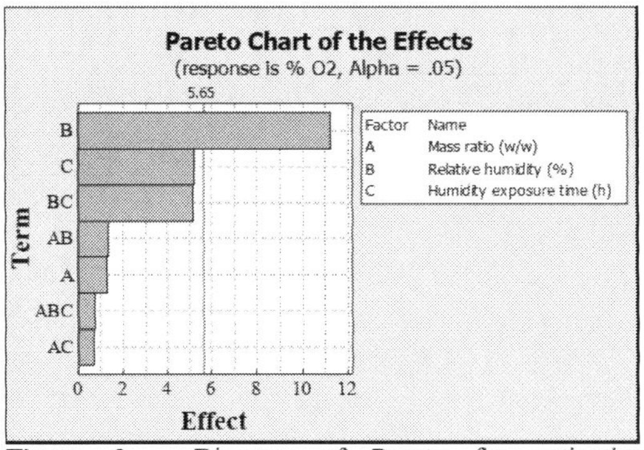

Figure 3 – Diagram of Pareto for activation experimental design.

Given that performance and in order to decide about the proper relationship between erythorbate sodium and ferrous sulphate, the cost of raw materials with a commercial value of sodium erythorbate about 6 USD / kg and ferrous sulphate 2 USD / kg, was evaluated. Each mass ratio to produce 1 kg of mixture was analyzed and we obtained the following results (Table 6).

Table 6 - Costs of two different ratios of erythorbate and sulphate.

Ratio (w/w) Erythorbate/sulphate	Sodium erythorbate	Ferrous sulphate	Total price
0,39	0,28 kg	0,72 kg	3,12 USD/kg
2,00	0,67 kg	0,33 kg	4,68 USD/kg

Therefore it is economically and technically coherent to use a ratio of 0.39 erythorbate/iron sulfate (w/w).

Absorption of oxygen mixtures with encapsulated sodium erythorbate

Encapsulation has a retarding effect on the ability to absorb oxygen from the mixture of sodium erythorbate and iron sulfate (Figure 4); similar percentages are achieved in five times as long.

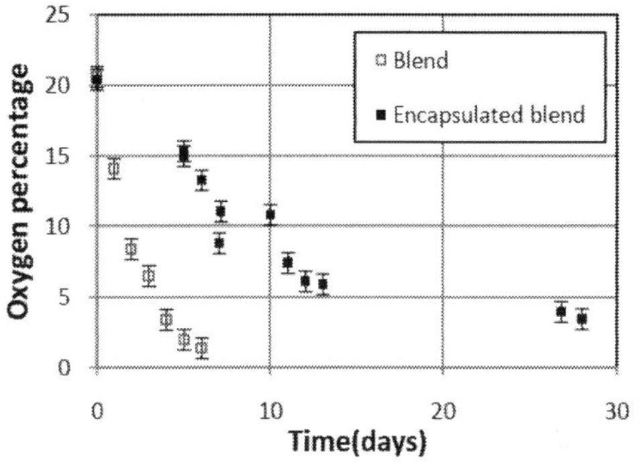

Figure 4 – Capacity of oxygen absorption of the blend erythorbate/sulphate (□) and the blend encapsulated (■).

Kinetics

The best fit of data for oxygen absorption by the erythorbate and ferrous sulphate mixture were achieved with a model of order 0.9, as shown in Figure 5 and obtained a rate constant $0.0543\frac{\%}{\text{day }\%^{0.9}}$.

Therefore, the kinetic model follows the equation (3)

$$\frac{dC}{dt} = kC^{0.9} \qquad (3)$$

In the case of the encapsulated mixture (Figure 6), the reaction order increased to 1.84 but the rate constant decreased an order of magnitude to a value of $0.0092\frac{\%}{day\ \%^{1.84}}$.

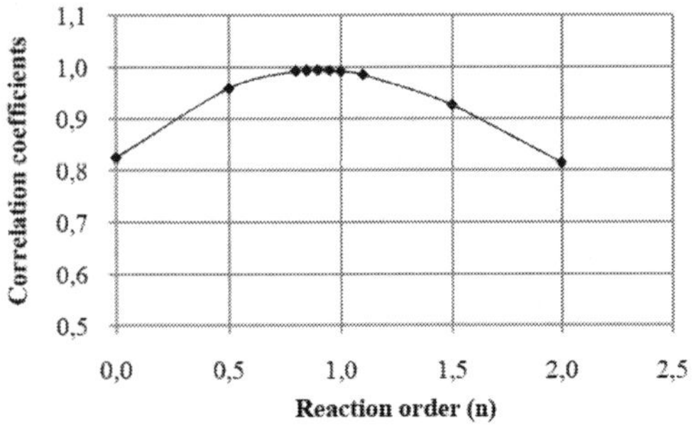

Figure 5 - Adjust of experimental data of oxygen absorption of erythorbate/ferrous sulphate to different reaction orders.

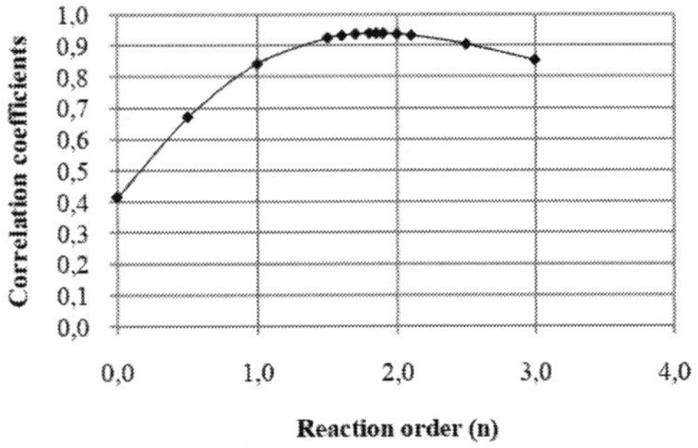

Figure 6- Adjust of experimental data of oxygen absorption of erythorbate/sulphate encapsulated to different reaction orders.

Viscoelastic properties

The viscoelastic behavior of sodium erythorbate and low density polyethylene mixture is shown in Figure 7.

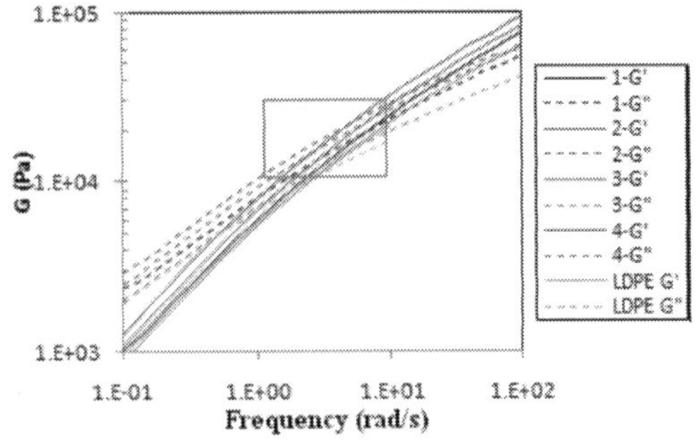

Figure 7 - LDPE + Sodium erythorbate at 160 ° C. Characterization in parallel plate rheometer, frequency sweep. (Experiments numbers are the L4 array in table 4)

Behavior of viscosity under shear loads

In order to inquire about the influence of particles and sodium erythorbate SiO_2, a robust design of experiments by the method of G. Taguchi was conducted (Taguchi, 2005).

Table 7 - Control parameters for robust design in the study of shear loads.

	1	2
Temperature (°C)	140	150
Particle	SiO_2	Sodium erythorbate
%wt.	2,5%	7,5%
ω	50rpm	30rpm

The averages of the shear load measured under the conditions of the control matrix are shown in the table 8 where can be seen that the SiO_2 is the particle that generates greater resistance under shear load.

Table 8- Average torques for robust design in the study of shear loads.

	1	2
Temperature (°C)	13,00 Nm	11,33 Nm
Particle	**13,00 Nm**	10,67 Nm
%wt.	10,67 Nm	11,67 Nm
ω	12,83 Nm	12,67 Nm

Behavior of viscosity under normal loads

Figure 8 presents a comparison between the stress results of pure LDPE and sodium erythorbate + LDPE at 135°C. It evidences how even with added particles the behavior is still according to power law, characteristic of a pseudo plastic fluid.

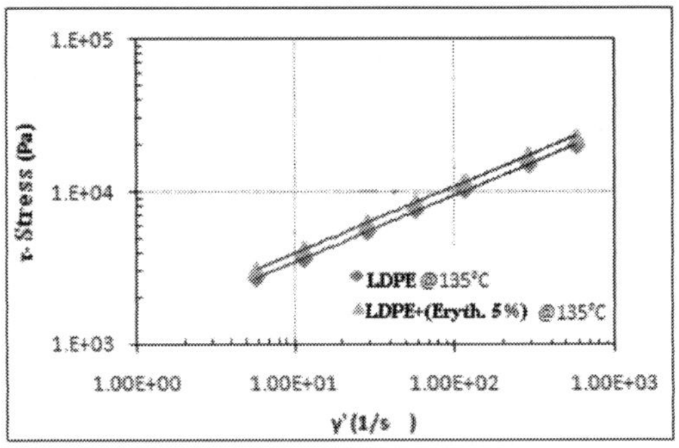

Figure 3 - Stress of LDPE pure and blended with sodium erythorbate.

Conclusions

Amongst the levels of the factors affecting the activation, the best conditions for improving the absorption capacity of the mixture of sodium erythorbate/iron sulfate are a weight ratio 0.39w/w exposed for 120 minutes to an environment of 90%RH, being able to decrease the exposure time to 30 minutes without significantly effects over the scavenging properties of materials.

The values of fractional reaction order of active system with and without encapsulation are attributed to the presence of parallel and intermediate reactions (hydrolysis, isomerism) for the catalyzed oxidation of sodium erythorbate.

Comparing the kinetic values with commercial oxygen scavengers, there is a difference with respect the order of the reaction, while the commercial types are only of first order, in this research it was possible make a model with a more specific order (0,9 and 1,8) even with a simple measurement method.
The rheological analysis reproduces the most relevant implications of the dispersion of the active particles related with the extrusion process. This behavior was studied under three methods. First method: When the polymer is exposed to normal loads, the fact that the viscous behavior of LDPE is still pseudo plastic with active particles, and not a dilatant behavior that is typical of fluids with solid particles dispersed, can be acknowledged. Second method: According with the behavior under shear loads, the SiO_2 showed the highest increase in viscosity. The viscosity is limited to an exponential mathematical model based on the percentage by weight of solid particles added to the matrix of LDPE. Third method: the analysis of the viscoelastic response of the material showed that sodium erythorbate raises the values of G than a given $\dot{\gamma}$. The rheological values obtained of the polyolefin with the particles dispersed suggest a strong potential for implementation in layered packaging structures.
The oxygen scavenger capacity of sodium erythorbate was studied, a simple model to predict the diminishing of the oxygen in time was proposed and a preliminary processability study was begun; still is pending the kinetic analysis of the masterbatch and its characterization as a packaging material.

References

Brinker, C. J., Ashley, C. S., Bhatia, R., & Singh, A. K. *Patent No. US 6,495,352 B1*. United States. (2002).

Brody, A. *Active Packaging: Beyond Barries.* West chester, PA: Packaging Strategies. (2002).

Day., B. P.. *Active Packaging of Food. Smart Packaging Technologies for Fast Moving Consumer Goods.* (J. K. Butler, Ed.) John Wiley & Sons, Ltd. (2008)

Emilia Frydrych, Z. F.. Oxygen scavengers for packing system based on zeolite adsorbed organic compounds. *Zeolites to Porous MOF Materials – the 40th Anniversary of International Zeolite Conference.* R. Xu, Z. Gao, J. Chen and W. Yan. (2007)

Food and Drug Administration. Department of Health and Human Services. (2001, April 01). Title 21. Chapter I. *Part 133.169, 133.173,133.179,136.110,155.200,169.115,169.140,169.150,172.814,173.340,175.300,176.170,176.200,184.1063,184.1400* . United States of America: Code of federal regulations.

Giorgia Spigno, D. M. Stabilità termica dell'acido ascorbico e delléritorbato di sodio. *Industrie Alimentari* , 538-545. (1999).

Jeane Santos de Rosa, R. L. Desenvolvimento de um método de análise de vitamina C em alimentos por cromatrografia líquido de alta eficiência e exclusao iônica. *Ciênc. Tecnol. Aliment, Campinas* , 837-846. (2007).

M. Ozdemir, J. D. Active Food Packaging Technologies. *Critical Reviews in Food Science and Nutrition* , 185-193. (2004).

P. Suppakul, J. M. (2003). Active Packaging Technologies with an Emphasis on Antimicrobial Packaging and its Applications. *Journal of food science* , 408-420.

Rogers, B. D. Use of oxygen scavengers in flexible packaging--an overview and case studies. *Paper, Film and Foil Converter* , 80. (2002).

Taguchi, G. *Taguchi's quality engineering handbook.* (2005).

SOLID-STATE PROPERTY PREDICTIONS FOR ORIENTED GLASSY POLYMERS USING A COMBINED MELT-SOLID CONSTITUTIVE MODEL

D.S.A. De Focatiis[1]* and C.P. Buckley[2]

[1] *Department of Chemical and Environmental Engineering, University of Nottingham, Nottingham, UK – davide.defocatiis@nottingham.ac.uk*
[2] *Department of Engineering Science, University of Oxford, Oxford, UK – paul.buckley@eng.ox.ac.uk*

Solid-state properties of processed polymers can be strongly influenced by the frozen-in orientation resulting from molecular flow during processing. The precise degree of molecular orientation present is a complex function of the non-linear rheology and process parameters, and the rheological behaviour itself is dependent on the molecular structure of the polymer. The interaction of all of these aspects needs to be addressed in order to achieve genuine predictive capability of solid-state properties for processed polymers. The authors have recently proposed a new constitutive model for glassy polymers applicable to both melt and solid states. The model ascribes two contributions to the free energy: perturbation of (A) the inter-atomic potentials, and of (B) the conformational entropy of the entanglement network. Within the model (A) consists of a multi-mode glassy constitutive model proposed previously in our Oxford laboratory, and (B) consists of a multi-mode set of ROLIEPOLY equations first proposed for non-linear rheology of linear entangled polymer melts. The molecular awareness is thus present in (B), parameterized *only* through the linear viscoelastic spectrum and through the number of entanglements per molecule, and hence the associated Rouse and reptation times. In this paper the model is applied to predictions of the effects of orientation on solid-state properties, and compared to experiments on oriented specimens representing a range of typical processing conditions. Numerical predictions and experiments are performed on two grades of linear monodisperse polystyrene with varying degrees of frozen-in orientation. The properties investigated are the optical birefringence, the stress required to initiate crazing in oriented material, and the large strain mechanical response of oriented material below the glass transition. The comparisons demonstrate the wide applicability of this molecularly aware melt-solid constitutive model to the prediction of multiple solid-state properties of processed polymer products.

Introduction

The ease with which thermoplastic polymers can be quickly and economically melt-processed into complex parts is imperative to their widespread use. The process of driving polymer melts through narrow channels followed by rapid cooling almost inevitably results in some degree of frozen-in orientation. Indeed, many plastic products rely on orientation to achieve their desired function and mechanical properties. The presence of this orientation, whether intended or not, affects a wide range of solid-state properties of processed polymers in such a way as to make predictions of properties a challenging but necessary task.

The degree of orientation present within a particular product is a complex function of its rheology and the process parameters employed (Ward, 1975). The rheological behaviour itself is intrinsically linked to the molecular weight and its distribution, and to the polymer architecture. Consequently, as industrial polymer processing engineers have long known, mechanical properties of practical importance such as yield stress, large deformation response, craze resistance and fracture toughness in any given direction are highly sensitive to the grade of polymer employed and to the flow history encountered during processing (De Francesco and Duckett, 2004; Embery et al., 2007). There is an undeniable engineering need to achieve a predictive capability, for use in optimising the design of polymer products and processes.

A constitutive model able to predict solid-state properties of processed polymers needs to be applicable to both melt and solid states. Melt flow stretches polymer molecules within an entanglement network, and the structure of the polymer (molecular weight and architecture) governs the timescales for relaxation of orientation processes. Subsequent solid-state deformation continues these orientation processes in parallel with more short-range effects such as segmental diffusion and structural evolution.

A substantial proportion of published works on constitutive modelling of solid polymers can trace their origins to the one-dimensional model of Haward and Thackray (1968). The Haward-Thackray model ascribes two contributions to the free energy arising from (a) perturbation of inter-atomic potentials, relaxed by segmental flow, and (b) perturbation of conformational entropy of the entangled molecular network, represented in the glassy state as a crosslinked network.

Many authors have used this approach for modelling the response of initially isotropic polymers below T_g (Boyce et al., 1988; Buckley and Jones, 1995; Tervoort et al., 1998). However, there is growing recognition that, in order to achieve quantitative fits to data, entanglement densities have had to be employed that are physically inconsistent with those evaluated from melt rheology (Govaert et al., 2008). Some phenomenological approaches have been proposed to address relaxation of the entanglement network (Dooling et al., 2002; Palm et al., 2006), but none of these are truly applicable to melt deformations, nor to the prediction of changes resulting from molecular structure.

Accurate predictions of the non-linear flow behaviour of entangled monodisperse linear polymer melts has been achieved through the model of McLeish and coworkers (Graham et al., 2003). A simplified model more applicable to large scale computation, known as the ROuse model for LInear Entangled POLYmers, or ROLIEPOLY model, was subsequently developed (Likhtman and Graham, 2003) and validated within a Lagrangian flow solver for complex highly elastic flows (Bent et al., 2003; Collis et al., 2005). The use of such a rheological model to account for the conformational behaviour of polymers *within* a Haward-Thackray framework for solid-state deformation addresses a number of of the previous inconsistencies.

The constitutive model recently proposed by the authors (De Focatiis et al., 2010) is such a hybrid: it is a glass-melt model consisting of a set of ROLIEPOLY equations that govern the conformational entropy elasticity of the polymer and its relaxation by tube diffusion in the melt, and a multi-mode glassy constitutive model for deformation of PS developed previously in the Oxford laboratory (Wu and Buckley, 2004).

It will be shown that this hybrid model makes considerable progress in the prediction of solid-state properties of processed polymers. Predictions from the constitutive model are compared to experiments performed on a range of linear monodisperse polystyrenes. Orientation is introduced through three procedures designed to explore separately the effects of draw temperature, stress-relaxation prior to quenching, and draw ratio. The solid-state properties of oriented material investigated here are optical birefringence, craze initiation stress parallel to the direction of orientation, and subsequent sub-T_g large deformation response.

Experimental

Materials

The materials used in this study were two samples of monodisperse linear atactic polystyrene (PS) (PDI = M_w/M_n < 1.15), synthesised by living anionic polymerization and provided by Dr Lian Hutchings of the University of Durham. Molar mass measurements were performed by size exclusion chromatography on a Viscotek TDA 302 machine with refractive index, viscosity and light scattering detectors, and molar mass dstributions are illustrated in Figure 1. The polymers are referred to as AF (M_w = 262kg/mol) and AG (M_w = 518kg/mol) for consistency with previous publications on these materials (De Focatiis and Buckley, 2008; De Focatiis et al., 2008).

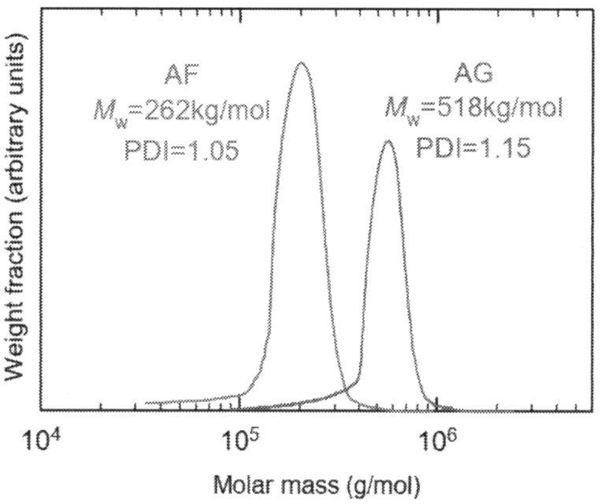

Figure 1 – Molar mass distributions of the two monodisperse atactic polystyrenes used in this study measured by SEC.

Rheometry

Linear melt rheology in shear was characterised for both materials using an Ares L2 rheometer with 10mm parallel plates by Dr John Embery of the University of Leeds. A temperature range of between 130°C and 230°C was used. The curves of G' and G" versus log(frequency) were shifted using time-temperature superposition to provide a single master curve for each polymer at a reference temperature of 170°C. The master curves are shown in Figure 2.

Non-linear rheology was characterised for both materials by Dr J. Embery of the University of Leeds using an Ares L2 rheometer with 10mm diameter cone and plate in transient shear, and with a Sentmanat extensional rheometer fixture in extension. A temperature of 170°C was used, with applied strain rates ranging from 0.01 s^{-1} to 6 s^{-1} in shear, and 0.06 s^{-1} to 10 s^{-1} in extension.

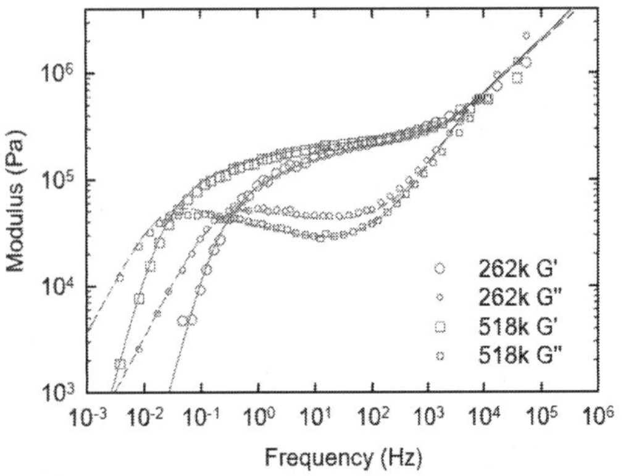

Figure 2 – Storage moduli (G' - large symbols) and loss moduli (G'' - small symbols) obtained from linear viscoelastic shear melt rheology, shifted to 170°C using time-temperature superposition, for the two polymers used in this study. Also shown are the moduli obtained from of the Likhtman-McLeish quantitative theory used to obtain the material parameters G_e, M_e and τ_e (G' - solid lines and G'' – dashed lines).

Specimen production

Isotropic bars of PS were formed by compression moulding polymer granules within a cavity mould in a hydraulic press heated to 170°C. In order to ensure a repeatable surface finish (essential for crazing studies) fresh sheets of disposable aluminium foil were used to line the surfaces of the mould. A detailed account including the recorded temperature history of such mouldings is available elsewhere (De Focatiis and Buckley, 2008).

Molecular orientation was introduced by melt-drawing the prismatic bars in an Instron 4204 testing machine fitted with an environmental chamber at a range of temperatures T around and above T_g, at a constant crosshead velocity corresponding to a nominal strain rate of 0.02 s^{-1}. The bars were gripped using pneumatic grippers and stretched uniaxially to a range of draw ratios λ between 2 and 7, and then quench cooled to below T_g, using a cold spray (cooling rate approximately 15°C/s). In some experiments a dwell time t was allowed at the end of stretching, for isothermal stress-relaxation to occur at constant grip displacement, before quenching. Strain was measured using an Instron non-contact video extensometer tracking black marks applied on the sample using a water-based ink. Temperature was monitored throughout the experiments using two thermocouples positioned close to the bars. The tensile load was recorded throughout the orientation and dwell process. The oriented bars were then stored at room temperature prior to re-drawing.

Three procedures were followed in order to explore systematically the separate roles of the three variables T, t and λ,:

Procedure I, varying T
Materials AF and AG were hot-drawn at a range of temperatures T from 105 to 135°C, to a fixed draw ratio $\lambda = 3$ at a constant nominal strain rate of 0.02 s^{-1}, and immediately quenched using the cold spray.

Procedure II, varying t
Material AF was hot-drawn at a temperature T =120°C, to a fixed draw ratio $\lambda = 3$ at a constant nominal strain rate of 0.02 s^{-1}, and a range of dwell times t from 1 s to 10^6 s were allowed prior to quenching, to allow relaxation to take place at the draw temperature.

Procedure III, varying λ
Material AG was drawn to a range of draw ratios λ from 2 to 7 at a constant nominal strain rate of 0.02 s^{-1} at a temperature T = 130°C, and immediately quenched using the cold spray.

The production of oriented specimens was challenging because of the limited availability of monodisperse materials precluding repetitions of the tests. In additon, polystyrene is prone to brittle fracture (below T_g) and to inhomogeneous deformation zones, or necks. The range of drawing conditions reported here indicate the range in which uniform specimens were able to be both produced and retested in the glassy state without such problems.

It is helpful to consider the various orientation procedures in terms of either the Weissenberg numbers associated with strain-rate and the various relaxation times of the polymer, or the equivalent Deborah numbers associated with dwell time and relaxation times. For example, $Wi_d^{AF} = \tau_d^{AF}\dot{\varepsilon}$ is the Weissenberg number associated with the disengagement, or reptation time τ_d^{AF} of polymer AF subjected to a strain rate $\dot{\varepsilon}$. A Weissenberg number much greater than unity indicates that relaxation does not occur during stretching, on the length scale associated with the corresponding relaxation time. Similarly, a Deborah number much greater than unity indicates that relaxation does not occur during the dwell time on the associated length scale.

Measurement of birefringence

The relative retardation R between the optic axes in the plane of the oriented bars was measured using an Olympus BH2 transmission optical microscope fitted with polarising optics and a Berek rotary compensator. Measurements were made at room temperature in white light. The optical birefringence $\Delta n = n_{\parallel} - n_{\perp}$ was then computed from $\Delta n = R / d$ where d is the specimen thickness. Measurements were taken on the oriented specimens prior to subsequent sub-T_g deformation and to creep crazing experiments.

Figure 3 shows the birefringence as a function of the experimentally recorded stress just prior to the application of the cold spray divided by absolute draw temperature, for both materials and all orientation procedures, illustrating the deviation from a linear stress-optical rule at high degrees of orientation.

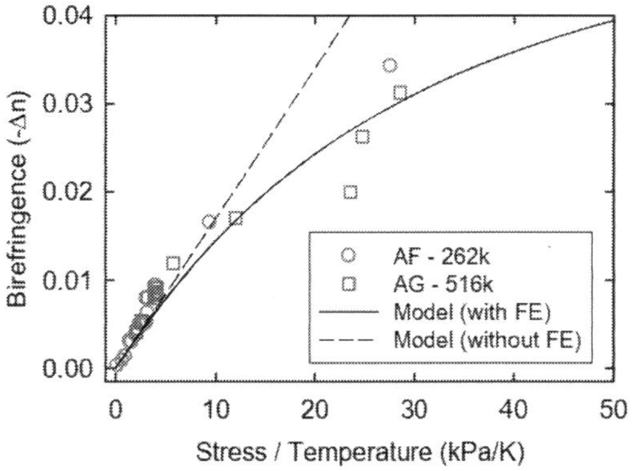

Figure 3 - Measurements of optical birefringence plotted against the experimentally recorded stress just prior to the application of the cold spray divided by absolute draw temperature, for both materials and all three procedures (symbols); constitutive model prediction of birefringence for the same range of conditions assuming a stress-optical coefficient $C = -4.5\times10^{-9}$ Pa^{-1}, with and without accounting for finite chain extensibility (lines).

Figure 4 illustrates measurements of birefringence as a function of draw temperature for melt-drawing performed according to Procedure I. Figure 5 illustrates measurements of birefringence as a function of post-drawing dwell time for melt-drawing performed according to Procedure II. Figure 6 illustrates measurements of birefringence as a function of draw ratio for melt-drawing performed according to Procedure III.

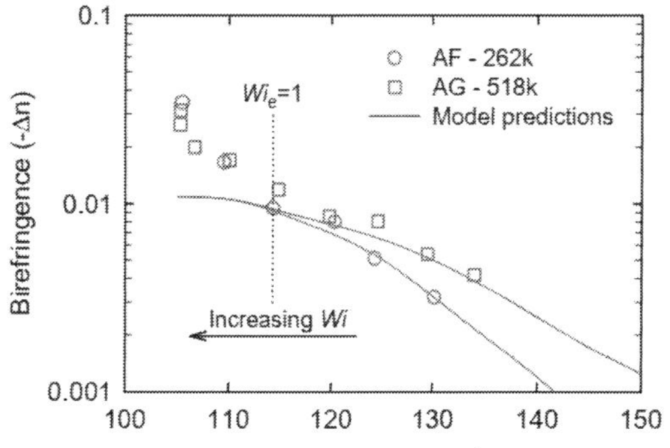

Figure 4 - Measurements of birefringence as a function of hot-drawing temperature T for melt-drawing performed according to Procedure I (symbols). Simulation predictions of birefringence computed using the constitutive model (lines). Also illustrated is the temperature at which the τ_e - based Weissenberg number is equal to unity.

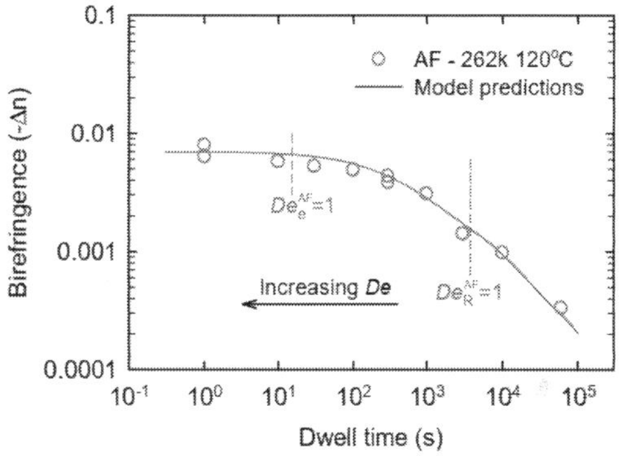

Figure 5 - Measurements of birefringence as a function of post-drawing dwell time t for melt-drawing performed according to Procedure II (symbols). Simulation predictions of birefringence computed using the constitutive model (line). Also illustrated are the times at which the τ_e and τ_R - based Deborah numbers are equal to unity.

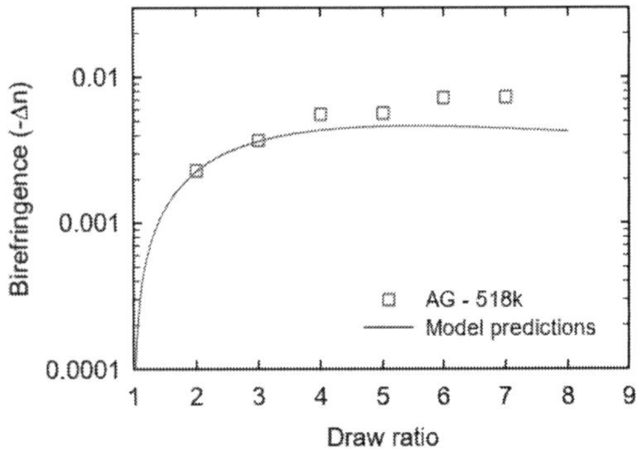

Figure 6 - Measurements of birefringence as a function of draw ratio λ for melt-drawing performed according to Procedure III (symbols). Simulation predictions of birefringence computed using the constitutive model (line).

Measurement of crazing stress

Miniature beam specimens with approximate dimensions 7 mm x 2 mm x 0.3 mm were cut from the oriented bars using a custom-made jig with parallel-sided single-bevelled blades. All miniature beam specimens were produced with the beam axis aligned with the orientation direction, since the oriented material was too brittle for specimens to be cut in the transverse direction. Specimens were soaked in analytical reagent grade diethylene glycol (DEG) for a minimum of 1 day prior to testing, after which no further solvent uptake could be measured. DEG is a mild crazing agent for PS, and was used in order to ensure the growth of fine, stable crazes required in the miniature specimens.

Crazing stress was measured in 3-point bending, using a miniature custom-built flexural creep rig, designed to accept miniature beam specimens cut from the oriented bars. A thorough account of the specimen preparation technique, the technique for measurement of visible craze initiation stress in the miniature specimens, and its use and validation against results for larger specimens, was provided in a previous publication (De Focatiis and Buckley, 2008) and only a condensed account is given here.

Each test consisted of subjecting the beam specimen, simply supported at its ends and still soaked in DEG, to a constant central load for a specified creep time. During this time the beam exhibited creep under the load, and, if the maximum stress was sufficient, crazes formed perpendicular to its axis. The crazes initiated at the surface where the tensile stress was highest. After unloading, the beam was removed from the rig and the surface was wiped clean of DEG before it was placed onto an angled support under a low power optical microscope. Transmitted light inclined at a small angle (e.g. 10°) to the specimen surface normal was then used to make the crazes visible. Within 1 hour of crazing, 5 measurements of the length of the crazed region were taken on each specimen across the width of the beam. The measurements were averaged to obtain the mean crazed length l_c. A minimum of four beam specimens were tested for each polymer under each orientation condition, although further repetitions were performed where sufficient material was available. In the present experiments, all creep tests were carried out at room temperature and creep times of 300 s were used throughout.

The elementary theory of bending of slender beams was employed to calculate the axial tensile stress at the surface of the beam at the extremities of the crazed region: i.e. at the points where crazes were just initiated and became visible under the conditions of the experiment:

$$\sigma_c = \frac{3P(l_s - l_c)}{2bd^2} \quad (1)$$

where P is the applied load, l_s is the support spacing, b the breadth and d the depth of the beam. In the results that follow, the mean values of σ_c are plotted, together with error bars representing 2 standard errors for each PS sample under each condition.

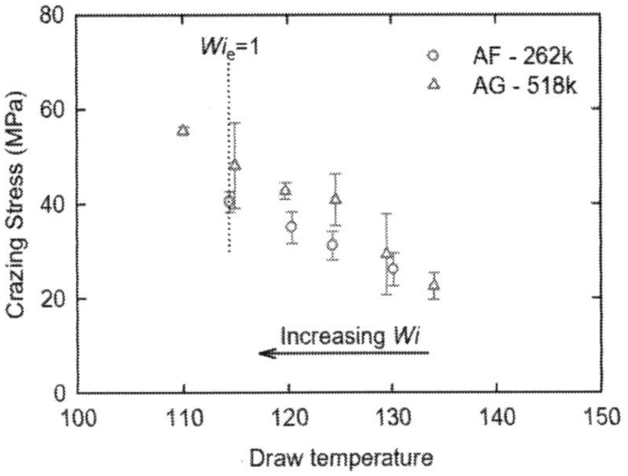

Figure 7 – Measurements of crazing stress as a function of draw temperature for specimens cut from melt-drawn tapes oriented according to Procedure I.

Measurement of sub-T_g mechanical response to large deformations

The oriented bars were subsequently tested in uniaxial tension in the same Instron testing machine as used for the orientation process, with the environmental chamber at a fixed temperature of 96±1°C, and a constant crosshead velocity corresponding to a nominal strain rate of $0.001 s^{-1}$. This temperature was chosen in the narrow temperature window between T_g and the temperature at which some of the bars became too brittle to test. The oriented bars were tested until failure, or until visible necking ensued. Strain was measured using an Instron non-contact video extensometer.

Figure 8 illustrates measured stresses as a function of nominal strains for bars of polymer AG oriented according to Procedure I. Similar measurements were made on polymer AF (not shown).

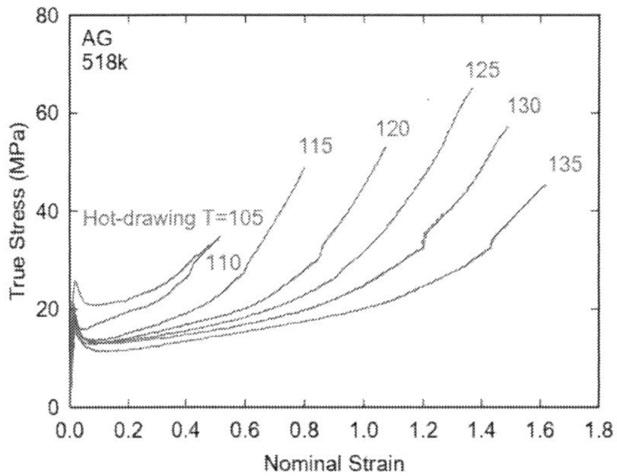

Figure 8 – Measured stress as a function of nominal strain for bars of polymer AG oriented according to Procedure I at different draw temperatures and redrawn just below T_g, illustrating the effect of draw temperature.

Constitutive model

Basis

The proposed model is a fully three-dimensional constitutive model for amorphous polymers, combining an existing version of the multi-mode Oxford Glass-Rubber (GR) constitutive model (Wu and Buckley, 2004) with the ROLIEPOLY constitutive equations of Likhtman and coworkers (Likhtman and Graham, 2003). The combined model is based on the assumption that in a deformed amorphous polymer free energy is stored through two mechanisms: (a) perturbation of interatomic potentials, or bond-stretching, and (b) changes in the supra-entanglement molecular conformational entropy; and that these two contributions to the free energy are additive.

The constitutive model describes the material response to a deformation gradient tensor $\boldsymbol{F}$ in terms of the Cauchy stress tensor $\boldsymbol{\sigma}$. $\boldsymbol{F}$ is separated into volumetric and deviatoric parts, given respectively by

$$J = \det \boldsymbol{F} \quad \text{and} \quad \bar{\boldsymbol{F}} = J^{-\frac{1}{3}} \boldsymbol{F}. \tag{2}$$

The stress tensor $\boldsymbol{\sigma}$ is similarly separated into volumetric and deviatoric parts, given by

$$\sigma_{\rm m} = K \ln J = \frac{1}{3} \mathrm{tr}\boldsymbol{\sigma} \quad \text{and} \quad \boldsymbol{S} = \boldsymbol{\sigma} - \sigma_{\rm m}\mathbf{I} \tag{3}$$

where K is the bulk modulus.

From the assumption of additivity of the bond-stretching and conformational free energies, the total deviatoric stress can be expressed as the sum of a bond-stretching stress $\boldsymbol{S}^{\rm b}$ and a conformational stress $\boldsymbol{S}^{\rm c}$

$$\boldsymbol{S} = \boldsymbol{S}^{\rm b} + \boldsymbol{S}^{\rm c}. \tag{4}$$

Bond-stretching stress

The bond-stretching part of the model mirrors our previous approach in the modelling of glassy polymers and only a brief treatment will be given here. The reader is advised to consult previous papers (Buckley et al., 2004; Buckley and Jones, 1995; De Focatiis et al., 2010; Wu and Buckley, 2004) for a more detailed discussion. The deviatoric rate of deformation tensor can be computed from the isochoric velocity gradient $\bar{\boldsymbol{D}}$ can be computed from the isochoric velocity gradient tensor $\bar{\boldsymbol{L}}$, given by

$$\bar{\boldsymbol{L}} = \dot{\bar{\boldsymbol{F}}}\bar{\boldsymbol{F}}^{-1}, \quad \bar{\boldsymbol{D}} = \frac{\bar{\boldsymbol{L}} + \bar{\boldsymbol{L}}^{\rm T}}{2} \tag{5}$$

where $\dot{\bar{\boldsymbol{F}}}$ is the time derivative of $\bar{\boldsymbol{F}}$. The deviatoric rate of deformation $\bar{\boldsymbol{D}}$ is the sum of a linear elastic bond stretching part, and a part from the viscous flow of molecular segments. Following Wu and Buckley (Wu and Buckley, 2004) we use a multi-mode discrete spectrum of M modes

$$\begin{aligned} \bar{\boldsymbol{D}} &= \frac{\hat{\boldsymbol{S}}_j^{\rm b}}{2G_{\rm b}} + \frac{\boldsymbol{S}_j^{\rm b}}{2G_{\rm b}\tau_j} \\ \boldsymbol{S}^{\rm b} &= \sum_{j=1}^{M} \upsilon_j \boldsymbol{S}_j^{\rm b}, \quad \sum_{j=1}^{M} \upsilon_j = 1 \end{aligned} \tag{6}$$

where $\hat{\boldsymbol{S}}_j^{\rm b}$ is is an objective, co-rotational, rate of change of the jth bond-stretching stress $\boldsymbol{S}_j^{\rm b}$ (implemented as the Jaumann rate), $G_{\rm b}$ is the bond-stretching shear modulus, υ_j represents the relative weighting of each mode, and τ_j is the relaxation time associated with the jth glassy mode. The glassy relaxation time is referred back to an unstressed relaxation time $\tau_{j,0}^*$ at a reference temperature T^* and structural state $T_{\rm f}^*$ through shift factors for temperature, structure and stress respectively

$$\tau_j = a_T a_s a_{\sigma,j} \tau^*_{j,0} \quad (7)$$

where

$$a_T = \exp\left[\frac{\Delta H}{R}\left(\frac{1}{T} - \frac{1}{T^*}\right)\right]$$
$$a_s = \exp\left[\frac{C}{T_f - T_\infty} - \frac{C}{T_f^* - T_\infty}\right] \quad (8)$$
$$a_\sigma = \frac{V_s \tau^b_{oct,j}}{2RT} \frac{\exp\left(-\frac{V_p \sigma_m}{RT}\right)}{\sinh\left(\frac{V_s \tau^b_{oct,j}}{2RT}\right)}$$

The material constants are as employed by Wu and Buckley (2004): ΔH is the activation enthalpy, T_∞ is the Vogel temperature, C is the Cohen Turnbull constant, V_s and V_p are the shear and pressure activation volumes, and $\tau^b_{oct,j}$ is the octahedral shear stress acting on the jth mode.

In the present paper, structural evolution, visible during mechanical deformation below T_g as a yield peak and subsequent strain-softening, is modelled through a semi-empirical expression connecting the evolution of T_f with viscoplastic strain (Buckley et al., 2004)

$$T_f = T_{f,0} + \Delta T_f \left[1 - \exp\left(-\frac{\bar{\varepsilon}^v}{\varepsilon_0^v}\right)\right] \quad (9)$$

where $T_{f,0}$ and ΔT_f are the initial fictive temperature and the increase in fictive temperature due to plastic strain, ε_0^v is a material parameter, and $\bar{\varepsilon}^v$ is the equivalent viscoplastic strain invariant.

Conformational stress

The conformational stress is computed via the ROLIEPOLY constitutive model, a simplified form of a full microscopic theory of linear entangled polymer melts. The ROLIEPOLY model has also been described in detail previously (Likhtman and Graham, 2003) and only a brief treatment will be given here. Since the present study explores the applicability of the model to semi-solid states with high degrees of stretch and limited relaxation, it is necessary to employ a version of RP equation with finite chain extensibility. In this study we do not consider the effects of convective constraint release.

Following Likhtman and Graham (2003), we use a spectrum of N discrete ROLIEPOLY modes to represent the conformational stress, and compute a stress in each mode from its corresponding orientation tensor $\boldsymbol{T}_k$. The orientation tensor is governed by the ROLIEPOLY differential equation, modified to allow for finite extensibility

$$\dot{\boldsymbol{T}}_k = \bar{\boldsymbol{L}} \cdot \boldsymbol{T}_k + \boldsymbol{T}_k \cdot \bar{\boldsymbol{L}}^T - \frac{1}{\tau_{d,k}}\left(\boldsymbol{T}_k - \lambda_k^2 \mathbf{I}\right) + -\frac{2}{\tau_{R,k}}\left(\frac{F(\lambda_k) - 1}{\lambda_k}\right)\boldsymbol{T}_k \quad (10)$$

where $\tau_{R,k}$ and $\tau_{d,k}$ are the Rouse and reptation times associated with the kth mode, and $\lambda_k = \sqrt{\frac{1}{3}\mathrm{tr}(\boldsymbol{T}_k)}$ is the chain stretch associated with the kth mode. $F(\lambda_k)$ is given by

$$F(\lambda_k) = \frac{\lambda_{max}}{3}\left(\frac{\lambda_{max}^2 - 1}{\lambda_{max}^2 - 1/3}\right)\mathcal{L}^{-1}\left(\frac{\lambda_k}{\lambda_{max}}\right) \quad (11)$$

where $\mathcal{L}^{-1}$ is the inverse Langevin function. After integration of equation (10) to obtain the orientation tensor, the stress in the kth mode is calculated from

$$\boldsymbol{\sigma}_k^c = G_e\left[\left(\frac{F(\lambda_k)}{\lambda_k}\right)\boldsymbol{T}_k - \mathbf{I}\right]$$
$$\boldsymbol{S}_k^c = \boldsymbol{\sigma}_k^c - \frac{\mathbf{I}}{3}\mathrm{tr}(\boldsymbol{\sigma}_k^c), \quad \boldsymbol{S}^c = \sum_{k=1}^{M} \upsilon_k \boldsymbol{S}_k^c, \quad (12)$$

where $\boldsymbol{\sigma}_k^c$ and $\boldsymbol{S}_k^c$ are the full and deviatoric conformational stresses of the kth mode, G_e is the entanglement modulus, and υ_k represents the relative volume fraction of the kth RP mode. The relaxation times are referred back to reference times $\tau^*_{R,j}$ and $\tau^*_{d,j}$ at temperature T^* and structure T_f^* through shift factors a_T for temperature and a_{struct} for structure, $\tau_{R,j} = a_T a_s \tau^*_{R,j}$ and $\tau_{d,j} = a_T a_s \tau^*_{d,j}$, as defined earlier in equation (8).

In order to characterise intrinsic material properties of linear PS melts, we used the optimiser Reptate (Ramirez and Likhtman, 2007) and the microscopic theory of linear polymer melts of Likhtman and McLeish (2000) applied to linear rheological data in shear simultaneously on the two materials. The material parameters G_e, the entanglement modulus, and M_e and τ_e, the molar mass and Rouse time of one entanglement length respectively, were established by treating them as variables for a best fit to the theory, following Likhtman and McLeish (2002). The parameter c_v was fixed at 1. The values of the parameters obtained are: $G_e = 317.9\text{kPa}$, $M_e = 13.14\text{kg/mol}$ and $\tau_e = 0.000697\text{s}$ at the reference temperature T^*=170°C. The longest (whole molecule) mode Rouse time and reptation time were computed following Collis et al. (2005) as

$$\tau^*_{R,1}=Z^2\tau_e$$
$$\tau^*_{d,1}=3\left(1-\frac{2.38}{Z^{0.5}}+\frac{4.17}{Z}-\frac{1.55}{Z^{1.5}}\right)Z^3\tau_e \tag{13}$$

where $Z=M/M_e$ represents the number of entanglements. The linear viscoelastic spectra obtained for the two materials are shown on Figure 2.

The finite extensibility (FE) of the chains can be derived from molecular theory using an equivalent Kuhn chain between entanglements, following Wagner (2006), and is given by

$$\lambda_{max}=\sqrt{\frac{n_b}{C_\infty}}\sin(\theta_b/2) \tag{14}$$

where n_b is the number of bonds between entanglements, $C_\infty=10$ is the characteristic ratio for PS, and θ_b is the carbon-carbon backbone bond angle, which for PS is 109.28°. Using the value of M_e obtained in the present work gives $\lambda_{max}=4.09$; this value was used in all the simulations reported here.

Constitutive model parameters

In order to determine a discrete relaxation spectrum fitting the linear viscoelastic data, the procedure employed was as follows. Starting with the mode with the longest relaxation time, $\tau_{d,1}$, calculated from equations (13) and (14), one mode of τ_d was assigned per decade of time/frequency to cover the range of data of interest. The relative weights of the modes were calculated using a custom-built optimiser written in Matlab to minimise the rms error between the logarithms of the values of G' and G'' as obtained by experiment, and as calculated from

$$G'=\sum_{i=1}^{M+N-1}G_i\frac{\omega^2(\tau_{d,i})^2}{1+\omega^2(\tau_{d,i})^2}$$
$$G''=\sum_{i=1}^{M+N-1}G_i\frac{\omega\tau_{d,i}}{1+\omega^2(\tau_{d,i})^2}. \tag{15}$$

Since the ROLIEPOLY constitutive model is based on entanglement physics, the following condition was imposed:

$$\sum_{k=1}^{N}G_k=G_e. \tag{16}$$

Hence relative weights $\upsilon_k=G_k/G_e$ were attributed to RP modes until the sum of the modes reached the entanglement modulus G_e, and relative weights $\upsilon_j=G_j/G_b$ were attributed to the remaining (bond-stretching) modes, until the sum of the modes reached the bond-stretch modulus $G_b=\sum_{j=1}^{M}G_j$. This procedure is illustrated for material AG in Figure 9. The same procedure was used to obtain a discrete spectrum for material AF (not shown). One mode occurring at $\omega\approx 1/\tau_e$ was partitioned between RP and glassy modes.

For each mode, the $\tau_{d,k}$ values are separated by a decade of time, while the corresponding $\tau_{R,k}$ values are calculated from the value of Z_k corresponding to that mode, by first solving

$$\tau_{d,k}=3\left(1-\frac{2.38}{Z_k^{0.5}}+\frac{4.17}{Z_k}-\frac{1.55}{Z_k^{1.5}}\right)Z_k^3\tau_e \tag{17}$$

for Z_k and then using the corresponding value of Z_k in $\tau_{R,k}=Z_k^{\,2}\tau_e$.

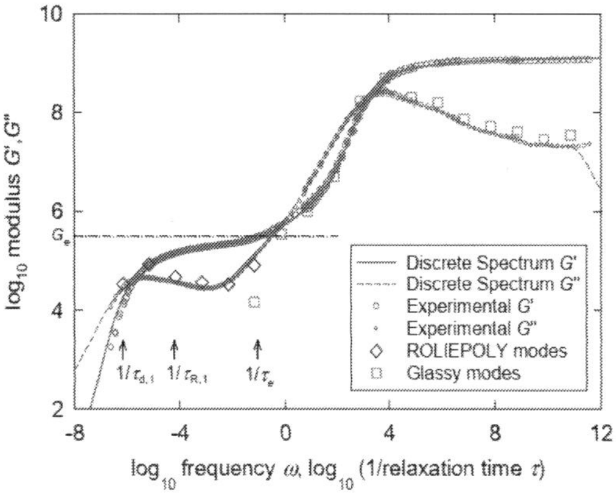

Figure 9 - Storage and loss moduli obtained from linear viscoelastic shear rheology (for $\log_{10}\omega<1$) and from dynamic mechanical analysis (for $\log_{10}\omega>1$) for polymer AG as a function of frequency (circles). Also shown are the storage and loss moduli calculated analytically from the 18-mode spectrum using equation (15) (solid line and dashed line respectively). The 6 G_k components of the 6-mode RP spectrum are shown as diamonds plotted versus the associated $1/\tau_{d,k}$. The 13 G_j components of the 13-mode glassy spectrum are shown as squares plotted versus the associated $1/\tau_j$. Also illustrated are the positions of the inverses of the fundamental relaxation times $1/\tau_d$, $1/\tau_R$ and $1/\tau_e$ and the value of G_e.

For the bond-stretch part of the model, where possible the material parameters follow the work on PS of Wu and Buckley (2004). In order to reflect the different thermal history between the present specimens and those of Wu and Buckley, appropriate values of $T_{f,0}$, ΔT_f and ε_0^v have been selected for the present data.

Results and Discussion

Preliminary results

As a first test of the constitutive model, melt rheology was simulated in the nonlinear viscoelastic regime and its predictions were compared with experiment. The model simulations are overlaid on the experimental data for polymer AG in Figure 10, and illustrate very good agreement with the data.

The constitutive model was also used to replicate successfully the simulations of Wu and Buckley (2004) of glassy-state compression yield and flow of isotropic specimens of PS at a range of temperatures below T_g (not shown). Simulations of melt rheology and isotropic glassy-state yield and flow provide a test of the individual parts of the model essentially in isolation, since in the melt viscoelastic relaxation times are much shorter than the timescale of the experiment, and deep in the glass the stress is dominated by glassy modes and relaxation of the conformational network is essentially frozen by the long relaxation times.

The model was then applied to the experimental conditions outlined in Procedures I, II and III. This involved simulating the following steps:

1. The melt-stretching process at a nominal strain rate of $\dot{\varepsilon} = 0.02\ \mathrm{s}^{-1}$ at the prescribed temperature T to the prescribed draw ratio λ.
2. Holding for the required dwell time t (assumed to be 1 s for cases where no dwell was specifically imposed).
3. A rapid quench, approximated by a decrease in temperature at a constant rate of 15°C/s down to a temperature of 0°C.
4. Unloading of the stress (simulated by a rapid contraction at a rate of $\dot{\varepsilon} = -1\ \mathrm{s}^{-1}$; the strain required to achieve unloading is between 0.1% and 0.3% depending on the drawing conditions).

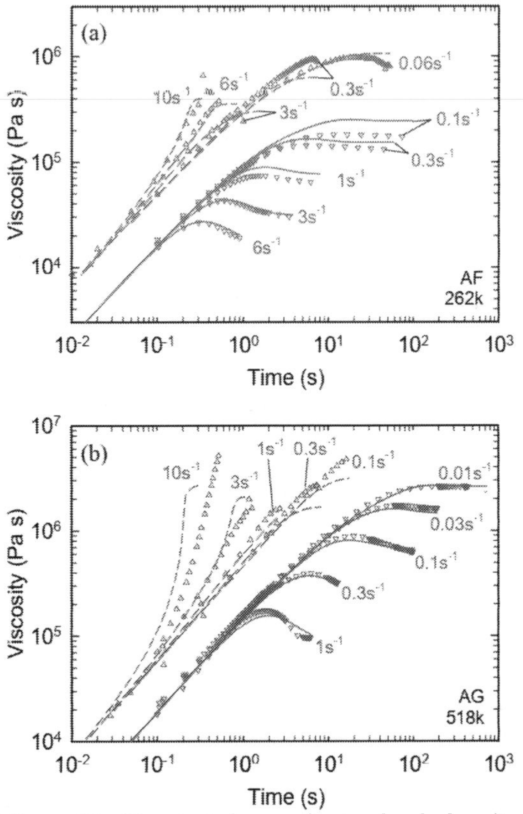

Figure 10 - Transient shear and extensional viscosity measured at 170°C for (a) polymer AF at a range of rates from 0. 1 s^{-1} to 6 s^{-1} in shear, and 0.06 s^{-1} to 10 s^{-1} in extension; and (b) polymer AG at a range of rates from 0.01 s^{-1} to 1 s^{-1} in shear, and 0.1 s^{-1} to 10 s^{-1} in extension. Also shown are simulations from the constitutive model.

Information from the constitutive model was then used to predict optical birefringence, and to predict changes in craze initiation on the oriented materials. Sub-T_g deformation of melt-drawn material required the following additional simulation steps:

5. Reheating to the glassy-state draw temperature of 96°C (simulated by a temperature rise at a constant rate of 0.8°C/s for 120 s).
6. Holding at 96°C for a time of 600 s (intended to simulate the time taken for the environmental chamber to acclimatise).
7. Uniaxial stretching at 96°C at a nominal strain rate of $\dot{\varepsilon} = 0.001\lambda\ \mathrm{s}^{-1}$ (intended to simulate the stretching of oriented material at a nominal strain rate of $\dot{\varepsilon} = 0.001\ \mathrm{s}^{-1}$ calculated using its new length after the hot drawing).

Birefringence predictions

In a phenyl polymer such as polystyrene, optical birefringence can arise from both bond stretching distortion (i.e. changes in intersegmental distances, primarily in the carbon-carbon backbone) and from changes in the molecular conformation (originating primarily from the angle of the phenyl side group relative to the main chain), each with its own associated stress-optical coefficient. The magnitude of the distortional coefficient is several orders smaller than that of the conformational birefringence, and can safely be neglected in the experiments of this paper.

In melts, where the distribution of the end-to-end vectors of the chain segments is Gaussian and the the stress is proportional to the bond orientation average, birefringence can be computed from the ROLIEPOLY model for a given stress state by straight-forward use of the stress-optical rule. Finite chain extensibility and the use of the inverse Langevin function in equations (10) and (11) lead to a non-linear force-distance relationship at large chain extensions. The resulting effect is that for an increasing degree of orientation birefringence saturates while stress continues to grow. This can be seen in the experimental measurements of stress prior to freezing plotted against birefringence in Figure 3. Conformational birefringence comes from orientation of chain segments, and hence can be described using the eigenvalues of tensor $\boldsymbol{T}_j$. For example, for uniaxial deformation in the 3 direction,

$$\Delta n = \left(n_1 - n_3\right) = C\sum_{j=1}^{m} G_j\left(\boldsymbol{T}_{j,1} - \boldsymbol{T}_{j,3}\right) \qquad (18)$$

where $\boldsymbol{T}_{j,1}$ and $\boldsymbol{T}_{j,3}$ are the eigenvalues of the jth orientation tensor associated with directions 1 and 3. Simulations are performed with the computed spectra for each material and test condition, and the computed values of birefringence are shown in Figures 3, 4, 5 and 6. A standard literature value of $C = -4.5\times10^{-9}$ Pa^{-1} was used throughout the computations of birefringence (Luap et al., 2006).

Figure 3 illustrates the need of finite chain extensibility in simulating experiments with significant degrees of orientation, characteristic of the solid state, by illustrating simulations both with and without finite extensibility. In the simulations in this paper, finite chain extensibility was used throughout.

Figures 4, 5 and 6 compare the predictions of birefringence from the model with the measured values for melt-orientation procedures I, II and III. In Figure 4 the model accurately predicts birefringence for a wide range of hot-drawing conditions for both materials at $Wi_e < 1$, but underpredicts the birefringence for $Wi_e > 1$. Since the constitutive model was constrained to contain only modes contributing to the entanglement modulus (see eq. (16)), it is natural to expect the model to only be able to capture orientation on a length-scale greater than an entanglement. When $Wi_e > 1$, stretching is taking place on a sub-entanglement length scale, and this is currently not intended to be accounted for by the model.

Figure 5 illustrates that the model is able to account for the effect of stress-relaxation remarkably accurately for a wide range of dwell times. In this case, $Wi_e < 1$ during stretching. Figure 6 illustrates procedure III, stretching to different draw ratios followed by immediate quench. Here the hot-drawing is performed at $Wi_e < 1$ and application of the model produces good predictions of birefringence for moderate stretch, up to $\lambda \sim 3$. For $\lambda \geq 4$ the model underpredicts the birefringence. One possible explanation for this discrepancy is that, although $Wi_e < 1$ during stretching, when chains are significantly stretched on a longer length scale, this stretch feeds back into the shorter length scales. Chains may experience some stretch on a sub-entanglement length scale as a result of the significant stretch present on longer length scales. There is at present no way to account for interactions between length scales in the model.

In summary, the constitutive model is appropriate for accurate predictions of locked-in birefringence in oriented monodisperse materials, provided that there is no significant sub-entanglement level chain stretch. In most contexts this equates to $Wi_e < 1$, but care must be taken in the cases where there are large stretches, as these could induce sub-entanglement stretch through interactions between length scales. Typical polymer processing operations do not normally introduce significant amounts of sub-entanglement level chain stretch, and thus the model, with its limitations, is applicable in this context.

Craze initiation predictions

Kramer's model of craze fibril growth (Kramer, 1983), proposed for crazing in isotropic polystyrene, states that the macroscopic craze stress σ_c^{iso} is given by

$$\sigma_c^{iso} = \frac{8\Gamma}{\beta D_0} \qquad (19)$$

where Γ is the surface energy for craze formation, D_0 is the fibril spacing, and β is a proportionality constant less than 1. Maestrini and Kramer (1991) extended this model to oriented PS by observing that neither Γ nor D_0 change with orientation, but that the presence of orientation leaves the polymer in a state of self-stress, with a conformational so-called back stress balanced by equal and opposite glassy stress. They proposed the form

$$\sigma_c - \sigma_c^{iso} = \frac{\sigma_{conf}}{\beta} \quad (20)$$

where σ_{conf} is the conformational stress. In order to compare our experimental measurements to equation (20), the birefringence was recorded individually for each crazing specimen. A value of conformational stress was then obtained by using the stress-birefringence relationship shown in Figure 3. Isotropic values of crazing stress were taken from De Focatiis et al. (2008).

Figure 11 shows a plot of $\sigma_c - \sigma_c^{iso}$ against σ_{conf} using the individual experimental measurements from procedure I shown in Figure 7. Linear regression is performed only on the data points for which $Wi_e < 1$. This restriction is imposed since the model does not presently account for sub-entanglement stretch. The regression line, shown in Figure 11, has a gradient of 15.58 ± 0.62. This equates to a value of Kramer's β factor of 0.064 ± 0.005.

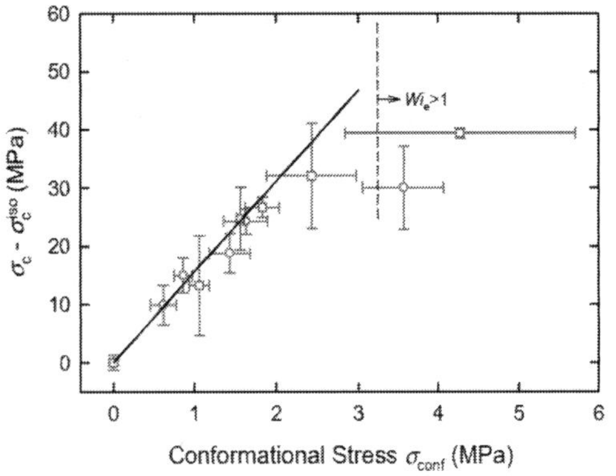

Figure 11 - Measurements of the increment in crazing stress relative to the isotropic value as a function of the conformational stress (computed from the birefringence using the master curve in Figure 3 for experimental measurements). Drawing was performed according to Procedure I. Also shown is the portion of the data set for which $Wi_e > 1$. Linear relationship predicted by Kramer's theory using β = 0.064 (solid line).

Using this value of β and equation (20) together with isotropic craze stress values, the constitutive model may be used to predict changes in crazing stress resulting from process-induced orientation. To the authors' knowledge, this is the first time a value of β has been obtained for polystyrene from experimental data.

Sub-T_g large deformation mechanical response predictions

The constitutive model was used for simulating the subsequent glassy-state mechanical response of melt-drawn tapes, as described previously, by simulating steps 1-7. In the simulation results shown below, only results from the final step are shown. The "nominal strain" during the uniaxial stretching of the oriented material shown in the simulations was computed from the continuous measure of nominal strain e used in all the process as $(e+1)/\lambda - 1$. This was to account for the fact that the measurement of strain in the experiments is computed relative to the new length of the bars after hot drawing and prior to glassy-state stretching.

Figure 12 shows constitutive model simulations of stresses versus nominal strain for the experimental conditions of Procedure I, for material AG. The plot is intended to simulate the experimental measurements shown in Figure 8 and is shown on identical axes to illustrate the differences between experiment and simulation. Qualitatively the simulation encompasses all the correct features of the deformation: a yield stress followed by a lower flow stress, followed by different degrees of strain hardening that depend on the melt history. Quantitatively, however, it is apparent that the simulation underpredicts yield stress levels and the onset of strain hardening when simulating material melt-drawn close to T_g. Similar conclusions can be drawn by comparison of experiments and simulations of polymer AF (not shown).

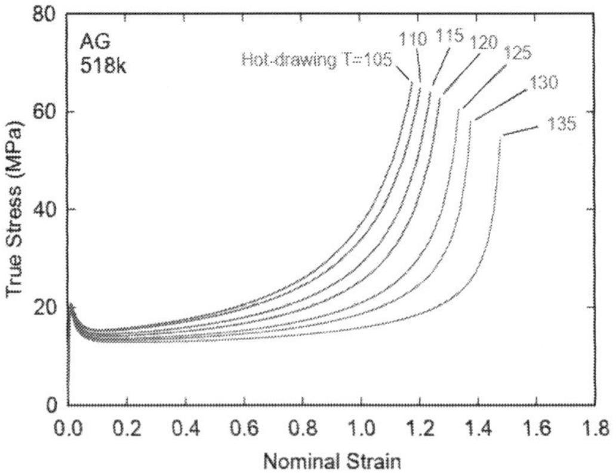

Figure 12 - Constitutive model simulations of glassy-state uniaxial testing of material melt-stretched according to procedure I, at a range of temperatures T = 105-135°C followed by immediate quenching, for material AG.

To some extent, this discrepancy is expected, for the same reasons that account for the model's inability to capture the highest degrees of orientation. Although the model accounts for orientation for a wide range of length scales, it lacks the correct physics when dealing with sub-entanglement orientation.

In order to illustrate how the constitutive model is able to capture not only the change in the onset of strain hardening for different draw ratios, but also the effect of structure (through molecular weight effects), one additional procedure is illustrated. Both materials were drawn to a range of draw ratios λ from 2 to 4 at a constant nominal strain rate 0.02 s^{-1} at a temperature T = 115°C. A dwell time t of 300 s was allowed prior to quenching. Sub-T_g restretching of oriented tapes was performed following the conditions of step (7) as described previously. Experimental measurements and constitutive model simulations are shown in Figure 13.

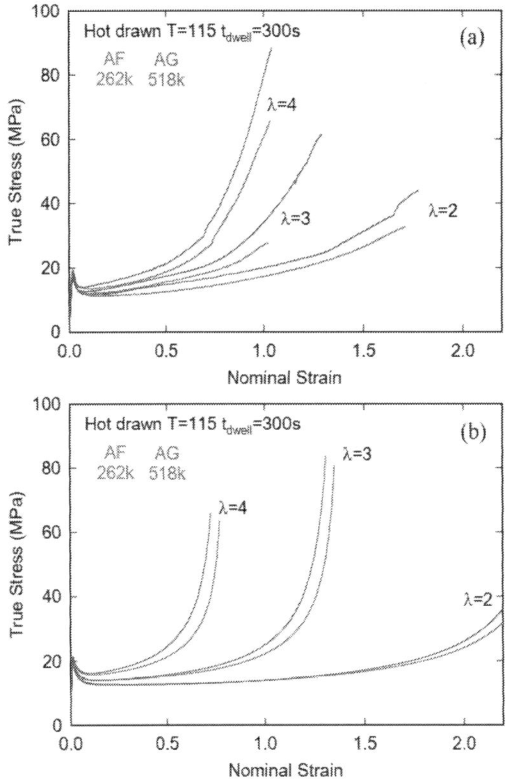

Figure 13 – (a) True stress plotted against nominal strain measured during glassy-state uniaxial testing of bars melt-stretched at a temperature of T = 115°C followed by a dwell time of 300 s prior to quenching, for draw ratios $\lambda = 2$, $\lambda = 3$ and $\lambda = 4$. (b) Constitutive model simulations for the same conditions.

Conclusions

This study has presented experimental measurements of the effects of orientation on solid-state properties of two grades of monodisperse linear polystyrene. Orientation has been introduced through three procedures designed to explore the effects of draw temperature, stress-relaxation prior to quenching, and draw ratio. Measurements of birefringence were performed on all specimens. The effects of orientation were recorded through changes in craze initiation stress and in the large strain sub-T_g mechanical response. The data have been compared to predictions made with a new framework for constitutive modelling of processed oriented polymers based on parallel coupling of two existing melt-state and glassy-state models. The new combined glass-melt model is parameterised *only* through its full linear viscoelastic spectrum, known molecular parameters of polystyrene, and additional parameters needed for capturing viscoplastic deformation of a glass.

When the model is assessed against results of non-linear rheological experiments in the melt, it is found to be quantitatively successful in capturing the material rheology. The model is also successfully used to predict values of frozen-in birefringence using a standard literature value of stress-optical coefficient. The conformational stress provided by the model was used to predict the rise in craze initiation stress that results from frozen-in orientation, using Maestrini and Kramer's (1991) model of crazing in oriented materials.

The model was also used to simulate the effects of processing conditions on the large deformation mechanical response in the glassy-state. The model captures qualitatively all of the features seen in the experimental results, and also has the advantage of a molecular basis for the prediction of orientation. There are, however, quantitative shortcomings in the current model's predictions, which are predominant when the time-temperature conditions of melt-stretching lead to molecular orientation on a sub-entanglement length-scale. The model presented here contains no representation of intrinsic anisotropy on this length-scale.

Acknowledgements

The authors wish to acknowledge the assistance of Dr L. Hutchings of Durham University in the supply and characterisation of the polystyrenes; Mr D. Groves and Dr J. Embery of Leeds University for the DMA and rheological measurements; Dr J. Ramirez and Prof. A.E. Likhtman for the Reptate toolbox; and Dr D.J. Read of Leeds University and Dr R. Graham of Nottingham University for helpful discussions. This work was supported by the UK EPSRC grant number GR/T11845/01.

References

Bent, J., et al., "Neutron-mapping polymer flow: scattering, flow visualisation, and molecular flow", Science, **301**, 2691-1695 (2003)

Boyce, M.C., et al., "Large Inelastic Deformation of Glassy-Polymers .1. Rate Dependent Constitutive Model", Mech. Mater., **7** (1), 15-33 (1988)

Buckley, C.P., et al., "Deformation of thermosetting resins at impact rates of strain. Part 2: constitutive model with rejuvenation", J. Mech. Phys. Solids, **52** (10), 2355-2377 (2004)

Buckley, C.P. and D.C. Jones, "Glass-Rubber Constitutive Model for Amorphous Polymers near the Glass-Transition", Polymer, **36** (17), 3301-3312 (1995)

Collis, M.W., et al., "Constriction flows of monodisperse linear entangled polymers: Multiscale modeling and flow visualization", J. Rheol., **49** (2), 501-522 (2005)

De Focatiis, D.S.A. and C.P. Buckley, "Determination of craze initiation stress in very small polymer specimens", Polym. Test., **27** (2), 136-145 (2008)

De Focatiis, D.S.A., et al., "Roles of Chain Length, Architecture and Time in the Initiation of Visible Crazes in Polystyrene", Macromolecules, **41**, 4484-4491 (2008)

De Focatiis, D.S.A., et al., "Large deformations in oriented polymer glasses: experimental study and a new glass-melt constitutive model", J. Polym. Sci. Part B: Pol. Phys., in press (2010)

Dooling, P.J., et al., "Hot-drawing of poly(methyl methacrylate) and simulation using a glass-rubber constitutive model", Polymer, **43** (8), 2451-2465 (2002)

Govaert, L.E., et al., "Does the Strain Hardening Modulus of Glassy Polymers Scale with the Flow Stress?" J. Polym. Sci. Part B: Polym. Phys., **46** (22), 2475-2481 (2008)

Graham, R.S., et al., "Microscopic theory of linear, entangled polymer chains under rapid deformation including chain stretch and convective constraint release", J. Rheol., **47** (5), 1171-1200 (2003)

Haward, R.N. and G. Thackray, "The use of a mathematical model to describe isothermal stress-strain curves in glassy thermoplastics", Proc. Roy. Soc. Lond. A Mat., **302**, 453-472 (1968)

Kramer, E.J., "Microscopic and molecular fundamentals of crazing", Adv. Polym. Sci., **52/53**, 1-56 (1983)

Likhtman, A.E. and R.S. Graham, "Simple constitutive equation for linear polymer melts derived from molecular theory: Rolie-Poly equation", J. Non-Newton. Fluid. Mech., **114**, 1-12 (2003)

Likhtman, A.E. and T.C.B. McLeish, "Quantitative theory for linear dynamics of linear entangled polymers", Macromolecules, **35**, 6332-6343 (2002)

Likhtman, A.E., et al., "Microscopic theory for the fast flow of polymer melts", Phys. Rev. Lett., **85**, 4550-4553 (2000)

Luap, C., et al., "Limit of validity of the stress-optical rule for polystyrene melts: Influence of polydispersity", J. Non-Newton. Fluid Mech., **138** (2-3), 197-203 (2006)

Maestrini, C. and E.J. Kramer, "Craze structure and stability in oriented polystyrene", Polymer, **32** (4), 609-618 (1991)

Palm, G., et al., "Large strain mechanical behavior of poly(methyl methacrylate) (PMMA) near the glass transition temperature", J. Eng. Mater. Technol., **128** (4), 559-563 (2006)

Ramirez, J. and A.E. Likhtman. Rheology of Entangled Polymers: Toolbox for the Analysis of Theory and Experiments (Reptate). 2007; Available at http://www.reptate.com.

Tervoort, T.A., et al., "A constitutive equation for the elasto-viscoplastic deformation of glassy polymers", Mech. Time Dependent Mater., **1**, 269-291 (1998)

Wagner, M., "Nonlinear rheology of polymer melts: a new perspective on finite chain extensibility effects", Korea-Aust. Rheol. J., **18** (4), 199-207 (2006)

Ward, I.M.: Structure and Properties of Oriented Polymers, Applied Science Publishers Ltd, Cambridge (1975)

Wu, J.J. and C.P. Buckley, "Plastic deformation of glassy polystyrene: A unified model of yield and the role of chain length", J. Polym. Sci. Part B: Pol. Phys., **42** (11), 2027-2040 (2004)

Analysis of Microstructural Deformation of HDPE Film during Stretching

F. Sadeghi and A. Ajji*

CREPEC, Chemical Engineering Department, Ecole Polytechnique, C.P. 6079, Succ. Centre- Ville, Montreal, QC, H3C 3A7 Canada

Abstract

The HDPE film was produced by film blowing process and a lamellar crystalline morphology was obtained applying proper process conditions. A sample of the film was annealed to thicken the lamellae and remove the defects. The film samples (annealed and non-annealed) were stretched by a tensile stage machine inserted inside an SEM field emission microscope. The structure deformation during stretching was observed and micrographs were snapped online as stretching was carried out. The stress-strain curves were extracted and discussed accompanied by morphology deformation. The non annealed sample showed void formation during the initial stage of stretching. At high draw ratios, lamellae were broken down and reoriented in machine direction to form elongated thin fibrils. For the second sample that was fully annealed, initial stretching separated the lamellae but further stretching did not rotate and reorient the lamellae. Stretching of this sample caused intense tie chain drawing leading to fibril crystallization (stress induced crystallization).

Introduction

Polyethylene (PE) is considered to be the most widely used polyolefin in the market. Due to its large application field, this polymer has attracted considerable attention in research and development. Physical and mechanical properties of this polymer in its final form such as film or pipe are strongly related to the microstructure that is developed during its processing and consisted of two phases: amorphous and crystalline. Crystalline phase is composed of packed folded chains that form lamellae blocks. The lamellae blocks are connected to each other by tie chains. Tension loading of such sample results in microstructural deformation. Exploring of this deformation could be a lead in understanding the role of initial precursor film microstructure in morphology evolution and also in designing loading conditions. ??

Crystal and amorphous phases in the film show different arrangements and orientations depending on the preparation process. Most of the polyethylene products in their final shape have oriented structure that means there is a preferred direction for formation of crystal and amorphous phases. Orientation causes the change of properties from one direction to another direction [3]. The orientation either is desired or is unavoidable. The degree of orientation can be calculated quantitatively and directly influence mechanical properties of material [3]. The arrangement and shape of crystal-lamellae phases defines the initial microstructure. In blown film extrusion, applying a proper draw ratio and cooling condition right after die, can result in formation of a stacked lamellar morphology oriented in MD direction [4].

Tensile test is a common way to analyze the mechanical behaviour of the sample in response to an applied stress. As stress is applied the microstructure begins to deform. The deformation can be tracked by recording deviations in the course of stress-strain graph accompanied by online surface observation. Depending on the initial microstructure, the stress-strain course (during stretching) and therefore the mechanism of microstructure deformation would be different [5].

Wade Adams et al. [6] studied the morphology change in high density polyethylene (HDPE) films having initial lamellar morphology. They showed lamellae separation and transformation into a final fibrillar structure. The HDPE films were stretched at 120°C. They performed direct visualization during the stretching for the annealed and non annealed samples. They found a larger voiding area when the annealed sample was stretched. For non-annealed sample a big portion of energy was spent to shear apart the lamellae at a lower strain level (50% vs 300% for annealed sample). In general the lamellar structure was not longer recognizable at 100% deformation.

The process to track the mechanism of microstructure deformation from both macroscopic and microscopic views gives valuable information that can be used to analyze and predict failure mechanism. The results therefore can be employed to design loading and study the effect of the process to obtain the desired product. In this work a tensile stage was inserted in a field emission scanning electron microscope. Snapping high resolution pictures online with analyzing stress-strain course give us a more precise investigation of morphology evolution with stretching.

Experimental Material and Film preparation

HDPE 6200 from Petromont Company was used in this study. HDPE 6200 with a MFR of 0.33 (190°C/2.16 kg) is a medium molecular weight grade and is usually used for food film packaging.

Blown film process was employed to produce film samples. The key processing parameters for blown film were: 1) the cooling conditions; 2) draw down ratio (DDR), which is the ratio of take-up speed to the average velocity of the extrudate at the exit of the die and represents the amount of stretching in the machine direction; 3) blow-up ratio (BUR), which is the ratio of the bubble diameter to the diameter of the die and 4) die temperature. The film sample was produced under following process conditions: Normal cooling conditions with blowing air, DDR=40, BUR=1, final film thickness of 28micron and die temperature of 220°C. The film produced with these conditions considered sample1. The same film was annealed at 110°C for 20minute to obtain the sample 2.

Film characterization

Fourier transform infrared spectroscopy (FTIR):

For FTIR measurements, a Nicolet Magna 860 FTIR instrument from Thermo Electron Corp. (DTGS detector, resolution 2 cm^{-1}, accumulation of 128 scans) was used. The beam was polarized by means of a Spectra-Tech zinc selenide wire grid polarizer from Thermo Electron Corp. The measurement is based on the absorption of infrared light at certain frequencies corresponding to the vibration modes of atomic groups present within the molecule.

Morphology:
A field emission scanning electron microscope (FE-SEM- Hitachi S4700) was employed for the observation of the surface of films. This microscope provides high resolution of 2.5 nm at a low accelerating voltage of 1 kV and high resolution of 1.5 nm at 15 kV with magnification from 20× to 500kx. A tensile stage was built in a way to stretch the sample inside the field emission microscope. That would enable us to directly track the morphology change during stretching as snapping of high quality SEM micrographs was carried out. In the fixture, one side was fixed and the other side was stretched in variable speeds. The initial gap was 7mm and the width of the sample was 6mm. The speed of stretching was 50 mm/min.
The samples were etched with a solution mix of phosphoric and sulfuric acid to obtain a better phase clearance in morphology.

DSC:
Differential scanning calorimetry (DSC) tests were carried out on a TA Instruments Q1000, with a heating rate of 10 °C/min.

Results and discussion

The morphology obtained for the HDPE film was almost a stacked lamellae oriented in machine direction. As stretching was applied, SEM micrographs were snapped and some are presented in Fig.1 and 2.The micrographs of sample1 for three draw ratios (DR) of 0.1,0.3 and 4 are shown in the Fig.1. The produced film was annealed for 10 minutes at 110°C to obtain the sample 2. The morphology of this sample for three draw ratios (DR) of 0.1, 0.9 and 2.5 are also depicted in Fig.2. The corresponding stress-strain curves are presented in Fig.3, Lamellae twisting (specially at the ends) that is observed in HDPE film morphology is already observed and addressed by Zhang et al. [4]. Lamellae twisting is more obvious for HDPE with narrower molecular weight distribution and is intensified when a medium draw down ratio was applied in the film blowing process. For the non-annealed sample at early stage of drawing the tie chains are elongated up to their limit (elastic zone), then tearing of chains creates voids and deformation begins as stress transforms to crystal blocks. Further

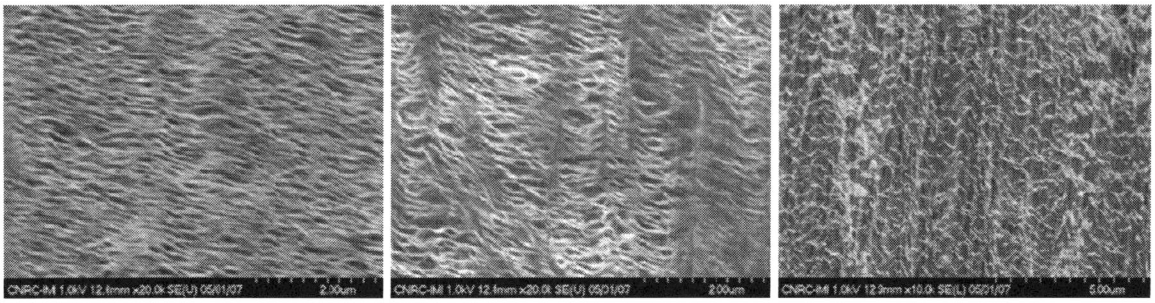

Fig.1 SEM micrographs of HD sample film taken from DR of 0.1, 0.3 and 4 from left to right respectively.

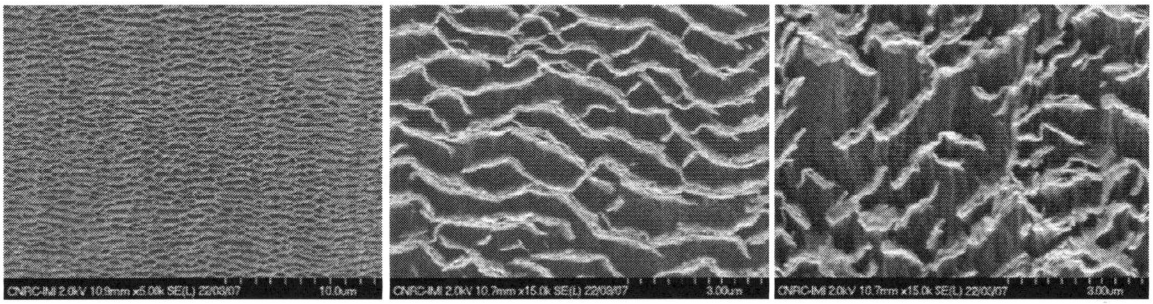

Fig.2 SEM micrographs of HDA sample film taken from DR of 0.1, 0.9 and 2.5 from left to right respectively.

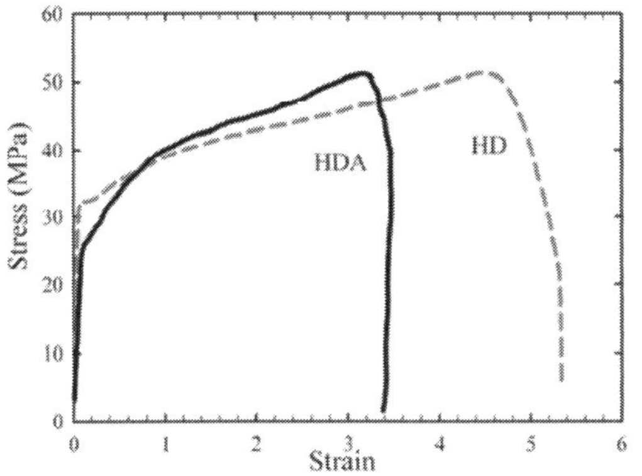

Fig.3. Stress-strain graph for the samples stretched at speed of 50mm/min.

stretching deforms the lamellae that include rotation, breaking down and reorientation of lamellae in long fibrils as it is observed in Fig.1. Annealing is a widely studied matter and has been covered specially in our previous works for polypropylene [2]. For the annealed sample, the lamellae are thick and so are difficult to rotate. Even the lamellae crystalline structure is not appropriate to create voids. It is previously shown that a certain orientation is needed for the precursor film to develop a microporous structure consist of stable pores [7].

The most significant effect of annealing beside of lamellae thickening is orientation improvement. The orientation of the samples was calculated from FTIR spectrum [2]. The results have been demonstrated in Table1. It is believed that low orientation of lamellae is the main reason for preventing of sustainable pores formation for the annealed sample. In this case most of the stretching energy is spent to separate the lamellae and in MD direction while the main body of lamellae block are preserved and just got get away from each other. At high draw ratios the lamellae are partially broken apart and some skin detachment occurred. However they are still resistant to rotation that could account for intense strain hardening behaviour observed in Fig.3.

Table1. Orientation functions along MD for the precursor and final stretched films.

HD	f_a	f_b	f_c
HD	0.04	-0.35	0.31
HD-S	-0.39	-0.31	0.71
HDA	0.04	-0.47	0.43
HDA-S	-0.42	-0.38	0.80

Since the orientation of the initial microstructure is not very high (as a result of the process conditions), the lamellae separation and voids distribution is not very uniform as it is seen in Fig.1. Over DR=2,the lamellae begin to break apart, reorient (rotate) and recrystallize in the form of elongated fibrils. From DR=3, the initial lamellae are almost deformed and a fibril network is produced, over that DR a more uniform fibril network is produced and the connecting lamellae are getting thinner while the population of fibrils increases.

The orientation of c-axis along MD improves as the sample is stretched in tensile stage. Orientation improvement corresponds to a better crystal orientation in fibril network that is result of the aligning of the broken crystal parts in machine direction and also drawing of tie chains and crystallization of them in fibril structures. In FTIR spectrum shown in Fig.4, the absorbance intensity for both 720 and 730 cm^{-1} decrease with stretching, however for 730 cm^{-1} (a-axis) there is a big fall for 0 polarization that leads to orientation drop of f_a from 0.04 to -0.39 as mentioned in Table 1. That could be interpreted that the orientation of c-axis is built with strong contribution of a-axis rotation, while the change in b-axis is relatively smaller.

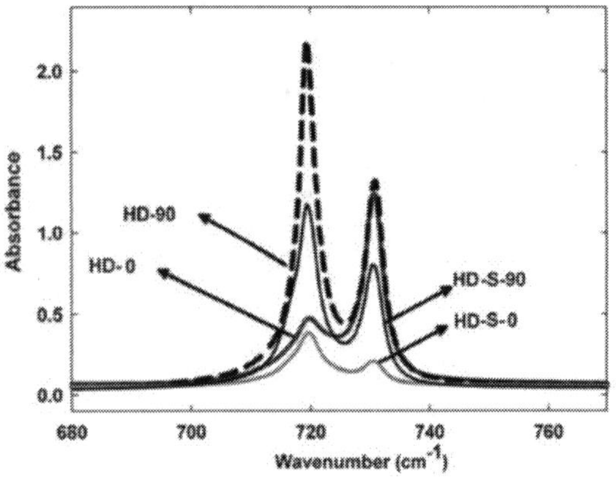

Fig.4. FTIR spectrum for HD samples.

To examine the effect of stretching on DSC spectrum, the result of experiment for the stretched sample of HD (the sample was taken at failure point of tensile) are presented in Fig. 5.

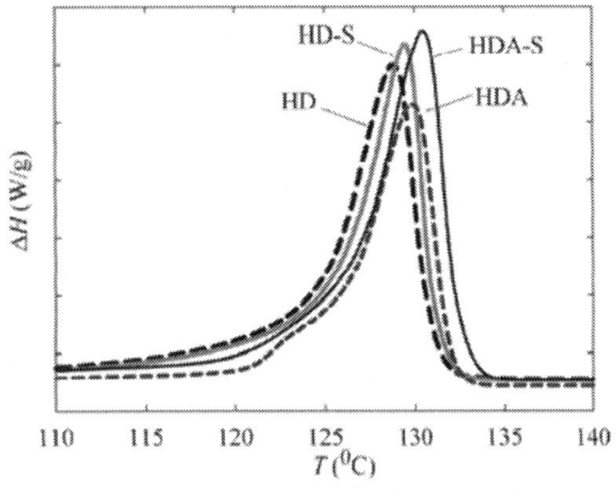

Fig.5. DSC thermograms of HD samples.

A shift in melting point is observed as a result of structure transformation to fibrils due to stretching and a single peak attributed to fibrils is appeared. In fact it can be predicted that as stretching approach failure point, the lamellae are transformed to fibrils and the contribution of lamellae to DSC peak will be limited to very tiny lamellae remained on the formed fibrils

That increase of melting point for HDA as a result of lamellae thickening (that is because of annealing) has been addressed in Ref.7. There is also a slight shoulder emerging at lower temperature for the highly annealed sample (HDA) that was also observed for polypropylene [8]. Stretching causes formation of more and thicker fibrils that results in shifting of melting point peak.

Conclusions

In this work microstructure deformation for HDPE film was investigated using a SEM field emission microscope equipped with a tensile stage. The deformation in lamellar crystalline structure was recorded and analyzed. The stress-strain curves were extracted and discussed with morphology deformation. The non annealed sample showed void formation during initial stage of stretching. In high degree of stretching the lamellae were broken down and reoriented in machine direction to form elongated thin fibrils. In the case of high annealing conditions thick lamellae were generated. The thick lamellae were difficult to break apart and reorient. Stretching of tie chains (in this sample) in the space between the lamellae was more pronounced and resulted in fibril crystallization that was represented by intense stress induced crystallization.

Acknowledgements

Financial support from NSERC (Natural Science and Engineering Research Council of Canada) and from FQRNT (Fonds Québécois de Recherche en Nature et Technologies) is gratefully acknowledged.

References

1. Sadeghi F., Ajji A., Carreau PJ. J. Plastic Film and Sheeting 2005; 21, 199-217.

2. Sadeghi F., Ajji A., Carreau PJ. Polym. Eng. Sci. 2007;47:1170-1178.

3. Ward IM., Structure and properties of oriented polymers, London: Chapman and Hall 1997.

4. Zhang X. M., Elkoun S., Ajji A. and Huneault M.A. Polymer 2004; 45: 217-229.

5. Drummond K. M., Shanks R. A. and Cser F. J. Appl. Polym. Sci. 2002; 83: 777–784.

6. W. Wade Adams, D. Yang and Edwin L. Thomas, J. Materials Sci. 1986; 21: 2239-2253

7. Yu T. H., PhD thesis, Virginia Polytechnic Institute and State University, 1996.

8. Sadeghi F., Ajji A., Carreau PJ. J. Membrane Sci. 2007; 292 : 62-71.

STRUCTURE FORMATION AND CRYSTAL ORIENTATION DURING SHEAR-INDUCED NONISOTHERMAL CRYSTALLIZATION OF POLYPROPYLENE NANOCOMPOSITES WITH EXFOLIATED MONTMORILLONITE

E. Szkudlarek[1], S. A. E. Boyer[2,3], J.-M. Haudin[2]*and E. Piorkowska[1]

[1]*Polish Academy of Sciences, Centre of Molecular and Macromolecular Studies, Lodz, Poland – eszkudl@cbmm.lodz.pl; epiorkow@bilbo.cbmm.lodz.pl*
[2]*MINES ParisTech, Centre de Mise en Forme des Matériaux, CNRS UMR 7635, Sophia Antipolis, France – Severine.Boyer@univ-bpclermont.fr; Jean-Marc.Haudin@mines-paristech.fr*
[3]*MINES Douai, Département Technologie des Polymères et Composites & Ingénierie Mécanique, Douai, France – Severine.Boyer@mines-douai.fr*

Abstract - Shear-induced nonisothermal crystallization in iPP based nanocomposites with organo-modified montmorillonite was followed by the light depolarization technique. Prior to the crystallization, samples were sheared in different temperature ranges during a cooling at 10 °C min^{-1}. Structure of the solidified specimens was investigated by scanning electron microscopy and X-ray diffraction. Strong enhancement of the nucleation and crystallization after shearing was observed in the compatibilized nanocomposites. Clay exfoliation was found to accelerate strongly shear-induced nucleation and overall crystallization kinetics. However, the sheared samples exhibited only a weak crystalline orientation that probably resulted from a small population of oriented crystals that formed due to shear-induced orientation of iPP chains, and served as nuclei for further nearly isotropic growth. This is in agreement with the recorded crystallization kinetics.

Introduction

Organo-modified montmorillonite (o-MMT) is widely used to prepare nanocomposites with a polymer matrix. A number of studies have considered its effects on matrix crystallization. It appears that nucleating activity of o-MMT in isotactic polypropylene (iPP) is rather weak in quiescent conditions. Svoboda et al. (2002) observed the iPP nucleation enhanced by o-MMT tactoids rather than by exfoliated platelets. Pozsgay et al. (2002) concluded that nucleation did not occur on the surfaces but rather in the interlayers of o-MMT particles. Conversely, crystallization can be considerably accelerated by flow if o-MMT is compatibilized with maleic anhydride grafted polypropylene (PP-MA) (Somwangthanaroj et al., 2003; Nowacki et al., 2004; Rozanski et al., 2009).

In our previous work (Rozanski et al., 2009) performed in isothermal conditions, strong enhancement of the nucleation and acceleration of the overall crystallization kinetics after shearing were observed in the compatibilized nanocomposites with exfoliated clay. However, the sheared samples exhibited only a weak orientation of α phase with (040) crystallographic planes parallel to shearing direction. This was interpreted in terms of a small population of oriented crystals that formed due to shear-induced orientation of iPP chains and served as nuclei for further nearly isotropic growth. In order to interpret this shear-sensitivity, we propose here to extend the work by studying the influence of shear on crystallization of iPP-based nanocomposites with organo-modified montmorillonite **under non-isothermal conditions**. Shear-induced nonisothermal crystallization will be followed *in-situ* by light-depolarizing microscopy.

Experimental

Materials and Preparation of Nanocomposites

The studies utilized one iPP produced by BASF, Germany: Novolen 1100H (PPH), with MFI of 1.8 g/10 min (230 °C, 2.16 kg), M_w= 4.0 x 10^5 g/mol, M_w/M_n =5. The o-MMT used was octadecyl amine modified Nanomer I30P, produced by Nanocor, USA, having particle size within the range of 16-20 μm. Maleic anhydride grafted iPP, Polybond 3150 (PB) with 0.5% of MA groups (MFI=50 g/10 min at 190 °C) manufactured by Uniroyal Chemical, was used as a compatibilizer.

A PPH/PB/MMT nanocomposite was prepared by two-step blending in a Brabender mixer at 200 °C for 20 min at the processing speed of 60 rpm. First, o-MMT was mixed with PB in the weight proportion of 3:10, and, in the second step, an appropriate amount of iPP was added to obtain 3 wt.% concentration of o-MMT in composites. The resulting PB content in the nanocomposites was 10 wt.%.

To have reference materials, PPH was also mixed with 11 wt.% of PB (PPH/PB) and with 3 wt.% of o-MMT (PPH/MMMT) for 20 min at 200 °C and 60 rpm.

Nonisothermal Flow-Induced Crystallization

180-190 μm thick specimens were prepared by compression and then sheared in the apparatus used in our previous study (Rozanski et al., 2009), which consists of a polarizing light microscope and a cell containing two heated glass-rails positioned on a microscope stage. The specimen was placed between the glass-rails, the entire sandwich was heated up to 225 °C for 3 min and then cooled down at the rate of 10 °C min^{-1}. During cooling, a simple shear was applied by means of a displacement of the lower rail, with different shearing programs (Fig. 1): shearing at 10, 20 or 40 s^{-1} for 10 s, starting at 163, 153 or 143 °C.

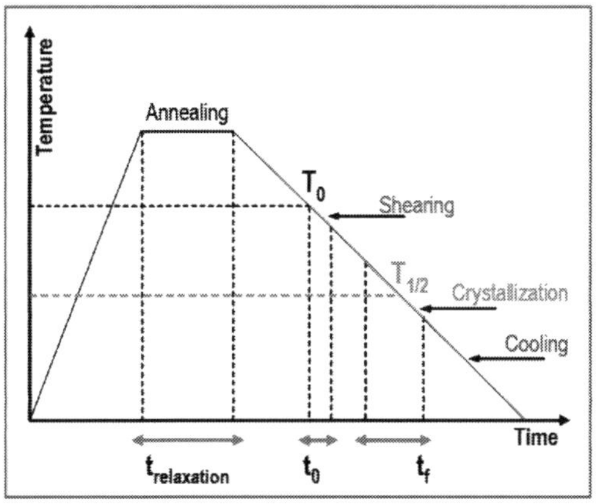

Figure 1 – Thermo-mechanical treatment.

Post-shearing crystallization was followed by polarized light microscopy (PLM). The light depolarization technique was also employed. The intensity of transmitted depolarized light, I, was measured to follow the overall crystallization kinetics. The conversion degree as a function of time t was calculated according to the equation:

$$\alpha(t) = [I(t)-I_0]/[I_e-I_0] \qquad (1)$$

where I_0 and I_e denote the intensity of transmitted depolarized light prior to crystallization and after completion of crystallization, respectively. After quenching to room temperature the samples were carefully removed from the shearing device for further studies.

Characterization

To study clay exfoliation 70 nm thick sections were microtomed from nanocomposite samples with an ultramicrotome and examined by transmission electron microscopy (TEM).
The microstructure of the samples was also studied by scanning electron microscopy (SEM) on etched microtomed sections. The specimens were etched for 1-2 h at room temperature with a mixture containing 0.7 wt.% of $KMNO_4$, dissolved in a 5:4:1 v/v mixture of 95% sulfuric acid, 85% phosphoric acid and distilled water, respectively.

The crystalline orientation was evaluated by wide-angle X-ray diffraction (WAXD) using imaging plates for recording the diffraction patterns.

Results and Discussion

Clay Exfoliation

In PPH/MMT, the exfoliation of o-MMT is poor with relatively large clay particles (Fig. 2A.). On the contrary, compatibilized nanocomposites prepared by two-step mixing exhibit numerous exfoliated clay platelets (Fig. 2B.).

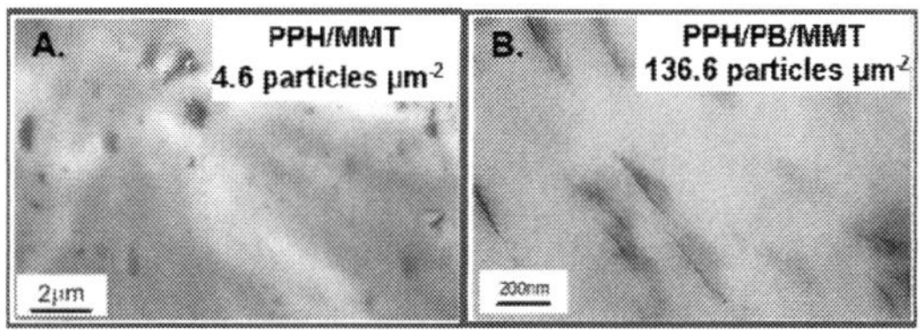

Figure 2 – TEM micrographs of **A.** PPH/MMT and **B.** PPH/PB/MMT (Rozanski et al., 2009).

Crystallization during Cooling. Influence of Flow

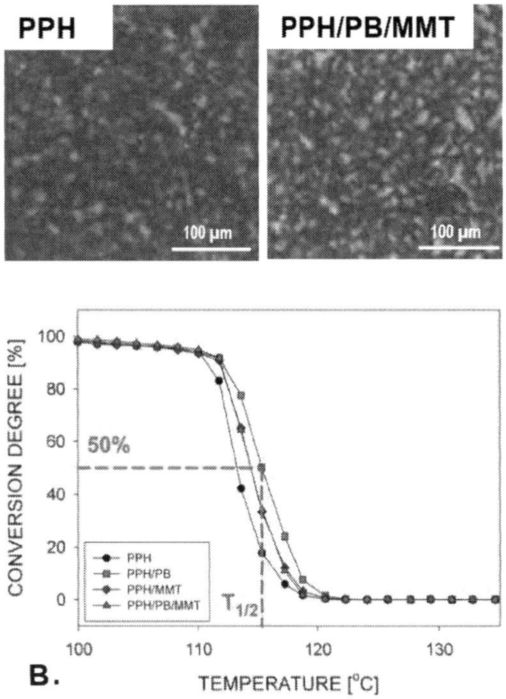

Figure 3 – **A.** PLM micrographs of PPH and PPH/PB/MMT and **B.** Overall crystallization kinetics of the four products in quiescent conditions.

In quiescent conditions, there are only weak differences in the crystallization behavior of the materials studied (Fig. 3). When a shear is applied, the differences are more important. For the experiments described in Fig. 4B., the curves for PPH/PB and PPH/MMT are superposed. With respect to the other ones, the curve for PPH/PB/MMT is quite typical:

- Crystallization starts at higher temperature.
- During 70 % of the transformation, the conversion rate is higher (steeper slope).
- Finally, the kinetics is slowed down and the curve tends to those of other materials.

As a consequence, a fine grain structure is obtained (Fig. 4A.).

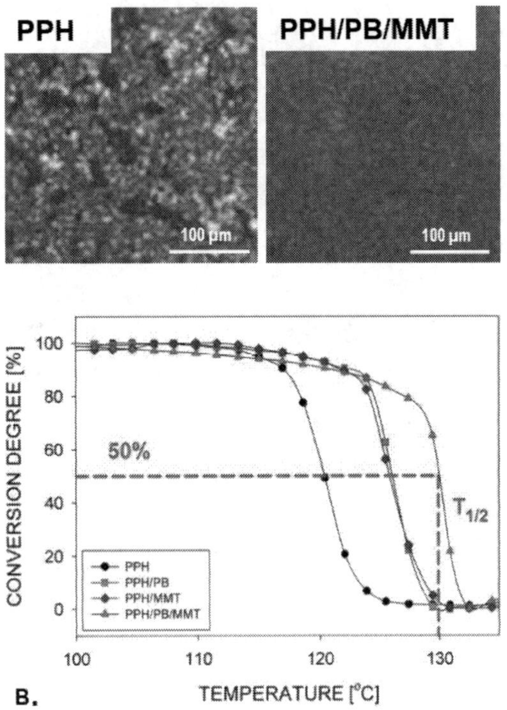

Figure 4 – **A.** PLM micrographs of PPH and PPH/PB/MMT and **B.** Overall crystallization kinetics of the four products: shearing from 143 °C, shear rate 40 s^{-1}, shearing time 10 s.

A good way to quantify the differences between the materials and to analyze the effects of shearing conditions is to use the half-crystallization temperature, i.e., the temperature at which the conversion degree is 50 % (Figs. 3B. and 4B.). The results are displayed in Figs 5 and 6.

Concerning the products, it appears that:

- Flow has a weak effect on PPH crystallization.
- Shear has a strong effect on the crystallization of the compatibilized nanocomposite. $T_{1/2}$ can reach 130 °C whereas it is about 114.5 °C in quiescent conditions.

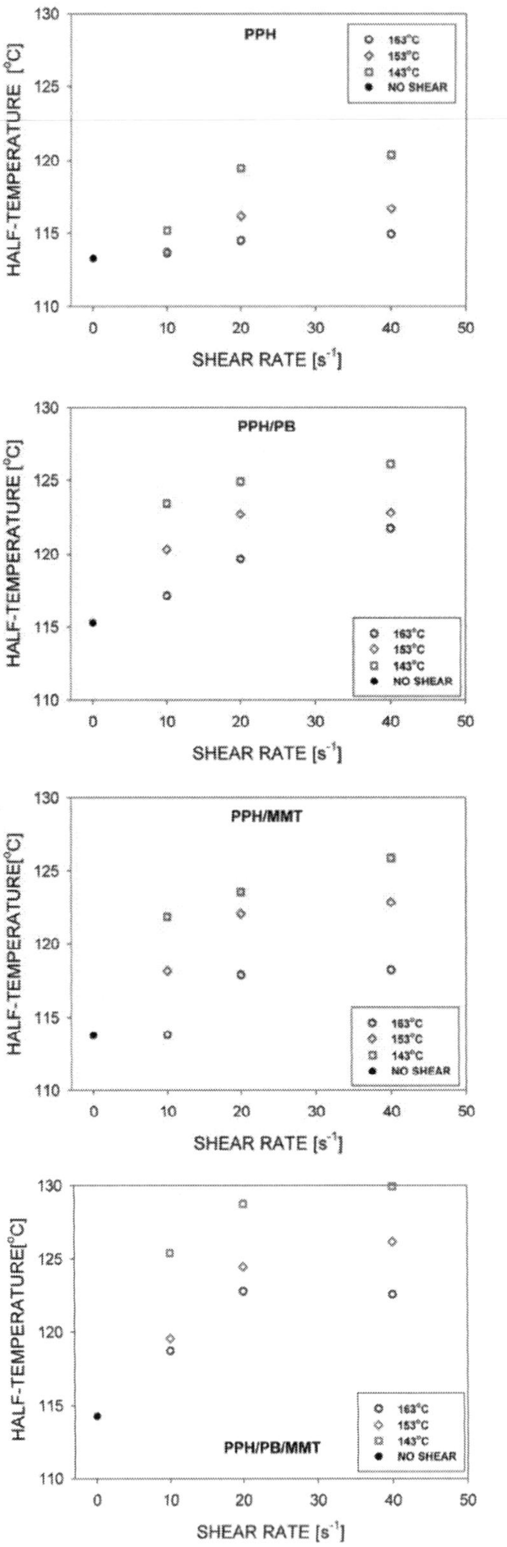

Figure 5 – Half-crystallization temperature plotted against shear rate.

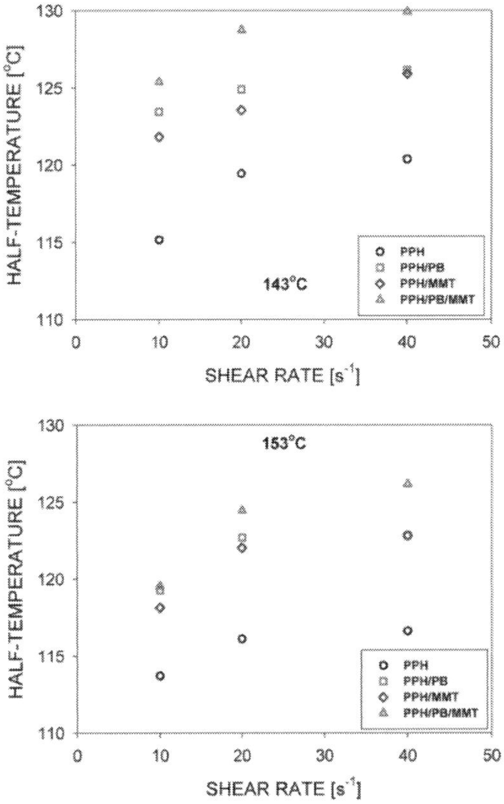

Figure 6 – Half-crystallization temperature against shear rate for shearing from 143 and 153 °C.

- The situation of PPH/PB and PPH/MMT is intermediate, some small differences between the two materials being observed as a function of the processing conditions.

For the shearing conditions:

- $T_{1/2}$ increases with shear rate.
- The most important influence is that of temperature. The decrease of temperature considerably enhances the crystallization kinetics.

SEM Observations

SEM micrographs confirm the results obtained by optical microscopy and provide additional information. The spherulitic morphology of the neat polymer is not affected by shearing (Figs. 7A. and B.). Conversely, a fine grain structure is observed in the sheared nanocomposite (Figs. 7C. and D.). It consists of polycrystalline aggregates analogous to those observed in isothermal conditions (Rozanski et al., 2009).

Crystalline orientation

Under static conditions, all the specimens can be considered as isotropic (Fig. 8D.). Flow does not create any significant orientation in PPH. In the same way, no orientation is generated in all the specimens when shear is applied from 163 °C (Fig. 8A.).

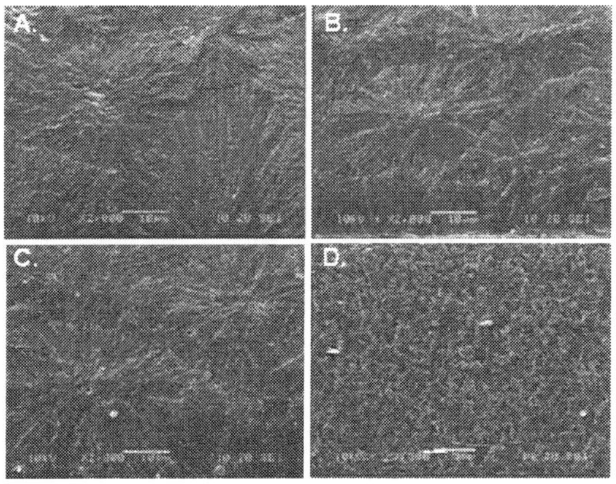

Figure 7 – SEM micrographs: PPH **A.** static conditions and **B.** shearing from 143 °C, shear rate 40 s^{-1}, duration 10 s; PPH/PB/MMMT **C.** static conditions and **D.** shearing from 143 °C, shear rate 40 s^{-1}, 10 s.

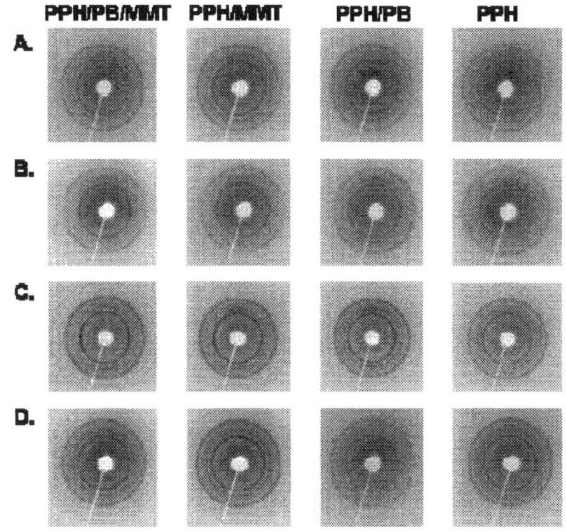

Figure 8 – WAXD 2D patterns of neat PPH and nanocomposites with **A.** shearing from 163 °C, shear rate 40 s^{-1}, shearing time10 s, **B.** shearing from 143 °C, 40 s^{-1}, 10 s, **C.** shearing from 143 °C, 10 s^{-1}, 10 s and **D.** static conditions. Shear direction horizontal.

In the other samples (Figs. 8B. and C., PPH/PB/MMT, PPH/MMT, PPH/PB), the same type of texture is observed:

- Reflections from (110) crystallographic planes parallel and perpendicular to shear direction.
- Reflections from (040) planes parallel to shear direction.

This is characteristic of a bimodal crystalline texture (Andersen and Carr, 1975), i.e., two populations of crystals with their **c** or **a*** axis along the flow direction, respectively (**a*** is the axis of the reciprocal lattice related to **a**). This is usually interpreted in terms of two types of lamellae: classical radial lamellae and tangential ones obtained by epitaxial growth, which is a particular feature of polypropylene crystallization.

Conclusions

The present work confirms and extends our previous conclusions deduced from isothermal crystallizations. Shear greatly accelerates the overall crystallization kinetics in the compatibilized nanocomposites as compared to that in the composite of iPP without compatibilizer and in the blend of iPP with the compatibilizer. This is clearly shown by using half-crystallization temperature as indicator, and attributed to an enhancement of nucleation observed by optical microscopy and SEM. Comparison of the materials based on the same iPP (PPH) but differing in the clay exfoliation level (PPH/MMT and PPH/PB/MMT) demonstrates that clay exfoliation and dispersion play a crucial role in shear-induced enhancement of the nucleation of crystallization.

As for isothermal experiments, X-ray studies evidenced a weak iPP crystalline orientation in sheared samples of compatibilized nanocomposites. It was concluded that this weak orientation in the sheared samples most probably resulted from a small population of well oriented shear-induced crystals that served as nuclei for further growth and accelerated the crystallization. The kinetic curves obtained in the present investigation support this interpretation, since they indicate a two-step crystallization process. Nevertheless, in the present work, a weak crystalline orientation is also detected in the composite without compatibilizer and in the blend of iPP with the compatibilizer, which is consistent with some enhancement of their crystallization kinetics.

Among processing parameters, the temperature range in which shear is applied plays a major role. The role of these different parameters (temperature, shear rate, shearing time) could be further discussed using a unifying concept, such as the specific work introduced by Janeschitz-Kriegl (2006).

Acknowledgements

The authors acknowledge Polish Academy of Sciences and CNRS for financial support through the contracts:
- PAN/CNRS Collaboration Project n°18261 (2008) "Shear-induced crystallization and structure of polymer based nanocomposites".
- PAN/CNRS PICS Project n°4702 (2009-2011) "Crystallization under shear and interface behaviors of polymers and polymer nanocomposites".

References

Andersen, P.G., Carr, S.H., "Formation of Bimodal Crystal Textures in Polypropylene", J. Mater. Sci., **10**, 870-888 (1975)

Janeschitz-Kriegl, H., "Phases of Flow-Induced Crystallization of i-PP: How Remote Pieces of the Puzzle Appear to Fit", Macromolecules, **39**, 4448-4454 (2006)

Nowacki R., Monasse B., Piorkowska E., Galeski A., Haudin J.M., "Spherulite Nucleation in Isotactic Polypropylene Based Nanocomposites with Montmorillonite under Shear", Polymer, **45**, 4877-4892 (2004)

Pozsgay A., Frater T., Papp L., Sajo I., Pukanszky B., "Nucleating Effect of Montmorillonite Nanoparticles in Polypropylene", J. Macromol. Sci., Part B-Phys., **41**, 1249-1265 (2002)

Rozanski A., Monasse B., Szkudlarek E., Pawlak A., Piorkowska E., Galeski A., Haudin J.M., "Shear-Induced Crystallization of Isotactic Polypropylene Based Nanocomposites with Montmorillonite", Eur. Polym. J., **45**, 88-101 (2009)

Somwangthanaroj A., Lee E.C., Solomon M.J., "Early Stage Quiescent and Flow-Induced Crystallization of Intercalated Polypropylene Nanocomposites by Time-Resolved Light Scattering", Macromolecules, **36**, 2333-2342 (2003)

Svoboda P., Zeng C., Wang H., Lee J., Tomasko D.L., "Morphology and Mechanical Properties of Polypropylene/Organoclay Nanocomposites", J. Appl. Polym. Sci., **85**, 1562-1570 (2002)

How to Nucleate Polymers

Gerrit W.M. Peters, Luigi Balzano*

EIndhoven University of Technology - Eindhoven - Netherlands

The effect of the thermo-mechanical history (deformation, pressure and cooling rate) on morphology and structure of semi-crystalline polymers is discussed, in particular for polyethylene (PE), isotactic polypropylene (iPP) and propylene/ethylene random copolymers (RaCo) and some of their blends with common additives like DMDBS and NA11. It is shown that the nucleation stage is dominating the final structure. In more detail, we will discuss the effect of: - pressure and cooling rate on the specific volume of iPP, iPP-DMDBS blends, iPP-NA11 blends and RaCo; - 'strong' deformations on the crystallization of bimodal PE, iPP+additives and RaCo, including the effect of shear deformations on the formation of beta phase polypropylene; - fast coolings on the structure of iPP, RaCo and their blends with additives; The main conclusion is that the effects process conditions, if well understood, can be used to act as good or even better as added nucleation agents

Assessment of the effect of nanoclay compounding in the morphology of PP mouldings produced with hybrid moulds

Antonio Sergio Pouzada (1), Alessandra Fortuna Neves (2), Mariana Azevedo Silva(1), Gean Victor Salmoria (2), Carlos Henrique Ahrens (2)*

(1) IPC--Institute for Polymers and Composites - Guimaraes - Portugal,
(2) CIMJECT – Dept Engenharia Mecânica, Universidade Federal de Santa Catarina - Santa Catarina - Brazil

The concept of hybrid mould combines the conventional techniques of mould manufacturing and Rapid Prototyping and Rapid Tooling, resorting to non-conventional materials for producing moulding blocks, e.g., epoxy resin composites (Pouzada 2009). These materials in the hybrid mould affect the morphology of the mouldings due to difference of thermal conductivities of steel and the epoxy composites (Martinho, Bartolo et al. 2009). The thermomechanical environment imposed during the injection moulding creates temperature and shear rate variations along the volume of the part, which cause the build up of density and crystallinity fields, according to the distance from the surface layer of the mould and from the gate (Brito, Cunha et al. 1991). For these moulds short steel fibre (SSF)/epoxy composites were considered adequate for improving the performance of moulding cores (Sabino-Netto, Salmoria et al. 2008). Nanoclays acts as an effective heterogeneous nucleating agent increasing the nucleating speed and the crystallization, and decreases the spherulite size (Wang, Xiao et al. 2005). This paper reports on a study on the influence of nanoclays in an injection moulding grade of polypropylene. The mouldings are a 3D part of nominal thickness of 2 mm and volume of 11 cm3 made of a mixture of polypropylene and nanoclay masterbatch. They were produced in a hybrid mould with epoxy/SSF composite core and steel cavity. In the microstructure evaluation the density and crystallinity were determined across the thickness of the part at a location midway between the gate and the end of fill. The skin-layer morphology as a function of the mould temperature, injection speed and nanoclay concentration was also observed and suggested that in some conditions a peculiar five-layer structure develops instead of the more common three-layer. Microhardness test were made to infer on the surface property effects of the nanoclay and also the correlation with the crystallinity data.

A STUDY ON THE ROLE OF NANOCLAY ON CORE-SHELL MORPHOLOGY DEVELOPMENT TERNARY BLEND NANOCOMPOSITES

Hadi Mohammadigoushki[1], Nima Mostofi[1] and Hossein Nazockdast[1*]

[1]Department of Polymer Engineering and Color Technology, Amirkabir University of Technology, 424 Hafez Ave, Tehran,Iran- Hadikhan64@yahoo.com, N_Mostofi@aut.ac.ir, Nazdast@aut.ac.ir

Abstract - Effect of the presence of organoclay on morphology development and microstructure of ternary polymer blend nanocomposites was studied by employing rheological as well as thermodynamical approaches. The ternary blend samples based on PA6/PP/PS with the same blend ratios and two types of surface modified montmorillonite (organoclay) with equal concentration but varying in surface modifier (intercalant) chemical functionality (Cloisite 15A and 30B) were considered. All samples were prepared by melt intercalation process in an internal mixer with using different feeding orders. From the melt linear viscoelastic results and comparing them with those predicted by Palierne model and SEM micrographs, it was found that, while presence of organoclay with greater intercalant polarity (Cloisite 30B) is mostly dispersed in PA6 matrix did not change original core-shell morphology of the samples, the organoclay with less polar intercalant (Cloisite 15A) changed the core-shell type morphology to the matrix-disperse. It was demonstrated that the results of nonlinear viscoelastic measurements can provide more reliable evidences to distinguish the core-shell from matrix-disperse morphology. These results were in good agreement with the results predicted based on Spreading Coefficient, Relative Interfacial Energy (RIE) and Young's Eq.

Introduction

During the last 2 decades a considerable attention has been directed towards polymer ternary blend systems, due to flexibility in controlling their morphology and therefore obtaining desirable physical and mechanical properties (Hemmati et. al 2001a, 2001b; Mohammadigoushki et. al 2009; Mostofi et. al 2009). Although a wide range of morphologies can be obtained for ternary blends, the core-shell morphology has received the most attraction. It has been shown that the thermodynamical parameters in particular spreading coefficient and relative interfacial energy (RIE) play most significant role in determining the morphology development and therefore the whole set of properties of this group of polymer blends.

On the other hand polymer blend nanocomposites have received special attention due to their enhanced physical and mechanical properties (Ren and Krishnamoorti 2003; Ren and Silva 2000; Gilman et. al. 2000; Gopakomar et. al 2002; Hotta and Paul 2004). The aim of the present work was to study the role of organoclay on interfacial interactions between the phases and morphology development of the ternary blend systems.

Experimental

Material

A commercial grades of PA6 (Akulon F 136C) purchased from DSM Co., an isotactic PP (HP500H) supplied by Arak Petrochemical Co. and PS (GPPS 1540) supplied by Tabriz Petrochemical Co., were used as blend components. Two different organically modified montmorillonite (namely Cloisite 15A and 30B) used, were purchased from Southern Clay Products.

Blend Preparation

Various PA6/PP/PS ternary blends and their nanocomposites with and without compatibilizers with composition given in Table 1 were considered. All samples were prepared by melt blending in an internal mixer (Brabender W50) at temperature of 260oC and rotor speed of 60rpm.

WAXD Experiments

The WAXD experiments were conducted using a Philips X' pert by Cu-Kα ray with the wave length of 0.154nm.

Rheological Measurements

The melt linear and nonlinear rheological measurements were preformed on the samples by using a rheometric mechanical spectrometer (Phaar Physica UDS200). The measurements were conducted at temperatures 260°C and 200°C, using a parallel plate fixture with diameter of 25mm and a constant gap of 1mm.

Surface Tension Measurements

The surface tension of the samples was calculated from the results of contact angle measurements using three appropriate solvents as: Diiode Methane, Water, Formamide.

Morphology

Morphology of the blend samples was studied by Scanning Electron Microscopy (SEM). Examination was performed on the samples that were cryofractured in liquid nitrogen and coated with gold prior to observation. In order to achieve better contrast for ternary blends the PS phase was etched by Chloroform.

Results and Discussion

WAXD Studies

Fig. 1 compares the WAXD patterns obtained for the PA6/PP/PS/nanoclay ternary blend nanocomposites. The corresponding calculated d-spacing are summarized in Table 2. As it can be seen the extent of intercalation obtained for Cloisite 30B organoclay is much greater than that of Cloisite 15A due to greater affinity of Cloisite 30B towards PA6 matrix.

Morphology Studies

Fig. 2 shows the SEM micrograph of PA6/PP/PS ternary blend system. As it can be noticed this sample exhibits a core-shell type morphology in which PP phase is encapsulated by the PS shell dispersed in the PA6 matrix. Similar results have been reported and confirmed by thermodynamic approach in our previous works (Mostofi et. al. 2009, Mohammadigoushki et. al 2009)

Fig. 3 compares the morphology of two ternary blend samples containing different type of organoclay. It can be noticed that, while the nanoclay with less polar surface modifier (Cloisite 15A) changed the morphology from core-shell to matrix-disperse, the nanoclay containing more polar surface modifier (Cloisite 30B) exhibits the same core-shell morphology as that of the ternary blend. In the case of sample containing Cloisite 30B the organoclay may mostly be located inside the PA6 phase and PA6/PS interface. As will be shown by thermodynamical calculation, this is in the favor of core-shell morphology formation. However, in the Cloisite 15A containing sample, the organoclay can preferentially be located in PP phase leading to increase in the compatibility of the PP and PA6; a process which is thermodynamically in favor of matrix-disperse morphology formation. This is evidenced by the viscoelastic results shown in Fig. 4, which shows much stronger low frequency solid body response for the ternary blend samples containing Cloisite 30B compared to that of 15A containing, ternary samples.

Rheological Studies

By comparing the results shown in figure 5, one may notice that the ternary blend samples exhibits stronger low frequency nonterminal behavior than that of binary blend samples. This can be considered as an indication of core-shell morphology developed in these ternary blend systems.

In order to provide more information about morphology and microstructure development of the samples, nonlinear viscoelastic tests measurements were also preformed on the samples.

Fig. 6 compares the results of startup flow test on the ternary and binary blend samples. As it can be seen the ternary blend sample in contrary to binary blend samples, which after stress overshoot quickly reaches to steady state, the ternary blend samples exhibits a stress undershoot which approaches to steady state after a mild stress growth. The undershoot and mild stress growth observed for ternary blend can be considered as an indication of a time dependent core shell morphology development.

The results of similar experiments obtained for ternary blend nanocomposite samples are shown in Fig. 7. By comparing these results one may clearly notice that while the samples containing Cloisite 30B shows a pronounced stress undershoot similar to that of ternary blend samples, the Cloisite 15A containing sample exhibit a behavior to binary blend samples. These results are in good agreement with the morphological observations discussed above.

Morphology Predictions Using Thermodynamic Calculations

Table 3 summarizes the values of surface tension of the blend components. The interfacial tension of each polymer pair calculated based on geometric-mean equation (Wu 1982a), are given in Table 4. Having considered these data the morphology of the system can be predicted using Harkin's spreading coefficient and Relative Interfacial Energy (RIE) approach as follows:

$$\lambda_{B/C}=\gamma_{AC}-\gamma_{AB}-\gamma_{BC} \quad \textbf{Eq. 1a}$$

$$\lambda_{C/B}=\gamma_{AB}-\gamma_{AC}-\gamma_{BC} \quad \textbf{Eq. 1b}$$

$$(\sum A_i\gamma_{ij})_{B+C}=(4\pi)^{1/3}[n_B^{1/3}x^{2/3}\gamma_{AB}+n_C^{1/3}\gamma_{AC}](3V_C)^{2/3} \quad \textbf{Eq. 2a}$$

$$(\sum A_i\gamma_{ij})_{B/C}=(4\pi)^{1/3}[n_B^{1/3}(1+x)^{2/3}\gamma_{AB}+n_C^{1/3}\gamma_{AC}](3V_C)^{2/3} \quad \textbf{Eq. 2b}$$

$$(\sum A_i\gamma_{ij})_{C/B}=(4\pi)^{1/3}[n_B^{1/3}x^{2/3}\gamma_{BC}+n_C^{1/3}(1+x)^{2/3}\gamma_{AC}](3V_C)^{2/3} \quad \textbf{Eq. 2c}$$

$$x=\frac{V_B}{V_C}$$

According to thermodynamic predictions in Table 5, the PA6/PP/PS system should have core-shell morphology in which the PP droplet should be encapsulated by PS skin dispersed in PA matrix; results which are in good agreement with experimental results. The partitioning of the nanoclay was predicted in terms of wetting factor using Young's Eq, (Ashabi et. al. 2008) and the results are summarized in Table 6.

From these results one can notice that Cloisite 30B is preferentially located in PA6, while the Cloisite 15A is mainly located in the interface. These results support our suggestion on the location of two different organoclay in our ternary blend nanocomposite system.

Table 1: Composition of the blends

Code	%PA6	%PP	%PS	%Cloisite 30B	%Cloisite 15A
B1	90	10	--	--	--
B2	90	--	10	--	--
B3	--	90	10	--	--
B4	--	10	90	--	--
B5	70	--	30	--	--
B6	70	30		--	--
T1	70	15	15	--	--
N1	70	15	15	3	--
N2	70	15	15	--	3

Table 2: gallery distances calculated from the WAXD patterns for Nanoclay and ternary nanocomposites

Sample	$2\Theta_1$	$2\Theta_2$	$2\Theta_3$	D_1(Å)	D_2(Å)	D_3(Å)
Cloisite 15A	3.41	4.5	8.24	25.9	19.61	10.71
Cloisite 30B	4.77	---	---	18.51	---	---
Ternary+15A	1.25	---	---	70.58	---	---
Ternary+30B	2.69	4.85	5.57	32.8	18.2	15.85

Table 3: Values of Surface Tension for Polymers

Polymer	T (°C)	γ (mN/m)	γ^p (mN/m)	γ^d (m N/m)	Ref.
PP	230	17.62	0.37	17.25	Ren and Krishnamoorti (2003)
PS	25	40.7	6.1	34.5	Wu 1982b
PA6	230	38.37	13.2	25.17	Ren and Krishnamoorti (2003)
PMMA	140	33.2	3.59	29.61	Ren and Krishnamoorti (2003)
Cloisite 15A	25	26.8	5.9	20.9	Measured
Cloisite 30B	25	40.43	15.78	24.65	Measured

Table 4: The interfacial tension of the polymer pairs

Polymer Pair	Experimental Values (mN/m)	Calculated (mN/m)
PP/PS (260°C)	2.71 (Macaúbas and Demarquette)	3.86
PA6/PS (260°C)	9.8	3.56
PA6/PP (260°C)	24.3	9.77
PP/PS (200°C)	---	5
PA6/Cloisite 15A (260°C)	---	1.303
PA6/Cloisite 30B (260°C)	0.21	0.149
PP/Cloisite 15A (260°C)	2.77	3.855
PP/Cloisite 30B (260°C)	---	12.6
PS/Cloisite 15A (260°C)	---	1.236
PS/Cloisite 30B (260°C)	---	6.925

Table 5: values of spreading coeffiecient and Relative Interfacial Energy calculated for ternary blends

Sample	$\lambda_{PP/PS}$	$\lambda_{PS/PP}$	RIE (PP+PS)	RIE (PP/PS)	RIE (PS/PP)
PA6/PP/PS	-17.21	11.79	34.1	41.28	18.26
PA6/PP/PS + Cloisite 30B	-11.11	3.99	17.29	23.57	11.59
PP/PMMA/PS	$\lambda_{PMMA/PS}$	$\lambda_{PS/PMMA}$	**RIE (PMMA+PS)**	**RIE (PMMA/PS)**	**RIE (PS/PMMA)**
	-3.51	0.13	61.5	52.1	62.5

Table 6: Values of wetting factor calculated using Young's Eq.

Blend	ω_{PA6}(30B)	ω_{PA6}(15A)	ω_{PP}(15A)	Location
PA6/PP	1.275>1	---	---	PA6
PA6/PS	1.726>1	---	---	PA6
PP/PS	---	---	-0.068>-1	Interface
PA6/PP	---	0.261<1	---	Interface
PA6/PS	---	-0.019<1	---	Interface

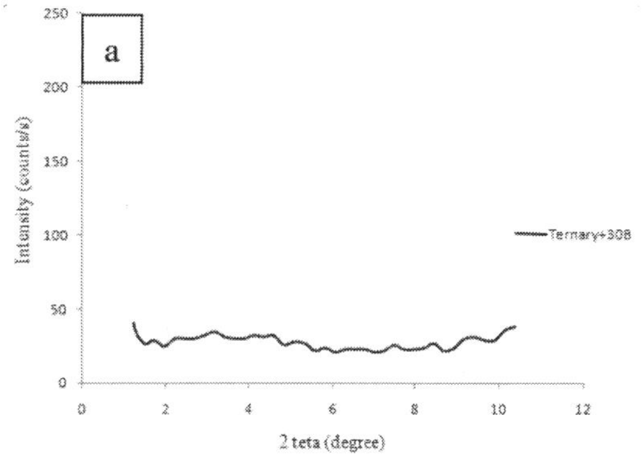

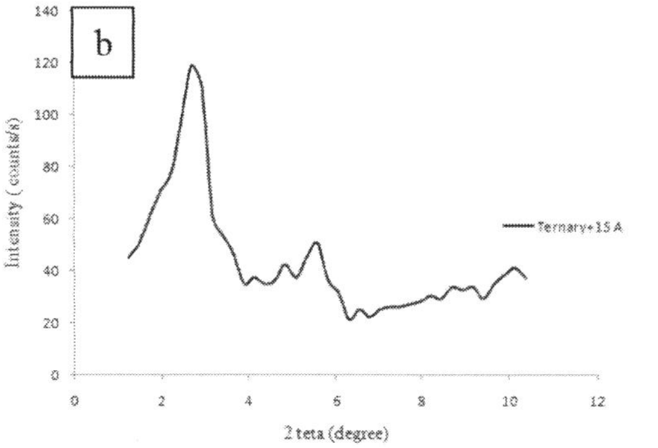

Fig. 1: WAXD pattern of ternary nanocomposite blends based on PA6/PP/PS-Organoclay a)Ternary + 30B, b) Ternary + 15A

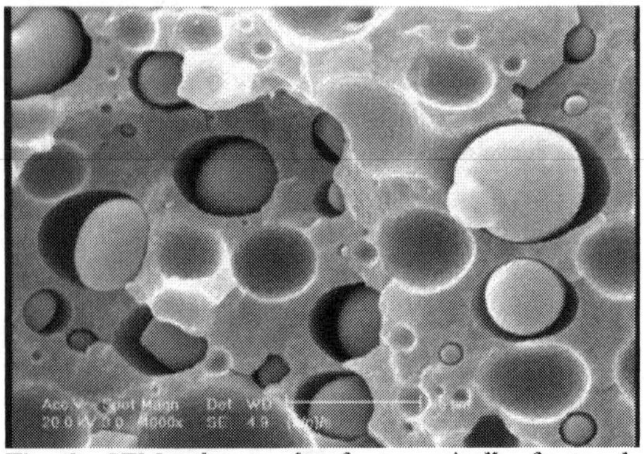

Fig. 2: SEM micrograph of cryogenically fractured surface of PA6/PP/PS ternary blend, the PS phase is selectively ethced by chlroform.

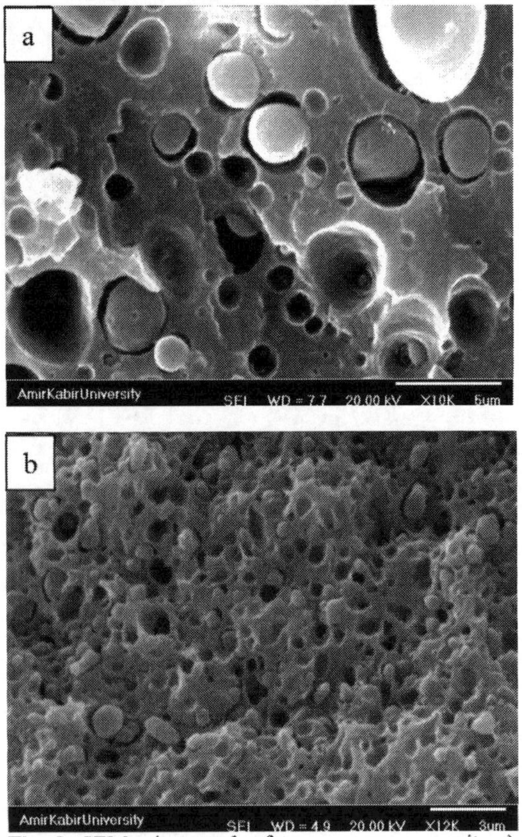

Fig. 3: SEM micrograph of ternary nanocomposite a) PA6/PP/PS-Organoclay (30B), b) PA6/PP/PS-Organoclay (15A)

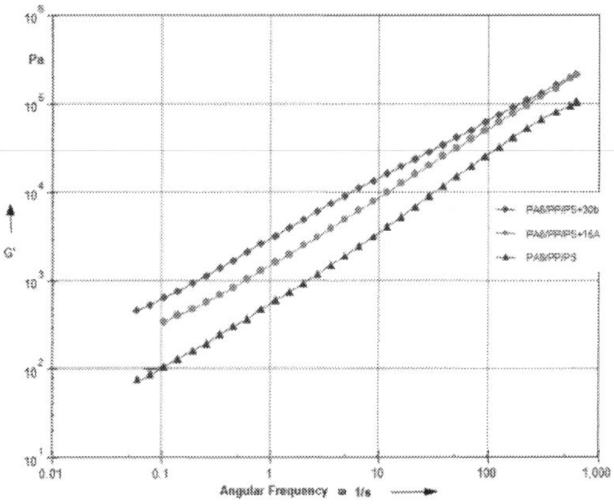

Fig. 4: Storage modulus of the ternary samples.

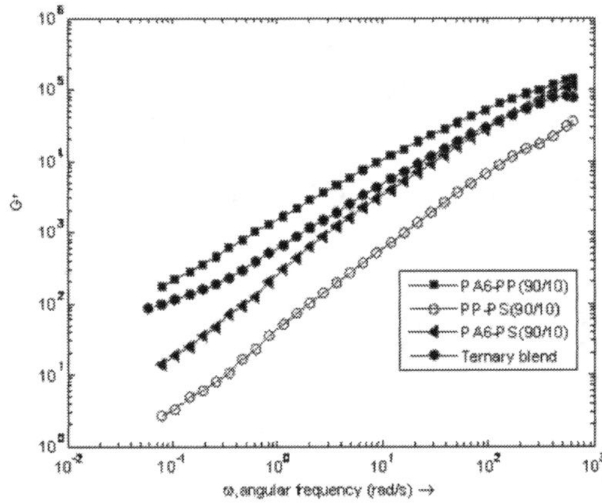

Fig. 5: Storage modulus of binary and ternary blend of PA6, PP and PS. The compositions of the blends are shown on the graph.

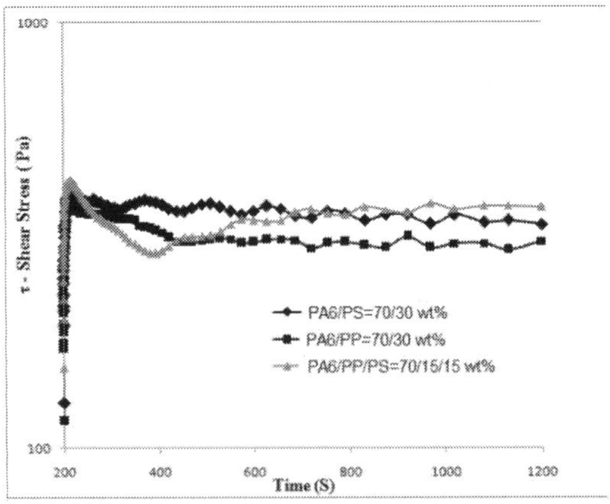

Fig. 6: Startup flow test for different samples, the sample compositions are shown on the figure.

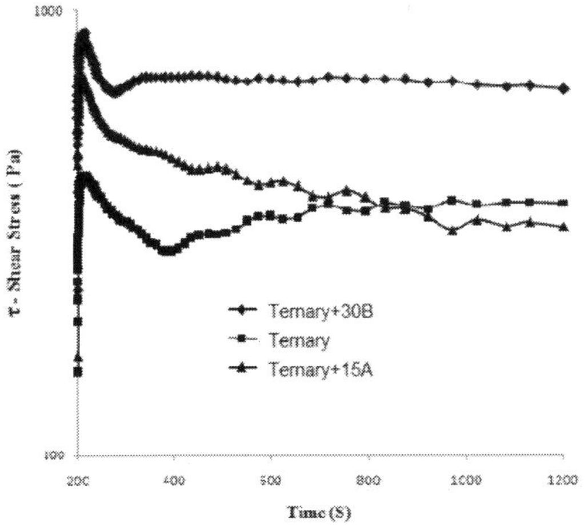

Fig. 7: Startup flow test for different samples, the sample compositions are shown on the figure.

Conclusions

It was demonstrated that incorporation of organoclay into a ternary blend systems can have different influence on morphology development, in particular, the core-shell formation of ternary blend systems.

In the systems in which organoclay is more compatible with matrix and shell and therefore forcing the organoclay to be located in the matrix and interface the core shell morphology remains almost intact, while for the system with greater compatibility between organoclay and two dispersed phases, the morphology will change from core-shell to matrix-disperse. It was also shown that the results of rheological measurements in conjunction with thermodynamical predictions based on spreading coefficient and RIE can provide valuable information to understand the morphology development of ternary blend nanocomposite system.

References

As'habi L., Jafari S. H., Baghaei B., Khonakdar H. A., Pötschke P., Böhme F., "Structural analysis of multicomponent nanoclay-containing polymer blends through simple model systems", Polymer 49, 2119-2126 (2008)

Gilman J. W., Jakson C. L., Morgan A. B., Harris R., "Flammability Properties of Polymer−Layered-Silicate Nanocomposites. Polypropylene and Polystyrene Nanocomposites" Chem. Mater, 12, 1866-1873(2000)

Gopakumar TG, Lee JA, Kontopoulou M, Parent JS., "Influence of clay exfoliation on the physical properties of montmorillonite/polyethylene composites", Polymer, 43, 5483-5491 (2002)

Hemmati M., Nazockdast H., Shariat panahi H., "Study on morphology of ternary polymer blends. I. Effects of melt viscosity and interfacial interaction", J. Appl. Polym. Sci., 82, 1129-1137 (2001)

Hemmati M., Nazockdast H., Shariat panahi H., "Study on morphology of ternary polymer blends. II. Effect of composition", J. Appl. Polym. Sci. 82, 1138-1146 (2001)

Hotta S. and Paul D. R., "Nanocomposites formed from linear low density polyethylene and organoclays" Polymer, 45, 7639-7654 (2004)

Macaúbas P. H. P., Demarquette N. R. "Morphologies and interfacial tensions of immiscible polypropylene/polystyrene blends modified with triblock copolymers" Polymer, 42, 2543-2554 (2001)

Mohammadigoushki H., nazockdast H. and Mostofi N., "Morphology Development and Melt Linear Viscoelastic Properties of PA6/PP/PS ternary blend systems", J. Elastomers and Plastics, 41, 339-351 (2009)

Mostofi N., Nazockdast H. and Mohammadigoushki H., "Study on Morphology and Viscoelastic Properties of PP/PET/SEBS Ternary Blend and their Fibers", J. Appl. Polym. Sci, 114, 3737-3743 (2009)

Reignier J. and Favis B. D., "Control of the Subinclusion Microstructure in HDPE/PS/PMMA Ternary Blends", Macromolecules, 33, 6998-7008 (2000)

Ren J and Silva A. S., "Linear Viscoelasticity of Disordered Polystyrene−Polyisoprene Block Copolymer Based Layered-Silicate Nanocomposites" Krishnamoorti R. Macromolecules, 33, 3739-3746 (2000)

Ren J. and Krishnamoorti R. "Nonlinear Viscoelastic Properties of Layered-Silicate-Based Intercalated Nanocomposites" Macromolecules, 36, 4443-4451 (2003)

Wu S., "Chapter 3 Interfacial and Surface Tensions of Polymer Melts and Liquids", in Polymer Interface and Adhesion, Wu S., Marcel Dekker, New York, p. 103 (1982)

Wu S., "Chapter 3 Interfacial and Surface Tensions of Polymer Melts and Liquids", in Polymer Interface and Adhesion, Wu S., Marcel Dekker, New York, p. 89 (1982)

ENHANCEMENT IN TEAR PROPERTIES OF SINGLE-SITE CATALYZED LLDPE BLOWN FILMS AT HIGHER DRAW-DOWN

Nitin Borse*, Norman Aubee, Paul Tas
NOVA Chemicals Technical Center, 3620-32 Street NE Calgary T1Y 6G7
borsen@novachem.com, aubeend@novachem.cem, taspa@novachem.com

In most applications, packaging films are required to possess high tear resistance. In this study we used conventional Ziegler-Natta catalyzed linear low density polyethylene (Z-N LLDPE) resins and single site catalyzed sLLDPE resins. The films produced at different draw-down conditions were analyzed by wide-angle X-ray scattering (WAXS), optical microscopy and Raman spectrometry to estimate crystalline orientation, optical birefringence and the film crystallinity. For the conventional Z-N LLDPE blown films it was observed that the anisotropy in tear properties was higher at higher draw-down ratios. The films blown at higher draw-down with lower blow-up ratio and at thinner gauges showed higher TD (Transverse Direction) tear strength but the MD (Machine Direction) tear strength was reduced. Unlike conventional Z-N LLDPE resins, the sLLDPE resins showed an increase in TD as well as MD tear properties when blown at higher draw-down ratios. We observed that in all the LLDPE films the TD tear strength correlated with the crystalline phase orientation, whereas MD tear strength correlated with the overall orientation which was the sum of both the crystalline phase and the amorphous phase orientations. The Z-N LLDPE blown films produced at higher draw-down ratio showed higher overall polymer chain orientation in the MD, as estimated by the birefringence analysis. The sLLDPE blown films on the other hand, when produced at higher draw-down, showed higher overall orientation in the TD. This anomalous orientation behavior in the sLLDPE blown films resulted in unusually higher MD and TD tear properties in these films at higher draw down ratios.

Introduction

Linear Low Density Polyethylene (LLDPE) resins are mostly used in film applications. Although different film market segments have different performance requirements, superior tear, tensile and dart impact strength are always desirable. Packaging films are required to possess high tear resistance in most applications. The tear resistance of the films, along with the other physical properties, depends upon the molecular architecture, microstructure and molecular orientation. It has been recognized (Chen et al., 2006; Krishnaswami and Sukhadia, 2000) that film performance strongly depends upon the orientation of both the crystalline phase and the amorphous chains.

Chen et al. (2006) fabricated LLDPE films at different conditions of blow up ratio, die gap, and frost line height. The White-Spruiell orientation factors of crystal unit cells, amorphous chains, and Herman's orientation factors of lamellae for the films were determined from wide-angle X-ray scattering (WAXS) pole figures, birefringence and small angle X-ray scattering (SAXS). They found that in typical Z-N LLDPE films the amorphous chains were aligned preferentially along the MD. The results show that at higher draw down the TD tear strength was higher, but the MD tear strength and dart impact strength were reduced. Krishnaswamy and Sukhadia (2000) found that the MD tear strength of LLDPE blown films was higher when non-crystalline chains were close to equi-biaxial in the plane of the film, while the TD tear strength was higher when the crystalline lamellae were relatively straight and close to the TD. Krishnaswamy and Lamborn (2000) explained the differences in the tensile properties of LLDPE blown films between the MD and TD in terms of lamellar orientation. They used various LLDPE resins polymerized using Ziegler-Natta, metallocene and chromium oxide based catalysts that were blown into films at similar processing conditions. Based on molecular orientation and microscopic investigations, the LLDPE blown film microstructure was described as a parallel array of lamellae with their long axes along the TD-normal plane. The morphology was typically Keller-Machin I type. The correlation between the Elmendorf tear resistance and tensile yield characteristics was noticed.

There have been attempts to relate the orientation, morphology and properties of the blown polyethylene films to the MD/TD stress balance at the frost line (Choi et al., 1982; Kwack and Han, 1988) and draw down ratio (Patel et.al., 1994). Zang et al. (2004) studied oriented structure and anisotropy properties of polyethylene blown films. The Z-N LLDPE films with octene comonomer showed higher tear strength than those produced using butene comonomer. The butene comonomer Z-N LLDPE films showed higher anisotropy in MD/TD tear strength. However in both the films the MD tear strength was reduced at high draw-down ratio.

The method first described by Herman and Platzek (1939) to measure orientation in fibres was determination of birefringence by optical retardation in a polarized light microscopy. The birefringence is defined as the difference in the refractive indices between two perpendicular directions. In semi-crystalline polymers, both crystalline and amorphous phases possess different orientation levels. The measured birefringence is the sum of the contributions from both these phases. The crystalline structure of polyethylene was first determined by Bunn (1939) using a long chain branched polyethylene (LDPE) on the basis of wide-angle X-ray diffraction studies. Polyethylene is found to crystallize into an orthorhombic unit cell with dimensions of *a*, *b* and *c*-axes as 7.41, 4.94 and 2.54 Å respectively. The crystal growth is seen in the direction of *b*-axis, and *c*-axis is the polymer chain axis. The crystalline phase orientation is usually analyzed by wide-angle X-ray scattering (WAXS) and then constructing pole figures (Choi et al., 1982). This allows a complete representation of the distribution of polymer chains and crystallographic axes. White and Spruiell (1981) developed the representation of a series of second moment biaxial orientation functions, which are useful for characterizing molecular orientation in films and sheets. White and Cakmak (1988) showed that this type of representation of crystalline orientation functions could be used to predict the properties of blown films. Figure 1 shows the White-Spruiell triangular representation of orientation of crystalline axes in a film. By knowing the relative orientation of all three axes (namely *a*-, *b*- and *c*-axis) of a unit crystal, the overall crystalline orientation in the film could be visualized.

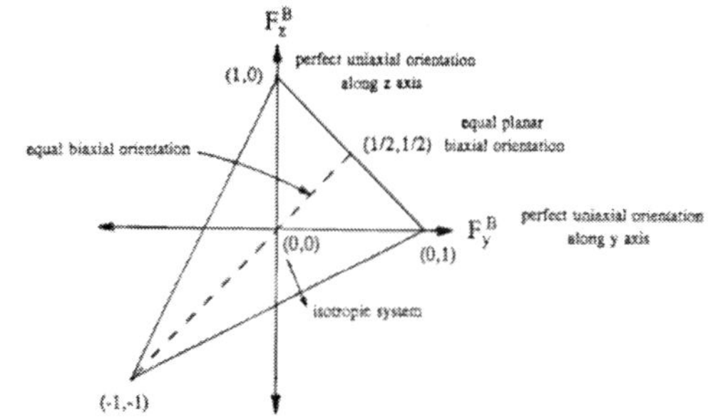

Figure 1: Orientation triangle for White-Spruiell orientation factors (Pazur and Prud'homme, 1996)

The amount of amorphous orientation in the blown films can be estimated by assuming the simple two phase model for polyethylene (Choi et al., 1982; Pazur and Prud'homme, 1996; Chen et al. 2006). In this case, the total birefringence can be represented by,

$$\Delta_{MN} = n_M - n_N$$
$$= w[\Delta_{bc}^{oc} f_{c,MD}^{c} + \Delta_{ab}^{oc} f_{a,MD}^{c}] + (1-w)\Delta^{oa} f_{am,MD} \quad \text{(Equation 1)}$$

$$\Delta_{TN} = n_T - n_N$$
$$= w[\Delta_{bc}^{oc} f_{c,TD}^{c} + \Delta_{ab}^{oc} f_{a,TD}^{c}] + (1-w)\Delta^{oa} f_{am,TD} \quad \text{(Equation 2)}$$

where,

w : film crystallinity (by Raman spectroscopy),

n_M , n_N , n_T : refractive indices in Machine, Normal and Transverse direction of film,

f_c^c , f_a^c : crystalline orientation functions of *c* and *a*-axis of PE unit crystal (by GADDS pole figure analysis),

Δ_{MN}, Δ_{TN} are MD/TD and TD/ND Birefringence,

$f_{am,MD}$ and $f_{am,TD}$ are MD and TD amorphous orientation function,

$\Delta_{bc}^{oc}, \Delta_{ab}^{oc}$ are intrinsic birefringence values of PE crystals in respective direction

$\Delta_{bc}^{oc} = 0.056, \ \Delta_{ab}^{oc} = -0.005$

Δ^{oa} intrinsic birefringence of amorphous phase $= 0.058$

Subtracting equation 2 from equation 1, leads to the equation in terms of MD/TD birefringence Δ_{MT}, which contains the term showing the difference in the amorphous orientation between MD and TD.

$$\Delta_{MT} = w[\Delta_{bc}^{oc}(f_{c,MD}^{c} - f_{c,TD}^{c}) + \Delta_{ab}^{oc}(f_{a,MD}^{c} - f_{a,TD}^{c})] + (1-w)\Delta^{oa}(f_{am,MD} - f_{am,TD}) \quad \text{(Equation 3)}$$

Using equation 3 the difference in MD and TD amorphous phase orientation functions ($f_{am,MD} - f_{am,TD}$) can be calculated by knowing MD/TD birefringence.

Choi et al. (1982) showed that MD/TD, MD/ND and TD/ND birefringence of polyethylene blown films increased with increase in draw down ratio at constant blowup ratio. The crystalline orientation functions for the HDPE blown films were interpreted by using pole figure (White and Spruiell, 1981) data. The difference in MD and TD *c*-axis functions correlated well with the drawdown ratio as well as with the MD/TD stress difference at the frost line. Krishnaswami and Sukhadia (2000) observed that the extent of *a*-axis orientation along the plane of the film was a function of the molecular characteristics. LLDPEs based on Ziegler-Natta catalysts and chromium oxide catalysts displayed higher degrees of crystalline orientation relative to the LLDPEs based on metallocene catalysts.

Zang et al. (2004) observed that for 1.0 MI (gm/10 min) LLDPEs produced with Z-N and single site catalysis, the morphologies displayed spherulite-like superstructures and relatively random lamellar arrangements. Row nucleated morphology of Shish-Kebab type was not achieved even though the highest possible DDR and very low frost-line were used. They attributed this to the narrow molecular weight distribution of the LLDPEs they used. The *a*-axis oriented towards MD with higher orientation at higher DDR. The *b*-axis oriented perpendicular to MD with decreasing orientation at high DDR. The *c*-axis showed MD orientation at low DDR but showed negligible orientation at high DDR. They concluded that these were characteristics of transcrystalline morphology. The transcrystalline morphology is formed with the preferential nucleation at surface of the film and crystal growth in the thickness direction. The single-site catalyzed LLDPE showed the lowest level of orientation and promoted transcrystalline morphology. They reported that the shorter relaxation time determined that the oriented melt of LLDPEs under stress could relax back to a low orientation isotropic state before crystallization started. This means that the lower the relaxation time, the lower the orientation of molecular chains and lesser is the chance to form row nucleated structure.

In this study we prepared sLLDPE, Z-N LLDPE, mLLDPE (metallocene catalyzed LLDPE) and super-hexene LLDPE films at low and high draw- down conditions by varying blow up ratio, film processing speed and film gauge. The films were tested for MD and TD tear strength. The crystalline orientations were analyzed for White-Spruiell orientation factors of crystal unit cells by wide-angle X-ray diffraction (WXRD). The film crystallinity was analyzed by Raman spectroscopy. Polarized optical microscopy technique was used to estimate MD/TD birefringence of the films. The amorphous phase orientation was estimated from the crystalline orientation, birefringence and the crystallinity of LLDPE films. The unique tear behavior was observed in sLLDPE films, where both the MD and the TD tear strength were significantly higher at high draw ratio. The tear properties of LLDPE films are explained in terms of the orientation of molecular chains in crystalline and amorphous phases.

Experimental

Materials

Three sLLDPE, two Z-N LLDPE, one m-LLDPE and one super-hexene co-polymer LLDPE resins were blown into films at low and high draw-down conditions. Tables 1 and 2 gives the materials and material characteristics of the resins used.

Table 1: Materials

Product	Reference Name in This Paper	Co-monomer
SURPASS FPs016-C	sLLDPE1	Octene
SURPASS FPr018-D	sLLDPE2	Octene
SURPASS FPs117-C	sLLDPE3	Octene
SCLAIR FP020-C	Z-N LLDPE1	Octene
SCLAIR FP120-C	Z-N LLDPE2	Octene
COMP mLLDPE	mLLDPE	Hexene
NOVAPOL TD-9022-D	Sup hex LLDPE	Hexene

Table 2: Material Characteristics

	Melt Index (gm/10 min)	**Density** (g/cm^3)	**Relaxation Time*** **@190°C** (sec)
sLLDPE1	0.65	0.9160	0.1144
sLLDPE2	0.86	0.9180	0.0878
sLLDPE3	0.92	0.9170	0.0918
Z-N LLDPE1	0.67	0.9210	0.2373
Z-N LLDPE2	0.93	0.9200	0.1885
mLLDPE	1.18	0.918	0.0188
Sup hex LLDPE	0.64	0.9195	0.2276

* Relaxation time is calculated by fitting DMA data to Ellis model for polymer melt rheology

Experimental Procedure

All films were produced on a Macro 8" blown film line, equipped with a general purpose 3.5" single screw with barrier design and length to diameter ratio 30. A general-purpose 8" spiral die with die gap 50 mil was used. The cooling unit consisted of a dual lip air ring in combination with internal bubble cooling (IBC).

Table 3: Blown film processing conditions

	Film gauge (mil)	**Blow-Up Ratio**	**Line Speed** (ft/min)	**Draw-Down Ratio**
Low DDR	1.5	3:1	130	10.4
High DDR	0.75	2:1	390	30.5

All the films produced were tested for Elmendorf tear strength as per ASTM D1922 in machine direction (MD) and transverse direction (TD).

The film crystallinity was determined by Raman spectrometry using Renishaw inVia Raman spectrometer with air cooled Argon ion 514.5 nm Laser.

Birefringence analysis of all the films was done as per ASTM D4093-95 using Olympus BX51 microscope with U-CBR1 compensator for measuring the birefringence.

Crystalline orientation in the films was analyzed using the pole figure from the wide angle X-ray diffraction. The pole figure method utilizes the General Area Detector Diffraction System (GADDS) manufactured by Bruker AXS Inc. operating in the wide angle X-ray diffraction mode. The X-ray source was generated using a copper anode which emits characteristic wavelength lines of Kα at 1.54 nm with nickel filter. The X-ray generator was set at 40 kV and 40 mA. The sample to detector distance was fixed at 4.90 cm. The films were rotated 180° at 5° intervals. The diffraction pattern is then integrated for the (200) and (020) reflections. The results for the Hermans and White-Spruiell orientation functions obtained with the GADDS were represented in the form of White-Spruiell triangle plots as shown in Figure 1.

Results and Discussion

Film Crystallinity, Birefringence and Tear strength

Table 4 shows the results of crystallinity of the films produced. Raman spectrometry gives the three-phase analysis which includes interphase as an intermediate phase between the crystalline and amorphous phases. In the present study, half of the interphase was taken as the part of the crystalline phase and the other half as part of the amorphous phase. In all the films, crystallinity decreased slightly as draw down ratio increased. Z-N LLDPE films showed slightly higher crystallinity than did the sLLDPE films.

Table 5 gives the results of birefringence analysis and the MD and the TD tear strength. The positive sign of the birefringence indicates that the orientation of the polymer chains in the film is in the MD, and the negative sign indicates that it is in the TD. In all the sLLDPE films birefringence at higher DDR increases in the TD. In the Z-N LLDPE films on the other hand birefringence at higher DDR increases in the MD. The mLLDPE films show birefringence in the MD at low DDR and close to zero at high DDR. The superhexene LLDPE films already show high TD birefringence even at low DDR, which further goes to the TD at high DDR.

Table 4: Crystallinity of films by Raman spectrometry

Resin Type	DDR	Raman film Crystallinity (%)
sLLDPE 1	Low	37.8
	High	36.7
sLLDPE 2	Low	39.2
	High	37.7
sLLDPE 3	Low	38.7
	High	36.6
Z-N LLDPE 1	Low	42.0
	High	40.2
Z-N LLDPE 2	Low	41.3
	High	39.4
mLLDPE	Low	40.4
	High	39.4
Sup hex LLDPE	Low	38.8
	High	36.3

Table 5: Birefringence, and MD and TD Tear

Resin Type	DDR	Birefringence (MD – TD) (X 10^{-4})	MD Tear (gm/mil)	TD Tear (gm/mil)
sLLDPE 1	Low	– 4.63	272	445
	High	– 7.83	783	658
sLLDPE 2	Low	– 1.3	313	455
	High	– 5.6	481	691
sLLDPE 3	Low	0	313	502
	High	– 3.83	595	753
Z-N LLDPE 1	Low	– 11.7	288	636
	High	+ 16.33	116	917
Z-N LLDPE 2	Low	– 6.47	327	593
	High	– 1.65	313	982
mLLDPE	Low	+ 5.63	311	383
	High	– 0.61	307	492
Sup hex LLDPE	Low	– 7.31	456	592
	High	– 13.13	350	888

In all the films the TD tear strength always increased with DDR. Only in the case of sLLDPE resins did the MD tear strength significantly increase at high DDR. In all other LLDPE films MD tear strength decreased at high DDR. It is interesting to see the correlation between birefringence and the MD tear strength (Figure 2). Small and large data points are shown in Figure 2 and the subsequent figures to denote the two films prepared from each resin. The small data point represents the film produced at low draw-down ratio (10.4) and the large data point represents the film produced at high draw-down (30.5). The general trend observed from Figure 2 is that when the birefringence values increase towards TD, the MD tear strength increases. All the sLLDPE films show higher TD birefringence at high DDR, which results in higher MD tear strength. The Z-N LLDPE films when produced at higher DDR show their birefringence values increasing towards MD, which results in decreasing MD tear strength.

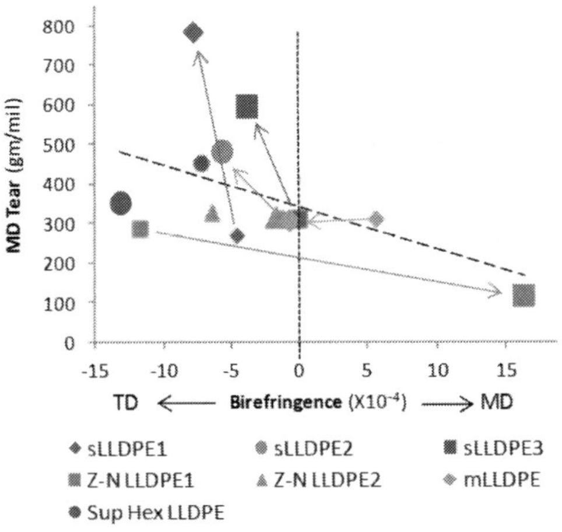

Figure 2: Relation between MD tear strength and film birefringence (MD–TD)
Small data point: low DDR (10.4);
Large data point: high DDR (30.5)

Crystalline Orientation

In figures 3 and 4 the difference between the MD and the TD orientation function is taken as the measure of orientation, the positive values indicate MD orientation and the negative values indicate TD orientation. The higher draw-down increases *a*-axis orientation in the MD and *b*-axis simultaneously shows increased orientation in the TD. For all the films *a*-axis always showed orientation in the MD and *b*-axis in the TD. This means the crystal growth direction, which is the *b*-axis, is always in the TD. As seen in Figure 4, the polymer chain axis (*c*-axis) shows MD orientation in most films. As the draw-down ratio increases, *c*-axis MD orientation decreases. In the case of Z-N LLDPE2, the *c*-axis orientation is in the TD at higher draw-down ratio. In this study the *c*-axis, which is polymer chain axis of a unit crystal is taken as a measure of crystalline orientation.

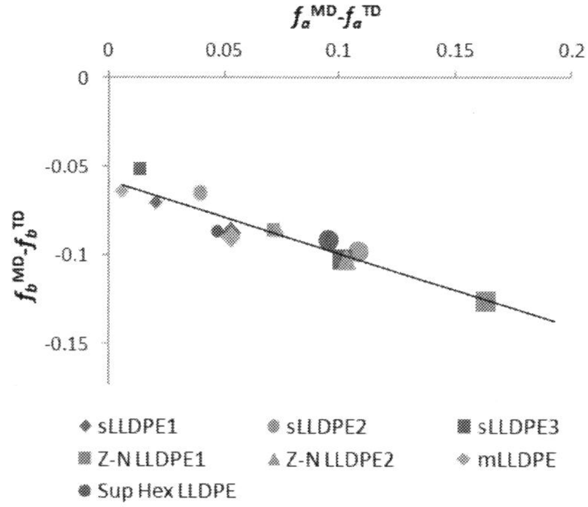

Figure 3: Orientations of *a*- and *b*- axes of unit crystal in the LLDPE blown films
Small data point: low DDR (10.4);
Large data point: high DDR (30.5)

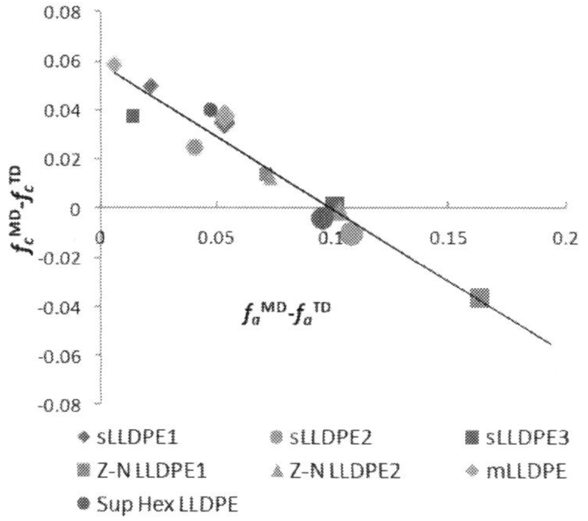

Figure 4: Orientation of *a*- and *c*- axes of unit crystal in LLDPE films
Small data point: low DDR (10.4);
Large data point: high DDR (30.5)

Amorphous and Crystalline Chain Orientation

Using equation 3 the difference in MD and TD amorphous phase orientation functions ($f_{am,MD}$ – $f_{am,TD}$) was calculated. Figure 5 represents the amorphous chain orientation, *c*-axis orientation and birefringence values plotted together for all the LLDPE films.

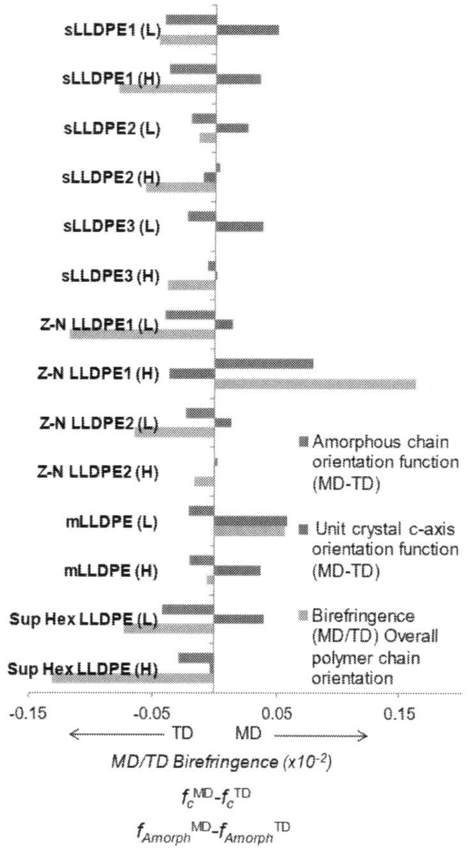

Figure 5: Amorphous chain orientation, crystalline chain *c*-axis orientation and birefringence of LLDPE films. (L)-low DDR film; (H)-high DDR film

In all the films it is seen that the amorphous chain orientation at high DDR shifts towards the MD and the crystalline chain axis orientation shifts towards TD. The birefringence is the sum of these two contributions. In sLLDPE films this sum, which represents overall polymer chain orientation, increases towards TD at high DDR. In Z-N LLDPEs at high DDR this sum shifts towards the MD. As seen in Figure 2, the sum of the contributions of the crystalline and the amorphous chains ultimately determine the MD tear strength of LLDPE films. In the case of mLLDPE at high DDR the birefringence value is close to zero and the MD tear is not significantly affected. In the case of superhexene LLDPE the birefringence at low DDR is already in the TD, which imparts good MD tear to these films at low DDR. Further increase in birefringence towards TD at high DDR may not affect the MD tear properties.

The TD tear of LLDPE films showed good correlation with the orientation of unit crystal axes, namely a-axis and c-axis. Figures 6 and 7 show these plots. The high DDR resulted in higher a-axis orientation towards MD. The higher the MD orientation of a-axis higher was the TD tear strength. The increase in the MD orientation of a-axis resulted in decrease in the c-axis MD orientation.

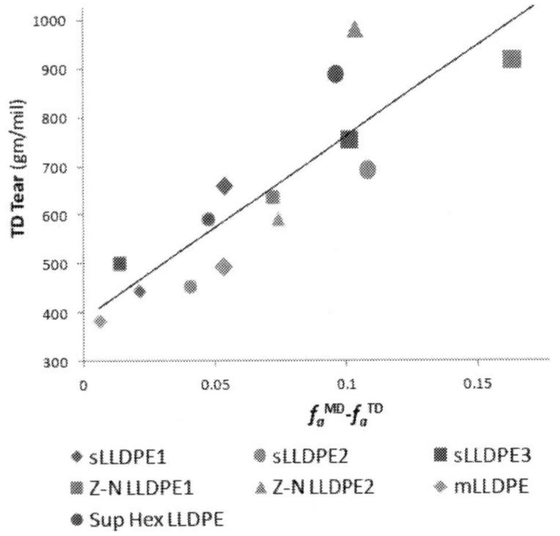

Figure 6: Relation between TD tear strength and the orientation a-axis of unit crystal
Small data point: low DDR (10.4);
Large data point: high DDR (30.5)

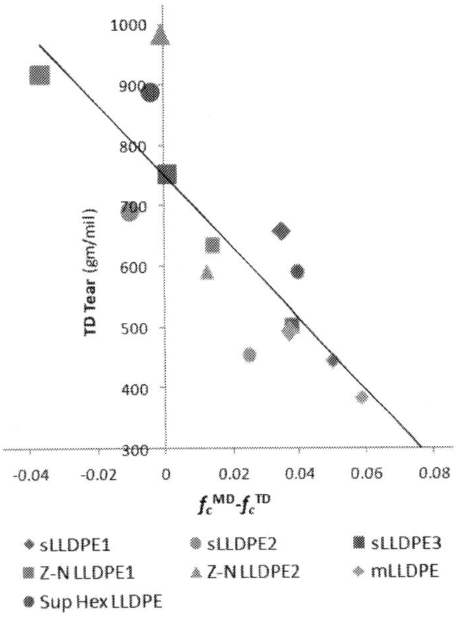

Figure 7: Relation between TD tear strength and the orientation c-axis of unit crystal
Small data point: low DDR (10.4);
Large data point: high DDR (30.5)

The unique tear properties of sLLDPE films could be attributed to the longer side chains of octene copolymer and the narrow molecular weight distribution. Zang et al. (2004) compared tear resistance of LLDPE, LDPE and HDPE films. The HDPE showed high tear resistance in TD and extremely low resistance in MD due to the separated row nucleated structure. In the LDPE film the twisted lamellae from adjacent row nuclei are strongly connected, which is responsible for the high MD tear. LLDPEs showed a relatively balanced tear resistance since less oriented localized spherulite-like structure was observed.

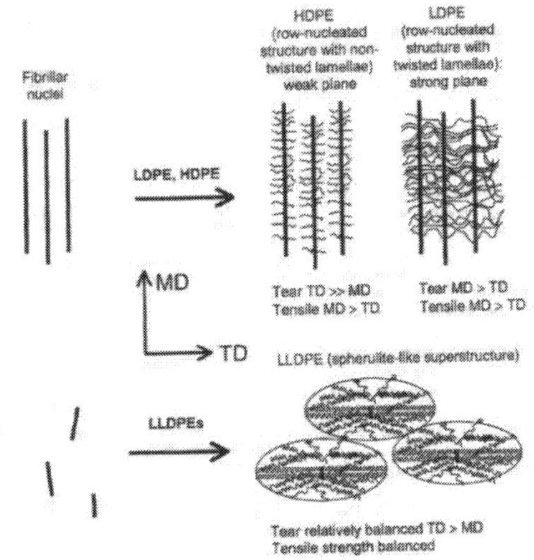

Figure 8: Schematic of morphological development and structure-tear resistance relationship for LDPE, LLDPE and HDPE (Zang et al. 2004)

In the case of sLLDPEs, due to their narrow molecular weight distribution there are few longer than average chains available to form the row nuclei. At high DDR the molecular chains in the MD are under stretch and may not be available for chain folding to form lamellar crystalline structure. The chains in the TD on the other hand are somewhat relaxed and the crystal growth proceeds in the TD. As the lamellae grow in length they get bent towards MD to form a sheaf-like structure. This type of morphology in high DDR sLLDPE films could result in high MD and TD tear strength. The Z-N LLDPE, which is broader in molecular weight in comparison to the sLLDPEs, may still have some longer than average chains to initiate row nucleated structure, and the unique increase in

MD and TD tear properties which are seen in sLLDPE films may not be apparent in Z-N LLDPE films.

Conclusions

In LLDPE films we observed that MD tear strength correlated with the overall orientation of the polymer chains in the film plane, which included the orientation in the amorphous phase and the crystalline phase. The overall polymer chain orientation was evaluated by analyzing the films for birefringence using optical microscopy. The TD tear strength of the films depended strongly on the crystalline phase orientation as indicated by *a*- and *c*- axes orientations of the unit crystal of polyethylene.

We observed that when sLLDPE films were blown at high draw-down conditions, the overall polymer chain orientation increased in the TD. This resulted in higher MD tear strength. The Z-N LLDPE resin films when blown at higher draw-down ratios showed higher overall orientation in the MD, which resulted in lower MD tear strength in these films. All the LLDPE films at high DDR showed higher *a*-axis orientation in the MD, which resulted in higher TD tear strength of LLDPE films at higher DDR. The anomalous and counter intuitive enhancement in the MD and TD tear properties of sLLDPE films at high DDR are attributed to their unique resin architecture and morphology.

References

Bunn C. W., "The crystal structure of long-chain normal paraffin hydrocarbons. The shape of the <CH2 group", Trans. Faraday Soc., **35**, 482-491 (1939).

Chen H. Y., Bishop M. T., Lands B. G., Chum S. P., "Orientation and Property Correlation for LLDPE Blown Films", J. Appl. Polym. Sci., **101**, 898-907 (2006).

Choi K-J., Spruiell J. E., White J. L., "Orientation and Morphology of High Density Polyethylene Films Produced by Tubular Blowing Method and its Relationship to Process Conditions", J. Appl. Polym. Sci., **20**, 27-47 (1982).

Hermans P. H. and Platzek P., Kolloid Z, **88**, 68-72 (1939).

Kwack T. H., Han C. D., "Development of Crystalline Structure during Tubular Film Blowing of Low Density Polyethylene", J. Appl. Polym. Sci., **35**, 363-389 (1988).

Krishnaswamy R. K., Lamborn M. J., "Tensile Properties of Linear Low Density Polyethylene (LLDPE) Blown Films", Polym. Eng. Sci., **40**(11), 2385-2396 (2000).

Krishnaswamy R. K., Sukhadia A. M., "Orientation Characteristics of LLDPE Blown Films and their Implications on Elmendorf Tear Performance", Polymer, **41**, 9205-9217 (2000).

Patel R. M., Butler T. I., Walton K. L., Knight G. W., Polym. Eng. Sci., **34**(19), 1506-1514 (1994).

Pazur R. J.,, and Prud'homme R. E., "X-ray Pole Figure and Small Angle Scattering Measurements on Tubular Blown Low-Density Poly(ethylene) Films", Macromolecules, **29**(1), 119-128 (1996).

White J. L. and Spruiell J. E., "Specification of Biaxial Orientation in Amorphous and Crystalline Polymers", Polym. Eng. Sci., **21**(13), 859-868 (1981).

White J. L. and Cakmak M., "Orientation, Crystallization, and Haze development in Tubular Film Extrusion", Adv. Polym. Tech., **8**(1), 27-61 (1988).

Zang X. M., Elkoun S., Aji A., Huneault M. A., "Oriented Structure and Anisotropy Properties of Polymer Blown Films: HDPE, LLDPE and LDPE", Polymer, **45**, 217-229 (2004).

SIX TYPES OF SPHERULITE MORPHOLOGIES PACKED WITH POLYMORPHIC CRYSTALS IN POLY(HEPTAMETHYLENE TEREPHTHALATE)

K. C. Yen* and E. M. Woo

[1]*Department of Chemical Engineering, University of Cheng Kung, Tainan, Taiwan* – liangbow0110@yahoo.com.tw

Six types of spherulite morphologies packed with polymorphic crystals and their growth kinetics in melt-crystallized poly(heptamethylene terephthalate) (PHepT) were characterized using polarized-light optical microscopy (POM), Fourier transformed infrared microspectrometry (micro-FTIR), differential scanning calorimetry (DSC) and atomic force microscopy (AFM). Two maximum melting temperatures (T_{max}), a higher 150°C and a lower 110°C, were used to melt the initially crystallized PHepT crystals of either α- or β-crystal. By heating to the higher T_{max}=150°C erasing all prior nuclei and quenched to T_c, the crystallized PHepT exhibits the ring-banded and ringless spherulites packed of only the β-crystal at T_c=50~65°C and other temperatures, respectively. Residual nuclei types (α- or β-crystal) are influential on subsequent morphologies upon crystallization of PHepT. On the other hand, by heating to the lower T_{max}=110°C with trace of α-crystal nuclei and quenching to T_c, tiny crystals and small spherulites packed solely of the α-crystal are produced in melt-crystallized PHepT. However, by heating to T_{max} (110°C) with trace of β-nuclei, PHepT exhibits six types of spherulites (Ring Type-I, -II, -III, Maltese-cross Type-1, -2, and -3). Thermal and spectroscopic analyses conclude that these six types of spherulites are solely packed of either α- or β-crystal form. Ring Type-I, Maltese-cross Type-1 and -3 spherulites are packed of the sole β-crystal, while Ring Type-II, -III and Maltese-cross Type-2 spherulites are attributed to the sole α-crystal. The crystallization regime transition temperatures of the α- and β-crystal from regime III to II occur at 60 and 63°C, respectively, which are both independent of the spherulite morphological changes.

Introduction

Semicrystalline polymers exhibiting polymorphism and different types of spherulite morphology have been studied, such as in poly(vinylidene fluoride) (PVDF),[1-3] poly(1,4-butylene adipate) (PBA),[4-6] poly(hexamethylene terephthalate) (PHT)[7-9] and isotactic poly(1-butene) (i-PBu-1).[10-13] Polymorphic behavior in PVDF, PBA, or PHT is dependent on crystallization temperature (T_c); besides, the spherulite morphology in these three polymers depends on not only T_c but also polymorphism.[1,2,4-9]

PHT, an aryl-polyester with six methylene groups between two terephthalate groups, exhibits complex polymorphic behavior upon melt crystallization. PHT contains both the α- and β-crystal when melt-crystallized at T_c lower than 140°C, whereas when T_c is higher than 140°C, PHT contains only the β-crystal. Upon crystallization at T_c=100°C, PHT possesses mixed α- and β-crystal cells, and PHT can be crystallized into mixed morphologies composed of normal Maltese-cross spherulites and dendrites. An earlier investigation points out that there is a correlation between the spherulite morphology and the polymorphism in PHT.[7] The normal Maltese-cross spherulites correspond to the α-crystal and the dendritic morphology is associated with the β-crystal. Apart from the effect of the crystallization temperature, polymorphic behavior in semicrystalline polymers is also influenced by crystallization condition, such as cold crystallization, melt crystallization, solvent-induced crystallization or presence of nucleating agent. Isotactic poly-1-butene (iPB-1) is a polymorphous semicrystalline polyolefin exhibiting four types of crystals, form I, form I', form II and form III. Form I' is crystallized from a certain dilute solution or in thin film at elevated temperature (150°C, 5 days).[8] Form II is crystallized in melt crystallization under normal pressure and packed into the normal Maltese-cross spherulites. Form III is also crystallized from the dilute solution similar to the crystallization condition of form I'; besides, a recently published paper[8] indicates that form III can be obtained together with the more common form II and I' from a melt in presence of amyl acetate. Form I, the most stable crystal form in iPB-1, is produced through transformation from Form-II when kept at ambient temperature for long time.[8] These four types of crystals show different spherulite morphology corresponding to each crystal form, in which the spherulites of Form I and II are ringless and those of Form III are ring-banded.[8]

Polymers mentioned above are all polymorphous and exhibit different spherulite morphology corresponding to each crystal form. In this study, poly(heptamethylene terephthalate) (PHepT) served as a model polymer to go further on investigations of the spherulite morphology of a polymorphous polymer. PHepT exhibits polymorphism (α- and β-crystal) in melt crystallization depending the crystallization temperature (T_c) and maximum melting temperature (T_{max}).[15-17] When heated at a lower T_{max} (110°C) containing some traces of α- or β-nuclei and quenched to T_c, the crystallized PHepT shows the sole α-crystal or the mixed α+β crystal, respectively. In contrast, when heated at a higher T_{max} (150°C) and quenched to T_c, the crystallized PHepT is packed of the sole β- and α-crystal at T_c higher than 35°C and lower than 25°C, respectively. Although the polymorphic behavior in PHepT has been well analyzed and proven,[15] the corresponding spherulite morphology in association with each of these two crystal forms (α- or β-crystal) has yet to be characterized. In this study, the spherulitic morphology and growth kinetics in melt-crystallized PHepT packed of either of α- or β-crystal or both crystal cells were characterized using polarized optical microscopy (POM), atomic force microscopy (AFM), Fourier transformed infrared microspectrometry (micro-FTIR) and differential scanning calorimetry (DSC) in order to reveal the exact natures of multiple types of spherulites packed with polymorphic crystal cells.

Experimental

Materials

Poly(heptamethylene terephthalate) (PHepT) was not commercially available and was synthesized in-house using a catalyst (butyl titanate) by following the method described earlier in the literature.[18] Characterizations showed basic physical data for PHepT, whose T_g = -1.6°C, T_m = 96°C, M_w = 37,500 g/mol (gel permeation chromatography, GPC), and polydispersity index (PDI) =1.7 (GPC).

PHepT was heated to a maximum temperature (T_{max}), held to ensure equilibrium, then quenched rapidly to a crystallization temperature (T_c) for various times to ensure full crystallinity. Two maximum melting temperatures (T_{max}), 150°C and 110°C, were used to melt the initially crystallized PHepT crystals of either α- or β-crystal. PHepT is polymorphic and samples could be prepared to contain solely α or β type crystal, prior to being subjected to heating to the prescribed T_{max}. The higher T_{max} (150°C) was capable of erasing all prior nuclei, and upon crystallization at T_c higher than 35°C, only β-crystal was found.[15] Samples treated at T_{max}=150°C are coded as **PHepT-150**. The other T_{max} (110°C) would permit presence of trace nuclei of either α- or β-crystal, depending on the crystal forms prior to heating to T_{max}. PHepT originally of β-form was heated at T_{max}=110°C so that they contained β-form nuclei at this T_{max}; upon crystallization at any T_c by quenching from T_{max}=110°C would lead to crystals of both α- and β-forms. PHepT samples treated at T_{max}=110°C and crystallized with β-nuclei are coded as **PHepT-110β**. On the other hand, PHepT originally of α-form was heated at same T_{max}=110°C so that they contained α-form nuclei at this T_{max}; upon crystallization at any T_c by quenching from T_{max}=110°C would lead to only α-crystal. Samples treated at T_{max}=110°C and crystallized with α-nuclei are coded as **PHepT-110α** These crystallization procedures were performed either in DSC cells or on microscopic hot stages, both with precision temperature control, depending on characterization to be followed.

Apparatus

Differential scanning calorimetry (DSC).

DSC measurements were made in a Perkin-Elmer DSC-Diamond equipped with a mechanical intracooler under dry nitrogen purge. Temperature and heat flow calibrations at different heating rates were done using indium and zinc. Considerably slow heating rates of 0.1°C/min and 0.5°C/min were used to completely separate the overlapped melting peaks of the α- and the β-crystal.

Fourier-transform infrared micro spectroscopy (micro-FTIR).

Micro-FTIR measurements were performed at a resolution of 2 cm^{-1} on a Varian 600 UMA microscope with a Varian 3100 FT-IR spectrometer equipped with an MCT detector. The lateral resolution of micro-FTIR is 11 μm × 11μm. Samples for micro-IR measurements were thin PHepT films of 4~5 μm thickness, which were prepared by dissolving the polymer in chloroform (~2 wt%) and cast onto KBr pellets at ambient temperature. The films on KBr pellets were dried in vacuum at 40°C for 2 days to remove the residual solvent.

Polarizing-light optical microscopy (POM).

A Nikon Optiphot-2 polarized-light optical microscope (POM) equipped with a charge-coupled device (CCD) digital camera and a Linkam THMS-6000 microscopic heating stage with TP-92 temperature programmer was used to observe the crystal morphology of the isothermally crystallized samples. The CCD equipped with automated image software was used to capture the morphology of the isothermally crystallized samples at preset time intervals. Thin-films specimens for POM were sandwiched between two glass slides.

Results and Discussion

PHepT-110β is crystallized into both the α- and β-crystal at all T_c's. [15] It was thus of interests to compare the spherulite morphology in melt-crystallized PHepT packed of the mixed α- and β-crystal. At T_c=40°C, PHepT-110β is crystallized into two types of spherulites where one is ringless spherulite with a lower birefringence (named as Maltese-cross Type-1, M1 Type) and the other is ringless spherulite with a higher birefringence (named as Maltese-cross Type-2, M2 Type).

PHepT is a polymorphic polymer packed of the α-, β-crystal, or both crystals of various fractions depending on T_{max} and T_c, which further complicates the characterization on origins of the different types of spherulites in melt-crystallized PHepT. In a previous study,[15] the individual peak-melting temperatures of the α- and β-crystal have been determined, where the peak-melting temperature of the α-crystal is 90°C and that of the β-crystal is 96°C upon scanning at 10°C/min. Similarly, in this study, difference in the peak-melting temperature of the α- and β-crystals was used to analyze the correlations between the type of spherulites and crystal forms therein in the spherulites. **Figure 1** shows DSC thermograms upon scanning to (A) 110°C and (B) 101°C (scanning rate=0.1°C/min) of thin-film PHepT-110β samples (coated on micro-glass slide) crystallized at 40°C exhibiting M1 and M2 spherulites, which are labeled previously in Fig. 5. The corresponding POM micrograph of the spherulite morphology in sample-(A) at 40°C is shown as inset, where the darker one is M1 and the brighter one is M2 spherulite as defined earlier. Upon scanning from T_c (40°C) to 110°C, the DSC trace-(A) shows two melting peaks at 98°C and 104°C, which are attributed to the α- and β-crystal, respectively. The lower scanning rate (0.1°C/min) leads to the shifts in the melting temperatures of the α- and β-crystal to higher temperatures compared to those scanned with scanning rate=10°C/min. DSC trace-(B) shows one melting peak at 98°C, which is attributed to the melting of the α-crystal upon heating from T_c to 101°C. After heating to 101°C, the thin-film sample was quenched to the original T_c (=40°C) and then allowed to re-crystallize again for 1 hr, prior to POM characterization. The corresponding POM micrograph of the spherulite morphology of the 40°C-recrystallized sample is shown as inset, and only M1 spherulites and tiny crystals are observed. It indicates that upon heating to 101°C to melt the α-crystal, M2 spherulites melt. In addition, upon re-crystallization at the original T_c, tiny crystals are produced because 101°C is not high enough to erase all traces of the nuclei and thus leads to the heterogeneous nucleation and the crystallization of tiny crystals. Thus, a correlation can be concluded that M1 and M2 spherulites are attributed to the α- and β-crystals, respectively.

At T_c=50~70°C, PHepT-110β is crystallized into two types of ringed spherulites where one is with zig-zag bands (named as Ring Type-I, RI Type) and the other is irregular ring-banded (named as Ring Type-II, RII Type).

Thermal analyses on the RI and RII spherulites were performed to identify their corresponding polymorphic crystals or lamellae. **Figure 2** shows DSC thermograms for thin-film PHepT-110β samples (sample coated as thin films and thermally treated on micro-glass slide) crystallized at 65°C exhibiting RI and RII spherulites (as labeled previously) scanned to (A) 110°C and (B) 101°C (scanning rate=0.5°C/min). The corresponding POM micrograph of the spherulite morphology in sample-(A) at 65°C is shown as inset showing RI and RII spherulites. Upon scanning from T_c (65°C) to 110°C, the DSC trace-(A) shows two melting peaks at 96°C and 102°C, attributed to the α- and β-crystal, respectively. The lower scanning rate (0.5°C/min) leads to up-shifting of the melting temperatures of the α- and β-crystal to higher temperatures in comparison to those scanned with rate of 10°C/min. DSC trace-(B) shows one melting peak at 96°C, attributed to melting of the α-crystal upon heating from T_c to 101°C. After heating to 101°C, the thin-film sample was quenched to the original T_c (=65°C) and then allowed to re-crystallize again for 1 hr, prior to POM characterization. The corresponding POM micrograph of the spherulite morphology of the 65°C-recrystallized sample is shown as inset, and only RI spherulites and tiny crystals are observed. Upon heating to 101°C to melt the α-crystal, RII spherulites melt. In addition, after melting at 101°C and upon re-crystallization at the original T_c, tiny crystals are produced because 101°C is not high enough to erase all traces of the nuclei and thus these nuclei lead to heterogeneous nucleation into tiny crystals upon recrystallization at T_c. Thus, it can be concluded that RI and RII spherulites are packed by the α- and β-crystals, respectively.

At T_c=75~80°C, PHepT-110β is crystallized into one type of ringless spherulites (named as Maltese-cross Type-3, M3 Type) and one type of double-ring patterns (named as Ring Type-III, RIII Type). Correlations between the polymorphic behavior and RIII as well as M3 spherulites in PHepT-110β were analyzed by using Fourier-transformed infrared micro-spectrometry (micro-FTIR), because of the difficulty in resolving the overlapped melting peaks for the α- and β-crystals crystallized at this temperature range (75~80°C) upon DSC scanning. Micro-FTIR provides a direct determination of the absorption spectra of the individual spherulite of micrometer scale. A previous study[15] has shown that the individual characteristic absorption bands correspond to the α- and β-crystal of PHepT. The corresponding absorption bands of the α-crystal are at 954, 1005, 1308 and 1476 cm^{-1}, and those of the β-crystal are at 950 and 1468 cm^{-1}.

Figure 3 shows the micro-FTIR spectra in the range of (A) 900-1200 cm^{-1} and (B) 1250-1600 cm^{-1} of RIII and M3 spherulites of PHepT-110β. The spectra of RIII spherulite show the absorption bands at 954, 1005, 1308 and 1476 cm^{-1} which are associated with the α-crystal of PHepT. The spectra of the M3 spherulite show the absorption bands at 950 and 1468 cm^{-1} which are associated with the β-crystal of PHepT. Thus, RIII and M3 spherulites are attributed to the α- and β-crystal, respectively. From the discussions above, the spherulite morphology of melt-crystallized PHepT depends on not only the crystallization temperature but also polymorphic behavior.

Conclusions

Two maximum melting temperatures (T_{max}), 150°C and 110°C, were used to melt the initially crystallized PHepT crystals of either α- or β-crystal. The higher T_{max} (150°C) was capable of erasing all prior nuclei, while the lower T_{max} (110°C) would permit presence of trace nuclei of either α- or β-crystal. Residual nuclei types (α- or β-crystal), whose presence was determined by T_{max}, are influential on the final spherulite and polymorphism morphologies of PHepT upon subsequent crystallization at T_c after quenching from T_{max}. When heated to the lower T_{max}=110°C that allows some residual α-crystal nuclei, upon quenching to T_c to crystallize, PHepT develops only tiny crystal spherulites packed solely of the α-crystal. However, if the initial crystal nuclei is of the β-crystal type, the same T_{max} (110°C) leads to crystallization of PHepT at various T_c's to exhibit as many as six types of different types of spherulites (Ring Type-I, -II and -III, and Maltese-cross Type-1, -2 and -3) packed solely of either α- or β-crystals at various T_c.

Polymorphism distributes differently in the six different types of spherulites in PHepT. Ring Type-I, Maltese-cross Type-1 and -3 spherulites are packed of only the β-crystal, whereas Ring Type-II, -III and Maltese-cross Type-2 spherulites are packed of only the α-crystal cell. The spherulitic growth rates of the α-crystal at all T_c's are slightly lower than those of the β-crystal. Regime transition temperatures from regime III to regime II of the α- and β-spherulites are at 60 and 63°C, respectively, at which temperatures no evident morphological changes are observed with regime transition behavior of both the α- and β-crystal.

Figure 1. DSC thermograms of thin-film samples of PHepT-110β exhibiting Maltese-cross Type-1 and -2 spherulites scanned to (A) 110°C and (B) 101°C (scanning rate=0.1°C/min). Inset-(top): Spherulite morphology of sample-(A) at 40°C. Inset-(bottom): morphology of Sample-(B) scanned to 101°C then re-crystallized at 40°C. Scale bar = 20 μm.

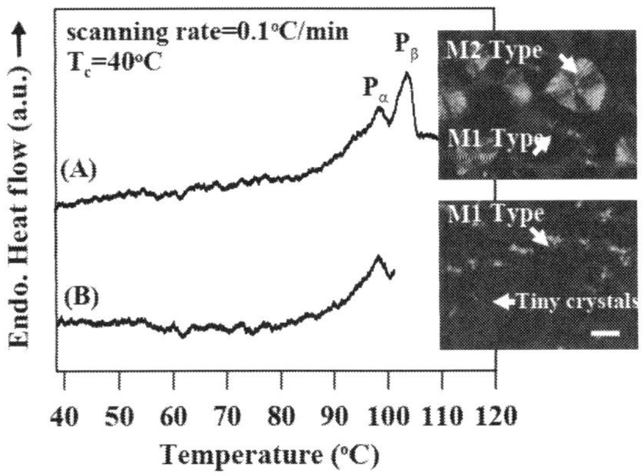

Figure 2. DSC thermograms of thin-film samples of PHepT-110β exhibiting Ring Type-I and -II spherulites scanned to (A) 110°C and (B) 101°C (scanning rate=0.5°C/min). Inset-(top):morphology of Sample-(A) at 65°C. Inset-(bottom): morphology of Sample-(B) scanned to 101°C then re-crystallized at 65°C. Scale bar = 20 μm

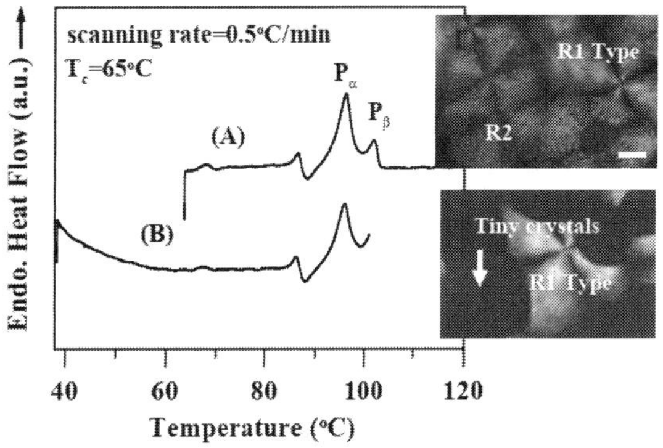

Figure 3. Micro-FTIR spectra of Ring Type-III and Maltese-cross Type-3 spherulites in range of (A) 900~1200 cm^{-1} and (B) 1300~1600 cm^{-1}.

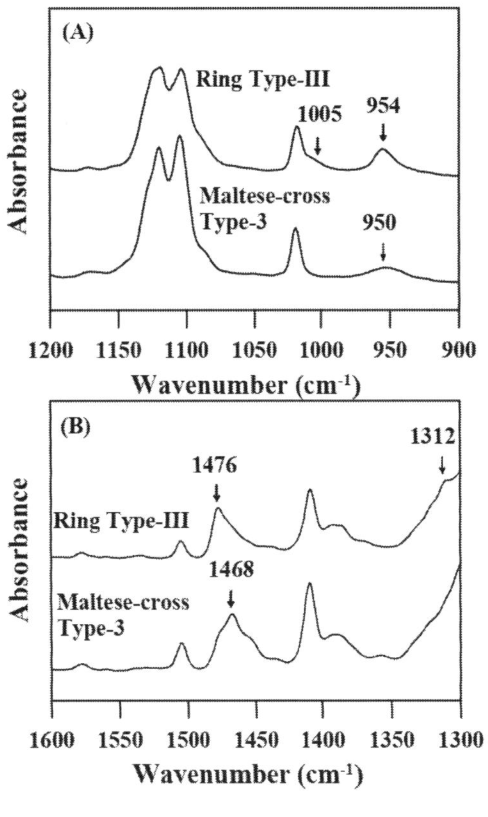

Acknowledgements (Times New Roman 10, bold)

This work has been financially supported by a basic research grant (NSC-95 2221 E006 183) for three consecutive years from Taiwan's *National Science Council* (NSC), to which the authors express their gratitude.

References

1. Gregorio, Jr. R., Cestari, M., "Effect of crystallization temperature on the crystalline phase content and morphology of poly(vinylidene fluoride)", J. Polym. Sci. Polym. Phys. **32**, 859-870 (1994)
2. Kressler, J., Schafer, R., Thomann, R., Imaging of semicrystalline polymers and polymer blends by FT-IR microspectroscopy. Appl. Spectro. **52**, 1269-1273 (1998).
3. Gregorio, Jr. R., Capitâo, R. C., Morphology and phase transition of high melt temperature crystallized poly(vinylidene fluoride). J. Mater. Sci. **35**, 299-306 (2000).
4. Wu, M. C., Woo, E. M., Effects of α-form or β-form nuclei on polymorphic crystalline morphology of poly(butylene adipate). Polym. Int. **54**, 1681-1688 (2005).
5. Woo, E. M., Yen, K. C., Wu, M. C., Analysis of multiple melting behavior of spherulites comprising ring-band shell/ringless core in polymorphic poly(butylene adipate). J. Polym. Sci. Polym. Phys. **46**, 892-899 (2008).
6. Zhao, L., Wang, X., Li, L., Gan, Z., Structural analysis of poly(butylene adipate) banded spherulite from their biodegradation behavior. Polymer **48**, 6152-6161 (2000).
7. Ghosh, A. K., Woo, E. M., Sun, Y. S., Lee, L. T., Wu, M. C., Characterization and analyses on complex melting, polymorphism, and crystal phases in melt-crystallized poly(hexamethylene terephthalate). Macromolecules **38**, 4780-4790 (2000).
8. Hall, I. H., Ibrahim, B. A., The structure and properties of poly(hexamethylene terephthalate): 1. The preparation, morphology and unit cells of three allomorphs. Polymer **23**, 805-816, 1982.
9. Palmer, A., Poulin-Dandurand, S., Revd, J. F., Brisse, F., Poly(hexamethylene terephthalate)—I. Interpretation of its diffraction patterns. Eur. Polym. J. **20**, 783(1984).
10. Lotz, B., Thierry, A., Spherulite morphology of form III isotactic poly(1-butene). Macromolecular **36**, 286-290 (2003).
11. Marigo, A., Marega, C., Cecchin, G., Collina, G., Ferrara, G., Phase transition II→I in isotactic poly-1-butene: wide- and small-angle X-ray scattering measurements. Eur. Polym. J. **36**, 131-136 (2000).
12. Di Lorenzo, M. L., Righetti, M. C., The three-phase structure of isotactic poly(1-butene). Polymer 49, 1323-1331 (2008).
13. Jiang, S., Duan, Y., Li, L., Yan, D., Chen, E., Yan, S., An AFM study on the structure and melting behavior of melt-crystallized isotactic poly(1-butene). Polymer **45**, 6365-6374 (2004).
14. Yen, K. C., Woo, E. M., Thermal, spectroscopy, and morphological studies on polymorphic crystals in poly(heptamethylene terephthalate). Polymer **50**, 662-669 (2009).
15. Gilbert, M., Hybart, F., Effect of chemical structure on crystallization rates and melting of polymers: Part 1. Aromatic polyesters. Polymer **13**, 327-332 (1972).

Morphology/ Block Ratio Correlation and Their effect on Viscoelastic Behavior of Anionic synthesized Styrene –Isoprene-Styrene(SIS) tri-Block Copolymer

Faramarz Afshar Taromi, Mitra Vasei Nasrabadi*

Amirkabir University of Technology - Tehran - Iran

In this study efforts have been made to synthesize number of block-copolymers of styrene and isoprene monomers via anionic polymerization technique using naphthalene sodium as a bi-functional initiator. All these polymers were prepared by two step sequential addition of isoprene and styrene respectively. All the samples have been characterizes and their viscoelastic and thermal properties have been investigated. Prepared thermoplastic elsatomer block copolymers have been precipitated using methanol and then have been dissolved in THF and aged to undergo phase separation. Rheomechanical measurement (RMS) has been used to investigate the melt viscoelastic behavior of prepared samples. The data obtained by means of RMS showed the change in viscoelastic properties are directly resulting from change in side block/ center block molecular weight ratio. Dependence of these properties on the block size has been discussed and the AFM holograms have been compared to illustrate the relation between microstructure and rheomechanical properties of prepared copolymers. The scanning electron microscopy analysis has been carried out on the samples to investigate the effect of block size and styrene to isoperene block ratio on developed microstructure of each sample. Finally to study the thermal stability of the samples, thermal gravimetric analysis (TGA) and differential scanning calorimetry (DSC) have been used and the change in thermal stability and the Tg of each block have been followed by increasing the block size ratio of the synthesized block copolymers.

COMPOUND EYE MICROLENS ARRAY FABRICATION USING POLYMER FLOW INTO MICRO CHANNELS

Byung Guk Park[1], Kiwoon Choi[1], Han Sup Lee[2]*

[1]*Department of Advanced Fiber Engineering, Inha University, Incheon, Republic of Korea, - bksy82@nate.com; kiwoon80@gmail.com*
[2]*Faculty of Engineering, Inha University, Incheon, Republic of Korea,-hslee@inha.ac.kr*

Abstract - The animal eye is a remarkable imaging device, with many attractive design features and new technical tools. Compound eyes allow arthropods to obtain visual information about their environment in a small eye volume with a large field of view, at the cost of comparatively low spatial resolution. This paper describes an effective method for fabrication of compound eye array with controlled surface profiles, using a novel method. The various silicon masters of micro-channels arrays are prepared using photolithography technique and the film (PC, polycarbonate, glass transition temperature 150℃, refractive index 1.59) is used to fabricate the compound eye. Using the method developed in this work, array structures of compound eye having many spherical smaller convex and concave microlens array structure on top of larger spherical convex and concave microlens array structure have been successfully fabricated. The surface profile of compound eye microlens array was analyzed using the optical microscopy(OM), scanning electron microscope(SEM), atomic force microscope(AFM) and confocal microscopy(CM).

Introduction

The animal eye is a remarkable imaging device, with many attractive design features and new technical tools. Compound eyes allow arthropods to obtain visual information about their environment in a small eye volume with a large field of view, at the cost of comparatively low spatial resolution.[1,2] Microlens array (MLA) and artificial compound eye array can be used in many area including the flat panel displays, optical communication, and photodiode system to control the convergence, divergence, and reflection of the visible light.[3]

In this work, the MLAs with convex and concave surface profile have been fabricated by forcing the flow of the polymer melt into the micro channel array. Furthermore, the compound eye MLA has been also prepared by controlling the flow temperature and pressure during the process. The three dimensional surface profile of the MLA has been characterized with SEM, confocal microscopy, AFM, and optical microscopy.

Experimental

Materials

Two different master molds have been prepared using the conventional photolithographic technique and silicon wafers. (hole diameter 100 ㎛, depth 200 ㎛, pitch 150 ㎛), (hole diameter 2 ㎛, depth 10 ㎛, pitch 3 ㎛) MLA and compound eye MLA have been fabricated with polycarbonate(PC, glass transition temperature 150℃, refractive index 1.59).

Methods

The process to fabricate the convex MLA is shown schematically in Scheme 1.

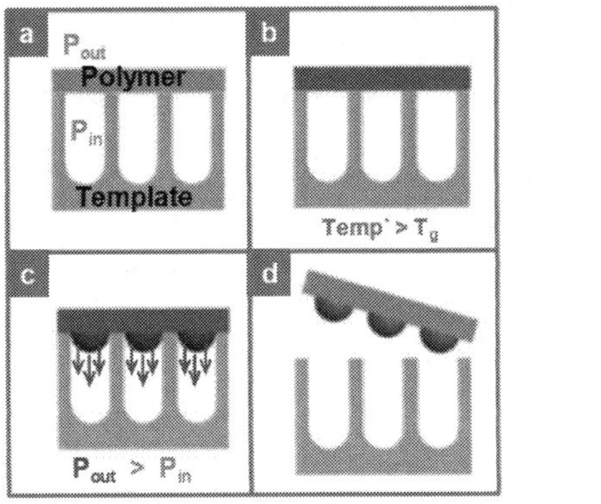

Scheme 1. Process to make convex MLA (a) vacuum applied (b) heating (c) vacuum release and suction (d) separation step.

Initially, the PC film is attached to the master mold and the setup is set under vacuum in the vacuum chamber. The setup is then heated to above PC's glass transition temperature and the polymer film is pressed against the master mold to seal the vacuum state in the pore. By purging the vacuum chamber while maintaining the vacuum state in the pore, the molten PC is forced to flow into the micro channels due to the pressure difference between inside and outside the pore. Finally, the entire setup is cooled down to room temperature to obtain the convex MLA structure which is shown in Fig. 1. Slight variation of the process in Scheme 1 can be used to prepare concave MLA and compound eye MLA.

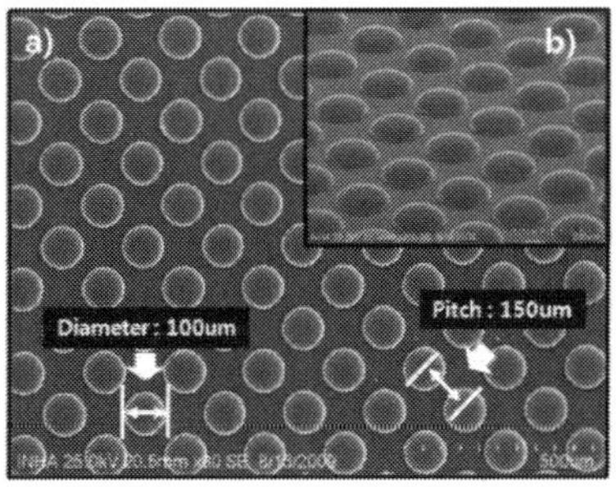

Fig. 1 (a) Surface SEM image of the convex MLA. (b) Tilt view (70°) is shown in inset.

Results and Discussion

As shown in Fig.1, the convex MLA film was successfully fabricated on the PC surface. The uniformity over the large area has been confirmed and each lens has very low roughness (less than few nm). The 3D topographical profiles of the MLA and compound eye MLA can be obtained from the confocal microscopic data. (Fig.2)

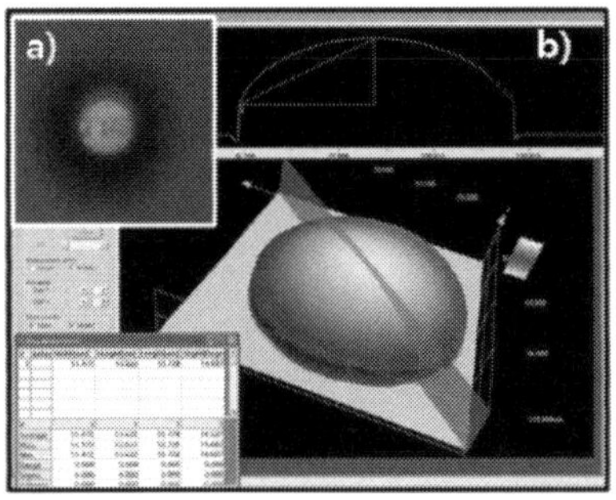

Fig. 2 (a) OM micrograph showing the focused image ("PNS") by a microlens. (b) Confocal microscopic 3D image and surface profile of a microlens.

The relation between the radius of curvature (R), sag height (h), diameter of the lens, and focal length of the convex lens can be expressed with a set of equations in Eq. 1,2 and refractive index (n) of polymer material used. [4]

$$R = \frac{D^2 + 4h^2}{8h} \quad (1) \qquad f = \frac{R}{n-1} \quad (2)$$

The advantage of the process in Scheme 1 is that the focal length and sag height can be tailored by controlling the process parameters such as temperature of the film, pressure difference, and time to allow the polymer flow in the micro channels. The effect of the process temperature on the focal length and radius curvature is shown in Fig. 3. It is noted that the focal length of each lens in MLA can be controlled to any value between 97 and 167 μm for the conditions used.

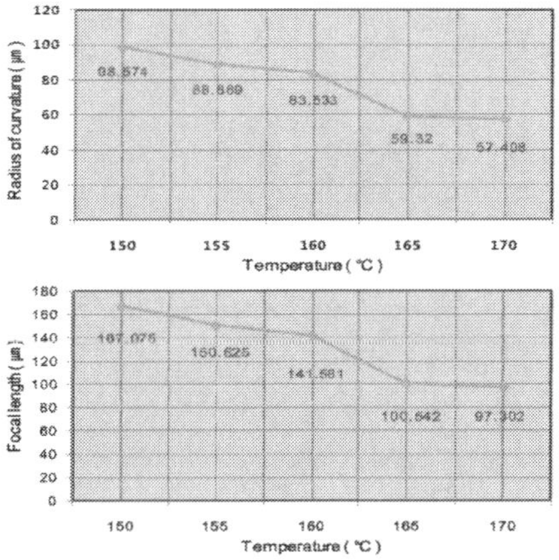

Fig. 3 Dependence of radius of curvature, focal length of convex MLA on processing temperature

Shown in Fig.4 is concave MLA prepared by slightly modifying the process in Scheme 1. The radius of curvature and focal length of concave MLA can be also controlled by varying the process parameters.

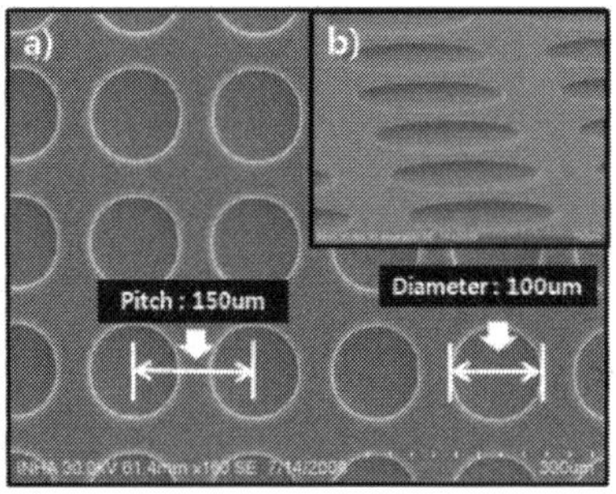

Fig. 4 (a) Surface SEM image of the concave MLA (b) 70° tilt view

If the process in Scheme 1 is applied to the PC film with micro lens already on it, it is also possible to form the smaller lens on the curved surface of the larger lens. The "lens on lens" (convex on convex) structure formed in this work is shown in Fig. 5 which resembles the compound eye structure generally observed with many arthropods.

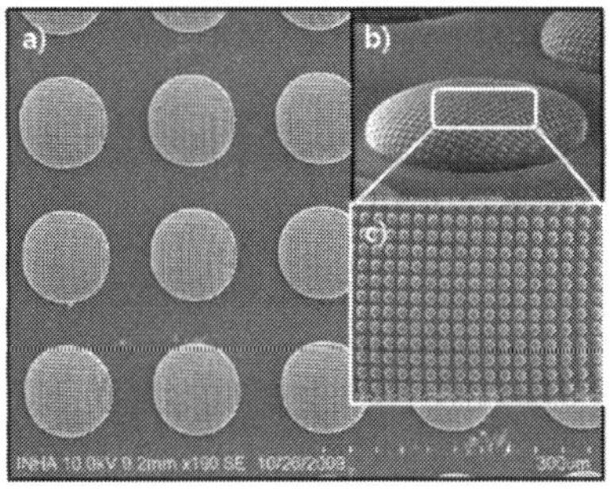

Fig. 5 (a) Surface SEM image of the compound eye (convex + convex) MLA (b) 70° tilt view (c) enlarged image of small lenses

The object image ("PNS") was viewed through the tiny lenses in the compound eye MLA and microscope objective lens and the results are shown in Fig. 6. It is clear from Fig.6 that each lens forms the image of "PNS" at the focal point of the lens which is again viewed at the focal point of the microscope objective lens.

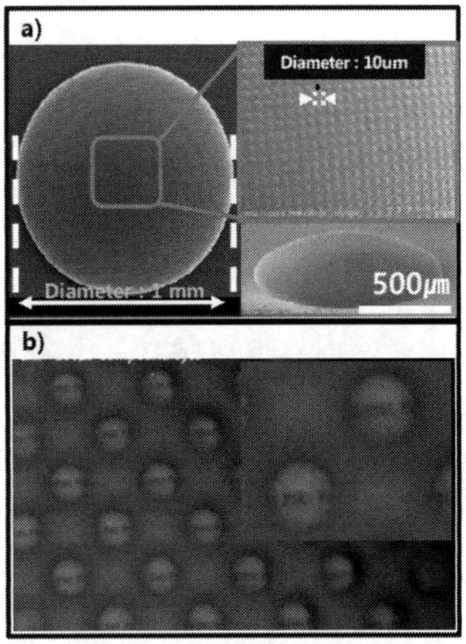

Fig. 6 (a) Surface SEM image of the compound eye (convex + concave) MLA (b) OM micrograph showing the focused image ("PNS") by lenses in a compound eye MLA

Conclusions

New process has been developed to fabricate the polymeric MLA with either convex or concave curvature. It was also demonstrated that the various optical parameters such as focal length, radius of curvature can be tailored by controlling the process parameters such as temperature, pressure difference, processing time.

"Lens on lens" MLA resembling the compound eye structure frequently observed in nature has been also fabricated. It was confirmed that the smaller lenses on larger lens surface act as an independent optical lens, forming the image at their focal points clearly. Furthermore, the curvature (either convex or concave) in compound eye MLA can be selected, resulting in four different combinations of compound eye MLA.

Acknowledgements

This work was supported by the 2010 Research Instrument and Its Key Technology Development Project sponsored by the Korea Basic Science Institute and 2009 Project for the Technique Innovation in the Knowledge Economy, which was sponsored by the Ministry of Knowledge Economy, the City of Incheon, and Inha Univeristy in the Republic of Korea.

References

[1] L.P. Lee, R. Szema, "Inspirations from Biological, Optics for Advanced Photonic Systems", Science, **310**,1148-1150 (2005)

[2] K.H. Jeong, J. K., et al., "Biologically Inspired Artificial Compound Eyes", Science, **312**, 557-561 (2006)

[3] J.J. KIM, S. C., et al., "Micropatterned Single Lens for Wide-Angle Light-Emitting Diodes", Optics Letters, **35**, 823-825 (2005)

[4] S. Sinzinger and J. Jahns : Microoptics, Wiley VCH, Weinheim, Germany (1999)

New understanding of unexpected β--crystal formation of sheared, beta-nucleated isotactic polypropylene

Zhong-Ming Li (1), Yan-Hui Chen (1), Yan Wang (1), Yimin Mao (2), Benjamin S. Hsiao (2)*

(1) Sichuan University - Sichuan - China,
(2) Dept. of Chemistry, Stony Brook University, Stony Brook, - New York - USA

Shear and β-nucleating agent can separately, efficiently induce β-form formation of isotactic polypropylene (iPP). When shear and β-nucleating agent coexist in iPP melt, the nucleation and growth process of α and β-crystals are characterized by in situ synchrotron wide-angle X-ray diffraction and in situ small-angle X-ray scattering for the first time. Unexpectedly, combination of shear and β-nucleating agent would suppress the growth of β-crystals, and then decrease the content of β-crystals. Moreover, the higher shear rate, the less β-crystals. After careful observing on WAXD and SAXS results, we proposed that the reduction of β-crystals is not because shear destroys the structure of β-nucleating agent, and the oriented α-row nuclei induced by shear go against the formation of β-crystals, but because shear makes the melt oriented, subsequently, the oriented molecular chains in the melt array along the shear direction meanwhile the oriented α-row nuclei form, which changes the conformation of molecular chains, as a result, it is hard for molecular chains with changed conformation to fit into the β-crystals lattice on the surface of β-nucleating agent, finally leads to the decrease of β-crystals. Our results suggest that the mechanism of the growth inhibition of β-crystals is responsible for suppression of β-crystals of sheared, β-nucleated iPP. Based on these results, a two-stage strategy was adopted to manipulate the hierarchic structure of β-nucleated iPP during injection molding, where initially, flow-induced crystallization takes place to form shish-kebabs in the outer layer of the sample, subsequently, β-nucleating agent induced nucleation leads to high β-crystal fraction in the core layer of the sample. Accordingly, the resultant iPP samples show superb strength and toughness. Our attempt is a good example for shear induced crystallization applying to factual polymer processing, and also opens up a door towards achieving high-performance polymer materials.

NANOSTRUCTURE EVOLUTION OF THERMOPLASTIC POLYURETHANES UNDER UNIAXIAL DEFORMATION MONITORED BY SAXS

Achim Frick[1], Marzena Mikoszek[1], Norbert Stribeck[2], Morteza Ganjaee Sari[2] and Ahmad Zeinolebadi[2*]

[1]*Institute of Polymer Science and Processing, Aalen University, 73430 Aalen, Germany – Achim.Frick@htw-aalen.de, Marzena.Mikoszek@htw-aalen.de*
[2]*Institute of Technical and Macromolecular Chemistry, University of Hamburg, Bundesstraße, 20146 Hamburg, Germany – Norbert.Stribeck@desy.de, Ganjaee@chemie.uni-hamburg.de, * Zeinoleb@chemie.uni-hamburg.de*

Thermoplastic polyurethanes (TPU) are applied as shape-memory materials, medical devices, piezoresistive sensor elements, etc. Their mechanical properties strongly depend on their nanostructure (soft-segment matrix and hard domains) that varies as a function of processing conditions and load. Materials of different uniaxial orientation and nanostructure have been prepared by varying the melt temperature in an injection molding process. The materials have been subjected to different types of uniaxial mechanical load (continuous tensile test, stress relaxation, retraction), while the evolution of the nanostructure has been monitored by two-dimensional small-angle X-ray scattering. Nanostructure parameters have been evaluated by the chord distribution function (CDF) method. For example, the nanoscopic strain is determined and compared to the macroscopic strain. After releasing the stress, different relaxation mechanism are observed. Data evaluation methods are discussed briefly. Aided by the image processing language PVWAVE, hundreds of SAXS patterns have been processed automatically to yield nanostructure parameters describing size and arrangement of the hard domains. Ultimately, the variation of nanostructure parameters is discussed in relation to the changing macroscopic load.

Introduction

Thermoplastic polyurethane elastomers (TPUs) are a versatile class of polymers with increasing applications as shape-memory materials, piezoresistive sensor elements, medical devices and many others (De and Gaymans, 2009). Thanks to their abundant chemistry TPUs can be formed with tailored properties and morphologies. The mechanical properties of TPUs strongly depend on their morphology, namely shape, size and orientation of phase-separated hard segment domains. Soft segments usually form amorphous regions with elastomeric properties while hard segments act as physical cross-links. Hard segment domains can form an amorphous rigid phase or crystallize according to their concentration and processing conditions (Waletzsko et al., 2009). During deformation TPUs go through different reversible or irreversible microstructural variations and after releasing the stress soft segment and hard segment domains show different relaxation mechanisms (Koerner et al., 2008).
We study microstructural variations of thermoplastic polyurethanes by small angle x-ray scattering (SAXS) to shed light on the structure-property relationships.

Experimental

Materials. Commercial grade polyurethane (soft segments: linear polycaprolactone diol, chain extender: HQEE and Hard segments: MDI) with shore A of 95 is used. The samples are injected molded at 205, 215 and 235 °C and labeled as TPU205, TPU215 and TPU235, respectively.

Figure 1 Simultaneous recording of SAXS pattern and mechanical data. The red cross on the sample shows the position of the beam path

Stretching set-up. Parallel fiducial marks are imprinted on the sample by a rubber stamp. The marks are a reference to calculate the deformation of the sample. The sample is mounted on the tensile machine so that the beam passes through its midsection, Fig. 1. The machine performs symmetric drawing. Signals from load cell and transducer are recorded during the experiment. The sample is monitored by a video-camera, and video frames are grabbed and stored together with the experimental data. The machine is operated at a cross-bar speed of 2 mm/min. Measurement of the true macroscopic elongation

$\varepsilon = l - l_0 / l_0$ is computed from the initial distance, l_0, between fiducial marks and the respective actual distance, l. The details of the method have been explored in our previous works (Stribeck et al., 2010 (a); Zeinolebadi and Stribeck 2010). In addition, the TPU235 sample is strained up to ca. 220% and then kept constant at this strain. Stress relaxation behavior is assessed by monitoring reduction of the stress with time.

SAXS measurement. SAXS is performed at the synchrotron beamline A2 at HASYLAB, Hamburg, Germany. The wavelength of the X-ray beam is 0.15 nm, and the sample-detector distance is 2488 mm. Scattering patterns are collected by a two-dimensional position sensitive marccd 165 detector (mar research, Norderstedt, Germany) operated in 1024 × 1024 pixel mode (pixel size: 158.2×158.2 m2). During the deformation experiments, scattering patterns are recorded every 60 s with an exposure time of 50 s.

Light Microscopy. Samples are cut using a cryo-microtome at -70 °C. Morphology of the sample is investigated by a transmission light microscope (Zeiss Axioplan 2).

Results and Discussion

Figure 2 shows true stress-strain curves for samples processed at different melt temperatures. All samples show typical behavior of elastomers with an elastic region at very small strains followed by a rubbry plateau and finally a strain hardening regime. The samples processed at lower melt temperatures show higher elastic modulus infered from the slope of the curves at small strains, narrower rubbery plateau and slightly steeper strain hardening effect. This is attributed to the higher degree of phase separation and higher amount of segregated hard segments at lower melt temperatures. This can be also understood in anology to the behavior of cured elastomers with increasing cross-link density.

In the stress relaxation curve of TPU235 (Fig. 3) three regimes can be distinguished; i) A rather sudden drop in stress, ca. 10 % of stress is relaxed in the first 30 seconds, ii) A steep reduction regime until 80 min after stopping deformation and iii) a very slow decline regime.

SAXS patterns and corresponding CDFs of different TPU samples are presented in Figs. 4-6. All samples are oriented in their initial state. This orientation is caused during injection molding process. However, the meridional peaks are weaker and broader for undeformed TPU235 in comparison with the samples molded at lower temperature. The lower intensity of the peaks is due to less developed phase separation and is in line with the mechanical data. The lower orientation of TPU235 can also be due to lower melt viscosity at higher temperature. As the samples are deformed, the meridional peaks become weaker and also an equatorial streak develops. The decrease of peak intensity indicates deterioration of hard segment domains. The development of the equatorial streak can be due to fibrillation of soft segment phase or growth of microvoids and crazes.

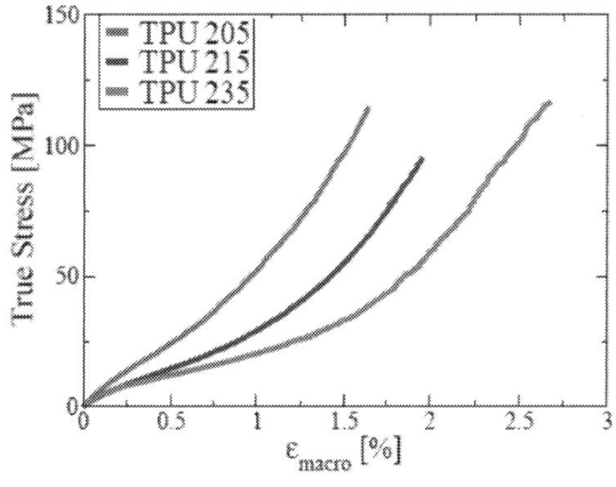

Figure 2 True stress – Strain curves of TPU samples

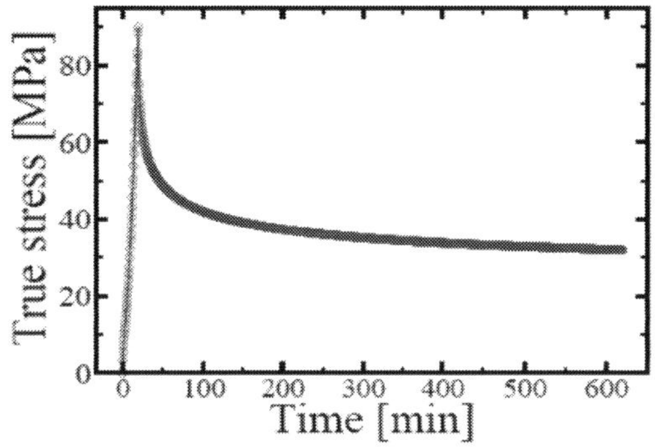

Figure 3 Stress relaxation curve of TPU235

Figure 7 shows SAXS patterns of TPU235 at different stages of relaxation. Peak shape and position do not vary during stress relaxation but peak intensity increases. This shows that the amorphous phase is responsible for the stress relaxation and the main mechanism is irreversible flow of the soft segment chains.

After having kept the sample for 12 hours at constant strain, the sample is slowly retracted to its initial length. By pushing the sample to its original length a 4-point pattern replaces the 2-point meridional peaks, Fig. 8. This is cuased by shearing mechanism due to local torque applied by the soft segment chains (Waletzsko et al., 2009). Figure 9 shows the schematic relation between orientation of the structural moieties and the position of peaks in the SAXS pattern. Thus, macroscopic relaxation of the sample appears to cause additional loss of orientation accompanied by formation of a preferred orientation of the hard domains to a set angle with respect to the axis of the material. Another explanation of the observation were lateral arrangement of the hard domains in a macrolattice (Fronk and Wilke 1985).

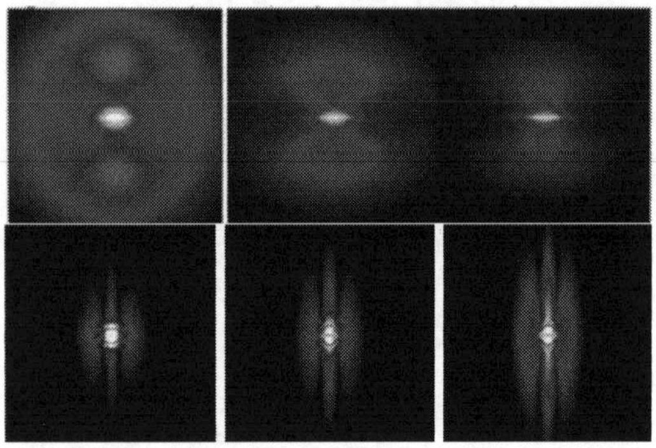

(a) (b) (c)

Figure 4 SAXS patterns (1st row) and corresponding CDFs (2nd row) of TPU205 sample: (a) initial state, (b) half maximum strain (c) maximum strain

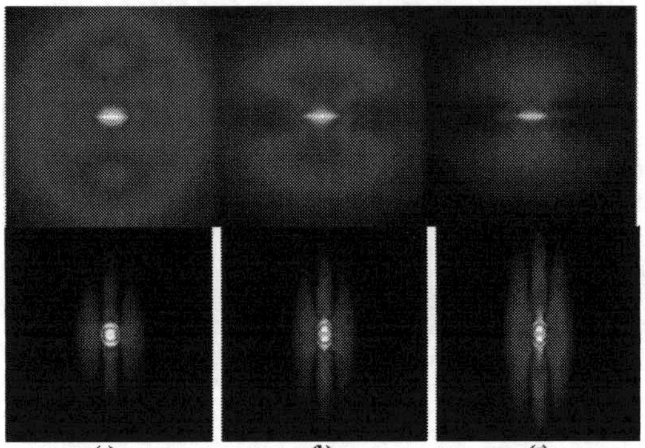

(a) (b) (c)

Figure 5 SAXS patterns (1st row) and corresponding CDFs (2nd row) of TPU215 sample: (a) initial state, (b) half maximum strain (c) maximum strain

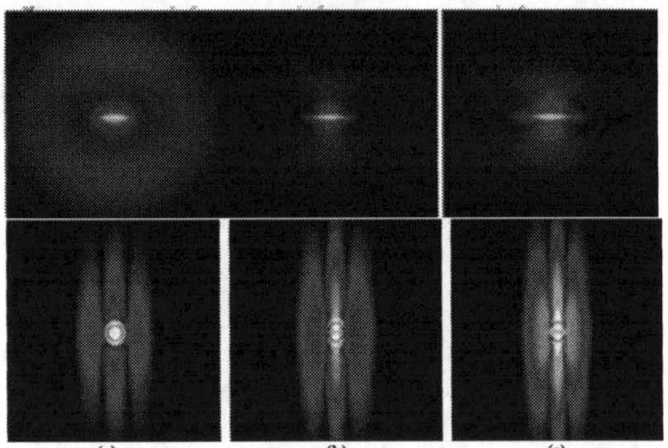

(a) (b) (c)

Figure 6 SAXS patterns (1st row) and corresponding CDFs (2nd row) of TPU235 sample: (a) initial state, (b) half maximum strain (c) maximum strain

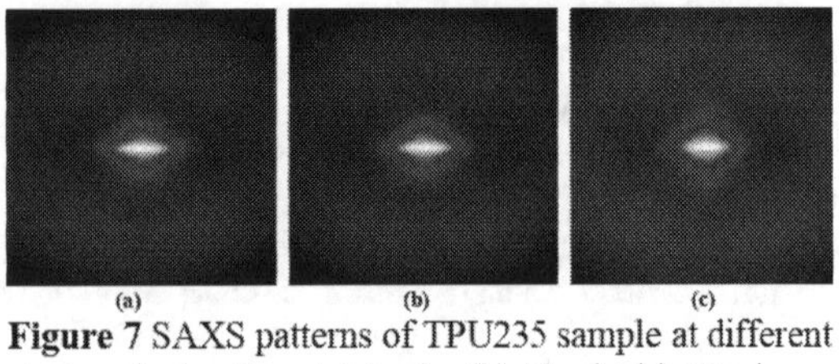

(a) (b) (c)

Figure 7 SAXS patterns of TPU235 sample at different stages of retraction: (a) 1 min, (b) 10 min (c) 20 min

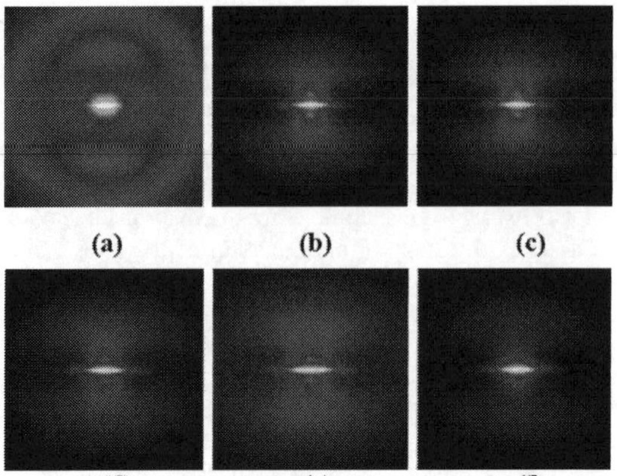

(a) (b) (c)

(d) (e) (f)

Figure 8 SAXS pattern of TPUX sample at different stages of relaxation. a) initial state, b) maximum stress, c) 1min relaxation, d) 30 min relaxation, e) 80 min relaxation, f) 230 min relaxation

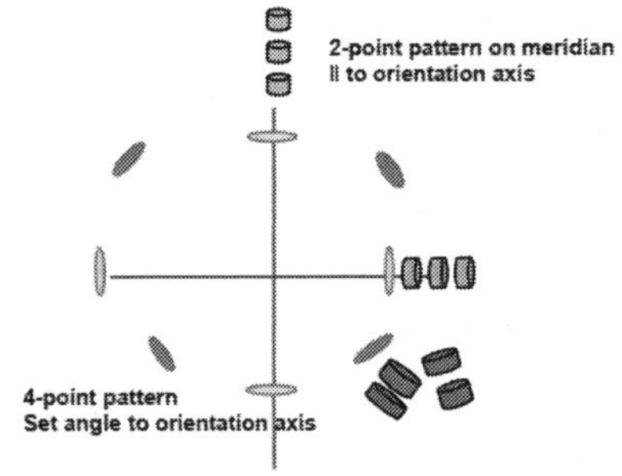

Figure 9 Schematic map of SAXS pattern of samples with different angles with respect to the orientation axis

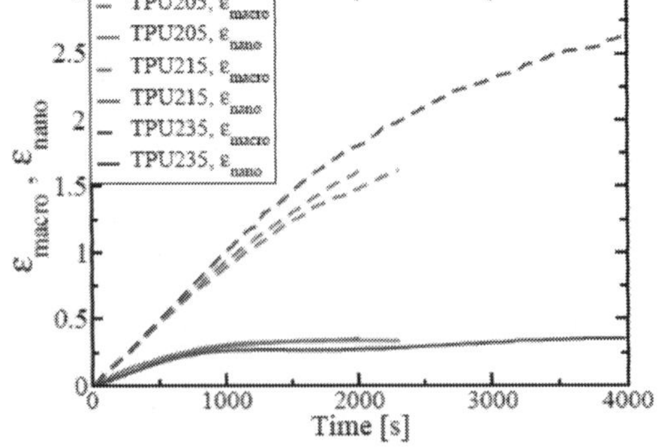

Figure 10 Macroscopic and nanoscopic strain of different TPU samples

In order to obtain more information about the nanoscopic variations, long periods of structural moieties are extracted from peak positions in CDFs. It is seen that the nanoscopic strain is quite smaller than the macroscopic strain, Fig. 10. We have observed this behavior in other materials such as high density polyethylene matrixes reinforced by polyamide microfibrils at large deformations (Stribeck et al., 2010 (b)). This shows that hard segment domains form colonies embeded in large amorphous regions (Fig. 11) and inhibit deformation of the small amorphous regions between them. Thus deformation mainly takes place in the amorphous phase far away from hard segment colonies.

During stress relaxation the nanoscopic strain decreases with time indicating that the amorphous phase outside colonies is mainly responsible for stress relaxation. As this free amorphous phase flows, the stress reduces and the strain imposed on hard segment colonies is released. Hence, the relaxation behavior can be modeled by considering a standard linear solid element, Fig. 12.

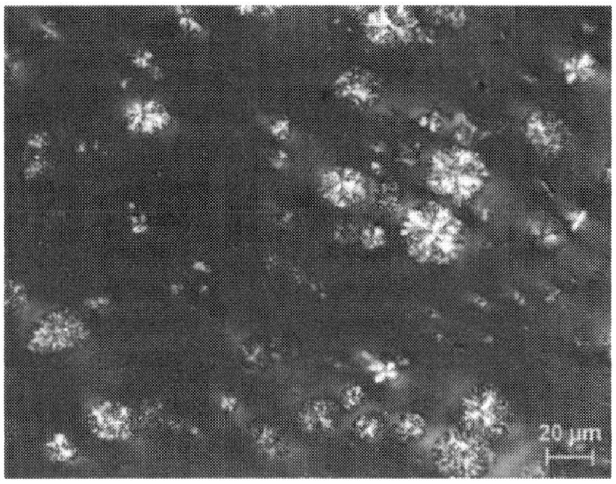

Figure 11 Transmission light microscopy of TPU205. White regions show hard segment colonies

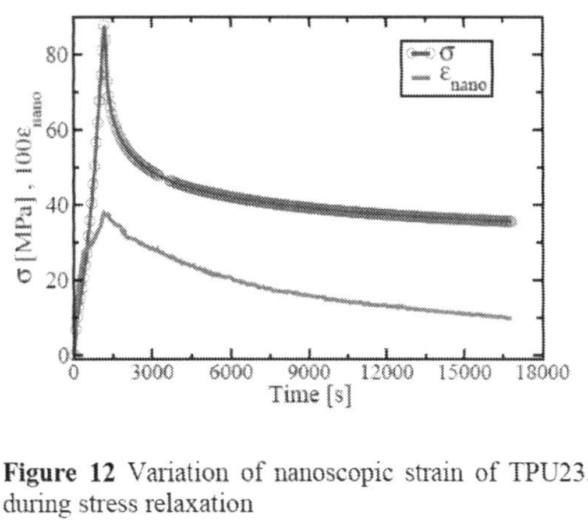

Figure 12 Variation of nanoscopic strain of TPU235 during stress relaxation

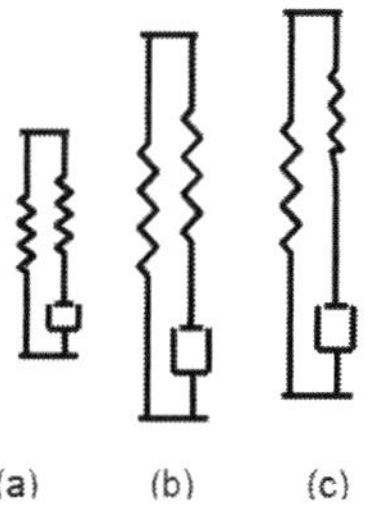

Figure 12 A model for stress relaxation behavior of TPU samples. a) initial state, b) strained, c) stress relaxation

Conclusions

Monitoring nanoscopic structural variations during mechanical deformation improves our understanding of performance of thermoplastic polyurethanes. All samples are oriented in the initial state. During deformation the hard segment domains are deteriorated. The nanoscopic strain is smaller than the macroscopic one indicating that hard domains form colonies and deformation mainly take place in the amorphous phase outside the colonies. Stress relaxation occurs by viscous flow of the amorphous phase and can be modeled considering a standard linear solid element.

Acknowledgements

The author would like to thank the Hamburg synchrotron radiation laboratory (HASYLAB) for granting the beam time. This work has been partially supported by the 7th framework program of the European Union (Project NANOTOUGH FP7_NMP_2007_LARGE)

References

De, D., Gaymans, RJ., " Transient Microstructure of Low Hard Segment Thermoplastic Polyurethane under Uniaxial Deformation", Macromol. Mater. Eng., **294**, 405-413 (2009)

Fronk, W., Wilke, W., "Small angle scattering of partially oriented polymers: model calculations with monoclinic macrolattice" Colloid Polym Sci (1985), **263**, 97-108.

Koerner, H., et al., "Transient Microstructure of Low Hard Segment Thermoplastic Polyurethane under Uniaxial Deformation" Macromolecules, **41**, 4709-4716 (2008)

(a) Stribeck, N., et al., "Nanostructure and Mechanical Properties Studied During Dynamical Straining of Microfibrillar Reinforced HDPE/PA Blends" J Polym Sci Part B Polym Phys (2010), **48(3)**, 237-250

(b) Stribeck, N., et al., "Nanostructure and fatigue studied in load cycling experiments of microfibrillar reinforced HDPE/PA blends", manuscript under preparation.

Waletzko, RS., et al., "Role of Increased Crystallinity in Deformation-Induced Structure of Segmented Thermoplastic Polyurethane Elastomers with PEO and PEO-PPO-PEO Soft Segments and HDI Hard Segments", Macromolecules, **42**, 2041-2053 (2009)

Zeinolebadi, A., Stribeck, N., J Phys Conf Ser, submitted for publication (2010)

Influence of extrusion conditions on the blend morphology: In-line monitoring by small angle light scattering

Bernd Steinhoff, Hans Kothe, Felix Simon, Ingo Alig*

Deutsches Kunststoff-Institut - Darmstadt - Germany

Typical domain sizes in immiscible polymer blends to exhibit optimum mechanical performance (e.g. high impact strength) are between 100 nm and 1 micron. This length scale can be probed by in-line small angle light scattering [1-4]. The objective of this study was to understand the relationship between blend morphology (e.g. mean domain size and aspect ratio), processing conditions (e.g. shear rate) and material related parameters (e.g. melt viscosity). The advantage of in-line measurements for this purpose is that changes in morphology due to changes in the extrusion conditions can be detected instantaneously in the scattering pattern. For our experiments a slit die with sapphire windows and an optical detection system was developed [3] and attached to the outlet of a pilot plant twin screw extruder. The extrusion experiments were performed on model blends of SAN/PMMA and PS/PC, and on technically relevant PP/POE blends (POE: polyolefine elastomer). For each set of processing conditions morphology parameters (e.g. diameter of droplets and strings, length of strings) were extracted from the scattering patterns using home made software. For comparison, samples from the extruded melt strands (quenched) were taken to analyze the morphology by scanning electron microscopy. From these results universal relationships between morphology parameters (e.g. numerical ratio of droplets and strings vs. capillary number) were established. By using in-line light scattering in a pilot plant environment it was demonstrated, that the method is suited for industrial process control.

[1] Li S, Migler K B, Hobbie E K, Kramer H, Han C C, Amis E J, J. Polym. Sci.: Part B: Polymer Physics 35 (1997) 2935
[2] Schlatter G, Serra C, Bouquey M, Muller R, Terrisse J, Polym. Eng. Sci 42 (2002) 1965
[3] Alig I Jahresbericht des DKI (ISBN 1618-0062) 2001 pp 40-45; 2008 pp 59-63.
[4] Lellinger D, Steinhoff B, Alig I, PPS-21 Conference Proceedings (ISBN 3-86010-784-4) P13.5 (2005) 19-23

Analysis, comparison and interpretation of instrumented Charpy and puncture impact test results of brittle to ductile transition behaviour of PP-compounds

Sybille Simon (1), Dieter Huber (1), Zoltan Major (2), Reinhold W. Lang (3)*

(1) Borealis Polyolefine gmbH - upper Austria - Austria,
(2) Institute of Polymer Product Engineering - Upper Austria - Austria
(3) Institute of Polymeric Materials and Testing - Upper Austria - Austria

Impact behaviour is one of the basic parameters for material classification. This parameter defines the ability of material to resist against crack propagation and furthermore break or failure. For polymers this resistance means the absorption of the impact energy through deformation which defines the toughness of a material. Different pp-compounds were tested and analysed in order to evaluate different influences on the breaking mechanisms of different material compositions more en detail. Samples were injection moulded and compression moulded for evaluation of the influence of processing conditions. There are different kinds of impacts possible and therefore different methods for determination of the impact behaviour are in use. Two methods were analysed more en detail: Instrumented Charpy test, where the stress state is uniaxial and instrumented puncture impact test, where the stress state is biaxial. The latter becomes more and more important as a biaxial stress state is more relevant for further applications. Instrumented Charpy tests were performed on notched and unnotched specimens. Force- deformation curves at different temperatures and the further analysis of these curves gives an improved insight of the material response. The evaluation of the deformation and energy offers the possibility to interpret this transition regime. An overview of the evaluated transition regimes and the consequences will be given, furthermore different influences as notching, material, test performance and evaluation methods will be discussed more en detail. It could be found that not only the transition temperature but also the behaviour in the transition regime is different and could be used for improved material characterisation. The specimen failure modes were analysed in addition and compared with the evaluated results. Furthermore puncture impact tests were performed with different notch geometries in order to evaluate the influence of an initial defect on the mate.

NUCLEATION OF POLYOLEFINS: THEORY AND PRACTICE

Markus Gahleitner*, Christelle Grein, Saeid Kheirandish and Johannes Wolfschwenger

Borealis Polyolefine GmbH, Innovation Headquarters Linz, St. Peterstr. 25, 4021 Linz,Austria
markus.gahleitner@borealisgroup.com

Abstract – From the physical principles of crystal nucleation in polymers to practical examples of polypropylene nucleation an overview of different aspects will be presented. The overall efficiency of nucleation results from a combination of epitaxy quality and dispersion quality, while for practical purposes also other factors like toxicity, thermal stability, discoloration and price need to be considered. Differences between the two physically distinct classes of nucleating agents for iPP and PE, namely particulate systems with a classical dispersion mechanism and soluble systems with a solution/recrystallization sequence for dispersion will be presented. A brief review of recent developments in the understanding of the complex interaction between nucleation, polymer parameters and processing phenomena will be given, leading to detailed results on mechanical and optical consequences of heterogeneous nucleation.

Physical principles of crystal nucleation

As soon as crystallinity had been identified as essential factor for the thermo-mechanical behaviour of important mass polymers like polyethylene and polypropylene, interest in ways to control crystallization developed both in industry and academia. One of the earliest cornerstones in systematizing data on polymer crystallization is probably the work of Van Krevelen [1], who used isotactic polystyrene as an ideal and slowly crystallizing model system for developing the concept of separation between nucleus formation and growth in polymer crystallisation, thus at the same time opening the door for intentional and controlled heterogeneous nucleation.

In case of isotactic polypropylene (iPP), the situation for nucleation is especially advantageous. Moderately slow crystal growth at large undercoolings goes together with the practical absence of sporadic nucleation, rather the presence of a thermal activation process of pre-existing (dormant) nuclei inherent to the polymer even in absence of external impurities, as clearly outlined by Janeschitz-Kriegl [2]. In comparison to that, the situation for polyethylene (PE) with its very fast growth and very limited undercooling is much less favourable.

The early systematic work on polymer nucleation can be summarized as largely based on "trial and error", involving the scanning of a wide range of substance classes as for example done by Binsbergen [3]. Only with the fundamental work in the group of Lotz in Strasbourg, France, the development of an integrated theory of epitaxial interaction between nucleating agent and polymer was possible. A first example of a successful proof of this concept for PE was given in 1981 [4]; later, a generalization to other polymers including polymer/polymer interaction and autoepitaxy was achieved [5]. According to Lotz' concept, "lattice matching" as shown in figure 1 for the interaction between iPP and poly[vinyl cyclohexane) is required for epitactic crystallization and, consequently for nucleation. This concept can even explain selectivity for the β-modification of iPP and multiple nucleation for both the α- and the γ-modification.

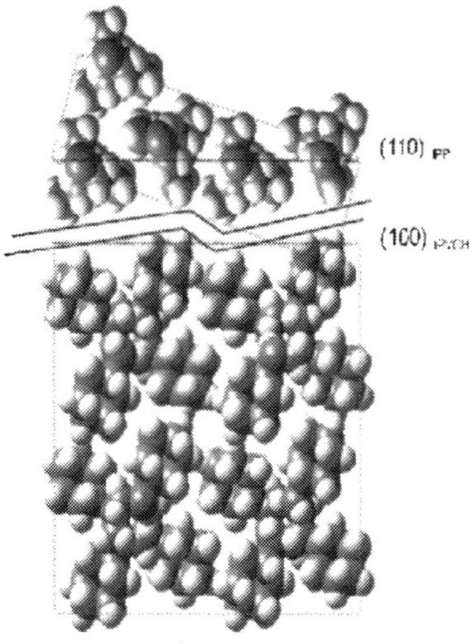

Figure 1 – Topographical interactions between iPP and PVCH contact faces; the PVCH and iPP helices have matching corrugations with a ≈22Å periodicity (from ref. [6])

In addition to that, the dispersibility of a nucleating agent is an essential prerequisite for practical use. Here, two physically distinct classes of nucleating agents for iPP exist: particulate systems with a classical dispersion mechanism, for which particle size and polarity difference to polymer are decisive, examples including talc, sodium benzoate, organophosphates etc. [7], and soluble systems with a sequence of dissolution upon heating and recrystallization upon cooling defining the degree of dispersion. In the latter case, solubility (again, of course, a question of polarity difference) and the resulting crystal shape are of prime importance for the efficiency with examples like sorbitol derivatives and benzene-trisamides [8].

Nucleating agents, even "unintentional" ones like organic pigments, need to be differentiated from solid particles inducing flow-induced crystallization on their surface – like glass fibres [9], minerals and especially nanofillers. These can be differentiated by quiescent crystallization experiments, where such particles do not show a measurable increase of the crystallization temperature (T_c) in a standard DSC experiment.

Nucleating agents for α- and γ-polypropylene

Inorganic nucleating agents

Before the introduction of nanocomposites, talc, wollastonite and mica were the three main groups of nucleating agents used industrially as fillers for polypropylene [10,11]. Nucleation efficiency and the subsequent reinforcement effect of these minerals are known to depend on a number of parameters, including concentration, filler surface treatment, particle size and distribution and particle shape. Taking T_c as reference, nucleating efficiency of the three commonly used mineral fillers can be sorted as: Talc > Wollastonite > Mica and it has been shown that addition of 5 wt% talc can lead to as much as 40% increase in nucleating efficiency [12].

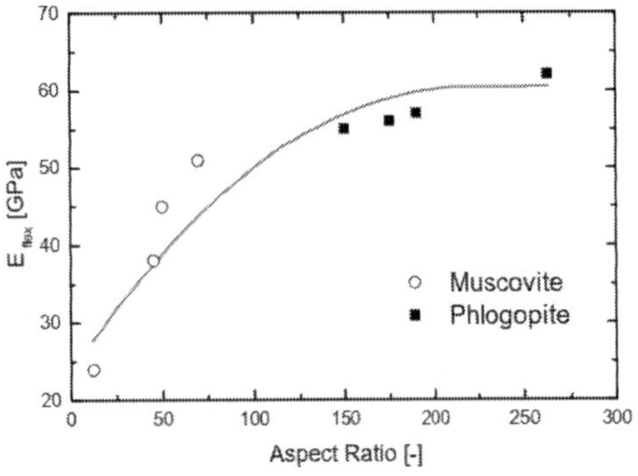

Figure 2 – Dependence of flexural modulus of a mica filled PP on particle aspect ratio (data from ref. [13])

In practice, however, particle orientation after processing is the determining factor for overall mechanical performance. Since inorganic nucleating agents for PP have been used almost exclusively in injection molded parts, the critical role of particle orientation and aspect ratio should be taken into account. A study on a series of mica-filled iPP's containing the same amount of mica with different aspect ratios has shown that the flexural modulus of composites can vary significantly only upon changing the aspect ratio between 10 and 267, as shown in figure 2 [13].

Despite their price advantage and availability, the main limiting factor with inorganic nucleating agents is their detrimental contribution to the impact behaviour. The transition from ductile to brittle fracture mechanism in filled polypropylene is generally controlled by a sharp increase in the number of initiation sites. This aspect is best explained by considering two separate mechanisms: stress-localization due to the presence of filler and existence of lower spherulite diameters arising from nucleation [14].

Organic nucleating agents

This by far biggest class of α/γ-nucleants for iPP can again be sub-divided into three categories: Particulate nucleating agents like benzoates and organophosphates, soluble nucleating agents like sorbitols and trisamides, and polymeric nucleating agents. Next to dispersion quality, lattice matching functions as the determining factor in efficiency of organic NA's. Recent experimental [15] and simulation [16] studies on varieties of available chemicals (sorbitols and amides among others) have shown that small structural changes can significantly affect the nucleating efficiency in PP.

The biggest family of particulate nucleating agents are salt crystals, for which Beck [17] has suggested in 1967 that carboxylic acid salts can be envisioned as suitable nucleating agents for PP as they consist of an organic (compatibilizing or solubilizing) and a polar (insolubilizing) group. Two tentative conclusions were drawn on the relationship between molecular structure and nucleation efficiency:
1. Sodium salts are among the most effective salts and,
2. Aromatic organic groups are preferred to aliphatic ones, as the existence of unsaturated double-bonds in the aliphatic chain can further diminish crystallinity, due probably to a decrease in chain flexibility.

In addition to benzoates and substituted benzoates, aromatic organophosphates are among the best investigated substance classes. Two extensive studies by Zhang et. al. report the performance of NA'a based on phosphoric acid and its derivatives [18] and its various metal salts with Na, Li, K, Ca, Mg, Zn and Al as metal substitute [19]. They have showed that crystallization and mechanical properties are improved by the first three monovalent metal salts (like the commercial type Adekastab NA-11) are more effective in comparison to the rest. More comprehensive studies of epitaxy of organic salts can offer more insight into the functioning mechanism of these nucleants [20];

figure 3 gives an example for lattice matching in this case.

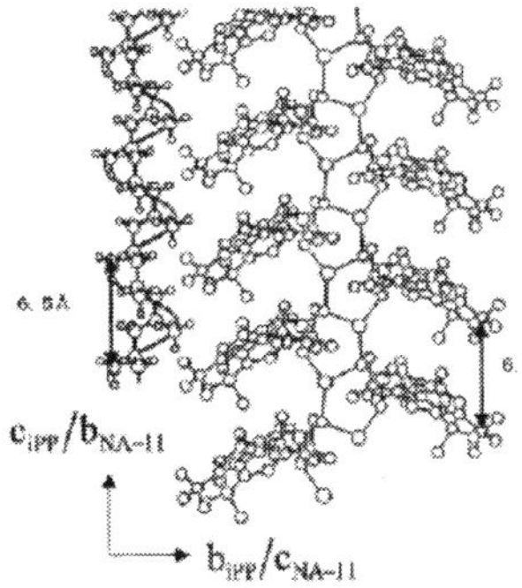

Figure 3 – Structural matching between iPP and NA-11 molecules (from Ref. [20])

Unlike salt-based NA's, sorbitol-based NA's are known to melt at the application temperature of the base resins, forming a fibrillar network in the melt that is soluble in PP, leading to fine, homogeneous dispersions with minimized scattering of light. Already in the 90's, the existence of a network texture between spherulites was observed for systems of dibenzylidene sorbitol (DBS) dissolved in ethylene glycol, whereas low DBS concentrations lead to a clear gel having two mesophases [21]. By increasing the DBS concentration a sharp drop, followed by a moderate one, in light transmission was seen. A sudden increase in mechanical strength at around the same concentrations was also observed. Both phenomena are attributed to the gelation and subsequent overlapping of spherical domains at increasing DBS concentrations. This phenomenon is unique to DBS as it is able to form a network into the solvent and maintain it, whereas other chiral gelators with a similar structure would eventually precipitate in the form of crystals in the solvent.

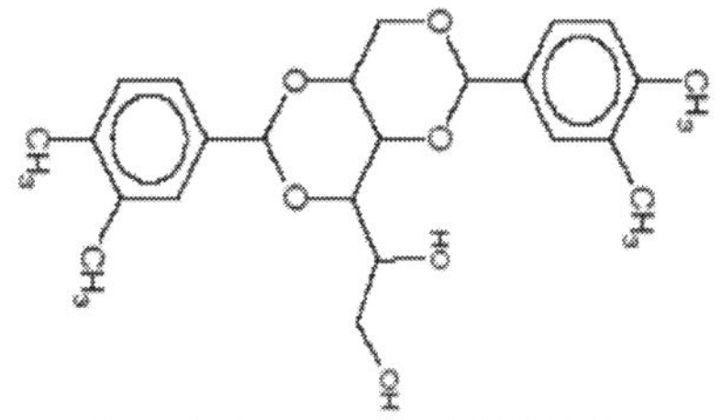

Figure 4 – Chemical structure of DMDBS

Increasing the degree of substitution on the benzyl groups of DBS results in even better nucleating efficiency, like in the case of bis(2,3-dimethyl benzyl)sorbitol (DMDBS, commercially available as Millad 3988 and presented in figure 4). Three sorbitol-based NA's with wide-spread commercial applications, DBS, MDBS and DMDBS, comprise the most relevant compounds within this group. The fundamental studies on the interesting features of this family of chemicals have been performed by producing gels by dissolving DBS in a controlled medium [21, 22]. Using microscopy on solutions with DBS concentrations as low as 1%, theses studies showed evidence of "local order", with a combination of stable, isotropic spherulites and a metastable mesophase between the spherulites. Own experimental studies using temperature sweeps in a dynamic rheometer show the network formation before the onset of crystallization for the combination of DMDBS with a PP random copolymer (see figure 5).

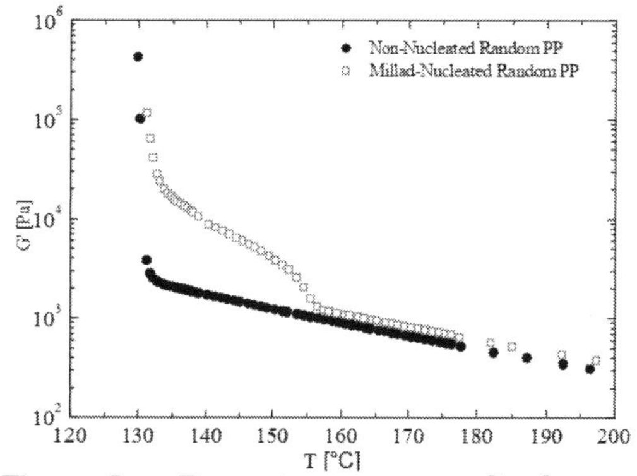

Figure 5 – Temperature sweep results for non-nucleated and Millad-nucleated PP-R

Nucleating agents for β-polypropylene

Among the three known crystalline structures (α, β, γ) of isotactic polypropylene (PP), the β-modification iscertainly the most fascinating one; an extensive review about the properties and performance of this modification has been prepared by Grein in 2004 [23]. The mostly used nucleation agents are γ-quinacridone (Permanent Red E3B) [24, 25], calcium pimelate or suberate [26, 27] and N,N-dicyclohexyl-2,6-naphthalene dicarboxamide produced by New Japan Chemicals - better known as NJ Star NU-100 - [28].

While the α-modification is monoclinic, β-PP crystallizes in a hexagonal form. It belongs to a class of polymers which displays a "frustrated" structure, referring to the fact that the helices in the unit cells have different azimutal settings [29].

The main reason for the interest in β-nucleated PP results from its significantly increased toughness, which it even exhibits at low temperatures. In comparison to other ways of impact modification for PP, β-nucleation causes only a rather limited loss in stiffness (modulus), making this a very attractive option for product design. On top of the applied type of nucleating agent, polymer type and processing conditions will determine the efficiency of this measure, the average molecular weight being the predominant factor in the case of homopolymers [23] as demonstrated in figure 6.

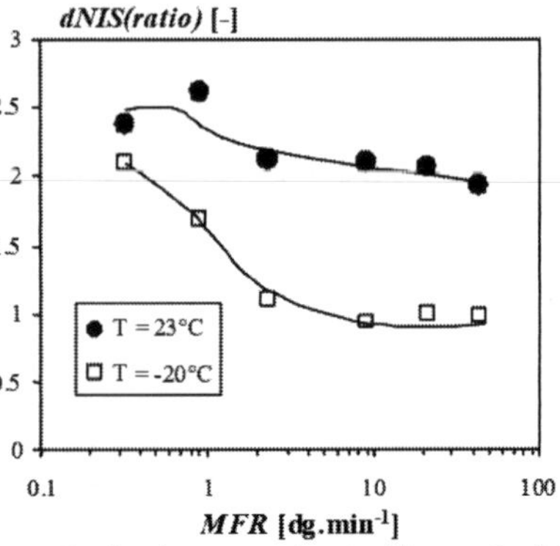

Figure 6 – Ratios between the double notched impact strengths of β-nucleated and non-nucleated grades in between MFR 0.3 and 40 dg.min^{-1} at room temperature and -20°C [23]

The combined effect at ambient and sub-zero temperatures results from a shift of the brittle/ductile transition temperature, which can reach the level of 50°C. While polydispersity has a limited effect on impact strength changes, disturbances in the chain like in random copolymers will clearly reduce the effect (see figure 7). This phenomenon, which results in a lower tendency to the formation of the β-modification at identical nucleation, has been reported before e.g. by Fujiyama [24] and Varga [30]; it can be explained by a combination of thermodynamic and kinetic effects.

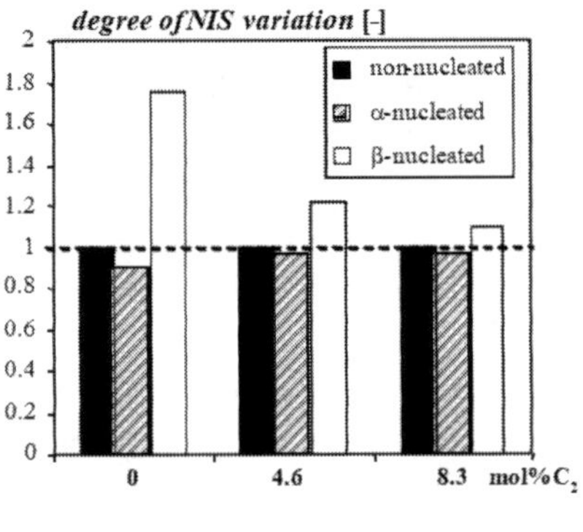

Figure 7 – Relative Charpy notched Impact Strengths (*NIS*) of non-nucleated, α-nucleated and β-nucleated versions of a series of PP grades with varying C_2 content (MFR 2 in all cases)

The development of the β-modification is controlled by the relative crystallization thermodynamics and kinetics of the stable α-modification and of the smectic phase towards the metastable β-phase. For PP homopolymers, it is generally accepted that under isothermal conditions, the α-phase grows more rapidly at temperatures below 105 and above 140°C than its counterpart, which in turn is more prone to develop in between these two temperatures in the presence of selective β-promoters [31]. While it is difficult to match these requirements to practically applicable processing conditions, the selection of the right type and amount of nucleating agent is easier; in any case a critical saturation concentration has to be exceeded for obtaining highly β-crystalline materials (see figure 8).

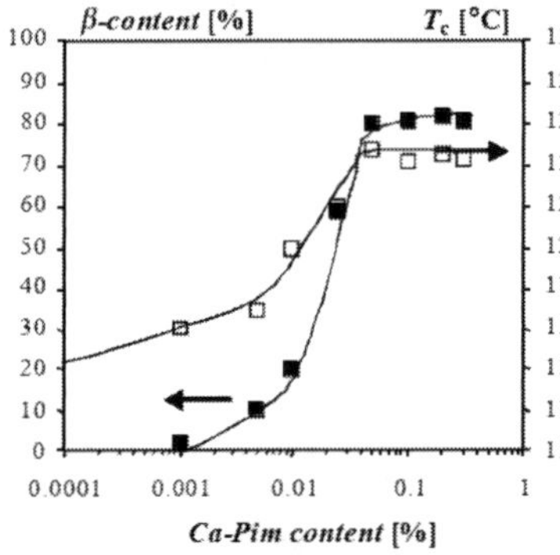

Figure 8 – Evolution of the amount of β-phase and the crystallization temperature Tc (both from DSC) as a function of the amount of calcium pimelate

The fact that conflicting data have been reported before for the interaction of the β-modification with elastomers (impact modifiers) and fillers or reinforcing fibers in PP results from complex dependencies. For PP/EPR-systems, morphology plays an important role in determining the outcome of β-nucleation [32], while for fillers positive effects can only be achieved with practically non-nucleating minerals like calcium carbonate [33]; otherwise the predominant a-nucleating effect of the filler – like in case of talc – will prevent the development of relevant amounts of β-modification.

Effects of α-crystal nucleation

Mechanical effects of α-nucleation

An important contribution to the mechanical performance of polyolefins is the correlation of modulus to not only the overall crystallinity but also the lamellar thickness in the system. The latter is correlated to the crystallization temperature according to the Gibbs-Thompson equation, resulting in a general correlation also for the case of PP as demonstrated in figure 9, where for three types of polymer independent but parallel dependencies were achieved. As shown by Pukánszky et al. [34] nucleation contributes significantly to the performance via this relation. The use of nucleating agents to increase the modulus and related properties like creep resistance or heat deflection temperature was the first major target of nucleation in the early years of development, while quickly recognizing the advantage of parallel positive consequences on impact strength and transparency.

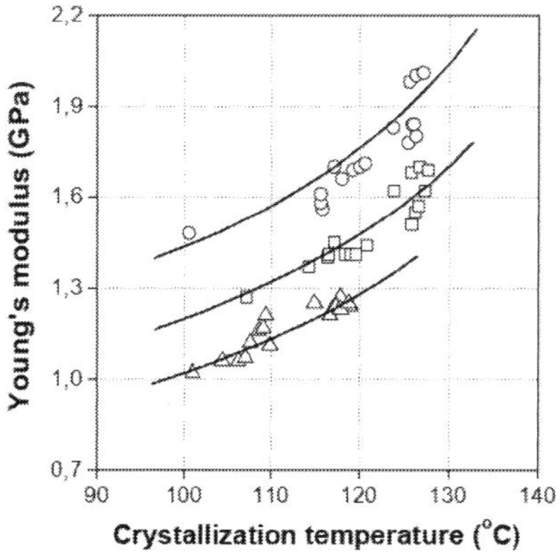

Figure 9 – Dependence of the tensile modulus (Young's modulus) and crystallization temperature of non-nucleated and differently nucleated versions of three different PP types: ○ – homopolymer, □ – impact copolymer, Δ – random copolymer (data from ref. [34])

Details of the respective mechanical effects will depend both on the type of polypropylene and the type of nucleant. In the most simple case of PP homopolymers, the effect of the average molecular weight (like the weight average molecular weight M_w, roughly correlated to the MFR) and the broadness and shape of the molecular weight distribution MWD has been studied [35, 36]. Roughly speaking, M_w will affect or even decide impact strength effects of nucleation, while the relative modulus changes can rather be related to the polydispersity of the MWD as expressed for example by the ratio between M_w and M_n.

For PP random copolymers a differentiated view is required: In case of the predominant class of propylene-ethylene copolymers mostly the comonomer content [37] is decisive for the nucleation effect and the resulting expression of properties. As a rule, reduced crystallinity will also reduce nucleation effects on mechanics, as the relative amount of crystallisable material reduces. When compared to homopolymers, especially the optical performance (transparency and haze) as well as the impact strength are affected differently by α-nucleation [38].

For the wide and increasingly important class of impact-modified heterophasic PPs including an elastomer (reactor-made ethylene-propylene impact copolymers or extruder-made blends) the situation becomes even more complex. The effects of phase composition and morphology are a versatile design tool for such materials, being heightened and made more complicated in case of specific nucleation [39, 40]. Even if only the crystalline PP matrix is affected by nucleation, quite contrary effects on the mechanical profile are possible, depending on the respective base polymer structure, with special relevance for molecular weight and morphology of the elastomer phase (like EPR). A typical example for the M_w(EPR)-effect (expressed as intrinsic viscosity of the xylene soluble content) can be seen in figure 10; details of such correlations extending to shifts in the brittle/ductile transition temperature have been described recently [32].

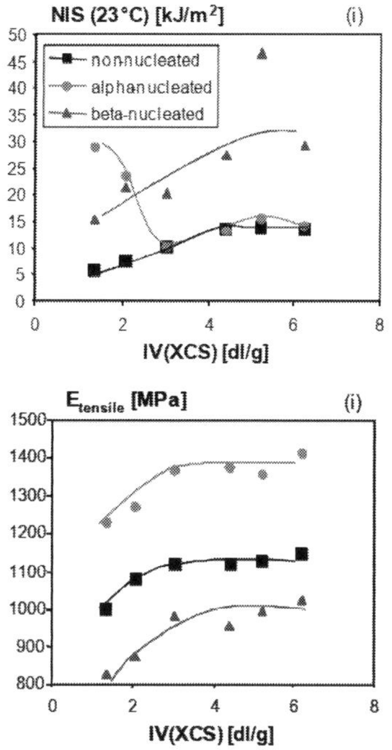

Figure 10 – Impact toughness at 23 °C (top) and tensile modulus (bottom) for a series of non-nucleated, α-nucleated and β-nucleated heterophasic PP copolymers

Optical effects of α-nucleation

Basically every nucleating agent results in a higher number of nuclei by some orders of magnitude, resulting in a decreased spherulite size and improved transparency. When it comes to "clarification" of polypropylene (i.e. when highest transparency is needed) then the soluble nucleating agents are clearly the preferred substances [8, 41]. The mechanism has so far best been studied for sorbitol derivatives [42]; a possible explanation for the final superior transparency of sorbitol-nucleated mixtures is the combination of the interpenetrating network of the clarifier with the locally twisted structure of the sorbitol-fibrils resulting from the "butterfly" structure of the molecule, in case of injection moulding possibly supported by the reduced presence of flow-induced superstructures (skin layers, as discussed in [38]).

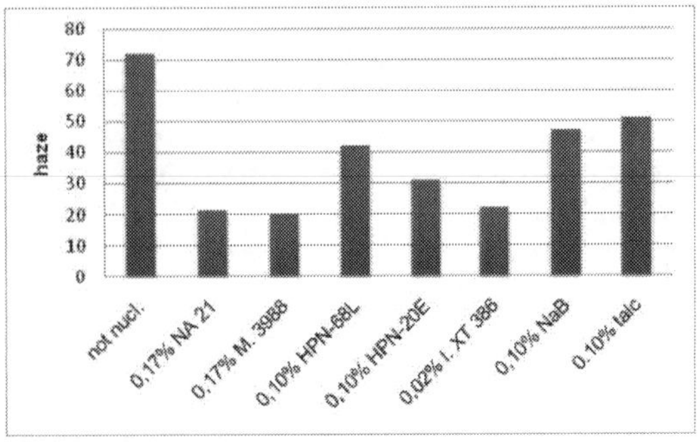

Figure 11 – Transparency of injection moulded plaques (2 mm thickness), (C3/C2 random copolymer, MFR (230°C/2,16kg) 1,5 g/10min)

In figure 11 the efficiency in terms of haze of different nucleating agents/clarifiers in an ethylene/propylene random copolymer is demonstrated. Only a limited number of particulate nucleating agents like one organophosphate (Adekastab NA-21E) and poly(vinyl cyclohexane) [43] are capable of achieving similar levels of transparency as the soluble types. Next to the polymer type, also processing effects play a predominant role here, making comparisons between films and injection moulded parts very difficult.

Post-crystallization effects of nucleation

Numerous studies regarding controlled aging or annealing of PP at different temperatures for various times have been performed in the past [44, 45]. In detail, attention has to be paid to the combined effects of physical ageing and post-crystallization, which both are affected by three main factors: the nature of the polymer (essentially by the chain regularity), the presence of nucleating agents and processing conditions.

The tendency to both post-crystallization and resulting changes in mechanical properties is clearly increased by both chain defects – e.g. in random copolymers – and by quenching part of the polymer into mesomorphic and amorphous state – e.g. at very high cooling rates. Especially in the case of PP films [46], this makes the application of highly transparent random copolymers for applications requiring thermal post-treatment like pasteurization or sterilization processes difficult. While stiffness increase can be foreseen and compensated, both the loss of transparency and the toughness reduction are critical in many applications like medical or food packaging. Selective α-nucleation can offer a way to stabilize films against such effects, reducing post-crystallization effects significantly even in highly transparent random copolymers [47]. As figure 12 shows, both the polymer selection and the type of nucleating agent will affect the final result critically.

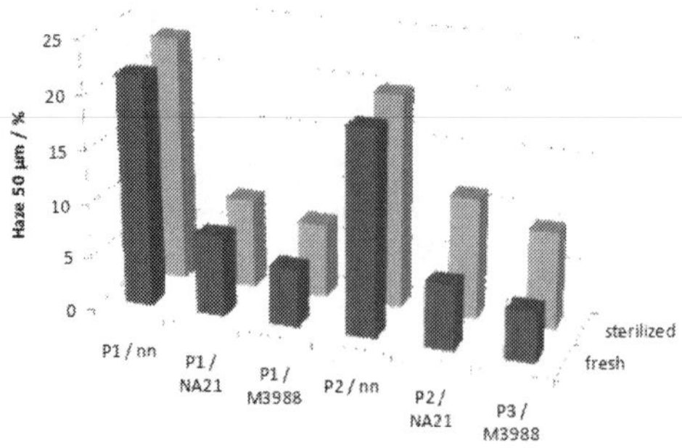

Figure 12 – Nucleation and sterilization (120°C/30min) effects on haze (turbidity) of PP cast films of 50 µm thickness (chill roll temperature 50°C) for two different C_3/C_2 random copolymers with similar C_2 content and different randomness

Conclusions

Despite a wealth of both experimental and computational studies on the interaction between polypropylenes and nucleating agents it is still difficult to draw general conclusions. Predicting the action of a specific (organic) molecule as nucleating agent is possible to a certain extent based on polarity studies (defining dispersibility) and the crystal structure (defining lattice matching). The practical usefulness and efficiency will, however, depend on far more factors, and for different target applications different "ideal" combinations of specific polymers and nucleating agents have been identified.

References

1 – Boon, J,; Challa, G.; Van Krevelen, D.W.: J.Polym.Sci.A-2 6, p. 1835-1851 (1968)
2 – Janeschitz-Kriegl H.: Coll.Polym.Sci. 275, p. 1121-1135 (1997)
3 – Binsbergen, F.L.: Polymer 5, p. 253-267 (1970)
4 – Wittmann, J.C.; Lotz B.: J.Polym.Sci.: Polym.Phys.Ed. 19, p. 1837-1851 (1981)
5 – Wittmann, J.C.; Lotz B.: Prog.Polym.Sci. 15, p. 909-948 (1990)
6 – Alcazar, D.; Ruan, J.; Thierry, A.; Lotz, B.: Macromolecules 39, p. 1008-1018 (2006)
7 – Gahleitner, M.; Wolfschwenger, J.; Bernreitner, K.; Neißl, W.; Bachner, C.: J.Appl.Polym.Sci. 61, p. 649-656 (1996)
8 – Kristiansen, M.; Werner, M.; Tervoort, T.; Smith, P.; Blomenhofer, M.; Schmidt, H.-W.: Macromolecules 36, p. 5150-5156 (2003)
9 – Varga, J.; Karger-Kocsis, J.: Compos.Sci.Tech. 48, p. 191-198 (1993)
10 – Ferrage, E.; Martin, F.; Boudet, A.; Petit, S.; Fourty, G.; Jouffret, F.; Micoud, P.; de Parseval, P.; Salvi,S.; Bourgerette, C.; Ferret, J.; Saint-Gerard, Y.;

Buratto, S. and Fortune, J.P.:, J.Mat.Sci. 37, p.1561-1573 (2002)
11 – Vu-Khanh, T.; Sanschagrin, B. and Fisa, B.: Polym.Compos. 6, p.249-256 (1985)
12 – de Medeiros, E.S,; Tocchetto,R.S.; de Carvalho, L.H.; Santos, I.M.G. and Souza, A.G.: J.Therm.Anal.Calor. 66, p.523-531 (2001)
13 – Lusis, J.; Woodhams, R.T. and Xanthos, M.: Polym.Eng.Sci.13, p.139-145 (1973)
14 – Schultz, J.M.: Polym.Eng.Sci. 24, p.770-785 (1984)
15 –Blomenhofer, M.; Ganzleben, S.; Hanft, D.; Schmidt, H.W.; Kristiansen, M.; Smith, P.; Stoll, K.; Mäder, D.; Hoffmann, K.: Macromolecules 38, p. 3688-3695 (2005)
16 – Nagarajan, K.; Myerson, A.: Cryst.GrowthDesign 1, p. 131-142 (2001)
17 – Beck, H.N.; J.Appl.Polym.Sci. 11, p. 673-685 (1967)
18 – Zhang, G.P.; Xin, Z.; Yu, J.Y.; Gui, Q.D.; Wang, S.Y.: J.Macromol.Sci:B.Phys. B42, p. 467-478 (2003)
19 – Zhang, Y.F.; Z Xin, Z.: J.Appl.Polym.Sci. 100, p. 4868-4874 (2006)
20 – Nagasawa, S.; Fujimori, A.; Masuko, T.; Iguchi, M.: Polymer 46, p. 5241-5250 (2005)
21 – Yamasaki, A.; Tsutsumi, H.: Bull.Chem.Soc.Jpn. 67, p. 906-911 (1994)
22 – Yamasaki, A.; Tsutsumi, H.: Bull.Chem.Soc.Jpn. 67, p. 2053-2056 (1994)
23 – Grein, C.: Adv.Polym.Sci. 188, p. 43-104 (2005)
24 – Leugering, H.J.: Makromol.Chem.109, p.204-212 (1967)
25 – Fujiyama, M.: Intern.Polym.Proc. 11, p. 271-279 (1996)
26 – Li, J.X.; Cheung, W.L.: J.Vinyl. Addit.Technol. 3, p. 151-162 (1997)
27 – Varga, J.; Mudra, I.; Ehrenstein, G.W.: J.Appl.Polym.Sci. 74, p. 2357-2364 (1999)
28 – Marco, C,; Gomez, M.A.; Ellis, G.; Arribas, J.M.: J.Appl.Polym.Sci. 86, p. 531-538 (2002)
29 – Stocker, W.; Schumacher, M.; Graff, S.; Thierry, A.; Wittmann, J.C.; Lotz, B.: Macromolecules 31, p. 807-814 (1998)
30 – Varga, J.; Schulek-Toth, F.: J.Therm.Anal. 47, p. 941-975 (1996)
31 – Varga, J. : J.Macromol.Sci.Phys. 41, p. 1121-1129 (2002)
32 – Grein, C. ; Gahleitner, M.: eXPR.Polym.Lett. 2, p. 392-397 (2008)
33 – Tjong, S.C.; Li, R.K.Y, ; Cheung, T. : Polym.Eng.Sci. 37, p. 166-172 (1997)
34 – Pukánszky, B., Mudra, I., Staniek, P.: J.Vinyl. Addit.Technol. 3, p. 53-57 (1997)
35 – Gahleitner, M., Wolfschwenger, J., Bernreitner, K., Neißl, W., Bachner, C.: J.Appl.Polym.Sci. 61, p. 649-657 (1996)
36 – Avella, M., dell'Erba, R., Martuscelli, E., Ragosta, G.: Polymer 34, p. 2951-2960 (1993)
37 – Spruiell, J.E., Lu, F.M., Ding, Z., Richeson, G.: Proc. SPE ANTEC '96, p. 2305-2311 (1996)
38 – Gahleitner, M., Wolfschwenger, J., Fiebig, J., Neißl, W.: Macromol.Symp. 185, p. 77-87 (2002)
39 – Kotek, J., Scudla, J., Slouf, M., Raab, M.: J.Appl.Polym.Sci. 103, p. 3539-3546 (2007)
40 – Fanegas, N., Gómez, M.A., Jiménez, I., Marco, C., Garcia-Martínez, J.M., Ellis, G.: Polym.Eng.Sci. 48, p. 80-87 (2008)
41 – Bernland, K,; Tervoort, T.; Smith, P.: Polymer 50, p. 2460-2464 (2009)
42 – Balzano, L.; Rastogi, S.; Peters, G.W.M.: Macromolecules 41, p. 399-408 (2008)
43 – Menyhárd, A., Gahleitner, M., Varga, J., Bernreitner, K., Jääskeläinen, P., Øysæd, H., Pukánszky, B.: Eur.Polym.J. 45, p. 3138-3148 (2009)
44 – Piccarolo, S.: Polymer 47, p. 5610-5622 (2006)
45 – Gahleitner, M.; Fiebig, J.; Wolfschwenger, J.; Dreiling, G. and Paulik, C.: J.Macromol.Sci.-Phys. B41, p. 833-849 (2002)
46 – Macauley, N.J.; Harkin-Jones, E.M.A. and Murphy, W.R.: Polym.Eng.Sci. 38, p. 662-70 (1998)
47 – Gahleitner, M.; Wolfschwenger, J.; Grein, C. Blell, R. and Koch, T.: Proc. of ADDCON World Conf. (RAPRA) 2008, Barcelona/ES, October 15-16 (Session 6, paper 15)

PET Nanocomposites based on carbon nanotubes as matrix for continuous glass fibre composites

M. Aurilia (1), L. Sorrentino (1),(2),, F. Velotta (2), S. Iannace (1),(2)*

(1)IMAST, Technological District on Polymeric and Composite Materials Engineering and Structures, P. E. Fermi 1 - loc. Granatello, Portici 80055, (NA), Italy,
(2) Institute for Composite and Biomedical Materials, National Research Council of Italy, P. E. Fermi 1 - loc. Granatello, Portici 80055, (NA), Italy

Continuous glass fibre composites prepared with polyethylenetherephthalate (PET) matrix, reinforced by multiwalled carbon nanotubes, have been produced in order to obtain higher mechanical properties composites. The carbon nanotubes were dispersed in the PET at nanoscale level through a melt-mixing process, as detected by both optical and electronic microscopy analyses. DMA results have shown that the storage modulus of PET nanocomposites increased compared to neat PET in the 30-200°C temperature range. The continuous glass fibre composites have been manufactured by means of the film stacking technique. Optical images of composite polished surfaces showed that the PET nanocomposites thoroughly impregnated the glass fibres. Flexural mechanical properties of the composites based on PET nanocomposites matrices increased at both room and high temperatures with respect to the neat PET composite. SEM images of the fractured surfaces of the tested glass fibres composites showed a good adhesion between nanocomposite matrices and glass fibres. The carbon nanotubes addition to the PET matrix resulted in improved Mechanical properties improvement of the PET, due to, successfully resulted in mechanical properties improvement of continuous glass fibre composites.

MICRO-FIBRILLATION OF POLYMER BLENDS DURING LAMINATING-MULTIPLYING EXTRUSION AND ITS EFFECT ON THEIR PROPERTIES

Guo Shaoyun *, Shen Jiabin, Li Ting , Sun Xiaojie, Li Jiang

The State Key Lab.of Polymer Materials Engineering,Polymer Research Institute of Sichuan University - Chengdu 610065, Sichuan – China

nic7702@scu.edu.cn

Three kinds of polymeric blends: nylon 6 (PA6)/isotactic polypropylene (iPP), carbon black (CB)-PA1010/iPP, and ethylene vinyl alcohol copolymer (EVOH)/propylene-ethylene compolymer (PPE) are prepared by applying an assembly of laminating-multiplying elements (LMEs) connecting with an extruder. The morphological observations illustrate that the LME may provide biaxial stretching force on the melt leading to the orientation of dispersed phases. As their concentrations are less than 10%, the fibrillation effect is more distinct along flowing direction. However, as the concentrations increase, the morphologies of those dispersed phases change from one-dimension fibers to two-dimension sheets gradually. Meanwhile, it has been found that the LME may also act as a blender to make the distribution of those dispersed particles become more even. The tensile, electrical or barrier properties are investigated respectively and demonstrate the development of morphologies in LMEs. In the PA6/iPP system, as the LME numbers increase to nine, both of the tensile strengh and elongation at break of iPP are enhanced more than 100%, respectively. In the CB-PA1010/iPP system, when seven LMEs are applied, a conductive network is formed by the oriented CB-nylon phase, resulting in about 5-order decrease in volume resistivity and at least 10 times increase in elongation at break compared with those results without LMEs. In the EVOH/PPE system, it is found that with increasing LMEs numbers, the EVOH phase tends to be extended biaxially, which induces the decrease of permeability coefficient about 2 orders accompanied by the increase of elongation at break more than 27 times compared to conventional blending system. Results suggest that the laminating-multiplying extrusion system is a useful tool to obtain polymeric composites with specific morphologies and properties.

Introduction

Recently, in situ microfibrillar composites (MFCs) are developed by Fakirov S et al through blending polyolefins with other polymers, especially engineering polymers like polyamide (PA), poly(ethylene terephthalate) (PET) [4-7]. Comparing with the inorganic fibre- and LCP-reinforced composites, the MFCs offer many potential advantages, such as no mineral additives, ability for recycling, reduction in weight, easy processing, no need for extra compatibilizers, high cost performance, etc [8]. The preparing methods of MFCs can generally be divided into two kinds: one is "cold drawing method" [5, 9, 10], in which the microfibrils of high-melting-temperature component are formed through solid state cold stretching after melt blending process; the other is "hot drawing method" [11, 14], in which the extrudate is stretched in melt state so that the microfibrils of high-melting-temperature component can be acquired, and then quenched quickly to preserve the microfibrillar morphology. The external stretching force is prerequisite for these two methods with the different processing temperatures. This leads to the introduction of additional equipment to produce the stretching force. In contrast, the utilization of internal shearing and elongational forces produced when polymer melt flows through the exit die should provide an economic and simple method to microfibrillize the dispersed engineering polymers. Unfortunately, the conventional extruder cannot bring large enough internal shearing and elongational forces for the microfibrillation. Therefore, the research and development of new style die will be a profound work in the field of in situ microfibrillar composites.

Recently, a novel extrusion die, which has an assembly of laminating-multiplying elements (LMEs), has been designed in our Lab. When the polymeric melt flows through this die, the laminating-multiplying process in each LME can induce stronger shearing and elongational forces on the melt, which may not only lead the dispersed phase to be oriented along the flow direction, but also make the distribution of those particles become more even. In this work, nylon 6 (PA6)/isotactic polypropylene (iPP), carbon black (CB)-PA1010/iPP, and ethylene vinyl alcohol copolymer (EVOH)/propylene-ethylene compolymer (PPE) are prepared by an extruder, which combines with a series of LMEs designed by our Lab at die. The morphological development, with increasing the number of LMEs, is examined by scanning electron microscope (SEM) to indicate its effect on the crystalline behavior, dynamic rheological properties and tensile properties.

Experimental (Times New Roman 10, bold)

PA6 , PA1010, EVOH, carbon black and PP are dried in an oven at 105 °C and 85 °C for 12 hour respectively, and mixed together in a certain weight ratio in PA6/iPP, carbon black (CB)-PA1010/iPP, and EVOH/PPE blends in a high-speed mixer prior to processing. The mixed granules are extruded from the experimental system as illustrated in Figure 1. The melt goes through the extruder, connector, laminating-multiplying elements (LMEs), and forms a sheet about 1.5 mm thick and 40 mm wide. Figure 2 shows the flowing behavior of the melt in LME, where the melt is sliced into left and right sections by a divider, and then recombines vertically. An assembly of n LMEs can make the melt experience this course n times.

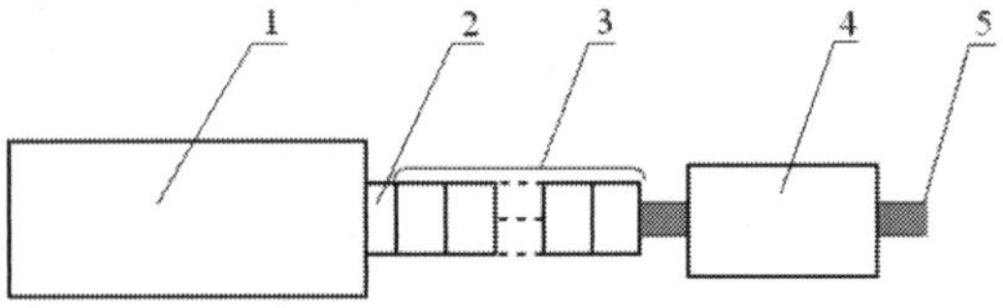

Fig.1 Schematic of experimental system: 1-single screw extruder; 2-connector; 3- laminating-multiplying; 4-water cooling block; 5-sample

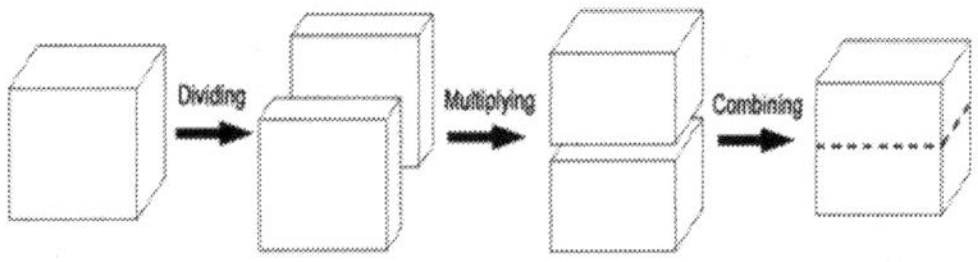

Fig.2 Schematic of a laminating-multiplying element (LME)

Results and Discussion

Figure 3 shows the effect of LMEs on the morphology of dispersed phase in PP/PA6 blend. The samples are immersed in hot xylene for 2 hours to etch iPP matrix away. It is shown that by introducing more LMEs, the PA6 particles are elongated obviously accompanied by the decrease in diameter. When the number of LMEs reaches nine **(Figure 3d)**, it can be found that the specimen exhibits well-defined microfibrillar morphology.

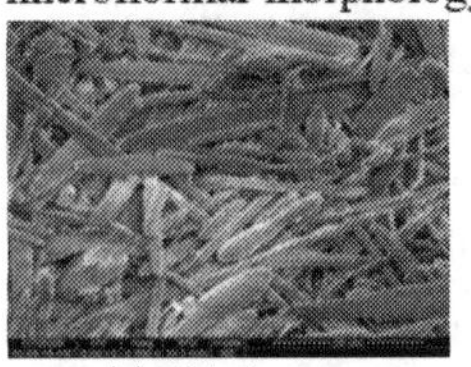

(a) 5%-0

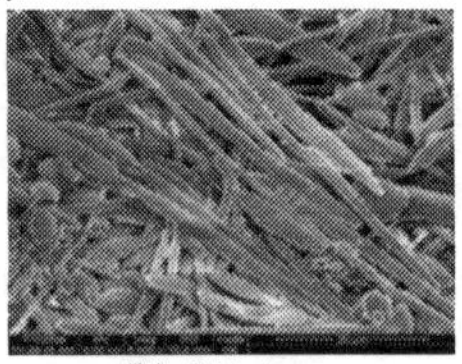

(b) 5%-1

(c) 5%-3

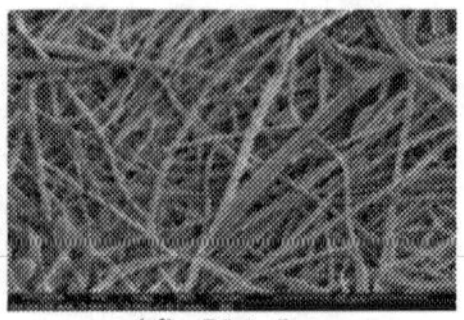

(d) 5%-9

Fig.3 SEM images of PA6/iPP blends with 5 wt% PA6 prepared with 0, 1, 3 and 9 LMEs

Figure 4(a) shows the dependence of the yield strength and strength at break of neat iPP, PA6 and PA6/iPP blends on the number of LMEs. It can be found that PP, PA6 and their composites deform in a ductile manner with necking and have approximately the same yield strength (about 34 MPa). However, as no LME is applied, the strength of the composite at break is distinctly less than that of iPP, which may be related to the poor interfacial adhesion between phases. With increasing the number of LMEs, the strength is enhanced obviously, from 19 MPa to 47 MPa. Similarly, the elongation at break of the material with nine LMEs is increased more than 100% compared to the value without LMEs, as illustrated in Figure 12(b).

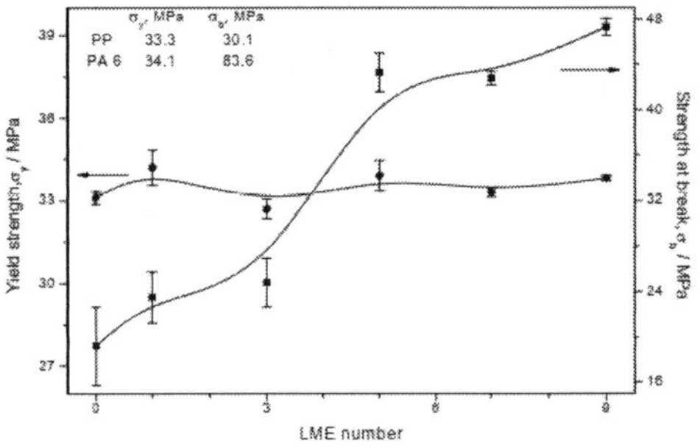

(a)

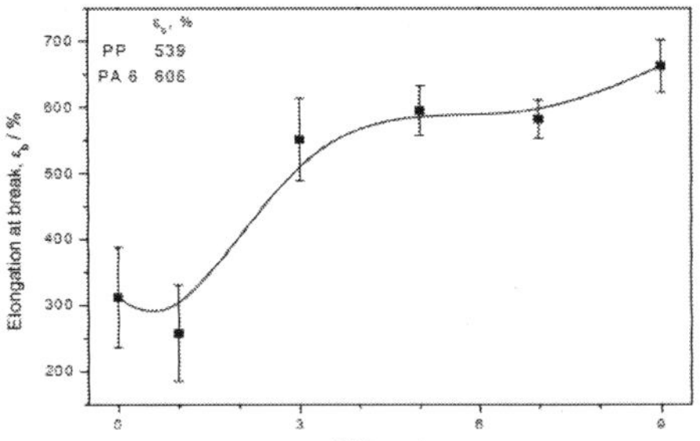

(b)

Fig.4 Effect of the number of LMEs on the tensile properties of PA6/iPP blends

Figure 5 shows the effect of LMEs on the morphology of dispersed phase in iPP/(PA1010+CB) (90/10) blends. It is also shown that by introducing more LMEs, the PA6 particles are elongated obviously accompanied by the decrease in diameter. when seven LMEs are applied, a conductive network is formed by the oriented CB-nylon phase, resulting in about 5-order decrease in volume resistivity (Fig.6) and at least 10 times increase in elongation at break compared with those results without LMEs.

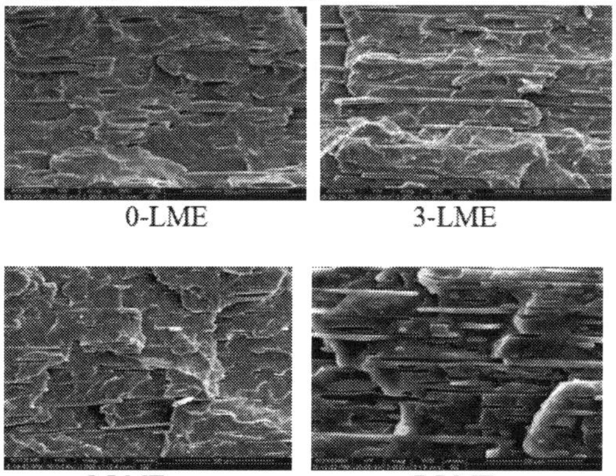

Fig.5 SEM images of iPP/(PA1010+CB)(90/10) blends

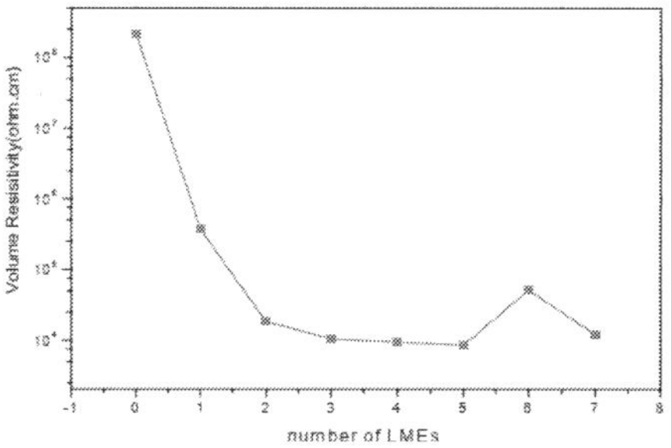

Fig.6 The effect of number of LMEs on the volume resistivity of iPP/PA1010+CB (along flowing direction)

Conclusions

As the polymeric melt flows through a LME, the dividing-multiplying process takes place and induces stronger shearing force. The scanning electron micrographs illustrate that with increasing the number of LMEs, the dispersed particles tend to be fibrillated along flowing direction. Compared with the sample prepared without LME, the strength and elongation at break of the sample get grealy enhanced. Results suggest that the laminating-multiplying extrusion system is a useful tool to obtain polymeric composites with specific morphologies and properties.

Acknowledgements

The authors are grateful to the Special Funds for Major State Basic Research Projects of China (2005CB623800), National Natural Science Foundation of China (50933004, 50773047,) for financial support of this work.

References (Times New Roman 10, bold)

[1] Shields R J, Bhattacharyya D, Fakirov S. Composites A, 39, 940 (2008).

[2] Friedrich K, Evstatiev M, Fakirov S, Evstatiev O, Ishii M, Harrass M. Comp Sci Technol, 65, 107 (2005).

[3] Evstatiev M, Nicolov N. Polymer, 37, 4455 (1996).

[4] Fakirov S, Evstatiev M, Petrovich S. Macromolecules, 26, 5219 (1993).

[5] Fakirov S, Bhattacharyya D, Shields R J. Colloid Surface A, 313-314, 2(2008)..

[6] Jayanarayanan K, Thomas S, Joseph K. Composites A, 39, 164 (2008).

[7] Taepaiboon P, Junkasem J, Dangtungee R, Amornsakchai T, Supaphol P. J Appl Polym Sci, 102, 1173 (2006).

[8] Li Z, Yang M, Xie B, Feng J, Huang R. Polym Eng Sci, 43, 615 (2003).

[9] Monticciolo A, Cassagnau P, Michel A. Polym Eng Sci, 38, 1882 (1998).

MELT MIXING STUDIES OF MISCIBLE AND IMMISCIBLE DRUG-POLYMER SYSTEMS

Jin Uk Ha and Marino Xanthos*

Department of Chemical, Biological, and Pharmaceutical Engineering, New Jersey Institute of Technology, Newark, NJ, 07102, USA - xanthos@njit.edu

In this study, examples of melt batch or continuous mixing for pharmaceutical applications are presented. Particularly, drug-polymer miscible and immiscible systems are compared in order to understand and improve the drug dissolution rate and the apparent solubility. The miscibility of the system was preliminarily determined by comparing the solubility parameters of drugs and polymer. Indomethacin and Diclofenac were selected as examples of drug-polymer miscible and immiscible systems, respectively, and water soluble Eudragit® E100 acrylic copolymer was used as the drug excipient. A novel manufacturing method involving the dispersion of anionic nanoclays containing intercalated drugs into their interlayer space is studied in order to improve the bioavailability of the miscible and immiscible drug polymer systems. Furthermore, the effects on drug stability due to the presence of the anionic nanoclay and the acrylic copolymer were studied. Drug miscibility and morphology in the polymer matrix was determined by XRD and SEM analysis. FTIR, TGA, DSC, and UV-Vis were used for further analysis.

Introduction

Recently, the study of producing pharmaceutical products through hot melt extrusion (HME) is receiving great attention from the pharmaceutical and polymer processing industries, and academia. It has been reported that HME, in addition to being a continuous operation, can provide many advantages compared to conventional methods, such as increasing the dissolution rate and thereby increasing bioavailability of drugs with poor solubility, and elimination of solvent usage [1, 2].

Liu et al [1], recently, studied the effects of extrusion process parameters on the dissolution behavior of an Active Pharmaceutical Ingredient (API) in Eudragit® EPO. They reported that processing parameters are significantly affecting the state of the drug in the polymer matrix and its miscibility with the polymer, thereby affecting dissolution behaviors.

Nollenberger et al. [3] reported that the drug felodipine in the Eudragit® E matrix improved the dissolution of the drug in the acidic environment. This is because the polymer dissolved rapidly and formed a polymer-rich phase in which the drug molecules could dissolve, thus resulting in increased aqueous drug solubility.

By comparing the solubility parameters (δ) of polymers and APIs one can approximate their miscibility. Among other criteria for predicting the miscibility of the system, it is apparent that the closer the values of the solubility parameters of the polymer and the API, the higher the tendency to be miscible. Greenhalgh et al. [4] showed that for differences in solubility parameters of less than 7.0 $MPa^{1/2}$, in the case of drug and polymer system, materials are considered to be miscible.

The objective of this research is the understanding of the differences between drug-polymer miscible and immiscible systems in terms of morphology, rheology and bioavailability. Furthermore, a novel method for enhancing the bioavailability of drug-polymer immiscible system by incorporating an anionic nanoclay is introduced.

Experimental

Materials

Butyl/ methyl methacrylate, dimethylamino-ethyl methacrylate terpolymer (1:1:2), Eudragit® E 100 donated from Evonik Industries (Piscataway, NJ), was used as excipient in a formulation containg an API and an API modified clay. Eudragit® E 100 is soluble in the gastric fluid up to the pH 5.0. This polymer has a reported T_g of 48°C and its maximum processing temperature is 220 °C.

The API for this study is sodium diclofenac (DIK-Na^+) purchased from Spectrum Chemical & Laboratory Products (Gardena, CA). DIK-Na^+ is a white powder with molecular weight of 296.14 g/mol and melting point (T_m) of 284 °C. The reported water solubility of DIK-Na^+ at pH 1 is reported as 1.2 mg/L [5]. DIK-Na^+ is considered as anionic drug since it contains carboxylic group in its molecular structure.

DIK-Na$^+$ is selected for the non miscible drug-polymer system since this drug has a higher difference in solubility parameter than 7.0 MPa$^{1/2}$ when compared to Eudragit® E100 (Table 1).

Another API used for the miscible drug-polymer system is indomethacin (IND), purchased from Spectrum Chemical & Laboratory Products (Gardena, CA). IND is a non steroidal anti inflammatory drug which is commonly used to reduce fever, pain, and swelling. IND is a white powder with molecular weight of 357.7 g/mol and melting point (T_m) of 162 °C. This drug has very poor water solubility reported as 4.0 - 8.8 μg/mL [6]. IND can be also considered as anionic drug due to its carboxylic group in its molecular structure. IND was selected for the miscible drug polymer system because IND has a lower difference in solubility parameter than 7.0 MPa$^{1/2}$ compared to Eudragit® E100 (Table 1).

Hydrotalcite, the anionic clay with carbonate as the exchangeable anion, was provided by Sasol Germany. It has a double layered metal hydroxide structure consisting of magnesium and aluminum hydroxide octahedrally interconnected via the edge. The weight ratio of Al_2O_3 : MgO is reported as 38 : 62. The chemical formula is $Mg_4Al_2(OH)_{12}CO_3\,nH_2O$ with basal spacing of 0.77 nm. The reported anionic exchange capacity (AEC) of this clay is 340 meq /100g.

Table 1. Calculated Solubility parameters*

Polymers	Eudragit ® E100	DIK-Na$^+$	IND
Solubility parameters, δ (MPa)$^{0.5}$	19.7	27.9	24.5

*Fedor's method

Preparation of Drug Polymer Composites

Eudragit® E100 blends containing 4 wt% and 15 wt% DIK-Na$^+$, and 10 wt% of DIKClay (containing 4wt% DIK) were prepared by compounding in a Brabender batch mixer (PL2000, C. W. Brabender) at 50 rpm and 130 °C for 6 minutes. IND 5wt% and 15wt% were also compounded with Eudragit® E100 and processing conditions are the same as DIK polymer composites. After melt mixing, the compounds were pressed for 50 sec at 130 °C into thin films (2 mm) for further analyses.

Modification of nanoclays with DIK-Na$^+$ (DIK/Clay)

Anionic exchange of calcined hydrotalcite after removal the carbonate anions and water was carried out with methanol (200 ml) and de-ionized water (50 ml) solutions containing 2X the stochiometric amounts of DIK-Na$^+$ based on the AEC of the HT. The solution pH was set to 6.0 by adding 0.01N of HCl. The reaction was carried out at 60 °C for 2 days in a closed reactor. After the reaction, the modified clays were filtered under vacuum, washed with a methanol/water mixture and then separated from the solution by centrifuging. These washing procedures were repeated several times in order to remove the residual DIK-Na$^+$ from the clay surface. The modified clays were then dried at room temperature for 24 hours and dried again at 80 °C for 24 hours under vacuum.

Characterization

FTIR spectra of all the samples were obtained using a Spectrum One FTIR Spectrometer® (Perkin-Elmer Instruments). Thermal stability was studied by thermogravimetric analysis, TGA (TA instruments' QA 50 thermogravimetric analyzer), at a heating rate of 10 °C /min under N_2 atmosphere. X-ray diffraction (XRD) with a Philips PW 3040 diffractometer was used to in order to determine the miscibility of the API in the polymer matrix, and calculate the interlayer space of unmodified and modified clays from diffraction angles. The fracture surface of compounds was examined by Scanning Electron Microscopy (LEO 1530 VP Emission SEM). The surface elements were identified by Energy Dispersive X-ray (LEO 1530 VP Emission SEM). The DIK loading in the CHT interlayer space was analyzed by calculating the amount of carbon and nitrogen through Elemental Analysis (QTI intertek, Whitehouse, NJ). An optical microscope (Carl Zeiss Universal Research Microscope) was used to characterize the miscibility of APIs in the polymer matrix. The drug dissolution was studied by a Distek dissolution system 2100A with a Distek temperature control system TCS 0200 according to USP dissolution Apparatus II with the paddle rotation speed of 50 rpm at 36.5 °C. The dissolution medium was 1L of pH 1.2 HCl and 7.4 phosphate buffer solution. The test fluid was collected at predetermined time interval, filtered through a 0.45 μm filter and then analyzed at 276 nm and 318 nm for DIK and IND, respectively by an UV-Vis spectrophotometer (Evolution 60, Thermo Scientific

Results

Figure 1 shows the XRD results of DIK-Na$^+$, IND, Eudragit® E100 and Eudragit® E100 containing APIs in its matrix. Both APIs showed strong crystalline peaks by contrast to the amorphous Eudragit E100. As predicted from the calculations of solubility parameters, $DIK\text{-}Na^+$ was not dissolved in the polymer matrix. Therefore, the crystallinity of the $DIK\text{-}Na^+$ was still observable (Figure 1 (c)). On the other hand, IND did not show any crystalline peaks as indicated in figure 1(b). This suggests that IND was dissolved during melt mixing and hence was in an amorphous state

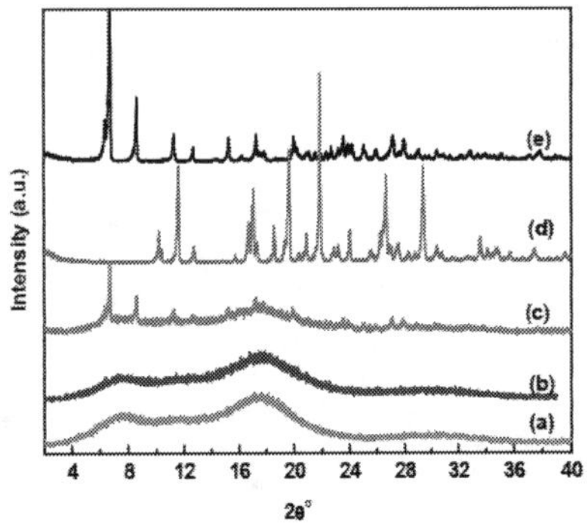

Figure 1 XRD results of (a) Eudragit E100, (b) Eudragit E100/15%IND, (c) Eudragit E100/15%DIK, (d) IND, and (e) DIK.

The miscibility of APIs with the polymer was also determined by polarized microscopy (PM). By this method, the crystalline drugs are appearing as much brighter objects than the background matrix while dissolved and amorphous drug do not show any characteristic brightness. Figure 2 shows PM pictures of $DIK\text{-}Na^+$ and IND composites. As shown in Figure 2, $DIK\text{-}Na^+$ in the polymer matrix shows bright particles which indicate that $DIK\text{-}Na^+$ is dispersed in Eudragit E100. However, IND composites did not show any particles since IND was dissolved in the polymer matrix, in agreement with the XRD results.

Figure 2. Images from polarized optical micro scopic (left) Eudragit E100/15wt% IND, and (right) Eudragit E100/15wt% $DIK\text{-}Na^+$.

Thermal property changes of the API composites were observed by DSC. Table 2 lists the Tg of pure polymer and polymer containing APIs. Polymer composites containing increasing amounts of $DIK\text{-}Na^+$ showed a somewhat increased Tg. On the other hand, the Tg of the IND composites is reduced as more IND was compounded with the polymer. Reduced Tg of miscible drug -polymer systems has also been observed in other research [1].

The viscosity of the pure polymer and API composites was evaluated by RMS. As it was expected from the DSC results, the API - polymer immiscible system showed increased viscosity while the miscible system showed reduced viscosity (not shown in this article). Therefore, it can be assumed that the drug dissolved in the polymer matrix functions as a plasticizer while the dispersed drug acts as a filler.

Table 2 T_g of Polymer and API composites

Materials	**Tg (°C)**
Eudragit E100	42.8
Eudragit E100/4wt% DIK	45.9
Eudragit E100/15wt% DIK	46.3
Eudragit E100/5wt% IND	41.3
Eudragit E100/15wt% IND	36.4

Intercalating a drug into the nanoclay interlayer space, and then compounding with the polymer can be a new method that can increase the bioavailability of the drug-polymer immiscible system. Figure 3 shows the dissolution profiles of the DIK/Polymers (physically mixed and melt mixed) and DIK/Clay/Polymer systems. As shown in figure 5, the DIK/Clay/Polymer system achieved the fastest dissolution rate and the highest solubility. Among the reasons; the hydrophilic clay helps to increase the wettability of the drug on its surface or at its interlayer space. Furthermore, the API/Clay particles are much smaller than the API particles in the polymer matrix. Therefore, reduced size of the drug affected the dissolution profiles [6]. It should be noted that the HT clay is attacked and dissolved under acidic conditions. Since $DIK\text{-}Na^+$ has an extremely low solubility at the low pH, the increased solubility of all species is mainly due to the surfactant effect from the polymer.

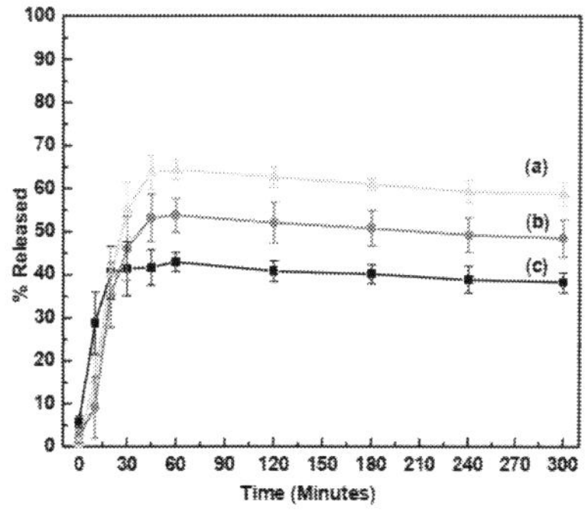

Figure 3 Dissolution results of (a) Eudragit® E100/DIK/Clay (b) Eudragit® E100/DIK-Na^+ (4 wt%), and (c) Eudragit® E100 physically mixed with DIK-Na^+.

Conclusions

In this study, the differences between drug-polymer miscible and immiscible systems in terms of morphology, rheology and bioavailability. Furthermore, a novel method for enhancing the bioavailability of drug-polymer immiscible system by incorporating an anionic nanoclay is introduced.
The miscible systems are preferable in certain applications since drug acts as a plasticizer thereby reducing T_g and viscosity. Immiscible systems showed somewhat increased T_g and viscosity. API incorporated with the clay can increase the bioavailability and dissolution rate of the API.

References

[1]. Liu, H. et al., "Effects of extrusion process parameters on the dissolution behavior of indomethacin in Eudragit® E PO solid dispersions" *Int. J. Pharm.* **383,** 161–169 (2010).
[2]. Crowley, M.M. et al., "Pharmaceutical applications of hot-melt extrusion: part I" *Drug Dev. Ind. Pharm.* **33**, 909-926 (2007).
[3]. Nollenberger, K. et al., "Pair distribution function X-ray analysis explains dissolution characteristics of felodipine melt extrusion products" *J. Pharm. Sci.*, **98**, 1476-1486 (2009).
[4]. Greenhalgh, D. J. et al., "Solubility parameters as predictors of miscibility in solid dispersions" *J. Pharm. Sci.,* **88**, 1182-1190 (1999).
[5]. Kincl, M. et al., "Study of Physicochemical Prameters Affecting the Release of Diclofenac Sodium from Lipophilic Matrix Tablets" *Acta Chim. Slov.*, **51**, 409–425 (2004).
[6]. Zhu, Y. et al., "Controlled release of a poorly water-soluble drug from hot-melt extrudates containing acrylic polymers" Drug Dev. Ind. Pharm., **32**, 569-583 (2006).
[7]. Noyes, A.A., Whitney, W.R., "The rate of solution of solid substances in their own solutions." J. Am. Chem. Soc. **19**, 930–934 (1897).

PLYWOOD STRUCTURE OF INJECTION-MOLDED POLYPROPYLENE

M. Yamaguchi[1]*, Y. Irie[1], H. Hagihara[1]

[1]*School of Materials Science, Japan Advanced Institute of Science and Technology, Japan – m_yama@jaist.ac.jp*
1-1 Asahidai, Nomi, Ishikawa 923-1292 Japan

Abstract - New method to control the molecular orientation is proposed for injection-molded products composed of isotactic polypropylene (PP) containing a β modification nucleating agent. The plaque obtained at specific mixing and processing conditions shows MD orientation in the skin layer and TD orientation in the core, which is revealed by birefringence measurements. Further, α monoclinic form is dominant in the skin layer, whereas only beta trigonal form is detected in the core layer. The intercross phenomenon of molecular orientation prohibits the crack propagation from one surface to the other at impact test, and thus leads to enhanced toughness. Since β crystalline form of PP is responsible for ductile property due to energy absorption mechanism during yield process, the plaque shows high level of impact strength even without rubbery dispersion.

Introduction

Plywood is composed of piled wooden sheets in which adjacent plies have their wood grain at right angles to each other for marked strength, resistance of cracks, shrinkage and twisting. Since most polymer products show molecular orientation to some degree, the same concept will be applicable. However, the difficulty of assembling process, leading to poor cost performance, prevents from prevailing the method.

In our previous papers, the effect of addition of *N,N'*-dicyclohexyl-2,6-naphthalenedicarboxamide, known as a high performance β-nucleating agent, on the crystallization of PP and the mechanical properties of the extruded sheets was studied.[1,2] The nucleating agent is dissolved into a molten PP and appears as needle crystals during cooling process. In the flow field, the needle crystals orient to the applied flow direction. Further, PP chains grow perpendicular to the long axis of the needle crystals, *i.e.*, perpendicular to the flow direction at crystallization. As a result, PP molecules orient perpendicular to the applied flow direction, which leads to extraordinary mechanical anisotropy. Further, it was clarified that the processing condition affects the molecular orientation and crystalline form to a great extent. For example, β form crystals are generated dominantly in the specific temperature range, i.e., 100-140 °C.[3-6] Further, the mixing temperature with PP has to be higher than 250 °C in order to dissolve the nucleating agent.[1] If the nucleating agent is undissolved (remains as spherical shape), the obtained sheet loses the anomalous molecular orientation because there is no needle crystals of the nucleating agent in a molten PP.

In this study, injection-molded products of the compound were prepared. Since there is a huge difference in flow field and cooling condition between skin and core layers, unique molecular orientation was obtained at specific processing condition. Furthermore, failure behavior was also studied in detail.

Experimental

Materials and Processing Condition

PP and *N,N'*-dicyclohexyl-2,6-naphthalene dicarboxamide, NU-100, (New Japan Chemical, NJ Star^{TR} NU-100) were melt-mixed with a small amount of thermal stabilizer by a twin-screw extruder at 280 °C. The amount of the nucleating agent was 500 ppm. The extruded strands were immediately quenched in a water bath and then cut by a strand cutter. Injection-molded plaques with a thickness of 1.0 and 2.0 mm were prepared by an injection-molding machine at the following conditions; Barrel temp. 200 °C, Mold temp. T_M either 80 or 120 °C, Injection time 20 sec., and Cooling time 30 sec. It should be mentioned that it is impossible to obtain a plaque without the nucleating agent when T_M = 120 °C because of prolonged crystallization of PP.

Measurements

Orientation birefringence was measured by an optical microscope. Film specimens with 10 μm thickness were cut out from the plaque by an ultra microtome at -100 °C parallel to the MD-TD plane.

Wide-angle X-ray diffraction (WAXD) measurements were performed at room temperature using a powder X-ray diffractmeter by refractive mode. Samples were mounted directly into the diffractmeter. The experiments were carried out using CuKα radiation operating at 40 kV and 30 mA at a scanning rate of 1°/min over 2θ (Bragg angle) range from 10° to 30°. In this study, the surface of the plaque was measured at first. Then, the surface area was removed by grinding the plaque in water, and the grinded product was mounted again to investigate the crystalline form of the inside of the plaque.

Impact tests were carried out by a Dupont impact tester (Yasuda Seiki Seisakusyo, Dupont Impact Tester) at room temperature following JIS K 5600. Various weighs were fallen onto the punch placed on the samples.

Results and Discussion

Crystallization behavior of PP containing NU-100 is observed by an optical microscope equipped with a hot stage at a cooling process from 260 °C at a rate of 5 °C/min. Although nothing is detected under crossed polars at 260 °C, the nucleating agent appears around at 220 °C as needle crystals. Further, the crystallization of PP is observed around at 135 °C, which occurs on the surface of the needle crystals of NU-100.

Crystalline form of the injection-molded plaques is studied by WAXD measurements. Figure 1(a) shows WAXD profile of the surface of the plaque with 1.0 mm thickness. As seen in the figure, the peak ascribed to β trigonal crystals, i.e., $2\theta = 16.1°$, is significantly weak, especially for the pure PP. On the contrary, α monoclinic form, that is the most stable crystalline form for PP, is detected dominantly. This is reasonable because β form is generated at specific cooling conditions such as temperature (100 – 140 °C)[3-6] and flow field.[7,8] The intense flow field would lead to α form crystalization at the current processing condition. On the contrary, β form is prevailing in the core layer (0.5 mm from the surface) for the sample containing NU-100.

Molecular orientation is evaluated by the birefringence measurement employing thin sliced films cut out from the plaque for PP containing NU-100, obtained at T_M = 80 °C. Figure 2 shows the optical micrographs under crossed polars inserting a full-wave plate that adds 530 nanometers to an optical path. As shown in the figure, the thin slice at the surface is blue, demonstrating that MD is the fast direction, i.e., direction of molecular orientation. This would be attributed to flow induced crystallization as observed for most injection-molded products.[9-12] On the other hand, the slice in the core layer (0.4 mm deep from the surface) is yellow. It indicates that TD is the direction of molecular orientation in the core, which is completely perpendicular to the applied flow field. As similar to our preceding experiments,[1,2] PP crystallizes on the needle crystals of NU-100, which orient to the flow direction in the core layer. Since the peculiar growth direction of PP crystals on the nucleating agent, as illustrated in Figure 3, c-axis of PP orients perpendicular to MD. The values of oriengtation birefringence in Figure 2 are measured by a tilting compensator and found to be 1.0 x 10^2 (left) and -1.0 x 10^2 (right).

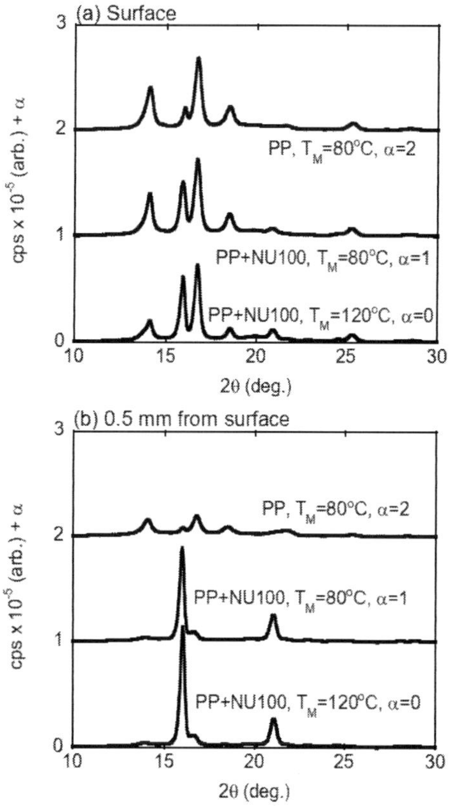

Fig.1 WAXD profiles for the injection-molded plaque with 1 mm thickness.

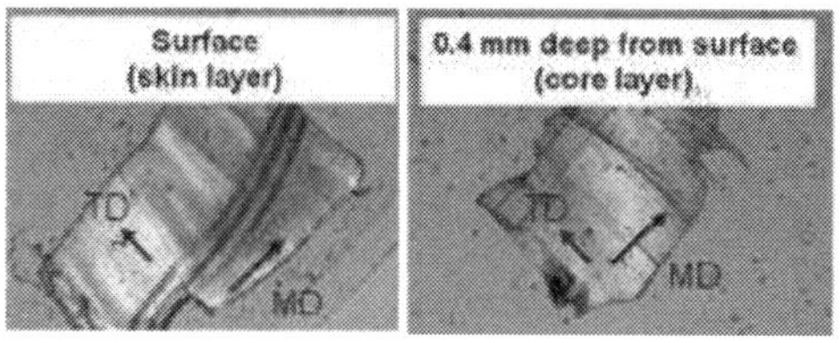

Fig.2 Optical micrographs under crossed polars inserting a full-wave plate for the slices (MD-TD plane) from the plaque with 1 mm thickness for PP containing NU-100 (T_M = 80 °C); (left) surface and (right) 0.4 mm deep from the surface.

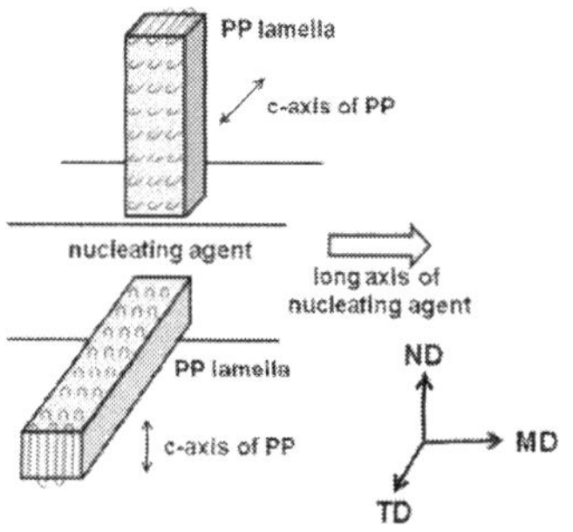

Fig.3 Schematic illustration of the core layer for injection-molded plaque of PP containing NU-100.

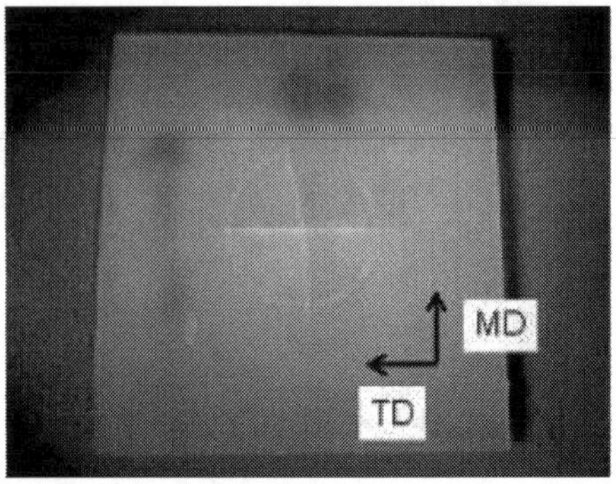

Fig.4 Sample specimen after Dupont impact test. The plaque containing NU-100 was obtained at T_M = 120 °C. The thickness of the plaque is 2 mm and the impact energy was 1.5 J.

The plywood structure of the injection-molded plaques is responsible for the enhanced toughness. Further, it is well known that β crystalline structure provides marked toughness due to the energy dissipation mechanism during yield process, where the phase transformation from β to α occurs.[13-16] Consequently, the obtained plaque having plywood structure shows considerably high level of toughness. For example, Dupont impact strength of the plaque with 1 mm thickness for PP containing NU-100, obtained at T_M = 80 °C, is approximately 1.5 J, whereas that of the pure PP is lower than 0.6 J at room temperature. Furthermore, the plywood structure provides the extraordinary failure behaviour, which is pronounced for the plaque with 2.0 mm thickness, as shown in Figure 4. The crack propagation, initiating near the surface, is prohibited inside the plaque because of the abrupt change of the molecular orientation. The complicated crack propagation will be benefits for various applications, because it changes the shape of the broken pieces

Conclusions

The effect of the nucleating agent, *N,N'*-dicyclohexyl-2,6-naphthalenedicarboxamide, on the crystallization of PP and the mechanical properties of injection-molded plaques are studied. The nucleating agent is dissolved into a molten PP and appears as needle crystals during cooling process. In the flow field, the needle crystals orient to the applied flow direction by hydrodynamic force. Further, during crystallization, PP chains grow perpendicular to the long axis of the needle crystals, *i.e.*, perpendicular to the flow direction, which leads to extraordinary mechanical anisotropy. Such type of anomalous molecular orientation is attained in the core layer of the injection-molded plaque. On the contrary, the skin layer shows MD orientation as similar to conventional products, which will be attributed to the flow induced crystallization. Finally, the plywood structure of the injection-molded PP is responsible for the marked toughness as well as complicated crack propagation.

Acknowledgement

The authors would like to express their gratitude to the members of New Japan Chemical for their valuable advices and discussion.

References

1. Yamaguchi, M., Fukui, T., Okamoto, K., Sasaki, S., Uchiyama, Y., Ueoka, C., *Polymer*, **50**, 1497-1504 (2009).
2. Uchiyama, Y., Iwasaki, S., Ueoka, C., Fukui, T., Okamoto, K., Yamaguchi, M., *J Polym Sci Polym Phys Ed*, **47**, 424-433 (2009).
3. Varga, J., *J Thermal Analysis*, **35**, 1891-1912 (1989).
4. Lotz, B., Wittmann, J. C., Lovinger A. J., *Polymer*, **37**, 4979-4992 (1996).
5. Lotz, B., *Polymer*, **39**, 4561-4567 (1998).
6. Varga, J., *J Macromol Sci*, **41**, 1121-1171 (2002).
7. Scudla, J., Eichhorn, K., Raab, M., Schmidt, P., Jehnichen, D., Häuβler, L., *Macromol Symp*, **184**, 371-387 (2002).
8. Scudla, J., Raab, M., Eichhorn, K., Strachota, A., *Polymer*, **44**, 4655-4664 (2003).
9. Sprague, B. J., *J Macromol Sci Phys*, **B8**, 157-187 (1973).
10. Tan, V., Gogos, C. G., *Polym Eng Sci*, **16**, 512-525 (1976).
11. Yamaguchi, M., Suzuki, K., Miyata, H., *J Polym Sci Polym Phys Ed*, **37**, 701-713 (1999).
12. Tenma, M., Yamaguchi, M., *Polym Eng Sci*, **47**, 1441-1446 (2007).
13. Karger-Kocsis, J., Varga, J., *J Appl Polym Sci*, **62**, 291-300 (1996).
14. Karger-Kocsis, J., *Polym Eng Sci*, **36**, 203-210 (1996).
15. Nezbedova, E., Pospisil, V., Bohaty, P., Vlach, B., *Macromol Symp*, **170**, 349-357 (2001).
16. Bohaty, P.; Vlach, B.; Seidler, S.; Koch, T.; Nezbedova, E., *J Macromol Sci B*, **41**, 657-669 (2002).

EFFECT OF PARTICLE PRESENCE AND FLOW AT PROCESSING SPEEDS ON POLYPROPYLENE CRYSTALLIZATION

Michelle D'Haese[1], Florentin Langouche[2] and Peter Van Puyvelde[1]*

[1] *Department of Chemical Engineering, Leuven Materials Research Centre MRC, Katholieke Universiteit Leuven, W. De Croylaan 46, 3001 Leuven, Belgium – peter.vanpuyvelde@cit.kuleuven.be*
[2] *Solvay Central Laboratory, Neder Over Heembeek, Rue de Ransbeek 310, 1120 Brussels, Belgium*

Abstract – The crystallization of mixtures of polypropylene and zinc oxide after shear from low rates to processing speeds was studied by means of birefringence measurements using a specially designed sandwich-type flow-cell. At the lower end of the investigated shear rate range, there are pronounced differences between the crystallization behavior of the mixtures and the pure polymer. These differences correspond to differences in nucleating efficiency of the particles, as determined by the Lotz efficiency scale. As the shear rate increases, though, the crystallization behavior of the mixtures and the pure polymer becomes increasingly similar. The transition to the highly oriented shish-kebab structure still occurs at a somewhat lower shear rate as the nucleating efficiency of the particles in the mixture is higher, but towards the higher end of the shear rate range, the differences in crystallization behavior of the mixtures and pure polymer have disappeared. This suggests that particle presence dominates the crystallization behavior at low shear conditions, but as shear intensity increases, particle presence gradually becomes less important, until the crystallization behavior is dominated entirely by flow.

Introduction

Both application of flow and addition of particles influence the solidification of semi-crystalline polymers, and hence the morphology and properties of a plastic product (Gahleitner et al., 2002; Housmans et al., 2009a). The effect of flow history on polymer crystallization has been studied extensively in the past decades (Janeschitz-Kriegl & Eder, 2007; Kornfield et al., 2002; Kumaraswamy, 2005). Flow accelerates crystallization by enhancing nucleation and sufficiently intense flow can cause a morphological transition from isotropic spherulitic to highly oriented 'shish-kebab' structures (Kumaraswamy, 2005). Similarly, the influence of particle presence on the (quiescent) crystallization of polymers has been the subject of many studies (Pukanszky, 2005; Pukanszky & Moczo, 2004; Tang et al., 2004; Vesely & Ronca, 2001). Depending on their characteristics, certain particles can accelerate the crystallization process by enhancing heterogeneous nucleation in semi-crystalline polymers (Pukanszky & Moczo, 2004; Tang et al., 2004; Vesely & Ronca, 2001).

The combined influence of flow and particles, however, has not been studied systematically yet. Some single-point studies involving rather intense flow have been carried out (Byelov et al., 2008; Kelarakis et al., 2006; Nowacki et al., 2004), while other studies have focused on obtaining a larger range of shear rates, under rheometrically well-controlled flow conditions but limited to low or moderate shear intensities (Lagasse & Maxwell, 1976; Naudy et al., 2007; Rozanski et al., 2009; Somwangthanaroj et al., 2003). Neither type of study, though, is able to provide a full overview of the contributions of flow and particle presence to the polymer crystallization.

In order to access a shear rate range up to processing speeds, some research groups have used setups involving extrusion, fiber drawing or injection molding (Jerschow & Janeschitz-Kriegl, 1997; Lu & Spruiell, 1993; Spruiell et al., 1996). These setups lead to complex flow histories, and it is very difficult to distinguish the exact influences of different parameters, both flow and particle related.

This study aims to perform a systematic study of the influence of particles on flow-induced polymer crystallization, using a specially designed shear cell, able to apply simple shear flow from low rates up to processing speeds (Baert et al., 2006; Langouche, 2006).

Experimental

Materials. An isotactic polypropylene, obtained from Borealis GMBH, and zinc oxide particles with average diameters of 35 nm, 200 nm, 500 nm and 1 μm, obtained from Umicore Group Zinc Chemicals, have been used in this study. Mixtures of the polypropylene with 1 wt% (approx. 0.16 vol%) of each zinc oxide were made by means of a twin-screw mini-extruder (DSM-Research, The Netherlands).

Methods. To determine the nucleating efficiency of the particles, DSC experiments have been performed with a DSC Q2000 (Universal V4.3A TA Instruments). The flow-induced crystallization has been studied by means of birefringence measurements. These experiments were performed using an optical train consisting of a laser modulated by means of a rotating half wave plate (OAM module, TA Instruments) and at the other side of the shear cell, a circular analyzer and a photo detector. The optical train was combined with a sandwich type sliding-plate shear cell, developed at the Solvay Central Laboratory (Langouche, 2006). The setup allows achieving deformations up to a few hundred shear units, which is in the order of the shear rates realized in the outer layers of injection molded products. The sliding plate concept ensures a constant shear rate throughout the sample. More detailed information about the shear cell and the signal processing can be found in reference (Langouche, 2006).

The temperature protocol used consists of an annealing step at 220°C. Then, the sample is cooled to the crystallization temperature, 145°C, at a rate of 5°C/min. The shearing protocol used consists of a shear step applied at the moment the crystallization temperature is reached. All shear steps are applied at a strain of approximately 100 units and the shear rates vary between 20 and 420 s^{-1}.

Results and Discussion

Before considering the combined influence of particle presence and flow application on crystallization, both aspects have to be assessed separately.

The nucleation by the different zinc oxides was characterized by means of the Lotz efficiency scale for polymer nucleating agents (Fillon et al., 1994), which references a certain polymer/particle mixture to both a lower and an upper limit of nucleation. For the mixtures used in this study, the nucleating efficiencies of the midrange particles (35 nm, 200 nm and 500 nm) are quite similar: they vary between 4-9%. For the largest particles (1 µm) though the nucleating efficiency is 22%. This is contrary to what would be expected, namely that the smaller particles, having more contact surface with the polymer, would nucleate more. This deviation is probably caused by differences in the particle properties (shape, surface irregularities, slight differences in chemical composition etc.).

In order to summarize the influence of shear rate (at constant strain) on the crystallization of the pure polymer, a characteristic value for its crystallization kinetics is defined. The characteristic time of crystallization at a certain shear rate is obtained from the birefringence measurements during crystallization after a shear step at that shear rate. Here, the time to the minimum of the intensity signal, one of the signals measured during birefringence experiments, is chosen. This characteristic time corresponds to the occurrence of impingement of the growing crystallites (Ding & Spruiell, 1996).

The resulting data for the pure polymer are presented in Figure 1. They show an acceleration of the crystallization kinetics at the lower end of the shear rate range, and another acceleration at its higher end. The data further suggest that a transition region in the form of a plateau is present between the two accelerations. Previous studies have reported the presence of the first acceleration and a plateau (Kumaraswamy et al., 1999; Somani et al., 2000; Vleeshouwers & Meijer, 1996), followed by another acceleration (Baert et al., 2006; Housmans et al., 2009b; Langouche, 2006), although in other studies the transitional plateau was absent (Vega et al., 2009). In all studies though, the transition region was shown to indicate a change in morphology from spherulite-dominated to shish-kebab- dominated (in this study, through the appearance of an upturn in the birefringence during flow (Kornfield et al., 2002; Kumaraswamy et al., 2000)). The precise mechanism behind this transition remains to be clarified.

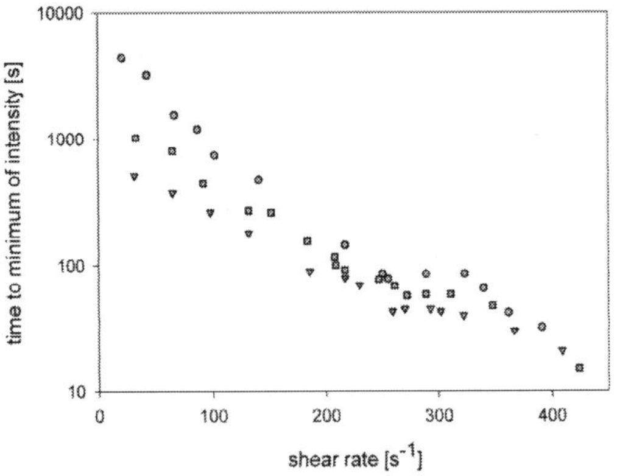

Figure 1 – Dependence of the characteristic crystallization time (time to minimum of intensity) on shear rate at a strain of approx. 100 units for pure polypropylene (circle) and for polypropylene with 1wt% 200 nm zinc oxide (square) and with 1wt% 1 µm zinc oxide (downward triangle)

The combined influence of particle presence and flow application on the polypropylene/zinc oxide mixtures can again be expressed in terms of characteristic time of crystallization versus shear rate, which is shown for two of the mixtures, together with the data for the pure polymer, in Figure 1. The data of the mixtures show the same trends, two accelerations and the suggestion of a transitional plateau, as the data of the pure polymer. Naturally, in view of the heterogeneous nucleation by the particles, the crystallization of the mixtures is faster than for the pure polymer- at least, this is the case in the low shear rate region, where the influence of the shear flow is still limited. These low shear data show clear link between the crystallization kinetics and the nucleating efficiency. However, it can be seen from Figure 1 that as the shear rate is increased, the differences between the mixtures and the pure polymer diminishes. Moreover, the differences in steepness of the slope of the first acceleration show that the effect of shear flow is less drastic as the quiescent nucleation density of the material is higher. This behavior corresponds with previous observations (Jerschow & Janeschitz-Kriegl, 1997; Lagasse & Maxwell, 1976; Lu & Spruiell, 1993; Naudy et al., 2007; Somwangthanaroj et al., 2003; Spruiell et al., 1996).

The differences between the mixtures and the pure polymer are still significant at the level of the transition region, and from the birefringence measurements during flow it can be concluded that the transition to shish-kebab structures occurs at a somewhat lower shear rate as the nucleating efficiency of the particles in the mixture is higher. But at the highest shear rates of the investigated range, the kinetics of all materials are the same. Thus, at the highest shear rates, the crystallization behavior becomes independent of the presence of particles, regardless of their nucleating efficiency, and is entirely dominated by the influence of flow.

Conclusions

This paper presents a systematic study of the influence of simple shear flow, from low to processing rates, on the crystallization behavior of polypropylene with micro- to nanosized particles. First, the separate influences of particle presence and shear flow were determined. The Lotz efficiency scale is shown to be a good instrument to characterize the influence of particle presence. As to the influence of flow, a characteristic value derived from birefringence measurements, time to minimum of intensity, is used to visualize the dependence of the kinetics on shear rate (at constant strain). Two accelerations of the crystallization kinetics are visible: one at low to moderate shear rates, and another at the highest shear rates investigated. In the transition region between these accelerations, the data suggest the presence of a plateau. Finally, the combined influence of particle presence and flow application is studied. In the low shear rate region the differences between the pure polymer and the different mixtures with zinc oxide are pronounced and in accordance with nucleating efficiency, rather than particle diameter. The difference though diminishes steadily with increasing shear rate. In the high shear rate region, the crystallization kinetics is entirely dominated by the influence of shear flow, independent of the particles and their nucleating efficiency.

Acknowledgements

The authors thank Borealis GMBH and Umicore Group, for providing the materials used in this study, and Solvay, for allowing the use of their facilities at Solvay Central Laboratory.

References

Baert, J., Van Puyvelde, P. , Langouche, F., "Flow-induced crystallization of PB-1: From the low shear rate region up to processing rates", Macromolecules, **39**, 9215-9222 (2006)

Byelov, D., Panine, P., Remerie, K., Biemond, E., Alfonso, G.C. , de Jeu, W.H., "Crystallization under shear in isotactic polypropylene containing nucleators", Polymer, **49**, 3076-3083 (2008)

Ding, Z.M. , Spruiell, J.E., "An experimental method for studying nonisothermal crystallization of polymers at very high cooling rates", J. Polym. Sci., Part B: Polym. Phys., **34**, 2783-2804 (1996)

Fillon, B., Thierry, A., Lotz, B. , Wittmann, J.C., "Efficiency Scale for Polymer Nucleating-Agents", J. Therm. Anal. Calorimetry, **42**, 721-731 (1994)

Gahleitner, M., Wolfschwenger, J., Fiebig, J. , Neissl, W., "Influence of nucleants on the formation of shear-induced structures in polypropylene", Macromol. Symp., **185**, 77-87 (2002)

Housmans, J.W., Gahleitner, M., Peters, G.W.M. , Meijer, H.E.H., "Structure-property relations in molded, nucleated isotactic polypropylene", Polymer, **50**, 2304-2319 (2009a)

Housmans, J.W., Steenbakkers, R.J.A., Roozemond, P.C., Peters, G.W.M. , Meijer, H.E.H., "Saturation of Pointlike Nuclei and the Transition to Oriented Structures in Flow-Induced Crystallization of Isotactic Polypropylene", Macromolecules, **42**, 5728-5740 (2009b)

Janeschitz-Kriegl, H. , Eder, G., "Shear induced crystallization, a relaxation phenomenon in polymer melts: A re-collection", J. Macromol. Sci., Part B: Phys., **46**, 591-601 (2007)

Jerschow, P. , Janeschitz-Kriegl, H., "The role of long molecules and nucleating agents in shear induced crystallization of isotactic polypropylenes", Int. Polym. Process., **12**, 72-77 (1997)

Kelarakis, A., Yoon, K., Sics, I., Somani, R.H., Chen, X.M., Hsiao, B.S. , Chu, B., "Shear-induced orientation and structure development in isotactic polypropylene melt containing modified carbon nanofibers", J. Macromol. Sci., Part B: Phys., **45**, 247-261 (2006)

Kornfield, J.A., Kumaraswamy, G. , Issaian, A.M., "Recent advances in understanding flow effects on polymer crystallization", Ind. Eng. Chem. Res., **41**, 6383-6392 (2002)
Kumaraswamy, G., "Crystallization of polymers from stressed melts", J. Macromol. Sci., Polym. Rev., **C45**, 375-397 (2005)
Kumaraswamy, G., Issaian, A.M. , Kornfield, J.A., "Shear-enhanced crystallization in isotactic polypropylene. 1. Correspondence between in situ rheo-optics and ex situ structure determination", Macromolecules, **32**, 7537-7547 (1999)
Kumaraswamy, G., Verma, R.K., Issaian, A.M., Wang, P., Kornfield, J.A., Yeh, F., Hsiao, B.S. , Olley, R.H., "Shear-enhanced crystallization in isotactic polypropylene Part 2. Analysis of the formation of the oriented "skin"", Polymer, **41**, 8931-8940 (2000)
Lagasse, R.R. , Maxwell, B., "Experimental-Study of Kinetics of Polymer Crystallization During Shear-Flow", Polym. Eng. Sci., **16**, 189-199 (1976)
Langouche, F., "Orientation development during shear flow-induced crystallization of i-PP", Macromolecules, **39**, 2568-2573 (2006)
Lu, F.M. , Spruiell, J.E., "The Role of Crystallization Kinetics in the Development of the Structure and Properties of Polypropylene Filaments", J. Appl. Polym. Sci., **49**, 623-631 (1993)
Naudy, S., David, L., Rochas, C. , Fulchiron, R., "Shear induced crystallization of poly(m-xylylene adipamide) with and without nucleating additives", Polymer, **48**, 3273-3285 (2007)
Nowacki, R., Monasse, B., Piorkowska, E., Galeski, A. , Haudin, J.M., "Spherulite nucleation in isotactic polypropylene based nanocomposites with montmorillonite under shear", Polymer, **45**, 4877-4892 (2004)
Pukanszky, B., "Interfaces and interphases in multicomponent materials: past, present, future", Eur. Polym. J., **41**, 645-662 (2005)
Pukanszky, B. , Moczo, J., "Morphology and properties of particulate filled polymers", Macromol. Symp., **214**, 115-134 (2004)
Rozanski, A., Monasse, B., Szkudlarek, E., Pawlak, A., Piorkowska, E., Galeski, A. , Haudin, J.M., "Shear-induced crystallization of isotactic polypropylene based nanocomposites with montmorillonite", Eur. Polym. J., **45**, 88-101 (2009)
Somani, R.H., Hsiao, B.S., Nogales, A., Srinivas, S., Tsou, A.H., Sics, I., Balta-Calleja, F.J. , Ezquerra, T.A., "Structure development during shear flow-induced crystallization of i-PP: In-situ small-angle X-ray scattering study", Macromolecules, **33**, 9385-9394 (2000)
Somwangthanaroj, A., Lee, E.C. , Solomon, M.J., "Early stage quiescent and flow-induced crystallization of intercalated polypropylene nanocomposites by time-resolved light scattering", Macromolecules, **36**, 2333-2342 (2003)
Spruiell, J.E., Lu, F.M., Ding, Z. , Richeson, G., "The influence of isotacticity, ethylene comonomer content, and nucleating agent additions on the structure and properties of melt-spun isotactic polypropylene filaments", J. Appl. Polym. Sci., **62**, 1965-1975 (1996)
Tang, J.G., Wang, Y., Liu, H.Y. , Belfiore, L.A., "Effects of organic nucleating agents and zinc oxide nanoparticles on isotactic polypropylene crystallization", Polymer, **45**, 2081-2091 (2004)
Vega, J.F., Hristova, D.G. , Peters, G.W.M., "Flow induced crystallization regimes and rheology of isotactic polypropylene: Effects of molecular architecture", J. Therm. Anal. Calorimetry, accepted (2009)
Vesely, D. , Ronca, G., "Epitaxial nucleation of crystallization at polymer-filler interfaces", J. Microsc. (Oxford, U.K.), **201**, 137-143 (2001)
Vleeshouwers, S. , Meijer, H.E.H., "A rheological study of shear induced crystallization", Rheol. Acta, **35**, 391-399 (1996)

FABRICATION OF ULTRAHIGH-DENSITY ARRAY OF NANOPATTERNS BY AFM LITHOGRAPHY AT ROOM TEMPERATURE

Ara Jo, and Jin Kon Kim*

National Creative Research Initiative Center for Block Copolymer Self-Assembly,
Department of Chemical Engineering and Polymer Research Institute,
Pohang University of Science and Technology, Pohang, Kyungbuk 790-784, Korea
**jkkim@postech.ac.kr*

Data-storage media based on scanning probes offer very high data densities, but existing approaches require the probe to be heated, which increases the energy expenditure and fabrication complexity. Here we demonstrate the writing (data-storage density of 1 Tb/in^2), reading and erasure of nanoscopic indentations without heating either the scanning probe tip or the substrate (Jo *et al.* 2009). Because of baroplastic properties of polystyrene-*block*-poly(n-pentyl methacrylate) copolymer (PS-*b*-PnPMA) which enable the nanophase transition at relatively lower pressure and temperature, nanopatterns were easily fabricated by the indentation at room temperature. The use of a pressure-based phase-change memory at room temperature may expedite the development of next-generation ultrahigh density data-storage media.

Introduction

SPM-based lithography, such as scanning tunneling microscope and atomic force microscope (AFM) lithography, has proven an effective means to fabricate an ultrahigh density array of nanopatterns for next-generation data storage media. For AFM-based lithography, nanopatterns are generated by the mechanical contact between a nano-sized tip and a polymeric film surface. Each of the nanopatterns is easily transferred to "1" or "0" for the binary digits. IBM fabricated the nanopatterns on poly(methyl methacrylate) (PMMA) or poly(styrene-*r*-benzocyclobutene) based on a thermomechanical recording system called as the Millipede (Vettiger *et al.* 2000, 2002; Pantazi *et al.* 2008). The fabrication of the nanopatterns on PMMA was achieved by using heated AFM tip with ~ 350 °C,

Here, we introduce a novel concept for the fabrication of nanopatterns on a polymer film by AFM without using a heated tip. The nanopatterns are generated by the AFM tip at room temperature on the film and these nanopatterns retain their original shapes at room temperature. This process is only possible by using a specific block copolymer film exhibiting baroplastic property that enables one to process at a relatively lower pressure and temperature by microphase transition (Ruzette *et al.* 2003; Gonzalez-Leon *et al.* 2003; Lee *et al.* 2006).

Experimental

Fabrication of nanopatterns on a baroplastic block copolymer thin film.

A symmetric PS-*b*-PnPMA, having a styrene volume fraction of 0.5 and a number-average molecular weight (M_n) of 54,000, was prepared by anion polymerization. The lamellar spacing (L_o) of PS-*b*-PnPMA was 22 nm measured by synchrotron small angle X-ray scattering.

Nanopatterns are generated by indentation of the AFM tip to the polymer film at room temperature. We used three kinds of AFM tips: AR5-NCHR, AR5T–NCHR, and SSS–NCHR tip (purchased from Nanosensors™). Nanopatterns were prepared by AFM (SPM: XE-100, PSIA) with a silicon cantilever having a force constant of 42 N/m.

Characterization of Nanopatterns.

Height images and height profiles of the fabricated nanopatterns are obtained by using an atomic force microscope (AFM: XE100, PSIA). The cross-sectional image of PS-*b*-PnPMA film was observed by high resolution transmission electron microscope (HR-TEM: Jeol-1200 EXII). For this purpose, the film was stained with ruthenium tetroxide (RuO_4) for 45 minutes; thus the dark region appearing in TEM images represents the PS phase. The patterned region was carefully cut by using the focused ion beam (FIB: SMI 3050, Seiko) and transferred on a TEM grid.

Piezoelectrical read system.

We employed the lead zirconate titanate (PZT) sensor to read the nanopatterns (Nam *et al.* 2003, 2005). As the PZT cantilever scans across the patterned sample, the cantilever deflects according to the topography of the sample, causing stress on the PZT film. Then the charge is self-generated on the surface of the capacitor without the applied voltage, as the electrical domain of piezoelectric material is changed by the stress, causing a variation in the surface charge. This generated charge is amplified by a charge amplifier and the amplified signals are compared to initial setting values by using the comparator. By this comparison, the data are determined to be either "0" or "1".

Results and Discussion

The schematic of the fabrication of the nanopatterns with an ultrahigh density by using AFM tip is shown in Fig. 1a-d. Fig. 1e –h show AFM height images (each inset is the magnified image) and height profiles of nanopatterns prepared by the indentation at room temperature with three different AFM tips. Pattern density in Fig. 1e is 170 Gb/in^2. Once a high aspect ratio tip with a tilted angle was used, the pattern density was increased to 260 Gb/in^2. The increased pattern density is attributed to the fact that this AFM tip can contact at a right angle (90°) to the substrate. If we carefully controlled the indentation force for a tilted angle tip, a pattern density up to 465 Gb/in^2 was generated. Furthermore, when another sharpened silicon tip was used, the pattern density was increased to 1.03 Tb/in^2, as shown in Fig. 1h. The above results indicate that a pattern density higher than the present value would be achieved once a more sharpened AFM tip with smaller diameter and higher aspect ratio, for instance, carbon nanotube AFM tip, is employed.

The nanopatterns by the indentation of AFM tips were formed due to the microphase transition of the lamellar microdomains to disordered state. To confirm whether the disordered state indeed occurred during the AFM indentation, we investigated the cross-sectional TEM image of the PS-*b*-PnPMA film. The formation of disordered regions is because the pressure at these regions is higher than the critical pressure to induce the microphase transition into disordered state. Higher pressure was achieved due to the smaller size of the AFM tip even though the indentation force is as small as of the order of nN. We consider the reason why the disordered region cannot return again to the lamellar microdomains after the AFM tip is detached from the PS-*b*-PnPMA film. This is simply because the T_g of the disordered state for PS-*b*-PnPMA was ~70 °C, higher than room temperature. Namely, the disordered structure achieved by the indentation of the AFM tip becomes quickly the froze-in structure at room temperature when the force (or the pressure) was removed suddenly from the film.

Once nanopatterns are formed, these should be read electrically. For this purpose, we employed a piezoelectric sensing method where piezoelectric charges are generated from a lead zirconate titanate film sensor. The piezoelectric charge outputs exactly respond to the depth profiles of the fabricated nanopatterns; thus the nanopatterns are easily converted to either "0" or "1" as binary digits.

Finally, the erase and rewrite capability of the nanopatterns is very essential for the storage medium. For this purpose, we repeated write-read-erase-rewrite operations on the same area of the film. Once this patterned film was simply placed for 2 s on a heating plate maintained at 120 °C, the nanopatterns were completely erased and the film became flat. We also found that during the bulk erasing process, the disordered state was completely transformed into the well-ordered lamellar microdomains, essentially the same morphology of the initial film. Repeated ten times writing and bulk erasing processes indicate that the nanopattern arrays are duplicated without any damage to the film surface.

We also performed the local erasing of the individual bit (or line) by using the method in ref. (Altebaeume *et al.* 2008). However, we did not use a heated AFM tip to erase the pattern. We show the erasing of one bit and one line at room temperature.

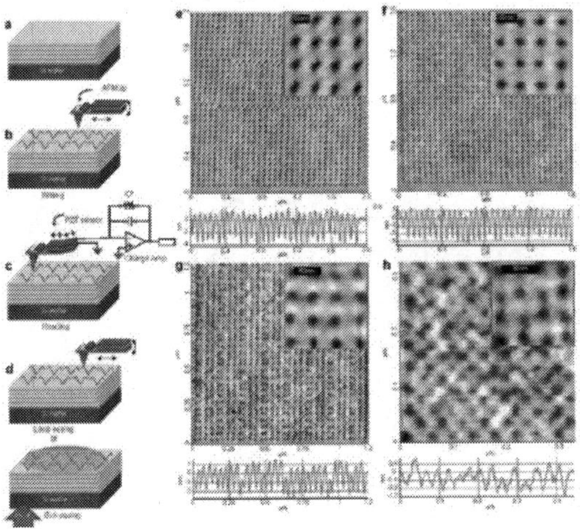

Figure 1 (a-d) Schematic of the fabrication of a nanopatterned surface with an ultrahigh density array of nanoscopic indentations. (e-h) AFM height images of nanopatterns prepared by the indentation of three kinds of the AFM tips.

Conclusions

In this study, we fabricated a nanopatterned surface with an ultrahigh density array of nanoscopic indentations on PS-*b*-PnPMA thin film by AFM tip at room temperature. We clearly demonstrated, via cross-sectional transmission electron microscopy (TEM) image that the phase transition from microdomain structures to disordered state indeed occurred during the indentation of AFM tip on the film. We also transformed the fabricated nanopatterns into electrical signals by using a piezoelectric sensor. Finally, local and bulk erasing of the nanopatterns was successfully carried out. The generated nanopatterns with an ultrahigh density could be used as next-generation storage medium.

Acknowledgements

This work was supported by the National Creative Research Initiative Program supported by the National Research Foundation of Korea (NRF) and the second stage of the BK 21 program of Korea.

References

1. Altebaeumer, T. *et al.* "Nanoscale shape-memory function in highly cross-linked polymers", *Nano Lett.* **8**(12), 4398-4403 (2008).
2. Gonzalez-Leon, J. A., Acar, M. H., Ryu, S. W., Ruzette, A. V. G. & Mayes, A. M. "Low-temperature processing of 'baroplastics' by pressure-induced flow", *Nature* **426**, 424-428 (2003).
3. Jo, A., Joo, W., Jin, W-H., Nam, H., and Kim, J. K. "Ultrahigh-density phase-change data storage without the use of heating", *Nature Nanotechnology,* **4**, 727-731 (2009).
4. Lee, D. H., Kim, H. J. & Kim, J. K. "Closed-loop phase behavior and baroplasticity of deuterated polystyrene-*block*-poly(n-pentyl methacrylate) copolymer investigated by SANS and FTIR spectroscopy", *Macromol. Symp.* **240**, 123-129 (2006).
5. Nam, H. J. *et al.* "Integrated nitride cantilever array with Si heaters and piezoelectric detectors for nano-data-storage application", *IEEE MEMS* **2005**, 247-250 (2005).
6. Nam, H. J. *et al.* "Silicon nitride cantilever array integrated with silicon heaters and piezoelectric detectors for probe-based data storage", *Sens. Actuators A* **134**, 329-333 (2007).
7. Pantazi, A. *et al.* "Probe-based ultrahigh-density storage technology", *IBM J. RES. & DEV.* **52**, 493-511 (2008).
8. Ruzette, A. V. G., Pollard, A. M., Russell, T. P. & Hammouda, B. "Pressure effects on the phase behavior of styrene/*n*-alkyl methacrylate block copolymers", *Macromolecules* **36**, 3351-3356 (2003).
9. Vettiger, P. *et al.* "The "Millipede"-more than one thousand tips for future AFM data storage", *IBM J. Res. & Dev.* **44**, 323-340 (2000).
10. Vettiger, P. *et al.* "The "Millipede"-nanotechnology entering data storage", *IEEE Trans. Nanotechnol.* **1**, 39-55 (2002).

Structure development during manufacture of BOPP

Basargekar Rajeev S, Preschilla Nisha

Structure development during manufacture of BOPP Rajeev Basargekar, Nisha Preschilla, Swayajith Sahdevan, Raghvendra Singh, Rajendra Patankar * Reliance Technology Group-Polymers, Reliance Industries Limited, V.N. Purav Marg, Chembur, Mumbai 400 071 Email: rajeev_basargekar@ril.com, Ph: (O) +91+22 67677337, (M) +91 +9322596646 Fax: +91+22 67677080 Abstract One of the major applications of polypropylene is Biaxially Oriented Polypropylene (BOPP) films. The grades of PP suitable for manufacture of BOPP are distinctly different from those used for moulding and tape extrusion. The architecture of such grades is dictated not only by the properties desired in the BOPP film but by the processability on the BOPP manufacturing lines also. The ever expanding capacities of BOPP manufacture in terms of the line speeds and film width make the requirements more critical. BOPP manufacture has three stages of casting on a chilled roll, uniaxial orientation in machine direction (MDO) and orientation in traverse direction (TDO) perpendicular to machine direction. The study of structure development during these stages is useful to give important insights on the suitability of grade architecture. In the present work the samples from three stages from a commercial line were studied for morphology and orientation by various techniques like Optical Polarizing microscopy, FTIR, DSC, X-Ray diffraction etc. The observation indicated conversion of spherulitic to oriented lamellar morphology and a remarkable rise in crystallinity as well as orientation during the TDO stage of BOPP manufacture. The Crystalline morphology developed during the first stage of casting on chilled roll in terms of absolute crystallinity as well as crystallite size distribution and also the possibility of rise in oriented crystalline morphology were observed to be favourable features for PP grades suitable for uninterrupted stretching during the BOPP process. Keywords: Biaxially Oriented Polypropylene, Structure Development in Processing, Morphology, Orientation, Crystallite Size Distribution.

Phase transformation in extruded poly(1-butene)

*Kristyna Montagova**

Tomas Bata University in Zlín - The Czech Republic - The Czech Republic

Poly(1-butene) (PB-1) is generally known as a material in which the crystalline phase structure undergoes transition within a time. Consequently, mechanical properties of processed PB-1 parts are dramatically changed what can complicate PB-1 applicability. In this work, transformation of phase structure and mechanical properties in extruded PB-1 was studied. Two commercially available grades of PB-1 homopolymer (extrusion and injection molding type) and a PB-1 copolymer were used. Extruded tapes were exposed during the transformation to various constant temperatures (-22, 5, 22, 40, and 60 °C). The evolution of structure and mechanical properties was followed by wide-angle X-ray scattering and tensile testing. The highest rate of transformation was found by specimens annealed at 5 and 22° C. PB-1 with lower melt flow index transforms faster than the PB-1 type intended for injection molding. However, the fastest transformation rate was found in PB-1 copolymer. Tensile modulus unambiguously follows the transformation extent; the stiffness rises monotonically with increasing content of form I.

FORMATION OF SHEAR-INDUCED PRECURSORS IN IPP CONTAINING NUCLEATING AGENTS

Andrew Phillips[1]*, Peng-wei Zhu[1] and Graham Edward[1]

[1]*Department of Materials Engineering, CRC for Polymers, Monash University, VIC 3800, Australia. Andrew.Phillips@eng.monash.edu.au, (Ph) +613 9905 4928, (Fax) +613 9905 4940*

The addition of nucleating agents complicate the prediction of morphology during processing of isotactic polypropylene (iPP). In this work, the formation of shear-induced precursors to crystallisation in a continuously sheared melt of iPP held at 175 °C containing two nucleants with different aspect ratios was examined by in-situ rheo-SAXS/WAXS and rheo-FTIR. Sharp equatorial streaks were observed in the SAXS patterns and a coil to helix transition was observed in FTIR spectra in all samples above a critical shear rate. However, no associated crystalline scattering was observed. These results are consistent with the formation of non-crystalline shear-induced precursors aligned along the flow direction. The critical shear rate required to induce the formation of the precursors was decreased by the presence of both nucleants compared to pure iPP. The high aspect ratio nucleant resulted in a greater number of crystal precursors, which then had a greater alignment along the flow direction when compared to the low aspect ratio nucleant. These results suggest that nucleant shape can affect the development of morphology in sheared iPP.

Introduction

Shear flows, such as those found in injection moulding, are well known to create molecular extension and orientation in polymer melts. This leads to significant changes to crystallisation kinetics, and the development of morphology. For example, shear flow can change a spherulite structure which forms under quiescent conditions, to an oriented one containing the molecular assemblies termed 'shish-kebabs'. The shish-kebab structure is a polymer crystal assembly, which consists of chain folded lamellae (kebabs) held together by a core of highly extended chains (shish). Preceding the formation of shish must be some form of shear-induced oriented precursor structure which is associated with the transition from coiled to stretched chains that occurs above a critical shear rate (De-Gennes, 1974). Much of the evidence for the precursor structure comes from scattering experiments where sharp equatorial streaks are observed at small scattering angles (Somani et al., 2005, Kumaraswamy, 2005; Kimata et al., 2007) as well as birefringence measurments (Seki et al., 2002; Fernandez-Ballester et al., 2008). More recently, FTIR experiments show the formation of precursors is also associated with a molecular coil to helix transition (Geng et al., 2009; Xinyuan et al., 2000). However, as yet only few authors have presented combined FTIR and scattering measurements.

The formation of precursors is complicated by the addition of foreign particle such as nucleating agents and colourants. Previous studies by our group on injection moulded isotactic polypropylene (iPP) containing sodium benzoate (Zhu et al., 2005; Zhu et al., 2006) and two colourants (Zhu et al., 2008) showed their presence decreased the shear intensity required to form the shish-kebab structure. More recently experiments by Byelov et al. (2008) showed that the shape of the nucleant may be important, with anisotropic nucleants leading to an increase in the number of nucleating sites for crystal formation compared to isotopic nucleant agents. These results show foreign particles influence the formation of precursors. However, the mechanism by which this occurs still remains a mystery.

This paper reports the results of recent small and wide angle x-ray scattering (rheo-SAXS/WAXS) and rheo-FTIR experiments on iPP containing two nucleants with widely different aspect ratios. The aim was to determine the influence of nucleant aspect ratio on the formation of the shear-induced precursor structure under continuous shear conditions. The shear temperature was chosen (175 °C) to be above the lamellae melting point to avoid artefacts from crystallisation, and long shear times (30 s) were used to allow investigation of precursor formation during shear flow. It will be shown that the high aspect ratio nucleant was more effective at forming oriented shear-induced precursors compared to the low aspect ratio nucleant.

Experimental

The polymer used was an iPP homopolymer manufactured by the Borealis group (HD601CF). This grade was chosen as it has been extensively investigated by our group and contains almost no additives. In particular it contains no calcium stearate which has recently been shown to influence the formation of the crystal precursor structure (Byelov et al., 2007). The average weight molecular weight (Mw) and the average number molecular weight (Mn) were

367 000 and 74 000, respectively. Various loadings of two nucleating agents, sodium benzoate (SB) and bicyclo[2.2.1]heptane-2,3-dicarboxylic acid (HPN), were compounded into the iPP matrix in a Haake Rheocord 90 batch mixer. The nucleant morphology is shown in Figure 1. SB is highly elongated. Standard techniques were employed to determine the average length (2.7 μm), width/thickness (0.3 μm), and aspect ratio (10) of SB. HPN was found to have a cubic morphology which was roughly isotropic, with dimension of 0.5 μm and aspect ratio near 1.

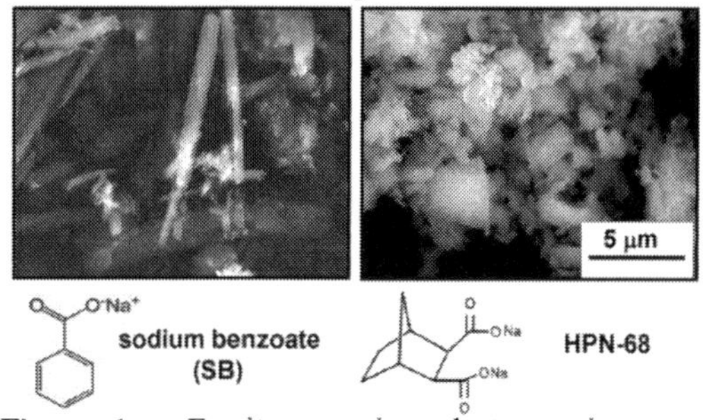

Figure 1 – Ex-situ scanning electron microscope images of the two nucleants.

A Linkam CSS450 shear cell was used to control the thermal–shear profile, Figure 2. The CSS450 is a parallel plate shearing device in which a small aperture in the base and lid allows observation of the polymer in-situ. Samples were heated to 220 °C and compressed slowly to the working distance (x-ray = 500 μm; FTIR = 250 μm). The compressed disks were held at the melting temperature for 5 minutes to erase the samples previous thermal history. Samples were then cooled to the shear temperature (175 °C) and held there for 5 minutes to allow the sample to equilibrate. A series of controlled continuous shears of increasing shear rate between 0 – 80 s^{-1} was then applied to the melt. The step shears were applied for 30 s with a 60 s relaxation time between each successive shear.

Simultaneous 2D rheo-SAXS/WAXS experiments were performed on the SAXS/WAXS beamline at the Australian Synchrotron. The set up allowed for in-situ investigations with a considerable amount of the scattering intensity around the azimuth ($\phi \sim 80°$) in the WAXS region (Polaris CCD 2D detector) to be captured whilst simultaneously capturing complete SAXS patterns (MAR CCD 2D detector). Exposure times were 1 s with successive exposures being 4.8 s apart. Details of the apparatus have been reported in a previous paper (Phillips et al., 2010). Data analysis was performed with the aid of the SAXS15iD program.

FTIR is sensitive to the conformation and local environment of the iPP molecules. In particular, information can be drawn about the length of the iPP helix (Geng et al., 2009; Xinyuan et al., 2000). The bands at 1100 cm^{-1}, 998 cm^{-1}, 841 cm^{-1} and 1220 cm^{-1} correspond to helical lengths with monomers of 6, 10, 12 and 14 respectively. The CSS450 was adapted for use with IR radiation by fitting it with 1 mm thick ZnSe windows. Rheo-FTIR experiments were performed on a Perkin-Elmer GX FTIR over a wavenumber range of 2000-500 cm^{-1}, and resolution of 4 cm^{-1}. An empty cell was used as the background.

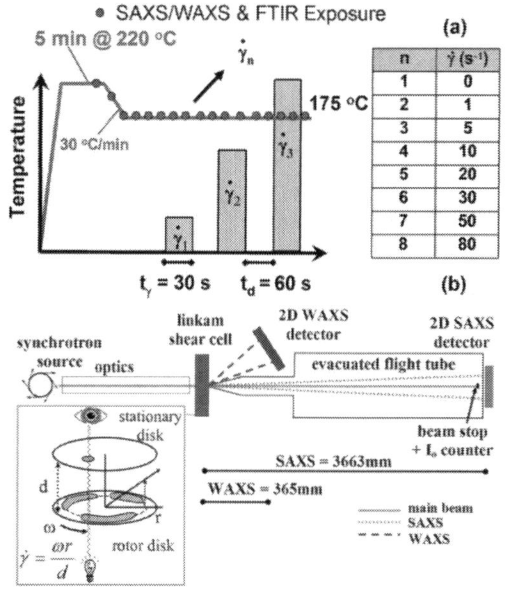

Figure 2 – Schematic diagrams of (a) the thermal-shear profile and (b) the simultaneous 2D WAXS/SAXS setup.

Results & Discussion

Influence of nucleants on the quiescent melt

In-situ 2D SAXS patterns of pure iPP and iPP containing the two nucleants in the quiescent melt at 175 °C are shown in Figure 3a. In all patterns the scattering profile was isotopic. The addition of both nucleants led to an increase in the overall amount of scattering. In particular, the addition of the HPN nucleant lead to a significant increase in the amount of scattering compared to pure iPP and iPP containing SB. This increase in scattering intensity could be due to the effect of the nucleant particles themselves or the effect of the nucleants on the structure of the melt. To further investigate this, in-situ rheo-FTIR spectra of the quiescent melt were obtained, Figure 4a. In pure iPP no absorbance could be found at 841 cm^{-1} indicating that no long helices were present. The addition of both nucleants lead to the occurrence of the 841 cm^{-1} peak. The addition of HPN also lead to the occurrence of the 1220 cm^{-1} peak, however pure HPN also asorbs at this frequency. This indicates that both nucleants could promote the coil to helix transition without shear. It is unclear at this stage the mechanism by which this occurs. However, it may be that an appropriate nucleant surface can stabilise the conformation of shorter iPP helices allowing them to reach a greater length than would otherwise be obtainable in the melt.

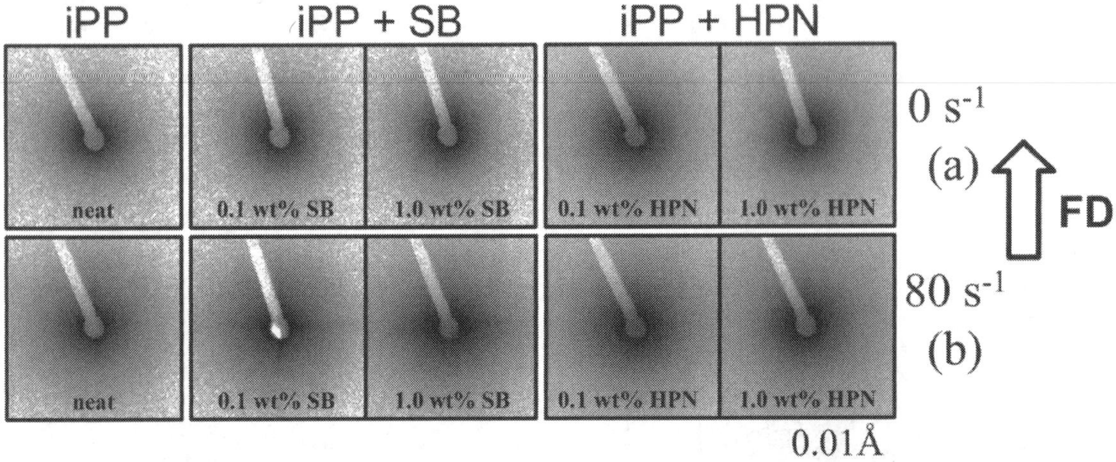

Figure 3 – In-situ small angle x-ray scattering patterns showing the influence of the two nucleants on (a) the quiescent melt and (b) the continuously sheared melt held at 175 °C.

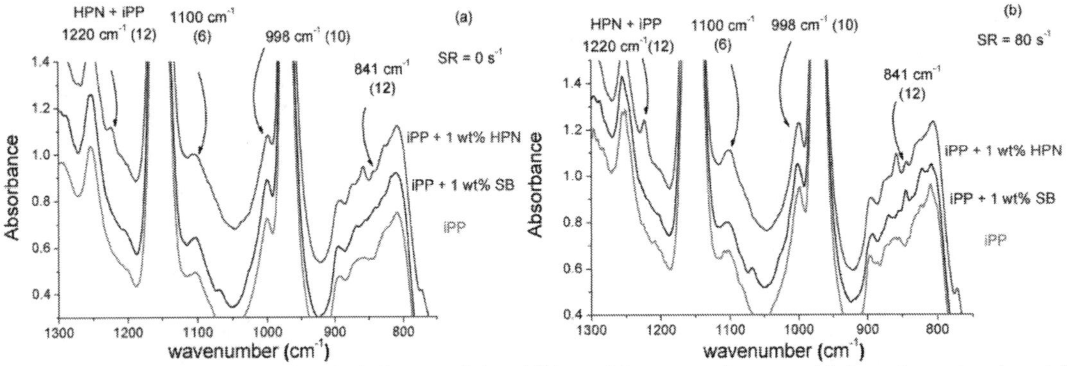

Figure 4 – (a) FTIR spectra showing the influence of the addition of the two nucleants on (a) the quiescent melt and (b) the continuously sheared melt held at 175 °C. The number in the brackets is the number of monomers in the iPP helix.

Influence of nucleants on the sheared melt

Figure 3b shows the last frame before the end of a continuous shear of shear rate 80 s^{-1} for 30 s at 175 °C. Sharp equatorial scattering was observed in all samples above a critical shear rate. This is consistent with the formation of oriented crystal precursors aligned along the flow direction. Recent studies by Balzano et al. (2009) found crystals formed during short fast shears in super-cooled iPP at a shear temperature of 145 °C. However, in these experiments no crystalline scattering could be discerned in the WAXS profiles. This suggests that the precursors at this temperature are primarily non-crystalline entities. Corresponding FTIR spectra to the SAXS results shown in Figure 3b are provided in Figure 4a. A critical shear rate above which an increase in the absorbance of the 998 cm^{-1} and 841 cm^{-1} peaks compared to their quiescent state was also observed. This suggests a shear induced coil to helix transition is associated with the formation of the precursors. Using criteria proposed by Doi et al. (1988), the critical persistence length for a shear-induced isotropic to nematic transition in iPP has been found to be 11 monomer units (Geng et al., 2009; Li and de Jeu, 2004). Therefore in combination with the SAXS results the shear-induced occurrence of the 841 cm^{-1}, which corresponds to helices greater than 12 monomer units long, indicates that structure of the precursors may be liquid crystalline in origin.

The addition of both nucleants significantly influenced the formation of the precursor structure. In the SAXS patterns the sharpness of the equatorial streak can be interpreted as the level of alignment of the precursors along FD while its intensity can be related to precursor density. The addition of SB increased both the sharpness and intensity while the addition of HPN appeared to reduce the sharpness and only mildly increase the intensity of the equatorial peak when compared to pure iPP. The critical shear rate identified in the SAXS patterns and the FTIR spectra was also investigated. The results are presented in Table 1 along with the nucleation efficiency of the nucleants at the various loading calculated using the method proposed by Fillon et al. (1993). Both nucleating agents decreased the critical shear rate for precursor formation and coil to helix transition compared to pure iPP. SB was slightly more effective at reducing the two critical shear rates than HPN even though its nucleation efficiency was less.

Table 1 – The critical shear rate for the occurrence of equatorial scattering in the SAXS patterns and the 841 cm^{-1} (helix of 12 monomers) and 998 cm^{-1} (helix of 10 monomers) peaks in the FTIR spectra.

Material	Loading (wt%)	Nucleation Efficiency (%)	SAXS (s^{-1})	998 cm^{-1} (s^{-1})	841 cm^{-1} (s^{-1})
iPP	0	0	30	30	50
iPP + SB	0.1	11	5	5	10
	1	30	1	5	10
iPP + HPN	0.1	49	10	10	30
	1	59	5	5	5

These interesting results can be explained by the effect of the nucleant on the local flow of iPP molecules near the nucleant particles. Assuming the velocity of the particle is lower than the velocity of melt ($V_{particle}<V_{melt}$) then the presence of the nucleants would result in an increase in the local flow rate near the particle, amplifying the shear field. Additionally, the flow constriction imposed by the particles would result in a greater elongational component to the shear flow (Hwang et al., 2006). Both would lead to the observed decrease in the critical shear rate required to form the precursor structure. Compared to high aspect ratio particles, low aspect ratio particles are known to tumble considerably more in shear fields (Zettner and Yoda, 2001). This leads to increased vortices which resulting in more uneven flow near the particles (Majhi and Vasudevaiah, 1982). The uneven flow field near the low aspect ratio nucleant is then thought to lead to the observed higher critical shear rate for precursor formation as well as their reduced alignment along the flow direction compared to the high aspect ratio nucleant.

Conclusions

Two nucleating agents with different aspect ratios were added to an iPP matrix and the formation of crystal precursor structures was investigated in-situ by simultaneous rheo-SAXS/WAXS and rheo-FTIR. In the quiescent melt the addition of nucleating agents induced some iPP molecules to undergo a coil to helix transition. In the continuously sheared melt oriented crystal precursor structures were formed in all samples above a critical shear rate. The critical shear rate for precursors formation was reduced in the presence of both nucleating agents compared to the pure iPP. The high aspect ratio nucleant reduced the critical shear rate more than the low aspect ratio nucleant. These results were explained by what happens to the flow of material around the nucleant particle.

Acknowledgements

This work was performed at the SAXS/WAXS beamline at the Australian Synchrotron, Victoria, Australia. The authors would also like to acknowledge George Theodossiou at the Monash Centre for Electron Microscopy (MCEM) for his invaluable assistance in preparing the SEM images.

References

Balzano, L., Rastogi, S. & Peters, G. W. M. Crystallization and Precursors during Fast Short-Term Shear. Macromolecules, **42,** 2088-2092 (2009).

Byelov, D., Panine, P. & De Jeu, W. H. Shear-Induced Smectic Order in Isotactic Polypropylene Revisited. Macromolecules, **40,** 288-289 (2007).

Byelov, D., Panine, P., Remerie, K., Biemond, E., Alfonso, G. C. & De Jeu, W. H. Crystallization under shear in isotactic polypropylene containing nucleators. Polymer, **49,** 3076-3083 (2008).

De-Gennes, P. G. Coil-stretch transition of dilute flexible polymers under ultrahigh velocity gradients. The Journal of Chemical Physics, **60,** 5030-5042 (1974).

Fernandez-Ballester, L., Gough, T., Meneau, F., Bras, W., Ania, F., Balta-Calleja, F. J. & Kornfield, J. A. Simultaneous birefringence, small- and wide-angle X-ray scattering to detect precursors and characterize morphology development during flow-induced crystallization of polymers. Journal of Synchrotron Radiation, **15,** 185-190 (2008).

Fillon, B., Lotz, B., Thierry, A. & Wittmann, J. C. Self-nucleation and enhanced nucleation of polymers. Definition of a convenient calorimetric efficiency scale and evaluation of nucleating additives in isotactic polypropylene (alpha phase). Journal of Polymer Science Part B: Polymer Physics, **31,** 1395-1405 (1993).

Geng, Y., Wang, G., Cong, Y., Bai, L., Li, L. & Yang, C. Shear-Induced Nucleation and Growth of Long Helices in Supercooled Isotactic Polypropylene. Macromolecules, **42,** 4751-4757 (2009).

Hwang, W. R., Peters, G. W. M., Hulsen, M. A. & Meijer, H. E. H. Modeling of Flow-Induced Crystallization of Particle-Filled Polymers. Macromolecules, **39,** 8389-8398 (2006).

Kimata, S., Sakurai, T., Nozue, Y., Kasahara, T., Yamaguchi, N., Karino, T., Shibayama, M. & Kornfield, J. A. Molecular Basis of the Shish-Kebab Morphology in Polymer Crystallization. Science, **316,** 1014-1017 (2007).

Kumaraswamy, G. Crystallization of Polymers from Stressed Melts. Polymer Reviews, **45,** 375 - 397 (2005).

Li, L. & De Jeu, W. H. Shear-Induced Smectic Ordering in the Melt of Isotactic Polypropylene. Physical Review Letters, **92,** 075506 (2004).

Majhi, S. & Vasudevaiah, M. Flow separation in a viscous parabolic shear past a sphere. Acta Mechanica, **45,** 233-249 (1982).

Doi, M., Shimada, T. & Okano. Concentration fluctuation of stiff polymers. II. Dynamical structure factor of rod-like polymers in the isotropic phase. The Journal of Chemical Physics, **88,** 4070-4075 (1988).

Phillips, A., Zhu, P. & Edward, G. Polystyrene as a versatile nucleating agent for polypropylene. Polymer, **51,** 1599-1607 (2010).

Seki, M., Thurman, D. W., Oberhauser, J. P. & Kornfield, J. A. Shear-Mediated Crystallization of Isotactic Polypropylene: Shear-Mediated Crystallization of Isotactic Polypropylene: The Role of Long Chain–Long Chain Overlap. Macromolecules, **35,** 2583-2594 (2002).

Somani, R. H., Yang, L., Zhu, L. & Hsiao, B. S. Flow-induced shish-kebab precursor structures in entangled polymer melts. Polymer, **46,** 8587-8623 (2005).

Xinyuan, Z., Deyue, Y., Hongxi, Y. & Pingfang, Z. In situ FTIR spectroscopic study of the regularity bands and partial-order melts of isotactic poly(propylene). Macromolecular Rapid Communications, **21,** 354-357 (2000).

Zettner, C. M. & Yoda, M. Moderate-aspect-ratio elliptical cylinders in simple shear with inertia. Journal of Fluid Mechanics, **442,** 241-266 (2001).

Zhu, P., Tung, J., Edward, G. & Nichols, L. Effects of Different Colorants on Morphological Development of Sheared Isotactic Polypropylene: A Study Using Synchrotron Wide-Angle Scattering. Journal of Applied Physics, **103,** 124906 (2008).

Zhu, P, Phillips, A., Tung, J. & Edward, G. Orientation distribution of sheared isotactic polypropylene plates through thickness in the presence of sodium benzoate. Journal of Applied Physics, **97,** 104908 (2005).

Zhu, P, Tung, J., Phillips, A. & Edward, G. Morphological development of oriented isotactic polypropylene in the presence of a nucleating agent. Macromolecules, **39,** 1821-1831 (2006).

STUDY OF LAYERED SILICATE CLAYS AS A LOW CONCENTRATION ADDITIVE FOR POLYPROPYLENE

N. Preschilla[1*], S. Shyamroy[1], M. Saroop[1], Z. Ahmad[1], A. Biswas[1], J. Bellare[2]

[1]*Reliance Technology Group, Reliance Industries Limited, Reliance Corporate Park,, Ghansoli, Thane-Belapur Road, Navi Mumbai - 400701, India – nisha.preschilla@ril.com; subarna.shyamroy@ril.com; madhumita.saroop@ril.com; zubair.ahmad@ril.com; amit.biswas@ril.com*

[2]*Department of Chemical Engineering, Indian Institute of Technology-Bombay, Powai, Mumabi – 400076, India - jb@iitb.ac.in*

Abstract - There are many reports pointing to the effectiveness of nano-clays as functional fillers for Polypropylene (PP). These studies have mostly been carried out at high loadings of nano-clay, of more than 1%. It is not known what outcome nano-clay can bring, at an additive-like loading in the ppm or sub-percentage level. In this study, two classes of PP-nanoclay compositions were characterized, with the objective to understand the effect of nanoclay on morphology, when incorporated at low concentrations. In the fist case, nano-clay was added in PP at ppm levels, in combination with two commercial nucleating agents - an inorganic salt of a cyclic aromatic phosphinate and bis-(3,4-dimethylbenzylidene) sorbitol (DMDBS). Substitution of a considerable fraction of organic nucleating agents with organically modified inorganic nano-clay resulted in a unique synergy between the two, in the nucleation of PP. Synergistic improvement of mold shrinkage anisotropy of PP was also achieved. A wide range of microscopy and scattering techniques at different length scales was used to gain insight into such synergistic behavior. A hybrid network consisting of exfoliated organo-clay platelets and organic nucleating agent molecules could be proposed, which is more stable and stiffer than the network of nucleating agent alone. In the second case, effect of nanoclay at less than 0.5% loading was explored in oriented tape products, towards attainment of high tenacity. Polarized light microscopy studies revealed a highly aligned and oriented structure of PP, composed of fine, dense crystallites, collectively forming the macro-scale structure of oriented crystallite strands. In-depth transmission electron microscopy and phase-contrast microscopy studies showed novel morphologies; consisting of exfoliated, aligned nanoclay platelets leading to such an uncommon nucleation behavior, and subsequently to improved mechanical and optical performance.

Introduction

Organically modified layered silicate clay or organo-clay is the most used nano-filler for the preparation of polymer nanocomposites. Full exfoliation and homogeneous dispersion of clay platelets throughout the matrix polymer has been widely reported to result in remarkably improved engineering performance of these nanocomposites (Ray, S.S. , Okamoto M., 2003; LeBaron, P.C., 1999, Hussain et.al. 2006). Preferentially aligned, lamellar and intercalated layers is also advantageous to specific functions, such as barrier to moisture and gas, flame retardance, abrasion resistance etc. Because the surface areas of these clays are very large, a small percentage of the clays when fully exfoliated and dispersed, saturate the matrix polymer and forms the basis for a unique two-way performance granted by this nano-scale material, i.e., both as a structural and functional modifier. Therefore, when added to semi-crystalline materials such as Polypropylene (PP), it can act as an effective nucleating agent, while also imparting other beneficial property improvements.

Nucleation and crystallization behavior of polypropylene (PP) have been widely researched over several decades (Zweifel and Amis, 2001; Pasquini, 2005). Numerous nucleating agents have been explored with the objective of increasing the onset of crystallization temperature and crystallization rates while cooling, since this will lead to reduction of cycle times involved in the production of PP based articles. In order to acheive this, heterogeneous nucleating agents are added to PP, usually at small dosage of 0.1-0.3% (Libster et al.2007). In addition to reduced cycle times, nucleation phenomenon leads to other advantageous changes in properties; such as improved mechanical performance. Reduced crystallite sizes as a result of nucleation, can also lead to reduced haze, and consequently improved clarity in certain cases.

A large range of organic and inorganic materials have been demonstrated as very effective nucleating agents for PP. Commonly reported inorganic nucleating agents for PP are as sodium benzoate, phosphoric acid, alkali salts and alkaline-earth salts, aluminum and titanium salts (Libster et al.2007, Jang et.al., 2001, Nagasawa et.al., 2005), or fillers such as talc or $CaCO_3$. Such inorganic materials do not melt or dissolve in PP, and hence are called "dispersion type" nucleating agents. Talc and $CaCO_3$ have been used at reinforcement levels, i.e, high loadings, and therefore, in addition to nucleation they act as reinforcing agents (Menczel and Varga, 1983). Among the organic group of nucleating agents are certain sorbitol derivatives and salts which have been classified as "dissolution type" nucleating agents (Sherpard et. al.,1997; Nagarajan, 2000). Sorbitol-based nucleating agents are typically organo-gelators, which form a stable, thermo-reversible, three-dimensional, fibrillar gel network with a large surface area, which serves as nucleating sites for PP (Lipp et.al., 2006).

In addition to the commonly used nucleating agents mentioned above, it is also known that numerous nano-scale materials nucleate PP. The most commonly cited ones are organo-clays, single-walled carbon nanotubes, multi-walled carbon nanotubes, nano-silica, nano-alumina, nano-$CaCO_3$, nano-ZnO and elastomeric nano-particles. Studies with nano-scale materials have mostly been carried out with the objective of imparting specific functions offered by the nanofiller to PP, and therefore they have been added at the typical concentrations of that of functional fillers, i.e., of more than 1%. However, owing to the high surface area of nanofillers, it has been noted that addtion of nanofillers invariably leads to improved nucleation and crystallization of PP. This points to a very attractive possibility that the nanofillers could act as a replacement for expensive nucleating agents, at very low loadings. Some of the nano-materials have been explored at relatively low concentrations, of about 0.5-0.8% (Bhattacharyya et.al., 2003) but issues were reported in acheiving good dispersion.

In this study, effect of organoclay at an additive-like loading in the range of 0.01 – 0.2 weight percent (100 – 2000 ppm) on the properties of PP is explored, especially in bringing about synergistic performance, in combination with other additives in the formulation. In the first case, organoclay was added in combination with commercial nucleating agent and the nucleation and crystallization behaviors were explored. Detailed studies were done to understand the influence on thermal and mechanical properties, and the crystalline structure of polypropylene. In addition, explorations were carried out towards obtaining reduced shrinkage anisotropy for injection molded PP parts, when nanoclay is added at low concentrations. With common organic nucleating agents, the value of shrinkage anisotropy is generally observed to be about 0.90. High performance nucleating agents, which are specially designed molecules intented for perfectly isotropic crystals inside molten PP matrix, are very costly. Effect of organoclay in this regard has not been widely studied; there have been only a few reports on the dimensional stability of organoclay based polymer nanocomposites (Revilla-Dıaz et.al., 2007), but employing high organoclay loading of more than 4% by weight. Therefore, properties of PP formulations with organo-clay in combination with organic nucleating agent were evaluated, to assess the possibility of forming a hybrid nucleating network composed of organo-clay platelets and the fibrillar network of nucleator molecules. Detailed analysis of the compositions was conducted using various characterization methods to get insights into the structural changes that happen in the presence of the hybrid network.

In the second case, addition of nanoclay into PP at less than 0.5% loading was explored towards attainment of high tenacity, for applications such as tapes and films. Formulations and processing conditions were optimized with the intention of generating unique morphologies with highly aligned clay platelets, which is favorable for improved tensile performance. Polarized light microscopy and thermal analysis were carried out to study nucleation patterns and spherulite growth in such products. In-depth transmission electron microscopy and phase-contrast microscopy studies were conducted to understand these novel morphologies.

Experimental

Materials:

A polypropylene random copolymer grade, manufactured by Reliance Industries Limited, India, was used for the investigations of nanoclay as a low concentration nucleating additive. This PP grade contained 3% ethylene content in the backbone and had a melt flow index (MFI) of 10g/10min. Organo-clay used in this work was supplied by Vesta Organano Private Limited, India. Appropriate amounts of maleic anhydride-grafted PP (MA-PP) compatibilizer was added to the formulations. Four different combinations of a conventional nucleating agent and organo-clay were used, with total loading of 0.2 weight percent, at organo-clay/ nucleating agent ratios of 1:0, 1:3, 1:1 and 0:1. Two nucleating agents were explored in this work, and were were 3,4-dimethyl dibenzylidene acetal of sorbitol (NA-1) and a cyclic aromatic phosphinate salt (NA-2), respectively. They were added separately to a stock of PP powder from the reactor, and stabilized with an appropriate stabilizer package.

In the second study involving organo-clay as a low concentration functional additive, a homopolymer grade PP with MFI of 3 g/10min was used. Organo-clay was added to PP at loadings of 0.012%, 0.25%, 0.4%, 0.6% and 0.8%. Pre-mixed formulations were compounded and extruded. Final tapes were prepared through a small scale tape extrusion line, and were oriented through the standard solid state stretching conditions used for tape applications. Three stretch ratios of 3.5, 4.5 and 5.5 were used.

Methods:

Polarized light microscopy (PLM) studies were carried out on an Olympus microscope, equipped with a Linkam THMS600 hot stage. Samples were isothermally crystallized at specific crystallization temperature in the cooling cycle, and spherulite formation was studied. Samples were heated to 200°C and held for 10 minutes, and cooled at 5°C/min to set crystallization temperature of 120°C for complete crystallization.

Thermal characterization was performed using Differential Scanning Calorimeter (DSC) from TA Instruments Model Q200. Samples of 4-6 mg were heated to 200°C at a rate of 10°C/minute in nitrogen atmosphere to study the melting endotherms. Cooling cycle was carried out at the rate of 10°C/minute to record the crystallization exotherm.

Crystallographic structure and lamellar morphology were studied by wide and small angle X-ray techniques, respectively. Simultaneous small and wide angle X-ray scattering data were measured at room temperature from samples of 100 microns thickness in a SAXSess (Anton Paar, Austria) system with Ni filtered CuK_α radiation source. Intensity profiles were obtained with a slit collimated compact Kratky camera and recorded with a two dimensional imaging plate. Two dimensional scattering data was linearly averaged to get intensity I(q) vs. scattering vector q(nm^{-1}), where $q = 4\pi/\lambda(\sin\theta)$; 2θ is the scattering divergence angle and λ is the CuK_α X-ray wavelength. TEM studies were performed on a FEI Tecnai 12 TEM operated at 120kV. For TEM analysis, 80nm sections were prepared at -60°C on a Leica UCT ultra-microtome and were stained with ruthenium tetroxide.

Results and Discussion

Nanoclay as a Synergistic Nucleating Additive

Fig. 1(a) – (h) show the PLM images of non-nucleated PP and for samples with different organo-clay/nucleating agent ratios for the two nucleating agents. Well-formed large spherulites, with a mean size of more than 200microns were observed for non-nucleated PP, as expected. These spherulites were characterized by the typical cross-hatched morphology of non-nucleated PP. Fig (1(b) – (e)) shows PLM images for the samples having both organoclay and the nucleating agent, at organoclay/NA-1 at ratios of 1:0, 1:3, 1:1 and 0:1. In comparison to the non-nucleated PP, great reduction in spherulite size was observed for samples with combination of NA-1 and organo-clay. These samples show a disordered distribution of crystals of sizes less than 1 micron (estimated from images recorded at higher magnifications), for both organo-clay/NA-1 ratios of 1:1 (Fig. 1(c)) and 1:3 (Fig. 1(d)). This morphology was comparable to that seen for samples with nucleating agent NA-1 alone in the formulation (Fig. 1(e)). A similar organo-clay/nucleating agent synergy was also seen for the the second nucleating agent NA-2, and PLM images are shown in Fig 1(f)-(h).

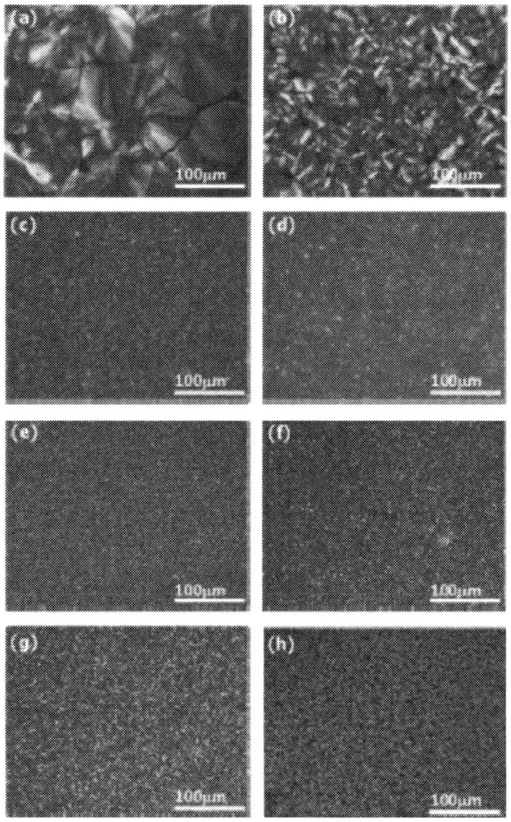

Figure 1: Polarized light microscopy images of (a) Non-nucleated PP (b) PP with 0.2 weight percent organoclay; and comparison to samples with organo-clay/NA-1 ratios of (c) 1:1 (d) 1:3 (e) 0:1 and organo-clay/NA-2 at ratios of (f) 1:1 (g) 1:3 and (h) 0:1.

Figure 2 shows comparison of melting and crystallization of non-nucleated PP with samples containing the two nucleating agents at different organo-clay/ nucleating agent ratios. DSC thermograms show a strong dependence of melting and crystallization characteristics on organo-clay/NA-1 ratio, similar to the trends seen in PLM. In the heating cycle, non-nucleated PP showed presence of two distinct melting peaks in the range of 145 - 155°C. This double endotherm has been attributed to two different types of lamellae in α-PP. Thinner transverse lamellae melt at lower temperatures, while the higher-temperature endotherm is due to the melting of the radial lamellae, which are usually thicker, and agrees with the well formed spherulitic structure generally seen in non-nucleated PP.

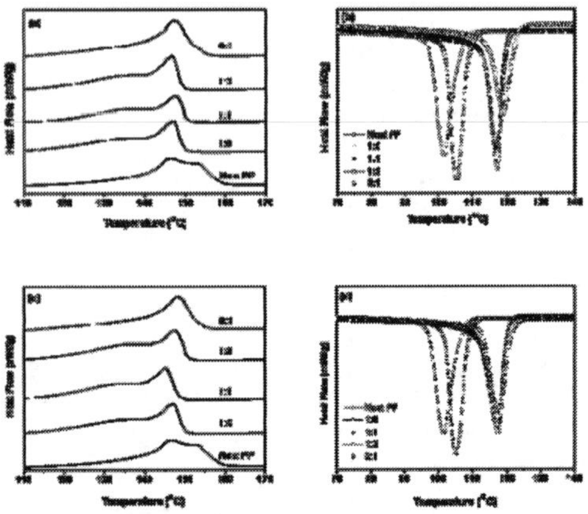

Figure 2: DSC curves showing (a) melting and (b) crystallization curves for non-nucleated PP, along with samples with different organo-clay/NA-1 ratios. (c) and (d) are the respective curves for samples with NA-2 nucleating agent.

For the sample with organo-clay/NA-1 ratios of 1:0, 1:1 and 1:3, a prominent melting peak was observed at about 147°C, matching that of samples with NA-1 alone i.e, 0:1. A broad shoulder was seen at lower temperatures in the range 125 – 140°C in these samples and is more obvious in samples with both organoclay and the nucleating agent. This melting peak has been attributed to the γ-phase of PP, consequent of different degree of the imperfection of crystals during crystallization. Presence of this sub-endothermal peak at lower temperatures has also been attributed to β crystal phase of PP in small concentrations. However, formation of β-PP can be ruled out, as they are usually formed only under stress or when specific β-nucleating agents are added in the formulation.

Degree of nucleation efficiency of NA-1, organo-clay or the combination manifested very well in the cooling cycle as well, and is given in Figure 2(b). Addition of organo-clay/NA-1 combination or NA-1 alone to PP resulted in increased crystallization temperature, to about 117°C, notably enhanced from that of 101°C for non-nucleated PP. Further, crystallization peaks were sharper and narrower for samples with both organo-clay and NA-1. Another important parameter, the supercooling ΔT, also showed strong trends on nucleation efficiency, depending on the organo-clay/NA-1 ratio. ΔT provides an indication to differences in free energies of melt and crystal phases and explains the nucleation effects seen in these samples. ΔT for non-nucleated PP was 46°C, while ΔT for samples with both organo-clay and NA-1, at ratios of 1:1 and 1:3 was found to be 30°C, comparable to that of the ΔT for the sample with only NA-1. ΔT was higher for the sample with organoclay alone, at a value of 38°C. Important parameters estimated from DSC analysis, i.e., the temperature of crystallization onset ($T_{c\text{-}onset}$), peak crystallization temperature ($T_{c\text{-}peak}$), peak melting temperature T_m and the supercooling ΔT ($T_m - T_{c\text{-}peak}$) are given in Table 1, along with the relative γ / α polymorphs in the formulation, estimated from area under the respective melting peaks.

Wide and small angle X-ray scattering studies were carried out to explore changes in crystallographic and lamellar structures of PP, as a function of organo-clay/nucleating agent ratios. Samples with either NA-1 or organo-clay showed primarily α-structure, with a considerable fraction of the γ-polytype. Appearance of another γ peak corresponding to (113) plane, at about 2θ of 15 degrees is also seen for samples with either NA-1 or organoclay. In the case of samples with synergistic combination of both organo-clay and NA-1, a prominent increase in the intensity of peaks corresponding to the γ fraction could be seen, and agreed well with the DSC observations.

One dimensional small angle X-ray scattering (SAXS) spectrum in the range of scattering vector, q, from 0.056–28 nm^{-1} was recorded for samples with different combinations of nucleating agent NA-1 and neat organo-clay. No clear peaks corresponding to clay basal layers were seen in the SAXS spectra, and the SAXS profile steadily decreased with q. A plot of one-dimensional, integrated correlation function determined by a modified Porod law approach was plotted for the samples with NA-1, and is given in Figure 3, along with the correlation function for neat organo-clay, estimated using an integrated correlation function (Rasheed, A. S. A. et.al, 2009).

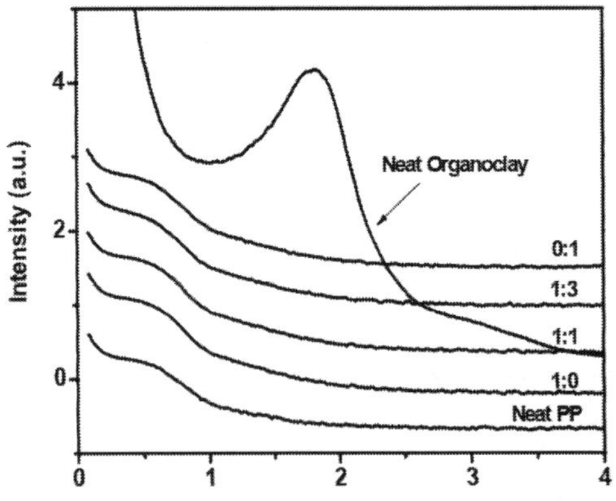

Figure 3: Small angle X-ray scattering spectra in the 2θ range of 0.08-4 degrees that covers the lamellar response and organo-clay response region, for non-nucleated PP and neat organo-clay; and comparison to samples with different organo-clay / NA-1 ratios

Given the high sensitivity of these integrated correlation functions, any signal corresponding to presence of organo-clay stacks was not observed, and it may be concluded that clay platelets are completely exfoliated in all the samples. TEM studies further confirmed this; organo-clay layers were found to be well dispersed in the sample as individual platelets, and aggregates were not seen in any of the samples. It could be concluded that complete exfoliation of organo-clay at the additive level loadings could be obtained in PP matrix. This is not unexpected, as that is about an order of magnitude lower than the estimated critical concentration of 0.0125 volume fraction for the clays used in this work for complete exfoliation in a polymer (Hussain, F. et.al., 2006).

Efforts to image any network formation of organic molecules in these samples with TEM were not successful, even with appropriate staining with ruthenium tetroxide. In the case of our samples, difficulty in imaging organo-clay and the organic molecular network together could be due to the large imaging contrast between the inorganic clay platelets and the organic molecules that has only a small fraction of phenyl component, and staining therefore does not sufficiently decrease the relative contrast between these two components. However, presence of larger fibrillar bundles or agglomerates of NA-1 could be easily detected in TEM, absence of which can imply formation of individual fibrils. Based on such an assumption, and combining the structural parameters determined from TEM and SAXS, a network between organo-clay platelets and nucleator fibrils was postulated. The proposed network structure in the case of synergistic combination of orgno-clay and nucleating agent is a hybrid 1D-2D network, connecting well exfoliated organo-modified clay 2D platelets and 1D nanofibrils of the nucleating agent. A schematic of this hybrid network is given in Figure 4.

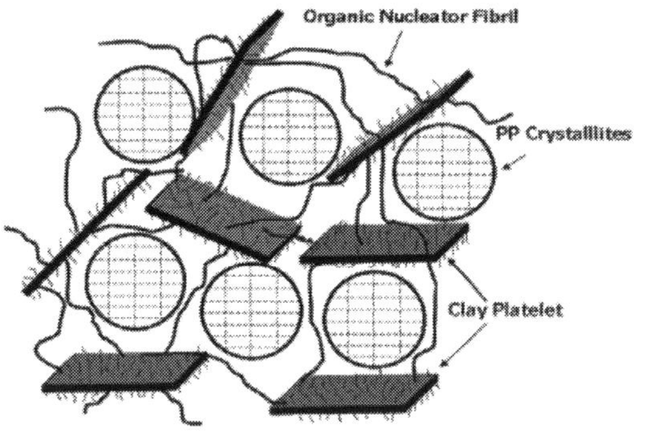

Figure 4. Schematic of the organo-modifed clay platelet - nucleating agent fibril network and the inter-network crystallites of PP.

Such a structure forms a rigid and highly stable architecture, offering high surface area for heterogeneous nucleation of PP. Since the clay platelet thickness is about 1nm – about an order lesser than fibril diameters, they could offer equivalent or larger surface area nucleation sites, coupled to the nucleating agent fibril network. Such a network also reduces PP chain mobility as a result of the narrow space available between clay platelets and nanofibrils, therefore leading to formation of smaller and less ordered crystals, mostly defective as a result of chain interception, explaining the high concentration of γ-polymorph in these samples.

Nanoclay as a Low Concentration Functional Additive

In the second study, suitability of nanoclay was explored as a functional additive to PP, with the objective to increase the tenacity of tape products, without loss in elasticity. Among the range of additive concentrations studied, best enhancement of tensile strength of about 20% was observed, for samples with 0.12% concentration of nanoclay, as compared to neat PP. Loss in elongation was not seen for this sample. At higher loadings of nanoclay, tensile strength was not seen into improve significantly with reference to unfilled PP.

DSC studies showed very interesting trends of nucleation and crystallization behavior of these nano-composites. At a given stretch ratio, higher degree of nucleation efficiency was observed for samples with 0.12% of nanoclay, as indicated from the increased crystallization temperature and subsequently, decreased ΔT, the temperature of super-cooling. However, at higher concentrations of nanoclay, the nucleation efficiency was found be either similar to or poorer compared to neat PP. Studies conducted to understand changes in crystallinity of the compositions showed that with increase in stretch ratio at a constant filler content, the crystallinity increased. This is due to the stretch induced orientation mechanism commonly seen in such oriented PP products (Saroop, M. et.al., 2008). Interestingly, at a given stretch ratio, an increase in orientation was seen for samples with 0.12% of nanoclay. Orientation of the crystalline fractions of PP was estimated through an FT-IR based method (Kissin and Drawer, 1983), and showed maximum orientation for the sample with 0.125% nanoclay.

Detailed analysis was carried out to understand the effect of nanoclay concentration on the structure of PP, and in turn to the distinctive improvement of properties seen at lower concentrations of nanoclay. PLM studies showed that in compounded samples, nanoclay increase the nucleation efficiency at all concentrations. However, the lowest crystallite size was seen for samples with 0.12% nanoclay, which showed a morphology consisting of densely packed tiny crystals, compared to well-formed spherulites of about hundred microns in size for neat PP. Crystallization studies of tape samples showed a very interesting microstructure, as given in Figure 5. All samples showed highly aligned strands consisting of lined up spherulites. For the neat PP sample, the spherulites were seen to be larger, and the strands were thicker. Samples with 0.12% nanoclay showed large number of thin strands made of tinier crystals. At higher concentrations, the size of the crystals increased, and consequently the number density of strands was seen to decrease. These results are summarised in Figure 5. It is known that smaller crystals are easier to orient, and such morphology is desired for oriented tape or film applications.

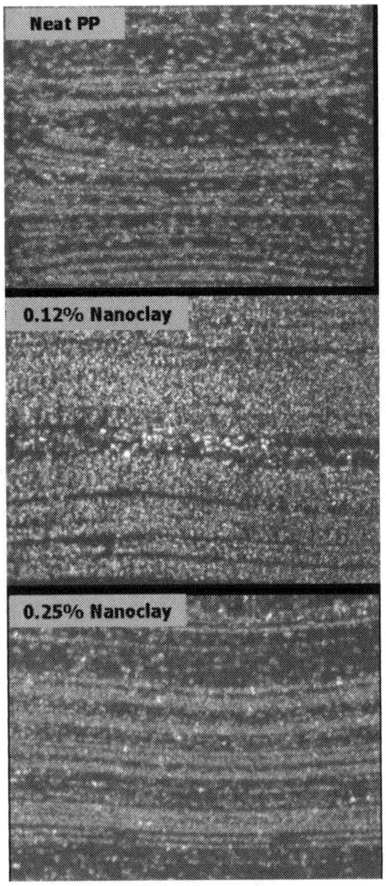

Figure 5: PLM images showing a comparison crystalline morphology for neat PP and samples with nanoclay at 0.12% and 0.25% concentrations.

Correlating the mechanical and thermal performance with PLM results, it is evident that nanoclay added at low concentrations behaves very differently than that at higher concentrations. This could be due to the differences in dispersion of nanoclay, which plays a crucial role in the final properties observed. In order to understand this, TEM studies were carried out.

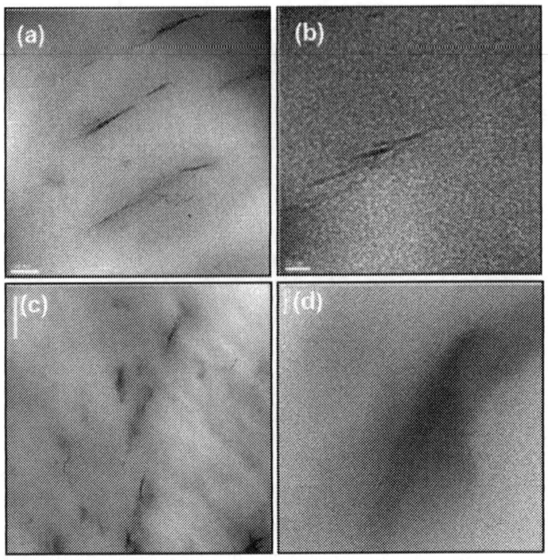

Figure 6. TEM images of PP-nanoclay samples with (a) 0.12% nanoclay and (b) 0.25% nanoclay. (c) and (d) are images recorded at higher magnification for the respective samples.

TEM results shown in Figure 6 readily provided insight into the dispersion of nanoclay. Samples with 0.12% of nanoclay showed that nanoclay was well exfoliated, almost into individual layers on 2-3 layers. In addition, they were seen to be highly aligned along the extrusion direction. Comparison to samples with higher concentration, say, the sample with 0.25% nanoclay, showed presence of larger tactoids, indicating poorer levels of exfoliation. Further, they showed lesser degree of alignment along extrusion direction, unlike the case of sample with 0.12% organo-clay.

TEM observations could be correlated very well to the crystallization behavior seen in PLM studies. Since the samples with low concentration show good level of exfoliation of nanoclay, they act as high surface area nucleating sites in larger numbers, leading to growth of smaller crystals. Large clusters of nanoclay does not provide as efficient nucleation performance. Presence of large clusters of nanoclay also adversely affects the mechanical properties, as observed. It may be also noted that the morphologies seen in the case of lower concentrations are ideal for achieving better barrier performance and optical clarity.

Conclusions

Organically modified layered silicate clay was shown to impart interesting performance properties to PP, both as a structural as well as a functional modifier, at additive level concentrations. Two case studies showed that nanoclay acts a high performance additive, when added at low concentrations of less than 0.5% in PP. In the first case, same nucleation efficiency of PP as with commercial, organic nucleating agents could be retained in synergistic combinations with organo-clay, which is otherwise a poorer nucleating agent by itself at similar loadings. Replacement of a substantial fraction of nucleating agent with organo-clay has significant potential to achieve multiple favourable properties. Synergy of organo-clay with both NA-1 and NA-2 in the nucleation of PP were proven with PLM and DSC studies, and further insights were obtained with wide and small angle X-ray scattering, and TEM analysis. Synergistic improvement of mold shrinkage anisotropy of PP towards highly isotropic values closer to unity was also another property of these formulations worth a lot of significance as anisotropy gives rise to warpage in the molded parts over a period of time leading to high part rejection. Nanoclay could also be utilized at low concentrations for tape and film applications, in obtaining high tenacity products. Acheivement of high level of exfoliation at concentration less than 0.25% led to a morphology of highly aligned platelets, and therefore simultaneously improved nucleation as well as tensile properties.

Acknowledgements

Authors are thankful for the TEM work of PP nanoclay tapes performed in the EPIC facility of NUANCE Center at Northwestern University. Assistance of Shobha Ramagiri at Indian Institute of Technology, Mumbai, for microscopy work is also gratefully acknowledged.

References

Bhattacharyya, A. R.. Sreekumar, T.V., Liu, T., Kumar, S., Ericson, L. M., Hauge, R. H., Smalley, R. E., Polymer, **44**, 2373 (2003)

Hussain, F., Hojjati, M., Okamoto, M., Gorga, R. E., J. Comp. Mater. **40**, 1511 (2006).

Jang, G. -S., Cho W. -J., Ha, C. -S., J Polym Sc., Part B: Polym Phys., **39**, 1001(2001)

Kissin, Y.V., Drawer, P.W., Journal of Polymer Science: Polymer Physics Edition **21**, 2085 (1983).

LeBaron, P.C., Wang, C., and Pinnavaia T., J Appl. Clay Sc. **15**, 11 (1999)

Libster, D., Aserin. A.., Garti, N., Polym Adv Technol., **18**, 685 (2007).

Lipp, J., Shuster, M., Terry, A. E., Cohen, Y., Langmuir, **22**, 6398 (2006).

Menczel, J., Varga, J. J., Therm. Anal. Cal., **28**, 161 (1983).

Nagasawa, S., Fujimori, A., Masuko T., Iguchi, M., Polymer, **46**, 5241, (2005).

Nagarajan, K., Levon K., Myerson, A. S., J. Therm. Anal. Cal., **59**, 497 (2000).

Pasquini, N.: Polypropylene Handbook; 2nd ed., Hanser: Munich (2005).

Ray, S.S., Okamoto M., Prog. Polym. Sci. **28**, 1539 (2003).

Revilla-Dıaz , R., Sanchez-Valde, S., Lopez-Campos, F., Medellın-Rodrıguez ,F.J., Lopez-Quintanilla, M.L., Macromol. Mater. Eng., **292**, 762 (2007).

Rasheed, A. S. A., Preschilla, N., Sivalingam, G., Tyagi, S., Biswas, A., Bellare, J. R., J. Nanosc. Nanotech., **9**, 4948 (2009).

Saroop, M., Preschilla, N., Biswas, A., Ramagiri, S.V., Bellare, J., and Mathur, G.N., PPS 24, Italy

Sherpard, T. A., Delsorbo, C. R., Louth, M., Walborn, J. L., Normann, D. A., Noel, G., Harvey, N. G., Spontak, R. J., J Polym Sc., Part B: Polym Phys., **35**, 2617 (1997).

Zweifel, H. et.al.: Plastics Additives handbook; 3rd ed., Hanser: Munich, 2001.

Influence of the molecular weight and styrene to butadiene ratio on morphology, phase behaviour and mechanical response of S-(S/B)-S triblock copolymers with high PS-content

*Martin Ganss (1), Mahendra Thunga (1), Konrad Knoll (2), Roland Weidisch (1)**

(1) Institute of Materials Science - Jena - Germany
(2) BASF-AG - Ludwigshafen - Germany

The influence of the molecular weight and styrene to butadiene ratio in the random middle block of linear polystyrene-block-(styrene-random-butadiene)-block-polystyrene triblock copolymers (20-60-20, 30-40-30) on the mechanical and fracture behaviour were studied via tensile testing and essential-work-of-fracture concept. TEM and SAXS investigations revealed a change in morphology (lamellae, perforated lamellae, weakly segregated co-continuous phase, not phase separated) with variation of the molecular weight and styrene to butadiene ratio in the random middle S/B block due to the alteration in the combined parameter chieffN where chi denotes the effective interaction parameter and N is the copolymer degree of polymerization. Generally the increase of the styrene content in the random middle S/B block leads to an increase in the Young´s modulus and yield stress of the block copolymers. However the state of phase separation strongly affects the strain at break and tensile toughness where a transition from ductile to brittle behaviour can be observed with a change from phase separated to weakly/not phase separated morphology. The drop in toughness with respect to the state of phase separation is reaffirmed from fracture investigations based on the essential-work-of-fracture approach. The contribution generally should show new possibilities of optimising the toughness of block copolymers with high PS-content and therewith connected limits of the concept.

MELT AND COLD CRYSTALLIZATION KINETICS OF PLA: A COMPREHENSIVE APPROACH

F. De Santis, R. Pantani, G. Titomanlio*

Polymer Technology Group – www.PolymerTechnology.it
Department of Chemical and Food Engineering, University of Salerno, Fisciano SA, Italy
fedesantis@unisa.it; rpantani@unisa.it; gtitomanlio@unisa.it

Abstract - Polylactide (PLA) is a biopolymer that behaves quite similarly to polyolefins and can be shaped into plastic products by standard processing methods, such as injection molding and extrusion. The physical and mechanical properties of PLA are largely dependent on its crystallinity and morphology. Although PLA has many desirable properties, its crystallization rate is very slow; thus the parts obtained by usual processing operations of this material could be completely amorphous. A commercial grade PLA was characterized using differential scanning calorimetry under both isothermal and non-isothermal conditions, and the experimental data were analyzed evaluating the crystallization kinetics constants for melt and cold isothermal crystallization at some temperatures. The crystallization evolution was described on the basis of nucleation density and growth rate, being the growth rate and nucleation density functions of the crystallization temperature and minimum cooling temperature, respectively. Temperature dependencies were described by the classical Lauritzen-Hoffman equation. Resulting Kolmogoroff Avrami model describes main features of both the melt and cold crystallization experimental results.

Introduction

The crystallization is a key process affecting the physical properties of polymers. The crystallization behavior of biodegradable polyesters has been extensively investigated in recent decades. Generally, the physical properties of thermoplastic polymers such as thermal, mechanical properties, and biodegradability (of biodegradable polymers) are considerably influenced by the crystalline structure and morphology that can be manipulated by changing the crystallization conditions. Accordingly, studies on the relationships between structure, morphology, and properties are of fundamental importance for controlling the final properties of polymeric materials.
Poly(lactic acid) (PLA) is considered as one of the most promising bioplastics because it's a raw material. L-lactic acid or D-lactic acid, can be efficiently produced by fermentation from renewable resources such as starchy materials and sugars. PLA can be chemically synthesized either by the condensation polymerization of lactic acid or by the ring-opening polymerization of lactide. Lactide (or lactic acid) is a chiral molecule and has two optically active isomers: L-lactic acid and D-lactic acid. The polymerization of optically pure monomers leads to the formation of either stereoregular poly(L-lactide) (PLLA) or poly(D-lactide) (PDLA). Both stereoregular PLLA or PDLA are semicrystalline, with a melting temperature of about 175 °C and a glass transition temperature, Tg, of about 60 °C.
Although, PLLA (or PDLA) has been known for more than a century, it has become of commercial interest only in recent years, because of its biodegradability. However, PLLA has very bad processability because of slow crystallization rate under supercooled state (Nam et al, 2006).
Stereocomplexation enhances the mechanical properties, the thermal-resistance, and the hydrolysis-resistance of PLA based materials. These improvements arise from a peculiarly strong interaction between L-lactyl unit sequences and D-lactyl unit sequences. More compact side-by-side crystallization between the two enantiomeric polymers results in a much higher melting point. Recently the melting point of PLLA/PDLA stereocomplex has been reported to be around 230 °C, which is about 50 °C higher than that of either lactide homopolymer.
This material has attracted much attention from both academic and practical viewpoints in the last two decades (Tsuji, 2005).
Enhancement of overall crystallization rate of PLLA is a matter of concern when PLLA is utilized for various industrial applications. Numerous studies have been carried out for improving the crystallization rate of PLLA by mixing it with nucleating agents such as talc. Besides study on addition of an inorganic nucleating agent to enhance crystallization rate of PLLA, it also reported that crystallization rate of PLLA increased with addition of poly(D-lactic acid), PDLA.
When the asymmetric blend of PLLA and PDLA is subjected to heat up to 200 °C, all the homocrystals are in the molten state whereas the stereocomplex crystallites stay unmelted because of its high melting temperature. Intensive studies mainly on the crystallization kinetics have been performed by means of calorimetry. The stereocomplex crystal acts as a nucleating agent. However, the crystallization behavior

of PLA seems not to have been examined sufficiently in the scientific literature.

Our study mainly focus on the crystallization kinetics of a commercial PLA, being a L-LA and D-LA copolymer. The aim of this work is to clarify the mechanism of crystallization of PLA, discussing on the basis of the detailed analysis the morphology from the molten state as well as from the glass state of PLA.

Experimental

Material

The poly(lactic acid) supplied by NatureWorks is a commercial grade (PLA 2002D) comprising about 4.5 % D-LA comonomer (Li and Huneault, 2007). PLA polymer 2002D is specifically designed for extrusion/thermoforming applications, with a melt flow index equal to 6. The molecular weight distribution was determined by a size exclusion chromatography, with M_n=145000 and M_w=235000.

The pellets were predried at 65 °C for 8 h in a vacuum oven, before any characterization, to prevent degradation.

Differential scanning calorimetry (DSC)

A differential scanning calorimeter DSC 822™ from Mettler Toledo Inc. was used for determination and measurement of the transition behavior of PLA. The calibration of the temperature was done with the extrapolated onset temperature of the phase transition of indium. The samples with a weight of about 10 mg were put into an aluminum pan and hermetically sealed. All the experiments were conducted in nitrogen (flow rate 50 mL/min), in order to prevent oxidative degradation.

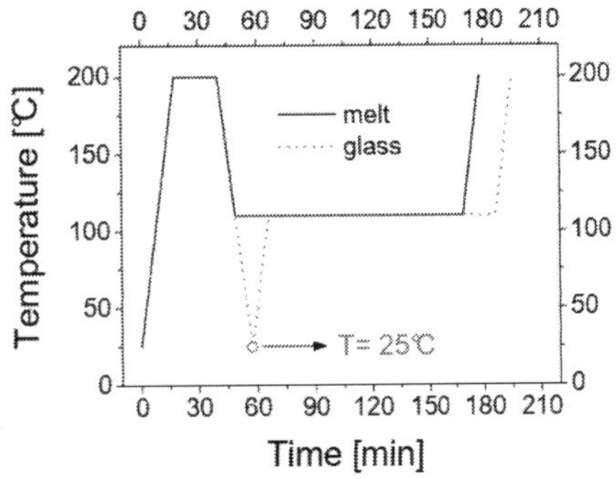

Figure 1 – Experimental protocols for isothermal crystallization from the molten and the solid state

Results and Discussion

Isothermal crystallization

The crystallization kinetics was investigated starting from both the melt amorphous material and the glassy amorphous material. Both conditions were obtained after holding the sample at 200 °C for 5 minutes in order to erase the effect of previous thermomechanical history. Preliminary test were performed with temperature holding sample temperature up to 220 °C, or at 200 °C with times up to 30 minutes to test the material behavior; the "couple" 200 °C for 5 minutes was selected as standard procedure to obtain an amorphous melt, erasing destroying thermomechanical history.

In order to study melt isothermal crystallization, the sample was then cooled from 200 °C directly to isothermal test temperature (crystallization temperature), with a cooling rate of 10 K/min.

Whereas in order to study isothermal cold crystallization, the samples were cooled down to 25 °C, with a cooling rate of 10 K/min, and subsequently heated to isothermal test temperature, with a heating rate of 10 K/min. Both experimental protocols, are sketched in Figure 1.

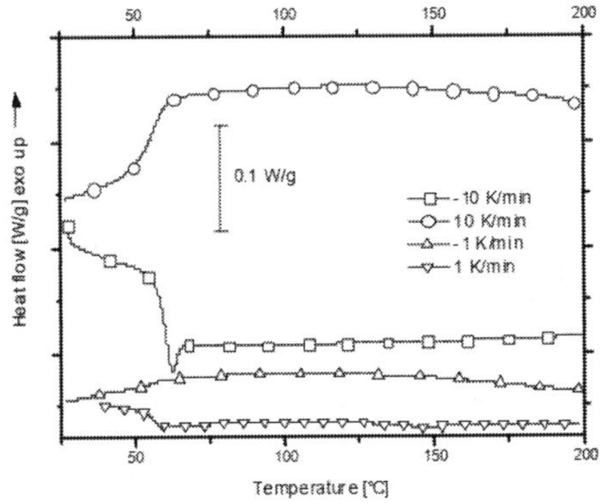

Figure 2 – DSC curves during cooling and heating at 1 K/min and 10 K/min

The cooling and the heating rate were previously checked to be appropriate, i.e. preventing any crystallization phenomena during temperature change before reaching the isothermal test temperature. For example cooling and heating at 10 K/min the initial amorphous sample doesn't undergo to any transformation, as can be evinced analyzing a calorimetric test, in Figure 2, with a such experimental protocol.

Isothermal crystallization kinetics were investigated at different temperatures each 5 K within the range 95 °C to 125°C. Even if isothermal crystallization kinetics were explored in the temperature range in which crystallization times were minima, the peak associated to crystallization revealed by calorimetric tests is often so slow to build up that thermal analysis results difficult to be interpreted.

The results of the isothermal step at 110 °C and 120 °C, from the molten and the glassy state, show that crystallization develops on time scale of the order of 10^4 s. In the case of lower or higher temperature during isothermal test the heat flow is almost flat.

The uncertainty related to these tests, for example in the selection of the correct baseline, was got around by performing several tests at the same temperature with increasing times, analyzing the subsequent melting by heating up at 10 K/min the partially crystallized sample. This procedure is obviously long, and at each isothermal temperature the time is increased until crystallization at that temperature is complete. This protocol lets to overcome the difficulties related to the slow crystallization of this material, increasing significantly the confidence to the experimental data, as explained in the results.

Crystallization kinetics were found always faster starting from the glassy state than from the molten state at each isothermal test temperature.

In order to evaluate the kinetic behavior all the isothermal tests, performed from partial to complete crystallization, or rather the subsequent melting peaks were analyzed, both from the glassy state and from the molten state.

Time integration of the melting peaks, obtained during heating ramps performed at 10 K/min after isothermal crystallization, lets to evaluate indirectly the enthalpy associated to each crystallization time.

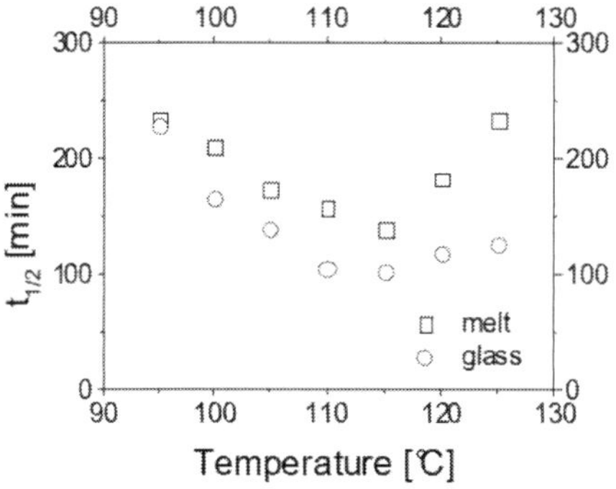

Figure 3 – Crystallization half time vs temperature

The crystallization half time at each temperature, as reported in Figure 3, were evaluated checking, if possible, the coherence with crystallization peak during isothermal step.

Further isothermal tests: self nucleation

The material is further investigated, using an experimental protocol, especially designed to evaluate the effect of self nucleation at different temperatures. Starting from the hypothesis that the growth rate is only function of the crystallization temperature, the difference in between the crystallization starting either from the molten state or from the glassy state has to be ascribed to the difference in the nucleation density evidentely due to the nuclei formed during the cooling to 25 °C.

In particular to evidence this self nucleation, the effect of the minimum cooling temperature was explored by a protocol sketched in Figure 4, with the isothermal step kept constant, i.e. 120 minutes.

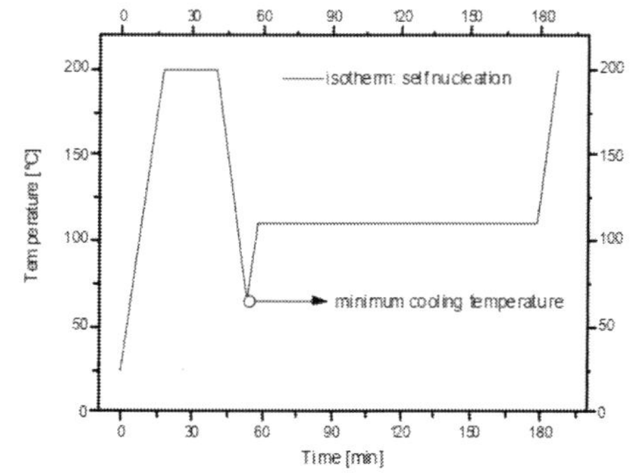

Figure 4 – Experimental protocols for isothermal crystallization changing minimum cooling temperature

In order to monitor the crystallization evolution the isothermal step was always followed by a heating step (10 K/min) to check the melting.

This protocol was planned so that the effect of the minimum temperature on nucleation could be clear, if the formed nuclei were stable, while the crystallization temperature and time were unchanged.

In order to monitor the crystallization evolution the isothermal step was always followed by a heating step (10 K/min) to check the melting.

This protocol was planned so that the effect of the minimum temperature on nucleation could be clear, if the formed nuclei were stable, while the crystallization temperature and time were unchanged.

The same protocol was repeated for different crystallization temperatures, always keeping the isothermal time equal to 120 minutes, and similarly changing the minimum cooling temperature, as sketched in Figure 4. The result is summarized in Figure 5, where the enthalpy of the melting peak subsequent the isothermal crystallization, at 100, 105 and 110 °C, is reported as function of the minimum cooling temperature.

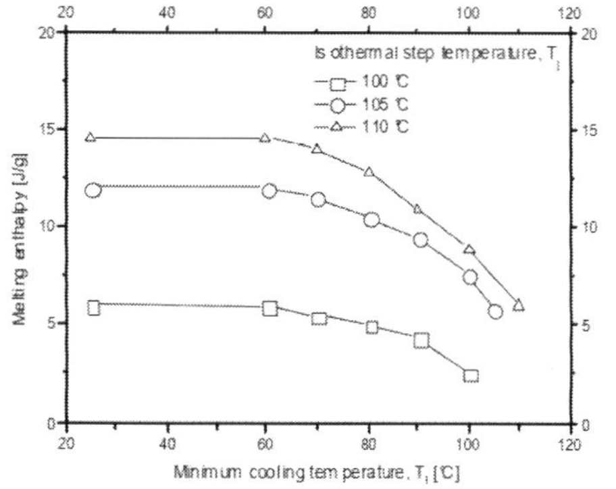

Figure 5 – Melting enthalpy vs minimum cooling rate (isothermal crystallization time = 120 minutes)

The increase of melting enthalpy of partially crystallized samples, at each temperature, has to be ascribed to the nuclei formed going to the minimum cooling temperature, because the following crystallization is performed isothermally, with constant growth rate.

Crystallization kinetics was evaluated using the Avrami model, through nucleation and growth rate, describing both the melt and cold crystallization experimental results in Figure 6.

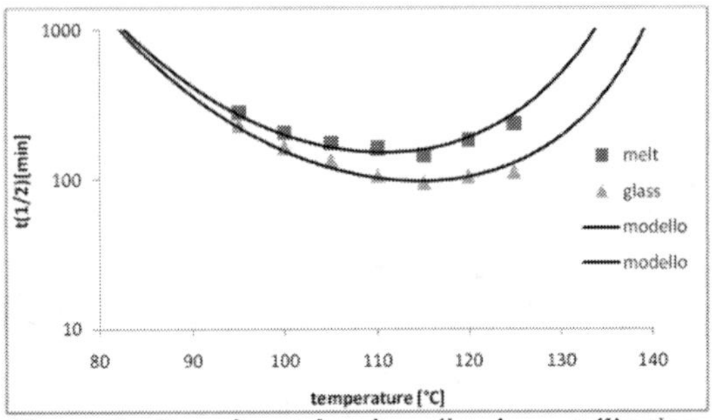

Figure 6 – Experimental and predicted, crystallization half time vs temperature, from the molten and the glassy state

These morphological characterizations were modeled using the classical Lauritzen-Hoffman theory.

Conclusions

A commercial grade PLA was characterized using differential scanning calorimetry under both isothermal and non-isothermal conditions, and the experimental data were analyzed evaluating the crystallization kinetics constants for melt and cold isothermal crystallization at some temperatures.

A peculiar experimental protocol was adopted, varying each time the minimum temperature before the isothermal step, so that nucleation and growth rate were evaluated in a adequate range of temperature.

These morphological characterizations were modeled using the classical Lauritzen-Hoffman theory, which assumes that the free energy barrier associated with nucleation has an energetic origin, estimating the dependence on temperature using physical parameters derived from the linear Hoffman-Weeks plot.

Crystallization kinetics was evaluated using the Avrami model, through nucleation and growth rate, describing both the melt and cold crystallization experimental results.

These characterization lets to describe crystallization kinetics on morphological basis, and can be used for any thermal protocol, to define process parameter to tailor material properties.

References

Nam J. Y., Okamoto M., Okamoto, Nakano H., Usuki M., Matsda A., M. Polymer, **47**, 1340 (2006)

Tsuji H., Macromol. Biosci., **5**, 569–597 (2005)

Hongbo Li, Huneault Michel A. Antec 2615-2618 (2007)

ADDITIVE ASSISTED LASER SINTERING OF POLYETHERETHERKETONE: INFLUENCE OF PROCESS PARAMETERS AND ADDITIVES ON MORPHOLOGY DEVELOPMENT

M. Kroh[1]* and P. Eyerer[2]

[1]*Institute for Polymer Technology (IKT), University of Stuttgart, Germany – michael.kroh@ikt.uni-stuttgart.de*
[2]*Fraunhofer Institute for Chemical Technology (ICT); Germany – peter.eyerer@ict.fraunhofer.de*

Selective laser sintering of polymers is a common rapid prototyping technique that has been developed over the last four decades. New approaches have been done to develop the laser sintering process itself and new materials have been qualified. Additive assisted laser sintering uses a Nd:YAG-laser and NIR sensitive additives like carbon black to heat the polymer particles indirectly. Due to that a more homogenous structure and a limited material damage by laser treatment is expected. This study focuses on the influence of the addition of carbon black and processing parameters on the structure formation in laser sintered specimen. The morphology was investigated by optical light microscopy, transmission electron microscopy (TEM), synchrotron WAXD experiments and correlated to mechanical properties. Isothermal and non-isothermal treatment within the laser sintering process was simulated by a hot stage and the development of the crystalline structure was detected by WAXD.

Due to the addition of carbon black and laser radiation an influence of the crystalline structure was detected. 'Tunnels of carbon black' around the sintered polymer particles were found which induce trans-crystalline areas at the border of the polymer particles. Additional tempering within the laser sintering process improved the stress-strain behavior of the manufactured samples. Due to tempering different lamellar structures were observed by TEM-analysis. WAXD experiments indicated no significant influence of the additive concentration on the crystalline structure.

Introduction

Rapid Prototyping (RP) is the designation for a group of primary shaping technologies without the need for building costly moulds known e.g. from injection moulding, so called *solid free-form manufacturing*. The basic principle of all RP-techniques is a layer wise building process. 3D-CAD-models are transformed into the *surface tessellation language* (STL). In a next step these STL-models are sliced into thin layers of typical 0.1 to 0.2 mm. In a laser sintering machine these layer informations are scanned by a laser onto a prepared powder layer. So the powder layer starts to melt at the scanned geometry. After that a new powder layer is wiped on the powder bed. That process is sequentially repeated until the initial designed model is built.

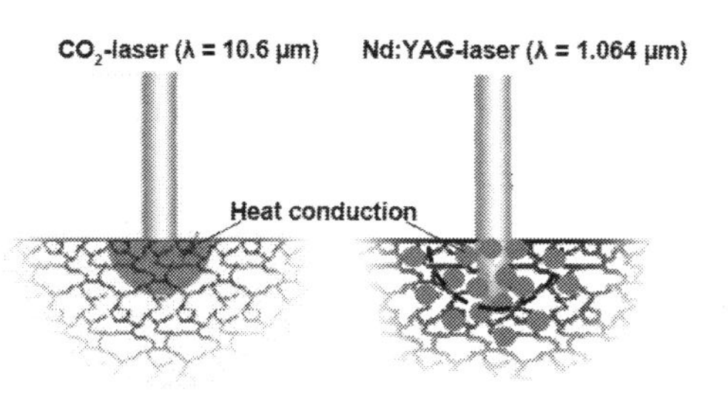

Figure 1 – Illustration of conventional laser sintering with CO_2-laser and the additive assisted laser sintering with Nd:YAG-laser

The presented study deals with the additive assisted laser sintering process of polyetheretherketone (PEEK) using carbon black as NIR absorber to enable an adjustable optical absorption behavior of the polymer powder (Kroh 2007, 2008). Due to its specific wavelength of λ = 1.064 µm the Nd:YAG-laser penetrates into the polymer powder and heats the incorporated absorber particles, which leads to a volume like heat source within the powder layer as illustrated in figure 1.

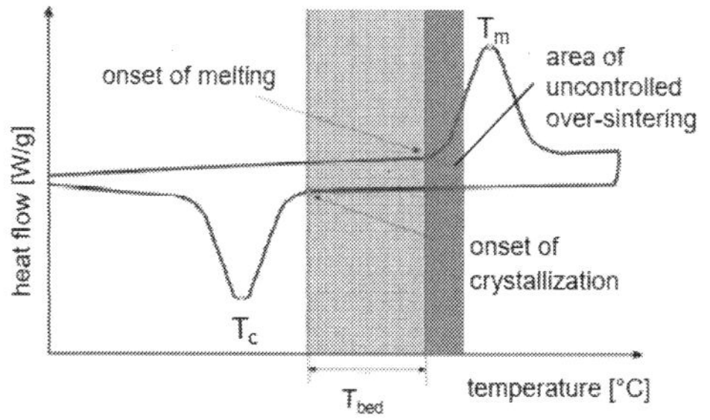

Figure 2 – DSC-graph of PEEK with processing range of the powder bed temperature within the quasi-isothermal laser sintering process

To avoid the so called 'curl' effect (Alscher 2000) the powder bed is held at a temperature between the melting point and above the onset of crystallization point during the sintering process, as illustrated by a qualitative DSC-graph in figure 2. Usually, the building process runs for several hours depending on part size and resolution. To avoid oxidation and thermal degradation the building chamber is flooded with nitrogen.

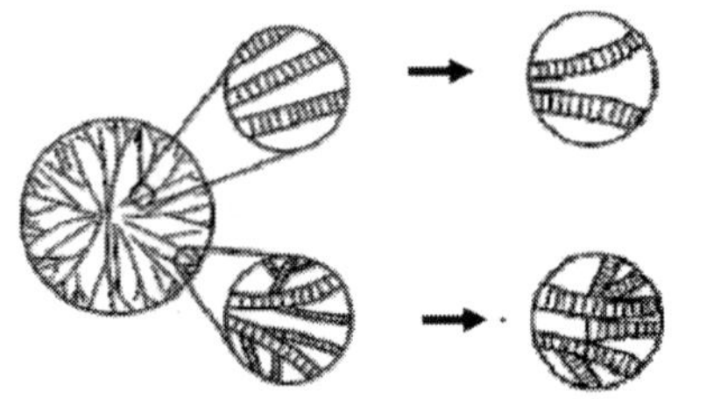

Figure 3 – Schematic illustration of lamellar polymorphism of PEEK as a result of different thermal treatment according to Ko and Woo (1996): A) heating with rates > 3 K/min B) slow heating or tempering

During that building process the powder and the molten polymer is more or less tempered. This iso-thermal treatment leads to different changes in material properties. As reported (Ko and Woo, 1996, Jonas et al 1995, Lee 1987, Marand and Prasad, 1992) PEEK shows multiple minor endothermic peaks in addition to the main melting peak if annealed above T_c or slowly cooled (< 3 K/min). The remelting behavior was associated with lamellae of multiple thicknesses.

Experimental

For all experiments commercially available PEEK™450PF (PEEK) polymer powder from Victrex Plc. was used. An overview of the properties is given in table 1. As NIR-absorbing additive Cabot Vulcan XC72R (Worlee Chemie) was mixed with varying mass-% into the polymer powder. The powder was sieved to an average particle size of 60 μm and a maximum diameter of 90 μm. Particle size distribution was determined by laser diffraction (Mastersizer 2000, Malvern instruments).

Table 1- Properties of PEEK™450PF

melting point	343 °C
continuous operating temperature	up to 260°C
tensile strength (23 °C)	97 MPa
Young's modulus	3600 MPa

Transmission electron microscopy (TEM)

The lamellar structure of the PEEK samples was examined using a transmission electron microscope (Zeiss EM-10C/CR) at an accelerating voltage of 80kV. The specimens have been contrasted with RuO_4-steam. Laser sintered and reference samples were cut by an ultra microtome (Reichert Jung, Ultracut E, FC 4 D) with a diamond knife into thin slices of 80 nm.

Differential scanning calorimetry (DSC)

For DSC measurments a Mettler Toledo DSC 821 (STAR sofware V9.10) was used. The sample masses were ca. 8 mg. For isothermal measuring the samples were heated with 80 K/min to 380 °C and held for 5 minutes. The samples were then cooled down to 350 °C with a rate of 20 K/min and then slowly cooled down to 330 °C and 320 °C, respectively with a rate of 10 K/min and held isothermal. For dynamic measuring the samples were heated from RT up to 380 °C at 10 K/min.

Solution viscometry

To investigate a thermal damage of the sintering process the viscosity number was determined according to DIN EN ISO 307. PEEK specimens have been dissolved in 96 % sulphuric acid for 96 h at 23 °C.

Wide-angle X-ray diffraction (WAXD)

WAXD experiments were carried out at the SCD beamline of the synchroton radiation facility (ANKA in Karlsruhe, Germany). Using a fixed photon energy (E) of 12.4 keV, which corresponds to a wavelength of $\lambda = 0.1$ nm.

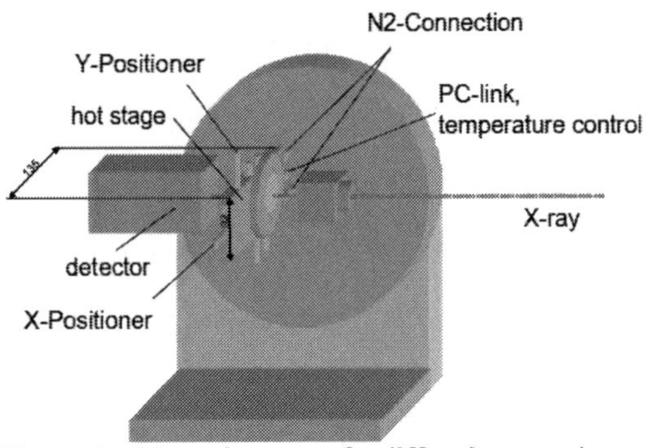

Figure 4 – Measuring setup for diffraction experiments with a hot stage (Linkam TMS94)

Powder samples have been molten at 380 °C and tempered for 5 minutes. The samples were then cooled down to 350 °C with a rate of 20 K/min and then slowly cooled down to 330 °C with a rate of 10 K/min and held isothermal for 4 hours.

Preliminary tests

Main objective of preliminary tests was to specify a processing window and the determination of an optimal powder bed temperature. According to that a basic determination was done of the optical penetration depth and crystallization behavior due to the quasi-isothermal laser sintering.

Hot stage - Near infrared spectroscopy (IR-ATR)

The temperature dependence of optical penetration depth l_a of the powder mixtures has been measured with near infrared spectroscopy in combination with a hot stage device (Linkam TMS94). Due to the high diffuse reflexion within and at the surface of the powder sample an integration sphere (Zeiss OFR D/8°) was used. The powder surface has been exposed with a beam angle of 0° and a measuring distance of 20 mm. The powder samples were scanned with a heating rate of 20 K/min to 400 °C.

According to Lambert-Beer law the optical penetration depth has been determined as plotted in figure 5 over carbon black content. The measured intensity maintained constant until a temperature of 330 °C was reached. Above 330 °C the powder starts to melt and leads to a fast reduction in diffuse reflexion. For temperatures below 330 °C, which represent the powder bed before laser treatment, a maximum optical penetration depth of 400 µm for PEEK powder with 0.1 mass-% of carbon black was detected. Higher concentrations lead to penetration depths of less than 180 µm which is not sufficient for a stable connection between the sintered layers.

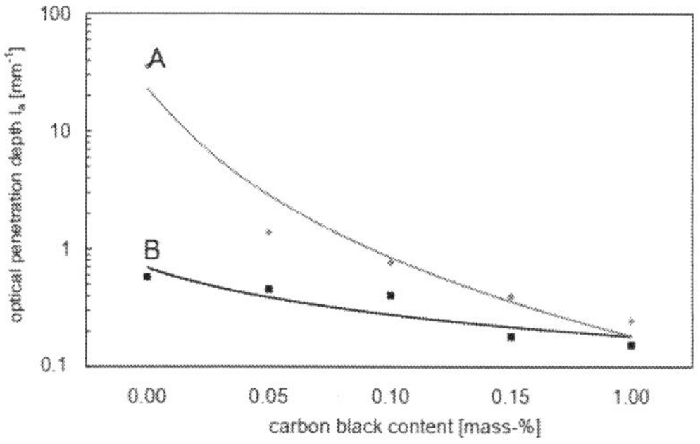

Figure 5 – Optical penetration depth of PEEK powder in dependence of carbon black content: A) T_{bed} = 380 °C, B) $T_{bed} \leq$ 330 °C

There is a well known effect of an accelerated crystallization due to fillers or additives like carbon black or at boundary surfaces. To determine the dependence of additive concentration on the time of crystallization DSC experiments have been performed. Isothermal measuring at 330 °C showed no significant crystallization behavior for powders with 1 mass-% and 0.1 mass-% carbon black after 2 hours. Within the laser sintering process new powder layers are wiped onto the surface of the molten area which can cause a short temperature drop of approximately 10 K. So the isothermal temperature for DSC was set to 320 °C. As shown in figure 6 the crystallization starts earlier with decreasing isothermal temperature, but showed no significant influence of the carbon black content. All samples crystallize after approximately 30 minutes. It is supposed that the nucleating effect of carbon black is superimposed by the crystallization effect on surfaces at the powder particles.

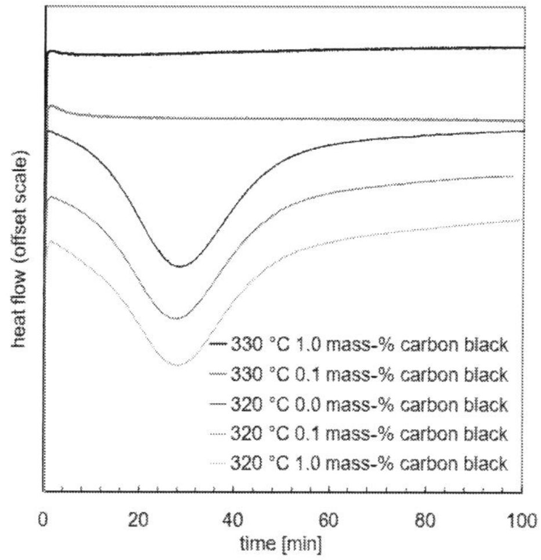

Figure 6 – DSC-graph of PEEK depending on isothermal crystallization temperature and carbon black content

Due to that the powder bed temperature in the SLS-process has to be chosen higher than 320 °C to enable a manageable processing window for sintering parts or tensile bars.

Single layer samples have been laser sintered for further determination of a processing window for manufacturing of PEEK-parts with an aspired layer thickness of 150 µm $\leq \delta \leq$ 180µm. Therefore a sintering depth of ca. 250 µm is essential to get a good connection to the previously sintered layer.

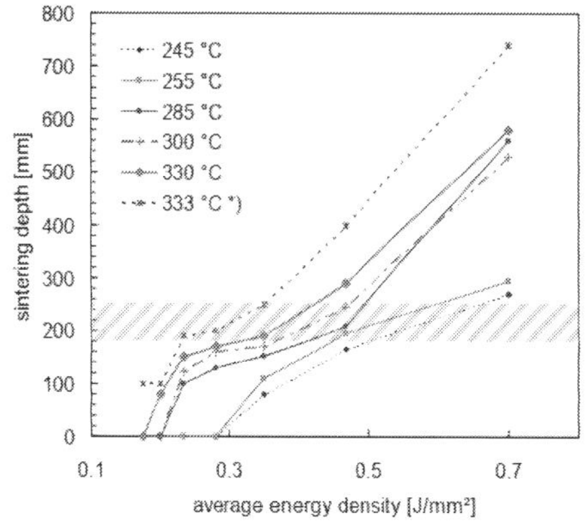

Figure 7 – Measured sintering depths of single layer samples in dependene of average energy density and powder bed temperature; hatched area marks necessary sintering depth; *) at $T_{bed} \geq$ 333 °C uncontrolled over-sintering occurs

Uncontrolled over-sintering occurs at a powder bed temperature of $T_{bed} \geq$ 333 °C. Due to that and with respect to DSC measurements the powder bed temperature is limited to a range of 320 °C to 330 °C. So tensile bars according to DIN EN ISO 527-2 have been laser sintered at a powder bed temperature of T_{bed} = 330 °C

Reference samples
To determine the influence of laser treatment on the sintered parts, reference samples have been produced by DSC to compare thermal treatment and by heating press to compare mechanical and morphological properties. The samples in the heating press have been sintered without pressing the powder due to the pressureless sintering characteristic of the SLS process (Alscher, 2000). All specimens have been heated up to 380 °C with 80 K/min and held for 5 minutes. For tensile tests plates have been sintered and tempered for 30 minutes at 330 °C, which is equivalent to the processing time in the laser sintering process. From these plates tensile bars have been machined.
To compare longer thermal treatment, powder specimens have been tempered additionally at 330 °C for 4 hours in nitrogen atmosphere.

Laser sintering of tensile bars
The RP-Lab at IKT offers a laboratory scaled laser sintering machine (Kroh 2008), which enables a manual variation of almost all processing parameters due to a modular design. Main item of this SLS-machine is a Nd:YAG-laser system (ACSYS Lasertechnik GmbH, Kornwestheim, Germany) with a nominal laser power of $P_{L,nom}$ = 70 W and a scanning head (Scangine 10, Scanlab) with a maximum scan speed of $v_{s,max}$ = 6000 mm/s. For the measuring of the powder bed temperature special thermocouples were designed which allow besides a sufficient measuring rate (2 Hz) a space-resolved measuring at the powder bed surface.
Tensile bars have been manufactured at a powder bed temperature of T_{bed} = 330 °C (see preliminary tests) and a powder layer thickness of δ_s = 180 µm, (s. figure 8). An additional tempering from 0 to 80 seconds was performed after each laser sintered layer.

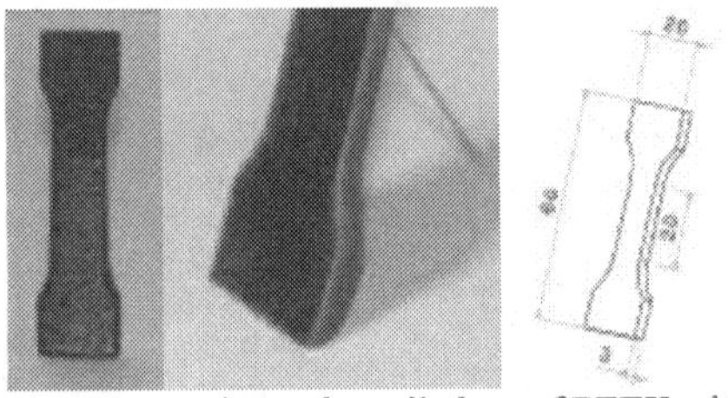

Figure 8 – Laser sintered tensile bars of PEEK with 0.1 mass-% Carbon black, T_{bed} = 330 °C, hatch distance 0.02 mm, scan speed v_s = 4500 mm/s

Tensile tests have been performed with a Zwick universal test machine 1476 according to DIN EN ISO 527 with a testing speed of 1 mm/min

Results and Discussion

Tensile tests
The stress-strain diagram of the sintered specimen in figure 9 showed a significant difference compared to injection molded tensile bars. Except for a nearly equal Young's -modulus of 3400 MPa of the laser sintered parts and the reference sample, the elongation at break indicates a highly brittle behavior.

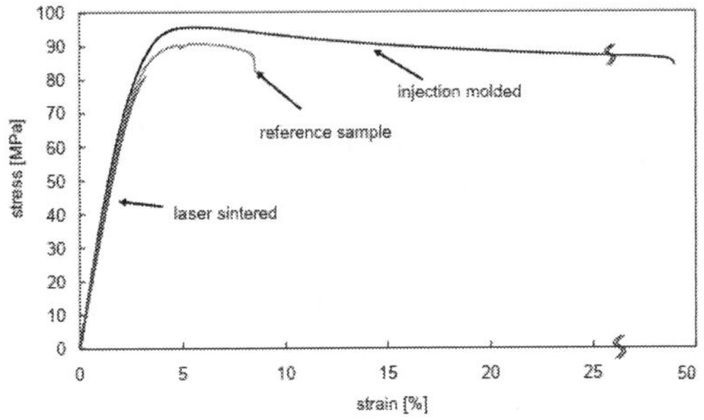

Figure 9 – Tensile tests of PEEK with 0.1 mass-% carbon black, comparison of different manufacturing techniques

It is supposed that this behavior is caused by a multiple notch effect of the powder particle as marked in figure 10. Due to these 'multiple welding' lines a plastic deformation is inhibited.

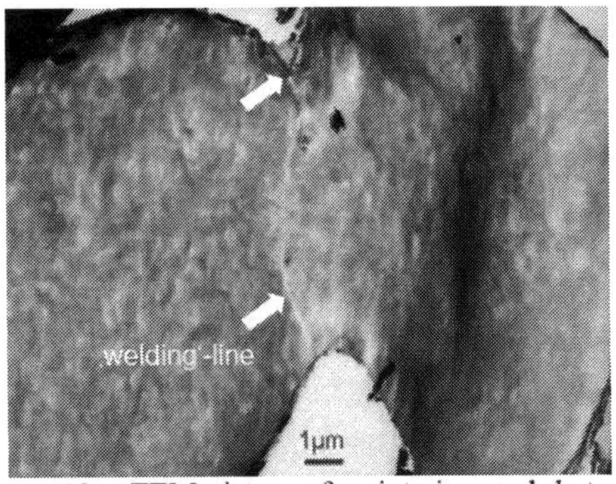

Figure 10 – TEM-picture of a sintering neck between two PEEK-particles.

Additional tempering after each scanning leads to an increased elongation at break as shown in figure 11. There was a threefold higher elongation at break whereas tensile strength was improved with a factor of 2.5. This improvement was observed until a tempering time of 60 seconds; longer tempering intervals showed no further improvement and lead to a reduced tensile strength and elongation.

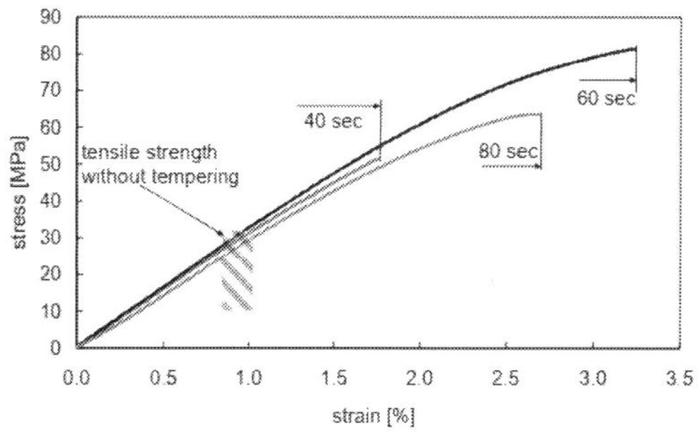

Figure 11 – Influence of the tempering time on the stress-strain behavior of PEEK tensile bars

Figure 12 shows cross-sections of tensile bars without tempering and with a tempering interval of 60 seconds. It is obviously that particles have been deformed by temperature treatment and it is supposed that 'welding lines' became less dominant in tensile testing.

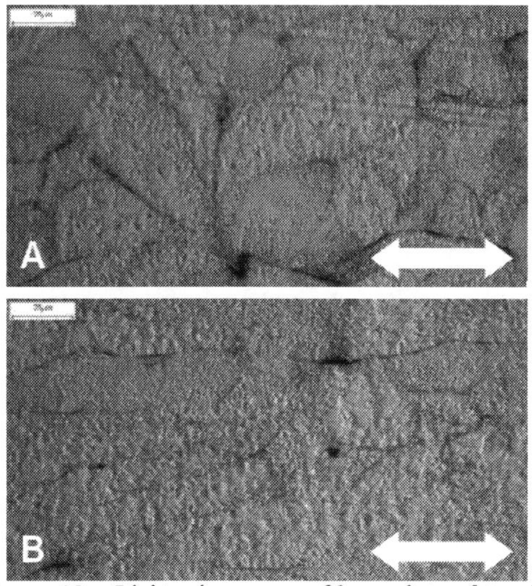

Figure 12 – Light microscopy of laser sintered tensile bars without (A) and with a tempering interval of 60 seconds (B), arrows indicate direction of tension

Differential scanning calorimetry
DSC results, as seen in figure 13, showed a secondary endothermic melting peak at 301 °C and a small peak at 320 °C of the laser sintered specimens as compared to initial powder.
As shown by Ko (Ko et al., 2006) the second endothermic peak was correlated with the tempering temperature. In the DSC graph above a difference was observed of 20 K and 10 K, respectively. To prove the influence of tempering time and carbon black content further DSC-runs have been performed as shown in figure 14.

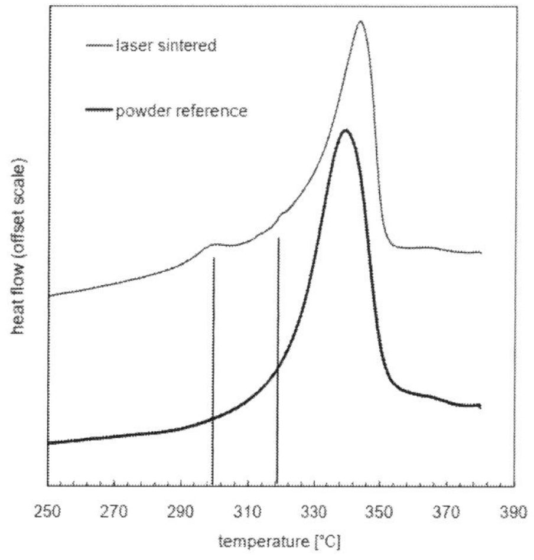

Figure 13 – DSC-graph (10 K/min) of PEEK before and after laser sintering (quasi-isothermal at a powder bed temperature of 330 °C, time for the building process 30 minutes)

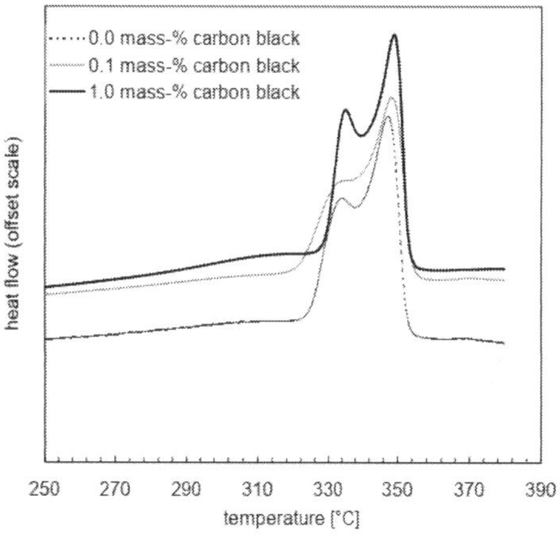

Figure 14 – DSC scan (10 K/min) of PEEK after tempering for 4 hours at 330 °C

For all specimens the second melting peak was observed at 330 °C, so there was no significant influence of the carbon black content on the double melting behaviour.

The calorimetric behavior of the laser sintered parts indicates an altered morphological structure or inhomogenity in the powder bed temperature. So the samples were investigated as follows by optical light microscopy, TEM and WAXD.

Optical light microscopy showed two different types of crystalline structure.

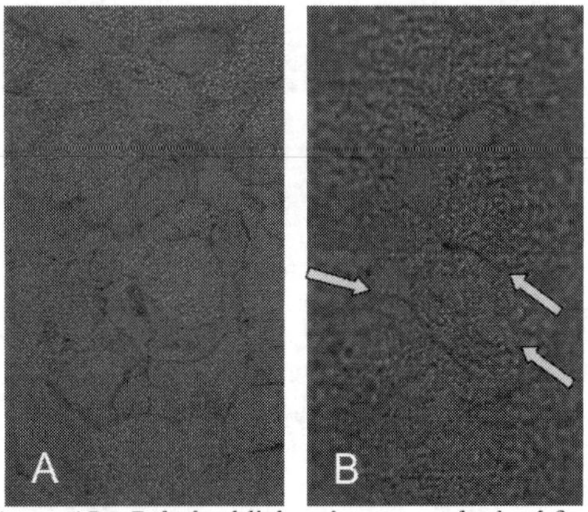

Figure 15 – Polarized light microscopy obtained from cross-sections of tensile bars of PEEK with 0.1 mass-% carbon black; A) reference, B) laser sintered

As shown in figure 15 and 16 by cross-sections of tensile bars carbon black channels occur in all sintered and laser sintered specimens. Spherulitic structures are similar but smaller to these reported by Marand (Marand and Prasad, 1992). Larger spherulitic structures are hindered by the multiple contact surfaces of the powder particles. Pressure less sintered reference samples showed no evidence of an altered crystalline structure. In contrast to that the laser sintered specimens showed a trans-crystalline like structure starting at the border of the polymer particles.

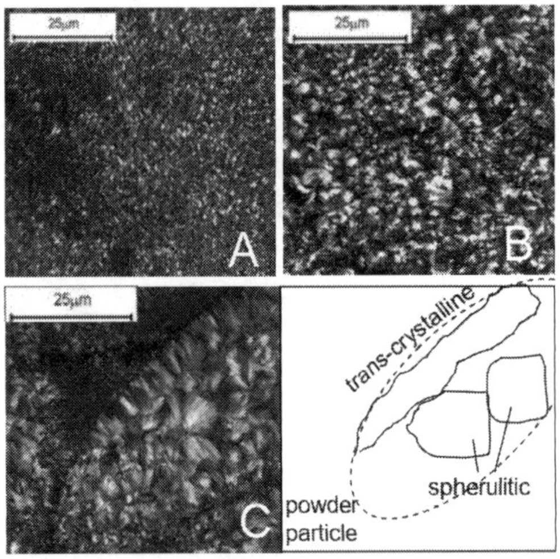

Figure 16 – Polarized light microscopy optained from cross-sections of laser sintered tensile bars of PEEK, A) origin powder, B) with 0.1 mass-% carbon black, C) laser sintered with 0.1 mass-% carbon black

TEM-investigations in figure 17 and 18 showed two different types of lamellar structures, which can be correlated to DSC investigations above and results presented by Ko and Woo (1996). In laser sintered samples a thickening of the investigated lamellae could be observed.

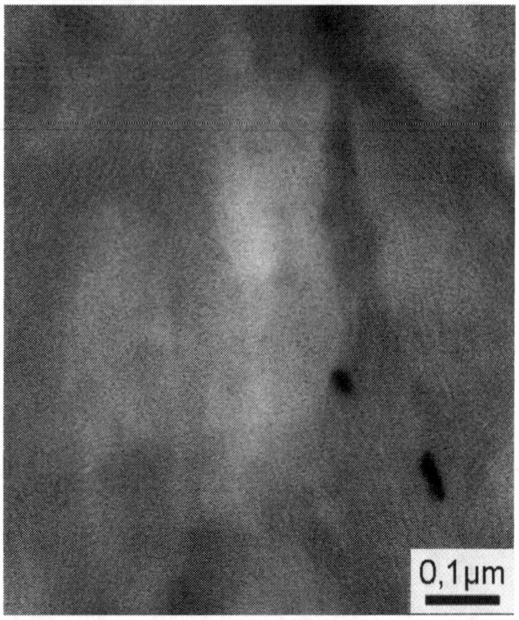

Figure 17 – TEM-picture of reference specimen, PEEK with 0.1 mass-% carbon black

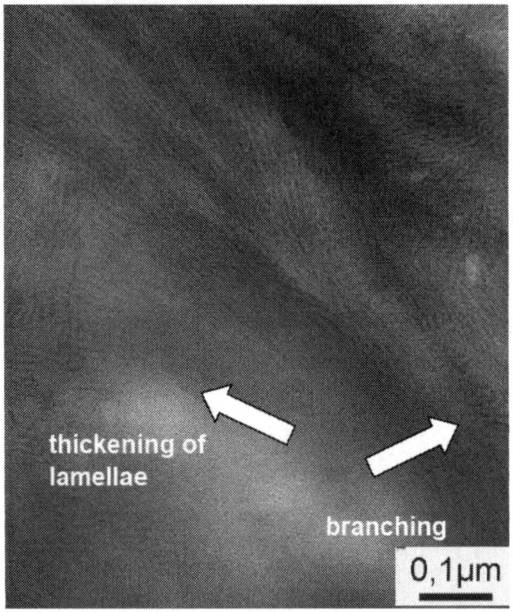

Figure 18 – TEM-picture of laser sintered specimen, PEEK with 0.1 mass-% carbon black

WAXD experiments

To investigate the influence of laser treatment, carbon black content and tempering on crystal structures WAXD experiments have been performed. As seen in figure 19 slightly increased peaks of the main reflexes with increasing tempering time were observed and indicate a higher degree of crystallization or crystal perfection.

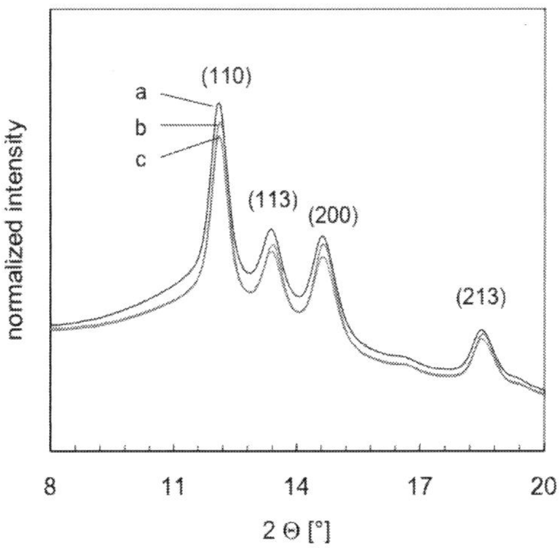

Figure 19 – WAXD scans of laser sintered PEEK samples: additional tempering after each sintered layer of a) 60 seconds, b) 40 seconds, c) 20 seconds

Figure 20 shows the results of diffraction experiments after an isothermal treatment of 4 hours. A slight displacement of the three main reflexes of the sample with 1 mass-% carbon black was observed compared to these with 0.1 mass-% or no carbon black.

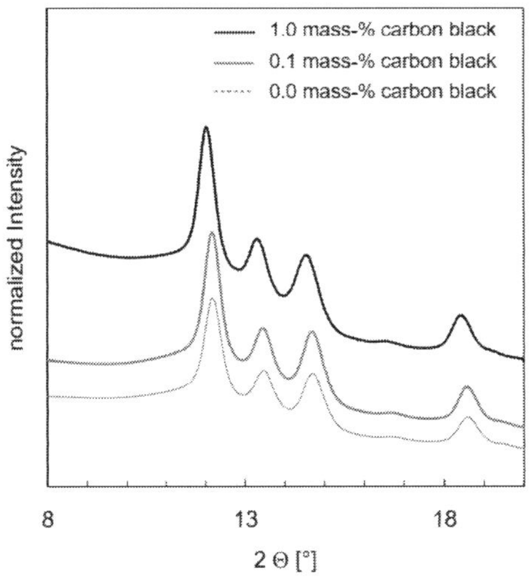

Figure 20 – WAXD scan of PEEK with varying carbon black content; tempered for 4 hours at 330 °C

Solution viscometry

Compared to the reference powder the viscosity number was reduced due to the addition of carbon black as seen in figure 21. In a second step the PEEK-powder with 0.1 mass-% carbon black showed no significant change due to the laser sintering process, so it is supposed that the average molar mass does not change in a siginificant manner within the additive assisted laser sintering and additional tempering.

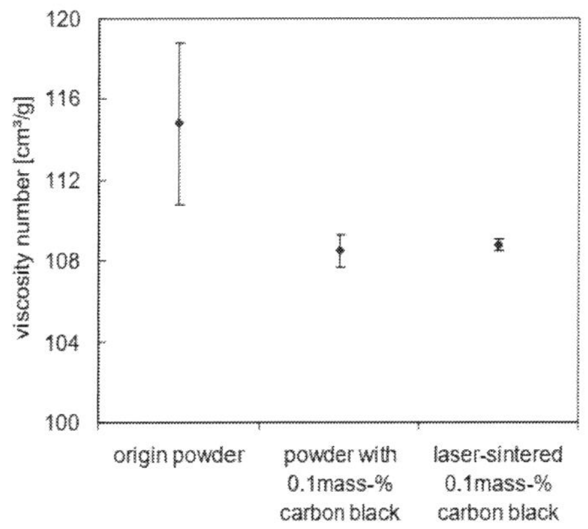

Figure 21 – Viscosity number of PEEK-powder, dissolved in sulfuric acid 96 % for 96 hours at 23 °C

Conclusions

Due to additional quasi-isothermal tempering of every sintered layer within the additive assisted laser sintering process an improved stress strain behavior of the investigated tensile bars was observed.

It is supposed that the improved stress-strain behavior is ascribed to an altered morphological structure due to thermal treatment. A changed morphology was indicated by calorimetric measurements due to a double melting in the laser sintered parts. Similar results were reported by Jonas et al. (1995). He indicates an increasing in Young's - modulus over annealing time which was correlated to an increasing crystal density and lamellar thickening.

It was shown by TEM-investigations that the isothermal treatments lead to a thickening of lamellar structures. Moreover it was indicated that trans-crystalline structures were caused by laser treatment. It is supposed that these structures are induced by the laser radiated carbon black channels due to a short glowing followed by a short overheating of the polymer material. This over heating may lead to some locally bordered breaking of polymer chains. Based on solution viscometry it can be assumed that there is no further damaging of the polymer by Nd:YAG-laser radiation. An influence on crystal structure or unit cell of the PEEK material were also supposed due to isothermal diffraction measurements, but further investigations have to be done.

Acknowledgements

The authors want to thank ACSYS Lasertechnik GmbH for their support and cooperation with in the research project. Further thank is extended to the technicians at IKT S. Osterloh, U. Fritz, U. Müller, H. Klink and A. Bendfeld from ICT, Pfinztal

References

Alscher, G., "Das Verhalten teilkristalliner Thermoplaste beim Lasersintern", PhD Thesis University of Essen (2000)

Ko, T. Y., Woo, E. N., "Changes and distribution of lamellae in the spherulites of poly(ether ether ketone) upon stepwise crystallization", Polymer, 37, 1167-1175 (1996)

Lee, Y., Porter, R.S., "Double-Melting Behaviour of Poly(ether ether ketone)" Macromolecules, 20, 1336-1341 (1987)

Jonas, A.M., Russell, T.P., Yoon, D.Y., "Synchroton X-ray Scattering Studies of Crystallization of Poly(ether-ether-ketone) from the Glass and Structural Changes during Subsequent Heating-Cooling Processes", Macromolecules, 28, 8491-8503 (1995),

Kroh, M., "Development of a modified processing technique for selective laser sintering of high temperature resistant polymers with a Nd:YAG-laser", PPS-24, Conf. Proc., paper S17-342 (2008)

Kroh, M., Woicke, N., Eyerer, P., "Lasersintern von Hochleistungspolymeren mittels Nd:YAG-Laser", in Entwicklung und Erprobung innovativer Produkte - Rapid Prototyping, Bertsche, B., Bullinger, H.-J., (Ed.). Springer, Berlin, p.430-443 (2007).

Marand, H. Prasad, A., "On the Observation of a New Morphology in Poly(arylene ether ether ketone). A Further Examination of the Double Endothermic Behavior of Poly(arylene ether ether ketone)", Macromolecules, 25, 1731-1736 (1992)

Towards the elaboration of new polymer blend morphologies with a new type of mixer-reactor device based on elongational flows

Rene Muller, Michel Bouquey, Julien Bonnet, Badi Triki*

University of Strasbourg - LIPHT - Alsace - France

Mixing is a key step in almost every polymer processing operation, affecting material properties, processability and cost. But, until now, some polymer blends like thermoplastics/thermoset blends,or some polymer composites like epoxy/NTC are very difficult to elaborate with the classical mixing devices. The present study describes a new approach for the mixing process allowing overcoming some well-known difficulties in this domain like mixing two fluids with very different viscosities or introducing highly volatile components into a polymer matrix. Thus, we designed a new mixing device, later called Ram-stat MiXer (RMX) in which the material to be mixed is alternatively pushed by two hydraulically driven pistons from one mixing chamber to another through a static mixing element. In the present study, the mixing element was a single small diameter die. Due to the continuous reciprocal motion of the pistons the material is subjected to both shear- and strong elongational flows in the contraction and expansion zones. RMX offers a number of potential advantages over conventional rotational mixers, one of the most significant being its high distributive and dispersive mixing efficiency. Moreover, RMX is characterized by: water- and air-tightness, the possibility of moulding or sampling specimens on-line, and easy handling of reactive systems. Its technical performances were evaluated by following the evolution of morphology of model blends under different experimental conditions. More specifically, we describe the influence of: the number of mixing cycles, the geometry of the mixing element, the flow rate (controlled by the speed of the pistons), and finally the viscosity ratio between matrix and dispersed phase. Results on PS/PMMA blends and reactive blends based on functionalized polyolefin/diamine oligomer systems will be presented. Finally, we comment on the contribution of elongational flow to the organoclay exfoliation process in a polypropylene matrix.

MORPHOLOGY AND MECHANICAL PROPERTIES OF PHYSICAL BLENDS OF P(L)LA AND PBS BIODEGRADABLE POLYESTERS

V. Berthé[1,2], A. Bergeret[1*], L. Ferry[1], J. C. Benezet[1]

[1] *Centre des Matériaux de Grande Diffusion (CMGD), Ecole des Mines d'Alès, 6, Avenue de Clavières, 30319 Alès cedex, France – Anne.Bergeret@mines-ales.fr; Laurent.Ferry@mines-ales.fr; Jean-Charles.Benezet@mines-ales.fr*
[2] *MateriaNova asbl, Zoning Industriel de Ghislenghien, Rue des Foudriers, 1, 7822 Ghislenghien Belgium – Vincent.Berthe@materianova.be*

Abstract - Two biodegradable polyesters, poly-L-(lactic acid) (P(L)LA) and poly(butylene succinate) (PBS) were melt-blended using a twin-screw extruder and subsequently injection moulded. The mechanical properties showed significant improvements in elongation at break over that of individual polymers while impact strength was improved as compared to pure P(L)LA. Mechanical models taking into account blend crystallinities were used to predict blend Young's moduli. Melting point depression indicated thermodynamic interactions between the blend constituents, while immiscibility could be inferred from Microscopy observations and Differential Scanning Calorimetry. Dynamic Mechanical Thermal Analysis showed all blends had two T_gs at positions close to those of the pure components. However, a shift in the tan(δ) peak position suggested partial miscibility of PBS in P(L)LA rich phase. This phenomenon was maximal for blends with high PBS contents. Melt miscibility was also highlighted by melt viscosity for blends having high PBS contents.

Introduction

Recently, a new type of aliphatic copolyester poly(butylene succinate-co-lactate) _PBSL_, (trade name: GS Pla) was developed and commercialized. Although the conventional poly(butylene succinate) _PBS_ is chain-extended by hexamethylene di-isocyanate, PBSL does not contain the isocyanate-based unit. The monomers for the synthesis of PBS are succinic acid, 1,4-butanediol, and L-lactic acid, all of which can be derived from renewable plant-resources since succinic acid can be prepared by fermentation of starch, and then be converted to 1,4-butanediol by hydrogenation. This work focuses on the blending of two commercial polyesters PBSL and polylactide _P(L)LA_ in order to evaluate blend impact strength / Young moduli ratio over the one of the pure components. For those two polyesters were melt-blended using a twin-screw extruder and subsequently injection moulded. Mechanical properties and morphology are reported. Interactions between amorphous phases, crystalline phases and in the melt are studied using DSC, DMTA and ARES. The mechanical properties showed significant improvements in elongation at break over that of individual polymers while impact strength was improved as compared to pure P(L)LA. Melting point depression indicated thermodynamic interactions between the blend constituents, while immiscibility could be inferred from Microscopy observations, Differential Scanning Calorimetry. Dynamic Mechanical Thermal Analysis showed all blends had two T_gs at positions close to the pure components. However, a shift in the tan(δ) peak position suggested partial miscibility of PBSL in P(L)LA rich phase. This phenomenon was maximal for blends with high PBSL contents. Melt miscibility was also highlighted by melt viscosity for blends having high PBSL contents.

Experimental

Materials

Materials were P(L)LA (Teramac TE-7000, Unitika Co. Mn = 123500 Da, polydispersity index I = 1.72 (SEC, THF, T=25 °C), T_g = 61 °C and T_m = 174 °C) and PBS (GS Pla AZ-type, lactate unit ca. 3 mol%, melt flow rate (190 °C, 2.16 kg) 25 g/10 min, specific gravity 1.26) that was supplied from Mitsubishi Chemical Corporation (Japan). Drying of pellets was performed for 8 hours at 50 °C under reduced pressure before use. Samples are respectively called by their weight fractions e.g. P(L)LA/PBS 90/10 and weight proportions are varied by steps of 10 %.

Melt Compounding

A twin-screw extruder (Clextral 900 mm) with co-rotating screws was used for melt blending. Cooling was performed using a water bath. The experimental conditions are temperature 170 °C for P(L)LA, 165 °C for all blends and 145 °C for PBS, screw speed 250 rpm, and feeding rate 4 kg/h. The extrudate was pelletized and injection-moulded to obtain tensile bars. The injection moulding was performed using a Sandretto / 95 t. Temperature was set at 170 °C for

P(L)LA and all blends and 155 °C for PBS. Hold time was as follows: 1.5 s for P(L)LA and blends ranging from 10 wt% PBS to 60 wt% PBS and 40 s for blends ranging from 70 wt% PBS to pure PBS.

Differential Scanning Calorimetry (DSC)

DSC experiments were performed on Perkin Elmer (Pyris Diamond) equipment, calibrated with indium and zinc standards and data were recorded at a heating rate of 10 °C/min. Sample weights were 6–8 mg, and they were analyzed in standard aluminum DSC pans. Prior to the measurement, the baseline was established with two empty pans. Thermal degradations were prevented by circulating nitrogen gas around the samples during analysis. The samples were heated from -90 to 190 °C. First scans were studied in order to measure sample crystallinities and relate them to mechanical properties. Three samples taken from the middle of three different bars where tested. With the maximum of the endothermic peak detected on heating scans being taken as the melting temperature, the peak area was used to calculate the enthalpy of melting using Pyris 8.0 software. The crystallinities of P(L)LA and PBS, respectively noted χc(P(L)LA) and χc(PBS) were calculated using the following Eq.1:

$$\chi_c(\%) = \frac{(\Delta H_m - \Delta H_{cc})}{\omega\ \Delta H_m^0} \quad (1)$$

where ΔH_m, $\Delta H_m°$ and ΔH_{cc} (J/g of polymer) and ω, are respectively the melting enthalpy, the melting enthalpy of infinite thickness crystal, the cold crystallisation enthalpy and the weight fraction of each component. For P(L)LA, $\Delta H_m°$(P(L)LA) = 106 J/g (Sarasua et al., 1998). The theoretical melting enthalpy for PBS was calculated on the basis of the group contribution method proposed by Van Krevelen, $\Delta H_m°$(PBS) = 110.3 J/g (Hoffman et al., 1976).

Tensile Properties

Mechanical properties were measured after conditioning the samples at room temperature under standard atmospheric conditions (22–24 °C) for 72 hours before testing. Maximum tensile strength and elongation at break was measured according to the ISO 527 standard (crosshead speed: 50 mm/min) using a Zwick Z100 tensile testing machine (load cell range 0-10 kN). The values recorded were the average of at least five specimens. Young's modulus was also recorded using ISO 527 standard (crosshead speed: 1 mm/min). Unnotched Charpy impact strength was determined using a Zwick apparatus according the ISO 179 standard (2 J, 4 J and 7.5 J). Testing was done on specimens of 80×10×4 mm^3. Mechanical properties were measured after conditioning the samples at room temperature under standard atmospheric conditions (22–24 °C) for 72 hours before testing.

Environmental scanning electron microscopy (ESEM)

Blend morphologies were examined with a QUANTA 200 FEG (FEI Company) environmental scanning electron microscope. Injection-moulded bars were fractured in liquid nitrogen and stuck to aluminium stubs. In order to prevent P(L)LA degradation by the electron beam, samples were coated with carbon using a Carbon Evaporation Device CED030 (Balzers). Once the pressure was obtained (P = 10-2 mbar) the coating was carried out for 15 minutes. The width of the carbon layer obtained was around 15 nm. Then the samples were dried for 8 hours under reduced pressure at 50 °C. ESEM experimental observation conditions were very low pressure (around 10^{-7} Torr), voltage 2-5 kV and work distance 8 mm.

Dynamic Mechanical Thermal Analysis (DMTA)

The dynamic mechanical properties were studied with a VA 815 Metravib RDS apparatus with a heating rate of 3 °C/min, over a temperature range from -100 to 120 °C. The samples (40×10×4 mm^3) were tested using an imposed frequency of 5Hz and an oscillation amplitude of 5-10 µm in the tensile/compression mode. The temperature dependence of oscillatory tensile parameters in the solid state, such as tensile storage modulus E', loss modulus E" and loss factor tan(δ), were measured as a function of temperature, the maximum of the tan(δ) peak being related to the main mechanical relaxation (α-relaxation related to glass-transition temperature T_g). A duplicate test was performed for each sample.

Rheological properties in molten state

The linear and non-linear shear rheological behaviours of P(L)LA/PBS melts were investigated by an advanced rheology expended system (ARES). The strain-controlled rheometer (ARES) had a torque transducer capable of measurements over the range of 2–2000 g.cm^{-1}. Tests were performed with a 25 mm diameter plate at 170 °C. The gap was set at 1 mm. Experiments were run on dried moulded tensile. Different rheological measurements were carried out:

- Ageing of materials at constant frequency 6.28 rad.s^{-1} (2π /s)
- Dynamic strain sweep tests to confirm the linearity of viscoelastic region up to 100 % strain, at 10 rad.s-1 frequency.
- Dynamic frequency sweep tests over a frequency range of 0.2 – 20 rad.s^{-1}
- Steady shear rate sweep tests over a shear rate rage of 0.02 – 20 s^{-1}

Ageing of materials at 170 °C was conclusive for all samples, even over 30 min of tests. This also showed that the drying conditions were good enough to prevent any degradation that may have occurred because of remaining traces of water. Linear viscoelastic responses of blends were determined by measuring their storage and loss shear moduli, respectively G' and G", in order to verify that tests were carried out in a linear viscoelastic region. P(L)LA, PBS and their blends were found to be within their linear viscoelastic range, for up to 5 % strain.

Results and Discussion

Mechanical properties: elongation at break and impact strength

The tensile elongation at break and impact strength of P(L)LA/PBS blends are given in Fig. 1. Blend impact strengths increased with PBS content in the blend: bars did not break above 30 wt% PBS in the blends. Both polymers had low initial elongation at break, 7 % for P(L)LA and 25 % for PBS. Blend elongation at break increased as a function of PBS content in the blend.

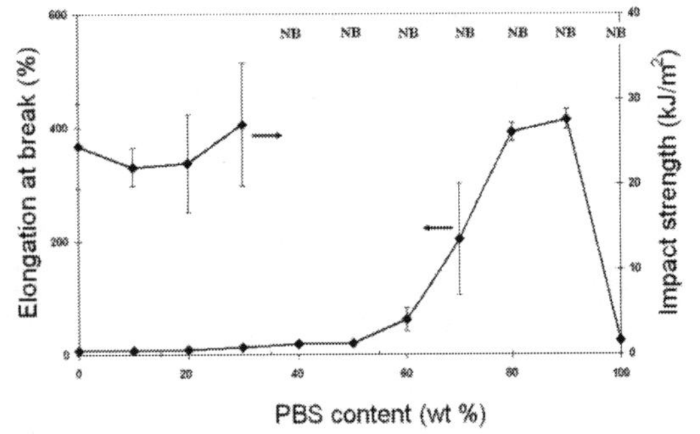

Figure 1 – Tensile properties of P(L)LA/PBS blends. Elongation at break and impact strength as a function of blend composition (NB stands for specimens that did not break on the impact test)

The elongation at break of the blends was quite similar to that of neat P(L)LA for compositions up to 40 wt% PBS and then, starting at 60 wt% PBS, increased with PBS content to 400 % for P(L)LA/PBS 90/10, which was well above pure PBS. For neat P(L)LA and blends under 40 wt% PBS no yield phenomenon was observed and samples broke rapidly after yielding (brittle fracture). For neat PBS, specimens broke without necking and stress increased with the strain up to the yield stress. PBS low elongation at break may originate from a higher crystallinity compared to that in blends, where PBS is the continuous phase. But since χ_c(PBS) in pure PBS was not significantly different from that of P(L)LA/PBS 90/10, 80/10 and 70/30, which had high elongation at break, other explanations need to be found to account for this observation.

Morphology

Environmental scanning electron microscopy was used to investigate the microstructure of P(L)LA/PBS blends. A good degree of mixing between the polymer phases is observed. Under 50 wt% PBS, microstructure of blends was characterized by relatively small spherical inclusions of what was assumed to be PBS in P(L)LA matrix. The particles had diameters increasing with PBS content ranging from 0.3 µm for P(L)LA/PBS 90/10 to 0.9 µm for P(L)LA/PBS 70/30 and appeared to be isolated from each other by the matrix for all pictures except for P(L)LA/PBS 50/50 which showed complex phase behaviours. Above 50 wt% PBS, inclusions of what is assumed to be P(L)LA in PBS were observed. Spherical P(L)LA droplets with smooth surface were in a continuous phase of PBS. The particle size (i.e. mean diameter) decreased with PBS content from 0.6 µm for P(L)LA/PBS 30/70 to 0.3 µm for P(L)LA/PBS 10/90. A traditional morphology of immiscible blends was observed since for all blends, interfaces between spherical inclusions and the matrix appeared to be sharp, which proved there was little interfacial adhesion between the two phases. This corroborates the results obtained by DSC and DMTA. The fact that a great level of mixture was obtained for this blend system even though those two polyesters were not miscible was consistent with the fact that P(L)LA and PBS have fairly close thermal expansion coefficients and a difference in solubility parameters of less than one, like in the case of P(L)LA/poly-ε-caprolactone (PεCL) blends (Park et al.,1992).

Dynamic Mechanical Thermal Analysis (DMTA)

One of the most common criteria to evaluate miscibility or phase separation of a polymer pairs is that of the T_g. In miscible blends, an intimate mixture is obtained at molecular level and therefore a single T_g is observed at an intermediate temperature of the pure components. On the other hand, when polymer blends exhibit two separate T_gs at the same positions as that of pure components, an immiscible system is envisaged. Intermediate examples belong to partially miscible systems, where two T_gs are observed and one, or both T_gs shift toward one another (Lopez Rodriguez et al., 2006). Yet, completely immiscible blend can also result in T_g variations. For instance, morphology of semi-crystalline polymer blends were reported increases in the T_g of polystyrene (the higher T_g of the blend), in polystyrene / polyethylene and polystyrene / polypropylene blends (Thirtha et al., 2006). In such cases, variations were of the order of 5 °C. Finally, neat homopolymers chain mobility can also be impacted by crystallinity. For P(L)LA in the case of an increase in crystallinity of 40%, a 5°C reduction in T_g was reported (Saiter et al., 2007). For pure PBS, the storage modulus E' decreases around -20 °C which was ascribed to glass transition, and at 110 °C due to melting. For pure P(L)LA the storage modulus E' fell off sharply around 60 °C due to glass transition and then increased again at 100 °C, which could be ascribed to cold crystallization. As for PBS, E' decreased moderately with temperature before decreasing sharply around 110 °C due to melting. From those observations it was concluded DMTA compared well with DSC results, even though heating speeds were different. In the case of the blends, the enhancement of E' due to the cold-crystallization of P(L)LA was shifted to lower temperatures by blending with PBS. This phenomenon indicated that PBS accelerated the cold crystallization of P(L)LA as pointed out by DSC and in previous works (Shibata et al., 2006; Wuk Park et al., 2002; Yokohara et al.,2008). Fig. 2 and 3 summarize the T_g values calculated from the tan(δ) peak maxima. Results showed two tan(δ) peaks, indicating blends were not miscible in the amorphous phase which may have originated either from the semi-crystalline nature of the components or/and the entropy restrictions of mixing those two high molecular weight polymers. On the other hand, previous studies (Wuk Park et al., 2002) demonstrated, using the Flory-Huggins theory, that P(L)LA and PBS are thermodynamically miscible and that P(L)LA and PBS can form a compatible mixture in the melt state, thus partial miscibility between P(L)LA and PBS in the amorphous phase can be expected.

Interestingly, T_g(P(L)LA) was dependent of the blend ratio, indicating that mutual dissolution may have taken place to some extent in the amorphous phase, in P(L)LA rich phases. For blends based on P(L)LA, T_g(P(L)LA) decreased from 69 °C, for neat P(L)LA, to 64 °C, for P(L)LA/PBS 50/50. Then, when phase inversion occurred at the equimolar composition of the blend (between 50 and 60 wt% PBS based on weight compositions) T_g(P(L)LA) reached a plateau. This observation suggested that although P(L)LA and PBS phase separated, an amount of amorphous PBS might have been dissolved in P(L)LA rich phase leading to a depression of the T_g. Partial miscibility of PBS in P(L)LA rich phase can be quantified by using the Fox equation (Eq. 2).

$$\frac{1}{T_{gb}} = \frac{\omega_{PBS}}{T_{g\,PBS}} + \frac{(1-\omega_{PBS})}{T_{g\,P(L)LA}} \qquad (2)$$

Here T_gb is the T_g of P(L)LA-rich phase in the blend, ωPBS accounts for the weight fraction of PBS into P(L)LA rich phases. Introducing the T_g values measured by DMTA (cf. Fig. 3) gave an approximation ranging from 2 wt% PBS miscible in P(L)LA rich phase (for P(L)LA/PBS 80/20) to about 4 wt% PBS (for P(L)LA/PBS 50/50). Then, above 50 wt% PBS in the blend the predicted mass fraction of PBS in P(L)LA rich phases remained constant. The trend of T_g(P(L)LA in P(L)LA/PBS blends very well match other studies (Lopez Rodriguez et al., 2006). In P(L)LA/PεCL blends (Lopez Rodriguez et al., 2006) T_g(P(L)LA) decreased with PεCL composition, as long as PεCL was the dispersed phase and then, above phase inversion, T_g(P(L)LA) remained constant. Using Fox equation, the authors ascribed the 7 °C T_g(PL)LA decrease to a miscibility of PεCL in P(L)LA rich phase of approximately 1 wt% PεCL, which was slightly lower than in the case of PBS. Those results are consistent with the total solubility parameters calculated from Hoy's group contributions of molar attraction constant method based on the classical method for Hansen solubility parameters (Van Krevelen, 1990).

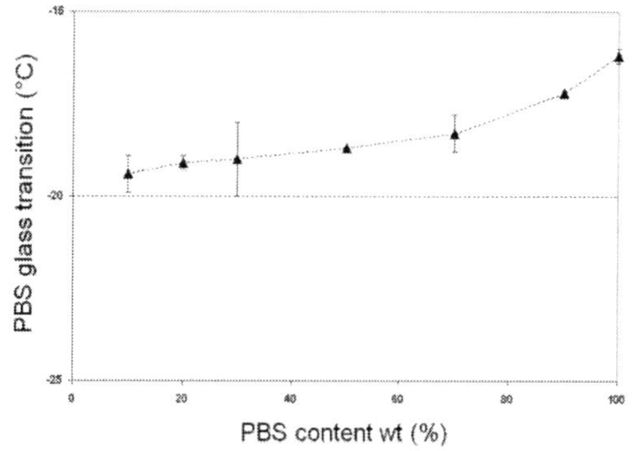

Figure 2 –T_g of PBS (determined by DMTA)

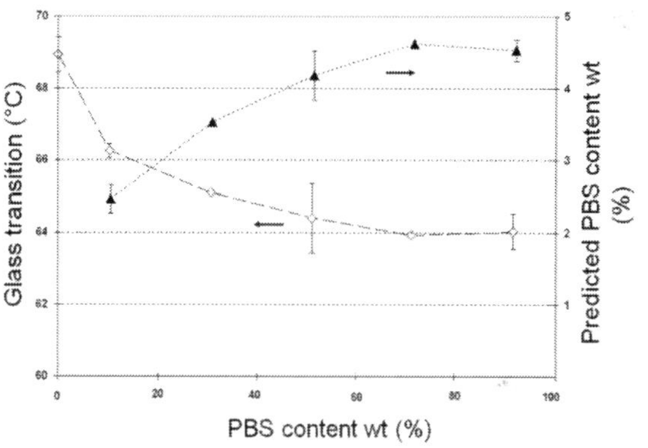

Figure 3 –T_g of P(L)LA (◊) in blends (determined by DMTA) and predicted PBS content in P(L)LA rich phase using Fox equation (▲), as a function of blend composition

Eq. 3 gives the calculation of the total solubility parameter based on Hoy's method.

$$\delta_t = \frac{F_t}{V} + \frac{277 \times \Delta_T^P}{0.5 \times V} \qquad (3)$$

With δ_t the total solubility parameter, V the molar volume of the structural unit of the polymer, F_t the molar attraction function and Δ_T^P the Lyderson correction for non-ideality. Results (see table 1) indicated that P(L)LA total solubility parameter 21.3 $(J/cm^3)^{1/2}$ was slightly closer from that of PBS 21.4 $(J/cm^3)^{1/2}$ than that of PεCL 20.4 $(J/cm^3)^{1/2}$, thus this accounted for the fact that PBS solubility in P(L)LA amorphous phases was greater than that of PεCL.

T_g(PBS) slightly decreased from -16 °C for pure PBS to -19 °C for P(L)LA/PBS 90/10 thus showing that mutual dissolution with P(L)LA did not take place in PBS rich phases. In fact, T_g variations, in the case of PBS, might originate from changes in blend morphology and crystallinity. Indeed polymer blend crystallintiy and morphology impact chain mobility by reducing it (Saiter et al., 2007; Thirtha et al., 2006), thus increasing T_g.

Table 1 – Total solubility parameters of P(L)LA, PBS and PεCL calculated from tables (Van Krevelen, 1990) using Hoy's group contribution method

Additives molar functions	P(L)LA	PBS*	PεCl
F_t	1119,5	2821,565	1985
V	54,8	137,4	101,5
Δ_T^P	0,08	0,21	0,15
δ_t $(J/cm^3)^{1/2}$	21.3	21.4	20.4

(*) PBS poly(butylenes-co- succinate-co-lactate) is based on 3 mol% lactate (Mitomo et al., 2005).

Consequently, the decrease of T_g(PBS), with increasing amount of P(L)LA in the blends may be explained by observing that both P(L)LA and PBS crytallinities decreased with P(L)LA content in the blend. DSC data showed that T_g(PBS) followed the trend of blend crystallinities by decreasing with P(L)LA content. The variation in T_g(PBS) was not as significant as the one of T_g(P(L)LA) and no sign of partial solubility of P(L)LA in PBS rich phase could be drawn from these data.

Rheological properties in molten state (ARES)

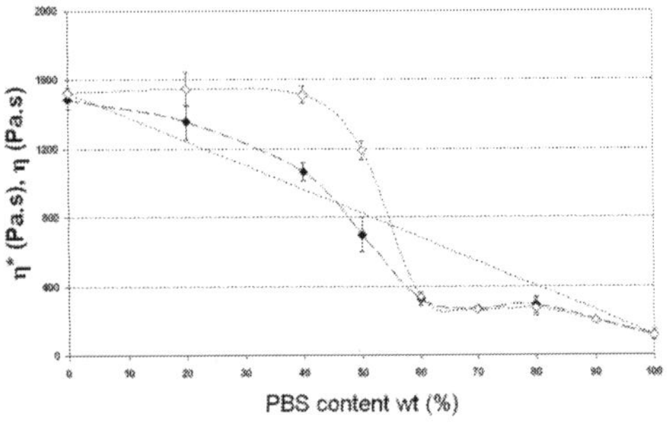

Figure 4 – Viscosity versus PBS content for P(L)LA/PBS blends at 170°C, (♦) complex viscosity at 0.05 rad/s; (◊) shear viscosity at $0.05s^{-1}$

Reports (Utracki, 1989) on polymer blend rheological behaviours by considering their thermodynamics and structures showed that polymer blends can be divided into four different groups. Plots of viscosities, at constant rates or frequencies, versus blend compositions make it possible to classify blend systems into those groups. Utracki (Utracki, 1983) underlined that this method should be based on zero shear viscosities while the second best criterion could be viscosity at constant shear stress. However, when neither of these two parameters would be available other rheological dependencies on composition could be used. The four different groups of blends are listed below: blends showing viscosities that follow the log-additivity rule, ($\log \eta_i = \phi_i \log \eta_i$ where η is the viscosity and ϕ is the blend volume fraction), blends having viscosities higher than the log-additivity rule called Positive Deviation Blends (PDB), those where the opposite effect is observed named Negative Deviation Blends (NDB), and the remaining mixed-behaviour systems called Positive-Negative Deviation Blends (PNDB). Generally, blend having viscosities following the linear log-additivity rule are considered to be miscible systems (Tayyab et al., 2007). To PDB belong the homogenous miscible blends and the heterogeneous blends with strong inter-domain interactions that could be either compatible or completely immiscible. To NDB belong heterogeneous blends where the interactions are weak (either miscible or immiscible). Finally, intermediate systems belong to PNDB type, which are heterogeneous blends in which there is a concentration-dependent transition of structure. Examples of Positive-Negative Deviation Blends (PNDB) are reported in literature e.g. high-density polyethylene/polystyrene (Han et al., 1975), polyethylene terephthalate/polyamide6-6 (Utracki, 1983) and poly(methyl methacrylate)/polystyrene (Utracki, Sammut, 1990). Those systems are characterized by a shift of viscosity (from PDB to NDB) at concentrations related to phase inversion (Utracki, 1989).

Plots of the storage modulus versus the loss modulus of P(L)LA, PBS and their blends at 170 °C show that for P(L)LA/PBS 80/20, 60/40 and 50/50, G' presented a shoulder peak related to shear thinning behaviour in the low frequency region, proving the existence of long relaxation time. Besides, dynamic complex viscosities of blends as a function of frequency at 170 °C of P(L)LA, PBS and P(L)LA/PBS 40/60, 30/70 and 20/80 exhibited a Newtonian behaviour in the investigated frequency range while P(L)LA/PBS 80/20, 60/40 and 50/50 showed a strong thinning behaviour at low frequencies. For all frequencies, P(L)LA complex viscosity was higher than that of PBS and blends. Fig. 4 shows P(L)LA, PBS and blend complex viscosities at 170 °C, at a fixed frequency of 0.05 rad/s. P(L)LA complex viscosity was higher than that of PBS, and complex viscosity followed approximately a PNDB behaviour: P(L)LA/PBS 80/20, 60/40 and 50/50

exhibited values higher than the log-additivity line while blends over 50 wt% PBS were under this line. In addition, for high PBS content, complex viscosities were very close from the log-additivity rule thus suggesting miscibility for P(L)LA/PBS 20/80 and 10/90. These results indicated P(L)LA and PBS were immiscible for most blend compositions while revealing potential miscibility in the melt for high PBS contents.
Dependence of steady shear rate viscosity on shear at 170 °C for neat polymers and their blends shows neat polymers, P(L)LA/PBS 20/80 and 10/90 present a plateau at low shear rate indicating zero-shear viscosities. The shear viscosities of all blend decreased with increase in shear rate, exhibiting shear thinning behaviour. Those shear thinning behaviours were particularly intense for P(L)LA/PBS 80/20, 60/40, 50/50 and 40/60 thus causing the crossover of the viscosity curves of blends. Similar behaviours have been reported for aliphatic polyesters such as PεCL (Kister et al., 1998), PBS and P(L)LA (Bhatia et al., 2007). Fig. 4 shows P(L)LA, PBS and blend shear viscosities at 170 °C, at a fixed rate of 0.05 s^{-1}. P(L)LA shear viscosity was higher than that of PBS and all other blends. Like for complex viscosity, shear viscosity followed approximately a PNDB behaviour, even though experimental points didn't perfectly fit in with theory (Utracki, 1983). Again, for blends having high PBS content, melt miscibility was inferred since shear viscosities were very close from the log-additivity line. This intimated that at 170 °C, P(L)LA and PBS were miscible in P(L)LA/PBS 20/80 and 10/90, while for all other samples melt immiscibility could be inferred.

The results showed that the compositional dependence of both complex viscosity and zero shear viscosity of PL(L)A/PBS blends deviated greatly from linear log-additivity. When PBS was dispersed in a P(L)LA continuous phase, the blends showed positive deviation, while for blends with P(L)LA dispersed in a PBS matrix, negative deviation was generated. Thus, around the equimolar composition of the blend (between 50 and 60 wt% PBS based on weight compositions) a shift of viscosity (from PDB to NDB) is reported. Most blend complex viscosity and shear viscosity did not lie between that of the constituents. However, miscibility in the melt was suggested for blends having high PBS content. Then, a parallel can be drawn between DMTA and ARES results, since DMTA results also indicated P(L)LA/PBS blend system was possibly partially miscible in P(L)LA rich phases and that blends having the highest amount of PBS showed the largest interactions. Still, those results are not consistent with the findings of Bhatia. They reported a PDB behaviour for P(L)LA/PBS (Bhatia et al., 2007) and highlighted miscibility for blends having PBS contents lower than 20 wt%. Both the study of Bhatia and this study used extrusion blending on comparable high molecular weight polymers. Those discrepancies may be attributed to the complex P(L)LA/PBS system and its multiphase morphology (Wuk Park et al., 2002).

Conclusions

The effects of processing blends of two biodegradable polyesters have been studied. The mechanical properties can be controlled by selectively formulating a blend composition for a specific end use. Starting from pure P(L)LA, impact strength sharply increased at 30 wt% PBS while elongation at break increased at 60 wt% PBS. Interestingly, the best elongation at break was obtained for 90 wt% PBS while pure PBS had an unexpected low elongation at break which could not be accounted for.
Microscopy revealed typical immiscible morphologies while a great level of mixture was observed. Two distinct melting temperatures, corresponding to the melting point of the individual polymers was measured by DSC, which indicated that these blends did not undergo co-crystallization. Small shifts in both melting temperatures revealed the existence of thermodynamic interactions between P(L)LA and PBS.
Further, there were distinct double peaks in the dynamic mechanical spectra ascribed to the glass transition of each component thus intimating P(L)LA and PBS were not miscible in the amorphous phase either. Small shifts in the T_g(P(L)LA) were believed to be an indication of interactions between the two polymers. Using Fox equation a maximum of 4 wt% PBS miscibility in P(L)LA rich phase was estimated for blends having high PBS contents. Still, mutual dissolution was not observed since T_g(PBS) did not increase with P(L)LA content.
Interactions into the melt were measured by melt viscosity revealing a PNDB behaviour for P(L)LA/PBS blend system. Blends with high PBS content were very close from the log-additivity rule, thus indicating potential melt miscibility.
In conclusion, further work should be performed on those materials in order to evaluate interchange reactions. Eventually it would also be of interest to optimize the blending conditions possibly using reactive compatibilizers as long as those blends would remain biodegradable even if chemically modified.

Acknowledgements

The authors wish to thank Mr M. Longerey for his generous help, and gratefully acknowledge Dr. A.S. Caro and Prof. P. Gaudon, for the discussions and their help with interpretations. The authors are also indebted to Mr. J.M. Taulemesse for the SEM micrographs. The authors gratefully acknowledge the Agence Nationale de la Recherche (ANR) that financially supported this work.

References

Bhatia A., Gupta R.K., Bhattacharya S.N., Choi H.J., "Compatibility of Biodegradable Poly (lactic acid) (PLA) and Poly (butylene succinate) (PBS) Blends for Packaging Application", Korea-Aust. Rheol. J., **19**(3), 125-131 (2007)

Han C.D., Kim Y.W, "Dispersed Two-Phase Flow of Viscoelastic Polymeric Melts in a Circular Tube", Trans. Soc. Rheol., **19**(2), 245-269 (1975)

Hoffman J.D., Davis G.T., Lauritzen I., "Chapter 7 The rate of crystallization of linear polymers with chain folding", in Treatise on solid state chemistry, Hanny NB, Editor, vol. 3. New York: Plenum Press, (1976)

Kister G., Cassanas G., Vert M., "Effects of morphology, conformation and configuration on the IR and Raman spectra of various poly(lactic acid)s", Polymer, **3**(2), 267-273 (1998)

López-Rodríguez N., López-Arraiza A., Meaurio E., Sarasua J.R., "Crystallization, morphology, and mechanical behavior of polylactide/poly(ε-caprolactone) blends", Polym. Eng. Sci., **46**(9), 1299-1308 (2006)

Mitomo H., Kaneda A., Quynh T.M., Nagasawa N., Yoshii F., "Improvement of heat stability of poly(L-lactic acid) by radiation-induced crosslinking", Polymer, **46**(13), 4695-4703 (2005)

Park T.G., Cohen S., Langer R., "Poly(L-lactic acid)/pluronic blends: Characterization of phase separation behavior, degradation, and morphology and use as protein-releasing matrices", Macromol., **25**(1), 116-122 (1992)

Saiter A., Delpouve N., Dargent E., Saiter J.M., "Cooperative rearranging region size determination by temperature modulated DSC in semi-crystalline poly(l-lactide acid)", Europ. Polym. J., **43**(11), 4675-4682 (2007)

Sarasua J.R., Prud'homme R.E., Wisniewski M., LeBorgne A., Spassky N., "Crystallization and Melting Behavior of Polylactides", Macromol., **31**(12), 3895-3905 (1998)

Shibata M., Inoue Y., Miyoshi M., "Mechanical properties, morphology, and crystallization behavior of blends of poly(l-lactide) with poly(butylene succinate-co-l-lactate) and poly(butylene succinate)", Polymer, **47**(10), 3557-3564 (2006)

Tayyab H., Ibnelwaleed A.H., "Melt miscibility and solid-state properties of metallocene LLDPE blends with HDPE: Influence of MM of LLDPE", J. Cent. South Univ. T., **14**(1), 183-185 (2007)

Thirtha V., Lehman R., Nosker T., "Morphological effects on glass transition behavior in selected immiscible blends of amorphous and semicrystalline polymers", Polymer, **47**(15), 5392-5401 (2006)

Utracki L.A., Sammut P., "On the uniaxial extensional flow of polystyrene/polyethylene blends", Polym. Eng. Sci., **30**(17), 1019-1026 (1990)

Utracki L.A., "Melt flow of polymer blends", Polym. Eng. Sci., **23**(11), 602-609 (1983)

Utracki L.A., Polymer alloys and blends, 1st edition, Hanser Publishers, New York (1989).

Van Krevelen D.W., "Chapter 5 Calorimetric properties", in Properties of polymers, 3rd edition Amsterdam: Elsevier, (1990)

Wuk Park J., Soon Im S., "Phase behavior and morphology in blends of poly(L-lactic acid) and poly(butylene succinate)", J. Apll. Polym. Sci., **86**(3), 647-655 (2002)

Yokohara T., Yamaguchi M. "Structure and properties for biomass-based polyester blends of PLA and PBS", Eur. Polym. J., **44**(3), 677-685 (2008)

MULTIPLE SHAPE-MEMORY EFFECT POLYMERS ON THE BASIS OF PEROXIDE CROSSLINKED BINARY AND TERNARY POLYETHYLENE BLENDS

H.-J. Radusch*, I. S. Kolesov

Center of Engineering Sciences, University Halle-Wittenberg, Halle, Germany
hans-joachim.radusch@iw.uni-halle.de

In our prior investigation was shown that shape-memory polymers can be successfully produced on the basis of peroxidic crosslinked blends of ethylene-1-octene copolymers and high density polyethylene using the melting temperature of the specific crystalline phases as switching temperature for starting recovery, and the covalent network in the amorphous phase as the essential element of driving recovery forces. Continuative investigations showed that binary and ternary blends from different branched ethylene-1-octene copolymers and/or linear high density polyethylene as well as polycyclooctene can be created possessing a multiple shape-memory behavior. The shape-memory behavior of the blends investigated was evaluated by the characteristic parameters "shape-memory recovery strain", "strain fixity ratio" and "recovery strain rate" in dependence on temperature. The shape-memory experiments have been carried out in tensile mode at different programming temperatures at constant programming strain. In dependence on branching degree of the ethylene-1-octene copolymer used and the cross-linking degree generated by the peroxide the single ethylene-1-octene copolymers and high density polyethylene exhibit a very good shape-memory behavior with high values of strain fixity ratio and strain recovery ratio of up to 99% at suitable programming temperatures. On the basis of these results binary and ternary blends were generated for creation of the multiple shape-memory effect. Multiple shape-memory behavior appeared at sequential stepwise application of good adjusted values of programming deformation and programming temperature for each blend component and crystalline phase, respectively, i.e., multi-step programming. Polycyclooctene as a further blend component was involved into the investigation to achieve a multiple shape-memory effect only with one-step programming. Furthermore a comparison between chemical crosslinking by peroxide and crosslinking by electron irradiation was performed. The results will be discussed in connection to the blend morphology.

Introduction

The shape-memory (SM) effect of materials was detected approximately 60 years ago [1]. Materials exhibit a shape memory effect if they can be deformed and fixed into a temporary shape and recover their original permanent shape only by the influence of external stimuli like heat, light, magnetic field, etc. Today, the most widely used shape-memory material is the Ni–Ti alloy "Nitinol ®" [2]. Intensive research in the field of polymeric shape memory materials started in the beginning of the 80th of the last century [3]. Currently, several research groups are intensively engaged in research of SM polymer materials generation as well as characterization [4-8]. The fields of use of SM materials are quite wide and innovative, thus very different applications became known, e.g. for heat-shrinkable tubing and films [3], packaging industry, automotive or aerospace [6]. SM polymers have also been developed for biomedical applications, e.g. for stents and microtubing or suture materials [8]. For this purpose only biocompatible polymers come into consideration. Preconditions for proper appearance of the thermal-induced SM effect are either the existence of a stable physical (entanglement) or a covalent molecular network, and glass or phase transition (e.g. melting) at matching temperatures, i.e. the transition temperature T_{trans}. Good results for effective working SM polymers were mainly achieved if the SM polymer shows a pronounced phase character caused by block-copolymer basis or crystalline-amorphous structure of the material.

Multiple SM polymers are of specific interest because of the opportunity to undergo stepwise temperature controlled shape changes, and thus several technical functions become possible. As multiple SM material a material is named if after e.g. a thermal stimulus the change from a first shape to a second and from there to a third etc. becomes able. Such a multiple SM effect is possible only on the basis of specific phase morphology of the polymer system used. Normally multiple SM polymers need multi-step programming, but recently Behl et al. have shown [9] that the triple-shape capability with good SM characteristics can be initialized only by conventional one-step programming for complex multiphase (AB) polymer networks, which were created on the basis of poly(ε-caprolactone) and poly (cyclohexyl methacrylate) with suitable compositions.

Unfortunately, in the most cases effective SM polymers have to be especially synthesized, and therefore they are rather expensive. Thus, a big interest exists to create easy to produce and cheap SM polymer materials.

In our investigations was shown that shape-memory polymers can be successfully produced on the basis of peroxidic crosslinked ethylene-1-octene copolymers (EOC) and high density polyethylene (HDPE) using the melting temperature of the specific crystalline phases as switching temperature for initiating recovery, and the covalent network in the amorphous phase as the essential element of driving recovery forces. Binary and ternary blends from different types of polyethylene were produced and characterized in regard to their SM behavior. Crosslinking of the amorphous phase of the polyethylenes was performed by peroxides and alternatively also by electron irradiation. Multiple SM effect could be created based on the heterogeneous crystalline phase structure of the polyethylene blends. Additionally and alternatively to PE polycyclooctene (PCO) was used for improving the SM effect of the polyolefinic blends.

Experimental

Materials

Short-chain branched polyethylenes (SCBPE) with approx. 30 and 60 hexyl branches per 1000 C (EOC30 and EOC60, respectively) as well as high density polyethylene (HDPE) produced by Dow Chemical were used. Polycyclooctene (PCO) by Evonik was involved in present investigation instead of EOC60. Polycyclooctene is thermodynamically incompatible with polyethylenes and has a low melt viscosity. In Table 1 relevant parameters of these materials are displayed.

Table 1 - Parameters of polymer materials used.

Polymer	Melting Temp. T_m [°C]	Density ρ [kg/m^3]	Molecular Mass $\overline{M}_w$ [kg/mol]	Polydis-persity $\overline{M}_w/\overline{M}_n$	Branching Degree $[CH_3/1000C]$
HDPE	132	955	-	-	-
EOC30	96	900	64	2.3	~ 30
EOC60	59	870	87	2.5	~ 60
PCO	54	910	90	5.2	-

As crosslinking agent for chemical crosslinking 2,5-dimethyl-2,5-di-(tert. butylperoxy)-hexane (DHBP) (Evonik) was employed. The use of DHBP guaranteed a high efficiency and a good processability of the chosen systems due to the availability of two peroxy groups and the high thermal stability at before the initiation of reaction. For SM sample preparation the used polymer pellets were impregnated with a nominal DHBP concentration of 1, 2, 3 and 4 wt%. DHBP.

Melt mixing of DHBP impregnated polymers in the compositions as shown in Table 2 was performed using a single-screw extruder (Brabender) with the barrel temperature of 130°C whereby the homogenization of DHBP was achieved too. As reference sample a cross-linked HDPE was prepared and investigated subsequently.

Table 2 - Composition of SM polymer samples.

Sample designation	Composition			
	HDPE [%]	EOC30 [%]	EOC60 [%]	PCO [%]
100HDPE	100	-	-	-
50HDPE/50EOC30	50	50	-	-
33HDPE/33EOC30/34EOC60	33	33	34	-
10HDPE/45EOC30/45EOC60	10	45	45	-
10HDPE/25EOC30/65EOC60	10	25	65	-
50EOC30/50EOC60	-	50	50	-
30EOC30/70EOC60	-	30	70	-
50 EOC30/50 PCO	-	50	-	50
20 HDPE/40EOC30/40 PCO	20	40	-	40

The true DHBP content after compound preparation was determined by thermogravimetric analysis (TGA, *Mettler-Toledo*). The extrudates were further processed to films with thickness of 1 and 2 mm and cross-linked at a temperature of 190°C. Cross-linking time and temperature were optimized using a vulcameter „Elastograph"(Göttfert). For comparison, beside chemical crosslinking the PCO containing blends were alternatively crosslinked by electron irradiation by means of an electron beam source. Blend sheets were impacted with an irradiation dose of 200 kGy at room temperature under nitrogen atmosphere.

Network characterization

The crosslink density ν_c representing the density of network chains is connected with $\overline{M}_c$ and the material density ρ at measurement temperature by the simple equation $\nu_c = \overline{M}_c/\rho$. The average molecular mass of the links, i.e., the network chains between two neighboring network nodes $\overline{M}_c$, were calculated by means of the *Flory-Rehner* equation [10] on the basis of swelling measurements. The swelling measurements were performed at the samples stored in xylene at 110 °C after at least 10 h using the gel of crosslinked polymer obtained after extraction of the soluble fraction. The ν_c and $\overline{M}_c$ values for the crosslinked polymers were also evaluated from stress-strain diagrams ascertained at 120°C on the basis of the *Mooney-Rivlin* equation [11,12]. Hence, in the text, tables and legends below, the mean ν_c values calculated on the basis of both swelling and tensile tests are given.

Tensile tests

The tensile tests in the temperature range of SM samples programming (120°C and 140°C) have been carried out using a mechanical testing machine Zwick 1425 (Zwick/Roell) with heating chamber and a load cell 10 N at a crosshead speed of 50 mm/min. Shouldered test bars were used with the cross-section area 2.0×2.0 mm^2.

Thermal analysis

Thermal analysis was performed with a DSC 820 (Mettler-Toledo) at a heating and cooling rate of 20 and 10 K/min, respectively.

Shape-memory behavior

For the characterization of the shape-memory behavior, at first, the samples were treated under a specific temperature-deformation program. The samples were programmed by a strain of $\varepsilon_{pl} \leq 100$ % during a dwell period of 120 s at the programming temperature T_{p1}, and cooled down to T_{p2} with an average cooling rate of approx. 11 K/min in a constant deformed state. Before loading to the next programming strain ε_{pi} the sample temperature was kept constant for 10 min. After the last programming step the specimen was cooled down to 10°C at the programming strain of 100%, thermal equilibrated for 10 min and finally unloaded. The recovery strain was measured not before a delay of 10 min at 25°C (ε_v) and then in the course of a heating run with a rate of 2 K/min and 'zero' stress of 70 Pa (ε_{rec}(T)). The SM investigations were carried out in tensile mode using a mechanical spectrometer with the measuring head Mark III (Rheometric Scientific). The ability of SM polymers to fixation the strain and to regeneration the shape is characterized by the strain fixity ratio (R_f) and the strain recovery ratio (R_r), respectively, which are defined according to Lendlein and Kelch [4] as

$$R_f = \frac{\varepsilon_v}{\varepsilon_{pr}} \qquad R_r = \frac{\varepsilon_{pr} - \varepsilon_{rec,m}}{\varepsilon_{pr}} \qquad (1)$$

where ε_{pr} is the strain caused by programming, ε_v is the strain that remains after programming, cooling, unloading and recovery at the lowest temperature of experiment of specimen and $\varepsilon_{rec,m}$ is the residual strain that remains after thermal-induced recovery (shrinkage) at maximum temperature of experiment. This means that in the case of ideal SM behavior the relations $\varepsilon_v = \varepsilon_{pr}$ and $\varepsilon_{rec,m} = 0$ must be valid.

If strain recovery is recorded at constant heating rate, the relation between temperature and measuring time is linear. In this connection the kinetics of SM recovery can be described by means of the recovery rate, i.e. the first derivative of recovery strain (ε_{rec}(T)) with respect to the time ($d\varepsilon_{rec}$(T)/dt). For characterization of the SM recovery kinetics the absolute peak value of recovery rate $|d\varepsilon_{rec}(T)/dt|_{max}$ has been employed. The switching temperature (T_{sw}), as a further important characteristic of SME, was estimated as the temperature, which corresponds to the peak of recovery rate [13,14].

Results and Discussion

Network parameters ν_c and $\overline{M}_c$ were determined on the basis of the stress-strain diagrams and the *Mooney-Rivlin* equation for uniaxial tension as well as from swelling measurements by means of the *Flory-Rehner* equation. The fitted values of *Mooney-Rivlin* equation coefficients C_1 and C_2 as well as calculated values of ν_c and M_c for the polyethylene types investigated are collected in Table 3. The ν_c values at a temperature of 140 °C change not essentially and varies approximately from $0.8 \cdot 10^{-10}$ mol/m³ for HDPE and $1.5 \cdot 10^{-10}$ to $1.7 \cdot 10^{-10}$ mol/m³ for blends at 140°C, respectively. A small decrease of ν_c and consequently increase of $\overline{M}_c$ values was found for the reduced temperature of 120°C [13].

Beside the network characteristics of the amorphous phase, which are responsible for the development of recovery forces, the characteristics of the crystalline phases are of essential impact for the control of the switching temperature for initiating recovery. Table 4 and 5 represent the thermal characteristics of the crystalline phases of the generated SM polymers determined in heating and cooling run, respectively.

Table 3 - Network parameters ν_c and $\overline{M}_c$ of cross-linked HDPE and HDPE/EOC30/EOC60 blends.

Blend Composition	$2C_1$ [MPa]	$2C_2$ [MPa]	T [°C]	$\nu_c \cdot 10^{10}$ [mol/m³]	$\overline{M}_c$ [kg/mol]
100HDPE	0.135	1.386	140	0.79	10.093
50HDPE/50EOC30	0.266	0.879	140	1.55	5.140
33HDPE/33EOC30/34EOC60	0.292	0.602	140	1.70	4.682
10HDPE/45EOC30/45EOC60	0.267	0.468	140	1.55	5.119
50EOC30/50EOC60	0.253	0.428	140	1.48	5.391
10HDPE/25EOC30/65EOC60	0.252	0.418	140	1.47	5.419
30EOC30/70EOC60	0.249	0.359	140	1.45	5.480

Table 4 - Melting temperatures T_m of crystalline phases in SM polymers and total melting enthalpies $\Delta H_{m\Sigma}$

Blend Composition	Melting			
	$T_{m\,HDPE}$ [°C]	$T_{m\,EOC30}$ [°C]	$T_{m\,EOC60}$ ($T_{m\,PCO}$) [°C]	$\Delta H_{m\Sigma}$ [kJ/kg]
100HDPE	124.3	-	-	139
50HDPE/50EOC30	116.1	1)	1)	108
33HDPE/33EOC30/34EOC60	114.0	1)	1)	86
10HDPE/45EOC30/45EOC60	~114 2) ~105 3)	~ 91 3) 82.9	~53 2)	74
50EOC30/50EOC60	-	85.0	~68 2) ~35 2)	68
10HDPE/25EOC30/65EOC60	~115 3) 108.3	~94 2) ~79 2)	~61 3) ~30 2)	65
30EOC30/70EOC60	-	82.7	64.7 ~35 2)	64
50 HDPE/50 PCO	126*	-*	(40.3)*	114.7*
20 HDPE/40EOC30/40 PCO	120.8*	90.3*	(40,6)*	100.9*

1) peaks or shoulders assignable to EOC30 or EOC60 phase not to identify in thermogram; 2) weak and flat shoulder; 3) distinct shoulder. * electron irradiation crosslinked samples

Generally, as displayed in Table 4 and 5 all blends, especially ternary blends demonstrate *multiple* behavior for both melting and crystallization. In some ternary blends the HDPE phase crystallizes only at 97.7° and 100.3°C, respectively, because the crystallization enthalpy amounts just approx. 5 and 10 % of the expected value, respectively [13].

Table 5 - Crystallization temperatures T_c of crystalline phases in SM polymers and total crystallization enthalpies $\Delta H_{c\Sigma}$

Blend Composition	Crystallization			
	$T_{c\,HDPE}$ [°C]	$T_{c\,EOC30}$ [°C]	$T_{c\,EOC60}$ ($T_{c\,PCO}$) [°C]	$\Delta H_{c\Sigma}$ [kJ/kg]
100HDPE	108.0	-	-	139
50HDPE/50EOC30	100.3	~71	-	109
33HDPE/33EOC30/34EOC60	97.7	84.7	~20 2)	89
10HDPE/45EOC30/45EOC60	97.7	77.2	~45 3)	73
50EOC30/50EOC60	-	62.8	~46 3)	68
10HDPE/25EOC30/65EOC60	100.3	74.3	55.7 40.3	68
30EOC30/70EOC60	-	58.2	~39 3)	63
50 HDPE/50 PCO	113.3*	-*	(24.3)*	116.2*
20 HDPE/40EOC30/40 PCO	109.6*	77*	(33)*	98.5*

1) peaks or shoulders assignable to EOC30 or EOC60 phase not to identify in thermogram; 2) weak and flat shoulder; 3) distinct shoulder. * electron irradiation crosslinked samples

Moreover, local cross-linking degree in the HDPE phase is lower than the average cross-linking degree of blends as given in Table 3. Obviously, the main part of the HDPE phase crystallizes together with the EOC30 phase at significantly lower temperature. Interestingly, the melting of HDPE phase in the ternary blends 10HDPE/45EOC30/ 45EOC60 and 10HDPE/ 25EOC30/65EOC60 occurs also stepwise in two stages. As the temporary shapes are fixed with different crystalline domains, for a sharp stepwise recovery, melting in a narrow temperature range for each component of blend is required.

Shape-memory behavior

The temperature dependences of SM recovery strain and recovery strain rate of HDPE and HDPE/EOC30/EOC60 blends and their components having networks with similar cross-linking degree are demonstrated in Figure 1. Here, only the first cycles of SM tests are demonstrated. The sharp step of SM recovery strain and high, well separated peaks of SM recovery rate could be generated by melting of the phase with sufficient high crystallinity and relatively perfect crystallites, which show strong and well separated DSC peaks. For explicit appearance of SM effect the availability of stored sufficient high viscoelastic forces is as important as melting behavior. These viscoelastic forces caused by network deformation whilst programming must be efficiently stored by the crystalline structure formed in the actual temperature range; i.e. the crystalline network has to hold during programming achieved deformation after unloading. As expected, the blends having a higher average crystallinity and well resolved melting peaks show the high strain fixity ratio R_f and SM recovery rate $-d\varepsilon/dt$. Correspondingly, these parameters increase

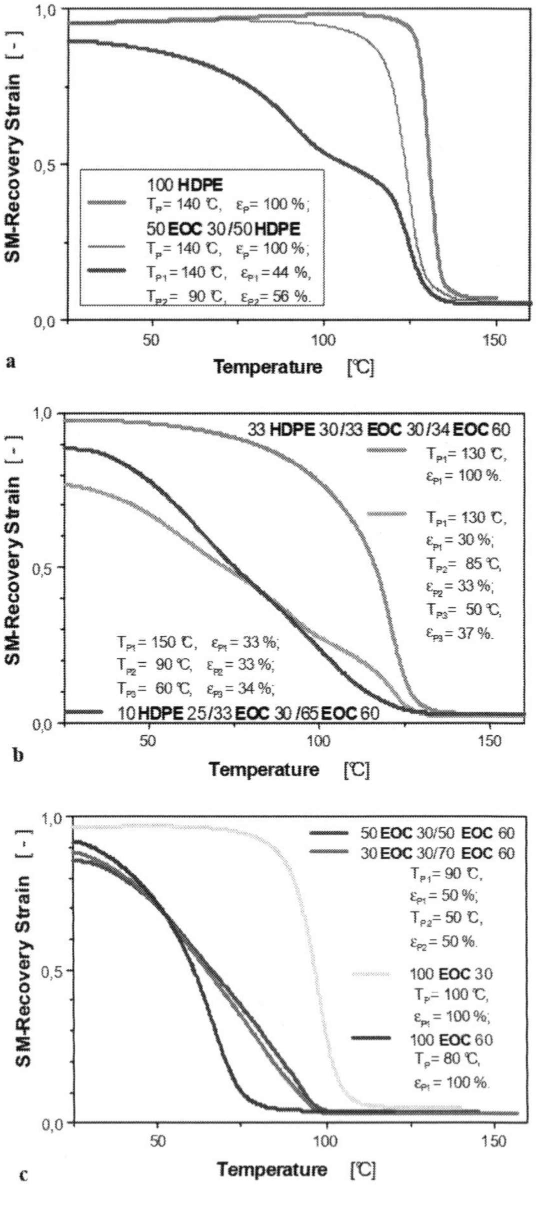

Figure 1 - Effect of composition and programming conditions on the temperature dependence of SM recovery strain of HDPE and HDPE/EOC30/EOC60 blends.

with increasing HDPE (and partly EOC30) content in blends. The one-step programming of all blend types results in appearance of only one step and accordingly one peak in temperature dependences of SM recovery strain and SM recovery rate, respectively, i.e. in dual-shape behavior.

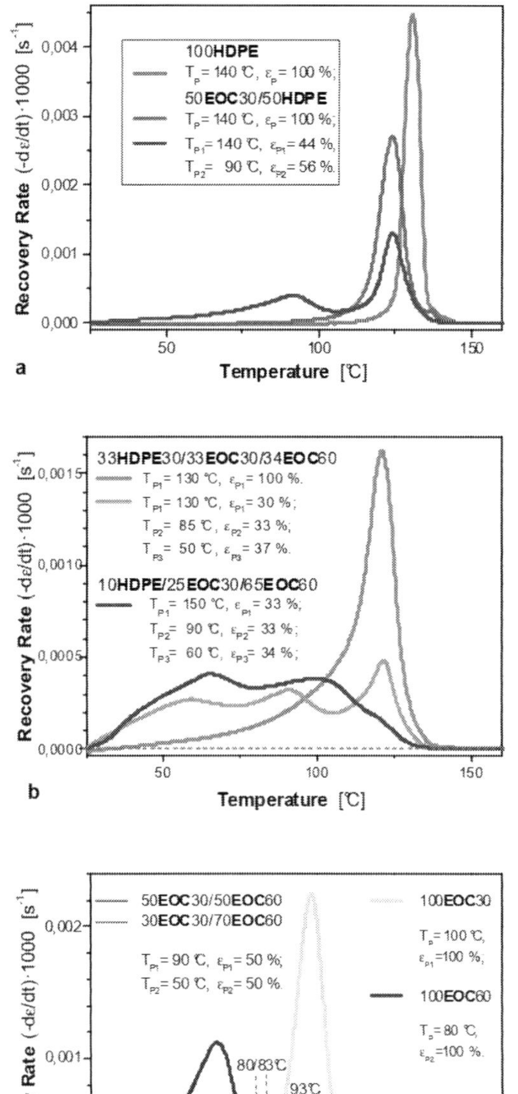

Figure 2 - Effect of composition and programming conditions on the temperature dependence of SM recovery strain rate of HDPE and HDPE/EOC30/EOC60 blends.

The temperatures for two- or three-step programming were chosen allowing for the temperatures of a quiescent crystallization of present crystal populations as characterized by means of DSC. Compared to the specimens arranged by one-step programming the same blends after two- or three-step programming show significant lower R_f and $-d\varepsilon/dt$ values and particularly peak values $[-d\varepsilon/dt]_{max}$ at the melting temperatures of components [13].

Disadvantages of above described multi-step programming besides the technical complexity of procedure and in many cases the very diffuse and indistinct steps and peaks of SM recovery strain and rate, respectively, are the relatively low values of strain fixing ratio R_f and strain recovery ratio R_r. This undesired effect by multi-step programming is a consequence of the drastic increase of irreversible deformation part caused by plastic deformation of already crystallized phase/phases in the second and third programming step.

As discussed above for the blends of linear and short-chain branched polyethylenes the multiple SM behavior cannot be achieved by one-step programming. One reason for this behavior is seen in the high chemical affinity of the blend partners in molted state and correspondingly by insufficiently segregated crystalline phases of different thermal stability that was also indicated by the results of DSC investigations. In this connection the polycyclooctene, which is thermodynamically incompatible with polyethylenes and has a low melt viscosity, was involved in recent investigation instead of EOC60. Already Liu et al. [15] have shown that the single PCO cross-linked by dicumyle peroxide (DCP) exhibits a SM behavior with high performance.

Figure 3 demonstrates the SM behavior of a binary and a ternary PE/PCO blend. The melt mixed blend specimens were subsequently cross-linked by peroxide (DHBP) as well as by electron radiation (ER) alternatively, and programmed by one-step process. It is interesting to see that the chemical crosslinked PCO containing blends show not a significant better multiple SM behavior in comparison to the HDPE/EOC blends. But blends, which were cross-linked by electron irradiation, show a more pronounced appearance of multiple SM behavior and higher R_f values compared to blends cross-linked by peroxide. These differences can be ascribing to the strong decrease of crystallinity and melting temperature in the case of peroxidic cross-linking of PCO. Preliminary investigation point out that PCO owning the lowest melt temperature and viscosity compared to the other components forms the matrix of the blends during mixing at the given compositions. The PCO matrix does not constrain the SM recovery of the blend system after melting of the crystalline phase / phases.

The SM blends cross-linked by electron irradiaton offer high values of strain fixing and strain recovery ratios of 95 to 99 %.

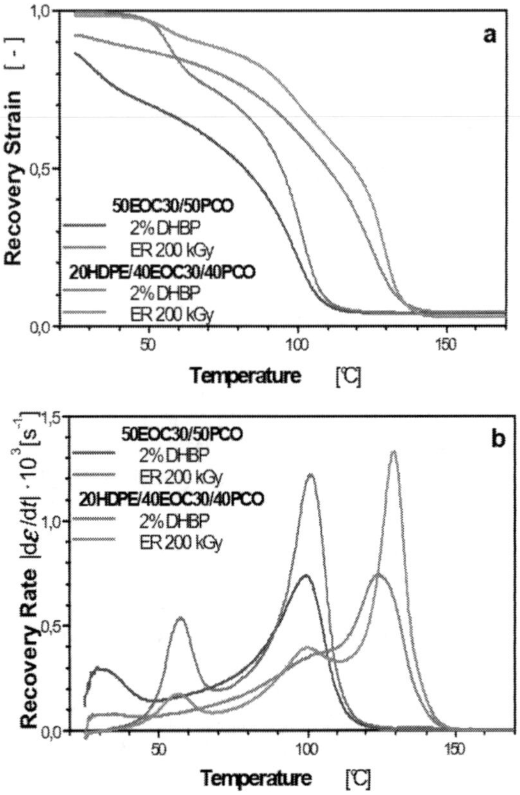

Figure 3 - Comparison of SM recovery strain (**a**) and SM recovery rate (**b**) of binary and ternary blends (50EOC30/50PCO and 20HDPE/40EOC30/40PCO) cross-linked by peroxide (2% DHBP) or electron irradiation (200 kGy) as function of temperature. One-step programming of the specimens at 120°C.

Conclusions

The finding of the investigation of EOC/HDPE blends show that multiple SM behavior appears only at consequent stepwise application of convenient programming strain and temperature. However, only for 50HDPE/50EOC30 blend, which shows especially fine disperse and distinctly segregated phase morphology the temperature dependence of SM recovery strain demonstrates two pronounced steps. All the other binary and ternary EOC/HDPE blends exhibit the triple or quadruple shape-memory effect only as very diffuse steps in the SM recovery strain curves and as rather small superimposed peaks of SM recovery rate vs. temperature. Obviously that is caused by multiple crystallization and melting behavior of these blends in correlation to the melt-mixing and crosslinking procedure. Using electron irradiation for blend crosslinking a pronounced improvement of the SM behavior as well as multiple SM effects can be generated.

Acknowledgements

The authors wish to thank the German Research Federation (DFG) for financial support. We thank the Leibniz Institute of Polymer Research (G. Heinrich, U. Wagenknecht and U. Gohs) for electron irradiation experiments.

References

[1] Chang, L.C., Read, T.A., "Plastic deformation and diffusion-less phase changes in metals - the gold-cadmium beta phase", Trans AIME, **189**, 47-52(1951)

[2] Buehler, W.J., Gilfrich, J.W., Wiley, R.C., "Effects of low temperature phase changes on the mechanical properties of alloys near composition TiNi", J. Appl. Phys. **34**, 1475-1477 (1963)

[3] Kleinheins, G., Starkl, W., Nuffer, K., „Besonderheiten bei der Qualitätssicherung strahlungsvernetzter Wärmeschrumpferzeugnisse", Kunststoffe **74**, 445–449, (1984)

[4] Lendlein, A., Kelch, S., "Shape-memory polymers", Angewandte Chemie, Int. Edition, **41**, 2034-2057 (2002)

[5] Liu, C., Qin, H., Mather, P.T. "Review of progress in shape-memory polymers" J. Mater. Chem. **17**, 1543–1558 (2007)

[6] Atli, B., Gandhi, F., Karst, G. "Thermo-mechanical characterization of shape memory polymers", J. Intel. Mat. Syst. and Struct., **20**, 87-95 (2009)

[7] Tao Xie, T. "Tunable polymer multi-shape memory effect", Nature, **464**, 267-270 (2010)

[8] Lendlein, A., Langer, R., "Biodegradable, elastic shape memory polymers for potential biomedical applications", Science, **296**, 1673-1676 (2002)

[9] Behl, M., Bellin, I., Kelch, S., Wagermaier, W., Lendlein, A., "One-step process for creating triple-shape capability of AB polymer networks", Adv. Func. Mater., **19**, 102–108, (2009)

[10] Flory, P.J., Rehner, J., "Statistical mechanics of cross-linked polymer networks. I and II" J. Phys.Chem., **11**, 512-520 and 521-526 (1943)

[11] Hoffmann, M., „Der Einfluß von Verhakungen auf das Dehnungsverhalten von vernetzten Polymeren", Kolloid-Zeitschrift und Zeitschrift für Polymere, **250**,197–206 (1972)

[12] Heinrich, G., Straube, E., Helmis, G., Molekularstatistische Theorien Polymerer Netzwerke (Review), Acta Polymerica, **31**, 275-286 (1980)

[13] Kolesov, I.S., Radusch, H.-J., Multiple shape-memory behavior and thermal-mechanical properties of peroxide cross-linked blends of linear and short-chain branched polyethylenes eXPRESS Polym. Lett., **2**, 461-473 (2008)

[14] Kolesov, I.S., Kratz, K., Lendlein, A., Radusch, H.-J., "Kinetics and dynamics of thermally-induced shape-memory behavior of crosslinked short-chain branched polyethylenes", Polymer, **50**, 5490-5498 (2009)

[15] Liu, C., Chun, S.B., Mather, P.T., Zheng, L., Haley, E.H., Coughlin, E.B., "Chemically cross-linked polycyclooctene: Synthesis, characterization, and shape memory behavior", Macromolecules, **35**, 9868-9874 (2002)

EFFECT OF ELASTOMER TYPE ON THE CRYSTALLINITY AND PHYSICAL PROPERTIES OF PHB OR PHB-CO-HV / ELASTOMER BLENDS

Gauthier Catherine [1*], Calvao Patricia [2], Chenal Jean-Marc [1], Demarquette Nicole [2], Cavaillé Jean-Yves [1]

[(1)] *MATEIS, INSA de Lyon - FRANCE - FRANCE, catherine.gauthier@insa-lyon.fr, jean-marc.chenal@insa-lyon.fr, jean-yves.cavaille@insa-lyon.fr*

[(2)] *Departement of Metallurgical and Materials Engineering, University of Sao Paulo - BRAZIL – BRAZIL nick@usp.br*

In this work, the effect of the addition of a rubbery phase on the properties of two bacterial polyesters namely Poly(3-hydroxybutyrate) (PHB) and its copolymer with hydroxyvalerate (HV) (PH*co*BV) was studied. Two different rubber were chosen either (i) ethylene-propylene-diene terpolymer (EPDM) or (ii) alternatively poly(vinyl butyral) (PVB). The change of PHB or PHBcoHV / rubber blends processing way on the crystallinity, the crystalline structure and physical properties of the composites were investigated. In both PHB and PHBcoHV blends, the main role of EPDM is its nucleating effect evidenced by a decrease of the crystallization temperature and an increase of the crystallinity with the higher EPDM ratio regardless of the processing route. While EPDM has a weak effect on the PHB (or PHBcoHV) glass transition temperature, PVB induces an outstanding decrease of this temperature due to its plasticizer component that swells the PHB amorphous phase. For both EPDM and PVB, 90/10 is the optimal PHB/rubber ratio to obatin a compromise between impact properties improvement and modulus decrease.

Introduction

Poly(3-hydroxybutyrate) (PHB) and its copolymer with hydroxyvalerate (HV) (PH*co*BV) are bacterial polyesters presenting the advantages of biodegradability and biocompatibility with large potential applications. The properties of PHB are often compared to those of polypropylene[1]; however, PHB is both stiffer and more brittle and shows a rapid thermal degradation and a rather narrow window of processability. In order to improve its processability and decrease its brittleness, copolymers containing valerate units have been synthesized to obtain poly[3-hydroxybutyrate)-*co*-(3-hydroxyvalerate)] (PHB*co*HV). Another alternative to cope with the brittleness of PHB is the obtention of blends of PHB with plasticizers or other polymers. The aim of the study is : (i) to obtain rubber toughened PHB or PHB*co*HV by conventional processing methods (ii) to address the effect of the addition of a rubbery phase either ethylene-propylene-diene terpolymer (EPDM) or polyvinyl butyral (PVB), according to the way of processing, on the blend microstructure, on the crystallization and thermomechanical properties.

Experimental

Materials and processing

PHB (M_n = 106,000 and M_w=163,567) and PHB*co*HV with 6 mol% of hydroxyvalerate HV (M_n = 190,000 and M_w=270,000) were supplied in powder form by "PHB Industrial – Biocycle". EPDM (Keltan 3200A) with 48,8% of ethylene was cryogenically grinded before processing. PVB is a random copolymer of vinyl butyral, vinyl alcohol and vinyl acetate whose properties can be controlled by the percentage of each monomer within the chains but also by the addition of plasticizer. The PVB used here (Saflex RB41) consists in a blend of PVB with 33 wt% of plasticizer (triethylene glycol di-2-ethylhexanoate). Blends of PHB/EPDM and PHB*co*HV/EPDM were prepared in different weight ratio (90/10, 80/20, 70/30) using an internal mixer Haake PolyLab 900 at a processing temperature of 170°C for 10 min and a rotor speed of 50 rpm followed by fast cooling in water. Then, the blends were cooled in liquid nitrogen and milled using a grinding mill. Sheets were prepared from the batch by compression molding at 195°C for 5 min and water quenching. These samples will be hereafter referred to "pressed samples". Another processing route has also been used to produce PHB blends, hereafter called "injected samples". PHB, in a powder form, was extruded in a twin screw extruder Haake Polylab 900/Rheomix PTW16, pelletized and further extruded with PVB or with EPDM with different weight ratio (90/10, 80/20, 70/30). The extrusion process was performed with a rotation speed of the screws of 100 rpm in the 160-170°C temperature range. The pellets of PHB/EPDM and PHB/PVB blends were molded into tensile tests bars (ASTM- ASTM D 638-98) using a Demag Ergotech pro 35-115 injection machine at 125 Bar of injection pressure. The molding temperature was set at 50°C and the barrel zone temperature ranged from 165 to 175°C. SEC experiments showed that the molecular weight of pressed PHB samples (M_n=57300 and M_w=122900 $g.mol^{-1}$) is lower than that of injected PHB samples

(M_n=72200 and M_w=151600 g.mol^{-1}), indicating that the degradation process is more pronounced for pressed sample.

Experimental techniques

The morphology of the blends was characterized by scanning electron microscopy (SEM) using a Philips, model XL 30 microscope. Differential scanning calorimetry was performed on a Perkin Elmer-Pyris Diamond differential scanning calorimeter (DSC) under nitrogen atmosphere. Crystallization kinetic was analyzed from crystallization exotherm recorded during the second heating step. Wide and Small Angle X-Ray Scattering (WAXS and SAXS) experiments were conducted on a device equipped with a copper rotating anode (λ=1.54Å), Gobel's mirrors collimation system and a 2D detector (Princeton instrument SCX2D). Solvent extraction experiments were conducted to detect co-continuity in PHB/EPDM and PHB/PVB blends. The EPDM was dissolved for 30 days at room temperature using toluene as solvent, and PVB was dissolved in ethanol at 80°C for 8 hours. Dynamic mechanical analysis (DMA) was performed on injected samples using an inverted torsion pendulum developed at MATEIS laboratory (INSA-Lyon). Tensile tests were carried out at room temperature using an Instron MTS1/ME tensile tester at speed of 0.1mm/min at room temperature. Izod impact tests were performed using a Tinius Olsen (IT 504) apparatus according to ASTM D-256. The experiments were conducted for unnotched samples at room temperature with a 2.82J pendulum. At least five samples for each blend composition and for neat PHB were tested for mechanical tests.

Results and Discussion

PHB/rubber blends have been successfully obtained by thermo-mechanical processing methods instead of solvent route. Considering the chosen route (pressed or injected), the obtained morphologies are different. MEB observations reveal a droplet dispersion morphology type for both pressed PHB*co*HV/EPDM and PHB/EPDM blends. The characteristics sizes of the rubber domains are around a few microns. Conversely, P(3HB)/EPDM injected samples display a non-uniform morphology with domains of large sizes (typical size ~25 μm vs ~2μm for pressed samples). Moreover, solvent extraction experiments show that a certain amount of co-continuity is present in EPDM/PHB injected samples, especially at high rubber ratio (Figure 1). P(3HB)/PVB injected blends show a droplet dispersion morphology type for all compositions (typical size ~2μm).

It is well known that the final morphology of a immiscible blend is dependent of several factors such as: processing conditions, composition, rheological parameters (viscosity and elasticity ratios) of the individual components and interfacial tension between the phases. From both rheological measurements and interfacial tension estimation, a smaller size of domains in PHB/PVB compared to PHB/EPDM injected blends is expected.

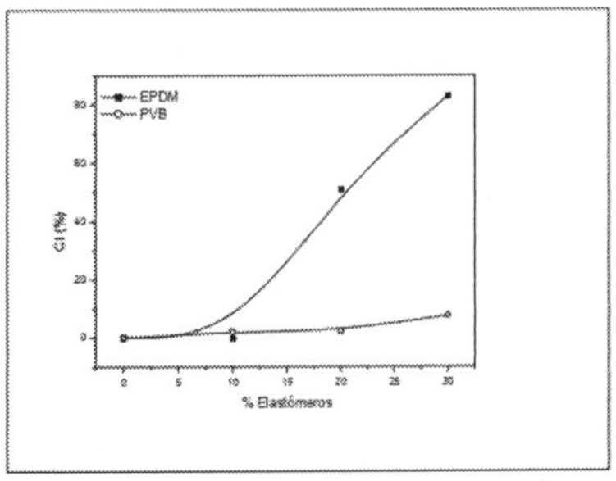

Figure 1 – Co-continuity index (CI) versus elastomer content (%) for EPDM/PHB and PVB/PHB injected samples.

In both PHB and PHB*co*HV blends, the main role of EPDM is its nucleating effect evidenced by a decrease of the crystallization temperature and an increase of the crystallinity with the higher EPDM ratio regardless of the processing route. While EPDM has a weak effect on the PHB (or PHB*co*HV) glass transition temperature, PVB induces an outstanding decrease of this temperature due to its plasticizer component that swells the PHB amorphous phase. This plasticizer appears very efficient to lower PHB glass transition temperature, and its influence is also observed on G" versus temperature curves (Figure 2).

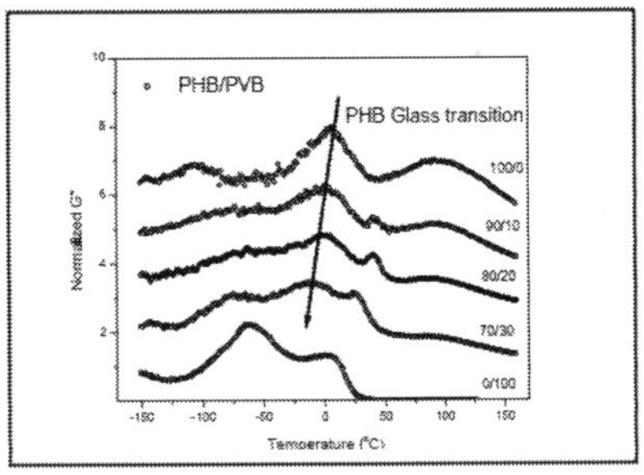

Figure 2 – Loss modulus (G'') versus temperature for PHB/PVB injected samples.

The tensile and impact properties obtained for P(3HB) and its blends with EPDM and PVB in different concentrations are reported in Table 1. It can be seen that the increase of elastomer concentration in P(3HB) matrix decreases its tensile strength and tensile modulus but increases the elongation at break and the impact strength.

	Tensile strength (MPa)	Modulus (MPa)	Strain at break (%)	Impact strength (J/m)
PHB	26,8 ± 0,5	2590 ± 20	1,5 ± 0,1	74 ± 5
PHB/ EPDM				
90/10	24,4 ± 0,7	2200 ± 50	1,7 ± 0,1	104 ± 5
80/20	19,0 ± 0,3	1800 ± 30	1,8 ± 0,1	106 ± 5
70/30	13,8 ± 0,3	1400 ± 30	2,2 ± 0,2	104 ± 5
PHB/ PVB				
90/10	25,3 ± 0,2	2150 ± 60	1,8 ± 0,1	93 ± 5
80/20	18,0 ± 0,5	1600 ± 30	1,8 ± 0,3	91 ± 5
70/30	16,0 ± 0,7	1090 ± 30	3,7 ± 0,5	83 ± 5

Table 1 – Mechanical characteristics of PPHB/EPDM and PHB/PVB injected samples

Taking into account that the improvement of impact strength is not influenced by the amount of rubber, it can be concluded that 90/10 is an optimal ratio for PHB/EPDM and PHB/PVB blends. Even if the PHB/PVB blends display droplet dispersion morphology with small domains, EPDM appears to be a more efficient impact modifier than PVB, probably because of its lower glass transition.

Conclusions

This work demonstrates that is possible to process PHB / rubber blends by conventional processing methods. EPDM The presence of the EPDM phase is found to modify the PHB crystallinity through a nucleating effect. In the case of PVB, a plasticizing effect is observed. In both cases, it can be seen that the increase of elastomer concentration in P(3HB) matrix decreases its tensile strength and tensile modulus but increases the elongation at break and the impact strength. EPDM appears to be a more efficient impact modifier that PVB. Future work might be to investigate the role of the plasticizer of PVB in PHB.

Acknowledgements

We would like to thank FAPESP, CNPq e CAPES for financial support.

References

1. Hocking PJ, Marchessault RH. Chemistry and Technology of Biodegradable Polymers. Blackie Academic & Professional. 1994

The role of interfacial elasticity on the rheological behavior of polymer blends

Joao Maia (1), Jorge Silva (2), Ana Vera Nóbrega (2), Paula Moldenaers (3)*

(1) Case Western Reserve University - Ohio - USA
(2) University of Minho - Guimarães - Portugal, (3) KU Leuven - Leuven - Belgium

In this work, the influence of interfacial modification on the rheological behavior of polymer blends was investigated. First, commercial blends of polypropylene and ethylene-propylene copolymer, PA6/EPM, non-compatibilized and compatibilized with EPM grafted with maleic anhydride, with high viscosity and elasticity ratios were studied. The observation of the morphology frozen during extension tests seemed to indicate that the droplets of the highly compatibilized blends are slightly deformed unlike those of the non-compatibilized ones. Stress relaxation experiments, both in shear and extension, showed that the introduction of the compatibilizer leads to the appearance of a second very long relaxation mechanism, the question then being what is its cause. In order to better understand the relative importance of relaxation of the droplets and relaxation of the interfaces in this kind of compatibilized blends, model high viscosity ratio blends of optically transparent materials, poly(methylmethacrylate), PMMA, polystyrene, PS, and oxazoline-modified polystyrene, PSOX were prepared. This allows SALS to be used in dilute blends to complement the rheometrical characterization and thus evaluate the deformation of the droplets in transient shear flows. In these blends there is also an additional longer relaxation time present but no changes in the morphology can be readily observed. In addition, the blend with functionalized PS shows a droplet deformation in the vorticity direction, which is an indication of an increase in the elasticity of the interface. Thus, our findings seem to indicate that the main factor behind the increase in relaxation time stress relaxation at and in the interfaces and not droplet deformation and recovery. These results were confirmed when PP/EVOH blends, non-compatibilized and compatibilized with Na+ were tested. In this case the elasticity of the interface is negligible and compatibilization did not alter the relaxation behavior.

EFFECTS OF MISCIBLE POLYMERIC DILUENT ON SPHERULITES OF STEREOCOMPLEXING POLY(L-LACTIC ACID) AND POLY(D-LACTIC ACID)

Ling Chang[1] and Eamor. M. Woo[2]*

[1]*Department of Chemical Engineering, National Cheng Kung University, Tainan, 701-01, Taiwan-n3897107@mail.ncku.edu.tw*
[2]*Department of Chemical Engineering, National Cheng Kung University, Tainan, 701-01, Taiwan-emwoo@mail.ncku.edu.tw*

Abstract
Effects of poly(3-hydroxybutyrate) (PHB) on crystalline morphology of stereocomplexing capacity of poly(*L*- and *D*-lactic acid) (PLLA and PDLA) were studied by differential scanning calorimetry (DSC), polarizing-light optical microscopy (POM), atomic force microscopy (AFM) and wide-angle X-ray diffraction (WAXD). When crystallized at high T_c (130°C or above), morphology transition of stereocomplexed PLA (sc-PLA) occurs from original well-rounded Maltese-cross spherulites to dendritic form in blends of high PHB contents (50 wt% or higher), where PHB acts as an amorphous species. Microscopy characterizations show that morphology sc-PLA in PHB/sc-PLA blends crystallized at T_c=170°C no longer retain original complexed Maltese-cross well-rounded spherulites; instead, the spherulites are disintegrated and restructured into two types of dendrites: (1) edge-on feather-like dendrites and (2) flat-on wedge-like crystal plates by growing along different directions and exhibiting different optical brightness. Despite of spherulite disintegration, WAXD result shows however that these two PHB-induced dendrites still retain the original unit cells of complexes, and thus these two new dendrites are still sc-PLA.
Keyword: stereocomplex, poly(*L*-lactic acid), poly(*D*-lactic acid); poly(3-hydroxybutyrate); dendrite.

Introduction

Semicrystalline polylactides (PLA's) exists in three isomeric forms, such as *L*-lactic acid (PLLA), *D*-lactic acid (PDLA), and racemic-lactic acid (PDLLA), which display a wide variety of properties. Stereocomplexation can occur between some polymers, which is based on interactions of stereoselective van der Waals forces, such as L-configured with D-configured PLA.[1-3] The typical morphology of stereocomplexed PLA (sc-PLA) spherulite is of a Maltese-cross pattern for melt-crystallized polymer, whereas the sc-PLA spherulites may become disordered and less distinct under non-equimolar conditions.[2-3] As a sc-PLA complex is being packed by cooling from a mixture in the melt state, properties of PLA complexes may be influenced by various parameters, such as molecular weight, mixing ratios, and crystallization temperature (T_c) and time (t_c).[2-3] As PLLA/PDLA mixtures are further blended with diluents, effects of above parameters and additionally interaction strength of PLAs with the diluents, on final complex morphology of PLA complexes have yet to be fully understood.

The miscibility and phase behavior of blends of poly(3-hydroxybutyrate) (PHB) with PLLA is strongly dependent on the molecular weights of either of constituents. It is known that PHB can be miscible with low-molecular-weight PLLA (M_w < 18,000) in the melt at 200°C over the entire compositions, whereas PHB blends with high-molecular-weight PLLA (M_w > 18,000) show phase separation.[4-7] Two types of spherulitic morphologies, respectively related to the crystal forms of PHB and PLLA, are observed when PLLA/PHB blends are cooled from the melt.[4]

Although blends of PHB with PLLA have been widely investigated;[4-7] however, studies have yet to be directed to investigation of influence of miscible diluents co-crystallizing with the complexes on the crystalline morphology of sc-PLA structures. It was of great interest to probe whether the stably complexed Maltese-cross spherulites in sc-PLA might be altered or disrupted when sc-PLA crystallized in presence of miscible diluents or polymers, such as PHB. In blends kept at T_c of PLAs, PHB is molten and remains noncrystallizable in a melt state, and thus PHB acts as an amorphous/molten diluent throughout crystallization of sc-PLA being packed into complexes from the liquid PHB/sc-PLA mixtures. It was of interest to investigate effects of PHB contents on miscibilty and morphology of isothermally crystallized complexed sc-PLA system. Influencing factors of the crystallization conditions including PHB compositions and crystallization temperature, on spherulitic morphologies of sc-PLA in blends were discussed.

Experimental

Thermal analysis was carried out with a differential scanning calorimetry (DSC) (Diamond, Perkin-Elmer

Corp.), under a dry/inert nitrogen atmosphere. The melted samples were kept at 230°C for 1 min for completely removing prior thermal history of blends, and immediately placed in liquid nitrogen to quench into an amorphous state for T_g measurements. For analysis of blends, the procedure for isothermal crystallization experiments was as follows: all samples were heated to 230°C for 1 min and immediately quenched to the desired crystallization temperature, T_c. For T_g measurements, the PHB/sc-PLA blends were heated to 230°C for 1 min (above T_m's of either sc-PLA or PHB), immediately immersed in liquid nitrogen to an amorphous state. In such rapid quenching condition, sc-PLA or PHB crystallites could not form during rapid cooling, and their glassy states were preserved for T_g characterization.

The spherulitic morphologies were observed on a Nikon Optiphot-2 polarizing-light optical microscope (POM) equipped with a Nikon CCD digital camera. Non-polarizing light optical microscopy (OM) was also used for better discerning the topographic shapes of spherulites. Samples were first melted on a hot stage at 230°C for 1 min, and then quickly transferred to hot stage (Linkam THMS-6000) equilibrated at the desired crystallization temperature. To further investigate the morphologies of blends, samples were also examined using atomic-force microscopy (AFM) (Caliber, Veeco-DI Corp., USA) with a tapping mode. Silicon tips with a resonance frequency of ~280 KHz were used, and the scan rate was 0.3 Hz. All samples were deposited on substrates of glass slides with an open face up, and were characterized using a set-point amplitude ratio of 0.8.

Wide-angle X-ray diffraction (WAXD) patterns for neat PHB, neat sc-PLA and blends were characterized with X-ray instrument (Shimadzu/XRD-6000) with a CuK_α radiation (30kV and 40mA). The scanning angle (2θ) covered a range between 10 and 30° at a rate of 2°/min.

Results and Discussion

Figure 1 shows the DSC thermograms of melt-quenched sc-PLA blend with various PHB contents. It has been found that neat sc-PLA exhibits a glass temperature (T_g) around 40°C, and a melting temperature (T_m) at about 221°C with ΔH_f of 84.9 J/g, which agree with literature.[1-3] Fig. 1 (A) shows that a single and composition-dependent T_g (onset position indicated with arrow marks), and this provides that the sc-PLA and PHB are miscible in the melt. Dependence of T_g on composition relationship for PHB/sc-PLA blend was fitted with the Gordon-Taylor equation[8]: $T_{g,blend} = (w_1T_{g,1}+w_2kT_{g,2})/(w_1+kw_2)$, with k = 0.43 as shown in Fig. 1 (B).

Figure 2 shows the morphologies of neat and blended sc-PLA with various weight fractions of PHB (70/30-30/70) crystallized at 130°C. In blends held at these T_c's, PHB is in an amorphous molten state and acts as a diluent throughout crystallization of sc-PLA. The spherulites of neat sc-PLA crystallized at 130°C are typical Maltese-cross patterns as shown in Fig. 2-(a), whereas those of sc-PLA in blend tend to formation of highly dendritic crystals. For the PHB-rich compositions (i.e., 50/50-70/30 blends), Fig. 2-(c) and (d) show dendritic sc-PLA spherulitic textures with disruption of Maltese-cross patterns. In addition to PHB composition, others factors, such as crystallization temperature (T_c), may also influence the morphologic formation of stereocomplexes PLA in blends. **Figure 3** shows POM and OM images of sc-PLA with PHB (i.e., 50/50) melt-crystallized at various temperatures (T_c= (a) 150°C and (b) 170°C). The POM images of sc-PLA in blends with PHB crystallized at 150°C (Fig. 3-a) show dendritic textures with disruption of Maltese-cross patterns, whereas those of sc-PLA in blends melt-crystallized at 170°C, show coexistence of feather-like and wedge-like spherulites (Fig. 3-b). At an even higher T_c=170°C, the dendrites can be classified into two types, indicated with dotted- and solid-line arrows, respectively. Upon crystallization of the 50/50 blend at 150°C or above, original circular Maltese-cross spherulites no longer exist. Instead, the Maltese-cross pattern typical of the sc-PLA spherulites disappears, and is now replaced with dendritic bundles, which may be ascribed to presence of PHB molecules dissolving and reorienting the stereocomplexed spherulites into dendrites. In Fig. 3-(c) and (d), POM micrographs for the enlarged areas of sc-PLA in blends crystallized at T_c=170°C show feather-like patterns with higher brightness and wedge-like patterns with lower brightness, respectively. In Fig. 3-(c) and (d), nonpolarizing OM micrographs for the enlarged areas of feather-like and wedge-like crystals, respectively, are shown jointly for comparison with the interested areas as illustrated in POM photos. The OM image (Fig. 3-d) further reveals that the wedge-like dendrites display a coarser texture than the feather-like ones. The POM and DSC results suggest that sc-PLA spherulites with dendritic textures contain less stable (or less perfect) lamellae in crystalline structure.

To explain the nature of morphologic changes of sc-PLA in blends, the wide-angle X-ray measurements on blends isothermally crystallized at various temperatures were analyzed. **Figure 4** shows WAXD profiles of (a) neat sc-PLA, (b) PHB/sc-PLA blends = 50/ 50 and (c) neat PHB, isothermally crystallized at indicated T_c's. Neat sc-PLA shows diffraction peaks at 2θ = 11.5°, 20.5°, and 23.5°, which are characteristic of sc-PLA crystallized with a triclinic unit cell of dimension: a = b = 1.50 nm, c = 0.823 nm, $\alpha = \beta = 90°$, $\gamma = 120°$.[9] Two main characteristic diffraction peaks located at 2θ = 12.9°, and 16.4° are assigned to the neat PHB crystal, which is in agreement with the literature results.[10-11] The blended sc-PLA exhibits the same diffraction peaks as the neat sc-PLA, indicating that the unit cells of sc-PLA in blends are not altered by addition of PHB.

As discussed earlier, POM graphs in Fig. 3 for the PHB/sc-PLA blends (50/50) crystallized at 170°C show coexisting feather-like and wedge-like dendrites/spherulites with bright and dark brightness, respectively. These two different morphologies grown from the disintegrated complexes of the PHB/sc-PLA blend only occurs at high T_c temperatures, suggesting that mobility of molten PHB chains in contacting with sc-PLA is the main factor for the morphology transition from sc-PLA complex spherulites to dendrites. The lamellar patterns of these two different dendrites were further analyzed in details using AFM. **Figure 5** shows AFM height images for PHB/sc-PLA blends (50/50) crystallized at 170°C, revealing height profiles for two different structures in the PHB-disintegrated sc-PLA spherulites. The height image of Fig. 5 (A) shows that feather-like and wedge-like dendrites are different in heights. The height profile of the wedge-like dendrite is much higher than that of the feather-like one as shown in the bottom plot of Fig. 5 (A). The enlarged areas of the square-boxed regions (b) and (c) of Fig. 5 (A) are shown in Fig. 5 (B) and (C), corresponding to the wedge-like and feather-like dendrites, respectively. The arrow mark in Fig. 5 (B) and (C) indicates the radial direction of spherulite growth. Obviously, the wedge-like and feather-like dendrites are of two distinctly different crystal types: platelet and long-stripe morphologies, respectively. Fig. 5 (B) shows that the wedge-like dendrite of sc-PLA is composed of platelet crystals; however, the feather-like dendrite is made up of long-stripe lamellae.

Figure 6 shows AFM phase images of the enlarged area of the wedge-like and feather-like dendrites (see Fig. 5 (B) and (C)) in the PHB/sc-PLA blend (50/50) crystallized at 170°C. Fig. 6 (A) shows the phase image of wedge-like dendrites, revealing wedge-like dendrites with multi-layered platelet morphology. The inset in Fig. 6 (A) shows the enlarged area of the square-boxed region of Fig. 6 (A), indicating that the lamellar thickness of the wedge-like dendrite, measured from the cross-sectional phase image, is ca. 16 nm and the direction of platelet crystal growth is almost perpendicular to the radial direction of spherulites growth. For the feather-like dendrite, the inset in Fig. 6 (B) shows that the lamellar thickness of the feather-like dendrite is ca. ~13 nm and the growth direction of long stripes of lamellae is almost parallel to the radial direction of spherulites growth. The results of AFM characterization show that the feather-like (bright area) and wedge-like (dark area) dendrites mainly consist of edge-on long-stripe lamellae and flat-on platelet lamellae, respectively.

Note that the AFM morphology for the platelet lamella in the wedge-like dendrites of the PHB/sc-PLA blend is quite similar to the flat-on lamella in the dark bands of ring-banded spherulites of PLLA crystallized between 125-130°C.[12] More strikingly, the AFM morphology for the long stripes of lamella in the feather-like dendrites of the PHB/sc-PLA blend is quite similar to the edge-on lamella in the bright bands of ring-banded spherulites of PLLA crystallized between 125-130°C.[12] The fact demonstrates that the crystal species of flat-on and edge-on lamellae can exist side-by-side and not only as concentric alternating rings. Note also the crystal species in the feather-like and wedge-like dendrites grow simultaneously from a common center, and they are not inter-related by continuous spiraling from one to another. The feather-like dendrite is filled with thin stripes of lamellar bundles, while the wedge-like dendrite is composed of discrete crystal platelet intersected by amorphous interfaces. The crystal species in two dendrites are not continuous inter-related but grow side-by-side. The morphologies of crystal species in these two dendrites demonstrate that edge-on and flat-on lamellae can exist by simultaneous or alternating growth, and not necessarily are created by spiraling from one lamella orientation to another. Therefore, edge-on and flat-on lamellae, discrete and not inter-related, can also be aligned in concentric fashion. This postulation can be an alternative mechanism to the conventional assumption that the lamellae continuously and rhythmically from flat-on to edge-on orientation to create alternating dark/bright bands, respectively.

Conclusions

The PHB/sc-PLA blend system shows a single composition-dependent T_g in all composition, indicating miscibility and interactions between PHB and sc-PLA, which leads to final morphology of the sc-PLA complexes is altered. PHB, via intimate miscibility interactions with both PLLA and PDLA, acts as an amorphous diluent component to disintegrate the originally well-rounded sc-PLA spherulites into dendrites of two distinctly different types. The final morphology in the blends finally exhibit coexisting dual types of dendritic spherulites: feather-like and wedge-like patterns, which totally replace the original well-rounded Maltese-cross spherulites. Various microscopy characterizations show that orientation of these two crystals of edge-on and flat-on lamellae, respectively, exhibit different brightness and birefringence. This fact also provides interesting and useful clues that the flat-on and edge-on lamellar orientation in PLLA/PDLA complexes can be independently and discretely packed into separate crystal entities, and are not necessarily inter-connected in an alternating twisting pattern. Despite of alteration of spherulites into dendrites by PHB, the WAXD result indicates that the unit cells of sc-PLA in PHB/sc-PLA blends are not altered by addition of PHB.

Acknowledgements

This work has been financially supported by basic research grants (NSC-95 2221 E006 183 and NSC-97-2221-E006-034-MY3) in three consecutive years from Taiwan's National Science Council (NSC), to which the authors express their gratitude.

References

(1)Tsuji,H.,"Poly(lactide) Stereocomplexes: Formation, Structure, Properties, Degradation, and Applications", Macromol. Biosci., **5**, 569-597 **(2005)**

(2) Tsuji, H., Ikada, Y., "Stereocomplex Formation between Enantiomeric Poly(lactic acids). 9. Stereocomplexation from the Melt", Macromolecules, **26**, 6918-6926 **(1993)**

(3) Zhang, J., "Infrared Spectroscopic Study of $CH_3 \cdots O=C$ Interaction during Poly(L-lactide)/Poly(D-lactide) Stereocomplex Formation", Macromolecules, **38**, 1822 **(2005)**

(4) Blümm, E., Owen, A. J., "Miscibiliy, crystallization and melting of poly(3-hydroxybutyrate)/poly(L-lactide)", Polymer, **36**, 4077-4081 **(1995)**

(5) Koymama, N., Doi, Y., "Miscibility of binary blends of poly[(R)-3-hydroxybutyric acid] and poly[(S)-lactic acid]", Polymer, **38**, 1589-1593 **(1997)**

(6) Park, J. W., et al., "Uniaxial Drawing and Mechanical Properties of Poly[(*R*)-3-hydroxybutyrate] /Poly(l-lactic acid) Blends", Biomacromolecules, 5 ,1557-1566 **(2004)**

(7) Furukawa, T., et al., "Structure, Dispersibility, and Crystallinity of Poly(hydroxybutyrate)/Poly(L-lactic acid) Blends Studied by FT-IR Microspectroscopy and Differential Scanning Calorimetry", Macromolecules, **38**, 6445-6454 **(2005)**

(8) Gordon, M.; Taylor, J. S. J. Appl. Chem., **1952**, 2, 493.

(9) Sawai, D., et al., "Isothermal crystallization kinetics and spherulitic morphology of poly(4-hydroxybutyric acid-alt-glycolic acid)", J. Polym. Sci. Part B: Polym. Phys., **45**, 2632-2639 **(2007)**

(10) Gazzano, M., et al., "X-ray investigation on melt-crystallized bacterial poly(3-hydroxybutyrate)", Macromol. Chem. Phys, **198**, 71-80 **(1997)**

(11) Canetti, M., et al., "Influence of the morphology and of the supermolecular structure on the enzymatic degradation of bacterial poly(3-hydroxybutyrate)", Polymer, **40**, 2587-2594 **(1999)**

(12) Nurkhamidah S. M.S. thesis, "Spherulitic Rings and Cracks in Poly(L-lactic acid) and Its Blends with Poly(3-hydroxybutyric acid)", Dept. Chem. Eng., National Cheng Kung University, Tainan, Taiwan, **(2009)**.

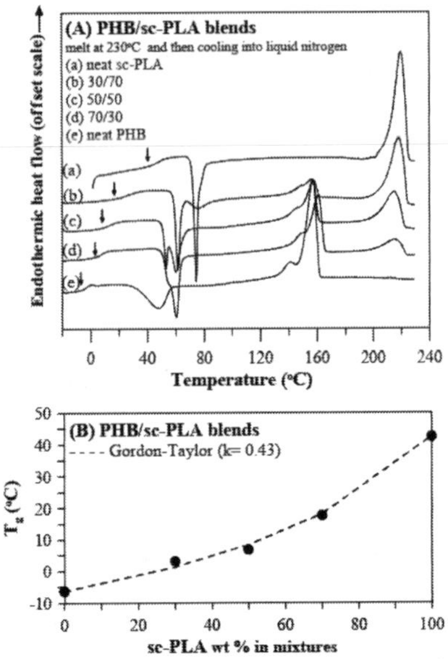

Figure 1. (A) DSC thermograms of PHB/sc-PLA blends; (B) plot of blend's single T_g's vs. composition.

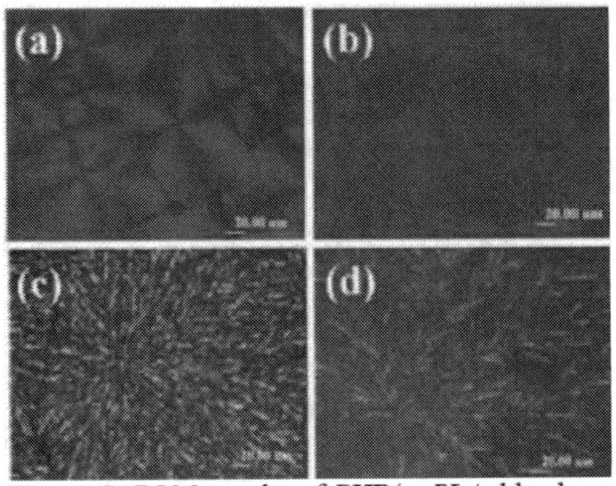

Figure 2. POM graphs of PHB/sc-PLA blends melt-crystallized at 130°C: (a) neat sc-PLA, (b) 70/30, (c) 50/50, and (d) 30/70.

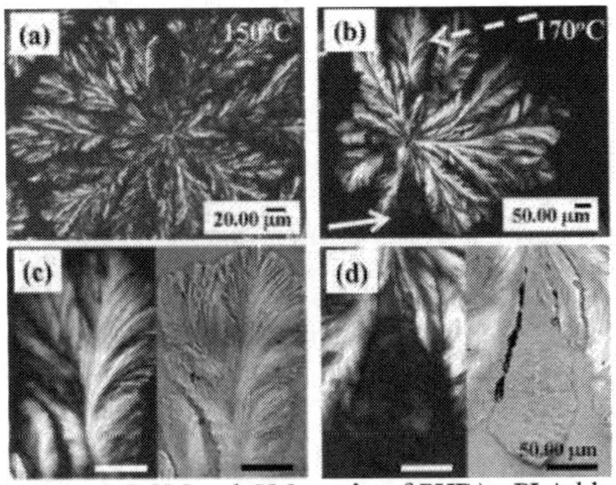

Figure 3. POM and OM graphs of PHB/sc-PLA blend (50/50) melt-crystallized at various temperatures: (a) 150°C, (b) 170°C; enlarged micrographs showing (c) feather-like, and (d) wedge-like lamella (indicated with dotted and solid-line arrows, respectively).

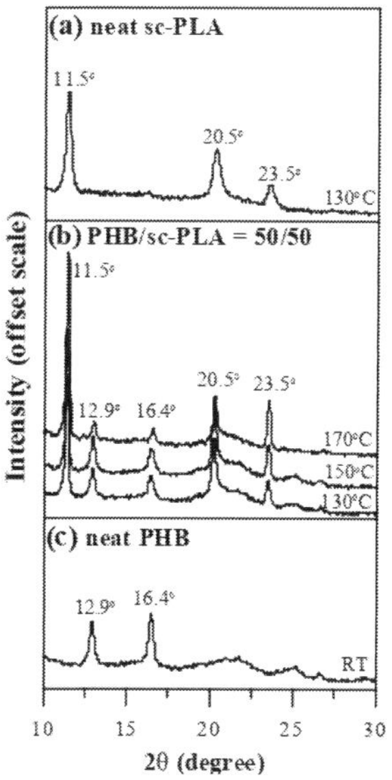

Figure 4. WAXD profiles of (a) neat sc PLA, (b) PHB/sc PLA blend (50/50) at three T_c's, and (c) neat PHB, isothermally crystallized at T_c's

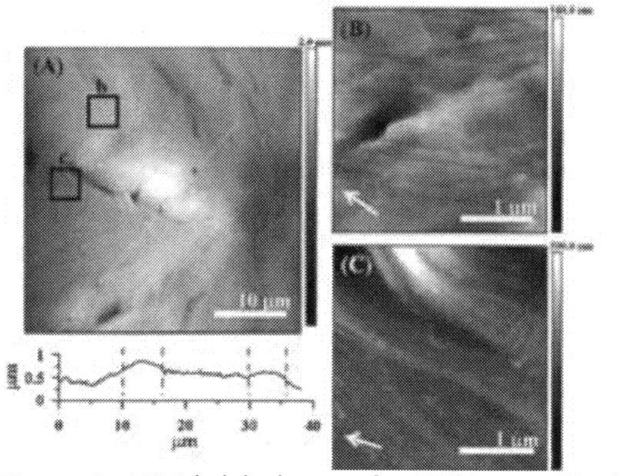

Figure 5. AFM height images for 170°C-crystallized PHB/sc-PLA blend (50/50). Graphs-(B) and (C) are the enlarged areas of the square-boxed regions (b) and (c) in Fig. 5 (A).

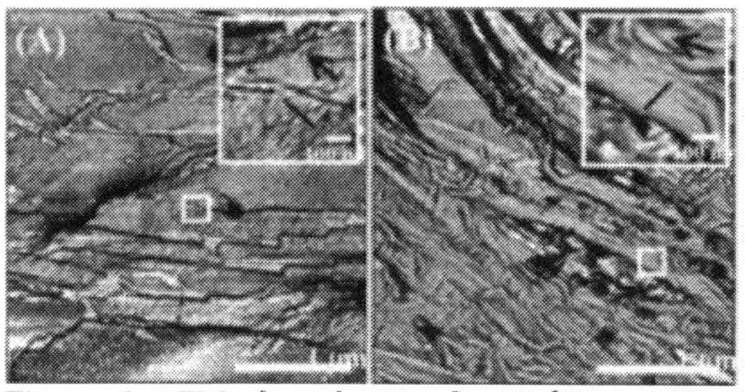

Figure 6. AFM phase images for 170°C-crystallized PHB/sc-PLA blend (50/50). Insets: the enlarged areas of square-boxed region in Fig. 6 (A) and (B).

COMPATIBILIZATION OF POLYAMIDE 6 AND POLYLACTIC ACID BASED RESIN FOR MEDICAL APPLICATIONS BY POLYALKENYL-POLY-MALEIC-ANHYDRIDE BASED DERIVATES

N. Miskolczi*[a], Cs. Varga[a], H. Szakács[a], V. Sedlařík[b,d], T. Kitano[b], A. Gregorova[c], M. Mozetic[d] and U. Cvelbar[d]

[a] University of Pannonia, Faculty of Engineering, Institute of Chemical Engineering and Process Engineering, Department of Hydrocarbon & Coal Processing, H-8200, Veszprém, Egyetem u. 10, Hungary

[b] Tomas Bata University, Faculty of Technology, Polymer Centre, Zlín, Nám. T.G. Masaryka 275, 76272 Zlín, Czech Republic

[c] Institute of Wood Science and Technology. University of Natural Resources and Applied Life Science, Vienna, Peter Jordan Strasse 82, A-1190 Vienna, Austria

[d] Jozef Stefan Institute, Jamova cesta 39, 1000 Ljubljana Slovenia

e-mail: mnorbert@almos.uni-pannon.hu

Thermoplastic blending of different polymers have been studied widely for many years because this process allows obtaining a new material with specific properties, which are, in some cases, even better that those of the individual homopolymers. It is a consequence of increasing material quality and functionality, which can be observed especially in area of medical devices production. From this point of view polyamide 6 (PA6) and polylactic acid (PLA) represent perspective group of materials applicable in medicine. PA6 has been used in various medical applications such as sutures, plastic items. Polyamide has good hardness, impact strength and abrasion resistance. Nether PA6 nor PLA do not carry the combination of excellent mechanical both properties and desired bioactivity. The combination of PA6 and PLA could bring the blend with good mechanical properties and improved bioactivity with regards to biomedical application. The problem is that phase separation effect and the loss of material properties can be arisen in PA6/PLA mixtures due to differ in physical form and in chemical composition of them. Therefore compatibilizing agents have been used. The aim of this work is to study and describe in detail the compatibilization effect of newly developed polyalkenyl-poly-maleic-anhydride coupling agents on resulting material properties (mechanical, structural, rheological) of the polymer blends based on PA and PLA co-polyester (Bioflex®). The main attention was paid to achievement of better chemical and physical interaction between the polymers. The results reveal the significant role of the compatibilisers in the PA6/Bioflex systems, which was confirmed by scanning electron microscopy (homogeneity improvement) as well as mechanical properties testing (e.g. 21 % increase in Charry impact strength). Fractured faces of the composites with and without compatibilizers have been studied by SEM. The micrographs showed the degree of homogeneity of the different compositions.

Blends of poly(ethylene terephthalate) with a maleic anhydride grafted Styrene-Ethylene-co-Butylene-Styrene

Hadi Gilvari, Ahmad Aref Azar*

Amirkabir University of Technology - Tehran - Iran

Styrene-Ethylene-co-Butylene-styrene (SEBS) was modified by the grafting reaction of maleic anhydride (MA) with the aid of dicumyl peroxide (DCP) as initiator in a Brabender mixer and the effect of the concentration of MA and initiator, as well as the processing conditions, such as temperature and rotor speed, on the grafting yield and gel content were examined. The amount of grafted MA groups on the rubber is determined by chemical titration and the degree of grafting is measured by infrared analysis. Then several home-made and commercially available SEBS grafted with maleic anhydride (SEBS-g-MA) were blended with poly (ethylene terephthalate) (PET). Both binary and ternary blends (consisting of a blend of SEBS and SEBS-g-MA) were examined and the mechanical and morphological properties of the resulting blends were evaluated by SEM, tensile and impact tests. Calorimetric studies indicated that the presence of a compatibilizer had no appreciable effect on the crystallization behavior of the blends. Particle size was strongly dependent on the functionality of the graft copolymer: the higher the functionality, the smaller the particle size and the narrower the particle size distribution. The finest phase morphologies having submicron particle sizes were obtained with commercial grade of SEBS elasomer, functionalized with 1.84 wt% maleic anhydride grafted onto the ethylene-butylene midblock. The presence of only a small amount of MA (0.12 wt% anhydride in compatibilizer) in compatibilizer is sufficient to induce a pronounced reduction of the dispersed phase particle size.

PROPERTIES OF POLY (LACTIC ACID) AND THERMOPLASTIC STARCH BLENDS

S. RONASI[1], S. HOPPE[1], G. H. HU[1, 2], J.L. SIX[3]

[1]Laboratoire Réactions et Génie des Procédés, CNRS-Nancy Université, 1 rue Grandville, BP 20451, 54001 Nancy, France – sara.ronasi@ensic.inpl-nancy.fr;sandrine.hoppe@ensic.inpl-nancy.fr; hu@ensic.inpl-nancy.fr
[2]Institut Universitaire de France, Maison des Universities, 103 Boulevard Saint-Michel, 75005 Paris,France
[3]Laboratoire de Chimie-Physique Macromoléculaire (LCPM), UMR CNRS-INPL 7568, Nancy University, Groupe ENSIC, 1 rue Grandville, BP 20451, 54001 Nancy cedex, France– Jean-Luc.Six@ensic.inpl-nancy.fr

Abstract -Blending starch with poly (lactic acid) (PLA) is one of the most promising efforts because starch is an abundant and cheap biopolymer and PLA is biodegradable with good mechanical properties. Different plasticizers are used and compared in order to best plasticize the starch. Polymer blends of PLA and thermoplastic wheat starch at various ratios are prepared by using a laboratory-scale twin screw extruder and compression molding. The thermal and mechanical properties as well as the morphology of the blends are studied. Starch and PLA are immiscible polymers. Copolymer of Amylose-g-PLA is prepared ex-situ in order to improve the properties of the blends. The grafting polymerization of PLA onto the amylose is done in three steps. In the first step, in order to limit and control the number of grafting chains, the amylose is partially silylated. The second step is the ring opening polymerization of lactide onto the remaining OH groups of the amylose and the last step is to deprotect the silylether groups. The number and chain length of the grafted PLA are varied. Three different structures of amylose-g-PLA are synthesized: (a) limited number but long chains of PLA grafts; (b) higher number but shorter chains of PLA grafts (the number and the length of the grafts are chosen in a way to have the same mass fraction of PLA in these two copolymers); (c) high grafting percentage of amylose and long chains of PLA.

Introduction

The development of synthetic polymers using monomers from natural resources provides a new direction to develop biodegradable polymers from renewable resources. One of the most promising polymers in this regard is PLA, because it is made from agricultural products and is readily biodegradable.

Blending starch with PLA is a route to reduce raw material costs, energy consumption and also to increase its rate of biodegradation.

Starch is one of the most promising natural polymers because of its inherent biodegradability, overwhelming abundance and its annual renewal (Blanshard, 1987). Native starch is usually in a granular form in which the polymer molecules are hydrogen bonded and aligned radially. Starch alone in a dry state is not considered a true thermoplastic in the sense that it would be degraded upon heating rather than melt on its own. However, in the presence of a plasticizer, starch would behave as a thermoplastic (Simmons and Thomas, 1995)

The common plasticizers are usually water, glycerol or other polyols (Sorbitol, PEG,...) (Poutanen, 1996). Glycerol and sorbitol are polyhydric alcohols that are miscible with starch and are good plasticizers to starch (Yu et al., 1998; Lawton and Fanta, 1994; Arvanitoyannis et al., 1998). Some hydroxyl groups of glycerol and sorbitol were expected to interact with the ester group from PLA in some degree, such a hydrogen bonding (Ke and Sun, 2001).

Other components such as those containing nitrogen (Urea, ammonium derived, amines,...) can be used. (Kazuo et al., 1998; Stepto and Tomka, 1987). These plasticized starches are commonly called "thermoplastic starches" or TPS. (Shogren, 1993).

Glycerol appears to be a less effective plasticizer than water (Forssell et al. 1997) However, Perry and Donald (2000) have shown that gelatinization proceeds even when no water is present: there is no chemical requirement for water, the only requirement being

mobility, induced by some low molecular weight plasticizer.
However, without water the temperature for the onset of gelatinization by glycerol increased (Nashed et al., 2003; Tan et al., 2004). Upon heating, starch granules become solvated, with the amorphous regions of the plasticized starch granule. This transformation of granular starch into a thermoplastic-like material allows it to be processed in a similar fashion to conventional polymers. At glycerol contents in the TPS lower than 10-15%, the plasticization leads to reduced mechanical properties (Lourdin et al., 1997). A more effective plasticization can be obtained at higher glycerol contents, in a range where glycerol demixing can occur in the thermoplastic starch. Thermoplastic starch itself is characterized by moisture sensitivity and weak mechanical properties as compared with other thermoplastic polymers. Blends of other polymers with thermoplastic starch represent an important route to overcome these limitations (Sarazin et al., 2008).
Martin and Ave´rous (2001) were the first to study TPS/PLA blends. They observed that the modulus of the blends decreased when the TPS fraction increased. In comparison to the pure TPS and PLA materials, all the blends showed a marked decrease in the elongation at break, tensile strength and impact resistance. They explained the poor mechanical properties by the lack of affinity between the two phases.
The type and quantity of the plasticizer influence the properties of the blends of TPS with PLA. Different plasticization strategies are compared in this work.
Polymer compatibility is an important factor in the development of polymer blends. Compatibility is critical in governing the morphology of the blend and, therefore, the final-state physical and mechanical properties of the system. One of the major governing factors that control the compatibility in polymer blends is interfacial tension (Vanoene, 1972).
One of the ways of improving the compatibility is the addition of a compatibilizer. The copolymer amylose-g-PLA with different structures is synthesized by a three step grafting strategy (Ouhib et al., 2009).

Experimental

Materials
PLA was purchased from Naturplast (France) the weight-average molecular weight was about 160,000~220,000 Da. The concentration of D-(—)-isomer was 12.0±1.0%. Wheat starch (10% moisture) was obtained from Sigma-Aldrich. The amylose and amylopectine contents were, respectively, 25 and 75%. The glycerol (purity) and D-Sorbitol were supplied by Aldrich.
Native amylose from potatoes was purchased from Fluka, N,O-bis-(trimethylsilyl) acetamide (BSA) and stannous octoate (SnOct2) were purchased from Sigma-Aldrich.

Samples Preparation
Thermoplastic starch was prepared according to the following procedure. Native starch was weighted and introduced in a turbo-mixer. Glycerol was then added and dispersed at high speed to obtain a homogeneous dispersion. The mixture was then placed in a vented oven at 170°C for 45 min and occasionally stirred, allowing vaporization of water and diffusion of glycerol into the starch granule. After cooling, the proper amount of water and sorbitol was added to the mixture that was once more dispersed at high speed. The powder was then extruded in a 15 cm^3 compounding system (DSM research; Netherlands) at 165°C for 5 min and a rotation speed of 60 rpm. The extruder has a screw length of 150 mm, a L/D of 18, and a net capacity of 15 cm^3. To obtain the desired specimen samples for various measurements and analysis, the molten composite materials were transferred after extrusion in a preheated cylinder to a mini-injection-molder, which was preset with the desired temperature (165°C). The mold temperature is 40°C. Injection molded samples were placed in sealed polyethylene bags in order to prevent moisture absorption.

	Water (%)	Glycerol (%)	Citric Acid (%)	Sorbitol (%)
TPS1	12	18	--	--
TPS2	12	18	3	--
TPS3	--	30	--	--
TPS4	--	15	--	15
TPS5	--	--	--	30

Table 1. Composition of different TPS

Scanning electron microscopy
Microstructures of the fractured samples were observed by using a scanning electron microscope (SEM) (JSM-6490-LV-JEOL-Japan). The surface of the samples was gold-plated.
In order to observe the starch phase, the sample was stored for 24h in a chloroform solvent to remove the PLA phase. For blends containing 20%TPS, the starch phase was removed by 5h storage in HCl (1N). The "skeleton" removed after the extraction is dried under vacuum for one day at 40°C and was then observed by SEM.

Mechanical properties of blends
Samples were prepared by injection molding. Tensile testing was done on mechanical tensile tester (Instron-Appolor). A crosshead speed of 10 mm/min is used. 5 samples of each blend composition were tested after one week period conditioning in sealed polyethylene bags. Results from 5 specimens were averaged.

Wide-Angle X-ray Diffraction
X-ray diffraction analysis was performed in a MiniFlex II-Rigaku diffractometer operating at 30 kV, 15 mA

and over a scattering angle (2θ) range of 10°-30° at 2° per minute.

Contact angle measurement

Contact angle measurements were carried out with a OCA 15 goniometre. A droplet was deposited onto the surface of a dumbbell form samples prepared by injection for traction test.

Water and diidomethane were used as test liquids because of their different polarities. The evolution of the droplet shape was recorded during 80s. A CCD video camera and image analysis software were used to determine the contact angle evolution.

The theoretical work of adhesion between polymers in solid state (w_{12}) which is the work required to reversibly separate two bulk phases can be calculated by contact angle measurement (1). First of all, the surface energies of solid materials are determined via contact angle measurement using the sessile drop method and two probe liquids. Wu's method (1982), practically adapted to our low energy systems, allow the calculation of the polar (γ^p) and dispersive component (γ^d) of the surface energies (2).

The harmonic mean equation is used to calculate the theoretical work of adhesion between two polymers TPS and PLA.

$$W_{12} = \frac{4\gamma_1^d \gamma_2^d}{\gamma_1^d + \gamma_2^d} + \frac{4\gamma_1^p \gamma_2^p}{\gamma_1^p + \gamma_2^p} \quad (1)$$

$$\gamma_L(1+\cos\theta) = 2\sqrt{\gamma_S^d}\sqrt{\gamma_L^d} + 2\sqrt{\gamma_S^p}\sqrt{\gamma_L^p} \quad (2)$$

	Surface tension (γ_L) $mJ.m^{-2}$	Disperse part (γ_L^d) $mJ.m^{-2}$	Polar part (γ_L^p) $mJ.m^{-2}$
Diiodmethane	50,8	48,5	2,3
water	72,8	26	46,8

Table 2 - Test liquids used for the determination of the surface free energy of solids

Results and Discussion

Starch plasticization

The force exercised by the microcompounder for the extrusion of different types of TPS (6 gram of each sample) is shown in Figure 1.

The force is an indication of the viscosity of the polymer. TPS1 and TPS5 have the highest viscosities. The high viscosity of TPS1 can be explained by the evaporation of water during the extrusion. The sample after the extrusion contains 1% of water. TPS5 contains no water and glycerol but 30% of sorbitol. The melting temperature of sorbitol is 90°c and it has less mobility than water and glycerol (bigger molecules) and so higher viscosity during the extrusion.

The low viscosity of TPS2 is due to the acid hydrolyze of starch in presence of citric acid.

Plasticization of starch is the disruption of granular starch and the transformation of the semi-crystalline granule into a homogenous, rather amorphous material with the destruction of hydrogen bonds between the macromolecules.

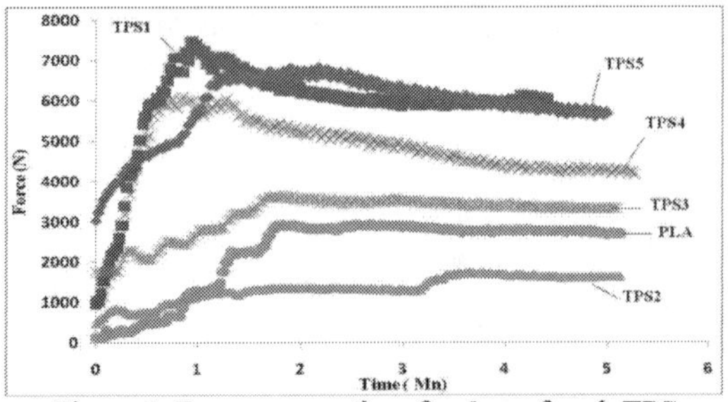

Figure 1. Force vursus time for 6 gr of each TPS sample

The extrusion conditions used for the plasticization of starch are adequate for disruption of the granular starch. The fragile fracture surface analysis obtained by SEM showed a smooth surface, typical of homogeneous amorphous materials (Figure 2).

The mechanical properties (tensile strength, elongation at break) of different TPS samples were shown in figure 3. A brittle characteristic is seen in TPS1 and TPS2. TPS3 and TPS4 exhibit a "plastic behavior'' plateau. Introduction of 30% glycerol led to a decrease of the tensile strength but the elongation at break increased. Generally one of the primary functions of a plasticizer is to improve the elongation at break and toughness; however this has to be achieved at the expense of tensile strength. Therefore this tendency is the same as a typical effect of plasticizer.

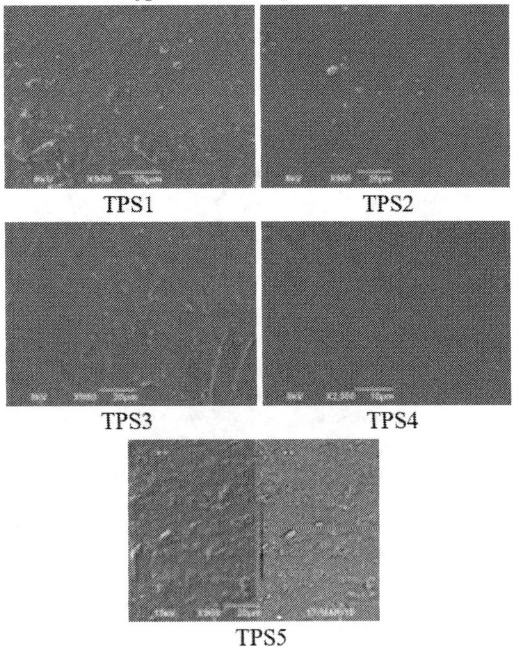

Figure 2. SEM Images of different TPS

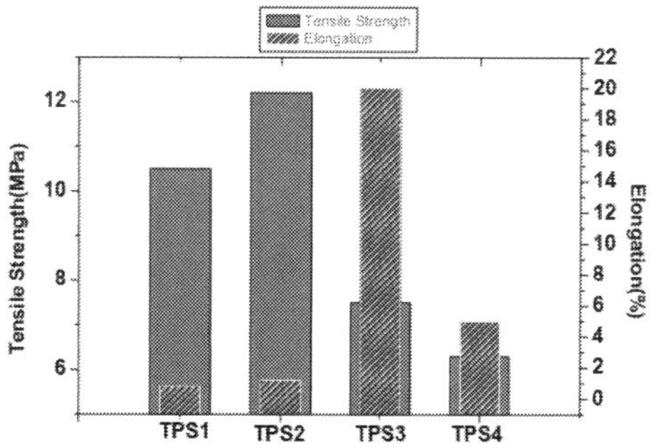

Figure 3. Comparison of mechanical properties of different TPS

X ray diffraction patterns of different TPS shows a typical V_a type patterns (Figure 4). V_a-type diffraction pattern of starch appears by a transformation from the A-type under extrusion at conditions of high temperature and pressure. For pure wheat starch, single peaks at 2θ of 14.88 and 22.54 and double peaks at 16.88 and 17.78 were observed (Figure 4). This pattern correspond to the typical A type X-ray diffraction pattern of wheat starch.

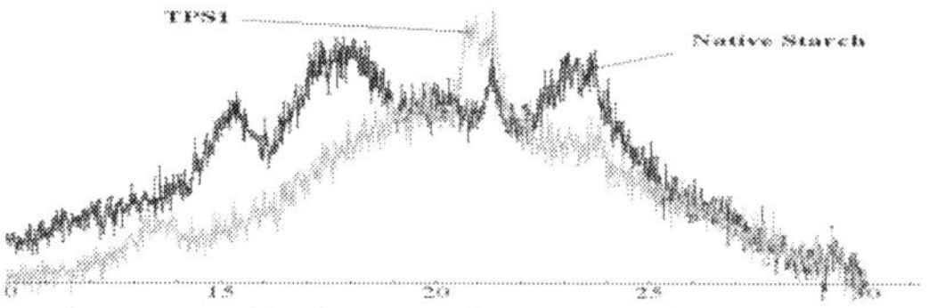

Figure 4. In black TPS1 after one week and in gray native starch (X-axe is 2θ)

All TPS samples have a V_a type pattern (Figure 5).The peaks of V_a pattern have been related to the presence of lipids in wheat starch granules, which have the ability to form complexes with amylose. The size of starch crystalline domains in TPS can typically be in order of 15-30 A°. (Rodriguez-Gonzalez et al., 2004) For TPS, the areas of the peaks which correspond to the crystalline region decreased in comparison with pure starch, which led to the decrease of crystallinity.
Citric Acid in TPS2 increases the crystallinity of TPS. Citric acid decrease the molecular weight of starch by hydrolyses acid (Wang et al., 2007). Shorter chains of starch would move easier and the crystallization is enhanced.

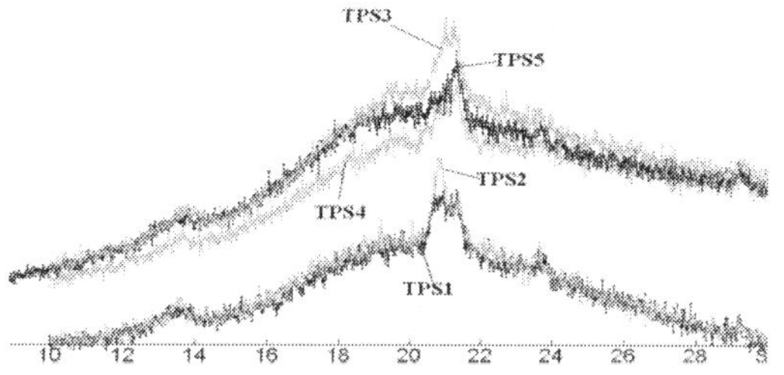

Figure 5. Different types of TPS after extrusion (X-axe is 2θ)

Polymer Blends

Plasticization and blending of starch with PLA must be done in two steps. A one step extrusion of PLA and starch with the introduction of plasticizers simultaneously in the extruder doesn't provide necessary shear for the fracture of the granule surface. As it can be seen on figure 6, many starch granules were obviously not plasticized.
So a two step extrusion is chosen. Starch is first plasticized as it has been explained earlier and after extrusion is ground in powder. PLA is dried for 5 hours at 40°C under vacuum in order to eliminate the water. Different TPS/PLA blends are prepared and introduced in the extruder at 165°C for 5 minutes and a screw speed of 60 rpm. In the two steps blending, the granules of starch are disappeared as it is shown in Figure 7.

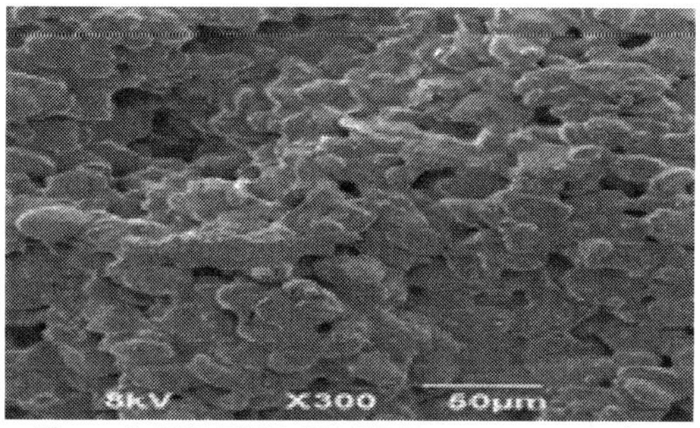

Figure 6. 50% TPS1-50%PLA- one step extrusion

Figure 7. 50%TPS1-50%PLA- two step extrusion

The tensile strength and elongation at break of 40/60; 50/50 and 60/40 TPS/PLA blends are compared on Figures 8 and 9. Compared with the mechanical properties of pure PLA (60MPa the tensile strength and 7% of elongation), those of TPS/PLA blend decreased dramatically, especially the tensile strength, which could be ascribed to the poor interfacial adhesion between two polymer phases.
It is expected that coarse blend dispersions will lead to premature failure because the larger dispersed particles act as defects that initiate cracks in the material. The improved dispersion in the blends is therefore expected to enhance the material ductility but the magnitude of this improvement is points to a change in fracture behavior similar to that observed in rubber-toughened materials. Improvement in solid-state adhesion

between the TPS and PLA is another factor that may improve the elongation at break. Even though the TPS is the most ductile component of the blend, the elongation at berak decreases with the TPS content. This is further indication that the presence of the TPS phase is changing the fracture mechanics of the material leading to a synergistic effect, (Huneault and Li, 2007).

Addition of 3% Citric Acid improves the tensile strength of the blends slightly. This improvement is due to the ester linkage between citric acid and starch and therefore the modification of the interfacial adhesion between TPS and PLA. The ester linkage between citric acid on starch (1740 cm^{-1}) is shown on Figure 10.

TPS3 contains higher residual plasticizer (30%glycerol) than TPS1 (18%glycerol) because of total evaporation of water during extrusion but these two have similar tensile strengths. This can be explained by a better dispersion of the TPS in the matrix in TPS3 as it can be seen in the Figure 11. The morphology analysis couldn't be done on the blends of TPS2/PLA because these blends are very sensitive to the extraction of PLA by chloroform.

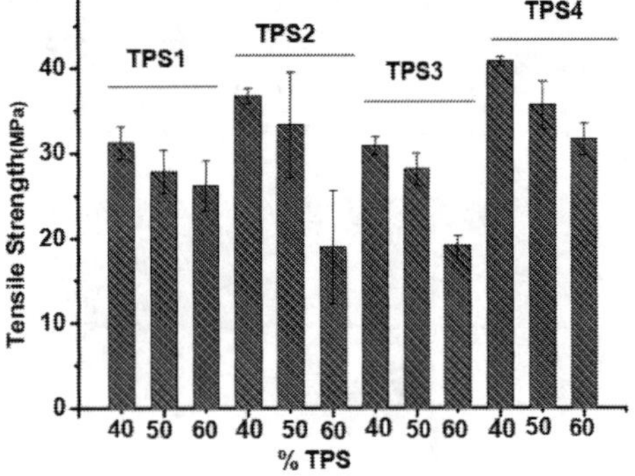

Figure 8. Tensile strength(Mpa) of 40/60, 50/50 and 60/40 TPS/PLA blends

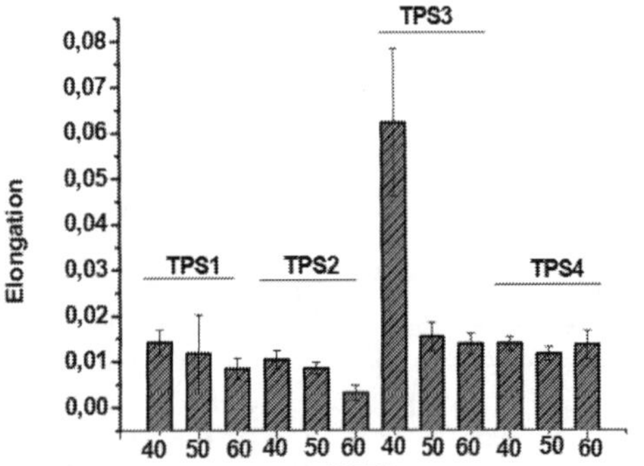

Figure 9. Elongation at break of 40/60, 50/50 and 60/40 TPS/PLA blends

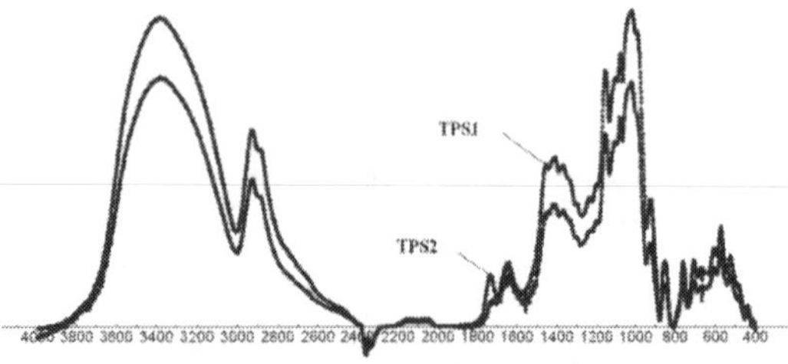

Figure 10. Influence of citric acid on the Starch (X-axe is wave number (cm^{-1}))

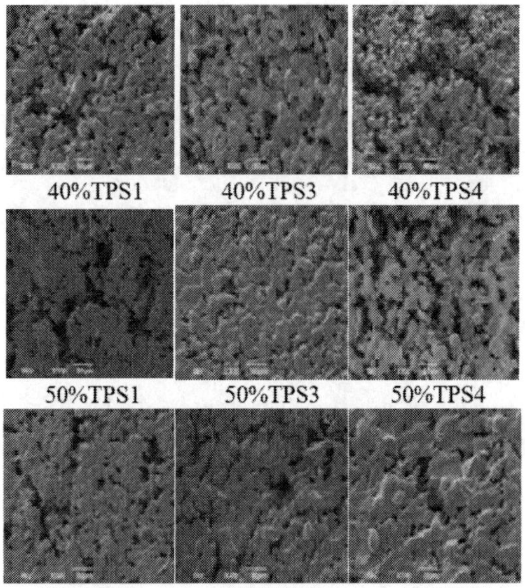

Figure 11. SEM images of 40-50-60% of TPS after extraction of PLA with chloroform

TPS4 shows the highest tensile strength comparing to the other TPS/PLA blends. Sorbitol seems to improve the adhesion of PLA and starch. The theoretical adhesion force calculated between different TPS and PLA pairs (Figure 12) shows stronger interaction between TPS4 and PLA.

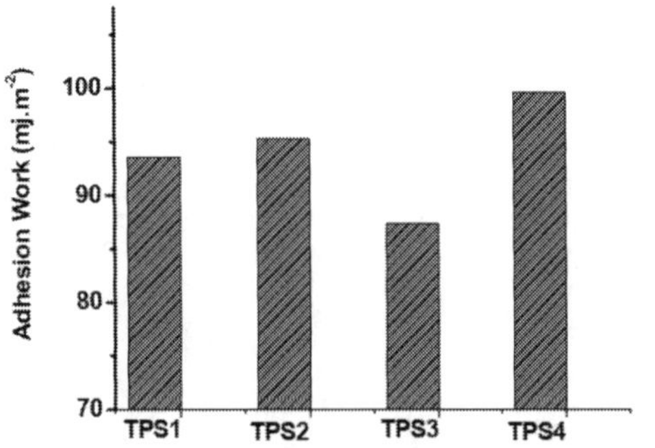

Figure 12. Adhesion work of TPS and PLA

As it was observed in Figure 11, we are in the co-continuity region of TPS for 40-60% of TPS and study

of the improvement of the dispersion by addition of a compatibilizer can hardly be observed. So lower quantities of TPS are chosen. TPS1 have a high viscosity during the extrusion and the water present during the extrusion might contribute to hydrolyze and degradation of PLA. This degradation might be more important in TPS2 because of the citric acid.
TPS3, TPS4 and TPS5 are chosen for study of the influence of compatibilizer.
Blends of TPS/PLA containing 20% TPS and 80%PLA are prepared with the same blending procedure used earlier.
In Figure 13, the mechanical properties of these blends are compared. By replacing 15% of glycerol of TPS3 (30%glycerol) by 15% sorbitol (TPS4), the tensile strength is increased by 30% and it approaches the tensile strength of the pure PLA. Figure 14 shows the FTIR spectra of the blends of TPS/PLA. For TPS5 we can observe a higher peak at 2924 cm^{-1}. This peak corresponds to the stretching of CH_3 group of PLA. Good dispersion and mechanical properties of TPS/PLA when sorbitol is used as plasticizer can be explained by interaction of sorbitol with PLA.
The elongation at break is decreased dramatically by adding TPS to the blend. TPS3 has the best elongation at break but it is less than the elongation of pure PLA.

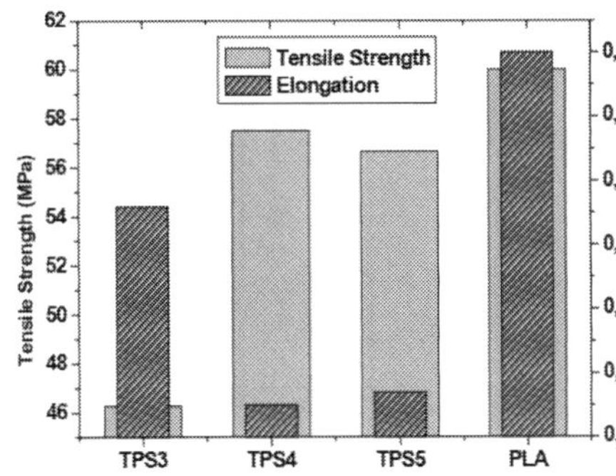

Figure 13. Tensile strength and elongation at berak of 80/20 PLA/TPS blends

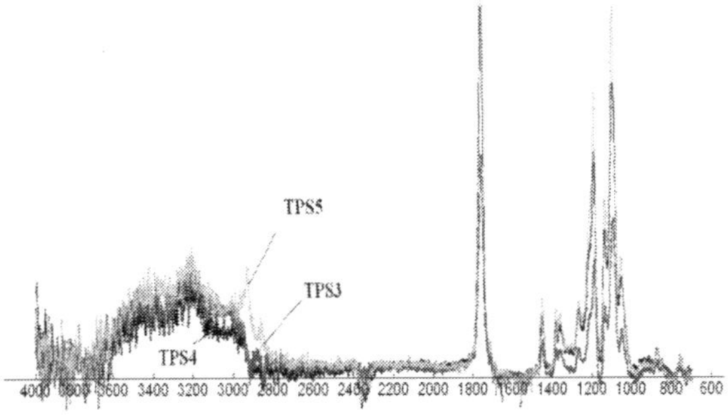

Figure 14. FTIR spectra of 80/20 PLA/TPS blends (X-axe is wave number (cm^{-1}))

SEM images of blends of 80/20 of PLA/TPS (Figure 15) demonstrate the efficiency of sorbitol for decreasing the size of the dispersed phase.

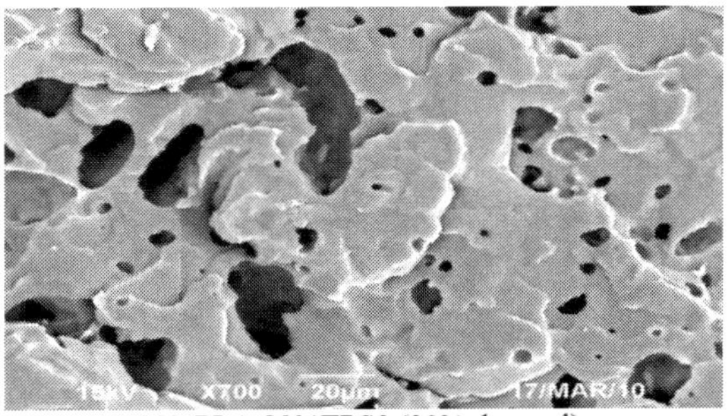

80%PLA-20%TPS3 (30%glycerol)

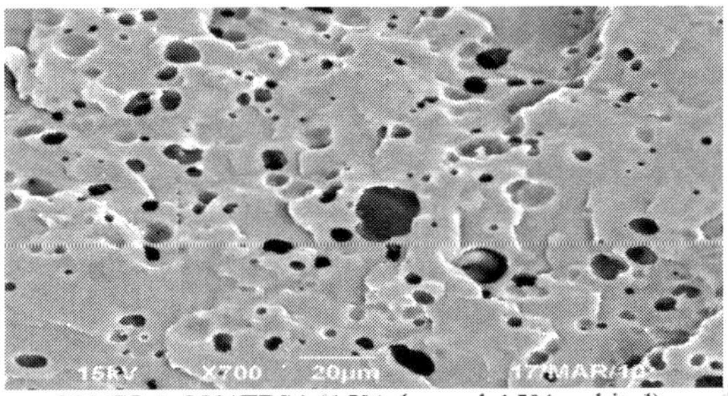

80%PLA-20%TPS4 (15%glycerol-15%sorbitol)

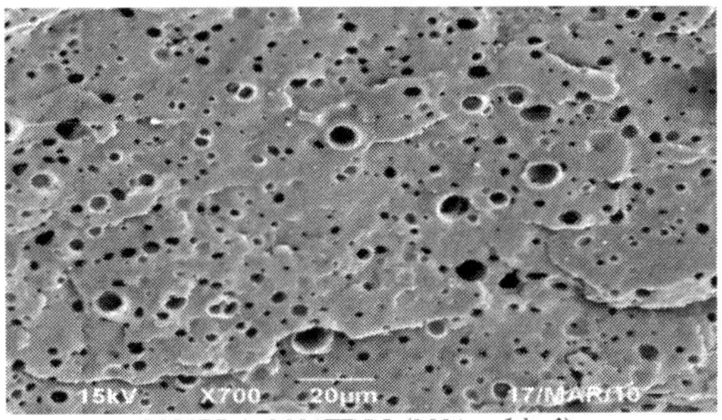

80%PLA-20%TPS5 (30%sorbitol)

Figure 15. SEM images of 80/20 blends of PLA/TPS

Copolymer Amylose-g-PLA

The method used in this work to improve the compatibility between PLA and TPS is the addition of copolymer Amylose-g-PLA.
This copolymer amylose-g-PLA is synthesized ex-situ by the procedure explained by Ouhib et al. (2008).
In order to control the structure, the copolymer is synthesized through a three-steps procedure (Figure 16): (i) partial silylation of the hydroxyl groups carried by amylose chains with N,O-bis-(trimethylsilyl) acetamide (BSA) as a silylation agent ; (ii) ring-opening polymerization (ROP) of D,L-lactide initiated from all the remaining hydroxyl groups on the partially silylated polysaccharide; and (iii) silylether deprotection under mild conditions.
Interface modification in polymer blends is expected to decrease interfacial tension in the melt phase leading to a smaller segregation scale and to improve interfacial adhesion in the solid state.
The stable location of the copolymer at the interface essentially depends on its molecular weight (Israels et

al., 1995; Matos, et al., 1995), architecture, chemical composition (Cigana et al., 1996), as well as the number of blocks (Cigana and Favis, 1998).

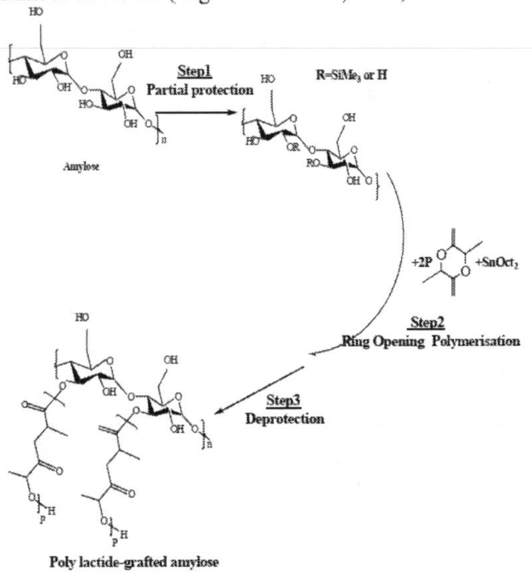

Figure 16. Schema of the three step polymerization of Amylose-g-PLA

In order to study the effect of number and chain length of the grafted PLA, three different structures have been synthesized (Table 3): 1.limited number but long chains of PLA grafts (run1); 2.higher numbers but shorter chains of PLA grafts (run2); 3.High grafting percentage of amylose and long chains of PLA (run3). The number and the length of the grafts in run1 and 2 are chosen in a way that the mass fraction of PLA in these copolymers is the same (F_{PLA}=0.67%).
In Table 3, *t (%)* is the silylation yield of the amylose (the number of trimethylsilylether groups per 100 OH functions) and *Y* is the number of PLA grafts per 100 glucose units. DP_n is the average degree of polymerization of each PLA graft, expressed in lactyle unit (molecular weight 72 g/mol). F_{PLA} is the PLA weight fraction in the silylated copolymers.

Run	t(%)	DS	Y	DP_n	Mn_{graft}	F_{PLA}
1	*90*	*2.7*	*30*	*15*	1080	0.68
2	*96*	*2.9*	*12*	*35*	2520	0.66
3	*83*	*2.95*	*51*	*74*	5400	0.94

Table 3. Polylactide-grafted silylated amylose parameters

Conclusions

The blends of thermoplastic starch and PLA are prepared. The efficiency of the plasticization procedure was observed by SEM images. All of the plasticizers (glycerol, Citric Acid and sorbitol) are able to destroy the granule structure of the starch. The blends of PLA and TPS plasticized by citric acid and sorbitol show the highest tensile strengths. Citric acid makes ester bonds with starch and improves its interaction with PLA and sorbitol makes hydrogen bonds with PLA. The study of morphology of these blends illustrates the co-continuity of starch and PLA for 40-60 % TPS. In this region the study of the improvement of the dispersion by addition of a compatibilizer can hardly be observed. The blends of 80/20 PLA/TPS are prepared. Finer dispersion of TPS in matrix is observed when sorbitol is used as plasticizer.
A three step procedure is used for the grafting reaction of PLA on amylose. This will enable us to study the role of the architecture of amylose-g-PLA on its compatibilizing efficiency in the starch/PLA blends.

References

Arvanitoyannis, I., et al., "Edible films made from hydroxypropyl starch and gelatin and plasticized by polyols and water", Carbohydrate Polymers, **36**(2/3),105-119(1998), *doi:10.1002/(SICI)1099-0488 (19960715)34:9<1691::AID-POLB18>3.0.CO;2-2.*

Blanshard, J.M.V., "Starch granule structure and function: a physiochemical approach". In: Starch: properties and potential, Galliard, T. (Ed.) Chichester: Wiley Publishers, p. 16–54 (1987).

Cigana, P., Favis, B.D., "The relative efficacy of diblock and triblock copolymers for a polystyrene/ethylene-propylene rubber interface" Polymer; **39**(15), 3373–8(1998), *doi: 10.1016/S0032-3861(97)10041-6.*

Cigana, P., et al. , "Diblock copolymers as emulsifying agents in polymer blends: Influence of molecular weight, architecture, and chemical composition" J Polym Sci B Polym Phys; **34**(9) ,1691–700(1996), *doi:10.1002/(SICI)1099-0488 (19960715) 34:9 <1691::AID-POLB18>3.0.CO;2-2.*

Forssell, P.M., et al., "Phase and glass transition behavior of concentrated barley starch-glycerol-water mixtures, a model for thermoplastic starch", Carbohydr Polym, **34**, 275-82 (1997), *doi: 10.1016/S0144-8617(97)00133-1.*

Huneault, M., Li, H., "Morphology and properties of compatibilized polylactide/thermoplastic starch blends", Polymer, 48, 270-280(2007), *doi:10.1016/j.Polymer. 2006.11.023.*

Israels, R., et al., "Compatibilizing A/B blends with AB diblock copolymers: effect of copolymer molecular weight", J. Chem Phys, **102**(20) ,8149–57(1995).

Kazuo, O., et al., "Studies on the retrogradation and structural properties of waxy potato starch", Bulletin of Chemical Society in Japanese, **71**, 1095-1100(1998).

Ke.T, Sun.X., " Thermal and Mechanical properties of poly(lactic acid) and Starch Blends with various plasticizers",American Society of Agricultural Engineers., **44**(4),945-953 (2001).

Lawton,J.W. , Fanta, G.F. , "Glycerol-plasticized films prepared from starch-poly(vinyl alcohol) mixtures: effect of poly(ethylene-co-acrylic acid)", Carbohydrate polymers, **23**(4), 275-80(1994), *doi:10.1016/0144-8617(94)90190-2.*

Lourdin, D., et al., "Antiplasticization' in starch-glycerol films?", J Appl Polym Sci,**63**,1047-53(1997)
Matos, M., et al., Polymer, **36**(20),3899–907(1995).

Martin, O., Ave´rous, L., "Poly(lactic acid): plasticization and properties of biodegradable multiphase systems", Polymer,**42**(6),209-19(2001), *doi:10.1016/S0032-3861(01)00086-6.*

Nashed, G., et al., "The plasticization effect of glycerol and water on the gelatinization of wheat starch" Starch/Starke; **55** (3-4), *doi: 131-7(2003), doi:10.1002/star.200390027.*

Ouhib, R., et al., "Biodegradable amylose-g-PLA glycopolymers from renewable resources", Carbohydrate Polymers, **77**(1), 32-40(2009), *doi: 10.1016/j.carbpol.2008.11.038.*

Perry, P.A., Donald, A.M., "The role of plasticization in starch granule assembly", Biomacromolecules, **1**,424-32 (2000), *doi: 10.1021/bm0055145.*

Poutanen, K., Forssell, P., "Modification of starch properties with plasticizers", Trends in Polymer Science, **4**, 128–132(1996).

Rodriguez-Gonzalez, F.Jw., "Rheological and thermal properties of thermoplastic starch with high glycerol content " ,Carbohydrate polymers,**58**,139-147(2004), *doi:10.1016/j.carbpol.2004.06.002.*

Sarazin, P., et al., "Binary and ternary blends of polylactide, polycaprolactone and thermoplastic starch", Polymer **49**, 599-609(2008), *doi: 10.1016/j.polymer.2007.11.029.*

Shogren, R. L., "Effect of moisture-content on the melting and subsequent physical aging of cornstarch" Carbohydrate Polymers, **19**(2), 83–90(1992), *doi:10.1016/0144-8617(92)90117-9.*

Simmons,S.,Thomas, E.L., "Structural characteristic of biodegradable thermoplastic starch/poly(ethylene-vinyl alcohol) blends", J.Appl. Polym.Sci.,**58**(12):2259-2285(1995), *doi: 10.1002/app.1995.070581215.*

Stepto, R.F.T., Tomka, I., "Injection molding of natural hydrophilic polymers in the presence of water", Chimia, **41**(3), 76-81(1987).

Tan, I., et al., "Investigation of the starch gelatinization phenomena in water-glycerol systems: Application of modulated temperature differential scanning calorimetry", Carbohydrate Polymer, **58**(2), 191-204(2004), *doi:10.1016/j.carbpol.2004.06.038.*

Vanoene, H., "Modes of dispersion of viscoelastic fluids in flow", Journal of Colloid and Interface Science,**40**(3),448(1972),*doi:10.1016/0021-9797(72) 90355-4.*

Wang, N., et al., Influence of citric acid on the properties of glycerol-plasticized dry starch(DTPS) and DTPS/Poly(lactic acid) blends" Starch/starke, **59**,409-417(2007), *doi: 10.1002/star.200700617.*

Wu, S. :Polymer Interface and Adhesion,Marcel Dekker Publishers, U.S.A.(1982).

Yu, J., et al., "A study on the properties of starch/glycerin blends", Starch/Staerke,**50**(6),246-250(1998).

Enhanced ionic conductivity in PEO/PMMA glassy miscible blends: Role of nano-confinement of minority component chains

Taheri Qazvini Nader

AC Impedance Spectroscopy was used to investigate the ionic conductivity of solution cast poly(ethylene oxide) (PEO) / poly(methyl methacrylate) (PMMA) blends doped with Lithium perchlorate. According to FTIR Spectroscopy, there is no trace of water and solvent left in samples. Based on DSC and DMTA graphs, all systems are fully miscible and glassy at room temperature. Conductivity measurements indicate that at low PEO contents (below overlap weight fraction w*), ionic conductivities are almost low. This could be due to nearly distant PEO chains in blend, which means ion transportation cannot be performed adequately. However, at weight fractions well above w*, a significant increase in ionic conductivity was observed. This enhanced ionic conductivity mimics the PEO segmental relaxation in rigid PMMA matrix, which can be attributed to the accelerated motions of confined PEO chains in PMMA matrix. At PEO content higher than 20 wt%, the conductivity measured at room temperature drops due to crystallization of PEO. However by increasing temperature to temperatures well above the melting point of PEO, a sudden increase of conductivity was observed which was attributed to phase transition from crystalline to amorphous state. The results indicate that some PEO/PMMA blends with well enough PEO content, which are structurally solid, can be considered as an interesting candidate for usage as solid-state electrolytes in Lithium batteries.

DISSOLUTION AND DIFFUSION OF NANOFILLED THERMOPLASTIC YARNS IN EPOXY RESINS

D.Dumont*, D.Daoust, M.Sclavons, J. Devaux

Institute of Condensed Matter and Nanosciences - Bio & Soft Matter (IMCN/BSMA), Université catholique de Louvain, Croix du Sud, 1 bte 4 ; Bât. Boltzmann (a2) B-1348 Louvain-la-Neuve, Belgium – david.dumont@uclouvain.be; daniel.daoust@uclouvain.be; michel.sclavons@uclouvain.be; jacques.devaux@uclouvain.be

In order to improve the performances of composites manufactured using modern liquid moulding technologies like Resin Transfer Moulding (RTM), a method to disperse simultaneously a soluble amorphous thermoplastic and nanofillers in epoxy matrices during processing is proposed. The concept consists in a nanofilled thermoplastic added to the dry composite preform in the form of yarns, veils or films. The dissolution of the thermoplastic element in the resin is completed during the temperature increase following the RTM injection stage. After thermoplastic and nanofiller diffusion, reaction induced phase separation generates the final microstructure. Unfunctionalized multiwall carbon nanotubes (MWNTs) were initially dispersed in poly(ether sulfone) (PES) and poly(hydroxyether of bisphenol A) (phenoxy) using an extrusion process. The pristine polymers and the nanocomposites were subsequently drawn to manufacture monofilaments. Dissolution of the thermoplastic yarns was measured in the HexFlow RTM6 resin widely used in the aeronautical industry and in alternative epoxy resin formulations. Polymer and MWNT concentration profiles were assessed using Raman spectroscopy after diffusion in the resin. The importance of thermoplastic type and resin composition on the dissolution kinetics and the diffusion extent is evidenced. The influence of the residual composition gradients on the local morphologies is explored by electron microscopy. For the nanocomposite yarns, the extent of MWNT diffusion demonstrates that soluble thermoplastics are adapted for the MWNT delivery in epoxy resins.

Introduction

The performances of fiber-reinforced polymer composites based on highly crosslinked epoxy networks combined with the low weight of the resin justify their outstanding use for aerospace applications and structures requiring high mechanical, thermal and chemical resistances. However, properties like fracture toughness or thermal and electrical conductivities remain important drawbacks of such thermosetting matrices. Therefore, incorporation of a thermoplastic or nanofillers in the epoxy resin has been proposed to improve these weak points.

Many attempts have been made to improve resistance to failure due to interlaminar crack propagation in laminated fibre-reinforced composites by modification of the thermosetting resin matrix with a thermoplastic. Blends of epoxy resins with various amounts of high-performance engineering thermoplastics contributing to the toughening of the cured network have been widely investigated (Hodgkin et al., 1998). A composition dependent microstructure is generated for the initially homogenous blends during the cure of the resin as a result of reaction induced phase separation. Unlike rubber toughening agents, thermoplastic additives that have a high glass transition temperature (T_g) and ductility can enhance the fracture resistance of the thermosetting resin formulation without compromising the thermomechanical stability of the epoxy network (Bonnet et al., 2001; Bucknall and Partridge, 1983; Cho et al., 1993; Hourston et al., 1997; Martinez et al., 2000; Mimura et al., 2000; Min and Kim, 1999; Park et al., 2007; Raghava, 1988) and are for that reason adapted to the severe requirements of sectors like the aerospace industry.

Liquid composite moulding technologies like Resin Transfer Moulding (RTM) or Liquid Resin Infusion (LRI) processes and variants have been emerging as cost effective alternatives to prepreg-autoclave to manufacture composite parts. Toughening of prepregs using a thermoplastic blended with the epoxy resin (Fernandez et al. 2004; Hayes and Seferis, 1998; Murakami et al. 2000; Turmel and Partridge, 1997) has demonstrated potential benefits in terms of interlaminar toughness improvement. However, transposition of these toughening methods to an injectable resin for a liquid infusion process is difficult because of the great viscosity increase of the system. Indeed, high molecular weight thermoplastics hinder infiltration throughout the fabric and preform impregnation which are essential for a final structure free of voids. To maintain a low resin viscosity, reinforcement of composite materials can be achieved by the introduction of solid thermoplastic into the preform without modification of the processing conditions. Thermoplastic additives being adapted to dissolve in the curable resin and enabling processing via liquid moulding process have been considered in the form of yarns (Beier et al., 2008, 2009; Carter et al., 2004; Thanomsilp and Hogg, 2005), films (An et al., 2002; Naffakh et al., 2006; Yun et al., 2004), membranes (Mortimer and Cawse, 2007) or non woven mats (Beier et al., 2009; LoFaro et al., 2006). These studies have highlighted enhanced performance in mode II shear cracking, the dominant failure mode for the initial formation of delamination during impact. This technology, providing a localized toughening of the composite, can additionally stabilize the dry preform to improve its handling.

Similarly to thermoplastic additives, carbon nanotubes have demonstrated their potential to improve significantly properties of epoxy resins like fracture toughness or electrical conductivity at very low content for optimized dispersion in the polymer matrix (Godara et al., 2009, Gojny et al., 2005, Sandler, 2003, Yu, 2008). Hence, the work reported in this paper specially examines the possibility to exploit the solubility of thermoplastics in epoxide-diamine formulations to deliver unmodified MWNTs in an epoxy resin. MWNT-thermoplastic nanocomposite yarns based on PES and phenoxy amorphous polymer are selected as vectors to incorporate simultaneously the polymer and the nanofiller tougheners in the epoxy resin. Due to its high T_g, no degradation of thermomechanical properties of the PES-modified epoxy resin is expected. In spite of its lower T_g, phenoxy has demonstrated ability to improve fracture toughness in two-phase morphology blends without significant reduction of other properties of the epoxy network (Siddhamali, 2000; Teng and Chang, 1996).

In a first time, the dispersion state of the MWNTs is evaluated in the melt processed thermoplastic nanocomposites. Subsequently, the thermal dissolution of a thermoplastic yarn is followed in the Hexflow RTM6 resin formulation qualified for aeronautic applications and in alternative resin compositions in order to define thermoplastic-epoxy resin associations suitable to achieve complete yarn dissolution before hindering by the crosslinking reaction of the resin. Thereafter, concentration profiles of thermoplastics and MWNTs are assessed for nanocomposites yarns following diffusion in the reactive resins. Finally, local microstructures expected after phase separation are discussed as a function of the residual gradients of composition in the blend.

Experimental

Materials

The HexFlow RTM6 resin from Hexcel, developed to fulfil the requirements of the aerospace industries in RTM, was tested as well as alternative epoxide-diamine resin formulations. The RTM6 resin is based upon tetraglycidyl-4,4'-methylenedianiline (TGMDA) and a blend of aromatic diamines. The alternative epoxy resin compositions contain a liquid diamine hardener, diethyltoluenediamine (DETDA, Lonzacure DETDA 80 from Lonza, isomer mixture composed by 77–81% 3,5-diethyltoluene-2,4-diamine and 18–22% 3,5-diethyltoluene-2,6-diamine) and the tetraepoxide TGMDA (Araldite MY 721 from Huntsman) or a triepoxide, triglycidyl para-aminophenol (TGAP, Araldite MY 0510 from Huntsman). The stoichiometry ratio of the epoxy resins TGAP-DETDA and TGMDA-DETDA was one in each case. A high and a low T_g amorphous thermoplastic were used, respectively poly(ether sulfone) (PES, Radel A-200 from Solvay Advanced Polymers) and, poly(hydroxyether of bisphenol A) (phenoxy, PKHP200 from InChem). The chemical structures of the above-mentioned resin monomers and thermoplastics are listed in Table 1.

Table 1 – Resin components and thermoplastics

EPOXIDES	Tetraglycidyl -4,4'-methylene dianiline (TGMDA) 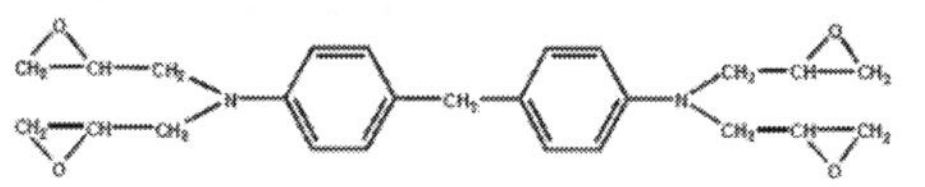 Triglycidyl para-aminophenol (TGAP) 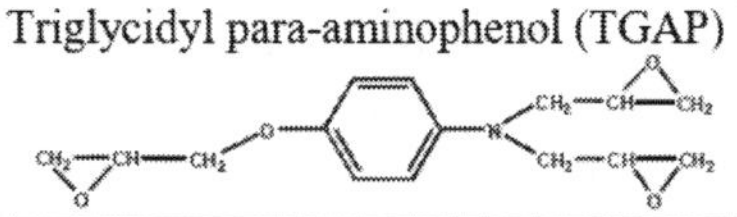
DIAMINES	Diethyltoluenediamine (DETDA) 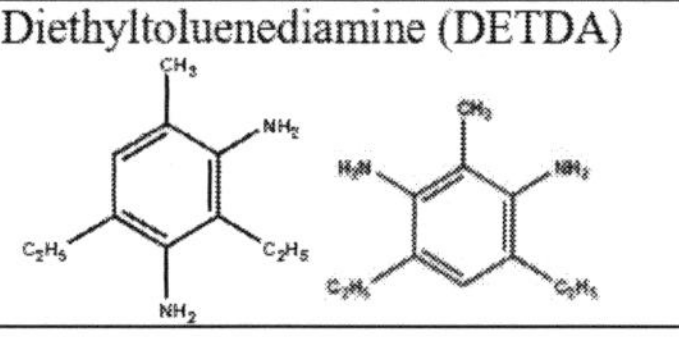
THERMOPLASTICS	Poly(ether sulfone) (PES) 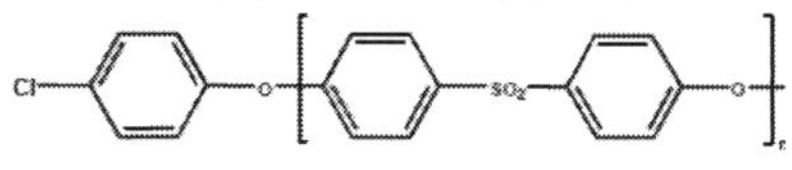 Poly(hydroxyether of bisphenol A) (phenoxy) 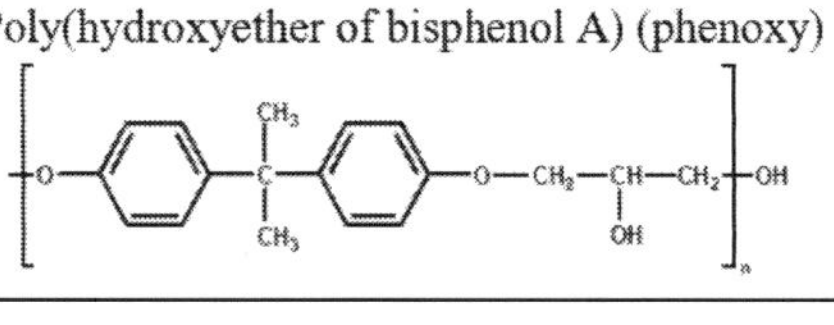

Polymer yarn manufacturing and characterisation

The thermoplastic yarns were drawn thanks to a DSM Xplore Fibre Spin Line. The thermoplastic, prior dried during 12 hours under vacuum, was initially extruded in a corotating twin-screw batch-type mini-extruder (DSM 15cm^3) at 220 °C for phenoxy and 340 °C for PES. The residence time was 5 minutes and the rotation speed of the screws was 150rpm. Subsequently, the winding unit allows the processing of monofilaments with diameters controlled by the winding speed and torque, the thermoplastic flow and the die dimensions (5 or 0.5mm diameter). Nanocomposites yarns containing 1, 3 and 5 wt% MWNTs Nanocyl7000 from Nanocyl were produced using the same conditions.

The MWNT dispersion was examined by means of transmission electron microscopy (TEM) for samples extruded and yarn specimens. Ultrathin sections of the nanocomposites of approximately 95 nm in thickness were cut at ambient temperature. The micrographs were imaged with a LEO 922 transmission electron microscope operating at 200 kV.

Shear rheology was performed on an ARES rheometer from Rheometric Scientific on compression molded samples. A parallel plate geometry was used with a diameter of 25mm and a gap of 1mm. Measurements were carried out under nitrogen atmosphere at a strain rate of 10% with an angular frequency range varying from 215 to 0,1 rad/s.

Dissolution of the yarns and diffusion in epoxy resins

The thermal dissolution of the thermoplastic yarns in the epoxy resins was followed in the case of a single 100 μm diameter thermoplastic fibre placed in 50 mg resin between two microscope glass slides. The sample was fitted into a Mettler FP82HT hot stage monitored by a Mettler FP90 central processor and continuously observed with an Olympus BX51 optical microscope connected to a ColorviewI camera. The yarn diameter was measured until its complete dissolution or constant values in the case of a partial dissolution.

To determine thermoplastic concentration profiles resulting from the dissolution and diffusion of a thermoplastic yarn in the epoxide-diamine formulation, the samples were removed from the hot stage after gelification of the resin and phase separation for analysis by Raman spectroscopy. Linear Raman imaging allows similarly the evaluation of MWNT diffusion. Spectra were acquired at intervals of 10 μm, perpendicularly to the initial yarn direction thanks to the motorized sample support of the confocal laser Raman spectrometer DXR SmartRaman from Thermo Scientific operating in the microscopic mode. The laser was focused between the two microscope glass slides, in the plane of the yarn center. Recorded spectra were the average of 5 spectra accumulated for 60 s. The incident laser excitation wavelength was 780 nm. The pinhole aperture was set to 50 μm in combination with a 50 times magnification objective. A high resolution grating was used in combination with a 2048 pixel charge-coupled device (CDD) detector.

Microstructure of the thermoplastic-epoxy resin blends

TEM and scanning electron microscopy (SEM) techniques are used to relate the composition gradients remaining after diffusion to the microstructure locally expected as a result of phase separation in the blend. On the one hand, the interphase morphology generated consequently to the interdiffusion between an epoxy resin and a thermoplastic yarn was observed and on the other hand, the microstructure dependence on the thermoplastic concentration was evidenced for blends with homogenous compositions. For the latter, samples were prepared by thermal dissolution of the thermoplastic powder in the epoxide component of the resin formulation at 140°C under stirring during 1 hour and by further addition of the diamine hardener. SEM analyses were performed on fractured surfaces of the cured samples using a LEO 982 (Zeiss) with prior etching of the thermoplastic phase to reveal the microstructure. The PES and phenoxy extractions were achieved respectively in methylene chloride and tetrahydrofuran during 24 hours.

Results and Discussion

Since thermoplastics are investigated as vector to disperse unfunctionalised MWTNs in epoxy resin, the phenoxy and PES nanocomposites have to be evaluated so as to determine if the melt processing conditions are adequate to disperse homogenously the carbon nanotubes in the thermoplastic matrix. Representative TEM micrographs of phenoxy nanocomposites containing 1 and 5 wt% MWNTs are presented in Figure 1 (a and b). A fine and homogeneous dispersion with individualized nanotubes randomly oriented is observed in the two cases, indicating that the selected process conditions without severe shearing allow to overcome the strong attractive forces resulting from entanglements and high Van der Waals forces between native MWNTs.

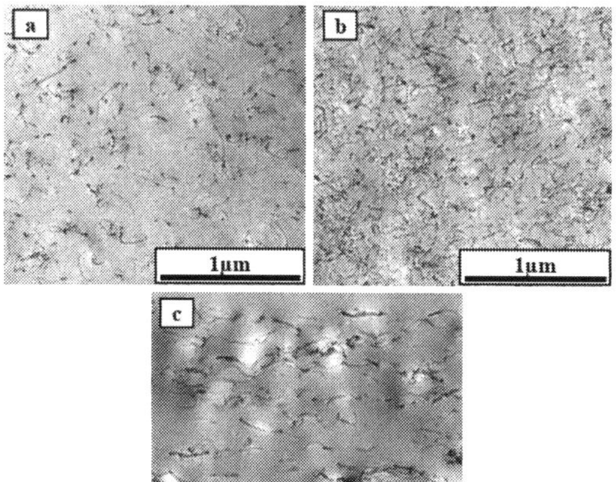

Figure 1 – TEM micrographs of the MWNT dispersion in phenoxy (a) 1wt% MWNTs extruded (b) 5wt% MWNTs extruded (c) yarn with 3wt% MWNTs (cut parallel to the yarn axis)

After spinning and winding of the extruded samples to produce nanocomposite yarns (Figure 1c), the fine distribution is conserved but MWNTs are oriented in the direction of the stretching, parallel to the yarn axis. The same results were observed for the PES nanocomposites.

The good dispersion state of MWNTs in the phenoxy and the formation of a percolated structure can be evidenced by the rheological analysis of the polymer nanocomposites. The frequency dependence of the shear storage modulus *G'* is shown in Figure 2 for a temperature of 150°C. At high frequencies, the sensitivity of response to the filler concentration is weak, indicating that the short-range polymer dynamics is not influenced by the carbon nanotubes. At low frequency, the storage modulus increases in presence of the MWNTs, particularly for high MWNT contents. A transition from a liquid-like viscoelastic behaviour to a solid-like or gel-like behaviour takes place and G' becomes nearly independent of the frequency. For 3wt% MWNTs, a maximum of tan δ and the absence of a Newtonian plateau for viscosity were also observed. These characteristics are typical of the formation of nanotube network-like structures with restriction of the long-range motion of polymer chains.

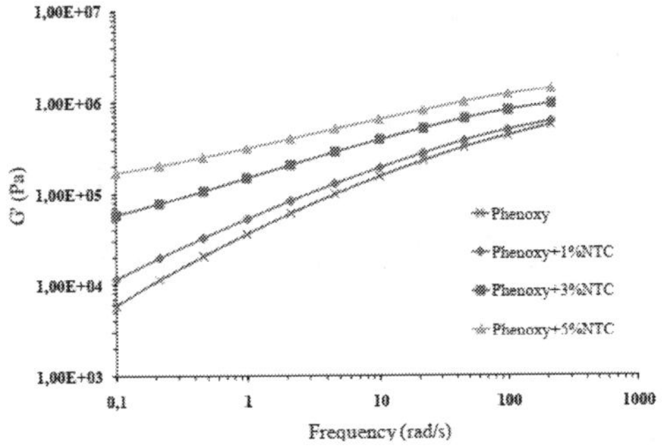

Figure 2 – Storage modulus as a function of angular frequency for neat phenoxy and nanocomposites with different MWNT contents (1, 3 and 5 wt%) at 150°C

Moreover, the Van Gurp-Palmen plot (phase angle δ as function of the absolute value of the complex modulus $|G^*|$), presented in Figure 7, allows the identification of the 'rheological percolation threshold'. For the neat matrix, the curves will approach a phase angle of 90° at low complex moduli whereas when 3 or 5 wt% MWNTs are added to the polymer, a significant decrease in phase angle is observed for low complex moduli. This is the evidence of the change of the nanocomposite structure from a viscous fluid behaviour to an elastic solid behaviour when carbon nanotube content increase. The transition is observed at the same MWNT concentration as in the *G'* plot and is attributed to the formation of a three-dimensional structure with distance between two carbon nanotubes being less than twice the gyration radius of the phenoxy macromolecules for which the nanotubes can hinder the motion of the polymer. It can be concluded that rheological percolation threshold exists at 150°C between 1 and 3wt% MWNTs for the phenoxy nanocomposites processed under the selected conditions and is in the same interval of concentration for PES nanocomposites at 260°C.

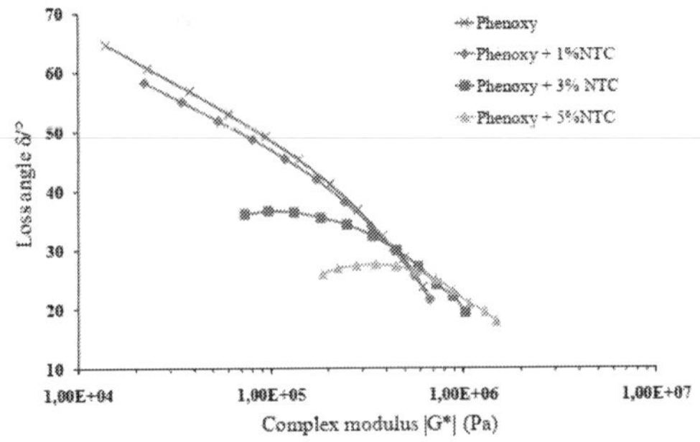

Figure 3 – Phase angle δ as a function of absolute value of the complex modulus $|G^*|$ for neat phenoxy and nanocomposites with different nanotube contents (1, 3 and 5 wt%) at 150°C (van Gurp-Palmen plot)

In order to guarantee a complete dissolution of nanocomposite yarns distributed in an epoxy resin, the selection of convenient epoxy resin-thermoplastic associations is primordial. From a mechanistic point of view, the thermal dissolution of thermoplastic yarns should occur through the diffusion of the resin monomer mixture into the thermoplastic and dissolution of the latter. After the initial interdiffusion regime, the dissolution process becomes inhibited by crosslinking and thermoplastic concentration effects. On the one hand, the molecular weight increase of the epoxy resin drops off the mobility of the diffusing molecules and on the other hand, diffusing species alter the conditions for further interdiffusion. Therefore, due to the competition between the cure reaction and the thermoplastic yarn dissolution, the temperature profile and the resin reactivity must also be considered as well as the yarn diameter and the global thermoplastic concentration characterized by the yarn distribution in the composite perform.

An experimental system based on a single yarn (diameter of 100 µm) in an epoxy resin was proposed to study the influence of the system composition on the yarn dissolution during an isothermal heating. In the case of a PES yarn in the RTM6 resin at 140°C, the dissolution is slow and the molecular weight increase of the reactive epoxy resin formulation becomes so large as to prevent further diffusion before complete solubilisation of the 100µm diameter yarn, as reported on Figure 4. The initial swelling of the yarn, observed before the diameter decrease, can be explained by the penetration of low molecular weight monomers into PES, inducing an elongated configuration of the thermoplastic chains, being faster than the yarn dissolution resulting from the disentanglement and the diffusion of these macromolecules. Moreover, it can be establish that the DETDA diamine and the TGAP triepoxide contribute to the improvement of the PES dissolution kinetics in the TGMDA-DETDA and the TGAP-DETDA alternative resin formulations in comparison to RTM6. Indeed, the 100µm PES yarn is completely dissolved in the TGAP-DETDA resin during the first minutes of the isotherm heating. In addition to the strong effect of the epoxy resin composition, the molecular structure of the thermoplastic reveal an obvious influence on the yarn dissolution in epoxy resins. In order to achieve complete yarn dissolution in the RTM6 resin, phenoxy can be chosen as substitute to PES. Although both thermoplastics have close medium molecular weights, the dissolution kinetics of phenoxy is considerably improved compared to PES.

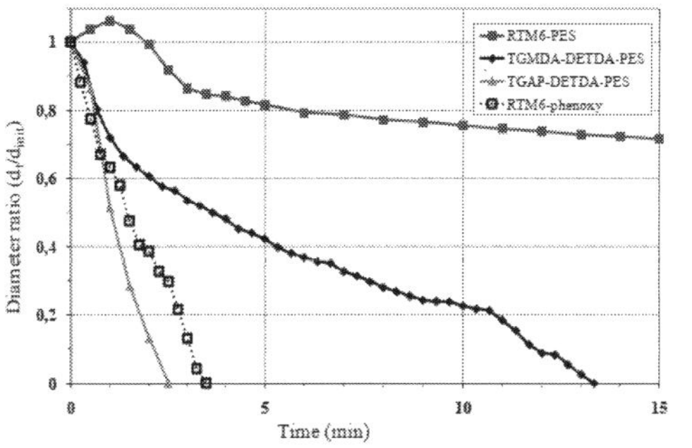

Figure 4 – Dissolution of a PES yarn in the RTM6, TGMDA-DETDA and TGAP-DETDA resins and phenoxy yarn in the RTM6 resin at 140°C

Further information concerning the thermoplastic distribution in the epoxy network after the yarn dissolution and the diffusion of the disentangled macromolecules is provided by Raman spectroscopy. Profiles illustrating the thermoplastic diffusion extent can be evaluated by computing the ratio between the intensity of a thermoplastic characteristic peak (1111 cm^{-1} for phenoxy and 1148 cm^{-1} for PES) and an epoxide reference peak (795 cm^{-1} or 1190 cm^{-1} for TGMDA and 825 cm^{-1} for TGAP) for all the spectra acquired at different distances from the initial location of the soluble yarn. The interphase of the samples was analyzed after dissolution of the thermoplastic yarn and further diffusion until gelification where no more evolution of the local blend composition was observed. Linear baseline subtraction was applied to remove the fluorescence background from the spectra. Thanks to the limited overlapping between the reference peaks, this comparative study can be achieved without decomposition treatment. Figure 5 shows the composition gradient after isothermal dissolution and diffusion of a 100µm PES yarn at 120°C. The diffusion distance is higher in the TGAP-DETDA resin than in the TGMDA-DETDA although its higher reactivity induces a shorter diffusion period. The earlier molecular weight increase in the TGAP-DETDA epoxy network due to the crosslinking reaction is counterweighed by the faster dissolution of the PES yarn and the improved relative diffusion of the thermoplastic macromolecules during the initial stages of the heating process in this resin. While the dissolution of a phenoxy yarn is drastically accelerated compared to a PES equivalent, due to stronger affinity of the resin components for phenoxy, the increase of diffusion extent observed for phenoxy compared to PES was less pronounced. Indeed, the diffusion coefficient of the phenoxy macromolecules having an average molecular weight close to the PES one remains limited compared to low molecular weight resin precursors.

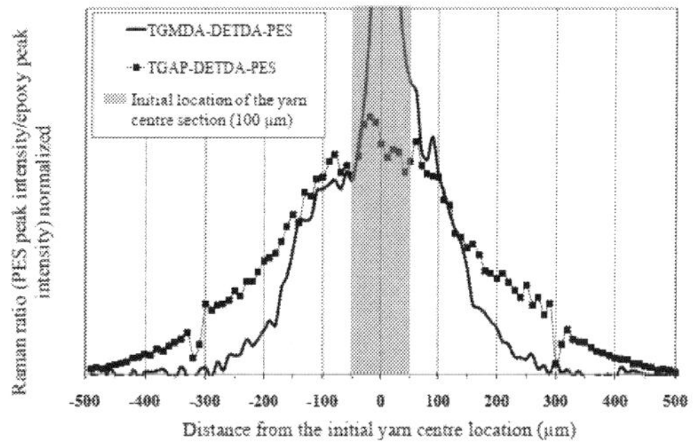

Figure 5 – Diffusion profiles of a PES yarn in TGMDA-DETDA and TGAP-DETDA at 120°C

The diffusion profiles presented in this paper are based on a qualitative exploitation of the Raman spectra. Nevertheless, a calibration was achieved thanks to samples of homogenous composition prepared by dissolution of thermoplastic concentrations comprised between 0 and 40wt% in the epoxy resins in order to quantify the local thermoplastic concentrations resulting from diffusion. The etched fracture surfaces of the calibration samples were subsequently observed by SEM (Figure 6).

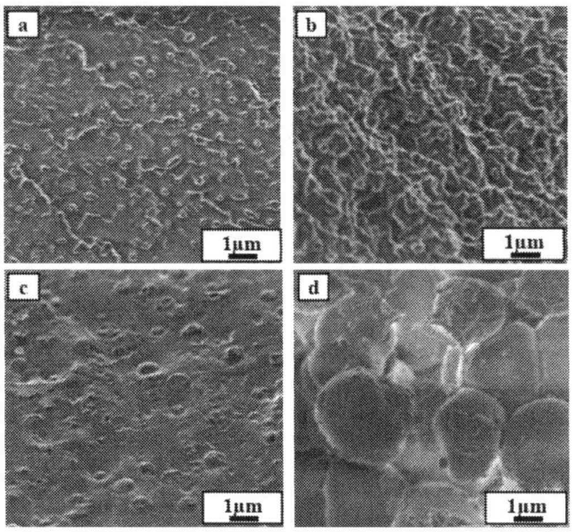

Figure 6 – SEM micrographs of the etched fracture surface of phenoxy in the RTM6 resin [(a) sea-island, 20wt% phenoxy (b) phase inversion, 40wt% phenoxy)] and PES in TGAP-DETDA resin [(c) sea-island, 10wt% PES (d) phase inversion, 20wt% PES]

Critical composition is comprised between 30 and 40wt% of phenoxy in the RTM6 resin and is located at lower PES concentration, between 15 and 20wt%, in the TGAP-DETDA resin. As the thermoplastic concentration assessed after diffusion remains locally higher than the critical composition, microstructures going from a sea (epoxy resin-rich phase)-island (thermoplastic-rich nodules) morphology to phase inverted morphology with the thermoplastic forming the continuous phase are expected. Figure 7 confirms this result for a PES yarn in the TGMDA-DETDA resin where the presence of an interphase with a wide morphology spectrum related to the composition gradient is clearly evidenced.

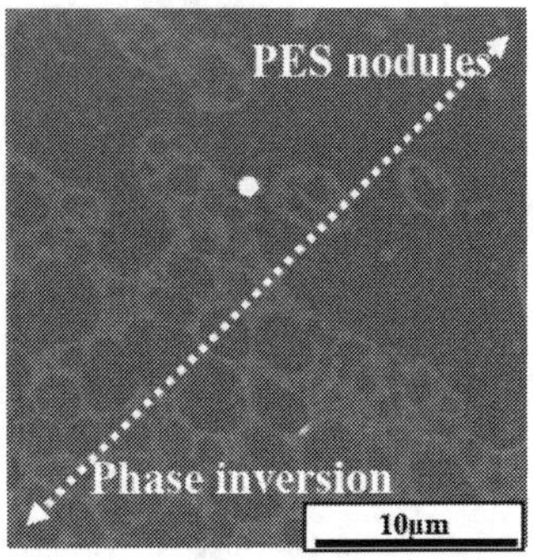

Figure 7 – Microstructure spectrum of the interphase between a PES yarn and the TGMDA-DETDA resin after dissolution and diffusion

The diffusion of MWNTs, initially dispersed in a nanocomposite yarn, was also investigated to demonstrate that soluble thermoplastic can act as vector for their delivery in epoxy resins. The MWNT distribution in the epoxy resin-thermoplastic blend can be determined thanks to the intense Raman peak centered at 1311cm^{-1}, characteristic of carbon nanotubes. Figure 8 suggests that the MWNT distribution depends on the simultaneous thermoplastic diffusion for a phenoxy yarn filled with 1wt% MWNTs since the normalized carbon nanotubes and phenoxy diffusion profiles have similar shapes.

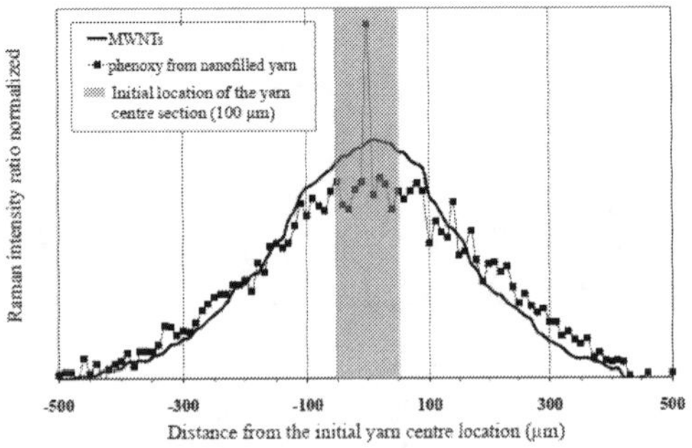

Figure 8 – Diffusion profiles of MWNTs and phenoxy from a phenoxy-1wt%MWNT nanocomposite in the TGMDA-DETDA resin at 120°C

The large diffusion distance confirms that MWNTs can be introduced quite homogeneously in epoxy resin via nanocomposite yarns being soluble in the epoxy resin without further mechanical or chemical dispersion process required. Moreover, the TEM micrograph of 20wt% phenoxy nanocomposite (3wt% MWNTs)-RTM6 resin blend presented in Figure 13 shows that all nanotubes seems to be located in the continuous epoxy-rich phase after phase separation. Consequently, the epoxy matrix should benefit from both carbon nanotubes and thermoplastic nodules tougheners. For this sample, a large amount of MWNTs agglomerates were formed during dissolution of the nanocomposite in the resin under stirring. However, this drawback is not observed for static dissolution and diffusion of the nanocomposite yarns in the epoxy resin.

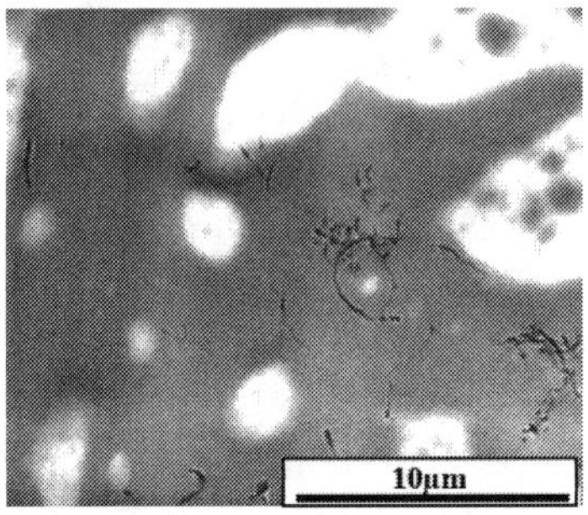

Figure 9 – TEM micrograph of 20wt% phenoxy nanocomposite (3wt% MWNTs) in the RTM6 resin

Conclusions

The potential to use soluble thermoplastic yarns as vectors to disperse MWNTs in epoxy resins without chemical modification of the nanotubes has been demonstrated in this work. The concept developed should be transposable for liquid moulding process of composite parts. Fine dispersions of the MWNTs in PES and phenoxy nanocomposites were generated by melt processing with rheological percolation thresholds comprised between 1 and 3wt% of MWNTs. Whereas phenoxy is soluble in the RTM6 resin, alternative resin formulations are proposed to dissolve PES yarns. The MWNTs distribution as well as the residual composition gradient of the blend, and consequently its microstructure spectrum, depends on thermoplastic diffusion in the epoxy resin.

Acknowledgements

The authors want to be grateful to FRIA (Fonds pour le Recherche dans l'Industrie et l'Agriculture, FRS-FNRS, Belgium) and "Région Wallonne" in the frame of the Plan Marshall program for the financial support. They acknowledge InChem for providing the phenoxy polymer, Lonza for the Lonzacure DETDA80 and Nanocyl for the carbon nanotubes.

References

-An, X., Ji, S., Tang, B., Zhang, Z., Yi, X. S., "Toughness improvement of carbon laminates by periodic interleaving thin thermoplastic films", J. Mater. Sci. Lett., **21**, 1763-1765 (2002)

-Beier, U., Wolff-Fabris, F., Fischer, F., Sandler, J. K. W., Altstädt, V., Hulder, G., Schmachtenberg, E., Spanner, H., Weimer, C., Roser, T., Buchs, W., "Mechanical performance of carbon fibre-reinforced composites based on preforms stitched with innovative low-melting temperature and matrix soluble thermoplastic yarns", Compos. Part A: Appl. Sci. and Manufact., **39**, 1572-1581 (2008)

-Beier, U. et al. "Mechanical performance of carbon fibre-reinforced composites based on stitched and bindered performs", Composites: Part A, **40**, 1756-63 (2009)

-Bonnet, A., Lestriez, B., Pascault, J. P., Sautereau, H., "Intractable high Tg thermoplastics processed with epoxy resin: Interfacial adhesion and mechanical properties of the cured blends", J. Polym. Sci. Part B: Polym. Phys., **39**, 363-373 (2001)

-Bucknall, C. B., Partridge, I. K., "Phase separation in epoxy resins containing polyethersulphone", Polymer, **24**, 639-644 (1983)

-U.S. Patent 2004/0041128 A1 (2004), Carter, J. T., LoFaro, C., Maskell, R. K., McGrail, P. T.

-Cho, J. B., Hwang, J. W., Cho, K., An, J. H., Park, C. E., "Effects of morphology on toughening of tetrafunctional epoxy resins with poly(ether imide)", Polymer, **34**, 4832-4836 (1993)

-Fernandez, B., Arbelaiz, A., Diaz, E., Mondragon, I., "Influence of polyethersulfone modification of a tetrafunctional epoxy matrix on the fracture behavior of composite laminates based on woven carbon fibres", Polym. Compos., **25**, 480-488 (2004)

-Godara, A., Mezzo, L., Luizi, F., Warrier, A., Lomov, S. V., van Vuure, A. W., Gorbatikh, L., Moldenaers, P., Verpoest, I., "Influence of carbon nanotube reinforcement on the processing and the mechanical behaviour of carbon fiber/epoxy composites", Carbon, **47**, 2914-2923 (2009)

-Gojny, F. H., Wichmann, M. H. G., Fiedler, B., Schulte, K., "Influence of different carbon nanotubes on the mechanical properties of epoxy matrix composites - A comparative study", Compos. Sci. Technol., **65**, 2300-2313 (2005)

-Hayes, B. S., Seferis, J. C., "Variable temperature cure polyetherimide epoxy-based prepreg systems", Polym. Eng. Sci., **38**, 357-370 (1998)

-Hodgkin, J. H., Simon, G. P., Varley, R. J., "Thermoplastic toughening of epoxy resins: a critical review", Polym. Adv. Technol., **9**, 3-10 (1998)

-Hourston, D. J., Lane, J. M., Zhang, H. X., "Toughening of Epoxy Resins with Thermoplastics: 3. An Investigation into the Effects of Composition on the Properties of Epoxy Resin Blends", Polym. Int., **42**, 349-355 (1997)

-U.S. Patent 2006/0252334 A1 (2006), LoFaro, C., Abusafieh, A., Webb, W. E., Doyle, M.

-Martinez, I., Martin, M. D., Eceiza, A., Oyanguren, P., Mondragon, I., "Phase separation in polysulfone-modified epoxy mixtures. Relationships between curing conditions, morphology and ultimate behavior", Polymer, **41**, 1027-1035 (2000)

-Mimura, K., Ito, H., Fujioka, H., "Improvement of thermal and mechanical properties by control of morphologies in PES-modified epoxy resins", Polymer, **41**, 4451-4459 (2000)

-Min, H. S., Kim, S. C., "Fracture toughness of polysulfone/epoxy semi-IPN with morphology spectrum", Polym. Bull., **42**, 221-227 (1999)

-PCT Int. Appl. Publ. WO 2007/110617 A1 (2007), Mortimer, S., Cawse, J. L.

-Murakami, A., Shonaike, G. O., Ooishi, K., Watanabe, O., Takezawa, M., "Fracture toughness of PEI modified epoxy resin CFRP composites", J. Reinf. Plast. Compos., **19**, 137-151 (2000)

-Naffakh, M., Dumon, M., Gérard, J. F., "Study of a reactive epoxy-amine resin enabling in situ dissolution of thermoplastic films during RTM for toughening compos.", Compos. Sci. Technol., **66**, 1376-84 (2006)

-Park, J. M., Kim, D. S., Kong, J. W., Kim, S. J., Jang, J. H., Kim, M., Kim, W., Devries, K. L., "Interfacial evaluation and self-sensing on residual stress and microfailure of toughened carbon fibre/epoxy-AT-PEI composites", Compos. Part B:Eng., **38**, 833-846 (2007)

-Raghava, R. S., "Development and characterization of thermosettingting-thermoplastic polymer blends for applications in damage-tolerant composites", J. Polym. Sci.: Polym. Phys. Ed., **26**, 65-81 (1988)

-Sandler, J. K. W., Kirk, J. E., Kinloch, I. A., Shaffer, M. S. P., Windle, A. H., "Ultra-low electrical percolation threshold in carbon-nanotube-epoxy composites", Polymer, **44**, 5893-5899 (2003)

-Siddhamali, S. K., Kyu, T., "Toughening of thermosetting/thermoplastic composites via reaction-induced phase separation: Epoxy/phenoxy blends", J. Appl. Polym. Sci., **77**, 1257-1268 (2000)

-Teng, K. C., Chang, F. C.,"Single-phase and multiple-phase thermoplastic/thermosetting polyblends: 2. Morphologies and mechanical properties of phenoxy/ epoxy blends", Polymer, **37**, 2385-2394 (1996)

-Thanomsilp, C., Hogg, P. J., "Interlaminar fracture toughness of hybrid compos itesbased on commingled yarn fabrics", Compos. Sci. Technol., **65**, 1547-63 (2005)

-Turmel, D. J. P., Partridge, I. K., "Heterogeneous phase separation around fibres in epoxy/PEI blends and its effect on composite delamination resistance", Compos. Sci. Technol., **57**, 1001-1007 (1997)

-Yu, N., Zhang, Z. H., He, S. Y., "Fracture toughness and fatigue life of MWCNT/epoxy composites", Mater. Sci. Eng., **494**, 380-384(2008)

-Yun, N. G., Won, Y. G., Kim, S. C., "Toughening of carbon fibre/epoxy composite by inserting polysulfone film to form morphology spectrum", Polymer, **45**, 6953-6958 (2004)

ROLE OF RUBBER CAVITAION IN TOUGHENING MECHANISMS OF EPOXY RESINS

Ehsan R. Mafi*, Morteza Ebrahimi, Hamed Vakili Tahami, Hossein Yahyaei

Polymer Engineering Department, Amirkabir University of Technology, 15875-4413 Tehran, Iran; ehsanrmafi@yahoo.com; ebrahimi@aut.ac.ir; hamed@hvakili.com; yahyaeihossein@yahoo.com

Abstract - The occurrence of rubber particle cavitation in rubber modified epoxy resins has generated some controversy in the literature and role of cavitation resistance in toughening mechanisms has remained an enigma. In this study, hydroxyl terminated polybutadiene (HTPB) and synthesized epoxy terminated polybutadiene (ETPB) have been used to toughen an epoxy resin. FTIR and titration techniques were used to monitor the ETPB synthesis. Mean fracture toughness, K_{IC}, of 0.65 $MNm^{-3/2}$ was measured for unmodified epoxy polymer. The measured toughness was increased by the addition of HTPB to 1.23 $MNm^{-3/2}$ and ETPB to 1.80 $MNm^{-3/2}$. The cavitation zone of the HTPB modified sample was bigger than that of ETPB modified sample which showed the higher cavitation resistance of ETPB particles which was due to crosslinking of rubber particles, confirmed by dynamic mechanical analysis. SEM studies showed volumetric expansion of rubbery phase in slow crack growth region of both samples which proved the plastic void growth. The results showed that for this specific case, elevated cavitation resistance of ETPB does not suppress plastic void growth. The optical microscopy of crack tips showed little shear yielding. Which was due to rationally large particle size of ETPB and HTPB domains in epoxy matrix and resulted high inter particle distance which suppresses the shear yielding mechanism. SEM micrographs of fracture surface of HTPB modified samples showed No tail formation while pinning tails were present in ETPB modified samples. In ETPB particles the crack front bows when reaches to rubber particle due to its high cavitation resistance. The results showed that higher cavitation resistance activates crack pinning process in ETPB modified samples. Deflection and forking of the main crack into secondary cracks was also observed in ETPB modified samples while these phenomena were not present in HTPM modified samples.

Introduction

Rubber toughening of epoxy resins has been extensively studied by several researchers and several toughening mechanisms have been proposed. Rubber cavitation and induced shear deformation in the form of dilatational void growth and shear banding between the particles are now generally accepted as the primary toughening mechanism of rubber modified epoxies [Yee and Du (2000)]. Several secondary toughening mechanisms such as plastic void growth [Lee and Yee (2000); Kawaguchi and Pearson (2003); Kawaguchi and Taylor (2002); Johnsen et al. (2007)] rubber particle bridging [Kunz-Douglass et al. (1980); Kunz and Beaumont (1981)], crack deflection [Yee and Du (2000); Sue (1991)], crack bifurcation [Yee and Du (2000); Sue (1991)] and crack pinning [Sue (1991)] may also contribute to the elevated toughness of modified epoxies.

However, there still are controversial discussions about the role of rubber particle cavitation on toughening of epoxy resins [Mafi and Ebrahimi (2008); Bagheri et al. (2009)]. Two hypotheses exist in regard of cavitation [Bagheri and Pearson (1996)], The first hypothesis claims that cavitation phenomena is independent of shear banding and only contributes to onset of plastic dilatation which might occur either before or after shear yielding. But the second hypothesis claims that cavitation resistance influences the size of process zone.

Bagheri and Pearson [Bagheri and Pearson (1996)] showed that voids at zero cavitation resistance can induce massive shear yielding in low cross-linked matrix. They concluded that cavitation resistance of rubbery phase did not play any role in toughening mechanism of epoxy resin. However, later Mafi and Ebrahimi [Mafi and Ebrahimi (2008)] showed that higher cavitation resistance of core-shell rubber particles of 120 nm size, enhances the fracture toughness of epoxy resins. They supported the hypothesis that greater cavitation resistance of core-shell rubber particles leads into greater build up of strain energy before cavitation so the cavitation of rubber particles enables the formation of shear band to proceed faster and results in bigger process zone.

However the role of cavitation resistance of rubbery phase with relatively big particle size has not been studied to our best knowledge. In this study, hydroxyl terminated polybutadiene (HTPB) and synthesized epoxy terminated polybutadiene (ETPB) have been used to toughen an epoxy resin. The cavitation resistance of produced blends was different while they had almost the same dispersion morphology. Toughening mechanisms have been studied to better elucidate the role of cavitation resistance.

Experimental

Materials

The materials used in this investigation consisted of a diglycidyl ether of bisphenol-A (DGEBA) based epoxy pre-polymer. The epoxide equivalent weight (EEW) of the epoxy resin was 190 g/equiv and it was supplied by Huntsman. A cycloaliphatic amine curing agent (Epikure F-205 from Huntsman) with a number of amine groups corresponding to 104 g/equiv was used as the hardener. Hydroxyl terminated polybutadiene (HTPB) with the number average molecular weight of 3400 g/mol was purchased from Zibo Qilong Chemical Industry Co. Ltd. The lab grade of maleic-anhydride (M.A) and triphenylphosphine (T.P.P) were purchased from Merck Chemicals. All materials were used as received.

Synthesis of Epoxy Terminated Polybutadiene (ETPB)

Synthesis of epoxy terminated polybutadiene was carried out in two steps. In the first step, carboxyl terminated polybutadiene (CTPB) was synthesized by reacting HTPB with maleic anhydride through a ring opening addition reaction at 80°C. This reaction was performed in the presence of epoxy resin to control the viscosity build up during the reaction. The reaction was carried out in a three necked 250 cc glass reactor which was equipped with a mechanical stirrer and blanketed with nitrogen. The reaction time was about 24 hours to ensure complete reaction of reagents. In the next step, ETPB was synthesized by reacting the resulted CTPB with an excess of epoxy resin (i.e. 75 wt %) in the presence of 0.2 wt% of triphenylphosphine as catalyst. The reaction was performed at 80°C and under vacuum (12 mmHg) for 24 hours.

Characterization of Synthesized Resins

The HTPB, CTPB and ETPB were characterized qualitatively by FT-IR spectroscopic technique. The FT-IR spectra of samples were taken on a Bomem FT-IR spectrometer (MB series; ABB Bomem, Inc., Quebec, Canada) at a 4 cm-1 resolution averaged over 20 scans. In addition, the acid value of synthesized CTPB and ETPB were determined according to ASTM D1639. The epoxy equivalent weight (EEW) of ETPB was also determined according to ASTM D1652.

Sample preparation

Two sets of samples containing 10 phr of HTPB or ETPB were prepared as described below. For preparing the test specimens, the epoxy resin was first degassed for 1 hour at 5mbar in a vacuum oven. After that, 10 phr of HTPB (considering total weight of epoxy and hardener) was added to the epoxy resin and was mixed under vacuum for 30 minutes at 1300 rpm. As ETPB samples contained some epoxy resin, the required amount of epoxy resin was added to the synthesized ETPB samples and was mixed for 30 minutes at 1300 rpm.

The stoichiometric amount of hardener was then added to the mixture and was mixed under vacuum. The mixture was poured into a pre-heated Teflon coated mold. The mold was placed in an air circulating oven at 80°C for 3 hours followed by 1 hour of post curing at 120°C. In the next step, the mold was removed from the oven and allowed to cool gradually to room temperature. The mold contained cavities with three and four point bending test specimen shapes (according to ASTM D5045) so no further machining of the specimens was required. By using this method, the risk of micro cracking of the samples during machining was eliminated.

Test methods

The fracture toughness was measured according to ASTM D5045-96 protocol. Critical strain energy release rate, K_{IC}, was determined using a single edge notched type specimen (6.5*12*80mm) in a three point bending (SEN3PB) geometry with a span of 50mm. The samples were notched and then pre-cracked by striking a razor blade, previously chilled in liquid nitrogen with rubber hammer. The ratio between the final crack length and the specimen width was held in a range between 0.4 and 0.6. The tests were performed using an Instron 5566 test machine. The crosshead speed was 1mm/min. A minimum of 6 specimens were tested to ensure accuracy. The value of the fracture energy, G_{IC}, was calculated from knowledge of the values of K_{IC} and E using the relationship:

$$G_{IC}=K_{IC}2(1-v^2)/E$$

Where E is the modulus of elasticity and v is the poisson's ratio taken to be 0.35 [Yee and Du (2000)].

Dynamic mechanical data were obtained using a Tritron DMTA equipped with a tension fixture. Shear moduli were measured in oscillating bending, at a frequency of 1Hz and strain amplitude of 1%. The samples were heated from -150°C to 50°C above Tg using 5°C/min ramp rate.

Double edge notch four point bending (DEN4PB) samples were used to investigate the crack tip. The bars with dimensions of 125mm*12mm*6.5mm were notched with a milling tool. The distance between notches was 2.25mm. Razor blade was inserted into the notches and two identical cracks were introduced to the samples. The samples were broken in DEN4PB geometry with crosshead speed of 1 mm/min. The damage zone around the survived DEN4PB crack was cut along the thickness direction into two halves. The plane-strain core region was prepared for transmission optical microscopy (TOM). In TOM investigations, thin sections of rubber modified samples were obtained by polishing the samples. The thin sections were then examined, using a Leica DMR optical microscope.

The fracture surface of the broken specimens was studied using scanning electron microscopy (SEM). The SEM examinations were carried out by a Philips scanning electron microscope at an accelerating voltage of 30kV. Samples were coated with a thin layer of gold to reduce any charge build up on the surface.
Tensile properties were measured according to ASTM D638 guideline. Type I dog bone specimens were tested using a screw driven Instron testing frame at crosshead speed of 5mm/min. A minimum of 5 specimens were tested. Tensile test results were used to check if the dimensions of the specimens provide the plane strain constraint criteria (ASTM D5054).

Results and Discussion

Characterization of synthesized components

The amount of carboxyl groups of the initial mixture of HTPB and maleic anhydride, the resulting CTPB and also ETPB were determined by titration method. It was found that the acid value of the CTPB was about 10 mg KOH/g which is close to acid value of the initial reaction mixture (i.e. 12 mg KOH/g). The comparison of the acid value of ETPB (i.e. almost zero mg KOH/g) with that of CTPB shows that all of the carboxyl functionalities of CTPB have reacted with epoxide groups of epoxy resin.
Also, the epoxy equivalent weight (EEW) of the resulting ETPB was determined to be 268 g/equiv which was in good agreement with the theoretically calculated EEW (i.e. 255 g/equiv).
Figure 1 shows the FTIR spectra of the HTPB, CTPB and ETPB. The spectrum of HTPB displays a characteristic absorption peak at 3500 cm-1 (peak A) due to hydroxyl stretching vibrations. An absorption bands at 1733 cm-1 (peak B), which can be ascribed to the carboxyl group of CTPB, confirmed the formation of carbonyl groups by reaction of maleic anhydride with hydroxyl groups of HTPB. As can be seen the spectrum of the ETPB sample is similar to that of the neat epoxy resin due to the high concentration of epoxy resin in the ETPB sample. The absorption bands at 913 and 839 cm−1, in the spectra of ETPB indicate oxirane group of epoxy resin.
The dispersion morphology of rubber phase in epoxy matrix was characterized by SEM. Figure 2 and 3 show the SEM micrographs of HTPB and ETPB modified samples respectively. As can be seen, the particle size of rubbery domains in both cases are nearly the same but dispersion of HTPB particles is more uniform. The findings of present study contradict the results of Soare et al. [Fabio et al. (2003)] which reported that epoxy end capping of HTPB chains results in considerably smaller particle size.

Fracture Energy

As described earlier, the stress intensity factor of neat and modified samples was measured by SEN3PB method according to ASTM D5045-98 protocol. The results of SEN3PB and tensile tests are summarized in Table 1. A mean fracture toughness, K_{IC}, of 0.65 $MNm^{-3/2}$ was measured for unmodified epoxy polymer. The measured toughness was increased by the addition of HTPB to 1.23 $MNm^{-3/2}$ and ETPB to 1.80 $MN^{m-3/2}$.
The fracture surface of unmodified epoxy polymer in slow crack-growth region is shown in Figure 4. As can be seen, the fracture surface is smooth and featureless and shows no plastic deformation which is typical of a brittle thermoset polymer [Kawaguchi and Taylor (2002); Johnsen et al. (2007)]. The observations agree well with the low measured toughness of the material.
One might attribute the variation in fracture energy of HTPB and ETPB samples to their degree of adhesion to the matrix but it has been shown experimentally that when the second phase consists of micron size rubber particles, the interfacial bonding has only a modest effect on the fracture properties of samples [Bagheri et al. (2009); Huang et al. (1993)].
The fracture surface of HTPB and ETPB modified samples in slow crack growth region are shown in Figure 5 and 6. Figure 5 clearly shows internal cavitation and tearing of the HTPB particles. No debonding or protrusion of rubber particles was observed in HTPB modified samples, while in ETPB modified specimen a combination of cavitation, debonding and rubber particle pull out was observed. The smaller size of cavities and debonding of rubber particles in ETPB particles modified samples suggest that cavitation resistance of ETPB particles might be higher than that of HTPB particles so further investigations were carried out to study this hypothesis. Four point bending specimens of HTPB and ETPB modified samples with exactly the same pre-cracks were broken in DEN4PB geometry with crosshead speed of 1 mm/min and the survival pre-crack was polished to its mid plane. The cavitation zone of the samples was visible with naked eyes due to the scattering of visible light. The cavitation zone of HTPB and ETPB modified samples are shown in Figure 7. As can be clearly seen, the cavitation zone of the HTPB modified sample is bigger than that of ETPB modified sample which shows the higher cavitation resistance of ETPB specimen. As the ETPB chains were end capped with oxirane groups, they could react with amine hardener so cross linking of the rubbery phase may occur which can increase the cavitation resistance of rubbery phase.
DMTA test was also utilized to investigate this hypothesis. The tan δ and storage modulus of neat and rubber modified samples are presented in Figures 8 and 9. The neat epoxy resin system showed a β peak at around -60°C. Addition of HTPB and ETPB did not result in a second low temperature peak (associated with the α transition of the rubbery phase) but essentially enhanced the magnitude of the β peak. However even if we could differentiate the peaks, it was not easy to judge the variations in rubber crosslink density according to their glass transition temperatures due to thermal shrinkage stresses. Triaxial thermal shrinkage stress develops in the dispersed domain phase on cooling since the coefficient of thermal

expansion of the rubbery state is larger than that of glassy state. The domains are constrained by glassy epoxy matrix. The glass transition of the rubber changes [Manzione et al. (1981)]. The rubbery region plateau of the specimens shows the decreased crosslink density of ETPB modified samples. This result suggests that a part of hardener was reacted with rubbery phase so a smaller amount of hardener reacted with the epoxy resin and therefore the crosslink density of the ETPB modified sample is decreased.

Toughening mechanisms
Huang and Kinloch proposed that fracture energy of a particle modified polymer may be expressed as [Huang and Kinloch (1992)]:

$G_{IC}=G_{ICU}+\Psi$

Where G_{ICU} represent the fracture energy of unmodified epoxy and Ψ represent the overall toughening mechanisms. These toughening contributions have been ascribed to a range of processes such as shear yielding, plastic void growth, crack bridging, crack pining, crack deflection and crack bifurcation. The following discussion considers these processes in more detail with specific consideration to role of cavitation resistance.

Plastic void growth
Triaxial stress-state ahead of crack tip exerts dilatational forces to rubbery particles. This dilatational force causes failure and void formation either internally in the rubbery particles (cavitation) or at the particle-matrix interface (debonding). Once the void is formed, it grows and dissipates energy [15]. As the poisson ratio of rubber is close to 0.5, its bulk modulus tends to infinity. This demonstrates that a rubber particle is a very rigid body when subjected to triaxial stresses and will be resistant to volumetric deformation and dilatation. As rigid rubbery phase prohibit large scale plastic dilatation in the matrix, the void formation in rubber modified epoxies is essential prerequisite for plastic void growth. Epoxy polymers strain soften after yielding which facilitates plastic void growth but they strain harder again [Haward (1975)]. So if the cavitation or debonding occurs after strain hardening the plastic void growth will be suppressed [Huang and Kinloch (1992)]. Thus the necessary criterion for plastic void growth is that rubbery particles cavitate or debond and do so before strain hardening. This criterion suggests that lower cavitation resistance might facilitate plastic void growth. Figure 10 and 11 compares the fast and slow crack growth regions of HTPB and ETPB modified specimens. Both figures show volumetric expansion of rubbery phase in slow crack growth region which proves the plastic void growth. The results show that for this specific case, elevated cavitation resistance of ETPB does not suppress plastic void growth.

Cavitation and induced shear yielding
In this mechanism, the rubber particles cavitate because of triaxial tension ahead of the crack tip. This cavitation reveals the plane-strain constraint ahead of the crack tip and allows the stress concentration associated with cavitated particle to activate shear yielding in the matrix [Yee and Du (2000)]. The role of rubber particle cavitation resistance on shear yielding of the matrix is discussed in our previous paper [Mafi and Ebrahimi (2008)]. It was shown that core shell rubber particles with higher cavitation resistance can enhance the shear yielding of matrix. In present study, the TOM micrographs of crack tips showed little shear yielding (Figure 12 & 13). This is because of large particle size of ETPB and HTPB domain in epoxy matrix. At constant volume fraction of modifier, large particle size leads in to higher inter-particle distance. Several researchers have reported the importance of the inter-particle distance in brittle to tough transition [Bagheri and Pearson (1996); Bagheri and Pearson (1995); Lovell et al. (1993)]. Different points of view exist in regard of the role of inter-particle distance. The findings of Bagheri and Pearson [Bagheri and Pearson (1996)] in rubber and hollow particle modified epoxies support the hypothesis that variation of inter-particle distance causes a transition from plane strain to plane stress state. They claimed that small enough inter-particle distance is required to cause plane stress condition [Bagheri and Pearson (1996)]. Apparently large particle size of rubber particles in this study did not provide strong plane stress condition ahead of crack tip so little shear yielding happened. Results showed that the role of cavitation resistance cannot be discussed when plane stress condition is not formed. The TOM images also show no crack bridging of rubber particles.

Cavitation and tearing of rubber particles
Two main reasons exist for cavitation of rubber particles. First triaxial thermal shrinkage stress which develops in the dispersed domain phase on cooling [Manzione et al. (1981)] and second triaxial stress field ahead of crack tip combined with high bulk modulus rubbery phase [Yee and Du (2000); Huang and Kinloch (1992)].
One may suppose that higher energy required for cavity formation in ETPB modified specimen stand for higher toughness. Bucknall et al. [Bucknal et al. (1994)] showed that cavitation itself could not be regarded as an important energy absorbing process and it could happen from comparatively small energy interchanges taking place within the rubber particles [Bucknal et al. (1994)]. Johnsen et al. [Johnsen et al. (2007)] also claimed that debonding process is generally considered to absorb little energy. So the energy of rubber cavitation or debonding cannot be regarded as the reason of the fracture toughness differences observed in ETPB and HTPB modified samples.

Crack pinning

In crack pinning mechanism, the crack begins to propagate within the material, the crack front bows out between the second phase dispersion whilst still remaining pinned at all positions where it has encountered the second phase particles [Kinloch and Young (1983)]. Crack pinning arises from interactions between the moving crack front and the second phase dispersion.

Theoretically crack pinning happens when the modifier particles are larger than the crack opening displacement [Johnsen et al. (2007)] under plane-strain conditions, the crack-opening displacement can be calculated by using this relationship [Kinloch and Young (1983)]

$\delta tc = G_{IC} / 6y$

Where G_{IC} is the fracture toughness and y is the yielding point. For modified samples with HTPB and ETPB the value of the crack opening displacement calculated are 1.17 μm and 2.8 μm respectively. It is clear that particles are bigger than the crack-opening displacement and are likely to cause crack pinning. In order to determine if the crack pinning mechanism was operative, SEM investigation at the fracture surface was conducted. The typical crack pinning feature is the formation of tails behind the rubber particles [Sue (1991)] which is due to the meeting of crack faces in different planes after bowing. Figure 5 and 10 show the SEM micrographs of fracture surface of HTPB modified sample. No tail can be observed in HTPB modified sample. Due to the high roughness of ETPB modified sample in slow crack growth region, it is difficult to judge the presence of pinning feature. Figure 14 show the SEM micrographs of ETPB modified sample in fast crack-growth region. Pinning tails can clearly be seen in fast crack growth region of ETPB modified sample. As the cavitation resistance of HTPB particles is low, the crack can propagate through the HTPB particles and no bowing happens. However, in ETPB particles due to the strong interface because of the formation of covalent bonds of rubber particles with matrix and higher cohesive strength because of crosslinking, the crack front bows when reaches to rubber particle. The results showed that higher cavitation resistance activates crack pinning process in ETPB modified samples.

Crack deflection and crack bifurcation

In these mechanisms, rubber particles cause the main crack to fork into many secondary cracks (bifurcation) or deviates the main crack off the principal plane of propagate (deflection).

Crack bifurcation disturbs the local stress intensity of the main crack into secondary cracks and crack deflection increases the area of the crack surface furthermore [Yee and Du (2000)]. Both mechanisms induce a mixed I and II mode of crack propagation. As the consumed energy in mode II of fracture is higher than mode I [Dieter (1988)], this mixed mode of crack propagation enhances toughness. Figure 15 shows the crack deflection mechanism of ETPB modified specimen. Deviating the crack from its principal plane also causes surface roughness which can be seen in Figure 6. Forking of the main crack into secondary cracks in ETPB modified sample can be seen in Figure 16. Due to the strong interface and high cavitation resistance of ETPB particles, the crack could not pass through particles so crack deflection and crack bifurcation process can be observed while these phenomena was not observed in HTPB modified samples due to the passing the crack through the rubber particles.

Results show that higher cavitation resistance enhanced the toughness of ETPB modified samples by inducing crack bifurcation and crack deflection process.

Table 1: tensile and fracure properties of neat and modifieds samples

	Elasticity Modulus (GPa)	Yield stress (Gpa)	$K_{IC}(MNm^{-3/2})$	$G_{IC}(J/m^2)$
Neat	2.97	91	0.65± 7%	125± 7%
HTPB Modified	2.42	78	1.23± 10%	548± 10%
ETPB Modified	2.28	74	1.80± 10%	1246±10%

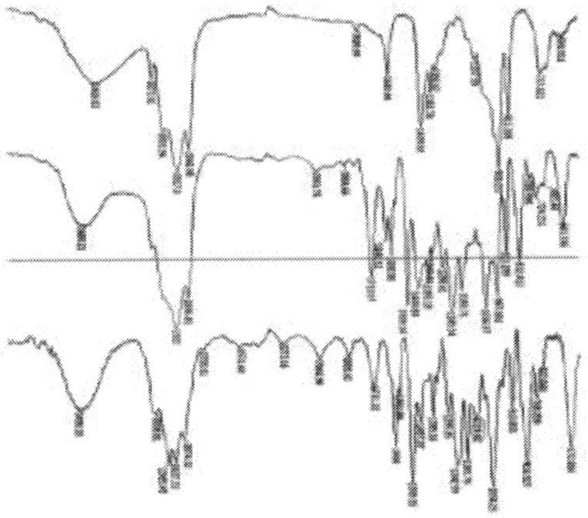

Figure 1- FTIR spectra of the a) HTPB, b) CTPB and c) ETPB

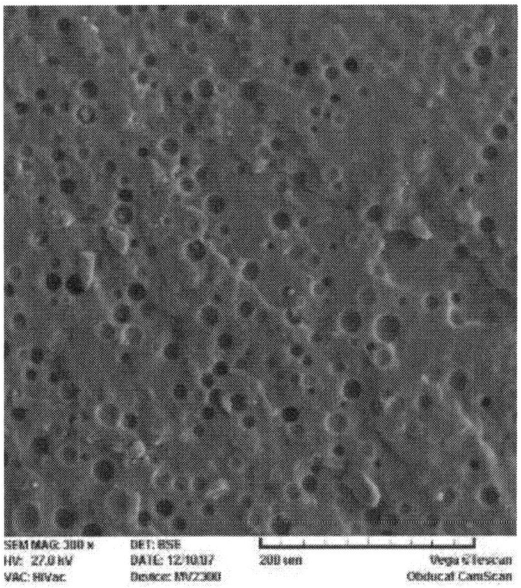

Figure 2: SEM micrograph of HTPB modified sample

Figure 4: The fracture surface of unmodified epoxy polymer in slow crack-growth region

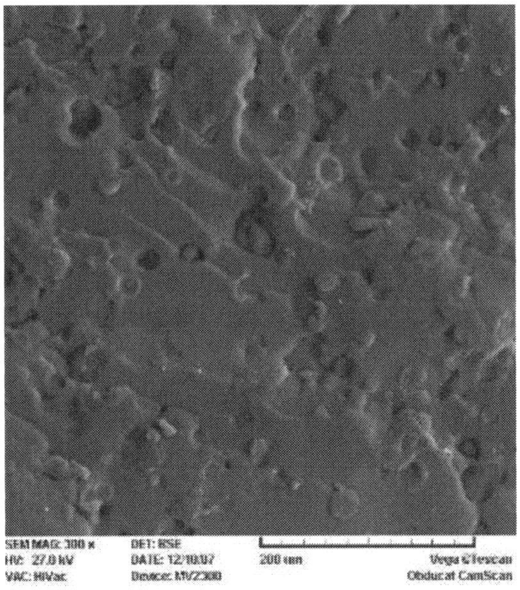

Figure 3: SEM micrograph of ETPB modified sample

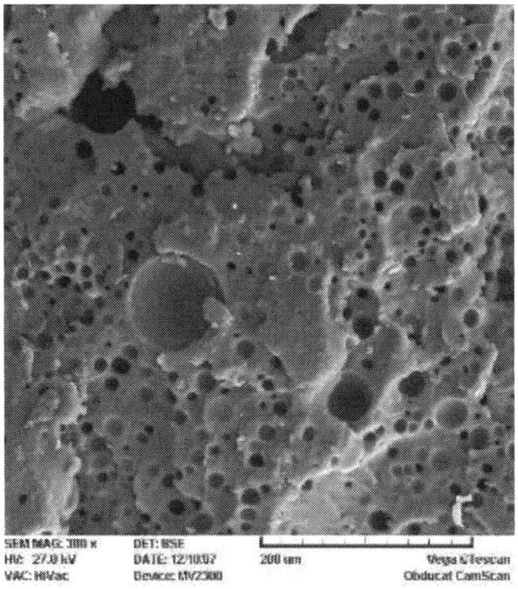

Figure 5: The fracture surface of HTPB modified sample in slow crack growth region

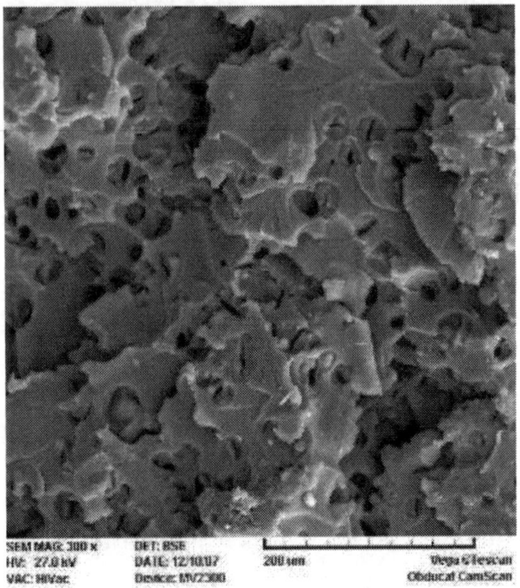

Figure 6 The fracture surface of ETPB modified sample in slow crack growth region

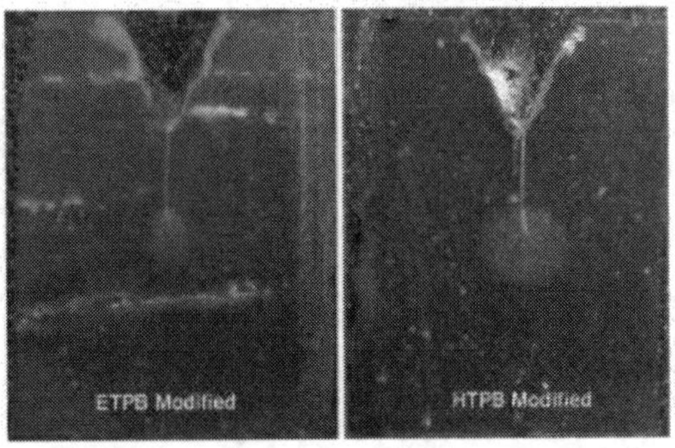

Figure 7: The cavitation zone of HTPB and ETPB modified samples

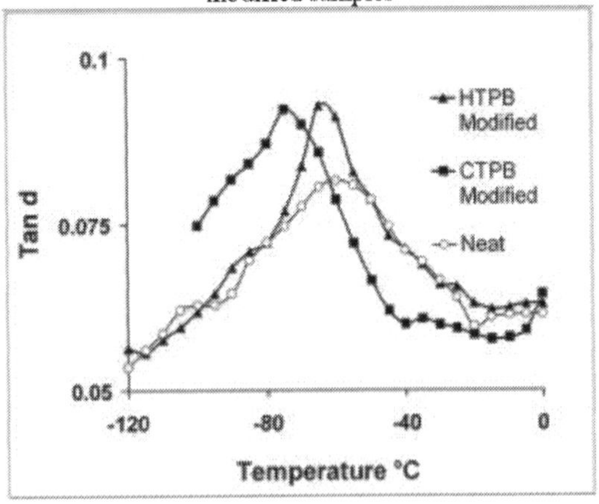

Figure 8: Tan δ of neat and rubber modified samples

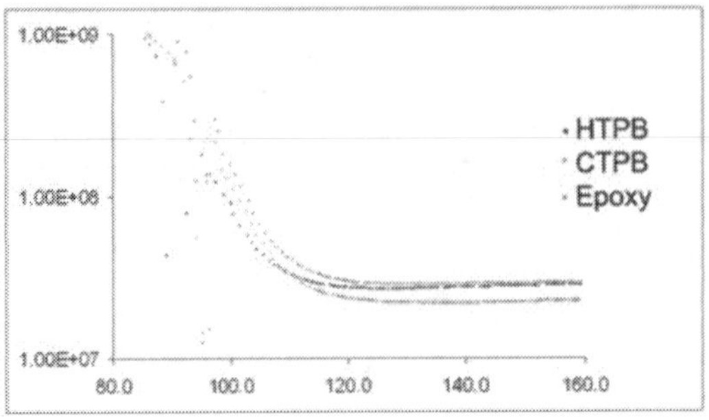

Figure 9: Storage modulus of neat and rubber modified samples

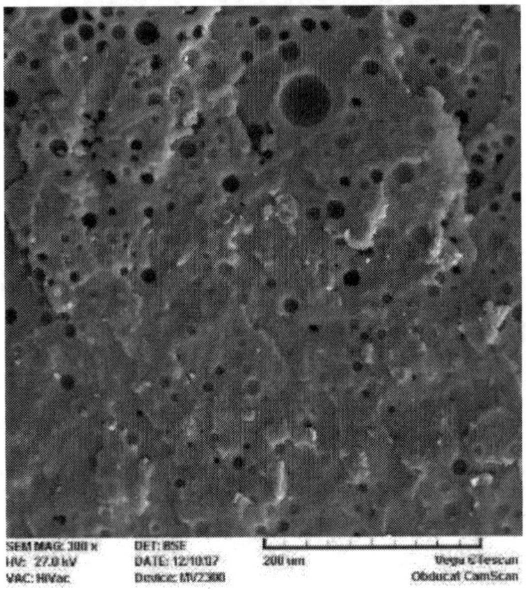

Figure 10: fast and slow crack growth regions of HTPB modified samples

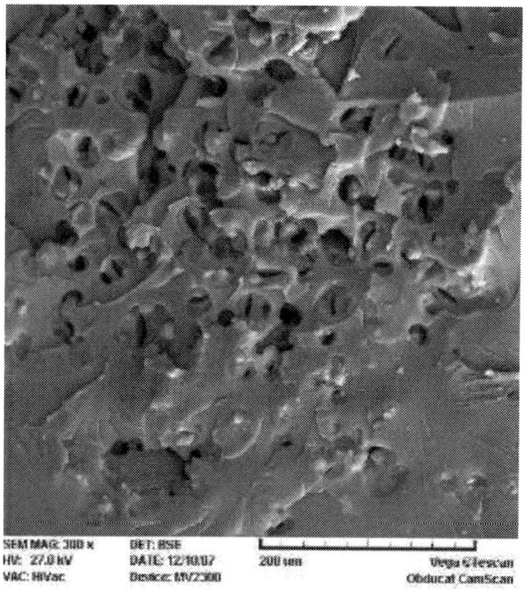

Figure 11: fast and slow crack growth regions of ETPB modified samples

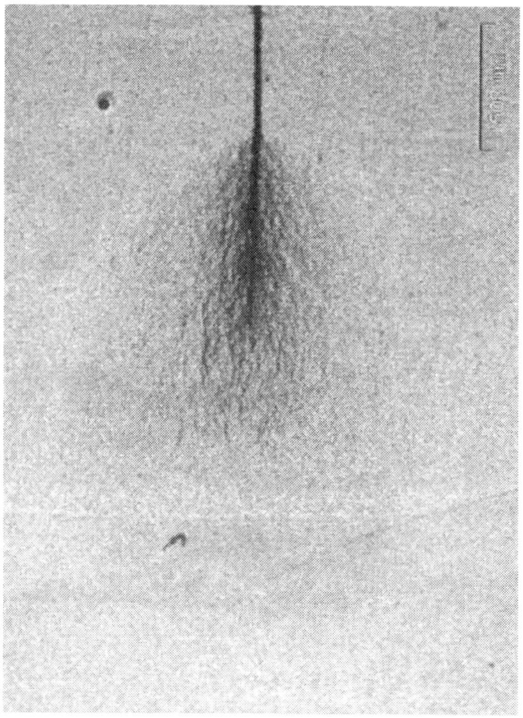

Figure 12: The crack tip TOM micrograph of HTPB modified sample

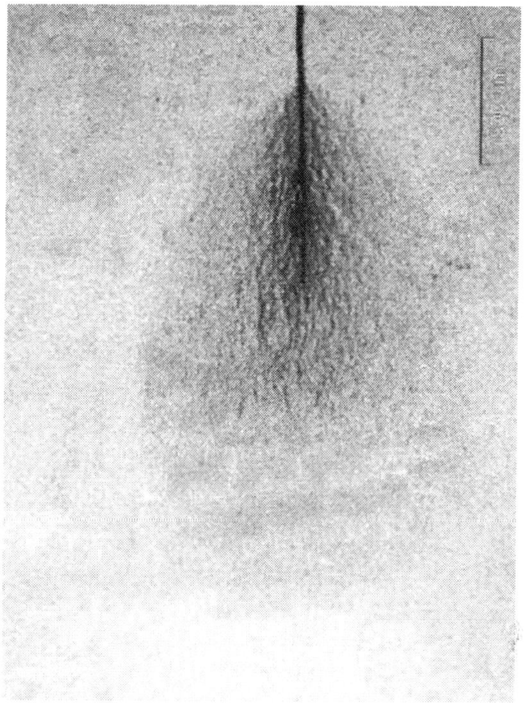

Figure 13: The crack tip TOM micrograph of ETPB modified sample

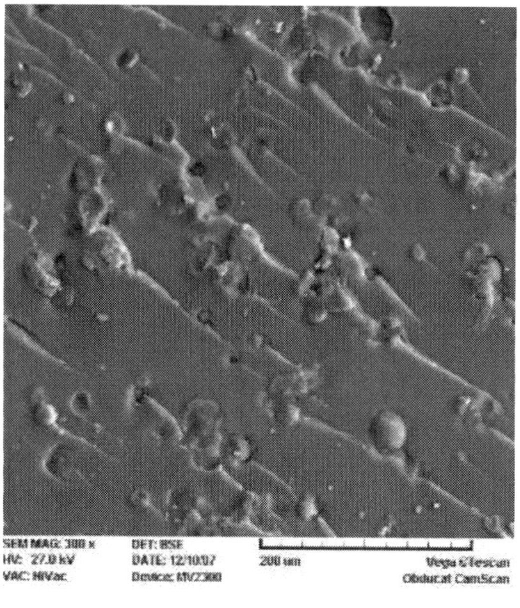

Figure 14: SEM micrograph of ETPB modified sample

in fast crack-growth region

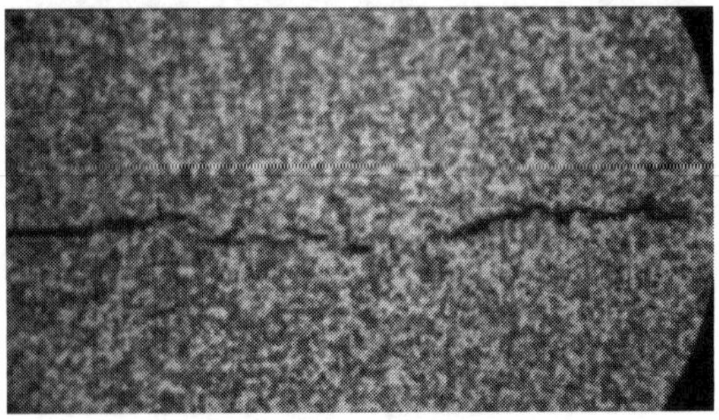

Figure 15: crack deflection mechanism of ETPB modified specimen

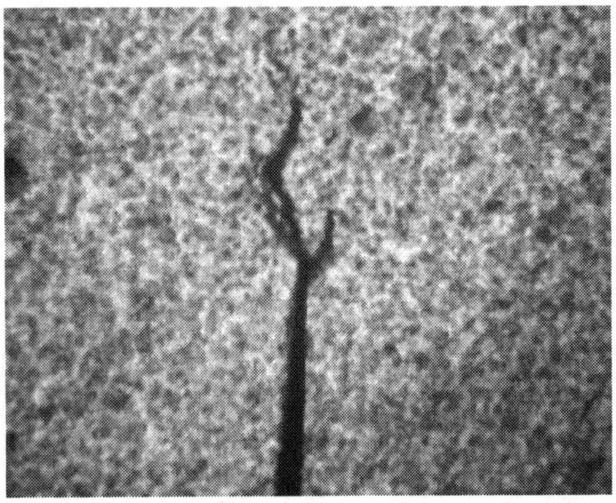

Figure 16: crack bifurcation mechanism of ETPB modified specimen

Conclusions

Hydroxyl terminated polybutadiene (HTPB) and synthesized epoxy terminated polybutadiene (ETPB) have been used to toughen an epoxy resin. The cavitation resistance of produced blends was different while they had almost the same dispersion morphology. The toughness of modified samples increased by the addition of liquid rubber and the toughening efficiency of ETPB was more significant.

The cavitation zone of the HTPB modified sample was bigger than that of ETPB modified sample which showed the higher cavitation resistance of ETPB specimen which was confirmed by DMTA analysis.

Both sets of samples showed volumetric expansion of rubbery phase in slow crack growth region which proved the plastic void growth mechanism. The results showed that for this specific case, elevated cavitation resistance does not suppress plastic void groeth. Little shear yielding was observed in samples due to large particle size of ETPB and HTPB domain in epoxy matrix.

As the cavitation resistance of HTPB particles was low, the crack could propagate through the HTPB particles and no crack pinning happened. However, in ETPB particles due to the strong interface because of the formation of covalent bonds of rubber particles with matrix and higher cohesive strength, the crack front bowed when reached to rubber particles. The results showed that higher cavitation resistance activates crack pinning process modified samples. Crack deflection and crack bifurcation processes were also observed in ETPB modified samples due to same reason Results showed that higher cavitation resistance enhanced the toughness of ETPB modified samples by inducing crack bifurcation, crack deflection and crack pinning processes.

References

Bagheri. R., Marouf. B. T., Pearson. R. A., "Rubber-Toughened Epoxies: A Critical Review", Polymer Reviews, **49**, 201-225 (2009)

Bagheri. R., Pearson. R. A., "Role of particle cavitation in rubber-toughened epoxies: 1. Microvoid toughening", Polymer, **37**, 4529-4538 (1996)

Bagheri. R., Pearson. R.A., "The use of microvoids to toughen polymers", Polymer, **36**, 4883-4885 (1995)

Bucknall. C. B., Karpodinis. A., Zhang. X. C., "A model for particle cavitation in rubber-toughened plastics", J. Mater. Sci., **29**, 3377-3383 (1994)

Dieter. G. E.: Mechanical Metallurgy, Mc Graw-Hill Book Company, London (1988)

Fabio L., Barcia., Thiago. P., Amaral., Bluma. Soares. B. G., "Synthesis and properties of epoxy resin modified with epoxy-terminated liquid polybutadiene", Polymer, **44**, 5811-5819 (2003)

Haward. R. N.: The physics of glassy polymers, Applied Science publishers, London, (1975)

Huang. Y., Kinloch. A. J., Bertsch. R., Sibert. A. R., "Particle-matrix interfacial bonding," In Toughened Plastics, Advances in Chemistry Series 233; Riew. C. K., Kinloch. A. J. (Eds), ACS, Washington, p. 189 (1993)

Huang. Y., Kinloch. A. J., "The role of plastic void growth in the fracture of rubber-toughened epoxy polymers", J. Mater. Sci. Lett., **11**, 484-487 (1992)

Johnsen. B.B., Kinloch. A.J., Mohammed. R.D., Taylor. A.C., Sprenger. S.," Toughening mechanisms of nanoparticle-modified epoxy polymers", Polymer, **48**, 530-541 (2007)

Kawaguchi. T, Pearson. R. A., "The effect of particle–matrix adhesion on the mechanical behavior of glass filled epoxies. Part 2. A study on fracture toughness", Polymer, **44**, 4239-4247 (2003)

Kinloch. A. J., Taylor. A. C., "The toughening of cyanate-ester polymers Part I Physical modification using particles, fibres and woven-mats", J. Mater. Sci., **37**, 433-460 (2002)

Kinloch. A. J., Young. R. J.: Fracture behaviour of polymers, Applied Science Publishers, London (1983)

Kunz-Douglass. S., Beaumont. P. W. R., Ashby. M. F., "A model for the toughness of epoxy-rubber particulate composites", J. Mater. Sci., **15**, 1109-1123 (1980)

Kunz. S. C., Beaumont. P. W. R., "Low-temperature behaviour of epoxy-rubber particulate composites", J. Mater. Sci., **16**, 3141-3152 (1981)

Lee. J., Yee. A. F., "Fracture of glass bead/epoxy composites: on micro-mechanical deformations" , Polymer, **41**, 8363-8373 (2000)

Lovell. P.A., McDonald. J., Saunders. D.E.J., Young. R.J., "Studies of rubber-toughened poly(methyl methacrylate): 1. Preparation and thermal properties of blends of poly(methyl methacrylate) with multiple-layer toughening particles", Polymer, **34**, 61-69 (1993)

Mafi. E. R., Ebrahimi. M., "Role of core-shell rubber particle cavitation resistance on toughenability of epoxy resins", Polym. Eng. Sci., **48**, 1376-1380 (2008)

Manzione. L. T., Gillham. J. K., McPherson. C. A., "Rubber-modified epoxies. I. Transitions and morphology", J. Appl. Polym. Sci., **26**, 889-905 (1981)

Sue. H. J., "Study of rubber-modified brittle epoxy systems. Part II: Toughening mechanisms under mode-I fracture", Polym. Eng. Sci., **31**, 275-288 (1991)

Yee. A. F., Du. J., "Toughening of Epoxies", in Polymer Blends Vol2, Paul. D. R., Bucknall. C. B. (Eds.), John Wiley & Sons. Inc., New York, p. 225 (2000)

POLYMER BLEND ELECTRON BEAM IRRADIATED RHEOLOGICAL BEHAVIOR

Z. Ahmadi[1*], F. Afshar-Taromi[2], H. Nazokdast[2]

[1] *Polymer Engineering Department, Islamic azad university of Mahshahr (IAUM), Mahshahr, Iran, z.ahmadi@mahshahriau.ac.ir*
[2] *Polymer Engineering Faculty, Amirkabir University of Technology, Tehran, Iran*

Abstract-Morphology and interfacial adhesion between phases are among the most important parameters effecting the properties and behavior of multiphase polymer blend. One of the method used for obtaining multiphase polymer blends with ordered structure is freezing the concentration fluctuations structure developed at different stages of spinodal decomposition process in polymer blends which can be stabilize by imposing electron beam radiation crosslinking process. The blend exhibits a lower critical solution temperature (LCST) thermodynamic behavior, i.e. phase separation upon heating. The combination of dissimilarity provides the necessary prerequisites for a unique and informative study of radiation effect on rheological behavior of polymer blend. The main objective of present work was to study the rheological behavior of irradiated blend samples at temperature above their LCST. The melt flow behavior of samples showed that irradiated samples with gel content up to 40% can flow with a power law like melt behavior, indicating a matrix/disperse type morphology for these samples. From the results of relaxation time distribution ($H(\lambda)$) is was found that the samples which were irradiated in their single and two phase states exhibit a two phase behavior at a temperature which is above their LCST. The melt viscoelastic properties of the irradiated samples showed that the irradiation can induced copolymerization and/or crosslinking between two phases whose extent was found to be increased with increasing the radiation dose.

Introduction

The rheology of multiphase polymer blends has long attracted the interest of the scientific and industrial community, due to the interesting physical problems and the ever increasing technology application [1].

One method to be used for producing multiphase polymer blend is to freeze and stabilize the microstructure formed during spinodal decomposition process of the lower critical solution temperature (LCST) blend system [2].

As rheological measurement of this material must be perform at temperature above their LCST rheological behavior expect to be similar to those two phase polymer system.

In many cases the induced morphology is not stabilized enough for further required melt process and therefore needs to use an appropriate method to fix morphology. The electron beam (EB) radiation of two phase polymer blend in their solid state has been used as one of the most successful method for morphology stabilization [2, 3]. In previous paper [2] we were shown that crosslinking via electron beam irradiation of the dispersed phase is a promising rout to stabilize morphology.

The rheological behavior of miscible and partially miscible polymer blends has been investigated extensively [4,5]. But a little work was reported about electron beam radiation effect on rheological behavior of polymer blends.Van Gisbergen et.al studied the effect of electron beam on melt rheology of PS/LDPE [6] and PP/EPDM [7,8] blends.

Fahrlander et al. [9] observed a power law behavior at the low frequency range on study of rheological properties of heterogeneous dibenzylidene sorbitol network formed by temperature in PPO matrices.

Polystyrene (PS) and Poly vinyl methyl ether (PVME) are a rather unique pair of polymers in that they are chemically dissimilar and also dissimilar in their response to irradiation. In the other hand PS/PVME blend exhibits the LCST behavior [3] i.e., phase separation upon heating, also PS is inert and PVME is radiation crosslinkable phase.

To the authors' knowledge, this is the first detailed report of the rheological behavior of electron beam irradiated PS/PVME.

In this work we consider the effect of electron beam irradiation on the rheological behavior of irradiated both miscible and immiscible polymer blend in the two phase state induced by temperature elevation from one phase into the spinodal region. Miscible and two phase polymer blends samples were imposed to electron beam at different doses. Rheological behavior of miscible and two phase irradiated blend were study in a temperature upper than LCST on two phase region by using The Rheometrics Dynamic Spectrometer (RMS). Results shown copolymerization and/or crosslinking between phases, on the phase separated blend samples are most effective phenomena on melt viscoelastic behavior and relaxation time of samples.

Experimental

Commercial grade PVME obtained from BASF Co. and PS supplied by Scientific Polymer Product Co. with specification listed in the Table 1 were used.

Table 1 Specification of the polymers

Polymer	name	Supplier	Mw	Density (g/cm^3)	code
PS	PS25	S.P.P	250000	1.050	P1
PVME	Lutonal M40	BASF	87000	1.057	P2

For rheological study the two sets samples with constant composition 30/70 PS/PVME was prepared in toluene as cosolvent. First set which is miscible blend which are denoted by A and second set is two phase blend which prepare via spinodal decomposition process and denoted by B (figure 1), based on cloud point curve described in elsewhere [2].The solution samples were cast on Teflon sheet and let it to be for 24 hrs at room temperature then placed the samples in 675 mmHg vacuum oven for 24 hrs in order to remove residual solvent. In order to study the morphology, the samples prepared at a time within the phase separation period were examined by optical microscopy (Nikon model SMZ-1500).The samples, both in single and two phase state were irradiated in air at a room temperature by using the Rhdotoron TT200 (Belgium) with specification given in Table 2. The samples were imposed to irradiation of 50, 100 and 150 kGy at room temperature.

Table 2 Rhodotoron TT200 specification

Energy range	5 - 10 Mev
Beam power range	35 - 70 Kw
Scanning range	30 – 100 cm
Total power consumption	< 300 kw
Electron gun average current	0 – 10 mA
Resolution	± 50 µA
Energy dispersion	± 300 kev

For gel fraction measurement, the irradiated samples were extracted by THF by means of Soxhlet extractor for 24 hrs at ambient temperature. The insoluble fraction was dried at 80 °C for 3 hrs. The gel fraction was then calculated using equation (1):

$$Gel\ fraction = 1 - \frac{w_2}{w_1}$$

where w1 is the initial weight of the sample, and w2 is weight of the insoluble residue.

Dynamic storage, loss modulus and complex viscosity were measured as a function of angular frequency ranging from 0.1 to 1000 Sec-1 and relaxation time distribution was measured as a function of time ranging 10-3 to 10 Sec. at constant temperature by using a Rheometrics Dynamic Spectrometer (Model UDS200) in the oscillatory mode with parallel plate fixture (25 mm diameter).

Results and Discussion

Figure 2 represent the micrograph of the phase separated sample indicating that sample exhibit two phase type morphology consisting of phase dispersed PVME in PS matrix at a temperature above LCST.

Figure 3 show that the gel content of the blend samples increase with increasing radiation dose. It is also seen that trend of increasing the gel content with dose varies for different blends samples depending on morphology and the extent of phase separation and interdiffusion of the samples.

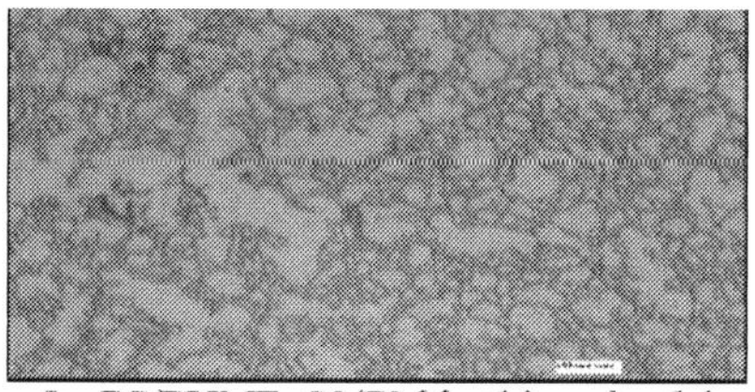

Figure 2 PS/PVME 30/70 blend in spinodal region (T = 170 °C, t = 120 sec)

Figure 4 shows the results of complex viscosity η*, storage modulus G' and loss modulus G" as a function of frequency for the irradiated samples. By considering these results one may notice that even the samples containing 40% of gel content shows a power law type flow behavior. These results suggest that the gels formed between two phases by irradiation may preferentially migrate into the PS phase at rheological measuring temperature (170 °C) which is above the PS/PVME blend LCST. The results of relaxation time distribution (H (λ)) obtained for blend samples "B" irradiated in different dose are shown in figure 5. By comparing these results one may notice that by increasing irradiation dose the maximum relaxation time distribution H (λ) corresponding to the PVME phase is shifted towards the higher relaxation time, in expense of reducing the extent of the PS relaxation time (H (λ)) peak, such that at dose 150 kGy only one single H (λ) peak with broad distribution is obtained. imilar results obtained for blend samples "A" are also shown in figure 6.

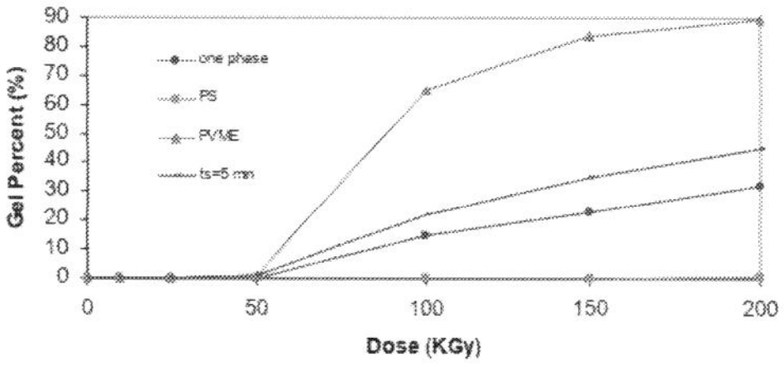

Figure 3 Gel fraction data for irradiated PS, PVME and blends

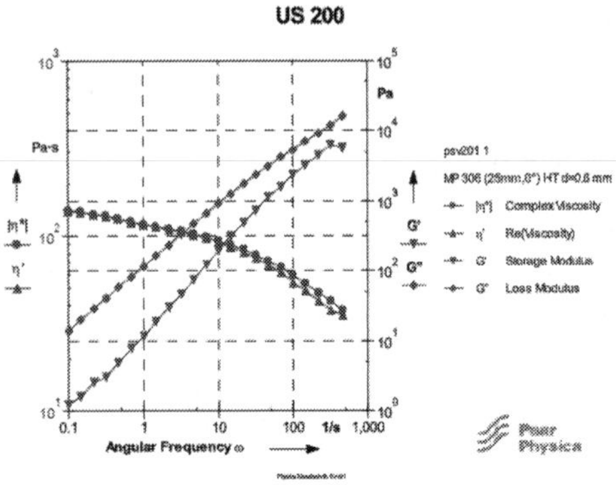

(a)

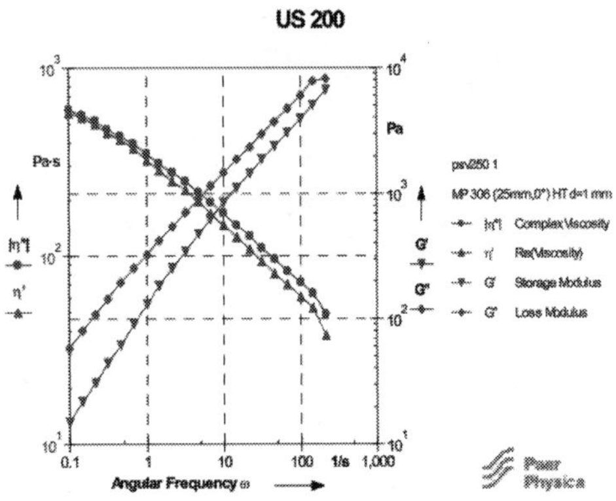

(b)

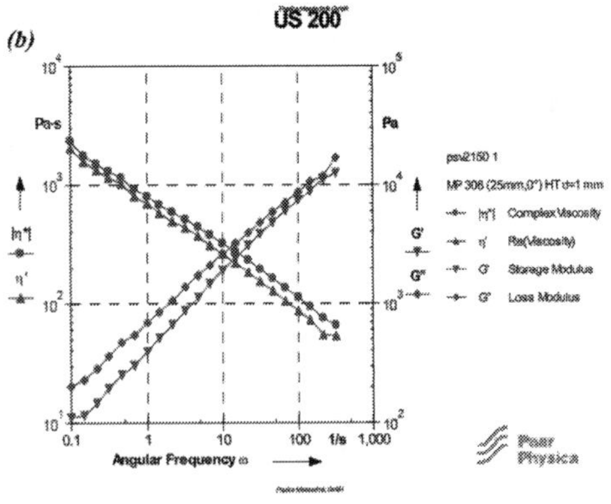

(c)

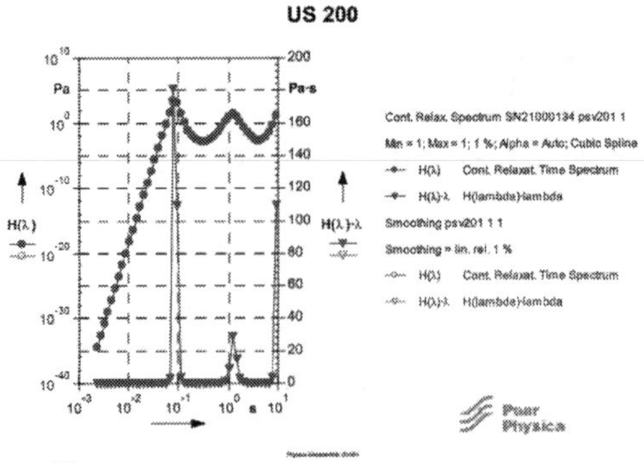

(a)

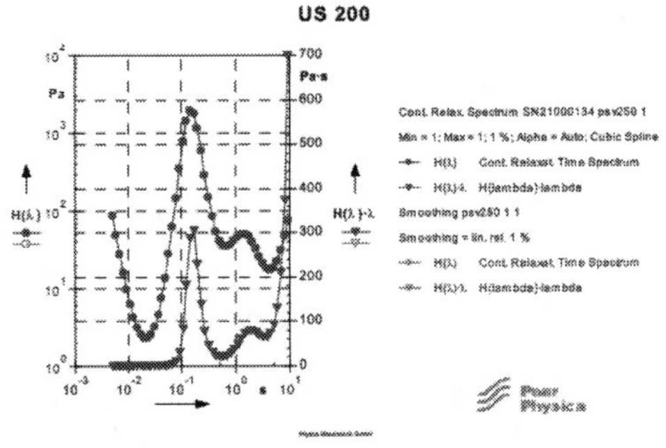

(b)

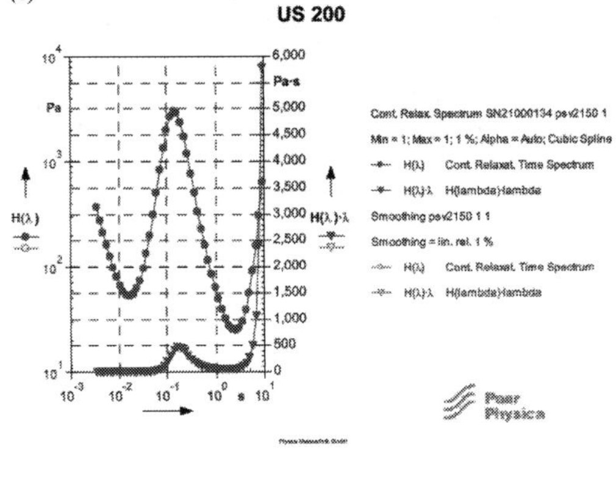

(c)

Figure 4 Radiation effect on the viscoelastic behavior of blend measured at (a) 0 kGy , (b) 50 and (c) 150 kGy

Figure5 Radiation effect on the relaxation time distribution of blend "B" measured at (a)0 kGy, (b)50 and (c)150 kGy

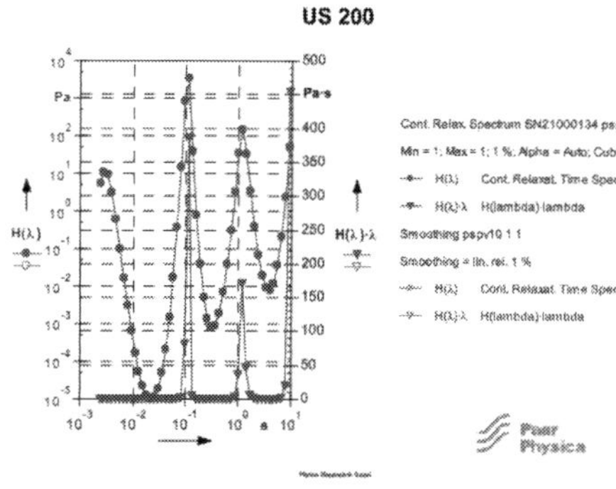

(a)

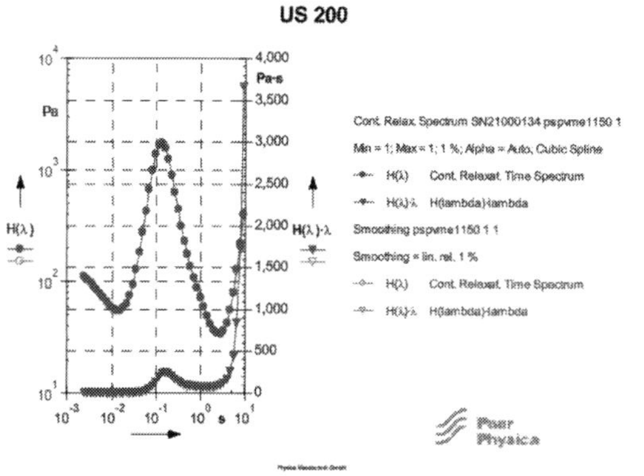

(b)

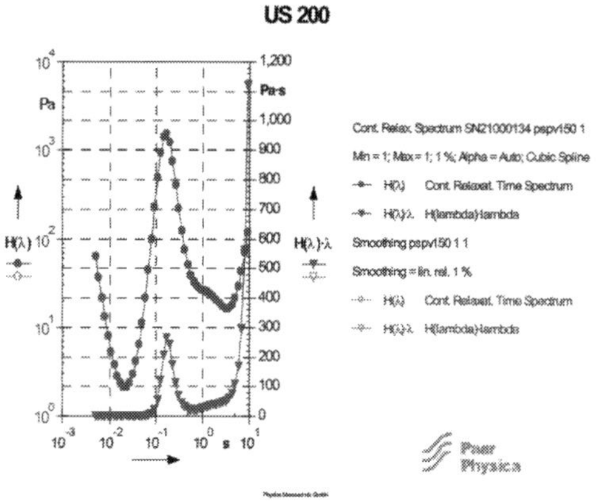

(c)

Figure 6 Radiation effect on the relaxation time distribution of blend "A" measured at (a)0 kGy, (b)50 and (c)150 kGy

Conclusions

The melt flow behavior of samples showed that irradiated samples with gel content up to 40% can flow with a power law like melt behavior, indicating a matrix/disperse type morphology for these samples. From the results of relaxation time distribution (H (λ)) is was found that the PS/PVME samples which were irradiated in their single and two phase states exhibit a two phase behavior at temperature 170 oC which is above their LCST. The melt viscoelastic properties of the irradiated samples showed that the irradiation can induced copolymerization and/or crosslinking between two phases whose extent was found to be increased with increasing the radiation dose.

References

Utracki L.A., Polymer Alloys and Blends, Hanser Publishers, 20 (1990).

Ahmadi Z., Afshar-taromi F., Mohammadi N., Nazockdast H., Haji-saeid S.M., Effect of electron beam radiation on morphology and properties of PS/PVME two phase blends, Iran. Poly. J., in press.

Araki T., Tran-Cong Q., Shibayama M., Structure and properties of multiphase polymeric materials, Marcel Dekker, NewYork, 40 (1998).

Kim J.K., Lee H.H., Son H.W., Phase behavior and rheology of PS/PS and PS/PVME blend systems, Macromolecules, 31, 8566-8578(1998).

Kapnistos M., Hinrichs A., Vassopoulos D., Anastasiadia S.H., Stammer A., Wolf B.A., Rhelogy of a lower critical solution temperature binary polymer blend in the homogeneous, phase separated and transitional regimes, Macromolecules, 29, 7155-7163 (1996).

Van Gisbergen J.G.M, Meijer H.E.M., Influence of electron beam irradiation on the microrheology of incompatible polymer blends, J.Rheol., 35, 63-87 (1991).

Van Gisbergen J.G.M, Hoeben W.F.L.M., Meijer H.E.M., Melt rheology of electron beam irradiated blends of PP/EPDM, Poly. Eng. Sci., 31, 1539-1544 (1991).

Van Gisbergen J.G.M, Meijer H.E.M., Lemstra P.J., Structure polymer blends: processing of polypropylene/EPDM blends: controlled rheology and morphology fixation via electron beam irradiation, Polymer, 30, 2153-2157 (1989).

Fahrlander M., Fuchs K., Friedrich Chr., Rheological properties of dibenzylidene sorbitol networks in polypropylene oxide matrices, J.Rheol., 44, 1103-1119 (2000).

Utracki L.A., On the viscosity-concentration dependence of immiscible polymer blends, J.Rheol., 35, 1615-1637 (1991).

Graebling D., Froelich D., Muller R., Viscoelastic properties of polydimethylsiloxane-polyoxyethylene blends in the melt, J.Rheol., 33, 1283-1291(1989).

Graebling D., Muller R., Rheological behavior of polydimethylsiloxane / polyoxyethylene blends in the melt emulsion model of two viscoelastic liquids, J.Rheol., 34, 193-205 (1990).
Borisenkova E.K., Kulichikhin V.G., Plate N.A., Generalized characteristic of polymer blend melt viscosity, Rheol. Acta, 30, 581-584 (1991).
Mani S., Malone F., Henning Winter H., Influence of phase separation on the linear viscoelastic behavior of miscible polymer blend, J.Rheol., 36, 1625-1649 (1992).

TUNING THE LOCALIZATION OF FUNCTIONALIZED MWNT IN REACTIVE PC/SAN BLENDS

M. Gültner, A. Göldel, and P. Pötschke*

*Leibniz Institute of Polymer Research Dresden (IPF), Hohe Str. 6, 01069 Dresden, Germany, – *poe@ipfdd.de*

Abstract - The influence of a reactive component (RC, N-phenylmaleimid styrene maleic anhydride) on the blend morphology, the localization of functionalized MWNT, and the electrical resistivity of the MWNT filled blend systems of polycarbonate (PC) and poly(styrene-co-acrylonitrile) (SAN) was investigated. The systems of PC, SAN, amino-functionalized MWNT (Nanocyl® 3152), and the RC were melt mixed in a DSM Xplore microcompounder. The RC containing maleic anhydride groups is miscible with the SAN and is assumed to enhance its polarity and surface tension. On the other hand it has the potential to form chemical bonds towards the functionalized MWNT.

By analyzing the blend systems with and without the RC co-continuous structures as well as the localization of the MWNT in one of the blend phases could be shown. The influence of the reactive component on the morphology of the blend systems was studied at different ratios of PC and SAN, whereas for all blends with 40 wt% SAN and 60 wt% PC co-continuous morphologies were found. In non-modified PC/SAN blends the MWNT are localized in the PC phase independently from the mixing procedure used, even if the carbon nanotubes were first mixed into SAN and then blended with PC. However, the modification of SAN using RC can change its properties in such a way, that the MWNT localize in the SAN-RC phase. This was found as well for mixing CNT first in SAN-RC, in PC, or adding all components together as well as adding RC to the ready PC/SAN/MWNT blends.

The localization of the MWNT after addition of the RC depends on the concentration of the MWNT and the RC. By adapting that ratio and the mixing strategy, the localization of the carbon nanotubes can be tuned. The investigations have shown that the MWNT once coupled with the RC remain in the SAN-RC Phase. Thus, the chemical reaction seems to be the driving force for localization of the MWNT in SAN-RC.

Introduction

Blending polymers is an economic way to produce new custom-made polymeric materials with outstanding properties. The characteristics of the employed polymers are combined, in some cases even synergistic effects are generated. The blend morphology influences significantly the properties of the materials. Another option to produce materials with outstanding properties is to add functional fillers into polymeric matrices. The incorporation of multiwalled carbon nanotubes (MWNT) in insulating polymer blends can improve their electrical proporties. When using a co-continuous blend a lower filler concentration compared to a composite with a one-phase-structure is required for electrical percolation, if an electrical conductive filler percolates and localizes in one of the phases (double percolation). This concept of douple percolation was first described by Sumita et al. (1991) using immiscible polymer blends with carbon black. The concept was later also adapted to other conductive fillers, e. g. to MWNT. As an example Göldel et al. (2009) described the double percolation concept in co-continuous polycarbonate (PC) / poly(styrene-co-acrylonitrile) (SAN) blends and discussed the driving force for the selective localisation of the MWNT in the PC phase. The PC/SAN blend was taken as a model system of the more complex PC/poly(acrylonitrile butadiene styrene) blend. This commercial blend was among the two most important polymer blend systems. The PC/SAN model blend has a reduced complexity, in addition this blend system show a high compatibility between the two phases.

Based on that blend system this paper investigates the influence of the addition of a reactive component (N-phenylmaleimid styrene maleic anhydride) on the blend morphology and the localization of amino-functionalized MWNT.

The aim of using amino-functionalized MWNT is that these MWNT can couple with the maleic anhydride group of the reactive component which is fully miscible with the SAN phase of the blend system. Thus, a fixation of the nanotubes in the SAN phase could be achieved.

Experimental

Materials

For extrusion of blends and composites polycarbonate Makrolon® 2600 (Bayer MaterialScience AG, Leverkusen, Germany) and poly(styrene-co-acrylonitrile) Luran® 358N (BASF AG, Ludwigshafen, Germany) were used. As MWNT material short thin amino-functionalized nanotubes Nanocyl® 3152

(Nanocyl S.A., Sambreville, Belgium) with a purity of 95% were used. As reactive component (RC) the N-phenylmaleimid styrene maleic anhydride (DENKI KAGAKU KOGYO KABUSHIKI KAISHA, Tokyo, Japan) was added to the SAN. This reactive component contains maleic anhydride groups and DSC studies confirmed that it is fully miscible with the SAN. RC is assumed to enhance the polarity and surface energy of SAN. On the other hand, it has the potential to form reactive bonds towards the modified nanotubes.

Preparation

All materials were dried under vacuum at 80°C for 12 h. The composites and blend systems of PC, SAN, carbon nanotubes, and the reactive component were melt mixed in a 15 cm^3 microcompounder (DSM Xplore, Geleen, The Netherlands) operating at 260 °C, 100 rpm and a mixing time of 5 min.

PC and SAN without or with RC were blended in the proportion 60:40 wt% or 60:20:20 wt%, respectively, to get blends with a co-continuous morphology. 0.5 wt% MWNT were incorporated in the polymer blends in different mixing sequences with or without the reactive component. On the one hand, the MWNT were pre-compounded in PC or SAN and in a second step these pre-compounds were melt mixed with the respective blend phase. On the other hand, all components were melt compounded in one step. The filler content in the blends was 0.5 wt% MWNT.

In addition, RC was added in a second step in PS/SAN/MWNT blends by varying the RC amount (20 %, 2 %, 0.2 % of the SAN replaced by RC).

For additional investigations, also 5 wt% MWNT loading was used in blends containing RC, and these blends were afterwards diluted to contain 0.5 wt% MWNT.

Morphological characterization

The morphology of the MWNT filled blend systems was characterized on thin sections using a light microscope. The localization of MWNT in such blends was analyzed by a transmission electron microscope (TEM, LIBRA 200 MC) on ultrathin sections without staining. In addition an Ultra Plus scanning electron microscope (SEM, Carl Zeiss SMT AG, Oberkochen, Germany) was used to investigate cut samples of extruded strands, in which the PC phase was selectively etched by hydrolysing using a solution of sodium hydroxide (30 wt%) at 105°C (see Dong et al. (1993)).

Results and Discussion

The blend systems with and without reactive component were analyzed concerning their morphology. In PC/SAN blends with and without RC a co-continuous structure as well as the selective localization of the MWNT in one phase of the blend was achieved independent on the mixing sequence.

The phase localization of the MWNT was proven using SEM and TEM. The investigations showed, that in blends containing 0.5 wt% MWNT without the reactive component the MWNT were always localized selectively in the PC phase (Figure 1A). This is analogous to findings on PC/SAN blends with 2 wt% Baytubes® C150 HP by Gödel et al. (2009).

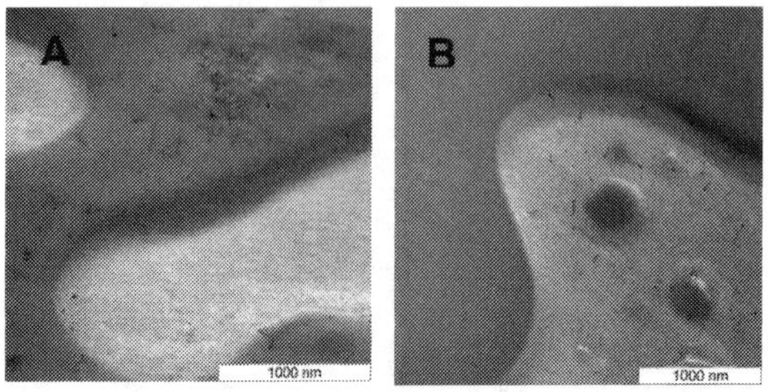

Figure 1 – TEM image on microtomed PC/SAN/0.5 wt% MWNT nanocomposite blends (PC phase is visible as unruffled surface, SAN and SAN-RC phase as rough areas): (A) without reactive component illustrating MWNT localized in the PC phase and (B) after replacing 20 wt% of SAN by RC showing MWNT in the SAN-RC phase

In blends containing the reactive component, the MWNT were localized in the SAN-RC phase (Figure 1B). MWNT localization was also found independently on the mixing sequence. Using blends of PC/SAN/MWNT with localization of MWNT within PC, the addition of the reactive component (20 wt%) in a second mixing step moved the nanotube localization towards SAN-RC illustrating the efficiency of RC as a "tuning additive".

The reason of the change in localization without reactive component illustrating MWNT could be either the chemical coupling and interaction of amino-functionalized MWNT and the maleic anhydride groups of RC, or interface effects by changing surface energy and polarity of SAN due to the addition of RC favourable for better wetting of nanotubes in the PC-RC phase.

In order to clarify these effects, RC was added to ready blends varying the amount of RC (20 %, 2 %, 0.2 %) in SAN

It was found that also after the addition of 2 % RC the MWNT (0.5 wt%) moved during the second mixing step from the PC to the SAN RC phase. However, the addition of 0.2 % RC is not sufficient to change the localization of 0.5 wt% MWNT. At this low RC concentration the MWNT remain in the PC phase (Figure 2).

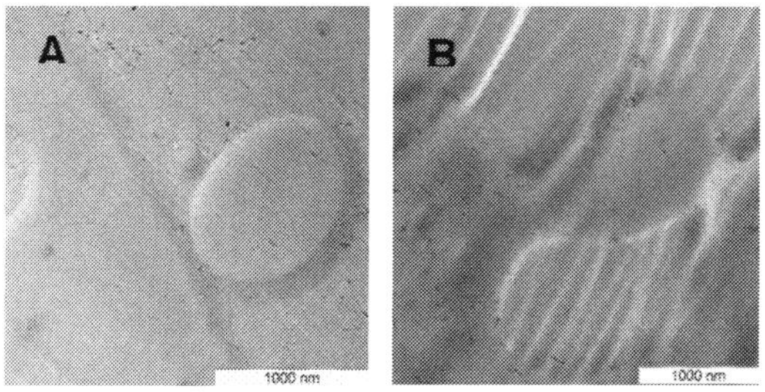

Figure 2 – TEM image on microtomed nanocomposite SAN/PC/0.5wt% MWNT blends (PC phase is visible as unruffled surface, SAN and SAN-RC phase as rough area): (A) - after replacing 2 wt% of SAN by RC illustrating MWNT localized in the SAN-RC phase and (B) after replacing 0.2 wt% of SAN by RC (B) showing MWNT remaining in the PC phase

Using a higher amount of MWNT, namely 5 wt% in the blends, already an amount of 2 wt% RC is not enough to localize the MWNT in SAN-RC and MWNT localize in both phases. When using 20% RC 5 wt% MWNT are exclusively localized in SAN.
Obviously, a certain ratio between RC concentration in SAN and MWNT-NH_2 is needed to achieve complete localization within SAN-RC, which is between 0.4:1 and 4:1.
Interestingly, when diluting a blend of PC/SAN-RC/MWNT having 5 wt% MWNT with MWNT localization in both phases using additional PC and SAN to get a blend with 0.5 wt% MWNT and 0.2 % RC, the nanotubes stay in the SAN-RC and PC phase, whereas when preparing that blend composition directly, a selectively localization in PC was found.
This localization difference indicates that the mixing sequence influence the MWNT localization. MWNT which once got in contact with the reactive component remain in the SAN-RC phase. In conclusion, the chemical reaction or strong interactions of RC and the amino-functionalized MWNT seem to be the driving force for the localization change rather than changes in interfacial energy relationships.

Conclusions

The investigations showed that the addition of a reactive component changes the localization of functionalized MWNT within an immiscible PC/SAN blend. Whereas in non-modified blends the MWNT are localized in the PC phase due to interfacial effects (Gödel et al., 2009), the modification of the SAN phase by the reactive component changes its properties in such a way, that the MWNT localize in that phase.
In summary, the localization of the MWNT after addition of the RC depends on the concentration ratio between the MWNT and the RC. By adapting that ratio and the mixing strategy, the localization of the nanotubes can be tuned. The chemical reaction or interactions of RC and the amino-functionalized MWNT seem to be the driving force for the localization change.

Acknowledgements

The authors thank the company DENKA (DENKI KAGAKU KOGYO KABUSHIKI KAISHA, Tokyo, Japan) for providing the reactive component.

References

Dong L., Greco R., Orsello G., "Polycarbonat/acrylonitril-butadiene-styrene blends: 1. Complementary etching techniques for morphology observations", Polymer, **34** (7), 1375-1382 (1993)

Göldel A., Kasaliwal G., Pötschke P., "Selective Localization and Migration of Multiwalled Carbon Nanotubes in Blends of Polycarbonate and Poly(styrene-acrylonitrile)", Macromol. Rapid Commun., , **30** (6), 423-429 (2009)

Sumita M., Sakata K., Asai S., Miyasaka K., Nakagawa H., "Dispersion of fillers and electrical-conductivity of polymer blends filled with carbon-black", Polym. Bulletin, **25**, 265-271 (1991)

MECHANICAL, THERMAL AND RHEOLOGICAL CHARACTERIZATION OF POLYPROPYLENE/ETHYLENE VINYL ACETATE BLENDS

Alothman, Othman. Y. (1,2)*, Elleithy, Rabeh H (1,2), Al-Zahrani, Saeed M.(1,2)

1 Department of Chemical Engineering, King Saud University - Riyadh - Saudi Arabia.
2 Center of Excellence for Research in Engineering Materials, King Saud University - Riyadh - Saudi Arabia.

Abstract - Polymer blending is usually performed to improve specific properties of a polymer. Various blends of PP/EVA were extruded at different extrusion temperatures and speeds. In this work, mechanical, thermal, and rheological characterization of PP/EVA blends is presented. The main finding is that the extrusion conditions have no significant effects on the properties of the blends. However, the blending ratio was found to be of high importance on the mechanical energy of blends. Data is to be discussed in details.

Introduction

Characterization of polymeric materials is of great importance to optimize processing conditions. The flow behavior of polymers depends on the molecular characteristics, flow geometry, and processing conditions such as temperature, shear rate, time of flow, etc. Basic information about the influence of the different parameters on viscosity, processing, elasticity, and extrudate characteristics will be a guideline in the selection of proper polymer under a given set of processing conditions.
Blending of two or more polymers to produce new materials is common in industry to achieve specific combinations of physical properties, processing characteristics, and cost. For example, Linear Low Density Polyethylene blends (LDPE/EVA) have good mechanical strength, processability, impact strength and insulation properties.
Arsac et al investigated the rheology of EVA with various molecular weights and VA contents (Arsac, Carrot et al. 1999). They found that the molecular weight together with branching remain the major parameter that influences the rheological behavior in the molten state and that the chemical composition has only a secondary effect.
Gupta et al studied EVA/Polypropylene blends, with various VA contents, over a temperature range of 210-230 0C (Gupta, Ratnam et al. 1992). They found no significant effect of VA content on the viscosity of the blend melts. They also find that EVA increases the impact strength of PP (Gupta, Ratnam et al. 1992). Increasing EVA was found to decreases the melt viscosity of the LLDPE/EVA blends (Moly, Oommen et al. 2002).
Kontopoulou et al investigated Blends of Ethylene-Vinyl Acetate (EVA) copolymer with a metallocene catalyzed elastomeric ethylene-□-olefin copolymer (EBC) (Kontopoulou, Huang et al. 2003). They found that adding EVA increased the shear thinning and elastic behavior of the blend, which improved the blend's processebility during film blowing. Although the blends were immiscible, they were mechanically compatible (Kontopoulou, Huang et al. 2003). Kontopoulou and Huang, also, found that EVA and EBC were more compatible, which resulted in blends with improved end-use properties (Kontopoulou and Huang 2004).

Experimental

Materials

The materials used in this part of the project were:

1. Polypropylene supplied by SABIC, (PP570P).
2. Ethylene Vinyl Acetate containing 27 wt% VA, supplied by ExxonMobil Chemicals, (Escorene™ Ultra FL 00328) It is referred to as EVA328. Its melt index is 3 g/10 min.
3. Ethylene Vinyl Acetate containing 6.5 wt% VA, supplied by ExxonMobil Chemicals, (Escorene™ Ultra FL00206) It is referred to as EVA328. Its melt index is 2.5 g/10 min.

Experiments

Blending and Sampling: Blends were manually prepared according to the desired ratios. Extrusion mixing, melting, and pelletizing were made by a twin screw extruder. Processing conditions were two temperatures (220 °C and 220 °C) and two screw RPM (12 and 24). The pellets were injection molded to testing samples.
Characterization: rheological properties were characterized using Melt Flow Indexer, (MFI) using 7.16 kg weight. Differential Scanning Calorimetry was used for thermal characterization at a rate of 5 °C/s. Unversal tensil machine and Izod Impact tester were used to measure the tensile stress, elongation and impact.

Results and Discussion

The characterization of the first blend and its variations were conducted. This includes rheological, thermal, and mechanical properties.

Rheological Properties

The flowability of the blends was compared by obtaining the Melt Flow index (MFI) for each blends under process conditions defined earlier. The load used in the MFI tests were 7.16 kg. Table 1 shows the values of blends of EVA/PP at various blending ratios and process conditions.

Table 1: Values of MFI of the Blends.

Load: 7.16	Temperature	200		220	
	RPM	12	24	12	24
EVA328	5%	69	93	96	69
	20%	91	88	89	81

To examine the effect of each variable, namely processing temperature, screw speed (RPM) and ratio of lend, the method of Taguchi was used. This analysis shows that the process conditions (processing temperature and screw speed) have no significant effect on the MFI. Expectedly, however, the blending ratio affected the MFI of the blends, as shown in Figure 1.

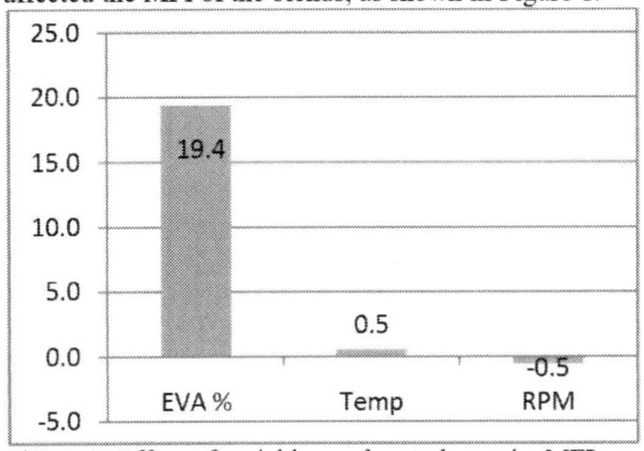

Figure 1: Effect of variables under study on the MFI.

No conclusive effect of any variable on the parameters of the Power-Law model, as shown in Table 2. Moreover, **Error! Reference source not found.** shows that the shear rate-viscosity curves at different conditions are overlapping, indicating no effect on viscosity. Similar trends were found for PP/EVA206 blends.

Table 2: Parameters of Power-Law Model at Various Conditions.

EVA %	RPM	Temperature	n	m
20%	12	200	0.54	4141.0
	24		0.55	3817.0
	12	220	0.53	4291.0
	24		0.56	3708.0
5%	12	200	0.54	4140.0
	24		0.48	6417.0
	12	220	0.55	3601.0
	24		0.51	5145.0

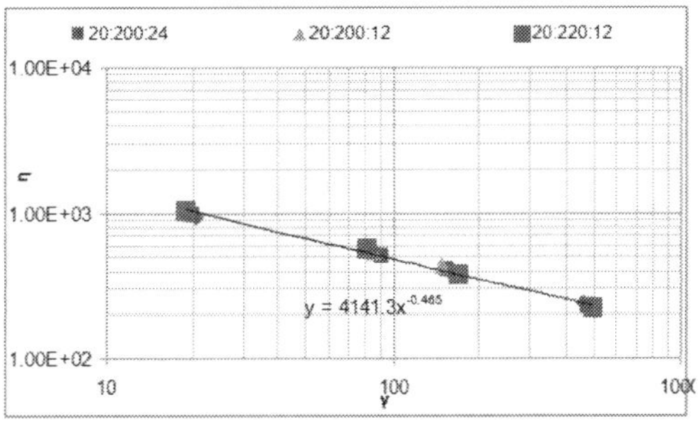

Figure 2: Shear rate-Viscosity Curves at Different Conditions.

Thermal Properties

The melting temperatures for PP/EVA328 and PP/EVA206 blends were obtained using the DSC technique. The thermal curves of PP/EVA328 and PP/EVA206 blends are shown in Figure 3 and Figure 4, respectively. Adding EVA into PP shifted the melting temperature of the blends slightly into the left.

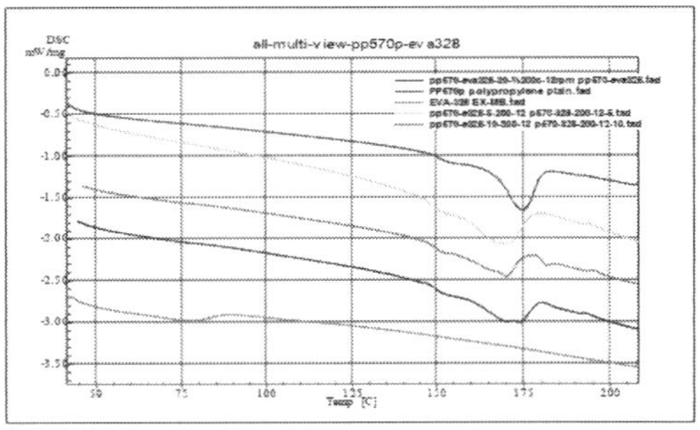

Figure 3: DSC Thermal curves of PP, EVA328 and Their Blends.

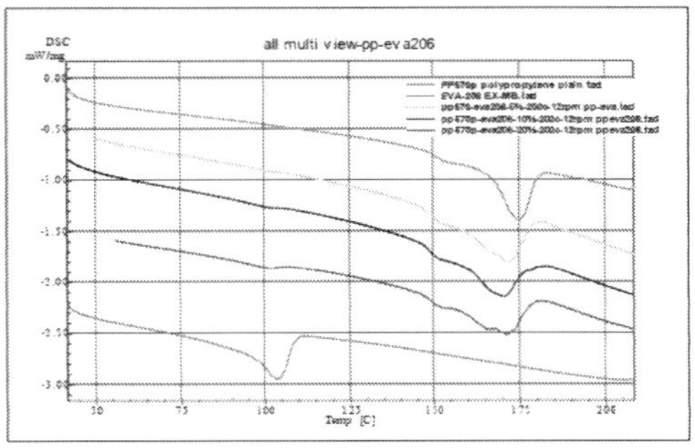

Figure 4: DSC Thermal curves of PP, EVA206 and Their Blends.

The effect of the variables studied on the melt temperature of PP/EVA206 is shown in Figure 5. In general, these results show no conclusive effect of all three variables on the melting temperatures of the blends. This might be due to that the percentage of the EVA is low. Thus, blends of higher percentages of EVA will be tested later. Another issue related to the thermal subject is the possibility of degradation during processing. Closer investigation of degradation occurrence is worth considering. Compatibility limitations may explain some multiple peaks observed in Figure 3 and Figure 4. Thus, closer examination of miscibility of the two polymers is of great importance.

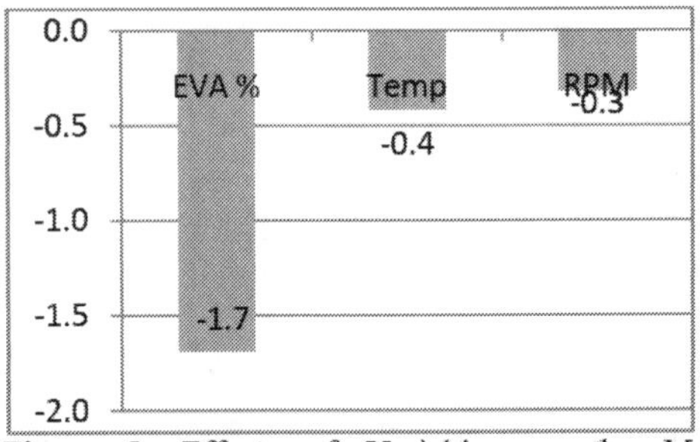

Figure 5: Effects of Variables on the Melting Temperature of the PP/EVA206 Blends.

Mechanical Properties

The mechanical properties, including Izod impact, tensile stress and elongation, for PP/EVA328 were studied. Anova analysis of the three variables (blending ratio, processing temperature and screw speed) suggested that processing conditions have no significant effect on the mechanical properties as shown in Figure 6. However, EVA was found to significantly improve the mechanical impact of PP as shown in Figure 7. The effect of the processing conditions and EVA content is illustrated by Figure 8.

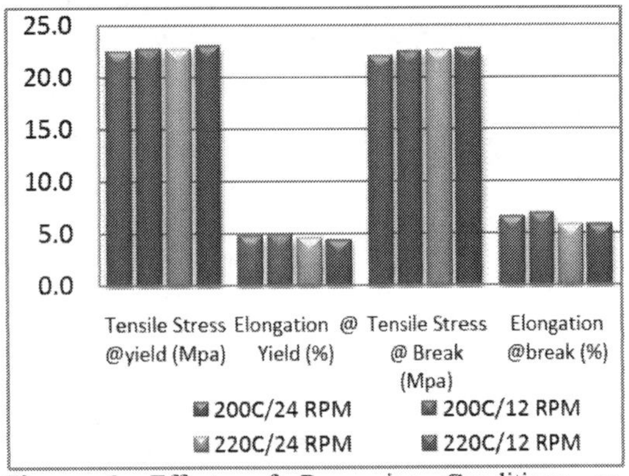

Figure 6: Effects of Processing Conditions on Mechanical Properties of a 10% EVA328 Blend.

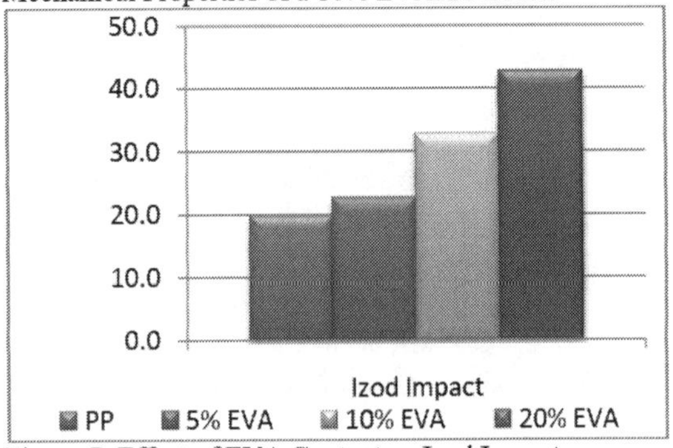

Figure 7: Effect of EVA Content on Izod Impact.

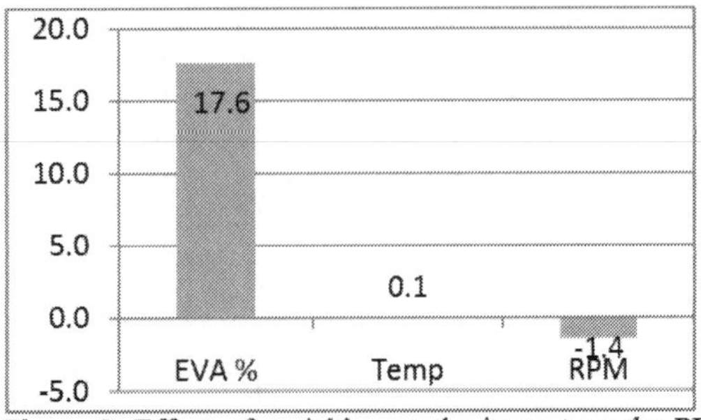

Figure 8: Effect of variables on the impact on the PP/ EVA328 blend.

Conclusions

The main conclusion remarks of this study are:

1. The three variables under study have shown no significant effect on the rheological properties (including flow curves) of all blends of PP/EVA328 and PP/EVA206. However, more investigation is planned including viscoelastic properties to look in more details at the dynamic rheological behavior of PP/EVA blends.

2. The thermal curves have shown a small shift of the melting temperature from the pure PP melting temperature as the EVA content increased. Multiple peaks were observed in some cases, which raised concerns about miscibility (or compatibility) of the two polymers.

3. The addition of EVA into PP improved significantly the impact of PP, as a main result of the cristallinity decrease. Moreover, cristallinity drop resulted in a drop in the PP tensile stress and an increase in its elongation.

4. Generally speaking, the results have clearly shown that the processing conditions under considerations (process temperatures of 200 0C and 220 0C and screw speeds of 12 and 24 RPM) have no significant influences on the properties under investigation.

Acknowledgements

The authors of this report would like to thanks the following:

1. Center of Excellence Research in Engineering Materials, CEREM, for its financial sponsorship of this project
2. SABIC Polymer Research Chair, SPRC, for using its processing and characterization facilities.
3. SABIC Technical Service Lab, TSL, for helping us with the mechanical properties testing.
4. ExxonMobil Chemicals for providing the EVA grades.
5. Technicians and students who have been working with us in this project.

References

Arsac, A., C. Carrot, et al. (1999). "Rheological Characterization of Ethylene Vinyl AcetateCopolymers." J. Appli. Polym. Sci. 74: 2625.

Gupta, A. K., B. K. Ratnam, et al. (1992). "Impact Toughening of Polypropylene by Ethylene Vinyl Actate Copolymer." Journal of Applied Polymer Science 45: 1303.

Gupta, A. K., B. K. Ratnam, et al. (1992). "Melt-Rheological Properties of PP/ EVA Blend." Journal of Applied Polymer Science 46: 281.

Kontopoulou, M. and L. C. Huang (2004). "Rheology, Structure, and Properties of Ethylene–Vinyl Acetate/Metallocene-Catalyzed Ethylene– -Olefin Copolymer Blends." J. Appli. Polym. Sci. 94: 881.

Kontopoulou, M., L. C. Huang, et al. (2003). "Binary blends of EVA and metallocene-catalyzed ethylene--olefin copolymers and their film properties." Adv. in Polym. Tech. 22: 209.

Moly, K. A., Z. Oommen, et al. (2002). "Melt rheology and morphology of LLDPE/EVA blends: Effect of blend ratio, compatibilization, and dynamic crosslinking." Journal of Applied Polymer Science 86(13): 3210-3225.

DISSIPATIVE PARTICAL DYNAMICS STUDIES ON INTERFACE PROPERTIES OF POLYMER BLENDS

Guo-Hua HU[1*, 2], Jing-Gang GAI[1, 3], Hui-Lin LI[3], Sandrine HOPPE[1]

1. *Laboratory of Reactions and Process Engineering, CNRS, Nancy Université-INPL-ENSIC, 1 rue Grandville BP 20451, 54001 Nancy, France – hu@ensic.inpl-nancy.fr;sandrine.hoppe@ensic.inpl-nancy.fr*
2. *Institut Universitaire de France, Maison des Universités, 103 Boulevard Saint-Michel, 75005 Paris, France*
3. *State Key Laboratory of Polymer Materials Engineering, Polymer Research Institute of Sichuan University, Chengdu, Sichuan 610065, People' Republic of China – gaijinggang@scu.edu.cn; nic7703@scu.edu.cn*

Abstract: Dissipative particle dynamics (DPD) is used to analyze the interface between polydisperse polymers by considering the simplest molten polydisperse system: LPE(long PE)/SPE(small PE)/LPP(long PP)/SPP(small PP). DPD results show that both LPE and LPP prefer to be in their own rich phases. Both SPE and SPP, however, are not only located in their own rich phases, but also in the neighboring phases. Besides, there is a significant accumulation of small chain at the interface. DPD results also show that the interfacial tension is lowered by the presence of small chains and decreases remarkably with increasing the concentration of the small chains, which is consistent with the density distributions of all components at the interface. The physical reason of the accumulation might be attributed to the fact that having more small chains at the interface has the effect of lowering the interfacial tension, and small chains play a surfactant role.

Introduction

When two immiscible homopolymers are mixed, they often separate into two distinct phases and the properties of the blends are controlled to a large extent by interfacial characteristics between the phases. Therefore it is very important to fine-control the interfaces properties of such polymer systems. Computer simulations have provided valuable microscopic insights into the interfacial behaviors of the immiscible molecules, such as the interface thickness and the molecular orientation at the interface.

Dissipative particle dynamics (DPD), which is a mesoscopic simulation technique, was invented in the early 1990s by Hoogerbrugge and Koelman [1]. As an example of application, Groot and Warren [2] calculated the interfacial tension between incompatible systems with the DPD model. They found that DPD simulation results can be used to quantitatively predict properties of real systems. However, studies with DPD on the interfacial behavior of polymer blends with polydispersity are scarce. Real polymer samples are polydisperse. It is therefore of interest to investigate the effect of polydispersity on interfacial properties with DPD.

Experimental

The DPD simulations of the LPE/SPE/LPP/SPP blends were performed in a cell of size 30×30×30, with the bead density ρ=3. Periodic boundary conditions were applied and the shear flow was simulated by means of Lees-Edwards boundary conditions. For convenience, the particle mass m, and k_BT were all taken as unit. The time step Δt was taken as 0.05 [3], and adjacent particles in the polymer chain interacted via a linear spring with a harmonic spring constant of 4.0, according to Groot and Liu [4 - 6]. Moreover the friction coefficient γ was chosen as 4.5 [6].

Results and Discussion

Fig. 1 show the density profiles of the LPE/SPE/LPP/SPP blends with different compositions. Both PE and LPP prefer to be in their own rich phases. Both SPE and SPP, however, are not only located in their own rich phases, but also in the neighboring ones.

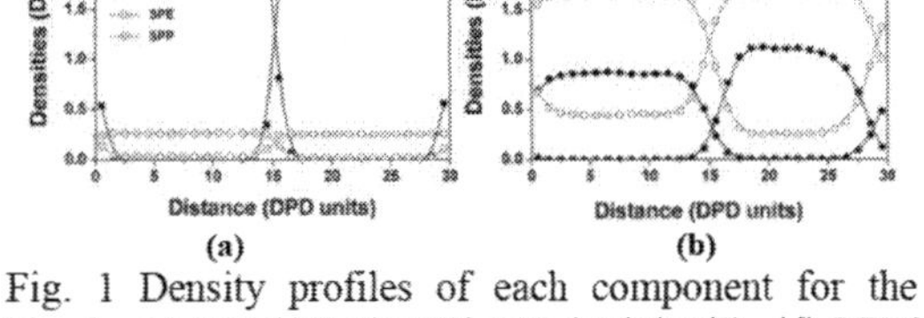

Fig. 1 Density profiles of each component for the blends: (a) LPE/ SPE/ LPP/ SPP (45/5/45/5); (d) LPE/ SPE/ LPP/ SPP (15/35/15/35)

Besides, there is a significant accumulation of small chains at the interface as indicated in Fig. 2 This accumulation of the small chains at the interface pushes the long chains further away from the interface with increasing the concentration of the small chains. One interesting quantity is the volume fraction of small chains, ϕ_{SPE} + ϕ_{SPP}. We have represented in Fig. 2 the small-chain excess

$$\varepsilon(x) = \phi_{SPE}(x) + \phi_{SPP}(x) - \overline{\phi_{SPE}} - \overline{\phi_{SPP}} \qquad (1)$$

as a function of x for different polydisperse systems.

Our calculations indicate that the function of $\varepsilon(x)$ has a positive maximum at the interface (x = 0 and 15). For example, both $\varepsilon(0)$ and $\varepsilon(15)$ are about 0.075 for the LPE/SPE/LPP/SPP (15/35/15/35) blend. Thus, at the interface, 22% of the monomers belong to long chains and 78% to small ones. Fig. 2 also indicates that the small-chain excess at the interface decreases when the concentrations of small chains as the amounts of SPE and SPP decrease.

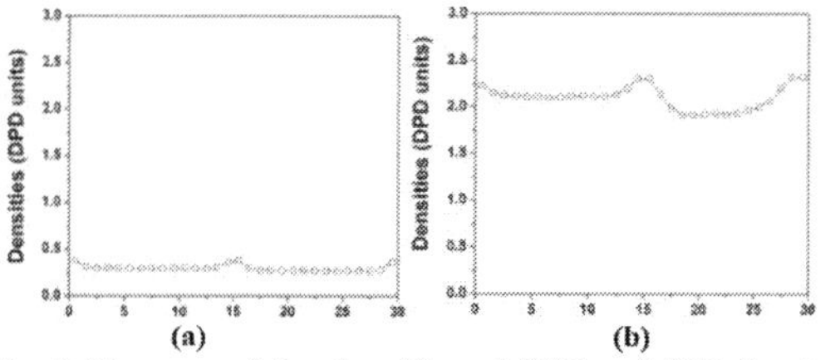

(a) (b)

Fig. 2 The sum of the densities of SPE and SPP for the blends: (a) LPE/SPE/LPP/SPP (45/5/45/5); (d) LPE/SPE/LPP/SPP (15/35/15/35)

The physical reason of the accumulation might be attributed to the fact that having more small chains at the interface has the effect of lowering the interfacial tension. Besides, it is clear that the interface thickness depends strongly upon the concentration of small chains. As the concentration of small chains increases, the interface thickness of the blend increases dramatically (Fig. 3).

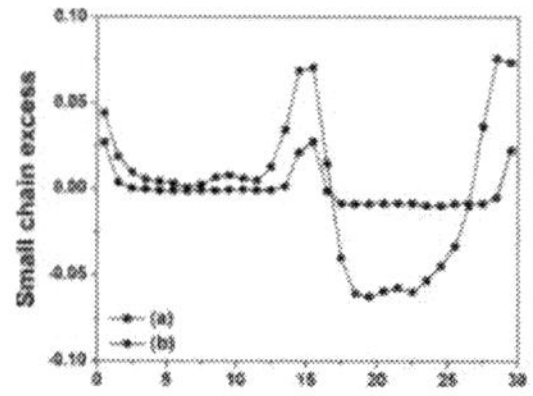

Fig. 3 The small chain excess (SPE and SPP) for the blends: (a) LPE/ SPE/ LPP/ SPP (15/35/15/35); (b) LPE/ SPE/ LPP/ SPP (45/5/45/5)

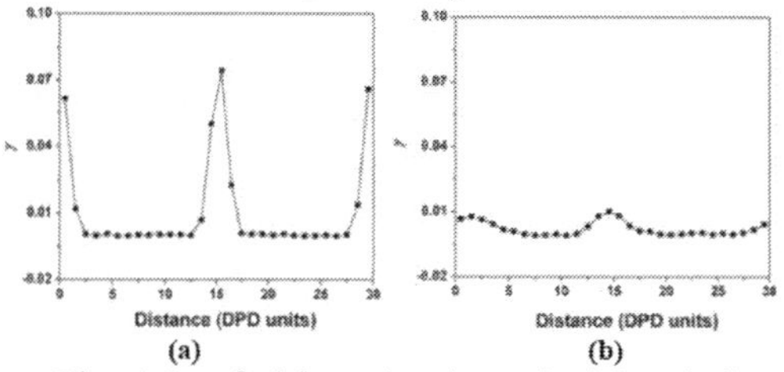

(a) (b)

Fig. 4 Interfacial tension in each slab calculated from Eqs. (1) and (4.4) for: (a) LPE/ SPE/ LPP/ SPP (45/5/45/5); (b) LPE/ SPE/ LPP/ SPP (15/35/15/35)

Fig. 4 shows that the interfacial tension is lowered by the presence of small chains and decreases remarkably with increasing the concentration of the small chains, which is consistent with the density distributions of all components at the interface. Small chains in fact play a surfactant role.

Conclusions

DPD is used to analyze the interface between polydisperse polymers by considering the simplest molten polydisperse system: LPE/SPE/LPP/SPP. DPD result show that both LPE and LPP prefer to be in their own rich phases. Both SPE and SPP, however, are not only located in their own rich phases, but also in the neighboring ones. Besides, there is a significant accumulation of small chain at the interface. The physical reason of the accumulation might be attributed to the fact that having more small chains at the interface has the effect of lowering the interfacial tension, and small chains play a surfactant role.

Acknowledgements

This work was supported by the National Basic Research Program of China (2005CB623800).

References

1. Koelman, J. M. V. A., Hoogerbrugge, P. J., "Dynamic simulation of hard sphere suspensions under steady shear", Europhys. Lett., 21, 363-368 (1993).
2. Groot, R. D., Warren, P. B., "Dissipative particle dynamics: Bridging the gap between atomistic and mesoscopic simulation", J. Chem. Phys., 107, 4423-4435 (1997).
3. Liu, D. H., Zhong, C. L., "Cooperative Self-Assembly of Nanoparticle Mixtures in Lamellar Diblock Copolymers: A Dissipative Particle Dynamics Study", Macromol. Rapid. Commun., 27, 458-462 (2006)
4. Groot, R. D., "Electrostatic interactions in dissipative particle dynamics - simulation of polyelectrolytes and anionic surfactants", J. Chem. Phys., 118, 11265-11277 (2003)
5. Özen, A. S., Sen, U., Atilgan, C., "Complete mapping of the morphologies of some linear and graft fluorinated co-oligomers in an aprotic solvent by dissipative particle dynamics", J. Chem. Phys., 124, 064905-1-064905-9 (2006)
6. Liu, D. H., Zhong, C. L., "Multicompartment micelles formed from star-dendritic triblock copolymers in selective solvents: A dissipative particle dynamics study", Polymer 49, 1407-1413 (2008)

SYNTHESIS OF GRAFT INTERPENETRATING POLYMER NETWORKS BASED ON EPOXY AND URETHANE. STUDY OF MECHANICAL AND MORPHOLOGICAL PROPERTIES

M. Ghafghazi[1*], M. Esfandeh[1], J.Morshedian

[1]*Department of Composite Materials, Iran Polymer & Petrochemical Institute, Tehran, Iran-mehdighafghazi@gmail.com*

Abstract- This paper describes the preparation of Epoxy/Urethane (EP/PU) graft interpenetrating polymer networks (g-IPNs) and investigates the effect of EP/PU weight ratio on the mechanical, morphological properties of the IPN system. Here, g-IPN was prepared by thorough mixing of an isocyanate-terminated urethane prepolymer with an epoxy resin followed by simultaneous curing of the resins. Polytetra hydrofuranate (PTHF), molecular weights (M_w) 1000 g/gmol, was used to prepare urethane prepolymers. EP/PU weight ratios were 75/25, 50/50, 30/70 and 15/85. Disappearance of epoxide and isocyanate functional groups was followed by Fourier Transform Infrared spectroscopy (FT–IR), showing curing of the resins. Tensile measurements and Scanning Electron Microscopy (SEM) were used to study thermal, mechanical and morphological properties of the prepared systems. The best mechanical properties were obtained at EP/PU weight ratio 75/25 which also shows a fine and uniformly dispersed morphology. Moreover, at this ratio, with increasing PTHF M_w in the urethane prepolymer, the mechanical properties were improved.

Introduction

The interpenetrating polymer networks (IPNs) are a unique type of polyblend which define as a mixture of two or more crosslinked polymer networks that have partial or total physical interlocking between them [1–3]. In fact, formation of IPNs is the only way of intimately combining crosslinked polymers [4]. If a high degree of intermixing is intended, the Simultaneous IPN (SIN), in situ mixing and polymerization of monomers or prepolymers, is preferred. This is due to the compatibility of the monomeric mixture, which is much higher than that of a polymeric mixture [5].

The entanglements in IPNs must be of a permanent nature and are made so by homocrosslinking of the two or more polymers [4]. These entanglements of multiple crosslinked polymers lead to forced miscibility compared to simple blends and the resulting IPN materials exhibit excellent size stability and a possibly synergistic combination of the properties of their components [6-8]. These make IPNs interesting in several applications [9,10].

In fact, g-IPNs possess high tensile modulus and strength and low elongation at break (ε_b) due to the existence of covalent chemical bonding between the networks [11]. The properties of IPN materials are determined by the kinetics of formation of the growing networks and by the thermodynamics of mixing of the constituent polymers [12].

Epoxy resins are widely used as the matrices of high-performance composite materials because of their stiffness, chemical resistance, and high temperature stability. However, they suffer from a low fracture strength and inherent brittleness [13]. To meet the required end-use performance, they need to be modified with either thermoplastics or elastomers. The IPNs using a rubbery polyurethane phase have been considered for such applications [14]. Polyurethanes are a versatile group of polymers that can be used in a wide spectrum of polymeric products, including foams, elastomers, coating resins, adhesives, and fibers. However, their relative low thermal resistance and strength [15], high water absorption and poor heat resistance limit their applications [16]. Therefore, some researchers have proposed the change of its phase structure by blending with one or two other polymers to improve its properties [13].

The object of this research is to prepare EP/PU graft interpenetrating polymer networks and to investigate the effect of Epoxy/Urethane weight ratio and urethane's prepolymer molecular weight on the mechanical and morphological properties of the IPN systems. The experimental design is in such a way to optimize the epoxy/urethane weight ratio in which the synergistic effect in properties is observed and then at this ratio the effect of PTHF molecular weight in urethane prepolymer is investigated. The properties of the prepared systems will be studied by FT-IR, tensile test,SEM and Wide-Angle X-ray Diffraction (WAXD).

Experimental

Materials

Epoxy resin Epon 828 DGEBA (Diglycidyl Ether of Bisphenol A, EMM 184-190g) was supplied by Shell Chemicals Co. (USA). Epoxy hardener, Aradur 917, was obtained from Huntsman Advanced Materials (Switzerland). TDI (Toluene Diisocyanate, an 80/20 mixture of 2,4/2,6 isomers), Dibutyltin dilaurate (urethane catalyst), tributylamine (epoxy catalyst) and glycerol (urethane cross-linker) were Merck grade. PTHF (Polytetra hydrofuranate, M_w= 1000, 2000 and 3000 g/gmol) was supplied by Arak Petrochemical Co. (Iran). Glass fiber tissue AEMS 20gr/m2 was purchased from Tishan Co. (China).

PTHF, glycerol and epoxy resin were dried at 80ºC in vacuum for 24h before use. Other raw materials were used without further purification.

Instrument

FTIR spectra were recorded on an Equinox 55 FTIR spectrophotometer (Bruker Co. Germany). The WAXD studies of the samples were carried out using a SIEMENS D5000 X-ray diffractometer (Germany) and under CuKα radiation. Scanning Electron Microscopy (SEM) images were prepared from the cryogenically fractured (in liquid N2) surface of the samples, using a S360 (Cambridge Co. England). Tensile properties of the specimens were determined with a tensile testing machine STM-150 (Santam Co. Iran).

Synthesis of Polyurethane Prepolymer

The isocyanate-terminated polyurethane prepolymer was prepared by reacting TDI with PTHF. For this purpose, a weighed amount of TDI was placed in a round-bottomed, three-necked flask, equipped with a mechanical stirrer and dry nitrogen inlet, and heated to 60°C. While stirring vigorously, a predetermined amount of PTHF (NCO:OH ratio was 2:1), with a certain M_w, was added slowly to the flask. The isocyanate content was continuously determined according to ASTM D 2572. The end of the reaction was regarded as the point in which the isocyanate content (%NCO) reaches its theoretical value [11].

Synthesis of Epoxy/Urethane Graft IPN

For the synthesis of EP/PU g-IPN, a simultaneous synthesis method was used. Here, a predetermined amount of epoxy precursor and its hardener (100:90) was thoroughly mixed with urethane prepolymer at room temperature and under dry nitrogen atmosphere. Then the urethane cross linker (glycerol), epoxy catalyst and finally urethane catalyst were added to the mixture. The mixing was further continued for about 15min. The mixture was then degassed under vacuum for about 30min and cured in an oven at 120°C for 6h.
To prepare the specimens for tensile testing, a composite lamina was prepared by full impregnating a glass fiber tissue with the resin mixture (96 %wt resin). The samples were then cured at 120°C for 6h.

Results and Discussion

In Figure 1 FTIR spectra for the mixture of epoxy resin with urethane prepolymer prior to curing (Figure 1a) and after formation of the g-IPN (Figure 1b) are shown. As it can be seen, in the mixture of two resin systems, the absorption bands for –NCO and epoxide group are appeared at 2270 cm^{-1} and 915 cm^{-1}, respectively. However, these bands are disappeared as the mixture is heated up to cure temperature of the individual systems; an indication of formation of the corresponding networks and completion of the curing reactions.

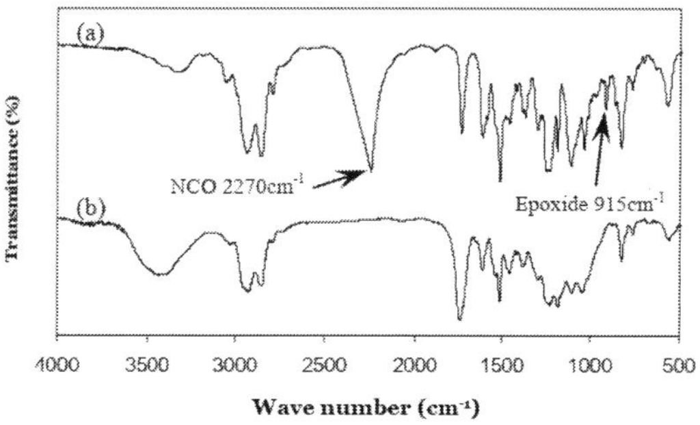

Figure 1. FTIR spectra of (a) mixture of epoxy resin and NCO terminated urethane prepolymer, 50/50 wt%, prior to curing, (b) cured system.

Mechanical and Morphological Properties

The results of mechanical properties for IPN systems at various EP/PU weight ratios are shown in Figures 2 and 3. As it is shown, there is an ascending order for the tensile modulus and strength with increasing EP/PU weight ratio (Figure 2) but the trend for the tensile strength is reversed at epoxy weight ratio above 75 (wt%). For ε_b and toughness (calculated from the area under stress-strain curve) a decrease is observed up to 50 (wt%) of epoxy but beyond that an increase is seen with a maximum at 75 (wt%) of epoxy. The initial increase in tensile modulus and strength and the reduction in toughness and ε_b are attributed to the increase in epoxy percentage which has a higher tensile modulus and strength and a lower toughness and ε_b than the PU component.
The distinct increase in tensile strength and modulus at EP/PU weight ratio 75/25 can be due to the a good interpenetration of epoxy and urethane chains which improves their compatibility, fills the defects existing in the single polymers, and increases the effective crosslinking density. Also, the increase in interpenetrating surface area leads to a more energy absorption by PU phase. Moreover, PU particles embedded in epoxy phase acts as stress concentration points which can initiate crazes and stop its development, thus toughness and ε_b amount increases [17]. Base on these results, a synergistic increase in tensile properties for EP/PU g-IPN 75/25 is observed. At this ratio, EP/PU g-IPN has a ε_b 8.85% and a tensile strength of 49.56 MPa (compared with 1.88% and 35.74MPa for pure epoxy). Considering these facts, this ratio seems to be the optimized one, and therefore, in the course of the research the effect of PTHF molecular weight was only studied at this ratio.
The microstructure of the polymer blend is also important in determining its physical and mechanical properties. In fact, the morphological changes and also the size of the dispersed phase can be responsible for the enhancement of the mechanical properties [18].
In Figure 4 SEM micrographs of g-IPNs at various EP/PU weight ratios are compared to the micrographs

of pure individual systems, i.e. epoxy and PU. The IPN system with EP/PU 75/25 (wt%) showed a homogeneous and finely dispersed PU phase in epoxy matrix (Figure 4b). At EP/PU weight ratio 15/85 again dispersed phase morphology is observed but the size of the dispersed phase is much bigger than that of with EP/PU weight ratio 75/25. Therefore, it is reasonable to have an improvement in properties at EP/PU g-IPN 75/25 (wt%) [19]. For EP/PU g-IPN 50/50 (wt%), a co-continuous morphology with a dispersed phase is seen. Comparing this with EP/PU g-IPN 30/70 (wt%), which shows an almost co-continuous morphology, the tensile modulus and strength are increased but the changes in toughness and ε_b are not that significant.

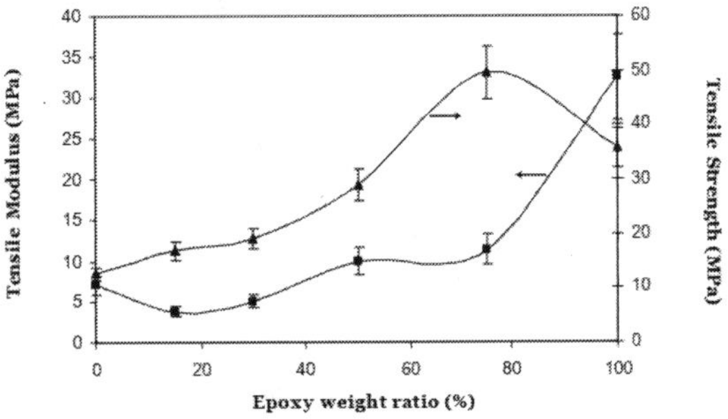

Figure 2. Tensile modulus and strength vs. epoxy weight ratio for g-IPNs, (■ Tensile modulus, ▲ Tensile strength).

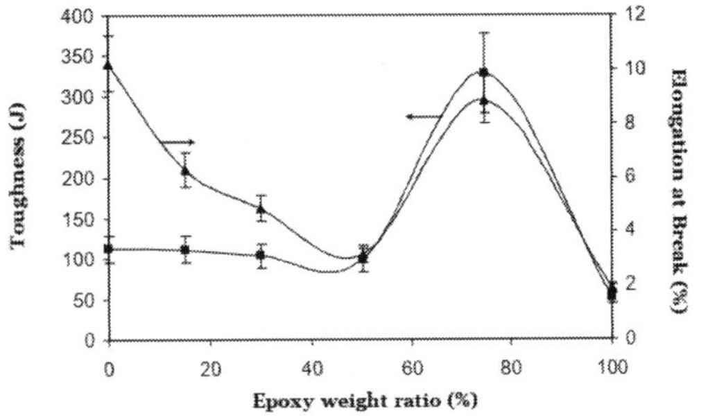

Figure 3. Toughness and elongation at break vs. epoxy weight ratio for g-IPNs, (■ Toughness, ▲ Elongation at break).

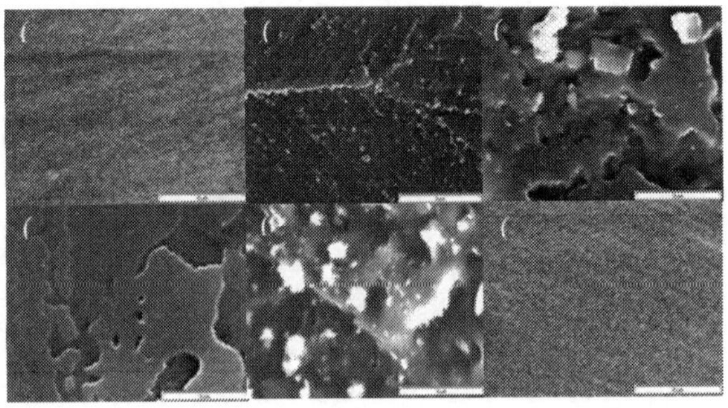

Figure 4. SEM micrographs for g-IPNs with various EP/PU weight ratios: (a) pure epoxy, (b) g-IPN 75/25, (c) g-IPN 50/50, (d) g-IPN 30/70, (e) g-IPN 15/85, (f) pure PU1000

Conclusions

Epoxy/Urethane graft interpenetrating polymer networks were prepared by a simultaneous polymerization method and the effect of Epoxy/Urethane weight ratio on the mechanical and morphological properties of the IPN system were studied.

The tensile modulus and strength of the IPN systems were increased with increasing EP/PU weight ratio but the trend for the tensile strength was reversed at epoxy ratios above 75 (wt%). For the elongation at break and toughness a decrease was observed up to 50 (wt%) of epoxy but beyond that an increase was seen with a maximum at 75 (wt%) of epoxy.

SEM results for g-IPN at EP/PU 75/25 showed good dispersion of PU phase in epoxy matrix. For EP/PU 15/85, the size of the dispersed phase was much bigger than that of with EP/PU weight ratio 75/25. For EP/PU 50/50, a co-continuous morphology with a dispersed phase was seen. Comparing this with EP/PU 30/70, which showed an almost co-continuous morphology, the tensile modulus and strength were increased but the changes in toughness and ε_b were not that significant.

References

1- Ma, C. C. M., et al., "Interpenetrating Polymer Networks of Polyurethane and furfuryl alcohol, I: Morphology and mechanical properties", Die Angewandte Makromoleculare Chemie., **214**, 39-56, (1994)

2- Kim, S. C., et al., "Polyurethane Interpenetrating Polymer Networks. II. Density and Glass Transition Behavior of Polyurethane-Poly (methyl methacrylate) and polyurethane-Polystyrene IPN's", Macromolecules, **9**, 263, (1976)

3- Jeevananda, T., Siddaramaiah., "Synthesis and Characterization of Polyaniline Filled PU/PMMA IPNs", Eur. Polym. J., **39**, 569-578, (2003)

4- Frisch, K. C., Klempner, D., "Recent advances in interpenetrating polymer networks", Polym Eng Sci., **22**, 1143-1152, (1982)

5- Kim, J. H., Kim, S. C., "Analysis of reaction injection molding process of polyurethane-unsaturated polyester blends. Part II: Mechanical properties", Polym. Eng. Sci., **27**, 1252-1257, (1987)

6- Sugimoto, H., et al., "Preparation and properties of urethane acrylate-epoxy interpenetrating polymer networks containing silica nanoparticles", Polymer Bulletin, **57**, 975-982, (2006)

7- Zhang, H., et al., "Synthesis and characterization of nanocomposites of silicon dioxide and polyurethane and epoxy resin
interpenetrating network", Polym. Int., **52**, 1493-1497, (2003)

8- Hsieha, K. H., et al., "Graft interpenetrating polymer networks of urethane-modified bismaleimide and epoxy (I): mechanical behavior and morphology", Polymer, **42**, 2491-2500, (2001)

9- Zhu, Y., et al., "Influence of MMT content on the microstructure and miscibility for PU/EP IPN nanocomposites by positron ", Phys. Stat. Sol.(c)4, **10**, 3780-3784, (2007)
10- Chen, L., Chen, S., "Latex interpenetrating networks based on polyurethane, polyacrylate and epoxy resin", Prog. Org. Coat., **49**, 252-258, (2004)
11- Cassidy, E. F., et al., "Two-component interpenetrating polymer networks (IPNs) from polyurethanes and epoxies. I. Studies on full IPN, pseudo-IPN, and graft polymer alloys of polyurethanes and epoxies", J. Polym. Sci. Part A, Polym. Chem., **22**, 1839-1850, (1984)
12- Karabanova, L. V., et al., "Phase separation in the polyurethane/poly(2-hydroxyethyl methacrylate) semi-interpenetrating polymer networks synthesized by different ways", Polym. Eng. Sci., **48**, 588-597, (2008)
13- Zhang, Y., Hourston, D. J., "Rigid interpenetrating polymer network foams prepared from a rosin-based polyurethane and an epoxy resin", J. Appl. Polym. Sci., **69**, 271-281 (1998)
14- Raymond, M. P., Bui, V. T., "Epoxy/castor oil graft interpenetrating polymer networks", J. Appl. Polym. Sci., **70**, 1649-1659, (1998)
15- Hourston, D. J., Schafer, F. U., "Poly(ether urethane)/poly(ethyl methacrylate) interpenetrating polymer networks: morphology, phase continuity and mechanical properties as a function of composition", Polymer, **37**, 3521-3530, (1996)
16- Jia, Q., et al., "Synthesis, characterization and properties of organoclay-modified polyurethane/epoxy interpenetrating polymer network nanocomposites", Polym. Int., **55**, 257-264, (2006)
17- Li, Y., Mao, S., "Study on the properties and application of epoxy resin/polyurethane semi-interpenetrating polymer networks", J. Appl. Polym. Sci., **61**, 2059-2063, (1996)
18- Pearson, R. A., Yee, A. F., "Influence of particle size and particle size distribution on toughening mechanisms in rubber-modified epoxies", J. Mat. Sci., **26**, 3828-3844, (1991)

A New Theory to Predict Interfacial Tension of Polymer Pairs in Compressible Condition of the Interface

*Hadi Ramezani-Dakhel, Hamid Garmabi**

(1) Department of Polymer Engineering and Color Technology,
Amirkabir University of Technology - Tehran - Iran

Interfacial tension is a significant factor that determines the miscibility of the components, the blend morphology and its mechanical behavior. In this study, a new theory combined with a simple free energy model was employed to predict the interfacial tension of polymer melts. Indeed, a new relation for compressible condition of the interface was derived in order to predict the interfacial tension of polymer pairs melt state. The new theory predicts both value and temperature dependence of interfacial tension, although no adjustable parameter is used in calculating interfacial tension. The whole required parameters can be obtained from the literature or may be calculated theoretically. The results were compared well with the predictions of previous theories (Broseta et al. theory which is a modified version of Helfand and Tagami theory) and also with experimental data reported in the literature. The prediction ability of the temperature dependency of interfacial tension for some polymer pairs using the new theory was demonstrated through comparison with reported experimental data.

TOUGHENING OF PA 6 BY FINE DISPERSED NANOSIZED PA 6/POLYETHER BLOCK COPOLYMER PARTICLES

S. Geier[1]* and M. Poindl[1]

[1]Institute of Polymer Technology, University of Stuttgart, Germany – simon.geier@ikt.uni-stuttgart.de; Institute of Polymer Technology, University of Stuttgart, Germany – marcus.poindl@ikt.uni-stuttgart.de.

This study focuses on a novel approach of toughening PA 6 by applying an anionic polymerized PA 6/polyether block copolymer (PA 6 - copo) as an impact modifier. PA 6/PA 6 - copo blends were prepared by melt compounding using ground PA 6 - copo plates of adjustable polyether soft phase (10 to 40 mass-%) followed by injection moulding of test samples. Mechanical studies show a significant enhancement of the notched Izod impact strength from about 5 to 41 kJ/m² at a blend with an overall polyether content of 10 mass-%. Moreover it was found that both, the elastic modulus as well as the tensile strength only decreases moderately about 18 and 24 %, respectively. Additionally, phase dispersion and chain mobility were examined by means of dynamic mechanical thermal analysis (DMTA). The occurrence of two characteristic glass transition temperatures indicates the chemical immiscibility of polyether and PA 6 within the copolymer. In case of the blend the retention of this pattern implies the compatibility of PA 6 and PA 6 - copo. AFM reveals fine dispersed soft phases with diameters smaller than 100 nm. Due to the high molecular affinity of both materials no interfacial tensions are recognizable. Furthermore, TEM proves that the original network - like structures of the PA 6 - copo are maintained in the blend. Interference between these structures and the lamella formation of PA 6 was not observed. Lamellae seem to grow through or by the PA 6 - copo soft domains.
Finally, the influence of the polyether content of the PA 6 - copo on the toughening effect of PA 6/PA 6 - copo blends was investigated. Here the use of PA6 - copo incorporated 30 mass-% polyether shows the best balance between stiffness and ductile properties of the blends.

Introduction

Polyamide 6 (PA 6) is one of the most important engineering plastics widely used in industrial applications such as radiator fans and air-intake manifolds of automobiles because of their remarkable combination of mechanical, thermal and chemical properties as well as processability (Elsner, 2008). However, PA 6 is susceptible to moisture absorption and notch-sensitive. PA 6 exhibits a high resistance to crack initiation, however in the presence of a notch the resistance to crack propagation is very poor (Tjong and Bao, 2005). With respect to the glass transition temperature of PA 6 (about 53 °C), the chain segments in the amorphous phase are frozen at room or lower temperature. As a consequence, the toughness properties under impact loading are rather limited and often do not meet the requirements in many applications.
To improve the impact behaviour of PA 6, an elastomeric second phase either by compounding or heterophasic copolymerization (synthesis) has to be regularly dispersed in the PA 6 homopolymer matrix. The rubbery phase interrupts the fracture propagation and lowers the stress concentration during deformation. This energy absorbing mechanism results in a significant enhancement of the toughness, also at low temperatures (Grellmann and Seidler, 1998).
Besides the chemical compatibility of the components, concentration, geometry and structure of the impact modifier play a decisive role on the resulting properties (Elias, 2001).
Among the synthesis routes, the anionic polymerization of caprolactam together with end functionalized polyether prepolymers followed by reaction injection moulding (RIM) leads to a PA 6/polyether block copolymer (Gabbert et al., 1983). The morphology of the PA 6/polyether block copolymers shows the existence of a non-crystallizable polyether (soft) phase incorporated in the crystallizable PA 6 (hard) matrix which yields in a material toughening even at temperatures below the glass transition of PA 6 (Göschel and Geier, 2009).
In contrast, the addition of impact modifiers during melt compounding describes an easy way in toughening of PA 6 and is of large industrial interest. Here, commonly used impact modifiers are various types of rubbers or copolymers (Tanrattanakul, 2008). Generally, these materials are grafted with maleic anhydride (MA) before blending with PA 6 to improve the compatibility between matrix and modifier. The functional group of the compatibilizer (MA) can react with the end groups of PA 6 resulting in a fine dispersion and good filler matrix bonding of the blends.
By avoiding the use of compatibilizer, PA 6/polyether block copolymer particles have been applied as impact modifier during melt compounding. The present study discusses the toughening effect of

PA - copos with respect to structure formation in the blend.

Experimental

Materials

PA 6 - polyether block copolymer plates with a thickness of about 2 to 4 mm as prepared by an anionic polymerization followed by RIM processing have been used. The polyether content was chosen to 10, 20, 30 and 40 mass-%, which corresponds to the commercial Nyrim® 1000 (N1000), Nyrim® 2000 (N2000), Nyrim® 3000 (N3000) and Nyrim® 4000 (N4000) from BrüggemanChemical, Heilbronn. Recently, mechanical and morphological properties of these materials have been discussed in detail by Göschel and Geier (2009). Before melt compounding the semi - finished plates were cooled down in liquid nitrogen and ground.

PA 6 (Ultramid® B40) with a molar mass of M_n = 33,000 g/mol was kindly provided from BASF AG, Germany.

Processing

PA 6/PA 6 - copo blends were prepared by melt compounding using a Coperion ZSK 26, co - rotating, intermeshing twin screw extruder (D = 26 mm, L/D = 40). The screw speed was set up to 250 rpm with a total feed rate of 10 kg/h at a maximum barrel temperature of 260 °C. Prior to each processing step, all materials were dried in a vacuum oven at 80 °C for 24 h.

Compounded pellets were injection moulded into standard tensile bars using an Arburg 220 M injection moulding machine. The injection pressure was 1600 bar at a mould temperature of 80 °C.

Characterization

All mechanical characterizations were conducted at standard atmosphere, 23 °C and 50 % relative humidity (DIN EN ISO 291). Before testing samples were stored at 80 °C in a vacuum oven for 72 h.

Notched Izod impact strength measurements were carried out in accordance to DIN EN ISO 180 on a Zwick Z5 impact testing machine. The samples (80 x 10 x 4 mm³) were tested edgewise with the notch - type A.

Tensile tests were performed in accordance to DIN EN ISO 527 on a Zwick 1476 universal testing machine, equipped with an automatic travel sensor "mulitsens". A traverse speed of 50 mm/min at an initial sample length of 50 mm was chosen. Young's Moduli were determined at a traverse speed of 1 mm/min.

Dynamic mechanical thermal analysis (DMTA) were performed with a relaxation spectrometer Rheometrics RDS II from - 100 °C to 200 °C, at a ramp rate of 5 K/min, a sinusoidal shear strain amplitude of 0.15 % and a vibration frequency of 1 Hz. As test samples the parallel area between the shoulders of the tensile bar were used (50 x 10 x 4 mm³).

The TEM samples were embedded in epoxy resin and cured at 23 °C over night, cut, stained with RuO_4 vapour at 23 °C for 16 h, rinsed by distilled water and dried. Thereafter, thin slices of 80 to 100 nm were sectioned at - 170 °C using the ultramicrotome Reichert-Jung, Ultracut E/FC4D with a diamond knife. The TEM studies were performed at a Zeiss EM - 10 at 80 kV.

AFM measurements were carried out with an Explorer-TopoMetrix System (ATOS) in non-contact mode. Samples were also cut at - 170 °C using the ultramicrotome.

Results and Discussion

Figure 1 shows the screw design of the compounding process of the PA 6/PA 6 - copo blends. Both, PA 6 and PA 6 - copo were fed in the extruder feed port located at zone 1. Due to the high affinity of the materials no other additives like compatibilizer were necessary during processing.

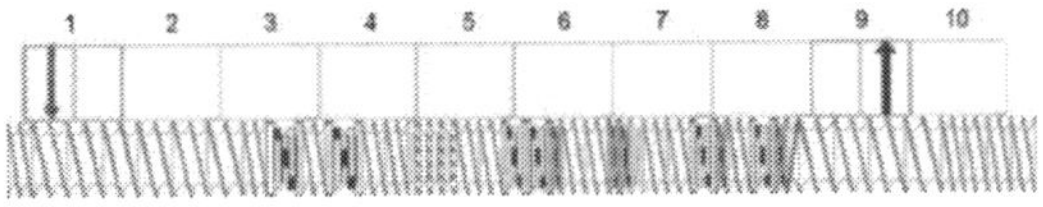

Figure 1 - Screw design

As a first approach PA 6 was compounded with 25 mass-% N4000 to obtain a blend with an overall polyether content of 10 mass-% which is comparable to N1000. Table 1 demonstrates the mechanical properties of the blend in comparison to PA 6, N4000 and N1000. Blending PA 6 with N4000 increased the notched impact strength by more than 500 %, whereas the impact properties of N1000 are only about 40 % higher than the PA 6 values. Simultaneously, stiffness properties of the blend, evident in the Young's modulus and yield stress only decrease moderately about 18 and 24 %, respectively.

Table 1 - Mechanical properties

material	Young's modulus [GPa]	yield stress [MPa]	impact strength [kJ/m²]
PA 6	2.76	83	6
PA 6 + 25 mass-% N4000	2.27	62.9	41.4
N1000	2.42	68.6	9.6
N4000	0.5	-	no break

Morphological analysis indicates fine dispersed nanosized soft domains. Figure 2 represents an AFM - picture (phase mode) of a PA 6 blend with 25 mass-% N4000 where particle diameters smaller than 250 nm are visible. Furthermore no interfacial tensions in terms of small cracks or displacements are

recognizable. This concludes the existence of perfect matrix filler bonding.
The question whether the polyether phase itself or, alternatively, the Nyrim® particle is responsible for the toughening effect in PA 6 reveals figure 3. The TEM - picture exhibits the preservation of the N4000 morphology within the soft domain. By means of stained thin cuts the typical polyether network of the N4000 structure becomes visible. Furthermore an influence of the soft particle on the lamellar morphology of the PA 6 matrix seems to be negligible. Lamellae grow trough or by the PA 6 - copo domains. This behaviour can be directly attributed to the PA 6 segments incorporated in the N4000 particles and verifies the assumption of a perfect filler matrix bonding.

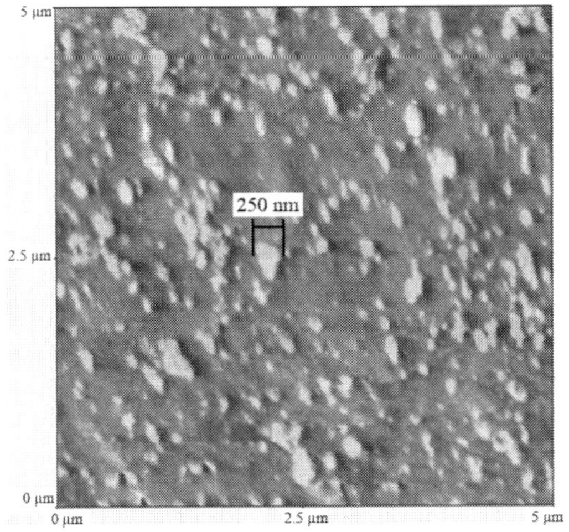

Figure 2 - AFM pircture of PA6/N4000 blend

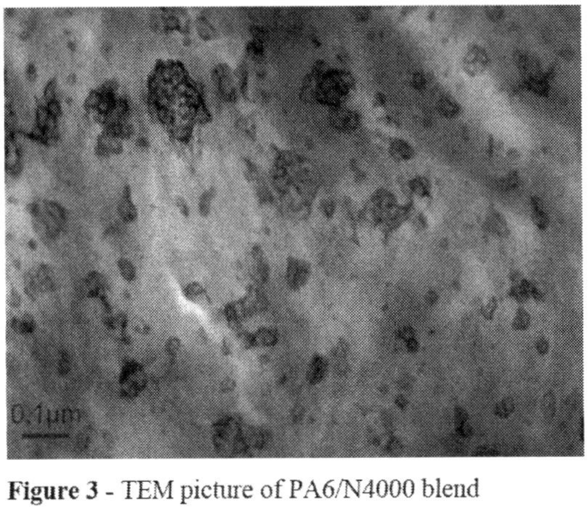

Figure 3 - TEM picture of PA6/N4000 blend

Table 2 - Correlation between polyether and filler content

[mass-%]	N3000	N4000	N3000
filler content	**25**	**25**	33.3
polyether content	7.5	**10**	**10**

Since both materials, N4000 and N3000 reveal a network - like polyether structure their influence of the toughening effect on PA 6 was investigated (Göschel and Geier, 2009). Therefore two kinds of blends have been provided. On the one hand the filler content of 33.3 mass-% N3000 and 25 mass-% N4000 represent a constant overall polyether content of 10 mass-%. On the other hand a constant filler content of 25 mass-% of both materials was compared (s. table 2).
Figure 4 shows the influence of a constant polyether and filler content, respectively on the stiffness/toughness properties of the blend. Again, the responsible toughening effect is due to the filler rather than the overall polyether content.
Hence, grinding and shear forces during melt processing cannot degrade Nyrim structure, but break it up to nanosized particles. The original network structure is maintained.
At the same filler content N3000 shows similar or a slightly better performance as an impact modifier than N4000.

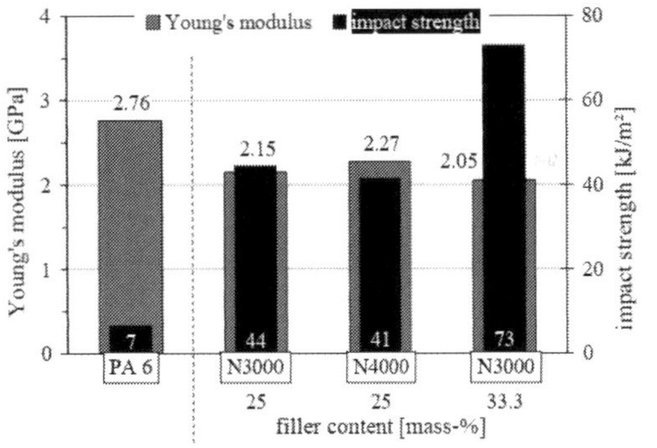

Figure 4 - Influence of constant polyether and filler content on stiffness/toughness properties

In order to examine molecular mobility and phase inversion of the blends' morphology dynamic mechanical thermal analysis have been carried out. Because of thermodynamic reasons most polymers are unmiscible on a molecular level resulting in multi - phase systems. As a consequence domains of one polymer are presented in the matrix of the second polymer.
Figure 5 and 6 display the loss factor tan δ as well as the storage modulus G' of PA 6, N3000 and a PA 6/N3000 blend with 33.3 mass-% N3000 over a temperature range from -100 °C to 200 °C.

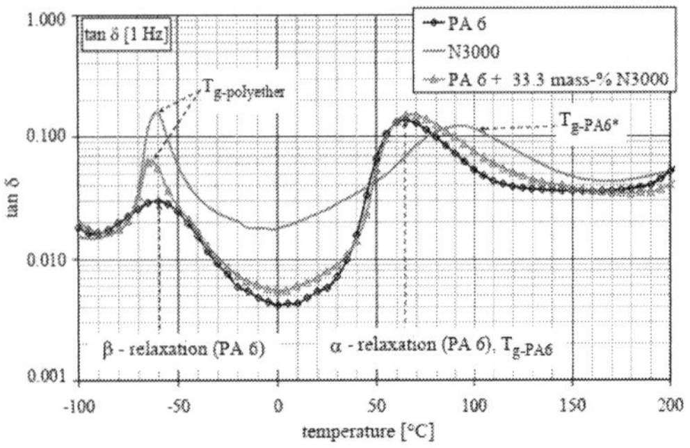

Figure 5 - Loss factor tan δ of PA 6, N3000 and PA 6/N3000 blend

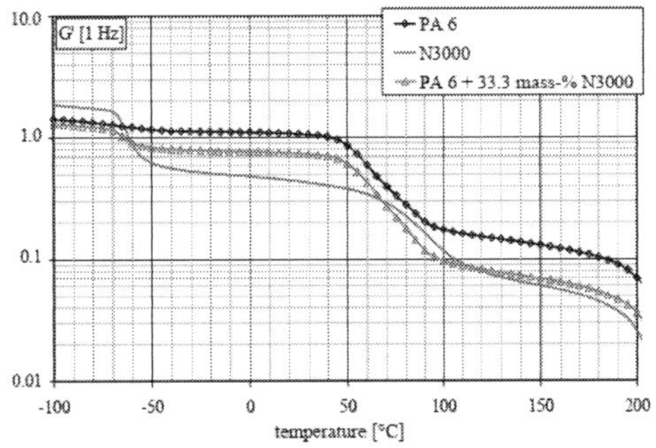

Figure 6 - Storage modulus G' of PA 6, N3000 and PA 6/N3000 blend

Besides the glass transition temperature ($T_{g\text{-PA6}}$) of the amorphous phase at about 65 °C (α - relaxation) PA 6 reveals two secondary relaxation temperatures in the low temperature range (β - and γ - relaxation). In the range of -90 to 0 °C the β - relaxation represents an increasing mobility of weak bonded carbon amide groups. Whereas at the γ - relaxation (< -90 °C, not shown in figure 5) a rising motion of CH_2 - sequences, comparable with polyethylene, is discussed (McCrum et al., 1991).

In terms of the α - relaxation rising energy supply increases the chain mobility which leads to an overcoming of secondary binding forces in disorder amorphous regions. As a result chain segments up to 50 atoms are able to move together which causes a significant drop of G'.

Regarding N3000 two pronounced glass transitions are recognizable. This indicates the nonexistent compatibility of the single components of the copolymer (Utracki, 1989). By statistically alternating monomer sequences of PA 6 and polyether within the N3000 their properties are maintained. The polyether phase reveals a relatively narrow developed peak at about - 60 °C ($T_{g\text{-polyether}}$). The sharpness and intensity of the peak and the drop in G' at this temperature reflects the amorphous structure of the polyether. Although the occurrence is situated in the same temperature range this peak formation differs clearly from the β - relaxation of PA 6.

The PA 6 - phase within the PA 6 - copo reveals a relatively broad $T_{g\text{-PA6*}}$ above room temperature (about 90 °C). Because of the cross - linking this peak is broader than the one of virgin PA 6. Further the polyether co - phase yields a reduction of the stiffness (G') in a temperature rage over $T_{g\text{-polyether}}$ (-60 °C) (Tsui and Johnson, 1995).

Blending of PA 6 and N3000 leads to a new morphology where only two significant relaxation patterns are visible. One occurs at low temperature and one above room temperature, respectively.

The presence of one $T_{g\text{-PA6}}$ for the PA 6 phases within the blend shows again that both components are compatible. In case of immiscibility both characteristical transition temperatures would be preserved. Hence the consequence is a blend with fine dispersion.

Furthermore the relaxation peak at low temperatures is clearly assigned to the polyether phase ($T_{g\text{-polyether}}$). In comparison with the β - relaxation of PA 6 this peak of the blend is considerably sharper. The lower peak area reflects the reduction in overall polyether content within the blend.

As shown in figure 7, the ductility enhances with increasing N3000 content. On the contrary yield strength decreases whereas at a filler content of 50 mass-% a pronounced formation vanishes. Here the material shows a rubber like behaviour. Moreover at filler content of 33.3 mass-% the characteristic necking behaviour of PA 6 changes.

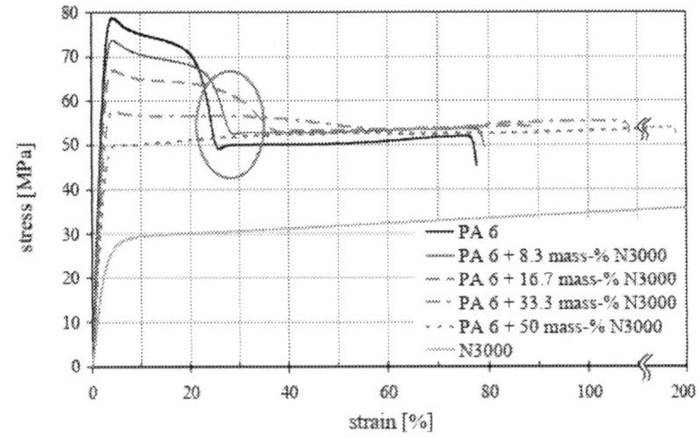

Figure 7 - Stress - strain curves of PA 6/N3000 blends

Figure 8 illustrates the mechanical properties of PA 6/N3000 blends with different N3000 content. It is shown that with increasing filler content stiffness properties in terms of Young's modulus decreases constantly from 2.8 GPa (PA 6) to 1.7 GPa in case of the blend with 50 mass-% N3000. The toughening performance is represented by the impact strength values. Compared to the Young's modulus a significant step between a filler content of 16.7 and 33.3 mass-% is evident. Subsequently the slope to 50 mass-% is reduced again. Also the impact strength value of virgin

N3000 lies below the value of the blend with 33.3 mass-% N3000.

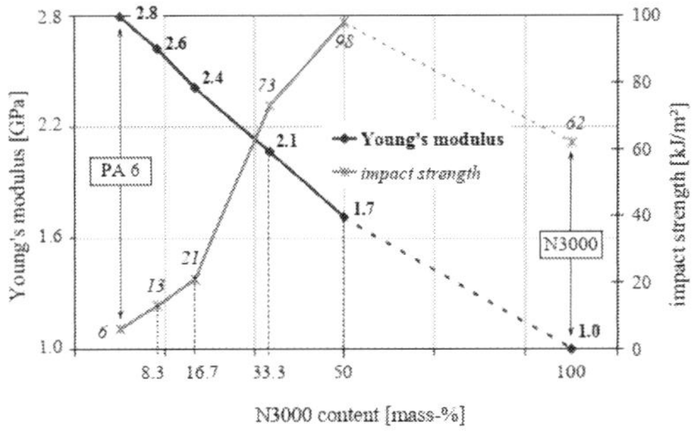

Figure 8 - Young's modulus and impact strength depending on N3000 content

Conclusions

The toughness behaviour of PA 6 has been enhanced significantly by the use of an PA 6/polyether block copolymer as impact modifier. With respect to AFM and TEM measurements the resulting morphology shows nanodispersed PA 6 - copo particles incoporated in the PA 6 matrix. TEM - analysis shows that the network like N3000 structure within this soft domains is preserved. Hence the PA 6 phase of the copolymer is completly compatible with the PA 6 matrix and results in a perfect filler matrix bonding. This fact is also confirm with DMTA where two diffrent PA 6 glass transions of the virgin materials merge to one in case of the PA 6/N3000 blend. With respect to the toughening effect of PA 6 - copo it was shown that filler amount plays a more important role than the polyether content of the filler itself. At a filler content of 25 mass-% no significant differences between N3000 and N4000 as impact modifier were observed. This effect can be derived from the morphology of both N4000 and N3000 which shows a cross - linked polyether structure. Here, at a filler content of 50 mass-% of N3000 the impact strength value is about 60 % higher than of the N3000 itself. Also the Young's modulus of the blend is 70 % higher than of the N3000.

Thus, the application of PA 6/polyether block copolymers as impact modifier demonstrates a promising route in optimizing the toughness behaviour of PA 6.

In further studies the presented results have been used for micromechanical modelling the effect of nanodispersed impact modification on the mechanical behaviour of PA 6 (Huang et al., 2010). Here, the aim is to get an input for optimization the morphologies of both, the filler as well as the blend.

Acknowledgements

The authors gratefully acknowledge the financial support by the German Research Foundation (DFG) for the funding in the project GO 579/11-1 and the project SCHM 746/74-1. Acknowledged is the material supply of BrüggemannChemical, Heilbronn, Germany.

References

Elsner, P., "Chapter 2.2.1.1 Polyamide", in Kunststoffe - Eigenschaften und Anwendungen, Elsner, P., Eyerer, P., Hirth, T. (Ed.), Springer Publishers, Berlin, p. 638-478 (2008).

Göschel, U., et al., "Nanodispersed impact modifieres in highly crystalline PA 6", PPS 25 Tech. Papers, (2009).

Grellman, W., Seidler, S.: Deformation und Bruchverhalten von Kunststoffen, Springer Puplishers, Berlin (1998).

Gabbert, J., et al., "Reinforced nylon 6 block copolymers", Polym. Comp., **4**, 196-199 (1983).

Elias, H.-G.: Makromoleküle, Band 2: Physikalische Strukturen und Eigenschaften, WILEY - VCH Puplishers, Weinheim (2001).

Huang, J., et al., "Micromechanical modelling of Young's modulus of semi-crystalline polyamide 6 (PA 6) and elastomer particle-modified PA 6", Comp. Mat. Sci., special issue IWCMM-19, (2010).

Langer, B.: Bruchmechanische Bewertung von Polyamid - Werkstoffen, Logos Puplisher, Berlin (1998).

McCrum, N. G., et al.: Anelastic and dielectric effects in polymeric solids, Dover Puplications, New York (1991).

Tanrattanakul, V., N., et al., "Rubber toughening of nylon 6 with epoxidized natural rubber", Polym. Test., **27** , 794-800 (2008).

Tjong, S. C. and Bao, S. P., "Impact fracture toughness of polyamide-6/montmorillonite nanocomposites toughened with a maleated styrene/ethylene butylene/styrene elastomer", J. Polym. Sci.: Part B, **43**, 585-595 (2005).

Tsui, S.-W. and Johnson, A. F., "Thermal behaviour of nylon 6-poly(ether-esteramide) block copolymers", J. Mat. Sci., **30**, 5967-5972 (1995).

Utracki, L. A.:. Polymer alloys and blends: Thermodynamics and rheology, Hanser Publishers, Munich (1989).

UNDERSTANDING PLA CRYSTALLIZATION IN COMPLEX PROCESSING OPERATIONS: INTERRELATION BETWEEN PLASTICIZATION, CHAIN BRANCHING AND FOAMING.

Michel A. Huneault[1*], Mihaila Mihai[2], Basil D. Favis[3]

[1]*Université de Sherbrooke, Chemical and Biotechnological engineering department, 2500 Boul. de l'Université, Sherbrooke (Québec) J1K 2R1, Canada; michel.huneault@*usherbrooke.ca
[2]*Industrial Materials Institute (NRC), 75 de Mortagne, Boucherville, QC J4B 6Y4; mihaela.mihai@*cnrc-nrc.gc.ca
[3]*CREPEC - Chemical Engineering Department, École Polytechnique de Montréal, Qc, Canada; basil.favis@*polymtl.ca

Abstract –Pure PLA crystallizes too slowly to develop significant crystallinity upon the rapid cooling typical of most extrusion or injection molding operations. In response to this fact, nucleants and plasticizers can be added to respectively boost the crystalline nucleation density and to improve chain mobility during the crystallization process. Such formulations can be crystallized in injection molds using high mold temperatures and relatively long cycle time. The foaming process is complex from a rheological and crystallization behavior point of view. The viscosity of the polymer increases rapidly due to the phase separation of the blowing agent. Shear forces are substituted by biaxial stresses as the material exits the extrusion die and enters the expansion phase. The crystallization occurs in this complex flow environment where the release of the blowing agent from the polymer drastically changes the T_g and crystallization rate of the polymer. Therefore it is not surprising that strong interactions between the foaming process, foam quality and crystallization behavior were found. This presentation will discuss and review the crystallization behavior of PLA in complex processing operation and will use extrusion foaming as a means to better illustrate this topic.

Introduction

PLA is one of the few biobased compostable materials that are readily available. Foams are often regarded as materials that are difficult to recycle. Because of this, the foaming of PLA has attracted immediate attention for replacement of PS foams. One interesting feature of PLA is that its crystallinity can be controlled by adjusting the content of its two optically active isomers, L-LA and D-LA. The PLLA crystallinity can attain around 45%. This crystallinity level decreases with D-LA content as the D-LA units acts as defects in the PLLA crystals. Therefore, it is possible to control the maximum crystalline level as well as the crystallization rate by playing with the L:D ratio. The crystallization rate can be further modified by adding plasticizer, nucleants (Li et al. 2007) or by adding a second polymer phase (Li et al. 2008). It is also possible to rapidly increase PLA crystallinity levels by subjecting the material to biaxial stresses at a temperature above its T_g (Chapleau et al, 2008).

The foaming process is interesting from a PLA crystallization point of view because it combines many features that can promote the crystallization. First, the blowing agent dissolved in the polymer can act as a powerful plasticizer, the extrusion temperatures are typically close the temperature window where the highest crystallization rates are encountered. Biaxial stretching is present during the expension phase and could also lead to crystallinity enhancement. Research on this complex topic has generated very useful knowledge on the crystallization of PLA in processing conditions and under conditions of large plasticizer concentration. It will be reviewed and compared to results obtained in other processing operations in the presentation.

Experimental

Three different PLA grades, supplied by NatureWorks, were used in this study. Their D-LA content was 2, 4 and 10% (grades 4032D, 2002D and8302D respectively). A chain-extender (CE), CesaExtend OMAN698493 from Clariant Additive Masterbatches, was used to create branched PLA. It is an epoxy / styrene / acrylic copolymer, provided in masterbatch form in a PLA carrier.

A Leistritz 34 mm co-rotating twin-screw extruder comprising 12 barrel zones was used to prepare the PLA foams. CO_2 was used as the blowing agent. It was pumped at mid extruder and solubilised in the polymer. As the blowing agent concentration increased, it was possible to produce the foam at lower extrusion temperature due to T_g depression and viscosity reduction. The extruder was operated at a constant screw rotation speed of 150 rpm. A capillary die with a 2 mm diameter was used to produce foamed rods.

Results and Discussion

Crystallinity control in PLA foams is of primary practical importance to increase heat resistance to acceptable levels. Therefore, the PLA grade must be selected with a sufficient L-LA purity to enable crystallization. On the other hand, foaming crystallizable polymers is more challenging because they need to be processed at a higher temperature relative to their glass transition. In the case of PLA, it was also shown to be very sensitive to foam cell rupture leading to a high open-cell content (Mihai et al, 2007) Surprisingly however when foaming crystallizable PLA grades, best foam morphologies are obtained in a very narrow processing window that lead to simultaneous crystallinity development and foam expansion. For example, Figure 1 and 2 compare foams obtained with at 5 and 9% CO2 content produced at their minimum extrusion foaming temperature (i.e. torque-limited conditions). At 5% content, the viscosity of the material is not sufficiently reduced by the plasticization effect to extrude the material in its crystallization temperature range. The resulting materials are very poorly foamed due to a lack of expansion and gas losses through open cells. At 9% CO2, extrusion temperature can be significantly lowered and the foams are produced in the 100-120oC range where the PLA crystallization rate is maximum. The resulting foams have very low densities and a fully developed crystalline structure. More details on this topic can be found in recently published paper (Mihai et al. 2009)

A second issue with PLA foaming is the rheology. Because of their linear structure and high rigidity, PLA chains cannot entangle as readily as other commonly foamed material leading to a lack of elasticity or melt strength that is detrimental for foam stabilization. This can be corrected through a reactive processing strategy where low molecular weight reactive copolymers are added to branch PLA chains through their acid chain ends. A commercial epoxy-based copolymer was used for this purpose with significant success in terms of rheological modification. This was easily achieved during batch mixing tests where reaction time can be adjusted at will. More importantly however, it was shown that similar extent of reaction could also be obtained during in situ reaction in the extrusion-foaming process. Rheological assessment of the macromolecular structure was obtained through elongational rheometry as shown in Figure 2. A series of blends with different chain extender concentration were produced, characterized in terms of their rheology and foamed. The complete results are reported elsewhere (Mihai et al., 2010). The strain-hardening phenomenon, typical of branched polymer structures was completely absent in pure PLA samples and was clearly accentuated as the reactive modifier content was increased. Surprisingly, this wasn't accompanied by significant changes in the foaming behavior. This was interpreted as an additional proof of the superseding role of crystallization on the foam cell stabilization. It was also postulated that crystal nuclei formation may slightly precede foam expansion and play an important role on the foam nucleation itself.

The presentation will review and examine these complex interrelations between crystallinity development, chemical branching of PLA and rheology and will examine the consequence of these interrelations on the complex and challenging process of making low-density foams. The presented phenomena encompass the technological aspects of foaming since they enable to understand the main challenges and characteristics of PLA processing.

a)
b)
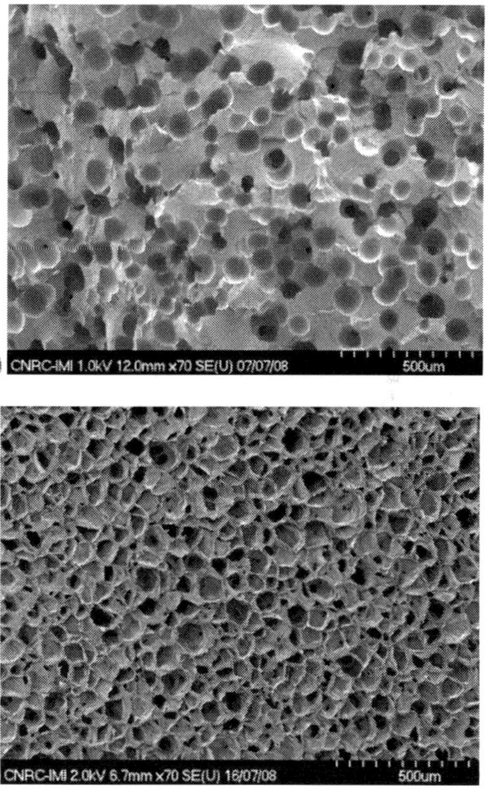

Figure 1 – Morphology of PLA foams. Micrograph a) shows a material foamed 5% CO2 where no crystallinity development and limited expansion occurred. By contrast, micrograph b) shows a fully expanded and fully cristalline PLA foam produced with 9% CO2.

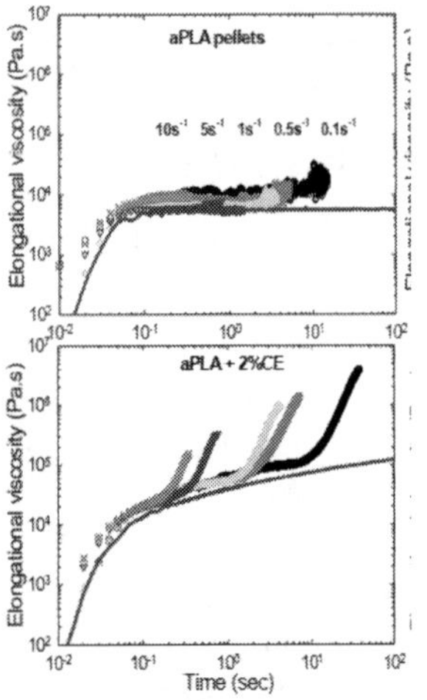

Figure 2. Elongational data obtained without chain-extender (upper graph) and with 2% chain extender (lower graph) for an amorphous PLA grade (8302). Similar strain-hardening was obtained with chain-extended crystallizable PLA grade.

Conclusions

The study of PLA crystallization in the foaming process illustrates well the complexity (and beauty) of the interaction between thermodynamics, rheology and crystallization all interlinked in a single polymer processing operation. The presence of a high plasticizer concentration (i.e. blowing agent) concentration in the polymer, the low extrusion temperature and the potential nucleating effect of CO_2 clusters promote PLA crystalline nucleation prior to die exit and foam expansion. The presence of these crystalline nuclei in turn significantly enhances the foam uniformity and expansion ratio. Complete crystallization can be attained when the conditions above are combined leading to good quality highly crystalline PLA foams.

References

Li H., Huneault M.A., "Effect of Nucleation and Plasticization on the Crystallization of Poly(Lactic Acid)", Polymer 48, 6855-6866 (2007).

Li H., Huneault M.A., "Crystallization in PLA/Thermoplastic Starch Blends", International Polymer Processing, 23, 412-418 (2008).

Chapleau N., Huneault M. A., H. Li, "TPS/PLA biaxial stretching, International Polymer Processing", 22, 402-409 (2007).

M. Mihai, M.A. Huneault, B.D. Favis, H. Li, "Extrusion foaming of semi-crystalline PLA and PLA/thermoplastic starch blends", Macromolecular Bioscience, 7(7), 907-920 (2007).

Mihai, M., Huneault, M.A., Favis, B.D., "Crystallinity development in cellular poly(lactic acid) in the presence of supercritical carbon dioxide", Journal of Applied Polymer Science, 113 (5), 2920-2932 (2009).

Mihai, M., Huneault, M.A., Favis, B.D. "Rheology and extrusion foaming of chain-branched poly(lactic acid)", Polymer Engineering and Science, 50 (3), 629-642 (2010).

FOAMING OF COMPATIBILIZED POLYMER BLENDS USING ORGANIC LIQUIDS AS BLOWING AGENTS

V. Altstädt[1*], P. Gutmann[1], F. Wolff Fabris[1], A.H.E. Müller[2]

[1]*Department of Polymer Engineering, University of Bayreuth, Universitätsstrasse 30, 95447 Bayreuth, Germany – altstaedt@uni-bayreuth.de;* peter.gutmann@uni-bayreuth.de; *felipe.wolff-fabris@uni-bayreuth.de*
[2] *Department of Macromolecular Chemistry II, University of Bayreuth, Bayreuth, Germany – axel.mueller@uni-bayreuth.de*

Immiscible blend systems are commonly compatibilized by block copolymers to improve the interfacial adhesion and the mechanical properties. The compatibilization by block copolymers not only affects the interfacial adhesion and the mechanical properties but also changes the foaming behavior of the blend system. The aim of the study is to systematically investigate the influence of compatibilization on the diffusion and solubility of the blowing agent in the compatibilized blend system as well as the influence on the foaming behavior and the final foam structure. An immiscible blend of 60 wt% poly(2,6-dimethyl-1,4-phenylene ether) (PPE) and 40 wt% poly(styrene-co-acrylonitrile) compatibilized with different amounts of a polystyrene-*block*-polybutylene-*block*-poly(methyl methacrylate) (SBM) triblock terpolymer is selected as polymeric material. As the properties of the compatibilized blend system additionally influence the foaming behavior, first the morphology and the rheological properties of the PPE/SAN/SBM blends are investigated. Two organic liquids, an alcohol and an alkane, are used as physical blowing agents. The diffusivity and solubility of the organic liquids in the blend systems are characterized by absorption measurements at a temperature near the boiling point of the blowing agent. In order to obtain cellular materials, the saturated materials are foamed by a batch-foaming process at three different temperatures for 30 sec. Subsequently, the properties of the compatibilized blend systems and the investigated transport properties, as measured by the absorption experiments, are systematically correlated to foam density analyzed using the buoyancy method. In addition, the final foam structure is characterized using scanning electron microscopy.

Introduction

Polymeric foams offer a wide range of beneficial properties as a thermal and acoustical insulation, improved energy absorption, cushioning effect or less materials usage. As result of these beneficial properties, polymeric foams are widely used in industrial applications (aerospace industry, packaging, furniture, building and construction or shock and sound attenuation) (Klempner and Frisch, 1991; Gendron, 2005). To satisfy steadily increasing demands on the property profile of polymeric foams, novel and innovative approaches have to be developed.

Combining blending and foaming can be identified as a promising approach to fulfill these increasing demands. In addition to the property profile of the compact material (Paul and Bucknall, 2000; Utracki, 2002), blending of polymers can adjust the important properties for foaming as the rheological poperties, diffusion, solubility or cell nucleation (Gendron, 2005; Harrats, 2006; Paul and Bucknall, 2000;).

However, combining blending and foaming not only offers great chances but also poses challenges. Regarding immiscible polymer blends, the difference in processing window as the difference in viscosity and glass transition temperature can inhibit foam expansion (Ruckdäschel et al., 2008). In addition, immiscible blend systems have to be compatibilized as result of their low interfacial adhesion (Datta and Lohse, 1996; Harrats, 2006; Robenson, 2007). Therefore, block copolymers are commonly used as compatibilization agents due to a selective miscibility of the blocks with either blend component (Datta and Lohse, 1996; Auschra and Stadler, 1993; Brown et al., 1996).

Foaming a PPE/SAN blend system compatibilized with a SBM triblock terpolymer using carbon dioxide as blowing agent, Ruckdäschel observed a finer foam structure with cell sizes in the nm range (Ruckdäschel, 2009). This phenomenon was explained by the finer blend morphology as result of the compatibilization leading to an improved cell nucleation. But the high viscous PPE-phase restricted the foam expansion as it shows a significant higher complex viscosity and glass transition temperature compared to the SAN-phase.

In this study the influence of two different blowing agents (ethanol and n-pentane) on the foaming behavior of a PPE/SAN blend system compatibilized by SBM is investigated. First the blend properties are characterized via shear rheological measurements and transmission electron micrographs. The blowing agent diffusivity and solubility as function of the SBM-content is analyzed by absorption experiments at a temperature near the boiling point of the blowing agent. To provide cellular materials, the saturated blend systems are batch-foamed at three different temperatures for 30 sec. To evaluate the foaming behavior, the foam density is measured using the buoyancy method while the foam structure is evaluated by scanning electron micrographs. Finally the blend properties are correlated with the results of the absorption experiments and the observed foaming behavior.

Experimental

Materials and blend systems
An immiscible blend of 60 wt% poly(2,6-dimethyl-1,4-phenylene ether) (PPE PX 100F, Mitsubishi Engineering Plastics Europe) (T_g = 216 °C, density = 1,065 g/cm^3) and 40 wt% poly(styrene-co-acrylonitrile) (SAN VLL 19, BASF SE) (T_g = 114 °C, density = 1,07 g/cm^3) is used in this work. This blend system is compatibilized by 0, 5, 10 and 20 wt% of a symmetric polystyrene – *block* – polybutylene – *block* – poly(methyl methacrylate) (SBM) triblock terpolymer (wt%(PS) = 32; wt%(PB) = 35; wt%(PMMA) = 33; M_n(SBM) = 89 kg/mol) synthesized at the department of macromolecular chemistry II, University of Bayreuth. To each blend system 0.1 wt% of a stabilizer (Irganox 1010, Ciba) is added. Before melt-compounding at a maximum temperature of 260 °C using a Brabender DSE 20/40 co-rotating twin screw extruder with a length–to–diameter ratio of 29.5, the materials are dry blended.
As blowing agents an alcohol (ethanol, boiling point: 78 °C, vapor pressure at 20 °C: 5.8 kPa) and an alkane (n-pentane, boiling point: 36 °C, vapor pressure at 20 °C: 56.2 kPa) are taken.

Sample preparation
The blend systems are pulverized under liquid nitrogen using a cryogenic mill (Pulverisette 14, Fritsch). Afterwards, disc like samples with a diameter of 25 mm and a thickness of 1 mm for shear rheological measurements and quadrate plates with a length of 120 mm and a thickness of 1 mm for the absorption experiments and batch-foaming are compression molded at 260 °C. The quadrate plates are further cut in 15 mm x 15 mm x 1 mm samples

Blend characterization
In order to investigate the shear rheological properties of the blend systems dynamic shear measurements were performed using a Rheometric Scientific RDA III in the plate-plate configuration (plate diameter 25 mm, gap 1 mm). First, strain sweep measurements at different temperatures are performed to determine the linear viscoelastic region. In a second step, dynamic temperature range tests in the viscoelastic region are performed to evaluate the complex viscosity as function of the temperature and the content of the SBM triblock terpolymer.
To investigate the morphology of the non-foamed blend systems, transmission electron micrographs are prepared using a transmission electron microscope from Zeiss type 902 with an acceleration voltage of 80 kV. Ultrathin samples are cut out of the plates for the absorption measurements using a microtome Ultracut E from Leica. These ultrathin samples are stained with osmiumtetroxide (OsO_4) for 30 sec as well as with rutheniumtetroxide (RuO_4) for 30 min. Due to this staining process the PPE-phase appears dark, the SAN-phase bright while the PB-domains of the SBM triblock terpolymer are black.

Absorption measurements
Absorption measurements of the blowing agents ethanol and n-pentane in the blend systems are performed based on the method described by Bernardo et al. (Bernardo and Vesely, 2007, 2008). Gastight glass bottles are filled with the liquid blowing agents, where one sample of the blends is placed. Prior to the absorption experiments the samples are vacuum-dried at 60 °C for at least 48 hours. The bottles including the liquid blowing agent and one sample are heated up to 60 °C in the case of ethanol and to 20 °C (room temperature) in the case of n-pentane. After defined times the mass uptake of the samples are measured using a microbalance (MX 5, Mettler Toledo). The temperatures are chosen as the blowing agents are still in the liquid phase but the temperature is near the boiling point of the organic liquid, leading to a simplified handling at an improved diffusivity. As the glass bottles are gastight and closed at normal pressure, the saturation pressure is set by the vapor pressure of the blowing agent at the saturation temperatures (ethanol at 60 °C: 46.7 kPa; n-pentane at 20 °C: 56.2 kPa).

Batch-foaming
The batch-foaming experiments were performed at 160, 190 and 220 °C at a foaming time of 30 sec. The samples are saturated either with ethanol for 35 days at 60 °C or with n-pentane for 56 days at 20 °C. The saturation temperature is chosen near the boiling point of the blowing agent. For the foaming the saturated samples were foamed in an oil bath and immediately cooled down in a water bath to stabilise the foam morphology.

Foam evaluation
The density of the foamed blend systems is measured using the buoyancy method. Each foamed blend system is measured three times and the average value is calculated.
To evaluate the foam structure of the blend systems scanning electron micrographs of the core region of the foams are prepared, using a scanning electron microscope, Joel IC 848, with an acceleration voltage of 15 kV. For sample preparation the foamed blends are fractured and the fracture surface is coated with a gold layer of around 12 nm.

Results and Discussion

Blend properties

The blend morphology as well as the rheological properties have a significant influence on the foaming behavior of immiscible blend systems, especially on cell nucleation, growth and stabilization (Gendron, 2005; Ruckdäschel, 2009). Analyzing the complex viscosity of the PPE/SAN 60/40 blend system as function of the temperature (Figure 1), first, a slight and steady increase with decreasing temperature can be seen. At a temperature in the range of the glass transition temperature of the PPE the slope of the complex viscosity is significantly increased. Compatibilizing the immiscible blend system with 5 wt% SBM no significant change in complex viscosity can be seen. Further increasing the SBM content, a similar behavior can be observed for elevated temperatures while the increase in complex viscosity slope is shifted to lower temperatures. This phenomenon can be explained by the reduction in glass transition temperature as result of the miscibility of the high viscous PPE-phase with the low viscous PS-block of the SBM triblock terpolymer (Everaert et al., 1999; Göldel et al., 2008; Ruckdäschel, 2009). Therefore, less restriction of the foam expansion due to the high viscous PPE-phase can be expected (Ruckdäschel et al., 2008).

The influence of the SBM triblock terpolymer on the blend morphology is investigated by transmission electron microscopy (Figure 2). The morphology of the non-compatibilized PPE/SAN blend system is shown in Figure 1a. A relative coarse morphology can be observed for the non-compatibilized blend while the addition of SBM leads to a finer morphology by reducing the phase size and increase the interfacial area (Figure 2b – c). As the SBM triblock terpolymer is located in the interfacial area between the PPE- and the SAN phase (Figure 3) more interfacial area is necessary by increasing the SBM content (Ruckdäschel, 2009).

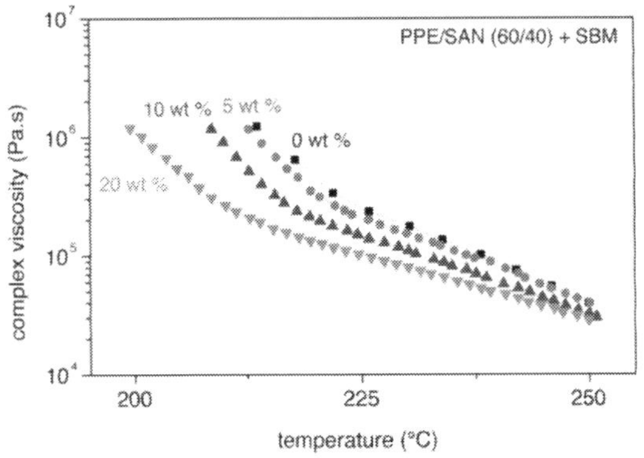

Figure 1 - Complex viscosity as function of the temperature of the blend systems compatibilized with SBM.

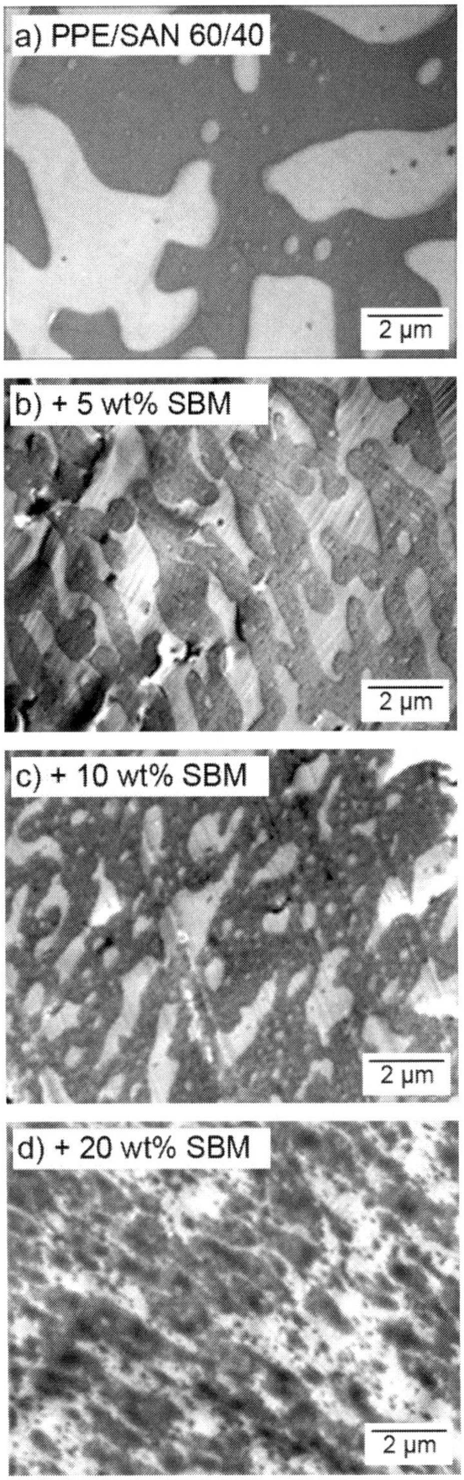

Figure 2 – Morphology of the PPE/SAN 60/40 blend system compatibilized with a) 0 wt%, b) 5 wt%, c) 10 wt% and d) 20 wt% SBM.
(PPE – dark, SAN – bright, PB from SBM – black)

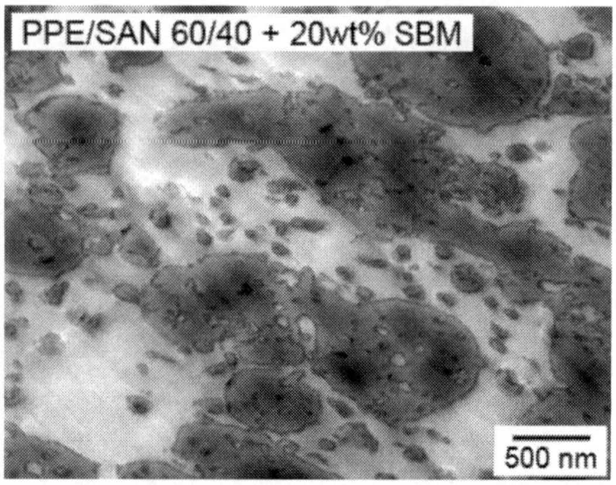

Figure 3 – Transmission electron micrograph of the PPE/SAN 60/40 blend system compatibilized with 20 wt% SBM. (PPE – dark, SAN – bright, PB from SBM – black)

Blowing agent absorption

Regarding immiscible blend systems the blowing agent absorption in the blend phases as well as the blend morphology have a significant influence on the solubility and diffusivity of the system (Paul and Bucknall, 2000). To evaluate the influence of the compatibilization on the blowing agent absorption, first the neat blend phases have to be analyzed. The blowing agent absorption in the neat polymers is listed in Table 1. While PPE and SAN show good ethanol absorption, only less amount of n-pentane is soluble in SAN and PPE stress cracked after 7 days and a n-pentane absorption of 18 wt%.

As result of this absorption behavior in the neat polymers, a nearly linear ethanol absorption could be observed in former studies in the PPE/SAN blend systems with no significant influence of the blend morphology. In contrast, the blend morphology was found as key factor for n-pentane absorption in the immiscible PPE/SAN blends (Gutmann, et al. 2009).

Analyzing the influence of the compatibilization using a SBM triblock terpolymer a similar behavior can be observed. As the blend morphology showed no influence on the ethanol absorption in the non-compatibilized blend systems the finer morphology as result of compatibilization also has no significant influence on ethanol absorption. Similar results could also be found using carbon dioxide as blowing agent (Ruckdäschel, 2009). A dramatic improve of the n-pentane absorption can be observed comparing the non-compatibilized and the compatibilized PPE/SAN blends. This phenomenon can be explained by the finer blend morphology and the PB-phase in the PPE/SAN interphase improving diffusivity.

Table 1 – Ethanol and n-pentane absorption of the neat Polymers.

	PPE	SAN
ethanol absorption (wt%)	9.0	21.5
n-pentane absorption (wt%)	18.0 (7 days)	0.2

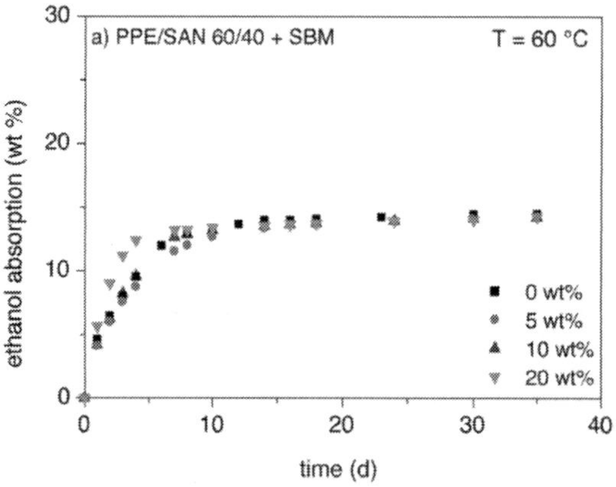

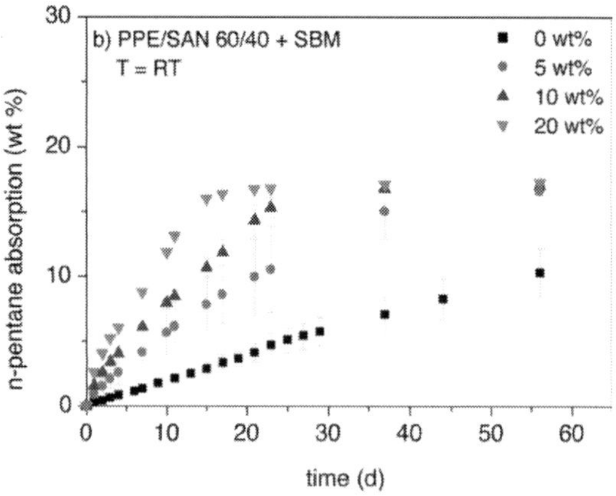

Figure 4 - Absorption of a) ethanol and b) n-pentane in the blend systems at a temperature near the boiling point of the blowing agent.

Ethanol as blowing agent
First, the influence of the SBM compatibilization on the foaming behavior of the immiscible PPE/SAN blend system using ethanol as blowing agent is analyzed. The foam densities of the compatibilized blends as function of the SBM content are highlighted in Figure 5. Comparing the foam densities of the PPE/SAN blend system and the compatibilized blend containing 5 wt% of SBM, similar foam densities can be observed for all investigated foaming temperatures. Elevated SBM contents of 10 and 20 wt% in the immiscible PPE/SAN blend result in an increasing foam density. As no significant differences in ethanol absorption of the non-compatibilized compared to the compatibilized blend systems could be observed the increase in foam density can directly be correlated to the change in rheological properties and the foam stabilization. While the immiscible blend and the compatibilized blend system containing 5 wt% SBM show similar complex viscosity (Figure 1) also a similar foaming behavior can be found. An increase in SBM content leads to a decrease of the complex viscosity and therefore cell coalescence and collapse can occur earlier resulting in an increase in foam density.
These phenomena can also be detected analyzing the foam structure (Figure 6). Interestingly, the finer blend morphology leading to a finer foam structure using carbon dioxide as blowing agent (Ruckdäschel, 2009) seems to have no impact on the foam structure using ethanol as blowing agent. Instead, the smaller cells next to the large cells visible by analyzing the foam morphology of the immiscible blend system are disappearing evaluating the compatibilized blend. Further increasing the SBM content leads to collapsing foam structures. The different foaming behavior can be correlated to the blowing agent properties. As organic liquids show slower diffusivity and therefore additionally longer plasticization, less nucleation occurs using organic liquids as blowing agents and the cells have more time to grow. Therefore, lower foam densities compared to carbon dioxide as blowing agent are achieved.

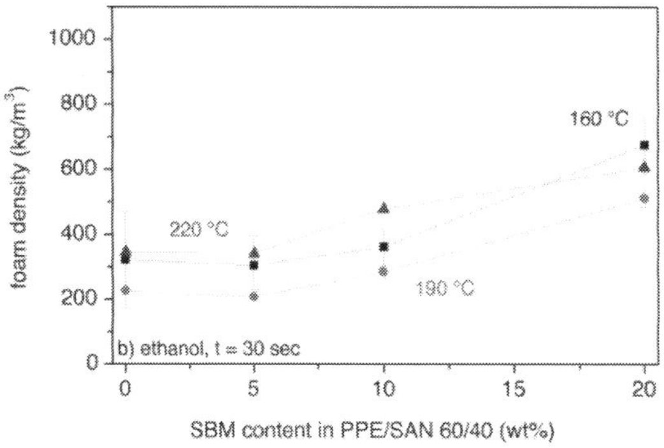

Figure 5 - Foam densities of the immiscible blend system as function of the SBM content using ethanol as blowing agent.

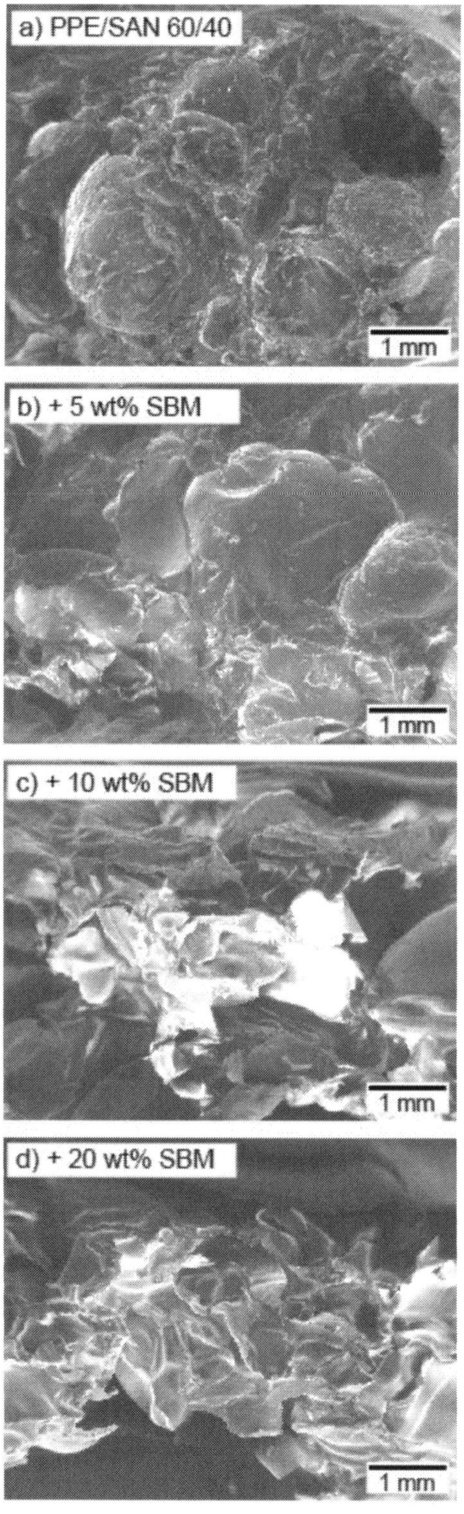

Figure 6 - Scanning electron micrographs of the foam structures of the immiscible and compatibilized blend systems using ethanol as blowing agent at a foaming temperature of 190 °C and a foaming time of 30 sec.

n-Pentane as blowing agent

In the case of n-pentane as blowing agent, the effect of the SBM compatibilization on the foam density of the immiscible blend system scales with the foaming temperature. Less influence of the SBM content on the final foam density can only be seen for 160 °C, while increasing the foaming temperature to 190 and 220 °C the foam density is increasing with increasing SBM content. Similar to the compatibilized and foamed blend systems using ethanol as blowing agent, the increase in foam density can be related to the viscosity decrease leading to a collapsing foam structure.

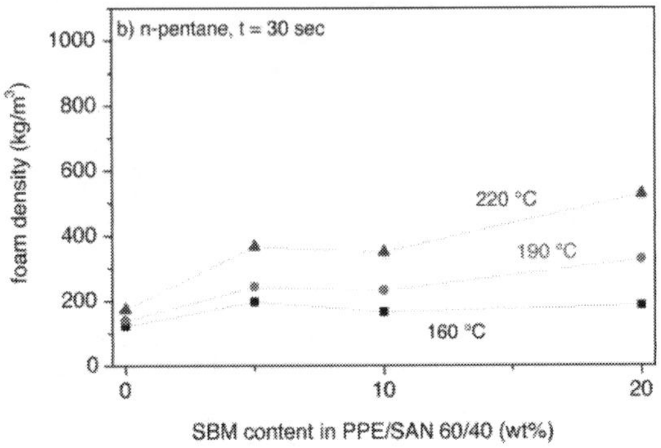

Figure 7 - Foam densities of the immiscible blend systems as function of the SBM content using n-pentane as blowing agent.

The scanning electron micrographs of the blend systems foamed at 190 °C as function of the SBM content are highlighted in Figure 8. Analyzing the noncompatibilized PPE/SAN 60/40 blend sytem a homogeneous foam structure with cell sizes in the range of 1.5 mm can be observed. The addition of 5 wt% SBM leads to a reduction in cell size below 1 mm and a slight increase in foam density. By adding the SBM triblock terpolymer a finer blend morphology could be achieved (Figure 2), improving the blowing agent diffusivity and the blowing agent content in the blend system (Figure 4b). As the blend morphology, the diffusivity and the blowing agent content are key factors for cell nucleation (Gendron, 2005), improving these factors result in a finer foam structure (Figure 8b). The finer foam structure can be detected for all three investigated SBM contents. As result of the decrease in complex viscosity by increasing the SBM content, stabilization of the foam structure is reduced and further cell coalescence and collapse occurs.

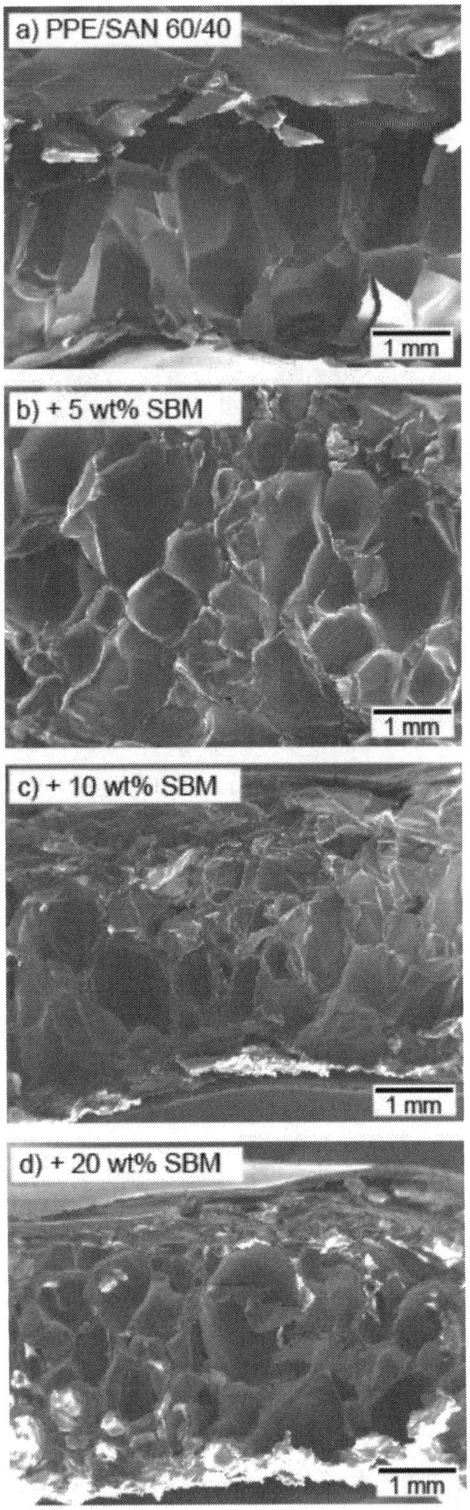

Figure 8 - Scanning electron micrographs of the foam structures of the immiscible and compatibilized blend systems using n-pentane as blowing agent at a foaming temperature of 190 °C and a foaming time of 30 sec.

Conclusions

The detailed investigations described have shown the influence of blend compatibilization on the properties relevant for foaming and the foaming behavior using organic liquids as blowing agents.
The addition of SBM to the immiscible PPE/SAN 60/40 blend system lead to a finer blend morphology, beneficial for cell nucleation and a reduction in complex viscosity.
Regarding ethanol as blowing agent no significant influence of the SBM content and the finer blend morphology on the transport properties could be observed. In contrast, the finer blend morphology could be detected as key factor for the transport properties of n-pentane in the blend system. This could be related to the blowing agent absorption in the neat polymers. While ethanol shows good solubility in both polymers, PPE and SAN, only low n-pentane absorption in SAN could be observed. Therefore, SAN act as a barrier material for n-pentane.
In contrast to the use of carbon dioxide as blowing agent, the finer blend morphology seems to have no influence on the foaming behavior and the foam structure using ethanol as blowing agent. However, the finer blend morphology improved the transport properties of n-pentane in the blend system and a finer foam structure could be detected.
In both cases, the use of ethanol as well as n-pentane as blowing agent, the decrease in complex viscosity lead to an increase in foam density as result of cell coalescence and collapse.

Acknowledgements

This work has been supported by the German Research Foundation (DFG) within SFB 481 (project A10). The polymers were kindly provided by BASF SE, Ludwigshafen and MEP Europe, Düsseldorf. We further acknowledge Susanne Edinger (Department of Macromolecular Chemistry II) for the synthesis of the SBM and Anne Lang and Carmen Kunert (Department of Polymer Engineering) for the experimental support.

References

Auschra, C., Stadler, R., "Polymer Alloys Based on Poly(2,6–dimethyl-1,4-phenylene ether) and Poly(styrene-co-acrylonitrile) Using Poly(styrene-b-(ethylene-co-butylene)-b-methyl methacrylate) Triblock Copolymers as Compatibilizer" Macromolecules, **26**, 6364-6377 (1993)

Bernardo, G., Vesely, D., "Equilibrium Solubility of alcohols in polystyrene attained by controlled diffusion" Eur. Polym. J., **43**, 938-948 (2007)

Bernardo, G., Vesely, D., "Solubility of Alkanes in a Polystyrene Matrix" J. Appl. Polym. Sci., **110**, 2393-2398 (2008)

Brown, H.R., Krappe, U., Stadler, R., "Effect of ABC Triblock Copolymers with an Elastomeric Midblock on the Adhesion between Immiscible Polymers", Macromolecules, **29**, 6582-6588 (1996)

Datta, S., Lohse, D.J.: Polymeric Compatibilizers: Use and Benefits in Polymer Blends. Hanser Publishers, Munich (1996)

Everaert, V., Aerts, G., Groeninckx, G., "Phase morphology development in immiscible PP/(PS/PPE) blends influence of the melt-viscosity ratio and blend composition" Polymer, **40**, 6627-6644 (1999)

Gendron, R.: Thermoplastic foam processing. Principles and Development, CRC Press, Boca Raton (2005)

Göldel, A., Ruckdäschel, H., Müller, A.H.E., Pötschke, P., Altstädt, V. "Controlling the phase morphology of immiscible poly(2,6-dimethyl-1,4-phenylene ether) / poly(styrene-co-acrylonitrile) blends via addition of polystyrene" e-Polymers, **no. 151** (2008)

Gutmann, P., Hildebrandt, K., Altstädt, V., Müller, A.H.E., "Organic liquids as blowing agents to foam an immiscible blend system" PPS-25, Goa, India (2009)

Harrats, C., Thomas, S., Groeninckx, G.: Micro- and Nanostructured Multiphase Polymer Blend Systems. Phase Morphology and Interfaces, CRC Press, Boca Raton (2006)

Klempner, D., Frisch, K.C.: Polymeric Foams, Hanser Publishers, Munich (1991)

Paul, D.R., Bucknall, C.B.: Polymer Blends, John Wiley & Sons, New York (2000)

Robeson, L.M.: Polymer Blends. A Comprehensive Review, Hanser Publishers, Munich (2007)

Ruckdäschel, H.: Micro- and Nanostructured Polymer Blends – Processing, Properties and Foaming Behavior. ISBN: 3-930400-34-0 (2009)

Ruckdäschel, H., Rausch, J., Sandler, J.K.W., Altstädt, V., Schmalz, H., Müller, A.H.E., "Correlation of the melt rheological properties with the foaming behavior of immiscible blends of Poly(2,6-dimethyl-1,4-phenylene ether) and Poly(styrene-co-acrylonitrile)", Polym. Eng. Sci., **48**, 2111-2125 (2008)

Utracki, L.A.: Polymer Blends Handbook, Kluwer Academic, New York (2002)

Foaming Behavior of PP/EPDM Blends with Carbon Dioxide

Han-Xiong Huang, Gui-Long Zhong*

South China University of Technology - Guangdong - CHINA

Foaming of PP/EPDM blends was carried out in a batch process at rubbery state (152°C) and molten state (172°C) of PP, respectively. Results showed that PP/EPDM blends exhibit much dramatically improved foamability than pure PP at both rubbery and molten states of PP. Cells spread as "a string of beads" and orient in the direction of the long diameter of the spherical EPDM droplet in foamed blends prepared at rubbery state of PP. Cell size and density increase as droplet size and density for foamed blends prepared at molten state of PP.

CHARACTERIZATION OF FOAM CELL DENSITY WITH PRINCIPLE OF CRITICAL BUBBLE LATTICE

Piyapong Buahom and Surat Areerat*

School of Chemical Engineering, Faculty of Engineering, King Mongkut's Institute of Technology Ladkrabang, Bangkok, 10520, Thailand – kasurat@kmitl.ac.th

In this study, models for estimating the cell density of isotropic polymeric foams using the surface cell density were developed. The basic morphological unit cell for these models is a gas-filled pentagonal dodecahedral cell cavity. The critical bubble lattice model was introduced to associate the packing structure of the pentagonal dodecahedral cells with a face-centered cubic (FCC) packing structure, and the cell density was estimated. This model was then used to investigate the effect of non-perfect cell cross-sections as found in the micrographs of cell densities of foams. Two cases were examined: first, the ideal case of a perfect cross-section (FCC-PF), and second, the actual case of a non-perfect cross-section (FCC-NPF). The plot of cell density versus surface cell density from recently published data show the effect of a non-perfect cross-section on the estimation of cell density within an FCC lattice with perfect and non-perfect cell cross-sections. The percentage average absolute deviation (AAD) showed that the non-perfect cell cross-section model reliably predicts the cell density of isotropic low density foam with an AAD of 39.2%. The perfect cell cross-section model reliably predicts the cell density of isotropic high density foam with an AAD of 57.2%. These new models are more reliable than the conventional model, which give AADs 59.4% for low density foams and 62.4% for high density foams. Furthermore, the isotropic foam index was introduced as a parameter to indicate the isotropic foam behavior.

Introduction

Polymeric foamed materials have been the subject of many studies and developments. The cell density is an important parameter which refers to the number of cells in a unit volume and is related to the mechanical properties of foams in many different applications [1]. In addition, the cell density is also used to approximate the nucleus density in foaming processes [2]. Foaming processes produce gas bubbles dispersed in the polymer resins. In isotropic systems, such as injection molded foams, their properties are uniform in all directions. At present, the cell density of isotropic polymeric foams are estimated using a simple averaging equation, where the surface cell density (N_S) represents the number of cells (n) in the area of interest (A) from a micrograph, as shown in Equation (1).

$$N_S = \frac{n}{A} \tag{1}$$

The characterization of surface cell density was reported in a study by R. Gosselin and D. Rodrigue [1]. The cell density was estimated by assuming that the foams were isotropic with a uniform distribution of spherical bubbles in all directions. In this case, a foam's cell density (N_F) can be estimated by using the surface cell density [2].

$$N_F = (N_S)^{\frac{3}{2}} \tag{2}$$

This equation was developed by a simple average method and is widely used as a conventional model for both closed-cell and open-cell foams.

In addition, a material balance-based model which includes the expansion ratio and average cell size can be used to characterize cell density. The average cell size (d) is measured directly from a micrograph, and the expansion ratio (ϕ) is the ratio of the densities of the polymer resin to the foam. In this case, the foam's cell density is estimated using Equation (3) [1].

$$N_{F,HD} = \frac{6}{\pi(d)^3}\left(1 - \frac{1}{\phi}\right) \tag{3}$$

To date, the true cell densities of foams are still not well-estimated, with the material balance-based method being the most reliable. Both the material balance-based and conventional models give similar results for high density, isotropic polymeric foams [1]; however, the different models do not give similar results for low-density isotropic foams.

This study aims to develop physical and mathematical models to estimate the cell density of both isotropic high density and isotropic low density closed-cell foams. The critical bubble lattice model was developed to associate the similar cell structure of foams and atomic packing structures while considering perfect and non-perfect cell cross-section images to estimate a foam's cell density. This new model and its useful estimation range were compared with experiments from previous studies [6-11]. In addition, the appropriate application and reliability of the newly-developed model and conventional models are discussed.

Methodology of Models Integration

In this study, three previous physical models were examined: the atomic packing model [3], the 3D foam structure model [4], and the cell cross-sectional windows model [5]. These models were integrated to develop a mathematical model to estimate the cell density of polymeric foam, the methodology of which is shown in Figure 1.

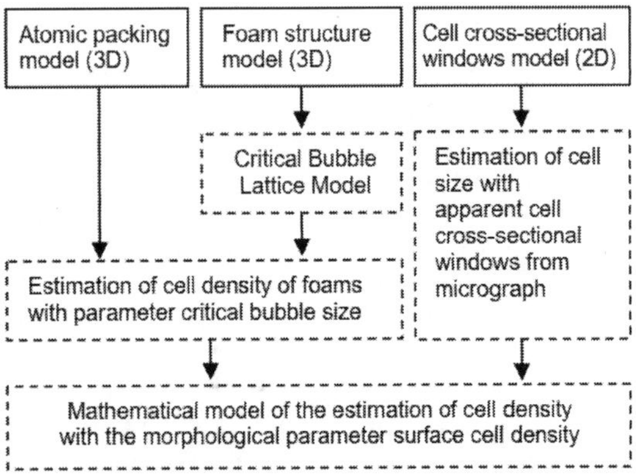

Figure 1 – Models integration methodology for the development of polymeric foam cell density estimation model (solid-bordered blocks refer to existing models and dashed-bordered blocks refer to newly developed components).

The 3D structures of foams are currently modeled for a variety of reasons which include the following: to describe cell morphology such as bubble shapes and average sizes, to study transport phenomena in foaming processes, and to analyze the mechanical properties of foamed materials with a finite element method. In this study, a comparison between an atomic packing structure and a 3D foam structure using cell structure and cell shape was performed. The critical bubble lattice concept was introduced to associate the pentagonal dodecahedral cells of a foam structure with the face-centered cubic (FCC) atomic packing structure. Their similarity was used to extend the material balance-based cell density model to various cell structures.

Principles of the Critical Bubble Lattice Model

Concept of the Critical Bubble Lattice

The critical bubble lattice model was developed to describe a uniform distribution of gas bubbles in a periodic structure, and the similarity between the critical bubble lattice and an atomic packed structure was studied. A critical bubble is assumed to be the largest spherical gas bubble that could fully fill a cell cavity (Figure 3B). The geometric properties of atomic packing were applied to estimate the properties of the critical bubble lattice; for instance, the void fraction of the critical bubbles equals the atomic packing factor (APF), and the cell density of both the critical bubble lattice and an atomic packed structure are equal. The similarities between the atomic packing model and the critical bubble lattice model are shown in Table 1.

Table 1 – Similarities between an atomic packed structure and the critical bubble lattice.

Property	Atomic packed structure	Critical bubble lattice
Packing	Spherical solid atoms in void space	Spherical gas bubbles in polymeric resin
Volume fraction	APF, volume fraction of solid spheres in unit cell and total volume of unit cell	Void Fraction, fraction of spherical bubble in unit cell and total volume of unit cell
Cell density	Number of solid spheres per unit volume	Number of critical bubble spheres per unit volume

In brief, the critical bubble lattice explains the periodic packing structure of the critical bubbles, similar to an atomic packed structure. This structure may be simple cubic (SC), body-centered cubic (BCC) or face-centered cubic (FCC) packing [3]. The cell cavities of low density isotropic polymeric foams are typically pentagonal dodecahedral in shape; thus, the typical structures of the critical bubble lattice are packed similarly to the FCC atomic packing structure, as shown in Figure 2.

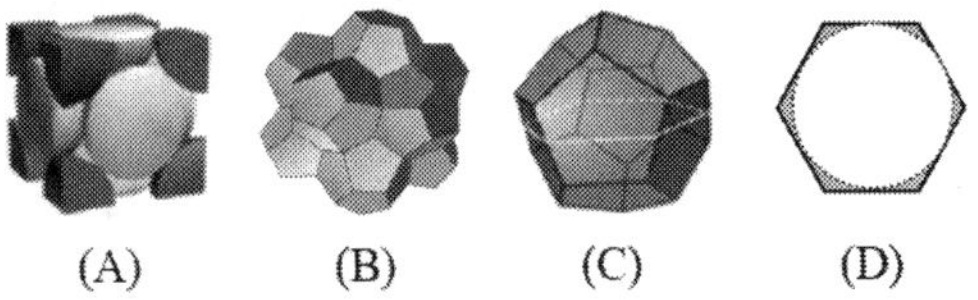

Figure 2 – Schematic of (A) FCC atomic packing structure, (B) pentagonal dodecahedron foam structure, (C) pentagonal dodecahedral cell and its perfect cross-section and (D) perfect cross-sectional window and its critical bubble.

Identical Cell Density at the Same Cell Size

The concept of the critical bubble lattice introduces a useful principle for characterizing the cell density independent of bubble shape or size: the critical bubbles signify only the cell size, as shown in Figure 3. Furthermore, the cell density is identical at the same cell size, independent of the bubbles inside. The cell density of any uniform or non-uniform gas bubbles inside the same cell structure are similar to the cell density the of critical bubble lattice ($N_{F,CR}$), as expressed in Equation (4).

$$N_F = N_{F,CR} \tag{4}$$

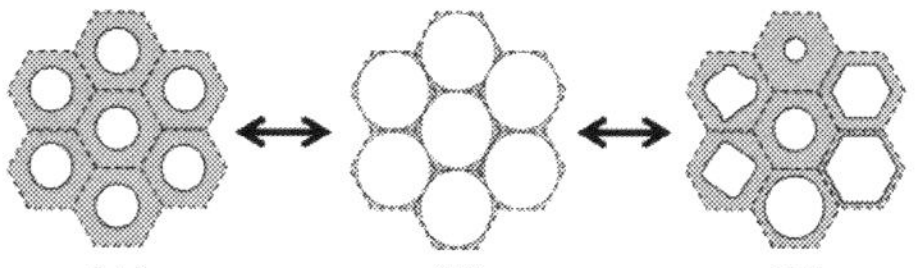

Figure 3 – Identical hexagonal cells and the gas bubbles contained therein: (A) uniform spherical bubbles, (B) critical bubbles, (C) non-uniform gas bubbles.

The Critical Bubble as a Criterion for the Classification between High Density and Low Density Foams

Consider a critical bubble as the theoretical limit of spherical expansion, as shown in Figure 4. In the bubble growth step of the foaming process, the bubble nuclei grow gradually in all directions as spheres before transforming into polyhedrons. The critical bubble is a useful parameter to distinguish between high density and low density foams as spherical and polyhedral bubble foams, respectively, thus explicitly introducing the definitions of high density and low density foams. High density foams have void fractions less than that of the critical bubble, and low density foams have void fractions greater than that of the critical bubble.

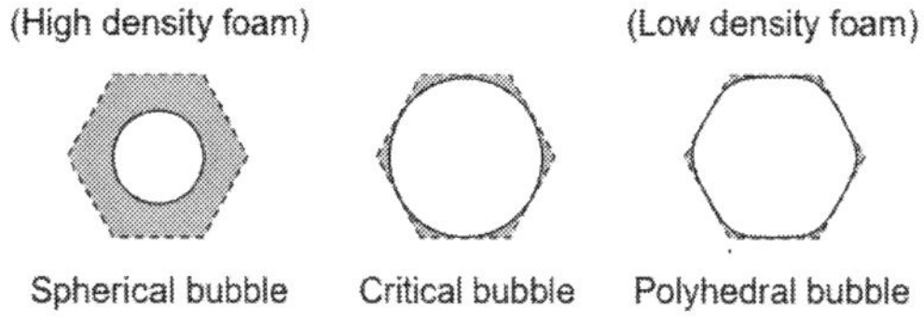

Figure 4 – Various bubble shapes at the same cell size.

The theoretical limit of spherical expansion can be investigated by expansion parameters such as void fraction (ε) or expansion ratio (ϕ) of the critical bubble lattice, as shown in Table 2.

Table 2 – Various shapes of bubbles in the same cells and their parameters.

Bubble shape	Cell size	Equivalent gas-filled spherical bubble	Measured bubble size	Void fraction	Expansion ratio
Spherical Bubble	D_C	$d^*<d$	$d < D_C$	$\varepsilon < \varepsilon_{CR}$	$\phi < \phi_{CR}$
Critical Bubble	D_C	$d^*=d$	$d = D_C$	$\varepsilon = \varepsilon_{CR}$	$\phi = \phi_{CR}$
Polyhedral Bubble	D_C	$d^*>d$	$d = D_C$	$\varepsilon > \varepsilon_{CR}$	$\phi > \phi_{CR}$

In practice, void fractions were taken from the APFs of atomic packed structures, while expansion ratios were calculated with these void fractions. The void fraction represents the ratio of gas bubble volume (V_G) and bulk volume of expanded foams (V_T), and the expansion ratio is the ratio of the bulk volume of the expanded foam and the volume of the polymer resin (V_P), as expressed in Equations (5) and (6), respectively.

$$\varepsilon = \frac{V_G}{V_T} \quad (5)$$

$$\phi = \frac{V_T}{V_P} \quad (6)$$

In addition, the total volume of foam equals the volume of the polymer plus the volume of gas bubbles, as in Equation (7).

$$V_T = V_P + V_G \quad (7)$$

Consequently, combining Equations (5) to (7) gives the expansion ratio as a function of the void fraction, as in Equation (8).

$$\phi = \frac{1}{1-\varepsilon} \quad (8)$$

The theoretical limit of expansion in terms critical void fraction (ε_{CR}) or critical expansion ratio (ϕ_{CR}) of various cell structures are shown in Table 3.

Table 3 – Various cell structures and their theoretical limit of spherical expansion parameters [3].

Lattice	Shape of Unit Cell	ε_{CR} (-)	ϕ_{CR} (-)
SC	Cubic	0.5236	2.0991
BCC	Tetrakaidecahedron	0.6802	3.1270
FCC	Pentagonal dodecahedron	0.7404	3.8521

Material Balance Based Cell Density using the Principle of the Critical Bubble Lattice

Fundamentals of Material Balance-Based Cell Density

The cell density refers to the number of cells (n_{cell}) in a unit volume of foam, expressed as follows:

$$N_F = \frac{n_{cell}}{V_T} \quad (9)$$

Because the number of cells equals the total volume of gas bubbles divided by the volume of a unit bubble (V_{bb}):

$$n_{cell} = \frac{V_G}{V_{bb}} \quad (10)$$

Equation (9) can be written as Equation (11).

$$N_F = \frac{1}{V_T}\left(\frac{V_G}{V_{bb}}\right) \quad (11)$$

Substituting V_G from Equation (7) into (11) gives Equation (12).

$$N_F = \frac{1}{V_{bb}}\left(\frac{V_T - V_P}{V_T}\right) \quad (12)$$

Combining Equations (6) and (12) leads to the general form of the material balance-based cell density model, as shown in Equation (13).

$$N_F = \frac{1}{V_{bb}}\left(1-\frac{1}{\phi}\right) \quad (13)$$

The material balance-based cell density model can also be written as a function of the void fraction, as shown in Equation (14).

$$N_F = \frac{\varepsilon}{V_{bb}} \quad (14)$$

Material Balance-Based Cell Density Model with Critical Bubble Lattice for Generic Foams

In order to determine the material balance-based cell density of various foams, both spherical and polyhedral bubbles must be examined. The combination of the principle of the critical bubble lattice and the material balance-based cell density in Equations (4) and (13) is expressed in Equation (15).

$$N_F = \frac{6}{\pi (D_C)^3}\left(1 - \frac{1}{\phi_{CR}}\right) \quad (15)$$

In Equation (15), the value of the critical expansion ratio was obtained from Table 3, depending upon the packing structure. Therefore, the general form of the cell density of atomic packed structures with a new parameter, the periodic factor (α), is given in Equation (16).

$$N_F = \frac{\alpha}{(D_C)^3} \quad (16)$$

Various atomic packed structures and their periodic factors are summarized in Table 4.

Table 4 – Various atomic packed structures and their geometric parameters.

Packed Structure	Shape of unit cell	Periodic factor, α (-)
SC	Cubic	1
BCC	Tetrakaidecahedron	1.2990
FCC	Pentagonal dodecahedron	1.4142

It is important to note that it is not possible to measure the cell size (D_C) directly from the micrographs. For this reason, the relationships between cell size and measurable parameters such as surface cell density or average cell cross-sectional window area were studied.

Investigation of the Relationship between Cell Size and Average Cross-Sectional Cell Window Area

Generally, the morphologies of polymeric foams observed in micrographs displayed non-perfect cross-sections with a non-uniform cell size and random cell structures, as shown in Figure 5.

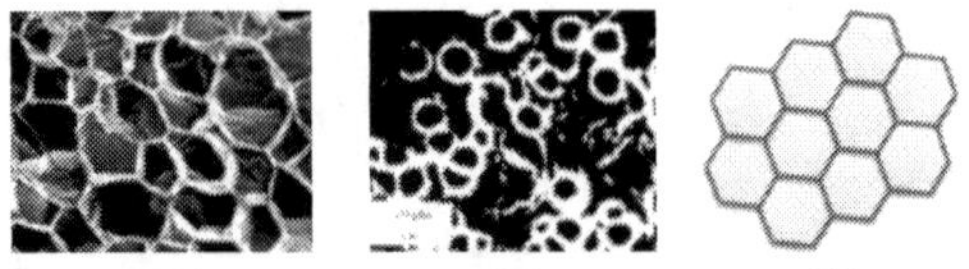

(A) (B) (C)

Figure 5 – Cell cross-sections of (A) low density foams [6] and (B) high density foams from micrographs [11]; (C) perfect hexagonal cell windows.

The relationship between cell size and cell window area were studied in two cases: perfect and non-perfect cell cross-sections. For the perfect case, cell windows are hexagonal and fully cover their critical bubbles (Figure 6A). In reality, non-perfect cell windows are observed, which are usually pentagonal, hexagonal or square and are variously distorted (Figure 6B).

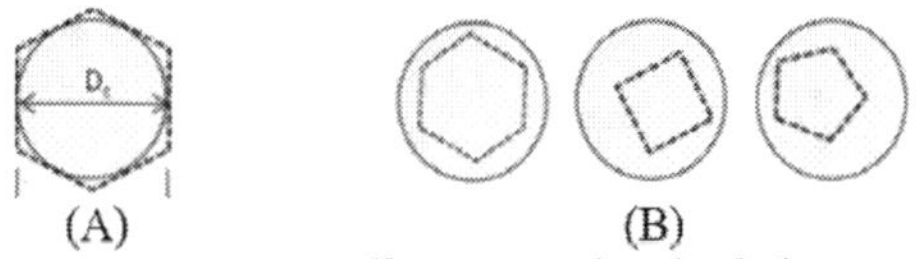

(A) (B)

Figure 6 - Apparent cell cross-sectional windows and their critical bubbles for (A) a perfect cell cross-section; (B) non-perfect cell cross-sections.

The average cell cross-sectional window area ($\bar{A}$), which refers to the overall area of interest (A) divided by the number of cells (n), is expressed in Equation (17).

$$\bar{A} = \frac{A}{n} \quad (17)$$

Perfect Cell Cross-Section

To evaluate the relationship between cell size (D_C) and average cross-sectional cell window area ($\bar{A}$), six combined triangles of a hexagonal cell were considered, as shown in Figure 7.

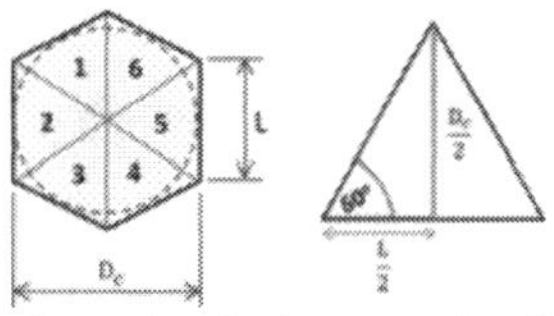

Figure 7 – Schematic of a hexagonal cell window and its six combined equilateral triangles.

The relationship between D_C and L was obtained by trigonometry from the equilateral triangle in Figure 7.

$$\tan 60^\circ = \frac{D_C/2}{L/2} \quad (18)$$

$$L = \frac{D_C}{\sqrt{3}} \quad (19)$$

The area of the hexagonal window equal the sum of the area of six combined equilateral triangles, as expressed below.

$$\bar{A} = 6\left(\frac{1}{2}\right)\left(\frac{D_C}{2}\right)\left(\frac{D_C}{\sqrt{3}}\right) \quad (20)$$

Finally, the mathematical relationship between the cell size and the average cross-sectional cell window area for the perfect cross-section model is expressed as follows:

$$D_C = 1.0746\left(\bar{A}\right)^{\frac{1}{2}} \quad (21)$$

Non-Perfect Cell Cross-Section

In a previous study, K. M. Lewis *et al.* [4] presented a reliable mathematical relationship between the observed cell cross-sectional window area and the average cell size, which has allowed the realization of useful correlations for low density closed cell foams. The cell windows were usually pentagonal, hexagonal or square and were variously distorted, and all the cells were assumed to be dodecahedral. The average cell diameter was calculated by the following formula:

$$D_C = 2.1354\left(\bar{A}\right)^{\frac{1}{2}} \quad (22)$$

The General Form of Mathematical Relationship between Cell Size and Average Cell Window Area

To simplify the analysis of the relationship between cell size and cell window area of perfect or non-perfect cell cross-sectional windows as measured from a micrograph, the mathematical relationship between cell size and average cell window area can be expressed with the apparent factor (κ), as in following equation:

$$D_C = \kappa \left(\overline{A}\right)^{\frac{1}{2}} \tag{23}$$

Model Formulation

The General Form of the Model

The estimation of cell density using the material balance-based model with the cell size in Equation (16) and the estimation of cell size with average cell window area in Equation (23) were combined to investigate the general form of the mathematical relationship between cell density and surface cell density. A new correcting factor, the isotropic factor (φ), was used, as expressed in Equation (24).

$$N_F = \varphi (N_S)^{\frac{3}{2}} \tag{24}$$

Where,

$$\varphi = \frac{\alpha}{\kappa^3} \tag{25}$$

Cell Density of Low Density Foams

The model developed in Equation (24) was used to investigate the effect of non-perfect cell cross-sectional windows on the cell density of foams in two cases: first, based on the FCC lattice with perfect cross-sectional windows (FCC-PF), and second, based on the FCC lattice with non-perfect cross-sectional windows (FCC-NPF). Each model and its parameters, as summarized and calculated from Equations (16), (21), (22) and (25), are shown in Table 5.

Table 5 – Conventional and newly developed models and their parameters.

Model	Periodic factor (α)	Apparent factor (κ)	Isotropic Factor (φ)
Conventional model	-	-	1
FCC-PF	1.4142	1.0746	1.1398
FCC-NPF	1.4142	2.1354	0.1453

Cell Density of High Density Foams

In the case of high density foams, the rearrangement of the random distribution of the spherical bubbles into a periodic pattern was considered to simplify the analysis. The method of analysis was considered to be different for high density and low density foams. In reality, bubbles grow from the nuclei dispersed in the polymer matrix until contact with neighboring bubbles causes their transformation from spherical bubbles into polyhedral cavities. The rearrangement of random nuclei (Figure 8A) into a periodic pattern was performed before the bubble growth step (Figure 8B-8D), with the assumption that no deflation or collapse of the bubbles occurs. For this assumption, the number of nuclei and the number of cells should be equal; in other words, the cell densities of random dispersed spherical bubbles and periodic structured bubbles inside the same cavities of interest should be equal. The cells are typically pentagonal dodecahedral cavities and the periodic pattern of the nuclei and the bubbles is the FCC lattice. Two-dimensional hexagonal cells can be used to describe cell boundaries, similar to the FCC-PF model.

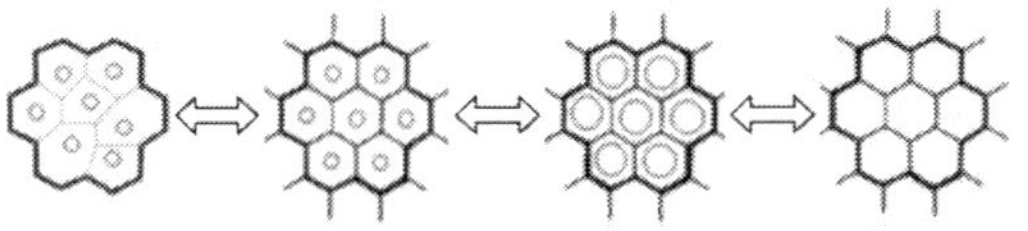

Figure 8 – Distribution of gas bubbles in the same cell cavities of interest, (A) random dispersion of spherical bubbles, (B) rearrangement into periodic lattice cavities of 2D hexagonal cells, (C) spherical bubble growth, and (D) polyhedral bubbles completely filling the cell cavities.

Isotropic Foam Index

The isotropic foam index (η) was defined to indicate a deviation from the isotropic foam behavior, given as the logarithm of the ratio of the apparent isotropic factor (φ) and the standard isotropic factor (φ_o), as shown in Equation (26).

$$\eta = \log\left(\frac{\varphi}{\varphi_o}\right) \tag{26}$$

In the case of isotropic foams, the isotropic foam index is equal to zero. The standard isotropic factor φ_o in Equation (26) is dependent on the foam structure, which affects the appearance of the foam morphology on a micrograph. This factor refers to the isotropic behavior of various foam structures.

Combining Equations (24) and (26) gives the following:

$$N_F = 10^{\eta} \varphi_o (N_S)^{\frac{3}{2}} \tag{27}$$

Equation (27) expresses the relationship between the cell density and the measured surface cell density, the isotropic factor, which depends on the shape of the gas bubbles, and the isotropic foam index, which corrects the effect of non-isotropic foam as determined by the plot of cell density and surface cell density.

The standard cell density (N_F^o) is the cell density of isotropic foam with an isotropic foam index equal to zero, expressed as follows:

$$N_F^o = \varphi_o (N_S)^{\frac{3}{2}} \tag{28}$$

In practice, the isotropic foam index can be evaluated by combining Equations (27) and (28), obtained by taking the logarithm of material balance-based cell density in Equation (24) divided by the standard cell density in Equation (28), as shown in Equation (29).

$$\eta = \log\left(\frac{N_F}{N_F^{\circ}}\right) \tag{29}$$

Discussion

The Critical Bubble Model as the Theoretical Limit of High Density Foam

In recent studies, many physical and mathematical models for either high density of low density foams have been presented; however, no criterion exists for the classification of foam as either high or low density. For this reason, the critical bubble criterion was used to identify the limit of spherical expansion, or the limit of high density foam, which depends upon the periodic pattern of the critical bubble lattice. Since the basic unit cell is typically a pentagonal dodecahedral cavity in an FCC lattice, the theoretical expansion limit of a high density foam yields a void fraction less than 0.74 and an expansion ratio less than 3.85. Many models for high density foams can be used if their expansion parameters are less than those given by the expansion limit.

Reliability and Scope of the Material Balance-Based Cell Density Estimation using the Critical Bubble Lattice

At present, the true cell density of polymeric foams cannot be quantified. The most reliable predictive model is the material balance-based model. High density foams can use directly using Equation (13) as shown in Equation (3). For low density foam, the material balance-based model can be used by modifying it in terms of the critical bubble lattice, as shown in Equation (13). The critical bubble concept, as illustrated in Figure 3, can be applied not only to high density and low density closed-cell foams, but also to open-cell foams by considering them as low density closed-cell foams with a pseudo-wall thickness. Therefore, the material balance-based model with a critical bubble lattice can be applied to all foam morphologies, independent of bubble shape or size by analyzing the cell boundaries.

FCC Lattice as the Maximum Limit of Estimation

The FCC lattice (α=1.415) was considered to be the maximum value for cell density estimation compared with previous experimental data [6-11]. Figure 9 illustrates the actual periodic lattice structures of both conventional and microcellular foams and their agreement with the model of the foams having typically pentagonal dodecahedral cells packed in an FCC lattice. The good fit of experimental data to the model indicates that the FCC lattice can be applied to these foams, although some of the experimental data falls below the line for the FCC model. The periodic factor α refers to the periodic structure of foams. Because the FCC lattice is the densest packed structure of equally-sized spheres, it is the theoretical upper limit for the estimation of cell density of isotropic foams. Thus, $\alpha \leq 1.414$ is the theoretical limit condition for cell density estimation in Equation (16).

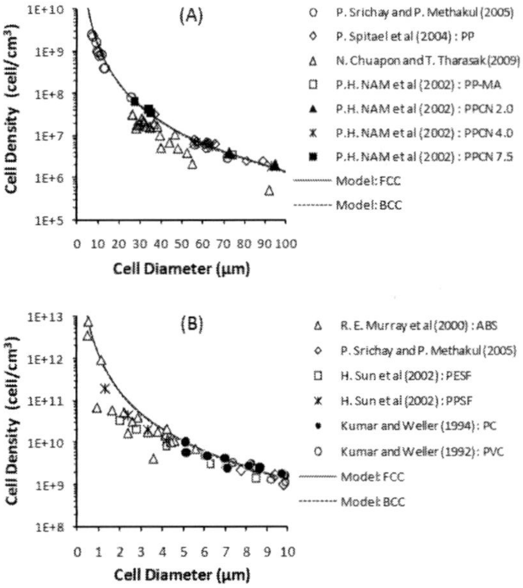

Figure 9 – Relationship between cell density and average cell size of (A) conventional foams (B) microcellular foams.

Range of Estimation of Cell Density for Isotropic Foam

To investigate the cell density of polymeric foams using Equation (24), cell cross-sectional windows must be considered, which may be perfect (κ=1.0746) or non-perfect (κ=2.1354). In the case of perfect cross-sections with pentagonal dodecahedral cells (φ=1.1398) arranged on an FCC lattice, the calculation results in the maximum cell density. Because of uncertainty in cell cross-sectional windows which may either be perfect or non-perfect, the value of apparent factor can fall in the range $1.0746 \leq \kappa \leq 2.1354$ (-). Assuming pentagonal dodecahedral cells on an FCC lattice, calculation result from equation (25) may appear within $0.1453 \leq \varphi \leq 1.1398$ (-), as shown in Figure 10. Using the same cell size, the calculated cell density of an FCC lattice with a perfect cross-section window (upper estimation limit) is about ten times greater than the cell density calculated with a non-perfect cross-section window (lower estimation limit).

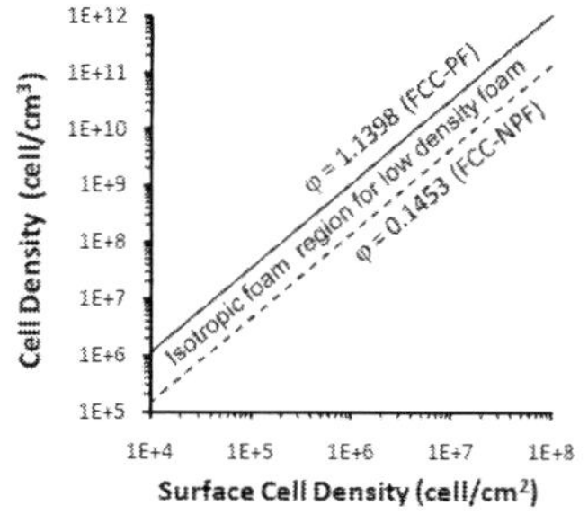

Figure 10 – Surface cell density versus cell density and theoretical isotropic region for low density foams.

Effect of Non-Perfect Cross-Section of Low Density Foams on Cell Density Estimation

The FCC-PF ideal model has the maximum value of the area of hexagonal windows, while the FCC-NPF model has a smaller value of the area of hexagonal windows. The calculated cell densities that ignore this effect are overestimated. Thus, the true cell density should be between the values calculated for the FCC-PF and FCC-NPF models. This hypothesis was investigated by comparing it with the experimental data from previous studies [6-11], which estimated the cell density using the material balance-based model. Most of the experimental data fell in between the two boundaries, as shown in Figure 11. For conventional foams (10-100 μm cell size), some of the experimental data deviated from the assumed range of estimation because of non-isotropic effects caused by extrusion. For microcellular foams (<10 μm cell size), the small cells are highly isotropic and most of the experimental data fell within the range of estimation; only a few experimental data fell outside the range of estimation.

The deviation of cell cross-section window area appeared due to the effect of deviation in measuring the surface cell density, as shown in Figure 6. Some unobservable cross-sectional area (A^*) did not appear. As a result, the true cell density ($N^*_{F,LD}$), as shown in Equation (30), was less than that calculated using the FCC lattice with perfect cross-section windows model.

$$N^*_{F,LD} = 1.1398\left(\frac{n}{A + A^*}\right)^{\frac{3}{2}} \quad (30)$$

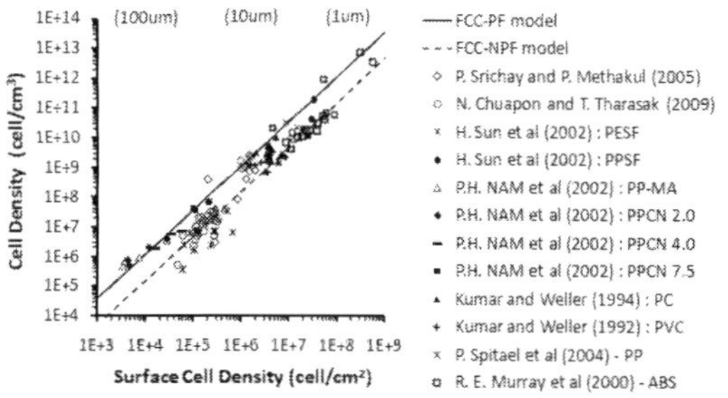

Figure 11 – Relationship between surface cell density and cell density of low density foams.

The percent average absolute deviation (%AAD) was used to measure accuracy of the models, calculated by the following formula:

$$\%AAD = \left|\frac{x_{exp} - x_{model}}{x_{exp}}\right| \times 100 \quad (31)$$

where x_{exp} is the experimental data and x_{model} is the calculation of the model. As a result, the percentage average absolute deviation shows that the FCC-NPF model is reasonable to predict the cell density of low density isotropic foams (AAD=39.2%), which is more reliable than either the conventional model (AAD=59.4%) or the FCC-PF model (AAD=64.4%).

Effect of Non-Observable Cell Boundaries of High Density Foams on Cell Density Estimation

For high density foams, a comparison between the cell density from previous experimental data [11] and the estimation result of both the conventional and developed models are shown in Figure 12. It can be seen that the cell densities measured in all of the previous experiments are higher than those calculated using the FCC-PF model, which is the theoretical maximum. Because of variations in the measured surface cell density due to improper identification of cell window boundaries, it is difficult to identify cell boundaries in high density foam, resulting in a lower apparent cell count than the true value. Some unobservable cells (n^*) did not appear in the area of interest on a micrograph, and therefore the cell count is lower than the true value ($n+n^*$), the surface cell density is estimated too low, and the estimated cell density (N_F) is less than the true cell density ($N^*_{F,HD}$), as in Equation (32).

$$N^*_{F,HD} = 1.1398\left(\frac{n+n^*}{A}\right)^{\frac{3}{2}} \quad (32)$$

Thus, the true cell density should be greater than the calculated result using the FCC-PF model ($N_F<N_F^*$). A comparison of the conventional and the two newly developed models indicate that the FCC-PF model (AAD=57.2%) gives the lowest deviation from the experimental data, while the conventional model (AAD=62.4%) and the FCC-NPF model (AAD=94.5%) deviate more from the experimental data. Therefore, the FCC-PF model appropriately predicts the cell density of an isotropic high density foam.

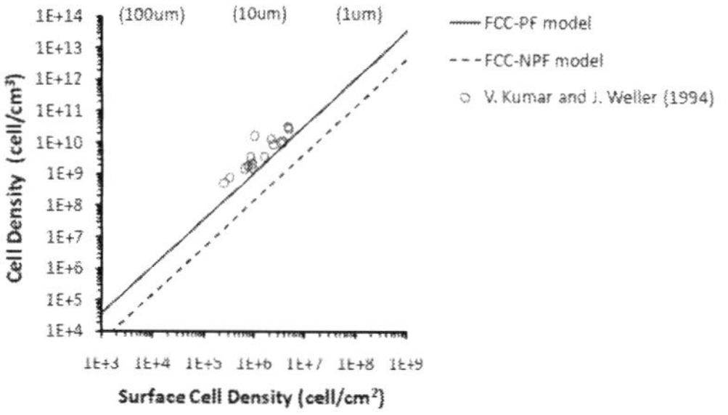

Figure 12 – Relationship between surface cell density and cell density of high density foams.

Standard Isotropic Factor

In practice, the apparent isotropic factor φ can be calculated fitting to the graph of surface cell density versus cell density. In the previous section, the isotropic foam index was introduced and calculated using Equation (29). The values of the isotropic factor for both spherical and polyhedral bubbles were discussed. Isotropic high density foams were found to be modeled well using the FCC-PF model, and the standard isotropic factor ($\varphi_{o,HD}$) is 1.1398. On the other hand, the FCC-NPF model appropriately predicts the cell density of the isotropic low density foams, and the standard isotropic factor ($\varphi_{o,LD}$) is 0.1453.

Testing of Isotropic Foam Behavior – Investigation of the Isotropic Foam Index

In Figure 10, the plot of cell density and surface cell density is a very useful tool to determine whether the foam in question is isotropic or non-isotropic, or how it deviates from isotropic foams. The isotropic foam index (η) has been introduced to describe this behavior. The value of the isotropic index is evaluated using Equation (29), and depends on the standard isotropic cell density (N_F^o) from Equation (28) and the true cell density (N_F), which is the material balance-based cell density. In the previous section, the standard isotropic factor of high density foam was found to be 1.1398, and isotropic microcellular foam was found to deviate from the FCC-PF model. The range of isotropic behavior between the model and the experiment was calculated. The result indicates that isotropic low density foams behave in a manner that is between the FCC-NPF and FCC-PF models, and the isotropic foam index of low density foams should be within the range of 0 and 0.895.

Conclusion

The newly-developed critical bubble model applies the equation used to estimate the cell density of the atomic packing model to the periodic structure of foams. In order to consider the variation of the apparent cell cross-sectional windows from the perfect case, another model was investigated. An estimation of the possible cell density is within the range of those predicted by the FCC-NPF and FCC-PF models ($0.1453<\varphi<1.1398$). The FCC-NPF model reasonably predicts the cell density of isotropic low density microcellular foams (cell size <10μm). The FCC-PF model appropriately predicts the cell density of isotropic high density foams. These newly-developed models are more reliable than the conventional model used for conventional foams, which ignores the variation of non-perfect cross-sectional windows. Standard isotropic factors of high density and low density foams are 1.1398 and 0.1453, respectively. In order to characterize the isotropic foam behavior, the isotropic foam index of isotropic low density foams is within the range of 0 and 0.895.

Reference

[1] Gosselin, R., Jerschow, P., 'Cell morphology analysis of high density polymer foams', Polymer Testing, **24**, 1027-1035 (2005)

[2] Kumar, V., Suh, N.P., 'A process for making microcellular thermoplastic parts', Polymer Engineering and Science, **30**, 1323-1329 (1990)

[3] Callister, W. D.: Materials Science and Engineering: An Introduction, 7th edition, John Wiley & Sons, New York (2006).

[4] Lewis et al., 'An image analysis method for cell-size and cell-size distribution measurement in rigid foams', Journal of Cellular Plastics, **32**, 235–257 (1996)

[5] Kraynik, A.M. et al., 'Foam micromechanics: Structure and rheology of foams, emulsions, and cellular solids', Proc. of the School on Foams, Emulsions and Cellular Materials, 259-286 (1999)

[6] Van der Burg, M.W.D. et al, 'A numerical study of large deformations of low-density elastomeric open-cell foam', Mechanics of Materials, **30**,125-140 (1998)

[7] Sun, H. et al., 'Microcellular foams from poly-ethersulfone and polyphenylsulfone preparation and mechanical properties', European Polymer Journal, **38**, 2373-2381 (2002)

[8] Murray R. E. et al., 'Solid-state microcellular acrylonitrile-butadyne-styrene foams', Cellular Polymers, **19**, 413-426 (2000)

[9] Srijan P., Methakul P.: Microcellular bio-gradable plastic foam from tapioca starch, Special Project, King Mongkut's Institute of Technology Ladkrabang, Bangkok (2005)

[10] Chuapon N., Tharasak T.: Studying of parameters for preparation of microcellular foam, Special Project, King Mongkut's Institute of Technology Ladkrabang, Bangkok (2009)

[11] Kumar, V., Weller, J. E., 'Production of microcellular polycarbonate using carbon dioxide for bubble nucleation,' ASME Journal of Engineering for Industry, **116**, 413-420 (1994)

Study of Injection Molding with Expandable Thermoplastic Microspheres

Jun Peng[1], Jian Wang[2], Yottha Srithep[3], Maf Ahmad[4] Lih-Sheng Turng[3], Xiang-Fang Peng[1*]*
[1]National Engineering Research Center of Novel Equipment for Polymer Processing
South China University of Technology, Guangzhou 510640, China
[2]College of Mechanical and Electrical Engineering, Beijing University of Chemical Technology, Beijing, 100029
[3]Polymer Engineering Center, University of Wisconsin-Madison, Madison, Wisconsin 53706
[4]Expancel Inc. Duluth, GA 30096

Abstract

Injection molding with expandable thermoplastic microspheres (ETM) containing blowing agents is capable of fabricating light weight, dimensionally stable plastic parts while using less material and energy. This paper presents the study of injection molded parts with ETM having better surface quality than conventional foamed parts and tailored mechanical properties. The tensile properties of injection molded samples—low density polyethylene (LDPE) and polylactic acid (PLA)—with variable ETM weight ratios have been investigated. The microcell morphology and cell density in the fracture cross-sectional surfaces were characterized by using scanning electron microscope (SEM) and an image analysis tool (UTHSCSA Image Tool). The surface characteristics were evaluated with a 2D surface roughness analyzer. As shown by the testing results, the cell microstructures—such as cell size, cell density, and a layered structure with a foamed core sandwiched by skin layers—play an important role in the surface quality and mechanical properties. A smaller cell diameter and a thicker skin layer help to improve the surface roughness and tensile properties of the injection molded parts with ETM. In addition, an appropriate microsphere content has a positive effect on cell microstructure and weight reduction, whereas too high a concentration of microspheres adversely affects the tensile properties and surface roughness.

Keywords: expandable thermoplastic microspheres, injection molding, surface roughness, cell microstructure

Introduction

Microcellular injection molding blends supercritical fluid (SCF, usually nitrogen or carbon dioxide) with the polymer melt in the machine barrel to create a single-phase solution for injection molding processes [1-4]. This promising technology enables an injection molded process with lower temperatures and pressures, leading to a reduction of clamp tonnage requirement, cycle time, and energy consumption [5, 6]. However, the benefits of this technology also come with some limitations, notably a rough surface finish with swirls and sliver scratches. For applications with strict part appearance requirements, the surface swirl marks of the molded part limit the adoption of microcellular injection molding unless the surface quality can be improved with other techniques.

In order to eliminate the surface defects of microcellular injection molded components, several approaches previously used for structural foam molding to improve surface quality have been employed to improve the surface quality of the molded parts. These approaches include vapor polishing, hot/cold mold cycling, expanding mold, Teflon film coating, co-injection molding, gas counter-pressure, and so on. The relatively rough surface characteristics of microcellular injection molded parts are mainly due to gas coming out of the polymer/gas solution and being trapped at the part–mold interface. Therefore, it is plausible to improve the surface quality via controlling or containing the emitted gas from the polymer/gas solution during mold filling. In fact, expandable microspheres could be effective in producing microcellular structures to reduce material usage while maintaining the mechanical properties without machine retrofitting. Most importantly, the surface quality of the molded parts can be improved if limited or no gas escapes from the expandable microspheres to cause the surface swirls. In this study, we compounded LDPE and PLA with expandable thermoplastic microspheres to study the microstructures, surface quality, and mechanical properties of the resulting foamed parts.

Experimental Procedure

Melt Processing

A commercial grade low density polyethylene (LDPE) was used in this study, together with commercially available, expandable thermoplastic microspheres, 950 DU 120, from Expancel Inc., as the blowing agent additive. The LDPE homopolymer is a

* Corresponding Author: turng@engr.wisc.edu, pmxfpeng@scut.edu.cn

regular injection molding grade with a melt index of 25 g/10 min. Polylactic acid (PLA, NatureWorksTM 3001D), a lactic acid copolymer based on 1.5% d-lactic acid and 98.5% l-lactic acid, was obtained in pellet form. It has a specific gravity of 1.24 and a melt flow index of around 15 g/10 min (190 °C/2.16 kg). The commercial blowing agent, a hydrocarbon encapsulated in hollow expandable thermoplastic microspheres (ETM), is able to expand up to four times its original diameter when being subjected to heat (160 to 200°C). In this study, 1.5, 5, and 8.5 wt% of ETM for LDPE and 2.0, 5.0, and 8.0 wt% of ETM for PLA were compounded. The injection molded samples are referred to as LDPE–1.5, LDPE–5, LDPE–8.5, PLA–2.0, PLA–5.0 and PLA–8.0, respectively.

The injection molding experiments were carried out to produce standard ASTM-D638 tensile bars on an ARBURG-320S injection molding machine. The typical weight reduction of microcellular injection molded samples in this study was approximately 10 to 20% compared to conventional solid injection molded samples. The nozzle temperatures were 210°C for LDPE and 200°C for PLA, with a coolant (mold) temperature of 15 °C for both. Solid and foamed tensile bars were molded at the processing conditions indicated in Tables 1 and 2. Thirty samples were collected during the course of each molding trial after discarding the first 15 samples. Additional trials were conducted for solid neat LDPE parts.

Table 1. Injection molding parameters of solid LDPE and foamed parts.

	LDPE	LDPE–1.5	LDPE–5.0	LDPE–8.5
Injection Speed (ccm/s)	30	70	70	70
Pack/hold Pressure (bar)	10	0	0	0
Pack/hold time (s)	8	0	0	0
Shot Volume (ccm)	20.0	19	18	17.5
Average Part Weight (g)	7.1	6.55	6.05	5.55

Table 2. Injection molding parameters of solid PLA and foamed parts.

	PLA	PLA –2.0	PLA –5.0	PLA –8.0
Injection Speed (ccm/s)	30	70	70	70
Pack/hold Pressure (bar)	10	0	0	0
Pack/hold time (s)	8	0	0	0
Shot Volume (ccm)	20.9	19.5	18.5	17.8
Average Part Weight (g)	10.6	9.45	8.75	8.2

Testing Techniques

Two-dimensional surface roughness (2D-SR) measurements were taken via a surface roughness analyzer (Surfan Alyzer 4000). This instrument amplifies the deflection of the stylus and calculates the surface roughness parameter, Ra, from the reading. Ra is the arithmetic mean of the departures of the profile from the mean line and is used here to quantify the roughness of the samples. The stylus traversed 2.5 mm in the melt flow direction on the sample surface at a fixed rate of 0.25 mm/s. The testing location is shown in Fig. 1. Fifteen samples were analyzed for each molding trial condition; then the mean and standard deviation values were calculated.

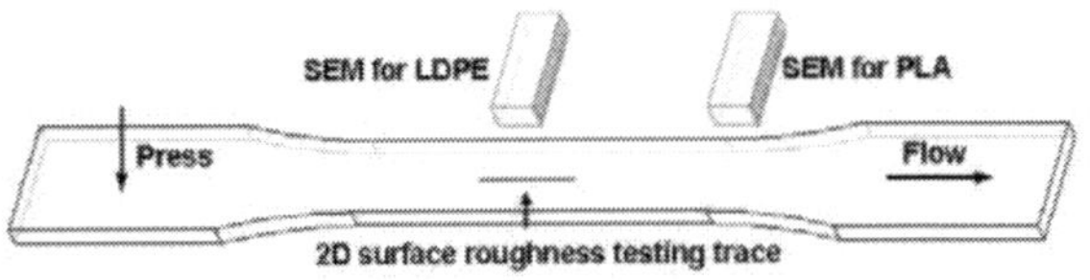

Figure1. Schematic of surface roughness trace and SEM sample location.

The static tensile properties (modulus, strength, toughness, and strain-at-break) were measured at room temperature with a 50 KN load cell on an Instron Model 3369 tensile tester. The cross-head speed was set at 50 mm/min for solid and foamed LDPE samples and 20 mm/min for solid and foamed PLA samples, which are more brittle. All tests were carried out according to the ASTM D638-03. Five samples of each molding trials were tested and the average results were reported.

The morphologies (porous structures) were observed by a scanning electron microscope (SEM; Model LEO 1530, JEOL, Japan), operated at an accelerating voltage of 3 kV. The samples were cryogenic fractured with liquid nitrogen. The average cell sizes, cell density, and skin layer thickness were analyzed using an image analysis tool (UTHSCSA Image Tool). The cell density was calculated using the following formula

$$\text{Cell Density} = \left(\frac{N}{L^2}\right)^{\frac{3}{2}} \quad (3)$$

where N is the number of cells and L is the linear length of the area.

Results and Discussion

Surface Roughness

The two-dimensional (2D) surface roughness was characterized at the middle location of the solid and foamed injection molded tensile bars along the flow direction. The arithmetic surface roughness was averaged across 15 samples with different amounts of ETM and is plotted in Figs. 2 and 3. It is evident that solid samples without ETM (i.e., neat LDPE or PLA resin) exhibit the lowest surface roughness. Surface qualities decrease gradually with the addition of ETM. In comparison with solid parts, the surface roughness of LDPE–8.5 and PLA–8.0 are almost two times rougher. However, they are still smooth and acceptable to the naked eye. Recall that the relatively rougher surface characteristics for microcellular injection molded parts are mainly because of the gas diffusing out from polymer/gas solution and being trapped within the limited surface space in the mold cavity. The more physical or chemical blowing agents that were mixed into the LDPE or PLA matrix, the more gas that is possibly emitted from the ruptured microspheres and trapped, thus leading to a worse surface quality. The maximum weight reduction achieved by this study was 20% for both LDPE and PLA samples, which would result in significant material cost savings for mass-produced parts.

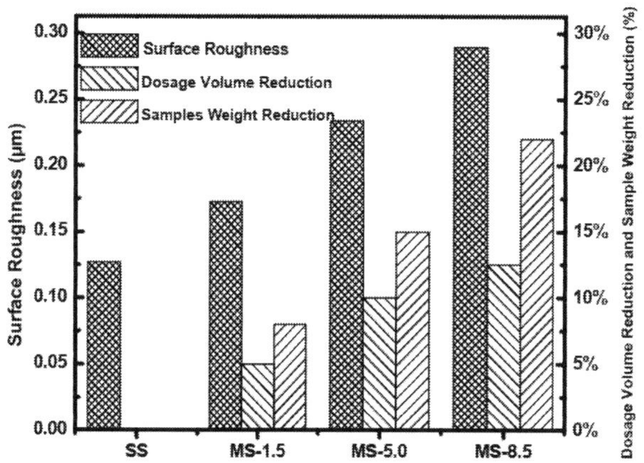

Figure 2. Two-dimensional surface roughness of samples molded with an injection speed of 70 ccm/s, a mold temperature of 15 °C, and four different dosage (shot) volumes. SS denotes solid injection molded tensile bars and MS means LDPE/ETM samples with 1.5, 5.0, and 8.5 weight ratios of ETM.

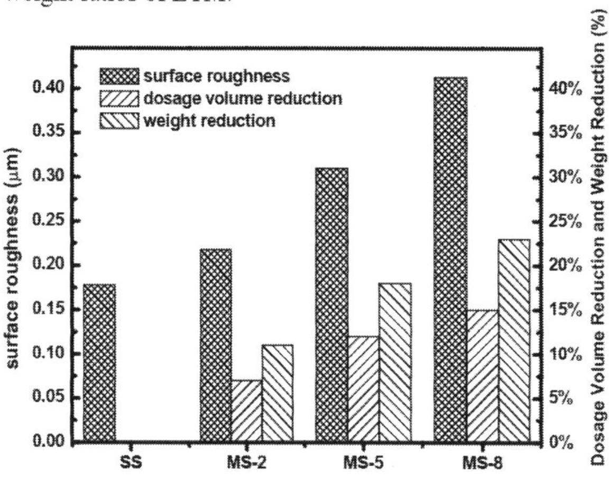

Figure 3. Two-dimensional surface roughness of samples molded with an injection speed of 70 ccm/s, a mold temperature of 15°C, and four different dosage (shot) volumes. SS denotes solid injection molded tensile bars and MS means LDPE/ETM samples with 2.0, 5.0, and 8.0 weight ratios of ETM.

It is known that the part weight reduction depends on the shot volume (dosage volume) reduction, which is equal to the volumetric amount of polymer melt injected into the mold for the tensile bar sample containing runners and sprue. When the changes of density due to the presence of ETM, temperature, and pressure inside the barrel are neglected, the dosage (shot) volume reductions should equal the part weight reduction of the samples. Figures 2 and 3 show that the samples weight reduction (without runners and sprue) has a higher value (greater reduction) than dosage volume at different ETM weight contents. This variation results from the possible expansion of ETM within the injection molding barrel and more significant volume expansion within the cavity as dosage volume decreases. The part weight reduction is also reflected by the different layered structures and cell characteristics in the molded foamed samples, as shown in Figs. 6 and 7. The outer layer (skin layer) has a high density with few voids, while most of the cells concentrate at the center (core layer). Since the core layer has a lower density in comparison with the outer layer, the thickness variation of these two layers affects the part weight.

Tensile Properties

The representative stress–strain curves of solid LDPE and LDPE/ETM, as well as solid PLA and PLA/ETM foamed samples, were featured in Figs. 4 and 5. Tables 3 and 4 tabulate the values of Young's modulus and the ultimate strength of those samples. As can be seen in Fig. 4 and Table 3, the ultimate strength of the solid LDPE samples was higher than the LDPE/ETM foamed tensile bars. The reductions in strain at break and ultimate strength observed with ETM were attributed to bubbles in the center of the samples, which are equivalent to the reduction in the effective cross section area. The degree of reduction increased with increasing ratio of ETM. For PLA and PLA/ETM parts, the ultimate strength rapidly decreases from solid to foamed samples as shown in Fig. 5 and Table 4. Comparisons between Figs. 4 and 5 show that the degree of reduction of PLA is more significant than that of LDPE, probably resulting from different original ultimate strengths of the solid samples.

Table 3. Specific mechanical properties of solid LDPE and LDPE/ETM.

	SS	MS-1.5	MS-5.0	MS-8.5
Young's Modulus (psi)	20218.1	21200.3	19674.6	19402.5
Ultimate Strength (psi)	1276.8	1140.6	912.1	807.1
Strain at Break	1.45	1.15	0.55	0.33

Table 4. Specific mechanical properties of solid PLA and PLA/ETM.

	SS	MS-2.0	MS-5.0	MS-8.0
Young's Modulus (psi)	321234.6	272113.3	238878.8	222643.0

Ultimate Strength (psi)	9958.9	5025.0	3890.0	3621.9
Strain at Break	0.07	0.28	0.36	0.39

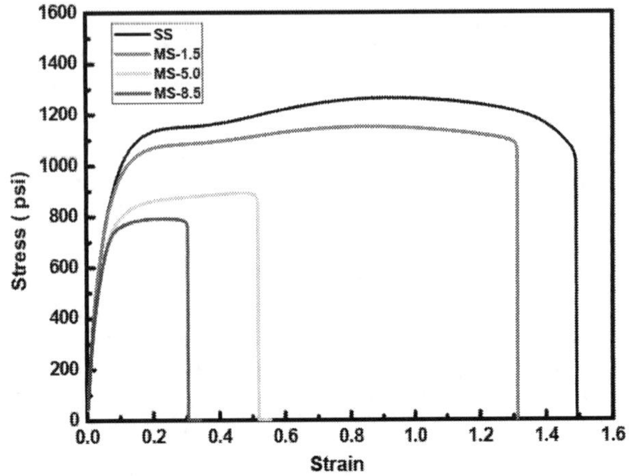

Figure 4. The mechanical properties of LDPE/ETM (MS–1.5, MS–5.0, MS–8.5) injection molded under different ETM concentrations in comparison with solid sample (SS).

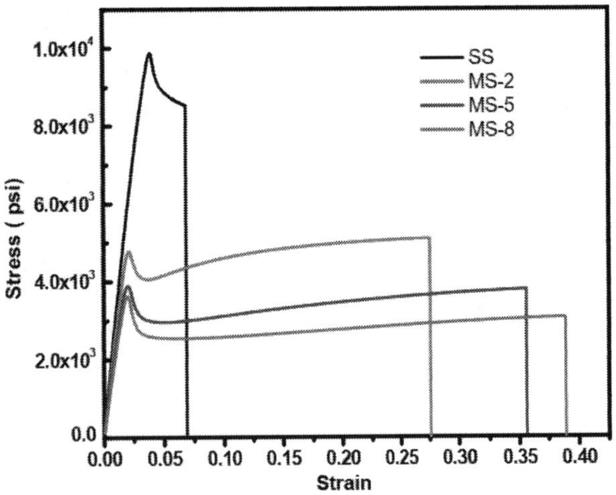

Figure 5. The mechanical properties of PLA/ETM (MS–2.0, MS–5.0 and MS–8.0) injection molded under different ETM concentrations in comparison with solid sample (SS).

Interestingly, the changes in strain at break showed the opposite trends with the addition of variable ETM in LDPE and PLA. That is, the strain at break monotonically increases for PLA with ETM and monotonically decreases for LDPE with ETM. The reason for this opposite trend warrants further study.

As shown in Table 3, the differences in Young's modulus of LDPE/ETM and their solid counterparts were small, approximately less than ±5%. The cause of the different Young's moduli is due to the different layered structure of the LDPE/ETM specimens; namely, the ratio of the outer skin layer to the core inner layer. Compared to the inner layer that composes mostly cells, the out layer has fewer and smaller cells, resulting in a higher density. Thus, the skin layer plays a more significant role in contributing to the specific modulus along the flow direction than the center portion. More specifically, the Young's modulus basically depends on the thickness and density of the outer layer rather than the thickness alone. With the addition of ETM, although the thickness of the outer layers decreased, the reduction in specific modulus only dropped by less than 5% for LDPE (solid LDPE samples can be considered as a single outer skin layer).

However, as a result of different material characteristics, the values and changing trends of Young's modulus of PLA is different from that of LDPE. Although the outer layer has smaller and fewer cells, the voids seem to be detrimental to the mechanical properties (i.e. Young's modulus) of more brittle PLA foamed parts.

Morphology of the Fractured Surface and Microcells

To investigate the effects of ETM on the LDPE matrix foamed samples, the morphology of the injection molded tensile bars was examined using scanning electron microscopy (SEM). For LDPE and LDPE/ETM, the cross-sectional fractural surface (as seen by SEM) locates the fracture at the middle point of tensile bar along the flow direction, while for the PLA and PLA/ETM specimens, SEM images found fractures closer to the end of the molded tensile bars (cf. Fig. 1). This is because for the tensile tests, the fracture of tensile bars with ETM takes place almost at the midpoint for LDPE and near the end of the part for foamed PLA. The location of the fracture point suggests that the cell size distribution is more uniform for LDPE samples whereas for PLA samples, the cells become larger towards the end of the molded parts.

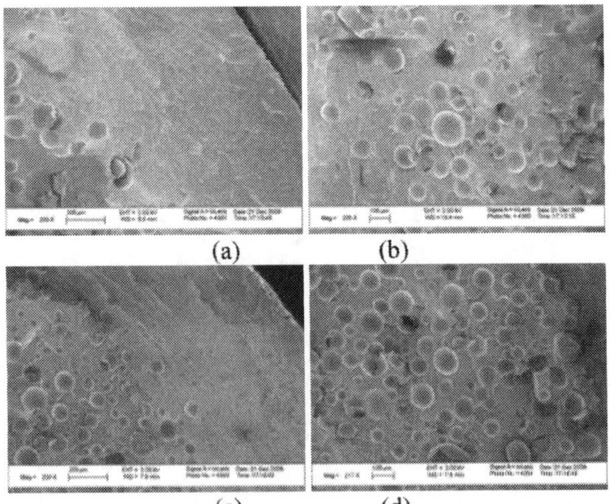

(a) (b)

(c) (d)

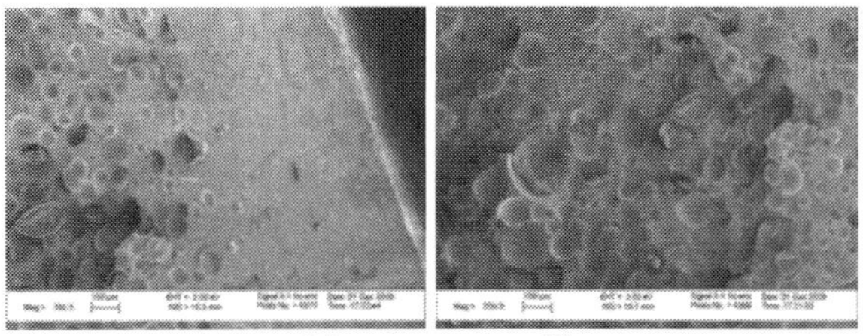

(e) (f)

Figure 6. SEM of fracture surface structure in cross-section direction of LDPE/ETM injection molded tensile bars: (A) outer layer of MS–1.5, (B) center layer of MS–1.5, (C) outer layer of MS–5.0, (D) center layer of MS–5.0, (E) outer layer of MS–8.5, and (F) center layer of MS–8.5.

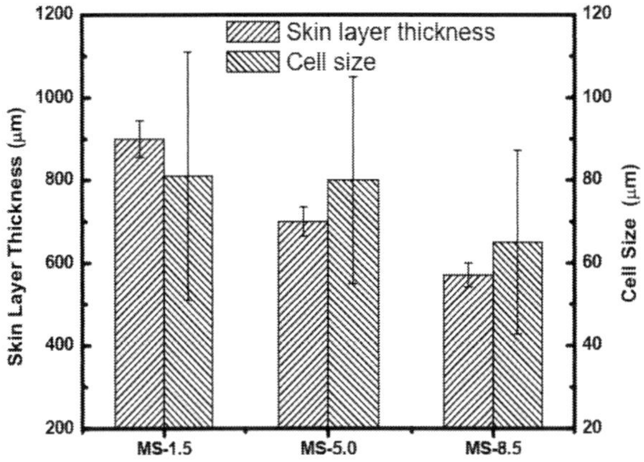

(a)

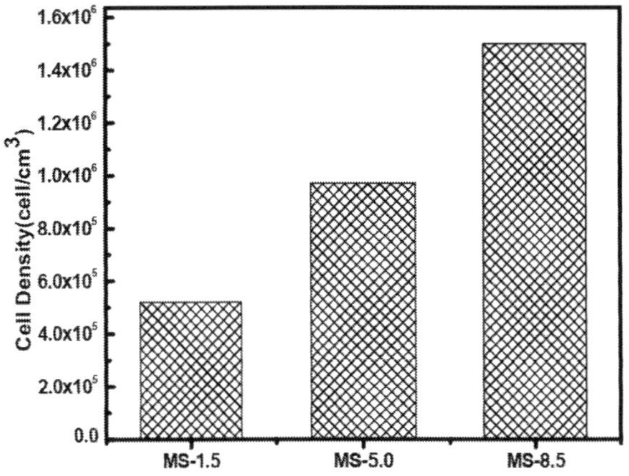

(b)

Figure 7. SEM of LDPE/ETM foamed samples were analyzed to report a representative value for (a) the skin layer thickness, the average cell size, and the standard deviation, and (b) the cell density.

The microcells and microstructures were influenced by the existence of ETM. This is confirmed by the experiments via examination of the SEM micrographs as shown in Figs. 6 and 8 and plotted graphically in Figs. 7 and 9. Two distinct layer structures were observed: the outer skin layer and the inner layer with most of cells. The skin layer possesses a higher density resulting from few cells. Recall that the cells in the center effectively lead to a decrease in cross-sectional area. The dense skin layer plays a primary role in the tensile properties and the effect varies for LDPE and PLA: (1) the lower ultimate strengths of both LDPE and PLA foamed samples are attributed to the reduction of the outer layer thickness and increasing amount of the microcells; (2) the cells in the center layers cause an opposite trend in strain at break with variable ETM contents; and (3) the Young's modulus of solid and foamed LDPE samples varies only slightly (less than 5%), while PLA samples show a drastic reduction in Young's modulus with increasing ETM weight ratio.

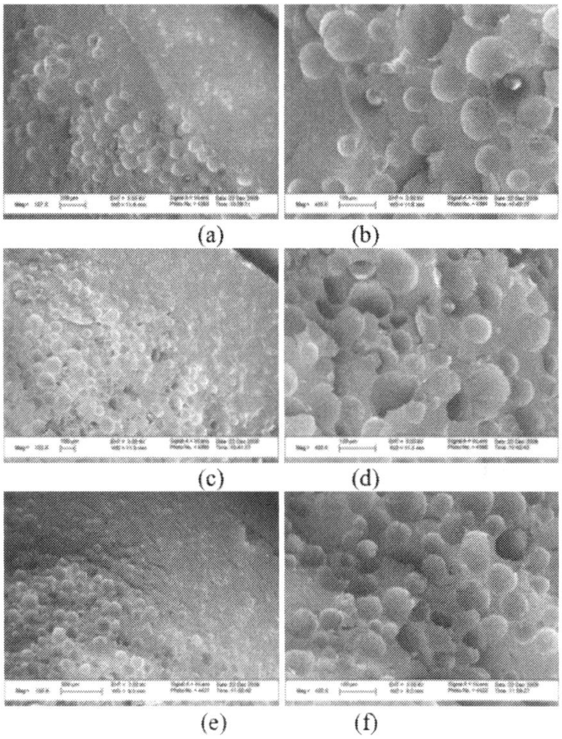

(a) (b) (c) (d) (e) (f)

Figure 8. SEM of fracture surface structure in cross-section direction of LDPE/ETM injection molded tensile bars: (A) outer layer of MS–2.0, (B) center layer of MS–2.0, (C) outer layer of MS–5.0, (D) center layer of MS–5.0, (E) outer layer of MS–8.0, and (F) center layer of MS–8.0.

Figures 7 and 9 show the cell size and cell density of foamed LDPE samples, which indicate a general trend toward decreasing cell size and increasing cell density with increasing ETM content. The cell diameters are similar for LDPE/ETM and PLA/ETM. When the ETM weight ratio is approximately 2 wt% and 5wt%, the cell size maintains at 80 μm, which is the fully expanded volume. When the ratio is increased to more than 8 wt%, the cell size drops to 65 μm, suggesting that they are not fully expanded. Note that average cell size deviations are somewhat large with different ETM concentrations. This is because the SEM observations were taken on fractured cross-sectional surfaces. It is reasonable to expect that the fracture might have propagated through the largest cells or cut through the center of the cells where the void size is the largest.

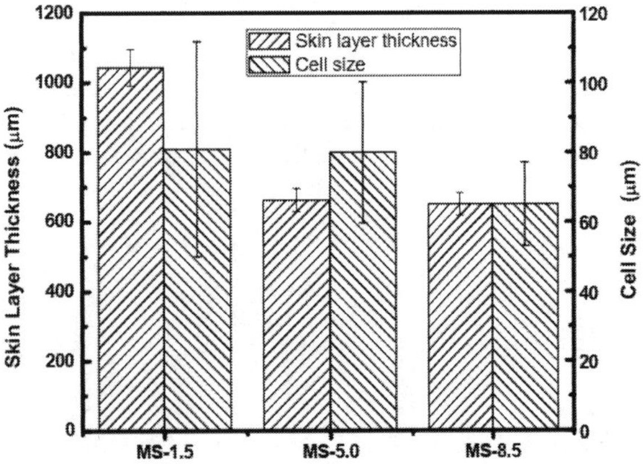

(a)

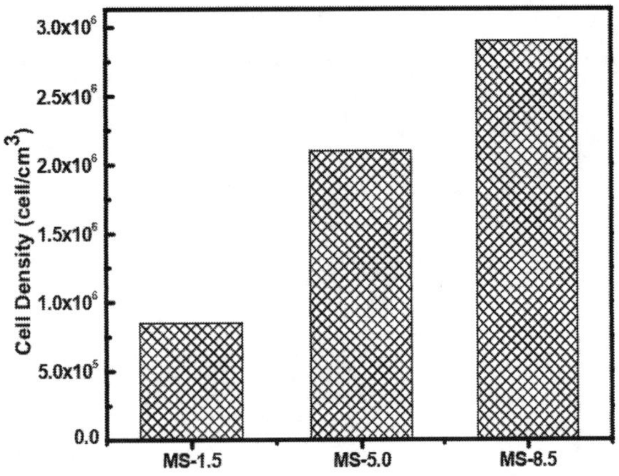

(b)

Figure 9. SEM of PLA/ETM foamed samples was analyzed to report a representative value for (a) the skin layer thickness, the average cell size, and the standard deviation, and (b) the cell density.

It was speculated that the thickness of the outer skin layer could be closely related to the viscosity of the matrix material. At a low concentration of ETM, 1.5 to 2 wt%, PLA/ETM has a larger skin layer due to the relatively higher viscosity. With an increasing ETM weight ratio, the viscosity of LDPE and PLA decreases gradually. Even with an equal quantity, the ETM has a greater influence on the viscosity of the PLA matrix. This is why the cell density of PLA/ETM is slightly higher than its LDPE counterparts.

Conclusions

Expandable thermoplastic microspheres enable part weight reduction and maintain dimensional stability of injection molded parts. Compared with microcellular injection molding technology, the expandable microspheres are able to improve the surface quality while retaining some mechanical properties by controlling the microstructures, such as cell size, cell density and skin layer thickness, through the content of ETM.

Acknowledgements

The authors would like to thank the Expancel Company for generously donating the materials used in this study. The authors also especially wish to acknowledge the China Scholarship Council for their financial support of the first author.

References

1 Kramschuster, A. and Turng, L. S., Journal of Biomedical Materials Research: Part B - Applied Biomaterials, in press, 2009.
2 Kramschuster, A. and Turng, L. S., Biofoams 2009, September 26-28, 2009, Niagara Falls, Canada.
3 Kramschuster, A. and Turng, L. S., Polymer Processing Society Annual Meeting (PPS-24), Salerno, Italy, June 15-19, 2008.
4 Kramschuster, A. and Turng, L. S., Biofoams 2007, Capri, Italy, September 26-28, 2007.
5 Royer, J. R., Gay, Y. J., Desimone, J. M. and Khan, S. A. (2000J. Polym. Sci.: part B- Polym. Phys., 38(23): 3168.
6 Kwang, C., Manke, C.W. and Gulari, E. (2001). Ind. Eng. Chem. res., 40(14): 3048.

INCREASING THE MELT STRENGTH OF LINEAR PP FOR FOAMING APPLICATIONS BY USING COAGENTS

Y. Zhang[1], M. Kontopoulou[1]*, J.S. Parent[1] and C.B. Park[2]

[1]*Department of Chemical Engineering, Queen's University, Kingston, Ontario, K7L 3N6, Canada – ying.zhang@chee.queensu.ca; marianna.kontopoulou@chee.queensu.ca; parent@chee.queensu.ca*
[2]*Microcellular Plastic Manufacturing Laboratory, Department of Mechanical and Industrial Engineering, University of Toronto, Toronto, Ontario, M5S 3G8, Canada - park@mie.utoronto.ca*

Reactive modification of linear PP by radical mediated grafting in the melt state is investigated as a means of increasing the melt strength of linear PP to improve its foam processing. A tri-functional coagent, trimethylolpropane trimethacrylate (TMPTMA) is used to introduce long chain branching to a linear PP homopolymer. Rheological characterization is conducted to relate the rheological properties to the modified PP chain architecture. The foaming behaviour of modified and original PP is investigated using a compression moulding approach, with azodicabonamide (AZ) as a foaming agent.

Introduction

Although linear polypropylene (L-PP) has many desirable attributes, it is challenging to process on a commercial scale in processes such as thermoforming, blow molding and foaming, which involve extensional flow. This has been attributed to its low melt strength at processing temperatures (Burt, 1978). Branched polypropylene (B-PP) has been shown to offer higher expansion ratio, alleviate cell coalescence and provide a more homogeneous cell structure compared to L-PP. Park and co-workers performed intensive studies on the development of extruded PP foams having high expansion ratios and a fine cell structure (Park and Cheung, 1997; Park et al., 1998; Naguib et al., 2001; Naguib et al., 2003). Extremely large volume expansion ratios, up to 10 times could be achieved by foaming a B-PP under optimized conditions. In comparison the volume expansion of L-PP was found to be much lower because of severe cell coalescence.

Among the methods available to increase the melt strength of commercial L-PP, solvent-free reactive processing, which employs peroxide initiated chain scission and co-agent assisted crosslinking is a particularly attractive methodology (Parent et al, 2009). This single stage process involves simultaneous PP fragmentation and crosslinking, the balance of which controls the length, frequency and distribution of branches. A general scheme of such radical post-modification of PP and a recent review on this topic can be found in Passaglia et al. (2009). The melt state rheology of co-agent treated PP have been studied under steady shear (Kim and Kim, 1993; Wang et al., 1996), oscillatory shear (Nam et al., 2005; Borsig et al., 2008) and extensional deformation (Nam et al., 2005; Parent et al., 2009). A wide range of coagent systems have been proven effective of generating LCB characteristics from a linear PP homopolymer. The efficiency of different systems of co-agents to introduce LCB as opposed to crosslinking and degradation of PP homopolymer was compared by Borsig et al. (2008) and Parent et al. (2009). In particular Tian et al. (2006 b) showed that complex viscosity curves of selected pentaerythritol triacrylate (PETA) modified samples progressively changed shape with increasing PETA content: the transition from a Newtonian plateau to the shear thinning regime was shifted to lower frequencies, thus indicating the presence of LCB. Their Cole-Cole plots were also sensitive to the different LCB amounts. Similar trends in the complex viscosity curves, accompanied with the appearance of longer relaxation modes were also reported by Parent et al. (2009) for triallyl and triacrylic systems.

Even though co-agent modified PPs have been well-characterized in the literature, their foaming behaviour has not been investigated extensively (Nam et al., 2005). The objectives of this work are to prepare, characterize and foam PP formulations containing a tri-functional coagent, trimethylolpropane trimethacrylate (TMPTMA).

Experimental

Materials

Trimethylolpropane trimethacrylate (TMPTMA, 98%, Singma-Aldrich) and dicumyl peroxide (DCP, 98%, Sigma-Aldrich) were used as received. ESCORENE™ PP 1042, a linear polypropylene homopolymer (L-PP) manufactured by ExxonMobil (MFR = 1.9 g/10 min, under 2.16 kg load at 230°C was used as the starting material. PF 611, a high melt strength branched polypropylene homopolymer (B-PP) manufactured by LyondellBasell (MFI = 30 g/10 min, under 2.16 kg load at 230°C) was chosen as a commercially available comparison material. 0.1 phr of Irganox B225 (CibaGeigy) was used as an antioxidant to prevent samples from thermal degradation during further theological characterization and foaming processes. Azodicarbonamide (AZ, Sigma-Aldrich Canada) was used as the chemical blowing agent.

Sample preparation

All blends were prepared using a Haake Polylab Rheocord torque rheometer equipped with a mixing chamber with a pair of rotors at 170 °C, 50 rpm in a nitrogen blanket. The formulations and sample designation are listed in Table 1. PP powder was pre-blended with DCP and TMPTMA using acetone as a solvent. After the acetone was evaporated, the PP powder was charged into the mixing chamber and compounded for 10 min to ensure complete decomposition of DCP and reaction of TMPTMA. Subsequently Irganox B225 was added and the compounding continued for another 2 minutes. 1 phr AZ was charged into the blend and mixed for 4 more minutes. A steady state torque reading was obtained for each formulation. The L-PP and B-PP grades were also processed under identical conditions, to ensure consistent thermal history.

A Carver® manual hydraulic press was used to make PP foams by compression moulding. The polymer melt was compression moulded inside a frame mold between two copper plates at 200 °C under 5 tons for 3 minutes, followed by the release of pressure, upon which expansion of the foam took place.

Table 1. Sample designation and formulations

Designation	DCP (phr)	TMPTMA (μmol/g PP)
L-PP	-	-
L-PP/D	0.1	-
L-PP/L	0.1	45
L-PP/H	0.1	90
B-PP	-	-

Testing

Compression moulded discs of 25 mm in diameter at 180 °C were used for rheological characterization. A controlled stress rheometer, ViscoTech by Reologica, with parallel plate fixtures was used in the oscillatory mode at a gap of 1 mm to measure linear viscoelastic properties of prepared PP samples as a function of angular frequency (ω) at 180 °C. Stress sweeps were performed to ensure that all measurements were within the linear viscoelasticity regime.

Foamed PP samples were characterized based on their expansion ratio and cell morphology. The density of the foams was measured by an electronic densimeter (Mirage MD-200S, Mirage Trading Co., Ltd) based on the water displacement method. The volume expansion ratio (Φ) was calculated as the ratio of bulk density of the pure polyolefin material (ρ_p) to the bulk density of the foam sample (ρ_f):

$$\Phi = \frac{\rho_p}{\rho_f} \qquad (1)$$

Foams were subjected to cryogenic fracture for SEM imaging. A JEOL JSM-840 SEM was used to characterize the cell morphology of foams. SigmaScan Pro® 5.0 (SPSS Inc.) was utilized to measure the average cell size of each sample. The cell density of the foam with respect to the unfoamed polymer volume (N) was calculated by the following equation:

$$N = \left(\frac{N_o}{A}\right)^{3/2} \Phi \qquad (2)$$

where N_0 is the number of cells in a specified area A, and Φ is the volume expansion ratio.The shape factor f was evaluated as a measure of the extent of spherity of the sample as follows:

$$f = \frac{4\pi v}{l^2} \qquad (3)$$

where v is the cell area in the planar profile of a 3D cell and l is total edge length in the planar profile of a 3D cell. Values of f closer to 1 indicate a more spherical sample.

Differential scanning calorimetry (DSC) was conducted on all mPEs to determine their crystallization temperature (T_c), enthalpy of crystallization (ΔH_c) and melt temperature (T_m) using a TA Instrument DSC Q 100. Weighed samples were sealed in aluminum hermetic pans and analyzed between 40 and 210 °C at a heating/cooling rate of 10 °C/min. After the first heating, each sample was held isothermally at 210 °C for 2 minute before cooling, followed by a second heating step.

Results and Discussion

Small amplitude oscillatory shear rheology

The storage modulus and complex viscosity of all samples as a function of the angular frequency (ω) are presented in Figure 1 (a) and (b) respectively. Reaction by DCP caused a significant decrease in the storage modulus and complex viscosity values compared to the neat L-PP grade. This is because of degradation reactions caused by the fragmentation of tertiary alkyl macroradicals. As a result, the complex viscosity showed a Newtonian plateau at a much lower value, indicating a lower molecular weight and narrower molecular weight distribution (Tzoganakis et al. (1988)). TMPTMA causes the formation of secondary macroradicals, leading to grafting/recombination reactions (Hu et al., 1996). The resulting chain extension and/or branching resulted in increased values of the complex viscosity of the polymer melts compared to their degraded counterparts, as well as in the re-appearance of shear thinning.

Additional indication of presence of branching in the coagent modified samples can be drawn from inspection of the storage modulus curves. Contrary to the degraded sample, which shows terminal flow behaviour, (G' scaling with ω^2) addition of TMPTMA, resulted in a deviation from terminal flow and an increase in G' at low frequencies. The higher G' values suggest the presence of a longer relaxation mechanism for the coagent treated PP, which can be attributed to the branching. The non-terminal behavior of TMPTMA treated PP can also be illustrated in Cole-Cole plots and compared to the curves of the commercial B-PP. (Figure 2). Whereas the Cole-Cole plots of L-PP (not shown in Figure 2) and L-PP/D are shaped as semicircles, the onset of an upturn is observed for the coagent modified PPs, which is indicative of the presence of branching (Tian et al. 2006 b) Moreover, the radius of the semi-circle curves for the latter samples is larger than that of the degraded PP, but lower than that the unmodified PP indicating that degradation was partially offset by the generation of branching in the presence of the coagent.

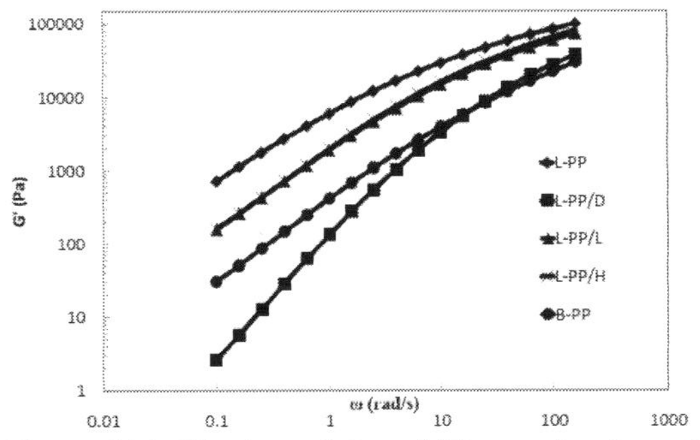

Figure 1(a). Elastic modulus of PP as a function of frequency at 180 °C.

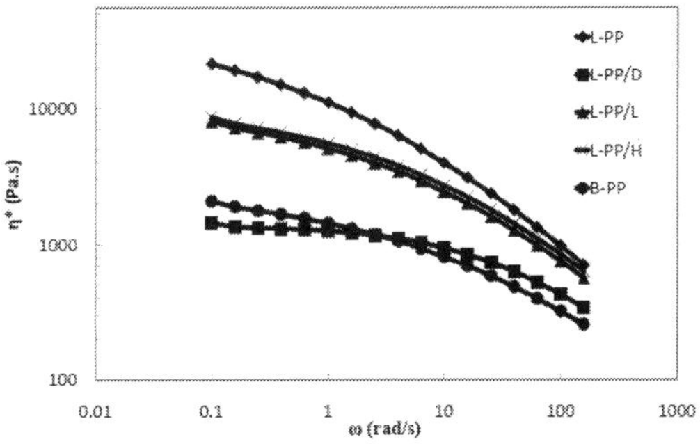

Figure 1(b). Complex viscosity of PP as a function of frequency at 180 °C.

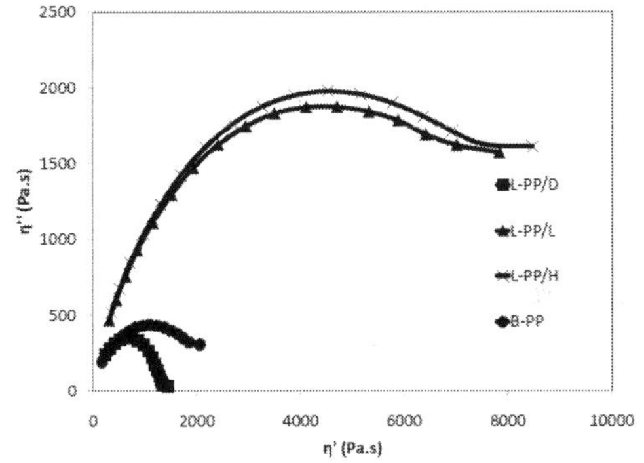

Figure 2. Cole-Cole plots at 180 °C.

It is noted that the viscoelastic behavior of TMPTMA treated PP does not vary significantly with the amount of TMPTMA used. Finally by comparing the complex viscosity, storage modulus and Cole-Cole plots of the commercial B-PP material to those of the L-PP and its derivatives, it can be concluded qualitatively that B-PP has sparse long chain branching with a molecular weight that is comparable with L-PP/D.

Thermal properties

The degradation, recombination, and grafting reactions that take place simultaneously in these systems result in changes in the molecular weight and molecular weight distribution, as well as in the chain architecture. The melting and crystallization points, T_m and T_c respectively, as well as the ΔH_c for all the derivatives are shown in Table 2. The T_c of the coagent modified samples is almost 10°C higher than the L-PP and its degraded counterpart. ΔH_c also increases upon coagent modification. The slightly higher T_c and ΔH_c observed in L-PP/D can be explained by its lower molecular weight; a lower molecular weight polymer generally crystallizes more readily due to higher bulk chain mobility. The higher values of T_c and ΔH_c observed in TMPTMA treated L-PP can be attributed to the combined effects of the decrease in molecular weight caused by degradation and branching introduced by coagent. It is known that in polyethylene the crystallization temperature and crystallinity generally decrease as short or long branches are incorporated along the backbone of a linear polymer, because of the disruption of ordered structure. An opposite effect has been found in branched PPs (McCallum et al., 2007). The side chains may tend to pack into crystal domains themselves, forming new sites for nucleation and thus increasing the overall bulk crystallization rate and crystallinity (Tian et al. (2006 a)).

Table 2. Thermal properties of PP samples.

Sample Name	T_m (°C)	T_c (°C)	ΔH_c (J/g)
L-PP	165.48	113.20	80.25
L-PP/D	162.83	114.42	85.87
L-PP/L	165.02	124.00	86.55
L-PP/H	164.39	122.98	91.52
B-PP	163.27	125.01	94.95

Foam expansion ratio and cell morphology

The cell morphology of the foamed samples showed different trends, depending on the degree of chemical modification (**Figure 3**). The foams made from neat L-PP had a non-uniform morphology, featuring large fluctuation in cell sizes and distortion of the cells. (**Figure 3** (a)). After degradation by DCP, the foam exhibited a coarser cell structure with an increase in the average cell diameter from 79 to 137 μm (Figure 4). Upon addition of TMPTMA the average cell diameter dropped to a value comparable to that of the initial neat L-PP. What is noteworthy however is that the cell diameter distribution was sensitive to chemical modification, i.e. the amount of LCB. As TMPTMA amount increased, the cell size distribution of the samples became significantly narrower (**Figure 5**). Lastly the cell distribution of the B-PP was narrower than the L-PP, but with a higher average cell diameter (**Figure 4**). This results in a smaller overall number of cells for the B-PP sample compared to the L-PP, as shown in **Figure 6** which shows the cell density of all foamed samples.

A qualitative presentation of the shape of the cells is presented in **Figure 7** in the form of shape factor. The shape factor of neat L-PP was the lowest signifying a larger deviation from a spherical shape, followed by L-PP/L. The highest shape factor value was achieved by L-PP/H and B-PP.

The observations noted in Figures 3-7 are likely due to complex synergistic effects between molecular weight and degree of LCB and their effects on rheological and thermal properties as explained below.

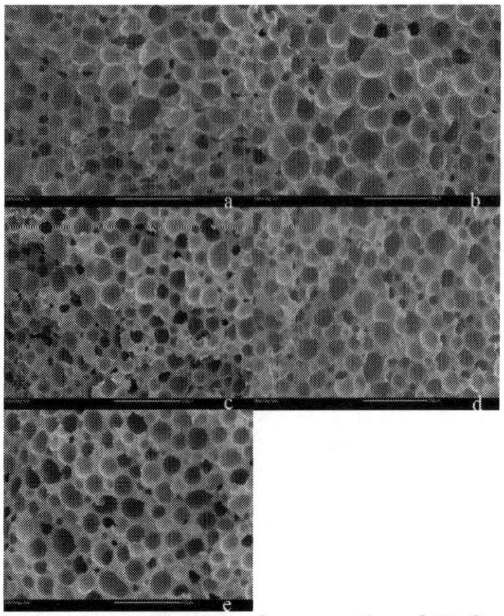

Figure 3. SEM images of cross-section of PP foams: (a) L-PP, (b) L-PP/D, (c) L-PP/L, (d) L-PP/H, (e) B-PP. Scale bars in the images represent 500 μm.

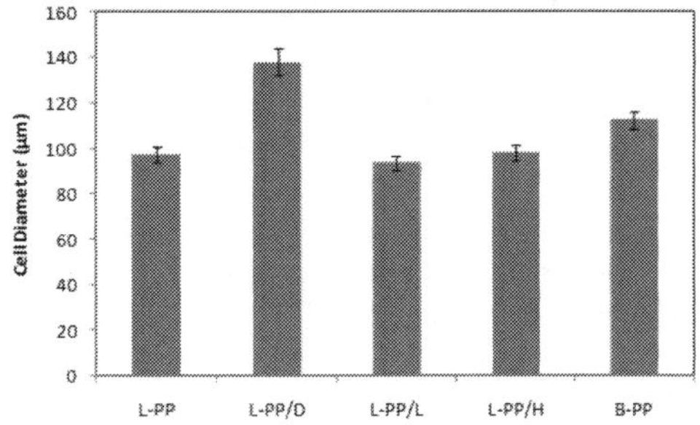

Figure 4. Average diameter of cells for the foamed PP samples. The error bars in the graph represent the standard error.

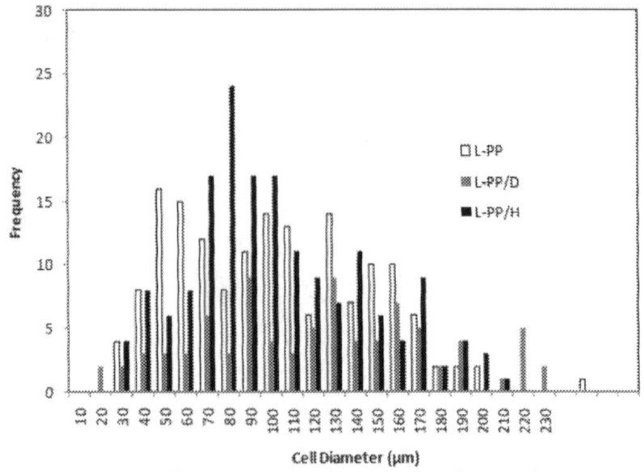

Figure 5. Cell diameter distribution for foamed PPs.

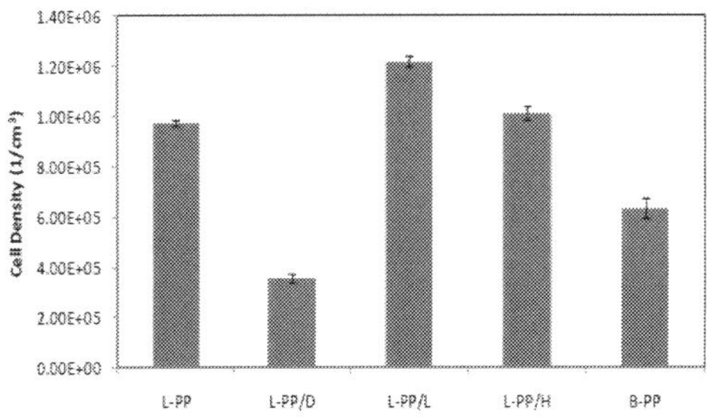

Figure 6. Cell density of PP foams

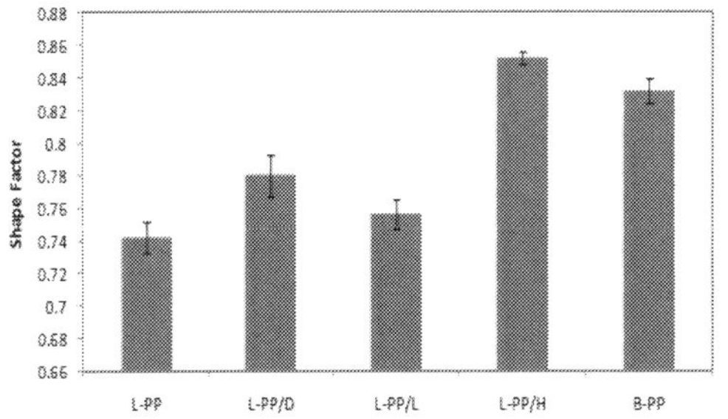

Figure 7. Shape factor of the cells. The error bars represent the standard deviation.

The B-PP and L-PP/D materials have the lowest molecular weight and lowest viscosity overall. This results in a faster rate of cell growth, and consequently higher value of the average cell diameter (**Figure 4**). Since the amount of AZ available was the same for all samples, this resulted in a smaller value of the cell density for these samples (**Figure 6**). However these samples are very different in terms of their extensional behaviour, with B-PP showing strain hardening (McCallum et al., 2007). This results in higher melt strength and a reduced propensity for coalescence. Additionally, its crystallization point is much higher, meaning that the cells have a reduced opportunity to coalesce because the melt solidifies sooner. These two effects likely result in a more uniform cell size distribution for the B-PP material and a shape factor closer to one.

On the other hand, L-PP and its derivatives have higher viscosities, leading to lower rates of cell growth, and a smaller average cell size. L-PP does not exhibit strain hardening because of its linear nature, therefore it will be more prone to coalescence, compared to its coagent modified derivatives, which display a more uniform cell size distribution, at the highest coagent level (**Figure 5**). Here once again, both the favourable rheological properties, as well as the higher crystallization point would act toward suppressing coalescence.

In spite of the differences in cell size and distribution, the overall expansion ratio was not significantly sensitive to the differences in structure (**Figure 8**). The overall expansion ratio reflects the average properties, therefore it is not sensitive to the cell distribution and shape. It is noteworthy that out of all the samples the lowest expansion ratio was observed for B-PP, which generally had less and larger cells, with the rest of the samples being comparable.

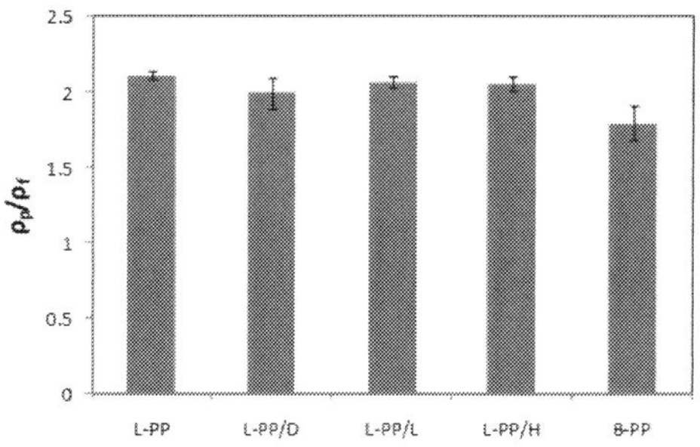

Figure 8. Expansion ratio of PP foams.

Conclusions

Coagent modified PP had a more uniform and spherical cell structure compared to the linear and degraded counterparts. The average cell diameter, average cell diameter distribution and the shape factor were sensitive to the amount of TMPTMA used to modify L-PP. The best combination of foam properties was achieved at the highest amount of coagent studied. These findings were associated to the rheological and thermal characteristics of the polymers.

Acknowledgements

The authors would like to thank NSERC and the Auto-21 –NCE for funding this project.

References

Burt, J. G., "The Elements of Expansion of Thermoplastics Part II", J. Cell. Plast. 14, 341-345 (1978)

Borsig, E., Duin, M. van, Gotsis, A. D., and Picchioni, F., "Long chain branching on linear polypropylene by solid state reactions", Eur. Polym. J. 44, 200-212 (2008)

Hu, G. H., Flat, J. J., and Lambla, M., "Free radical grafting of monomers onto polymers by reactive extrusion: principles and applications", In: Reactive modifiers for polymers, Al-Malaika, S. (Ed.), Blackie Academic and Professional Chapman and Hall, London, p. 1-77, (1996)

Kim, B. K. and Kim, K. J., "Crosslinking of polypropylene by peroxide and multifunctional monomer during reactive extrusion", Adv. Polym. Tech. 12, 263-269 (1993)

McCallum, T.J., Kontopoulou, M., Park, C.B., Muliawan, E.B. and Hatzikiriakos, S.G., "The

rheological and physical properties of linear and branched polypropylene blends", Polym. Eng. Sci. 47, 1133-1140, (2007)
Naguib, H. E., Xu, J. X., Park, C. B., Hesse, A., Panzer, U. and Reichelt, N., "Effects of blending of branched and linear polypropylene resins on foaming", SPE ANTEC Tech. Papers, 1623-1630 (2001)
Naguib, H. E., Park, C. B. , Wang, J., Mukhopadhyay, A. K. and Reichelt, N., "Effect of recycling on the rheological properties and foaming behaviors of branched PP", Cell. Polym. **22**, 1-22 (2003)
Nam, G. J., Yoo, J. H., and Lee, J. W., "Effect of Long-Chain Branches of Polypropylene on Rheological Properties and Foam-Extrusion Performances", J. Appl. Polym. Sci. 96 1793-1800 (2005)
Parent, J. S., Bodsworth, A., Sengupta, S. S., Kontopoulou, M., Chaudhary, B. I., Poche, D., and Cousteaux, S., "Structure-rheology relationships of long-chain branched polypropylene: Comparative analysis of acrylic and allylic coagent chemistry", Polymer 50, 85-94 (2009)
Park, C. B. and Cheung, L. K., "A study of cell nucleation in the extrusion of polypropylene foams", Polym. Eng. Sci. 37, 1-10 (1997)
Park, C. B., Cheung, L. K. ,and Song, S. W., "The effect of talc on cell nucleation in extrusion foam processing of polypropylenes with CO_2 and isopentane", Cell. Polym. **17**, 221-251 (1998)
Passaglia, E., Coiai, S., and Augier, S., "Control of Macromolecular architecture during the reactive functionalization in the melt state of olefin polymers", Prog. Polym. Sci. 34, 911-947 (2009)
Tian, J., Yu, W., and Zhou, C., "Crystallization kinetics of linear and long-chain branched polypropylene", J. Macromol. Sci. B. 45 969-985 (2006 a)
Tian, J., Yu, W., and Zhou, C., "The preparation and rheology characterization of long chain branching polypropylene", Polymer 47 7962-7969 (2006 b)
Tzonganakis, C., Vlachopoulos, J., and Hamielec, A. E., "Production of controlled-rheology polypropylene resins by peroxide promoted degradation during extrusion", Ploym. Eng. Sci. 28 170-180 (1988)
Wang, X., Tzoganakis, C., and Rempel, G. L., "Chemical modification of polypropylene with peroxide/pentaerythritol triacrylate by reactive extrusion", J. Appl. Polym. Sci. 61 1395-1404 (1996)

COMPUTER MODELING OF PERCOLATION THRESHOLD IN ELECTRICALLY CONDUCTIVE CARBON FIBER POLYMER COMPOSITE FOAMS

G. H. Motlagh[1*], A. Foroozani[1], M. R. Thompson[2], A. N. Hrymak[3]

[1]*Department of Chemical Engineering, Faculty of Engineering, University of Tehran, Tehran, Iran- ghmotlagh@ut.ac.ir, foroozani_b@yahoo.com*
[2]*Department of Chemical Engineering, McMaster University, Ontario, Canada- mthomps@mcmaster.ca*
[3]*Dept. of Chemical & Biochemical Engineering, The University of Western Ontario, Ontario, Canada- ahrymak@uwo.ca*

Abstract - Electrically conductive polymer composites are a rapid growing market with versatile applications in electronic, energy and transportation industries. Foaming can be used as a tool to control distribution and orientation of conductive particles while it reduces product weight and usually cost. Foaming is also beneficial by reducing melt viscosity leading to improved processability during processing and molding of highly filled composites. Although a lot of studies have been carried out to model percolation threshold and electrical conductivity of the solid polymer composites, the modeling of these properties for the polymer composite foams is an interesting subject for further exploration. The aim of this work is to model percolation threshold and electrical conductivity of short carbon fiber polymer composite foams. The procedure is based on computer programming in MATLAB. The fiber is assumed as a hard core surrounded by a soft shell which is permeable to transfer of electrons. Fibers and cells are randomly distributed in the space with applicable controlling constrains. The number of dispersed fibers in the polymer matrix containing bubbles is continually increased and the program checks for the formation of a continuous path which spans the considered geometry. The effects of microstructural factors such as fiber aspect ratio, porosity and bubble size have been studied. The sample size has been varied to adequately determine its effect on percolation threshold. The obtained results have been compared with the experimental values to verify the model. The model predicted an increase in the effective percolation threshold via foaming, which was more considerable at lower fiber aspect ratio and smaller bubble size.

Introduction

The use of electrically conductive polymer composites for a variety of applications such as EMI shielding, bipolar plates, actuators, sensors and battery electrodes has shown to be a rapid growing market in recent years. The desired conductivity is commonly achieved by the addition of conductive fillers such as carbon black, carbon fiber, carbon nanotube and graphite to a polymer matrix (Wei et al., 2007; Huang, 2002; Strumpler and Glatz-reichenbach, 1999; Motlagh et al., 2007b). To create conductive paths through the insulating matrix material the amount of filler must be above its percolation threshold which usually makes the material difficult to process and the final product brittle. On the other hand foaming can reduce melt viscosity to improve processability and enhance properties such as impact resistance while reducing density and usually cost. In addition foaming can be used to control orientation and distribution of conductive particles in the composite (Motlagh et al., 2007a). Therefore there are valuable advantages to foaming of electrically conductive polymer composites. There is little experimental research on this group of materials which makes them very potent to explore (Thompson et al., 2010; Motlagh et al., 2007a). Although many attempts have considered the modeling of percolation threshold and electrical conductivity of solid polymer composites (Wang and Ogale, 1993; Ogale and Wang 1993; Dani and Ogale 1996; Ma and Goa, 2008; Munson-McGee, 1991; Weber and Kamal, 1997; Zhang and Yi, 2008) but for foam polymer composites this area is quite fresh to explore. The aim of this work is to model the percolation threshold for short carbon fiber polymer composite foams. The effects of a series of microstructural factors will be evaluated.

Modeling Approach

MATLAB environment was used to carry out the simulation process. Conductive fibers with diameter of

0.775 and two aspect ratios of 8 and 16 were considered for the simulation. A cube with certain length, named as sample size, was assumed as the control volume to create a composite with random distribution of fibers and bubbles. First, spherical bubbles with uniform size were distributed randomly throughout the 3D cube. Then fibers were randomly placed into the cube containing bubbles with the constraint that the fibers could not go through the bubbles. A soft shell with a thickness δ=0.1125 was considered around the hard core of the fibers. Within the soft shell electrons can jump or tunnel through the material. Therefore two fibers which were at a distance less than 2δ were considered as in contact. The number of fibers was continually increased and their connectivity was checked by considering body-to-body, end-to-body and end-to-end contacts until a cluster of connected fibers could span the cube length. At this point the number of fibers was used to determine the percolation threshold of the composite. Figure 1 shows one of the composites created by this procedure.

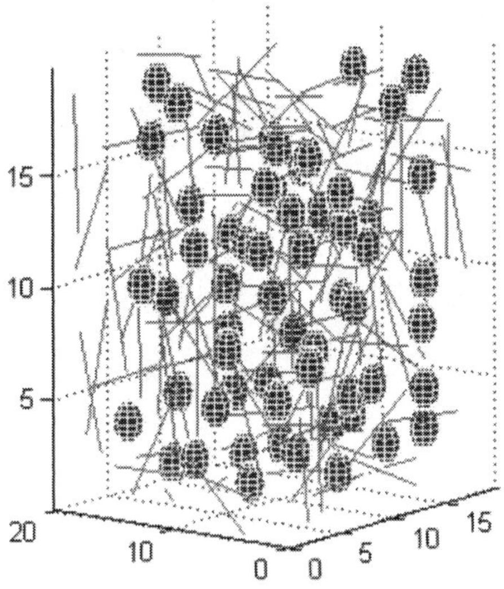

Figure 1- Morphology of a foam composite containing randomly distributed fibers and buubles produced by the model in MATLAB.

Results and Discussion

The value of the (effective) percolation threshold was calculated in the program based on the amount of the matrix. It means that the volume of bubbles has been excluded during the calculation of the percolation thresholds throughout this paepr. Figure 2 shows the percolation threshold versus the number of runs for the MATLAB code. It is seen that the percolation threshold obtained by a small number of runs is not reproducible. By increasing the number of runs and using their average, the result become more reproducible. Therefore at least the average of 100 runs

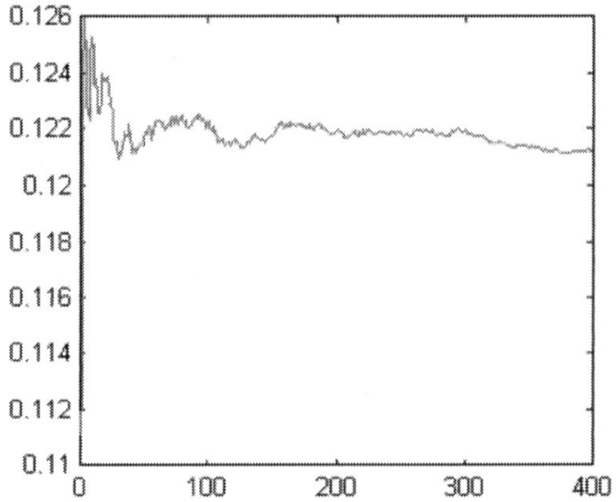

Figure 2- Average percolation threshold versus the number of runs for the MATLAB code.

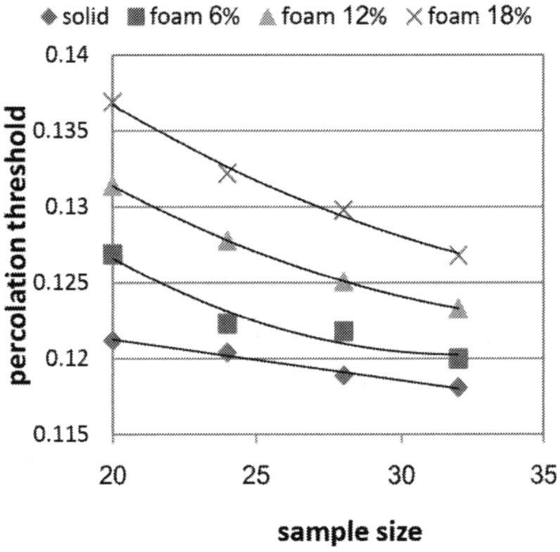

Figure 3- Average percolation threshold of solid and foam composites: fiber aspect ratio=8 and bubble radius=2.5

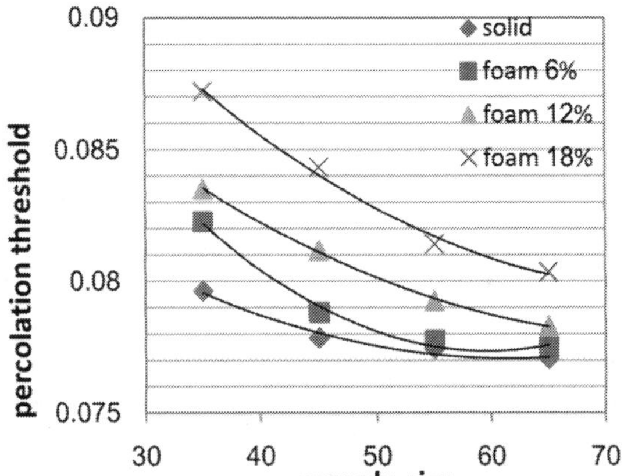

Figure 4- Average percolation threshold of solid and foam composites: fiber aspect ratio=8 and bubble radius=2.5

(for most cases 400 runs) was used to report a percolation threshold.
Figure 3 shows the results for the composite containing fibers with aspect ratio of 8 and bubble radius of 2.5. The obtained percolation threshold for the solid composite approaches 0.115 for a very large sample size. The percolation threshold for a short fiber composite with aspect ratio of 8 reported by Bigg (1979) is also 0.115. This comparison indicates the good accuracy of the developed code in this work.
Figure 3 shows that by introducing bubbles to the solid composie, percolation threshold of the composite increases. The increase is larger for the smaller sample sizes. Therefore a solid conductive short fiber composite can become nonconductive via foaming at a lower void content as the size of the sample decreases and become closer to the length of the fibers. For the composite containing fibers with higher aspect ratio of 16 Figure 4 shows generally the same trend as Figure3. For the solid composite the percolation threshold at a large sample size approaches 0.077 which is in agreement with the value (0.076) reported by Bigg (1979).
The experimental work by Motlagh (2008) on COC/carbon fiber composites containing 15 vol% fiber, with an aspect ratio of about 16 showed that foaming by 20% reduces the bulk electrical conductivity by a factor of 5. Using the universal percolation model shown below, the critical exponent (t) of 3 and the percolation thresholds of 0.077 and 0.08 for the solid and foam composites (Figure 4), respectively, one can calculate a reduction by a factor of 7 in the electrical conductivity via 20% foaming. Therefore the prediction is in rough agreement with the experimental value of 5. This can be accounted as a nondirect verification of the model. However further experimental work is needed to determine the percolation thresholds of the foam composites to directly evaluate the model accuracy. Universal percolation model is as follows:

$$\sigma_C = \sigma_F (\varphi - \varphi_{Crit})^t$$

Where, σ_c and σ_F are the electrical conductivity of the composite and fibers, respectively, and φ_{crit} is the fiber percolation threshold and φ is the fiber volume content in the composite.

By comparing Figure 3 and 4 it is deduced that increase in percolation threshold with foaming is less considerable for the composites with longer fibers.

As seen in Figures 3 and 4 by increasing sample size the percolation threshold decreases. Moreover the rate of decrease increases with the foam percent. The infinte size percolation threshold can be obtained by (Stauffer, 2003):

$$|p_{ave} - p_c| \sim \Delta$$

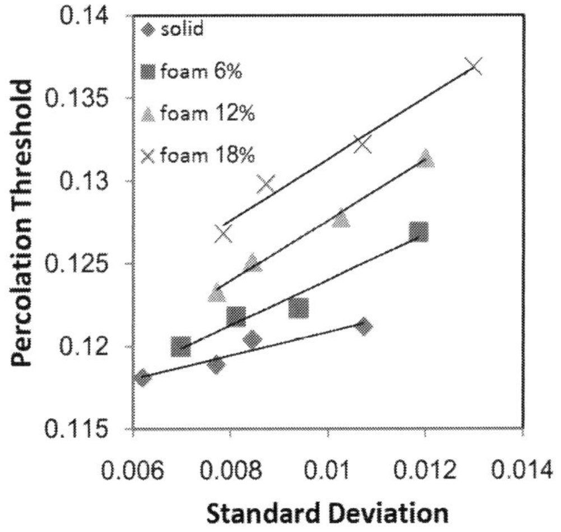

Figure 5- Average percolation threshold (p_{ave}) versus standard deviation: fiber aspect ratio=8

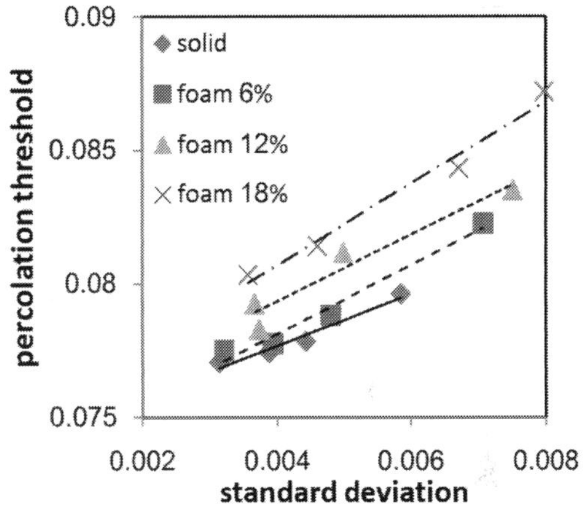

Figure 6- Average percolation threshold (p_{ave}) versus standard deviation: fiber aspect ratio=16

Table 1- Percolation thresholds of infinite size samples obtained from intercepts in Figures 5 and 6

Composite type	Fiber aspect ratio 8	Fiber aspect ratio 16
	Infinite size percolation threshold	
Solid	0.113	0.073
6% foam	0.110	0.072
12% foam	0.109	0.074
18% foam	0.112	0.074

where, p_{ave} and Δ are the average value and standard deviation of the percolation thresholds obtained from the multiple runs of the program for a finite size sample, respectively, and p_c is the infinite sample sizepercolation threshold. By plotting p_{ave} against standard deviation for several sample sizes and fitting the best line through the points (Figures 5 and 6), the intercept will be p_c. Table 1 depicts the intercept values obtained from Figures 5 and 6. As seen the percolation threshold (based on matrix volume) for an infinite size sample does not change considerably via foaming.

The effect of bubblel size on the percolation threshold for the composites is shown in Figure 7 and 8. At constnt void contant, for the composites containing fibers with an aspect ratio of 8, a decrease in bubble size from 2.5 to 1.5 causes a slight increase in the percolation threshold (Figure 7). In Figure 8 the bubble size does not show a clear effect on the system including fibers with higher aspect ratio of 16.

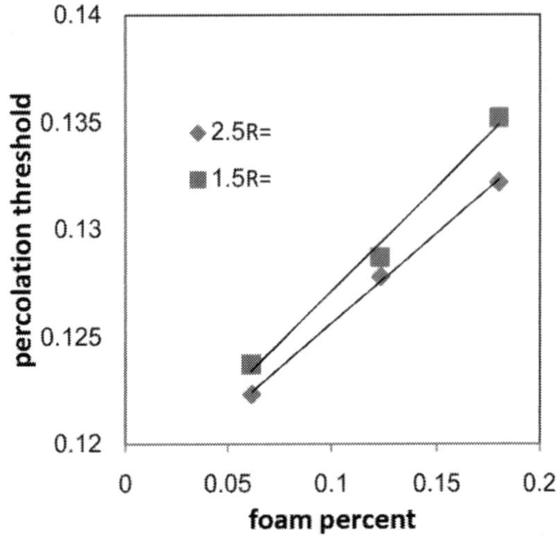

Figure 7- The effect of bubble size on percolation threshold: fiber aspect ratio 8 and sample size 24.

Conclusions

The predicted percolation thresholds for the conductive short fiber composites obtained from the computer programming in this work are in very good agreement with the previous works for the solid composites. The model accuracy to predict the percolation threshold of fiber composite foams was indirectly verified by using the universal percolation model along with the electrical conductivity data of short carbon fiber composites. The model shows that the effective percolation threshold increases with an increase in the level of foaming for the samples with a finite size. The percolation thresholds of the infinite size samples did not show a considerable change via foaming. At constant foaming level, with a decrease in the bubble size, the model predicted a slight increase in the percolation threshold of the foam composites. This effect reduces as the fiber aspect ratio increases.

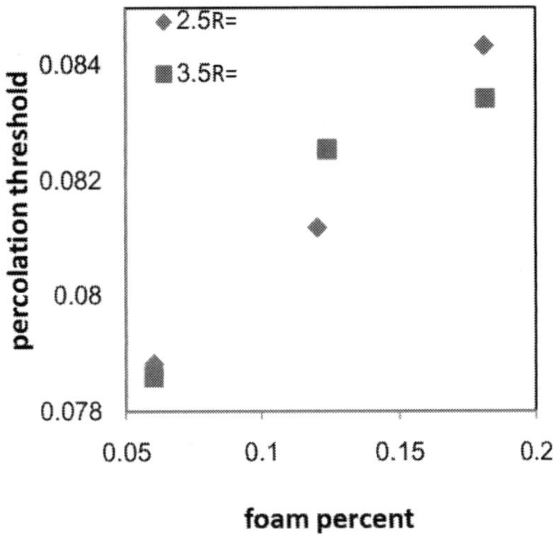

Figure 8- The effect of bubble size on percolation threshold: fiber aspect ratio 16 and sample size 45.

References

Bigg, D. M.," Mechanical, Thermal and Electrical of Metal Filled Polymer Composites", Polym. Eng. Sci., 19, 1188-1192 (1979)

Dani, A., Ogale, A. A., "Electrical Percolation Behavior of Short-Fiber Composites: Experimental Characterization and Modeling", Composites Sci. Technol., 56, 911-920 (1996)

Huang, J., "Carbon Black Filled Conducting Polymers and Polymer Blends", Adv. Polym. Technol., 21 , 299-313 (2002)

Ma, H. M., Goa, X. L., "A Three-Dimensional Monte Carlo Model for Electrically Conductive Polymer Matrix Composites Filled with Curved Fibers ", Polymer, 49, 4230-4238 (2008)

[a]Motlagh, G. H., et al., "Improved Through-Plane Electrical Conductivity in a Carbon-Filled Thermoplastic via Foaming", Polym. Eng. Sci., 48, 687-696 (2007)

[b]Motlagh, G. H., et al., "Properties of a Carbon Filled Cyclic Olefin Copolymer", J Polym. Sci.: Part B: Polym. Phys., 45, 1808-1820 (2007)

Motlagh, G. H.: Electrically Conductive Amorphous Thermoplastic Composites, PhD Thesis, McMaster University, Hamilton, Ontario, Canada (2008)

Munson-McGee, S. H., "Estimation of Critical Concentration in an Anisotropic Percolation network", Physical Review B, 43, 33314-3336 (1991)

Ogale, A. A., Wang, S. F., "Simulation of the Percolation Behavior of Quasi- and Transversely Isotropic Short-Fiber Composites With a Continuum Model", Composites Sci. Technol., 46, 379-388 (1993)

Stauffer, D., Aharony, A.: Introduction to Percolation Theory, 2nd Revised Edition, Taylor & Francis (2003)

Strumpler, R., Glatz-reichenbach, J., "Conducting Polymer Composites", J Electroceramics, 3:4,329-346 (1999)

Thompson, M. R., et al., "Multiple Percolation in a Carbon Filled Polymer Composites via Foaming", J App. Polym. Sci., 115, 646-654 (2010)

Wang, S. F.,Ogale, A. A., "Continuum Space Simulation and Experimental Characterization of Electrical Percolation Behavior of Particulate Composites", Composites Sci. Technol., 46, 93-103 (1993)

Weber, M., Kamal, M. R., "Estimation of the Volume Resistivity of Electrically Conductive Composites", polymer composites, 18, 711-725 (1997)

Wei, Z., et al., "Carbon Based Conductive Polymer Composites", J Mater. Sci.,42, 3408-3418 (2007)

Zhang, T., Yi, Y. B., "Monte Carlo Simulation of Effective Electrical Conductivity in Short-fiber Composites", J Applied Phys., 103, 014910 (2008)

OBSERVATION OF 3D FORM MORPHOLOGIES OF PP AND HIPS IN MICROCELLULAR INJECTION MOLDING USING X-RAY CT

T. YAMADA [1] and *H. YOKOI* [2]

[1] *Saitama Industrial Technology Center,Saitama,Japan– yamada.takehiro@pref.saitama.lg.jp*
[2] *Institute of Industrial Scinece ,The University of Tokyo, Tokyo,Japan – hiyokoi@iis.u-tokyo.ac.jp*

The present paper discusses the results of observing and comparing the molding conditions (maximum pressure of polymer in the cavity) and 3D foam morphologies of PP and HIPS in microcellular injection molding using X-ray CT. The common morphologies shared by HIPS and PP were found to be having a silver streak layer (skin layer 1), having no foaming layer (skin layer 2) and foaming layer (core layer). When the maximum cavity pressure Pmax was 5MPa, the foam morphologies in HIPS and PP became quite similar. When Pmax exceeded 15MPa, many microcells (d < 50μm) were found to coexist with the morphology of HIPS in PP (a crystalline polymer). When Pmax was 30MPa, formation of "icicle-shaped long cells" from the boundary area between skin layer 2 and the core layer to the internal layers was seen only for PP, and this shape widened the cross-sectional diameter of the cells toward the middle.

Introduction

Microcellular injection molding using supercritical fluid is drawing attention for advantages such as weight reduction and high-precision products. In the future, both high impact polystyrene HIPS (an amorphous polymer) and PP (a crystalline polymer), which are broadly used for motor parts etc., are expected to be utilized for molding applying this technology[1]. Thus investigations on foam morphologies affecting the quality of foam products are important.

In the past, SEM observations of fractures were carried out to observe the morphology of molded products[2,3,4]. In recent years, observations of detailed cell distribution and shape using X-ray CT are being attempted[5,6]. However such trials have only been reported for extrusion, and not for microcellular injection molding. In this study, the relationship between molding conditions (maximum pressure of polymer in the cavity) and 3D morphologies was compared between HIPS and PP through detailed investigations using X-ray CT technology.

Experimental

Figure 1 shows the cavity shape. The cavity shape was rectangular and a side gate was adopted to investigate the basic foam morphology. Polymer pressure was measured with a pressure transducer at A shown in Figure 1 (type 9221, ϕ3.5mm, Kistler Japan Co., Ltd.). The injection molding machine used was J85ELⅢ-110H-Mucell with a mold clamping force of 833kN and screw diameter of ϕ32mm (Japan Steel Works, Ltd.). Table 1 shows the molding conditions and PP and HIPS grades.The foaming agent used was scf-N_2. The most suitable concentration of the foaming agent was provided for each polymer to ensure microcellular formation and to ensure that the polymer

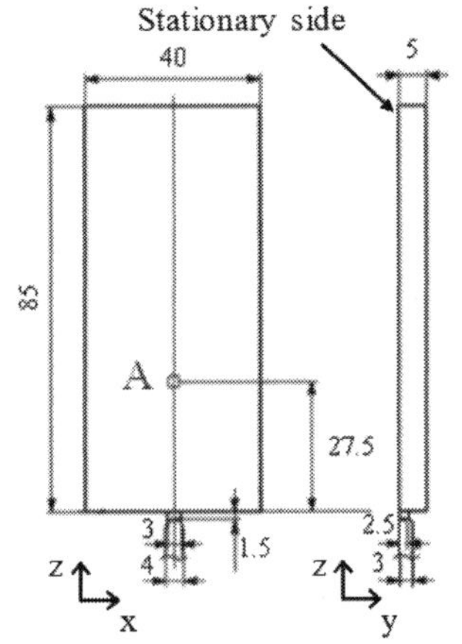

Figure 1 Cavity shape (unit:mm)

Table 1 Experimental conditions

Resin	HIPS M220 (Japan Polystyrene Inc.)	PP MH4 (Japan Polypropylene Corp.)
Resin temperature (°C)	220-215-215-215-215-200-190-170(N-H)	200-200-190-180-180-200-190-180(N-H)
Mold temperature (°C)	30	
Maximum cavity pressure Pmax at A (MPa)	5/15/30	5/14/30
Injection rate (cm^3/s)	40.2	
Holding pressure (MPa)	0	
Back pressure (MPa)	15	
Cooling time (s)	110	
N_2 concentration (wt%)	0.3	0.4

and gas were not separated. The injection volume was adjusted according to the molding conditions, and the maximum pressure (Pmax) at A was set at 5MPa, 15MPa, and 30MPa in the compression process.

The morphologies of one part of the molded sample (width direction x; 2.3mm, flow direction z; 1.3mm, thickness direction y; 2.3mm) around A were investigated using micro-focus X-ray CT (SMX-130CT SV, Shimadzu Corp.). Cell shapes and

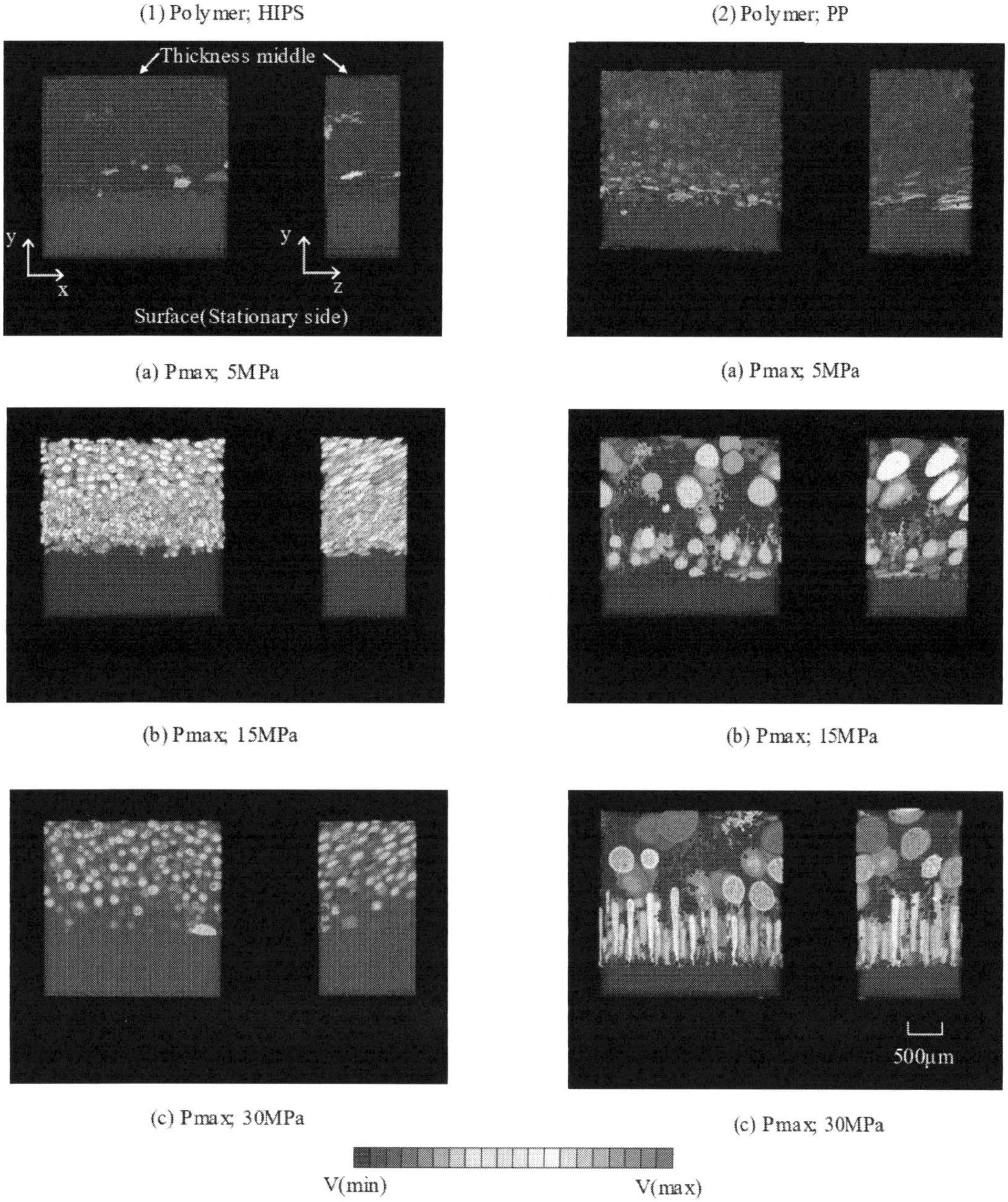

Figure 2 X-ray micro-tomography images from surface (stationary side) to middle of thickness at each Pmax (Injection rate;40.2cm^3/s, Cavity position;A)

distributions obtained by X-ray CT were analyzed using the software VG Studio Max1.2.1 (Nihon Visual Science, Inc).

Results

Figure 2 shows the 3D micro-tomography images of HIPS and PP at each Pmax condition. A color scale was assigned to the cell volume in each micro-tomography image (Left side; perpendicular section to the flow, Right side; cross section along the flow direction in Figure 2). Cells were colored and relatively displayed. The characteristic layer morphologies in the thickness direction are shown for each condition. The common morphologies shared by the two polymers at each condition were as follows. The surface layer (skin layer 1) had silver streaks and the inner layer was composed of micro cells (d: Cell diameter < 30μm) and had a thickness of 80μm. Inside this, a nonfoam layer (skin layer 2) was formed, and there was a foam layer (core layer) with cells at the center. Only the cell morphology of the core layer changed according to the change in Pmax. When Pmax increased, the average cell volume of both polymers increased, while the cell density decreased. Next, the changes in the core layer were observed for both polymers at various conditions. When Pmax was 5MPa, the core layer of HIPS showed distribution of micro cells over the whole area, and large cells with high aspect ratio in the flow direction at the boundary area between skin layer 2 and core layer. When Pmax was 15MPa, the core layer was found to be a multi-layer structure where the vicinity of its surface had a high cell density composed of cells of small volume, while its central layer had low cell density composed of cells with large volume. When Pmax was 30MPa, the core layer had a few large cells distributed over the whole area. When Pmax was 5MPa, the core layers of PP and HIPS were quite similar to each other except the absolute values of cell volume and layer thickness. When Pmax was 15MPa, HIPS and PP shared the same basic structure, but differed in that the PP core layer had micro cells distributed at the center (d < 50μm). When Pmax was 30MPa, the PP core layer differed significantly from that of HIPS. "Icicle-shaped long cells" were observed in the thickness direction at the boundary area between skin layer 2 and core layer. These "icicle-shaped cells" were touching the skin layer, and grew towards the inside layer, widening their own cross-sectional diameters. Large cells and numerous micro cells (d < 50μm) were found to coexist at the middle area of the core layer.

Discussion

When Pmax is low, the cell state in the filling process is maintained even in the compression process, resulting in the formation of the foam morphology of the molded product. On the other hand, when Pmax is high, cells disappear during the compression process, and new cells are generated during the decompression and cooling processes, resulting in the formation of the foam morphology of the molded product. Because the cell distributions of PP and HIPS are almost similar in the filling process, their foam morphologies are also quite similar. On the other hand, when new cells are generated in the decompression and cooling processes, the difference in the glass transition temperature (Tg) and whether crystallization occurs in the cooling process cause the difference in morphology between HIPS and PP. Because the temperature of amorphous polymer is below Tg in the latter half of the decompression and cooling process, cells which were formed around the same time in the early stage grow in the same way, resulting in the uniformity of their cell diameters. On the other hand, because the Tg of PP is much lower than the mold temperature throughout the decompression and cooling process, cells are generated continuously in the whole prosess. The cells generated in the early stage of the process grow considerably, while in the latter stage, numerous microcells were formed due to the effects of crystalline nucleus and the increase in the viscosity of the polymer surrounding the cells. For this reason, many microcells coexist in PP. When Pmax was 30MPa, nucleation and cell growth are restrained in the decompression and cooling processes. Because the cooling rate slows down from skin layer 2 towards the surface to the middle section, volumetric shrinkage is thought to occur with the crystallization process in the latter stage of the cooling process. For this reason, the cells near skin layer 2 grow toward the inner layer to make up for the volumetric shrinkage. Cells are thought to become the icicle shape at the boundary area between skin layer 2 and core layer.

Conclusions

The present paper discusses the correlation between molding conditions (maximum pressure of polymer in the cavity) and 3D foam morphologies of PP and HIPS in microcellular injection molding through comparative investigations using X-ray CT. The following results were obtained:

1. The common morphologies shared by HIPS and PP were found to be having a silver streak layer (skin layer 1), having no foaming layer (skin layer 2) and foaming layer (core layer).
2. When Pmax was 5MPa, the foam morphologies of HIPS and PP became quite similar. When Pmax exceeded 15MPa, many micro cells (d < 50μm) were found to coexist with the morphology of HIPS in PP (a crystalline polymer).
3. When Pmax was 30MPa, formation of "icicle-shaped long cells" from the boundary area between skin layer 2 and the core layer to the internal layers was seen only for PP, and this shape widened the cross-sectional diameter of the cells toward the middle.

References

1. Ogawa,J., et al. "Development of Door Module Carrier Using Foam Molding Technology", Mazda Technical Review,27,21-25(2009)
2. Yuan,M. "Effects of Nano- and Micro-fillers and Processing Parameters on Injection-Molded Microcellular Composites", Polymer Engineering & Sience,45,6,773-788(2005)
3. Huang,H.X.,Wang,J.K. "Equipment development and experimental investigation on the cellular structure of microcellular injection molded parts", Polymer Testing,27,4,513-519(2008)
4. Yamada,T., et al. "Correlation between Molding Conditions and Foam Morphology in Microcellular Injection Molding", seikei-kakou, 21, 10, 633-639(2009)
5. Cho,K.Y.,Rizvi,S.S.H. "3D microstructure of supercritical fluid extrudates I", Food Res Int, 42, 5-6,595-602(2009)
6. Cho,K.Y.,Rizvi,S.S.H. "3D Microstructure of supercritical fluid extrudates. II: Food Res Int, 42, 5-6,603-611(2009)

The influence of polymer nanocomposite rheology on foam cell structure

Sati N Bhattacharya, Rahul K Gupta*

(1) RMIT University - Victoria - Australia

Polymer clay nanocomposite foams were prepared in a batch process. The effect of polymer melt rheology on foam cell morphology was modelled using the PTT rheological model. A finite difference technique was used to generate simulation results using the rheological model. The PTT model was chosen since this constitutive equation incorporates shear as well as extensional rheology in its form. Subsequently the dynamic and shear rheological properties of the nanocomposites were studied using the ARES rheometer.The experimentally determined relaxation spectrum obtained from the dynamic data was used to determine the effect of extensional viscosity and dynamic viscosity on the nanocomposite sample. The simulation results show that the polymer melt viscoelasticity has a domineering influence in the generation of fine cells. More over the results reveal that the use of a branched polymer leads to smaller cell sizes.
Keywords: Nanocomposites, Shear Rheology, PTT Model, Nano-cellular Foam.

EXTRUSION OF HYDROSILYLATED POLYPROPYLENE WITH SUPERCRITICAL CARBON DIOXIDE

Mercy P. Bulsari, Costas Tzoganakis, Alex Penlidis*

University of Waterloo - Ontario - Canada

Polypropylene (PP) is a commodity polymer with balanced bulk physical and mechanical properties, used in a wide range of applications. On the other hand, polymer surface properties play an important role in determining wettability and adhesion, friction and wear, gloss and scratch resistance, paintability and printability. Hence, it is of interest to find a way to chemically alter PP to affect such properties. This would further open a path to the formation of interesting copolymers, and the production of compatibilizers for inorganic fillers and polymer blends. In the present work, such a modification of PP has been attempted through hydrosilylation reactions. Hydrosilylated PP or silicone-modified PP has been investigated in a twin-screw extrusion process. This process involved two steps: i) peroxide initiated degradation, which leads to the formation of double bonds, and ii) the addition of silanes to such bonds. Detailed description of hydrosilylation of PP has been discussed in Bulsari M et al. (2008). Extrusion of silicone-modified PP with supercritical carbon dioxide was studied in a twin-screw extruder. Silicone-modified PPs are expected to exhibit better foaming behavior compared to conventional PP due to increased solubility of supercritical CO_2 into the silicone component of the material and hence supercritical carbon dioxide was used in this reactive extrusion system. Rheological measurements were carried out by using a wedge die, showing the effect of the concentration of the supercritical carbon dioxide on the shear viscosity of the silicone-modified PP blends. Results revealed that incorporation of supercritical carbon dioxide led to a reduction in shear viscosity of the silicone-modified PP material. Foaming behavior was evaluated through scanning electron microscopy and it was found that the supercritical carbon dioxide resulted in significant change in the foaming structure of the silicone-modified PP blends.

DYNAMIC MECHANICAL BEHAVIOUR OF POLYSTYRENE BEAD FOAMS

M.I.Stumpf[1]*, F.Fischer and V.Altstädt[1]

[1] *Department of Polymer Engineering, University of Bayreuth, Bayreuth, Germany*
* *marieluise.stumpf@uni-bayreuth.de*

Since their development in the 1950's, bead foams based on expandable polystyrene (EPS) have been widely used as packaging materials and for insulation products. The main disadvantage of EPS is the brittleness, leading to a reduced service life. The material shows two common failure mechanisms, such as brittle cracking under impact and cell buckling under permanent vibrations. Further alternatives like expanded polypropylene (EPP), expanded polyethylene (EPE), and EPS modified with copolymers or high impact modifiers, have been investigated and showed an improved mechanical behavior, however, the processing of such materials is rather difficult. Therefore, the fundamental understanding of the fracture mechanisms of EPS is of major relevance.
Parts made from these materials show a rather special composition, as they are made of single beads, which have a closed-cell foam structure and show a thin compact layer as shell. In the case of EPS, the beads are pre-foamed to settle the density and afterwards moulded to parts. Therefore, the parts consist of foamed beads with a percentage of interfaces, which can be varied using different bead sizes, and a compact layer outside the part. Furthermore, there are small cavities between the beads that can act as voids when a crack propagates through the material.
The aim of the study is to investigate the influence of density, compact layer and bead size on the fracture behavior of moulded EPS parts. Static and dynamic stepwise loading tests have been performed under compression and three-point flexural. Subsequently, the fracture surfaces are investigated by scanning electron microscopy to reconstruct the failure mechanisms under different loadings. Results show that the density plays the most important role on the mechanical properties, nevertheless, the thickness of the compact layer as well as the bead size can be used to tailor the properties of bead foam parts.

Introduction

Polystyrene bead foams are widely used in packaging and insulation industries, as well as for protective equipment like helmets (Stupak et al., 1991, Mills and Kang, 1994). The foamed material itself is known since more than fifty years, and the wide range of densities and material varieties offers a broad application range. EPS is used as a low-density protective material in packaging applications, in thermal insulation applications for the low thermal conductivity and in electrical insulation applications for their high dielectric properties. The material is also used for the low stiffness in cushioning applications as mentioned before (Bureau and Gendron, 2003).
Expandable polystyrene (EPS) consists of single beads that are pre-expanded in a first step. Therefore the material is heated up above glass-transition so the entrapped n-pentane (normally around 5-7 %) leads to expansion of the PS. After a selected time (normally around 8-24 h) of interim storage with the atmosphere for pressure equalization, the beads can be moulded (Rosacci and Shivkumar, 2003).
In this step the pre-expanded beads are fused with steam in an aluminum mould with a certain steam pressure. Therefore, the pre-expanded beads are heated up a second time above glass-transition. The remaining blowing agent leads to further expansion and fusion of the beads, thus forming a pattern of the mould cavity (Rosacci and Shivkumar, 2003). The quality of the bead fusion depends on several factors, like remaining n-pentane content, mould temperature, steam pressure and pre-treatment of the surface. To facilitate the pre-expansion process and to obtain no agglomerates the beads are normally covered with a stearate.
A main disadvantage of EPS is the brittle fracture behaviour of the material during service life, especially regarding the applications like protective equipment and packaging. While looking at the usual applications, there can be identified two different requirements: impact, which is a problem while using the material in protection applications or as packaging material and transportation phenomena like permanent vibrations while stacking the packages on top of the other.
The use of single pre-foamed beads leads to interfaces in the foamed part. Furthermore, the moulding leads to a compact layer around the part, as can be seen in Figure 1.
For the reason of investigations on the influence of the interfaces of the single beads in the foamed parts, another foamed material, extruded polystyrene (XPS) is used. The material is also often used in the case of insulation applications like EPS, but is manufactured by a foam extrusion process. The material used in this

study shows a similar cell structure and density to EPS but has no interface structure due to the missing beads.

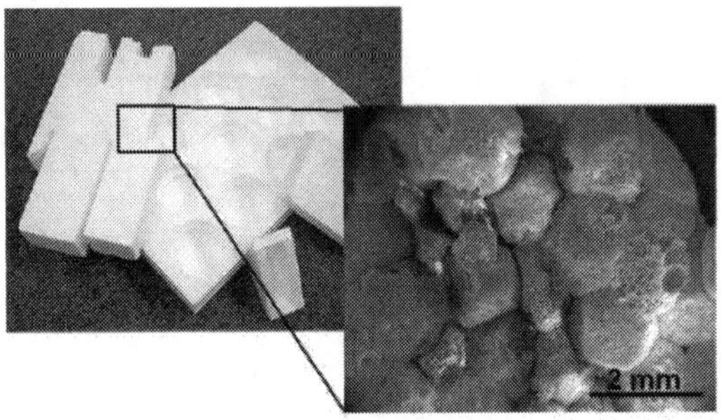

Figure 1 Part made of EPS and SEM picture of the material. The material consists of single beads with a foamed core and a compact surface around the whole part.

The moulded EPS-parts show a density gradient to the middle of the part that has to be taken in consideration if mechanical data is evaluated. For the beads fused with steam, there are also irregular small cavities between the beads, which can act as voids, and lead to early failure. During cooling of the part the steam diffuses out through these voids (Mills and Gilchrist, 2007).
In this work, quasi-static and dynamic properties of particle foams were assessed. For quasi-static tests, compression and three-point flexural tests were performed for three different densities and two bead sizes. For the three-point flexural tests, the influence of the compact layer was also investigated. In the case of dynamic testing, three-point flexural tests under stepwise increasing loading were carried out.

EXPERIMENTAL

Material and sample preparation
EPS discs with three different densities, 15 g/l, 25 g/l and 80 g/l and two different bead sizes (0.7-0.9 and 0.9-1.3 mm relating to the compact beads) were used to evaluate the influence of the density and the bead size on the mechanical properties. All samples were prepared using the same moulding parameters and moulding machine. The discs were moulded on a Erlenbach moulding machine with a steam pressure of 0.8 bar.
As a reference, samples made of XPS, a commercially available material under the brand name Styrodur from BASF SE, with a density of 25 g/l were also used for the dynamic tests.
The specimens for the mechanical tests were taken out of discs with a diameter of 300 mm and a thickness of 55 mm. Moreover, specimens with and without compact layer (with CL, without CL) were prepared to investigate its influence on the properties of the part. Except as noted otherwise, the compact layer is on the compression side of the samples.

Table 1 Sample matrix

Bead size (mm)	Density g/l without compact layer (without CL)			Density g/l with compact layer (with CL)		
0.7-0.9	15	25	80	15	25	80
0.9-1.3	15	25	80	15	25	80
XPS		25				

For further investigations on the effect of the compact surface, specimen with different abrasion depths of the compact layer (up to 7 mm, which means complete remotion of the compact layer) were prepared by keeping the thickness of the specimen constant. These samples were also tested by three-point bending.

Morphology & Density distribution
The morphology of the samples was investigated by scanning electron microscopy. The fracture surfaces of the tested specimen were therefore coated with a thin layer of gold to prevent charge build-up and examined with a Jeol SEM under an acceleration of 15 kV.
Two discs of each density were tested on the homogeneity of the density distribution. Therefore the discs were cut in three layers, the layers cut in eleven cubes per layer and the density of each cube was defined precisely by calculating the density out of the volume and the weight of the cubes.

Quasi-static testing
For the quasi-static and the dynamic mechanical measurements, three-point bending and compression test specimen were prepared by sawing the specimen out of the discs. Compression tests were carried out according to DIN 53421 on a universal testing machine Zwick Z050 with a load cell of 1 kN. Five cubes per material with an edge length of 50 mm were compressed with a deformation rate of 5 mm/min related to the original height of the specimen.
Three-point flexural tests were carried out using a Zwicki Z2.5 according to DIN 53423. Five test specimens were sawed with a thickness of 20 mm, a width of 25 mm and a length of 120 mm. The displacement was 100 mm. For the quasi-static testing, three densities and two varying bead sizes were used to investigate the influence of density and interface fraction of the samples. Furthermore the influence of the compact layer was investigated by preparation samples with varying abrasion depths.

Fatigue testing
The dynamic stepwise increasing load tests under bending were carried out with a servohydraulic test machine with a 10 kN servo Cylinder and a 200 N load cell at room temperature. The load ratio was R=0.1 with a sinusoidal waveform. A frequency of 5 Hz was used. The displacement was 100 mm, the sample geometry similar to the quasi-static three-point flexural loading tests. Three specimens per material were investigated. The maximum dynamic strength of the

sample was determined under stepwise increasing loading until fracture of the sample occured or the end of the load cycles was achieved. The loading pattern can be seen in Figure 9.

By continuous monitoring of several characteristics (minimum strain, dynamic modulus, damping) during the test the gradually fatigue of the material was quantified. The length of time of such tests is typically several hours.

RESULTS & DISCUSSION

Morphology & Density distribution

The main differences of the cellular structure of the materials, regarding the bead size and inhomogeneity of the cells, can be observed in the SEM-pictures. Figure 2 shows the structure of the parts based on two different bead sizes. The size is related to the compact, non-foamed beads. The size of the foamed beads can be up to 5 mm.

The bead fusion depends on moulding time, moulding pressure and blowing agent content and is characterized as the ability to cleave or remove a single bead (Mills and Kang, 1994). Figure 2 shows a high percentage of cellular fracture, what indicates a high-quality bead fusion for both bead sizes. Furthermore, the material shows no cavities what also indicates a good bead fusion. Nevertheless, there are sporadic big cells in the single beads, which also can act as imperfections. These inhomogeneities can be particularly seen for the specimen prepared with smaller bead sizes.

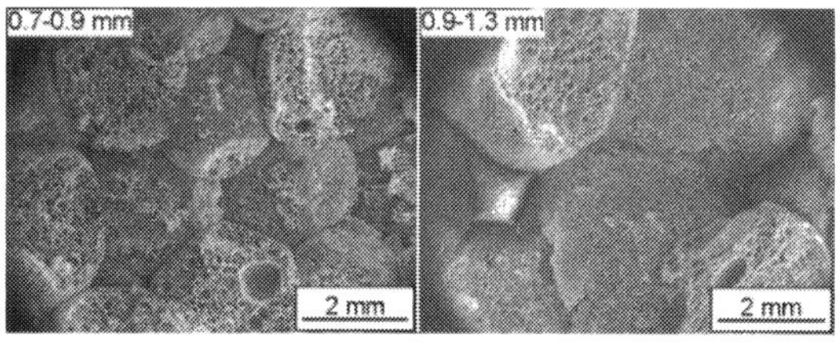

Figure 2 Foamed EPS with compact bead size 0.7 – 0.9 mm and 0.9 – 1.3 mm. Both have a density of 25 g/l.

The density investigations on the 25 g/l disc showed a high reproducibility for all discs. The discs used have a thickness of 55 mm and a diameter of 300 mm. The average real density of the tested disc shown here exemplary is 25.16 g/l. The material shows a maximal density deviation of 2.78 %. These values were controlled and verified with further determinations on other discs. It could be also seen, that there is a density gradient from the layers to the center of the disc. The specimens used for further investigations were prepared precisely from the middle of the disc.

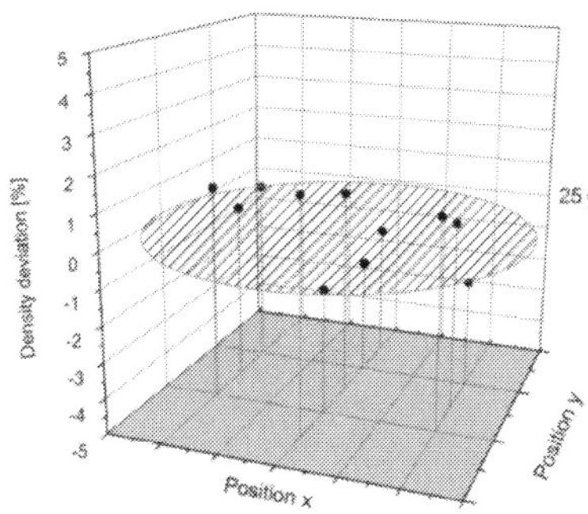

Figure 3 Density deviation of an exemplary disc with a density of 25 g/l and a bead size of 0.7-0.9 mm.

The lower density in the core in comparison to the skin layers of the moulded part is due to the manufacturing process of the material. The steam diffuses from the outside of the beads to the core and first compacts the surface. Therefore less fusion is achieved in the center. The materials with the density of 80 g/l show variations up to 4.35 %. There were no significant changes of the density regarding a change of the bead sizes.

Quasi-static testing

The density of the material shows a gradient between the surface layer and the core as can be seen in Figure 3. Depending on the density of the material, there are variations up to 4.35 % for the density 25 g/l.

Table 2 Density deviation of the sample matrix

Bead size (mm)	Real density g/l			Deviation		
0.7-0.9	13,91	25,16	77,03	1,22	2,78	3,79
0.9-1.3	13,79	24,37	81,61	2,83	4,35	3,42

Therefore, the density of each specimen was measured and compared to the results. To achieve a better comparability, the values were normalized.

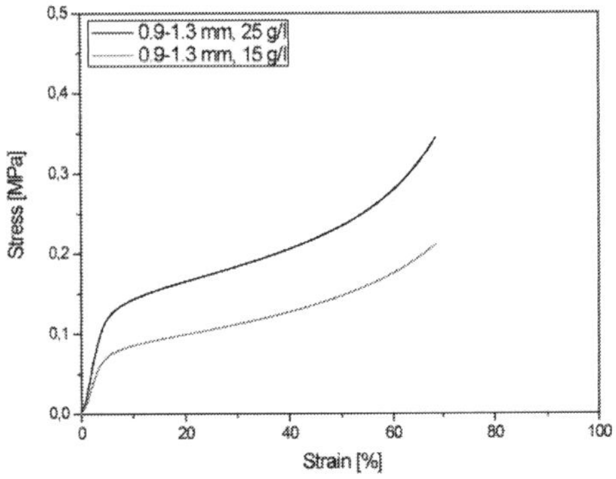

Figure 4 Mastercurve out of five measurements of the compression test of one bead sizes and two different densities.

The mastercurves in Figure 4 are the average of five single measured curves. As can be seen in Figure 4, the stress is at the beginning highly ascending until a strain of 4 % is obtained. Afterwards the curve shows a more moderate progression. In this area, the cell struts are compressed and further compression leads to compaction of the material. With further increasing compressive strain, the material shows a higher resistance for the reason of now being a nearly compact material.

As it can be seen in Figure 5, the specific compression modulus shows a clear dependence to the density. As expected, an increase of the density, results in an increase of the specific modulus.

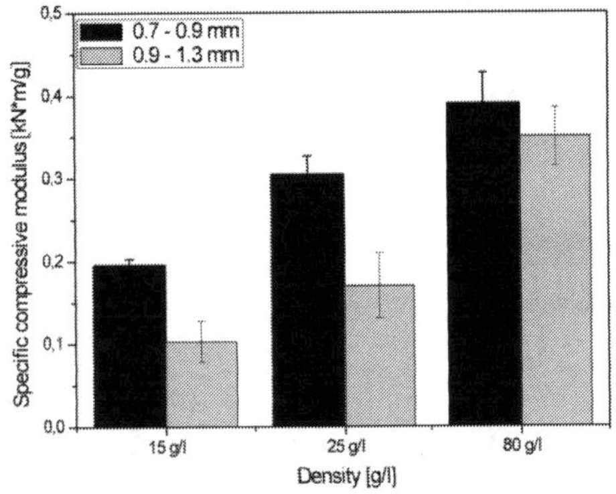

Figure 5 Specific compression modulus of three densities and two bead sizes.

The material with smaller beads and therefore a higher number of compact surfaces shows a higher resistance against compressive deformation compared to the material with the larger beads. This effect can also be seen in the compression modulus (Figure 5) and is due to the higher amount of beads in the specimen.

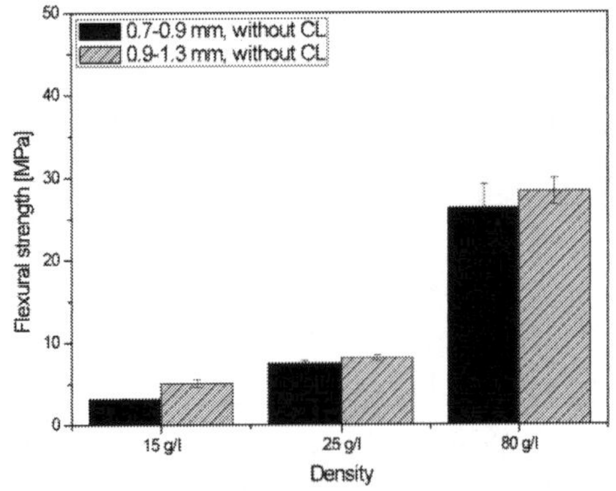

Figure 6 Influence of bead size on the flexural modulus

The influence of the bead size on the specific flexural modulus is shown in Figure 6. The smaller beads show a slightly decrease on the flexural modulus. This is in contrast to the results of the compression test (see Figure 5). A possible reason is the higher percentage of interfaces between the beads, which are less resistant to tension than to compression (Moosa and Mills, 1998). Furthermore, the presence of big cells in the single beads is another possible reason as they act as imperfections.

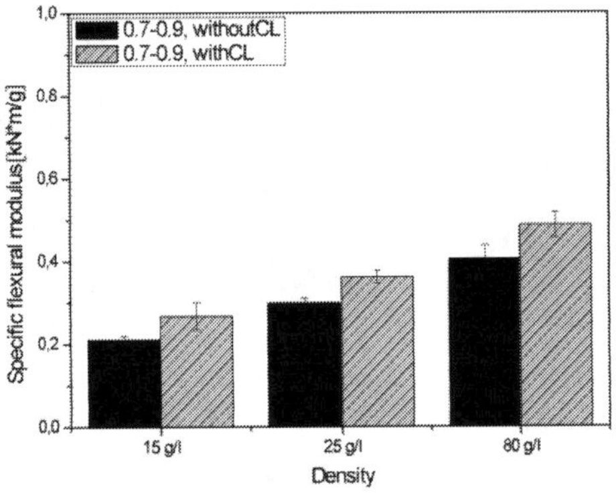

Figure 7 Influence of three densities and the compact layer on the specific flexural modulus of EPS. The compact layer increases the specific flexural modulus slightly.

As can be seen in Figure 7, there is a slight difference for the specific flexural modulus between samples with and without compact layer. To investigate the influence of the depth of the compact layer, further measurements with different abrasion depths of the surface were carried out (Figure 8) by keeping the thickness of the specimen constant.

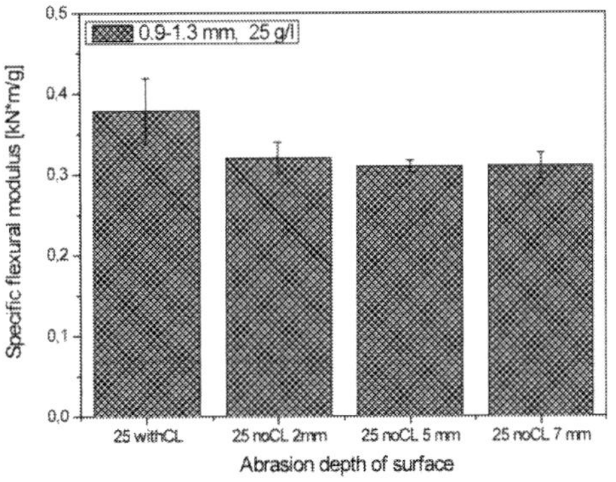

Figure 8 Influence of surface depth on specific flexural modulus

The variation of the abrasion depth shows an influence up to 2 mm depth. Between 5 and 7 mm, the material shows no more differences in the specific flexural modulus. The presence of the complete compact layer shows the highest specific flexural modulus. These results suggest therefore that the compact layer has a thickness between 2-4 mm. The exact thickness of the compact layer is currently under investigated and will be discussed in future works.

Fatigue testing

The stepwise increasing load test was carried out with a loading pattern as seen in Figure 9. The specimen was constantly loaded with 20 % of the static flexural strength and the loading pattern was increased up to 80 %. Between the loading steps, partial unloading for 1000 cycles was carried out.

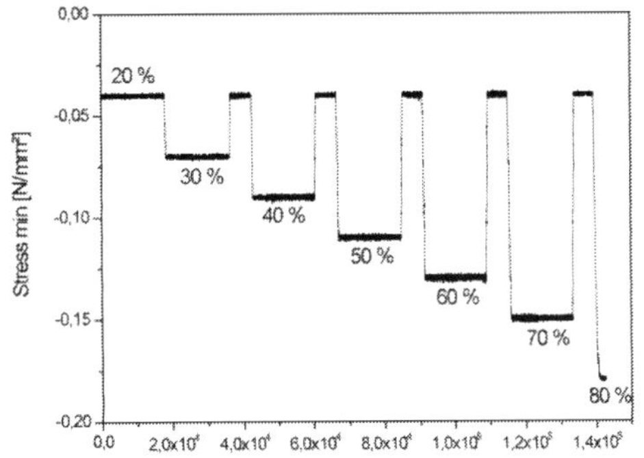

Figure 9 Loading pattern as a function of the number N of cycles for the stepwise increasing load test (SILT) methodology. The stress amplitude was incrementally increased in steps of 10 % after unloading for 1000 cyles.

The specimens with a density of 25 g/l without compact layer but different bead sizes showed no failure up to the end of the loading cycles. Compared to the same material density but different bead size and a further extruded material XPS as reference, it can be seen that an increase in the bead size leads to higher dynamic creep. The lowest dynamic creep occurs for the extruded material.

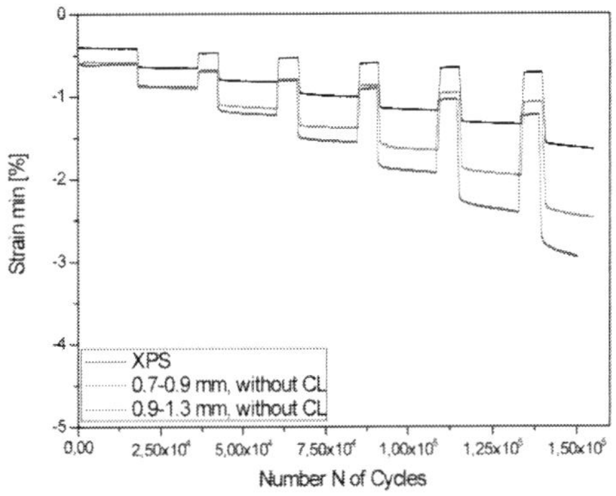

Figure 10 Minimal strain as a function of number N of cycles for EPS with a density of 25 g/l and two bead sizes.

The comparison of samples with and without compact layer shows a clear influence of the existing compact layer, as can be seen in Figure 11. The abrasion of the compact layer leads to a significant lower minimal strain and higher dynamic creep.

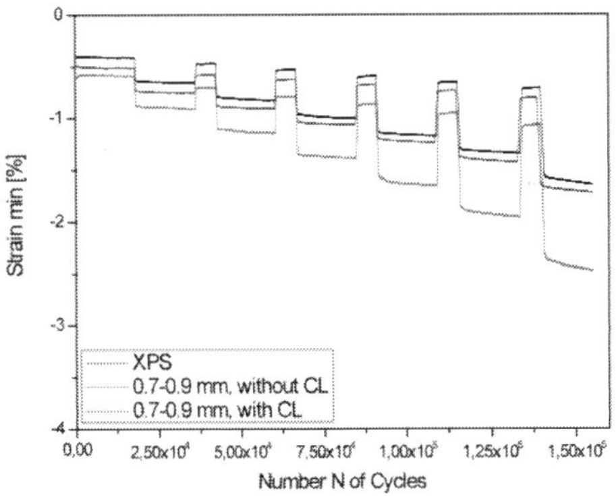

Figure 11 Minimal strain as a function of number N of cycles for EPS and XPS with a density of 25 g/l and a bead size of 0.7 – 0.9 mm.

Regarding both bead sizes and the existence of the compact layer, it can be clearly seen, that smaller beads lead to lower dynamic creep as well as an existing compact layer (Figure 12).

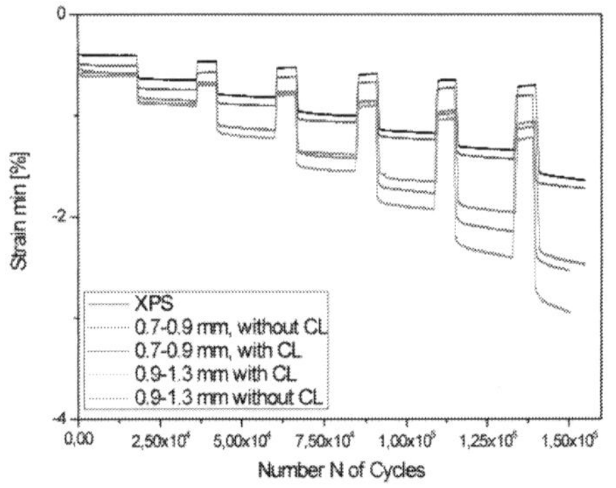

Figure 12 Minimal strain as a function of number N of cycles for EPS and XPS with a density of 25 g/l and two bead sizes as well as an existance of the compact layer.

The comparison of the minimal strain shows relative large differences between the bead sizes (Figure 13). Therefore, the minimal strain at the end of each loading cycle was determined. Especially the material with 0.9-1.3 mm beads and without compact layer shows a stronger decrease in dynamic strain with accessing number N of cycles.

The extruded polystyrene and the smaller bead size (0.7-0.9 mm) with existing compact layer show a lower decrease in minimal strain.

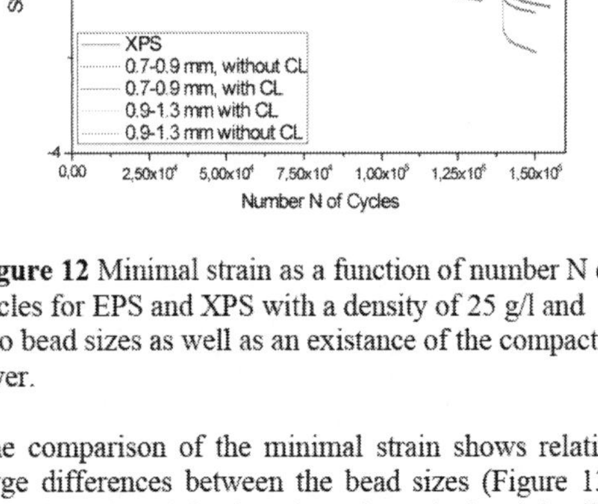

Figure 13 Fraction of the flexural strength (minimal strain). The data points are determined at the end of each loding cycle.

Due to the dynamic creep during testint in the materials the strain is increasing despite a constant stress amplitude in the relaxation cycles (10 % of static flexural strength). This effect can be seen for all materials (Figure 10 and Figure 11), especially for the bead size of 0.7-0.9 mm with compact layer.

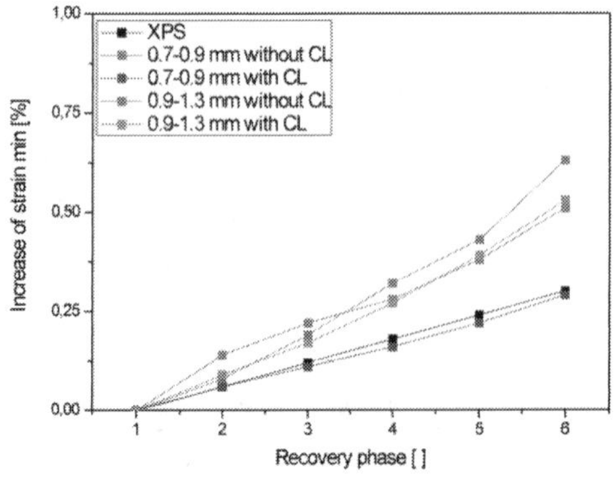

Figure 14 Strain at relaxation in comparison with the particular recovery phases.

CONCLUSION

The dependence of the bead size of expandable polystyrene shows clear differences according to type of loading. In quasi-static compression load, the higher amount of interfaces while using smaller beads improves the resistance against compression strain. The material showed compaction and no cell fracture during loading. The stress-strain curves show similar curve progressions for the different densities.

In contrast, the three point flexural tests show the opposite behaviour for the varying bead size. In this case the detached existing big cells act as imperfections for the crack. Furthermore, the influence of the compact layer was investigated by varying the abrasion depth while keeping the thickness constant. It could be illustrated that the influence of the compact layer is to a depth of 2 mm.

Both investigations showed a high influence of the density on the mechanical properties.

The stepwise increasing load test showed dynamic creep of the material. The compact layer, as well as the smaller size of beads contribute to improve the fatigue behaviour. For the materials with the same density and an existing compact layer, it was observed that the layer acts positively on the dynamic testing. Futhermore, the extruded polystyrene shows in comparison to the other materials the lowest dynamic creep.

In both static and dynamic flexural loading, specimens failed by crack propagation on the tension side of the beam.

Therefore, it can be pointed out that the use of EPS can be adjusted by using different densities and bead sizes, which depends on the particular application. Furthermore the compact layer plays a major role on the mechanical behaviour and acts favourably regarding flexural loading. The use of smaller beads can be beneficial especially for applications in the area of protective equipment, where the material is submitted to compression loads.

These trials show the importance of further investigations on the fracture mechanical properties to further identify the failure mechanisms.

ACKNOWLEDGEMENTS

The authors acknowledge the support from BASF SE by kindly providing the materials.

REFERENCES

Journal: P.R. Stupak, W.O. Frye, and J.A. Donovan, "The effect of bead fusion on the energy absorption of polystyrene foam. Part I: Fracture toughness" Journal of cellular plastics, 1991. **27**: p. 484-505.

*Journal:*N.J. Mills and P. Kang, "The effect of water immersion on the mechanical properties of polystyrene bead foam used in soft shell cycle helmets" Journal of Cellular Plastics, 1994. **30**(3): p. 196-222.

*Journal:*M.N. Bureau and R. Gendron, "Mechanical-morphology relationship of PS foams" Journal of cellular plastics, 2003. **39**(5): p. 353-367.

*Journal:*J. Rossacci and S. Shivkumar, Bead fusion in polystyrene foams. J. Mater. Sci., 2003. **38**(2): p. 201-206.

*Journal:*N.J. Mills and A. Gilchrist, "Properties of bonded-polypropylene-bead foams: data and modelling" J. Mater. Sci., 2007. **42**: p. 3177-3189.

*Journal:*A.S.I. Moosa and N.J. Mills, "Analysis of bend tests on polystyrene bead foams" Polymer Testing, 1998. **17**(5): p. 357-378.

NANOCELLULAR FOAMS OF IMMISCIBLE AND MISCIBLE PEEK/PEI POLYMER BLENDS

M. Ohshima[1], T. Nemoto[2], and J. Takagi[2]

[1]*Material Process Engineering Division, Dept. of Chemical Engineering, Kyoto University, Kyoto 615-8514, Japan*
oshima@cheme.kyoto-u.ac.jp
2 Research & Development Laboratory Mitsubishi Plastic Inc. Shiga 526-8660, Japan

Abstract - Plastic foam with a nano-scale cell structure was prepared from poly(ether ether ketone) (PEEK)/para-diamine poly(ether imide) (p-PEI) as well as PEEK/meta-diamine poly(ether imide) (m-PEI) blends by a temperature quench foaming method with CO_2. The difference in chemical configuration between the m- and p-PEI gave rise to a prominent change in the higher-order blend morphology and cell structure of their foams. The bubble nucleation site and bubble size were controlled by templating the morphology of the PEEK/p-PEI blend, which shows an immiscible and unique strip-patterned crystalline morphology. The properties influenced by the immiscibility of the PEEK/p-PEI blend were investigated using SEM, thermal analysis and rheology and compared with the properties of the miscible PEEK/m-PEI blend. The bubble size and location were highly controlled in the PEI disperse domain that was aligned between the PEEK crystalline layers in the PEEK/p-PEI blend.

Introduction

In recent years, there has been great interest in nano-structured materials. Nanoscale porous structured polymer material is attracting attention because of its many applications. Polymer foams can be used to prepare these nanoscale porous polymeric materials (or nanocellular foam). Previous efforts have focused on creating microscale porous polymeric foams (or microcellular foam). The microcellular technique realized cellular structures that were c.a. 10 μm in cell diameter; unfortunately, it could not yield smaller structures with diameters of 50 – 500 nm. Subsequent studies have thus concentrated on investigating methods to reduce the cell size down to the nanoscale level[1,2]. We have also been working on nanocellular foams and proposed a method of nanocellular foam preparation[3-5]: it is to use nanoscale-ordered spherical morphology of polymer blends or alloys as a template for bubble nucleation and growth and to make solubility of foaming agent higher and viscoelasticity of disperse domain lower than those in the matrix polymer for localizing bubble nucleation and suppressing the bubble growth in disperse domain.

In this study, we investigated the use of nanocellular foam of an immiscible polymer blend to reduce the cell size furthermore and to control the bubble location. The polymer blend used was a high Tg-polymer blend, i.e., the polyether ether ketone (PEEK) and poly(ether imide) (PEI) blend. We used PEEK as the matrix polymer to exploit its crystalline nature to control elasticity and PEI as the disperse phase to localize bubble nucleation. In this study, a PEI, whose chemical structure for the diamine component comprises para-phenylenediamine, was used to prepare an immiscible PEEK/PEI polymer blend. The PEEK/para-diamine PEI (PEEK/p-PEI) blend shows unique sea-island morphology, in which the PEI composes the strip-pattern disperse domain. Utilizing the unique morphology of PEEK/p-PEI, nanocellular foam having multi-layers of bubbles smaller than 100 nm in diameter was prepared.

Experimental

A. Material

Poly(ether ether ketone) (PEEK: PEEK450G) received from VICTREX was used as in its original condition. The weight-average molecular weight of PEEK450G was 4.0 $\times 10^4$ g/mol. Two different types of poly(ether imide) (PEI), of which the chemical configuration for the diamine component was composed of either meta or para-phenylenediamine, were used in this study: one was Ultem1000, which is made up of meta-diamine and hereafter called m-PEI, and the other was UltemCRS5001, which is made up of

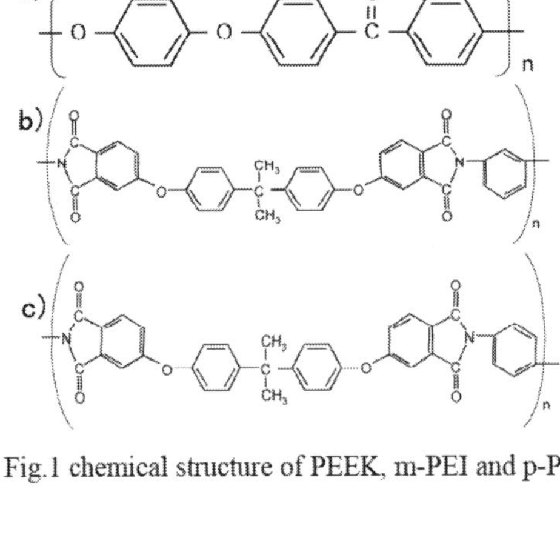

Fig.1 chemical structure of PEEK, m-PEI and p-PEI

para-diamine and hereafter called p-PEI. The weight-average molecular weight of Ultem1000 was 3.2 $\times10^4$ g/mol and that of UltemCRS5001 was 3.7 $\times10^4$. The chemical structures of PEEK, m-PEI and p-PEI are shown in Fig. 1 a), b) and c), respectively. Carbon dioxide 99.99 % in purity was purchased from Showa Tansan CORP and used as a blowing agent.

B. Melt Blend and Film Preparation

A single-screw extruder was used at a screw rotating speed of 30 rpm to blend PEEK and PEI at various weight fractions. The barrel temperature was set at 380 °C. Without drawing down and stretching at the die, films about 250 μm in thickness were also prepared and used as a reference to observe the effect of the draw down ratio on cell morphology. The four different PEEK/PEI blend films were prepared from m-PEI and p-PEI by varying the composition of PEI at two levels, 40 and 60 wt%.

C. Morphological Observation

The morphology of the blend polymers was observed by a transmission electron microscope (TEM: JEM-1200EX, JEOL) after the following treatment: the sample was coated with epoxy (EPSON 828) and cured at room temperature with 10 wt% triethylenetetramine for 24 hrs. It was then sliced by an ultra microtome Reichart Ultracut S (Leica) with diamond knives. The cell morphologies of the polymer foams were observed with a Hitachi S-4500 (Hitachi High-Technologies Corporation) scanning electron microscope (SEM) after the samples were coated with approximately 10 nm gold. RUO4 was used to provide sufficient phase contrast in morphology.

D. Sorption Experiments

CO_2, 99.9% in purity, was used for sorption measurements as well as for foaming experiments. For sorption measurements, the sample was loaded in a high-pressure autoclave and pressurized by CO_2 at 20 MPa while keeping the temperature at 40 °C. Five different sorption times (1, 2, 5, 8, and 15 hrs) were examined, and the weight gains, i.e., CO_2 uptakes, were measured after each sorption time interval just after releasing the pressure without foaming.

E. Dynamic Storage Modulus and Loss Tangent Measurement

The dynamic storage modulus (E') and loss tangent (tan δ) were determined by the dual-level bending mode of dynamic mechanical analysis (DMA) (DVA-200). A small-amplitude bending oscillation (10 Hz) was applied in DMA to PEEK, PEIs and their blends with a 10.0 mm gap. The temperature was increased at a rate of 3 °C min^{-1} from -50 to 400 °C. The test pieces were prepared to be approximately 3.5 – 4.5 mm wide and 40 mm long.

E. Foaming Experiments

In this study, the temperature quench method was employed. The temperature quench method is a two-stage procedure: In the first stage, the polymer is placed in a high-pressure vessel. A foaming agent gas, CO_2 , is introduced into the vessel and is dissolved into the polymer. After depressurization of the vessel at low temperature, that is, a temperature normally slightly below the T_g of the polymer, the polymer is taken out of the vessel and brought to atmospheric pressure without foaming. During the second stage, bubble nucleation and growth are induced by heating. The polymer sample was prepared to be 100 mm long and 100 mm wide and placed in the high-pressure vessel. The vessel was kept at 40 °C and pressurized by CO_2 up to 20 MPa. CO_2 was dissolved into the sample for 1 hr while the temperature was maintained at 40 °C. After an equilibrium state was established, the pressure in the vessel was released at the rate of about 1 MPa/s without foaming the sample. Then, the sample was rapidly immersed in an air oven where the temperature could be controlled at three different levels (150, 180, and 200 °C).

F. Characterization of Cell Structure

The cell diameter (D) was calculated from SEM micrographs. The cell density (N_f), that is, the number of cells per cubic centimeter of foam, was calculated by:

$$N_f = \left(\frac{nM^2}{A}\right)^{3/2} \qquad (1)$$

where n is the number of cells observed in a SEM micrograph, A is the area of the micrograph (in cm^2) where the number of bubbles is counted, and M is the magnification factor.

The density of the sample (d) was calculated by

$$d = \frac{m_{gas}}{m_{water}} \times \rho_{water} \qquad (2)$$

where ρ_{water} is the density of the distilled water at room temperature, and m_{gas} and m_{water} are the masses of the sample in atmosphere and in distilled water, respectively.

RESULTS AND DISCUSSION

A. Cast sheet of neat samples

Figure 1 shows the measured dynamic storage modulus (E') and the loss tangent (tan δ) of neat PEEK cast, m-PEI and p-PEI sheets. The storage modulus of PEEK rapidly depressed when the temperature exceeded 150 °C. At around 165 °C, it reached its minimum value, started increasing and finally leveled around 2.0 x10^8 Pa. The thermal analysis indicates that the T_g of PEEK cast film is 144 °C and its cold crystallization temperature is 174 °C. Thus, the minimum value for E' occurred at around 165 °C. The thermal property data also indicated that the crystallinity of the PEEK cast sheet was initially about

7.7% (=10/130*100) and became 30.7% after the cold crystallization. The crystallinity was calculated assuming that the heat of fusion of a perfect crystal is 130 J/g . m-PEI and p-PEI are amorphous, and their T_g values are 215 and 225 °C, respectively. The dynamic storage moduli of m-PEI and p-PEI decreased monotonically when the temperature exceeded their T_g values.

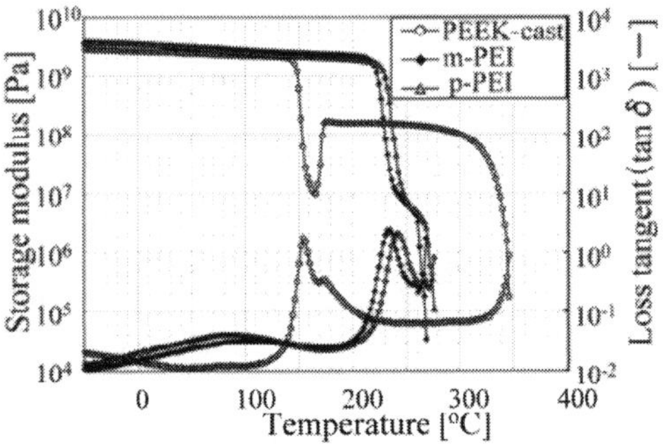

Fig.1 Rheological data of PEEK, m-PEI and p-PEI

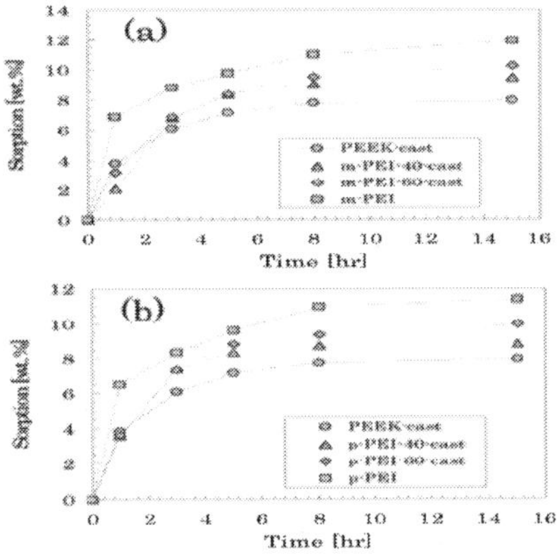

Fig.2 CO_2 sorption curves at 40 oC

Figure 2 shows the sorption curves of CO_2 in PEEK-cast, m-PEI, p-PEI and their blends. The mass uptake, which converges to a steady-state value over the sorption time period, was used to measure the CO_2 solubility in the polymers. The solubility of CO_2 in PEI was larger than that in PEEK at 40 °C. The diffusivity of CO_2 in PEI, i.e., the mass uptake rate, was also larger than that in PEEK. No significant difference in the solubility of CO_2 in m-PEI and p-PEI. In all the PEEK/PEI blends, the solubility of CO_2 increased as the PEI content increased.

Figure 3 shows the SEM micrographs of the foamed, PEEK, m-PEI, and p-PEI neat polymer samples at two different magnifications. Foaming of neat polymers was conducted using a temperature quench method at 200 °C. The cell structures of m-PEI and p-PEI were similar; only their void sizes were slightly different. The void size of m-PEI was smaller, about 80 nm, whereas that of p-PEI was about 100 nm. The uniformity of cell structure was lowest for the PEEK-

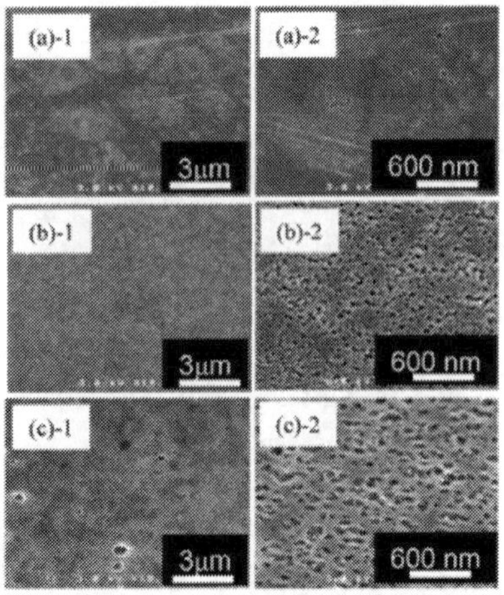

Fig. 3 SEM micrographs of foamed PEEK (a), m-PEI(b) and p-PEI(c)

cast film, where both foamed and unfoamed domains were present. The low degree of uniformity in the cell structure of foamed PEEK is due to the presence of the crystalline phase in PEEK.

B. Cast Sheet of Polymer Blends

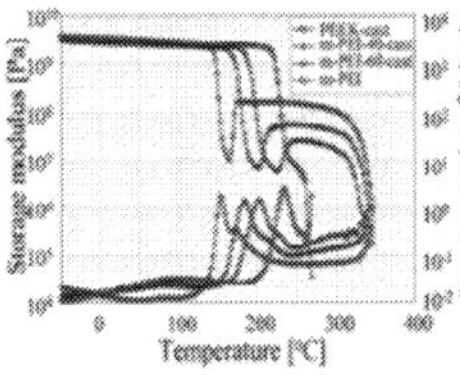
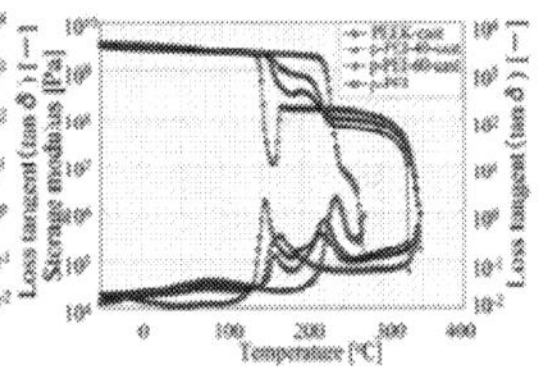

Fig.4 E'&tan δ of m-PEI Fig.5 E'&tan δ of p-PEI

Figures 4 and 5 illustrate the rheological behaviors of the PEEK/PEI blend cast sheets, i.e., the PEEK/m-PEI and PEEK/p-PEI cast sheets. In **Figure 5**, the tan δ versus temperature curves of the PEEK/p-PEI blends showed double peaks while the curves for the PEEK/m-PEI blends showed a single peak (Fig. 4); one of the peaks is at around 165 °C, which corresponds to the T_g of PEEK and the other peak, associated with the T_g of p-PEI, is at 240 °C. The temperatures at which these two peaks were found were shifted from T_g of neat PEEK and that of p-PEI by addition of PEI. However they were not changed by the increase of p-PEI content. The existence of two peaks was maintained even when the p-PEI contents are changed in the blend. This indicates that the PEEK and p-PEI were immiscible.

Figure 6 shows TEM images of both PEEK/m-PEI and PEEK/p-PEI blends. The TEM image clearly shows that PEEK and m-PEI are miscible while PEEK and p-PEI are immiscible. The amorphous domain of PEEK was dyed and is seen as black domains in the TEM image. We ascertained a sea-and-island morphology only in the PEEK/p-PEI blends. When the p-PEI blend ratio was 40%, the p-PEI composed the disperse domains. However, when the PEI blend ratio exceeded 60%, both PEEK and p-PEI formed the continuous phase. It can be said that having meta- and

Fig. 6 Blend morphologies of (a) PEEK/m-PEI-40-cast, (b) PEEK/m-PEI-60-cast, (c) PEEK/p-PEI-40-cast and (d) PEEK/p-PEI-60-cast.

para–diamine in PEI induces a large difference in miscibility as well as in blend morphology.

Figure 7 shows the SEM micrographs of the cross-sectional areas of foamed PEEK/m-PEI and PEEK/p-PEI blends. Just as with the foaming of the neat polymers, foaming of these blend polymers was conducted at 200 °C after dissolving in CO_2 at 40 °C. The average cell size was between about 1 - 6 μm for the PEEK/m-PEI blend with 40% PEI content and between 0.5 and 3 μm for the PEEK/m-PEI with 60% PEI content. Comparing the foams of neat PEEK and the foams of neat PEI (Fig. 3) with the foamed blend polymers (Fig. 7), it could be noticed that the cell sizes in the foamed blends are larger than those of foamed neat polymers. The cell size of the foamed PEEK/p-PEI blend was about 80-150 nm for 40% PEI content and 100 nm for 60% PEI. The cell structures of the PEEK/p-PEI blends were found to be strongly associated with their blend morphologies. The alignment of bubble location increased as the p-PEI ratio increased. This indicated that the bubbles nucleated in the aligned p-PEI domains.

Figure 8 shows TEM micrographs of the cross-sectional area of a PEEK/p-PEI cast. Bubbles

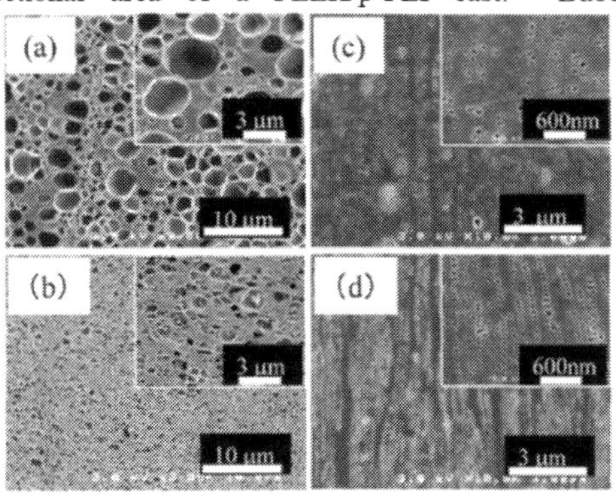

Fig. 7 SEM images of a cross sectional area of (a) m-PEI-40-cast, (b) m-PEI-60-cast ,(c) p-PEI-40-cast and (d) p-PEI-60 foamed at 200°C.

(spherical voids) were observed in PEI domains (black region) in foams. Only the p-PEI domains were foamed. Looking at the bubble location in p-PEI domains in foamed cast sheets, it was observed that the bubbles existed uniformly in the p-PEI domain of the cast sheet foams. These TEM micrographs clearly showed that the bubble nucleation site and bubble size were controlled by templating the morphology of the PEEK/p-PEI blend and utilizing the semi-crystalline polymer, PEEK, as a matrix polymer to increase the stiffness and to suppress bubble growth in the p-PEI domains.

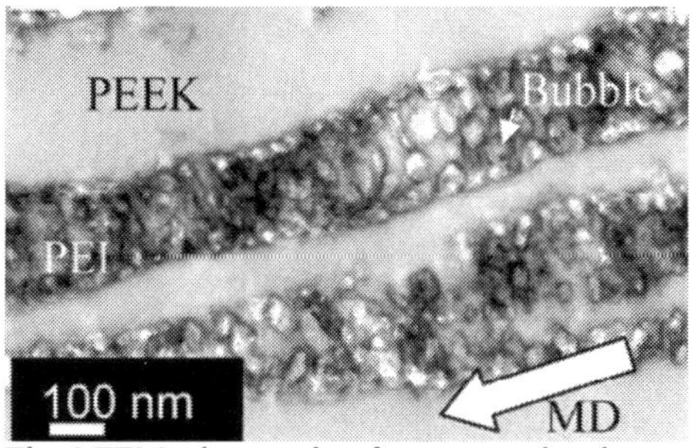

Fig. 8 TEM micrographs of a cross-sectional area of p-PEI-40 cast film foamed at 200 °C . (MD: machine direction).

CONCLUSION

Nanocellular foam was prepared by the CO_2 temperature quench method from a high T_g engineering plastic blend, i.e., the poly(ether ether ketone) (PEEK) and poly(ether imide) (PEI) blends. This study showed that a small difference in the structure of the diamine in PEI induced a large difference in the blend morphology and cell structure of the blends' foams. The PEEK/m-diamine PEI blends were completely miscible for all blend ratios, and the PEEK/p-diamine PEI blends were immiscible for all blend ratios. By exploiting the immiscible blend morphology as a template and distinguishing the CO_2 solubility of the amorphous polymer domain from that of the crystalline domain, the bubble nucleation site and growth can be controlled. The cell size was reduced to 100 nm and the bubble location was highly controlled in the p-PEI domain, aligned between the PEEK layers.

REFERENCE

1. B. Krause, H.J.P. Sijbesma, P. Munuklu, N.F.A. van der Vegt, M. Wessling, *Macromolecules*, **34**(25) 8792(2001)
2. B. Krause, K. Diekamann, N.F.A. van der Vegt, M. Wessling, *Macromolecules*, **35**(5) 1738(2002)
3. T. Ohtsuka, K. Taki, M. Ohshima, *Macromol. Mater. Eng.*, **293**(1), 78(2008)
4. T. Nemoto, J. Takagi, M. Ohshima, *Macromol. Mater. Eng.*, **293**(7),574(2008)
5. T. Nemoto, J. Takagi, M. Ohshima, *Macromol. Mater. Eng.*, **293**(12) 991(2008)

NANOCOMPOSITE PREPARATION BY A FOAMING PROCESS

Youngwook P. Seo[1], Kyounghwan Oh[1], Yongsok Seo[1*], Soon Man Hong[2]

[1]Department of Materials Science and Engineering, College of Engineering, Seoul National University, Seoul, Korea – ysseo@snu.ac.kr
[2] Hybrid Materials Research Center, Korea Institute of Science and Technology, Seoul, Korea

Abstract - We present an investigation of the dispersion and exfoliation of the nanoclay particles in polypropylene copolymer matrix. Supercritical carbon dioxide (sc CO_2) and subcritical carbon dioxide (sub CO_2) were used as a foaming agent. Small angle X-ray diffraction showed that full exfoliation of the clay particles were obtained. Similar results were observed when sub CO_2 was used. The clay particles provide more nucleating sites for the foaming of polymer. Expansion of the adsorbed CO_2 facilitates the exfoliation of the clay platelets, thus, sc CO_2 at lower temperature is more efficient for uniform foaming cell production.

Introduction

In order to overcome the difficulties in PP foaming, supercritical CO_2 (sc CO_2) has been used since 1990s over gaseous CO_2 because sc CO_2 has much higher solubility and diffusivity in polymers.[1] Besides the improvement in various physical properties including the mechanical properties for the matrix polymers, the fact that nanoclay particles can be a nucleator for foaming has stimulated researchers to apply the supercritical CO_2 foaming process and nanocomposite formation for the preparation of microcellular forms with better physical properties which have great potential for a variety of applications such as packaging, insulator, structural components for automotive and furniture industries. In this study, we explore further the role of nanoclay particles in the foaming of PP/clay nanocomposites. The nanoclay particles were doing dual role as a nucleating agent for CO_2 gas or fluid and polypropylene molecules. We tried to figure out the relation between the cell generation and the processing variables.

Description of the theoretical model.

Here, we briefly present the basic equations from our earlier report.[2] In the Ozawa equation, the Avrami equation is expressed using a cooling rate

$$\ln[-\ln(1-x_v(T)_U)] = \ln K(T) - n \ln U \qquad (1)$$

where $x_v(T)$ is the volume fraction of the polymer transformed at a temperature T and the cooling rate U, and K(T) is the so-called cooling function, which only varies as a function of the temperature. Since the Ozawa equation is based on the volume fraction of the crystallites, conversion of the weight fraction of the polymer, $x_w(T)$, to volume fraction of the polymer, $x_v(T)$, is needed. This can be easily done by using the density of the amorphous phase and the density of the crystallized phase. As suggested by the theory, a linear dependence between ln K(T) and the temperature T is assumed, $\ln K(T) = aT + b$. When the temperature reaches the peak of the exothermal curve, T_{max}, for a given cooling rate, the first and the second derivatives of the curve with respect to the temperature should be zero. Using equation (1) with this condition, a linear relation between T_{max} and lnU can be obtained, i.e., $n \ln U = aT_{max} + b - \ln[-\ln(1-x_v(T_{max})_U)]$. Therefore, Eq.(1) can be rewritten as

$$\ln[-\ln(1-x_v(T)_U)] = a(T - T_{max}) + \ln[-\ln(1-x_v(T_{max})_U)] \ (2)$$

Hence the value of the parameter a can be estimated from the slope of a plot of $\ln[-\ln(1-x_v(T)_U)]$ against $T - T_{max}$. Also, plotting T_{max} versus ln U gives a straight line whose slope is n/a and intercept is $(\ln[-\ln(1-x_v(T_{max})_U)] - b)/a$; thus, all the parameter values can be determined without resorting to any numerical process.

Experimentals

Polypropylene terpolymer (poly(propylene-co-ethylene-co-1-butene), with a composition of 94.5wt% of polypropylene unit, 3wt% of ethylene unit, and 2.5wt% 1-butene unit, was obtained from Honam Petrochemicals Co. (Korea). Foamed terpolymer was prepared by injecting out the blowing-agent submerged terpolymer particle in an autoclave through a nozzle. In order to aid the compatibilization between PP terpolymer and the annoclay particles, small amount of maleic anhydride grafted polypropylene (MAPP) was added. Surface modified monmorillinite (Cloisite 20A) was purchased from Southern Clay products Inc. For the master batch preparation, MAPP and clay were premixed in the weight ratio of 6:4 in the internal mixer (HAAKE) at 210°C, 50 RPM for ten minutes. Terpolymer and premade master batch were mixed in a twin screw extruder (Prism) and extruded. Weight ratio between the matrix polymer and the master batch compound was controlled to set the clay amount s 0.2 wt% and 1wt%.

Results and Discussion

Small angle X-ray diffraction (SAXD) patterns for pristine organoclay 20A, two sc CO_2 foamed nanocomposites are illustrated in Figure 1 (0.2 wt% of clay)). Unfoamed samples show a diffraction peak of Cloisite 20A. The peak of unfoamed sample is shifted to lower angle compared to that of clay, which indicates that unfoamed samples have intercalated morphology, i.e., some expansion of the clay gallery. This expansion could be attributed to the diffusion of CO_2 and compatibilizer polymer chains into the clay galleries. On the other hand, SAXD patterns of all foamed samples show no peaks, which indicate full exfoliation of the nanoclay particles. The samples foamed under the subcritical condition show no peak at all. This means the foaming by CO_2 caused the exfoliation. The polymer chains are not in the fully melt state in this condition. This implicates that further smearing of polymer molecules into the gallery was not the cause of exfoliation or most of the exfoliation was done by the CO_2 molecules adsorption and growth by later pressure release.

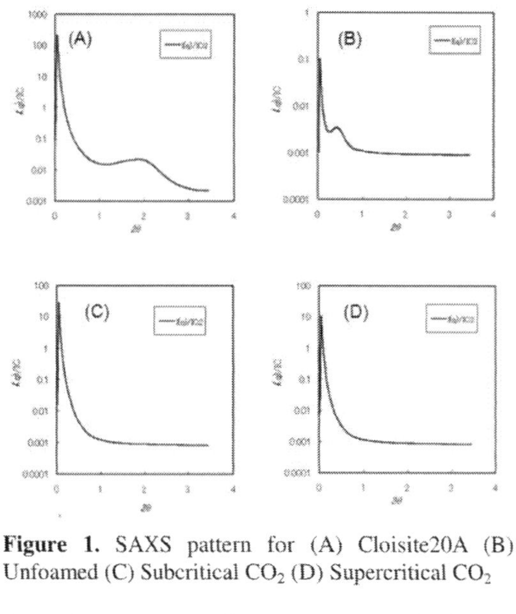

Figure 1. SAXS pattern for (A) Cloisite20A (B) Unfoamed (C) Subcritical CO_2 (D) Supercritical CO_2

Figure 2 shows the cell morphology of the foamed structure of PP and three composite samples. The pure PP foam has but smaller than composite foamed one. All the foaming temperatures were lower than the melting temperature of PP at the applied pressure, hence the crystal lamellae later split by the cell growth to form the walls of the cell. Nanocomposites show uniform cell size . More uniform and bigger cell size for the composite foamed by sub CO_2 implies that large amount of CO_2 molecules smears into the clay galleries which grows later as each cells. Foamed one at low temperature shows smaller cells than the one foamed at high temperature. Smaller cell for the one at low temperature is ascribed to the increased modulus whereas for the composite, more fluid molecules smear into the clay gallery which serves as the nucleating sites during the foaming process. This shows that the cell growth is very sensitive to temperature change as well as the CO_2 pressure.

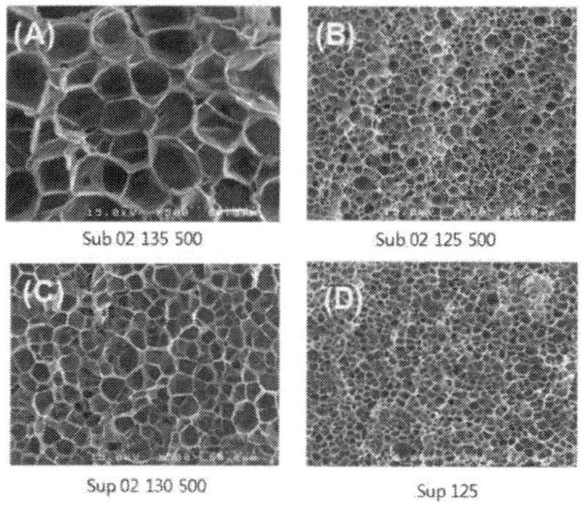

Figure 2. SEM photos of the fractured surfaces (A) Sub CO_2 (135°C foaming), (B)Sub CO_2 (125°C foaming), (C) Sup CO_2 (130°C foaming), (D) PP Sup CO_2 (125°C foaming).

Conclusions

The effect of foaming by CO_2 under pressure on the dispersion and exfoliation of nanoclay particles in PP copolymer was investigated. The SAXD analyses showed that the foaming process facilitates dispersion of clay particles. Fully exfoliated structures were obtained for all foamed samples. Clay layers provide more heterogeneous nucleating sites for foaming. Nanocomposites of better dispersion of clay platelet produces more cells and more uniform structure. More details will be presented at the conference.

Acknowledgements

Support from SNU CTL Undergraduate Research Program as well as ITRC through ERC program (RIAM 2010-0001609) is greatly appreaciated

References

[1] Q. T. Nguyen, D. G. Baird *Polymer* 48, 6923 (2007)

[2] Y. Seo, T. Kang, S. Hong, H. Choi *Polymer* 48, 3844 (20007)

INVESTIGATION OF DIMENSIONAL STABILITY THROUGH FOAMING OF MICRO INJECTION MOLDED COMPONENTS

M. Mahmoodi[1], S. S. Park[1]* and G. Rizvi[2]

[1] *Micro Engineering, Dynamics and Automation Laboratory (MEDAL), Department of Mechanical and Manufacturing Engineering, University of Calgary, Calgary, Alberta, Canada – mahmoom@ucalgary.ca, *simon.park@ucalgary.ca*
[2] *Faculty of Engineering and Applied Science, University of Ontario Institute of Technology, Oshawa, Ontario, Canada – ghaus.rizvi@uoit.ca*

Abstract - The trend towards the miniaturization of parts has created a strong demand for the manufacturing of miniature precision-molded plastic components, due to their high productivity and cost-effectiveness. The objective of this study is the investigation of the feasibility of foaming for micro injection molded components using chemical blowing agents and wood fibers. In order to investigate the foamed miniature molded parts, a miniature micro mold was designed and manufactured using a micro milling machine. The flow pattern of the thin molded components was investigated using different materials (pure linear low density polyethylene – LLDPE, foamed LLDPE, and wood fiber LLDPE composite), and it was found that the viscosity significantly affected the flow patterns. Dimensional accuracy and shrinkage of the molded parts was measured using various gauges. Controlling the cell size, cell density and distribution was especially difficult for thin components.

Introduction

The miniaturization of components with three-dimensional (3D) surfaces with high aspect ratios is becoming increasingly important for various industries, and this trend is expected to continue to increase. Traditionally, micro components have been fabricated based on semi-conductor processing techniques, where silicon materials are photo-etched through chemical and dry processes. However, these methods are slow and limited to a few silicon-based materials and essentially planar geometries.

With the advent of miniaturized systems for biomedical applications, the fabrication of polymeric and composite components has received a great deal of attention. Micro biomedical systems present unique advantages over conventional macro analysis, such as fast responses, laminar flow characteristics, low sample consumption, portability, and point of care (POC) (Chin et al., 2007). Some of the key criteria for miniature biomedical devices are biocompatibility, low cost and disposability (Chae et al., 2005); and, micro injection processes are better suited to meet these criteria than the photolithographic processes. In micro injection molding, identical structures can be reproduced quickly and cost-effectively once accurate molds have been fabricated. In particular, the production of lab-on-chips (LOCs), which are used for handling extremely small fluid volumes down to less than picoliters, have been reported recently (Chien, 2006; Jung et al., 2007).

Unlike macro injection molding processes, micro molding processes face several challenges that require a paradigm shift from macro processes. These issues stem from the miniaturization of components, such as premature solidification, shrinkage, parting line, and handling (Bibber, 2004; Heckele and Schomburg, 2004). Among these challenges, dimensional accuracy is one of critical factors for micro fluidic or LOC devices, since the mating and bonding between upper and lower plates can cause leakage and imprecise measurements that can result from warpage and poor dimensional accuracy. Most of the filling of micro features is completed in the packing phase. Shrinkage can be reduced by increasing the holding pressure and time; however, as the holding pressure is increased, the residual stress and minute shrinkage in the micro scale causes big challenges.

In this study, we examine the feasibility of micro injection molding through the use of the foaming technique and dispersion of wood fibers in the polymer matrix. A miniature mold is designed and fabricated to investigate the feasibility. Since gas fills the interstitial sites between polymer molecules, it effectively reduces the viscosity (Kwag et al., 1999; Kwag et al., 2001; Royer, 2000) and the glass transition temperature of the polymer melt (Kwag et al., 2001; Suh, 1996; Zhang and Handa, 1997). This enables the material to be processed at much lower pressure and temperature levels (Suh, 1996).

The lower pressures experienced with foam injection molding also correlate to a reduction in residual stresses in the part, increasing the dimensional stability. Along with the decrease in residual stresses, the internal pressure arising from foaming eliminates sink marks and further improves the dimensional stability of the molded parts (Kramschuster, 2005).

To minimize modification of the existing micro injection molding setup, we employed chemical blowing agents (CBAs) that were evenly mixed with the polymer resin. The wood fiber polymer resins were prepared using a twin-screw extrusion process and a pelletizer. The flow characteristics were examined with variations in pressure and mold temperature for pure, foamed and wood fiber polymers. The shrinkage analysis of micro components was performed through the comparison of the in-flow and cross-flow shrinkage of the molded parts. The dimensional accuracy was measured using an optical microscope and a surface profilometer.

Experimental

Experimental Setup

The micro molds were fabricated based on the micro milling process. A micro injection molding machine (Boy 12A) was used to injection mold the specimens. The screw diameter of the machine used was 18 mm with a length/diameter (L/D) ratio of 20. The mold base was made out of mold steel, and the mold inserts were fabricated from Aluminum 7075. An electrical heating system was applied for controlling the mold temperature. Figure 1 depicts the experimental setup.

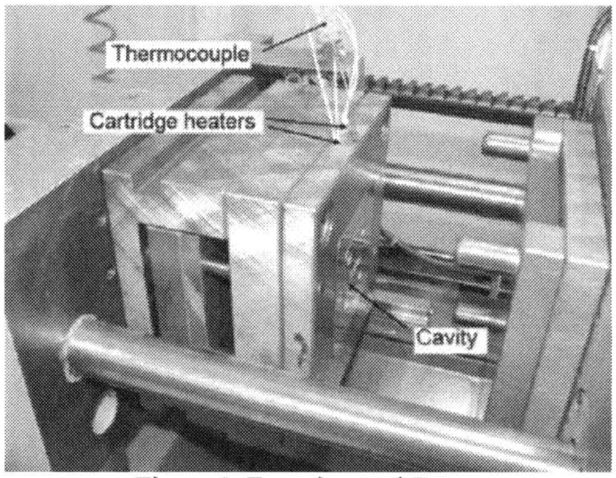

Figure 1. Experimental Setup

To measure the surface profile of the micro parts, a surface profilometer (Mitutoyo Surftest SJ201P) was utilized. Photos of the micro parts were taken using an optical microscope (Nikon Optiphot), and their flow behavior was then analyzed using ImageJ^{TM} software.

To investigate the dimensional stability of the molded specimens, a new mold was designed similar to the ASTM D955-Type D_2, which is a standard test method for determining the cross-flow and in-flow shrinkage behavior of materials in injection molding processes. In the designing procedure, the dimensions of the cavities were scaled down by a factor of 10. To control the mold temperature, two cartridge heaters were used. Figure 2 shows the schematic and pictorial views of the designed mold with cavity dimensions of 6×6×0.2 mm. An ultra-precision micro machining centre (KERN Micro 2522, Germany) was used to fabricate the molds with miniature tungsten carbide (WC) flat end mills (500 µm diameters) at 60,000 rev/min. The surface of the molds were polished with 4 µm grit-size diamond pastes using a rotary polisher.

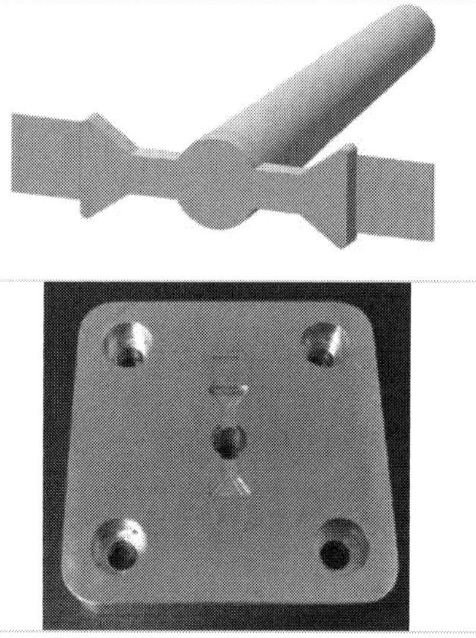

Figure 2. Miniature test mold based on ASTM D955

Materials

In this study, linear low density polyethylene (LLDPE) was used as the polymeric material. This material has a high melt flow index (MFI), which makes it suitable for thin wall molding applications. The main characteristics of the polymer LLDPE (Nova Chemicals Sclair 2316) are a density of 0.931 g/cm^3 and MFI of 72g/10 min at 190°C and 2.16 kg. Also, Celogen OT was used as the CBA for producing micro cells inside the polymeric matrix. This chemical foaming agent generates nitrogen gas as it reaches its decomposition temperature.

An electrical blender was used to mix the CBA with the polymer granules. To prepare the wood-plastic material, a twin-screw extruder was used. For compounding wood fibers (2 wt. %) and LLDPE resin, a 3% coupling agent was used.

Procedure

To investigate the effects of the process parameters on the flow behavior and dimensional accuracy of the molded parts, the injection pressure, holding pressure, and mold temperature were varied. The temperature of the plasticizing barrel ranged from 160 to 180°C, and the nozzle temperature was set at 170°C.

In micro molding, the cavity temperature plays a critical role and should be kept at a high enough temperature to postpone solidification of the gate. This parameter can also affect the thickness of the skin layer in the filling stage and ease the flowing of the melt in the mold cavity. The cavity temperature was varied at

25, 35, 45 and 55°C. The injection and holding pressures were also varied: 20, 25, 30 and 40 bar. For each experiment, the holding pressure was set to the same as the injection pressure.

Results and Discussion

In micro injection molding, process parameters can greatly affect the filling quality of the micro molded parts. To investigate the influence of material type, mold temperature and injection pressure on filline, and the flow behavior of thin wall micro molding, several injection molding experiments were conducted using pure LLDPE, CBA-LLDPE and wood-fiber composite. Adding wood particles and CBAs to a polymer can change its viscosity and flow behavior, which may affect the filling quality of the parts in thin wall micro molding.

In the following sections, the effects of process parameters (mold temperature and injection pressure) on the filling behavior of the thin wall molded parts are presented. The discussions and assumptions associated with the investigations are reported.

Flow Behavior of Thin Micro Molded Components

To investigate the effects of the process parameters, such as injection pressure, holding pressure and mold temperature of the injection molding machine on forming of micro parts (200 μm thickness), a series of experimental tests were performed using different materials. In particular, the melt flow and partial fills were investigated based on the micro mold.

Figure 3 depicts forming schematics of micro parts during the filling of two cavities under different injection pressures and mold temperatures. As can be seen in Figure 3, the melt temperature played an important role on the flow behavior. For all the tested materials, the right section of the melt flow tended to move faster; and, consequently, an asymmetric flow pattern can be observed. However, by increasing the mold temperature, the melt front tended to be more symmetric; and, due to the lower heat loss between the melt and the mold wall, higher filling was achieved. This trend was observable for all the injected materials.

For foamed specimens, the higher filling ratio was obtainable under the same injection pressure, since the viscosity of the melt was lowered, due to the addition of the CBA. For foamed micro molded samples, the flow pattern seemed to be more symmetric compared with pure LLDPE micro parts. In constrast, for the wood-plastic samples, which had a higher melt viscosity due to the wood particles dispersed in the LLDPE matrix, a lower filling ratio was observable compared to pure LLDPE. In addition, the melt front of the wood-plastic composite samples showed more asymmetry under the same injection pressure compared with pure LLDPE samples. Therefore, the viscosity played a significant role in the filling and melts flow patterns.

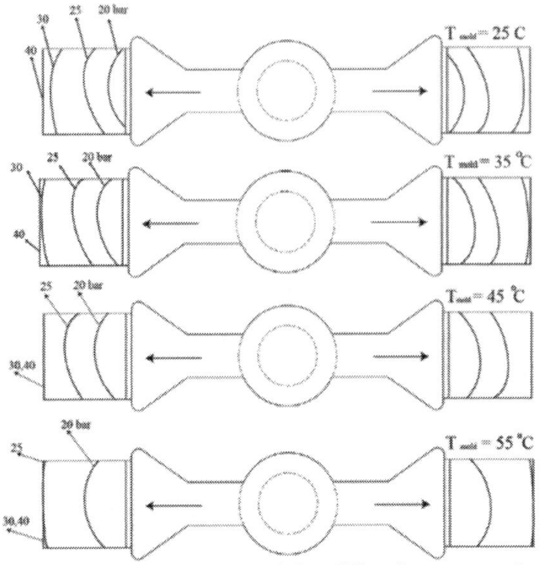

Figure 3. Flow patterns of shot filling for a two-cavity mold under different molding conditions for pure LLDPE

Figure 4 shows optical images of pure LLDPE, CBA-LLDPE and wood-plastic molded micro parts, which were taken by an optical inspection microscope. The flow front of the pure LLDPE material showed more uniformity compared with that of the wood-plastic material, which can be explained by the presence of the wood particles inside the polymer matrix. Furthermore, wood particles were homogenously dispersed within the LLDPE, demonstrating the effectiveness of the twin-screw extruder used to mix wood particles with polymer resin.

To analyze the flow behavior and filling ratio of the thin wall samples, three different points of the front edge were adopted for measurement. The average of the three points was selected as the filling length, as represented in Figure 5.

Figure 6 shows the measurement results of the filling ratio for the micro parts under different injection pressures and melt temperatures. As can be seen, an increased mold temperature had a significant effect on the filling ratio of the micro molded parts by lowering the heat conduction between the melt and the mold wall. The effect of the mold temperature was more significant at lower injection pressures. At higher mold temperatures, the thickness of the solidified layer decreased, which improved the fluidity of the melt. For wood-plastic composite, due to the higher viscosity, a lower filling ratio was achievable at the same injection/holding pressure, compared with pure LLDPE and CBA-LLDPE material.

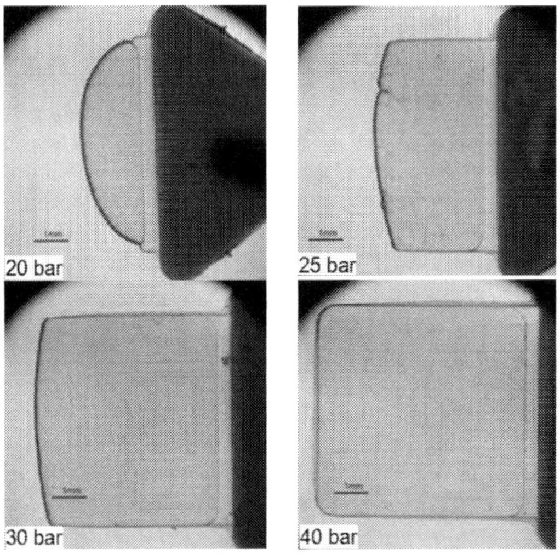

(a) Pure LLDPE

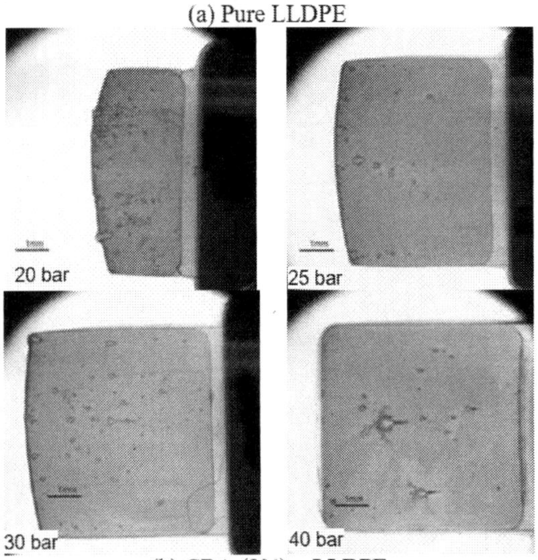

(b) CBA (2%) – LLDPE

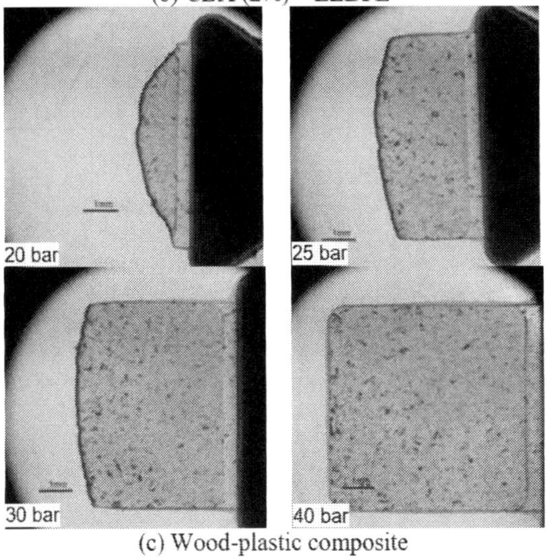

(c) Wood-plastic composite

Figure 4. Microscopic images of specimens molded at 25°C and different injection and holding pressures

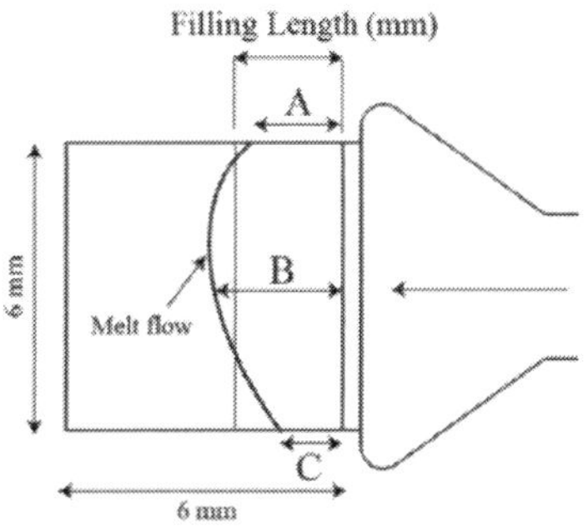

Figure 5. The measured point to obtain the filling length

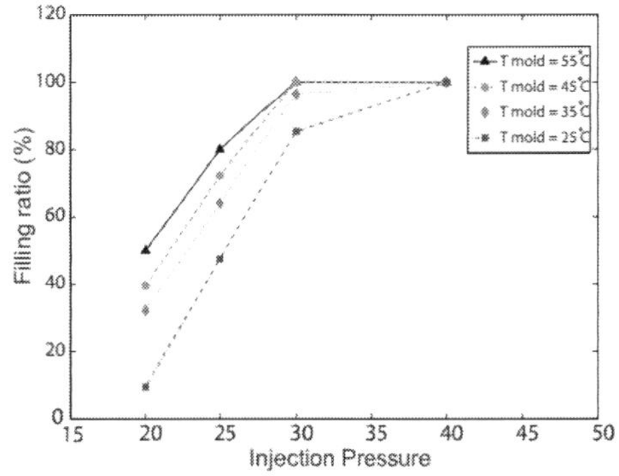

(a) Pure LLDPE

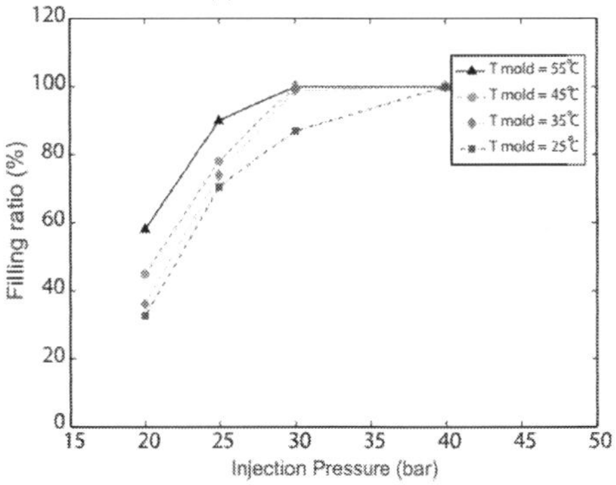

(b) CBA - LLDPE

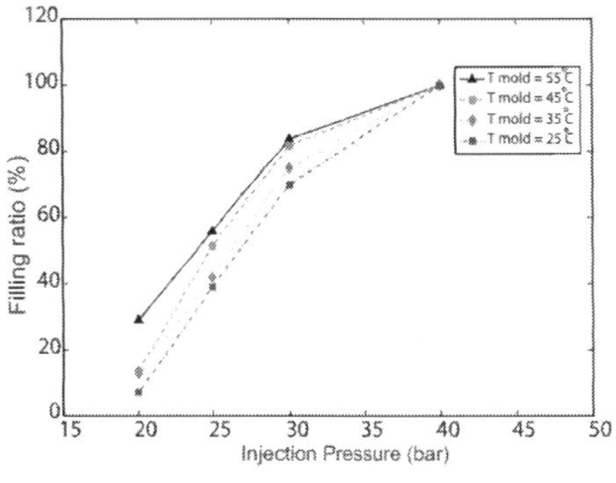

(c) Woodfiber composite

Figure 6. Influence of mold temperatures and injection pressures on the average filling ratio of molded parts

In-Flow and Cross-Flow Shrinkage Analysis and Surface Profile

Micro injection molded components are mostly used in cases where high dimensional stability is needed. The surface profile of the micro parts can also significantly affect the applicability of these components. The in-flow shrinkage was investigated through an optical analysis for the thin micro injection molded parts at a pressure of 40 bar with a 25°C mold temperature. The results show that, for the pure LLDPE and CBA-LLDPE materials, the in-flow shrinkage was less than 0.3%; whereas, for the wood fiber composite material, this value was about 0.8%. This may have been due to the large size of the wood particles dispersed within the polymeric matrix compared to the small size of the micro parts (Figure 7a). Furthermore, the cross-flow shrinkage was approximately 0.1% for all the micro molded samples. Figure 7b shows a scanning electron microscope (SEM) picture of a wood fiber in the LLDPE microstructure.

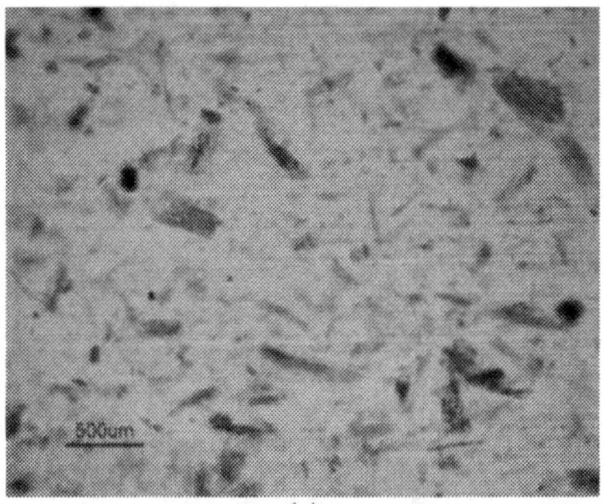

(a)

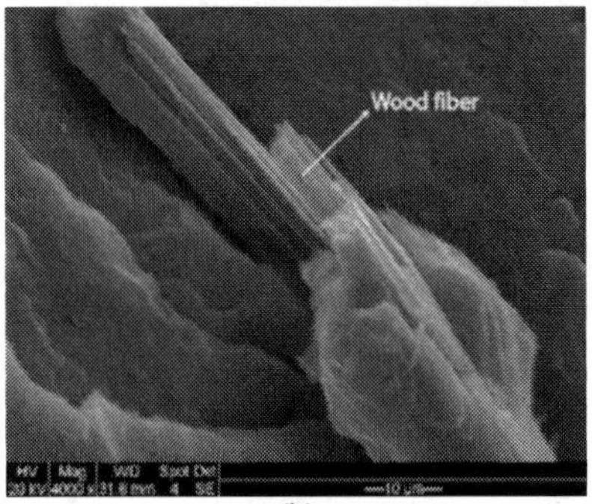

(b)

Figure 7. (a) Dispersion and (b) SEM image of wood particle within the LLDPE matrix

In thin wall micro molding, however, the shrinkage was not as much as the shrinkage in conventional injection molding processes. This is due to the low thickness of the micro parts, which increased the surface area to volume ratio, promoting an even cooling. The part thickness had a dominant role on the shrinkage behavior of the molded parts in the injection molding processes. In addition, since LLDPE is a semi-crystalline polymer with very low thickness, the cooling time is very short; therefore, the crystalline microstructure cannot be developed, which causes minimal shrinkage (Fischer, 2003).

The surface profile of the micro molded parts was measured using a surface profilometer on 5 mm of the surface perpendicular to the flow direction. The results showed that thin wall micro samples molded with pure LLDPE and CBA-LLDPE materials had a smoother surface than the wood-plastic composite, as shown in Figure 8. This is due to the existence of the wood particles in the microstructure and surface of the composite. In addition, the shrinkage of the molded samples with pure LLDPE and CBA-LLDPE was less than 6 µm; whereas, this value for wood-plastic composite was approximately 10 µm, which shows that shrinkage of the wood plastic samples was higher than that for the other two materials in the thickness direction. Compared with conventional injection molding parts, thin micro injection molded components showed less shrinkage in all directions, due to the small size of the micro parts and the high heat conduction between the polymer melt and the mold.

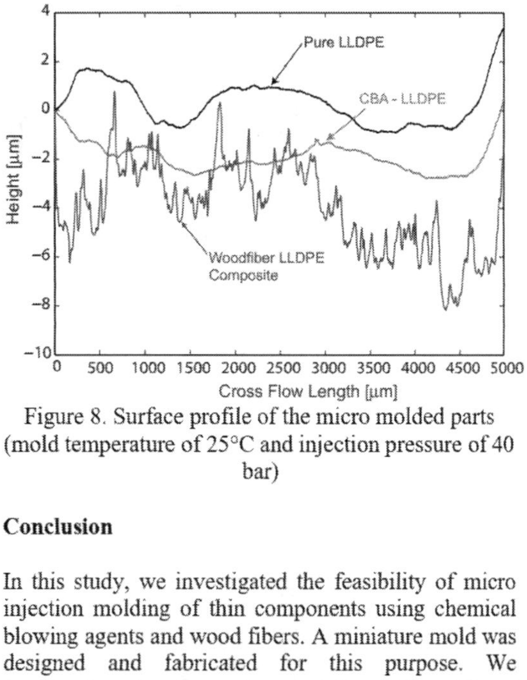

Figure 8. Surface profile of the micro molded parts (mold temperature of 25°C and injection pressure of 40 bar)

Conclusion

In this study, we investigated the feasibility of micro injection molding of thin components using chemical blowing agents and wood fibers. A miniature mold was designed and fabricated for this purpose. We investigated the flow characteristics with different operating conditions. The viscosity decreased when the polymeric melt was mixed with a CBA. We observed that the viscosity was an important parameter for flow characteristics in thin micro parts. In addition, foaming of very thin micro parts was found to be very challenging, since CBA-based foamings do not produce micro scale voids. Due to the large surface area to volume ratio of thin micro parts, the shrinkage was not significant, especially for the foamed and pure LLDPE samples. The surface profiles of molded parts were investigated. It was found that the wood fiber composites showed rough surfaces; whereas, foamed and pure LLDPE parts showed comparable surface roughness.

Acknowledgements

This research was supported by Auto21 and NSERC. The authors thank American Wood Fibers (AWF) for supplying wood fibers, Lion Copolymer Geismar Co. for CBAs, and Dr. Naguib from University of Toronto and Dr. Wood-Adams from Concordia University, for their support.

References

Bibber, D. M., "Micro Molding Challenges", SPE ANTEC Tech. Papers, 3703-3711 (2004).

Chae, J., et al., "Investigation of Micro-Cutting Operations," Int. J. Mach. Tools Manuf, **46**, 313-332 (2006).

Chien, R. D., "Micromolding of Biochip Devices Designed with Microchannels", Senors and Actuators A, **128**, 238-247 (2006).

Chin, C. D., et al., Lab Chip, "Lab-on-a-chip Devices for Global Health: Past Studies and Future Opportunities", **7**, 41-57 (2007).

Fischer, J. M.: Handbook of Molded Parts Shrinkage and warpage, William Andrew Pub., Plastic Design Library (2003).

Heckele, M., Schomburg, W.K., , "Review on Micro Molding of Thermoplastic Polymers", J. Micromech. Microeng., **14**, R1-14 (2004).

Jung, W. C., et al., "Micro Machining of Injection Mold Inserts for Fluidic Channel of Polymeric Biochips", Sensors, **7**, 1643-1654 (2007).

Kramschuster, A., et al., "Quantitative Study of Shrinkage and Warpage Behavior for Microcellular and Conventional Injection Molding", Polym. Eng. Sci., **45**, 1408-1418 (2005).

Kwag, C., et al., "Rheology of Molten Polystyrene with Dissolved Supercritical and Near-Critical Gases", J. Polym. Sci., Part B: Polym. Phys., **37**, 2771-2781 (1999).

Kwag, C., et al., "Effects of Dissolved Gas on Viscoelastic Scaling and Glass Transition Temperature of Polystyrene Melts", Ind. Eng. Chem. Res., **40**, 3048-3052 (2001).

Royer, J. R., et al., "High-Pressure Rheology of Polystyrene Melts Plasticized with CO_2: Experimental Measurement and Predictive Scaling Relationships", J. Polym. Sci., Part B: Polym. Phys., **38**, 3168-3180 (2000).

Suh, N.P., "Microcellular Plastics," in Innovation in Polymer Processing-Molding, J.F. Stevenson, ed., Hanser Publishers, Munich, **93** (1996).

Zhang, Z., Handa, Y. P., "In Situ Study of Plasticization of Polymers by High-Pressure Gases", J. Polym. Sci., Part B: Polym. Phys., **36**, 977-982 (1997).

DEVELOPING WATER RESISTANT STARCH POLYMERS

P.J.Halley[1*], C.Chaleat[2], R.Truss[3], M.Gidley[4], A.K.Shrestha[5], R.G.Gilbert[6], W-C Liu[7], A.Roberts[8], N.McCaffrey[9]

[1] *AIBN/School of Chemical Engineering, The University of Queensland QLD 4072 Australia – p.halley@uq.edu.au*
[2] *AIBN, The University of Queensland– c.chaleat@uq.edu.au*
[3] *School of Chemical Engineering, The University of Queensland– r.truss@uq.edu.au*
[4] *Centre for Nutrition and Food Science, The University of Queensland -mike.gidley@uq.edu.au*
[5] *Centre for Nutrition and Food Science, The University of Queensland -a.shrestha@uq.edu.au*
[6] *Centre for Nutrition and Food Science, The University of Queensland -b.gilbert@uq.edu.au*
[7] *Centre for Nutrition and Food Science, The University of Queensland -weichen.liu1@uqconnect.edu.au*
[8] *Plantic Technologies Ltd– annr@plantic.com.au*
[9] *Plantic Technologies Ltd– nickm@plantic.com.au*

Abstract - This paper summarises the development of novel thermoplastic starch polymers (TPS), that are aimed to enable TPS's to be able to be used in wider-ranging application areas. Specifically this work will examine the importance of understanding related fundamental structure-property-degradation relationships, developing novel analytical techniqies? for structure-processing-property development (repetititive) and developing novel platforms for water resistantTPS. It is anticipated a water resistant, fully biodegradable range of TPS's will open new markets for TPS materials

Introduction

Thermoplastic starch-based renewable plastics are of significant current interest in the plastics industry to reduce carbon footprints and minimize plastics waste. However, traditionally most thermoplastic starch polymers suffer from low water resistance and loss of mechanical properties when in contact with water. This paper reviews recent developments of biodegradable thermoplastic starch polymers at The University of Queensland. Specifically we will examine the fundamental interrelationships between structure-processing-properties-degradation, novel analytical technique development and new developments in water resistant starch polymers for novel thermoplastic starch polymers that are processible on conventional polymer processing equipment.

Results and Discussion

Fundamentals

Our labs have developed novel starch polymers for thermoforming, film blowing and injection molding processes (McGlashan (2003), Halley (2006)) that have been used commercially by Plantic Technologies Ltd. This research and development work went hand in hand with underlying fundamental research which enabled us to more deeply understand the fundamental structure-property-processing-performance-biodegradation relationships of the systems, and hence develop new materials for different applications. For example current fundamental studies [Russo et al. (2009a-b), in our laboratory have focused on understanding structure-processing-property relationships (effects of relative humidity and temperature on ultimate mechanical properties – (Chaleat (2008a-b))) and structure-processing-biodegradation relationships (diffusion of water in thermoplastic starch blends – (Russo (2009a-b))). Initial conclusions are that modulus, yield stress and crack energy is highly affected by relative humidity, and that water diffusion, enzymatic breakdown and biodegradation is closely related to starch blend structure and plasticisers.

Novel Analytical Techniques

Alongside the fundamental development research we have developed novel characterization tools to allow us to better understand underlying structures of the starch. For example Tan (2007) developed an improved method to analyze the 13C NMR spectra of starches, which considers the contribution of the V-type conformation and the nature of the amorphous component. Here the spectra of the starch are separated into amorphous and ordered subspectra and the ordered subspectra are further analysed to determine the presence of both V-type single helices and B-type double helices. Then relative proportions of amorphous, single, and double-helical conformations are estimated by apportioning intensity of C1 peak areas between conformational types on the basis of ordered and amorphous subspectra of the starch. Quantitative analysis shows that the V-type single-helical component increases with amylose content of starches.

Most recently work by Liu (2010) has developed a novel size-exclusion chromatography (SEC) technique for analysing molecular size changes during extrusion for a range of modified maize starches with varying amylose contents. For example Figure 1 shows the molecular size distributions of starches of varying amylose contents, as they are collected along the extruder.

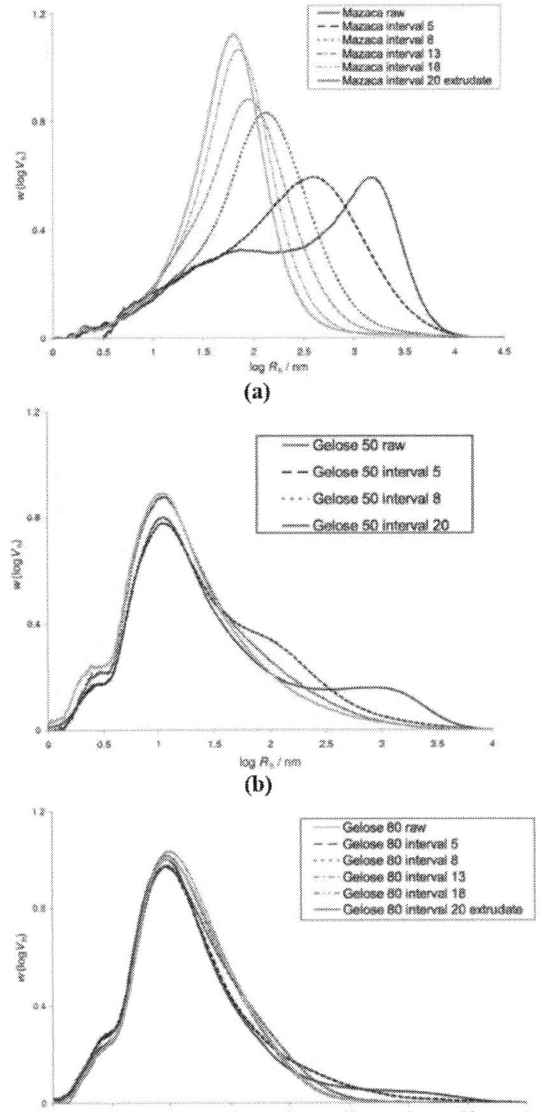

Figure 1 – Molecular mass distribution profiles for (a) 20% amylose, (b) 50% amylose and (c) 80% amylose starch as a function of section (distance along) of the extruder. [Liu (2010)]

Analysis by of both whole and debranched samples subject to various levels of extrusion showed that the extrusion degradation process involved preferential cleaving of larger molecules, while causing the size distribution to narrow and converge toward a maximum stable size. This is analogous to a similar effect of shear degradation of droplets in emulsions. It was also found that high branching density and short branch length were associated with higher susceptibility to shear degradation.

Improving Water Resistance of TPS

Current work is examining development of water resistant TPS systems, via a new project with Plantic Technologies, funded via the Cooperative Research Centre (CRC) for Polymers. This work will involve similar links between fundamental structure-processing-property-biodegradation relationships and novel characterization tools as outlined above, but of course incorporate new formulations, plasticisers and extrusion conditions to maximize gelatinization and melting, maintaining key property requirements such as mechanical and barrier properties, and maintaining biodegradability and low carbon footprint.

Conclusions

Clearly thermoplastic starch based polymers offer a very attractive low cost base for new biodegradable polymers due to their low material cost and ability to be processed on conventional plastic processing equipment. The engineering of more enhanced structures and properties into this low cost base materials will continue to be the main technological drive into the future. This development will best be undertaken by interweaving fundamental understanding of structure-processing-properties-degradation interrelationships, development of novel analytical tools to illucidate these relationships and new developments in water resistant TPS and blends for novel thermoplastic starch platforms.

Acknowledgements

Thanks to the CRC Food Packaging, Australian Research Council, Plantic Technologies Ltd and the CRC Polymers for funding.

References

Chaléat, C.M. et al., "Properties of a plasticised starch blend. Part 1. Influence of moisture content on fracture properties", Carbohydrate Polymers, **71**(4), 535-543 (2008a).

Chaleat CM et al. , Properties of a plasticised starch blend - Part 2: Influence of strain rate, temperature and moisture on the tensile yield behaviour, Carbohydrate Polymers, **74** (3) 366-371 (2008b).

Halley, P et al., "Biodegradable plastics for injection molded applications", 28th RACI Australiasian Polymer Symposium, Rotorua, 5-9 February 2006, p C4/1 (2006).

Liu W-C, et al. "Mechanism of Degradation of Starch, a Highly Branched Polymer, during Extrusion", Macromolecules, **43** (6), 2855–2864 (2010)

McGlashan, S et al., "Scale-up of Biodegradable Films" International Polymer Processing Society Meeting PPS19, July 9-12, Melbourne, CD (2003)

Russo MAL, et al., "The enzymatic hydrolysis of starch-based PVOH and polyol plasticised blends", Carbohydrate Polymers **77** (3) 442-448 (2009a)

Russo, MAL. et al., "The anaerobic degradability of thermoplastic starch: Polyvinyl alcohol blends: Potential biodegradable food packaging materials", Bioresource Technology **100** 1705–1710 (2009b).

Tan, I, et al., "A Method for Estimating the Nature and Relative Proportions of Amorphous, Single, and Double-Helical Components in Starch Granules by 13C CP/MAS NMR", Biomacromolecules **8**(3) 885-891 (2007).

A NEW GENERATION OF NANO-BIOCOMPOSITES BASED ON RENEWABLE RESOURCES

E. Pollet, V. Martino, L. Avérous*

LIPHT-ECPM, EAc (CNRS) 4379, Université de Strasbourg, 25 rue Becquerel, 67087 Strasbourg Cedex 2 – France
Email : luc.averous@unistra.fr

Nano-biocomposites are a new generation of nanocomposites, elaborated by the association of biopolymers with nanofillers (mainly nanoclays). The biopolymers/nanoclays materials are a valid answer to produce low cost, highly competitive and pioneering environmentally friendly materials. Moreover, since the citizens are more and more concerned about sustainable development and since the global need and the price of the fossil resources increase, these materials now represent an interesting option to replace conventional fossil resources based plastics. These materials present a wide range of applications, such as packaging, agriculture devices, or even for biomedical applications. This presentation shows main results obtained with different nano-biocomposites, which are based on various matrices (only bio-matrices based on plasticized starches are presented in this paper). The elaboration of these different multiphase systems based on starch are presented with their main properties (thermal, mechanical behavior ...) and the corresponding nano-structures.

INTRODUCTION

It is increasingly being realized that the use of long-lasting polymers for short-lived applications (packaging, catering, surgery, hygiene...) is not entirely adequate. This is not justified when increased concern exists about the preservation of ecological systems.

The potential of biodegradable and renewable polymers obtained from agro-resources (e.g., the polysaccharides) has long been recognized. But until now, agro-polymers such as starch are largely and mainly used in food industry applications and have not found extensive applications in e.g., the packaging industry, to replace conventional plastic materials.

Native starch can be transformed in thermoplastic-like polymers after destructuration and plasticization processes. The most common non-volatile plasticizers are polyols, like glycerol. Plasticized starch (PLS) presents strong attributes (i) a total compostability without toxic residues, (ii) the renewability of the resource and (iii) the processability of this material using conventional plastic processing. Besides, compared to synthetic thermoplastic, PLS is a rather cheap material. Compared to fossil resources, the price of the starch resources stays stable and tends to decrease, with the cereal over-production on the world market. PLS can be easily processed with a large range of process such as extrusion, injection molding, thermoforming ... According to the formulation (plasticizer, filler ...), PLS shows a wide properties range. But unfortunately, PLS presents some issues such as great moisture sensitivity, rather weak mechanical properties compared to synthetic polymers and, an important post-processing aging. For instance, mechanical properties vary several weeks after the processing. To overcome these issues, during the last decades two main strategies have been carried out. Starch modification (chemical way) has been developed since the first half of the XXth century and we can find a large literature on this topic. It is well known that starch esterification (e.g., by acetylation) improves the water resistance. But, this strategy presents some strong limitations such as the toxicity and the diversity of the by-products obtained during the chemical reactions. Besides, the chemical reactions present some incidences on the reduction of the polysaccharide molecular weight. Such products do not fulfill the requirements to substitute PLS for material applications. Therefore, since one or two decades another strategy has been developed. It is based on the association of PLS with other biodegradable compounds such as biodegradable polymers, agro-fillers to obtain compostable multiphase materials. We can obtain different structures and a large range of properties. This approach induces different problematics linked to the interfaces, to the continuity between the phases, to the compatibility between these materials ...

PLASTICIZED STARCH-BASED MULTIPHASE SYSTEMS.

Different systems can be obtained:

- **Plasticized starch -based blends**

Blending is the easier process to associate together different polymers. Blending provides a powerful route to obtain materials with improved property/cost performances. This approach is cheaper than the development of new polymers. Besides, blends are also commonly used as models to test the compatibility between different polymeric phases, because a blend presents a great surface at the interface compared e.g., to multilayer structures.

PLS has been widely used in blends with other polymers [1]. A lot of patents have been published on this topic. These considerable research efforts have led to starch-based blends being commercialized e.g., Mater-Bi [2] from Novamont (Italy). To produce these commercial blends, starch is blended with non-biodegradable polymers (polyolefins) or with biodegradable polyesters (e.g., polycaprolactone). Applications concern packaging, disposable cutlery, gardening, leisure, hygiene ...

In the past, blends with synthetic polymers, such as PE, were developed leading to non-fully biodegradable materials. These controversial materials, because presented as biodegradable, have been named "bio-fragmentable". To maintain the compostability feature, different biodegradable blends were developed. A great number of patents have been published on this topic. We find some associations of PLS with agro-polymers such as proteins or pectins. But, most of the researches are focused on the blending of plasticized starch (PLS) with biodegradable polyesters: polycaprolactone (PCL) [3-9], polyesteramide (PEA) [10], polyhydroxybutyrate-co-hydroxyvalerate (PHBV) [11], polybutylene succinate-adipate (PBSA) [12-13], poly(butylene adipate-co-terephtalate) (PBAT) [13] or polylactic acid (PLA) [14]. These commercially available polyesters show some interesting and reproducible properties such as a more hydrophobic character, lower water permeability and some improved mechanical properties, compared to PLS. According to the polyester nature, we can obtain different levels of compatibility with the starchy phase, which can be given by e.g., the theoretical work of adhesion. On agreement with mechanical properties (tensile tests results) [13], we can see that the best results are obtained with PEA.

According to the starch/polyester ratio and to the nature of the polyester (in connection with the compatibility level), we obtained co-continuous or granular structures.

To improve both the compatibility between main phases and the final material performances, different compatibilisation strategies are generally developed. These strategies could imply the addition of a compound, the compatibiliser, which can be obtained by modification of at least one polymer initially present in the blend. For polyester/starch compatibilisation, the literature presents different routes, which can be classified into three main groups [1]:

(i) Starch or polyester functionalisations with e.g., maleic anhydride (polyester) or urethane functions (starch), [16, 4]

(ii) the starch-polyester reticulation with a coupling agent such as e.g., peroxides,

(iii) the use of copolymers e.g., polysaccharide-grafted PLA.

Results show improvements for the interphase continuity, observed by MEB and by DSC, and for the mechanical properties. One of the most efficient approaches seems to be the use of tailored copolymers. But, such an approach need to strictly control the copolymer structure e.g., with long lateral branches [15]. In fact, too short branches induce a reduced compatibility, which is linked to a lack of entanglements between the compatibiliser and e.g., the polyester phase

- **Plasticized starch-based multilayers**

Compared to blends, multilayer structures present some strong attributes. Moisture sensitivity is not fully addressed in a blend because of starch phase distribution close to the surface. Development of moisture-resistant starch-based products shall be undertaken through multilayers, allowing for the preparation of sandwich-type structures with PLS as the central layer and the hydrophobic biodegradable component as the surface outer layers. On this topic, some patents were published. For instance, Bastioli et al. [16] (1996) proposed different processes (coextrusion, casting and hot melt techniques) to protect starch-based materials with wax layers. Another alternative is SiOx or carbon deposit at the surface (e.g., by plasma treatment, see below).

Coextrusion seems the best option since it offers the advantages of being a one-step, continuous and versatile process. Multilayer coextrusion has been widely used in the past decades to combine the properties of two or more polymers into one single multilayered structure. However, some problems inherent to the multiphasic nature of the flow are likely to occur during coextrusion operations, such as non-uniform layer distribution, encapsulation, and interfacial instabilities, which are critical since they directly affect the quality and functionality of the multilayer products. By optimising the process conditions, we can decrease this issue. Despite the number and diversity of studies on multilayer flows and stability, only few articles and patents reported the use of plasticized starch and polyester in coextrusion processes. Different stratified structures were processed by coextrusion and studied: with PCL [17-18], PBSA [17-18], PEA [19], PLA [17-18], or with PBAT [18]. We can produce e.g., trays for food applications.

- **Plasticized starch-based Biocomposites**

In the literature, different types of fibres or microfibrils were tested in association with plasticized starch [20-23] such as microfibrils from potato pulp, bleached leafwood fibres, fibres from bleached eucalyptus pulp, flax and jute fibres ... The different authors have shown high compatibility between both polysaccharides. After mixing with cellulose and ligno-cellulose fibers, as other authors [20-23], we have found high improvements of the performance of PLS-based material such as (i) higher modulus, (ii) reduced water sensitivity due to fiber-matrix interactions and to the higher hydrophobic character of the cellulose, linked to its high cristallinity, (iii) higher thermal resistance, due to the transition shift and an increase of the rubber plateau, (iiii) reduced post processing ageing, due to the formation of a 3D network between the different carbohydrates, through hydrogen bonds. This network induces a stabilization of the material and a strong reduction of the post-processing aging.

RECENT DEVELOPMENTS ON NANO-BIOCOMPOSITES BASED ON STARCH

This paper highlights more particularly nano-biocomposites, which are obtained by introducing nano-fillers into a starchy matrix

Nano-biocomposites

Organoclays, mainly montmorillonites (MMT) more or less modified with different type of surfactant have been tested.

On a recent patent based on PLS-montmorillonites [24], Fischer and Fischer (2001) present clay-based compounds as processing aids, which decrease the viscosity during extrusion step. Besides, the mineral compounds act to prevent plasticizer lost during ageing and then, improves the PLS stability. On PLS-biodegradable polyester blends, McGlashan and Halley (2003) [25] have shown that nanocomposites tend to improve the mechanical properties (tensile and impact tests), the barrier properties and the transparency by cristallinity modification. These improvements have been confirmed by others authors [26-28] on PLS materials with "natural" MMT (Na-MMT).

In recent communications [29-35], we have shown that we can improve the d-spacing (distance between the layers) of plasticized wheat starch using natural co-product from sea resources (seaweed) as surfactant and then, we can obtained exfoliated materials with improved mechanical properties, largely superior than we can obtain with Na-MMT materials.

These nanohybrid materials mainly differ from conventional nanocomposites due to their hydrophilic character. Consequently, only microcomposite morphologies are reached with the dispersion of the conventional commercially available hydrophobic organo-modified nanoclays, such as Cloisite (quaternary ammonium-based MMT. Because of the high thermal sensitivity of the carbohydrate chains, solvent intercalation process is often required or used for the nano-biocomposites elaboration. This strategy allows an efficient nanoclay exfoliation in starchy matrices. Nevertheless, since this process is cost and time prohibitive, it may be not adapted for an industrial scale-up.

The melt blending elaboration process appears as a better option to produce polysaccharide-based nano-biocomposites. According to the reported studies, the plasticizers (polyols such as, glycerol or sorbitol) greatly affect the nano-hybrid morphology with the establishment of strong hydrogen interactions with the nanofillers, which perturb the clay platelets intercalation/exfoliation process. Consequently, these nano-hybrids display mainly intercalated structures at high plasticizer content. To overcome this limitation and reach exfoliation, an organo-modification of the clay surface with hydrophilic surfactants, such as carbohydrate surfactants (cationic starch), has been proposed and tested to modify the nanofiller polarity and thus the clay/matrix affinity. Such nano-hybrid materials display improved mechanical reinforcement, higher thermal stability, and lower moisture sensitivity. Besides, new starch-based nano-biocomposites have been recently elaborated from needle-shaped nanofillers (sepiolite) [35]. Compared to more conventional layered silicates, they present substantial differences, namely their aspect ratio and their surface properties (silanol groups located at the nanofiller surface) and thus could permit the elaboration of innovative polysaccharide-based nano-hybrid materials with extended properties.

CONCLUSION

This paper tried to show the high potential presented by multiphase materials based on renewable resources and more particularly, by plasticized starch-based materials. Because of its low cost, its renewable and biodegradable character, starch seems to be nowadays an important challenger for the development of future agro-based products. Face to petrochemical polymers (fossil resources), plasticized starch should be a future good competitor, when petroleum resources will drastically decrease. Besides, these agro-materials are totally on agreement with the emergent concept of sustainable development.

Compared to the starch chemical modifications, which present some strong drawbacks, the association of plasticized starch with nano-fillers is an interesting alternative to overcome main starch issues and to fulfill some requirements on different fields.

These polysaccharides/nanoclays materials are a valid answer to produce low cost, highly competitive and pioneering environmentally friendly materials. Moreover, since the citizens are more and more concerned about sustainable development and since the global need and the price of the fossil resources increase, such agro-based materials now represent an interesting option to replace conventional fossil resources based plastics. Thanks to the agro-polymers intrinsic characteristics and improved properties brought by the nanofillers, these materials could present a wide range of applications, such as packaging, agriculture devices, or even for biomedical applications with some developments based on the biocompatibility of these materials.

Perspectives of such a topic could be (i) on the use of "new" types of starch e.g., starch with low amylopectin content, new botanical resources, or genetically modified starch and (ii) on the use of new technologies and recent approaches based for instance, on advanced nanotechnology.

Acknowledgements
The author wants to thank the following persons which have been involved since 1996, into this thematic:
From URCA-INRA Reims–France: Olivier Martin, Emmanuelle Schwach, Laurent Belard, Patrice Dole and Christophe Fringant. From LIPHT-ECPM Strasbourg–France: Frederic Chivrac.
Some collaboration as A/Pr. Peter Halley and Rowan Truss (UQ – Australia), or O. Gueguen and S. Ahzi (IMFS-Strasbourg/France)

REFERENCES

1- L. Avérous J. Macromol. Sci.-Part C. Polymer Reviews 2004, 3, 231.
2- C. Bastioli Polym. Deg. Stab. 1998, 59, 263.
3- R. Mani; J.Tang; M. Bhattacharya Macromol Rapid Commun. 1998, 19, 283
4- W. Amass; A. Amass; B. Tighe Polym Int 1998; 47, 89.
5- L. Avérous; L. Moro; P. Dole; C. Fringant, Polymer 2000, 41, 4157.
6- O. Myllymäki; P. Myllärinen; P. Forssell; T. Suortti; K. Läteenkorva; R. Ahvenaien; K. Poutanen Packag Technol Sci 1998, 11, 265.
7- S.J. Huang; M.F. Koening; M. Huang, in Biodegradable polymers and packaging; C. Ching; D.L. Kaplan; E.L. Thomas, Eds; Technomic publication: Basel, 1993; 97-110.
8- M.F. Koenig; S.J. Huang Polymer 1995, 36, 1877.
9- M. Vikman; S.H.D. Hulleman; M. Van Der Zee; P. Myllarinen; H. Feil J. Appl. Polym. Sci. 1999, 74, 2494.
10- L. Avérous; N. Fauconnier; L. Moro; C. Fringant J. of Applied Polymer Science 2000, 76, 1117.
11- H. Verhoogt; N. St-Pierre; F.S. Truchon; B.A. Ramsay; B.D. Favis; J.A. Ramsay Can J Microbiol 1995, 41, 323.
12- J.A. Ratto; P.J. Stenhouse; M. Auerbach; J. Mitchell; R. Farrell Polymer 1999, 40, 6777.
13- L. Avérous; C. Fringant Polym Eng. Sci. 2001, 41, 727.
14- O. Martin; L. Avérous Polymer 2001, 42, 6237.
15- E. Schwach, PhD Thesis, University of Reims Champagne-Ardenne, France, 2004
16- C.Bastioli; G. Romano; M. Scarati; M. Toscin, U.S. Patent 5,512,378; 1996.
17- L. Wang; R.L. Shogren; C. Carriere J. Appl. Polym. Sci. 2000, 40, 499
18- O. Martin; E. Schwach; L. Avérous; Y. Couturier Starch/Starke 2001, 53, 372.
19- O. Martin; L. Avérous Journal of Applied Polymer Science 2002, 86, 2586.
20- A. Dufresne; M.R. Vignon Macromolecules 1998, 31, 2693.
21- L. Avérous; C. Fringant; L. Moro Polymer 2001, 42, 6571.
22- A.A.S. Curvelo; A.J.F. De Carvalho; J.A.M. Agnelli Polymer 2001, 45, 183.
23- M. Wollerdorfer; H. Bader Industrial Crops and Products 1998, 8, 105.
24- H.R. Fischer; S. Fischer, Patent WO 01/68762, 2001
25- S. McGlashan; P.J. Halley Polym Inter 2003, 52, 1767.
26- H.M. Park Macromol. Mater. Eng. 2002, 287, 553.
27- H.M Park J. of Mat. Sci 2003, 38, 909.
28- M.F Huang; J.G. Yu; X.F. Ma Polymer 2004, 45, 7017.
29- F. Chivrac, O. Gueguen, E. Pollet, S. Ahzi, A. Makradi, L. Avérous. Acta Biomaterialia 2008, 4, 1707
30- F. Chivrac, E. Pollet, M. Schmutz, L. Avérous. Biomacromolecules 2008, 9, 896-900
31- F. Chivrac, E. Pollet, L. Avérous. Materials Science & Engineering R 2009, 67, 1-17
32- F. Chivrac, E. Pollet, L. Avérous. Polymers for Advanced Technologies 2010, DOI: 10.1002/pat.1468.
33- F. Chivrac, O. Gueguen, E. Pollet, L. Averous, S. Ahzi, S. Belouettar. Journal of Nanoscience and Nanotechnology 2010, 10, 2949–2955
34- F. Chivrac, E. Pollet, P. Dole, L. Avérous. Carbohydrate Polymers 2010, 79, 941–947
35- F. Chivrac, E. Pollet, L. Avérous. Carbohydrate Polymers 2010, 80, 145–153

DEVELOPMENT OF MODIFIED POLYLACTIDE (PLA)

H. Kishimoto[1], A. Takenaka[1], H. Moriwaka[2] and H. Enomoto[2]

[1] *Kao Corporation, Wakayama, Japan – kishimoto.hiroaki@kao.co.jp*
[2] *Kao Corporation, Tokyo, Japan*

Abstract - In recent years, polylactide (PLA) and other bio-based plastics have been attracting much attention for environment problems. PLA is hard and good transparent resin, but the application of PLA has been restricted because of its poor flexibility, fragile, unavailable in high temperature and low productivity. In this report, we have developed modified PLA resin for extrusion sheet and injection moldings based on "Technology of Nano-Modification for Polymer" such as control of softening, control of PLA crystallization in nano size, and modification with another component. Due to these original technologies, we have achieved that physical properties and moldability are almost equal to propylene (PP) and acrylonitrile-butadiene-styrene (ABS) resin. As a result, we developed two type of modified PLA. First is clear and soft PLA for extrusion molding, second is high moldability PLA for injection molding. Now, we have applied these modified PLA as alternative plastics of PP and ABS to stationery, packaging, convenience goods, electrical appliance and so on. The performance and the technology will be presented.

1. Introduction

Because of its characteristics of being, "lightweight, strong and inexpensive", oil-derived, all-purpose plastic is used in a variety of applications, such as electrical goods, vehicles, stationery, housing material, commodities, and packaging. As the raw material of plastic is oil, problems have arisen, such as exhaustion of resources, pollution and contributing to global warming. Because of this, the use of plastic made from plants including polylactide ("PLA"), has been gaining attention. As PLA is a plant-derived resin, there is no increase in carbon dioxide, the resources are sustainable, and PLA returns to the soil as it is biodegradable. As PLA is an environment-conscious resin whose industrial production has been established, the application of PLA for general consumer goods and durable material has been discussed.

PLA is a rigid resin with excellent transparency; however, the purposes of PLA is limited because it is rigid, fragile, has low heat resistance and is nonproductive. To become more widespread, the physicality and productivity of PLA, needs to be improved. To improve its heat resistance, which is one of the issues to be solved, crystallization of PLA is effective. It takes about 10 min. to crystallize unmodified PLA at 110°C which limits practical productivity. In addition, transparency which is important in the field of extruded sheeting significantly decreases with crystallization.[1, 2] This study focused on modification using additive agents in the field of extrusion molding and injection molding and developed modified PLA resin. This study emphasized the harmonic content with the environment and used only pure PLA resin, not alloying it with oil-derived resin which has been recently considered.

Extruded sheeting is used mainly for packaging containers, such as food containers, stationery, blister packs or clear cartons where transparency is necessary. When unmodified, or the existing modified PLA is used for stationery or clear cartons, application of bend stress, bending or punching processing is often conducted. Major problems occur with the unmodified PLA sheeting and the existing modified PLA sheeting, such as; blushing when PLA sheeting is bent or damages when PLA is punched, because it is too rigid or too weak. When there is a hinge segment on a blister pack, cracks occur even with only a very small amount of opening and closing movements. The heat resistance of PLA is low and it deforms at 50°C degree or higher. When PLA is used for food containers, it deforms during storage or transport of the food and it cannot be used for storage of food with temperatures of 50°C or higher. We have developed "Softness control technology" and "Nano-crystalcontrol technology", creating sheeting that is flexible, impact strength and heat resistant, equal to or surpassing, those of polypropylene ("PP"), and will be mass produced with extrusion molding while maintaining transparency.[2, 3]

In the field of injection molding, PLA is used for case of home electrical appliances and commodities. Mechanical property such as heat resistance (amorphous) or impact resistance of unmodified PLA is significantly inferior to that of all-purpose resin like PP and ABS, and unmodified PLA has not been practical for common use. As mentioned above, it requires a longer time and a higher mold temperature, compared to all-purpose resin, to crystallize PLA for the improvement of heat resistance. There are many production issues that the molding cycle is unpractical, the existing equipment cannot be used (due to the capabilities of the mold temperature controller) or deformation occurs when PLA is removed from the mold as the temperature of PLA is high. We have developed "High speed crystallization technology with common temperature molding" aiming at 30°C for the mold temperature and at 30 sec. of molding time, which is productivity equivalent to that of ABS resin or all-purpose resin.[5]

2. Experimental

2-1. Extruded Sheeting

2-1-1. Compound Production

Modified PLA pellets were made by mixing PLA with a variety of additive agents, melting and kneading the mixture at 190°C with a corotating twin-screw extruder. The pellets that were produced were dehumidified and dried at 80°C for 5 hours.

2-1-2. Sheeting Production

The pellets were caught by ferrotype plates, with a spacer of 0.5mm thick inserted. The pellets were pressed in a press molding machine for 3min. with residual heat and 1 min of pressurization at 190°C. Then, the sheeting that was produced was rapidly cooled to glass transition point (Tg) or lower. Then, to make the crystallization sheeting, annealing was given to the sheeting in a press machine at a temperature of 80°C for 30 sec. For unmodified PLA, annealing was given to the sheeting in a press machine for 10 minutes at a temperature of 120°C.

2-1-3. Evaluation of Mechanical Property

For mechanical property, a tensile test was performed with Tensilon, to evaluate flexibility using the modulus and elongation at break. Haze values were measured using a haze meter to evaluate transparency. For heat resistance, storage modulus was obtained using a dynamic viscoelasticity measurement under the conditions of 2°C/min. for a range from -20 to 160°C and frequency of 10Hz.

2-1-4. Observation of Crystallization Dimension

The surface of the crystallization sheeting produced above was observed with an atomic force microscope ("AFM").

2-1-5. Extruded Sheeting Production

For the developed formulation, sheeting was made from the pellets using a T-die extraction molding machine at a width of 250 mm. The 1st roll was rapidly cooled so that the roll would be amorphous and was contacted with the 2nd roll, heated to 80°C, to make crystallized sheeting, in-line.

2-2. Injection Molding

2-2-1. Compound Production

The compound was made in the same manner as that of 2-1-1.

2-2-2. Evaluation of Crystallization Speed

Half crystallization time obtained by a differential scanning calorimetry ("DSC") was used to evaluate crystallization speed. The pellets were weighed to 7 to 8mg, melted for 5 min. at 200°C and then, the temperature was rapidly cooled to each crystallization temperature (T°C) at the speed of 500°C/min. Half time of when crystal is saturated, was set to be the half crystallization time (t1/2) at the T°C. The shorter the half crystallization time indicates a faster crystallization speed and favorable moldability. If crystallization is completed during rapid-cool down at a speed of 500°C/min. and it is impossible to measure time, the time is set to t1/2=0 indicating that the speed of crystallization is significantly high.

2-2-3. Evaluation of Moldability (productivity)

Moldability when the mold temperature and molding times were changed, was evaluated using an injection molding machine and test piece mold. If crystallization is well advanced, moldings can be removed without deformation; therefore, moldable time and temperature were evaluated.

2-2-4. Measurement of Relative Crystallinity

The pellets were weighed to 7 to 8 mg and were enclosed in an aluminum pan. The 1st Run temperature was raised from 20°C to 200 °C at a temperature rising speed of 20 °C/min. using DSC and the temperature was kept at 200 °C for 5 min. Then, the temperature was decreased from 200 °C to 20°C at a temperature descendent speed of -20 ° C/min. and temperature was kept at 20 °C for 1 min. On the 2nd Run, temperature was raised from 20°C to 200 °C at 20 ° C/min. The absolute value ⊿Hcc of cool crystallization enthalpy of PLA, observed in the 1st RUN and crystalline melting enthalpy ⊿Hm observed in the 2nd RUN, were obtained. The relative crystallinity was obtained using those values and the formula below.

Relative crystallinity = {((⊿Hm-⊿Hcc)/ ⊿Hmx100)}

2-2-5. Machine Physicality

A three-point bend test, Izod impact test, and measurement of deflection temperatures under load were performed, using the test pieces molded in the above process.

3. Results and Discussion

3-1. Extruded sheeting

To improve flexibility and impact resistance, we have developed two key technologies. The first one is "Softness control technology" which softens the amorphous portion of PLA. The second one is "Nano-crystalcontrol technology" to control nanocrystal, which promptly forms fine crystals and improves heat resistance with the synergy effect of the modifier, while maintaining transparency.

To demonstrate the introduction effects of the technology of crystal control, the crystal dimension of PLA was analyzed using an AFM. It took 10 min. to crystallize the unmodified PLA sheeting at 120°C and spherocrystals of 0.5 μm were observed. On the other hand, introduction of the developed technology of nanocrystal control, succeeded in forming fine crystals at 80°C within 30sec., the average dimension being about 16 nm, while maintaining transparency (Fig. 1).

To elevate heat resistance, the temperature spectrum of storage modulus of transparent modified PLA sheeting previously produced, and of PP sheeting, was compared (Fig. 2). Heat resistance, equivalent to or exceeding that of PP, was achieved by crystallization.

Aiming at mass production of transparent extruded sheeting, in-line crystallization of T-die extrusion molding was discussed (Fig. 3). The 1st roll was rapidly quenched and was contacted with the 2nd roll, annealed to 80°C for 15 sec. It was found that this saturates crystallization and transparent extruded sheeting can be mass-produced. Transparency and flexibility equivalent to or exceeding those of PP were obtained. Impact resistance and elongation at break, were also excellent (Table 1). It was confirmed that the transparent extruded sheeting has other characteristics, blushing resistance to bending, scratch resistance, resistance to chemicals and glossiness, all equivalent to or exceeding those of PP.

Stamping, UV offset printing, ruled line folding, punching and adhesive quality, were studied for secondary processability, which is important for transparent packaging containers and stationery. By selecting proper ink, each printing under the same conditions of PP can be conducted and the results of the tape peeling test and scratch test were acceptable. For ruled line folding, optimization of the processing conditions for extruded sheeting was necessary; however this was not a major problem. For improvement of the adhesive quality, the extruded sheeting was adhered using hot-melt and welding.

Thermoforming moldability which is important to food containers and blister packs was evaluated. Amorphous sheeting of a thickness of 400μm was produced using a T-die extrusion molding machine and pressure forming of the sheeting was evaluated together with HIPS and A-PET sheeting. Table 2 shows evaluation results of moldability and thermoforming products. With the "Nano-crystallization technology", molding cycles were the same as those of oil-derived resin and the relative crystallinity of PLA reached 100%. Transparency was acceptable, although it did not reach that of A-PET, and impact resistance was excellent. As crystallization of PLA was completed, heat resistance was superior to that of A-PET and the possibility was indicated that the sheeting could be applied to food containers, which can be heated in a microwave oven as with PP. Targeting stationery and packaging materials, sheeting produced with extrusion molding has been taken off the market.

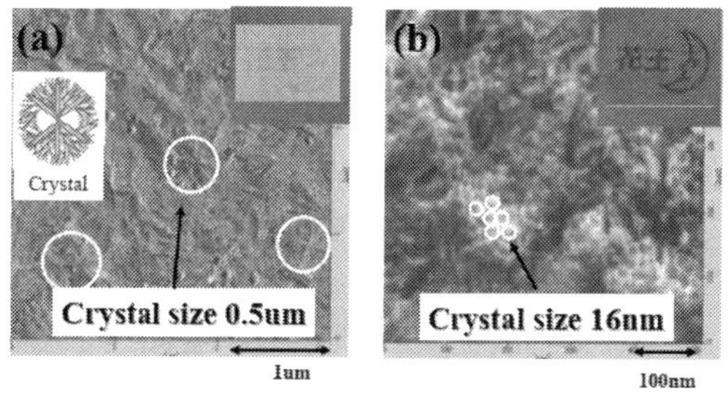

Figure 1 – AFM image of unmodified PLA sheeting with spherocrystals of 0.5 μm (a) and modified PLA sheeting using "Nano-crystalcontrol technology" with fine crystals of 16 nm (b).

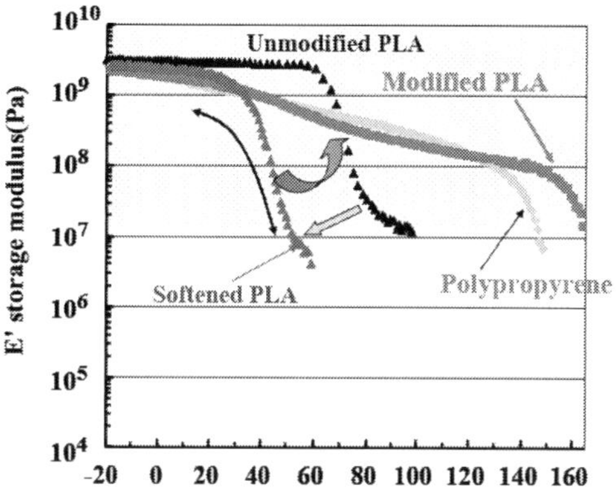

Figure 2 – Viscoelastic spectrum of PLA and PP.

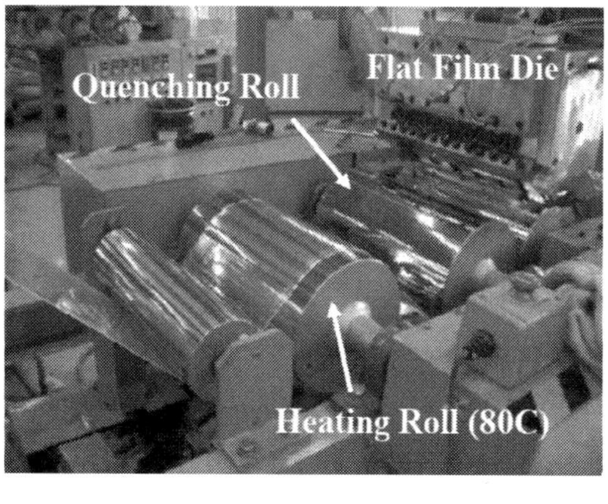

Figure 3 – Mass production of transparent extruded sheeting, in-line crystallizationusing of a T-die extraction molding machine at a width of 250 mm.

Table 1 – Mechanical property of modified and unmodified PLA sheeting and PP sheeting.

	unit	Modified PLA sheeting	PP sheeting	Unmodified PLA sheeting
Thickness	μm	200		
Haze value	%	5	33	2
Tensile modulus	GPa	1.4	1.2	2.8
Tensile strength	MPa	37	34	58
Impact strength (Dupont method)	J/mm	12.8	3	0.1
Elongation at break	%	300	100	<5
Blushing resistance to bending	-	○	△	×

Table 2 – Evaluation of moldability and thermoforming products of modified PLA, HIPS and A-PET sheeting.

	unit	Modified PLA sheeting	HIPS sheeting	A-PET sheeting
Thickness	μm	400		
Moldability				
Draw down	-	○	○	○
Molding time	sec	3	4	4
Mechanical property				
Haze value	%	5	83	2
Tensile modulus	GPa	2.7	2.5	2.5
Impact strength (Dupont method)	J/mm	18	10	18
Heat resistance (80°C for 4 hours)	-	○	○	×
Relative crystallinity	%	100	-	-

3-2. Injection Molding

Fig. 4 shows half crystallization time at each crystallization temperature of modified PLA and all-purpose resin, PP. It reveals that the crystallization speed of unmodified PLA is very slow as the half crystallization time ranges from 600 sec. to 650 sec at temperatures of 100°C to 120 °C. It indicates that the crystallization speed of PP is significantly faster from high temperature ranges to low temperature ranges. To enable injection molding with a common mold temperature as seen in PP, it is necessary to achieve the same crystallization speed as that of PP.
Addition of crystallization promoter "A" achieves high speed crystallization at high temperature ranges (Fig. 4 "A" agent). Then, addition of crystallization promoter "B" achieves high speed crystallization at middle and low temperature ranges (Fig. 4 "B" agent). Furthermore, the use of the crystallization promoter "A" and "B" together achieves high speed crystallization at wide ranges of temperature equivalent to those of PP, all-purpose resin (Fig.4 "A" agent + "B" agent).
Injection moldability (productivity) of the compound which showed crystallization speed equivalent to that of PP in the evaluation of half crystallization time was evaluated (Table 3). Crystallization rapidly advanced at 30°C of mold temperature and for 30 sec. of molding time, which are the same as those of all-purpose resin, and molding was performed without deformation. The measurement results show that the relative crystallinity of unmodified PLA is low and crystallization is not completed, but the relative crystallinity of the above modified PLA is 100%, which means that crystallization is completed and molding is rapidly performed. We succeeded in developing the desired technology of common mold temperature and high speed molding, which will realize productivity equal to that of PP and ABS.
Introduction of above technology to control crystallization and to soften PLA, led to success in developing modified PLA which is in no way inferior to all-purpose oil-derived resin. Table 4 shows the characteristics of ECOLA S-2010 and ECOLA S-2020 which have been launched as a commercial grade product. ECOLA S-2010 has a flexural modulus equivalent to that of PP and Izod impact strength exceeding that of PP. We are aiming at it becoming an alternative to PP in the field of all-purpose injection. ECOLA S-2020 is ABS alternative grade, having a high flexural modulus, deflection temperature under

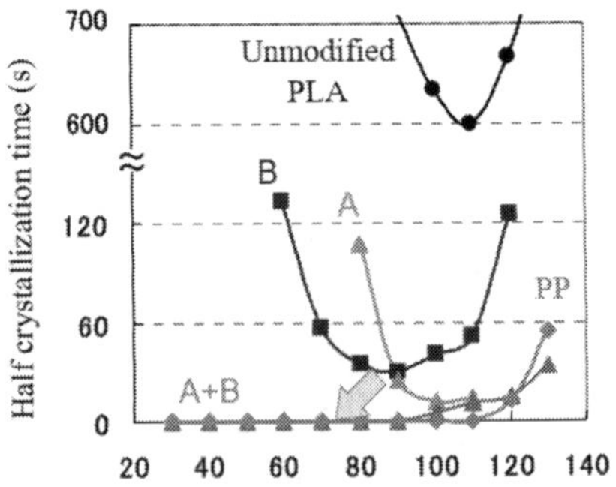

Figure 4 – Half crystallization time as a function of crystallization temperature of modified and unmodified PLA and PP compound.

Table 3 – Evaluation of moldability and relative crystallinity of PLA, PP and ABS compound.

	unit	Unmodified PLA	Modified PLA (for extrusion)		Modified PLA (Fig.4 A+B	PP	ABS
Mold temperature	°C	120	80	30	30	30	30
Molding time	sec	600	60	30	30	30	30
Relative crystallinity	%	100	100	70	100	100	-

Table 4 – Mechanical property of modified PLA "ECOLA" for injection molding.

			unit	ECOLA S-2010	ECOLA S-2020	ECOLA S-103X
Material property	MFR(190°C/2.16kgf)		g/10min	5.1	2.2	5.2
	Mold shrinkage	MD	-	1.19	0.44	1.08
		TD	-	1.21	0.78	1.04
Mechanical property	Flexural modulus		GPa	1.9	2.4	2.1
	Flexural strength		MPa	41	45	43
	Deflection at break		%	Non break	Non break	Non break
	Izod impact strength		J/m	80	105	67
Thermal property	DTUL	0.45 MPa	°C	132	140	125
		1.81 MPa	°C	-	70	-
	Flame resistnce(UL94 V)		-	-	-	V-2(1.8mm)
Moldability	Molding time/temperature		sec/°C	30/80	30/80	30/80
Biobased content (ASTM D6866)			%	89	79	83

load and Izod impact strength equivalent to those of ABS resin. A flame resistance formula is added, for the developed of case for information home electrical appliances. Flame retardants, such as phosphorous, nitrogen system, and inorganic system flame retardants, except for halogenated flame retardants were discussed. After discussion of a variety of flame retardants, we developed ECOLA S-103X where a non-halogenated flame retardant, which does not have PLA decomposition nor bleed out and has the necessary flame resistance, is added (Table 4).[6] ECOLA S-103X of a thickness of 1.8 mm achieved flame resistance V-2 while moldability was maintained. As resins at all grades use PLA only, the vegetation and environmental properties are favorable.

4. Conclusions

In the field of extruded sheeting, we established and developed the technology necessary for alternatives to all-purpose flexible resins such as PP with the synergy effects of modifiers. For mechanical property, modified PLA resin for extruded sheeting with qualities exceeding PP, was developed by using nanocrystallization, adding excellent transparency, heat resistance and stable flexibility. Drastic improvement of crystallization speed enabled in-line crystallization and established the industrial production of ECOLA sheeting. Also, secondary processability for thermoforming, printing adequacy, clear cartons and stationery is acceptable. The extruded sheeting can also be applied to packaging containers. We will continue improving the developed core technology and expand the use of extruded sheeting for an alternative to other all-purpose resins.

In the field of injection molding, essential technology for alternatives to all-purpose resins such as PP or ABS was established with the synergy effect of 2 kinds of crystallization promoters and developed PLA resin, moldable at 30°C of mold temperature and for 30 sec. of molding time, having excellent productivity. We developed ECOLA S-2010 which performs equivalent to that of PP, S-2020 which shows rigidity and toughness equivalent to those of ABS resin and S-103X, where flame resistance is added. In the future, we will improve the element technology necessary for case of information home electronic appliances and develop technology which will reduce costs so that PLA resin will be used in more costly commodities.

References

1. Inomata I., Plastics Age, **54**, [12], 81 (2008)
2. Kishimoto H., Plastics Age, **55**, [12], 70 (2009)
3. Takenaka A., et al., "Development of modified poly lactic acid resin for transparent extruded sheets", JSPP Symposium. Papers, 141–142 (2008)
4. Hashimoto R., et al., "Development of poly lactide resin for secondary processing", JSPP Symposium. Papers, 183–184 (2009)
5. Kishimoto H., et al., "Development of modified poly lactic acid resin for injection molding", JSPP Symposium. Papers, 143-144 (2008)
6. Kishimoto H., et al., "Development of flame resistant polylactide resin for injection molding", JSPP Symposium. Papers, 181-182 (2009)

Melt Processed Polyhdroxyalkanoates' Behavior in the Marine Environment

Jo Ann Ratto (1), Christopher Thellen (2), Robert Whitehouse (3), Carl Wirsen (4)*

(1) U.S. Army Natick Soldie Research, Development and Engineering Center - MA - USA, (2) U.S. Army Natick Soldier Research, Development and Engineering Center - MA - USA, (3) Telles - MA - USA, (4) Woods Hole Oceanographic Institution - MA - USA

Polyhydroxyalkanoate (PHA), a versatile family of polymers from renewable resources, is melt processable for a variety of applications. Extruded pellets, films and foams as well as extrusion coated paper and injection molded items were produced and then characterized in the marine environment for biodegradation behavior. A tierd testing approach was used where the samples were first evaluated using the ASTM 6691 method to determine the amount of carbon dioxide evolved as a function of time. These samples were tested using inoculums of either a known consortium of microorganisms or natural seawater with its indigenous microflora, supplemented with nitrogen and phosphorous nutrients . These PHA samples typically demonstrated 80 percent mineralization within 30 days. Since these materials met the minimum 30 percent mineralization requirement of the ASTM 7081 specification, they were eligible candidates for the tier two level of testing. These experiments involved weight loss experiments in both static laboratory and dynamic aquarium incubations as a function of time. Typical results exhibited over 90 percent weight loss in 45 days in a static experiment with controlled temperatures and 50 percent weight loss in 100 days for the dynamic test where nutritive and seasonal effects influence the weight loss results. Tier three level of testing exposed these samples in situ on coastal moorings that would replicate a more typical real life environment. Tier three testing of these samples demonstrated rates of weight loss that were even slower than aquarium incubations; however, the samples still had over 20 percent weight loss after 60 days. Overall, the PHA polymers, despite the melt processing method used to form these items, exhibited exceptionally high rates of biodegradation in the marine environment in comparison to other polymer samples studied with this same tierd testing approach. With marine debris issues today, this family of polymers can potentially replace other petroleum based polymers and be an alternative for many plastic products.

MAJOR ENHANCEMENTS IN CRYSTALLIZABILITY AND HIGH-TEMPERATURE MECHANICAL ROBUSTNESS OF POLY(LACTIC ACID): PRODUCTION OF NOVEL HYBRIDS OF PLA AND CELLULOSE VIA SOLID-STATE SHEAR PULVERIZATION

J.M. Torkelson[1,2,*], Philip J. Brunner[2], and John R. Dorgan[3]

[1]*Department of Chemical and Biological Engineering, Northwestern University, Evanston, Illinois 60208 USA – j-torkelson@northwestern.edu*
[2]*Department of Materials Science and Engineering, Northwestern University, Evanston, Illinois 60208 USA – philipbrunner2008@u.northwestern.edu*
[3]*Department of Chemical Engineering, Colorado School of Mines, Golden, Colorado 80401 USA – jdorgan@mines.edu*

Abstract - Biobased and biodegradable polymer blends or hybrids consisting of poly(lactic acid) (PLA) and 1-25 wt% microcrystalline cellulose (MCC) are created via a novel, industrially scalable process called solid-state shear pulverization (SSSP). The SSSP process may be followed by some extent of melt mixing (MM) achieved by the melt processing needed to form a final product. Scanning electron microscopy reveals that samples made by SSSP only and a two-step SSSP/MM process can result in relatively uniform MCC dispersion, with the SSSP/MM process yielding MCC particle sizes ranging from several microns in diameter down to the nanoscale. In contrast, almost no dispersion is achieved when MM is used alone without any SSSP processing, resulting in MCC particle diameters in the hundreds of microns. The high level of dispersion as well as the excellent wetting of the MCC filler by the PLA matris leads to major enhancements in crystallization kinetics and other material properties. For example, with a 99/1 wt% PLA/MCC hybrid made by the two step SSSP/MM process, a 38% crystallinity is achieved during non-isothermal crystallization from the melt state (10 K/min cooling rate), and a crystallization half-time of 1.7 min is achieved during isothermal crystallization at 378 K. In contrast, neat PLA achieves less than 1% crystallinity under similar thermal histories. Dynamic mechanical analysis also reveals that PLA/MCC samples that were processed by SSSP alone can yield nearly full theoretical crystallinity after only 3 minutes of cold crystallization time at 378 K. Thus, the temperature at which PLA maintains some mechanical robustness can be easily increased from roughly 333 K (the PLA glass transition temperature) to 390 K or higher. The increased crystallinity achieved in the PLA/MCC hybrids made by SSSP or SSSP/MM also results in increases in other selected mechanical properties and major decreases in oxygen permeation. Well-matched refractive indices also lead to good PLA/MCC film transparency.

Introduction

Biobased and biodegradable polymers have been extensively studied as environmentally friendly, sustainable alternatives to synthetic polymers derived from petroleum feedstocks [1,2]. Among the family of biodegradable polymers, poly(lactic acid) (PLA), also called polylactide, has many advantages. For example, PLA can be completely derived from renewable resources such as corn, sugar, and milk byproducts and is biocompatible and readily commercially available [3-5]. Moreover, PLA is less expensive than most other biodegradable polymers and, in some cases, has good mechanical properties. However, technological application of PLA has been limited because of several factors, including its low toughness and thermal stability and its low crystallinity in the as-processed state. This low crystallinity is associated with the grudgingly slow crystallization rate of PLA, which makes application in processes such as injection molding impractical when the desired product is semi-crystalline rather than amorphous PLA. Overcoming the technological challenge associated with the very slow crystallization rate is essential for the full commercial potential of PLA to be realized.

Here we demonstrate that this challenge can be met by making hybrids or blends or composites of PLA with low levels of cellulose. In particular, we mix 1 - 5 wt% micro-crystalline cellulose (MCC) with PLA via a continuous, industrially scalable process known as solid-state shear pulverization (SSSP) which can then be followed by melt mixing (MM). This two-step SSSP/MM process yields a fine MCC dispersion in PLA, with MCC particles ranging from the nanoscale to several microns in diameter. Relative to neat PLA, such blends result in factor of 20 reductions in PLA crystallization half-times, crystallinity levels (~38%) in the as-processed state near the maximum attainable (40-45%) for PLA, and improvements in both Young's modulus and yield stress.

Experimental

Materials. Linear PLA was from Cargill-Dow Polymers (now NatureWorks) with with M_n = 61,000 g/mol and M_w = 95,000 g/mol as determined by gel universal calibration. Micro-crystalline cellulose (MCC) was kindly donated by FMC Biopolymer.

Pulverization. The SSSP apparatus is a modified twin-screw extruder made by Berstorff. The main difference between the pulverizer and a conventional twin-screw melt extruder is that the barrels are cooled below room temperature using a -7 °C cooling medium. Further descriptions of the SSSP apparatus are given in refs. 6-8. Here, polymer/MCC hybrids were first hand-mixed in a plastic bag prior to being fed at a controlled rate into the pulverizer. A mild screw design involving two forward bi-lobe elements in the mixing zone and seven tri-lobe elements in the pulverization zone was used to pulverize the materials at screw speeds of 200 to 300 rpm.

Melt Mixing. After SSSP, some of the pulverized PLA product was melt mixed for 2 min in an Atlas Laboratory mini-max mixer at 200 °C, with three small steel ball bearings added to ensure relatively intense mixing [9]. Collected samples were placed in a liquid N_2 bath to freeze the microstructure.

Characterization. Crystallization behavior was characterized using a Mettler Toledo DSC822. Mechanical properties were measured with a Sintech 20G. The micromorphologies of fractured surfaces of the polymer-MCC composites were characterized by scanning electron microscopy (Leo 1525) employing a backscattered electron detector. Surfaces were sputter coated with 5 nm of Au/Pd and examined at 5 kV accelerating voltage.

Results and Discussion

Nonisothermal crystallization curves obtained via differential scanning calorimetry are shown in Figure 1 for samples of neat PLA (as-received pellet), pulverized neat PLA, and 99/1 wt% PLA/MCC composites made by SSSP alone and by the two-step SSSP/MM process. Related data are provided in Table 1. No exothermic crystallization peak is evident in the cooling scan of neat PLA, indicating that neat PLA does not crystallize (within experimental error) when cooled at a rate of 10 °C/min from the melt state. For neat PLA subjected to SSSP, a barely perceptible exothermic peak is evident in the cooling cycle, which we estimate to be indicative of 3% crystallinity. We hypothesize that the pulverization process leads to an enhanced dispersion of heterogeneous nucleation agents present in the PLA, and thereby to a greater development of crystallinity relative to a PLA sample never subjected to SSSP.

In contrast to the neat PLA, the sample made by addition of 1 wt% MCC to PLA via SSSP yields a substantial exothermic peak in the 100-120 °C temperature range, consistent with the development of 20% crystallinity of the PLA during the cooling cycle.

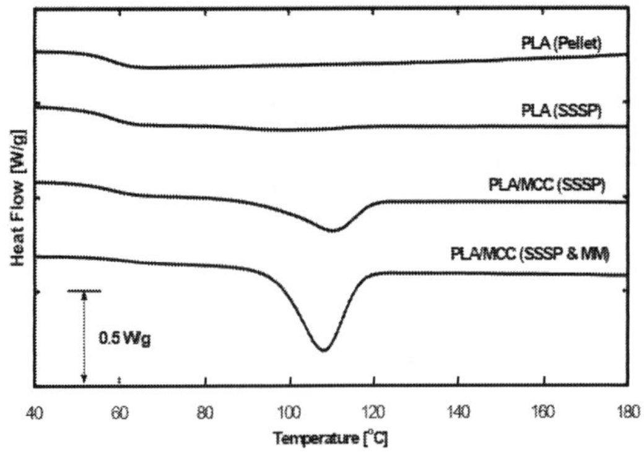

Figure 1. Non-isothermal crystallization curves for PLA pellets, pulverized PLA, and 99/1 wt% PLA/MCC composites made by SSSP only and the two-step SSSP/MM procedure. Data were obtained via differential scanning calorimetry by heating samples from 40 °C to 200 °C and then cooling to 40 °C at a rate of 10 °C /min, and repeating the heating and cooling cycle, with data taken during the second cooling cycle.

Table 1. Crystallization Results from DSC Analysis of PLA and 99/1 wt% PLA/MCC Composites

Sample	$X_{c,n}$ (%)[1]	$\tau_{1/2}$ (min)[2]	$X_{c,i}$ (%)[2]
PLA (Pellet)	<1	30	<1
PLA (SSSP)	3	4.1	27
PLA/MCC (SSSP)	20	3.1	29
PLA/MCC (SSSP & MM)	38	1.6	34

[1]$X_{c,n}$ = crystallinity achieved during non-isothermal cooling from melt at 10 °C/min; [2]$\tau_{1/2}$ = isothermal crystallization half-time at 105 °C; [3]$X_{c,i}$ = crystallinity obtained during the isothermal hold at 105 °C.

An even greater level of crystallinity (38%) is obtained upon cooling a 99/1 wt% PLA/MCC sample made by SSSP followed by MM; this 38% crystallinity is close to the 40-45% maximum level of crystallinity obtained in PLA subjected to long-term isothermal crystallization. These results are consistent with the hypothesis that the interfacial regions between PLA and MCC serve as highly effective nucleation sites for PLA crystallization.

We used differential scanning calorimetry to characterize the isothermal crystallization behaviors of PLA and PLA/MCC composites at 105 °C. As indicated in Table 1, we obtained a crystallization half-time for neat PLA of 30 min, in agreement with the range of 17-45 min reported in the literature [10-13]. The crystallization half-times are dramatically decreased in the samples subjected to SSSP, with the neat PLA and 99/1 wt% PLA/MCC composite having 4.1 min and 3.1 min half-times, respectively. An even greater reduction in crystallization half-time, at 1.6 min nearly a factor of 20 smaller than the value for neat PLA, is obtained in the 99/1 wt% PLA/MCC composite made by the two-step SSSP/MM procedure. Thus, while SSSP processing alone yields major changes in the crystallizability and crystallization kinetics of PLA and 99/1 wt% PLA/MCC composites, SSSP followed by MM results in even greater changes.

The further enhancement in crystallization caused by melt mixing the already pulverized 99/1 wt% PLA/MCC composite is likely associated with the greater dispersion of the MCC in PLA. As shown in Figure 2, in a 95/5 wt% PLA/MCC composite, SSSP processing alone yields MCC particles with ~10-20 micron length scales while the two-step SSSP/MM process results in MCC particle sizes ranging from several microns in diameter down to the nanoscale.

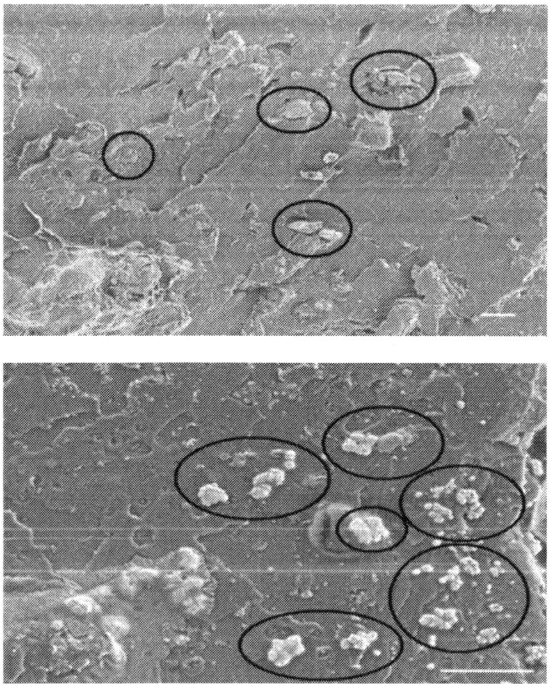

Figure 2. SEM images of 95/5 wt% PLA/MCC composites made by SSSP only (upper) and by the two-step SSSP/MM process (lower). The size bars represent length scales of 20 μm (upper) and 10 μm (lower), respectively.

We also characterized the tensile properties of PLA and 99/1 wt% PLA/MCC composites. In all cases, SSSP processing and inclusion of 1 wt% MCC leads to property improvements. Relative to neat PLA, Young's modulus (2.67 +/- 0.06 GPa in neat PLA) increases by 9%, 7%, and 15% in neat PLA processed by SSSP, in 99/1 wt% PLA/MCC processed by SSSP, and in 99/1 wt% PLA/MCC subjected to the two-step SSSP/MCC procedure, respectively. Similarly, the yield stress (61.2 +/- 1.9 MPa in neat PLA) increases by 5%, 9%, and 12% in neat PLA processed by SSSP, in 99/1 wt% PLA/MCC processed by SSSP, and in 99/1 wt% PLA/MCC subjected to the two-step SSSP/MCC procedure, respectively.

As indicated by the results described above, there are many positive outcomes associated with the SSSP-based addition of low levels of biobased MCC to biobased polymers (e.g., PLA). The application of SSSP processing may contribute in significant ways to making sustainable, green engineering of polymers synonymous with property improvement.

References

1. Vink, E.T.H.; Rabago, K.R.; Glassner, D.A.; Gruber, P.R. *Polym. Degrad. Stabil.* **2003**, *80*, 403.
2. Huda, M.S.; Mohanty, A.K.; Drzal, L.T.; Schut, E.; Misra, M. *J. Mater. Sci.* **2005**, *40*, 4221.
3. Braun, B.; Dorgan, J.R.; Knauss, D.M. *J. Polym. Environ.* **2006**, *14*, 49.
4. Sobkowicz, M.J.; Dorgan, J.R.; Gneshin, K.W.; Herring, A.M.; McKinnon, J.T. *J. Polym. Environ.* **2008**, *16*, 131.
5. Zhang, J.; Tsuji, H.; Noda, I.; Ozaki, Y. *J. Phys. Chem. B* **2004**, *108*, 11520.
6. Furgiuele, N.; Lebovitz, A.H.; Khait, K.; Torkelson, J.M. *Macromolecules* **2000**, *33*, 225.
7. Tao, Y.; Lebovitz, A.H.; Torkelson, J.M. *Polymer* **2005,** *46*, 475.
8. Walker, A.M.; Tao, Y.; Torkelson, J.M. *Polymer* **2007**, *48*, 1066.
9. Maric, M.; Macosko, C.W. *Polym. Eng. Sci.* **2001**, *41*, 118.
10. Brandrup, J.; Immergut, E.H.; Grulke, E.A.; Eds. *Polymer Handbook*, Wiley: New York, 1999.
11. Schmidt, S.C.; Hillmyer, M.A. *J. Polym. Sci., Part B: Polym. Phys.* **2001**, *39*, 300.
12. Anderson, K.S.; Hillmyer, M.A. *Polymer* **2006**, *47*, 2030.
13. Kolstad, J.J. *J. Appl. Polym. Sci.* **1996**, *62*, 1079.
14. Wakabayashi, K.; Pierre, C.; Dikin, D.A.; Ruoff, R.S.; Ramanathan, T.; Brinson, L.C.; Torkelson, J.M. *Macromolecules* **2008**, *41*, 1905.
15. Masuda, J.; Torkelson, J.M. *Macromolecules* **2008,** *41*, 5974.

PROPERTIES OF THERMOPLASTIC POTATO STARCH AND ITS BLENDS

A.N. Hrymak [1*], S. Falah-Toosi [2], Q. Liu [3]

(1) Department of Chemical and Biochemical Engineering, University of Western Ontario, London, Ontario, Canada – ahrymak@uwo.ca
(2) Department of Chemical Engineering, McMaster University, Hamilton, Ontario, Canada,
(3) Guelph Food Research Centre, Agriculture and Agri-Food Canada, Guelph, Ontario, Canada - Qiang.Liu@agr.gc.ca

Abstract - Potato starch granules were plasticized with different amounts of glycerol and water. Differential scanning calorimetry (DSC) was used to investigate the effect of the plasticizers on gelatinization temperature of dried potato starch. By increasing the amount of glycerol, gelatinization peaks in TPS shift to higher temperatures. While increasing the amount of water leads to a decrease in gelatinization peak temperatures. In order to mimic one form of extrusion processing, glycerol was added to the blend of starch and water at various time steps. At this point, the formulation of 60/ 30/10 (starch/glycerol/water) wt% was selected for subsequent runs. Blends were processed in micro compounder extruder (extruder DSM xplore) to get fully gelatinized starch. Mechanical properties and thermal properties were studied for TPS blends. Also, the influence of aging over time on the mechanical properties of thermoplastic potato starch has been monitored. Poly lactic acid (PLA) was added to the blend to improve the mechanical properties of TPS. In order to investigate the interfacial modification of the blend through MA, free radical-initiated grafting of maleic anhydride (MA) onto the PLA backbone and then mixing the PLAgMA with TPS was performed. The extent of maleation for grafted samples was determined by titration. Before adding PLAgMA to TPS matrix, the residual initiator was decomposed by washing and heating the PLA grafted samples. To improve the mechanical properties and water uptake behavior of potato thermoplastic starch, the addition of sisal fibers as reinforcement materials to TPS and TPS/PLA matrices was investigated. The effects of fibers on both matrices were studied through tensile tests, contact angle, and water uptake. As the last step, in order to increase the water resistance of the samples, all the blends including TPS, TPS/PLA, TPS/sisal fibers, and TPS/PLA/sisal fiber

RELEVANCE OF LAB-SCALE CONICAL TWIN-SCREW EXTRUDER FOR THERMOPLASTIC STARCH/PLA BLENDS RHEOLOGY STUDY

E. Chabrat[123*], A. Rouilly[12], P. Evon[12], A. Longieras[3] and L. Rigal[12]

*1. Université de Toulouse, INPT, LCA (Laboratoire de Chimie Agro-industrielle), ENSIACET, 4 Allée Emile Monso, 31432 Toulouse Cedex 4, France - *elodie.chabrat@ensiacet.fr*
2. INRA, LCA (Laboratoire de Chimie Agro-industrielle)
*3. Vegeplast, Parc des Pyrénées, 2 rue de Troumouse, 65420 IBOS, France -*ec@vegeplast.com*

During the last decade, thermoplastic starch (TPS) has been studied more and more, alone and in blends. One key property of TPS is its viscosity. A lab-scale conical twin-screw extruder has been used to process polymers or blends of polymers at a small scale (7 cm^3), and to measure their viscosity by pressure loss in a backflow channel. TPS has been tested at different temperatures (100-180°C) and for a 100-900s^{-1} shear rate range. One surprising result is the viscosity evolution of starch/glycerol blends (supposed to have become TPS before the rheology measurements) with temperature. An increase is observed between 120 and 140°C and then it decreases between 140 and 160°C. Therefore plasticization may happen only from 140°C. Blends of TPS with PLA and with different additives have also been studied. The additives are introduced in a second time, ten minutes after the TPS/PLA blend recirculation. Their influence on the blend viscosity is instantaneous. A viscosity increase after the additive introduction can be a clue for an improvement of the blend compatibilisation, as observed with CMC. Thanks to this micro-compounder, viscosity and behaviour of polymers can be evaluated easily even for TPS which needs shear to flow. Moreover, the solidified sample collected in the backflow channel at the end of the experiment can be used to test different properties like mechanical or thermal (DMTA, DSC) ones. Using this kind of lab-scale extruder represents then a good opportunity for preview screening studies.

Introduction

During the last decade, thermoplastic starch (TPS) has been studied more and more, alone (Stepto, 2009) and in blends (Ke and Sun, 2001). Indeed, starch is a renewable and biodegradable resource which can be transformed into a thermoplastic material by using plasticizers, shear and temperature. This product is very promising, in spite of its sensibility to water and its poor mechanical properties. It is well-known that starch plasticization can be realised in a twin screw extruder (Martin et al. 2003). Twin-screw extruders range widely from lab-scale (g/min) to industrial-scale (Tons/min).

Among the smallest twin-screw extruders, one can find the Thermo-Haake Minilab (Thermo Scientific, Haake, Germany). As this instrument contains co-rotating twin screws, it can represent the performance of a co-rotating extruder (industrial scale) (Moghaddam et al. 2009). The Minilab operates at speeds between 10 and 360 rpm, corresponding to shear rates between 35s^{-1} and 1248s^{-1} in the geometry controlled back flow channel and its heating capacity and maximum temperature are 800W and 450°C respectively. The Minilab has a maximum feed capacity of 7 cm^3.

The backflow (recirculating) channel in the Minilab allows its use as a mixing reactor with controlled recirculation rather than direct extrusion through the die. In addition to melt processing operations, the Minilab can also be used as a rheometer by using a pair of pressure transducers in the backflow channel. An approximate value for the melt viscosity is calculated automatically by the Minilab software using the pressure loss between these two sensors.

This apparatus can thus be used to study new blends or new products. After a screening at small scale, the interesting blends can be processed on an industrial twin-screw extruder with new parameters like temperature profile (and not just one temperature), choice of the screw configuration…

In this study, starch has first been plasticized with glycerol. Variable factors were the temperature, the screw speed (shear rate) and the glycerin rate. Then, thermoplastic starch has been blended with a biodegradable polyester, the polylactic acid (PLA). This blend already studied in several previous works is not miscible and needs to be compatibilized. The last part of this work describes a method to visualize the additive impact on the viscosity.

Experimental

Materials

Wheat starch was supplied by Gers Farine (France). Starch plasticizer was glycerol (purity 99%) supplied by Gaches Chimie (France). The PLA (Ingeo 3251D was supplied by Natureworks (USA).
The three tested additives were CMC (carboxymethylcellulose), which is a Blanose 7M grade (Aqualon), succinic acid (Sigma-Aldrich) and xanthan gum (IMCD).

TPS fabrication

Starch was hand-blended with glycerol and then introduced in a lab-scale twin-screw extruder, the Minilab (Thermo Scientific Haake, Germany) (Figure 1).

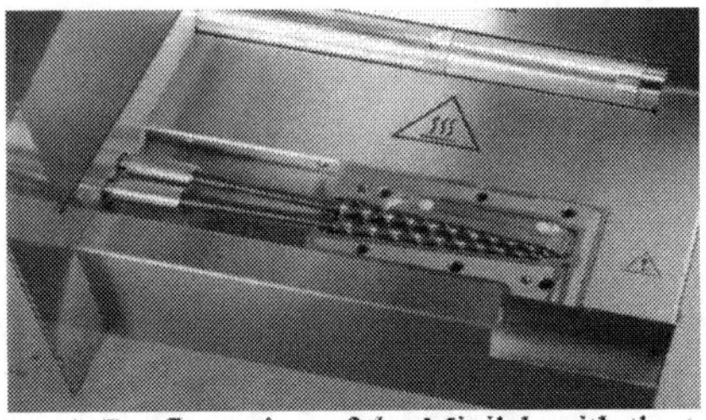

Figure 1 Configuration of the Minilab with the twin screws and the pressure transducers

In our case, the experiments were performed at 120-160°C and screw speeds between 50 and 250rpm. The sample was processed in the Minilab in recirculating mode for 8 minutes (stabilization time) before measuring the viscosity for 20 minutes, i.e. 28 minutes of recirculation. To study the influence of glycerol rate, temperature and shear rate, a design of experiment (DOE) was made as described in Table 1.

Table 1 - Design of experiment for the starch plasticization

Experiment number	Temperature (°C)	Screw speed (rpm)	Glycerol content (%)
1	160	150	30
2	120	150	30
3	150	237	30
4	130	63	30
5	150	63	30
6	130	237	30
7	150	179	34.1
8	130	121	25.9
9	150	121	25.9
10	140	208	25.9
11	130	179	34.1
12	140	92	34.1
13	140	150	30
14	140	150	30
15	140	150	30

Characterization

Viscosity was directly measured with the lab-scale extruder thanks to its pressure transducers. It was calculated as described in Equation 1 and measured every minute for 20 minutes. Finally, these viscosities were averaged. The average is the product viscosity given in the results part.

Equation 1 Viscosity formula

$$\eta = \frac{\tau}{\gamma}$$

$$\text{With } \tau = (\frac{h}{2\Delta l}) \times \Delta P$$

$$\text{and } \gamma = (\frac{6}{wh^2}) \times 8.10^{-7} n$$

Viscosity (Pa.s) : η

Shear stress (Pa) : τ and shear rate (s^{-1}) : $\dot{\gamma}$

Depth of the flow channel : h = 1.5mm
Width of the flow channel : w = 10mm
Distance between the transducers : Δl = 64mm
Pressure drop (Pa) : ΔP
Screw speed (s^{-1}) : n

Extruded samples were collected from the backflow channel for analytical purposes. All experiments were performed in triplicates. These samples were 70mm long, 8mm large and 1mm thick. They were conditioned at 25°C and 60%HR for stabilization for one week prior to testing.
The tensile testing was carried out at a rate of 5mm/min according to ASTM D638.

Dynamic mechanical analyses (DMA) of the samples were performed with a Tritec 2000 DMA (Triton Technology, UK) over the temperature range -100 to 200°C with a scanning rate of 2°C/min. The amplitude and frequency were maintained constant at 25µm and 1 and 10Hz, respectively.

Thermal study was also performed on a Pyris 1 differential scanning calorimeter (DSC) (Perkin-Elmer). The measurement cells were purged with dry nitrogen. The temperature was calibrated by use of indium (T_f = 156.6°C). The pans used for this study were airtight steel capsules with an 0-ring. They withstand high pressure (40bar) and are therefore well adapted for studying weakly hydrated biopolymers at high temperatures. The sample weighed approximately 20mg. The measurements were taken during the first scan, at 20°C/min between 40 and 200°C.

Results and Discussion

Recirculation in the Minilab

To ensure that the matter correctly recirculates in the lab-scale extruder, the following experiment was made: A starch/glycerol blend was introduced in the lab-scale extruder at 140°C with a screw speed of 150rpm. After 5 minutes, a small quantity of a dye (erythrosine) was introduced (t=0s) and the experiment was stopped after 30s, 1min, 2min and 5min.

The color of the sample in the backflow channel is representative of the circulation of the matter. On Figure 2, one can find the photographs of the 4 samples (4 different times of experiment). After 30s (top-left), the dye has only reached the middle of the channel. After one minute, the dye has colored almost all the backflow channel except for the extreme sides. After 2 minutes, the whole backflow channel is colored, the dye had circulated everywhere in the channel. After 5 minutes, the sample is darker, meaning that the dye had already circulated several times in the backflow channel. Hence, the matter does effectively recirculate in the lab-scale extruder.

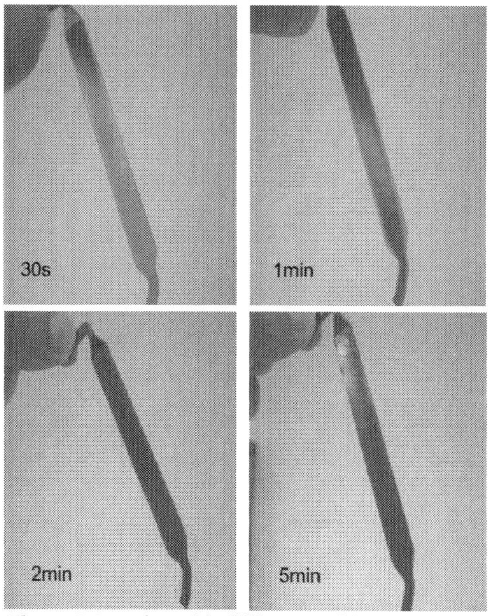

Figure 2 Samples with different residence time of the dye (30s, 1min, 2min, 5min)

Plasticization of starch

DOE results are represented on figures 3 to 5 (response surface). The temperature has only a small effect on the viscosity comparing to glycerol rate and screw speed. A screw speed increase (i.e. shear rate increase) leads to a decrease of the viscosity whatever the temperature and the glycerol rate, which is consistent with the shear-thinning behavior of TPS. A glycerol rate increase also leads to a viscosity decrease whatever the temperature or the screw speed.

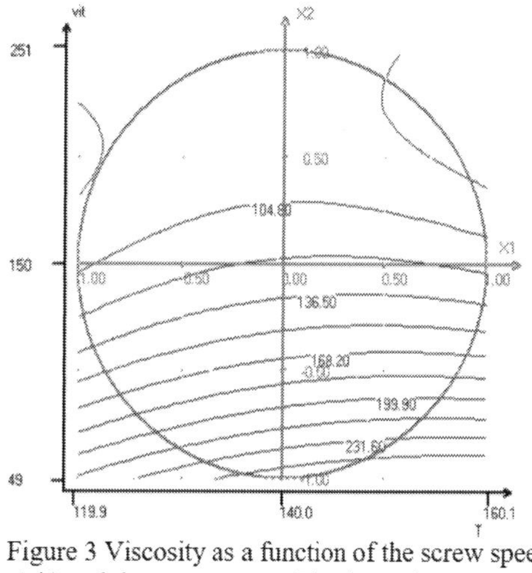

Figure 3 Viscosity as a function of the screw speed (vit) and the temperature (T), glycerol rate = 30%

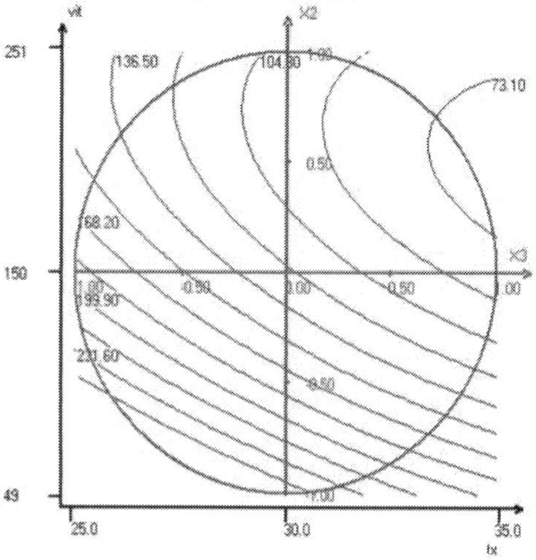

Figure 4 Viscosity as a function of the screw speed (vit) and the glycerol rate (tx), T fixed = 140°C

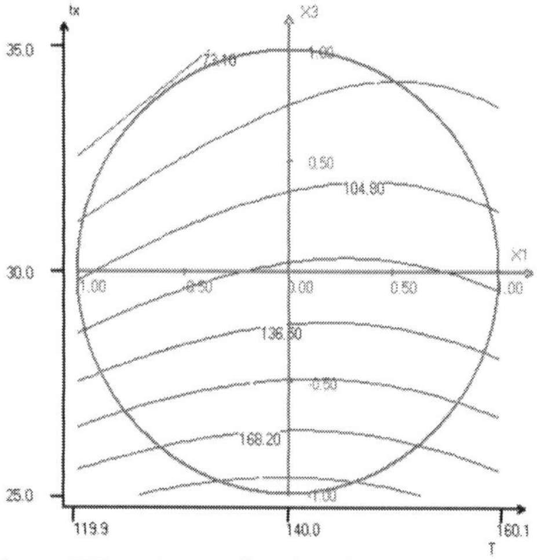

Figure 5 Viscosity as a function of the temperature (T) and the glycerol rate (tx), screw speed fixed = 150rpm

Even if the temperature has a small effect on the viscosity, at high screw speed a viscosity maximum is observed around 140°C as represented on Figure 6. This viscosity maximum could be explained by a transformation happening at 140°C. After 140°C, the viscosity decreases with the temperature which is a classical behaviour for a thermoplastic. Following this hypothesis, this means that before 140°C, the destructuration (and/or plasticization) of starch may not be complete. This idea was confirmed with thermal analysis (Figure 7). On the DSC graph, a peak is observed around 170-180°C for native starch. For a manual blend of starch and glycerol, this peak is divided into two peaks, around 175°C and around 140°C. This second peak was still observed after extrusion at 120°C, whereas it disappeared after extrusion at 140°C. This peak may be representative of the native structure and disappears totally after complete destructuration (and/or plasticization). Van Soest et al. (1996) have already studied the melting of potato starch with low and high amounts of glycerol and have observed similar behavior.

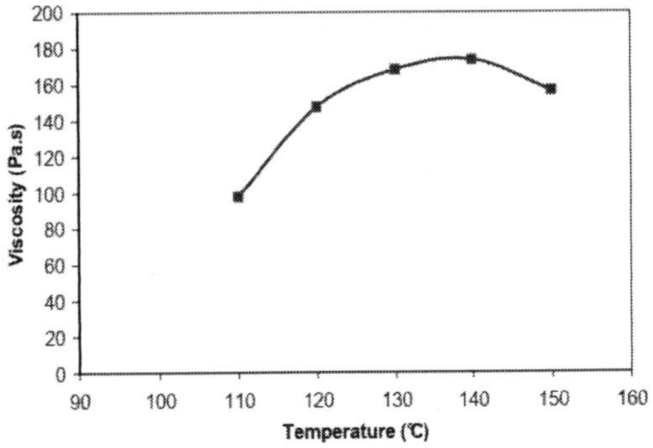

Figure 6 Viscosity of a starch/glycerol blend (75/25) at $622s^{-1}$ (175rpm) for different temperatures

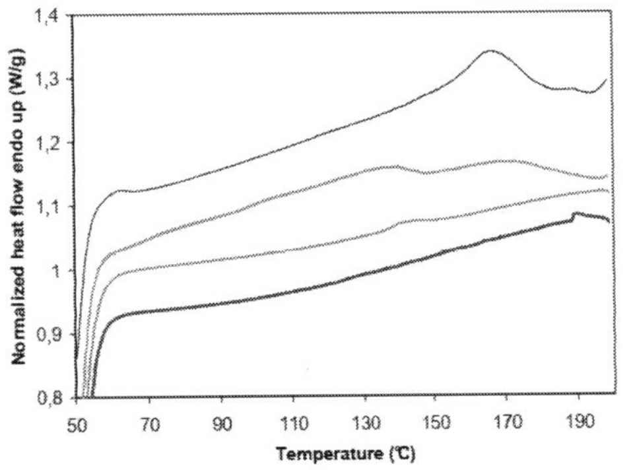

Figure 7 Thermal properties of starch and its blends with glycerol (from top to bottom : a-native starch, b-starch + glycerol non extruded, c-starch + glycerol extruded at 120°C, 175rpm for 7min, d-starch + glycerol extruded at 140°C, 50rpm for 20min)

Blending with PLA

Blends of thermoplastic starch and PLA were also made in the lab-scale extruder (Figure 8). At 140°C, TPS and PLA viscosities intersected at $500s^{-1}$, meaning that at lower shear rates, PLA is more viscous than thermoplastic starch and reversely at higher shear rates. This is interesting because it means that at $500s^{-1}$, the two products have the same viscosity, which is a perfect condition for blending. What is also interesting on this graph is the blend viscosities. Whatever the proportions, it is always lower than the viscosity of the two components. This result can be an indication of a bad blend compatibility where the two phases repel each other and the blend is more fluid. Opposingly, when a blend is homogeneous, the viscosity of the blend is higher than the viscosity of the components because of interactions between the two phases.

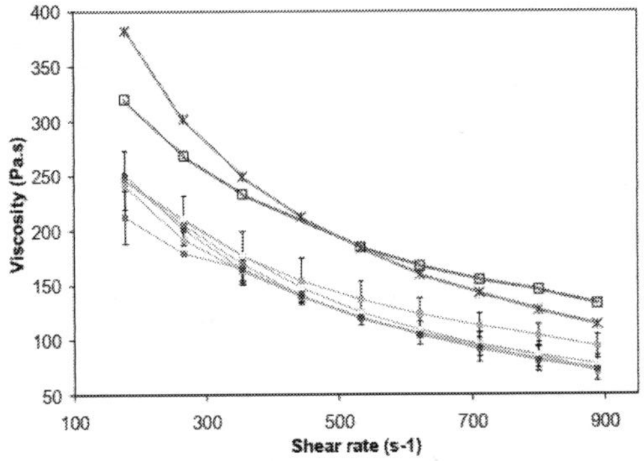

Figure 8 Flow curves for different TPS/PLA blends at 140°C – (□ : TPS, * : PLA, other curves : blends TPS/PLA from 95/5 to 50/50)

Method to study additives influence

Here is presented a simple method to study the additive influence on a blend. In each experiment, the beginning was the same. Starch, glycerol and PLA were extruded for 18min (8min of stabilization and 10min of measure). After these 18min, the additive was introduced in the extruder and the viscosity was measured for 10min (Figure 9).

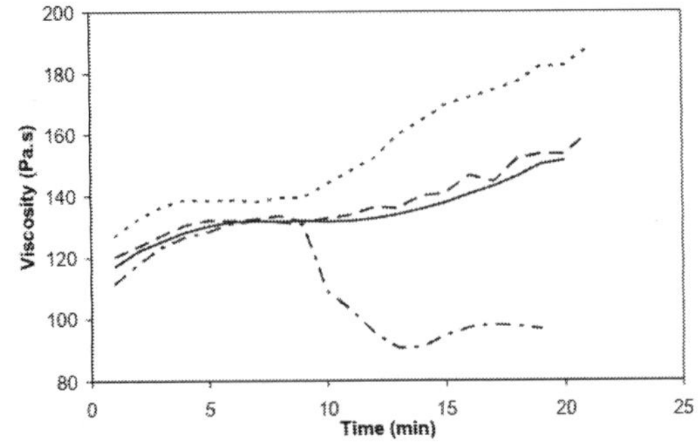

Figure 9 Evolution of the viscosity with the introduction of an addtive at mid-experiment (continuous curve : no additive, point curve : CMC, dash curve : xanthan gum and point & dash curve : succinic acid)

The continuous line is the reference curve (without mid-experiment add). Some additives lower the viscosity (succinic acid - dash and point curve), one can imagine maybe a lack of affinity between the components or a degradation of some of them (acidic degradation). Some additives have no influence on the viscosity (xanthan gum - dash curve). These additives have maybe no influence on the blend. The other additives increase the viscosity (CMC - point curve). The increase can be simply explained by a thickening effect of some products or by a compatibilisation of the blend. A viscosity increase of the blend can be a clue for a blend compatibilisation improvement.

Conclusions

Starch plasticization, starch blending with PLA and even with compatibilizers was possible in the lab-scale extruder. Moreover, instantaneous viscosity measurements provided useful information about the matter state in the extruder.

Acknowledgements

We would like to thank Maria Konstantakopoulou for her help for the experiments and the French Environment and Energy Management Agency (ADEME) for financial help.

References

Ke, T. & Sun, X., "Effects of moisture content and heat treatment on the physical properties of starch and poly(lactic acid) blends", *Journal of Applied Polymer Science*, 81(12), 3069-3082 (2001)

Martin, O., Averous, L. & Della Valle, G., "In-line determination of plasticized wheat starch viscoelastic behavior: impact of processing", *Carbohydrate Polymers*, 53(2), 169-182 (2003)

Moghaddam, L., Martin, D., Halley, P. & Fredericks, P., "Vibrational spectroscopic studies of laboratory scale polymer melt processing: Application to a thermoplastic polyurethane nanocomposite", *Vibrational Spectroscopy*, 51, 86-92 (2009)

Stepto, R.,. "Thermoplastic Starch", *Macromolecular Symposia*, 279(1), 163-168 (2009)

Van Soest, J.J.G., Bezemer, R.C., de Wit, D. & Vliegenthart, J.F.G., "Influence of glycerol on the melting of potato starch", *Industrial Crops and Products* 5, 1-9 (1996)

Calculation of the Activation Energy for Self Diffusion and Molecular Healing in PLA Films

The bonding of semicrystalline polylactic acid (PLA) films was investigated for a broad range of contact times and temperatures above the glass transition point in a lap shear joint geometry using an impulse welding system. The goal was to gain insight into molecular diffusion across the weld interface.

Molecular diffusion occurs during plastics welding of two polymer interfaces that are brought into contact above their glass transition temperature (T_g). At this temperature, the polymer chains diffuse across the contact area until the interface is indistinguishable from the bulk material. During the process, the mechanical strength of the bond increases until it reaches the bulk material strength [1,2,3]. The parameters of such "autohesion" processes are contact time, interfacial temperature, and pressure. In order to re-establish the bulk material strength at polymer interfaces or the virgin material strength, the chains need to diffuse a distance to sufficient to heal the interface. The mechanism by which autohesion occurs can be attributed to the two characteristics of high molecular weight polymers, which are a random chain and flexible macromolecules that are able to move within the bulk material. Schematically shown in Figure 1, the autohesion process for an amorphous polymer assumes no bulk flow. At the initial contact in Figure 1 (a) the surfaces match each other. After some contact time above the T_g some of the polymer chains penetrate the surfaces of the adjacent parts Figure 1(b), which is due to free chain movement that results from increased molecular free volume at the elevated temperatures. With increasing contact time the penetration depth of the diffusing polymer chains, as well as the number of chains crossing the interface increases, resulting in more entanglements and higher interfacial strength. In Figure 1(c) at a contact time of $t=t_\infty$ (time infinity) the healing process can be considered complete as the interface disappears and strength, as well as morphology are not distinguishable from the bulk material [4]. Some argue that sufficient diffusion occurs when the center of mass of a molecule has diffused a distance of approximately its radius of gyration to achieve full adhesion or complete healing [5].

It is assumed that the interface structure and the shear strength (degree of autohesion) at any time during the welding is proportional to the fourth root of the weld time t_w [5].

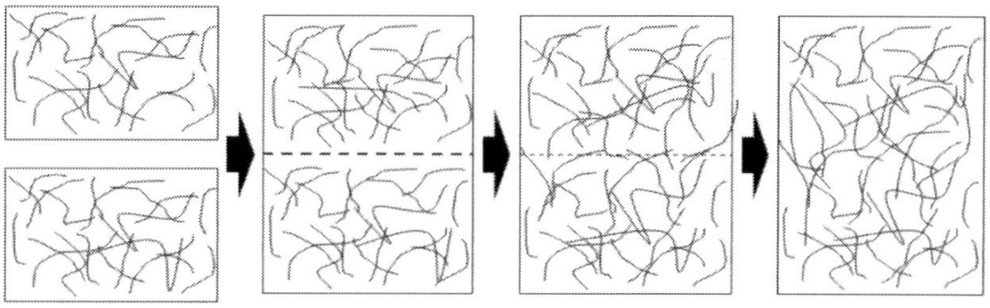

Figure 1 Physical picture of autohesion

The time dependency of the interface strength during processing, such as welding, can be expressed as the ratio of a cohesive strength at a weld time t_w and the cohesive strength when the strength becomes independent of the processing time at t_∞ [6] is defined in Eq. (1).

$$D_{Au} = \frac{A_u(t_w, T)}{A_{u,\infty}(t_\infty, T)} \tag{1}$$

The strength of autohesion A_u depends on the contact time of the interfaces, t_w and the temperature of the material, T. However, Jud [6] reported that the strength of autohesion can be represented by an Arrhenius law of diffusion $D(T)$:

$$D(T) = D_0 \exp\left(-\frac{E_a}{RT}\right) \tag{2}$$

where R is the universal gas constant (8.3144 J mol^{-1}K^{-1}), T is the absolute temperature, E_a is the activation energy of diffusion and D_0 is a diffusion constant. The degree of autohesion in Eq. (2) can be written as

$$D_{Au} = D_{Au_0} + K(T) * t_w^{\frac{1}{4}} \tag{3}$$

where D_{Au0} is initial degree of autohesion. In this case, $K(T)$ is a temperature dependent parameter that represents the temperature dependent slope of D_{Au} when it is plotted as a function of the fourth root of the weld time, and D_{Au0} is the intersection with the ordinate of the cartesian coordinate system. In more detail, $K(T)$ is the product of a proportionality constant K_0 times the self diffusion coefficient of the Arrhenius equation:

$$K(T) = K_0 * \exp\left(-\frac{E_a}{RT}\right) \tag{4}$$

With equation (3) and (4) the degree of autohesion D_{Au} can be written as

$$D_{Au} = D_{Au_0} + K_0 * \exp\left(-\frac{E_a}{RT}\right) * t_w^{\frac{1}{4}} \tag{5}$$

By plotting the natural logarithm of $K(T)$ as a function of the inverse temperature, the slope of the graph is the activation energy of diffusion and the intersection of the graph with the ordinate is the proportionality factor K_0, as it can be inferred from Eq. (8).

$$\ln(K(T)) = \ln(K_0) - \frac{E_a}{RT} \tag{6}$$

The sample dimensions followed the ASTM D 638 [7] procedure shown in Figure 2. The PLA used in this study was biaxially oriented film (Evlon®) provided by Bi-Ax International Inc. and manufactured from a Nature Works.

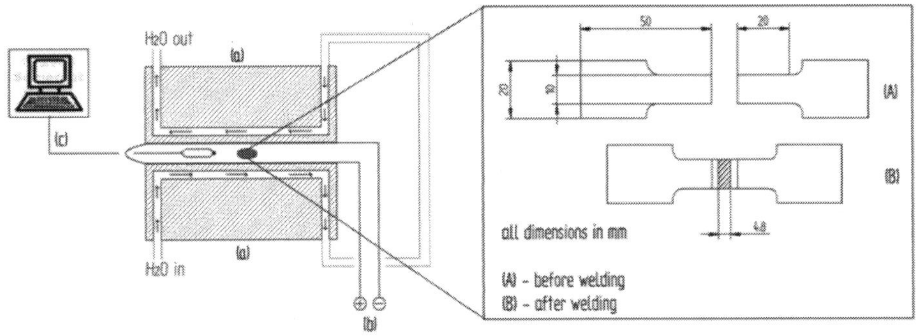

(A) (B)

Figure 2 (A) Setup for impulse welding: (a) welding head, (b) Nichrome heating wire connected to a DC power supply, (c) digital data aquisition system; (B) sample dimensions

The welding head is shown in Figure 2 (A).The head consist of a heating element that are resistively heat by DC electrical current which maintains the temperature at a constant value until the welding process is completed. The temperature during the welding was acquired with a J-type thermocouple (TC) which was connected to a digital data acquisition system that works with a sampling rate of 1000 Hz. The TC was placed in at the weld interface in a second sacrificial sample. The heating time was defined to be welding time t_w or time electrical current was applied to the heating elements. Because of the heat and pressure during the welding process, the polymer chains at the faying surfaces started moving with some of the chain segments crossing the interface, creating a sufficiently strong bond that can be mechanically support a load. The failure test were done using an Instron model 4500 load frame at a crosshead rate of 10 mm/min. Six duplicate samples were produced and tested for each set of experimental parameters. The failure load is report as an apparent shear stress which was calculated as the applied force at failure divided by the overlap area, assuming a uniform stress distribution in the overlap with no stress concentrations. The weld pressure on the overlap was approximately 0.33 MPa, to provide a satisfactory contact between the two surfaces.

The degree of autohesion as a function of the ¼ root of time in Eq.(5) is shown in **Figure 3**.

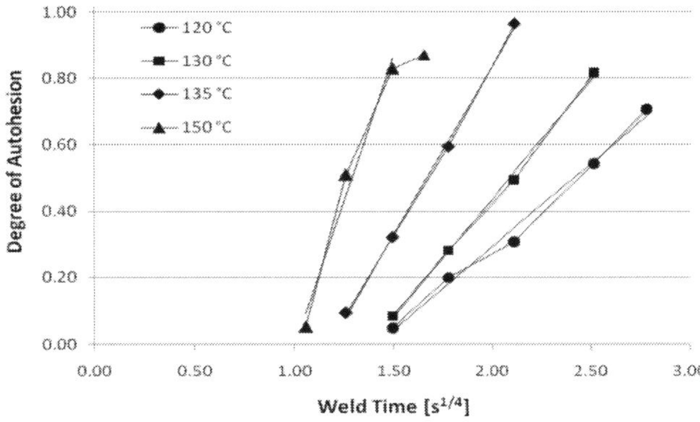

Figure 3 Degree of autohesion as a function of the fourth root of weld time for 0.1 mm thick PLA films

The trendlines which represent the slope of the linear parts of the plotted graphs for the various temperatures, which are in good agreement with published data on the linear development of weld strength with the fourth root of weld time [4,6,7]. Thus, it is assumed to be the temperature dependent *K(T)* of Eq. 4.

By plotting the natural logarithms of the slopes of the trend lines in **Figure 3** as a function of the inverse of the welding temperature, it is possible to calculate the activation energy E_a of Eq. 6, and the extension of this plot to the intersection with the ordinate will represent the natural logarithm of the material parameter K_0.

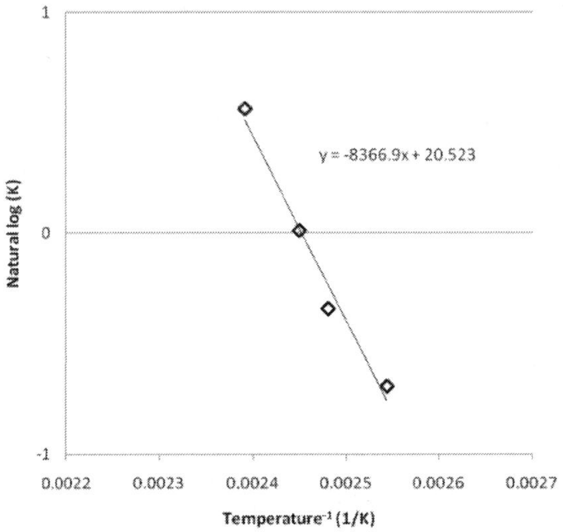

With a slope of ($-E_a/R$ = -8366 K) and an intersection at $\log_{10}(K(T))$ = 20.523 the activation energy E_a is calculated as 69,561 J mol^{-1}, and $K_0 = 8.18*10^8$ $s^{-1/4}$.

Thus, in summary, PLA can be fully welded with impulse and the autohesion process can be modelled based on fundamental processes.

---

1 Kim H. K., Wool R.P., Macromolecules, Vol. **16**, pp. 1115-1120, (1983)

2 Wool R.P., O'Connor K.M., J. Appl. Phys., Vol. **52**, #10 (1981)

3 Lee I.W., Springer G.S., J. Compos. Mater. Vol. 21 (1987)

4 Dara P.H., Loos A.C., Thermoplastic Matrix Composites Processing Model, Virginia Polytechnic Institute and State University (1985)

5 Boiko Y.M., Prud'homme R.E., Macromolecules, Vol. **31**, pp. 6620-6626, (1998)

6 Jud K., Kausch H.H., Williams J.G., Journal of Materials Science, Vol. 16, pp. 204-210 (1981)

7 ASTM D 638 - 08 Standard Test Method for Tensile Properties of Plastics, American Society for Testing and Materials

MORPHOLOGY DEVELOPMENT AND INTERFACIAL INTERACTIONS IN POLYCAPROLACTONE/THERMOPLASTIC STARCH BLENDS

B. D. Favis* and G. Li

CREPEC, Department of Chemical Engineering, Ecole Polytechnique de Montreal, P.O. Box 6079 Station Centre-Ville, Montreal, QC, Canada H3C 3A7

Abstract – The The coalescence, continuity development, dynamic and static mechanical properties, as well as interfacial interactions, were studied for polycaprolactone/ thermoplastic starch blends. These materials, prepared by a one-step extrusion process, demonstrate the features of a highly interacting system and the tensile mechanical properties demonstrate exceptional ductility at very high levels of thermoplastic starch without any added interfacial modifier.

Introduction

Starch materials have attracted considerable attention in the bioplastics field due to their abundance, biodegradability and low cost. The mixing of conventional polymers with native unplasticized starch blends always leads to brittle materials. In that case the starch component behaves as a solid filler. St Pierre *et al.*[1] carried out an investigation on thermoplastic starch/polyethylene blends and demonstrated that dispersed phase/matrix morphology control protocols could be applied to this blend. In a later work, B. D. Favis *al.* developed an effective one-step melt processing technique and controlled the level of continuity of the TPS phase [2]. This is achieved through the application of heat and in the presence of water and/or another plasticizer for starch. The advantage of TPS is that it can be processed on typical melt-processing equipment.

It is well known in polymer blends that the morphology control of the respective phases is a key factor to achieve desired material properties. The phases in a polymer blend can be structured in droplet, fiber, laminated, and co-continuous form. Phase coalescence is an important phenomenon affecting the final morphology of polymer blends and has been studied in some detail in recent years. Typically, coalescence can be divided into two categories: dynamic and static processes. The dynamic case is a flow-induced coalescence process during polymer blending with the final morphology being governed by a balance of particle breakup and coalescence. Static coalescence is a quiescent process whereby the phase or phases are coarsened over time at high temperature. Co-continuous systems, in particular can demonstrate very elevated levels of coalescence during static annealing [3].

PCL is a synthetic polymer having unique properties that make it attractive for biomaterials applications. It has excellent biodegradability and biocompatibility properties. However, PCL is still more expensive than conventional plastics, and the degradation rate is not completely satisfactory in some instances. The blending of PCL with TPS could provide a potential route towards more economic fully biodegradable materials. A number of studies on PCL/TPS blends have been carried out [4]. Shin et al. reported large diameters of the TPS dispersed phase in blends of PCL and thermoplastic starch plasticized with 20% glycerol. Matzinos et al. reported on the obtention of a completely plasticized thermoplastic starch and a morphology with a relatively fine and uniform dispersion of TPS phase within the PCL matrix after a three-times extrusion protocol. The mechanical and thermal properties of PCL/TPS with various moisture and glycerol contents were investigated by Averous et al.. Results indicated low compatibility between PCL and TPS according to DMA and DSC measurements. Sarazin et al. showed that the addition of small amounts of polycaprolactone to a blend of poly(lactic acid) and thermoplastic starch significantly improved the ductility and impact strength of the blends. Recently Shin et al. studied blends of polycaprolactone with thermoplastic starch and showed that the chemical modification of starch with maleic anhydride led to improvements in interfacial adhesion and processability. To date, very little work has been done on the detailed morphology development and morphology control in PCL/TPS blends.

The objective of this study is to carry out a detailed study on the morphology development in PCL/TPS blends with high plasticizer content prepared using a one-step process. This work will closely examine the coalescence and continuity development in these systems.

Experimental

Two commercial grades of polycaprolactone CAPA6500 and CAPA6800, with different molecular weights were used in this work. Native starch was obtained from ADM/Ogilvie and is composed of 25%

amylose and 75% amylopectin. The plasticizers used were water and pure glycerol (SIMCO Chemical products Inc. 99.5%).

Starch granules were gelatinized, plasticized with glycerol and water, and blended with PCL in a one-step extrusion process. For the composition study, the entire ranges of PCL/TPS36 blend composition from 100/0 to 0/100 in steps of 10 wt% were prepared. For the investigation of the effect of viscosity ratio on blend morphology, two kinds of PCL with different molecular weights and TPS with different glycerol contents were blended;

The coalescence, continuity development, dynamic and static mechanical properties, as well as interfacial interactions, were detail studied by rheological measurements, SEM, quiescent annealing test and Fourier transform infrared spectroscopy (FTIR).

Results and Discussion

The rheological properties of the neat PCL and TPS as a function of frequency are measured. The TPS with high glycerol concentration exhibits the rheological behavior of a typical gel and is characterized by a storage modulus (G') which is larger than the loss modulus (G") over the entire frequency sweep (Fig 1a). This behavior has been observed by other authors and is generally explained by the presence of an elastic network embedded in a softer matrix, i.e. the existence of a protein network; remaining crystalline structure in the samples; or strong hydrogen bonding [5]. In the present case, no crystalline structure in this highly plasticzed TPS was found as determined by X-ray diffraction and very little protein exists in the starch as well. The gel-like behavior is most likely the result of hydrogen bonding between the starch and plasticizer and could also be due to high levels of entanglement of starch molecules in the melt state creating a type of pseudo-crosslinking effect. Note that the elasticity is significantly higher than the complex viscosity over a wide range of shear rates. Also, the TPS is significantly more elastic than PCL (not shown here). The gel-like properties of TPS and its elastic nature will tend to make it more difficult to deform/disintegrate as a dispersed phase in a polymer blend.

The complex viscosity (η^*) of TPS follows a power-law behavior and there is no Newtonian plateau for the viscosity curve within the measured region of frequency range (Fig 1a), another characteristic of gel-like behavior. Thus, the zero shear viscosity of TPS cannot be obtained. The extreme shear thinning behaviors have been associated with a melt yield stress. The complex viscosities of PCL1 and PCL2 were measured at temperatures of 110 and 130 °C, respectively. PCL1 and PCL2 demonstrate shear thinning behavior in the high frequency range from 0.5 to 500 rad/s (Figure 1b).

In the present work, two kinds of PCL with different molecular weights and two kinds of TPS with various glycerol concentrations were chosen to study the influence of viscosity ratio on the blend morphology of strands after twin-screw extrusion. Figure 2 demonstrates the SEM image of PCL/TPS (70/30 weight %) blends at various viscosity ratios. The results show that the viscosity ratio, in the range studied, has virtually no influence on dispersed phase diameter. Although previous studies have clearly shown that the viscosity ratio has an important influence on phase size for immiscible blends, compatibilized blends often show significantly less dependence on viscosity ratio. Another potential explanation for the low dependence on viscosity ratio could be related to the presence of strong elongational flow fields present during twin-screw extrusion. It is well known that elongation flow is much more effective in droplet breakup than shear flow.

The SEM images of PCL1/TSP36 blends indicate that TPS exists as dispersed droplets in a PCL matrix up to a TPS concentration of 30 wt% (not shown here). Image analysis of these droplets indicate that the *dn* and *dv* are around 1.4 and 1.8 microns, respectively (Fig. 3). Further increasing the TPS concentration to 40 wt%, the co-existance of elongated structures (fiber-like) and droplets of TPS with a similar diameter are observed. Finally, when the TPS content increases up to 50 wt%, a solely fibrillar TPS is attained and the aspect ratio increases to about 25. It is of particular interest to note that the diameter of TPS droplets and fibers are independent of composition right up to 45 vol% TPS. Further increasing the TPS36 concentration above 54 vol%, results in dramatically increased TPS fiber diameters. The region of dual-phase continuity appears to exist in the concentration range of 54 to 65 vol% TPS. TPS appears to resist deformation/disintegration, but once deformed forms stable fibers.

Veenstra et al. investigated the morphology of blends of polystyrene and poly(ether-ester) thermoplastic elastomer [6]. When the processing temperature was chosen below the block copolymer order to disorder transition, the flow curves of poly(ether ester) did not show a Newtonian plateau in viscosity. Shear thinning behaviour over the entire range of measured shear rate was found. In that case, physical crosslinks of crystalline structures are present in the melt state and a melt yield stress was estimated in the range between 500 and 3800 Pa. These structures were found to limit and even stop the breakup and retraction behaviour of poly(ether ester) threads and formed stable elongated structures.

In the present work, the rheology of TPS demonstrates a gel behaviour and no Newtonian plateau in viscosity was obtained during melting processing (Fig. 1). It indicates the existence of strong interactions within the TPS that can be attributed to hydrogen bonding. The presence of physical crosslinks in the melt state leads to highly elastic properties and a melt yield stress, which also have a stabilizing effect on elongated structure. Thus, this highly elastic state in the TPS typically resists deformation, but once deformed forms stable fibers. It is possible that the elongated structure can also be attributed to twin-screw extrusion processing which provides a significant elongational flow component. It is well known that the deformation of a droplet to produce a fibril is more easily obtained by subjecting the fluid to an extensional flow.

As a comparison, PCL2/TPS36 blends were prepared and the morphologies are studied. As observed for the PCL1/TPS36 blends, the dispersed TPS droplet/fiber diameters are independent of concentration.

The continuity of TPS/PCL2 via gravimetric solvent extraction is also shown in Figure 4 in order to examine the influence of a more viscous PCL on continuity development. In that case it can be seen that the percolation thresholds of both dispersed PCL2 and dispersed TPS36 are shifted to higher concentration. This is an unexpected shift since in typical thermoplastic blends, a lower viscosity ratio blend typically results in a more readily deformed dispersed phase and thus typically displays a lower percolation threshold. The opposite is observed here. This result is likely a reflection of the fact that both the elastic nature and internal hydrogen bonding in TPS give it the character of a pseudo-partially-crosslinked material. In this case the dispersed TPS is quite difficult to breakup, but once deformed forms stable fibers with less of a tendency to breakup than typical thermoplastics. Crosslinked materials classically demonstrate asymmetric continuity diagrams. In the case of TPS36/PCL2, since the more viscous PCL2 has virtually no additional effect on the deformation of dispersed TPS, the only effect is that the more viscous PCL2 actually retards the coalescence required for TPS droplets to transit to stable TPS fibers. Clearly these TPS/PCL blends do not respect the classical empirical description of the relationship of viscosity ratio and co-continuity composition.

The morphologies of injection moulded samples were obtained (not shown here). The results show the small effect of concentration based coalescence. The change in processing condition did lead to significantly different particle sizes. This indicates that changes in processing techniques, with their different balances of shear and elongation flow fields, can have an effect on TPS coalescence.

As an immiscible blend is annealed in the melt state, the phase size grows as a function of annealing time and temperature indicating that significant coarsening effects are taking place. This coarsening is strongly affected by the state of the interface with compatibilized blends showing significantly less coalescence and more stable morphologies. A 25 vol% TPS36 in PCL1 blend with droplet/matrix morphology was used to study the TPS coalescence at annealing temperatures of 110 and 150 °C. The results clearly show that the TPS particle size in the blend does not change with annealing time and temperature indicating no coalescence whatsoever. In a second quiescent annealing experiment, a 43 vol% TPS in PCL1 blend (50/50 by weight), was also studied. However even at this concentration no further coarsening was observed even after long times of annealing. As a comparison, a completely immiscible HDPE/TPS36 blend (50/50 by weight) was also subjected to quiescent annealing to study the coarsening of TPS. In this case strong coalescence effects are observed for the TPS phase with the average TPS36 phase diameter increasing from 6.0 to 14.9 μm when annealing time was increased from 0 min to 60 min. These results strongly suggest the presence of compatibilization effects between TPS and PCL.

DMTA results for TPS, PCL and their blends are presented in Figure 5. For 100% TPS, the storage modulus falls in two steps, the first between -60 and -20 °C, and the second between -10 and 45 °C. This corresponds to the tan δ peaks for 100% TPS at -38.3 and 28.0 °C, respectively. The high temperature relaxation (T_α) is attributed to the glass transition of the starch-rich phase, whereas the low temperature (T_β) arises from the glycerol-rich phase owing to the phase separation of starch and glycerol at high glycerol concentration as reported previously. The T_g of PCL from the tan δ peak is -38.5 °C which overlaps with the T_β of TPS.

The storage moduli of PCL/TPS blends demonstrate two main tendancies which can also be used to determine the boundary of the co-continuity interval. At TPS concentrations less than and equal to 50 wt%, the moduli drop down at the melting point of PCL indicating PCL is the continuous phase. On the other hand, the 80 and 90 wt% TPS blends with PCL demonstrate a moduli which decreases dramatically following the starch-rich phase transition at 0 °C indicating TPS is the continuous phase. This effectively corresponds to the boundaries of phase inversion as determined by solvent gravimetry in the previous section.

It can be seen from Figure 5 that the T_β of PCL1/TPS36 blends decreases with increasing TPS36 concentration. This effect can be attributed to the further phase separation between starch and glycerol upon blending with PCL. Recent work from this laboratory, in fact, has shown that a thin glycerol-rich layer forms at the interface between TPS and polymer in TPS/polymer blends [7]. It is also interesting to note that there is a significant shift in the T_β peak to higher temperatures when the concentration is increased from 70-80 wt% TPS. This corresponds to the phase inversion of PCL and TPS as reported earlier in the paper and is related to the onset of formation of a TPS matrix phase and PCL dispersed phase.

The T_α peak for 100% TPS36 is 28.0 °C. However, PCL/TPS blends with 50 and 70 wt% TPS have a TPS T_α peak which only appears as a diffuse shoulder at -10 and -8 °C. This is a peak shift in excess of 30 C°. At TPS contents of 30 wt% and below, the TPS T_α peak disappears entirely. These results strongly imply interactions between PCL and the starch-rich phase of TPS. The interaction is likely to be cause by the hydrogen bonding interactions between the carbonyl groups of PCL and hydroxyl groups on starch.

The FTIR spectra of PCL1, TPS36 and their blend at 50/50 wt%. PCL1 had a strong carbonyl stretching absorption at a wave number of 1720 cm^{-1}, which shifts to 1724 cm^{-1} after blending with TPS36. This shift in the stretching vibration frequency of carbonyl groups suggests a hydrogen bonding interaction between the carbonyl groups of PCL and the hydroxyl groups of TPS. Similar results have been obtained with poly(propylene carbonate)/thermoplastic starch and poly(hydroxyl ether of bisphenol A)/polycaprolcatone blends.

The tensile properties of PCL1, TPS36 and their blends were studied as well. The stress at yield and modulus decrease continuously with increasing TPS concentrations. The remarkable aspect of the results in this mechanical property data set are the elongation at break values. The elongation at break of PCL/TPS36 blends over a very wide concentration range exhibit outstanding ductility. Samples with a TPS36 content of 50 wt% and less did not even break during the tensile test (max value set at 1000%). Even PCL/TPS blends with 80 and 90 wt% of TPS36 showed elongations at break of 450 and 550%, respectively. The key to these properties appears to be the high level of plasticization of the starch. It is likely that the high plasticizer concentrations used here increase the mobility of the starch chains and thus promote a high level of specific interactions between the PCL and starch. This excellent interfacial interactions result in the exceptional ductilities observed above.

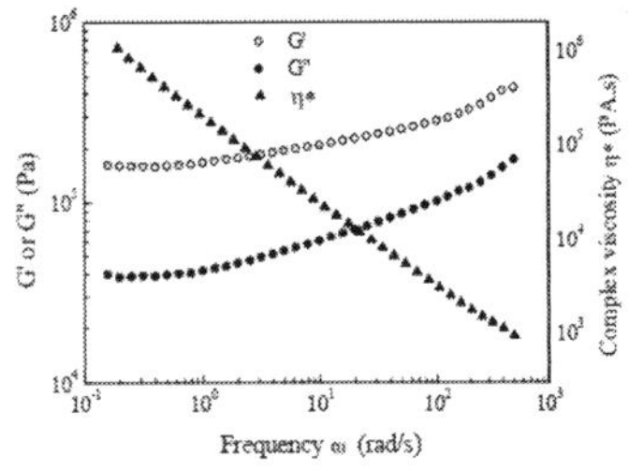

a

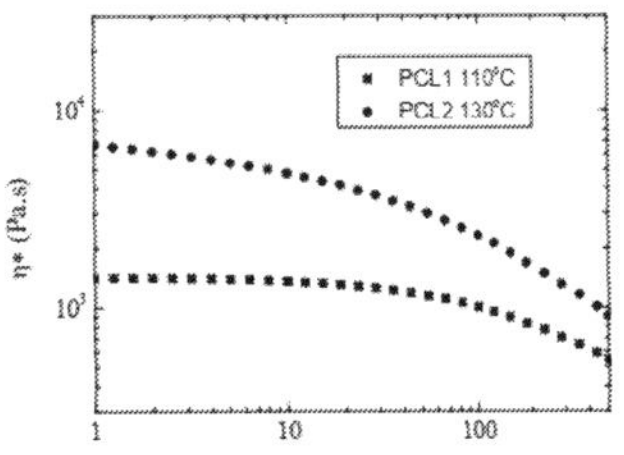

b

Figure 1. Rheological properties of a) TPS36 at 110°C and complex viscosity at 110°C and 130 °C for PCL.

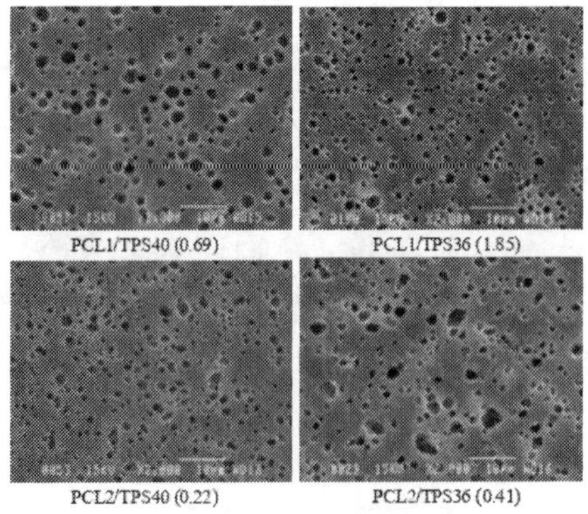

Figure 2. Effect of viscosity ratio on the morphology of PCL/TPS36 blends (70/30 weight %). The number in brackets indicates the viscosity ratio of dispersed phase to matrix.

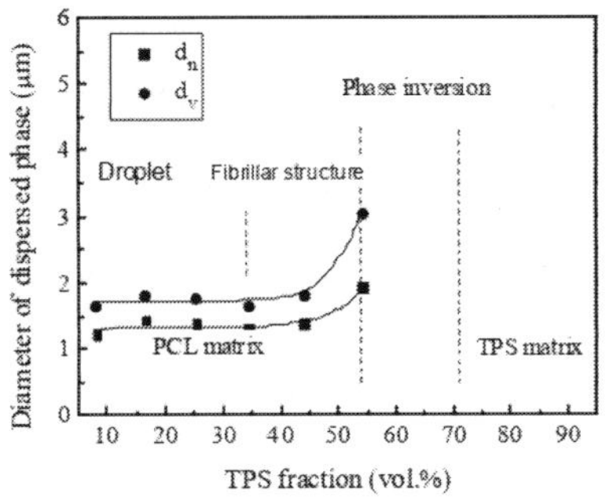

Figure 3. Diameter of phases as a function of TPS concentration obtained from the image analysis of PCL1/TPS36 extruded strands

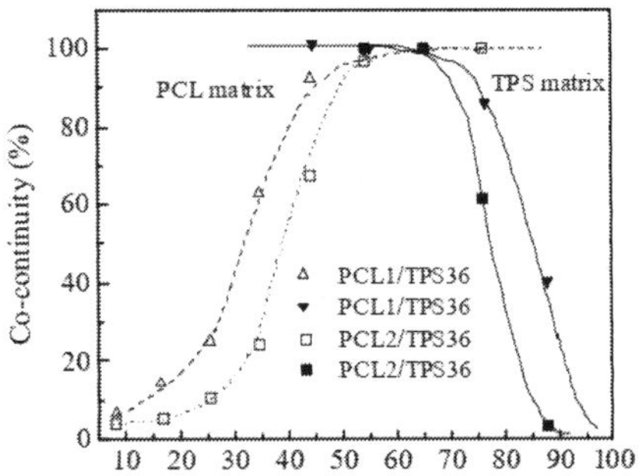

Figure 4. Phase continuity development of PCL/TPS36 blends using the gravimetric solvent extraction technique.

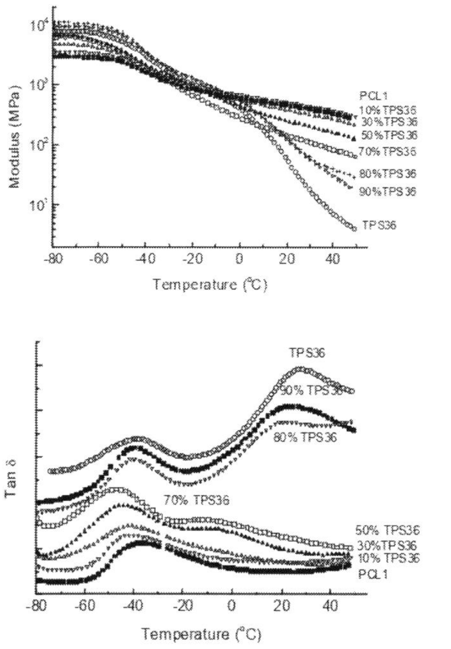

Figure 5. DMTA results for PCL, TPS and their blends. Concentrations are in weight %.

Conclusion

In this work a detailed study of the morphology development in thermoplastic starch/polycaprolactone blends, prepared using a one-step extrusion process has been undertaken. These highly plasticized blends, without any added interfacial modifier, display remarkably low levels of coalescence and exceptional mechanical properties. The phase size of the dispersed thermoplastic starch, after twin-screw extrusion, is virtually independent of TPS composition and TPS droplets transit to stable fiber-like phases of identical phase size above 30% TPS. Although the TPS is quite insensitive to concentration-based coalescence, it displays some coalescence sensitivity to changes in processing technique (compounding vs. injection). A study of percent continuity vs TPS concentration, as determined by solvent gravimetry, indicate the region of co-continuity to be 55-65 vol% TPS. This is an asymmetric phase inversion region and indicates a strong tendency of polycaprolactone to encapsulate the thermoplastic starch, a behaviour similar to polymer blends containing partially crosslinked rubbers. The preferential encapsulation of PCL about TPS is especially evident in the continuity development region at high TPS concentration where 10 wt% of PCL1 (the lower viscosity PCL) displays 40% continuity and 20 wt% PCL1 displays 80% continuity, a remarkable result. It is interesting to note that the midpoint value of phase inversion is identical irrespective of the viscosity of PCL used. The coalescence, as observed by static annealing experiments, show that the TPS phase at both 30 and 50 wt% TPS in PCL displays no coalescence after 60 minutes of annealing. A comparative study of static annealing of TPS/polyethylene on the other hand shows significant coalescence. These results strongly point to the likely presence of specific interactions between the TPS and the PCL. Dynamic mechanical analysis confirms the region of dual-phase continuity and also strongly indicates a specific interaction between PCL and TPS. FTIR results show the presence of a hydrogen bonding interaction between the carbonyl groups of PCL and the hydroxyl groups on starch. It is likely that the high plasticizer concentrations used here increase the mobility of the starch chains and thus promote a high level of specific interactions between the PCL and starch. Mechanical properties display extremely high elongations at break, even at high TPS concentrations, typical of those observed for highly compatibilized immiscible polymer blends.

References

1.St-Pierre, N. et al., "Processing and Characterization of Thermoplastic Starch/Polyethylene Blends", Polymer, **38**, 647-655(1997).
2.US Patent 6,605,657 (2003), Favis, B. D. et al..
3.Yuan, Z. H. and Favis, B. D., "Coarsening of Immiscible Co-Continuous Blends During Quiescent Annealing", AIChE J., **51**, 271-280 (2005)
4.Averous, L. et al., "Properties of Thermoplastic Blends: Starch-Polycaprolactone", Polymer, **41**, 4157-4167 (2000).
5.Wilhelm, H. M. et al., "Dynamic Rheological Properties of Yam Starch/Hectorite Composite Gels", Polymer International, **54**, 814-822 (2005).
6.Veenstra, H. et al, "Stability of Co-continuous Polystyrene/ Poly(ether–ester) Blends in Shear Flow", Polymer, **40**, 5223-5226 (1999).
7.Taguet, A. et al., "Interface/Morphology Relationships in Polymer Blends with Thermoplastic Starch", Polymer, **50**, 5733-5743 (2009).

SURFACE MODIFICATION OF UN- AND HIGHLY FILLED PLA FILMS USING CORONA TREATMENT

S. Laske[1]*, E. Ruef[2], C. Holzer[1]

[1] *Institute of Plastics Processing, University of Leoben, Franz Josef Strasse 18, 8700 Leoben, Austria – stephan.laske@unileoben.ac.at; clemens.holzer@unileoben.ac.at*
[2] *Polymer Competence Center Leoben GmbH, Roseggerstrasse 12, 8700 Leoben, Austria – elisabeth.ruef@stud.unileoben.ac.at*

Polylactid acid (PLA) is a biodegradable, multifunctional, aliphatic, thermoplastic polyester based on renewable resources. At this stage, PLA mainly replaces polyolefins (PP, PE) or polystyrene (PS) in packaging applications. The wettability for printing or bonding is an important factor, but the wettability of PLA is – typically for most polymers – insufficient. One possibility to improve it is to increase the surface energy using corona discharge treatment. This conditioning is based on the principle, that emitted electrons from the cathode of the discharging station have enough energy to create radicals on the surface. As a consequence, functional groups are formed at the polymer surface, for example hydroxides, ketones or carbon acids and esters, which are partly very polar and therefore enhance the surface energy.
This study deals with the wettability improvement of un- and highly filled PLA films by corona discharge with the main focus on improving the printability. The aging behaviour (loss of printability) and the ideal dismissal of treated PLA are studied as well. Single layer films unfilled respectively filled with 40wt% calcium carbonate were extruded and treated in-line with four different dosage levels. All samples were conditioned at room temperature in aluminum foil to protect the surface from dust, oil and especially fingerprints, so that the measurements were not falsified. The surface energy was periodically measured after 0, 24, 168, 336 and 672 hours with a contact angle goniometer.
The results show, that corona discharge increases the surface energy of PLA significantly and above the important threshold for printability. The aging shows an exponential decrease of surface energy over time and reaching a stable value already after 3 weeks. Furthermore, the influence of calcium carbonate particles on the surface energy and the ideal discharge dosage is in evidence.

Introduction

The problem with finishing of polymeric films by means of printing, laminating or bonding lies in the low surface energy of polymers. For being wettable, the surface energy of the polymer must be higher then the surface tension of the ink or adhesives.
The surface energy of polymer films can be increased by surface modifications like the corona treatment. This method is used in a broad range of polymer processing techniques. The problem with corona treatment is puncture, occurring when air is between the film and the roll of the corona device. This harms the film in a way that preprocessing can be hindered.
In this work, the influence of corona treatment of un- and highly filled PLA (polylactic acid) films is examined. The surface tension of films treated with different dosages is measured over storage time using a contact angle goniometer.

Experimental

Surface tension vs. surface energy

The wettability of a solid material, in this case a polymer film, is determined by the surface energy. The surface energy is the energy, which is required for breaking up the chemical bondings, when a new surface is formed per unit area with the SI-unit J/m^2. The surface energy can not be measured directly, as it is not possible to create a new surface, without expending energy for other processes (e.g. deformation of the solid) [1, 2].
One way for making the surface energy respectively the wettability of solid materials measurable, is the correlation with the surface energy of liquids which is equivalent to their surface tension. This can be done by using specific liquids with known surface tensions by measuring the contact angle between the liquid and the solid material [1, 2].

Surface tension measurements

In this work, the surface tension is measured using a contact angel measurement device (Figure 1) and specific nonpolar liquids (water, diiodemethane, and ethylen-glycole) [6].

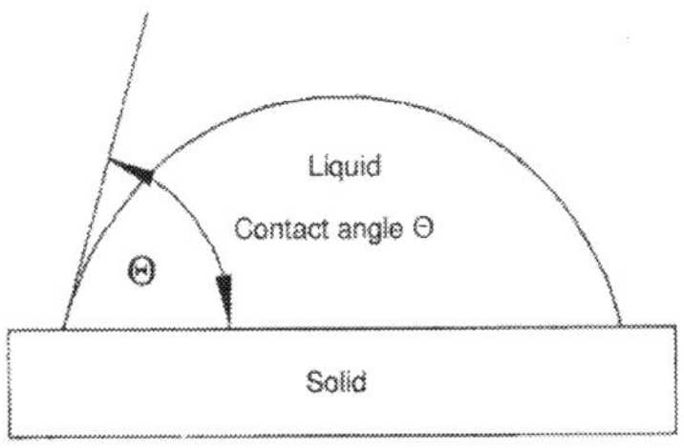

Figure 1 - Contact angle measurement [6]

The evaluation of the results is done using the OWRK (Owens, Wendt, Rabel and Kaelble) method. This technique delivers the disperse as well as the polar part of the surface tension. Assuming that only uniform interactions can be active between the phases (solid – liquid), only one part of the surface tension (in this case the disperse part) can interact between the nonpolar liquid and the polar polymer films [3, 4].

$$\frac{(1+\cos[(\theta)\sigma_l])}{2\sqrt{\sigma_l^d}} = \sqrt{\sigma_s^p}\sqrt{\frac{\sigma_l^p}{\sigma_l^d}} + \sqrt{\sigma_s^d} \quad \text{(Eq. 1)}$$

σ_s^d = solid dispers surface tension
σ_s^p = solid polar surface tension
σ_l^d = liquid dispers surface tension
σ_l^p = liquid polar surface tension

Using equation 1, the known values of σ_l, σ_l^d and σ_l^p and the measured contact angle a straight line can be drawn in a diagram which delivers the disperse and polar part of the surface energy of the solid.
Using

$$x = \sqrt{\frac{\sigma_l^p}{\sigma_l^d}} \text{ and } y = \frac{(1+\cos[(\theta)\sigma_l])}{2\sqrt{\sigma_l^d}}$$

delivers

$$k = \sqrt{\sigma_s^p} = \text{polar surface energy}$$

$$d = \sqrt{\sigma_s^d} = \text{disperse surface energy}$$

The processed films were measured right after production and at 5 different dates (Table 1), to evaluate ageing by measuring the decrease of surface tension over time.

Table 1 - Contact angle measurement time plan

Nr.	1	2	3	4	5	6
Time [h]	0	24	168	336	504	672
Time [d]	0	1	7	14	21	28

Production of PLA-films
Single layer PLA-films were processed using a single screw extruder E20 T-H from Dr. Collin Company, Germany, in combination with a flat film die. The films had a thickness of 120 µm and a width of 180 mm. The highly filled films contained 40wt% of calcium carbonate. The haul-off speed was 1.1 m/min. Corona treatment is some kind of plasma treatment, where the film is exposed an electric discharge. Thereby, high-voltage is applied between a grounded steel-roll and an isolated electrode (Figure 2) [6].

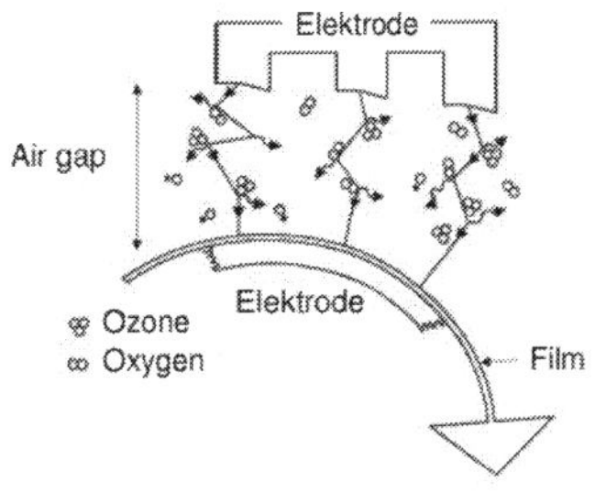

Figure 2 - Corona discharge technique [6]

The continuous discharge leads to various chemical and physical reactions on the film, creating radicals on the surface and, as a consequence, functional groups like hydroxides, ketones or carbon acids, which are partly very polar and therefore increase the surface energy.
The applied energy can be calculated using equation 2 where P_{sum} is the total power (W), P_{spez} is the specific power (W min / m^2), v is the line speed (m / min) and l is the deckle width (m).

$$P_{spez} = \frac{P_{sum}}{v * l} \quad \text{(Eq. 2)}$$

The corona treatment was done in-line using a discharging station ASOEH-11V100E from Ahlbrandt Systems, Germany. Four different dosages were applied (Table 2).

Table 2 - Corona treatment parameters

	Reference	Corona 1	Corona 2	Corona 3	Corona 4
Adjustment [W]	0	0.6	0.8	1.15	1.5
Dosage [W/min]	0	3.33	4.44	6.39	8.33

Results and Discussion
The threshold for printability of PLA-films is around 41 mN/m and marked as a bar in figures 3 and 4. Corona 4 with the highest dosage of 8.33 W/min delivers very diffuse and non-reproducible results caused by a deterioration of the material. Therefore it is excluded from the charts.

Unfilled PLA-films
Table 3 shows the measured values and Figure 3 the approximation of the results of the contact angle measurements for the surface tension of the unfilled

PLA films treated with 3 different dosages and the reference.

Table 3 - Results of the unfilled films

Time [h]	Surface tension (mN/m)	Corona 0 (mN/m)	Corona 1 (mN/m)	Corona 2 (mN/m)	Corona 3 (mN/m)
0	disperse	30.90	31.18	31.14	31.53
	polar	7.42	12.51	12.34	10.63
	total	**38.32**	**43.69**	**43.48**	**42.16**
24	disperse	30.90	31.18	31.14	31.53
	polar	8.15	10.67	11.08	8.32
	total	**39.05**	**41.85**	**42.22**	**39.85**
168	disperse	30.90	31.18	31.14	31.53
	polar	7.21	10.35	11.03	9.16
	total	**38.11**	**41.53**	**42.17**	**40.69**
336	disperse	30.90	31.18	31.14	31.53
	polar	8.62	9.94	9.66	8.38
	total	**39.52**	**41.12**	**40.80**	**39.91**
504	disperse	30.90	31.18	31.14	31.53
	polar	8.86	10.44	9.68	9.12
	total	**39.76**	**41.62**	**40.82**	**40.65**
672	disperse	30.90	31.18	31.14	31.53
	polar	7.89	9.76	9.90	8.19
	total	**38.79**	**40.94**	**41.04**	**39.72**

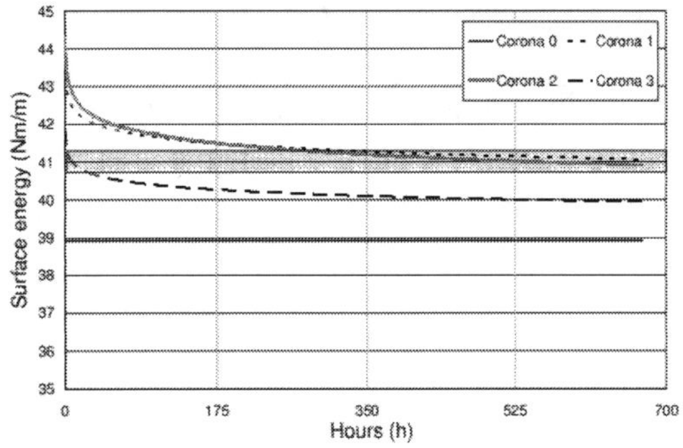

Figure 3 - Results of the unfilled films

It can be seen, that the treatment of all three Corona dosages increases the surface tension above the threshold of around 41 mN/m, directly after processing. The results of Corona 3 are to low, to keep that level for more than some hours, which can be enough for in-line but is insufficient for off-line printing methods. This can be explained by the initiating material deterioration. Corona 1 and 2 deliver good results, keeping a sufficient high surface tension for at least 3 weeks. As Corona 1 and 2 accomplish almost the same results, Corona 1 should be preferred, because the material load is lower.

Highly filled PLA-films

Table 4 shows the measured values and Figure 4 the approximation of the results of the contact angle measurements and the resulting values for the surface tension of the highly filled PLA films treated with 3 different dosages and the reference.

Table 4 - Results of the highly filled films

Time [h]	Surface tension (mN/m)	Corona 0 (mN/m)	Corona 1 (mN/m)	Corona 2 (mN/m)	Corona 3 (mN/m)
0	disperse	30.45	30.03	30.30	30.24
	polar	7.66	13.89	12.38	12.00
	total	**38.11**	**43.92**	**42.68**	**42.24**
24	disperse	30.45	30.03	30.30	30.24
	polar	6.83	11.83	11.14	9.62
	total	**37.28**	**41.86**	**41.44**	**40.16**
168	disperse	30.45	30.03	30.30	30.24
	polar	7.52	9.85	10.15	9.72
	total	**37.97**	**39.88**	**40.45**	**39.96**
336	disperse	30.45	30.03	30.30	30.24
	polar	6.57	10.31	9.21	9.24
	total	**37.02**	**40.34**	**39.51**	**39.48**
504	disperse	30.45	30.03	30.30	30.24
	polar	7.51	9.24	11.47	9.35
	total	**37.96**	**39.27**	**41.77**	**39.59**
672	disperse	30.45	30.03	30.30	30.24
	polar	7.83	8.48	10.84	9.53
	total	**38.28**	**38.51**	**41.14**	**39.77**

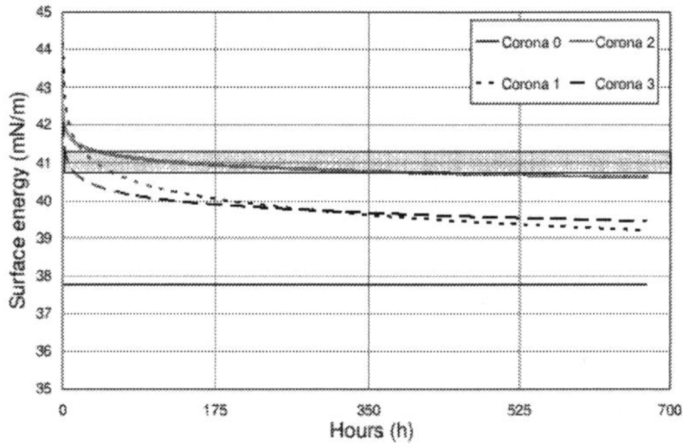

Figure 4 - Results of the highly filled films

The values of the highly filled PLA-films are generally lower then that of unfilled PLA. This is because of the surface modificated calcium carbonate. The used modification reduces the surface energy of the calcium carbonate particles for a better miscibility with the polymer. According to this, the overall surface energy respectively surface tension of the filled compound is as well reduced compared to the unfilled PLA.

An unexpected fact is, that with Corona 1 after a few hours the surface tension is lower than with Corona 2, which can be explained with the lower sensitivity of the compound versus the discharge treatment because of the filler. In that case Corona 1 lasts, same as Corona 3, only for some hours over the threshold, while Corona 2 last for at least 5 days. With Corona 3, the reason for the low results is again an initiating material deterioration. This results show, that for high-filled PLA-films Corona technique is not necessarily the preferential treatment, but may be beneficiary in some cases.

Conclusions

Corona treatment is a very easy and simple method for increasing the surface energy respectively tension of un and highly filled PLA-films. Attention should be paid especially on the correct modification parameters (e.g. line speed, dosage, ...) and the material formulation, because wrong treatment settings could on the one hand result in almost no improvement and on the other hand destroy the material. As there is a smooth transition between good and bad results, correct settings are even more important. Processing films filled with calcium carbonate the process parameters must be changed to higher values, because the untreated values of surface tension are already lower and the compound sensitivity versus the discharge is much lower compared with unfilled PLA. Generally spoken, treating PLA-films with Corona technique leads to a better wettability and to a sufficient printability, which is durable even over weeks, so off-line as well as in-line printing is possible. Only with highly filled films, a more proper technique should be evaluated.

Acknowledgements

This research work was performed at the Polymer Competence Center Leoben GmbH (PCCL, Austria) within the framework of the Kplus-program of the Austrian Ministry of Traffic, Innovation and Technology with contributions by the University of Leoben. The PCCL is funded by the Austrian Government and the State Governments of Styria and Upper Austria.

References

[1] Godderidge, A., „Kontaktwinkelmessung zur Bestimmung der Oberflächenspannung von Festkörpern“, Bachelor thesis, University of Leoben (1996)

[2] Hlawacek, G., „Kontaktwinkelmessung“, Master thesis, University of Leoben (2005)

[3] Alexander, M., „Untersuchungen über die Benetzbarkeit von Kunststoff-Folien durch Druckfarbe“, Westdeutscher Verlag, Munich (1971)

[4] Kunststoffinstitut Lüdenscheid „Lackierung von Kunststoffbauteilen“, hand book (2004)

[5] Czeslik, C., Seemann, H. and Winter, R., „Basiswissen Physikalische Chemie“, B.G. Teubner Verlag, Wiesbaden (2007)

[6] Wratschko G., „Herstellung und Charakterisierung von hochgefüllten Folien aus thermoplastischem Polyester“, Master thesis, University of Leoben (2005)

BIO-BASED COMPOSITES FROM LIGNIN, TANNIN AND SISAL FIBERS

E. C. Ramires, and E. Frollini*

Instituto de Química de São Carlos, Universidade de São Paulo, C P 780, 13 560 970, São Carlos, São Paulo, Brazil, ecramires@iqsc.usp.br; elisabete@iqsc.usp.br

In this study, sisal fibers were used in the proportion range of 30-70% (w/w) as a reinforcement of composites prepared using phenolic-type thermoset resins based on lignin and tannin. The materials were widely characterized e.g. by ^{1}H and ^{13}C (pre-polymers), SEM, thermal analyses, water absorption test. The Izod impact strength increased up to 40% (w/w) of sisal fibers in lignin-based matrix composites and 50% in tannin-based matrix composites. The SEM images showed good fiber/matrix interface in these composites, which also had high storage modulus. The composites reinforced with large amount of sisal fibers also showed good fiber/matrix interface however, fiber agglomeration caused a decrease in the Izod impact strength values when compared to the composites reinforced with 40 and 50% of sisal fibers. Overall, the results demonstrated that composites with good properties may be prepared using high proportions of materials obtained from renewable sources.

Introduction

The use of lignocellulosic materials as replacement for synthetic fibers in the production of composites has gained significant importance as they are obtained from renewable sources, they are biodegradable, they present low density, and they are inexpensive (Frollini et al., 2004; Ramires et al., 2010). Among the natural fibers, sisal can be highlighted due to its high mechanical strength and other physical characteristics related to its high cellulose content (Megiatto et al., 2009, 2010; Mu et al., 2009). The sisal plant (Agave sisalana) is native from Mexico, but nowadays Brazil is the biggest worldwide producer of this fiber (FAO, 2009).

The hydrophilic and polar nature of lignocellulosic fibers and the hydrophobic characteristic of many thermoplastic matrices can impair the properties of the composites, due to the lack of adhesion and non-uniform dispersion of the fibers in the matrix (John and Thomas, 2008; Nicolai et al., 2008). When a thermoset matrix like phenolic resin is used, the lack-of-adhesion can be considerably minimized by favorable interaction between polar hydroxyl groups of the phenolic matrix and the hydrophilic fibers, improving the properties of the obtained composite (Ramires et al., 2010; Trindade et al., 2004).

The phenolic thermosets resins are widely known and used due to their excellent properties, such as dimensional and thermal stability, and chemical resistance (Ku et al., 2008). These resins are basically produced by the reaction of phenol and formaldehyde, which are obtained in a large-scale from non-renewable sources. Therefore, the substitution of these reagents by equivalents obtained from non-fossil sources is an interesting alternative from both economical and environmental perspectives (Frollini et al., 2008; Paiva and Frollini, 2006; Ramires et al., 2010).

Tannins are natural products present in several types of plants, and have a large amount of phenolic rings in their structure (Mosiewicki et al., 2007). Condensed tannins (Figure 1A) are the most suitable for the production of phenolic resins, due to the presence of phenolic rings di-substituted with hydroxyl groups in their structure. These hydroxyl groups make these rings activated to react with formaldehyde, owing to the electron donor effect of the hydroxyl groups (Vázquez et al., 2006; Tondi et al., 2009).

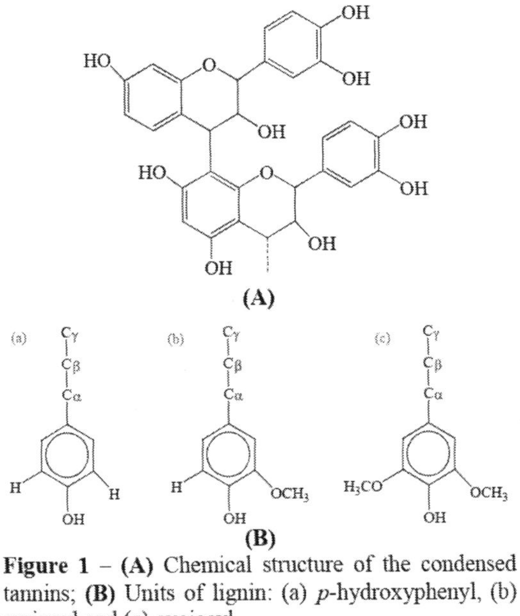

Figure 1 – **(A)** Chemical structure of the condensed tannins; **(B)** Units of lignin: (a) *p*-hydroxyphenyl, (b) syringyl and (c) guaiacyl.

Lignin is a biomacromolecule that is highly branched and it has a wide variety of functional groups. The main units in lignin are guaiacyl (G), syringyl (S) and *p*-hydroxyphenyl (H) units, which differ by the presence or absence of ortho-methyl groups in the aromatic rings (Figure 1B). The phenolic aromatic rings present in lignin enable this biomacromolecule to substitute the phenol in preparation of phenolic resins (Paiva and Frollini, 2006; Ramires and Frollini, 2008).
In the present study, the phenolic-type thermoset resins were prepared using lignin and condensed tannin as substitutes of phenol. These resins were used to prepare composites reinforced with sisal fibers in the proportion range of 30-70% (w/w).

Experimental

Lignin-formaldehyde pre-polymer synthesis
The pre-polymer, in which the phenol was replaced by lignin (kindly supplied by Dedini SA, Piracicaba, SP, Brazil), was synthesized using formaldehyde, lignin and potassium hydroxide (1.38:1.0:0.075, w/w). The lignin was dispersed in the aqueous formaldehyde solution under mechanical stirring for 15 min at room temperature. Potassium hydroxide was then added and mixed for 45 min. After that, the mixture was heated at 97°C, under mechanical stirring, for 2 h. The obtained solution was cooled to room temperature, neutralized with concentrated hydrochloric acid and the water was eliminated under reduced pressure.

Tannin-phenolic pre-polymer synthesis
The pre-polymer, containing 40% w/w of tannin (kindly supplied by Tanac, Montenegro, RS, Brazil) replacing the phenol, was synthesized using formaldehyde, phenol, tannin and potassium hydroxide (1.38:0.6:0.4:0.06, w/w). First, tannin, formaldehyde and potassium hydroxide were mixed under mechanical stirring, at 40°C for 30 min. After that, phenol was added and mixed under mechanical stirring, at 40°C for 1h. Then, the solution was cooled to room temperature, neutralized with concentrated hydrochloric acid and the water was eliminated under reduced pressure.

Composites preparation
The composites were prepared by mixing the lignin-formaldehyde or tannin-phenolic resins with resorcinol, the curing accelerator (10:1 w/w). Then the sisal fibers (kindly supplied by Lwarcel, Lençóis Paulista, SP, Brazil) were added to the resin with resorcinol and submitted to mechanical stirring for 20 min (JVJ mixer, Pardinho, SP, Brazil), in order to obtain an optimum impregnation of the fibers by the resin. The sisal fibers (3 cm of length and randomly distributed) were previously dried for 4 h in an air-circulated stove at 105°C to remove moisture. Compression molding was carried out in a mold (300 x 140 x 5 mm) under pressure. The molding cure cycles were based on previous studies (Frollini et al., 2008; Paiva and Frollini, 2006; Ramires and Frollini, 2008; Razera and Frollini, 2004) and were different for each resin:
Lignin-formaldehyde cycle: (50°C/1.5h/0ton; 65°C/1h/2ton; 80°C/1h/16ton; 95°C/0.5h/16ton; 105°C/0.5h/16ton; 115°C/1h/16ton; 125°C/2h/16ton).
Tannin-phenolic cycle: (40°C/0.5h/0ton; 50°C/1h/10ton; 85°C/1.5h/16ton; 125°C/2h/16ton; 150°C/2h/16ton).

Pre-polymer characterization
The ^{1}H NMR (Nuclear Magnetic Resonance) spectra of lignin-formaldehyde resin were recorded on a Bruker AC-250 spectrometer at 300 K, in 5 mm tubes, using DMSO-d6 as solvent, with 1024 scans. The ^{1}H spectra of tannin-phenolic resin were recorded on a Bruker AC-200 spectrometer at 300 K, in 5 mm tubes, using DMSO-d6 as solvent, with 512 and 20000 scans, respectively. The chemical shifts were referenced to the residual signal of DMSO-d6, δ= 2.5 ppm.

Composites characterization
The Izod impact strength was assessed using Izod impact CEAST Resil 25 equipment. Twenty unnotched samples were cut from each plate and shaped according to ASTM D256 (63.5 mm length × 12.7 mm width × 4.5 mm thickness). The impact tests were carried out at room temperature with an impact speed of 4 m s^{-1} and incident energy of 5.5 J.
The Scanning Electron Microscopy (SEM) images were taken with a Zeiss-Leica apparatus, Model 440, with an electron acceleration of 20 kV. Fractured samples were covered with a thin layer of gold in a sputter-coating system prior to analysis.
The Dynamic Mechanical Thermoanalysis (DMTA) was carried out using a DMA thermal analyzer, model 2980 from TA Instruments, operating with the 3-point bending clamp (flexural mode), oscillation amplitude of 20 μm, frequency of 1 Hz, heating rate of 2°C min^{-1}, and temperature range of 30-230°C. The composites samples dimensions were 64 mm of length × 12 mm of width × 3.2 mm of thickness.
The Thermogravimetric (TG) analyses were carried out using Shimadzu equipment, Model TGA-50. The samples (8–10 mg) were placed in platinum pans and heated from 25°C to 800°C at the rate of 10°C min^{-1} under a synthetic air atmosphere (flow of 20 mL min^{-1}).
The Differential Scanning Carlorimetry (DSC) analyses were performed using a Shimadzu instrument, Model DSC-50. The samples (6–8 mg) were placed in aluminum seal pans and heated from 50°C to 400°C at the rate of 10°C min^{-1} under a synthetic air atmosphere (flow of 20 mL min^{-1}).
The water absorption analyses were performed according to the ASTM D570. The samples (76.2 mm × 25.4 mm × 3.2 mm) were immersed in distilled water at room temperature. After certain intervals, they were removed from water, put in a piece of dry cloth (only to remove the excess of water), and weighed on a high-precision balance. The percentage of mass increase,

due to the water content, was calculated by the weight difference.

Results and Discussion

The NMR analyses were carried out to confirm the occurrence of the reactions generating the phenolic-type thermoset resins based on lignin and tannin.

In the ^1H NMR spectrum of the lignin-formaldehyde resin (Figure not shown), the peak characteristic of the protons of the methoxyl groups (OCH_3) was observed at 3.59 ppm and the peak characteristic of aromatic protons was observed at 6.58 ppm. The peak at 8.5 ppm was related to residual formaldehyde. It is important to notice that formaldehyde reacts in the next step, being incorporated in the cross-linked structure generated in the cure stage. The peaks at 4.6 and 4.7 ppm are due to the introduction of CH_2 groups in lignin, as hydroxymethyl groups and methylene bridges, respectively, confirming the occurrence of the reaction generating the resin.

In the ^1H spectrum of the tannin-phenolic resin (Figure not shown) the intense peak at 4.7 ppm is characteristic of protons of hydroxymethyl groups linked to aromatic rings. In the region of aromatic protons it was observed two sets of peaks, between 6.7 and 7.1 ppm, corresponding to higher chemical shifts when compared to the tannin. These peaks are related to the aromatic protons of tannin and to the phenol hydroxymethylated rings linked by methylene bridges, generating the tannin-phenolic resin.

The phenolic-type thermosetting resins based on lignin and tannin were used to prepare composites reinforced with sisal fibers in the proportion range of 30-70% (w/w). Figure 2 shows the results of the Izod impact strength test of the tannin-phenolic and lignin-formaldehyde composites reinforced in the range of 30 to 70% w/w by sisal fibers.

The thermoset, without fiber reinforcement, presented an impact strength value of about 15 Jm^{-1} (Ramires and Frollini, 2008). The results of the Izod impact strength test of the composites showed that the presence of fibers greatly improved this property. The fibers were able to distribute the tension applied to the whole matrix efficiently, improving the impact strength of the material (Figure 2). The increase in the proportion of the sisal fibers used as reinforcement led to an increase in the Izod impact strength up to 40% (w/w) of sisal fibers in lignin-based matrix composites (Figure 2B) and 50% in tannin-based matrix composites (Figure 2A). Generally, composites prepared from thermoset matrix reinforced with vegetable fibers show improvement in their mechanical properties with the increase in the proportion of fibers used as reinforcement. However, after a limit, which depends on both the matrix and the type of vegetable fiber used, a decrease in the mechanical properties can occur by increasing the amount of fiber. This happens due to the higher fiber-fiber contact that can occur, in addition to incomplete impregnation of the fiber by the polymeric matrix.

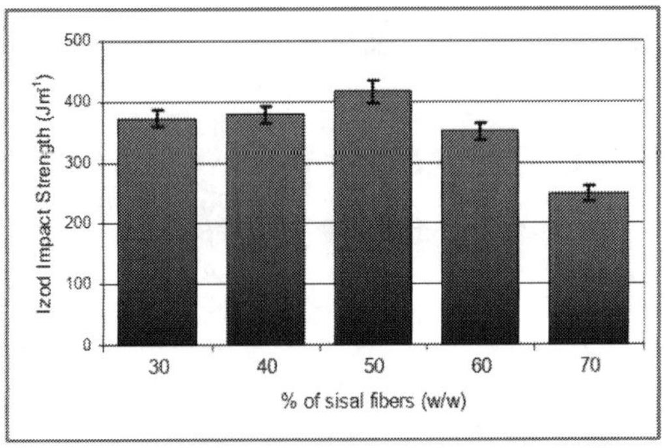

(A)

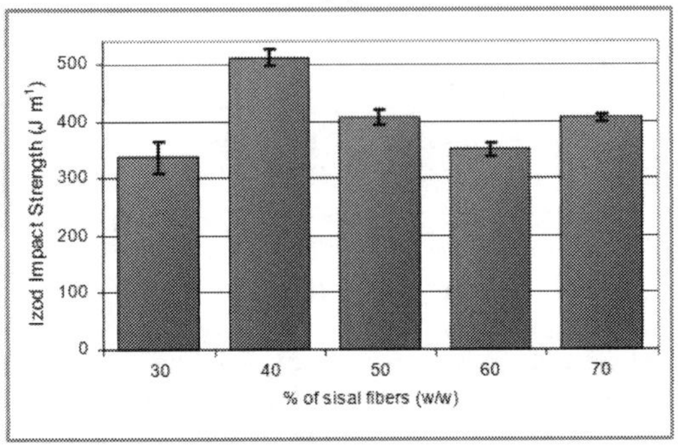

(B)

Figure 2 – Izod impact strength and standard deviation of the composites: **(A)** tannin-phenolic **(B)** lignin-formaldehyde.

The lignin-formaldehyde and tannin-phenolic composites reinforced with sisal fiber showed Izod impact strength values higher than that of the corresponding phenolic (phenol-formaldehyde) composite (Figure 3). This result demonstrates that the substitution of phenol by lignin and tannin is feasible, leading to materials with improved mechanical properties when compared to that of phenolic composites, currently used in industrial scale.

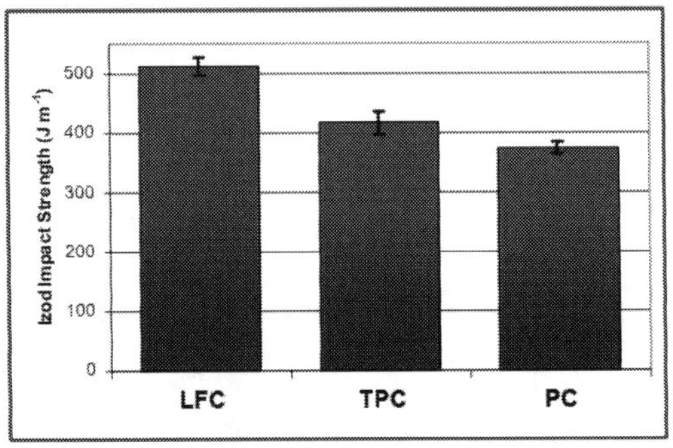

Figure 3 – Izod impact strength and standard deviation of the lignin-formaldehyde (LFC), tannin-phenolic (TPC) and phenolic (phenol-formaldehyde) composites reinforced with sisal fibers.

Figure 4 shows the SEM images of impact fractured surfaces of the composites reinforced with sisal fibers.

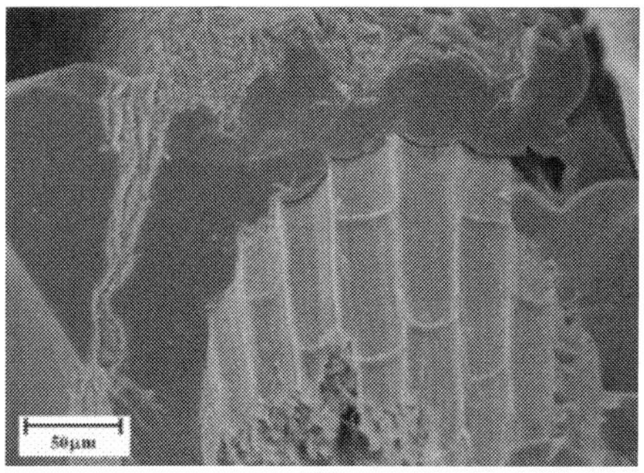

(A)

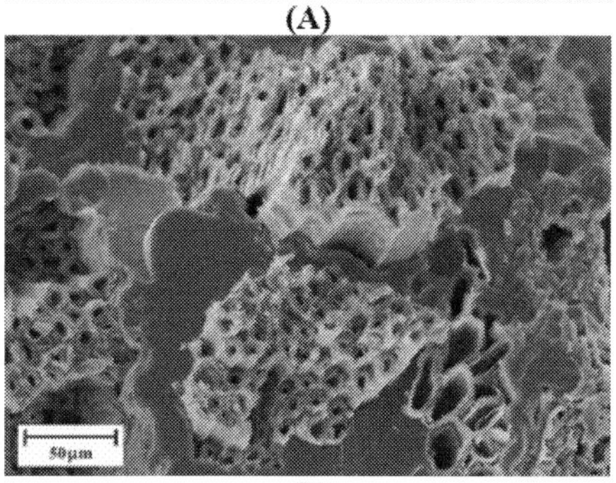

(B)

(C)

(D)

Figure 4 – SEM images of impact fractured surfaces of the composites: **(A)** tannin-phenolic/50% of sisal fiber; **(B)** tannin-phenolic/60% of sisal fiber; **(C)** lignin-formaldehyde/40% of sisal fiber; **(D)** lignin-formaldehyde/60% of sisal fiber.

The composites reinforced with 30, 40 and 50% (w/w) of sisal fibers showed good adhesion at the fiber/matrix interface, since the fibers remained adhered to the matrix after the impact test (Figure 4A and 4C). In these composites, the sisal fibers were covered and filled by the matrix, which prevents the tips of the fibers from acting as tension concentrators. The composites reinforced with higher amount of sisal fibers (Figure 4B and 4D) also showed good fiber/matrix interface. However, fiber agglomeration caused a decrease in the Izod impact strength values when compared to the composites reinforced with 40 and 50% of sisal fibers.

Figure 5 shows the curves of storage modulus (E') *versus* temperature of the tannin-phenolic and lignin-formaldehyde composites.

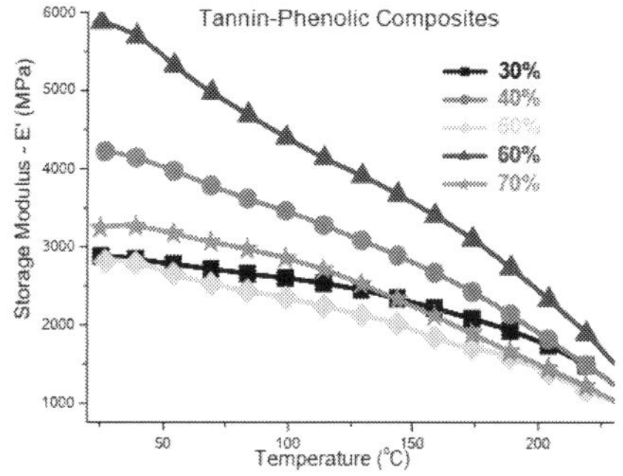

(A)

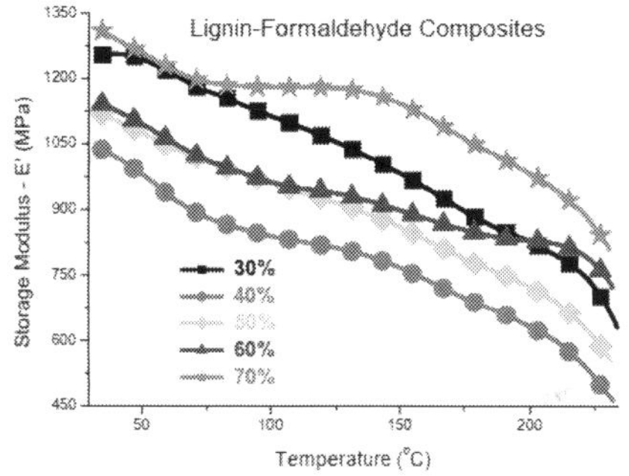

(B)

Figure 5 – DMTA curves of the storage modulus (E') *versus* temperature of the composites reinforced with sisal fibers: **(A)** tannin-phenolic **(B)** lignin-formaldehyde.

The storage moduli of the composites increased with increasing proportion of the sisal fiber. The sisal fiber has high cellulose content (64%) and crystallinity (61%) and the modulus of the fiber is superior to those of tannin-phenol and lignin-formaldehyde matrices. Hence, the increase in the proportion of fibers led to a proportional increase in the storage moduli of the composites. Two exceptions to this behavior were observed. The composite tannin-phenolic reinforced with 70% of the sisal fiber showed smaller storage modulus than the expected, if considered only the effect of hardening caused by the sisal fibers. The low value of E' in the composite reinforced with 70% of fiber is probably due to the dissipation of energy in the transfer of load through the interface. This dissipation of energy occurs due to the weak fiber/matrix interface in this composite, caused by the difficulty in impregnating such a high amount of fiber.

Another exception was observed for the lignin-formaldehyde composite reinforced with 30% of sisal fiber. The high value of E' shown by this composite was probably due to the higher crosslinking of the matrix in this composite. The smaller amount of fiber the greater amount of matrix, which favors the occurrence of bonds between the chain segments of the matrix.

Figure 6 shows the first derivative of TG (dTG) curves of the tannin-phenolic and lignin-formaldehyde composite reinforced with 40, 50 and 70% of sisal fiber. The other composites showed similar thermal behavior (curves not shown).

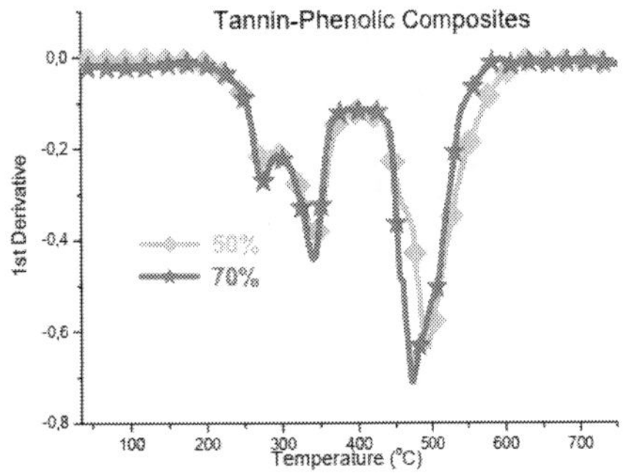

(A)

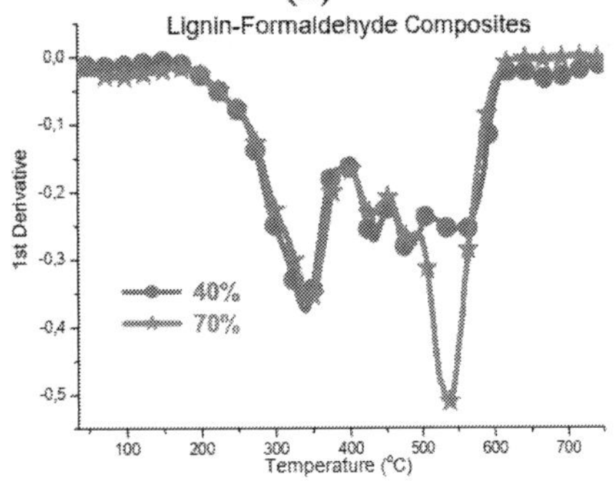

(B)

Figure 6 – dTG curves of the composites reinforced with sisal fibers: **(A)** tannin-phenolic **(B)** lignin-formaldehyde.

In the dTG curves of the composites (Figure 6) around 275°C, it is observed a peak of mass loss (a shoulder in the lignin-formaldehyde composites) probably due to the volatilization of water generated in stages of residual cure occurred during the scanning, in addition to the decomposition of hemicellulose of the sisal fibers (Ramires et al., 2010). At 350°C the peak of mass loss is related to the beginning of the thermal decomposition of the cellulose present in the sisal fibers. The last stage of mass loss begins around 400°C (peak-maximum around 500°C) and it is related to both the decomposition of the polymeric chains and the aromatic rings of the matrix, in addition to the decomposition of the lignin present in the sisal fibers. In the lignin-formaldehyde this peak of mass loss is also related to the decomposition of the lignin present in the matrix, justifying its enlargement.

Figure 7 shows the DSC curves of the tannin-phenolic and lignin-formaldehyde composite reinforced with 40, 50 and 70% of sisal fiber. The other composites showed similar thermal behavior (curves not shown).

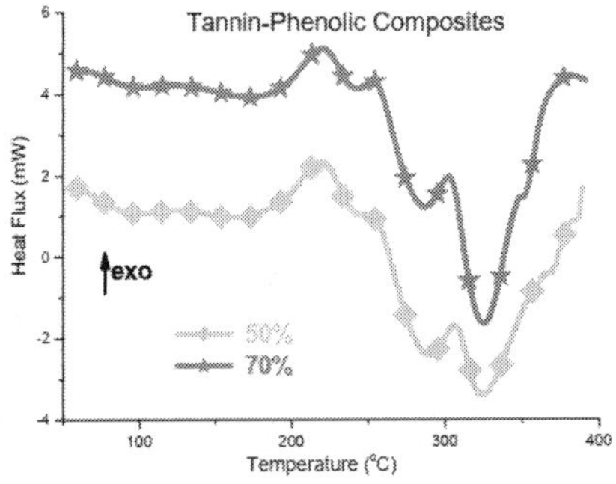

(A)

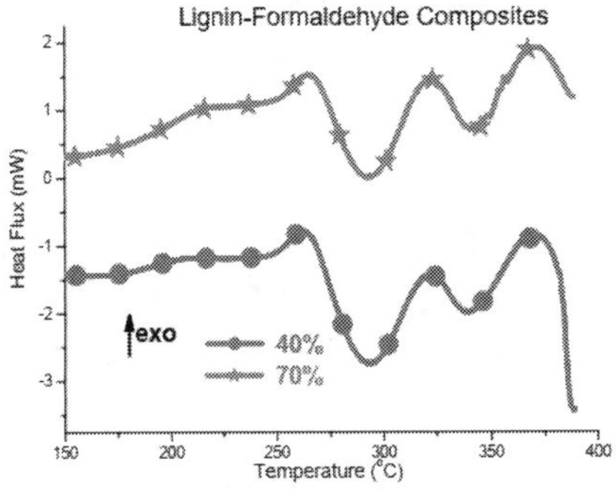

(B)

Figure 7 – DSC curves of the composites reinforced with sisal fibers: **(A)** tannin-phenolic **(B)** lignin-formaldehyde.

Around 220°C, in tannin-phenolic composites (Figure 7A) and 250°C in lignin-formaldehyde composites (Figure 7B) it is observed an exothermic peak related to the residual cure in addition to the decomposition of the hemicellulose. Around 300°C, it is observed an endothermic peak due to the release of volatiles generated by oxidative process, in addition to the decomposition of the cellulose present in the sisal fibers. The decomposition reactions are exothermic, but as at the same time the release of volatiles occurs (endothermic event), the observed peak can be exothermic or endothermic, depending on the balance between the decomposition reactions and the release of volatiles.

The composites were also analyzed by the water absorption test (figures not shown). It is important to emphasize that the conditions considered in this test were intentionally considerably more aggressive than those found where this kind of material is used, aiming to showing the maximum absorption that could occur under unfavorable conditions (Hoareau et al., 2006).

As the sisal fibers have hydrophilic character, the increase in the proportion of the used fiber as reinforcement generally intensifies the process of water absorption in the composites, increasing the water absorbed by the composites. This behavior was observed for the tannin-phenolic composites, the water absorption increased with the increased proportion of sisal fibers. Thus, for these composites, the hydrophilic nature of the sisal fibers, that facilitates the interact ion with the water molecules, was the mainly responsible for the water absorption. In the fiber composition, the hemicellulose is considered the main responsible for the water absorption, although non-crystalline cellulose and lignin also play an important role in this process (Paiva and Frollini, 2006).

In the lignin-formaldehyde composites, the water absorption did not increase with the increased proportion of sisal fibers used as reinforcement in the composites. All the composites showed close values of water absorption. This result is very important, because it shows the high affinity between fiber and matrix and, additionally, makes viable the use of high proportion of fibers, without increasing the amount of absorbed water.

The speed of diffusion of the water molecules inside the composites can be related to the diffusion coefficient (D), which can be obtained from Equation (1):

$$\frac{M_t}{M_\infty} = \frac{4}{L}\left(\frac{D}{\pi}\right)^{1/2} t^{1/2} \qquad (1)$$

where M_t is the water content at a time t; M_∞ is the water content in equilibrium and L is the thickness of the sample.

Figure 8 shows the values of the diffusion coefficient (D) of the tannin-phenolic composites.

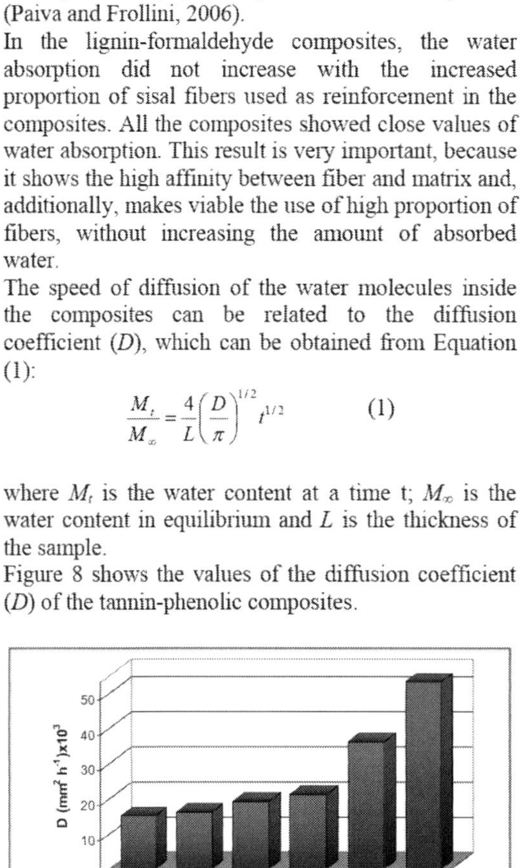

Figure 8 – Values of diffusion coefficient (D) of the tannin-phenolic thermoset and composites reinforced with sisal fibers.

The composites reinforced in the range of 30 to 50% of sisal fiber showed values of the diffusion coefficient (D) close to that of thermoset (without fiber reinforcement) (Figure 8). The tannin-phenolic matrix has a high proportion of hydroxyl groups (Figure 1A) that can interact with water molecules via hydrogen bonds. Thus, the introduction of sisal fibers up to 50% w/w did not change the rate of diffusion of the water molecules inside the composites.

In the composites reinforced with 60 and 70% of sisal fibers, especially for the latter, it was observed a significant increase in the values of D (Figure 8). This considerable increase was due to the presence of flaws, mainly in the fiber/matrix interface, which behave as "free spaces" for the diffusion of water molecules, accelerating the diffusion inside the composites.

Figure 9 shows the values of the diffusion coefficient (D) of the lignin-formaldehyde composites reinforced with sisal fibers.

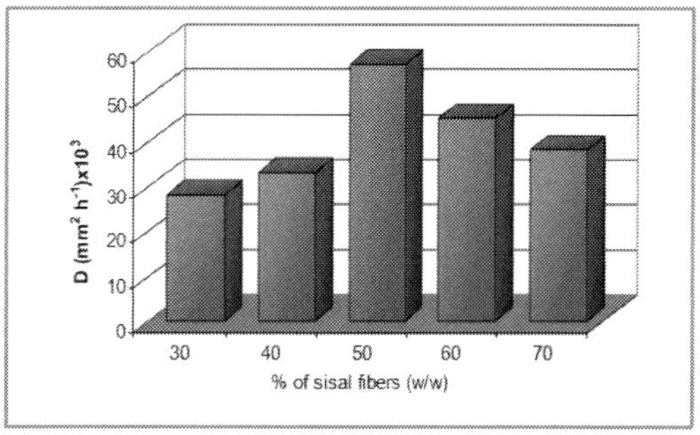

Figure 9 – Values of diffusion coefficient (D) of the lignin-formaldehyde composites reinforced with sisal fibers.

The increase in the proportion of sisal fibers used as reinforcement from 30 to 50% (w/w) resulted in increased values of the diffusion coefficient (D). As the internal channels of fibers, which are the more hydrophilic components, were filled by the resin, the water diffusion through these channels was difficult, and the diffusion occurred mainly through the matrix. Due to the intense fiber/matrix interaction, the matrix had a smaller amount of free polar groups to interact with water, via hydrogen bonds. This justifies the increase in the diffusion coefficient, since the development of hydrogen bonds between the matrix and water slows down the diffusion of these molecules through the material, then decreasing D. The composites reinforced with 60 and 70% (w/w) sisal fibers showed a decrease in the value of D. Such a decrease is due to the higher amount of fibers, which makes the impregnation of the fibers by the matrix more difficult. In these composites, the fibers were less protected by the matrix and could interact with water molecules, establishing hydrogen bonds, which slows down the diffusion of water molecules into the composites. However, these values of D were higher than that of the composite reinforced with 30% of sisal fiber. These results favor the composite reinforced with 70% of fiber, when compared to the composite reinforced with 30% of fiber, because they have close values of impact strength and the former has a higher content of raw materials from renewable source.

Conclusions

The lignin and condensed tannin have a high potential to be used as substitutes of phenol in the preparation of phenolic-type resins, aiming at the production of lignocellulosic fiber-reinforced bio-based composites. The results demonstrated that composites with good properties may be prepared using high proportions of materials obtained from renewable sources, i.e., lignin, tannin and sisal fibers can be used to obtain added-value materials.

Acknowledgements

The authors gratefully acknowledge FAPESP (The State of São Paulo Research Foundation, Brazil) for the fellowship for E.C.R., and for financial support. E.F. is grateful to CNPq (National Research Council, Brazil) for the research productivity fellowship and financial support.

References

B.R. Patent PI0801091 (2008) Frollini, E., Ramires, E. C., Oliveira, F. B., Barbosa Júnior, V.

FAO-Food and Agriculture Organization of the United Nations "Overview of the Various Alternative Uses of Sisal", www.fao.org (2009)

Frollini, E. et al., "Chapter 12 Lignophenolic and phenolic resins as matrix materials", in Natural Fibers, Plastic and Composites, Wallenberger, F. T., Weston, N. (Ed.), Kluwer Academic, USA, p. 193–219 (2004).

Hoareau, W., Oliveira, F. B., Stephane, G., Frollini, E., Castellan, A., "Fiberboards based on sugarcane bagasse lignin and fibers", Macromol. Mater. Eng., **291**, 829–839 (2006)

John, M.J., Thomas, S., "Biofibres and biocomposites", Carbohydr. Polym., **71**, 343–364 (2008)

Ku, H., Rogers, D., Davey, R., Cardona, F., Trada, M. "Fracture toughness of phenol formaldehyde composites: pilot study", J. Mater. Eng. Perform., **17**, 85–90 (2008)

Megiatto Jr., J. D., Silva, C., Ramires, E. C., Frollini, E., "Thermoset matrix reinforced with sisal fibers: effect of the cure cycle on the properties of the biobased composite", Polym. Test., **28**, 793-800 (2009)

Megiatto Jr., J. D., Ramires, E. C., Frollini, E., "Phenolic matrices and sisal fibers modified with hydroxyl terminated polybutadiene rubber: impact strength, water absorption, and morphological aspects of thermoset and composites", Ind. Crops Prod., **31**, 178–184 (2010)

Mosiewicki, M. A., Aranguren, M. I., Curvelo, A. A. S., Borrajo, J., "Effect of natural rubber on wood-reinforced tannin composites", J. Appl. Polym. Sci., **105**, 1825–1832 (2007)

Mu, Q., Wei, C., Feng, S., "Studies on mechanical properties of sisal fiber/phenol formaldehyde resin in-situ composites", Polym. Compos., **30**, 131–137 (2009)

Nicolai, F. N. P., Botaro, V. R., Cunha Lins, V. F. "Effect of saline degradation on the mechanical properties of vinyl ester matrix composites reinforced with glass and natural fibers", J. Appl. Polym. Sci., **108**, 2494–2502 (2008)

Paiva, J. M. F., Frollini E., "Unmodified and modified surface sisal fibers as reinforcement of phenolic and lignophenolic matrices composites: thermal analyses of fibers and composites", Macromol. Mater. Eng., **291**, 405–417 (2006)

Ramires, E. C., Frollini, E., "Renewable resources as reinforcement of lignophenolic and tanninphenolic matrices", 9th Conference of Science and Wood Industries. 1–8 (2008)

Ramires, E. C., Megiatto Jr, J. D., Gardrat, C., Castellan, A., Frollini, E., "Biobased composites from glyoxal-phenolic resins and sisal fibers", Bioresour. Technol., **101**, 1998–2006, (2010)

Razera, I. A. T., Frollini E., "Composites based on jute fibers and phenolic matrices: properties of fibers and composites", J. Appl. Polym. Sci., **91**, 1077–1085 (2004)

Tondi, G., Pizzi, A., "Tannin-based rigid foams: characterization and modification", Ind. Crops Prod., **29**, 356–363 (2009)

Trindade, W. G., Hoareau, W., Razera, I. A. T., Ruggiero, R., Frollini, E., Castellan, A., "Phenolic thermoset matrix reinforced with sugar cane bagasse fibers: Attempt to develop a new fiber surface chemical modification involving formation of quinones followed by reaction with furfuryl alcohol", Macromol. Mater. Eng., **289**, 728–736 (2004)

Vázquez, G., González-Álvarez, J., Antorrena G., "Curing of a phenol-formaldehyde-tannin adhesive in the presence of wood. Analysis by differential scanning calorimetry", J. Therm. Anal. Calorim., **84**, 651–654 (2006)

PREPARATION AND CHARACTERIZATION OF COMPOSITES FROM THERMOPLASTIC STARCH, BIOPOLYMERS AND LIGNOCELLULOSIC FIBERS

Quevedo, T. T.[1], Viana, H. M.[2*], Bergereth, A.[3], Benezet, J. C.[4]

[1]FAENG/CUFSA - Santo André – Brazil, – taquevedo@gmail.com
*[*2]FAENG/CUFSA - – Av. Principe de Gales, 821 – 09060-650 - Santo André – Brazil, – hviana@fsa.br*
[3]École dês Mines d'Alès – Alès –France – anne.bergeret@mines-ales.fr
[4]École dês Mines d'Alès – Alès – France – jean-charles.benezet@mines-ales.fr

Abstract

Population and economy growth worldwide have caused an enormous increase in waste production. To minimize this problem, some alternatives are being studied, such as, for instance, recycling of plastic materials and substitution of conventional plastics by biodegradable polymers *(Heller et al., 2003; Kim, 2008)*. Biopolymer materials derived from renewable agricultural resources have focused a great interest in the last few years. In this sense, many different thermoplastic biopolymers have been developed and commercially produced using current levels of technology *(Lörck et al., 2000)*. By means of a extrusion process, blends composed of polylactide (PLA), poly(ε-caprolactone) (PCL) and thermoplastic starch was prepared to conquer the major short comings (brittle and high price) of PLA. The thermoplastic starch was prepared by an extrusion process with glycerol as a plasticizer, native and acetylated starches were used. During extrusion, starch granules are exposed to high temperature and shear and undergo structural changes such as gelatinization, melting and fractionation. The ability to process starch and the resulting physical properties depend on the extent of structural changes of the starch. During gelatinization, starch molecules are released from the granule structure *(Gomez & Aguilera, 1983, 1984)*. The addition of a plasticizer to gelatinized starch allows free starch molecules to behave in a similar fashion to common thermoplastic synthetic polymers. By the addition of PCL, PLA was tuned from rigid to ductile but its tensile strength was also reduced. In this work, the thermal, rheological and mechanical properties of these blends were analyzed using different techniques including tensile tests, DSC, TG, and SEM. All these results were used to choose the bests blends to incorporate lignocellulosic fiber.
Keywords: Polymeric blends, thermoplastic starch, lignocellulosic fibers, natural composites.

Introduction

As people's growing concern for environmental protection, biodegradable materials have attracted considerable attention as interesting sustainable plastics. Starch is a naturally occurring biopolymer that can be extracted from many crops including corn, wheat, rice, potato and so on *(Tester & Karkalas, 2002)*. As an inexpensive and renewable source, it has been used as filler (*Chandra & Rustgi, 1997; Griffin, 1977; Tang, Zou, Xiong, & Tang, 2008a)* for environmentally friendly plastics for about three decades. The main renewable sources of biopolymers are proteins, polysaccharides and lipids. Proteins are thermoplastic heteropolymers of both polar and non polar amino acids that are able to form numerous intermolecular linkages, and undergo different interactions, yielding a wide range of potential functional properties. Moreover, plant proteins are inexpensive, renewable and abundant raw materials *(Irissin-Mangata et al., 2001; Song and Zheng, 2008)*.
The most popular and important biodegradable polymers are aliphatic polyesters [e.g. polylactide (PLA), poly(ε-caprolactone) (PCL), polyethylene oxide (PEO), poly(3-hydroxybutyrate) (PHB) and polyglycolicacid] and thermoplastic proteins.
Poly(ε-caprolactone) is a semi-crystalline polymer that shows a high degree of miscibility with a number of different polymers. Poly(ε-caprolactone)(PCL) is an important biodegradable material in degradable packaging applications *(Lepoittevin et al., 2002)*. Poly(ε-caprolactone) is a biodegradable polymer and as such it is of great importance also in drug delivery systems and biomedical applications such as tissue engineering. It is used in constructing scaffolds including composites with inorganic particles, fibers, or sintered ceramic matrices. In all these applications, physical properties such as crystallinity, morphology, microstructure, and mechanical properties play important roles. The degree of crystallinity of the polymer, for example, determines the rate of its degradation.
The biodegradable PLA, which is a hydrophobic and semi crystalline polyester and a renewable material that can be utilized by microbes within 30–40 days *(Hakkarainen, Karlsson, & Albertsson, 2000; Itavaara, Karjomaa, & Selin, 2002)*, had attracted increasing attention as a candidate for us in industrial applications

since it possessed excellent mechanical properties (especially in tensile strength and modulus) and was produced with corn or other starches as the starting raw material. With these advantages, PLA was a versatile material with applications in the medical, textile and packaging felds, but its brittleness and high price in comparison to petroleum-based thermoplastics limited its applications. Blending of PLA with TPS is a good way to balance the cost-effective issue and get a new material that has good performances. However, previous studies have shown that TPS/PLA blend have rather poor mechanical properties due to the poor interfacial affinity of starch and PLA *(Ke & Sun, 2000; Kim, Chin, Yoon, Kim, & Jung, 1998; Ren, Liu, & Ren, 2007; Wang, Yu, Chang, & Ma, 2008)*.

However, both PLA and PCL have some disadvantages that restrict their practical applications. For instance, PLA has high strength and melting temperature, but it degrades relatively slow and is hard and brittle. PCL has high flexibility and relatively good degradation rate, but its strength is relatively low, and the 60°C melting point is too low for many applications. Obviously, PLA and PCL have strong complementary performance.

A way of processing protein-based biomaterials is the mechanical method, or thermoplastic processing, which consists of mixing proteins and plasticizer to obtain a dough-like material *(Attenburrow et al., 1999; Jerez et al., 2007; Song and Zheng, 2008)*. The use of plasticizers reduces the intermolecular forces and increases the mobility of polymeric chains, thereby improving the flexibility and the extensibility of the material *(Gao et al., 2006)*.

Therefore, bioplastics can be processed using existing plastics processing machinery, including thermoforming, various types of injection moulding, compression-moulding, extrusion (films and fibers), and extrusion coating and lamination *(Kim, 2008; Kumar and Singh, 2008; Kunanopparat et al., 2008; Kadokawa et al., 2009)*.

Starch is a naturally occurring biopolymer that can be extracted from many crops including corn, wheat, rice, potato and so on *(Tester & Karkalas, 2002)*. As an inexpensive and renewable source, it has been used as filler *(Chandra & Rustgi, 1997; Griffin, 1977; Tang, Zou, Xiong, & Tang, 2008a)* for environmentally friendly plastics for about three decades. However, native starch has to be modified in order to be melt-processed as a thermoplastic because the melting temperature (Tm) of pure dry starch is close to 220–240°C and the onset temperature of starch degradation is around 220°C *(Sarazin, Li, Orts, & Favis, 2008; Souza & Andrade, 2002; Stepto, 2003)*.

In order to obtain thermal plastic starch (TPS), both physically and chemically methods of modification of starch have been widely studied. Physical modification of starch can be prepared by extrusion using various plasticizers *(Da Roz, Carvalho, Gandini, & Curvelo, 2006; Forssell, Mikkila, Suortti, Seppala, & Poutanen, 1996; Graaf, Karman, & Janssen, 2003; Ma, Yu, & Wan, 2006)*. Water *(Teixeira, Da Roz, Carvalho, & Curvelo, 2007)* and glycerol *(Tang, Alavi, & Herald, 2008b)* are two of the most widely used plasticizers. The role of plasticizers is to destructurize granular starch by breaking hydrogen bonds between the starch macromolecules, accompanying with a partial depolymerization of starch backbone. Chemical modification of starch such as esterification *(Raquez et al., 2008)* is the other well known methods to get TPS.

Acetylated starch that can be obtained through changing part of the hydroxyl groups into acetyl groups has been extensively studied over the last two decades *(Reinisch, Radics, & Roatsch, 1995; Wang & Wang, 2002)*. Although various methods have been tried to modify native starch in the last few years, it is still difficult to completely overcome its moisture absorption nature and poor mechanical properties of the final products.

Development of such materials has not only been a great motivating factor for materials scientists, but also an important provider of opportunities to improve the standard of living of people around the world. This is true, since many of the renewable materials mostly based on agricultural materials, as a source of raw materials particularly to plastic industries could generate a non-food source of economic development for farming and rural areas in developing countries, which have greater potential in these materials even though major thrust for their use has been given by the industrialized countries. One such material is lignocellulosic fiber such as jute, sisal, pineapple, curauá, sugar cane (bagasse) etc, whose extraction is an important process that determines the properties of the fibers, can generate rural jobs since these fibers have established their potential as reinforcing fillers in many polymers and several products based on these are produced commercially and are being used.

The lignocellulosic fibers have some unique attributes, such as being less abrasive to tooling and not causing as many respiratory problems for workers. Furthermore, because they are inexpensive and have load-bearing potential, use of natural fiber based composites has spread to various sectors, including aircraft, construction, grain and fruit storage and footwear. In addition, some European countries, where greater environmental consciousness exists, have already established the use of Lignocellulosic Fiber Incorporated Polymer Composites (LCFIPC) in automotive applications. Traditional engineering thermoplastics markets, such as those for exterior auto components, noise shields and door panels, or as "wood ceramic" for electromagnetic shielding, have been effectively taken over by the new commodity polymers and their composites due to their improved performance. In addition, many of lignocellulosic materials, such as rice and wheat straw, sugarcane bagasse and rice and coffee husks, which were either used as household fuels or sometimes burnt *in situ* in the fields as a means of disposal, have also been tried as fillers/reinforcements in polymer or cement to develop new materials for the construction industry.

A number of advanced technologies are being applied to bioplastics to provide added value, including active packaging technology, natural fiber reinforcements, nanotechnology, and innovative product design *(Kumar and Singh, 2008; Kunanopparat et al., 2008; Kadokawa et al., 2009).*

The increasing focus on eco-friendliness, such as the 7 European Union's directive that all new vehicles must use 95% recyclable materials to achieve the "end of life" required by 2015 and about 85% of this is recoverable through reuse or mechanical recycling and about 10% through energy recovery or thermal recycling has led to greater impetus for the development of commercially viable biodegradable composites. It is therefore not surprising to note that many leading plastics companies in U.S. and other western countries are in the forefront in developing biodegradable materials from natural materials, including "secondary resources", to mitigate environmental impacts. Also, it is reported that by 2020, about 10% of the basic chemical building blocks will come from renewable plant resources, which is expected to rise to 50% by 2050 as hurdles to their use are reduced.

Extrusion of starch-based products is common in the food industry *(Gomez & Aguilera, 1983, 1984).* In the polymer field, interest in this biopolymer has been recently renewed due to its abundance, low-cost, biodegradability and the possibility of processing using conventional polymer processing equipment.

Polymer blending is an important way to obtain new materials that can meet different needs. Blending of TPS with other polymers represents an important route to overcome the limitations of TPS.

The physicochemical and rheological characteristics of extruded starch products have been studied from the point of view of polymer science *(Aichholzer & Fritz, 1998; Della Valle, Buleon, Carreau, Lavoie, & Vergnes, 1998; Della Valle, Vergnes, & Tayeb, 1992; Willett, Jasberg, & Swanson, 1995; Willett, Millard, & Jasberg, 1998).*

Experimental

Materials

The following materials were obtained from commercial suppliers and used as received. The potato starch was supplied by Roquette and glycerol was supplied by Fluka. Part of the potato starch was acetylated. The polylactide (PLA, NatureWorks 7000D), (purity > 98% and density = 1,24g/cm^3), was obtained from Cargill-Dow. The poly(ε-caprolactone) (PCL, CAPA 6800), with a molar mass of 80,000g/mol, (purity > 99%), was obtained from Solvay Chemicals, Inc. All materials were used with no further purification

Thermoplastic Starch (TPS) preparation

TPS can be obtained by processing granular starch at low water or other plasticizer content using thermal and mechanical forces (Myllarinen, Partanen, Seppala, & Forssell, 2002). In this work, TPS was prepared by a twin-screw extruder in the following proportions: 75wt% of starch and 25wt% of glycerol. The thermal profile was 30-60-70-70-70-70-75-75-75-80°C (zones 1–10, respectively), the screw speed was 250rpm and feeder rate was 0,95kg/h. Two different TPS were produced: one with native (TPS_N) and another with acetylated starch (TPS_A). The material was extruded and granulated.

Preparation of PLA/TPS_N and PLA/TPS_A blends

PLA/TPS blends were prepared by the melt blending method using a twin-screw extruder. Before blending, PLA was dried in a vacuum oven at 50°C for 6h. The mass ratios of PLA to TPS were chosen as 80/20, 70/30, 50/50, 30/70 and 20/80. $PLA^xTPS_N^y$ and $PLA^xTPS_A^y$ were used as the symbols for the PLA/TPS blends, where x and y indicated mass percentages of PLA and TPS, respectively. A determined amount of PLA and TPS was put together into the feeder to extrude in a twin-screw extruder under the conditions that the screw speed and the blending temperature were kept at 250-300rpm and 30–140°C, respectively. The material was extruded and granulated.

Preparation of PCL/TPS_N and PCL/TPS_A blends

PCL/TPS blends were prepared by the melt blending method using a twin-screw extruder. Before blending, PLA was dried in a vacuum oven at 50°C for 6h. The mass ratios of PCL to TPS were chosen as 80/20, 70/30, 50/50, 30/70 and 20/80. $PCL^xTPS_N^y$ and $PCL^xTPS_A^y$ were used as the symbols for the PCL/TPS blends, where x and y indicated mass percentages of PCL and TPS, respectively. A determined amount of PCL and TPS was put together into the feeder to extrude in a twin-screw extruder under the conditions that the screw speed and the blending temperature were kept at 250-300rpm and 50–140°C, respectively. The material was extruded and granulated.

Preparation of $PCL^{70}TPS_A^{30}$/Fiber and $PCL^{30}TPS_A^{70}$/Fiber

$PCL^xTPS_N^y$/Fiber blends were prepared by the melt blending method using a twin-screw extruder. Before blending, the binary blends were dried in a vacuum oven at 50°C for 6h. The mass ratio of binary blends to TPS was chosen as 85/15. $PCL^{70}TPS_A^{30}/SC^{15}$ and $PCL^{30}TPS_A^{70}/SC^{15}$ were used as the symbols for the binary blends/fiber blends. A determined amount of binary blends and sugar cane fiber was put together into the feeder to extrude in a twin-screw extruder under the conditions that the screw speed and the blending temperature were kept at 250-300rpm and 50–140°C, respectively. The material was extruded and granulated.

Preparation of $PLA^{70}TPS_A^{30}$/Fiber and $PLA^{30}TPS_A^{70}$/Fiber

$PLA^xTPS_N^y$/Fiber blends were prepared by the melt blending method using a twin-screw extruder. Before blending, the binary blends were dried in a vacuum oven at 50°C for 6h. The mass ratio of binary blends to TPS was chosen as 85/15. $PLA^{70}TPS_A^{30}/SC^{15}$ and $PLA^{30}TPS_A^{70}/SC^{15}$ were used as the symbols for the binary blends/fiber blends. A determined amount of binary blends and sugar cane fiber was put together into the feeder to extrude in a twin-screw extruder under the conditions that the screw speed and the blending temperature were kept at 250-300rpm and 30–140°C, respectively. The material was extruded and granulated.

Injection Moulding

Injection moulding was accomplished in a Sandretto 95T injection moulding machine. Subsequently, the granulate blends were made into standard specimens for characterization and mechanical analysis. The binary blends follow the following processing conditions: barrel temperature 125°C-155°C and mould temperature 30°C. The ternary blends follow the following processing conditions: barrel temperature 125°C-155°C and mould temperature 30°C. All samples were conditioned at 40°C for at least 5 days before testing.

Mechanical Properties

All the mechanical property measurements were performed at room temperature on blends specimens. The tensile properties were determined with a Zwick equipment model Z010 universal testing machine using specimens type dumb-bell, according to the ISO 527 standard. The load cell had a capacity of 10 KN, at a rate of 50mm/min. The data was averaged of 5 specimens. The average and standard deviations of the tensile strength, elongation at break and Young's modulus were determined for each formulation.

Differential Scanning Calorimetry (DSC)

Samples were taken from the blends and conditioned at 40°C for at least 5 days before testing. The thermal properties were measured using a Differential Scanning Calorimeter Pyris 1 from Perkin–Elmer, under a nitrogen atmosphere, at a heating rate of 10°C/min. Glass transition temperatures *(Tg)*, higher thermal transitions, such as fusion and crystallization were scanned from -20°C to 200°C. The samples were first heated from -20°C to 200°C to eliminate their thermal history, and then cooled to -20°C and immediately reheated to 200°C. The thermal transitions were calculated from the second heating cycle.

Scanning Electron Microscopy (SEM)

A FEI Quanta 200 FEG scanning electron microscopy was used to study the morphology of blends and to measure the starch phase size in the polymer matrix. The tensile fractured surfaces of the blends were fixed in a metallic support and coated with graphite, using a sample metallizer (BAL-TEC CED 030) and observed by SEM, under different tensions according to each sample.

Impact Resistance

To assess the impact resistance of these blends were performed impact tests with specimen size of 80x10x4mm, in a Zwick 5102 equipment with a pendulum of 4J.

Results and Discussion

Thermoplastic Starch (TPS) preparation

Initially, the last heating zones of the extruder were heated above 100°C. The extruded material had a puffed appearance, reducing the temperatures of all heating zones this effect was minimized. The maximum temperature reached was 80°C. The material was extruded and granulated.

Preparation of binary and ternary blends

The extrusion temperatures were adjusted in each case, each of the blends required different conditions of processing temperature. During the extrusion process, it was necessary to be alert to increases of the pressure, mainly in blends with higher amount of thermoplastic starch.

Injection Moulding

After extrusion, specimens were injected in sufficient quantity to make all characterization and mechanical analysis. The blends showed a narrow range of processing due to the low thermal stability, requiring high shear rates to make possible the plasticizing and the moulding. However, it was necessary to care that the processing temperature does not degrade the material, reducing its molecular weight and therefore their mechanical properties.

Mechanical Properties

The tensile strength, elongation at break and modulus of binary and ternary blends are shown in Figure 1. When a small amount of compatibilizer was added to the binary blends, increases in both tensile strength and elongation at break were observed over the whole composition range. The increases could be ascribed to the poor interfacial adhesion between TPS and polyesters being improved after the compatibilizer was added. For the non-compatibilized blends, there was a decrease in the tensile strength of the blends with increasing PLA or PCL content. The modulus for these blends decrease gradually with increasing PLA or PCL content. This is because PLA and PCL are flexible polyesters, as its content increasing the blends became much softer, thus resulting in a decrease in modulus values. The compatibilized blends showed higher modulus values comparing to the non-compatibilized blends. Compatibilized blends have better interfacial adhesion between the blends components, thus resulting in a relatively compact phase structure and an increase in modulus values.

The result shows that the neat PLA gives a high tensile strength at break and a low elongation at break, exhibiting that PLA is a high modulus and brittle material. On the contrary, the neat PCL is a ductile material with a relative low tensile at break value and a high elongation at break value. It is clear that the addition of PCL makes blends a substantial reduction in tensile at break but an increase en elongation at break value as their PCL content increases.

Figure 1a shows that when the amount of TPS_A or TPS_N increases there is a reduction of the elongation at break. As the PLA has a higher tensile strength, reducing the content of this polymer in the blends leads to a reduction in tensile strength. The tensile strength is stable with the increase of the amount of TPS_A and it is possible to observe that with the increase of the amount of TPS_N there is an increase in tensile strength values. There is a sharp increase in the values of tensile strength in blends with more than 80% of TPS_N. For modulus, the results show that when the amount of TPS_A or TPS_N increases there is a reduction of the modulus values.

In Figure 1b it is possible to observe that when the amount of TPS_A increases there is a reduction of the elongation at break. Over 70% of TPS_A or TPS_N there is an increase in the value of elongation at break. When the amount of TPS_A increases there is a reduction of the tensile strength and there is an increase of these values in blends with up to 70% of TPS_N, above this, it shows a sharp drop in these values. For modulus, the results show that when the amount of TPS_A increases up to 30%, there is an increase of the modulus values, after this amount, it shows a sharp drop in these values up to 70% of TPS_A, after this, the values increase again. It can be observed that when the amount of TPS_N increases, there is an increase in modulus values.

Figure 1c shows that with the incorporation of sugar cane fibers in binary blends of PCL there is a little change in the value of elongation at break and a reduction of the value of tensile strength when comparing to binary blends. For binary blends of PLA, there is a sharp drop in the value of elongation at break and an increase of the value of tensile strength, comparing to binary blends. For both types of ternary blends, it is possible to observe a sharp increase of the value of modulus.

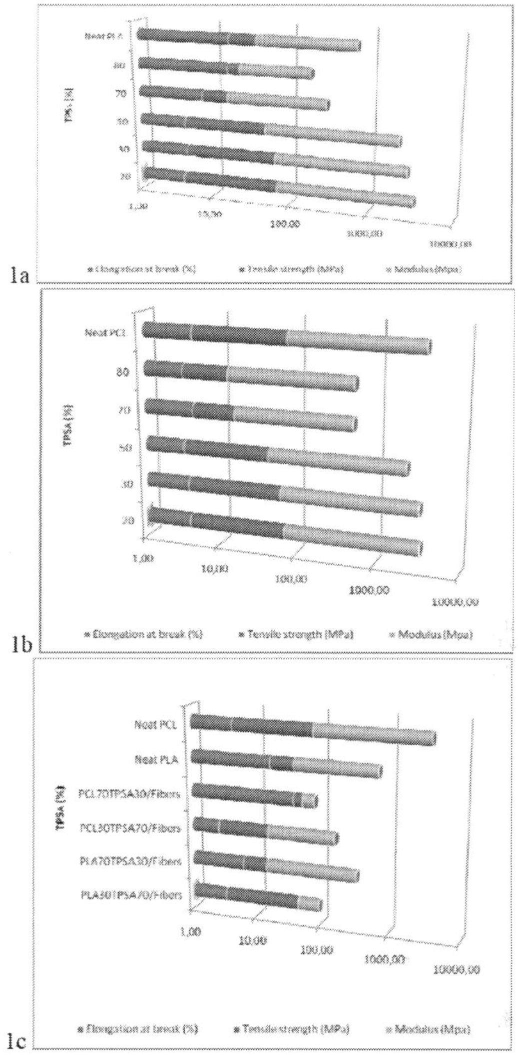

Figure 1 Mechanical analysis: 1a) Mechanical analysis for binary blends using PLA; 1b) Mechanical analysis for binary blends using PLA and 1c) Mechanical analysis for ternary blends

Interfacial interaction between the fillers and matrix was an important factor affecting he mechanical properties of the composites. Thus, theoretical tensile strength of the composites was modeled for the cases of adhesion and no adhesion between the filler particles and matrix.

Differential Scanning Calorimetry (DSC)

The ability to determine transition temperatures and enthalpies makes DSC a valuable tool in producing phase diagrams for various chemical systems.

Table 1 shows all the results. The weight loss can be attributed to the out of water, previously absorbed by the potatoes, when exposed to the environment.

Table 1: DSC results

Blend	Tg (°C)	Hot Tc (°C) (during cooling step)	Cold Tc (°C) (during heating step): difficult to see	Tf (°C)
PLA20/TPSA80	58,77	x	116,34	150,93
PLA30/TPSA70	57,75	x	125,30	149,54
PLA50/TPSA50	63,04			147,40
PLA70/TPSA30	62,38			148,75
PLA80/TPSA20	62,75			152,52
PLA20/TPSN80	59,95	x	x	148,85
PLA30/TPSN70	59.96	x	x	149.89
PLA50/TPSN50	61,12			146,34
PLA70/TPSN30	57,95			148,19
PLA80/TPSN20	59.10			103,18
PCL20/TPSA80	x	23,84	x	58,28
PCL30/TPSA70	x	22,51	x	58,74
PCL50/TPSA50	x			63,91
PCL70/TPSA30	x			63,22
PCL80/TPSA20	x			65,55
PCL20/TPSN80	x	22,74	x	58,43
PCL30/TPSN70	x	22,11	x	57,78
PCL50/TPSN50	x	27,04	x	61,86
PCL70/TPSN30	x	27,57	x	58,19
PCL80/TPSN20	x	27,06	x	58,92
PLA30TPSA70/Fiber				147,94
PLA70TPSA30/Fiber	57,28			150,28
PCL30TPSA70/Fiber	x	25,90	x	58,86
PCL70TPSA30/Fiber	x			63,42
Neat PLA	60,27	x	129,25	153,02
Neat PCL	-60,12	28,75	x	60,19

Scanning Electron Microscopy (SEM)

The morphology of the blends has been analyzed by SEM and Figure 2 shows the starch granules distributing in the polymer matrix and Figure 3 shows de fibers distributing in the binary blends. In this study, PLA^X and PCL^X forms the polymer matrix for Figure 2, whereas the starch is the dispersed phase and $PLA^XTPS_N^{\ Y}$, $PLA^XTPS_A^{\ Y}$, $PCL^XTPS_N^{\ Y}$ and $PCL^XTPS_A^{\ Y}$ forms the matrix for Figure 3, whereas the fibers are the dispersed phase. Good dispersion of starch and fibers, effective wetting of starch by the polymer and fibers by the binary blends and strong interfacial adhesion between two phases or more are required to obtain satisfactory mechanical properties. By examining the morphologies shown in Figure 2, it can be seen that large starch phases domains could be found for the nom-compatibilized blends, after compatibilized, the starch granules were mostly melted and formed a continuous phase with the polymer matrix.

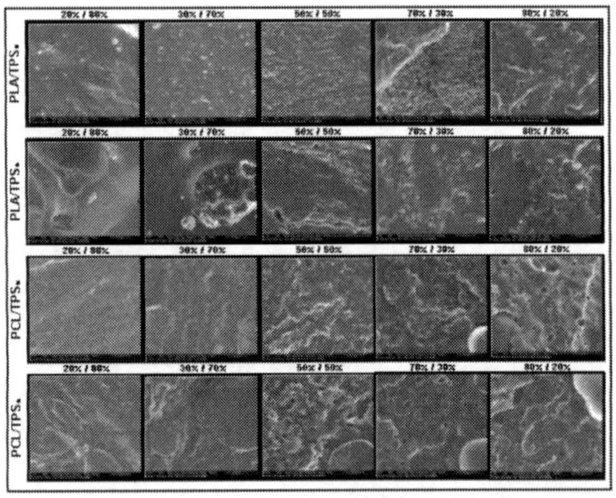

Figure 2 SEM micrographs for binary blends

By examining the morphologies shown in Figure 2, it can be seen that blends with 80% of TPS have good miscibility, very small starch phases and better dispersion for $PLA^{20}TPS_A^{\ 80}$ and $PCL^{20}TPS_N^{\ 80}$ blends. For blends with 70% of TPS, it can be observed a very flat surface with large phases of starch for $PLA^{30}TPS_A^{\ 70}$ blend, the outlines of the specimens are quite different from the center for $PLA^{30}TPS_N^{\ 70}$ blend and large phases of starch for $PCL^{30}TPS_A^{\ 70}$ blend. It can be seen that blends with 50% of TPS have a bad miscibility and starch appears to be fragile for $PLA^{50}TPS_N^{\ 50}$, nodules and risks like a half-moon, large and small phases of starch for $PLA^{50}TPS_A^{\ 50}$ and presence of scaling for $PCL^{50}TPS_N^{\ 50}$. For blends with 30% of TPS, it can be observed big fragility, difficult to analyze because it degrades rapidly with the applied tension (0,53torr), good miscibility and high porosity for $PLA^{70}TPS_N^{\ 30}$ blend; a little difference between binary blends of PCL with natural and acetylated starch; many bubbles, problems of miscibility and less degradation than binary blends with PLA for $PCL^{70}TPS_A^{\ 30}$. It can be seen that blends with 20% of TPS have a high speed of degradation, good miscibility and very compactness for $PLA^{80}TPS_N^{\ 20}$ blend; bigger phases and better dispersion for $PLA^{80}TPS_A^{\ 20}$ blend when compared to $PLA^{70}TPS_A^{\ 30}$ blend; there is a little of immiscible material, fusion phenomenon, failures of ligament and little phases of starch for $PCL^{80}TPS_N^{\ 20}$; smaller amount of bubbles compared to $PLA^{80}TPS_A^{\ 20}$ blend, existence of small dark spots and little phases of starch.

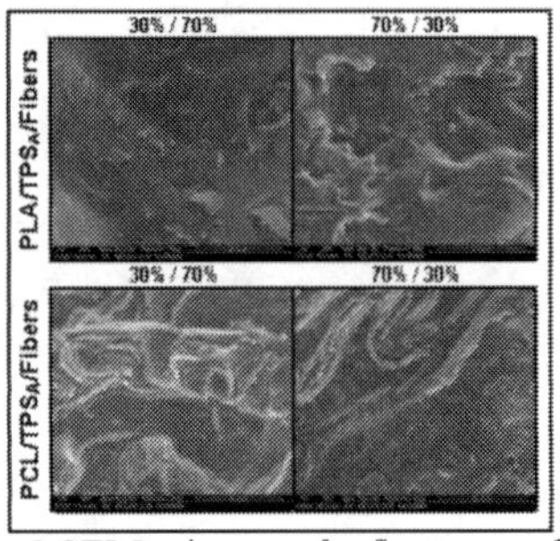

Figure 3 SEM micrographs for ternary blends

By examining the morphologies shown in Figure 3, it can be seen that the ternary blends have bad miscibility, small fibers phases for ternary blends using PCL and major fibers phases for ternary blends using PLA. It can be observed a better dispersion for ternary blends using PLA.

Impact Resistance

Figure 4 shows the quantitative results – energy needed to fracture the material - for the binary blends. Fig. 8a shows the results for binary blends with PLA, with TPS_A or TPS_N. Good results are achieved with PLA/TPS_A blends, it is possible to observe that when the amount of TPS_A increases there is an increase of the impact resistance, but under 70%, this value decreases. Fig. 8b shows the results for binary blends with PCL, with TPS_A or TPS_N. Good results are achieved with PCL/TPS_N blends, it is possible to observe that the specimens don't break with a pendulum of 4J.

4a

4b

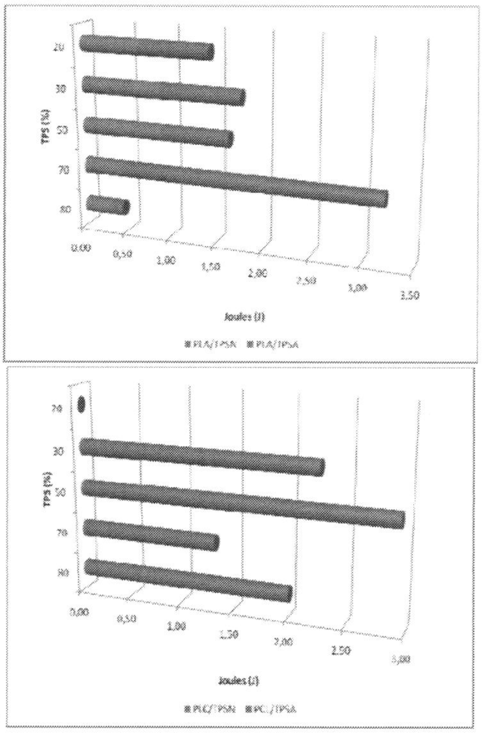

Figure 4 - Results of impact tests: 4a) Results of impact tests of binary blends with PLA, 4b) Results of impact tests of binary blends with PCL

Conclusions

An advantage of a polymeric blend is to obtain materials with new properties and/or improved. These properties can be: stiffness, impact resistance, toughness, dimensional stability at high temperatures, weather resistance and others. There are no major problems related to the extrusion of potato starch with glycerol, the only inconvenience was the puffed formation due to high temperature of the heating zones. The difficulty of processability of PLA and PCL was attenuated by the fact that we work with the resin previously dried in a vacuum oven. The DSC tests showed that the binary blends exhibited two separate Tms, demonstrating that PLA or PCL and TPS formed a partial miscible blend.

Acknowledgements

This work was supported by École des Mines d'Alès (France) and FAENG - Faculdade de Engenharia Eng° Celso Daniel/CUFSA and authors are grateful to them. They also express their appreciation to the reviewers for their constructive and meticulous assessment of the manuscript.

References

Aichholzer, W., & Fritz, H-G. (1998). Rheological characterization of thermoplastic starch materials. Starch/Starke, 50, 77–83.

Chandra, R., & Rustgi, R. (1997). Biodegradation of maleated linear low-density polyethylene and starch blends. Polymer Degradation and Stability, 56(2), 185–202.

Da Roz, A.L., Carvalho, A.J.F., Gandini, A., & Curvelo, A.A.S. (2006). The effect of plasticizers on the thermoplastic starch compositions obtained by melt processing. Carbohydrate Polymers, 63, 417–424.

Della Valle, G., Buleon, A., Carreau, P.J., Lavoie, P.-A., & Vergnes, B. (1998). Relationship between structure and viscoelastic behavior of plasticized starch. Journal of Rheology, 42 (3), 507–525.

Della Valle, G., Vergnes, B., & Tayeb, J. (1992). Mesure de la viscosité d'amidons fondus faible menthydratés dans um nouveau rhéomètreen ligne. Entropie, 169, 59–63.

Forssell, P., Mikkila, J., Suortti, T., Seppala, J., & Poutanen, K. (1996). Plasticization of barley starch with glycerol and water. Journal of Macromolecular Science – Pure and Applied Chemistry, 33 (5), 703–715.

Graaf, R., Karman, A., & Janssen, L.(2003). Material properties and glass transition temperatures of different thermoplastic starches after extrusion processing. Starch, 55 (22), 80–86.

Griffin, G.J.L. (1977). Synthetic resin sheet material. US Patent Office, Pat.No.4021388.

Gomez, M.H., & Aguilera, J.M. (1983). Changes in the starch fraction during extrusion —cooking of corn. Journal of Food Science, 48, 378.

Gomez, M.H., & Aguilera, J.M. (1984). Aphysico chemical model for extrusion of corn starch. Journal of Food Science, 49, 40.

Hakkarainen, M., Karlsson, S., & Albertsson,A.C. (2000). Rapid biodegradation of polylactide by mixed culture of compost micro-organisms-low molecular weight products and matrix changes. Polymer, 41 (7), 2331–2338.

Heller J, Pharma AP, Domb AJ.Adv Drug Deliv Rev 2003; 55: 445e6.

Itavaara, M., Karjomaa, S., & Selin, J. (2002). Biodegradation of polylactide in aerobic and anaerobic thermophilic conditions. Chemosphere, 46 (6), 879–885.

Ke, T., & Sun, X. (2000). Physical properties of poly(lactic acid) and starch composites with various blending ratios. Cereal Chemistry, 77 (6), 761–768.

Kim, S.H., Chin, I., Yoon, J., Kim, S.H., & Jung, J. (1998). Mechanical properties of biodegradable blends of poly(L-lacticacid) and starch. Korea Polymer Journal, 6 (5), 422–427.

Ma, X.F., Yu, J.G., & Wan, J.J. (2006).Urea and ethanol amine as a mixed plasticizer for thermoplastic starch. Carbohydrate Polymers, 64, 267–273.

Raquez, J.M., Nabar, Y., Srinivasan, M., Shin, B.Y., Narayanb, R., & Dubois, P. (2008). Maleated thermoplastic starch by reactive extrusion. Carbohydrate Polymers, 74 (2), 159–169.

Reinisch, G., Radics, U., & Roatsch, B. (1995). Efficient syntheses for starch acetates. Angewante Makromolecular Chemistry, 233, 113–120.

Ren, J., Liu, Z.C., & Ren, T.B. (2007). Mechanical and thermal properties of poly(lactic acid)/starch/mont morillonite biodegradable blends. Polymers & Polymer Composites, 15 (8), 597–602.

Sarazin, P., Li, G., Orts, W.J., & Favis, B.D. (2008). Binary and ternary blends of polylactide, polycaprolactone and thermoplastic starch. Polymer, 49 (2), 599–609.

Souza, R., & Andrade, C. (2002). Investigation of the gelatinization and extrusion processes of corn starch. Advances in Polymer Technology, 21 (1), 17–24.

Stepto, R. (2003). The processing of starch as a thermoplastic. Macromolecular Symposia, 201 (1), 201–212.

Tang, X.Z., Alavi, S., & Herald, T.J. (2008b). Effects of plasticizers on the structure and properties of starch – clay nanocomposite films. Carbohydrate Polymers, 74,552–558.

Tang, S.W., Zou, P., Xiong, H.G., & Tang, H.L. (2008a). Effect of nano-SiO2 on the performance of starch/polyvinyl alcohol blend films. Carbohydrate Polymers, 72, 521–526.

Teixeira, E.M., DaRoz, A.L., Carvalho, A.J.F., & Curvelo, A.A.S. (2007). The effect of glycerol/sugar/water and sugar/water mixtures on the plasticization of thermoplastic cassava starch. Carbohydrate Polymers, 69, 619–624.

Tester, R.F., & Karkalas, J. (2002). The effects of environmental conditions on the structural features and physico-chemical properties of starches. Starch, 53 (10), 513–519.

Wang, Y.J., & Wang, L. (2002). Characterization of acetylated waxy maize starches prepared under catalysis by different alkali and alkaline-earth hydroxides. Starch, 54 (1), 25–30.

Wang, N., Yu, J.G., Chang, P.R., & Ma, X.F. (2008). Influence of formamide and water on the properties of thermoplastic starch/poly(lacticacid) blends. Carbohydrate Polymers, 71 (1), 109–118.

Willett, J.L., Jasberg, B.K., & Swanson, C.L. (1995). Rheology of thermoplastic starch: Effects of temperature, moisture content,and additives on melt viscosity. Polymer Engineering Science, 35 (2), 202–210.

Willett, J.L., Millard, M.M., & Jasberg, B.K. (1998). Extrusion of waxy maize starch: melt rheology and molecular weight degradation of amylopectin. European Polymer Journal, 10, 1477–1487.

PREDICTION OF THE GAS PERMEABILITY PROPERTY OF POLY (LACTIC ACID)/ POLY (BUTYLENE SUCCINATE) (PLA/PBS) NANOCOMPOSITES

Rahul K Gupta*, Sati N. Bhattacharya and Deeptangshu Chaudhary[1]
Rheology and Materials Processing Centre,
School of Civil, Environmental and Chemical Engineering,
RMIT University, 124 Latrobe Street,Melbourne,Vic 3001 Australia
Rahul.gupta@rmit.edu.au, Sati.bhattacharya@rmit.edu.au
[1]Department of Chemical Engineering, Curtin University of Technology, Perth, Australia
D.Chaudhary@curtin.edu.au

Abstract

PLA/PBS/Cloisite 30BX (organically modified MMT) clay nancomposites were prepared by using simple melt extrusion method. Composition of PLA and PBS polymers were fixed as 80% and 20% by weight for all the nanocomposites. Nanocomposites containing 1, 3, 5, 7 and 10 wt% of C30BX contents were prepared to investigate the effect of clay content on morphological, rheological and thermal and gas barrier properties of these nanocomposites. Gas permeability, morphological, rheological, thermal properties of nanocomposites prepared with the mixing of poly (lactic acid) (PLA), poly (butylene succinate) (PBS) and organically modified layered silicate were investigated. From morphological studies using wide angle X-ray diffraction (WAXD) and transmission electron microscopy (TEM), nanocomposite having 1 wt% of clay was considered to attain mixed morphology of intercalated and exfoliated structure, while some clusters or agglomerated particles were detected for nanocomposites having 3 and more than 3 wt% of clay content. Gas barrier properties were measured for oxygen and water vapours. Oxygen barrier property improved significantly as compared to water vapour. A model based on gas barrier was used for the validation of oxygen relative permeabilities of PLA/PBS/C30BX nanocomposites.

Keywords: Nanocomposites, poly (lactic acid) (PLA), poly (butylene succinate) (PBS), barrier properties, gas barrier model

1. INTRODUCTION

Several researchers (Garlotta, 2001, Fujimaki, 1998, Bhatia et al. (2009, 2010) have studied biodegradable polymers which are derived from renewable and petroleum resources such as poly (L-lactide) (PLA), poly (butylene succinate) (PBS), poly (butylene adipate/terephthalate) (PBAT), poly (ethylene terephthlate/succinate) (PETS), poly (ε- caprolactone) (PCL) etc. Poly (L-lactide) (PLA), a feedstock of corn rather than petroleum, has been recently used for packaging material. Packaging material for food should provide sufficient barriers against water vapour to prevent the food degradation, atmospheric gases to prevent the oxidation of food material, and volatile organic compounds to preserve the aromas and the flavours of the food. (Oliveira et al., 2004). In food packaging, mass transfer both from the food to the polymer and vice-versa is responsible for the contamination of food and thus causes degradation of the packaging material. Since, PLA is produced from the lactide monomer which comes from lactic acid and is a non-toxic substance readily available in human body. Thus, food contamination can be neglected for PLA packaging material. Therefore, it can be stated that the determination of permeability of oxygen, water vapour and aromas in polymers used for food packaging material are important parameters.

PLA, thermoplastic polyester has good mechanical properties, thermal plasticity and biocompatibility. Thus, it can be easily fabricated. However, due to the ring-opening polymerization of lactide monomer is cost demanding and its brittleness; limit the commercial use of PLA polymer. Therefore, blending of PLA with another biodegradable polyester polymer is necessary so that the resultant blend should overcome the limitations of PLA and will also add good qualities of the added polymer and will reduce its cost. In this study, other polymer selected to blend PLA with is poly (butylene succinate) (PBS). PBS is chemically synthesized by the polycondensation of 1, 4-butanediol with succinic acid. (Bhari et al., 1998). PBS has high flexibility, excellent impact strength, thermal and chemical resistance (Doi and Fukuda, 1994). PBS has excellent processability, so it will also make PLA processability easy.

2. LITERATURE REVIEW

Theoretical models were developed for predicting properties of composite materials based on the properties of the pure components and the morphology of the composite. These theories enable a simple route for evaluation of the individual contribution of components properties such as matrix type, volume fraction, filler aspect ratio, filler orientation, filler distribution etc, Enhancements in various properties are well-known in polymer-silicate nanocomposites; their main dependence is on the distribution of the silicate layers in the polymer matrices. However, main factors behind the improvement in gas barrier properties are not yet fully understood. As clays are crystalline materials, they are believed to increase the barrier properties by creating a maze or "tortuous path" that restricts the progress of the gas molecules to pass through the polymer matrices (Nielsen, 1967).

According to filled polymer systems theory developed by Nielsen (1967), if the fillers are impenetrable to a diffusing gas or liquid molecule, then the diffusing molecule must go around the filler particles. He introduced tortuosity factor (τ) which is defined as the ratio of the actual distance (d') that a penetrant must travel to the shortest distance (d) that it would have travelled in the absence of the layered silicate. Nielson added that thin plates like fillers having large aspect ratio (L/2W) (where L is the length of the filler particle and W is the thickness of the filler plates) can dramatically reduce the permeability only if the particles are oriented in such a way that their flat surfaces are parallel to the surfaces of the film. Nielsen also stressed that incomplete filler dispersion, voids, and only partial orientation of the particles should result in reverse affect in permeability.

Tortuosity factor (τ) is expressed as:

$$\tau = \frac{d'}{d} = 1 + \frac{L}{2W}\phi s$$

(1); where Φ_s is the volume fraction of the sheets.

The effect of tortuosity on the permeability can be expressed as:

$$\frac{Ps}{Pp} = \frac{1-\phi s}{\tau}$$

(2); where P_s and P_p represent the permeabilities of the nanocomposite and pure polymer, respectively. Bharadwaj (2001) invoked a simple model developed earlier by Nielsen correlating between the sheet length, concentration, relative orientation, and state of aggregation of the filler in the polymer matrix. Bharadwaj (2001) predicted from equation (2) that the relative permeability is a function of the silicate sheet length. Bharadwaj (2001) modified the tortuosity factor to include the orientation order (S), and thus the relative permeability changed to following equation (3):

$$\frac{Ps}{Pp} = \frac{1-\phi s}{1 + \frac{L}{2W}\phi s\left(\frac{2}{3}\right)\left(S + \frac{1}{2}\right)} \quad (3)$$

The above expression reduces to equation (2) when S = 1 (planar arrangement i.e. maximum tortuosity) and provides approximately the permeability of the pure polymer when S = -1/2 (orthogonal arrangement). Further, Bharadwaj (2001) concluded that incorporating sheets with L > 500 nm with random orientation is more beneficial for barrier property as the case where the sheets were aligned perpendicular to the diffusing path.

Exfoliation appears to be the critical factor in determining the maximum performance of polymer nancomposites for barrier applications. He explained that, at L = 1000 nm, the relative permeability is ~ 0.05 for W = 1 nm and 0.2 for W = 7 nm. While for L = 500 nm, relative permeability changes from 0.1 to 0.3 for W = 1 to 7 nm, thus underscoring the need for complete exfoliation of the silicate sheets.

Both Nielsen (1967) and Bharadwaj (2001) models are based on the arguments based on the tortuosity factor. Further, equations related to the permeability and tortuosity were developed to understand the diffusion of small molecules in conventional composites, however, they do extremely fit in predicting and relating the experimental results for the relative permeability in PLS nanocomposites. The main assumption of above model is that the sheets are placed such that sheet normal is coincident with the direction of diffusion (S = 1). Clearly, this arrangement results in the highest tortuosity. Any deviation from this arrangement would lead to deterioration in the barrier properties. Also, a sheet like morphology (W = 1) is particularly efficient at maximizing the path length of the penetrant to travel due to the large length-to-width ratio compared to spheres or cubes.

In this study, PLA/PBS/C30BX nanocomposites having fixed composition of PLA and PBS as 80 wt% and 20 wt% with various clay loadings were prepared. Various aspects of selected morphological, oxygen and water vapour permeability properties of the nanocomposites were compared to the polymer blend without clay. In particulate, this study mainly aimed to obtain an extensive understanding of the gas barrier properties of biodegradable polymer nanocomposites using a model based on gas barrier.

3. EXPERIMENTAL MATERIALS AND METHODS

3.1 Materials and nanocomposite preparation

Both the polymers used in this study have been supplied by Ire Chem. Co., Korea. PLA grade 4032D-biaxially oriented films grade (Nature Works) produced by Cargill Dow LLC was used in this study. And Poly (butylene succinate) (PBS) polymer grade G4460, another biodegradable polyester was used to blend with PLA polymer.

Densities of PLA and PBS are 1240 kg/m^3 and 1230 kg/m^3, respectively. PLA has crystalline melting point range between 150-170°C, while PBS has melting range between 95°-114.7°C.

In this work, experimental grade of Cloisite 30B organoclay has been used. Cloisite C30B grade is SCPX 3016, however in this paper this clay has been referred as C30BX. It was supplied by Southern Clay Products (SCP) Inc (USA. All the PLA/PBS/C30BX nancomposites have been prepared by melt compounding the polymers with C30BX clay at 180°C using Brabender twin-screw extruder at 40 rpm. The pellets of both PLA and PBS were dried in a vacuum oven prior to processing at a temperature of 50°C for 48 hours. For the determination of oxygen and water vapour transmission rates, film samples of approximately 0.3 mm thickness were used. All the samples including morphological and gas barrier property evaluations were obtained by using compression moulding machine at 200°C and pressure of 40 kN for 5 minutes.

3.2 Characterisation Methods

Oxygen permeability was measured according to ASTM D3985 standard, with MOCON Ox-Tran 2/21 instrument manufactured by Modern Controls Inc. (MN). The values of oxygen permeability were obtained at 23°C and 0% RH. The exposed area and thickness were 50 cm^2 and 300 μm, respectively. A mixture of 97 % N_2 with 3 % H_2 was used as carrier gas and 99.9 % oxygen was used as test gas. The film samples were placed in two test (permeability) cells. An average of four samples were analysed and their mean value has been reported in this study.

The water vapour transmission rate (WVTR) of each PLA/PBS/C30BX nanocomposite sample was measured using a Modern Controls Inc. (MOCON) Permatran-W1 (Minneapolis, MN, USA), according to ASTM standard F 1249 at 100% Relative Humidity (RH) and 37.8°C temperature. Exposure area for the WVTR measurements was 50 cm^2 and thickness of the film sample was 300 μm. The flow rate of the air during the test was set to 60 cm^3/min. After conditioning, actual water vapour transmission (WVTR) was noted usually after 6-8 hours. An average of five values has been reported in this work after normalizing the source values according to the thickness of the sample.

4. RESULTS AND DISCUSSION

All the nanocomposite films showed better oxygen barrier than pure PLA films as well as PLA/PBS blend films (Figure 1). From all the nanocomposites, nanocomposite having 3 wt% of C30BX content exhibited the highest reduction in oxygen permeability of approximately 26%. XRD and TEM results indicated exfoliated or disordered intercalated morphology occurred at 1 and 3 wt% of clay nanocomposites. That is why improvement in 3 wt% of nanocomposite seems to be maximum. As the clay content was increased, aggregation of the clay particles formed which gave less improvement in oxygen barrier property. Research has shown that the degree of tortuosity also reflects geometrical influences like shape and state of exfoliation/intercalation of platelets and their orientation in the polymer matrix. For instance, SinhaRay et al.(2002) found not much significant decrease in oxygen permeability for compression moulded PLA nanocomposites samples. These authors concluded that the planar orientation of the dispersed clay particles in the compression moulded films did not increase the tortuosity of the matrix adequately..

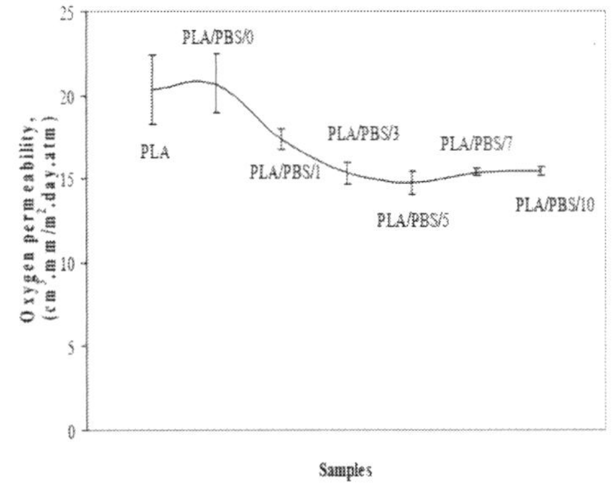

Figure 1: Effect of clay (C30BX) on oxygen permeability of blend and nanocomposites of different clay loadings

More importantly, it has been an interesting point to note that even though the oxygen permeability did not decrease significantly at higher clay content (7 and 10 wt%) in PLA/PBS/C30BX nanocomposites, it was far less to be considered in comparison to PLA/PBS (80/20) blend. This implies that due to the presence of impermeable crystalline clay particles restriction in the passage of oxygen occurred. However, due to the increased degree of aggregation with the clay concentration supported by the TEM micrographs, restriction by the clay particles in the passage of oxygen decreased.
Krook et al.(2001) attributed two reasons for the relatively modest improvement of the barrier properties in their nanocomposites, firstly, the void-content, which provided means for non-diffusion transport through the film and secondly, the non-uniformity of the clay particles. The relative improvement of the barrier properties of PLA/PBS/C30BX nanocomposites compression moulding samples was probably because of the combination of the increased crystallinity due to the presence of clay and the small decrease in void density.

4.1 Water Vapour Permeability:
PLA/PBS/C30BX nanocomposites exhibited modest improvement in water vapour barrier properties relative to that of the neat PLA and PLA/PBS blend films. Water vapour transmission rates (WVTR) of all the samples are shown in Figure 2. Nielsen stated that water penetrates montmorillonite layered silicate surface, form water clusters in the nanocomposites, and does not allow water to pass thorough quickly, and thus the diffusivity and the overall transport of water across the films is decreased. In PLA/PBS/C30BX nanocomposites water vapour permeation rate was not much reduced as in the case of oxygen permeation rate due to the hydrophilic nature of the polymers. When PBS polymer was added in PLA polymer in the ratio of 80/20 wt% (PLA/PBS), the water vapour transmission rate got increased. During the transportation of water through the PLA/PBS films, water first gets absorbed and once saturation level is attained, water started to diffuse through the film and transportation takes place. PLA and PBS are highly hydrophilic, thus once water get absorbed in the film, it allowed water to evaporate to the other end of the film where air flows, has less moisture and absorbs water vapour from the moistened cotton pad in the opposite end. All the nanocomposites showed higher values of water transmission rates in comparison to that of neat PLA polymer but with the increase in C30BX in the binary blend, the water vapour transmission rates decreased. However, the decrease in water vapour transmission rate was very modest. C30BX also acted as impermeable crystallites which decreased the amount of amorphous material through which water vapour could diffuse.

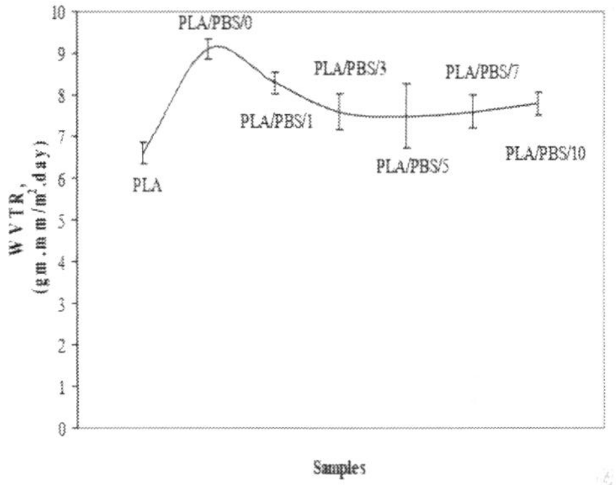

Figure 2: Water vapour transmission rates (WVTR) of PLA/PBS (80/20) blend and their nanocomposites

The polymer crystalline regions are considered to be impermeable to small molecules such as gases, which can lead to modifications in the final properties. It has been analysed that the presence of the filler (clay) may nucleate crystallization, which lead to a change in morphology i.e. degree of crystallization. These can have an indirect effect on the transport properties. The significant drop in permeability with increased crystallinity is usually explained in two ways. Firstly, inclusion of impermeable crystallites decreases the amount of amorphous material through which the permeant can diffuse. Secondly, impermeable crystallites increase the tortuosities of the transport path. As shown in our previous publication, although MDSC measurements indicated that the polymer crystallinity in the nanocomposites was not affected by the presence of fillers, all the nanocomposites have lower oxygen and water vapour permeabilities than that of the pure polymer blend (80/20, PLA/PBS). This improvement in the gas barrier properties can be attributed to the achievement of the mixed morphology suggested by XRD and TEM results and decrease in void density. Gas barrier properties have improved due to good distribution of clay in both the polymers in the bulk phase.

4.2 Model validation of Gas Permeability:

As discussed earlier, according to the Bharadwaj model the permeability of the nanocomposite (Ps) is related to the permeability of the pure polymer (Pp) and the volume fraction of the sheets (Φs), length (L) and width (W) of the sheets (clay particles) as shown in equation (3).

The width of the basic structural unit of montmorillonite is well defined and is approximately given as W ~ 1 nm; however, the length shows large distribution ranging between 100 nm to several microns depending upon the concentration of the filler in the polymer. Effective length and width of the sheets of the nanocomposites can change rapidly through aggregation.

Importantly, two main reasons made the estimation of sheet length quite complicated. Firstly, on the micrograph images of the nanocomposites only the projections of the particle transect were seen. Therefore, the real dimensions could not be estimated. Secondly, the features appearing on the micrograph images were not straight; therefore a computer assisted analysis method was required.(2004) A semi-automated image analysis was conducted and image analysis method has been used to estimate the sheet length of the clay particles. The volume fractions of the sheets in nanocomposites were calculated from the weight fraction using the density of C30BX.

In the case of nanocomposites, contrast could also arise due to the atomic number within the materials. Regions of high atomic number appear dark; while of low atomic number appear light. However, to naked eye, this discrimination between the greyish polymeric regions and the dark lines corresponding to the aluminosilicate platelets are relatively easy. In semi-automated image analysis method, conversion of the original TEM photomicrograph into black and white image is necessary.

A semi-automated method (Fornes and Paul, 2003) consists of firstly, manually tracing the dispersed platelets and the agglomerates on the transparent film overlaid on the micrograph image print using a black permanent pen. The isolated platelets were considered as single platelets. The resulting transparent film was then electronically scanned and converted into a greyscale tagged-image film format (TIFF) image file. The micrograph image and its scanned image obtained by manually tracing the particles are presented in Figure 3. This greyscale image can then be converted into black and white or binary image by using image analysis software (Scion Image Beta 4.02, Scion Corporation, USA). This software easily convert TIFF file into a binary image and creates a duplicate of the image, length of the traced particles were measured and exported into a separate file. From this separate file, the sheet lengths of the particles were used to obtain histograms of various nanocomposites using Minitab 15 software.

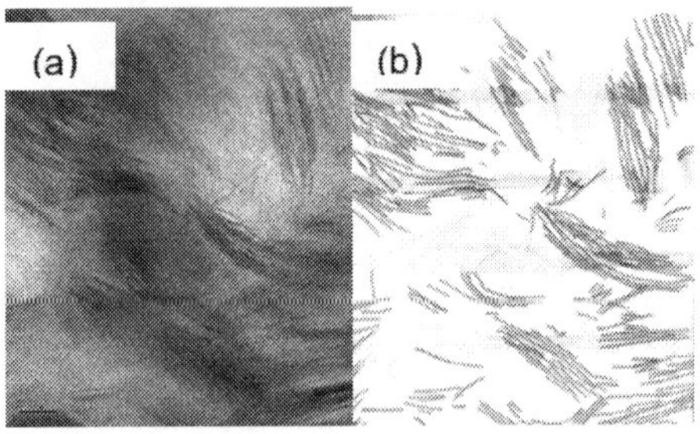

Figure 3: TEM (a) and its scanned image (b) of film obtained by manually tracing the particles by superimposing on micrograph

By comparing the relative permeabilities of oxygen obtained from the experiments and from the model, it can be easily seen that the experimental data very well validate the Bharadwaj model for nanocomposites having lower concentration of C30BX. 1 wt% of nanocomposite has the shortest sheet length and with the increase in clay loadings sheet length increased. At low concentrations of clay contents, clay particles were dispersed well as compared to the nanocomposites of higher content of C30BX. Increase in sheet length also means that at higher concentrations intercalation of sheets occurred. From TEM images, it was seen that as concentration of clay increased agglomerates and clay clusters formed. Therefore, effective width of the clay sheets also increased. The increased width included polymer layer thickness is called as effective width. Instead of 1 nm, it becomes 9 or 11 nm or more, depending upon the thickness of the clay clusters. However, here W ~ 1 nm and S = 1 (planar arrangement of silicate layers to the path of gas molecule to diffuse) has been taken as assumption while doing calculations using model based on gas barrier.

Figure 4 showed that experimental data validated the Bharadwaj model up to sheet length approximately 80 nm. After that at high sheet lengths more and more deviation from the Bharadwaj model was observed. The reason behind this may be that certain assumptions like good dispersion of clay particles in the polymer matrix, S =1 and W ~ 1 nm has been considered in this model. Nanocomposites having clay concentration more than 3 wt%, intercalation occurred. Therefore, due to agglomeration/intercalation, width of the sheet can not remain as 1 nm. At high concentrations width becomes effective width, which depends on the stacks in sheet layers.

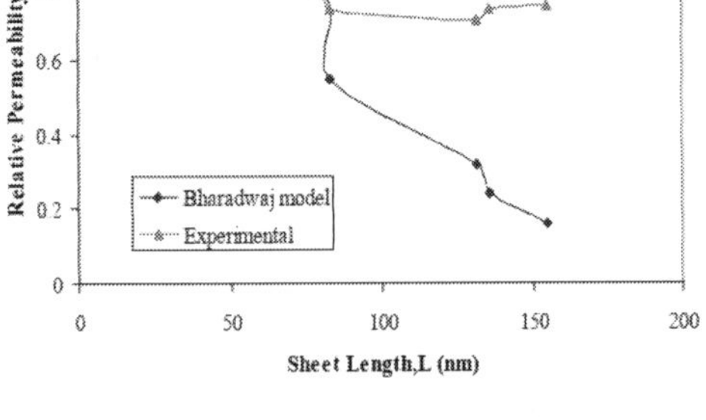

Figure 4: Validation of the Bharadwaj model for PLA/PBS (80/20) blend and their nanocomposites

The tendency of the silicate sheets to agglomerate was indeed expected to become more severe as the clay concentration increases because of the higher probability of collision between particles at higher concentration. Due to the change in width of the silicate sheet, aspect ratio (L/2W) also changed, thus high enhancement in gas barrier was not noted and so as the validation of the Bharadwaj model.

5. CONCLUSIONS

Improvement in oxygen permeability was significantly noticed for PLA/PBS/C30BX nanocomposites with the addition of C30BX, however, water vapour permeation rate showed modest improvements in this system. Due to the planar orientation of the dispersed clay particles, tortuosity path for the gas molecules did not increase significantly and so were the gas barrier properties. PLA/PBS/C30BX nanocomposites validated the Bharadwaj model only up to 3 wt% of clay content and with the higher concentration of C30BX nanocomposites , due to the formation of clusters and agglomerates, deviation in the Bharadwaj model was observed as effective width increased from 1 nm to 5-7 nm..

6. REFERENCES

Bharadwaj, R. K. *Macromolecules* **34**, 9189, 2001.

Bhari, K.;Mitomo, H.;Enjoji, T.;Yoshii, F.;Makuuchi, K. *Polymer Degradation and Stability* **62**, 551, 1998.

Bhatia, A., Gupta, Rahul K., Bhattacharya, S.N. and Choi, H.J , *International Polym. Processing*, **XXV**, 5-14 (2010).

Bhatia, A., Gupta, Rahul K., Bhattacharya, S.N. and Choi, H.J., *J App. Polym Sci.*, 114 (5), 2577 - 3342 (2009).

Doi, Y.;Fukuda, K. *Studies in Polymer Science* **12**, 627, 1994.

Fornes, T. D.;Paul, D. R. *Polymer* **44**, 4993, 2003

Fujimaki, T. *Polymer Degradation and Stability* **59**, 209, 1998.

Garlotta, D. *Journal of Polymers and the Environment* **9,** 63, 2001.

Krook, M.;Albertsson, A.-C.;Gedde, U. W.;Hedenqvist, M. S. *Polymer Engineering & Science* **42**, 1238, 2002.

Nielsen, L. E. *Journals of Macromolecular Science, Part A* **1**, 929, 1967.

Oliveira, N. S.;Oliveira, J.;Gomes, T.;Ferreira, A.;Dorgan, J.;Marrucho, I. M. *Fluid Phase Equilibra* **222-23**, 352, 2004.

SinhaRay, S.;Yamada, K.;Ogami, A.;Okamoto, M.;Ueda, K. *Macromolecular Rapid Communications* **23**, 943, 2002.

Vincent, M. PhD Thesis, In Department of Chemical Engineering; McGill University, Montreal, QC, Canada: 2004.

NEW RENEWABLE BIOCOMPOSITES OF DIMER FATTY ACID-BASED POLYAMIDES: THERMAL PHYSICAL AND MECHANICAL PROPERTIES

E. Hablot[1]*, R. Matadi[2], S. Ahzi[2], L. Avérous[1]

1 Laboratoire d'Ingénierie des Polymères pour les Hautes Technologies, ECPM-LIPHT, EAc (CNRS) 4379, Université de Strasbourg, 25, rue Becquerel 67087 Strasbourg Cedex2, France – elodie.hablot@ulp.u-strasbg.fr, luc.averous@unistra.fr

[2] Institut de Mécanique des Fluides et des Solides; IMFS, Université de Strasbourg, 2, rue Boussingault, 67000 Strasbourg, France - Rodrigue.Matadi-boumbimba@imfs.u-strasbg.fr, said.ahzi@imfs.u-strasbg.fr

Nowadays, replacing petroleum-based materials with renewable resources is a major concern in terms of both economical and environmental viewpoints. In good agreement with this emergent of concept of sustainable development, this work deals with the study of innovative green polyamides (DAPA) based on rapeseed oil-dimer fatty acid reinforced with cellulose fibres (CF). Thermal, morphological and mechanical properties of DAPA and its biocomposites were examined. With the increase of fibre content, the biocomposites exhibited a high increase in glass transition temperature, which can be attributed to carbonyl (DAPA) and hydroxyl (CP) groups' interactions, reducing the polymer mobility. In correlation with these results, tensile tests revealed that the Young modulus and the yield stress highly increased with increasing CF concentration. Halpin-Tsai and Eyring's micromechanically-models were found to predict successfully the Young modulus and the yield stress respectively, of DAPA and DAPA/Cellulose biocomposites. These different results have clearly highlighted the attractive properties of DAPA-based biocomposites and have demonstrated that these materials are good alternatives to conventional composites.

Introduction

The use of renewable materials to produce existing chemicals or to create original products is the subject of growing number of academic and industrial research projects. As an example, vegetable oils are expected to be an ideal alternative chemical feedstock, since they are produced in abundance throughout the world. Moreover, they have the particularity to contain lots of active chemical units that can be used for polymerization as bio-based monomers or building blocks (Biermann, 2000; Guner, 2006; Khot, 2001; Meier, 2007; Nayak, 2000; Hablot, 2008).

Bio-based polyamides (excluding proteins and polypeptides) remain little-developed. One easy route to produce bio-based polyamides is to use dimers of fatty acids, as presented in Figure 1.

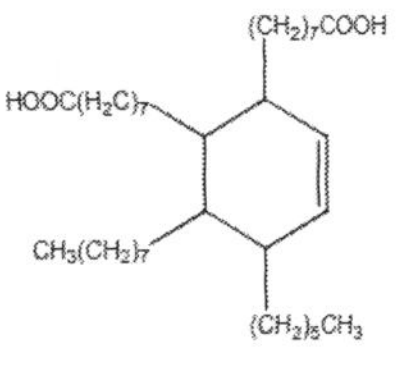

Figure 1 – Fatty dimer acid structure

They are well-known and commercially available products and thus are good candidates to elaborate bio-based thermoplastic polyamides Cavus, 2006). Dimer fatty acid based-polyamides (DAPA), due to their special molecular structure based on the dimer acid, show relatively good performance. But they are still far from conventional polyamide thermoplastics properties (Brandrup, 1999). Then, an effective approach to improve the DAPA behaviour is the incorporation of fibres into the matrix. In order to successfully meet the environment and recycling problems, agro-based fibres motivate more and more different industrial sectors (e.g. automotive) to replace e.g., common fibreglass. Because of their relative cheapness, their low density and ability to recycle (Bledzki, 1999; Eichhorn, 2001), cellulosic fibres seems to be good candidates to reinforce DAPA and to elaborate biocomposites essentially based on renewable materials, on agreement with a sustainable development. In this work, we propose the study of the thermal and mechanical properties of DAPA and DAPA biocomposites (DAPAC). Mechanical models were also developed for modelling the Young modulus and the yield stress of DAPA and DAPAC.

Experimental

Materials

Dimer acid used in the synthesis is supplied by Prolea (France). It is a yellowish, viscous liquid at room temperature with the dimer, trimer and monomer content at 96.6, 2.8 and 0.6%wt, respectively. 1,6-diaminohexane was purchased from Sigma Aldrich. All chemicals are used as received.

Pure cellulose fibres (CF), issued from leafwood, are used in this study. This product, Arbocel B400, is supplied by JRS (Rosenberg, Germany). According to this supplier, they show an average length and thickness of 900 and 20μm, respectively. In a previous paper, Amash et al Amash, 2000) have presented some characteristics of these fibres. According to these authors, the density of the fibres is 1.50 and the cellulose content is greater than 99.5%.

Polyamide synthesis

The polyamide (DAPA) is synthesized by condensation polymerization. Dimer acid is charged in a 1L five-necked round-bottom flask equipped with mechanical stirrer, a thermometer, a nitrogen inlet, a Dean-Stark apparatus and a dropping funnel. Dimer acid is heated at 343K under stirring and nitrogen atmosphere. 1,6-diaminohexane equivalent to total acid equivalent is added by using dropping funnel during 10 min. After 1,6-diaminohexane addition, the mixture is heated gradually to 453K within 3 hours. The completion of the polymerization is checked by amine and acid titration. The reaction is stopped after the consumption of more than 99% of the acid functions. The final acid-based thermoplastic polyamide is then discharged from the flask. DAPA is yellowish, transparent and flexible at ambient temperature with a molecular weight of 14 000 g/mol, a glass transition temperature of 263K and a fusion temperature of 354K.

Biocomposites processing

Prior to blending, fibres and DAPA are dried in an air-circulating oven at 383K for 1h and 343K for 1h, respectively. DAPA and varying amounts of CF are directly added in the feeding zone of an internal mixer (Counter-rotating mixer Rheocord 9000, Haake, USA) at 373K, with a rotation speed of 50 rpm and a processing time not exceeding 8 min, in order to avoid DAPA degradation. Four biocomposites (DAPAC) are prepared with 5, 10, 15 and 20%wt of CF content to form DAPAC5, DAPA10, DAPA15 and DAPAC20, respectively. After melt processing, the composites materials are compression-moulded, to obtain plates, with a hot press at 403K applying 200 MPa pressure for 5 min. The moulded specimens are then quenched between two steel plates for 10 min.

Optical micrographs of thin sheets of DAPAC20 showed very good fibre dispersion into the DAPA matrix.

Characterization

DSC measurements were performed on DSC 2910 TA Instruments under airflow. To determine the thermal characteristics, the samples (5-10 mg placed in an aluminium pan) were heated up to 423K at 10K/min. Then, the temperature was maintained for 3 min to get rid of thermal history. Then, samples were cooled to 233K at 10K/min and heated again to 473K at 10K/min. The glass transition temperature T_g and the melting temperature T_m were determined on the second heating scan.

Tensile tests were carried out with an Instron tensile testing machine (model 4204, USA) at different temperatures ranging from 298 to 333K, and different strain rates ranging from $10^{-4}s^{-1}$ to $1s^{-1}$. For tests conducted at elevated temperatures, the test samples were held at those temperatures for 40 min to ensure equilibration of the samples. The yield stress in this work was defined in the traditional sense for polymers as the first maximum stress on the material attained right after elastic response, on the stress-strain response. Even if the tensile tests are conducted at temperatures around T_g and beyond, the real plastic deformation is controlled with tensile loading and unloading tests (see Figure 2).

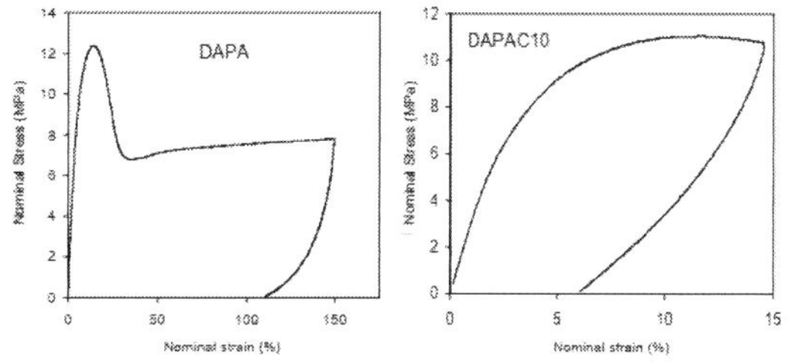

Figure 2 – Loading and unloading curves of DAPA and DAPAC10 at strain rate of $10^{-2}s^{-1}$ and at 298 K

Results and Discussion

Main thermal characteristics of DAPA and DAPAC are summarized in Table 1. The main gaps are observed on T_g values. Increasing amounts of CF results in a significant increase in T_g, from 263K to 270K for DAPA and DAPAC20, respectively. This trend may be explained with the carbonyl (DAPA) and hydroxyl (CF) groups interactions, which reduce the polymer mobility. The DAPA-based biocomposites do not show any significant variation in T_m, in agreement with previous studies based on these cellulose fibres (Averous, 2001).

Table 1 – Main DSC results

Sample	T_g (K)	T_m (°C)
DAPA	263	354
DAPA5%	269	353
DAPA10%	268	354
DAPA-15%	272	353
DAPA-20%	270	352

The stress-strain response, under tensile loading of DAPA and DAPAC at 298K, represented in Figure 3, showed significant increase in the Young's modulus and yield stress of DAPAC. These latters increase from

181 MPa (without CF) to 707 MPa (with 20%wt of CF) and from 11.4 (without CF) to 19.1 MPa (with 20wt% of CF), respectively. The Young's modulus and yield stress increases are probably due to a good dispersion of the fillers and good interfacial interactions between the fibres and the matrix. Parallel, the decreases of strain at break (from 600% without CF to 10% with 20%wt of CF) are related to the areas of stresses concentration caused by the cellulose fibre. In fact because of the modulus gap between cellulose and the matrix, stresses concentration are located at the fillers/matrix interfaces. Under tensile loading, crazes and cracks can be initiated at those weak points.

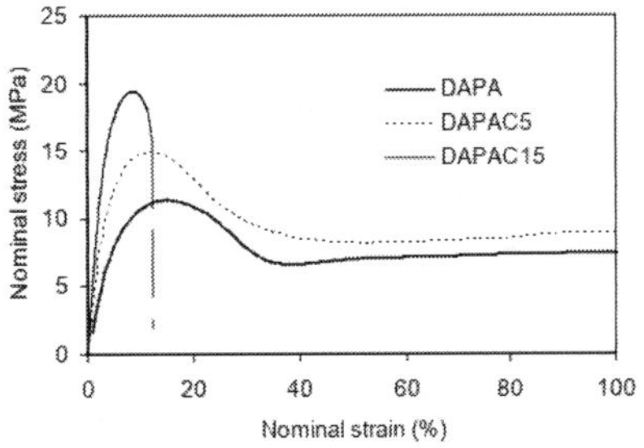

Figure 3 – Stress-strain curves of pure DAPA and DAPAC with different

Young modulus modelling of DAPAC

Considering the DAPA/cellulose composites as a two phase material, the Young's modulus can be estimated by the contribution of the two phases. Models predictions of Young's modulus of DAPAC were investigated using the classical bounds theory of Voigt (Voigt, 1928) (upper bound) and Reuss (Reuss, 1929) (lower bound) and Halpin-Tsai equation (Halpin, 1976), especially appropriated for short fibres biocomposites. The results are presented in Figure 4 (Hablot, 2010a).

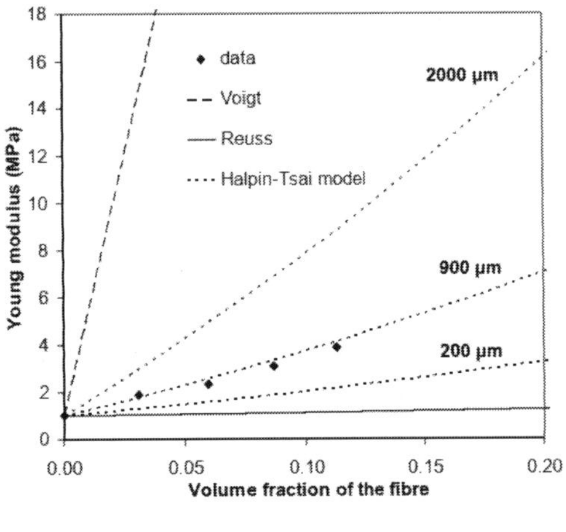

Figure 4 – Experimental modulus values compared to theoretical predictions obtained with different models (Voigt, Reuss, and Halpin-Tsai) (Hablot, 2010a)

As expected the experimental data lie between the upper and the lower bound models. The Voigt bound overestimate the Young's modulus, while Reuss bound underestimate the Young's modulus. But compared to Voigt prediction, the experimental results are closer to Reuss predictions, due to the great contrast between the Young's modulus of pure DAPA and the cellulose. Figure 4 also depicted the Halpin-Tsai model results for different fibre dimensions. One can show that the Halpin-Tsai model provide a good fit of experimental data for a fibre dimension of 900 µm long. These different models assumed that the interface is an ideal attachment without slip or chemical reactions, and the fibres are well dispersed in the matrix. But the studied systems don't show a perfect adhesion between the DAPA matrix and the cellulose fibres. The best fit of experimental results suggests that the enhancement of the Young's modulus is more related to the good fibres dispersion than strong interactions between the fibres and DAPA matrix. But this model couldn't be a perfect model for the full description of complex physics and chemistry controlling the behaviour of DAPA/cellulose composite. But the Halpin-Tsai model helped to obtain a good estimation of stiffness in the case of short fibre composites (Tucker, 1999).

Yield stress modelling of DAPAC

The temperature and strain rate dependence of yield behaviour of various polymers can be described by the following cooperative model of Richeton (Richeton, 2005; Bauwens-Crowet, 1969; Robertson, 1966; Roetling, 1965) where the yield stress σ_y is expressed below T_g in the Equation 1.

$$\frac{\sigma_y}{T} = \frac{\sigma_i(0) - m \cdot T}{T} + \frac{2k}{V_{eff}} \sinh^{-1}\left(\frac{\dot{\varepsilon}}{\dot{\varepsilon}_0 \exp\left(-\frac{\Delta H_{eff}}{kT}\right)}\right)^{1/n} \quad (1)$$

T is the absolute temperature, k the Boltzmann constant, V_{eff} the effective activation volume, ΔH_{eff} the effective activation energy, $\dot{\varepsilon}$ the strain rate, $\dot{\varepsilon}_0$ a pre-exponential constant, $\sigma_i(0)$ the internal stress at 0 K. m is a material parameter equal to $\sigma_i(0)/T^*$, T^* being the compensation temperature and n describes the cooperative character of the yield process (Richeton, 2005).

By several calculations, the model parameters of DAPA and DAPAC were determined and are presented in Table 2 (Hablot, 2010b). The values reported mentioned a slight increase of the activation energy with the cellulose content, and a slight decrease of the activation volume with increasing cellulose content. These results are in agreement with the experimental results where the yield stress increases with the cellulose concentration.

On a previous publication, we have shown that the T_g increases with the fibres addition (Amash, 2000). this observation correlated to the increase of Yield Stress validate the fact that the addition of cellulose fibres leads to a reduction of the DAPA mobility in agreement with the behaviour of most of the cellulose-based composites (Bledzki, 1999).

Table 2 – Variation of effective parameters with the cellulose content

Cellulose content (wt %)	Volume fraction	ΔH_{eff} (kJ/mol)	V_{eff} (m^3)
0	0	139	2.50E-27
5	3.17E-3	144	2.32E-27
15	9.55E-3	154	2.25E-27

The cooperative model predictions of the yield stress are in good agreement with the experimental results.

Conclusions

This study has stressed the very interesting properties of bio-based polyamides from rapeseed oil based-dimer acid (DAPA) reinforced with cellulose fibres. Tensile test have been carried out on DAPA and DAPA/Cellulose composites. The experimental results have shown that the cellulose addition in DAPA polymer matrix led to a high increase of its T_g and Young's modulus and yield stress, mainly due carbonyl (DAPA) and hydroxyl (CF) groups' interactions. Good description of the experimental data was obtained with adapted models. In terms of tensile tests results, these biocomposites show similar behaviour to conventional reinforced thermoplatics, with a high increase of Young modulus and a drastic decrease of the elongation at break. Halpin-Tsai model seems to be an accurate and simple answer to evaluate the elastic modulus evolution. Considering the yield stress, the cooperative model of Richeton showed that the activation energy increased slightly with the cellulose content and the activation volume decrease with increasing cellulose content. As a perspective, the cooperative model used in this work could be implemented in the formulation of large deformation of DAPAC or others cellulose composite materials.
The association of this innovative bio-based polymer with cellulose fillers is a good answer to overcome primary issues with green polymers, i.e. the material cost and the mechanical properties. Then, the association we have elaborated can be used in different fields such as automobile or coating industry. The key-point is that these materials are in very good agreement with the concept of sustainable development.

References

Amash, A., Zugenmaier, P. "Morphology and properties of isotropic and oriented samples of cellulose fibre-polypropylene composites", Polymer, **41**(4), 1589-1596 (2000)

Averous, L. et al. "Plasticized starch-cellulose interactions in polysaccharide composites", Polymer, **42**(15), 6565-6572 (2001)

Bauwens-Crowet, C. et al. "Tensile yield-stress behavior of glassy polymers" J. Poly. Sci. : Part A-2, **7**, 735-742 (1969).

Biermann, U. et al. "New syntheses with oils and fats as renewable raw materials for the chemical industry", Angew. Chem. In. Ed., **39**(13), 2207-2224 (2000)

Bledzki, A. K., Gassan, J. Composites reinforced with cellulose based fibres. Progr. Polym. Sci., **24**(2), 221-274 (1999)

Brandrup, J. et al.: Polymer Handbook, 4th Edition, Wiley-Interscience. New York (1999)

Cavus, S., Gurkaynak, M. A. "Influence of monofunctional reactants on the physical properties of dimer acid-based polyamides", Polym. Adv. Technol., **17**(1), 30-36 (2006)

Eichhorn, S. J. et al. "Review: Current international research into cellulosic fibres and composites", J. Mater. Sci., **36**(9), 2107-2131 (2001)

Guner, F. S. et al. "Polymers from triglyceride oils", Prog. Polym. Sci., **31**(7), 633-670 (2006)

Hablot, E. et al. "Polyurethanes Based on Castor Oil: Kinetics, Chemical, Mechanical and Thermal Properties", Macromol. Mater. Eng., **293**(11), 922-929 (2008)

Hablot, E. et al. "Renewable biocomposites of dimer fatty acid-based polyamides with cellulose fibres: Thermal, physical and mechanical properties", Comp. Sci. Technol., **70**(3), 504-509 (2010)

Hablot, E. et al. "Yield behaviour of renewable biocomposites of dimer fatty acid-based polyamides with cellulose fibres", Comp. Sci. Technol., **70**(3), 525-529 (2010)

Halpin, J., Kardos, J. "The Halpin-Tsai equations : a review", Polym. Eng. Sci., **16**(5), 344-352 (1976)

Khot, S. N. et al. "Development and application of triglyceride-based polymers and composites", J. of Appl. Polym. Sci., **82**(3), 703-723 (2001)

Meier, M. A. R. et al. "Plant oil renewable resources as green alternatives in polymer science", Chem. Soc. Rev., **36**(11), 1788-1802. (2007).

Nayak, P. L. "Natural oil-based polymers: Opportunities and challenges", J. Macromol. Sci., Rev. Macromol. Chem. Phys., **C40**(1), 1-21 (2000)

Reuss, A., Angew, Z. "Berechnung der fliessgrenze von Mishkristallen auf Grund der Plastizitatsbeding fur Einkristalle", Appl. Math. Mech., **9**, 49 (1929)

Richeton, J. et al. "A formulation of the cooperative model for the yield stress of amorphous polymers for a wide range of strain rates and temperatures", Polymer, **46**(16), 6035-6043 (2005)

Robertson, R. E. "Theory for the plasticity of glassy polymers", J. Chem. Phys., **44**, 3950-3956 (1966)

Roetling, J. A. "Yield stress behaviour of polymethylmethacrylate", Polymer, **6**, 311-317 (1965)

Tucker, C. L., Liang, E. "Stiffness predictions for unidirectional short-fiber composites: Review and evaluation", Compos. Sci. Technol., **59**(5), 655-671 (1999)

Voigt, W.: Lehrbruch des Krystalphysik, Berlin, (1928).

THERMAL STABILIZATION, ENHANCED PROCESSABILITY AND PROPERTIES OF BIODEGRADABLE POLYMERS

Shaked Yanir [(1)], Kenig Samuel [(1)*], Dodiuk Hanna [(1)]

[1]Department of Plastics Engineeing, Shenkar College of Engineering and Design, Ramat-Gan, Israel. samkenig@shenkar.ac.il

The use of biodegradable polymers, which are based on aliphatic polyesters, is limited due to their thermal degradation and low re-crystallization rate from the melt during processing. Moreover, these aliphatic polyesters have inherent low toughness as the result of large and non-uniform crystal stracture. To remedy these shortcomings the effects of hyperbranched polyesters (HBP) and plasticizers were studied in the case of the biodegradable poly-hydroxyalkaonoates (PHA). Experimental results have shown that a synergistic effect takes place when a liquid HBP has been combined with a plasticizer based on citrate oil (CTP) having a combined content of only 5%. In the optimized case the molecular weight of the processed PHA containing the HBP and the CTP increased by close to 20% compared to the processed neat PHA. This was attributed to the reduction in viscous heating of the composite phase separated melt. Furthermore, the re-crystallization rate constant increased by 200 fold while the total crystallinity was raised from 20% for the neat PHA to more than 40% of the composite PHA while crystallization temperature increased from 65°C to 90°C, respectively. This was attributed to heterogeneous nucleation of the HBP and CTP phase. As for the mechanical properties the elongation to break was enhanced from the low level of 12% for the neat PHA to the level of 26% in the case of the composite PHA. This increase in toughness was accompanied with a moderate decrease of the modulus from 740MPa to 570MPa. The formation of a separated phase containing HBP and CTP and its effect on the plasticization, nucleation and mechanical properties were studied and verified by analytical means.

Introduction

Poly-Hydroxyalkanoates, PHAs, are considered as one of the most promising technologies of bio-plastics. PHAs are thermoplastic, semi-crystalline, aliphatic polyesters produced by bacteria and fully biodegradable [Lemoigne, 1925]. Poly-Hydroxy-Butyrate, PHB, a common PHA, and its copolymers, are bio-plastics comparable to Polypropylene, Polyethylene, and even Polystyrene. However, with no modifications PHB possess inherent weaknesses like: limited thermal stability, slow recrystallization and, poor toughness that limit its use in many applications were biodegradation and sustainability can be an advantage.

PHB tends to recrystallize slowly from the melt, mainly because of low nucleation density rather than slow crystal growth rate [Withey and Hay,1999, Kai and Inoue, 2005] PHB recrystallization is strongly dependent on temperature [Braham et al, 1984]. Furthermore, when melt processed above its melting point PHB thermally degrades through a random chain scission reaction. The degradation proceeds by forming a six-membered ring intermediate, leading to chain scission through a β-elimination process. As the processing temperature increases the rate of degradation increases exponentially, following an Arrhenius type behavior [Grassie and Murray, 1984]. Furthermore, under extensive shear, local viscous heating can initiate degradation as well. When exposed to temperatures just above its melting temperature, the molecular weight of

Limited thermal stability combined with prolonged exposure to heat, as a result of the relatively slow crystallization rate, lead to enhanced loss in molecular weight of the PHB. The loss in molecular weight results in low melt strength and inferior solid mechanical properties [Holmes 1985, Melik and Schechtman, 1995,]. In previous work [Shaked et al 2008, 2009] had indicated the effect of nano sized modifiers on the rheological and thermal behavior of PHB. Modifiers, such as Hyperbranched polymers (HBP) and Poly-Oligomeric-Silsesquioxane (POSS) were showed to accelerate nucleation and improve thermal stability of PHB.

In this work, P3HB-co-4HB copolymer modified with liquid HBP was studied with respect to its thermal stability, recrystallization from the melt, and mechanical properties.

Experimental

Materials and Preparations

Liquid HBP (Perstorp Boltorn, H2004, Sweden), was studied as nano modifier for P3HB-co-4HB copolymer, (Telles, Metabolix, USA). Structures of PHB copolymer and HBP are described in **Figures 1&2**. HBP was compounded with PHB using a counter rotating intermeshing twin screw extruder (Brabender®) in concentrations of up to 5wt% of HBP in PHB. The PHB polymer was in the form of dense powder and was dried prior to compounding in a desiccant drier at

80°C over night. HBP was in a form of a viscous paste and used as received.

HBP was introduced into the PHB in two different ways. First, in a binary blend consisting of PHB and HBP. Second, in ternary blends, where the HBP was first dissolved in Citrate oil plasticizer (CTP) and then introduced into the PHB. Control blends of PHB and CTP were also prepared for comparison.

A reverse temperature profile of 185°C 185°C 165°C was used for PHB-HBP compounding. Screw speed was kept constant for all compounds.

Table 1, the compositions used in this study.

COMPOUND	*MODIFIER*	*MODIFIER WT%*
PHB-0-0	-	0
PHB-H2004-1	HBP-H2004	1
PHB-H2004-5	HBP-H2004	5
PHB-CTP-1	Citrate Oil	1
PHB-CTP-5	Citrate Oil	5
PHB-H2004CTP-14	HBP-H2004	1
	Citrate Oil	4
PHB-H2004CTP-23	HBP-H2004	2
	Citrate Oil	3
PHB-H2004CTP-32	HBP-H2004	3
	Citrate Oil	2
PHB-H2004CTP-41	HBP-H2004	4
	Citrate Oil	1

Analysis

Isothermal and non-isothermal crystallization behavior were examined using Differential Scanning Calorimeter (DSC Q200, TA Instruments). DSC results were fitted to the Avrami crystallization model. Molecular weights were obtained from Gel Permeation Chromatography (GPC) using size exclusion, refractive index in chloroform, relative to a standard polystyrene. Tensile mechanical properties were evaluated using extensometer (Instron IX).

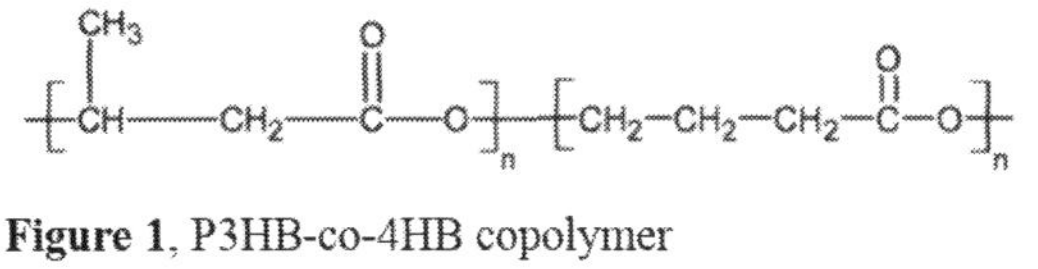

Figure 1, P3HB-co-4HB copolymer

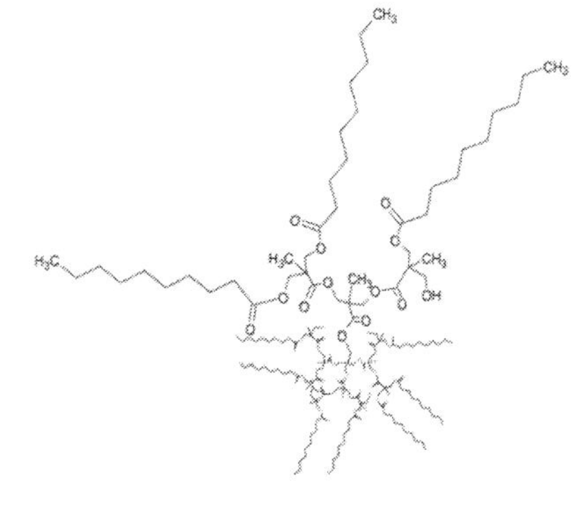

Figure 2, Modified 2nd generation, liquid, Hyperbranched Polyester. A single branch is enlarged for clarification

Results and Discussion

Figure 3, is showing the relative molecular weight retained after processing of the different PHB-HBP compositions. One can see that all compounds exhibited better molecular weight retention, compared to PHB. Whilst the combination of 4% HBP dissolved in 1% CTP lead to a molecular weight retention, higher by close to 20% compared to the neat PHB. This can be attributed to the reduction of viscous heating as a result of the lubricating effect of the vicous liquid HBP and plasticizer.

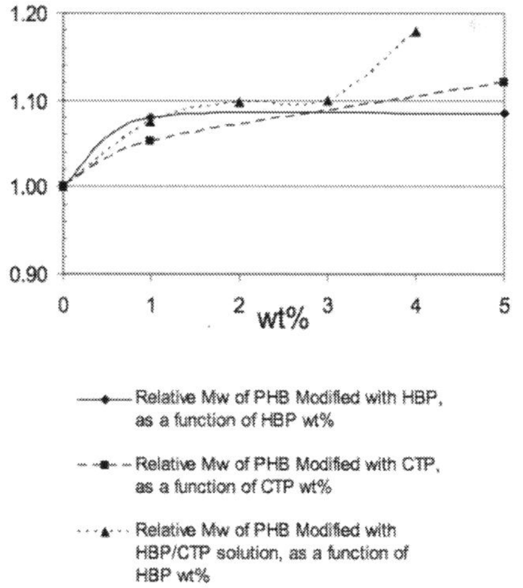

Figure 3, Relative Mw after processing of PHB compounds

Significant increase in the rate of recrystallization was observed with the addition of HBP. It can be seen in **Figure 4**, that 5wt% of HBP and 3wt% HBP carried by 2wt% of CTP lead to almost 2 orders of magnitude increase in the overall rate of recrystallization. 4wt%

of HBP carried by 1wt%, co-modifier (CTP) led to an approximate ~200 fold increase of the recrystallization rate. The "S" shape of the curve describing the rate of recrystallization of PHB modified by a solution of HBP and CTP, may offer explanation to the contribution of each component to the nucleation process. In solutions consisting of less HBP than CTP, the effect of HBP as a nucleating agent is suppressed by the co-modifier, CTP. Once the HBP becomes the major modifier (3wt% and up) an enhanced nucleation effect is observed. The Citrate oil, lubricates and eases the mobility of the HBP, and weakens the intermolecular forces between the HBP molecules. As a result, HBP polar groups are more prone to interact with the polyester chain, thus initiating crystallization at a higher rate than HBP alone.

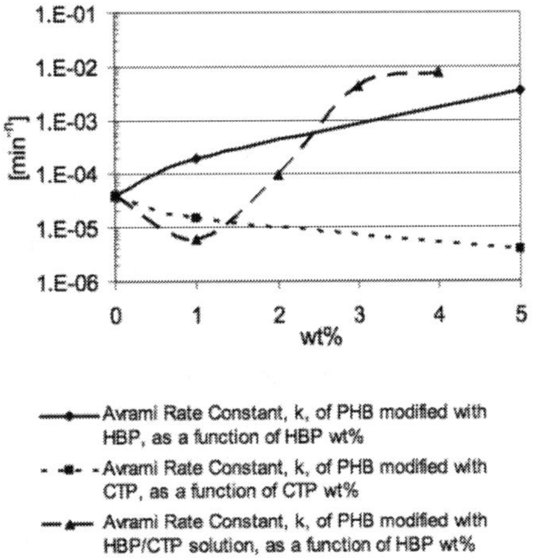

Figure 4, Overall Avrami rate constant, k, for recrystallization of the modified PHB compounds. Under Isothermal conditions @ 120°C

The crystallization temperature of the modified PHB under non-isothermal conditions also increased with the addition of HBP, **Figure 5**. The addition of CTP slows down the crystallization, and decreases Tc, while HBP increases Tc when present as the only modifier or as a major component carried by the co-modifier, CTP. The "S" shape change in Tc is further manifested in the degree of crystallization results, **Figure 6**. The correlation between the rate of recrystallization, Tc, and the degree of crystallinity is well established. Adding the solution of HBP/CTP into the PHB, where HBP is the major component and CTP is the co-modifier, lead to a rate increase of 2 orders of magnitude, 25°C increase in Tc, and almost twice the crystal phase content. Adding the solution of HBP/CTP into the PHB, where HBP is the major component and CTP is the co-modifier, lead to a rate increase of 2 orders of magnitude, 25°C increase in Tc, and almost twice the crystal phase content.

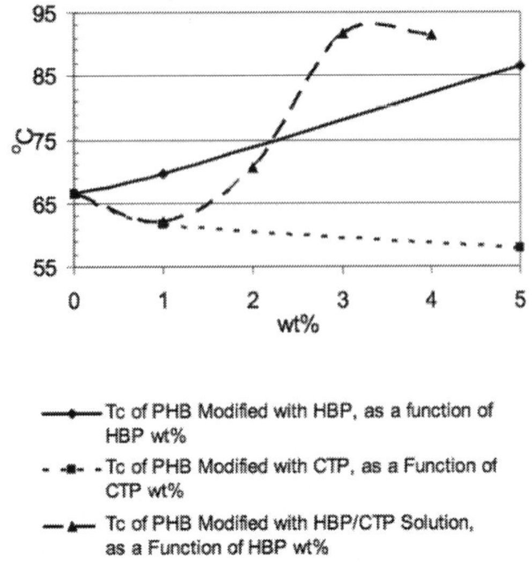

Figure 5, Nonisothermal recrystallization temperature of PHB and modified PHB. Data generated at a cooling rate of 10°/min

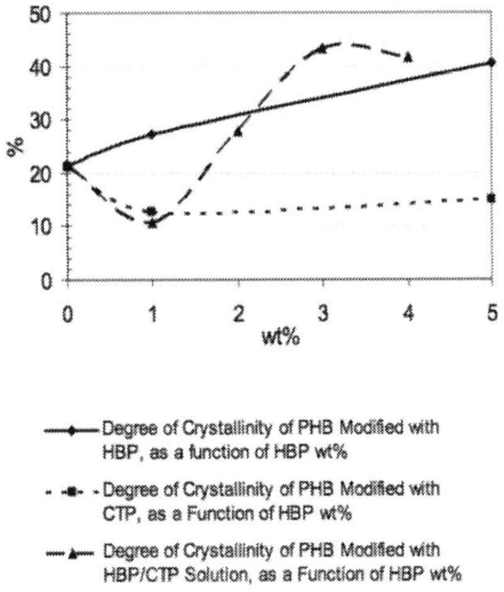

Figure 6, % crystallinity achieved under non-isothermal conditions of PHB modified with HBP, CTP, and HBP/CTP solutions. Cooling rate of 10°/min

Increased nucleation effects may be attributed to strong intermolecular forces between PHB and the HBP due to the high concentration of hydroxyl groups in the outer shell of the HBP. The hydroxyl groups may, form a complex that leads to heterogeneous nucleation. The immobilized chains then start to grow crystals at a high rate. Phase separation of PHB and HBP blends was presented in previous work [Shaked et al, 2009] hence nucleation is thought to be heterogeneous.

The addition of both HBP and CTP lead to a reduction in Young's modulus. When the solution of HBP and CTP was added to PHB, an initial reduction was observed followed by a leveling of the modulus, **Figure 7**. At 1wt% CTP mixed with 4wt% HBP (PHB-H2004CTP-14) the modulus is similar to that obtained with 1wt% CTP alone. Further increase in HBP in CTP results in deviation from the cases of neat CTP and neat HBP. The different trends for the H2004CTP line can be a result of the higher degree of crystallinity. The increase in the degree of crystallinity is not being manifested in Young's modulus for the PHB-H2004 series. This may be due to the heterophasic system obtained comprising of fairly large liquid domain.

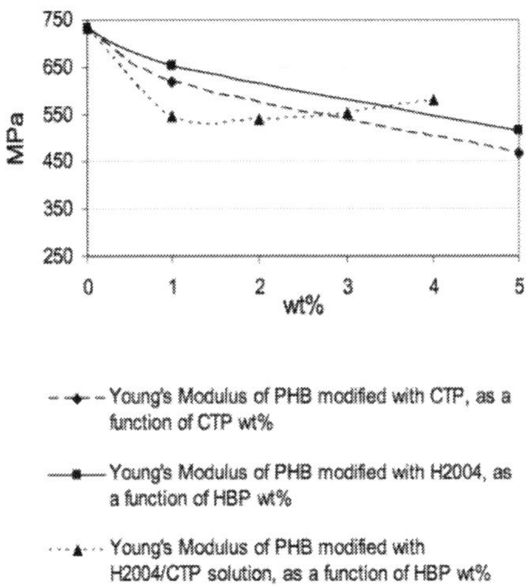

Figure 7, Young's Modulus of PHB modified with HBP, CTP and HBP/CTP solutions

The most significant and important effect on the mechanical properties of the PHB modification is the toughening effect. Blends modified with HBP/CTP solution, rich in HBP, had higher elongation at break compared to PHB modified with either HBP or CTP alone. Maximum performance was achieved with 3wt % HBP and 2wt% CTP, where the elongation at break increased by ~115% and the energy to break by ~106%. **Figure 8,** shows the stress Strain curves of some selected modified PHB blends. A yield behavior appears in the PHB modified with HBP/CTP solution and HBP, while the neat PHB appears to have a brittle behavior.

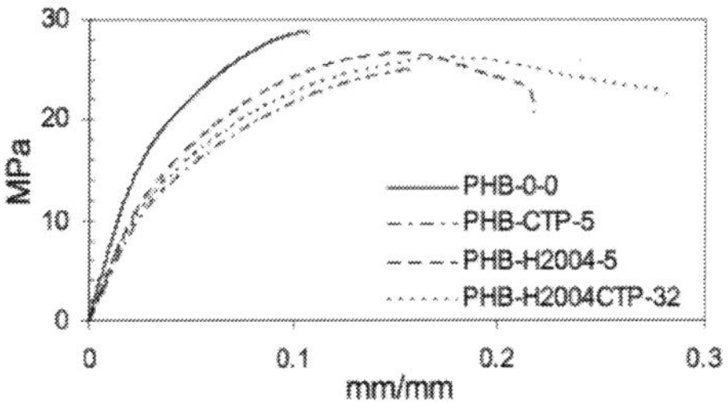

Figure 8, Stress Strain curves of selected PHB compositions

Conclusions

Liquid HBP polyesters and CTP were studied as modifiers for poly Hydroxybutyrate copolymer. All blends retained higher molecular weight during processing, compared to the neat PHB (having the same thermal history). A solution of HBP and CTP at a ratio of 4:1, lead to Mw higher by 18% compared to the neat PHB. It is assumed that reduction in viscous dissipation by plasticization is the dominant factor for the phenomena. The liquid HBP modified PHB had crystallized at a much higher rate under isothermal conditions and at a higher temperature and to a higher degree of crystallinity non-isothermally. Nucleation is attributed to strong intermolecular forces due to high concentration of hydroxyl groups in the outer shell of the HBP. The hydroxyl groups assume to form a complex with the polymer chains that leads to heterogeneous nucleation. Nucleation effect is more profound with the presence of CTP as a co-modifier. Mechanical performance, namely, toughness, of the PHB was greatly improved with the introduction of HBP into PHB.

Liquid HBP accompanied by a plasticizer, were shown to improve some of the major drawbacks in PHB inherent characteristics - thermal instability, slow recrystallization, and brittleness.

References

1. Grassie, N. and E. J. Murray, The Thermal Degradation of Poly(-(D)-Beta-Hydroxybutyric Acid): Part 2 - Changes in Molecular Weigh. Polymer Degradation and Stability, 1984. 6: p. 95-103
2. Holmes, P. A., Applications of PHB - a Microbially Produced Biodegradable Thermoplastic. Physics in Technology, 1985. 16(1): p. 32-36
3. Kai, W., Y. He, and Y. Inoue, Fast Crystallization of Poly(3-Hydroxybutyrate) and Poly(3-Hydroxybutyrate-Co-3-Hydroxyvalerate) with Talc and Boron Nitride as Nucleating Agents. Polymer International, 2005. 54(5): p. 780-789
4. M. Lemoigne, Annales De L'institute Pasteur, 39, (1925)
5. Melik, D. H. and L. A. Schechtman, Biopolyester Melt Behavior by Torque Rheometry. Polymer

Engineering and Science, 1995. 35(22): p. 1795(12).
6. Withey, R. E. and J. N. Hay, The Effect of Seeding on the Crystallisation of Poly(Hydroxybutyrate), and Co-Poly(Hydroxybutyrate-Co-Valerate). Polymer, 1999. 40(18): p. 5147-5152
7. Y. Shaked, H. Dodiuk, S. Kenig, C. Schwier, S. McCarthy, Thermal Stabilization of Bioderadable poly-Hydroxy-Butyrate (PHB) in Melt Extrusion; Part 1: Poly-Oligomerics-Silsesquioxane (POSS). Society of Plastics Engineers, ANTEC 2008, Conference Proceedings
8. Y. Shaked, H. Dodiuk, S. Kenig, S. McCarthy, The Effect of Hyperbranched Polymers on Processing and Thermal Stability of Biodegradable Polyesters. Polymer Engineering and Science, 2009, 49 (3): p.559-566

3D FINITE ELEMENT SOLUTION OF THE FLOW IN SINGLE AND TWIN-SCREW EXTRUDERS

Hétu Jean-François [1]*, Ilinca Florin [2]

[1] *Industrial Materials Institute, National Research Council – Quebec, Canada – jean-francois.hetu@cnrc-nrc.gc.ca*
[2] *Industrial Materials Institute, National Research Council – Quebec, Canada – florin.ilinca@cnrc-nrc.gc.ca*

Abstract - This work is aimed at the numerical modeling the flow inside screw extruders. Important efforts has been devoted during the last decade to the numerical modeling of the injection molding process with more and more emphasis being given to three-dimensional modeling. However, little attention was given to the 3D solution of the flow inside the plasticization unit. Here we present applications of an immersed boundary finite element method that is capable of solving the flow in the presence of complex non-stationary solid boundaries. The method is first validated against the solution obtained on a body-conforming grid for a single screw extruder and then applied to a twin-screw mixer. The time dependent single screw problem can also be solved in a rotating reference frame for which a steady state solution can be obtained. This allows the evaluation of time integration errors of the moving immersed interface algorithm. For instance the flow is considered isothermal and the material behaves as a Generalized Newtonian fluid. Because the viscosity depends on the shear rate, solutions will be shown for various rotation velocities of the screw and compared with solutions obtained on body-conforming grids. The method is shown to give very accurate results and opens the way for in-depth investigation of the material behavior in extruders.

Introduction

In injection molding and extrusion, polymers are plasticized (melted by viscous generated heating) inside screw mixers and then pushed into the cavity or die to form the desired part. Important efforts has been devoted during the last decade to the numerical modeling of the injection molding process with more and more emphasis being given to three-dimensional modeling (Kim and Turng, 2004; Ilinca and Hétu, 2001, 2003). However, little attention was given to the 3D solution of the flow inside the plasticization unit mostly because of the difficulty of the task. A survey of various attempts to solve a 2D analogy of the flow inside twin screw extruders is presented by Kajiwara et al. (1996) and Bravo et al. (2000). An effort in modeling the 2D flow in a single C-shaped chamber of a counterrotating twin screw extruder is reported by Speur et al. (1987). Kajiwara et al. (1996) developed a 3D numerical technique to simulate fluid flow in the case of full flight conveying screw elements. Their simulations were however restricted to one section of the analysis domain, located in the nip region, and relied upon periodic boundary conditions. Bertrand et al. (2003) present a fictitious domain method based on a mesh refinement technique overlapping a single reference mesh. The mesh is locally adapted to enrich the discretization in small gaps and applications are shown for 2D flow simulation in twin-screw extruders. The main hurdle when modeling the flow inside twin-screw extruders is the permanent change in the computational domain determined by the rotation of the screws. Various techniques can be used to deal with the time dependent flow domain: global remeshing, addition/deletion of elements based on a reference mesh, fictitious domain or immersed boundary methods. Remeshing and mesh adaptation techniques are used generally in conjunction with a solution algorithm based on body-conforming grids. While most industrial solvers use body conforming meshes, there is increasing interest for algorithms using non conforming grids. They are known under the names: immersed boundary, immersed interface, embedded mesh and fictitious domain methods. For simplicity, we will use in the present work the immersed boundary (IB) term to identify a non body-conforming method. These techniques all rely on the use of fixed grids with the discretization being performed on a single domain enclosing both fluid and solid regions. It is understood that solid surfaces will intersect the mesh between grid points and that they will not be formally represented by the grid (i.e. by no-slip surfaces at the fluid-solid interface). These methods trade the geometrical complexity of time dependant domains for the simplicity of simpler geometries using relatively uniform meshes. This necessitates abandoning the classical implementation of boundary conditions for a more complex one.

The original IB method was introduced by Peskin (1972). It introduces additional source terms distributed through the solid region to indirectly enforce the proper boundary conditions on the solid-fluid interface. IB methods received particular attention in recent years. Their applications cover a broad spectrum of fluid dynamics problems from flow through heart devices (Peskin, 1972, 1977), to the interaction of body movement and fluid flow (Mittal and Iaccarino, 2005), modeling anguilliform swimming (Tyson et al., 2008) and 3-D parachute simulation (Kim and Peskin, 2009). Reviews of IB methods may be found in (Mittal and Iaccarino, 2005) and (Margnat and Morinière, 2009). Most of these developments were made using a finite difference framework. IB methods using finite elements were developed more recently (Glowinski et al., 1994) and an overview can be found in (Ilinca and Hétu, 2010a).

In most IB methods, boundary conditions on immersed surfaces are handled either accurately by using dynamic data structures to add/remove grid points as needed or in an approximate way by imposing the boundary conditions to the grid point closest to the surface or through least-squares. Using a combinaison of local modifications of element basis functions and element level condensation of added degrees of freedom, our recently proposed approach (Ilinca and Hétu, 2010a) achieves the level of accuracy of cut cell dynamic node addition techniques without their drawbacks (increased CPU time and costly dynamic data structures). This approach was extended to moving fluid/solid interfaces in (Ilinca and Hétu, 2010b).

In this paper, the isothermal flow inside twin-screw extruders is solved using the 3D finite element IB method developed in (Ilinca and Hétu, 2010a, 2010b). The solution is interpolated using linear elements and time integration is done by an implicit Euler scheme. The immersed boundary is represented using a time dependent level-set function using the same linear interpolation functions that are used to solve the flow problem. For low Reynolds number, isothermal, viscous flows, the solution does not depend upon time and the proposed approach can be verified against solutions on body-conforming grids. For high velocity flows, in which convection transport is important, the proposed IB method is shown to predict adequately the flow structures produced by the continuous repositioning of the screw surfaces.

The paper is organized as follows. First the model problem and the associated finite element formulation is presented. The IB formulation and the procedure to impose the boundary conditions on moving solid surfaces are discussed briefly in section 3. Section 4 illustrates the performance of the present IB method for the 3-D flow inside screw extruders. The paper ends with conclusions.

Model Equations

Flow equations

The flow of viscous incompressible fluids is described by the Navier-Stokes equations

$$\rho\left(\frac{\partial \mathbf{u}}{\partial t}+\mathbf{u}\cdot\nabla\mathbf{u}\right)=-\nabla p+\nabla\cdot\left[\eta(\nabla\mathbf{u}+\nabla\mathbf{u}^T)\right]+\mathbf{f},$$

$$-\nabla\cdot\mathbf{u}=0,$$

where ρ is the fluid density, t denotes time, u is the velocity, p is the pressure, is the dynamic viscosity and f is a volumetric force vector.
The interface between the fluid and solid regions is defined using a level-set function ψ, defined as a signed distance function from the immersed interface:

$$\psi(\mathbf{x})=\begin{cases} d(\mathbf{x},\mathbf{x}_i), & \mathbf{x} \text{ in the fluid region,}\\ 0, & \mathbf{x} \text{ on the fluid/solid interface,}\\ -d(\mathbf{x},\mathbf{x}_i), & \mathbf{x} \text{ in the solid region,}\end{cases}$$

where $d(\mathbf{x},\mathbf{x}_i)$ is the distance between the point P($\mathbf{x}$) and the fluid/solid interface. Hence, points in the fluid region will take on positive values of ψ, whereas ψ for points in the solid region will be negative.

Boundary conditions

The boundary conditions associated to the momentum-continuity equations are

$$\mathbf{u}=\mathbf{U}_d(\mathbf{x}), \quad \text{for } \mathbf{x}\in\Gamma_D,$$

$$\eta(\nabla\mathbf{u}+\nabla\mathbf{u}^T)\cdot\hat{\mathbf{n}}-p\hat{\mathbf{n}}=\mathbf{t}(\mathbf{x}), \quad \text{for } \mathbf{x}\in\Gamma_t,$$

where Γ_D is the portion of the fluid boundary where Dirichlet conditions are imposed, and **t** is the traction imposed on the remaining fluid boundary. Dirichlet boundary conditions are also imposed at the interface between fluid and solid regions. Because this interface is not represented by the finite element discretization, a special procedure is used to enforce velocity boundary conditions on this surface. The approach is discussed in the following section on the Immersed Boundary Method.

The finite element formulation

In this work, the momentum-continuity equations are discretized using a GLS (Galerkin Least-Squares) method with linear continuous shape functions for both velocity and pressure. See Ilinca et al., 1997 for further details.

The Immersed Boundary Method

In the proposed approach, the computational domain is, at least, the volume in which the fluid may flow for all times within the studied system. It includes moving mechanical parts that are considered a solid region as it often simplifies mesh generation. The mesh is intersected by the fluid/solid interface at points located along element edges and we consider those points as additional degrees of freedom in the finite element formulation. A special treatment is then applied to solid region and interface degrees of freedom. In

Figure 1 the mesh is shown by the black continuous lines, the interface is indicated by the discontinuous line and the additional interface nodes are indicated by the square symbols. The proposed technique was implemented for 3-D meshes but for the ease of the presentation only a 2-D case is illustrated in Figure 1.

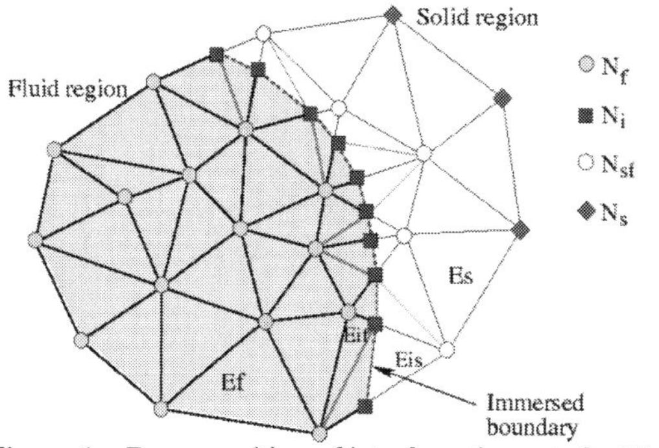

Figure 1 – Decomposition of interface elements for IB method.

The fluid/solid interface is defined by the level-set function and therefore the additional interface nodes are easily determined as being those for which $\psi=0$. Moreover, because the level-set function is interpolated using linear shape functions, the intersection between the interface and a tetrahedral element is a plane, either a triangle or a quadrilateral. Elements cut by the interface are therefore decomposed into sub-elements which are either entirely in the fluid or entirely in the solid region. The additional degrees of freedom associated to the nodes on the interface would normally determine a modification of the global matrix resulting from the finite element equations. In such a case, the implementation is more challenging and would result in an increase in computational cost inherent to dynamic data structures and renumbering. The present approach does not need the explicit addition of the degrees of freedom associated to those nodes. The degrees of freedom associated to the interface nodes are eliminated at the element level by using the Dirichlet boundary conditions for the velocity and by static condensation for the pressure. Hence, the pressure interpolation is taken continuous except for the element faces cut by the interface where it is discontinuous. The procedure results in the local enrichment of the pressure discretization. For more details regarding the implementation of the IB method the reader may consult (Ilinca and Hétu, 2010a, 2010b).

Numerical Application

In this section the IB solution algorithm is used to solve the flow inside extruders. Two distinct flow regimes are investigated. For Stokes flows, characterized by negligible inertia compared to viscous forces, the solution does not depend upon time (i.e., for each position of the screws a steady state is reached). Therefore such analysis can also be conducted on body-conforming grids and the accuracy of the IB method can be assessed against that of a standard finite element method. For convective flow problems (Reynolds number Re > 1) the solution is time dependent and the IB algorithm should consider the continuous change in the screw position.

Steady state flow inside a single screw mixer

This problem is inspired from extrusion and injection molding applications in which plastic pellets are mixed in a screw plasticizing unit prior to being injected into a mold. The flow is solved around a single screw rotating at a constant angular speed ω inside a fixed barrel of radius R. The screw profile, illustrated in Figure 2 has an inner radius $R_i = 0.6R$, outer radius $R_o = 0.98R$ and pitch $L = R$. The profile has a trapezoidal shape with round corners of radius $r = 0.05R$.

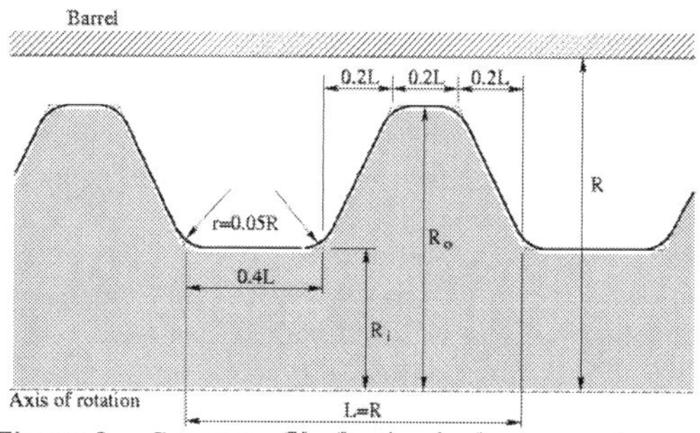

Figure 2 – Screw profile for the single screw mixer.

A non-dimensional solution is obtained by defining reference values for the variables describing the geometry and the flow. The reference length L_0 is taken equal to the barrel radius R and the reference velocity is given by $U_0 = \omega R$. The advection forces are considered very small when compared to the viscous forces and therefore are neglected. Equations corresponding to the Stokes flow are therefore solved in the following dimensionless form

$$0 = -\tilde{\nabla}\tilde{p} + \tilde{\nabla}\cdot\left[\tilde{\eta}(\tilde{\nabla}\tilde{\mathbf{u}} + \tilde{\nabla}\tilde{\mathbf{u}}^T)\right],$$
$$-\tilde{\nabla}\cdot\tilde{\mathbf{u}} = 0,$$

where the dimensionless variables are defined as:

$$\tilde{\nabla} = \frac{\partial}{\partial\tilde{x}}\mathbf{i} + \frac{\partial}{\partial\tilde{y}}\mathbf{j} + \frac{\partial}{\partial\tilde{z}}\mathbf{k},$$
$$\tilde{x} = x / L_0, \quad \tilde{y} = y / L_0, \quad \tilde{z} = z / L_0$$
$$\tilde{\mathbf{u}} = \mathbf{u} / U_0$$
$$\tilde{\eta} = \eta / \eta_0, \qquad \eta_0 = \eta(\gamma_0),$$
$$\gamma_0 = U_0 / L_0$$
$$\tilde{p} = p / p_0, \quad p_0 = \eta_0 U_0 / L_0$$

The material is considered as behaving as a generalized Newtonian fluid with the viscosity described by the Cross model:

$$\eta(\dot{\gamma}) = \frac{\eta_r}{1 + \left(\frac{\eta_r\dot{\gamma}}{\tau^*}\right)^{1-n}}$$

where model constants take on values corresponding to polystyrene at 200°C (Ilinca and Hétu, 2001): $n = 0.26$, $\tau^* = 2.5 \cdot 10^4 Pa$, $\eta_r = 3.841 \cdot 10^3 Pa \cdot s$.

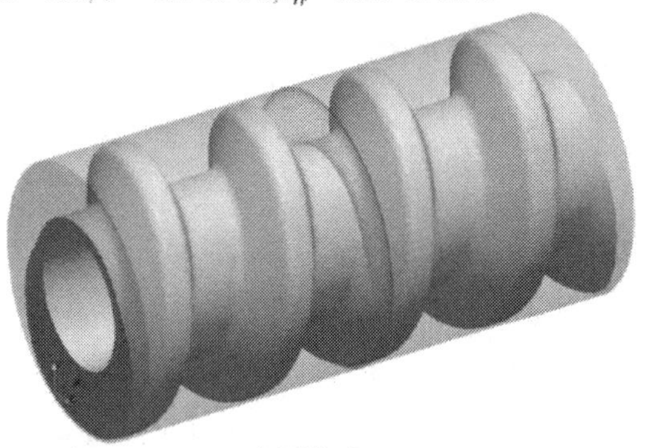

(a) 3D view

Figure 3 – Computational domain and screw geometry.

Numerical simulations were carried out considering a total screw length $L_s = 4L$. The computational domain spans from $x = 0$ to $x = 4L$ and from $r = 0.5R$ to $r = R$, where $r = (y^2+z^2)^{1/2}$. Figure 3 shows the computational domain, with the fluid region shown in transparent gray. The mesh has a total of 344,544 nodes and 1,658,880 tetrahedral elements.

The following boundary conditions are imposed:

- Non-slip boundary conditions on the barrel surface, $u = v = w = 0$;
- Imposed axial velocity $u = 0$ and zero-traction condition for the v and w velocity components at $x = 4L$;
- Zero-traction condition for all velocity components at $x = 0$. This condition will set the zero pressure level on this surface;
- Screw velocity $\mathbf{u} = \boldsymbol{\omega} \times \mathbf{r}$ for points inside the screw. Screw rotation speed is $\boldsymbol{\omega} = -\omega\mathbf{i}$, which gives $u = 0$, $v = \omega z$, $w = -\omega y$.

The apparent velocity of the screw profile is $\mathbf{u}_a = -L\omega / (2\pi)\mathbf{i}$, which means that the profile is moving in the negative direction of the x-axis by a distance equal to the screw pitch L in a time interval equal to the rotation period $T = 2\pi/\omega$.

Computations were carried out using the IB method and also by a standard finite element method on a body conforming mesh having 337,695 nodes and 1,835,873 elements. Velocity solutions for different rotation speeds are compared in Figure 4 for the y-component (v) and in Figure 5 for the z-component (w) in a section at $x = 2L$. In each case the velocity is plotted along the thick black line shown in the left hand side picture. Results indicate that the agreement with the body-conforming solution is very good for all rotation speeds. Note that the solution depends on the rotation speed due to the shear thinning viscosity of the material.

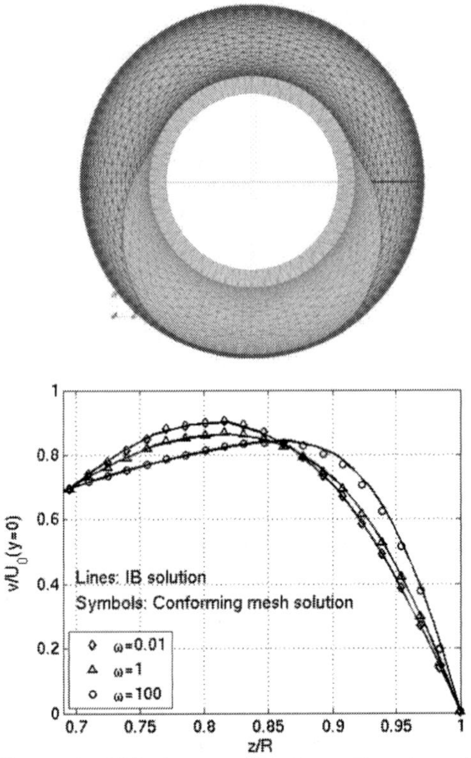

Figure 4 – Velocity v profile at $x = 2L$ and $y = 0$ for three different angular velocities.

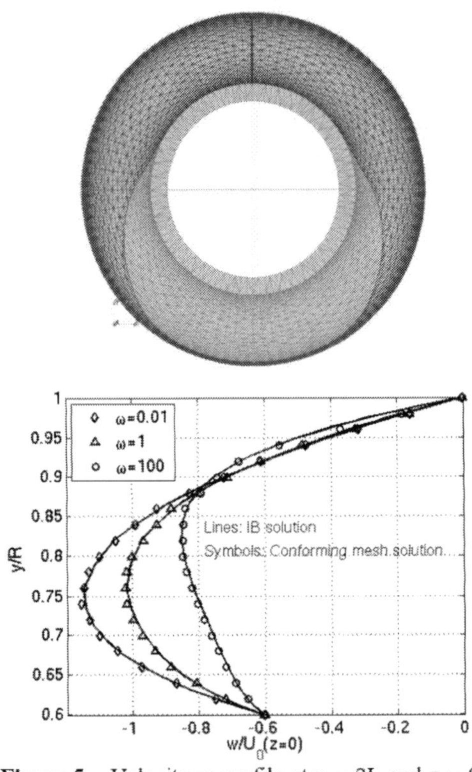

Figure 5 – Velocity w profile at $x = 2L$ and $z = 0$ for three different angular velocities.

The pressure distribution on the surface of the screw for $\omega = 100s^{-1}$ is shown in Figure 6. The dimensionless pressure $\tilde{p} = p / p_0$ is shown. Here again we can observe that the IB solution agrees very well with the solution computed on a body-conforming grid. Because the apparent movement of the screw is towards the left, the pressure is higher on the left side of the screw. The variation along the surface of the barrel at y = 0, z = R of the pressure/velocity ratio for various rotation speeds is shown in Figure 7. The magnitude of the pressure/velocity ratio decreases when the rotation speed increases as a result of a decrease in the viscosity. The dependence of the reference viscosity with the rotation speed is shown in Figure 8(a), while the variation of the mean pressure gradient along the screw axis is shown in Figure 8(b). At low rotation speed ($\omega < 0.01s^{-1}$) the viscosity is practically constant and the dimensionless pressure gradient changes little with the rotation speed. Because the maximum shear rate is about 50 times higher than the rotation speed, $\gamma_{max} = \omega R/(R-R_0) = 50\omega$, shear thinning effects are apparent for $\omega > 0.01s^{-1}$. At very high rotation speeds ($\omega > 1000s^{-1}$) the dimensionless pressure gradient changes very little as the material behaves like a power law fluid. However, because in this range of flow conditions the reference pressure decreases with increasing speed as seen in Figure 8(a), the absolute value of the pressure gradient continues to decrease when the screw rotation speed is increased.

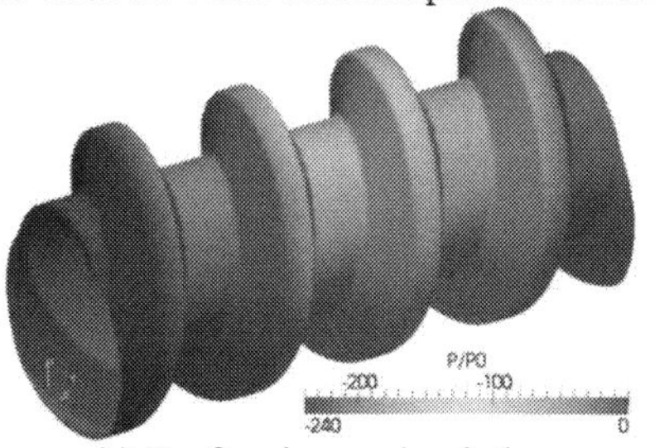

(a) Conforming mesh solution

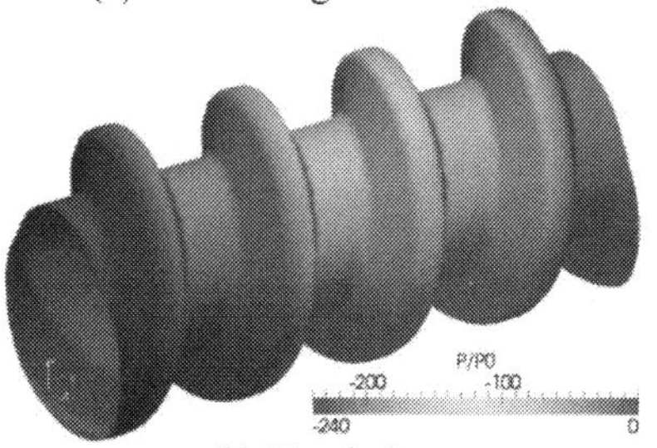

(b) IB solution

Figure 6 – Pressure distribution on the screw surface for $\omega = 100s^{-1}$

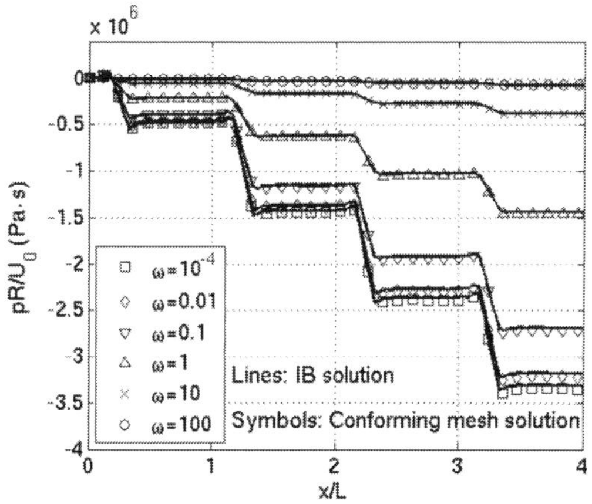

Figure 7 – Pressure along the barrel at y = 0, z = R.

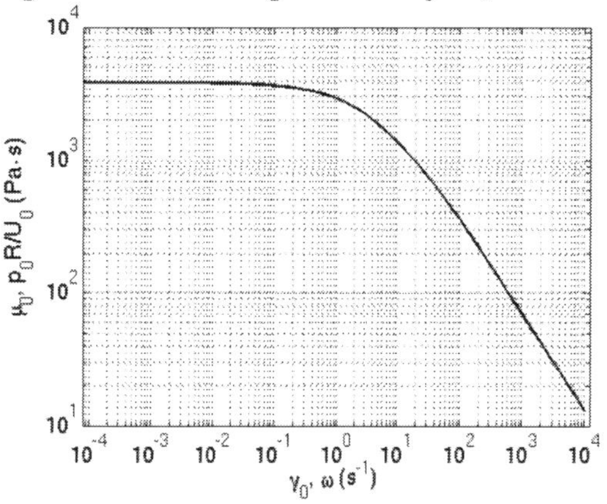

(a) Reference pressure (viscosity)

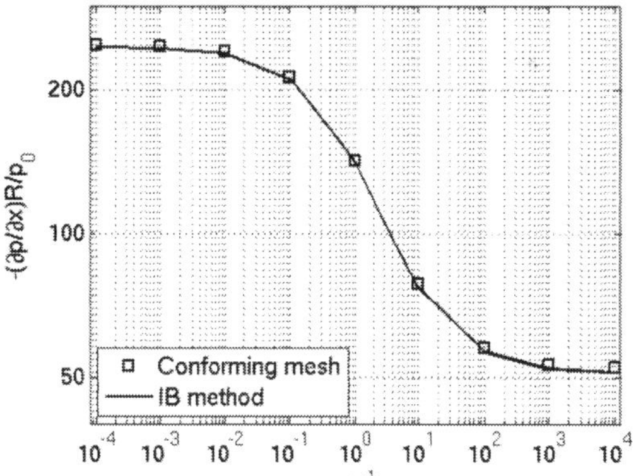

(b) Dimensionless pressure gradient

Figure 8 – Pressure drop as function of the rotation speed.

Steady-state flow inside a co-rotating twin-screw mixer

In this application, we consider a co-rotating twin screw mixing device in which each screw has the same geometry as the one defined in the previous example. This problem can again be solved on a body-conforming mesh in the case of a Stokes flow because the solution is not affected by the inertial terms.

However, as shown for the single screw case, the IB method is general and can be applied to both steady state and transient solutions involving moving fluid/solid interfaces.

Figure 9 – Computational domain for the twin-screw mixer.

The screws are located at a distance $\Delta z = 1.6R$ (measured between their axis of rotation) and the computational domain has a length Ls = 4L. Figure 9 shows the computational domain, with the fluid region shown in transparent gray. A cross-section of the grid at x = 2L is also shown. The mesh has a total of 678,669 nodes and 3,264,000 tetrahedral elements. Both screws rotate in the same direction at angular speed ω. Solutions are computed using the same non-dimensional form of the equations and boundary conditions as for the single screw case.

Figure 10 shows the y-component of the velocity at x = 2L and y = 0, while Figure 11 shows the z-component of the velocity at x = L and z = 0 (see thicker black lines on the left side pictures for the locations where the velocities are plotted). As can be seen, the agreement with the reference solution obtained on a body-conforming grid is very good. Figure 11 indicates that the flow exchange between the two sides (left and right screws) is high and increases as the rotation speed increases. For rotation speeds ω > 1s-1, the maximum velocity along the vertical line at x=2L and z=0 is as high as 6 times the reference velocity. Small differences are observed between the IB and body-conforming grid solutions in the region of high velocity, where the velocity gradient is also highest. This may be due to the fact that the IB grid is slightly more refined in this region of the flow.

The pressure distribution on the surface of the twin-screw extruder for ω = 100s-1 is shown in Figure 12. Here again we can observe that the IB solution agrees very well with the solution computed on a body-conforming grid.

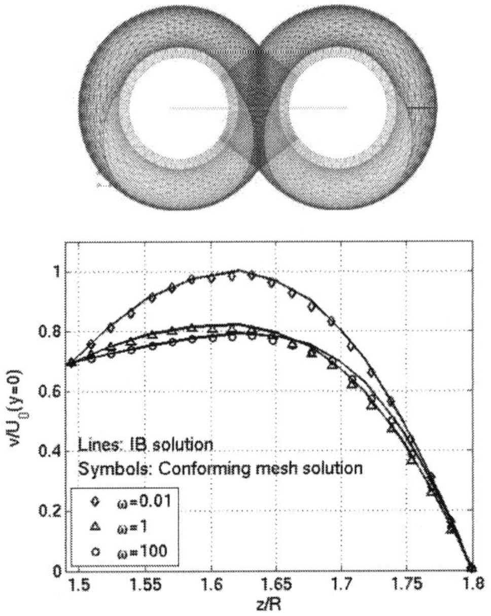

Figure 10 – Velocity v profile at x=2L and y=0 (thick black line).

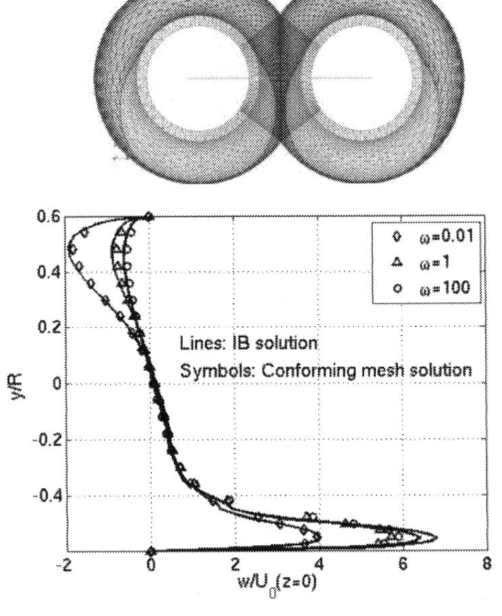

Figure 11 – Velocity w profile at x=2L and z=0 (thick black line).

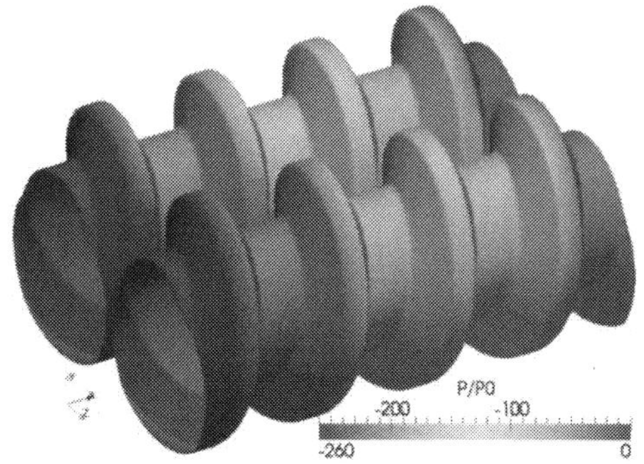

(a) Conforming mesh solution

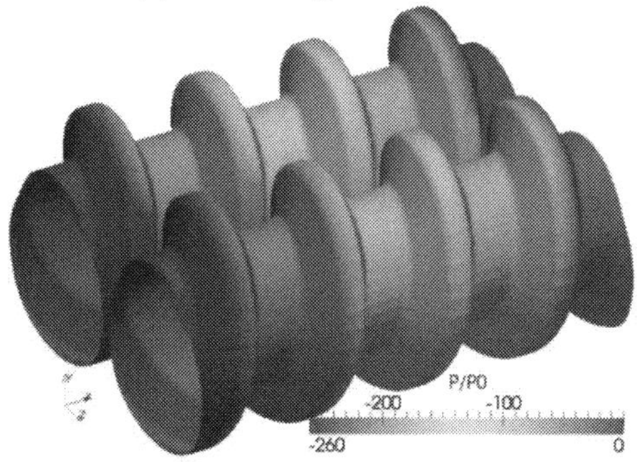

(b) IB solution

Figure 12 – Pressure distribution on the screw surface for ω=100s-1

Transient flow inside a co-rotating twin-screw mixer

The single screw problem was also solved for transient, high velocity conditions for which the convective terms are important. In this case the solution depends with time and simulations were carried out for constant viscosity corresponding to a Reynolds number $Re=\rho U_0 R/\eta=100$. In the presence of moving interfaces, as in this case, the computational grid remains the same during the course of a simulation but the level set function defining the screw geometry is re-evaluated at each time step. That means ψ is a function of time such as the $\psi(x,t) = 0$ iso-surface represents the screw surface at time t. A first computation was carried out using a time step increment $\Delta t = T/40$, with T being the time period of the screw rotation, and then by reducing successively Δt by a factor of two: $\Delta t = T/80$, T/160, T/320 and T/640. This problem cannot be solved on a body-conforming grid without extensive remeshing work and therefore the solution for the smallest time step (T/640) was used as a reference to which the other solutions are compared.

The position of the two screws at different phases of rotation (T/4, T/2, 3T/4 and T) is shown in Figure 13 for x = 2L. The corresponding velocity profile along the horizontal (y = 0) and vertical (z = 0) sections indicated in Figure 13 by the thick black lines is presented in Figure 14. Results correspond to the solution computed with $\Delta t = T/40$. The gap size for the horizontal section (see Figure 14(a)) varies between 0.02R and 0.4R. The fluid velocity is given by the screw velocity at the left hand side extremity of the respective section and is zero at the barrel surface (z = 1.8R). The vertical section, on the barrel symmetry plane (z = 0), spans from y = -0.6R to y = 0.6R, at both extremities the velocity being zero (barrel surface). The highest velocity for this section is observed for t = T/2 and t = 3T/2 when the section is not obstructed by the screws. At t = T/4 and t = 3T/4 part of this vertical section is occupied by the screws and the flow velocities are smaller.

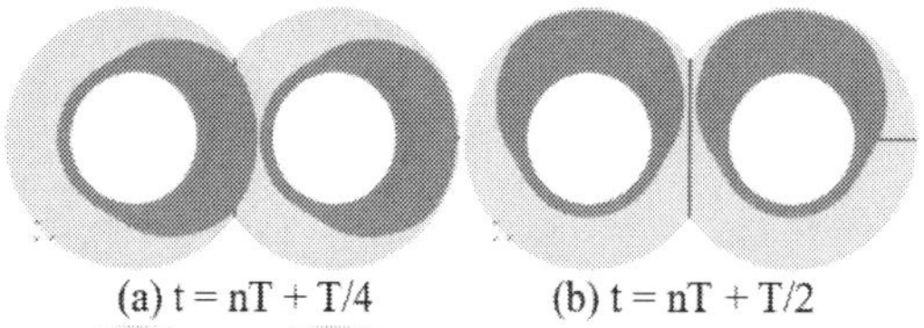

(a) t = nT + T/4 (b) t = nT + T/2

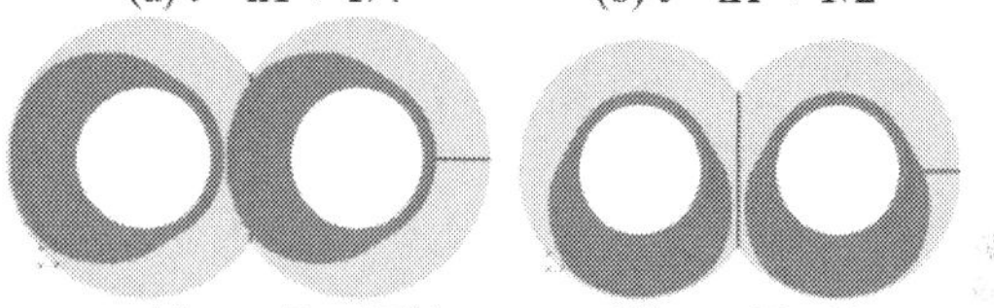

(c) t = nT + 3T/4 (d) t = nT + T

Figure 13 – Screws position at different times.

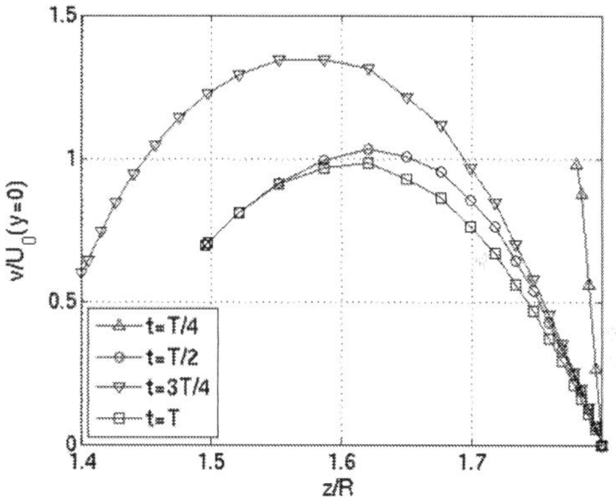

(a) v-velocity at y = 0

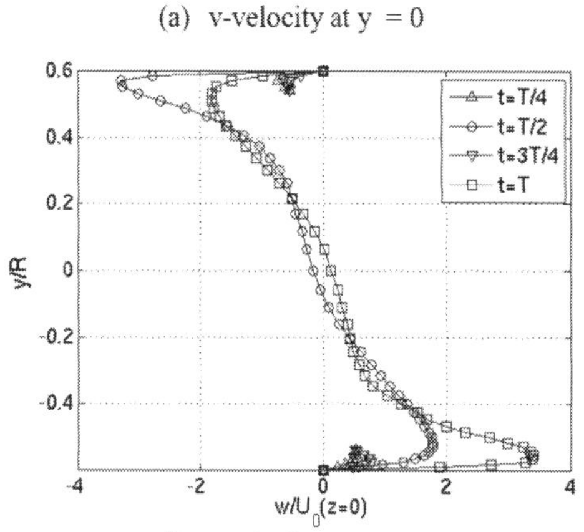

(b) w-velocity at z = 0

Figure 14 – Velocity profile for various screw positions

The IB solutions at t = (n+1)T (screws position as shown in Figure 13(d)) for different time step increments are presented in Figure 15 and Figure 16. Figure 15 shows the y-component of the velocity (a) and the error with respect to the reference solution (b) at x = 2L and y = 0. Figure 16 shows the z-component of the velocity (a) and the error with respect to the reference solution (b) at x = 2L and z = 0. The reference solution is the solution with the smallest time step (Δt = T/640). Results indicate that the solution improves when decreasing the time step increment as expected. This can be also seen in Figure 17 presenting the mean error of the solution for the two cross-sections considered in Figure 15 and Figure 16. The slope of the curve log(Error) - log(Δt) is very close to the theoretical value of 1 of a first-order Euler implicit scheme. This indicates that the IB method describes accurately the change in the flow domain determined by the rotation of the screws and that the procedure for imposing the boundary conditions on the surface of the screw perform well.

Conclusions

An IB finite element method that accurately imposes the boundary conditions on the fluid/solid interface is applied to the solution of single- and twin-screw extruders. The procedure consists of incorporating into the grid the points where the mesh intersects the boundary. The degrees of freedom associated with the additional grid points are then eliminated either because the velocity is known or by static condensation in the case of the pressure.

The three-dimensional IB finite element method was used to solve the flow inside a single-screw mixer and the solution was compared with the one obtained on a body-conforming grid. The IB solution is as accurate as the one computed on a body conforming grid. For a non-Newtonian fluid, the pressure drop along the screw axis depends on the rotation speed. Significant changes were also observed in the velocity distribution when changing the rotation speed.

The flow inside a co-rotating twin-screw extruder is more complex. For a Stokes flow a reference solution can however be computed on a body-conforming grid. The IB method performs again very well and the accuracy is similar to the one observed for the body-conforming grid solution. For transient flow at Re=100 the solution improves when the time step decreases.

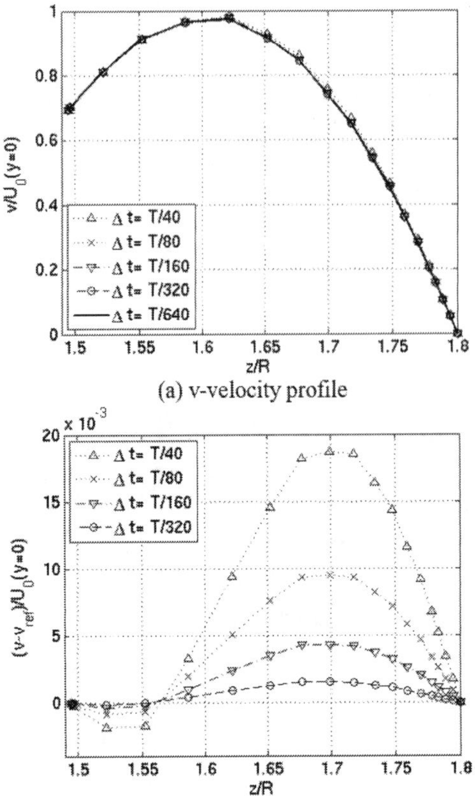

(a) v-velocity profile

(b) Error with respect to the reference solution

Figure 15 – Y Velocity component v at x = 2L, y = 0.

References

1. Bertrand, F., et al., "Adaptive finite element simulations of fluid flow in twin-screw extruders", Comp. Chem. Eng., 27, 491-500 (2003)
2. Bravo, V.L., et al., "Numerical simulation of pressure and velocity profiles in kneading elements of a co-rotating twin screw extruder", Polym. Eng. Sc., 40(2), 525-541 (2000)
3. Franca, L.P. and Frey, S.L., "Stabilized finite element methods: II. The incompressible Navier-Stokes equations", Comp. Methods Appl. Mech. Eng., 99(2-3), 209-233 (1992)
4. Glowinski, et al., "A fictitious domain method for external incompressible viscous flows modeled by Navier-Stokes equations", Comput. Methods Appl. Mech. Engrg., 112, 133-148 (1994)
5. Ilinca, F. and Hétu, J.-F., "A finite element immersed boundary method for fluid flow around rigid objects", Int. J. for Num. Methods in Fluids, early view (electronic), (2010a)
6. Ilinca, F. and Hétu, J.-F., "A finite element immersed boundary method for fluid flow around moving objects", Computers & Fluids, submitted (2010b)
7. Ilinca, F. and Hétu, J.-F., "Three-dimensional filling and post-filling simulation of polymer injection moulding", Int. Polym. Process., 16, 291-301 (2001)
8. Ilinca, F. and Hétu, J.-F., "Three-dimensional finite element solution of gas-assisted injection moulding", Int. J. Numer. Methods Eng., 53, 2002-2017 (2003)
9. Ilinca, F., et al., "An adaptive finite element method for a two-equation turbulence model in wall-bounded flows", Int. J. Numer. Methods Fluids, 24, 101-120 (1997)

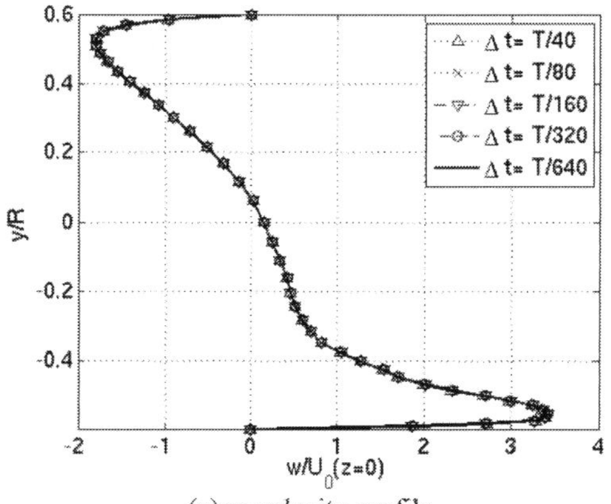

(a) w-velocity profile

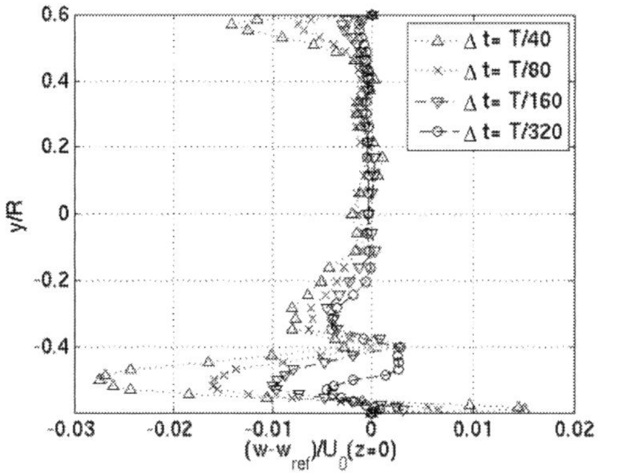

(b) Error with respect to the reference solution

Figure 16 – Z Velocity component w at x = 2L, y = 0.

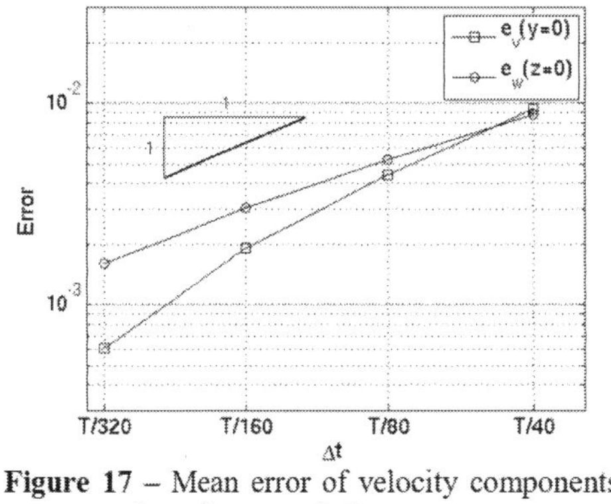

Figure 17 – Mean error of velocity components with respect to the reference solution.

10. Kajiwara, et al., "Numerical study of twin-screw extruders by three-dimensional flow analysis - Development of analysis technique and evaluation of mixing performance for full flight screws", Polym. Eng. Sc., 36(16), 2142-2152 (1996)

11. Kim, S.-W. and Turng, L.-S., "Developments of three-dimensional computer-aided engineering simulation for injection moulding", Modelling Simul. Mater. Sci. Eng., 12, S151-S173 (2004)

12. Kim, Y. and Peskin, C.S., "3-D parachute simulation by the immersed boundary method", Computers & Fluids, 38, 1080-1090 (2009)

13. Margnat, F. and Morinière, V., "Behaviour of an immersed boundary method in unsteady flows over sharp-edged bodies", Computers & Fluids, 38, 1065-1079 (2009)

14. Mittal, R. and Iaccarino, G., "Immersed boundary methods", Annu. Rev. Fluid Mech., 37, 239-261 (2005)

15. Peskin, C.S., "Flow patterns around heart valves: a numerical method", J. Comput. Phys., 10, 252-271 (1972)

16. Peskin, C.S., "Numerical analysis of blood flow in the heart", J. Comput. Phys., 25, 220-252 (1977)

17. Speur, J.A., et al., "Flow patterns in calender gap of a counterrotating twin screw extruder", Adv. Polym. Technology, 7, 39-48 (1987)

18. Tezduyar, T.E., et al., "Incompressible flow using stabilized bilinear and linear equal-order-interpolation velocity-pressure elements." Research Report, Univ. of Minnesota/Supercomputer Inst. (1990)

19. Tyson, R., et al., "Modelling anguilliform swimming at intermediate Reynolds number: A review and a novel extension of immersed boundary method applications." Comput. Methods Appl. Mech. Engrg., 197, 2105-2118 (2008)

PREDICTION OF SWELL PHENOMENA USING ADAPTIVE LOCAL MESH REFINEMENT AND WEIGHT OPTIMIZATION PROCEDURE IN EXTRUSION BLOW MOLDING PROCESS

Z. Benrabah*, F. Thibault, R. DiRaddo

Industrial Materials Institute – NRC – 75 de Mortagne Blvd., Boucherville, Québec J4B 6Y4
zohir.benrabah@cnrc-nrc.gc.ca; francis.thibault@cnrc-nrc.gc.ca; robert.diraddo@cnrc-nrc.gc.ca

Abstract: Extrudate swell is an observable and common phenomenon in the polymer extrusion industry. A better understanding of this phenomenon will ultimately lead to improvements in the prediction of the extrusion process, such as the optimization of both die design and processing parameters. Consequently, this paper presents an adaptive local refinement procedure combined with a robust numerical simulation tool for parison formation and blowing, so as to ensure a more accurate prediction of the entire blow molding process with an optimal computational time. A numerical validation, in terms of thickness distribution of the final blow molded part, is performed. Results obtained using a locally refined mesh are compared to those obtained using a uniform refined mesh. The gain, in terms of part weight and CPU time, using adaptive local mesh refinement and optimization procedures are presented for an automotive plastic fuel tank (PFT) during weight optimization iterations.

Introduction

Extrusion blow molding is extensively used in the manufacturing industry for various commercial applications including complex, hollow, automotive parts. The first stage of this process is parison formation, whereby a molten thermoplastic is extruded from an annular die. The hollow parison is then deformed by applying a pressure during the inflation stage. The part is then cooled and ejected out of the mold. The most critical stage of this process is parison formation, as its shape, diameter and thickness profile strongly affect the final part's thickness distribution and ultimately its mechanical performance. On the other hand, the parison's geometry is governed by the deformation of the molten thermoplastic after exiting the die. In particular, the swell and sag phenomena cause variation in parison dimensions between the time the melt emerges from the die and the moment of inflation. Extrudate swell is qualitatively explained as the release of elastic energy stored in the polymer die flow after its rapid restructuring at the die exit [1]. The sag, however, results from tensile stresses generated by gravity acting on the parison while it hangs from the die. The degree of swell and sag is controlled by die design, resin characteristics, and processing parameters such as melt temperature, suspension time and parison length. In general, due to the viscoelastic fluid behavior of molten polymer, anything that increases the elastic response of the melt will lead to greater swell. Therefore, it is highly important to be able to predict the swell of an annular polymer melt under the effect of sag during extrusion. Two main methods can be identified to achieve this goal: semi-empirical method and numerical simulation method. By using the numerical method, the flow stresses for viscoelastic fluids is computed based on two different approaches as the polymer melt is considered either a Newtonian fluid [2] or non-Newtonian fluid [3]. In the first approach, the flow kinematics is derived using a generalized Newtonian model which is then used as input to calculate the flow stress distribution using an appropriate viscoelastic constitutive equation [4]. In the second approach, namely the direct approach, the viscoelastic behavior of the material is taken into account from the beginning to calculate the flow kinematics and thereafter the corresponding flow stresses [5]. As reported by [6], Baaijens [7] compared these methods and concluded that there was a good agreement between the two, although the former was more cost effective.

The finite element software BlowParison, developed at IMI, is based on the first approach previously discussed. This software has been successfully used in the past to predict the parison formation, accounting for swell, sag and non-isothermal effects, for several industrial parts. It is based on a hybrid approach that couples fluid mechanics to represent the die flow, solid mechanics to represent the parison behavior outside the die, and a phenomenological model to capture the die geometry effect. This approach circumvents instability issues encountered by traditional fluid mechanics alone, especially at high Weissenberg numbers. The fundamental of this approach is largely discussed in the references [8-10].

Membrane theory and material incompressibility assumptions are considered. The die channel and the parison wall are discretized by Constant Strain triangle T3 (CST) plane finite element. Due to strong mechanical and geometrical nonlinearities including instabilities, the quasi-static approach has been adopted and the fully Newton-Raphson incremental procedure is used to solve the fully coupled non linear set of equations raised from both fluid mechanics and solid mechanics.

The work presented in this paper has two aims: first, to integrate a design optimization modulus developed at IMI, BlowDesign, to manipulate processing parameters such as the die gap opening profile and flow rate in order to optimize the parison weight; and second, to integrate an adaptive local refinement procedure in order to ensure a more accurate prediction of the entire blow molding process, with an optimal computational time.

Theoretical Background

The actual swell of an annular parison under the effect of the sag can be described using two swelling expressions, namely, diameter swell (B1) and thickness swell (B2) defined as:

$$B_1 = \frac{D_p}{D_0} \qquad (1)$$

$$B_2 = \frac{h_p}{h_0} \qquad (2)$$

where D_p is the outer parison diameter, D_0 is the outer die diameter, h_p is the parison thickness and h_0 is the die gap [11]. It is well known that the parison swell is a viscoelastic phenomena and the maximum extrudate swell is related to recovery strain [12]. The Weissenberg number is a quantity that is mostly used to represent the recoverable strain and is defined as follows:

$$WE = \frac{Q \cdot \bar{\tau}}{A_0 \cdot h_0} \qquad (3)$$

where Q is the volumetric flow rate, A_0 is the flow channel area at the die exit, and $\bar{\tau}$ is the Maxwell relaxation time represented by the reciprocal of the crossover frequency, where the storage modulus (G') and the loss modulus (G'') coincide in dynamic shear tests (G' = G''), according to the Maxwell viscoelastic model [13].

Kinematical Flow in the Die

The flow inside the die was predicted by using IMI's ParMesh software. This software is based on the generalized Hele-Shaw model developed for lubrication problem for a generalized Newtonian fluid under non isothermal conditions [14]. Assuming Carreau model type behavior, the viscosity of the melt is given by the following expression:

$$\eta = \eta_0 a_T \left[1 + \left(\lambda \cdot a_T \cdot \dot{\gamma}\right)^2\right]^{\frac{n-1}{2}} \qquad (4)$$

where η is the viscosity of the polymer melt at the processing temperature, η_0 is the zero-shear Newtonian viscosity at a reference temperature, $\dot{\gamma}$ is the shear rate, n is the power-law index for the shear-thinning zone, λ is a characteristic time representing the transition between the Newtonian and shear-thinning zones, and a_T is the temperature shift factor according to the WLF shift function [15]:

$$\ln(a_T) = -\frac{C_1(T - T_{ref})}{C_2 + T - T_{ref}} \qquad (5)$$

where C_1 and C_2 are material parameters, T is the fluid temperature, and T_{ref} is a reference temperature.

Initial Stress after Exiting the Die

The overall stress components are evaluated for each element by the integration of the deformation history through the die. However, the entire deformation history of a material element is derived from the results of the previous flow model (i.e., the kinematics and temperature field of the flow). The material elements are then tracked along their specific streamlines to obtain the required strain history. These stresses, which are considered initial stresses, are then removed at the moment the element emerges from the die. Therefore, the stress relaxation of the molten extrudate after exiting the die is, in turn, predicted based on the K-BKZ viscoelastic model [6].

Formulation of the Solid Mechanics Approach

Based on the principal of stationary potential energy, we know that the total energy, π, of a deformable, semi-solid extrudate subjected to external loads, tends to a minimum with regards to the displacement, and is expressed as [16]:

$$\delta\Pi = \delta\Pi_{Int} + \delta W_{ext} = 0 \qquad (6)$$

where $\delta\Pi_{Int}$ is the variation of strain energy of the deformable body and δW_{ext} is the variation of the work of external forces. In the Lagrangian formulation, the variation of the strain energy can be expressed as a function of the 2nd deviatoric Piola-Kirchoff stress tensor S^d and the conjugated green-Lagrange strain tensor E as follows:

$$\delta\Pi_{Int} = \int_{V_0} S^d_{ij} \delta E_{ij} dV \qquad (7)$$

where V_0 is the initial volume of the undeformed body.

For the constitutive behaviour, the viscoelastic integral model of K-BKZ is used to describe the viscoelastic deformations that occur during polymer forming processes; i.e., stress (S^d) – strain (E) relationship of equation (7). This model is well known to adequately represent polymer behaviour at the molten state [17]. It gives the true stresses σ at time t as a function of the strain history according to:

$$\boldsymbol{\sigma}(t) = -p\mathbf{I} + \frac{1}{1-\theta}\int_{-\infty}^{t} M(t-\tau)\cdot h(\mathrm{I}_1,\mathrm{I}_2)\cdot\left[\mathbf{C}^{-1}(t,\tau)+\theta\mathbf{C}(t,\tau)\right]d\tau \quad (8)$$

where p is the hydrostatic pressure, $M(t-\tau)$ is the memory function, $h(I_1,I_2)$ is the damping function , and $C(t,\tau)$ and $C^{-1}(t,\tau)$ are the Cauchy and Finger strain tensors, respectively [17]. The memory function $M(t-\tau)$ is defined by:

$$M(t-\tau) = \sum_k \frac{g_k}{\tau_k}\exp\left(-\frac{t-\tau}{\tau_k}\right) \quad (9)$$

where g_k, τ_k are the materials relaxation modulus and time respectively. However, τ is considered here as pseudo time.

Weight Optimization Procedure

In this section we highlight the main components of the BlowDesign optimization software developed at IMI. The software manipulates the design variables or processing parameters such as die gap opening profiles, die geometry and flowrate, in order to minimize parison weight subject to process constraints such as extrusion time and parison length. At each optimization iteration, BlowDesign manipulates the design variables to minimize the objective function, which is defined as the PFT weight subject to a minimum inflated parison thickness or part thickness constraint (T_{min}). To target this objective, we propose the following optimization scheme [18]:

Objective function:

$$\min F(\mathrm{X}) = weight = \frac{\sum_{i=1}^{nElem} \rho T_i}{nElem} \quad (10)$$

subject to

$$L_p = cte \quad (11a)$$

$$T_i > T_{min} \quad (11b)$$

$$X_{min} < X < X_{max} \quad (11c)$$

by manipulating

$$X = \{\Theta_i(t), \Delta Gap(\theta), Q, P_b\} \quad (11d)$$

where L_p, T_i, $\Delta Gap(\theta)$, Q, P_b and $\Theta_i(t)$ are the targeted parison length , the nodal parison thickness, the circumferential die gap, the flowrate, the pre-blow pressure and the programming point profile respectively. In the plastic fuel tank industry, it is well accepted that the PFT minimum target thickness value should be around 3.0 mm. So, each time the processing parameters are updated, the thickness distribution on the PFT will be modified accordingly.

Refinement criteria and Refinement Procedure

The major problem encountered when linear triangular elements are used in geometrical discretization is that they remain flat through the finite element mesh. Indeed, given the very large deformations that these elements undergo, this does not perfectly simulate the evolution of the parison in high curvature areas of complex parts. To address this issue, a simple refinement criterion coupled with robust and recursive refinement methods is highly desirable [19]. Therefore, for simplicity, we referred to a geometric criterion for deciding whether an element should be subdivided or not. When we decide to remesh the parison, for each k consecutive optimization iteration, each membrane element is tested. To proceed, we first select all the elements sharing a common edge with current element and then the average normal of the elements group is calculated. Thereafter, the angle between the normal of each group element and the average normal of the elements group is considered. If the element angle is greater than a prescribed angle limit, set by the user, the corresponding element is subdivided; otherwise it remains unchanged for subsequent iterations until the next remeshing test. A list of elements that should be subdivided is prepared. This list is sorted in order of increasing length of the longest element edge. The subdivision starts with the last element (i.e. the one with the longest edge), and is continued recursively until the list is empty. The subdivision is carried out by bisecting the longest element edges in the list. The details and criticism of this procedure can be found in [20].

Simulation Results and Discussion

To illustrate the prediction of the swell using the optimization procedure and refinement technique, a Kautex Company PFT tank will be used. Fuel tank walls are typically multilayer (i.e., made up of different material layers), however, as this is not the scope of the present study, only a monolayer HDPE blow-molded tank is considered. Table 1 summarizes the material properties.

Table 1: Relaxation times and relaxation strengths

Maxwell relaxation spectrum	
g_k(Pa)	τ_k (s)
2.08593E-001	7.14000E-003
7.38500E-002	4.99790E-002
3.03100E-002	3.49854E-001
1.06100E-002	2.44898E+000
3.03400E-003	1.71429E+001
1.11900E-003	1.20000E+002
WLF constants	
T_{ref} (°C)	150
C_1	4.063
C_2	106.01

Numerical comparisons, in terms of thickness distribution of the final blow molded part and computational times (CPU), are performed on the two cases studies; one using local mesh refinement and the other using a uniform refined mesh.

The die mesh and the initial parison mesh, without swell and sag effects, for a given set of die gap openings and programming points, are obtained from IMI's ParMesh software. The parison mesh contained 127 600 T3 (CST) elements for the uniform refined case, and 55 929 T3 (CST) elements for the adaptive local refined case. The parison formation, accounting for swell and sag as well as non isothermal effects, was subsequently predicted using the BlowParison software as depicted in Figure 1.

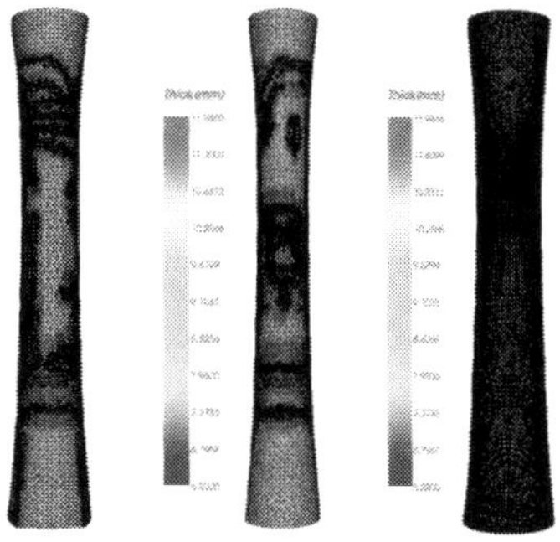

Figure 1: Front and back views of parison with sag and swell effect for the local refined mesh and uniform refined mesh.

It is important to note here, that the axial variation in the thickness profile is obtained after 6 optimization iterations using the BlowDesign Software. At each iteration, BlowDesign manipulates the design variables, or processing parameters, in order to minimize the objective function, which is defined as the PFT weight subject to a minimum part thickness constraint (T_{min}) for a given parison length constraint (L_p) after swell and sag effects. On the other hand, the adaptive refinement procedure is performed at every second consecutive optimization iteration specified by the designer. Therefore, the local refined mesh presented in Figure 1 is obtained after 3 local refinement procedures. As seen from Figure 1, the final parison length (L_p ~ 2100 mm) for the two refined meshes are very similar.

In Figure 2, we illustrate the die gap evolution of the Vertical Wall Distribution System (VWDS) during the optimization iteration in order to minimize the parison thickness subject to a minimum part thickness constraint.

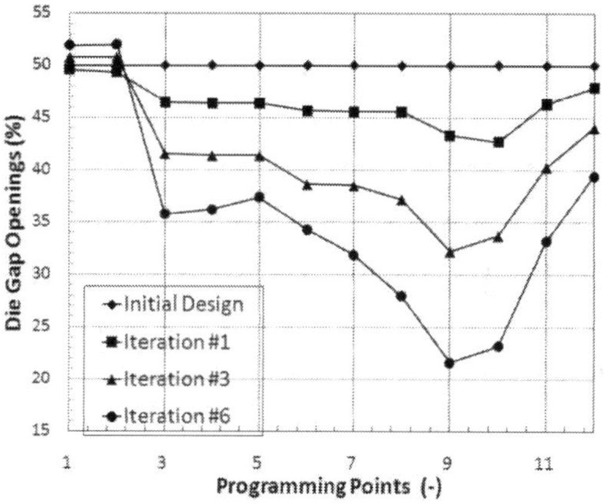

Figure 2: Die gap evolution of the Vertical Wall Distribution System during the optimization iteration.

An overview of the final part thickness distribution, with adaptive local refinement, is presented in Figure 3. It can be seen that the minimum thickness constraint (T_{min} = 3.0 mm) is well respected over the entire PFT wall. In terms of weight, however, the initial design weight has been reduced from 21.7 kg down to 17.6 kg (~ 20% reduction) after 6 optimization iterations.

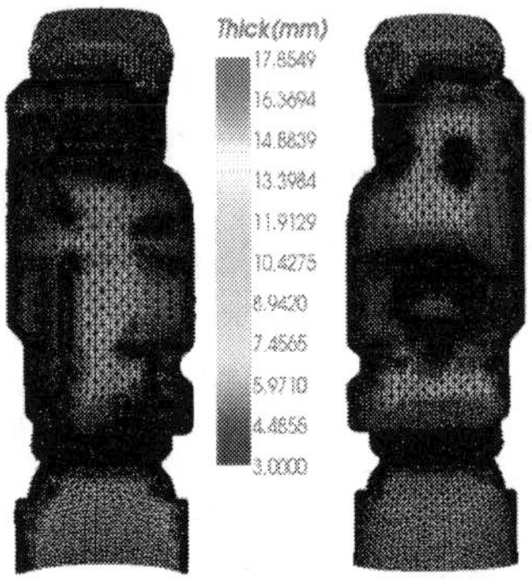

Figure 3: Front and back views of thickness distribution on the inflated final part using the adaptive local refined mesh.

Figures 4a and b compare the simulation thickness distribution of the final blow moulded part, obtained using uniform and locally refined meshes.

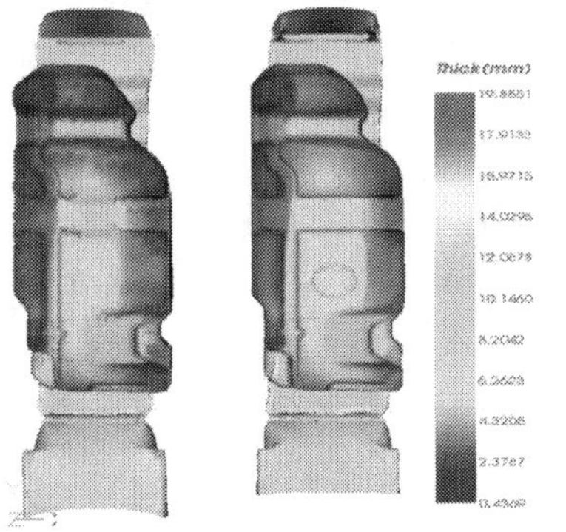

(a)

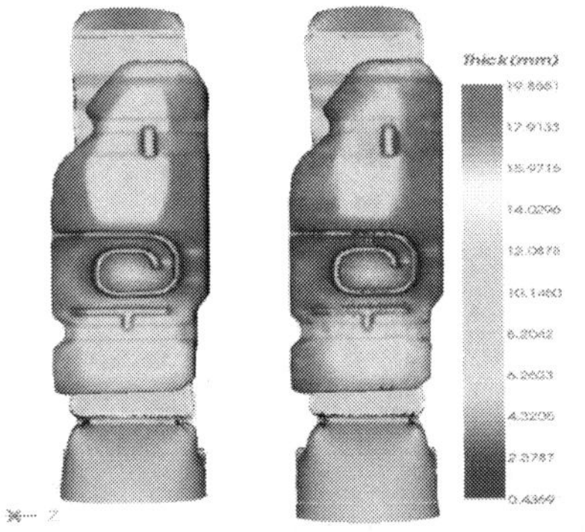

(b)

Figure 4: (a) Front view of inflated final part obtained using the uniform and locally refined meshes respectively. (b) Back view of inflated final part obtained using the uniform and locally refined meshes respectively.

The differences, in terms of thickness distribution for the front and back side, between the two cases studied are presented in Figure 5. As it can be seen, the maximum difference is less than 5 mm and is observed in the high curvature zones for both the front and back side of the PFT. Another area where the difference is important has been observed on the back side of the PFT. This area is flat and meshed with relatively large T3 membrane elements (refer to Figure 3). Hence, the difference in the thickness distribution could be explained by the difference in the element mesh size used in this area (i.e., uniform and locally refined mesh, respectively).

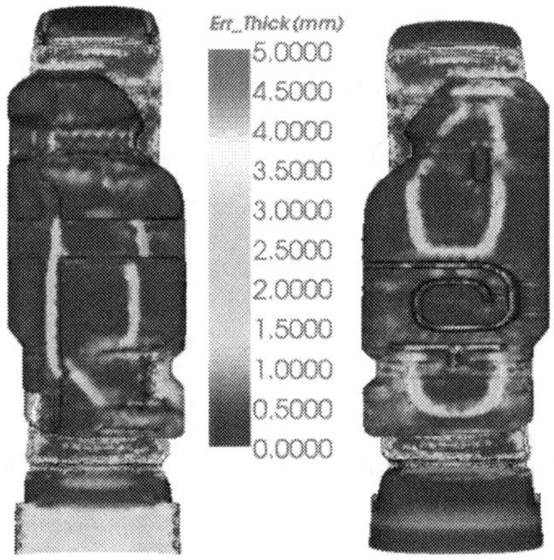

Figure 5: Front and back views of inflated final part thickness distribution difference between the two refined meshes.

The computational times on a single PC with Intel Quad Core Processor Q9450 (2.66 GHz) for the two meshes are summarized in Table 2.

Table 2: The computational times (CPU) in minutes

	Uniform Mesh	Refined Mesh
ParMesh	100	37
BlowParison	172	66
Total	272	103

Given the results shown in Figures 4 and 5, we could say that the simulation results obtained with a local refined mesh are as accurate as those obtained with a uniform refined mesh, but with a fraction (~1/3) of the computational cost.

Conclusion

In this work, an adaptive local refinement procedure was presented, combined with a robust numerical simulation/optimization tool for parison formation and blowing. Its accuracy, in terms of thickness distribution, has been analyzed. It can be concluded that the computational results obtained with a local refined mesh are as accurate as a uniform refined mesh, at a fraction of the computational cost. This work also demonstrated the capability of the modeling tools in predicting the parison shape after emerging from the die as well as the final part thickness distribution while optimizing weight and computational times. Future work will consider the possibility of replacing the 2.5D membrane approach with a 3D prediction of the viscoelastic fluid flow kinematics and parison formation, to better capture the die geometry effect.

Acknowledgment

The authors would like to acknowledge the contribution of IMI's SIGBLOW/SIGFORM group members for their continuous feedback concerning the EBM and Optimization software improvements. Special thanks to Anna Bardetti for her contribution.

References

1. Tanner R.I., Engineering Rheology, Clarendon Press, 1985.
2. Mitsoulis E., Heng, F.L. Rheologica Acta, 1987 (26), 414-417.
3. Seo Y., J. Appl. Polym Sci,1990 (4), 25-3
4. Douven L., Towards the Computation of Properties of Injection Moulded Product, Ph.D. Thesis, Technische Universiteit Eindhoven, Netherlands (1991).
5. Erchiqui F., Godard F., Ben Yedder, J. Polym. Eng., 2004 24 (6), 537-555.
6. Yousefi A. M., Collins P., Chang S., DiRaddo R., Polym. Eng. Sci., 2007 47 (1), 1-13.
7. Baaijens F.T.P, Rheo. Acta, 1991 30 (3), 284-299
8. Yousefi A. M., Doelder J.D., Rainville M.A, Koppi K., Polym. Eng. Sci., 2009 49 (2), 251-263.
9. Yousefi A. M., Atsbha H., Polym. Eng. Sci., 2009 49 (2), 229-239.
10. Yousefi, A.M., A Combined Fluid-Mechanics and Solid-Mechanics for Swell Prediction, Technical Report IMI-2002-93585-CNRC, 2002.
11. Garcia-Rejon A., Dealy J.M., Swell of Extrudate from an Annular Die, Polym. Eng. Sci., 1982 22 (3), 158-165.
12. R.J. Koopmans, *Polym. Eng. Sci.*, **32** , 1750-1754
13. R.J. Koopmans, *Polym. Eng. Sci.*, **32** , 1755-1764
14. Tucker C.L. III, "Computer Modeling for Polymer Processing – Fundamentals", C. H. Verlag Publishers (1989).
15. Williams M.L., Landel R.F., and Ferry J.D., Am. Chem J. Soc., 77, 3701-3711 (1955).
16. Bonet, J., Wood R.D., Nonlinear Continuum Mechanics for Finite Element Analysis, Cambridge University Press, (1997).
17. Laroche D., Kabanemi K., Pecora L., DiRaddo R., Polym. Eng. Sci., 1999 39 (7), 1223-1233.
18. Thibault F, Mir H., Duan Y., Diraddo R., Atshba H., To be submitted to Polym. Eng. Sci.(2010).
19. Nambiar R.V., Valera R. S., Lawrence, K. L., Morgan R.B., Amil, D., Int. J. Num. Meth. Eng., 1993, 36, 499-509.
20. Verron E., Contribution Expérimentale et Numérique aux Procédés de Moulage par soufflage et Thermoformage, Ph.D. Thèses, École Centrale de Nantes, 1997.

Suspension flow modeling for general geometries

*Jeff Morris**

CCNY - NY - USA

A frame-invariant model for the flow of concentrated dispersions is developed. This model develops the nonlinear rheology of the suspension and relates it to the migration of the particles and fluid. The relation to nonequilibrium osmotic pressure concepts will be discussed to provide insight to the fundamental basis for unique suspension behavior. Application of the model to geometries including abrupt contractions or expansions, eccentric cylinder flow, and flow in and past a cavity illustrate the predictions made. These are generally in agreement with existing data but point to the need for more detailed experimentally-based understanding of the flow of dispersions in complex geometries. Experimental consideration of flow past immersed bodies illustrates the key importance of stagnation point, corner flows, and closed streamline regions, and the status of these issues in flow of suspensions will be discussed in terms of the present model status. The implementation of the modeling approach into a purpose-built flow solver as well as a commercial flow solver will be illustrated to indicate the relative ease of the approach.

ORINTATIONAL BEHAVIOR OF THE BINARY CARBONACEOUS MESOPHASE MIXTURES UNDER THE EXTENSIONAL FLOW

M. Golmohammadi[1], A.D. Rey[1]

[1] *Department of Chemical Engineering, McGill University, Montreal, Quebec, Canada-*
mojdeh.golmohammadi@mail.mcgill.ca, alejandro.rey@mcgill.ca

Abstract - A Maier-Saupe model for binary mixtures of uniaxial discotic nematogens under extensional flow is formulated to describe phase ordering in carbonacoues mesophases used as precursors in the carbon fiber melt spinning process. The orientational structure and the thermodynamic phase diagram of these liquid crystalline materials depend on (i) the intrinsic properties of the mixture, i.e. molecular weight difference and the interaction parameter, (ii) the operating properties, i.e. concentration and temperature and (iii) deformation rates (Deborah De number). The results obtained from our pervious study [Golmohammadi et al., L.C., 2009] shows that depending on the intrinsic properties of the system, two types of uniaxial nematic mixtures arise: (i) non-ideal mixtures with a minimum in its Nemtic to Isotropic (NI) transition temperature, and (ii) ideal mixtures with a monotonic trend of the NI transition temperature as a function of concentration. In the current computational work we study the effect of extensional flow on the phase transitions in non-ideal mixtures. Predictions of fiber microstruture as a function of thermodynamic and rheological parameters are discussed.

Introduction (Times New Roman 10, bold)

Carbonaceous mesophases (CMs), are discotic nematic liquid crystalline (DNLC) mixtures obtained from petroleum pitches and synthetic naphthalene precursors (Golmohammadi and Rey, 2008). Manufacturing of high performance carbon fibers by melt-spinning process is one of the major applications of these precursor materials. Carbon fiber spinning process involves unixial extensional flow of liquid crystalline polydisperse CM precursors. As a result, the final structure of the fiber which determines its final application depends not only on the precursor chemistry and the operating temperature, but also, on the uniaxial extensional flow which is applied during the process.

This paper seeks to extend the thermodynamics of carbonaceous mesophase binary mixtures (Golmohammadi and Rey, 2009) to non-equilibrium states produced by an imposed uniaxial extensional flow.

An ideal crystalline solid (Figure 1a) has a perfect orientational order and a perfect positional order. On the other hand an isotropic (completely random) liquid (Figure 1c) lacks any kind of ordering, either orientational or positional. Between these two extremes lies a nematic liquid crystal that has a partially orientational order without any positional order (Figure1b) (Golmohammadi and Rey, 2008).

In nematic liquid crystals (Gupta and Rey, 2005) (NLCs), the long molecular axes (**u** direction in Fig. 2) are preferably oriented along a particular direction called the director **n** (Figure 2). In discotic nematic liquid crystals (DNLCs) the director **n** is perpendicular to the long axis of discotic molecules (Figure 2). In rod-like nematic liquid crystals (RNLCs) **n** is parallel to the long axis of the molecules (Figure 2). As mentioned above CMs are discotic nematic liquid crystalline (DNLC) mixtures. The type of the structure and degree of ordering in CM fibers come from: (1) the intrinsic properties of the CM raw material, molecular weight difference of the components within the CM mixtures and the interaction between the components, (2) the concentration of the components within the mixture and the operating temperature and (3) the flow rate. These variables determine the ultimate physical and mechanical properties of the fiber and its final application and hence it is essential to understand the fundamental physics to predict their properties. In this work, the Maier-Saupe liquid crystal model adapted to discotic mesogens (Golmohammadi and Rey, 2009) is extended to incorporate the flow effects on the degree of ordering.

This theory is widely used to describe the thermodynamics of nematic liquid crystals (Golmohammadi and Rey, 2008). This mean field theory gives the temperature-dependence of the molecular orientation as well as macroscopic entropic aspects in mesogenic materials. It predicts the values of the experimentally measured scalar order parameters very well (Chandrasekhar, 1992), and hence has been applied to different nematic liquid crystalline systems and can be adjusted to their mixtures (Golmohammadi and Rey, 2009; Golmohammadi and Rey, 2008). For binary mixtures of uniaxial mesogens the mixture quadrupolar order parameter is related to that of the individual components (Golmohammadi and Rey, 2009):

$$\mathbf{Q}_{mix} \equiv \mathbf{Q} = m_1\mathbf{Q}_1 + m_2\mathbf{Q}_2 \quad (1)$$

where m_i is the mole fraction of ith component. $\mathbf{Q}_1$ and $\mathbf{Q}_2$ are the quadrupolar tensor order parameters of each component:

$$\begin{cases} \mathbf{Q}_1 = S_1(\mathbf{n}_1\mathbf{n}_1 - \frac{\mathbf{I}}{3}) + \frac{P_1}{3}(\mathbf{m}_1\mathbf{m}_1 - \mathbf{l}_1\mathbf{l}_1) \\ \mathbf{Q}_2 = S_2(\mathbf{n}_2\mathbf{n}_2 - \frac{\mathbf{I}}{3}) + \frac{P_2}{3}(\mathbf{m}_2\mathbf{m}_2 - \mathbf{l}_2\mathbf{l}_2) \end{cases} \quad (2)$$

where $\{S_i\ ;\ i=1,2$ and $P_i\ ;\ i=1,2\ \}$ are the pure component scalar order parameters and $\{n_i\ ;\ i=1,2,\ m_i\ ;\ i=1,2$ and $l_i\ ;\ i=1,2\ \}$ are the orientation unit vectors; each orientation triad is orthogonal: $\mathbf{n}_1 \cdot \mathbf{m}_1 = \mathbf{n}_1 \cdot \mathbf{l}_1 = \mathbf{n}_2 \cdot \mathbf{m}_2 = \mathbf{n}_2 \cdot \mathbf{l}_2 = 0$. Using eqns.(2) we find the scalar order parameters (S_{mix}, P_{mix}) for the mixture:

$$\begin{cases} S_{mix} = \frac{3}{2}\left(m_1\mathbf{Q}_1 : \mathbf{nn} + (1-m_1)\mathbf{Q}_2 : \mathbf{nn}\right) \\ P_{mix} = \frac{3}{2}\left(m_1\mathbf{Q}_1 + (1-m_1)\mathbf{Q}_2\right) : (\mathbf{mm} - \mathbf{ll}) \end{cases} \quad (3)$$

The ordering states of each species are coupled and are also dependent on the temperature and the dilution, so that we can expect different orientation and the thermal behavior for different compositions of the mixtures.

Theory and Simulation

Details of the extension of Maier-Saupe model to a binary mixture of NLCs is given in our pervious studies (Golmohammadi and Rey, 2009; Golmohammadi and Rey, 2008). Here we briefly present the main features of the model that are necessary to obtain the ordering and thermo-rheological behavior of the components within the mixture.

The Helmholtz free energy per unit mole of the homogeneous mixture A is:

$$A(\mathbf{Q}_1, \mathbf{Q}_2) = -N_A\left(E_{mix} + k_B T \ln Z\right) \quad (5)$$

where N_A, k_B, T are the Avogadro's number, Boltzmann's constant and temperature respectively, and E_{mix}, S_{mix}, and Z are the internal energy , the orientational entropy, and the partition function of the mixture per molecule respectively. (Golmohammadi and Rey, 2009; Golmohammadi and Rey, 2008). For a homogeneous binary mixture, the internal energy E_{mix} per molecule, due to self (1-1, 2-2) and cross (1-2) species interactions, is given by the sum of three contributions:

$$E_{mix}(\mathbf{Q}) = -\frac{3}{4}W_{11}\mathbf{Q}_1 : \mathbf{Q}_1 - \frac{3}{4}W_{22}\mathbf{Q}_2 : \mathbf{Q}_2 - \frac{3}{2}W_{12}\mathbf{Q}_1 : \mathbf{Q}_2 \quad (6)$$

The composition and molecular weight-dependent interaction parameters $\{W_{11}, W_{22}, W_{12}\}$ are as follows:

$$W_{ii} = m_i^2 U_{ii}\ ;\ W_{12} = m_1 m_2 U_{12};\quad i=11,22 \quad (7)$$

where

$$U_{ii} = \bar{U}_{ii}\,\frac{Mw_i}{m_1 Mw_1 + m_2 Mw_2};\quad U_{12} = \bar{U}_{12} \quad (8)$$

and M_{wi} denotes the molecular weight of the ith component, m_i its mole fraction, and $\{\bar{U}_{11}, \bar{U}_{22}, \bar{U}_{12}\}$ are the bare interaction parameters. The mixture partition function Z is factorized as:

$$Z = Z_1^{m_1} Z_2^{m_2} = \left(\int e^{-\Phi_1/m_1 k_B T - H_1/k_B T} d\mathbf{u}_1\right)^{m_1} \left(\int e^{-\Phi_2/m_2 k_B T - H_2/k_B T} d\mathbf{u}_2\right)^{m_2} \quad (9)$$

With Φ_i being the molecular potential and Hi are the flow potential of the ith component wich are given with the following expressions:

$$\Phi_1 = \frac{\partial E_{mix}}{\partial \mathbf{Q}_1} : \left(\mathbf{u}_1\mathbf{u}_1 - \frac{\mathbf{I}}{3}\right) = -\frac{3}{2}\left(W_{11}\mathbf{Q}_1 + W_{12}\mathbf{Q}_2\right) : \left(\mathbf{u}_1\mathbf{u}_1 - \frac{\mathbf{I}}{3}\right)$$

$$\frac{H_i}{k_B T} = -\frac{3\dot{\varepsilon}}{4\bar{D}_{ri}}(\mathbf{u}.\mathbf{e}_z)^2 R \quad (10)$$

where $\dot{\varepsilon}$ is the steady rate of uniaxial extension, $\bar{D}_{ri}$ is the rotational diffusivity of the ith component, **u** is the molecular unit normal, e_z the flow (extension) direction, and R is the rheological effective shape factor which is negative for disk-like molecules (Golmohammadi and Rey, 2010). By replacing the expression for D_r of the discotic liquid crystalline molecules (Golmohammadi and Rey, 2010) the following dimensionless flow potential is obtained:

$$\frac{H_1}{k_B T} = -\frac{3}{4}(\mathbf{u}.\mathbf{e}_z)^2 RDe\left(1-\frac{3}{2}\mathbf{Q}_{ext}:\mathbf{Q}_{ext}\right)^2 \frac{\left(\frac{m_1}{\mathrm{m}_1 Mw_1 + \mathrm{m}_2 Mw_2}\right)^2}{\left(\left(\frac{m_1}{\mathrm{m}_1 Mw_1 + \mathrm{m}_2 Mw_2}\right)_{ref}\right)^2}$$

$$\frac{M_{w1}^{\frac{3}{p}}}{\left(M_{w1}^{\frac{3}{p}}\right)_{ref}} \frac{\left(m_1 M_{w1}^{\frac{3}{p}} + (1-m_1) M_{w1}^{\frac{1}{p}} M_{w2}^{\frac{1}{p}} \frac{\left(M_{w1}^{\frac{1}{p}} + M_{w2}^{\frac{1}{p}}\right)}{2}\right)^2}{\left(\left(m_1 M_{w1}^{\frac{3}{p}} + (1-m_1) M_{w1}^{\frac{1}{p}} M_{w2}^{\frac{1}{p}} \frac{\left(M_{w1}^{\frac{1}{p}} + M_{w2}^{\frac{1}{p}}\right)}{2}\right)_{ref}\right)^2} \quad (11)$$

where *De* is the dimensionless Deborah number given by :

$$De = \frac{\dot{\varepsilon}}{Dr_{ref}} = \frac{\dot{\varepsilon}}{CkT_{ref}} \left(\left(\frac{m_1}{\mathrm{m}_1 Mw_1 + \mathrm{m}_2 Mw_2}\right)_{ref}\right)^2 \left(M_{w1}^{\frac{3}{p}}\right)_{ref} \times$$

$$\left(\left(m_1 M_{w1}^{\frac{3}{p}} + (1-m_1) M_{w1}^{\frac{1}{p}} M_{w2}^{\frac{1}{p}} \frac{\left(M_{w1}^{\frac{1}{p}} + M_{w2}^{\frac{1}{p}}\right)}{2}\right)_{ref}\right)^2 \quad (12)$$

For steady uniaxial extensional flow along the "z" direction, the two director triads are proved to be congruent (Golmohammadi and Rey, 2010),

$$\mathbf{n} = \mathbf{n}_1 = \mathbf{n}_2; \mathbf{m} = \mathbf{m}_1 = \mathbf{m}_2; \delta_z = -\mathbf{l}_1 = -\mathbf{l}_2$$

Since the compression plane normal to the stretching direction is isotropic the orientation of the **n-m** axes within this plane is irrelevant. Scaling the extended Helmholtz free energy *A* (eqn.(12)) with the bare interaction parameter $\overline{U}_{11}$, minimizing the resulting dimensionless free energy with respect to $\mathbf{Q_1}$, $\mathbf{Q_2}$, and double-contracting the tensorial equations with nn and mm we find the governing equations for the scalar order parameters S_1,S_2,P_1,P_2:

$$(a): m_1\varphi_1 S_1 + \sqrt{m_1\varphi_1 m_2\varphi_2} L_1 S_2 - T_r \sum \frac{m_i}{Z_i}\frac{\partial Z_i}{\partial S_1} = 0$$

$$(b): m_2\varphi_2 L_2 S_2 + \sqrt{m_1\varphi_1 m_2\varphi_2} L_1 S_1 - T_r \sum \frac{m_i}{Z_i}\frac{\partial Z_i}{\partial S_2} = 0$$

$$(c): \frac{1}{3} m_1\varphi_1 P_1 + \sqrt{m_1\varphi_1 m_2\varphi_2} L_1 \frac{P_2}{3} - T_r \sum \frac{m_i}{Z_i}\frac{\partial Z_i}{\partial P_1} = 0$$

$$(d): \frac{1}{3} m_2\varphi_2 L_2 P_2 + \sqrt{m_1\varphi_1 m_2\varphi_2} L_1 \frac{P_1}{3} - T_r \sum \frac{m_i}{Z_i}\frac{\partial Z_i}{\partial P_2} = 0 \quad (11)$$

The thermodynamics of the mixture is defined by the dimensionless temperature Tr and two effective mole fractions as follows:

$$T_r = \frac{k_B T}{\overline{U}_{11}} \quad ; \quad \varphi_i = \frac{m_i Mw_i}{\sum m_i Mw_i} \quad (13)$$

The two material parameters of the model are:

$$L_1 = \frac{\overline{U}_{12}}{\overline{U}_{11}}; \quad L_2 = \frac{\overline{U}_{22}}{\overline{U}_{11}} \quad (14)$$

The material parameters (L_1, L_2) are functions of the species' molecular weights. To approximate real CMs we fix the M_w of the first and the second species as 1400 and 1000, respectively (Hurt and Hu, 1999). The concentration m_1 and m_2 are also fixed as 0.7 and 0.3. By using the molecular weight dependency of the transition temperatures and the material parameter L2 (Golmohammadi and Rey, 2009; Golmohammadi and Rey, 2008), we obtain T_{NI}, T_{N2} and L_2 as 900 K, 600 K and 0.6667. Since the molecular weight dependency of L_1 is not suggested by actual data1, we use:

$$L_1 = \beta L_2 \quad (15)$$

where β is the interaction coefficient and is positive for mixtures of similar molecules (disk/disk). In this study we fix β as 0.1. This combination of ΔMw, interaction parameter and concentration results in a non-ideal mixture above its critical concentration. (Golmohammadi and Rey, 2009)

The present thermodynamic model is given by the four nonlinear integral equations (eqns.(11)); the solution vector consists of the four scalar order parameters (S_1,S_2, P_1,P_2); the two material parameters are β and ΔMw; the concentration is also fixed; the thermorheological phase diagram is obtained by sweeping over temperature T_r and flow rate in terms of deborah *De* number. Equations (11) are solved by the Newton-Raphson method, using an eighth order Simpson integration method. Stability, accuracy, and convergence were ensured using standard methods (Huebner et al., 2001).

Results and Discussion (Times New Roman 10, bold)

The six classes of steady state solutions to equations (11) using the sign and the ordering of the eigenvalues of $\mathbf{Q_1}$ and $\mathbf{Q_2}$ and the orientation of the dominant eigenvector with respect to the extension direction are:

N_U^+ : Uniaxial Nematic with $\mu_n > \mu_l = \mu_m$

N_U^- : Uniaxial Nematic with $\mu_n < \mu_l = \mu_m$

N_B^+ : Biaxial Nematic with $\mu_n \rangle \mu_m \rangle \mu_l$

PN_U^+ : Uniaxial Paranematic with $\mu_n > \mu_l = \mu_m$

PN_U^- : Uniaxial Paranematic with $\mu_l < \mu_n = \mu_m$

PN_B^- : Biaxial Paranematic with $\mu_n \rangle \mu_m \rangle \mu_l$

where the superscripts refer to the sign of the largest absolute eigenvalue and the subscript refers to the state (*U*: uniaxial, *B*: biaxial) . If there are two identical eigenvalues the state is uniaxial, otherwise it is biaxial. The corresponding ellipsoidal structures are prolate, oblate and scalene. Each radii of the ellipsoid refers to one eigenvalue. We use standard nomenclature: prolate is a rugby ball shaped ellipsoid with two identical eigenvalues; oblate is a disk-shaped ellipsoid with two identical eigenvalues and a scalene ellipsoid represents a structure with three different eigenvalues.

Figure 3 shows the phase diagram of the binary system obtained as the solutions to equations (13). The lower Mw component ("2") is always uniaxial while the high Mw component "1" may be uniaxial or biaxial; below when referring to biaxial state it is understood to refer to the higher Mw component "1". The phase diagram contains four regions separated by four distinct transition lines. Each region is characterized by a phase and a structural state of the two components. In each region, the structure on the left corresponds to the first component and the one on the right corresponds to the second component. The

Region A : $N_{U1}^+ / N_{U2}^+ \Rightarrow prolate \perp / prolate \perp$

Region B : $N_{B1}^+ / N_{U2}^- \Rightarrow scalene \perp / oblate \perp$

Region C : $PN_{B1}^- / PN_{U2}^+ \Rightarrow scalene \parallel / prolate \perp$

Region D : $PN_{U1}^- / PN_{U2}^+ \Rightarrow oblate \parallel / prolate \perp$

regions are:

where the subscript i=1, 2 refers to the component number and $\perp$ and $\parallel$ refer to the orientation perpendicular and parallel to the flow direction. The four transition lines that define the four regions are:

(1) $N_{UB,PO}$: Nematic Uniaxial-Biaxial and Prolate-Oblate transition line which seperates region *A* from region *B*(horizental dashed line)

(2) *DNPN* : Discontineous Nematic-Paranematic transition line which seperates regions *A* and *B* from regions *C* and *D*(dotted and full line with superposed dotts

(3) *NPN* : Nematic-Paranematic transition line which seperates region *B* from region *C* (full line)

(4) PN_{BU} : Paranematic Biaxial-Uniaxial transion line which seperates region *C* from region *D*(dashed line)

The *DNPN* line that describes the discontinuous transition between nematic and paranematic states is itself divided into two segments:

(i) $0 < De < De_{c-} = 0.05$: here the discontinuous transition is between uniaxial nematic/ uniaxial nematic (*A*) and uniaxial paranematic/uniaxial paranematic (*D*) or biaxial paranematic/uniaxial paranematic (*C*).

(ii) $De_{c-} = 0.05 < De < De_{c+} = 1.1$: here the discontinuous transition is between biaxial nematic (*B*) and biaxial paranematic/uniaxial paranematic (*C*). For $De > De_{c+} = 1.1$ the nematic/paranematic transition (full line) is always continuous.

The phase diagram of the pure system has three regions (figure 4): at low temperatures (region N_U^+) there is thermodynamic and flow cooperation. Thermodynamics favors the uniaxial orientation N_U^+ while extensional flow also favors orientation along **n** with n along the compression plane and hence there is no competing effect. At higher temperatures and higher *De*, the thermodynamic uniaxial alignment along **n** now competes with the biaxial flow alignment producing the biaxial nematic state (region N_B^+) with **n** along the compression plane. Further temperature increase eventually leads to the biaxial nematic/ uniaxial paranematic transition. In the paranematic state (region PN_B^-) the oblate state has the unique axis (**l** vector) along the extension direction; here the competition of thermodynamics that prefers an isotropic state and the flow effect that seeks molecular alignment along the compression plane results in the oblate state.

According to figure (3) the dominant component ("1") in the mixture essentially follows the standard behavior of the pure mesophase (Fig. 4) with a small perturbation which is the formation of the biaxial paranematic (PN_{B1}^-, region C). The second low molar mass component, however, behaves differently. Although at low temperatures it shows the standard phase state which is uniaxial nematic N_{U2}^+, at medium and high temperatures and also at high *De* numbers (*B*, *C* and *D* regions) it adopts different orientational and structural states. Extension affects the molecules of the second low molar mass component differently. For instance, in region *D*, flow tends to align the molecules of species "1" in the compression (**n-m**) plane to form a parallel oblate but it aligns the molecules of the second component in a prolate $\perp$ structure. This phenomenon is due to the coupled viscoelastic forces generated within the mixture.

To clarify how different regions on the phase diagram form we discuss the ordering and structural behavior of the component within the mixture by imposing the extensional flow on the homogeneous mixture and by increasing the dimensionless temperature T_r at De= 0.7 which creats all four regions.

Figure 5 shows the scalar order parameters $(S_1, S_2, S_{mix}, P_1, P_2, P_{mix})$ for $De_{c-} = 0.05 < De = 0.7 < De_{c+} = 1.1$, corresponding to discontinuous transition between nematic and paranematic states. By increasing the dimensionless temperature from zero four regions *A*, *B*, *C* and *D* appear. In region *A*, (like in the pervious case, Fig.4) the stable phases are both uniaxial nematic, N_U^+, and the corresponding structures are both prolate $\perp$. By increasing temperature at this *De*, a transition within the nematic phase takes place and region *B* is reached. In region *B* the stable phase for the first component is biaxial nemtaic, N_{B1}^+, its scalar order parameter is positive and its biaxiality is significant. There are three distinct eigenvalues, $\mu_{n1} \rangle \mu_{m1} \rangle \mu_{l1}$ with μ_{n1} being positive and the largest value; the corresponding structure is scalene $\perp$. For the second component, however, the stable phase is uniaxial nemtic, N_{U2}^-, with negative scalar order parameter and two identical eigenvalues, $\mu_{n2} < \mu_{l2} = \mu_{m2}$, which denote a perpendicular oblate state. By further increasing the temperature, a nematic to paranematic discontinuous transition takes place and region *C* emerges. In region *C* the stable phase for the first component is uniaxial paranematic, PN_{U1}^-, its scalar order parameter is positive and its biaxiality is significant. There are three distinct eigenvalues, $\mu_{n1} > \mu_{m1} > \mu_{l1}$ with μ_{l1} being negative and of largest value, and the corresponding structure is the parallel scalene. For the second component, however, the phase is uniaxial paranematic, PN_{U2}^+, with two identical eigenvalues, $\mu_{n2} > \mu_{l2} = \mu_{m2}$, denoting a prolate $\perp$. By increasing temperature in the paranematic phase at this De a continuous transition within the paranematic phase takes place and region *D* is reached. In region *D* the stable phases are uniaxial paranematic, PN_{U1}^-, for the first component and uniaxial paranematic, PN_{U2}^-, for the second component. The corresponding structures are an oblate || and prolate $\perp$, respectively.

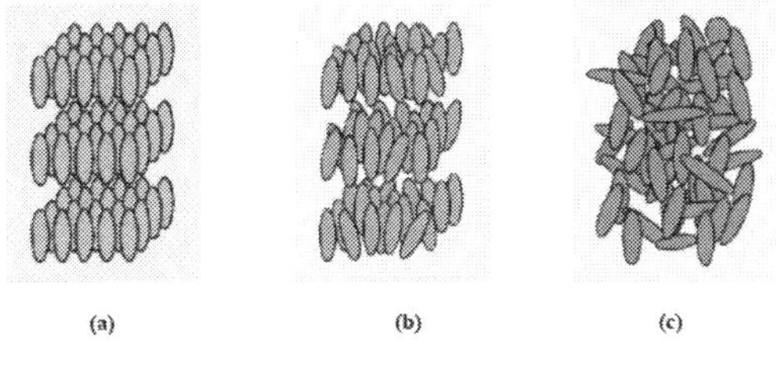

Fig. 1. The comparison of three states of the material: a perfect crystalline solid with both positional and orientational order (a), a liquid crystal with positional disorder and partial orientational order (b) and an isotropic liquid with positional and orientational disorder (c). (Golmohammadi and Rey, 2008)

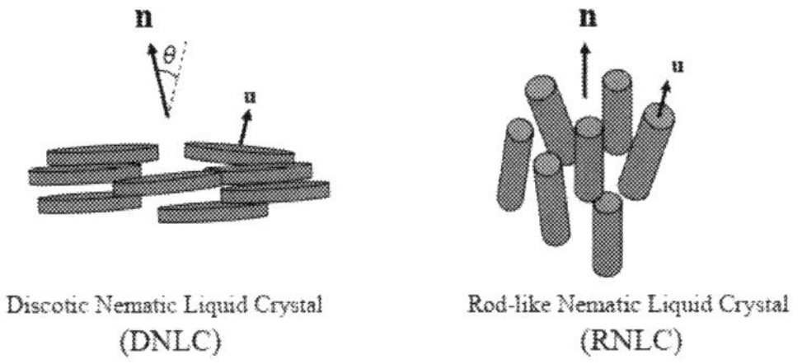

Fig. 2. Schematic of nematic liquid crystals (NLCs), the director **n** is the average of the molecular orientation **u**, and its classification into discotic and rod-like molecules. In discotic nematic liquid crystals the director **n** (average orientation) is perpendicular to the long axis of the molecules; however, in rod-like nematic liquid crystals the director **n** is parallel to the long axis of the molecules. (Golmohammadi and Rey, 2008)

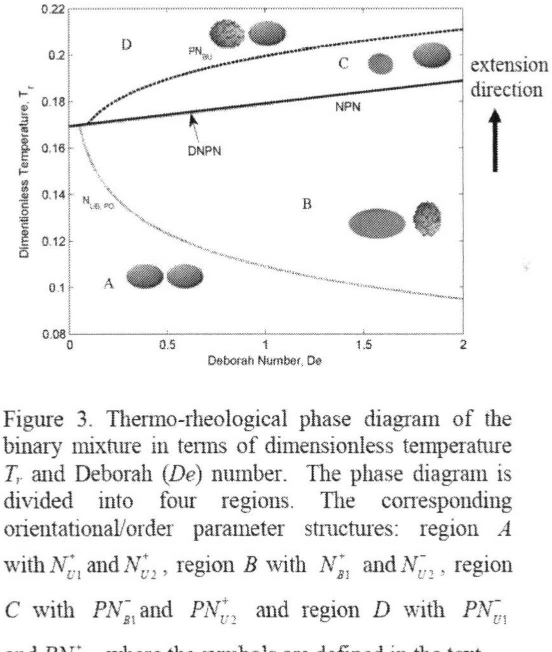

Figure 3. Thermo-rheological phase diagram of the binary mixture in terms of dimensionless temperature T_r and Deborah (*De*) number. The phase diagram is divided into four regions. The corresponding orientational/order parameter structures: region *A* with N_{U1}^+ and N_{U2}^+, region *B* with N_{B1}^+ and N_{U2}^-, region *C* with PN_{B1}^- and PN_{U2}^+ and region *D* with PN_{U1}^- and PN_{U2}^+, where the symbols are defined in the text.

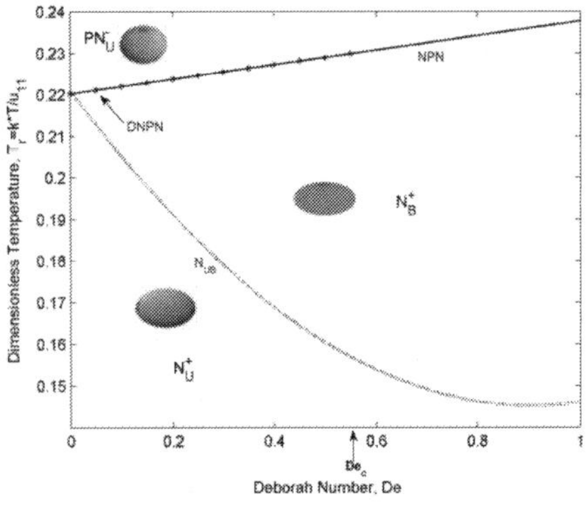

Figure 4. Thermo-rheological phase diagram of a single discotic nematogen with its corresponding orientational/order parameter structures adapted from (Golmohammadi and Rey, 2010): N_U^+ at low temperatures, N_B^+ at medium temperatures and *De* numbers and PN_B^- at high temperatures. Comparing Figs. 3 and 4 we see that the high M_w species follows essentially the pure species response but that low molar mass component exhibits new behavior.

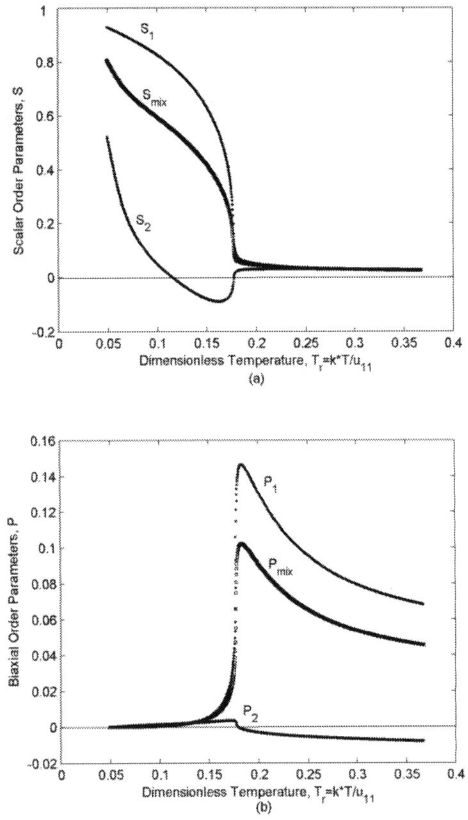

Figure 5. Scalar order parameters (*S* and *P*) (a, b) of the components as a function of dimensionless temperature for *De=0.7*. As temperature increases at this *De*, regions *A*, *B*, *C* and *D* of the phase diagram are traversed.

Conclusions (Times New Roman 10, bold)

This paper uses the extended Maier-Saupe model for binary mixtures of discotic nematogenes to investigate the effect of steady uniaxial extensional flow on the orientational structure and molecular order of carbonaceous mesophase mixtures.

This work shows that mixing high and lower molecular weight nematogenic species introduces asymmetric coupling effects, such that the higher molecular weight component changes the flow-induced structures of the lower molecular weight component, but that the converse is not true.

The main difference with the flow-response of the pure system is the presence of a direct uniaxial nematic/paranematic transition for 0<De<0.3 and the presence of temperature/*De* region with a biaxial paranematic state (Region *C* in Fig.3). The thermorheological behaviour of the low molecular weight component in a binary mixture is starkly different than for the pure system.

The deviations from the pure species response to extensional flow are due to coupling effects. The present work provides a foundation in mixing to manipulate extensional flow-induced structures in carbon fiber-forming processes.

Acknowledgements (Times New Roman 10, bold)

This research is supported by a grant from the American Chemical Society-Petroleum Research Fund (PRF). M.G. is thankful to the Eugene Ulmer Lamothe (EUL) Scholarship Fund (Department of Chemical Engineering, McGill University). A.D.R. wishes to acknowledge his collaboration with Profs A. Ogale and Dan Edie, and the Center for Advanced Fibers and Films on the rheology of carbonaceous mesophases.

References (Times New Roman 10, bold)

Insert text here (Times New Roman 10, single line spacing, justified paragraph, 2 columns).
References should be cited in the text using the following formats:

Journal:

Hurt R. H., Hu Y., "Thermodynamics of carbonaceous mesophase", Carbon, 37(2) 281-292 (1999).

Golmohammadi M., Rey A. D., "Thermodynamic modelling of carbonaceous mesophase mixtures", Liquid Crystals, 36.1, 75-92 (2009).

Golmohammadi M., Rey A. D., "Entropic Behavior of Binary Carbonaceous Mesophases", Entropy 10.3, 183-199(2008).

Gupta, G., Rey A.D., "Texture Rules for Concentrated Filled Nematics", Phys. Rev. Lett. 95, 127802:1-127802:4 (2005).

Golmohammadi M., Rey A. D., "Structure and Phase Transitions of Carbonaceous Mesophase Binary Mixtures Under Uniaxial Extensional Flow", Submitted to the Journal of Non-Newtonian Fluid Mechanics.

Book:
Chandrasekhar S., Liquid crystals. 2nd Ed.; Cambridge University Press: Cambridge [England] ; New York, NY, USA, (1992).

Huebner, K.H, D.L. Dewhirst, D.E. Smith, T.G.: Byrom (2001), The Finite Element Methods for Engineers, 4th Ed.; John Wiley & Sons, INC.: New York, NY, USA (2001).

ISSUES IN THE MODELING OF PASTE EXTRUSION

E. Mitsoulis[1*], Th. Zisis[1], S.G. Hatzikiriakos[2]

[1]*School of Mining Engineering & Metallurgy, National Technical University of Athens (NTUA)*
Zografou 157-80, Athens, GREECE, mitsouli@metal.ntua.gr, zisis@metal.ntua.gr
[2]*Dept. Chemical and Biological Engineering, University of British Columbia(UBC), Vancouver, Canada –* hatzikir@apsc.ubc.ca

Abstract - In a recent article (*P. Patil, J. Feng, S.G. Hatzikiriakos, Constitutive modelling and flow simulation of polytetrafluoroethylene paste extrusion, J. Non-Newtonian Fluid Mech., 139, 2006, 44-53*), a rheological constitutive equation was proposed to model the incompressible flow of poly-tetra-fluoro-ethylene (PTFE) paste and its structural development (fibrillation between the PTFE particles) during flow. The constitutive model was a combination of shear-thinning and shear-thickening viscosity terms, depending on a structural parameter, ξ, which obeyed a convective-transport equation. In the latter, a non-objective flow type parameter, ψ, was employed, which depended on the magnitudes of the rate-of-strain and vorticity tensors. In the present work an objective flow type parameter is used instead (*R.G. Larson, Flows of constant stretch history for polymeric materials with power-law distributions of relaxation times, Rheol. Acta, 24, 1985, 443-449*). Despite its complexity and the severe slip at the wall, good convergence has been obtained even for very high apparent shear rates. New results are shown as contours of the ξ-parameter in the flow field as well as contours of the new objective flow type parameter, ψ^*, and compared with previous results. It is shown that the corrected model predicts higher structural levels (more fibrils) in PTFE paste flow through extrusion dies and that the new results are more consistent with experimental observations.

Introduction

In a recent article Patil et al. [1] proposed a rheological constitutive equation to model PTFE paste flow and its structural development in tapered axisymmetric dies. The authors performed steady-state flow simulations, which were governed by the conservation equations of mass and momentum under isothermal, incompressible conditions. A linear slip law at the wall was also implemented due to the presence of the lubricant. The constitutive model proposed was a combination of a shear-thinning and a shear-thickening viscosity terms, linearly connected by a structural parameter, ξ, defined as the percentage of fibrillated domains in the paste. This parameter obeyed a convective-transport equation (see below for details of the set of equations). The paste flow was assumed to be incompressible. However, significant amounts of air voids exist in the paste due to low lubricant concentrations. More recent work by Mitsoulis and Hatzikiriakos [2] re-examined the PTFE paste flow problem by taking into account the significant compressibility of the paste and by implementing a slip law based on the consistent normal-to-the-surface unit vector. New results were shown as contours of the ξ-parameter in the flow field in order to study in detail the effect of compressibility on the structure formation and flow kinematics of paste. These results offered a better understanding of the flow behaviour of PTFE in flow through extrusion dies.

A problem with the previous models [1,2] arises from using a non-objective flow type parameter, ψ, that represents the strength of the flow. This measure of flow type was originally introduced by Giesekus [3] and later used by Fuller and Leal for homogeneous flows [4]. It is given by:

$$\psi = \frac{|\dot{\gamma}| - |\omega|}{|\dot{\gamma}| + |\omega|}, \quad (1)$$

where $|\dot{\gamma}|$ and $|\omega|$ are the magnitudes of the rate-of-strain tensor $\dot{\gamma} = \nabla u + \nabla u^T$ and the vorticity tensor $\omega = \nabla u - \nabla u^T$, respectively, u is the velocity vector, ∇u is the velocity gradient tensor, and ∇u^T is its transpose. As explained in [3,4], $\psi = -1$ in pure rotational flow, $\psi = 0$ in pure shear flow, and $\psi = 1$ in pure elongational flow. In a mixed flow, such as flow through an extrusion die considered here, $-1 \leq \psi \leq 1$. Since values of ψ less than zero correspond to weak flows, and since such flows do not contribute to fibril formation (structure development), Patil et al. [1] were setting all negative values of ψ equal to 0 in their calculations (no structure development).

Astarita [5] pointed out the necessity for a frame-invariant generalization of eq. (1), and this was also explained in detail by Singh and Leal [6] for the case of flows generated by a co-rotating two-roll mill. The new definition of the objective structural parameter, ψ', is given by:

$$\psi' = \frac{|\dot{\gamma}| - |\Omega|}{|\dot{\gamma}| + |\Omega|}, \quad (2)$$

where $|\Omega|$ is the magnitude of the relative rate-of-rotation tensor, $\boldsymbol{\Omega}$, which is the difference between two tensors, the vorticity tensor, $\boldsymbol{\omega}$, and the antisymmetric tensor, $\boldsymbol{W}$, of the rate of rotation of the rate-of-strain tensor, $\dot{\boldsymbol{\gamma}}$. The definition of $\boldsymbol{W}$ is given by [5]:

$$\frac{\mathrm{D}\boldsymbol{e}_i}{\mathrm{D}t} = \boldsymbol{W} \cdot \boldsymbol{e}_i, \tag{3}$$

where D/Dt is the substantial time derivative and $\boldsymbol{e}_i$ are the proper vectors of $\dot{\boldsymbol{\gamma}}$ (i.e., unit vectors along the principal axes of $\dot{\boldsymbol{\gamma}}$). Similarly, ψ' varies between −1 and +1, and in flow simulations the negative values of ψ' should be set arbitrarily equal to zero in order to exclude weak flows that do not contribute to structure development [1,2].

Deriving an objective flow type parameter is clearly a challenge, as it involves the determination of angular velocity of eigenvectors. Some authors [7,8] have worked out such modifications, but it is certainly not a trivial matter. Larson [9] has put forward a frame-invariant flow strength parameter, S_f, defined by:

$$S_f = \frac{2\left(tr\boldsymbol{D}^2\right)^2}{tr\overset{\circ}{\boldsymbol{D}}{}^2}, \tag{4}$$

where $\overset{\circ}{\boldsymbol{D}}$ is the Jaumann time derivative of the rate-of-deformation tensor, $\boldsymbol{D}$, defined as $2\boldsymbol{D} \equiv \dot{\gamma}$, i.e., the time derivative of $\boldsymbol{D}$ with respect to a frame that rotates with the angular velocity of the fluid element. The parameter S_f ranges from 0, for pure rotational flow, to 1, for simple shear flow, to infinite, for pure elongational flow. To make it readily usable in the flow model, S_f can be accordingly normalized to result in a new objective flow type parameter. It is worthwhile to mention that the above flow type or flow strength parameters have been used in the past, particularly in polymeric flows that involve mixing by Manas-Zloczower and co-workers [10-14]. On the other hand, the non-objective ψ-parameter was primarily used due to its ease of computation, while the much more difficult to calculate S_f-parameter has been used in [13].

It is the purpose of the present work to re-examine the flow of PTFE paste in extrusion dies by using a consistent and objective flow type parameter, based on S_f. The new flow strength parameter and the model are then formulated accordingly. The results are compared with those obtained with the use of the non-objective flow type parameter, ψ, reported in Ref. [2].

Governing Equations and Rheological Modeling

We consider the conservation equations of mass and momentum for compressible fluids under isothermal, creeping, steady flow conditions. These are written as [2]:

$$\boldsymbol{u} \cdot \nabla\rho + \rho\left(\nabla \cdot \boldsymbol{u}\right) = 0, \tag{5}$$

$$0 = -\nabla p + \nabla \cdot \boldsymbol{\tau}, \tag{6}$$

where ρ is the density, $\boldsymbol{u}$ is the velocity vector, p is the pressure and $\boldsymbol{\tau}$ is the extra stress tensor. For a compressible fluid, pressure and density are connected as a first approximation through a simple linear thermodynamic equation of state [2]:

$$\rho = \rho_0\left(1 + \beta p\right), \tag{7}$$

where β is the isothermal compressibility with the density to be ρ_0 at reference pressure p_0 (=0). The viscous stresses are given for inelastic non-Newtonian compressible fluids by the relation [2]:

$$\boldsymbol{\tau} = \eta\left(|\dot{\gamma}|\right)\left(\dot{\boldsymbol{\gamma}} - \frac{2}{3}\left(\nabla \cdot \boldsymbol{u}\right)\boldsymbol{I}\right), \tag{8}$$

where $\eta\left(|\dot{\gamma}|\right)$ is the apparent non-Newtonian viscosity, which is a function of the magnitude $\bar{\bar{\dot{\gamma}}}$ of the rate-of-strain tensor, and is given by:

$$|\dot{\gamma}| = \sqrt{\frac{1}{2} II_{\dot{\gamma}}} = \left(\frac{1}{2}\left(\dot{\boldsymbol{\gamma}} : \dot{\boldsymbol{\gamma}}\right)\right)^{1/2}, \tag{9}$$

where $II_{\dot{\gamma}}$ is the second invariant of $\bar{\bar{\dot{\gamma}}}$. The tensor $\boldsymbol{I}$ in eq. (8) is the unit tensor.

In the previous works [1,2], a simple linear combination of shear-thinning and shear-thickening viscosities was assumed for PTFE paste, depending on a structural parameter, ξ:

$$\eta = (1-\xi)\eta_1 + \xi\eta_2. \tag{10}$$

In the above, η_1 and η_2 are the shear-thinning and shear-thickening viscosities, respectively, that are expressed by a Carreau model [1,2]:

$$\eta_i = \eta_{0,i}\left[1 + \left(\lambda_i|\dot{\gamma}|\right)^2\right]^{(n_i-1)/2}, \tag{11}$$

where i = 1 refers to shear-thinning (n_1 < 1) and i = 2 refers to shear-thickening (n_2 > 1). The model parameters are the zero-shear-rate viscosity, η_{0i}, the power-law index, n_i, and the time constant, λ_i. Their values are listed in Table 1.

Table 1 – Constants used in the simulations for PTFE paste extrusion.

Parameter	Value
$\eta_{0,1}$	0.004 MPa·s
n_1	0.5 (-)
λ_1	0.3 s
$\eta_{0,2}$	0.0016 MPa·s
n_2	1.3 (-)
λ_2	1.0 s
β	0.02 MPa^{-1}
ρ	1600 kg/m^3
b_{sl}	1.92 m/MPa s

For the structural parameter, ξ, a convective-transport equation (CTE) was introduced in [1] based on arguments of creation and destruction of fibrils (see [1] for details). The CTE is written as:

$$\boldsymbol{u} \cdot \nabla\xi = f - g, \tag{12}$$

where f and g denote the rate of creation and elimination of fibrillated domains in the paste. These functions are given by:

$$f(|\dot{\gamma}|,\zeta) = 2|\dot{\gamma}|\sqrt{\zeta(1-\zeta)}, \tag{13a}$$

$$g(|\dot{\gamma}|,\xi) = |\dot{\gamma}|\xi, \tag{13b}$$

where the proportionality constants in the above functions proposed by Patil et al. [1] were assumed to be 1 in order to bound ξ between 0 and 1; the parameter ζ is related to S_f, according to Larson [9] as follows. For a planar flow of constant stretch history, there is a frame in which

$$\nabla \boldsymbol{u} = |\dot{\gamma}|\begin{pmatrix} 0 & \zeta \\ 1-\zeta & 0 \end{pmatrix}, \tag{14}$$

where ζ is related to S_f by:

$$S_f = \frac{1}{(1-2\zeta)^2}. \tag{15}$$

This corresponds to weak flows for $0 \le S_f \le 1$ and strong flows for $S_f > 1$. The relationship between ζ and S_f (see eq. 15) leads us to define a parameter ψ^* as:

$$\psi^* \equiv 4\zeta(1-\zeta) = \frac{S_f - 1}{S_f}. \tag{16}$$

The value of ψ^* is bounded between 0 and 1 for strong flows. This is similar to ψ and ψ', which also take values between 0 and 1 for strong flows (of main interest in the present work).

We can still define ψ^* in terms of S_f according to eq. (16), and therefore it retains the significance of a flow type parameter. As Brunn [15] has pointed out, most definitions of flow type parameters have their basis in steady 2-D constant stretching history flows, and it is tempting to use them for general steady flow in the absence of a better alternative. As also pointed out, the Larson flow type parameter, S_f, and therefore ψ^*, satisfy Tanner's and Huilgol's suggestion of strong and weak flows [16,17].

Therefore the CTE can be written now as:

$$\boldsymbol{u}\cdot\nabla\xi = |\dot{\gamma}|\left(\sqrt{\psi^*} - \xi\right). \tag{17}$$

Obviously from the above form of CTE, ψ^* and ξ are bound between 0 and 1.

The above rheological model is introduced into the conservation of momentum (eq. 6) and closes the system of equations. Boundary conditions are necessary for the solution of the above system of equations. Figure 1 shows the solution domain and boundary conditions for the contraction geometry. Because of symmetry only one half of the flow domain is considered, as was done in Ref. [2].

Also, severe slip occurs at the die wall due to the lubricant [1,2]. Due to slip at the wall (boundary EDSC in Fig. 1), the experimental data have been fitted with a linear slip law, which takes the usual form [2]:

$$u_{sl} = \beta_{sl}\tau_w, \tag{18}$$

where u_{sl} is the slip velocity, τ_w is the shear stress at the die wall, β_{sl} is the slip coefficient. For the 2-D axisymmetric problem, the above law means that the tangential velocity, u_t, is given by the slip law, while the normal velocity, u_n, is set to zero, i.e.,

$$\beta_{sl}(\boldsymbol{tn}:\boldsymbol{\tau}) = \boldsymbol{t}\cdot\boldsymbol{u} = u_t, \quad u_n = \boldsymbol{n}\cdot\boldsymbol{u} = 0 \tag{19}$$

where $\bar{n}$ is the unit outward normal vector to a surface and $\bar{t}$ is the tangential unit vector in the direction of flow.

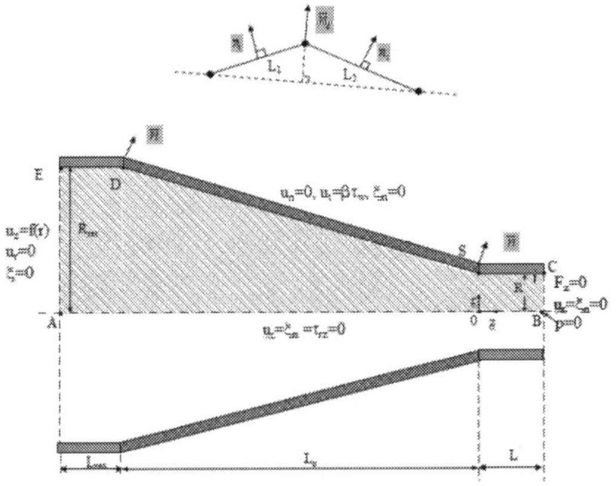

Figure 1 – Flow domain and boundary conditions. The shaded part is the solution domain due to symmetry. In the upper section the consistent normal to the surface is shown as implemented for the slip boundary condition.

All lengths are scaled by the die radius R, all velocities by the average velocity U at the die exit defined as $U = Q/\pi R^2$, all pressures and stresses by $\eta_{0,1}\lambda_1^{n_1-1}(U/R)^{n_1}$, and the density by ρ_0. Then the dimensionless numbers which control the flow phenomena are [2]:

(a) for slip at the wall, the slip parameter, B_{sl}, given by:

$$B_{sl} = \frac{\beta_{sl}}{U}\frac{\eta_{0,1}}{\lambda_1^{1-n_1}}\left(\frac{U}{R}\right)^n. \tag{20}$$

For the present conditions (see Table 2), $1.2 < B_{sl} < 2.6$, which corresponds to severe slip;

(b) for compressibility, the compressibility parameter, B, given by:

$$B = \beta\frac{\eta_{0,1}}{\lambda_1^{1-n_1}}\left(\frac{U}{R}\right)^n. \tag{21}$$

For the present conditions (see Table 2), $0.003 < B < 0.007$, which is substantial for the material at hand.

Table 2 – Dimensionless parameters for the range of simulations for PTFE paste extrusion at 35°C.

Apparent shear rate $\dot{\gamma}_A$ (s^{-1})	Slip Parameter B_{sl} *(eq. 20)*	Compressibility Parameter B (eq. 21)
1875	2.550	0.00316
2812	2.082	0.00387
3750	1.803	0.00447
4688	1.613	0.00500
5869	1.441	0.00559
8305	1.212	0.00666

Method of Solution

The numerical solution is obtained with the Finite Element Method (FEM), employing as primary variables the two velocities, pressure, and structural parameter (*u-v-p-ξ formulation*) as explained in Ref. [2]. We use Lagrangian quadrilateral elements with biquadratic interpolation for the velocities and the structural parameter, and bilinear interpolation for the pressures. The densities, appearing in the compressible conservation equations, are substituted by the equation of state (eq. 7) through the pressures and their linear interpolation per element.

An important extra feature of the governing equations is the handling of the purely convective-transport equation (eq. 17) with the new ψ^*–parameter. Since the velocities are interpolated quadratically, their first derivatives (the rates-of-strain) are interpolated linearly, and their second derivatives needed in the definition of ψ^* are constant per element. As pointed out by Yao and Manas-Zloczower [13], "high density mesh designs are required in order to minimize the numerical error", which in their case were prohibitive for most of the calculations with the objective ψ^*–parameter. This was borne out by our calculations. Despite dense meshes and 2×2 subdivisions of each parent element for the CTE, when no slip was assumed, divergence occurred. However, with the presence of severe slip which brings down the strain rates substantially, convergence was always guaranteed, and it was fast and monotonic.

The finite element meshes used in the simulations are shown in Fig. 2 for the 2α=30° geometry (both the full domain and a blown-up section of it near the die entry). No smoothing of the boundaries was necessary, in contrast to what was done in Refs. [1,2]. The entry has been set at $-140R$ and the exit at $+40R$ (L/D=20). The less dense mesh M1 has 1700 elements, 7161 nodes, and 22644 unknown degrees of freedom (DOF) with 21 points in the radial direction at entry. A more dense mesh M2 has 6800 elements (resulting by subdividing M1 into 2×2 sub-elements) is used to guarantee results, which are mesh-independent. Note that for the ξ-solution of the convective-transport equation each parent element is further subdivided into 2×2 sub-elements according to the SUPG formulation.

Another feature of the solution is the proper imposition of a slip boundary condition along the solid die walls. There are two points of sharp change in the geometry at the beginning and the end of the conical section (points D and S, respectively, in Fig. 1). The consistent method of Engelman et al. [18] and Page et al. [19] for the incompressible case was modified to take into account the density changes due to compressibility assumption. Namely, the following formula, based on mass conservation, was used for the consistent normal, $\boldsymbol{n}_c$:

$$\boldsymbol{n}_c = \frac{\rho_1 L_1 \boldsymbol{n}_1 + \rho_2 L_2 \boldsymbol{n}_2}{\rho_c (L_1 + L_2)}, \quad (22)$$

where the symbols are defined in Fig. 1 (inset) for element sides 1 and 2, and common node C. The density ρ_c is the arithmetic average of the densities ρ_1 and ρ_2, which are themselves averages of the densities at the element nodal points for each side. Obviously, the consistent normal, and hence the local angle, depend on the element sizes that share the problem nodes (D and S in Fig. 1). We found that the correct implementation of the slip boundary condition was essential for obeying the mass balance (no loss of mass) and for obtaining good and convergent results either with sharp corners or with rounded ones.

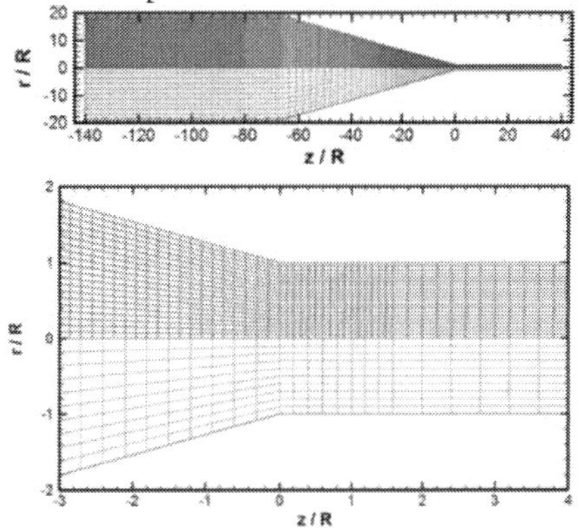

Figure 2 – View of the finite element meshes used in the simulations (upper half mesh M2 with 6800 elements, lower half mesh M1 with 1700 elements). Die design with 2α=30°, RR=352, L/D=20.

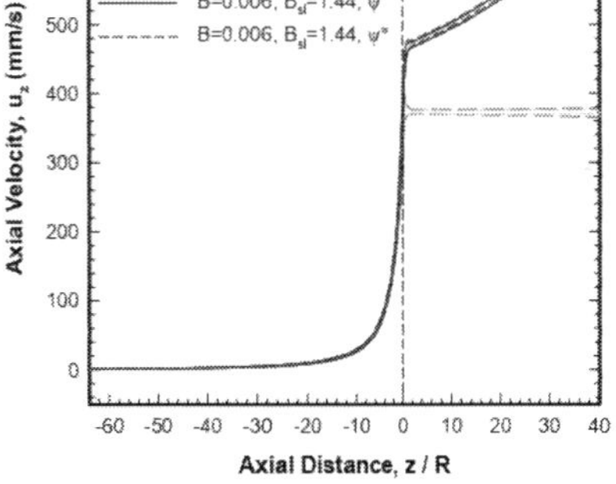

Figure 3 – Axial velocity distribution for PTFE paste extrusion at 35°C using the parameters listed in Table 1. Effects of compressibility and slip for the different flow type parameters.

Results and Discussion

Axial distributions in the flow field

Figure 3 shows the axial distribution of the axial velocity for the cases studied. Surprisingly, there are no significant differences in the results coming from using the non-objective, ψ, and the objective, ψ^*, flow type parameters, meaning that objectivity does not affect much the kinematics of the flow. For the case of incompressibility and slip (B=0, B_{sl}=1.44), the velocity rises to values of about 370 mm/s at the wall and 380 mm/s at the centreline in the die region. Thus, slip is substantial, as there is a small but discernible difference between centreline and wall velocities. For the case of both compressibility and slip, the velocity rises significantly higher in the die region, reaching at the exit 660 mm/s at the wall and 680 mm/s at the centreline. Thus, compressibility raises the velocity profiles as the material proceeds towards the exit, considerably. This is due to the decompression of the material as it approaches the exit where the pressure decreases substantially. This additional acceleration causes extension of the paste in the die that influences the ξ-parameter. Any small discrepancies with our previous work [2] in the numerical values are due to using here no-smoothing or rounding of the corners in the boundaries.

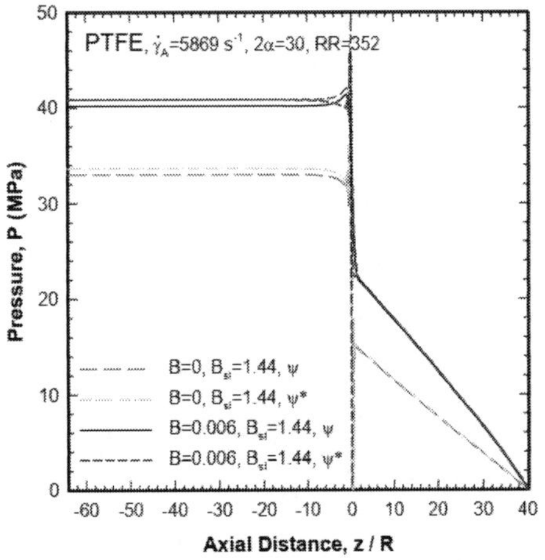

Figure 4 – Axial pressure distribution along the wall for PTFE paste extrusion at 35°C using the parameters listed in Table 1. Effects of compressibility and slip for the different flow type parameters.

Figure 4 shows the axial distribution of pressure for the cases studied. Again, there are no significant differences in the results coming from using the non-objective, ψ, and the objective, ψ^*, flow type parameters. If anything, objectivity serves to increase slightly the pressure (by 1.5%). Thus, for incompressibility and slip (B=0, B_{sl}=1.44), the pressure increases from 33 MPa to 33.5 MPa at entry. For the case of both compressibility and slip, the pressure is now higher, reaching about 41 MPa at entry. Thus, compressibility raises the pressure by about 20% compared with the incompressible case, a significant contribution.

Figure 5 depicts the axial distribution of the structural parameter, ξ, at the centreline and the wall for the cases studied. There are now considerable differences in the behaviour of ξ for all cases. The main differences with before (using the non-objective ψ-parameter), are the elevated values of ξ when using the objective ψ^*-parameter. At the centreline the structural parameter rises continuously from the entry to reach the maximum value of 1 in the conical region (pure elongational flow) before entry to the die, and then continues at that level towards the exit. This behaviour is the same for all the cases. It can be traced to the exponential increase of the elongational rate dv_z/dz, which according to the CTE give values of ξ=1. These values are then carried to the exit due to the convective nature of CTE. At the wall the structural parameter shows more pronounced differences for the two cases. First, we observe that in the tapered wall region, there are virtually no differences by assuming either compressibility or no. This is not the case, however, along the die wall, where compressibility serves to increase the structure levels markedly. The effect of the objective ψ^*–parameter is to raise rapidly the value of ξ in the conical region from 0 to values close to 1, and then as the material enters the die to decrease it somewhat and keep it constant along the die wall. This is in marked difference with the results obtained from using the non-objective ψ-parameter, where upon entry into the die, ξ decreases exponentially, to below 0.1 for incompressible conditions and to about 0.3 for compressible conditions (due to additional acceleration in the die). The new results, which show ξ slightly rising inside the die, mean that shear flow does not alter the structure of the material, a prediction that is in agreement with the experimental findings [1,20]. This effect of objectivity seems so far to be an important finding in the present work, i.e., it renders the model predictions to be in qualitative agreement with experimental observations.

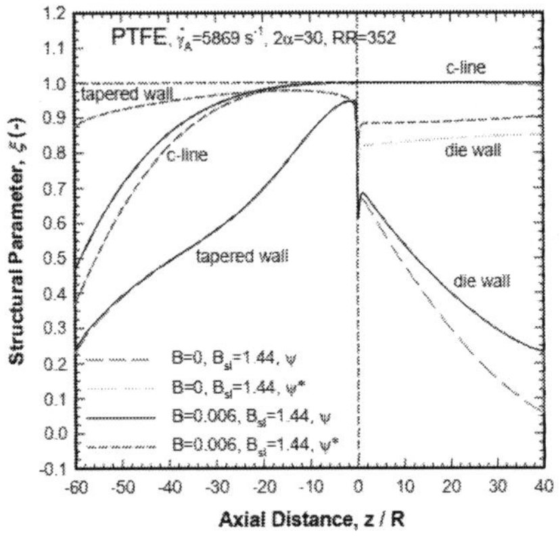

Figure 5 – Axial distribution for the structural parameter, ξ, along the centreline and the wall for PTFE paste extrusion at 35°C using the parameters listed in Table 1. Effects of compressibility and slip for the different flow type parameters.

Effect of flow and design parameters

The effect of the apparent shear rate for a given geometry (2α=30°) is shown in Fig. 6 for the objective parameter, ψ^*. The ξ-parameter increases in the tapered wall for all flow rates, but the decrease in the die is a strong function of apparent shear rate. A lower apparent shear-rate flow causes a stronger flow at the wall, i.e., slip is higher at lower apparent shear rate (see Table 2), although dv_z/dz is smaller. The structural parameter increases with apparent shear rate at the exit, which is again in agreement with experimental observations [1,20]. The increase of ξ towards the exit is due to acceleration in the die that is caused by the decompression of the compressible paste and by the low strength of the shear flow.

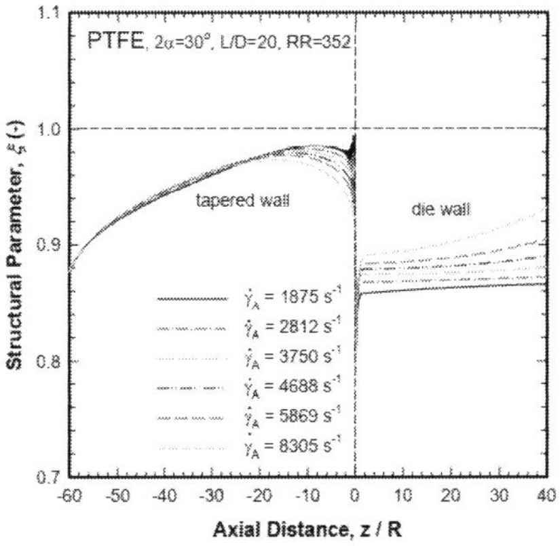

Figure 6 – The effect of apparent shear rate, $\dot{\gamma}_A$, on the structural parameter, ξ, using the objective ψ*-parameter. Axial distribution of ξ along the wall for PTFE paste extrusion at 35°C using the parameters listed in Table 1. Effects of compressibility and slip included.

The effect of the die entrance angle on the structural parameter ξ is depicted in Fig. 7 for the objective parameter, ψ*. We present results for the reference case of compressibility and slip at the apparent shear rate of $\dot{\gamma}_A$ = 1875 s^{-1}. According to Table 2, under these conditions B_{sl}=2.55 and B=0.003. The results are shown along the centreline (Fig. 7a) and the wall (Fig. 7b), respectively. At the centreline and in all cases of different angles ranging from 8° to 90°, ξ reaches 1 at the die entry and remains approximately constant. The shorter the conical die the sharper the increase in the value of ξ, which grows according to the elongational component of the flow. At the wall, the objective ψ*-parameter gives high values (Fig. 7a), with ξ increasing rapidly up to the die entry, but its maximum never reaches 1. Then after entry to the die, ξ follows a small decrease or increase depending on the angle, although for all practical purposes it may be taken that ξ remains constant, a prediction that is in agreement with experimental observations (explained above as well). On the contrary, use of the non-objective parameter ψ, caused ξ to decrease significantly in the die, which is in disagreement with experimental observations. This was found to be the case in all calculations with ψ (significant decrease of ξ in the die), and therefore no results using the non-objective ψ-parameter will be presented here.

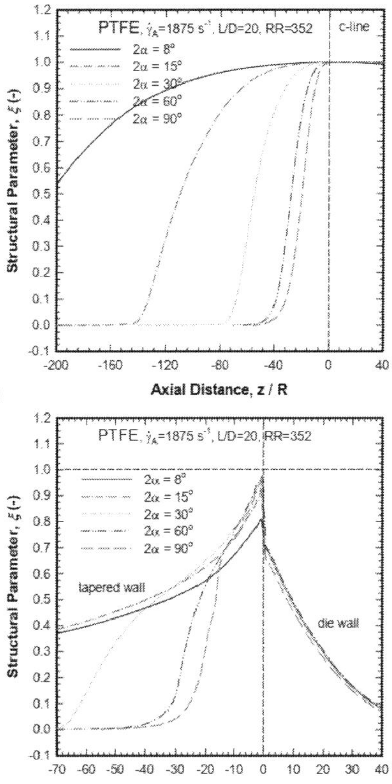

(a)

(b)

Figure 7 – The effect of die entrance angle, 2α, on the structural parameter, ξ, using the objective ψ*-parameter. Axial distribution of ξ along (a) the centreline, and (b) the wall for PTFE paste extrusion at 35°C using the parameters listed in Table 1. Effects of compressibility and slip included (B=0.003, B_{sl}=2.55).

Finally, the effect of L/D on the structural parameter for a given geometry (2α=90°) and apparent flow rate ($\dot{\gamma}_A$ = 1875 s^{-1}) is shown in Fig. 8. The objective parameter ψ* produces again much higher values, between 0.8 and 1 in the die land, while the non-objective ψ gave rapid exponential decays that are experimentally unreasonable (not presented here). A longer die allows the material to relax and to reach lower values of ξ. This has been observed experimentally that for long dies a small increase in the final strength of the samples has been observed [1,20].

Conclusions

The present results establish the behaviour of the structural parameter, ξ (structure formation), when employing an objective flow type parameter, ψ*, in the presence of compressibility and slip, in paste extrusion of PTFE resins through dies. The recently proposed model by Patil et al., [1] was solved for different cases of extrusion rate and die geometry, with emphasis on the effects of the objective flow type parameter, ψ*, as opposed to its easier non-objective counterpart, ψ, used in many cases before [1-2,10-14]. While the results for velocity and pressure are not affected appreciably, this is not the case for the ξ-parameter. The ξ-parameter is now much higher (closer to 1) in the die land for all cases. It depends much less than before (non-objective ψ) on die length, contraction angle, and of course on the flow rate, with higher flow rates and longer dies offering more structure in the material in paste extrusion. These predictions are consistent with experimental observations [1,20].

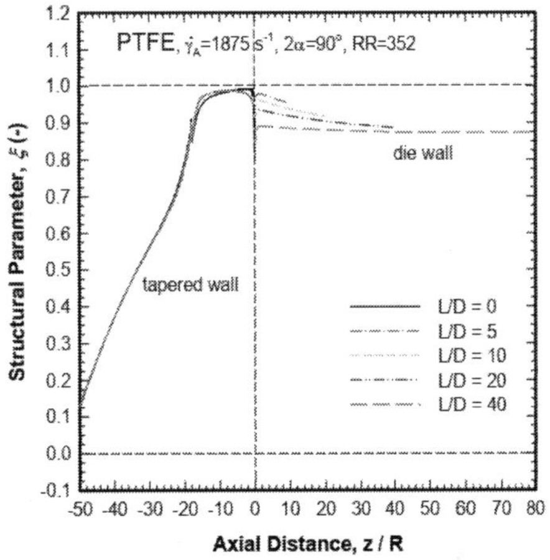

Figure 8 – The effect of die length, L/D, on the structural parameter, ξ, using the objective ψ^*-parameter. Axial distribution of ξ along the wall for PTFE paste extrusion at 35°C using the parameters listed in Table 1. Effects of compressibility and slip included.

Acknowledgements

Financial assistance was provided by the Natural Sciences and Engineering Research Council (NSERC) of Canada, and the National Technical University of Athens (NTUA), Greece, in the form of a basic research project, code-named "PEBE 2009-2011". This assistance is gratefully acknowledged.

References

1. P.D. Patil; J.J. Feng; S.G. Hatzikiriakos *J. Non-Newtonian Fluid Mech.* 2006, *139*, 44.
2. E. Mitsoulis; S.G. Hatzikiriakos *J. Non-Newtonian Fluid Mech.* 2009, *157*, 26.
3. H. Giesekus *Rheol. Acta* 1962, *2*, 112.
4. G.G. Fuller; L.G. Leal *Rheol. Acta* 1980, *19*, 580.
5. G. Astarita *J. Non-Newtonian Fluid Mech.* 1979, *6*, 69.
6. P. Singh; L.G. Leal *J. Rheol.* 1994, *38*, 485.
7. E. Ryssel; P.O. Brunn *J. Non-Newtonian Fluid Mech.* 1999, *85*, 11.
8. G. Mompean; L. Thais *J. Non-Newtonian Fluid Mech.* 2007, *145*, 41.
9. R.G. Larson *Rheol. Acta* 1985, *24*, 443.
10. C. Wang; I. Manas-Zloczower *Intern. Polym. Proc.* 1994, *9*, 46.
11. H.-H. Yang; I. Manas-Zloczower *Intern. Polym. Proc.* 1994, *9*, 291.
12. C. Wang; I. Manas-Zloczower *Intern. Polym. Proc.* 1996, *11*, 115.
13. C.-H. Yao; I. Manas-Zloczower *Intern. Polym. Proc.* 1997, *12*, 92.
14. H.F. Cheng; I. Manas-Zloczower *Polym. Eng. Sci.* 1997, *37*, 1082.
15. P.O. Brunn *Rheol. Acta* 2006, *46*, 171.
16. R.I. Tanner; R.R. Huilgol *Rheol. Acta* 1975, *14*, 959.
17. R.R. Huilgol *J. Non-Newtonian Fluid Mech.* 1980, *7*, 91.
18. M.S. Engelman; R.L. Sani; P.M. Gresho *Int. J. Numer. Meth. Fluids* 1982, *2*, 225.
19. M. Page; M.-I. Farinas; A. Garon *Commun. Numer. Meth. Engng.* 1997, *13*, 949.
20. I. Ochoa; S.G. Hatzikiriakos *Powder Technology* 2005, *153*, 108.

THERMAL STRESSES AND BIREFRINGENCE IN TUBES AND MULTI-LAYERED POLYMER SLABS: THEORY AND EXPERIMENT

Avraam I Isayev, Antonio J Carrillo, Nam Hyun Kim*

The University of Akron - Akron, OH - USA

Free quenching of tubes of PC and PS and three-layered slabs of PS-PC-PS and PC-PS-PC from various initial melt temperatures to below the glass transition temperature was carried out. The thermal residual stress and birefringence distributions along the thickness direction of the tubes and slabs were simulated and the birefringence was measured. Simulations were based on the linear viscoelastic and photoviscoelastic constitutive equations and the first-order rate equation for volume relaxation. The relaxation modulus and strain-optical coefficient functions of polymers as a function of time and temperature were used in the simulations. Comparisons indicated a reasonable agreement between the simulated and measured data.

NUMERICAL ANALYSIS OF THE MELTING PROCESS FOR BARRIER-FLIGHTED SINGLE-SCREW EXTRUDERS USING SCREW ROTATION PHYSICS

Gregory A. Campbell[1] *and Mark A. Spalding*[2]

[1] *Castle Research Associates, 1145 Mason Bay Road, Jonesport, ME 04649 swcgac@maineline.net*
[2] *The Dow Chemical Company, Midland, MI 48667 maspalding@Dow.com*

Abstract

The focus of this investigation was to develop a new model for the melting dynamics of the solid bed for barrier melting sections in single-screw extruders. This analysis is based on the screw rotation melting model presented earlier for conventional screws. In this paper we present a new physical concept for melting in a barrier screw for single-screw extruders that is consistent with more rapid loss of solid material in the cross-channel direction when compared to conventional screw melting discussed previously. We developed a dynamic melting model and then simulated this first-order theory. The simulation qualitatively fits the literature observations for melting in a barrier-flighted melting section of a screw.

Introduction

Barrier screws have been used commercially since they were first developed in the 1960s (Maillefer, 1962; Lacher, 1966). Since then, the design of the section has varied in many ways, all with the goal of increasing the melting capacity of the screw and decreasing the extrudate temperature. If the melting capacity of a screw can be increased the screw can be operated at higher screw speeds and thus higher rates. Typically, a well designed barrier screw can provide an increase in the rate of about 20 to 30% over a well designed conventional screw. The design features of a barrier section include the location or position downstream where the barrier section starts, the entry geometry, the lead length of the main flight and the barrier flight, the barrier flight undercut from the main flight, and the geometry of the exit of the section. The proper specification of these design features can be difficult to design while producing high rates, low extrudate temperatures, and without degradation of the resin.

Experiments for studying the single-screw extrusion melting mechanism were first carried out by Maddock (1959) and then by Street (1961). The melting mechanism was observed by performing Maddock solidification experiments. For these experiments, the extruder was fed with a mixture of natural and colored (1-3%) polymer. When steady state was achieved, screw rotation was stopped, and full cooling was applied to the barrel, solidifying the polymer in the channels. Usually, the original conditions existing in the extruder were retained. By examining the cross sections of the solidified polymer after taking it out of the screw channel, a clear picture of the melting behavior was obtained, as shown in Figure 1 for a conventional melting section.

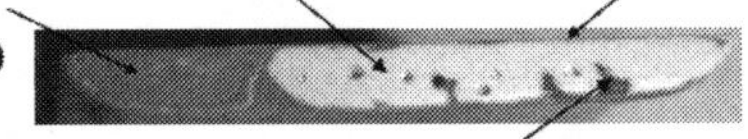

Figure 1. Photograph of resin solidified in a conventional melting section after a Maddock solidification experiment.

Maddock and Street's famous solidification experiments revealed that melting in many situations takes place in a specific order. The experiments showed that after melting began, there was a continuous solid bed and a melt film over the inner barrel surface. Later the solid bed was completely surrounded by melt. Farther downstream, a melt pool developed between the pushing flight and the solid bed. The solid bed disappeared gradually all the way down the channel (Maddock, 1959). Tadmor's research group (1966, 1978) observed similar melting phenomena. They studied many different materials which included low density polyethylene (LDPE), high density polyethylene (HDPE), acrylonitrile butadiene styrene (ABS), rigid polyvinyl chloride (PVC), and polypropylene (PP) resins using different operating conditions and extruder geometries. The results were interpreted to demonstrate that the Maddock melting mechanism prevails in quite a large range of melting processes with most resins. Other investigators' experiments led to other melting phenomena, such as the acceleration of the solid bed, break up of the solid bed, and a melt pool existing by the trailing flight (Agassant, 1991). Maddock first proposed the physical description of the melting process in 1959. In his description, the energy for the melting process came from two sources: energy conduction from the barrel and viscous dissipation from the film at the inner barrel surface. Beginning from 1966 when Tadmor (1966) set up the first mathematical model based on the Maddock melting mechanism, he proposed a reorganizing solid bed that led to a focus on the cross-channel direction (x-dimension) as rate controlling in the melting process. That is, the rate of melting was defined as the ratio of the solid bed width to the width of the channel.

Many other investigators continued working on improving the melting model in later years. Numerous papers have since been written which refine the Tadmor model; all are based on the hypothesis that the solid bed reorganizes as it melts leading to the bed being consumed in the cross-channel direction until it is completely melted. We will in this paper re-examine the assumptions in the literature models in order to address the classical melting data. The classical melting data originally presented by Tadmor and Klein (1978) for PE resin were examined using digital dimensional analysis (Campbell and Tang, 2004; Campbell et al., 2003), and the melting results for three specific runs are shown in Figure 2. These investigations were extended to other types of screw configurations. It was proposed that there was deceleration of the bed (Ingen, Housz, and Meijer, 1981) in a Barr barrier screw (Barr, 1972). A new model (Elbirli et al., 1983) adapted the historical single-screw melting model to a barrier melting section.

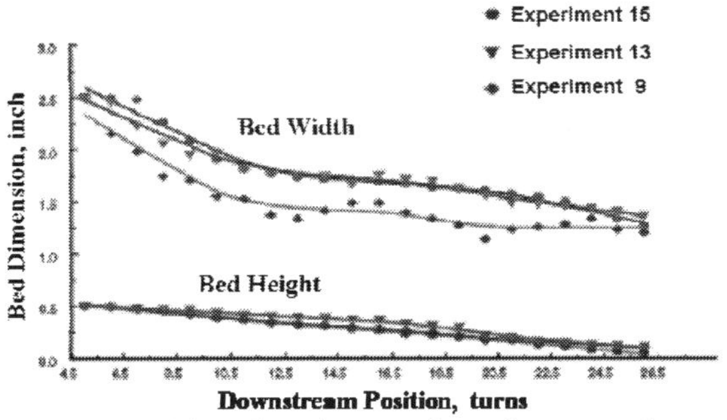

Figure 2. Melting data from Tadmor for PE resins (Campbell and Tang, 2004; Campbell et al., 2003). At a downstream position of 25 turns, the solids were completely melted and thus the bed width and bed height were zero.

In essentially all of the data sets found in in the literature (Tadmor and Klein, 1978), the melting results resemble the data shown in Figure 2. The data analysis is of course most difficult near the end of melting, and in essentially all cases the fraction of the bed remaining in the cross-channel direction is much greater than in the thickness direction. Also, in general the last third of the x direction material appears to melt in the last turn of the channel, creating nearly a step function in the melting rate with respect to the axial distance, as shown in Figure 2. Step functions, however, do not in general occur in processes that are heat conduction dominated. To reiterate, the digital analysis conclusively demonstrated that the y direction thickness always goes to zero before the x direction width (Campbell and Tang, 2004; Campbell et al., 2003). This observation led us to the development of a new physical model for single-screw melting that is based on these new observations (Campbell et al., 2009). Two different styles of barrier section geometries were studied here. Both geometries use a melting model that is an extension of the melting model developed for a conventional transition section. For the first geometry, a barrier flight was positioned at the start of the section at the pushing flight, and it was positioned with a lead length that was larger than that for the primary flight; i.e., a transverse barrier flight. This design caused the solids channel to decrease evenly in width in the downstream direction. The solids channel depth also decreased with the downstream direction. The transverse barrier flight merged with the trailing flight at the exit of the barrier section (Elbirli et al., 1983). The second geometry used a barrier flight that had the same lead length as the primary flight maintaining a solids channel with a constant width. The depth of the solids channel decreased in the downstream direction (Barr, 1972).

Theory Development

The new barrier section melting concept presented here was developed based on recognizing that the rate limiting melting dynamics for solid bed consumption was more complex than for melting in a conventional screw. Thus, we used a diagram that focused on the four melt films surrounding the solid bed to set the foundation for the development of the physics of melting in a barrier melting section, as shown in Figure 3.

Consistent with the results presented in Figure 2, separate melting dynamics were defined for the four melt films as originally defined by Pearson (1976), and as depicted in Figure 3 for the solids (or melting) channel; C is the film between the solid bed and the barrel wall, D is the film at the screw root (or core) polymer interface, E is the film at the trailing side of the primary flight, and B is the film on the pushing side of the barrier flight. The melted polymer flows over the barrier flight and into the melt conveying channel forming the melt pool.

We use screw rotation velocities in this analysis because the process is energy dissipation dominated (Campbell et al., 2001a, 2001b, 1992a, 1992b, 1996, 2008, 2009) and dissipation is not frame indifferent. We start with the analysis developed classically (Tadmor and Klein, 1978). One major difference is that the barrel is not moving so the velocities that are used in the simulation are adjusted to account for screw rotation dissipation. The diagram of this adaptation of the Tadmor analysis is shown in Figure 4 for zone C. We start by unwrapping the screw into x, y, and z coordinates. The barrel is hot as depicted by the red color and its velocity is zero. The velocity gradient in the z direction goes from V_{sz}, the solid bed velocity in the downstream directions (z direction), to 0 at the barrel interface. A similar diagram was developed for the other three fluid interfaces around the solid bed, remembering that for these liquid interfaces the metal velocity is the screw velocity, as shown in Figure 5.

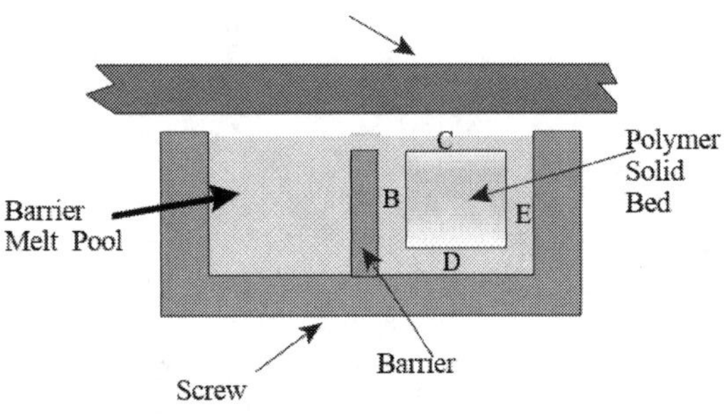

Figure 3. Schematic for the zones of the new barrier screw melting concept.

A major difference in this analysis is that we retain V_{sy} as a variable because this allows us to calculate the change in bed thickness in our re-analysis of the literature data, as shown in Figure 2. This provides the ability to evaluate the effect of the channel geometry on the rate of solid loss in both the cross-channel direction (x-direction) and the solid thickness direction (y-direction).

The nomenclature used here follows that used in the literature (Tadmor and Klein, 1978) as much as possible. Where deviations occur, a full explanation will be presented. A nomenclature list is presented at the end of the paper.

The development that follows is very similar to that found in our previous paper (Campbell and Spalding, 2009). The velocity of the solid bed in the z direction, V_{sz}, is as defined by Tadmor and Klein (1978). The velocity of the solid bed in the x direction, V_{sx}, due to screw rotation is as follows:

$$V_{sx} = \frac{\pi N D}{60}\sin\theta \tag{1}$$

$$\tan\theta = \frac{\pi L}{D} \tag{2}$$

where N is the screw rotation speed in rpm, D is the diameter at the surface of the solid bed, θ is the helix angle at diameter D, and L is the lead length of the flight.

We calculated the thickness of the melt film, δ_C, at film C using similar assumptions used by Tadmor and Klein (1978) and thus solve their equation (5-23) for $V_{sy,C}$, the velocity of the loss of the solid bed due to melting at δ_C:

$$V_{sy,C} = \frac{\left[\frac{k_m}{\delta_c}(T_b - T_m) + \frac{\mu V_j^2}{2\delta_c}\right]}{\rho_s\lambda + \left[\rho_s c_s (T_m - T_s)\right]} \tag{3}$$

where k_m is the thermal conductivity of the melt, T_b is the barrel temperature, T_m is the melting temperature of the resin, m is the shear viscosity, V_j is the vectorial velocity in the film (discussed later), r_s is the density of the compacted solid bed, l is the heat of fusion, c_s is the heat capacity, and T_s is the temperature of the solid bed. The term $V_{sy,C}$ here is equivalent to Tadmor and Klein's V_{sy}. Equation (3) is simply an energy balance for the melting process. Melting velocities at the other three interfaces $V_{sy,D}$, $V_{sx,B}$, and $V_{sx,E}$ are essentially identical to that for Equation (3) except that the vectorial velocity used corresponds to the specific film. These equations were omitted here.

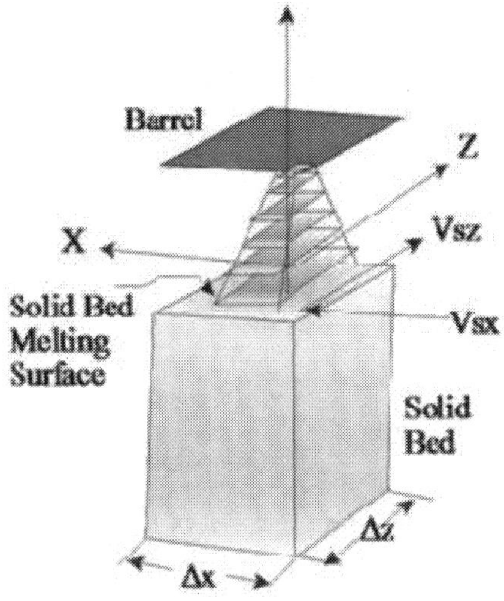

Figure 4. Melting differential element for a rotating screw for the melt film in zone C.

The melting velocities of the solid bed interfaces are the foundation of this new approach. The mathematical equality to calculate the rate of loss of the solid at the interface next to the barrel is found in Equation (4).

$$\left.\frac{dY}{dt}\right|_{\delta c} = V_{sy,C} \tag{4}$$

where Y is the local thickness of the solid bed and t is the independent time variable.

The melt film thickness for zone C, d_c, is determined using Equation (5). Note that V_{sx} is essentially equal to the traditional V_{bx} as specified by Tadmor and Klein (1978) because of the small value of the film thickness.

$$\delta_c = \left\{ \frac{\left[\left[2k_m(T_b - T_m) + \mu V_j^2\right] X/\alpha\right]}{V_{sx}\rho_m\left[c_s(T_m - T_s) + \lambda\right]} \right\}^{\frac{1}{2}} \quad (5)$$

where X is the local width of the solid bed and a is a fitting factor. Equation (5) is a mass balance for the zone C film and is essentially identical to Tadmor and Klein's equation (5-31). The flow equation is a first-order approximation since it neglects flows due to pressure gradients. Ideally, the value of a is set to 2 so as to provide an average thickness for the film in the cross-channel direction. The value of a was set to 0.5 for the simulation here to force the melting lengths to be equivalent to those in the literature.

The film thickness, d_e, for zone E was determined using a similar mass balance equation as follows:

$$\delta_e = \left\{ \frac{\left[\left[2k_m(T_b - T_m) + \mu V_j^2\right] Z/\alpha\right]}{V_{sz}\rho_m\left[c_s(T_m - T_s) + \lambda\right]} \right\}^{\frac{1}{2}} \quad (6)$$

where Z is downstream position of the solid bed. The melt film thicknesses, d_b and d_d for zones B and D were found by difference:

$$\delta_b = W(z) - X - \delta_e \quad (7)$$

$$\delta_d = H(z) - Y - \delta_c \quad (8)$$

where $W(z)$ is the width of barrier screw solids channel as a function of z and $H(z)$ is the local depth of the channel as a function of z. Both $W(z)$ and $H(z)$ are fixed by the geometry of the screw.

The boundary conditions for the velocity at the screw root interfaces are shown in Figure 5. The x and z component velocities of the screw are calculated as follows and they are used as the boundary conditions at the screw surface:

$$Vscrew\_x = \frac{\pi N D_s}{60}\sin\theta_s \quad (9)$$

$$Vscrew\_z = -\frac{\pi N D_s}{60}\cos\theta_s \quad (10)$$

$$\tan\theta_s = \frac{\pi L}{D_s} \quad (11)$$

where $Vscrew\_x$ is the velocity of the screw root in the x direction, $Vscrew\_z$ is velocity of the screw root in the z direction, D_s is the diameter at the screw root, q_s is the helix angle at the screw root. The direction of $Vscrew\_z$ is in the opposite direction of V_{sz}.

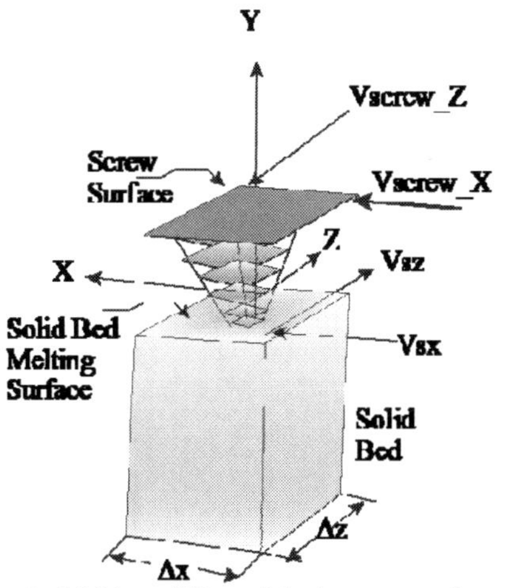

Figure 5. Melting differential element at the screw surfaces for a rotating screw reference frame.

The vectorial velocities for the four polymer films that account for the heat transfer and dissipative melting are as follows:

$$V_j\big|_C = \left(V_{sx}^2 + V_{sz}^2\right)^{1/2} \quad (12)$$

$$V_j\big|_D = \left|V_{screw\_z}\right| + V_{sz} \quad (13)$$

$$V_j\big|_E = \left|V_{screw\_z}\right| + V_{sz} \quad (14)$$

$$V_j\big|_B = \left|V_{screw\_z}\right| + V_{sz} \quad (15)$$

Note that the z velocity at the barrier flight is assumed to be the average of the z velocity at the screw root and flight tip.

The new concepts presented here remove the literature assumption that the solid bed reorganizes, and it allows melting at all solid bed interfaces. The classical Tadmor model allows melting at only a single interface; i.e., as specified by the melting velocity V_{sy} in zone C. The model presented here allows melting at all four interfaces: two in the y-direction and two in the x-direction. This facilitates determining the effect of the barrier flight on the solids melting rate.

We calculated the δ_d in film D by subtracting from the local channel depth, H, the δ_c at C and the remaining bed thickness Y_{remain}. Y_{remain} is the original bed thickness minus the loss predicted by the solid bed melting velocities, the two $V_{sy,C}$ and $V_{sy,D}$ equations at both the zone C and D fluid and solid bed surfaces. An artistic representation of the solid bed in a barrier section with a transverse barrier flight channel is shown in Figure 6. We depict the four films as the solid bed is consumed in the melting section of the screw.

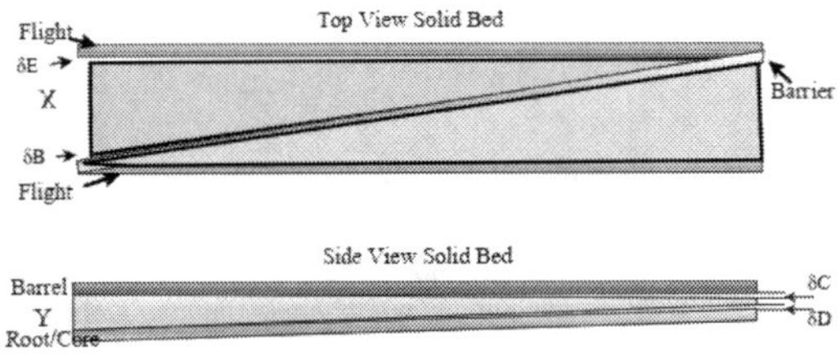

Figure 6. Qualitative shapes of x and y bed dimensions and melt film thicknesses for a barrier screw model with a transverse barrier flight. The solids channel is shown in green while the blue is the melt conveying channel.

The calculations for the x-direction melting are done in a parallel manor. First evaluate the melting at the δ_e solid interface, the surface adjacent to the trailing flight, and then calculate the size of δ_b at zone B, normally described as the melt pool (conventional melting section), in a similar manner as was done for film D. For the barrier section with the transverse barrier flight, the barrier flight position is moving closer to the trailing flight in the downstream direction, and the solids channel width is decreasing. Thus, the zone B thickness would be expected to be substantially smaller as the solids move down channel. Essentially all of the experimental data show an increase in the thickness of δ_e as the polymer solid bed progresses from the initial melting film to the end of melting at or near the end of the compression zone. We have incorporated this observation in our theory. We at this time do not address solid bed breakup since solids cannot go over the barrier flight into the melt conveying channel.

Results and Discussion

We are now in a position to determine if the proposed theory will produce results that are consistent with the observations of the literature data shown in Figure 2 and the expectation of a more rapid loss in the x direction for the barrier melting section case.

This new theory requires the solution of five coupled differential equations where x, y, and z are the local coordinates of the solid bed:

$$\left.\frac{dY}{dt}\right|_{\delta c} = V_{sy,C} \tag{16}$$

$$\left.\frac{dY}{dt}\right|_{\delta d} = V_{sy,D} \tag{17}$$

$$\left.\frac{dX}{dt}\right|_{\delta e} = V_{sx,E} \tag{18}$$

$$\left.\frac{dX}{dt}\right|_{\delta b} = V_{sx,B} \tag{19}$$

$$\frac{dZ}{dt} = V_{sz} \tag{20}$$

The first four of these equations were developed using physics and the analysis that led to Equations (3) through (6). We used a differential equation solver to obtain the results of the initial value problem.

Typical simulation results are presented in Figure 7 for a conventional screw. As shown by this figure, the Y thickness goes to zero at a down channel position of 250 cm into the melting section. We see that the change in the bed width (x direction) leaves about 60% of the original bed width remaining just at the end of melting. The end of melting is defined as when the y direction thickness of the solid bed goes to zero. This simulation was performed using a PE resin with a shear viscosity of 880 Pa-s.

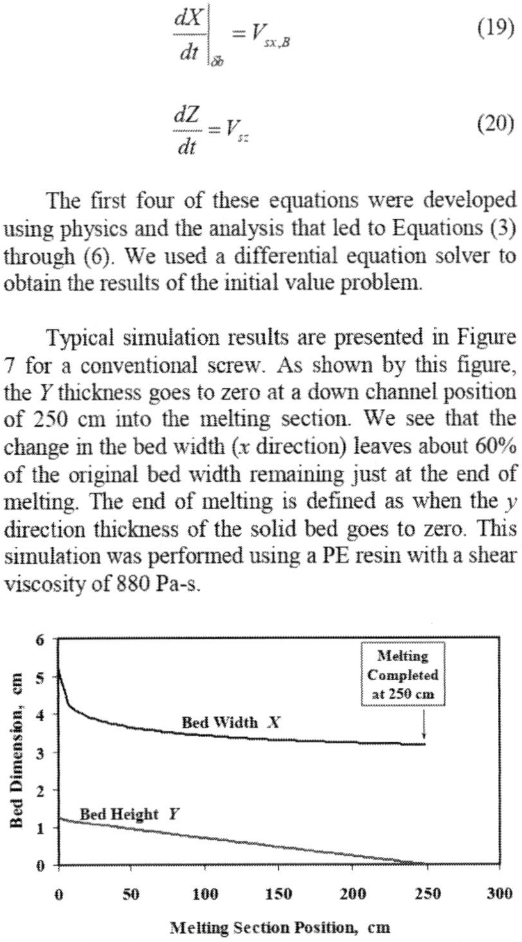

Figure 7. Simulation results using Example 5.1 (Tadmor and Klein, 1978) data. Melting was completed at 250 cm downstream from the start of melting for a resin with a shear viscosity of 880 Pa-s (Campbell and Spalding, 2009).

Now that it has been established that the model predicts the melting behavior demonstrated in Figure 2, the case of a barrier screw is evaluated. The focus of this first order simulation is on the physics that occur on the solid bed side of the barrier section. It is assumed that once the solid melts that the fluid polymer is carried across the barrier flight to the melt conveying channel.

The melting dynamics for a barrier section with a transverse barrier flight are presented in Figure 8. These initial results demonstrate interesting differences during the melting in a conventional screw with a constant taper for the depth and a constant channel width melting section, and a barrier section with similar characteristics except the channel width is decreasing linearly as the solid is transported from the entrance of the melting section to the end of the barrier section. The most dramatic observation is that the x direction bed thickness decreased much more rapidly in the transverse barrier case when compared with the conventional screw, as shown in Figure 8. There is also a more rapid decrease in solid thickness in the y direction at the end of melting in the barrier screw. Another observation for this set of simulation experiments is that it takes less down channel distance to complete the melting for the barrier section than for the conventional transition section. For these similar cases, melting was completed at 180 cm downstream for the transverse barrier section while the complete melting occurred at 200 cm for the conventional screw.

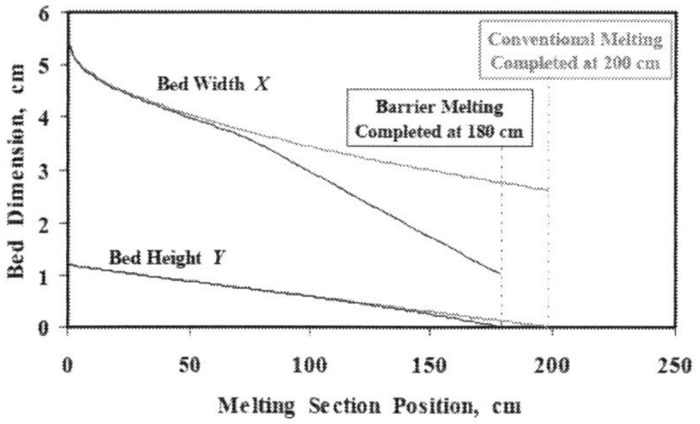

Figure 8. Comparison of melting dynamics for conventional melting channel and a transverse barrier melting channel. The conventional channel is in red while the barrier melting section is in blue.

A similar simulation was performed using a barrier melting channel with a barrier flight that is positioned parallel to the primary flight; i.e., the width of the melting channel is constant with the downstream direction. The melting dynamics for this case are shown in Figure 9. For this geometry, the solids channel at the entry to the barrier section is about 60% the width of the conventional melting channel. The smaller width of this parallel flighted barrier section is caused by the abrupt addition of the constant width melt conveying channel; i.e., the melt channel occupies the remaining 40%. This smaller width is apparent in Figure 9 at a melting section position of 0 cm. The velocity of the solid bed for the barrier case is 40% higher than that for the conventional screw in order to maintain the same flow rate. The melting dynamics presented in Figure 9 are similar to those for the conventional screw. That is, at the end of melting the thickness in the y direction goes to zero while the width of the solid bed is about 40% of the original width. Moreover, this barrier section geometry required a melting length of 235 cm to complete as compared to 200 cm for the conventional screw.

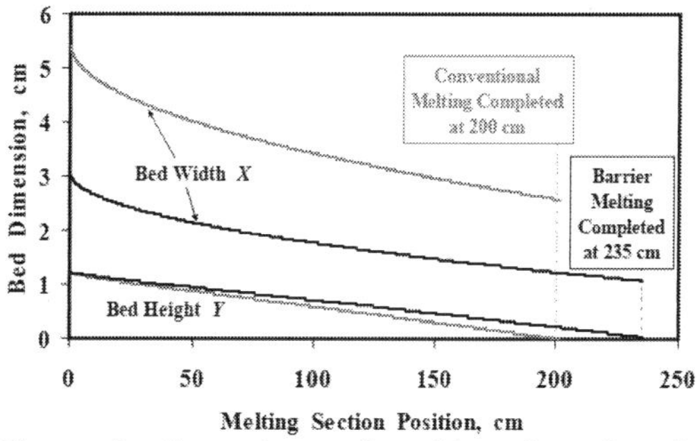

Figure 9. Comparison of melting dynamics for conventional melting channel and a barrier section with the barrier flight parallel to the primary flight. The conventional channel is in red while the barrier melting section is in black.

The simulations suggest that conventional screws may have a higher or lower melting capacity, measured by the distance traveled down the screw, than barrier-flighted melting sections dependent on the barrier design. However, in both designs the melting time is less than the conventional screw. Commercial practice, however, shows that barrier screws can operate at rates 20 to 30% higher than conventional screws. The combined data suggest that barrier screws can be operated at higher rates via higher specific rates by maintaining the solid bed intact. The higher specific rates also allow the extruder to discharge at lower temperatures. Conventional screws will discharge solid polymer fragments into the extrudate when the melting capacity is exceeded due to solid bed breakup, especially for high screw speeds (Spalding and Hyun, 2003). The barrier section entry position and design are also known to be problematic (Hyun et al., 1995; Spalding, 2004). That is, if the design of the entry section and position downstream for the screw start are not correct, the entry section can cause a restriction and limit the rate of the extruder.

The results presented here are encouraging but only qualitative and have been produced using this first-order model. We will be continuing to refine this model and intend to evaluate its quantitative predictions in the future. Current limitations of the model are the use of a constant viscosity function independent of temperature and shear-rate. Also, the dynamic local temperature of the barrel and screw (Campbell and Spalding, 2009) must be incorporated into the model; they are currently set as constants. An enhanced model for the film thickness at both the barrel and screw surfaces will also be added to the current model along with flows induced by pressure gradients.

Conclusions

A new model is proposed for the melting of the solid polymer bed for screws with barrier melting sections. The model alters the heat transfer and melting velocities originally proposed by Tadmor and Klein (1978). The model releases three of the original literature assumptions: a) elimination of the V_{sy} melting velocity in the y direction of the bed, b) the concept that the bed rearranges, and c) barrel rotation dissipation dynamics. This required recasting the heat conduction and energy dissipation analysis using screw velocities and extending the material balances to all four bed solid interfaces at the fluid films. This model for barrier melting sections suggests that by placing the barrier flight in the transition section the rate of melting increases leading to a shorter time to complete melting for both barrier types. This is consistent with both the experimental and theory presented previously (Campbell et al., 2009) where the rate controlling step is the loss of the bed in the y direction. The model predicts the dynamics of melting previously reported (Campbell et al., 2003, 2004) as determined by digital dimensional analysis of the solids remaining in the bed, as shown in Figure 2 for a conventional screw. The model predicts that it may take a longer melting section to melt all of the polymer using a barrier screw due to the increased bed velocity, but since solids cannot, however, get over the barrier flight no solids would be expected in the melt exiting the extruder at normal screw speeds.

Nomenclature

c_s heat capacity of the solid.
D diameter at the surface of the solid bed.
D_s diameter at the screw root.
H local channel depth.
k_m thermal conductivity of the melt.
L lead length of the flight.
N screw rotation speed in rpm.
T independent time variable.
T_b barrel temperature.
T_m melting temperature of the resin.
T_s temperature of the solid bed.
V_j vectorial velocity in the films.
V_{bx} velocity of the barrel in the x direction for a barrel rotation reference frame (not used).
V_{sx} velocity of the solid bed in the x direction.
$V_{sx,B}$ velocity of the solid bed consumption in the thickness direction at the zone B interface (x direction).
$V_{sx,E}$ velocity of the solid bed consumption in the thickness direction at the zone E interface (x direction).
V_{sy} velocity of the solid bed consumption in the thickness direction at the zone C film (Tadmor).
$V_{sy,C}$ velocity of the solid bed consumption in the thickness direction at the zone C interface (y direction).
$V_{sy,D}$ velocity of the solid bed consumption in the thickness direction at the zone D interface (y direction).
V_{sz} velocity of the solid bed in the z direction.
V_{screw_x} velocity of the screw root in the x direction.
V_{screw_z} velocity of the screw root in the z direction.
W width of the channel perpendicular to the flight.
X width of the solid bed.
x independent variable in the cross-channel direction.
Y thickness of the solid bed in the y direction.
Y_{remain} original bed thickness minus the loss predicted by solid bed melting.
y independent variable in the channel-depth direction.
Z downstream position of the solid bed.
z independent variable in the down-channel direction.
α fitting factor.
θ helix angle at diameter D.
θ_s helix angle at the screw root.
σ_m density of the molten polymer.
σ_s density of the compacted solid bed.
λ heat of fusion.
δ_b thickness of melt film B.
δ_c thickness of melt film C.
δ_d thickness of melt film D.
δ_e thickness of melt film E.
μ shear viscosity.

Acknowledgments

We would like to thank The Dow Chemical Company, Becton Dickinson, Eastman Chemical, XALOY, American Leistritz, 3M, and Pechiney. Much of this work was initiated in the Polymer Extrusion Consortium at Clarkson University.

References

Agassant, F. , Avenas, P. , Carreau, P.J., and Sergent, J.Ph., "Polymer Processing: Principles and Modeling," Oxford University Press, New York, 1991.

Barr, R.A. "Extruder, Or Extruder-Like Melting Apparatus," US Patent 3,698,541 (1972).

Campbell, G.A., Sweeney, P.A. and Felton, J.N. "Experimental investigation of the Drag Flow Assumption in Extruder Analysis," *Polym. Eng. Sci.*, **32**, 1765 (1992a).

Campbell, G.A., Sweeney, P.A., and Felton, J.N., "Analysis of an Alternative Extruder Screw Pump Design," *Intern. Polym. Proc.*, **7**, 240 (1992b).

Campbell, G.A, Sweeney, P.A., Dontula, N., and Wang C., "Frame Indifference: Fluid Flow in Single Screw Pumps and Extruders," *Intern. Polym. Proc.*, **11**, 3 (1996).

Campbell, G.A., Cheng, H., Wang, C., Bullwinkel, M. and te-Riele, M. A, "Temperature Rise in Single Screw Pump-Extruders," *SPE-ANTEC Tech, Papers,* **47**,152 (2001a).

Campbell, G.A., Cheng, H., Wang, C., Bullwinkel, M. and te-Riele, M.A., "An Investigation of Flow Rate and Viscous Dissipation in the Single Screw Pump-Extruder," *Intern. Polym. Proc.*, **16**, 323 (2001b).

Campbell, G.A., Tang, Z., Wang, C., and Bullwinkel, M., "Some New Observations Relative to Melting in Single Screw Extruders," *SPE-ANTEC Tech. Papers*, **49**, 213 (2003).

Campbell G.A. and Tang, Z., "Solid Bed Melting In Single Screw Extruders - An Alternative First Order Mechanism," *SPE-ANTEC Tech. Papers*, **50**, 162 (2004).

Campbell, G.A., Spalding, M.A, and Carlson, F., "Prediction of Melt Temperature Rise in Single-Screw Pump Extruders," *SPE-ANTEC Tech. Papers*, **54**, 267 (2008).

Campbell, G.A., Spalding, M.A, and Tang, Z. "An Alternative Analysis of Single-Screw Melting," *SPE-ANTEC Tech. Papers*, **55**, 147 (2009).

Elbirli, B., Lindt, J.T. ,Gottgetreu, S.R., and Baba, S.M., "Mathematical Modeling of Melting of Polymers in Barrier-Screw Extruders," *Polym. Eng. Sci.*, **23**, 86 (1983).

Hyun, K.S., Spalding, M.A., and Powers, J., "Elimination of a Restriction at the Entrance of Barrier Flighted Extruder Screw Sections," *SPE-ANTEC Tech. Papers*, **41**, 293 (1995).

Ingen Housz, J.T. and Meijer, E.H., "The Melting Performance of Single-Screw Extruders," *Polym. Eng. Sci.*, **21**, 352(1981).

Lacher, F.K. "Extruder," US Patent 3,271,819 (1966).

Maddock, B.H., "A Visual Analysis of Flow and Mixing in Extruder Screws," *SPE J.*, **15**, 383 (1959).

Maillefer, C. Swiss Patent 363,149 (1962).

Street, L.F. "Extrusion Plastification," *Int. Plast. Eng.*, **1**, 289 (1961).

Pearson, J.R.A., Halmos, A.L., and Shapiro, J., "Melting in Single Screw Extruders," *Polymer*, **17**, 905 (1976).

Spalding, M.A. and Hyun, K.S., "Troubleshooting Mixing Problems in Single-Screw Extruders," *SPE-ANTEC Tech. Papers*, **49**, 229 (2003).

Spalding, M.A., "Metering Channel Flows and Troubleshooting Single-Screw Extruders," *SPE-ANTEC Tech. Papers*, **50**, 329 (2004).

Tadmor, Z., "Fundamentals of Plasticating Extrusion. I. A Theoretical Model for Melting," *Polym. Eng. Sci.*, **6**,185(1966).

Tadmor, Z. and Klein, I., "Engineering Principles of Plasticating Extrusion," Robert E. Krieger publishing company, New York, 1978.

A NEW TYPE OF MIXER BASED ON ELONGATIONAL FLOW: NUMERICAL SIMULATION AND FLOW VISUALIZATION

C. Loux[1]*, R. Muller[1], G. Bouchet[2], M. Bouquey[1]

[1] *Laboratoire d'Ingénierie des Polymères et des Hautes Technologies (LIPHT), 25 rue Becquerel, 67200 STRASBOURG; cyril.loux@ecpm.u-strasbg.fr*
[2] *Institut de Mécanique des Fluides et des Solides (IMFS), 2 rue Boussingault, 67000 STRASBOURG*

Abstract - A new lab-scale mixing device based on an original concept was built and tested. This device presents important technical features like tightness to liquids and gases, the possibility of direct moulding of specimens after mixing and easy handling of reactive systems. Compared to existing laboratory mixers, the flow in the mixer is characterized by a high contribution of elongational flow. Morphological data on model PS/PMMA blend systems proved the high distributive and dispersive mixing efficiency as compared to a classical rotational batch mixer. To better understand the origin of distributive mixing in the system, finite element numerical simulations were carried out together with an experimental study for visualization of the flow. For this purpose, experiments were carried out at room temperature in a transparent system with the same geometry. The fluids were either Newtonian or viscoelastic polymer solutions containing fluoresceine and the flow was visualized with a linear laser beam. For the numerical simulations, we used the passive scalar method and more recently the area tensor method. The first conclusions which came out from the simulations and visualization experiments are the following: In creeping low conditions (Re below 0.001) the viscosity gradient in the chambers due to viscous dissipation in the mixing element leads to flow irreversibilities and eventually distributive mixing. At inertial flow conditions (Re around 10) and negligible viscous dissipation, downstream vortices appear leading to irreversible flow and mixing. The geometry of the mixing element and in particular the contraction angle between the chamber and the static mixing element was shown to have a significant influence on mixing efficiency. A quasi-Newtonian model which allows taking into account strain-hardening effects in elongational flow was used for the simulation of viscoelastic fluids.

Introduction

The blending of two polymers has been the subject of fairly extensive studies. This topic directly concerns the polymer processing industry which is constantly looking for new methods to obtain materials with improved properties.
The usual mixing equipments, such as internal mixers or extruders, mainly generate shear flows and their efficiency is limited, especially when a high viscosity phase is to be dispersed in a low viscosity matrix.
Several devices, designed to generate elongational flows which are assumed to be more efficient for dispersive mixing, have been described in the littérature. In particular, Meller et al. (1999, 2002a, 2002b), studied the deformation and breakup of dispersed droplets in molten polymer blends of different viscosities. They used a capillary rheometer equipped with dies having different entry profiles and showed that the mixing efficiency in the converging flow zone was dependent both on the shape of the convergence and on the flow rate.

The use of convergent (and divergent) elongational flow to enhance the efficiency of mixing in an industrial process has been limited up to now, mainly due to the high pressure drops which arise if several convergent/divergent flow units are associated in series for instance at the exit of a continuously operated extruder (Meller et al. (1999) and Tokihisa et al. (2006)). On the other hand, the flow between two opposite chambers through a small diameter die was first described by Hausman (1957 and 1960) and later used by Westofer (1961) to measure the pressure dependence of the viscosity of polymer melt. Mackley et al. (1995), adapted this geometry to design their so-called "multipass rheometer" in which the influence of the number of passes through the central die on the rheological properties can be studied. Recently, Son (2009) has shown that the concept of the multi-pass rheometer can be adapted to design a batch mixer in which an unlimited number of convergent/divergent flows can be applied to the material to be mixed.

In the present paper, we describe a new mixing device (Terrisse (2007)) based on the geometry of Mackley's multi-pass rheometer and whose design was significantly modified with respect to Son's microcompounder, thus allowing original operation modes. The efficiency of our device as a mixer was tested on model immiscible polystyrene/polymethylmethacrylate (PS/PMMA) blends for which easy and accurate morphology characterizations by TEM are possible. A finite element numerical simulation of the flow within the mixer was carried out with the objective of explaining the high distributive and dispersive mixing capability of the device.

Principle

The recently built device, later called RMX (elongational flow Reactor and MiXer), is schematically represented in Figure 1. The material to be mixed is alternately pushed from one cylindrical chamber to the other through a central static mixing element by two reciprocally moving pistons. For all results presented in this paper, the mixing element is composed of a small diameter cylindrical die. The convergent and divergent elongational flows at the entrance and exit of the die are expected to contribute significantly to dispersive mixing. The following original features characterize the RMX device:

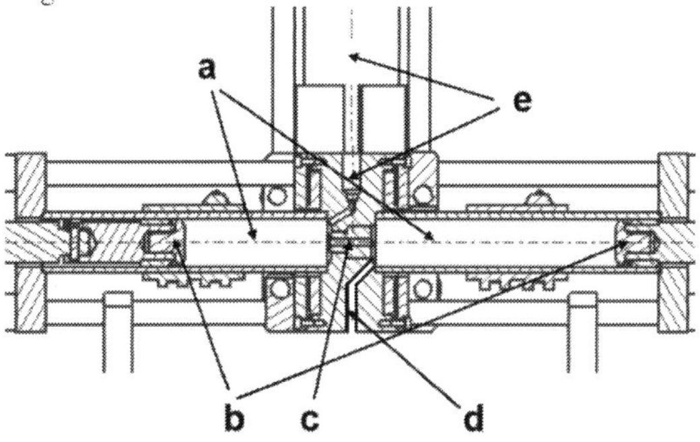

Figure 1 – Schematic drawing of the mixing device. a: mixing chambers, b: pistons, c: mixing element, d: feeding unit and feeding channel, e: outlet channel (sampling or molding)

Feeding of components

In addition to the central channel which connects the chambers (mixing channel), the mixing element is fitted with three additional channels, one for the feeding of the components to be mixed, one for the outlet of the mixed material

The feeding of a component into the mixer is schematically represented in Figure 2a: the material flows through the feeding channel into the chamber on the left side, a small counter pressure is applied on the corresponding piston which moves back until the total volume of material is reached. At the same time, the piston in the opposite chamber is in a fixed position thus preventing the material to flow to the right through the mixing channel.

However, it is also possible to feed sequentially several components: two for instance mutually reactive components A and B can be independently fed into the mixer in the following way: component A is introduced first following step 2a. It is then transferred into the right chamber through the mixing channel by the displacement of the left piston. Component B is the introduced according to step 2a, the left piston moving back again and the right piston remaining in a fixed position thus preventing component B from flowing into the left chamber, and minimizing the area of the A/B interface before the beginning of the mixing step.

Mixing

Figure 2b represents schematically the reciprocal displacement of the pistons during the mixing step. At this time, both the feeding and the outlet channels are closed. The volume of material inside the mixer is adjusted by the initial positions of the pistons and can be varied in the range [10-100 cm^3]. The pistons are hydraulically driven by a servovalve at controlled speed in the range [0.18 – 180 mm/s] which for 3cm diameter pistons corresponds to volume flow rates in the range [0,125 – 125 cm^3/s]. The maximum pressure in the upstream chamber is limited by the power of the hydraulic station and is therefore a function of the flow rate. Typically at the highest flow rate, the upstream pressure can reach around 150 bars. A mixing sequence is defined by the piston speed (or flow rate) and the number of cycles, and an experiment can be composed of several consecutive sequences at different flow rates. The pressure in one of the chambers is continuously measured during the mixing sequence by a pressure transducer (Dynisco range of pressure [0-700 bars]), connected to this chamber through the mixing element. Since the flow is reversed during each cycle, only one pressure transducer is required to measure the pressure drop between the two chambers: As a matter of fact, the transducer measures alternately the upstream and the downstream pressure at each flow reversal, so that the pressure drop between the chambers is merely the difference of both pressure measurements.

Outlet of the mixed material.

At the end of the mixing step, all material is first pushed into the right chamber connected with the outlet channel. As shown in Figure 2c, the outlet channel can be used to take samples at the end of an experiment (or even during an experiment between two mixing sequences), or can be connected to a mould. During moulding, the left piston is in a fixed position and the right piston pushes the material into the mould. During this step the device is working like an injection machine. Unlike for other types of microcompounders (Twin-screw or Brabender type), no additional step (like recovering the melt, placing it into a mould, reheating and moulding) is therefore required to mould specimens of specified shape after mixing.

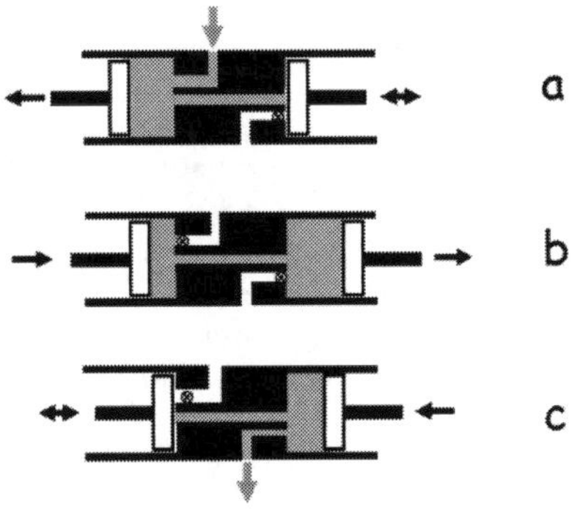

Figure 2 – Schematic operation of the mixer

It is important to notice that the central part of the mixing element is removable and therefore the shape and diameter of the central mixing channel can be easily varied.

Preparation of blends

Two blend compositions were investigated in the present study: PS/PMMA 10/90 in weight for which the ratio of dispersed phase over matrix viscosity is equal to p=0.25 (low viscosity droplets dispersed in higher viscosity matrix) and PS/PMMA 90/10 in weight for which the ratio of dispersed phase over matrix viscosity is equal to p=4 (high viscosity droplets dispersed in lower viscosity matrix). The components were filled into the mixers in the form of a dry blend of pellets.
The morphologies of blends obtained on the RMX device with two different mixing éléments (long die and short die geometries) were compared to those obtained with an internal batch mixer (Haake Rheomix 600 equipped with counter-rotating roller-type blades). The initial temperature at the beginning of the mixing step was set to 210°C for all tests. Depending on the mixing conditions, a temperature rise due to viscous heating was observed which could reach about 20°C.

Morphological characterization

At the end of the mixing step, the material was cooled down to room temperature. Transmission electron microscopy (TEM) was used (Hitachi H7500 operated at 60 kV) to characterize the particle size distribution within the blends. 50 and 80 nm cuts were prepared with a Leica EM Ultracut 6 microtome equipped with diamond knives. For the system PS/PMMA, the phases are clearly identified in TEM, the PS phase appearing in dark and the PMMA in white. For each sample, at least 3 cuts taken at several positions were analyzed and the particle size distribution was determined using image analysis (Aphelion software) on micrographs corresponding to 40 μm x 40 μm domains. The number- and volume-average particle diameters values, Dn and Dv, were only discussed on a comparative basis as a function of mixing conditions, and were therefore directly determined from these micrographs without any correction like that of Schwartz–Saltikov which is often used for polymer blends40.

Comparison of morphologies of blends obtained on RMX-LDME and on Haake Rheomix

Figure 3 shows the values of Dn and Dv as a function of the specific energy input for PS/PMMA 90/10 blends obtained in the RMX with the long die mixing element at a piston speed of v=10 mm/s in the Haake Rheomix internal batch mixer at 100 rpm. It is found that at equivalent energy input, a finer dispersion is obtained with the RMX, which can be attributed to the higher elongational flow component. It should be recalled that for the blend considered here, the ratio of dispersed phase over matrix viscosity p is equal to 4, and that for this value, the critical capillary number in simple shear becomes infinite [Grace], thus confirming the importance of the elongational flow component in the break up process.

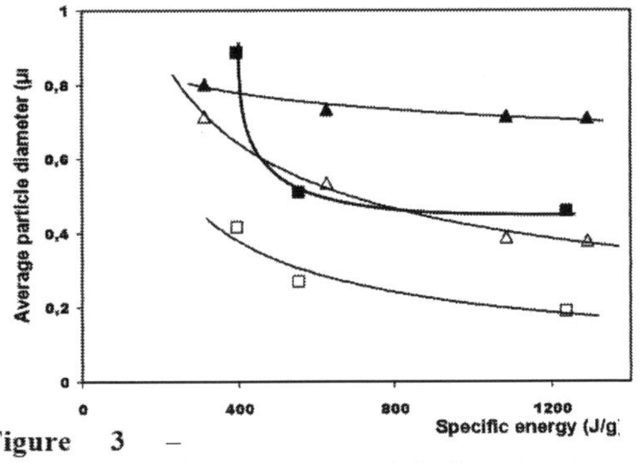

Figure 3 – Number and volume average particle diameters in mm versus specific energy input in J/g. PS/PMMA 90/10, v=10 mm/s. □ : Dn long die ME; ■ : Dv long die ME; Δ : Dn Haake Rheomix; ▲ : Dv Haake Rheomix.

Influence of the number of cycles

Whatever the blend composition (10/90 or 90/10), the geometry of mixing element or the value of the piston speed, a minimum number of about 20 cycles was necessary to obtain a dispersion-type morphology. As seen in Figure 4a corresponding to a PS/PMMA 10/90 blend obtained with the long-die mixing element and a piston speed of v=10 mm/s, after 10 cycles a complex morphology with large PS domains containing themselves PS inclusions is observed. After 40 cycles (Figure 4b), a dispersion-type morphology is reached for which the average particle diameters can be determined.

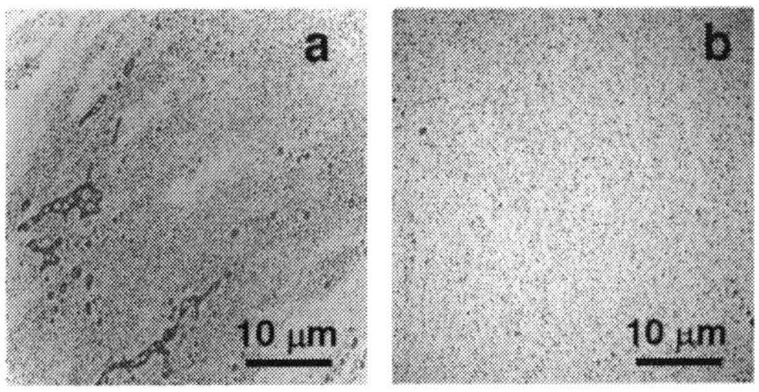

Figure 4 – TEM micrographs for two different numbers of cycles. PS/PMMA 10/90: v=10 mm/s, a: N=10, b: N-40

A general observation for all mixing conditions is a rapid initial decrease of particle size with number of cycles or energy input. Most often the average particle values reach a plateau value indicating a steady-state blend morphology (Figure 5). On the other hand, small particles are present from the early steps of the mixing process and increasing the number of cycles leads to the progressive break-up of the largest particles and to a decrease of the particle size polydispersity.

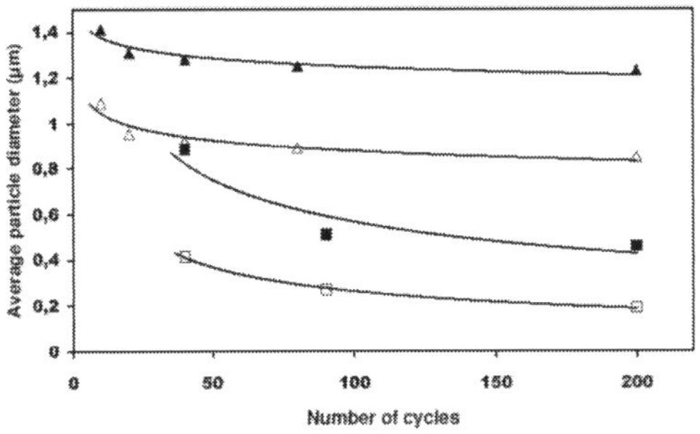

Figure 5 – Number and volume average particle diameters in mm versus number of cycles. PS/PMMA 90/10, v= 10 mm/s. □ : Dn long die ME; ■ : Dv long die ME; Δ : Dn short die ME; ▲ : Dv short die ME.

Transparent prototype

To visualize the flow inside the RMX mixer, a transparent prototype device was built. Product is illuminated with an argon laser. A camera is installed on a perpendicular plane to the laser cut.

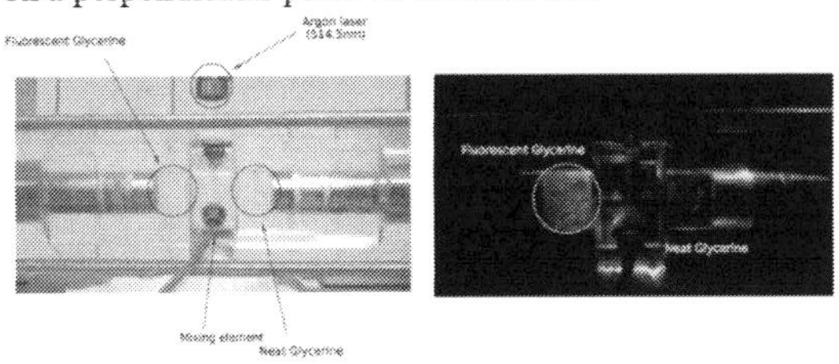

Figure 6 – Transparent protoype: view from camera

Products use for newtonian validation experiments is glycerine. At low Reynolds numbers (Re<0.1), the flow is a stokes and it is reversible so no mix appear. At inertial flows (Re>0.1), reversibility is broken due to recirculation in divergent regions, at the outlet of the mixing element.

For the viscoelastic approach, we worked with polyvinylpyrrolidone. In this case, vortices are present in the convergent region, even at low Reynolds numbers. In divergent regions, there is a striation of the material (Figure 7).

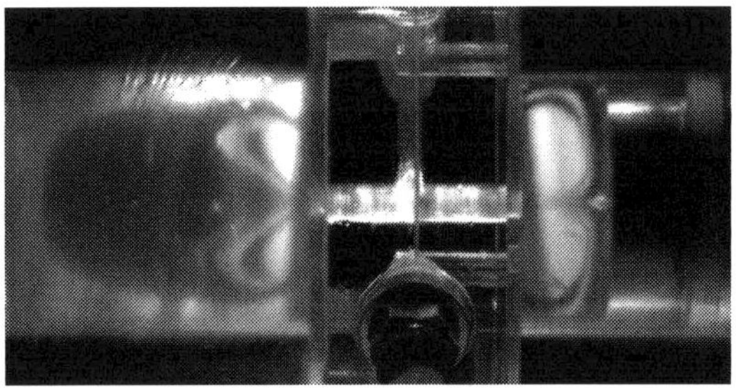

Figure 7 – Non-Newtonian flow (Re=0.00001). Pistons go from right to left.

Numerical study

A numerical simulation of the flow within the RMX mixer was carried out with the objective of characterizing the relative importance of elongational and shear flows. The flow was assumed to be axisymmetric and the volume within both chambers and in the mixing element was discretized using a structured mesh. A non-slip boundary condition was imposed on the walls and surface of the pistons and the pistons move with a constant velocity parallel to the axis. The total volume is conserved during the simulation. To solve this problem, we use a finite volume code which allows solving the continuity and Navier-Stokes equations for an incompressible fluid. From the viscosity versus shear rate curves, the parameters of a Carreau-type equation could be identified for each polymer and taken into account in the simulation by assuming that the material follows a generalized Newtonian equation of state (Bird, Wiest, 1995). Since the geometry is in motion, remeshing of the volume is required at each time step. The solver compresses the cells on one side and expands them on the other side. The mesh displacement is taken into account in the calculations using the ALE method (simple interpolation).

We also performed numerical simulations taking into account the temperature-dependence of the viscosity (through an Arrhenius-type equation (eq.1)).

$$\mu = \mu_0 . exp\left(\frac{E_a}{R}\left(\frac{1}{T} - \frac{1}{T_0}\right)\right) \qquad (1)$$

When the piston velocity and the pressure drop increase, viscous dissipation and temperature rise are increased (Figure 8) and the viscosity decrease is more important (Figure 9).

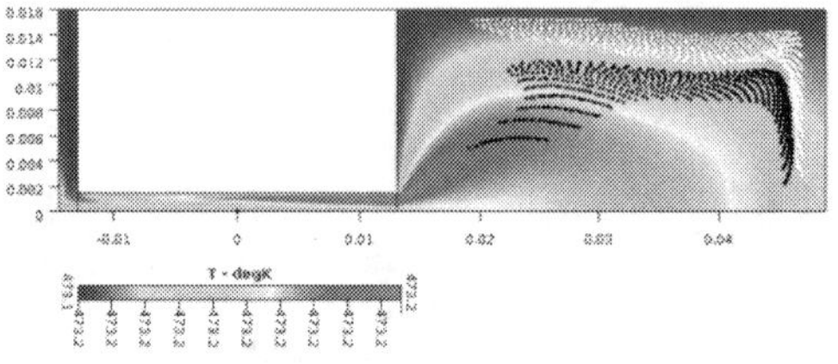

Figure 8 – Increased of temperature. Pistons go from left to right.

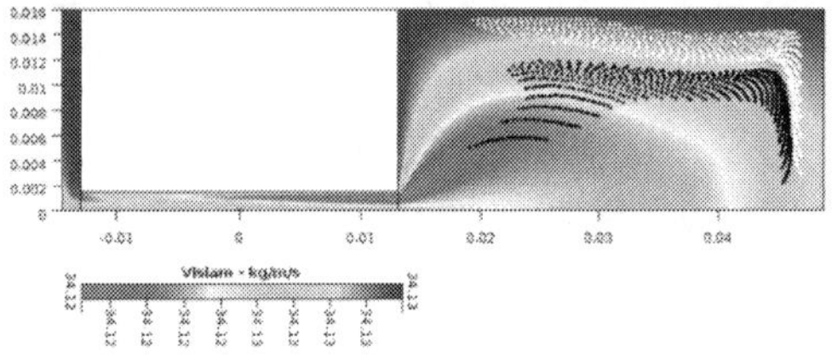

Figure 9 – Decreased of viscosity. Pistons go from left to right.

Whereas the deformation experienced by a fluid particle is mainly simple shear in the canal where the x-component of the velocity is much higher than the y-component, the flow tends to be closer to pure elongation in the convergent and the divergent sections of the mixing chambers. To visualize the mixing, we used a passive scalar (represented by a grid of points). In the Newtonian case at low Reynolds numbers (1.10^{-3}), the grid almost exactly returns to its initial shape after one complete cycle (no mixing). A difference was observed in the case where we solved simultaneously the energy equation. As a consequence of viscous dissipation, the size of the corner vortices increases, and the grid becomes stretched (Figure 10). Even if vortices are present, it seems that they are not due to inertial effects but rather to viscous dissipation.

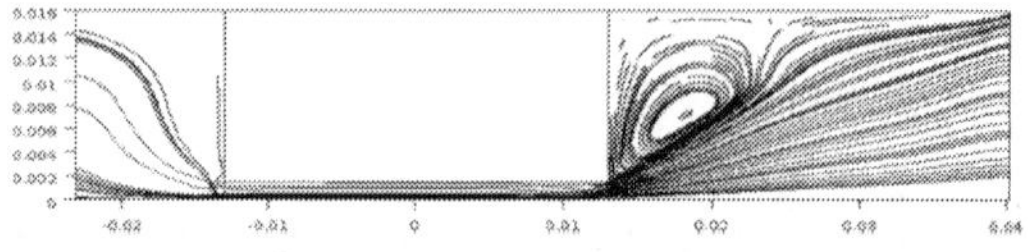

Figure 10 – Effects of viscous dissipation

The flow geometry can be characterized by a scalar parameter χ (eq.2) defined by Astarita :

$$\chi = 2\frac{|\Omega_{rel}|}{|\Omega_{rel}| + |D|} \qquad (2)$$

where D is the magnitude of the strain rate tensor and Ω_{rel} is the magnitude of the relative rate of rotation tensor $\underline{\Omega}$-$\underline{W}$. $\underline{\Omega}$ is the vorticity tensor and $\underline{W}$ is the tensor that gives the rate of rotation of the eigenvectors of $\underline{D}$. With this definition, χ is an objective quantity whose values are in the range [0,2]. χ=0 corresponds to pure elongational flow, χ=1 to simple shear flow and χ=2 to rigid body motion.

Figure 11 shows a mapping of Astarita's parameter χ in the volume of the mixer. As expected, it is seen that the flow is nearly elongational in the contraction and expansion zones close to the mixing element, and purely simple shear in the die of the mixing element.

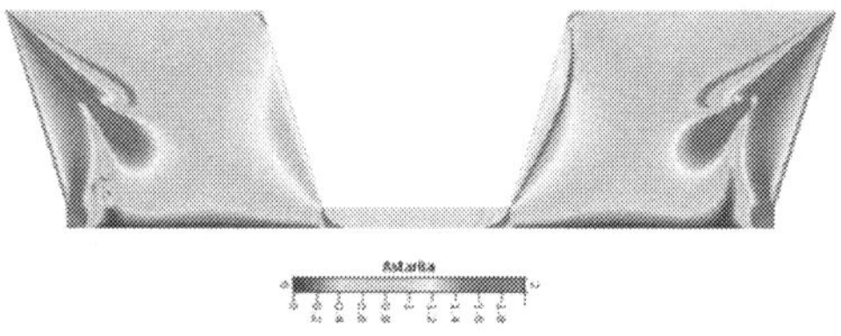

Figure 11 – Mapping of flow by Astarita's parameter

Once the velocity field has been determined, one can also estimate the mapping of elongational strain rate in the convergent and divergent zones. The values along the axis of symmetry are shown in Figure 12 for a piston velocity of 1 cm/s (flow rate equal to 7 cm^3/s). Very high values of the elongational strain rate are found close to the mixing element but the corresponding residence times are very short due to increasing fluid velocity. Due to strong gradients of strain rate, it will be difficult to compare the particle sizes obtained in our mixer to data obtained either experimentally by Grace (1982) or theoretically by Taylor in steady flow.

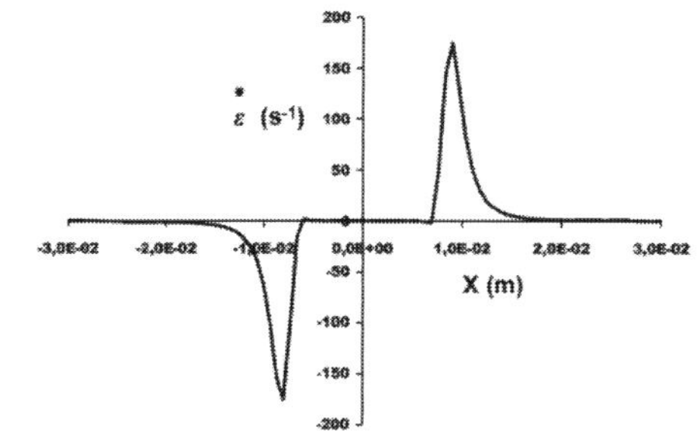

Figure 12 – Strain rate along symmetry axis

We then used a quasi Newtonian model (eq.3) - which is a function of χ parameter - which is able to take into account strain-hardening effects in elongation. Very good results were obtained with this model proposed by Ryssel and Brunn (1999) which presents the advantage of being simple to implement in a commercial code.

$$\mu = f(\chi)\mu_s + (1 - f(\chi))\,\mu_e \qquad (3)$$

Moreover, computation times remain reasonable compared to differential or integral viscoelastic models. In fact, elongation in this geometry has different effects on material. On expansion region, elongation tends to remain flow particles in their initial geometry. There is a relaxtion of the fluid. We implemented the quasi-Newtonian in upstream regions and a carreau law downstream. Numerical flow solved shows good approximation of experiments physical effects. There is the creation of vortices upstream the mixing channel (in circle in Figure 13) and a presence of striation downstream (in circle in Figure 14).

Figure 13 – Vortices in the upstream channel (pistons go from right to left)

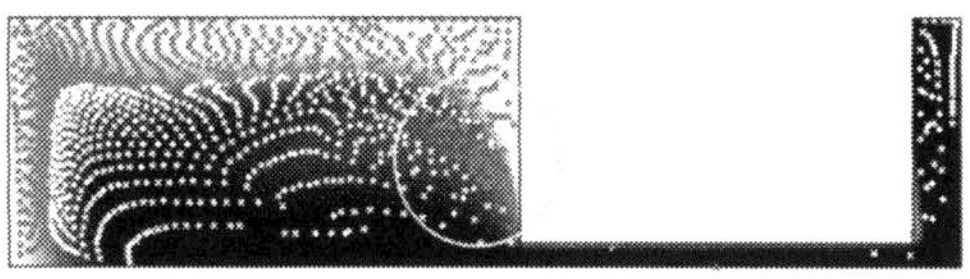

Figure 14 – Striation of the material downstream (pistons go from right to left)

A further objective of the numerical simulation is to explain the high distributive mixing efficiency of the RMX device during the first cycles of the mixing step, which was also observed by Son (2009). This issue will be the topic of a forthcoming paper and will not be further discussed here.

Conclusions

The results of the present paper demonstrate the efficiency of RMX as compared to a commercial lab-scale mixer. Its efficiency for dispersive mixing was attributed to the combination of elongational flow in the convergent zone and shear flow in the die of the mixing element. The size distribution of dispersed domains obtained in RMX are typically reduced by a factor of 2 with respect to blends obtained for the same specific energy input on a Haake Rheomix 600 mixer.

A still open issue which is currently investigated by numerical simulation is the efficiency of RMX for distributive mixing. Several sources of flow irreversibility during the cyclic mixing process were identified even at very low Reynolds numbers and these results will be published in a separate paper.

Finally, it should be pointed out that due to the possibility to work under controlled atmosphere and high pressures, RMX has a broad field of potential applications. It can in particular be operated as a chemical batch reactor, either for polymer modification with volatile reactants or even for polymerization. Moreover, the design of the mixing element allows continuous feeding and outlet, which opens the possibility of use as a stirred continuous reactor.

References

Journal: Bird, R. B.; Wiest, J. M., Annual Review of Fluid Mechanics 27, 169 1995
Journal: Grace, H. P., Chem. Eng. Commun.1982.
Journal: Hausman, J. M.: 1960
Journal: Hausman, J. M., United States of America: 1957.
Journal: Mackley, M. R.; Marshall, R. T. J.; Smeulders, J., Journal of Rheology 39, 1293 1995.
Journal: Son, Y., Journal of Applied Polymer Science 112, 609 2009.
Journal: Meller, M.; Luciani, A.; Manson, J. A. E., International Polymer Processing 14, 221 1999
Journal: Meller, M.; Luciani, A.; Sarioglu, A.; Manson, J. A. E., Polymer Engineering and Science 42, 611 2002
Journal: Meller, M.; Luciani, A.; Manson, J. A. E., Polymer Engineering and Science 42, 634 2002.
Journal: Ryssel E. and Brunn P.-O. « Flow of a quasi-newtonian fluid through a planar contraction » J. Non-Newtonian
Fluid Mech., **85**, 11–27, (1999)
Journal: Taylor, G. I., Proc. Roy. Soc. A1381932.
Journal: Terrisse, J.; Bouquey, M.; Muller, R., France: 2007.

Journal: Tokihisa, M.; Yakemoto, K.; Sakai, T.; Utracki, L. A.; Sepehr, M.; Li, J.; Simard, Y., Polymer Engineering and Science 46, 1040 2006.
Journal: Westofer, R. E., Soc. Plast. Eng. Trans.1961.

The effect of wall slip in 4:1 contraction flows of a PTT fluid

*Luis Lima Ferras (1), Femando Taveres Pinho (2), Olga Sousa Carneiro (1), Joao Miguel Nobrega (1)**

(1) University of Minho - Guimarães - Portugal, (2) University of Porto - Porto - Portugal

The phenomenon of wall slip has been investigated for several years by the scientific community, being most of the published work related to experimental measurements performed with the objective to evaluate the effect of slip in the velocity distribution, pressure drop or the onset of flow instabilities. Recently, the research team involved in this work has implemented in a numerical modelling code based in the Finite Volume Method the capability to model flows with the wall slip boundary condition. This code was already used to study the effect of the wall slip boundary condition on flows of Newtonian fluids in 4:1 contractions [1], using a implicit approach for the implementation of the boundary condition. In the present work the previous study is extended to model flows of viscoelastic fluids, being the code used to perform a systematic investigation of the effect of slip in the flow of a PTT fluid in 4:1 contraction, with the objective to better understand some of the phenomena related to the vortex dimensions and shape.

MODELING CHEMICALLY BLOWN FOAM EXPANSION USING THE FINITE ELEMENT METHOD

Rekha R. Rao*, Lisa A. Mondy, Mathew C. Celina, Douglas B. Adolf, Jamie M. Kropka, and Edward M. Russick[1]

[1]*Sandia National Laboratories, PO Box 5800, Albuquerque, NM 87185-0836, USA–* rrrao@sandia.gov; lamondy@sandia.gov; mccelin@sandia.gov; jmkropk@sandia.gov; emrussi@sandia.gov;

We are developing production-level foam process models for polyurethanes and other chemically blown foams. These foams have at least two simultaneous competing reactions, the polymerization of the reactive resins and carbon dioxide gas formation that creates the foamed material. The kinetics of the resin cure were evaluated using a dry polyurethane precursor, as water is essential to carbon dioxide formation, in order to separate the polymerization and blowing reactions, which both consume primary isocyanate groups. The kinetics of the foaming reaction was measured by tracking volume evolution with time. Rheology was performed on the dry polyurethane system in order to quantify the effects of cure on the continuous phase viscosity. Rheology of the foam was carried out to understand the effect of gas fraction on the flow of polymer. A finite element model was developed combining the equations of motion, an energy balance, a density model to represent the foam blowing reaction, and a rate equation for the polymerization reaction. This was combined with a level set method to track the location of the free surface as it evolves in time. The viscosity of the foam is dependent on degree of cure, temperature and gas fraction as the foam evolves from a bubbly liquid to a jammed system of bubbles. The three-dimensional model is compared to experimental flow visualization data in a complex geometry involving flow through an aluminum mold with serpentine features.

Introduction

Polyurethane foam is used for a variety of applications from foam cushions to encapsulants of sensitive electronic components (Figure 1). Since encapsulation is our application of interest, we seek to understand the dynamics of polyurethane formation as it fills a mold by developing continuum-level finite element models of the process, informed by careful experimental measurements.

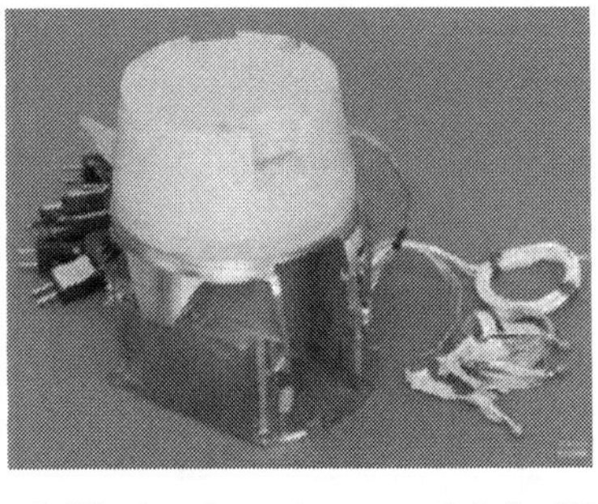

Figure 1. Electronic part encapsulated with PMDI.

Polyurethanes are chemically blown foams that begin as mixtures of polyisocyanate, polyol, water, surfactant and catalyst. Four competing reactions occur simultaneously, the most important of which are the polymerization reaction and the foaming reaction. Polymerization occurs via condensation reaction of the polyisocyanate and the polyol to form a crosslinked polyurethane. The foaming reaction occurs between the polyisocyanate and the water, forming carbamic acid which then decomposes to form carbon dioxide and an amine group [Tesser et al, 2003]. The carbon dioxide is a gas, which is responsible for blowing the polymer into a foam.

Since both primary reactions use polyisocyanate, separating these two reactions for the purpose of understanding the details of their reaction kinetics and rheology is intrinsically challenging. For this reason, micro-attenuated total reflection (ATR) infrared spectroscopy (IR) measurements were undertaken to obtain spectral signatures from the dried non-foaming system, where only the urethane curing reaction occurs. These were then compared to IR measurements on the foam system, where both foaming and curing reactions occur in parallel. Figure 2 shows the typical IR-spectra for two different temperatures, basically before and after cure. The peak at 1218 cm^{-1} represents the pure curing reaction, as it relates to the C=O-O-R linkage (i.e the ester side of the urethane) between the polyol and urethane group. Unfortunately, no similar specific peak exists for the foaming reaction, first of all since the carbamic acid generating CO2 is a transient species, and secondly since after decomposition the resulting urea linkages (carbonyl and amide groups) overlap with the urethanes in the spectrum. Therefore, carbon dioxide production rates must be inferred from measurements of volume with time.

Once the reaction kinetics of polymerization have been understood at an adequate level, the extent of reaction can be correlated with time-dependent rheological measurements of the dry system. The rheology of the foam is also measured and compared to correlations based on bubble volume fraction to develop a closed form solution for viscosity of the polyurethane as a function of temperature, degree of cure, and gas volume fraction.

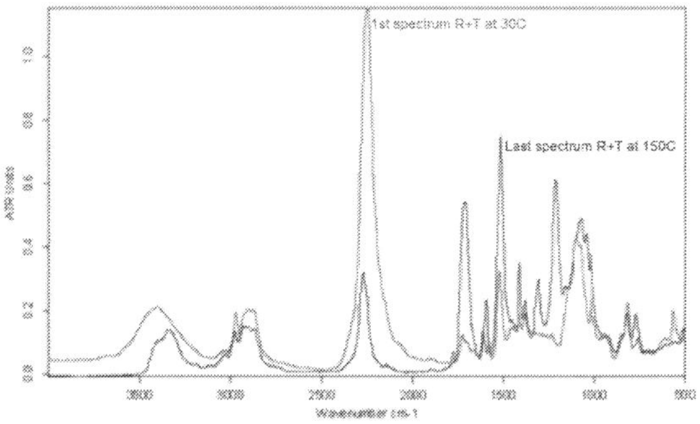

Figure 2. IR spectra for polyurethane curing. Peak 1218cm^{-1} represents pure curing reactions: polyol-isocyanate urethane reactions.

The continuum-modeling capability is based on a finite element discretization of the self-expansion that occurs during foam blowing; coupling the equations of motion with the movement of the free surface represented using the level set method [Sethian 1999]. This continuum model uses a homogenized description of foam, which does not include the gas in the foam cells explicitly but instead tracks an evolving density equation. The current model is an engineering design tool that will allow us to study processing variables such as temperature, as well as the effect of gate and vent locations for optimizing the foam encapsulation process. Comparison of the model against validation experiments is also shown in the results section.

Experimental

Materials and Mixing Methods

The system studied is a polymeric methylene diphenyl diisocyanate (pMDI) system, termed pMDI-4 formulation consisting of an R-component (resin) and a T-component (isocyanate). Specifically the R-component is comprised of Voranol 490 polyol [Dow Chemcial Co., Midland, MI] (100g), water (2.4g), DC-197 surfactant [Air Products and Chemicals, Inc., Allentown, PA] (3.0g), and 33LV catalyst [Air Products and Chemicals, Inc.] (0.2g). The T-component contains PAPI 27 isocyanate [Dow Chemical Co.] (100g) and Voranol 490 polyol (2.6g) as a slightly condensed prepolymer. The R- and T-components are mixed in a 37.5% to 62.5% ratio to produce a 4 lb/ft^3 density foam in an unconfined geometry (i.e., free-rise). The materials are hand-mixed for 0.5-1 minute using a wooden spatula at room temperature, and then the mixture is injected or poured into a mold or container to be foamed. The mold is then placed into a 70°C oven where foaming and curing occur simultaneously.

ATR IR Spectroscopy to Determine Reaction Kinetics

Micro-ATR was used to probe the kinetics of polyol isocyanate and related reactions at a range of temperatures. We also examined whether significant differences in spectral features exist that could be used to decouple the polyol cure and water based foaming reaction which competitively consume the isocyanate.

Micro-ATR experiments were conducted using a heated diamond micro-ATR attachment in a Bruker Equinox 55 spectrometer with a DTGS (ID301/8) detector operating at RT. Spectra were collected as an average of 64 scans at 4 cm^{-1} resolution. For rapid cure conditions at the higher temperatures (>50°C), we used 16 scans to better capture early reaction kinetics. Repetitive loop macros were used to acquire sequential spectra over multiple hours. The observable spectral features are related to the dominant reactions, namely consumption of the isocyanate at 2260 cm^{-1}, the formation of carbonyls either as urethanes (1723 cm^{-1}) or ureas close to 1700 cm^{-1}, the amide II band at 1521 cm^{-1} and 1510 cm^{-1}, a band at 1308 cm^{-1} and the "urethane ester linkage" at 1218 cm^{-1}. Since the spectra are heavily convoluted, it is not possible to integrate specific band areas for all of these bands. This creates significant constraints for the quantitative analysis of spectral data. While a collective carbonyl band area could be established using a range from 1650-1750 cm^{-1} for integration, all other peak changes were quantified using peak height instead. The spectra also show a decrease of hydroxyls at approximately 3400 cm^{-1} and an increase of NH absorption at 3320 cm^{-1}. Neither one of these broader bands is suitable for simple quantification.

A standard sample mass of two grams was mixed, as discussed above, for every run and then placed on the preheated temperature-controlled plate for spectral acquisition. The first spectrum is generated after approximately 64 seconds for 64 scans and 16 seconds for 16 scans. For all dried system experiments, four small beakers filled with anhydrous calcium sulfate were placed around the sample and covered with a beaker, and the sample was covered with a small strip of aluminum foil to minimize any diffusion of water into the sample. Wet foam experiments were also covered with a closed container.

For the pMDI foam under investigation, the isocyanate reactions on a molar basis split into 86% for the polyol addition to form urethanes, and roughly 14% for the isocyanate water foaming reaction. In principle, approximately 14% of urea formation from an amine isocyanate addition reaction should be observable as a follow-up of the CO_2 producing carbamic acid decomposition.

Two resin systems were evaluated: The basic foam system consisting of a standard foam system in which the polyol component contains a pre-mixed surfactant, water and catalyst, and a "dried" system from which the water had been removed from the polyol component using common laboratory-grade zeolites as a drying agent. In principle, a comparison of the spectral features between a dried non-foaming system, and a water-containing polyurethane foam system should reveal spectral variations in the nature and quantity of the underlying reactions. The dried system should exclusively display a dominant urethane formation, while the wet system should involve a combination of urethane and urea products. However, both systems will be further convoluted by subsequent allophanate and biuret reactions when isocyanates are used in slight excess. This mix of reactions implies that a deconvolution of the foaming and urethane cure reactions based on IR spectral features alone is intrinsically difficult. A complicating issue is that the carbonyl group of the urea does not display a specific isolated spectral band, but only appears at a slightly lower wave-number than the urethane carbonyl. Deconvolution of the urethane and urea carbonyls, particularly at low concentrations of urea is prohibitive, unless chemometrics and sophisticated algorithms are applied. Such approaches were not the goal of this overview study.

To determine curing rates in the full water-containing foam system, IR data were obtained for the 1218 cm^{-1} band height as a function of time and temperature (Figure 3).

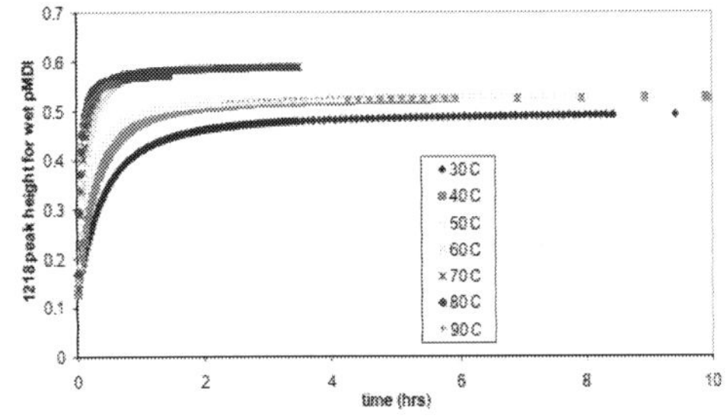

Figure 3. Peak height as a function of time for the 1218 cm^{-1} peak and the wet pMDI material. Isothermal tests were carried out for various temperatures ranging from 30°C to 90°C.

The peak height from Figure 3 can be normalized by the maximum peak height (~0.6cm^{-1}), to obtain the extent of reaction as a function of time and temperature. Shifting the various temperature data to obtain a master curve (Figure 4) gives shift factors as a function of temperature; this can then be used to determine the activation energy for the Arrhenius rate constant.

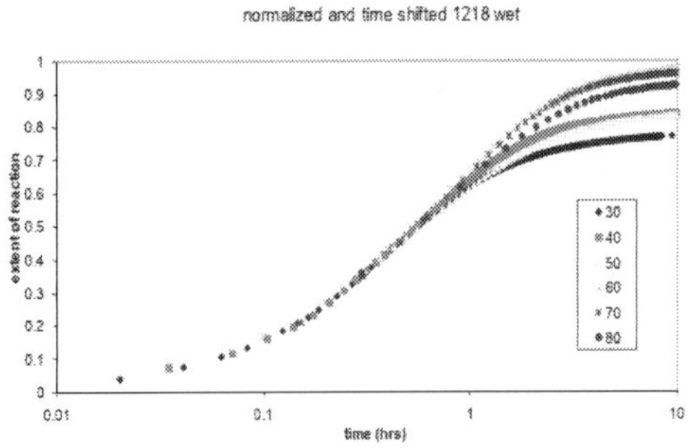

Figure 4. Shifted extent of reaction for isothermal tests carried out for various temperatures ranging from 30°C to 80°C. The shift to a master curve works well up until 1-2 hours.

Figure 5 shows a plot of the natural logarithm of the shift factors versus the inverse of temperature in Kelvin. The slope of this linear plot gives the activation energy for the rate constant.

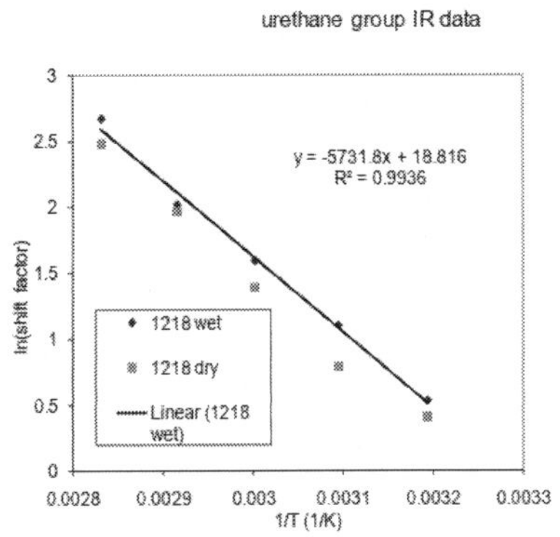

Figure 5. The natural log of the shift factor plotted versus the reciprocal temperature in Kelvin, gives the activation energy for the Arrhenius rate constant for the polymerization reaction. Here the data are plotted for the wet and dry foam, demonstrating that the blowing reaction does not greatly affect the activation energy of the urethane cure.

Once the extent of reaction and activation energy are determined, the reaction kinetics (e.g. rate constant and order of reaction) can be obtained by fitting the equation below to the numerically differenced data:

$$\frac{d\xi_{cure}}{dt} = k_0 e^{\Delta E/RT} \left(1-\xi_{cure}\right)^n \qquad (1)$$

Where k_o is the rate coefficient, ξ_{cure} is the extent of reaction, t is time, T is the temperature in Kelvin, ΔE is the activation energy, R is the universal gas constant, and n is the reaction rate. The results comparing the fit to the data for extent and rate of reaction are given in Figures 6 and 7 respectively.

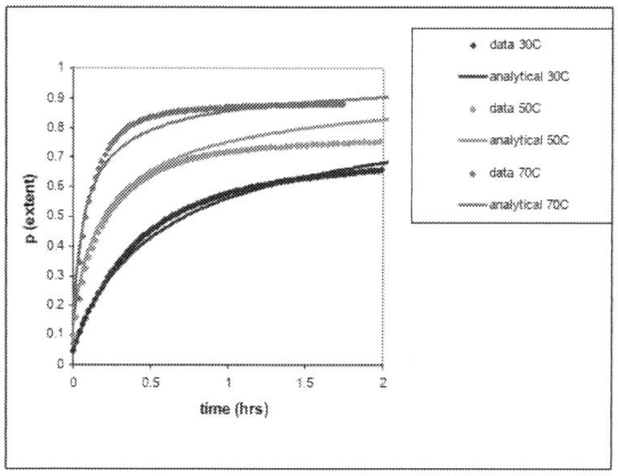

Figure 6. The extent of reaction compared to the numerical fit from equation 1 for three different temperatures.

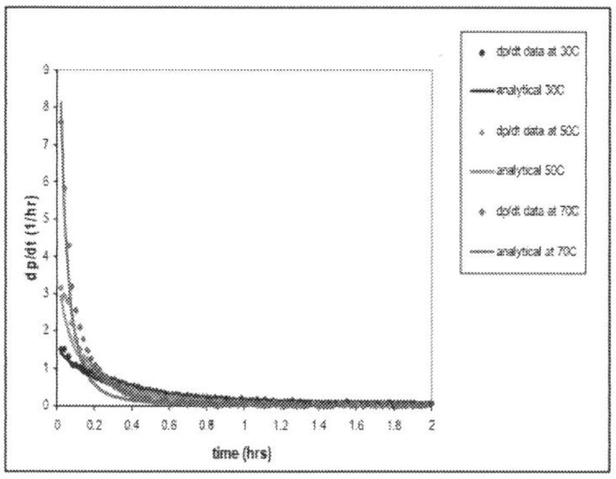

Figure 7. The rate of reaction compared to the numerical fit from equation 1 for three different temperatures.

The order of the reaction is 2.75, which is higher than values of 1-2 seen in the literature [Bikard et al., 2007] and consistent with polymerization reaction following a fractal dimension [Martin et al., 1989]. The rate coefficient, k_o, is found to be 2.96×10^8 1/hr while the activation energy is ~ 2.89 kcal/mol. The heat of reaction is roughly 57.4 cal/g.

Reaction Kinetics of Gas Generation Reaction

As discussed previously, the kinetics of the gas production could not be measured via IR, so we used a simpler method to understand the carbon dioxide formation. Here we observe the height evolution with time, and infer the gas formation by the increase in volume. This method is somewhat inadequate as it cannot account for all the carbon dioxide produced as some is lost to the atmosphere through the free surface and does not produce an increase in volume, and some remains supersaturated in solution and does not nucleate to form a gas bubble. The volume increase as a function of time is shown in Figure 8 for three different temperatures assuming the lost volume is negligible.

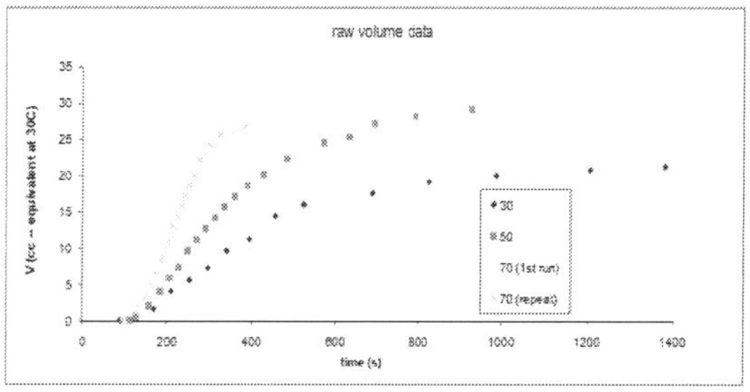

Figure 8. Volume evolution from carbon dioxide formation for three different temperatures inferred from height versus time experiments.

From Figure 8, we can see that the lost volume becomes more significant with temperature, probably due to the fact that the viscosity is lower and the degree of supersaturation is higher, so less gas is trapped in the polymer. As seen previously for the curing reaction, we can convert volume to extent of reaction based on maximum volume produced from stoichiometric considerations (~30 cc) and obtain an extent of reaction. This can be shifted to obtain a master curve (Figure 9).

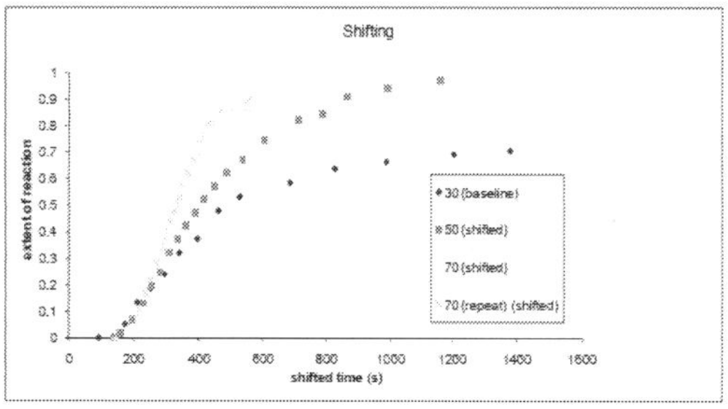

Figure 9. Extent of reaction for carbon dioxide formation for three different temperatures inferred from height versus time experiments. Note that the green points are a repeat for 70°C.

Because the shift did not work well for times greater than 300s and the foaming reaction goes on for significantly longer, we decided to pursue a simpler approach and track the density as a function of time and temperature following the work of Seo *et al.*, [2003]. The density would then be fit to an empirical model that would be used in the finite element model. For the range of the temperatures tested, as the temperature increases the foam rises faster. By considering the data at the various temperatures, one can include the effects of temperature.

$$\rho = (\rho_{initial} - \rho_{final}) \exp\left[\frac{-(t - t_{initial})}{\frac{A}{T} - B}\right] + \rho_{final} \qquad (2)$$

Figure 10 documents the density changes at three oven temperatures, assuming that the initial density was 1.4 g/cm^3. Also shown in Figure 10 is a fit to preliminary data in the form of equation (2) to take into account the effects of temperature based on the nominal oven temperature only. Here, ρ_{final} was taken to be 0.09 g/cm^3 and T is the temperature in Kelvin. Fitting equation (2) to the data in Figure 10 resulted in values for the parameters, A and B, of $3.0\text{x}10^5$ K s and 810 s, respectively, yielding time constants for the exponential change in the foam density of 64.6 s at 70 °C.

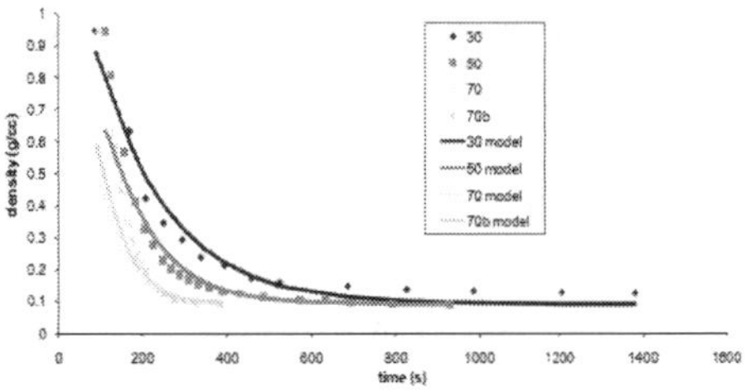

Figure 10. Foam density as a function of time and temperature.

Once we develop a better experimental procedure for determining the total amount of carbon dioxide produced, we can work on producing a more satisfying kinetic approach to gas generation.

Rheology of Continuous Phase and Foam

Foam rheological properties are complex, exhibiting effects of changing microstructure as bubble growth or breakage occurs due to coalescence or shear. However, here we will assume that the gross foam rheology can be approximated with an apparent viscosity and slip at the wall. The suspension/emulsion literature has clearly shown that the effects of the continuous phase are separable from the discontinuous particle, emulsion, or gas bubble phases [Prud'homme and Khan, 1996]. For this reason, we have chosen to separate the viscosity into two parts dependent on 1) continuous phase epoxy properties, dependent on extent of reaction and temperature, 2) gas bubble volume fraction. We assume these components are multiplicative and can be decoupled.

$$\eta = \eta_{cure}\eta_{\varphi} \tag{3}$$

For the first term, small strain oscillatory rheological measurements were made for the continuous phase material without the water as a function of time. These measurements we carried out in a Rheometrics ARES rheometer with parallel plate geometry. Results are given in Figure 11 for the storage and loss modulus at three temperatures.

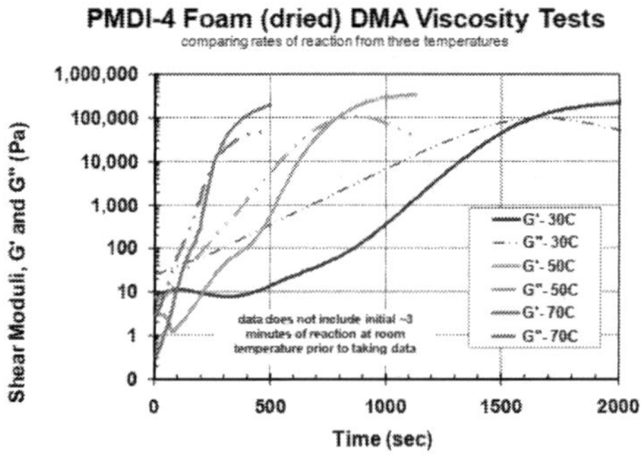

Figure 11. Storage and loss modulus for dry polyurethane at 30°C, 50°C, and 70°C. The cross over point of G' and G'' gives the gel point and gel time of the polymer.

For operating temperatures far above the current glass transition temperature of the reacting epoxy, the temperature dependence can be modeled accurately by an Arrhenius relationship [Ferry, 1980]. Dynamic percolation theory predicts a dependence of the Newtonian viscosity on extent of reaction with the form

$$\eta_{epoxy} = \eta_{00}\exp\left(\frac{E_a}{RT}\right)\left(\frac{\xi_c^b - \xi^b}{\xi_c^b}\right)^q \tag{4}$$

where ξ_c is the gel point, E_a is the activation energy, and η_{00} is the uncured viscosity at a reference temperature $T_{0,}$ and b and q are exponents for the model, with b being positive and q being negative [Martin et al., 1989]. The parameters we use to populate this model are given in Table 1.

Table 1. Curing Viscosity Model Parameters

Curing Viscosity Model Parameter	Value
Uncured viscosity at reference T, η_{00}	$2.7\text{x}10^{-9}$ Pa s
Activation energy, E_a	6.4 kcal/mol
Extent of reaction at gel point, ξ_c	0.46
Curing viscosity exponent, p	-2.0
Curing viscosity exponent, b	1.0

The data are compared to equation (4) with these parameters in Figure 12.

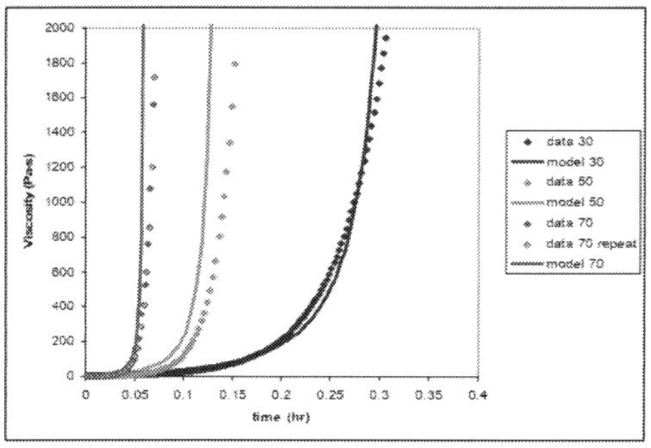

Figure 12. Dry polyurethane at 30°C, 50°C, and 70°C plotted with viscosity model.

We expected the second term in equation (3), the foam viscosity η_ϕ, to be a strong function of the gas volume fraction ϕ_g and, at least at the zero shear rate limit, to follow the Taylor-Mooney form derived from emulsion experiments, extrapolating the discontinuous phase viscosity to zero [Taylor 1934, Mooney 1951, Prud'homme and Khan 1996]:

$$\eta_\phi = \eta_0 \exp\left(\frac{\phi_g}{1-\phi_g}\right) \quad (5)$$

To determine if this was a satisfactory approximation to the foam effective viscosity, we tested our polyurethane foam as it was expanding in a shear rheometer using the parallel plate geometry. Because the foam was expanding during the test, the foam escaped out the side of the parallel plates and, therefore, the volume of the sample did not change, but the density of the sample did. Both quantities are needed for the interpretation of the viscosity measurement. We used a temperature ramp in the rheometer that mimicked that of the free-rise experiments. Knowing that the viscosity of the foam would be sensitive to cell breakage, we tested a shear rate as low as possible given the resolution of the torque sensor in the rheometer. We assumed that the shear rate in the geometries of most interest, under the conditions of self expansion, would indeed be very low. Figure 13 shows the results of the rheological experiments compared to the theoretical correlation of Taylor-Mooney.

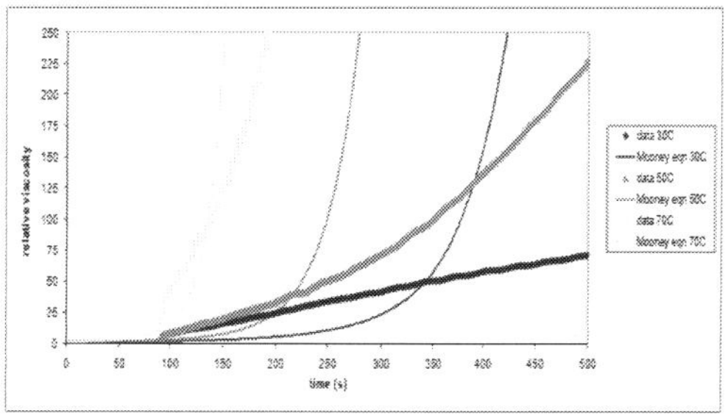

Figure 13. Relative viscosity as a function of time and temperature from rheological measurements on pMDI-4 and Taylor-Mooney Correlation gives the gas volume fraction dependence of the foam.

Equations and Numerical Method

Here we encapsulate the complex physics of nucleation and boiling into a time- and temperature-dependent density function (as discussed in the experimental section), which decreases in time creating volume and velocity through the continuity equation. The continuity equation is written to emphasize the change in density as the source of foam velocity generation, where v is the mass-averaged velocity and ρ is the foam density.

$$\nabla \bullet v = -\frac{1}{\rho}\left(\frac{\partial \rho}{\partial t} + v \bullet \nabla \rho\right) \quad (6)$$

Conservation of momentum takes into account gradients in the fluid stress, τ, and pressure, p, as well as gravitational effects. Note that gravity is applied to the homogenized foam material and does not take into account the buoyancy differences between the epoxy and the gas bubbles.

$$\rho\frac{\partial v}{\partial t} + \rho v \bullet \nabla v = \nabla \bullet \underline{\underline{\tau}} - \nabla p + \rho \underline{g} \quad (7)$$

The stress tensor has a generalized Newtonian shear viscosity, in addition to a generalized Newtonian bulk viscosity. The bulk viscosity is associated with the fact that the divergence of the velocity field is non-zero and we have a dilatational flow [Bird et al., 1960]. The bulk viscosity term produce only normal stresses and not shear stresses.

$$\underline{\underline{\tau}} = \eta(\nabla v + \nabla v^t)) - (\frac{2}{3}\eta - \kappa)(\nabla \bullet v)\underline{\underline{I}} \quad (8)$$

The shear viscosity, η, can be a function of temperature, cure, and gas fraction as discussed in the viscosity section. An expression for the bulk viscosity κ for non-dilute foams was developed by Kraynik *et al.* [2005] who used 2D boundary integral calculations and

3D ALE free surface calculations of expansion in a Kelvin cell. Their numerical results demonstrated that the following formula was appropriate for non-dilute systems.

$$\kappa = \frac{4}{3}\eta_{polymer}\frac{(1-\phi_g)}{\phi_g}. \qquad (9)$$

The energy in the system must also be tracked to determine the temperature variations from various sources including oven heating and exothermic polymerization reactions. The energy equation has a variable heat capacity, C_p, and thermal conductivity, k, both of which depend on the gas volume fraction.

$$\rho C_p \frac{\partial T}{\partial t} + \rho C_p v \bullet \nabla T = \nabla \bullet (k \nabla T) + S_{rxn} \qquad (10)$$

The equations of motion are discretized with the finite element method. An Eulerian level set method is used to track the location of the free surface as a function of time [Sethian, 1999]. Dohrmann-Bochev pressure stabilization is used to reduce the condition number of the matrix, allow for equal order interpolation, and enable the use of Krylov-based iterative solvers [Dohrmann and Bochev, 2004]. Further details of the modeling approach and equations, the numerical methods used and the finite element implementation can be found in a report by Rao et al. [2008].

Results and Discussion

Preliminary Modeling Results and Model Validation

Flow visualization experiments were carried out in a complex model containing expansions and contractions and a serpentine area near the vent. The mold and the finite element mesh based on it are shown in Figure 14.

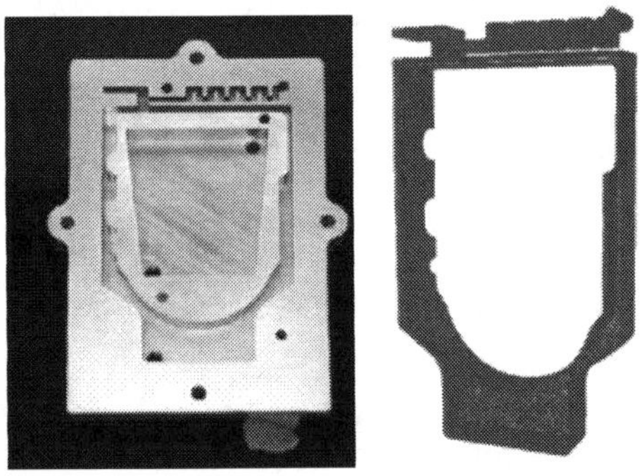

Figure 14. Kansas City mold (left) and finite element mesh (right).

Results from the flow visualization experiments are seen in Figure 15. The material is injected into the mold as an incompressible fluid. Over time the gas generation increases the volume of material in the mold, until it filled at 279s.

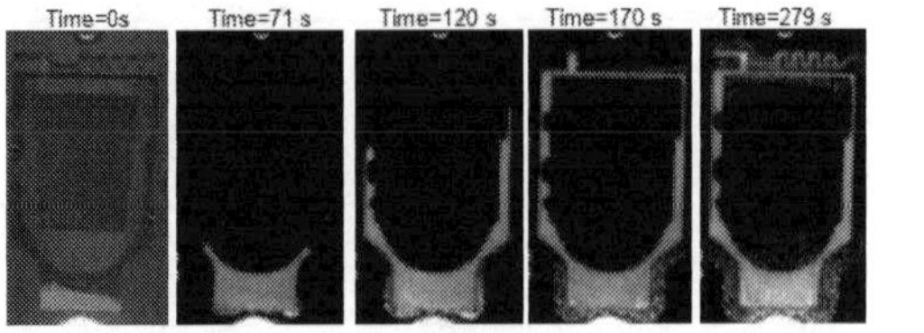

Figure 15. Flow visualization experiments for pMDI-4 filling the Kansas City Mold.

Preliminary modeling results from the level set/finite element model are given in Figure 16. Numerical difficulties with the full viscosity model occurred, necessitating the use of a simpler, constant viscosity model for these preliminary results. From previous work [Rao et al., 2008], it has been determined that the form of the viscosity does not affect the location of the free surface as a function of time. This is solely a function of the density model, which is both one of the advantages and drawbacks of the model.

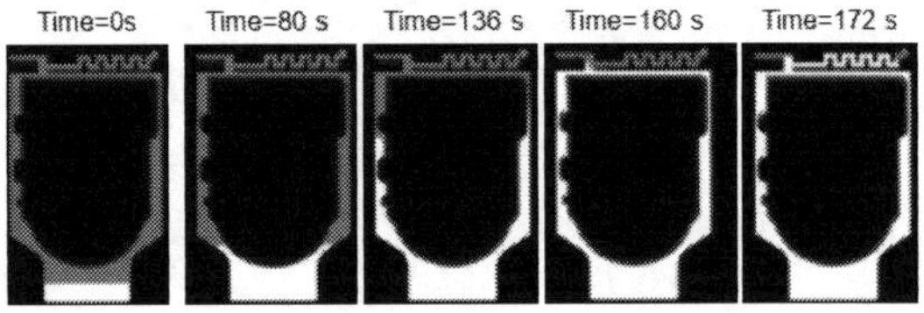

Figure 16. Level set simulation results showing foam expansion in the finite element mold.

Comparing the experimental results with the simulation, we can see that the model agrees well with the experiment in terms of the shape of the free surface, but the timescales are different for the model and the real polyurethane. The model first lags behind the experiment substantially. At late times, once the mold is nearly full, *e.g.* after 160s, the model becomes faster than the experiment. The simple density model does not seem to be able to capture the almost asymptotic expansion that happens at late times. We hope that the more kinetic-based approach we are working on will be able to account for this phenomenon.

Conclusions and Future Work

We have presented IR-spectroscopy data to better elucidate the polymerization of polyurethane, and have fit the data to an extent of reaction equation indicating a reaction order greater than 2. The foam blowing reaction is studied by tracking the volume increase with time, though these experiments are less definitive than the IR since some of the carbon dioxide is lost to the atmosphere. We are still working on determining an experimental technique that will allow us to track all the gas produced. A viscosity model was obtained that included the effects of temperature, degree of polymerization and bubble volume fraction.

A numerical model has been developed based on a finite element discretization and a level set method to determine the location of the free surface as the foam expands. The model includes the kinetics of polymerization, but uses a simplified Newtonian viscosity. The model agrees qualitatively with the experimental results, but fills the mold much faster than the real polyurethane. For future work, we plan on refining the model and addressing the numerical difficulties that prevent us from using the full viscosity model.

Acknowledgements

The authors would like to acknowledge the previous efforts on ARIA code development in support of foam expansion models from David Noble, Harry Moffat, and Pat Notz. The Kansas City mold mesh was made by James Mahoney. This work was supported by the Laboratory Directed Research and Development program at Sandia National Laboratories. Sandia is a multiprogram laboratory operated by Sandia Corporation, a Lockheed Martin Company, for the United States Department of Energy's National Nuclear Security Administration under contract DE-AC04-94AL85000.

References

J. Bikard, J. Bruchon, T. Coupez & L. Silva, *Colloids Surfaces A: Physicochem. Eng. Aspects* **309**: 49-63, 2007.

R. B. Bird, W.E. Stewart, E.N. Lightfoot, *Transport Phenomena*, John Wiley and Sons, New York, USA, 1960.

C. R. Dohrmann and P. B. Bochev, "A stabilized finite element method for the Stokes problem based on polynomial pressure projections," *Int. J. Num. Meth. Fluids*, **46**, 183–201, 2004.

J. D. Ferry, *Viscoelastic Properties of Polymers*, p. 290, Wiley, NY, USA, 1980.

A. M. Kraynik, T. A. Baer, M. Martinez, Sandia National Laboratories, Personal Communication, Aug. 2005.

J. P. Martin, D. B. Adolf, and J. P. Wilcoxon, *Physical Review A* **39**: 1325, 1989.

Mooney, M., "The viscosity of a concentrated suspension of spherical particles," *J. Colloid Sci.*, **6**, 162-170 (1951).

Prud'homme, R. K. and Khan, S. A., "Experimental Results on Foam Rheology," in *Foams, Theory, Measurements and Applications,* Prud'homme and Khan (Eds.), Marcel Dekker, Inc. New York, p. 234 (1996).

R.R. Rao, L.A. Mondy, H.K. Moffat, D.R. Noble, P.K. Notz, D.B. Adolf, T.A. Baer, *Foam Process Models*, SAND2008-6094, Albuquerque, NM, September, 2008.

D. Seo, J. R. Youn, C.L.I. Tucker *Int. J. Numer. Meth. Fluids*, **42**, 1105, 2003.

J. A. Sethian. *Level Set Methods and Fast Marching Methods*, volume 3 of Cambridge Monographs on Applied and Computational Mathematics. Cambridge University Press, New York, USA, 2nd edition, 1999.

Taylor, G. I., " The formation of emulsions in definable fields of flow," *Proc. Roy. Soc. Ser. A.* **146**, 0501-0523 (1934).

R. Tesser, M. Di Serio, A. Sclafani, E. Santacesaria, "Modeling of Polyurethane Foam Formation," *J. Appl. Polymer Sci.*, **92**, 1875, 2004.

VAPOR SORPTION ISOTHERM BY DIFFERENT BIOPOLYMERS AND POLYMERS

V. V. Ugrozov

Moscow State University of Food Productions 125080 Moscow, Volokolamskoy highway, 11, Problem science-research laboratory (PSRL) "Physical and chemical bases of processing of vegetative raw material", Department of Higher and Applied Mathematics, E-mail: vugr@rambler.ru

Introduction

Process sorption of gases and vapor by polymers and biopolymers represents the big scientifically-practical interest and it is intensively investigated. It is known that depending on a temperature a polymer can be either in glassy (T<Tg) or the rubbery (T>Tg) states (where Tg – glass transition temperature). Now a number of the equations of isotherms vapor sorption by amorphous polymers are offered. So the famous isotherm of dual mode sorption model well describes the given process if polymer, is in glassy state or by isotherm Flory-Huggins if it is in a rubbery state. However till now there is no the uniform isotherm, allowing to describe process of sorption vapor by the amorphous polymer which is being any specified conditions. The basic purpose of the given work is determination of such isotherm.

Model sorption

As is well known the free volume of glassy polymer includes two types of "holes".
Holes of the first type arise owing to thermal mobility of various fragments of polymer chains.
The second type holes are immobile microvoids which arise owing to nonequilibrium of structures of a polymer matrix.

In given work was been suggested model (Ugrozov 2010), in which hypothesizes that two types of dynamic clusters are formed in the course of reversible vapor sorption kinetics in glassy polymer.

Clusters of the first type consist of the first type holes each of which filled by the vapor molecule. During of sorption kinetics the sizes of these clusters can increase or decrease owing to attachment to them or detachment from them one molecule of vapor (fig.1a).

Clusters of the second type are formed from microvoids and holes of first type which were filled by the vapor molecule. It is assumed that at each of microvoids has the active site on which the initial vapor molecule is adsorbed with rate constant - k_{1a} or desorbed with rate constant - k_{1d}. The subsequent cluster formation is carried out by the vapor molecule attachment or its disintegration into the vapor molecule and the less size cluster (fig. 1b).

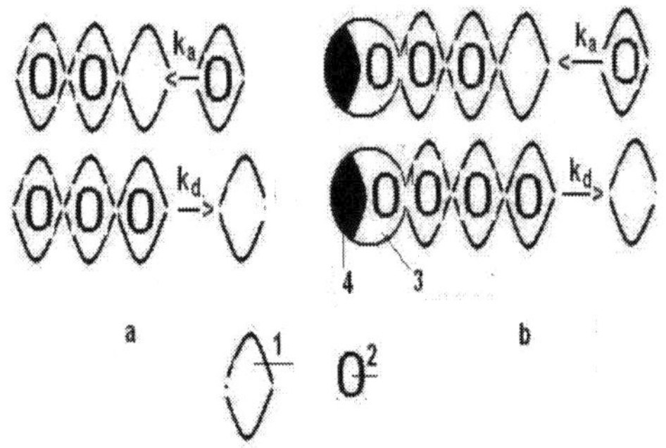

Fig.1. Schematic illustration of formation of the clusters: a – first and b- second types. 1- migration hole ; 2 – molecule of vapor; 3 - microvoid 4 – active adsorption site. ka (kd) – constants of rate aggregation (fragmentation) of vapor molecules.

Results and Discussion

The systems of the differential equations describing kinetics of reversible vapor sorption by glassy polymers are obtained. It is shown that the equation of the vapor sorption isotherm by polymer, can be described as follows

$$a=(a_{1\infty}+a_{2\infty})\left[\frac{\beta x}{(1-\beta x)}+\lambda\frac{(C-\beta)x}{(1+(C-\beta)x)}\right] \quad (1)$$

where $a_{1\infty}$ + $a_{2\infty}$ - Langmuir's capacity of two types clusters, λ - parameter of state the amorphous polymer ($\lambda \approx 1$ - the glassy state ; $\lambda = 0$ - the rubbery state), $x = p/p_s$, p -pressure, p_s - vapor saturation pressure.

It is found out that the equation (1) well describes experimental data of vapor sorption by the glassy and rubbery polymers (fig.2-3). It has been shown that isotherm (1) may be used for description of water sorption by different biopolymers: starches, starch-containing products, crups (Ugrozov et.al.2007a, 2007b, 2008c) (fig.4) and natural polysaccharides (Ugrozov et.al. 2008) in a wide humidity range. This work was supported by the Russian Foundation for Basic Research (project no.08-08-00832-a).

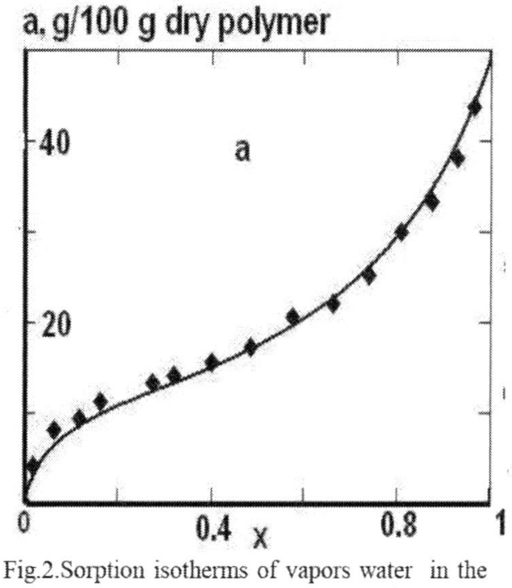

Fig.2.Sorption isotherms of vapors water in the sulfonated polyimide copolymers at 50° C.(Guo et.al.2004). Solid lines is eq.(1) and the dots refer to the experimental data.

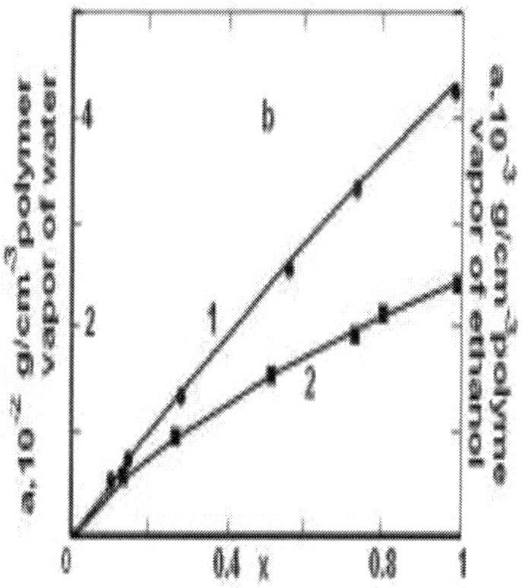

Fig.3.Sorption isotherms of vapors water -1 and ethanol - 2 in cellulose acetate at 25° C (Hiroshi et.al.1995). Solid lines is eq.(1) and the dots refer to the experimental data.

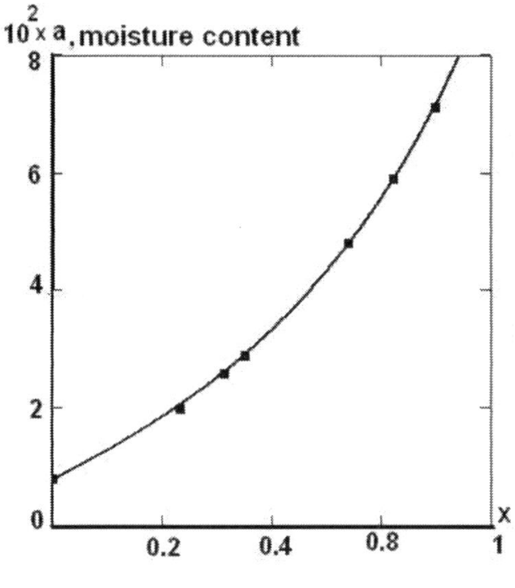

Fig.5.Vapor sorption isotherms by rubbery polymers water in nylon 6,6 at 35°C (Loong et.al.1999); Solid lines is eq.(1) and the dots refer to the experimental data.

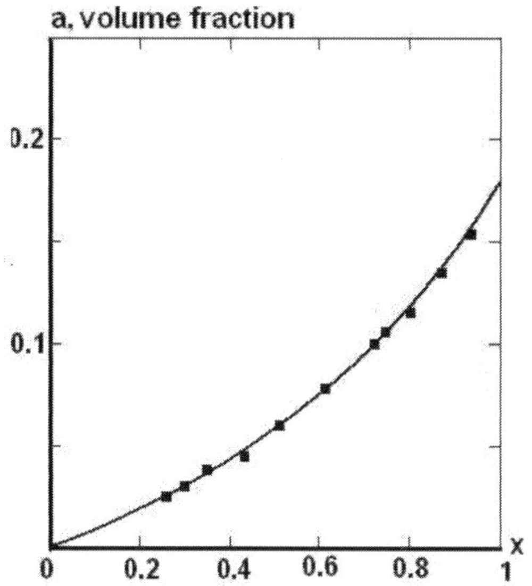

Fig.6.Vapor sorption isotherms of ethanol in polar elastomer at 30°C (Jonquieres et.al. 1998) Solid lines is eq.(1) and the dots refer to the experimental data.

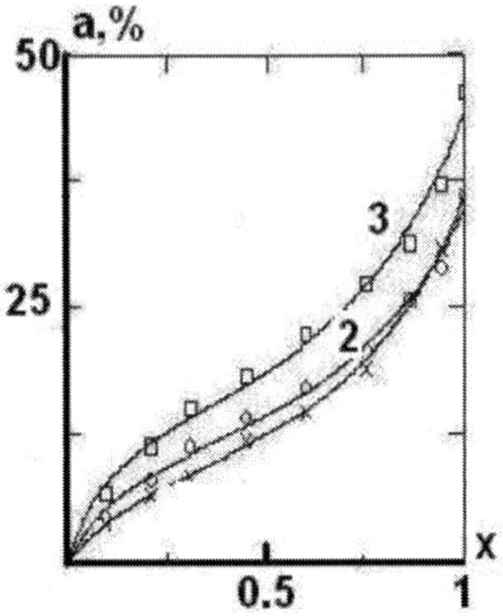

Fig.7.Isotherms of water vapor sorption by grains of different starches: 1- wheat, 2- corn, and 3- potato. Solid lines are eq (1) and the dots refer to the experimental data.

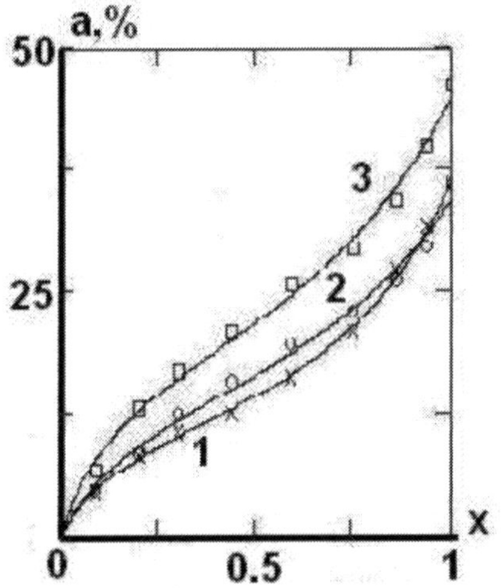

Fig.8. Isotherms of water vapor desorption by grains of different starches: 1- wheat, 2- corn, and 3- potato. Solid lines are eq (1) and the dots refer to the experimental data.

References.

1. Guo X., Fang J.H., Kazuhino T., Hidetoshi K., Kenichi O. " Synthesis and properties of novel sulfonated polyimides from 2,2′-bis(4-aminophenoxy)biphenyl-5,5′-disulfonic acid." J. Polym. Sci. Polym. Chem. Ed **.42.** p.1432 (2004)
2. Hiroshi O.,Kazuhiko R.,Tadashi U. " Sorption and permeation of water and ethanol vapors in poly(vinylchloride) membrane." J. Membr. Sci.. **103**.,.31-35, (1995)
3..Jonquieres A., Perrin L., Arnold.S., Lochon. P. "Comparison of UNIQUAC with related models for modelling vapour sorption in polar materials."» J. Membr. Sci., **150,** .125-134 (1998).
4. Loong T.L., Britt I.J., Marvin A.T. "Sorption and transport of water vapor in nylon 6,6 film" J. Appl. Polym. Sci. 1999.v.71.p..197.
5. Ugrozov V.V., Filippov, A.N., and Sidorenko, Yu.I. «Theoretical description of the hygroscopicity of hydrophilic biopolymers and their mixtures. » Colloid J. (Russia), **69,** 256-270 (2007)
6. Ugrozov V.V., Filippov, A.N., and Sidorenko, Yu.I. . «On a mathematical description of the isotherm of water vapor sorption on grains of various cereals.». J.of Phys. Chem. (Russia) **70**, 383-386 (2007)
7. Ugrozov V.V., Shebershneva N.N., Filippov A.N., Sidorenko Y.I. « Sorption and Desorption of Water Vapor by Grains of Native Starch of Some Crops» . Colloid J. . (Russia)**,** **70**, No.3, 366–371 (2008)
8. Ugrozov V.V., Artamonova S. D. , Sharnina F. F., Ivshin V. P, Grunin L. Yu. , Kataeva L. I. «Water vapor sorption by chitin-containing materials.» Colloid J. (Russia), **70**, No.6, 780–783 (2008)
9. Ugrozov V.V. « *Cluster model of kinetics of vapor sorption* by amorphyty polymers», Colloid J. (Russia), **73**, .1-9 (2010)

IMPROVED PERFORMANCE AND PRODUCT QUALITY BY AUTOMATED GEOMETRY OPTIMIZATION OF EXTRUSION DIES

Prof. Dr.-Ing. J. Wortberg[1], Dipl.-Ing. K. Saul[1]*

[1] *Universtity of Duisburg-Essen, Institute for Product Engineering, Engineering Design and Plastics Machinery, Duisburg, Germany – kenny.saul@uni-due.de*

Based on previous works, an enhanced method for automatic geometry optimization is topic of current research activities. In continuation to the automatic optimization of simple flow channel geometries like 90° elbows, this project comprises the optimization of more complex geometries like spiral mandrel dies. The main ambition of this research activity is to reduce the time of development and to simplify the design of complex extrusion dies. Using different software systems the developed process meshes the geometry of an initial CAD-model, calculates the flow behavior, interprets the results and changes the CAD-model autonomously until the optimization criteria defined before are fulfilled.

The process is capable to optimize the performance of the spiral mandrel die due to an embedded ability to vary important construction parameters like the channel width, gap width, number of spirals as well as the helix- or wrap angle autonomously. Optimization criteria can be (e.g.) a low pressure loss, homogeneous velocity distribution or even an improved purge behavior. Manufacturing restrictions can be considered and the combination of different optimization criteria can be taken into account to prevent non-realistic design proposals. In first optimization runs the method showed a close match with analytically calculated solutions. This geometry proposal will be manufactured and validated in a blown film test-bench.

Introduction

Extrusion dies are the main components of every extrusion-system. The requirements on these systems are versatile. On the one hand the die has to realize a high product quality, but on the other hand the purge behavior has to be very well and the manufacturing costs for the die should be as low as possible. Processing of different polymer-materials or different operation points shall also be possible so that the whole system can be used for different purposes. These requirements complicate the design of extrusion dies more and more.

In the production of tubes, pipes, preforms, hoses and similar products in engineering plastics these days nearly exclusively spiral-mandrel- or flat-spiral-dies are used because of their advantages in the avoidance of weld-lines in contrast to other die-geometries. A direct comparison between spiral-mandrel-die (SMD) and flat-spiral-die shows that the SMD has advantages in flow distribution but has disadvantages in a high pressure loss [Malekzadeh et al. 2008]. Manufacturing costs and installation space are further disadvantages of SMDs.

Without using numerical simulation systems, the designer at first suggests the outlined diameter of the mandrel to fulfill the essential technological requirements. Further he suggests, according to his experience, the dimension of the spiral channel and the thickness of the slot, whereby the spiral channel is chosen so that its depth decreases uniformly. The form of the slot is chosen in such a way that the surface of the mandrel forms either a cylinder or a cone. This chosen form is used for the calculation of the pressure and the volume flow rates distribution [Vlcek et al. 84]. To calculate the design of the die, the volume flow rates are observed in different spiral-segments. The volume flow coming into the spiral is divided into a volume flow into the gap and another volume flow out of the spiral-segment (dx).

$$\dot{V}_1(x) = \dot{V}_2(x) + \dot{V}_3(x) \qquad \text{Eq.: 1}$$

V_1 (x) = volume flow into the spiral-segment
V_2 (x) = volume flow into the gap
V_3 (x) = volume flow out of the spiral-segment

using the power law (Ostwald – de Waele)

$$\tau^m = \eta \cdot \dot{\gamma} \qquad \text{Eq.: 2}$$

τ = shear stress
η = viscosity
m = material parameter
$\dot{\gamma}$ = shear rate

follows:

$$\dot{V}_2(x)=\frac{1}{\eta}\cdot K_2'(x)\cdot\Delta p^m(x) \qquad \text{Eq.: 3}$$

$$\dot{V}_3(x)=\frac{1}{\eta}\cdot K_3'(x)\cdot\Delta p^m(x) \qquad \text{Eq.: 4}$$

K = parameter of channel geometry
p = pressure
x = channel length

from geometry:

$$\Delta p_2(x)=\frac{\Delta p_3(x)}{\partial x}\cdot\frac{t}{\sin\varphi} \qquad \text{Eq.: 5}$$

φ = helix angle of the spiral
t = distance of two spirals

with rectangular spiral-geometry:

$$\frac{\dot{V}_2(x)}{\dot{V}_3(x)}=\frac{\partial x}{B(x)}\cdot\left[\frac{H_2(x)}{H_3(x)}\right]^{2+m}\cdot\left[\frac{t}{L_2(x)\sin\varphi}\right]^{m}\cdot\frac{1}{f_p} \qquad \text{Eq.: 6}$$

B = channel width
H_2 = gap height
H_3 = channel height

and

$$\frac{\dot{V}_2(x)}{\dot{V}_3(x)}=\frac{1}{\frac{\dot{V}_3(x)}{\dot{V}_2(x)}+1} \qquad \text{Eq.: 7}$$

Using above equations follows with V_k = Volume flow in a spiral:

$$\dot{V}_1(x)=\dot{V}_k-\int_0^x\left[\frac{\dot{V}_2(x)}{\partial x}\right]dx=\dot{V}_k-\int_0^x\left[\frac{\dot{V}_2(x)}{\dot{V}_1(x)}\right]\cdot\frac{\dot{V}_1(x)}{\partial x}dx$$

Eq.: 8

with

$$\dot{V}_2(x)=\dot{V}_1(x)\cdot\left[\frac{\dot{V}_2(x)}{\dot{V}_1(x)}\right] \qquad \text{Eq.: 9}$$

and for the flow in the gap:

$$[\dot{V}_2(x)]_B=\{[\dot{V}_2(x)]_A+[\dot{V}_1(x)]_B\}\left\{\frac{\dot{V}_2(x)}{\dot{V}_1(x)}\right\}_B \qquad \text{Eq.: 10}$$

[Michaeli 1979]

The main goal of the design of a SMD is a homogeneous distribution of the velocity at the outlet of the die. Inhomogeneous distribution would lead to different wall thicknesses in the product and consequently to low product-quality. It is obvious that this calculation method can only be an approximation and that it is very interesting to substitute this method by numerical simulation methods.

State of the art in die design

Three-dimensional simulation of fluid flows seemed to be improvident not many years ago: "*Three-dimensional FEA calculations are too costly on account of the high time outlay involved in structuring the die geometry (network generation) and the as yet still inadequate computing capacity of personal computers*" [Limper 1998]. The appearance of high-performance computation systems in the last years changed this situation significantly, so that Computational Fluid Dynamic (CFD) has become more and more interesting.

Today, the die-design with powerful 2D or mostly 3D-Computer-Aided-Design-systems (CAD) is state of the art. Modern numerical fluid-simulation-systems are able to use these 3D-design models as an initial model and after preprocessing and creating the finite element or finite volume mesh to calculate the flow-behavior. To interpret the solution, the CFD-engineer uses color-contour-plots, 2-dimensional or 3-dimensional diagrams or scalar values. Thus the interpretation-quality, the amount of time for interpretation and the deviated design-changes for geometry optimization depends on the experience of the designer.

Characteristic for such optimization processes is an iterative procedure in which the designer starts with an initial geometry, calculates and interprets the results of this geometry and changes the geometry to start this process from the beginning. In each iteration step the designer has to vary design-parameters of the die-geometry manually and has to preprocess, solve and postprocess the whole simulation run. It is obvious that this procedure can last for quite long but currently there is no method available which allows a completely mathematical calculation of the die geometry for extrusion- or profile-dies [Kaul 2004].
Another difficulty for the designer is to avoid geometry proposals that would lead to high manufacturing costs or which are even impossible to manufacture with standard-manufacturing-tools. Resulting from this, the designer has to be experienced in the manufacturing of such extrusion dies.

Therefore an automated procedure for optimization of an extrusion die is presented, based on previous works, as shown in [Wortberg et al. 08][Wortberg et al. 09].

Geometry-Design

A 3D-CAD-model of a spiral-mandrel-die is used for this optimization-study. According to [Cretu 2008] this model comprises currently three degrees of freedom – the numbers of spirals (in a range between 4 and 12), the starting width of the spirals and the depth of the spirals. In addition to these degrees of freedom currently the starting gap-width, the further trend of the gap width and the angle of the spirals is integrated into the CAD-model.

Another important point in the 3D-CAD-model is the integration of manufacturing knowledge. Impossible geometries are detected and blocked, so that these design-proposals will not be used for the CFD-analysis. So called User Defined Functions (UDF) are used to integrate knowledge into the CAD-system. An UDF is a function in the CAD-system in which different consecutive program-intern processes are executed sequentially. Additionally this UDF is able to execute clauses to check and steer the functions/features.

The approved degrees of freedom are used by the optimization-algorithm to vary the CAD-geometry within predefined constraints. Figures 1 to 3 show different design proposals of the optimization-system.

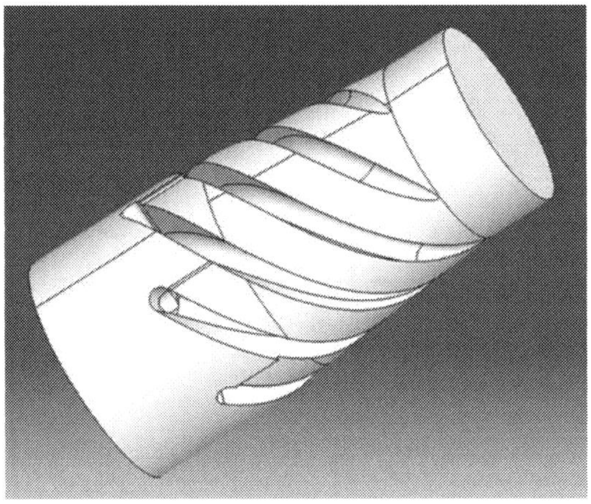

Figure 1 – Optimization Progress - Design-Proposal 1

Figure 1 shows the initial-model for the optimization process. This spiral-mandrel-die has 6 spirals.

The second design-proposal of the optimization run is shown in Figure 2. The width of the spirals is increased.

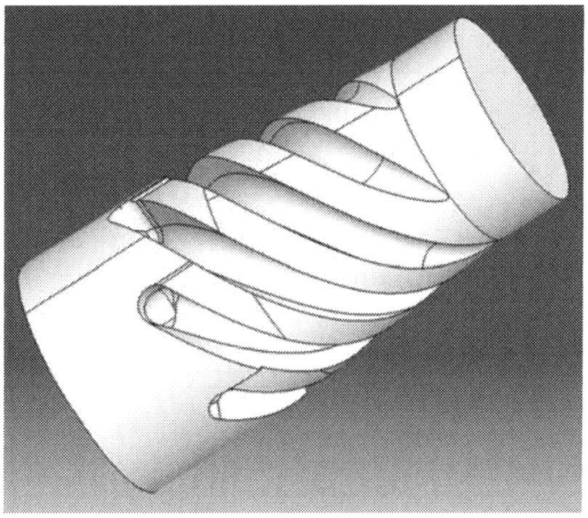

Figure 2 – Optimization Progress - Design-Proposal 2

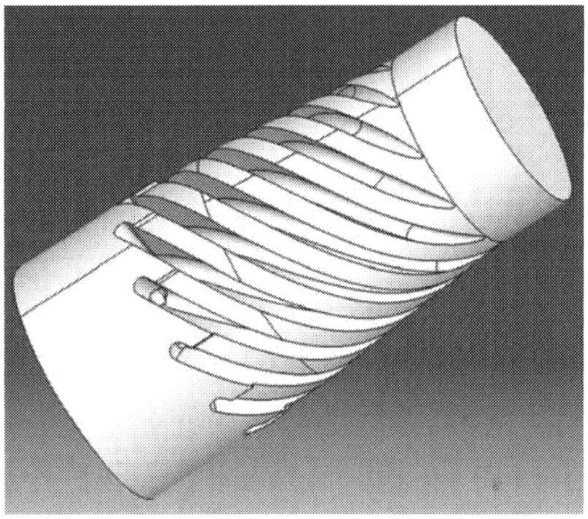

Figure 3 – Optimization Progress - Design-Proposal 4

Figure 3 shows the final spiral-mandrel-die design. The number of spirals is increased to 12 and the channel width is reduced.

Computational Fluid Dynamics

Each geometry-proposal of the optimization system is prepared for the CFD-simulation. Therefore the fluid-volume (the negative) is extracted from the die geometry (the positive). This fluid-volume is discretized with tetraeder and boundary-layer method (wedges near the wall).

The extraction of the fluid volume and the discretization are completely automated by software-intern system-macros, so that no user-interaction is needed.

The predefined discretization-parameters have to be selected intelligently. On the one hand the discretization has to be very fine, so that the discretization has no further influence on the solution quality; on the other hand the discretization has to be coarse to gain minimum solution time.

To be sure about the quality of the meshing-parameters, several convergency-studies have been solved.

Figure 4 shows the discretized 3D-CAD-Model of the extracted flow volume. This model is discretized using the boundary layer method combined with tetraeder-meshing-method to increase the accuracy of the numerical solution.

Boundary Layers increase the accuracy of the mesh near the wall. This method is often used when the flow is influenced mainly in this area. In extrusion systems the polymer sticks to the wall, so that there are high gradients of the shear rate in the area near the wall. To achieve a good resolution of the melt flow in the cylindrical gap at the outlet, it is important to have at least 9 to 11 cells across the height of the channel.

The green arrows show the zones where the polymer-melt enters the fluid-volume and where the fluid-volume is left (boundary conditions). This preprocessing is done fully automated by program-intern system-macros.

Figure 4 – Discretized 3D-Geometry-Model (Fluid-Volume)

In figure 5 another perspective of the discretized fluid-volume is shown. The melt-flow-inlet areas (six drillings for the melt – negative model of these drillings because of extraction of fluid volume) are clearly visible.

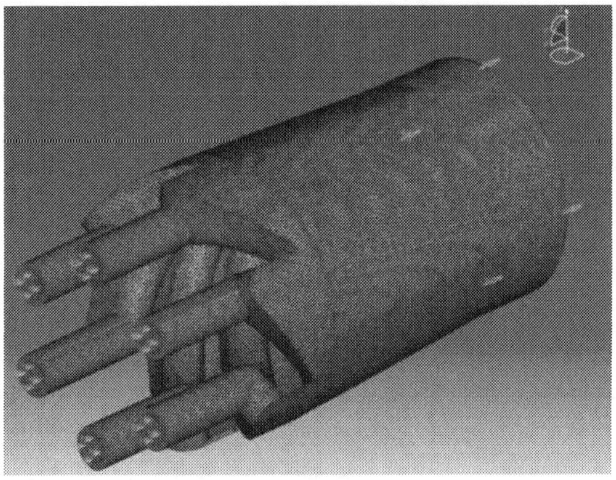

Figure 5 – Discretized 3D-Geometry-Model (Fluid-Volume) – second perspective

The discretized fluid-volume is transferred to the CFD-Preprocessor. The preprocessing is executed automatically by using a batch-file. After completing this, the solution-process is started, resulting in report files. These report files are used for the interpretation of the design-proposal by the optimization algorithm.

The interpretation of the quality of the design-proposal can be executed with different foci. Scalar values like the maximal pressure loss or the maximal wall shear stress within the geometry are easy to evaluate. But in most cases it is more important to get knowledge about distributions of values in different (predefined) zones of the geometry.

The velocity distribution at the outlet of the die geometry is such an interesting value. This distribution is responsible for the wall thickness in the product and so it is essentially for the product quality. The use of value tables as an optimization goal is aligned with vastly increase in calculation time, so that it is necessary to reduce this table. Therefore the positive and negative aberration of the average velocity is calculated and the minimization of this value is used as an optimization goal.

Other interesting values can be the distribution of the temperature, the absence of areas with low wall shear stresses or even the whole distribution of the wall shear stresses because of their essential influence for the purge behavior of the whole die geometry [Niemeier 06].

For interpreting the melt-distribution in the spiral-mandrel-system it would be very nice to have knowledge about the distribution-angle of the polymer-melt coming from one spiral-mandrel. At present this method is investigated at the IPE and the automation of this interpretation method is tested.

Figure 6 shows the velocity distribution of one of the first design-proposals. It is obvious that there are various peaks in this distribution which will result in a bad product-quality. The two different scatter plots in the diagram are the velocity in different distances from the channel wall. While the velocity near the wall decreases because of the no-slip-condition for the wall-boundaries, the velocity in the center of the channel has the highest value. The x-axis shows the position over the outlet diameter of the die.

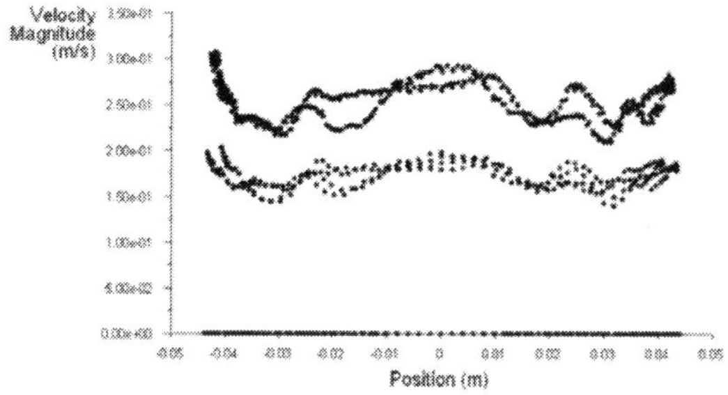

Figure 6 – Velocity profile at the outlet of a spiral-mandrel-die, one of the first optimization runs

Figure 7 and 8 show the velocity distribution at the outlet of the die in further optimization runs. It is apparent, that the difference between maximum and minimum value decreases. The final geometry-proposal in figure 10 has a very smooth velocity distribution.

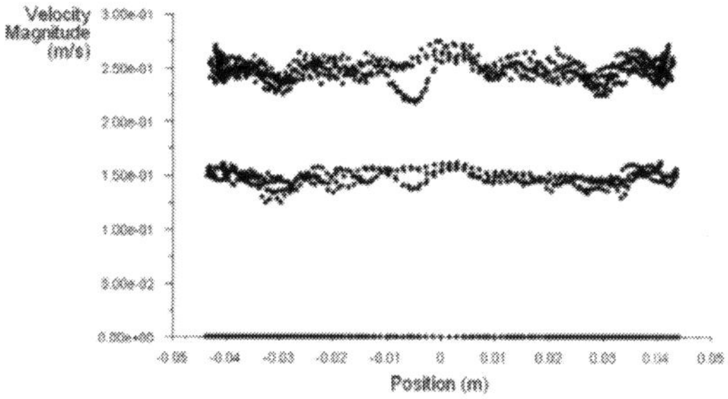

Figure 7 – Velocity profile at the outlet of a spiral mandrel die, median optimization run

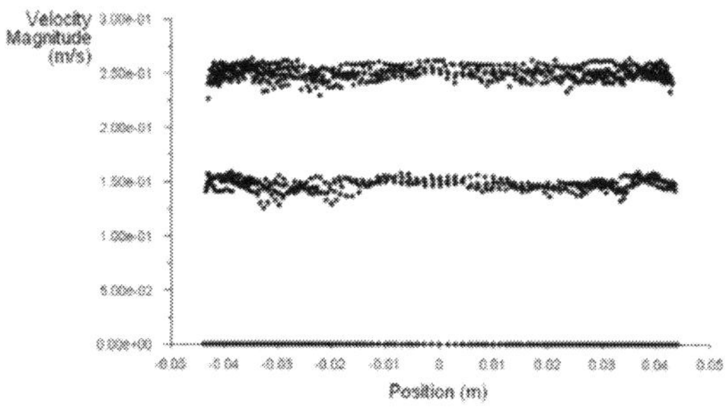

Figure 8 – Velocity profile at the outlet of a spiral mandrel die, last optimization run

Another interesting method to interpret the result is to show the distribution of the material over the thickness of the outcoming material. A main function of the spiral mandrel die is to avoid weld-lines. So it can be very interesting to get knowledge about the distribution of the material coming from different spirals. This interpretation method is investigated in present work and shows good accordance to practical measurements. The automation of this method will be a future step to increase the power of the optimization-system.

Figure 9 shows the graphical analysis of the weld-line avoidance. Blue points in the diagram represents the material coming from the first spiral, red triangle represent the material coming from the second spiral. On the y-axis the thickness of each film is shown over the angle of the cylindrical sector. The black curve shows a freehand sketch to gain a mathematical description of the distribution. In this example the distribution is stretched across an angle of 60° degrees.

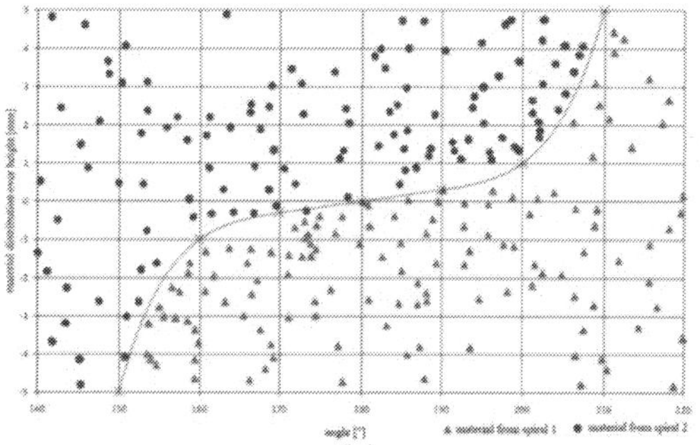

Figure 9 – Currently investigated method to interpret the avoidance of weld-lines, different die-geometry than shown in the pictures before

Figure 10 and figure 11 show pathline-graphics of spiral mandrel dies which are used to create the diagram shown in figure 9.

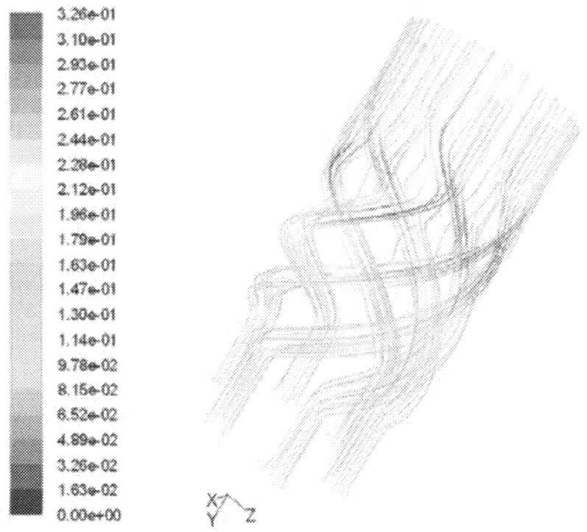

Figure 10 – Pathline graphic to illustrate the flow path in a 6-spiral mandrel-die

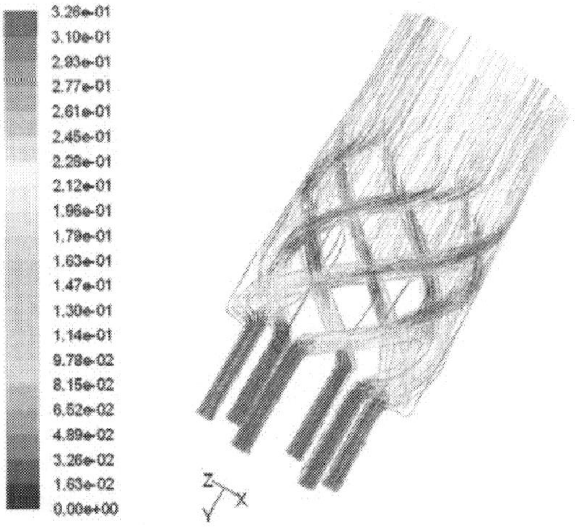

Figure 11 – Pathline graphic to illustrate the flow path in a 6-spiral mandrel-die with decreased channel width.

Conclusion

The design of SMD is a complex process in which different aspects have to be considered. The die has to ensure a constant high product quality with absence of weld-lines and homogeneous distribution of velocity and temperature combined with low pressure loss. Fast purge-behavior is another key-factor for die design and is mainly influenced by the absence of zones with low shear-stress.

Since powerful computation hardware has been developed, 3D-CAD-systems and finite element/volume method has become more and more efficient in everyday optimization processes.

Therefore the automated combination of 3D-CAD and 3D-CFD is the next consequential step to reach high quality die-design with low development-time.

Using the example of an easy 3-dimensional spiral-mandrel die with a starting number of three degrees of freedom, it is shown that this method performs well.

Next steps are, to increase the number of degrees of freedom and to implement the manufacturing restrictions into the CAD-system. The investigation of additional ways for interpreting the results – even in an automated process – has started yet. Further it is necessary to increase the robustness of the 3D-CAD-system – some design parameter changes lead to aborts in the optimization-run.

Acknowledgements

This project is supported by:

German National Science Foundation
Deutsche Forschungsgemeinschaft, DFG

Title: Integrative Interconnection of Design, Calculation and Simulation for Dimensioning a Flow Channel in Polymer Processing

Project-No. WO-302-37-2, KO1620-9-2

Deutsche Forschungsgemeinschaft DFG

and by

NRW Ziel-2 Program
CheK.NRW – Chemie und Kunststofftechnik
ERDF – European Regional Development Fund

Title: Computer-Aided Optimization of Extrusion-Dies in Consideration of Manufacturing Restrictions

Project-No. 280809302

References

Journal: Malekzadeh, M., Goharpey, F., Foudazi, R., "Flow Analysis of Flat Spiral Dies and Comparison with Cylindrical Spiral Mandrel Dies", Int. Polym. Process., XXIII, 38-46 (2008)

Journal: Vlcek, J., Kral, V., Kouba, K. "The calculation of the form of a spiral mandrel", Plastics and Rubber Processing and Applications 4 (1984), 309-315

Book: Michaeli, W.: Extrusion Dies for Plastics and Rubber, 1st Edition, Hanser Publishers, Munich (1979)

Proceedings: Limper, A., Stieglitz, H., "Recent Research on Mathematical Design of Spiral Mandrel Dies", SPE ANTEC Tech. Papers, 309-313, 1998

Proceedings: Wortberg, J., Köhler, P. Saul, K., Lupa, N. "Integrative Interconnection of Design, Calculation and Simulation for Dimensioning a Flow Channel in Polymer Processing", Proceedings of Polymer Processing Society 24th, Page II. 142, ISBN 88-7897-024-7

Proceedings: Wortberg, J., Köhler, P. Saul, K., Lupa, N. "Integrative Interconnection of Design, Calculation and Simulation for Dimensioning a Flow Channel in Polymer Processing", Proceedings of Polymer Processing Society,

Dissertation: Niemeier, H. "Analyse und Optimierung wandnaher Schmelzeströmungen in Extrusionswerkzeugen für die Kunststoffverarbeitung", Shaker Verlag Aachen 2006, ISBN 3-8322-5385-8

Dissertation: Kaul S. "Computer Based optimization strategies for profile extrusion dies", 2004, ISBN 3-86130-492-9

Dissertation: Cretu I. M. "Analyse, Auslegung und Optimierung von Wendelverteilerwerkzeugen", 2008, Shaker Verlag Aachen, ISBN 978-8322-7705-5

THE MESOSCOPIC CONSTITUTIVE EQUATIONS FOR POLYMERIC FLUIDS AND SOME EXAMPLES OF FLOWS

Gusev Alexsei [(1)], Afonin Grigory [(1)], Tretyakov Ilya [(1)], Pyshnogray Grigory [(1)*]

[(1)] *Department of Mathematics, Altai State Technical University, Barnaul, Altai, Russia – pyshnogray@mail.ru*

ABSTRACT

Constitutive equations for melts and concentrated solutions of linear polymers are derived as consequences of dynamics of a separate macromolecule. The model is investigated for viscometric flows. It was shown that the model gives a good description of non-linear effects of simple polymer flows: viscosity anomalies, first and second normal stresses, non-steady shear stresses. A simple rheological equation of state (RES) which can be chosen as an initial approximation in formulating such a sequence of RES are obtained and studied. In this work, RES is extended to the case of allowance for the additional corrections caused by intrinsic viscosity and the delayed interaction of a macromolecule with its environment. Realization of this approach involves consequent solution of two problems: formulation of the equations of dynamics for a macromolecule and transition from the formulated equations to RES. The resulting equations can be recommended as a first approximation in constructing a sequence of RES. Comparison this approach with another authors are given. Two cases of steady-state flow between unlimited parallel planes under the action of a constant pressure gradient are considered and the same constitutive equations allow us to expand calculations also on the process of extension of the jet after the leaving of the die. Considering the processes of stretching, which occur at the lower temperatures, one has to take into account the possible process of crystallization of polymer.

Introduction

Rheological equation of state of linear polymers solutions and melts obtained with using methods of statistical mechanics and thermodynamics for description and interpretations of nonlinear effects. The base of this equations of Brown dynamics with entered non Markovian anisotropic noise have been formulated and has been shown, that introduction of the certain kind non Markovian fluctuations leads to occurrence at a macrolevel of intermediate scale of length which is meaningful radius of a tube as in theory De Gennes, or lengths of a chain between entanglements as at Graessley.

At the description of process of polymeric film formation the account of that is necessary, that the received film is cooled and exposed to stretching. As these processes occur simultaneously at their mathematical modelling the joint decision of the equations for stresses and heat conduction is necessary. In work for a finding of the established stress at stretching it has been used rheological model which parameters are functions of temperature. At thermal calculation it was supposed, that thickness of a film is small. Also in work calculation of non-uniform current of a nonlinear viscoelastic liquid between parallel planes under action of constant pressure gradient is executed. Comparison of results of calculation of a velocity with experimental data is executed. It allows draw a conclusion, that in the considered case of Pouselle flow, the rheological model describes not parabolic profile of velocity between parallel plates that does not contradict experimental data.

Film casting process

In this work the problem of mathematical description of film casting process which schematic image is presented according to [1–3] on Fig. 1 is considered, whence it is visible that at first polymer melt is pressed through die and after expression the received film is reeled up on a drum. As a result of movement of a film from die to a drum there is a material cooling, change of its width and a thickness. During this moment the film is exposed to a stretching and there is so-called «an effect of a neck». All these processes occur simultaneously and consequently at their mathematical modelling the joint decision of the equations for stresses and heat conduction is necessary. In work for a finding of the established pressure at a stretching has been used modified rheological model Vinogradov-Pokrovskii which parameters are known functions of temperature. At thermal calculation it was considered that the thickness of a film is small enough and consequently it is possible to assume that the temperature on thickness direction of a film is constant.

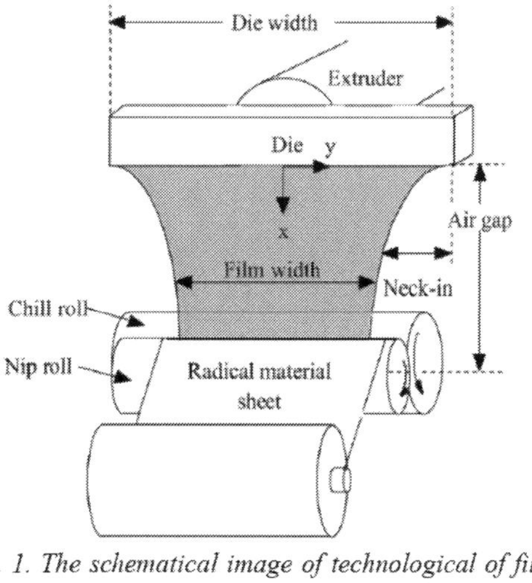

Fig. 1. The schematical image of technological of film casting process[1-3].

For the decision of a problem of a stretching and cooling of a film after an exit from die will use rheological equation of state [4]:

$$\sigma_{ik} = -p\delta_{ik} + 3\frac{\eta_0}{\tau_0}a_{ik};$$
$$\frac{d}{dt}a_{ik} - v_{ij}a_{jk} - v_{kj}a_{ji} + \frac{1+(\kappa-\beta)I}{\tau_0}a_{ik} = \frac{2}{3}\gamma_{ik} - 3\frac{\beta}{\tau_0}a_{ij}a_{jk}, \tag{1}$$

Where σ_{ik} - stress tensor; p – pressure; η_0 and τ_0 – initial values of shear viscosity and relaxation time; $v_{ik} = \frac{\partial v_i}{\partial x_i}$ – tensor gradients of velocity; a_{ik} - symmetric second rank tensor of anisotropy; $I = a_{ii}$ – the first invariant anisotropy tensor a_{ik}; $\gamma_{ik} = \frac{1}{2}(v_{ik} + v_{ki})$ – simmetric tensor gradients of velocity; κ and β – the phenomenological parameters of model considering in the equations of dynamics of macromolecule the sizes and the form of macromolecular coil.

Then in Cartesian co-ordinates the system of the equations of dynamics looks like:

$$\frac{\partial v_x}{\partial x} + \frac{\partial v_y}{\partial y} + \frac{\partial v_z}{\partial z} = 0,$$
$$\rho\left(\frac{\partial v_x}{\partial t} + v_x\frac{\partial v_x}{\partial x} + v_y\frac{\partial v_x}{\partial y} + v_z\frac{\partial v_x}{\partial z}\right) = \frac{\partial\sigma_{11}}{\partial x} + \frac{\partial\sigma_{12}}{\partial y} + \frac{\partial\sigma_{13}}{\partial z}, \tag{2}$$
$$\rho\left(\frac{\partial v_y}{\partial t} + v_x\frac{\partial v_y}{\partial x} + v_y\frac{\partial v_y}{\partial y} + v_z\frac{\partial v_y}{\partial z}\right) = \frac{\partial\sigma_{21}}{\partial x} + \frac{\partial\sigma_{22}}{\partial y} + \frac{\partial\sigma_{23}}{\partial z},$$
$$\rho\left(\frac{\partial v_z}{\partial t} + v_x\frac{\partial v_z}{\partial x} + v_y\frac{\partial v_z}{\partial y} + v_z\frac{\partial v_z}{\partial z}\right) = \frac{\partial\sigma_{31}}{\partial x} + \frac{\partial\sigma_{32}}{\partial y} + \frac{\partial\sigma_{33}}{\partial z},$$

Where, v_i – melt velocity along axes, and accordingly; ρ – density. For non-isothermal case the system (1, 2) should be added by the equation of carrying over of energy [5, 6]:

$$\rho C_V\left(\frac{\partial T}{\partial t} + v_x\frac{\partial T}{\partial x} + v_y\frac{\partial T}{\partial y} + v_z\frac{\partial T}{\partial z}\right) =$$
$$= \frac{\partial}{\partial x}\left(\lambda\frac{\partial T}{\partial x}\right) + \frac{\partial}{\partial y}\left(\lambda\frac{\partial T}{\partial y}\right) + \frac{\partial}{\partial z}\left(\lambda\frac{\partial T}{\partial z}\right) - \frac{\mu}{h}(T - T_0), \tag{3}$$

Where C_V - a specific thermal capacity at constant volume; T – film temperature, T_0 – an ambient temperature, λ – heat conductivity factor; μ – heat exchange factor, $h = \frac{S}{p}$ – the relation of the area of section of a film to section perimeter ($h = \frac{ab}{2a+2b} = \frac{b}{2}$, for thin film $a >> b$). Last composed in (3) considers film cooling through its surface. We will notice that in (3) the stream of heat arising at nonzero gradients of speed is not considered.

Let's consider the decision of a stationary problem when in motionless system of co-ordinates there are no sizes time-dependent. The beginning of co-ordinates we will place in the middle of an exhaust outlet, an Ox axis we will direct along movement of a film and we will search for the decision of system depending only from a variable (1-3)

$$v_x = v_x(x); \qquad T = T(x).$$

If to address to the experimental data [7] received for various samples at identical degree of a stretching we will see that various polymeric samples have various width. It does not correspond to a mode uni-axial elongation.

This effect can be considered if to assume that in an axis direction Oz operates additional stretching pressure that leads to following expression for tensor gradients of velocity.

As a results stress tensor looks like:

$$\begin{bmatrix}\sigma_{11} & 0 & 0\\ 0 & 0 & 0\\ 0 & 0 & \sigma_{33}\end{bmatrix}$$, here σ_{11} – elongation pressure, σ_{33} – additional pressure. (4)

Thus speed of change of width will differ from speed of change of its thickness and tensor gradients of velocity will look

like: $$v_{ij} = \begin{bmatrix}u'(x) & 0 & 0\\ 0 & -\frac{1+\alpha}{2}u'(x) & 0\\ 0 & 0 & -\frac{1-\alpha}{2}u'(x)\end{bmatrix}, \tag{5}$$

Here α is parameter of elongation anisotropy.

In this case it is possible to find dependence between speed of a film u (x) both width and thickness of a film. As from definition: $v_{22} = \frac{\partial u_y}{\partial y}$, то

$\frac{\partial u_y}{\partial y} = -\frac{(1+\alpha)}{2} u'(x)$, obtained

$U_y(x,y) = -y\frac{(1+\alpha)}{2} u'(x)$, if $y=a(x)$ and

$U_y = \frac{dy}{dt} = \frac{da(x)}{dt} = a'(x)\cdot\frac{dx}{dt} = u(x)a'(x)$, or

$u(x)a'(x) = -a(x)\cdot\frac{1+\alpha}{2}u'(x)$.

$$\frac{a'(x)}{a(x)} = -\frac{1-\alpha}{2}\frac{u'(x)}{u(x)}$$

Or after integration we receive expression for width of a film:

$$a(x) = a(0)\left(\frac{u(x)}{v_0}\right)^{-\frac{1+\alpha}{2}} \qquad (6)$$

And similar distribution for a thickness:

$$b(x) = b(0)\left(\frac{u(x)}{v_0}\right)^{-\frac{1-\alpha}{2}} \qquad (7)$$

In a stationary case, for the description of an anisotropic stretching in one-dimensional approach from the equations (1-3) taking into account (5), (7) we will receive:

$$\rho u\frac{du}{dx} = 3\frac{d}{dx}\left(\frac{\eta_0}{\tau_0}(a_{11}-a_{22})\right)$$

$$u\frac{da_{11}}{dx} - 2\frac{du}{dx}a_{11} + \frac{1+(\kappa-\beta)(a_{11}+a_{22}+a_{33})}{\tau_0}a_{11} = \frac{2du}{3dx} - 3\beta a_{11}^2$$

$$u\frac{da_{22}}{dx} - (1+\alpha)\frac{du}{dx}a_{22} + \frac{1+(\kappa-\beta)(a_{11}+a_{22}+a_{33})}{\tau_0}a_{22} = -\frac{1+\alpha}{3}\frac{du}{dx} - 3\beta a_{22}^2$$

(8)

$$u\frac{da_{33}}{dx} - (1-\alpha)\frac{du}{dx}a_{33} + \frac{1+(\kappa-\beta)(a_{11}+a_{22}+a_{33})}{\tau_0}a_{33} = -\frac{1-\alpha}{3}\frac{du}{dx} - 3\beta a_{33}^2$$

$$\rho C_v u\frac{dT}{dx} = \lambda\frac{d^2T}{dx^2} - \frac{\mu}{h}(T-T_0)$$

It is necessary to add these equations with boundary conditions for speed and temperature. In a case u (x) these conditions look like: $v_x(0) = v_0$; $v_x(l) = kv_0$, where k - factor of a stretching of a film.

For temperature we will consider boundary conditions: the film temperature at contact with a cooling roll coincides with the established temperature of the roll and is equal to an ambient temperature: $T(0) = T_1; T(l) = T_0$. Here T_1 - is melt temperature on an exit from die.

For initial viscosity and initial relaxation time used expressions: $\eta_0(T) = \eta_0(T_1)e^{\frac{\Delta H}{R}\left(\frac{1}{T+273}-\frac{1}{T_1}\right)}$,

$\tau_0(T) = \frac{\eta_0(T)}{nk(T+273)}$, here n – number of macromolecules in unit of volume, k – constant Boltzmann, R – universal gas constant, $\Delta H/R$ – energy of activation.

Let's consider now results of modelling of process. To have full idea about behaviour of investigated system we will enter into consideration dimensionless characteristic numbers: **Prandtl** number (approximating the ratio of momentum diffusivity (kinematic viscosity) and thermal diffusivity); **Nusselt** number (the ratio of convective to conductive heat transfer across the boundary); **Reynolds** number (gives a measure of the ratio of inertial forces to viscous forces) and **Weissenberg** number (the ratio of the relaxation time of the fluid and a specific process time) and let's consider their influence on system decisions.

The greatest interest for practice the temperature and width of a film represent velocity of flow in various points,

From drawings it is visible that at small **Pr** numbers there is a fast cooling of a film and the temperature leaves on the final value equal to environment temperature. At great **Pr** numbers of cooling of a film practically does not occur, what not influence on flows.

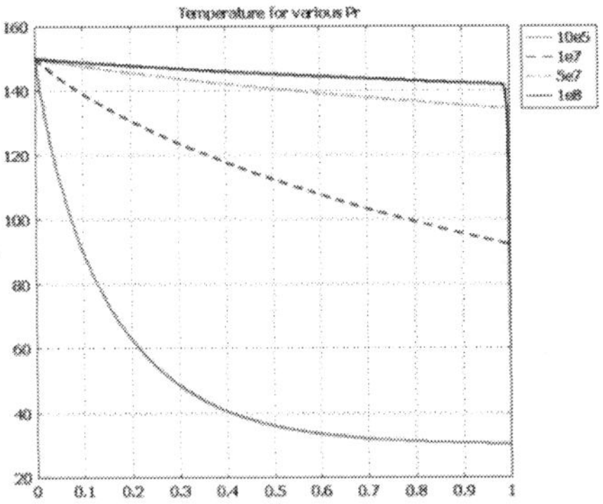

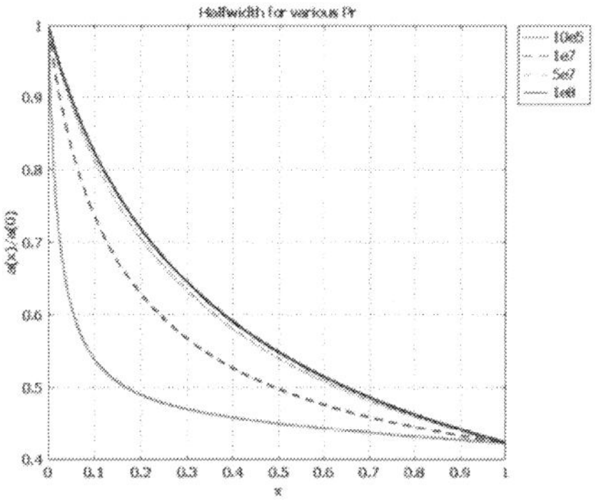

Fig. 2.

To number Nu growth also there is faster cooling of a film that influences its form (dependence of width on distance).

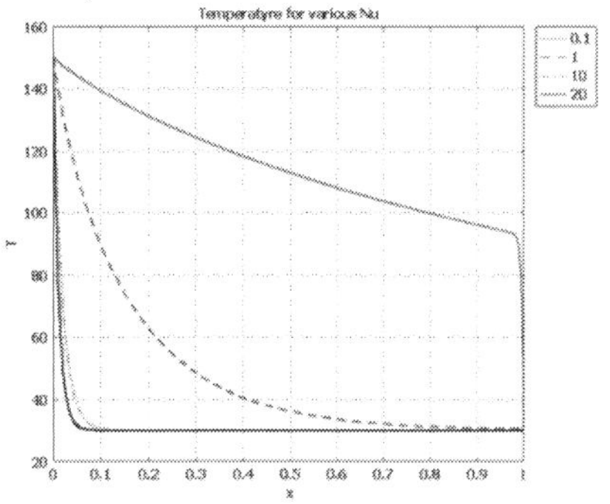

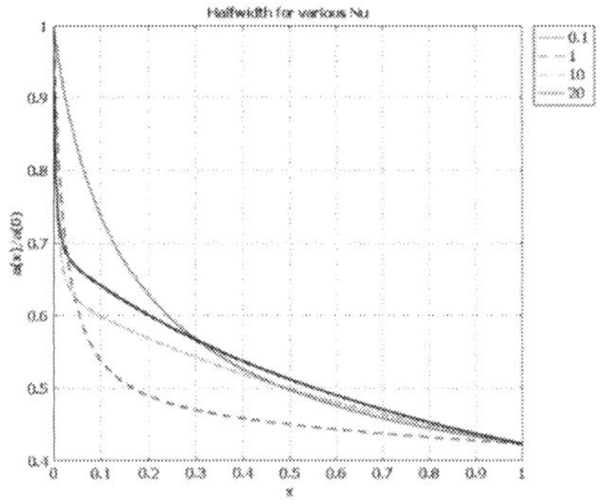

Fig. 3.

The **Re** number increase on the contrary leads to that film is cooled worse.

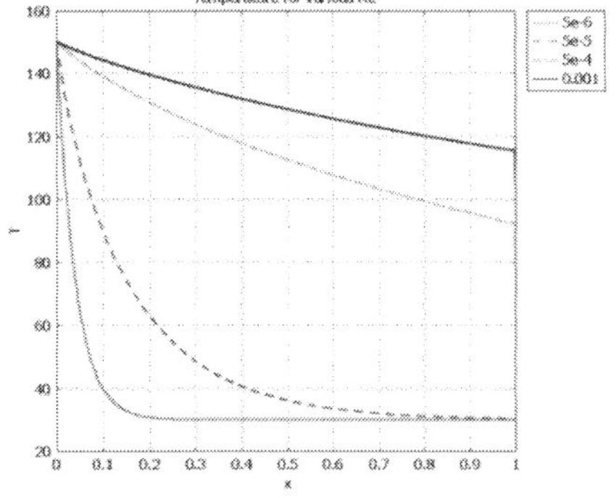

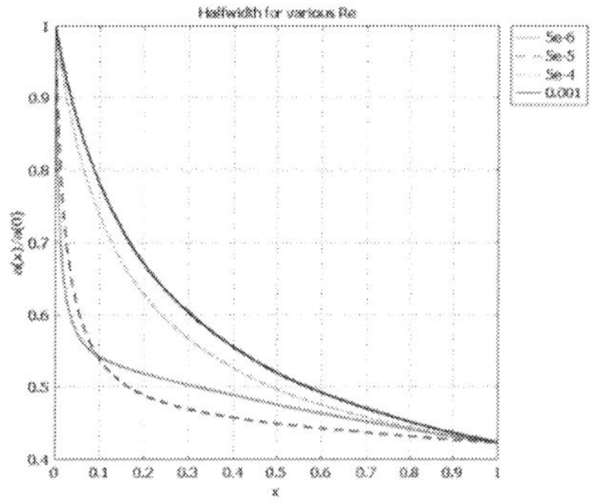

Fig. 4.

The most essential made impact of parameter of anisotropy of a stretching. Apparently from drawing with growth α the width of a film increases.

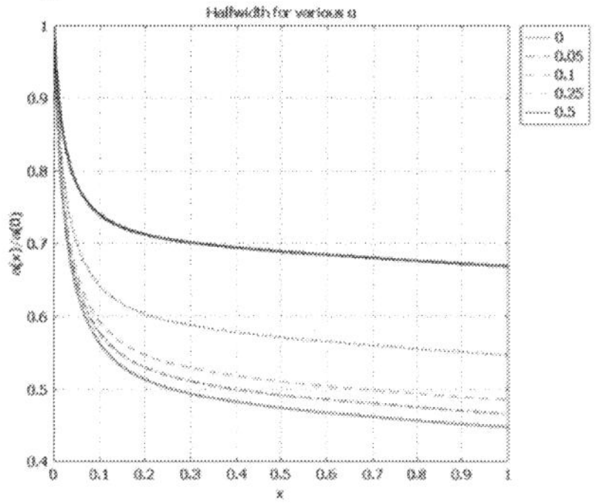

Fig. 5.

On a following slide comparison with experimental data from work is shown. It is possible to notice the good consent of the theory and experiment that confirms the made assumptions. Here k - factor of a stretching of a film.

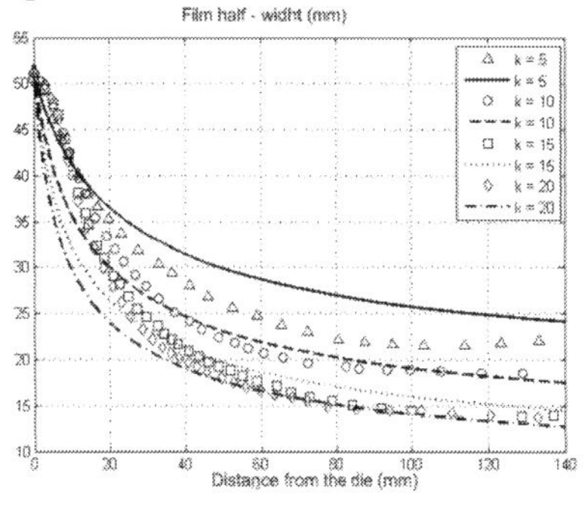

Fig. 6. Comparison with experimental data

From the picture it is visible that on this dependence there is a site of constant width, and the basic change of width of a film occurs on distance of 20 %-30 % from an exit, it is so-called effect of occurrence of "neck" which would not be observed without film cooling.

Possibility of use modified rheological is shown model Vinogradov-Pokrovskii for the description of linear polymers melts flows in various modes of deformation.

The system of the equations of dynamics is written down in one-dimensional approach, at the account heat exchange. Comparison with experimental data available in the literature for film half-width is spent and adequacy of the applied approach is shown.
Thus, it is possible to draw a conclusion that rheological model (1, 2) at the account of heat exchange (3) consistently describes observed at formation of polymeric films effects [1–3], and the received dependences making stress tensor from distance to an exit from die can be used at modelling of process of crystallisation polymer melt under the influence of a stretching and cooling.

The flow between parallel planes

Shall consider a problem of motion of the non-linear viscoelastic liquid, governing by the established constitutive equations, in a gap between parallel planes under constant pressure gradient. The system of the equations consists of the continuity equation and the equation of dynamics, which, in the Cartesian system of coordinates, are written (2).
Due to the form of the die, we can consider the flow in the channel of the land as a flow between two parallel planes with distance h.
One can chose the beginning of coordinates at one of the wall of the die, directing the axis Ox along the stream, the axis Oy perpendicular to the planes, and the axis Oz perpendicular to the axes Ox and Oy. Due to the symmetry of the problem, the solution, that is the distribution of velocity vx, vy, vz and orientation axx, axy, ... does not depend on co-ordinate z and the system of the equations (2) is reducing to:

$$\frac{\partial V_x}{\partial x}+\frac{\partial V_y}{\partial y}=0,$$

$$\rho\left(\frac{\partial V_x}{\partial t}+V_x\frac{\partial V_x}{\partial x}+V_y\frac{\partial V_x}{\partial y}\right)=\frac{\partial \sigma_{xx}}{\partial x}+\frac{\partial \sigma_{xy}}{\partial y},$$

$$\rho\left(\frac{\partial V_y}{\partial t}+V_x\frac{\partial V_y}{\partial x}+V_y\frac{\partial V_y}{\partial y}\right)=\frac{\partial \sigma_{xy}}{\partial x}+\frac{\partial \sigma_{yy}}{\partial y},$$

$$\sigma_{xx}=-p+3\frac{\eta_0}{\tau_0}a_{xx},\sigma_{xy}=3\frac{\eta_0}{\tau_0}a_{xy},\sigma_{yy}=-p+3\frac{\eta_0}{\tau_0}a_{yy},$$

$$\frac{\partial a_{xx}}{\partial t}+V_x\frac{\partial a_{xx}}{\partial x}+V_y\frac{\partial a_{xx}}{\partial y}-2a_{xx}\frac{\partial V_x}{\partial x}-2a_{xy}\frac{\partial V_x}{\partial y}+\frac{1+(\kappa-\beta)I}{\tau_0}a_{xx}=$$
$$=\frac{2}{3}\frac{\partial V_x}{\partial x}-3\frac{\beta}{\tau_0}\left(a_{xx}^2+a_{xy}^2\right),$$

$$\frac{\partial a_{xy}}{\partial t}+V_x\frac{\partial a_{xy}}{\partial x}+V_y\frac{\partial a_{xy}}{\partial y}-a_{xy}\frac{\partial V_x}{\partial x}-a_{xy}\frac{\partial V_y}{\partial y}-a_{xx}\frac{\partial V_y}{\partial x}-a_{yy}\frac{\partial V_x}{\partial y}+$$
$$+\frac{1+(\kappa-\beta)I}{\tau_0}a_{xy}=\frac{1}{3}\left(\frac{\partial V_x}{\partial y}+\frac{\partial V_y}{\partial x}\right)-3\frac{\beta}{\tau_0}a_{xy}\left(a_{xx}+a_{yy}\right),$$

$$\frac{\partial a_{yy}}{\partial t}+V_x\frac{\partial a_{yy}}{\partial x}+V_y\frac{\partial a_{yy}}{\partial y}-2a_{xy}\frac{\partial V_y}{\partial x}-2a_{yy}\frac{\partial V_y}{\partial y}+\frac{1+(\kappa-\beta)I}{\tau_0}a_{yy}=$$
$$=\frac{2}{3}\frac{\partial V_y}{\partial y}-3\frac{\beta}{\tau_0}\left(a_{xy}^2+a_{yy}^2\right)$$

(9)

The system of the equations (9) describes the two-dimensional unsteady flows of polymeric medium. Further, the solution of equations can be considered to be independent on the variable x , and we consider that the gradient of pressure is constant, that is $\frac{\partial p}{\partial x}=-A$

Then we shall have:

$$\frac{\partial V_y}{\partial y}=0,$$

$$\rho\left(\frac{\partial V_x}{\partial t}+V_y\frac{\partial V_x}{\partial y}\right)=-\frac{\partial p}{\partial x}+3\frac{\eta_0}{\tau_0}\frac{\partial a_{xy}}{\partial y},$$

$$\rho\left(\frac{\partial V_y}{\partial t}+V_y\frac{\partial V_y}{\partial y}\right)=-\frac{\partial p}{\partial y}+3\frac{\eta_0}{\tau_0}\frac{\partial a_{yy}}{\partial y},$$

$$\frac{\partial a_{xx}}{\partial t}+V_y\frac{\partial a_{xx}}{\partial y}-2a_{xy}\frac{\partial V_x}{\partial y}+\frac{1+(\kappa-\beta)I}{\tau_0}a_{xx}=-3\frac{\beta}{\tau_0}\left(a_{xx}^2+a_{xy}^2\right),$$

$$\frac{\partial a_{xy}}{\partial t}+V_y\frac{\partial a_{xy}}{\partial y}-a_{xy}\frac{\partial V_y}{\partial y}-a_{yy}\frac{\partial V_x}{\partial y}+\frac{1+(\kappa-\beta)I}{\tau_0}a_{xy}=$$
$$=\frac{1}{3}\frac{\partial V_x}{\partial y}-3\frac{\beta}{\tau_0}a_{xy}\left(a_{xx}+a_{yy}\right),$$

$$\frac{\partial a_{yy}}{\partial t}+V_y\frac{\partial a_{yy}}{\partial y}-2a_{yy}\frac{\partial V_y}{\partial y}+\frac{1+(\kappa-\beta)I}{\tau_0}a_{yy}=\frac{2}{3}\frac{\partial V_y}{\partial y}-3\frac{\beta}{\tau_0}\left(a_{xy}^2+a_{yy}^2\right)$$

(10)

Note that it follows from equation (9), that $V_y(y)$ is a linear function of y, but so as, due to the boundary conditions $V_y(0)=V_y(h)=0$, one has $V_y(y)=0$, and the equation of continuity is satisfied identically.
If consider a steady-state case for which the above equations (8) can be reduced to more simple ones:

$$3\frac{\eta_0}{\tau_0}\frac{\partial a_{xy}}{\partial y}=\frac{\partial p}{\partial x},$$

$$3\frac{\eta_0}{\tau_0}\frac{\partial a_{yy}}{\partial y}=\frac{\partial p}{\partial y},$$

$$a_{xx}=\frac{\tau_0}{1+(\kappa-\beta)I}\left(2a_{xy}\frac{\partial V_x}{\partial y}-3\frac{\beta}{\tau_0}\left(a_{xx}^2+a_{xy}^2\right)\right),$$

$$a_{xy}=\frac{\tau_0}{1+(\kappa-\beta)I}\left(\left(a_{yy}+\frac{1}{3}\right)\frac{\partial V_x}{\partial y}-3\frac{\beta}{\tau_0}a_{xy}\left(a_{xx}+a_{yy}\right)\right),$$

$$a_{yy}=-3\frac{\beta}{1+(\kappa-\beta)I}\left(a_{xy}^2+a_{yy}^2\right)$$

(11)

We shall look for solution of the system (11) by the method of consecutive approximation, considering terms of the first order with respect to small parameters κ and β. Then the system of the equations for the zero approximation will become:

$$\frac{\partial a_{xy}}{\partial y}=-\frac{1}{3}\frac{\tau_0}{\eta_0}A,\quad a_{xx}=2\tau_0 a_{xy}\frac{\partial V_x}{\partial y},\quad a_{xy}=\frac{1}{3}\tau_0\frac{\partial V_x}{\partial y},$$

$$a_{yy}=0.$$

From here, we shall find, replacing the partional derivatives with the full ones:

$$\frac{1}{3}\tau_0\frac{d^2}{dy^2}V_x=-\frac{\tau_0}{3}\frac{A}{\eta_0};\frac{d^2}{dy^2}V_x=-\frac{A}{\eta_0};V_x(y)=-\frac{1}{2}\frac{A}{\eta_0}y^2+ay+b.$$

where the constants a and b ought to found from the boundary conditions: $V_x(0)=V_x(h)=0$
so that in the zero approximation, one has the solution

$$V_x(y)=-\frac{1}{2}\frac{A}{\eta_0}(y-h)y=\frac{1}{2}\frac{A}{\eta_0}(h-y)y;$$

$$\frac{dV_x}{dy}\frac{1}{2}\frac{A}{\eta_0}(h-2y);$$

$$a_{xy}=\frac{1}{3}\tau_0\frac{A}{2\eta_0}(h-2y)=\frac{\tau_0 A}{6\eta_0}(h-2y);$$

$$a_{xx}=\frac{1}{6}\tau_0^2\frac{A^2}{2\eta_0^2}(h-2y)^2.$$

This is a well-know Poiseuille law for viscous liquid. Having introduced the symbol $\overline{A}=\frac{A\tau_0}{\eta_0}$, the above solution can be rewritten as

$$V_x(y)=\frac{\overline{A}}{2\tau_0}(h-y)y;\ a_{xy}=\frac{\overline{A}}{6}(h-2y);\quad a_{xx}=\frac{A^2}{6}(h-2y)^2;\ a_{yy}=0.$$

The next step is a calculation of the amendments of the first order with respect to κ and β. In this case, using (11) we are receiving:

$$I=a_{xx}+a_{yy}=\frac{1}{6}\overline{A}^2(h-2y)^2;$$

$$\tau=\frac{\tau_0}{1+(\kappa-\beta)I}=\tau_0\left(1-(\kappa-\beta)\frac{1}{6}\overline{A}^2(h-2y)^2\right);$$

$$a_{xx}=2\tau_0\left(1-(\kappa-\beta)\frac{1}{6}\overline{A}^2(h-2y)^2\right)\frac{dV_x}{dy}a_{xy}-$$
$$3\beta\left(\frac{1}{36}\overline{A}^4(h-2y)^4+\frac{1}{36}\overline{A}^2(h-2y)^2\right);$$

$$a_{yy}=-\beta\frac{1}{12}\overline{A}^2(h-2y)^2;$$

$$a_{xy}=\tau_0\left(1-(\kappa-\beta)\frac{1}{6}\overline{A}^2(h-2y)^2\right)\frac{dV_x}{dy}\left(\frac{1}{3}-\beta\frac{1}{12}\overline{A}^2(h-2y)^2\right)-$$
$$-\beta\frac{1}{12}\overline{A}^3(h-2y)^3.$$

(12)

The last expression can be rewritten in the approximation of the first order with respect to parameters κ and β:

$$a_{xy}=\tau_0\frac{dV_x}{dy}\left(\frac{1}{3}-(\kappa-\beta)\frac{1}{6\cdot 3}\overline{A}^2(h-2y)^2-\frac{\beta}{12}\overline{A}^2(h-2y)^2\right)-$$
$$-\beta\frac{1}{12}\overline{A}^3(h-2y)^3.$$

(13)

With help of this expression, we are receiving from equation (10)

$$-\frac{1}{3}\overline{A}=\tau_0\frac{d^2V_x}{dy^2}\left(\frac{1}{3}-\left(\frac{\kappa}{18}+\frac{\beta}{36}\right)\overline{A}^2(h-2y)^2\right)+\tau_0\frac{dV_x}{dy}\left(\frac{2\kappa}{9}+\frac{\beta}{9}\right)\overline{A}^2(h-2y)-$$
$$-\beta\frac{1}{12}\overline{A}^3 3(h-2y)^2(-2).$$

Having substituted here $\frac{dV_x}{dy}$ in the zero approximation on κ and β, we have

$$-\frac{1}{3}\overline{A}=\tau_0\frac{d^2V_x}{dy^2}\left(\frac{1}{3}-\left(\frac{\kappa}{18}+\frac{\beta}{36}\right)\overline{A}^2(h-2y)^2\right)+\frac{\overline{A}^3}{2}\left(\frac{2\kappa}{9}+\frac{\beta}{9}\right)(h-2y)^2+\frac{\beta}{2}\overline{A}^3(h-2y)^2.$$

or

$$\tau_0\frac{d^2V_x}{dy^2}\left(\frac{1}{3}-\left(\frac{\kappa}{18}+\frac{\beta}{36}\right)\overline{A}^2(h-2y)^2\right)=-\frac{1}{3}\overline{A}-\overline{A}^3\left(\frac{\kappa}{9}+\frac{5\beta}{9}\right)(h-2y)^2.$$

Multiplying both parts of this expression by 3, we find:

$$\tau_0\frac{d^2V_x}{dy^2}=-\frac{\overline{A}+\overline{A}^3\left(\frac{\kappa}{3}+\frac{5\beta}{3}\right)(h-2y)^2}{1-\left(\frac{\kappa}{6}+\frac{\beta}{12}\right)\overline{A}^2(h-2y)^2}.$$

Or, keeping terms of the first order with respect to κ and β,

$$\tau_0\frac{d^2V_x}{dy^2}=-\left(\overline{A}+\overline{A}^3\left(\frac{\kappa}{3}+\frac{5\beta}{3}+\frac{\kappa}{6}+\frac{\beta}{12}\right)(h-2y)^2\right);$$

$$\tau_0\frac{d^2V_x}{dy^2}=-\left(\overline{A}+\overline{A}^3\left(\frac{\kappa}{2}+\frac{7\beta}{4}\right)(h-2y)^2\right).$$

(14)

One can integrate equation (12), taking the boundary conditions $V_x(0)=V_x(h)=0$
into account, and get:

$$V_x(y)=\frac{\overline{A}}{2\tau_0}y(h-y)+\frac{\overline{A}^3}{6\tau_0}\left(\frac{\kappa}{2}+\frac{7\beta}{4}\right)y(h-y)(h^2-2hy+2y^2)$$

or

$$V_x(y)=\frac{\overline{A}}{2\tau_0}y(h-y)+\frac{\overline{A}^3}{6\tau_0}\left(\frac{\kappa}{2}+\frac{7\beta}{4}\right)y(h-y)((h-y)^2+y^2)$$

(15)

Note that that the gradient of pressure in the perpendicular direction is different from zero, as one can see from equations (10) and (13),

$$\frac{\partial p}{\partial y} = \beta A(h-2y) \neq 0. \qquad (16)$$

There are no flow in this direction inside the channel, but after the exit from the channel, it gives extention of the jet.

To calculate the volume of liquid, which is coming through unit of the length of the gap in unit of time, that, is the specific volume, or the output flow volume rate Q, one has to integrate expression (13) with respect to the variable y from 0 to h and obtain:

$$Q = \int_0^h V_x(y)dy = \frac{\overline{A}}{12\tau_0}h^3 + \frac{\overline{A}^3}{60\tau_0}\left(\frac{\kappa}{2}+\frac{7\beta}{4}\right)h^5. \qquad (17)$$

The reversion of this formula determines the pressure gradient as function of Q, which allows us to calculate, with help of equation (15), distribution of velocities in the channel of the die at different values of the specific volume Q. The plots of dependencies are depicted in Fig. 7. One can see that the increase in the specific volume Q provokes the deviation of velocities from the Poiseuille distribution, corresponding to viscous liquid. This statement is demonstrated in Fig. 9, where the ratio of velocity of flow to its maximum value is shown. The calculations show, that the distribution of velocities changes non-essentially when Q>10.

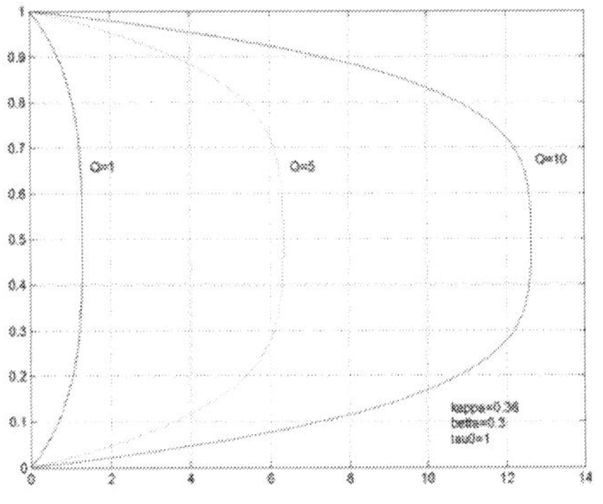

Fig. 7. The distribution of velocities in the channel of the die at different values of the specific volume Q.

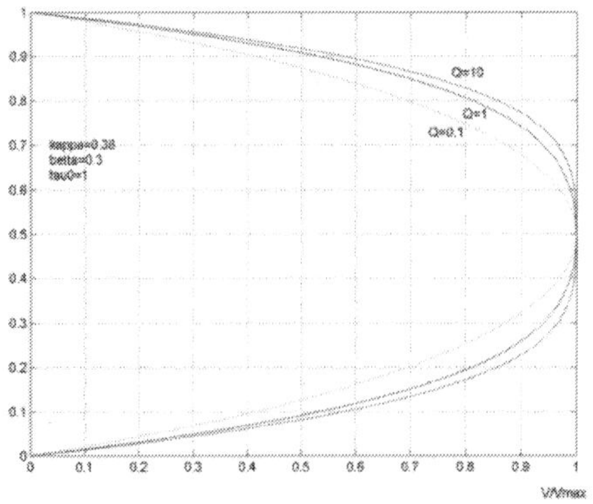

Fig. 8. The distribution of velocities in the channel of the die at the different values of the specific volume Q.

We have investigated also effects of the parameters of the rheological model on the distribution of the velocity. Figure 9 demonstrate the influence of relaxation time τ_0 on the distribution of the velocities in the channel, when the specific volume Q is constant.

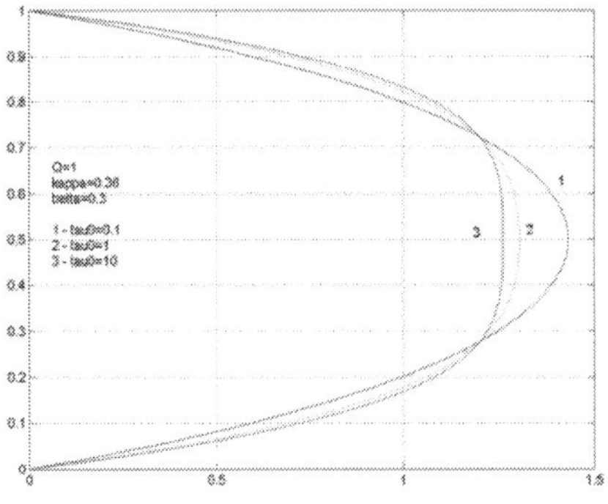

Fig. 9. The distribution of velocities in the channel of the die at the different values of relaxation time of the flowing system.

To test the applicability of the results to polymer liquid, we use the measurement of velocity distribution by [6]. The measurements were fulfilled inside the channel with cross-section 1 by 14 mm^2, the height of the channel was 50 mm. The velocity distribution was measured at 10 mm from the outlet of the channel. The results of comparison of empirical data with theoretical ones, calculated specially for this case, are shown in Figure 10.

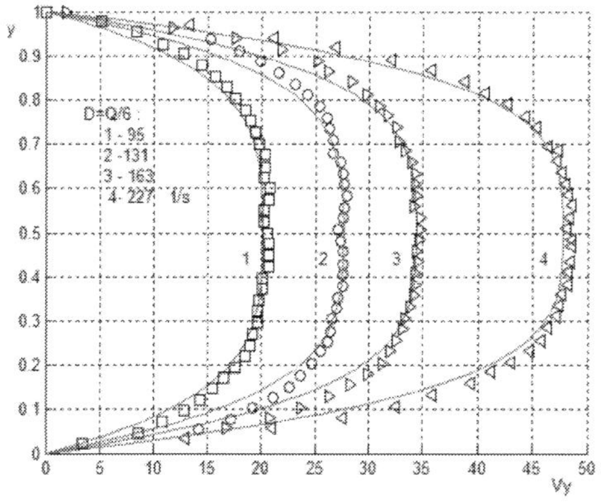

Fig. 10. Comparison of calculated and measured velocity distribution.

Conclusions

Possibility of use modified rheological is shown model Vinogradov-Pokrovskii for the description of linear polymers melts flows in various modes of deformation. The system of the equations of dynamics is written down in one-dimensional approach, at the account heat exchange. Comparison with experimental data available in the literature for film half-width is spent and adequacy of the applied approach is shown.

Thus, it is possible to draw a conclusion that rheological model (1, 2) at the account of heat exchange (3) consistently describes observed at formation of polymeric films effects [1–3], and the received dependences making stress tensor from distance to an exit from die can be used at modelling of process of crystallisation polymer melt under the influence of a stretching and cooling.

Comparison of results of calculation of a velocity with experimental data is executed. It allows to draw a conclusion, that in the considered case of Pouselle flow, the rheological model describes not parabolic profile of velocity between parallel plates that does not contradict experimental data.

Acknowledgements

The authors greatly appreciate professor Vladimir Pokrovskii for his interest to this work and useful discussions and acknowledge financial support of the Russian Foundation for Basic Research (RFBR) under grant 10-01-00293.

References

1. D. Silagy, Y. Demay, J.F. Agassant Stationary and stability analysis of the film casting process // J. Non-Newtonian Fluid Mech., 79, 1998, p. 563-583.

2. T. Kajiwara, M. Yamamura, T. Asahina Relationship between neck-in phenomena and rheological properties in film casting // Journal of the Society of Rheology, Japan, 2006, Vol. 34, № 2, p. 97 – 102.

3. H. Ito, M. Doi, T. Isaki, M. Takeo, K. Yagi 2D flow analysis of film casting process // Journal of the Society of Rheology, Japan, 2003, Vol. 31, № 3, p. 149 – 155.

4. G.V. Pyshnograi, V.N. Pokrovskii, Yu.G. Yanovsky, Yu.N. Karnet, I.F. Obraztsov Constitutive equation of non-linear viscoelastic (polymer) media in zeroth approximation by parameter of molecular theory and conclusions for shear and extension, Physics { Doklady 39 (1994) 889 - 892.

5. G.V. Pyshnograi, A.S. Gusev, V.N. Pokrovskii Constitutive equations for weakly entangled linear polymers// J. Non-Newtonian Fluid Mech.

6. E. Wassner, M. Schmidt, H. Munstedt Entry flow of a low-density-polyethylene melt into a slit die: An experimental study by laser-Doppler velocimetry // J. Rheol., 1999, 43(6), p. 1339-1353.

7. C. W. Seay, D. G. Baird Sparse Long-chain Branching's Effect on the Film-casting Behavior of PE // International Polymer Processing, Issue 2009/05, 2009, p. 428-438.

EXPERIMENTAL AND NUMERICAL STUDY OF TRANSCRIPTABILITY OF MICRO STRUCTURED SURFACE IN INJECTION MOLDING

J. G. Lee[1], M. Rha[1], T. G. Kang[3] and T. H. Kwon[1,2]*

[1]Department of Mechanical Engineering, Pohang Univerisity of Science and Technology, San 31 Hyojadong, Namgu, Pohang, 790-784, Republic of Korea – gulee@postech.ac.kr; woomoo98@postech.ac.kr; thkwon@postech.ac.kr
[2] Department of Integrative Bioscience and Biotechnology, Pohang Univerisity of Science and Technology, San 31 Hyojadong, Namgu, Pohang, 790-784, Republic of Korea – thkwon@postech.ac.kr
[3] School of Aerospace and Mechanical Engineering, Korea Aerospace University, 100 Hanggongdae-gil Hwajeon-dong, Goyang, Gyeonggi, 412-791, Republic of Korea – tgkang@kau.ac.kr

Experimental and numerical studies are carried out in order to investigate the transcriptability of the micro structures in injection molding. For this purpose, a mold insert is fabricated which has micro pole patterns. Mold insert has a ground plate with the thickness of 1.5mm, and has many micro poles over the ground plate. Height of the micro pole is 100 μm, and the aspect ratio is one. As the melt flows, melt inside the micro pole located near the gate starts to cool down. When the ground plate is completely filled, the melt front away from the gate is relatively hot. Thus melt inside the micro pole far from the gate is able to flow further. Consequently, the transcriptability of the micro pole is dependent on the relative position of micro pole to the gate. Numerical analysis system based on the domain decomposition method has been developed to verify micro structure filling phenomenon. Developed numerical system is composed of 2.5D program for ground plate and 3D program for micro structures. Further investigation on the effect of processing conditions, such as flow rate, mold wall temperature and melt temperature, on the transcriptability has been performed. The most dominant factors which affect the transcriptability of the micro pole are flow rate and mold wall temperature. High flow rate and high mold wall temperature enhance the transcriptability of pole structure.

Introduction

In general, injection molding is known to be one of the most important mass production technologies due to its high productivity with many industrial applications. In these days, fabrication of polymeric microstructures has attracted many researchers' attentions to manufacture microfluidic devices, lab-on-a-chip devices, micro lens arrays, back light unit for LCD panels, etc. (Attia 2009; Giboz, et al., 2007; Heckele and Schomburg 2004). Injection molding is found to be one of the most efficient methods to realize such various products with microstructures. The injection molding technique used in such applications is nowadays termed as micro injection molding.

Products manufactured by micro injection molding can be classified into two categories: one with molded products less than few cubic centimeters in volume (Zhao, et al., 2003), namely micro part injection molding, and the other with microstructures on a surface of the conventional sized products (Yu, et al., 2002) which could be named as micro structure injection molding.

The micro injection molding filling process is not so much different from the conventional filling except the fact that small volume of polymeric material should fill the cavity with less cooling during the filling process in a shorter period of time compared to the conventional one.

However, as for the latter case, the filling process involves a little more complexity since the molten polymer tends to first fill the major part of the product, the so-called ground plate (Figure 1), and then the microstructures. It is well known that the microstructures are sometimes incompletely filled (i.e. short shot), which results in poor replication quality. One wants to prevent such poor replication for precision injection molded products. In this regard, it is quite important to understand how the molten polymer fills the cavity with microstructures, which is closely related to the transcriptability issue in micro structure injection molding.

The present study limits the focus on the latter case of micro injection molding and investigates the transcriptability of micro pole structures on a flat surface both numerically and experimentally. There are many devices having micro structured surfaces, and injection molding is the most powerful method in commercialization of such devices. However, it is still difficult to make a complete filling of micro structures all over the region. This study includes mold insert fabrication, accompanying injection molding experiment and numerical simulation.

In numerical analysis, the computational domain will be decomposed into two parts; the ground plate part and the micro structure part (Lee, et al., Online published). The ground plate part and the micro structure part will be numerically simulated via 2.5D and 3D numerical systems, respectively. The experiment is performed under the same processing condition and geometry that are used in numerical analysis. Based on the numerical and experimental results, processing conditions can be controlled to enhance the transcriptability of micro structures.

Experimental

Mold insert fabrication

In order to make a micro structured product, micro mold insert is mostly fabricated by MEMS process and installed inside the mold cavity (Kim, et al., 2006). To carry out this study, mold inserts have been fabricated as shown in Figure 2. Each micro structure has the same values of length(l), width(w), height(h) and spacing(s) specified by value of 100μm. Mold base is designed to include a square ground plate having thickness of 1.5mm and width of 35mm.

The micro pole patterns with shape described above is obtained via SU-8 photolithography process, followed by Nickel (Ni) electroplating. The backside of the electroplated Ni plate is lapped. The residual Si, Cr and Au layers can be easily removed by chemical etching process. However, the removal of SU-8 is not easy due to the chemical and thermal stability of SU-8 (Dentinger, et al., 2002; Kim, et al., 2006). Moreover, SU-8 is placed inside of Ni mold insert, making removal process even more difficult. PR ashing process is used to remove residual SU-8 structures. The processing condition and etch rate are as follows: power 1000W, pressure 520mT, temperature 150°C, O_2 100 sccm, CF_4 7 sccm, 10 μm/min.

Injection molding

There are many control parameters in the injection molding which affect the final quality and transcriptability of the product. In the present study, three control parameters such as injection speed, melt temperature and mold temperature are considered as major parameters. Packing pressure and packing time are not controlled in this study, because packing procedure (Lee, et al., Online published) has little effect on the filling of micro structures. These processing conditions are varied to investigate the effects on the transcriptability of micro poles. Detailed processing conditions are listed in Table 1. The values in the first column are chosen as the reference values, from which one parameter is varied while the other parameters are fixed as indicated in the second column. The injection molding experiment will be done with a full-electrical injection molding machine (SE-50D, Sumitomo). Polymethyl methacrylate (PMMA; IF870S, LG MMA) is used as a molding material.

Measurement

The heights of the injection molded micro poles are measured using a surface profiler (NanoScan, NanoSystems) in ten separate positions as shown in Figure 3. For convenience, each pole is named as Pole1~Pole10.

Numerical Analysis

Physical modeling

The numerical simulation for the ground plate part is performed with traditional 2.5D numerical analysis system including both the filling stage and the packing stage (Chiang, et al., 1980; Hieber and Shen, 1980). The thickness of the ground plate part is relatively small compared with its width. Therefore, the adoption of Hele-Shaw approximation (Hieber and Shen, 1980) is reasonable and there is no trouble in using 2.5D numerical analysis system. 2.5D simulation employs finite element scheme for the momentum equation in planar direction and finite different scheme for the energy equation in the thickness direction. Pressure is evaluated in equation below, where S is a fluidity of molten polymer.

$$\frac{\partial}{\partial x}\left(S\frac{\partial p}{\partial x}\right)+\frac{\partial}{\partial y}\left(S\frac{\partial p}{\partial y}\right)=0$$

$$S \equiv \int_0^b \frac{z^2}{\eta}dz$$

For the analysis of micro structures, 3D numerical analysis is carried out with various boundary conditions. The behavior of molten polymer is assumed to be an incompressible non-Newtonian Stokes flow. Corresponding continuity, momentum and energy equations are shown below

$$\nabla \cdot u = 0$$

$$-\nabla p + \nabla \cdot \tau = 0$$

$$\rho c_p\left(\frac{\partial T}{\partial t}+u\cdot\nabla T\right)=\nabla\cdot k\nabla T+\mu\dot{\gamma}'^2$$

The constitutive equation of Stokes flow can be modeled as a generalized-Newtonian fluid model. The generalized Newtonian fluid model can be written as

$$\tau = 2\mu D\sigma = -pI + 2\eta D$$

, where η is the viscosity of the fluid, D is the rate-of-deformation tensor, p is the pressure and I is the identity tensor. The rate-of-deformation tensor is defined by

$$D=\frac{1}{2}\left(\nabla u+(\nabla u)^T\right)$$

Boundary conditions for 3D

The analysis data, such as time-varying pressure, temperature and velocity, of the ground plate are imposed on the micro structures as boundary conditions. Two different approaches of domain decomposition methods are used to simulate the micro pole filling. One is to compute the micro pole region only (Approach1), where pressure and temperature history obtained from 2.5D analysis are imposed on the inlet of micro poles. The other approach has both micro pole and the corresponding ground plate as a computational domain (Approach2). In this case, velocity profile, pressure and temperature are adopted as boundary conditions. These boundary conditions are shown in Figure 4.

The time-varying velocity and pressure are prescribed

at the inlet Γ_g, Γ_i, respectively. On the symmetric boundary Γ_s, the tangential traction force and relative normal velocity are imposed as zero. The velocity is evaluated during the simulation and pressure is prescribed at the outlet, as is at Γ_i. For the other boundaries, boundary conditions are treated using the adjustable Robin boundary condition for the velocity and traction force. These boundary conditions may be summarized as follows:

$$\begin{array}{ll} u_n = \bar{u}, u_s = 0 & \text{on } \Gamma_g \\ p = \bar{p}, u_s = 0 & \text{on } \Gamma_i \\ u_n = 0, t_s = 0 & \text{on } \Gamma_s \\ p = \bar{p} & \text{on } \Gamma_o \\ u_n = 0, au_s + t_s = 0 & \text{on } \Gamma_w \\ au_n + t_n = 0, au_s + t_s = 0 & \text{on } \Gamma_v \end{array}$$

, where Γ_g, Γ_i, Γ_s, Γ_o, Γ_w, and Γ_v are the gate region, inlet region, symmetric region, outlet region, mold wall and air vent, respectively. The Robin penalty parameter, which is a function of pseudo concentration, a is defined as

$$a = a(F) = \begin{cases} \text{large number} & \text{if } F > 0.5 \text{ (polymer region)} \\ 0 & \text{otherwise (air region)} \end{cases}$$

Viscosity model

The Cross-WLF (Williams-Ladel-Ferry) model is employed as the viscosity model in simulation for a precise description of polymeric flow. It may be noted that the Cross-WLF model shows good agreement with the material behavior of amorphous polymer near the glass transition temperature.

$$\mu(\dot{\gamma}, T, p) = \frac{\mu_0}{1 + \left(\frac{\mu_0 \dot{\gamma}}{\tau^*}\right)}$$

τ^* is the shear stress at the transition between Newtonian and power law behavior, μ_0 is a zero-shear-rate viscosity described by the WLF model:

$$\mu_0(T,p) = D_1 \exp\left(-\frac{A_1(T - T^*)}{A_2 + T - T^*}\right)$$

$$A_2(p) = \tilde{A}_2 + D_3 p$$

$$T^*(p) = D_2 + D_3 p$$

Material constants for the viscosity model used in this study are listed in Table 2.

Results and Discussion

The time history of the injection molded micro pole with reference processing condition is obtained by measuring series of short shot samples in Figure 5. One can observe that most of micro pole filling takes place during the filling stage. The portion of fillings in the filling stage to the final height of a pole varies from 75% to 95%. In this regard, only filling history of the 2.5D analysis is used in the following 3D numerical simulation. Although the additional filling in the packing stage is mush less, it has relatively greater influence on the pole which is far from the gate (Pole5). The heights of micro poles, injection molded with varying processing conditions, are shown in Figure 6. Generally, as shown in the figure, micro poles which are far from the gate are filled better with the exception of Pole1 which is very close to the gate. Lee, et al. (Online published) had explained already such phenomenon via race tracking and hesitation in the injection molding of micro rib structures. As hot polymeric melt flows into the mold cavity, micro structure close to the gate cools down and the frozen layer of the polymer becomes thicker. Thus hot melt heated by viscous heating is supplied to the micro pole far from the gate. As for micro poles which are very close to the gate (Pole1), hot melt is continuously injected through the gate and micro poles can be filled better.

The most affecting processing parameter is injection speed as shown in Figure 6 (a). In the case of 30 cm^3/s, Pole5 and Pole10 show greater values of height than the micro mold insert. It seems that the micro pole has been stretched during the demolding process. Though high temperature of the melt and the wall enhance the filling of micro pole, they have less influence than the injection speed with the given range of variance. Since the mold wall temperature is set to be lower than the glass transition temperature in this study, more investigation is needed to find out the filling behavior near the glass transition temperature.

A typical pressure and temperature history of the injection molding, which is used as boundary value in 3D analysis, is shown in Figure 7. Packing stage starts around 0.12 sec and only filling data is used in this study as mentioned before.

A result of final filling state of the Approach1, which has a computational domain of micro pole alone and analyzed with reference processing condition, is shown in Figure 8. The heights of micro poles are presented in Figure 9. It may be mentioned that the height of micro pole tends to be smaller as it gets far away from the gate. This may come from the less pressure imposed on the inlet region. Additionally, it is thought that a little higher melt temperature at the pole far from the gate has a negligible effect on the filling of micro pole.

As for the second approach, which includes the micro pole and the corresponding ground plate as a computational domain (Figure 10), the shape of melt front is quite different with that of Approach1, does not verify the tendency shown in the experiment.

Table 1 – Reference processing condition and variations

Processing parameter	Reference	Variations
Injection speed (cm^3/s)	15	7.5, 30
Packing pressure (MPa)	60	-
Pacing time (s)	2	-
Melt temperature (°C)	230	210, 250
Wall temperature (°C)	60	40, 80

Table 2 – Cross WLF viscosity model coefficients for the PMMA

n	2.5450×10^{-1}
τ	7.9120×10^{4} Pa
D_1	4.3800×10^{18} Pa·sec
D_2	3.7715×10^{2} K
D_3	0.0
A_1	4.7208×10^{1} K
A_2	5.1600×10^{1} K

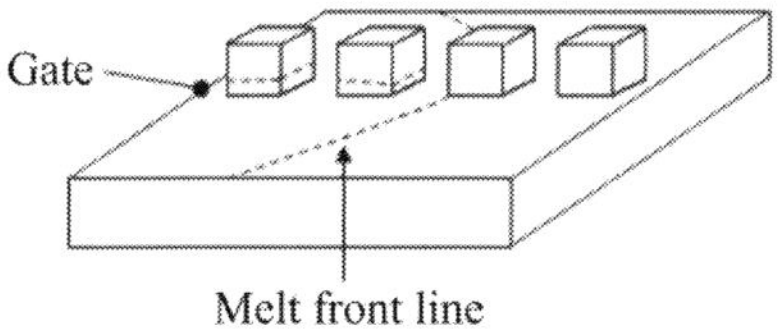

Figure 1 – Typical filling pattern of the micro structured product

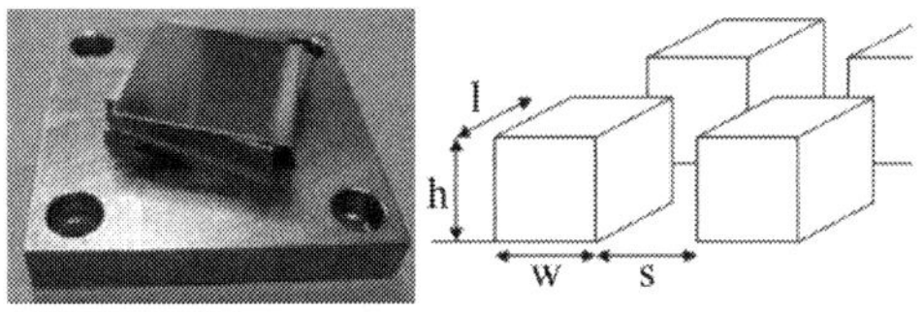

Figure 2 – Micro pole mold insert having 100μm dimple array on the surface and schematic view of the micro poles

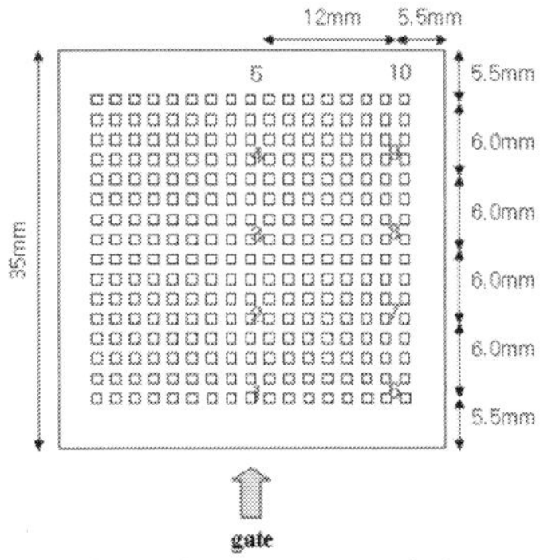

Figure 3 – Schematic geometry of the ground plate and ten measured points

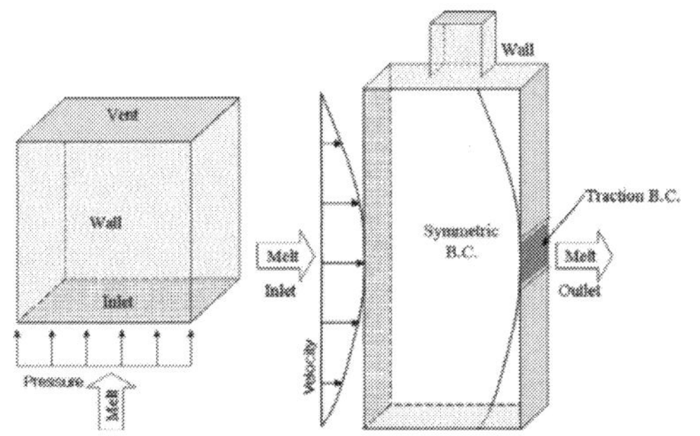

Figure 4 – Two numerical approaches to analyze the micro pole filling

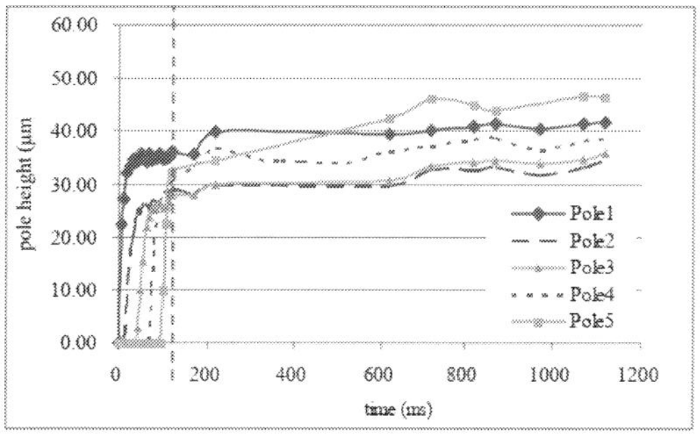

Figure 5 – The history of the 100μm pole height during the filling and packing process (red vertical dot line indicates the end of filling i.e. switch over)

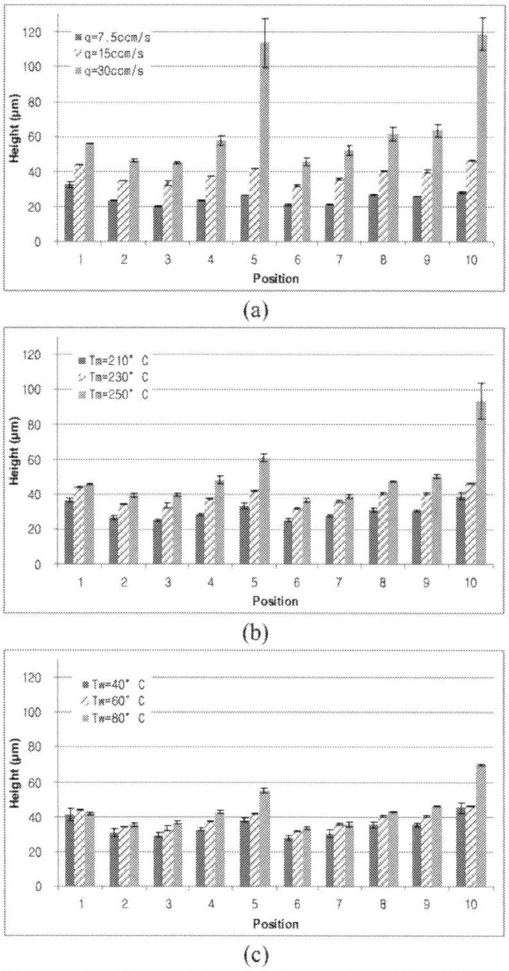

Figure 6 – The height of micro poles with different processing conditions (a) injection speed, (b) melt temperature, (c) wall temperature

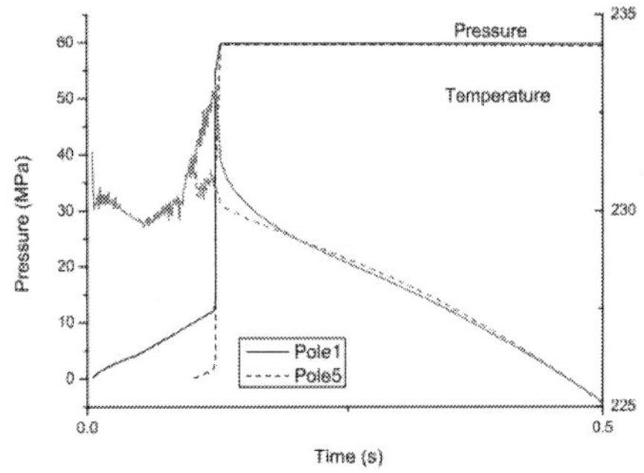

Figure 7 – Pressure and temperature history of Pole1 and Pole5 in a ground plate

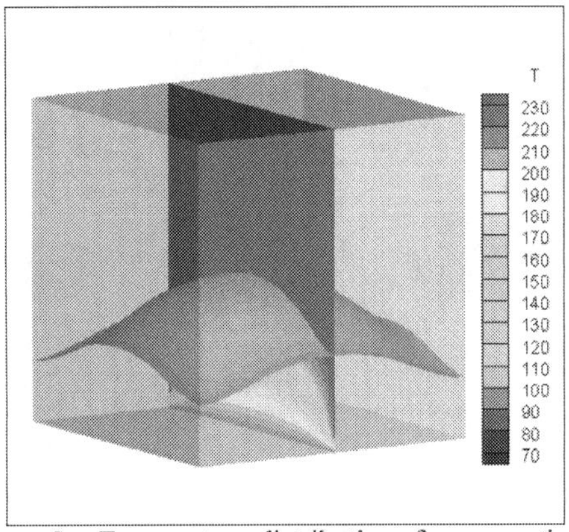

Figure 8 – Temperature distribution of cross section at the center and the melt front at the end of filling (Approach1, Pole1 and reference processing condition)

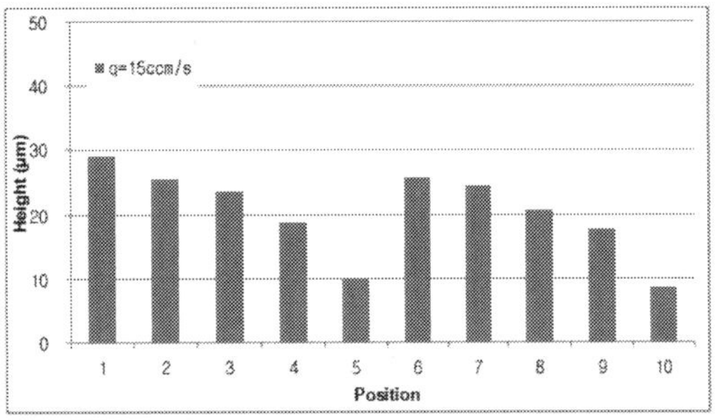

Figure 9 – Numerical analysis result with reference processing condition (Approach1)

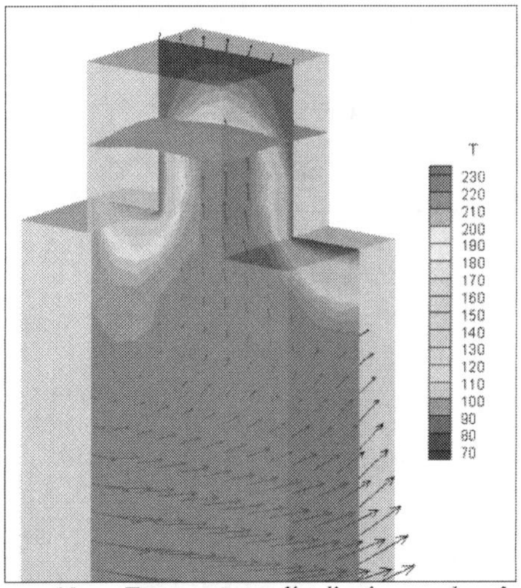

Figure 10 – Temperature distribution and velocity vector of cross section at the center and the melt front at the end of filling (Approach2, Pole1 and reference processing condition)

Conclusions

Experimental had been carried out to investigate the filling of micro pole structures. Higher injection speed, melt temperature and wall temperature enhanced the filling of micro poles. The micro poles close to the gate and far from the gate were filled better than those in the intermediate region. Numerical simulation also had been attempted to analyze the injection molding of micro pole structures. However, it did not show the variation of filled height depending on the position of micro pole.

Acknowledgements

This research was supported by Basic Science Research Program through the National Research Foundation of Korea (NRF) funded by the Ministry of Education, Science and Technology (2009-0083905) and "Development of Intelligent Robot Technology for Total Clinical System based on Bio Technology" under the Development of Next-Generation New Technology program (10024733) of the Korean Ministry of Knowledge Economy.

References

Attia, Usama M., Marson, S., Alcock, J.R., "Micro-injection Moulding of Polymer Microfluidic Devices", Microfluid. Nanofluid., 7, 1-28 (2009)

Chiang, H.H., Hieber, C.A., Wang, K.K., "A Unified Simulation of the Filling and Postfilling Stages in Injection Molding. Part I: Formulation", Polym. Eng. Sci., 31, 116-124 (1991)

Dentinger, P. M., Clift, W.M., Goods, S.H., "Removal of SU-8 Photoresist for Thick Film Applications", Microelectron. Eng., 61-62, 993-1000 (2002)

Giboz, J., Copponnex, T., Mélé, P., "Microinjection Molding of Thermoplastic Polymers: a review", J. Micromech. Microeng., 17, R96-R109 (2007)

Heckele, M., Schomburg, W.K., "Review on Micro Molding of Thermoplastic Polymers", J. Micromech. Microeng., 14, R1-R14 (2004)

Hieber, C.A., Shen, S.F., "A Finite-element/Finite-difference Simulation of the Injection-molding Filling Process", J. Non-Newton. Fluid, 7, 1-32 (1980)

Kim, S., Yang, H., Kim, K., Lim, Y.T., Pyo, H.-B., "Study of SU-8 to Make a Ni Master-mold:Adhesion, sidewall profile, and removal", Electrophoresis, 27, 3284-3296 (2006)

Lee, J.G., Lee, B.-K., Kang, T.G., Kwon, T.H., "Experimental and Numerical Investigation of Injection Molding with Micro Rib Patterns", Polym. Eng. Sci., Online published

Yu, L., Koh, C.G., Lee, L.J., Koelling, K.W., "Experimental Investigation and Numerical Simulation of Injection Molding with Microfeatures", Polym. Eng. Sci., 42, 871-888 (2002)

Zhao, J., Mayes, R.H., Chen, G., Xie, H., Chan, P.S., "Effects of Process Parameters on the Micro Molding Process", Polym. Eng. Sci., 43, 1542-1554 (2003)

DETERMINATION OF THE TRANSITION TEMPERATURE AT DIFFERENT COOLING RATES AND ITS INFLUENCE ON PREDICTION OF SHRINKAGE AND WARPAGE IN INJECTION MOULDING SIMULATION

T. Lucyshyn[1]*, G. Knapp[1] and C. Holzer[1]

[1]*Institute of Plastics Processing, University of Leoben, Leoben, Austria – thomas.lucyshyn@unileoben.ac.at; gilbert.knapp@stud.unileoben.ac.at; Clemens.holzer@unileoben.ac.at*

In injection moulding simulation the phase change from melt to solid state is usually simplified by using a so called transition temperature (also known as no-flow-temperature). Above this temperature the polymer is assumed to behave like a fluid and below this temperature it is regarded as a frozen solid without any flow velocity. A common method to determine this material parameter is the Differential Scanning Calorimetry (DSC) with a cooling rate of typically 20 K/min. This cooling rate is much lower compared to the injection moulding process. It can be expected that the way of determining the transition temperature has an influence on the prediction of shrinkage and warpage with commercial injection moulding simulation tools.
In the present work the transition temperature of several amorphous and semi-crystalline polymers was determined using DSC-runs at different cooling rates up to 100 K/min. The dependence of the transition temperature is described as a function of cooling rate. There was a weak influence of the cooling rate on the transition temperatures of the investigated amorphous polymers, but the transition temperatures of the semi-crystalline polymers were significantly shifted to lower temperatures with increasing cooling rate.
The obtained transition temperatures of the selected semi-crystalline polymers were then used in injection moulding simulations with the commercial software package Autodesk Moldflow Insight to calculate the shrinkage and warpage of box-shaped test parts. The test parts were injection moulded and the dimensions of these boxes were determined using an optical 3D-scanner. Finally, the simulation results were compared with the experimental values of the injection moulded boxes. The results showed a strong influence of the transition temperatures on simulation results of the 3D model and a very low influence for the 2.5D model. Generally the simulation results of the 3D models matched better with experimental values. Transition temperatures at higher cooling rates even improved the 3D simulation results for several dimensions.

Introduction

The computer simulation of shrinkage and warpage of injection moulded parts is a difficult task. Commercial software packages offer such simulations, but the accuracy of the results is not always sufficient. Regarding the highly complex material behaviour of thermoplastic polymers it is clear that the determination of material data is an important need for the injection moulding simulation. Especially a correct modeling of the phase change from melt to solid state is a major challenge. Although there are numerous research groups investigating the phase transition phenomena (for example [3,4,5,8,9]) up to now the typical approach in simulation is the rather simple concept of the no-flow or transition temperature which defines the limit between melt and solid. As soon as the temperature of the polymer is below the transition temperature, it is defined as a solid without flow velocity. This temperature is either determined on a high capillary rheometer or nowadays Autodesk Moldflow Insight (AMI) uses the Differential Scanning Calorimetry (DSC) at a cooling rate of 20 K/min [6]. As cooling rates in the injection moulding process are much higher (app. 1000 K/min) and it is well known that the cooling rate has an influence on the DSC curves ([1,2]), the target of this research work was to determine the transition temperatures from DSC curves obtained at different cooling rates using a commercially available DSC measuring device. Standard DSC measurement devices offer only limited cooling rates up to 100 K/min and therefore a mathematical approximation of the transition temperature as a function of the cooling rate should be found in order to be able to extrapolate the transition temperature to higher cooling rates. The influence of the varying transition temperatures on the simulation of shrinkage and warpage with AMI was the final aim of these investigations.

Experimental

Determination of transition temperature

The transition temperature was determined using a DSC 1 device by Mettler-Toledo International Inc. (USA), which is commercially available. A small sample of the polymer is prepared and put into a temperature controlled furnace chamber, together with a reference sample (figure 1). The furnace is heated to the desired starting temperature and then cooled down at defined cooling rates. The heat flow difference between the polymer and the reference sample to obtain the same temperatures is measured as a function of temperature.

It has to be mentioned that the applied cooling rate is not constant over the whole temperature range, but rather decreases with lower temperatures (according to the decreasing temperature gradient between the sample and the cooling agent). Figure 2 shows the cooling rate that can be achieved with the DSC 1 device as a function of temperature. In principle, this behaviour is similar to the conditions in an injection mould, where the cooling rate also decreases with lower temperatures of the injection moulded part.

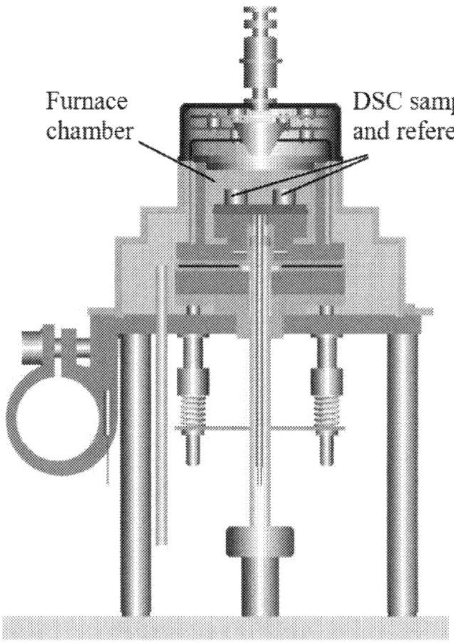

Figure 1 – Mettler-Toledo DSC 1 device for determination of the transition temperature [7].

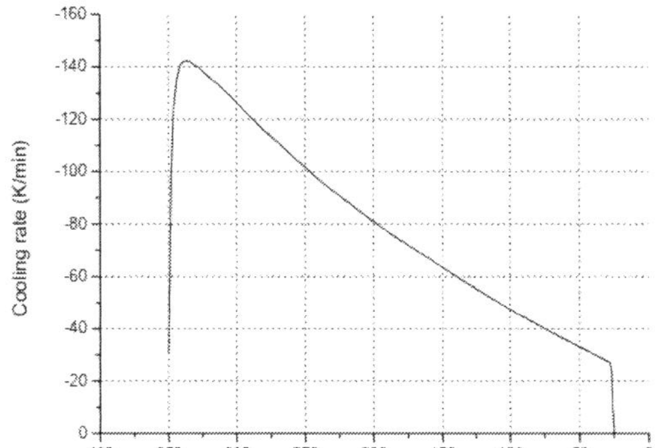

Figure 2 – Cooling rate of the DSC 1 as a function of temperature, measured without polymer sample.

In order to find out the influence of cooling rate on the transition temperature of different thermoplastic materials, two amorphous and two semi-crystalline polymers were investigated. The material grades are summarized in table 1.

Table 1 – Investigated material grades

Grade name	Supplier	Morphology
PS 454 C	BASF, Germany	amorphous
ABS Urtal M122	Polimeri Europa, Italy	amorphous
PP Hostacom BR 735 G	Basell Polyolefins, Netherlands	semi-crystalline
PC/PET Stapron E EM 605	DSM Engineering Plastics, Netherlands	semi-crystalline

The determination of the transition temperatures from the heat flow functions obtained by the DSC-measurements is explained in the following section. For semi-crystalline polymers the transition temperature was obtained at the intersection of the tangent to the heat flow curve above the crystallization peak and the tangent to the heat flow curve in the steep section of the crystallization peak (fig. 3).

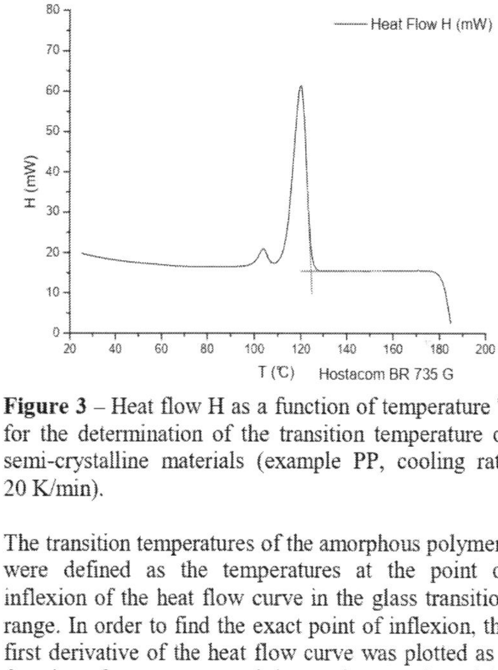

Figure 3 – Heat flow H as a function of temperature T for the determination of the transition temperature of semi-crystalline materials (example PP, cooling rate 20 K/min).

The transition temperatures of the amorphous polymers were defined as the temperatures at the point of inflexion of the heat flow curve in the glass transition range. In order to find the exact point of inflexion, the first derivative of the heat flow curve was plotted as a function of temperature and the maximum value of the first derivative in the glass transition region marks the transition temperature (fig. 4).

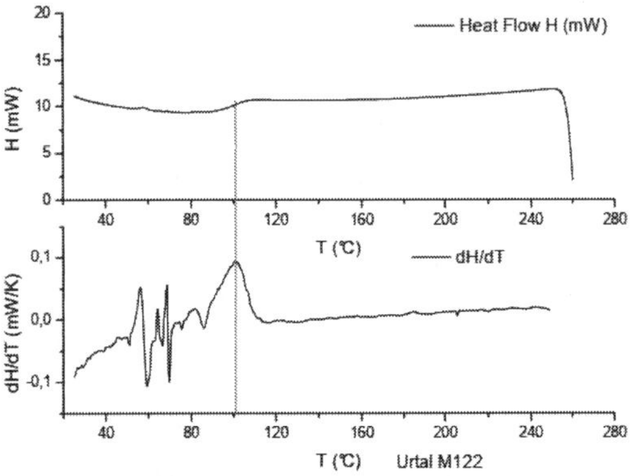

Figure 4 – Heat flow H and first derivative dH/dT of heat flow as a function of temperature T for the determination of the transition temperature of amorphous materials (example ABS, cooling rate 20 K/min).

Injection moulding experiments

In order to be able to compare the simulation results with real parts, several test parts were produced on a fully electric injection moulding machine (Battenfeld 1000 CD-SE, Austria). The basic geometry was a simple box with the dimensions 100 mm x 100 mm x 40 mm (CAD-model of the box cf. fig. 5). Using different mould inserts, the wall thickness of the bottom and of the side walls of the box can be varied (1 mm, 1,8 mm and 3 mm). The mould is equipped with a hot runner and gate is at the centre of the box. The process settings of the injection moulding machine were recorded and used for the simulation.

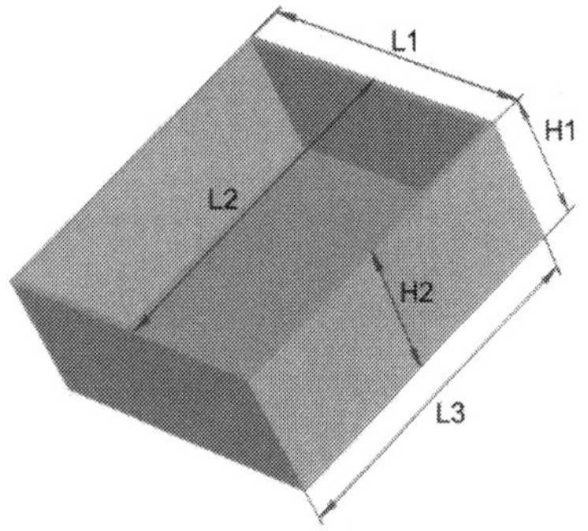

Figure 5 – Model of the box, which was used for simulation and injection moulding experiments with the chosen dimensions for the comparison of simulation and experiment.

Measuring of the injection moulded parts

After producing the parts, the complete geometry was measured using an optical 3D-scanner available at the Polymer Competence Center Leoben, Austria (fig. 6).

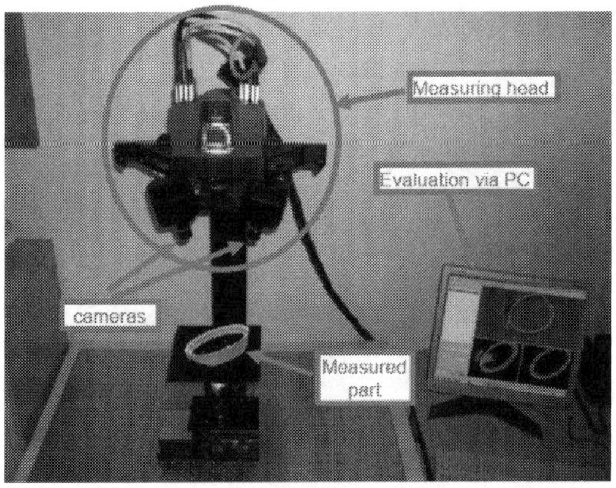

Figure 6 – Optical 3D-scanner ATOS II SO from GOM (Germany) for the measuring of the injection moulded parts.

The system ATOS II SO from GOM (Gesellschaft fuer optische Messtechnik, Germany) consists of a projector and two digital cameras with high resolution. The projector projects a strip pattern onto the measured part, which is recorded by the two cameras. The measuring principle is similar to the human three-dimensional viewing capability. Using the two photos of the part from two different viewing angles, the evaluation software calculates the three-dimensional shape of the part surface. By attaching reference points to the part, photos from different sides can be overlaid to compose the complete 3D-surface geometry which can be used to compare the real part geometry with the shrinkage and warpage results obtained by simulation. For an easier evaluation, a few characteristic dimensions of the box were chosen for comparison (fig. 5): The four side lengths at the rim of the box were averaged to the dimension L1, the two distances of opposite side walls at the rim of the box were averaged to the dimension L2 and the average value of the four lengths at the bottom of the box was defined as the dimension L3. Furthermore, the dimension H1 is the mean value of the four heights of the side walls at the corners and finally H2 is the average value of the height in the middle of the four side walls.

Simulation with Autodesk Moldflow Insight 2010 (AMI)

Using the process settings from the experiments, the test parts were simulated with AMI, including the cooling and runner system. The simulations were performed using the transition temperatures obtained from DSC-measurements at different cooling rates. Finally, the calculated deformations of the dimensions L1, L2, L3, H1 and H2 were evaluated. The deformation is defined as the relative difference between the cavity and final part dimensions.

Results and Discussion

Transition temperatures at different cooling rates

As the recommended cooling rate for the determination of the transition temperature according to Moldflow is 20 K/min, this cooling rate was used as a reference.
For the amorphous materials, no significant influence of the cooling rate on the transition temperature was observed. Figure 7 shows the heat flow curves of PS 454 C as a function of temperature at cooling rates of 20 K/min and 40 K/min. The transition region (marked with the blue circle) remains almost unchanged.

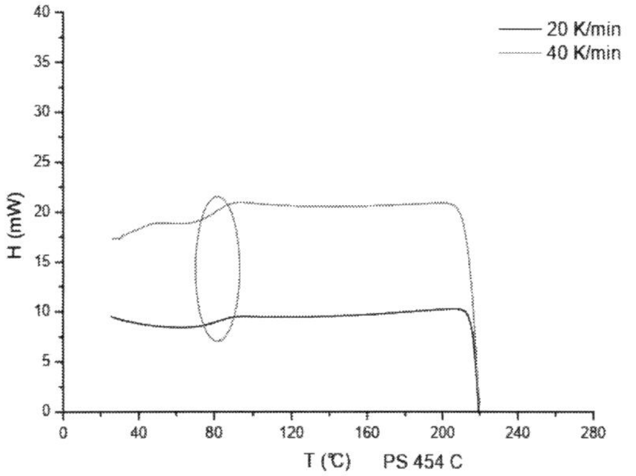

Figure 7 – Heat flow as a function of temperature at two different cooling rates for PS 454 C.

The results for the second amorphous polymer (ABS Urtal M122) are very similar. As for the polystyrene, there is no remarkable shift in the transition region (fig. 8).

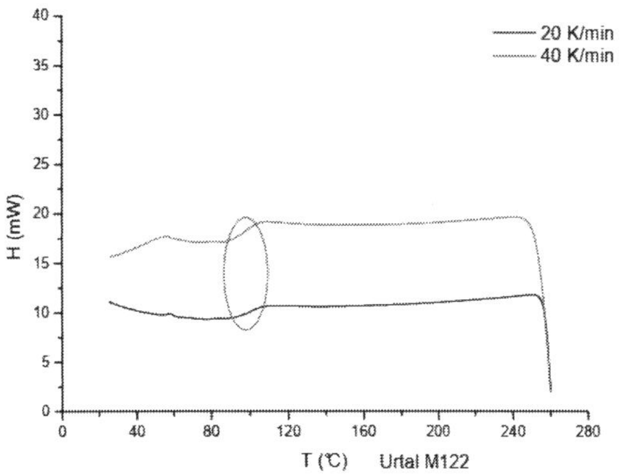

Figure 8 – Heat flow as a function of temperature at two different cooling rates for ABS Urtal M122.

As a consequence of these findings it can be assumed that the determination of the transition temperature at a standard cooling rate of 20 K/min is sufficient. Therefore no computer simulations were performed with the amorphous polymers.

As expected, the heat flow curves of the investigated semi-crystalline polymers showed a significant sensitivity to the cooling rate of the DSC-measurements. This behaviour has already been described well in several publications. Target of the actual research was the determination of the transition temperature as a function of cooling rate with a commercially available testing device. In principle, the maximum cooling rate achievable with the DSC 1 should be 100 K/min. During the investigations it turned out that reasonable results for the investigated materials could only be achieved up to about 70 K/min. Figure 9 shows the heat flow curves of the polypropylene Hostacom BR 735 G as a function of temperature and at different cooling rates (5, 10, 20, 40 and 50 K/min). The shift of the transition temperature to lower temperatures with increasing cooling rates can clearly be seen.

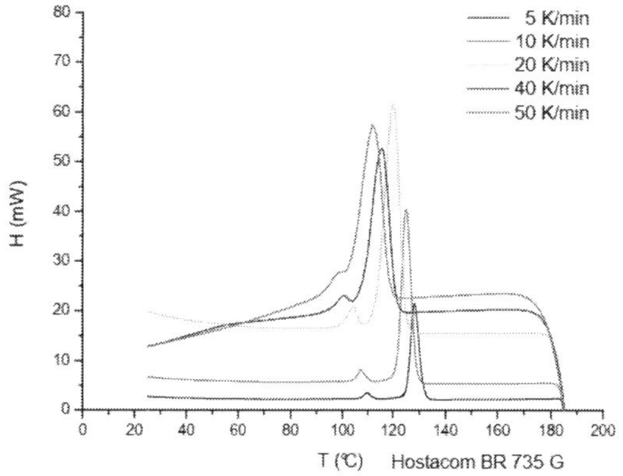

Figure 9 – Heat flow as a function of temperature at different cooling rates for PP Hostacom BR 735 G.

For the PC/PET blend with mainly semi-crystalline character Stapron E EM 605 cooling rates up to 70 K/min could be achieved. The heat flow curves at the various cooling rates are illustrated in fig. 10. There is also a significant dependence of the transition region on the cooling rate.

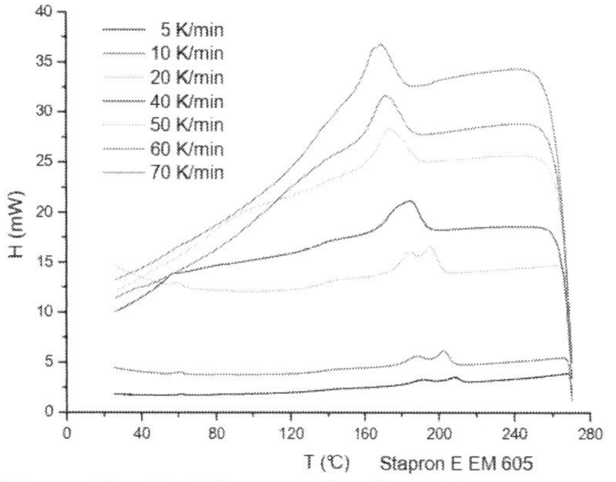

Figure 10 – Heat flow as a function of temperature at different cooling rates for PC/PET Stapron E EM 605.

In order to be able to extrapolate the results to higher cooling rates, a simple mathematical function was found to describe the transition temperature (in K) as a function of cooling rate (equation 1):

$$T_{trans} = a \cdot \left(\frac{\partial T}{\partial t} \right)^b \quad (1)$$

with

T_{trans} Transition temperature (K),

$\left(\frac{\partial T}{\partial t} \right)$ Cooling rate (K/min)

a Material parameter ($min^b K^{(1-b)}$)

b Material parameter (-)

The material parameters a and b for the two materials are summarized in table 2.

Table 2 – Material parameters and correlation coefficient R^2 for the approximation of the transition temperature as a function of cooling rate.

Parameter	PP Hostacom BR 735 G	PC/PET Stapron E EM 605
a ($min^b K^{(1-b)}$)	415,4	512,1
b (-)	-0,0148	-0,0266
Correlation coefficient R^2	0,983	0,94

This function shows good agreement with the experimental values of the two materials. The transition temperature as a function of cooling rate for PP Hostacom BR 735 G is shown in figure 11 and the results for PC/PET Stapron E EM 605 can be seen in figure 12.

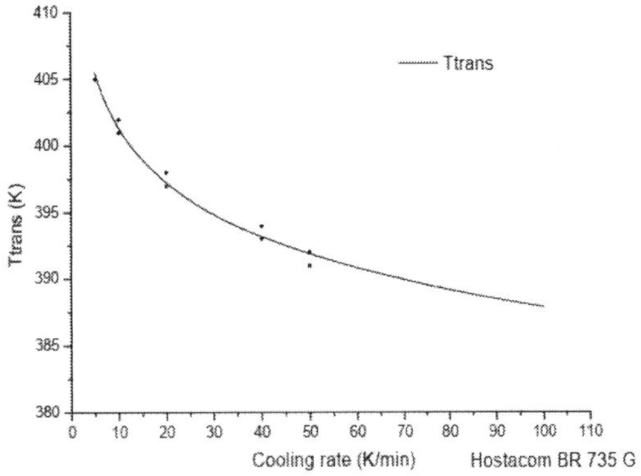

Figure 11 – Transition temperature as a function of cooling rate for PP Hostacom BR 735 G (experimental values and data fit).

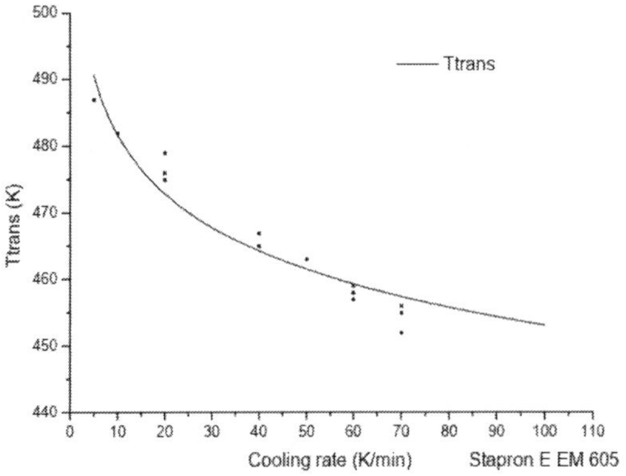

Figure 12 – Transition temperature as a function of cooling rate for PC/PET Stapron E EM 605 (experimental values and data fit).

Shrinkage and warpage simulation results

In the following figures the warpage simulation results performed with the transition temperatures obtained at different cooling rates are compared with experimental results. The simulations in AMI were done with 3D as well as 2.5D models. Figure 13 shows the results of the 3D calculation for the PP Hostacom BR 735 G. The predicted shrinkage decreases with increasing cooling rates. For some dimensions (L1 and L3) the transition temperature at higher cooling rates improves the simulation results.

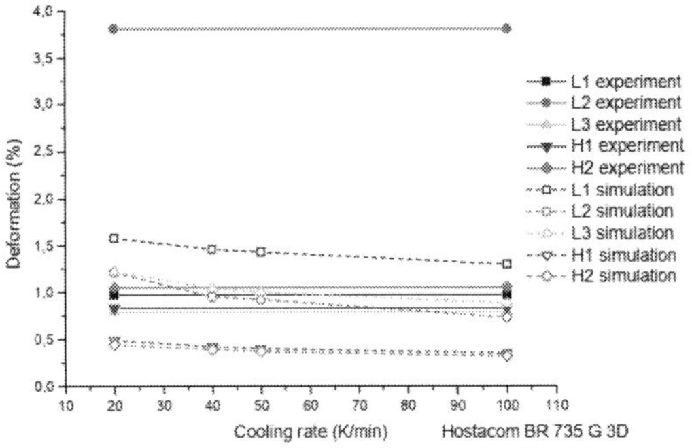

Figure 13 – Warpage simulation results for Hostacom BR 735 G (box with 1 mm wall thickness, 3D model).

Compared to the 3D model, the influence of the transition temperature on the warpage simulation results for PP Hostacom BR 735 G was much lower in the 2.5D model (figure 14). The results obtained with the 2.5D model are generally too low compared to the experimental data.

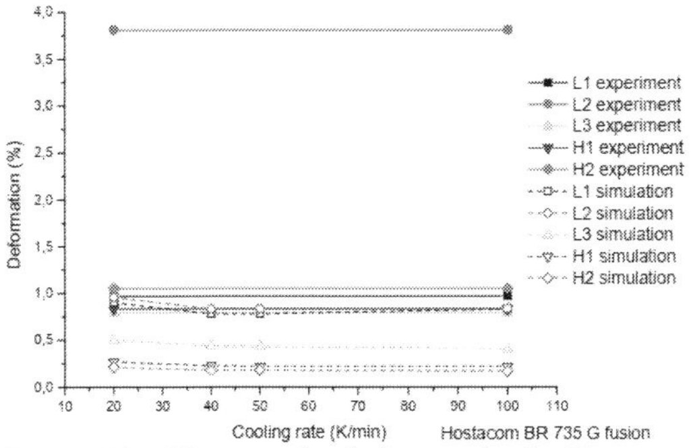

Figure 14 – Warpage simulation results for Hostacom BR 735 G (box with 1 mm wall thickness, 2.5D model).

Similar tendencies were found for the PC/PET Stapron E EM 605. Using the 3D model, transition temperatures at higher cooling rates improved the simulation results (figure 15). The 2.5D, once again, did not show much influence of the transition temperatures on the warpage simulation results (figure 16).

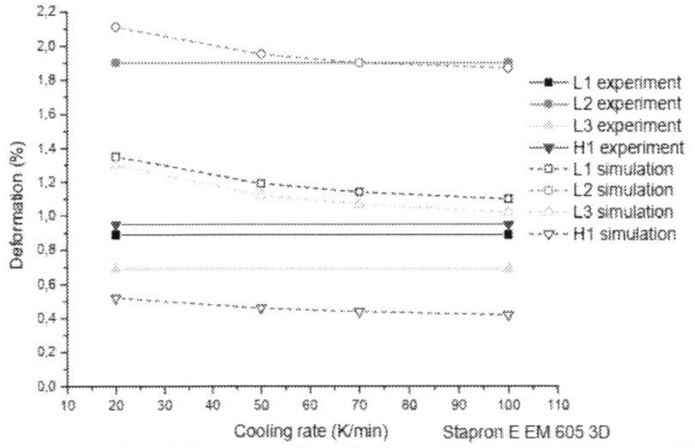

Figure 15 – Warpage simulation results for Stapron E EM 605 (box with 1,8 mm wall thickness, 3D model).

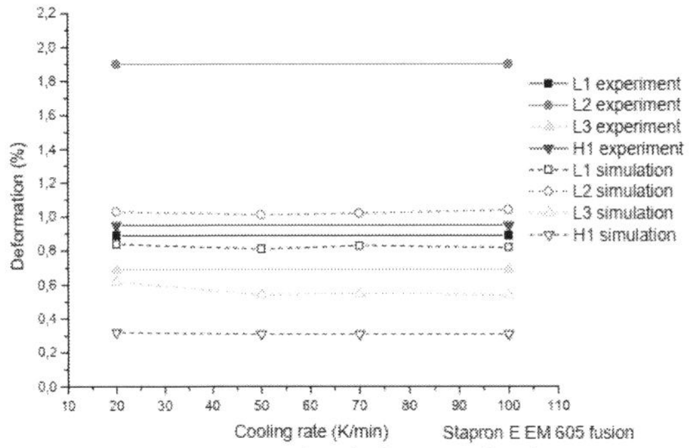

Figure 16 – Warpage simulation results for Stapron E EM 605 (box with 1,8 mm wall thickness, 2.5D model).

Conclusions

In this paper the influence of cooling rate on the determination of the so called transition temperature with DSC measurement was investigated for two amorphous and two semi-crystalline materials. The amorphous polymers hardly showed any sensitivity of the transition temperature to the cooling rate. Contrary, the cooling rate had a strong influence on the transition temperatures of the two semi-crystalline polymers. For the latter, a mathematical function was found to describe the dependence of the transition temperature on the cooling rate in order to be able to extrapolate to higher rates which are not accessible by standard DSC-measurements.
In the next step box shaped test parts of the two semi-crystalline materials were injection moulded and simulated using the commercial injection moulding software Autodesk Moldflow Insight 2010. The influence of transition temperatures obtained at different cooling rates on the warpage simulation results was analysed and the predicted values were compared to the experimental values which had been measured using a 3D optical scanner. The calculations were performed using 3D as well as 2.5D models. The results showed a strong influence of the transition temperatures on simulation results of the 3D model and a very low influence for the 2.5D model. Generally the simulation results of the 3D models matched better with experimental values. Transition temperatures at higher cooling rates even improved the 3D simulation results for several dimensions.
Based on these results further research will be performed also taking into account the different curves of the specific heat at different cooling rates.

Acknowledgements

The research work of this paper was performed in cooperation with the Polymer Competence Center Leoben GmbH (PCCL, Austria) within the framework of the COMET-program of the Austrian Ministry of Traffic, Innovation and Technology. The PCCL is funded by the Austrian Government and the State Governments of Styria and Upper Austria.

References

[1] Adamovsky, S.; Minakov, A; Schick, C.: Scanning microcalorimetry at high cooling rates. Thermochim Acta, 403, 55-63 (2003).
[2] De Santis, F.; Adamovsky, S.; Titomanlio, G.; Schick, C.: Scanning nanocalorimetry at high cooling rate of isotactic polypropylene. Macromolecules, 39, 2562-2567 (2006).
[3] Eder, G.; Janeschitz-Kriegl, H.: Crystallization. Material Science and Technology, Vol 18, Processing of Polymers. 269-342, Wiley-VCH, New York (1997).
[4] Huang, T.; Kamal, M.R.: Morphological modeling of polymer solidification, Polymer Eng. Sci., 40, 1796-1808 (2000).
[5] Janeschitz-Kriegl, H.: Crystallization Modalities in Polymer Melt Processing – Fundamental Aspects of Structure Formation. SpringerWienNewYork (2010).

[6] Kennedy, P.: Practical and Scientific Aspects of Injection Molding Simulation. PhD thesis, Technical University of Eindhoven (2008).

[7] N.N.: DSC 1 product information brochure, Mettler-Toledo AG, Schwerzenbach (2007).

[8] Pantani, R.; Coccorullo, I.; Speranza, V. Titomanlio, G.: Modeling of morphology evolution in the injection molding process of thermoplastic polymers. Prog. Polymer Sci., 30, 1185-1222 (2005).

[9] Zuidema, H.: Flow Induced Crystallinity of Polymers, Applications to Injection Molding. PhD thesis, Technical University of Eindhoven (2000).

EXTENDED FINITE ELEMENT METHOD FOR FLOW INSIDE TWIN-SCREW EXTRUDERS

A. Sarhangi Fard, M.A. Hulsen, H.E.H Meijer, P.D. Anderson*

Materials Technology, Eindhoven University of Technology, Eindhoven, Netherlands
a.sarhangi.fard@tue.nl, m.a.hulsen@tue.nl, h.e.h.meijer@tue.nl, p.d.anderson@tue.nl

Determining the velocity field in twin-screw extruders operating under realistic conditions is still very challenging despite all progress made in simulations during last years. There are two main difficulties that arise when solving the balance equations. The first is that we have to deal with moving geometries, the screws, in a fixed barrel, while the gap widths are extremely small and change in time during rotation. Obtaining a velocity field and its derivative very close to the moving screws is not trivial, but essential for further particle tracking analysis. The second challenge is to deal with the viscosity that is a function of position, shear rate and temperature, and can change orders of magnitude over the height of the channel. Here we present an extended finite element method (XFEM) to model non-Newtonian Stokes flow inside the twin-screw extruder and demonstrate its accuracy and efficiency by systematically refining the mesh and compare results with those of boundary-fitted calculations. Two-dimensional cross-sectional velocity and pressure profiles are compared with full three-dimensional simulations. In addition we discuss strategies to speed up particle tracking in these moving grid simulations without compromizing accuracy. Residence time distributions are compared for various screw designs.

Introduction

Simulation of flow behaviour in twin screw extruders (TSE), has gained great endeavors in the past decades due to change of geometry with screw rotation and the presence of narrow gap regions. The fictitious domain method (FDM) is used to avoid re-meshing. But the accuracy of FDM is not sufficient, particularly in the gap regions (Fig. 1) with their typical high shear rates. For mixing analysis the accuracy of velocity field in the highly shear rate regions is essential.

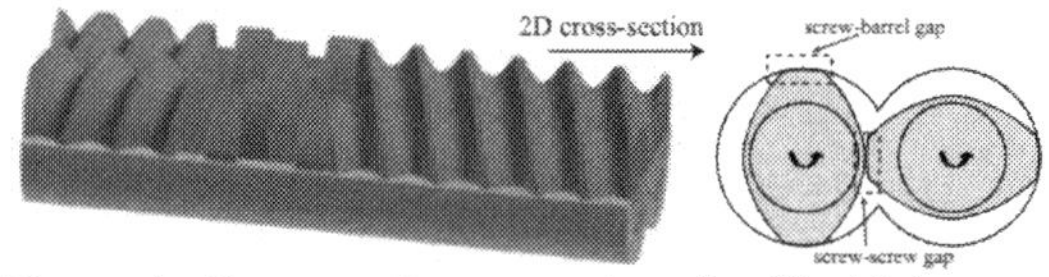

Figure 1. Cross section geometry of self-whiping co rotating twin screw extruders with narrow gaps between screw-screw and screw-barrel.

In this work we present a new strategy for non-conforming mesh refinement method based on FDM for rectangular Taylor-Hood elements and its adaptation for complex geometries with internal moving parts and narrow gap regions. Next step we interpret the XFEM enrichment scheme using virtual degrees of freedoms and the method is applied to Stokes flow inside a 2D cross section of the twin screw extruder. We compare the results of non-conforming mesh refinement in FDM with those of XFEM to explore the effect of discontinuities of physical variables, specially for the complex geometries with different flow regimes and high shear rate regions. Finally we apply XFEM also for the full 3D geometry of twin screw extruders.

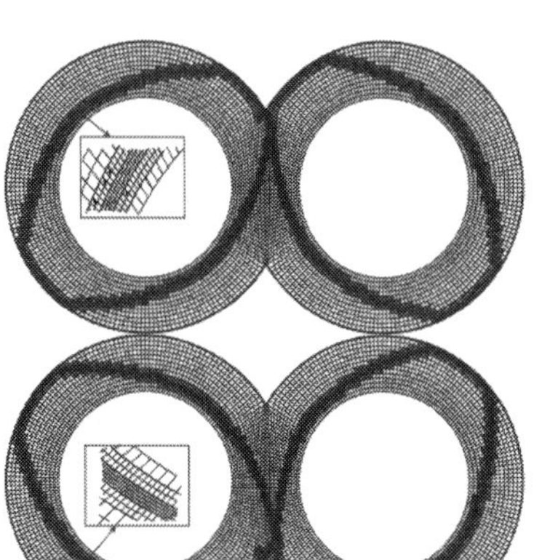

Figure 2. Example of a finite element mesh used for FDM using different mesh refinement schemes.

Numerical methods

A isothermal and incompressible fluid is modelled in the flow domain where, since the Reynolds number of polymeric melts is very low, only Stokes flow is considered. The equations of conservation of mass and momentum read:

$$\nabla . v = 0 \tag{1}$$

$$-\nabla p + \nabla . \tau = 0 \tag{2}$$

where v is the velocity vector and p is the hydrostatic pressure; τ is the extra stress tensor. The equations are solved fully coupled using a finite element discretization method. We use a Carreau model as the generalized Newtonian constitutive equation. Using FDM, the velocity of screw is imposed as an extra constraint during solving equations (Glowinski et al. 1994). The accuracy of the FDM depends on the number of elements intersecting with moving boundaries. This accuracy becomes important in gap regions. To obtain adequate accuracy in these regions, the size of elements of the finite element mesh are decreased by subdividing them into small sub-elements. The non-conforming mesh refinement technique is based on one fixed reference mesh where the quality of the refined mesh is further preserved (Figure 2). To ensure continuity across non-conforming regions, a Lagrangian multiplier is imposed as a constraint (Ben Belgacem 1999). In the FDM intersected elements exist that contain parts which are inside the fluid while the remainder is inside the rigid body. As a result there is continuity in field variables such as a velocity, pressure across these intersected elements. We used XFEM to decouple the fluid and internal moving rigid bodies (Gerstenberger et al. 2008). With XFEM the discontinuities of physical variables are captured using virtual degrees of freedom as enrichment and by applying a special integration scheme over the elements intersected by the rigid body (Figure 3).

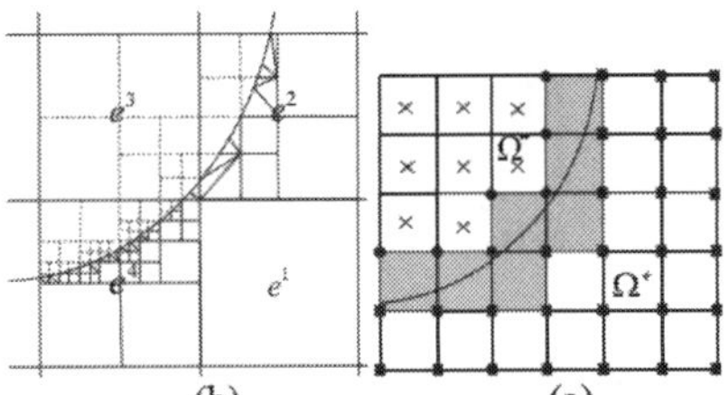

Figure 3. XFEM: (a) Non-intersected elements fully inside Ω^+ are fully integrated, intersected elements (gray color) are integrated on the inside part (Ω^+) and the elements (×) fully outside (Ω^-) are ignored. (b) Subdivision of elements for integration.

Results and Discussion

The cross-section of a twin screw extruder is selected as a test case. We compared non-conforming FDM and XFEM results with those applying a boundary fitted mesh (Figure 4). In the gap region the XFEM shows a much more accurate result than FDM and it predicts the high value of velocity in y direction the same as the BF result. Next we apply our XFEM method on the full 3D geometry of a TSE (Figure 5).

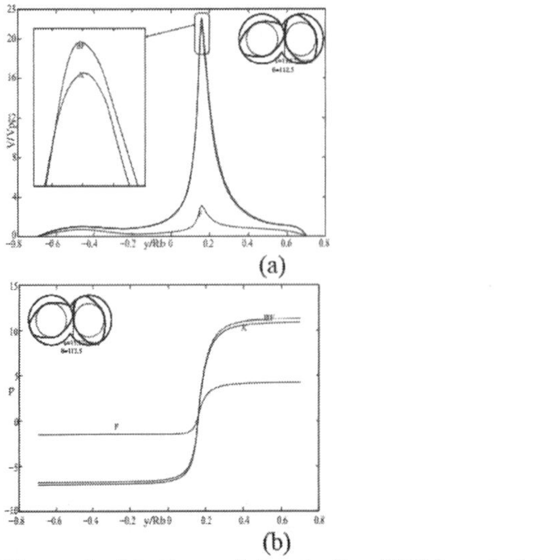

Figure 4. (a) Tangential velocity (V/Vp) and (b) pressure (p) as a function of y/Rb over the line x = 15.177 mm, XFEM (X), boundary fitted (BF) and FDM with high order refinement (F), Rb is barrel radius.

From the Eulerian velocity field $\mathbf{v}(\boldsymbol{x}, t)$ the particle path of $\boldsymbol{X}$ is given by the numerical solution of:

$$d\mathbf{x}/dt = \mathbf{v}(\boldsymbol{x},t), \quad \boldsymbol{x}=\boldsymbol{X}_0 \text{ at } t=t_0 \tag{3}$$

A total 9900 particles were initially located to cover a complete inlet cross-section of the extruder at z=0 mm. particles are tracked until they exit the extruder. Residence time distributions are obtained based on the calculation of particle tracers. Figure 6 shows snapshots of particles position at different times.

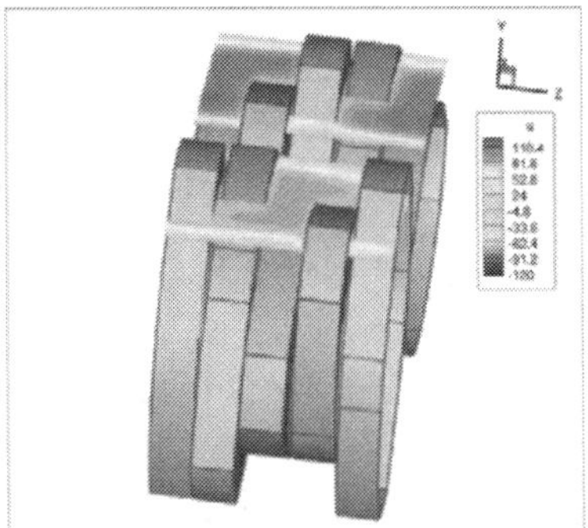

Kneading element with 45 staggering angle

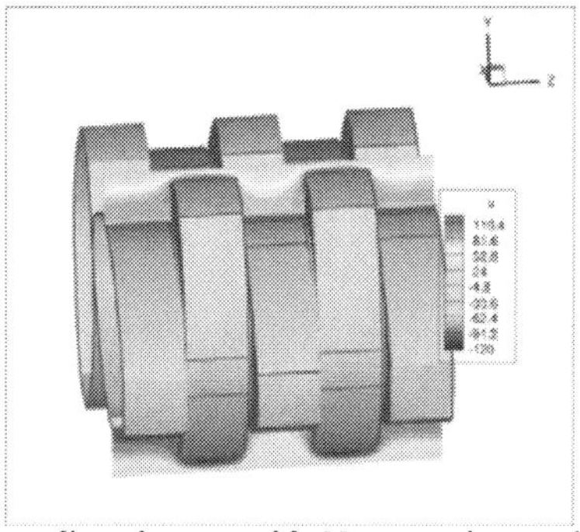

Kneading element with 90 staggering angle

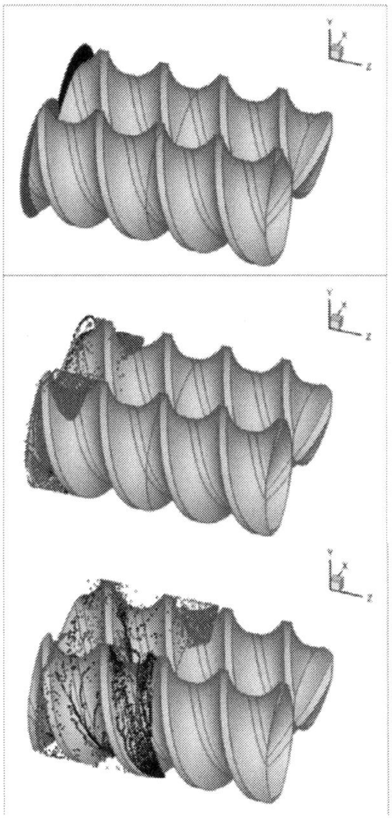

Figure 6. Particle tracking in a conveying element

Figure 7 shows a RTD curve for a conveying element with pitch size 30 mm and screw speed of 80 RPM. The screw consists of two conveying elements (L=60 mm).

Conclusions

Modified techniques based on mortar elements, which are FDM and XFEM, are investigated in their

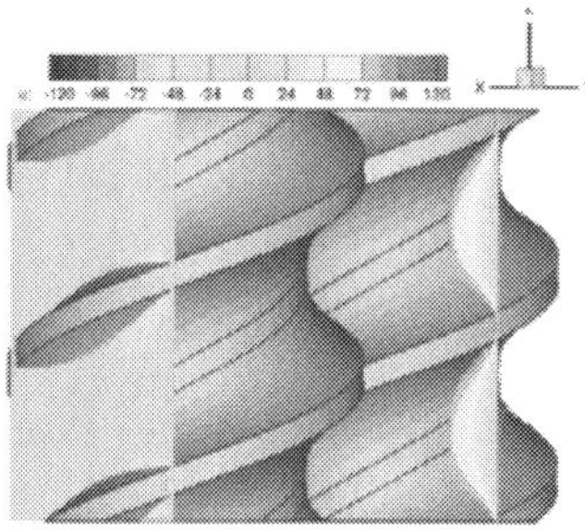

forward conveying element

Figure 5. Contour plot of velocity at x direction for 3D screw elements.

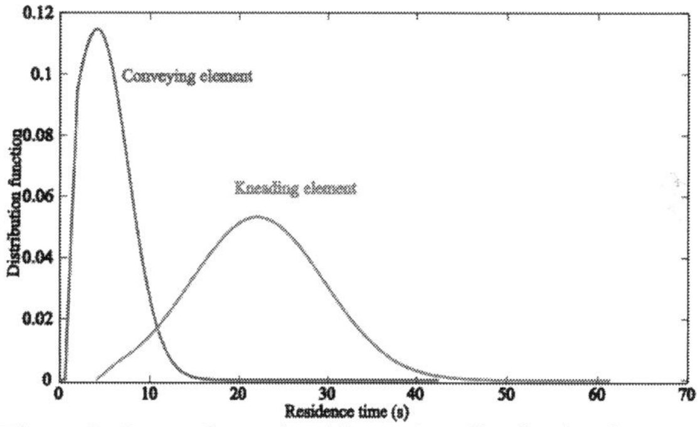

Figure 7. Comparison of residence time distribution for conveying element and kneading element (90^0 staggering angle)

capability to simulate the fluid flow inside a 3D geometry of a TSE with narrow gaps. The accurately calculated velocity field is accordingly used to start mixing analysis by calculating the residence time distribution.

Acknowledgements

ASF thanks FP6 program PEPTFLOW for financial support.

References

Ben Belgacem, F., "The mortar finite element method with Lagrange multipliers", Numerische Mathematik, **84**, 173–197 (1999)

Gerstenberger, A., Wall, W.A., "An extended finite element method/Lagrange multiplier based approach for fluid-structure interaction", Comput. Methods Applied Mech. & Eng., **197,** 1699–1714 (2008)

Glowinski, R., et al., "A fictitious domain method for Dirichlet problem and applications", Comput. Methods Applied Mech. & Eng., **111,** 283–303 (1994)

MODELING OF REACTIVE EXTRUSION: A NEW METHOD

Ionut BANU[1], Jean-Pierre PUAUX[2], Grigore BOZGA[3], Iosif NAGY[3]

(1) Petroleum Institute, Abu Dhabi, United Arab Emirates; bionut@pi.ac.ae
(2) IMP, UMR 5223, Laboratoire des Matériaux Polymères et Biomatériaux, Université Lyon 1, 69622 Villeurbanne cedex (France); jean-pierre.puaux@univ-lyon1.fr
(3) University "Politehnica" Bucharest, 313 Spl. Independentei, sector 6, 060042 Bucharest, Romania; g_bozga@chim.upb.ro

Abstract – The flow and mixing of melted material in the twin extruders are frequently described by classical flow models (axial dispersion model or different versions of compartment type models). The validation of these flow models and the evaluation of their intrinsic parameters are usually based on residence time distribution (RTD) measurements using tracer concentration signals. However, such experimental measurements are not always straightforwardly available, due to different technical reasons (most frequently the lack of adequate tracer and its measurement sensor). In such a case an intermediary approach can be applied in the reactive extrusion modeling, based on both experimental measurements and the use of flow simulation software. In this work we present such an approach, with application to the modeling and simulation of the dilactide polymerization by reactive extrusion. The flow and mixing of the polymerization mixture in the screw channels is modeled using a commercial simulation software. This software computes the filling ratios of the screw sections, as well as the local residence times. The accuracy of these predictions was proved by RTD measurements on melted polypropylene flow. In the dilactide polymerization modeling, the mass balance equations are written considering both axial dispersion model and a series of ideal flow zones (the filled sections were approximated as perfectly mixed whereas the conveying sections as plug flow zones). The comparison of calculated dilactide monomer conversion and polymer molecular weight versus measured data shows a good adequacy, proving the correctness of the approach.

Introduction

The complexity of the flow in twin-screw extruders, as well as the large number of parameters and interrelated variables that affect the flow and mixing in these apparatus, make the reactive extrusion process difficult to understand, control, and optimize. As underlined by Vergnes (Vergnes et al. 1998), two different approaches can be distinguished in the modeling of flow in twin-screw extruders. The first one uses a local description of the flow velocities in the extruder channels, by momentum balance equations. This approach is, theoretically, the most accurate one, but rather complex, which leads to difficulties regarding the accurate identification of the transport properties of the material, identification of the boundary conditions or difficulties in numerical calculations. The second approach is based on a global description of the entire flow process, from the feed to the exit section of the extruder, and, consequently, is a simpler one (Zagal et al., 2005). This second approach was often used in the published literature to describe the flow and mixing in twin screw extruders (Poulesquen, 2003 a & b) It is based on a global description of the flow characteristics along the screw channels by using the well known concept of flow models, in chemical engineering, using RTD. The information needed in this purpose is obtained from experimental residence distribution (RTD) measurements, performed by using appropriate tracer techniques. The main advantage of this global approach is to avoid the difficulties in modeling the flow in the extruder screw, a complex geometry apparatus. Nevertheless, the parameters of models don't have always a physical signification, and the number of parameters could be a difficulty encountered in fitting the experimental data.

In order to develop a mathematical model for the reactive extrusion process following this global approach, two sets of data are to be known: firstly the kinetic data of the polymerization process and secondly the details of the flow, including mean residence times and filling ratios in each section of the screw. These last data cannot be always obtained experimentally. There are cases when the simulation is the only method allowing a relatively accurate description of the flow. One of the difficulties is to know as precisely as possible the viscosity of the melt, assuming that the viscosity changes drastically during the process.

A flow simulation software yields the position and lengths of the filled and conveying sections, the filling ratios, and the local residence times, so long we have the values of the viscosity.

From these data, two classical flow models have been used: firstly the axial dispersion model, and secondly the compartment type model consisting in a series of

plug flow and mixing sections deduced from the simulation results.
This modeling approach was developed and applied to the bulk polymerization of l-lactide, starting from l-dilactide, yielding polylactide (PLA).
For this polymerization system the necessary RTD data were not directly measured due to the lack of an appropriate tracer permitting to work in the presence of melted polylactide mixture. Instead, the RTD measurements were replaced by a theoretical evaluation of the RTD of L-lactide polymerization mixture, performed by flow simulation with the software Ludovic®, a specialized package for the calculus of material flow in twin-screw extruders. In addition, even if the measurement of RTD for polylactide mixture was not possible, some experimental RTD measurements were performed on the same extruder and screw structures used in the polymerization study. In this purpose we used melted polypropylene as flowing material and a fluorescent tracer technique developed by Cassagnau (Cassagnau, 1991). This allowed to obtain a minimum information concerning the material flow in this apparatus. The results of this experimental study were used to prove the ability of Ludovic® simulator to predict the RTD curves.

Experimental

Extrusion experiments

Experimental studies have been carried out using the screw profile below fig. 1.

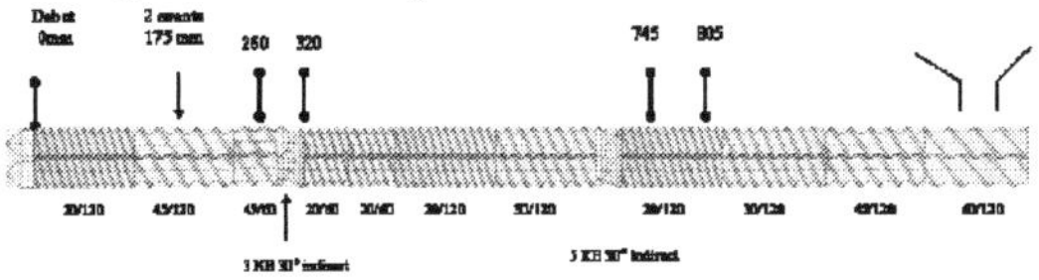

Fig 1. Screw profile

We performed the experiments of polymerization described elsewhere (Puaux et al. 2007, Banu et al. 2010). For these experiments, Stannous Octanoate and Triphenylphosphine have been used as initiators.
The kinetic data are given in the table below, $\overline{k_i}$ and E_i being the preexponential and activation energy of the initiation step, $\overline{k_p}$ and E_p of the propagation step, $\overline{k_{tm}}$ and E_{tm} of the termination transfer to monomer, and $\overline{k_{ts}}$ and $\overline{E_{ts}}$ of the termination to solvent, respectively.

Table 1. Kinetic parameters

$\overline{k_i}$, $l.mol^{-1}\min^{-1}$	$\overline{k_p}$, $l.mol^{-1}\min^{-1}$	$\overline{k_{tm}}$, $l.mol^{-1}\min^{-1}$	$\overline{k_{ts}}$, $l.mol^{-1}\min^{-1}$
1.279x10^{1}	5.53 x10^{1}	6.99 x10^{-2}	2.05 x10^{1}
E_i, $J\cdot mole^{-1}$	E_p, $J\cdot mole^{-1}$	E_{tm}, $J\cdot mole^{-1}$	E_{ts}, $J\cdot mole^{-1}$
5.46 x10^{4}	9.11 x10^{4}	7.65 x10^{4}	1.06 x10^{5}

Simulation of the flow

This simulation is done using the Ludovic® software (Vergnes, 1998). Some simulation results are presented in the text below.

Simulation of the reactive extrusion

The modeling of the L-lactide extrusion process is performed by using two approaches, one based on the RTD simulated curves using an axial dispersion model and another one approximating the TSE as a succession of ideal flow zones (compartment model).

Axial dispersion model (ADM)
Based on Ludovic calculated RTD curves, an axial dispersion model can be adapted to describe the flow in the twin screw extruder and to predict the polymer characteristics at the extruder exit.

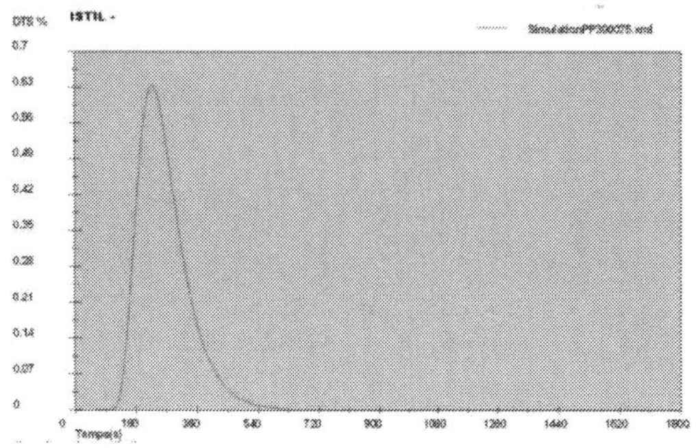

Fig. 1. Example of residence time computed using Ludovic®

The Peclet number (estimated from RTD curves) and the mean residence time values calculated in this approach are presented in the following table 2.

Table 2. ADM parameters calculated from Ludovic simulated RTD

Extruder mass flow rate , [kg/h]	Screw speed, [rpm]	Pe values	Mean residence time, [min]
0.75	50	44.2	7.54
0.75	200	30.0	5.84
0.75	300	28.2	5.67
1.25	50	57.6	5.38
1.25	200	32.3	5.00
1.25	300	30.4	3.61
1.5	50	68.2	4.82
1.5	200	35.0	3.25

This allows us to determine the parameters needed in order to simulate a reaction using an axial dispersion flow model, whose basic differential equation and the boundary conditions are:

$$\frac{1}{P_e}\frac{d^2C}{dX^2}+\frac{dC}{dX}-t_s v_{R,C}=0;\ X=\frac{z}{L};\ Pe=\frac{u\cdot L}{D_L}$$

$$B.C.1:\quad X=0,\quad C_0-C=-\frac{1}{P_e}\frac{dC}{dX}$$

$$B.C.2:\quad X=1,\quad \frac{dC}{\partial X}=0$$

The boundary conditions are of the "close" type. The C variable can be initiator concentration (I), monomer concentration (M), 0th, 1st and 2nd order moments of live (λ_0, λ_1, λ_2) and dead (μ_0, μ_1, μ_2) polymer concentration distributions as well as impurities concentration (S). The reaction rates appearing in the balance equations are substituted by:

$$v_{R,C_I}=k_i IM$$

$$v_{R,C_M}=k_i IM+k_p M\lambda_0+k_{tm}M\lambda_0$$

$$v_{R,\lambda_0}=-k_i IM$$

$$v_{R,\lambda_1}=-(k_i IM+k_p M\lambda_0+k_{tm}M(\lambda_0-\lambda_1)+k_{ts}S(\lambda_0-\lambda_1))$$

$$v_{R,\lambda_2}=-(k_i IM+2k_p M\lambda_1+k_p M\lambda_0+k_{tm}M(\lambda_0-\lambda_2)+k_{ts}S(\lambda_0-\lambda_2))$$

$$v_{R,\mu_0}=-(k_{tm}M\lambda_0+k_{ts}S\lambda_0)$$

$$v_{R,\mu_1}=-(k_{tm}M\lambda_1+k_{ts}S\lambda_1)$$

$$v_{R,\mu_2}=-(k_{tm}M\lambda_2+k_{ts}S\lambda_2)$$

$$v_{R,C_S}=-k_{ts}S\lambda_0$$

For details see Banu (Banu, 2009) Using the process kinetics parameters given in Table 1, as well as the flow parameters in Table 2, one can proceed to simulation of reactive extrusion with axial dispersion model, based on the Ludovic® predictions.

The number-average molecular weight of the dead and live polymers are calculated based on the distribution moments of 0th and 1st order, following the relation:

$$Mn=(MW)\frac{\mu_1+\lambda_1}{\mu_0+\lambda_0}$$

with MW – the monomer molecular weight.

Compartment model

Having the succession of ideal plug flow and perfectly mixed zones, as determined from the Ludovic® simulation and the kinetic model, the calculation can be done from the hopper to the die, starting from the entry conditions. The regions with higher filling ratios (degree of mixing) and residence time (kneading discs sections) are treated as perfectly mixed flow regions whereas the transport sections, with lower intensity of mixing (smallest filling ratio) and lower residence time, are treated as plug-flow regions. In this way, the conversion and the average weights of the polymer along the extruder are computed.

The mass balance for the ith plug-flow zone (with the residence time $t\in(0,t_s)$, t_s – mean residence time in ith region) is given by:

$$\frac{dC_i}{dt}=-v_{R,i}$$

The mass balance for the ith perfectly mixed flow region is:

$$C^{i-1}=C^i-t_s^i v_{R,i}$$

The problem arising in this case is the calculation of the unknown concentrations and moments, C^i, at the exit of the perfectly mixed zone, knowing the inlet values C^{i-1}. In this purpose, a Newton-Raphson procedure was implemented.

The screw profile, the viscosity (might be different in each section), the screw speed, and the flow rate have to be known in order to run the software.

The viscosity data for PLA were taken from the published literature (Cooper-White and McKay, 1999) and fitted on a Carreau-Yasuda viscosity model. The calculations were performed approximating a constant averaged viscosity along the TSE. The weight-average molecular weights for the polymers are generally higher than 50000 kg/kmole. Two viscosity models for 40000 and 130000 kg/kmole weight-average molecular weights were implemented.

The parameters η_0, λ, m and a of the Carreau-Yasuda equation $\eta = \eta_0 \left(1+(\lambda\dot{\gamma})^a\right)^{\frac{m-1}{a}}$ were estimated by using the viscosity data published by Cooper-White and Mackay (Cooper-White and McKay, 1999) . The calculated parameters are given in Table 3 and a comparison between the predicted and published values of polymer viscosity is presented in figure 2. Further, considering the range of polymer molecular weight values, the reactive extrusion calculations will be performed by the viscosity model 1.

Table 3. Estimated values for the parameters of the Carreau-Yasuda viscosity model

Model	η_0	λ	a	m	M_w
1	1.029×10^2	7.658×10^{-3}	2.018×10^0	7.711×10^{-1}	40000
2	7.118×10^3	1.133×10^{-1}	9.033×10^{-1}	5.100×10^{-1}	130000
3	9.525×10^5	1.081×10^1	3.843×10^{-1}	3.582×10^{-1}	360000

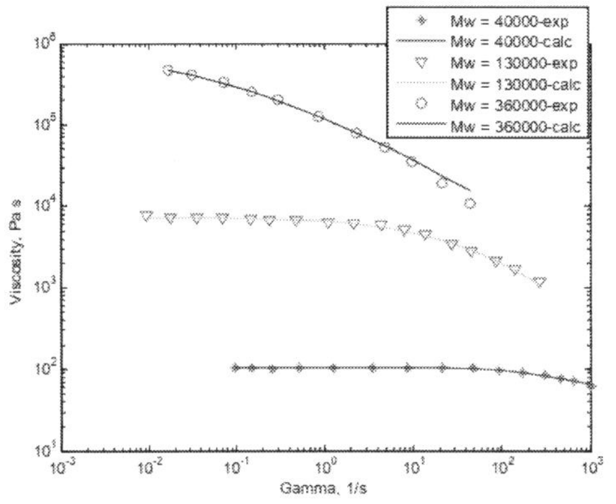

Figure 2. Published and calculated viscosity values at 200°C

Results and Discussion

Several local and global residence times predicted by Ludovic® are presented in the following figures for the screw profile in fig 3. A constant temperature of 200°C was considered along the extruder barrel, accordingly with the experimental operating conditions. The red curve shows the local residence times, while the blue curve is the cumulative one.

The Ludovic® simulations result was an argument for the insensitivity of the residence time with the weight average molecular weight of the PLA and consequently with the viscosity model implemented in the calculations.

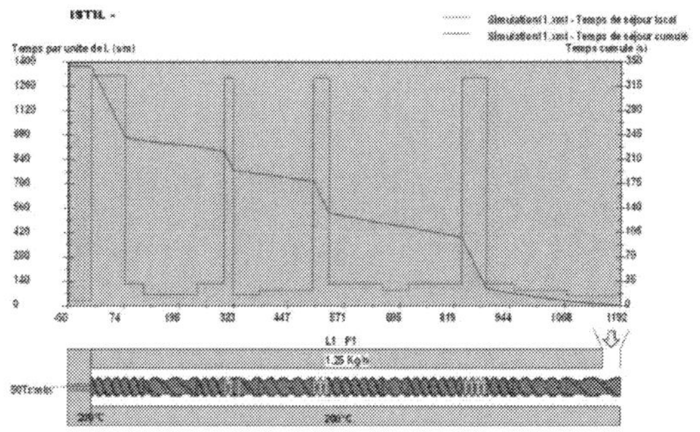

Fig. 3. Example of residence time computed using Ludovic®

The filling ratio is relatively high only in the first section of kneading discs and before the die (Figure 4). Further, the reactive extrusion modeling will be performed using the viscosity model. Ludovic® simulations for PLA reactive extrusion were performed for the representative values of the flow rate and screw speeds.

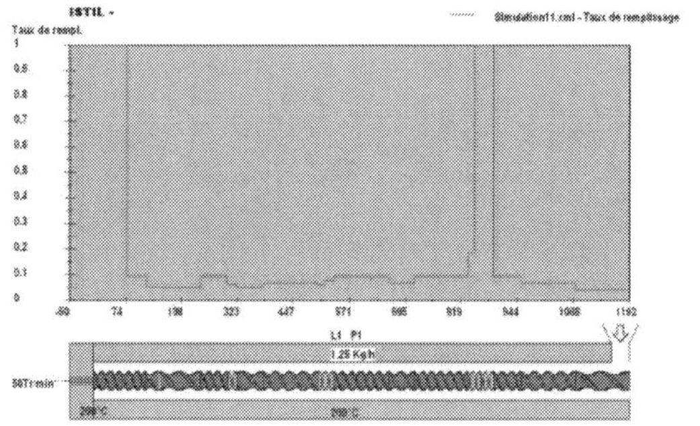

Fig. 4. Example of filling ratio computed using Ludovic®

The data presented in Table 4 are extracted by reading the local and global residence time profiles given by Ludovic® simulations.

Table.4. The parameters for 1.25 kg/h, 50 rpm, screw 1) and viscosity model

Zone #	Type	Residence time,[s]	Zone length, [mm]
Zone 1	Plug-flow	27	300
Zone 2	Perfectly mixed	73	-
Zone 3	Plug-flow	35	300
Zone 4	Perfectly mixed	46	-
Zone 5	Plug-flow	14	177
Zone 6	Perfectly mixed	28	-
Zone 7	Plug-flow	20	223
Zone 8	Perfectly mixed	100	-
Zone 9	Plug-flow	2	50

The mass balance of the monomer for the ith perfectly mixed zone assuming equal inlet and outlet volume flow-rates,

$$x^{i-1} = x^i - t_s^i v_{R,i}$$

Where x_i represents the concentration of the species x in the ith region.
The expressions of the reaction rates involved in the equations are given by relations, presented in the previous section. From these equations, one can successively calculate the concentrations at the exit of the perfectly mixed region, given the feed concentrations and the mean residence times t_s^i. This is performed by solving the nonlinear algebraic system of mass balance equations particularized for a series of perfectly mixed zones, by the Newton-Raphson method.
The mass balance for the ith plug flow (PF) region is given by:

$$\frac{dx_i}{dt} = -v_{R,i}, \quad t = \frac{z}{u_{z,i}}$$

Where: x_i - the concentration of the species x in the ith zone;
$u_{z,i}$ - the flow velocity in the ith zone;
$v_{R,i}$ - the consumption rate of the x_i species;
The integration of the PF model was performed with a procedure based on a Runge-Kutta fifth order algorithm implemented in MATLAB® programming package. An example of simulation will be given for a mass flow-rate of 1.25 kg/h, a screw speed of 50 rpm, using the results obtained for the first viscosity model. Using the flow parameters specified in the table 2 and the kinetic constants presented in a previous paragraph, the behavior of the reactive extrusion process was simulated. The results in terms of monomer conversion and number-average molecular weights profiles are presented in fig. 5 and 6.
A monomer conversion of about 80% and a number-average molecular weight of 71000 kg/kmole are obtained at the extruder exit.
One example of simulation in terms of monomer conversion and number-average molecular weights profiles along the extruder is presented in figures 5 and 6 below. A comparison with the experimental results is shown in the following table 5.

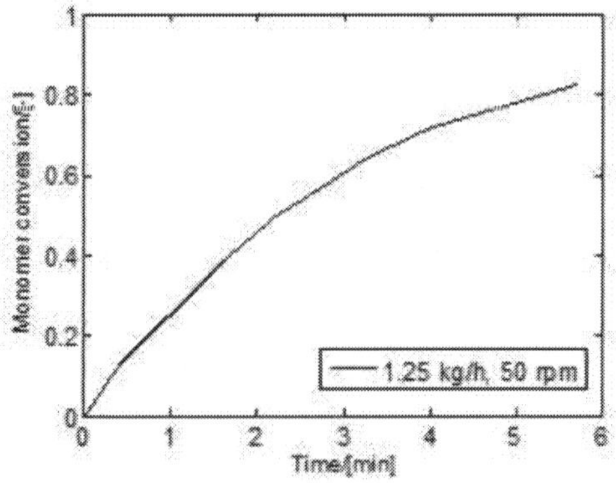

Fig. 5. Monomer conversion for a M/I of 4500 (initiator $SnOct_2$/TFF, flow rate 0.75 kg/h, screw speed 50 rpm)

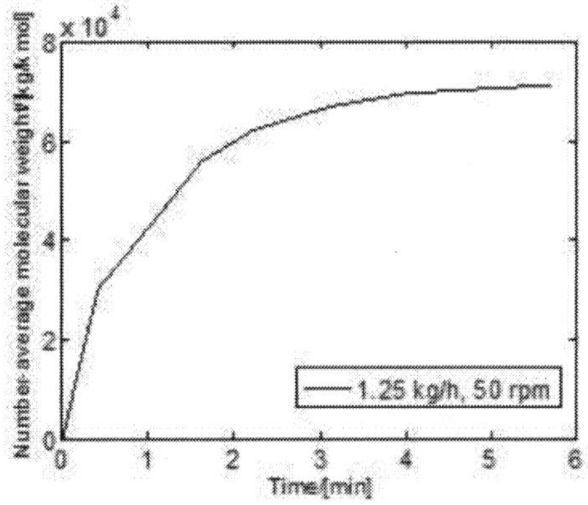

Figure 6. Mn for a M/I of 4500 (initiator $SnOct_2$/TFF, flow rate 0.75 kg/h, screw speed 50 rpm)

Table 5. Experimental and predicted polymer characteristics for M/I = 4500 (initiator $SnOct_2$/TPP)

Mass flow rate, [kg/h]	Screw speed, rpm	Monomer conversion,[-] (experimental)	Monomer conversion, [-] (model)
0.75	50	0.80	0.65
0.75	200	0.75	0.56
0.75	300	0.70	0.55
1.25	50	0.73	0.53
1.25	200	0.65	0.49
1.25	300	0.60	0.40
1.5	50	0.70	0.49
1.5	200	0.50	0.37

Mass flow rate, [kg/h]	Screw speed, rpm	Number-average molecular weight, [kg/kmol] (experimental)	Number-average molecular weight, [kg/kmol] (model)
0.75	50	95000	90000
0.75	200	90000	86000
0.75	300	72000	85000
1.25	50	81000	85000
1.25	200	73000	83000
1.25	300	60000	79000
1.5	50	75000	84000
1.5	200	67000	76000

Conclusions

It can be seen from the above data that the convergence between calculated and experimental data is acceptable in this case.

The work presented a modeling and simulation study of the L-lactide polymerization by reactive extrusion. The flow and mixing of the PLA reaction mixture in the screw channels were computeded by simulation using the flow simulation package Ludovic®. The capability of this software to describe the flow of melts in a twin screw extruder was proved by an experimental test, using polypropylene as flowing material. An acceptable concordance was obtained between the measured RTD characteristics and the calculated ones by Ludovic package, for a screw structure used in L-lactide polymerization experiments.

Based on the RTD features predicted by Ludovic package for the L-lactide polymerization mixture, the flow in the TSE was described by two flow models: (i) axial dispersion model and (ii) a compartment model. Based on these flow models the reactive extrusion process was simulated for a set of experimental operating conditions. An acceptable concordance between the calculated and experimentally measured polilactide characteristics was obtained, thus demonstrating the correctness of this modeling approach.

Considering the dimension of the modeling study, the number of parameters involved, the accuracy of extruder operating parameters measurements and control as well as the approximations used in the extruder flow model, the results are acceptable. The complexity of such model is an argument: we took into account all possible parameters and all the effects. A validation using other systems will be an interesting test.

Acknowledgements

We acknowledge the French Ministry of Foreign Affairs and the Region Rhône-Alpes for scholarships, and the National University Research Council of Romania for a research grant.

References

Banu I., Puaux J. P., Bozga G., Nagy I., "Modeling of L-lactide Polymerization by reactive extrusion", Macromolecular Symposia, 289, 108-118 (2010)

Banu I., PhD thesis, "Modeling and optimization of tubular polymerization reactors", Bucharest and Lyon (2009)

Cassagnau, P., C. Mijangos, and A. Michel, "An ultraviolet method for the determination of the residence time distribution ina twin screw extruder." Polymer Engineering and Science, **31** (11), 772–778 (1991).

Cooper-White J. J., MacKay M. E., "Rheological properties of polylactides. Effect of molecular weight and temperature on the viscoelasticity of poly(L-lactic acid), J. of Polymer Science, Part B: Polymer Physics, 37 (15) 1803-1814 (1999)

Poulesquen, A. and Vergnes B., "A Study of Residence Time Distribution in Co-Rotating Twin-Screw Extruders. Part I: Theoretical Modeling." Polymer Engineering and Science, **43** (12), 1841–1848 (2003)

Poulesquen, A., and Vergnes B., "A Study of Residence Time Distribution in Co-Rotating Twin-Screw Extruders. Part II: Experimental Validation." Polymer Engineering and Science, **43** (12), 849-1862 (2003)

Puaux J. P., Banu I., Bozga G., Nagy I., "A study of L-lactide Ring-Opening Polymerization Kinetics", Macromolecular Symposia, 259, 218-226 (2007)

Vergnes B., Valle G. D. and Delamare L., "A Global Computer Software for Polymers Flows in Corotating Twin Screw Extruders", Polymer Engineering and Science, 38 (11), 1781-1792 (1998)

Zagal, A., Vivaldo-Lima E., and Manero O., "A Mathematical Model for the Reactive Extrusion of Methyl Methacrylate in a Co-rotating Twin-Screw Extruder" Ind. Eng. Chem. Res. **44** (26), 9805-9817 (2005)

ANALYSIS OF A SPLIT AND RECOMBINE MIXER FOR POLYMER MELTS

F.R. Phelan Jr.*, D. Moon and K.B. Migler

Polymers Division, National Institute of Standards and Technology (NIST), Gaithersburg, MD 20899 – frederick.phelan@nist.gov; doyoung.moon@nist.gov; kalman.migler@nist.gov

Abstract - Moon and Migler[1] recently developed a novel microfluidic mixer for high temperature polymer melts based on the use of split and recombine flow channels formed from the stacking of stainless steel shims (50 μm thickness) that uses only 20 μL of material per sample. The device produces an exponential 2^n multilayer domain morphology (where n is the number of units in series) in a direction either perpendicular or parallel to the plane of the shims. For immiscible systems, mixing is produced by reducing layer thickness progressively closer to the molecular length scale of the components. The device also has potential as an alternative to coextrusion for the production of multilayer laminates in the case of limited material quantity. Flow in the mixer is analyzed using two techniques. First, multiphase flow equations based on a level-set model are solved using the finite element method (FEM) to track the interface between the immiscible phases. In a second technique, flow modeling is combined with multi-species particle tracking to produce Poincare diagrams of the mixing in the outflow cross-sections of each serial unit (these results will be presented at the conference). Results for the two techniques are compared with ideally expected results and suggestions for future work are discussed. The two major challenges to producing uniform laminates for immiscible components using the shim based flow channel design are generating equal splits, and recombining the flow streams without producing rotational skew of the laminate layers.

Introduction[1]

The mixing of polymer melts is most commonly accomplished using extrusion with either single or twin-screw extruders due to the high viscosities of the materials and the elevated temperatures required to process them[2]. However, there is an increasing need to process melt materials in small volumes at microfluidic size scales in order to reduce sample size, process rare/limited quantity materials, as well as for the economic incentive of avoiding the high costs associated with complex dies and industrial scale equipment.

A number of techniques have been developed for effectively mixing aqueous solutions at the microscale[3,4], and are generally divided into two categories, active and passive mixing. A technique that lends itself well to the processing of more viscous polymer melts is a passive technique called split and recombine (SAR) mixing[5-8]. The idea of SAR mixing is akin to a mathematical model called a Baker's Map[9] and is shown in Figure 1. In each step, the flow stream is split lengthwise in a direction perpendicular to the gradient direction between components, and then recombined in a manner so as to double the number of laminate layers. The technique produces an exponentially increasing 2^n multilayer domain morphology, where n is the number of units in series. For miscible systems, mixing is produced by substantially lowering the length scale for molecular diffusion in the gradient direction between components. For immiscible systems such as polymer melts, mixing is effectively produced by reducing layer thickness down (or close to) the molecular length scale and freezing in the structure. For melts, the technique also has potential as an alternative to coextrusion for the production of multilayer laminates in cases where the laminate length scale remains larger than the molecular scale.

Moon and Migler[1] recently developed a novel microfluidic mixer for high temperature polymer melts based on the use of split and recombine flow channels formed from the stacking of stainless steel shims. The shims have a thickness of 50 μm, and can be designed to form laminates in a direction either parallel or perpendicular to the plane of the shims. The device uses only 20 μL of material per sample The arrangement of the shim structures in the context of the Multi-sample Microfluidic Rheometer[2] (MMR) used to

[1] Official contribution of the National Institute of Standards and Technology; not subject to copyright in the United States.

[2] Identification of a commercial product is made only to facilitate reproducibility and to adequately describe

control the flow of the polymer streams is shown in Figure 2a; Figure 2b coarsely details the etching in the shim structures that is used to create the flow channels. An example of a flow channel formed from the arrangement for a perpendicular shim stacking geometry[3] is shown in Figure 3a. Details of the inlet region and the points at which the flow stream is split and recombined are show in Figure 3b.

In what follows, we show some example results for the modeling of two immiscible fluids in the perpendicular stacking SAR mixer geometry shown in Figure 3. The results show that the ability to produce uniform laminate layers is highly dependent on the geometry, and that care must be taken to eliminate non-uniform splitting and rotational skew of the laminate layers upon recombination. These effects are particularly important in immiscible systems, which do not have the benefit of diffusion to smooth over non-uniformities.

Modeling

The governing equations used to model the flow are the unsteady Navier–Stokes equations together with the incompressible condition and the level-set equation[10]

$$\rho\frac{\partial \underline{v}}{\partial t}+\rho\underline{v}\cdot\nabla\underline{v}=-\nabla p+\eta\nabla^2\underline{v}-\sigma\kappa\delta_\varepsilon(\phi)\nabla\phi \quad (1.1)$$

$$\nabla\cdot\underline{v}=0 \quad (1.2)$$

$$\frac{\partial\phi}{\partial t}+\underline{v}\cdot\nabla\phi=\gamma\left(\varepsilon\nabla\phi-\phi(1-\phi)\frac{\nabla\phi}{|\nabla\phi|}\right) \quad (1.3)$$

where $\underline{v}$ is the velocity, p is the pressure, ϕ is the level set parameter, ρ and η are the density and viscosity, respectively, σ is the surface tension, ε and κ are the interface thickness and curvature, respectively, γ is a level set parameter called the phase velocity, and $\delta_\varepsilon(\phi)$ is an interface delta function. The level set parameter, ϕ, varies smoothly between [0,1], with $\phi=0$ denoting a composition of 100% component 2, and $\phi=1$ denoting 100% of component 1. Numerical solutions were obtained using the commercial finite-element package COMSOL MultiPhysics (COMSOL MP)[11]. Solutions are obtained by setting boundary conditions for $\phi=0$ and $\phi=1$ at the appropriate injection points, with an initial condition of $\phi=1$ throughout the domain. The system is then integrated in time until the second component has propagated throughout the system and steady-state is achieved.

Results and Discussion

Some typical results for the mixing of two immiscible fluid components are shown in Figures 4-6. Figure 4 shows results for the level set parameter at various cross-sections in the first section of the device, which encompasses two split-recombine units. The Figure shows the two different fluids (denoted by red-blue) impinge in the initial T-section and then are progressively processed into an increasing number of layers.

Finer details of the results are depicted in Figures 5-6. Figure 5 shows results for the level set parameter in the cross-sections midway along the left and right side of the curved section after the first split. Ideally, the thicknesses of the red and blue fluids would be equal in size, with a small multi-color interface section denoting the transition from one fluid to another. However, it is evident that in left section after the split, the ratio of red:blue is about 2:1, while in the right section the same ratio is the inverse at 1:2. This difference is accounted for by the fact that the fluid towards the inside of the channel encounters a much longer flow path as it splits to either the left or the right. Rounding the flow channel at these points rather than the sharp transition present in the initial design is a possible remedy to come closer to a 1:1 split ratio.

Figure 6 shows results for the laminate produced in the first recombination section. While the laminate shows four layers as expected, due to the non-uniform split previously shown in Fig. 5, the layers are not equal in size as would be ideal, but the inner layers are much thinner than the outer layers. However, it is also evident that in addition to being non-uniform, the layers also have a rotational skew and do not maintain an orientation perpendicular to the plane of the shims. The reason for the rotational skew can be traced to the way the fluids streams impinge to recombine as shown in Figure 7. In the first recombine section, the fluid on the left comes in through the bottom shim, while the fluid on the right comes in through the upper shim. While the offset is small, it is enough to produce a net rotation in the laminate layers during the impingement process. This rotational skew has a progressive effect, and is a possible explanation for the formation of cylindrical type domains in experiments using the perpendicular shim design[1].

Further details of the flow simulation, and what can be learned towards building more effective split-recombine mix units using a shim based design will be discussed at the conference.

---

procedure. In no case does it imply endorsement by the National Institute of Standards and Technology (NIST) or that it is necessarily the best product for the procedure.

[3] That is, the device produces layers that are perpendicular to the plane of the shims.

Conclusions

Flow simulation has been used to model the flow of two immiscible fluids in SAR mixer geometry based on the design of Moon and Milger[1]. The results show that the ability to produce uniform laminate layers is highly dependent on the geometry, and that care must be taken to eliminate non-uniform splitting and rotational skew of the laminate layers upon recombination.

References

1. Moon, D. & Migler, Kalman B. ANTEC 2009: Annual Technical Conference Proceedings, Society of Plastics Engineers. 1838-1840 (2009).
2. Tadmor, Z. & Gogos, C.G. *Principles of Polymer Processing*. (Wiley-Interscience: 2006).
3. Nguyen, N. & Wu, Z. Micromixers—a review. *J. Micromech. Microeng.* **15**, R1-R16 (2005).
4. Hessel, V., Löwe, H. & Schönfeld, F. Micromixers--a review on passive and active mixing principles. *Chemical Engineering Science* **60**, 2479-2501
5. Tofteberg, T., Skolimowski, M., Andreassen, E. & Geschke, O. A novel passive micromixer: lamination in a planar channel system. *MICROFLUIDICS AND NANOFLUIDICS* **8**, 209-215 (2010).
6. Amarie, D., Glazier, J.A. & Jacobson, S.C. Compact Microfluidic Structures for Generating Spatial and Temporal Gradients. *Analytical Chemistry* **79**, 9471-9477 (2007).
7. Ohkawa, K., Nakamoto, T., Izuka, Y., Hirata, Y. & Inoue, Y. Flow and mixing characteristics of [sigma]-type plate static mixer with splitting and inverse recombination. *Chemical Engineering Research and Design* **86**, 1447-1453 (2008).
8. Hardt, S., Pennemann, H. & Schonfeld, F. Theoretical and experimental characterization of a low-Reynolds number split-and-recombine mixer. *MICROFLUIDICS AND NANOFLUIDICS* **2**, 237-248 (2006).
9. Ottino, J.M. *The Kinematics of Mixing: Stretching, Chaos, and Transport*. (Cambridge University Press: 1989).
10. Sethian, J.A. & Smereka, P. LEVEL SET METHODS FOR FLUID INTERFACES. *Annu. Rev. Fluid. Mech.* **35**, 341-372 (2003).
11. *COMSOL MP User's Guide*. (COMSOL AB: Burlington, MA, 2009).

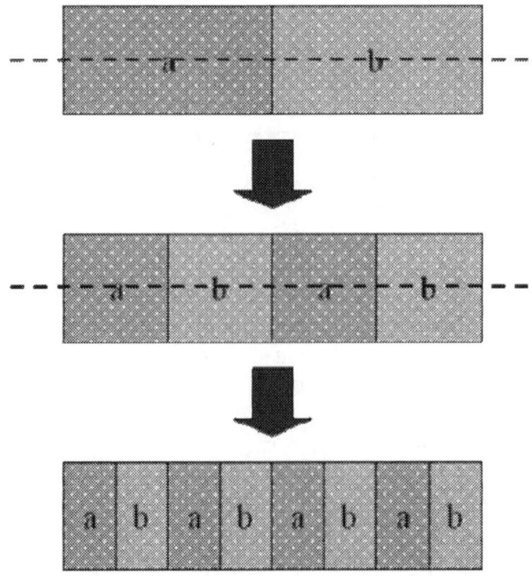

Figure 1 – In SAR mixing, the flow stream is split lengthwise in a direction perpendicular to the gradient direction between components, and then recombined in a manner so as to double the number of laminate layers. Repetition creates an exponentially increasing multi-layer domain morphology.

Figure 2 – a) The arrangement of the shim structures in the context of the Multi-sample Microfluidic Rheometer (MMR) used to control the flow of the polymer streams[1].

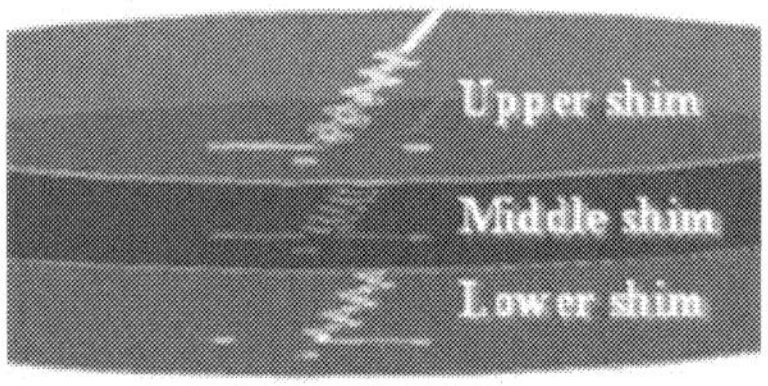

Figure 2 – b) Coarse details of the etching in the shim structures that is used to create the flow channel.

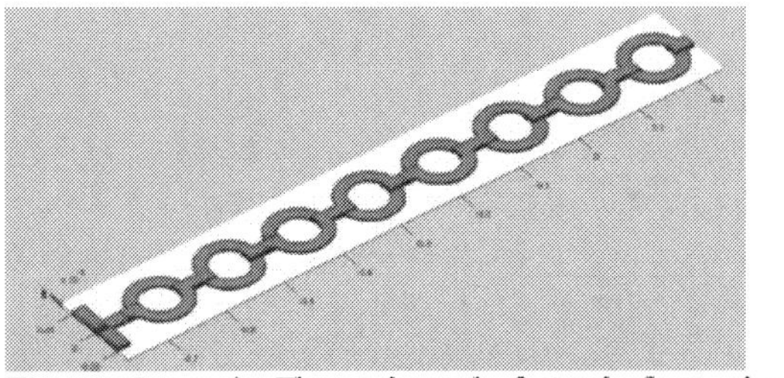

Figure 3 – a) Flow channel formed from the arrangement for a perpendicular shim stacking geometry.

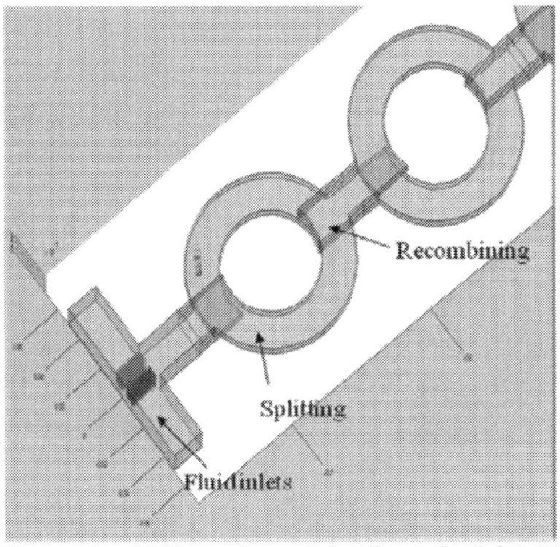

Figure 3 – b) Flow channel details at the inlet. Two components merge at the inlet T-section, and then are repeatedly split/recombined in the areas indicated.

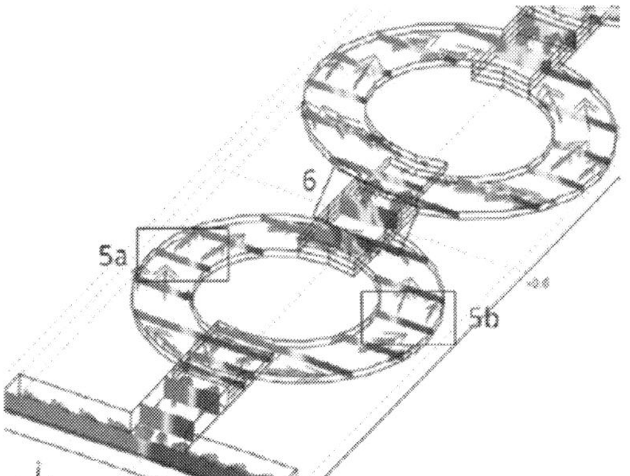

Figure 4 – Results for the level set parameter at various cross-sections in the first section of the device. The two different fluids are denoted by red and blue; other colors indicate a gradient region between phases. The location of the detail pictures shown in Figures 5-6 are also indicated.

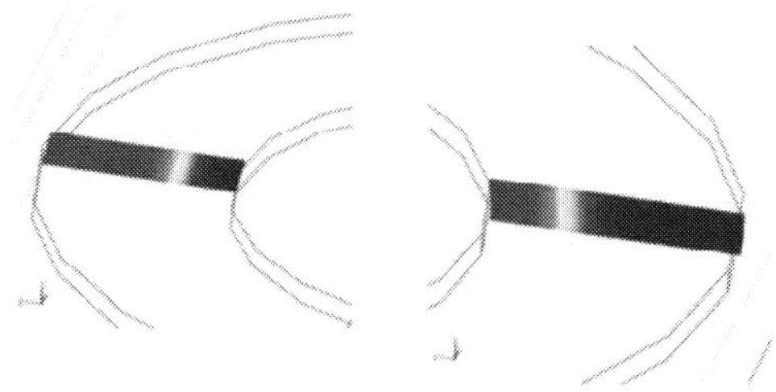

Figure 5 – Detail for the level set parameter in the cross-sections midway along the left and right side of the curved section after the first split.

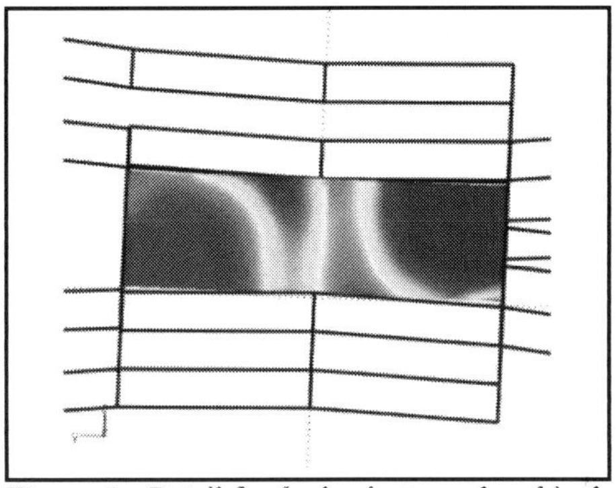

Figure 6 – Detail for the laminate produced in the first recombination section.

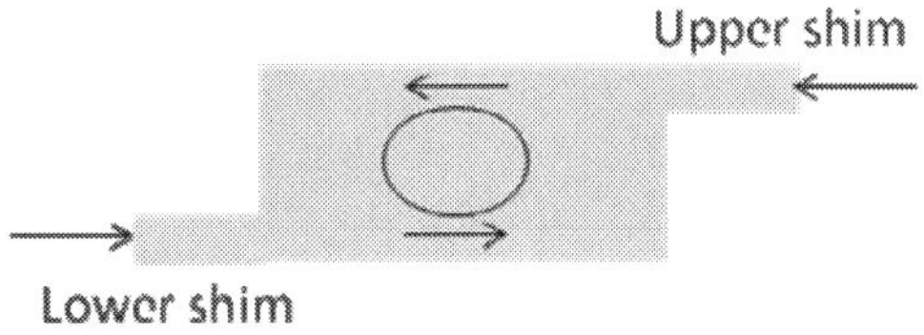

Figure 7 – Cross-section of the fluid impingement region in the first recombination section.

Preliminary Efforts in the Simulation of Molding of a Polypropylene Melt Reinforced with Long Glass Fibers using Transient Shear Rheology

Ortman, K.C.[1], Neeraj Agarwal[1], D.G. Baird[1] and P. Wapperom[2], Jeffrey Giacomin[3]

Chemical Engineering Department[1], Virginia Tech, Blacksburg, VA 24061
Mathematics Department[2], Virginia Tech, Blacksburg, VA 24061
Mechanical Engineering Department[3], University of Madison-Wisconsin, Madison, WI 53706

Abstract

The purpose of this research is to understand the transient fiber orientation and the associated rheology of long glass fiber (> 1mm) reinforced polypropylene, in a well-defined simple shear flow, to extend the results and knowledge gained from these fundamental experiments to the use of simulating (more complex) molding processes. Specifically, we are interested in associating the rheological behavior of glass fiber reinforced polypropylene with the transient evolution of fiber orientation in simple shear flow in an effort to ultimately model fiber orientation in complex flow. A sliding plate rheometer was designed to measure stress growth in the startup and cessation of steady shear flow. Results were confirmed by independent measurements on another sliding plate rheometer[13]. Two fiber orientation models were investigated to predict the transient orientation of the long glass fiber system. One model, the Folgar-Tucker model[2,3] (FT), has been particularly useful for short glass fiber[1] systems and was used in this paper to assess its performance with long glass fibers. A second fiber orientation model[8], one that accounts for the flexibility of long fibers as opposed to rigid rod models commonly used for short fibers, was investigated and results were also compared with experimentally measured values of orientation. The accuracy of these models, when used with the stress tensor predictions of Lipscomb[12], was evaluated by comparing the model predictions against experimental stress growth data. Samples consisting of 10% wt. glass fiber in polypropylene with an average fiber length of 4 mm were prepared with random initial orientation and were sheared at different rates. Model predictions showed that fiber flexibility has the effect of retarding transient fiber orientation but provides poor rheological predictions with the chosen stress model. Additionally, it was shown that the predictions of the Folgar-Tucker model are not able to capture the dynamics of neither the orientation evolution nor the stress growth evolution that was measured experimentally.

Experimental Methods

In this research, the transient startup shear rheology of highly concentrated long glass fibers, in a polypropylene melt, was studied in a sliding plate rheometer. Ten weight percent, 12 mm long glass fibers pellets were extruded and re-pelletized to provide homogeneity. Extrusion does cause fiber breakage. The resulting number average fiber length after extrusion was found to be approximately 4 mm. This material was then compression molded to construct samples of known initial orientation that can be rectilinearly sheared (Figure 1). In this research, samples were subjected to a simple shear flow (Equation 1).

$$v_x(y) = \dot{\gamma}\, y \tag{1}$$

The transient shear rheology was measured using a shear stress transducer. Fiber orientation, within this research, was experimentally measured using the method proposed by Leeds[6]. This method refers to optically analyzing metallographically polished (microtomed) sample slices (in a plane of interest) and determining the tangential orientation from the resulting ellipse.

Fiber Orientation Models

Both the Folgar-Tucker[2] model and Bead-Rod model, suggested by Strautins and Latz[7], were used with the stress predicting Libscomb model to fit model parameters that are most consistent with both the measured orientation evolution and the transient rheology. The accuracy of the models will be discussed. The reader is referred to the literature for model explanations.

Folgar-Tucker Model

The Folgar-Tucker model[2], which has seen much use since its introduction in the 1980's, was developed for short glass fibers in the concentrated regime. Here, the term concentrated is used to describe the ability for significant fiber-fiber interactions to occur. The model has been developed to extend Jeffery's model, introduced in the 1920's, by introducing an isotropic rotary diffusion term. This term, called the Folgar-Tucker term, inhibits the steady state alignment of the fibers from being fully oriented (Equation 2).

$$\frac{D\underline{\underline{A}}}{Dt} = \overbrace{\underline{\underline{A}}\cdot\underline{\underline{\kappa}}^{T} + \underline{\underline{\kappa}}\cdot\underline{\underline{A}} - [(\underline{\underline{\kappa}}+\underline{\underline{\kappa}}^{T}):\underline{\underline{A}}]\underline{\underline{A}}}^{\text{Generalized Jeffery for Fibers}} + \overbrace{2C_1 II_D(\underline{\underline{\delta}} - 3\underline{\underline{A}})}^{\text{Folgar-Tucker Term}} \qquad (2)$$

Within this model, D/Dt is the material derivative, $\underline{\underline{\kappa}}^{T}$ is the velocity gradient tensor whose i,j[th] component is defined as $\partial v_j / \partial x_i$, and C_I is the isotropic rotary diffusion term, II_D is the second invariant of the rate of strain tensor ($\underline{\underline{D}}$) defined as $\underline{\underline{D}} = \frac{1}{2}\left(\underline{\underline{\kappa}} + \underline{\underline{\kappa}}^{T}\right)$, and $\underline{\underline{\delta}}$ is the unit tensor. For the simulation of this model, the quadratic closure approximation is used to decouple to the fourth order orientation tensor (Equation 3).

$$\underline{\underline{A}}_4 = \underline{\underline{A}}\,\underline{\underline{A}} \text{ (quadratic closure approx.)} \qquad (3)$$

Because our fibers are long, we wish to explore a model that attempts to account for flexibility.

Bead Rod Model

The model being referred to here was published in 2007 by Strautins and Latz[8]. This Bead Rod model (BR) was published as a continuum fiber model for dilute solutions that provides a first approximation to flexibility. This is accomplished by modeling a fiber as two rods connected by a pivot allowing bead (Figure 2). The model constructs two rigid segments with length l_B that are allowed to slightly pivot about the connecting bead, with some restorative spring rigidity k. Both $\vec{p}$ and $\vec{q}$ are unit vectors that represent the orientation of the corresponding fiber segments, with respect to the center bead. This model makes use of two orientation tensors ($\underline{\underline{A}}$ and $\underline{\underline{B}}$) that represent the second moment of the orientation distribution function with respect to the unit vectors, $\vec{p}$ and $\vec{q}$, in the following manner (Equations 4 and 5).

$$\underline{\underline{A}} = \int \vec{p}\vec{p}\;\psi(\vec{p},\vec{q})\,d\vec{p}\,d\vec{q} \qquad (4)$$

$$\underline{\underline{B}} = \int \vec{p}\vec{q}\;\psi(\vec{p},\vec{q})\,d\vec{p}\,d\vec{q} \qquad (5)$$

In simple shear flow, the model's equations are reduced significantly due to the non-existence of second order velocity derivatives. Additionally, we can adapt the isotropic rotary diffusion term of Folgar and Tucker in an ad-hoc manner to probe concentrated suspensions in the following model (Equation 6 and 7).

$$\frac{D\underline{\underline{A}}}{Dt} = \underline{\underline{A}} \cdot \underline{\underline{\kappa}}^T + \underline{\underline{\kappa}} \cdot \underline{\underline{A}} - [(\underline{\underline{\kappa}} + \underline{\underline{\kappa}}^T) : \underline{\underline{A}}]\underline{\underline{A}} - 2k[\underline{\underline{B}} - \underline{\underline{A}}\ tr(\underline{\underline{B}})] - 6C_I II_D\left(\underline{\underline{A}} - \frac{1}{3}\underline{\underline{\delta}}\right) \quad (6)$$

$$\frac{D\underline{\underline{B}}}{Dt} = \underline{\underline{B}} \cdot \underline{\underline{\kappa}}^T + \underline{\underline{\kappa}} \cdot \underline{\underline{B}} - [(\underline{\underline{\kappa}} + \underline{\underline{\kappa}}^T) : \underline{\underline{A}}]\underline{\underline{B}} - 2k[\underline{\underline{A}} - \underline{\underline{B}}\ tr(\underline{\underline{B}})] - 6C_I II_D\left(\underline{\underline{B}} - \frac{tr(\underline{\underline{B}})}{3}\underline{\underline{\delta}}\right) \quad (7)$$

The physical interpretation of the $\underline{\underline{A}}$ tensor is that it describes the orientation tensor of a given rod, either $\vec{p}$ or $\vec{q}$. The $\underline{\underline{B}}$ tensor, on the other hand, describes the end-to-end distance between the two rods at any given time and thus, in simple flows where no second order velocity gradients exist, behaves as a time delay for the model fiber to "open" (provided it has some initial bending to it). To take advantage of this time delayed effect, we wish to phenomenologically choose to use an orientation tensor constructed from the dimensionless end-to-end vector (Equation 8), (Figure 3). This choice will give us a little more control over the dynamics of the orientation evolution.

$$\vec{r} = \vec{p} - \vec{q} \quad (8)$$

Using this definition of the end-to-end vector, an analogous definition of the orientation tensor (constructed from the end-to-end vector) can be formed by taking the second moment of this vector with respect to the orientation tensor (Equation 9).

$$\underline{\underline{rr}} = \int \vec{r}\vec{r}\ \psi(\vec{p},\vec{q},t)d\vec{p}d\vec{q} \quad (9)$$

This form of the orientation tensor is not fully useful because it has not yet been normalized. Hence, the normalized form of this tensor is obtained by dividing the tensor by the magnitude of the end-to-end vector. This is equivalent to normalizing the tensor with respect the trace of this tensor to form (Equation 10).

$$\underline{\underline{R}} = \frac{\underline{\underline{rr}}}{tr(\underline{\underline{rr}})} \text{(normalized end-to-end tensor)} \quad (10)$$

Libscomb Stress Model

The Libscomb stress model was used in this research to relate the transient fiber orientation within the melt to the transient viscosity behavior of the suspension. The quadratic closure approximation was again used. The orientation representation that we have chosen to use in this model was of the end-to-end form (Equation 10). With this adaptation, the Libscomb model[12] for the transient shear viscosity (η^+) becomes (Equation 11).

$$\eta^+ = \eta_s + c_1\varphi\eta_s + 2\varphi\eta_s N R_{12}^2 \quad (11)$$

In this expression, η_s is the neat matrix viscosity, φ is the volume fraction of fibers, c_I is a coefficient responsible for the viscosity enhancement due to the presence of fibers, N is a parameter associated with the enhancement due to fiber orientation, and R_{12} is the 1-2 component of the orientation tensor. It is worth mentioning, for the case of Folgar-Tucker, that the orientation tensor ($\underline{\underline{A}}$) of a rigid rod is the same as its end-to-end orientation ($\underline{\underline{R}}$) because a rigid rod does not bend. For concentrated suspensions, both c_I and N were used as fit parameters for this stress model.

Results

Extruded long glass fiber strands (with a post-extruded average fiber length of 4 mm) were given an initially random planar orientation and sheared for 70 strain units at shear rates of 0.4, 1.0, and 4.0 sec^{-1}. The transient rheologies of the fiber suspensions are shown in (Figure 4). As one can see, the shear stress overshoot are larger at lower shear rates. Likewise, shear thinning is also visible. These effects have been reported in literature and have been said to be due to long fiber flocculation[14]. At higher shear rates, the aggregates are more easily broken and hence lower peaks and steady state viscosities are realized. The corresponding orientation evolution at 1.0 sec^{-1} is depicted in (Figure 5 and 6).

Using this orientation information, the orientation model parameters can be fit. Initial conditions of the orientation models for planar random fibers are $A_{11}=A_{33}=1/2$, $A_{ij\neq11,33}=0$, and $B_{ij}=0$. These conditions infer that the end-to-end orientation tensor has the initial conditions of $R_{11}=R_{33}=1/2$, $R_{ij\neq11,33}=0$. One disadvantage of the Folgar-Tucker model is that the orientation kinetics cannot be slowed down. This is untrue for the relatively new 'Reduced Strain Closure' model of Tucker, et al.[15], but this form will not be evaluated in this paper. Using the original form of the Folgar-Tucker model, the isotropic rotary diffusion coefficient (C_I) may be fit to obtain the largest strain value of the A_{11} component of the orientation tensor given in (Figure 5). Ideally, this value should conform to the steady state value of the transient orientation, but the data at 70 strain units shown in (Figure 5) is the closest value available to the true steady state orientation. This same value will also used in the Bead Rod model, but k may be varied to either slow or speed up the orientation kinetics, in this situation. It is very important to reiterate the distinction between these two models again. The Bead Rod model was used in this paper to slow orientation kinetics by giving the fibers a random in-plane orientation, identical to that used by the Folgar-Tucker model, but also with a random initial bending angle (not found in rigid rod theory such as the Folgar-Tucker model). This allows the restorative spring rigidity k to be varied and behave as a time delayed orientation parameter, as mentioned before. The isotropic rotary diffusion term was found to be $C_I = 0.038$ to asymptotically approach the largest strain value of orientation available. Using a least squared error method, the bending rigidity term of the Bead-Rod model was determined to be $k = 0.012$. Using these parameters, the orientation model predictions can be compared with the experimentally determined values (Figure 7 and 8).

As one can see from (Figures 7 and 8), the Folgar-Tucker model over predicts the rate of orientation. The Bead-Rod model, on the other hand, approaches the predictions of the Folgar-Tucker more slowly. One important difference is apparent in (Figure 8). The models' values of the 1-2 component of the orientation tensor converge in (Figure 8), but the Bead-Rod model shows no overshoot. This is especially apparent when the orientation models are used in combination with the Libscomb model to predict the transient stress growth (Figure 9).

For the 10% wt. glass fiber systems of interest, the volume fraction of fiber is $\varphi=0.038$, and the neat polyproplene matrix viscosity is $\eta_s=560$ Pa*sec at a shear rate of 1.0 sec^{-1}. The stress model parameters (c_I and N), in combination with the orientation models and their respective model parameters already determined from the transient orientation data, were also determined by using a minimized error approach. Despite the small viscosity overshoot predicted by using the Folgar-Tucker model (in combination with the stress model), it is not able to capture the magnitude of the overshoot. The Bead Rod model, on the other hand, does not show even a slight overshoot prediction and overall produces a rather trivial and useless viscosity profile. Additionally, when the Bead Rod model, and its parameters that were determined from one set of data, are used to predict the orientation evolution and corresponding rheology at a shear rate of 4.0 sec^{-1}, it produces results that are further incorrect. Specifically, the orientation versus strain profile is predicted to be even slower than what was fit at 1.0 sec^{-1}. This is because the delay in orientation is based on time and not strain, thus effectively retarding the orientation evolution even more drastically. This is incorrect because, although not shown in this paper, experimentally we see fibers align more quickly at higher shear rates. Again, this retardation completely eliminates the prediction of a viscosity overshoot and hence provides a particularly poor understanding of the rheology. It is apparent from this treatment that neither of these orientation models can be used with the stress model chosen to accurately relate the transient rheology with the underlying fiber orientation.

Summary and Conclusion

An orientation model that tries to account for flexibility was used with the stress predicting Libscomb model to try to link the rheological behavior of long glass fiber reinforced polypropylene to its fiber orientation microstructure. Additionally, the much more widely acknowledged Folgar-Tucker model was used as a standard for comparison. Orientation model parameters were fit from transient orientation data, and stress model parameters were fit from simple shear flow rheology, both at a shear rate of 1.0 sec^{-1}. It was shown that the orientation and stress models used in this paper are not fully consistent with experimentally determined values. This is apparent in the lack of ability of either model to accurately capture the dynamics of the measured transient rheology. Notably, the Bead Rod model does not show an overshoot in the A_{12} component with increased flexibility, and thus provides poor rheology predictions with the chosen stress model. Additionally, when the Bead Rod model is used in the manner describe by this paper, it predicts slower orientation kinetics at higher shear rates, which is in contradiction with experiment. The fundamental goal of this research is to accurately predict complex flow processes involving long glass fibers, but a better combination of orientation and stress model may still be needed.

Acknowledgments

Financial of NSF: DMI-0853537

Figure Appendix:

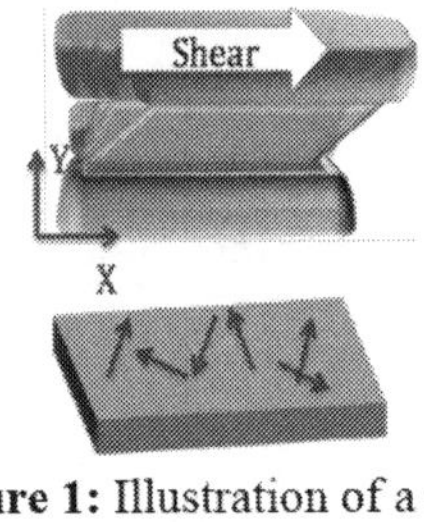

Figure 1: Illustration of a compression molded sample containing planar random initial fiber orientation (below) in a sliding plate apparatus (above).

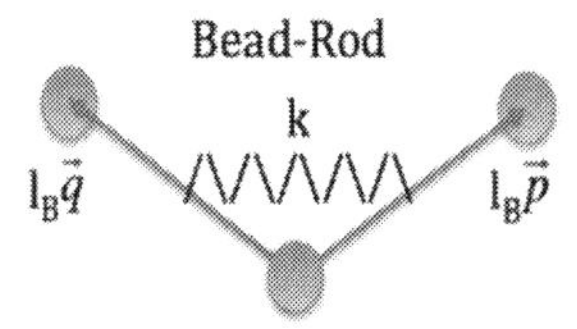

Figure 2: Fiber model, with segment length l_B, allowing semi-flexibility. The segment orientations are denoted by unit vectors $\vec{p}$ and $\vec{q}$, and resist bending by some spring constant k.

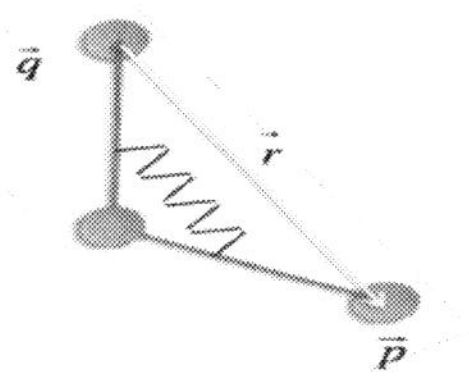

Figure 3: Illustration of the dimensionless end-to-end vector of a bead-rod model fiber.

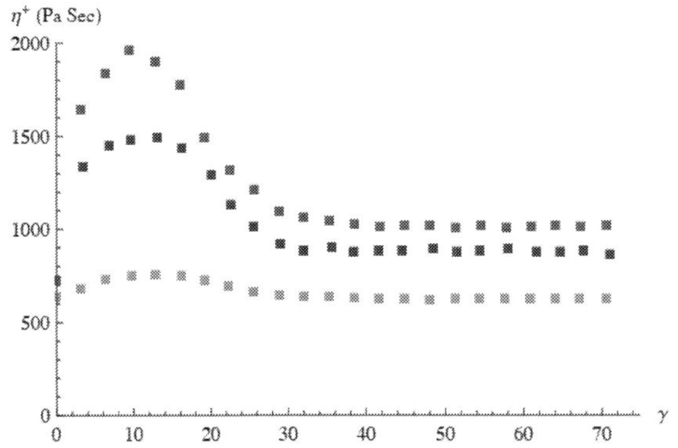

Figure 4: Experimentally measured transient viscosity vs. strain for 10% long glass fibers initially oriented randomly in the shear plane, at various shear rates [■=0.4 sec^{-1}, ■=1.0 sec^{-1}, ■=4.0 sec^{-1}]

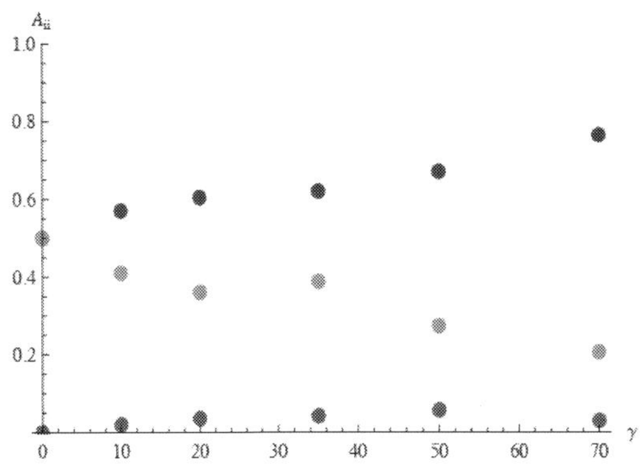

Figure 5: Experimentally determined data for the orientation evolution of the trace components of the orientation tensor versus strain at a shear rate of 1.0 sec^{-1}. System is 10% wt. glass fibers initially random in shear plane. [Data=Points, ●=A_{11}, ●=A_{22}, ●=A_{33}]

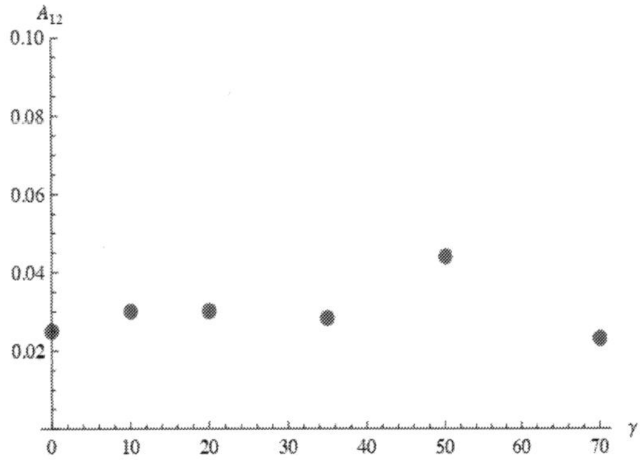

Figure 6: Experimentally determined data for the orientation evolution of the 1-2 component of the orientation tensor versus strain at a shear rate of 1.0 sec^{-1}. System is 10% wt. glass fibers initially random in shear plane. [Data=Points, ●=A_{12}]

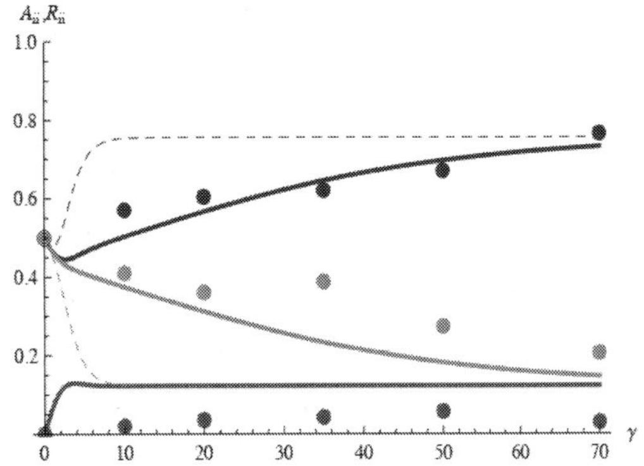

Figure 7: Experimentally determined data and model predictions for the orientation evolution of the trace components of the orientation tensor (A_{ii} for Folgar-Tucker and Data, and R_{ii} for the Bead Rod model) versus strain at a shear rate of 1.0 sec^{-1}. System is 10% wt. glass fibers initially random in shear plane. C_I =0.038, k =0.012. [Data=Points, FT=Dashed, BR=Solid, ●=A_{11},R_{11}, ●=A_{22},R_{22}, ●=A_{33},R_{33}]

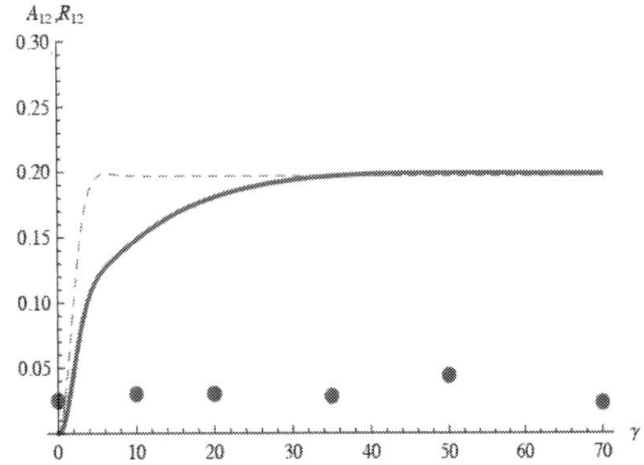

Figure 8: Experimentally determined data and model predictions for the orientation evolution of the 1-2 component of the orientation tensor (A_{12} for Folgar-Tucker and Data, and R_{12} for the Bead Rod model) versus strain at a shear rate of 1.0 sec^{-1}. System is 10% wt. glass fibers initially random in shear plane. C_I =0.038, k =0.012. [Data=Points, FT=Dashed, BR=Solid, ●=A_{12},R_{12}]

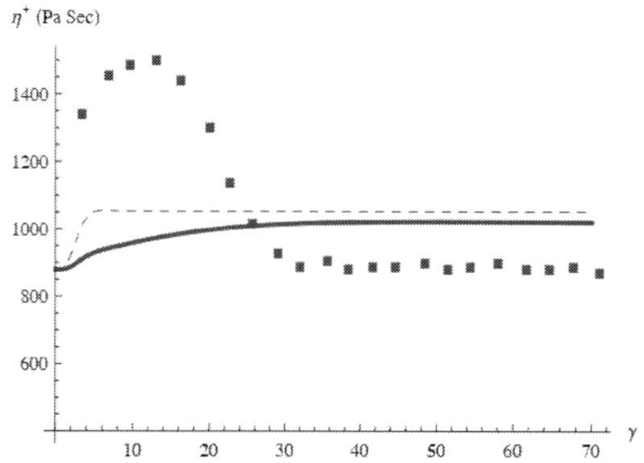

Figure 9: Experimental transient viscosity (η^+) and model fits (using the orientation models in combination with the stress Libscomb model) for fibers initially random in plane and sheared at 1.0 sec^{-1}. [■=1.0 sec^{-1}, Data=Points, FT=Dashed, BR=Solid]. [Model Parameters for FT: C_I=0.038, c1=15, N=105. Model Parameters for BR: C_I=0.038, k=0.012, c_l=15, N=85].

References

1. Eberle, A., et al., *J. Rheol.* **53**, 685-706 (2008).
2. Advani, S.G., and Tucker III, C.L., *J. Rheol.* **31**, 751-84 (1987).
3. Advani S.G. and Tucker III C.L., *J. Rheol.* **34,** 367-386 (1990).
4. Baird D. G. and Collias, D. I., Polymer Processing, Boston: Butterworth-Heinemann, 1998.
5. Folgar, F. P. and Tucker III, C. L., *J. Reinf. Plast. Comp.* **3**, 98-119 (1984).
6. Hine, P.J., Davidson, N., Duckett, R.A., *Poly. Comp.* **17**, 720–729 (1996).
7. Bay R. S. and Tucker III, C. L., *Polym Eng Sci.* **32**, 240-253 (1992).
8. Strautins, U. and Latz,A., *Rheol. Acta.* **46**. 1057-1064 (2007).
9. Smith, W. and Hashemi, J., Foundations of Material Science and Engineeering. *McGraw Hill* (2006)
10. Junke, X. and Costeux, S., *Rheo. Acta.* **46**, 815-824 (2007).
11. Velez-Garcia, G., Ortman K., *ICOR Conf. Proc.* **1027**, 42-44 (2008).
12. Lipscomb, G., et al., *J. Non-Newt. Fluid. Mech.* **26**, 297–325 (1988).
13. Kolitawong, C. and Giacomin, A.J., J. Non-Newt. Fluid. Mech. **102**, 71-96 (2002).
14. Keshtkar, M., et al., J. Rheol., 2009. 53: p. 631-650.
15. Wang, J., O'Gara, J., and Tucker III, C. L., *J. Rheol.* **52**(5), 1179-1200 (2008).

Key Words: Long glass fiber rheology, long glass fiber orientation, sliding plate rheometer, flexible fiber, composite, glass reinforced thermoplastic

PERMEABILITY OF FOAMS WITH TANGENTIALLY-MOBILE INTERFACES

I.J. Galea[1] and M. Loewenberg[2]
Department of Chemical Engineering, Yale University, New Haven, CT 06520-8286-
[1] *ivan.galea@aya.yale.edu*, [2] *michael.loewenberg@yale.edu*

The permeability of foams with tangentially-mobile interfaces is analyzed. Two complementary regimes are considered: (1) foams with surface tension gradients (Marangoni stresses) associated with an insoluble non-diffusing adsorbed surfactant, and (2) foams that have negligible tangential stresses on the gas-liquid interface. It is shown that Marangoni stresses do not enforce no-slip boundary conditions at the gas-liquid interface and that the resulting permeability is much larger than that expected based on this erroneous assumption, provided that the surface viscosity is low enough to retain the tangential mobility of the interface. Boundary integral simulations for the three-dimensional flow at the tetrahedral junction of four Plateau borders are used to determine the permeability of a foam under the complementary conditions where the gas-liquid interface exerts no tangential stresses on the liquid in the Plateau borders.

Introduction

The foam-drainage equation [1, 2] is frequently used to predict the macroscopic dynamics of wet foams, including buoyancy-driven forced drainage and capillary-driven free-drainage. Its derivation follows from a Darcy's-law description for porous media augmented by Plateau's laws which describe the equilibrium microstructure of a foam under the assumption that the microstructure is only slightly perturbed from local equilibrium during the dynamical process of interest. According to Plateau's laws, the liquid resides in a continuous network of channels and nodes, each channel surrounded by three bubbles, and each node surrounded by four bubbles and connected to four channels, as illustrated in Figure 1. The problem simplifies further for foams with low liquid content because of the separation of length scales that occurs in this regime. Despite the foregoing simplifications, macroscopic descriptions of foam dynamics is not well understood largely due largely to uncertainty regarding the permeability of a foam.[3, 4, 5]

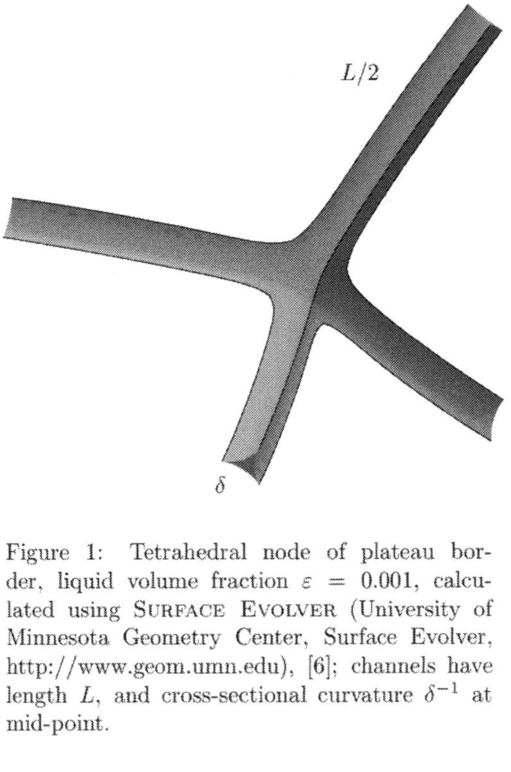

Figure 1: Tetrahedral node of plateau border, liquid volume fraction $\varepsilon = 0.001$, calculated using SURFACE EVOLVER (University of Minnesota Geometry Center, Surface Evolver, http://www.geom.umn.edu), [6]; channels have length L, and cross-sectional curvature δ^{-1} at mid-point.

Leonard and Lemlich (1965) derived the first model for the permeability of a foam, considering unidirectional flow in a straight channel with three concave constant-curvature sides, as depicted in Figure 2. In their model, the resistance to flow at the nodes of the Plateau borders was neglected because it was assumed to be small compared to the resistance to flow in the long slender channels under low liquid-content conditions. The effects of surfactants were incorporated through a surface shear viscosity with no-slip boundary conditions recovered for large values of this parameter. However, Marangoni stresses (surface tension gradients), and thus surfactant conservation, were ignored. As a result, the predicted permeability diverges in the absence of interfacial viscosity.

Recent models of foam drainage [8, 9, 10] still rely on the original framework developed by Leonard and Lemlich (1965), including the ad hoc assumption that the velocity vanishes at the three cusps of the channel cross-section (cf. Figure 2) despite evidence to the contrary [11, 12, 13, 14]

The classical understanding of Marangoni stresses is that they simply enforce no-slip boundary conditions, unless they are dissipated by surface diffusion or by the exchange of surfactant between the interface and the bulk-phase liquid. This picture holds for the axisymmetric case of a channel with a circular cross-section[15] but it is incorrect for the non-circular cross-section of a Plateau border.

Under some conditions, and for some foams, tangential stresses do not develop on the gas-liquid interface. In this case, the nearly straight, constant cross-section channels of the Plateau borders offer negligible hydrodynamic resistance and the controlling resistance is associated with the three-dimensional flow in the nodes of the Plateau borders.[16] Widely differing permeabilities have been reported in this "node-dominated" regime.[17, 18, 19, 20, 21] Numerical computations of the permeability in this regime have been elusive because of numerical difficulties in resolving the three-dimensional flow in the node of a Plateau border with free-surface boundary conditions.[22]

Formulation

Herein, we consider the permeability of a foam with an insoluble non-diffusing surfactant adsorbed on the gas-liquid interface, rigorously incorporating the effect of Marangoni stresses, and the complementary case of a foam with negligible tangential stresses on the gas-liquid interface. The first case is a reasonable limit for certain large surfactant molecules (e.g., proteins); the second case is relevant for certain non-aqueous foams and aqueous foams with small-molecule surfactants that rapidly adsorb and desorb from the gas-liquid interface, thereby dissipating the convectively-generated fluctuations in surfactant concentration that give rise to Marangoni stresses.

Low-liquid-fraction conditions, $\varepsilon \ll 1$, are assumed throughout. Accordingly, the channels of the Plateau border network are long and slender, and distinct from the four-way junctions (nodes) where channels meet. Under these conditions, the liquid volume fraction has the expansion

$$\varepsilon = c_1 (\delta/L)^2 + c_2 (\delta/L)^3 + O(\delta/L)^4 , \qquad (1)$$

where L is the average length of the channels and δ is the curvature of the channel cross-sections as defined in Figures 1 and 2. The first and second terms in expansion (1) represent, respectively, the volume of the channels and nodes of the liquid-filled Plateau borders. The coefficient of the first term, c_1, depends analytically on the underlying dry-foam lattice.

We shall assume that capillary stresses dominate buoyancy stresses,

$$GL \ll \sigma_0/L , \qquad (2)$$

where G is the magnitude of the pressure gradient or body force acting on the liquid phase, σ_0 is the surface tension corresponding to the equilibrium surfactant concentration on the gas-liquid interfaces Γ_0. Accordingly, the equilibrium foam microstructure is not appreciably distorted by the fluid motion and is thus described by Plateau's laws.

Insoluble Non-diffusing Surfactant

Under the assumption that Marangoni stresses balance the pressure gradient in the channels of the Plateau borders, variations of surface tension $\delta\sigma$ scale as

$$\delta\sigma/L \sim GL , \qquad (3)$$

Combining Eqs. (2) and (3), we obtain

$$\delta\sigma/\sigma_0 \ll 1 , \qquad (4)$$

from which it follows that the interfacial surfactant concentration Γ is only slightly perturbed by the flow

$$\delta\Gamma/\Gamma_0 \ll 1 , \qquad (5)$$

provided that the surfactant elasticity

$$E_0 = -\left(\frac{\partial \log \sigma}{\partial \log \Gamma}\right)_{\Gamma=\Gamma_0} , \qquad (6)$$

is not too small. The exceptional cases $E_0 \ll 1$ correspond to dilute surfactant, or phase changes in the adsorbed surfactant layer.

Condition (5) corresponds to the regime of incompressible surfactant.[23, 24] Accordingly, the equation for surfactant conservation on the gas-liquid interface reduces to the interfacial incompressibility constraint

$$\nabla_s \cdot \mathbf{u}_s = 0\,, \tag{7}$$

where $\mathbf{u}_s$ is the fluid velocity on the interface, and ∇_s is the surface divergence operator. Here, we have neglected surface diffusion and the flux of surfactant between the bulk-phase fluid and the interface. Note that the elasticity (6) does not enter the surfactant conservation equation (7) in the same way that an equation of state relating the density and pressure does not enter the mass conservation equation for an incompressible fluid.

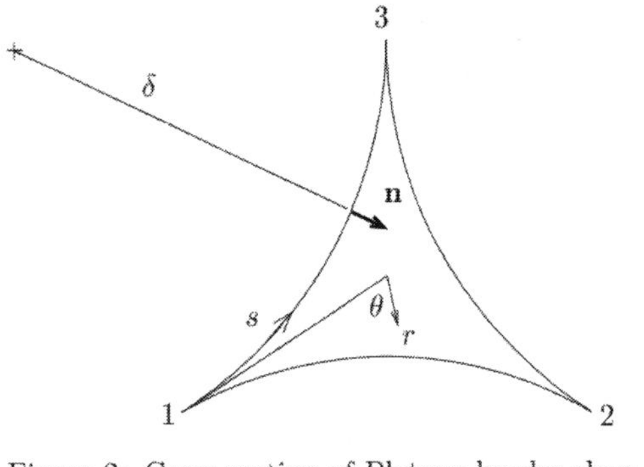

Figure 2: Cross-section of Plateau border channel showing cylindrical coordinate system (r, θ, z) with $r = 0$ at center of channel, z-axis aligned with channel, and cusps $1, 2, 3$ at $\theta = 0, 2\pi/3, 4\pi/3$. Concave sides have inward normal vector $\mathbf{n}$, constant curvature δ^{-1}, and arclength coordinate s measured from cusp as indicated.

To compute the permeability of a foam with an adsorbed surfactant, we consider unidirectional flow in a channel with cross-section and coordinates defined in Figure 2. Accordingly, the fluid velocity has the form $\mathbf{u} = u(r, \theta)\mathbf{e}_z$, where $\mathbf{e}_z$ is a unit vector in the z-direction, and $u(r, \theta)$ is governed by

$$\mu \nabla^2 u = G\,, \tag{8}$$

where μ is the fluid viscosity and the Laplacian contains only derivatives in the r- and θ-directions.

Marangoni stresses are enforced via the boundary condition,

$$\mu \frac{\partial u}{\partial n} = \sigma_z + \mu_s \frac{\partial^2 u}{\partial s^2}\,, \tag{9}$$

where the normal and tangential directions are as indicated in Figure 2. Here, σ_z is the Marangoni stress, and μ_s is the surface shear viscosity. The Marangoni stress is obtained by the force balance

$$C\sigma_z = GA\,, \tag{10}$$

where $C = \delta\pi$ and $A = \left(\sqrt{3} - \pi/2\right)\delta^2$ are the perimeter and cross-sectional area of the channel. Under the assumption that there is negligible momentum exchange between the fluid in the channel and the films (bubble faces) that emanate from the cusps of the channel, we impose the condition

$$\mu_s \frac{\partial u}{\partial s} = 0 \quad \text{at} \quad \theta = 0\,, \frac{2\pi}{3}\,, \frac{4\pi}{3}\,. \tag{11}$$

The velocity is not required to vanish at the cusps.

According to Eqs. (10)-(11), the applied pressure gradient is balanced by Marangoni stresses; no momentum is transported to the bubble faces. This formulation contrasts with the previous models based on Leonard and Lemlich (1965) (discussed above), which ignore Marangoni stresses and assume that the pressure gradient is balanced by forces at the edges of the bubble faces (i.e., at cusps).

For unidirectional flow, the surfactant conservation equation (7) reduces to

$$\oint_C u(s)ds = 0\,, \tag{12}$$

where the integration is performed around the perimeter of the channel. Note that this constraint does not force the velocity to vanish on the interface; it would force the velocity to vanish on the interface only if the channel cross-section were circular.

Eqs.(8)-(12) form a closed set of equations. Given the three-fold symmetry of the channel cross-section, the velocity has the form

$$u = \frac{Gr^2}{4\mu} + \sum_{k=0}^{\infty} a_k r^{3k} \cos k\theta\,, \tag{13}$$

which satisfies Eqs.(8) and (11). The coefficients a_k ($k > 0$) are obtained by truncating the sum at $k = N$ and satisfying boundary condition (9) at a set of equally-spaced collocation points, $0 \leq s \leq \delta\pi/6$; the constant a_0 is determined from the surfactant conservation constraint (12). The results depend on the interfacial mobility parameter,

$$m = \left(\frac{\mu A^{1/2}}{\mu_s}\right)^{1/2}. \tag{14}$$

Near the cusps, we have the local lubrication solution

$$\frac{\mu u}{GA} \sim \frac{1}{2\pi}\log\left(r_k^2 A^{-1} + m^{-2}\right) \tag{15}$$

for $r_k A^{-1/2} \ll 1$, where r_k is the distance from cusp k ($k = 1, 2, 3$). The result indicates that the fluid velocity is logarithmically singular at the cusps for $\mu_s = 0$ ($m \to \infty$). In this case, the fluid velocity is obtained by a modified procedure. For $m \to \infty$, the velocity has the form

$$u = \frac{GA}{\mu\pi}\sum_{k=1}^{3}\log\frac{r_k}{A^{1/2}} + \frac{Gr^2}{4\mu} + \sum_{k=0}^{\infty} a_k\, r^{3k}\cos k\theta\,. \tag{16}$$

This form also satisfies the unidirectional flow equation (8). The coefficients a_k are obtained by satisfying boundary condition (9) (with $\mu_s = 0$) and the surfactant conservation equation (12), as described above.

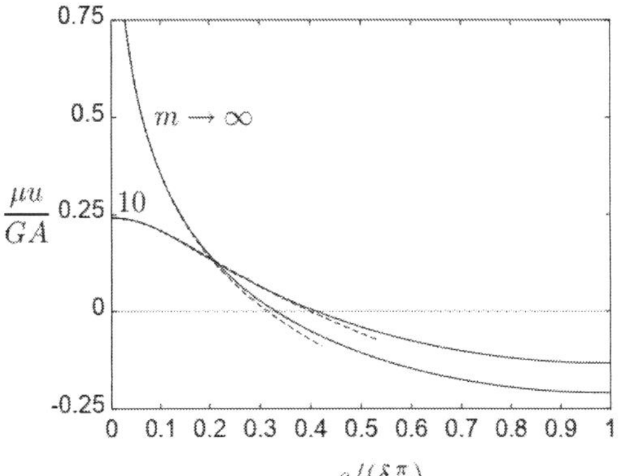

Figure 3: Velocity profile on interface of channel (solid curves); lubrication solution (15) (dashed curves); surface mobility parameter, m, as indicated.

Velocity profiles on the channel interface are depicted in Figure 3 for two cases corresponding to zero and finite surface viscosity (i.e., $m \to \infty$, and $m = 10$). The predicted counter-current interfacial flow has been observed in experiments.[11, 14]

The permeability is obtained by integrating the velocity over the cross-sectional area of the channel,

$$k = \frac{\mu}{GA}\int_A u(r,\theta)\,dA\,. \tag{17}$$

The dependence of the permeability on the interfacial mobility parameter, is shown in Figure 4. For $m \gg 1$, we obtain

$$kA^{-1} = \bar{k}_0 - 12^{3/4}\pi^{-1}m^{-1}\,, \tag{18}$$

where $\bar{k}_0 = 0.3633$ is the result corresponding to $\mu_s = 0$ which is approximately eighteen times larger than that corresponding to no-slip boundary conditions.

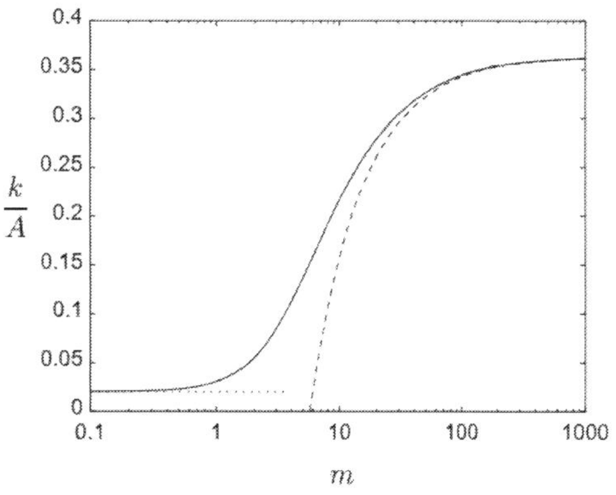

Figure 4: Permeability for a foam with an adsorbed insoluble non-diffusing surfactant as a function of interfacial mobility (solid curve), high mobility formula (18) (dashed curve), no-slip boundary conditions $k/A = 0.02010$ [7] (dotted line).

Stress-Free Interface

Here, we consider the complementary case of a foam with free-surface boundary conditions, where the interface exerts no tangential stress on the fluid. In this case, the flow in the channels is characterized by plug flow and the work done on the fluid by the applied pressure gradient is primarily dissipated by the three-dimensional flow in the nodes of the Plateau border network.

We analyzed this problem using boundary integral simulations for creeping flow through a single tetrahedral node, as depicted in Figure 1. For a tetrahedral node, the liquid volume fraction is related to the geometry by relation (1) with $c_1 = (2 - \pi/\sqrt{3})$. The boundary conditions for our simulations were given by

$$\mathbf{u} \cdot \mathbf{n} = 0\,, \quad \mathbf{f} \cdot (\mathbf{I} - \mathbf{nn}) = 0\,, \qquad (19)$$

on the gas-liquid interfaces, and by

$$\mathbf{u} \cdot (\mathbf{I} - \mathbf{nn}) = 0\,, \quad \mathbf{f} \cdot \mathbf{n} = \mathbf{G} \cdot \mathbf{x} \qquad (20)$$

on the exposed mid-channel cross-sections as shown in Figure 1, where $\mathbf{x}$ is the position vector and the pressure gradient $\mathbf{G}$ is aligned with one of the channels. The SURFACE EVOLVER[6] was used to generate triangulated meshes for a range of liquid volume fractions $.004 \le \varepsilon \le .02$. For a given mesh, our simulations have $O(N^2)$ accuracy, where N is the number of boundary elements; using $N = 3584$, we estimate the relative accuracy of the results presented in Figure 5 at $\pm$ 1%. Here, we define the permeability by Eq. (17) with the integration over the cross-section of the inlet channel. Extrapolating the results as shown in the inset of Figure 5, we obtain

$$kL^{-2} = \bar{k}_0(\delta/L) + \bar{k}_1(\delta/L)^3 + O(\delta/L)^4\,, \quad (21)$$

where $\bar{k}_0 = .08836$ and $\bar{k}_1 = -.1135$.

The only known attempt to compute the permeability of a foam with free-surface boundary conditions is reported by Cox et. al. (2001) who obtained the estimate $\bar{k}_0 \approx .025$ using commercial numerical software (Fluent). However it is difficult to compare their result with ours because they did not report the value(s) of δ/L on which their result is based, and they considered a high-Reynolds-number flow in the Plateau borders[1] whereas we consider Stokes flow.

[1]Cox et. al. (2001) used a Reynolds number, $\mathrm{Re} = u_i\delta/\nu$, given by $\mathrm{Re} = 200$, where u_i is the velocity entering the Plateau border and ν is the kinematic viscosity of the liquid.

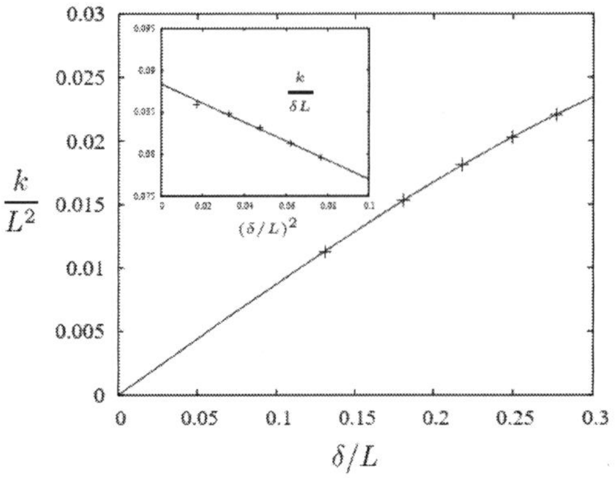

Figure 5: Permeability for a foam with free-surface boundary conditions as a function of liquid fraction, δ/L; boundary integral simulations (+), extrapolation (21) (solid curves).

References

[1] G Verbist, D Weaire, and AM Kraynik. The foam drainage equation. *J. Phys. Condensed Matter*, 8:3715–3731, 1996.

[2] P. M. Kruglyakov, S. I. Karakashev, A. V. Nguyen, and N. G. Vilkova. Foam drainage. *Curr. Opinion Colloid Interface Sci.*, 13:163–170, 2008.

[3] HA Stone, SA Koehler, S Hilgenfeldt, and M Durand. Perspectives on foam drainage and the influence of interfacial rheology. *J. Phys. Condensed Matter*, 15:S283–S290, 2003.

[4] Qicheng Sun, Lianghui Tan, and Guangqian Wang. Liquid foam drainage: An overview. *Int. J. Modern Phys. B*, 22:2333–2354, 2008.

[5] E. Lorenceau, N. Louvet, F. Rouyer, and O. Pitois. Permeability of aqueous foams. *Eur. Phys. J. E*, 28:293–304, 2009.

[6] KA Brakke. The surface evolver. *Exp. Math.*, 1:141–165, 1992.

[7] RA Leonard and R Lemlich. A study of interstitial liquid flow in foam .I. theoretical model and application to foam fractionation. *AICHE J.*, 11:18–25, 1965.

[8] AV Nguyen. Liquid drainage in single Plateau borders of foam. *J. Colloid Interface Sci.*, 249:194–199, 2002.

[9] SA Koehler, S Hilgenfeldt, and HA Stone. Foam drainage on the microscale - I. Model-

ing flow through single Plateau borders. *J. Colloid Interface Sci.*, 276:420–438, 2004.

[10] O Pitois, C Fritz, and M Vignes-Adler. Hydrodynamic resistance of a single foam channel. *Colloids Surfaces A*, 261:109–114, 2005.

[11] SA Koehler, S Hilgenfeldt, ER Weeks, and HA Stone. Drainage of single Plateau borders: Direct observation of rigid and mobile interfaces. *Phys. Rev. E*, 66, 2002.

[12] A Saint-Jalmes. Physical chemistry in foam drainage and coarsening. *Soft Matter*, 2:836–849, 2006.

[13] O Pitois, C Fritz, and M Vignes-Adler. Liquid drainage through aqueous foam: study of the flow on the bubble scale. *J. Colloid Interface Sci.*, 282:458–465, 2005.

[14] F. Rouyer, E. Lorenceau, and O. Pitois. Film junction effect on foam drainage. *Colloids Surfaces A*, 324:234–236, 2008.

[15] M Durand and D Langevin. Physiochemical approach to the theory of foam drainage. *Eur. Phys. J. E*, 7:35–44, 2002.

[16] SA Koehler, S Hilgenfeldt, and HA Stone. Liquid flow through aqueous foams: The node-dominated foam drainage equation. *Phys. Rev. Lett.*, 82:4232–4235, 1999.

[17] SA Koehler, S Hilgenfeldt, and HA Stone. A generalized view of foam drainage: Experiment and theory. *Langmuir*, 16:6327–6341, 2000.

[18] V Carrier, S Destouesse, and A Colin. Foam drainage: A film contribution? *Phys. Rev. E*, 65:0614041–0614049, 2002.

[19] SJ Neethling, HT Lee, and JJ Cilliers. A foam drainage equation generalized for all liquid contents. *J. Phys. Condensed Matter*, 14:331–342, 2002.

[20] A Saint-Jalmes, Y Zhang, and D Langevin. Quantitative description of foam drainage: Transitions with surface mobility. *Eur. Phys. J. E*, 15:53–60, 2004.

[21] O. Pitois, N. Louvet, E. Lorenceau, and F. Rouyer. Node contribution to the permeability of liquid foams. *J. Colloid Interface Sci.*, 322:675–677, 2008.

[22] SJ Cox, G Bradley, S Hutzler, and D Weaire. Vertex corrections in the theory of foam drainage. *J. Phys. Condensed Matter*, 13:4863–4869, 2001.

[23] J. Bławzdziewicz, E. Wajnryb, and M. Loewenberg. Hydrodynamic interactions and collision efficiencies of spherical drops covered with an incompressible surfactant film. *J. Fluid Mech.*, 395:29–59, 1999.

[24] J. Bławzdziewicz, V. Cristini, and M. Loewenberg. Stokes flow in the presence of a planar interface covered with incompressible surfactant. *Phys. Fluids*, 11:251–258, 1999.

A hyperelastic damage model for continuous fiber reinforced thermoplastic composites: formulation and numerical simulations

Hicham Chaouki, Augustin Gakwaya, Fouad Erchiqui, Mario Fafard*

Department of Civil Engineering - Laval, Quebec - Canada

In this work an orthotropic hyperelastic damage model is presented for modeling and simulation of continuous fiber reinforced thermoplastic composites behavior. For this purpose, the free energy function is additively decomposed into three components: 1) a component taking into account the matrix response; 2) a component taking account of the fibers behaviors and 3) a component describing the matrix-fibers interactions. A continuum damage formulation is used by associating to each free energy function component an isotropic internal variable describing the irreversible aspect of the damage evolution of the material constituent considered. Thus, the response of a reinforced thermoplastic composite is obtained through the response of each constituent with its damage evolution for the matrix and the fibers separately, and for the matrix-fibers interactions. Parameters laws were obtained by inverse identification from experimental results of uniaxial tractions and cyclic loading performed at 0 and 45 degrees of Twintex fiber directions, a woven thermoplastic composite, and at a temperature higher than the matrix fusion temperature. A user material subroutine VUMAT for an explicit dynamic analysis has been developed and implemented in Abaqus software. In order to evaluate the constitutive law predictive capability, finite element simulations of thermostamping process of a woven thermoplastic composite were performed with the four nodes S4R thick shell element. Results show that the model provides interesting physical responses compared to an orthotropic hyperelastic model without damage. This constitutes a first step before making a benchmark with experimental results that will be performed.

A CALIBRATED INTEGRATIVE SIMULATION MODEL FOR BLOWN FILM EXTRUSION

M. Bussmann[1], M. Janas[1], J. Wortberg[1]*

[1]*Institute of Product Engineering, engineering design and plastic machinery, University Duisburg-Essen, Duisburg, Germany – markus.bussmann@uni-due.de*

Increasing the capacity of a blown film line is the major task to gain a more efficient production plant. Still the film cooling system is the device which sets the upper limit for the whole plant. Developments in the recent past have shown that thanks to computational blown film-calculation models it is theoretically possible to optimize or even redesign cooling systems.
This essay deals with the development of such an integrative computational model, to predict the bubble shape in the tube formation zone, taking into account the interaction of the cooling air flow and the molten polymer. It is based on an iteration process which contains the linkage between the bubble deformation and its interaction with the air flow. Thus the contour calculation algorithm based on the rheological model of Phan-Thien and Tanner has been linked with a CFD-Program in an iteration circle.
To gain a tool which is suitable for industrial usage, this contour calculation model is not founded on previous rheological experiments. To avoid this, a calibration of the contour calculation procedure is invented to fit the necessary polymer data by means of an experimentally determined set of bubble contours. Furthermore a simple approach is introduced taking into account the rheological elongation viscosity depending on the stretching directions.
In comparison to experimental determined bubble shapes it could be outlined, that using this kind of calibration varying process states can be represented quite well.

Introduction

One of the most important processes to produce high quality plastic films is the blown film extrusion. The cooling system is the device, which sets the upper limit for the whole production line. Recent developments have shown, that it is possible to increase the capability of the cooling systems using air guiding systems. This system forces the cooling air volume to remain longer at the bubble surface and use it more efficiently. With altered cooling conditions and thus changing pressure profiles on the bubble surface, the free deforming polymer tube reacts in an unknown way. To make the design and optimization process of such cooling systems more efficient, it is necessary to predict the bubble's reactions. Thus it is obvious to simulate the whole stretching procedure in interaction with the cooling system.
Since the 70ties models were developed, which should have the ability to represent the bubble formation zone. Due to the increasing complexity limitations in contour description these model have not made their way into any industrial usage. A simulation to optimize or design a completely new cooling system makes only sense, when it is capable to generate results with a reasonable amount of experimental afford, as *Vlachopoulos 2008* motioned.

Aims

The aim of this work is to develop such a simulation tool. Therefore, an existing rheological based contour calculation model is used to design an integrative process model. The difficulty is that the complete tube formation zone is not geometrically fixed and completely freely deformable. To gain a realistic film, reaction parameters have to be determined which allow the model to represent an equilibrium process condition. Furthermore the simulation model should be designed in such a way that it can be calibrated to varying process conditions, polymers and different plants. To achieve this, reasonable calibration parameters have to be established. measured

State of the art contour calculation

Researches in the recent past led to very complex and extensive modelling of the polymer behaviour within the tube formation zone. These detailed models led to the question of applicability of such tools in industrial usage.
Most publications deal with the rheological characteristics and the phase change of the polymer to get hold of the bubble shape. But it is well-known that the cooling conditions and thus the rheological behaviour are strictly related to all other influences on the film. Especially the pressure profile has a major effect.
So far the approach which has been made was to serve the process model with as much measured parameters as possible, to gain a most realistic bubble image.
An exact rheological measurement of the polymer behaviour in generalized elongation is only limited.

Polymer behaviour in elongation

The film stretching process is mainly affected by the biaxial elongation process. Referring to the trouton rule, which describes the connection between the material parameters for incompressible media in shear and elongation, the elongation viscosity corresponds to three times the shear viscosity. (see equation 1)

$$\eta_\varepsilon = \begin{pmatrix} 2\eta_0\dot{\varepsilon} & 0 & 0 \\ 0 & \eta_0\dot{\varepsilon} & 0 \\ 0 & 0 & -\eta_0\dot{\varepsilon} \end{pmatrix} = 3\cdot\eta_0 \qquad \text{Eq. 1}$$

This relation is only valid for low elongation rates, above the elongation viscosity increases with the elongation rate. After reaching its maximum as shown in figure 1 the viscosity decreases. In addition to this behaviour the zero shear viscosity in equivalent biaxial elongation - an equal stretching in all direction - increases. A six times higher deformation resistance according to equation 2 is assumed.

$$\eta_\varepsilon = \begin{pmatrix} 2\eta_0\dot{\varepsilon} & 0 & 0 \\ 0 & 2\eta_0\dot{\varepsilon} & 0 \\ 0 & 0 & -4\eta_0\dot{\varepsilon} \end{pmatrix} = 6\cdot\eta_0 \qquad \text{Eq. 2}$$

Furthermore, the rheological elongation characteristics are highly dependent on the molecular structure of the polymer, its side branches and molecular weight. (*Laun 2005*)
Depending on the inner structure, phenomenons different elongation viscosities are detectable. As for example *Münstedt et. al. 2005* have presented characteristics of increasing but also decreasing viscosity at higher elongation rates can be appear.

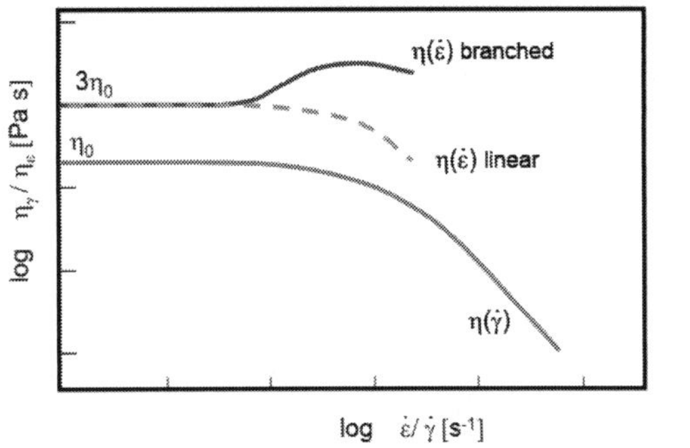

Figure 1 – Relation between shear and elongation viscosity depending on the molecular structure

Such properties are crucial for the film stretching process given the fact that the elongation viscosity and thus the melt rigidity are extremely important for the appearance of the bubble. A suitable numerical model for this material behaviour can only be set up when it is possible to approximate the material behaviour in a suitable way. This means for the film blowing process that the melt characteristic in generalized biaxial elongation has to be presented correctly. Up to now there is no existing model which is capable to represent those properties entirely.

Due to these reasons an extension of the rheological model has been made to take into account the rheological characteristics in each elongation direction. This is essential to describe the anisotropic behaviour of the polymer caused by the molecular orientation.

An integrative process model
The aim of this work was to set up a model which is able to display and predict various film contours. This can only be done by forcing the model to effectuate an equilibrium state. To gain this, all process parameters have to been considered as connected. This done because it could be proved that changing only a single parameter is not reasonable, respectively meaningful. The alteration of one parameter abrogates virtually the equilibrium state of the whole tube formation process, as shown schematically in figure 2.

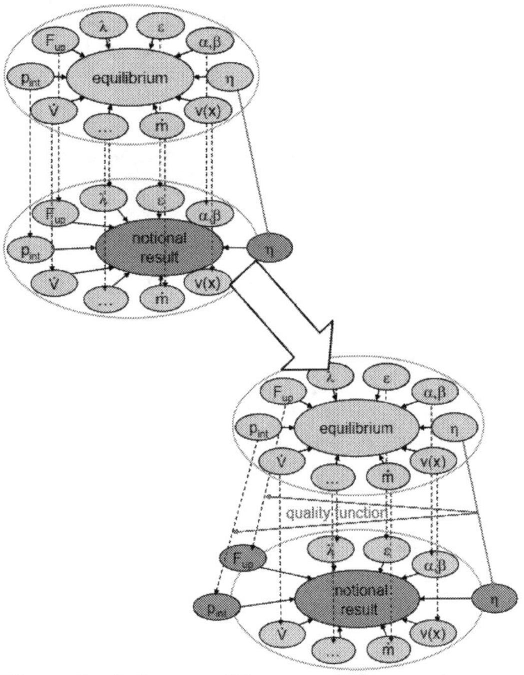

Figure 2 – Influence of the connected parameters representing the equilibrium state

By changing only one parameter detached form all the others, one gains a purely numerical solution for the film contour. The gained result represents an unbalanced state. This provisional result can only be corrected by changing the other connected values. Due to this parameters have to be determined which are allowed to be altered to regain the equilibrium state.
Based on these facts an integrative process model has been developed. The integrative model is based on two primary calculation modules as shown in figure 3.

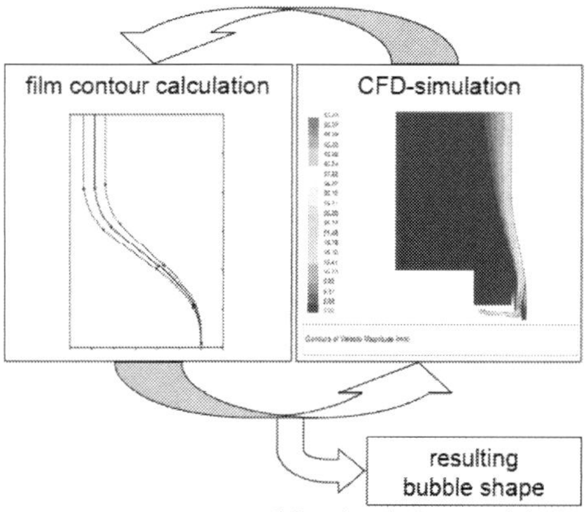

Figure 3 – Structure of the simulation model

Fundamental for a representation of the transient tube formation zone is a true-to-life description of the film contour by a numerical model. In this model all influences; mechanical and rheological are integrated.
It is a fact that the polymer stretching process is mainly temperature dependent. To assure a best possible description of the thermodynamically and flow induced procedures, the calculation of the heat transfer is accomplished by a CFD-analysis. As used in previous publication the CFD-program Fluent is chosen to be the second module,(*i.e. Zahng Lafleur 2006*). The two modules interact with each other in an iteration loop. This loop is performed until a resulting end contour is determined. To calculate the quasi-stationary state for each iteration step an inner iteration loop is necessary. Aim of this loop is to determine the best possible contour by the given boundary condition. During this loop the predefined process parameters are iterated to regain their balance. By means of a quality-function the parameters are altered. Thus it is assured, that in each iteration step the optimum for the film contour representation is found.

Model for the film contour calculation
It is necessary to use a film calculation model which determines reliably and true-to-life the film contour. Based on the equations established by *Pearson and Petrie 1970* (PP) a numerical model for the film contour calculation has been set up. Preexaminations of different types of rheological models have shown, that the material description of the *Phan-Thien and Tanner 1977* (PTT) model achieved best results. Due to this process model, film contour could be reliable determined.

To describe the film contour depending on the stresses caused by the elongation of the polymer the PP-model is used. This model is based on the membrane theory for shells. The bubble is considered to by axis symmetric and the whole process steady state. The thickness of the film is regarded to be small relative to the radius of the bubble. (see figure 4)

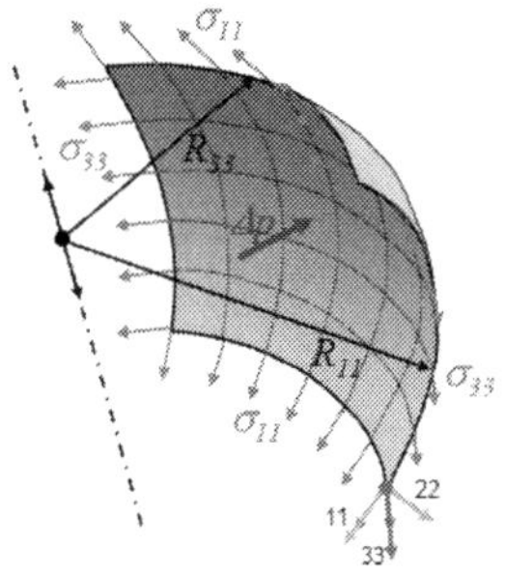

Figure 4 – Membrane element with its two radii (R_{11} R_{33})

Each film is under influence of the take up force and the pressure difference between the inside and the outside of the bubble.
Assuming this, an equation for the force balance for each element can be written as followed.

$$\frac{\Delta p}{h} = \frac{\sigma_{11}}{R_{11}} + \frac{\sigma_{33}}{R_{33}} \qquad \text{Eq. 3}$$

Thanks to the CFD-analysis, the pressure on the bubble outside can be considered as depending on the extrusion height.

$$\Delta p = p_{\text{int}} - \left(p - p_{atm}\right) \qquad \text{Eq. 4}$$

The two radii for the meridian (R_{33}) and extrusion (R_{11}) direction can be described using the following relationships.

$$R_{11} = \frac{-1}{\dfrac{d^2 r}{dx^2}\cos^3(\theta)} \qquad \text{Eq. 5}$$

$$R_{33} = \frac{r}{\cos(\theta)} \qquad \text{Eq. 6}$$

$$\cos(\theta) = \frac{1}{\sqrt{1+\left(\dfrac{dr}{dx}\right)^2}} \qquad \text{Eq. 7}$$

Converting these equations with respect to r", a second order nonlinear differential equation is obtained.

$$r'' = \frac{\left(r'^2+1\right)\cdot\left(h\,\sigma_{33} - \Delta p\, r\sqrt{r'^2+1}\right)}{h\, r\, \sigma_{11}} \qquad \text{Eq. 8}$$

Taking into account the mass conservation throughout the whole process the film thickness can be calculated.

$$\frac{dh}{dx} = -\left(\frac{\dot{v}}{v} + \frac{\dot{r}}{r} + \frac{\dot{\rho}}{\rho}\right)\cdot h \qquad \text{Eq. 9}$$

Rheological description

The rheological model introduced by Phan-Thien and Tanner is based on an enhancement of the Maxwell-model. It allows describing the nonlinear behaviour of the melt in elongation and in shear. Therefore the model is enhanced by two values. The dimensionless parameter ε influences the nonlinear behaviour in elongation, when the as well dimensionless parameter ξ on the other hand determines the nonlinear behaviour in shear flow. Both parameters are fixed values for the calculation and are only defined by the material properties of the polymer. Suggesting that the Polymer is not sheared in the region of bubble formation ξ is set to zero.

Further material properties are used, such as the viscosity depending on the temperature or the relaxation time. Depending on measured properties and the heat transfer condition established by the CFD-simulation, stresses can be calculated.

$$\frac{d\tau_{11}}{dx} = \frac{2\eta_{11}\frac{dv}{dx}}{\lambda v} - \frac{\tau_{11} e^{\left(\frac{\varepsilon\lambda}{\eta_{11}}(\tau_{11}+\tau_{22}+\tau_{33})\right)}}{\lambda v \cos\theta} + \frac{2(1-\xi)\tau_{11}\frac{dv}{dx}}{v}$$

Eq. 10

$$\frac{d\tau_{22}}{dx} = 2\eta_{22}\frac{1}{h\lambda}\frac{dh}{dx} - \frac{\tau_{22} e^{\left(\frac{\varepsilon\lambda}{\eta_{22}}(\tau_{11}+\tau_{22}+\tau_{33})\right)}}{\lambda v \cos} + 2(1-\xi)\tau_{22}\frac{1}{h}\frac{dh}{dx}$$

Eq. 11

$$\frac{d\tau_{33}}{dx} = 2\eta_{33}\frac{1}{r\lambda}\frac{dr}{dx} - \frac{\tau_{33} e^{\left(\frac{\varepsilon\lambda}{\eta_{33}}(\tau_{11}+\tau_{22}+\tau_{33})\right)}}{\lambda v \cos} + 2(1-\xi)\tau_{33}\frac{1}{r}\frac{dr}{dx}$$

Eq. 12

As mentioned before the rheological model by PTT is enhanced by a simple approach to take into account the anisotropic material properties. The elongation process of the polymer is highly dependent on the elongation viscosity characteristics. These are fairly good represented by the PTT-model. During the biaxial elongation process, viscosities will occur which are elongation rate and strain direction dependent. Due to the structure of the linear chains, anisotropic viscosities arise which in addition are dependent on each other. Considering a generalized elongation process is affected by this macromolecular orientation within the polymer.

Figure 5 displays the influence of film blowing process on the inner polymer structure. The final product has the same film thickness only the BUR is changed.

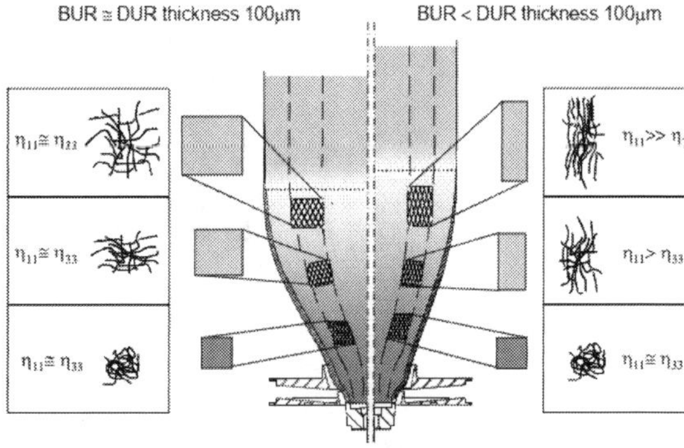

Figure 5 – Elongation-affected molecular structure with its dependant viscosities. Schematically shown at two states with similar film thickness

This figure shows why it is necessary to use different viscosities in the stretching direction. Due to entanglements, the resistances with respect to the deformation are dependent on each other. The assumption of a generalized three times or even six times higher elongation viscosity can not cope with this case. It can be assumed that the viscosities in circumferential and extrusion direction differ form each other.

According to this, a direction depending modification of the zero viscosity (η_{22}, η_{33}) has been introduced. This is done by a simple factorisation and thus a transformation of the viscosities on the basis of the zero viscosity in extrusion direction.

$$\eta_{22} = \alpha \cdot \eta_{11}$$

Eq. 13

$$\eta_{33} = \beta \cdot \eta_{11}$$

Eq. 14

Han and Park 1975 motioned that kind of polymer properties alteration on basis of the elongation conditions.

Initial conditions

To solve the second order differential equation of the contour initial condition at the die exit are required. The calculation is performed beginning at the die exit up to the freeze line height. Therefore the initial values for the start conditions (x=0) are used.

Required are the values for the start radius $r(0)$ and the derivative of the radius $r'(0)$. The origin of the contour is similar to the radius at the die exit.

Beside those initial values for contour calculation the PTT-model requires initial values for the stress calculation. In this particular case the initial stresses at the die exit have to be known. According to *Muslet et. al. 2004* the stresses can be calculated by means of the force balance at the die exit using the following equation.

$$\sigma_{11,0} = \frac{F_{up} - \pi \Delta\bar{p}\left(r_{frost}^2 - r_0^2\right)}{2\pi r_0 h_0 \cos(\theta)}$$

Eq. 15

$$\sigma_{33,0} = \frac{r_0}{h_0 \cos(\theta)} \Delta p_{start} \qquad \text{Eq. 16}$$

Definition of calibration and iteration parameters

It is assumed that the whole process model can be calibrated with regards to the polymer. Therefore material parameters are valid for the calibration. These are especially η, λ, ε and ξ for the fundamental characterization of the polymer. With regards to the film stretching process the sheering within this zone is limited. Therefore the constant in the PTT-model to describe the nonlinear polymers behaviour in shear is set zero ($\xi = 0$), according to *Kahn and Larson 1987*.

Due to the fact, that the tube formation zone is freely deformable, the bubble will always tend to gain an equilibrium. Thus the stresses in the stretching process calculated by the rheological model have to be at the equilibrium with all other mechanical process parameters. In this process model it is requested, that the equilibrium has to exist in every iteration step. In an inner iteration loop parameters are optimized by means of a quality function to gain these conditions. Based on the interacting influences on the bubble shape, parameters have been predefined which are valid for the optimization.

As aforementioned each process is defined by varying elongation conditions with respect to the elongation direction. Due to this, two further constants (α, β) have been implemented in the model to alter the direction depending viscosity. Furthermore the initial conditions at the die exit are unknown and thus as well set as iteration parameters. Those can be calculated using the process values F^*_{up} and Δp_{start}.

The figure shows the inner iteration loop to calculate the film contour. It contains six coupled equations, which have to be solved simultaneously.

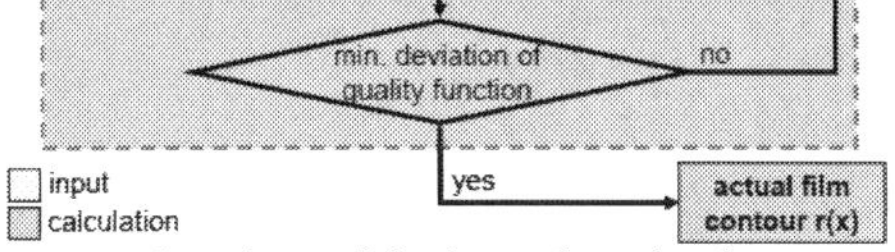

Figure 6 – Flowchart of the inner iteration loop

The criterion to stop the iteration loop is defined by the deviation of the quality function. The quality function displays the rate how exact the calculated contour fits the predefined targets. In the generalized case to calculate an unknown bubble shape the targets are the blow up ratio (BUR), bubble shape, the film thickness at the frost line as well as the frost line height itself. Based on these values the quality function calculates a scalar quality value. Goal of the iteration loop is to minimize this value. The result is a film contour with its set of iteration parameters. For the whole model setup the software Matlab has been used.

CFD-simulation

Besides the contour calculation, the correct representation of the cooling airflow is essential to describe the film blowing process. Thus the CFD-simulation program fluent has been used to calculate the heat transfer conditions.

According to the model *Wolf 1995* established and *Spirgatis 2004* extended, a simulation domain has been set up. The CFD-domain is automatically built by so a called journal-file using the actual film contour and the film thickness. The film contour sets a boarder in the CFD-domain.

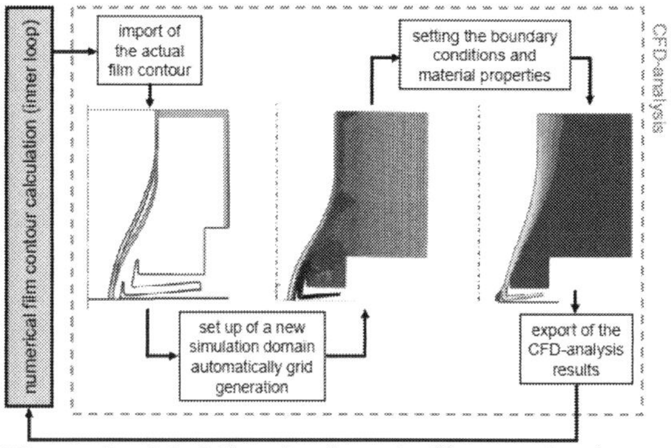

Figure 7 – Modeling the dynamic film reaction in a CFD-simulation

In each iteration the film contour may change dramatically caused by the boundary conditions. Therefore it is necessary to create new grid with each step. Figure 7 shows the whole iteration loop in combination with the CFD-simulation

As already mentioned by *Sidiropoulos and Vlachopoulos 2005* it is crucial for the contour calculation to know the exact temperature profile. Because of highly different cooling conditions on the outside and on the inside of the bubble a temperature gradient exists in the film thickness. Due to the low heat transfer coefficient of polymers this gradient can not be neglected. Within the CFD-simulation the material parameters have been taken into account to calculate a representative temperature profile. Analysis have shown that a minimum amount of six cells have to be used in film thickness to gain a temperature profile which is accurate enough. These temperature profiles were compared to data determined by a thermographic imaging system.

Experimental

To achieve a verification and calibration of the before presented model, it is crucial to compare the results with the real process. A detailed analysis of the cooling

and stretching process sets the basis to determine the necessary data.
To give evidence of the capability of this model, two process states have been detected experimentally.

Table 1 – Data of the process states

BUR	$\dot{m}$ [kg/h]	h_{frost} [µm]	v_{frost} [m/min]	$\dot{V}$ [m³/h]	p_{int} [Pa]	x_{frost} [m]
2	35	100	10	600,92	98	0,532
2	35	100	10	803,05	98	0,415

Both states were chosen in a way to include all important influences on the film formation. The main aim was to display the influence of the air guiding system, which is capable to causes a Venturi-effect leading to an increased cooling capacity. This is gained by simply altering the cooling air flow volume. All other conditions have been kept constant. Figure 8 shows both states. in addition the air flow guiding system display schematically. One represents an airflow volume (AV) of 600 m³/h and the other of 800 m³/h.

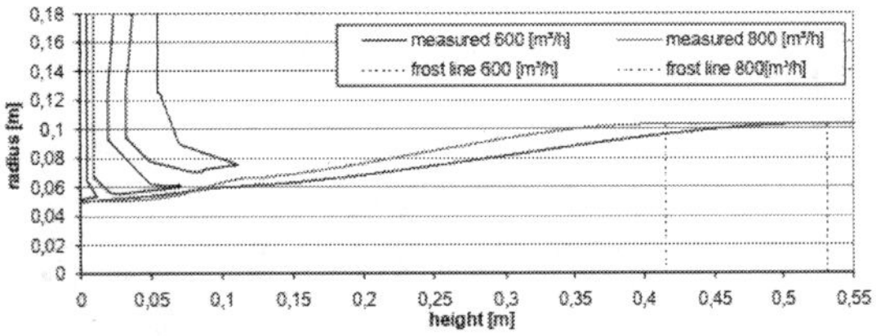

Figure 8 – Process states for the verification

It is evident, that a higher cooling airflow leads to a so called Venturi-effect at the upper airing lip. In this region a suction pressure occurs and the film is bulging. Due to this the space between the lip and the film narrows, which leads to a higher air velocity. In combination with this acceleration of the airflow the heat transfer is increased, as it can be seen in Figure 9.

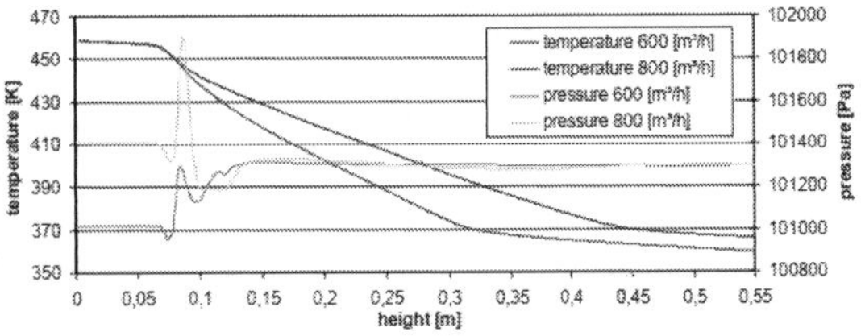

Figure 9 – CFD-results of the temperature- and pressure profile

The measured values were the film contour, the freeze line height, the air velocity, the film velocity, the air temperature and the pressure on the inside of the bubble. An accurate measurement of the film contour is extremely important because the film contour represents the sum of all parameters as a final result.

The simulation model for the verification

This essay deals with the verification of the process model. In further publications an extended calibration basing on a calibrations space and furthermore a calculation of unknown process condition will be presented.

Dealing with the verification the simulation model reduces to the inner iteration loop. It is not necessary to include the CFD-simulation in the loop, because it is only accomplished once.

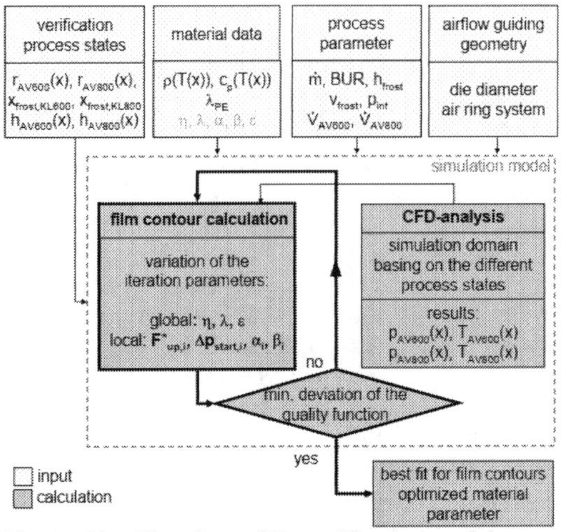

Figure 10 – Flowchart of the verification model

By means of the experimentally determined contours two CFD-analyses are accomplished to gain the film temperature and the pressure profile as shown in figure 9. Within the Matlab-model the material parameters (η, α, β, ε), as well as the initial conditions ($\sigma_{11,0}$, $\sigma_{33,0}$) are iterated simultaneously.
As global iteration parameter, i. e. for both states in union, the viscosity, the relaxation-time as well as the constants for the PTT-model were chosen. Due to the different elongation conditions in machine and circumferential direction it is expected that the elongation viscosities differ. Thus a difference in α and β is allowed.
The cooling airflow volume varies in both states and hence the pressure profile on the bubble surface. Therefore the initial condition which follow F^*_{up} und Δp_{start} can not be regarded as global. They are set as local parameters which are individual for each process.

Table 2 – Iteration parameters

globale iteration parameter:	η, λ, ε
lokale iteration parameter:	α, β, F^*_{up}, Δp_{start}

The quality function for this verification has been designed to gain the best suited representation of both process conditions. The main criterion is the absolute deviation of each reference contour and in addition the deviation of their derivatives. To achieve better convergence of the optimization the film thickness and the BUR were implemented into the quality function.

Results and Discussion

Figure 11 shows the capability of simulation model to represent the dynamic tube formation zone. A very good description of the film contour could be achieved. It is possible to depict nearly the whole contour, except for the lower region in the airing about 0,1 m. Aside form this, this region displays the extrapolation of the contour, which is done by using a few known points. The experimentally detected film contour above the air guiding system as well as the BUR and the film thickness are represented entirely.

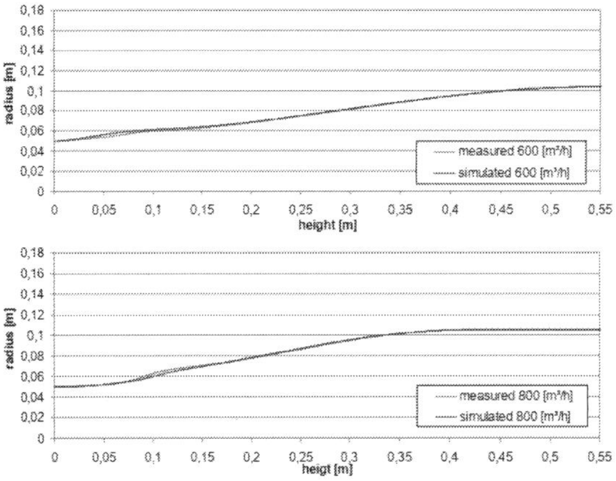

Figure 11 – Calculated film contours in comparison to the measured contours.

Material parameters determined by the model are in a realistic range and are comparable to real polymer data.

Table 3 – Determined material data

process state	η_{11} [Pa s]	λ [s]	ε [-]	F^*_{up} [-]	p_{start} [Pa]	α [-]	β [-]
BUR2-h100-KL600	42648,17	0,896	0,038	0,605	952,013	0,274	0,649
BUR2-h100-KL800				0,84	26,358	0,223	0,627

Table 3 shows the determined material data. A value smaller than 1 is determined for β. Hence the viscosity in circumferential direction (η_{33}) is predicted thinner than the one in the machine direction (η_{11}). Whereas α for the viscosity in film thickness is predicted as ~0,25. Due to this value the stresses τ_{22} in the PTT-model are regarded low.

Conclusions

It could be outlined, that a combination of CFD-simulation with a rheological based film calculation model has been successfully linked into an appropriated process model. By using the CFD-simulation the real life pressure profile and its influence on the bubble appearance could be included into this the model. Only this allowed to consider phenomenons such as the Venturi-effect. Hence the influence of an airflow guiding system on the whole tube formation process could be incorporated. In addition this enables to consider the increased heat transfer due to the Venutri-effect, which is highly responsible for the film contour.

Thanks to the extended rheological model the quality of the film representation could be increased. Two constants were introduced to improve the description of the polymer behaviour in elongation. These constants are allowed to be iterated due to different elongation rates in varying process states. Thus the anisotropic material behaviour with its different molecular orientations respectively viscosities can be incorporated.

Furthermore the unknown stresses induced by the flow conditions in the die were declared as process state dependent parameters. The usage of the unknown material parameters enabled to calibrate the model for these two states. As shown by the calculated film contours, this calibrated integrative model is capable of representing the film blowing process quite well.

In forthcoming publications the calibration of the process model in an extended calibration space will be presented. This space includes varying BURs and film thicknesses. Basing on this calibrated model it is possible to predict unknown film contours. Starting form a fully known state this model is capable to calculate user-defined states iteratively.

Acknowledgements

This project was supported by the German National Science Foundation (Deutsche Forschungsgemeinschaft, DFG), project-nr. WO 302/28-3.

References

Journal:

Han, C. D., Park, J.Y., "Studies on Blown Film Extrusion. I. Experimental Determination of Elongational Viscosity", Journal of Applied Polymer Science Vol. 19, page 3257-3276 (1975)

Kahn, S. A., Larson, R. G., "Comparison of Simple Constitutive Equations for Polymer Melts in Shear and Biaxial and Uniaxial Extensions" Journal of Rheology Vol. 31(3), page 207-234 (1987)

Münstedt, H., et. al., "Correlation between rheological behaviour in uniaxial elongation and film blowing properties of various poly-ethylenes", Rehologica Acta, page 14-22 (2005)

Muslet ,et. al., "Computational simulation of the film blowing process incorporating crystalliziation and viscoelasticity", Journal of Rheology Vol. 48(3), page 525-550 (2004)

Pearson, J.R.A., Petrie, C.J.S., "The flow of a tubular film I: Formal mathematical rep-resentation", Journal Fluid Mechanic, Vol. 40 Part 1, page 1-19 (1970)

Pearson, J.R.A., Petrie, C.J.S., "The flow of a tubular film II: Interpretation of the model and discussion of solutions", Journal Fluid Mechanic, Vol. 42 Part 3, page 609-625 (1970)

Phan- Thien, N., Tanner, R.I.J., "A new constitutive equation derived from network theory", Journal of Non- Newtonian Fluid Mechanic Vol. 2, page 352-364 (1977)

Sidiropoulos, V., Vlachopoulos, J., "Temperature Gradients in Blown Film Bubbles", Advances in Polymer Technology Vol. 24, page 83-90 (2005)

Book:
Laun, H.-M., "Industrial Rheology of Polymer Melts", lecture notes, University of Dortmund (2005)

Spirgatis, J., "Untersuchungen des Einflusses des instationären konvektiven Wärmeübergangs bei der Folienherstellung auf die Produktqualität", Dissertation an der Universität Duisburg-Essen (2004)

Wolf, D., „Analyse von Kühlluftsystemen zur Kühlleistungssteigerung und Verbesserung der Folienqualität bei der Herstellung von Blasfolien", Dissertation an der Universität-GH Essen (1995)

Proceedings:
Vlachopoulos, J., Sidiropoulos, V., "Rethinking the Mathematical Modeling of Blown Film Extrusion", Polymer Processing Society Conference, (2008)

Zahng, Z., Lafleur, P.G., "The effect of cooling air aerodynamics on bubble instability in blown film", Annual Technical Conference ANTEC (2006)

CHARACTERISATION OF LOCAL RESIDENCE TIME DISTRIBUTIONS IN CO-ROTATING TWIN SCREW EXTRUDERS, USING POSITRON EMISSION PARTICLE TRACKING FLOW VISUALISATION TECHNOLOGY - PEPTFLOW -

J. Diemert [1*], C. Chilles [2], J. Colbert [2], T. Miri [3], A. Ingram [3]

[1] *Fraunhofer Institute for Chemical Technology, Pfinztal (Berghausen), Germany – jan.diemert@ict.fraunhofer.de*
[2] *Smithers Rapra Technology Ltd,. Shawbury, United Kingdon – cchiles@rapra.net*
[3] *School of Chemical Engineering, University of Birmingham, Birmingham, United Kingdon – a.ingram@bham.ac.uk*

Co-rotating twin-screw extruders are widely used compounding machines. They are mainly configured based on extensive experience and iterative approaches to optimise output and composite quality. The visualisation technology developed in the EU-project PEPTFlow allows visualisation of composite flow in twin-screw extruders under realistic processing conditions by tracking radioactive tracer particles in the polymer melt, using a specially developed camera system. This new approach allows polymer flow to be studied in different screw elements and screw configurations under realistic compounding conditions at normal temperatures and melt pressures. The paper presents the latest developments in the camera systems as well as the different ways to use and interpret the results. Detailed analysis of residence times and residence time distributions for standard compound screw elements, like kneading discs, conveying elements and reverse elements are presented.

Introduction

The compounding of thermoplastic materials on an industrial scale is mainly carried out in co-rotational twin-screw extruders (TSE) that were specifically designed to offer high throughput and good mixing capabilities. Due to the inherent flexibility in the machine design of TSEs, where barrel segments, screw elements and dosing points can be varied, it is possible to adapt this machine to the varying demands of different material systems and processing requirements. Typical adaptations are the modification of the screw profiles tailoring the amount of mechanical mixing, residence time and pressure levels. These adaptations are mainly carried out based on extensive experience and some basic calculations and scale-up rules. Simulation of TSE processes often suffer from the lack of precise material data describing the material behaviour, especially during the important mixing phases (Kohlgrueber and Wiedmann, 2008).
PEPTFlow offers a new way to study flow within TSE under realistic processing conditions by following the movement of a radioactive trace incorporated into the compounded material. Using the results of PEPTFlow, local residence times can be analyzed, flow velocities can be studied and consequently screw profiles can be optimized.
The paper presents the PEPTFlow technology, the ways the data can be analysed and selected results from standard processing conditions.

Experimental

Positron Emission Particle Tracking (PEPT)
Positron emission particle tracking is a non-invasive technique, developed at the University of Birmingham, for following the motion of a radioactive tracer particle in three dimensions within non-transparent systems. It is derived from the medical technique Positron Emission Tomography and uses the same imaging equipment. The set-up for the standard PEPT camera, shown in figure 1, consists of a pair of parallel detectors, a radioactive tracer particle and an algorithm for the calculation of the particle location. The tracer particle is labelled with a radionuclide which decays by $\beta+$ decay resulting in the emission of a positron. Each positron rapidly annihilates with an electron, producing a pair of back-to-back 511keV γ-photons which are then detected by the position-sensitive detectors. Triangulation of successive photon pairs enables calculation of particle location. The frequency and accuracy of tracer location depends on the active strength of the tracer and the mass of material that must be penetrated by the γ-photons. Typically a tracer can be located to within 1mm at speeds up to 1ms-1. For more detail see Hawkesworth et al, 1991; Parker et al, 1993; Parker et al, 1997; Parker et al, 2002; Fan et al, 2006 and Parker and Fan et al, 2008.

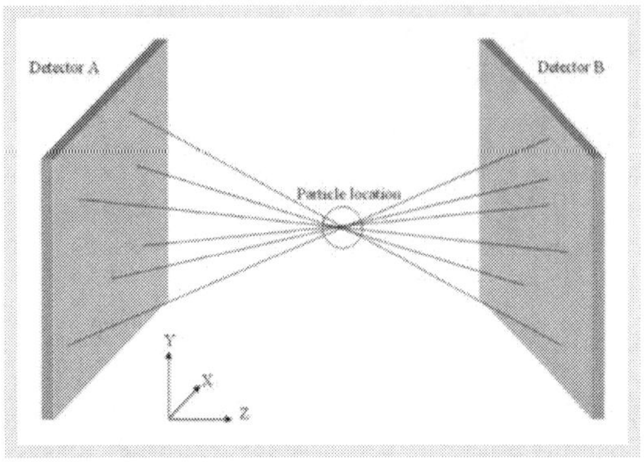

Figure 1: Schematic representation of PEPT

The technique has been employed in numerous applications extensively reported in literature, predominantly in granular systems due to the particulate nature of the tracer. Seville et al (2009) presented an extended summary of case studies and applications for PEPT. For example, Jones and Bridgwater (1998) employed PEPT to study motion and mixing in a ploughshare mixer and observed how particles move between inter-plough regions. Ding et al (2001) studied the rolling mode for a rotating drum and characterisation of the stochastic motion in the active region. Some work has also been carried out on paste systems and liquid systems (Fishwick et al, 2003). Wildman et al (1999) observed plug flow inside the barrel and die land region during ram extrusion of ceramic paste through analysis of particle velocity distribution. Fishwick et al (2003) used PEPT to infer liquid-particle slip velocity in multiphase stirred reactor vessels.

The PEPTFlow camera and extruder

The recent trend for UK medical PET centres to replace existing PET scanners with new dual function PET/CT scanners has enabled the Positron Imaging Centre at Birmingham to acquire a number of redundant CTI ECAT 931 and 951 cameras. The cameras are essentially modular in design. The detecting elements are individual blocks of bismuth germinate (BGO) scintillating crystal arranged in groups of four on a board with associated electronics. The camera comprises two rings of 16 of these so called "buckets" of four detector blocks (128 blocks in total) that can be dismounted and rearranged as desired. In fact it is also possible to separate the detector blocks a short distance from the buckets. This gives much increased flexibility in terms of matching the camera to the geometry of the system under investigation. Work started at the University of Birmingham a few years ago to develop a methodology for building modular PEPT cameras from the components of the 931 and more recently the 951 PET cameras (Parker et al, 2008; Leadbeater et al, 2007 and 2008). Since then, configurations of the camera have been used to study flow in a pilot scale high pressure fluidised bed and circulation in an industrial fluidised bed at BP's (now Ineos's) site in Hull (Ingram et al, 2007). The camera flexibility also means it can be "focussed" on a small field of view, with consequently high data rate and accuracy, or strategically deployed around larger systems to give a discontinuous field of view. This is ideal for PEPTFlow since it has enabled the camera to be positioned in order to surround the field of view.

The experimental set-up was based on a Leistritz Micro 27 mm twin-screw extruder that was modified to provide a PEPT window in the extruder barrel. Studies showed that the thick steel of the extruder barrel was effectively impenetrable to gamma photons. For PEPT to work, a section of barrel with reduced wall mass was therefore required. This was provided by a surface hardened aluminium insert with nominal wall thickness of 25mm. The length of the exposed aluminium section, 110mm, gave a field of view of approximately 90mm. Other construction materials were considered and tested (such as ceramic), but were not deemed suitable. The set-up of the PEPTFlow window section of the extruder is illustrated in figure 2.

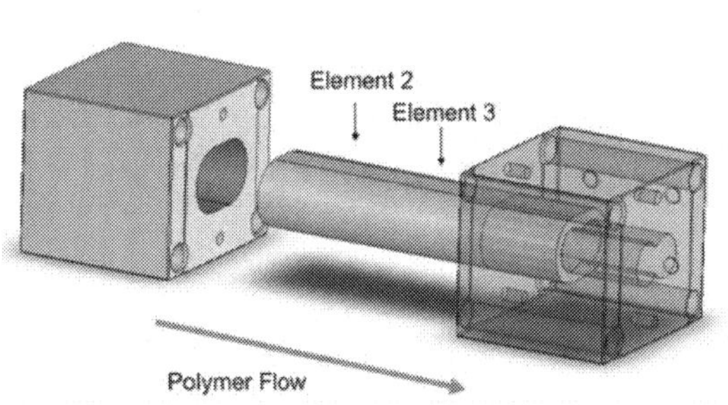

Figure 2: Barrel inliner and screw element numbering

The flexible design of the Leistritz 27mm extruder allowed the gearbox with the attached barrel to be moved forward on the machine bed, freeing additional space around the barrel, allowing the better application of the ring type camera system (see figure 3)

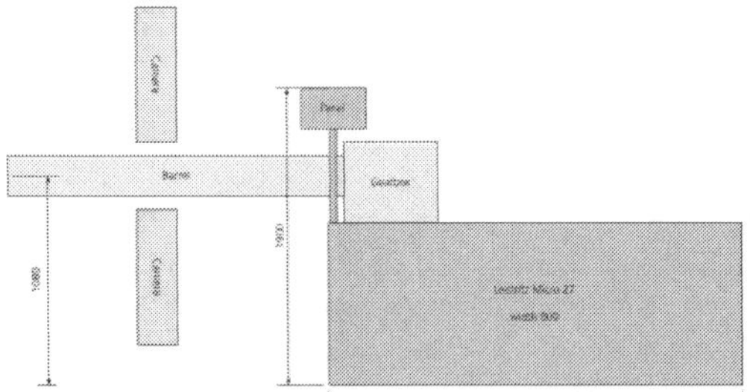

Figure 3: Set-up of the extrusion line and camera position

In addition to the modification of the base extruder, a refeeding system had to be developed to refeed the PEPT-particle after it had flowed through the extruder. This was necessary due to the limited number of particles that could be manufactured in one day, increasing the data output for one particle. With the refeeding systems an average of 5-10 runs were possible using one particle.

Figure 4 illustrates the ring camera mounted around the barrel and the downstream extruder die and waterbatch.

Figure 4: PEPTFlow camera mounted on the lab extruder

A comprehensive experimental plan was designed to maximise the range of conditions investigated within the measurement time available. This is discussed in Section 2.

PEPTFlow data evaluation

Positron Emission Particle Tracking generates trajectory data initially as a list of co-ordinates of the paired gamma photons detected as the tracer moves through the field of view. Electronic circuitry within the camera, termed "coincidence boards", pairs the photons according to detection time. Each detected pair is known as an "event" and the line connecting them is termed a line of response (LOR). Events are generated and particle locations are obtained from triangulation of the lines of response collected over a period of a few milliseconds. The data however comprises a mixture of true and so-called "corrupted" events. The latter arise from photon scattering or invalid pairing, and reduce the certainty of location measurement. It is not possible to determine a priori which pairs are true and which are corrupted (although a proportion of photons are sufficiently scattered that they can be eliminated because the consequent energy loss is detectable and measured by the camera). A statistical algorithm is therefore used to home in on the most convergent point of the lines by iteratively removing outliers.

Based on the delivered X, Y, Z over time coordinates for an individual run, several data analyses and interpretations can be carried out:

- 3D visualisation of the trajectory, enabling visual interpretation and investigation of special flow phenomena.
- Calculation and visualisation of different velocities and accelerations based on the position data.
- Statistical interpretation of the results of several runs of one experiment to determine average velocities, residence times and local residence time distributions.

Results and Discussion

Materials and processing conditions

All experiments investigating the influence of different processing conditions on the residence times were carried out using the same PP-Polymer with an MFI of 70. This comparably low viscosity polymer was chosen to minimise the stress on the PEPT-Particle. The temperature profile in the extruder was set to 220-240°C.

The screw set-up of the upstream elements was also kept constant to guarantee constant and comparable melt quality. The screw set-up of the upstream elements was chosen to plasticise the polymer gently without too much shear stress, in order to apply the least possible stress to the PEPT-particle which was fed with the polymer, consequently passing the melting section.

For the interpretation of especially the residence time graphs it is important to note the numbering of the elements underneath the PEPT-window as illustrated in figure 2. In the discussion of the residence time only the elements fully covered by the PEPT-window (element 2 and 3) are discussed.

Effect of throughput in kneading disc sections

The following section discusses the influence of different processing conditions on standard kneading discs.

<u>Screw configuration</u>

A screw set-up with a 60° and a 90° kneading disc was chosen for this study. A 30° kneading disc was in front of this set-up, which was followed by a simple 30 mm pitch conveying element. The whole set-up is illustrated in figure 5. The 60° and 90° kneading discs are therefore both fully covered by the PEPT window.

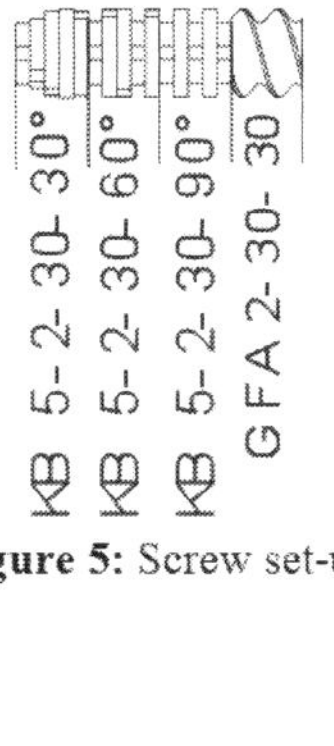

Figure 5: Screw set-up for experiments 5 and 6

Processing conditions

In order to investigate the effect of throughput on the residence time and the residence time distribution, two different processing conditions were chosen (table 1).

Table 1: Conditions of experiments 5 and 6

	Experiment 5	Experiment 6
Temperature, °C	220-240	220-240
Screw speed, 1/min	200	200
Feedrate, kg/h	4.8	9.6

Effect on residence time

The effect on the residence time distribution of the change in processing conditions between experiments 5 and 6 is illustrated in the three following graphs 6-8.

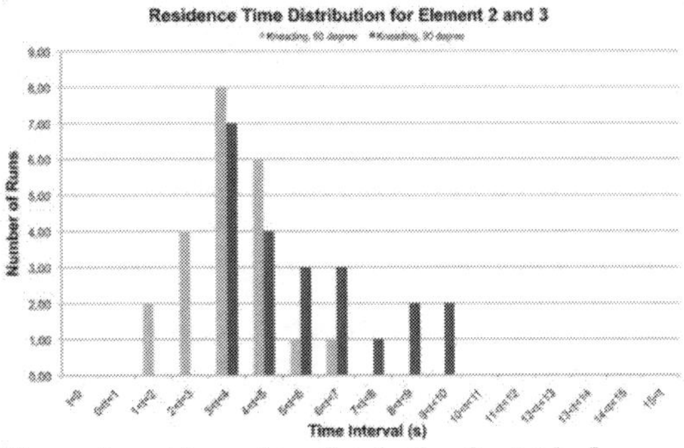

Figure 6: residence time distribution for 4.8 kg/h

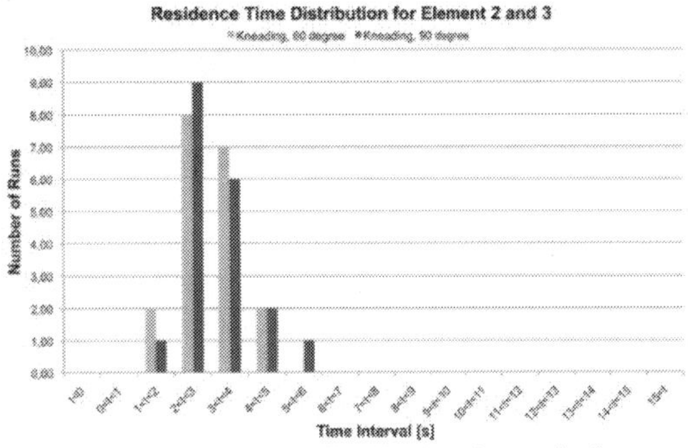

Figure 7: residence time distribution for 9.6 kg/h

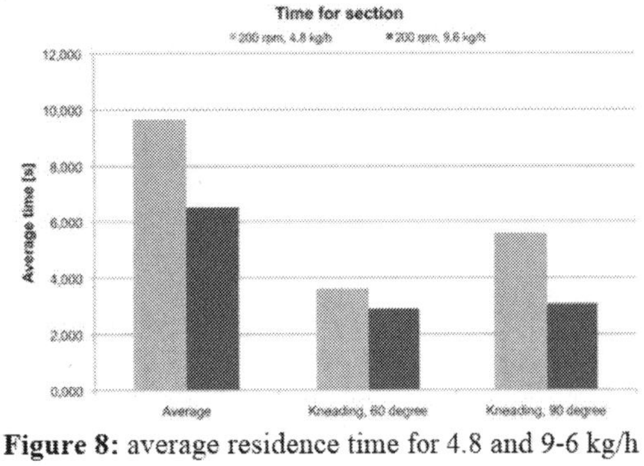

Figure 8: average residence time for 4.8 and 9-6 kg/h

The change in throughput from 4.8 to 9.6 kg/h while keeping the other processing conditions constant has a quite substantial effect on the residence time distribution. Figures 6 and 7 clearly show a different residence time distribution for the two elements, which is also reflected in the average residence times shown in figure 8. The 60° kneading disc shows a narrower distribution and a shorter average residence time at 4.8 kg/h then the 90° kneading disc. The fill level can explain this large difference in residence time for these two kneading discs. The 60° kneading disc has a small but positive conveying capacity. At such a low throughput this results in the 60° kneading disc not being 100% full.

At 9.6 kg/h the situation changes quite substantially. The average residence time for the 60° discs is reduced by about 25%, whereas the 90° disc shows around a 50% residence time reduction. The later was expected because of the fully filled state, where the material flow is a direct function of the material volume flow through the screw geometry. Most interestingly the situation in the 60° kneading disc changes completely and the average residence time as well as the residence time distribution is now very similar to the 90° kneading disc. This can be explained by the 60° element now being fully filled and therefore the residence time is only a function of the material flow, which is now identical in both elements.

It is important to note that this was not expected to this extent. The change in throughput narrowed the residence time distribution for both kneading discs. This clearly reflects a decrease in distributive mixing efficiency of kneading discs at higher output rates or degree of fill.

Effect in conveying elements

After looking in detail at the effect of throughput on residence time distribution in kneading discs, the following section describes the effect of changing the throughput in a conveying section.

Screw configuration

A screw set-up with a 40mm and a 30mm conveying element was chosen for this study. A 40mm conveying element was in front of this set-up, which was followed by a simple 30 mm pitch-conveying element. The whole set-up is illustrated in figure 9.

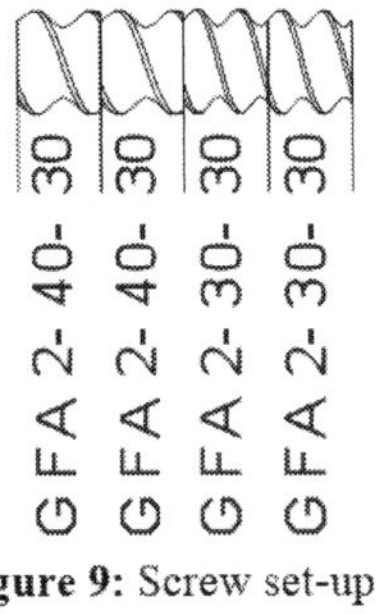

Figure 9: Screw set-up for experiment 9 and 12

Processing conditions

For the study of the effect of throughput on the residence time and the residence time distribution two different processing conditions were chosen (table 2)

Table 2: Conditions of experiments 9 and 12

	Experiment 9	Experiment 12
Temperature, °C	220-240	220-240
Screw speed, 1/min	80	80
Feedrate, kg/h	4.8	9.6

Effect on residence time

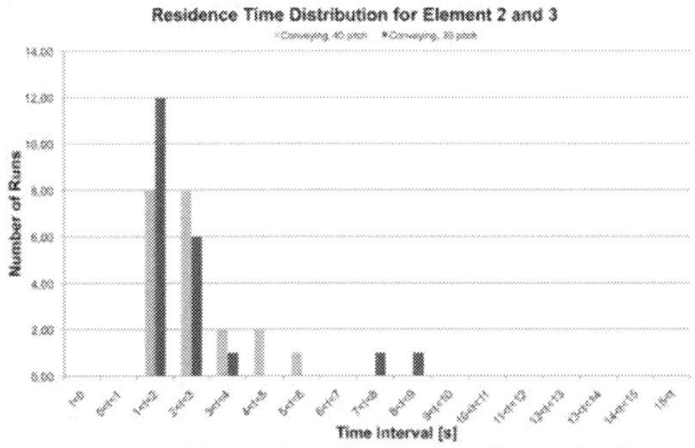

Figure 10: residence time distribution for 4.8 kg/h

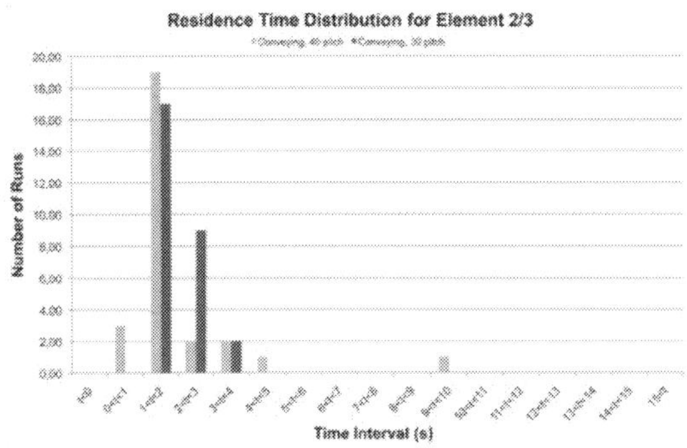

Figure 11: residence time distribution for 9.6 kg/h

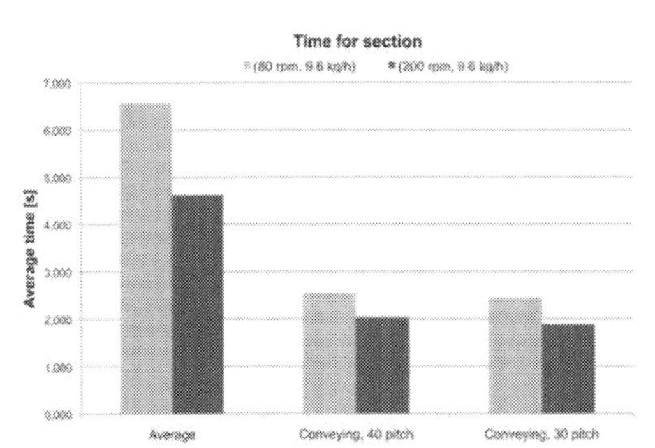

Figure 12: average residence time for 4.8 and 9-6 kg/h

By comparing the effect of changing the throughput in a conveying section to the effect the same change in processing conditions has in a kneading disc section, it is firstly obvious that the average residence time for the conveying section is significantly shorter and the residence time distribution is a lot narrower for the conveying elements. There are some quite slow passages, as can be seen in figures 10 and 11, but the residence time distribution for both conditions shows a sharp peak and short passage times. In absolute values, the average residence time for the conveying element is less than 50% of the time for the 90o kneading discLooking at the effect of changing the throughput for the conveying element section, it can be noted that the average residence time, as well as the shape of the residence time distribution, does not change significantly by increasing the throughput from 4.8 kg/h to 9.6 kg/h. The explanation for this is that the elements are not 100% full for the 4.8 kg/h condition or, most probably, for the 9.6 kg/h condition. There is therefore enough conveying capacity left to take up the additional material. Interestingly, the average residence time decreases for the conveying element when the throughput is increased, although these elements are not 100% full. Looking at the shape of the residence time distribution and taking into account that the peak passage time is between 1 and 2 seconds for both conditions, then this perhaps not directly anticipated phenomena can mainly be explained by a significant reduction in long passages for the higher throughputs.

Effect of back-pressure on residence times in conveying elements

Screw configuration

For this study a screw set-up was used that contains a reverse element at the end of the PEPT-window section (here the 4th element, see figure 13). This element produces a strong back-pressure, guaranteeing that at least the conveying element in front of the reverse element is fully filled with melt. The reverse element has a pitch of only 20mm and the pressure build capacity of the reverse element is higher than the pressure build-up generated by the 30mm pitch conveying element in front of it. So it can be assumed that the fully filled section is also covering part of the second element underneath the PEPT-window (see figure 2).

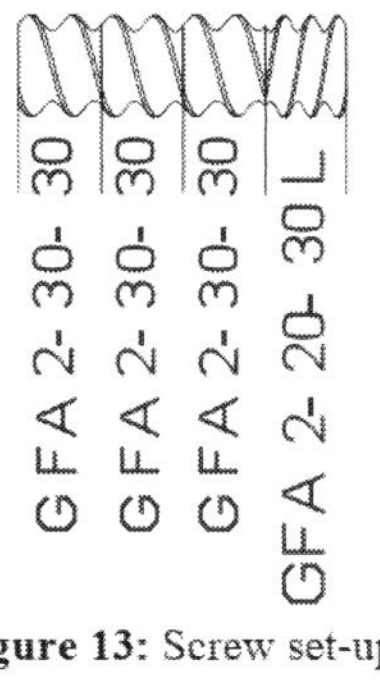

Figure 13: Screw set-up for experiment 14

Processing conditions

For the study of the effect of back-pressure on the residence time and the residence time distribution, the following processing conditions were chosen:

Table 3: Conditions of experiments 14

	Experiment 14
Temperature, °C	220-240
Screw speed, 1/min	80
Feedrate, kg/h	4.8

PEPTFlow partner Scientific Computer Consultant carried out a 3D simulation of this screw set-up under the above processing conditions. The resulting melt pressure is illustrated in figure 14. It can be observed that the pressure directly in front of the reverse element reaches almost 80 bars, with pressure of >40 bars going back at least one screw element.

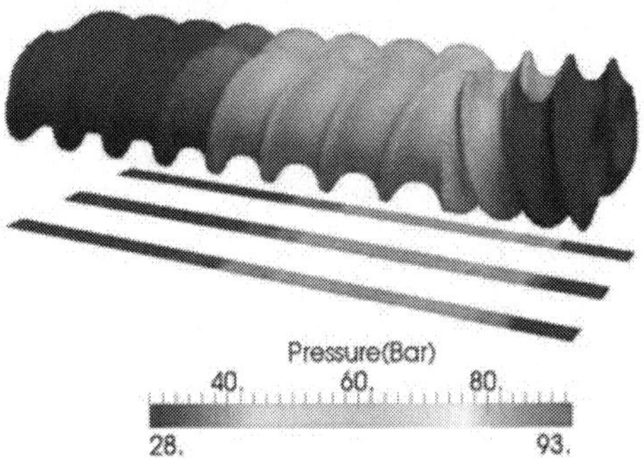

Figure 14: 3D simulation of the pressure in front of the reverse element.

Effect on residence time

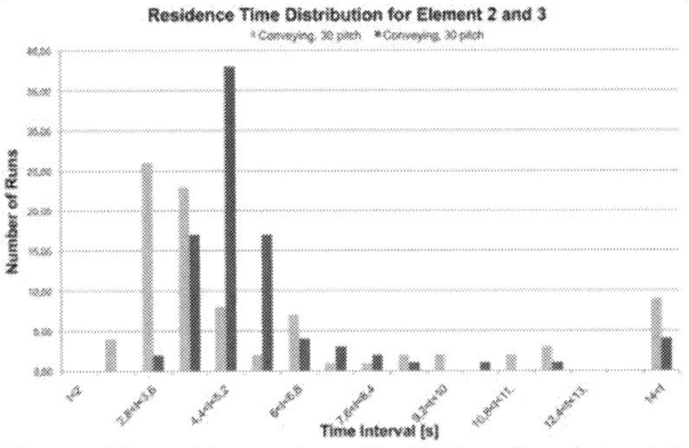

Figure 15: residence time distribution for element 2 and 3

Figure 15 illustrated substantial differences in the residence time and residence time distribution for the two monitored conveying elements underneath the PEPT-window. Element 2 can be assumed not to be fully filled, whereas element 3 can be regarded as being fully filled due to the back-pressure.

The pressure level within the conveying elements has a very strong influence on the residence time distribution. Element 2, which is freely conveying the material in most of its length shows substantially shorter residence times than element 3. The peak of residence times shifts from 2.8-3.6 to 4.4-5.2 seconds by pressurising the conveying element with the reverse element. On the other hand the general shape of the main peak of the residence time distribution has not changed.

An important finding was made within the PEPTFlow-project concerning the influence of back-pressure on the number of very long passage times or residence times. Looking at the right-hand side of figure 15 it is very obvious that the number of these long passages is reduced drastically in element 3, which has to build up pressure. This behaviour was not expected at first sight. The most probable explanation of the reduction in long passages can be found if we consider the areas of the screws and barrels where these long residence times can occur. They mainly occur due to the melt film between the screw and barrel and on the screw surface. This melt film is not moving fast and has no substantial exchange with the main material stream.

Pressurising these sections results in two phenomena, that consequently reduce the residence time:

- Due to the pressure build up the screws are a pushed a little further towards the barrel. This reduces the clearance and consequently reduces quantity of material trapped in the melt film
- Perhaps more important is that the back-pressure increases the leakage flow over the screw tip and between the screws. This intensive material flow through these small gaps washes away the melt film, resulting in a much more intense exchange of the melt film with the main material flow.

Effect of back-pressure on residence times in kneading discs

Screw configuration

The following screw configuration (see figure 16) was used for the trials with kneading discs in front of a reverse element to guarantee a 100% fill.

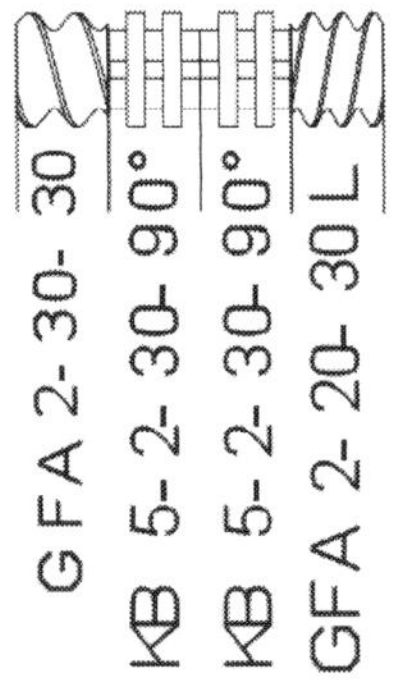

Figure 16: Screw set-up for experiment 15

Processing conditions

Table 4: Conditions of experiments 15

	Experiment 15
Temperature, °C	220-240
Screw speed, 1/min	80
Feedrate, kg/h	4.8

Similar to the conveying section PEPTFlow partner SCC has carried out a 3D simulation of this screw set-up under the mentioned processing conditions. The resulting melt pressure is illustrated in figure 17. It can be observed that the pressure directly in front of the first kneading disc element reaches almost 70 bar with a slight decrease towards the end of the kneading disc section and a steep decrease over the reverse element.

Figure 17: 3D simulation of the pressure in the screw profile

Effect on residence time

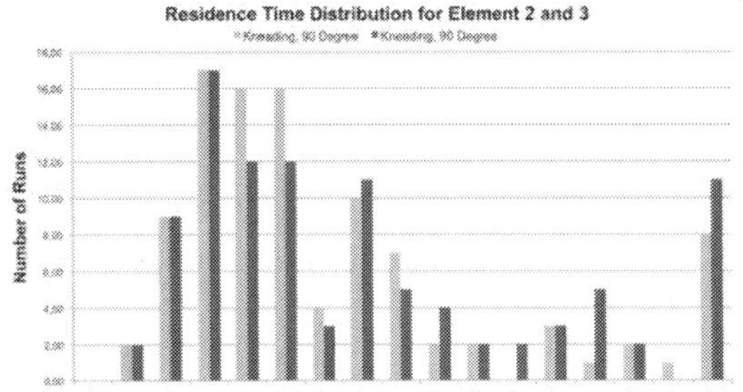

Figure 18: Residence time distribution for a 100% fully filled kneading disc section under back-pressure.

In the kneading disc section that is fully filled under these processing conditions the effect of the pressure on the residence time and residence time distribution is negligible (see figure 18). This was assumed, as the back-pressure only increases the static pressure in this area.

It should be noted that there are a substantial amount of very slow passages that are summarised in the category 14plus seconds on the right-hand side of the graph. They reflect the good axial mixing capacities of kneading discs but they can also be a problem with temperature and shear sensitive materials.

Observed flow phenomena - Sticking to the screw

The following paragraph briefly summarises certain observed flow conditions that show interesting or unexpected flow phenomena. As some of them can have a strong influence on residence time it is important to know that these conditions and phenomena like those presented can occur.

During experiments with the following screw set-up (figure 19)

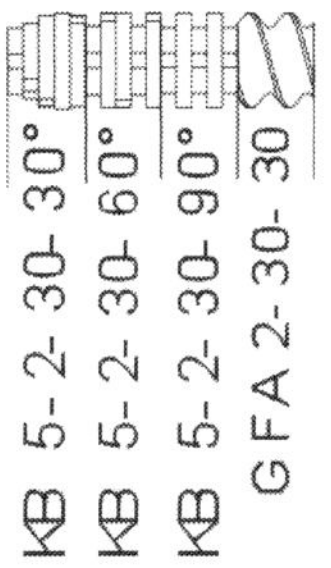

Figure 19: Screw set-up for experiment 3

and the following processing conditions (table 5), using the standard PP material a flow path was observed where the particle rotated for several revolutions on one screw.

Table 5: Conditions of experiments 3

	Experiment 3
Temperature, °C	220-240
Screw speed, 1/min	80
Feedrate, kg/h	4.8

As illustrated in figure 20 the particle rotated for 11 revolutions on the right screw.

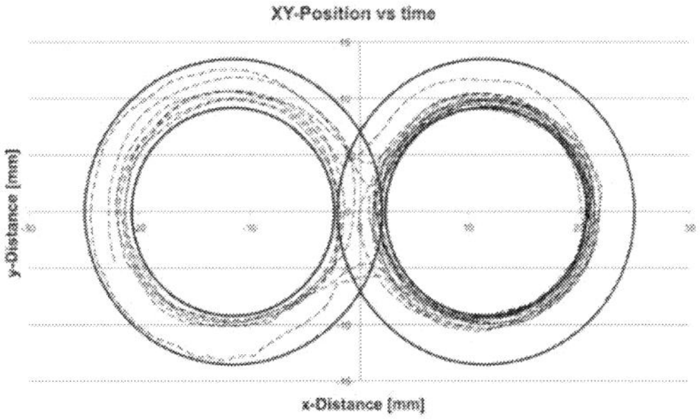

Figure 20: Particle sticking to the right screw

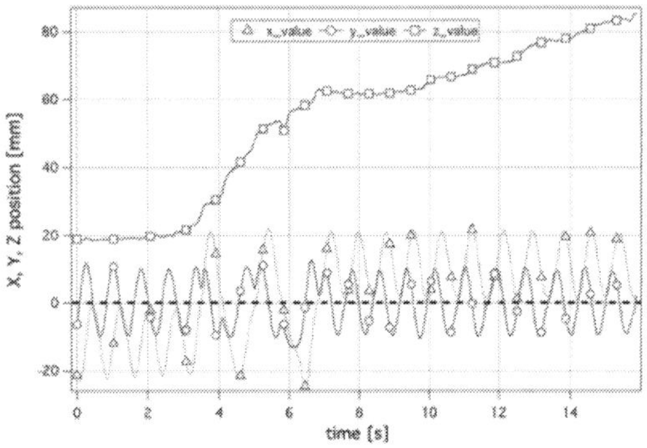

Figure 21: XYZ-coordinates of a particle stuck on one screw.

As illustrated in figure 21 the particle did not move in the z direction for almost 5 revolutions (7-10 seconds) and then moved slowly 10mm forward during the next 6 revolutions.

This condition has led to a passage time that is almost twice as long as the average passage time for this element.

Such a flow condition can be a problem with very temperature sensitive material or in reactive processing, where narrow residence time distributions are needed.

Observed flow phenomena - Sticking to the barrel

During the same experiments the following, additional flow phenomena was observed.

As illustrated in figure 22 the particle travelled slowly but constantly along in the figure of eight shape of the inner barrel surface for one complete revolution.

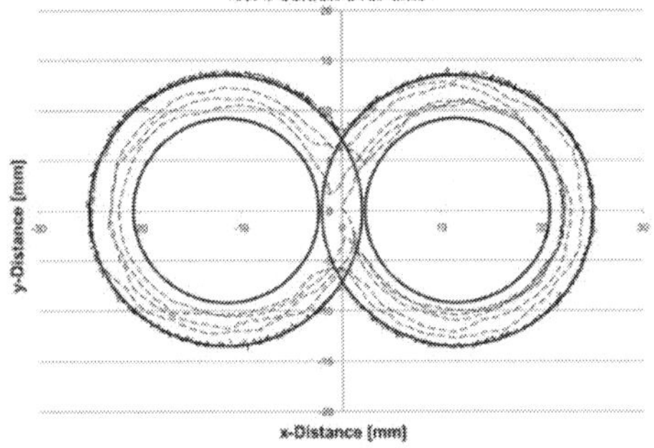

Figure 22: Flow path of a particle stuck in the melt film.

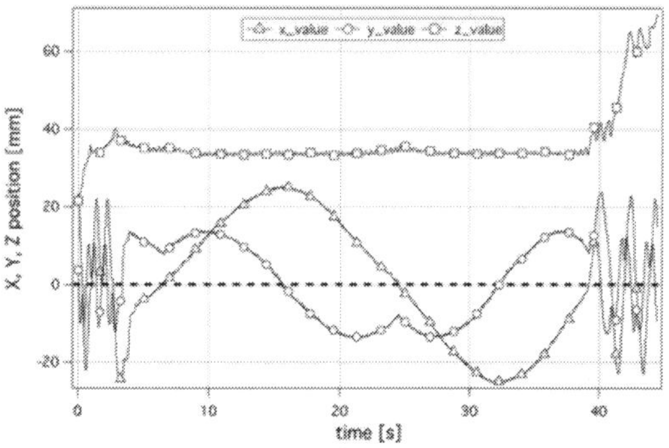

Figure 23: XYZ-coordinates of a particle stuck on one screw.

As illustrated in figure 23 the particle did not move in the z direction while it was stuck in the melt film: it was moving in a circumferential direction.

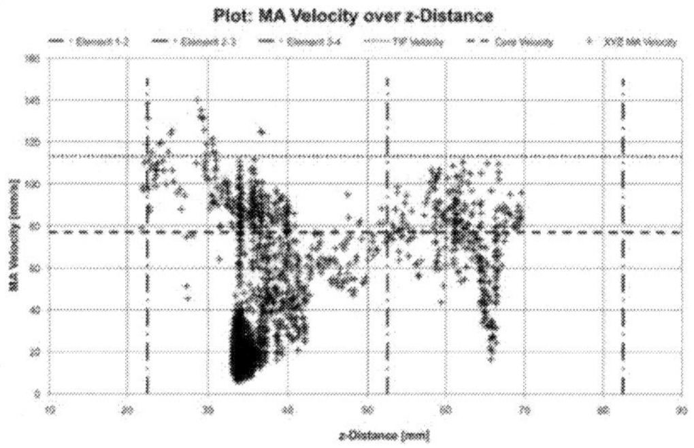

Figure 24: Velocity profile of a particle stuck in the melt film.

In the velocity profile illustrated in figure 24 it can furthermore be seen that the speed of the particle was in a very low range of around mm/s while it was stuck in the melt film at an axial distance of z=33-35mm.

Then at an observation time of almost 40 seconds the particle was suddenly washed away from the barrel surface and accelerated back to normal particle speeds of about 60-100 mm/s.

Again such flow conditions can be a problem with very temperature sensitive materials or in reactive processing, where narrow residence time distributions are needed.

Conclusions

PEPTFlow is a completely new characterisation technology for polymer flow within polymer processing machines. It offers insights into the machine that have not been available before. PEPTFlow allows monitoring of polymer flow under realistic processing conditions using real polymers at high temperatures and high pressures, and monitoring of flow locally in individual screw elements, making it possible to identify regions with very long residence time. PEPTFlow can be improved. Due to the current limitation on positional frequency and some necessary smoothing to avoid excessive noise, it is presently only reasonable to monitor screw speeds of up to 300 rpm. Only one particle can be traced in an individual run. Therefore only 50-60 trajectories can be taken in one day

Most of the results obtained in extensive studies have confirmed expectations and experimental data obtained before. On the other hand, non-standard flow conditions occurred more often than expected during the experiments.

Acknowledgement

The European Commission within the 6th framework program under contract Coll-CT-2006-030191 funded the work leading to the set-up of the extruder in Birmingham and the characterization of different processes.

For further project information see:

www.peptflow.com

Furthermore the company Leistritz Extrusionstechnik GmbH supported the work.

References

Ding, Y. L.; Seville, J. P. K.; Forster, R.; Parker, D. J. (2001): Solids motion in rolling mode rotating drums operated at low to medium rotational speeds. *Chemical Engineering Science*, vol. 56, no. 5, pp. 1769-1780.

Fan, X.; Parker, D.J.; Smith, M.D. (2006). Labelling a single particle for positron emission particle tracking using direct activation and ion-exchange techniques. *Nuclear Instruments and Methods in Physics Research Section A: Accelerators, Spectrometers, Detectors and Associated Equipment, vol. 562, no.1, pp. 345-350.*

Fishwick, R.P, Winterbottom, J.M., Stitt, E.H. (2003), Explaining mass transfer observations in multiphase stirredreactors: particle-liquid slip velocity measurements using PEPT, *Catalysis Today,* 79–80, 195–202.

Hawkesworth, M. R.; Parker, D. J.; Fowles, P.;Crilly, J. F.; Jefferies, N. L.; Jonkers, G. (1991) Nonmedical applications of a positron camera. *Nuclear Instruments and Methods in Physics Research Section A: Accelerators, Spectrometers, Detectors and Associated Equipment*, vol. 310, no. 1-2, Pages 423-434.

Ingram, A.; Hausard, M.; Fan, X.; Parker, D.J.; Seville, J. P. K.; Finn, N. (2007) Portable positron emission particle tracking (PEPT) for industrial use, 12th International Conference on Fluidization. *New Horizons in Fluidization Engineering, ECI Engineering Conference International Symposium Series*, paper 60.

Jones, J. R.; Bridgwater, J. (1998): A case study of particle mixing in a ploughshare mixer using Positron Emission Particle Tracking. *International Journal of Mineral Processing*, vol. 53, no. 1-2, pp. 29-38.

Kohlgrüber, K.; Wiedmann, W. (2008): Co-Rotating Twin-Screw Extruders, 1st edition, Carl Hanser Verlag, Munich

Leadbeater, T.W. and Parker, D.J., A Positron Camera with Flexible Geometry for the Study of Industrial Processes, 5th World Congress on Industrial Process Tomography, Bergen, Norway, 2007.

Leadbeater, T.W. and Parker, D.J., A High Speed PC-Based Data Acquisition & Control System for Positron Imaging, 8th International Conference on Position Sensitive Detectors, Glasgow, Scotland, 2008. (Accepted for publication in Nucl. Inst. Meth. A.)

Parker, D.J., Broadbent, C.J., Fowles, P., Hawkesworth, M.R. and McNeil, P., Positron Emission Particle Tracking - A technique for studying flow within engineering equipment. (1993), Nucl. Inst. Meth., A326:592–607.

Parker, D. J.; Allen, D. A. ; Benton, D. M. ; Fowles, P. ; McNeil, P. A.; Tan, Min; Beynon, T. D. (1997). Developments in particle tracking using the Birmingham Positron Camera. *Nuclear Instruments and Methods in Physics Research Section A: Accelerators, Spectrometers, Detectors and Associated Equipment*, vol. 392, no. 1-3, 21 June 1997, pp. 421-426.

Parker, D. J.; Forster, R. N.; Fowles, P.; Takhar, P. S. (2002) Positron emission particle tracking using the new Birmingham positron camera. *Nuclear Instruments and Methods in Physics Research Section A: Accelerators, Spectrometers, Detectors and Associated Equipment*, vol. 477, no. 1-3, pp. 540-545.

Parker, D.J., Leadbeater, T.W., Fan, X, Hausard, M.N., Ingram, A and Yang, Z, Positron imaging techniques for process engineering: recent developments at Birmingham, Meas. Sci. Technol. 19 (2008) 094004 (10pp).

Parker, D.J. and Fan, X, Positron emission particle tracking—Application and labelling techniques, Particuology, Volume 6, Issue 1, February 2008, Pages 16-23.

Seville, J.P.K.; Ingram, A.; Fan, X.; Parker, D.J. (2009): Chapter 4 Positron Emission Imaging in Chemical Engineering. *Advances in Chemical Engineering*, vol. 37, pp. 149-178.

Wildman, R. D.; Blackburn, S.; Benton, D. M.; McNeil, P. A.; Parker, D. J. (1999): Investigation of paste flow using positron emission particle tracking. *Powder Technology*, vol. 103, no. 3, pp. 220-229.

Measurement of Melt Temperature Profile in Injection Moulding

Kelly Adrian, Mulvaney-Johnson Leigh, Brown Elaine, Coates Phil

The injection moulding process is dependent upon the repeatable production of moulded parts, which itself depends largely on the consistent preparation of a melt volume prior to injection. Melt temperature is a particularly important variable as it influences rheology during injection, cooling rate, crystallisation dynamics and final part properties. Melt temperature is however difficult to measure in-process without adversely affecting the melt flow. Here, melt temperature has been measured using a sensor consisting of a grid of thermocouple wires located in front of the nozzle, across which melt flows during injection. Radial melt temperature profiles have been recorded during a series of air-shots in an injection moulding machine with a screw diameter of 22mm. The stability of melt temperature profile along the length of the shot is reported, in addition to the effect of various process variables such as set barrel temperature, injection velocity, plastication screw speed and back pressure. Such measurements offer the potential to optimise moulding conditions and screw design to produce a homogeneous melt reservoir.

Compact and flexible optical sensor designed for on-line morphology monitoring

René Muller, Michel Bouquey, Rodrigue Matadi, Badi Triki, Laurent Jourdainne*

University of Strasbourg ECPM - 25 rue Becquerel 67200 Strasbourg - France

A new in line optical sensor based on light scattering dedicated to real-time monitoring during processing of polymers blends and polymers nanocomposites has been developed. The sensor is designed to monitor the quality of a polymer melt or the dispersion of nanofillers in a polymeric matrix during the process of extrusion. This sensor presents three openings: two for light sensor setups (one for transmission measurements and the other for 90°-scattered light measurements) and one for the incident laser beam. The experimental validation of this optical sensor have been realized on two polymers: poly(methylmethacrylate) (PMMA) and polypropylene (PP), and on the nanocomposites obtained by the melt mixing of theses polymers matrices with modified montmorillonite (MMT) clays. Validation of experiments proved the good results in terms of signal repeatability and sensitivity to organoclay particle concentration and structures of nanocomposites respectively. Keywords: optical sensor, transmitted light, scattered light, polymers, nanocomposites, melt mixing

Development of New Measuring Mold with Sensor-installed Rotary Cylinder for Measuring Plane Distributions of Melt Temperature and Pressure

Hidetoshi Yokoi[1*],Hirohito Kai[1]

[1]*Institute of Industrial Science*
The University of Tokyo, Tokyo 153-8505, Japan
hiyokoi@iis.u-tokyo.ac.jp

Abstract

In injection molding, the analysis of resin flow behaviour at the branched portion of runners of various shapes is a very important task for realizing the filling balance of a multi-cavity mold. In this study, we first developed a measuring block comprised of a rotary cylindrical block inserted with temperature and pressure sensors whose positions can be changed easily, and realized highly functional measurement at any resolution and range by motor drive. Using this measuring block, we measured the grid surfaces at the branched portion of a Y-shaped isometric runner and the changes in temperature and pressure distributions. Furthermore, we also evaluated the filling imbalance inside the cavity from visualization images of resin flow behaviour observed using a glass-inserted mold, and reviewed the correlation with the measured temperature and pressure distributions. As a result, we found that at the wall continuing from the main runner, the higher the injection rate, the more conspicuous was the rise in resin temperature due to shear heating. In addition, the temperature was also seen to drop due to the decrease in shear heating effect at the expanded portion of the flow path. Comparisons of the filling behaviour of each cavity extracted from the visualization images and temperature measurement results confirmed correlation between the asymmetrical filling behaviour balance in each cavity and temperature distribution along the width of the runner.

1. Introduction

Multi-cavity molding is applied widely in injection molding. To ensure good filling balance in a multi-cavity mold, the most important task is to analyze resin filling behaviour at the branched portion of runners of various shapes. In previous studies, some of the authors conducted correlation analysis of temperature distribution and filling balance from visualization results of resin flow patterns inside the cavity, and results of temperature distribution inside the runner and cavity using an infrared thermometer built inside a movable mold [1]. However, this movable thermometer block must be moved manually, which means measurement of multi points would require enormous work. In addition, because the block can only be moved in the horizontal direction, there is a need to change the mold for two-dimensional measurements when the vertical direction position of sensors is changed, making large area high precision observations difficult.

In this study, the authors therefore developed a measuring block comprised of a rotary cylindrical block inserted with a sensor whose position can be changed easily to realize high performance measurements at any resolution and observed area by driving the block with a motor.

Using this measuring block, we measured the grid surfaces at the branched portion of a Y-shaped isometric runner and the changes in temperature and pressure distributions. Furthermore, we also evaluated the filling imbalance inside the cavity from visualization images of resin flow behaviour observed using a glass-inserted mold, and reviewed the correlation with the measured temperature and pressure distributions. The results are reported below.

2. Outline of Experimental System

2. 1 Mold Base

Figure 1 (a) shows the structures of the measuring block and the mold base developed in this study, and **Fig. 1** (b) shows their external view. In the experimental system comprising of the measuring block and mold base, the stationary side of the mold base has a 210mm×222mm square hole that is fitted with a 180mm×155mm measuring block and guide block for horizontal movement. The measuring block has a rotary cylindrical block (rotary core) which is rotated by a different mechanism. The horizontal position of the measuring block is positioned and fixed by the stepping motor (Oriental Motor Co., Ltd./CRK543APB-H100) controlling the ball screw positioned horizontally, while its vertical position is fixed according to the height of the guide block.

2.2 Rotary System

The measuring block has a built-in rotary cylindrical block (rotary core) with a diameter of 125mm that is connected to a worm wheel. The rotary core has five temperature sensors and five pressure ones which are arranged at equal intervals from the center of rotation taking into account sensor size and measuring efficiency. **Fig. 2** shows the

layout of the sensors on the rotary core. The temperature sensors are made up of a straight infrared irradiation temperature probe (Futaba Corporation /EPSSZT-04.0×030 N150) and high response infrared thermometer (Tokyo Seiko Co. Ltd. /PRH-OK001SHA). The pressure sensors used are piezoelectric pressure sensors of 1.8mm in probe diameter (6006B, Priamus System Technologies AG). Rotation of the rotary core is controlled by a stepping motor (Oriental Motor Co., Ltd./CRK513APB-H100) via the worm gear which is linked to the worm wheel.

This enables sensor positions to be changed up and down easily. By combining the horizontal movement of the measuring block, the sensors can also be moved to any position inside the movable range. **Fig. 3** shows how the stepping motor is connected to the wheel via the worm gear (back of the measuring block). The sensor cables are pulled outside the mold via the groove at the base of the rotary core.

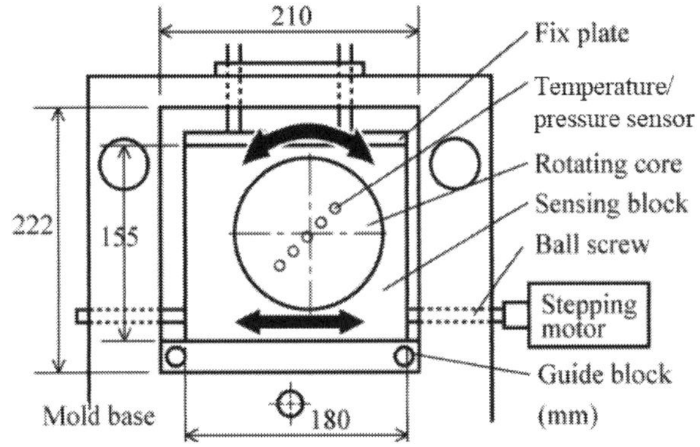

(a) Schematic structure of measuring unit

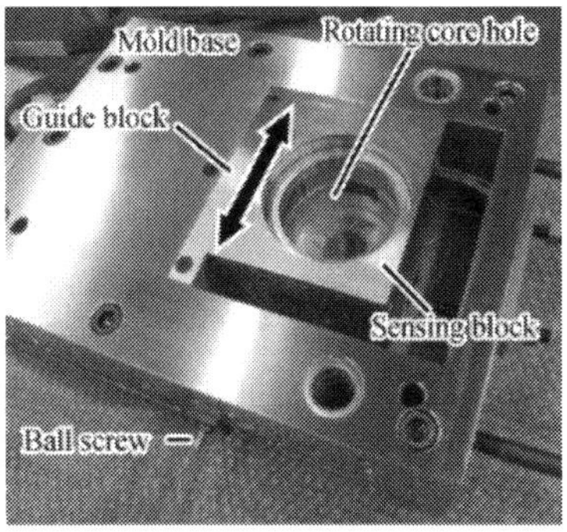

(b) Appearance of measuring unit

Fig. 1 Basic structure of mold-base and measuring unit

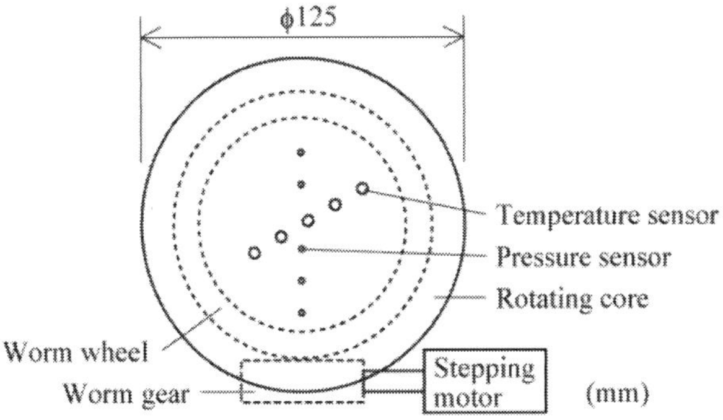

Sensor position from the core center; 3, 9, 15, 21, 27 (mm)

Fig.2 Set-up location of temperature and pressure sensors

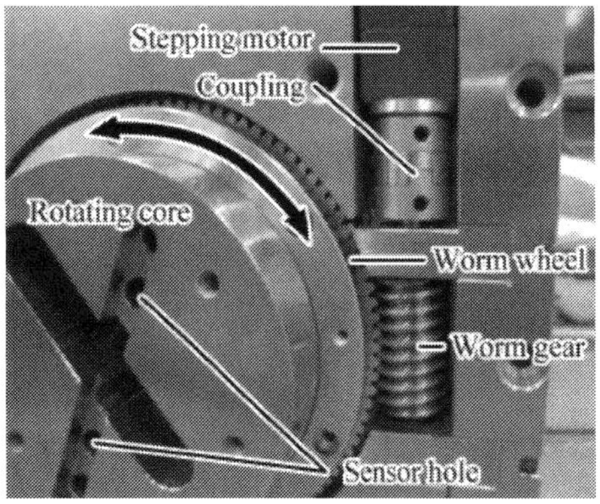

Fig. 3 Picture of rotating unit (Bottom side)

3. Experimental Method

3.1 Temperature Distribution Measurement Experiments

Figure 4 shows the Y-shaped isometric runner and cavity used in this experiment. In the temperature and pressure measurements, two sets of sensors positioned 21 mm and 27 mm from the rotary center respectively were used taking into account the measuring area. The resin used was PP (Prime Polymer Co., Ltd. / J-3054) and the molding machine used was the FANAC ROBOSHOT α-100iA LINEAR. **Table 1** shows the molding conditions.

Regarding the branched portion of the Y-shaped isometric runner, temperature distribution was measured at a total of 289 points (289 shots for each injection rate) located at a pitch of 0.7 mm in the horizontal and vertical directions, and pressure distribution was also measured at a total of 169 points at a pitch of 1.4mm, then two dimensional distribution images were composed by the measuring data. **Fig.5** shows the measured points.

3.2 Visualization Experiments

To investigate the correlation between temperature/pressure distribution and filling behaviour, we used a glass-inserted mold to visualize (1) the resin filling behaviour in each cavity located after Y-shaped runner and (2) the resin filling behaviour at the branched portion of a Y-shaped isometric runner. For (1), we used a high speed video camera (HS4540, Kodak) for videoing behaviour, and recorded images at 4,500 fps. For (2), we used a high speed video camera (MEMRECAM fx k4, nac Image Technology) and recorded images at 40,000 fps. In the observation of the resin filling behavior in each cavity (1), we measured the length for which the resin touches both walls extending to the right and left sides of the gate of each cavity. Specifically, as shown in **Fig.4** (b), taking the left side cavity wall extending 270 degrees left and right from each gate entrance to be P and the right side cavity wall to be Q, we obtained the lengths Lp and Lq of the areas for which P and Q contact the resin respectively. Taking the difference [Lp - Lq] to be α, we quantitatively evaluated the change in α accompanying the screw stroke. On the other hand, in the observation of the resin filling behaviour at the branched portion (2), we mixed tracers into the resin ($0.2mm^2$ aluminium fragment) and multi-synthesized the observed images to extract the streamline of the resin flow.

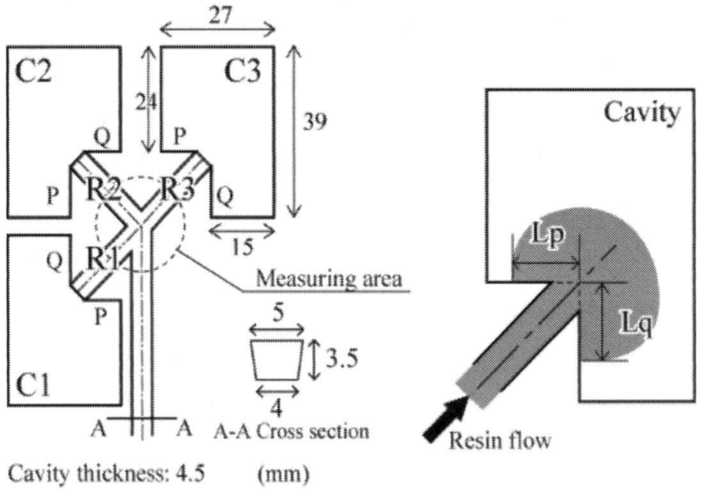

(a) Cavity shape (b) Definition of Lp & Lq

Fig. 4 Shape of runner and cavity

Table 1 Molding conditions

Screw speed (mm/s)	159	1274
Injection rate (cm^3/s)	50	400
Resin	PP / J-3054 Prime Polymer Co., Ltd	
Cylinder temperature* (℃)	225 - 235 - 235 - 215 - 55	
Mold temperature* (℃)	50	

*Set-up value

4. Measurement Results and Discussion

4.1 Measurement of Temperature Distribution

Figures 6 and **7** show changes in the temperature distribution at each screw position S at the injection rates of 50 and 400cm^3/s. At each injection rate, no significant changes are seen in the temperature distribution immediately after resin passes the branched portion of a Y-shaped isometric runner (**Fig.6, 7**(a)), but temperature was seen to rise in the paths to R1 and R3 from the main runner after filling. In **Fig.6, 7**(d), the temperature distribution along the width of each runner in **Fig.5** a and b (3 mm below the corner) is compared in **Fig. 8**. Here, the measured points are numbered from the left in the flow direction. Compared to low injection rates, temperature rise of about 30 degrees was found at high injection rates. Temperature peak was generated near the left wall for R1 and near the right wall for R3. This phenomenon was especially seen clearly at high injection rates, probably due to the rise in resin temperature near the main runner wall as a result of shear heating phenomenon generated by resin filling, and the flow of hot melt along the R1 and R3 walls continuing from the main runner wall. On the other hand, at R2, temperature distribution was more or less flat or a slightly convex shape, which is thought to be due to the flow into the R2 of the resin in the middle area of main runner where temperature gradient is small. Compared to the center of the main runner, resin temperature was found to be consistently low at the center of R2 at the downstream, and this tendency was also found at high injection rates where there are no cooling effects of the wall.

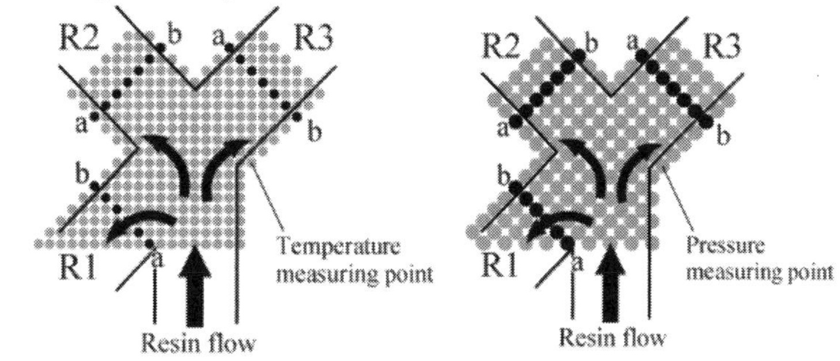

(a) Temperature measuring points (b) Pressure measuring points

Fig. 5 Measuring points at split runner area

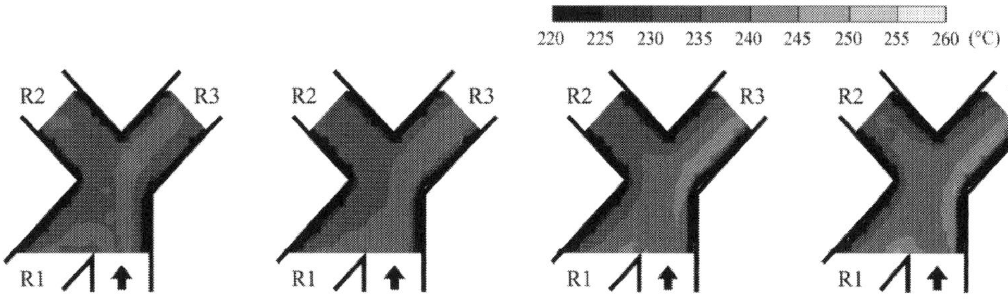

(a) S = 48mm (171ms) (b) S = 42mm (209ms) (c) S = 34mm (260ms) (d) S = 26mm (311ms)

Fig.6 Measuring results showing the change in temperature distribution (Inj. rate 50cm^3/s)

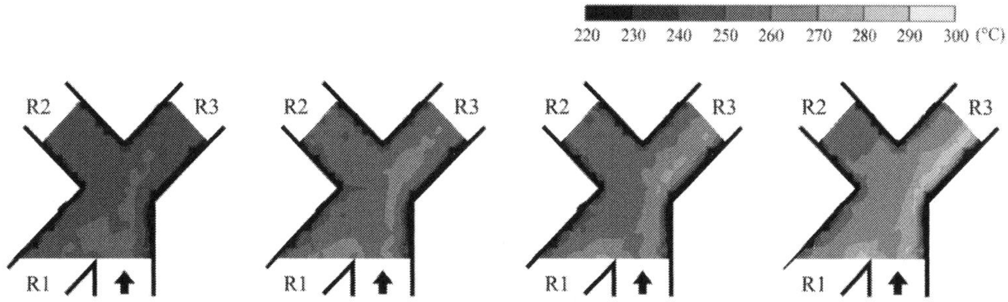

(a) S = 45mm (29ms) (b) S = 40mm (34ms) (c) S = 36mm (37ms) (d) S = 30mm (42ms)

Fig.7 Measuring results showing the change in temperature distribution (Inj. rate 400cm^3/s)

Figure 9 compares the changes in the cavity filling process with screw stroke at the injection rates of 50 and 400cm^3/s taken from the visualization images. The vertical axis is α (Lp - Lq) and the horizontal axis is the screw stroke β, after the screw position at which flow to C1 starts is set to 0. At both injection rates, the left and right filling balance of each cavity was conspicuously different especially at C1 and C3. Filling preceded at P at C1 and at Q at C3. Furthermore, the higher the injection rate, the greater was this difference. These results correspond to those in **Fig. 8**, suggesting that the temperature difference along the runner width directly affects the filling behavior in each cavity.

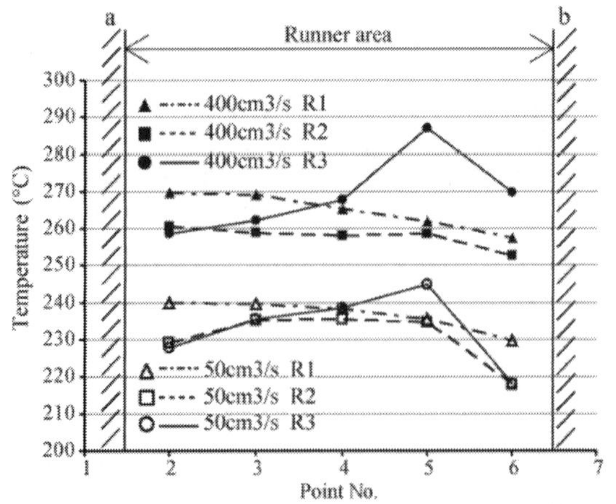

Fig.8 Temperature profiles across each runner width in Fig. 6 (d) & Fig. 7 (d)

4.2 Measurement of Pressure Distribution

Figures 10 and **11** show changes in the pressure distribution at each screw position (S) at the injection rates of 50 and 400 cm^3/s. A sudden drop in pressure was observed from the high pressure area of the runner to the branched portion at both injection rates. Looking at the isobaric line shape of the high pressure area protruded to the branched portion, at the injection rate of 400 cm^3/s, the shape is vertically long compared to the shape at the injection rate of 50cm^3/s, stretching directly from the resin flow direction. This suggests that the isobaric line shape corresponds to the inertial resin flow at high injection rates. At side 'c' of the upper wall of R1, a unique pressure distribution where the pressure at the wall temporarily increases can be seen.

Inside the runner after branching, flat pressure distribution without huge peaks in the width direction was seen. Although temperature drop from the upstream runner to the branched runner was observed as described in the previous section,

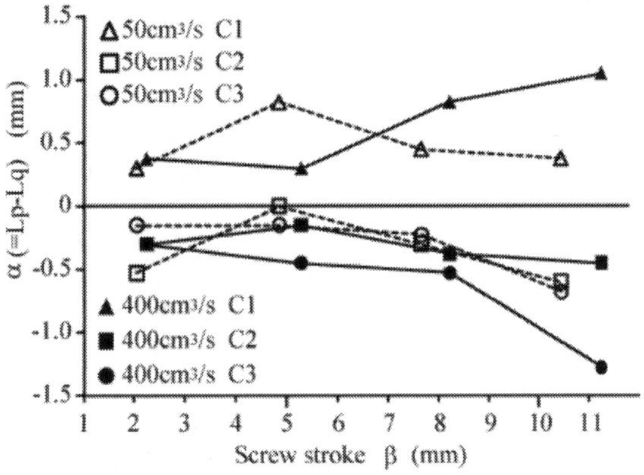

Fig.9 Relationship between contact length difference α and screw stroke β

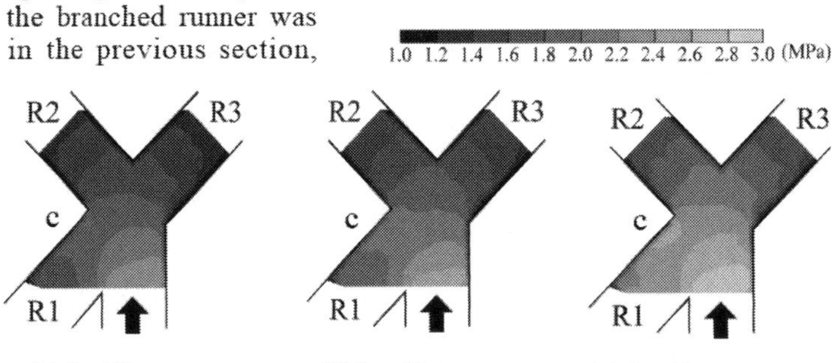

(a) S = 39 mm (b) S = 31 mm (c) S = 23 mm

Fig.10 Measuring results showing the change in pressure distribution (Inj. rate 50cm^3/s)

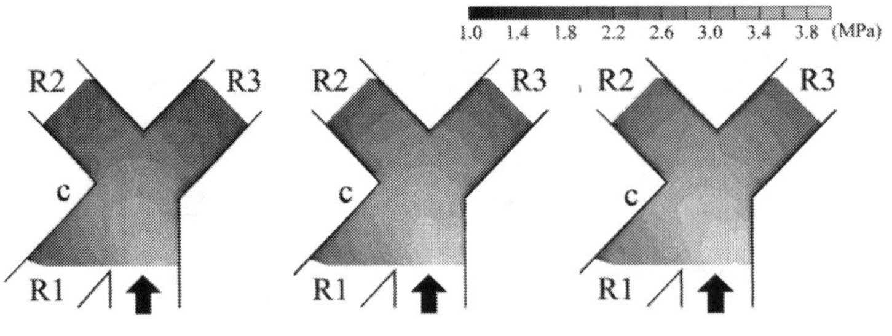

(a) S = 39 mm (b) S = 31 mm (c) S = 23 mm

Fig.11 Measuring results showing the change in pressure distribution (Inj. rate 400cm^3/s)

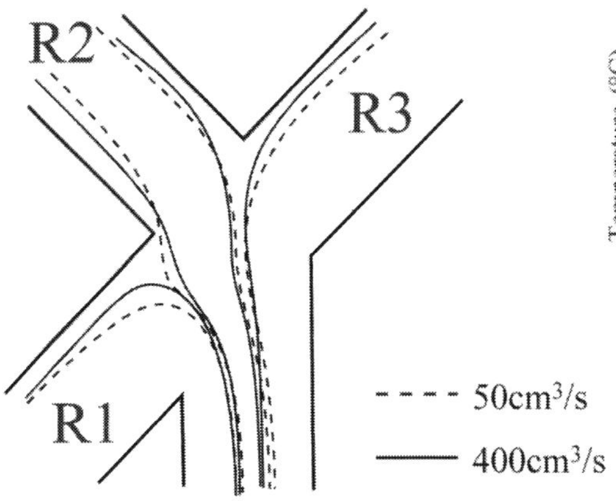

Fig.12 Streamline at Y-shaped runner split zone

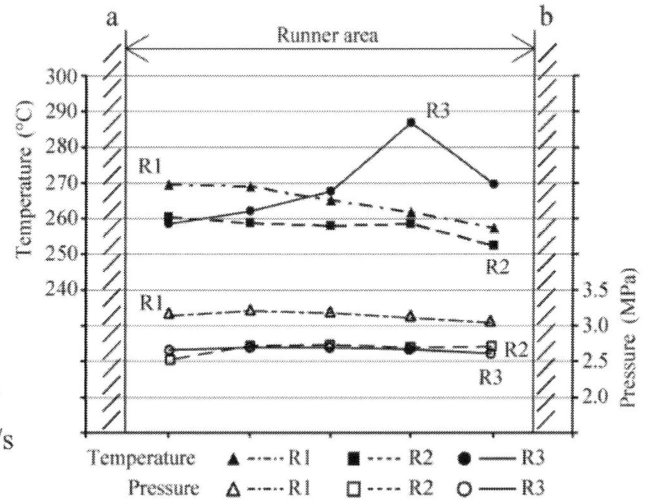

Fig.13 Temperature & pressure profiles across each runner width (Inj. rate 400cm^3/s)

pressure loss during this time was small at less than 1MPa. Such temperature drop is thought to be caused by reduced shear heating effect near the cavity walls due to the sudden decrease in the flow rate at the branched runner, instead of adiabatic expansion and cooling factors. **Fig. 12** compares the streamline at the branched area at the injection rates of 50 and 400 cm^3/s extracted from the tracer behavior of the visualized images. At both injection rates, the center of flow shifts to the left at the branched portion from the main runner to R1, resulting in the formation of a dividing ridge of resin flowing to R1 and R2 at the right wall 'c' of R1. Compared to the streamline at 50 cm^3/s, the straight flow from the main runner is seen to intensify at 400 cm^3/s. These observation results correspond with the generation of the isobaric line from the main runner to the branched portion, and to the formation of the unique high pressure area at 'c' wall where the streamline interferes with the wall.

Depending upon the pressure distribution data in **Fig.11** (c), pressure distribution curves along each runner width in **Fig.5** (b) 'a' and 'b' (downstream position 3 mm from the corner) are extracted and compared with each other in **Fig. 13.** The figure also shows the temperature distribution reported in **Fig.8**. Here, the left side wall to the flow direction is displayed as 'a' and the right side as 'b'. In the previous section, it was confirmed that the melt front is more advanced in the cavity filling process on the elongated side of the runner wall (near wall 'a' in R1 and near wall 'b' in R3), where the temperature peak is reached due to the shear heating effect. When the melt temperature becomes higher due to shear heating, melt viscosity drops. Under the uniform pressure distribution along the runner width, melt flows at relatively higher speeds along the low viscosity layer due to shear heating than along the high viscosity layer in the vicinity. This results in the difference in flow velocity distribution in the runner width direction, and also in cavity width direction. The imbalance of cavity filling is thus thought to occur due to the formation of asymmetrical melt flow velocity along the runner width, which directly reflects the diverging velocity distribution of melt injected into the cavity.

5. Conclusions

The authors developed a measuring mold in which a measuring block is installed with a rotary core embedded with multipoint sensors. By moving the measuring block with a horizontal drive mechanism, temperature and pressure at any position in the movable area can be measured. Using a measuring system based on a rotary cylindrical block, we measured the temperature and pressure of an area of an isometric runner branching into three directions from the main runner in the shape of the letter Y. The results obtained are summarized below.

(1) At the wall continuing from the main runner, the rise in resin temperature due to shear heating effect was greater with increasing injection rate. It was also found that the runner flow resin temperature was the lowest at the central runner among the three-branched runners.

(2) As a result of comparing the filling behaviour in each cavity with the temperature measurement results extracted from visualization images, a clear correlation between the asymmetrical filling behaviour inside each cavity and temperature distribution along the runner width was verified.

(3) A more or less flat pressure distribution was observed in respect to the width in the runner after the branched portion at both higher and lower injection rates. When the injection rate increases, the intervals between the isobaric lines decrease, resulting in the rise in pressure gradient.

(4) Compared to the results of pressure distribution measurement and visualization of the streamline inside the branched portion, an area with rising pressure is formed on the wall of the dividing ridge where the streamline divides the branched area. In addition, corresponding to the shift in the streamline from the upstream runner, the high pressure distribution area protruding from the upstream runner moves to the side at low injection rates, and in a straight direction at high injection rates.

(5) The results of measuring the resin temperature and pressure distribution showed that a high temperature low viscosity area is formed due to shear heating inside the runner, along the walls continuing from the main runner. Because the resin at the high temperature layer along the runner width flows at a higher velocity, an asymmetrical filling phenomenon is thought to be generated in the cavity.

Acknowledgements

The authors would like to thank Mr. Yoshinori Kanetoh of our laboratory for his support in this research and to Futaba Corporation for providing the temperature sensors. This research was conducted by the FY2009 "Ultimate Injection Molding Technology" Research Committee of the Foundation for the Promotion of Industrial Science.

References

[1] K. Jiang Y. Kanetoh, H. Yokoi: PPS 2009 Europe/Africa Regional Meeting, 59 (2009)

In-line monitoring along the process chain: Compounding, extrusion and mould injection

Ingo Alig, Bernd Steinhoff, Dirk Lellinger, Hans Kothe, Torsten Finnberg*

Deutsches Kusnststoff-Institut - Darmstadt - Germany

Since the final properties of plastic parts depend strongly on the processing conditions, we extended or activities from "melt monitoring" [1-4] to monitoring of the final properties of the plastic parts [5]. Different spectroscopic and scattering techniques and methods for other physical properties are compared regarding their potential for different process applications [1-5]. In addition to information on chemical composition and state of the process, the in situ detection of morphology (e.g. in polymer composites and polymer blends) is considered. Optical (UV/VIS, NIR) and ultrasonic attenuation spectra recorded during extrusion were found to be sensitive to chemical composition as well as size and degree of dispersion of micro or nanofillers [1-3]. In-line light scattering allows detection of process-induced structure formation in polymers blends during compounding. By using conductivity sensors [3,4], the influence of extrusion conditions on electrical conductivity was monitored for carbon black and carbon nanotubes. In-line conductivity measurements during mould injection demonstrated the influence of injection velocity, mould and melt temperature on the electrical conductivity of plastic parts of carbon nanotube-polymer compounds [5]. Ultrasonic measurements were used to detect the time-dependent growth of the crystalline boundary during mould injection and validation of FEM simulations. More recently we developed methods to monitor electrostatic properties and crystalline morphology during flat foil extrusion. [1] Alig I., Steinhoff B., Lellinger D., Meas. Sci. Technol. (submitted) [2] Alig, I., Fischer, D., Lellinger, D., Steinhoff, B, Macromol. Symp. 230 (2005) 51 [3] Wang Y., Steinhoff B., Brinkmann Ch., Alig I., Polymer 49 (2008) 1257. [4] Alig I, Lellinger D, Engel M, Skipa T, Pötschke, Polymer 49 (2008)1902. [5] D. Lellinger, D. Xu, A. Ohneiser, T. Skipa, I. Alig, Phys. Stat. Sol. (b) 245 (2008) 2268.

INJECTION MOLDING PROCESS CONTROL USING 2D MODEL PREDICTIVE ITERATIVE LEARNING ALGORITHM

J. Shi[1], Y. Yang[2] and F. Gao[2*]

[1] *Dept. of Chemical & Biochemical Engineering, School of Chemistry & Chemical Engineering, Xiamen University, Xiamen, China – jshi@xmu.edu.cn*

[2] *Dept. of Chemical & Biomolecular Engineering, Hong Kong Univ. of Sci. & Tech., Hong Kong, China – keyy@ust.hk; kefgao@ust.hk*

Abstract - Injection molding is a typical batch process. The control of such a process must maintain certain key process variables to follow predetermined profiles repetitively and accurately to ensure a good part quality. Unlike traditional feedback control algorithm that only utilizes current cycle's response, the information of past controls can be explored for improvement of the current and future cycles. In this paper, an algorithm that combines the iterative learning control (ILC) and the model predictive control (MPC) systematically has been adopted to control the injection molding process. Experimental results of packing pressure control have proved the effectiveness of this 2D model predictive-iterative learning method.

Introduction:

Injection molding is a typical batch process. It mainly consists of injection (filling), packing-holding, and cooling (plastication) three stages that be performed in defined order. The completion of this sequence of stages creates one or more product(s). The process must be repeated to produce more products. The control of such a batch process, unlike that of the continuous processes, involves keeping the defined sequence of operations. At the same time, the control system must also control the key process variables to follow certain profiles repetitively and accurately to ensure a good part quality. Like any manufacturing process, injection molding is affected by various disturbances. These disturbances can come from the materials, machine itself, environments and human interferences. Most disturbances cause the process to drift slowly over cycles instead of abrupt changes. For example, mold temperature may increase from cycle to cycle during the warm-up stage as a result of the continuous molding operation. The dynamics of injection molding process, thus not only changes along the time direction within one cycle, but also varies along the cycle direction. The traditional feedback that only uses the past information of the current cycle is therefore not sufficient for injection molding process. The information of past controls shall be explored for improvement of the current and future cycles.

Iterative learning control (ILC) is motivated to mimic human learning process. It is originally developed for the manipulation of industrial robots, in which it is required to repeat a given task with high precision (Uchiyama, 1978; Arimoto, 1984). By using the repetitive nature of the processes, ILC progressively and iteratively improves the control accuracy along two time dimensions for control input, that is the batch direction from cycle to cycle, and the elapsed time during one cycle from step to step. This two-dimensional learning results in advantages over the conventional real-time feedback control techniques where only one time dimensional input/output information are feed-backed for control decision along the time direction. The conventional ILC scheme, using only the information of previous cycles for control input refinement, is in fact an time-wise open-loop feed-forward control for each cycle that does not have the ability to guarantee control performance along time within the cycle. For this reason, a real-time feedback control is often combined with the conventional ILC, commonly referred as feedback feed-forward ILC, to improve control performances not only along time but also along cycle (Xu, 1998). Previous works of such combination (Amann, 1996; Owens, 2000; Lee, 2000; Chin, 2004) did not provide a unified and robust design algorithm.

Injection molding process controlled by ILC (Havlicsek and Alleyne, 1999; Shi et al., 2001) can be viewed as a two-dimensional (2D) control system, where the dynamics behavior along the time is determined by the process and the feedback control, whereas the ILC law introduces the dynamics along the cycle. Shi *et al.* (2001; 2006; 2007) extended robust control concept to such 2D systems with uncertainties, resulting in an integrated design of a robust real-time feedback control within the cycle and an ILC law from cycle to cycle. It can be seen from their results that model and design a feedback feed-forward ILC system as a 2D system can not only result in a united design of feedback control and cycle-wise ILC but also allow analyzing and optimizing the resulted ILC system along time and cycle indices.

Generalized Predictive Control (GPC) proposed by Clarke et al. (1987a; 1987b) is one of the most popular process control strategy which has been widely adopted in industries with good performance and good robustness. GPC has also been applied to injection molding control with good results (Yang and Gao, 2000). It can be expected that combining GPC as the feedback control with feed-forward ILC will result in good control performances along both the time and cycle. To achieve this, in this paper, a 2D ILC scheme is developed based on the GPC infrastructure for injection molding process control. The resulted 2D model predictive iterative learning control (MPILC) algorithm is introduced briefly in the following control algorithm section first, and then applied to the injection molding process control experimentally using the packing pressure as an example to demonstrate the effectiveness of the proposed control scheme.

2D MPILC Algorithm

Due to the space limitation, the control algorithm is introduced briefly here. The detailed derivation and analysis were given in reference (Shi et al., 2006; Shi et al. 2007).

An SISO batch process can be described by the following Controlled Auto-Regressive Integrated Moving-Average (CARIMA) model

$$\Sigma_P: \quad A(q_t^{-1})y_k(t) = B(q_t^{-1})\Delta_t\left(u_k(t)\right) + w_k(t) \qquad (1)$$

$$t = 0,1,\ldots,T; \quad k = 1,2,\ldots$$

where t and k represent the discrete-time and cycle/batch index, respectively, T is the time duration of each cycle, $u_k(t)$, $y_k(t)$ and $w_k(t)$ are the input, output and disturbance of the process at time t in the kth cycle, respectively, q_t^{-1} indicate the time-wise unit backward-shift operator, $A(q_t^{-1})$ and $B(q_t^{-1})$ are both operator polynomials

$$A(q_t^{-1}) = 1 + a_1 q_t^{-1} + a_2 q_t^{-2} + \cdots + a_n q_t^{-n} \qquad (2)$$

$$B(q_t^{-1}) = b_1 q_t^{-1} + b_2 q_t^{-2} + \cdots + b_m q_t^{-m} \qquad (3)$$

and Δ_t represents the time-wise backward difference operator, i.e. $\Delta_t\left(f(t,k)\right) = f(t,k) - f(t-1,k)$.

For the above repetitive process, introduce an ILC law with the form

$$\Sigma_{ILC}: \quad u_k(t) = u_{k-1}(t) + u_k(t-1) - u_{k-1}(t-1) + r_k(t) \qquad (4)$$

$$u_0(t) = 0, \quad t = -1,0,1,\ldots,T$$

where $r_k(t)$ is referred as the updating law to be determined online based on the MPC philosophy, and $u_0(t)$ is the initial profile of iteration. Let q_k^{-1} represents the cycle-wise unit backward-shift operator, $u_k(t)$ and $r_k(t)$ have the following relationship,

$$u_k(t) = \frac{1}{(1-q_k^{-1})} \cdot \frac{1}{(1-q_t^{-1})} \cdot r_k(t) \qquad (5)$$

which is a 2D system with 2D integral transformation. Control law (4) can be expressed equivalently as

$$\Delta_t\left(u_k(t)\right) = \Delta_t\left(u_{k-1}(t)\right) + r_k(t) \qquad (6)$$

or

$$\Delta_k\left(u_k(t)\right) = \Delta_k\left(u_k(t-1)\right) + r_k(t) \qquad (7)$$

where Δ_k represents the cycle-wise backward difference operator, i.e. $\Delta_k\left(f(t,k)\right) = f(t,k) - f(t,k-1)$.

Substituting (7) into model (1) leads to the following 2D system:

$$\Sigma_{2D-P}: \quad A(q_t^{-1})y_k(t) = A(q_t^{-1})y_{k-1}(t) + B(q_t^{-1})r_k(t) + \Delta_k\left(w_k(t)\right) \qquad (8)$$

where $r_k(t)$, $y_k(t)$ and $\Delta_k\left(w_k(t)\right)$ represent the input, output and disturbance of the system, respectively.

To ensure smooth operations both along the time and cycle directions, the following quadratic cost function is defined as the general control performance index over one cycle for the 2D system (8)

$$J(t,k,n_1,n_2) = \begin{pmatrix} \hat{\boldsymbol{e}}_{k|k}(|_{t+n_1}^{t+1} \mid t) \\ \Delta_t\left(\boldsymbol{u}_k(|_{t+n_2-1}^{t})\right) \\ \Delta_k\left(\boldsymbol{u}_k(|_{t+n_2-1}^{t})\right) \end{pmatrix}^T \begin{pmatrix} \boldsymbol{Q} & \boldsymbol{0} & \boldsymbol{0} \\ \boldsymbol{0} & \boldsymbol{S} & \boldsymbol{0} \\ \boldsymbol{0} & \boldsymbol{0} & \boldsymbol{T} \end{pmatrix} \begin{pmatrix} \hat{\boldsymbol{e}}_{k|k}(|_{t+n_1}^{t+1} \mid t) \\ \Delta_t\left(\boldsymbol{u}_k(|_{t+n_2-1}^{t})\right) \\ \Delta_k\left(\boldsymbol{u}_k(|_{t+n_2-1}^{t})\right) \end{pmatrix} + \boldsymbol{r}_k^T(|_{t+n_2-1}^{t})\boldsymbol{R}\boldsymbol{r}_k(|_{t+n_2-1}^{t})$$

$$= \hat{\boldsymbol{X}}_{k|k}^T(t)\bar{\boldsymbol{Q}}\hat{\boldsymbol{X}}_{k|k}(t) + \boldsymbol{r}_{k+l}^T(|_{t+n_2-1}^{t})\boldsymbol{R}\boldsymbol{r}_{k+l}(|_{t+n_2-1}^{t}) \qquad (9)$$

where the notations

$$\boldsymbol{f}_k\left(|_{t_2}^{t_1}\right) = \left(f_k(t_1) \quad f_k(t_1+1) \quad \cdots \quad f_k(t_2)\right) \qquad f \in \{y,u,r,e\} \qquad (10)$$

and $\hat{\boldsymbol{f}}_{k|k}\left(|_{t+n_1}^{t+1} \mid t\right)$ indicates the prediction vector of k th cycle over the time-wise prediction horizon based on the history information before time t of k th cycle, and

$$\hat{\boldsymbol{e}}_{k|k}(|_{t+n_1}^{t+1} \mid t) = \boldsymbol{y}_r(|_{t+n_1}^{t+1}) - \hat{\boldsymbol{y}}_{k|k}(|_{t+n_1}^{t+1} \mid t) \qquad (11)$$

$$\hat{\boldsymbol{X}}_{k|k}(t) = \begin{pmatrix} \hat{\boldsymbol{e}}_{k|k}(|_{t+n_1}^{t+1} \mid t) \\ \Delta_t\left(\boldsymbol{u}_k(|_{t+n_2-1}^{t})\right) \\ \Delta_k\left(\boldsymbol{u}_k(|_{t+n_2-1}^{t})\right) \end{pmatrix} \qquad (12)$$

$$\boldsymbol{Q} = diag\{\eta(1),\ldots,\eta(n_1)\} \qquad (13)$$

$$\boldsymbol{R} = diag\{\alpha(0),\ldots,\alpha(n_2-1)\} \qquad (14)$$

$$\boldsymbol{S} = diag\{\beta(0),\ldots,\beta(n_2-1)\} \qquad (15)$$

$$\boldsymbol{T} = diag\{\gamma(0),\ldots,\gamma(n_2-1)\} \qquad (16)$$

$$\bar{\boldsymbol{Q}} = diag\{\boldsymbol{Q},\boldsymbol{S},\boldsymbol{T}\} \qquad (17)$$

By the procedure proposed in reference[10][11], the best prediction model for the 2D system (8) can be formulated as follows:

$$\hat{\boldsymbol{y}}_{k|k}\left(|_{t+n_1}^{t+1} \mid t\right) = \boldsymbol{G}\boldsymbol{r}_k\left(|_{t+n_2-1}^{t}\right) + \boldsymbol{y}_{k-1}\left(|_{t+n_1}^{t+1}\right) + \boldsymbol{F}_k(t) \qquad (18)$$

the detailed formulation of matrix $\boldsymbol{G}$ and $\boldsymbol{F}_k(t)$ can be found in reference [10][11]

To minimize cost function (11), the following relationships between variables $\Delta_t\left(\boldsymbol{u}_k(|_{t+n_2-1}^{t})\right)$, $\Delta_k\left(\boldsymbol{u}_k(|_{t+n_2-1}^{t})\right)$ and updating variable $\boldsymbol{r}_k(|_{t+n_2-1}^{t})$ are required

$$\Delta_t\left(\boldsymbol{u}_k(|_{t+n_2-1}^{t})\right) = \boldsymbol{r}_k(|_{t+n_2-1}^{t}) + \Delta_t\left(\boldsymbol{u}_{k-1}(|_{t+n_2-1}^{t})\right) \qquad (19)$$

$$\boldsymbol{r}_k(|_{t+n_2-1}^{t}) = \boldsymbol{H}\Delta_k\left(\boldsymbol{u}_k(|_{t+n_2-1}^{t})\right) - \Delta_k\left(u_k(t-1)\right)\Big|_{n_2} \qquad (20)$$

where

$$
\boldsymbol{H}=\begin{pmatrix}1 & 0 & 0 & \cdots & 0\\ -1 & 1 & 0 & \cdots & 0\\ 0 & -1 & 1 & \cdots & 0\\ \vdots & \vdots & \vdots & \ddots & \vdots\\ 0 & 0 & 0 & \cdots & 1\end{pmatrix}_{n_2\times n_2},\quad \Delta_k\left(u_k(t-1)\right)\Big|_{n_2}=\begin{pmatrix}\Delta_k\left(u_k(t-1)\right)\\ 0\\ 0\\ \vdots\\ 0\end{pmatrix}_{n_2\times 1}
$$

Let $\boldsymbol{V}=\boldsymbol{H}^{-1}$, then

$$
\Delta_t\left(\boldsymbol{u}_k(|^{t}_{t+n_2-1})\right)=\boldsymbol{V}\boldsymbol{r}_k(|^{t}_{t+n_2-1})+\boldsymbol{V}\Delta_k\left(u_k(t-1)\right)\Big|_{n_2} \qquad (22)
$$

Together with (18)(19)(22), the following generalized 2D prediction model is obtained

$$
\hat{\boldsymbol{X}}_{k|k}(t)=\bar{\boldsymbol{G}}\boldsymbol{r}_k(|^{t}_{t+n_2-1})+\bar{\boldsymbol{A}}\boldsymbol{X}_{k-1}(t)+\boldsymbol{W}_k(t) \qquad (23)
$$

where

$$
\bar{\boldsymbol{G}}=\begin{pmatrix}-\boldsymbol{G}\\ \boldsymbol{I}\\ \boldsymbol{V}\end{pmatrix},\ \bar{\boldsymbol{A}}=\begin{pmatrix}\boldsymbol{I} & \boldsymbol{0} & \boldsymbol{0}\\ \boldsymbol{0} & \boldsymbol{I} & \boldsymbol{0}\\ \boldsymbol{0} & \boldsymbol{0} & \boldsymbol{0}\end{pmatrix} \qquad (24)
$$

$$
\boldsymbol{X}_{k-1}(t)=\begin{pmatrix}\boldsymbol{e}_{k-1}(|^{t+1}_{t+n_1})\\ \Delta_t\left(\boldsymbol{u}_{k-1}(|^{t}_{t+n_2-1})\right)\\ \Delta_k\left(\boldsymbol{u}_{k-1}(|^{t}_{t+n_2-1})\right)\end{pmatrix},\ \boldsymbol{W}_k(t)=\begin{pmatrix}-\boldsymbol{F}_k(t)\\ \boldsymbol{0}\\ \boldsymbol{V}\Delta_k\left(u_k(t-1)\right)\Big|_{n_2}\end{pmatrix} \qquad (25)
$$

$$
\boldsymbol{e}_{k-1}(|^{t+1}_{t+n_1})=\boldsymbol{y}_r(|^{t+1}_{t+n_1})-\boldsymbol{y}_{k-1}(|^{t+1}_{t+n_1}) \qquad (26)
$$

Based on the above prediction model, it results from optimization algorithm that the quadratic cost function (11) is minimized by the following optimal control law

$$
\boldsymbol{r}_k^*(|^{t}_{t+n_2-1})=-\left(\boldsymbol{R}+\bar{\boldsymbol{G}}^T\bar{\boldsymbol{Q}}\bar{\boldsymbol{G}}\right)^{-1}\bar{\boldsymbol{G}}^T\bar{\boldsymbol{Q}}\left(\bar{\boldsymbol{A}}\boldsymbol{X}_{k-1}(t)+\boldsymbol{W}_k(t)\right) \qquad (27)
$$

It follows from the definitions (9)(12)(24) that

$$
\begin{aligned}
\boldsymbol{r}_k^*(|^{t}_{t+n_2-1})&=\left(\boldsymbol{R}+\boldsymbol{G}^T\boldsymbol{Q}\boldsymbol{G}+\boldsymbol{S}+\boldsymbol{V}^T\boldsymbol{T}\boldsymbol{V}\right)^{-1}\boldsymbol{G}^T\boldsymbol{Q}\left(\boldsymbol{e}_{k-1}(|^{t+1}_{t+n_1})-\boldsymbol{F}_k(t)\right)\\
&-\left(\boldsymbol{R}+\boldsymbol{G}^T\boldsymbol{Q}\boldsymbol{G}+\boldsymbol{S}+\boldsymbol{V}^T\boldsymbol{T}\boldsymbol{V}\right)^{-1}\boldsymbol{S}\Delta_t\left(\boldsymbol{u}_{k-1}(|^{t}_{t+n_2-1})\right)\\
&-\left(\boldsymbol{R}+\boldsymbol{G}^T\boldsymbol{Q}\boldsymbol{G}+\boldsymbol{S}+\boldsymbol{V}^T\boldsymbol{T}\boldsymbol{V}\right)^{-1}\boldsymbol{V}^T\boldsymbol{T}\boldsymbol{V}\Delta_k\left(u_k(t-1)\right)\Big|_{n_2}
\end{aligned} \qquad (28)
$$

Now, let $\boldsymbol{K}_1$ and $\boldsymbol{K}_2$ be the first rows of matrices $\left(\boldsymbol{R}+\boldsymbol{G}^T\boldsymbol{Q}\boldsymbol{G}+\boldsymbol{S}+\boldsymbol{V}^T\boldsymbol{T}\boldsymbol{V}\right)^{-1}\boldsymbol{G}^T\boldsymbol{Q}$ and $-\left(\boldsymbol{R}+\boldsymbol{G}^T\boldsymbol{Q}\boldsymbol{G}+\boldsymbol{S}+\boldsymbol{V}^T\boldsymbol{T}\boldsymbol{V}\right)^{-1}\boldsymbol{S}$, respectively, and K_3 be the up-left element of matrix $-\left(\boldsymbol{R}+\boldsymbol{G}^T\boldsymbol{Q}\boldsymbol{G}+\boldsymbol{S}+\boldsymbol{V}^T\boldsymbol{T}\boldsymbol{V}\right)^{-1}\boldsymbol{V}^T\boldsymbol{T}\boldsymbol{V}$, then, the 2D-MPILC scheme is given as follows

$$
\begin{aligned}
\Sigma_{2D-GPILC}:\quad \Delta_t\left(u_k(t)\right)&=\Delta_t\left(u_{k-1}(t)\right)+\boldsymbol{K}_1\left(\boldsymbol{e}_{k-1}(|^{t+1}_{t+n_1})-\boldsymbol{F}_k(t)\right)\\
&+\boldsymbol{K}_2\Delta_t\left(\boldsymbol{u}_{k-1}(|^{t}_{t+n_2-1})\right)+K_3\Delta_k\left(u_k(t-1)\right)
\end{aligned} \qquad (29)
$$

Experimental Setup

All the experiments were conducted on a Chen Hsong reciprocating-screw injection molding machine, model number JM88MKIII, located in the Advanced Materials Lab at HKUST. The machine clamping tonnage is 88 tons with a maximum shot weight of 128 g.

A KISTLER 6171BA piezo-electric pressure transducer is mounted on the injection (21) nozzle to measure the nozzle pressure, as shown in Figure 1.

The control system is developed on a real-time multitasking operating system platform. A set of real-time programs is developed in house to perform the data acquisition, machine sequence control, and key process variables control. The hardware system is shown in Figure 2, including the power supply system, transducers, data acquisition system, and the control computer.

A flat mold is used for all the experiments, and the material is high-density polyethylene.

Results and Discussion

Nozzle packing pressure

Packing is an important stage in the injection molding process. Too much packing could result in highly stressed parts and mold flush. It may also cause ejection problems. Insufficient packing may cause poor surfaces, sink marks and uneven shrinkage. Proper packing control is critical for producing molded parts with good mechanical properties, dimensional stability, and satisfactory surface characteristics. The degree of packing is indicated well by the cavity pressure, which is determined by nozzle pressure, cavity geometry, and mold temperatures. However, control of cavity pressure requires each mold to be fitted with a pressure transducer. The hardware expenses plus the tooling and maintenance cost make it difficult to be accepted by industry. Nozzle pressure, a process variable close to the mold cavity, can better indicate the degree of packing. Unlike a cavity pressure transducer, a nozzle pressure transducer installed at an injection nozzle can be used directly for different molds. It is therefore adopted in this paper to demonstrate the performance of proposed 2D MPILC.

Nozzle Pressure Prediction Model

The nonlinear and time-varying characteristics of nozzle pressure have been studied extensively in references (Yang and Gao, 1998; Yang and Gao, 1999). Since the analysis of the nozzle pressure through first-principle model is difficult, a data based first-order input-output model is established using the open-loop test results, as follows:

$$
(1-0.8596q_t^{-1})y_k(t)=1.408u_k(t)+w_k(t) \qquad (30)
$$

Figure 3 gives the comparison between the actual nozzle packing pressure response and the output of the above model. It is shown that the pressure response of the first cycle is different from the other cycles, which is caused by the material leakage during the machine initial heat-up. This difference may cause initial problem for the iterative learning. The discrepancy between the actual response and the model prediction also clearly indicates that the model has significant mismatch.

2D MPILC Design and Experimental Test

The 2D MPILC is designed using the design parameters listed in Table 1. The resulted controller parameters are also listed in Table 1. The controller has been tested on the machine experimentally. To perform a thorough test over a wide range of operating conditions, the injection velocity that affects the initial status of packing and nozzle pressure settings are changed to combine 5 operating conditions. All the five conditions are conducted continuously in 107 cycles. The cycle number and the corresponding operating conditions are listed in Table 2. The experimental results of these five conditions are shown in Figures 4 to 8, respectively, where Figures 4-8 (a) shows the nozzle pressure responses and Figures 4-8(b) plots the corresponding proportional valve opening.

It is shown in Figures 4 to 8 that in all the five conditions, the control can converge within 1 to 2 cycles, and keep the good performance with the continuous molding cycles. Even though the nozzle pressure dynamics change for different operating conditions, and initial uncertainties are existing in all the condition changes, a good and consistent control result can be achieved using the proposed 2D MPILC controller.

Conclusions

The injection molding process controlled by an ILC is transformed to a 2D system. A 2D MPILC controller is then applied to provide a unified and robust control along both the time direction and the cycle direction. The controller is demonstrated experimentally using the nozzle pressure as an example. It is shown that even though the prediction model has significant mismatch, the controller designed still can achieve good control results. Different operating conditions have been tested, and the results show that the control can converge within 1 to 2 cycles, and keep good results in the following cycles. These results prove the robustness of the proposed controller.

References:

Amann N., et al., "Iterative learning control using optimal feedback and feedforward actions", *Int. J. Control*, Vol. 65(2), p277-293 (1996)

Arimoto S., et al., "Bettering operation of robots by learning", *Journal of Robotic Systems,* Vol.1(2), p123-140 (1984)

Chin I., et al., "A Two-Stage Iterative Learning and Batch Control Technique with Independent Disturbance Rejection Capability", *Automatica*, **40**(11), p1913-1922 (2004)

Clarke D.W., et al., "Generalized predictive control – Part I. The basic algorithm", *Automatica*, **23**(2), p137-148 (1987)

Clarke D.W., et al., "Generalized predictive control – Part II. Extensions and interpretations", *Automatica*, **23**(2), p149-160 (1987)

Havlicsek H., Alleyne A., "Nonlinear control of an electrohydraulic injection molding machine via iterative adaptive learning", *IEEE/ASME Transactions on Mechatronics*, **4**, p312-323 (1999)

Lee J. H., et al., "Model-based iterative learning control with a quadratic criterion for time-varying linear systems", *Automatica*, **36**, p641-657 (2000)

Owens D. H., et al., "Analysis of linear iterative learning control schemes – a 2D system/repetitive processes approach", *Multidimensional Systems and Signal Processing*, **11**, p125-177 (2000)

Shi, J., F. Gao, and T. Wu, "Robust Iterative Learning Control with Applications to Injection Molding Process", *Chemical Engineering Science*, **56**, 7025. (2001)

Shi, J., F. Gao, and T. Wu, "Single-cycle and Multi-cycle Generalized 2D Model Predictive Iterative Learning Control (2D-GPILC) Schemes for Batch processes". *Journal Of Process Control*, **17**(9): p. 715-727 (2007).

Shi, J., F. Gao, and T. Wu. "2D Model Predictive Iterative Learning Control Schemes for Batch Processes". *IFAC International Symposium on Advanced Control of Chemical Processes.* Gramado, Brazil.(2006)

Uchiyama M., "Formation of high-speed motion pattern of a mechanical arm by trial", *Transactions of SICE (Society for Implementation and Control Engineers) (in Japanese)*, **14**, 706 (1978)

Xu J.-X., Bien Z. Z., "The frontiers of iterative learning control", *Iterative Learning Control: Analysis, Design, Integration and Application.* Kluwer Academic Publishers, Boston/Dordrecht/London, p9-35 (1998)

Yang Y., Gao F., "Adaptive Control of the Filling Velocity of Thermoplastics Injection Molding", *Control Engineering Practice*, **8**, 1285 (2000)

Yang, Y., Gao, F., "Adaptive Control of Nozzle Melt Packing Pressure", *Journal of Intelligent Material Systems and Structures*, **9**, 1046. (1998)

Yang, Y., Gao, F., "Cycle-to-cycle and Within-cycle Adaptive Control of Nozzle Pressure During Packing-holding for Thermoplastic Injection Molding", *Polymer Engineering & Science.*, **39**, 2042. (1999)

Table 1: List of controller parameters and design parameters

Design Parameters	Controller Parameters
$n_1 = n_2 = 5$	$K_1' = (0.0013\ 0.0024\ 0.0033\ 0.0040\ 0.0046)$
$\alpha(j) = 1000,\ j = 0,1,2,3,4$	$K_2' = (0\ 9.3091e\text{-}004\ 6.2538e\text{-}004\ 3.4665e\text{-}004\ 1.2717e\text{-}004)$
$\beta(0)=0,\ \beta(j) = 20,\ j = 1,2,3,4$	$K_3' = Null$
$\gamma(j) \equiv 0,\ j = 0,1,2,3,4$	$K_4' = (0.0383\ \text{-}0.0539)$
$\eta(i) = 1,\ i = 1,2,3,4,5$	$K_5' = 0$

Table 2: List of experimental test operating conditions

Condition Num.	Cycle Num.	Operating Conditions (set points) (*MT: Mold Temperature; BT: Barrel Temperature; IV: Injection Velocity; NP: Nozzle Pressure*)
1	1~42	*MT*:25°C; *BT*:200°C; *IV*:15mm/s; *NP*:350bar
2	43~61	*MT*:25°C; *BT*:200°C; *IV*:25mm/s; *NP*:350bar
3	62~76	*MT*:25°C; *BT*:200°C; *IV*:15mm/s; *NP*:425bar
4	78~91	*MT*:25°C; *BT*:200°C; *IV*:20mm/s; *NP*:425bar
5	92~107	*MT*:25°C; *BT*:200°C; *IV*:25mm/s; *NP*:425bar

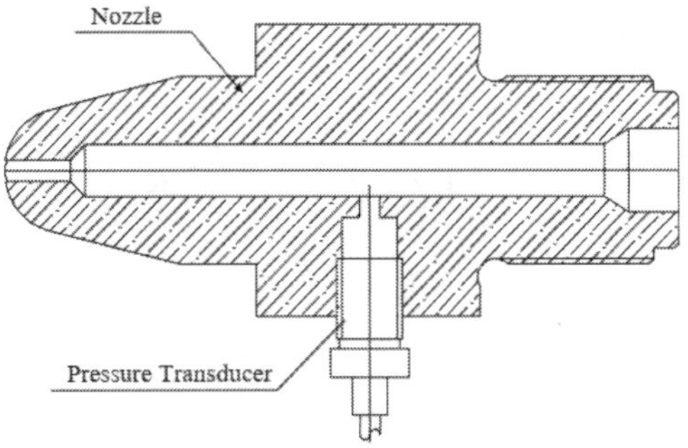

Figure 1: Installation of nozzle pressure transducer

Figure 2: IMM control system in HKUST

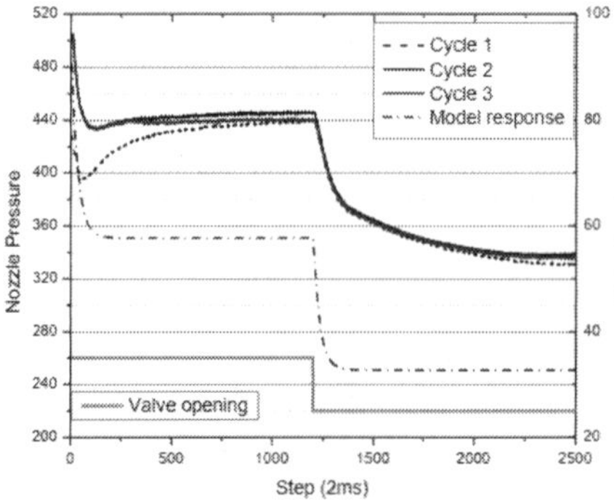

Figure 3: Comparison of the actual nozzle packing pressure response and the output of the model

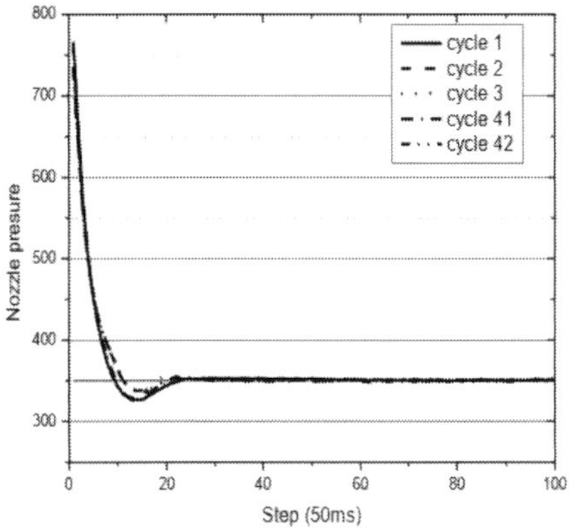

(a)

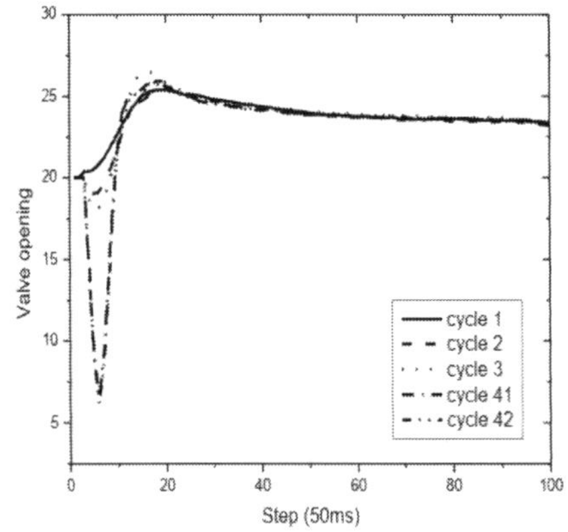

(b)

Figure 4: Experimental control results, condition 1. (a) nozzle pressure measurements and (b) corresponding control valve opening

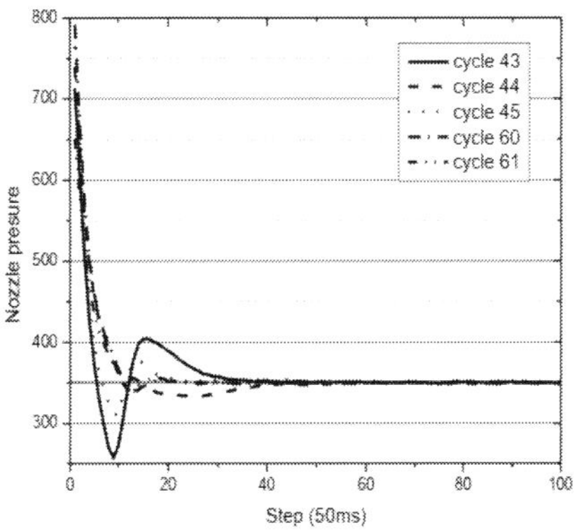

(a)

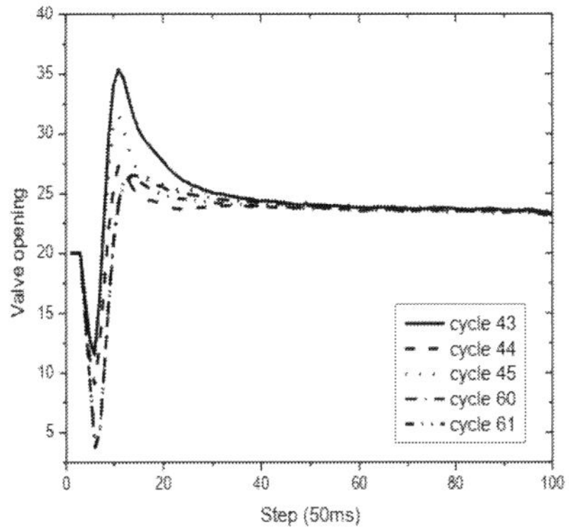

(b)

Figure 5: Experimental control results, condition 2. (a) nozzle pressure measurements and (b) corresponding control valve opening

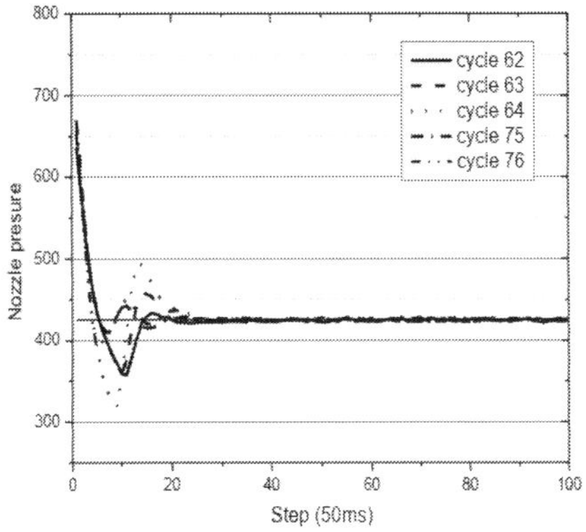

(a)

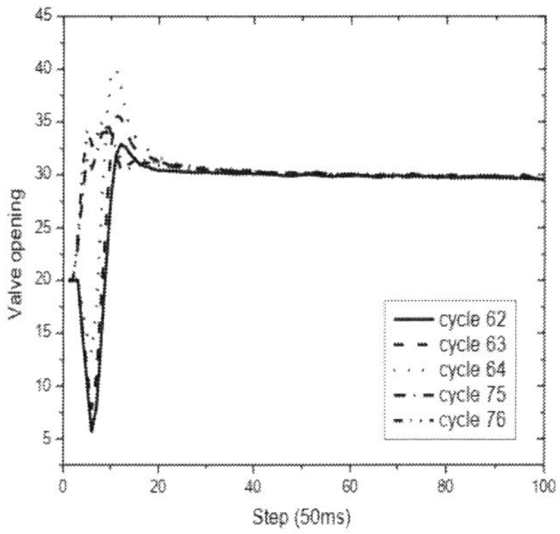

(b)

Figure 6: Experimental control results, condition 3. (a) nozzle pressure measurements and (b) corresponding control valve opening

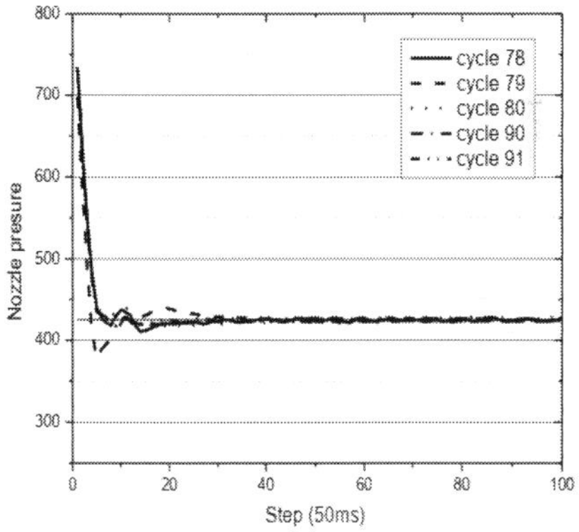

(a)

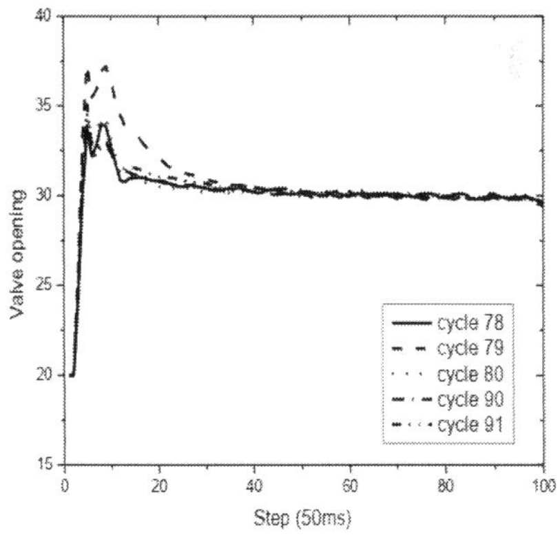

(b)

Figure 7: Experimental control results, condition 4. (a) nozzle pressure measurements and (b) corresponding control valve opening

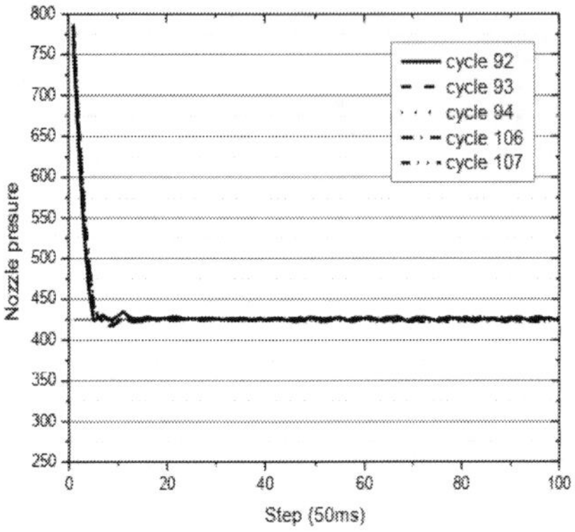

(a)

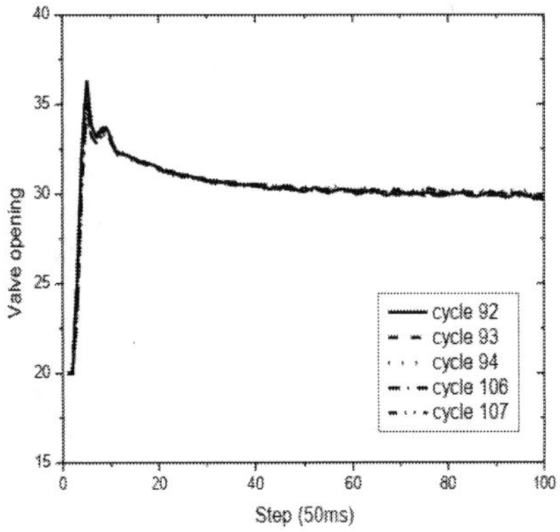

(b)

Figure 8: Experimental control results, condition 5. (a) nozzle pressure measurements and (b) corresponding control valve opening

FEASIBILITY ANALYSIS OF AN IN-MOLD MULTIVARIATE SENSOR

David O. Kazmer[1*], Stephen P. Johnston[1], and Robert X. Gao[2]

[1] *University of Massachusetts Lowell, 1 University Avenue, Lowell, MA 01854; david_kazmer@uml.edu; stephen_johnston@uml.edu*
[2]*University of Connecticut, 191 Auditorium Road, Unit 3139, Storrs, CT 06269; rgao@engr.uconn.edu*

Abstract: The design of a multivariate sensor is described for the measurement of melt temperature, melt pressure, melt velocity, melt viscosity, and mold temperature. Melt pressure and temperature are respectively obtained through the incorporation of a piezo-ceramic element and infrared photodetector within the sensor head. Melt velocity is derived from the initial response of the melt temperature as the polymer melt flows across the sensor's lens. The apparent melt viscosity is then derived based on the melt velocity and the time derivative of the increasing melt pressure given the cavity thickness. The feasibility of the envisioned sensor are analyzed using an instrumented mold including piezoelectric pressure sensors, infrared pyrometer, and thermocouples. Several predictive models of part weight are developed using multiple regression of data obtained from a design of experiments to evaluate the efficacy of the multivariate sensor. The results indicate a correlation coefficient, R^2, of 0.79 for a model based on the machine settings, 0.80 for a model based on a cavity pressure sensor, 0.90 for a model based on the multivariate sensor, and 0.98 for a non-linear model based on the multivariate sensor. The implication is that multiple orthogonal sensing streams with high fidelity models are necessary to provide automatic quality assurance sufficient for fully automated plastics manufacturing.

Introduction

Continuing advances in machine and control system designs have been made such that response and repeatability of the machinery have improved dramatically (Agrawal, Pandelidis et al. 1987; Schulx 2005; Giboz, Copponnex et al. 2007). While such advances in machine control are important, a fundamental difficulty in polymer processing is the transient and distributed state of the polymer melt. Four key states – pressure, temperature, velocity, and viscosity – are known to govern the quality of manufactured products through the constitutive viscoelastic behavior of the polymer being processed (Mann 1974; Dininger 1994; Macosko 1994; Coates and Speight 1995; Dubay, Bell et al. 1997; Tanner 2000; Malkin and Isayev 2005). Yet, the vast majority of production machinery uses only conventional machine sensors for closed-loop control according to the machine setpoints. Plastics manufacturers typically use this machine-centric approach with set control limits hoping that their setup is "good enough." As a result, the *in-situ* states of the polymer are neither accurately observed nor properly controlled, leading to suboptimal processes, low yields, and environmental waste.

Striving to remain competitive and economically viable, plastics manufacturers are increasingly relying on fully automated systems for fault diagnosis, quality control, and materials handling (Aeppel 2002). Such system designs must operate in an on-line, in-process fashion for effective polymer process control and optimization (Kazmer 2009). Unfortunately, every plastics manufacturing application is unique with respect to the mold geometry, material properties, and processing conditions. The ideal sensor suite is therefore unique with respect to the number, type, position, and use of signals to provide the best process observability at lowest cost. Numerous instrumentation companies have developed sensing and control philosophies specific to the plastics industry based on measurement of pressure and/or temperature [ref]. However, no single control strategy or system design has been universally successful and defective components are continuing to be manufactured in high volume production (Kazmer 2006).

Accordingly, there is an urgent need for more advanced and intelligent in-mold sensors that can provide comprehensive, real-time state feedback of the process internal to the mold where the product is formed. The measurement and analysis of a complete set of process states remains one of the most significant barriers to the success of more advanced polymer processing control systems.

Related Research

While nearly all plastics manufacturing machinery has sensors, the instrumentation suite varies widely with respect to type and specification. Furthermore, instrumentation for *in situ* monitoring and control of the polymer states is nearly non-existent. It has been estimated that less than 1% of processes have been instrumented with in-mold sensors (Groleau 2002), and that most of the molds which are instrumented typically have only in-mold thermocouples or pressure transducers. As such, the quantity and quality of information gleaned from these sensors is often inadequate.

More advanced sensing platforms have been developed though few are yet common in industry use. Infrared melt pyrometers have existed for some forty years (Orroth 1974). Infrared pyrometers provide faster rise times and greater sensitivity for measurement of melt temperature than thermocouples but their expense, fragility, and calibration remain barriers in most applications. Optical methods have also been invented to measure orientation by fluorescence (Bur and Thomas 1997; Thomas and Bur 1999) as well as in-mold shrinkage by optical interferometry (Thomas and Bur 1999). In-mold part shrinkage sensing has also been accomplished by examination of the strain recovery in a sensor initially loaded by the melt pressure (Panchal and Kazmer 2009). Such energy harvesting from the polymer melt for wireless sensor excitation has also been explored for dual pressure and temperature measurement (Gao, Fan et al. 2008). Remote monitoring of the polymer melt location during mold filling has also been enabled by ultrasonic scanning across the mold cavity; similar techniques can also indicate cooling, solidification, and shrinkage (Wang, Cao et al. 1997; Franca, Jen et al. 2000; Edwards and Thomas 2001; Michaeli and Starke 2005).

Sensor data must be processed for automatic interpretation of the process behavior and determination of any corrective control actions. Most practitioners rely on simple statistics such as the average, minimum, maximum, and duration for process representation. More detailed data features can be calculated based on more advanced signal processing techniques (Kazmer and Westerdale 2009). After a process is investigated according to a design of experiments or other techniques, empirical models are derived on an application-specific basis to estimate the process outputs, such as part weight or dimensions, as a function of the data features or machine settings. Many such studies have been performed using neural networks (Woll, Cooper et al. 1996), response surface methods (Kazmer, Lotti et al. 2004), fuzzy logic (Chiang 2007), principal component analysis (Lu, Gao et al. 2004; Kazmer, Westerdale et al. 2008), and other modeling approaches (Chen and Turng 2007).

Incorporation of mechanistic process simulations within on-line quality control systems has been attempted (Kazmer, Nageri et al. 2006; Johnston, Kazmer et al. 2009). However, their success has been limited given the numerous assumptions required for simulation and the lack of detailed geometry and constitutive material models on a real time basis. For these reasons, we currently believe that the best approach for determination of product quality is the acquisition of orthogonal multivariate data streams from one or more sensors coupled with extensive development and validation of high fidelity models on an application specific basis. The next section suggests the design of such a multivariate sensor.

Instrumentation

For research related to on-line process simulation, an ASTM test mold was extensively instrumented with five piezoelectric pressure transducers (Priamus 6001A) at critical junctions in the mold, two type N in-mold thermocouples (Primaus 4001A) having exposed junctions located at the end of flow, and an infrared melt pyrometer (Omega OS1562) opposite one of the in-mold thermocouples (Kazmer, Westerdale et al. 2008; Kazmer and Westerdale 2009). All experimentation was performed on an all-electric molding machine (50 ton Ferromatic Milacron) having a 25 mm diameter screw. This machine was extensively instrumented with other pressure, temperature, and displacement transducers not central to this study (Johnston 2008). All process data from both the machine and in-mold sensors were acquired at 2 ms intervals using a data acquisition system (Priamus eDart).

Figure 1 provides the cavity geometry, sensor locations, and typical traces obtained for a polypropylene resin with a 2 s fill time; the nominal thickness of the mold cavities was 3.175 mm. The data provides many interesting insights into both the polymer behavior. For example, the early filling of the plaque cavity at 1 s produces a large pressure spike in the injection and runner pressure as the polymer compresses and redirects down the other runner primaries. As another example, the pressure in the tensile bar cavity is the first to decay given its smaller gate and distance from the sprue. This data indicates that the total injection forward time could be reduced from 12.5 to 9.5 s with no decrease in the density of the molded parts.

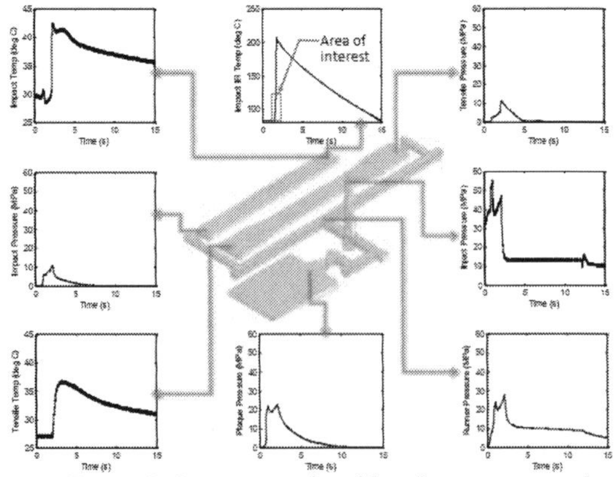

Figure 1: Instrumented mold with process signals

The data of Figure 1 provides some interesting insights into the behavior of the process sensors as well. In-mold, type N thermocouples are being increasingly used for detection of the arrival of the polymer melt at the end of the mold cavity. The response time of these thermocouples is quite good, on the order of 0.05 s, though they may sometimes provide a false signal due to the compression and heating of the air of the polymer melt in the closed cavity ahead; such a behavior is observed at a time of 1 s in the upper left graph providing the thermocouple output located at the end of the impact bar cavity. By comparison, the infrared pyrometer located across the cavity provides an even faster rise time than the thermocouple as well as much better responsivity. The reason for the reduced range of the thermocouple is the large thermal mass of the surrounding mold, such that the thermocouple junction does not approach the polymer melt temperature. As such, in-mold thermocouples largely provide a relative value and not an absolute indicator of the melt temperature.

Upon further analysis of data collected under different processing conditions, it was observed that the rise time of the infrared pyrometer signal is dependent on the velocity of the polymer melt crossing the pyrometer's lens. This behavior is explicitly plotted in Figure 2. Processes having a faster injection speed yielded a rise time significantly less than processes with a slower injection speed. The underlying reason is that the infrared pyrometer senses the irradiation from the polymer melt. If only a portion of the lens is in contact with the polymer, then the amount of emitted radiation is reduced and the "temperature" sensed by the pyrometer will be proportionally less.

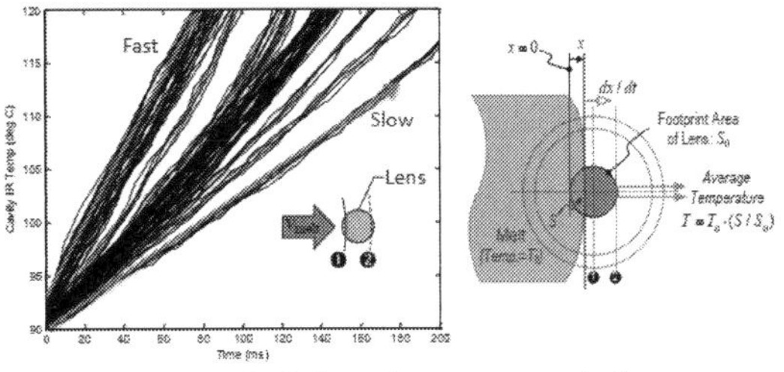

Figure 2: Infrared temperature behavior

The observed rise time of the temperature signal will be a function of 1) the rise time of the infrared detector, 2) the diameter of the lens, and 3) the velocity of the polymer melt. Fortunately, advancements in IR detectors have led to fast response time (1 to 10^{-2} μs), making it possible to capture the temporal variation of the fast-changing temperature profile during the injection stage of a molding cycle where the melt front is flowing over the surface of the sapphire lens. Specifically, the observed temperature sensed by the pyrometer can be expressed as a function of the melt front location, x, and melt velocity v_x:

$$\frac{dT}{dt} = \frac{2r}{S_0} \cdot T_0 \sqrt{1-(1-\frac{x}{r})^2} v_x \qquad (1)$$

where T_0 is the melt temperature, and S_0 and r are the footprint area and radius of the IR sensor lens, respectively. Accordingly, the maximum temperature ramp rate occurs when the melt front crosses the middle of the sapphire lens at point ❶ of Figure 2, where $x = r$:

$$v_x = \frac{S_0}{T_0 \cdot 2r} \cdot \left(\frac{dT}{dt}\right)_{\max} \qquad (2)$$

Equation (2) indicates that the maximum temperature ramp rate can be approximated as a linear function of the melt velocity, while the coefficient can be determined from the designed or measured parameters S_0, r, and T_0. Hence, if the sampling rate for the infrared detector is sufficient to capture the maximum time derivative of the melt temperature, the melt front velocity can be determined analytically.

Figure 3 plots the transient temperature sensed by the pyrometer as well as its time derivative. It can be observed that the sensed temperature follows an "S" type curve associated with the area of the lens covered by the flowing polymer. Once the lens is fully covered, the temperature peaks and then slowly decays with the cooling of the polymer. The melt temperature's time derivative is also plotted, and occurs at the inflection point of the "S" curve as the polymer crosses the centerline of the lens. Figure 3 provides data for two different melt velocities: it can be observed that the temperature rate of change is higher for the higher melt velocity. We theorize that they are proportional.

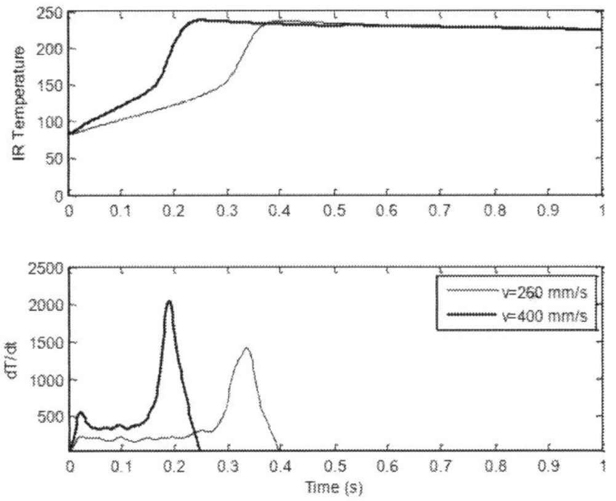

Figure 3: Melt temperature dynamics

The possibility of inferring melt velocity from melt temperature suggests a multivariate sensor design as shown in Figure 4. In the envisioned design, an infrared photodetector is housed behind an optically transparent lens. One or more piezoelectric (PZT) ceramic rings surround the photodetector, support the lens, and measure the melt pressure. An application specific integrated circuit (ASIC) can be provided to perform local signal processing and provide process data in a digital format, thereby reducing requirements on the process controller while increasing the signal to noise ratio. A thermocouple or resistance temperature detector can be incorporated within the ASIC as already common in semiconductor design. Since the sensor is housed within the cooled mold, the PZT, photodetector, and ASIC can all operate safely within their specified temperature operating ranges.

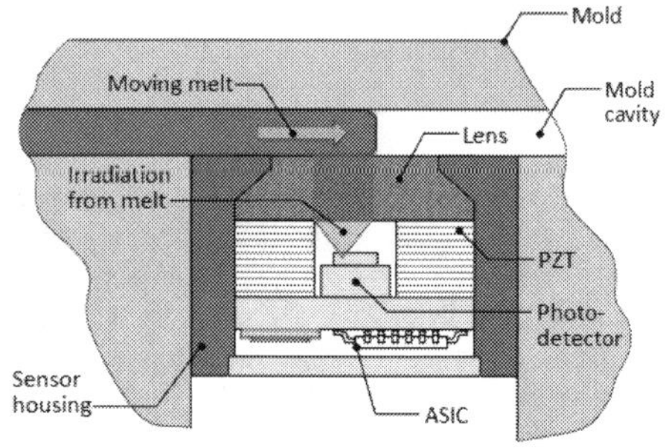

Figure 4: Envisioned multivariate sensor

The fusion of melt pressure, temperature, and velocity at a single point within the mold also supports the inferencing of the apparent melt viscosity. The polymer's shear viscosity is a critical indicator of the polymer morphology (Bicerano 2002), and so is also a determinant of the residual stress distribution and resulting part quality (Wang, Xiao et al. 2003). Given the measured melt velocity, v_x, and melt pressure, P, the apparent melt viscosity, μ, can be inferred assuming a Newtonian fluid (Malkin and Isayev 2005):

$$\tau = \mu\dot{\gamma} \rightarrow \frac{dP}{dx} = \frac{12\mu v_x}{H^2} \qquad (3)$$

where H is the known thickness across the mold cavity. Since the velocity, v_x, is the time derivative of the melt position with respect to time, application of the chain rule provides:

$$\mu = \frac{H^2}{12 v_x}\frac{dP}{dt} \qquad (4)$$

where dP/dt is the time derivative of the observed melt pressure as the polymer melt crosses the centerline of the sensor. While this sensor design currently exists only in concept, its efficacy can be evaluated using the instrumented molding cell depicted in Figure 1. The next section validates the measurement of melt velocity and apparent viscosity by sensor fusion, after which the implications of multivariate sensing are discussed.

Results

A 21 run central composite design of experiments with four factors at three levels was conducted for a semi-crystalline polypropylene (PP, Braskem CP201) as well as an amorphous high impact polystyrene (HIPS, American Polymers 550-11). Both materials were processed according to the conditions listed in Table 1; these specific settings were selected to investigate a diverse set of conditions that all resulted in acceptable moldings without defects. The results of the two materials were extremely similar and so only the results for PP are presented herein.

The instantaneous melt velocity in the mold was calculated using the response from the infrared pyrometer according to equation (2). Additionally, the average melt velocity across the cavity was calculated as:

$$\bar{v}_x = \frac{x}{t_{IR} - t_P} \qquad (5)$$

where x is the linear distance between the infrared pyrometer and pressure transducer located in the impact bar cavity of Figure 1, t_{IR} is the time at which the temperature sensed by the pyrometer exceeded 100 C, and t_P is the time at which the pressure sensed by the pressure transducer exceeded 0.5 MPa. These threshold levels were selected to represent the minimum conditions by which the presence of the melt at the sensors could be reliably indicated.

Table 1: Central composites design of experiments

Run	Injection Velocity (mm/s)	Mold Coolant Temperature (°C)	Barrel Temperature (°C)	Cooling Time (s)
1	60	20	210	20
2	100	20	210	20
3	60	50	210	20
4	100	20	250	20
5	60	50	250	20
6	100	50	250	20
7	60	20	210	30
8	60	50	210	30
9	100	50	210	30
10	60	20	250	30
11	100	20	250	30
12	100	50	250	30
13	60	35	230	25
14	100	35	230	25
15	80	20	230	25
16	80	50	230	25
17	80	35	210	25
18	80	35	250	25
19	80	35	230	20
20	80	35	230	30
21	80	35	230	25

The instantaneous melt velocity at the IR pyrometer and the average melt velocity between the two in-mold sensors are plotted in Figure 5. It is observed that as the ram velocity increases, the melt velocity also increases; it is not possible to accurately predict the in-mold melt velocity directly from the ram velocity data since the polymer melt is compressible, the mold has multiple cavities, and the area of the melt front varies as a function of time.

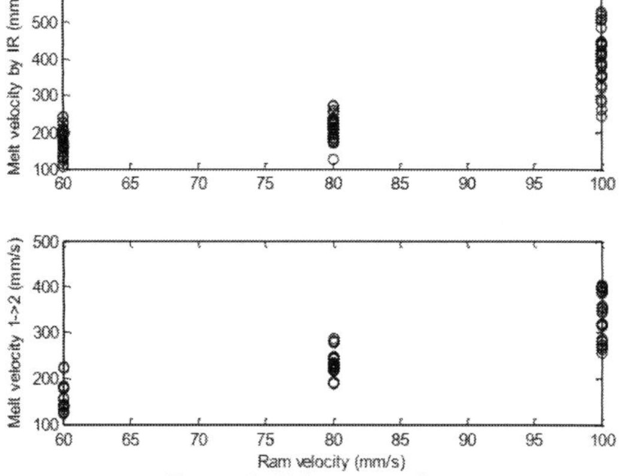

Figure 5: Velocity verification

There is a surprising amount of variation between the instantaneous and average melt velocities compared to the ram velocity. The primary reason is that changes in melt temperature will change the rheology of the material. Decreases in temperature will increase the viscosity and melt pressure required to fill the mold cavity. As a result, the in-mold melt velocity will be a function of not only ram velocity but other process conditions resulting in the variance from a straight line fit as observed in Figure 5.

Figure 6 supports the direct comparison of the instantaneous and average melt velocity measurements. While there is clearly a correlation, there is again a surprising amount of variation between the instantaneous and average melt velocities within a single cavity. There are two likely reasons, both of which are related to the IR pyrometer being located at the end of the mold cavity. First, inspection of the process data provided in Figure 1 indicates that the plaque and tensile bar cavities are filling in parallel with the impact cavity. As such, the completed filling of the plaque cavity, for instance, will increase the velocity of the melt while the other two cavities are filling. Second, the molding machine was set according to a short shot method so that the velocity-controlled filling stage transitions to the pressure-controlled packing stage when the mold cavities were 95% full. Accordingly, the instantaneous melt velocity at the IR pyrometer could be expected to be a function of not only the ram velocity, but also the melt temperature and packing pressure. These suppositions are supported by the fact the that the range of the instantaneous melt velocity $\in$ [105, 530 mm/s] exceeds the range of the average melt velocity $\in$ [120, 410 mm/s].

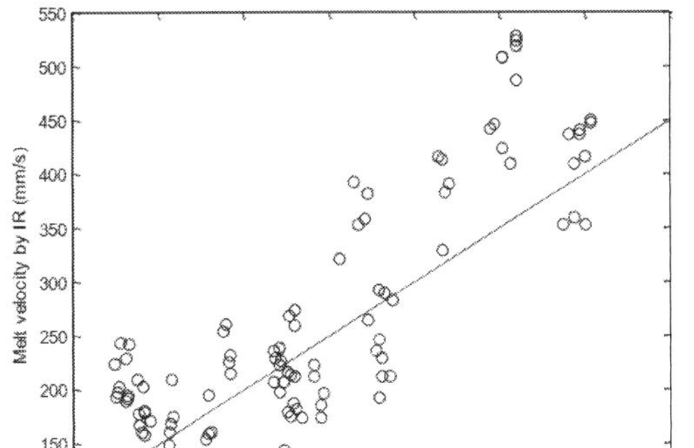

Figure 6: Comparison of in-mold velocity measurements

Figure 7 plots the melt viscosity calculated from the in-mold pressure and velocity data according to equation (4). Since the pressure and temperature transducers are not currently co-located, the time derivative of the pressure was taken when the polymer melt first passed over the pressure transducer, i.e. when the sensed melt pressure exceeded 0.5 MPa. The average melt velocity across the mold cavity was also chosen as v_x in equation 4 to provide a representative value for the calculation of the apparent viscosity in the mold cavity. The results plotted in Figure 7 match the expected rheological behavior of the polymer including a) shear thinning behavior causing a reduction in the viscosity at increasing melt velocities, and b) non-isothermal behavior providing a decrease in the melt viscosity at increasing melt temperatures. A Newtonian plateau is not exhibited or expected given the narrow range of shear rates around 100 s^{-1} that were investigated. Still, there is surprisingly little variation in the behavior of the apparent melt viscosity as calculated with the use of the in-mold sensor data.

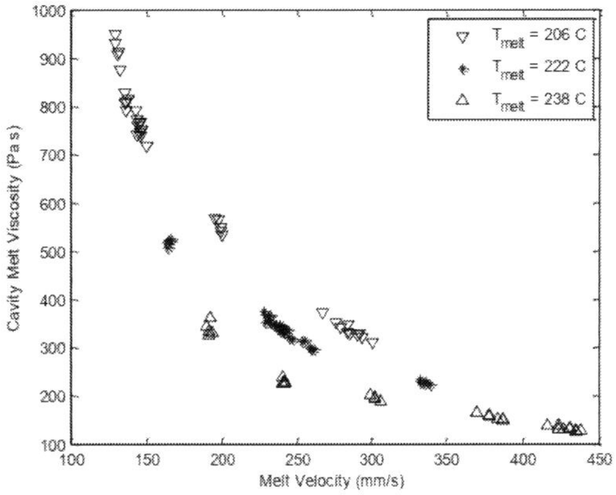

Figure 7: Viscosity inferencing from pressure and velocity data

Discussion

To evaluate the efficacy of different instrumentation strategies, a second design of experiments was conducted as specified in Table 2. This 2^3 full factorial design was performed with the same PP (Braskem CP201) to characterize the effect of injection velocity, mold temperature, and melt temperature on part weight. Two additional runs were added at an intermediate injection velocity and different melt temperatures to provide some curvature information. After reaching steady state for each run condition, 25 molding cycles were consecutively performed in a fully automatic mode.

Table 2: Design of experiments for weight predictions

Run	Barrel Temperature (°C)	Coolant Temperature (°C)	Injection Velocity (mm/s)
1	210	25	20
2	210	25	40
3	210	45	20
4	210	45	40
5	230	45	40
6	230	45	20
7	230	25	20
8	230	25	40
9	230	45	30
10	210	25	30

Most modern plastics machinery provide some form of rudimentary statistical process control by which alarm limits can be set for individual process settings such as barrel temperatures, injection pressures, etc. However, plastics manufacturers are increasingly using more sophisticated statistical modeling techniques for quality assurance and process control since such models can provide significantly better predictive capability necessary for tight tolerance applications. Among the most common modeling techniques is the use of multiple regression whereby a product quality attribute (such as a dimension) can be predicted as a function of the values of other independent variables (such as machine settings).

Perhaps the most common regression modeling in industry practice is performed to predict the product quality as a function of the machine settings. Such modeling is often performed with design of experiments to provide the main effects used for tuning and optimization of the machine settings. For this study, the measured weights, w, were regressed against the machine settings provided in Table 2.The resulting model is:

$$\begin{array}{l} w = 28.15 - 0.011 \cdot T_B \\ +0.0089 \cdot T_C + 0.042 \cdot v_R \end{array} ; \quad R^2 = 0.794 \qquad (6)$$

where T_B, T_C, and v_R are the specified barrel temperature, mold coolant temperature, and ram velocity. The coefficients indicate that the weight predominantly increases with decreasing barrel temperature and increasing ram velocity. The resulting predictions are plotted as a function of the observed weights in Figure 8. With a correlation coefficient, R^2, of 0.794, the regression model provides a reasonable prediction of the part weight as a function of the machine settings. However, the primary shortcoming with this modeling approach is that it considers only the machine settings; the model is deterministic and fails to consider any cycle to cycle variation. The ten groups observed in Figure 8 correspond to the ten run settings of Table 2. As such, this model may be helpful for process setup but is not useful for quality assurance on a cycle by cycle basis.

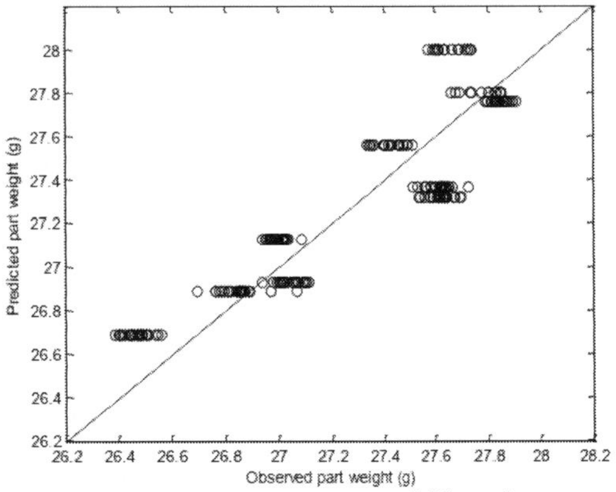

Figure 8: Regression model based on machine settings

Plastics manufacturers seeking a simple solution for on-line quality control often base their models on data from one or more cavity pressure transducers incorporated into the mold. Two common data features used for quality assurance include the peak cavity pressure, P_C, as well as the integral of the cavity pressure, P_I. The authors' industry experience indicates that the former is often the primary predictor of quality in thin-walled applications while the latter is the primary predictor in thicker walled applications. In any case, a mold-centric instrumentation methodology would support the prediction of part quality based on these two measures. The corresponding model for this study is:

$$\begin{array}{l} w = 23.65 + 0.229 \cdot P_C \\ +0.0089 \cdot P_I \end{array} ; R^2 = 0.804 \qquad (7)$$

The minimum, median, and maximum values for the peak and integral pressure from the pressure transducer in the impact cavity used in this model are provided in Table 3. The coefficients indicate that increasing the peak and integral pressures increase the part weight as more polymer is forced into the mold cavity.

Table 3: Statistics of independent variables

Eq	State	Description	Min	Mean	Max
6	T_B	Metering zone barrel temperature (C)	210	230	230
6	T_C	Mold coolant Temperature (C)	25	45	45
6	v_R	Ram velocity (mm/s)	20	30	40
7, 8, 9	P_C	Peak in-mold melt pressure (MPa)	12.5	14.9	17.4
7, 8, 9	P_I	Integral of in-mold pressure (MPa s)	14.9	18.9	92.8
8, 9	T_P	Sensed polymer melt temperature (C)	171.1	190.9	196.7
8, 9	v_P	Sensed in-mold melt velocity (mm/s)	98.9	148.3	199.4
8, 9	μ_P	Apparent in-mold melt viscosity (Pa s)	42.1	65.9	348.3
8, 9	T_M	Sensed in-mold melt temperature	32.0	49.5	50.8

The correlations between the observed and predicted weights according to equation (7) are plotted in Figure 9. The results indicate that the overall predictive capabilities of the models presented in equations (6) and (7) are quite similar. Yet, the predictive model based on the pressure data, equation (7), provides two specific advantages. First, the data is cycle specific which means that a quality prediction can be made on a cycle by cycle basis for quality assurance. Second, no knowledge of the machine settings is required to make a prediction, which simplifies the system design for the end-user. However, the correlation coefficient, R^2, remains too low to provide much quality assurance. Of specific concern is the lack of modeling sensitivity exhibited in Figure 9 for the range of observed part weights $\in$ [26.3, 27.1 g] for which the predictive model suggests a uniform weight of 26.7 g. This lack of sensitivity indicates that the use of cavity pressure data according to models like equation (7) will often fail to include other factors that are significant determinants of the manufactured product quality.

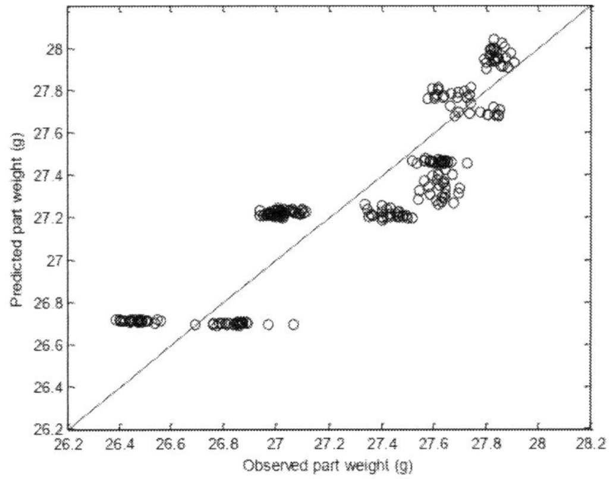

Figure 9: Regression model based on cavity pressure data

Focusing now on the envisioned multivariate sensor, it is possible to model the part weight as a function of the polymer melt temperature, T_P, mold temperature, T_M, melt velocity, v_M, apparent polymer melt viscosity, μ_P, peak cavity pressure, P_C, and integral cavity pressure, P_I, as sensed on a cycle by cycle basis. The resulting multivariate model is:

$$\begin{aligned} w = 41.57 - 0.066 \cdot T_P + 0.01 \cdot T_M \\ +0.015 \cdot v_P - 0.001 \cdot \mu_P \quad ; \quad R^2 = 0.914 \\ -0.297 \cdot P_C + 0.01 \cdot P_I \end{aligned} \tag{8}$$

The minimum, median, and maximum values for the values of these independent variables are also provided in Table 3. A comparison between this model's weight predictions and the observed weights is provided by the data plotted in Figure 10.

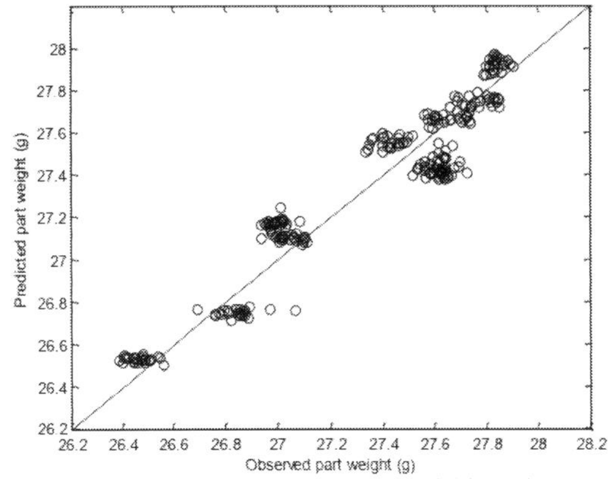

Figure 10: Regression model based on multivariate data

The incorporation of temperature, velocity, and viscosity information has significantly increased the predictive capability as quantified by the correlation coefficient, R^2, of 0.91. The ten clusters corresponding to the run conditions of Table 2 can still be observed, and the predictive model still lacks sensitivity as indicated by the horizontal predicted weights for some of the clusters. The standard error, taken as the mean of the absolute values of the residuals, is equal to 0.11 g or 0.4% of the mean weight. This level of predictive capability may be acceptable for commodity applications with loose specifications, but would often be insufficient for applications having tighter specifications. If the envisioned multivariate sensor is to enable automatic quality assurance, then the predictive capability must be higher.

The relatively low predictive capability from the multivariate data was surprising given the amount of information available from the in-mold sensors: intuition and experience dictated that the polymer states in the mold cavity should determine part quality, and there were no states left to measure. We hypothesized that the linear form of the models did not match the true topology of the process behavior. Accordingly, we inspected the significance of the first order interaction and pure quadratic terms. Of the 21 higher order terms, three were found to be statistically significant at the 95% confidence level: melt temperature with viscosity, melt temperature with the integral of pressure, and the square of the integral of pressure. The resulting model is:

$$\begin{aligned} w = 55.89 - 0.15 \cdot T_P + 0.009 \cdot T_M \\ +0.021 \cdot v_P + 0.011 \cdot \mu_P + 0.066 \cdot P_C \quad ; \quad R^2 = 0.978 \\ -1.63 \cdot P_I - 5.3 \cdot 10^{-5} \cdot T_P \cdot \mu_P + \\ 0.0082 \cdot T_P \cdot P_I + 6.05 \cdot 10^{-6} \cdot P_I^2 \end{aligned} \tag{9}$$

It is informative to compare the coefficients of the various models. While the magnitude of the coefficients is dependent upon the range and significance of the independent variables, the coefficients should be consistent across the various models. Interestingly, the coefficients for the sensed polymer melt temperature, mold temperature, and polymer melt velocity in equation (9) closely match with those for the barrel temperature, coolant temperature, and ram velocity in equation (6). This matching suggests that the machine settings can be deduced on a cycle by cycle basis from the in-mold multivariate data. The peak and integral pressure coefficients of equation (9) also closely follow those of equation (7) when adjusted for the additional higher order terms.

The weight predictions are plotted against the weight observations in Figure 11. While the clusters from the DOE run are still apparent, the multivariate model now has sufficient fidelity to predict some of the intra-run variation. The standard error has been reduced to 0.05g or 0.18% of the nominal part weight. By comparison, the standard deviation of the weight for parts molded across 25 consecutive molding cycles at the same run conditions is 0.21 g. The model still lacks sensitivity for one of the clusters corresponding to run 8 of Table 2 with observed part weights $\in$ [27.3, 27.5 g]. Still, the authors suggest that the described multivariate sensing and modeling approach may be sufficient for automatic quality assurance across a diverse set of plastics molding applications.

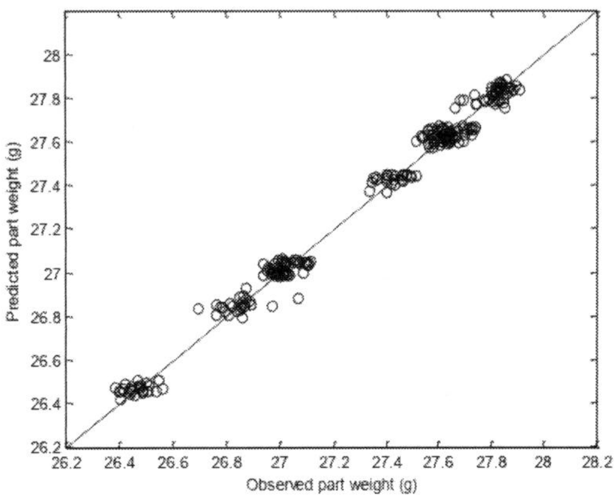

Figure 11: Regression model based on multivariate data with three higher order terms

Conclusions

Plastics manufacturers are striving to implement simple quality assurance systems that are uniformly operable across different applications. Suppliers of instrumentation and auxiliary quality control systems have differentiated themselves in part by the type of sensors they sell and the modeling philosophy they endorse. Accordingly, plastics manufacturers can be uncertain as to the instrumentation strategy that should be adopted.

The results of the current study indicate that one type of process data, be it either pressure or temperature, will often be insufficient for automatic quality assurance. The combination of pressure and temperature data is clearly preferable, since the two states are orthogonal and together enable more robust models that are sensitive to process variation. The move to infrared temperature sensors is clearly preferable, since they rely on irradiation rather than heat conduction and so provide much greater temperature sensitivity. Once an infrared photodetector is incorporated into an in-mold sensor, the calculation of instantaneous melt velocity and apparent melt viscosity can be readily implemented as described.

The additional multivariate process data provides vital information, but a surprising result of this study is that careful modeling beyond conventional linear regression is necessary to provide sufficient model fidelity. Specifically, there are clearly higher order and potential nonlinear effects that relate the polymer states to the product quality attributes. We believe that the mechanistic modeling of the process physics can provide important clues as to the existence and forms of these missing terms. In the absence of such knowledge, however, routine inspection of first order interactions and quadratic effects can provide the needed modeling fidelity on an application-specific basis.

Advancements in machine control technology are changing the architecture of manufacturing operations on both the process and plant levels. At the process level, machine controllers are becoming more open and able to accommodate diverse sensing and analysis methodologies. At the plant level, machine controllers are increasingly networked to provide better verification and translation of molding applications between the various machines. The development and implementation of multivariate sensors could be highly enabling for those plastics manufacturers whose costs are driven not by the purchase cost of their instrumentation but rather by the opportunity costs related to lower productivity and higher defect levels associated with suboptimal manufacturing processes.

Acknowledgements

The authors would like to thank Priamus System Technologies for the use of their sensors and data acquisition system.

Bibliography

Aeppel, T. (2002). Workers Not Included. The Wall Street Journal. New York: B1.

Agrawal, A. R., I. O. Pandelidis, et al. (1987). "Injection-Molding Process Control - a Review." Polymer Engineering and Science **27**(18): 1345-1357.

Bicerano, J. (2002). Prediction of polymer properties, CRC.

Bur, A. J. and C. L. Thomas (1997). "Fluorescence monitoring of polymer injection molding: Model development." Polymer Engineering and Science **37**(9): 1430-1441.

Chen, Z. and L. S. Turng (2007). "Injection molding quality control by integrating weight feedback into a cascade closed-loop control system." Polymer Engineering & Science **47**(6).

Chiang, K. T. (2007). "The optimal process conditions of an injection-molded thermoplastic part with a thin shell feature using grey-fuzzy logic: A case study on machining the PC/ABS cell phone shell." Materials and Design **28**(6): 1851-1860.

Coates, P. D. and R. G. Speight (1995). "Towards intelligent process control of injection moulding of polymers." Proceedings of the Institution of Mechanical Engineers, Part B: Journal of Engineering Manufacture **209**(B5): 357-367.

Dininger, J. (1994). Three critical measurements on injection molding processes. IEEE Industry Applications Society Annual Meeting.

Dubay, R., A. C. Bell, et al. (1997). "Control of plastic melt temperature: A multiple input multiple output model predictive approach." Polymer Engineering and Science **37**(9): 1550-1563.

Edwards, R. and C. Thomas (2001). "On-line measurement of polymer orientation using ultrasonic technology." Polymer Engineering and Science **41**(9): 1644.

Franca, D. R., C. K. Jen, et al. (2000). "Ultrasonic in-line monitoring of polymer extrusion." Polymer Engineering and Science **40**(1): 82-94.

Gao, R., Z. Fan, et al. (2008). "Injection molding process monitoring using a self-energized dual-parameter sensor." CIRP Annals – Manufacturing Technology **57**: 389-393.

Giboz, J., T. Copponnex, et al. (2007). "Microinjection molding of thermoplastic polymers: a review." Journal of Micromechanics and Microengineering **17**(6): 96.

Groleau, R. J. (2002). Best practices with in-mold sensors. Society of Plastics Engineers Regional Technical Conference, Erie, PA.

Johnston, S., D. Kazmer, et al. (2009). "On-line simulation-based process control for injection molding." Polymer Engineering and Science **49**(12): 2482-2491.

Johnston, S. P. (2008). Instrumentation, Analysis, and On-Line Simulation for Improved Process Control of Injection Molding. Plastics Engineering. Lowell, MA, Univ. Mass. Lowell**:** 121.

Kazmer, D. (2006). Precision Process Control of Precision Injection Molding. Precision Injection Molding: Process, Materials, and Applications J. Greener and R. Wimberger-Friedl. Munich, Carl Hanser Verlag**:** 265-297.

Kazmer, D. (2009). Plastics Manufacturing Systems Engineering. Munich, Carl Hanser Verlag.

Kazmer, D., C. Lotti, et al. (2004). "Tuning and Control of Dimensional Consistency in Molded Products." Advances in Polymer Technology **23**(3): 167-175.

Kazmer, D. and S. Westerdale (2009). "A model-based methodology for on-line quality control." Int J Adv Manuf Technol **42**(3): 280-292.

Kazmer, D. O., R. Nageri, et al. (2006). "Validation of three on-line flow simulations for injection molding." Polymer Engineering and Science **46**(3): 274-288.

Kazmer, D. O., S. Westerdale, et al. (2008). "A Comparison of Statistical Process Control (SPC) and On-Line Multivariate Analyses (MVA) for Injection Molding." International Polymer Processing **23**(5): 447-458.

Lu, N., F. Gao, et al. (2004). "PCA-Based Modeling and On-Line Monitoring Strategy for Uneven-Length Batch Processes." Industrial & Engineering Chemistry Research **43**: 3343-3352.

Macosko, C. W. (1994). Rheology : Principles, Measurements, and Applications. New York, NY, VCH.

Malkin, A. Y. and A. I. Isayev (2005). Rheology: Concepts, Methods, and Applications, ChemTec Publishing.

Mann, J. W. (1974). "Process Parameter Control: the Key to Optimization." Plastics Engineering **30**(1): 25-27.

Michaeli, W. and C. Starke (2005). "Ultrasonic investigations of the thermoplastics injection moulding process." Polymer Testing **24**(2): 205-209.

Orroth, S. A. J. (1974). Infrared melt temperature measurements in injection molding of thermoplastics. United States -- Massachusetts, University of Massachusetts Lowell. **M.S.:** 111.

Panchal, R. R. and D. O. Kazmer (2009). In-mold shrinkage monitoring sensor for injection molding. SPE Annual Technical Conference.

Schulx, H. (2005). Characteristics of Modern Manufacturing Techniques. Seventh International Conference on Advanced Manufacturing Systems and Technology, Springer-Verlag New York Inc.

Tanner, R. I. (2000). Engineering rheology. Oxford ; New York, Oxford University Press.

Thomas, C. L. and A. J. Bur (1999). "In-Situ Monitoring of Product Shrinkage During Injection Molding Using an Optical Sensor." Polymer Eng Sci **39**(9): 1619.

Thomas, C. L. and A. J. Bur (1999). "Optical monitoring of polypropylene injection molding." Polymer Engineering & Science **39**(7): 1291-1302.

Wang, H., B. Cao, et al. (1997). "On-line ultrasonic monitoring of the injection molding process." Polymer Engineering and Science(USA) **37**(2): 363-376.

Wang, Y., Y. Xiao, et al. (2003). "The morphology and mechanical properties of dynamic packing injection molded PP/PS blends." POLYMER **44**(5): 1469-1480.

Woll, S. L. B., D. J. Cooper, et al. (1996). "Online pattern-based part quality monitoring of the injection molding process." Polymer Engineering and Science **36**(11): 1477-1488.

PHOTOINDUCED CREEP, SHAPE CHANGE AND STRESS RELAXATION IN CROSSLINKED POLYMERS

W D Cook[1]*, F Chen[1], Q D Nghiem[1], C Morhoff[1], J Sun[1], S Thang[2], TF Scott[3], CN Bowman[3]

[1] *Department of Materials Engineering, Monash University, Clayton, Victoria, Australia*
- wayne.cook@eng.monash.edu.au; fei.chen@eng.monash.edu.au; dat.nghiem@eng.monash.edu.au; cornelis.moorhoff@eng.monash.edu.au; jiazeng.sun@sci.monash.edu.au
[2] *CSIRO Molecular and Health Technologies, Clayton, Victoria, Australia*
- san.thang@csiro.au
[3] *Department of Chemical and Biological Engineering, University of Colorado, Boulder, Colorado, USA*
- timothy.scott@colorado.edu; christopher.bowman@colorado.edu

Abstract

There is considerable interest in the area of stimuli responsive polymers which can change their shape in response to an external stimulus such as heat, magnetic/electric fields, pH and irradiation. A related area is where the polymer undergoes creep or stress relaxation only under the action of the stimulus such as light. The main limitations of these methods are their need of specific UV radiation to cause the photo-plastic effect, the long irradiation periods required and the low quantum yields (i.e. ≤ 1 chain scission per photon).

As an alternative to the existing systems, thiol-ene polymer networks containing a free radical photoinitiator and allylic dithioether groups in the network strands were previously shown to undergo creep and stress relaxation behavior during irradiation. The mechanism for this process is attack by a thiol radical on the allylic dithioether followed by its β-cleavage and reformation of a new network strand with a regenerated allylic dithioether.

In this work we show photoplastic behaviour in a range of networks formed from the thio-ene addition of a tetrathiol and a divinyl ether, the cationic photopolymerization of a divinyl ether or of a diepoxide, the anhydride curing of a diepoxide and the copolymerization of a diepoxide and a tetrathiol.

Introduction

Thermosets are crosslinked polymers with the glass transition temperature above room temperature and so can provide thermal resistance which is often highly desirable in a polymeric material. However, the same crosslinked structure that is the origin of many of the desirable mechanical and thermal properties results in the polymer shape and properties generally being fixed upon polymerization, requiring degradation of the crosslinks to make them reshapeable. Additionally, the polymerization process is generally associated with shrinkage as monomer is converted into polymer. In combination with the formation of a network structure, this densification can lead to significant development of stress inhomogeneities in these systems.

Recently (Scott et al, 2005; Cook et al 2008), we have shown that thiol-ene polymer networks containing a free radical photoinitiator and allylic dithioether groups in the network strands are able to undergo creep and stress relaxation behavior during irradiation. The mechanism for this process is attack by a thiol radical on the allylic dithioether followed by its β-cleavage and reformation of a new network strand with a regenerated allylic dithioether (see Scheme 1).

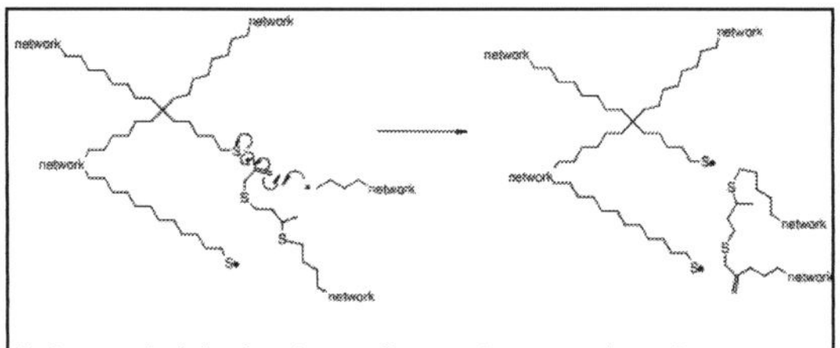

Scheme 1. Mechanism of creep/stress relaxation

In contrast to previously studied systems, this type of photo-induced creep and stress relaxation should be applicable to a wide range of network polymers and it is this aspect that this paper addresses.

Experimental

Scheme 2 shows the monomers used in the study. The PETMP/TEGDVE/MDTVE thiol-ene systems were polymerized by a visible light photoinitiator system (camphorquinone/tetramethyl aniline) while the ADTDE, ADTDBDE or ADTDNEGDE/PETMP diepoxy-thiol or anhydride systems were copolymerized using a teriary amine catalyst (dimethyl benzylamine). After network formation, the photoplastic behaviour was obtained by photodegrading free radical UV-photoinitiators present in the networks (usually benzoin ethyl ether).

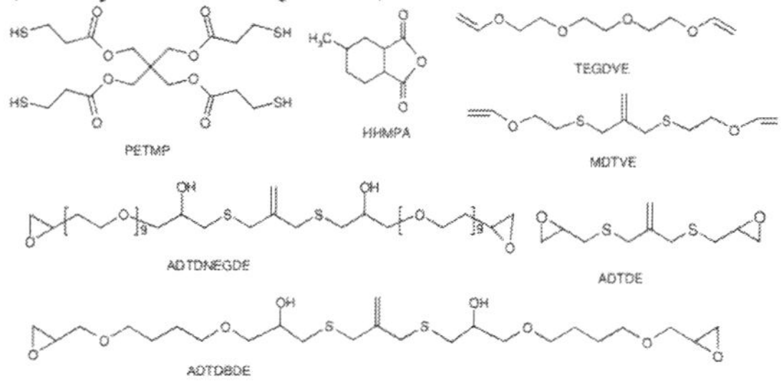

Scheme 2. Structures of the monomers used.

Results and Discussion

Previous studies [Scott et al., 2005] of photo-induced stress relaxation employed a cyclic monomer containing the allylic dithioether unit, MDTO to copolymerize with PETMP/TEGDVE and form a thiol-ene network. A major disadvantage with the PETMP/TEGDVE/MDTO system is that MDTO reduces the network crosslink density. To counter this problem, we synthesized MDTVE which is a divinyl ether containing the allylic dithioether divinyl ether [Cook et al., 2008]. In contrast to MDTO, the two vinyl ether units of MDTVE can copolymerize with two thiol groups of PETMP molecules thus forming a network strand with a β-cleavable allylic dithioether unit so that theoretically the level of photo-plasticity can be varied depending on the ratio of TEGDVE to MTDVE. Fig 1 shows the photoplasticity of this system. The extent of stress relaxation is lees than that observed previously [Scott et al., 2005] in the PETMP/TEGDVE/MDTO system and may suggest that for extensive photoplasticity to occur, the network strands need to be sufficiently long to be able to reorient and attack an adjacent allylic group during β-scission.

Photoplasticity experiments have also been undertaken on other network systems. Figure 2 shows the photo-induced stress relaxation behaviour of the network formed from PETMP and the relatively short diepoxy ADTDE. Again this system undergoes limited stress relaxation during irradiation. In contrast the network formed from PETMP and a longer chain diepoxy, ADTDNEGDE, containing the allylic dithioether unit (Figure 3) undergoes substantial stress relaxation. This suggests that this mechanism of photoplasticity requires sufficiently long chains during β-scission to be able to reorient and attack an adjacent allylic group. Additional photoplastic monomers are under study.

Conclusions

In this paper, we have shown the generality of β-scission of allylic dithio ether network strands for generating photoplasticity in crosslinked polymers.

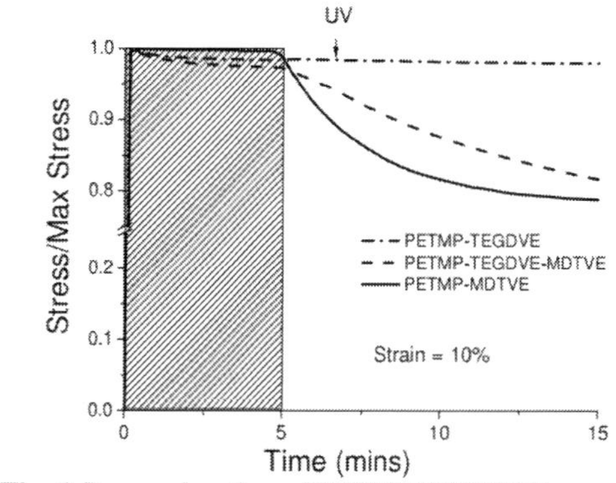

Fig. 1 Stress relaxation of PETMP/TEGDVE/ MDTVE networks with 1 wt% BEE as radical source during exposure to UV radiation (un-shaded regions).

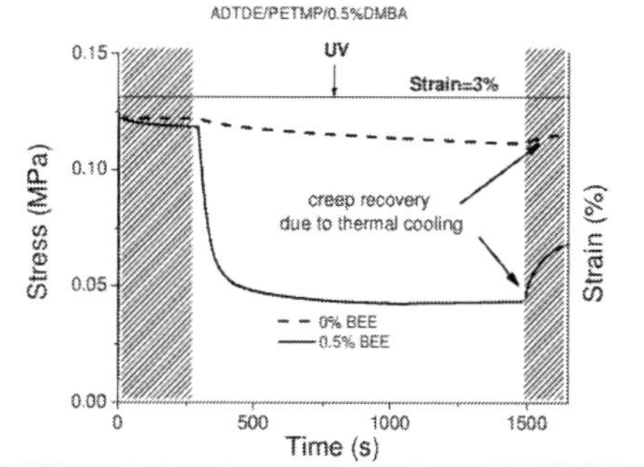

Fig. 2 Photo-induced stress relaxation of PETMP/ ADTDE/DMBA catalyst with differing levels of BEE as radical source during exposure to UV radiation.

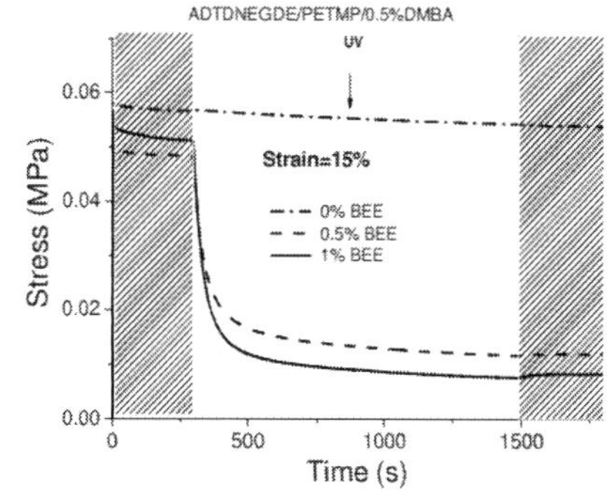

Fig. 3 Photo-induced stress relaxation of PETMP/ ADTDNEGDE/DMBA catalyst with differing levels of BEE during exposure to UV radiation.

Acknowledgements

The authors wish to acknowledge the support of ARC DP0877382 in this work.

References (Times New Roman 10, bold)

Scott, T. F., *et al.*, "Photo-induced plasticity in crosslinked polymers", Science, **308**, 1615-1617 (2005)

Cook W. D., *et al.*, "Photo-polymerization kinetics, photo-rheology and photo-plasticity of thiol-ene-allylic sulfide networks", Polym. Int., **57**, 469-478, (2008)

SYNTHESIS OF BIODEGRADABLE COPOLYMERS

I. Moura[1], A. V. Machado[1]*, R. Nogueira[2] and Véronique Bounor-Legare[3]

[1]*IPC – Instituto de Polímeros e Compósitos, University of Minho, 4800-058 Guimarães, Portugal – isabelm@dep.uminho.pt, avm@dep.uminho.pt*
[2] *Centro de Engenharia Biológica, University of Minho, Campus de Gualtar, 4710-057 Braga, Portugal – regina@deb.uminho.pt*
3 Université de Lyon, Lyon, F-69003, France, Université de Lyon 1, Lyon, F-69003, France, CNRS UMR5223, Ingénierie des Matériaux Polymères : Laboratoire des Matériaux Polymères et Biomatériaux, F-69622 Villeurbanne, France - bounor@univ-lyon1.fr

Abstract - This work aimed to prepare biodegradable copolymers based on ethylene vinyl acetate and polylactic acid using transesterification reactions. The materials were characterized by elemental analysis, infrared spectroscopy, rheology, scanning electron microscopy and gel permeation chromatography.
Selective extractions for all samples were made, and the results indicate that a maximum of 25% of EVA-g-PLA copolymer was synthesized by this method. Biodegradation tests were carried out using the standard ISO 14851 (1999) which specifies a method for determining the biochemical oxygen demand in a closed respirometer.
This procedure allowed the synthesizes biodegradable copolymers with mechanical properties similar to conventional materials.

Introduction

In recent years the importance of biodegradable polymers increased significantly because of the growing awareness of society and the depletation of natural oil reserves [1, 2]. However, synthetic polymers are more effective than the natural ones, due to the best properties (mechanical, thermal, UV resistance, etc.). Many attempts have been made to combinate the best properties of both by grafting natural polymers onto synthetic ones.
One important challenge would be to develop biodegradable materials based on copolymers of ethyl vinyl acetate (EVA) and polylactic acid (PLA), which it is cheap, biodegradable and readily available. PLA can be obtained from renewable agricultural sources, from the fermentation of crop like corn starch and sugar feed stocks [3].
An important feature of polyesters blends is its potential for transesterification. During the reaction, chemical modification depends on the degree of substitution. The extent of grafting reaction depends on the composition of the reaction mixture and on the reaction conditions. Such reactions are expected to induce morphological changes of blends. Usually, transesterification reactions only occur in a presence of a catalyst. Among others, $Ti(OPr)_4$ are well known to promote transesterification reactions. Thus, in the present work, biodegradable graft copolymers were synthesized from the transesterification reactions between EVA and PLA using titanium propoxide ($Ti(OPr)_4$) as catalyst. The produced materials were characterized by several techniques, such as, X-ray, GPC, SEM, rheology, DSC, TGA. Biodegradation tests were carried out using the standard ISO 14851.
The results show that this method allows to synthesized biodegradable copolymers with mechanical properties similar to conventional polymers.

Experimental

Materials
Ethylene vinyl acetate 28 % of vinyl acetate(28 % VA) (Escorene Ultra Lot. 61E466) supplied from Exxon was used as non-biodegradable polymer and polylactic acid (PLA) supplied by Aldrich was used as a biodegradable polyester. Titanium propoxide ($Ti(OPr)_4$, used as transesterification catalyst, was purchased from Aldrich was. The compositions of the prepared samples are shown in Table 1.

Table 1: Composition of the samples

Sample	EVA (wt %)	PLA (wt %)	$Ti(OPr)_4$ (wt %)
EVAPLA0	60.0	40.0	0.0
EVAPLA1	59.5	38.6	1.9
EVAPLA2	59.5	39.5	0.9
EVAPLA3	59.5	42.0	0.7
EVAPLA4	59.5	40.0	0.5

Compounding
The pellets of both polymers were dried in a vacuum oven at 60°C for 24 hours before use.Copolymers of EVA and PLA were prepared in a Haake batch mixer

(Rheocord 90; volume 50 ml), equipped with two rotors running in a counter-rotating way. The rotor speed was 50 rpm and the set temperature was 160 °C. The copolymers were prepared using the following sequence: first the EVA pellets were introduced into the hot mixer, after melting, PLA and catalyst were added. Titanium propoxide was collected and carried to the mixer inside the syringe in an argon atmosphere, to prevent oxidation. After 5 min, when a constant torque was reached, the rotors were stopped and the total sample was removed.

Material characterization

After fracture of the samples in liquid nitrogen and gold plating, the morphology of the blends was studied using a FEI Quanta 400 Scanning Electron Microscope Gel permeation chromatography (GPC) was also performed to characterize the grafting reaction, in order to monitor the variation of molar weight during sample preparation. The number average molecular weight (M_n), the weight average molecular weight (M_w), were determined by GPC analysis, in $CHCl_3$, using a Waters 150-CV gel permeation chromatography apparatus. The flow rate was 1ml/min and a calibration curve was previously obtained using polystyrene standards with narrow molecular weight distribution.

Biodegradability of the materials was evaluated using the biochemical oxygen demand method (BOD). The only available carbon source for microorganisms in the prepared middle is present in the polymeric material.

Results and Discussion

Figure 1 shows, as example of the morphology of the samples, SEM micrographs of EVA/PLA blend without (a) and with catalyst (b). In Figure 1a it is possible to observe that the adhesion between PLA and EVA is not good, which results from the low compatibility between both polymers. However, when titanium propoxide is added (Figure1b), the particle size decreases and the interfacial adhesion increases. This supports the occurrence of reaction between EVA and PLA.

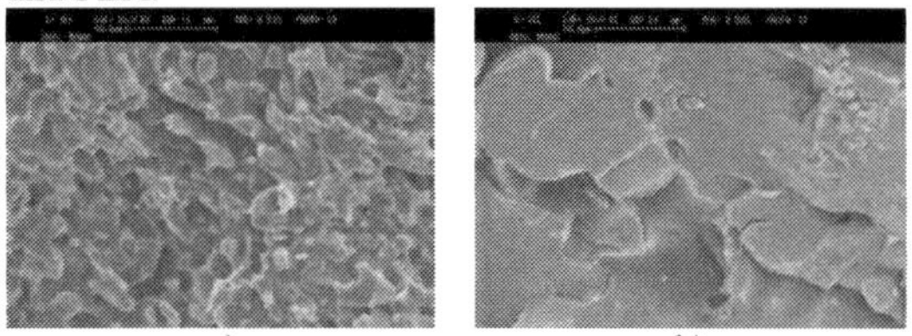

a) b)

Figure 1 – (a) EVAPLA0 (b) EVAPLA1

As shown in Figure 2, the extent of the transesterification can be evaluated by changes in the molecular weight. From Figure 2, curves of samples differs in magnitude and elution time, and decrease for the physical blend (EVAPLA0) to the copolymers, but among the copolymers it is possible to verify that with an increase of the catalyst amount the molar weight decreases. For EVAPLA1 and EVAPLA2, the peaks of copolymer and homopolymer can be distinguished in the cromathograms. These results suggest that copolymers were accomplished in both cases. Contrarily, for EVAPLA3 and EVAPLA4, it is possible to observe a single peak, the chromatograms are very similar to the one obtained for the physical blend (EVAPLA0).

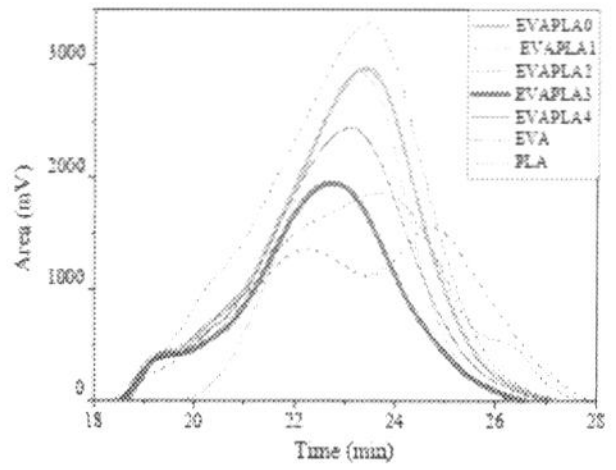

Figure 2 - GPC analysis of EVA, PLA and prepared samples.

Biodegradability values were expressed as the amount of O_2 consumed during sample biodegradation divided by their theoretical oxygen demand (ThOD) and are presented in Figure 3. The EVAPLA0 sample is the one that exhibits higher biodegradability. Comparing the biodegradability of EVAPLA0 and copolymers, and the SEM results, a possible explanation is that the particles size is higher in the case of the blend, which contributes to the formation of void spaces and consequently, enhance the capability of microorganisms to get inside the material, which increases biodegradation. Copolymers biodegradability increases with the increase of the amount of titanium propoxide, which is related with the low molar mass obtained.

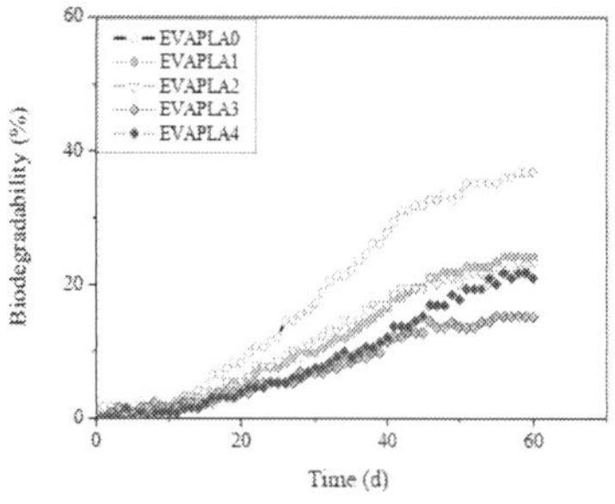

Figure3. Percentage of biodegradation of the copolymers and blend according to ISO 14851(1999)

Conclusions

The results obtained show that polymers with enhanced biodegradability can be obtained by transesterification reactions of synthetic/biodegradable polymers. The biodegradability results indicated that biodegradability increases as the catalyst amount increases. Among the various materials studied, EVAPLA1 shows the lower molar weight, and consequently are most biodegradable.

Acknowledgements

The authors are grateful to FCT for funding the project.

References

[1] Schurz J., "Trends in Polymer Science – A bright future for cellulose", Prog Polym Sci., **24**, 481-483 (1999).
[2] Scott G., "'Green" Polymers",. Polym. Degrad. Stab., **68**, 1-7 (2000).
[3] Lunt J., "Large-scale production, properties and commercial applications of polylactic acid polymers", Polym. Degrad. Stab., **59**, 145-152 (1998).

A REACTIVE EXTRUSION PROCESS FOR THE SYNTHESIS AND PROCESSABILITY OF POLYAMIDE GRAFT COPOLYMERS

Guo-Hua HU[1*, 2], Cai-Liang ZhANG[3], Lian-Fang FENG[3*], Sandrine HOPPE[1]

1 *Laboratory of Reactions and Process Engineering, CNRS, Nancy Université-INPL-ENSIC, 1 rue Grandville BP 20451, 54001 Nancy, France – hu@ensic.inpl-nancy.fr;sandrine.hoppe@ensic.inpl-nancy.fr*

2 *Institut Universitaire de France, Maison des Universités, 103 Boulevard Saint-Michel, 75005 Paris, France*

3 *State Key Laboratory of Chemical Engineering, Department of chemical and biological engineering, Zhejiang University, 310027 Hangzhou, China – zhangclzju@hotmail.com; fenglf@zju.edu.cn*

Abstract: A "grafting from" approach is used to synthesize graft copolymers of polyamide 6 (PA6) and polystyrene (PS) using a random copolymer of styrene (St) and 3-isopropenyl-α, α-dimethylbenzene isocyanate (TMI), PS-co-TMI. The latter activates the polymerization of ε-caprolactam (CL) in the presence of sodium ε-caprolactam (NaCL) as an anionic catalyst. When the graft copolymer is subjected to high temperature and high shear, the graft chains degrade, resulting in a decrease in content of graft chains in the graft copolymer. As the volume fraction of the graft chains decreases, the interface curvature toward graft chain increases, which drives the graft copolymer to self-assemble into various types of morphologies.

Key words: anionic polymerization; graft copolymer; polystyrene; polyamide 6

Introduction

Graft copolymers are a class of nonlinear copolymers and composed of a backbone to which one or more side chains are connected through covalent bonds, which have been used as emulsifiers of polymer blends [Pitsikalis et al. 1998]. However, many existing chemical routes cannot be followed easily or in an economically feasible fashion to synthesize architecturally well-defined graft copolymers. The one that has been frequently practiced is the so-called in-situ compatibilization. It consists in reacting two premade functional polymers. Nevertheless, this method hardly leads to the formation of pure copolymers when the functional polymers are immiscible. Hu et al. (2002, 2005) had reported on a novel method to synthesize pure PP-g-PA6 graft copolymers by activated anionic polymerization of ε-caprolactam (CL) using sodium ε-caprolactam (NaCL) as catalyst and a 3-isopropenyl-α, α-dimethylbenzene isocyanate (TMI) bearing polypropylene (PP) as activator (PP-g-TMI). However, PP-g-TMI was soluble only in some solvents at temperatures above room temperature. We used a random copolymer of styrene (St) and TMI (PS-co-TMI) as an activator to replace PP-g-TMI (zhang et al. 2008). The resulting copolymer would be composed of a PS backbone and PA6 grafts. It was much easier to characterize the PS-co-TMI/CL/NaCL polymerization system than the PP-g-TMI/CL /NaCL one before and after the polymerization. The phase behavior of graft copolymer under the polymer processing conditions was investigated.

Experimental

The free radical copolymerization of styrene (St) and TMI was carried out in a one-litter glass reactor. The reactor was purged by nitrogen and its temperature was controlled at 80 °C by circulating water. The polymerization lasted 24 h. The resulting product was precipitated twice in methanol, filtered and then dried in a vacuum at 80 °C for 24 h. The anionic polymerization of CL from the PS-co-TMI in the presence of catalyst was carried out using a Haake torque rheometer. The temperature was set at 230 °C. Samples were taken out from the mixing chamber after prescribed mixing time.

Results and Discussion

After the polymerization, the as-polymerized product was purified by successive extraction in methanol under reflux, tetrahydrofuran under reflux and formic acid at room temperature. The architectures of the graft copolymer from PS-co-TMI/CL/ Cat (30/70/5) are shown Figure 1A.

It was well-known that PA6 is prone to degradation when subjected to high temperature and high shear while PS is relatively stable below 250 °C. In other words, the degradation of PS-g-PA6 results mainly from the rupture of graft chains. The molecular architecture, composition and molar mass of the backbone and grafts chains for the copolymer from the PS-co-TMI/CL/ Cat (30/70/5) system before and after treatment by static annealing, Haake mixing and formic acid are shown in Figure 1.

TEM micrographs of the phase morphologies of the PS-g-PA6 copolymer after various treatments are shown in Figure 2. The PA6 appears black and PS white on the TEM micrographs.

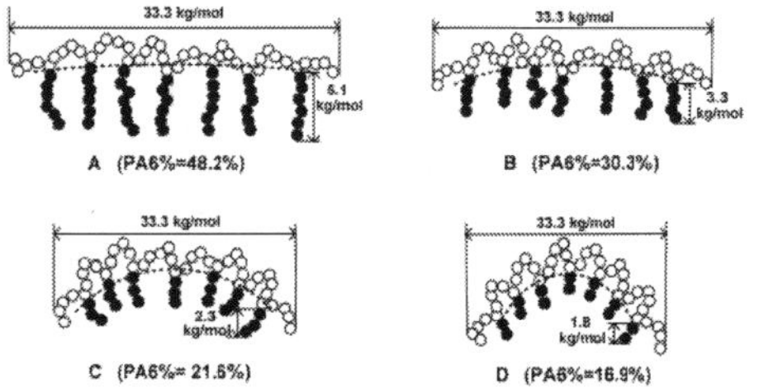

Figure 1 Schematic of the molecular architectures, composition (based on the volume fraction) and molar masses of PS-g-PA6 graft copolymers by various treatments. A: without treatment; B: Formic acid treatment 12 h at 30 °C; C: Formic acid treatment 12 h at 30 °C after static annealing for 2 h at 230 °C; D: Formic acid treatment 12 at 30 °C after Haake mixing for 10 min at 230 °C.

From Figure 2 (A and B), the morphologies of the pure PS-g-PA6 graft copolymer before and after the treatment in formic acid are similar and are cylindrical. When it was subjected to static annealing or Haake mixing, its morphology changed into a two phase structure, especially the one by Haake mixing (Figure 2 C and E). This mainly results from the aggregation of the degraded PA6 graft chains. The dissolution of the two phase structures in formic acid dissolution, the most of the free PA6 chains are removed, as shown in Figure 2 (D and F). At the same time, it should also be noted that their morphologies after formic acid treatment are different from that of the original pure PS-g-PA6. The PS-g-pA6 obtained by static annealing treatment displays wormlike micelles while that by Haake mixing treatment exhibits ordered spheres. The above results indicate that the PS-g-PA6 graft copolymer undergoes structural and morphological changes under high temperature and high shear conditions.

The graft copolymer subjected to Haake annealing treatment then purification in formic acid contained 16.9% PA6, the most asymmetric volume fraction and the corresponding morphology is spherical because the curvature towards to PA6 was the highest (Figure 1D). As the content of graft chain increases, the volume available for the graft chain increases as one moves form the concave to the convex side of a curved interface, leading to a smaller mean curvature, and ultimately, a flat interface (almost zero mean curvature), as shown in Figure 1. Therefore, as the volume fraction difference between the grafts and the backbone decreases, so does the interfacial curvature across the sequence of observed morphologies: spheres, wormlike micelle, cylinders.

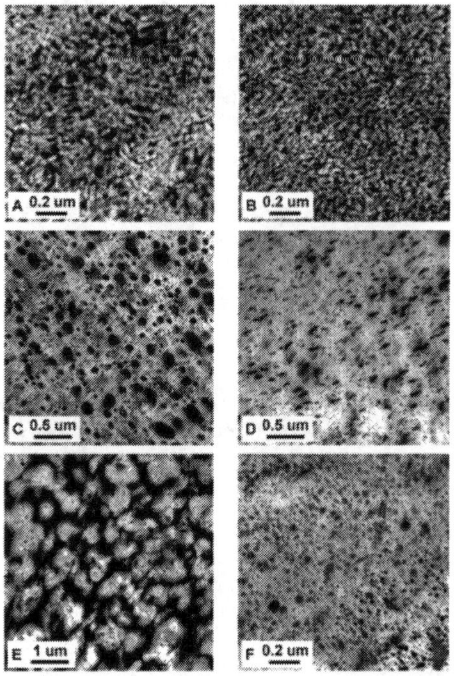

Figure 2 TEM micrographs of the phase morphologies of various PS-g-PA6 by different treatments. A: Pure PS-g-PA6; B: Formic acid treatment 12 h at 30 °C; C: Static annealing for 2 h at 230 °C; D: Formic acid treatment 12 h at 30 °C after static annealing for 2 h at 230 °C; E: Haake mixing for 10 min at 230 °C; F: Formic acid treatment 12 at 30 °C after Haake mixing for 10 min at 230 °C.

Conclusions

Graft copolymers with PS as backbone and PA6 as grafts were synthesized by the anionic polymerization of ε-caprolectam (CL) onto PS-co-TMI in the presence of Cat. The PA6 grafts undergo degradation under high temperature and high shear conditions, leading to different types of morpholgies.

References

Hu, G. H., et al., "Rate of the activated anionic polymerization of ε-caprolactam onto an isocyanate bearing polypropylene in the melt", Polymer, **46**, 4562-4570 (2005).

Pitsikalis, P., et al., "Nonlinear block copolymer architectures", Adv. Polym. Sci., **135**, 1-137. (1998).

Zhang C. L., "Grafting of polyamide 6 by the anionic polymerization of ε-caprolactam from an isocyanate bearing polystyrene backbone", J. polym. Sci. Part A: polym. Chem., **46**, 4766-4776 (2008).

ADVANCEMENTS IN EFFICIENCY AND QUALITY OF THERMOSETTING RESINS BY USING MICROWAVE-CURING

P. Lechner[1]*, S. Wiebel[1], H. Haupt[1], C. Berger[1]

[1] *Zentrum für Konstruktionswerkstoffe Staatliche Materialprüfungsanstalt Darmstadt, Fachgebiet und Institut für Werkstoffkunde, Technische Universität Darmstadt, Darmstadt, Germany –*

lechner@mpa-ifw.tu-darmstadt.de; wiebel@mpa-ifw.tu-darmstadt.de; haupt@mpa-ifw.tu-darmstadt.de; berger@mpa-ifw.tu-darmstadt.de

Abstract - Thermosetting resins are often used for electrical insulation applications because of their excellent thermal, electrical and mechanical properties. To ensure these properties an adequate degree of cross-linked polymer chains in the material is needed. With conventional methods using convective heat transfer or infrared radiation curing time depends strongly on the wall-thickness and increases extremely for thick parts because of the absorption of radiation and the low thermal conductivity of thermosetting resins. This has a large influence in production costs and energy consumption and is a competitive disadvantage of thermosetting materials compared to other materials.
The use of microwave power is an encouraging alternative for curing thermosets. The internal friction of oscillating polar groups in a microwave field generates heat simultaneously in the whole irradiated volume of the material. With this homogeneous heating distortion and residual stress can be reduced. The fact that uncured polar groups interact with the microwave field is another advantage of this method because during the curing-process the material can not be heated unregulated. Thermal damage due to overheating can be prevented.
The handling of potential inhomogeneous electrical field distribution in the microwave chamber that causes inhomogeneous heat distribution is the main challenge to realize this curing method.
Initial trials with two-component casting resins show, that there is a high potential in the microwave curing method. The tests were conducted on various epoxy and polyurethane compounds. Test samples in a microwave field cure much faster than comparative samples cured conventionally with convective heat flow methods.
Analytical verifications with Differential Scanning Calorimetry show that the thermosetting resins are nearly completely cross-linked after the microwave curing-process. A post-curing reaction was not be detected.

Introduction

Thermosetting resins are used widely in electrical insulation technology. The casting of electronics serves as protection against humidity, environmental influences, shock and vibration. The mechanical and electrical properties of the resin have to fulfil special requirements. Variations in curing or cross-linkage may minimize the insulation effect.
These materials are mostly formed on the basis of epoxy resins, phenolic resins, polyurethane resins and unsaturated polyester resins (Domininghaus 1993).
The good electrical and mechanical properties and thermal endurance are attributed to, among other factors, the high degree of cross-linked monomers during the molding process to a non-meltable macromolecular compound which is typical for thermosetting materials.
In order to ensure sufficient cross-linkage, long cycle times are often necessary, which restricts the competitiveness.

The curing process of reaction resins is started either by direct heat development, e. g. by ultraviolet light, or indirect heat input via convective heat flow and heat conduction. Process temperatures for the curing of thermosetting materials typically range from 70 °C to 180 °C (Feher 2009).
A disadvantage of the currently used methods is the low achievable layer and wall thicknesses.
During the cross-linkage process, the heat penetrates through the surface in the material for curing, therefore outer layers link faster, while the core of the part is still unlinked (Fig. 1). The consequences are internal stress in the material and long process cycles (Peters 2002).
Apart from the long cycle times, the high energy costs are a main factor which influences production efficiency (Schubert and Gruenewald 1983).
In the course of discussions about the emission of carbon dioxide, it is especially important for manufacturing companies to focus on energy-efficient production technologies and shorter cycle times in the future (Chen et al. 1993, Feher 2009).

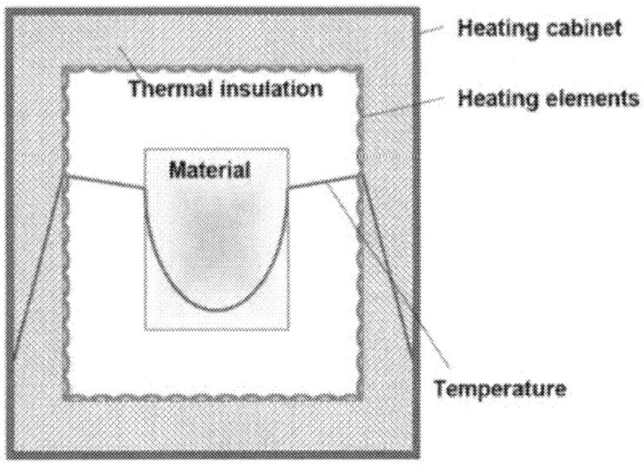

Figure 1 Temperature in case of heating with convective methods (Fischer et al. 2007)

By means of innovative curing methods, e.g. of microwave technology, it is possible to make the processing of thermosetting materials more economically.
The aim of this study is to show the benefits, but also the difficulties of curing thermosetting resins by means of microwaves. A general feasibility is given by the chemical structure of the investigated materials.

Experimental

Curing by means of Microwave

Microwave heating is a dielectric heating process. This means that heat is generated in non-conducting or just low electrical conducting materials, e.g. electrical insulation materials, by the influence of a high-frequency alternating electromagnetic field (Scheel et al. 2008). A basic prerequisite for the dissipation of microwave energy is a polar molecular structure, as present in non-linked thermosetting materials (Emmerich 2008).

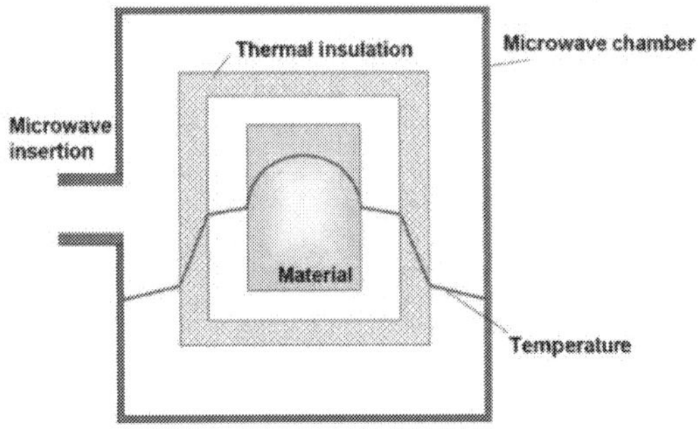

Figure 2 Temperature when heating by microwave (Fischer et al. 2007)

The specific frequency range of microwaves extends between 1 GHz and 300 GHz (Scheel et al. 2008, IEV).

In contrast to the currently used curing methods, microwaves penetrate much deeper into the object to be heated because of their large wavelengths (Fischer et al. 2007). The electromagnetic energy of the microwave is converted directly into heat in the inside of the material. The heat arises simultaneously over the entire volume; therefore, the process is also called a volumetric heating process (see Fig. 2).
The heating procedure takes place at the molecular level (Emmerich 2008). In an electromagnetic alternating field, forces are exerted on the molecules of a dielectric material. The result is a polarization of the material. Existing dipoles, e.g. hydroxylgroups or molecular dipoles align themselves in the field. Orientation polarization appears and contributes to the heating of the material (Fig. 3). Another effect which may contribute to the heating by microwave is the ion conductivity of the uncured thermosetting materials.

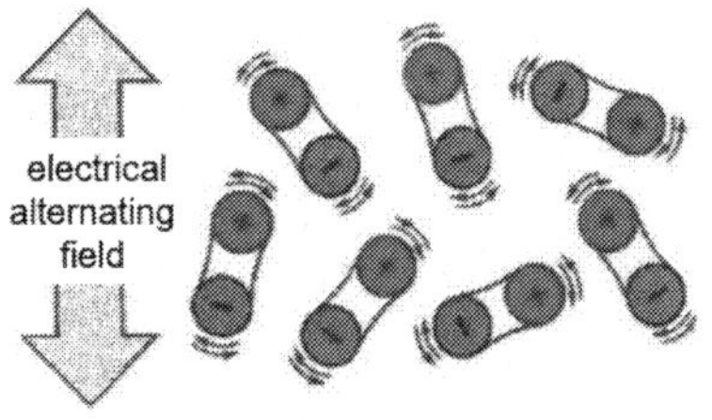

Figure 3 Orientation polarization in electrical alternating field (Scheel et al. 2008)

The rotation and vibration of the dipoles and polar molecules in the ever-changing electrical field produce intermolecular friction effects in the inside of the material, the resulting kinetic energy is converted into heat (Habenicht 2006). Since dipoles are difficult to move in the molecule unit, the dielectric loss angle δ is large according to amount and a lot of heat is generated. The dielectric loss angle δ depends on the dielectric loss index ε'', the imaginary part of the complex permittivity $\varepsilon_r = \varepsilon' - j\varepsilon''$, of the material. It is a degree of how much energy of the acting electric field is converted into heat. If there is a material with a high dielectric loss index ε'', the penetrated microwave energy is converted into thermal energy faster than in case of materials with a low dielectric loss index.
Therefore the absorption of microwave energy of a material is directly depending on the dielectric loss index ε'' (Buchenhorst 2005).
The extent to which the material absorbs depends on the frequency as well as on the temperature (Emmerich 2008). But in the curing process of thermosetting materials, the increasing degree of cross-linking of the material contributes to the absorption capacity (Mijovic

and Wijaya 1990). This should be considered when curing thermosetting compounds and can also be specifically used to influence curing conditions.
The absorbed microwave energy relating to the volume is equal to the heat generated in a volume element (see formula 1) (Scheel et al. 2008). This power density is dependent on the electric field power of the microwave field E, the frequency f and the dielectric loss index ε'' of the heated material. The dielectric loss index is built by the product of the permittivity and the tangent of the dielectric loss angle δ. Both variables are dependent on the frequency and temperature.

Absorbed microwave power with respect to volume [W/m³]:

$$p = 2\pi \cdot f \cdot \varepsilon_0 \cdot \varepsilon_r' \cdot \tan\delta \cdot E^2 = 2\pi \cdot f \cdot \varepsilon_0 \cdot \varepsilon_r'' \cdot E^2 \quad (1)$$

The fact that diffraction and refraction of the wave occur when microwaves penetrate in a material is aggravating. According to geometry, power concentrations at corners and edges and in certain areas of the inner volume of goods to be heated may occur.
Because of these complex relations between the different variables, calculations are just possible for simple idealized arrangements, the interpretation of the heating chambers for microwave technology relies heavily on experience and experiments.

Microwave chamber

A commercial microwave oven with a 2.45 GHz magnetron is used for the first experiments. The maximum power of the microwave is 950 W. The average introduced power can be varied by means of pulsed coupling times.
The field distribution of the microwave chamber is not completely homogeneous.

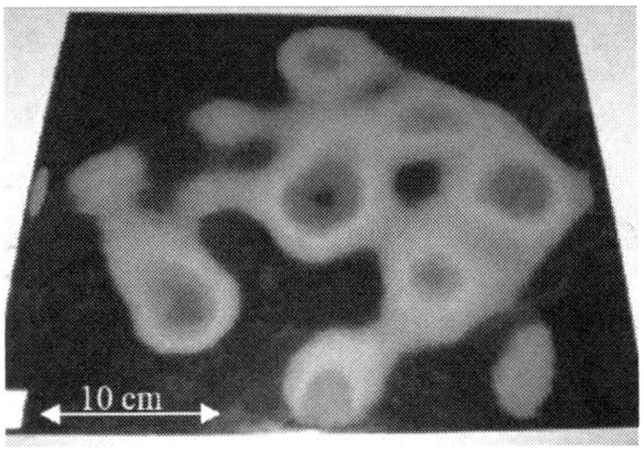

Figure 4 Distribution of the electrical field within the microwave chamber

By means of the temperature distribution in the chamber, the electrical field distribution and the occurrence of hot and cold spots can be made visible. To record temperature distribution, films are used on the base of temperature sensitive liquid crystals. An image of such a film after microwave irradiation (15 sec, 860 W) can be seen in Fig. 4. The used film reacts on temperatures between 25 °C and 30 °C. The illustrated film shows the temperature distribution in the interior of the microwave chamber that is directly depending on the electrical field distribution of the microwaves.
To obtain comparable and reproducible results, the samples to be cured are positioned in the same domain of the microwave chamber.

Materials

Several polyurethane potting compounds and epoxy resins are examined. All materials are two-component potting compounds, which are formulated of a resin-hardener mixture. The materials were all mixed according to the proportion agreed on in the data sheet.
The results of a polyurethane potting compound as well as those of an epoxy resin potting compound are shown as representatives of the respective material group.

Temperature of the components and the mixture

The single components of the respective potting compounds and the mixture of their respective components are placed separately into ceramic crucibles in the microwave chamber. The sample was taken due to defined time intervals and the temperature of the component was promptly recorded using an infrared thermometer. The measuring times were after 3 min, 5 min and 10 min and at 280 W, 670 W and 860 W microwave power.
The study aims to incorporate the different coupling of the microwave to the respective components and the resulting heat input into the mixtures due to microwave power.

Results and Discussion

The temperature analysis of the individual components with an infrared thermometer shows that the absorbed energy from the microwave varies a lot between the two components of any two-component resin. This effect can be shown for both, the epoxy and the polyurethane resin.
The temperature of the polyurethane components with a maximum of about 250 °C at 860 W microwave power is higher than that of the epoxy resin components (see Fig. 6). Both epoxy resin components reach a maximum temperature of about 200 °C at 860 W microwave power as well (see Fig. 6).

Phenomenologically, after 3 min in the microwave chamber, a mild bubble formation and slight evaporation can be seen on the resin component and as well on the hardener component of the epoxy resin. The bubble formation and the vaporization rate are increased after 5 min (see Fig. 5).

The resin component of the epoxy is substantially reduced after 10 minutes at 860 W microwave power and has increased in viscosity.

a)  b) 

Figure 5 a) EP resin component after 5 min at 860W
b) EP hardener component after 5 min at 860W

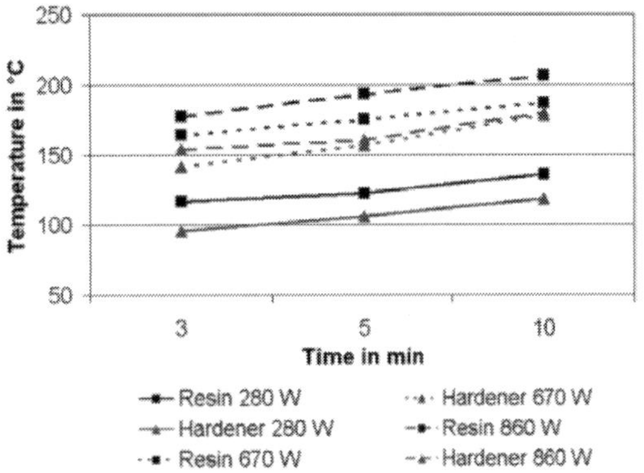

Figure 6 Temperature of the single EP components at 280 W, 670 W and 860 W.

The temperature analysis of the EP resin-hardener mixture shows that the temperature level of the mixture is at nearly the same range than the single components during microwave curing (Fig. 7). For the epoxy resin it is noticeable that the increase of temperature after 5 min decelerates. This can be founded to the advancing cross-linking process and the resulting decrement of functional groups that can be influenced by the microwave field.

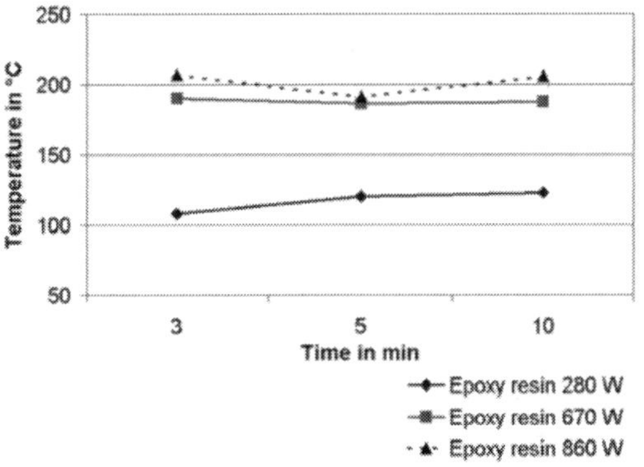

Figure 7 Temperature of the EP component mixture at 280 W, 670 W and 860 W.

The temperature record of the polyurethane components is also varying. Thus, at first, it comes to a slight formation of fumes and bubbles in both components, which get stronger in the course of time after 10 minutes more in the microwave chamber (Fig 8a).

The hardener component shows strong vaporization after 3 and 5 minutes in the microwave chamber. After 10 minutes under microwave influence, the microwave absorption is so strong that there is a thermal degradation with a strong resulting foaming of the component. The volume of the component has extremely increased and the color of the component has turned dark (see Fig. 8b).

The measurement shows that the general temperature level of the PUR components after microwave heating is higher than for the epoxy resin. Temperatures of more than 250 °C are measured. The located drop in temperature of the polyurethane hardener component after long curing times and at high power levels can be explained by the increased volume (Fig. 9). The temperature measurement by means of the infrared thermometer is more difficult and the cooling process after the microwave influence is faster because of the changed relation of surface and volume in the sample.

a) 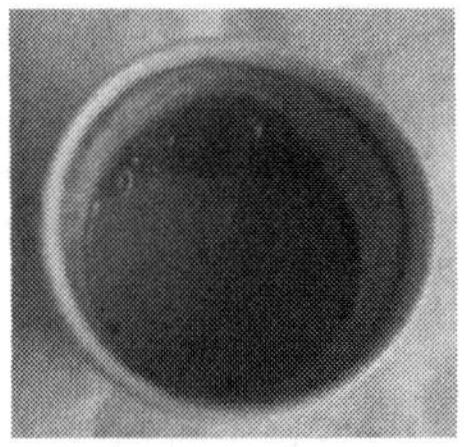 b) 

Figure 8 a) PUR resin component after 10min at 860W
b) PUR hardener component after 10 min at 860 W

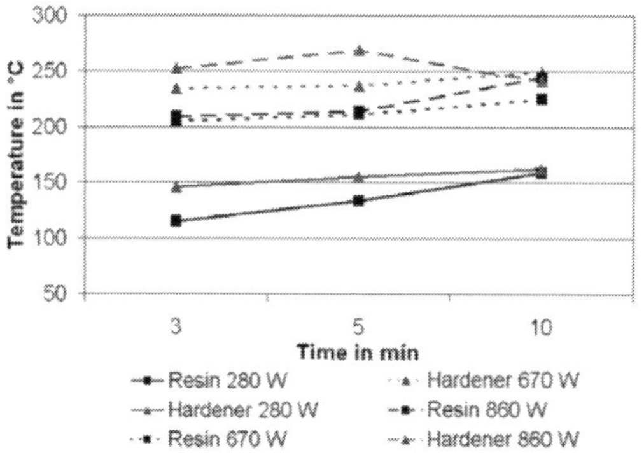

Figure 9 Temperature of the single PUR components at 280 W, 670 W and 860 W

In contrast to the measured temperatures of the single components it is conspicuous that the temperature of the PUR compound is on a lower level of just 200 °C after 10 min at 860 W (Fig. 10). This is nearly the same level than for the epoxy resin. The difference between the measured values is that for the tested PUR resin the temperature does not decrease after 5 min. This leads to the fact that the curing of the PUR resin proceeds with lower reaction rate than for the epoxy resin. But the difference between the temperature levels of the single components and the mixture indicate that it would be meaningful to use microwave energy for the curing of the testes PUR resin just at low energy levels to avoid the overheating of single components during the curing process.

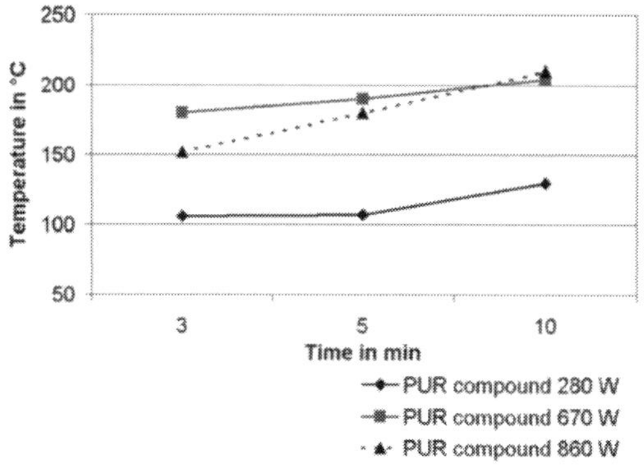

Figure 10 Temperature of the PUR compound at 280 W, 670 W and 860 W.

In Fig. 11, epoxy resin samples are shown, which were cross sectioned after curing with a sharp blade in order to be able to make a statement about the cross-linking of the entire volume. The curing of epoxy resin potting compounds with microwaves seems to allow a significant time saving, in order to visually exhibit a relatively cured looking condition.

A bubble formation may be seen if the microwave irradiation is too long (see Fig. 11 b)). Small bubbles may be seen in Fig. 11 a) as well.

Fig. 11 b) clearly shows the volumetric heat generation in the interior of the resin to be cured, the bubble formation is stronger at the core of the prepared sample than in peripheral areas.

a)
b)
c)

Figure 11 a) EP (5 min at 380 W)
b) EP (10 min at 280 W)
c) EP regularly cured (60 min at 100 °C)

Directly after the curing process, the polyurethane potting compound samples were also divided (see Fig. 12 a) – 12 c)). Even with this material, a significant bubble formation can be seen (Fig. 12 a).

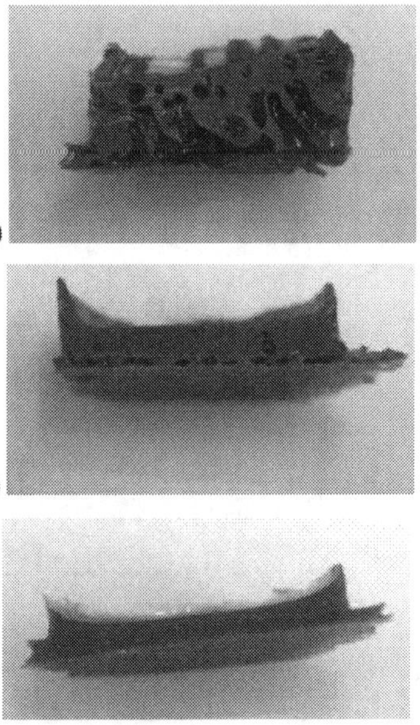

a)

b)

c)

Figure 12 a) PUR (2 min at 860 W)
b) PUR (5 min at 480 W)
c) PUR convective curing (150 min, 80°C)

DSC-analysis

For the analytical determination of the curing degree of the manufactured samples, calorimetric measurements were additionally carried out by means of Differential Scanning Calorimetry (DSC) (Ehrenstein et al. 2004). To characterize the curing process by comparing conventional and microwave curing, samples that are visually similar and have no bubbles were measured first. While comparing the measurements, one searched for non-linked sections within the various cured samples in this way. The heating-up profile used for this purpose consists of two linear heating cycles from ambient temperature up to 250 °C at 10 K/min and a cooling cycle of 10 K/min in between.

In the first heating-up phase, still non-linked parts react. The comparison of the exothermic peaks in the enthalpy of the curve can thus conclude, based on the deviation of the curve between the first and the second heating-up cycle, on the degree of cross linking of the sample.

Table 1 shows the enthalpy values of three EP samples witch are cured under different conditions. Sample 1 is an uncured specimen that cures just because of the heat input during the DSC measurement and shows an obvious exothermal enthalpy of 147 J/g with an onset at 55 °C. The other samples were cured before the DSC analysis. One is conventionally cured in a heating cabinet for 60 min at 100 °C. The other sample is cured in a microwave oven for 10 min at 200 W. Both samples do not show a measurable exothermal peak which concludes that both are nearly completely cured. No further cross link process can be detected.

Table 1 Measured enthalpies of different EP samples

Status of curing	Exothermal enthalpy
uncured	147 J/g
60 min at 100 °C (oven)	not measurable
10 min at 200 W (microwave)	not measurable

In the same way DSC studies of some PUR samples were made. Table 2 shows the results of three PUR samples. One is a uncured sample; one is cured in a heating cabinet for 150 min at 80 °C and the third is cured in the microwave oven for 5 min at 480 W. For the uncured sample a resulting exothermal enthalpy of 47 J/g with an onset at 76 °C can be detected. As expected the heating cabinet cured sample has no detectable enthalpy peak. But for the microwave cured sample a slight exothermic curve of 21 J/g with an onset at 76 °C during the first heating phase.

Table 2 Measured enthalpies of different PUR samples

Status of curing	Exothermal enthalpy
uncured	47 J/g
150 min at 80 °C (oven)	not measurable
5 min at 480 W (microwave)	21 J/g

Despite the visual similarity to the conventionally cured sample, a difference can be seen in the measurement here. It is reasonable to assume that the DSC measurement leads to a post-curing and resulting post-cross-linking of the material and the exposure that time and power of the microwave have not been sufficient for a complete curing of the material.

Comparing the sample cured in a microwave oven to a conventionally cured sample, the tiny bubbles in the interior of the material volume can be noticed. Despite the bubble formation, the material is still not fully cross-linked. One reason for the incomplete cross-linking could be that a part of the curing component thermally decomposes in the process because of to much application of energy, and thus it is no longer available for the curing process.

Another reason could be that the curing process is not completely finished and the operation of microwave should be longer and probably more sensitive.

For this reason, further optimization is necessary in the process to ensure the stoichiometry of the cross-linking components.

Conclusions

The current work shows that thermosetting casting resins from the group of epoxy resins and polyurethane resins cure and cross-link with the aid of microwaves.
The experiments also show that the required time and energy costs are considerably lower than those when using conventional curing methods by means of convectional heat flow.
The temperature absorption of individual components is very different. During the curing of a two-component potting compound there is risk that one component heats faster and absorbs substantially more energy than the other. Due to excessive temperature formation in the interior, it is possible that a component thermally decomposes before the cross-linking process. A fully cured material can no longer be generated. The properties of the cured material are thus significantly affected.
The formation of bubbles in the inside of the material being cured is an indication of overheating during the curing process.
The uniformity of the microwave field is crucial for the homogeneous curing of thick-layered components and layers with large surfaces. As shown, the field in a conventional microwave is very inhomogeneous. Considerable hot and cold spots can be seen.
In this way, a further focus for future work is set. Once this prerequisite is created, the curing process for different materials should be optimized. The next step is to get a homogenous distribution of the microwave energy in the chamber. This can be realised by a special constructed chamber with special coupling variants. Once this disadvantage of the used microwave curing technology is solved specific process parameters for the curing of a large number of thermosetting potting compounds can be developed.

Acknowledgements

This research is financially supported by the „Arbeitsgemeinschaft industrieller Forschungs-vereinigungen (AiF)“ as part of the project „Schnelle und eigenspannungsfreie Aushärtung von Beschichtungsstoffen, duroplastischen Harzen und rieselfähigen Duroplasten für die Elektro-Isoliertechnik mittels Mikrowellen“ in cooperation with the Faculty of Computer Science and Electrical Engineering IEF / IGS of the University of Rostock.

References

Buchenhorst, D., “Anwendung von Radiowellen zur Erwärmung von Adsorbenzien und Katalysatoren“, Martin-Luther-Universität Halle-Wittenberg, Halle (Saale), (2005).

Chen, M.; Siochi, E. J.; Ward, T.C.; McGrath, J. E., “Basic Ideas of Microwave Processing of Polymers“, Polymer Engineering and science, MID-September, Vol. 33, No. 17, (1993).

Debye, P., “Polare Molekeln”, 8th Ed., Hirzel, Leipzig, (1929).

Domininghaus, H., “Plastics for Engineers: Materials, Properties, Applications”, Hanser Gardner Publications, (1993).

Ehrenstein, G. W.; Riedel, G.; Trawiel, P., “Thermal Analysis of Plastics: Theory and Practice“, Hanser Gardner Publications, Munich, (2004).

Emmerich, R., “Mikrowellentechnik“, ICT Fraunhofer Institut für Chemische Technologie, Pfinztal, (2008).

Feher, L. E., “Energy Efficient Microwave Systems – Materials Processing Technologies for Avionic, Mobility and Environmental Applications“, Springer-Verlag, Berlin, Heidelberg, (2009).

Fischer, B.; Lampke, Th.; Walter, G.; Wielage, B., “Mikrowellenangepasstes Sintern von Ferriten“, Materialwissenschaft und Werkstofftechnik, Vol. 38, No. 10, Wiley-VCH Verlag, Weinheim, (2007).

Habenicht, G., “Kleben – Grundlagen, Technologien, Anwendungen“, VDI-book, 5th Ed., Springer, Berlin New York, (2006).

International Electrotechnical Commission, Electropedia, reference “microwave”, EN 713-06-03, Genf, (2010).

Mijovic, J.; Wijaya, J., “Review of Cure of Polymers and Composites by Microwave Energy“, Polymer Composites, Vol. 11, Nr. 3, New York, (1990).

Peters, N.; Nunge, S.; Geldermann, J.; Renz, O., “Bericht über Beste Verfügbare Techniken (BVT) im Bereich der Lack- und Klebstoffverarbeitung in Deutschland“, Deutsch-Französisches Institut für Umweltforschung (DFIU), Karlsruhe, (2002).

Scheel, W.; Wittke, K.; Nowottnick, M., “Niedrigtemperaturmontage hochintegrierter elektronischer Baugruppen durch selektrive Mikrowellenerwärmung MICROFLOW“, Dr. Marcus S. Detert, Templin, (2008).

Schubert, H.; Grünewald, T., "Wärmequellenverteilung in Modellkörpern bei der Mikrowellen-Erhitzung", Chemie Ingenieur Technik, Vol. 55, No. 7, p. 553 – 555 (1983).

Stoye, D.; Freitag, W., "Resins for Coatings: Chemistry, Properties and Applications", Hanser Gardner Publications, Munich, Vienna, (1996).

MELT STRENGTHENING OF POLY (LACTIC ACID) THROUGH BRANCHING BY REACTIVE EXTRUSION WITH EPOXY-FUNCTIONALIZED CHAINS

Y-M. CORRE, A. MAAZOUZ*, J. DUCHET, J. REIGNIER

Université de Lyon, F-69003, Lyon, France, INSA Lyon, F-69621, Villeurbanne, France
CNRS, UMR5223, Ingénierie des Matériaux Polymères. Abderrahim.maazouz@insa-lyon.fr

ABSTRACT: Poly (lactic acid) is an industrially mature, bio-sourced and biodegradable polymer. However, current applications of this eco-friendly material are limited as a result of its poor melt properties. One of the keys to extending its processing field is melt strengthening and the present investigation considers chain branching as a valuable solution for reaching such an objective. A commercially available additive consisting of epoxy-functionalized PLA chains was employed as a chain brancher, and reactive extrusion runs were performed with various amounts of additive and at several processing temperatures. The efficiency and kinetics of the reactions were revealed by monitoring the melt pressure and analyzing the molecular weight distribution. Dynamical and extensional rheology confirmed the melt strengthening for the branched PLA samples.

Main Conclusions

Poly (lactic acid), PLA, is a polymer that possesses many physical characteristics rendering it suitable for replacing commodity polymers [Vert et al. (1995)]. Beside it being biocompatible and available from renewable resources, PLA also demonstrates a high strength, thermoplastic processability, good crease-retention, grease and oil resistance and an excellent aroma barrier [Lunt (1998), Wehrenberg et al. (1981), Drumright et al. (2000), Lipinsky et al. (1986)]. The PLA has an elevated modulus and strength, but has a brittle nature and lack in toughness. These drawbacks are especially disadvantageous in the film extrusion industry. However, there are means of improving the flexibility and toughness in PLA, for instance by blending with a plasticizer.

Previous laboratory studies have mainly been focused on modifying the resin through branching reactions, polymer blending and charge filling [Zhou et al. (2007), Palade et al. (2001), Lamnawar et al. (2006), Villalobos et al. (2006)]. There are also numerous reports on blending PLA with plasticizers for biomedical [Henry et al. (1985), Yolles et al. (1976, 1978)], as well as for film applications [Ljungberg et al. (2002, 2003, 2005), Martin et al. (2003)]. When synthesizing PLA from corn starch, its molecular weight distribution is generally not well controlled and the polydispersity of common commercial grades is often quite broad. This specificity, in addition to degradation problems [Labrecque et al. (2007)] arising from PLA's polyester nature, puts the studies of this material apart from those of classical resins.

Few papers have been dedicated to the rheological behavior of melt strengthening through chain extension of poly (lactic acid). A prerequisite in order to allow a larger processing window.

PLA was modified through reactive extrusion with this chain brancher and the resultant material was characterized with regard to its molecular weight distribution (MWD), as well as by means of rheological analysis.

An optimized processing temperature was identified as a key parameter to obtain the desired efficiency of the reaction, since this would reduce the competition between chain extension and thermal chain scission.

Chain branching drastically modified the structure of PLA by shifting the molecular weight distribution from a monomodal to a bimodal population, and this new chain distribution had an enlarging effect on the relaxation time spectrum. Moreover, the melt rheology of the materials was influenced by these molecular couplings. On the one hand, the low shear viscosity could be drastically improved by increasing the content of the branching agent, and on the other hand, converging values were obtained for high shear rates. The melt thus obviously became stronger with an increasing epoxide content.

The thermal properties of the modified PLA materials were found to diverge from those of neat PLA. For instance, the crystallization nucleation was significantly increased when branching reactions were performed. This phenomenon induced a kinetic improvement but influenced the final crystallinity, which was found to become lowered when the branching level was increased. To finish, the branched materials featured a broader crystallization temperature window as compared to that of the neat PLA.

References

Arraiza A. L., Verdu J., Colin X., I. Polym. Proc. XXII, **5,** 389-394 (2007).

Drumright R. E., Henton D. E., Adv. Mater., **12**, (23), 1841 - 1846 (2000).

Dorgan J.R., J. S. W., J. Rheol., 43, **(5),** 1141-1155 (1999).

Henry, E. W.; Chiu T.-H.; Nyilas, E.; Brushart, T. M.; Dikkes, P.; Sidman, R. L., Exp. Neurol., 90, **(3),** 652-676 (1985).

Labrecque L. V., R. A. K., V. Davé, R. A. Gross, S. P. McCarthy, J. Appl. Polym. Sci., **66,** (8), 1507-1513 (1997).

Lamnawar K, Maazouz A, Rheol. Acta, **45**, 411-424 (2006).

Ljungberg N., B. W., T J. Appl. Polym. Sci., **86,** (5), 1227-1234 (2002).

Ljungberg, N.; Wesslén, B., Polymer, **44**, **(25),** 7679-7688 (2003).

Ljungberg, N.; Wesslen, B., Biomacromolecules, **6**, (3), 1789-1796 (2005).

Lipinsky E.S., R. G. S., Chem. Eng. Prog., **82**, 26–32 (1986).

Lunt, J., Polym. Degrad. Stab., **59**, (1-3), 145-152 (1998).

Martin, O.; Avérous, L., Polymer, **42**, (14), 6209-6219 (2001).

Palade L-J, H. J. L., Dorgan J. R., Macromolecules, **34**, 1384-1390 (2001).

Vert, M.; Schwarch, G.; Coudane, J. J. Macromol. Sci., Part A. Pure Appl. Chem., **32**, 787-796 (1995).

Villalobos M., Greeley T., Turco G., Deeter G., Energy, **31**, 3227–3234 (2006).

Wehrenberg, R. H., Mater. Eng., **94** (3), 63-66 (1981).

Yolles S., T. L., Ward L., Boettner F., Parenteral Drug Assoc., **30**, (6), 306-312 (1976).

Yolles, S., J. Parenteral Drug Assoc., **32**, (4), 188 (1978).

Zhou Z. F., G. Q. H., Xu W. B, Ren, F. M. eXPRESS Polym. Lett., **1**, (11), 734–739 (2007).

CHAIN EXTENSION / BRANCHING OF THERMOPLASTIC POLYETHER-ESTER ELASTOMER BY REACTIVE EXTRUSION

Byeong Joon Jeong[1*], Hunsoo Kim[1], Jonggun Park[1], Eon Seok Lee[1], Marino Xanthos[2]

[1]*Tech Center, LG Chemical Ltd. - Daejeon – Korea*
bjjunge@lgchem.com; hanskim@lgchem.com; pjgun@lgchem.com; esleea@lgchem.com

[2]*Otto H. York Department of Chemical, Biological and Pharmaceutical Engineering,*
New Jersey Institute of Technology, NJ, USA
xanthos@njit.edu

Abstract

Chain extension and branching of a thermoplastic polyether-ester elastomer (TPEE) with two reactive agents by reactive compounding was carried out. The commercial TPEE used in this study was a step-polymerized copolymer between a polybutylene terephthalate (as hard segments) and a polyoxy-alkylene glycol (as soft segments). Commercial reactive agents were a polycarbodiimide and an ethylene / glycidyl methacrylate / vinyl acetate terpolymer. The reactivities of the two additives were compared by rheological measurements of the modified products. The reactively modified products showed improved cell homogeneity of the foamed samples.

Introduction

Thermoplastic polyether-ester elastomers (TPEE) have superior performance and ease of processing as compared with thermoset rubber. Furthermore, the strength of TPEE, which is combined with flexibility at a wide range of temperature, is known as an advantage over usual engineering plastics. Owing to these, the applications of TPEE become wider, especially in the automotive industry. For example, TPEE is replacing the traditional rubber, in the application of automotive suspension joints, where resistance to fatigue and shock absorption are required.
Depending on its application, TPEE is usually melt-mixed with a rigid polyester (e.g. polybutylene terephthalate) to control its hardness. During compounding, TPEE would degrade by hydrolytic and thermooxidative degradation reactions, which result in its reduced viscosity. The detailed degradation behavior during compounding is summarized elsewhere [1]. In order to compensate for the viscosity reduction during compounding, reactive modifiers were introduced for chain extension/branching of the TPEE molecules. Furthermore, enhanced viscoelasticity could be obtained by adjusting the concentration of the reactive modifiers.
The objective of this study is to investigate the reactive chain extension/branching of TPEE by introducing a modifier with multiple polycarbodiimide and with glycidyl functional groups. The extent of reactive modification is investigated by rhelogical analyses of the modified products and extrusion foaming with a chemical blowing agent.

Experimental

Materials: A commercial TPEE, produced by LG Chemical Ltd., was chosen. The TPEE (grade name: Keyflex BT 2140D) was an extrusion grade with MFI of 5 / 10 min at 230 °C/2.16 kg. The resin was pre-dried overnight at 90 °C before the experiments. Two reactive modifiers were selected. One was multi-functional polycarbodiimide (named as Modifier-1 in this study), and the other was ethylene/glycidyl methacrylate/vinyl acetate terpolymer (Modifier-2). The concentration of the GMA comonomer in the Modifier-2 was known to be 12%. Modifier-1 was a fine powder, while Modifier-2 was in the form of regular size pellets. Azodicarbonamide (ADCA) was used in extrusion foaming as a chemical blowing agent. The onset temperature of nitrogen release was expected at 205 °C.

Reactive Processing: Reactive extrusion was performed using a twin screw extruder (ZSK 25, L/D: 44, Coperion). Set temperatures of the barrels and the die were between 235 ~ 250 °C. The screw speed and the feed rate were constant for all the samples as 300 rpm and 10 kg/hr, respectively. The extrudate from a 2.5mm rod die was cooled in a water bath, and then pelletized.

Rheological characterization: A capillary rheometer was used for the viscosity measurements with a 1.0mm diameter die (L/D of 30) at 230 °C. The Rabinowitsch correction was applied on each measurement to calculate the true shear rate and viscosity.

Extrusion Foaming: A lab-scale single screw extruder (D: 19 mm, L/D: 33, Haake Rheomex) was used for foam extrusion. A static mixer and a rod die (D: 1 mm, L/D: 3) were attached to the extruder. The chemical blowing agent (CBA) was premixed with the PBT resin using a small batch mixer, and then fed into the hopper

Results and Discussion

The carbodiimide functional group of the Modifier-1 and the glycidyl functional group of the Modifier-2 are expected to react with the carboxyl end group of TPEE preferentially. Schematics of these reactions can be found in ref [2] and [3], respectively.

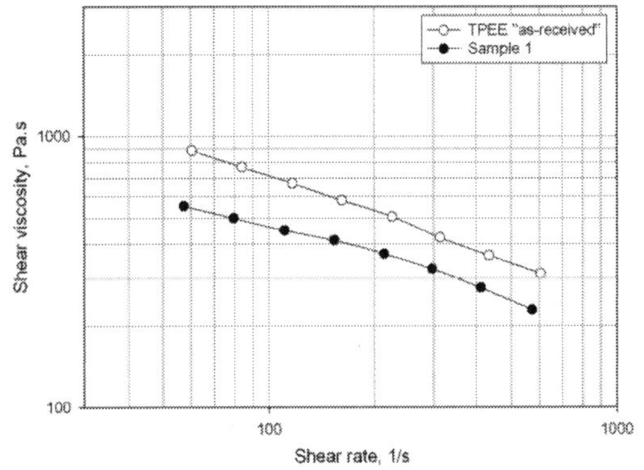

Figure 1. Viscosity reduction before and after extrusion processing at 300 rpm, 10kg/hr

Figure 1 shows the effect of the degradation of the TPEE during melt processing at the given operational conditions (300 rpm, 10 kg/hr). The minimum residence time under the operational conditions measured by using a tracer (colored pellets) was about 52 seconds. The viscosity of sample 1, which was made by extrusion without any modifier, showed a reduced viscosity after extrusion as compared with that of the "as-received" TPEE. This indicates that hydrolytic and/or thermooxidative degradation occurred, in spite of the fact that a certain amount of antioxidant was premixed with the TPEE before extrusion.
Shear viscosities, which were increased proportionally to the concentration of Modifier-1, are shown in Figure 2. The concentrations of Modifier-1 in samples 2, 3, and 4 are 0.2, 0.4 and 0.6 phr, respectively. Note that the viscosity of the "as-received" TPEE (Figure 1) lies between the viscosity curves of the Samples 3 and 4 in Figure 2. This means that Modifier-1 could be used as an anti-degradation agent when a constant fluidity (e.g. Melt Flow Index) after compounding was required.

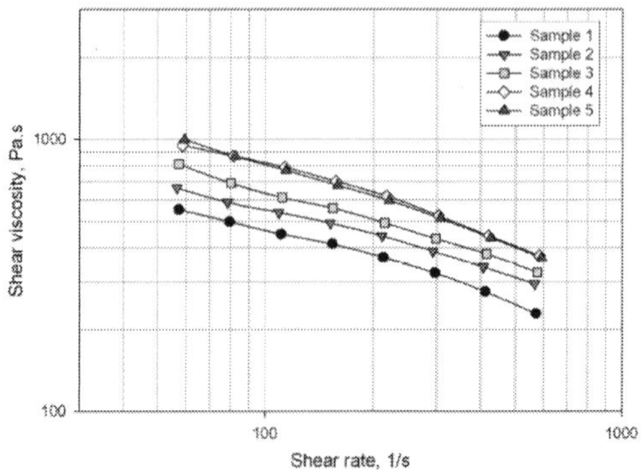

Figure 2. Effect of type and amount of modifiers on shear viscosity/shear rate curves

A series of experiments with Modifier-2 was carried out in order to compare the reactivity of the two modifiers. It was found that five times higher concentration of Modifier-2, as compared with Modifier-1, was required to achieve a similar level of viscosity increase. Figure 2 shows a similar viscosity curve of sample 5 (the sample with 3.0 phr of Modifier-2) as compared with that of sample 4 (the sample with 0.6 phr of Modifier-1).
Jeong and Xanthos [4] made a branched polyester from a linear polyester by reactive modification with a tri-functional modifier. They observed an early onset of the shear thinning behavior with their branched sample and different slopes (the Power law indexes) of the viscosity curves between the linear and the branched polyesters. In Figure 2, there exists no considerable change of the slopes of the viscosity curves. It appears that the structures of the TPEE molecules were not changed significantly (e.g. from linear to branched) in this study. However, it should be noted that the measured range of the shear rates in Figure 2 is relatively narrow; thus, further rheological measurements in a wider range of shear rates (esp. at lower shear rates) is required for definite conclusions on the structural changes of the modified samples.
Modifier-2 was favored in terms of mixing, since the required concentration of Modifier-2 was relatively high (similarly to a master batch) enough to make a homogeneous premix and also prevent the fluctuation of the feed concentration at low modifier concentration.. However, the main chains of Modifier-2 are very flexible (mainly polyethylene based), thus, a high concentration of Modifier-2 may adversely affect properties of the product such as thermal resistance (low heat distortion temperature). A combination of the two modifiers would be recommended, considering the ease of processing and the desirable rheological and thermal properties.

The densities and the blow-up ratio (the diameter of the sample divided by that of the die) of the foamed samples are shown in Table 1. Note that the solid density of the unmodified TPEE is 1.14 g/cc. Unexpectedly, the density reduction of sample 4 was relatively lower than that of sample 5.

Table 1. Characteristics of foamed sample

Sample No	Foam density, g/cc (std deviation)	Blow-up ratio (std deviation)
Sample 1	0.75 (0.05)	1.65 (0.06)
Sample 4	0.71 (0.02)	1.93 (0.03)
Sample 5	0.66 (0.03)	1.98 (0.02)

The cell morphologies of the fractured surfaces of the foamed samples are shown in Figures 3 - 5. As anticipated, relatively large and uneven cell sizes, presumably due to the frequent coalescence of the cells, are observed with sample 1 (Figure 3). A more even cell morphology is found with sample 5 (Figure 5). However, the difference of the cell morphologies between samples 1 and 4 is not so significant as between samples 1 and 5. This is presumably due to 1) the effect of the soft segments of the backbone of Modifier-2, and/or 2) a sign of the different structures (and viscoelastic behavior) by the reactive modification between samples 4 and 5, although further rheological investigations are required.

Conclusions

Chain extended/branched TPEE could be made by reactive extrusion with two different types of modifiers. The rheology of the reactive extrusion product could be controlled. Different viscoelastic behavior was observed in the foamed samples depending on the type of modifiers.

References

1. R.K. Adams et al., "Thermoplastic Polyether Ester Elastomers", in Thermoplastic Elastomers, G. Holden et al. (Ed.), Hanser Publishers, Munich, p.191-225 (1996)
2. M. Xanthos, M. Young, G. Karayannidis and D. Bikiaris, *Polym. Eng. Sci.*, 41, 643 (2001)
3. R. Jerome and C. Pagnoulle, "Key Role of Structural Features of Compatibilizing Polymer Additives in Reactive Blending", in Reactive Polymer Blending, W.E. Baker et al. (Ed.), Hanser Publishers, Munich, p. 82-112 (2001)
4. B. Jeong and M. Xanthos, *Polym. Eng. Sci.*, 47, 244 (2007)

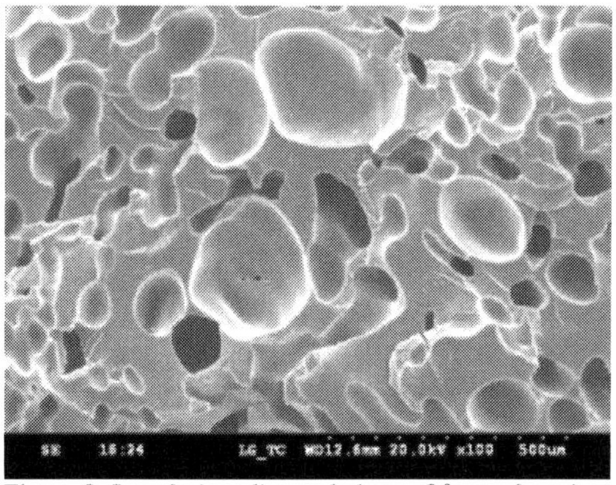

Figure 3. Sample 1: cell morphology of foamed product

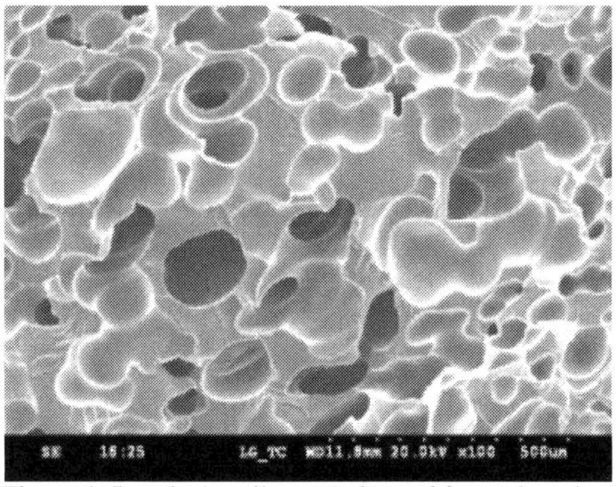

Figure 4. Sample 4: cell morphology of foamed product

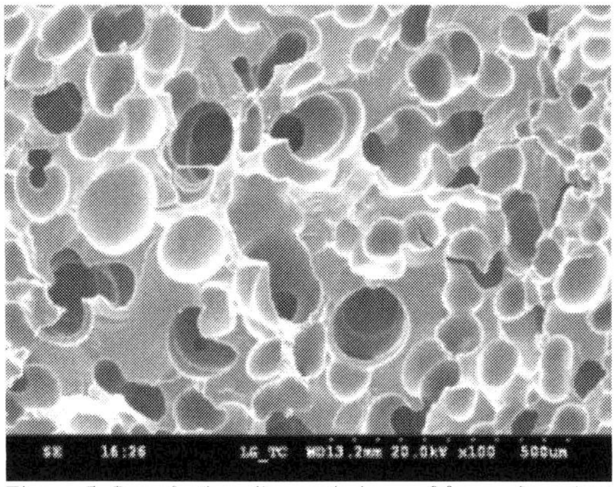

Figure 5. Sample 5: cell morphology of foamed product

IN SITU SYNTHESIS OF TITANIUM DIOXIDE IN A MOLTEN POLYPROPYLENE MATRIX BY HYDROLYSIS-CONDENSATION REACTIONS IN A TWIN SCREW EXTRUDER

W. Bahloul[1,2], V. Bounor-Legaré[2], F. Melis[2], P. Cassagnau[1,2]*, O. Oddes[3], B. Vergnes[3]

[1] *Université Lyon 1, 69622 Villeurbanne (France), philippe.cassagnau@univ-lyon1.fr*
[2]*CNRS, UMR 5223, Ingénierie des Matériaux Polymères, Laboratoire Matériaux Polymères et Biomatériaux, 69622 Villeurbanne (France)*
[3] *MINES ParisTech, CEMEF, UMR CNRS 7635, Sophia Antipolis (France)*

Abstract – The objective of this work was to develop an original method for the preparation in molten condition of PP/TiO_2 nanocomposites. As a breakthrough of nanocomposite preparation, the in-situ formation of melt-processable nanofillers offers new opportunities to design multiphase polymer blends with new property profiles. Consequently, in order to favour and to control the dispersion of the nanofillers within the polymeric matrix, a sol-gel approach for in-situ synthesis of the TiO_2 inorganic phase was developed. It was shown that the size and the quality of particles dispersion at the nano-scale depend strongly on the processing conditions. According to observations from small-angle X-ray scattering (SAXS) and transmission electron microscopy (TEM), these new nanocomposites have an average size of the TiO_2 phase around 20 nm and a fractal structure. Furthermore the molten viscoelastic properties show a nano-structuration of the materials. Comparatively, usual dispersion methods (melt mixing of TiO_2 particles) provide nanoparticles dispersion at the micro-scale with agglomeration phenomena. Finally, to simulate the development of the sol-gel reaction along the extruder, the Ludovic© software, previously developed at CEMEF, was used. As a result, this modelling approach allowed us to better understand and to optimize the processing conditions in order to obtain at the die exit a controlled conversion rate at the maximum output conditions.

Introduction

Nanocomposites materials obtained by melt blending have been intensively described (Vaia et al., 1993; Vaia et al., 1997; Liu et al., 1999; Dennis et al., 2001; Lertwimolnun and Vergnes, 2006). In that case, melt processing conditions have an important influence on the nature and properties of the nanocomposite materials. However, due to the poor compatibility of hydrophilic nanoparticles with conventional hydrophobic polymer matrices, such as polyolefins, nanodispersion is difficult to reach without particle agglomerations. As a consequence, an original way to overcome these problems is the in-situ synthesis of the inorganic filler in the molten polymer matrix by reactive extrusion. This synthesis is based on hydrolysis-condensations reactions, called classically "sol-gel", of metal oxide precursors (Dou et al., 2008). This approach has been largely developed for coating applications (Chen et al., 1999; Barlier et al., 2007) allowing in mild conditions to create multifunctional organic-inorganic materials.

The transposition of this chemistry to the reactive extrusion consists in using an extruder (single or twin-screw) as a continuous chemical reactor (Xanthos, 1992; Cassagnau et al., 2007). Besides the classical functions of a screw extruder (solid conveying, melting, mixing, pumping), additional functions are involved, such as the development and the control of a chemical reaction (sol-gel reaction). The objective of the present work is first to demonstrate the potentiality of the reactive extrusion for synthesizing PP/TiO_2 nanocomposites. Secondly, we will characterize the TiO_2 filler dispersion in the PP matrix over a wide length range through the combination of small angle X-ray scattering (SAXS), transmission electron microscopy (TEM) and linear viscoelasticty in molten conditions. Finally, the effect of processing parameters, such as screw speed, feed rate and concentration of precursor, upon the titanium n-butoxide hydrolysis-condensation reactions conversion will be investigated by using numerical simulations.

Experimental

Materials

Polypropylene (PP) HP500N was kindly supplied by Basell company. The titanium n-butoxide 97% from ABCR was used as inorganic precursor. This monomer has the following characteristics: molar mass M = 340g.mol^{-1}, density ρ = 0.99 and boiling temperature: T_b = 350°C. The titanium dioxide anatase LO-CR-K was used as pre-formed particle reference and was supplied from Sachtleben.

Nanocomposite preparation

The nanocomposite materials were prepared using a co-rotating twin screw extruder (Leistritz LSM 30-34, screw diameter D = 34 mm, length to diameter ratio L/D = 34.5). A specific screw profile was designed for the in-situ synthesis of titanium dioxide in molten PP matrix. 10, 20 and 30 wt% of titanium n-butoxide were mixed with molten PP in order to achieve TiO_2 concentrations of 2.9, 6 and 9.3 wt% respectively in PP/TiO_2 nanocomposites. For a comparative purpose, anatase TiO_2 particles were mixed with molten PP in the same twin screw extruder under the same processing conditions. Three samples were prepared at TiO_2 concentrations of 3, 6 and 9 wt%.

Results and Discussion

The morphology of the in situ PP/TiO_2 nanocomposite samples has been explored by SAXS and TEM microscopy. First of all, it must be mentioned that SAXS test was performed at high temperature (200°C) above the melting temperature of the PP matrix. Figure 1 shows the variation (bi-logarithmic plot) of the scattering intensity I versus the scattering vector q for I-2.9TiO_2 and I-9.3TiO_2 samples. The measured data exhibit a $q^{-1.5}$ and $q^{-3.5}$ power law scattering in the low q region for the two studied samples. These results have been argued to be linked to the fractal nature of scattering objects in the literature (Wijnen et al., 1991).

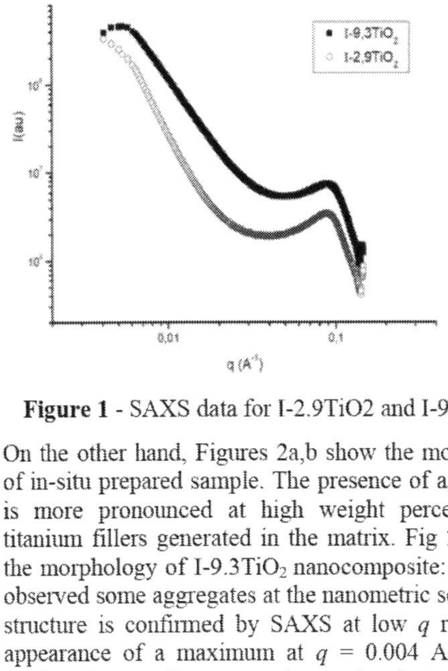

Figure 1 - SAXS data for I-2.9TiO2 and I-9.3TiO2

On the other hand, Figures 2a,b show the morphology of in-situ prepared sample. The presence of aggregates is more pronounced at high weight percentage of titanium fillers generated in the matrix. Fig 2b shows the morphology of I-9.3TiO_2 nanocomposite: it can be observed some aggregates at the nanometric scale. This structure is confirmed by SAXS at low q range, the appearance of a maximum at q = 0.004 A^{-1} can be attributed to a distance or interaction between some aggregates.

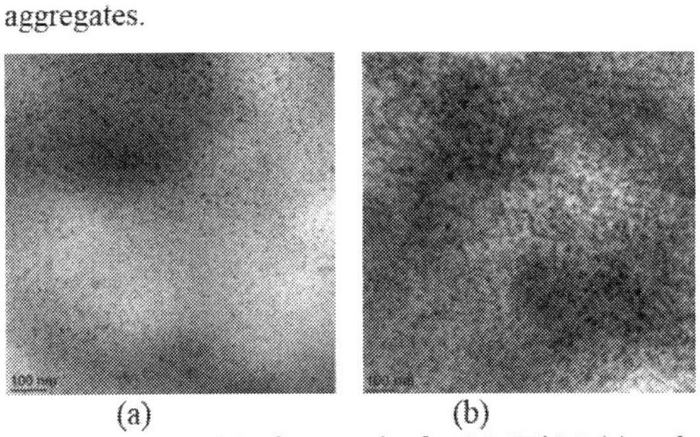

(a) (b)

Figure 2 – TEM micrographs for I-2.9TiO2 (a) and I-9.3TiO2 (b)

Concerning rheological properties, the viscoelastic behaviour of PP/TiO_2 composites prepared by melt blending of TiO_2 (anatase) in polypropylene matrix and the in-situ method via sol-gel reaction was compared. As shown in Figure 3, the simple dispersion of titanium dioxide (anatase) with 9 wt% does not considerably affect the linear viscoelastic behaviour of the PP matrix: the reinforcement by the TiO_2 micro-particles is weak. This result means that there are no interparticle interactions between the micro-size TiO_2 particles. On the opposite, the sample prepared by in situ method shows that the terminal relaxation zone (variation of the storage modulus G') of the nanocomposite has been strongly altered. For 9.3 wt% of in-situ generated TiO_2 particles, the viscoelastic behavior does not show any terminal flow zone and the elastic character of this suspension becomes predominant at low frequencies with the appearance of a secondary plateau. Actually, this nanocomposite shows a solid-like behavior response which includes a non–terminal zone of relaxation.

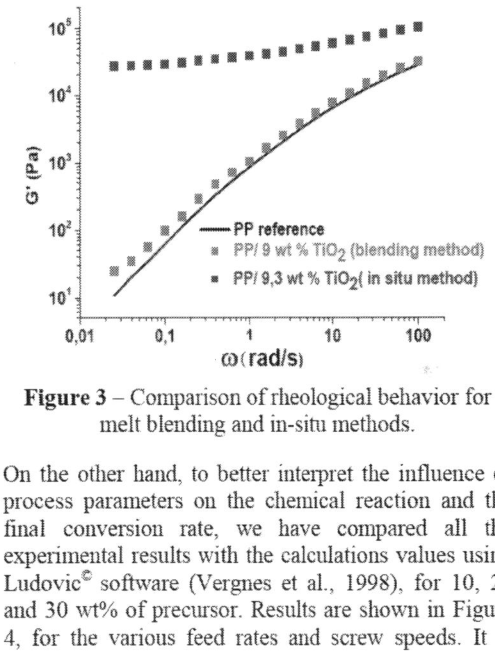

Figure 3 – Comparison of rheological behavior for melt blending and in-situ methods.

On the other hand, to better interpret the influence of process parameters on the chemical reaction and the final conversion rate, we have compared all the experimental results with the calculations values using Ludovic© software (Vergnes et al., 1998), for 10, 20 and 30 wt% of precursor. Results are shown in Figure 4, for the various feed rates and screw speeds. It is clearly shown that the prediction for 10 and 20 % are satisfactory, leading to a correct value at ± 15%, whatever screw speed and feed rate. However, the calculations at 30 % provide an important over-estimation of the conversion rate. We can imagine that the corresponding kinetics is too fast. Even if based on experimental values, it is probably because this kinetic has been established in a model medium (squalane of low viscosity) and is no more valid in a high viscous polymer. We can also imagine that the dispersion of a low viscous liquid in a very viscous polymer is more difficult at 30 wt%, leading to a slower mechanism which is not taken into account in the model applied.

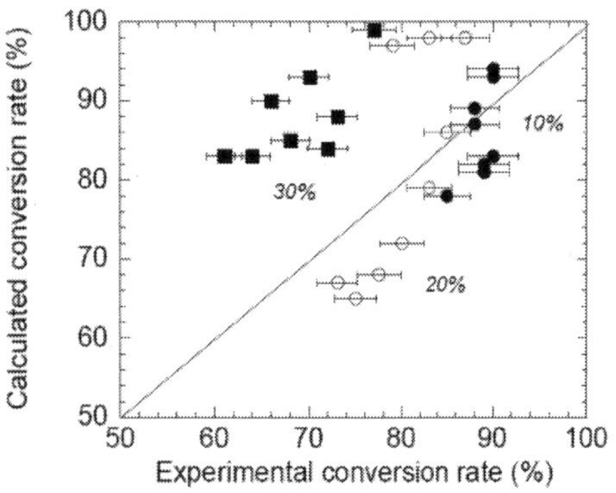

Figure 4 - General comparison between experimental and computed conversion rates

Conclusion

An original method based on the in-situ generation of TiO_2 inorganic particles in a polypropylene matrix was applied. SAXS measurements confirmed and enriched the morphological information that has been obtained via TEM observation.

Rheological experiments have shown that melt polymers filled with in-situ titanium dioxide particles exhibit a modification of viscoelastic properties. The rheological behaviour observed highlights a continuous distribution of relaxation times and a weak law dependence of storage at low frequencies appear for in-situ prepared materials.

Finally, the effects of processing conditions on the development of the sol-gel reaction have also been investigated by using numerical simulations. The comparison between experimental and theoretical results shows that this simulation approach is relevant to predict the conversion of the titanium n-butoxide as titanium dioxide precursor through hydrolysis-condensations reactions in molten PP.

Acknowledgements

This work was carried out during Multihybrid project, supported by the European Commission in the frame of the 6th PCRD.

References

Barlier, V., Bounor-Legare, V., Alcouffe, P., Boiteux, G., Davenas, J., ''Formation of TiO2 domains in poly(9-vinylcarbazole) thin film by hydrolysis-condensation of a metal alkoxide'', Thin Solid Films., **16**, 6328-6331 (2007).

Cassagnau P., Bounor-Legaré V, Fenouillot F, "Reactive processing of thermoplastic polymers: A review of fundamental aspects" Intern. Polym. Proc., **22**, 218-258, (2007).

Chen, J., Chareonsak, R., Puengpipat, V., Marturunkakul, S., "Organic/inorganic composite materials for coating applications", J. Appl. Polym. Sci., **74**, 1341-1346 (1999).

Dennis, R., Hunter, L., Chang, D., White, L., Cho, W., Paul, R., "Effectt of melt processing conditions on the extent of exfoliation in organoclay-based nanocomposites", Polymer, **41**, 9513-9522 (2001).

Dou, Q., Zhu, X., Peter, K., Demco, E., Moller, M., Melian, C., "Preparation of polypropylene/silica composites by in-situ sol–gel processing using hyperbranched polyethoxysiloxane", J. Sol-Gel Sci. Technol., **48**, 51-60 (2008).

Lertwimolnun, W., Vergnes, B., "Effect of processing conditions on the formation of polypropylene /organoclay nanocomposites in a twin screw extruder" , Polym. Eng. Sci., **10**, 315-323 (2006).

Liu, L., Qi, Z., Zhu, J., "Studies on nylon 6/clay nanocomposites by melt-intercalation process", J. Appl. Polym. Sci., **71**, 1133 (1999).

Vaia, A., Hope, I., Giannelis, P., "Synthesis and properties of two-dimensional nanostructures by direct intercalation of polymer melts in layered silicates", Chem. Mater., **5**, 1694-1696 (1993).

Vaia, A., Giannelis, P., "Polymer melt intercalation in organically-modified layered silicates: model predictions and experiment", Macromol., **30**, 8000-8009 (1997).

Vergnes, B., Della Valle, G., Delamare, L., "A global computer software for polymer flows in corotating twin screw extruders", Polym. Eng. Sci., **38**, 1781-1792 (1998).

Wijnen, P., Beelen, M., Rummens, J., Saeijs, H., Vansanten, A., "Silica gel from waterglas: a SAXS study of the formation and aeging of fractal aggregates", J. Appl. Crystallogr, **24**, 759-764 (1991).

Xanthos, M., "Reactive Extrusion: Principles and Practice", Munich, Hanser. (1992).

POLYMER RHEOLOGY AND POLYMER MIXING 'ON A CHIP'

D. Moon, K. B. Migler*

Polymers Division, NIST, Gaithersburg, MD, USA – kalman.migler@nist.gov

We present a novel microfluidic rheometer and mixer for high temperature polymer melts. In the rheological mode, it is termed a multi-sample micro slit rheometer (MMR), capable of measurements over a broad range of temperatures, viscosities and shear rates. The instrument is inherently simple as the flow is generated by external gas pressure and the shear rate is measured through optical tracking of the flow front. In the current implementation, the required volume of each sample is approximately 20 micro-liter and we measure four samples simultaneously. In the mixing mode, it is a planar polymer micro-mixer (PPMM), essentially a static mixer with multiple consecutive units of splitting and recombination flow formed by stacking three stainless steel shims (50 μm thickness, 45 mm in diameter). The mixing results show an evolution of morphology for a polystyrene/polypropylene blend, cylindrical domain/matrix or the transition of a multilayer to a domain/matrix or a co-continuous morphology depending on the channel configuration and the number of mixing unit passed by. This mixer will be particularly useful not only as a polymer melt mixer but as a multilayer processor when a limited material quantity (only 1 or 2 pellets) is available. The microfluidic platform is based on flow through shim slits and is easily reconfigurable so that it can emulate other kind of polymer melt processing such as capillary rheometer or micro-injection molder. Taken together, the two modes form the basis for a "polymer processing lab on a chip.

Introduction

The modern polymer processing laboratory typically contains disparate instruments for various operations such as rheometry, mixing, injection molding and interfacial tension measurement. Modification and customization of these instruments to meet the demands of a particular application is costly and time intensive. The required sample sizes for these operations ranges from $(1 - 10^4)$ g. We propose to merge the multiple processing and rheology operations into a single platform that is inspired by microfluidics methods. We term the platform "Polymer Processing on a Chip" and describe here two key modes in which it can be used: rheometry and mixing. Advantages of this platform are the ability to structure blends, to work on independent samples in parallel, to ease the reconfiguration of flow geometry, and to work with small sample sizes (~50 mg).

The first mode that we have developed is melt and solution rheometry (Moon et al, 2008; Moon and Migler 2009a). Melt rheometry is essential to the polymers industry because it provides critical information regarding the processability of new materials and provides insights into a wide range of fundamental material properties. However, bulk rheological techniques are problematic in cases where material supply is limited, where large numbers of sample measurements are required and when rapid sample loading and cleaning procedures are needed. Two principle classes of rheometers have wide scale use in polymers research: capillary rheometers and rotational rheometers. These two classes of rheometers are thus not amenable to the high throughput approach being developed in many laboratories.

We call our rheological mode a Multi-sample Micro-slit Rheometer (MMR) - essentially a pressure-driven slit-rheometer. The use of pressure as the driving force allows for simple parallelization of the design as well as mechanical simplicity. The induced flow is measured by optically tracking the position of the flow front as the fluid fills the channel. In our current design, we require approximately 20 μL per sample and measure four samples simultaneously. The MMR offers the simplicity of capillary rheometry and a

Identification of a commercial product is made only to facilitate reproducibility and to adequately describe procedure. In no case does it imply endorsement by the National Institute of Standards and Technology (NIST) or that it is necessarily the best product for the procedure.

design that enables parallel measurements on limited quantity samples. The second mode for the polymer processing on a chip is the melt mixing of immiscible polymers (Moon and Migler, 2009b). Blending by melt mixing of immiscible polymers to create materials with superior properties is widely used in the polymer industry across multiple product lines. Significant research is devoted to development of new polymers, their blending strategy, the resulting structure and the ultimate properties. It is well known that specific structures can enhance application specific properties; for example, a multi-layer structure may be desired to reduce gas transport , a co-continuous structure may be desired for a polymer scaffold pre-cursor for tissue engineering and a domain/matrix structure is utilized to improve properties such as brittleness. There are two issues associated with the development of new immiscible blends materials that we address in the current work. The first is the difficulty in generating complex and targeted structures. The second is the frequent gap between the quantity of materials available during the development of new polymers and the quantity needed for typical laboratory scale mixers. In the present work, we present a new strategy – based on a planar polymer micro-mixer (PPMM) - to create targeted structures from immiscible polymers.

The PPMM differs from the MMR in that the shims may be stacked upon each other and are designed so that the different polymers can mix with each other as they traverse the channels. In the simplest case, the shims bring two polymer streams together in a T-junction. In the more general case, we utilize multiple stacked shims to achieve flexibility in the flow. The second major modification that we make for the MMR is to allow independent pressure control of each sample holder, which is utilized in our demonstration of active mixing.

Experimental

The primary design considerations were mechanical simplicity (no moving parts), high temperature capability (materials of construction are stainless steel, aluminum and sapphire), cleanability (no inaccessible cavities), multi-sample capability and optical access to the fluid. Figure 1 shows the cross-section of the main fixture when working in the rheological mode. The heart of the device is the steel shim (standard grade shim-stock) which contains four slits through which four independent fluids can flow. As mentioned above, in the mixing mode, multiple shims may be used and the flow is such that the flow channels intersect so as to create a mixing flow.

Figure 2 is a schematic of a typical shim setup for the mixing mode – the planar polymer micro-mixer. There

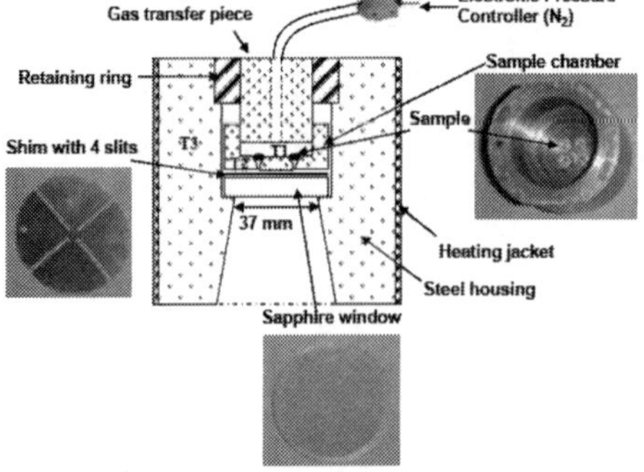

Figure 1: Schematic diagram of the polymer processing lab on a chip in the multi-sample micro-slit rheometer mode. Pressurized gas drives the fluid from the sample chamber through the slits in the shim. The flow front position is measured as a function of time.

are three shims which are stacked together. Note that there are two sample inputs which bring two dissimilar polymers together. The channels are cut so that the polymer can flow independently along any shim, while allowing for polymer to pass from shim to shim. In the implementation shown in Figure 2, it is designed to be a split and recombine mixer.

The temperature is controlled through a band heater operating in conjunction with a control thermocouple and PID temperature controller. For pressure control, nitrogen gas from the regulator of a standard laboratory cylinder flows into a programmable electronic pressure controller and then to the sample chamber. Upon application of pressure, various polymeric fluids flow down from the sample chamber and through the slits, displacing the air. The flow front is imaged by a high-end consumer video camera through a mirror (placed for geometrical convenience.)

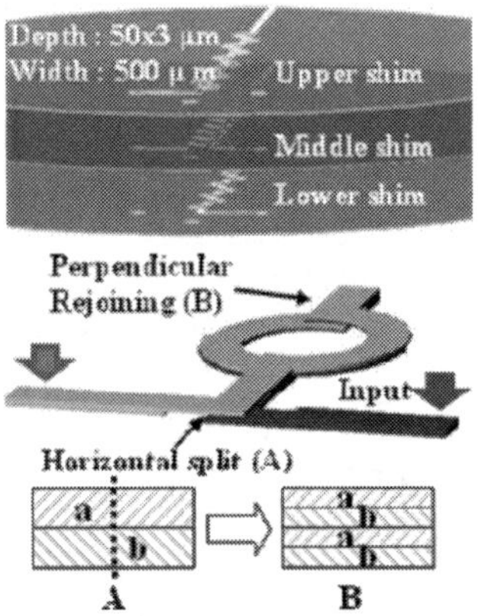

Figure 2. Split and Recombine (SAR) mixer. (top) Schematic of three stacked shims showing multiple mixing units. (Middle) Schematic of the sample inputs and the first mixing unit.. (Bottom) Idealized mixing structure.

To demonstrate the rheology mode, multiple materials are used to demonstrate the capabilities of the device. Two of them are NIST Standard Reference Materials (SRMs): a PIB solution (SRM2490, polyisobutylene dessolved in 2, 6, 10, 14-tetranmethlypentadecane) and a PDMS melt (SRM2491, polydimethlysiloxane). A third is a polystyrene (PS612, DOW Chemical), which is chosen to demonstrate the elevated temperature capabilities. Finally, standard viscosity oils were utilized. The measured viscosity from the MMR was compared to the reference data for the two SRMs and to the measured results of a capillary rheometer for the polystyrene.

The apparent viscosity can be described by Eq. (1):

$$\eta_a = \frac{\tau_w}{\dot{\gamma}_a} = \frac{\frac{h}{2}\left(\frac{\Delta P}{l}\right)}{\frac{6}{h}V_f} = \frac{h^2 \Delta P}{12}\frac{1}{lV_f} \quad \text{eq. 1}$$

where τ_w is the wall shear stress, $\dot{\gamma}_a$ is the apparent wall shear rate, V_f (= dl/dt) is flow velocity of the flow front, ΔP is the pressure drop, h is the slit height, and l is the length of sample in the channel. By measuring the flow velocity V_f at flow length l, the apparent viscosity can be easily obtained from Eq. (1). The analysis has a five step process: (1) flow length versus time; (2) conversion of data to flow velocity versus time; (3) conversion of data to apparent viscosity versus apparent shear rate; (4) Rabinowitsch procedure; (5) entrance corrections. In the case of the standard oils, a modified Bagley correction was utilized to obtain the capillarity.

To demonstrate the mixing mode, the polymer pair utilized in these experiments is polystyrene (PS) (Styron 666D, Dow Chemical) and polypropylene (PP) (Exceed 4062, Exxon) chosen in because their viscosities closely match each other over the shear rate range (0.1 to 200) s-1 in shear rate at 200 °C (Zhao and Macosko, 2002) and there is image contrast between the two. For the split and recombine experiments, a linearly increasing pressure profile (138 kPa/min) up to a maximum 6.9 MPa (at 50 min) at 200 °C was used in order to roughly maintain a constant averaged flow rate through the successive mixing elements.

Results and Discussion

Results from the multi-sample micro-scale rheometer are shown in Figure 3 in the case of the PDMS sample. The measurements from the MMR are shown in the open symbols. These values are in agreement with those from the certified values. The certified values were obtained by both steady shear and dynamic measurement in a rotational rheometer. Note that the minimum and maximum shear rate ranges covered by the MMR is greater than that for the corresponding rotational measurements. In these experiments, the shear rate range is intermediate between a rotational and a capillary rheometer.

In figure 4, we compile results from a range of polymers and fluids. Across the fluids, the shear rate range covers approximately five orders of magnitude while the viscosity range covers approximately eight orders of magnitude. Further, the temperature range covered in these experiments ranges from room temperature for the oils, the PIB and the PDMS, to 180 °C for the PS. In other experiments (not shown) we have measured polycarbonate at 270 °C. In the case of the oils, we must make corrections for the capillarity (the wicking effect) of the oil as it enters the channel. This process is done through a modification of the Bagley method. While time consuming, it also allows for a direct measure of the capillary pressure and for an indirect measure of the dynamic contact angle.

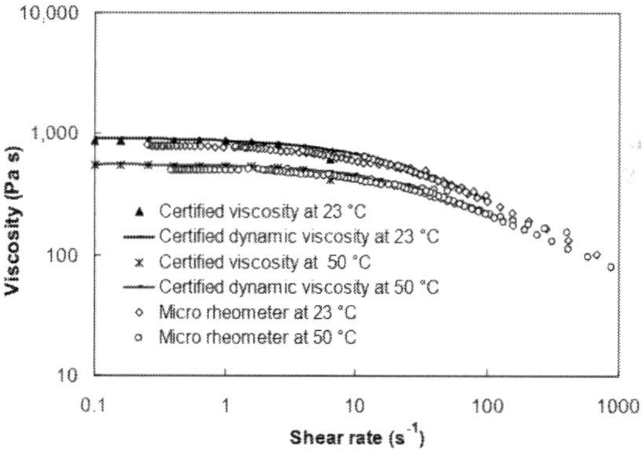

Figure 3. Comparison of the certified values of the PDMS standard reference material with those obtained by the multi-sample micro-slit rheometer.

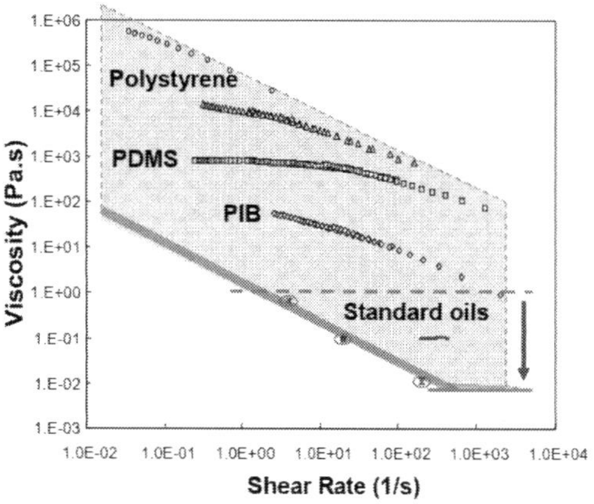

Figure 4: Demonstration of the range of shear rates, viscosities and temperatures that can be achieved with the multi-sample micro-scale rheometer.

Results from the planar polymer micro-mixer can be seen in Figure 5 which are described in terms of the distance down the channel in terms of the mixing unit number. SEM micrographs from six mixing units are shown in Fig. 6. Up until the 5th mixing unit, a multi-layer pattern is observed although the layer thicknesses are not uniform. At the 5th unit the morphology varies from multi-layer in the central region of the cross-section to curved structures at the left and right regions. At the 6th mixing unit, the multilayer pattern remains in the central region, but discrete domains are now visible on the sides. At the 8th and 10th unit, the morphology is changed completely to the domain/matrix or co-continuous pattern. This morphology change can be confirmed by the whitening of the flow as seen visually. The mixture becomes increasingly white

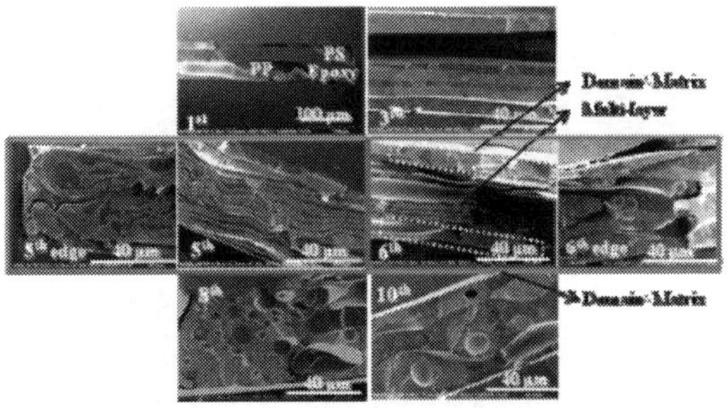

Figure 5. Split and recombine mixer in parallel multi-layer configuration. SEM images of the cross-sections of the PS/PP samples at the 1st, 3rd, 5th, 6th, 8th and 10th mixing units. The SEM reveals a breaking-up of structure from multilayer to domain/matrix morphology

at progressively larger *n* indicating the transitions from multi-layer to curved. Note that the distribution of the domain sizes of the mixture in the 8th and 10th sections is bimodal (many small domains and a few large domains). This is consistent with a previous observation that when the multilayer structure is broken during flow, a population of smaller particles is formed because of the Rayleigh instability and a second population of larger particles is formed by a relaxation mechanism[14] of multi-layer breakup and the corner effect as previously described.

Conclusions

In conclusion, we have demonstrated two modes of a new paradigm for a polymer laboratory that we call polymer processing on a chip. One can imagine a number of additional modes that could be carried out on this platform, such as reactive blending, compatibilization, and interfacial tension measurements and micro-injection molding. In order for this to become a reality, it will be useful to develop mechanical tests on the small part sizes that can be created with this platform.

References

Moon D., Migler, KB "Multi-sample micro-slit rheometry" Journal of Rheology, **52**, 1131-1142 (2008)

Moon D., Migler, Kalman B, "Measurement of dynamic capillary pressure and viscosity via the multi-sample micro-slit rheometer", Chemical Engineering Science **64,** 4537-4542 (2009a)

Moon D., Migler, Kalman B. ANTEC 2009: Annual Technical Conference Proceedings, Society of Plastics Engineers. 1838-1840 (2009b)

Zhao R, Macosko CW. "Slip at polymer-polymer interfaces: Rheological measurements on coextruded multilayers" , Journal of Rheology, **46**, 145-67 (2002)

STRUCTURES AND RHEOLOGY OF POLYMERS WITH LONG CHAIN BRANCHING

Wei Yu*, Lijuan Lou, Jianye Liu, Chixing Zhou
Advanced Rheology Institute, Department of Polymer Science and Engineering, Shanghai Jiao Tong University, Shanghai 200240, P. R. China – wyu@sjtu.edu.cn; loulijuan@sjtu.edu.cn; jyliu@sju.edu.cn; cxzhou@sjtu.edu.cn

Abstract - Polymers with low amount of long chain branching (LCB) exhibit very different properties from linear polymers. A lot of works have been done to understand the structures and rheology of LCB polymers, however, it is still difficult to characterize clearly the topological structures of polymer chains and the compositions of a LCB polymer if it is a mixture of different LCB polymers. In this work, LCB poly(lactic acid) (PLA) was successfully prepared and the topological structure of these LCB generated from functional groups reactions as well as free radical reactions were investigated thoroughly. Qualitative information on the branched structures could be readily obtained from linear viscoelasticity, nonlinear oscillatory shear experiments and strain hardening in elongational experiments. For quantitative information on chain structure, linear viscoelasticity combined with branch-on-branch (BOB) dynamic model was used to predict exact compositions and chain topology of the products.

Introduction

Introduction of long chain branching into polymer backbones has been shown to affect its properties substantially. Among all the possible properties that can be affected, rheological properties are very important not only due to its great impact on the elongation-related processing such as foaming and film blowing, but also due to its potentials in quantitatively determination of the structures of branched chains. Therefore, preparation and characterization of LCB polymers like polypropylene have attracted a lot of interests (Tian et al. 2006). On the other hand, it has been found that poly(lactic acid) is difficult to foaming due to its poor melt strength, although it has a large potential application and is completely biodegradable. In this work, we try to prepare the PLA with LCB by different methods, and most important, to quantitatively analyze the branched structures by rheology with the help of branch-on-branch model.

Experimental

The PLA used in this work was dried at 40°C for 24h under vacuum before mixing. Modification of the linear PLA took place in a Haake Rheocord90 mixer at 190°C and 210°C, respectively. At first, the rotation speed was set at 20rpm, then just after the pellets of PLA (65g) were totally melted, 0.5 wt % Irganox 1330 were added into each product as the antioxidants followed by 1min mixing to stabilize the products during the subsequent long chain branching reactions and rheological measurements. Pyromellitic dianhydride (PMDA) was added, then after 1min mixing, Triglycidyl isocyanurate (TGIC) were added into samples followed by 2min mixing to make the additives disperse as evenly as possible in PLA. At last, the rotation speed was changed to 60rpm until the torque curves reached its steady value. In order to evaluate the long chain branching effect of the functional group reaction, a comparison with the randomly branched sample by 0.2% Dicumyl peroxides (DCP) was treated under the same processing conditions. The formulations and the processing conditions for all the samples are summarized in Table 1.

The linear rheological tests were performed on a rotational rheometer (Gemini 200HR, Bohlin Instruments) with parallel plate geometry of 25mm in diameter and a gap of 0.9mm. The samples for rheological measurement were obtained by compressing molding into sheet (thickness ≈1mm) at 180°C under 12MPa. Small amplitude oscillatory shear (SAOS) was carried out for all samples. Frequencies sweep were performed with a given strain amplitude of 5%, which has been proved to lie in the linear regime. The nonlinear oscillatory shear experiments were performed on a strain-controlled ARES G2 rheometer (TA instrument) with parallel plate geometry of 25mm in diameter and a gap of 1mm. Tests were performed at 180°C with the frequency of 0.5Hz. For uniaxial elongational viscosity measurements, sample sheets were cut into pieces with width of 10 mm and length of 17 mm. The measurements were carried out on ARES rheometer (TA instrument) with the extensional viscosity fixture (EVF) at different strain rates.

Results and Discussion

The torque curve of Haake mixer can give a clear indication of coupling reaction happened during reactive mixing. It is shown in Figure 1 that the torque increases greatly during mixing for all samples, while it takes longer time for the torque to reach a steady value under lower processing temperature. Linear viscoelasticity measurement show quite slow relaxation process at low frequencies. This can also be illustrated in vGP plot as shown in Figure 2. It is clear the phase angle decrease monotonically with the increase of complex modulus for linear PLA, while a bump with local minimum has been shown for PLA samples modified with PMDA and TGIC. Such transition is not clearly seen in PLA modified with

DCP. This suggests although long chain branching might exists in all samples, they are different due to different reaction mechanisms. The existence of LCB can also be justified by the strain hardening behaviors in uniaxial elongation (Figure 3) and nonlinear rheological behaviors under medium amplitude oscillatory shear (not shown here).

Although qualitative analysis of LCB structure is convincing as shown above, it is still difficult to differentiate the structures in these polymers. The most difficult thing is that the LCB sample could be a mixer composed of chains with different topologies. Here, we try to use the Branch-On-Branch (BOB) model (Das and McLeish 2006) to fit the linear viscoelastic behavior of different samples. The result is shown in Figure 4 for P-P02-T04. A relative satisfying fitting suggests that this sample is composed of 60% linear chains with Mw=110kg/mol and PDI=1.2, 25% 3-arm star chains with Ma=110kg/mol and PDI=1.2, 15% two-generation Cayley tree-like structures with the molecular mass 50kg/mol for zero generation and 200kg/mol for 1st generation.

Conclusions

LCB PLAs were successfully prepared by different methods in this work, where the existence of LCB was verified by various rheological properties. Detailed information on exact compositions and chain topological structure of the products was also successfully obtained by fitting the branch-on-branch (BOB) dynamic model to the linear viscoelasticity.

Table 1–The processing conditions and the formulation of samples.

Sample Code	1330 (wt%)	PMDA (g)	TGIC (g)	T (°C)
P	0.5	-	-	190
P-P03-T04-190	0.5	0.3	0.4	190
P-P02-T04	0.5	0.2	0.4	210
P-P03-T04	0.5	0.3	0.4	210
P-P04-T04	0.5	0.4	0.4	210
P-DCP	0.5	-	-	190

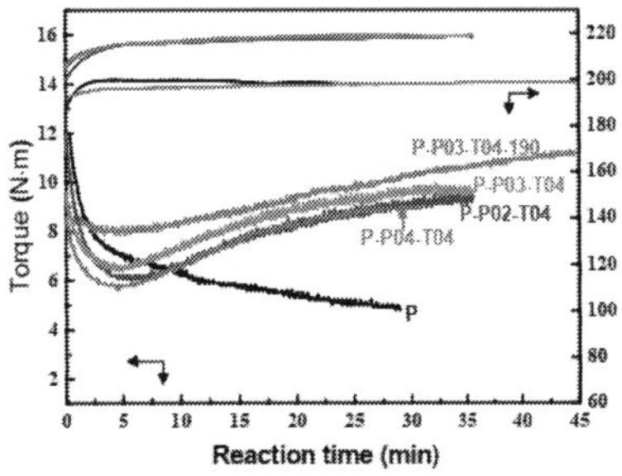

Figure 1-Torque and temperature evolution for all sample reactions.

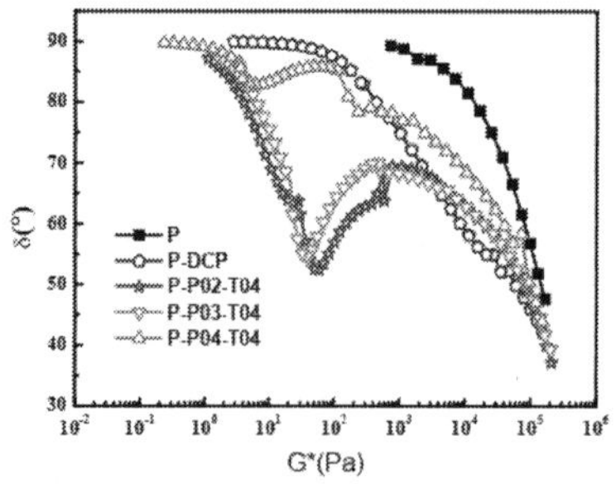

Figure 2-vGP plot for original PLA and modified products at 180°C.

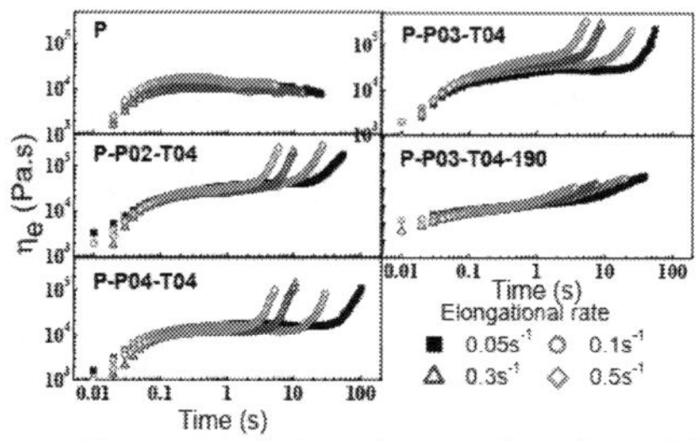

Figure 3-Elongational viscosity as a function of time at different elongational rates (0.05, 0.1, 0.3, 0.5s^{-1}) for original PLA and modified products at 160°C.

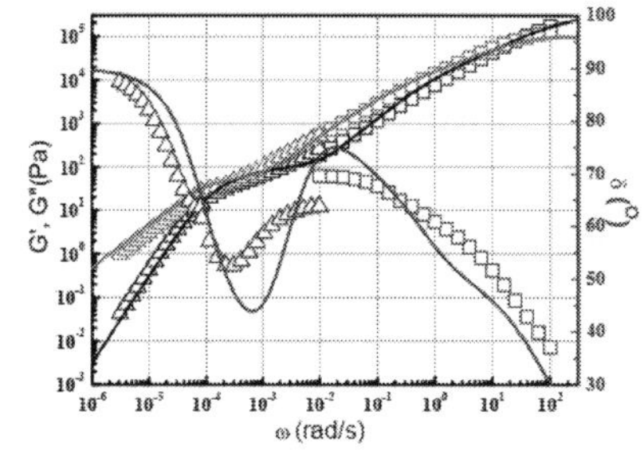

Figure 4-Linear viscoelastic spectra at 180°C for P-P02-T04. (Squares are the SAOS data, and triangles are the data interpreted from creep test; lines are the results of model fitting).

Acknowledgements

This work is financially supported by the Natural Science Foundation of China (NSFC) no. 50390095.

References

Das, C., McLeish, T. C. B., "Computational linear rheology of general branch-on-branch polymers" J. Rheol., 50, 207-234 (2006)

Tian, J. H., Yu, W., Zhou, C. X., "The preparation and rheology characterization of long chain branching polypropylene ", Polymer, 47, 7962-7969 (2006)

EFFECT OF POLYMER/FILLER INTERACTIONS ON THE RHEOLOGICAL PROPERTIES OF POLYOLEFIN ELASTOMER/NANOSILICA COMPOSITES

M. Bailly[1], M. Kontopoulou[1*] and K. El Mabrouk[2]

[1] *Department of Chemical Engineering, Queen's University, Kingston, ON, Canada K7L 3N6 – mathieu.bailly@chee.queensu.ca; kontopm@queensu.ca*
[2]*Institute of Nanomaterials and Nanotechnology, Rabat, Morocco – k.elmabrouk@inanotech.ma*

Nanocomposites were prepared using a silane-grafted polyolefin elastomer (POE) matrix reinforced with hydrophobic silica nanoparticles (oct-SiO_2). The composites were differentiated according to type of silane agent grafted on the matrix, either monofunctional (VTES) or bifunctional (VTEOS). The latter resulted in covalent bonds between the VTEOS functionality and the hydroxyl groups present at the surface of the particles, which generated strong polymer/filler interactions. TEM micrographs revealed a fractal-like composite structure. Rheological measurements in the melt state revealed significant differences, depending on the type of matrix used. Time-sweep experiments carried out in the LVE region showed that the storage modulus of POE-g-VTES-based composites increased by 53% after only 500s. On the contrary, the response of POE-g-VTEOS-based composites was more stable. It is suggested that the strong interactions occurring between POE-g-VTEOS and oct-SiO_2 prevented the particles from aggregating when subjected to small shear deformation. During stress-sweep experiments the POE-g-VTEOS-based composites exhibited a higher critical strain for the onset of non-linearity. Lastly, DMA measurements showed that more energy is dissipated during the glass transition in the composites with enhanced polymer/filler interactions.

Introduction

Colloidal and fumed silica suspensions have been used extensively in applications involving paints, foodstuffs, as well as emerging technologies such as photonics and microelectronics, due to their function as thickening and thixotropic agents in low molecular weight solvents (Khan and Zoeller, 1993 ; Raghavan and Khan, 1995). Silica particles have also successfully served as reinforcing agents for rubbers (Kraus 1965) and more recently for thermoplastic matrices (Zou et al., 2008).

Given the industrial importance of silica-containing thermoplastic composites, a lot of attention has been paid recently to their rheological properties. Dramatic increases in the storage moduli at low frequencies, leading to the appearance of a secondary plateau or a "solid-like behaviour" and a drop of the storage modulus upon increasing strain, also known as the "Payne effect", are commonly observed (Cassagnau, 2008).

The rheological response of polymer-based composites is influenced substantially by the extent of polymer/filler interactions, which in turn depends on the surface treatment of the particles, and the nature of the polymeric matrix. Strong polymer/filler interactions may occur between hydrophilic silica particles and a polar matrix or in the presence of bifunctional coupling agents that can form covalent bonds with functional groups present in the polymer matrix (Aranguren et al., 1992); Barus et al., 2009 ; Wu and Liao, 2003). This leads to the formation of an immobilized layer of "bound" polymer surrounding the filler particles. The presence of bound polymer has been frequently evoked in elastomer technology to explain the enhanced hydrodynamic effect, leading to an increase in the modulus when filler particles are added in elastomeric matrices (Wang, 1998).

Strong chemical interactions between the matrix and the fillers are crucial in obtaining improved mechanical properties of the resulting composites (Bailly and Kontopoulou, 2009 ; Bansal et al, 2006 ; Rong et al, 2006 ; Taniguchi et al, 2005), however the influence of chemical bonds on the rheological properties of the composites is still unclear. In this work we used a polyolefin elastomer matrix functionalized with reactive and non-reactive silanes, the former of which have the potential of establishing chemical bonds with the nanosilica particles. Based on detailed rheological characterization and imaging, we report the effect of chemical bonds on the structure of the composites and on their rheological response.

Experimental

Materials

Ethylene-octene copolymer polyolefin elastomer POE, trade name Engage 8130, density 0.864 $g.cm^{-3}$, MFR 1.0g/10 min at 190°C, copolymer content 42 wt%, was obtained from Dow Chemical. The melting point and crystallinity of this polymer, as measured by DSC are 67.1°C and 3.7%. Vinyltriethoxysilane (VTEOS, 97%, Aldrich Chemical Company Inc.), vinyltriethylsilane (VTES, 98%, Sigma-Aldrich) and dicumyl peroxide (DCP, 98%, Sigma Aldrich) were used as received. A hydrophobic surface modified nanosilica Aerosil®

R805 modified with octylsilane (oct-SiO_2) having a SSA of 150±25 m^2/g was used.

Grafting Procedure and Composite Preparation

POE pellets, 0.1wt% of DCP and 2wt% of VTEOS or VTES were mixed in a Haake PolyLab rheometer equipped with a Rheomix 610p mixing chamber and roller rotors at 80°C for 10 min at 60 rpm. Both VTEOS and VTES monomers were grafted onto POE to give POE-g-VTEOS and POE-g-VTES respectively. The mechanism of grafting of the silane monomers has been described elsewhere (Sengupta and Parent, 2006). After the completion of the grafting reaction, the filler was added; the total compounding time was 12 minutes. Nanosilica contents ranged from 4 to 12 wt%.

Characterization

FT-IR spectra of the functionalized POEs were obtained using a Nicolet Avatar 360 FTIR ESP instrument. The state of dispersion of the fillers was assessed by TEM imaging, using an FEI Tecnai 20 instrument. Samples were compression molded using a Carver press at 150°C and 10MPa for 1 min. The amount of bound polymer was determined by dissolving the composites into 40 mL of toluene at 80°C for 3 hrs to dissolve the polymer that was not bound to the filler. The gel containing the bound polymer and silica particles was dried under vacuum at 60°C overnight. The amount of bound polymer was determined using a TA Instruments Q500 series Thermogravimetric analysis instrument to heat samples under nitrogen atmosphere from 25°C to 700°C at a rate of 10°C/min.

Rheological characterization was carried out in a Reologica ViscoTech oscillatory rheometer using 20 mm parallel plate fixtures under nitrogen blanket. Stress sweep experiments were carried out from 1 to 10^4 Pa at a frequency of 0.1Hz and at a temperature of 200°C. Time sweeps were done at 1, 2 and 10% strain, at 0.1 rad/s and 190°C.

Dynamic mechanical properties were studied on compression-molded samples with a Rheometer Solid Analyzer RSA-II. The instrument was operated in the dual cantiliver configuration and flexion mode. The dynamic responses were tested from -100 to 20°C at a frequency of 1Hz and a heating ramp of 2°C/min.

Results

Morphology

Figure 1 displays the TEM images of the two different systems containing 12 wt% silica. Silica exists in an aggregated state with the size and shape of the aggregates varying slightly, depending on the type of matrix and particles used. At this loading the composites appear to be almost interconnected, signifying that this composition is at the vicinity of the percolation threshold.

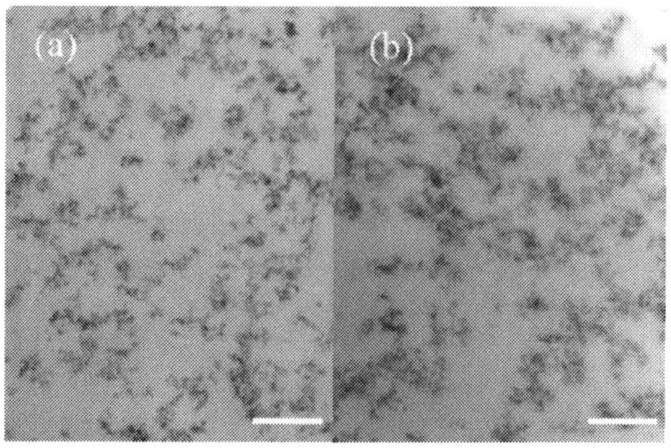

Figure 1. TEM images of composites containing 12wt% nanosilica. (a) POE-g-VTEOS/oct-SiO_2: (b) POE-g-VTES/oct-SiO_2. The white scale bar represents 500nm.

Image analysis showed that composites based on POE-g-VTEOS have a higher fraction of isolated particles and a smaller amount of large aggregates than composites based on POE-g-VTES, which display a higher fraction of large aggregates.

The improved dispersion of the oct-SiO_2 particles within the polymer matrix is most likely due to the disruption of the hydrogen bond formation between the silanol groups that cause aggregation. In the presence of the silane coating agent the specific component of the surface tension is lowered substantially, whereas the dispersive component becomes more significant, improving the affinity with the polyolefin matrix (Elias et al., 2008).

Dispersion is also facilitated by improving the stress transfer between polymer and fillers during compounding, thus breaking the aggregates apart. This can be achieved by establishing stronger polymer/filler interactions. The difference between VTEOS and VTES lies in the presence of an ethoxy group in VTEOS instead of an alkyl group in VTES. The hydroxyl groups present at the surface of the particles can react with the ethoxy group of VTEOS through a hydrolysis reaction (Bailly and Kontopoulou, 2009), thus establishing strong interfacial interactions between the polymer and fillers. However, in the absence of the ethoxy group in VTES, this reaction cannot take place.

Apart from the size distribution of the aggregates, image analysis clearly confirms that their aspect ratio strongly depends on their size; the bigger the aggregate, the less spherical it is. Aspect ratios below 0.2 are found for large aggregates above, implying that the aggregates do not have a spherical shape; rather they appear to have a fractal structure (Heinrich and Kluppel, 2002).

Bound polymer

Bound polymer characterizations are frequently used in rubber technology as a means to quantify polymer/filler interactions (Leblanc, 2002). The results of the bound polymer measurements are summarized in Table 1. Assuming an idealized case of spherical silica particles of density 2.2 $g.cm^{-3}$ and diameter of 12 nm surrounded by a bound polymer shell of density 0.864 $g.cm^{-3}$, the shell thickness can be calculated. Results are included in Table 1.

Table 1 Amount of bound polymer for composites containing 7wt% silica

	POE-g-VTEOS/oct-SiO_2	POE-g-VTES/oct-SiO_2
Bound polymer in wt%	7.0	3.5
Bound polymer per mass of silica	1.0	0.5
Shell thickness (nm)	2.9	1.7

As expected the POE-g-VTEOS-based composites exhibit higher amounts of bound polymer compared to POE-g-VTES-based materials, which is a possible indication of the existence of polymer that is covalently bound to the filler. The amount of bound polymer in POE-g-VTES-based composites probably corresponds to a residual bound polymer made of chains that have been physically adsorbed or trapped within the large particle aggregates (Wang, 1998 ; Yatsuyanagi et al, 2001). The presence of an immobilized "bound" polymer layer has been related to substantial increases in the rheological properties of filled polymers. The concept of an "effective volume fraction" which takes into account not only the volume occupied by the particles but also that of the rigid shell surrounding them, has been used to explain this increase.

Rheology

It is well-known that suspensions of fumed silica experience microstructural rearrangements during shear, leading to flocculation and formation of particle aggregates (Raghavan and Khan, 1995 ; Zhang and Archer, 2002). Functionalized polyolefins that are commonly used as nanocomposite matrices are also prone to time-dependent effects, because of the propensity for functional group associations (Lee et al., 2004). We therefore begin our investigation with a comparison of the time-dependent rheological response of the materials. Time-dependency was noted in the composite rheology at filler loadings above 5 wt%, signifying that a minimum amount of filler is required in order to observe a rearrangement and the corresponding increase in modulus, as pointed out previously by Romeo et al. (2008).

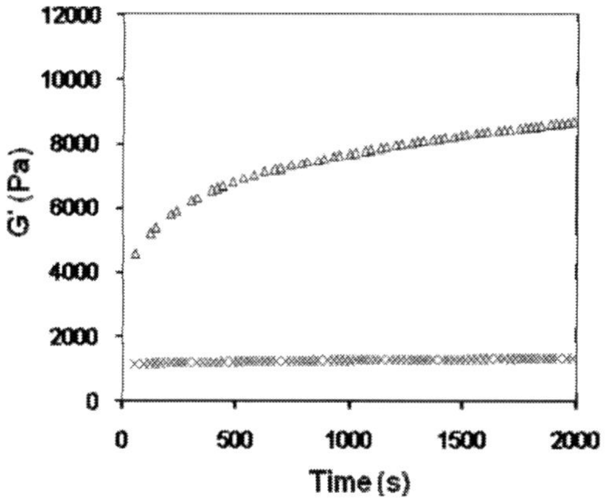

Figure 3. Time dependence of the elastic modulus, G', at 190°C under a 2% strain and 0.1Hz frequency. Silica content is 12wt%. Symbols are: (△) POE-g-VTES/oct-SiO_2; (X) POE-g-VTEOS/oct-SiO_2.

The pronounced increase in storage modulus seen for the POE-g-VTES based composites suggests that the silica particles have the propensity to associate and form aggregates during the time sweeps. The relative increase is more dramatic for the POE-g-VTES/ oct-SiO_2 composite, for which the value of G' becomes double compared to its initial value. An equilibrium value is not reached even after 3500s, although the majority of the change happens in the first 500s. This implies that the free silanol groups existing on the silica surface maintain their capability to associate through hydrogen bonding under the low strains imposed during the time sweep experiments, thereby leading to aggregation. The situation changes when the VTEOS-based matrix is used, since the silanol groups are now able to form covalent bonds with the ethoxysilane functionality. This limits their capacity to associate with each other and form aggregates, leading to a steadier rheological response, presumably attributed to a more stable morphology.

It is also noteworthy that the composites, especially the POE-g-VTES-based ones are very sensitive to pre-shearing and to the strain level imposed during the time sweeps, as shown in Figures 4(a) and (b).

Stress sweeps

Figure 5 presents the elastic modulus, G', as a function of strain for the two systems at a 12wt% silica loading obtained during the stress-sweep experiments.

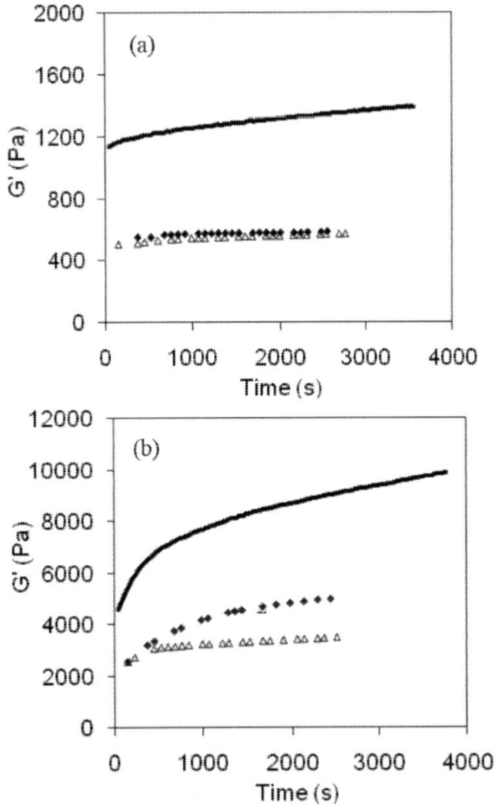

Figure 4. Elastic modulus dependence as a function of time: solid line, no preshear; (Δ) pre-sheared at 1000 Pa for 100 s, strain of 1% during time sweep; (♦) pre-sheared at 1000 Pa for 100 s, strain of 10% during time sweep. (a) POE-g-VTEOS/oct-SiO_2; (b) POE-g-VTES/oct-SiO_2.

The POE-g-VTES/oct-SiO_2 composite shows very strong strain dependence; the elastic modulus curve increases at moderate strains, and subsequently decreases. The initial increase may be due to "strain-induced" aggregation. This is in agreement with the findings from the time sweeps reported in Figures 3 and 4, where this particular composite seemed to be prone to aggregation. The critical strain for the onset of non-linearity is significantly lower for the POE-g-VTES compared to the POE-g-VTEOS-based composites, with the latter being more strain resistant.

In addition the POE-g-VTES based composites display a peak in their loss modulus vs. strain response –data not shown-, which is related to the energy dissipation produced by the disruption of extensive agglomeration (Cassagnau and Mélis, 2003 ; Galindo-Rosales et al., 2009 ; Paquien et al., 2005 ; Wang., 1998 ; Yziquel et al., 1999). On the contrary, the corresponding G″ curves of POE-g-VTEOS-based composites are flat, confirming the absence of significant breakdown of filler aggregates.

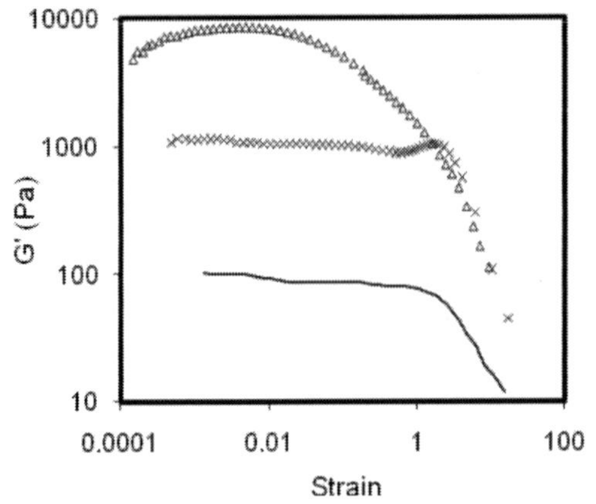

Figure 5. Stress sweeps for composites containing 12wt% nanosilica, at 200°C and 0.1Hz (0.62 rad/s). Symbols are: (Δ) POE-g-VTES/oct-SiO_2; (X) POE-g-VTEOS/oct-SiO_2. The solid line corresponds to the unfilled POE-g-VTEOS.

What is also noteworthy is the presence of a small peak in the elastic modulus vs. strain curves of the POE-g-VTEOS composites at very large strains (Fig. 5), right at the vicinity of the transition to non-linear behaviour. To our knowledge this has never been reported before, except from a similar observation by Cassagnau (Cassagnau, 2003) in silica-filled EVA/xylene mixtures. One possible explanation is stress-induced debonding of the polymer chains that are bound to the filler surface through covalent bonding.

Dynamic Mechanical Analysis

Figure 6 presents the storage and loss moduli as functions of temperature, obtained by DMA for a 7wt% oct-SiO_2 loading. Fig. 6 (a) reveals that although both of the systems possess the same storage modulus at temperatures below the glass transition, their behavior at temperatures around and above the glass transition differs, with the POE-g-VTEOS-based composite having a higher value of the elastic modulus. Similarly, as Fig. 6 (b) shows, the area under the E″ vs. T curve is higher for the POE-g-VTEOS based composites, suggesting that more energy is dissipated when the glass transition temperature is reached, at about -50°C. It should be noted that the curves corresponding the POE-g-VTES composite are very similar to the response of the neat polymers (not shown here), suggesting a minimal effect of the filler. It can be suggested therefore that the dynamic elastic and loss moduli are affected only when strong polymer/filler interactions are present. Further more these charts suggest a very slight increase in the glass transition temperature, T_g, for the POE-g-VTEOS composites.

Some studies have shown that the motion of macromolecules in the presence of nanoparticles is hindered when strong polymer/filler interactions exist, resulting in higher T_g (Wu and Liao, 2003). This is still a disputed topic, as some authors could not detect any shift in T_g, supporting the idea that polymer motion is not affected by filler reinforcement (Bogoslovov et al., 2008 ; Robertson et al, 2008).

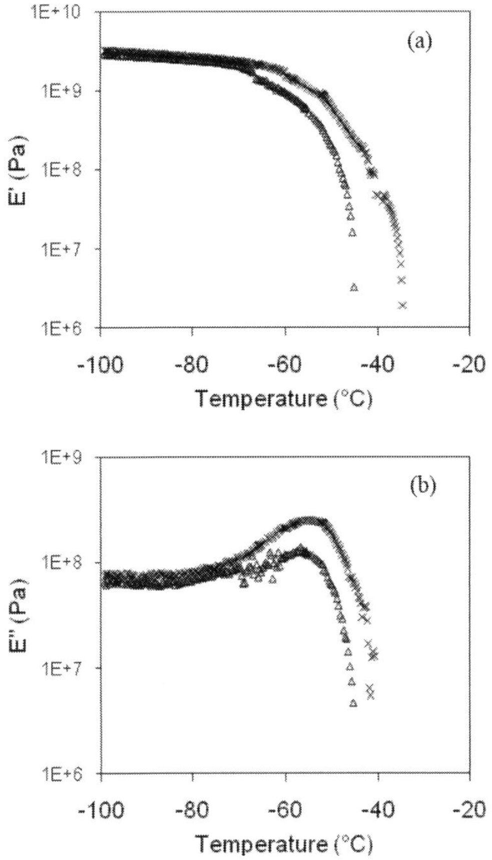

Figure 6. DMA curves at 1Hz. Symbols are: (△) POE-g-VTES/oct-SiO_2; (X) POE-g-VTEOS/oct-SiO_2. Silica content is 7wt%; (a) elastic modulus and (b) storage modulus.

Conclusions

Grafting reactive and non-reactive silanes onto the POE matrix enabled the comparison of two model systems, which are differentiated by the presence of polymer/filler interactions. A correlation was found between the amount of bound polymer and the presence of polymer/filler interactions when VTEOS, which is able to form covalent bonds with the hydroxyl groups of silica, is grafted onto the POE matrix. Both particle modification and grafting of VTEOS on the POE matrix contributed to the fine dispersion of the nanoparticles, as revealed through TEM imaging.

Time-sweep experiments showed that the composites are prone to aggregation, in the absence of chemical interactions between filler and hosting polymer. Stress-sweep experiments showed that VTEOS grafted composites are more capable of enduring high strains without significant network disruption, as their critical strains were much higher compared to VTES grafted composites. This behaviour was attributed to the presence of polymer/filler interactions that create linkages between the hosting polymer chains and the nanoparticles. Differences were also detected in the solid state DMA evaluations, which showed that a higher amount of energy is dissipated at the glass transition temperature when the VTEOS based composites are used.

Acknowledgements

This research has been supported by the Natural Sciences and Engineering Research Council of Canada (NSERC). The authors would like to thank Evonic Industries for donating the nanosilica samples used in this work. The contributions of Doug Holmyard and Andrea Liskova are also greatly appreciated.

References

Aranguren, M. I., Mora, E., Degroot, J. V. and Macosko, C. W., "Effect of Reinforcing Fillers on the Rheology of Polymer Melts", J. Rheol., **36,** 1165-1182 (1992)

Bailly, M., Kontopoulou, M., "Preparation and characterization of thermoplastic olefin/nanosilica composites using a silane-grafted polypropylene matrix", Polymer, **50,** 2472-2480 (2009)

Bansal, A., Yang, H., Li, C., Benicewicz, B. C., Kumar, S. K. and Schadler, L. S., "Controlling the thermomechanical properties of polymer nanocomposites by tailoring the polymer–particle interface", J. Polym. Sci. B Polym. Phys., **44,** 2944-2950 (2006)

Barus, S., Zanetti, M., Lazzari, M. and Costa, L., "Preparation of polymeric hybrid nanocomposites based on PE and nanosilica", Polymer, **50,** 2595-2600 (2009)

Bogoslovov, R. B., Rol, C. M., Ellis, A. R., Rall, A. M. and Robertson, C. G., "Effect of Silica Nanoparticles on the Local Segmental Dynamics in Poly(vinyl acetate)", Macromolecules, **41,** 1289-1296 (2008)

Cassagnau, P., "Melt rheology of organoclay and fumed silica nanocomposites", Polymer, **49,** 2183-2196 (2008)

Cassagnau, P., Mélis, F., "Non-linear viscoelastic behaviour and modulus recovery in silica filled polymers", Polymer, **44,** 6607-6615 (2003)

Cassagnau, P., "Payne effect and shear elasticity of silica-filled polymers in concentrated solutions and in molten state", Polymer, **44,** 2455-2462 (2003)

Elias, L., Fenouillot, F., Majesté, J. C., Alcouffe, P. and Cassagnau, P., "Immiscible polymer blends stabilized with nano-silica particles: Rheology and effective interfacial tension", Polymer, **49,** 4378-4385 (2008)

Galindo-Rosales, F. J., Rubio-Hernandez, F. J. and Velazquez-Navarro, J. F., "Shear-thickening behavior of AerosilA (R) R816 nanoparticles suspensions in polar organic liquids", Rheol. Acta, **48,** 699-708 (2009)

Heinrich G, Kluppel M Recent Advances in the Theory of Filler Networking in Elastomers. In: Anonymous Filled Elastomers Drug Delivery Systems. Berlin: Heidelberg, 2002. pp. 1-44

Khan, S. A., Zoeller, N. J., "Dynamic Rheological Behavior of Flocculated Fumed Silica Suspensions", J. Rheol., **37,** 1225-1235 (1993)

Kraus, G.: Reinforcement of elastomers,Anonymous , New York, (1965)

Leblanc, J. L., "Rubber-filler interactions and rheological properties in filled compounds", Prog. Polym. Sci., **27,** 627-687 (2002)

Lee, J. A., Kontopoulou, M. and Parent, J. S., "Time and shear dependent rheology of maleated polyethylene and its nanocomposites", Polymer, **45,** 6595-6600 (2004)

Paquien, J. N., Galy, J., Gerard, J. F. and Pouchelon, A., "Rheological studies of fumed silica-polydimethylsiloxane suspensions", Colloid Surf. A-Physicochem. Eng. Asp., **260,** 165-172 (2005)

Raghavan, S. R., Khan, S. A., "Shear-Induced Microstructural Changes in Flocculated Suspensions of Fumed Silica", J. Rheol., **39,** 1311-1325 (1995)

Robertson, C. G., Lin, C. J., Rackaitis, M. and Rol, C. M., "Influence of Particle Size and Polymer−Filler Coupling on Viscoelastic Glass Transition of Particle-Reinforced Polymers", Macromolecules, **41,** 2727-2731 (2008)

Romeo, G., Filippone, G., Fernández-Nieves, A., Russo, P. and Acierno, D., "Elasticity and dynamics of particle gels in non-Newtonian melts", Rheol Acta, **47,** 989-997 (2008)

Rong, M. Z., Zhang, M. Q. and Ruan, W. H., "Surface modification of nanoscale fillers for improving properties of polymer nanocomposites: a review", Mater. Sci. Technol., **22,** 787-796 (2006)

Sengupta, S. S., Parent, J. S., "Comparative analysis of

radical-mediated polyethylene modifications:

Vinyltriethoxysilane versus mercaptopropyltriethoxysilane addition", Polym. Eng. Sci., **46,** 480-485 (2006)

Taniguchi, Y., Shirai, K., Saitoh, H., Yamauchi, T. and Tsubokawa, N., "Postgrafting of vinyl polymers onto hyperbranched poly(amidoamine)-grafted nano-sized silica surface", Polymer, **46,** 2541-2547 (2005)

Wang, M. J., "Effect of polymer-filler and filler-filler interactions on dynamic properties of filled vulcanizates", Rubber Chem. Technol., **71,** 520-589 (1998)

Wu, C., Liao, H., "Modification of polyethylene–octene elastomer by silica through a sol–gel process", J Appl Polym Sci, **88,** 966-972 (2003)

Yatsuyanagi, F., Suzuki, N., Ito, M. and Kaidou, H., "Effects of secondary structure of fillers on the mechanical properties of silica filled rubber systems", Polymer, **42,** 9523-9529 (2001)

Yziquel, F., Carreau, P. J. and Tanguy, P. A., "Non-linear viscoelastic behavior of fumed silica suspensions", Rheol. Acta, **38,** 14-25 (1999)

Zhang, Q., Archer, L. A., "Poly(ethylene oxide)/Silica Nanocomposites: Structure and Rheology", Langmuir, **18,** 10435-10442 (2002)

Zou, H., Wu, S. and Shen, J., "Polymer/silica nanocomposites: Preparation, characterization, properties, and applications", Chem. Rev., **108,** 3893-3957 (2008)

MELT RHEOLOGY OF HYPERBRANCHED POLYSTYRENE SYNTHESIZED WITH MULTI-SITE MACROMONOMER

M. Sugimoto (1)*, T. Koizumi (1), T. Taniguchi (1)(3), K. Koyama (1), K. Saito (2), D. Nonokawa (2), T. Morita (2)

[1] *Graduate School of Science and Engineering, Yamagata University, Yonezawa, Japan – sugimoto@yz.yamagata-u.ac.jp*
[2] *Central Research Laboratories, DIC Corporation, Sakato 631, Sakura 285-8668, Japan*
[3] *Present address: Graduate School of Engineering, Kyoto University, Katsura, Kyoto, Japan*

Abstract - Melt rheological behaviors of hyperbranched polystyrene (PS) copolymerized by dendric macromonomer technique are presented. The time-temperature superposition principle was applicable to the hyperbranched PS. The branched PS showed slightly lower zero shear viscosity in comparison with linear PS regardless of a presence of a number of branches expected from the dendric macromonomer technique. Although the influence of use of multi-methacryloyl macromonomer in the polymerization process was marginal for linear viscoelastic regime, nonlinear shear and uniaxial elongational flows showed distinct differences between linear and branched PS. The strain dependence of the damping function became weak as increase of macromonomer content. The branched PS exhibited the growing elongational viscosity function comparing with linear PS. This prominent effect on the elongational flow behavior can be explained by the molecular architecture of the branched PS.

Introduction

It is well known that an introduction of long chain branch enhances elasticity, melt strength, elongational flow behavior of polymer melts. The elongational flow behavior is particularly important for the polymer processings which involves free surface flows e.g., thermoforming, film blowing, blow molding and foaming. As for polystyrene (PS) copolymerization with divinylbenzene (DVB) as a crosslinker has been reported to crosslink PS.(Antonietti and Rosenauer 1991, Campbell et al. 2005, Ourdouillie et al. 2001) The series polymerization method using DVB, however, needs utilization of large amount of crosslinker to enhance branching degree. It is very difficult to control the three-dimensional crosslinking reaction. Ferri and Lomellini(Ferri and Lomellini 1999) reported the uniaxial elongational viscosity of branched polystyrenes synthesized by copolymerization of styrene and DVB. The randomly branched PS showed the strong strain hardening under the elongational flow at 140 °C, though linear PS just followed the linear viscoelasticity of Trouton's rule. Hepperle et al. reported that nonlinear shear and elongational behavior of polystylenes grafted with various average number and molecular weight of grafted side chains.(Hepperle and Munstedt 2006, Hepperle et al. 2005) It is shown that the grafted branches in which the side chains have rather small molecular weight (ranging from the entanglement molecular weight M_e to $2M_e$) is still effective for the strain hardening in elongational flow.

In this work macromonomer was used to obtain hyperbranched polystyrene. We report rheological behaviors of hyperbranched PS melts: dynamic and step strain measurements in shear, and transient flow measurements in elongation.

Experimental

The dendric multi-methacryloyl macromonomer was used to synthesize hyperbranched PS. We used three samples; PS-A is linear polymer, PS-B and C were synthesized with the dendrimer type multi-methacryloyl macromonomer of 500 and 900 ppm, respectively. The bulk polymerization of PS-B and C were conducted in same way as that of PS-A except for use of the multi-methacryloyl macromonomer.

The rheological measurements were performed with ARES rheometer (TA instruments) for dynamic frequency sweep tests and transient elongational flow tests. Step-strain stress relaxation measurements were performed with MCR 301 rheometer (Anton Paar) with parallel-plates geometry (plate diameter of 25 mm). All the measurements were carried out under nitrogen atmosphere to suppress oxidative degradation of the samples.

To discuss effect of molecular weight and molecular architecture on the rheological behaviors, gel

Table 1. Molecular characterization of PS-A, B and C

	Macromonomer content / ppm	$\overline{Mw}$ / 10^4	$\overline{Mn}$ / 10^4	$\overline{Mw}/\overline{Mn}$	η_0 *1 /10^6 Pas	m *2
PS-A	0	26.3	14.3	1.8	1.58	-
PS-B	500	30.6	11.8	2.6	1.44	15
PS-C	900	44.5	13.2	3.4	1.16	80

*1: Zero-shear viscosity at 160 °C

*2: Number of branches at shoulder of the MWD

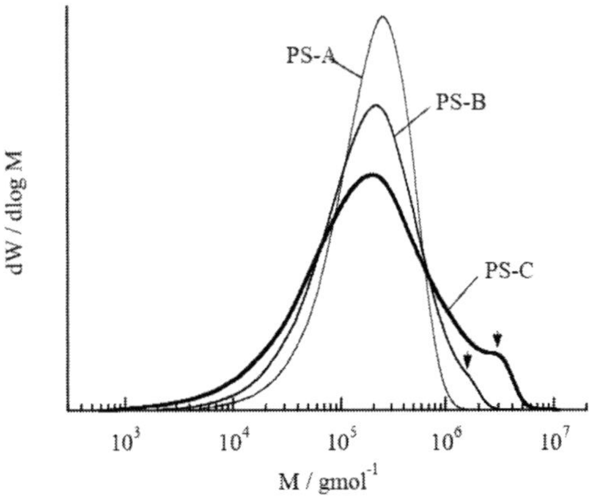

Fig. 1 Molecular weight distribution of PS-A, B and C.

permeation chromatography analysis coupled to an online multi-angle laser light scattering system (GPC-MALLS) was carried out.

Results and Discussion

The molecular weight distribution curves are also shown in Fig. 1. The molecular characteristics of PS-A~C are summarized in Table I. An addition of the dendrimer type multi-methacryloyl macromonomer of 500 ppm (PS-B) leads to a marked increase of high molecular weight chains, though the peaks of the curves showed little change in comparison with conventional PS-A. With increasing macromonomer content up to 900 ppm (PS-C) high molecular weight tail emerged in bimodal curve.

The isothermal dynamic viscoelastic measurements were carried out at 160 ~ 240 °C. The resulting storage modulus *G'* and loss modulus *G"* were horizontally shifted to the curve at reference temperature of 160 °C. Figure 3 shows *G'* and *G"* as

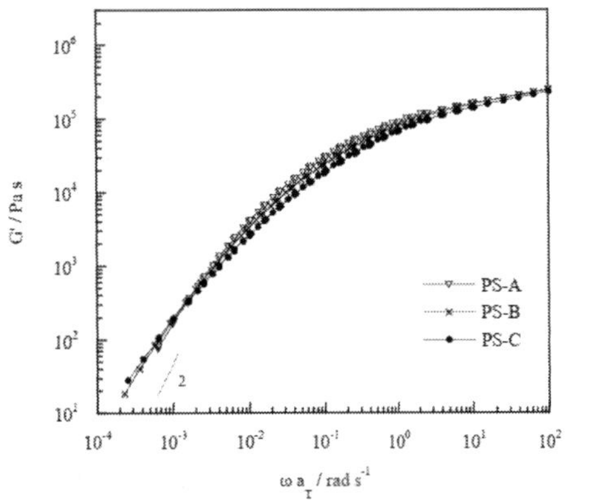

Fig. 2 Mastercurves of *G'* and *G"* as a function of angular frequency for PS- PS-A, B and C. The reference temperature T_0 is 160 °C.

a function of angular frequency for PS-A, B and C. The mastercurves reveal that *G'* decreases with increase of multi-methacryloyl macromonomer dose at intermediate frequency region. In the high frequency region which corresponds to the onset of segmental relaxation the dynamic modulus tends to converge on a single curve. In the low frequency region below 10^{-3} rads^{-1} *G'* indicates signs of a slight increase with macromonomer content, though PS-A shows terminal flow transition where slopes of log *G'* and log *G"* against log ω are 2 and 1, respectively. This suggests marginal change of relaxation time distribution before the terminal region. Thus the difference between linear PS and hyperbranched PS is not significant than expected from the polydispersity of molecular weight and a presence of high molecular weight tail.

Although the zero-shear viscosity of PS-A follows the well-known power-law relationship as a function of weight-average molecular weight, PS-B and C shows lower zero-shear viscosities than those of linear PS with same molecular weights. The rheological parameter is directly related to the molecular structure such as long-chain branching. The viscosity reduction and deviation from the power law relationship can be attributed to the reduced coil size due to a presence of the long-chain branching.

Conclusions

The hyperbranched PS was successfully polymerized with dendric multi-methacryloyl macromonomer which contains average active functionality of 8. The branched PS-B and C was obtained via same process with linear PS-A excepting the utilization of the macromonomer.

Oscillatory shear measurements were conducted at various temperatures and PS-A, B and C followed time-temperature superposition principle. The effect of use of multi-methacryloyl macromonomer in the polymerization process was marginal for linear viscoelastic regime. We will discuss the non-linear viscoelastic flow behavior under shear and elongational flow in more detail.

References

Antonietti M and Rosenauer C Properties of Fractal Divinylbenzene Microgels Macromolecules, 24, 3434-3442 (1991)

Campbell JD, Teymour F and Morbidelli M Production of hyperbranched polystyrene by high-temperature polymerization Macromolecules, 38, 752-760 (2005)

Ferri D and Lomellini P Melt rheology of randomly branched polystylenes J. Rheol. 43, 1355-1372 (1999)

Hepperle J and Munstedt H Rheological properties of branched polystyrenes, nonlinear shear and extensional behavior Rheol. Acta 45: 717-727 (2006)

Hepperle J, Munstedt H, Haug PK and Eisenbach CD Rheological properties of branched polystyrenes: linear viscoelastic behavior Rheol. Acta 45, 151-163 (2005)
Ourdouillie P, Chaumont P, Mechin F, Dumon M, Durand D and Nicolai T Characterization of randomly branched polymers formed by end-linking linear polystyrene using controlled free radical polymerization Macromolecules 34, 4109-4113 (2001)

Constitutive model for polymers in rubber like state for thermoforming and stretch blow moulding: How to use chain statistic

*Noëlle BILLON**

Mines-Paristech CEMEF - NA - France

An innovative concept of visco hyper elastic modelling for polymers based on revisited Edward Vilgis network models will be presented. The model is described at the level of chain statistic and considers the polymer to be an upon strain dependent network. This model allows reproducing strain rate and temperature dependences thanks to the use of WLF approach. Inelastic effects are introduced through the changes in the parameters of the initial network. Model is written in a full 3D thermo dynamical approach and allows thermo mechanical coupling. Efficiency will be discussed on the basis of tensile and shearing testing performed over the range of forming and using stain field measurements. Experimental data base includes loading - unloading loops in tension and in shear as well as loading - relaxation - unloading - anti relaxation tests. This innovative model can be introduced in numerical code and models complex behaviour with a small number of parameters each keeping a physical meaning.

RHEOLOGICAL MODEL FOR SHORT FIBER SUSPENSION IN POLYMERIC LIQUID

Jang Min Park and Tai Hun Kwon*

Department of Mechanical Engineering, Pohang University of Science and Technology, Pohang, Korea
thkwon@postech.ac.kr

Abstract - A new rheological model for a short fiber suspension in a polymeric liquid is proposed via irreversible thermodynamics approach. Particularly in this model, the anisotropic effect due to the fiber orientation is introduced in the kinematic equation of the polymer in a manner of positive entropy production. Fiber orientation kinematics is described by Folgar-Tucker model using fiber orientation tensor where invariant based optimal fitting closure approximation is employed. The final form of the stress tensor combines the viscoelasticity of polymeric matrix of Leonov model and the fiber contribution of Dinh-Armstrong model. Comprehensive investigations are carried out with the proposed model using steady and transient shear flows. The model shows a particular feature, which is different from the other models in the literature, that the slop of the material functions (shear viscosity and first normal stress difference in a steady shear flow) in a high shear rate regime is nearly independent of the fiber volume fraction and the fiber-polymer interaction. Basic features of the fiber suspension in polymeric liquid can be predicted by the proposed model, and also the model could fit the experimental data available in the literature very well.

Introduction

Short fiber reinforced thermoplastic is one of the most widely used composite materials in the relevant industries due to its enhanced thermo-mechanical property over the neat thermoplastic and a mass-producibility by injection and/or injection-compression molding processes [Advani 1994]. Since the short fiber reinforced thermoplastic is commonly processed in a melt state, its rheological property is of technical importance for a systematic control of the molding process.

The rheological behavior of the fiber suspensions in a Newtonian and non-Newtonian liquid has been a subject of considerable amount of research. An extensive review on the experimental observations was recently carried out by Eberle et al. (2008). As far as the rheological modeling of the short fiber suspension is concerned, most of the previous works was on the dilute suspension in the Newtonian matrix, and not much work has been carried out for a non-Newtonian viscoelastic matrix case [Petrie 1999].

Previously, a rheological equation for a fiber suspension in viscoelastic media was developed by Ait-Kadi and Grmela (1994) by combining a viscoelastic rheological equation of FENE-P model by Bird et al. (1980) and a suspension rheological equation by Dinh and Armstrong (1984). In particular, a simple modification to the Dinh-Armstrong model was made to include the non-Newtonian effect of the matrix in the drag force on the fiber. Later, more rigorous models were proposed by Azaiez (1996) and Ramazani et al. (1997) in which the viscoelastic part of the rheological equation was modified to take the anisotropic effect due to the fiber orientation into account. Azaiez (1996) modified Giesekus and FENE-P model with a hypothesis that the fiber orientation induces an anisotropic hydrodynamic drag force on the polymer chains which was described by a fiber orientation dependent friction tensor. Meanwhile, Ramazani et al. (1997) could develop rheological models for the suspension in the viscoelastic media via the unified approach of a general Poisson bracket formulation [Beris and Edwards 1994]. The fiber-polymer interaction was introduced in the fourth-order mobility tensor which depends on the polymer conformation and fiber orientation tensors. In those models by Azaiez (1996) and Ramazani et al. (1997), the strength of the fiber-polymer interaction was characterized by namely a fiber-polymer interaction parameter.

In the present study, a new rheological model for a fiber suspension in polymeric liquid is proposed according to the irreversible thermodynamics of viscoelastic deformation of polymer matrix. Particularly in the kinematic equation of the polymer, the anisotropic effect due to the fiber orientation is taken into account in a manner of a positive entropy production. The rheological behavior predicted by the present model has a particular feature, which is quite different from FENE-P based model by Azaiez (1996) and Ramazani et al. (1997), in that the slops of the material functions (shear viscosity and first normal stress difference in steady shear flows) in a high shear rate regime are nearly independent of the fiber volume fraction and the fiber-polymer interaction parameter.

In the following sections, the fundamental theories and model derivation procedure are described first. Then, the details of the rheological behavior of the proposed model are investigated with a steady and transient shear flows to understand the effects of various parameters and to confirm whether the model could represent the important features of short fiber suspended polymeric liquids.

Model development

Kinematics of fiber and polymer

The orientation state of a group of fibers can be described by a probability distribution function, $\Psi(\underline{p},\underline{r},t)$, which is defined such that a probability to find a fiber oriented in the direction of $\underline{p}$ within an angular range of $d\underline{p}$ at the position $\underline{r}$ at time t is $\Psi(\underline{p},\underline{r},t)d\underline{p}$ [Folgar and Tucker 1984]. In the practical viewpoint of the efficient computation, a fiber orientation tensor is more commonly used than the probability distribution function due to its compact representation [Advani and Tucker 1987]. The second- and fourth-order fiber orientation tensors are defined as follows:

$$\underline{\underline{a}}(\underline{r},t)\equiv\int \underline{p}\underline{p}\Psi(\underline{p},\underline{r},t)d\underline{p} \tag{1}$$

$$\underset{\sim}{a}(\underline{r},t)\equiv\int \underline{p}\underline{p}\underline{p}\underline{p}\Psi(\underline{p},\underline{r},t)d\underline{p} \tag{2}$$

The evolution equation of the second-order orientation tensor can be written as follows [Advani and Tucker 1987]:

$$\frac{D\underline{\underline{a}}}{Dt}=-\frac{1}{2}\left(\underline{\underline{\omega}}\cdot\underline{\underline{a}}-\underline{\underline{a}}\cdot\underline{\underline{\omega}}\right)+\frac{\lambda}{2}\left(\underline{\underline{\dot{\gamma}}}\cdot\underline{\underline{a}}+\underline{\underline{a}}\cdot\underline{\underline{\dot{\gamma}}}-2\underset{\sim}{a}:\underline{\underline{\dot{\gamma}}}\right)+2D_r\left(\underline{\underline{\delta}}-3\underline{\underline{a}}\right) \tag{3}$$

where D_r is a rotary diffusion for the fiber-fiber interactions. Folgar and Tucker (1984) suggested to use $D_r=C_I\dot{\gamma}$ where $\dot{\gamma}$ is a generalized shear rate and C_I is an empirically determined fiber interaction coefficient. Particularly for a fiber suspended in viscoelastic liquid, Ramazani et al. (2001) proposed a phenomenological model for C_I as follows:

$$C_I=\frac{C_I^*}{\frac{h}{L}\left[\mathrm{tr}\left(\underline{\underline{a}}\cdot\underline{\underline{c}}_e\right)\right]^n} \tag{4}$$

where h is average distance between neighboring fibers and L is fiber length; $\underline{\underline{c}}_e$ is a Finger strain tensor for the reversible deformation of polymer.

It migh be mentioned that, according to several experiments, fibers orient about ten time slower than the model of Eqn. (3) predicts. In this regard, Wang et al. (2008) recently developed, namely a strain reduced closure model, which could describe the slow orientation kinetics.

To solve Eqn. (3), one needs a closure approximation, which represents the fourth order orientation tensor ($\underset{\sim}{a}$) in terms of the second order orientation tensor ($\underline{\underline{a}}$).

There are many closure approximations in the literature based on different approaches [Advani and Tucker 1987; Cintra and Tucker 1995; Chung and Kwon 2002; Jack et al. (2009)]. In the present study, invariant-based optimal fitting (IBOF) closure approximation by Chung and Kwon (2002) is employed because of its accuracy and computational efficiency.

The kinematics of polymer is described according to Leonov's postulate [Leonov 1976]. The deformation history of the continuum media can be expressed by $\underline{x}=\underline{x}(t,\underline{X})$ where $\underline{x}$ and $\underline{X}$ is Eulerian and Lagrangian coordinate of material point, respectively. Leonov postulated that there is an unloaded state $\hat{\underline{X}}$ when the deformed material is released from the stresses, and the unloaded state can be related to the initial state as $\hat{\underline{X}}=\hat{\underline{X}}(t,\underline{X})$. Then, one can define a deformation gradient tensor ($\underline{\underline{F}}_e$) and Finger strain tensor ($\underline{\underline{c}}_e$) for the reversible deformation as follows:

$$\underline{\underline{F}}_e\equiv\left(\nabla_{\hat{\underline{X}}}\underline{x}\right)^T \tag{5}$$

$$\underline{\underline{c}}_e\equiv\underline{\underline{F}}_e\cdot\underline{\underline{F}}_e^T \tag{6}$$

The evolution equation for $\underline{\underline{c}}_e$ can be written as follows:

$$\overset{\triangledown}{\underline{\underline{c}}}_e=-\underline{\underline{e}}_p\cdot\underline{\underline{c}}_e-\underline{\underline{c}}_e\cdot\underline{\underline{e}}_p \tag{7}$$

where $\underline{\underline{e}}_p$ is a dissipative rate of strain tensor and the upper triangular indicates the upper-convective time derivative defined as follows:

$$\overset{\triangledown}{\underline{\underline{c}}}_e\equiv\frac{\partial\underline{\underline{c}}_e}{\partial t}+\underline{v}\cdot\nabla\underline{\underline{c}}_e-\nabla\underline{v}^T\cdot\underline{\underline{c}}_e-\underline{\underline{c}}_e\cdot\nabla\underline{v} \tag{8}$$

Now, it remains to specify $\underline{\underline{e}}_p$ in Eqn. (7) to complete the kinematic equation of the polymer as a part of the constitutive theory.

Stress tensor

The total stress tensor ($\underline{\underline{\sigma}}$) of the fiber suspended polymeric liquid can be written as follows:

$$\underline{\underline{\sigma}}=-p\underline{\underline{\delta}}+\underline{\underline{\sigma}}_s+\underline{\underline{\sigma}}_f+\underline{\underline{\sigma}}_e \tag{9}$$

where p is pressure, and $\underline{\underline{\sigma}}_s$, $\underline{\underline{\sigma}}_f$ and $\underline{\underline{\sigma}}_e$ is stress tensor due to solvent, fiber and polymer, respectively. For the solvent stress part, one can use following relation for Newtonian fluid:

$$\underline{\underline{\sigma}}_s=\eta_s\underline{\underline{\dot{\gamma}}} \tag{10}$$

where η_s is solvent viscosity. As for the fiber contribution to the total stress, Dinh-Armstrong model is employed which can be written as follows [Dinh and Armstrong 1984]:

$$\underline{\underline{\sigma}}_f=N_p\eta_m\underline{\underline{\dot{\gamma}}}:\underset{\sim}{a},\quad N_p=\frac{\pi n_f L^3}{12\ln(2h/D)} \tag{11}$$

where η_m is matrix viscosity, n_f is particle number density; D is fiber diameter. The stress contribution from the polymer can be written as follows [Leonov 1976]:

$$\underline{\underline{\sigma}}_e=2\frac{\partial W}{\partial I_1}\underline{\underline{c}}_e-2\frac{\partial W}{\partial I_2}\underline{\underline{c}}_e^{-1}, \tag{12}$$

where W is elastic potential, I_1 and I_2 is first and second invariant of $\underline{\underline{c}}_e$, respectively. In this study, a simple form of $W(I_1, I_2) = \mu(I_1 - 3)$ will be used for the elastic potential.

Dissipative rate of strain tensor

To complete the rheological constitutive equation, one needs an appropriate form for $\underline{\underline{e}}_p$ in the Eqn. (7). Previously, Leonov (1976) gave a general form of $\underline{\underline{e}}_p$ for the polymeric liquid via irreversible thermodynamics with a local equilibrium hypothesis. Particularly in this study, the fiber orientation tensor is introduced in $\underline{\underline{e}}_p$, so that the anisotropy of the fiber orientation state affects the dissipation process of the polymer.

According to Leonov (1976), $\underline{\underline{e}}_p$ for Maxwellian viscoelastic media can be written as follows:

$$\underline{\underline{e}}_p = \underset{\sim}{M} : \underline{\underline{\sigma}}_e' \tag{13}$$

where $\underline{\underline{\sigma}}_e' \equiv \underline{\underline{\sigma}}_e - \frac{1}{3}\left(\mathrm{tr}\,\underline{\underline{\sigma}}_e\right)\underline{\underline{\delta}}$, and $\underset{\sim}{M}$ is fourth-order kinetic tensor. In the present study, the anisotropic effect due to the fiber orientation is included in $\underset{\sim}{M}$ to take, namely the fiber-polymer coupling, into account via the following expression of the kinetic tensor:

$$\underset{\sim}{M}\left(\underline{\underline{c}}_e, \underline{\underline{a}}\right) = \alpha_1 \underset{\sim}{M}^{\text{Leonov}}\left(\underline{\underline{c}}_e\right) + \alpha_2 \underset{\sim}{M}^{\text{F-P}}\left(\underline{\underline{c}}_e, \underline{\underline{a}}\right) \tag{14}$$

$$\begin{aligned} M_{ijkl}^{\text{Leonov}} = {} & \frac{1}{2} m_1 \left(\delta_{ik}\delta_{jl} + \delta_{il}\delta_{jk} - \frac{2}{3}\delta_{ij}\delta_{kl}\right) \\ & + m_2 c_{ij}' c_{kl}' + m_3 c_{ij}' c_{kl}^{-1'} + m_4 c_{ij}^{-1'} c_{kl}' + m_5 c_{ij}^{-1'} c_{kl}^{-1'} \\ & + \frac{1}{4} m_6 \begin{pmatrix} c_{ik}'\delta_{jl} + c_{jk}'\delta_{il} - \frac{4}{3}\delta_{ij}c_{kl}' \\ + c_{il}'\delta_{jk} + c_{jl}'\delta_{ik} - \frac{4}{3}\delta_{kl}c_{ij}' \end{pmatrix} \\ & + \frac{1}{4} m_7 \begin{pmatrix} c_{ik}^{-1'}\delta_{jl} + c_{jk}^{-1'}\delta_{il} - \frac{4}{3}\delta_{ij}c_{kl}' \\ + c_{il}^{-1'}\delta_{jk} + c_{jl}^{-1'}\delta_{ik} - \frac{4}{3}\delta_{kl}c_{ij}^{-1'} \end{pmatrix} \end{aligned} \tag{15}$$

$$M_{ijkl}^{\text{F-P}} = n_1 a_{ij}' c_{kl}' + n_2 a_{ij}' c_{kl}^{-1'} + n_3 a_{ij}^{2'} c_{kl}' + n_4 a_{ij}^{2'} c_{kl}^{-1'} \tag{16}$$

where $c_{ij}^{-1} \equiv \left(\underline{\underline{c}}_e^{-1}\right)_{ij}$ and $a_{ij}^2 = \left(\underline{\underline{a}}^2\right)_{ij}$; m_i $(i = 1 \sim 7)$ is scalar function of invariants of $\underline{\underline{c}}_e$; n_i $(i = 1 \sim 4)$ is scalar function of joint invariants of $\underline{\underline{c}}_e$ and $\underline{\underline{a}}$. In Eqn. (14), α_1 and α_2 represents the each contribution of polymer and fiber-polymer coupling to the total kinetic tensor, respectively. In this study, $\alpha_1 = 1 - k\phi_f$ and $\alpha_2 = k\phi_f$ are used where ϕ_f is the fiber volume fraction and k is a fiber-polymer interaction parameter.

By substituting Eqns. (14), (15) and (16) into Eqn. (13), one can obtain the following form of the dissipative rate of strain tensor:

$$\underline{\underline{e}}_p = \alpha_1 \underline{\underline{e}}_p^{\text{Leonov}} + \alpha_2 \underline{\underline{e}}_p^{\text{F-P}} \tag{17}$$

$$\underline{\underline{e}}_p^{\text{Leonov}} = C_1\left(\underline{\underline{c}}_e - \frac{1}{3}I_1\underline{\underline{\delta}}\right) - C_2\left(\underline{\underline{c}}_e^{-1} - \frac{1}{3}I_2\underline{\underline{\delta}}\right) \tag{18}$$

$$\underline{\underline{e}}_p^{\text{F-P}} = B_1\left(\underline{\underline{a}} - \frac{1}{3}J_1\underline{\underline{\delta}}\right) - B_2\left(\underline{\underline{a}}^2 - \frac{1}{3}J_2\underline{\underline{\delta}}\right) \tag{19}$$

where C_i $(i = 1,2)$ is scalar function of invariant of $\underline{\underline{c}}_e$ and B_i $(i = 1,2)$ is scalar function of joint invariant of $\underline{\underline{c}}_e$ and $\underline{\underline{a}}$; $J_1 = \mathrm{tr}(\underline{\underline{a}}) = 1$ and $J_2 = \mathrm{tr}(\underline{\underline{a}}^2)$. As for C_i, a simple form of $C_1 = C_2 = C = \frac{1}{4\theta}$ is used according to previous work by Leonov (1976). As far as B_i is concerned, it is assumed that $B_1 = B_2 = B$ for simplicity of the model. It should be noted that the entropy production during the non-equalibrium process should be a positive quantity, which is namely Clausius-Duhem inequality. In this regard, following form of B is proposed in this study (details of the derivation procedure will be published in elsewhere):

$$B = \frac{1}{4\theta}\left[\mathrm{tr}\left\{\left(\underline{\underline{a}} - \underline{\underline{a}}^2\right)\cdot\underline{\underline{c}}_e\right\} - \frac{1}{3}(J_1 - J_2)I_1\right] \tag{20}$$

It should be noted that the present model could be easily extended to a multimode model via the same procedure mentioned above, and details will be published in elsewhere.

Rheology in steady and transient shear flows

Steady shear flow

Fig. 1 shows the effect of n (in Eqn. (4)) on the shear viscosity (η) for the suspension with ϕ_f=14 % and k=3; η of the neat polymer are also plotted for a comparison purpose. As n increases, η of the suspension tends to decrease particularly at a high shear rate regime. This behavior attributes to the decrease of C_I as n increases at the high shear rate regime. In the low shear rate regime, η is almost insensitive to the change of n, since $\mathrm{tr}(\underline{\underline{a}}\cdot\underline{\underline{c}}_e)$ approaches to unity as shear rate decreases. The first normal stress difference (N_1) is found to be insignificantly affected by the change of n.

Fig. 2 shows the effect of k on the rheological behavior of the model (with n=0.5). η increases with k in the low shear rate regime, and also the shear thinning tends to start appearing at the lower shear rate as k increases (Fig. 2(a)). In the high shear rate regime, all η curves with different values of k tend to merge to a single curve which is nearly parallel to that of the neat polymer, thus the effect of k is nearly negligible in the high shear rate regime. N_1 increases with k in the overall range of the shear rate, and their slops at the high shear rate regime are almost the same as that of the neat polymer (Fig. 2(b)).

Fig. 3 shows the effect of ϕ_f with n=0.5 and k=5. As ϕ_f increases, η increases in the overall range of the shear rate (Fig. 3(a)). Particularly, the increase is larger at the low shear rate regime than the high shear rate regime. Also the shear thinning tends to start appearing at the lower shear rate as ϕ_f increases. It might be noted that such a rheological behavior of the fiber suspension (in a polymeric matrix) is consistent with the experimentally observed one in the literature [Laun 1984; Chan et al. 1978; Eberle et al. 2008]. The shear thinning slop does not significantly change with the ϕ_f, thus the η curves are almost in parallel with each other at the high shear rate regime. The model also predicts the increase of N_1 in the overall range of the shear rate, and the curves for different fiber fractions are nearly in parallel with each other at the high shear rate regime (Fig. 3(b)).

The present rheological model is compared with the experimental data for the short fiber reinforced thermoplastic melts which was reported by Chan et al. (1978). Fig. 4 compares the model prediction (with two relaxation modes) and experimental results for the short fiber reinforced PS melt at 180°C. The material parameters for the model prediction is shown in Table 1. As far as η is concerned, the model can predict the experimental result with a high accuracy in the overall range of the shear rate with three fiber volume fraction cases: the shear rate at which the shear thinning appears, the shear thinning slop, and the viscosity increment due to the fibers volume fraction are in good agreements between the prediction and experiment. Meanwhile the model over-predicts N_1 than the experimental data, and this is mainly due to the over-prediction by the original Leonov model for the neat polymer melt. However, the increasing slop and overall shape of the first normal stress difference curve are well predicted by the model.

Transient shear flow

Fig. 5 shows the effect of n on a transient behavior of the model in terms of $\eta+$ as a function of the applied shear strain ($\gamma = \dot{\gamma}t$) with specific values of k=3 and ϕ_f=14 %. The shear rate of 10 s^{-1} is used since the effect of n is observable only at the high shear rate regime as discussed in the previous section. The neat polymer shows an overshoot of the shear viscosity which originates from the polymer kinematics being subject to the high shear rate flow. The fiber suspension also has the overshoot of the shear viscosity which, in this case, originates from the kinematics of both the fibers and the polymers. The magnitude of the viscosity overshoot of the fiber suspension is little affected by the change of n, while the steady state viscosity after the overshoot tends to decrease as n increases as expected from the results of Fig. 1.

Shown in the Fig. 6 is the effect of k on transient behavior of $\eta+$ with n 0.5 and ϕ_f=14 %. For these results, the shear rate of 0.1 s^{-1} is used since the effect of k is observable only in the low shear rate regime as discussed with Fig. 2a. The neat polymer has no viscosity overshoot in this case of the low shear rate flow, while the fiber suspension shows the viscosity overshoot which is purely due to the fiber kinematics. The $\eta+$ increases with k which is found to be mostly due to the polymer kinematics according to the analysis results.

Fig. 7 shows the effect of ϕ_f on the transient behavior of the model at the shear rate of 0.1 s^{-1}. $\eta+$ increases with ϕ_f and there are three contributions to the increase of the shear viscosity: one is N_p which increases with the fiber volume fraction; two other contributions are from the kinematics of the fiber and the polymer, respectively. In contrast to the other parameters of n and k, it is observed that ϕ_f affects both the polymer and fiber kinematics significantly in the low shear rate regime. According to the fiber interaction model of Eqn. (4), the larger ϕ_f results in the increase of C_I due to the decrease in h. Thus less aligned fiber orientation state is predicted as ϕ_f increases. As far as the polymer kinematics is concerned, the effect of ϕ_f is similar to that of k which is obviously due to the expressions of α_1 and α_2: as the fiber volume fraction or the fiber-polymer interaction parameter increases, the dissipation process of the polymer is more affected by the presence of the fibers.

Recently, transient rheology with a flow reverse is getting more attentions in the relevant research field [Eberle et al. 2009]. By introducing the flow reverse, one can obtain more reliable and reproducible experimental data for the transient rheology of the fiber suspensions since one can have rather consistent initial conditions of the fiber orientation state for each experiment. Fig. 8 shows the effect of reversed shear flow on the transient behavior of five different cases at the shear rate of 0.1 s^{-1}: one case is for a conventional start-up flow, while the other four cases undergo the backward (BWD) shear flow before the start-up of a consecutive forward (FWD) shear flow. Particularly in the flow reverse cases, a flow cessation is allowed in between BWD and FWD flows for 0~5 seconds, by which the stress contribution due to the polymers can be relaxed to some extent depending on the cessation time. Plotted in Fig. 8 are the transient behaviors at the start-up of the FWD shear flow. The flow reverse cases (BWD-FWD) show the overshoot of $\eta+$ and N_1+ at a larger shear strain than that of the conventional FWD only case, and also the peak value and the magnitude of the overshoots are rather smaller in the flow reverse cases than that of FWD only case. The effect of the flow cessation between BWD and FWD flows could be observed only at the early stage of the deformation as an initial condition effect. The shear viscosity has a negative value at the beginning of the start-up flow without the flow cessation. The negative shear viscosity is due to the polymers having a negative value of the shear stress which is developed during the

BWD flow. As more relaxation is allowed by increasing the flow cessation time, $\eta+$ at the beginning of the deformation grows in the positive direction until the stress contribution from the polymers is fully relaxed. With 5 seconds of the flow cessation, the negative shear stress due to the polymer is fully relaxed thus $\eta+$ has a positive value at the beginning of the start-up flow. As far as N_1+ is concerned, the flow reverse case without the stress relaxation begins with the positive value, which subsequently is followed by a significant undershoot to the negative value at the early stage of the shear deformation (Fig. 8b). As the flow cessation time increases, N_1+ at the beginning of the deformation reduces to the negative direction and the undershoot gets more blunt. Eventually with 5 seconds of the flow cessation, N_1+ has a negative value at the beginning of the deformation and the undershoot behavior is invisible. In N_1+ at the beginning of the start-up flow, there are two contributions: i) one is a negative contribution due to the fiber orientation state which is not affected by the flow cessation and ii) the other is a positive contribution due to the polymer which can be relaxed during the flow cessation. The undershoot behavior is a result of a recovery and subsequent growth of $(c_{11}-c_{22})$ during the start-up flow. When the polymer contribution is fully relaxed by 5 seconds of the flow cessation, only the negative N_1+ due to the fiber appears at the beginning of the shear deformation. In this regard, it might be suggested to introduce a sufficient relaxation time between BWD and FWD flows particularly for a fiber suspension in a nonlinear viscoelastic matrix so that one can identify only the fiber anisotropy effect in the transient stress growth using the flow reverse experiments.

Concluding Remarks

A new rheological model for a short fiber suspension in a polymeric liquid is derived from the irreversible thermodynamics approach. The kinematic relations of the fiber and polymer, respectively, are introduced first. The stress tensor is stated as a summation of each stress contribution from the Newtonian solvent, fibers and polymers by using Dinh-Armstrong model and Leonov model, respectively. In particular, a coupling effect between the fiber and polymer are introduced in the kinematic equation of the polymer by assuming that the anisotropy due to the fiber orientation affects the dissipation process of the polymer. The second law of thermodynamics is satisfied by using an appropriate form of a scalar function appearing in the irreversible (dissipative) deformation rate tensor in the proposed model. An additional material parameter, the fiber-polymer interaction parameter k, appears in the proposed model, and $k\phi_f$ characterizes the amount of anisotropic contribution of the fiber orientation to the polymer kinematics.

There are two material parameters of k and n which characterize the fiber-polymer coupling effect in the proposed model and each parameter has quite different effect on the rheological behavior: k affects shear viscosity only for a low shear rate regime while the effect of n on the shear viscosity is confined only for a high shear rate regime.

The fiber volume fraction affects both the shear viscosity and the first normal stress difference for overall range of the shear rate, and particularly the shear viscosity increment as fiber volume fraction increases is larger for the low shear rate. Also it is observed that the shear thinning tends to appear at a lower shear rate as k or fiber fraction increases.

According to the transient shear rheology, the shear rate dependency of the fiber kinematics results in a decrease of the shear viscosity particularly at the high shear rate regime as n increases. Also, the increase of the shear viscosity at the low shear rate regime as k increases is found to be the result of the polymer kinematics where the dissipation process of the polymer is more affected by the anisotropy of the fiber orientation as k increases.

It is also shown that the present model could fit the experimental data of steady shear viscosity and the first normal stress difference for fiber reinforced thermoplastic melt very well.

Acknowledgements

This research was supported partially by Basic Science Research Program through the National Research Foundation of Korea (NRF) funded by the Ministry of Education, Science and Technology (2009-0092817), and partially by a grant to MEMS Research Center for National Defense funded by Defense Acquisition Program Administration.

Table 1 – Material parameters used for fitting experimental data of 20 wt% and 40 wt% fiber reinforced PS.

θ_1 [s]	2.444×10^{-1}	C_I^*(20 wt%)	0.001
θ_2 [s]	6.299	C_I^*(40 wt%)	0.01
η_1 [Pa s]	9.836×10^{3}	k_1	2
η_2 [Pa s]	6.222×10^{4}	k_2	4.5
s	8.038×10^{-3}	n(20 wt%)	0.5
		n(40 wt%)	0.1

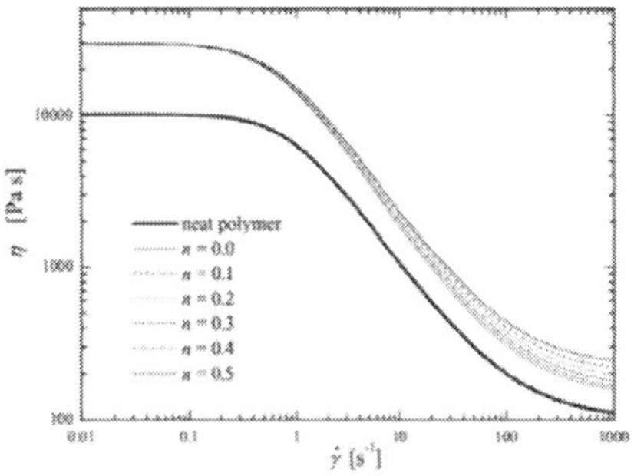

Figure 1 – Shear viscosity of neat polymer and fiber suspensions with different values of n.

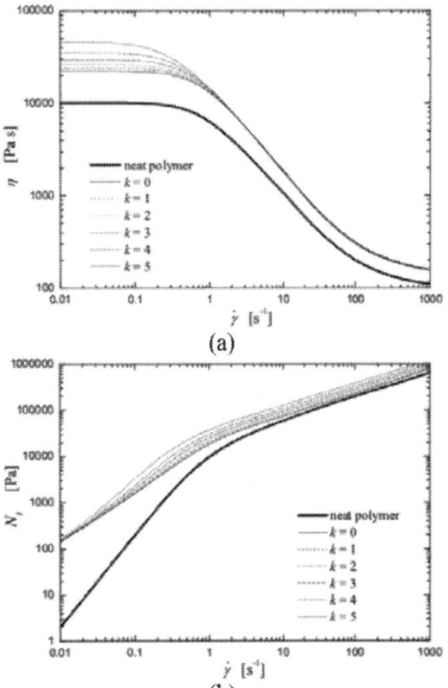

Figure 2 – (a) Shear viscosity and (b) the first normal stress difference of neat polymer and fiber suspension with different values of *k*.

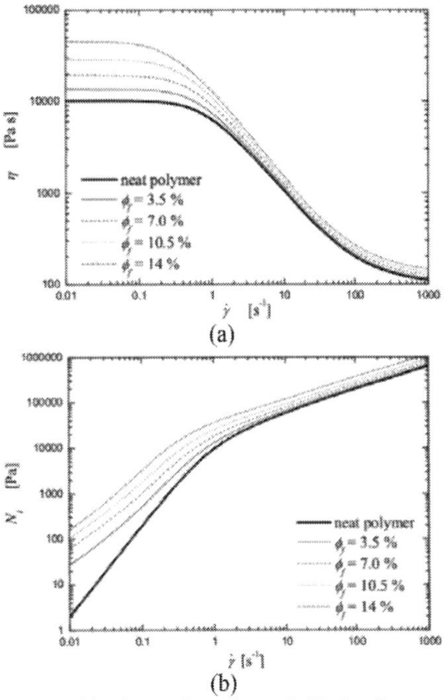

Figure 3 – (a) Shear viscosity and (b) the first normal stress difference of fiber suspensions for different ϕ_f with k=5.

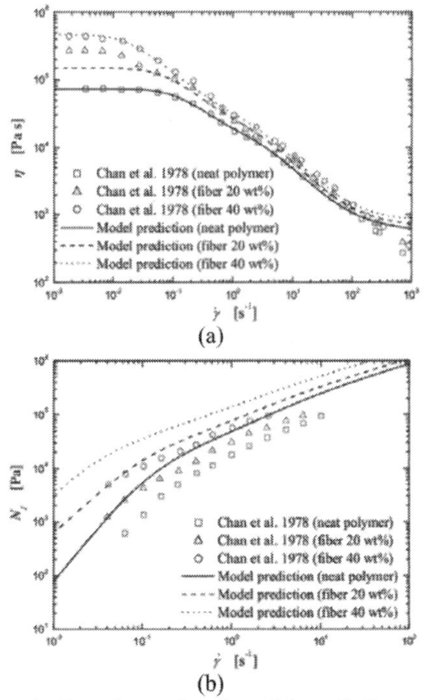

Figure 4 –Experimental and model prediction results for a fiber reinforced PS melt; (a) shear viscosity and (b) the first normal stress difference.

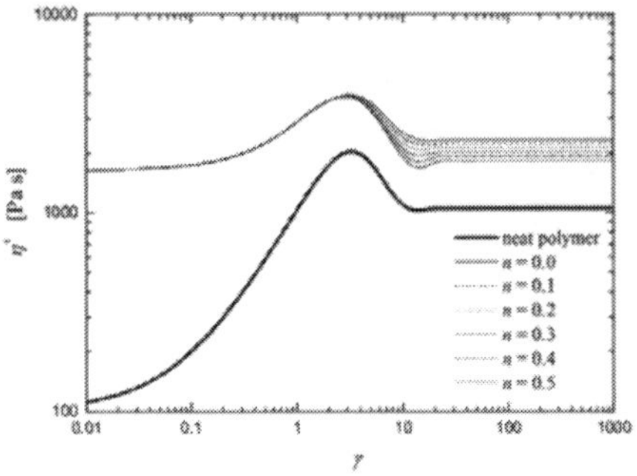

Figure 5 – Transient behaviors of shear viscosity growth at $\dot{\gamma}$=10 s^{-1} for different values of n with k=3 and ϕ_f =14 %.

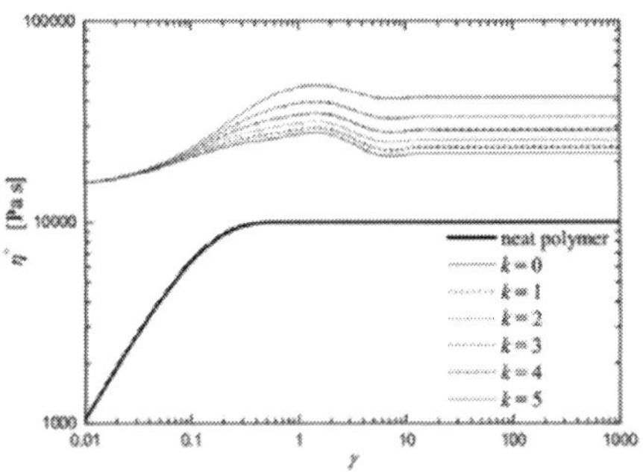

Figure 6 – Transient behaviors of shear viscosity growth at $\dot{\gamma}$=0.1 s^{-1} for different values of k with n=0.5 and ϕ_f =14 %.

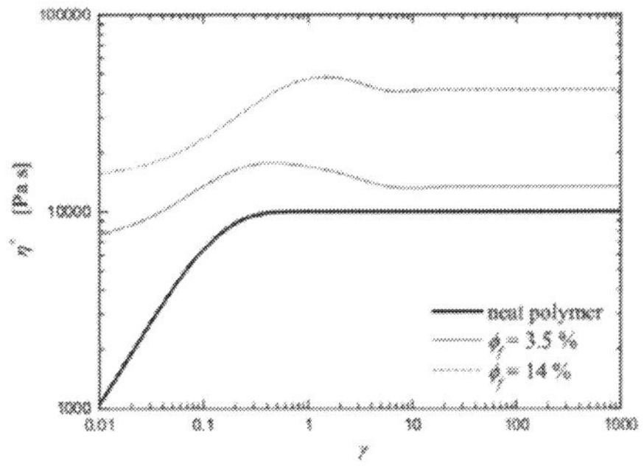

Figure 7 – Transient behaviors of shear viscosity growth at $\dot{\gamma}$ = 0.1 s^{-1} for two different values of ϕ_f.

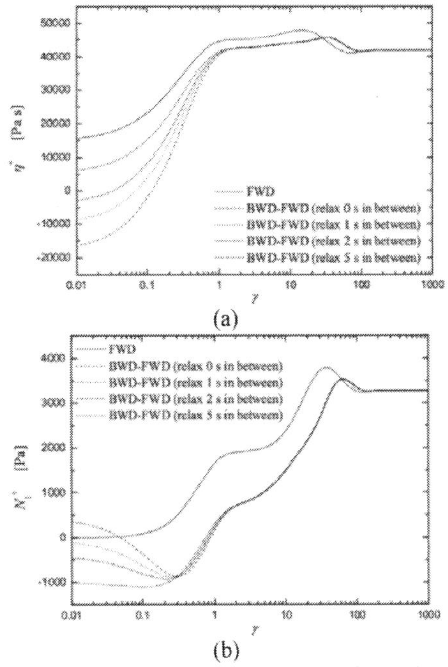

(a)

(b)

Figure 8 – Transient behaviors of (a) shear viscosity and (b) the first normal stress difference at $\dot{\gamma}$ = 0.1 s^{-1}; four samples undergo backward shear flow with a shear rate of 0.1 s^{-1} and subsequent relaxation of 0~5 seconds before the forward shear flow.

References

Advani, S. G.: Flow and rheology in polymer composites manufacturing, Elsevier Science B. V. (1994)

Advani, S. G. and C. L. Tucker "The use of tensors to describe and predict fiber orientation in short fiber composites," J. Rheol. **8** 751-784 (1987)

Ait-Kadi A. and M. Grmela "Modelling the rheological behavior of fibre suspensions in viscoelastic media," J. Non-Newtonian Fluid Mech. **53** 65-81 (1994)

Azaiez, J. "Constitutive equations for fiber suspensions in viscoelastic media," J. Non-Newtonian Fluid Mech. **66** 35-54 (1996)

Beris, A. N. and B. J. Edwards: Thermodynamics of flowing systems with internal microstructure, Oxford University Press, Inc. (1994)

Bird, R. B., P. J. Dotson and N. L. Johnson "Polymer solution rheology based on a finitely extensible bead-spring chain model," J. Non-Newtonian Fluid Mech. **7** 213-235 (1980)

Chan, Y., J. L. White and Y Oyanagi "A fundamental study of rheological properties of glass-fiber-reinforced polyethylene and polystyrene melts," J. Rheol. **22** 507-524 (1978)

Chung, D. H. and T. H. Kwon "Invariant-based optimal fitting closure approximation for the numerical prediction of flow-induced fiber orientation," J. Rheol. **46** 169-194 (2002)

Cintra, J. S. and C. L. Tucker "Orthotropic closure approximations for flow-induced fiber orientation," J. Rheol. 39 1095-1122 (1995)

Dinh, S. M. and R. C. Armstrong "A rheological equation of state for semiconcentrated fiber suspensions," J. Rheol. **28** 207-227 (1984)

Eberle, A. P. R., D. G. Baird and P. Wapperom "Rheology of non-Newtonian fluids containing glass fibers: A review of experimental literature," Ind. Eng. Chem. Res. **47** 3470-3488 (2008)

Eberle, A. P. R., D. G. Baird, P. Wapperom and G. M. Vélez-García "Obtaining reliable transient rheological data on concentrated short fiber suspensions using a rotational rheometer," J. Rheol. **53** 1049-1068 (2009)

Folgar, F. and C. L. Tucker "Orientation behavior of fibers in concentrated suspensions," J. Reinf. Plast. Compos. **3** 98-119 (1984)

Jack D. A., B. Schache and D. E. Smith "Neural network-based closure for modeling short-fiber suspensions," Polym. Compos. Doi: 10.1002/pc.20912 (2009)

Leonov, A. I. "Nonequilibrium thermodynamics and rheology of viscoelastic polymer media," Rheol. Acta **15** 85-98 (1976)

Petrie, C. J. S. "The rheology of fibre suspensions," J. Non-Newtonian Fluid Mech. **87** 369-402 (1999)

Ramazani, A., A. Ait-Kadi and M. Grmela "Rheological modelling of short fiber thermoplastic composites," J. Non-Newtonian Fluid Mech. **73** 241-260 (1997)

Ramazani, A., A. Ait-Kadi and M. Grmela "Rheology of fiber suspensions in viscoelastic media: Experiments and model predictions," J. Rheol. **45** 945-962 (2001)

Wang, J., J. F. O'Gara and C. L. Tucker "An objective model for slow orientation kinetics in concentrated fiber suspensions: Theory and rheological evidence,", J. Rheol. **52** 1179-1200 (2008)

RHEOLOGY AND MORPHOLOGY OF NANO-SILICA CONTAINING POLYPROPYLENE AND POLYPROPYLENE/LIQUID CRYSTALLINE POLYMER BLEND

R. Foudazi[1], N. Nazockdast[1]* and M.M. Salehi Barmi[1]

[1]*Department of Polymer Engineering, Amirkabir University of Technology, Tehran, Iran – foudazi@aut.ac.ir; nazdast@aut.ac.ir; msalehibarmi@aut.ac.ir*

Abstract - We studied the rheology and morphology of hydrophilic and hydrophobic silica containing Polypropylene (PP) and Polypropylene/Liquid crystalline polymer (LCP) blend. It was found that hydrophilic silica has higher tendency for aggregation and forms bigger aggregate size. The lower percolation threshold and the higher fractal dimension for hydrophilic silica are due to its stronger particle-particle interaction compared to hydrophobic one. While the hydrophobic silica has lower thickening capability and lower coalescence hindrance by aggregates, it locates at the interface of PP/LCP blend and results in smaller droplet size and higher elasticity of hybrid samples in comparison to hydrophilic silica. These results suggest that the hydrophobic silica has a compatibilization capability for PP/LCP blend, while the hydrophilic silica mostly works as a thickening agent and suppresses the coalescence.

Introduction

The presence of nanofillers in molten polymers results in significant change of viscoelastic properties. Consequently, melt linear viscoelastic behavior is a way generally used to assess the state of dispersion of nanocomposites directly in the melt state. During the past few years, intensive discussions have been hold on the rheology of polymer layered silicate nanocomposites whereas the rheology of usual nanocomposites filled with silica particles, and more particularly fumed silica, seems to have been received less attention. The particle-particle and particle-polymer interactions have a major impact on the rheological and reinforcement properties of fumed silica nanocomposites due to their large surface area of particles (Cassagnau, 2008; Aranguren et al., 1992; Khan and Zoeller, 1993). Fumed silica fillers have the potential for self-aggregation due to their fractal structure and high specific area, and hence they can form a network of connected or interacting particles in the molten polymer (Cassagnau, 2008; Aranguren et al., 1992; Khan and Zoeller, 1993; Yziquel et al., 1993).

The nanocomposites containing fumed silica show a solid-like behavior response at low deformation rates with following characteristics: a non-terminal zone of relaxation, apparent yield stress and a shear-thinning dependence on viscosity at high deformation rates (Cassagnau, 2008). This particular rheological behavior arises from the presence of a network structure. This network structure can be formed by partly adsorption of polymer chains on the filler surface and partly entanglement with neighboring ones (Aranguren et al., 1992), which give rise to a subset of stiffer chains extending from the interface, resulting in a dynamically stiffer matrix as well as a higher loss modulus due to an increase in hydrodynamic friction (Sternstein and Zhu, 2002). This finding was evidenced by viscoelastic experiments showing that the low frequency modulus of the composites decreases considerably when the particles are chemically treated with organo-silane. However, the density of adsorbed chain (bound chain) and their conformation at the filler surface are generally quite difficult to access (Aranguren et al., 1992; 1997).

Two mechanisms are presented in the literature in order to depict the solid-like behavior qualitatively: the particle-particle interaction is the dominant mechanism in fumed silica nanocomposites, while the particle-polymer interaction is the dominant one in colloidal silica nanocomposites at identical filler concentration. Obviously, these interactions are balanced in each nanocomposite systems by the silica surface treatment (chain grafting, silane modification) and the matrix characteristics. (Cassagnau, 2008)

Polymer blends comprising of a flexible matrix and small amount of liquid crystalline polymers (LCP) are of industrial and academic interests because the dispersed LCP phase can lead to easier processing and enhanced mechanical properties (e.g. Kim and Denn, 1992; Riise et al., 1999). Addition of compatibilizer has been found to improve the dispersion of the LCP phase and in some cases enhance the final mechanical properties of the blend systems. Using a compatibilizer – which is compatible or miscible with both phases – is the classical route to ensure adhesion between two phases. For example the addition of maleic anhydride grafted polypropylene as a compatibilizer to polypropylene/LCP blends was studied by Datta and Baird (1995) and O'Donnel and Baird (1995). The enhanced dispersion in compatibilized polymer blends is due to the reduction of interfacial tension and suppression of coalescence (Harrats et al., 2006).

Another less explored compatibilization method is that by use of inorganic solid particles (Lipatov, 1995; Zhang et al., 2004; Ray et al., 2004). Cassagnau and his co-workers studied the effect of both hydrophilic and hydrophobic nano-silica on PP/PS (Elias et al., 2007), and PP/EVA blends (Elias et al., 2008a). They showed that while the hydrophilic silica tends to locate inside PS phase in PP/PS blend, the hydrophobic silica preferentially concentrates at the interface and PP phase (Elias et al., 2007). For PP/EVA blend samples, it was found that the hydrophilic silica tends to confine in the EVA phase whereas hydrophobic one was located close to the PP/EVA interface in the EVA phase. For the latter case, no clear interphase of PP/silica/EVA has been observed (Elias et al., 2008a).
It has been shown that the hydrophobic nano-silica enhances the fibrillation process of LCP in PP/LCP blends (Lee et al., 2002). The thickening effect of nano-silica in polypropylene matrix was suggested to be the reason of this observation. Zhang et al. (2003) also studied the effect of hydrophobic and hydrophilic nano-silica on PP/LCP blends and reported that only hydrophobic type silica enhances the fibrillation of LCP phase.
In the first part of this work, the aggregation and rheological behavior of hydrophobic and hydrophilic silica filled PP are studied. Then in the second part, the effects of these nano-particles, as a new potential type of compatibilizers, on the rheology and morphology of PP/LCP blends are studied.

Experimental

Materials

An isotactic polypropylene, Moplen HP500H obtained from Arak petrochemicals, with melt flow rate (MFR) of 1.8 and a melting point of 165°C, was used as matrix. The dispersed phase was a thermotropic liquid crystalline polymer, Vectra A950 (from Hoechst-Celanese, now Ticona) which is a commercial random copolyester of 73% pHBA and 27% 4,6-HNA. The as-received pellets had a nominal melting point of 280°C. The nematic–isotropic transition is not known, since it lies above the decomposition temperature.
Fumed nano-silica, Aerosil 200 and Aerosil R816 abbreviated respectively as Silica and m-Silica (the "silica" term is used for both silica types), were kindly supplied by Degussa GmbH. The hydrophilic nature of the Aerosil 200 powder, which has a specific surface area (BET) of 200±25m^2/g and an average particle size of 12nm, is due to the presence of silanol groups on its surface. The Aerosil R816 is a hexadecylsilane surface treated hydrophobic nano-silica based on Aerosil 200. The hydrophobic surface treatment changes slightly the specific surface area (BET) to 190±20m^2/g. The nano-silica particles have a spherical surface and are free of pores.

Sample preparation

Two different series of samples were prepared: (1) PP/Silica and PP/m-Silica nanocomposites, and (2) PP/LCP/Silica and PP/LCP/m-Slica hybrid blends. Different concentration of silica nano-particles were used in these samples. All samples were prepared by melt blending in a Brabender internal mixer at 290°C and a rotor speed of 60rpm for approximately 6min (time to reach a constant torque). In order to inhibit the thermal degradation during melt blending and rheological measurements, 0.5wt.% Irganox 1076 and Irgafos 168 was added during melt mixing.

Morphological studies

The transmission electron microscopy (TEM), Philips EM 208 operating at 100kV was used to study the aggregate size and location of nano-silica in the samples. The samples were prepared by using cryo-microtoming (with a diamond knife) at -80°C to prevent any possible morphological changes during cutting.
CamScan model MV2300 scanning electron microscopy (SEM) was employed to study the morphology of blend samples. The blend samples were cryo-fractured in liquid nitrogen and the resulting fractured surfaces were coated with gold prior to observations by microscope operated at 25.0kV and under high vacuum.

Rheological measurements

The rheological measurements were performed with a stress/strain controlled rheometer, Paar Physica model UDS 200, equipped with parallel plate geometry with 25mm diameter and 1mm gap. The temperature was set at 290°C during experiments in which the LCP is in the nematic state. The frequency sweep tests were performed on samples in the pre-determined linear viscoelastic region.

Results and Discussion

Polypropylene nanocomposites

The TEM micrographs of polypropylene filled with hydrophilic and hydrophobic silica are shown in Figure 1.

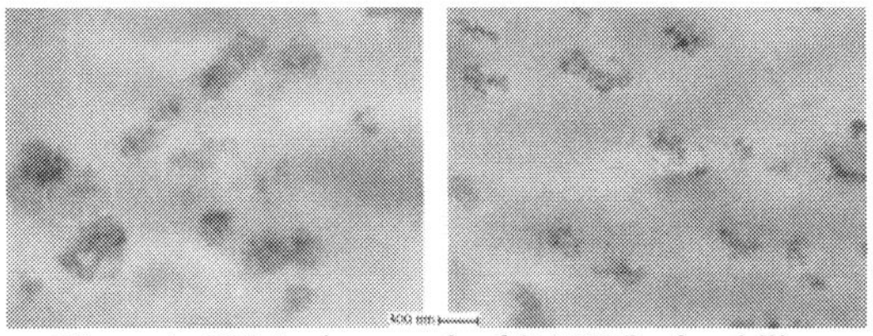

Figure 1: TEM micrograph of (A) PP/hydrophilic silica, and (B) PP/hydrophobic silica

The estimation of aggregate sizes performed for several aggregates in the TEM micrographs of samples shows that the average aggregate sizes in the PP matrix are about 230nm and 390nm for hydrophobic and hydrophilic silica filled samples, respectively. The observed higher extent of aggregation for hydrophilic silica is in agreement with the results reported by Aranguren et al. (1992) for nano-silica suspensions in PDMS.

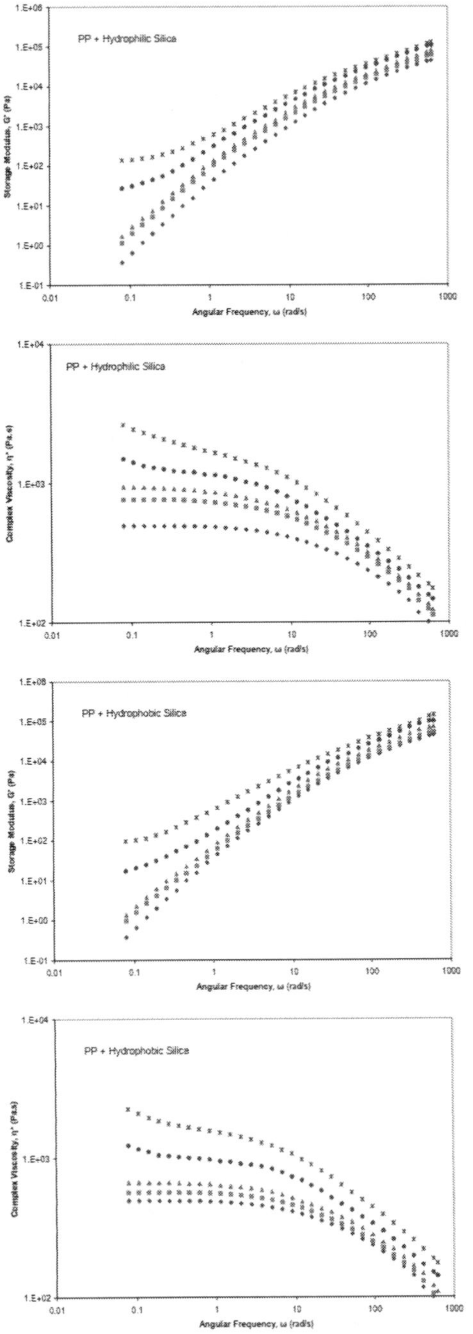

Figure 2: Storage modulus and complex viscosity versus angular frequency of polypropylene nanocomposites for 0% (♦), 1% (■), 1.2% (▲), 2.4% (●), and 3.6% (*) silica contents

The effect of nano-silica type and content on the melt linear viscoelastic behavior of the samples is shown in Figure 2. It is seen that the addition of nano-silica even at very low contents increases the viscosity and elasticity of the samples which is a characteristic of nano-structured material.

It can clearly be seen that with increasing silica content of samples, the non-terminal behavior in storage modulus and viscosity upturn at low frequencies become more pronounced. The observed behavior is an indication of three dimensional physical network formed in these nanocomposite samples. It should be noted that the hydrophilic silica shows higher increase in complex viscosity and storage modulus in comparison to hydrophobic one which is in agreement with higher extent of aggregation observed in microscopy study.

The results of the melt linear viscoelastic measurements can be used to assess the microstructure of the dispersions. In order to quantify the percolation threshold, the following percolation model can be used:

$$G' \sim (\varphi - \varphi_{per})^{\nu} \quad (1)$$

where ν is a power law exponent and φ_{per} is the percolation threshold volume fraction (Aranguren et al., 1992; Rueb and Zukoski, 1997). Vermant et al. (2007) used a linear-linear plot and approximated the φ_{per} by a linear regression of data points above the percolation. However, it seems that fitting the eq. (1) to find the φ_{per} as shown in Figure 3 is more reasonable. The slightly lower percolation threshold for hydrophilic silica filled samples is in agreement with its higher extent of aggregation seen in TEM micrographs. The percolation threshold volume fraction was found to be ~0.8% and ~0.9% for hydrophilic and hydrophobic silica respectively, which is in agreement with the results reported in the literature about 1% for nano-silica (Paquien et al., 2005).

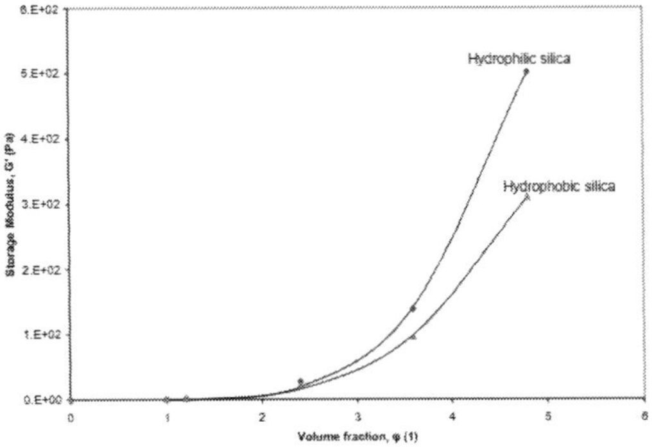

Figure 3: The fitting of eq. (1) for nano-silica filled samples

When the particulate networks are above the percolation threshold φ_{per}, the scaling concept of the elastic properties for fractal networks, developed by

Shih et al. (1990) can be used to quantitatively analyze the rheological data. Shih et al. considered the structure of a particle network as a collection of fractal aggregates that are elastically linked together. In this case, low frequency solid body response is dominated by the elasticity of the backbones of the aggregates. In this regime, the value of the plateau modulus is predicted to have a power-law dependence on volume fraction:

$$G'_p \sim \varphi^{(3+x)/(3-d_f)} \qquad (2)$$

where d_f is the fractal dimension of the aggregate network and x is an exponent that relates the particle volume fraction with aggregate size, i.e., x depends on the number of particles per aggregate. According to Potanin (1991) the ratio d_f /x can be considered as an invariant. He proposed d_f /x =3/2 for a three-dimensional network. Although the storage modulus plateau may not always be present in the experimental window, the low frequency storage modulus can be used to study this scaling behavior (Paquien et al., 2005; Vermant et al., 2004). Figure 4 shows the variations of the low frequency elastic modulus as functions of the solids content. For the hydrophilic silica, power-law exponent was found to be 4.1 which corresponds to fractal dimension $d_f \approx 2$. For the hydrophobic silica, exponent 3.9 is obtained corresponding to $d_f \approx 1.9$. The lower value of fractal index for the m-Silica containing samples can be explained by a structure which is less dense than that of the Silica ones. The obtained fractal dimensions correspond to chemically limited aggregation interaction type, and are in agreement with the results of Yziquel et al. (1999) and Khan et al. (1993).

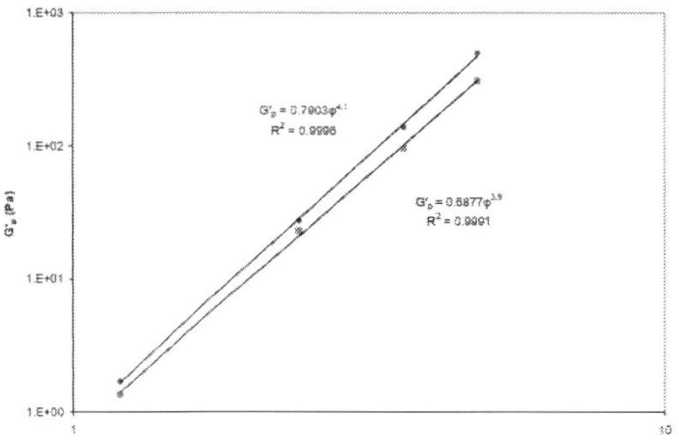

Figure 4: fitting the power-law model on the storage modulus versus volume fraction dependency

The lower percolation threshold and hence higher aggregation tendency of hydrophilic silica may be explained by its lower affinity with polypropylene and hydrogen bonding tendency between particles in comparison to hydrophobic silica. However, considering fundamental theories concerned with particle interactions in terms of electrostatic, Van der Wasls and steric forces (Israelachvili, 1992) can be more helpful. The electrostatic repulsion results from the overlapping of the electric double layer that develops at charged interfaces. In the simplest case, a repulsive force develops between the interfaces due to the entropic confinement of the counter-ions which neutralize a charged interface (Israelachvili, 1992). The neutral nature of polymeric media (the absence of ions) and uncharged interface of nano-silica implies insignificant electrostatic interaction. According to the microscopic theory by Hamaker, the Van der Waals interaction between two macroscopic bodies can be found by integration of Van der Waals interaction between molecules over all couples of molecules (Israelachvili, 1992). The presence of silanol groups in surface of hydrophilic silica compared to the hexadecyl chain in the surface of m-Silica results in stronger dipoles at the particle surface and hence stronger Van der Waals attraction between particles (Israelachvili, 1992). On the other hand, hexadecyl chains attached to the surface of hydrophobic silica cause higher degree of steric repulsion between particles of hydrophobic silica. Therefore, the higher degree of aggregation and denser aggregates of hydrophilic silica compared to hydrophobic one are expected.

For concentrated hard sphere suspensions, the Krieger and Dougherty (1959) equation is suitable to describe the relationship between the relative viscosity and the volume fraction of solid particles:

$$\eta_r = \frac{\eta}{\eta_m} = (1 - \frac{\varphi}{\varphi_m})^{-[\eta]\varphi_m} \qquad (3)$$

where η_r is the relative viscosity of filled polymer with respect to the matrix, φ_m is the maximum packing fraction, and $[\eta]$ is the intrinsic viscosity of suspensions. This equation has been used at both low and high shear rates of viscosity obtained from flow curve measurement (Chen et al., 2005), and for storage and loss moduli at low or high frequencies as suggested by Vermant et al. (2007). While the low frequency viscoelasticity mainly gives information about the aggregates and eventually the percolating structure formed by the aggregates, the high frequency viscoelasticity is dominated by the polymeric matrix contributions due to an increase of hydrodynamic contribution of matrix. In the present work, both low and high frequency complex viscosities of samples were fitted to the Krieger and Dougherty equation, as shown in Figure 5. It is seen that the model fits quite well, and consequently it can predict the relative viscosity at various volume fractions.

The results (summarized in Table 1) show that φ_m of hydrophilic silica is lower than that of hydrophobic one indicating that the former reaches the closest packing at lower volume fractions. Higher value of $[\eta]$ for hydrophilic silica filled samples is expected to be higher due to its higher thickening effect observed in Figure 2. Similar results are presented by Chen et al. (2005) for suspension of silica particles in mineral oil. While the maximum packing fraction (φ_m) was found to be close to the theoretical value (0.74) for the micro-particles, for the nano-silica filled samples the values were in the range of 0.07~0.22 that is comparable to our results (Chen et al. 2005). The lower φ_m of nano-particles was attributed to the formation of branched aggregates by Chen et al. (2005). Higher value of $[\eta]$ for hydrophilic silica corresponds to the higher aspect ratio of aggregates (Brenner, 1974) which could be the reason of lower φ_m value of this silica type.

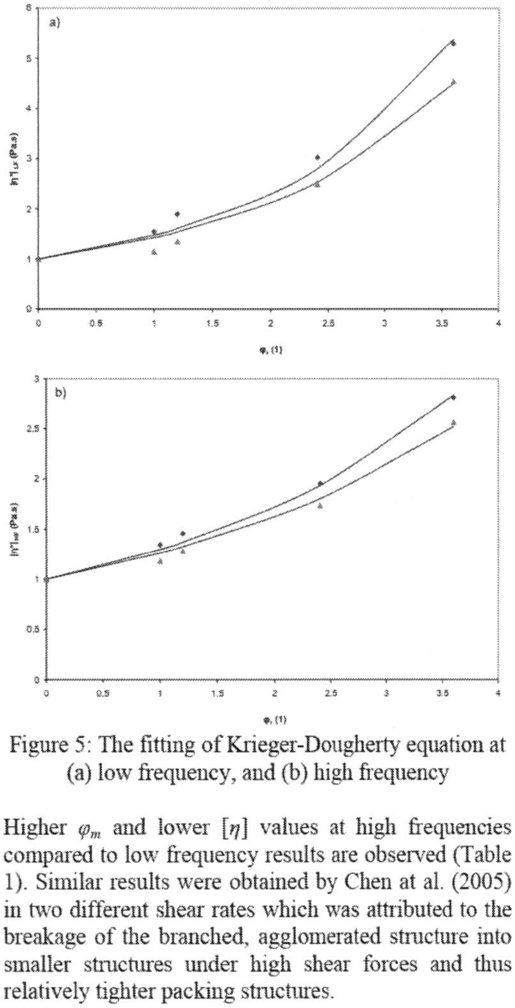

Figure 5: The fitting of Krieger-Dougherty equation at (a) low frequency, and (b) high frequency

Higher φ_m and lower $[\eta]$ values at high frequencies compared to low frequency results are observed (Table 1). Similar results were obtained by Chen at al. (2005) in two different shear rates which was attributed to the breakage of the branched, agglomerated structure into smaller structures under high shear forces and thus relatively tighter packing structures.

Table 1: The $[\eta]$ and φ_m obtained from Krieger-Dougherty equation

	$[\eta]_{LF}^{a}$	$\varphi_{m,LF}$	$[\eta]_{HF}^{b}$	$\varphi_{m,HF}$
Hydrophilic silica	37.1	0.094	25.1	0.140
Hydrophobic silica	34.0	0.103	22.6	0.155

a- LF index is corresponding to low frequency data
b- HF index is corresponding to data taken at high frequencies

While there is hardly breakage of aggregates in the small amplitude oscillatory shear flow compared to the shear forces in rotational flow, the higher φ_m and lower $[\eta]$ values at high frequencies can be attributed to the dominated hydrodynamic contributions of polymeric matrix and less contribution of three dimensional network between aggregates.

Blend nanocomposites

The PP/LCP blend with 85/15 weight fraction was selected to study the effect of nano-silica on its rheology and morphology. Three different weight fraction of silica (1, 2 and 3wt.%) was added during melt blending to this system. If we assume that silica selectively locates in the PP matrix (to be discussed afterwards), these weight fractions correspond to 0.5, 1 and 1.5 volume fractions of silica in PP phase.
The thermodynamic preference of filler to distribute selectively can be predicted by introducing a wetting coefficient, *Wa* (Sumita et a., 1991):

$$Wa = \frac{\gamma_{\text{filler-B}} - \gamma_{\text{filler-A}}}{\gamma_{\text{A-B}}} \qquad (4)$$

where $\gamma_{\text{filler-A}}$ and $\gamma_{\text{filler-B}}$ are the interfacial tensions between the filler and polymer A or B, respectively, and $\gamma_{\text{A-B}}$ is the interfacial tension between polymer A and B. If $Wa>1$, the filler is located within A-phase, if $-1<Wa<1$, the filler concentrates at the interface, and if $Wa<-1$, the filler is selectively distributed in the B-phase.
In the present study, the interfacial tension between PP and LCP was calculated according to harmonic mean equation (Wu, 1982):

$$\gamma_{\text{A-B}} = \gamma_{\text{A}} + \gamma_{\text{B}} - \frac{4\gamma_{\text{A}}^{d}\gamma_{\text{B}}^{d}}{\gamma_{\text{A}}^{d} + \gamma_{\text{B}}^{d}} - \frac{4\gamma_{\text{A}}^{p}\gamma_{\text{B}}^{p}}{\gamma_{\text{A}}^{p} + \gamma_{\text{B}}^{p}} \qquad (5)$$

where γ_A and γ_B are the surface energy of two polymers A and B, and the superscripts "*d*" and "*p*" denote the dispersion and polar components of the surface energy, respectively. By using this equation, the interfacial tension of PP and LCP phases is calculated 16.2mN/m.
In order to calculate the interfacial tensions between filler and polymers, the geometric mean equation of Wu (1982) which is suitable for systems composed of a high energy material such as nano-silica and a low-energy material such as a polymer can be used:

$$\gamma_{\text{A-B}} = \gamma_{\text{A}} + \gamma_{\text{B}} - 2\sqrt{\gamma_{\text{A}}^{d}\gamma_{\text{B}}^{d}} - 2\sqrt{\gamma_{\text{A}}^{p}\gamma_{\text{B}}^{p}} \qquad (6)$$

The values of the surface energy taken from literature are listed in Table 2.

Table 2: Surface tension and its components for polymers and two types of nano-silica

	γ	γ^d	γ^p	Ref.
PP	30.1[a]	30.1	0	Elias et al., 2007
LCP	43.8	27.7	16.1	Kim & Denn, 1992
Silica	80	29.4	50.6	Elias et al., 2007
m-Silica	81.7	72.3	9.5	Ou & Yu, 1994[b]

a- All values are presented in mN/m.
b- Values of nano-SiO2 with almost the same size and surface properties.

For the PP/LCP blend systems, *Wa* was found to be -2.53 for the hydrophilic silica particles and -0.45 for the hydrophobic one. Thus at thermodynamic equilibrium, the hydrophilic silica particles should be preferentially located in the LCP phase, while the hydrophobic silica should be concentrated at the interface.

The TEM micrographs of silica filled PP/LCP blend sample are shown in Fig. 6. While hydrophobic silica is located at the interface and in the matrix, hydrophilic silica is distributed in both phases. These results are in agreement with the experimental results reported by Zhang et al. (2003). By considering this result and those predicted by thermodynamic approach presented earlier, one may notice that there is some inconsistency between the experimental and theoretical results. This discrepancy can be explained in terms of the concentration effect, significance of diffusion of silica particles (Elias et al., 2008b) and/or temperature effects on interfacial properties.

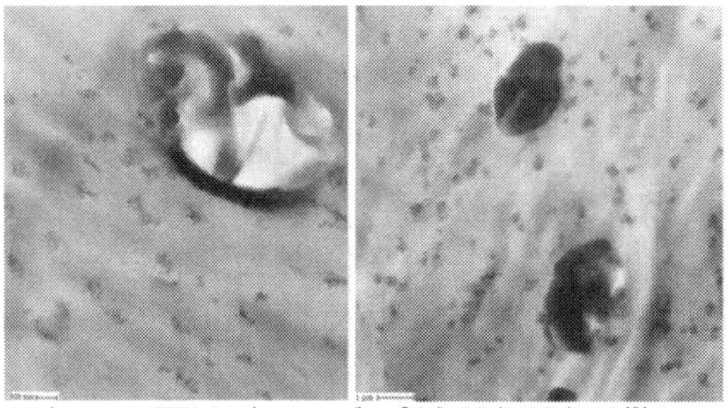

Figure 6: TEM micrograph of (a) PP/LCP/m-Silica, and (b) PP/LCP/Silica

The time required for a particle with radius r (in this case an aggregate) to diffuse a distance of its radius can be expressed as (Elias et al., 2008b):

$$t_D = \frac{r^2}{D_s} \quad (7)$$

where $D_s(\varphi)$ is the short time self-diffusion coefficient which reduces to the following equation for a dilute suspension of spherical droplets:

$$D_0 = \frac{k_B T}{6\pi\eta r} \quad (8)$$

In this equation, k_B is the Boltzmann constant, and η is the viscosity of media at temperature T and shear rate imposed on the sample. The viscosity of the melt blend was calculated at operational conditions of internal mixer by Bousmina et al. (1999) method (in which the shear rate was found to be about $52s^{-1}$). Thus, aggregates need 17~145min to diffuse a distance equal to their size. This evidences the inconsistency between the thermodynamic prediction and morphological observations.

On the other hand, if a spherical nano-particle acts as a compatibilizer, the effective concentration ratio of nano-silica to the LCP dispersed phase ($\varphi_{Silica}/\varphi_{LCP}$) can be estimated by assuming full-coverage of one droplet surface by silica particles. Through multiplying the volumetric ratio of one silica particle to one LCP droplet by number of particles required to cover the surface of one droplet, one can obtain the $\varphi_{Silica}/\varphi_{LCP}$ ratio as follows:

$$\frac{\varphi_{Silica}}{\varphi_{LCP}} = \frac{\frac{\text{surface of one LCP droplet}}{\text{coverage of one silica particle}} \times \frac{4}{3}\pi r^3}{\frac{4}{3}\pi R^3} = \frac{\frac{2\pi}{\sqrt{3}}(\frac{R}{r}+1)^2 \times r^3}{R^3} \approx \frac{2\pi}{\sqrt{3}} \cdot \frac{r}{R} \quad (9)$$

where φ, r and R are volume fraction, the radius of silica particle and LCP dispersed droplet size, respectively. This equation supports the idea stating that hydrophobic silica can act as a compatibilizer, if its concentration is about 0.1% that is much lower than concentrations used in PP/LCP hybrid blends. This is evidenced by TEM micrograph which shows a fraction of hydrophobic silica particles tends to locate in the PP phase after saturation of interface.

The comparison of aggregate sizes between the PP/LCP blends and PP nanocomposites does not show a considerable difference which may be due to the insignificant effect of LCP dispersed phase on aggregate size, calculation errors and/or the local characteristic of TEM images.

Typical SEM micrographs of PP/LCP hybrid blends are shown in Fig. 6. As it is clearly seen, the LCP minor phase in the form of nearly spherical droplets is distributed in PP matrix. For a quantitative analysis, the droplet size was determined by using image analysis. Typically, 300 particles were analyzed per sample and the number average radius $\overline{R}_N$ and volume average radius $\overline{R}_V$ of dispersed phase were calculated by assuming the dispersed phase as spherical particles by following equations:

$$\overline{R}_N = \frac{\sum n_i R_i}{\sum n_i} \quad (10)$$

$$\overline{R}_V = \frac{\sum n_i R_i^4}{\sum n_i R_i^3} \quad (11)$$

It should be noted that in the emulsion theories, the domain size is usually expressed as volume average radius. The polydispersity index, PDI, of droplet size distribution can be estimated in terms of $\overline{R}_V / \overline{R}_N$ ratio. For a monodisperse system, this ratio tends to unity and the values greater than one show the broader droplet size distributions.

Figure 6: The SEM micrographs of PP/LCP/silica hybrid samples with 3wt.%: (left) hydrophilic, (right) hydrophobic silica

The values of droplet size and polydispersity of LCP dispersed phase in hybrid samples are shown in Figure 7. These results show that the addition of nano-silica results in reduction of the droplet size during melt-blending, though their effect on the polydispersity is rather complicated to draw a clear conclusion.

Incorporation of nano-filler in a polymer blend sample can reduce the droplet size by three different mechanisms: (a) reducing the viscosity ratio by thickening the matrix (Zhang et al., 2003; Chen et al., 2007), (b) suppression the coalescence during mixing (Vermant et al., 2004; Zhang et al., 2005), and (c) compatibilization by concentrating at the interface (Elias et al., 2007; Zhang et al, 2005; Pisharath et al., 2006; Fenouillot et al., 2009). Note that the mechanisms (b) and (c) are somehow interrelated, because the presence of compatibilizer at the interface also results in suppression of coalescence (Harrats et al., 2006). Nano-silica is especially known as a thickening agent in the emulsion and suspension systems. From the above results, one realizes that while the hydrophilic silica has higher thickening capability, higher degree of droplet size reduction is observed for hydrophobic one. This suggests that other phenomena control the final droplet size of blend samples. In the case of PP/LCP blend, a thickening effect of hydrophobic nano-silica has also been reported (Zhang et al., 2003). However in our work, the existence of m-Silica at the interface of prepared PP/LCP blend, and weaker thickening effect and smaller resultant droplet size of m-Silica filled blend samples in comparison to hydrophilic silica filled ones suggest a compatibilization effect of m-Silica for this system.

From the three mechanisms mentioned above, both hydrophilic and hydrophobic silica can induce coalescence hindrance and thickening effect. The coalescence can be hindered by the presence of aggregates in the matrix or particles at the interface. The bigger aggregates of hydrophilic silica imply higher extent of coalescence hindrance in comparison to hydrophobic one. The results discussed earlier showed that hydrophilic silica has higher extent of thickening effect on the viscosity of polypropylene nanocomposites. On the other hand, small amount of hydrophilic silica in LCP droplets was observed which can affect the viscosity ratio. It can be suggested that the viscosity ratio has insignificant effect on the final droplet size, since for our blend samples this ratio is already much higher than the critical ratio of $\eta_d/\eta_m=3.5$ (see Figure 8) for droplet break-up. It seems that the presence of elongational flow field in the internal mixer (Manas-Zloczower and Tadmor, 1994) controls the droplet break-up process. Therefore, considering only coalescence hindrance and thickening mechanisms suggest smaller droplet size for hydrophilic silica containing blend samples. However, the morphological studies showed the opposite trend. Thus, the compatibilizing effect of hydrophobic silica can be the mechanism which controls the final morphology of PP/LCP/m-Silica hybrid samples.

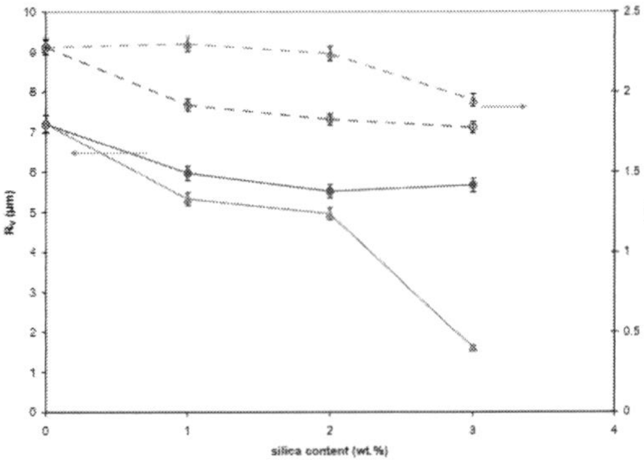

Figure 7: Droplet size and its distribution for PP/LCP blends with different: (▲, △) hydrophobic, and (•, ○) hydrophilic silica contents

If an additive acts as a thickening agent, it is expected to enhance the shear stress transfer particularly in low shear rate region – where the greater droplet size exists – and therefore it will lead to a narrower droplet size distribution. This reveals that the hydrophobic silica has not appreciable thickening effect on PP/LCP blend (at least in 0~2wt.% silica content range), because it does not affect the PDI in this range. This result supports the compatibilization effect of hydrophobic nano-silica which decreases the mean droplet size more than hydrophilic one. The reduced LCP droplet size in hydrophilic silica containing blend samples can be attributed to the thickening effect of this silica and hindrance effect of aggregates on coalescence process.

The abrupt reduction in LCP droplet size after 2wt.% for hydrophobic silica filled blend is matching the range of percolation volume fraction in the PP matrix. Thus, while small amount of silica is needed to cover one droplet, its volume fraction should be above the percolation threshold to overcome slow diffusion to the interface during mixing. In other words, the small diffusion coefficient may be conquered by increasing the collision probability above the percolation threshold. While the concentration of hydrophilic silica in 3wt.% containing PP/LCP is also above the percolation threshold, we don't see any further refinement in this blend sample. This supports the idea that the hydrophilic silica is not a potential compatibilizing agent for PP/LCP system. The decrease of PDI by increase in hydrophilic silica content shows that this type of silica works mostly as a thickening agent for PP/LCP blend.

Figure 8 shows the storage modulus, G′, and complex viscosity, η*, as a function of angular frequency for PP, LCP and PP/LCP blend samples. It is seen that LCP exhibits much greater viscosity and melt elasticity compared to PP matrix at whole frequency range. The storage modulus plateau of LCP at low frequencies is attributed to the solid body behavior resulted from the highly ordered nematic state close to the melting point.

The G' and η* of PP/LCP blend sample are compared with simple mixing rule and Palierne's model in Figure 8. The results show a negative deviation from simple mixing rule due to weak interfacial adhesion between PP and LCP phases. The Palierne's advanced model (Palierne, 1991) was found to be unable to predict the rheological behavior of LCP containing blend, something that has been reported in the literature (Riise et al., 1999). The pronounced non-terminal behavior of PP/LCP blend sample indicating longer relaxation time of dispersed droplets compared to Palierne's model (1991) prediction, could originate from internal structure of LCP droplets.

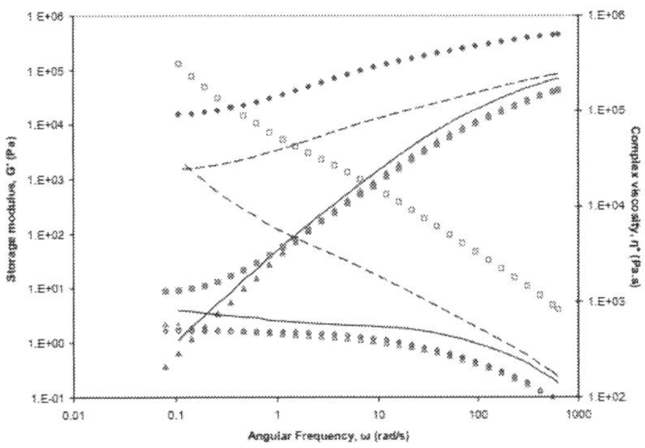

Figure 8: The rheological behavior of PP (▲, △), LCP (♦, ◊), PP/LCP blend (■, □), and the prediction of mixing rule (···) and Palierne's model (–). Filled symbol shows the storage modulus and open one shows the complex viscosity.

The linear viscoelastic properties of silica containing PP/LCP blend samples are shown in Figure 9. As seen in Figure 9, the extent of increase in G' and η* with increasing silica content is higher for hydrophobic silica compared to hydrophilic one. This supports the compatibilization of hydrophobic silica which has lower thickening ability and smaller aggregates to hinder the coalescence process. As it was shown in Figure 8, the Palierne's model could not be used to predict the rheological behavior of simple PP/LCP blend sample when the interfacial tension is known and vice versa. Similar behavior was found for silica containing blend samples. Vermant et al. (2004) suggested that for estimating the ratio of droplet size to interfacial tension for nano-silica containing blend, the interface contribution can be obtained by subtracting the matrix and dispersed phase contribution; and then the crossover frequency between G'_{INT} and G''_{INT} can be used to estimate this ratio. However, in the present work the observed negative deviation from mixing rule shows that both G'_{INT} and G''_{INT} are negative in the whole frequency range for all blend samples. Therefore, neither the Palierne's model is able to predict the rheological behavior of hybrid samples, nor the Vermant et al.'s method can be applied for estimation of interfacial tension between PP and LCP in the presence of nano-particles. It means that an advanced model or new experimental methods may be developed to describe this behavior.

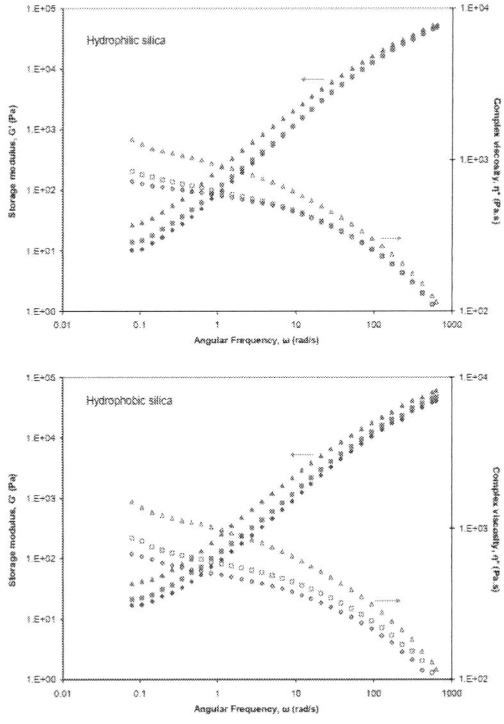

Figure 9: The rheological behavior of silica filled PP/LCP blend

Results in Figure 9 can be presented in terms of storage modulus variation of hybrid blend samples versus corresponding G' values of silica containing nanocomposites at low and high frequencies (Figure 10). It is seen that the presence of LCP phase in hydrophobic silica containing blend samples results in higher extent of elasticity at both low and high frequencies. The higher order of elasticity at high frequency, where the molecular scale contributions are significant, shows the enhanced stress transfer from matrix to dispersed phase at molecular scale. At low frequency, the contributions of interfacial energy of dispersed droplets (micro-scale contributions) are significant and thus the higher order of elasticity for hydrophobic silica containing blend samples suggest higher order of droplet deformation due to enhanced interfacial adhesion. Therefore, higher degree of compatibility between PP and LCP are achieved in the presence of hydrophobic silica compared to hydrophilic one.

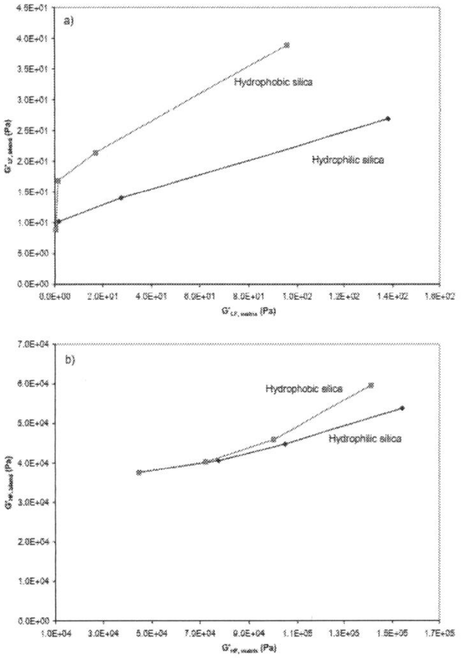

Figure 10: Variation of storage modulus of PP/LCP/silica samples versus storage modulus of PP/silica nanocomposites at (a) low frequency, and (b) high frequency

Conclusions

The rheology and morphology of nano-silica containing PP and PP/LCP blend were studied. The rheological results were used to calculate the percolation threshold, fractal index, and the intrinsic viscosity of PP/silica nanocomposites. It was found that hydrophobic silica has higher tendency for aggregation and forms bigger and denser aggregates. The lower percolation threshold and the higher fractal dimension for hydrophilic silica are due to its stronger particle-particle interaction rather than polymer-particle interaction compared to hydrophobic one. While the hydrophobic silica has lower thickening capability and lower coalescence hindrance caused by aggregates, it locates at the interface of PP/LCP blend and results in smaller droplet size and higher elasticity in comparison to hydrophilic silica. It was found that if the m-Silica content in PP/LCP is higher than its percolation threshold, a significant reduction in LCP droplet size is observed. The results suggest that the hydrophobic silica has a compatibilization capability for PP/LCP blend, while the hydrophilic silica mostly works as a thickening agent and suppresses the coalescence.

References

Aranguren M.I., Mora E., DeGroot J.V., Macosko C.W.: Effect of reinforcing fillers on the rheology of polymer melts, J. Rheol. 36 (1992) 1165-1182.

Aranguren M.I., Mora E., Macosko C.W., Compounding Fumed Silicas into Polydimethylsiloxane: Bound Rubber and Final Aggregate Size, J. Colloid Interface Sci. 195: 329-337 (1997)

Bousmina M., Ait-Kadi A., Faisant J.B.: Determination of shear rate and viscosity from batch mixer data, J. Rheol. 43 (1999) 415-433.

Brenner H., Int. J. Multiphase Flow, 1, 195 (1974)

Cassagnau Ph., Melt rheology of organoclay and fumed silica nanocomposites, Polymer 49: 2183-2196 (2008)

Chen S., Øye G., Sj□blom J.: Rheological Properties of Silica Particle Suspensions in Mineral Oil. J. Dispersion Sci. Tech. 26, 791-798 (2005)

Chen J., Chen P., Wu L., Zhang J., He J., Fibrillation of liquid crystalline polymer in polysulfone promoted by increased system elasticity via adding nano-silica, Polymer 48: 4242-4251 (2007)

Datta D., Baird D.G.: Compatibilization of thermoplastic composites based on blends of polypropylene with two liquid crystalline polymers, Polymer 36 (1995) 505-514.

Elias L., Fenouillot F., Majeste J.C., Cassagnau Ph.: Morphology and rheology of immiscible polymer blends filled with silica nanoparticles, Polymer 48 (2007) 6029-6040.

Elias L., Fenouillot F., Majesté J.C., Alcouffe P., Cassagnau Ph., Immiscible polymer blends stabilized with nano-silica particles: Rheology and effective interfacial tension, Polymer 49: 4378-4385 (2008a).

Elias L., Fenouillot F., Majeste J.C., Martin G., Cassagnau Ph.: Migration of nanosilica particles in polymer blends, J. Polym. Sci.: Part B: Polym. Phys. 46 (2008b) 1976-1983.

Fenouillot F., Cassagnau Ph., Majeste J.-C., Uneven distribution of nanoparticles in immiscible fluids: Morphology development in polymer blends, Polymer 50: 1333-1350 (2009).

Harrats C., Thomas S., Groeninckx G. (eds.), Micro- and Nanostructured Multiphase Polymer Blend Systems: Phase Morphology and Interfaces, CRC Press, Taylor & Francis Group, Boca Raton, 2006.

Israelachvili, J. N., Intermolecular and Surface Forces. Academic Press, London (1992)

Khan S.A., Zoeller N.J.: Dynamic rheological behavior of flocculated fumed silica suspensions. J Rheol 37:1225-1235 (1993)

Kim W.N., Denn M.M.: Properties of blends of a thermotropic liquid crystalline polymer with a flexible polymer (Vectra/PET), J. Rheol. 36 (1992) 1477-1498.
Lipatov Y.S.: Polymer reinforcement. Toronto, Canada: ChemTec Publishing; (1995).

Krieger, I.M. and Dougherty, T.J.: A mechanism for non-Newtonian flow in suspensions of rigid spheres. Trans. Soc. Rheol., 3: 137-152 (1959).

Lee M.W., Hu X., Yue C.Y., Li L., Tam K.C., Nakayama K.: Novel approach to fibrillation of LCP in an LCP/PP blend, J. Appl. Polym. Sci. 86 (2002) 2070-2078.

Manas-Zloczower I., Tadmor Z., eds., Mixing and Compounding of Polymers- Theory and Practice, Hanser, New York (1994)

O'Donnell H.J., Baird D.G.: In situ reinforcement of polypropylene with liquid-crystalline polymers : effect of maleic anhydride-grafter polypropylene, Polymer 36 (1995) 3113-3126.

Ou Y.C., Yu Z.Z.: Effects of alkylation of silica filler on rubber reinforcement, Rubber Chem. Technol. 67 (1994) 834–844.

Palierne J.F., Linear rheology of viscoelastic emulsions with interfacial tension. Rheol Acta 29: 204-214 (1991)

Paquien J.-N., Galy J., Gérard J.-F., Pouchelon A.: Rheological studies of fumed silica–polydimethylsiloxane suspensions, Colloids and Surfaces A: Physicochem. Eng. Aspects 260 165-172 (2005)

Pisharath S., Hu X., Wong S.C.: Rheology–morphology relationships in nylon–LCP hybrid composites, Composites Science and Technology 66 (2006) 2971–2979.

Potanin A.A.: On the mechanism of aggregation in shear flow of suspensions. J Colloid Interface Sci. 145:140-157(1991).

Ray S.S., Pouliot S., Bousmina M., Utracki L.A.: Role of organically modified layered silicate as an active interfacial modifier in immiscible polystyrene/polypropylene blends, Polymer 45 (2004) 8403–8413.

Riise B.L., Mikler N., Denn M.M.: Rheology of a liquid crystalline polymer dispersed in a flexible polymer matrix, J. Non-Newtonian Fluid Mech. 86 (1999) 3-14.

Rueb, C. J., C. F. Zukoski, "Viscoelastic properties of colloidal gels," J. Rheol. 41, 197-218 (1997).

Shih, W.H., Shih W.Y., Kim S.I., Aksay A.I., "Scaling behavior of the elastic properties of colloidal gels," Phys. Rev. A 42, 4772-4779 (1990).

Sternstein S.S., Ai-Jun Zhu, Reinforcement Mechanism of Nanofilled Polymer Melts As Elucidated by Nonlinear Viscoelastic Behavior, *Macromolecules* **2002,** *35,* 7262-7273

Sumita M., Sakata K., Asai S., Miyasaka K., Nakagawa H.: Dispersion of fillers and the electrical conductivity of polymer blends filled with carbon black, Polym. Bull. 25 (1991) 265-271.

Vermant J., Cioccolo G., Golapan Nair K., Moldenaers P., Coalescence suppression in model immiscible polymer blends by nano-sized colloidal particles, Rheol Acta 43: 529–538 (2004)

Vermant J., Ceccia S., Dolgovskij M.K., Maffettone P.L., Macosko C.W., Quantifying dispersion of layered nanocomposites via melt rheology, J. Rheol. 51, 429-450 (2007).

Wu S.: Polymer interface and adhesion, vol. 1, New York: Marcel Dekker (1982)

Yziquel F., Carreau P.J., Tanguy P.A.: Non-linear viscoelastic behavior of fumed silica suspensions, Rheol Acta 38: 14-25 (1999).

Zhang L., Tam K.C., Gan L.H., Yue C.Y., Lam Y.C., Hu X.: Effect of nano-silica filler on the rheological and morphological properties of polypropylene/liquid-crystalline polymer blends, J. Appl. Polym. Sci. 87 (2003) 1484-1492.

Zhang Q., Yang H., Fu Q.: Kinetics-controlled compatibilization of immiscible polypropylene/polystyrene blends using nano-SiO_2 particles, Polymer 45 (2004) 1913-1922.

Zhang B., Ding Y., Chen P., Liu C., Zhang J., He J., Hu G.-H., Fibrillation of thermotropic liquid crystalline polymer enhanced by nano-clay in nylon-6 matrix, Polymer 46: 5385-5395 (2005)

FLOW INDUCED CRYSTALLIZATION: THE EFFECT OF UNIAXIAL EXTENSION

Martin Sentmanat[1], Omar Delgadillo-Velázquez[2], and Savvas G. Hatzikiriakos[2*]

[1]*Xpansion Instruments, 1425 Glenoak Dr., Tallmadge, OH 44278, USA*
[2]*Department of Chemical and Biological Engineering, The University of British Columbia, Vancouver, Canada*

Abstract - In this paper the effect of uniaxial extension on the crystallization of a linear low-density polyethylene is examined by using rheometry coupled with differential scanning calorimetry (DSC). Uniaxial extension experiments were performed at temperatures below and above the peak melting point of the polyethylene in order to characterize its flow induced crystallization behavior at extensional rates relevant to processing. The degree of crystallinity of the stretched samples was quantified by DSC i.e. by analyzing the thermal behavior of samples after stretching. Analysis of the tensile strain hardening behavior very near the peak melt temperature revealed that crystallization depends on temperature, strain and strain rate. In addition, it was revealed that a very small window of temperatures spanning just 1-2 degrees Celsius can have a dramatic effect on polymer crystallization.

Introduction

Plastic products are manufactured by deforming the polymer at elevated temperatures into a desired shape. Examples include film blowing, film casting, fiber spinning and calendaring. Polymers used in these processes include most polyolefins i.e. linear-low density (LLDPE), low density (LDPE) and high-density (HDPE) polyethylenes. When transitioning from the molten state to the solid state during processing, these polymers undergo crystallization and as a result their final mechanical properties heavily depend on the conditions under which crystallization takes place (Mc Hugh, 1995; Eder et al., 1990; Kornfield et al., 2002; Janeschitz-Kriegl, 2003; Scelsi and Mackley 2008). In fact, the mechanical properties depend heavily on the thermal and deformation histories experienced by the polymer during the crystallization process. For instance in extruded fibers the crystals are oriented in the direction of extension, while in films formed by film blowing, the crystals are oriented in the plane of biaxial extension. In both these applications the orientation of the crystalline phase enhances the mechanical properties of the final product (Gelfer and Winter, 1999). There are examples where formation of crystalline phase might influence negatively the mechanical properties of final products, i.e. crystalline phase skin formation during injection molding.

Polymers such as polyethylene are always semi-crystalline in solid form (Dealy and Wissbrun, 1990). Their crystallization rates at temperatures below the melting point are very high, and it is not possible to cool the polymer rapidly enough to a temperature below the glass transition temperature without substantial crystallization taking place. Under quiescent conditions, the crystallization rate for polyethylenes is usually observed to be zero at the melting temperature and at the glass transition temperature with a maximum occurring at a temperature in between these two temperatures. However, under flow conditions at temperatures near the melt state the kinetics of crystallization might significantly change and it may become a function of deformation history depending on both strain and rate (Kitoko et al., 2003; Hidinata et al., 2006; Dai et al., 2006; Tanner and Qi, 2009; Tanner et al., 2009). The type of flow deformation can also play a significant role in affecting crystallization. For example, extensional flow (which is an inherently strong flow) causes molecules to orient and stretch in the direction of extension (as in the case of fiber spinning) facilitating the process of flow induced crystallization (Swartjes et al., 2003; Janeschitz-Kriegl, 2003; Kornfield et al., 2002: Stalbauer et al., 2004). Although shear deformations also influence the crystallization behavior of polyethylenes (Monasse, 1995; Vleeshouwers and Meijer , 1996; Alfonso and Scardigli, 1997; Koran and Dealy, 1999; Lagasse and Maxwell, 1976; Fortenly et al., 1995; Fernandez-Ballester et al., 2009), because these shear flow fields are not typically as strong as extensional flows, very large strains must be generated in order to observe shear flow induced crystallization.

The main focus of this paper is to study first the rheological behavior of a metallocene linear low density polyethylene at temperatures around the point of crystallization/melting and secondly the effect of extensional flows on its crystallization. Film blowing and calendaring are two flows that involve significant degrees of extensional flow elements and therefore becomes important to see how extension influences structure formation (crystalline phase formation) in polyethylenes. In addition the obtained data is very useful to evaluate important flow induced crystallization models (Doufas et al., 2000a; 200b; Tanner et al., 2009; and Tanner and Qi, 2009).

Experimental

A film grade metallocene linear low-density polyethylene (m-LLDPE) Exact 3128 from ExxonMobil was studied in this work with a specific interest in its flow induced crystallization behavior. This polymer has an MFI of 1.2, a molecular weight of about 80 kg/mol, and a polydispersity of about 2. Because of the higher degree of crystallization that can be achieved in the solid state, linear polymers are particularly sensitive to extensional flows very near the melt temperature and for this reason a m-LLDPE resin was selected for this study (Sentmanat and Hatzikiriakos, 2004, Sentmanat et al., 2005a).

A Shimadzu DSC-60 calorimeter was used to study the thermal behavior of the m-LLDPE. Measurements were made on samples of about 1-2 mg sealed in aluminum pans and nitrogen flow. The samples were heated from 30°C to 180°C, at a heating rate of 10°C/min, in order to determine the melting temperature (T_m) and heat of fusion (ΔH_m). The calorimeter was calibrated periodically for melting temperature and heat flow using Indium and Zinc as standards.

The linear viscoelastic (LVE) rheological properties of the polymer were characterized in small amplitude oscillatory shear using an Anton Paar rotational rheometer (MCR 501) equipped with circular platens 25 mm in diameter. The polymer was also characterized in simple extension using a model SER2-P SER Universal Testing Platform (Sentmanat, 2003, Sentmanat, 2004) fixture hosted on the MCR 501. Uniaxial extension experiments were run at various temperatures, strain rates and strains to examine their effect on the crystallization behavior of the m-LLDPE. In addition, tensile stress relaxation measurements were also performed on the polymer by stopping the uniaxial extension at various levels of strains. As will be discussed later, this type of experiment shows clearly the effect that the extensional strain plays in the crystallization of polymers. Finally tensile experiments were run at temperatures from ambient up to melting point in order to assess the importance of deformation behavior at temperatures near the peak melting point temperature, T_m. As will be further discussed, it is within this range of temperatures near T_m that most of the critical crystallization phenomena occur due to the combined effects of molecular mobility and the affinity of the molecule to crystallize. For example at temperatures well below T_m, molecules lose their flexibility or ability to orient and additional crystallization is difficult to achieve, unless significant deformation is imposed.

Melt rheology

Figure 1 contains the rheological master curves for linear viscoelastic shear moduli and the complex viscosity of the m-LLDPE at a reference temperature of 170°C. Small amplitude oscillatory shear experiments were carried out over a wide range of temperatures from 130°C to 230°C at 20°C increments while the time-temperature superposition shift factors were calculated from MATLAB. The superposition of the data was excellent illustrating the thermorheological simplicity of this m-LLDPE, an observation that is consistent with most linear polymers.

Flow induced crystallization (FIC) in uniaxial extension

Figure 2 depicts the melting thermographs of the m-LLDPE obtained from Differential Scanning Calorimetry (DSC) and shows that the onset of melting occurs at about 87°C with the peak melt temperature at 92°C. Figure 3 plots the engineering stress, F/A_o (instantaneous tensile force F divided by the initial cross sectional area of the sample) as a function of the Hencky strain (also known as true strain), ε_H, for several temperatures shown in Fig. 2 and for a given Hencky strain rate of $\dot{\varepsilon}_H = 1\,s^{-1}$. The top 3 curves from 23°C to 87°C correspond to the solid physical state of the polymer depicting classic plastic behavior of yielding, necking, and strain hardening to tensile rupture. It should be noted that to control the thermal history of all of the samples at a temperature near the melt state, each test specimen was loaded on the SER fixture and pre-soaked within the CTD oven of the MCR501 for 5 minutes at a temperature of 102°C in order to eliminate any residual crystallinity prior to cooling and stretching of the test specimen. Note that a higher pre-soak temperature was not used in order to avoid excessive melt sagging of the test sample during the pre-soak period on the SER fixture. Over a range of temperatures from 89°C to 93°C (within the peak melt trough of the DSC thermograph) the material is molten yet experiences strain hardening behavior with increasing strain as a result of the onset of flow induced crystallization (FIC). However at test temperatures beyond the DSC peak melt trough region (95°C and above) there is no evidence of FIC and as a result the engineering tensile stress keeps decreasing monotonically with increasing Hencky strain in a manner consistent with classic molten flow behavior of linear polymers. This series of experiments clearly depicts the significant roles that temperature and extensional deformation play in the crystallization of polymers.

Tensile stress growth coefficient, η_E^+

Figure 4 plots the tensile stress growth coefficient, η_E^+ of m-LLDPE as a function of Hencky strain at several Hencky strain rates ranging from $0.01s^{-1}$ to $30s^{-1}$ and three temperatures, namely 91°C, 93°C and 102°C. First at 102°C, the polymer is in the fully molten state (well above the melting point) and the tensile stress growth coefficient follows the LVE envelope that from LVE theory equals three times the linear viscoelastic shear stress growth coefficient, $3\eta^+$, and is typically described by the base curve on which all of the low strain transient extensional data for all Hencky strain rates tend to collapse. It is noted that there is no sign of strain hardening (defined as a positive stress deviation from the LVE envelope) in a manner consistent with the classic flow behavior of molten linear polymers (Dealy and Wissbrun, 1990). At 93°C, although the flow for a given strain rate initially follows along the LVE envelope, beyond a critical strain significant deviations from LVE behavior are observed despite being just 1°C above the peak melt temperature. At all Hencky strain rates (except the smallest of $0.01s^{-1}$), strong strain hardening is observed, which provides clear evidence of FIC. At a temperature of 91°C, just 1°C below the peak melt temperature, the flow for all of the strain rates again exhibits significant deviation from the LVE envelope but at an even earlier onset of critical strain as a result of FIC. Furthermore, the significant shift in the LVE envelope from 93°C to 91°C displays the inherent effect of temperature on the quiescent crystallization state within the bulk polymer which apparently serves to assist or "seed" the observed FIC behavior. In other words, the underlying crystallinity present in the bulk polymer due to quiescent crystalliazation prior to sample stretch appears to serve as "seed" crystallization sites within the polymer matrix brought on by the increased orientation and molecular chain stretch during uniaxial extension thereby resulting in the observed dramatic flow induced crystallization behavior.

These strain hardening effects can be seen more clearly if one plots the data of Figure 4 in terms of a relative deviation from LVE behavior as the ratio of the instantaneous tensile stress to the tensile stress from the LVE envelope, in other words, $\sigma_E / \sigma_{E-LVE}$. The data of Figure 4 for all three temperatures of 102°C, 93°C and 91°C are plotted in Figures 5, 6, and 7, respectively, in terms of relative strain hardening ratio as $\sigma_E / \sigma_{E-LVE}$ vs. the Hencky strain, ε_H. Figure 5 shows the data at 102°C where the polymer is 10°C above its peak melt temperature. The ratio $\sigma_E / \sigma_{E-LVE}$ remains around a value of 1 at all Hencky strain rates prior to sample failure in a manner consistent with the classic behavior of a linear polymer melt, indicating that FIC is absent. As shown in Figure 6, at 93°C (that is 1°C above the peak melting point) the effect of FIC is significant. The ratio $\sigma_E / \sigma_{E-LVE}$ increases well above a value of 1 (the definition of strain hardening) for all Hencky strain rates beyond a critical Hencky strain of about 2. As depicted in Figure 7, at 91°C (only 1°C below the peak melting point) the effect of FIC is even more pronounced. The ratio $\sigma_E / \sigma_{E-LVE}$ increases sharply beyond a critical Hencky strain just above 1. In general, the critical Hencky strain at the onset of strain hardening behavior appears to somewhat decrease with increasing Hencky strain rate. Furthermore, the discernable trends in the data from Figures 4, 6 and 7 appear to suggest that the effects of quiescent crystallization and FIC on the bulk polymer rheology are somewhat separable, which implies that for a given temperature condition from a potential constitutive modeling perspective the effects of quiescent crystallization can be modeled primarily as a function of time affecting the linear viscoelastic bulk rheological properties while the effects of FIC can be can be modeled primarily as a function of strain and rate thereby affecting the non-linear viscoelastic bulk rheological properties of a given polymer system.

Tensile stress relaxation after imposition of steady uniaxial extension

Figure 8 contains a plot of the engineering stress, F/A_o (instantaneous tensile force F divided by the initial cross sectional area of the sample) as a function of the Hencky strain, ε_H at a temperature of 91°C and Hencky strain rate of $\dot{\varepsilon}_H = 1\,s^{-1}$. As discussed before and witnessed by the data in Figure 8, at the early stages of stretching the polymer appears to remain in a molten state displaying little deviation from the LVE envelope, however at strains greater than 1, FIC crystallization appears evident from the inflection and sudden increase of the engineering tensile stress curve. In order to characterize the stress relaxation behavior of the polymer after the onset of FIC, cessation of extension experiments (in which the stretching deformation is suddenly stopped at a pre-determined strain) were performed on samples after a pre-determined strain imposition corresponding to various points along the curve in Figure 8, i.e. after 1 s when the polymer exhibits near-LVE behavior, after 1.5, 2, 2.5 and 3 s during the FIC process, and finally at a time 3.3 s immediately prior to the onset of sample rupture.

The results for these relaxation experiments are plotted in Figure 9 which depicts the instantaneous stress as a function of time. As described earlier, each test specimen was loaded on the SER fixture and pre-soaked within the CTD oven of the MCR501 for a period of 5 minutes at a temperature of 102°C in order to eliminate any residual crystallinity prior to cooling and stretching of the test specimen at the test temperature of 91°C. When the stretching is stopped at the early stages of the experiment (after 1s) the polymer is still in a molten state and stress relaxation occurs normally as the shown by exponential decay in the tensile stress. However, this is not the case upon the onset of FIC at cessation times greater than 1s. As seen from Figure 9, the tensile stress shows no such exponential decay but rather tends to approach an equilibrium stress as in the case of a rubber-like material. At higher cessation strains corresponding to the second characteristic peak in Figure 8, the sample appears to approach a limiting elastic behavior prior to the onset of sample rupture. Finally, at times beyond the second maximum (roughly for times greater than 3s), the relaxation becomes abnormal due to the onset of an extensional instability or necking failure that has been previously described in the literature with respect to a Considère criterion (McKinley and Hassager, 1999; Sentmanat et al., 2005b).

Conclusions

The extensional behavior of a m-LLDPE was characterized over a broad range of temperatures (well above, well below and around its peak melt temperature) and Hencky strain rates. It was shown that flow induced crystallization (FIC) occurs over a range of temperatures near the DSC peak melt temperature. Analysis of the tensile strain hardening behavior very near the peak melt temperature revealed a temperature & strain dependence on the FIC behavior. The tensile stress growth coefficient revealed that the critical FIC behavior of the m-LLDPE occurs in a significant way over a span of just 1°C above and below the peak melt temperature. Cessation of extension experiments very near the peak melt temperature revealed the retardation effect that FIC has on the stress relaxation behavior of the crystallizing melt. These results demonstrate the significant role that extensional deformation plays in the occurrence and degree of crystallization of polymer melts relevant to the polymer processing industry.

Acknowledgements

Financial assistance from the Natural Sciences and Engineering Research Council (NSERC) of Canada is gratefully acknowledged.

References

Alfonso GC, Scardigli P, (1997) Melt Memory Effects in Polymer Crystallization. Macrom Symp 118:323-328.

Dai S-C, Qi F, Tanner RI, (2006) Strain and Strain-rate formulation for flow-induced crystallization. Polym Eng Sci, 46:659-669.

Dealy JM, and Wissbrun KF, Melt Rheology and its Role in Plastics Processing – Theory and Applications, Van Nostrand Reinhold (1990).

Doufas AK, McHugh AJ, and Miller C, (2000a) Simulation of melt spinning including flow-induced crystallization Part I. Model development and predictions. J Non-Newtonian Fluid Mech 92:27–66.

Doufas AK, McHugh AJ, Miller C, and Immaneni A, (2000b) Simulation of melt spinning including flow-induced crystallization Part II. Quantitative comparisons with industrial spinline data. J. Non-Newtonian Fluid Mech. 92:81–103.

Eder G, Janeschitz-Kriegl H, S Liedauer S (1990) Crystallization processes in quiescent and moving polymer melts under heat transfer conditions. Prog Polym Sci 15:629-714.

Fernandez-Ballester L, Thurman DW, Kornfield JA, (2009) Real-time depth sectioning: Isolating the effect of stress on structure development in pressure-driven flow. J Rheol 53:1229-1254.

Gelfer MY and HH Winter (1999) Effect of Branch Distribution on Rheology of LLDPE during Early Stages of Crystallizattion. Macromolecules 32:8974-8981.

Hadinata C, Gabriel C, Ruellmann M, Kao N, Laun HM (2006) Shear-induced crystallization of PB-1 up to processing-releant shear rates. Rheol Acta 45:539-546.

Janeschitz-Kriegl H (2003) How to understand nucleation in crystallizing polymer melts under real processing conditions. Colloid Polym Sci 281:1157-1171.

Kitoko V, Keentok M, Tanner RI, (2003) Study of shear and elongational flow of solidifying polypropylene melt forlow flow deformation rates. Korea-Australia Rheol J 15:63-73.

Koran F, and JM Dealy (1999) A high Pressure Sliding Plate Rheometer for Polymer Melts. J Rheol 43:1279-1290.

Kornfield JA, Kumaraswamy G, Issaian AM (2002) Recent advances in understanding flow effects on polymer crystallization. Ind & Eng Chem Res 41:6383-6392.

Lagasse RR, and Maxwell B (1976) An experimental study of the Kinetics of Polymer Crystallization during Shear flow. Polym Eng Sci 16:189-199 (1976).

McHugh AJ, Flow induced crystallization in polymers, in J. Lyngaae-Jorgenson, K Sandergaard (Eds.) Rheophysics of Multiphase Polymeric Systems, Technomic Publishing Co., Lancaster, PA, pp.227-267, (1995).

McKinley GH, and Hassager O, (1999) The considère condition and rapid stretching of linear and branched polymer melts. J Rheol 43:1195-1212.

Monasse B, (1995) Nucleation and anisotropic crystalline growth of polyethylene under shear. J mater Sc, 30:5002-5012.

Sentmanat ML (2003) Dual windup extensional rheometer. US Patent 6,578,413.

Sentmanat ML (2004) Miniature universal testing platform: from extensional melt rheology to solid-state deformation behavior. Rheol Acta 43:657-669.

Sentmanat ML, SG Hatzikiriakos SG, (2004) Mechanism of gross melt fracture elimination in the extrusion of polyethylenes in the presence of boron nitride. Rheol Acta 43:624-633.

Sentmanat M, Muliawan EB and Hatzikiriakos SG, (2005a) Fingerprinting the Processing Behavior of Polyethylenes from Transient Extensional Flow and Peel Experiments in the Melt State. Rheol Acta 44:1-15.

Sentmanat M, Wang BN, McKinley GH, (2005b) Measuring the transient extensional rheology of polyethylene melts using the SER universal testing platform. J Rheol 49: 585-606

Scelsi L, Mackley MR (2008) Rheo-optic flow-induced crystallization of polypropylene and polyethylene within confined entry-exit flow geometries. Rheol Acta 47:895-908.

Stadlbauer M, Janeschitz-Kriegl H, Eder G, and Ratajski E, (2004) New extensional rheometer for creep flow at high tensile stress. Part II. Flow induced nucleation for the crystallization of iPP. J Rheol 48:631-639.

Swartjes FHM, Pters GWM, Rastogi S, Mejer HEH (2003) Stress induced crystallization in elongational flow. Intern Polym Proc 18:53-66.

Tanner RI, Hadimata C, and LeeWo D. (2009) Behaviour of a simple crystallization model in tube and channel flow. Rheol Acta 48:499-507.

Tanner RI and Qi F (2009) Stretching, shearing and solidification. Chem Eng Sci 64:4576-4579.

Vleeshouwers S, Meijer HEH, A rheological study of stress induced crystallization (1996). Rheol Acta 35:391-399.

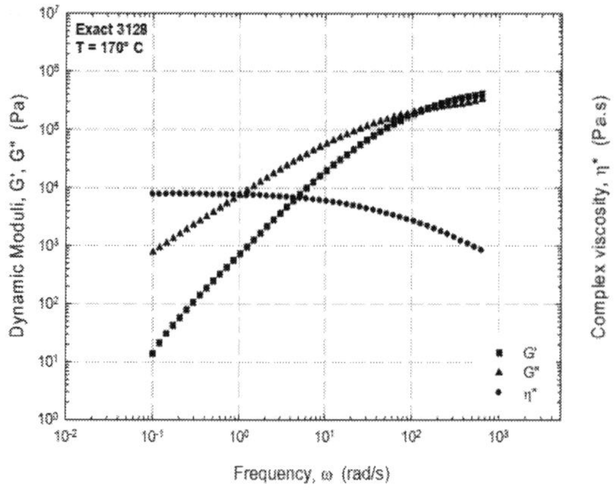

Figure 1: Linear viscoelastic moduli, G', G", and complex viscosity, $|\eta^*(\omega)|$ of m-LLDPE at 170°C.

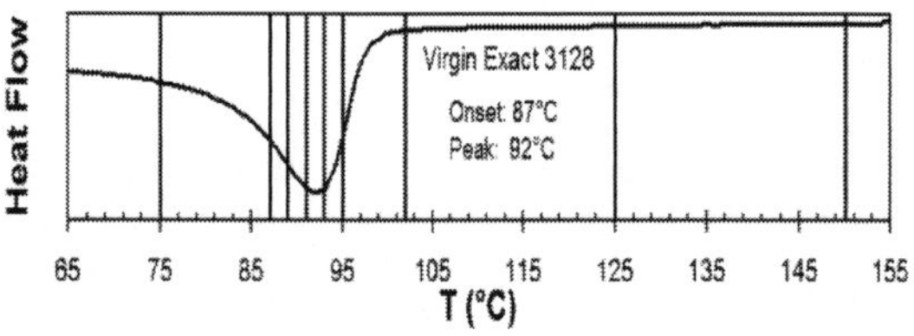

Figure 2: The thermograph (DSC) of the m-LLDPE showing the onset of melt occurs at 87°C with the melting peak at 92°C.

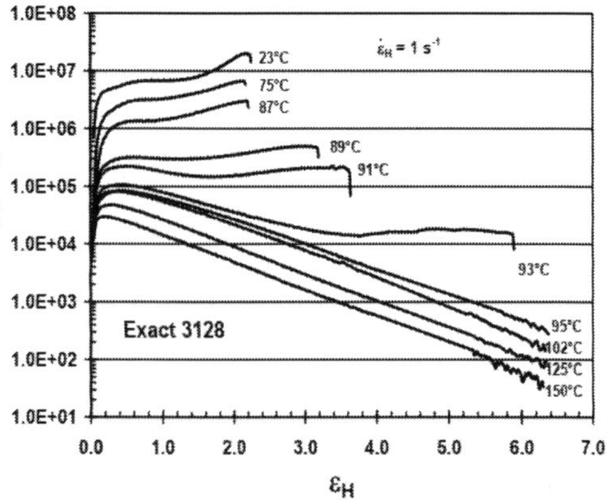

Figure 3: A series of steady Hencky strain rate uniaxial experiments for the m-LLDPE at various temperatures shown in Fig.2, well below, around and well above the peak melting point.

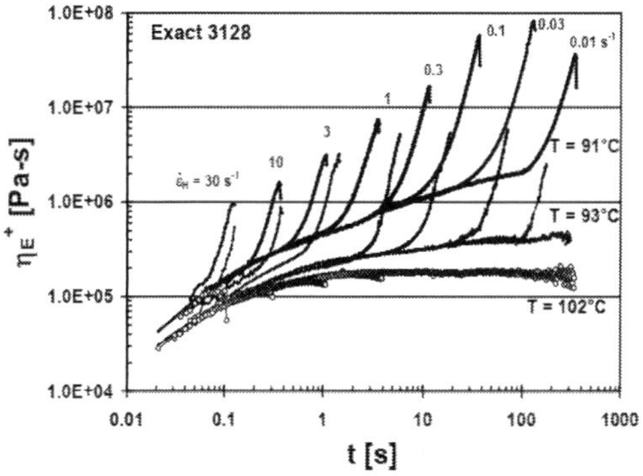

Figure 4: The tensile stress growth coefficient, η_E^+ of m-LLDPE at various temperatures (91°C, 93°C and 102°C) and Hencky strain rates ranging from 0.01 s^{-1} to 30 s^{-1}.

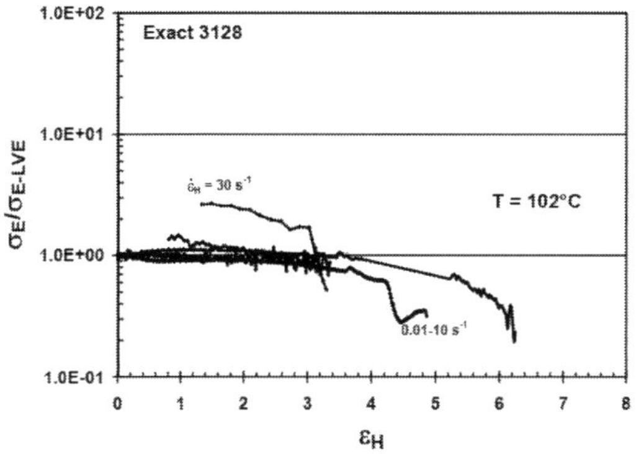

Figure 5: The tensile stress normalized by the tensile stress from LVE, that is $\sigma_E / \sigma_{E-LVE}$ as a function of the Hencky strain, ε_H at 102°C.

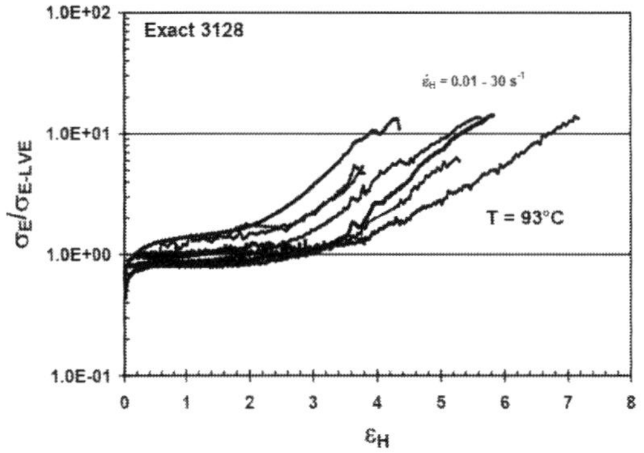

Figure 6: The tensile stress normalized by the tensile stress from LVE, that is $\sigma_E / \sigma_{E-LVE}$ as a function of the Hencky strain, ε_H at 93°C.

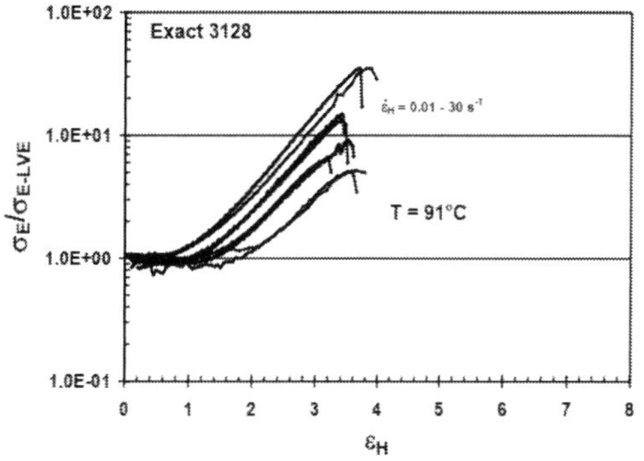

Figure 7: The tensile stress normalized by the tensile stress from LVE, that is $\sigma_E / \sigma_{E-LVE}$ as a function of the Hencky strain, ε_H at 91°C.

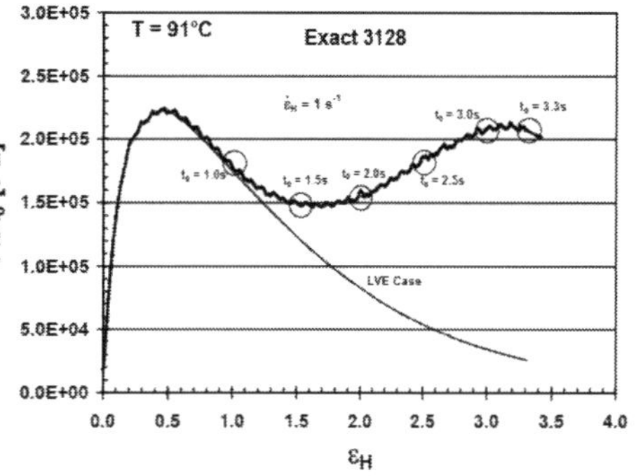

Figure 8: A constant Hencky strain rate ($\dot{\varepsilon}_H = 1 s^{-1}$) uniaxial experiment for the m-LLDPE at 91°C.

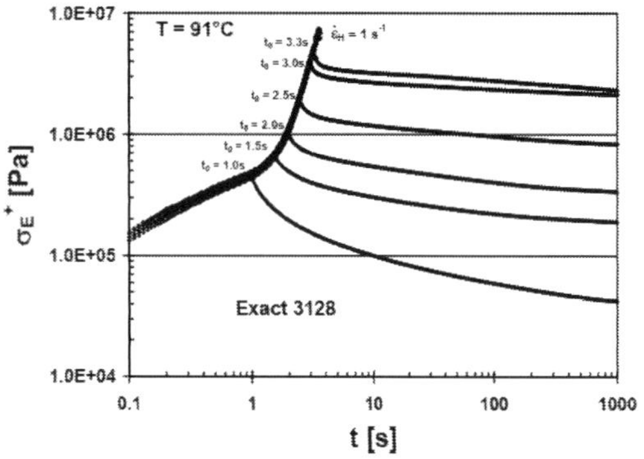

Figure 9: A series of tensile stress relaxation after imposition of a constant Hencky strain rate ($\dot{\varepsilon}_H = 1\ s^{-1}$) uniaxial experiments for the m-LLDPE at 91°C. Note that the tensile stress growth experiment was stopped at various times ranging from 1 to 3.5 s as indicated also in Figure 8.

EXPERIMENTAL STUDY OF THE INFLUENCE OF PRESHEARING ON THE YIELD STRESS OF CARBON BLACK FILLED RUBBER USING EXTRUSION RHEOMETRY

T. Hallmann[1]*, V. Schoeppner[1]

[1]*University of Paderborn, Kunststofftechnik Paderborn KTP, Paderborn, Germany - tobias.hallmann@ktp.upb.de*

Abstract - Filled rubber compounds exhibit specific flow phenomena like the existence of a yield stress, wall slippage and viscoelastic flow properties. In conventional extruders with deep flighted screws, the shear stress is lower than the yield stress in some areas of the screw. These areas remain unmixed and cold for a certain screw length as the rubber inside these areas heats up only through heat conduction instead of dissipation. Dependent on material properties like type of polymer, filler type, filler level and process parameters like temperature or preshearing history of the rubber, the existence and amount of yield stress varies. This experimental study investigates the influence of preshearing using extrusion rheometry.

By shearing filled rubber during processing, the filler network, whose origin is assigned to interactions between filler and elastomer, degrades and rheological properties like viscosity and yield stress change. The measurement of yield stress using extrusion rheometry is equivalent to the measurement using capillary rheometry. Using capillary rheometry, a piston is used to build up pressure in the filled barrel so the polymer melt flows out of the barrel. While discharging the barrel, the piston can be locked in position and the pressure in the barrel decreases to a certain level. The differential pressure between two pressure sensors can be interpreted as a wall shear stress and as the yield stress after the differential pressure is converged against a certain value. Using an extrusion rheometer, locking the piston is equivalent to stopping the extruder drive. By varying the screw speed before stopping the drive, the degree of shear deformation can be adjusted. The measured yield stresses at different rotational speed levels show the influence of preshearing on the yield stress.

Introduction

Most products made of rubber are processed during the production process by screw extruders or injection molding machines. The trend in rubber goods production is more and more towards cold feed extrusion and to process more and more products with a high level of viscosity. There are several flow phenomena like the existence of a yield stress, wall slippage and viscoelastic flow properties that have to be taken into account at these kind of processing techniques.

In the screw channel's cross section of a cold feed extruder some areas with a low shear stress exist, in which the material is not sheared sufficiently and as a result, the material remains cold and unmixed for a major part of the screw length. Figure 1 shows such a cross section of a cold feed extruder that has been fed with black and white compounds making the stream paths visible. After the extraction of the specimen, completely black and white areas can be detected in the channel, where no mixing takes place contrary to gray i.e. mixed areas.

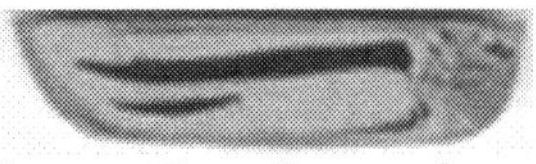

Figure 1 – Cross section of screw channel (Schramm, 2003)

The results of this effect are e.g. thermal inhomogeneities, which lead to different levels of die swell and different thermal expansion, consequently insufficient product quality.

The origin of this effect is assigned to the yield stress. The areas remain unmixed, in which the shear stress is below the yield stress. As the yield stress is a matter of the mass temperature, possible solutions to avoid this so-called "cold core" are preheating the compound with preheating mills and using warm feed extruders or cold feed pin type extruders.

A common mathematical model for yield stress fluids is the model stated by Bingham with the formulation

$$\tau = \tau_0 + K \cdot \dot{\gamma}^n$$

where K and n are power law parameters, τ is the shear stress, $\dot{\gamma}$ the shear rate and τ_0 the yield stress.

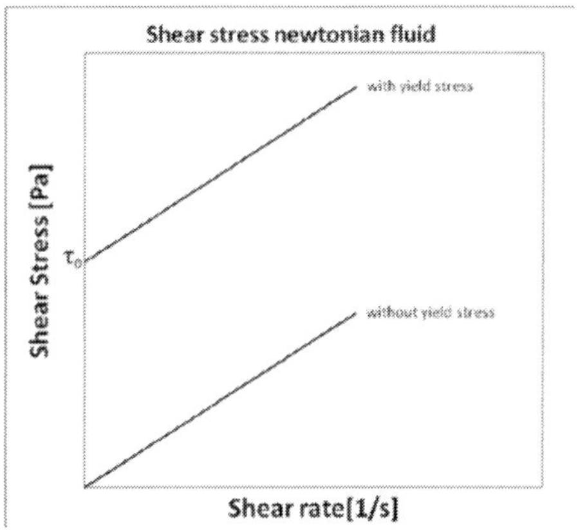

Figure 2 – Shear stress over shear rate for newtonian fluids with and without yield stress

Nevertheless the existence of a yield stress has been discussed lately in quite a number of papers by using several methods like sandwich rheometers or sliding cylinder rheometers (Barrès, 2000; Borgreve et. al, 1987, Muenstedt, 1981). These papers show that a universally valid statement about the existence of a yield stress for all rubber compounds cannot be given as different compounds show contradictory behaviors.

The yield stress in carbon black filled rubber compounds is attributed to a filler network developed between the single - usually nanoscopic - particles by van der Waals interactions as described in Ziegler (2004). As van der Waals interactions are a matter of particle size, a correlation of yield stress and particle size respectively carbon black type e.g. N110, N220 etc is assumed.
Further investigations have been done by Leblanc (2002).
It is well-known that after the rubber mixing process rheological properties evolve over a period of time to a final level as the filler network builds up and strengthens.
By shearing the compound the filler network be can temporarily and partially destroyed so that the remaining yield stress after shearing might be different at different levels of shearing.

The conclusion is that the shearing history needs to be considered for rheological characterizations of rubber compounds.
One method to measure the yield stress and vary the amount of shear stress is the use of an extrusion rheometer. The setup of the shear stress can easily be done by variation of rotational screw speeds.

Experimental
For the investigation of the influence of preshearing on the yield stress a NR/BR compound containing 75 phr NR, 25 phr BR and 55 phr carbon black (N110) was used and also further ingredients in small percentages. This type of carbon black has a high specific surface and a high effect on SHORE hardness (Häberlein, 2010).
For the measurement a "*Rubicon*" laboratory extruder with a screw length of 12 D and a screw diameter D = 32 mm was used. The rotational speeds of the extruder can be adjusted from 6 min^{-1} to 52 min^{-1}. Barrel and screw can be heated separately by two water-based temperature control units. A double flat channel die was attached to the extruder and also heated with water with a separate temperature control unit. The die is equipped with several pressure and temperature sensors. The sensors are connected to a data acquisition card in a personal computer and all pressure and temperature values are recorded twice per second. Pressure sensors were calibrated prior to each measurement.

Due to the position of the pressure sensors in the flat channel in a certain distance to the channel border, the entrance and the exit pressure drop do not need to be taken into account.

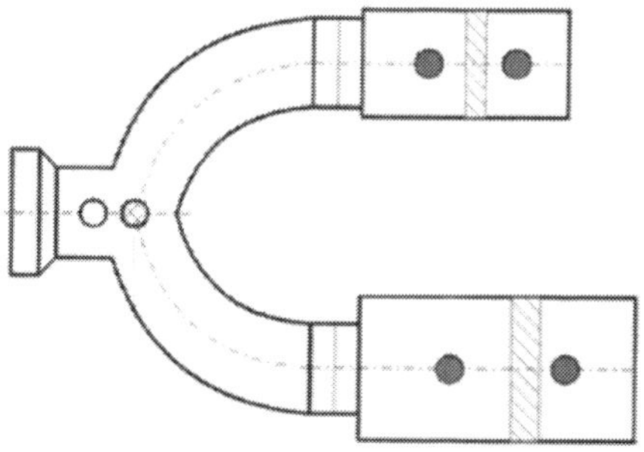

Figure 3 – Top view of flat channel die for yield stress measurement
Red dots = pressure transducers

During the extrusion of rubber through the die a pressure difference between the pressure transducers develops. The differential pressure between the two pressure transducers can easily be used to calculate the wall shear stress by

$$\tau_W = \frac{\Delta p}{L} \cdot \frac{H}{2}$$

where τ_W is the wall shear stress, Δp the differential pressure between the pressure transducers in the flat channel, L the length of the flat channel and H its height.

As Limper (1985) stated, one method to measure the yield stress using a capillary rheometer is to stop the piston and record the converging pressures. The wall shear stress at this point can be interpreted as the yield stress.
Similar to this procedure using the extrusion rheometer the extruder drive can be stopped and the pressure values before and after stopping the drive are recorded.

After stopping the extruder drive the absolute pressures and the differential pressure in the flat channel drop to a certain level so that the wall shear stress at this point can be interpreted as the yield stress.

In view of the fact that a higher rotational speed induces a higher mass temperature caused by the higher level of dissipation, the higher mass temperature implies a lower yield stress as showed in Limper (1985). As a result of this circumstance, the barrel and screw temperatures must be adjusted at the different rotational speed stages to maintain a constant mass temperature.
Figure 4 shows this pressure drop after shutdown.

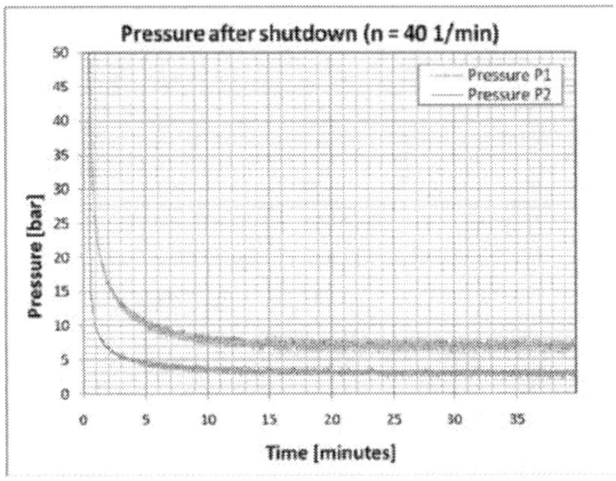

Figure 4 – Pressure over time after shutdown of extruder drive

As can be seen in Figure 4, the pressure drops to a certrain level after shutdown and does not fall any further after an operating point at a rotational speed of 40 min^{-1}. This becomes more clear in Figure 5 which shows a double logarithmic illustration of the pressure drop over time.

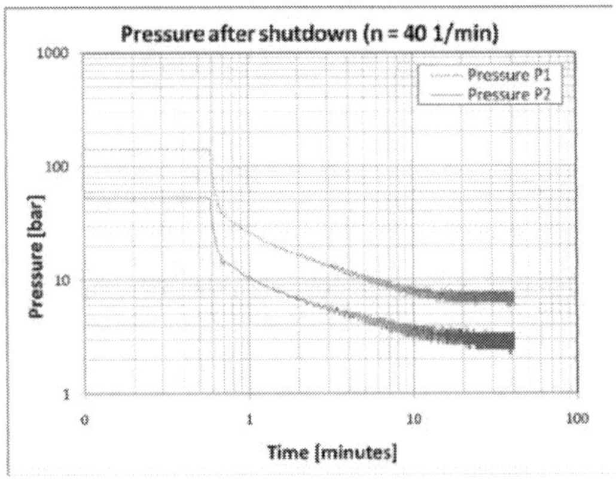

Figure 5 – Double logarithmic plot of pressure vs. time

The time till convergence varies over rotational speed and is large at low rotational speeds, which will be discussed in the next section. In order to investigate the repeatability of the measurement several measurements were taken at the same point of operation. Figure 6 shows that these repeated measurements lead to the same pressure drop curves with only very few deviations.

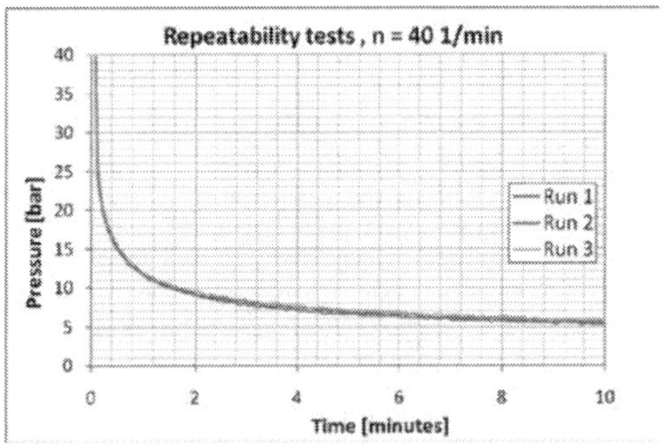

Figure 6 – Multiple tests of the same operating point to ensure repeatability

After the measurement the die was opened and the pressure values were checked without any rubber compound to ensure the pressure transducers again measure the ambient pressure.

Results and Discussion

As seen in Figure 4 and 5 the pressure, respectively the wall shear stress converges toward the yield stress after about 30 minutes. As this time varies over rotational speed, a convergence time of 40 minutes is assumed for all operating points in the following investigations.
In order to reduce the time required for the measurement investigations concerning a pressure coefficient

$$\frac{p_{10}}{p_{40}}$$

were done which simply means that the pressure value after 10 minutes is divided by the pressure value after 40 minutes or on the other hand the pressure value after 40 minutes can be calculated by the coefficient and the pressure value after 10 minutes.

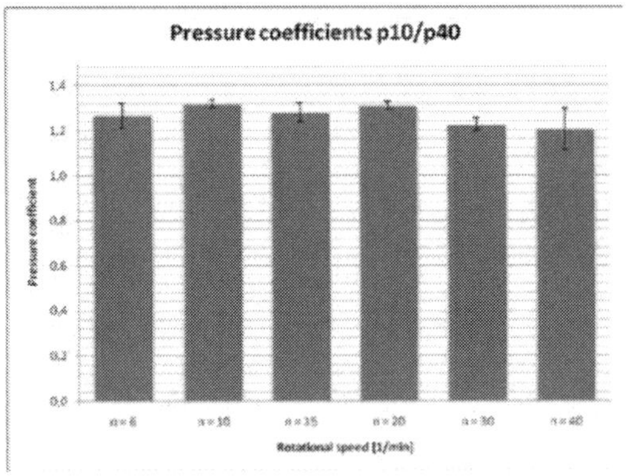

Figure 7 – Pressure coefficient p10/p40 for different rotational speeds

Figure 7 shows this pressure coefficient for all rotational speed levels investigated. As the deviations of the pressure coefficients are small, a measurement duration of 10 minutes was used for further investigations.

Taking a closer look at the yield stress after 40 minutes calculated by the pressure coefficient and the pressure curves during the first 10 minutes, it becomes obvious that the pressure drop velocity varies with the different rotational speed levels. According to the assumption of lower filler network strength with higher rotational speeds under the boundary condition of a constant mass temperature, a higher pressure drop velocity is expected for higher rotational speeds.

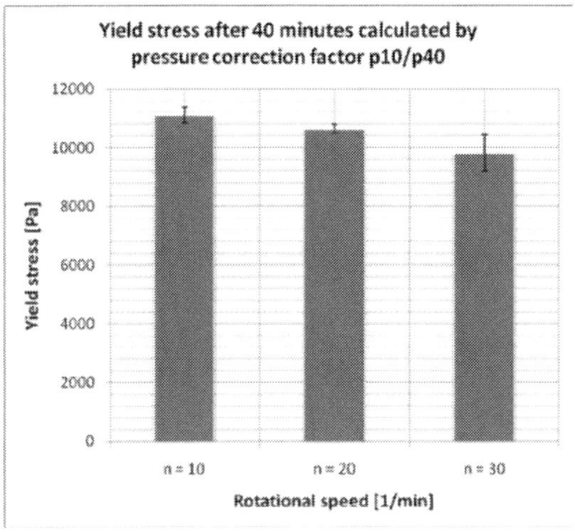

Figure 8 – Yield stress after 40 minutes calculated by pressure after 10 minutes and pressure correction factor

An approach to fit the pressure curves over the time t is given by the formulation of the dimensionless pressure

$$\frac{p-p_{\infty}}{p_0-p_{\infty}}=e^{(a\cdot t)^b}-1.$$

In case of assumed convergence time t = 40 minutes the pressure value p_{40} is used for the value of p_{∞}.

The pressure drop curves in figure 9 and 10 show that the dimensionless pressure over the time decreases faster at higher rotational speeds.
As the mass temperature in the die entrance was constant, the differences in dimensionless pressure curves and yield stress over rotational speed do not seem to be a result of the temperature dependency of the yield stress.

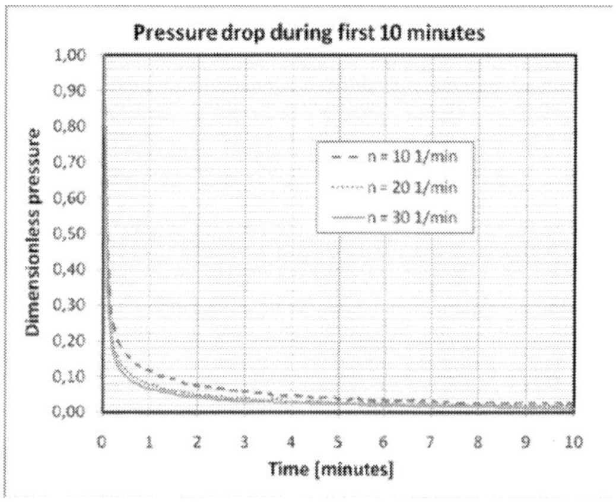

Figure 9 – Dimensionless pressure during the first 10 minutes of measurement for different rotational velocities

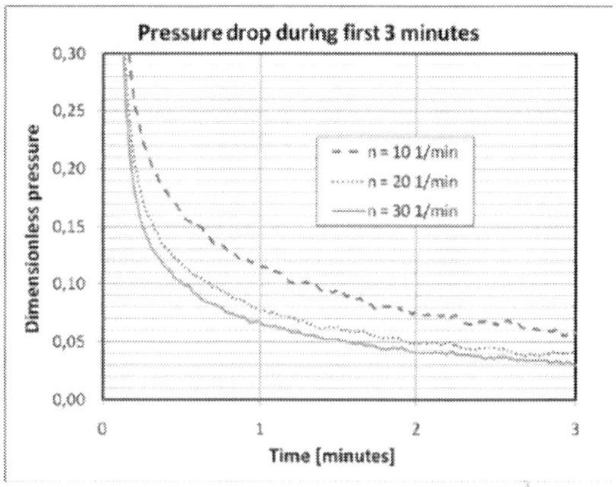

Figure 10 – Closer look at dimensionless pressure during the first 3 minutes of measurement for different rotational velocities

A regression of the dimensionless pressure curves with the equation above delivers for the coefficients a and b the values given in Table 1.

Table 1 – Coefficients a and b for regression of pressure drop curves

Rotational speed [1/min]	*a [1/s]*	*b*
10	60.659	-0.559
20	80.117	-0.598
30	86.267	-0.629

The coefficients a and b can be used as parameters for the speed of the pressure drop. These values over the rotational speed indicate an increase in the pressure drop speed at higher rotational velocities, as the curves show.
As it could be shown that the yield stresses and the pressure drop curves correlate to the rotational speed, a closer look at the preshearing history is important.

Throughput measurements can be used to calculate the average residence time t_{RT}. On the other hand, the shear rate $\dot{\gamma}$ can be calculated by the screw geometry by

$$\dot{\gamma} = \frac{v_{0z}}{H}$$

where v_{0z} represents the flow speed in channel direction and H the channel height.
With a shearing history factor f defined by

$$f = \dot{\gamma} \cdot t_{RT}$$

an almost constant shearing history factor was generated by using a constant screw length.
Figure 11 shows the yield stress after 40 minutes over the shearing history factor.

Figure 11 – Yield stress after 40 minutes over the shearing history factor

That the yield stresses and dimensionless pressure curves vary at almost constant shear history factor suggests the assumption that the shear rate is more important for the filler network degradation than the shearing history factor.

During the investigations it became clear that an essential problem in measuring the yield stress is the time till convergence. At high rotational speeds the measurement duration till convergence is less than one hour. At low rotational speeds the pressure values at the same time are higher compared to higher rotational speeds because of the slower pressure drop velocity as shown above. For low rotational speeds a convergence was not reached even after several hours of measurement.

Conclusions and Outlook

A yield stress can be detected in less than one hour during measurement after high rotational speeds. For low levels of rotational speeds a convergence for the pressure drop is not reached even after several hours of measurement but in fact that these low shear rates are not common in usual production processes this is not a big disadvantage. One problem of measuring for several hours is a varying ambient temperature that affects the die temperature and consequently the level of yield stress. In further investigations the die must be isolated from ambient influences in order to allow long term measurements of absolute yield stresses.

Nevertheless, it could be shown that under the boundary condition of a constant mass temperature the pressure drop velocity over the shear rate varies, which might be caused by the degraded filler network.
Other aspects to be investigated are influences of carbon black types on the yield stress and the time-dependent development of the yield stress after shearing.
Another aspect to be investigated is the influence of different shear history factors, which could be realized by different screw lengths.

Acknowledgements

We would like to thank Harburg-Freudenberger Maschinenbau GmbH for the kind support during the experimental studies.

References

Barrès, C., Leblanc, Jean L., "Recent developments in shear rheometry of uncured rubber compounds", Polymer Testing **19**, 177-191 (2000)

Borggreve, R.J.M., Gaymans, R.J., Schuijer, J., Ingen Housz, J.F.: "Brittle-tough transition in nylon-rubberblends: effect of rubber concentration and particle size", Polymer **28**, 1489-1496 (1987)

Häberlein, M.: http://chemie.fb2.fh-frankfurt.de/KAT/13Fuellstoffe.htm (2010)

Leblanc, Jean L: "Rubber-filler interactions and rheological properties in filled compounds", Progress in Polymer Science **27**, 627-687 (2002)

Limper, A.: PhD-Thesis, RWTH Aachen, 1985

Muenstedt, H.: "Rheology of Rubber-Modified Polymer Melts", Polymer Engineering and Science **21**, 259-270 (1981)

Nguyen, Q.D., Boger, D.V.:"Measuring the flow properties of yield stress fluids", Annu. Rev. Fluid Mech. **24**, 47-88 (1992)

Schramm, D.: PhD-Thesis, University of Paderborn, 2003

Ziegler, J.: PhD-Thesis, University of Hannover, 2004

Stability and breakup of confined threads

*Pieter JA Janssen (1), Patrick David Anderson (2)**

(1) University Wisconsin - MA - USA, (2) TU Eindhoven - NB - Netherlands

Microfluidic devices are often used to generate drops, but confined morphologies with at least one dimension much longer than the wall spacing also exist. An example of this are long threads, either generated by jetting, or coalescence of drops. We use a three-dimensional periodic boundary-integral method to investigate the stability of such threads. Our algorithm incorporates Green's function who exactly satisfy the no-slip condition at the walls. The periodic version makes use of the Hele-Shaw form formulation, that the wall-corrected Green's functions take in the far field. We primarily focus on the stability of threads versus small amplitude disturbances. At large wall separations, we obtain an excellent match with Tomotika's theory. With increasing the confinement ratio, the growth speed reduces, but much more in the direction perpendicular to the walls. Also the wavelength with the fastest growth speed changes slightly. We find that all threads will break up in the confinement, as long as the undeformed drop fits between the plates (confinement ratio < 1), although the breakup time increases significantly. One other aspect we show is the response of threads laying close to each other. Depending on the distance between the threads, waves on neighboring threads form in an in- or out-of-phase pattern. The crossover distance in unconfined situations is in agreement with experimental data and analytical results. We find that the confinement promotes out-of-phase behavior. Finally, we show the influence of shear. If the shear flow can convect distortions before they can grow, shear flow acts to stabilize the thread. Our results match quite well with a simple scaling argument for the critical capillary number, above which the threads are stable.

AGGREGATION OF CARBON NANOTUBES IN A SEMI-DILUTE SUSPENSION.

Ludovic Moreira[1], René Fulchiron[*1], Gérard Seytre[1], Sandrine Marceau[2], Philippe Dubois[2], Philippe Cassagnau[1]

[1] Université de Lyon, CNRS UMR5223, Ingénierie des Matériaux Polymères/LMPB, Université Lyon 1, F-69622 Villeurbanne, France – rene.fulchiron@univ-lyon1.fr, gerard.seytre@univ-lyon1.fr, Philippe.cassanau@univ-lyon1.fr
[2] Center of Innovation and Research in Materials & Polymers (CIRMAP), Laboratory of Polymeric and Composite Materials, University of Mons - UMONS, Place du Parc 20, B-7000 Mons (Belgium), philippe.dubois@umh.ac.be

Abstract :
Multiwall carbon nanotubes (MWNT) aggregation and de-aggregation mechanisms for semi-dilute suspensions (Φ<0.5 %) in low viscosity PDMS matrix were investigated. Both dynamic shear modulus and electrical conductivity measurements were performed. Especially, the effect of the total applied strain was scrutinized. First, from low strain measurements in the linear viscoelasticity domain, the complex shear modulus gives evidence for a second terminal relaxation zone at low frequencies which can be related to the Brownian motions of the dispersed nanotubes embedded by the surrounding ones. However, an additional slowing of the rotary diffusion for concentrations higher than 0.3% was obtained. This critical concentration is much lower than this deduced from the Doi-Edwards theory of rigid rod dynamics which predicts the transition between semi-dilute and the concentrated isotropic regimes near 2%. This deviation was explained by a clustering phenomenon of MWNTs. Therefore, the MWNTs aggregation mechanism was more deeply studied from simultaneous time resolved measurements of electrical conductivity and dynamic shear modulus for a concentration of nanotubes (Φ=0.2 %). The material was subjected to different strain magnitudes. Hence, after a non-linear compressive deformation, it was shown that the kinetics of MWNT aggregation strongly depends on the amplitude of the dynamic shear deformation. Actually, the kinetics of aggregation increases with increasing the deformation. However, above a critical deformation (60%), the equilibrium value of the storage modulus and electrical conductivity decreases with increasing the deformation. It was then concluded that low shear deformations enhance an aggregation mechanism, but these aggregates break down at high shear, forming small aggregates with less entanglements. Furthermore, this phenomenon of MWNT cluster aggregation was also highlighted by optical microscopy.

Introduction

Because of their large aspect ratio, Carbon Nanotubes (CNTs) can be viewed as one-dimensional nanomaterials. Moreover, they present very interesting properties like, for example, high electrical conductivity. These characteristics make them a very attractive object of study because, used as nanofillers, they can lead to interesting properties of nanocomposites even for very low CNT concentrations provided that they are suitably dispersed. In that field, the rheological behavior analysis can be very helpful to characterize the dispersion quality and the development of the eventual percolation network (Mitchell et al. 2002, Du et al. 2004, Hough et al. 2004, Pötschke et al. 2004, Zhang et al. 2006). Additionally, the electrical properties can also provide many information (Du et al. 2004, Pötschke et al. 2004, Zhang et al. 2006). Obviously, the studies are mainly focused on CNT concentrations near the percolation threshold, that is typically lower than 1 wt%. By the way, it can be added that the experimental statement of the percolation threshold concentration is still under discussion because it is very sensitive to the quality and the stability of the CNTs dispersion which is generally far from straightforward. Moreover, the detection of this threshold from rheological or from electrical measurements can lead to divergent results bringing controversies concerning the transfer mechanisms of the mechanical stress or the electrical carriers from one CNT to another.

In the present study, following earlier works (Marceau et al. 2009, Moreira et al. 2010), we focus on semi-dilute suspensions of CNTs in a relatively low molecular weight polydimethylsiloxane (PDMS) matrix. It was previously shown that the viscoelastic behavior of such suspensions deviates from the Doi-Edwards theory of rigid rods dynamics probably because of a clustering of the CNT. However, as it will be shown hereafter, this clustering phenomenon can take a very long time and be very dependant of the strain history of the material. The aim of the present work is to analyze this phenomenon using a coupled technique of viscoelastic and dielectric measurements.

Experimental

Nanocomposite preparation:

The nanocomposite (PDMS)/CNTs was prepared following the method developed by Beigbeder et al. (2008) based on untreated nanotubes. Indeed, for such a system, the CH-π interactions between the PDMS methyl groups and the π-electron rich surface of multiwall nanotubes (MWNT) are favorable to CNTs dispersion in PDMS at the level of individual CNTs. In the present study, multiwall carbon nanotubes with a diameter of d $\approx$10 nm, average length of L $\approx$1.5 µm (L/d~160) and a purity of mininimum 90%. Furthermore, they were directly used as masterbatch predispersed in PDMS (0.5 wt%) using a high shear industrial dispersion process and supplied under Biocyl 2000 trade name by Nanocyl SA (Sambreville, Belgium). Polydimethylsiloxane (PDMS) Sylgard 184-PartA (from Dow Corning, USA) was used as silicone sample. The number-average molar mass of this PDMS sample is Mn = 32 000 $g.mol^{-1}$. This molar mass is close to the critical molar mass for the viscosity Mc$\approx$30 000 $g.mol^{-1}$. The zero shear viscosity of this PDMS at room temperature (T=22 °C) is η_0= 5 Pa.s. Practically, CNTs and Sylgard 184-part A were mechanically mixed (1200 rpm) for 30 min at room temperature and used without any further mechanical treatment. For the present study, concentrations Φ of CNTs in PDMS of 0.1 and 0.2% were prepared. In terms of dilution regime for solutions of rigid rods (Doi and Edwards 1986), these concentrations correspond to the semi-dilute regime for which the number of CNTs per unit volume of solution is situated between $1/L^3$ and $1/dL^2$). In those conditions, the rods are considered not able to rotate freely anymore without any interaction from surrounding ones. However the excluded volume for interactions can be still neglected.

Rheological and electrical measurements:

The viscoelastic and electrical measurements were carried out on an ARES rheometer (TA Instruments) at room temperature (T $\approx$22°C). A home-made cell has been developed to carry out simultaneous rheological and dielectric measurements. The cell is constituted of isolated ring-plate electrodes connected to a frequency response analyzer HP4284. The upper and lower ring-plates have an inner diameter of 22 mm and an outer diameter of 25 mm. Thus, the strain difference between the inner and the outer limits of the cell is only of 14%. Compared with a conventional plate-plate geometry, such a ring-plate device allows us to apply a quasi-homogeneous shear rate and deformation throughout the sample volume. The dielectric parameters conductivity and capacitance were recorded at a frequency of 1kHz in steady or oscillatory shearing mode by using specific connections to the experimental setup. Note that such type of device has been already developed and successfully used by Alig et al. (2007).

Most of the rheological measurements were performed in an oscillatory shear mode (time or frequency sweep modes depending on the experiment) using the ring geometry. The specimens were prior stored at room temperature. A relatively thick film was then deposited on the ring-plate surface. Obviously, this positioning of the sample in the cell can not be achieved without a considerable deformation of the material. Moreover, it was then squeezed by a displacement of the upper ring down to 0.5 mm. Thus, the sample was initially submitted to an important deformation before the measurement which results in the break down of CNTs contacts as a posteriori checked. Moreover the further results will show that the recovering of a stability state in the material can take a very long time. Therefore, this initial state of the material is not the equilibrium but rather a "deconstructed" state. Then, to quantify the reformation of the CNTs network, time sweep experiments (combined electrical and dynamic measurements) at a mechanical frequency ω= $1 rad.s^{-1}$. Moreover as it will be shown, a strain may enhance this evolution towards the network recovery. Therefore, different types of experiment were performed like time sweeps after a steady shear treatment or with varying.

Optical microscopy:

In order to visualize the suspension evolution, some optical micrographs were taken using an optical microscope Orthoplan (Leitz). The sample was placed between glass slides, undergoing a relatively high strain because of the inherent squeezing. Then successive pictures were recorded along time.

Results and Discussion

In Figure 1 and 2, for Φ=0.1% and Φ=0.2% respectively, a measurement of the electrical conductivity σ evolution after the positioning of the sample in the cell is plotted. Then, after approximately 140 hours for the suspension with 0.1% of CNT and 50 hours for this with 0.2%, a steady shear of $1s^{-1}$ during 158 s was applied. Afterwards, another recovery measurement was programmed but with simultaneous σ and G' measurements (strain amplitudes of 80% for the suspension with Φ=0.1% and 25% for Φ=0.2%.)
Different comments can be drawn from these experiments. First, for both CNT concentrations at the experiment beginning, the measured conductivity is of 10^{-9} $S.m^{-1}$ which corresponds to the PDMS matrix conductivity. Thus, it can be concluded that effectively, the sample positioning in the cell breaks the CNT contacts hence collapsing the network. Moreover, this conductivity remains constant during ~60 h for Φ=0.1% before increasing whereas it grows up immediately for Φ=0.2%. Obviously, the conductivity increasing is due to the little by little recovering of the conductive network resulting from a mechanism of cluster aggregation. If the sample is left at rest, this

recovery takes a very long time for Φ=0.1% (more than 120 h) and slightly less time for Φ=0.2% (~50 h).

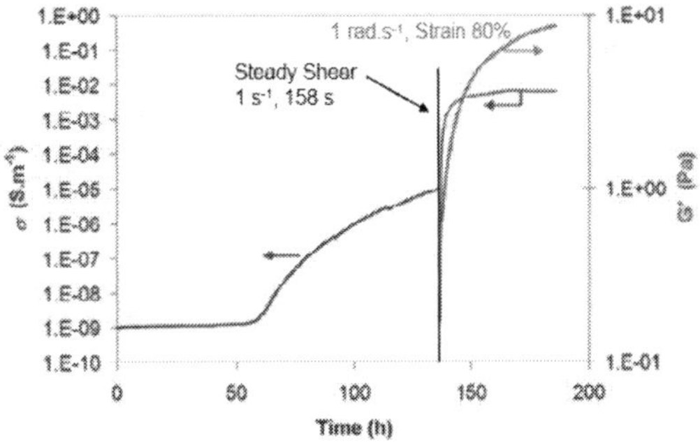

Figure 1 – Network recovery experiments for 0.1% CNT suspension. Electrical conductivity measurement followed by a steady shear of 1 s^{-1} during 158 s and a combined measurement of electrical conductivity and dynamic modulus (1 rad.s^{-1}, 80%).

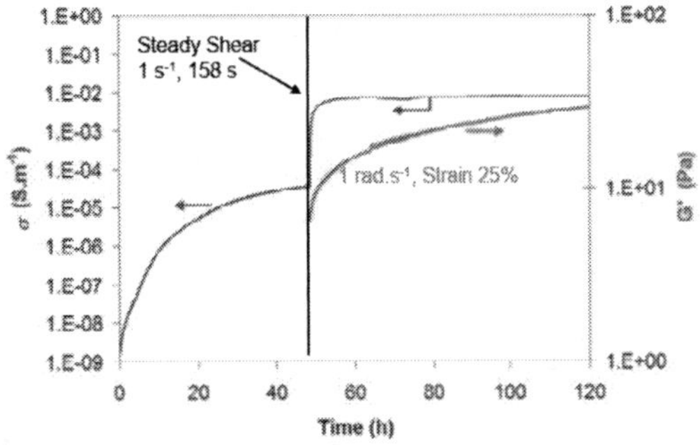

Figure 2 – Network recovery experiments for 0.2% CNT suspension. Electrical conductivity measurement followed by a steady shear of 1 s^{-1} during 158 s and a combined measurement of electrical conductivity and dynamic modulus (1 rad.s^{-1}, 25%).

The clustering phenomenon is also illustrated in Figure 3 where optical microscope pictures taken at different times after the sample preparation. Actually, the micrographs show no clear CNTs aggregates (some impurities may be spotted anyway) just after the sample preparation. However, after a period (at rest) of a few hours, the aggregation (clustering mechanism) of the CNTs is clearly visible. However, it can be pointed out that the scale of this clustering is relatively high since it can be evidenced by simple optical microscopy.

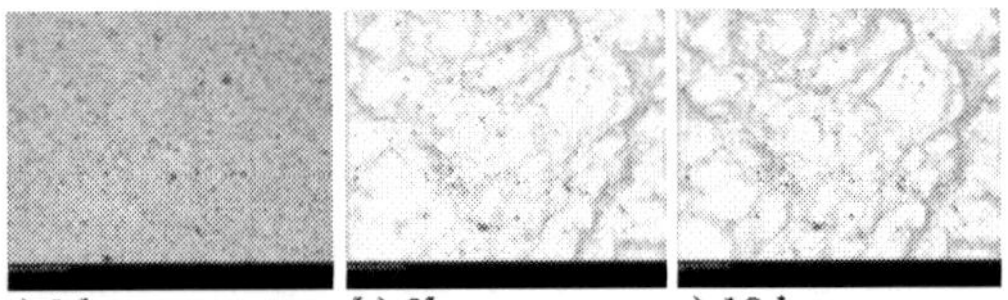

a) 0 h b) 9h c) 18 h

Figure 3 – Optical observation of the aggregation of CNTs in 0.2 % suspension. a): CNTs are well dispersed in PDMS matrix, b) and c): CNTs aggregate to build a network (Picture width: 600 μm)

After the recovery at rest, when applying a steady shear, the network is partially destroyed but interestingly, the subsequent recovery is much faster and the final conductivity level is higher than for the recovery at rest. Note that, for these second recovery measurements, the final conductivity is the same for both CNT concentrations. However, it must be reminded that during this second step of the experiment, the conductivity measurement is simultaneous to the dynamic modulus measurement. Thus, this measurement can not be considered totally at rest. However, even if cautions (not shown in the present paper) were taken to ensure that this measurement is carried out within the linear viscoelastic domain, an important doubt remains at this stage about the effect of the total oscillatory strain on the CNT network formation and evolution. However, the *G'* evolutions plotted in Figures 1 and 2, show that the elasticity recovery due the CNT network formation takes more time than the necessary time to reach a constant conductivity. Moreover, the variation magnitude of σ (6-7 decades) is much larger than this of *G'* (at the most 2 decades). It can be then admitted that the variation of the electrical conductivity is more sensitive to the variation of the number of nanotube-nanotube contacts than the modulus, at least at the frequency used. In other words, the electrical conductivity method is much more sensitive than the mechanical one to characterize the cluster-aggregation mechanism.

However, as mentioned before, the network formation due to the NTC aggregation is affected by the strain conditions. Thus, the dynamic modulus measurement by itself can influence this structuring and lead to the paradox that if the strain is to low, no reliable data is obtained because the stress is not enough to be measured and if it is too high, the data are not really characteristic of the network structuring at rest.

In order to more deeply evaluate this effect of the strain, different measurements were performed with varying oscillation shear strain for the 0.2% suspension. The results are gathered in Figure 4 and 5.

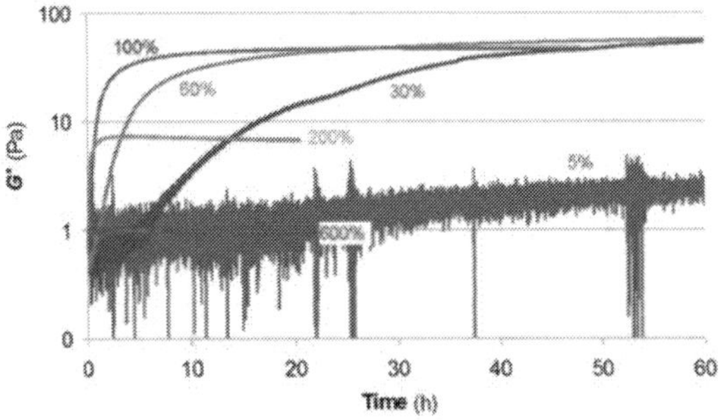

Figure 4 – Variation of storage modulus under different oscillation strains 5, 30, 60, 100, 200 and 600%.

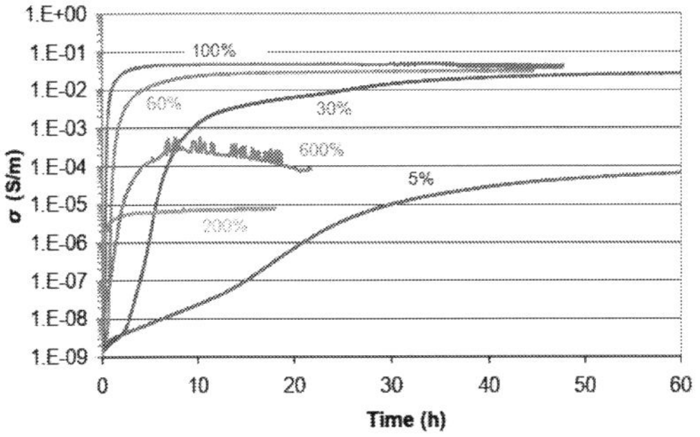

Figure 5 – Variation of electrical conductivity under different oscillation strains: 5, 30, 60, 100, 200 and 600% (experiments simultaneous to those of Figure 3)

Under small deformation (5%), Figure 4 shows a weak variation of the storage shear modulus G' which is extremely noisy due to this low value of the applied deformation. On the contrary, the conductivity of the suspension (Figure 5) increases progressively from 10^{-9} S.m^{-1}, corresponding to the PDMS matrix, to reach $\sigma \approx 10^{-4}$ S.m^{-1} after 48 h. This value is in the same order of magnitude as the conductivity of usual semiconductive nanocomposites. The electrical property of the suspension varies from nonconductive to conductive behavior whereas the weak variation of the complex shear modulus does not show any network formation at least for the used strain and frequency.
However, for a deformation of 60% the complex shear modulus and conductivity drastically increase with time. More precisely, the storage modulus increases within approximately 30 h from 0.5 to 60 Pa (two decades of variation) and the conductivity increases from10^{-9} S.m^{-1} to 4×10^{-2} S.m^{-1} (more than seven decades of variation). Moreover, the storage modulus crosses the loss modulus at $t\approx5$ h (results not shown here) indicating that the elastic behavior of the suspension becomes dominant compared with the viscous behavior. This crossover can be associated with the formation of a CNTs network under the dynamic shear force. According to our previous work (Marceau et al. 2009), this network elasticity resulting from the nanotubes interactions appears to be a temporary elasticity. In other words, this transition does not obey the sol-gel transition as it depends on the frequency used.
Furthermore, under higher deformation (200, 600%, the storage shear modulus and the conductivity rapidly increase but finally reach a much lower values (~ 1 Pa and ~ 10^{-4} S.m^{-1} respectively). Hence, these results show the existence of a critical deformation. Beyond this critical deformation, the variation of the storage modulus and conductivity remains still accelerated but the final values considerably decreases with increasing the deformation. From Figures 4 and 5, this critical strain appears to be at between 60% and 100%.
To check the critical value of the deformation, the famous effect of strain-amplitude dependence of the dynamic viscoelastic properties of filled polymers, generally referred as the Payne effect (Payne 1965) was studied. Figure 6 shows the variation of the reduced complex shear modulus and electrical conductivity versus the applied dynamic strain (ω=1 rad .s^{-1}). This experiment was performed just after the time sweep experiment under the deformation of 30%. As expected, beyond the critical strain of ~60%, both storage modulus and electrical conductivity decrease with increasing the strain.

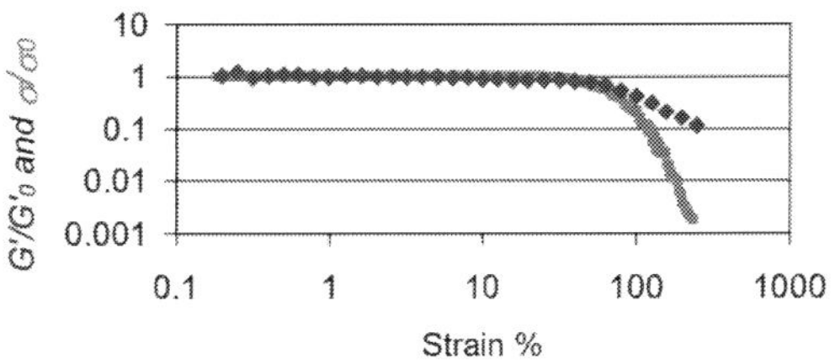

Figure 6 – Variation of normalized storage modulus and electrical conductivity versus the deformation. The CNTs network was previously built under a shear deformation of 30%.

Actually, the present results are in agreement with the predictions of the model of Ma et al. (2008) which shows that the clustering is enhanced by the strain if it remains moderate whereas the aggregates are broken for higher shear.
Finally, the cluster reconstruction after the positioning of the sample in the rheometer cell is clearly evidenced by Figure 7. Just after having placed the sample, the variation of the complex shear modulus (under 5% strain) shows a rheological behavior close to the viscoelastic behavior of the PDMS matrix because the CNTs network was destroyed. After aggregation process (here enhanced under 60% of strain) the viscoelastic behavior of the composite shows a solid-like behavior.

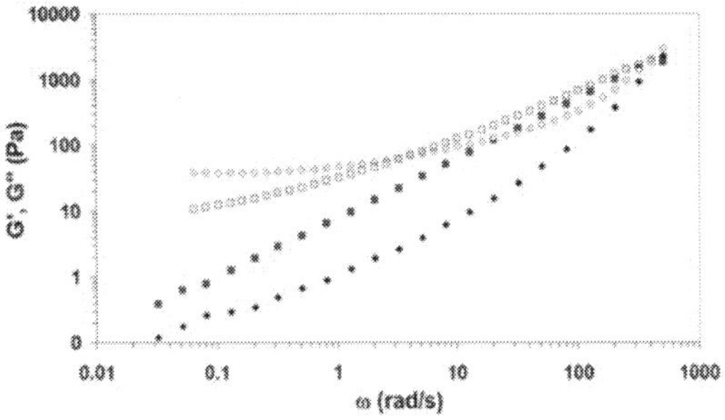

Figure 7 – Frequency dependence of the complex shear modulus of the PDMS/CNTs suspension (0.2 %). Filled symbol: just after the sample positioning in the cell (measured under a strain of 5%). Unfilled symbols: After the end the aggregation process (measured under a strain of 30%)

Conclusions

In the present study, coupled measurements of electrical conductivity and viscoelastic behavior were performed on suspensions of NTC in a relatively low molecular weight PDMS matrix. It has been shown that starting from suspensions in PDMS where the NTC are very well dispersed, a clustering is little by little established leading to a percolation which can be revealed by electrical conductivity measurements. Moreover, mechanical measurements can also highlight the network building but important cautions must be exercised because the applied strain needed for the measurements may drastically influence the network building by itself.

Furthermore, concerning the controversies about the percolation threshold detection, electrical measurements appears to be preferable since they do not influence the structuring mechanism contrary to mechanical measurements for which the strain can enhance or prevent the structuring depending if it is lower or higher than the critical one. Moreover, this critical value is possibly dependent on the concentration. Nevertheless, whatever the measuring method (electrical or mechanical), it is necessary to let the sample at rest after its positioning during a very long time to attain the equilibrium and it is far from indubitable that this condition is always fulfilled.

References

Alig, I.; Lellinger, D.; Dukin, S.; Potschke, P., "Conductivity spectroscopy on melt processed polypropylene-multiwalled carbon nanotube composites: Recovery after shear and crystallization" Polymer, **48**, 1020–1029 (2007).

Beigbeder A., Linares M., Devalckenaere M., Degee P., Claes M., Beljonne D., Lazzaroni R., Dubois P., "CH-pi interactions as the driving force for silicone-based nanocomposites with exceptional properties", Adv. Mater., **20**, 1003–1007 (2008)

Doi, M., Edwards, S. F.: The theory of polymer dynamics, Oxford Press: London, 1986.

Du F., Scogna R.C., Zhou W., Brand S., Fischer J.E., Winey K.I., "Nanotube Networks in Polymer Nanocomposites: Rheology and Electrical Conductivity", Macromolecules, **37,** p 9048-9055 (2004).

Hough L.A., Islam M.F., Janmey P.A., Yodh A.G., "Viscoelasticity of single wall carbon nanotube suspensions", Phys Rev Letters, **93**, 168102 (2004).

Marceau S., Dubois P., Fulchiron R., Cassagnau P. "Viscoelasticity of Brownian Carbon Nanotubes in PDMS Semidilute Regime" Macromolecules, **42,** 1433-1438 (2009)

Mitchell C.A., Bahr J.L., Arepalli S., Tour J.M., Krishnamoorti R, "Dispersion of Functionalized Carbon Nanotubes in Polystyrene", Macromolecules, **35**, 8825-8830 (2002)

Moreira L., Fulchiron R., Seytre G., Dubois P., Cassagnau P. "Aggregation of Carbon Nanotubes in Semidilute Suspension", Macromolecules **43**, 1467–1472 (2010).

Payne A.R., Chapter 3 in "Reinforcement of Elastomers", New York : Interscience, p.69-123 (1965).

Pötschke P., Abdel-Goad M., Alig I., Dudkin S., Lellinger D., "Rheological and dielectrical characterization of melt mixed polycarbonate-multiwalled carbon nanotube composites", Polymer, **45**, 8863-8870 (2004)

Zhang Q., Lippits D.R., Rastogi S., "Dispersion and rheological aspects of SWNTs in ultrahigh molecular weight polyethylene", Macromolecules, **39**, 658-666 (2006).

Ma, W. K. A.; Chinesta, F.; Ammar, A.; Mackley, M. R., ": Rheological modeling of carbon nanotube aggregate suspensions", J. Rheol. **52**, 1311–1330 (2008).

RHEOLOGICAL BEHAVIOUR AND RELAXATION PHENOMENA OF IMMISCIBLE BLENDS: APPLICATION OF CREEP

Vahid Shaayegan[1*], Paula Wood-Adams[1], Nicole Raymonde Demarquette[2] and Adriana M.C. Souza[2]

[1]*Department of Industrial and Mechanical Engineering, Concordia University, Montreal, Quebec, Canada – v_shaay@live.concordia.ca; woodadam@alcor.concordia.ca*
[2] *Department of Materials Engineering, University of Sao Paulo, Sao Paulo, Brazil – nick@usp.br; adrianacatelli@gmail.com*

Abstract - A new method, using creep measurements to study the long-time linear relaxation mechanisms in immiscible blends, was developed. Dynamic experiments were performed to analyze the shorter time relaxation processes. An experimental protocol based upon incomplete-creep/recovery was defined such that the morphological changes during flow were minimized. The creep and recovery measurements were used to detect the long-time portions of the weighted relaxation spectrum. Data from dynamic and creep measurements were combined using the method proposed by He *et al.* (2004) to construct an extended retardation spectrum. The extended retardation spectrum was then converted to an extended relaxation spectrum and other linear viscoelastic (LVE) material functions covering a broader viscoelastic window than either test on its own. This procedure was tested using a polypropylene/polystyrene blend. It was observed that the extended relaxation spectrum depicts well the long-time relaxation processes and differs from the spectrum obtained using only dynamic measurements at longer times.

Introduction

Linear viscoelastic properties of molten polymers provide valuable information about their structure, especially in the plateau or terminal zones (He *et al.*, 2004). One can get information on the molar mass and state of branching in the case of pure polymers (Wood-Adams and Dealy, 2000), on the interfacial tension between components of a blend (Bousmina, 1999; Graebling *et al.,* 1993a; Palierne, 1989) or on the state of exfoliation of clay in the case of clay containing nano-composites (Ren *et al.*, 2000). Small amplitude oscillatory shear experiments (SAOS) are normally used to study the linear viscoelastic behaviour of molten polymers (Yee *et al.*, 2007). However, the relaxation phenomena of branches or sheared droplets (in the case of immiscible blends) may occur at relatively long times that cannot be accessed when performing small amplitude oscillatory shear experiments. Dynamic measurements (SAOS) are not suitable to investigate the viscoelastic behaviour at long times (or very low frequencies) since the stress is normally too small to be measured or controlled precisely. Also, the most valuable information are obtainable at low frequencies for which the duration of the experiment is such that the samples may suffer thermal degradation during the test (Dealy and Larson , 2006; He *et al.*, 2004). Creep experiments, on the other hand, are useful to probe the long time behaviour of polymer melts. In creep experiments, low enough stresses are required to keep the sample within the linear viscoelastic limits. However, to reach steady-state flow, a large strain that may take the sample into nonlinear region is needed (Dealy and Larson, 2006; He *et al.,* 2004). Furthermore, the level of applied stress should be large enough to have accurate, noise-free results with reasonable resolution from the instrument. In addition, creep experiments are not accurate at very short times, because it takes time for the stress to reach its desired value during the start up. He *et al.* (2004) established a method to combine the data from dynamic and steady experiments. They converted the creep compliance to a continuous retardation spectrum using the NLREG technique (Honerkamp and Weese, 1993) and found a considerable time interval over which the two continuous retardation spectra from SAOS and creep experiment superpose. Using this procedure, they extended the time range over which the material can be studied.

In the case of polymer blends, linear viscoelasticity can be used as a tool to infer some information on the blend morphology, interfacial tension between its components and conformation of compatibilizer at the interface (Friedrich and Antonov, 2007). However, to use linear viscoelasticity to characterize polymer blends, one should ensure that almost no morphological changes happen during the experiments. Small amplitude oscillatory shear experiments (SAOS) are routinely used to analyze the linear viscoelastic behaviour of immiscible blends (Graebling and Muller, 1990; Macaubas and Demarquette, 1999; Velankar *et al.* 2004), and most studies have revealed that SAOS flow at small enough strain amplitudes does not change the morphology of blends provided the concentration of the dispersed phase is small enough (Calvão *et al.*

2005; Cavallo *et al.* 2003; Graebling *et al.* ,1993b; Vinckeier *et al.* ,1996), i.e. the deformations are within the linear viscoelastic regime. However, morphological modifications including droplet deformation, coalescence and break up are observed during most transient or steady rheological investigations. These micro-structural changes are the result of a competition between viscous forces, due to applied stress/deformation, tending to change the morphology, and interfacial forces that tend to retain the morphology (Ghodgaonkar and Sundararaj, 1996). Martin *et al.* (2000) showed that while there was no significant deformation in drops of high density polyethylene (HDPE) in polystyrene (PS) at low stresses and small strains, coalescence was observed if the flow was continued to a larger strain during creep experiments. At higher stresses, spherical droplets reshape to ellipsoids, orient in the flow direction and frequently break apart. Once these phenomena have occurred, the behaviour, by definition, is no longer linear. In order to capture the relaxation phenomena which occur at long times and cope with inevitable morphological changes during creep experiments, we developed a new experimental creep protocol at the limit of linear viscoelastic behaviour in this work. The proposed creep method involves the determination of the appropriate creep stress level and shearing time which keep the sample in the LVE regime while minimizing morphological changes. This method was tested on polypropylene/polystyrene (PP/PS) immiscible blends.

Experimental

Materials and sample preparation

Blends of PP/PS at two different concentrations, (90/10) and (70/30), were prepared in an extruder (Rheomix PTW-16) coupled to a torque rheometer (ThermoHaake PolyLab 900) with four heating zones, all set at 200° C and 100 rpm. The properties of the neat components are reported in Table 1. The pure components and the blend were compression moulded into discs using a Carver Laboratory Hydraulic Press at 200° C. All moulded PS samples as well as the blends were dried under vacuum before each experiment for 15 hours and 30 minutes at 110° C.

Table 1: Properties of pure components of blends

Name	Mw (g/mol)	MFI (g/10min)	Commercial name	Supplier
PP	340000	8	HM6100	Quattor
PS	N/A	4	158K	BASF

Rheological measurements

An MCR500 (Anton Paar US) rotational rheometer was used to perform dynamic and creep tests at 200° C under a nitrogen atmosphere. A parallel plate geometry with a diameter of 25 mm was used with a gap of 1mm which is large compared to the droplets' size. All samples were annealed in the rheometer at 200° C for 10 minutes before testing. Thermal stability tests were performed to ensure samples' stability and strain sweeps were carried out to determine the linear viscoelastic limits (LVE) for small amplitude oscillatory shear tests (SAOS). Frequency sweep experiments were carried out over four frequency decades based on the pre-determined strain magnitude. A detailed description of the creep experiments is provided below as the developed procedure is the key of obtaining reliable results.

Morphological investigations

The morphology of the blends was characterized prior to and after the rheological measurements. In the former case, the samples were annealed in the rheometer before observing the morphology. The samples were quenched *in situ* between parallel plates using liquid nitrogen at a cooling rate of about 4.25 °C/s to ensure that the morphology was captured unchanged. The samples were then fractured in liquid nitrogen. To see the morphological changes, the samples should be analysed in the flow direction; therefore the samples were fractured on a plane whose normal is in the radial direction. Furthermore, in parallel plate geometry, the shear rate is changing from a minimum at the center to a maximum at the edge, meaning that the morphology is expected to vary in the radial direction. To capture this effect, the samples were fractured at $3/4$ distance of the disk radius from the center (Martin *et al.*, 2000). To enhance the contrast between the phases, the fractured blends were etched for 10 hours in cyclohexane to dissolve the PS phase. The samples were then sputtered-coated with a thin layer of a gold/palladium alloy and observed with a scanning electron microscope (SEM), model JSM-840. Image analyses were done using SigmaPlot software to determine the polydispersity and volume average radius of the droplets.

Establishment of the creep method for blends

Preliminary creep tests were carried out to make sure that the stresses as well as the deformation time are small enough to keep the sample in the linear viscoelastic regime. Figure 1 shows the results of eight creep tests, each performed using a different specimen of PP/PS(90/10), at a stress of 5 Pa for 100 seconds. Clearly, the compliance function varies significantly from one sample to another, probably due to the sample-to-sample variation of morphology as indicated by the occurrence of the results in clusters.

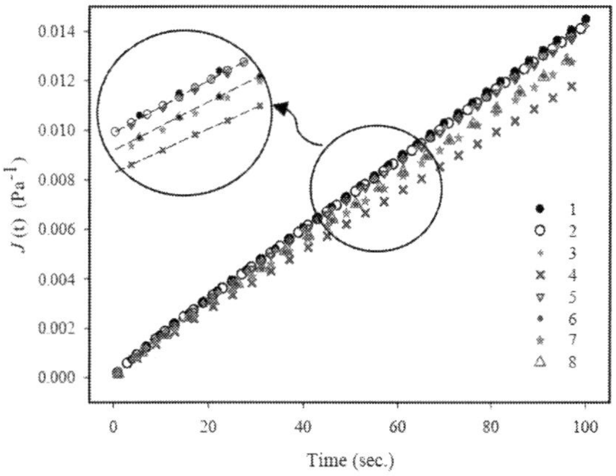

Figure 1: Creep compliance vs. time, at a stress of 5 Pa for 100 sec. on PP/PS(90/10) blend at 200° C

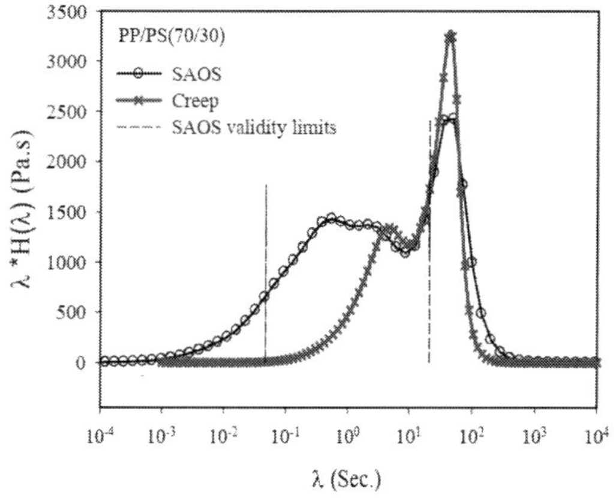

Figure 2: Weighted relaxation spectra for PP/PS(70/30) from SAOS and creep experiments

Due to this sample-to-sample variation, it was not possible to determine the appropriate stress magnitude to keep the behaviour in LVE regime from these data. To cope with the sample-to-sample variation, determination of linear viscoelastic regime was performed on the same specimen, i.e. sequential creep experiments were carried out using the same specimen at different stress while the appropriate relaxation time was allocated between two successive tests.

By omitting the sample-to-sample variation, the creep compliance functions at different stresses superposed, indicating that the variations in properties were due to the variation of morphology in different samples. We suspect such morphological variations are typical behaviour of immiscible blends; therefore *the same specimen* approach is the only possible method to determine the appropriate stress for linear viscoelastic creep experiments. The repeatability of the results was verified on at least three PP/PS(90/10) samples.

The amount of morphological modification is not only dependent on the stress magnitude, but also is a function of shearing time; therefore the effect of the time of deformation must be examined once the magnitude of the stress is chosen for the creep experiment. After the selection of the stress and deformation time with respect to the linear viscoelastic limits, five incomplete creep experiments followed by recovery were carried out on five different samples and the average values were taken to represent the overall behaviour of the blend.

Results and Discussion

Combination of Dynamic and Steady Results, Composite Spectra

Dynamic measurements as well as the creep results were converted to relaxation and retardation spectra using the NLREG software. Figure 2 compares the weighted relaxation spectra obtained from both experiments for PP/PS(70/30).

Clearly, the interface relaxation time is longer than the upper validity limit of the spectrum determined by SAOS. However, for this blend, the creep measurements characterize well the interface relaxation time and long-time parts of the spectrum. Figure 3 shows the corresponding retardation spectra inferred from SAOS and incomplete creep/recovery method. The vertical lines show the limits of validity for the SAOS measurements. There is a considerable time interval over which the two spectra overlap within the window for both experiments. Therefore a composite retardation spectrum was constructed using the short-time part of the retardation spectrum from the SAOS experiment and the long-time portion obtained from creep compliance shown in Figure 4.

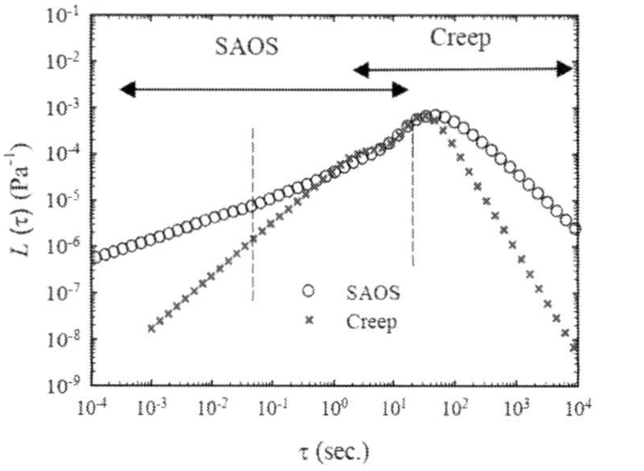

Figure 3: Retardation spectra from the SAOS and creep compliance for PP/PS(70/30) at 200° C

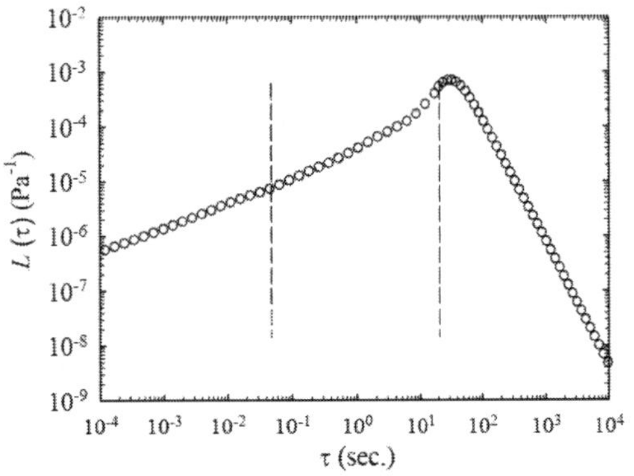

Figure 4: Composite retardation spectrum for PP/PS(70/30) at 200° C

The extended retardation spectrum was then used to calculate other material functions as well as a composite relaxation spectrum which is valid over a time span larger than the one obtained using only SAOS experiment. The obtained composite spectrum includes the information of both relaxation spectra from the SAOS experiments and creep tests. Figures 5 and 6 show the recalculation of dynamic moduli and complex viscosity from composite retardation spectrum as well as the extended relaxation spectrum for PP/PS(70/30) respectively.

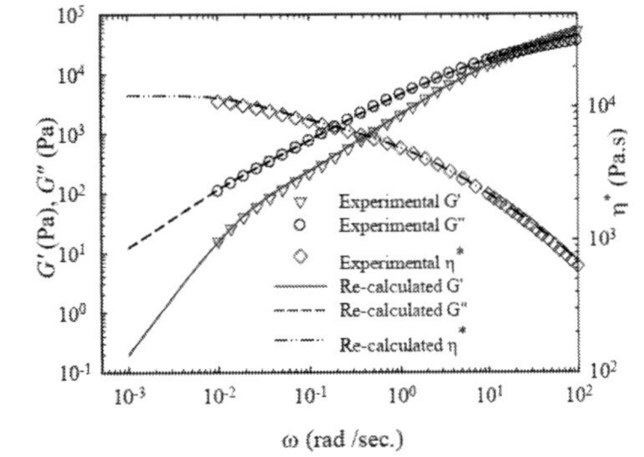

Figure 5: Recalculation of dynamic moduli and complex viscosity for PP/PS(70/30) at 200° C

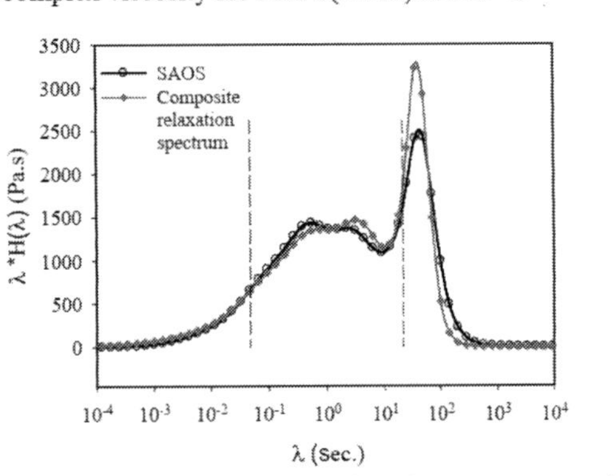

Figure 6: Composite relaxation spectrum for PP/PS(70/30) at 200° C

It can be seen that valuable information regarding the blend properties is accessible over very low frequencies. For instance, the zero shear viscosity, which is obtainable at the limits of very low frequencies, now can be determined as the complex viscosity curve levels off. In addition, the composite relaxation spectrum describes all relaxation processes precisely. From Figure 3, it can be seen that the retardation spectrum obtained from the incomplete creep/recovery method superposes that from the SAOS experiments over a reasonable time interval; hence the creep experiments performed on the immiscible blends were within the linear viscoelastic regime, meaning that the morphological modifications were negligible. The key procedure was the application of incomplete creep experiment followed by recovery and the use of Boltzmann superposition principle to recover the creep compliance while minimizing the strain applied to the sample.

To illustrate the importance of the application of the incomplete creep/recovery, a direct creep experiment was performed on the PP/PS(70/30) for 2000 seconds. Figure 7 shows the retardation spectra obtained from the direct creep experiment and the SAOS experiment.

It can be seen that there is only a 3 second overlap between two spectra and they lose the agreement well before the upper validity limit of the SAOS experiment; hence the retardation spectrum obtained from the direct creep experiment does not represent the LVE behaviour of the blend. This is due to significant coalescence that occurred during the direct creep experiment. Table 2 shows the changes in the volume average radius in incomplete creep/recovery and direct creep experiments for PP/PS(70/30).

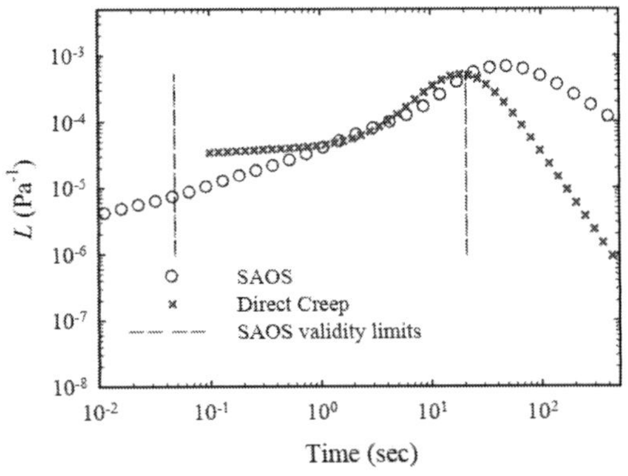

Figure 7: Retardation spectra from the SAOS and the direct creep experiment for PP/PS(70/30) at 200° C

Table 2: Changes in volume average radius due to the incomplete creep/recovery and direct creep experiments for PP/PS(70/30)

Experiment	Initial $\bar{R}_v(\mu m)$	Final $\bar{R}_v(\mu m)$
Incomplete creep/recovery	6.3 ± 0.6	4.6 ±0.6
Direct creep	6.3 ±1	11.9 ±1

From Table 2, it can be seen that while there was an 89% increase in the volume average radius of the dispersed phase for the PP/PS(70/30) due to 2000 seconds of creeping, almost no change in the morphology was observed using incomplete creep/recovery experiments . The large coalescence changed the overall rheological behaviour of the blend in the direct creep experiment. Such morphological modifications are clearly avoided by using incomplete creep followed by recovery.

Conclusions

The linear viscoelastic behavior of an immiscible blend of PP/PS was studied by means of creep experiments. Determination of the linear viscoelastic limit is possible by the proposed method of successive creep tests on *the same specimen*. The incomplete creep/recovery method based on Boltzmann superposition principle minimizes the morphological modifications, keeps the sample in LVE regime and can be used to study the long-time behavior of immiscible blends. The comparison between the weighted relaxation spectra from SAOS and creep experiments showed that while the SAOS tests are accurate at short times, creep experiments more accurately represent the relaxation processes at relatively long times, longer than the validity limit of SAOS tests. It is possible to extend the window of experimentally accessible LVE response of immiscible blends by combining SAOS and creep results to construct a composite retardation spectrum which contains the information of both experiments and can be used to calculate material functions and make an extended relaxation spectrum over a broader relaxation time range. The composite relaxation spectrum then can be used to analyze the relaxation processes of the interface in immiscible blends.

Acknowledgements

This work was supported by the NSERC discovery grants program. The authors are pleased to acknowledge FAPESP, CNPq and Kleber Vaccioli for the preparation of the blends.

References

Bousmina, M. "Rheology of polymer blends: Linear model for viscoelastic emulsions", Rheol Acta 38: 73-83 (1999)

Calvao P. S., Yee M., Demarquette N. R., "Effect of composition on the linear viscoelastic behavior and morphology of PMMA/PS and PMMA/PP blends" Polymer 46: 2610–2620 (2005)

Cavallo R., Guido S., Simeone M., "Drop deformation under small-amplitide oscillatory shear flow", Rheol Acta 42: 1-9 (2003)

Dealy J. M., Ronald G. Larson, "Chapter Structure and Rheology of Molten Polymers- From Structure to Flow Behavior and Back Again", Hanser Punlisher, Munich, p.(2006)

Friedrich C. Antonov YY, "Interfacial relaxation in polymer blends and gibbs elasticity", Macromolecules Vol. 40, no. 4, pp. 1283-1289 (2007)

Ghodgaonkar P. G. and Sundararaj U., "Prediction of dispersed phase drop diameter in polymer blends: The effect of elasticity", Polymer engineering and science, Vol. 36, No. 12, (1996)

Graebling D. and Muller R., "Rheological behaviour of polydimethylsiloxane/polyoxyethylene blends in the melt. Emulsion model of two viscoelastic liquids", J Rheol 34:193-205 (1990)

Graebling D., Muller R., and Palierne J.F., "Linear viscoelasticity of incompatible polymer blends in the melt in relation with interfacial properties", Journal de physique IV, colloque C7, supplement au Journal de physique III, Vol. 3, 1525-1534 (1993a)

Graebling D., R. Muller, and J.F. Palierne, "Linear viscoelastic behaviour of some incompatible polymer blends in the melt. Interpretation of data with a model of emulsion of viscoelastic liquids", Macromolecules, 26,320-329 (1993b)

He C., P. Wood-Adams, J. M Dealy, "Broad frequency range characterization of molten polymers", J. Rheol. 48(4), 711-724 July/August (2004)

Honerkamp J. and Weese J. "A nonlinear regularization method for the calculation of relaxation spectra", Rheol Acta 32:65-73 (1993)

Macaubas P. H.P., Demarquette N. R., "Interfacial tension, morphology and linear viscoelasticity behaviour of PP/PS blends", Polimeros, ciencia e tecnologia, 71-77 (1999)

Martin P., Carreau P.J., Favis B.D., R. Jerome, "Investigating the morphology/rheology interrelationships in immiscible polymer blends", J.Rheol.44, 569-583 (2000)

Palierne J.F. "Linear rheology of viscoelastic emulsions with interfacial tension", Rheol Acta 29:204-214 (1990)

Ren J., Silva AS. Krishnamoorti R, "Linear Viscoelasticity of Disordered Polystyrene–Polyisoprene Block Copolymer Based Layered-Silicate Nanocomposites", Macromolecules 33(10), pp 3739-3746 (2000)

Velankar S., Van Puyvelde P., Mewis J., and Moldenaers P., "Steady-shear rheological properties of model compatibilized blend", J. Rheol. 48(4), 725-744 (2004)

Vinckeier I., Moldenaer P., Mewis J., "Relationship between rheology and morphology of model blends in steady shear flow", J. Rheol, 40, 613- 631 (1996)

Wood-Adams PM, Dealy JM, "Inferring the Level of Long Chain Branching in mPE Using Rheological Data", Macromolecules, 33, 7481-7488 (2000)

Yee M., Calvao P. S., Demarquette N. R., "Rheological behaviour of poly(methyl methacrylate)/polystyrene (PMMA/PS) blends with the addition of PMMA-ran-PS", Rheol Acta, 46: 653-664 (2007)

THERMAL, MECHANICAL AND RHEOLOGICAL PROPERTIES OF HOMO HDPE SYNTHESIZED AT DIFFERENT POLYMERIZATION CONDITIONS

Mohammad A. Parvez[1], Ibnelwaleed A. Hussein[1*], Joao BP Soares[2], Mohammed A. Suleiman[1]

[1]Department of Chemical Engineering, King Fahd University of Petroleum and Minerals, Dhahran 31261, Saudi Arabia, ihussein@kfupm.edu.sa;maparvez@kfupm.edu.sa; smawwal@kfupm.edu.sa
[2]Department of Chemical Engineering, University of Waterloo, Waterloo ON N2L 6M2, Canada, jsoares@uwaterloo.ca

Abstract

Polymerization of linear high density polyethylene (HDPE) was carried out in a 300 ml semi-batch autoclave reactor equipped with a mass flowmeter and a temperature control unit. The polymerization was performed with a continuous ethylene flow to meet the desired ethylene pressure. Different combinations of polymerization conditions were used. Two levels of temperatures (60°C and 80°C); three levels of pressures of hydrogen (0, 2 and 5 atm) and two levels of polymerization times (0.5; 1 hr) were used. The synthesized polymers were characterized by GPC-IR for molecular weight and molecular weight distribution. Thermal properties of all linear HDPE were measured in TA Q1000 DSC for crystallinity, melting and crystallization peak temperatures. Mechanical properties such as tensile modulus, tensile strength and strain at break were measured using Instron testing machine. Rheological properties were measured in ARES rheometer. Thermal, mechanical and rheological properties of the different HDPE resins were correlated to their polymerization conditions. Results were analyzed to assess the impact of the different polymerization conditions on the thermal, mechanical and rheological properties of HDPE.

Introduction

Heterogeneous Ziegler-Natta catalysts are responsible for most of the industrial production of polyethylene and polypropylene. A unique feature of these catalysts is the presence of more than one active site type, making polyolefins with broad distributions of molecular weight (MWD), chemical composition distribution (CCD) and stereoregularity. These distributions influence the rheological and physical properties of polyolefins and are responsible for their performance and final applications.

Much work has been done by different researchers to improve the properties of polyethylene by blending or modifying the polymer. In particular, Hussein et al., 2003; Hussein, 2003, 2004, 2005; Hussein and Hameed 2005; Hameed and Hussein 2002, 2004.

The objective of this work aimed at relating thermal, mechanical and rheological properties to the reaction condition to predict the properties before synthesizing. The thermal and mechanical properties of the combined polymer fractions will provide tools for modeling and predicting the overall properties.

Experimental

Ethylene Polymerization

Polymerizations were carried out in a 300 ml semi-batch autoclave reactor equipped with a mass flowmeter and a temperature control unit consisting of a cooling coil and an electric heater. The polymerization temperature was maintained within ± 0.2°C of the set point. Prior to each reaction, the reactor was purged five times with nitrogen, then heated to 140°C under vacuum, and purged again to the set point temperature under nitrogen flow. A volume of 200 ml solvent and the triethyl aluminum activator were transferred to the reactor, and stirred for 5 minutes. Catalyst slurry with hexane was injected into the reactor and stirred for 10 minutes. The polymerization was performed with a continuous ethylene flow to meet the desired ethylene pressure under a stirring rate of 500 rpm. At the end of the polymerization time, the reactor was rapidly vented and the obtained polymer was precipitated in 200 ml ethanol, then filtered and dried under vacuum.

Thermal Characterization

Thermal properties of all linear HDPE were measured in TA Q1000 DSC for crystallinity, melting and crystallization peak temperatures. Three cycles were completed for each sample. The first heating is to remove the previous thermal history. The heating and cooling rate were 10°C/min 5°C/min respectively. Non-hermetic aluminum pans were used and 99.99 % pure nitrogen gas was purged during the DSC scanning.

Mechanical Testing

The tensile properties of all the samples were measured using Instron 5567 machine. The crosshead speed was 100 mm/min. The experiments were done according to the ASTM D-638 method.

Rheological testing

The powder samples were blended with 2500 ppm antioxidant (50% Irganox and 50% Irgafox) in Haake polydrive blender. The circular disc specimens were prepared using carver hydraulic press. The ARES rheometer was used in the dynamic mode to study linear rheological properties for each sample as a function frequency, ω. A 25 mm cone-and-plate was used. All tests were conducted at 200°C with nitrogen as the heating medium to avoid degradation during testing.

Results and Discussion

Effect of reaction conditions on thermal properties

The thermal analysis results showed that the reaction conditions (i.e pressure, temperature and reaction time) have direct influence on the thermal properties of the samples. Figure 1 shows the melting peak temperature has linear relationship with hydrogen pressure to the reactor. Increased the hydrogen pressure decreases the melting peak temperature. And it is also clear that the melting temperature increases with increasing ethylene pressure to reactor. However, the cooling peak temperature was not affected significantly by the reaction parameters. Cooling peak temperature were in the range of 117-120°C with slight variation due to the reaction conditions.

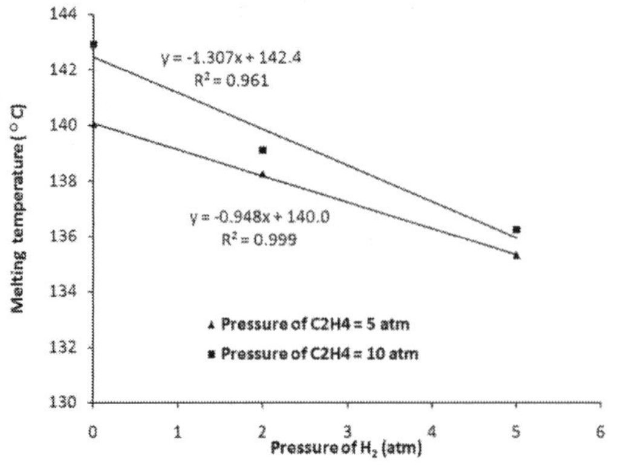

Figure 1: Melting peak temperature as function of hydrogen pressure to the reactor showing the effect of ethylene pressure to the reactor.

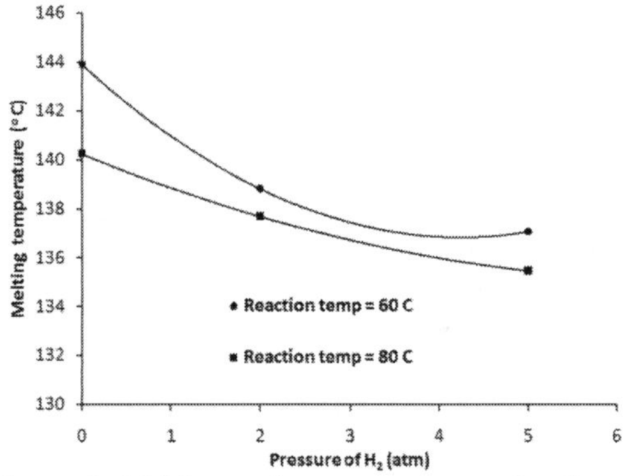

Figure 2: Melting peak temperature as function of hydrogen pressure to the reactor showing the effect of reaction temperature.

Figure 2 shows that the reaction temperature has also the similar effect on the thermal properties but the relation is not linear in this case. The increase in reaction temperature and hydrogen pressure decreases the melting peak temperatures.

The crystallinities were affected by the reaction parameters. The percent crystallinities for all the samples were in the range of 52-86% and 56-82% based on cooling and heating. Figure 3 shows that the increase in both hydrogen and ethylene pressure to the reactor increases the cooling crystallinities. However, the melting crystallinity first decreased and then increased with hydrogen pressure to the reactor.

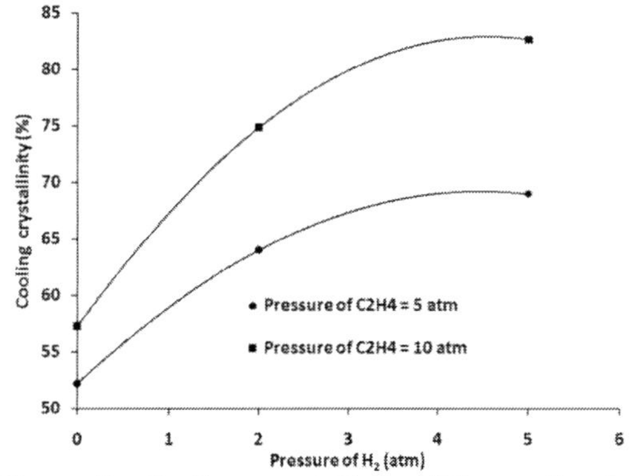

Figure 3: Cooling crystallinity (%) as function hydrogen pressure showing the effect of Ethylene pressure to the reactor.

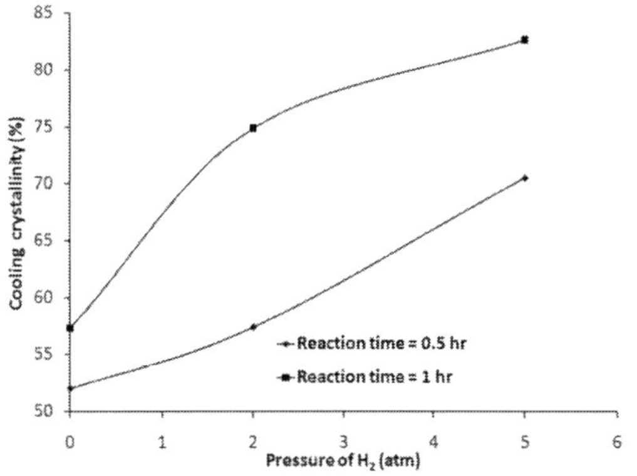

Figure 4: Cooling crystallinity (%) as function hydrogen pressure showing the effect of reaction time.

Figure 4 shows the effect of reaction time on the thermal properties. In general, the increase in reaction time increases the percent crystallinities. It has also the effect on melting and crystallization peak temperatures.

Effect of reaction conditions on mechanical properties

The tensile moduli for all the samples were in the range of 390-1020 MPa. It increases linearly with increase in hydrogen and ethylene pressure to the reactor as it can be seen from Figure 5. The Figure 6 shows that the reaction temperature also increases the tensile modulus. The increase in reaction time increases the tensile modulus. For 30 minutes increment in reaction time increase the modulus around 15%.

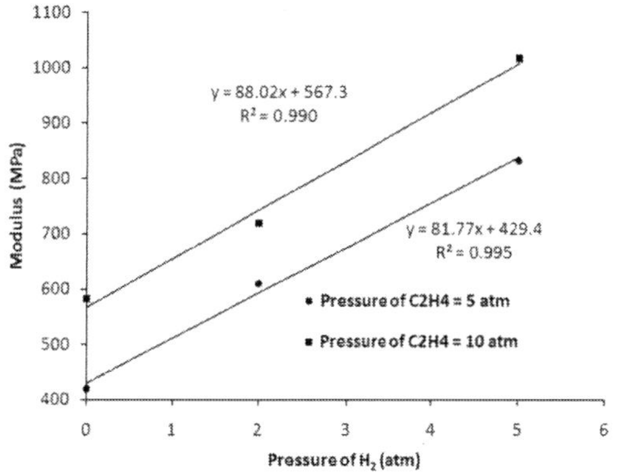

Figure 5: Tensile Modulii as function of hydrogen pressure showing the effect of ethylene pressure to the reactor.

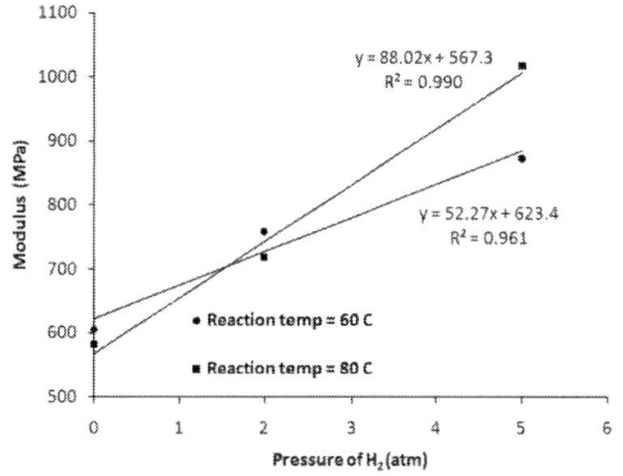

Figure 6: Tensile Modulii as function of hydrogen pressure showing the effect of reaction time.

Figure 7 represents the variation of tensile stress with ethylene and hydrogen pressure. The tensile stress increases with the increasing ethylene pressure to the reactor. However, hydrogen pressure has optimum point at 2 atm pressure and then decreases.

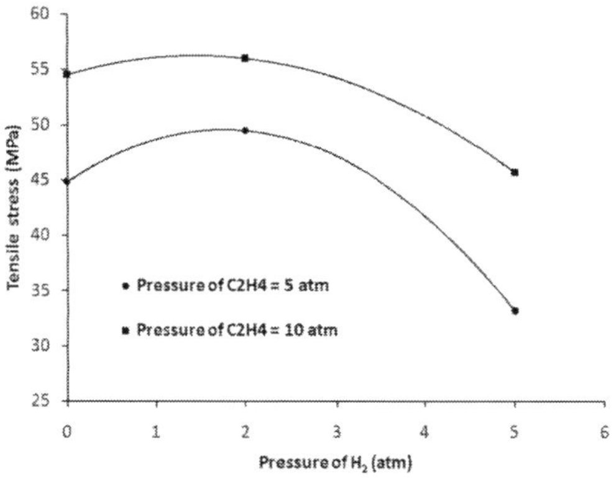

Figure 7: Tensile stress as function of hydrogen pressure showing the effect of ethylene pressure to the reactor.

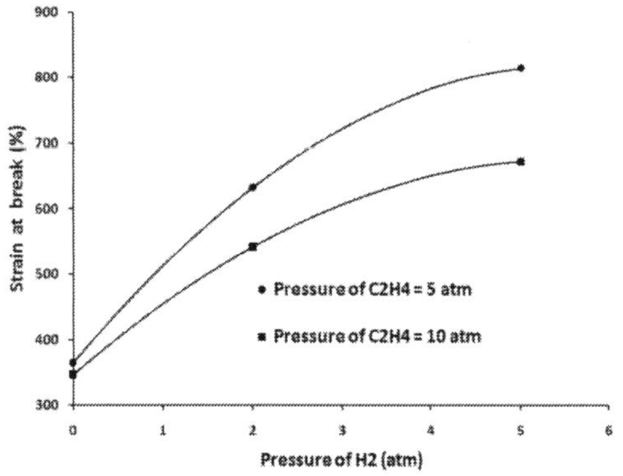

Figure 8: Strain at break as function of hydrogen pressure showing the effect of ethylene pressure to the reactor.

The ultimate breaking properties were shown to improve by increasing the hydrogen pressure to the reactor. Figure 8 shows that the strain at break is increasing with hydrogen pressure but reduces with ethylene pressure. The increase in reaction temperature from 60°C to 80°C increases the strain at break approximately 7%. Similarly a increase in reaction time from 0.5 hr to 1 hr increase the strain at break around 14%. However, the yield properties were not affected significantly by these reaction parameters.

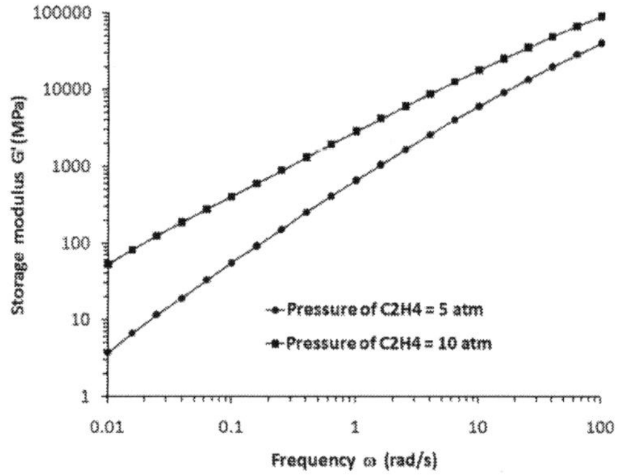

Figure 9: The results of frequency sweep of sample synthesized at two different level of ethylene pressure.

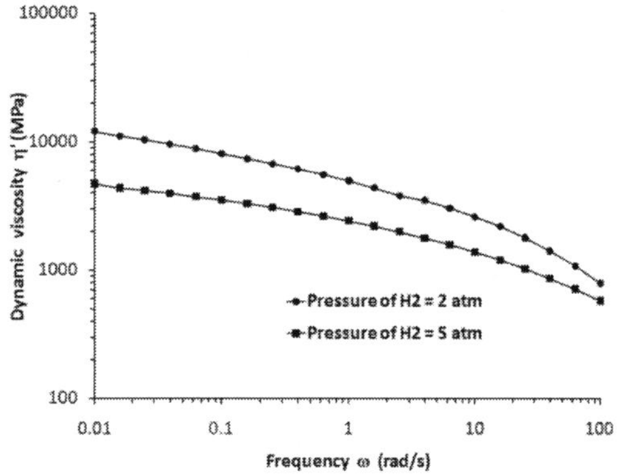

Figure 10: Dynamic viscosity versus frequency plot of two samples synthesized at two different level of hydrogen pressure.

Effect of reaction conditions on rheological properties

The rheological properties of all the samples were affected by the reaction conditions. The result shows that the sample synthesized at higher level of ethylene pressure has higher storage modulus than those synthesized at the lower level of ethylene pressure to the reactor. Figure 10 represents the dynamic viscosity versus frequency plot from it is clear that increase in hydrogen pressure decreases the dynamic viscosity. Figure 11 shows that increase in hydrogen pressure decreasing the storage modulus.

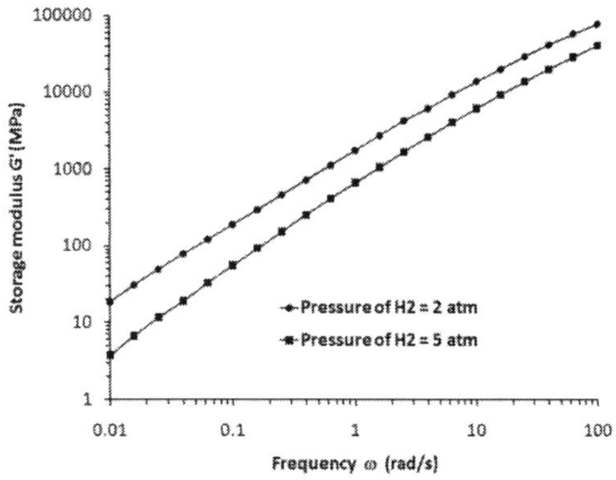

Figure 11: Storage modulus versus frequency plot of two samples synthesized at two different level of hydrogen pressure.

Conclusions

Both the hydrogen and ethylene pressure play an important role on the thermal, mechanical and rheological properties as it has been explained in the results and discussion part. The reaction time and reaction temperature also affect the final properties of the polymers. So, it is very important to control these parameter during polymerization. The end properties of these polymers were correlated to their reaction condition. It is possible to set the reaction condition to get the specific desired properties of HDPE.

Acknowledgements

This research project was supported by King Abdul Aziz City for Science and Technology (KACST) under project # AT-26-67. The authors would like to thank KFUPM for its support.

References

1. Ibnelwaleed A. Hussein, Tayyab Hameed, Basel F. Abu Sharkh, Khaled Mezghani, "Miscibility of hexene-LLDPE and LDPE Blends: Rheological Study of the Influence of Branch Content and Composition Distribution", Polymer 2003, 44(16), 4665-4672.

2. Ibnelwaleed A. Hussein, Tayyab Hameed, "Influence of Branching Characteristics on Thermal and Mechanical Properties of Ziegler-Natta and Metallocene LLDPE blends with LDPE", J. Applied Polymer Science, 97(6), 2488-2498, 2005\

3. Ibnelwaleed A. Hussein, Implications of melt Miscibility on Thermal and Mechanical Properties of metallocene LLDPE Blends with HDPE: Influence of Comonomer Type, Polymer International, 54(9), 1330-1336, 200.

4. Tayyab Hameed, Ibnelwaleed A. Hussein, "Rheological, Thermal and Mechanical Characterization of Metallocene LLDPE blends with HDPE: Influence of Mw", Polymer Journal, 38(11), 1-13, 2006.

Nonlinear Stress Relaxation Behavior of SiO_2/(Acrylic Polymer/Epoxy) Suspensions

K Koyama*, H Uematsu, M Sugimoto and Y Aoki

Department of Polymer Science and Engineering, Yamagata University, 4-3-16, Jonan, Yonezawa, Yamagata 992-8510, Japan- koyama@yz.yamagata-u.ac.jp

Nonlinear stress relaxation modulus $G(t,\gamma)$ was examined for the SiO_2 suspensions in a blend of acrylic polymer (AP) and epoxy (EP) with various SiO_2 volume fractions (ϕ) at various temperatures (T). AP/EP contained 70 vol% of EP. At $\phi \leq 30$ vol%, the SiO_2/(AP/EP) suspensions behaved as a viscoelastic liquid, and the time-strain separability, $G(t,\gamma) = G(t)h(\gamma)$, was applicable at long time. $h(\gamma)$ of the suspensions was more strongly dependent on γ than that of the medium (AP/EP). At $\phi = 35$ vol% and $T = 100$°C, and $\phi \geq 40$ vol%, the time-strain separability was not applicable. The suspensions exhibited a critical gel behavior at $\phi = 35$ vol% and $T = 100$°C characterized with a power-law relationship between $G(t)$ and t; $G(t) \propto t^{-n}$. The relaxation exponent n was estimated to be about 0.45, which was in good agreement with the result of linear dynamic viscoelasticity. This nonlinear stress relaxation behavior is attributable to strain-induced disruption of the network structure formed by the SiO_2 particles.

Introduction

The integrated circuit (IC) chips are three-dimensionally integrated on a wafer using an adhesive film in order to increase the capacity of memory without a change of size. Our adhesive film consists of three components: acrylic polymer (AP), epoxy resin (EP) and SiO_2 particles. It is very important to investigate the nonlinear rheological behavior of SiO_2/(AP/EP) because the adhesive film undergoes large strain at high strain rates in the actual processing. However the nonlinear rheology of this complex system has never been investigated. In this study, we focus on the effect of SiO_2 on larger deformation, nonlinear stress relaxation modulus $G(t,\gamma)$ and damping function $h(\gamma)$ of the SiO_2/(AP/EP) suspensions having various ϕ

Experimental

Materials

Blend of an acrylic polymer (AP) with linear chain and low molecular weight epoxy (EP) was used as suspending medium. Both AP and EP were provided from Hitachi Chemical Co. Ltd. The weight-averaged molecular weight, M_w, and heterogeneity index, M_w/M_n, of AP are 360,000 and 2.2, respectively, where M_n is the number-average molecular weight. EP consists of bisphenol f diglycidylether (L-epo), cresol novolac type epoxy resin (S-epo) and bisphenol A novolac (CA). The composition ratio of L-epo and CA is equal and that of S-epo is one third of L-epo. The content of EP in (AP/EP) was 70 vol%. Volume fractions of SiO_2 particles were 0, 20, 30, 35 and 40 vol%. The average diameter of SiO_2 was 0.5 μm. The surface of SiO_2 was treated with epoxy silane coupling agent to prevent agglomeration of SiO_2 particles.

Measurements

Step-shear stress relaxation measurements were carried out with a strain-controlled rheometer (ARES, TA Instruments) at several temperatures between 100 °C. A parallel-plate geometry with 8 mm diameter was used. The shear relaxation modulus $G(t,\gamma)$ was corrected with Soskey and Winter's equation [1]. The shear strain ranged from 0.01 to 8.0.

Results and Discussion

For the SiO_2/(AP/EP) suspensions with various particle concentrations, $\phi = 0$ and 30 vol %, Fig.1 shows the nonlinear relaxation modulus $G(t,\gamma)$ at 100 °C. To confirm the linearity for small γ, we also present the linear relaxation modulus $G(t)$ calculated from the G' and G'' data. The solid curves represent this $G(t)$ obtained with the method of Schwarzl[2]. For the medium $\phi = 0$ vol%, the $G(t,\gamma)$ curves for $\gamma < 0.5$ are very close to each other and agree with the linear $G(t)$ curve. This $G(t,\gamma)$ shows downward deviation from $G(t)$ at $\gamma > 0.5$. For the suspension having $\phi = 30$, the linear region is limited $\gamma < 0.1$.

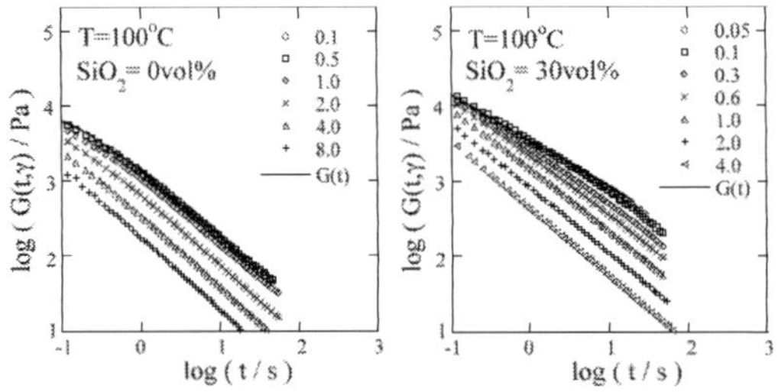

Fig. 1 Stress relaxation modulus $G(t,\gamma)$ at various strain γ obtained for SiO_2/(AP/EP) having ϕ =0 vol % and 30vol % at 100°C.

We evaluate the damping function for these samples as

$$h(\gamma) = G(t,\gamma)/G(t) \quad (t > \tau_k) \qquad (1)$$

The $G(t, \gamma)$ curves for $\phi = 0$ and 30 vol% at 100°C are shifted vertically (in the double-logarithmic scales) to be superposed on respective reference curves obtained for the smallest γ. The reference curve are in good agreement with the linear $G(t)$ curves at $t \geq \tau_k$. In other words, time-strain separability for medium (AP/EP) is valid. We tried the time-strain separability for suspensions. If τ_k is chosen as 5.5 ± 1.5s at 100°C in the suspension system, time-strain separability is approximately satisfied. Fig.2 (a) illustrates the damping function $h(\gamma)$ of the AP/EP and the suspensions. $h(\gamma)$ data for the AP/EP monotonically decrease with increasing of strain. On the other hand, strain dependency of $h(\gamma)$ becomes stronger with increasing SiO_2.

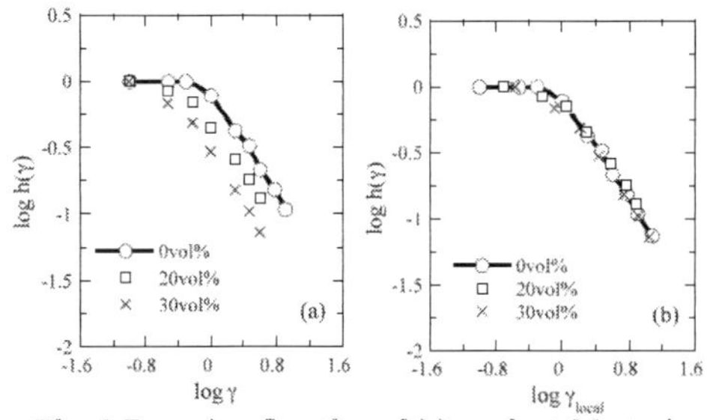

Fig. 2 Damping functions $h(\gamma)$ against (a) strain γ and (b) local strain γ_{local} of SiO_2/(AP/EP) suspensions having different SiO_2 volume fraction.

Aoki et al. have reported that strong damping was related to filler effect[3]. For well-dispersed suspension system, they discussed the strong damping in terms of local strain γ_{local} and external strain γ_{Ex}. To examine this expectation, we evaluated γ_{local} for the case of the simplest filler effect[4],

$$\gamma_{local} = \left[1 + 2.5\phi + 14.1\phi^2\right]\gamma_{Ex} \qquad (2)$$

where γ_{Ex} is the externally applied (nominal) strain, and ϕ is the volume fraction of the SiO_2 particles. The $h(\gamma)$ data of suspension are plotted against γ_{local} thus evaluated. The plots for various ϕ approximately agree with each other and coincide with that of AP/EP as shown in Fig.2 (b).

Fig.3 shows the nonlinear relaxation modulus $G(t,\gamma)$ of the SiO_2/(EP/AP) suspensions having ϕ = 35 and 40 vol % at 100°C. As seen in Fig.3, $G(t,\gamma)$ of suspensions agree with the linear relaxation modulus $G(t)$ calculated from the G' and G" data at $\gamma \leq 0.05$ and 0.01, respectively. This $G(t,\gamma)$ decreases with increasing γ, and the nonlinearity prevails. We should note that it was impossible to evaluate $h(\gamma)$ because time-strain separability could not be applicable for the suspensions.

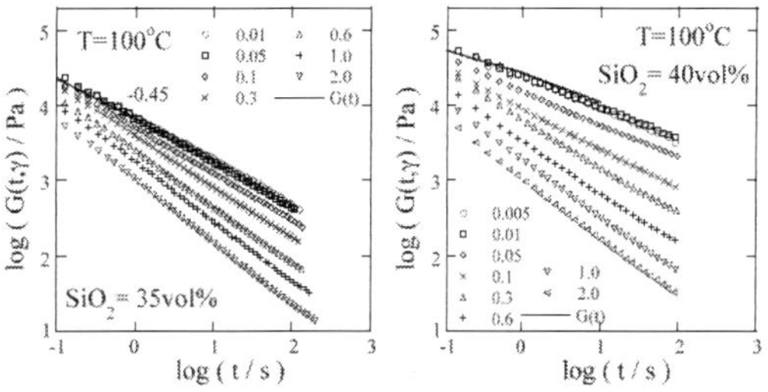

Fig. 3 Stress relaxation modulus $G(t,\gamma)$ at various strain γ obtained for SiO_2/(AP/EP) suspensions having ϕ =35 vol % and 40vol % at 100°C.

From the linear viscoelasticity, it is found that the SiO_2/(AP/EP) suspensions exhibited a liquid-like behavior at $\phi \leq 30$ vol%, the critical gel behavior at ϕ = 35 vol % at 100°C, and a solid-like behavior at $\phi \geq 40$ vol %. In critical gel state, it is known that the relaxation modulus $G(t)$ is expressed by a power law; $G(t) = St^{-n}$ [5]. Especially, the magnitude of n agrees with the critical gel exponent n at critical gel temperature. We found that the gradient of $G(t)$ is 0.45 in Fig. 3. It is suggested that critical gel behavior is exhibited in Fig. 3 because the slope of $G(t)$ in the double-logarithmic scales agrees with the critical gel exponent (n=0.45) calculated from linear viscoelasticity. Therefore, it is suggested that time-strain separability could not be satisfied because of strain-induced disruption of the network structure formed by the SiO_2 particles therein.

Conclusions

At $\phi \leq 30$ vol%, the time-strain separability, $G(t,\gamma) = G(t)h(\gamma)$, was applicable at long time. At $\phi \geq 35$ vol%, the SiO_2/(AP/EP) suspensions behaved as a critical gel and gel behavior, the time-strain separability was not applicable because gel network structure was broken with increasing strain

Acknowledgements

Authors are grateful to Dr. T. Inada, T. Iwakura and T Morimoto at Hitachi Chemical Co., Ltd. for support of this work

References

(1)Soskey P R, Winter H H, "Large Step Strain Experiments with Parallel-Disk Rotational Rheometers", J Rheol., **28**, 625-645 (1984)
(2)Schwarzl FR, "Numerical Calculation of Stress Relaxation Modulus from Dynamic Data for Linear Viscoelastic Materials", Rheol Acta, 14, 581-590 (1975)
(3)Aoki Y et al., "Nonlinear stress relaxation of ABS polymer in the molten state", Macromolecules, **32**, 3100-3107 (2001)
(4)Guth E, Gold O, "On the Hydrodynamic Theory of the Viscsoity of Suspension", Phys. Rev., **53**, 322 (1938)
(5)Ferry JD.: Viscoelastic Properties of Polymers, 3rd edition, Wiley, New York (1980)

STUDY OF POLYMER/CLAY NANOCOMPOSITE MORPHOLOGY: A RHEOLOGICAL AND ELECTRON MICROSCOPIC APPROACH

Saikat Banerjee[1], Mangala Joshi[2], Anup K. Ghosh[1*]

[1]*Centre for Polymer Sci. and Engg., Indian Institute of Technology, New Delhi - 110016, India-* *anupkghosh@gmail.com; saikat_tech@rediffmail.com*
[2]*Department of Textile Technology, Indian Institute of Technology, New Delhi -110016, India-* *mangalajoshi9@gmail.com*

Abstract

Understanding of the complex mechanism of dispersion and exfoliation of the clay tactoids may allow us to better control the final morphology and the homogeneity of clay nanocomposites and thus their macroscopic properties. Melt rheology is an important tool to characterize polymer–clay nanocomposites. In the present study, shear-induced clay dispersion in polypropylene matrix was ascertained from the changes of rheological properties under intermediate and large-amplitude oscillatory shear fields as well as by using transmission electron microscopy (TEM). PP-g-MA compatibilizer and Cloisite 15A were melt blended with polypropylene in a co-rotating twin screw extruder at selected screw speed and temperature profile. The effect of ratio of the compatibilizer to the organo-clay (1:1, 2:1, 3:1, 4:1 compositions) on the dispersion was studied. The dispersion behaviour of organoclay in the polymer matrix under oscillatory shear field was assessed through a change in complex viscosity and both the dynamic modulii. It was further supported by frequency dependent viscoelastic spectra, thermal analysis, dielectric spectroscopy and TEM. The main utility of the TEM methodology is to quantify the exfoliated or intercalated clay morphology with small number of layers with stacks. A new methodology was applied to determine the dispersion parameter, based on the measurement of the free-path spacing distance between the single clay sheets from the transmission electron microscopic images. Among all these compositions, the 3:1 compatibilizer to organo-clay ratio showed a better exfoliated morphology.

Introduction

The use of rheological techniques enables a real-time monitoring of process influences on the microstructure of materials over a short and medium period of time where the structure of materials exposed to the testing processes is not destroyed under small-strain amplitude. It has been extensively used in the study of nanocomposites in conjunction with basic characterization techniques like XRD, SEM, TEM etc. The advantages of rheological methods relative to the others are that the measurements can be performed in the molten state and different rheological methods can be used to study the response of the nanocomposites to both linear and non-linear deformation. The useful information that can be derived from dynamic rheology are storage modulus (G'), loss modulus (G"), damping behavior (tanδ) and complex viscosity (η*). These experiments give the idea about elastic or solid-like and viscous or liquid-like response as well as zero shear viscosity and relaxation behavior of the materials. These parameters are dependent on molecular structure of the material and hence dynamic rheological measurements help in understand the material property from molecular level. Dynamic rheological measurements are sensitive to size, shape and surface behavior of the fillers. When nanofillers are introduced in polymer matrix; the rheological behavior of nanocomposites becomes even more interesting, since this kind of reinforcements have been found to alter molecular mobility and crystallinity of the matrix [1]. Compared with the rheological behavior of the pure resin, polymer composites generally exhibit pronounced elastic properties and long relaxation times [2].

Experimental

Materials

The isotactic polypropylene (iPP, REPOL H110 MA, MFI~11) used in this study was commercially available from Reliance Industries Ltd, India. Maleic anhydride-grafted polypropylene (PP-g-MA, P-406) with a melt index of 24 gm/10 min at 190°C and 2.16 kg load was purchased from Pluss Polymer Ltd. India. The clay, Cloisite 15A, was procured from Southern Clay Products USA, which has a cation exchange capacity of 125 meq/100 gm, and was modified with dimethyl, hydrogenated tallow and quaternary ammonium.

Nanocomposite Processing

The melt mixing was done in a PRISM EUROLAB 16 co-rotating twin screw extruder having a high shear screw configuration for better mixing. The processing temperature profile was selected such as to ensure proper viscosity for the mixing while at the same time minimizing degradation of the polymer. Before melt processing, all the components were dried at 80 °C for 24 hrs. The feed rate was kept constant.

Characterization Techniques

The dynamic shear rheological measurements were carried out in Bohlin C-VOR rotational rheometer (Malvern Instruments Limited, UK). Parallel plate geometry with 25mm diameter and a 1mm gap was used in all tests. All the samples were initially tested using an amplitude sweep between 0.0001 to 10 strain units to determine the linear viscoelastic region (LVR). The frequency sweeps between 0.01 Hz to 100 Hz at 0.1 strain units were performed maintaining temperature at 200°C. For transmission electron microscopy (TEM) freshly cut glass knives with cutting edge of 45° were used to get the cryosections of 50 nm thickness by using a Leica Ultracut UCT microtome. JEOL-2100 electron microscope (Tokyo, Japan) having LaB6 filament and operating at an accelerating voltage of 200 kV was utilized to obtain the bright field images of the cryo-microtomed samples.

Results and Discussion

Rheological Analysis

Shear thinning component 'n' is a semi-quantitative measurement of nanodispersion which can be measured from power law equation, $\eta = A f^{n}$, where η= apparent viscosity, A= pre-exponential factor, f= oscillation frequency of the rheometer equivalent to shear rate [3]. At low shear rates the rheological response is most representative for the unperturbed platelet structure (Figure 1). Well exfoliated samples display a higher magnitude of n, while absence of exfoliation results in n close to zero (Table 1)

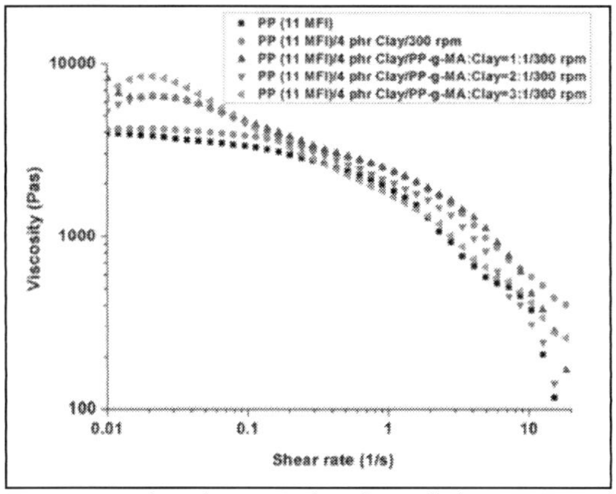

Figure 1. Viscosity as a function of shear rate and compatibilizer to clay ratio.

Figure 2 shows the elastic modulus G' as a function of the frequency f for the neat Polypropylene and for the Polypropylene/Cloisite 15A composites. G' is shown because it is the parameter that most reflects the rheological changes upon clay addition [2,4-5]. The neat polymer exhibits a viscous response with the G' scaling at low f in a manner characteristic of viscous fluids, i.e., with $G' \sim f^2$. Adding clay particles with an increase in compatibilizer to clay ratio causes a systematic increase in G', especially at low frequencies, where G' becomes less frequency-dependent. At the higher ratio, approaches a low-f plateau, which means that the sample relaxes very slowly at long time scales [2, 4]. G' approaches a low-f plateau, which means that the sample relaxes very slowly at long time scales. Such behavior is characteristic of a volume-filling "jammed" network of particles in the nanocomposite [2,6]. The higher compatibilizer to clay ratio leads to better exfoliation of the clay which forms a percolated network structure. With increasing frequency, more energy is lost due to friction between the PP chains and clays, which resulted in increase in G" (Figure 3).

Table 1. Shear thinning component and mechanical properties as function of compatibilizer to clay ratio.

Compositions	n, at low shear rate	Tensile Mod (MPa)	Flexural Mod (MPa)
PP (11 MFI)	-0.46154	229.38	1077.21
PP/ 4phr Clay	-0.32563	290.65	1255.21
PP/ 1:1 ratio	-0.31714	289.18	1373.85
PP/ 2:1 ratio	-0.34817	323.29	1485.17
PP/ 3:1 ratio	-0.41343	325.18	1445.54

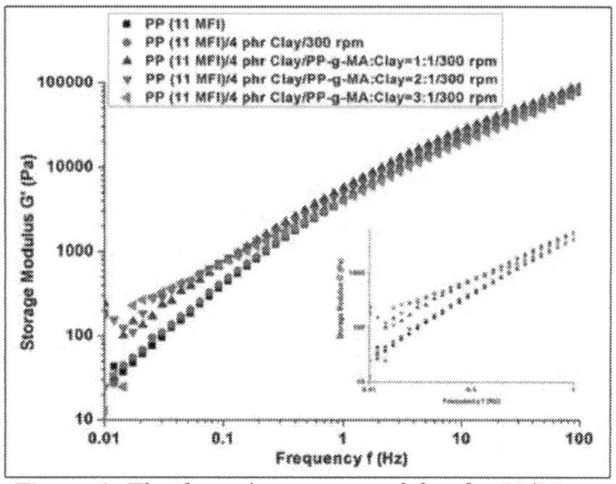

Figure 2. The dynamic storage modulus for PP/PP-g-MA/Clay samples.

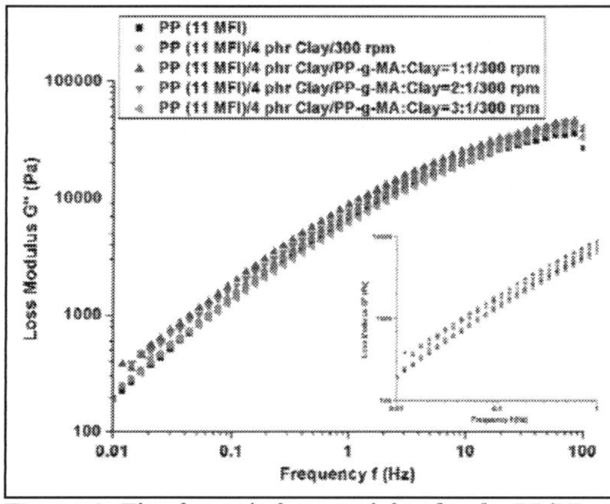

Figure 3. The dynamic loss modulus for the PP/PP-g-MA/Clay samples.

Transmission electron Microscopy

A new TEM methodology based on free-path spacing measurement between the single clay platelets was used to quantify the clay dispersion in clay/polymer nanocomposites [7]. The dispersion value $D_{0.1}$ was calculated from the distribution of the free-path spacing data using the equation:

$D_{0.1} = 1.1539 \times 10^{-2} + 7.5933 \times 10^{-2} (\mu/\sigma) + 6.6838 \times 10^{-4} (\mu/\sigma)^2 - 1.9169 \times 10^{-4} (\mu/\sigma)^3 + 3.9201 \times 10^{-6} ((\mu/\sigma)^4$,

where μ is the mean spacing , σ is the standard deviation. The $D_{0.1}$ parameter is dimensionless and is not related to the filler size, shape, aggregation, concentration, or the existence of additional filler. A higher $D_{0.1}$ value indicates more spacing data close to the mean μ and thus a better dispersion level.

Each micrograph was divided by using gridlines. The micrograph was divided into 100 samplings by using a 10x10 grid. The measurement scale bar was calibrated and the free-path distance x_i, the distance of the matrix, between the clay platelets, along the intercept lines was measured using a software. Calculation based on free-path distance data and histogram, sample mean $\mu = \sum_i x_i/N$ and standard deviation $\sigma = \sqrt{\sum_i (xi-x)^2/(N-1)}$ was calculated, where N is the sampling number (=100). Putting this μ and σ in the equation, the degree of exfoliation value $D_{0.1}$ (%) was estimated (Table 2). It was found that, if the value is below 4 % the morphology is immiscible. A value between 4 to 7 % shows intercalated structure and above which it is exfoliated. It is clear from Table 2, that a compatibilizer to clay ratio of 2:1shows a better exfoliated structure (Figure 4).

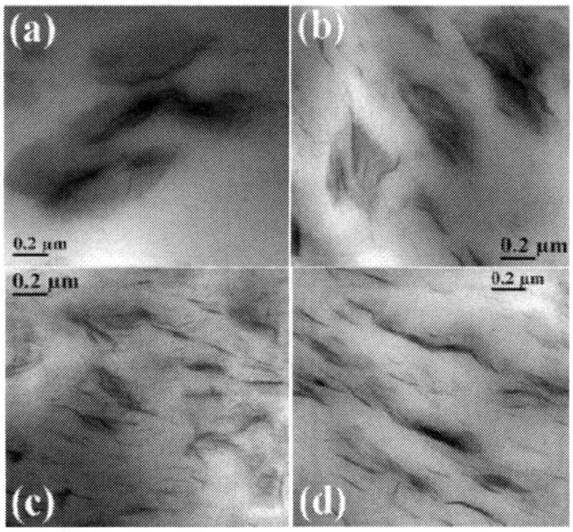

Figure 4. TEM images of PP (11 MFI)/PP-g-MA/Cloisite 15A mixture with a compatibilizer to clay ratio: (a) without compatibilizer, (b) 1:1, (c) 2:1 and (d) 3:1 at 12000x magnification.

Table 2. Morphology as function of compatibilizer to clay ratio.

PP-g-MA:Clay	μ (nm)	σ (nm)	$D_{0.1}$ (%)	Morphology
0	107.5	273.4	4.2	Intercalated
1:1	231.5	357.8	6.1	Intercalated/partially exfoliated
2:1	66.3	68.0	8.6	Well dispersed/exfoliated
3:1	28.6	25.6	9.7	Well exfoliated

Conclusions

In this work, a series of rheological responses of polypropylene/clay nanocomposite were mapped out to various compatibilizer to clay ratio and degrees of clay dispersion. The level of exfoliation was characterized by dielectric spectroscopy and TEM, was quantitatively correlated with a characteristic rheological response. A compatibilizer to clay ratio of 2:1 showed a better exfoliated morphology that was further supported by mechanical properties.

Acknowledgements

The authors thank Prof. Ratnamala Chatterjee of Indian Institute of Technology, Delhi, India for her cooperation in dielectric spectroscopic analysis. Financial support for this research by the Department of Science & Technology, Government of India and Reliance Industries Ltd India is gratefully acknowledged.

References

(1) Banerjee S., et al., Polymer Composites, accepted (2010).
(2) Geabling D.,et al., Macromolecules, **26**, 320 (1993).
(3) Durmus A., et al., Polymer, **48**, 4492 (2007).
(4) Giannelis, E. P.,et al., Adv. Polym. Sci., **138**, 107 (2000).
(5) Alexandre, M., Dubois P., Mater. Sci. Eng, **28,** 1 (1999).
(6) Ray S. S., Okamoto M., Prog. Polym. Sci., **28**, 1539 (2003).
(7) Luo Z.P., Koo J.H., Polymer, **49**, 1841 (2008).

Nonlinear Rheology of Phase Separating LCST Blends: Polystyrene/Poly (Vinyl Methyl Ether)

Jafar Khademzadeh Yeganeh, Fatemeh Goharpey, Reza Foudazi*

(1) Department of Polymer Engineering. Amirkabir University of Technology - Tehran - Iran

We investigated nonlinear rheology of different time dependant phase-separating morphologies for PS/PVME blends which a large difference between glass-transitions of PS and PVME (about 125oC) make this blend dynamically asymmetric. For this purpose phase diagram of PS/PVME blend, which exhibit lower critical solution temperature (LCST) was determined form the dynamic temperature sweep experiments. Phase contrast optical microscopy was employed to investigate morphological evolution of PS/PVME blends at various region of obtained phase diagram at constant temperature of 105 oC. At this temperature, depending on composition, in addition to the usual phase separation mechanisms (SD, NG) viscoelastic phase separation was observed, indicating interplay of thermodynamics and viscoelasticity in controlling phase separation behavior of PS/PVME blends. For sample at the PVME-rich binodal region of phase diagram, stress overshoot, observed in the shear start up experiments, was increasing function of phase separation time, which was attributed to the greater deformability of larger droplets. In contrast, for sample at unstable region, stress overshoot was decreasing function of phase separation time; and at early stages of spinodal decomposition a strong overshoot was observed which was attributed to the stretching and/or orienting of cocontinuous domains with a large interfacial area. The experimental results of stress growth experiments were compared with the predictions of Lacroix model. It was observed that, even without considering interfacial contribution, Lacroix model predicted much higher than experimental results. This discrepancy was attributed to the combination effects of interfacial slip and self-generated stresses induced in matrix phase. Furthermore, the validity of Doi-Ohta scaling relationships for different morphologies was studied.

Influence of thermo-rheological history on carbon nanotube composite melts: Combined rheo-electrical experiments

Ingo Alig, Dirk Lellinger, Tetyana Skipa*

Deutsches Kunststoff-Institut - Darmstadt - Germany

Although, carbon nanotube (CNT)-polymer composites are commercially available a major restraint in broad market acceptance is the wide variations in electrical conductivity as function of processing conditions. The tremendous influence of processing conditions on the electrical conductivity has been demonstrated by in-line measurements during extrusion [1,2] and mould injection [3]. In order to achieve a deeper understanding of the influence of thermo-rheological prehistory on electrical and mechanical properties systematic rheo-electrical experiments have been performed in the CNT-composite melts [4-6]. Electrical conductivity and complex shear modulus (G', G'') were measured simultaneously after transient shear, during quasi-steady shear and in the quiescent melt. The time-evolution of shear modulus, viscosity and conductivity under shear were found to depend strongly on the initial state of CNT agglomeration. Both, shear induced destruction and formation of the filler network was detected. Under steady flow the electrical and mechanical properties approach stationary values, representing a "dynamic equilibrium" of the filler network. Furthermore, is has been shown that the mechanisms for charge carrier transport and viscoelasticity are different. The findings were described by a model combining the interplay of destruction and reformation of the filler network under shear. The model has been applied for FEM simulation of mould injection. [1] Alig I., Lellinger D., Dudkin, S., Pötschke, P., Polymer 48 (2007) 1020. [2] Alig I., Lellinger D., Engel M., Skipa T., Pötschke, P., Polymer 49 (2008) 1902. [3] Lellinger, L., Xu D., T. Skipa, Alig I., Phys. Stat. Sol. (b) 245 (2008) 2268. [4] Alig I., T. Skipa, Lellinger, M. Engel, S., Pötschke, P., Phys. Stat. Sol. (b) 244 (2007) 4223. [5] Alig I., Skipa T., Lellinger D., Pötschke P. Polymer 49, (2008), 3524. [6] Skipa T., Lellinger D., Saphiannikova M., Alig I., Polymer 50 (2009), DOI: 0.1016/j.polymer.2009

SHEAR AND EXTENSIONAL RHEOLOGY OF CONTROLLED MICROSTRUCTURE POLY(LACTIDE) POLYMERS

Norhayani Othman[1,4], Alberto Acosta-Ramírez[2], Parisa Mehrkhodavandi[2], John R. Dorgan[3], Savvas G. Hatzikiriakos[1*]

[1]*Department of Chemical and Biological Engineering, The University of British Columbia, Vancouver, BC, Canada- nothman@chbe.ubc.ca; hatzikiriakos@apsc.ubc.ca*
[2]*Department of Chemistry, The University of British Columbia, Vancouver, BC, Canada- alberto710@chem.ubc.ca; mehr@chem.ubc.ca*
[3]*Department of Chemical and Biochemical Engineering, Colorado School of Mines, Golden, CO, USA- jdorgan@mines.edu*
[4]*Faculty of Chemical and Natural Resources Engineering, Universiti Teknologi Malaysia, Skudai 81310, Johor Bahru Malaysia- norhayani@fkkksa.utm.my*

Abstract - In this study, a series of controlled microstructure poly(lactide) polymers (PLAs) were synthesized by using novel ring-opening polymerization methods that utilize chiral main group catalysts for enantioselective polymerization of lactide. PLAs with different ratios of *L*- to *D*- enantiomeric monomers of 100:0, 75:25, 50:50, and 0:100 and molecular weight are investigated. The rheological study includes determination of zero shear viscosity and its relationship with the molecular weight, the relaxation spectrum and its relation with molecular weight characteristics, plateau modulus and other important rheological parameters that are helpful in predicting the linear viscoelasticity of PLA polymers. Emphasis is also placed on the uniaxial melt behavior of these polymers. At low temperatures, significant strain hardening is observed which gradually disappears with increase of temperature and decrease of Hencky strain rate. The origin of the strain hardening is investigated from (i) the point of view of flow induced-crystallization using DSC and XRD as PLAs are known to crystallize extremely slowly and (ii) from the point of view of polymer melt dynamics in terms of the relative magnitudes of the characteristic times of reptation and deformation. Consistent with previous studies, it is concluded that in spite of their linear structure, PLAs exhibit strain hardening which is not due to strain induced crystallization and it is solely due to the dynamics of molecular relaxation.

Introduction

Polymers from renewable resources have attracted an increasing amount of attention over the last decades, predominantly due to environmental concerns and the desire to reduce dependence on petroleum, since petroleum resources are finite. Poly(lactide) PLA is a biodegradable thermoplastic produced from corn and other renewable agricultural resources. Applications of PLA polymers are found in packaging such as films, bottles and food service products as well as in medical applications, where biodegradability and biocompatibility are desired (Stewart, 2008).

Lactide (the monomer for the production of PLA) is a chiral molecule that can exist as two different stereoisomers: *L*- and *D*- lactide the racemic mixture of which is referred to as *rac*-lactide, as well as the diastereomer *meso*-lactide. The ring-opening polymerization (ROP) of enantiopure *L*- and *D*- lactide produces completely isotactic PLAs, while the polymerization of *meso*-lactide gives syndiaotactic or heterotactic PLA. When *rac*-lactide is used as a monomer, atactic polymer can be obtained, and the challenge in the last few years has been to improve the polymer stereochemistry, thus controlling polymer crystallinity which is one of the most important factors in the properties of these materials (Cheng et al. 1999).

In this paper, a series of newly synthesized controlled microstructure PLAs from mixture of *L*- and *D*- lactide are studied (Douglas and Mehrkhodavandi, 2008). Simple shear and extensional rheology are used to determine interrelationships between molecular characteristics and rheological properties for comparison with previous studies (Palade at el. 2001: Dorgan et al. 2005). The extensional viscosity measurements for PLAs are presented over a broader range of Hencky strain rate and temperature than previously reported (Palade at el. 2001). The new data sheds additional light on the dynamics of this important class of polymers.

Experimental

PLAs of controlled molecular weight were synthesized via ring-opening polymerization methods by using a chiral Indium catalyst (Douglas and Mehrkhodavandi, 2008) with different ratios of *L*- to *D*- lactide of 100:0, 75:25, 50:50, and 0:100.

In a typical experiment, 2.0g of *rac*-lactide (0.014mol) was dissolved in 10mL of CH_2Cl_2 in a vial and stirred using a magnetic stir bar. To this solution 0.036g of indium complex (3.32×10^{-5}mol) in 5mL of CH_2Cl_2 was added. The reaction was allowed to proceed for 16h and then the reaction mixture was concentrated to dryness. The resulting polymeric material was dissolved in a minimum amount of CH_2Cl_2 and added to cold wet methanol (0°C, 7mL). The polymer precipitated out of solution, and was isolated by centrifugation. The supernatant was decanted off and the polymer was dried under high vacuum for 2h.

Shear measurements were performed using a MCR 501 rheometer, equipped with 8mm parallel plates. The dynamic measurements were carried out within the linear response at temperatures from 70-170°C. Tris(nonylphenyl)phosphate (TNPP) was added as a stabilizer against thermal degradation. The TNPP concentration used was 0.35%, as reported by Palade et al. (2001); weighed quantities of the polymers were dissolved in dry chloroform and dried in vacuo. Consistency of storage and loss modulus during the test indicated that sample degradation had not occurred during the testing time after addition of TNPP. Uniaxial extensional measurements were performed using SER-2 feature attached to MCR 501 rheometer. All the rheological measurements were performed under nitrogen atmosphere to minimize degradation of the polymer samples during testing. Cylindrical and rectangular samples with dimensions of 8mm in diameter and 16×50mm respectively were prepared by means of solution casting and compression moulding for the rheological measurements.

Table 1: PLA samples

Samples	%L/%D	10^3 M_w	10^3 M_n	T_g °C	T_m °C
L 41	100/0	40.9	19.5	55.9	169.4
L75D 51	75/25	70.9	35.1	60.1	-
L50D 51	50/50	53.9	24.9	52.7	-
L50D 71	50/50	71.0	43.6	51.5	-
L50D 91	50/50	98.4	46.9	52.6	-
L50D 22	50/50	246.0	160.8	52.6	-
D 12	0/100	109.3	49.2	57.2	170.7

Results and Discussion

Figure 1 shows the master linear viscoelastic modulus and dynamic viscosity of L50D 22 at a reference temperature of 150°C. The time-temperature superposition was applied to obtain the master curves. The superposition of the data was excellent pointing to the thermorheological simplicity of these linear polymers.

The plateau modulus, G_N^o was found to be 0.9MPa ±0.2 obtained from Van Gurp-Palmen plots (Trinkle et al. 2001). This value falls within the range reported by Dorgan et al. (2005) of 1.0MPa ±0.2 across all L/D compositions.

Figure 2 depicts the zero shear viscosity for the samples listed in Table 1. The relationship between the zero shear viscosity and molecular weight (determined by GPC using Polystyrene as a standard) can be described by $\log(\eta_o) = -13.8 + 3.4\log(M_w)$ at the reference temperature of 150°C. The validity of the power of 3.4 has also been reported by Dorgan et al. (2005) in their comprehensive study of a wide range L/D composition and molecular weights of PLAs.

The continuous line in Figure 1 represents the fitting of the parsimonious relaxation spectrum (PM) $\{G_i, \lambda_i\}$ calculated from equations 1 and 2.

$$G'(\omega) = \sum_{i=1}^{N} g_i \frac{(\omega\lambda_i)^2}{1+(\omega\lambda_i)^2} \tag{1}$$

$$G''(\omega) = \sum_{i=1}^{N} g_i \frac{\omega\lambda_i}{1+(\omega\lambda_i)^2} \tag{2}$$

Figure 3 plots the PM relaxation spectrum (5 modes) along with the Baumgaertel-Schausberger-Winter (BSW) continous relaxation spectrum for L50D 22 express by equation 3. It has been confirmed by Baumgaertel et al. (1990), and Hatzikiriakos et al. (2000) that BSW have shown an excellent agreement with the PM spectrum for linear and star polybutadiene polymers. This allows for an accurate representation of the linear viscoelastic relaxation with the smallest number of parameters.

$$H(\lambda) = n_e G_N^o \left[\left(\frac{\lambda}{\lambda_c} \right)^{-n_g} + \left(\frac{\lambda}{\lambda_{max}} \right)^{n_e} \right] \tag{3}$$

where λ_{max} is the longest relaxation time, n_e and n_g are the slopes of the spectrum in the entanglement and glass transition zones respectively, and λ_c is the crossover time to the glass transition. The values of the parameters for L50D 22 calculated from the fitting procedure (equation 3) are listed in the insert of Figure 3.

The molecular weight between entanglements, M_e, can be calculated from the plateau modulus utilizing Equation 4.

$$M_e = \frac{\rho_o RT}{G_N^o} \quad (4)$$

Where ρ_o is melt density (g/cm^3) at the chosen reference temperature. Using the relationship for the PLA density, (T in °C), reported by Witzke (1997).

$$\rho(T) = 1.2836\exp[-7.7(10^{-4})T]$$

the M_e of PLAs is found to be about 4400g/mol.

Figure 4a, b and c plot the tensile stress growth coefficient of PLA L50D 22 in steady uniaxial extension, at several extensional Hencky strain rates measured at 90, 110 and 130°C. The dashed lines represents the linear viscoelastic envelope (LVE) $3\eta^+(t)$, where $\eta^+(t)$ is the shear stress growth coefficient in a linear viscoelastic region. The coefficient $\eta^+(t)$ was calculated from the parsimonious relaxation spectrum $\{G_i, \lambda_i\}$.

At the lowest Hencky strain rate of 0.01 s^{-1}, the extensional data is superposed with the LVE indicating the consistency between the two sets of experimental data, namely shear and extensional. However, at higher values of the Hencky strain rate of 1 s^{-1} and 10 s^{-1}, a clear strain hardening effect can be observed (see Figures 4a and 4b). This strain hardening at high Hencky strain rate becomes less pronounced or completely disappears with increasing temperature (see Figure 4c).

Strain hardening for linear PLA polymers has been observed and reported by others (Palade et al. 2001; Yamane et al. 2004). In their study of extensional behavior of PLLA, Palade et al. (2001), have observed a strain hardening at a Hencky strain rate of 0.1 s^{-1} measured at the temperature of 180°C. Moreover, Yamane et al. (2004) observed a strain hardening in biaxial extensional flow at extensional rate of 0.1 to 0.006 s^{-1} for PLLA and PLLA/PDLA blends at 190°C. Such PLAs (100% L or 100% D homopolymer) are readily crystallizable and therefore since the temperature used was well above T_m, the hardening effect was solely attributed to the flow dynamics of the polymers.

In the present study, DSC and XRD were used to understand the strain hardening behavior that occurred at relatively low temperatures. These tests were performed on unstretched and stretched samples that were quickly quenched in order to freeze any crystallinity formed during stretching. No evidence of crystallinity or other supermolecular structure was detected in our samples by DSC and XRD. Therefore, it can be concluded that the strain hardening effect is solely due to the flow dynamics of the polymers.

It has been reported in the literature that the strain hardening behavior for polymers and in this case for linear polymers is exhibited when the rate of the deformation exceeds the rate of molecular relaxation (Kasehagen and Macosko, 1998). In other words, strain hardening always occurs when the characteristic time for molecular relaxation (longest relaxation time) is longer than the characteristic time of the deformation (inverse of Hencky strain rate).

The longest relaxation times obtained from Figure 3 for L50D 22 for the three temperatures are listed in Table 2. It can be seen that the strain hardening effect observed in this study occurs when the longest relaxation time is larger than the deformation time applied. For example at 130°C, the longest relaxation time (0.065s) is smaller than the characteristic times of Hencky strain rates from 0.01 to 10s^{-1} which are from 0.1 to 100 s and therefore strain hardening effects are not present. On the other hand at 90°C, the longest relaxation time is 7.63s, which is larger than most of the characteristic times of the deformation 1 and 0.1s for the Hencky strains of 1 and 10s^{-1} and therefore strain hardening is obtained. This is the case for most of the available data obtained for the other PLAs examined in this study.

Table 2: the longest relaxation times for L50D 22

Temperature (°C)	Longest relaxation time (s)
90	7.63
110	0.70
130	0.065

Conclusions

The melt rheological properties of PLA have been investigated for different optical compositions and molecular weights. The plateau modulus of PLAs was found to be 0.9 MPa±0.2. The PLAs melts have a molecular weight between entanglements of approximately 4400 g/mol that was determined from the scaling of zero-shear viscosity with molecular weight. The PLAs show a scaling of $\eta_o \propto M_w^{3.4}$. Furthermore it was demonstrated that strain hardening is possible to be obtained for these linear polymers for cases where the longest relaxation time is larger than the characteristic time of deformation.

Acknowledgements

The authors would like to acknowledge NSERC for the strategic grant (STPGP 364848). A scholarship granted by Ministry of High Education, Malaysia and Universiti Teknologi Malaysia to Norhayani Othman is also gratefully acknowledged.

References

Baumgaertel M, Schausberger A, Winter HH.,"The relaxation of polymers with linear flexible chains of uniform length", Rheologica Acta, 29(5), 400-8 (1990).

Cheng M, Attygalle AB, Lobkovsky EB, Coates GW., "Single-site catalysts for ring-opening polymerization: Synthesis of heterotactic poly(lactic acid) from rac-lactide", J Am Chem Soc., 15;121(49), 11583-4. (1999)

Dorgan JR, Janzen J, Clayton MP, Hait SB, Knauss DM., "Melt rheology of variable L-content poly(lactic acid)", J Rheol., 49(3), 607-619 (2005).

Douglas, A. F., Patrick, B. O., and Mehrkhodavandi P., "A Highly Active Chiral Indium Catalyst for Living Lactide Polymerization", Angew. Chem. Int. Ed., 47, 2290–2293, (2008).

Hatzikiriakos SG, Kapnistos M, Vlassopoulos D, Chevillard C, Winter HH, Roovers J., "Relaxation time spectra of star polymers", Rheologica Acta, 39(1), 38-43 (2000)

Kasehagen LJ, Macosko CW., "Nonlinear shear and extensional rheology of long-chain randomly branched polybutadiene", J Rheol, 42(6), 1303-1327 (1998)

Palade L, Lehermeier HJ, Dorgan JR., " Melt rheology of high L-content poly(lactic acid)", Macromolecules, 34(5), 1384-90 (2001)

Stewart. R., "Going Green: Eco-friendly materials and recycling on growth paths", Plastics Engineering. Society of Plastics Engineers, 64 (1), 17-22 (2008)

Trinkle S, Friedrich C., "Van Gurp-Palmen-plot: A way to characterize polydispersity of linear polymers", Rheologica Acta, 40(4), 322-328 (2001)

Witzke D.R., "Introduction to Properties, Engineering, and Prospects of Polylactide Polymers", Ph.D thesis, Chemical Engineering, Michigan State University, (1997)

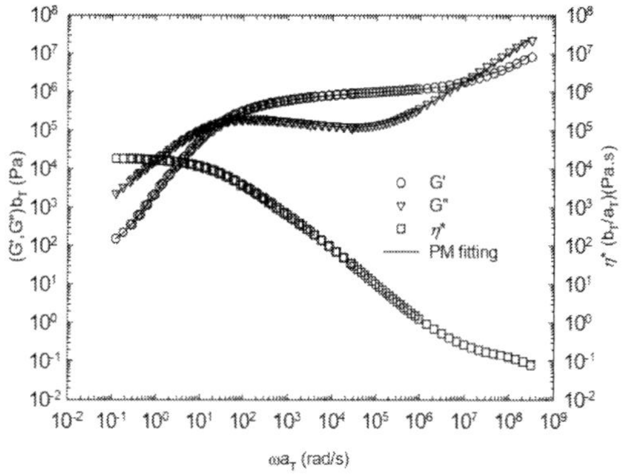

Figure 1: Master curve for L50D 22 measured by using a strain of 0.02 at a reference temperature of 150°C.

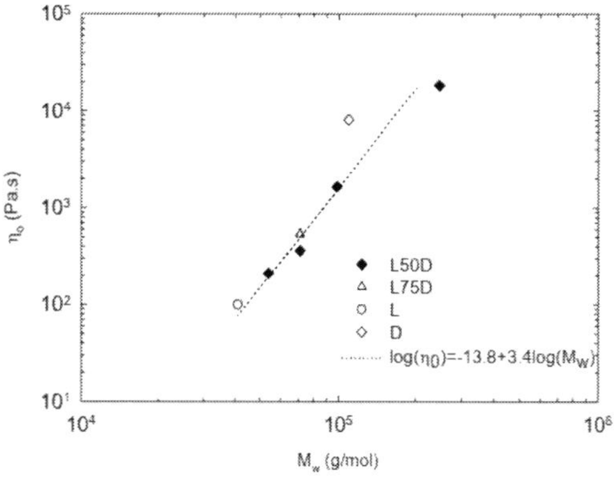

Figure 2: The scaling between the zero shear viscosity and molecular weight of PLAs at 150°C.

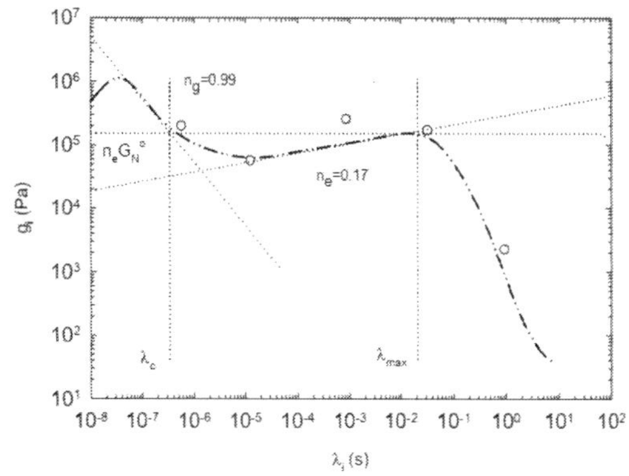

Figure 3: The PM along with a continuous representation of the relaxation spectrum of L50D 22 at 150°C.

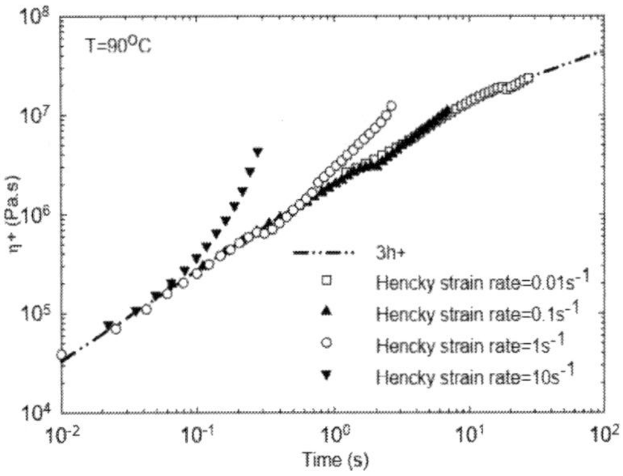

Figure 4a: Tensile stress growth coefficient as a function of time measured at various Hencky strain rates for L50D 22 at 90°C.

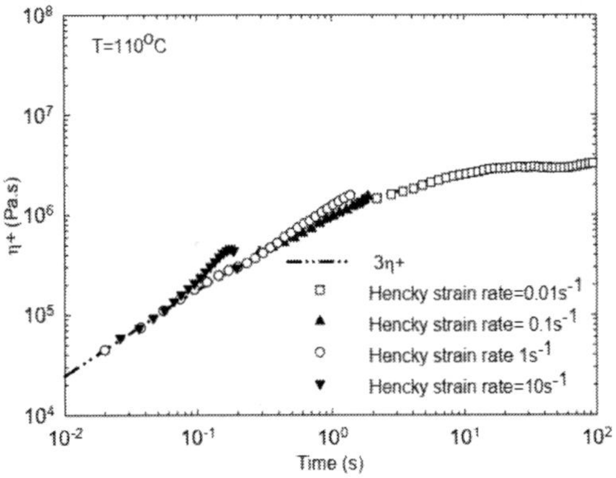

Figure 4b: Tensile stress growth coefficient as a function of time measured at various Hencky strain rates for L50D 22 at 110°C.

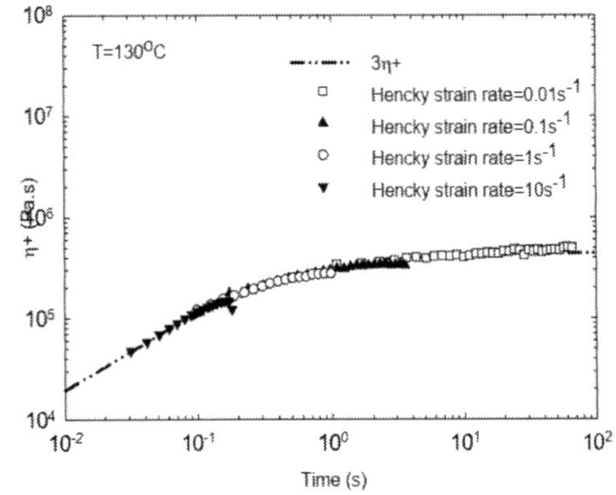

Figure 4c: Tensile stress growth coefficient as a function of time measured at various Hencky strain rates for L50D 22 at 130°C.

WALL SLIP AND MELT-FRACTURE OF POLYSTYRENE MELTS IN CAPILLARY FLOW

R. Komuro[1], K. Kobayashi[1], T. Taniguchi[2,3], M. Sugimoto[1] and K. Koyama[1*]

[1]*Graduate School of Science and Engineering, Yamagata University, Yonezawa, 992-8510, Japan, komuro@ckpss.yamagata-u.ac.jp, sugimoto@yamagata-u.ac.jp, koyama@yamagata-u.ac.jp,*
[2]*Department of Chemical Engineering, Graduate School of Engineering, Kyoto University, Kyoto 615-8510, Japan,*
[3]*CREST, JST, Kawaguchi, Saitama 332-0012, Japan, taniguch@cheme.kyoto-u.ac.jp*

Abstract - We investigated slip and unstable flow phenomena of polymer melts in capillaries by using three polystyrene samples with different molecular weights (M_w=192000, M_w=258000, and M_w=321000). The slip velocities are estimated by the Mooney method and the modified Mooney method. We found that the slip velocity increases and the critical slip stress $\sigma_c^{(S)}$ above which a slip starts to occur decreases with the temperature. We also observed the melt-fracture at above a critical melt fracture stress $\sigma_c^{(m)}$ higher than $\sigma_c^{(S)}$.

Introduction

In industrial processes of extrusion and injection molding, molten plastics experiences a flow with a high shear rate especially near a solid wall of die. In such situations, it has been known that a slippage sometimes takes place near the wall under a certain condition. So far many researchers have investigated slip phenomena, especially for polyethyrene (PE) using the Mooney, the modified Mooney method and so on. It has been reported by Hatzikiriakos (1991,1992) that for high density PE a slippage occurs at above a critical shear stress $\sigma_c^{(S)}$ and the obtained slip velocities V_s by the Mooney and the modified Mooney method satisfies the relation

$$V_s=a(\sigma_w)^m. \quad (1)$$

They have also reported that the exponent m is not a function of temperature, but the coefficient a increases with temperature. Such studies of slippage, however, concentrate on those of molten polyethylenes.

So, the aim of the present work is to investigate unstable flow phenomena such a slippage and melt fracture in polystyrene melts that have not been paid attention so much.

Experimental

We used three types of PS (PS1-3) that have different molecular weights (PS1: M_w=192,000, M_w/M_n=2.4, PS2: M_w=258,000, M_n=2.2, PS3: M_w=321,000, M_w/M_n=2.6). These samples are kindly supplied from Japan polystyrene inc. Since commercial samples usually contain mineral oils and it is considered that such mineral oils play important role in slip phenomenon, *i.e.*, work as lubricants, we used PS samples that do not contain any mineral oils to remove the effect of the oil. Using a capillary rheometer we measured the applied pressure P and volumetric flow rate Q. In order to reduce the effect of pressure loss at the entrance region to the capillary from the barrel, we used dies with an entry angle (2α=60°) and with an enough length L. We investigated on the slippage and unstable flow of PS1-3 to clarify the effect of molecular weight and temperature on the behavior of capillary flow of PS. All the experiments were performed at temperatures 180, 210, and 240°C to elucidate the temperature dependence of capillary flow behavior of PS.

we investigated whether a slippage takes place or not in polystyrene melts systems (PS1, PS2 and PS3). The shear stress σ_w at the wall and apparent shear rate $\dot{\gamma}_A$ are calculated from following equations:

$$\sigma_W = \frac{D}{4L}\Delta P \quad (2)$$

$$\dot{\gamma}_A = \frac{32Q}{\pi D^3} \quad (3)$$

The pressure difference ΔP between the pressures at the entry and the exit of die can be estimated through the Bagley correction where the applied pressure P are plotted as functions of the ratio of the length of capillary L to the diameter D, i.e., L/D by using at least three types of capillaries die and at various piston speeds. The equation for the shear rate eq. (3) is valid only for the Newtonian fluid in the absence of wall slip. If a slippage takes place at the wall, the Rabinowitsch correction cannot be applied to obtain the corrected shear rate for non-Newtonian fluid. Therefore, we used apparent shear rate $\dot{\gamma}_A$ analyze whether a slip takes place or not, and to estimate the slip velocity. The slip velocity V_s at the wall is estimated by

$$\dot{\gamma}_A = \frac{8V_S(T)}{D} + \frac{4}{3\sigma_W^3}\int_0^{\sigma_W^3} \dot{\gamma}(\sigma^3,T)d\sigma^3 \quad (5)$$

where T is temperature. From eq. (5), V_s is obtained by plotting apparent shear rate as a function of the inverse of diameter of die, $1/D$, i.e., the slope of the resultant graph corresponds to $8V_s$ (Mooney Method) .

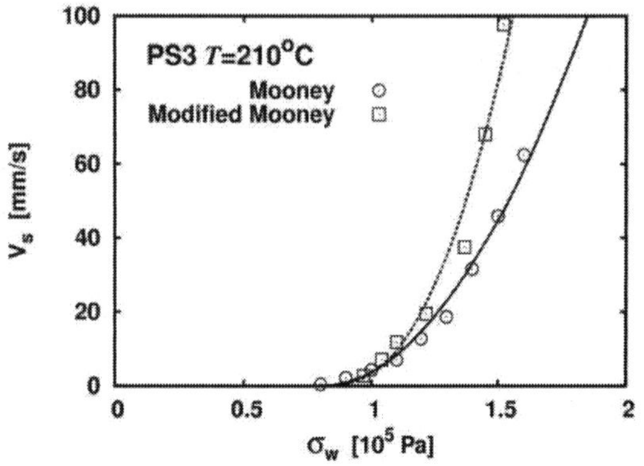

Figure 1 Comparison of the slip velocities evaluated by the Mooney method and the modified Mooney method for PS3 at 210 °C. The slip velocities are fitted by using eq.(11) and $\sigma_c^{(s)}$ =0.08 MPa. By the fitting, a=58.3 mm/s and m=1.99 in the Mooney method, and a=115 mm/s, $\sigma_c^{(s)}$=0.08 MPa, and m=2.53 in the modified Mooney method are obtained.

Furthermore, the slip velocity V_s can be estimated without performing experiments with various diameters of dies using the modified Mooney method. In the modified Mooney method, the polymeric fluid is assumed to be a power law fluid ($\sigma = K(\dot{\gamma})^n$), and the slip velocity V_s is expressed by the following equation :

$$V_S = \frac{D}{8}\left[\dot{\gamma}_A - \frac{4n}{3n+1}\left(\frac{\sigma_w}{K}\right)^{\frac{1}{n}}\right] \tag{6}$$

where K is the power law coefficient and n is the flow index. Firstly by performing the Bagley corrections, we evaluated the wall shear rate $\dot{\gamma}_A$. From performed measurements, the flow curve almost obeys a power law, but in the higher shear rate regiona deviation of the flow curve from a guide line that is obtained by a fitting of data to a power law function in a lower shear rate region. By the fitting of the flow curve data less than $\dot{\gamma}_A$ =180s^{-1} to the power law function, the flow index n and the power law factor K are found to be n=0.303 and K=0.0210MPa(sec)^n. In Fig. 1, we show the slip velocity that is obtained by the Mooney and the modified Mooney method. From the result of Fig. 1, the wall stress dependence of the slip velocity V_s can be expressed empirically by the following equation :

$$V_s(\sigma_W) = \begin{cases} a \cdot \left(\dfrac{\sigma_W - \sigma_c^{(s)}}{\sigma_c^{(s)}}\right)^m & \text{for} \quad \sigma_W \geq \sigma_W^{(s)} \\ 0 & \text{for} \quad \sigma_W < \sigma_W^{(s)} \end{cases} \tag{7}$$

where a is the slip coefficient. It is found from Fig. 1 that the slip velocity estimated by the modified Mooney method is a little bit higher than that evaluated by the Mooney method. The similar behavior is observed by Hatizikirialos et al(1992). They also found that the slip velocity decreases with the increase of L/D ratio, and discussed that the origin of L/D ratio dependence of the slip velocity comes from the errors related to the pressure loss at the entrance of die since in the modified Mooney method the Bagley correction is not applied to the estimation of stress. In our experiments, since L/D of die used in the modified Mooney method (L/D =20) is smaller than that in the Mooney method (L/D =25,33,50), it is considered that the slip velocity estimated by the modified Mooney method became larger than that by the Mooney method in similar fashion to the result of Hatizikirialos et al (1992). As for the critical slip stress, the both methods resulted in almost the same value $\sigma_c^{(s)}$=0.08 MPa.

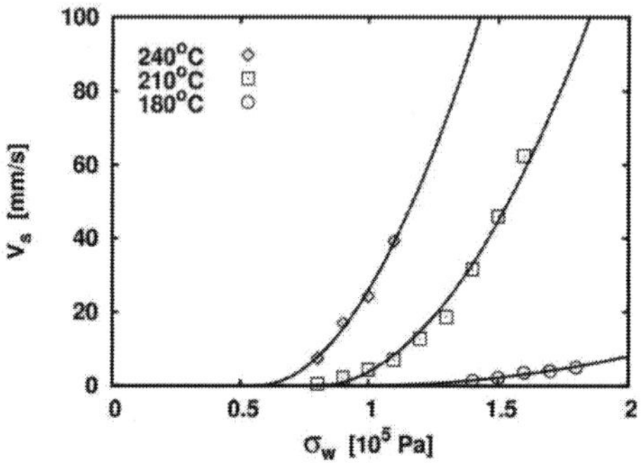

Figure 2 Slip velocities of PS-3 as a function of the wall shear stress σ_w at the temperatures 180, 210, 240 °C. The slip velocities are fitted by using eq.(7) and m=1.99 obtained by the analysis in the Mooney method. By the fitting, a=41.5mm/s and $\sigma_c^{(s)}$ =0.056MPa at 240 °C, a=58.3mm/s and $\sigma_c^{(s)}$ =0.08MPa at 210 °C and a=6.5mm/s and $\sigma_c^{(s)}$ =0.095MPa at 180 °C are obtained.

We show the temperature dependence of the slip velocity for PS3 in Fig.2. We also show the fitted lines to the data at T=180 and 240°C by eq. (7) with m=1.99 obtained by the fitting of data at T=210 °C in the Mooney method. By the fittings, we estimated the slip coefficient a and the critical slip stress $\sigma_c^{(s)}$ at T=180 and 240°C. As seen from the Fig.2, the data of slip velocity are nicely fitted by eq. (7) with the constant exponent m=1.99, which supports the result reported by Hatzikiriakos et al(1992) where m is not a function of temperature, but the coefficient a is an increasing function of temperature in similar fashion to the behavior of HDPE. The temperature dependence of a in PS3 is much larger than that in HDPE. From the results, the slip velocity increases and the critical slip stress $\sigma_c^{(s)}$ decreases with the temperature.

In order to examine the molecular weight dependence of slip phenomenon for polystyrene, we tried to obtain the slip velocity for PS1 and PS2 using the Mooney method. In PS1 and PS2, however, we could not observe slip phenomenon in the temperature range from 180 to 240 °C. We are considering that the shorter relaxation times in PS1 and PS2 makes the critical slip stresses in these systems higher ones which are much larger than our experimentally accessible stress or makes the slip velocities in these systems too small to estimate them by the Mooney methods.

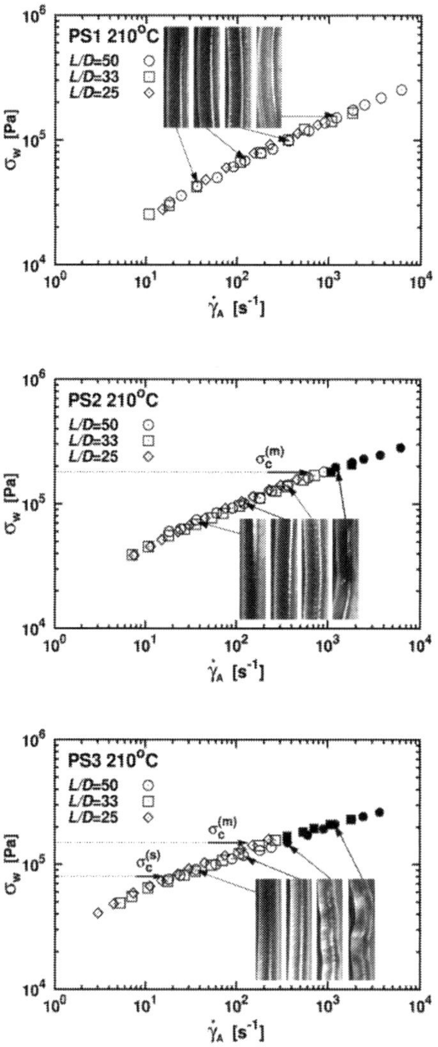

Figure 3 Flow curves and morphology of extrudates for PS1-3 at 210 °C. The morphology of extrudates obtained in *L*/*D*=50 at $\dot{\gamma}_A$=3.65x10^1, 1.22 x10^2, 3.65 x10^2, 1.22x10^3 s^{-1}.

At a shear rate region higher than the one where a slip takes place, on the other hand, we observed a melt-fracture. In Fig. 3, we show not only the flow curve of PS1-3 at 210 °C with capillaries of *D*=1, 1.5, 2 mm and *L*=50 mm but also the morphology of extrudates obtained at *L*/*D*=50 at $\dot{\gamma}_A$=3.65x10^1, 1.22 x10^2, 3.65 x10^2, 1.22x10^3 s^{-1}. The distortion of extrudate starts to appear from a critical shear rate and shear stress. Hereafter we will refer them to the critical melt-fracture shear rate $\dot{\gamma}_c^{(m)}$ and shear stress $\sigma_c^{(m)}$, respectively. At a stress higher than $\sigma_c^{(m)}$, the pressure and the shear stress oscillate between two values owing to an unstable flow. In Fig. 3, the filled symbol are used for data at a shear stress higher than the critical melt-fracture shear stress $\sigma_c^{(m)}$, because of the unstable pressure. This means that the critical stress and shear rate for the onset of melt-fracture are not sensitive to *L*/*D*, i.e., the oscillating melt-fracture originates from a flow instability in the barrel. It is now generally accepted that the melt-fracture is initiated at the entrance to a capillary die where the melt undergoes primarily uniaxial extension due to the contraction flow. The critical melt-fracture shear rate increases with temperature and molecular weight.

Conclusions

We investigated the slip phenomena and unstable flow phenomena in PS melts using a capillary rheometer. The slip velocities estimated by the modified Mooney method is nicely fitted by the eq. (7). The slip velocity obtained in the modified Mooney method is a little bit higher than that evaluated by the Mooney method in accordance with the result of Hatizikirialos et al(1992). On the other hand, as for the critical slip stress $\sigma_c^{(s)}$, the two methods result in almost the same value, $\sigma_c^{(s)}$=0.08MPa at *T*=210°C. Furthermore, the slip velocity increases and the critical slip stress $\sigma_c^{(s)}$ decreases with the temperature. We also found that the wall slip tends to easily occur with increase of the temperature and molecular weight. At a shear rate region higher than the one where a slip takes place, we observed a melt-fracture.

Acknowledgements

This work was supported by KAKENHI (Grant-in-Aid for Scientific Research) on Priority Area "Soft Matter Physics" from the Ministry of Education, Culture, Sports, Science and Technology of Japan.

References

Hatzikiriakos S. G., Dealy J. M., "Wall slip of molten high density polyethylene. I. Sliding plate rheometer studies", J. Rheo. **35**, 497-523 (1991).
Hatzikiriakos S. G., Dealy J. M. "Wall slip of molten high density polyethylene. II. Capillary rheometer studies", J. Rheo., **36**, 703-741 (1992).

COMPARATIVE ANALYSIS OF DAMPING OF LINEAR, POM-POM AND BRANCHED POLYMERS

A. A. Kharlamov[1]*, V. H. Rolón-Garrido[2] and P. Filip[1]

[1]*Institute of Hydrodynamics AS CR, Prague, Czech Republic – kharlamov@ih.cas.cz; filip@ih.cas.cz*
[2]*Chair of Polymer Engineering-Polymer Physics, Berlin Institute of Technology, Berlin, Germany – victor.h.rolongarrido@tu-berlin.de*

Abstract - Behavior of linear polymers is well characterized by Doi and Edwards (DE) theory and represents basis for describing any polymeric systems. Compared to linear polymers, the pom-pom polymers exhibit weaker damping for small strains, and collapse to DE damping for higher strains with branching point withdrawal. The branched polymers (combs and commercial LDPEs) exhibit weaker damping than the linear polymers, and do not show sudden branching point withdrawal. The branching points of those polymers are distributed randomly along backbone and produce less resistance to retraction than in pom-pom polymers. Thereafter, the branching points are, probably, withdrawn smoothly during deformation and the polymer backbone is subject to DE dynamics. This behavior brings in explanation of weaker damping of branched polymers.

Introduction

In the DE model (Doi and Edwards, 1986) the polymer strands between entanglements are treated as Hookean springs. A linear polymer molecule contacts the media of other polymer molecules only at entanglement points. After deformation it retracts and the distribution of polymer strands between entanglements is described by the DE model.
The distribution of the whole polymer molecules after retraction is, however, different from that of the strands between entanglements. We consider a system where the polymer backbones are subject to retraction by the DE model for linear molecules, but after retraction the whole molecule acts as a Hookean spring. This results in higher stress than that predicted by the DE model.
The behaviour considered is expected to be valid for branched polymers with randomly distributed branching points, e.g. combs and commercial polymers.
The predicted damping functions will be compared with the experiments with branched polymers.

Model

According to the DE model the stress after step deformation equals

$$\underline{\underline{\sigma}} = 3k_B T\nu \left\langle \frac{\underline{u}' \cdot \underline{\underline{F}}^{-1} \underline{u}' \cdot \underline{\underline{F}}^{-1}}{\left|\underline{u}' \cdot \underline{\underline{F}}^{-1}\right|} \right\rangle_0 \frac{1}{\left\langle \left|\underline{u}' \cdot \underline{\underline{F}}^{-1}\right| \right\rangle_0} . \quad (1)$$

If the whole molecules are considered as Hookean springs according to Kharlamov et al. (2010) the stress is expressed in the form

$$\underline{\underline{\sigma}} = 3k_B T\nu \frac{\left\langle \underline{u}' \cdot \underline{\underline{F}}^{-1} \underline{u}' \cdot \underline{\underline{F}}^{-1} \right\rangle_0}{\left\langle \left|\underline{u}' \cdot \underline{\underline{F}}^{-1}\right| \right\rangle_0} = k_B T\nu \frac{\underline{\underline{C}}^{-1}}{\left\langle \left|\underline{u}' \cdot \underline{\underline{F}}^{-1}\right| \right\rangle_0} , \quad (2)$$

where $\underline{\underline{C}}^{-1}$ is the Finger tensor.

Results and Discussion

The present model can describe polymeric systems for which retraction is subject to the tube model, but forces on a molecule are concentrated at its ends, as e.g. in the case of comb polymers. If during retraction the branching points are withdrawn smoothly, then retraction of a molecule backbone behaves similarly to retraction of a linear molecule.
Meanwhile, the branches fold themselves along the primitive path during retraction and produce more resistance to retraction at the ends of backbone than in the middle part. Thus the zone of connection of the backbone to the mean field of other chains shifts to the ends of the backbone and the backbone can be considered as a Hookean spring thus validating the present approach.
The example of non-smooth branching point withdrawal is found in Pom-Pom polymers, see e.g. Archer and Juliani (2004), or Archer and Varshney (1998). In a damping function in Archer and Varshney (1998) for shear strain less than a critical value of around 6.0 the damping was weaker than that of DE (see Fig.1), indicating stretch and tension in the backbone of molecule. At strain equal 6.0 the damping function dropped suddenly and approached the DE result, indicating branching point withdrawal.

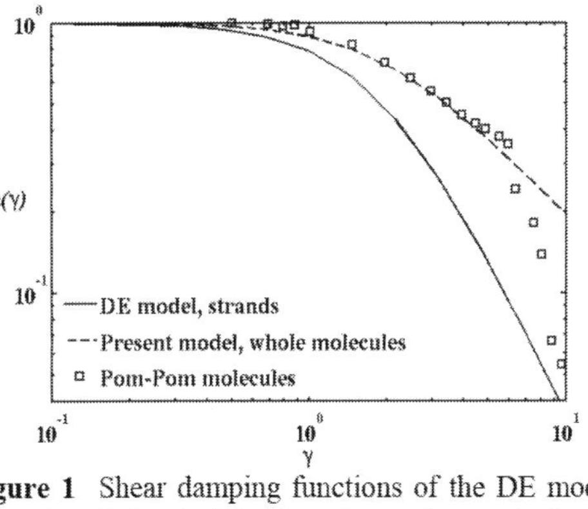

Figure 1 Shear damping functions of the DE model, present model and of the Pom-Pom polymers in Archer and Juliani (2004) and Archer and Varshney (1998).

In Archer and Juliani (2004) the damping function drop was observed prior to characteristic separability time. For comb polymers no such transition in damping is observed. Probably the branching points are withdrawn because they comprise fewer branches than those of Pom-Poms and hence are more easily withdrawn. Smoothness of branching points withdrawal in the combs is also stimulated by randomness of branching points distribution along the backbone.

A comparison of shear damping function of the present model with damping of branched polystyrene with a brush-like architecture and number average mass fraction of grafted chains Φ_{br} ca 0.5 (Hepperle and Münstedt, 2006) is presented in Fig.2. The polystyrenes with lower mass fraction of grafted chains, however, showed damping closer to that predicted by the DE model. The branched polybutadiene with Φ_{br} up to 0.39 (Kasehagen and Macosko, 1998) also exhibits weaker damping with an increasing mass fraction of branching content, the damping function of the most branched polymer used in Kasehagen and Macosko (1998) is shown in Fig.2.

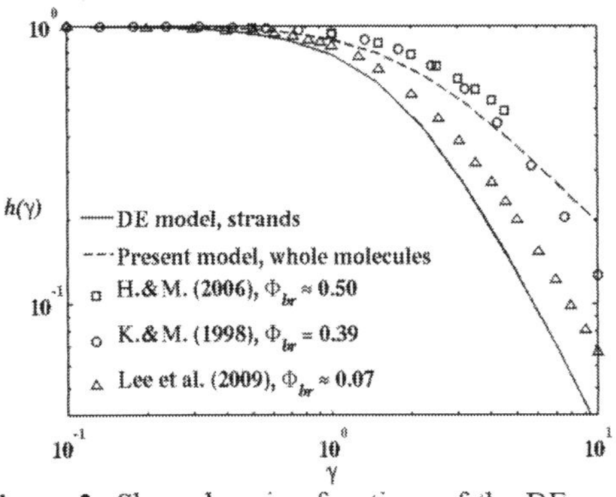

Figure 2 Shear damping functions of the DE model, present model and of the branched polymers in Hepperle and Münstedt (2006), Kasehagen and Macosko (1998), and Lee et al. (2009)

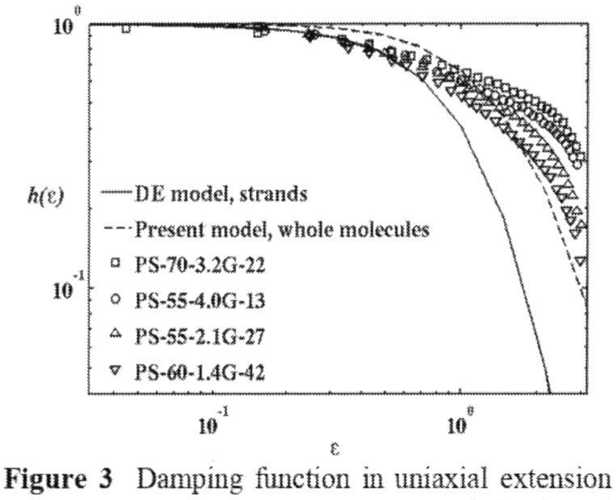

Figure 3 Damping function in uniaxial extension of the DE model, present model and of the branched comb polystyrenes in Hepperle and Münstedt (2006).

The damping function of branched polyisoprene with low mass fraction of branches, $\Phi_{br} \approx 0.07$ (Lee et al., 2009) shows slightly weaker damping than that predicted by the DE model, see Fig.2. This tendency can also be explained by the above mentioned arguments.

It is documented in Fig.2 that the present model satisfactorily predicts shear damping function for branched polymers when branches content is high enough.

A comparison of the damping function in uniaxial extension of polystyrene in Hepperle and Münstedt (2006) is presented in Fig.3. The description of the samples is presented using the nomenclature PS-*x*-pG-*y* with *x* describing the number average molar mass in kg/mol of polydisperse backbone, p represents the average number of grafted side chains per molecule randomly attached to the backbone, and *y* denotes the number average molar mass of side chains in kg/mol.

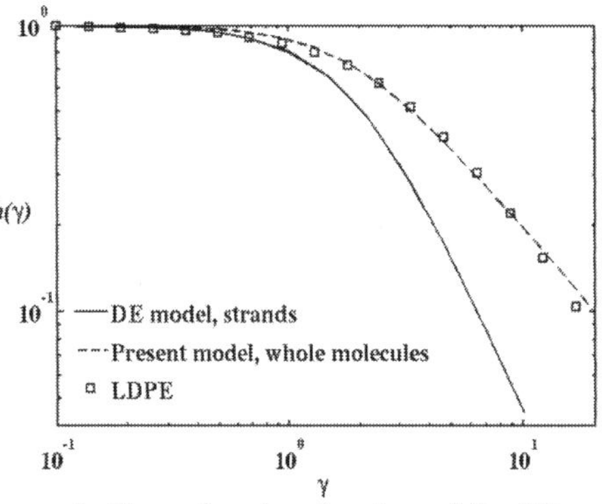

Figure 4 Shear damping function of the DE model, present model and of the LDPE in Laun (1978).

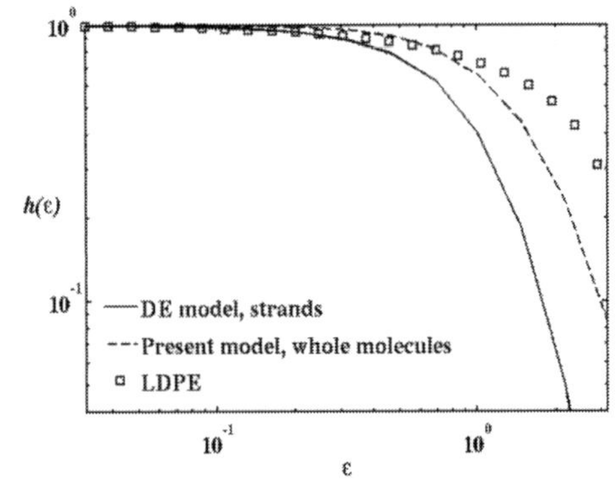

Figure 5 The damping functions in uniaxial extension for the DE model, present model and for LDPE in Wagner (1978).

Fig.4 shows shear damping function of LDPE (IUPAC A) in Laun (1978). The damping functions of other LDPEs – duPont Alathon 20 in Soskey and Winter (1984) and IUPAC X in Samurkas et al. (1989) are very close to the damping of LDPE in Laun (1978) and thus are not presented in Fig.4. The damping function of the present model fits the damping functions of the LDPEs very well.

Behavior of LDPE in elongation was studied by Wagner (1978). The experimental data are compared with the DE and present model predictions in Fig.5. In spite of the fact that the damping function of the present model is closer, an agreement is not as good as in the case of shear flow.

Conclusions

The stress tensor is constructed by treating the whole molecules as Hookean springs. The shear and uniaxial damping functions attain higher values than those in the DE model. The damping function predictions satisfactorily compare with damping of the highly branched commercial polyethylenes and comb polymers.

Acknowledgements

This work has been supported by the Grant Agency of the Czech Republic, Grant No. 103/09/2066 and by the German Science Foundation (DFG).

References

Archer, L. A., Juliani, "Linear and Nonlinear Viscoelasticity of Entangled Multiarm (Pom-Pom) Polymer Liquids", Macromolecules, **37**, 1076-1088 (2004)

Archer, L. A., Varshney, S. K., "Synthesis and Relaxation Dynamics of Multiarm Polybutadiene Melts", Macromolecules, **31**, 6348-6355 (1998)

Doi, M., Edwards, S. F.: The Theory of Polymer Dynamics, Clarendon Press, Oxford (1986)

Hepperle, J., Münstedt, H., "Rheological Properties of Branched Polystyrenes: Nonlinear Shear and Extensional Behavior", Rheol. Acta, **45**, 717-727 (2006)

Kasehagen, L. J., Macosko, C. W., "Nonlinear Shear and Extensional Rheology of Long-Chain Randomly Branched Polybutadiene", J. Rheol., **42**, 1303-1327 (1998)

Kharlamov, A. A. et al., "Probability Distribution Function of the Polymer End-to-End Molecule Vector after Retraction and its Application to Step Deformation", Macromol. Theory Simul., **19**, (2010), DOI: 10.1002/mats.200900071

Laun, H. M., "Description of the Non-Linear Shear Behaviour of a Low Density Polyethylene Melt by Means of an Experimentally Determined Strain Dependent Memory Function", Rheol. Acta, **17**, 1-15 (1978)

Lee, J. H. et al., "Damping Behavior of Entangled Comb Polymers: Experiment", Macromolecules, **42**, 1392-1399 (2009)

Samurkas, T. et al. "Strong Extensional and Shearing Flows of a Branched Polyethylene", J. Rheol., **33**, 559-578 (1989)

Soskey, P. R., Winter, H. H., "Large Step Shear Strain Experiments with Parallel-Disk Rotational Rheometers", J. Rheol., **28**, 625-645 (1984)

Wagner, M. H, "A Constitutive Analysis of Uniaxial Elongation Flow Data of a Low-Density Polyethylene Melt", J. Non-Newt. Fluid Mech., **4**, 39-55 (1978)

RHEOLOGY OF LIQUID CRYSTALLINE POLYMERS (LCPS)

Rahul K Gupta*, Sati N. Bhattacharya

Rheology and Materials Processing Centre,
School of Civil, Environmental and Chemical Engineering,
RMIT University, Melbourne, Vic 3001, Australia
rahul.gupta@rmit.edu.au; Sati.bhattacharya@rmit.edu.au

ABSTRACT

A liquid crystal polymer has a particular kind of structural orientation at one or two dimensional level. This structural orientation makes the rheological properties of Liquid Crystal Polymers (LCPs) often extraordinary and partially understood. Furthermore, these properties vary from one LCP to another, which result in an unpredictable behaviour. It has been reported that accurate measurements of the shear rheological properties of (LCPs) is critical to allow appropriate prediction, as it flows under pressure in a typical injection moulding machine. Thus, rheological properties of five thermotropic LCPs were investigated to accurately measure these properties. The complex viscosities of the five LCPs show a typical shear-shinning behaviour. As to the temperature effect, the complex viscosities give rise to different responses. A not so typical response of complex viscosity was found with grades 4179 and 4246. There was a decrease of viscosity as the LCPs were heated above the melting temperatures. Following this the viscosities levelled off and then increased gradually with further increasing of temperature. The abnormal temperature dependence of the complex viscosities can be explained by the nematic-isotropic transition, in which the rod-like molecules lost part of their orientation ordering and became partially isotropic with increase in temperature.

Keywords: Liquid Crystal Polymer, Shear Rheology, Shear Thinning

Introduction:

The rheology of the liquid crystalline polymer is a topic of practical importance because of their unique properties; these include low coefficients of thermal expansion, low viscosity in melting states, high modulus, low permeability to gases and chemical resistance (Acierno and Collyer, 1996, Baek et al, 1994, Chan et al, 2001). Rheological studies can provide effective means of improving the processes of LCPs or blends of LCPs with conventional polymers. There have been numerous studies on the rheological behavior of LCPs in the past. The main features observed may be summarized as follows:

- In the liquid crystalline phase, LCPs possess much lower viscosities than their isotropic counterparts (Chan, 2001);
- LCPs may show multiple melting behavior, and in the range of high temperature a transition from nematic phase to isotropic phase may occur, making the viscosity increase as the temperature increases (Hsieh, 1999)
- LCPs usually exhibit little or no die swell and sometimes a negative normal stress difference (Wilson and Baird, 1992);
- Some LCPs have a three-zone dependence of viscosity (Walker and Wagner, 1994)

To make more accurate predictions especially for injection moulding applications, we need to understand flow curves of these materials accurately. Once these properties are measured accurately then it will be easy to predict flow properties of such complex materials. There is a number of ways to understand flow curves; one is to see logarithmic plots of viscosity against shear rate; the others are viscosity against flow stress and flow stress against shear rate. These relationships can be obtained by measuring the pressure and flow rates in solutions for issotropics as well as temperature for themotropics. If we study these relationships we are able to make more accurate assumptions for rheological behavior.

It is important to understand the elastic properties of LCP. When combined with viscosity, it is referred to as viscoelasticity, which is measured in oscillatory experiments. In this work, shear rheology of five different grades of LCP materials are presented. The major emphasis of this work is to accurately measure shear rheology of these complex materials so that appropriate rheological equation can be used for flow predictions.

Experimental Materials and Techniques
Materials:
Five aromatic nematic thermotropic LCPs were provided by Autodesk, Melbourne, Australia. in the form of pellets. These materials are named by the production serial number as follows:

- 4046 (EXDAR NC-301BL)
- 4047 (UENO LCP 5050GM)
- 4179 (Vectra E950IRX)
- 4180 (Vectra E130i)
- 4247 (XYDAR M-450B)

In this paper different grades of LCPs will be referred by the four digit numbers e.g. 4046 for EXDAR NC-301BL.

Preparation of samples:
At first the LCP materials were dried at specified temperatures to achieve target moisture. These materials were then compression molded at the processing temperature as specified by the manufacturer to produce discs with a thickness of 2 mm and a diameter of 25 mm for shear rheology experiments.

Shear Rheology
The 2 mm thick compression molded samples were cut into 25 mm diameter discs. Shear rheological properties were studied using the Advanced Rheometrics Expansion System (ARES) in both complex and steady shear modes with parallel plate geometry. The tests were conducted using 25 mm diameter plates at 310°C to 390^{0}C for different LCP materials. All measurements were performed with a force transducer with a range of 0.2 to 200 g-cm torque. Prior to any test the zero-gap between the parallel plates was calibrated at the required temperature. In the measurements, the sample was placed between preheated fixtures and approximate 10 min were allowed to reach thermal equilibrium, then the gap was set. Complex frequency sweeps were conducted to determine the microstructure and dynamics of the materials. These tests were conducted at low strain amplitudes that were within the linear viscoelastic region and at frequencies ranging from 100 rad/s to 0.01 rad/s.

Results and Discussion
Storage and loss moduli for different LCPs were measured at several temperatures and shifted along the frequency axis to form master curves (not shown here). Most of the materials show storage moduli G' dominant loss moduli G" expect 4179. Thus majority of these samples show solid-like behavior, whereas only 4179 is still behaved as liquid-like. Sample 4180 has a cross over of G' and G" at 345°C. This might due to the glass filler which affect rheology behavior of this material at higher temperature.
Figures 1 to 5 present the magnitude of the measured complex viscosity for the material 4046, 4047, 4179, 4080, 4247, against various frequencies and different temperatures. The viscosity-frequency curves exhibit typical shear thinning and mostly all the curves are parallel for various temperatures. In the range of the experiments no constant viscosity plateau was found even with frequencies much lower than that depicted in the figures. Three states can be observed in the viscosity-temperature behaviors. At the first stage, with temperature increasing from the melting temperature, the complex viscosities decrease; at the second stage, the complex viscosities level off; and at the third stage the viscosities gradually increase with further increasing temperatures. This is illustrated for LCP samples of 4179 and 4247. An usual trend as that of other polymers, which has a decreased viscosity as temperature increased is best illustrated for sample 4180

The viscosities gradually increase with increasing temperatures. The dramatic increase of viscosity with increasing temperature is a typical behavior of LCPs. It can be well understood by the solid-nematic phase transition where the stiff rod-like molecules are aligned preferentially along the shear direction and the liquid is positionally disordered but orientationally ordered. The gradual increase of viscosities at higher temperatures can be attributed to the biphasic nature of these LCPs, that is, due to thermal activation, part of the rod-like molecules become a more viscous isotropic phase and the increase of temperature gives rise to increasing portion of isotropic phase at the expense of anisotropic phase (Gao 1996; Hsieh, 1999) The level of the nematic-isotropic transition is marginal for the material 4046 and 4047.

Modelling –preliminary results

The flow of liquids can be conceptually modelled as an activated process of driven molecular motion. The activation energy controlling the rate of molecular motion can often be related to the molecular structure of the fluid (Bird, 1960). The classical formulation for the temperature dependence of viscosity can be expressed as

$$\ln(\eta) = A_0 + \frac{\Delta G}{RT} \quad (1)$$

Where ΔG is the activation energy, R is the gas constant and T is the absolute temperature. For the tested materials the viscosities have a strong shear thinning character, thus A_0 should be a function of the frequency. Therefore, equation (1) can be shown as

$$\eta^* = K\omega^n EXP(\frac{\Delta G}{RT}) \quad (2)$$

Where K is a constant, ω is the angular frequency, n is the power law index.

This equation can be used to model complex viscosity of LCP 4046, 4047 and 4180. The corresponding values of the parameters obtained for these three LCPs are listed in table 1. Figures 6, 7 and 8 show the predicted complex viscosities respectively by using equation (2). Comparing with the experimental results, one can observe that this model predicts the shear-shinning behaviors with the best fit power law quite well. This model gives rise to the correct order of magnitude and the trend of either increased viscosity (4046 &4047) or decreased viscosity (4180) with the increasing temperature.

Table 1 Parameters of equation (2) obtained for LCP 4046, 4047 and 4180

Material	4046	4047	4180
K	5.4*E+26	3.8*E+6	7.7*E-18
n	-0.45	-0.7	-0.65
ΔG/R	-33352	-2946	31752

In order to take into account the abnormal temperature dependence of LCP materials, we assume that there is a critical temperature, T_c, above which the nematic-isotropic phase transition makes significant contribution to the activation energy such that the viscosity starts to increase with temperature. By allowing the activation energy to be a function of temperature, we propose the flowing modification of the equation (2),

$$\eta^* = K\omega^n EXP[\frac{\Delta G}{R(T-T_0)}(1+\frac{(T-T_0)^2}{(T_c-T_0)^2})]$$

(3)

At the temperature T_0 the viscosity becomes infinite. It seems reasonable that T_0 is taken as the lowest melting temperatures provided in Material Safety Data Sheet (MSDS). The resultant equation (3) satisfies the condition of minimum viscosity at the critical temperature, *i.e.,*

$$\frac{\partial|\eta^*|}{\partial T} = 0 \text{ at } T=T_c \quad (4)$$

By inspecting the experimental curves we can a priori choose the values of T_c. Then the parameters to be determined are K, n, and G_0. The set of independent data is two-dimensional. It consists of the experimental frequencies and temperatures. The corresponding values of the parameters obtained for two LCPs, 4179 and 4247, are listed in Table 2. Figures 9 and 10 show the predicted complex viscosities of the two LCPs, respectively, by using equation (3). Comparing with the experimental results, one can observe that this model predicts the shear-thinning behaviors quite well. As to the variation with temperature, this model gives rise to the correct order of magnitude and the trend of decreases followed by gradual increase of the complex viscosities with the increase of temperature.

Table 2 Parameters of the equation (3) obtained for LCP 4179 and 4246

Material	4179	4247
K	0.056	1.25*E-14
n	-0.25	-0.81
ΔG/R	-56.4	-424.6
T_0 (°C)	330	370
T_c (°C)	345	390

Conclusions

The rheological properties of five thermotropic LCPs were investigated to accurately measure the steady shear properties. The complex viscosities of the five LCPs show a typical shear-thinning behaviour. Due to the temperature effect, the complex viscosities give rise to different responses. There was a decrease of viscosity as the LCPs were heated above the melting temperatures. The viscosities levelled off and then increased gradually with further increasing of temperature. The abnormal temperature dependence of the complex viscosities can be explained by the nematic-isotropic transition, in which the rod-like molecules lost part of their orientation ordering and became partially isotropic.

Reference

Acierno, D. and Collyer, A.A., *Rheology and Processing of Liquid Crystal Polymers*, UK (1996)

Baek, S. G., Magda, J. J., Larrson, R. G. and Hudson, S. D. , Rheological differences among liquid-crystalline polymers. 2. Disappearance of negative N(1) in densely packed lyotropes and thermotropes. *J. Rheol.*, 503-1473 (1994)

Chan C.K., C. Whitehouse, P. Gao and C.K. Chai, Flow induced chain alignment and disentanglement as the viscosity reduction mechanism with TLCP/HDPE blends, *Polymer*, **42**, 7847-7856 (2001)

Cogswell, F. N. and Wissbrun, K. F., Rheology and processing of liquid crystal polymer melts. In *Rheology and Processing of Liquid Crystalline Polymers, eds.* D. Acierno and A. A. collier. London: Chapman & Hall, 86-134 (1996)

Donald, A., Windle, A., Hanna, S. *Liquid Crystalline Polymers*, UK (2006)

Hsieh, T.T., C. Tiu, G.P. Simon and R.Y. Wu, Rheology and miscibility of thermotropic liquid crystalline polymer blends, *J. Non-Newt. Fluid Mech.*, **86**, 15-35 (1999).

Gao, P., X.H. Lu and C.K. Chai, 1996, Rheology of low nematic transition temperature thermotropic liquid crystalline copolyester HBA/HQ/SA, *Polym. Eng. Sci.*, **36**, 2771-2780 (1996).

Walker, L. and N. Wagner, 1994, Rheology of region I flow in a lyotropic liquid-crystal polymer: the effects of defect texture, *J. Rheol.*, **38**, 1525-1547 (1994).

Wilson, T. S. and D.G. Baird, Transient elongational flow behavior of thermotropic liquid crystalline polymers, *J. Non-Newt. Fluid Mech.*, **44**, 85-112 (1992)

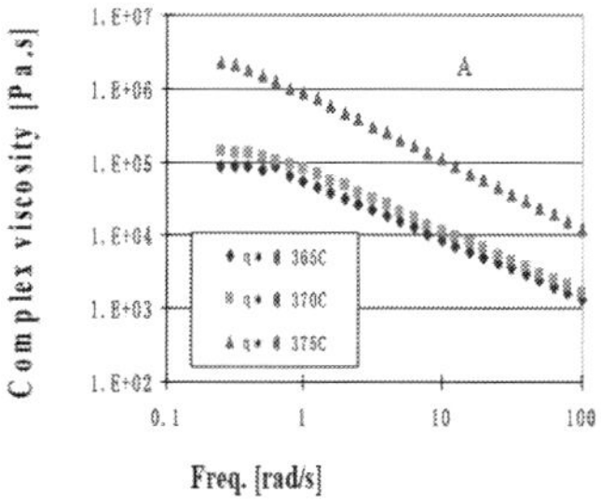

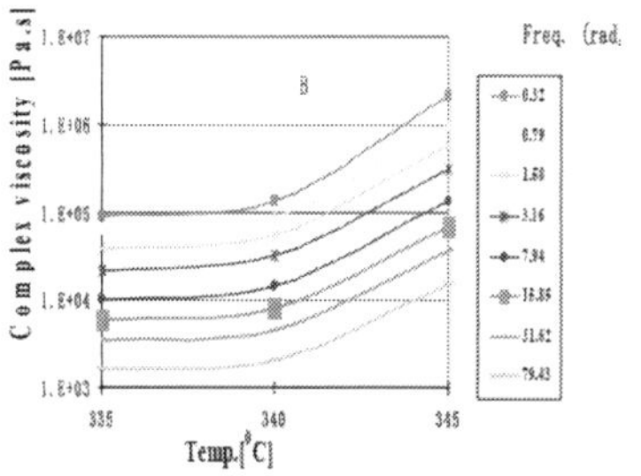

Figure 1: Complex viscosity as a function of frequency (A) and temperature (B) at different frequencies for LCP 4046.

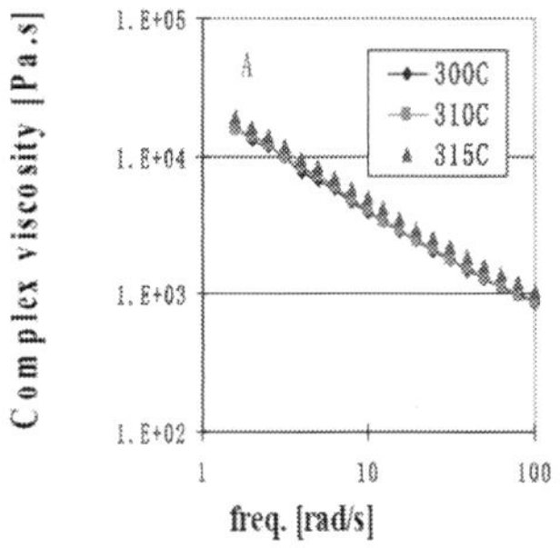

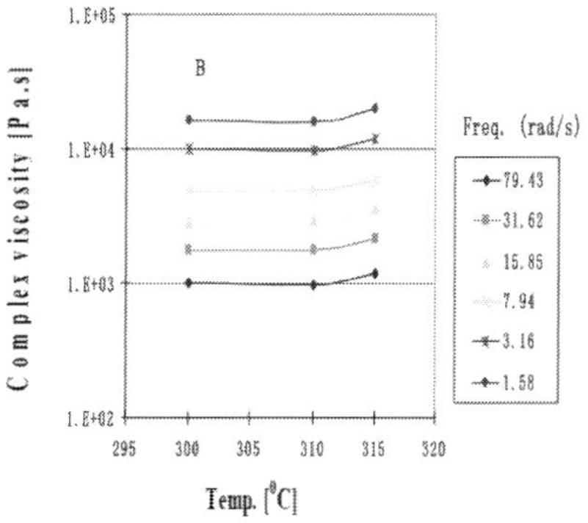

Figure 2: Complex viscosity as a function of frequency (A) and temperature (B) at different frequencies for LCP 4047.

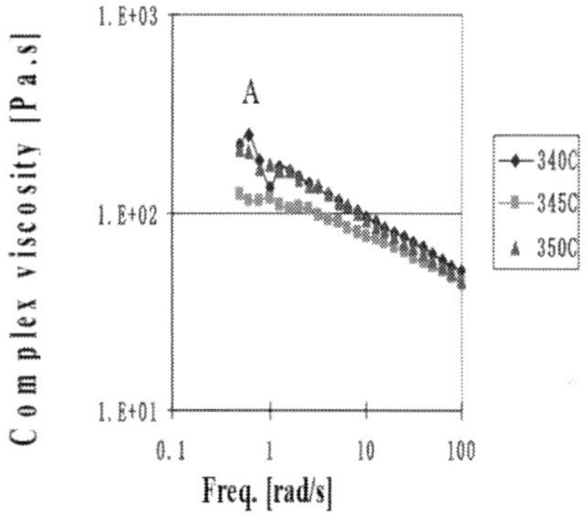

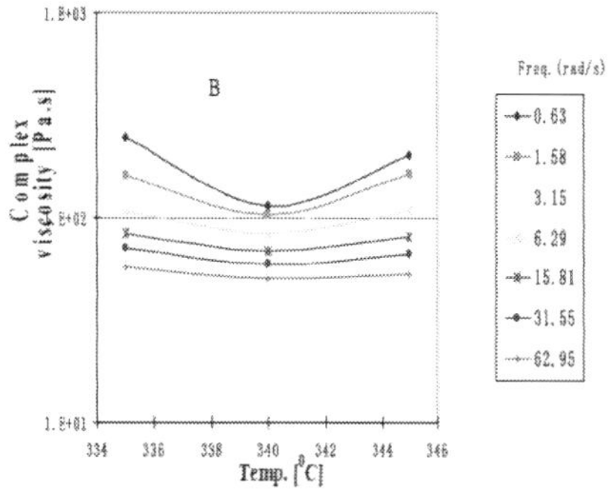

Figure 3: Complex viscosity as a function of frequency (A) and temperature (B) at different frequencies for LCP 4179

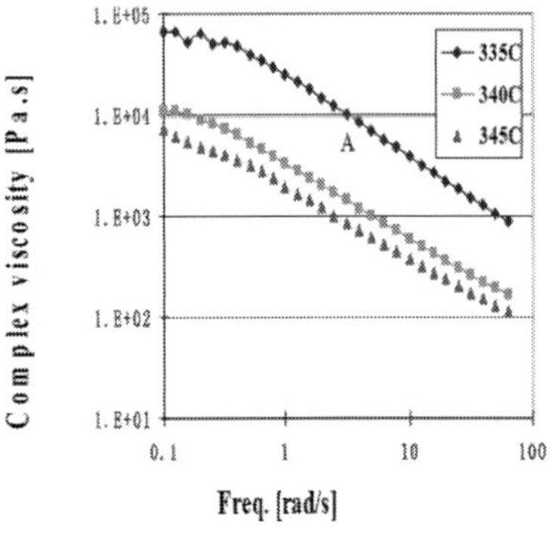

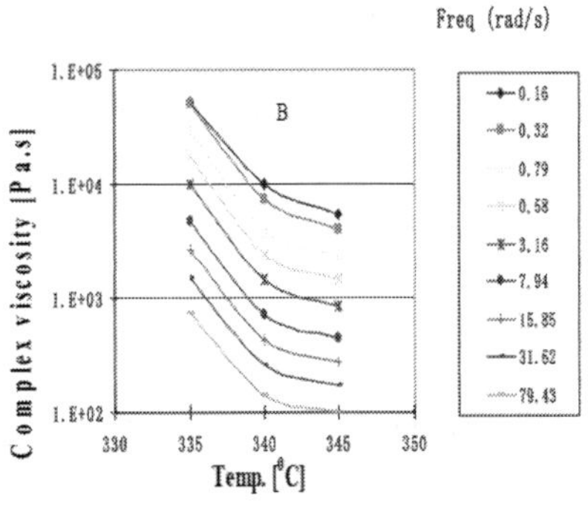

Figure 4: Complex viscosity as a function of frequency (A) and temperature (B) at different frequencies for LCP 4180

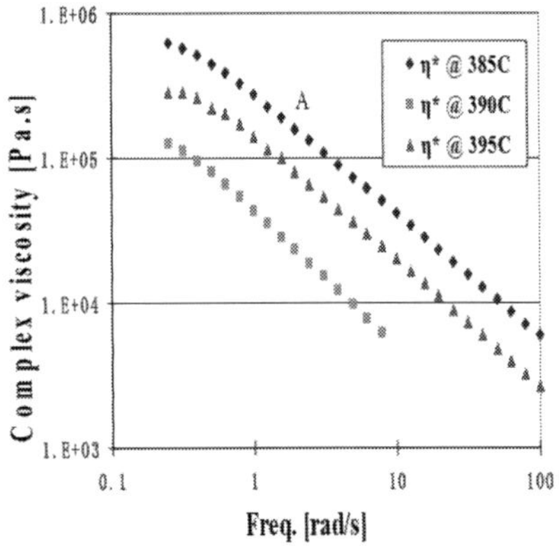

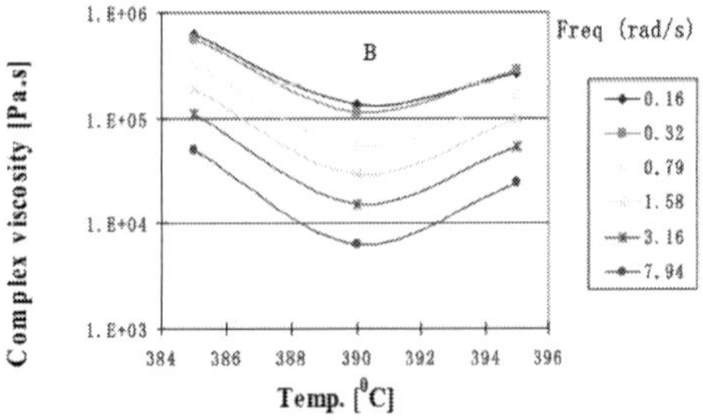

Figure 5: Complex viscosity as a function of frequency (A) and temperature (B) at different frequencies for LCP 4247

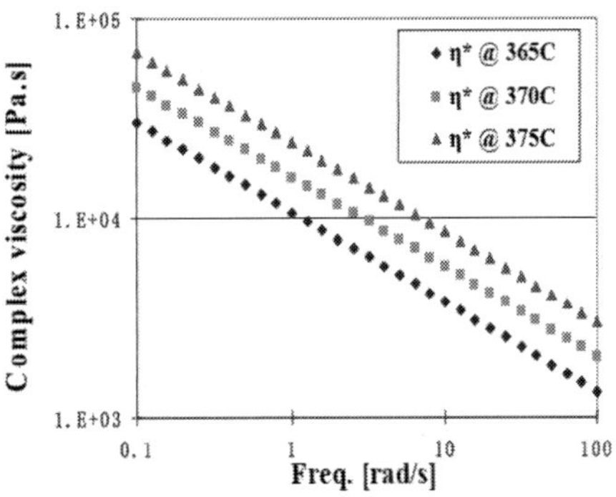

Figure 6: Prediction of Complex viscosity as a function
of frequency at different temperatures for LCP 4046

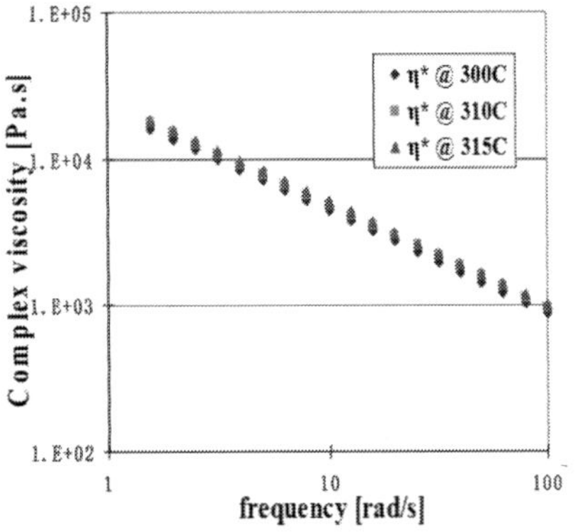

Figure 7: Prediction of Complex viscosity as a function of frequency at different temperatures for LCP 4047

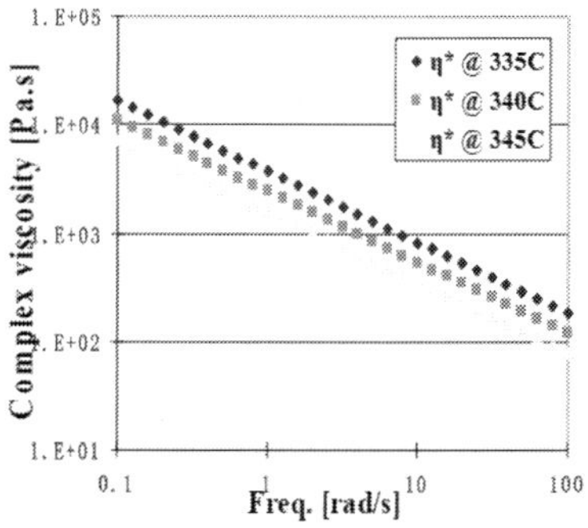

Figure 8: Prediction of Complex viscosity as a function
of frequency at different temperatures for LCP 4180

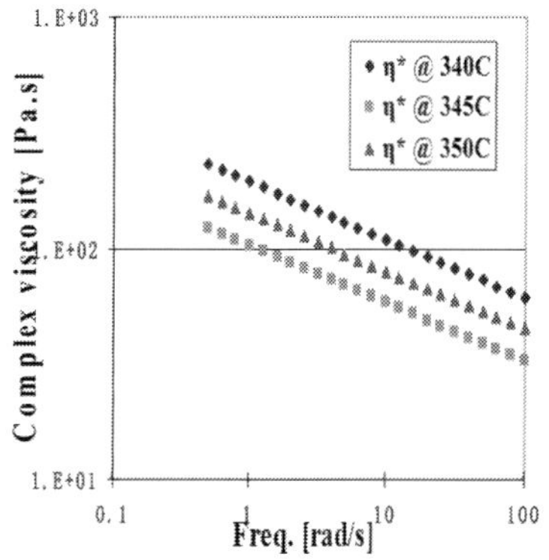

Figure 9: Prediction of Complex viscosity as a function of frequency at different temperatures for LCP 4179

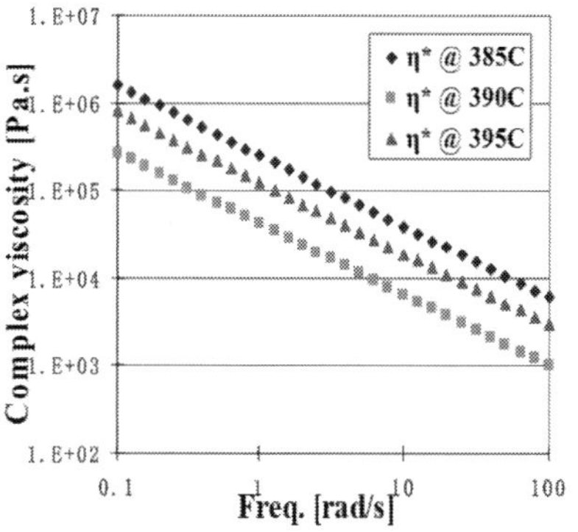

Figure 10: Prediction of Complex viscosity as a function
of frequency at different temperatures for LCP 4247

CHARACTERIZATION OF CLAY DISPERSION IN POLYMER NANOCOMPOSITES BY RHEOLOGICAL AND DIELECTRIC MEASUREMENTS

Hyung Tag Lim, Seong Han Lee, In Gook Hong, Seung Jong Lee, Kyung Hyun Ahn*

SchoolofChemicalandBiologicalEngineering,SeoulNationalUniversity,Gwanak-ro,Gwanak-gu,Seoul151-742,Korea lht0717@snu.ac.kr; begood59@snu.ac.kr; hik7459@snu.ac.kr; sjlee@snu.ac.kr; ahnnet@snu.ac.kr

Abstract - In this study, polymer /clay composites were investigated in terms of rheological and dielectric properties to define the heterogeneity of clay dispersion in polymer matrix. Clay dispersion was controlled by applying the electric field. The non-terminal behavior in the storage modulus at low frequency was gradually increased with increasing the strength of electric field. And the characteristic peak in X-ray diffraction also disappeared with electric field. Dielectric relaxation analysis showed that different mechanisms originate from different charge distributions of bound ions attached to the clay surfaces. When the clay dispersion increased, the electrode polarization increased in the real part of the complex dielectric (e') constant due to the contribution of ionic conduction. And the conductivity relaxation time decreased with increasing clay dispersion in imaginary part of the electric modulus (M''). The degree of symmetry in M'' was found to be related to conductivity relaxation time distribution. Distribution of conductivity relaxation time became narrow as the clay morphology changed from partially intercalated dispersion to exfoliation. Finally, we related clay dispersed morphology in TEM images with dielectric modulus.

Introduction

Over the past decade, polymer layered silicate (clay) nanocomposites gained rapidly increasing interest due to their mechanical reinforcement or formation of barrier structures which reduce the diffusive transport of gases or vapors. Much effort has been oriented to finding the combination of modified clay, resin and processing conditions that can yield exfoliated clay structure. So, the characterization of clay dispersion is very important to application of polymer clay nanocomposites. The evidences of clay dispersion are X-ray diffraction (XRD), rheological properties, TEM and so on. XRD can detect the periodically stacked layers of clay particles, i.e., the measurement of d001 spacing in the intercalated structure and little can be said about the spatial distribution of the clay particles. The rheological characterization may well be regarded as a sensitive tool to reveal the global state of particle dispersion in polymer nanocomposites. TEM is a strong method to observe directly the particle dispersion. And dielectric relaxation provides information of the heterogeneous structure.

In this study, polypropylene (PP) /clay composites were investigated by rheological and dielectric measurements to define heterogeneity of clay dispersion in each polymer matrix.

Experimental

PP (HP562T, PolyMirae Corp.) was used. Organically modified montmorillonite clay (C20A) was donated from Southern Clay Products Inc. Melt compounding was performed using an intermeshing, co-rotating twin-screw extruder at a screw speed of 100rpm. The weight percent of clay was 3wt%.

Rheometrics mechanical spectrometer (RMS 800) was used to assess the rheological performance of the composites. Two types of experiments were carried out: time sweep test and frequency sweep test. Time sweep test was performed at a frequency of 1 rad/sec and strain of 5%, which is in the linear region at 180°C. The input AC electric field was applied after 3minutes from the start of the test, and the rheological properties were monitored. The strength of electric field was fixed 1kV/mm at fixed AC frequency of 60 Hz. Frequency sweep test was performed at 3~5% strain under nitrogen atmosphere. The dielectric properties were measured with a frequency response analyzer solartron/Schlumberger FRA1260 connected to a buffer amplifier of variable gain (Chelsea dielectric interface).

Results and Discussion

Kim et al. (2003, 2006) applied electric field to construct the exfoliated structure of clays. The time sweep test shows the formation of the PP/clay nanocomposites when applying the electric field (AC electric field of 1kV/mm and 60Hz) in Figure 1. The storage modulus increases with time. As the strength of electric field increases, the non-terminal behavior and drastic increase in the storage modulus of PP/C20A at low frequency was found in Figure 2, and the disappearance of the characteristic peak in XRD after electric treatment and the morphological changes were observed in TEM.

The complex dielectric constant is a meaningful quantity under the condition that the electric field is an independent variable and the charge is a dependent variable. Considering the charge as the independent variable, conductivity relaxation effects can be suitably analyzed within the modulus formalism in terms of a dimensionless quantity M*. Analogous to mechanical relaxation, the 'electric modulus' M* is obtained from $M^*(\omega) = 1/\varepsilon^*(\omega) = M' + jM''$. Dielectric relaxation analysis showed that different mechanisms originate from different charge distributions of bound ions attached to the clay surfaces. As the clay dispersion improved, the relaxation peak of electric modulus shifted toward high frequency as in Figure 3. It means that the molecular structure has less flexibility and increases the value of d.c.-conductivity. And the degree of symmetry in M'' is related to conductivity relaxation time distribution. The distribution of conductivity relaxation time becomes narrow when the clay morphology changes from partially intercalated dispersion to exfoliation.

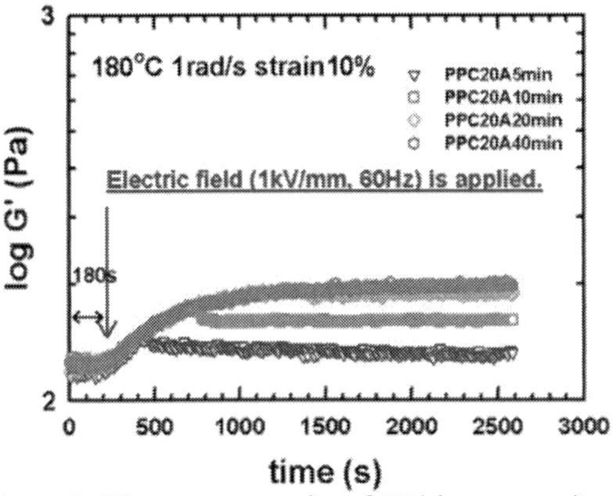

Figure 1. Time sweep results of PP/clay composites before and after applying 1 kV/mm a.c. 60 Hz electric field at 180°C.

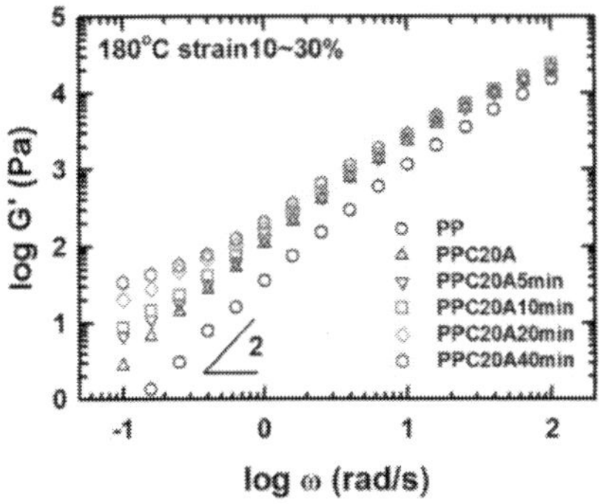

Figure 2. Frequency sweep results of PP/clay composites after time sweep tests.

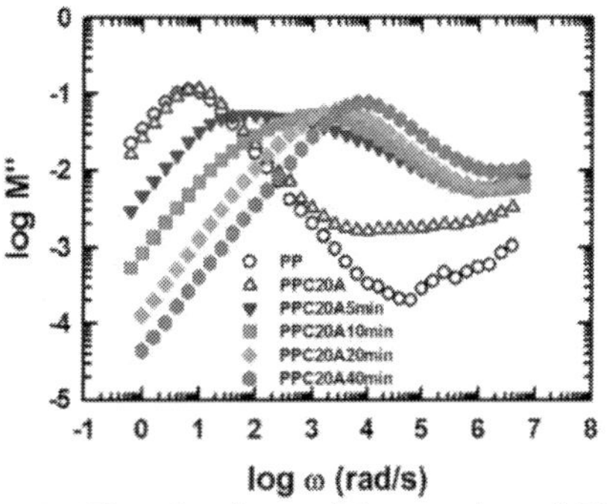

Figure 3. The electric modulus results of PP/clay composites after time sweep tests.

Conclusions

Based on rheology, XRD and TEM, the degree of clay dispersion could be controlled by changing the time of applying the electric field. And the frequency at peak and the degree of symmetry in M'' was found to beis related to clay dispersion .

Acknowledgements
This work was supported by the National Research Foundation of Korea(NRF) grant (No. 0458-20090039) funded by the Korea government(MEST)

References (Times New Roman 10, bold)

Kim et al., "Electrically Activated Poly(propylene) /Clay Nanocomposites", *Macromol. Rapid Commun.*, 24, 388-398 (2003)

Kim et al., "Microstructural evolution of electrically activated polypropylene/layered silicate nanocomposites investigated by in situ synchrotron wide-angle X-ray scattering and dielectric relaxation analysis", *Polymer*, 47, 5938-5945 (2006)

COMPOUNDING AND PROCESSING STUDIES OF SHORT ARAMID FIBER FILLED ELASTOMER COMPOUNDS

C. Hintze[1,2], S. Wiessner[1], U. Wagenknecht[1] and G. Heinrich[1]

[1] *Leibniz Institute of Polymer Research, Dresden, Germany – hintze@ipfdd.de; wiessner@ipfdd.de; wagenknecht@ipfdd.de; heinrich@ipfdd.de*
[2] *Dutch Polymer Institute (DPI), PO Box 902, 5600 AX Eindhoven, The Netherlands – info@polymers.nl*

Our studies are related to the investigation of controlled fiber orientation of aramid short cut fibers in elastomers that is induced during processing including its experimental characterization. Further objectives include the investigation of technological and material related parameters on fiber dispersion/distribution after compounding. Low fiber contents of Twaron® fibers with standard finish (low fiber-matrix adhesion) were used for this study to investigate the basic compounding influences on morphology and mechanical properties of a transparent EPDM model compound. We show how combined studies using optical microscopy, scanning electron microscopy, confocal scanning laser microscopy, dynamic mechanical thermal analysis and tensile testing are necessary to explore the new generation short-cut aramid fiber reinforced elastomer composites.

Introduction

In the past a wide variety of short cut fibers has been studied to reinforce elastomers. Natural fibers mainly cellulose, sisal, jute, silk, coir, coconut, bamboo and oil palm fibers have the advantage that they are renewable and sometimes show a better bonding to rubber. Their complicated fabrication and lower reinforcement factors are the disadvantages. Compared to natural fibers, synthetic fibers like nylon, polyester, glass, melamine, carbon and aramid exhibit continuous fabrication and therefore better tolerances in shape and uniformity. Nevertheless brittle fibers like carbon and glass are not so common for reinforcing elastomers, because of dramatic fiber breakage during processing. Due the lower fiber breakage and the considerable improvement of tribological properties of the resulting rubber composites there is a renewed interest in the application of short-cut aramid fibers in elastomers (Uchiyama, 2005; Wada, 1991, 1994a, 1994b) In particular the cut, chip and chunk resistance of truck tires can be enhanced, as reported by Datta et al. (2006). In addition, the hysteretic loss of aramid-fiber reinforced rubber is greatly reduced, which lowers the rolling resistance of the corresponding tires and, therefore, saves fuel and prolongs their durability. Further possible applications of short cut aramid fiber reinforced compounds are dynamically loaded rubber seals, diaphragms, engine mounts, rubber layers on counter-rotating rollers and rotating cylinders, transmission belts, conveyer belts, and hoses.
To obtain optimal performance of such short cut fibers in elastomers, the following two main governing goals should be reached: (i) a strong adhesion of the fiber to the matrix polymer to withstand shearing forces exerted on the fiber-matrix interface, and (ii) a controlled orientation of the short fibers in the matrix with respect to the application.
In the past, the use of short aramid fibers was hampered by the chemical inertness of the aramid surface, leading to a lack of interaction between fiber and elastomer, resulting in dispersion problems (Hintze, 2009) and insufficient fiber matrix adhesion (Orlob, 2009). The recent progress of Teijin Aramid BV in the development of surface modified aramid fibers might overcome these problems (Ahmad, 2005; Coffey, 2007; De Lange, 2001, 2007; Wennekes 2008).

Experimental

Compounds
An Ethylene Propylene Diene Rubber (EPDM, Keltan 8340A) based model compound with peroxide curing system (1.5 phr Perkadox® 14 40 MBR GR and 0.6 phr Actigran® 50) was selected to characterize the morphology of the chosen composite. The Twaron® short cut fibers supplied by Teijin Aramid BV were treated with a standard finish for processing reasons and had an average initial length of 3 mm.

Mixing
For the EPDM model compound the standard treated Twaron fibers with 3 mm length were incorporated via internal mixer (Haake Rheomix 600p) at 90°C with a rotor speed of 40 rpm. The following mixing sequence was found to be optimum. First 50% of the EPDM was kneaded for 2 min. After that the fibers were added with the rest of the matrix. This method was used for all experiments, since with other mixing sequences problems due to insufficient fiber wetting occurred.

The curatives were mixed in a separate step on two roll mill and the fibers were oriented that way. The influence of mixing parameter like rotor speed and mixing time on dispersion and fiber length was studied for variety of fiber contents (0, 1, 3, 5 phr).

Curing
Vulcanization was always carried out with hot press from Fortijne Grotnes BV, AC Vlaardingen, Holland at a pressure of 10 MPa and at a temperature of 170°C. For tensile testing and dynamic mechanical thermal analysis 2 mm thick sheets were pressed corresponding to optimum cure time (t90) of the compounds determined by rubber process analyzer (Scarabaeus SIS V50 from Scarabaeus Mess- und Produktionstechnik GmbH, Langgöns, Germany).

Fiber dispersion
The degree of fibre dispersion plays an important role for short fibre reinforced compounds, since agglomerates act as crack initiators under stress.
To get optimal dispersion different experiments were executed on internal mixer for the EPDM model compounds. Mixing sequences, screw speeds and dwell times were varied. Compounds were two roll milled to sheets with thickness of 0.5 mm. All agglomerates with a diameter over 3 mm were counted.

Fiber length distribution
To characterize the fiber length distribution after processing in the model compounds fluorescence microscopy was executed. A microscope in reflection mode (Axio Imager A1m von Carl Zeiss Jena GmbH, Germany) was used to study images of 15 µm thick films taken with magnification of 5x, which were later compiled with software (MethaMorph SPOT Insight) to 9 mm x 12 mm images.
Fiber length distribution was then determined by counting the length of minimum 200 fibers. Fibers touching the edge and fibers with diameter below the normal thickness were left out. The initial fiber length before processing was determined by measuring 100 random fiber bundles with a digital caliper. The following graph in the chapter results shows the added length of fibers of one length interval relative to overall added length.

Fiber orientation
To determine the fiber orientation of the model compounds, 2 mm thick S2 tensile bars were used. The parallel portion of the tensile bars (3.15 x 6.3 mm) was then scanned with a confocal laser scanning microscope (CLSM) before testing it with a tensile testing machine. The penetration depth of the laser was limited to ~300 µm. The image of one recorded plane was then characterized after method reported by Pegel et al. (2009) passing a binarization with threshold, autocorrelation by FD Math algorithm from ImageJ and elliptical fit for major axis to get the main orientation angle. This angle was then correlated to the secant modulus between 1 and 2% strain obtained from tensile testing.

Fiber matrix adhesion
The fiber-matrix interaction was investigated using scanning electron microscopy on the surface of fractured tensile bars. Therefore an Ultra 55 plus (Carl Zeiss SMT, Germany) scanning electron microscope with an acceleration voltage of 4kV was used. Before examination samples were sputtered with 2 nm thick layer of platinum.

Tensile testing
Mechanical properties were determined according to DIN 53504 with 5 S2 type tensile bars, a cross-head speed of 200 mm/min using a 1 kN load cell installed at tensile testing machine from Zwick GmbH, Ulm, Germany. These tensile bars obtained main fiber orientation in direction of applied tension. Secant modulus between 1 and 2% strain, modulus at 20%, 100% and 300% elongation, tensile strength at break and percentage elongation at break were recorded.

Dynamic Mechanical Thermal Analysis
The frequency–temperature dependent viscoelastic mechanical properties of the vulcanizates including storage, loss modulus and tan δ were measured in tension mode with the help of a dynamic mechanical analyzer, EPLEXOR 2000N (Gabo Qualimeter Testanlagen GmbH, Ahlden, Germany) using a 2 mm thick rectangular rubber vulcanizate block having an area of 15 mm × 8 mm. These specimens had a main fiber orientation in direction of the applied tension. The temperature scans were run from -80°C to 80°C at a heating rate of 2 K/min. The measurements were done with strain and frequency being set to 0.5% and 10 Hz, respectively dynamic. Static strain was set to 1% corresponding to a low elongation where the fibers were strained but not pulled out of the matrix.

Results and Discussion

Curing
Curing values like maximum and minimum torque, t10, t90 and scorch time can be observed in table below.

Table 1 – Curing values for EPDM model compound

	S' Min	S' Max	TC 10	TC 90	Scorch Time (TS 2)
	[dNm]	[dNm]	[min]	[min]	[min]
K8340 + 0phr	0,45	4,80	1,35	15,44	3,71
K8340 + 1phr	0,50	5,34	1,27	15,54	3,19
K8340 + 3phr	0,61	6,44	1,22	14,09	2,56
K8340 + 5phr	0,79	6,65	1,21	13,74	2,34

As expected the torque values rose with increasing fiber load, while the different curing times decreased.

This can be explained with faster heat transfer through the fiber-matrix interfaces into the matrix.

Fiber dispersion

The process of dispersion of fiber bundles consists of a mixture of erosion and rupture phenomena. To research the influence of the rotor speed induced shear stress on the dispersion different experiments were executed. Dwell time of the fiber was varied at constant screw speed, till no agglomerates were observable anymore in the milled sheets. It could be shown that time for optimum dispersion increased with increasing fiber loading, which can be seen in Figure 1.

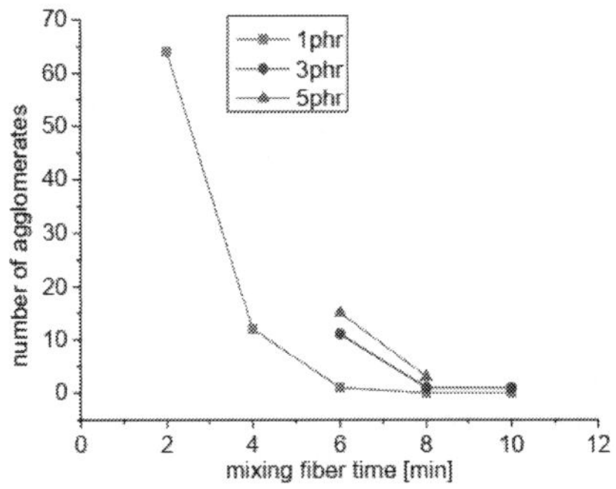

Figure 1 – Number of agglomerates versus mixing fiber time for different fiber loadings at constant rotor speed

Fiber length distribution

The initial fiber length showed some variation because of the cutting process, but the average value was 3 mm. Fiber length after processing was determined in terms of dwell time and fiber load. The important factor for the degree of fiber breakage was the total strain, the product of screw speed and fiber mixing time and therefore a probability how often a single fiber was going through the highest shear region. In Figure 2 this effect can be seen for a constant screw speed and two different fiber mixing times. The fiber length was also determined for different fiber loadings at constant screw speed and fiber mixing time (see Figure 3). The fiber length for 3 and 5 phr were slightly shifted to higher values in this case.

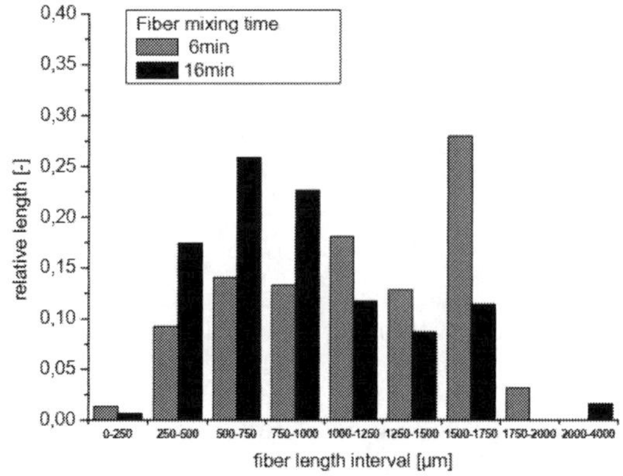

Figure 2 – Relative fiber length for constant fiber content of 1 phr and rotor speed of 40 rpm for varied fiber mixing times

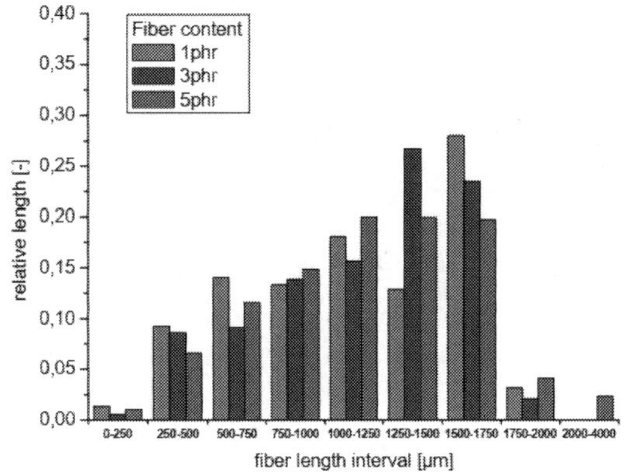

Figure 3 – Relative fiber length for constant fiber content of 1 phr mixing time, rotor speed of 40 rpm and fiber mixing time of 6 min

Fiber orientation

Assuming possible additional flow induced fiber orientation during compression molding, tensile bars were punched out in different distance from the overflow channels. A sketch of the numbered positions of the tensile bars can be seen in the figure below. Tensile bars number 1 till 4 had a distance of 5 mm to the each other or the left edge, while tensile bar 5 had a distance of 10 mm to the right edge. The arrow shows the original orientation of the fibers in the sheet obtained by two roll mill. DMTA stripe was punched out between tensile bars number 3 and 4.

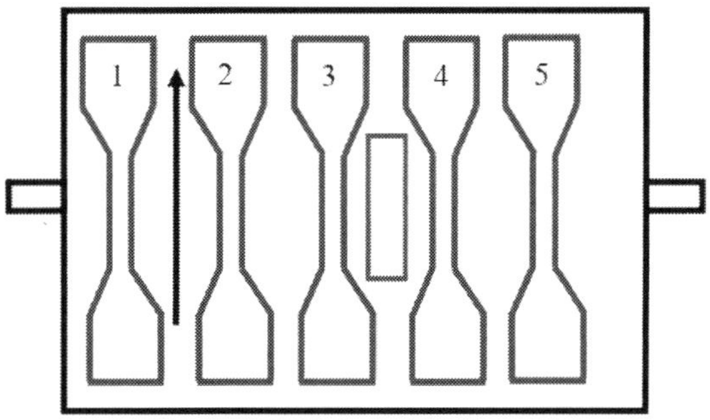

Figure 4 – Schematic sketch of positions for tensile bar and DMTA sampling

The results of the correlation between tensile testing results and angles received from the method described in the experimental part are shown in table 2. The angle obtained from CLSM describes hereby the deviation of the main fiber orientation to the main stress direction in the tensile bars under testing.

Table 2 – Tensile testing values correlation to orientation angle

	Angle	**Secant modulus 1-2%**	**Stress at 20%**	**Strain at break**
	°	MPa	MPa	%
K8340 + 0phr		**2,83**	**0,5**	**193,30**
Stan. Dev.		*0,21*	*0,01*	*24,61*
1		3,04	0,5	219,75
2		2,64	0,49	212,27
3		3,05	0,51	188,89
4		2,64	0,49	188,66
5		2,79	0,49	156,95
K8340 + 1phr		**3,81**	**0,73**	**175,15**
Stan. Dev.	--	*0,48*	*0,05*	*17,43*
1	68	3,47	0,67	176,51
2	53	3,45	0,69	149,31
3	--	3,48	0,71	195,82
4	--	4,46	0,79	169,58
5	26	4,2	0,78	184,50
K8340 + 3phr		**6,36**	**1,32**	**87,35**
Stan. Dev.	--	*0,81*	*0,15*	*18,19*
1	19	5,10	1,10	58,92
2	--	6,02	1,24	87,43
3	6	7,00	1,39	90,10
4	--	6,92	1,49	90,67
5	13	6,79	1,41	109,62
K8340 +5phr		**8,96**	**2,06**	**66,59**
Stan. Dev.	--	*0,94*	*0,26*	*8,29*
1	48	7,41	1,64	80,01
2	--	8,98	2,13	66,88
3	29	9,94	2,27	58,83
4	--	9,32	2,25	60,70
5	19	9,14	2,03	66,53

The tensile bars punched out of the middle of the sheet (away from press induced distortion flow of the overflow channels) obtained higher reinforcement values and lower strains at break then those taken from the edge. Especially the tensile bars from position 1 (with the lowest distance) were influenced significantly by this flow. The corresponding angles correlated more or less to the tensile testing results.

Fiber matrix adhesion

As expected the fiber matrix adhesion was low between the polar fibers and the non-polar matrix.
This could be seen by cavities observable in SEM images taken from the surfaces of fractured tensile test bars. If the fibers were oriented in direction of tension, these cavities were observable as holes (see upper part of Figure 5). If they were oriented perpendicular to the direction of tension they occurred as notches. Starting for a loading of 3 phr the fibers were creating entanglements, which led to higher reinforcement.

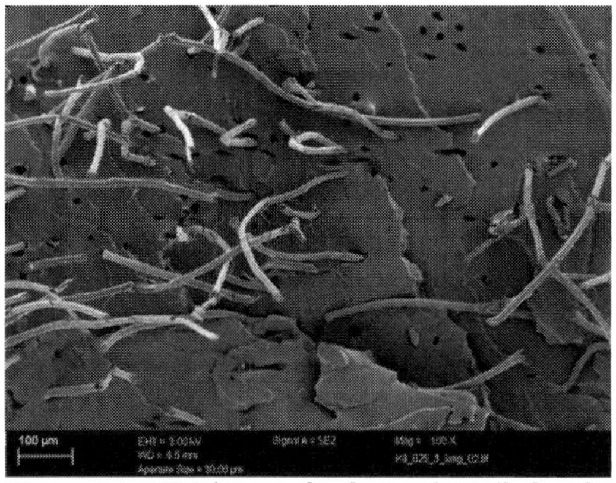

Figure 5 – SEM image of a fractured tensile bar for K8340 + 5phr with fibers oriented in direction of tensile testing

Tensile testing

Tensile testing values in fiber direction showed rising modulus and falling strain at break for increasing fiber loads (see table 2). The strain at break values were reduced to 50% for the compounds with 5 phr compared to the gum compound with 200%. The secant modulus between 1 and 2% strain was tripled compared to gum compound. The degree of reinforcement was mainly orientation- and fiber load dependent. The reinforcement from fiber-matrix interaction was taking place for strains under 50%.
This is observable especially for the stress-strain curve of the 1 phr sample (Figure 6). The slope is stiffer till a value of 50%, but follows the curve for the gum compound parallel afterwards, when the fibers-matrix interaction was lost. The influence of fiber shortening had no significant influence on the tensile testing results in this case like shown before in fiber distribution chapter.

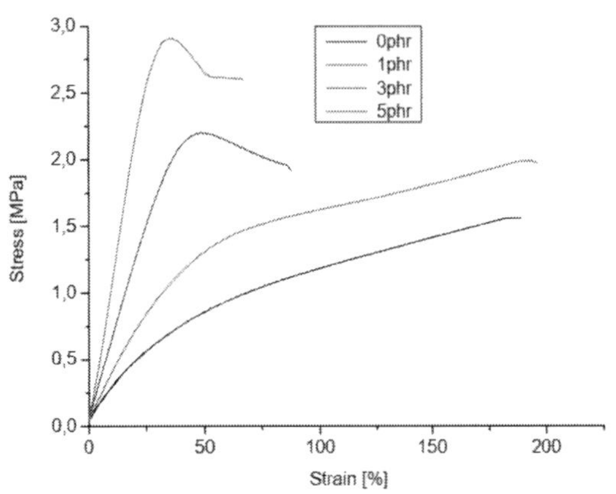

Figure 6 – Strain-Stress curves for different fiber loadings

Dynamic Mechanical Thermal Analysis

The DMTA measurements in fiber direction showed increasing storage modulus values with increasing fiber loading. Figure 7 shows also that the maximum of mechanical loss factor tan δ around T_G was decreased with increasing fiber load. Influence from fiber orientation and fiber length could be neglected, since these effects were discussed in the previous chapters.

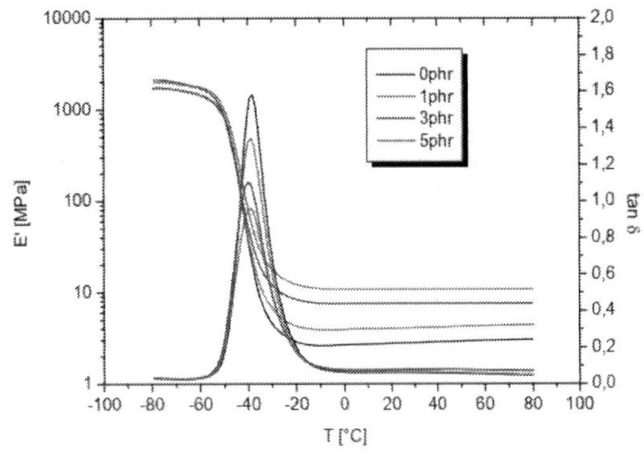

Figure 7 – Storage modulus E' and mechanical loss factor tan δ versus temperature for different fiber loadings

Conclusions

The influences of the compounding process on the morphology of standard treated Twaron® fibers with length of 3 mm in a transparent EPDM model system were investigated by optical microscopy, CLSM, SEM, tensile testing and DMTA. The used fibers had a low fiber matrix adhesion and were used for this study to investigate the basic compounding influences on morphology and mechanical properties of the composites.

It could be shown that the curing times decreased with increasing fiber load, while the torque values and the time needed for optimum dispersion went up.

The fiber length after processing in internal mixer was mainly influenced by the product of screw speed and fiber mixing time (total strain, which is the probability a single fiber is going through the highest shear region).

Fiber orientation was influenced by flow in mold through hot pressing near the overflow channels. The fiber-matrix interaction was as expected low, which could be confirmed by pull out holes through SEM.

Tensile testing values were mainly affected by orientation and fiber load. Higher fiber load and better orientation led like expected to increased modulus and lowered strain at break. Regarding the viscoelastic properties vulcanizates showed increasing storage modulus and decreasing tan δ peak values with increasing fiber loading.

Further studies will focus on the compounding influences of different mixing aggregates and surface treatments of fibers (adhesion enhanced) on the compounds and their resulting morphology and physical properties. In a later step this experience will be used to characterize carbon black filled industrial relevant composites and correlate their morphology and physical properties.

Acknowledgements

We gratefully acknowledge Teijin Aramid BV for providing Twaron® short cut fibers and DSM NV for providing of Keltan® K8340 und curatives. This work is part of the Research Program of the Dutch Polymer Institute DPI, Eindhoven, the Netherlands, project #664.

References

Ahmad et al. "Study of Fiber Surface Treatment on Reinforcement/Matrix Interaction in Twaron Fiber/ENR Composites", American Journal of Applied Sciences (Special Issue): 14-20 (2005)

Coffey et al. "Interfacial stress transfer in an aramid reinforced thermoplastic elastomer", Journal of Materials Science **42**, 8053–8061 (2007)

Datta R. N. "Reduced hysteresis in truck tread compounds by using aramid short fibres", Rubber Chemistry and Technology **79**, 26-38 (2006)

De Lange et al. "Characterization and micromechanical testing of the interphase of aramid-reinforced epoxy composites", Composites: Part A **32**, 31–342 (2001)

De Lange et al. "Controlled interfacial adhesion of Twaron aramid fibres in composites by the finish formulation", Composites Science and Technology **67**, 2027–2035 (2007)

Hintze et al. „Einfluss verfahrenstechnischer und stofflicher Parameter auf Morphologie und Anwendungseigenschaften aramidkurzfaserverstärkter Elastomercompounds" Technomer 2009, 21. Fachtagung über Verarbeitung und Anwendung von Polymeren, Elastomertechnik ETP5, CD-Rom Proceedings (2009)

Orlob I.; Reußmann T. "Hochleistungs-Elastomere auf Basis von aramidkurzfaserverstärktem Silikonkautschuk" Technomer 2009, 21. Fachtagung über Verarbeitung und Anwendung von Polymeren, Elastomertechnik ETV3, CD-Rom Proceedings (2009)

Pegel et al. "Spatial statistics of carbon nanotube polymer composites", Polymer **50**, 2123–2132 (2009)

Uchiyama et al. "Friction of Short-Fiber–Reinforced Rubber on Wet Surfaces", Journal of Applied Polymer Science **95**, 82–89 (2005)

Wada et al. "Dispersion and Orientation of Short fiber reinforced Rubber Composites", Kautschuk Gummi Kunststoffe **12**, 1142-1145 (1991)

Wada N., Uchiyama Y. "Effect of surface roughness and fibre diameter on the abrasive wear of short fibre reinforced rubber composites", International Polymer Science and technology **21**, 10 (1994)

Wada et al. "Friction of fibrious fillers and their orientations on the wear of short fibre reinforced rubber composites", International Polymer Science and technology **21**, 3, (1994)

Wennekes W. B., "Adhesion of RFL-treated cords to rubber: new insights into interfacial phenomena", Ph.D. thesis, University of Twente, Enschede, the Netherlands (2008)

MORPHOLOGY DEVELOPMENT DURING THE MIXING PROCESS OF CARBON BLACK AND SILICA FILLED SSBR/NR BLENDS AND RESULTING VISCOELASTIC PROPERTIES OF THE VULCANIZATES

S. Ilisch*, H.H. Le, D. Heidenreich, M. Hristov and H.-J. Radusch

Center of Engineering Sciences, Martin Luther University Halle-Wittenberg, Germany – sybill.ilisch@iw.uni-halle.de; hai.le.hong@iw.uni-halle.de; daniel.heidenreich@iw.uni-halle.de; hans-joachim.radusch@iw.uni-halle.de

For applications, where entropyelastic behavior is necessary, rubber is irreplaceable. Many rubber mixtures are produced in a discontinuous compounding process by means of internal mixers. The majority of technical used rubber materials are composites on the basis of rubber blends. The properties of such composites depend strongly on the morphology and localization of the filler in the blend phases. For carbon black and silica filled natural rubber (NR)/solution styrene butadiene rubber (SSBR) blend systems, the wetting, infiltration, dispersion and distribution of the filler during the mixing process were investigated by online measured electrical conductance test (OMEC), thermogravimetry and infrared spectroscopy on the rubber filler gel, as well as optical microscopy and atomic force microscopy in dependence on the mixing time. The knowledge about the morphology development during the mixing process is of great interest for tailoring the viscoelastic properties, which are in correlation with application-relevant properties of tires like wet skid resistance and rolling resistance.
In silica filled rubber compounds, an organosilane based coupling agent is used. During the compounding process, a chemical reaction between silane and silica is necessary for the improvement of viscoelastic properties of the vulcanizates. Furthermore, the time of adding of process oil influences the wetting behavior of the rubber. Therefore, the mixing regime of the rubber compounds is a key factor for the performance characteristics of the vulcanizates.

Introduction

Rubber compounds for tires are based on one or more polymers, reinforcing fillers, like carbon black or silica/silane, process oil, curing system and additives. A most fundamental step in processing rubber compounds is the incorporation of filler into the polymer by high-shear mixing procedures, like internal mixing or two-roll milling. In the case of silica filled rubbers, a chemical reaction between silica and silane during the mixing process takes part.
The quality of the mixtures is strongly dependent on an effective dispersion and uniform distribution of the filler throughout the polymer matrix. The properties of the gum depend in a high manner on the effort of this dispersion and distribution processes. Therefore different test procedures have been developed for the characterization of the filler dispersion. The properties of nanofiller dispersion may be broadly classified into three categories which are: agglomeration, networking, and phase distribution in polymer blends [1]. Filler agglomeration is the most negative aspect of dispersion and is also the most controllable in terms of good mixing procedure. The networking component of filler dispersion was originally defined by *Payne* [2]. The term refers to the tendency for the primary aggregates of carbon black to form a three-dimensional mesh-like configuration which is weakly held together by van der Waals bonding. The smaller the particle size of the nanofiller and the higher its loading, the more extensive the network and the greater its augmentation of low-strain dynamic modulus. Other aspects of networking include the occlusion of polymer between the aggregates in the network and the amount of polymer-filler linkages in conjunction with the filler-filler contacts. The occlusion of polymer by the carbon black during mixing has been described by *Boonstra* and *Medalia* [3]. The distribution of filler particles between the phases of an elastomer blend is affected by a number of factors, including unsaturation, viscosity, polarity, and mixing procedure. *Callan, Topcik* and *Ford* [4] showed for butyl/EPDM blends filled with carbon black (HAF (N 330)), that carbon black goes initially to the softer butyl phase of the blend, but with increasing carbon black content and higher shear forces the filler goes preferentially to the EPDM phase. *Hess, Scott* and *Callan* [5] firstly demonstrated the influence of the preferentially incorporation of filler in the phases of rubber blends on its mechanical properties. They found in 50/50 NR/BR blends filled with carbon black (ISAF (N 220)) a dependence of the phase selective CB localization on the total amount of CB in the blend. For a low CB amount of 10 phr the CB distribution in both phases is nearly homogeneous, but with increasing CB amount (40 phr) the CB localizes preferentially in the BR phase. For the 50/50 blend with 20 phr CB they showed the influence of viscosity: as BR viscosity increases, more CB locates in the NR.

Marsh et al. [6] showed for a 50/50 CR/NR blend the localization of carbon black in the interface or the CR phase, depending on the mixing regime. *Callan, Hess* and *Scott* [7] found a predominate localization of filler in one phase of binary blends (50/50 blends of NR with BR, SBR, CR, CIIR, NBR or EPDM), when large differences in unsaturation existed between the two elastomer components. In that case the filler was found in the highly unsaturated rubber. For 50/50 NR/SBR blends a strong dependence of the CB localization on the preparing conditions was shown. For solution masterbatches an equal CB distribution in both polymers was found, but upon mixing the carbon black located preferentially in the SBR phase. For masterbatches of NR/CB diluted with BR or SBR to get a loading of 40 phr CB (HAF-H (N 347)) a disposition of the CB in the NR phase was found. These experiments emphasized that the extent of transfer is very much related to factors such as rubber viscosity, heat and mechanical history, and total filler loading and polarity. A filler transfer can occur when the adsorptive capacity of the filler has not been fully utilized. This situation can exist if the masterbatch has minimum heat or mechanical history, or involves low molecular weight or low unsaturation elastomers. In 50/50 NR/SBR, NR/BR and SBR/BR blends with 60 phr N 220 no CB transfer was recorded when the first-stage masterbatches were mixed in Banbury for 5 min to a dump temperature of 149 °C. CB filled NR/SBR preblends indicated very significant preferential location in the SBR phase and in NR/BR preblends preferential BR location was indicated [8]. In SBR/BR blends which become miscible at higher mixing temperatures, it was found that carbon black tends to disperse preferentially in the polymer having higher molecular weight. For 50/50 NR/ESBR and 50/50 NR/SSBR blends 36 phr of N 339 were found in the NR (of a total amount of 45 phr). With increasing mixing time a consistent drop of CB loading in NR was found [9]. In 50/50 NR/BR blends with 50 phr N 550 it can be concluded that the CB does not have a preference for either the NR or BR phase. TEM images showed that NR is the dispersed and BR the continuous phase [10]. In silica filled EPDM/NBR blends [11] after sufficient mixing time the filler is preferentially located in the NBR phase (without silane).

From the beginning of rubber processing, the compounder has used some procedures to determine the quality of mixing. Especially surface inspection of stretched, cut or broken specimen was utilized. In 1956 *Leigh-Dugmore* [12] quantified the optical microscope analytical procedure to determine the percentage of carbon black dispersed below a certain agglomerate size (6 µm). His method was adopted by ASTM and is still widely used today. *Stumpe* and *Railsback* [13] used photographic technique in connection with a rating system for the characterization of the carbon black dispersion. Transmission electron microscope (TEM) procedures have also been applied extensively for filler dispersion analysis. *Von Ardenne* and *Beisher* [14] published the first electron micrographs of carbon black in rubber. The phase specific distribution of carbon black in rubber blends was initially investigated by phase contrast microscopy and TEM [4-8, 15- 19]. After the development of the atomic force microscopy (AFM) this technique was also applied for the investigation of filled rubber blends [20-24]. Differential scanning calorimetry (DSC) [10, 17, 25], differential pyrolysis [7-9], mechanical damping studies [26-30] and the analysis of the electrical conductivity [31-33] are further possibilities for the inspection of the phase specific distribution of filler in rubber blends. Most of these methods control the quality of the mixture after finishing the compounding process.

In 1983 *Jentsch* et al. [34] used the electrical conductance as control factor in an internal mixer. *Shiga* and *Furuta* [35] have determined changes in dispersion of carbon black agglomerates during mixing and proposed a concept of size reduction, which is consistent with the change in compound resistivity (onion model).

The method of controlling the electrical conductance during the mixing process was not used extensively because of the fact, that not only the dispersion and distribution influence the electrical conductance but also a series of technological and material factors. Furthermore, the stability of the conductivity sensor represents a problem regarding the high viscosity of the rubber melt.

Recently, we have developed a new method for the online characterization of the carbon black dispersion during the mixing process by means of electrical conductance [36-39]. This method was used to analyze the morphology development during the mixing process for a series of rubber [40] and carbon black [41], binary blends [42], thermoplastic vulcanizates (TPV) [43], thermoplastic melts [44, 45] and carbon nanotubes (CNT) [46].

Furthermore, we have generated a method for the description of the wetting behavior of the rubber to the filler and the quantification of the kinetics of the phase selective filler distribution in rubber blends [47, 48].

Experimental

Material

The rubbers used in the investigations are shown in *Tab. 1*; the fillers used are shown in *Tab. 2*.

Table 1 - Investigated rubbers.

Type	*Trade Name, Supplier*	*Remarks*
SSBR	SE SLR-4601, Dow Schkopau, Germany	21 wt.% styrene, 63 wt.% vinyl
NR	SMR 10, Astlett Rubber Inc., Malaysia	100 wt.% 1.4 cis

NR	SVR-L, Phuoc Hoa Rubber Company, Vietnam	100 wt.% 1.4 cis

Table 2 - Filler types, average diameter, structure (DBP: Dibutylphthalat, TEA: Triethanolamin) and specific surface area (NSA).

Type	*Supplier*	*Av. diam., nm*	*DBP/TEA, ml/100g*	*NSA, m^2/g*
CB Corax N 220	Evonik Degussa	21	114	117
Silica Ultrasil 7000 GR	Evonik Degussa	14	204	177

A basis recipe and general mixing regime for the carbon black filled compounds according *Tab. 3* and silica filled compounds according *Tab. 4* were applied. A total filler content of 50 parts per hundred parts of rubber (phr) was kept constant for all pre-investigations, that means the carbon black content was always above the percolation threshold (i.e. a carbon black content of approx. 32 wt% or 20 vol%, respectively).

Table 3 - Recipe and mixing regime for the carbon black filled compounds.

Ingredients	*Content, phr*	*Time of addition, min*
Rubber	100	0
ZnO	3	1
Stearic acid	1	1
Carbon black	50	2
Sulphur	2	2
Accelerator	1	2
T_A=50 °C, n=50 rpm, φ= 0.72, t=10 min		

Table 4 - Recipe and mixing regime for the silica filled compounds.

Ingredients	*Content, phr*	*Time of addition, min*
1st step		
Rubber	100 or 50/50	0 or 0 and 0,5
2/3 Silica	34	1
Silane	5	
1/3 silica	16	2,5
Processing oil	20	
ZnO	2,5	
Stearic acid	1	
Wax	1,5	
Antioxidant	2	
T_A=50/80/130/145 °C, n=60 rpm, φ= 0.72, t=11 min		
2nd step		
1st step batch	100	0
Sulfur	1,4	1,5
CBS	1,5	
DPG	1,5	
T_A=50 °C, n=50 rpm, φ= 0.72, t=5 min		

Processing and equipment for the online measured electrical conductance

The mixtures were produced in a laboratory internal mixer PolyLab System Rheocord 300p with a chamber volume of 72 ccm (Thermo Electron, Germany) - equipped with a combined temperature and electrical conductance sensor shown in *Fig. 1*.

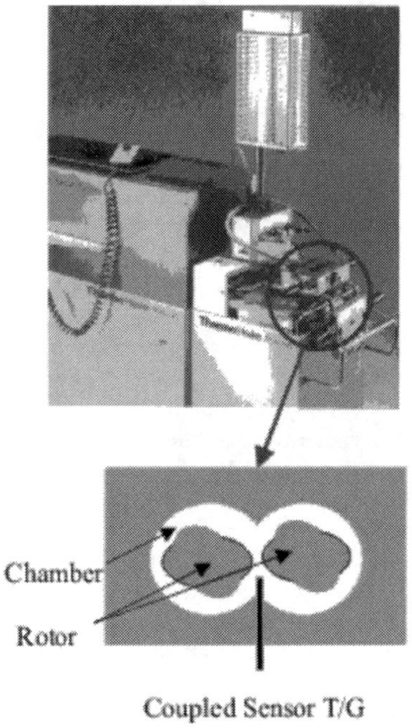

Figure 1 - Laboratory kneader with combined temperature / electrical conductance sensor.

The chamber wall acts as the second electrode. The sensor extends in a certain length into the chamber to realize a good contact to the material and to guaranty a sufficient stability against the appearing shear forces.

Characterization of the dispersion index (macrodispersion)

Optical microscopy is a rapid technique for evaluating the state of macrodispersion of filler in elastomers.

For the estimation of the dispersion index one produces gloss cuts by cutting thin, stretched samples by a razor blade at room temperature. The surface of the samples is checked by optical microscopy. If the surface of the cut contents filler agglomerates, the light is scattered and its area appears dark. With an image analysis program one can calculate the area of larger filler regions. The dispersion degree is assessed as the amount of the non-dispersed agglomerates with an average diameter larger than 6 µm. For this investigation a research microscope RM-DX (Leica, Germany) was used.

Characterization of the microdispersion
The dispersion and distribution of the filler in small lenght scale was analyzed by means of an AFM Q-Scope 250 (Quesant, USA) working in intermittend mode. The samples were cut in frozen state below the glass transition temperature with a microtom HM 360 (Microm, Germany), equipped with a cryo-chamber CN 30 by diamond knife to get preferably smooth surfaces for the AFM investigations.

Characterization of the wetting behavior, thermogravimetric analysis and infrared spectroscopy
Small pieces of the raw mixture (0.5 g) were stored in 100 ml solvent (cyclohexane, toluol) for seven days. The solvent was renewed after three days. Afterwards the resulting rubber-filler gel is dried until the mass stays constant. The rubber part in the gel L bound on the filler surface can be calculated by eq. 1.

$$L = \frac{m_2 - m_1 \cdot c_R}{m_2} = 1 - \frac{m_1}{m_2} \cdot c_R \quad (1)$$

The mass m_1 is corresponding to the rubber compound before extracting, and it is the sum of the mass of the undissolvable rubber part, the mass of the soluble rubber part and of the CB. m_2 is the mass of the rubber-filler gel, which is the sum of the undissolvable rubber part and the mass of the filler. c_R is the weight ratio of filler in the single rubber mixture or binary blends. This value is dimensionless and lies between 0 and 1. For determination of $L^{B(SBR)}$ and $L^{B(NR)}$ from $L^{B(SBR/NR)}$ a thermogravimetric analysis TGA / SDTA 851 (Mettler-Toledo, USA) was carried out to quantify the remaining rubber amount of each polymer blend phase [48] and by use of a Fourier transformed infrared (FTIR) spectrometer S2000 (Fa. Perkin Elmer) equipped with a diamond single ATR cell (Specac, Golden Gate) in the range of 4000-540 cm^{-1} [49].

Compression moulding and vulcanization
In front of viscoelastic testing, the raw mixtures were vulcanized in frame tools in a compression moulding machine to the time of vulcanization optimum.

Viscoelastic properties
For dynamic-mechanical tests we used the Qualimeter Eplexor 500 N (Gabo, Germany) in uniaxial mode.

Results and discussion

Carbon black filled rubber composites
The online measured electrical conductance G correlates directly with the dispersion and distribution of electrical conductive filler in the polymer matrix. *Fig. 1* shows, that the used SSBR faster than the NR disperses the carbon black (shift of the increase of G to longer mixing times). NR was masticated to the same viscosity like SSBR. The increase of the G-t-curve of the NR/SSBR blend (50/50) starts at the beginning of the mixing process at longer mixing times like the curve of the NR compound, after passing the local G-t-maximum it layes directly between the two single curves of SSBR and NR, which correlates to a nearly homogeneous wetting of the carbon black by both blend partners.

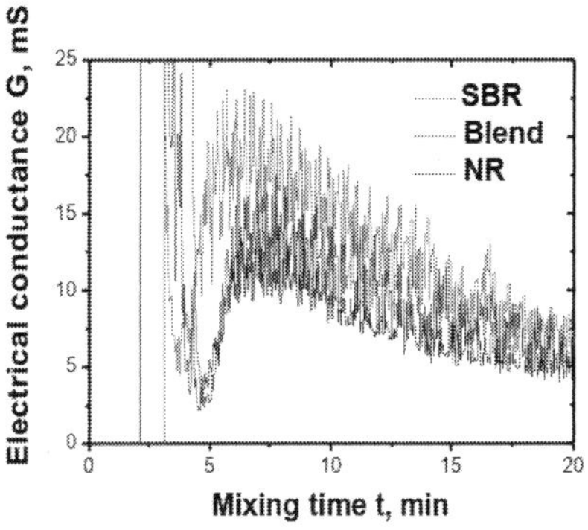

Figure 1 - Online measured electrical conductance versus mixing time for SSBR and NR compounds and 50/50 SSBR/NR blends filled with 50 phr N 220

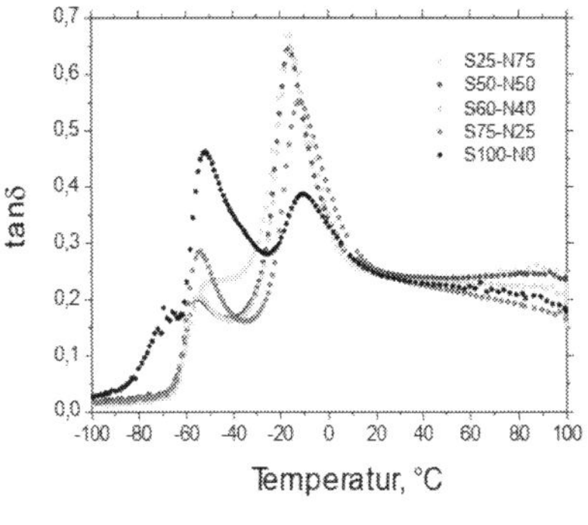

Figure 2 – Loss factor versus temperature for 50/50 NR/SSBR blends with 50 phr N 220 and varied localization of the CB in the blend phases (S25-N75: 25 % of CB in SSBR, 75 % in NR phase)

In systems with an inhomogeneous localization of the CB in both blend phases one can visualize sharp differences of the viscoelastic behavior by DMA (*Fig. 2*).

Silica filled rubber composites
For silica filled compounds the OMEC is only usable with moisture residues on the surface of the silica. The

characterization of the wetting behavior of the filler by the polymer chains was done like in the case of carbon black filled systems. The quantitative analysis of the phase selective wetting of the filler surface by the blend partner can be performed not only by the TGA/DTGA of the rubber-filler gel, but also by Fourier transformed infrared spectroscopy (FTIR).

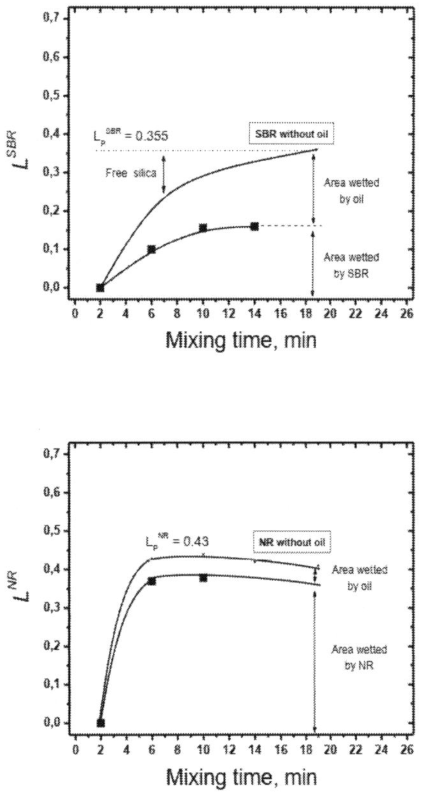

Figure 3 – Wetting behavior of single SSBR (above) and NR (below) mixtures

The *Fig. 3* clearly shows the wetting behavior of the single rubbers. NR occupies the free filler surface very fast in comparison with SSBR. The plateau in NR is reached at about 6 minutes mixing time against 10 minutes for SBR. The processing oil as low-molecular product is capable to wet the filler surface very fast and obviously the SSBR rubber layer formation is more affected by it. Therefore the oil addition affects the rubber-filler interaction negatively. Also of that NR forms thicker layer because of its flexible and longer polymer chains independently from the polarity.

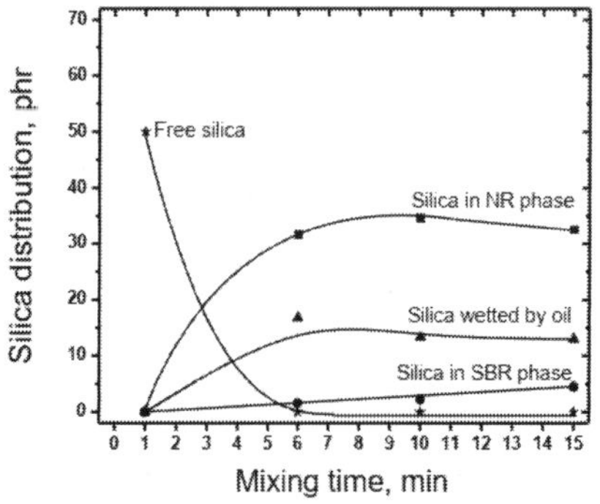

Figure 4 - Phase specific distribution of silica 50/50 SSBR/NR blend (with process oil)

Fig. 4 shows that NR wets the filler very fast. Until the 6th minute of the mixing process a great amount of silica is wetted by NR and the rest by the oil, because SSBR is not capable in high manner to compete NR. Respectively, no more free silica is observed after the 6th minute but not the entire filler surface is occupied by rubber phase. After 6th minute of the mixing SSBR because of its weak-polar nature starts to substitute the oil and the SSBR but very slow because the filler is not polar anymore. Responsible for this effect is the "oil shell" formed which prevents the migration of the filler thru phases. For comparison the kinetics of the distribution and migration of the filler in 50/50 SSBR/NR blend system without oil was investigated. In this system also no more free silica is observed after 6th minute but after this moment the filler starts intensively to migrate from NR to SSBR until even distribution is observed. After 14 minutes the filler is equally distributed in the phases and no more changes takes place. For comparison in the oil extended system only 5 phr from 50 phr are located in SSBR phase.

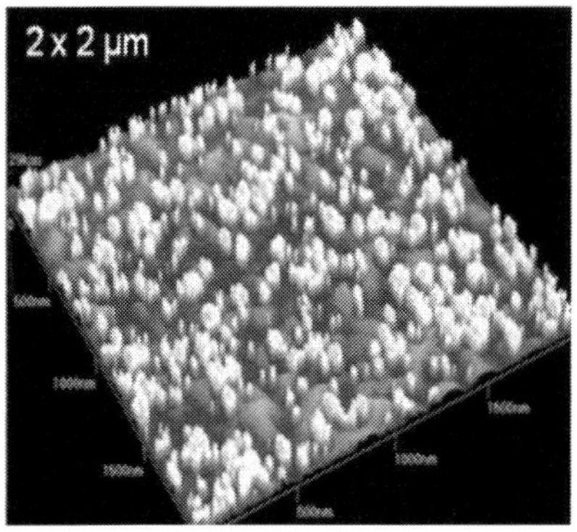

Figure 5 - AFM images of silica reinforced 50/50 SSBR/NR blend composites with process oil (above) and without oil (below).

Differences in the filler localization are also shown by AFM images (*Fig. 5*).

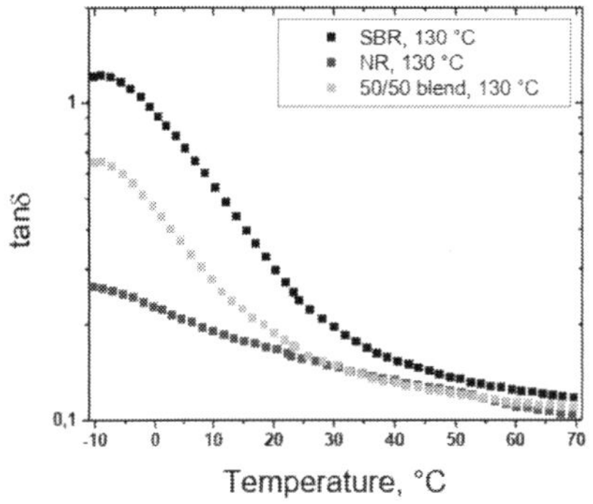

Figure 6 – Loss factor of single SSBR and NR mixtures and the 50/50 SSBR/NR blend, all filled with silica

The silane is capable to influence the dynamic properties too. It is widely popular that a low loss factor at high temperatures is in correlation with a low rolling resistance on one hand and high loss factor at low temperature indicates a high wet grip. In the blends SSBR with its relatively high glass transition temperature provides high loss factor at low temperatures. In the same moment NR is above its glass transition and works as continuous phase for the glassy SBR domains. It was described before that great filler amount is located in NR phase. At high working temperatures NR phase is responsible for the decreased loss factor.

Conclusions

Active fillers in rubber exhibit an high potential for the development of nanocomposites with enhanced properties. A key pre-condition for the successful use of such fillers is the control of dispersion mechanisms during the compounding process. Therefore the analysis of the electrical conductance, the knowledge of the wetting kinetics of the polymers and the kinetics of the selective phase distribution of the filler during the mixing process are helpful tools.

Acknowledgements

The authors thank the Federal Ministry of Education and Research (BMBF), the German Research Foundation (DFG) and the German Rubber Society (DKG) for the financial support of this work.

References

1. W.M. Hess *Rubber Chem. Technol.* 1991, *64*, 386.
2. A.R. Payne *J. Appl. Polym. Sci.* 1960, *3*, 127.
3. B.B. Boonstra, A.I. Medalia *Rubber Chem. Technol.* 1963, *36*, 115.
4. J.E. Callan, B. Topcik, F.P. Ford *Rubber World* 1965, *151*, 60.
5. W.M. Hess, C.E. Scott, J.E. Callan *Rubber Chem. Technol.* 1967, *40*, 371.

6 P.A. Marsh, A. Voet, L.D. Price *Rubber Chem. Technol.* 1967, *40*, 359.

7 J.E. Callan, W.M. Hess, C.E. Scott *Rubber Chem. Technol.* 1971, *44*, 814.

8 W.M. Hess, V.E. Chirico *Rubber Chem. Technol.* 1977, *50*, 301.

9 G.R. Cotton, L.J. Murphy *Rubber Chem. Technol.* 1988, *61*, 609.

10 J.M. Massie, R.C. Hirst, A.F. Halasa *Rubber Chem. Technol.* 1993, *66*, 276.

11 F. Boller, H. Geisler, R.H. Schuster *Kautsch. Gummi Kunstst.* 44, 1991, 1166.

12. C.H. Leigh-Dugmore *Rubber Chem. Technol.* 1956, *29*, 1303.
13. N.A. Stumpe, Jr., H.E. Railsback *Rubber World* 1964, *151*, 41.
14. M. von Ardenne, D. Beisher, *Rubber Chem. Technol.* 1941, *14*, 15.
15. P.A. Marsh, A. Voet, L.D. Price, T.J. Mullens *Rubber Chem.Technol.* 1968, *41*, 344.
16. P.A. Marsh, T.J. Mullens, L.D. Price *Rubber Chem. Technol.* 1970, *43*, 400.
17. A.K. Sircar, T.G. Lamond *Rubber Chem. Technol.* 1973, *46*, 178.
18. W.M. Hess, R.A. Swor, P.C. Vegvari *Kautsch. Gummi Kunstst.* 1985, *38*, 1114.
19. W.M. Hess, P.C. Vegvari, R.A. Swor *Rubber Chem. Technol.* 1985, *58*, 350.

20. A.A. Galuska, R. R. Poulter, K.O. Mc Elrath *Surface and Interface Analysis*, 1997, *25*, 418.
21. M. Ganter, R. Brandsch, Y. Thomann, T. Malner, G. Bar *Kautsch. Gummi Kunstst.* 1999, *52*, 717.
22. V. Herrmann, K. Unseld, H.B. Fuchs *Kautsch. Gummi Kunstst.* 2001, *54*, 453.
23. I.H. Jeon, H. Kim, S.G. Kim *Rubber Chem. Technol.* 2003, *76*, 1.
24. C.C. Wang, J.B. Donnet, T.K. Wang *Rubber Chem. Technol.* 2005, *78*, 17.
25. B.-L. Lee, C. Singleton *J. of Appl. Polym. Sci.* 1979, *24*, 2169.
26. A. Medalia *Rubber Chem. Technol.* 1978, 51 437.
27. S. Maiti, S.K. De, A.K. Bhowmick *Rubber Chem. Technol.* 1992, 65, 293.
28 M. Klüppel, R.H. Schuster, J. Schaper *Gummi Fasern Kunstst.* 1998, *51*, 508.
29. C. Sirisinha, J. Thunyarittikorn, S. Yartpakdee *Plastics Rubber Comp. Proc. Appl.* 1998, *27*, 373.
30 M. Klüppel, R.H. Schuster, J. Schaper *Rubber Chem. Technol.* 1999, *72*, 91.
31. Sircar, A. K. *Rubber Chem. Technol.* 1981, *54* 820.
32. A.Y. Zaikin, V.A. Nigmatullin *Kautsch. Gummi Kunstst.* 1994, *47*, 709.
33. B.G. Soares, F. Gubbels, R. Jérome, E. Vanlathem, R. Deltour *Rubber Chem. Technol.* 1997, *70*, 60.
34 J. Jentzsch, J. Spacek, H. Michael, J. Flohrer *Plaste und Kautsch.* 1983, *30*, 700.
35. S. Shiga, M. Furuta *Rubber Chem. Technol.* 1985, *58*, 1.
36. H.H. Le, S. Ilisch, B. Jakob, H.-J. Radusch *Kautsch. Gummi Kunstst.* 2003, *56*, 388.
37. H.H. Le, S. Ilisch, B. Jakob, H.-J. Radusch *Rubber Chem. Technol.* 2004, *77*, 147.
38. H.H. Le, S. Ilisch, I. Prodanova, H.-J. Radusch *Kautsch. Gummi Kunstst.* 2004, *57*, 355.
39. H.H. Le, I. Prodanova, S. Ilisch, H.-J. *Radusch Rubber Chem. Technol.* 2004, *77* 815.
40. H.H. Le, M. Tiwari, S. Ilisch, H.-J. Radusch *Kautsch. Gummi Kunstst.* 2005, *58*, 575.
41. H.H. Le, M. Tiwari, S. Ilisch, H.-J. Radusch *Rubber Chem. Technol.* 2006, *79*, 610.
42. H.H. Le, Z. Qamer, S. Ilisch, H.-J. Radusch *Rubber Chem. Technol.* 2006, *79*, 621.
43. H.H. Le, M. Tiwari, S. Ilisch, H-J. Radusch *Plastics Rubber Composites*, 2006, *35*, 410.
44. H.-J. Radusch, H.H. Le, S. Ilisch *Gummi Fasern Kunstst.* 2006, *59*, 46.
45. H.H. Le, S. Ilisch, A. Wutzler, H.-J. Radusch *Kautsch. Gummi Kunstst.* 2007, *60*, 295.
46. H.H. Le, G. Kasaliwal, S. Ilisch, H.-J. Radusch *Kautsch. Gummi Kunstst.*, 2009, 62, 326.
47. H.H. Le, S. Ilisch, G.R. Kasaliwal, H.-J. Radusch Kautsch. Gummi Kunstst. 2007, *60*, 241.
48 H.H. Le, S. Ilisch, G.R. Kasaliwal, H.-J. Radusch *Rubber Chem. Technol.*, 2008, 81, 767.
49 H.H. Le, S. Ilisch, D. Heidenreich, A. Wutzler, H.-J. Radusch, *Polymer Composites*, 2010, ID PC-09-0354.R1.

CONTINUOUS RUBBER DEVULCANIZATION USING SUPERCRITICAL CO2

Mohammad Meysami, Shuihan Zhu, Costas Tzoganakis*

University of Waterloo - Ontario - Canada

Continuous devulcanization of tire rubber crumb was performed using supercritical CO2 in an industrial scale twin screw extruder. A reasonably high throughput extrusion process has been developed and the effect of processing conditions has been studied. Crosslink density and percent of devulcanization of different samples are measured. Curing behaviour, tensile strength, and elongation at break of different compounds consisting of blends of virgin rubber with devulcanized crumb have also been evaluated.

INJECTION MOLDING OF THERMOPLASTIC/RUBBER-COMPOSITES

Prof. Dr.-Ing. Dr.-Ing. E.h. W. Michaeli, Dipl.-Ing. O. Grönlund*, Dipl.-Ing. T. Arping, Dipl.-Ing. K. Opdenwinkel, F. von Fragstein, M.Sc.

Institute of Plastics Processing (IKV) at RWTH Aachen University, Pontstraße 49, 52062 Aachen, Germany, groenlund@ikv.rwth-aachen.de

Facing increasing demands of the market and accretive efforts of productivity improvement the multi-component injection molding offers a substantial potential in developing and manufacturing technically ambitious and highly integrated plastic and elastomer products. The fundamental challenge in manufacturing multi-component moldings is to realize a sufficient bond strength between both components. To achieve high bond strengths it is desirable to have substance-to-substance bonds between the different materials. Typically thermoplastic/rubber-composites do not have the ability to build those substance-to-substance bonds. This leads to adhesion forces which are mostly too weak for manufacturing two-component parts with sufficient bond strengths.
Therefore the objective of current research is the design and validation of an entirely new process chain for molding various thermoplastic/rubber-composites. The desired bond strengths are ensured by using a plasma polymerized interlayer. The characteristics of plasma coatings can be varied systematically by the process determining parameters of the plasma process. The main parameters under investigation are type and flow rate of the process gases, the duration of the layer deposition and the applied electromagnetic power and process pressure. The constitutional idea is to gain the ability to set individual surface characteristics by varying those parameters for each material combination.
First results prove the possibility of directing the chemical composition as well as the physical properties of the plasma polymerised interlayer. Current investigations are now focused on the influence of certain characteristics on the bond strength between the thermoplastic and the rubber components.

Introduction

Multi-component molding is an established production technology within the plastics industry. Due to its high grade of automation it is offering the possibility to join different plastics easily, economically and quickly.
A great number of multi-component applications are rigid-soft combinations. Those composites are used for technical appliances as well as for consumer products. For applications facing high thermal loads or aggressive fluids the demands on the soft component can only be fulfilled by rubber.

Initial situation

Adhesion is defined as the resistance of a material combination against a disruptive load [Brinkmann 1996, Kuhmann 1998]. The adhesive strength σ_H is the inner force F_i required to override the adhesion related to the true contact surface A_W at break.

$$\sigma_H = \frac{F_i}{A_w} \qquad (1)$$

Since measurement of the inner force and the true contact surface is hardly possible the bond strength σ_V is used for comparing adhesive strengths. The bond strength is measured by destructive testing and defined as the quotient of the extraneous force F_a and the initial contact surface A_N.
Though the bond strength is a summary expression of

$$\sigma_V = \frac{F_a}{A_N} \qquad (2)$$

mechanical strength and only partially comparable to the adhesive strength which is based on several different adhesion phenomena it is well established as the decisive reference parameter [Bischof et. al. 1983].
Despite some good results in realization or improvement of this bond strength between thermoplastics and rubber the existing strategies unexceptionally implicate some severe handicaps. The development of special materials for one or both joining components is marked by high chemical and technological complexity and thereby by high costs. Another disadvantage of the chemical modification of the materials is the fact that many of the used substances and additives are not conform to the regulations for medical or consumer products. The usage of primer and/or bonding agents known from the coupling of rubber and metal is work extensive and mostly environmentally harmful.
Plasma treatment and plasma coating are offering an ecological and economical attractive alternative to the

existing methods for enhancing the bond strength of thermoplastic/rubber-composites.
The improvement of bond strength by *plasma treatment* of the contact surfaces is caused by an activation of the surface. The process gases get excited and ionized in the plasma and the emerging radicals, ions and free electrons hit the substrates' surface. They induce changes of the physical (i.e. topography) and chemical (i.e. functional groups) structure of the surface. But the success of this method strongly depends on the desired material combination and also the modifications achieved are temporary.
In opposition to this a plasma-polymerized coating is deposed on the substrate in *plasma coating* processes. This enables a durable chemical und physical modification of the original surface of the substrate.

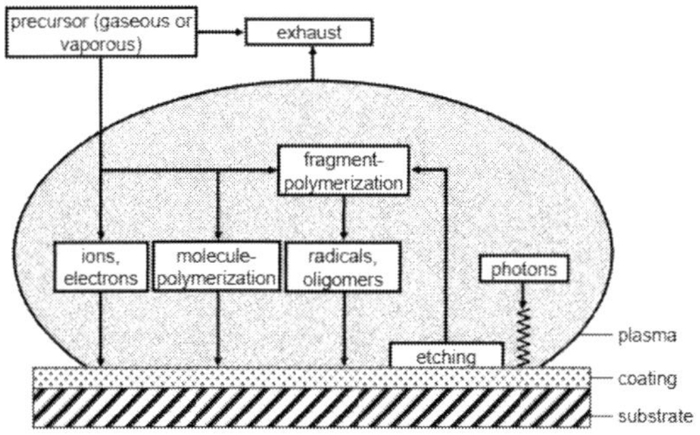

Figure 1: Principle of plasma coating processes

The applied monomers get fragmented and excited in the plasma and depose as a highly cross-linked coating on the parts surface. By inelastic collisions between electrons and molecules chemical bonds get broken up in the plasma and polymerization processes (molecule polymerization, fragment polymerization) take place. Contrarily to conventional polymerization processes, which are mostly based on molecule polymerization, is the plasma coating process marked by fragment polymerization. This fact and the other plasma specific properties like etching by excited molecules or high UV-radiation lead to surface characteristics which are strongly differing to the original surface. Plasma polymerized coatings are generally highly cross-linked, amorphous and of high density. The additional characteristics can vary in a wide range. So it is possible to adjust the coatings characteristics to the desired surface functionality as well as for optimal bonding to the substrate material.
By that the plasma coating offers the possibility to depose optimal coatings for the bonding of arbitrary materials in the injection molding process especially thermoplastics and rubber.

Objectives

In the future the task for multi-component injection molding will be the improvement of bond strengths of material combinations already in use and the realization of more powerful rigd/soft-composites with sufficient adhesion. Hence, the objective of actual research is the development and proving of an totally new process chain for manufacturing arbitrary thermoplastic/rubber-composites which bond strengths are based on custom tailored plasma polymerized coatings. For the hard component different technical thermoplastics and for the soft component different standard rubbers will be chosen as joining materials. Neither the thermoplastics nor the rubbers will be chemically modified. The adhesion should be exclusively gained by the plasma coating.
Based on the successful realization of sufficient bond strengths the influence of varying parameters of the plasma process (process gas, additive gases, time, power, pressure) on the bond strengths of different material combinations will be examined. By that a fundamental understanding of the process will be gained.
A further objective is to find the optimal coating characteristics (topography, chemical composition) for each of the investigated material combinations and to evaluate how the obtained results can be used for the joining of further arbitrary thermoplastic/rubber-composites.
The new process chain developed to achieve these objectives can be divided into three steps of processing.

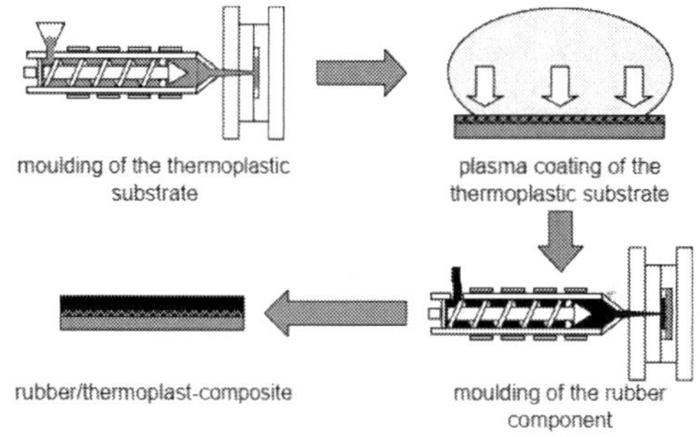

Figure 2: Process chain for the plasma aided manufacturing of thermoplastic/rubber-composites

The first step is the molding of the thermoplastic substrate. After that the demolded part is transferred into a low-pressure plasma reactor in which the plasma polymerized coating is deposed. In the last step the coated part is placed in the cavity and the rubber is overmolded.

Experimental series and results

Coating characteristics
The bond strength of composite-moldings depends decisively on the characteristics of the plasma polymerized coating in general and in particular on the surface topography and its chemical composition.
The diagrams in Figure 3 and Figure 4 illustrate the changes in surface topography in dependency on variations of power and pressure in the plasma process. With an increase of power at constant gas flow rates

the available power density per particle rises and is by that an influencing factor on plasma processes. An increase of power density causes an increase of electron density and by that an increase of excited particles [Inagaki 1996]. If there is enough power available an increase of pressure is also leading to an elevated number of excited particles.

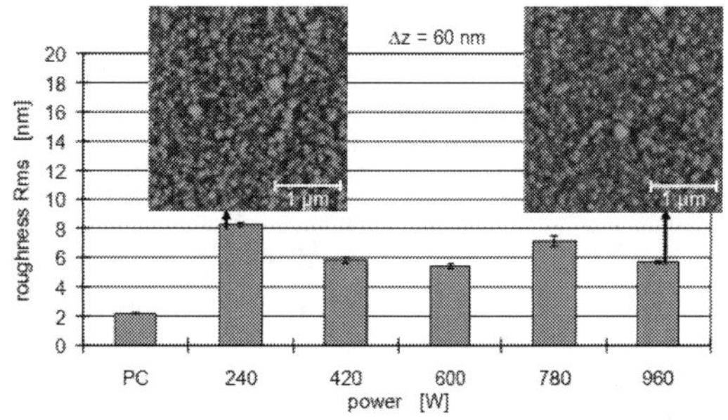

Figure 3: Surface topography for varied input powers

Firstly Figure 3 shows that the variation of power in the chosen span has no influence on the surface topography. But with all input pressures a very homogeneous, grained coating has been deposed. Analogous to the optical impression of the surface topography does the chemical composition not change distinctly within the experimental series. Yet differing to the substrate material PC do all the deposed coatings have a significantly higher fraction of oxygen and a lower fraction of carbon in it.

Figure 4 shows the influence of the process pressure on the topography of the deposed plasma polymerized coating. Contrarily to the experiments discussed before there is a clear relation between variations of the process pressure and the surface topography detectable. A decline of pressure leads to a finer surface structure and by that to a significantly lower roughness. The change in medial roughness between pressures from 0,2-0,8 mbar is about a factor of 10.

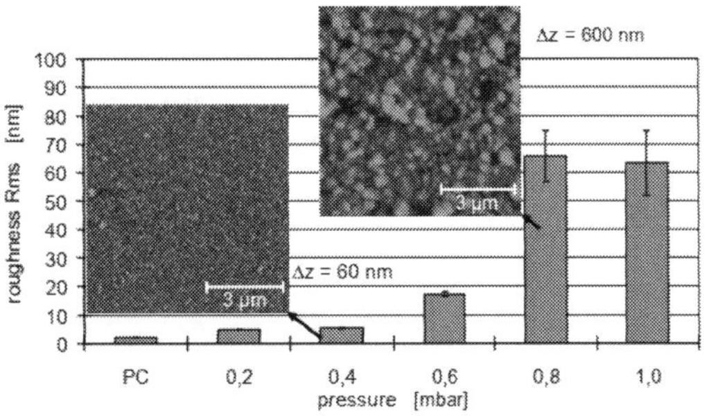

Figure 4: Surface topography against process pressures

In contrast the chemical composition of the surface is affected hardly by the pressure variation. Only a slight trend of a decreasing fraction of silicium with decreasing pressure is noticeable (Figure 5).

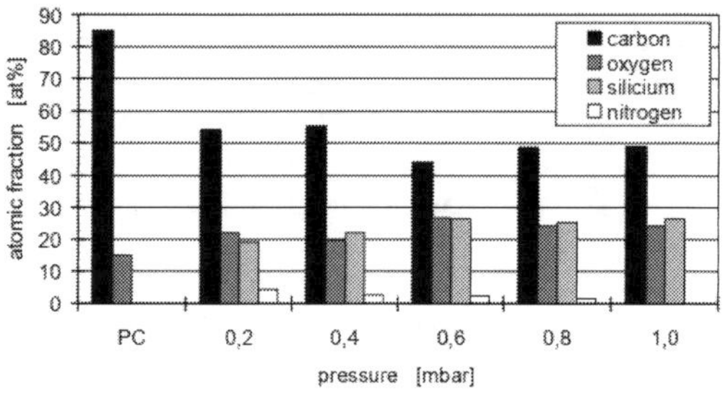

Figure 5: Chemical composition of the layers for varying process pressures

The discussed results display that an effective influencing or even direction of the chemical composition of the in the plasma process deposed coating is not achievable by a variation of process pressure or input power. Due to that a variation of the process gas was done. The dependency of the chemical composition on the gas flow of oxygen is shown in Figure 6. As reference the composition of the chosen substrate material PC is shown on the left.

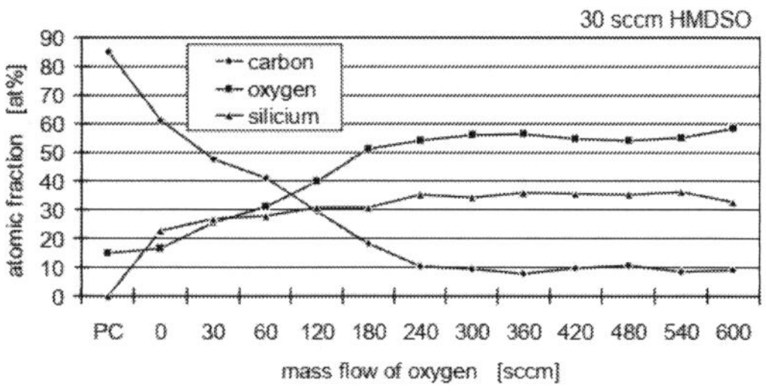

Figure 6: Chemical composition of the layers for varying compositions of the process gas

If the amount of oxygen in the process gas increases its fraction in the chemical composition of the plasma polymerized coating increases as well. Using pure HMDSO as process gas a fraction of 15 at% of oxygen is detectable in the coating. But this value can be risen to more than 50 at% with a process gas ratio of 8:1 oxygen/HDMSO. For a further increase of the oxygen fraction in the process gas are no more relevant changes in the surfaces' chemical composition detectable. The fraction of carbon in the deposed layer decreases to the same extent as the fraction of oxygen rises. It drops from about 60 at% to less than 10 at%. The amount of silicium in the coating is also elevated by an increase of the added oxygen. Starting with about 22 at% silicium for a pure HDMSO-plasma the silicium fraction raises for another 10 at% with an increase of oxygen addition. Like the curve of the oxygen fraction both of the described trends show no further changes for bigger oxygen/HMDSO ratios than 8:1.

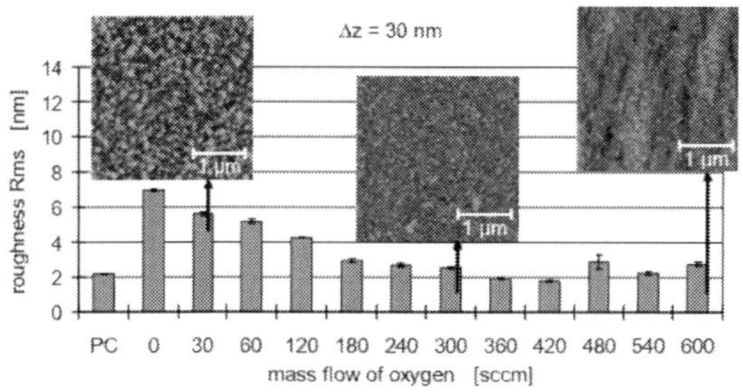

Figure 7: Surface topography for varying compositions of the process gas

Figure 7 shows the influence of the variation of the mass flow of oxygen against the medial surface roughness. A reduction of roughness is clearly visible up to a flow rate of 400 sccm. For higher amounts of oxygen in the process gas no further influence on the surface structure is noticeable.

Bond strength
All the experimental series and results discussed before are essential steps to achieve the objective of gaining sufficient bond strength between rubber and thermoplastics. Facing this challenge first experiments on the improvement of bond strength by a plasma polymerized coating were carried out. For different parameter variations in the plasma process adhesion tests for EPDM/PA composites were done (Figure 8)

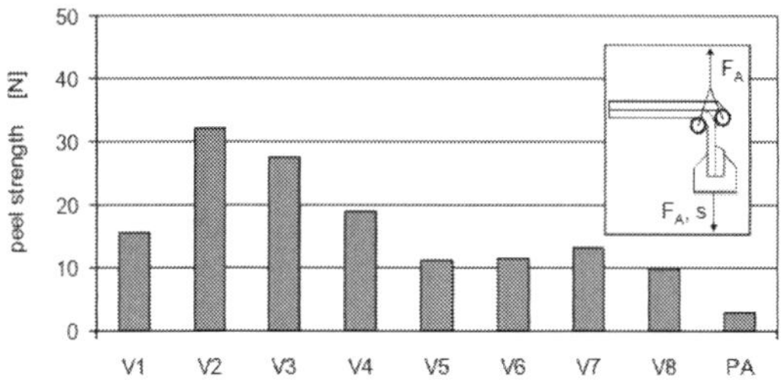

Figure 8: Peel test EPDM/PA-composites

Compared to the adhesion of EPDM on the uncoated PA-substrate shown at the right the potential of bond strength improvement by plasma polymerized coatings is apparent. In this experimental series the peel strength was risen by a factor of 10.

Conclusions and Outlook

The investigations carried out with the focus on the characterization and manipulation of the chemical composition and the topography of a plasma polymerized coating on a thermoplastic substrate prove that it is possible to direct the surface characteristics systematically. By varying the process parameters pressure and process gas of the plasma process the topography and the chemical composition of the deposed coating could be manipulated reproducibly.
Future investigations will have their focus on the realization and testing of bond strengths of different thermoplastic/rubber-composites. Composites of several technical thermoplastics and rubbers of frequent industrial use will be manufactured according to the process chain described. Afterwards the bond strengths of the rigid/soft-composite moldings will be analyzed in peel, shear and tensile tests.

Acknowledgements

The research project 16522 N of the Forschungsvereinigung Kunststoffverarbeitung was sponsored as part of the "Industrielle Gemeinschaftsforschung" (IGF) by the German Bundesministerium für Wirtschaft und Technologie (BMWi) through the AiF, to whom we extend our thanks.

References

Brinkmann, S.: Verbesserte Vorhersage der Verbundfestigkeiten von 2-Komponenten-Spritzgießbauteilen. Rheinisch-Westfälische Technische Hochschule Aachen, Dissertation, 1996

Bischof, C.; Bauer, A.; Leonhardt, H. W.: Haftfestigkeitsuntersuchungen an Metall-Polymer-Verbunden I – Theoretische und experimentelle Grundlagen – Stand und Grenzen der Prüfmethoden. Plaste und Kautschuk, 30 (1983) 1, S. 1-5

Inagaki, N.: Plasma surface modification and plasma polymerization. Technomic Publishing Company, 1996

Kuhmann, K.: Prozeß- und Materialeinflüsse beim Mehrkomponenten-Spritzgießen. Universität Erlangen-Nürnberg, Dissertation, 1998

MECHANICAL PROPERTIES OF NATURAL RUBBER WITH NANOMATRIX STRUCTURE

Seiichi Kawahara*, Yoshimasa Yamamoto, Keiichi Akabori

Department of Materials Science and Technology, Faculty of Engineering, Nagaoka University of Technology, Nagaoka 940-2188, Japan –kawahara@mst.nagaokaut.ac.jp

Morphology of natural rubber was observed by transmission electron microscopy. Nanomatrix of non-rubber components such as proteins and phospholipids was found to be inherently formed in natural rubber, in which natural rubber particles of about 0.5 μm in average diameter were dispersed. The nanomatrix of non-rubber components disappeared after deproteinization of natural rubber with urea. Strain at break of natural rubber increased after deproteinization, while stress at break did not change. When amount of the non-rubber components increased, the stress at break became significantly dependent upon the amount of non-rubber components. Viscoelastic properties of natural rubber were also dependent upon the nanomatrix of non-rubber components. Storage modulus of natural rubber increased significantly, when the amount of the non-rubber constituents increased.

INTRODUCTION

Nanomatrix structure is usually formed in natural rubber, in which natural rubber particles of about 1 μm in diameter are well dispersed in the nanomatrix of non-rubber components such as proteins, phospholipids and so forth. It may come from a layer structure of natural rubber particles present in latex. Since the particles are mainly covered with the proteins and phospholipids [1,2], after condensing the natural rubber particles, the nanomatrix is inherently formed in natural rubber, which is able to interact with *cis*-1,4 polyisoprene as a rubber hydrocarbon.

In the previous work, a linear *cis*-1,4 polyisoprene of natural rubber was proposed to consist of ω-terminal, two *trans*-1,4 isoprene units, about 1,000-3,000 *cis*-1,4 isoprene units, and α-terminal [3]. The α-terminal is composed of phospholipids linking to fatty acids, which play a role in branching structure of natural rubber. On the other hand, the ω-terminal of the rubber molecule in natural rubber was postulated to be a modified dimethylallyl group linking to a functional group, which is associated with the proteins to form crosslinking junctions through intermolecular hydrogen bonding formation [4]. In the previous work, the branching structure was associated with outstanding mechanical properties of natural rubber, such as tensile strength, tear strength, green strength and so forth, which was attributed to strain-induced crystallization of the rubber [5-7]. However, the mechanical properties did not change significantly after deproteinization of natural rubber. Therefore, it has been desired to find a dominant factor in the mechanical properties of natural rubber.

We investigated an effect of the nanomatrix on the properties of natural rubber with a model compound, prepared by graft copolymerization of styrene onto natural rubber particles after deproteinization of the rubber [8,9]. The stress at brake of the graft-copolymerized natural rubber with the nanomatrix structure was significantly higher than that of the graft-copolymerized natural rubber with island-matrix structure and it was quite similar to the stress at break of vulcanized natural rubber. Storage modulus of natural rubber increased by a factor of more than 10 times, even though loss tangent increased, especially at high frequency region. On the basis of the previous results, it is possible to postulate that the nanomatrix structure is a key factor to control mechanical properties of natural rubber. In order to prove the postulation, we have to relate the nanomatrix structure with the outstanding mechanical properties of natural rubber.

In the present work, we investigate a relationship between morphology and properties of natural rubber with nanomatrix of non-rubber constituents. Natural rubber containing various amounts of non-rubber constituents were prepared by changing a rotation speed at centrifugation. The morphology was observed by transmission electron microscopy. Changes in tensile at break and plateau modulus were associated with the amount of non-rubber constituents.

EXPERIMENTAL

Natural rubber latex used in this study was commercial high-ammonia NR (HANR) latex. The natural rubber latex was separated in particle size with a centrifugator. Rotations per minutes at centrifugation were 2000 (2k) rpm, 5000 (5k) rpm, 7000 (7k) rpm and 10000 (10k) rpm. After the centrifugation, serum fractions were recovered and as-cast serum films were prepared from the serum fractions. Deproteinization of natural rubber was carried out by incubation of natural rubber latex with 0.1 w/w% urea in the presence of 1 w/w% sodium dodecyl sulfate (SDS) at 303K followed by centrifugation at 10,000 g [10,11]. The cream fraction was re-dispersed in 1 w/w% SDS solution to make latex of 30 w/w% in dry rubber content (DRC) and it was washed twice by centrifugation to prepare deproteinized natural rubber (DPNR) latex. The DPNR latex was diluted with distilled water to adjust 30 w/w% DRC and concentration of SDS was adjusted to 0.1 w/w%.

Morphology of natural rubber and DPNR was observed with a transmission electron microscope (TEM), Hitachi H-800 at accelerating voltage of 100 kV. The ultra thin sections of the blends were prepared with an ultra-microtome equipped with a cryo kit (Leica ULTRA CUT UCT) at a temperature lower than glass transition temperature of natural rubber. The thin sections were stained with Tungstainic Phosphate, after annealing the section at 353 K for 30 min.

Dynamic mechanical properties were measured with a Rheometric Science ARES-4 at frequencies between 0.1 – 10 Hz and temperatures between 213 and 403 K. The tensile strength was measured with an Instron universal tester 3365, according to JIS-6251.

RESULTS and DISCUSSION

Table 1 shows dry rubber content (DRC) and nitrogen content (N%) of HANR, DPNR and serum rubbers. Values of the DRC and N% of HANR were 62.4 wt% and 0.229 wt%, respectively, which were reduced by deproteinization. Especially, the value of N% of DPNR was dramatically reduced to 0.013 wt%, as in the case of literature value [10,11]. The values of the DRC and N% of the serum rubber were dependent upon rotation speed of centrifugator; the larger the rotation speed, the lower is the value of DRC and the higher is the value of N%. The highest value of the N% was 2.285 wt%, which was about 10 times as high as that of HANR. This demonstrates that natural rubber containing a large amount of proteins (high-protein natural rubber) is prepared from HANR. Volume mean particle size of the rubbers of HANR, DPNR and the serum rubbers is also shown in Table 1. The volume

Table 1 Dry rubber content (DRC), volume mean particle diameter (<D>) and nitrogen content (N%) of the rubbers.

Sample	DRC / %	N %	<D> / μm
DPNR	25.3	0.013	0.383
HANR	62.4	0.229	0.437
Serum2k	25.3	0.482	0.425
Serum5k	15.6	0.971	0.356
Serum7k	11.9	1.305	0.243
Serum10k	7.0	2.285	0.193

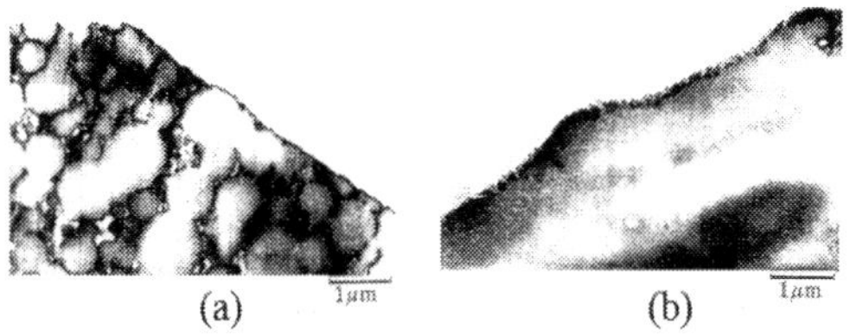

Figure 1 TEM images for HANR (a) and DPNR (b).

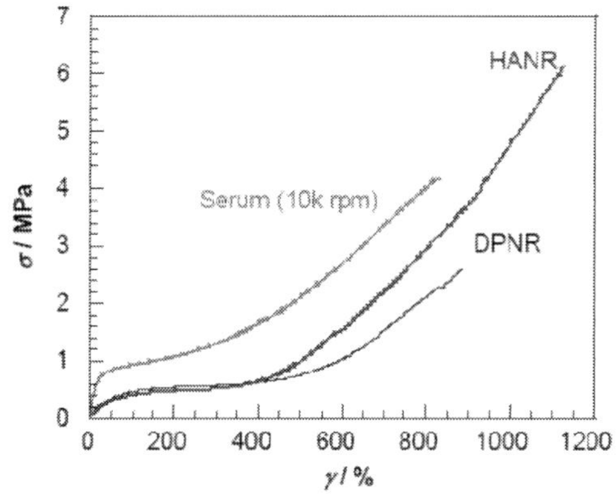

Figure 2 Stress-strain (s-s) curves for DPNR, HANR and the serum rubber (10k rpm).

mean particle size of HANR in the latex was similar to that of DPNR, i.e. about 0.5 μm. It was reduced by centrifugation. The value of the volume mean particle size of the serum rubber was lower than that of HANR by a factor of about one-half, after centrifugation at rotation speed of 10k rpm.

Figure 1 shows TEM images for HANR and DPNR. As for HANR, rubber particles of about 0.5 μm in diameter were found to be well dispersed in matrix of the non-rubber constituents such as the proteins and phospholipids. The phase separated structure was quite similar to the nanomatrix structure proposed in our previous work [8,9,12]. In contrast, for DPNR, no phase separated structure was observed in TEM image. The phase separated structure of natural rubber was not held after deproteinization due to the removal of the proteins.

Stress-strain (s-s) curves for DPNR, HANR and the serum rubber (10k rpm) are shown in Figure 2. The S-S curve for DPNR was distinguished from that for the serum rubber (10k rpm), whereas it was almost similar to that for HANR. Especially, at an initial stage, i.e. strain of 1, the value of stress of the serum rubber was twice as high as that of HANR and DPNR. In contrast, a value of stress at break of HANR was the highest among the samples. The large amount of the proteins may form weaker nanomatrix, as in the case of natural rubber with the nanomatrix of polystyrene reported in our previous work [8].

Figure 3 shows storage modulus (G') of DPNR, the serum rubber (10k rpm) and DPNR-graft-PS with the nanomatrix structure reported in our previous work [9]. Value of G' of serum rubber was

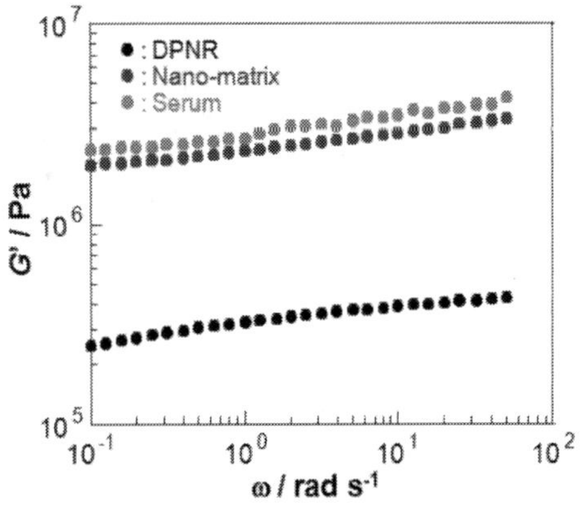

Figure 3 storage modulus (G') of DPNR, the serum rubber (10k rpm) and DPNR-graft-PS with the nanomatrix structure reported in our previous work [9].

quite higher than that of DPNR and it was similar to DPNR-graft-PS with the nanomatrix structure. Consequently, the nanomatrix structure played an important role in the properties of natural rubber.

In the previous work, the structure of natural rubber was investigated in molecular level [13]. The conclusion is the following. Natural rubber is a kind of long-chain telechelic polymer having peptide and phospholipid terminal groups. Both terminal groups are the origin of properties intrinsic to natural rubber. In the present work, the structure of natural rubber was investigated in nano and micro-meter levels. The nanomatrix structure was found to be quite important in the mechanical properties of the rubber.

CONCLUSION

Morphology of natural rubber was observed by transmission electron microscopy. It was found that natural rubber was a nanomatrix structured material consisting of natural rubber particles and non-rubber components. The stress at strain of 1 and the storage modulus increased dramatically, when the amount of non-rubber constituents increased. The nanomatrix structure was the origin of properties intrinsic to natural rubber.

Acknowledgement: This work was supported in part by a Grant-in-Aid (21655080) for Challenging Exploratory Research from Japan Society for the Promotion of Science and a Program for Developing the Supporting System for Global Multidisciplinary Engineering Establishment.

References

1. Koyama T, Steinbüchel A (2001) Biopolymers Volume 2 Polyisoprenoids, Wiley-VCH Verlag GmbH, Weinhein
2. Cornish K (2001) Photochemistry 57:1123
3. Tanaka Y, Mori M, Ute K, Hatada K (1990) Rubber Chem Technol 63:1
4. Tanaka Y, Kawahara S, Tangpakdee J (1997) Kautsch Gummi Kunst 50:6
5. Kawahara S, Kakubo T, Nishiyama N, Tanaka Y, Isono Y, Sakdapipanich JT (2000) J Appl Polym Sci 78:1510
6. Kawahara S, Kakubo T, Suzuki M, Tanaka Y (1999) Rubber Chem Technol 72:174
7. Kawazura T, Kawahara S, Isono Y (2005) J Appl Polym Sci 98:613
8. Kawahara S, Kawazura T, Sawada T, Isono Y (2003) Polymer 44:4527
9. Kawahara S, Yamamoto Y, Fujii S, Isono Y, Niihara K, Jinnai H, Nishioka H, Takaoka A (2008) Macromolecules 41:4510
10. Klinklai W, Saito T, Kawahara S, Tashiro K, Suzuki Y, Sakdapipanich JT, Isono Y (2004) J Appl Polym Sci 93:555
11. Kawahara S, Klinklai W, Kuroda H, Isono Y (2004) Polym Adv Technol 15:181
12. Pukkate N, Kitai T, Yamamoto Y, Kawazura T, Sakdapipanich J, Kawahara S (2007) 43:3208
13. Tanaka Y (1989) Prog Polym Sci 14:339

CROSSLINKING CONTROL OF PDMS RUBBER AT HIGH TEMPERATURES USING TEMPO NITROXIDE : EXPERIMENTAL STUDY AND RHEOLOGICAL MODELING

Skander Mani[1, 2], Philippe Cassagnau[2*], Mosto Bousmina[3] and Philippe Chaumont[2]

[1]*Department of Chemical Engineering, Laval University, Quebec G1V 0A6, Canada*
[2]*Université de Lyon, France, F-69003, Université de Lyon 1, France, F-69622, CNRS UMR5223, Ingénierie des Matériaux Polymères : Laboratoire des Matériaux Polymères et Biomatériaux, 15 Boulevard Latarjet, F-69622 Villeurbanne, France - philippe.cassagnau@univ-lyon1.fr*
[3]*Institute for Nanomaterials and Nanotechnology (INANOTECH), Hassan II Academy of Science and Technology, Rabat, Morocco*

Abstract

A novel composition using 2,2,6,6-tetramethylpiperidinyloxyl (TEMPO) and Dicumyl peroxide (DCP) for scorch delay and control of the final network topology of Polydimethyl-vinylmethyl-siloxane (vinyl-PDMS) at high temperatures has been proposed. The evolution of linear viscoelasticity during crosslinking reaction was carried out on parallel plate geometry rheometer. The rubber was crosslinked with different ratio of [TEMPO]/[DCP] in order to control scorch time which is defined as the time during which the PDMS rubber can be worked at a given temperature before crosslinking begins (i.e. the time at which the complex shear modulus suddenly increases). We showed that scorch delay increases according to the amount of TEMPO acting as an inhibitor. Nuclear Magnetic Resonance Spectroscopy (^{1}H-NMR) has been used to investigate the effect of TEMPO. This study reveals that the delayed action is the result of a TEMPO-grafted polymer action formed by reaction between TEMPO and PDMS macro-radicals. Furthermore, polymeric radicals are rapidly trapped by TEMPO before they are able to form crosslinks. In addition, differential scanning calorimeter (DSC) was used to characterize the crosslinking reaction. Specific heat data show two exothermic reactions. These reactions may be associated in one hand to the decomposition of DCP and TEMPO grafting in vinyl-PDMS, and on the other hand to the C-C covalent bonds creation. The DSC results indicate that the variation of scorch time with the [TEMPO]/[DCP] ratio is in reasonable agreement with the rheological results. On the other hand, an original method based on a relationship between the kinetic of the PDMS macro-radicals coupling [$R_{cc}(t)$] was derived from a developed fundamental kinetics model and the viscoelastic changes of the complex shear modulus ($G'(t)_\omega$ and $G''(t)_\omega$). The concentration variation of the active PDMS carbon-centered radicals [$R_p^\bullet(t)$]$_{act}$ and the [$R_{cc}(t)$] with reaction time were predicted using this kinetic model. Finally, the influence of the formulation such as the TEMPO concentration and temperature on viscoelastic variations are studied. As a result, the rheological modelling shows that this new method remarkably predicts the time variation of the complex shear modulus at any temperature and [TEMPO]/[DCP] ratio.

Introduction

Modeling of crosslinking process has recently received a great deal of attention (Yuxi et al. 2002). Indeed, chemists are interested in the crosslinking reactions with the intention to develop a fundamental kinetic model (Blaz and Matjaz 2008), materials scientists has been focused on the relationship between molecular structure and mechanical properties changes during the crosslinking process (Yuxi et al. 2004), and process engineers desire knowledge and control of the sol-gel transition (Friedrich et al. 1989). As crosslinking of polymer is a complex molecular phenomenon, the theoretical description of such phenomenon usually involves difficult mathematical treatment (Aranguren and Macosko 1988). Generally, the network growth in crosslinking process has been described in literature by statistical models (Shy and Eichinger 1986), kinetic models (Lopez et al. 2007) or computer simulation models in n-dimensional spaces (Kremer and Grest 1995). However, free radical crosslinking of rubber are a very complex chemical process and no known simulation techniques can directly investigate the physico-chemical properties changes of network at the molecular scale.

Polydimethylsiloxane (PDMS) rubbers are used for the production of seals in the automotive and aerospace industry, implants and devices for medical purpose and packaging in the food industry (Yoo et al. 2006). Commonly, a major curing mechanism frequently used for PDMS is the generation of polymer radicals through the use of organic peroxides that subsequently combine to form carbon-carbon bonds (Baquey 2005). Actually, the presence of vinyl functional groups in the polymer chain enables the free-radical crosslinking of PDMS (Ciullo and Hewitt 1999). However, free-radical crosslinking by peroxide suffers from premature crosslinking at high temperatures, which is called scorching (Dorn 1985). Robert (1997) first patented a process for grafting maleic anhydride onto a thermoplastic polymer in presence of a nitroxide such as TEMPO, the role of which is to avoid crosslinking during the grafting operation. Regarding crosslinking, Chaudhary et al. (2007) showed that the reaction of carbon-centered radicals with nitroxides and its derivatives can be a novel means for scorch time and cure control in peroxide crosslinking of polyethylene thermoplastic. We investigated in a previous work (Mani et al. 2009) the effect of TEMPO in the free-radical mechanism of PDMS rubber crosslinking initiated by dicumyl peroxide (DCP). The topological parameters of final networks such as the length of the network strands, functionality of cross-links, the amounts of entanglements and dangling chains have been quantitatively investigated and discussed. As a result, a remarkably scorch delay has been found with varying the molar ratio [TEMPO]/[DCP] in the range r =0 to 2.4. Furthermore, the characterization of the network features based on the phenomenological model of Langley (1968) and Dossin and Graessley (1979) provided that the control of the network topology can be achieved by using TEMPO nitroxide. Regarding viscoelastic properties, the equilibrium elastic modulus has been predicted only, whereas the modelling of the variation of the viscoelastic properties during crosslinking is of importance from a processing point of view.

The aim of the present work is to introduce a new simple modeling technique for direct investigation of the variation of viscoelastic properties during the randomly crosslinking process in presence of TEMPO. This new modelling aims to predict the variation of the complex shear modulus ($G'(t)_\omega$ and $G''(t)_\omega$) from the kinetic of the macro-radicals coupling. The present study is based on free radical crosslinking of PDMS controlled by TEMPO Nitroxide as used as model system in our previous work (Mani et al. 2009).

Experimental

PDMS: PDMS of high molar mass (PDMS gum) from ABCR was used. The number-average (M_n= *300,000 g.mol^{-1}*) and weight-average (M_w = *650,000 g.mol^{-1}*) were determined by size exclusion chromatography (SEC). This PDMS contains 0.2% mol of vinyl groups. The molar weight of molecular segment between two consecutive reactive sites, i.e. between two Vinyl sites is therefore: $M_0 = 13{,}000\ g.mol^{-1}$.

Free-Radical Crosslinking: Dicumyl peroxide (DCP, $M = 270\ g.mol^{-1}$) was used as the free-radical initiator of the crosslinking reaction and the nitroxide 2,2,6,6-tetramethylpiperidinyloxyl (TEMPO, $M = 156\ g.mol^{-1}$) was used as the inhibitor. Both products were purchased from Aldrich Chemicals and used without any further purification. All experiments were carried out with an identical concentration in DCP: $[DCP]_0$= 36 × 10^{-3} (mol.L^{-1}). The concentration of TEMPO was calculated in order to have the following molar ratios (r = [TEMPO]/[DCP]) r =1.2, 1.6, 1.8, 2 and 2.4) (Mani et al. 2009).

Dynamic Scanning Calorimetry (DSC): Differential scanning calorimeter equipment manufactured by TA Instruments (Q100 system), equipped with Sealed aluminium pans, was used to measure the heat of crosslinking reactions. The mass of the samples ranged from 20 to 23 mg. A sealed empty pan was used as a reference. The total heats of reactions were obtained from isothermal conditions (T=160°C) or non-isothermal method (Heating rate: $\dot{T} = 2.5°C.min^{-1}$). All experiments were performed under nitrogen purge.

Rheological measurements: The rheological experiments were carried out on a rheometrics mechanical spectrometer (RMS800) using a parallel-plate geometry (R=12.5 mm). The parallel plate system was preheated at the temperature of the experiment. Then, the sample was put between the plates once the temperature of regulation was reached. The rheological kinetics of cross-linking were carried out in real time at only one pulsation (ω=1 rad.s^{-1}) and at different temperatures T = 160, 180 and 200 °C. At the end of this crosslinking process, a frequency sweep experiment ($10^{-2}<\omega$ (rad s^{-1})$<10^2$) was performed on the same sample at the same temperature to determine the equilibrium modulus. In all experiments, sample response linearity with respect to strain amplitude was verified, and nitrogen gas was used to prevent thermal oxidation. In this type of experiment, we assumed that DCP and TEMPO were perfectly dispersed in molten polymer at a molecular scale (homogeneous conditions of reaction).

Results and Discussion

Rheology. The influence of the TEMPO concentration, via the ratio r=[TEMPO]/[DCP], on scorch time and crosslinking density of PDMS network is shown in Figure 1. From a qualitative point of view, Figure 1 (r=1.8 for illustrative example) shows that the crosslinking process is delayed by few minutes. The scorch time is defined in Figure 1 as the time at which the storage modulus suddenly increases. At 160 °C, the addition of TEMPO results in an increase of the scorch time ($t_{r,exp}$, see Table 1) from 7.2min (r=1.2) to 16.9 min (r= 2.0). Surprisingly, the crosslinking reaction is totally inhibited at higher amount of TEMPO (r=2.4). Although scorch mechanism has been reported in the literature (Alvarez-Grima et al. 2006), such quantitative experiments from TEMPO addition have been never shown in literature.

DSC Analysis. The rheology and isothermal DSC scan of (PDMS/DCP/TEMPO) results for r =1.2 are compared in Figure 2. Interestingly, the DSC kinetic allowed us to separate the own exothermic heat of C-C bonds creation (or quantum of energy of formed crosslinked bonds) in a complex system of the other reactions like the homolytic decomposition of the initiator (DCP) and its addition on the polymer chains.

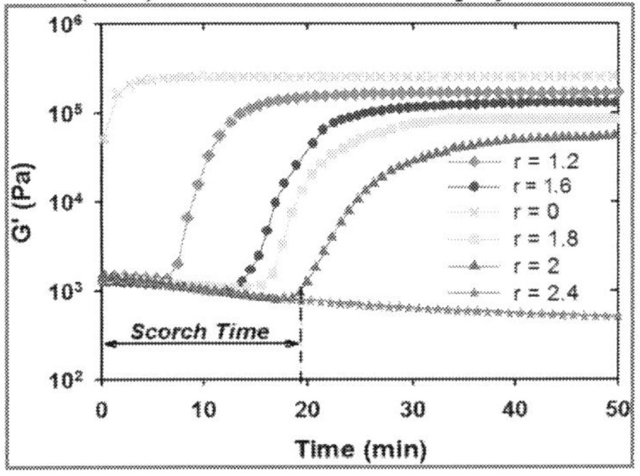

Figure 1: Variation of the storage modulus (r=0, 1.2, 1.6, 1.8, 2 and r=2.4). Definition of the scorch time. T=160°C.

Furthermore, both techniques are quite complementary for the kinetic study of this complex reactive system. At the earlier stages of the reaction, the rheology does not record any variation of the viscoelastic properties since TEMPO react with polymer radical and then embed the crosslinking reaction. On the contrary, DSC analysis shows a first exothermic peak of reaction which is spread out until the starting time (scorch time) of the complex modulus variation. This result confirms our last hypothesis: the first phase corresponds to the DCP decomposition, vinyl sites activation and the addition of nitroxyl radicals on the polymeric radicals (DSC exothermic peak and no variation of the complex shear modulus). The end of this inhibition phase is announced by both techniques: strong variation of the complex shear modulus and evidence of a second exothermic peak. Therefore, this second peak corresponds to the network formation through to the covalent bonds formation (or chemical crosslink) between the PDMS polymer chains.

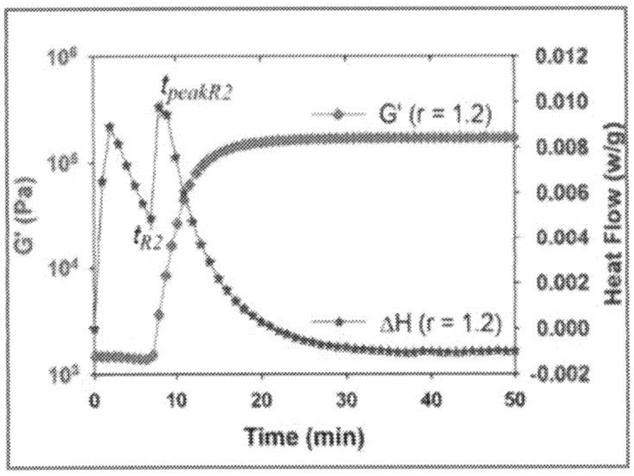

Figure 2: Comparison of the variation storage modulus and enthalpy of the reaction (r=1.2, T=160°C).

Rheo-Kinetics modeling. The main objective of this work is to predict the changes of viscoelastic properties of PDMS during a free-radical crosslinking process controlled by the addition of TEMPO. First we have established a kinetics model able to predict the the crosslink formation $[R_{cc}(t)]$. Then, rheo-kinetic modelling aims to predict the time variation of the complex shear modulus (storage $G'(t)_\omega$ and loss $G''(t)_\omega$ moduli). This can be achieved from the variation of the crosslinking bond formation $[R_{cc}(t)]$ derived from Equ 1.

$$[R_{cc}(t)]=\left\{\begin{array}{l}\left[2f[DCP]_0\left(exp(-k_d t_r)-exp(-k_d t)\right)\right]- \\ \left[\dfrac{2fk_d[DCP]_0\,exp(-k_d t)}{k_{cc}}\right]^{1/2}\times \\ \tanh\left[\left(2fk_{cc}k_d[DCP]_0\,exp(-k_d t)\right)^{1/2}(t-t_r)\right]\end{array}\right\} \quad (1)$$

The influence of the experimental conditions ($[DCP]_0$, $[N]_0$, α and T) on the cross-linking reaction kinetics and the network growth, has been studied at the molecular scale according to the developed kinetic model (Equ 1). As a result, the concentration variation of the crosslink covalent bonds $[R_{cc}(t)]$ with the reaction time (according to Equ 18 at T= 160°C) for r=1.8 is shown in Figure 3. During the inhibition stage, TEMPO inactive the primary PDMS macro-radicals and prevent the radical coupling $[R_{cc}(t < t_r)] = 0$. Therefore, if we accept that the TEMPO is completely consumed during the inhibition period and that the crosslinking reaction does not begin until the inhibitor is totally consumed, the bimolecular termination reaction starts but it is clearly slowed down in presence of higher TEMPO amount. In addition, Figure 3 shown comparison of the predicted storage and loss modulus with the rheometer data for r= [TEMPO]/[DCP]=1.8 at T=160°C. From a qualitative point of view, the viscoelastic variation of $G'(t)_\omega$ and $G''(t)_\omega$ was remarkably predicted by the rheo-kinetic model.

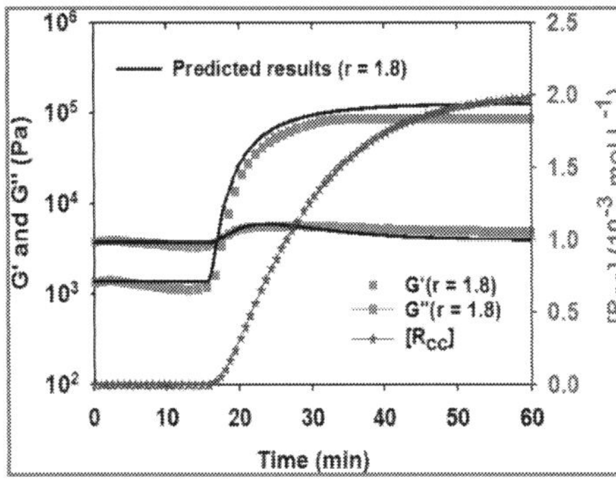

Figure 3: Modelling of the time variation of the complex shear modulus (r=1.8, T=160°C): Solid curves are from simulations and symbols are from experimental data. The concentration variation of the crosslink covalent bonds $[R_{cc}(t)]$ with the reaction time (according to Equ 18 at T= 160°C).

Moreover, Table 1 shows that initiator efficiency increases from 0.31 to 0.485 with $[N]_0$ (TEMPO concentration). This result is in agreement with the experiment results of Zhang and Ray (2002). Indeed, these authors proved that addition of stable radicals can boost the initiator efficiency.

Finally, the comparison in Table 1 of the predicted final $[R_{cc}]$; ie when the reaction is completed, with our last results of μ (density of chemical crosslink bonds) (Mani et al. 2009) computed by using the Pearson and Graessley model, shows very satisfactory agreement which validates our kinetic hypothesis.

Table 1. Comparison between the experimental and the calculated scorch time t_r. Efficiency of TEMPO and initiator. $[R_{cc}]$ is the total concentration of crosslinked bonds when the reaction is completed. Initial concentration of DCP: $[DCP]_0 = 36\times10^{-3}.mol.L^{-1}$ and T=160°C.

r $=[N]_0/[DCP]$	$t_{r,exp}$ (min)	$t_{r,cal}$ (min)	α =f/s	f	$[R_{cc}]$ *mol. m⁻³*	μ [a] *mol. m⁻³*
0	0	0	0.689	0.31	11.1	10.1
1.2	7.2	7.7	0.926	0.417	5.07	5.6
1.6	13	12.2	0.966	0.435	2.96	4
1.8	14.1	15.3	1.029	0.463	1.97	2.6
2.0	16.9	20.4	1.078	0.485	1.02	1.6

a) From (Mani et al. 2009)

Conclusions

In this study, a new rheological modeling method was developed to predict the various important modifications for PDMS network formation under free radical crosslinking reaction controlled by TEMPO addition. This new method is based on the relationship between the kinetic of the macro-radicals coupling $[R_{cc}(t)]$ derived from a developed fundamental kinetics model and the viscoelastic variation of the complex shear modulus ($G'(t)_\omega$ and $G''(t)_\omega$). The rheological modelling shows that this new method remarkably predicts the time variation of the complex shear modulus at any temperature and [TEMPO]/[DCP] ratio. This modelling can be then used as simulation to control the free radical crosslinking of PDMS rubber in presence of TEMPO.

Acknowledgements

The authors extended thanks to Agnès Crépet for conducting the ^{1}H-NMR measurements and Flavien Melis for his assisting in TGA-GC/MS analysis.

References

Aranguren, M. I., Macosko, C. W., "Modulus of Polybutadiene Networks Made by Hydrosilation Cross-Linking", Macromolecules, **21**, 2484-2491 (1988)

Baquey, G., "Decomposition of Di-*tert*-butyl Peroxide in Siloxane: An Approach of the Free Radical Cross-Linking of Silicones", Macromolecules, **38**, 9571-9583 (2005)

Blaz, L., Matjaz, K., "Simulation of Chemical Kinetics of Elastomer Crosslinking by Organic Peroxides", Polym. Eng. Sci., **49**, 60-72 (2008)

Chaudhary, B.I., et al., "Nitroxyls for Scorch Suppression, Cure Control, and Functionalization in Free-Radical Crosslinking of Polyethylene", J. Polym. Sci., **47**, 50-61 (2007)

Ciullo, P. A., Hewitt, N., "Chapter 1 Rubber", The Rubber Formulary, Noyes Publication, New York, p.17 (1999)

Dossin, L.M., Graessley, W.W., "Rubber Elasticity of Well-Characterized Polybutadiene Networks", Macromolecules, **12**, 123-130 (1979)

Dorn, M., "Modification of Molecular Weight and Flow Properties of Thermoplastics", Adv. Polym. Technol., **5**, 87-91 (1985)

E. P. Patent 0837080 (1997) Robert, P.M.

E. P Patent PCT/EP WO2006/100214 (2006) Alvarez-Grima, M.M., et al.

Friedrich, C., et al., "Evolution of time constants during sol-gel transition", Rheol. Acta., **28**, 535-539 (1989)

Kremer, K., Grest, G. S., "Chapter 4 Entanglement effects in polymer melts and networks", Monte Carlo and Molecular Dynamics Simulation in Polymer Science, Binder, K., Oxford University Press, New York, p. 194-271 (1995)

Langley, N.R., Ferry, J.D., "Dynamic Mechanical Properties of Cross-Linked Rubbers. VI. Poly (dimethylsiloxane) Networks", Macromolecules 1968, **1**, 353-358 (1968)

Lopez, L.M., et al.,"Modeling the Vulcanization Reaction of Silicone Rubber", Polym. Eng. Sci., **47**, 675-683 (2007)

Mani, S., et al., "Cross-Linking Control of PDMS Rubber at High Temperatures Using TEMPO Nitroxide", Macromolecules, **42**, 8460-8467 (2009)

Shy, L. Y.; Eichinger, B. E.,"Computer Simulations of Radiation-Cured Networks", Macromolecules, **19**, 2787-2793 (1986)

Yuxi, J., et al., "Investigation of computer-aided engineering of silicone rubber vulcanizing (II)", Polymer, **43**, 7515-7520 (2002)

Yuxi, J., et al., "Design of silicone rubber according to requirements based on the multi-objective optimization of chemical reactions", Acta. Materialia, **52**, 4153-4159 (2004)

Yoo, S.H., et al., "Mechanical and swelling properties of PDMS interpenetrating polymer networks", Polymer, **47**, 6226-6235 (2006)

Zhang, M., Ray, W. H., "Modeling of "Living" Free-Radical Polymerization Processes. I. Batch, Semibatch, and Continuous Tank Reactors", J. Appl. Polym. Sci., **86**, 1630-1662 (2002)

High Internal Phase Emulsions - Rheology and Microstructure in Various Flow Regimes

Marco Camesasca[1-2], William Hartt[2] and Ica Manas-Zloczower[1*]
1 Department of Macromolecular Science and Engineering, Case Western Reserve University, Cleveland (OH) USA
2 Corporate Engineering Technologies Lab, Procter & Gamble, Cincinnati (OH) USA
*ica.manas-zloczower@case.edu, Phone: +1(216)368-3596, Fax: +1(216)368-4202

Abstract
High internal phase emulsions (HIPEs) are colloidal emulsions for which the volume of the internal, dispersed phase is much greater than the volume of the continuous, matrix phase. For such systems above a critical volume ratio the rheological properties change substantially due to a change in the morphology and force balance in the system. In this work we present the mechanism of formation and morphology dependency on flow regimes for a HIPE system consisting of water droplets dispersed in an oil matrix. The rheological behavior of the system was tested in response to different shear rates and stress magnitudes in a cone-and-plate shear stress controlled rheometer and in a Couette and parallel-plate shear rate controlled rheometer. We have identified two different regimes of behavior. Above a critical shear rate the emulsion changes its microscopic structure which is also reflected by a different system response to shear stress. The experiments also reveal a transient stress response (initial peak/overshoot) under a constant shear rate applied. The system exhibits shear thinning and thixotropic behavior along with some non-linear response at short times.

Introduction: Concentrated emulsions are colloidal emulsions for which the volume of the internal, dispersed phase is much greater than the volume of the continuous, matrix phase. Concentrated emulsions have been studied theoretically and experimentally (Barry (1975), Mason, Bibette, Weitz (1995), Pons et al. (1993), Princen (1979, 1980, 1982, 1988), Princen, Kiss (1986, 1989), Quemada (1985, 1989), Reinelt, Kraynik (1993)). It is worthwhile noting that above a critical volume ratio the rheological properties change substantially due to a change in the morphology and force balance in the system. Particularly noteworthy is the existence of a critical yield stress below which the elastic behavior is predominant whereas above it, viscous properties dominate. Additionally due to the coexistence of two phases not in equilibrium, any external perturbation can destroy the system microstructure giving rise to effects such as coalescence or irreversible topological changes. Because of all these complex effects a comprehensive rheological description of these emulsions is difficult and such systems are mostly studied following a case-by-case approach.

Materials and Experimental Setup: Emulsions studied were concentrated water-in-oil (W/O) systems in the ratio 26:1 by weight without the complexity generated by adding the elastomeric silicone stabilizer. Specific composition: Specific composition: 3.7% wt silicone fluids including additional emulsifiers. The remaining 96.3% wt was an aqueous phase solution.

The experiments carried out were primarily stress vs. strain and stress vs time runs at different shear conditions: constant, on/off, incremental. The rheometers used are a strain controlled TA ARES with concentric cylinders (34 mm Couette) and a stress controlled Anton-Paar Physica MCR 501 with cone and plate (40 mm diameter). All rheological tests were conducted at room temperature and individual samples from the same batch were used for each run in order to minimize pre-shear effects that might change the emulsion microstructure.

Results: Figure 1 shows the microstructure of the emulsion as depicted by optical microscopy. The average droplet size is about 20 μm with some degree of polydispersity. In Figure 2 results for viscosity dynamics at different shear rates are presented. At low shear rates a large "overshoot" followed by a "relaxation tail" is observed. As the shear rate increases the plateau is reached in a shorter time.

In figure 3 the stress response over time is shown for two runs where the emulsion was sheared at 0.254 1/s for 240 s and let to recover for 60 s. The protocol was then repeated for a total of 6 runs. The overshoot is observed every time the emulsion is sheared after rest and pre-shear seems to affect only the long term stress response.

As the magnitude of the applied shear rate increases, the non-linear response of the emulsion becomes more visible and the response of the material is no longer predictable, especially the long term response. Surprisingly the overshoot is always present and it becomes more and more pronounced both in absolute and relative terms (with respect to the long term response) as pictured in figures 4 and 5. We hypothesize that the emulsion microstructure renders a stable rheological behavior up to a certain critical shear value over which the water droplets start deforming and eventually break therefore causing non-linear and unpredictable responses.

In figure 6 we present the dynamic stress behavior obtained for different applied shear rates ranging from 0.1 to 25.4 1/s. These results suggest the application of a double Hershel-Buckley model for fitting the stress responses (in the intervals from 0 to 2.5 1/s and from 2.5 to 25.4 1/s) (Figure 7). Fitting for the two different regimes returns unrelated yield stresses and characteristic constants, suggesting that the emulsion changes its microstructure. Nevertheless, the exponents are similar indicating the same sliding mechanism for different droplet size.

Conclusions: Concentrated water in oil emulsions are complex fluids presenting great challenges in their rheological behavior interpretation. In this work we studied the rheology of a typical skin-care emulsion system. We observed a short term behavior characterized by a peak in the stress response followed by a long term behavior that is time and shear dependent. The emulsion also presents two different regimes that can be well characterized by using a double Hershel-Buckley model. These results suggest that such systems exhibit, besides the typical yield stress point marking the shift between the elastic/solid to the plastic/liquid regime, a second transition due to a change in the microstructure under the applied shear.

Acknowledgements: We acknowledge the financial support of Procter & Gamble.

References:

Barry BW (1975) Adv Colloid Interface Sci 5:37–75
Mason TG, Bibette J, Weitz DA (1995) Physical Review Lett 75:2051–2054
Pons R, Erra P, Solans C, Ravey JC, Stebe MJ (1993) J Physical Chem 97:12320–12324
Princen HM (1979) J Colloid Interface Sci 71:55–66
Princen HM (1980) J Colloid Interface Sci 75:246–270
Princen HM (1982) J Colloid Interface Sci 91:160–175
Princen HM (1988) Langmuir 4:164–169
Princen HM, Kiss AD (1986) J Colloid Interface Sci 112:427–437
Princen HM, Kiss AD (1989) J Colloid Interface Sci 128:176–187
Quemada D (1985) J Theoretical Appl Mech special number, p 289–301
Quemada D (1989) Progress Colloid Polym Sci 79:112–119
Reinelt DA, Kraynik AM (1993) J Colloid Interface Sci 159:460–470

Figures:

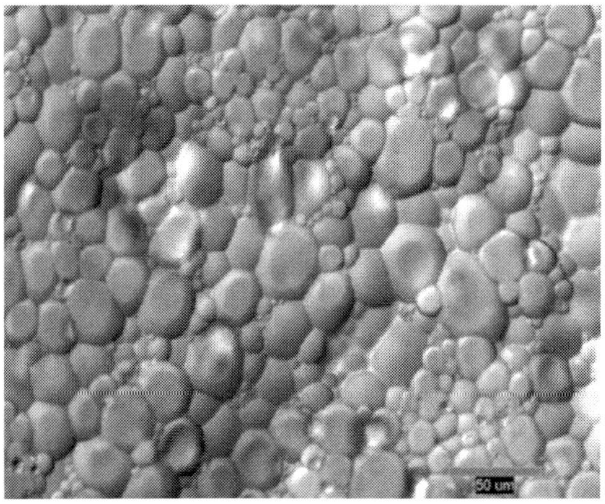

Figure 1 - Optical microscopy of concentrated W/O emulsion. Average droplet size is about 20 μm.

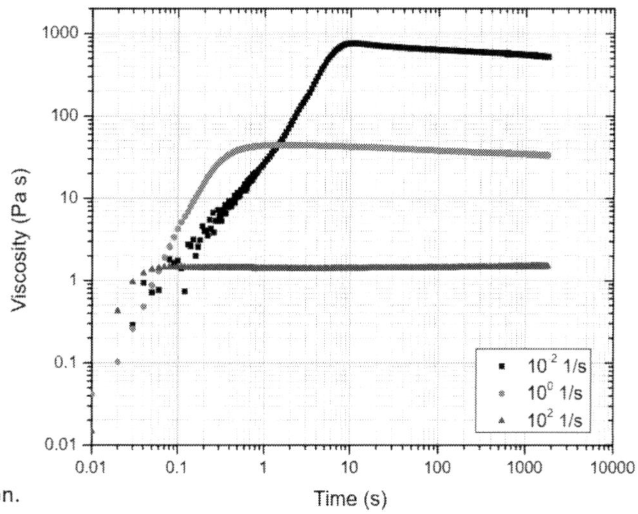

Figure 2-Viscosity dynamics for different shear rates.

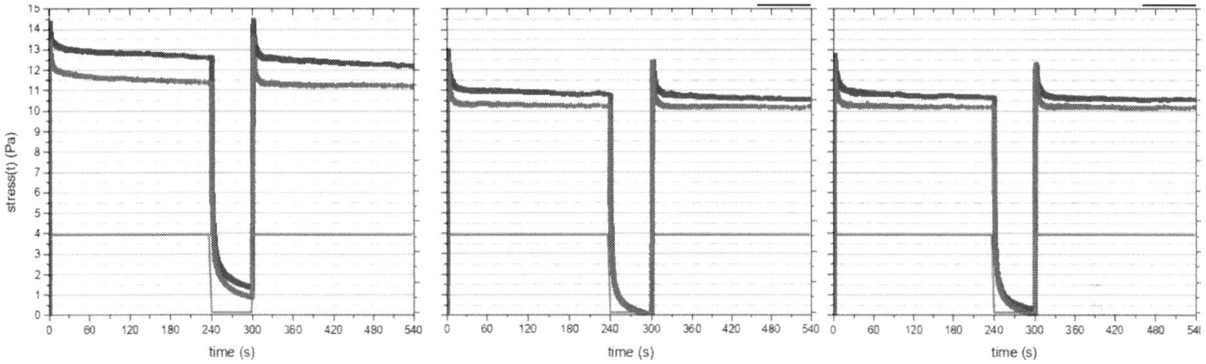

Figure 3 - Stress vs. time response for two samples (green and blue lines). A shear rate of 0.254 1/s (red line), is applied for 240 sec after which the sample is let to relax for 60 seconds.

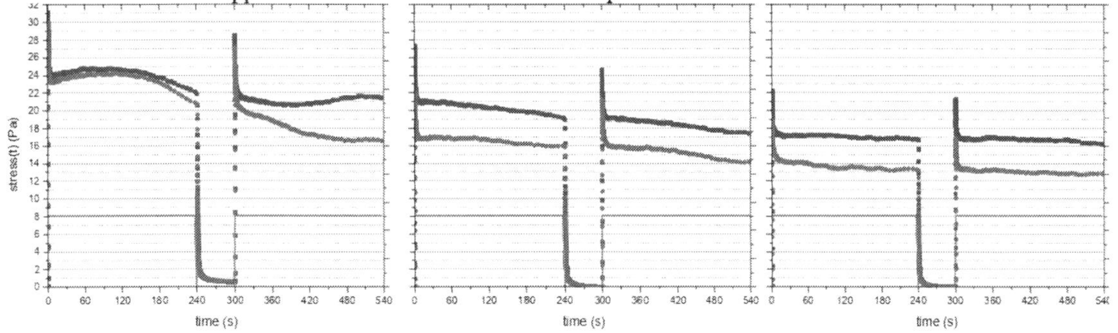

Figure 4 - Stress vs. time response for three samples (green and blue lines). A shear rate of 2.54 1/s (red line), is applied for 240 sec after which the sample is let to relax for 60 seconds.

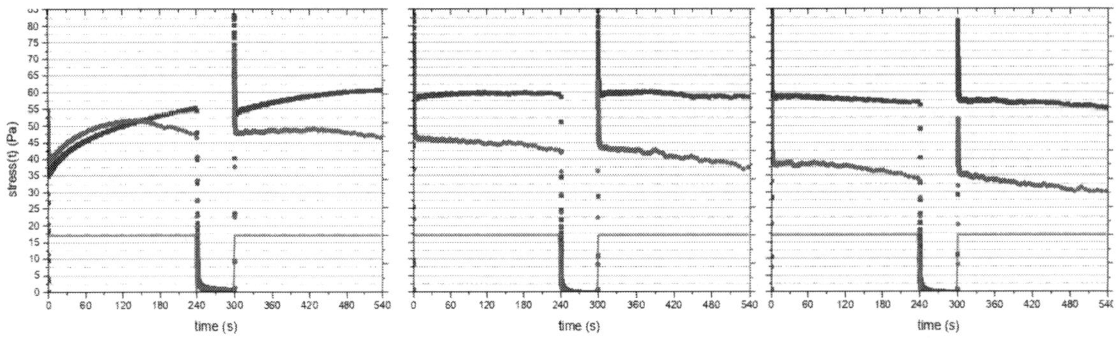

Figure 5 - Stress vs. time response for three samples (green and blue lines). A shear rate of 10 1/s (red line), is applied for 240 sec after which the sample is let to relax for 60 seconds.

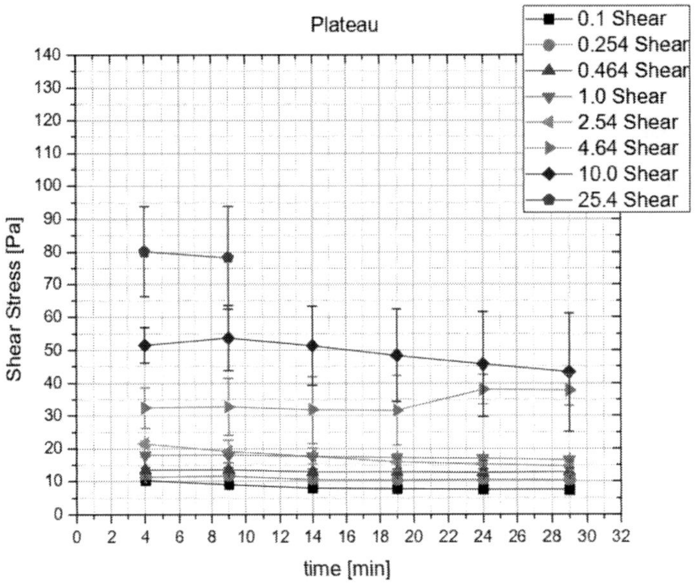

Figure 6 - Stress vs. time responses at various applied shear rate values.
The points represent the plateau (long term values).

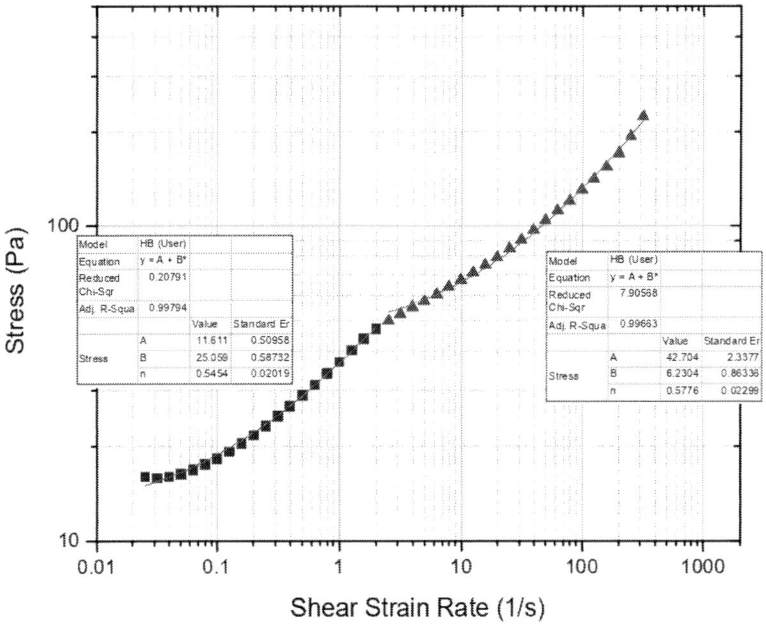

Figure 7 - Stress vs. shear for the long term responses. The continuous line represents a fitting using a double Hershel-Buckley model.

REACTIVE POLYMER BLENDING

Guo-Hua HU

1 Laboratory of Reactions and Process Engineering, CNRS, Nancy Université-INPL-ENSIC, 1 rue Grandville BP 20451, 54001 Nancy, France – hu@ensic.inpl-nancy.fr

2. Institut Universitaire de France, Maison des Universités, 103 Boulevard Saint-Michel, 75005 Paris, France

Abstract - This talk is dedicated to Prof. Chris Macosko on the occasion of his 65th birthday. The author met him for the first time in 1986 in Strasbourg, France. He was a visiting professor of the University Louis Pasteur – Institut Charles Sadron - EAHP while the author freshly joined the late Prof. Morand LAMBLA's group as a Ph.D. candidate. In this talk, the author will give a few research examples and recall a few moments to show how he has inspired me as a mentor, a scientist and a good friend and to express once again all my gratitude to him on this particular occasion.

Introduction

Prof. Chris Macosko's work has inspired the author's work in various topics. In this talk, the author will focus on reactive polymer blending.

Results and Discussion

Reactive polymer blending has become a very important contributor in the development of new polymer materials. It has undergone tremendous developments both in academia and industry during the last three decades or so.

In this talk, the author will show a few pieces of work on this subject. The author started to get involved in this topic in the second half of the 80s' of the 20th century when he was doing his thesis in the late Prof. Morand LAMBLA's group at the universty Louis Pasteur, Strasbourg, France (Hu, 1990). The following aspects will be addressed: (a) Reactions at polymer-polymer interfaces (Hu, 1995; Hu, 1998; Feng, 2004); (b) morphology development (Li, 2001; Cartier, 1999); (c) process (Sun, 1996; Zhang, 2009), and (d) Nanoblends (Hu, 1999).

Conclusions

Prof. Chris Macosko has significantly contributed to the fundamental understanding and practice of reactive polymer blending. His work has also inspired many young and senior researchers to get involved in this fascinating topic. As one of them, the author wishes to express his gratitude to him on the occasion of this special session dedicated to his 65th birthday.

Acknowledgements

The author is grateful to those (there are many) who have contributed to his work and his career.

References

Baker, W., et al., "Reactive polymer blending", Hanser publisher, Munich, Germany (2001).

Hu, G.H., "Etude des réactions d'échanges des copolymères d'esters acryliques et vinyliques à l' état fondu", Thèse de doctorat, Université Louis Pasteur de Strasbourg, France (March 28, 1990).

Hu, G.H., Lambla, M., "Chemical reactions between immiscible polymers in the melt: Transesterification of poly(ethylene-co-methyl acrylate) with mono-hydroxylated polystyrenes", J. Polym. Sci., Chem. Ed., 33, 97-107 (1995).

Hu, G.H., Kadri, I., "Modeling reactive blending: An experimental approach", J. Polym. Sci., Part B - Phys. Ed., Vol. 36, 2153-2163 (1998).

Feng, L.F., Hu, G.H., "Reaction kinetics at polymer-polymer interfaces under flow", AIChE.J.50, 2604-2614 (2004).

Li, H., Hu, G.H., "The early stage of the morphology development of immiscible polymer blends during melt blending: compatibilized vs. uncompatibilized blends", J. Polym. Sci., Part B: Polym. Phys. Ed., Vol. 39, 601-610 (2001).

Cartier, H., Hu, G.H., "Morphology development of in situ compatibilized semi-crystalline polymer blends in a co-rotating twin screw extruder", Polym. Eng. Sci., 39, 996-1013 (1999).

Sun, Y.J., et al., "In-situ compatibilisation of polypropylene and poly(butylene terephthalate) blends by one-step reactive extrusion", Polymer, 37, 4119-4127(1996).

Zhang, C.L., et al., "Residence time distribution: an old concept in chemical engineering and a new application in polymer processing", AIChE Journal, 55(1), 279-283 (2009).

Hu, G.H., et al., "Reactive extrusion: Toward nano-blends", Macromolecules, Vol.32, 4713 – 4718 (1999).

FLOW CHANNEL DESIGN FOR LAYER THICKNESS UNIFORMITY IN MULTILAYER COEXTRUSION

Patrick C. Lee[1], Hong Zheng[2], Christopher W. Macosko[3*], Patrick D. Anderson[4]

[1]*The Dow Chemical Company, Midland, Michigan, U.S.A – cdlee@dow.com*
[2] *The Dow Chemical Company, Shanghai, China – hzheng@dow.com*
[3]*Department of Chemical Engineering and Materials Science, University of Minnesota, Minneapolis, MN – macosko@umn.edu*
[4]*Eindhoven University of Technology, Netherlands- P.D.Anderson@tue.nl*

It is important to understand the effect of the flow channel shape on the final layer thickness distribution of multilayer products in coextrusion. These layers normally should have uniform thicknesses throughout the structure for optimal performance. However, layer thickness non-uniformities have been observed in many commercial coextruded structures. It is a very complex and challenging problem due to coupled material and processing variables. In this study, we present examples of non-uniform polystyrene (PS) and polypropylene (PP) multilayers produced using existing layer multiplication dies and demonstrate that layer thickness non-uniformity is closely related to the flow channel shape of the multiplication die. Two other designs are proposed and compared with the existing design using both experimental and flow simulation techniques using commercial multiphysics simulation software (COMSOL), ANSYS Polyflow. In addition, the mapping method is used to determine the layer thickness distribution during extrusion, using the obtained velocity field as input. The results demonstrate that by balancing the flow fields the modified designs yield more homogeneous layer distributions.

Introduction

It is of great importance to obtain uniform layer thickness in multilayer coextrusion processes to produce superior optical and barrier properties of the final products. Among the many factors which affect the final layer thickness distribution of multilayers, such as viscosity ratio, elasticity match, second normal stress differences, this work focuses on the effect of multiplication die channel design.

Not many studies on flow channel design optimization in a layer multiplication unit are available in the open literature. J.C. Van Der Hoeven and coworkers studied layer non-uniformity of a similar flow channel design and explained the non-uniformity in terms of velocity profile and volume flux of each layer. Their solution to improve uniformity was the addition of longer stabilization sections before and after the original multiplication unit (Van der Hoeven et al. 2001).

C.D. Mueller discussed on the effect of path length difference in a multiplication die on layer thickness distribution (Mueller, 1998). He improved the design by reducing the path length difference in the flow channels and experimentally verified the improvement using a 16 layer Polycarbonate/Poly(methyl methacrylate) system. The study demonstrated the importance of flow channel design on layer thickness distribution but lacked 3-D numerical simulation for design optimization. Here we use flow simulations as well as experiments to compare three die channel designs.

Experimental

Materials

We prepared alternately layered polypropylene (PP, PP4062, M_w=280 kg/mol and M_w/M_n=4.0, ExxonMobil Chemical Co.) and polystyrene (PS, Styron 666D with M_w=226 kg/mol and M_w/M_n=2.6, The Dow Chemical Company) multilayer samples using the setup shown in Figure 1. PS with 0.5 wt% black or white pigments were also used to study layer distribution. We assumed that this small amount of pigment did not change the PS rheology.

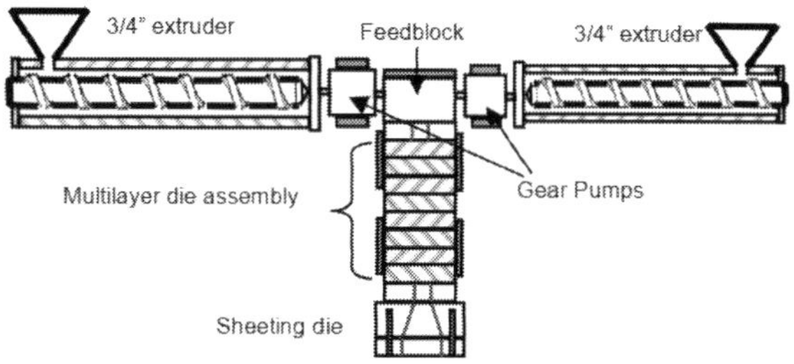

Figure 1. Multilayer coextrusion setup

A rotational rheometer (ARES, TA Instruments) and a sliding plate rheometer (SPR) were used to determine the viscosity and elasticity (G') of the two materials (Figure 2). Both viscosity and elasticity are well matched at 200°C, which is critical for layer thickness uniformity in multilayer coextrusion (Dooley, 2005).

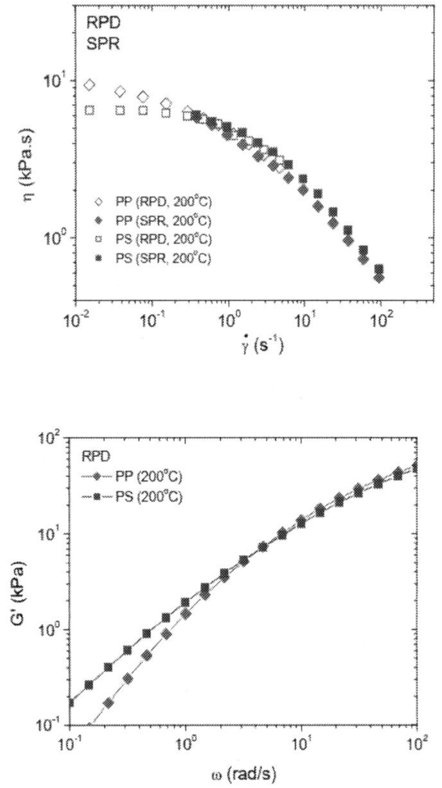

Figure 2. Shear viscosity and elasticity of PP and PS at 200 °C. Lines drawn through the data points to guide the eye.

Experimental Setups

In order to prepare multilayer samples, a coextrusion line with 0 to 4 layer multiplication dies has been set up (Fig. 1) (Dolgovskij, 2007; Zhao and Macosko, 2002). Two 19 mm single screw extruders were utilized to melt the polymers and deliver them to the gear pumps. The gear pumps (Zenith PEPII) were used to control constant volumetric flow rate of each polymer regardless of downstream pressure fluctuation in the extruders. The volumetric flow rates of both streams were kept constant (i.e., 50 vol%/50 vol%).

Two polymer melts are brought together in a feedblock and arranged into two (or more) layers. When the layered polymer flow enters a multiplication die, the melt is initially cut in half vertically, and then each half is compressed and then re-stretched to its original width, effectively doubling the number of layers each time through a multiplication die. By using n multiplication dies, the number of the total layers in the final sample when exiting the die will be 2^n times the number of initial layers. For this study, we produced 20, 80, 160 and 320 alternate layers of PP/PS by using a 20 layer feedblock and 0, 2, 3, and 4 multiplication dies, respectively. For PS (black and white pigments) experiments, 16 layer samples were produced using a 4 layer feedblock and 2 multiplication dies.

Results and Discussion

Layer Thickness Distribution of Design I (PS/PP)

Figure 3 shows the layer thickness distribution of PS/PP multilayers after 0, 1, 2 and 3 multiplication dies (Design I, Figure 4). Due to severe difference in the flow path length between left and right streams, each pass through a multiplication die produces a local maximum in the layer thickness distribution (i.e., a group of thicker layer sections). For example, a maximum starts to develop in the middle after 1 multiplication die (Fig. 3b, 80 layers) and develops further after 2 multiplication dies (Fig. 3c, 160 layers). After 3 multipliers two other maxima are observed beside the middle one (Fig. 3d, 320 layers), while a 20 layer case (without any multiplication die) shows no local maxima (Fig. 3a). Also, it is worthwhile to note that the layers close to the walls become relatively thicker as they pass through more multiplication dies.

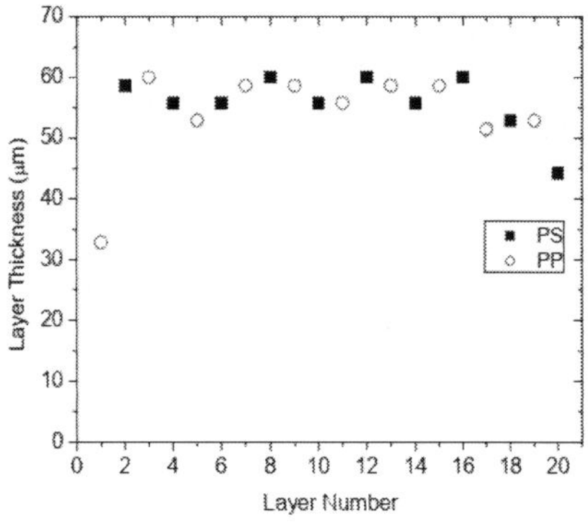

(a)

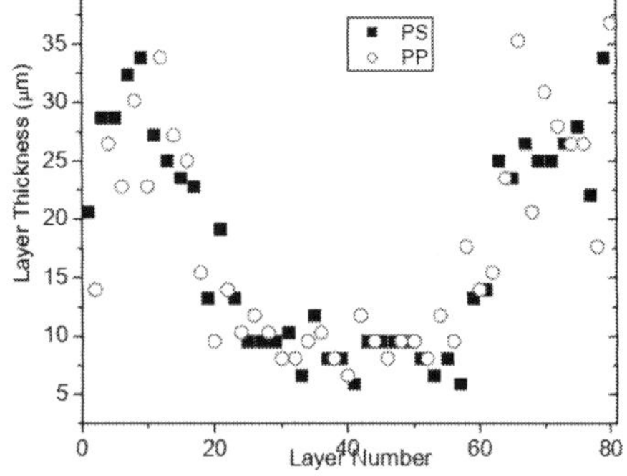

(b)

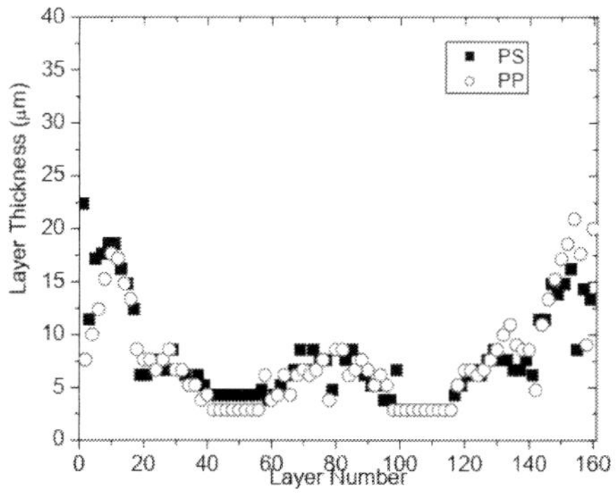

(c)

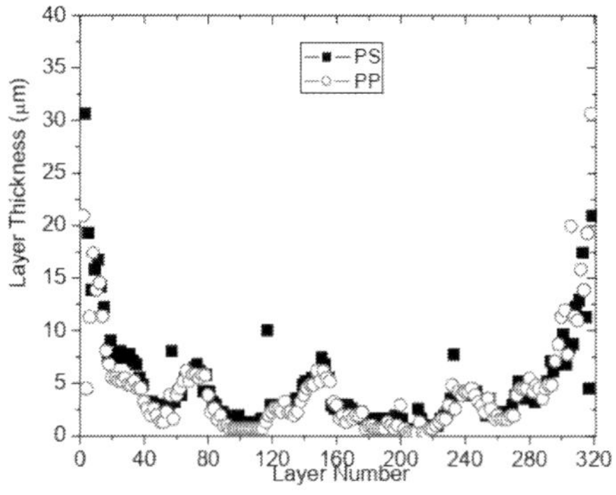

(d)

Figure 3. Layer thickness distribution of PP and PS multilayers: (a) 20 (without multiplication die), (b) 80 (1 MD), (c) 160 (2 MD), and (d) 320 layers (3 MD)

Path length difference is a key factor to determine the layer thickness distribution: the shorter the path length, the thinner the layers. Therefore, it is reasonable to minimize the path length difference.

Layer Thickness Distributions of Design I, II, and III (PS)

Figure 4 shows a 16 multilayer setup with 3 different channel designs: (a) Design I (path length difference = 10.5%), (b) Design II (5.8%), (c) Design III (4.9%). Each design starts with the 4 layer feedblock followed by clockwise and counter-clockwise multiplication dies. Figure 5 shows 16 layers produced for the three designs using black and white pigmented PS. It is obvious that Design III with the minimum path length difference produces the best layer thickness uniformity.

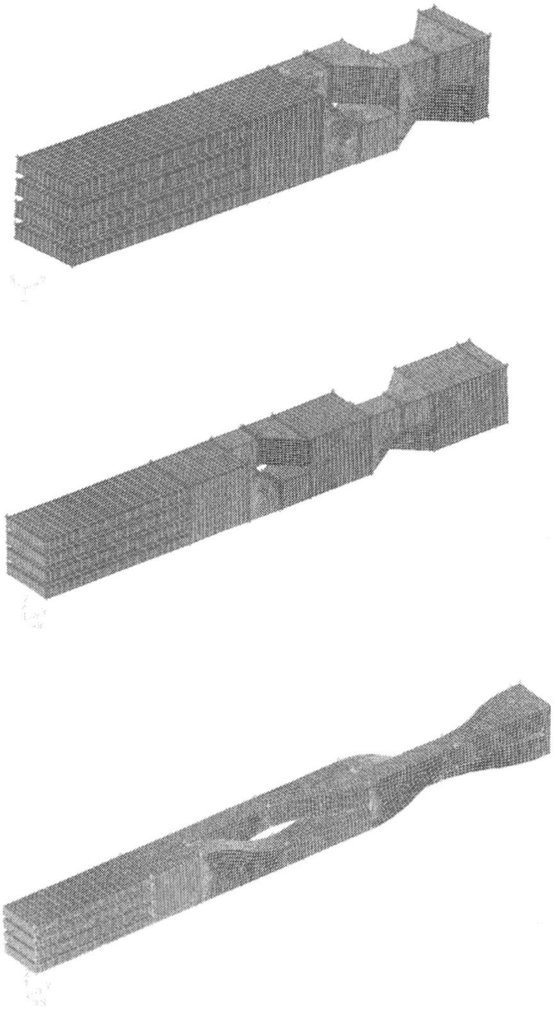

Figure 4. Three different channel designs: (a) Design I (path length difference = 10.5%), (b) Design II (5.8%), (c) Design III (4.9%). Flow is from left to right, from the four layer feedblock into two multiplication dies (2 MD). The first die rotates the flow clockwise, the second die rotates counterclockwise.

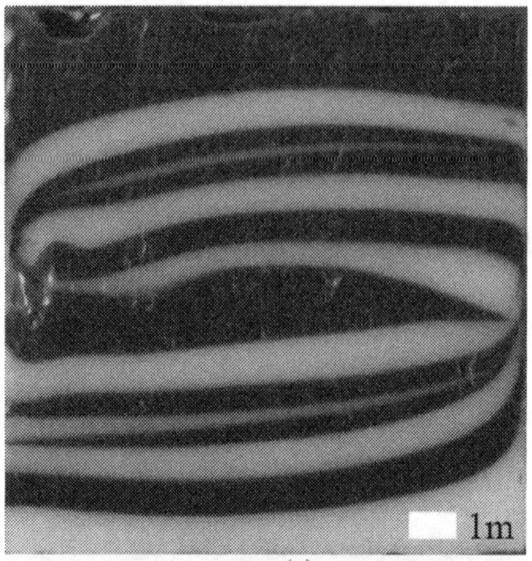

(a)

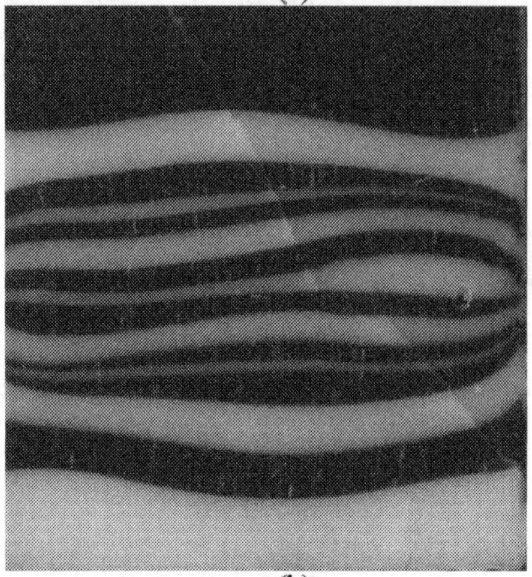

(b)

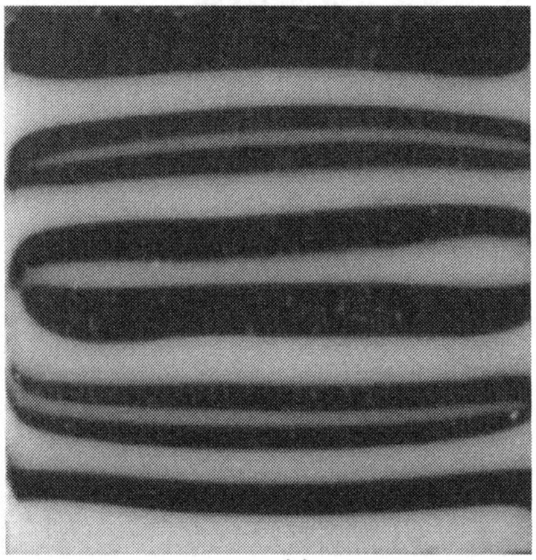

(c)

Figure 5. Optical micrographs after flow through the three different channel designs: (a) Design I, (b) Design II , (c) Design III. PS with 0.5% black or white pigment was feed in alternating inlet streams.

Flow through each design was simulated using COMSOL, Polyflow, and mapping method. The Carreau model (Macosko, 1994) was fit to the data of Figure 2 (η_o =5,210 Pa.s, n=0.295, and λ =0.18 s) and used for the simulations. The boundary conditions are: no slip on the wall, inlet volumetric flow rate = 36 g./min (666 mm^3/s), and fully-developed flow at the outlet. The unstructured tetrahedral elements were used for COMSOL while structured hexahedral elements for POLYFLOW. Only layers which correspond to the black pigmented layers of the experiments were traced with arrays of black particles.

To calculate the shape of layer at the outlet, the Navier-Stokes equations with its viscosity depending on shear rate were first solved using finite element codes in both softwares to simulate the polymer melt flow in the given channel. The calculated flow field includes the distribution of the velocity, pressure, shear rate and shear stress. The flow can be well investigated by visualize the movement of material particles along their trajectories. Based on the flow field, the trajectories of numerous material particles can be calculated. The calculation starts when a group of material particles were released on the inlet surface, representing the flow of the black pigmented layers. For the trajectory of each particle, the calculation ends as long as the particle reaches the exit of the flow domain or the lifetime reaches the set maximum value. The lifetime for this computation is set as 1,000 s. Any particle with a lifetime above this value is regarded to be very near the wall and cannot reach the exit. The distribution of material particles on the exit is shown in Figure 6, which well simulate the final shape of the current multi-layer co-extrusion.

The particle tracing results are shown in Figure 6.

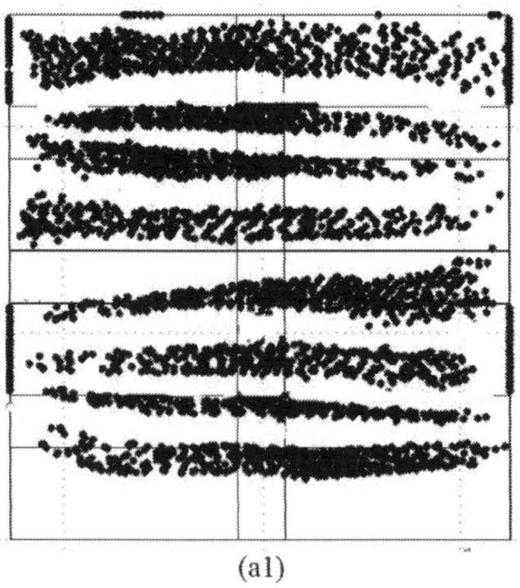

(a1)

(a2)

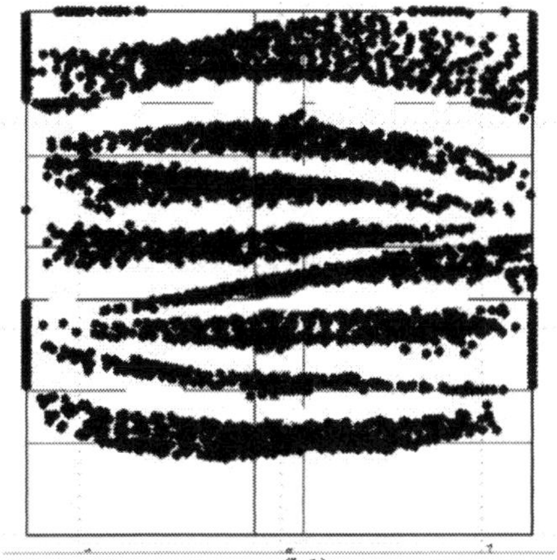

(b1)

(b2)

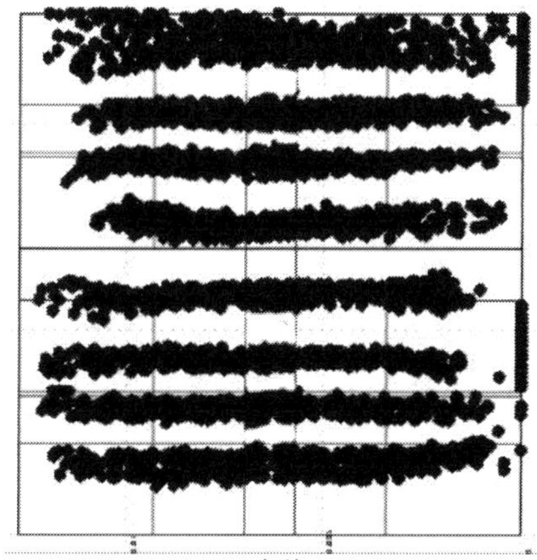

(c1)

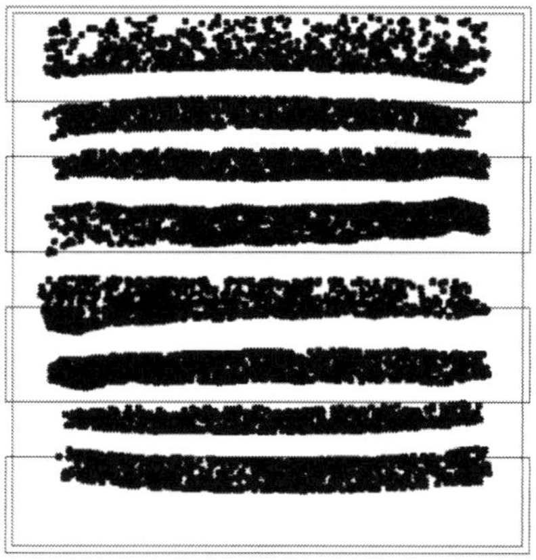

(c2)

Figure 6. Simulations of thickness distribution after 2 MD (16 layers): (a) Design I, (b) Design II , (c) Design III where 1, 2 are COMSOL and POLYFLOW outputs, respectively.

Figure 7 compares the experimental and simulation results. The error bars represent the degree of deviation of layer thickness across the width (3 measurements at ¼, ½, and ¾). It is demonstrated that layer thickness uniformities across both height and width of the cross-section are best for the design III with the smallest path length difference.

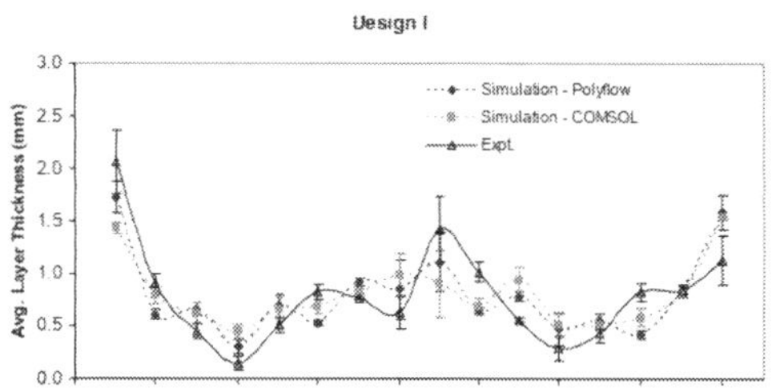

(a)

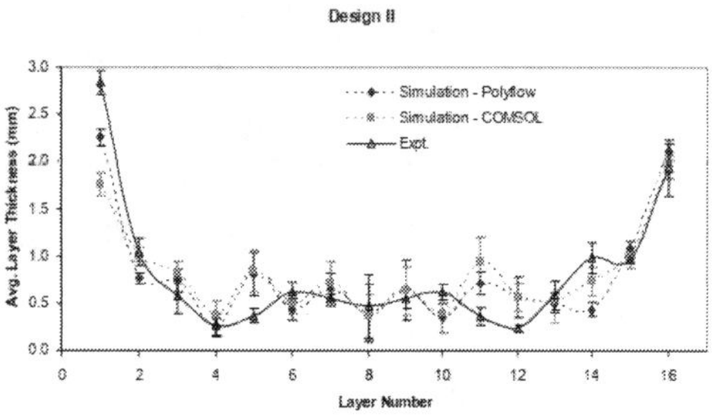

(b)

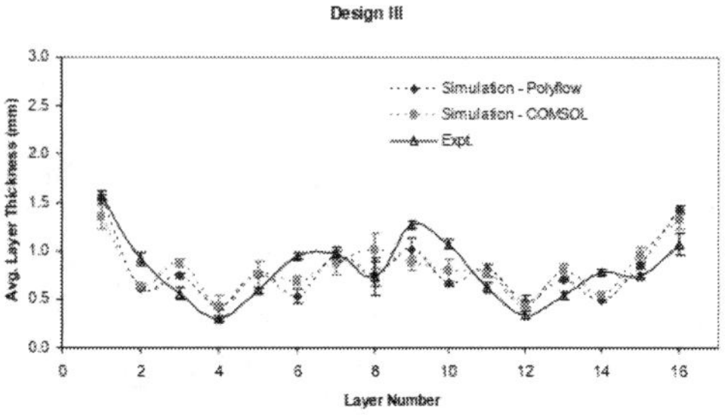

(c)

Figure 7. Comparison between experiments and simulation: (a) Design I, (b) Design II , (c) Design III

Conclusions

It is important to understand the effect of the flow channel shape on the final layer thickness distribution of multilayer products in coextrusion.

This study presents examples of non-uniform multilayers produced using our original layer multiplication dies with rheologically matched polystyrene (PS) and polypropylene (PP). We found that the layer thickness non-uniformity is closely related to the flow channel shape of the multiplication die. In order to reduce non-uniformity, two other die designs are proposed and compared with the existing design using both experimental and simulation techniques. Simulations and experiment are in reasonable agreement and show that the design with the least path length difference is the best to solve the layer thickness non-uniformity.

References

Dolgovskij, M. K., PhD, Univeristy of Minnesota (2007)

Dooley, J., "Coextrusion Instabilities," in Polymer Processing Instabilities: Control and Understanding," Hatzikiriakos, S. G. and K. B. Migler, (Ed.), Marcel Dekker, New York, p. 396 (2005).

Macosko, C. W.: Rheology: Principles, Measurements and Applications, Wiley, New York, p. 86 (1994).

Mueller, C. D., PhD Thesis, Case Western Reserve University (1998)

Van der Hoeven, J. C. Wimberger-Friedl, R. and Meijer, H. E. H., "Homogeneity of Multilayers Produced with a Static Mixer," Polym. Eng. Sci., **41**, 32-42 (2001)

Zhao, R., Macosko, C. W., "Slip at Polymer-polymer Interfaces: Rheological Measurements on Coextruded Multilayers," J. Rheol., **46**, 145-167 (2002).

MORPHOLOGY DEVELOPMENT AND PROPERTIES OF POLYMER/CLAY NANOCOMPOSITES

Uttandaraman Sundararaj*, G. Gelves and Sriram Lakshminarayan

Uttandaraman Sundararaj, G. Gelves and Sriram Lakshminarayan*
Department of Chemical and Petroleum Engineering
University of Calgary
Calgary, Alberta, CANADA
u.sundararaj@ucalagary.ca

Abstract - Layered silicate polymer nanocomposites have gained great importance in the recent years because they can greatly improve properties of polymers, much more significantly than fillers used in conventional composites. In this research, fluoroelastomer/clay nanocomposites were made for potential utilization in Progressive Cavity Pumps (PCPs) or other oilfield supplies operated at high temperatures. PCPs are an artificial lift method to pump oil from several thousand feet depths to the surface. The integral components of a PCP are the stator, which is usually made of elastomer, and a rotor, usually made of metal. The pumps are exposed to severe conditions during operation and the elastomer has to withstand high temperatures, high solids content, high viscous dissipation, etc. PCP's currently cannot be used for high temperature oil production because these temperature levels are not accessible using hydrocarbon elastomers. Rheological characterization was performed on these materials and the results were analyzed to understand the micro-/nano-structure. It was found that the clay intercalant played a key role in the intercalation/exfoliation of nanoclay and surprisingly, the initial concentration of intercalant in the clay had little effect on the final structure. The mechanical properties and rheological properties correlated well with the structural information obtained from XRD.

Introduction

Fluorocarbon elastomers (FKM), and more specifically, the copolymers and terpolymers synthesized from vinylidene fluoride (VDF), hexafluoropropylene (HFP) and tetrafluoroethylene (TFE) are widely used in automotive, chemical, petrochemical, aerospace and food processing industries, primarily as seals, gaskets, O-rings and hoses. The carbon-fluorine bonds in the backbone provide extremely high temperature resistance, and chemical and oil resistance. FKM compounds have limited low temperature flexibility and are not easy to process. These elastomers have moderate tensile strength at room temperature when compared to hydrocarbon rubbers and retain only a portion of it, when used at high temperatures.

There are continuing efforts to enhance properties of rubber compounds to extend their application to new areas and to improve their current performance. In the past decade, polymer clay nanocomposites have drawn significant attention due to the unique properties they exhibit. However, it has to be emphasized that most previous studies have been devoted to thermoplastics and thermosets, leaving elastomers highly unexplored. Some recent review articles have discussed the recent achievements in rubber/clay nanocomposites focussing on their preparation techniques and property enhancements.

In this study, fluorocarbon rubber is melt mixed with unmodified and organically modified nanoclays in an internal mixer. FKM compounds prepared with five different commercial nanoclays have been studied. Characterization of the uncured composites has been performed by dynamic rheology, X-ray diffraction (XRD) and transmission electron microscopy (TEM). In addition to comparing the reinforcement efficiency of different clays on FKM compounds, the plasticizing effect of organically modified nanoclays is presented.

Experimental

Materials

DyneonTM FPO 3741, a terpolymer of vinylidene fluoride (VDF), hexafluoropropylene (HFP) and tetrafluoroethylene (TFE) with fluorine content of 69.5%, specific gravity 1.89 and Mooney viscosity 38 ML (1+10) @ 121°C was obtained from Dyneon, 3M. Different grades of nanoclays viz. Cloisite® NA (untreated), 15A, 20A, 30B and 93A were obtained from Southern Clay Products.

FKM/Clay Composite Preparation

Clays were used in three different loading levels viz. 2.5, 5.0 and 10.0 phr. Mixing was carried out in a Haake Rheomix Series 600 with Banbury blades. FKM rubber was initially masticated for 2 min at 45 rpm and then clay was added in one part. Mixing was done for another 15 min at 45 rpm. Compound temperature was maintained at around 75°C during mixing. Fill factor was set at 0.70 and the thermal expansion was taken into consideration when calculating the mixing quantities.

X-Ray Diffraction and Rheology

For XRD and rheological characterization, samples of FKM/clay composites of 25 mm diameter and 2 mm thickness were prepared by compression molding at 90°C and 10 MPa for 2 minutes. X ray diffraction was performed by using a Rigaku diffractometer (CoKα, l=0.1789 nm) operated at room temperature in the range of 2-10° (2-12° except for NA clay and its composites). The detector (2θ) was rotated at twice the speed of sample (θ). Dynamic rheological characterization was done in a Rheometrics RMS800 with parallel plate fixture at a temperature of 180°C and angular frequencies from 10-2 to to 102s-1. Strain sweep experiments were carried out to find the linear viscoelastic region and then frequency sweep studies were performed within this region.

Results and Discussion

Rheological Characterization

The rheological response of polymer composites is sensitive to filler morphology and dispersion, and it has been used to characterize micro- and nano-composites. The effect of concentration of Cloisite® 20A, 15A, 93A and NA on the storage moduli (*G'*) of the composites in the frequency range of 10^{-2} to 10^{2}rad/s was studied. It is clear in all cases that as the filler content increases, so does *G'*. This is expected in rubber/filler compounds, when using reinforcing or semi-reinforcing fillers. However, an insignificant reinforcement effect is observed in compounds prepared with NA and 93A for the concentrations used. In the case of 20A and 15A, the response to increasing filler loading is significant with step increases in storage modulus. The data show that the dependence of *G'* with frequency decreases and leads to a lower slope in the terminal region (low frequency). A zero slope in the terminal frequency region is an indication of solid-like material and a decrease of slope suggests a gradual transition from liquid-like to solid-like behaviour, which is a strong indication of improved dispersion and/or enhanced intercalation of layered silicates by polymers$10^{2}s^{-1}$. Strain sweep experiments were carried out to find the linear viscoelastic region and then frequency sweep studies were performed within this region.

Morphology of FKM/clay nanocomposites

The XRD patterns of FKM compounds prepared with different concentrations of clay loading along with the diffractograms of nanoclays was studied. The peak positions were used to find the corresponding d-spacings using Bragg's law of diffraction. For the nanoclays used (NA, 15A, 20A, 30B, 93A) the d-spacings were found to be 1.10, 3.27, 2.55, 1.72, 2.62 nm respectively. The composites prepared from NA, 15A, 30B and 93A are not intercalated because the calculated d-spacings of the composites were not considerably higher that of the corresponding clays. This could be attributed to the inability of the polymer molecules to intercalate the silicate galleries of these nanoclays.

Conclusions

Fluoroelastomer layered silicate composites were prepared by melt mixing using an internal batch mixer. The morphology and rheological properties of the composites prepared using unmodified and organically-modified nanoclays at different concentrations were studied. A transition from liquid-like to pseudo solid-like behaviour was observed with the increase in clay concentration for composites prepared with Cloisite® 15A and 20A. The increase in storage modulus of composites prepared with different concentrations of Cloisite® 15A and 20A indicated significant clay dispersion and higher reinforcement than those prepared with Cloisite® NA, 30B and 93A. This was attributed to the reduction of the surface energy imparted to the clays by the presence of di(hydrogenated tallow-alkyl) dimethyl ammonium surfactant in Cloisite® 15A and 20A. XRD characterization of compounds prepared with different concentrations of Cloisite® 15A does not indicate a significant increase in the intergallery distance of the nanoclay in comparison to the 1.06 nm increase in d001 obtained for 20A compounds. This difference has been attributed to the presence of unbound surfactant in the silicate galleries that affects the mechanism of intercalation in 15A and 20A compounds.

Acknowledgements

The authors acknowledge financial support from NSERC.

NANOSTRUCTURATION OF EPOXY RESINS BY BLOCK COPOLYMERS: APPLICATION TO RTM-PROCESSED COMPOSITES

F. Gille, J. P. Pascault, J. F. Gérard, F. Lortie*

Université de Lyon, Lyon, F-69003, France; INSA de Lyon, IMP/LMM, Laboratoire des Matériaux Macromoléculaires, Villeurbanne, F-69621, France; CNRS, UMR 5223, Ingénierie des Matériaux Polymères, Villeurbanne, F-69621, France

Poly (styrene-*b*-butadiene-*b*-methyl methacrylate) (SBM) and poly (methyl methacrylate-*b*-n-butyl acrylate-*b*-methyl methacrylate) (MAM) copolymers were added to epoxy-based matrices and to continuous-fibers-reinforced epoxy-based composites. Crack resistance of the matrix and interlaminar fracture toughness of the composite were greatly improved by the presence of the block copolymers, as attested by K_{IC}, G_{IIC} and ILLS values. The mechanical properties appear to be strongly related to the nature and the morphology of the copolymers.

Introduction

Composites based on epoxy resin and continuous carbon fibers are widely used in the aeronautical industry for structural applications. However, these parts are very often suffering from a poor resistance to crack and delamination. Our group showed that it was possible to improve the mechanical properties of epoxy resins by adding poly (styrene-*b*-butadiene-*b*-methyl methacrylate) (SBM) or poly (methyl methacrylate-*b*-n-butyl acrylate-*b*-methyl methacrylate) (MAM) triblocks copolymers (Fine et al., 2005; Maiez-Tribut, 2007; Ritzenthaler et al., 2002, 2003). The resulting nanostructured samples remained transparent and the Tg of the matrix was only slightly affected by the block copolymers (BCP) content.

This paper aims at describing our strategy to transfer this approach to the fracture resistance improvement of continuous-fiber-reinforced epoxy composites manufactured by Resin Transfer Molding (RTM). This low pressure and low temperature process is based on the injection of a low viscosity thermosetting resin into a mold containing a fiber preform. Once the fibers fully impregnated, the resin curing takes place in the heated mold. BCP have to be integrated in the process steps so that they are homogeneously dispersed and organized in the final part. Since the dissolution of BCP in a thermoset precursor can result in a high increase of viscosity (incompatible with RTM injection), the BCP is rather powdered on the preform. An additional reason for adopting such a strategy is that delamination resistance of the cured parts is expected to be improved if BCP micelles are located at the fiber-matrix interface. Herein, we present the study of homogeneous BCP/resin blends before and after curing in order to select a BCP leading to a satisfactory resin toughness improvement. The BCP powder dissolution rate was then investigated in relation with the chemo-rheological behavior of the resin in order to evaluate its suitability with the RTM process. Finally, carbon-fiber-reinforced composites containing BCP were mechanically-evaluated in terms of fracture toughness.

Experimental

Materials:

The RTM resin is a mixture of two multifunctional epoxies and one diamine. The BCP are provided by Arkéma. $S_{28}{}^{12.7}B_{28}M_{44}$ (SBM) is a poly (styrene-*b*-butadiene-*b*-methyl methacrylate), the subscripts indicating the blocks weight percentages, the superscript indicating the PS molar mass in kg.mol^{-1}. $M_{31}A_{38}M_{31}$ (MAM31) and $M_{25}A_{50}M_{25}$ (MAM25) are poly (methyl methacrylate-*b*-n-butyl acrylate-*b*-methyl methacrylate), the subscripts indicating the blocks weight percentages, the number average molar masses being respectively 54 kg.mol^{-1} and 100 kg.mol^{-1}.

Mixtures preparation and polymerization:

BCP/resin blends were prepared at 130°C in a glass reactor under mechanical stirring. The polymerization steps are 1 hour at 130°C, a heating rate at 2°C/min up to 180°C and 2 hours at 180°C.

Composites manufacturing:

Carbon-fiber-based plies were powdered with BCP at 12 wt. %. Commercial powder Hexion was used as a reference. 0/90° preforms constituted of 8 plies were prepared and placed in a 4 mm thick RTM mold. The resin was injected at 10 cc/min. The polymerization sequence was 1 hour at 130°C, a heating rate at 2°C/min up to 180°C and 2 hours at 180°C. Composites qualities were controlled by ultrasound.

Rheological experiments:

Homogeneous non-reactive mixtures: To determine the order-disorder transition temperature T_{ODT} of BCP-resin solutions before polymerization, measurements were performed on a stress-controlled TA AR1000 rheometer equipped with a Peltier plate and 60 mm cone geometry. Rheological experiments were carried out versus temperature at 1 $rad.s^{-1}$ in the linear viscoelastic region following a temperature ramp from 30°C to 180°C at 2°$C.min^{-1}$. The T_{ODT} has been defined at the cross-over of G' and G".

BCP powder dissolution during resin polymerization: a 400 µm BCP disk was sandwiched between two layers of resin in a disposable 40 mm plate-plate geometry (the gap being of 1 mm). The measurements were performed on a strain-controlled TA ARES rheometer. The rheological experiments were carried out versus time at 1 $rad.s^{-1}$ in the linear viscoelastic region following the polymerization temperature sequence described above.

Transmission Electronic Microscopy (TEM):

Pictures were taken with a Philips CM 120 microscope under an acceleration voltage of 80 kV. 60 nm slices were cut at room temperature with an ultra microtome. No staining agent was used.

Mode I critical stress intensity factors K_{IC}:

Critical stress intensity factors K_{IC} were measured on notched samples in three-points bending mode at $10mm.min^{-1}$ on a MTS 2/M tensile machine. The ratio between the notch depth and the sample thickness lies between 0.45 and 0.55.

Interlaminar shear strength (ILLS):

Three-points bending tests at 0.5 $mm.min^{-1}$ with $L/h < 6$ were carried out on a MTS 2/M machine to determine first-crack energy of 0/90° carbon-reinforced composites.

Mode II critical strain energy release rate G_{IIC}:

The End Notched Flexure test (ENF) was applied to the continuous fiber composites and consists in the crack propagation in mode II and the determination of G_{IIC} following the recommendations of ESIS (Davies et al., 2001). The pre-crack was realized by inserting a 12.5 µm Tedlar (PVF) film during the manufacture of the 0/90° carbon-fiber reinforced composite. Tests were performed on a MTS 2/M tensile machine at 0.5 $mm.min^{-1}$.

Results and Discussion

Homogeneous BCP/resin mixtures

Epoxies/BCP blends containing 20 wt. % of each BCP (in the absence of hardener) were prepared. All the obtained solutions were transparent which indicates that no macrophase separation occurred. Depending on the BCP nature, the flow behaviors of the solutions were very different. To explain these differences, T_{ODT} were measured by rheology on a 30°C-180°C temperature range. T_{ODT} for epoxy/SBM lied above 180°C: the solution behaved as a gel (G'>G") at all temperatures between 30°C and 180°C. The transparency of the solutions gave evidence of the nanostructuration by BCP micelles with the PMMA block miscible in the epoxy while PS and PB blocks were segregated (Fine et al., 2005). T_{ODT} for epoxy/MAM31 was found at 55°C and T_{ODT} for epoxy/MAM25 at 105°C (**Figure 1**). Below T_{ODT}, BCP micelles were formed and resulted in a solid-like behavior for the solution (G'>G"). Above T_{ODT}, the BCP micelles were destroyed and as a consequence, the solution flowed (G">G'). A higher content of poly (n-butyl acrylate) and a higher molar mass can explain the improved thermal stability of MAM25 micelles compared to MAM31 ones.

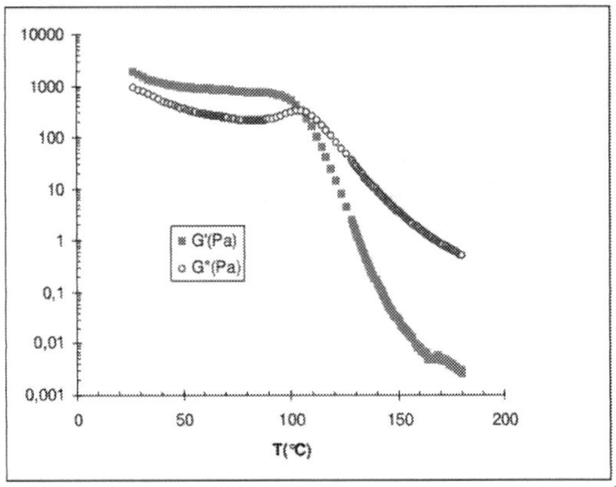

Figure 1 – G' and G" versus temperature at 1 $rad.s^{-1}$ for a epoxy/MAM25 solution with 20 wt. % of MAM25

Epoxies/amine/BCP homogeneous solutions containing 20 wt. % of BCP were then prepared and polymerized following the temperature sequence presented in the experimental part. Depending on the BCP and according to the measured T_{ODT}, the BCP is either organized when the polymerization starts at 130°C (SBM) or fully miscible (MAM31 and MAM25). As previously demonstrated, the key to reach a nanostructured morphology at the end of the curing is to keep a block miscible in the resin up to the end of the polymerization, while the other blocks segregate (Ritzenthaler et al., 2003). Epoxy-amine curing proceeds by a step polymerization and miscibility conditions are worsening upon curing since the polymer chains are growing. As a consequence, an initially miscible BCP may remain miscible (all blocks are miscible), nanoseparate (only one block remains miscible) or macroseparate (all the blocks segregate) once the polymerization is over. Morphologies of the cured blends were observed by TEM. In the case of SBM which was initially organized, the nanostructuration was clearly maintained at full conversion. Indeed, micelles constituted of PS-PB core and a miscible PMMA shell (D_n = 10-20 nm) were observed. (**Figure 2**)

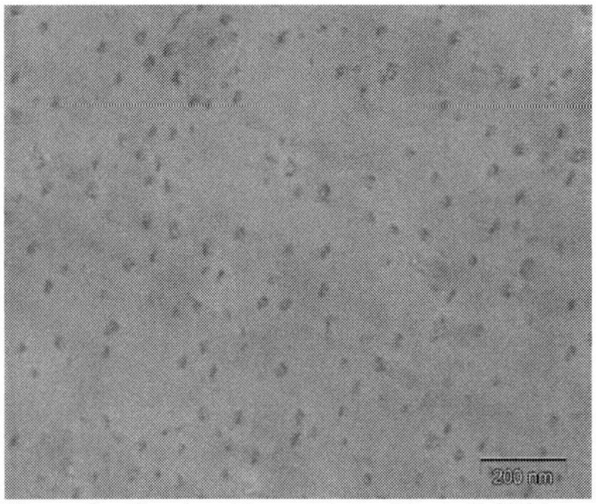

Figure 2 – Cured matrix containing 20 wt. % SBM

In the case of MAM31 which was initially fully miscible, a nanostructuration occurred during the curing. Micelles constituted of poly (n-butyl acrylate) core and a miscible PMMA shell (D_n = 10-20 nm) were observed (**Figure 3**).

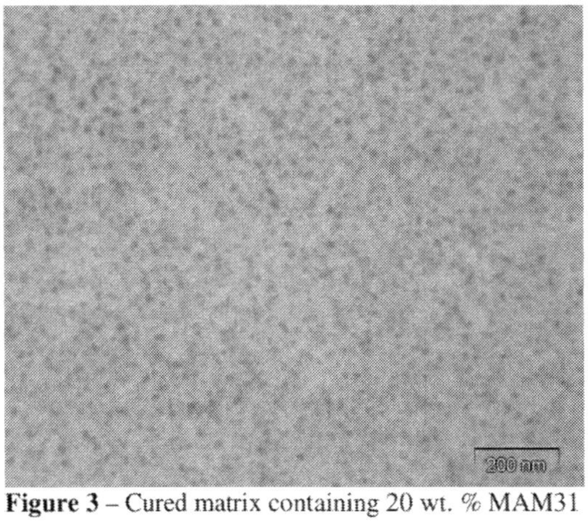

Figure 3 – Cured matrix containing 20 wt. % MAM31

In the case of MAM 25 which was also initially fully miscible, a nanostructuration occurred during the polymerization. A vermicular morphology is obtained, the dark domains being constituted of poly (n-butyl acrylate) (**Figure 4**).

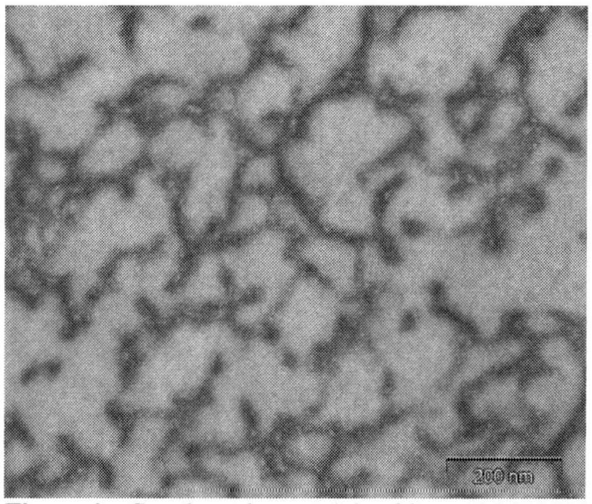

Figure 4 – Cured matrix containing 20 wt. % MAM25

The mechanical properties, especially toughness, are known to be related to the final morphology. To evaluate the crack propagation resistance of the matrices, critical stress intensity factors K_{IC} were measured for each blend (**Table 1**). The greatest improvement (165%) was observed for the matrix with MAM25 which exhibited vermicular morphology. In contrast, MAM31-based and SBM-based matrices which presented micellar morphologies, led to moderate toughness improvements (respectively of 54% and 33%). Hence, the vermicular morphology seems particularly suited to reinforcement.

	Neat	SBM	MAM31	MAM25
K_{IC} (MPa.m$^{1/2}$)	0.63 (±0.05)	0.84 (±0.13)	0.97 (±0.08)	1.67 (±0.23)

Table 1 – K_{IC} values for neat resin and cured matrices containing 20 wt. % of BCP

Simultaneous dissolution of BCP powder and resin curing

As explained previously, we chose to powder the BCP on the woven carbon-fiber preform instead of dissolving it in the injection resin to avoid any viscosity increase. When the resin flows inside the preform, the BCP has to solubilize quite rapidly without being washed out of the mold. The resin viscosity must remain low enough to impregnate the fibers and the BCP has to nanostructure before resin gelation. To evaluate the suitability of the different BCP to these process conditions, rheological tests have first been performed on a sandwich constituted of a 400 µm BCP disk between two layers of resin. This sample is supposed to mimic the behavior of BCP powders of 400 µm in presence of epoxy-amine. Complex viscosities were measured versus time for reactive solutions containing respectively 20 wt. % of SBM and 20 wt. % of MAM25 (**Figure 5**). In the case of SBM, η* increased regularly during the step at 130°C and did not reach any plateau when the temperature increased at 180°C. It means that the dissolution of this BCP is quite slow. In the case of

MAM 25, η* reached a plateau at 10 Pa.s after only 20 min indicating that the dissolution is very fast. T_{ODT} for a 20 wt. % MAM25 solution was 105°C, meaning that MAM25 cannot self-organize as micelles at 130°C. This may explain its fast dissolution and the relatively low final viscosity. For both BCP, the viscosity increases very rapidly after 90 min, corresponding to the gelation of the epoxy-amine system. By consequence, MAM25 was fully dissolved when the gelation occurred and thus appears as a very good choice for our RTM process.

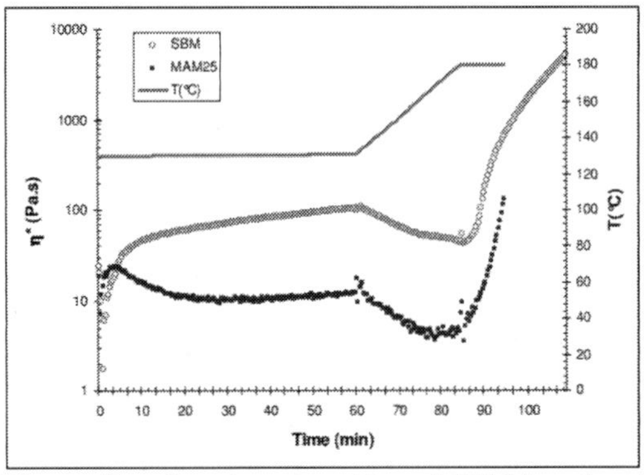

Figure 5 – Complex viscosities of SBM-based and MAM25-based reactive epoxy-amine solutions, measured at 1 $rad.s^{-1}$, according to the temperature sequence: 1 hour at 130°C, a heating rate at 2°C/min up to 180°C and 2 hours at 180°C

Composites interlaminar mechanical properties:

0/90° carbon-reinforced composites were manufactured according to the methodology described in the experimental part. SBM and MAM25 were used as preforms powders, while Hexion powder constituted the reference. Ultrasonic non-destructive control confirmed that the obtained parts were of good quality (no porosity, no dry area) and that the BCP dissolved homogeneously in the composite.
Interlaminar shear strengths (ILSS) were evaluated for each part by measuring the first-crack energy, as well as mode II critical strain energy release rates, G_{IIC} (**Table 2**).

	Reference	SBM	MAM25
E_{crack} (J)	1.54±0.09	2.18±0.04	2.77±0.14
G_{IIC} ($J.m^{-2}$)	3790±860	2280±220	7000±1350

Table 2 – First-crack energy and G_{IIC} for 0/90° carbon-fiber reinforced composites

First-crack energy is clearly improved in the presence of BCP: a 40 % increase is observed with SBM and a 80% increase with MAM25. The lower value for SBM may be partly explained by the fact that this BCP is partially dissolved in the matrix since its dissolution rate is slower as explained previously. G_{IIC} values confirmed the positive effect of MAM25 with an increase of 84% if compared with the reference. This improvement is consistent with the K_{IC} values measured on the matrix. It indicates that MAM25 is responsible for the greater toughness improvement both on matrix and composite. Unexpectedly, G_{IIC} of SBM-based composite was much lower than the value measured for the reference composite.

Conclusions

Poly (styrene-*b*-butadiene-*b*-methyl methacrylate) (SBM) and poly (methyl methacrylate-*b*-n-butyl acrylate-*b*-methyl methacrylate) (MAM) can nanostructure epoxy-amine matrices which are eligible to the RTM process. $S_{28}{}^{12.7}B_{28}M_{44}$ and $M_{31}A_{38}M_{31}$ were self-organized as spherical micelles while $M_{25}A_{50}M_{25}$ was organized as vermicular micelles. This last BCP led to the highest K_{IC} improvement (165% with a 20 wt. % BCP content) if compared with the neat matrix.
$S_{28}{}^{12.7}B_{28}M_{44}$ and $M_{25}A_{50}M_{25}$ dissolution rates as powders in relation with the resin curing were studied by rheology. It proved that $M_{25}A_{50}M_{25}$ is fully dissolved in the epoxy-amine precursors when the gelation occurs. On the contrary, $S_{28}{}^{12.7}B_{28}M_{44}$ is only partially dissolved at the gel point. This difference may be partly explained by the fact that $M_{25}A_{50}M_{25}$ is not organized as micelles at 20 wt. % in the epoxy-amine precursors (T_{ODT}=105°C) while $S_{28}{}^{12.7}B_{28}M_{44}$ is already self-organized for the same content.
0/90° carbon-reinforced composites were prepared by injecting the RTM resin on BCP-powdered preform (12 wt. % of $S_{28}{}^{12.7}B_{28}M_{44}$ or $M_{25}A_{50}M_{25}$). Homogeneous parts were obtained as shown by ultrasonic non-destructive control. Interlaminar mechanical properties were evaluated by measuring ILLS (first-crack energy) and G_{IIC}. Compared to the reference, $M_{25}A_{50}M_{25}$ afforded a 80% increase of the first-crack energy and a 84% increase of G_{IIC}, consistent with the toughness improvement observed on the unfilled matrix.

Acknowledgements

The authors thank Arkéma and EADS CCR for their financial support and scientific collaboration. Pr. Henry Sautereau is warmly thanked for his important contribution to the mechanical properties evaluations. Dr. Julien Bernard is thanked for his valuable suggestions aimed at improving the clarity of this manuscript.

References

Davies, P., Blackman, B., Brunner, A.J., "Fracture mechanics testing methods for polymers, adhesives and composites", Moore, D.R., Pavana,A., Williams, J.G. (Ed.), ESIS Publ. **28**, Elsevier, 307-334, 2001

Fine, T., Lortie, F., David, L., Pascault, J.P., "Structures and rheological properties of reactive solutions of block copolymers. Part I. Diblock copolymers in a reactive solvent", Polymer, **46**, 6605-6613 (2005)

Maeiz-Tribut S., Pascault, J.P., Soule, E.R, Borrajo, J., Williams, R.J.J, "Nanostructured epoxies based on the self-assembly of block copolymers: a new miscible block that can be tailored to different epoxy formulations", Macromolecules, 40, 1268-1273 (2007)

Ritzenthaler, S., Court, F., David, L., Girard-Reydet, E., Leibler, L., Pascault, J.P., "ABC triblock copolymers/epoxy-diamine blends. 1. Keys to achieve nanostructured thermosets", Macromolecules, **35**, 6245-6254 (2002)

Ritzenthaler, S., Court, F., Girard-Reydet, E., Leibler, L., Pascault, J.P., "ABC triblock copolymers/epoxy-diamine blends. 2. Parameters controlling the morphologies and properties", Macromolecules, **36**, 118-126 (2003)

NANOSTRUCTURED POLYMERS AND COMPOSITES FROM SELF-ASSEMBLY OF DEGRADABLE CHIRAL BLOCK COPOLYMERS AND THEIR APPLICATIONS

Rong-Ming Ho

Department of Chemical Engineering, National Tsing Hua University, Hsinchu, Taiwan –rmho@mx.nthu.edu.tw

Various nanostructures including sphere, cylinder, gyroid, lamellae, helix and core-shell cylinder with helical sense could be obtained in the self-assembly of PS-PLLA chiral block copolymer (BCP*). After the hydrolysis of PLLA blocks, mesoporous PS with ordered nanochannels could be fabricated, and treated as template for the following electroplating process or sol-gel reaction. By taking advantage of templated electroplating process, a straightforward and effective method for the fabrication of the patterned carbon nanotube (CNT) arrays can be achieved by electroplating Ni as catalysts for the growth of CNTs; the optical performance of field emission for CNT device might be achieved by finely modulating the configuration of patterned arrays. Furthermore, silica precursor mixture could be introduced into the PS template by pore-filling process. Well-defined nanocomposites with SiO_2 inorganic nanostructures orderly dispersed in polymeric matrix could be successfully obtained after the sol-gel process. Also, specific pore-filling process was adopted to increase the loading efficiency of sirolimus (a potent immunosuppressive drug) solution. After the pore-filling process, sirolimus-loaded cylindrical and lamellar nanoarrays could be obtained. A comparison with those of macroscale templates indicates that the developed nanoporous templates can successfully entrap the loaded drug in nanoscale pores, markedly increasing the duration of drug delivery.

Introduction

Various nanostructures including sphere, cylinder, gyroid, lamellae, helix and core-shell cylinder with helical sense could be obtained in the self-assembly of PS-PLLA chiral block copolymer (BCP*).[1-5] After the hydrolysis of PLLA blocks, mesoporous PS with ordered nanochannels could be fabricated, and treated as template for the following electroplating process or sol-gel reaction. By taking advantage of templated electroplating process, a straightforward and effective method for the fabrication of the patterned carbon nanotube (CNT) arrays can be achieved by electroplating Ni as catalysts for the growth of CNTs; the optical performance of field emission for CNT device might be achieved by finely modulating the configuration of patterned arrays.[6] Furthermore, silica precursor mixture could be introduced into the PS template by pore-filling process. Well-defined nanocomposites with SiO_2 inorganic nanostructures orderly dispersed in polymeric matrix could be successfully obtained after the sol-gel process.[7] Also, specific pore-filling process was adopted to increase the loading efficiency of sirolimus (a potent immunosuppressive drug) solution. After the pore-filling process, sirolimus-loaded cylindrical and lamellar nanoarrays could be obtained. A comparison with those of macroscale templates indicates that the developed nanoporous templates can successfully entrap the loaded drug in nanoscale pores, markedly increasing the duration of drug delivery.[8]

Results and Discussion

Chiral block copolymers (BCP*) comprising chiral entities, poly(styrene)-b-poly(L-lactide) (PS-PLLA), were designed to fabricate helical architectures (i.e., twisted morphologies) from self-assembly. A unique transmission electron microscopy (TEM) projection (Figure 1) for the self-assembled nanostructure of PS-PLLA with volume fraction of PLLA (f_{PLLA}^{v}) ~0.35 suggests a new helical phase (H*) whereas no such projection can be found in racemic BCPs (that is poly(styrene)-b-poly(racemic-lactide) (PS-PLA)). This H* phase with hexagonally packed, interdigitated PLLA helices in a PS matrix were directly visualized by electron tomography and identified as a P622 symmetry (Figure 2). The phase diagram of the PS-PLLA BCPs* was also established. Phase transitions from the H* phase to the stable cylinder and gyroid phases were found after long-time annealing, suggesting that the H* is a long-lived metastable phase. In contrast to racemic poly(styrene)-b-poly(D,L-lactide) BCPs, chiral interaction significantly enhances the incompatibility between achiral PS and chiral PLLA blocks in the PS-PLLA BCPs* and can be estimated through the determination of the interaction parameter. Furthermore, interesting crystallized morphology can be observed; various crystalline PS-PLLA nanostructures were obtained by simply controlling the crystallization temperature of PLLA at which crystalline nanohelices (PLLA crystallization directed by helical confined microdomain) and crystalline cylinders (phase transformation of helical nanostructure dictated by crystallization) occur while $T_{g,PS}>T_{c,PLLA}$ and $T_{c,PLLA}\leq T_{g,PS}$, respectively. For PS-PLLA with PLLA-rich fraction, a unique self-assembled nanostructure, a hexagonally packed core-shell cylinder phase of which the PS microdomains appear as shells and PLLA microdomains appear as matrix and cores can be found from the self-assembly of PS-PLLA with f_{PLLA}^{v}~0.65. Also, the helical microdomains can be hydrolyzed to generate helical nanochannels that can be used as templates. By exploiting the nanoreactor concept, reactions such as sol-gel process can be carried out within the helical nanochannels for manufacturing helical nanocomposites and inorganic objects as well. By taking advantage of templated electroplating process, a straightforward and effective method for the fabrication of the patterned carbon nanotube (CNT) arrays can be achieved by electroplating Ni as catalysts for the growth of CNTs; the optical performance of field emission for CNT device might be achieved by finely modulating the configuration of patterned arrays (Figure 3).. Furthermore, silica precursor mixture could be introduced into the PS template by pore-filling process. Well-defined nanocomposites with SiO_2 inorganic nanostructures orderly dispersed in polymeric matrix could be successfully obtained after the sol-gel process (Fig. 4). Also, specific pore-filling process was adopted to increase the loading efficiency of sirolimus (a potent immunosuppressive drug) solution. After the pore-filling process, sirolimus-loaded cylindrical and lamellar nanoarrays could be obtained. A comparison with those of macroscale templates indicates that the developed nanoporous templates can successfully entrap the loaded drug in nanoscale pores, markedly increasing the duration of drug delivery (Fig. 5).

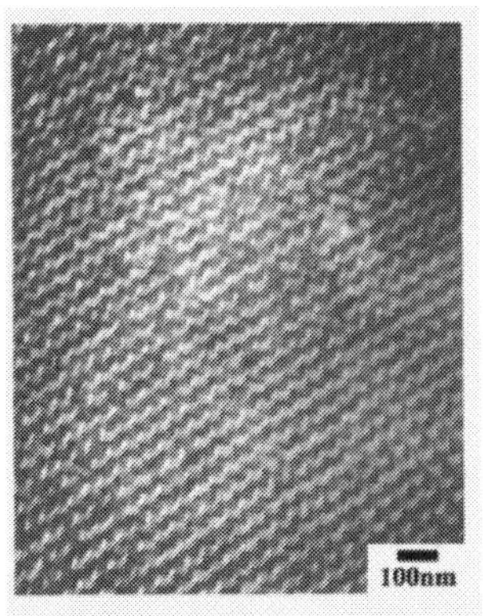

Figure 1 – TEM micrograph of PS280-PLLA127 (f_{PLLA}^{v}=0.35).

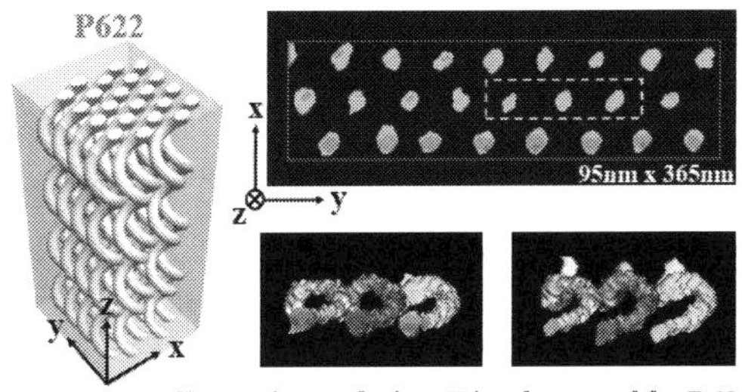

Figure 2 –Illustration of the H* phase with P622 symmetry and corresponding 3D TEM visualization with a dark matrix: cross-sectional view after binarization, and images viewed parallel to the helical central axes and slightly tilted from the helical central axes after binarization and segmentation, respectively (the domain within the box with the dashed perimeter in image viewed parallel to the helical central axes).

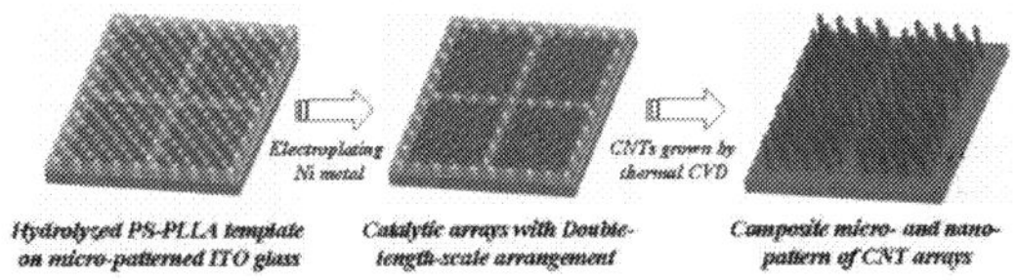

Figure 4 – Fabrication of double-length-scale patterns via lithography, block copolymer templation and electrodepostion.

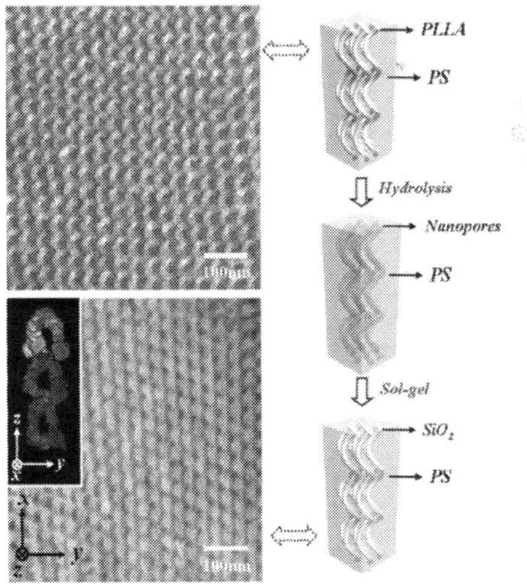

Figure 3 –Schematic illustration of the templation for generating well-defined helical nanocomposites and corresponding 2D TEM projection images and 3D TEM visualization.

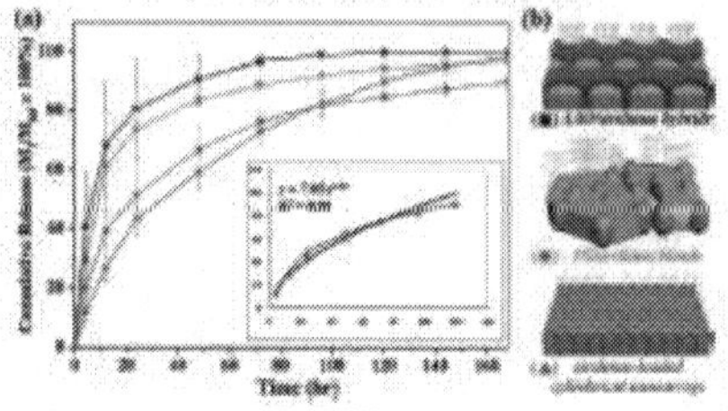

Figure 5 – Pore-filling nanoporous templates from degradable BCPs for nanoscale drug delivery.

Conclusions

Chiral block copolymers (BCP*) comprising chiral entities, poly(styrene)-b-poly(L-lactide) (PS-PLLA), were designed to fabricate helical architectures (i.e., twisted morphologies) from self-assembly. The phase diagram of the PS-PLLA BCPs* was also established. Various nanostructures including sphere, cylinder, gyroid, lamellae, helix and core-shell cylinder with helical sense could thus be obtained in the self-assembly of PS-PLLA chiral block copolymer (BCP*). After the hydrolysis of PLLA blocks, mesoporous PS with ordered nanochannels could be fabricated, and treated as template for the following electroplating process or sol-gel reaction. Consequently, a variety of nanostructured composites could be obtained in applications for functions and complexity.

References

1. Ho, R.-M.; Chiang, Y.-W.; Tsai, C.-C.; Lin, C.-C.; Ko, B.-T.; Huang, B.-H. *J. Am. Chem. Soc.* ***126***, 2704, (2004).
2. Ho, R.-M.; Chiang, Y.-W.; Chen, C.-K.; Wang, H.-W.; Hasegawa, H., Akasaka, S.; Thomas, E. L.; Burger, C., Hsiao, B. S. *J. Am. Chem. Soc.* ***131***, 18533, (2009).
3. Chiang, Y.-W.; Ho, R.-M.; Ko, B.-T.; Lin, C.-C. *Angew. Chem. Int. Ed.* ***44***, 7969 (2005).
4. Chiang, Y.-W.; Ho, R.-M.; Thomas, E. L.; Burger, C.; Hsiao, B. *Adv. Funct. Mater.* ***19***, 448 (2009).
5. Ho, R.-M.; Chen, C.-K.; Chiang, Y.-W.; Ko, B.-T.; Lin, C.-C. *Adv. Mater.* **18**, 2355 (2006).
6. Tseng, W.-H.; Chen, C.-K.; Chiang, Y.-W.; Ho R.-M. *J. Am. Chem. Soc.*, ***131***, 1356 (2009).
7. Tseng, Y.-T.; Tseng, W.-H.; Lin, C.-H.; Ho, R.-M. *Adv. Mater.*, **19**, 3584-3588 (2007)
8. Lo, K.-H.; Chen, M.-C.; Ho, R.-M.; Sung, H.-W. *ACS Nano*, ***3***, 2260.(2009).

Direct and Indirect Polymer-Polymer Interfacial Slip Measurements

HeeEon Park[1], Patrick C. Lee[2], and Chris W. Macosko[3]

[1]*Department of Chemical Engineering, McGill University, Montreal, Quebec, Canada*

[2]*The Dow Chemical Company, Midland, MI, USA*

[3]*Department of Chemical Engineering and Materials Science, University of Minnesota, Minneapolis, MN, USA*

Significant slip can occur during flow of two immiscible polymers due to reduced entanglements at their interface. The slip is of practical importance because of its effect on morphology and adhesion of these multi-phase materials, such as disordered two-phase blends and multilayer films. In this research, we are investigating the amount of polymer-polymer slip over a range of shear stresses using rheological (indirect method) and visualization measurements (direct method) of alternative multilayer samples. Interfacial slip was quantified using two polydisperse polymer pairs, of which each component has closely matched viscosity and elasticity but different miscibility: polypropylene (PP)/polystyrene (PS) and polyethylene (PE)/fluoropolymer (FP). PE/FP bilayer samples were also used to measure V_{slip} directly by imaging relative displacement between marks on the interface of each polymer. This direct method confirmed the indirect stress method. The nominal viscosity as well as shear normal stresses of the multilayer samples decreased with the number of layers. Both direct and indirect measurements confirmed that V_{slip} did not show a sudden increase at a critical shear stress as predicted by Brochard-Wyart and de Gennes [C. R. Acad. Sci., Ser. II: Mec., Phys., Chim., Sci. Terre l'Univers 317, 13 (1993)] but rather increased $V_{slip} \propto \tau^n$, where n = 3 for PS/PP and n = 6.2 up to 55 kPa and 1.8 at higher stresses for PE/FP. We extend our slip measurement study to monodisperse pairs of PS/poly(methyl methacrylate) (PMMA) and polybutadiene (PB)/polyisoprene (PI) to eliminate the effect of molecular weight distribution.

Heat capacity of high-density polyethylene at melt temperatures: Transitions at 210 and 230°C

I.A. Hussein[b] and M. C. Williams[a]

[a]Department of Chemical and Materials Engineering, University of Alberta
Edmonton, Alberta, Canada, T6G 2G6
mike.williams@ualberta.ca)

[b]Department of Chemical Engineering, King Fahd University of Petroleum and Minerals, Dhahran 31261, Saudi Arabia
(Email: ihussein@kfupm edu.sa)

Abstract

The first measurements of heat capacity of high-density polyethylene (HDPE) have been obtained for melts in the high-temperature region of 180 to 260°C. A heat-flow twin calorimeter was used. Abrupt transitions are noticed at 212 and 228°C, analogous to transitions seen earlier with rheological and surface tension measurements. Possible explanations for these phenomena are offered.

Key words: Polyethylene, HDPE melts, heat capacity at high temperature, transitions

MULTISCALE SIMULATION OF SOL-GEL SILICA POLYCONDENSATION IN DRYING FILMS

S.E. Rankin[1]* and X. Li[2]

[1] *Department of Chemical and Materials Engineering, University of Kentucky, Lexington, KY, USA – srankin@engr.uky.edu*
[2] *Medical College of Wisconsin, Milwaukee, WI, USA – xinli@uky.edu*

Beginning with the recursive method (Bailey et al. 1990) and progressing to the dynamic Monte Carlo (dMC) method (Hendrickson et al. 1994), C.W. Macosko and his collaborators provided the theoretical underpinnings needed to link the kinetics of siloxane-based sol-gel precursors to polymer structure and gelation. The recursive approach uses probability and statistics to link populations of differently connected building units to average properties (e.g. weight-averaged degree of polymerization) but is fundamentally limited by requiring the randomization of connections among the structural elements modeled. The dMC method also uses probability and statistics, but directly models the evolution of connections among of a finite set of monomers based on the relative rates of the polymerization reactions that occur. The dMC method was the first approach able to match both site-level reaction kinetics and the overall gelation conversion of a sol-gel silica system (Rankin et al. 2000) – primarily because this system is prone to extensive cyclization which, under typical reaction conditions, gives rise to a set of cage-like precursors to a gel (Ng et al. 1995). Here, a multiscale model is described that links the dMC simulation of a finite set of monomers to a finite difference model of diffusion and evaporation of solvent from a deposited coating. By including both first-shell substitution effects and cyclization reactions, the dMC model is able to realistically model the hydrolytic polycondensation behavior of precursors to silica-based sol-gel materials, such as tetraethoxysilane. This multiscale model is able to capture not only gradients across a film of gelation dynamics, but also gradients of structure due to different levels of cyclization that occur across the film with certain combinations of dimensionless parameters (Biot and Damköhler numbers).

Introduction

Silica derived by hydrolytic polycondensation of alkoxysilanes such as tetraethoxysilane has been of high interest since the 1980s. The process, sometimes known as "sol-gel" ceramic processing (Brinker and Scherer 1990), is amenable to a wide variety of novel forms and properties of silica because it can be carried out near ambient conditions, leading to precise control of the porosity and structure of the metal oxide by control of the conditions and incorporation of organic templates (such as surfactants (Ying et al. 1990)) and organic functionalized silanes (Stein et al. 2000).

Many of the potential applications of silica-based sol-gel materials take advantage of the relative ease with which they can form into thin films. Spin and dip coating are commonly used to form well-defined porous thin films for applications as diverse as sensing, membrane separations, and biomedical materials. (Sanchez et al. 2008) Despite these applications, large-scale production of sol-gel silica-based films will require being able to control and reduce defects such as crack formation, delamination, and uncontrolled porosity (Brinker and Scherer 1990).

Here, we briefly discuss our recent effort to extend dynamic Monte Carlo modeling of sol-gel silica formation to a multiscale model of thin film formation. We briefly describe the technique and provide an overview of key outcomes which will be discussed in more detail in forthcoming publications.

Modeling Approach

Developing a comprehensive model of sol-gel silica polycondensation has been an ongoing process since some of the detailed NMR studies by Assink and Kay (Assink and Kay 1988; Kay and Assink 1988) and Pouxviel et al. (1987). It became evident early on that the process does not follow the assumptions of random branching theory – notably by displaying strong first-shell substitution effects (Sanchez et al. 1996) and extensive cyclization (Rankin et al. 1998a). The substitution effects could reasonably be incorporated into the recursive method for statistically modeling network polymerization (Bailey et al. 1990) but the gelation conversion predicted even with the most severe substitution effect (50%) could not match the experimental value (82%). Isolated rings could increase the predicted gel conversion some but not enough (Kasehagen et al. 1996). However, more extensive cyclization with the formation of cage-like precursors prior to gelation was shown by Ng et al. (1995) to give a structural model capable of matching the experimental gel conversion.

With these considerations in mind, we have used a dynamic Monte Carlo (dMC) simulation technique based on the method described by Hendrickson et al. (1994). Many of the data structures described by Hendrickson et al. are used to track quantities such as lists of sites of different connectivity, numbers of different types of sites and bond blocks, and graphs of the growing polymers. The dMC simulation allows unlimited cyclization of three-silicon (6-atom) rings by unimolecular-like reactions of two-bond blocks, as described by Rankin et al. (2000). The primary kinetic features are:

Hydrolysis pseudoequilibrium is assumed for all sites, with equal equilibrium coefficients. This allows the hydrolysis reaction (eq. 1) to be treated as being at equilibrium, where the parameter χ describes the fraction of uncondensed groups that have been hydrolyzed. If the water content is high enough, χ is constant, so we have assume that this is true in our modeling

$$\equiv SiOR + H_2O \rightleftharpoons \equiv SiOH + ROH \qquad (1)$$

$$\chi = \frac{[\equiv SiOH]}{[\equiv SiOH]+[\equiv SiOR]} = \frac{K_h[H_2O]}{K_h[H_2O]+[ROH]} \qquad (2)$$

A negative first-shell substitution effect is assumed for condensation reactions between silicon sites. If Q_i represents a silicon site with i siloxane bonds attached to it (so a monomer is a Q_0 site, a chain end is a Q_1 site, etc.), the reaction rates for condensation reactions (eq. 3), R_{ij}^{bimol}, are given by eq. 4 where f represents the functionality of the monomer (in this case, 4). The rate coefficients follow the pattern that they drop by 10% if the connectivity of only one site changes, while they drop by 90% if the connectivity of both sites changes. In other words, $k_{01} = 0.9k_{00}$, $k_{02}=0.81k_{00}$, etc. and $k_{11}=0.1k_{00}$, $k_{22}=0.01k_{00}$, etc. (Rankin et al. 1998b).

$$Q_i + Q_j \xrightarrow{k_{ij}^*} Q_{i+1} + Q_{j+1} + H_2O \qquad (3)$$

$$R_{ij}^{bimol} = \begin{cases} \chi^2(f-i)(f-j)k_{ij}[Q_i][Q_j], & i \neq j \\ \frac{1}{2}\chi^2(f-i)(f-j)k_{ij}[Q_i][Q_j], & i = j \end{cases} \qquad (4)$$

Unlimited cyclization to form three-monomer (6-atom) rings is allows by counting 2-bond blocks with varying connectivities of the chain ends throughout the simulation (Rankin 2000). For a 2-bond block ending with Q_i and Q_j sites, which is denoted by $^2B_{ij}$, the rate of ring closure is assumed to be given by eq. 5.

$$^2R_{ij}^{cyc} = (4-i)(4-j)\chi^2 k_{3c(i,j)}[^2B_{ij}] \qquad (5)$$

The dMC simulation itself is conducted by starting with a population of 100,000 monomers. All concentrations required to calculate rates are updated continuously through the simulation. At each step, a reaction is first chosen randomly using the set of probabilities which are set proportional to the rates of all bimolecular (eq. 4) and cyclization (eq. 5) reactions. After selecting the reaction, the specific sites for reaction are chosen randomly from all types of sites that satisfy the criteria for the reaction – either having the right connectivity for bimolecular reaction, or being members of the correct type of bond block for cyclization. The matrices describing the molecular weight distribution, structure distribution and many other average properties are updated after the reaction. The time is also incremented using the continuous-time approach as described by Gillespie (1976), eq. 6.

$$\Delta t = \frac{\ln\left(1/r\right)}{\left(N/C_{Si}\right)\sum(R_{ij}+{}^2R_{ij}^{cyc})} \qquad (6)$$

where r = a random number selected from the interval (0,1), N is the total number of sites, and C_{Si} is the total silicon concentration, which comes from a finite difference simulation of a drying film.

The dMC model is coupled to a continuum model of diffusion and convection from the top surface of a thin film, as described in a recent paper from our group (Li and Rankin 2010). Basically, the continuum model uses the time steps from the dMC model to update the concentrations in a centered finite difference model of drying in a thin film. The film is assumed to initially be uniform. Only solvent diffusion is modeled, and the position of a particle of sol from its initial position is tracked. That sol particle marks the point in the film where the dMC simulation occurs, and the concentration at that point is used to determine the absolute rates of reaction and the time step in the dMC simulation. By tracking the behavior of several dMC simulations at different points in the film, it is possible to get an estimate of gradients of gel time, gel structure, etc. The multiscale modeling approach is illustrated in Fig. 1.

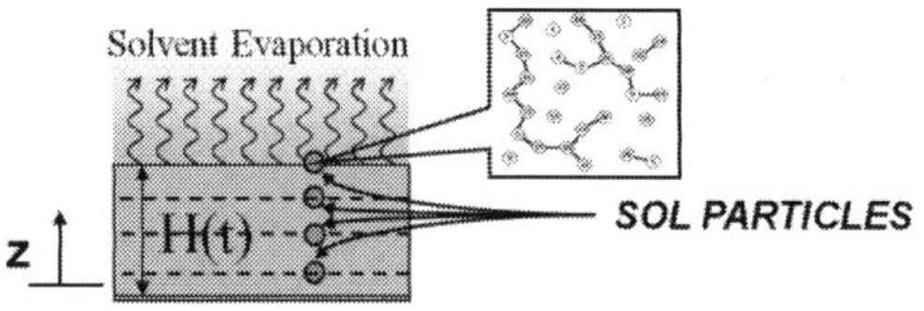

Figure 1. Schematic illustration of the multiscale model. Solvent drying and diffusion are modeled using a centered finite difference method to determine the concentration profile in the film, and sol particles are tracked. A dMC simulation is performed in each particle in the film.

The film is assumed to be in contact with one impenetrable substrate surface and one vapor interface, where solvent evaporates at a rate determined by the vapor-phase driving force. See Li and Rankin (2010) for full details and equations.

Three key dimensionless variables describe the behavior of this multiscale system. They are the Biot number (Bi), the Damköhler number (Da), and the cyclization tendency (κ), defined in eqs. 7-9.

$$Bi = \frac{k_g H_0 \hat{V}_1}{D_1} \quad (7)$$

$$Da = \frac{k_{00} C_{1,0} {H_0}^2}{D_1} \quad (8)$$

$$\kappa = \frac{k_{3c(1,1)}}{k_{11} C_{Si,0}} \quad (9)$$

where k_g = the vapor-phase mass transfer coefficient, H_0 = the initial film thickness, $\hat{V}_1$ = the molar volume of solvent, D_1 = the diffusivity of solvent, $C_{1,0}$ = the initial concentration of solvent, and $C_{Si,0}$ = the initial monomer concentration throughout the film.

Results and Discussion

With the model defined, calculations were performed for various values of the dimensionless parameters to survey the behavior of the system. The concentration field in which polymerization occurred varied depending on the value of Bi, but in all cases a drying time for the film, t_{dry}, was defined as the time at which the total amount of solvent was 1% of the initial quantity present. Fig. 2 shows an example of the evolution of the concentration profile in a film and the position of a dMC sol particle for a particular set of parameters.

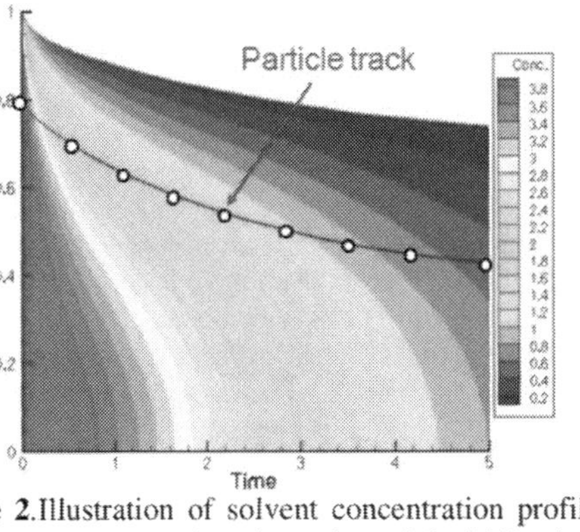

Figure 2. Illustration of solvent concentration profiles starting from $C_{1,0}$ = 4 and a sol particle 80% of the distance from the bottom of the film (z=0) to the top (z=1). For this calculation, Bi = 507. Time is scaled by D_1/H_0^2.

The primary outcome of interest from this multiscale method is to see how cyclization affects the formation of gradients in gelation and in structure across the film. These gradients arise naturally as a consequence of the concentration gradients illustrated in Figure 2. For instance, if we consider ideal polycondensation ($k_{00}=k_{01}=k_{02}=\ldots=k_{11}=k_{22}=\ldots$) with κ=0, Figure 3 illustrates how DP_w evolves with time at different positions in the film. The divergence of DP_w is defined to be the gel point, so this shows that for positions deeper in the film (smaller values of h_{p0}), gelation occurs later.

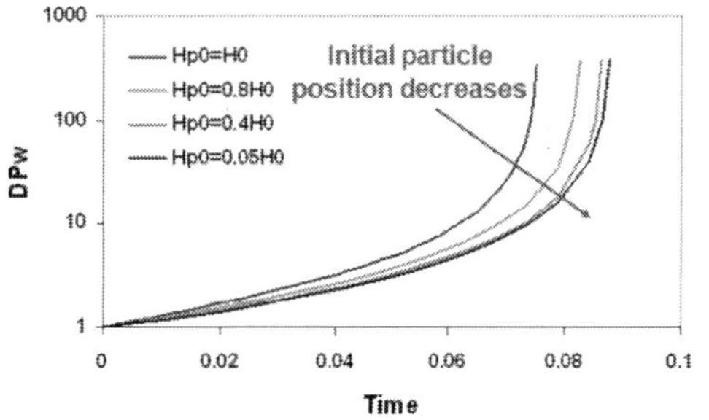

Figure 3. Evolution of DP_w with time for different initial particle positions. Ideal polycondensation was used for the dMC model to generate this plot. Bi=1030 and Da = 0.4.

Previous studies by Cairncross et al. (1996) using the kinetic-recursive approach (Bailey et al. 1990) indicated that depending on the values of Bi and Da, several phenomena could be observed in thin films. The first was gelation before drying, defined when the gel time at the base ($t_{gel,b}$) < 10% of t_{dry}. Gelation before drying, on the other hand, was defined as $t_{dry} < t_{gel,b}$. A third phenomenon was defined to encapsulate the possibility that a severe gradient in gel times could occur across the film relative to the rate of drying, and this is called "literal skinning", defined to be when $(t_{gel,b}-t_{gel,s})/t_{dry} > 0.1$ where $t_{gel,s}$ = the gel time at the surface of the film.

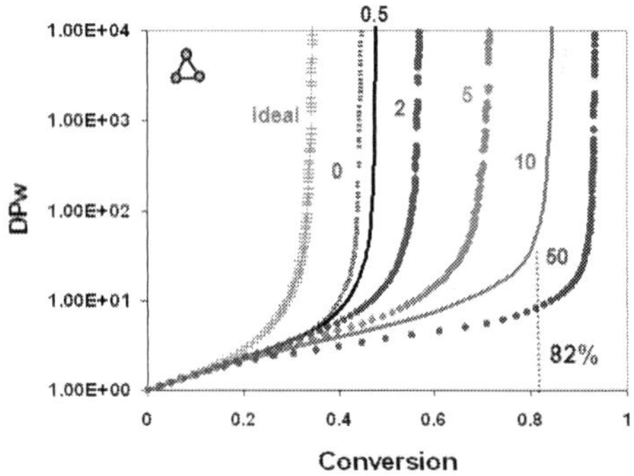

Figure 4. Evolution of DP_w with conversion for varying values of cyclization tendency. Here, Bi = 1000, Da = 60 and h_{p0} = 0.8. Divergence of DP_w indicates gelation.

We have confirmed using this multiscale model that similar boundaries for the gelation phenomena are found to Cairncross' report when only first-shell substitution effects (FSSE) are considered (Li and Rankin 2010). However, with FSSE, only the gel time changes with film position; because relative rates of reaction are always the same for an FSSE model (and are not concentration dependent), the gel conversion and structure of the gel remain constant throughout the film. With cyclization, gel conversion does depend significantly on the value of κ, and as shown in Figure 4, can vary from ~40% to over 90% (the experimental value is 82%) as κ increases. Along with this gel conversion change, the gel that forms goes from having no rings at all (κ=0) to being made of cage-like precursors.

Knowing that cyclization should not only strongly influence gelation rate but also gel structure, a "gelation regime map" (Cairncross et al. 1994) showing the transitions between different gelation and drying phenomena as a function of Bi and Da was prepared with κ = 5. The "literal skinning" region in the map (Figure 5) is significantly larger than in simulations without cyclization, suggesting that cyclization induces a larger variation in gel times across the film, consistent with the greater sensitivity expected for the mix of bimolecular and unimolecular rates in the model with cyclization. This is expected to increase the likelihood that coating defects will occur because in this region, a solid gel film occurs at the top surface well before one forms at the base. This gel can continue to cure while the underlying layer is liquid, leading either to wrinkling or cracking depending on the conditions.

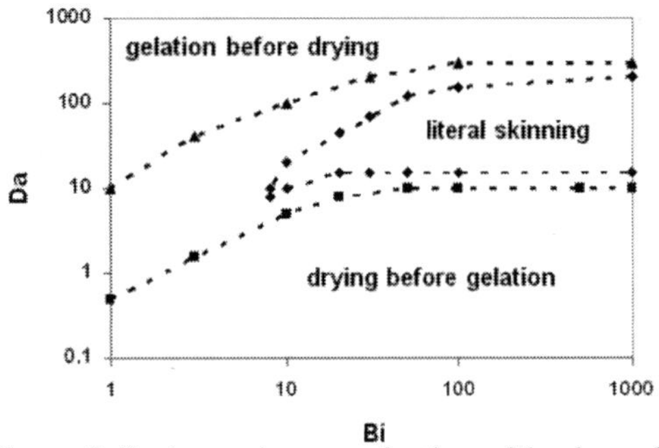

Figure 5. Drying regime map for the multiscale model with negative first-shell substitution effect and unlimited formation of three-silicon rings. The cyclization tendency κ = 5.

In addition to the kinetic effects of cyclization summarized in Figure 5, cyclization induces gradients in structure through the film. There are several ways to characterize these gradients, but one example is shown in Figure 6, which shows a contour plot of number-average cycle rank (*Crankn*) as a function of film position and time for conditions in the "literal skinning" region until gelation is reached. *Crankn* measures the average number of independent rings per molecule. As we can see, at the surface of the film, the cycle rank increases continuously as polymerization proceeds, but reaches a value of only 4 at the gel point. In contrast, cyclization continuous at the relatively low polymer concentration deeper in the film, and by the gel point at the base, *Crankn* at gelation is about 50% larger than at the film surface. This is a significant new prediction because a network with higher Crank would be expected to be more brittle due to the rigidity of polycyclic cage-like species and the small number of links between them needed to form a network. Thus, in addition to a kinetic variation of gel times through the film, our multiscale model suggests that a tougher gel (with few rings) would form at the surface than at the base of the film.

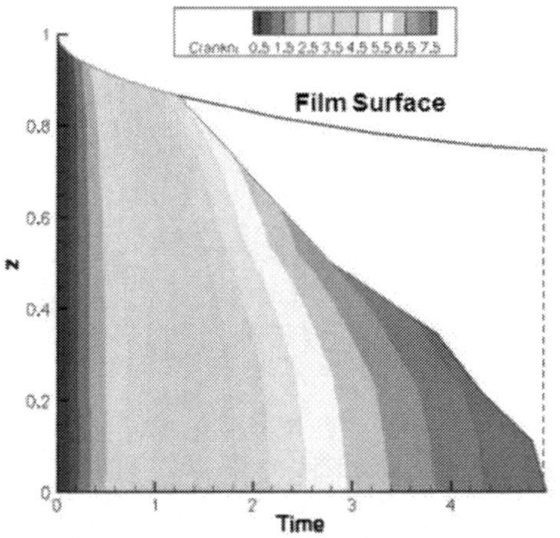

Figure 6. Variation of number-average cycle rank as a function of position (z) as predicted by this multiscale model. The contours end approximately at the gel point for each position in the film (where the model can no longer be used to describe the evolution of the film). Bi =800, Da = 80, and κ = 5.

Conclusions

A multiscale dynamic Monte Carlo / continuum model of sol-gel silica polycondensation with both first-shell substitution effects and extensive cyclization is described. In essence, this model used the dMC model of Rankin et al. (2000) and couples it to a finite difference model of film drying (Li and Rankin 2010). This model not only predicts a larger region of "literal skinning" (the conditions where a gel "skin" is likely to be found due to difference in gelation kinetics across the film) but also predicts that a gel structure gradient across the film should be found under conditions where this skinning occurs.

Acknowledgements

This report is based on work supported by the U.S. Department of Energy under grant numbers DE-FG02-03ER46033 and DE-FG02-07ER46375.

References

Assink, R.A., Kay, B.D., "Sol-gel kinetics I. Functional group kinetics", *J.Non-Cryst.Sol.*, **99**, 359 - 370 (1988)

Bailey, J.K., Macosko, C.W., Mecartney, M.L., "Modeling the gelation of silicon alkoxides", J. Non-Cryst. Solids, **125**, 208-223 (1990).

Brinker, C.J., Scherer, G.W.: Sol-gel Science: The Physics and Chemistry of Sol-Gel Processing, Academic Press, Boston (1990).

Cairncross, R.A., Francis, L.F., Scriven, L.E., "Predicting Drying in Coatings that React and Gel: Drying Regime Maps", AIChE J. **42**, 55 - 67 (1996).

Gillespie, D.T., "A General Method for Numerically Simulating the Stochastic Time Evolution of Coupled Chemical Reactions", J. Comp. Phys., **22**, 403-434 (1976).

Hendrickson, R., Macosko, C.W., Gupta, A.M., "Sol-Gel Polymerization: Monte-Carlo Simulation of Substitution Effects", J. Comput. Polym. Sci. **4**, 53-65 (1994).

Kasehagen, L.J., Rankin, S.E., McCormick, A.V., Macosko, C.W., "Modeling of First Shell Substitution Effects and Preferred Cyclization in Sol-Gel Polymerization", Macromolecules, **30**, 3921-3929 (1997).

Kay, B.D., Assink, R.A., "Sol-gel kinetics II. Chemical speciation modeling", J. Non-Cryst. Solids, **104**, 112-122 (1988).

Li, X., Rankin, S.E., "Multiscale Dynamic Monte Carlo / Continuum Model of Drying and Nonideal Polycondensation in Sol-gel Silica Films", AIChE J. in press (2010).

Ng, L.V., Thompson, P., Sanchez, J., Macosko, C.W., McCormick, A.V., "Formation of Cagelike Intermediates from Nonrandom Cyclization during Acid-Catalyzed Sol-Gel Polymerization of Tetraethyl Orthosilicate", Macromolecules, **28**, 6471-6176 (1995).

Pouxviel, J.C., Boilot, J.P., Beloeil, J.C., Lallemand, J.Y., "NMR study of the sol/gel polymerization", J. Non-Cryst. Solids, **89**, 345-360 (1987).

Rankin, S.E., Macosko, C.W., McCormick, A.V., "Importance of Cyclization during the COndensation of Hydrolyzed Alkoxysilanes", Chem. Mater., **10**, 2037-2040 (1998a).

Rankin, S.E., Macosko, C.W., McCormick, A.V., "Sol-Gel Polycondensation Kinetic Modeling: Methylethoxysilanes", AIChE J., **44**, 1141-1156 (1998b).

Rankin, S.E., Kasehagen, L.J., McCormick, A.V., Macosko, C.W., "Dynamic Monte Carlo Simulation of Gelation with extensive Cyclization", Macromolecules, **33**, 7639-7648 (2000).

Sanchez, C., Boissiere, C., Grosso, D., Laberty, C., Nicole, L., "Design, synthesis, and properties of inorganic and hybrid thin films having periodically organized nanoporosity", Chem. Mat. **20**, 682-737 (2008).

Sanchez, J., Rankin, S.E., McCormick, A.V., "^{29}Si NMR Kinetic Study of Tetraethoxysilane and Ethyl-Substituted Ethoxysilane Polymerization in Acidic conditions", Ind. Eng. Chem. Res., **34**, 117 - 129 (1996).

Stein, A., Melde, B.J., Schroden, R.C., "Hybrid inorganic-organic mesoporous silicates - Nanoscopic reactors coming of age", Adv. Mater., **12**, 1403-1419 (2000).

Ying, J.Y., Mehnert, C.P., Wong, M.S., "Synthesis and Applications of Supramolecular-Templated Mesoporous Materials", Angew. Chem. Int. Ed., **38**, 56-77 (1999).

PROCESS ANOMALIES WITH IMMISCIBLE BLENDS DURING EXTRUSION

Gibson L. Batch[1*], Patrick Lee[2], Carlos Lopez-Barron[2], Christopher W. Macosko[2]

1 3M Center, Building 218-1-1, Saint Paul, MN 55144 glbatch1@mmm.com
2 Department of Chemical Engineering and Materials Science, University of Minnesota, Minneapolis, Minnesota

Immiscible blends are not typically made into films due to their poor strength and lack of clarity. But if these limitations can be overcome, they do offer unique aesthetic, mechanical, and optical properties. We examine the morphology and physical properties of a PS/PP immiscible blend extruded into a film and drawing before the chill roll. At draw ratios greater than 7-to-1, we found a brittle-to-ductile transition in longitudinal direction at room temperature and an onset of longitudinal creep at elevated temperatures. To understand the nature of this transition, we characterized the films by WAXS, SEM, and DSC. We found a significant change in crystalline structure and particle size with increased draw ratios. A model was devised to illustrate how particle size reduction during drawing may result in the observed creep behavior.

Introduction

The morphology of immiscible blends are described as either dispersed droplets, stratified filaments, or a co-continuous interlocking network. In prior work, we found that elongational viscosity of a model PS/PP immiscible blend is greater than that of either component (permeably due to surface tension between the phases), whereas shear viscosity is lower than either component due to streamlining. Now we examine the properties of these blends when extruded and stretched into films.

To visualize co-continuous blends as a film, we examined a blend of fluorescently-tagged polystyrene and styrene-acrylonitrile copolymer using a confocal microscope. We compare the PS phase morphology after batch mixing to the morphology after pressing into an elongated channel. The elongated structure of an immiscible blend in films gives rise to anisotropic properties.

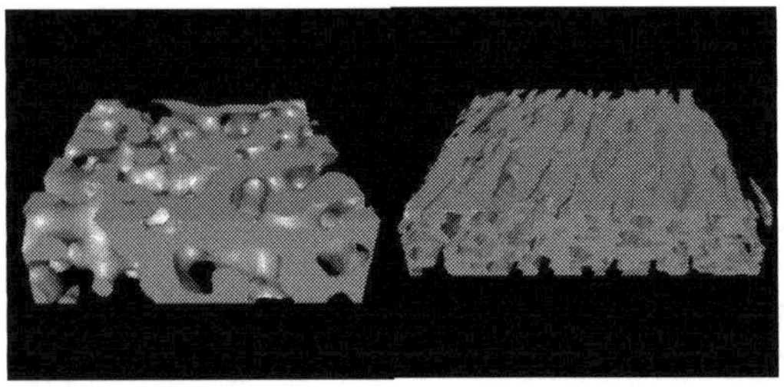

Experimental

Materials

A co-continuous blend was made by mixing equal weight fractions of polypropylene (PP4062, M_w = 280 g/mol, M_w/M_n=4.5, ExxonMobil Chemical Company) and polystyrene (Styron 666D, M_w=226 g/mol, M_w/M_n = 2.6, The Dow Chemical Company). These polymers were selected to have nearly the same shear viscosity at 200 C over a wide range of shear rates as shown below.

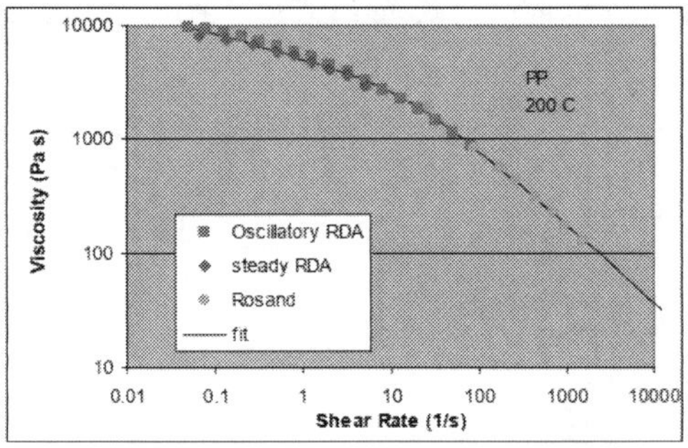

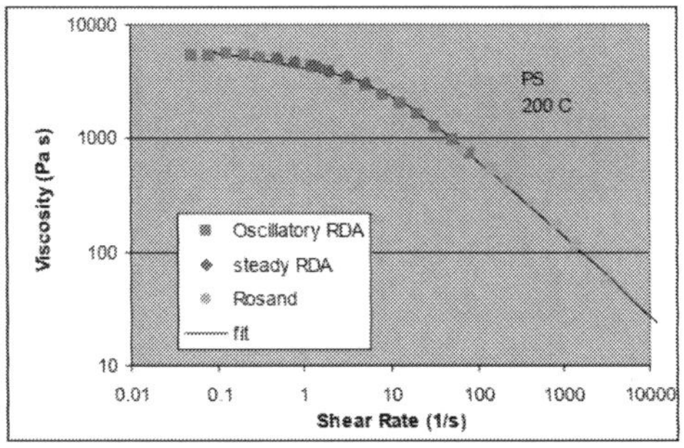

Extrusion

A PS/PP dry blend was fed into a 1.5 inch single screw extruder (Killion) with a 24:1 barrier melting screw and a medium work Maddock mixer near the end. The barrel temperature was 200 C. Extrusion was flood fed at 70 RPM (18 pph) and 16 RPM (5 pph). Additionally, samples were made at 160 RPM with a loss-in-weight feeder set to at 5 pph. This starved feed condition is 14 percent the projected throughput under flood feed conditions.

The film was made with a 6-inch coathanger die set to a nominal 25 mil gap. The casting roll was an 8-inch water-cooled chrome roll at room temperature. A rubber-coated nip roll was applied to the web to help stabilize the drawing about 2 inches away from the die exit. The casting roll velocity was varied from 5 to 40 fpm. By changing the roll velocity, we were able to obtain a wide range of draw ratios from 1:1 to 24:1.

Mechanical Properties
Tensile testing was performed at room temperature in both MD and TD directions using an Instron 5500R with a 4 inch dog bone tensile specimens. The crosshead speed was 1 in/min. Strain was measured by displacement of the grips, and stress from a 50 kg load cell.

Creep was measured at 125 C, a temperature above the T_g of the PS and below the melting point of the PP. Film samples were cut about 1 cm wide in the LD orientation and loaded with a 15 g dead weight. Creep measurements began immediately after the sample was heated to temperature. Total length was recorded as a function of time to determine strain.

Structural Characterization
Differential Scanning Calorimetry was performed on a TA Instruments 2920 DSC from -40 C to 220 C at 20 C/min, using a heat/cool/heat sequence. The glass transition temperature for PP and PS was determined upon heating, and the melting temperature and melting enthalpy were found upon both heating and cooling.

Reflection geometry WAXS data were collected by use of a Philips vertical diffractometer, copper K radiation, and proportional detector registry of the scattered radiation. The diffractometer is fitted with variable incident beam slits, fixed diffracted beam slits, and graphite diffracted beam monochromator. Scans were conducted from 5 to 35 degrees (2theta) using a 0.04 degree step size and 6 second dwell time. Transmission geometry WAXS data were collected by use of a Bruker GADDS microdiffractometer, copper K radiation, and HiStar 2D position sensitive detector registry of the scattered radiation. The 300 im pinhole collimated incident x-ray beam was directed normal to the sample plane at a sample to detector distance of 6 cm.

Scanning Electron Microscopy was performed in both in the machine direction (MD) and transverse direction (TD) of film samples. The pieces were dip-coated with epoxy and then cross-sections were prepared using an ultra-microtome at room temperature. The resulting cross-sections were examined Hitachi S-4700 field emission scanning electron microscope without deposition of a conductive coating. The samples were examined using compositional electron imaging (BSEI). Areas of higher average atomic number appear brighter in BSE images. One sample was re-faced and then exposed to RuO4 vapor for approximately 3 minutes to induce preferential staining of the PS. We found that the lighter phase is PS and the darker phase is PP.

Results and Discussion

Mechanical Properties
Properties were correlated with draw ratio between the die and chill roll, that is the ratio of chill roll draw velocity v and extrusion die velocity v_o,

$$DR = \frac{v}{v_o}$$

where v_o is found from the die gap H_o, die width W_o, and the mass per unit length of the extruded web (M/L).

$$v_o = \left(\frac{M}{L}\right)\frac{v}{\rho_{melt} W_o H_o}$$

The blend has a melt density of 0.85 g/cc.

The modulus and strength in both LD and TD were found not to be affected by screw speed or draw ratio. The TD elongation to break decreased with draw ratio. The MD elongation to break, however, displayed a remarkable brittle-to-ductile transition at draw ratios greater than 7:1.

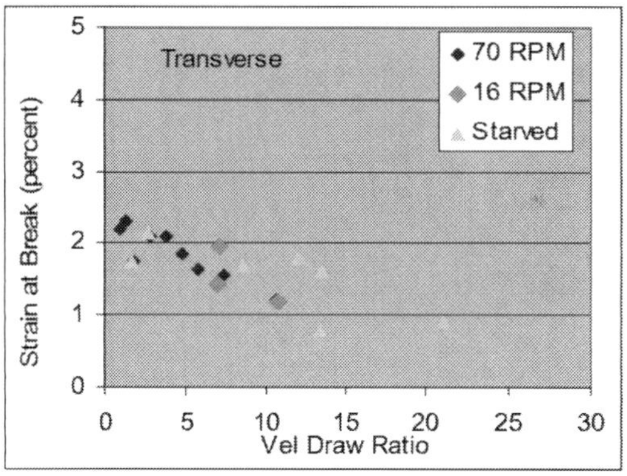

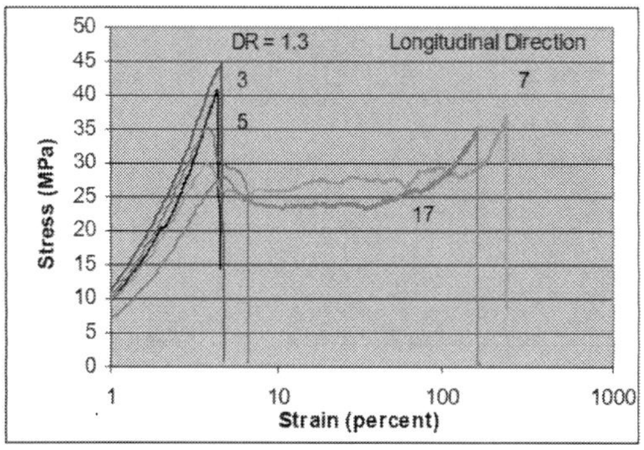

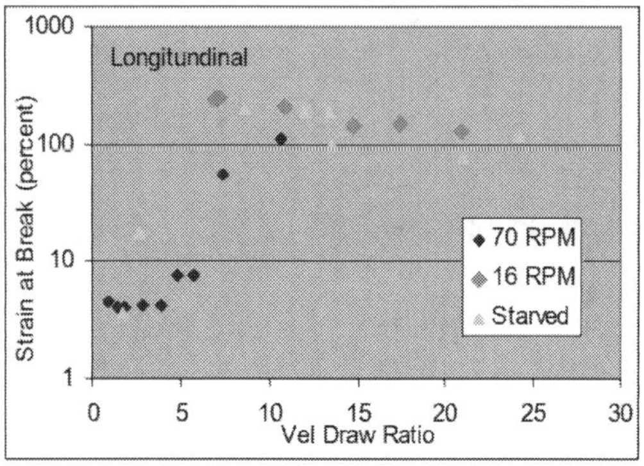

The creep experiments showed a remarkable increase in creep for films with a draw ratio greater than 7:1. Based on sample area (A, the product of thickness and width), we calculate the effective viscosity from a know stretch rate $\dot{\varepsilon}$.

$$\eta_{eff} = \frac{F}{A\dot{\varepsilon}}$$

This viscosity is 4 orders of magnitude higher than for the PS phase at 125 C, but decreases about 15X with increasing draw ratio.

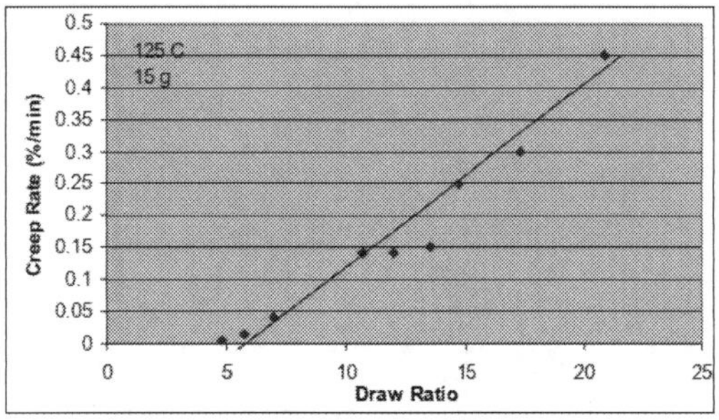

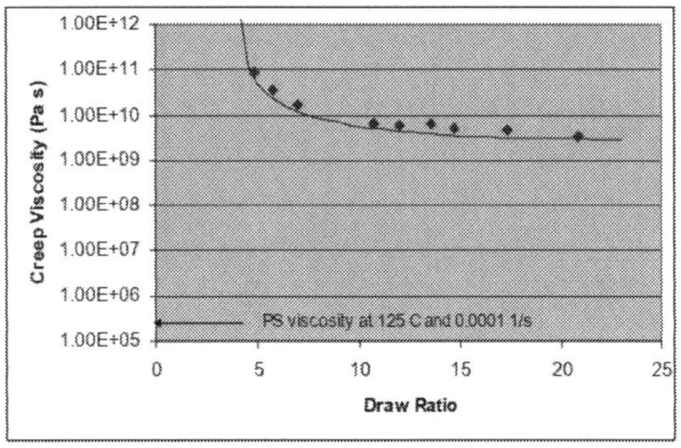

The brittle-to-ductile transition can be due to a number of factors, including changes in crystallinity, residual stress, and morphological differences in the films. We will examine these possible causes next.

Morphology

The micrographs from the film cross sections show that the continuous phase is PP (dark) and the discrete phase is PS (light). The thickness of either phase decreases with draw ratio but is insensitive to extrusion speed.

$$D_{PS} = 0.3\exp(-DR/20)$$

$$D_{Pp} = 0.65\exp(-DR/40)$$

At a draw ratio of 7:1, we have the PS phase of about 0.2 micron and the PP phase of about 0.55 micron.

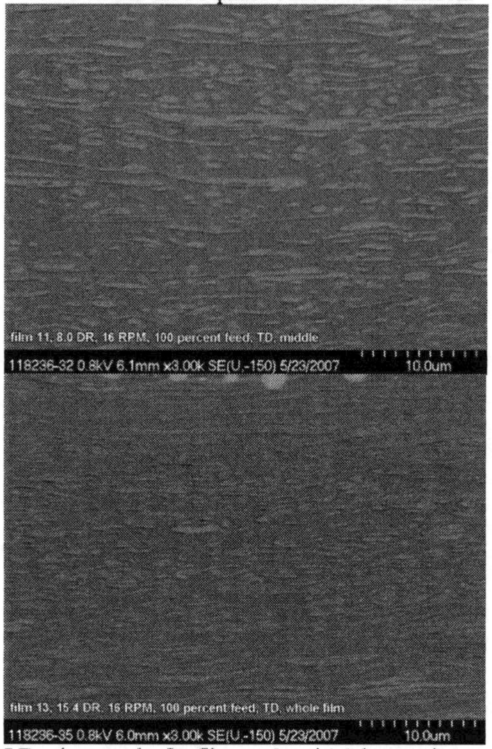

LD micrographs for films at 8 and 16 draw ratio.

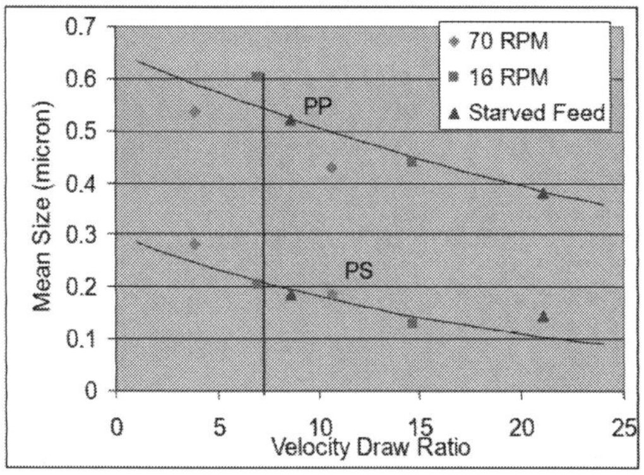

Cooling Rate Effects

Cooling rate will also affect the structure and properties of the blend to give the observed brittle-to-ductile transition. Neglecting the enthalpy of melting of PP and heat transfer to the rubber roll, we can estimate mean temperature upon cooling as

$$T = T_{cool} + (T_{melt} - T_{cool})\exp(-t/\tau_{cool})$$

$$\tau_{cool} = \frac{H^2 \rho_{solid} C_p}{k}$$

were thermal diffusivity ($k/\rho C_p$) of the blend is 0.0008 cm^2/s. We see that the cooling time is less than 1 second for most of the film studied.

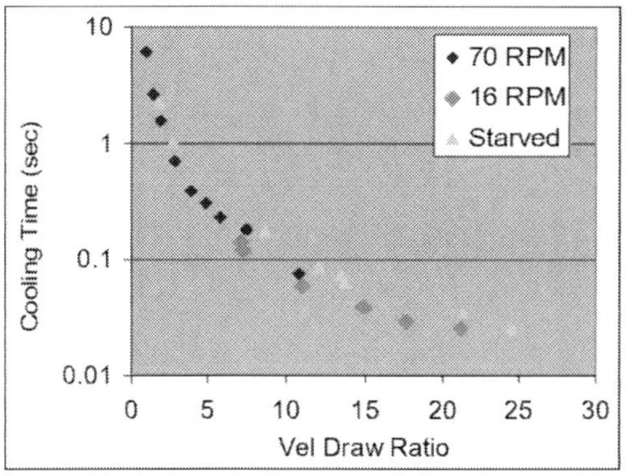

The degree of crystallinity X_c in the PP phase was determined by WAXS. We see that X_c correlates very well with cooling time. The curve was found by fitting the Avrami Equation

$$X_c = X_{max} \phi_{PP} \left(1 - \exp(-K t_{cool}^m)\right)$$

where X_{max} is known to be 0.63 for PP, volume fraction ϕ_{PP} is 0.56, and the Avrami exponent m is found to be 0.5, consistent with crystal growth that is simultaneously nucleated into rods with diffusion limitations. This condition seems reasonable given the orientation and rapid cooling of the film by the chill roll.

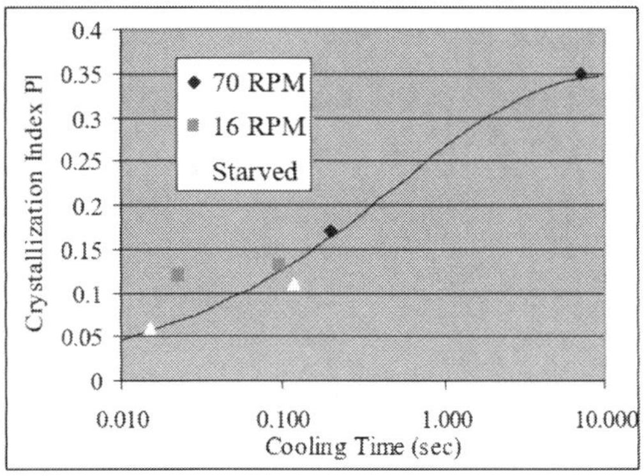

It is also noted from the WAXS results that mesomorphic (smectic) PP was observed for any film with a draw ratio larger than 7:1. The reduction of crystallinity and the dominance of mesomorphic PP can give rise to the ductile behavior at high draw ratios.

To probe changes in polystyrene, we observe with the DSC data the size of an exothermal peak at the glass transition temperature (100 C) upon the first heating. This exotherm results from free volume relaxation upon reheating of rapidly quenching films. No exotherm was detectable upon the second heating cycle in the DSC. It is possible that excess free volume may also have a role in the brittle-to-ductile transition.

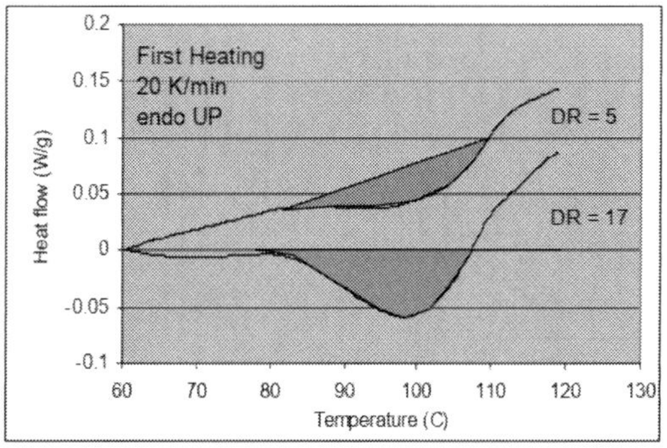

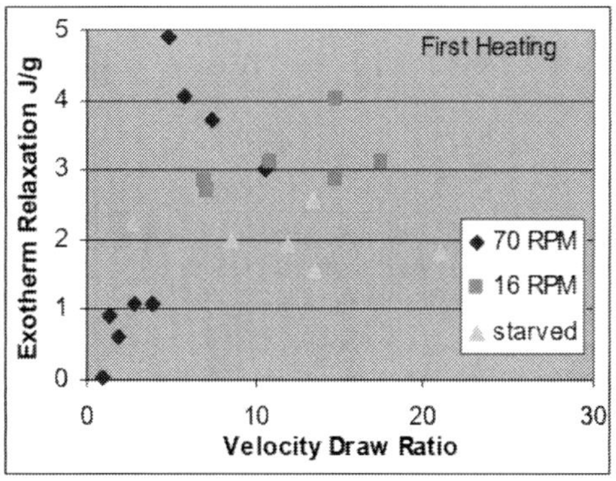

A Model of Creep Behavior

A model for creep considers the L/D ratio of rods a discrete phase separated by a matrix of a second material. The shear stress on the rod surface is

$$\tau_s = \eta_m \dot{\gamma}_m$$

where η_m is the matrix viscosity and the matrix shear rate is the total stretch rate multiplied by the mean length-to-thickness ratio of the overlapping rods.

$$\dot{\gamma}_m = \frac{\Delta v_d}{\delta_m} = \frac{\dot{\varepsilon} L_d}{2\delta_m}$$

The stretching force on the matrix is the sum of shear force on the dispersed phase and elongational force on the matrix,

$$F = \tau_s N_d A_d + 3\eta_m \dot{\varepsilon} A (1 - \phi_d)$$

where A_d ($=\pi\, \delta_d\, L_d$) is the mean surface area of the filaments and N_d is the number of filament per unit cross sectional area, found from the area fraction ϕ_d.

$$N_d = \frac{4A\phi_d}{\pi \delta_d^2}$$

Combining these expressions, we have

$$\eta_{eff} = \eta_m \left(f\phi_d + 3(1 - \phi_d)\right)$$

$$f = \frac{2\, L_d^2}{\delta_d\, \delta_m}$$

Ignoring the second term, we have

$$\eta_{eff} = \eta_m \frac{2L_d^2}{\delta_d \delta_m} \phi_d$$

where δ_d is the thickness of the PS phase, δ_m is the mean thickness of the PP phase, and L_d is the effective length of the PS rods. Fitting values of L_d to values of η_{eff}, we see a decreasing function with draw ratio. We need independently measure particle length for verification in further work.

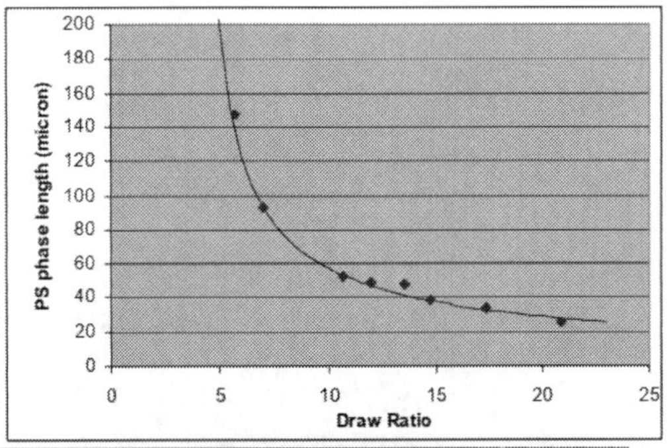

Conclusions

We have observed a room temperature brittle-to-ductile transition in extruded immiscible blend films with a draw ratio greater than 7:1. We also see an onset of high temperature creeping flow at similar films. Three potential causes of this are

1) the decrease in phase thickness and length,
2) the increase in cooling rates at high draw ratio results in a reduction crystalline structure, and
3) excess free volume in the amorphous phase.

At present, we have no clear evidence about which of these mechanisms may result in the observed ductility. In future work we will try to investigate this more by comparing properties of thin films of PP and PS only to the properties of the PP/PS blend.

Acknowledgements

This work was done under the IPrime Consortium as an Industrial Fellowship. DSC data was by Ron Dehn, extrusion was performed by Tom L. Wood, SEM images were by Mary Swierczek of 3M, and WAXS was by Myles Brostrom all of 3M.

VISCOELASTIC CHARACTERIZATION OF NOVEL THERMOPLASTIC POLYURETHANES

Rafael E. Camargo, Chris I Lindsay, Lifeng Wu and Tianshen Hu

Huntsman International LLC, CoreScience Group ,8600 Gosling Road, The Woodlands, TX 77381
rafael_e_camargo@huntsman.com

Abstract – Thermoplastic polyurethanes, TPUs, are a unique group of elastomers that can be tailored for a number of applications by suitable combination of the starting monomers and intermediates. In recent years, new concepts in TPUs have emerged using intermediates not normally used such as supramolecular structures or highly unique amine terminated chain extenders. These new TPUs offer unique property or processing advantages such as low melt viscosities or highly elastic behavior. In this paper, we have examined the viscoelastic response of two new TPU families using dynamic mechanical analysis. For the TPUs based on supramolecular architecture there exists a unique transition that leads to a rapid decrease in the melt viscosity as the supramolecular structures are broken up. This transition is likely caused by chain dissociation. This is confirmed by the unique results in a frequency sweep configuration. The relaxation behavior of the materials does not follow the classical time-temperature superposition behavior of conventional polymers. Dynamic mechanical thermal analysis of the new polyurethane-urea TPUs indicates a high degree of microphase separation characterized for a sharp soft segment glass transition and an extremely flat rubbery plateau. Most conventional TPUs are characterized by a continuous drop in storage modulus after the soft segment Tg. This behavior is caused by a distribution of hard domains that are poorly held together by weak chain-chain interactions. The presence of an amine chain extender introduces much stronger interactions via hydrogen bonding that are likely responsible for the stable value of the storage modulus in these new TPUs.

INTRODUCTION

Thermoplastic polyurethanes, TPUs are widely employed as high performance elastomers due to their unique combination of elasticity, mechanical performance and adhesive properties, as well as ease of processability in a variety of equipment [1]. Due to the broad formulation flexibility, the number of applications is very large and it includes injection molded parts, extruded film, extrusion profiles, cable and wire coating, slush molding and many others. In the broadest context, a TPU is the reaction product of three main types of monomers: a diisocyanate, aromatic or aliphatic, a short chain diol or extender, and an oligomeric higher molecular weight diol or polyol. Different products are typically the result of combining different materials in different proportions. For instance, if the polyol is a polyester diol, TPUs of good toughness and resistance to oil and many hydrocarbons can be obtained. One of the most common isocyanates used in the production of TPUs is 4,4'-diphenyl methane diisocyanate, MDI, while a common extender is 1,4 butane diol, BDO. The elastic behavior of TPUs, as it is commonly accepted, is derived from the two phase morphology that results from the incompatibility of the MDI-BDO sequences and the polyol chains. The MDI-BDO sequences, know as hard segments, tend to associate and form virtual crosslink points that interact with the polyol chains producing the elastomeric behavior. These materials are schematically represented in Figure 1.

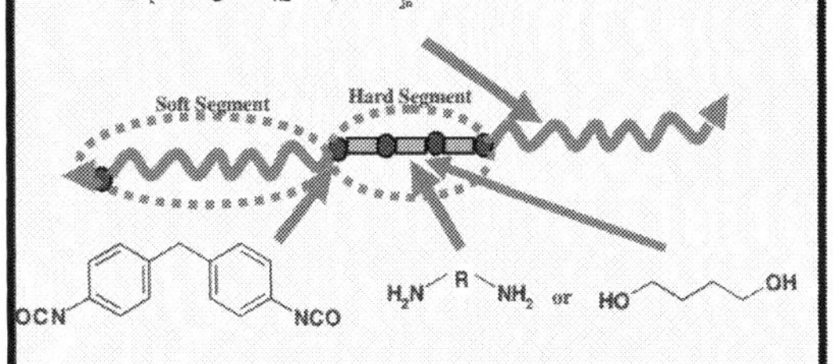

Figure 1: Schematic representation of an aromatic TPU elastomer

Two kinds of TPUs have been studied in this work: supramolecular TPUs and amine extended TPUs. Both materials are unique compared to the standard TPUs described earlier. Viscoelastic measurements are used to understand the behavior and unique structure of these materials.

EXPERIMENTAL

Initial rheological measurements were done in a stress-controlled AR-2000 rheometer (TA Instruments). Three types of tests were conducted: a temperature ramp at a constant frequency of 1 Hz. Relaxation behavior was characterized using a dynamic frequency sweep at multiple temperatures to develop a master curve using the time-temperature superposition principle[2]. Finally a time sweep at a given temperature was used to study the kinetics of crystallization of some of these materials.

RESULTS

a. Supramolecular-type TPUs:

The concept of the supramolecular TPU is best explained using the diagram shown in Figure 2. Certain links in the flexible chains can be disrupted by taking the material above a certain transition temperature. Upon cooling the bonds are reformed. A main objective of this study was to study the relaxation behavior of these unique materials and compare it to the one or more conventional TPUs.

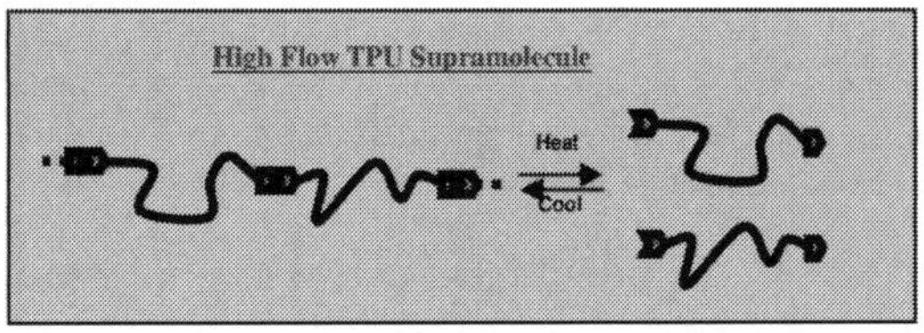

Figure 2: Principle of a supramolecular TPU

Figure 3 shows the temperature sweep, DMTA, of a supramolecular TPU and its conventional counterpart. Both materials show a major transition occurring at approximately 50°C. This is due to the melting of soft segment crystallites since the polyester used in these particular case are readily crystallizable. At approximately 100°C a second drop in storage modulus can be observed. This new transition is very pronounced in the supramolecular TPU. At about 160°C the storage modulus of the conventional TPU is almost two orders of magnitude larger. Melt viscosity measurements demonstrate that the viscosity of the conventional material is higher.

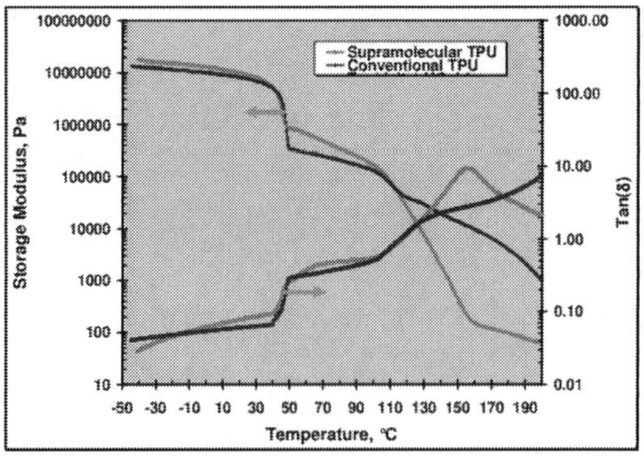

Figure 3: DMTA comparison of conventional and supramolecular TPU.

Figure 4 shows the relaxation behavior (G') of a conventional TPU using 150°C as the reference temperature. As expected, the time-temperature superposition allows the construction of a single curve over approximately 4 frequency decades. In contrast, Figure 5 shows the relaxation behavior for the supramolecular-type TPU. The relaxation behavior can not be explained by one single curve. This is caused by chain dissociation that is taking place over the test temperature range. The sample illustrated in this case had a different overall composition but the behavior was similar for all samples.

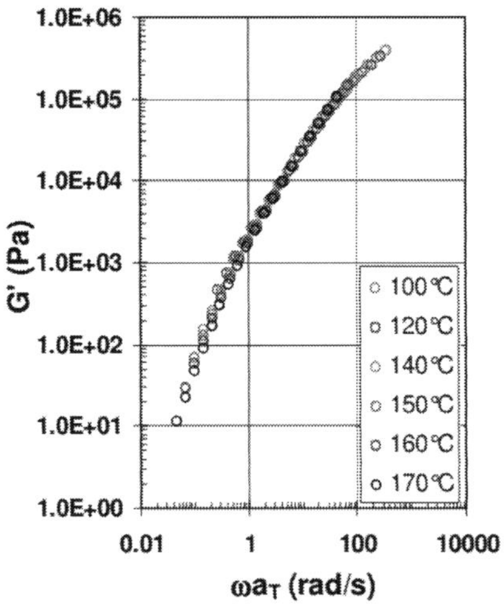

Figure 4: Relaxation behavior of a conventional TPU. Reference temperature is 150°C

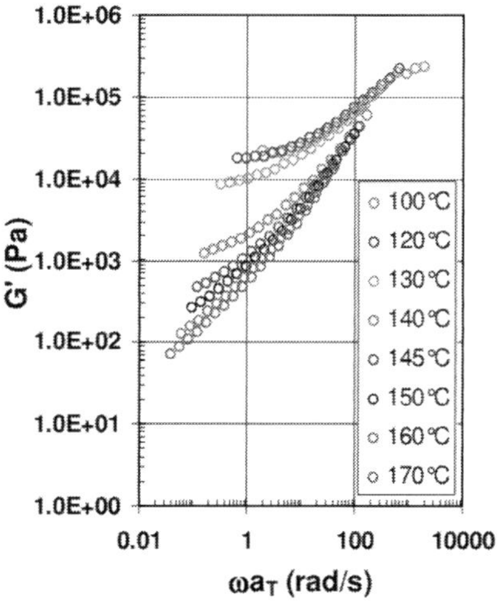

Figure 5: Relaxation behavior of a supramolecular TPU. Reference temperature is 150°C.

The solidification behavior was studied following the modulus development of a particular sample at a chosen annealing temperature. Samples were heated to 150° and rapidly cooled to the selected temperature. It was found, as shown in Figure 6 that the modulus development follows an Avrami type equation with the maximum rate of modulus development occurring at around 40°C for the sample shown.

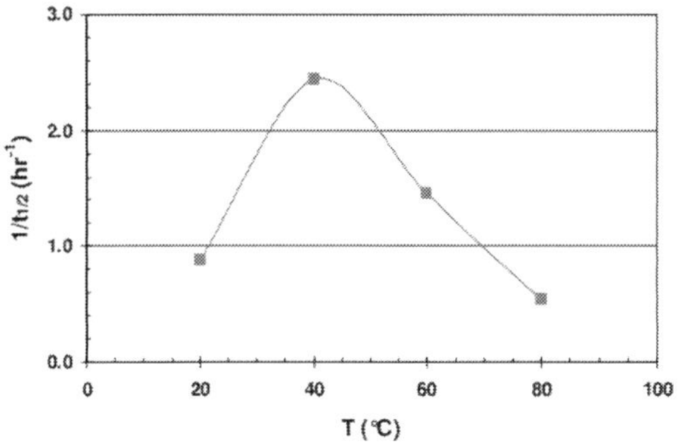

Figure 6: maximum rate of modulus development after heating to 150°C and rapidly cooling.

b. <u>Polyurea type TPU</u>

A second illustration of novel TPU concepts is the development of a polyurethane-urea TPU using a new proprietary diamine. While diamines and polyurea hard segments are known to provide very good properties, in practice the use of diamines in TPU manufacturing has been restricted by the fast reactivity of these materials. This reactivity leads to premature phase segregation and this results in a polymer of low molecular weight and poor properties. With the the new diamine chain extender it is possible to produce TPUs of high molecular weight, good mechanical properties and an excellent degree of phase segregation as demonstrated by the temperature ramp shown in Figure 7. This technology has enabled the production of TPU's of low modulus (and low hardness) without the use of plasticizers. As shown in Figure 7, when one compares these new TPUs with a conventional diol extended TPU, one can observe a much lower soft segment glass transition temperature, Tg, , a sharper transition and a much flatter rubbery plateau demonstrating the effectiveness of the polyurea soft microdomains to act as virtual crosslinks in these materials.

SUMMARY

In this work it has been shown using a number of viscolelastic measurements that, through the use of new intermediates it is possible to synthesize TPUs that behave significantly different from conventional TPUs.

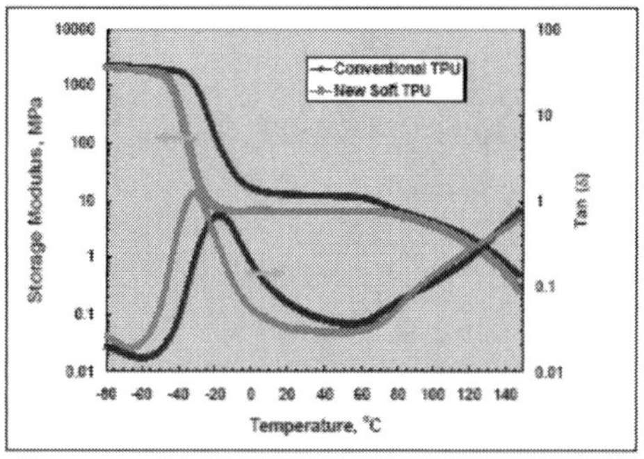

Figure 7: Comparison of a conventional TPU to a new polyurethane-urea TPU.

References:

[1] W. Meckel et al. , 1987, "Thermoplastic Polyurethane Elastomers", in *Thermoplastic Elastomers: A Comprehensive Review*, N. R. Legge, G. Holden, and H. E. Schroeder, eds., Hanser Publishers, New York.

[2] J. D. Ferry., 1980, *Viscoelastic Properties of Polymers*, Wiley, New York.

POLYMERIC MULTILAYER FILMS FOR MANUFACTURE OF A COMPOSITE DENTAL DEVICE

T. D. Jones[1]*, G. S. Deeb[1], N. Karim[1], S. Mitra[1], J. Yang[1], B. Craig[1], and D. Jacobs[1]

[1]3M Company, 3M Center, Bldg 260-2B-12, Saint Paul, MN, USA, 55116 – tdjones@mmm.com, gsdeeb1@mmm.com, nkarim@mmm.com, sbmitra@mmm.com, jyang3@mmm.com, bdcraig@mmm.com, dwjacobs@mmm.com

Recently a malleable, preformed, photocurable composite crown system has become available for the dental practitioner as a prosthesis to restore extensively damaged teeth. This single-visit crown (available commercially as Protemp™ Crown, 3M ESPE Dental Products) can be used for extended temporization, up to a year or more in the mouth. These crowns are manufactured using a clay-like highly-filled methacrylate-based dental composite with unusual wax-like rheology. These novel rheological characteristics are achieved through a combination of semi-crystalline oligomers and high surface area fillers. These same rheological characteristics which provide Protemp Crown's novel handling create unique challenges in the manufacture of a molded complex organic shape, specifically a replica tooth. In order to address these challenges, 3M has developed a family of multilayer polymer films for use as moldable release liners. These films, consisting of a thermoplastic skin and an elastomeric core, allow for high elongation with low recovery force. In turn, these properties permit the manufacture of malleable objects with accurate reproduction of complex organic features.

Introduction

Temporization is a critical element of restoring a tooth with a lab-fabricated crown, as is done for teeth with extensive damage or decay. Temporization creates an interim restoration, necessary for maintaining space between adjacent and opposing teeth and protecting the tooth during the time interval while a permanent restoration is prepared, typically of cast metal or ceramic (Christensen, 1996). In order to meet the dental practitioner's needs, materials for temporization need to be durable, but moreover very fast and easy to use.

Traditionally, temporization has been carried out by one of two methods: modifying and fitting a preformed metal alloy shell crown to a patient, or creating a more aesthetic replica by mixing a two-component system in a negative mold, or impression, of the tooth to be restored. Both of these methods suffer from shortcomings in practice. Due to their limited customizability, a large inventory of metal crowns must be maintained. Furthermore, metal alloy crowns are inherently unaesthetic. In contrast, two-component crowns require a smaller inventory of material, but require the patient to endure the creation of the needed negative mold. These materials are also limited in their mechanical properties, as they must be able to flow through a static mix tip under manual pressure to allow for the curing chemistry to take place.

Recently, 3M-ESPE has developed a novel preformed tooth-colored malleable composite crown, called Protemp Crown™, which can be easily shaped by the dentist for an individual patient and cured after customization. The composite material consists of a photocurable dimethacrylate resin matrix including a semicrystalline polycaprolactone dimethacrylate oligomer, with a filler mixture of fumed silica and a silica/zirconia micron-scale ceramic. These composite crowns have a clay-like consistency, enabling a much greater range of customization than can be achieved with an alloy crown. The dentist needs a much smaller inventory of crown sizes to cover clinical indications. The crowns do not require a negative mold of the tooth, and have superior post-cure properties to two-component systems.

The uncured state and rheological properties of this material significantly limit the conditions under which it can be processed. The crowns are formed by a molding process, wherein a conformable release liner covers the interior of the mold. This release liner must have high elongation in order to match the exterior dimensions of the crown, without forming perforations that could allow contact between the sticky uncured composite and the mold. It must also have low recovery stresses, in order to avoid deformation of the crown after molding is complete. The film must release cleanly from the mold itself after the part is formed. This set of requirements becomes particularly challenging in the case of tall, thin crowns, as is the case for anterior teeth i.e. maxillary (upper) and mandibular (lower) incisors.

While a single polymer was not found that could provide this mix of characteristics for anterior crowns, a range of films were designed to meet this need through a multilayer construction. Multilayer films have long been considered as a method of combining desirable properties from multiple materials (Schrenk, 1984), and are widely used today in the packaging industry. By combining the properties of a thermoplastic with an elastomer in a multilayer construction, the extensional properties, recovery properties and surface contact properties of the film can be adjusted to match the application need.

Experimental

Materials

A photocurable dental composite (3M, 2009) was prepared from 2,2-bis[4-(2-hydroxy-3-methacryl-oyloxypropoxy)phenyl]-propane (bisGMA, Aldrich), Cabosil™ M-5 fumed silica (Cabot Corp), a dimethacrylate functional polycaprolactone synthesized by reacting two equivalents of 2-isocyanatoethyl methacrylate with one equivalent of TONE-0230™ (Dow Chemical), and a silica-zirconia filler of average particle size 0.6 to 0.9 micrometers treated with 3-methacryloxypropyltrimethoxysilane.

Polymers used in this study to prepare core-shell films are summarized below in Table 1, and were obtained from the indicated manufacturers.

Table 1 - Table of Polymeric Film Materials

Material	*Supplier*	*Description*	*Placement in Construction*
Kraton G1657	Kraton™	S-EB-S block copolymer	core
Vector 4111	Dexco™	S-I-S block copolymer	core
Kraton 1106	Kraton	S-I-S block copolymer	core
Engage 8200G	Dow Chemical	Ultra low density polyethylene	core
PP 3445	Exxon™	Isotactic polypropylene	skin
PP 7C12N	Dow Chemical	PE-PP impact copolymer	skin
PE 4023	Exxon	Ethylene-based plastomer	skin
M6060	Equistar	High density polyethylene	skin
LD728.61	Exxon	Poly(ethylene-co-vinyl acetate), 19% vinyl acetate (EVA copolymer)	Comparative single layer film

Multilayer Film Extrusion

Multilayer films were prepared as A-B-A constructions, having a thermoplastic skin and an elastomeric core. Skin materials were conveyed through a single screw extruder (19 mm diameter, L/D of 32/1) having multiple temperature zones to a gear pump and then to the "A" and "C" channels of a Cloeren™ feedblock. Core materials were conveyed through a single screw extruder (32 mm diameter, L/D of 24/1) having multiple temperature zones to a gear pump and then to the "B" channel of the Cloeren feedblock. The multilayered flow stream passed through a single orifice 254 mm film die, drop cast onto a chill role, and collected. Temperatures and flow rates were adjusted to obtain the desired film thickness and core/skin ratio.

Test Methods

Film layer thickness was measured using an optical measuring method, available as LayerGauge™ from Davinor™ Corp. Film elongational properties were measured on an Instron™ tensile tester model 5500R, using variations on ASTM D882. A test method was prepared to mimic the extensional processes occurring during molding of an anterior crown. A film was extended to an engineering strain of 500% at 6.35 mm/s, followed by a hold at maximum extension of 10 s. The film then recovered in a mixed creep/stress relaxation mode at 4.23 mm/s to 0% strain, followed by a second stretch to 600% engineering strain, again at 4.23 mm/s. After a second hold for 10 s at maximum extension, the film again recovered in a mixed creep/stress relaxation mode at 4.23 mm/s to 0% strain. Properties of interest included the survival of the film to maximum extension at each stage, the tensile load at maximum extension, the extension (elastic recovery) at 2 N of tensile recovery load, and the extension (elastic recovery) at ~0 N of tensile recovery load (where 0 was defined as less than 0.044 N)

Paste Forming

In addition to measuring the tensile properties of the film, the success of the film in crown forming was also evaluated more directly. In a first test, a known volume of composite was forced against a sample film into a simple bullet-shaped mold, having a target height of ~11.5 mm, such that the dental composite was contained within a pocket formed of the sample film. The ratio of the final surface area of this pocket to the initial unstretched area of the film creating the pocket was 6.87:1. The film was then evaluated for punctures, tears, or perforations by visual examination. If a film passed this first test, a known volume of composite was created in a film pocket. This composite was then pressed into the shape of a maxillary incisor crown, having a target height of 13.1 mm. The ratio of the surface area of this crown to the initial unstretched surface area of the film was 6.49:1. Three to six crowns were prepared in each film and evaluated by visual inspection on both film condition and crown shape. Crowns were evaluated as Pass (no visual defects), Marginal (minor defects visible which did not render the crown unusable), or Fail (unacceptable gross defects). A second known volume of composite was then created in a film pocket, and pressed into the shape of a mandibular incisor crown. The ratio of the surface area of this crown to the initial unstretched surface area of the film was 8.38:1. Again, three to six crowns were prepared and evaluated in each film as described above.

Results and Discussion

Multiple combinations of core and skin layer thickness were evaluated for a number of polymer pairs, selected from the list of materials in Table 1. Illustrative examples are included in Table 2, following. Tensile properties of these films are described in Table 3.

Table 2 - Film Constructions

#	*Core*	*Thickness (μm)*	*Skin*	*Thickness (μm)*
1	Vector 4111	199	PP 7C12N	4.2
2	Kraton 1106	88.9	PP 3445	5.5
3	Vector 4111	141.6	PE 4023	18.1
4	Kraton G1657	39.7	PP 3445	12.6
5	Engage 8200G	66.3	PE M6060	6.4
6	Engage 8200G	75.1	PE M6060	12.5
7	Engage 8200G	66	PP7C12N	6
8	LD728.61 (single layer)	76.2	n/a	n/a

Table 3 – Film Extensional Properties

#	*Max Extension (%)*	*Max Load (N)*	*% recovery (2 N)*	*% recovery (~0 N)*
1	600%	16.1	81.2	94.7%
2	600%	15.3	9.3	77.0%
3	600%	15	63.8	91.0%
4	534%	26.1	n/a	n/a
5	600%	20.1	12.5	35.0%
6	600%	22	13.3	53.3%
7	566%	27.9	n/a	n/a
8	378%	37.8	n/a	n/a

Table 4 – Incisor Crown Forming Results

#	*Maxillary (Upper) Incisor*			*Mandibular (Lower) Incisor*		
	Film Survival	*Crown Quality*	*Height (mm)*	*Film Survival*	*Crown Quality*	*Height (mm)*
1	Pass	Fail	10.97	Pass	Fail	9.54
2	Pass	Pass	11.85	Fail	Fail	n/a
3	Pass	Fail	n/a	Fail	Fail	n/a
4	Fail	Fail	n/a	Fail	Fail	n/a
5	Pass	Pass	11.84	Pass	Pass	10.24
6	Marginal	Marginal	12.12	Pass	Pass	10.24
7	Marginal	Marginal	11.94	Fail	Fail	n/a
8	Fail	Fail	n/a	Fail	Fail	n/a

While a number of these films would appear to be promising candidates, the evaluation of the films in forming of the two challenging anterior crown shapes reveals the film requirements for this application are more extensive, as shown above in Table 4.

Previous experiments have demonstrated that a single-layer EVA copolymer film (sample #8) is capable of performing adequately for less challenging crown shapes, such as molar crowns. Traditional polyolefins, such as polyethylene or polypropylene, tend to tear during the forming process, despite their relatively high elongation to break. Tears in the film permit the uncured paste to contact the mold, leading to adhesion between the crown and mold and damage upon mold separation. Pure elastomers, such as Kraton 1106, or even VLDPE, such as Engage 8200G, have too much elastic recovery: although the films survive the crown forming process, after removal from the mold the residual stress in the film extensively deforms the uncured molded part. In addition, elastomers can themselves adhere to the mold, creating additional distortion of the formed part on mold separation.

By combining the properties of thermoplastic and elastomeric components in a multilayer construction, success in molding can be achieved. Placement of a thermoplastic in the A position on an A-B-A film enables successful mold separation. An elastomeric core prevents film tearing, but its recovery stresses can be balanced by the low recovery of the thermoplastic skin. Thus, while sample 4 demonstrates too much thermoplastic character and demonstrates insufficient elongation, samples 1 and 3 are too elastomeric, and excessively distort the crown. Samples 5 and 6 demonstrate an appropriate balance of thermoplastic and elastomeric character for this application.

Conclusions

The design of a mold release liner for creating an un-cured, malleable composite crown requires careful balancing of material properties, which has not been achieved in a single layer polymer film. By appropriate choice of core and skin layers and constructions, films can be achieved with a balance of extensional and recovery properties that permit the successful manufacture of these novel dental articles.

References

3M Company, "Protemp Crown Technical Product Profile" (2009).

Christensen, G. J.,"Provisional Restorations for Fixed Prosthodontics", J. A. D. A., **127**, 249–252 (1996)

Schrenk, W. J., "Multilayer Films", in Encyclopedia of Polymer Science and Engineering, **7**, p. 106 (1984)

Maleic Anhydride Copolymers via Controlled Radical Polymerization

B. Lessard[1] and M. Maric[1,*]

[1]*Department of Chemical Engineering, McGill University, Montreal, Canada – benoit.lessard@mail.mcgill.ca; milan.maric@mcgill.ca*
** corresponding author*

Abstract - Maleic anhydride functional copolymers have long been used as compatibilizing agents in reactive blending, as thickening agents and as adhesion promoters. Traditionally, such copolymers are produced by free radical polymerization and are characterized by broad molecular weight distributions and often limited microstructural control. The advent of controlled free radical polymerization processes such as nitroxide mediated polymerization (NMP) has enabled the possibility of synthesizing such polymers without excessive reagent purification or functional group protection while retaining the simplicity of conventional free radical processes. This enables not only control of molecular weight distribution essential for processing but also permits the production of block copolymers. These characteristics widen the possibilities of using such functional copolymers to develop much more tailored blend morphologies. Earlier, NMP was limited to being performed at relatively high temperatures (> 120 °C) using first-generation mediating compounds such as TEMPO. These relatively high temperatures made the attainment of strictly alternating sequences of styrene (ST) with maleic anhydride (MA) impossible for all feed compositions. Strictly alternating sequences can only be achieved at much lower temperatures ~ 80 °C. Recently developed alkoxyamines based on the SG1 family of nitroxides (BlocBuilder, Arkema) has permitted the polymerization of several monomers and polymerization at much lower temperatures that were not initially possible with TEMPO. We show that strictly alternating ST/MA sequences are possible by NMP using BlocBuilder if low temperatures of 80-90 °C are used. Indeed, block copolymers of poly(ST-*alt*-MA)-poly(ST) with narrow molecular weight distribution (polydispersity index ≈ 1.3) were formed from a feed of 20 mol% MA in a single step as confirmed by ^{13}C NMR measurements. Possibilities for using ST/MA copolymers produced via SG1-mediated NMP as compatibilizers are illustrated specifically using poly(ST-*alt*-MA)-poly(ST) copolymers or poly(ethylene-*ran*-butylene) rubber terminated with ST/MA sequences.

Introduction

Incorporation of maleic anhydride (MA) via copolymerization or grafting onto polymers has made such resins valuable for blend compatibilization and adhesion promoters (Jeon et al., 2004). Although possessing the requisite functionality, such resins are typically characterized by broad molecular weight distributions and a lack of microstructural control (i.e. precise placement of MA along the chain), with the exception of sophisticated strategies to place MA groups at a chain end (Cernohous et al., 1997). Recently, controlled radical polymerization (CRP) has enabled the synthesis of polymers with narrow molecular weight distributions and ability to control microstructure (Matyjaszewski and Davis, 2002). Further, functional groups such as MA can be directly incorporated into the resin without need of tedious protection strategies required previously for traditional controlled polymerizations.

It has long been known in conventional radical polymerization that MA copolymerizes with monomers such as styrene (ST) in an alternating fashion. Such an alternating copolymerization scheme in CRP is of interest since a one-step block copolymerization is feasible where the first block would consist of a poly(ST-*alt*-MA) block followed by a pure poly(ST) (Benoit et al., 2000). The production of such a block copolymer by CRP was presented although the ST-*alt*-MA sequence was not precisely alternating due to the high temperatures (120 °C) required for the particular variant of CRP used (nitroxide mediated polymerization – NMP). Of the CRP processes, reversible addition fragmentation transfer (RAFT) has been able to synthesize strictly alternating ST-MA sequences since RAFT can be performed at temperatures below which strictly alternating sequences can be obtained (< 80 °C) (Chernikova et al., 2003). However, alkoxyamine unimolecular initiators based on the SG1 family of nitroxides (BlocBuilder, Arkema) has expanded the possibilities for NMP applications. For example, polymerizations can be done at much lower temperatures (80-90 °C), thus making it possible to produce poly(ST-*alt*-MA)-*block*-poly(ST) in one step using BlocBuilder initiator (Lessard and Maric, 2010). The following study examines the copolymerization of ST with MA using BlocBuilder in addition to adding ST-MA sequences onto suitably BlocBuilder functionalized poly(ethylene-*ran*-butylene) liquid rubbers.

Experimental

Materials. A list of all materials for the ST/MA copolymerizations using 2-({*tert*-butyl[1-(diethoxyphosphoryl)-2,2-dimethylpropyl]amino}oxy)-2-methylpropanoic acid, also known as BlocBuilder (Arkema) (99%) {*tert*-butyl[1-(diethoxyphosphoryl)-2,2 dimethylpropyl]amino}oxidanyl, also known as SG1 (>85%) as the initiator and additional free nitroxide, respectively, can be found elsewhere (Lessard and Maric, 2010). Hydroxy terminated poly(ethylene-*co*-butylene) (PEB-$(OH)_2$, Kraton D2205) was obtained from Kraton Inc. while acryloyl chloride (98%) was acquired from Aldrich.

Synthesis of Poly(styrene-ran-maleic anhydride) (P(ST-ran-MA)) and Poly(styrene-ran-maleic anhydride)-block-Poly(styrene) (Poly(ST-ran-MA)-block-Poly(ST)) Copolymers. Details for the synthesis of the ST/MA copolymers can be found elsewhere (Lessard and Maric, 2010).

Synthesis of BlocBuilder terminated Poly(ethylene-co-butylene)-(BlocBuilder)$_2$ (PEB-(BlocBuilder)$_2$). The conversion of the hydroxy terminated PEB (PEB-$(OH)_2$, M_n = 6 500 kg·mol^{-1}) to a BlocBuilder terminated PEB (PEB-(BlocBuilder)$_2$) was done by an esterification followed by an intermolecular radical addition. First, PEB-$(OH)_2$ was dissolved in a 1L round bottom flask with purified methylene chloride (CH_2Cl_2) at room temperature under a nitrogen purge. Then, an equivalent amount of acryloyl chloride was injected. The reaction was allowed to continue under nitrogen purge for 24 h. The crude product was then precipitated in ether and dried under vacuum. Conversion of the hydroxyl end groups to vinyl groups was checked using ^{1}H NMR. The acrylate-terminated PEB was then re-dissolved in toluene and an excess of BlocBuilder was added and allowed to react at 100 °C for 6 h. The mixture was then cooled and precipitated in ether to remove the unreacted BlocBuilder. The remaining solvent was removed by rotary evaporation and the resulting product was dried in a vacuum oven for an additional 24 h at 70 °C. The conversion of the acrylate-terminated PEB to PEB-(BlocBuilder)$_2$ was evaluated using ^{1}H NMR. The PEB-(BlocBuilder)$_2$ had $\overline{M}_n$ = 7 500 kg·mol^{-1} as determined by GPC relative to linear poly(styrene) standards in THF at 35 °C.

Synthesis of Poly(ethylene-co-butylene)-block-Poly(styrene-ran-maleic anhydride) (PEB-b-P(ST-ran-MA) block copolymers. PEB-*b*-P(ST-*ran*-MA) with varying MA content were synthesized from PEB-(BlocBuilder)$_2$ macroinitiators. Typically 1 g of PEB-(BlocBuilder)$_2$ was dissolved in 15 g of 1,4 dioxane solvent at a concentration of in a 100 mL three-necked round bottom flask attached to a condenser/chiller unit. The flask was placed inside a heating mantle that sat on top of magnetic stirrer plate. The thermal well connected to the reactor was attached to the temperature controller. Then, the desired amounts of ST and MA monomer were added with a total mass of 15 g. After chilling the condenser with an ethylene glycol/water mixture at 5 °C, the reactor contents were stirred and bubbled with a nitrogen purge for 30 minutes. Then, the temperature was increased to 110 °C at a rate of about 8 °C·min^{-1}. When the temperature reached 105 °C, this was taken as the initial time for the reaction. Samples were taken periodically from the reactor and precipitated into methanol to isolate the copolymer product. The samples were washed repeatedly with methanol before being placed in a vacuum oven and dried overnight for 24 h at 70 °C. The chain extension was characterized by gel permeation chromatography (GPC) relative to linear poly(styrene) standards. The molar composition of MA in the copolymer, F_{MA}, was determined using ^{1}H NMR.

Characterization. Conversion was determined gravimetrically. Molecular weights were measured using gel permeation chromatography (GPC, Waters Breeze) with tetrahydrofuran (THF) as the mobile phase with a flow rate of 0.3 mL·min^{-1}. The GPC was equipped with three Waters Styragel HR columns (HR1 with molecular weight measurement range of 10^2 – 5 x 10^3 g·mol^{-1}, HR2 with molecular weight measurement range of 5 x 10^2 – 2 x 10^4 g·mol^{-1} and HR4 with molecular weight measurement range 5 x 10^3 – 6 x 10^5 g·mol^{-1} and a guard column was used. The columns were heated to 40 °C during the analysis. The molecular weights were determined by calibration with linear narrow molecular weight distribution poly(styrene) standards, and the GPC used the differential refractive index (RI 2410) detector. ^{13}C NMR (resonance frequency of 75 MHz) was performed in a 300 MHz Varian Mercury spectrometer under similar conditions to those previously reported by Ha (1999). The copolymer samples were dissolved in *d*-acetone in 5 mm diameter Up NMR tubes, and the ^{13}C NMR characterizations were performed using a flip angle of 30°, a relaxation time of 5 s, and suppression of the NOE.

Results and Discussion

The ST/MA copolymerizations controlled by BlocBuilder and additional SG1 free nitroxide were characterized by acceleration in polymerization rate as the MA feed content increased. Figure 1 shows the scaled conversion of ln[$(1-X)^{-1}$], where X = total monomer conversion, versus polymerization time t at 110 °C. From the slopes of these kinetic plots, an estimation of the product for the two key parameters governing the NMP of the copolymerization can be estimated. The plot of ln[$(1-X)^{-1}$] versus t provides an apparent rate constant of k_p[P·] where k_p is the propagation rate constant and [P·] is the concentration of macro-radicals. The apparent rate constant can get converted to k_pK where K is the equilibrium constant between the dormant chains capped by the SG1 groups whose concentration is denoted by [P-SG1] and active radical chains whose concentration is denoted by [P·].

$$K = \frac{[SG1][P^{\bullet}]}{[SG1 - P]} \quad (1)$$

Assuming that the initial concentration of free nitroxide is sufficiently high so that its concentration [SG1] does not change much (i.e. [SG1] ≈ $[SG1]_0$) and that irreversible termination reactions are sufficiently low (both of these assumptions are generally satisfied in the early stages of the polymerization) so that [P-SG1] ≈ $[BlocBuilder]_0$, then Equation 1 becomes:

$$k_p[P^{\bullet}] = k_pK\frac{[SG1-P]}{[SG1]} \cong k_pK\frac{[BlocBuilder]_0}{[SG1]_0} \quad (2).$$

$$\therefore k_pK \cong k_p[P^{\bullet}]r \quad where\ r = \frac{[SG1]_0}{[BlocBuilder]_0}$$

Values for k_pK are listed in Table 1 and indicate that increasing the feed concentration of MA dramatically accelerates the polymerization rate. An order of magnitude increase was observed compared to styrene homopolymerization. Chernikova et al. (2003) observed a similar acceleration for ST/MA copolymerizations using RAFT.

Table 2 indicates the molecular weight distribution data for the various ST/MA copolymers synthesized at 110 °C. All the molecular weight distributions remained narrow ($1.18 < \overline{M}_w/\overline{M}_n < 1.31$) despite some polymerizations being effectively complete after only a few minutes of polymerization. Further, alternating ST/MA sequences could be obtained for non-stoichiometric feeds when the polymerization temperature was 90 °C (Lessard and Maric, 2010). Stoichiometric feeds at higher polymerization temperatures resulted in alternating ST/MA sequences.

Table 1 – Molecular Characteristics of Poly(styrene-*ran*-maleic anhydride) Copolymers Synthesized Using BlocBuilder and SG1 at 110 °C in 1,4 Dioxane Solution.

Experiment ID[a)]	$f_{MA,0}$ [b)]	r [c)]	k_pK [d)] (s^{-1})
SMA-1	0.100	0.048	$(1.0\pm0.1) \times 10^{-5}$
SMA-2	0.198	0.049	$(9.6\pm1.7) \times 10^{-6}$
SMA-3	0.298	0.054	$(2.4\pm0.4) \times 10^{-5}$
SMA-4	0.398	0.043	$(2.2\pm0.4) \times 10^{-5}$
SMA-4.5	0.447	0.050	$(8.7\pm1.0) \times 10^{-5}$
SMA-5	0.489	0.054	$(9.6\pm0.2) \times 10^{-5}$
SMA-5.5	0.556	0.051	$(8.6\pm1.0) \times 10^{-5}$

[a]Experimental identification (ID) is given by SMA-X where SMA = styrene (S) and maleic anhydride (MA) and the number abbreviation refers to the experiment number. [b] $f_{MA,0}$ is the initial molar feed composition of maleic anhydride. [c]Initial molar concentration ratio of SG1 free nitroxide to BlocBuilder initiator = r = $[SG1]_0$/ $[BlocBuilder^{TM}]_0$. [d]$\langle k_p \rangle$ is the average propagation rate constant $\langle K \rangle$ is the average equilibrium constant. The error bars for the $\langle k_p \rangle \langle K \rangle$ are derived from the standard errors for the slopes from the kinetic plots shown in Figure 1.

Table 2 – Molecular Weight Characterization of Poly(styrene-*ran*-maleic anhydride) Copolymers Synthesized Using BlocBuilder and SG1 at 110 °C in 1,4 Dioxane Solution.

Expt. ID[a)]	$t_{polym.}$ (min)	X [b)]	$\overline{M}_n$ [c)] ($kg \cdot mol^{-1}$)	$\overline{M}_w/\overline{M}_n$ [c)]
SMA-1	120	0.54	12.1	1.18
SMA-2	155	0.67	19.1	1.23
SMA-3	45	0.74	14.1	1.28
SMA-4	25	0.53	15	1.27
SMA-4.5	12	0.65	15.4	1.33
SMA-5	9	0.70	15.5	1.31
SMA-5.5	6	0.72	13.3	1.22

[a]Experimental identification (ID) is given by SMA-X where SMA = styrene (S) and maleic anhydride (MA) and the number abbreviation refers to the experiment number. [b]X is the conversion determined gravimetrically. [c]Number average molecular weight ($\overline{M}_n$) and polydispersity index ($\overline{M}_w/\overline{M}_n$) obtained by using gel permeation chromatography (GPC) with linear poly(styrene) standards in tetrahydrofuran at 40 °C.

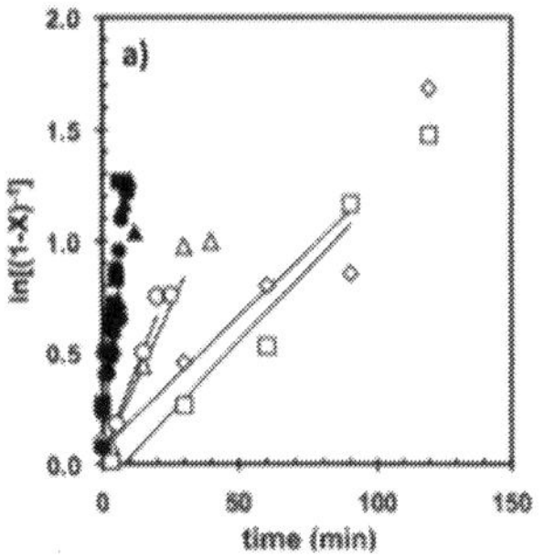

Figure 1 – Plot of $\ln[(1-X)^{-1}]$ (X = monomer conversion) versus polymerization time t for styrene/maleic anhydride (ST/MA) copolymerizations at 110 °C in 50 wt% 1,4 dioxane solution as a function of initial MA molar feed compositions $f_{MA,0}$ using a BlocBuilder/SG1 initiator/mediator system. The legend corresponds to the sample identification listed in Table 1 where $f_{MA,0}$ = 0.10 (SMA-1, □), $f_{MA,0}$ = 0.20 (SMA-2, ◇), $f_{MA,0}$ = 0.30 (SMA-3, △), $f_{MA,0}$ = 0.40 (SMA-4, ○), $f_{MA,0}$ = 0.45 (SMA-4.5, ▲), $f_{MA,0}$ = 0.49 (SMA-5, ●), and $f_{MA,0}$ = 0.55 (SMA-5.5, ■).

ST/MA copolymerizations were also initiated using PEB-(BlocBuilder)$_2$ macroinitiators at 110 °C in dioxane. Incorporation of BlocBuilder was clearly observed by ^1H NMR and chain extensions with pure styrene. The GPC chromatograms showing the PEB-(BlocBuilder)$_2$ initiator and the samples taken after styrene injection at various times are shown in Figure 2.

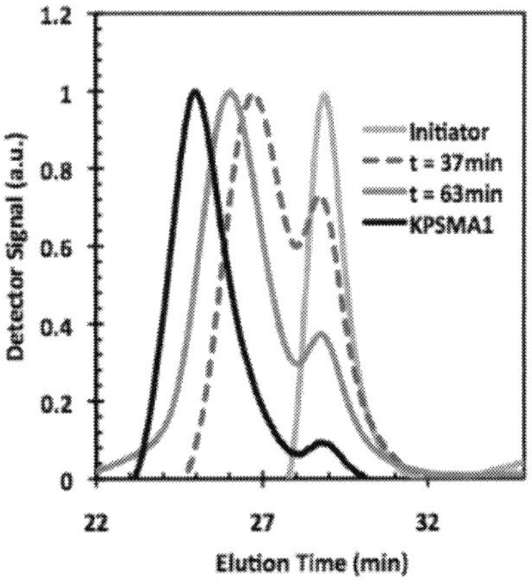

Figure 2 – Gel permeation chromatography (GPC) chromatograms of the PEB-(BlocBuilder)$_2$ macroinitiator shown at far right (solid, light grey line) and from samples taken after styrene was injected and polymerization at 110 °C at 37 minutes (dashed grey line), 63 minutes (solid, dark grey line) and final product after 191 minutes polymerization (solid black line).

The resulting product remains unimodal and there is only a small amount < 5% of unreacted presumable dead macroinitiator. The kinetic plot showing the effect of varying MA content in the feed is shown in Figure 3. The apparent rate constant $k_p[P\cdot] = 4.1 \times 10^{-5}$ s^{-1} is not dramatically different compared to that of a styrene homopolymerization at a similar temperature (Lessard, et al., 2007). However, when MA was incorporated with styrene into the feed initiated by PEB-(BlocBuilder)$_2$, the apparent rate constant dramatically increased in a similar fashion that when only BlocBuilder/SG1 were used as the initiator/mediator (Table 3). At a feed of 40 mol% MA, the solution rapidly became viscous at the reaction temperature of 110 °C as the monomer conversion increased rapidly to over 70% within a few minutes.

We are continuing our examination of ST/MA copolymerizations initiated by PEB-(BlocBuilder)$_2$ in terms of MA incorporation and determination of ST/MA sequence distribution by ^{13}C NMR. Further work is also being performed to examine if lower temperatures could also be used for this set of polymerizations. Regardless, using BlocBuilder/SG1 or a BlocBuilder terminated macroinitiator to initiate ST/MA copolymerizations yielded similar kinetic behavior.

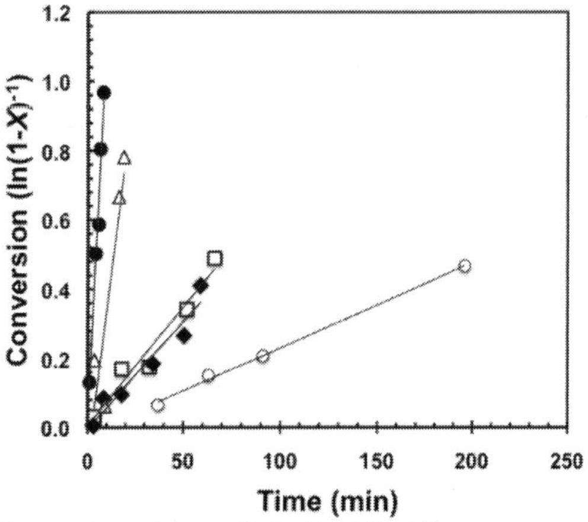

Figure 3 - Plot of $\ln[(1-X)^{-1}]$ (X = monomer conversion) versus polymerization time t for styrene/maleic anhydride (ST/MA) copolymerizations initiated by PEB-(BlocBuilder)$_2$ macroinitiator at 110 °C in 50 wt% 1,4 dioxane solution as a function of initial MA molar feed compositions $f_{MA,0}$. The legend corresponds to the sample identification listed in Table 3 where $f_{MA,0}$ = 0 (PEB-SMA-0, ○), $f_{MA,0}$ = 0.10 (PEB-SMA-1, □), $f_{MA,0}$ = 0.20 (PEB-SMA-2, ◆), $f_{MA,0}$ = 0.30 (PEB-SMA-3, △), $f_{MA,0}$ = 0.40 (PEB-SMA-4, ●).

Table 3 – Kinetic Parameters for Styrene/Maleic Anhydride Copolymerizations Initiated by a Poly(ethylene-*co*-butylene)-(BlocBuilder)$_2$ Macroinitiator at 110 °C in 50 wt % 1,4 Dioxane Solution.

Expt. ID[a]	$f_{MA,0}$ [b]	$k_p[P\cdot]$ [c] (s^{-1})
PEB-SMA-0	0	$(4.1\pm0.2) \times 10^{-5}$
PEB-SMA-1	0.100	$(1.1\pm0.2) \times 10^{-4}$
PEB-SMA-2	0.199	$(1.0\pm0.1) \times 10^{-4}$
PEB-SMA-3	0.301	$(7.4\pm2.7) \times 10^{-4}$
PEB-SMA-4	0.398	$(1.9\pm0.2) \times 10^{-3}$

[a]Expt. ID refers to the experiments where the Poly(ethylene-*co*-butylene)-(BlocBuilder)$_2$ (PEB-(BlocBuilder)$_2$) initiated various styrene/maleic anhydride (ST/MA) feeds. [b]$f_{MA,0}$ is the initial molar feed composition of with respect to maleic anhydride (MA). [c]Apparent rate constants $k_p[P\cdot]$ were obtained from the linear regions of the $\ln[(1-X)^{-1}]$ versus polymerization time t plots where X = total monomer conversion. The error bars are derived from the standard error in the slope of the linear fit.

Conclusions

Poly(ST-*ran*-MA) and PEB-(ST-*ran*-MA)$_2$ random and block copolymers could be synthesized using BlocBuilder/SG1 or BlocBuilder-terminated PEB initiators (PEB-(BlocBuilder)$_2$). Using BlocBuilder/SG1 as the initiator/mediator, strictly alternating ST/MA sequences can be obtained from any feed at temperatures < 90 °C while only equimolar ST/MA feeds at temperatures > 90 °C will provide strictly alternating ST/MA sequences. PEB-

$(BlocBuilder)_2$ was effective as a macroinitiator for ST/MA copolymerizations at 110 °C as very little unreacted macroinitiator remained for pure ST chain extensions. Dramatic increases in polymerization rates was observed as MA content in the feed increased, in much the same manner as for BlocBuilder/SG1 system.

Acknowledgements

This work was funded by an NSERC Discovery Grant (Grant No. 288125) and a Canada Foundation for Innovation (CFI) New Opportunities Grant. We thank Scott Schmidt and Noah Macy of Arkema Inc. for their help in providing the BlocBuilder initiator and SG1 free nitroxide.

References

Benoit, D., Hawker, C. J., Huang, E. E., Lin, Z., Russell, T. P. "One-step Formation of Functionalized Block Copolymers" Macromolecules, **33**, 1505-1507 (2000)

Cernohous, J.J., Macosko, C.W., Hoye, T.R. "Anionic Synthesis of Polymers Functionalized with a Terminal Anhydride Group", Macromolecules, **30**, 5213-5219 (1997)

Chernikova, E., Terpugova, P., Bui, C., Charleux, B. "Effect of Comonomer Composition on the Controlled Free-radical Copolymerization of Styrene and Maleic Anhydride by Reversible Addition-Fragmentation Chain Transfer (RAFT)", Polymer, **44**, 4101-4107 (2003)

Ha, N. T. H. "Determination of Triad Sequence Distribution of Copolymers of Maleic Anhydride and its Derivates with Donor Monomers by ^{13}C n.m.r. Spectroscopy" Polymer, **40**, 1081-1086 (1999)

Jeon, H. K., Feist, B. J., Koh, S. B., Chang, K., Macosko, C. W., Dion, R. P. "Reactively formed block and graft copolymers as compatibilizers for polyamide 66/PS blends" Polymer, **45**, 197-206 (2004)

Lessard, B., Graffe, A., Maric, M. "Styrene/*tert*-Butyl Acrylate Random Copolymers Synthesized by Nitroxide-Mediated Polymerization: Effect of Free Nitroxide on Kinetics and Copolymer Composition", Macromolecules, **40**, 9284-9292 (2007)

Lessard, B., Maric, M. "One-Step Poly(styrene-alt-maleic anhydride)-block-poly(styrene) Copolymers with Highly Alternating Styrene/Maleic Anhydride Sequences Are Possible by Nitroxide-Mediated Polymerization", Macromolecules, **43**, 879-885 (2010)

Matyjeszewski, K., Davis, T. P.: Handbook of Radical Polymerization, Wiley-Interscience, Hoboken, NJ (2002)

CONTRIBUTIONS OF JOHN VLACHOPOULOS TO MELT FRACTURE AND WALL SLIP LITERATURE

Savvas G. Hatzikiriakos*

Department of Chemical and Biological Engineering, The University of British Columbia, Vancouver, BC, Canada, hatzikir@interchange.ubc.ca

Abstract - Molten polymers exhibit different types of distortions when extruded at high shear rates. These phenomena are collectively known as melt fracture and include sharkskin or surface melt fracture, cyclic, stick-slip or oscillating and gross melt fracture. Furthermore, when the shear rates are high enough, molten polymers slip on solid interfaces and thus violate the classical no-slip boundary condition of Newtonian fluid mechanics. In this paper, the contributions of John Vlachopoulos to the literature of melt fracture and wall slip of molten polymers and their compounds are summarized starting from 1971 (Lidorikis and Vlachopoulos) to 2009 (Santi et al.).

Introduction

The instabilities that occur in the pressure driven extrusion of molten polymers are fascinating from the scientific perspective but troublesome and sometimes catastrophic from the industrial one. These are known as melt fracture phenomena and manifest themselves as distortions on the surface of emerging polymer profiles from extrusion dies.

Over the past 50 years, there has been a sustained interest in the understanding and control of these instabilities. They can occur in common extrusion operations for various polmers such as the manufacture of polymeric rods, tubes, sheets and wire coating. Over the years, processors have learned to work around these processing defects by a variety of means: slowing the manufacturing rate, using materials that process better but have reduced final mechanical properties or addition of processing additives. From the scientific point of view, the fundamental understanding of these instabilities remains as a major challenge despite much progress in the past decade. The primary motivation for continued research is the promise of a more rational design of materials and processes in order to enhance manufacturing efficiency.

Several papers have been written to analyze, quantify these phenomena and to devise ways of eliminating these phenomena. In this short paper, the contributions of professor John Vlachopoulos, known also as "JV" in the literature of melt fracture and wall slip are summarized.

Melt Fracture Phenomena

As discussed above the rate of production in many polymer processing operations including fiber spinning, film blowing, profile extrusion, and various coating flows, is limited by the onset of flow instabilities (Hatzikiriakos and Migler, 2004). In particular, in extrusion processes when the throughput exceeds a critical value, small amplitude periodic distortions appear on the surface of extrudates (surface melt fracture or sharkskin). At higher throughput rates more severe distortions of irregular form appear known as gross melt fracture. These two phenomena are independent and might occur simultaneously. The surface melt fracture phenomena are believed to originate in the land of the die next to the die exit, while gross melt fracture is initiated at the die entry.

In the early '70s the interest was limited to gross melt fracture and especially on the critical value of stress ratio or recoverable shear for its onset. Vlachopoulos and Lirodikis (1971) studied the gross melt fracture phenomena in the case of polystyrene and recognized the important effect of polydispersity. They found that the critical shear stress for the onset of melt fracture varies linearly with the inverse of molecular weight (Figure 1).

Vlachopoulos and Alam (1972) reported further on the gross melt fracture behaviour of polystyrene and found that the critical stress is of the order of 1 MPa. They also reported that the recoverable shear at the onset of melt fracture is proportional to the factor of $\overline{M}_z\overline{M}_{z+1}/\overline{M}_w^2$ recognizing the important effect of polydispersity (Figure 2).

Finally, Vlachopoulos and Chan (1977) reported that melt fracture in slit dies occurs at higher stresses in slits than in cylindrical dies and that the recoverable shear should always be used as a criterion for the initiation of melt fracture. This has been recognized ten years later by Kalika and Denn (1987).

In the late '70s with the increasing importance of linear-low density polyethylene (LLDPE), the sharkskin or surface melt fracture became a popular subject of study in academia and this is the case until today. New polymers such as metallocene polyolefins or biodegradable polyesters have developed over the years and while they exhibit better mechanical properties, unfortunately become more difficult to process. Kanev et al., (2009) have reported on such a case recently, studying the sharkskin melt fracture behaviour of biodegradable polylactic acid polymers.

Based on the above, one might think that gross melt fracture is less significant than surface melt fracture, because most polymers will exhibit sharkskin before the onset of gross melt fracture. This is the case for film application. However, in pelletization that happens at high shear rates, gross melt fracture is important.

Work of Adhesion, Wall Slip and Sharkskin Melt Fracture

Larrazabal et al., (2006a, 2006b) have recognized the importance of the work of adhesion that plays in sharkskin melt fracture phenomena and the wall slip of polymers such as HDPE and LLDPE. Their studies have important implications in industrial operations such as film blowing and profile extrusion. First, they studied the relationship between the critical shear stress for the onset of flow instabilities and the work of adhesion at the interface experimentally for a series of polyolefins on several steels, brasses, and coated steels. They have reported that the critical shear stress for the onset of melt fracture was considerably affected by the chemical nature of the die, an observation important in industry i.e. selection of the material of construction of the die to increase production. A linear relationship was identified between the critical shear stress for the onset of slip and the work of adhesion (Figure 3).

They have also reported that low-surface-energy materials (coated dies with fluoropolymers) were found to produce smooth extrudates by inducing slippage at the wall. Finally they concluded that steel, tungsten carbide, and brass surfaces promoted sharkskin defect at shear stresses above the first critical shear stress. Chrome and nickel plate dies have found to improve performance as well as the increase of the surface roughness of the dies. All these finding are extremely important in the industrial sector.

Surface tearing

Highly filled polymers such HDPE/wood composites exhibit a phenomenon known as extrudate surface tearing which resembles in some cases sharkskin melt fracture, although completely different (typical images are presented in Figure 4). This phenomenon of surface tearing is observed in highly filled materials and has become known in the '80s when such materials (highly-filled) were studied as possibilities to enhance the mechanical properties of polymers (Hristov et al., 2006). The Vlachopoulos group has contributed in this area significantly (Hristov et al., 2006; Hristov and Vlachopoulos, 2007; Santi et al., 2009). They have investigated the relation between the rheological properties and the wall slip phenomenon in capillary dies. Their results have shown considerable increase of storage modulus, dynamic and shear viscosity with increasing wood flour loading. It was also found that all wood filled composites did not obey the Cox–Merz rule possibly due to the existence of plasticity in such materials. The wall slip velocity also was found to depend on wood filler content and shear rate.

Generally, with increasing shear rate the slip velocity was found to sharply increase leading to plug-like flow. Finally, they have reported that the surface of the extrudates becomes smoother (decrease of surface tearing) with increase in shear rate and wood flour content, which means.

Conclusions

The contributions of JV in the literature of melt fracture and wall slip has been extremely important and significant. Important work has been published on sharkskin and gross melt fracture behaviour of various polymers, on the relationshipship between melt fracture sharkskin and work of adhesion that has implication in die selection and finally on the instabilities known as surface tearing that occurs in the case of highly filled materials such as HDPE/wood compounds.

References

Hatzikiriakos S.G., and K. Migler "Polymer Processing Instabilities : Control and Understanding" Marcel Deker ISBN 824 75386, (2004).

Hristov V., Elizabeth Takacs, and J. Vlachopoulos "Surface tearing and wall slip phenomena in extrusion of highly filled HDPE/Wood flour composites", Polym. Eng. Sci., 1204-1214 (2006).

Hristov V. and J. Vlachopoulos, "A study of viscoelasticity and extrudate distortions of wood polymer composites," Rheol Acta , **46**, 773–783 (2007).

Kalika D.S., M.M. Denn "Wall slip and extrudate distortion in linear low-density polyethylene" J. Rheology, **31**, 815-834 (1987)

Kanev D., E.Takacs and J. Vlachopoulos " Rheological evaluations and observations of extrusion instabilities of biodegradable polyesters' Intern. Polym. Proc., **22**, 395-401(2007).

Larrazabal H.J., A.N. Hrymak and J. Vlachopoulos, On the relationship between the work of adhesion and the critical shear stress for the onset of flow instabilities, Rheol. Acta, **45**, 705–715 (2006a).

Larrazabal H.J., A.N. Hrymak and J. Vlachopoulos, Effect of chemical and morphological conditions of the die wall on the extrusion of linear polyolefins", Intern. Polym. Proc., **XXI**, 132-140 (2006b).

Santi C.R., E. Hage Jr, J. Vlachopoulos, and C.A. Korea, "Rheology and processing of HDPE/Wood flour composites", **XXIV**, 346-353 (2009).

Vlachopoulos J., and S. Lidorikis, "Melt Fracture of Polystyrene," Polym. Eng. Sci., **11**, 1-5 (1971).

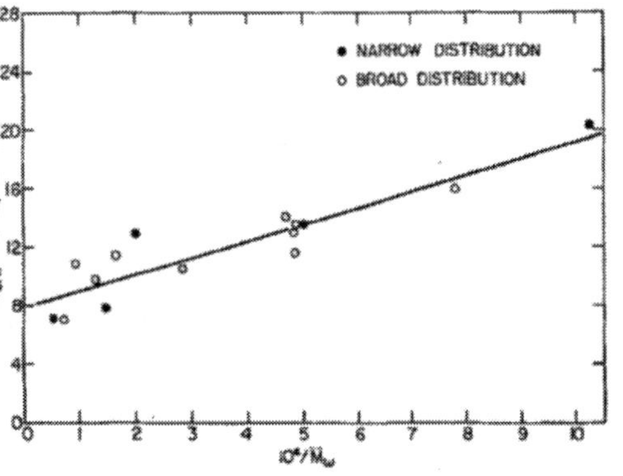

Figure 1: The critical shear stress vs. $1/M_w$ (from Vlachopoulos and Lidorikis, 1971).

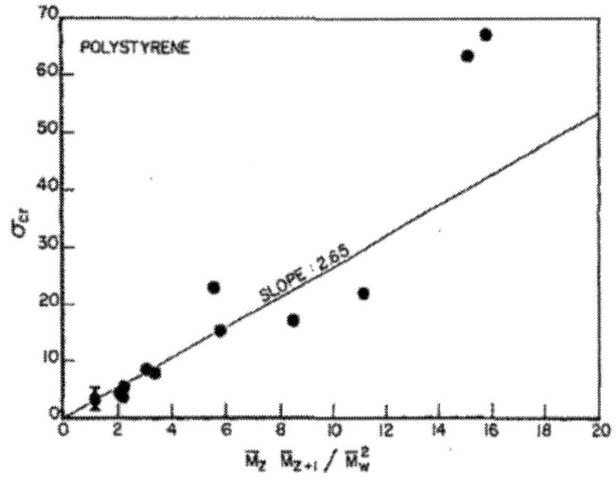

Figure 2: Critical recoverable shear strain at the wall (σ_{cr}) shear stress vs. $\bar{M}_z \bar{M}_{z+1} / \bar{M}_w^2$ (from Vlachopoulos and Alam, 1972).

Vlachopoulos J., and M. Alam, "Critical stress and recoverable shear for polymer melt fracture," Polym. Eng. Sci., **12**, 184-192 (1972).

Vlachopoulos J., and T.W. Chan, "A comparison of melt fracture initiation conditions in capillaries and slits," J. Applied Polym. Sci., **21**, 1177-1187 (1977).

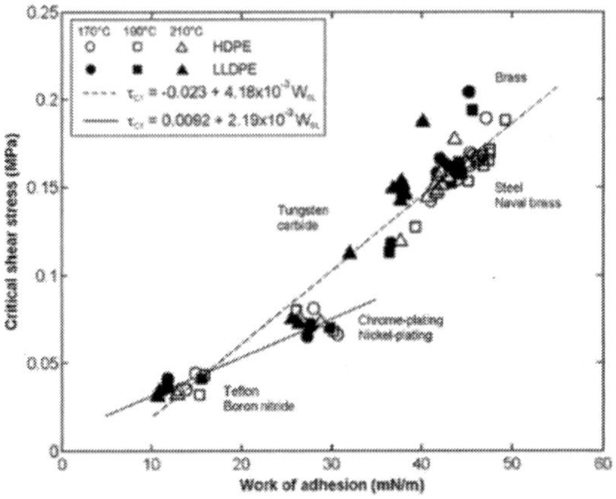

Figure 3: The critical shear stress for the onset of flow instabilities or wall slip as a function of the work of adhesion for a HDPE and a LLDPE at 170, 190, and 210°C (from Larrazabal et al, 2006b).

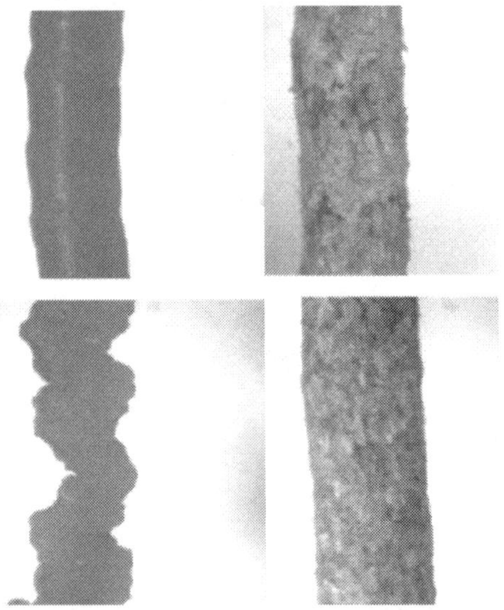

Figure 4: Left-typical images of extrudate distortions in virgin HDPE known as melt fracture. Right typical images of surface tearing in the extrusion of a highly filled HDPE (from Hristov et. al. 1986)

CALENDERING:
FROM JV'S EARLY WORK TO THE PRESENT

Evan Mitsoulis*

School of Mining Engineering & Metallurgy, National Technical University of Athens (NTUA)
Zografou 157-80, Athens, GREECE, mitsouli@metal.ntua.gr

Abstract - John Vlachopoulos (JV) started his polymer processing career with the process of calendering. In two landmark papers with Costas Kiparissides (PES, 1976; 1978) he introduced the Finite Element Method (FEM) to solve the governing equations of mass, momentum, and energy based on the Lubrication Approximation Theory (LAT). That early work was followed by the introduction of slip (with Andy Hrymak, PES, 1980). The first fully 2-D simulations for calendering PVC were carried out with the author (PES, 1985). In the intervening 25 years, other works have emerged, but our understanding has not been drastically improved since JV's early works.
Some new results have been obtained for pseudoplastic and viscoplastic fluids using the Herschel-Bulkley model. The emphasis was on finding possible differences with LAT regarding the attachment and detachment points (hence the domain length), and the extent and shape of yielded/unyielded regions. The results showed that while the former is well predicted by LAT, the latter is grossly overpredicted by LAT.
Presently, the transient problem is solved using the ALE-FEM formulation for moving free-boundary problems. The results are compared with the previous simulations for the steady-state and show a good agreement. The transient simulations capture the movement of the upstream and downstream free surfaces, and also provide the attachment and detachment points, which are unknown a priori. Finding these still remains the prevailing challenge in calendering.

Introduction

Calendering is a process used in a wide variety of industries for the production of rolled sheets or films of specific thickness and final appearance. The procedure has been theoretically introduced more than 50 years ago [1] and involves the feed from an infinite reservoir of a pair of co-rotating heated rolls (calenders) with a material to form a sheet. The case of calendering fed with a finite sheet of thickness H_f for the production of a final sheet with exit thickness H, as shown in Figure 1 [2], is a variant of the process. In calendering, the ratio $R/H_0 >> 1$ [O(1000)].

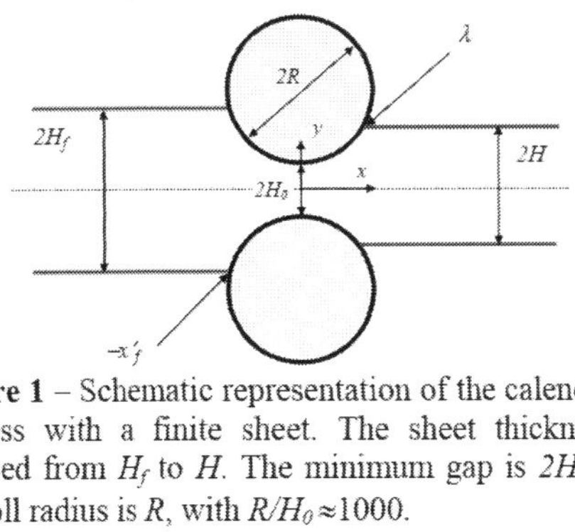

Figure 1 – Schematic representation of the calendering process with a finite sheet. The sheet thickness is reduced from H_f to H. The minimum gap is $2H_0$, and the roll radius is R, with $R/H_0 \approx 1000$.

The quality of the sheet or film produced is determined by the flow and heat transfer phenomena in the gap between two rotating rolls. It is thus of great importance to determine the velocity and temperature profiles as the melt is squeezed between the rolls, the pressure distribution on the roll surfaces, the roll-separating force and the torque and energy input. A realistic Newtonian and Bingham flow analysis by Gaskell [1] spearheaded further developments in modelling. McKelvey [3] reworked Gaskell's model for Newtonian flow and extended it to include non-Newtonian (i.e. shear-thinning) effects.
JV's first encounter with calendering occurred during his 1st sabbatical in 1974 in Germany, where some inaccuracies in McKelvey's [3] power consumption calculations were corrected by Ehrmann and Vlachopoulos [4]. The problem was then taken in earnest by Kiparissides and Vlachopoulos [5], who developed a finite element analysis for symmetric and asymmetric calendering (different roll speeds, different roll diameters). Viscous dissipation effects were included in finite difference solutions of the energy equation by Kiparissides and Vlachopoulos [6]. An isothermal model with slip was developed by Vlachopoulos and Hrymak [7] using the lubrication approximation and a Runge-Kutta solution.
The vast majority of the above works is based on the Lubrication Approximation Theory (LAT) of Reynolds. The textbook by Middleman [2] nicely summarises this and presents the most important findings up to 1977. Lifting this approximate assumption leads to a truly two-dimensional (2-D) analysis, as was done by Mitsoulis et al. [8] and Agassant and Espy [9]. These two works have shown

very interesting results with intricate patterns dominated by large vortices in the melt bank before the rolls. Zheng and Tanner [10] also solved the 2-D problem for Newtonian, power-law and viscoelastic fluids, with particular emphasis on finding the point where the fluid detaches from the rolls. They used a boundary element method and the criterion that the fluid detaches from the rolls at the point where the tangential traction becomes zero. Again, all works prior to 1990 have been summarized in the textbook by Agassant et al. [11]. More recently, Levine et al. [12] used a 2½-D approach to calculate the side spreading of the calendered sheet while employing the lubrication approximation for the flow through the nip. Mewes et al. [13] have done work on the simultaneous calculation of roll deformation and polymer flow, while the full three-dimensional (3-D) problem has been solved by Luther and Mewes [14] for power-law fluids.

A renewed effort has appeared recently to study viscoplastic materials in calendering and find out the effect of yield stress. A series of papers by Sofou and Mitsoulis [15-17] have used LAT and carried out full parametric studies for pseudoplastic and viscoplastic fluids obeying the Herschel-Bulkley model of viscoplasticity with or without slip at the wall. This model, upon appropriate simplifications, can be reduced to the Newtonian, power-law and Bingham plastic models. The results have shown interesting trends and yielded / unyielded regions. However, the latter are misleading as pointed out by Sofou and Mitsoulis [15], and by Mitsoulis [18] who used a fully 2-D analysis, and they are a direct consequence of using LAT and its inherent approximations.

It is, therefore, the purpose of the present paper to review the problem of calendering, to present the major results so far and to show new ones from time-dependent calculations. Some issues, which still remain unresolved, will be discussed for future work.

Mathematical Modeling

The problem at hand is shown schematically in Fig. 1, where a fluid is squeezed between two rotating rolls moving with a roll speed, U, and exits with a reduced thickness, $2H$, which is unknown a priori.

Many materials used in calendering are non-Newtonian, exhibiting either *pseudoplastic* (shear–thinning or –thickening) or *viscoplastic* (presence of a yield stress) behaviour (see, e.g., Bird et al. [19]). Models describing pseudoplastic behaviour include the power-law, Carreau, Cross, etc. Models describing viscoplastic behaviour include the Bingham, Herschel-Bulkley and Casson. The Herschel-Bulkley model has the advantage of reducing – with an appropriate choice of parameters – to the Bingham, power-law or Newtonian model. In simple shear flow it becomes:

$$\tau = \tau_y + K\dot{\gamma}^n, \qquad \text{for } |\tau| > \tau_y, \tag{1a}$$

$$\dot{\gamma} = 0, \qquad \text{for } |\tau| \le \tau_y, \tag{1b}$$

where τ is the shear stress, $\dot{\gamma}$ is the shear rate, τ_y is the yield stress, K is the consistency index, and n is the power-law index. Note that when $n = 1$ and $K = \mu$ (a constant), the Herschel-Bulkley model reduces to the Bingham model. When $\tau_y = 0$, the power-law model is recovered, and when $\tau_y = 0$ and $n = 1$, the Newtonian model is obtained.

It should be noted that in viscoplastic models, when the shear stress τ falls below τ_y, a solid structure is formed (unyielded). Also in viscoplasticity, a dimensionless yield stress can be defined by:

$$\tau_y^* = \frac{\tau_y}{\left(\frac{dP}{dx}\right)_0 H_0} = \frac{y_0}{H_0}, \tag{2}$$

where H_0 is a characteristic length (half the minimum gap between the rolls), $(dP/dx)_0$ is a pressure gradient evaluated at the nip, and y_0 is a yield line separating the yielded from the unyielded material. In all cases, the Newtonian fluid corresponds to $\tau_y^* = 0$. However, at the other extreme of an unyielded solid, $\tau_y^* \to 1$, a limiting value, where the material becomes all unyielded.

If the dimensionless yield stress is based on the kinematics through the roll speed U, it gives rise to the Bingham number, Bn, defined by:

$$Bn = \frac{\tau_y (2H_0)^n}{KU^n}. \tag{3}$$

Again, the Newtonian fluid corresponds to $Bn = 0$. However, at the other extreme of an unyielded solid, $Bn \to \infty$.

The flow is governed by the usual conservation equations of mass and momentum under isothermal creeping flow conditions for an incompressible viscous fluid. The constitutive equation that relates the stresses to the velocity gradients is the generalized Newtonian fluid. In full tensorial form the constitutive equation is written as:

$$\bar{\bar{\tau}} = \eta \bar{\bar{\dot{\gamma}}}, \tag{4}$$

where η is the apparent viscosity given for the Herschel-Bulkley-Papanastasiou model by

$$\eta = \left[K|\dot{\gamma}|^{n-1} + \frac{\tau_y\left(1-\exp(-m|\dot{\gamma}|)\right)}{|\dot{\gamma}|} \right], \tag{5}$$

and $|\dot{\gamma}|$ is the magnitude of the rate-of-strain tensor $\bar{\bar{\dot{\gamma}}} = \nabla\bar{v} + \nabla\bar{v}^T$, which is given by

$$|\dot{\gamma}| = \sqrt{\frac{1}{2} II_{\dot{\gamma}}} = \left[\frac{1}{2}\{\bar{\bar{\dot{\gamma}}} : \bar{\bar{\dot{\gamma}}}\}\right]^{1/2}, \tag{6}$$

where $II_{\dot{\gamma}}$ is the second invariant of $\bar{\bar{\dot{\gamma}}}$.

The stress growth exponent m is the regularization parameter introduced by Papanastasiou [20], and it has to be high enough to approximate the ideal viscoplastic model [20,21]. The dimensionless form of eq. (5) is written as

$$\eta = \left[|\dot{\gamma}|^{n-1} + \frac{Bn\left[1-\exp(-M|\dot{\gamma}|)\right]}{|\dot{\gamma}|} \right], \tag{7}$$

where the dimensionless stress growth exponent M is given by

$$M = \frac{mU}{H_0}. \quad (8)$$

As in our previous works [20,21], to track down yielded / unyielded regions, we shall define yielded and unyielded fluid regions by the following criteria:

yielded:

$$|\tau| = \sqrt{\frac{1}{2} II_\tau} = \left[\frac{1}{2}\{\bar{\bar{\tau}}:\bar{\bar{\tau}}\}\right]^{1/2} > \tau_y \quad (9a)$$

unyielded: $|\tau| \le \tau_y$ (9b)

In parametric studies for pseudoplastic models obeying the power law, the dimensionless parameter is the power-law index, n. For viscoplastic models obeying the Bingham model, the dimensionless parameter is the Bingham number, Bn. In the present simulations, the range of parameters is $0 \le n \le 2$, and $0 \le Bn \le 1000$.

Governing Equations (LAT)

LAT considers locally fully developed shear flow between the rolls. The conservation of momentum equation then gives

$$\frac{dP}{dx} = \frac{d\tau_{xy}}{dy}, \quad (10)$$

where $\tau_{xy} = \tau$ is the shear stress in the transverse direction. The shear stress is given by the Herschel-Bulkley model in the present work, i.e. eq. (1).

We integrate eq. (10) and apply the boundary conditions for symmetric calendering, i.e.

$$\tau_{xy} = 0 \text{ at } y = 0, \quad (11a)$$

$$u = U \text{ at } y = h(x). \quad (11b)$$

From the mathematical analysis, we know that there are two flow regions in the x-direction, one region near the nip and the exit, which has a negative pressure gradient and a negative velocity gradient, and another region away from the nip at the entrance with opposite signs. For viscoplastic materials, we know that there are also two flow regions in the y-direction, one where the material is unyielded ($y \le y_0$) and another where it is yielded ($y > y_0$). We also note that $\tau_y = |dP/dx| y_0$. Thus, we obtain:

$$u = \begin{cases} U + \frac{n}{(n+1)}\left(\frac{1}{K}\frac{dP}{dx}\right)^{\frac{1}{n}}\left[(y-y_0)^{\frac{n+1}{n}} - (h-y_0)^{\frac{n+1}{n}}\right], & y > y_0 \\ U - \frac{n}{(n+1)}\left(\frac{1}{K}\frac{dP}{dx}\right)^{\frac{1}{n}}(h-y_0)^{\frac{n+1}{n}}, & y \le y_0 \end{cases}$$

for $dP/dx > 0$, (12a)

$$u = \begin{cases} U - \frac{n}{(n+1)}\left(-\frac{1}{K}\frac{dP}{dx}\right)^{\frac{1}{n}}\left[(y-y_0)^{\frac{n+1}{n}} - (h-y_0)^{\frac{n+1}{n}}\right], & y > y_0 \\ U + \frac{n}{(n+1)}\left(-\frac{1}{K}\frac{dP}{dx}\right)^{\frac{1}{n}}(h-y_0)^{\frac{n+1}{n}}, & y \le y_0 \end{cases}$$

for $dP/dx < 0$. (12b)

Integration of the velocity profile gives the volumetric flow rate Q according to:

$$Q = 2Uh - \frac{2n[(n+1)h + ny_0]}{(n+1)(2n+1)}\left(\frac{1}{K}\frac{dP}{dx}\right)^{\frac{1}{n}}(h-y_0)^{\frac{n+1}{n}}$$

for $dP/dx > 0$, (13a)

$$Q = 2Uh + \frac{2n[(n+1)h + ny_0]}{(n+1)(2n+1)}\left(-\frac{1}{K}\frac{dP}{dx}\right)^{\frac{1}{n}}(h-y_0)^{\frac{n+1}{n}}$$

for $dP/dx < 0$. (13b)

In the case of the Bingham plastic model ($n = 1$), the above equation becomes:

$$Q = 2Uh - \frac{(2h+y_0)}{3\mu}\left(\frac{dP}{dx}\right)(h-y_0)^2 \quad \text{for all } dP/dx. \quad (14)$$

From eq. (12) the pressure gradient dP/dx can be found. In calendering the following dimensionless parameters are introduced [2]:

$$x' = \frac{x}{\sqrt{2RH_0}},\ h = H_0\left(1 + \frac{x^2}{2RH_0}\right),\ y' = \frac{y}{H_0},\ y_0' = \frac{y_0}{H_0},$$

$$P' = \frac{P}{K}\left(\frac{H_0}{U}\right)^n,\ \lambda^2 = \frac{Q}{2UH_0} - 1, \quad (15)$$

where λ is a dimensionless flow rate (or leave-off distance) and the rest of the symbols are defined in Fig. 1. After the appropriate manipulations, the following dimensionless pressure gradient is obtained:

$$\frac{dP'}{dx'} = -\left(\frac{2n+1}{n}\right)^n \sqrt{\frac{2R}{H_0}} \frac{(\lambda^2 - x'^2)|(\lambda^2 - x'^2)|^{n-1}}{\left[(1+x'^2) + \frac{n}{n+1}y_0'\right]^n} \frac{1}{(1+x'^2-y_0')^{n+1}} \quad (16)$$

where
$$y_0' = \tau_y^* \left|\frac{dP'}{dx'}\right| \Big/ \left|\frac{dP'}{dx'}\right|_0 \quad (17)$$

Equation (16) incorporates both cases: for $-\infty < x' < -\lambda$ the pressure gradient is positive, and for $-\lambda \le x' \le \lambda$ the pressure gradient is negative. Integration of the above equation (16) requires boundary conditions for the pressure gradient dP'/dx' and the pressure P' as well. In the case of an infinite reservoir, the standard conditions are the so-called *Swift conditions* [10]:

$$P' = dP'/dx' = 0, \text{ at } x' = \lambda_\infty, \quad (18a)$$

$$P' = 0, \text{ at } x' = -\infty. \quad (18b)$$

Then the pressure is given by the integral:

$$P' = \left(\frac{2n+1}{n}\right)^n \sqrt{\frac{2R}{H_0}} \int_{x'}^{\lambda_\infty} \frac{(\lambda_\infty^2 - x'^2)|(\lambda_\infty^2 - x'^2)|^{n-1}}{\left[(1+x'^2) + \frac{n}{n+1}y_0'\right]^n} \frac{1}{(1+x'^2-y_0')^{n+1}} dx' \quad (19)$$

The extrema of the pressure distribution occur at $x' = -\infty, \pm\lambda_\infty$. The dimensionless leave-off distance λ_∞ corresponding to an infinite reservoir can be found from the above equation knowing that $P'(x' = -\infty) = 0$. Therefore λ_∞ can be found from the relation:

$$0=\int_{-\infty}^{\lambda_\infty}\frac{\left(\lambda_\infty^2-x'^2\right)\left|\left(\lambda_\infty^2-x'^2\right)\right|^{n-1}}{\left[\left(1+x'^2\right)+\frac{n}{n+1}y_0'\right]^n}\frac{1}{\left(1+x'^2-y_0'\right)^{n+1}}dx'=\int_{-\infty}^{\lambda_\infty}I(n,\tau_y^*)dx' \qquad (20)$$

The above integral has no analytical solution for the general case of Herschel-Bulkley fluids. Therefore, a numerical solution must be found, based on some numerical algorithm for solving non-linear equations.

Sheet thickness

Once λ_∞ is found as a function of n and τ_y^*, then all other quantities of interest are readily available. The exiting sheet thickness H from an *infinite reservoir* is given by:

$$\frac{H}{H_0}=1+\lambda_\infty^2 . \qquad (21)$$

For the case of calendering sheets with a *finite thickness*, the leave-off distance λ_∞ is substituted by λ and $-\infty$ by $-x_f'$. The exiting sheet thickness H is given by:

$$\frac{H}{H_0}=1+\lambda^2 , \qquad (22)$$

while the thickness of the entering sheet H_f is entering the analysis according to the definition:

$$x_f'=\left(\frac{H_f}{H_0}-1\right)^{1/2} . \qquad (23)$$

Operating variables

The operating variables used in engineering calculations are also of interest [2]:

(i) the maximum pressure defined by:

$$\mathcal{P}(n,\tau_y^*)=\frac{P'_{max}}{\sqrt{2R/H_0}}=\left(\frac{2n+1}{n}\right)^n\int_{-\lambda}^{\lambda}I(n,\tau_y^*)dx' , \qquad (24)$$

(ii) the roll-separating force per unit width defined by:

$$\frac{F}{W}(n,\tau_y^*)=\int_{-x_f'}^{\lambda}P(x)dx=K\left(\frac{U}{H_0}\right)^n R\mathcal{F}(n,\tau_y^*) , \qquad (25a)$$

with $$\mathcal{F}(n,\tau_y^*)=2\left(\frac{2n+1}{n}\right)^n\int_{-x_f'}^{\lambda}\left[\int_{x'}^{\lambda}I(n,\tau_y^*)dx'\right]dx' , \qquad (25b)$$

(iii) the power input for both rolls defined by:

$$\dot{\mathcal{W}}(n,\tau_y^*)=2WU\int_{-x_f'}^{\lambda}\tau_{xy}\Big|_{y=h(x)}dx=WU^2K\left(\frac{U}{H_0}\right)^{n-1}\sqrt{\frac{R}{H_0}}\mathcal{E}(n,\tau_y^*) \qquad (26a)$$

with $$\mathcal{E}(n,\tau_y^*)=-2\sqrt{2}\left(\frac{2n+1}{n}\right)^n\int_{-x_f'}^{\lambda}I(n,\tau_y^*)\left(1+x'^2\right)dx' , \qquad (26b)$$

Results and Discussion

The first paper by JV [4] showed that small numerical errors in the attachment point $-x_f$* caused large errors in the power consumption. This becomes evident in Fig. 2, where the original figure from [4] is presented and it gives the correction to the coefficient f related to the power input (eq. 26a) as a function of the leave-off distance λ. This attention to detail established a genuine feel for the process of calendering, and what the numbers meant in meaningful calculations.

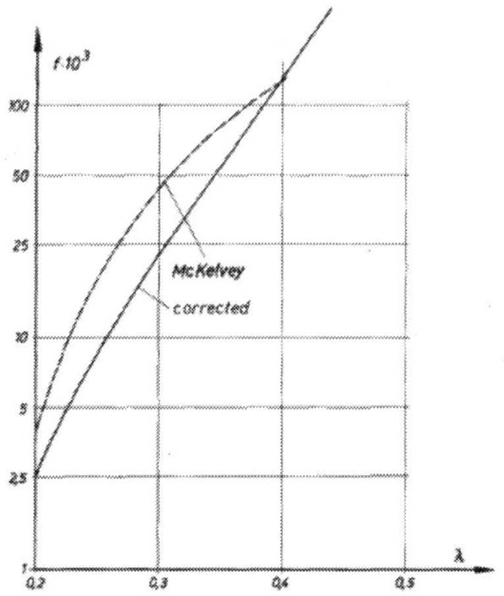

Figure 2 – Correction of the power consumption coefficient f as a function of the leave-off distance λ as found out by Ehrmann and Vlachopoulos [4] for calendering of Newtonian fluids.

The subsequent work by Kiparissides and Vlachopoulos [5] used the finite element method (FEM) to study the process. The original FEM grid from the computations is shown in Fig. 3 and it consists of 64 triangular elements and 157 nodes. Such grids at the time (1974) were only allowed due to memory limitations of the mainframe computers. The results for the pressure for Newtonian fluids showed that they were in good agreement with LAT only for low values of $\lambda \leq 0.3$, as shown in Fig. 4. This was due to the fact that although the grid was 2-D, only one velocity component u_x was used in the calculations due to memory limitations. Further results on non-isothermal calendering [6] took into account the viscous dissipation term in the energy equation and showed interesting non-monotonic temperature profiles close to the walls for the first time (Fig. 5).

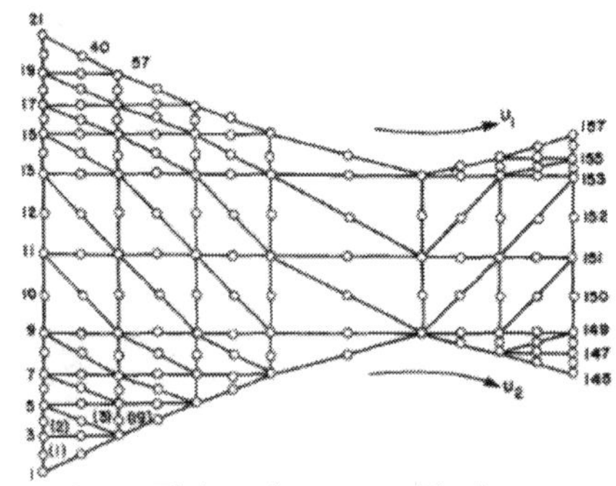

Figure 3 – Finite element grid for asymmetric calendering [5]. The grid consists of 64 triangular elements and 157 nodes.

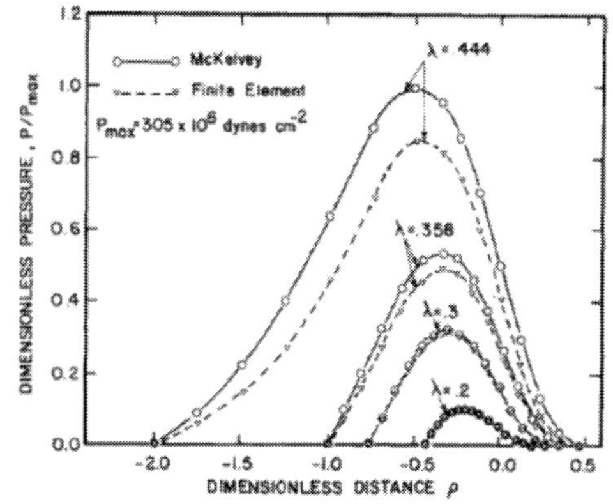

Figure 4 – Pressure distribution as a function of the dimensionless distance $\rho=x'$ in calendering of Newtonian fluids [5].

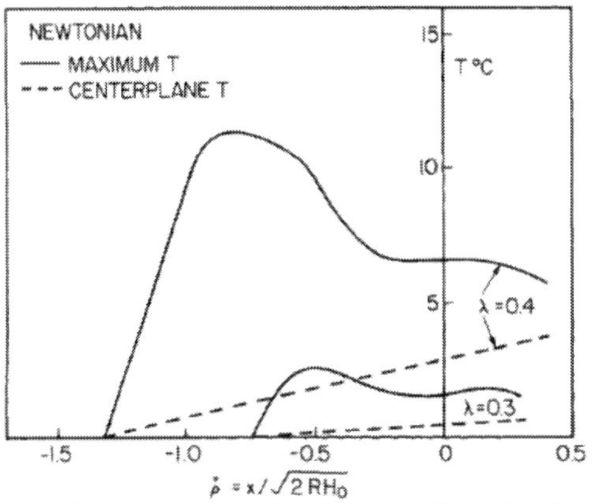

Figure 5 – Temperature development along the flow field in calendering of Newtonian fluids [6].

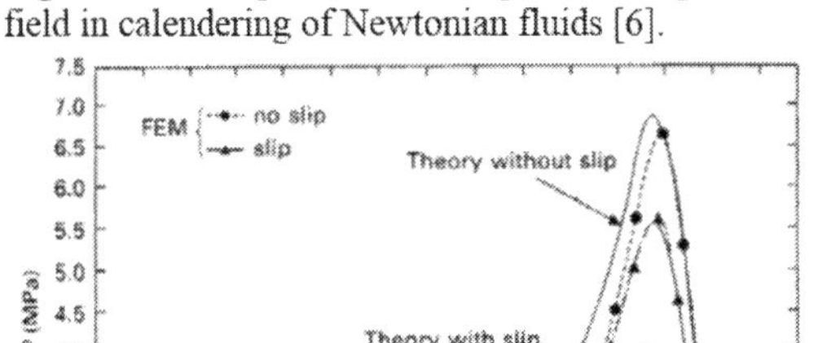

Figure 6 – Pressure distribution as a function of dimensionless distance taking into account slip at the wall for a rigid PVC melt. Continuous lines are from [7] with LAT, symbols are 2-D FEM results from [8].

JV's subsequent work was with Andy Hrymak (then an undergraduate student at McMaster University). The task was to simulate experimental data for a rigid PVC compound known to exhibit slip at the wall. In a landmark paper [7], the equations for power-law fluids were recast with slip and solved for with a 4th-order Runge-Kutta method, giving the results of Fig. 6 (continuous lines), which were later verified by the author with a full 2-D FEM solution (symbols) [8]. The paper by Vlachopoulos and Hrymak [7] was the only paper at the time that had comprehensive experimental data both for the resin and the process. It remains to this date perhaps the only well documented paper with experiments on calendering.

The full 2-D problem was first tackled by the author, then working as a graduate student for his PhD with JV, in [8]. JV's inspiration and guidance were unique in getting results for a fully 2-D, non-isothermal, pseudoplastic or viscoelastic FEM analysis with slip at the wall, and the determination of the free surface in the melt bank that had defied description up to that time. Results from that work appeared in 1985 in PES [8]. Reproduced here are the FEM grids for the velocities-pressure (u-v-p formulation) (Fig. 7a) and the temperature (T) and stream function (ψ) solution (Fig. 7b). The results for the streamlines and isotherms are shown in Figs. 8a and 8b, respectively. They showed for the first time the intricate patterns observed experimentally by Agassant and Espy [9] (Fig. 9).

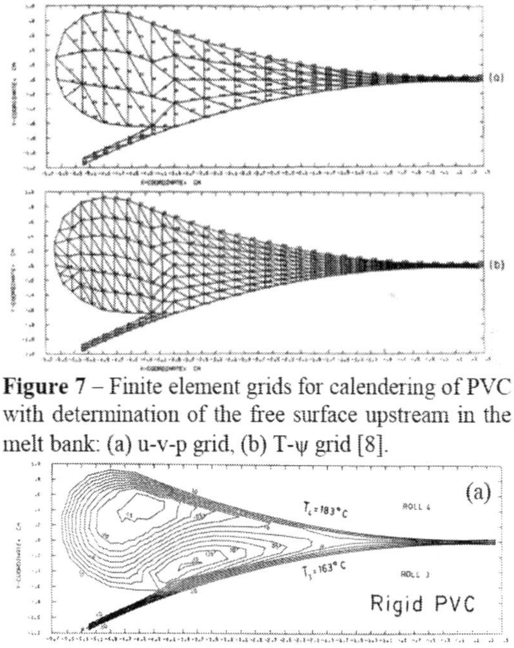

Figure 7 – Finite element grids for calendering of PVC with determination of the free surface upstream in the melt bank: (a) u-v-p grid, (b) T-ψ grid [8].

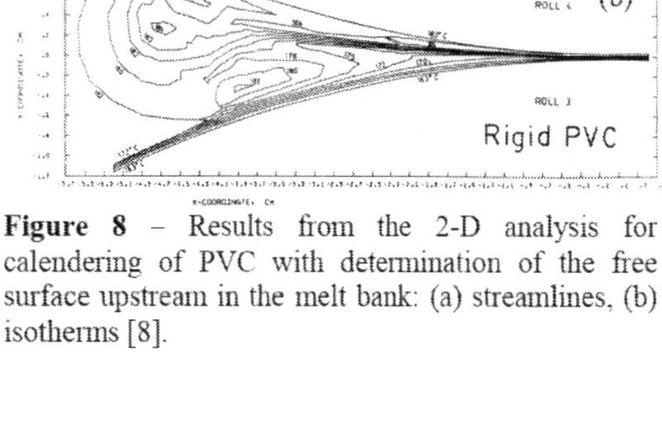

Figure 8 – Results from the 2-D analysis for calendering of PVC with determination of the free surface upstream in the melt bank: (a) streamlines, (b) isotherms [8].

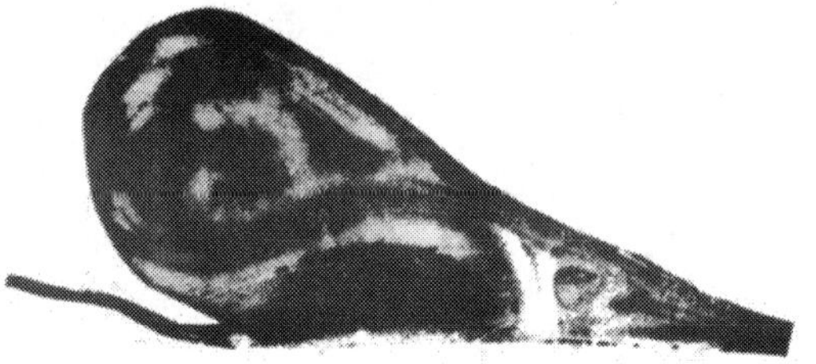

Figure 9 – Experimental results from calendering rigid PVC [9].

The 2-D results remained for a long time without any improvement or addition in the literature, as they seemed to have answered the major points of analysis in calendering. Viscoelastic results by Zheng and Tanner [10] (Fig. 10) showed that viscoelasticity did not alter the results appreciably, as also found earlier in [8]. However, that work pointed to an important issue, namely that of finding both detachment and attachment points without using the Swift condition (pressure and its derivative vanish at the detachment point, eq. 18).

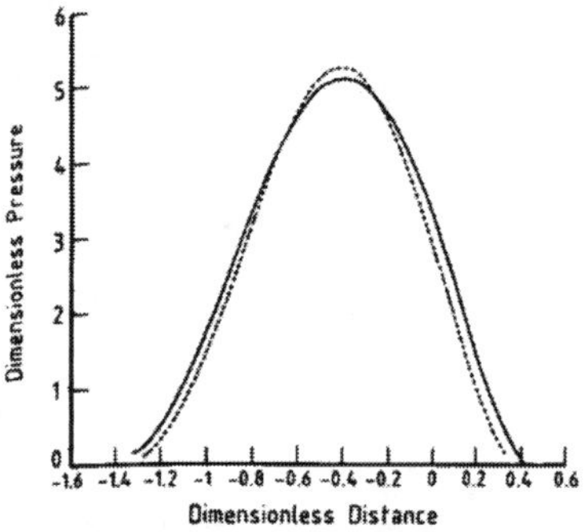

Figure 10 – The effect of viscoelasticity on the pressure distribution in calendering of a Newtonian fluid (dashed line) and a viscoelastic fluid obeying the PTT model at Weissenberg number Ws=1 [10].

It was only in the 2000's that new additions were made in the literature. They started with a 2½-D approach to calculate the side spreading of the calendered sheet while employing the lubrication approximation for the flow through the nip [12]. Typical results are shown in Fig. 11, where spreading of the sheet is more pronounced for a ratio W_0/H_0=10 than a ratio W_0/H_0=60, where W_0 is the incoming sheet width.

The full three-dimensional (3-D) problem for power-law fluids was tackled by Luther and Mewes [14], who used the commercial package POLYFLOW® and computed the upstream free surface in its full length. The results are indeed very impressive and exciting, as evidenced in Fig. 12. They reveal a three-dimensional spiral motion in the velocity field and a bending of the upstream free surface to accommodate a melt bank that is thickest at the middle and thinnest towards the ends of the rolls. If any criticism is made for an otherwise tremendous piece of work, it involves the difficulty of the solution, the lack of any information of how this was achieved (except that it was difficult), and the inability to reproduce these results by somebody else.

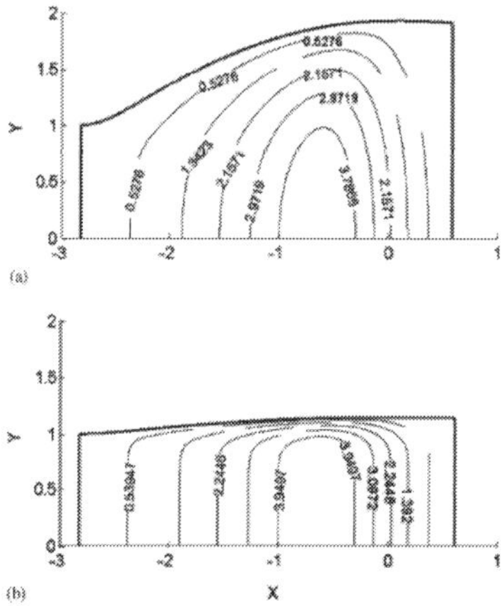

Figure 11 – Shape and pressure contours in calendering analysis of spreading sheets (2½-D approach): (a) W_0/H_0=10, (b) W_0/H_0=60 [12].

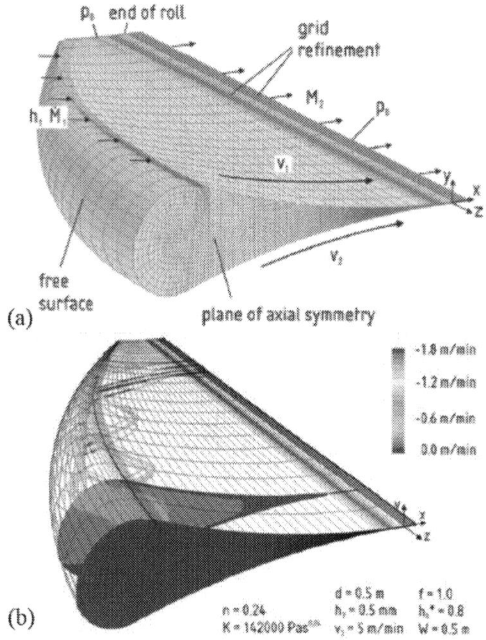

Figure 12 – 3-D computations of calendering power-law fluids: (a) FEM grid, (b) velocity field showing a spiral motion in the 3rd-direction [14].

Viscoplastic fluids

Starting in 2004 and in a series of papers, Sofou and Mitsoulis [15-17] examined the effect of viscoplasticity in calendering in full parametric studies of the power-law index, n, and the Bingham number, Bn. They used LAT and the equations presented above (eq. 12-26) to obtain solutions for three distinct cases: (a) feed from an infinite reservoir [16], (b) feed with a finite sheet [15], (c) both cases with slip at the wall [17]. Typical results for the pressure distribution of Bingham fluids is shown in Fig. 13, where viscoplasticity sharpens the pressure distributions and extends the flow domain.

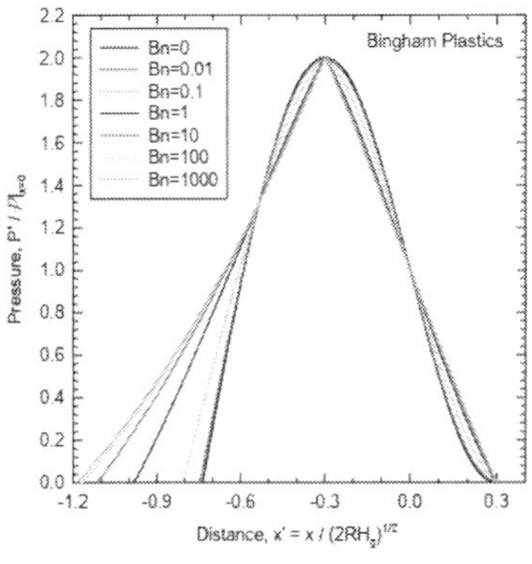

Figure 13 – Axial distribution of dimensionless pressure for different values of the Bingham number *Bn* for viscoplastic fluids obeying the Bingham model ($\lambda = 0.3$) [15].

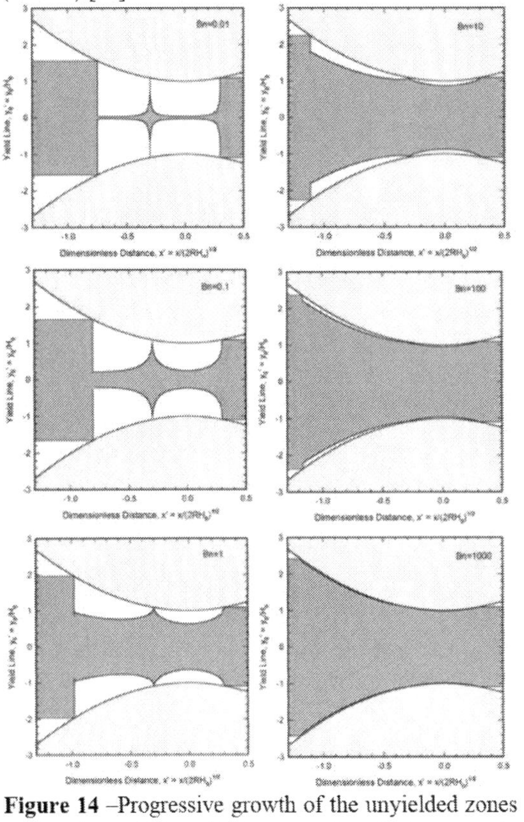

Figure 14 –Progressive growth of the unyielded zones as the dimensionless Bingham number *Bn* increases in calendering of viscoplastic materials obeying the Bingham plastic model. The shaded regions are unyielded. The exiting sheet thickness is fixed so that $\lambda = 0.3$ [15].

Figure 14 shows the progressive growth of yielded/unyielded regions for different *Bn* numbers and λ=0.3. For $Bn = 0$, we have purely viscous flow and the area is all yielded. At the other extreme, for $Bn \rightarrow \infty$, we have a purely plastic solid, and the area is all unyielded. Any other value of *Bn* between 0 and ∞ gives both yielded and unyielded regions. For the finite-sheet case shown in Fig. 14, the abrupt change of the unyielded region at the point of contact is due to the limitations of the lubrication approximation, which is not valid there. These results, although interesting and exciting, are erroneous because of the limitations of LAT for viscoplastic flows [15,16], namely the ever changing geometry in the flow field (presence of non-zero velocity gradients) does not allow for unyielded regions under the gap. However, LAT is good at giving correct pressures (Fig. 13) and sheet thicknesses, as shown in Fig. 15.

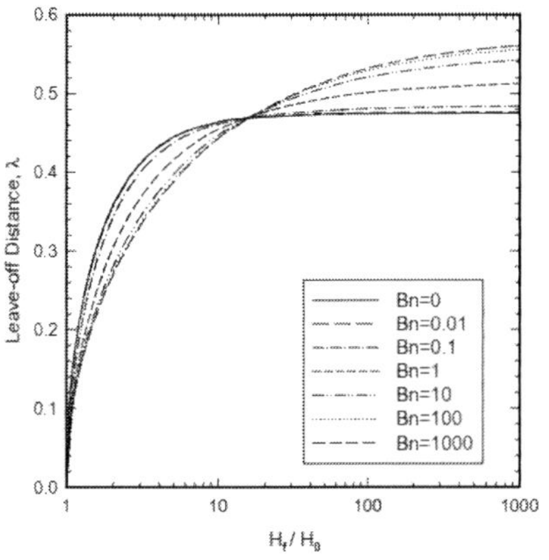

Figure 15 – Leave-off distance λ as a function of the entry thickness H_f/H_0 for viscoplastic fluids obeying the Bingham model for various values of *Bn*.

The results of Fig. 15 have been obtained for a wide range of entering sheet thickness ratios H_f/H_0, from the limiting case of $H_f/H_0 = 1$ (no thickness reduction) to the case of $H_f/H_0 = 1000$ (approaching a sheet coming from an infinite reservoir). For Bingham plastics, the results for the sheet thickness H/H_0 are obtained for various *Bn* values. The Newtonian values (for $Bn = 0$) of $\lambda_\infty = 0.475$ and $H/H_0 = 1.226$, corresponding to calendering from an infinite reservoir, are also obtained when $H_f/H_0 = 1000$, as expected. Figure 15 shows that for Newtonian fluids these values are attained for $H_f/H_0 \geq 20$, while Bingham plastics, like shear-thinning fluids, require even higher ratios as the degree of viscoplasticity increases (e.g., for $Bn = 1000$, $H_f/H_0 = 1000$ is not enough to attain the asymptotic values). It is interesting to note that for a certain value of $H_f/H_0 \approx 15.85$, the results do not depend on the *Bn* value (crossover points of Fig. 15). The corresponding values are $\lambda = 0.470$ and $H/H_0 = 1.221$. After this point, viscoplasticity increases the values. Thus, as was also the case with pseudoplastic fluids [2], it is no longer correct to say, without qualification, that the effect of non-Newtonian behaviour is to increase the sheet thickness.

2-D calculations have been pursued for viscoplastic fluids, obeying the Bingham-Papanastasiou model (eq. 7 with $n = 1$ and $M = 1000$) in a range of Bingham numbers Bn ($0 \leq Bn \leq 1000$). The extra feature in viscoplastic materials is the division of the flow field in yielded/unyielded regions by applying the criterion (9). Typical results from the simulations are shown in Figs. 16 and 17 (blown-up version of Fig. 16).

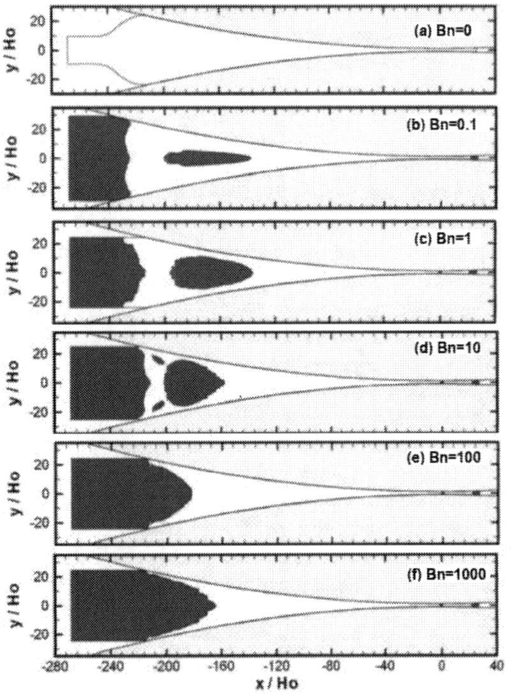

Figure 16 – Yielded/unyielded regions as a function of the Bingham number Bn for viscoplastic fluids obeying the Bingham-Papanastasiou model with $M = 1000$ [18].

Increasing the Bn number leads to a monotonic increase of the unyielded region (shaded solid dark). The yielded / unyielded regions are very interesting as shown in Fig. 16 for the whole flow field, and in blow-up in Fig. 17 near the nip region and exit. Most of the unyielded zones are evidenced in the entering and exiting sheets, which move as rigid plugs with a constant velocity (incidentally not the roll speed U). Unlike LAT predictions, the vast areas between the rolls are yielded, and only small unyielded islands appear in the nip around the centreline. However, the pressure distributions are not that much different from LAT and are not shown here. Hence, all other engineering quantities are well predicted by LAT.

Time-dependent simulations

Time-dependent numerical simulations for Newtonian and viscoplastic fluids were undertaken using the Arbitrary Lagrangian-Eulerian (ALE) formulation for moving free-boundary problems [22,23,24]. The reasoning behind it was that time-dependent simulations could provide the attachment and detachment points as part of the solution, as the free surface is allowed to move to accommodate the incoming and outgoing sheet.

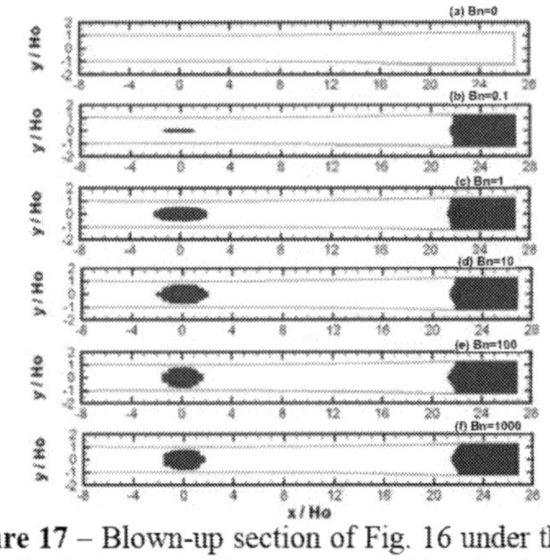

Figure 17 – Blown-up section of Fig. 16 under the nip in calendering of viscoplastic fluids obeying the Bingham-Papanastasiou model with $M = 1000$ [18].

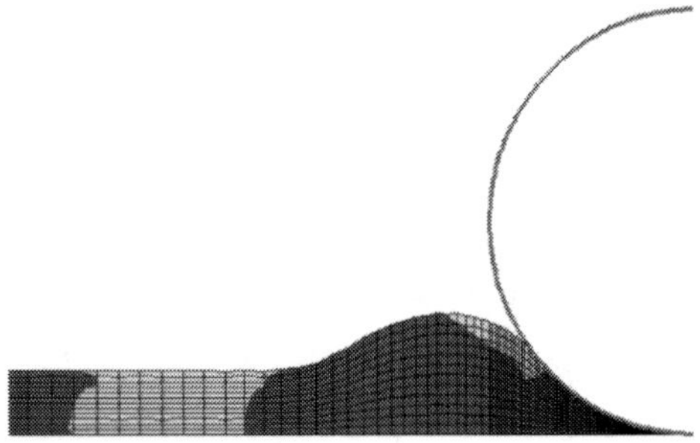

Figure 18 – Time-dependent ALE simulations of an incoming Newtonian fluid. The FEM grid deforms to accommodate the free surface.

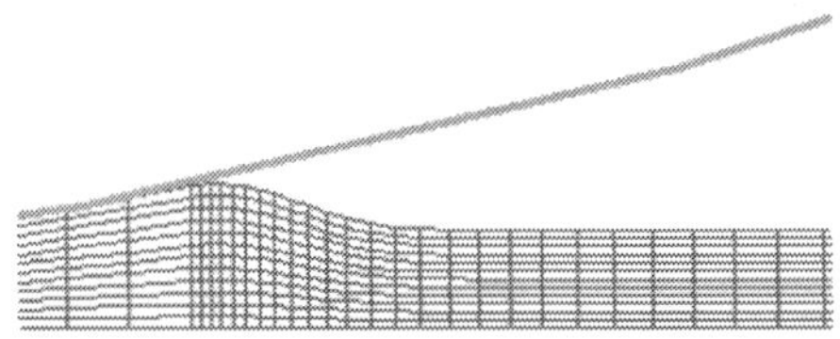

Figure 19 – Time-dependent ALE simulations of an outgoing Newtonian fluid. The FEM grid deforms to accommodate the free surface.

Typical results during the time evolution are shown in Fig. 18 for a Newtonian incoming sheet and in Fig. 19 for an outgoing Newtonian sheet. The FEM grid deforms to accommodate the free surface both upstream and downstream of the nip. The results for the Newtonian fluids reach their steady-state counterparts for a given set of conditions and are in good agreement with the LAT results. However, accurate prediction of attachment and detachment points needs further work for its resolution.

The runs for viscoplastic Bingham fluids at the time of writing are still in progress but they show a good promise. We expect to announce new results at the Conference.

Conclusions

John Vlachopoulos (JV) started his polymer processing career with the process of calendering in the early 70's. He introduced the Finite Element Method (FEM) to solve the governing equations of mass, momentum, and energy based on the Lubrication Approximation Theory (LAT). The 80's saw the introduction by JV of slip and the first fully 2-D simulations for calendering PVC under non-isothermal conditions with slip at the wall and the determination of the free surface. Viscoelasticity was also introduced through the CEF model. In the intervening 25 years, other works have emerged, but our understanding has not been drastically improved since JV's early works.

The only 3-D effort has been done by others (notably Luther [14] in Germany) and showed exciting new results, such as spiral flow and determination of the full melt bank. Some other new results have been obtained for pseudoplastic and viscoplastic fluids using the Herschel-Bulkley model. The emphasis was on finding possible differences with LAT regarding the attachment and detachment points (hence the domain length), and the extent and shape of yielded/unyielded regions. The results showed that while the former is well predicted by LAT, the latter is grossly overpredicted by LAT.

Presently, the transient problem is tackled using the ALE-FEM formulation for moving free-boundary problems. The transient simulations capture the movement of the upstream and downstream free surfaces, and also provide the attachment and detachment points, which are unknown a priori. A full investigation of the latter is still lacking and remains the prevailing challenge in calendering.

Acknowledgements

Financial support from the PEBE 2009-2011 project for basic research from NTUA is gratefully acknowledged. The author is indebted to his mentor, Prof. John Vlachopoulos, who over a period of 30 years – first as his PhD advisor and then as a colleague and above all a friend, had endless exciting and helpful discussions (scientific and otherwise) towards the solution of an exciting and rewarding problem among many others.

References

1. R.E. Gaskell *J. Appl. Mech.* 1950, *17*, 334.
2. S. Middleman *Fundamentals of Polymer Processing*, McGraw-Hill, New York, 1977.
3. J.M. McKelvey *Polymer Processing*, Wiley, New York, 1972.
4. G. Ehrmann; J. Vlachopoulos *Rheol. Acta* 1975, *14*, 761.
5. C. Kiparissides; J. Vlachopoulos *Polym. Eng. Sci.* 1976, *16*, 712.
6. C. Kiparissides; J. Vlachopoulos *Polym. Eng. Sci.* 1978, *18*, 210.
7. J. Vlachopoulos; A.N. Hrymak *Polym. Eng. Sci.* 1980, *20*, 725.
8. E. Mitsoulis; J. Vlachopoulos; F.A. Mirza *Polym. Eng. Sci.* 1985, *25*, 6.
9. J.F. Agassant; M. Espy *Polym. Eng. Sci.* 1985, *25*, 113.
10. R. Zheng; R.I. Tanner *J. Non-Newtonian Fluid Mech.* 1988, *28*, 149.
11. J.F. Agassant; P. Avenas; J.Ph. Sergent; P.J. Carreau *Polymer Processing: Principles and Modeling*, Munich, Hanser Publishers, 1991.
12. L. Levine; C.M. Corvalan; O.H. Campanella; M.R. Okos *Chem. Eng. Sci.* 2002, *57*, 643.
13. D. Mewes; S. Luther; K. Riest *Intern. Polym. Proc.* 2002, *17*, 339.
14. S. Luther; D. Mewes *Polym. Eng. Sci.* 2004, *44*, 1642.
15. S. Sofou; E. Mitsoulis *J. Plast. Film & Sheeting* 2004, *20*, 185.
16. S. Sofou; E. Mitsoulis *J. Polym. Eng.* 2004, *24*, 505.
17. E. Mitsoulis; S. Sofou *J. Appl. Mech.* 2006, *73*, 291.
18. E. Mitsoulis *J. Non-Newtonian Fluid Mech.* 2008, 154, 77.
19. R.B. Bird; G.C. Dai; B.J. Yarusso *Rev. Chem. Eng.* 1982, *1*, 1.
20. T.C. Papanastasiou *J. Rheol.* 1987, *31*, 385.
21. E. Mitsoulis; S.S. Abdali; N.C. Markatos *Can. J. Chem. Eng.* 1993, *71*, 147.
22. C.W. Hirt; A.A. Amsden; J.L. Cook *J. Comp. Phys.* 1974, 14, 227.
23. A. Alexandrou; E. Mitsoulis *J. Non-Newtonian Fluid Mech.* 2007, *146*, 114.
24. E. Mitsoulis; A. Alexandrou *Ann. Eur. Rheol. Conf. AERC'2009*, Cardiff, UK, 2009.

DESIGN AND PROPERTIES OF POLYMER MAGNETIC COMPOSITES

P. Saha[1, 2], N. Kazantseva[1]*, J. Vilcakova[1], R. Moucka[1], M. Omastova[3]

[1]Polymer Centre, Faculty of Technology, Tomas Bata University in Zlin, T.G. Masaryka 275, 76272 Zlin, Czech Republic – saha@ft.utb.cz; kazantseva@ft.utb.cz; vilcakova@ft.utb.cz; moucka@ft.utb.cz
[2]Medical Materials Research Centre, University Institute, Tomas Bata University in Zlin, Nad Ovcirnou 3685, 76001 Zlin, Czech Republic, saha@ft.utb.cz
[3]Polymer Institute, Slovak Academy of Sciences, Dubravska cesta 9, 842 36 Bratislava, Slovakia – upolmaom@savba.sk

**Polymer Centre, Faculty of Technology, Tomas Bata University in Zlin, T.G. Masaryka 275, 76272 Zlin, Czech Republic*

In the present paper we report on the specific features of the electromagnetic properties of polymer magnetic composites based on multifunctional filler that combine the properties of magnetic and conducting materials. The high-frequency magnetic permeability in such hybrid composites can be enhanced due to formation of a core-shell-like structure of a composite, in which ferrite particles are immersed into a conducting medium. Simultaneously, one can vary the dielectric properties of composite by changing the conductivity and thickness of "conducting shell".

Introduction:

New magnetic materials have received considerable attention, especially in studies of polymer composite materials with different types of ferromagnetic fillers. The principal feature of the frequency dependence of the complex permeability μ (magnetic spectra) of polymer magnetic composites (PMCs) is a significant decrease in μ at low frequencies (10^3-10^7 Hz) and a shift of the dispersion region of μ to higher frequencies (10^7-10^{10} Hz) compared with bulk ferromagnets. Thus, in a high-frequency region, the permeability of PMCs is greater than the permeability of bulk ferromagnets, as was reported in a number of works (Bespyatykh and Kazantseva, 2008; Kazantseva et al., 2004; Kazantseva et al., 2006; Nakamura et al., 1994; Slama et al., 2003). Because of these, PMCs have many useful applications, such as magnetic substrates for microwave antennas, microwave inductors, radio-absorbing materials, etc.
The change in the magnetic spectra of composites compared with those of bulk ferromagnets are associated with several physical mechanisms, primarily, with magnetic polarization of magnetic particles isolated by nonmagnetic polymer layer. The interphase polarization and the formation of effective magnetic charges/dipoles are responsible for the non-uniform distribution of magnetization over the bulk of the material. As a result, the effective magnetic field acting on a magnetic particle decreases by the value of the demagnetizing field, which is proportional to the demagnetization factor. Demagnetization factor takes into account the effect of the anisotropy of internal and external demagnetization factors. The anisotropy of internal demagnetization factors is associated with the structural inhomogeneity of the magnetic substance, whereas the anisotropy of external demagnetization factors is associated with the anisotropy of the sample shape. The structural inhomogeneity of PMC is determined by the structural inhomogeneity of the magnetic particles (distortion of chemical composition, inhomogeneous elastic stresses, pores, cracks, surface roughness, etc.), as well as by the structural inhomogeneity of the composite due to the mutual arrangement of magnetic particles in the composite, that depends on the size, shape, and concentration of the particles. Thus, in order to increase the μ, it is necessary to decrease the demagnetizing field within the particle and within the bulk of composite. The permeability is larger for PMC based on magnetic particles with lower demagnetization form-factor N_d. Thus μ is larger for fiber-based composites than for composites filled with spherical and flaky particles. When magnetic filler loading is increased, the effect of particle shape on μ becomes insignificant. For highly-filled PMCs, the external demagnetizing field dominates. The necessity to maintain the physical and mechanical properties of PMCs restricts the volume contents of the magnetic phase, which practically does not exceed 60 vol. %; therefore, one should seek new ways to further increase μ in PMCs.
The main challenge of this work is to design the microstructure of composites, which would have

remarkable effect on the permeability, and thus could be used for its control.

Experimental:

Materials

As composite matrix, commercially available non-filled urethane casting resin (UR 3420, Axson) was used. It consists of isocyanate and polyol components, which are mixed together in the weight ratio of 10:4 and cured at 80°C for at least 4 hours. The final polyurethane (PU) exhibits good dimensional stability and aging resistance. MnZn spinel-type ferrite powder with different particle size distributions (30–40 μm and 40–60 μm) was used as the main magnetic component of PMCs. The ferrite powder was prepared by mechanically grinding sintered MnZn ferrite cores (NM 3000, Ferropribor, Russia). This ferrite consists of 53.75 mol.% of Fe_2O_3, 26.10 mol.% of MnO, and 20.15 mol.% of ZnO. According to the product information, this ferrite is characterized by the initial permeability μ_i ~2700–3000, the Curie temperature T_C = 473 K, the conductivity σ_f = 2 S cm^{-1}, and the density ρ_f = 4.8 g cm^{-3}. MnZn ferrite was used for preparation of two types of magnetic fillers. First type was prepared by mixing of MnZn ferrite with different types of electrically conductive fillers: carbon black (CB), carbon fibres (CF), aluminium powder (Al) and polypyrrole (PPy). Second type was prepared by coating of ferrite particles by conducting polymers, namely polyaniline (PANI) or polypyrrole (Kazantseva et al., 2004; Kazantseva et al., 2006).

Methods

The complex permeability and permittivity of PMCs were studied by combining the impedance and resonant-perturbation methods. In the first instance, the measurements were performed in the frequency range from 1 MHz to 3 GHz by an Agilent E4991A Impedance/Material Analyser. In the latter case, measurements were carried out at discrete frequencies in the frequency range from 2 GHz to 10 GHz by resonant-perturbation method using a TE_{10n} mode rectangular waveguide with iris-coupled end plates.

Results and Discussion:

Figure 1 represents the magnetic spectra of composites filled with MnZn ferrite and its mixtures with different electrically conducting fillers (CB, CF, Al powder and PPy).

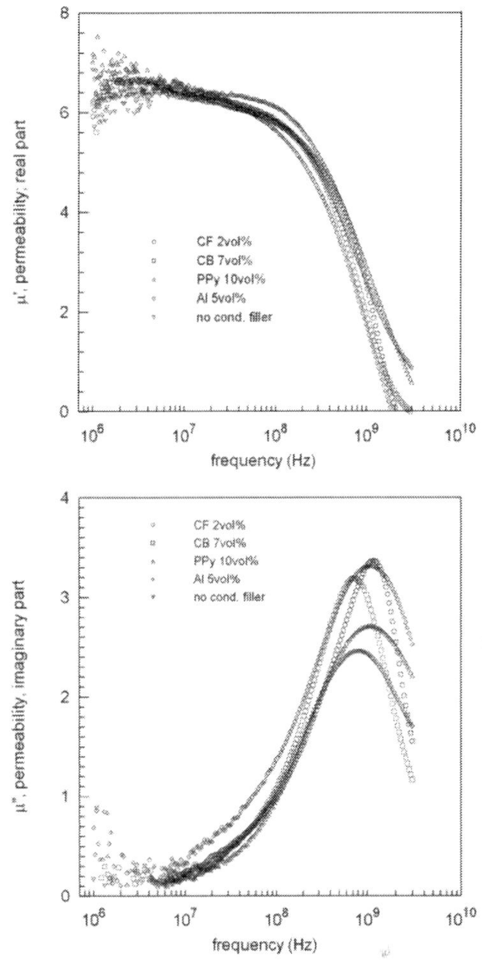

Figure 1 – Magnetic spectra of composites with 40 vol. % of MnZn ferrite and various conducting fillers. The content of conducting filler is given for each curve (vol.%).

The increase in the maximum of magnetic losses depends on the volume ratio between ferrite and conducting filler, which is in each case different. It was established that the increase in the magnetic losses in a composite occurs when the concentration of the conducting filler is higher than the electrical percolation threshold C_σ in the two-component system, namely polymer filled with conducting filler. For example, composites filled with the mixture of ferrite and carbon black show the increase in magnetic losses when the concentration of carbon black in a composite is 3.5 vol. %, which is above the C_σ (Fig. 2).

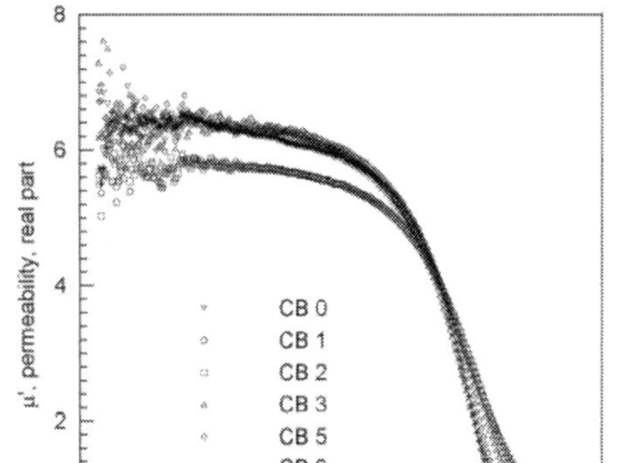

Figure 2 – Magnetic spectra of composites with 40 vol. % of MnZn ferrite and different content of carbon black. The content of carbon black (vol. %) is given for each curve.

The observed increase in magnetic losses is associated with the microstructure of the composites. In both composites (PU+MnZn ferrite and PU+MnZn ferrite+conducting filler), the continuous phase is formed by a polymer, and ferrite particles are distributed in it as inclusions. However, in the case of a composite with multicomponent filler, ferrite particles are embedded in a polymer matrix filled with conducting particles. Each ferrite particle is surrounded by a conducting medium and can be considered as a "core-shell-like" structure, where the core is a ferrite particle and the shell represents a conducting polymer layer that separates the ferrites particles (Fig. 3).

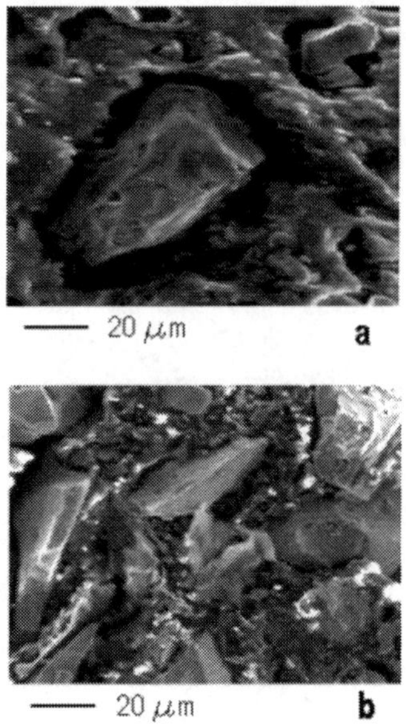

Figure 3 – SEM photographs of polymer composites:
(a) polyurethane with 40 vol. % of MnZn ferrite;
(b) polyurethane with 40 vol. % of MnZn ferrite and 5 vol. % of Al.

From the practical viewpoint, composites with multicomponent filler and "core-shell" structure have the following important property: the character of the frequency dispersion of the complex permittivity and the magnitude of the effective permittivity of a composite can be controlled by choosing the type and concentration of electrically conducting filler in the mixture (Fig. 4); this property may be crucial for the design of radio-absorbers on the basis of such composites.

Core-shell-like structure in a polymer composite is formed under the following conditions: (i) the concentration of ferrite particles in the composite is at least 40 vol. %; in the case of the MnZn ferrite, this concentration is higher than the magnetic percolation threshold in the two-component composite PU+MnZn ferrite; (ii) the size of ferrite particles is much greater than the size of the conducting filler particles, and the volume content of ferrite in the composite is much higher than that of electrically conducting filler; these two conditions ensure a "forced" arrangement (concentration) of conducting particles in the polymer interlayer between the ferrite particles; (iii) the volume contents of conducting particles in the composite is higher than the electrical percolation threshold C_{σ}, so that they form an infinite conducting cluster in the polymer.

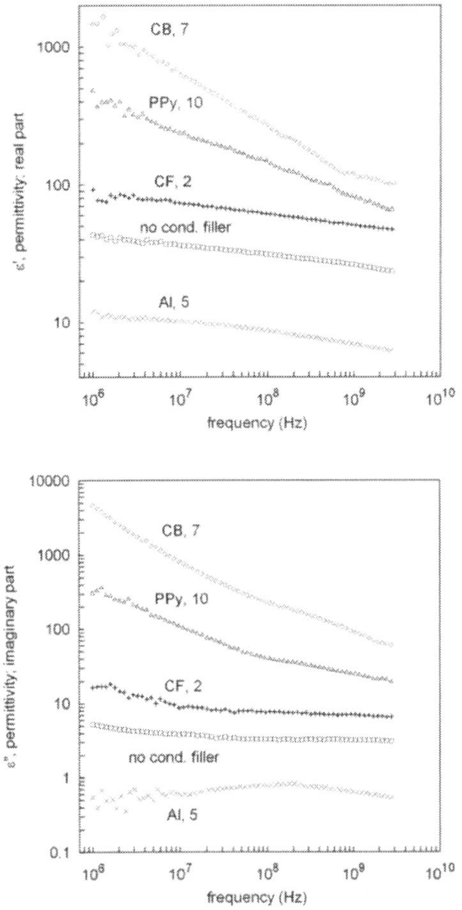

Figure 4 – Dielectric spectra of polymer composites filled with MnZn ferrite and different types of conducting filler.

Composites filled with multicomponent magnetic particles, namely with MnZn ferrite particles coated with nanolayers of conducting polymers (PANI, and PPy), possess the properties of core-like structure, and the μ of these composites exhibits behaviour similar to PMCs based on multicomponent filler. The main parameters of these particles that determine the electromagnetic properties of the composites are the particle size and the thickness and conductivity of the polymer shell. Figure 5 represents the magnetic spectra of composites based on polyurethane filled with MnZn ferrite particles (30–40 μm) coated with polypyrrole. According to thermogravimetric analysis, the concentrations of PPy in the composites were 10 and 15 wt. %. The magnetic spectra show an increase in μ at 15 wt. % of PPy. According to SEM data, this concentration corresponds to the formation of a continuous (defect-free) PPy layer on the surface of particles and the formation of particles with core–shell-like structure (Fig. 6).

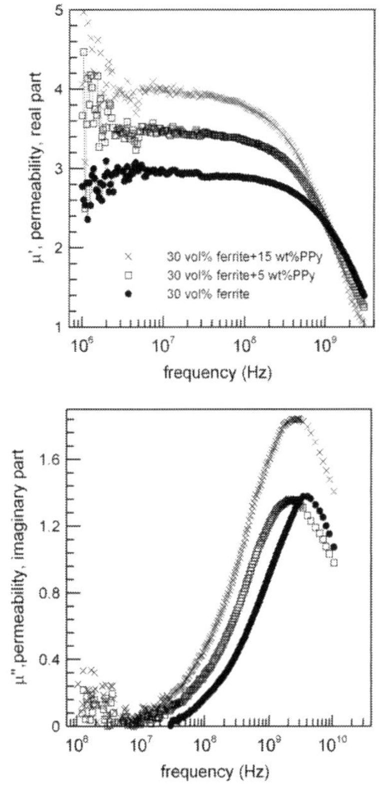

Figure 5 – Magnetic spectra of polyurethane composites containing 30 vol. % of MnZn ferrite particles (30-40-μm) coated with polypyrrole. The content of PPy in composite is given for each individual curve.

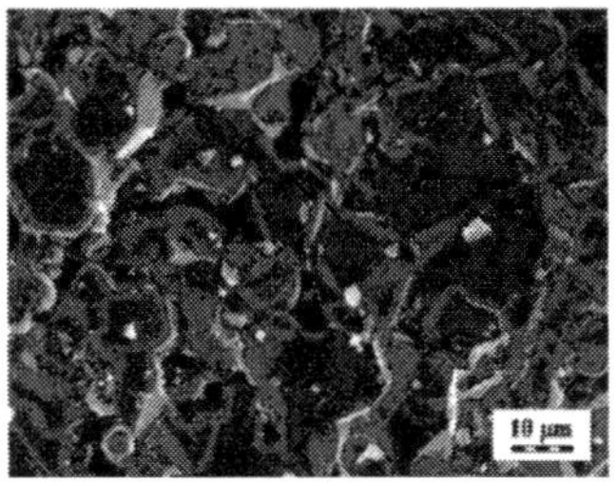

a

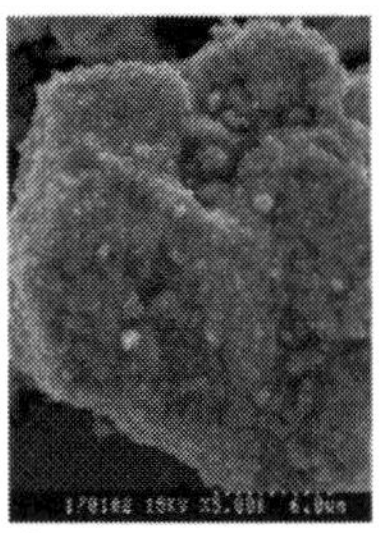

b

Figure 6 – The surface of MnZn ferrite particle before (a) and after (b) coating with polypyrrole (15 wt. %.).

Another factor that determines the μ of composites with multicomponent particles with core–shell-like structure is the conductivity of the shell. Figure 7 shows the magnetic spectra of composites based on polyurethane filled with MnZn ferrite particles (40-60 μm) coated with high-conductivity PANI. For comparison, the same figure represents the spectrum of a composite with pristine ferrite particles of the same size and concentration. The high-conductivity shell reduces the μ of the composite and leads to a significant shift in the resonance frequency f_r from MHz to GHz band. For example, while the composites filled with pristine particles of MnZn ferrite have $\mu' \sim 7$ and the maximum of magnetic losses of $\mu'' \sim 2.5$ at $f_r \sim 900$ MHz, the composites filled with MnZn ferrite particles coated with PANI film with conductivity of 4.4 S cm^{-1} have $\mu' \sim 3$ and the maximum of magnetic losses of $\mu'' \sim 1$ at $f_r \sim 3$ GHz. In addition, the dielectric properties of PMCs with multicomponent filler can be controlled due to the core-shell interfacial polarization (Moucka et al., 2008).

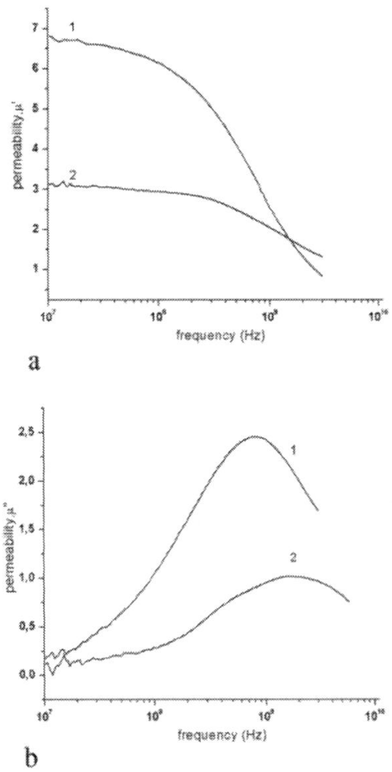

Figure 7 – The frequency dependence of the real (a) and imaginary (b) parts of the complex permeability of a polyurethane composite containing 50 vol. % of 40-60-μm MnZn ferrite particles; (1) no coating; (2) coated with polyaniline with $\sigma = 4.4$ S cm^{-1}.

In order to explain the observed behaviour of the permeability and the resonance frequency in a polymer magnetic composite with a core–shell-like structure, the effect of the conducting overlayer (shell) on the uniform ferromagnetic resonance in a spherical ferrite particle, and on the effective value of magnetic permeability μ of an ensemble of such particles embedded into a nonmagnetic matrix was analysed in (Bespyatykh and Kazantseva, 2008). Theoretical analysis shows that the effect of a conducting overlayer on μ is determined by a dimensionless quantity $\xi = \delta R / \Lambda^2$, where δ is the thickness of the conducting layer, R is the characteristic radius of a ferromagnetic particle, and $\Lambda = (2\pi \sigma \omega / c^2)^{-1/2}$ is the skin depth of the conducting layer (shell). Depending on the relation between the parameters δ, R, and Λ, the conducting layer manifests itself in the displacement of the ferromagnetic resonance frequency and (or) in the enhanced magnetic losses in the material.

Conclusions:

One of promising directions in the design of PMCs with high values of permeability in the radio-frequency and microwave bands is the use of multicomponent fillers and the formation of a core-shell-like structure of composite. A remarkable advantage of PMCs filled with spinal-type ferrite with high value of the initial permeability and low magnetic anisotropy fields is that they allow to control the permeability and make it possible to shift the region of permeability dispersion from MHz to GHz band by coating ferrite particles with conducting polymers (polyaniline or polypyrrole).

Acknowledgements:

The author thanks for the financial support of Ministry of Education, Youth and Sports of the Czech Republic (MSM 708 8532 101 and ME 883 Kontakt).

References:

Bespyatykh Yu. I., Kazantseva N. E., "Electromagnetic properties of hybrid composites", J. Commun. Technol. Electron., **53**, 143-154 (2008)
Kazantseva N. E., et al., "Magnetic behaviour of composites containing polyaniline-coated manganese-zinc ferrite", JMMM, **269**, 30-37 (2004)
Kazantseva N. E., et al., "Magnetic materials based on manganese–zinc ferrite with surface-organized polyaniline coating", JMMM, **301**, 155-165 (2006)
Moucka R., et al., "The influence of interfaces on the dielectric properties of MnZn-based hybrid polymer composites", J. Appl. Phys., **104**, 103718 (2008)
Nakamura T., et al., "Frequency dispersion of permeability in ferrite composite materials", JMMM, **138**, 319-328 (1994)
Slama J., et al., "Frequency dispersion of permeability in ferrite polymer composites" JMMM, **254-255**, 195-197 (2003)

FOUNTAIN FLOW REVISITED: WHAT NEW WE LEARNED SINCE JV'S WORK IN THE 80'S

Evan Mitsoulis*

School of Mining Engineering & Metallurgy, National Technical University of Athens (NTUA)
Zografou 157-80, Athens, GREECE, mitsouli@metal.ntua.gr

Abstract - John Vlachopoulos (JV) had a major impact in injection molding when he with his co-authors Harry Mavridis and Andy Hrymak were the first to accurately solve the fountain flow problem and show its detailed characteristics (PES, 1986). In a series of papers (AIChE J, 1987; 1988, JOR, 1986; 1988, etc.), they addressed the formation of V-shapes, the transient problem, and the viscoelastic non-isothermal problem. In the intervening 20 years, other works have emerged, but our understanding has not been drastically improved since JV's early works.
Just recently, the author revisited the problem. Using the original polar grids by Mavridis, the old steady-state results were reproduced and found to be remarkably accurate even with thousands of elements. New results have been obtained for pseudoplastic and viscoplastic fluids using the Herschel-Bulkley model in full parametric studies of the power-law index and the Bingham number. The effect of various fluid mechanics parameters has also been studied for Newtonian fluids. Viscoelasticity has been introduced with K-BKZ integral constitutive equation fitting rheological data for LDPE and HDPE melts. The emphasis was on finding the exact front shape and the front pressure correction (equivalent to exit or Couette correction in extrusion flows).

Introduction

Fountain flow is a well-known fluid-mechanics phenomenon present in all cases where a moving fluid displaces another immiscible fluid [1]. Within the context of non-Newtonian fluid mechanics, this type of flow is of interest in polymer processing, and in particular in the flow of polymer melts in injection-mold filling [2]. The theoretical problem has been adequately addressed in the mid-80's by a number of researchers [3-5], starting with the pioneering work of Mavridis et al. [3], who for the first time calculated correctly the front position. This is important because it dictates the behaviour of the fluid in the rest of the flow field. These early works dealt primarily with Newtonian fluids and showed intricate patterns developing when tracing particles, in agreement with experiments [3-6]. Mavridis et al. [3] showed in a sample run (for a power-law index of $n = 0.5$) that power-law fluids only slightly affect the position of the flow front. Beris [7] argued that fountain flow patterns (especially the observed V-shapes [8]) are independent of the model and form as a consequence of fluid-mechanics principles that the fluid obeys at the front. More recently, Bogaerds et al. [9] performed calculations with a viscoelastic model (the extended pom-pom [10]) and showed that the flow front is not affected appreciably by the elasticity of the polymer melt, and it does not deviate much from a semi-circle for a wide range of flow rates in a planar domain. All these works have shown that the results range from 0.8 to 1.05 of the radius of the semi-circle, with the lower values corresponding to the axisymmetric case of a Newtonian fluid, while shear-thinning of power-law fluids brings the results closer to a semi-circle front.
Although the problem is well understood from the physics and fluid mechanics points of view, it has become evident that numerically subtle differences exist for different fluid models and types of domain geometry (planar vs. axisymmetric). The problem of finding the free surface in fountain flow is not as easy as in the extrudate swell case [11], which is the prime example of free-surface flow in extrusion, and special care must be taken when constructing the domain discretization with finite elements, the method of choice for this problem. Powerful methods for tackling free-surface problems have been developed, such as the spine method [12], the arbitrary Lagrangian-Eulerian (ALE) method [13] and its variants [14], the volume-of-fluid (VOF) method [15], the level-set method [16], to name just a few. The rheology of the fluids, especially with highly nonlinear models, may also interact in unpredictable ways.
It is, therefore, the purpose of the present paper to revisit the problem of fountain flow for a 3-thronged approach: (a) a full parametric study of various fluid mechanics parameters for Newtonian fluids; (b) a full parametric study for power-law fluids (both shear-thinning and shear-thickening) and of viscoplastic fluids of the Bingham-Papanastasiou type; (c) a study of differences of viscoelastic polymer melts (LDPE and HDPE) for a wide range of apparent shear rates. The emphasis will be on providing detailed results both for the free surface location and the excess pressure in the system (front pressure correction or exit correction) as a function of the various parameters studied.

Mathematical Modeling

The problem at hand is shown schematically in Fig. 1, where a fluid is considered advancing with an average velocity, U, in a stationary frame of reference and thus the process is considered as steady-state [3].

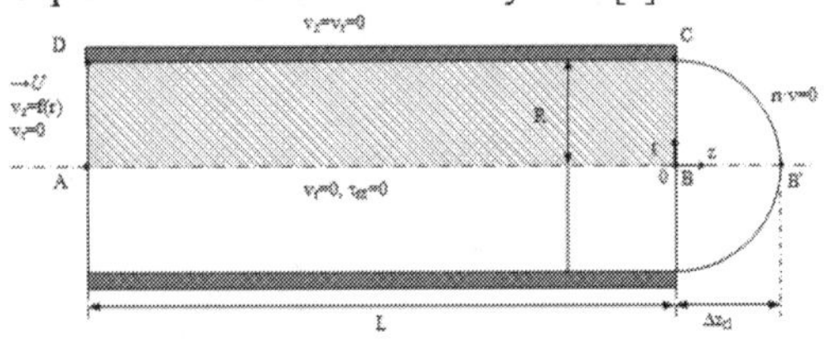

Figure 1 – Schematic diagram of flow domain in a stationary frame of reference and boundary conditions.

Many materials used in injection molding are non-Newtonian, exhibiting either *pseudoplastic* (shear–thinning or –thickening) or *viscoplastic* (presence of a yield stress) behaviour (see, e.g., Bird et al. [17]). Models describing pseudoplastic behaviour include the power-law, Carreau, Cross, etc. Models describing viscoplastic behaviour include the Bingham, Herschel-Bulkley and Casson. The Herschel-Bulkley model has the advantage of reducing – with an appropriate choice of parameters – to the Bingham, power-law or Newtonian model. In simple shear flow it takes the form [17]:

$$\tau = \tau_y + K\dot{\gamma}^n, \qquad \text{for } |\tau| > \tau_y, \qquad (1a)$$

$$\dot{\gamma} = 0, \qquad \text{for } |\tau| \leq \tau_y, \qquad (1b)$$

where τ is the shear stress, $\dot{\gamma}$ is the shear rate, τ_y is the yield stress, K is the consistency index, and n is the power-law index. Note that when $n = 1$ and $K = \mu$ (a constant), the Herschel-Bulkley model reduces to the Bingham model. When $\tau_y = 0$, the power-law model is recovered, and when $\tau_y = 0$ and $n = 1$, the Newtonian model is obtained.

It should be noted that in viscoplastic models, when the shear stress τ falls below τ_y, a solid structure is formed (unyielded). Also in viscoplasticity, a dimensionless yield stress can be defined by:

$$\tau_y^* = \frac{\tau_y}{\left(\frac{dP}{dx}\right)_0 H} = \frac{y_0}{H}, \qquad (2)$$

where H is a characteristic length (half the gap or the radius of the tube), $(dP/dx)_0$ is a pressure gradient for fully developed Poiseuille flow in a channel of length L without the fountain flow, and y_0 is a yield line separating the yielded from the unyielded material. In all cases, the Newtonian fluid corresponds to $\tau_y^* = 0$. However, at the other extreme of an unyielded solid, $\tau_y^* \rightarrow 3$ for planar or 4 for axisymmetric fully-developed Poiseuille flow.

If the dimensionless yield stress is based on the kinematics through the average velocity, it gives rise to the Bingham number, Bn, defined by:

$$Bn = \frac{\tau_y (2H)^n}{KU^n}. \qquad (3)$$

Again, the Newtonian fluid corresponds to $Bn = 0$. However, at the other extreme of an unyielded solid, $Bn \rightarrow \infty$.

The flow is governed by the usual conservation equations of mass and momentum under isothermal creeping flow conditions for an incompressible viscous fluid. The constitutive equation that relates the stresses to the velocity gradients is the generalized Newtonian fluid. In full tensorial form the constitutive equation is written as:

$$\bar{\bar{\tau}} = \eta \bar{\bar{\dot{\gamma}}}, \qquad (4)$$

where η is the apparent viscosity given for the Herschel-Bulkley-Papanastasiou model by

$$\eta = \left[K|\dot{\gamma}|^{n-1} + \frac{\tau_y\left(1-\exp(-m|\dot{\gamma}|)\right)}{|\dot{\gamma}|} \right], \qquad (5)$$

and $|\dot{\gamma}|$ is the magnitude of the rate-of-strain tensor $\bar{\bar{\dot{\gamma}}} = \nabla\bar{v} + \nabla\bar{v}^T$, which is given by

$$|\dot{\gamma}| = \sqrt{\frac{1}{2} II_{\dot{\gamma}}} = \left[\frac{1}{2}\{\bar{\bar{\dot{\gamma}}} : \bar{\bar{\dot{\gamma}}}\}\right]^{1/2}, \qquad (6)$$

where $II_{\dot{\gamma}}$ is the second invariant of $\bar{\bar{\dot{\gamma}}}$.

The stress growth exponent m is the regularization parameter introduced by Papanastasiou [18], and it has to be high enough to approximate the ideal viscoplastic model [19,20]. The dimensionless form of eq. (5) is written as

$$\eta = \left[|\dot{\gamma}|^{n-1} + \frac{Bn\left[1-\exp(-M|\dot{\gamma}|)\right]}{|\dot{\gamma}|} \right], \qquad (7)$$

where the dimensionless stress growth exponent M is given by

$$M = \frac{mU}{H}. \qquad (8)$$

As in our previous works [19,20], to track down yielded / unyielded regions, we shall define yielded and unyielded fluid regions by the following criteria:

yielded:

$$|\tau| = \sqrt{\frac{1}{2} II_{\tau}} = \left[\frac{1}{2}\{\bar{\bar{\tau}} : \bar{\bar{\tau}}\}\right]^{1/2} > \tau_y \qquad (9a)$$

unyielded: $\qquad |\tau| \leq \tau_y \qquad$ (9b)

In parametric studies for pseudoplastic models obeying the power law, the dimensionless parameter is the power-law index, n. For viscoplastic models obeying the Bingham model, the dimensionless parameter is the Bingham number, Bn. In the present simulations, the range of parameters is $n_{lim} < n \leq 2$, and $0 \leq Bn < Bn_{lim}$, where n_{lim} and Bn_{lim} are limiting numbers beyond which no convergence could be obtained.

Dimensionless Numbers

The relevant dimensionless numbers of the Navier-Stokes equations need a characteristic length, L, and a characteristic velocity, U. In the present case these are the channel half gap H (or radius R for axisymmetric flows), and the average incoming velocity. The following numbers are defined, arising from *field flow*:

(1) The Reynolds number, Re, is a measure of inertia forces over viscous forces:

$$Re = \frac{\rho UR}{\mu}. \quad (10)$$

When $Re = 0$, we have inertialess flow. In the current simulations, $0 \leq Re \leq 7$.

(2) The Stokes number, St, is a measure of gravity forces over viscous forces:

$$St = \frac{\rho g R^2}{\mu U}. \quad (11)$$

When $St = 0$, we have no gravity effects. In the current simulations, $0 \leq St \leq 40$.

(3) The compressibility coefficient, B, is a measure of fluid compressibility:

$$B = \frac{\beta \mu U}{R}. \quad (12)$$

When $B = 0$, we have incompressible flow. In the current simulations, $0 \leq B \leq 0.12$.

(4) The pressure-shift coefficient, B_p, is a measure of pressure-dependence of the fluid viscosity:

$$B_p = \frac{\beta_p \mu U}{R}. \quad (13)$$

When $B_p = 0$, we have no pressure-dependence of the viscosity. In the current simulations, $0 \leq B_p < 0.1$.

Also, the following dimensionless numbers arise due to *boundary flow*:

(5) The slip coefficient, B_{sl}, is a measure of fluid slip at the wall:

$$B_{sl} = \frac{\beta_{sl} \mu}{R}. \quad (14)$$

When $B_{sl} = 0$, we have no-slip conditions. When $B_{sl} \approx 1$, we have macroscopically obvious slip. In the current simulations, $0 \leq B_{sl} \leq 100$.

(6) The capillary number, Ca, is a measure of surface tension over viscous effects:

$$Ca = \frac{\mu U}{\gamma}. \quad (15)$$

When $Ca \to \infty$, viscous forces dominate, as is the case with very viscous fluids. When $Ca \to 0$, surface tension dominates, as is the case with very low-viscosity fluids. In the current simulations, $10000 \leq Ca < 0.02$.

Method of Solution

The constitutive equation for the pseudoplastic and viscoplastic fluids must be solved together with the conservation equations and appropriate boundary conditions. Figure 1 shows the solution domain and boundary conditions for the axisymmetric geometry. Because of symmetry, only one half of the flow domain is considered, as was done previously [3-6]. The boundary conditions are therefore (see Fig. 1):

- symmetry along AB' ($v_r = 0$, $\tau_{rz} = 0$),
- no slip along the walls CD ($v_z = v_r = 0$),
- along the inflow boundary DA, a fully developed Poiseuille velocity profile ($v_z(r) = f_1(r)$, $v_r = 0$), corresponding to an inlet average velocity U, which is obtained numerically by 1D-FEM solution (see below).
- along the free surface B'C, vanishing tangential and normal stresses ($(\bar{\bar{\sigma}} \cdot \bar{n}) \cdot \bar{t} = 0$, $(\bar{\bar{\sigma}} \cdot \bar{n}) \cdot \bar{n} = 0$) and no flow through the surface ($\bar{v} \cdot \bar{n} = 0$), where $\bar{n}$ and $\bar{t}$ are the normal and tangential vectors to the surface and $\bar{\bar{\sigma}} = -p\bar{\bar{I}} + \bar{\bar{\tau}}$ is the total stress.

Numerical experiments for setting a reference pressure showed that if it were set to zero at the singular point C, extra zig-zags appeared before and after point C. Also, setting the pressure level at the front centreline equal to 0 is unphysical, as there is a sub-pressure there (see below). The best solution is obtained by not specifying anywhere a reference pressure, as was done apparently in [3], although no mention is made of that.

All lengths are scaled with the radius R in axisymmetric geometry or with H in planar geometry, all velocities with the average velocity U, and all pressures and stresses with KU/R or KU/H, respectively.

The numerical solution is obtained with the Finite Element Method (FEM). It is important to note that when calculating a sensitive quantity such as the excess pressure loss in the system (usually called the Couette or exit correction, but here more appropriately called the front pressure correction), the imposition of an inlet boundary condition has to be accurate or compatible within the FEM context. Based on our previous experience with annular extrudate swell [20], the entrance fully-developed profile was obtained from a 1-D FEM with quadratic interpolation functions for the axial velocity (same order of interpolation as in the 2-D FEM and same discretization in the radial direction). Only then the 1-D and 2-D FEM results were in full agreement, and the flow field was indeed fully developed with parallel streamlines at the domain entry and vertical isobars. This is particularly true for the power-law model away from $n = 1$ (Newtonian) and for the viscoplastic Bingham-Papanastasiou model with the m–exponential term, which always needs a numerical solution [21].

For the domain discretization, special care is needed. Of course, any of the sophisticated free-surface methods developed over the years and mentioned in the Introduction can be used to solve the problem. However, we wanted to find the easiest way to solve this problem, as done originally by Mavridis et al. [3]. Therefore, we have opted for quadratic triangular elements and a polar grid, starting with an initial flow domain, which had a semi-circle at the flow front with the stagnation point now being B'. For this purpose, the original MACVIP program [26] was slightly modified to handle this polar grid arrangement, where now the spines are at a fixed angle ϕ. The new finite element meshes (designated as POLAR-M1 to M3 or for short PM1, PM2, PM3) are shown in Fig. 2 and their characteristics are given in Table 1. POLAR-M1 is the one used by Mavridis et al. [3] and is used here for verification of their results, while the other two meshes are based on our experience, regarding the domain length and the number of radial cuts needed for highly non-Newtonian fluids.

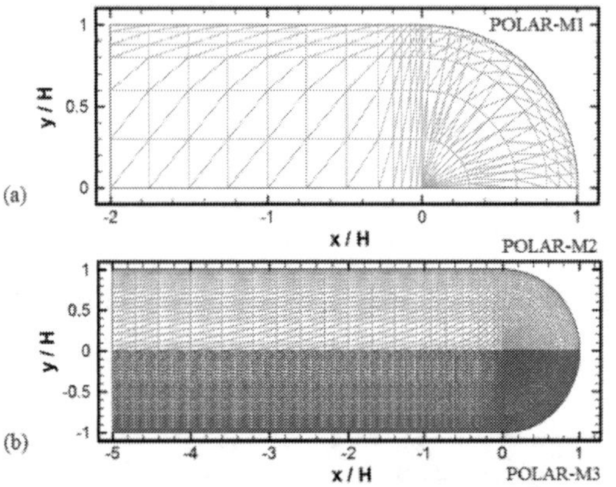

Figure 2 – Polar finite element meshes used in the current computations: (a) POLAR–M1 also used in [3], (b) upper half shows POLAR–M2, while lower half shows POLAR–M3. The meshes shown use a semi-circle as an initial guess for the flow front. Table 1 gives the various mesh characteristics.

Table 1 – Finite element mesh characteristics used in the simulations and Newtonian *planar* results for the centreline location of the flow front, χ, and the front pressure or exit correction, n_{ex}. POLAR corresponds to 6-node triangular elements. POLAR-M1 has been used in [3].

Mesh	*# FS nodes*	*# elem.*	*# nodes*	*# dof*	χ *(n=1)*	n_{ex} *(n=1)*
POLAR-M1	29	298	647	1371	0.893	0.0986
POLAR-M2	39	1541	3182	6983	0.895	0.0983
POLAR-M3	77	6202	12603	28005	0.895	0.0989

The free surface is still found by numerical integration according to the formula (e.g., for the axisymmetric case):

$$z_{new} = z_w + \int_R^r \frac{v_z - V_{B'}}{v_r} dr \, , \tag{16}$$

where z_{new} is the new position of the front and z_w is the exit position at point C (see Fig. 1). The numerical integration is carried out using Simpson's rule on the curve CB' using the velocity $V_{B'}$ but the new change in distance $dz = z_{new} - z_w$ is projected onto the spine to give a new spine radius $r+dr$, according to $dr = dz \sin\phi$. Thus, the spine angle ϕ is always kept constant to avoid distorted meshes. With this method, the results for the free surface were mesh-independent even with few elements, as will be shown below.

In all the above cases, the iterative scheme was direct substitution (Picard iterations) with a zero-order continuation in a parameter (*Re, St, Ca, n, Bn,* etc.). The criteria for termination of the iterative process were for both the norm-of-the-error and the norm-of-the-residuals < 10^{-3}, and for the maximum free surface change < 10^{-5}.

With the polar grid arrangement starting from a semi-circle and employing radial spines, we extended the lower range to $n_{lim} = 0.3$ (stretching it to 0.25 with a larger tolerance for planar flows and to 0.2 for axisymmetric ones). The number of iterations for the Newtonian case was around 8-9 (planar) and 16 (axisymmetric) with an under-relaxation factor ω = 0.5. Convergence was fast and good up to n = 2, while for shear-thinning fluids convergence was slower but still good down to n = 0.5. Below n = 0.5, the number of iterations increased substantially, and under-relaxation reached levels of ω = 0.05 (around n = 0.3). For viscoplastic fluids, the limiting Bn_{lim} number was around 10 for axisymmetric and around 4 for planar geometries. Convergence was fast and good up to Bn = 1, with the number of iterations around 10 and an under-relaxation factor for the vector of unknowns ω = 0.5. After that, the number of iterations increased substantially and it was necessary to decrease the under-relaxation factor down to ω = 0.01 near the limiting Bn–value, while the number of iterations ran into the thousands. The combination of the non-linear rheological model and the highly deformed free surface of the flow front proved to be a very difficult problem to solve.

Known tests with commercial software (namely FIDAP, which included the Newtonian case as an example) have not appeared in the literature, but do not seem to go below n=0.5 [27]. The reasons for this may be attributed to the very steep gradients for highly shear-thinning fluids and highly viscoplastic fluids, which tend to have a flat free surface as an extension of the wall, after which they obtain a precipitous perpendicular free surface down to the centreline. This is in sharp contrast with the extrudate swell problem, which has a free surface running parallel with the flow, and it can still give good solutions in the extremes of n = 0 and Bn = 1000 or higher [20].

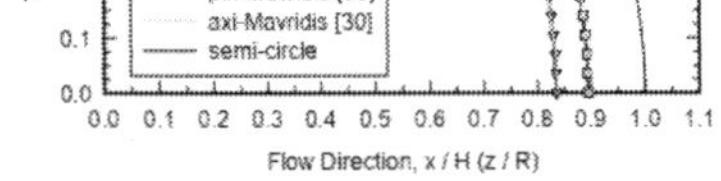

Figure 3 – Comparison of advancing front shapes for planar and axisymmetric flow (Newtonian fluid) [3,30].

Results and Discussion

Newtonian fluids

Before we embark on the non-Newtonian pseudoplastic and viscoplastic fluids, it is instructive to present the Newtonian results in some detail, in order to establish their accuracy. The results are given in terms of the dimensionless change in the location of the advancing front, χ, and of the dimensionless change in pressure drop over and above the fully developed values, n_{ex}. These are defined as follows:

(i) the dimensionless *centreline front location*, χ, defined by

$$\chi = \frac{\Delta z_{cl}}{R} \quad \text{or} \quad \chi = \frac{\Delta x_{cl}}{H}, \tag{17}$$

where $\Delta z_{cl} = z_{cl}-z_w$ is the change in the centreline position z_{cl} with respect to wall position z_w (distance BB′, see Fig. 1), and R is the tube radius; correspondingly for planar geometries, $\Delta x_{cl} = x_{cl}-x_w$ and H is the channel half gap.

(ii) the dimensionless *front pressure correction* or *exit correction*, n_{ex}, defined by

$$n_{ex} = \frac{\Delta P_w - \Delta P_0}{2\tau_w}, \tag{18}$$

where ΔP_w is the overall pressure drop in the system calculated at the wall, ΔP_0 is the pressure drop based on the fully developed flow in the tube (channel) without the fountain flow, and τ_w is the shear stress for fully developed Poiseuille flow at the tube (channel) wall.

The Newtonian values (for $n = 1$) are: $\chi = 0.895$ (planar) and $\chi = 0.835$ (axisymmetric). We note a difference of 6% in favour of the planar geometry (cf. the extrudate swell values of 18.63% for planar and 12.65% for axisymmetric geometries [28,29]). Thus, the maximum thrust forward of the advancing front at the centreline can be at most 89.5% (flat channel) and 83.5% (tubes) for Newtonian fluids. Note that with the previous non-polar meshes, we had found $\chi = 0.931$ (planar) and $\chi = 0.876$ (axisymmetric), so there is a distinct overestimation of the results with non-polar meshes (4-5%). As pointed out before, the fountain flow problem (the prime example of free-surface flow in injection molding) is more difficult than the extrudate swell problem (the prime example of free-surface flow in extrusion), because the free surface is perpendicular to the main flow direction, whereas in extrudate swell the free surface follows the main flow direction. Also, in the Newtonian extrudate swell the domain gets deformed in the order of 13-19%, while in fountain flow the domain is deformed in the order of 84-90%. Starting from a semi-circle, the deformations are then reduced to the same order as in the extrudate swell.

Figure 3 shows a comparison of the present results for the free surface (mesh POLAR-M2) between planar and axisymmetric flows for Newtonian fluids, as done by Mavridis et al. [3] (their Fig. 13), who used the less dense grid POLAR-M1 with 298 triangular elements. As mentioned above, the axisymmetric domain is further away from the semi-circle, followed by the planar case. Our results with the two polar meshes do not reveal any differences. Behrens et al. [5] have produced a table indicating the χ-values, which is updated here as Table 2. The reasons for the discrepancies among various methods are due to the mesh construction and its density, especially around the singular exit point C, as explained above.

Table 2 – Dimensionless centreline front location χ for Newtonian fluids.

Investigator	Geometry	Frame of Reference	χ
Mavridis et al. [3]	Planar	Moving-steady	0.90
Coyle et al. [4]	Planar	Moving-steady	0.93
Behrens et al. [5]	Planar	Moving-steady	0.94
Behrens et al. [5]	Planar	Fixed-transient	0.91
Bogaerds et al. [8]	Planar	Moving-steady	0.92
This work	Planar	Moving-steady	0.90
Mavridis et al. [3]	Axisym.	Moving-steady	0.83
Behrens et al. [5]	Axisym.	Moving-steady	0.82
Behrens et al. [5]	Axisym.	Fixed-transient	0.86
Behrens et al. [5]	Axisym.	Exp. result	0.83±0.04
This work	Axisym.	Moving-steady	0.84

Fluid Mechanics parameters

The calculations are pursued for various fluid mechanics parameters. Figure 4 shows results for different Re numbers ($0 \leq Re \leq 6.8$), while Fig. 5 shows results for different St^- numbers (gravity opposing the flow). It is seen that the front location greatly advances by inertia, while it is greatly reduced by gravity opposing the flow. The latter was also found in a sample run by Coyle et al. [4].

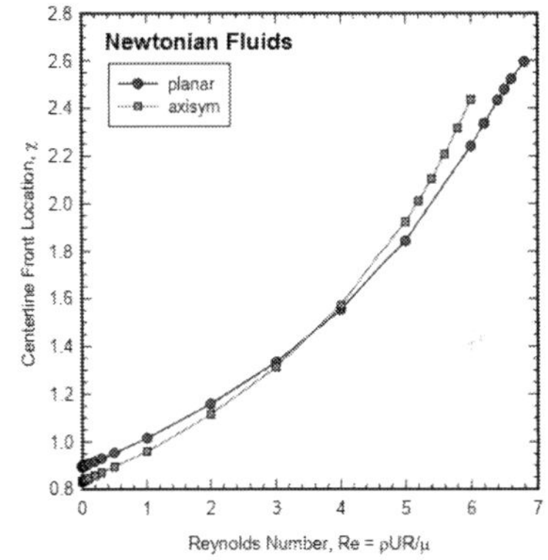

Figure 4 – Dimensionless centreline front location χ as a function of Re for Newtonian fluids.

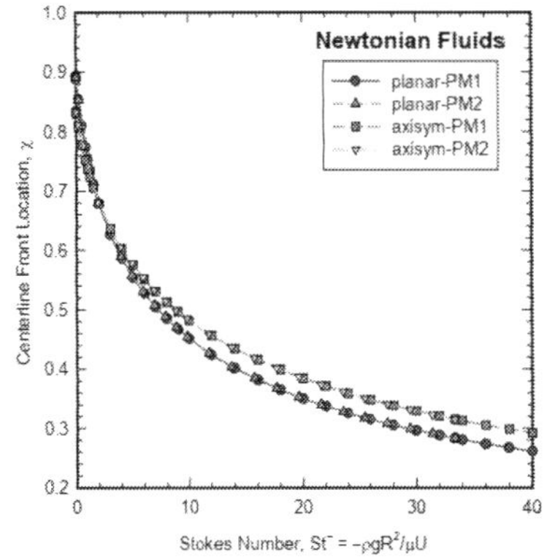

Figure 5 – Dimensionless centreline front location χ as a function of St^- (gravity opposing the flow) for Newtonian fluids. Results from polar meshes of Fig. 2 showing mesh independence.

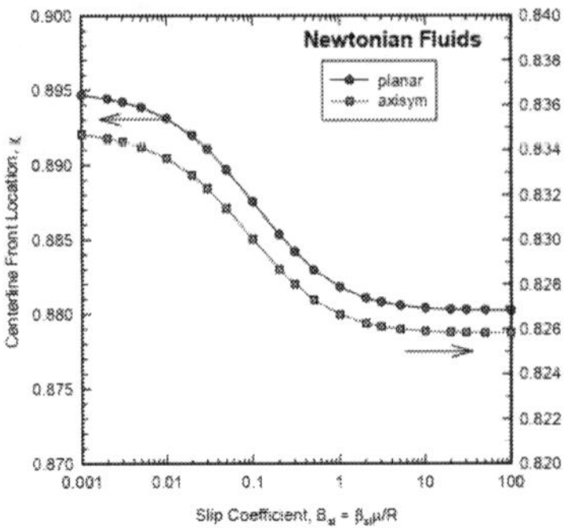

Figure 6 – Dimensionless centreline front location χ as a function of B_{sl} (slip at the wall) for Newtonian fluids.

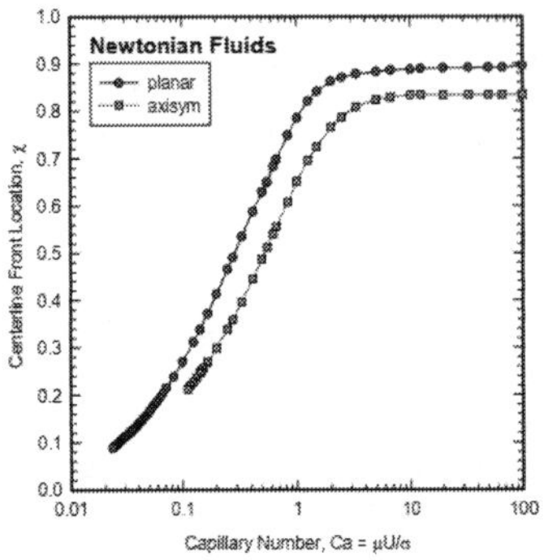

Figure 7 – Dimensionless centreline front location χ as a function of Ca for Newtonian fluids.

Figure 6 shows results for different B_{sl} numbers ($0 \le B_{sl} \le 100$), while Fig. 7 shows results for different Ca numbers ($0.01 < Ca \le 100$) (surface tension effects). It is seen that the front location is slightly reduced by slip, while it is greatly reduced by capillary forces. The full range of results for all fluid mechanics parameters can be found in a recent article by the author [31].

Pseudoplastic fluids

The calculations are pursued for pseudoplastic fluids, both shear-thinning ($n_{lim} \le n \le 1$) and shear-thickening ($1 \le n \le 2$). The Newtonian values are a starting point both for the shear-thinning ($n < 1$) and shear-thickening ($n > 1$) simulations.

The results for χ with the polar meshes POLAR–M1 (PM1) and POLAR–M2 (PM2) are given in Fig. 8. Mesh-dependence is much reduced and virtually non-existent between PM2 and PM3. Axisymmetric flows show a higher mesh-dependence, which is also more evident near the limiting n-values. Shear-thinning behaviour increases the χ–values, reaching for $n = 0.5$, $\chi = 1.007$ (planar) and $\chi = 0.845$ (axisymmetric). Shear-thickening decreases monotonically these values for the planar geometry, reaching for $n = 2$, $\chi = 0.877$, while it increases them for the axisymmetric geometry, reaching for $n = 2$, $\chi = 0.849$.

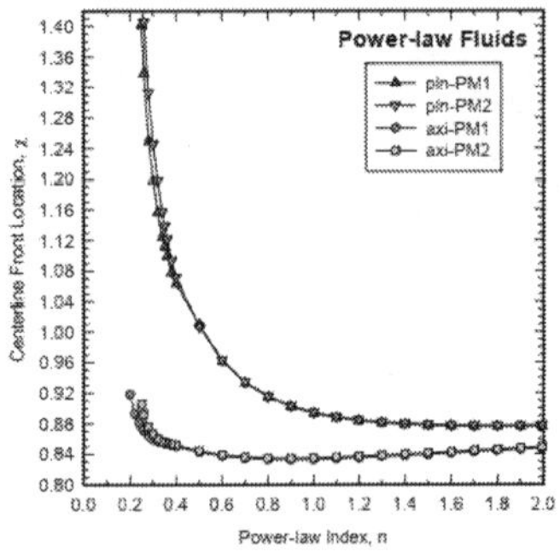

Figure 8 – Dimensionless centreline front location χ as a function of the power-law index n for pseudoplastic fluids obeying the power-law model. Results from polar meshes of Fig. 2 showing mesh independence.

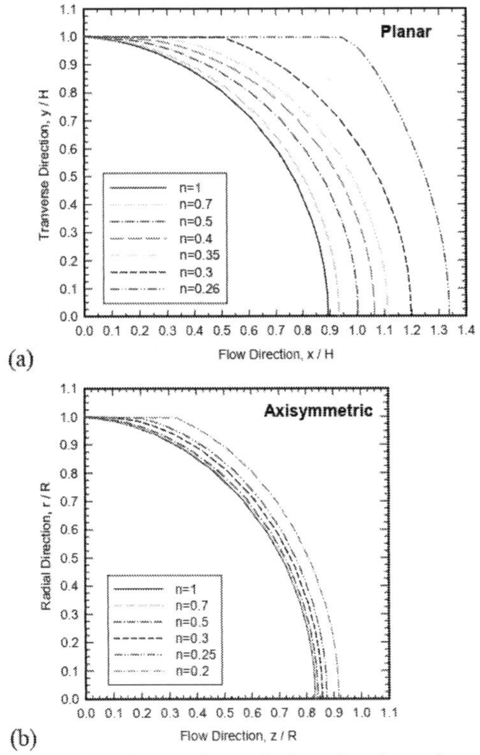

Figure 9 – Comparison of advancing front shapes for Newtonian and shear-thinning power-law fluids: (a) planar flow, (b) axisymmetric flow.

Figure 9a shows a comparison of the present results for the free surface (POLAR–M2) between a Newtonian ($n = 1$) and different power-law fluids for the planar geometry, as done by Mavridis et al. [3] (although they show only a comparison between a Newtonian and a power-law with $n = 0.5$, their Fig. 14). The power-law fluid with $n = 0.5$ matches the semi-circle as was also found out by Mavridis et al. [3]. However, all the interesting results appear after that, when increasing the shear-thinning character of the fluid leads to a dramatic increase in the shape of the flow front. This is in contrast to accepted knowledge that the flow front is basically a semi-circle for all types of fluids. The flow front seems to want to imitate the fully-developed Poiseuille profile, which from parabolic for Newtonian fluids tends to become plug-like for highly shear-thinning fluids. This trend is evident in the upper part of the profile after the die wall, where the fluid wants to continue moving at the same level as the wall, and then bends precipitously towards the centreline. This behaviour is the main reason for losing convergence below a limiting n-value, and making this a difficult problem to solve. Note that numerical tests with Carreau fluids of the same n-values did not perform any better than the power-law model as far as convergence was concerned.

The corresponding results for the axisymmetric case are shown in Fig. 9b. The trends are similar but the axisymmetric case has a much reduced front, and this allows for a lower limiting n-value, $n_{lim} = 0.2$, where it hardly exceeds $\chi = 0.92$.

The exit correction n_{ex} is obtained from the wall pressure values given by the simulations at the entry of the domain upstream according to eq. (18). This is a very sensitive quantity and reflects the adequacy of the domain length, the imposed entry profile, and the finite element mesh used. The results with the polar meshes show a good behaviour regarding mesh-independence, but also the inadequacy of the least dense POLAR–M1 for low values of n. These results are presented in Fig. 10. The Newtonian values are 0.098 (planar) and 0.162 (axisymmetric). The corresponding values from the extrudate swell problem are 0.147 (planar) and 0.227 (axisymmetric) [29].

The results with POLAR–M1 and an entry set at –2 are inadequate below a certain n-value (0.6 to 0.8, depending on geometry) and are discarded. The effect of pseudoplasticity is to increase the n_{ex}-values monotonically and substantially as one moves away from the shear-thickening behaviour into Newtonian and then shear-thinning. Similar results have been obtained for the extrudate swell problem for planar and axisymmetric geometries of pseudoplastic fluids [20].

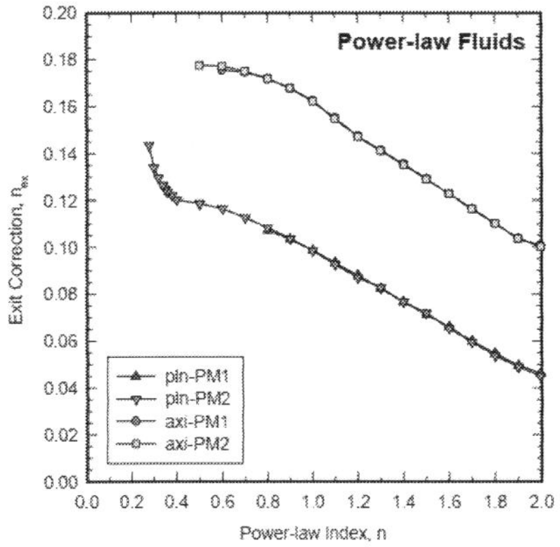

Figure 10 – Dimensionless front pressure or exit correction n_{ex} as a function of the power-law index n for pseudoplastic fluids obeying the power-law model.

Viscoplastic fluids

The calculations are then pursued for viscoplastic fluids, obeying the Bingham-Papanastasiou model (eq. 7 with n = 1 and M = 1000) in a range of Bingham numbers Bn ($0 \leq Bn \leq Bn_{lim}$). The extra feature in viscoplastic materials is the division of the flow field in yielded/unyielded regions by applying the criterion (9), as done in our previous work for extrudate swell from tubes, slits, and annuli [20]. Typical results from the simulations are shown in Fig. 11a for planar (Bn = 2, 3, 3.4, 3.6) and in Fig. 11b for axisymmetric geometries (Bn = 2, 4, 8, 10) top to bottom, respectively. Increasing the Bn number leads to a monotonic increase of the unyielded region (shaded solid dark). There is a distinct difference between planar and axisymmetric results. The planar results show unyielded regions right at the flow front, while axisymmetry prevents that, despite the fact that higher Bn numbers were achieved for the latter case. The lack of convergence for higher Bn numbers prevents us from observing the full phenomenon, where the unyielded regions inside the channel will eventually connect with the one at the flow front, as was the case in extrudate swell [20,32]. Generally speaking, for the same Bn, the planar flow has larger unyielded regions than the axisymmetric one, as also noted first in [32].

It must be borne in mind that the above results for the yielded/unyielded regions are a direct consequence of using the Papanastasiou regularization of the ideal Bingham model, which basically assumes a viscous response for all shear rates. As discussed in [20], strictly speaking the regularized model does not coincide with the ideal Bingham model. However, for all practical and numerical purposes, it yields results, which have physical significance for real viscoplastic fluids.

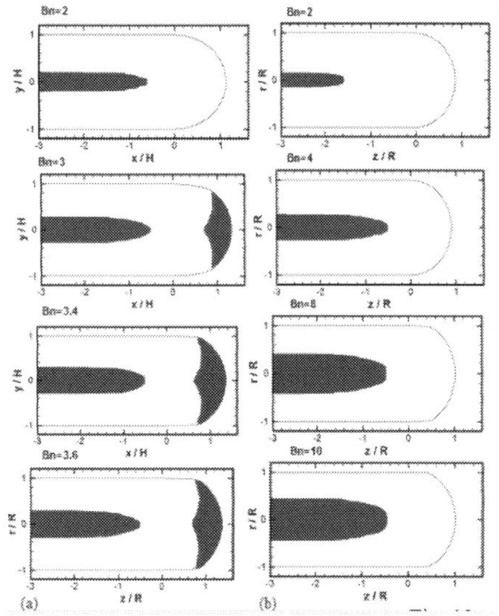

Figure 11 – Yielded/unyielded regions as a function of the Bingham number Bn for viscoplastic fluids obeying the Bingham-Papanastasiou model with M = 1000: (a) planar, (b) axisymmetric results.

The results for the centreline front location (χ) as a function of the Bingham number *Bn* are given in Fig. 12 for the two geometries. There is a monotonic increase for the planar case, which reaches high values for relatively low *Bn* numbers. This also leads to early loss of convergence (Bn_{lim} = 3.6). The axisymmetric case exhibits smaller values and a higher range of convergence (Bn_{lim} = 11.2). An interesting phenomenon is observed in the range 7 < *Bn* < 9, where a bump appears in the results. It is not clear why this happens, as it was observed with all polar meshes.

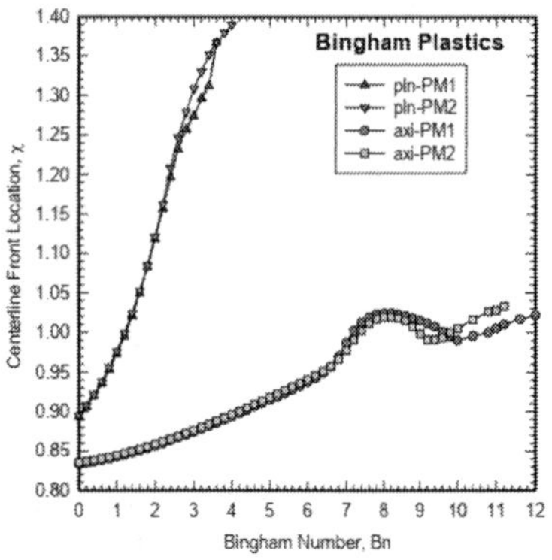

Figure 12 – Dimensionless centreline front location χ as a function of the Bingham number *Bn* for viscoplastic fluids obeying the Bingham-Papanastasiou model with M = 1000.

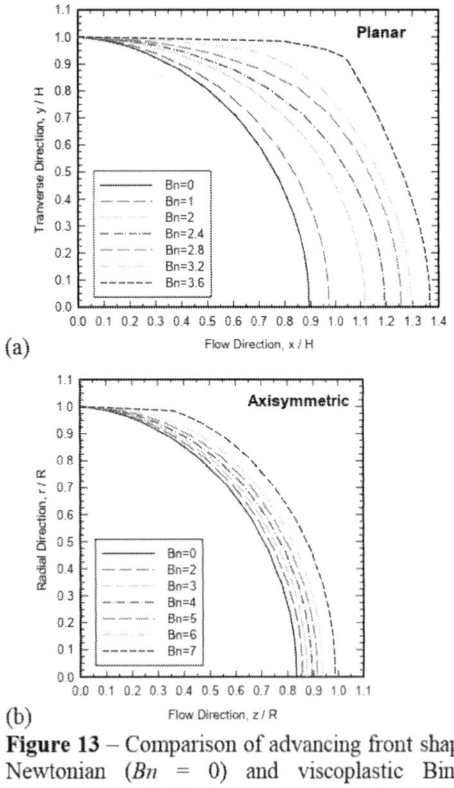

Figure 13 – Comparison of advancing front shapes for Newtonian (*Bn* = 0) and viscoplastic Bingham-Papanastasiou fluids (M = 1000): (a) planar flow, (b) axisymmetric flow.

Figure 13a shows a comparison of the present results for the free surface (POLAR–M2) between a Newtonian (n = 1) and different viscoplastic fluids for the planar geometry. The semi-circle is approached for *Bn* just above 1. As was the case with the power-law shear-thinning fluids, all the interesting results appear after that (*Bn* > 1), when increasing the viscoplastic character of the fluid leads to a dramatic increase in the shape of the flow front. The flow front again seems to want to imitate the fully-developed Poiseuille profile, which from parabolic for Newtonian fluids tends to become plug-like for highly viscoplastic fluids (equivalence of velocity profiles for $n \to 0$ and $Bn \to \infty$). This trend is evident in the upper part of the profile after the die wall, where the fluid wants to continue moving at the same level as the wall, and then bends precipitously towards the centreline. This behaviour is the main reason for losing convergence below a limiting *Bn*-value, and making this a difficult problem to solve.

The corresponding results for the axisymmetric case are shown in Fig. 13b. The trends are similar but the axisymmetric case has a much reduced front, and this allows for a higher limiting *Bn*-value, Bn_{lim} = 11.2, where it hardly exceeds χ = 1.

The results for the exit correction (n_{ex}) as a function of the Bingham number *Bn* are given in Fig. 14 for the two geometries. The exit correction rises monotonically for all cases, provided enough domain length is used (–5R). The results with POLAR–M1 are not reliable (except near the Newtonian solution) due to an inadequate smaller entry length (–2R), and are not shown here.

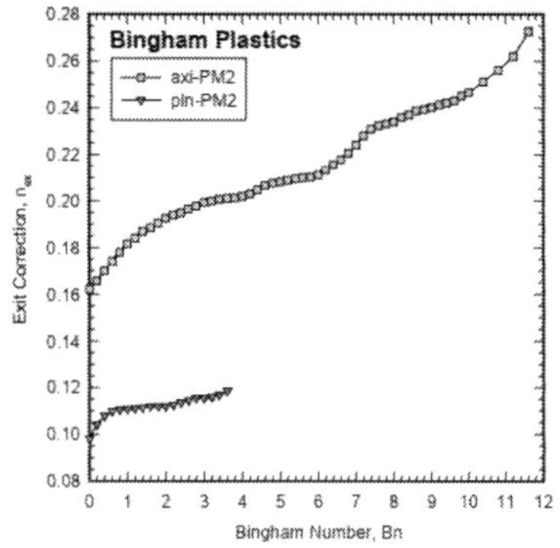

Figure 14 – Dimensionless exit correction n_{ex} as a function of the Bingham number *Bn* for viscoplastic fluids obeying the Bingham-Papanastasiou model with M = 1000.

Viscoelastic fluids

Viscoelastic numerical simulations for a branched LDPE melt and a linear HDPE melt have been done with the K-BKZ model for a wide range of apparent shear rates $\dot{\gamma}_a$ starting from 0.001 s^{-1} and reaching as high as 300 s^{-1}. Typical results (planar case) are shown in Fig. 15, where contours for several variables are given for the highest apparent shear rate $\dot{\gamma}_a$ = 300 s^{-1}.

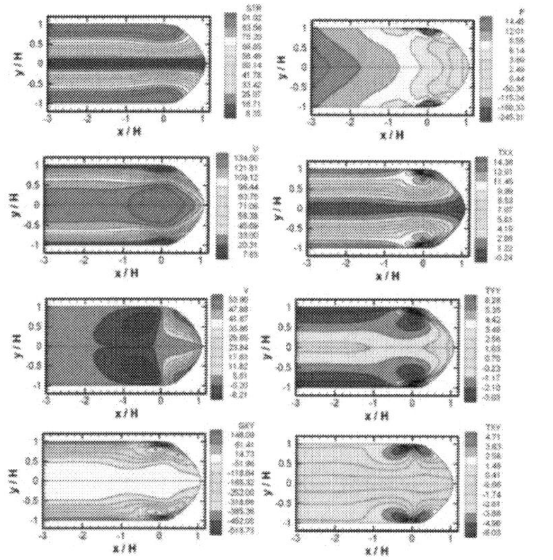

Figure 15 – Contours of kinematic and dynamic variables obtained from the planar simulations of fountain flow for the IUPAC-LDPE melt A at $\dot{\gamma}_a$ = 300 s^{-1}. Eleven (11) contour values are shown between maximum and minimum.

The kinematic variables are the stream function (PSI), the two velocities v_x (U) and v_y (V), and the shear rate $\dot{\gamma}_{xy}$ (GXY), while the dynamic variables are the pressure, p (P), and the 3 extra stresses, τ_{xx}, τ_{yy}, τ_{xy} (or TXX, TYY, TXY). The pressure and the stresses are given in dimensionless form, being divided by $\eta_0 U/H$. The velocities are divided by U. Interesting points to notice and in contrast with the Newtonian flow fields are the squeezing and bending of the streamlines near the exit due to the viscoelastic nature of the melt. Also the fully developed shear flow is reduced compared to the Newtonian. The melt now takes a full gap ($2H$) to rearrange itself near the exit, and produces a bullet-like profile with less roundness than in the Newtonian case. This behavior has been shown quite recently in viscoelastic transient simulations for a viscoelastic polymer solution [33], so it is not a unique feature of the LDPE melt. Because of the assumed non-zero second normal stress difference ($\theta \neq 0$), the isobars are not vertical lines but exhibit a parabolic profile between the walls. The maximum value of the pressure is at entry (not shown), while high negative pressures occur near the singularity, which is now much more pronounced. The shear rates for this case are indeed high, reaching values at the die wall > 500 s^{-1}. The stress contours show that the flow rearranges itself more than a gap from the exit, with most of the rearrangement occurring right at exit where the minima and maxima occur because of the exit singularity.
The front profiles are shown in Fig. 16a for the planar and in Fig. 16b for the axisymmetric geometry for different values of the apparent shear rate, $\dot{\gamma}_a$. These are compared with the base Newtonian case ($Ws=\lambda\dot{\gamma}_a=0$). It is seen that the curvature of the Newtonian fluid is lost and gives rise to a less rounded front shape for high apparent shear rates. For low apparent shear rates we observe a reduction of the profile compared to the Newtonian. Then the profile increases and surpasses the Newtonian, also losing its curvature in the process.

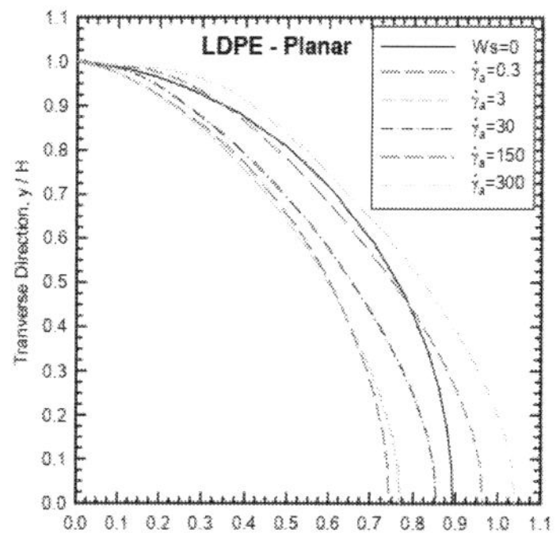

(a)

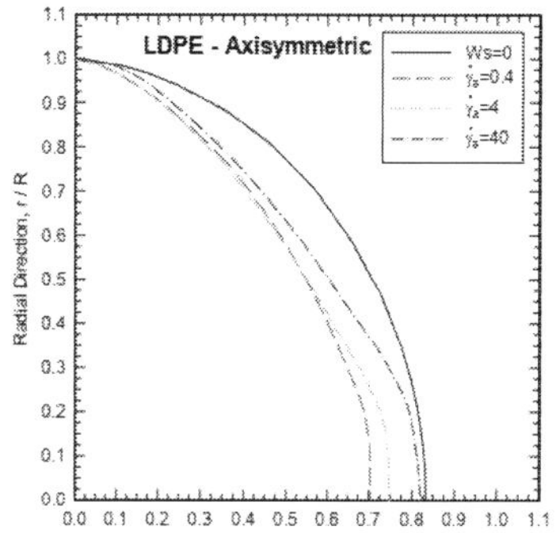

(b)

Figure 16 – Flow front profiles as a function of the apparent shear rate obtained from the simulations of fountain flow for the IUPAC-LDPE melt A.

The dependence of the centerline front location χ on the apparent shear rate $\dot{\gamma}_a$ is presented in Fig. 17a. The behavior of the front location is typical of polymer melts at low shear rates. Namely, starting near the Newtonian value of 89.5% of the half gap H at extremely low shear rates, it first decreases reaching a minimum of 74.5% around 0.3 s^{-1}. After that the present LDPE melt increases substantially reaching values of 104% at apparent shear rates of 300 s^{-1}. The same behavior is observed for both geometries (planar vs. axisymmetric) and with or without a second normal stress difference ($\theta \neq 0$ or $\theta = 0$, respectively). The axisymmetric results are lower than the planar ones in general, while the results with $\theta = -0.25$ are always higher than the ones with $\theta = 0$.
An important observation that can be made from these results is that the phenomenon is nonlinear. In the past, it has been assumed that fountain flow profiles are approximately those of a semicircle [1,2], and they do not change much either for different fluid models or different types of geometry. The current results show that those observations are generally true in a rough

way. The planar results for a 1-mode pom-pom model by Bogaerds et al. [9] show an increase of χ from 0.94 to a maximum of about 1.02 and then a decrease down to 0.98-0.99 within a *Ws*-range from 0.1 to 2.5. Therefore, different fluid models do play a role, the χ-variable depends for polymer melts on the flow rate, and the type of geometry is also an important parameter.

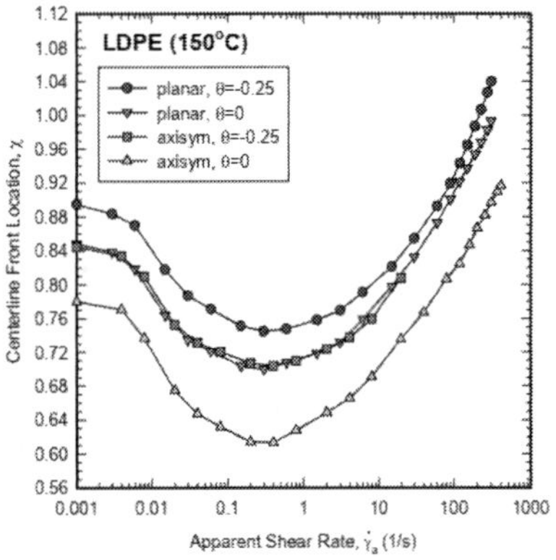

(a)

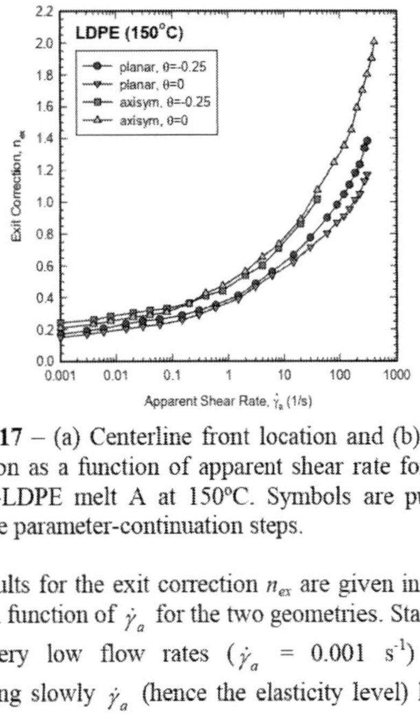

(b)

Figure 17 – (a) Centerline front location and (b) exit correction as a function of apparent shear rate for the IUPAC-LDPE melt A at 150°C. Symbols are put to show the parameter-continuation steps.

The results for the exit correction n_{ex} are given in Fig. 17b as a function of $\dot{\gamma}_a$ for the two geometries. Starting from very low flow rates ($\dot{\gamma}_a$ = 0.001 s^{-1}) and increasing slowly $\dot{\gamma}_a$ (hence the elasticity level) leads now to a monotonic increase of the excess pressure losses. The axisymmetric results are always higher than the planar ones. A negative θ-value gives always higher values than $\theta = 0$ for the planar case, but for the axisymmetric one this is true for the lower range of $\dot{\gamma}_a$ ≤ 0.1 s^{-1}, and the opposite occurs after that.

The viscoelastic numerical simulations for the linear HDPE melt have been carried out for a wide range of apparent shear rates $\dot{\gamma}_a$ starting from 0.001 s^{-1} and reaching as high as 450 s^{-1}. Typical results (planar case) are shown in Fig. 18, where contours are given for the highest apparent shear rate $\dot{\gamma}_a$ = 300 s^{-1}.

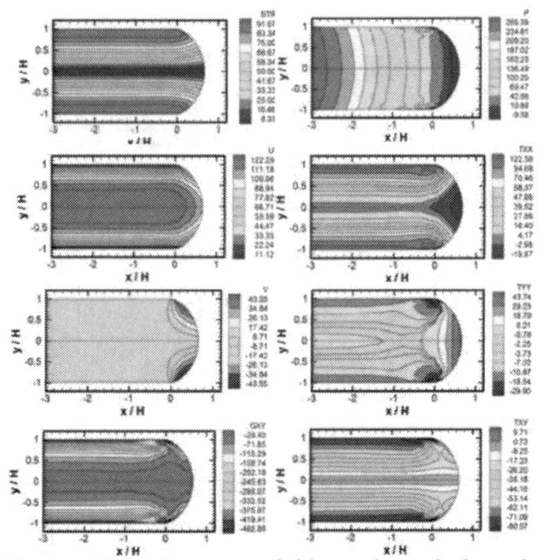

Figure 18 – Contours of kinematic and dynamic variables obtained from the planar simulations of fountain flow for the HDPE melt at $\dot{\gamma}_a$ = 300 s^{-1}. Eleven (11) contour values are shown between maximum and minimum.

The results are for the same variables as for the LDPE of Fig. 15 for easy comparison. Interesting points to notice and in contrast with the LDPE flow fields are the lack of squeezing and bending of the streamlines near the exit due to the much less elastic nature of the melt. The results resemble more those of a purely viscous fluid (like the Newtonian but taking into account the shear-thinning nature of the melt). Also the fully developed shear flow can be seen to occur up to one half-gap H from the exit. The flow front is much reduced compared to the Newtonian and presents none of the bullet-like profile evident with the LDPE melt. Again, as with the LDPE melt and because of the assumed nonzero second normal stress difference ($\theta \neq 0$), the isobars are not vertical lines but exhibit a parabolic profile between the walls. The maximum value of the pressure is at entry (not shown), while high negative pressures occur near the singularity, which is now less pronounced. The shear rates for this case are indeed high, reaching values at the die wall just below 500 s^{-1}. The stress contours show that the flow rearranges itself in less than a half gap from the exit, with most of the rearrangement occurring right at exit where the minima and maxima occur because of the exit singularity.

The front profiles are shown in Fig. 19a for the planar and in Fig. 19b for the axisymmetric geometry for different values of the apparent shear rate $\dot{\gamma}_a$. These are compared with the base Newtonian case (*Ws* = 0). It is seen that the curvature of the Newtonian fluid is now maintained but the profiles are progressively less extended as viscoelasticity increases, and this is true for both geometries. The axisymmetric geometry always gives less extended profiles than the planar one.

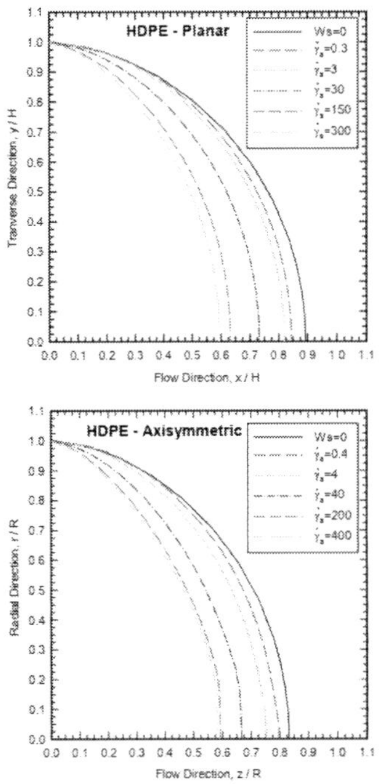

(a)

(b)

Figure 19 – Flow front profiles as a function of the apparent shear rate obtained from the simulations of fountain flow for the HDPE melt.

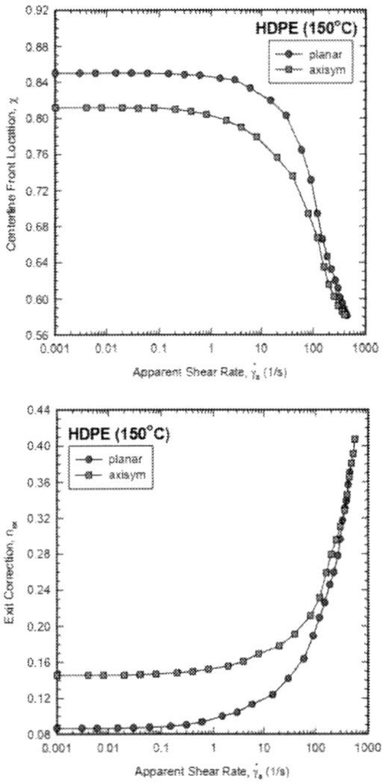

(a)

(b)

Figure 20 – (a) Centerline front location and (b) exit correction as a function of apparent shear rate for the HDPE melt at 150°C. Symbols are put to show the parameter-continuation steps.

The dependence of the centerline front location χ on the apparent shear rate $\dot{\gamma}_a$ is presented in Fig. 20a. The behavior of the front location is now in sharp contrast with that of the LDPE melt, in a sense that it monotonically decreases as the flow rate increases. The axisymmetric geometry always gives lower χ-values.

The results for the exit correction n_{ex} are given in Fig. 20b as a function of $\dot{\gamma}_a$ for the two geometries. As was the case with the Maxwell model, starting from very low flow rates ($\dot{\gamma}_a$ = 0.001 s^{-1}) and increasing slowly $\dot{\gamma}_a$ (hence the elasticity level) leads now to a monotonic increase of the excess pressure losses. This increase is much less pronounced than in the case of the LDPE melt, reaching values around 0.4 while in the former case they reached 2. Again, the axisymmetric results are always higher than the planar ones.

The differences in the results between LDPE and HDPE can be traced to their molecular structure (branched vs. linear), which is reflected in their rheology [34]. The LDPE is much more elastic than HDPE (as evidenced by the average relaxation times <λ>), and this gives rise to higher stresses and pressures, which in turn give much higher exit corrections. The pronounced elasticity of LDPE is responsible for the bullet-like profile of the flow front, which is present either with or without a second normal stress difference (θ-value). In contrast, HDPE produces reduced flow fronts, which get more so as the flow rate increases.

Conclusions

The finite element method has been used to derive numerical solutions for the dimensionless centreline front location and front pressure correction (or exit correction) in fountain flow of Newtonian, pseudoplastic, viscoplastic, and viscoelastic fluids for planar and axisymmetric domains. Polar grids were used for a better representation of the flow field, and the initial guess of a semi-circle is crucial to get accurate solutions.

Despite accepted knowledge that the flow front is close to a semi-circle for all types of fluids, this is not the case for highly non-Newtonian ones. In the case of power-law fluids, it was found that shear-thinning increases substantially the front location compared with the Newtonian value, while shear-thickening does not affect it appreciably. In the case of viscoplastic materials, the front location is also increased substantially as viscoplasticity increases. The non-Newtonian fluids tend to follow their fully-developed Poiseuille velocity profile in the development of the flow front. As this becomes plug-like in the extremes of non-Newtonian behaviour, the flow front tends to become highly extended from the die exit and flat. This is a reason for loss of convergence for low values of the power-law index, n, and high values of the Bingham number, Bn. Viscoelasticity serves to change the profiles depending on the viscoelastic nature of the polymer (LDPE vs. HDPE).

The results from the pressure gave rise to the exit correction, a sensitive quantity of the dynamic variables. The exit correction increases substantially with the non-Newtonian character of the fluids, both for shear-thinning, viscoplastic, and viscoelastic fluids. The problem of fountain flow (a prime example of free-surface flow in injection molding) was found to be much harder to solve than the corresponding problem of extrudate swell (a prime example of free-surface flow in extrusion), since the free surface in fountain flow is perpendicular to the main flow direction and deforms highly. The combination of non-Newtonian rheology and high deformations is still a difficult subject for routine computations.

Acknowledgements

Financial support from the European Union's Centres of Excellence consortium under the title of "Polymer Injection Advanced Molding" (PIAM) project and the PEBE 2009-2011 project for basic research from NTUA are gratefully acknowledged. The author is indebted to Harry Mavridis, Andy Hrymak, and John Vlachopoulos for many helpful hints towards the solution of a rather difficult but much rewarding problem.

References

1. W. Rose *Nature* 1961, *191*, 242.
2. Z. Tadmor; C.G. Gogos *Principles of Polymer Processing*, SPE Monograph Series, Wiley, New York, 1979.
3. H. Mavridis; A.N. Hrymak; J. Vlachopoulos *Polym. Eng. Sci.* 1986, *26*, 449.
4. D.J. Coyle; J.W. Blake; C.W. Macosko *AIChE J.* 1987, *33*, 1168.
5. R.A. Behrens; M.J. Crochet; C.D. Denson; A.B. Metzner *AIChE J.* 1987, *33*, 1178.
6. H. Mavridis; A.N. Hrymak; J. Vlachopoulos *AIChE J.* 1988, *34*, 403.
7. A.N. Beris *J. Rheol.* 1987, *31*, 121.
8. L.R. Schmidt *Polym. Eng. Sci.* 1974, *14*, 797.
9. A.C.B. Bogaerds; M.A. Hulsen; G.W.M. Peters; F.P.T. Baaijens *J. Rheol.* 2004, *48*, 765.
10. W.M.H. Verbeeten; G.W.M. Peters; F.P.T. Baaijens *J. Rheol.* 2001, *45*, 823.
11. R.E. Nickell; R.I. Tanner; B. Caswell *J. Fluid. Mech.* 1974, *65*, 189.
12. S.F. Kistler; L.E. Scriven *Int. J. Num. Meth. Fluids* 1984, *4*, 207.
13. C.W. Hirt; A.A. Amsden; J.L. Cook *J. Comp. Phys.* 1974, *14*, 227.
14. Y. Dimakopoulos; J. Tsamopoulos *J. Comp. Phys.* 2003, *192*, 494.
15. C.W. Hirt; B.D. Nichols *J. Comput. Phys.* 1981, *39*, 201.
16. S.J. Osher; J.A. Sethian *J. Comput. Phys.* 1988, *79*, 12.
17. R.B. Bird; G.C. Dai; B.J. Yarusso *Rev. Chem. Eng.* 1982, *1*, 1.
18. T.C. Papanastasiou *J. Rheol.* 1987, *31*, 385.
19. E. Mitsoulis; S.S. Abdali; N.C. Markatos *Can. J. Chem. Eng.* 1993, *71*, 147.
20. E. Mitsoulis *J. Non-Newtonian Fluid Mech.* 2007, *141*, 138.
21. Z. You; R.R. Huilgol; E. Mitsoulis *Intern. J. Eng. Sci.* 2008, *46*, 799.
22. E. Mitsoulis; A. Matsoukas *Proceedings of the Polymer Processing Society Annual Meeting,* PPS-21, Leipzig, Germany, June 2005, Paper SL7-9, pp. 1-16.
23. E. Mitsoulis *Proceedings of the 5th GRACM Congress on Computational Mechanics,* Eds. G. Georgiou, P. Papanastasiou, and M. Papadrakakis, University of Cyprus, Limassol, Cyprus, June 2005, pp. 645-652.
24. X.-L. Luo; E. Mitsoulis, *Int. J. Num. Meth. Fluids* 1990, *11*, 1015.
25. A. Hannachi; E. Mitsoulis *Adv. Polym. Technol.* 1993, *12*, 217.
26. E. Mitsoulis; J. Vlachopoulos *Adv. Polym. Technol.* 1984, *4*, 107.
27. A.N. Hrymak, private communication, 2009.
28. G.C. Georgiou; A.G. Boudouvis *Int. J. Numer. Meth. Fluids* 1999, 29, 363.
29. E. Taliadorou; G.C. Georgiou; E. Mitsoulis *Rheol. Acta* 2008, *47*, 49.
30. H. Mavridis *PhD Thesis*, Dept. Chem. Eng., McMaster University, Hamilton, ON, Canada, 1988.
31. E. Mitsoulis *AIChE J.*, in press, 2009, DOI 10.1002/aic.12038.
32. S.S. Abdali; E. Mitsoulis; N.C. Markatos *J. Rheol.* 1992, *36*, 389.
33. Y. Dimakopoulos; J. Tsamopoulos *J. Non-Newtonian Fluid Mech.* 2009, *159*, 95.
34. E. Mitsoulis *Intern. Polym. Proc.* 2009, *24*, 439.

NEW TONER TECHNOLOGIES FOR DIGITAL XEROGRAPHIC PRINTING

Enno E. Agur

Xerox Research Centre of Canada
2660 Speakman Drive, Mississauga, Ontario L5K 2L1 Canada

Digital printing, including high quality color printing, has evolved as a mainstream technology for imaging on paper and information sharing as people continue to have a high affinity for paper as a communication media. Laser xerography (or laser printing) is the technology that provides the highest image quality for color, as well as black and white image formation on paper. Xerographic toners are the "dry inks" that form the images on paper in laser printing.

Two advances in toner technology are described. First, a new toner resin and processes for making the toner resin are detailed. This technology enabled Xerox to launch its first monochrome (black and white) and color *digital* production printing products. The resin is a low cost polyester based polymer whose molecular properties are tailored by reactive extrusion. This process enables the formation of resins and toners with unique properties as accomplished by a controlled crosslinking reaction in the extruder resulting in specific product flow and gloss properties.

A further advancement is the development of Emulsion Aggregation (EA) toner. The EA technology is a nano-technology based process that creates micron size toner particles by controlled assembly from very small polymer particles in an aqueous system. The technology enables the formation of particles with controlled shape and building of complex particle structures in the particle bulk and on its surface using an energy efficient manufacturing process. These features result in toners that yield sharper image quality, higher reliability and require significantly less printing energy. Pages printed with EA toner use 40 to 50 percent less material than those printed with conventional toners. EA manufacturing is also 35 percent more energy efficient resulting in an overall energy savings of about 70 percent per printed page.

Introduction

At the beginning of the 21st century we are in the midst of an information explosion, specifically a digital information expansion encompassing a wide area from digital printing, computers, internet and mobile devices. Digital printing, including high quality color printing, has evolved in the last decade as a mainstream technology for imaging on paper and information sharing as people continue to have a high affinity for paper as a communication media. Laser xerography (or laser printing) is the technology that provides the highest image quality for color, as well as black and white image formation on paper. Applications include digital office printing, high volume production printing, graphic arts printing and the like.

Xerographic toners are the "dry inks" that form the images on paper in laser printing. Two advances in toner technology are described.

Reactive Extrusion of Specialty Polymers for Digital Xerographic Toners

For Xerox and Xerox customers to first succeed in the *digital* monochrome (black & white) and color laser production printing market, the toners in the new digital xerographic presses had to meet a large number of stringent property requirements. The compositions of toners are usually in the range 80 to 95 percent by weight polymer and so, the performance of these toners is substantially driven by that of their binder polymers. Therefore, it stands to reason that a key toner design activity was needed in the area of the toner polymer.

For the new digital xerographic presses, there were several requirements placed on the selection of a new binder polymer resin: (i) benchmark performance including fast melting at low temperature, high and low gloss depending on the printing application, etc., (ii) low cost, (iii) rapid delivery to market, and (iv) environmentally friendly manufacturing process. Incremental improvements to existing polymer resins already developed at Xerox were not proving to be successful for this new application. The answer lay in a polymer processing technology used for modifying specialty polymers – reactive extrusion.

Why reactive extrusion? It is a continuous process. Furthermore, it is a high viscosity reactor that enables efficient dispersive and distributive mixing. It is a broadly controlled reactor comprising of segregated and sequential processing, variable screw geometry and widely adjustable temperature, pressure and residence time. It is solvent-less, and therefore, environmental control is much easier than many polymerization processes that deal with solvents.

As stated above, many requirements and specifications were placed on the design and selection of the toner binder polymer. These are summarized in Fig. 1

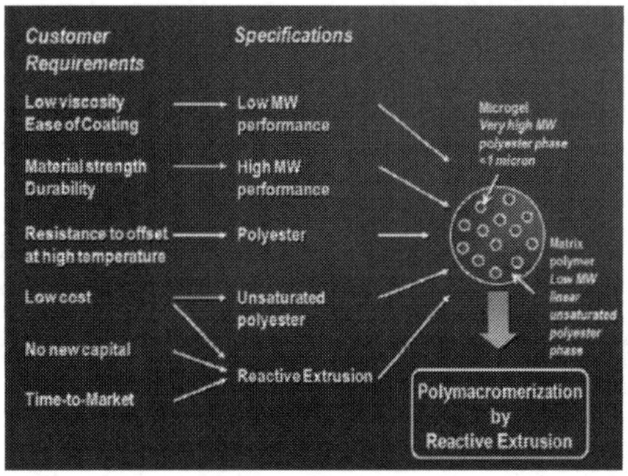

Fig. 1. Requirements and specifications for specialty polymer for xerographic digital toner application.

The end result was a partially crosslinked polyester resin obtained by reactive extrusion (REX). The details of this specialty polymer and the process for producing this material are as follows. Starting with a linear unsaturated polymer like unsaturated polyester and an initiator as shown in Fig. 2, the two are fed to a twin screw extruder where the polymer is melted and partial crosslinking (polymacromerization) is carried out under high shear. Control of the crosslinking reaction is facilitated by initiator concentration, polymer reactivity, extruder temperature, screw design. On exiting the extruder the product is a mixture of well dispersed microgels (less than 1 micron in size) in the linear polymer.

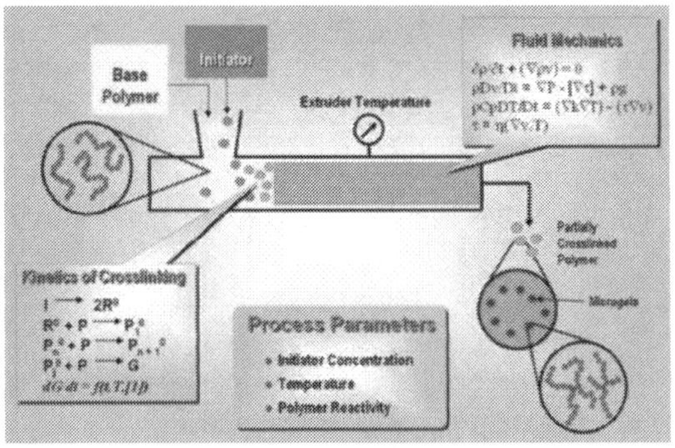

Fig. 2. Reactive extrusion of partially crosslinked polymer for digital xerographic toner application – polymacromerization.

The reaction mechanism for this process is shown in Fig. 3.

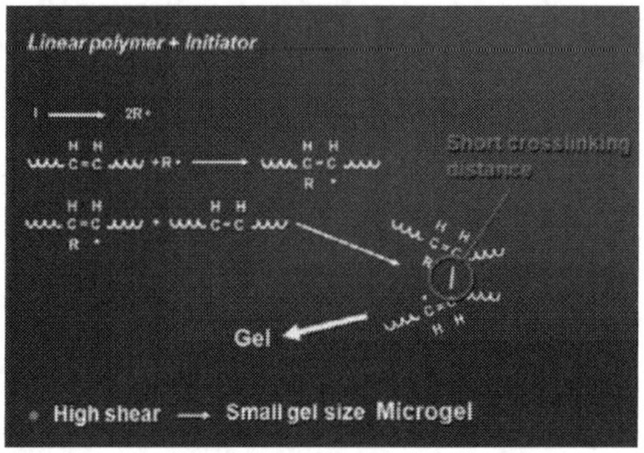

Fig. 3. Reaction mechanism for polymacromerization by reactive extrusion

It should be stressed, that for the formation of the microgels, the reaction must be carried out under high shear such as in the screw channels of a twin screw extruder. Shear stresses in more traditional reaction vessels used in polymerization or polymer modifications are much lower than in extruders resulting in much larger gel particles. A comparison of the two processes is shown in Fig. 4 as relating to the resulting molecular weight distributions.

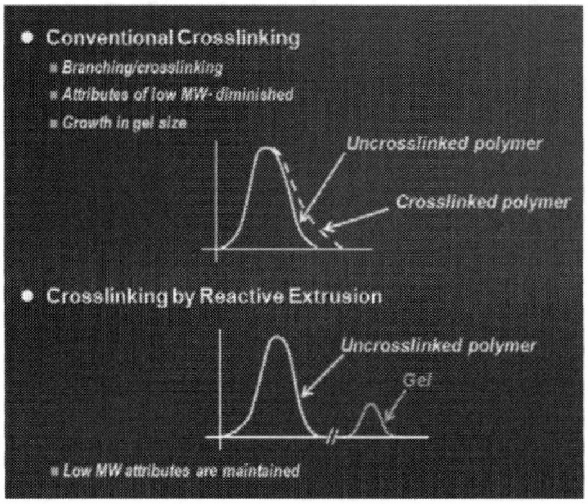

Fig. 4. Comparison of conventional crosslinking vs. reactive extrusion crosslinking.

By having the microgel dispersed in the linear resin, the resulting mixture portrays properties of both the linear resin AND crosslinked resin. Thus the low melting properties of the linear resin are maintained and not impacted by the presence of the microgel particles. The amount of microgel is controlled by the initiator concentration as shown in Fig. 5, extruder temperature, linear polymer reactivity, etc. Control of microgel concentration is critical for printed image gloss control. Increasing the gel content in the toner results in lowering of gloss levels.

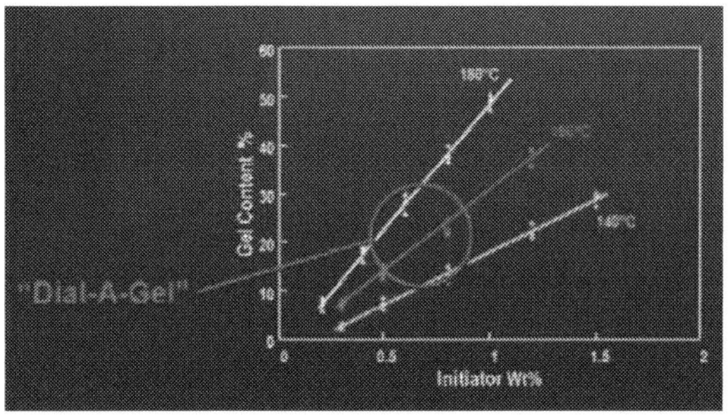

Fig. 5. Gel control in reactive extrusion process.

The end result of the invention and subsequent development was a toner resin with benchmark performance and cost enabling new digital presses, first for the monochrome printing market and later for the color market.

Details of the REX technology are described in a number of patents including Mahabadi et al. (1993, 1994, 1996).

Emulsion Aggregation (EA) Toner Technology

The conventional method of making xerographic toners is based on a melt mixing process (for example, extrusion) in which polymer resins are melted and blended with colorants and other additives . Following this the toner polymeric composite is cooled and then mechanically ground and classified in several steps to micron size particles. This method is very high in energy consumption, especially when targeting smaller than 10 microns toner particles required for a high image quality. Due to the grinding process, the toner particles are also of irregular shape and with a wide particle size distribution as shown in Fig. 6 which further affects print quality.

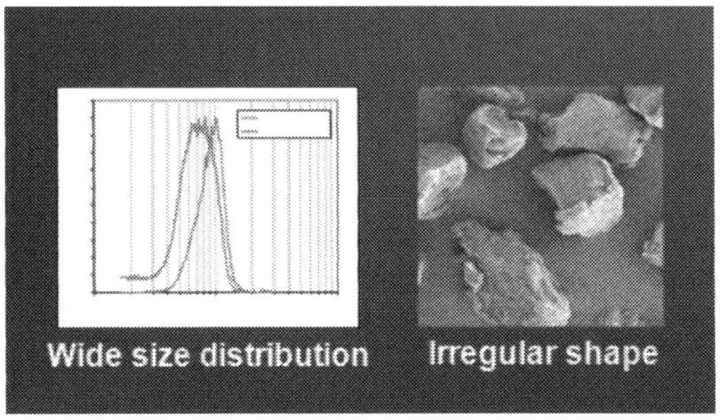

Fig. 6. Toner particle size distribution and shape from conventional toner process.

As the requirements dictated by the digital printing market for higher print quality and lower cost became more and more demanding, it became necessary to adopt a new and radically different approach to the manufacture of toners. The Emulsion Aggregation (EA) toner technology is our answer to meet these requirements, and represents a major advance in the chemistry and chemical processing of preparing toner materials. Toners such as EA and others that deviate fundamentally from conventionally melt mixed toners are often referred to as "chemical toners". The EA technology utilizes sophisticated chemical design and control based on nanotechnology methodology to generate micron sized particles in a bottoms-up approach from nanoscale polymer particles and other submicron size components by a controlled self-assembly as illustrated in Fig. 7.

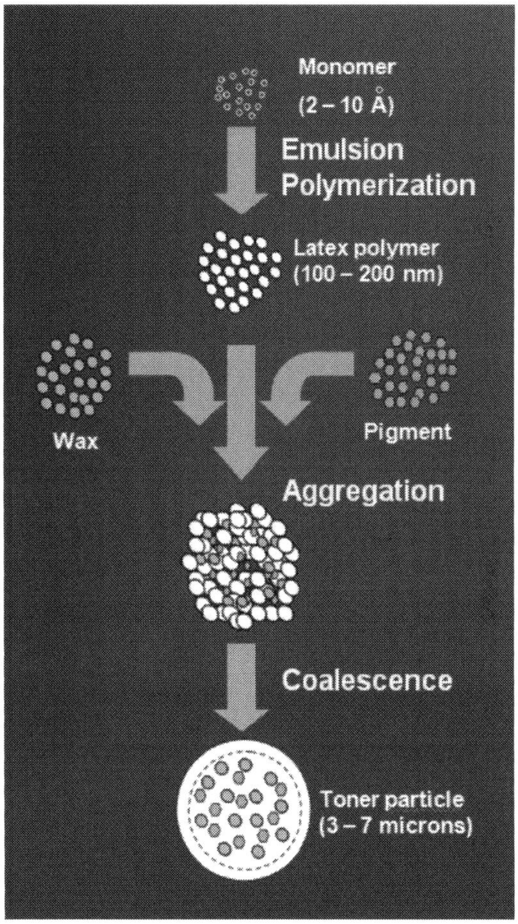

Fig. 7. EA toner process steps.

The first step in the EA process is an emulsion polymerization that generates nanoscale particles in water. It was required to invent a new formulation and polymerization conditions that produced the particles with desired properties, such as a relatively low molecular weight and a wide molecular weight distribution. This was achieved by a novel seeded semi-continuous emulsion with a special addition technique of the surfactant and chain transfer agent.

The second step in the process is the aggregation of the polymer emulsion (latex) particles with colorant (pigment) and wax particles in the form of aqueous dispersions to form rough aggregates of the desired size. It was necessary to identify the reagents and conditions that yielded the best control of this assembly process, as well as to find process parameters that allowed building particular structures (morphologies) of the particle. The end result was the ability to form 3 to 10 micron sized particles and achieve narrow particle size distributions at solids loadings and process times that makes the process economically feasible. From an engineering perspective, large scale process conditions had to be developed that allowed simultaneous control of (i) diffusion of various components, (ii) the electrostatic forces of particles and (iii) the shear characteristics of a highly complex solid-liquid system.

The third process step involves the coalescence of the particles to the targeted shape and desired surface properties. As in the aggregation step, control of the diffusion, electrostatic forces and viscosities of the components is crucial to the successful production of EA toner particles. The complex and precise "onion shell" particle structure that can be obtained by this process is illustrated in Fig. 8

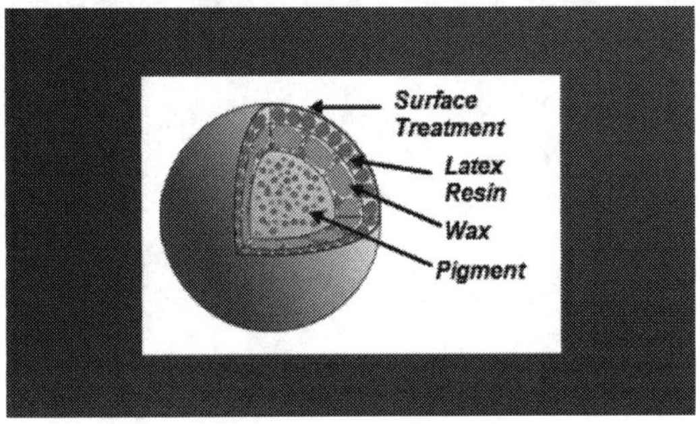

Fig. 8. EA toner particle with controlled structure.

By means of efficient control of the size, shape and structure of the EA particles which are essential for improved print quality, the result is also less toner usage, less toner waste and low energy printing. Typical toners with smoother and rounder surfaces and narrower particle size distributions from this process are shown in Fig. 9.

Thus, we have energy efficient EA toner manufacturing process. The features as discussed above in toners that yield sharper image quality, higher reliability and require significantly less printing energy. Pages printed with EA toner use 40 to 50 percent less material than those printed with conventional toners. EA manufacturing is also 35 percent more energy efficient resulting in an overall energy savings of about 70 percent per printed page.

Details of the EA toner technology are described in a number of patents including Chen et al. (2003), Vanbesien et al. (2008) and Patel et al. (2010).

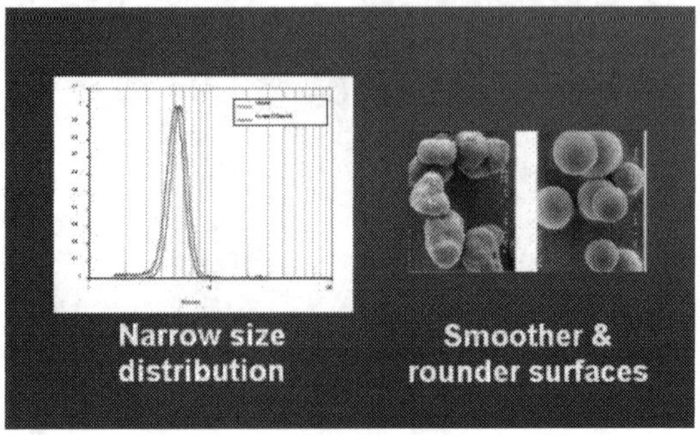

Fig. 9. Toner particle size distribution and shape from EA toner process.

Conclusions

Two advances in toner technology that serve the digital xerographic (laser) printing market are described above. The REX technology was the first developed to meet the requirements of higher print quality and lower cost. This was followed by a more radical approach of toner manufacturing, EA toner technology, which offers our customers many additional advantages including significantly more pages printed by using the same amount of toner and less energy, even higher print quality, environmentally friendly process, and the like.

Acknowledgements

The successful development and commercialization of the REX and EA toner technologies was made possible only by the close collaboration and teamwork of the many Xerox scientists, engineers and staff from our research, development and manufacturing organizations in Canada and U.S.A and with our partners at Fuji Xerox in Japan.

References

U.S. Patent 5,227,460 (1993), Mahabadi, H.K., Agur, E.E., et al.
U.S. Patent 5,352,556 (1994), Mahabadi, H.K., Agur, E.E., et al.
U.S. Patent 5,500,324 (1996), Mahabadi, H.K., Agur, E.E., et al.
U.S. Patent 6,503,680 (2003), Chen, A.K., et al.
U.S. Patent 7,413,842 (2008), Vanbesien, D., et al.
U.S. Patent 7,645,552 (2010), Patel, R.D., Agur, E.E., et al.

A GLOBAL MODELING PROGRAM FOR CO-ROTATING TWIN-SCREW EXTRUDERS

C. Teixeira, J. A. Covas and A. Gaspar-Cunha

I3N / IPC – Institute for Polymers and Composites, Department of Polymer Engineering, University of Minho, Guimarães, Portugal - denebe@dep.uminho.pt, agc@dep.uminho.pt, jcovas@dep.uminho.pt

Abstract - A global plasticating modeling software for twin screw extruders is presented. The program computes the evolution of pressure, average and maximum temperature, shear rate, viscosity, mechanical power consumption, degree of fill, residence time and mixing along a sequence of the various elements (conveying, left handed and kneading blocks) that are used to assemble a typical screw. Modeling is performed from hopper to die, considering solids conveying, melting and melt flow and takes into account the geometrical characteristics of each screw element. The effect of the operating conditions and screw geometry on the flow were compared with experimental data.

Introduction

Intermeshing co-rotating twin-screw extruders are commonly used by the polymer, pharmaceutics and food industries for processing, compounding and reactive extrusion operations, due to their construction flexibility, mixing capabilities, and output and residence time control.
However, flow and heat transfer inside these machines is quite complex, and is significantly affected when the screw profile is modified (e.g. by varying the number and/or the geometry of kneading blocks), or the operating conditions are changed (barrel temperatures, output, screw speed). Not surprisingly, modelling the flow in these machines has been attempted, either describing the overall process based on simplified flow assumptions, or describing in detail the complex 3D flow in specific mixing sections (Potente and Melish, 1996; Vergnes *et al.*, 1998; White *et al.*, 2001, Valette *et al*, 2008).
This work presents a modelling routine that takes into consideration the main process steps from hopper to die, by adopting individual models available in the literature and coupling them in a coherent way. The steps include: i) starve fed solids conveying, ii) non-isothermal solids conveying under pressure, iii) flow of a solid plug separated from the barrel by a melt film, iv) melting of a solid bed surrounded by melt films, v) melting of particles suspended in a melt, vi) non-isothermal melt conveying under pressure, and vi) melt conveying without pressure. The program computes the evolution along the screws of pressure, temperature, mechanical power consumption, shear rate, viscosity, residence time and filling ratio. The predictions are compared with experimental measurements. The response of the system to changes in the main process variables is investigated.

Process Modeling

Geometry

Given the geometrical complexity of the screws used in co-rotating twin-screw extruders, it is important to introduce simplifications than can avoid the need to perform expensive 3D computations. The following types of screw elements are considered:

i) Conveying, or right handed elements, having a positive helix angle and, consequently, high conveying capacity;
ii) Left handed elements, having a negative helix angle and, thus, restricting the flow;
iii) Kneading blocks, combining various kneading discs, with the possibility of creating different staggering angles. Blocks with positive staggering have good conveying capacity. Neutral or orthogonal blocks have no axial drag. Negative staggering angles can be highly restrictive, but they generate intensive mixing, and promote efficient heat.

Figure 1 shows a right handed screw section, using double flighted screw elements. The cross-section is converted into an equivalent rectangular channel with the same cross-sectional area (Booy, 1978; Rauwendal, 1986; White, 1990), fixing the height as $H_{max}/2$. Figure 2 depicts a kneading block and shows that the staggering angle forms two major flow channels, Q_C and Q_L, with positive and negative drag, respectively, both approximately rectangular but with different widths. In each channel, the flow is similar to that developing in right or left handed elements, (Booy, 1978; Vergnes *et al.*, 1998).

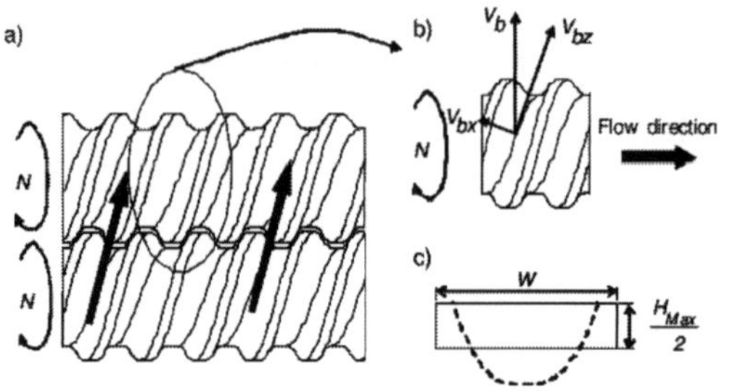

Figure 1- Geometry and flow in right-handed screw elements.

Individual Modeling Steps

The sequence and extension of each process step depends on the screw configuration and operating conditions. Figure 3 illustrates a typical series: 1) solids conveying without pressure, 2) solids conveying under pressure, 3) melting; 4) melt conveying under pressure and 5) melt conveying without pressure.
As soon as the material reaches the first restrictive element, pressure builds up and melting develops very quickly (White, 1990).
The corresponding individual models were linked together in a global model using coherent boundary conditions.

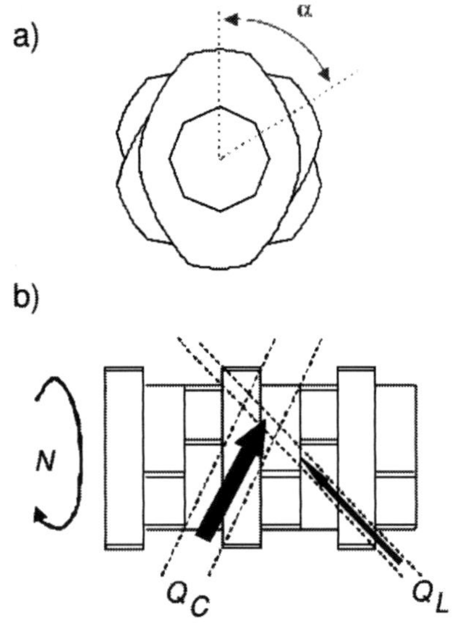

Figure 2- Geometry and flow in a kneading block.

Computer modeling

The global algorithm flowchart is presented in Fig. 4. The operating conditions (screw speed, output, barrel temperature profile), material properties and screw and die geometries are defined. Calculations progress from the screw entrance to the die exit.

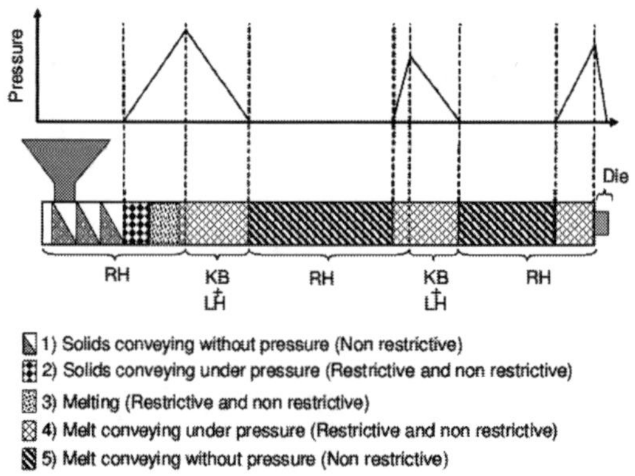

Figure 3- Process steps in a co-rotating twin-screw extruder (RH – Right handed elements; LH- Left)

The program starts by assessing the restrictive character of all screw elements. Then, flow calculations begin at the first restrictive element, an iterative procedure being used to identify the location upstream, L_0, where the channel becomes fully filled. Calculations can now be performed along small Δz channel increments. During flow along the first restrictive element solids conveying, delay, melting and melt conveying are included, while later only melt conveying becomes relevant.
This iterative process is repeated for every restrictive element. In the case of partially filled sections, only the temperature evolution is computed.

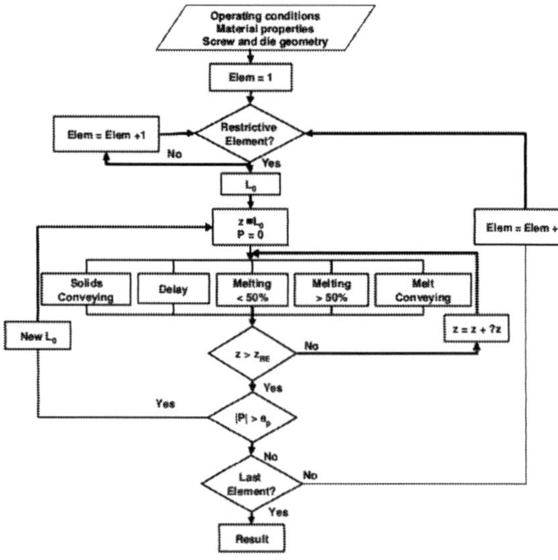

Figure 4- Modeling flowchart.

Case Study

In order to compare model predictions with experimental results, a laboratorial Leistritz LSM 30.34 co-rotating twin screw extruder was used. Figure 5 and Table 1 show the layout, screw configuration, and the location of sampling devices that were used to measure real melt temperatures (Carneiro *et al*, 2000). ISPLEN PP030 G1E, a polypropylene manufactured by Repsol, has the properties summarized in Table 2.

Table 1- Screw profile (a negative pitch indicates a left handed element and a KD-30° denotes a block of left handed kneading discs with a staggering angle of -30°).

Screw Element	**1**	**2**	**3**	**4**	**5**	**6**	**7**	**8**
Length (mm)	97.5	120	60	30	60	30	120	60
Pitch (mm)	45	30	30	30	KD -30°	60	30	20
Screw Element	**9**	**10**	**11**	**12**	**13**	**14**	**15**	**16**
Length (mm)	30	120	60	60	22.5	30	30	30
Pitch (mm)	KD -60°	60	45	30	KD 30°	-20	30	20

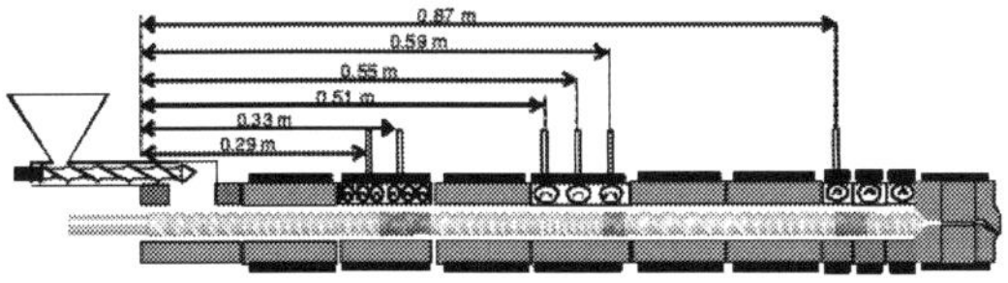

Figure 5- Geometrical description of twin-screw extruder and location of the pressure transducers.

Table 2- Polymer properties.

			PP (ISPLEN PP 030 G1E)	
Density	Solids	ρ_s	690.90	$kg.m^{-3}$
	Melt	ρ	902.00	
Thermal Conductivity	Solids	k_s	0.21	$W.m^{-1.o}C^{-1}$
	Melt	k_m	0.18	
Specific Heat	Solids	C_s	1881.92	$J.kg^{-1}$
	Melt	C_m	1974.55	
Melting	Heat	H	$89.49x10^3$	$J.kg^{-1}$
	Temp.	T_m	170	°C
Viscosity Carreau-Yasuda law		η_0	3041.48	$Pa.s$
		E/R	4023.29	K
		$\bar{\lambda}$	0.17	s
		a	1.82	
		n	0.35	
		T_0	493.15	K

The following parameters were measured experimentally: output, melt temperature (directly on the melt), pressure, fill ratio, mechanical power consumption and residence time distribution. The mechanical power consumption was estimated from the value of the current intensity measured (Rauwendal, 1986). The residence time was estimated through the use of a pigment as a tracer. For this purpose, material samples were collected at the die every 10 s, their pigment concentration being assessed.

Results

Figure 6 shows the computational and experimental pressure and temperature profiles along the extruder for a run with an output of 8kg/hr, a screw speed of 150rpm and a barrel temperature of 220ºC. As expected, the pressure starts to develop in the screw element before the location of the different restrictive elements (5, 9 and 13/14, see Table 1) and before the die. This is due to the flow constraint imposed by the presence of these types of elements. The maximum pressure is obtained at the beginning of the restrictive elements and at the beginning of the die. The temperature increases when the pressure is positive, due to viscosity dissipation and more efficiency heat conduction from the barrel. In partially filled channels the temperature decreases, since set value is lower.
The temperature increases sharply upon melting, and then mainly during flow through mixing zones, due to viscous dissipation, relaxing in partially-filled sections. generally, a good agreement is obtained between the two types of data, with the exception of the initial melt temperatures, that are overestimated. However, measurement of the temperature of a partially melted material is also cumbersome.

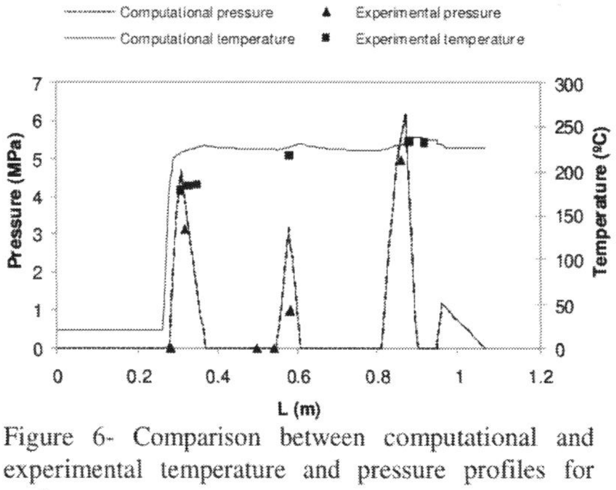

Figure 6- Comparison between computational and experimental temperature and pressure profiles for run1.

Figure 7 presents the local and cumulative residence times for the same run. The cumulative residence time calculated is comparable with the average total residence time measured. As expected, local residence times are higher at restrictive elements. The local residence time is very sensitive to the geometry of each screw element.

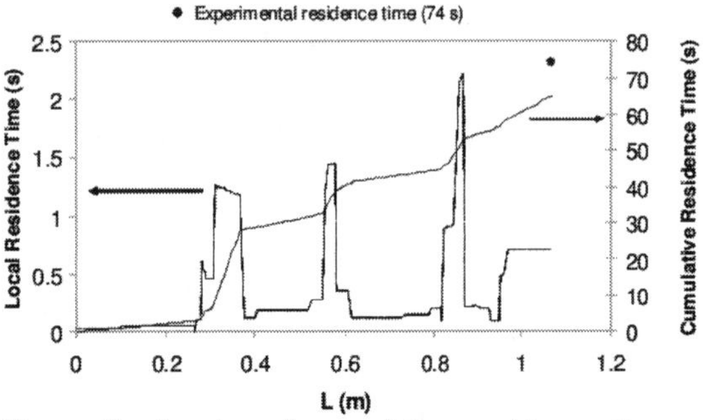

Figure 7- Local and cumulative residence time predicted along screws and die and residence time measured.

Conclusions

A global plasticating modeling software for twin screw extruders is presented. The computations were directly compared with experimental data, satisfactory agreement having been obtained. The program predicts correctly the influence of process parameters on the machine behavior.

Acknowledgements

The authors are grateful to Portuguese Fundação para a Ciência e Tecnologia for supporting this work under grant SFRH/BD//21921/2005.

References

M. L. Booy, "Geometry of Fully Wiped Twin-Screw Equipment", Polymer Engineering and Science, 18, p. 973-984, 1978.

O. S. Carneiro, J. A. Covas, B. Vergnes, Experimental and Theoretical Study of Twin Screw Extrusion of Polypropylene. J. Appl. Polym. Sci., **78**, 1419-1430, 2000.

H. Potente, U. Melisch, "Theoretical and Experimental Investigations of the Melting of Pellets in Co-Rotating Twin Screw Extruders", International Polymer Processing, XI, p. 101-108, 1996.

C. Rauwendaal, Polymer Extrusion, Hanser Publishers, Munich ,1986.

Valette, R., Vergnes, B.and Coupez, T., Int J Mat Forming, 1, 1131, (2008).

B. Vergnes, G. Della Valle, L. Delamare, "A Global Computer Software for Polymer Flows in Corotating Twin Screw Extruders", Polymer Engineering and Science, 38, p. 1781-1792, 1998.

J.L. White, "Twin Screw Extrusion; Technology and Principles", Hanser, Munich, 1990

J. L. White, B. Kim, S. Bawiskar, J. M. Keum, "Development of a Global Computer Software for Modular Self-Wiping Corotating Twin Screw Extruders", Polymer-Plastics Technology and Engineering, 40, p. 385-405, 2001.

MICROSTRUCTURE AND PROPERTIES OF VIBRATION AND LASER WELDED THERMOPLASTIC POLYMER JOINTS

Musa R. Kamal

Department of Chemical Engineering, McGill University,

Montreal, Canada, H3A 2B2 musa.kanal@mcgill.ca

ABSTRACT

The joining of polymers is an important step in the processing and integration of plastics products. In recent years, vibration welding and laser welding of thermoplastics have found growing use, especially in the production of automotive and aerospace components. Each process has comparative advantages and disadvantages. The presentation will give a brief description of each process, with emphasis on associated thermal and flow phenomena. We shall also report the results of studies to evaluate the impact of various factors on the microstructure and strength of the joints for various polymeric systems.

INTRODUCTION

Joining of thermoplastics parts can be achieved by chemical methods using adhesives, mechanical techniques using fasteners, or thermal methods such as welding. Welding involves heating the regions to be joined to create a localized melted zone that causes the molecules to diffuse at the molten interface, which subsequently produces a solid joint upon cooling. The common welding methods for plastics involve either direct heating as in hot plate welding [1], friction as in spin and vibration welding [2,3], or by laser transmission welding [4-6]. The need for effective welding techniques that are fast and precise has led to the expanding use of both vibration and laser transmission welding, which have shown growing use for joining various thermoplastic parts in automotive, electrical, electronics, and medical applications. In this discussion we provide some background and results regarding the microstructure and properties obtained with these two techniques.

Figure 1 illustrates the vibration welding (VW) process. In this process, frictional heat is generated to melt the surfaces of the joint interface between two parts, by pressing together the two parts while vibrating one of them in the plane of the joint. Virtually any kind of thermoplastic material can be used in this process. After 2-3 seconds, vibration is stopped at the exact required relative position of the two parts. Pressure is still applied briefly while the joined parts cool. A solid joint is achieved by allowing the molten interfacial film to solidify. Vibration welding is used for joining large parts that are practically impossible to be joined by other methods. The parts can be circular, rectangular or irregular in shape, as long as the weld joint is in a single plane. Most industrial machines operate at fixed weld frequencies of 120 Hz and 240 Hz, although higher frequencies are also employed. The amplitude of the vibratory motion is normally less than 5 mm (0.2 in.), and the weld time varies from 1 to 10 seconds, but typically it is 1 to 3 seconds. The solidification times usually range from 0.5 to 1.0 second. Total cycle times typically range from 6 to 15 seconds, resulting in 4 to 10 cycles per minute. At higher amplitudes (2 mm to 4 mm; 0.08 in. to 0.16 in.), lower frequencies (100 Hz) are used. On the other hand, higher frequencies (240 Hz) are used at lower amplitudes (0.7 mm to 1.8 mm; 0.03 in. to 0.07 in.). The small amplitudes are useful in applications, where a part has to be welded into a recessed cavity wherein the clearances may not be sufficient for large-amplitude vibration. Weld pressure varies widely (0.5-20 MPa; 72-2900 psi), although pressures at the lower end of this range are used.

Figure 2 illustrates the laser transmission welding (LTW) process. In laser transmission welding (LTW), a laser beam passes through a non-absorbing plastic part. It is absorbed at the contacting surface of a second part, that contains a small amount of an agent that absorbs the laser light (i.e. commonly carbon black), as illustrated in Figure 2. Most unmodified thermoplastics are relatively transparent to laser beams in the wavelength range between 800 and 1100 nm [7]. Good contact between the two parts ensures heating, by conduction, of the non-absorbing material at the interface. Diffusion in the molten material on both sides leads to formation of the weld upon cooling. The applicability of LTW depends significantly on the optical properties of the material to be welded, since these materials usually contain modifiers and/or pigments that influence their spectral transmitting/absorption properties. The effects of the optical properties of several polymers and additives on the transmitting, absorbing, and scattering properties of laser light have been extensively studied and characterized by Kagan and co-workers [8, 9].

VIBRATION WELDING JOINTS

The strength of the joint depends on the process employed. In each process, the shape, dimensions, and microstructure of the polymer within the joint depend on the processing conditions and the physical-chemical characteristics of the joint material. Figure 3 provides a typical polarized light micrograph (PLM) of a layer obtained by microtoming across the heat affected zone (HAZ) in a polyamide-6 (PA-6) joint [10]. It is obvious that the morphology in the HAZ is radically different from that in the bulk on both sides of the weld. Moreover, the morphology within the HAZ is variable, showing a symmetrical five morphological zone structure with varying distributions of crystalline structures. The microstructural characteristics and dimensions of each of these zones depend strongly on the thermo-mechanical history experienced by the material during the VW process and the physical-chemical properties of the material. Various researchers have used mathematical modeling and computer simulation to predict the thermo-mechanical history experienced by the material in the HAZ and to relate the observed microstructure to model predictions [11 - 13].

It should be emphasized that, beside the variation of weld microstructure across the thickness of VW joints, there are significant variations in the microstructure that can be observed across the width of the HAZ, due to melt outflow, as shown in Figure 4 [10]. These variations are most prominent near the open ends of HAZ, where significant changes in the flow pattern take place. These flow patterns depend on the applied pressure and the amplitude and frequency of the vibration cycle. The above variations in dimensions and morphology distributions within the HAZ and the properties of the joint show strong dependence on processing conditions, especially pressure and amplitude, as shown in Figure 5 [14]. It is interesting to note that weaker joints are obtained both at high pressure and high amplitude. Also, it appears that glass fiber reinforced (GFR) PA-6 vibration welded joints appear to be weaker than both the unreinforced joints and the GFR bulk, although the GFR bulk is usually stronger than the unreinforced bulk. It is also interesting to note that it is possible, at low pressure, to obtain unreinforced joints with the same strength as the unreinforced bulk.

LASER TRANSMISSION WELDING JOINTS

Figure 6 depicts a PLM micrograph of the weld structure in the laser affected zone (LAZ) in a PA-6 joint produced by laser transmission welding (LTW) [15]. It is important to note the differences in shape and dimensions between the LAZ and the HAZ. The LAZ exhibits a pseudo-elliptical shape with components in the absorbing and non-absorbing zones. The dimensions of the LAZ subzones in the laser absorbing and non-absorbing subzones are quite different. These characteristics depend on the laser power intensity, as shown in Figure 7 [15], and the material characteristics. However, as expected, the area of the subzone in the absorbing component is significantly larger than that in the non-absorbing component. Figure 8 shows the influence of laser power on the width of LAZ for various polymer joints: PA-6, GFR PA-6, and polycarbonates (PC) [15]. Recently, it has been shown that computer simulation of the process permits reasonable predictions of the shape and dimensions of the LAZ [17]. It should be noted that, in most instances, it is not possible to raise the laser power beyond certain limits, due to burning or degradation of the material. The power limits depend on the material.

As expected, the strength of LTW joints depends on the material, the laser power, and the characteristics of LAZ. Typical experimental data are shown for PA-6 in Figure 9 [16]. These results show that the joint strength increases with laser power, and it increases gradually until it levels off somewhat below the strength of the bulk material.

CLOSING REMARKS

Various factors influence the appropriate choice of a welding technique. Vibration and laser welding provide the ability to achieve precision joints, which are of high quality, at a fast rate.. However, both techniques have limitations with regard to the shapes and dimensions of the materials concerned. Laser welding generally requires a colored (commonly black) component, while mainly simple shapes can be used in vibration welding. In each case, further research is needed to permit optimal design and operation of corresponding welding systems. The key to success in these areas is the understanding and simulation of microstructure of the joints and its influence on joint quality.

ACKNOWLWDGEMENT

The author would like to express his appreciation for financial support from the AUTO 21 Network (Canada). The contributions of the following members of the Thermoplastics Joining Team have made important contributions to various aspects of the aboveresults: P. Bates, E. Zak, M. Chen, Y-M. Chung, M. Al-Wohoush, and J. Calderon.

REFERENCES

1. Stokes, V., "The Vibration and Hot-Tool Welding of Polyamides", Polymer Engineering and Science, 41, 1427-1439 (2001)
2. Potente, H., Uebbing, M., "The Effects of Moisture on the Vibration Welding of Polyamide", SPE ANTEC Proceedings, 1320-1326 (1994)
3. Froment, I., "Vibration Welding Nylon 6 and Nylon 66 – A Comparative Study", SPE ANTEC Proceedings, 1285-1289 (1995)
4. Kocheny, S., Zybko, J., "Three Approaches in Utilizing High Power Diode Laser to Join Thermoplastics", SPE ANTEC Proceedings, 1115-1119 (2002)
5. Kagan, V., "Innovations in Laser Welding of Thermoplastics: This Technology is Ready to be Commercialized", Tech. Paper 2002-01-2011, Automotive & Transportation Technology Congress & Exh., Paris, France (2002).
6. Prabhakaran, R., Kontopoulou, M., Zak, G., Bates, P., and Baylis, B., "Laser Transmission Welding of Glass Reinforced Nylon 6", Tech. Paper 2003-01-1134, SAE, (2003).
7. Grewell, D., Benatar, A., "Beam shaping with spatial modulators for laser micro-welding of plastics", SPE ANTEC Proceedings, 2279-2284 (2006)
8. Kagan, V., Bray, R.G. and Kuhn, W.P., "Laser Transmission Welding of Semi-Crystalline Thermoplastics-Part1: Optical Characterization of Nylon-Based Plastics", SPE ANTEC Proceedings, 1171–1181 (2000)
9. Kagan, V., "Innovations in Laser Welding Technology: State of the Art in Joining of Thermoplastics and Advances with Coloured Nylon for Automotive Applications
10. Chung, Y.-M, Kamal, M. R., "Morphology of PA 6 vibration welded joints and its effect on weld strength", Polymer Engineering and Science, 48(2), 240-248 (2008).
11. V. K. Stokes, "Vibration Welding of Thermoplastics. Part II: Analysis of the Welding Process," *Polym. Eng. Sci.*, **28**: 728-739 (1999)
12. Nonhof and Melchers, "Estimates for Process Conditions During the Vibration Welding of Thermoplastics", *Polym. Eng. Sci.*, **36**: 2018-2028 (1996
13. Young-Mi Chung, Martin E. Weber and Musa R. Kamal, "Prediction Of Temperature, Viscosity And Shear Stress During Steady State Of Vibation Welding Of Polyamide-6", SPE ANTEC 64, 2251 – 2256 (2006).
14. M. Kamal, Young-Mi Chung, R. Gomez, "Three-Dimensional Fiber Orientation in Vibration Welded Joints of Glass Fiber Reinforced Polyamide-6", Polymer Composites, 29(9), 954-963 (2008).

15. M.H. Al-Wohoush, S. Sanchez-Garcia, M. R. Kamal, SPE ANTEC 65, 2803 – 2807 (2007).
16. M. H. Al-Wohoush, M. R. Kamal, SPE ANTEC 66, "The microstructure and quality of laser-welded joints" 1844 – 1848 (2008).
17. M. Chen, G. Zak,P.J. bates, M. R. Kamal, M. Al-Wohoush, "Finite Element and Microstructure Analysis for Contour laserr Transmission Welding of Glass Fiber Reinforced Nylon-6", SPE ANTEC 67, 1295 – 1300 (2009).

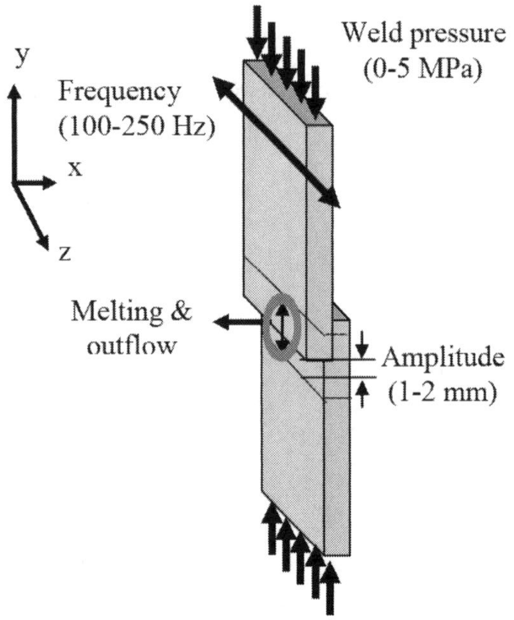

Figure 1. A schematic of the vibration welding process

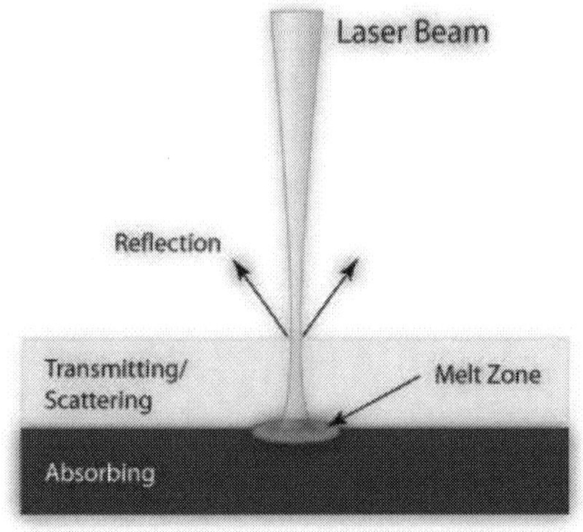

Figure 2. Principle of the laser transmission welding process

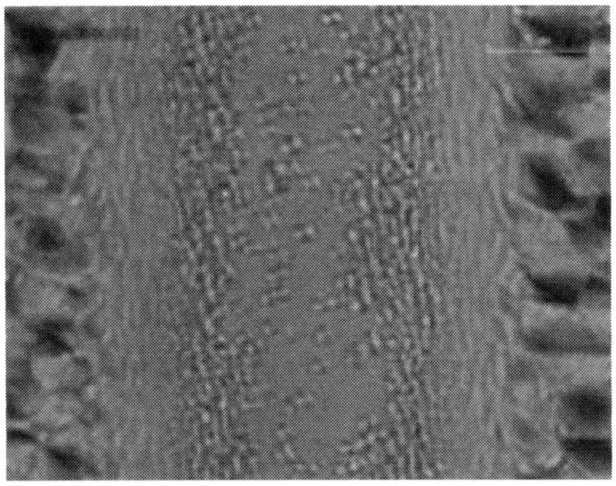

Figure 3. PLM micrographs of microtomed mid-section of vibration welded PA-6 joint

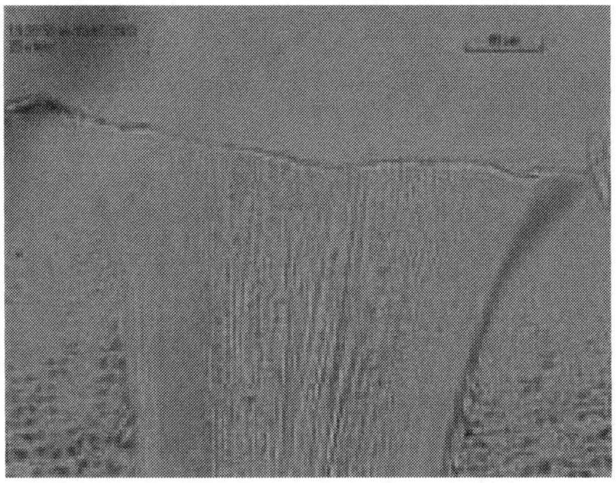

Figure 4. PLM micrographs of microtomed end section of PA- 6 joint

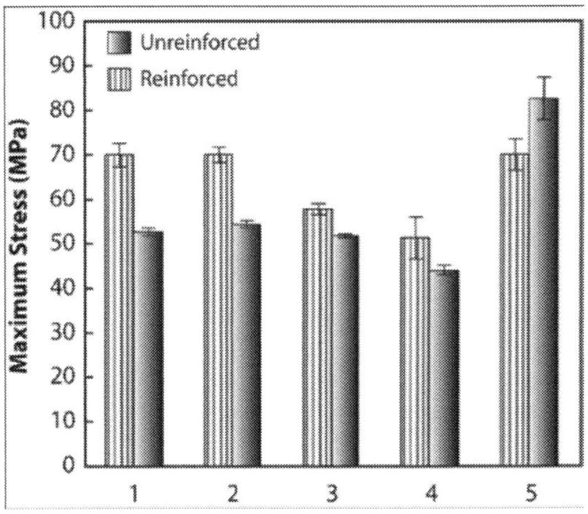

Figure 5. Tensile strength of VW joints of PA-6 ("P" Pa/"Amplitude" mm). (1) 0.6/1 (2) 0.6/1.78 (3) 5/1 (4) 5/1.78 (5) bulk.

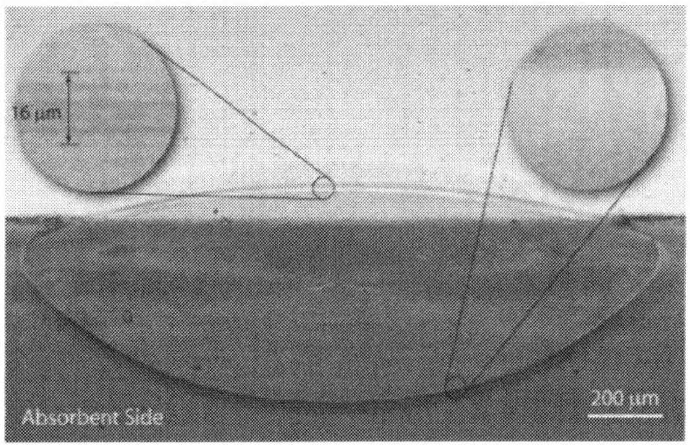

Figure 6. PLM image of laser welded section of PA-6 joint

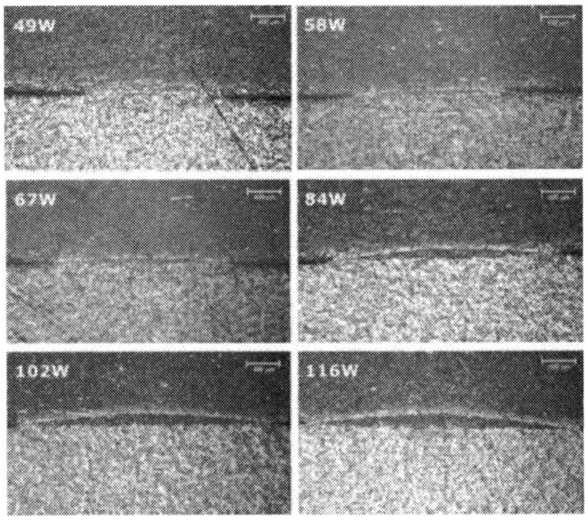

Figure 7. Microstructure of PA-6 welds at varying laser power

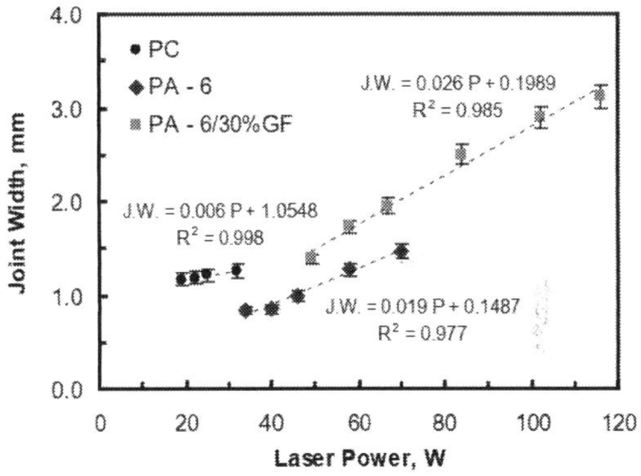

Fig. 8. Effect of laser power on joint width.

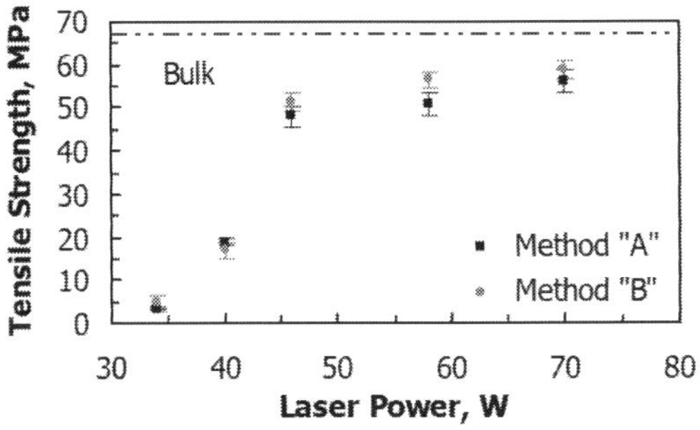

Figure 9. Joint strength vs. laser power

MICROINJECTION MOLDING – SIMULATION OF MOLD FILLING EFFECTS

A.N. Hrymak [(1)*], K. Apostolou [(2)], J. Chu [(3)], M.R. Kamal [(3)], S. Derdouri [(4)]

[1)] Department of Chemical and Biochemical Engineering, University of Western Ontario, London, Ontario, Canada – ahrymak@uwo.ca

[(2)] Faculty of Engineering, McMaster University, Hamilton, Ontario, Canada – apostol@mcmaster.ca

[(3)] Department of Chemical Engineering, McGill University, Montreal, Quebec, Canada - jingsong.chu@mcgill.ca, musa.kamal@mcgill.ca

(4) Industrial Materials Institute, National Research Council of Canada, Boucherville, Quebec, Canada - Salim.Derdouri@imi.cnrc-nrc.gc.ca

Abstract - Micron-sized plastic parts are the subject of increasing interest and have increasing functionality in well-established and emerging technologies. These parts are often manufactured through traditional molding techniques using either modified macro-scale or, preferably, specially designed micro-molding equipment. The high surface-to-volume ratio of micro-parts makes micro-molding significantly different from macro-molding as the effective control of heat transfer has a primary role in micro-sized operations. Many studies have been devoted to the analysis of the mechanical properties and structure of micro-sized parts. Comparisons of mechanical properties to those of macro-sized parts reveal that the micro-sized parts have reduced Young's modulus and often reduced yield stress because of different structural features as skin thickness and crystallinity. Analysis of the effect of processing parameters indicates the need for limiting heat transfer from the molded part to allow enough time for proper crystallization. This can be achieved through raised mold temperature or through innovative techniques as induction heating of the mold and low conductivity mold material. Injection speed appears to be another parameter that strongly influences product properties. A stronger link between the processing conditions and the material characteristics can be drawn through computational studies that provide the thermal history of the polymer as it fills the mold; however, there are limited such studies in the literature. The work of Gova and Lucchetta (2008) appears to be a complete three-dimensional computational study of micro-molding,but the study has some limitations. This presentation describes our modeling attempts on the filling stage of micro-injection molding of thermoplastics. The transient three-dimensional equations of mass, momentum, and energy conservation are solved to reveal the filling profile, temperature, viscosity, pressure, and velocity of the polymer as it fills the mold. The effect of processing conditions on the filling profile and polymer properties is examined. Comparison to filling on the macro-scale is also performed.

NANO-SCALE COMPOUNDING VIA TWIN SCREW EXTRUDERS

Charlie Martin

Leistritz, Somerville, New Jersey, USA – cmartin@alec-usa.com

Twin screw extrusion is a preferred manufacturing methodolgy for many polymer compounds. Complex formulations for high value-added products are being used for a variety of new applications, particularly in the medical, pharmaceutical and aerospace industries. The high cost and limited availability of the raw materials for these emerging compounds has resulted in the need to process very small amounts of prototype feed stocks to determine the viability of an extrusion process. Experimental data will be presented on a new extrusion device that can process as little as 20 to 100 gram micro-batches, or lower, during the development phase of these products. A potential scale-up path to production will also be defined.

Introduction

High speed, energy input (HSEI) twin screw extruders (TSE's) are the plastic industry's manufacturing methodology of choice for compounding, devolatilization and reactive extrusion. Laboratory scale TSE systems process as little as ½ kg/hr. of material. Production systems can be specified for 50,000+ kgs/hr. The end product is typically a pellet that is then fed into an injection molding machine or single screw extruder. It is also common to by-pass pelletization and to directly extrude a film/sheet, profile or other part directly from the HSEI TSE.

In recent years the HSEI TSE process has been embraced by the pharmaceutical industry to replace batch mixing operations to compound active pharmaceutical ingredients (API's) with polymers to create sophisticated dosage forms. Simultaneously, extensive development efforts to process wide ranging nano-particles for use in the aerospace, electronics and packaging industries has occurred. The commonality between API's and nano-particles is that in the early stages of development these materials can be very expensive and are only available in limited quantities.

Hence, a need arose to process very small samples (less than 100 grams) via the extrusion process. A review of basic twin screw theory precedes a description of how these new nano-scale hardware options are applied.

HSEI twin screw extruder theory and design basics:

HSEI twin screw extruders utilize segmented screws that are assembled on high torque shafts. Barrels are also modular and integrate internal bores for cooling. The motor inputs energy into the process via rotating screws. Segmented screws/barrels, in combination with the controlled pumping and wiping characteristics of the co-rotating, self wiping screws, allows screw/barrel geometries to be matched to the process tasks. Solids conveying and melting occurs in the first part of the process section. Screw elements for mixing and devolatilization are utilized as dictated by the process. Discharge elements then build and stabilize pressure. (Fig.1)

The free volume in the process section is directly related to the OD/ID ratio. The OD/ID ratio is defined by dividing the outside diameter (OD) by the inside diameter (ID) of each screw. Based on the use of a splined shaft, an OD/ID ratio of approximately 1.5/1 has generally been deemed to offer the best balance of torque and volume.

HSEI twin screw extruders are starve fed, with the output rate determined by the feeder(s), which meter pellets, liquids, powders and fibers into the process section. The HSEI TSE screw rpm is independent from the feed rate and is used to optimize compounding efficiencies. Because the pressure gradient is controlled, and zero for much of the process, materials can be introduced into downstream barrel sections to facilitate sequential feeding. (Fig.2)

Developments in nano-compounding:

A nano-16 twin screw extruder with 16 mm OD screws and a 1 mm flight depth has been designed to evaluate extrusion with as little as a 20 gram batch. Screws and barrels are segmented. A 1.2/1 OD/ID ratio results in a free volume of approximately 1 cc/diameter. Tri-lobal screw elements, as compared to bi-lobal elements, are specified due to geometric differences inherent with the smaller OD/ID ratio. The same gearbox is used for the nano-16 as for the Micro-18 mm HSEI TSE, which also allows an 18 mm process section to be mated to the gearbox for scale-up processing utilizing a 1.5/1 OD/ID ratio. (Fig.3)

To simulate the starve-fed process, a micro-plunger feeder (patent pending) was designed to meter small amounts of materials in virtually any form with near-full utilization of the batch. The micro-plunger is a piston, driven by a closed-loop Vector gearmotor, within a stainless steel tube that is filled with a 20 to 100 cc batch and mated to the bottom of the nano-16 feed barrel. The nano-16 feed screw elements are designed to convey the materials at a higher rate than is being delivered by the micro-plunger to facilitate starve-feeding. The combination of the nano-16 twin screw extruder and the micro-plunger feeder replicates the staging of unit operations and the shear imparting mechanisms of production scale twin screw extrusion equipment. (Fig.4)

Experimental

A series of trials were performed on the nano-16 processing hypromellose acetate succinate (HPMCAS), a pharmaceutical grade polymer, with 40% of a poorly soluble drug and a trace amount of FDA approved blue color pigment. The objective was to demonstrate the viability of the extrusion process utilizing a very small sample with minimal waste. The materials were pre-mixed and metered by the micro-plunger feeder. A 25 to 1 L/D process section was utilized. The screw design included flighted elements and kneading/shear inducing elements. An atmospheric vent and a low-volume strand die front-end attachment were used.

Tests were performed at different feed rates (2, 4, and 8 cc/min) and run conditions. A feed rate of 2 cc/min. translated to approximately 120 gms/hr., 4 cc/min. to 240 gms/hr., and 8 cc/min. to 480 gms/hr. The batch size selected for the pre-mix was 50 grams, of which 44 grams of the sample was collected and usable for evaluation purposes. Approximately 6 grams of material was lost as follows: 1 gram at the extruder/plunger interface, 2-3 grams on the screws, and 2 grams in the die/front end. The temperature profile and screw rpms were selected based upon experience with similar formulations. A PC based controls/data acquisition package allowed for detailed analysis of the run conditions. (Fig.5)

A strand was cooled in an air quenched annular chamber and cut into 1 mm pellets by a dual drive strand pelletizer.

Results and Discussion

The pellets were then milled into a powder and compressed into tablets for dissolution rate and solubility testing, which are common tests in the pharmaceutical industry. The results indicated the superiority of extrusion to process this poorly soluble API as compared to traditional "dry mixing" operations. The API was transformed from a crystalline to an amorphous structure during processing, and most of the drug was dissolved after 20 minutes. Without extrusion, this API might not have been a candidate for additional development. (Fig.6)

Subsequent tests were performed at 2 kgs/hr. (HSEI TSE with 18 mm screws dia.) and 6 kgs/hr. (HSEI TSE with 27 mm screws dia.) with similar results, confirming the viability of the nano-16 as an initial screening device. (note: approximately 40% of new drugs are poorly soluble, which is a barrier to usage)

Figure 1

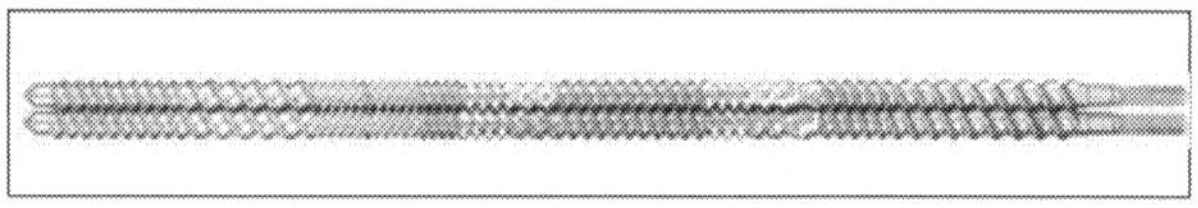

HSEI Co-rotating intermeshing twin screw extruder screw set

Figure 2

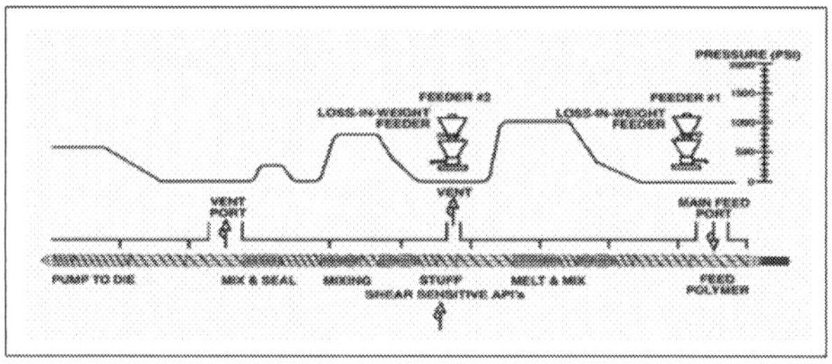

Pressure gradient in a HSEI twin screw extruder

Figure 3

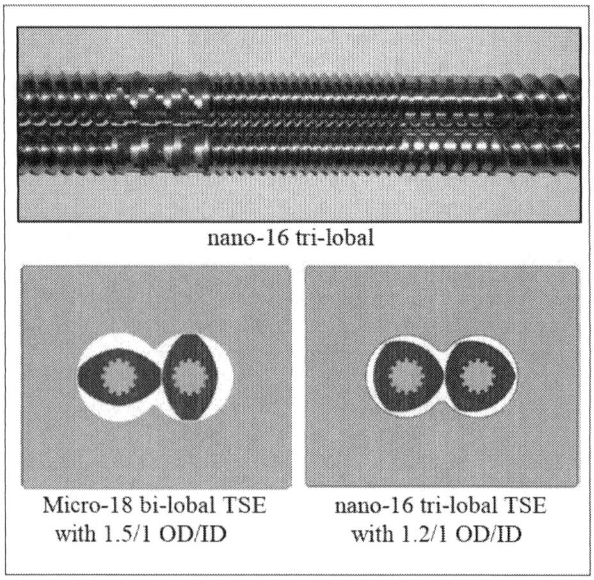

nano-16 tri-lobal

Micro-18 bi-lobal TSE with 1.5/1 OD/ID

nano-16 tri-lobal TSE with 1.2/1 OD/ID

Figure 4

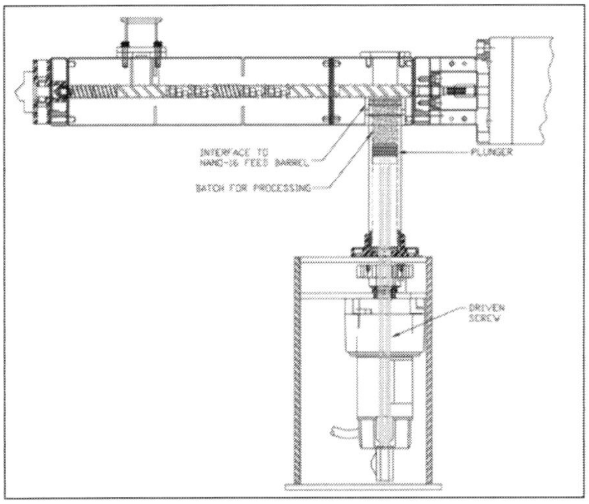

Micro-plunger feeder mated to a nano-16 twin screw extruder

Figure 5

Run Data

Runtime	Z1	Z2	Z3	Z4	Z5	A1	A2	A3	A4	A5
0:00:00	120	125	130	131	131	0	0	86	53	0
0:01:00	120	125	130	131	130	100	2	207	190	0.27
0:02:00	120	125	130	130	130	100	2	138	179	0.45
0:03:00	120	125	130	130	129	100	2	207	203	0.63
0:04:00	120	125	130	130	128	100	2	486	266	0.87
0:05:00	120	125	130	130	127	100	2	742	303	1.17
0:06:00	120	125	130	130	126	100	2	760	314	1.48
0:07:00	120	125	130	130	125	100	2	673	314	1.79
0:08:00	120	125	130	130	124	100	2	742	324	2.1
0:09:00	120	125	130	130	124	100	2	725	324	2.43
0:10:00	120	125	130	130	123	100	2	1019	385	2.79
0:11:00	120	125	130	130	123	100	2	1019	409	3.19
0:12:00	120	125	130	130	122	100	2	1036	401	3.6
0:13:00	120	125	130	130	122	100	2	932	374	3.98
0:14:00	120	125	130	130	122	100	2	1036	390	4.36
0:15:00	120	125	130	130	122	100	2	1088	398	4.77
0:16:00	120	125	130	130	121	100	2	1071	409	5.19
0:17:00	120	125	130	130	121	100	2	1209	443	5.62
0:18:00	120	125	130	130	121	100	2	1226	467	6.07
0:19:00	120	125	130	130	121	100	2	1140	427	6.51
0:20:00	120	125	130	130	121	100	2	1105	417	6.94
0:21:00	120	125	130	130	121	100	2	1157	459	7.37
0:22:00	120	125	130	130	121	100	2	1278	475	7.82
0:23:00	120	125	130	130	121	100	2	1364	496	8.31
0:24:00	120	125	130	130	121	100	2	1519	511	8.82
0:25:00	120	125	130	130	121	100	2	1347	525	9.32
0:26:00	120	125	130	130	121	100	2	1312	469	9.82
0:27:00	120	125	130	130	121	100	2	1450	562	10.31
0:28:00	120	125	130	130	121	100	2	1485	525	10.83
0:29:00	120	125	130	130	121	100	2	1416	506	11.35
0:30:00	120	125	130	130	121	100	2	1416	509	11.86
0:31:00	120	125	130	130	121	100	2	1381	546	12.35
0:32:00	120	125	130	130	121	100	2	1381	509	12.86
0:33:00	120	125	130	130	121	100	2	1329	485	13.36
0:34:00	120	125	130	130	121	100	0	1220	448	13.72

Z1 = Zone 1 Temperature; A1 = Screw Speed
Z2 = Zone 2 Temperature; A2 = Plunger Feeder cc/min
Z3 = Zone 3 Temperature; A3 = Pressure
Z4 = Zone 4 Temperature; A4 = Torque
Z5 = Melt Temp; A5 = TTQ

nano-16 Run Record

Figure 6

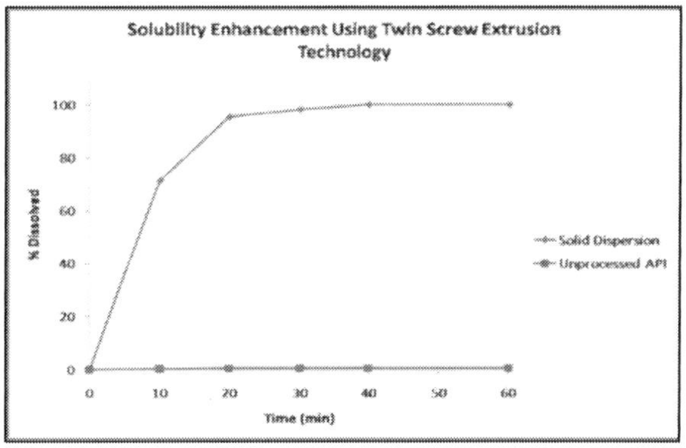

Solubility results from dissolution testing

Figure 7

ZSE-50 mm pharmaceutical production class HSEI TSE

Conclusions

Miracle-like materials are being synthesized for a variety of products in the pharmaceutical and other industries. For most commodity products the goal is to increase throughput rates, whereas during the inception phase of a products' life cycle the need is to downsize and minimize material usage. The use of the nano-16 system allows users to test early-stage materials, often only available in minute quantities, for extrusion viability to facilitate eventual integration into production. (Fig.7)

References

Charles Martin, 2010

Ghebre-Sellassie, I. and Martin, C., Pharmaceutical Extrusion Technology, Marcel-Dekker, New York (2003)

Martin, C.: "Continuous Mixing of Solid Dosage Forms via Hot-Melt Extrusion", Pharmaceutical Technology Magazine, October Issue (2008)

Todd, D.: Plastics Compounding: Equipment & Processing, Hanser/Gardner (1998)

Vlachopoulos, J. and Wagner, J.R. Jr.: The SPE Guide on Extrusion Technology and Troubleshooting, Society of Plastics Engineers (2001)

White, J.: Twin Screw Extrusion: Technology & Principles, New York: Hanser (1990)

The Effects of Additives on the Thermal Stability of Polylactic Acid

Marino Xanthos, Ku-Il Park, Quyuan Zhou*

NJ Institute of Technology - NJ - USA

The thermal stabiltiy of polylactic acid (PLA) is evaluated in the presence of additives including montmorillonite (MMT) reinforcing clays and plasticizing ionic liquids. PLA/MMT micro- and nanocomposites based on semicrystalline and amorphous polymers and unmodified or organomodified clays at 5wt% content were produced by melt mixing. During melt processing, thermomechanical degradation was more pronounced in the presence of all fillers, which apparently acted catalytically, but to different degrees. During isothermal degradation in air from 180°C to 200 oC, the thermal degradation rate constants of the amorphous PLA and its composites are lower than those of the semicrystalline PLA and its composites. Due to better filler dispersion in the polymer matrix, the thermal degradation rate constants of the nanocomposites in air are significantly lower than those of the unfilled polymers and their microcomposites. As per dynamic TGA data and thermal kinetic analysis, organomodified nanofillers have a complex effect on the polymer thermal stability, while unmodified fillers reduce polymer thermal stability. These TGA data and kinetic analysis results also support the findings that the thermal stability of the amorphous PLA and its composites is higher than that of the semicrystalline PLA and its composites, and the thermal stability of the nanocomposites is higher than that of the microcomposites. The effects of clays on the thermomechanical and thermooxidative stability of the amorphous PLA are compared with the degradative effects of two ionic liquids (IL) having different anions (decanoate, tetrafluoroborate) but with the same phosphonium based cation. Both ILs showed promising plasticizing/lubricating behavior in PLA. The blends containing the tetrafluoroborate anion IL showed higher thermal stability compared with the IL containing the decanoate anion as also confirmed from thermal degradation rate constants.

STRUCTURE AND PROPERTIES OF PLA NANOCOMPOSITES FILMS

Seyed H. Tabatabaei and Abdellah Ajji

Chemical Engineering Department, Ecole Polytechnique of Montreal, Montreal, QC, Canada

Abstract

In this study, we investigated the structure and some properties of multilayer films with nanoclay filled PLA and polyethylene (PE) obtained from the extrusion blowing process, pure PLA and filled with nanoclay cast films and fibers. The structure of the films is investigated using microscopy and X-ray diffraction. The properties that are studied are the mechanical properties in terms of modulus, strength and elongation for both films and fibers in addition to tear and haze for blown films. The effect of nanoclay content in the range of 1 to 7.5 wt% is investigated.

Introduction

Nanocomposites based on nano-layered silicates have attracted much attention during the past ten years because of their low cost, their ready availability, and their non-isometric structure derived from a high aspect ratio, which can maximize the reinforcing effect in terms of mechanical, thermal, and barrier properties (1).

Poly (lactic acid) (PLA) is one of the biodegradable polymers, used in fibers and films, and is formed from l-lactic acid produced from natural renewable sources such as corn. This polymer decomposes rapidly and completely in a typical compost environment which makes it an ideal replacement for non-degradable polymers in numerous applications. Due to its ideal combination of physical properties (high modulus, good film and fiber forming properties, good heat seal characteristics, barrier to flavor and aroma) and competitive costs, it is being investigated in a number of commodity applications including flexible packaging (films), fibers and containers (2).

Very few studies addressed the use of clay nanoparticles in polymer films and fibers, particularly from the molten state. The objective of this study is to take advantage of the reinforcing effect of the nanoparticles in films and fibers made from PLA.

Experimental

PLA 2002D, an extrusion grade, was purchased from Nature Works Company (previously Cargill Dow). A 10wt% master batch nanocomposite compound was produced in a twin-screw extruder using 30B clay from Southern Clay.

Blown and cast nanocomposite films were produced with the 2002D resin by diluting the master batch to concentrations of 1, 2.5, 5 and 7.5%. Blown films were made using a blow up ratio of 2 and draw down ratio of 15, and PLA and nanocomposites were sandwiched between two PE layers. The cast films were made using a cast film extrusion equipped with a die width of 20 cm with PLA and nanocomposites only. Some PLA and nanocomposites cast films were oriented in a biaxial stretcher and characterized as well.

The structure of the films was evaluated using field emission scanning electron microscopy (FE-SEM) and X-ray diffraction. The crystalline axes orientation factors were determined from wide-angle X-ray diffraction pole figures using a Bruker AXS X-rays goniometer equipped with a Hi-STAR two-dimensional area detector (3). The generator was set up at 40kV and 40mA and the copper K_α radiation (λ=1.542 Å) was selected using a graphite crystal monochromator. Sample to detector distance was fixed at 8 cm. Films or fiber samples were stacked to a thickness of about 2 mm in order to obtain enough accuracy in a reasonable time. The d-spacing of the (001) clay crystalline plane, indicative of the clay dispersion, was determined from the theta-2*theta plots superimposed on the two dimensional X-ray diagrams. The performance of the blown films was evaluated in terms of tensile mechanical properties; tear resistance and Haze according to ASTM standards.

Results and Discussion

Structure: X-ray results indicate that PLA in general did not crystallize in the films, due to the rapid cooling. For the fibers, some kind of superstructure can be observed in the 2D X-ray patterns of Figure 1. The results obtained for the *d*-spacing for clay concentration of 5 wt% was evaluated to about 3.15 nm at small angles and indicate some partial intercalation/exfoliation of the clay in the PNC (d spacing of 30B clay is 1.85 nm). Results on the orientation of the clay 001 axis in cast and blown nanocomposite films (normal to the clay platelets plane), shown in Figures 2, indicate that it is oriented in the normal direction, which is the thickness direction, indicating that the platelets are in the plane of the films, which is expected. The cast films nanocomposite was oriented uniaxially at 85 °C to a draw ratio of 4x and Figure 3 shows the distribution of the clay platelets after orientation. It is observed that the platelets normal shifts towards the TD-ND plane.

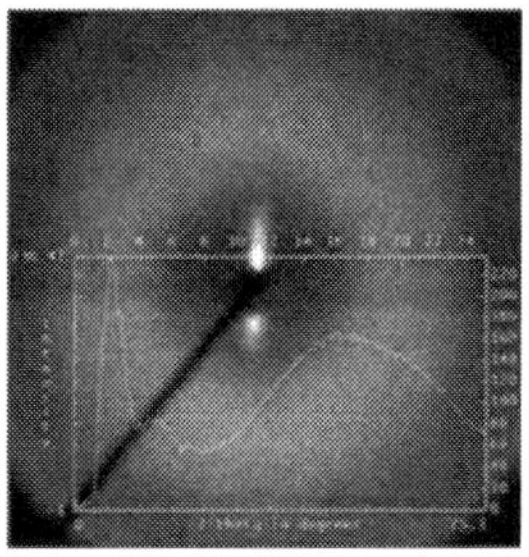

Fig. 1: 2D WAXD pattern for 2.5% clay cast film and intensity profile.

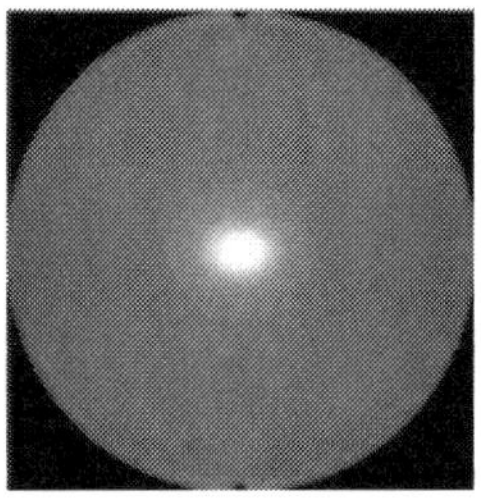

Fig. 2: X-ray Pole figure of the clay 001 crystalline reflection in PLA2002_5%30B cast film.

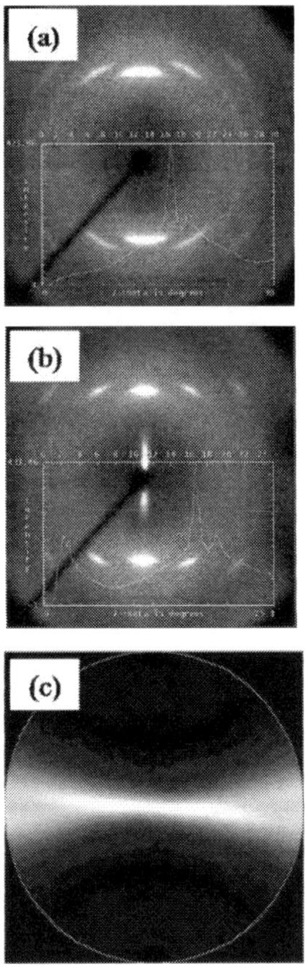

Fig. 3: 2D WAXD patterns of uniaxially oriented (a) pure PLA, (b) nanocomposite with 2.5 wt% Cloisite 30 B and (c) X ray pole figure for the clay 001 reflection. Draw ratio was 4 and stretching temperature 85 °C.

Tensile properties of various PE/PLA nanocomposite/PE blown films are shown in tables a, b and c in terms of tensile modulus, tensile strength and elongation at break respectively. Modulus increases about 30% for 7.5 wt% clay content in both machine and transverse directions. Strength decreases at low clay content, and then recovers at higher content to its original value. Elongation at break seems to improve significantly, particularly in the transverse direction for clay contents of 1, 2.5 and 5%, but decreases drastically at 7.5 wt%. Tear is improved by addition of clay nanoparticles and haze is unchanged except for the higher content of clay.

Table 1. Mechanical characterization of films.

a) Modulus

PE/PLA/PE	Thickness µm	MD Modulus MPa (deviation)	TD Modulus MPa (deviation)
33.3/33.3/33.3 +7.5%30B	25/25/25	1361 (147)	1334 (103)
33.3/33.3/33.3 +5%30B	25/25/25	1212 (144)	1192 (85)
33.3/33.3/33.3 +2.5%30B	25/25/25	1017 (73)	1095 (78)
33.3/33.3/33.3 +1%30B	25/25/25	1010 (100)	956 (32)
PE/PLA/PE 25/25/25	25/25/25	1105 (100)	1173 (46)
PLA 2002	13	2614 (584)	2341 (871)

b) Strength

PE/PLA/PE	Thickness µm	MD strength MPa (deviation)	TD strength MPa (deviation)
33.3/33.3/33.3 +7.5%30B	25/25/25	26.8 (2.5)	26.2 (2)
33.3/33.3/33.3 +5%30B	25/25/25	27.4 (3.4)	24.8 (1.8)
33.3/33.3/33.3 +2.5%30B	25/25/25	24.3 (1.1)	25.0 2.3
33.3/33.3/33.3 +1%30B	25/25/25	23.8 (2.9)	23.1 (0.8)
PE/PLA/PE 25/25/25	25/25/25	28.3 (2.5)	28.4 (0.7)
PLA 2002	13	53.5 (14.5)	44.0 (17.3)

c) Elongation

PE/PLA/PE	Thickness µm	MD Elong. % (deviation)	TD Elong. % (deviation)
33.3/33.3/33.3 +7.5%30B	25/25/25	237 (66)	25 22
33.3/33.3/33.3 +5%30B	25/25/25	85 (48)	374 (206)
33.3/33.3/33.3 +2.5%30B	25/25/25	121 (18)	343 (322)
33.3/33.3/33.3 +1%30B	25/25/25	137 (37)	668 (22)
PE/PLA/PE 25/25/25	25/25/25	130 (70)	40 (28)
PLA 2002	13	3.7 0.9	5.7 2.9

d) Tear resistance and haze

Content, PE/PLA/PE	MD Tear g/µm (deviation)	TD Tear g/µm (deviation)	Haze (deviation)
33.3/33.3/33.3 +7.5%30B	2.9 (1.0)	4.3 (1.3)	19.9 (1.6)
33.3/33.3/33.3 PLA+5%30B	2.5 (1.5)	4.1 (1.2)	16.9 (1.5)
33.3/33.3/33.3 +2.5%30B	3.0 (1.5)	4.7 (0.9)	16.3 (1.8)
33.3/33.3/33.3 +1%30B	3.2 (1.3)	4.7 (1.4)	17.9 (3.5)
PE/PLA/PE 25/25/25	0.4 (0.1)	1.6 (0.6)	18.2 (1.7)

References

1. Mosmann T. 1983. J. Immunol. Methods. 65:55-63.

2. P. B. Smith, A. Leugers, S. Kang, X. Yang and S. Ling Hsu, Macromol. Symp., 175, 81 (2001)

3. Cole, K.C. and A. Ajji, Chapter 3 in Solid Phase Processing of Polymers. Hanser Munich, 33-84 (2000).

MIRCOPOROUS MEMBRANES OBTAINED FROM POLYPROPYLENE/HIGH DENSITY POLYETHYLENE MULTILAYER FILMS BY STRETCHING

Seyed H. Tabatabaei, Pierre J. Carreau*, and Abdellah Ajji

CREPEC, Chemical Engineering Department, Ecole Polytechnique, C.P. 6079, Succ. Centre- Ville, Montreal, QC, H3C 3A7 Canada

Abstract

Polypropylene/high density polyethylene (PP/HDPE) multilayer films were prepared to develop microporous membranes using cast film extrusion followed by stretching. The effects of draw ratio (DR), cooling air flow rate (AFR), and annealing on the crystalline structure and orientation of the monolayer and components in the multilayer films were investigated using wide angle X-ray diffraction (WAXD) and Fourier transform infrared (FTIR). Scanning electron microscopy images of the membrane surface and cross-section obtained for the cold and hot stretch ratios of 55% and 75%, respectively, showed larger pores and higher porosity for the HDPE layer compared to the PP. As the level of the applied extension during cold stretching increased, the water vapor transmission rate (WVTR) of the HDPE monolayer improved while the effect was inversed for the PP monolayer. In addition, compared to the monolayer membranes, the multilayer ones showed smaller WVTR.

Introduction

Microporous membranes are commonly used in separation processes such as battery separators and medical applications to control the permeation rate of chemical components. Due to the wide range of chemical structures, optimum physical properties, and low cost of polymers, they are the best candidates for the fabrication of microporous membranes.
Commercially available lithium battery separators are made from polyolefins such as polypropylene (PP) and polyethylene (PE). These materials are compatible with the cell chemistry and can be used for many cycles without significant degradation in properties [1]. Lithium (Li) batteries will generate heat if accidentally overcharged. Separator shutdown is a useful safety feature for restricting thermal reactions in Li-batteries [1,2]. Shutdown occurs close to the melting temperature of the polymer, leading to pores collapse and restricting passage of current through the cell. PP separators melt around 160 °C whereas PE separators have shutdown temperature between 120 and 130 °C. If in a battery, the heat dissipation is slow, even after shutdown, the cell temperature may continue to increase before starting to cool [1]. Recently, manufacturers have started producing trilayer separators where a porous PE layer is sandwiched between two porous PP layers. In such a case, the PE layer has lower shutdown temperature while PP provides the mechanical stability at and above the shutdown temperature [1]. However, the dry-stretch process is based on the stretching of a polymer film containing a row-nucleated lamellar structure [3]. Three consecutive stages are carried out to obtain porous membranes by this technique: 1) creating a precursor film having a row-nucleated lamellar structure through shear and elongation-induced crystallization of the polymer having proper molecular weight and molecular weight distribution, 2) annealing the precursor film at a temperature near the melting point of the resin to remove imperfections in the crystalline phase and to increase lamellae thickness, and 3) stretching at low and high temperatures to create and enlarge pores, respectively [3,4].
In this study, we investigate the fabrication of microporous PP/HDPE/PP trilayer membranes using the cast film process.

Experimental

2.1 Materials

A commercial linear polypropylene (PP) and a commercial high density polyethylene (HDPE) were selected. PP5341E1 was supplied by ExxonMobil and had a melt flow rate (MFR) value of 0.8 g/10min (under ASTM D1238 conditions of 230 °C and 2.16 kg). HDPE 19A was provided by NOVA Chemicals and had an MFR value of 0.72 g/10min (under ASTM D1238 conditions of 190 °C and 2.16 kg). The main characteristics of the resins are shown in Table 1. The molecular weight (M_w) and polydispersity index (PDI) of the HDPE was supplied by company and that of the PP was measured using a GPC (Viscotek model 350) with 1,2,4-Trichlorobenzene (TCB) as a solvent at a column temperature of 140 °C. The melting point, T_m, and crystallization temperature, T_c, of the resins obtained from differential scanning calorimetry at a rate of 10 °C/min are also reported in Table 1.

Table 1. Main characteristics of neat polymers.

Resin code	Company	MFR	M_w (kg/mol)	M_w/M_n	T_m (°C)	T_c (°C)
PP5341E1	ExxonMobil	0.8 230 °C/2.16 kg	361	2.7	162	112
HDPE 19A	NOVA Chemicals	0.72 190 °C/2.16 kg	126	7.8	129	118

2.3 Film and membrane preparation

The cast films were prepared using an industrial multilayer cast film unit from Davis Standard Company (Pawcatuck, CT) equipped with a 2.8 mm opening and 122 cm width slit die and two cooling drums. The extrusion was carried out at 220 °C and the distance between the die exit to the nip roll was 15 cm. The die temperature was set at 220 °C and draw ratios of 60, 75, and 90 were applied. An air knife with dimensions of 3 mm opening and 130 cm width was mounted close to the die to provide air to the film surface right at the exit of the die. The variables of interest were draw ratio and amount of air flow rate. The films were produced under a chill roll temperature of 50 °C. The air cooling rates used were 1.2 and 12 L/s. These air cooling conditions are noted as: low air flow rate (L-AFR) and high air flow rate (H-AFR), respectively.

For the membrane fabrication, the precursor films with thickness, width, and length of 32 μm, 46 mm, and 64 mm, respectively, were used. The films were first annealed at 120 °C for 30 min and then cold and hot stretched at 25 °C and 120 °C, respectively. Both annealing and stretching were performed in an Instron tensile machine equipped with an environmental chamber. Drawing speeds of 500 mm/min and 25 mm/min were applied during cold and hot stretching steps, respectively.

2.4 Film and membrane characterization

2.4.2 Fourier transform infrared spectroscopy (FTIR): For FTIR measurements, a Nicolet Magna 860 FTIR instrument from Thermo Electron Corp. (DTGS detector, resolution 2 cm^{-1}, accumulation of 128 scans) was used. The beam was polarized by means of a Spectra-Tech zinc selenide wire grid polarizer from Thermo Electron Corp. The measurement is based on the absorption of infrared light at certain frequencies corresponding to the vibration modes of atomic groups present within the molecule. In addition, if a specific vibration is attributed to a specific phase, the orientation within that phase can be determined [5]. The details of the measurements for PP and PE can be found elsewhere [6].

2.4.3 X-ray diffraction: XRD measurement was carried out using a Bruker AXS X-ray goniometer equipped with a Hi-STAR two-dimensional area detector. The generator was set up at 40 kV and 40 mA and the copper CuKα radiation (λ = 1.542 A°) was selected using a graphite crystal monochromator. The sample to detector distance was fixed at 9.2 cm for wide angle diffraction and 28.2 cm for small angle X-ray scattering analysis. To get the maximum diffraction intensity several film layers were stacked together to obtain the total thickness of about 2 mm. The details of the measurements can be found in our previous work [6].

2.4.5 Morphology: A field emission scanning electron microscope (FE-SEM- Hitachi S4700) was employed for the observation of the etched precursor films and microporous membranes surfaces as well as cross-sections. This microscope provides high resolution of 2.5 nm at a low accelerating voltage of 1 kV and high resolution of 1.5 nm at 15 kV with magnification from 20x to 500kx.

2.4.6 Water vapor transmission: The permeability to water vapor was measured via a MOCON PERMATRAN-W Model 101K at room temperature. It is composed of three chambers: an upper chamber containing liquid water and separated from the center chamber by two porous films. Water vapor diffuses from the first film to fill the space between the films to reach 100% relative humidity (*RH*). The center chamber is separated from the lower one by the test film. The diffused vapor is swept away by N_2 gas to a relative humidity (*RH)* sensor.

Results and discussion

Figure 1 demonstrates the effect of air flow rate (AFR), draw ratio (DR), and annealing on the orientation features, in terms of $\cos^2(\phi)$ of the crystalline axes (i.e. *a*, *b*, and *c*) along MD, TD, and ND obtained from the Herman orientation function for the PP and HDPE single layers as well as the components in the multilayer film. As expected, the *c*-axis orientation characteristics along MD improve by increasing AFR and DR or by annealing. In addition, obviously, the *c*-axis alignment of the HDPE (both in the monolayer and multilayer) is significantly lower than that of the PP and also the *c*-axis orientation along MD of the PP and HDPE in the multilayer is lower than in the monolayer made under the same conditions. It is well known that the HDPE has much higher crystallinity and larger heat of fusion compared to the PP, resulting in a large heat release during its crystallization. This can explain lower orientation of the PP component in the trilayer than the PP monolayer.

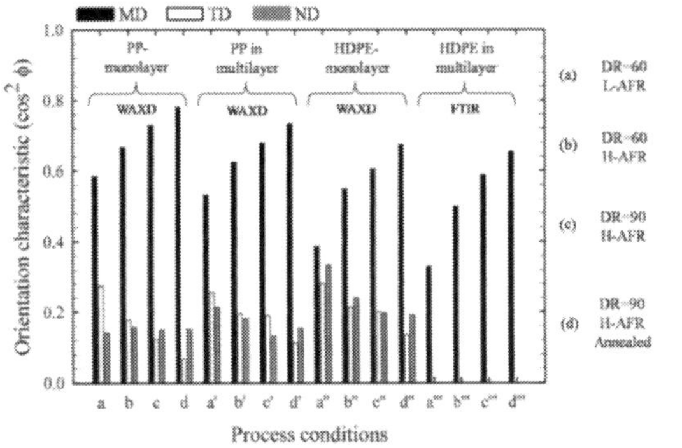

Fig. 1. Orientation characteristics as $\cos^2(\phi)$ of the *c*-axis along MD, TD, and ND for the films obtained under different DR, AFR, and annealing. Annealing was performed at 120 °C for 30 min.

Figure 2 presents SEM micrographs of the surface of the monolayer microporous membranes prepared at the cold and hot stretching extensions of 55% and 75%, respectively. The distribution of interlamellar tie chains in the precursor films may not be uniform [7]. Therefore, it is believed that the micropores are first developed in regions with a small amount of tie chains. It is clear that the size of the pores in the HDPE microporous membrane is much larger than in the PP membrane. The longer interlamellar microfibrils (bridges) in the HDPE porous membrane compared to the PP porous membrane are believed to be due to the longer tie chains in the precursor film of the former. It should be mentioned that the surface morphology of the PP/HDPE/PP membrane (not shown) was somewhat similar to the surface structure of the PP monolayer exhibited in Figure 2.

In general, blends of PP and HDPE are known to be immiscible systems. The interfacial morphology for the etched multilayer film is illustrated in Figure 3a. Some transcrystallization zone around the interface can be easily distinguished; they are the PE lamellae nucleated on the PP. In other words, crystallization of the PE overgrows at the interface. A transcrystalline layer is formed when a large number of nuclei are formed on an interface such that the crystallites are forced to grow normal to the interface and when a larger difference in crystallization temperature is present [8], which is the case between HDPE and PP (see Table 1). Figure 3b illustrates SEM micrographs of the cross-section of the multilayer porous membrane. Similar to the surface micrographs shown in Figure 2, it is clearly recognized that the HDPE layer has much larger pores than the PP layer. Additionally, a reasonable adhesion between the layers is seen, which can be explained by the transcrystallization and penetration of the HDPE lamellae into the PP layer observed in the cross-section image of the trilayer precursor film shown in Figure 3a.

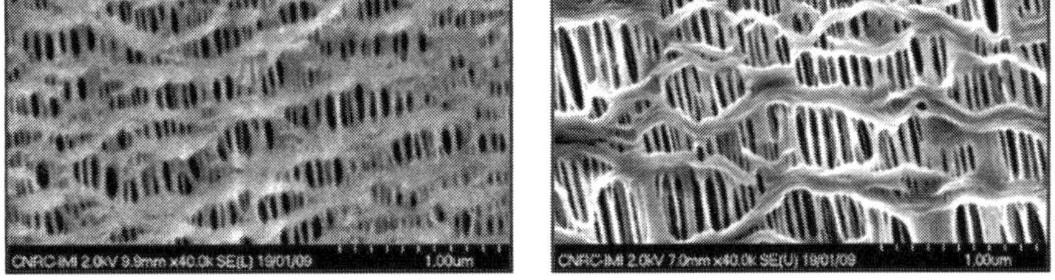

Fig. 2. SEM micrographs of the surface of microporous membranes (20 μm thick): a) PP and b) HDPE; DR=90, cold stretching of 55%, followed by hot stretching of 75%. MD ↑ and TD →

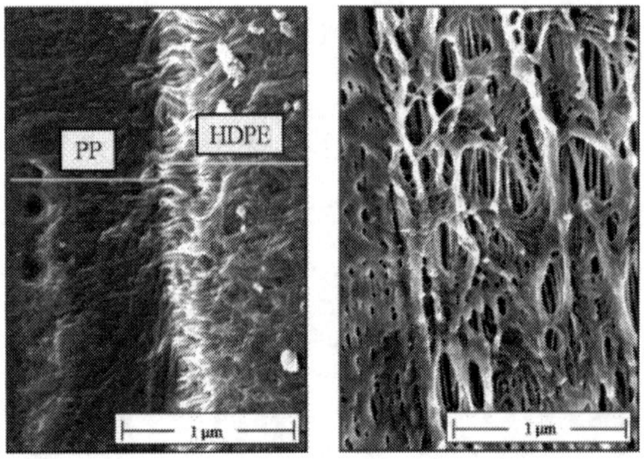

Fig. 3. SEM micrographs of the cross-section of trilayer precursor (left) and microporous membranes (right) (20 μm thick); DR=90, cold stretching of 55%, followed by hot stretching of 75%.

Our previous study [9] showed that the water vapor transmission rate (WVTR) of the cold stretched PP films increased as the stretch ratio increased up to 30% while further stretching resulted in a reduction in WVTR. To find the optimum cold stretching extension for the present PP as well as the HDPE, cold stretching was carried out under predetermined levels of extension while the amount of hot stretching was kept constant. Figure 4 reports the WVTR values normalized (multiplied) by the membrane thickness as a function of applied extension for the PP and HDPE monolayer porous membranes. It is obvious that 25% extension during cold stretching is not enough to initiate pores formation for the HDPE and PP. A monotonous increase in WVTR of the HDPE membranes was observed after further extension during cold stretching. In contrast, a significant enhancement in WVTR of the PP membranes was observed when 30% extension was applied while further stretching reduced the normalized WVTR. Additionally, it is clear that both the PP and HDPE have almost identical permeabilities at 55% cold drawing. Therefore, 55% was found to be the optimum cold stretching extension for the fabrication of the multilayer membranes. Finally, the trilayer microporous membranes obtained at 55% cold extension followed by 75% hot extension showed WVTR values about 30% lower than the monolayer PP and HDPE obtained under the same conditions. This could be due to the presence of the interface and lower orientation of the PP and HDPE components in the multilayer film than the monolayer ones.

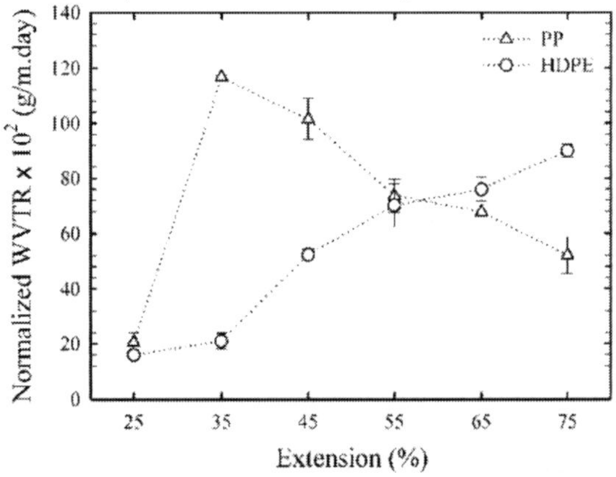

Fig. 4. Normalized water vapor permeability for PP and HDPE membranes as a function of extension during cold stretching at 25 °C; DR=90, H-AFR, hot stretching of 75%.

Conclusions

In this work, we have investigated the structure and performances of microporous membranes made from monolayer and trilayer films of the PP and HDPE. Significant effects of cooling air flow rate (AFR), draw ratio (DR), and annealing on the crystal orientation of the PP and HDPE monolayer films as well as the components in the multilayer one were observed. A transcrystallization of the HDPE lamellae penetrating into the PP at the interface of the multilayer films was observed and was explained by the difference in the resin crystallization temperature. Finally, by increasing the applied extension during cold stretching, water vapor transmission rate (WVTR) monotonically rose for the HDPE while for the PP, WVTR first increased significantly and then decreased

Acknowledgements

Financial support from NSERC (Natural Science and Engineering Research Council of Canada) and from FQRNT (Fonds Québécois de Recherche en Nature et Technologies) is gratefully acknowledged. We also acknowledge the large infrastructure grant received from the Canadian Foundation for Innovation

(Government of Canada and Province of Quebec), which allowed us to build the unique POLYNOV facility.

References

1. Venugopal G, Moore J, Howard J, Pendalwar S. J Power Sourc 1998;77:34-41.

2. Dahn JR, Fuller EW, Obrovac M, van Sacken U. Solid State Ionics 1994;69:265-270.

3. Sadeghi F. Developing of microporous polypropylene by stretching, Ph.D thesis, Ecole polytechnique de Montreal, 2007.

4. Johnson MB. Ph.D thesis, Virginia Polytechnic Institute and State University, 2000.

5. Sadeghi F, Ajji A, Carreau PJ. Polym Eng Sci 2007;47:1170-1178.

6. Tabatabaei SH, Carreau PJ, Ajji A. J Membr Sci 2009;345:148-159.

7. Yu TH. Ph.D thesis, Virginia Polytechnic Institute and State University, 1996.

8. Zhang XM, Elkoun S, Ajji A, Huneault MA. Polymer 2004;45:217-229.

9. Tabatabaei SH, Carreau PJ, Ajji A. J Membr Sci 2008;325:772-782.

EXPERIMENTAL AND NUMERICAL MODELING OF THE COEXTRUSION PROCESS

Joseph Dooley

The Dow Chemical Company, Midland, MI, USA – jdooley@dow.com

Coextrusion is a common method used for producing multilayer films and sheets. Coextrusion is a process in which two or more polymers are extruded and joined together in a feedblock or die to form a single structure with multiple layers. This technique allows the processor to combine the desirable properties of multiple polymers into one structure with enhanced performance characteristics. The coextrusion process has been widely used to produce multilayer sheet, blown film, cast film, tubing, wire coating, and profiles. This paper will discuss experimental programs as well as numerical modeling of the coextrusion process. The topics that will be covered include rheological measurements of coextruded structures as well as flow of coextruded structures through feedblocks and dies.

Introduction

Early multilayer products contained two or more layers. Before the development of coextrusion technology, individual films or sheets were produced and then laminated together to produce a product with the desired number of layers. The economics of producing multilayer structures was significantly improved by the breakthrough introduction of the coextrusion process.

There are many topics that have been covered in the literature related to coextrusion including viscous encapsulation (White et al., 1972, Southern and Ballman, 1973, Han, 1973, Everage, 1973, and Dooley and Rudolph, 2002) and interlayer instability (Khan and Han, 1977, Schrenk and Alfrey, 1978, Schrenk et al., 1978, Han et al., 1984, Khomami, 1990a, 1990b, and Ramanathan et al., 1996). In addition, there are a number of other areas in which information is needed in order to troubleshoot or design a coextrusion system. The three areas covered in this paper are the rheology of the coextruded structure as well as the geometry and flow through a feedblock and die. These three areas will be discussed in detail in the following sections.

Rheology of Coextruded Structures

A typical technique used to examine the rheology of a coextruded structure is to measure the rheology of the individual layers and assume that the rheology of the structure is some combination or average of those values. In order to improve on this methodology, a coextrusion line was modified with a rheology die to directly measure the rheology of coextruded structures.

The coextrusion line used in this study for determining the viscosity of coextruded structures consisted of two 31.75 mm (1.25 inch) diameter, 24:1 L/D single screw extruders. A schematic diagram of the extrusion line set-up is shown in Figure 1. The extruders fed individual gear pumps to ensure uniform flow of the polymer melts to the feedblock and rheology dies. The gear pumps were attached to a feedblock by transfer lines that contained variable depth thermocouples to ensure consistent and uniform temperatures from the extruders. A feedblock was designed to produce coextruded structures with the capability of varying the number of annular layers. This feedblock allows evaluation of various coextruded structures with different layer arrangements and thicknesses

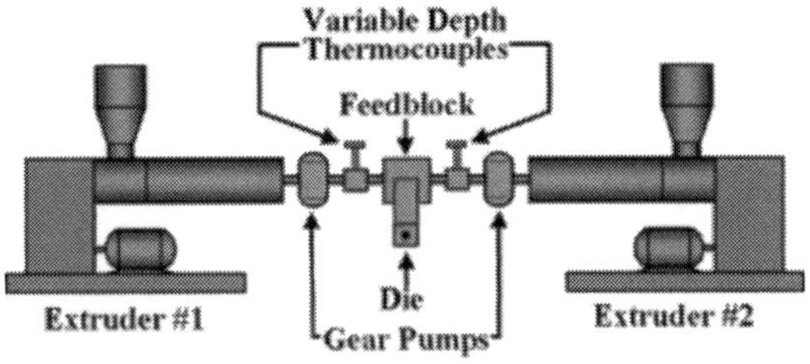

Figure 1 – Apparatus for measuring rheology of coextruded structures.

Attached to the exit of the feedblock was a rheology die that was fabricated with a circular cross section with a diameter of 7.9 mm (0.312 inch). This die had four pressure transducers spaced at 50.8 mm (2.0 inch) intervals down the length of the die. The rheology die is shown in Figure 2 with the pressure transducers and the heating jacket installed.

For the rheology experiments, the coextrusion line was run with the two extruders and gear pumps producing set flow rates until steady-state conditions had been reached. At this condition, the pressures from the four transducers were recorded along with the measured total polymer flow rate and temperature. This procedure was repeated at several different flow rates so that the viscosity of the monolayer or coextruded structures could be determined at different shear rates and/or layer ratios.

Figure 2 – Rheology die.

A typical set of experimental results for a 3 layered structure is shown in Figure 3. In this figure, Resins A and B are two low density polyethylene resins with melt indices of 8 and 2 dg/min, respectively.

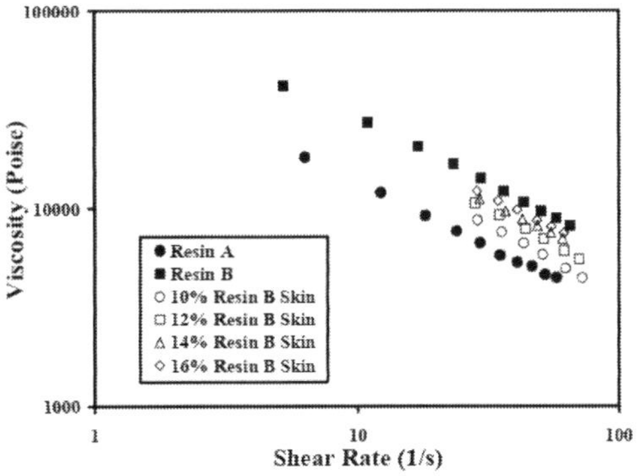

Figure 3 – Rheology data for individual components and for 3 layer coextruded encapsulated structures with various skin layer thicknesses of Resin B.

A numerical technique was developed to model the rheology of these types of structures. These models allowed prediction of the rheology of these structures as well as the pressure drop in the system and the velocity profiles.

Figure 4 shows an example of the velocity profiles calculated using these techniques. In this figure, Resins A and B are alternated in 1, 2, 3, and 4 layer structures. Note how different the velocity profiles are depending on the layer sequence.

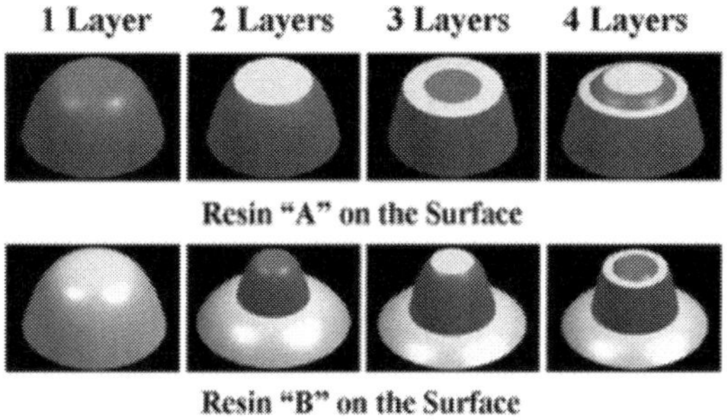

Figure 4 – Comparison of the velocity profiles of multilayer coextruded structures in a tube.

Flow of a Coextruded Structure through a Feedblock

The next area of interest in a coextrusion system is the feedblock. The geometry of the feedblock is one variable that can be modified to try to change the layer uniformity of the final coextruded structure.

A set of designed experiments was conducted on a commercial two layer coextrusion system to determine the effect of the feedblock shaping on the final sheet layer profile. Two ABS (Acrylonitrile-Butadiene-Styrene) type resins with similar viscosities were chosen for the study. The coextrusion line used consisted of a 114.3 mm (4.5 inch) diameter, 30:1 L/D single screw extruder for the substrate resin and a 63.5 mm (2.5 inch) diameter, 24:1 L/D single screw extruder for the cap resin. The sheet die was 1.42 meters (56 inches) wide. The feedblock was designed with inserts that would restrict the flow of the cap layer to the edges of the sheet. These inserts restricted the flow path widths to 34%, 67%, and 100% (control) of the available width.

Figure 5 shows the cap layer thickness distribution across the sheet width using the 3 different restricting inserts.

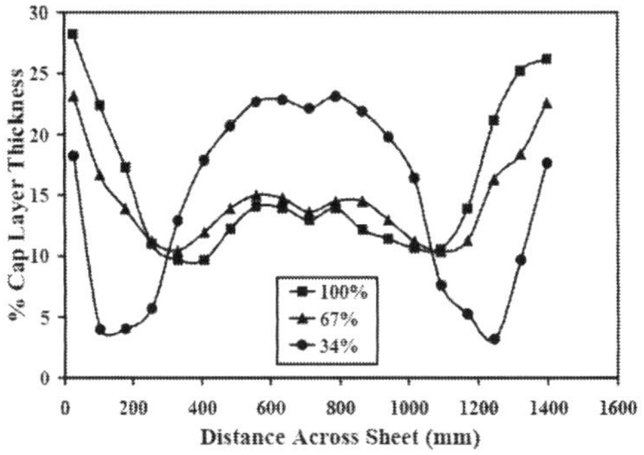

Figure 5 – Cap layer thickness for a two layer structure using different feedblock inserts.

This study used a Box-Behnkin type of designed experiment. The results were analyzed statistically and a model was developed that could be used to predict the most effective feedblock slot width for this system. Figure 6 shows a response surface analysis of the data. This statistical model had a correlation coefficient (R-squared) of 0.96 indicating that most of the variability in the data could be explained by the model.

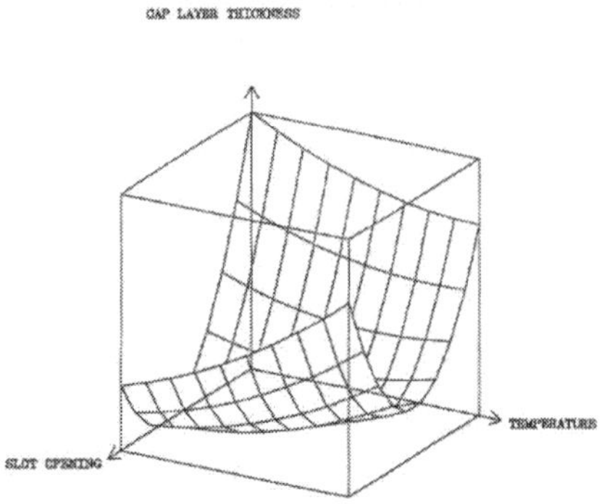

Figure 6 – Statistical analysis of the results of a designed experiment of processing and geometric variables in a feedblock.

A numerical model of the flow in a feedblock was subsequently developed to further develop our understanding of the effect of the feedblock geometry on the flow of the layers. Previous work on modeling feedblocks has been done by Perdikoulias et al. (1995). Figure 7 shows some of the results of the finite element model of the flow in a feedblock.

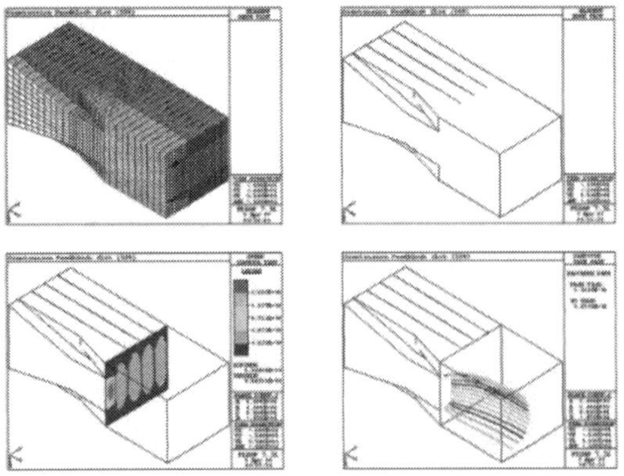

Figure 7 – Numerical simulation results for flow through a feedblock.

Figure 7 shows the geometry of the feedblock in the upper two images while the lower two show the velocity profile and particle flow paths. Note how restricting the flow of the cap layer in the insert affects both the velocity profile and particle flow paths.

Flow of a Coextruded Structure through a Die

The final area of interest in a coextrusion system is the die. The geometry of the die can substantially affect the layer uniformity of the final coextruded structure.

Figure 8 shows a photograph of coathanger style dies that were used to experimentally measure the flow distribution of different polymers through these die geometries.

Figure 8 – Experimental modular dies containing coathanger style distribution manifolds.

Finite element models were developed for these die geometries. Because of the symmetry of these dies, only one half of the flow domain was modeled. Figure 9 shows the finite element mesh developed for one of these geometries.

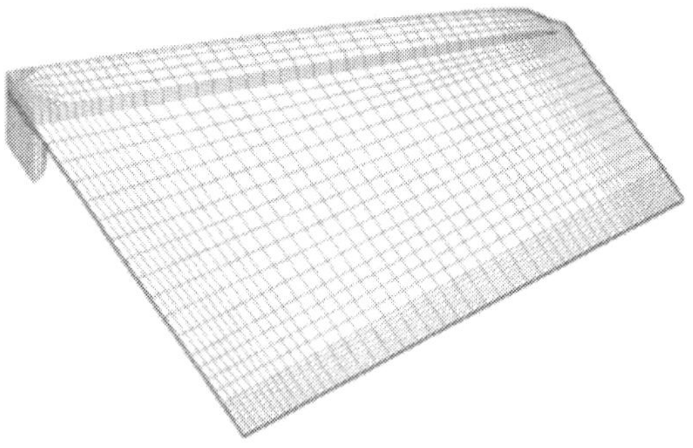

Figure 9 – Finite element mesh for a coathanger style die geometry.

Figure 10 shows some of the results of the finite element flow simulation. This figure uses particle path tracing to show the flow through the coathanger style die. This simulation was also used to predict the flow distribution at the exit of the die.

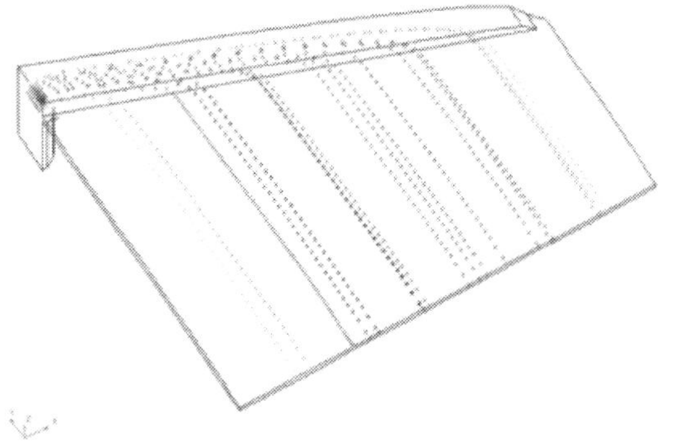

Figure 10 – Results of numerical simulation showing paths of massless particles flowing through the die.

In order to validate the numerical results, experimental measurements were made of the flow distribution from one of the dies at the die exit. The results of that experiment are shown in Figure 11.

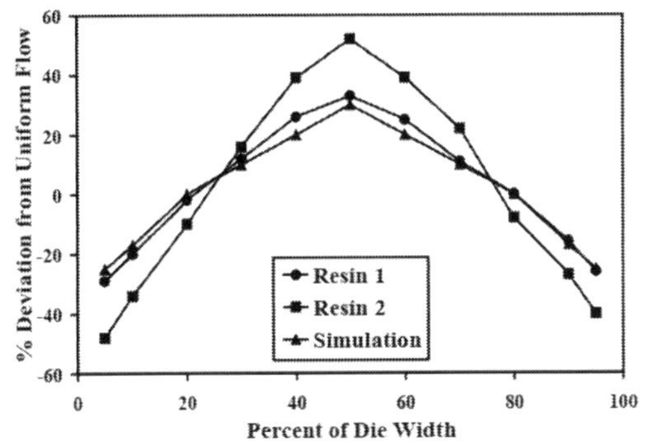

Figure 11 – Comparison of experimental and numerical flow distribution results for an elastic and an inelastic resin flowing through a coathanger die.

Figure 11 compares the numerical simulation results with experimental flow distribution results for two resins with very similar viscosities. This figure shows that the agreement between the numerical simulation results and the experimental results for Resin 1 is very good. However, the agreement is not as good for Resin 2, even though it has a similar viscosity to Resin 1.

The main difference between the two resins is that although they have similar viscosities, Resin 2 is more elastic than Resin 1. Since the numerical simulation is a viscous simulation alone (not a viscoelastic simulation), it appears to predict the results for the less elastic material better than for the more elastic material.

Since these experiments showed that elasticity is an important parameter even for monolithic structures, experiments were carried out to determine its importance in coextruded structures. In these experiments, a feedblock that produced a structure of 49 circular strands in a 7 by 7 matrix was used to feed an end-fed coathanger style die containing a rectangular distribution manifold. To ensure that there would be no viscosity differences, the same material was extruded in the fibers as in the matrix.

Figure 12 shows cross-sectional samples cut at set distances down the distribution manifold for a sample of a relatively inelastic polycarbonate resin. Note that all 49 strands are visible in the entry sample on the far left. As the flow progresses down the channel from left to right, the rows of strands flow uniformly out of the bottom of the manifold.

Figure 12 – Experimental results showing the flow patterns in the distribution manifold of a coathanger style die for a polycarbonate resin.

Figure 13 shows similar samples but for a relatively elastic polyethylene resin. This figure shows that the flow of this material down this manifold is significantly different than for the polycarbonate resin.

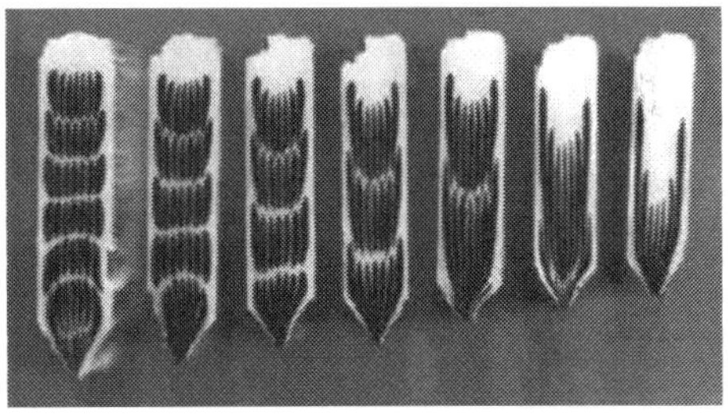

Figure 13 – Experimental results showing the flow patterns in the distribution manifold of a coathanger style die for a polyethylene resin.

The same polyethylene resin was run in this manifold but using a two-layer feedblock. This feedblock produces a two layer structure with a 20% cap layer. Once again the same materials were used in each layer. The results of this experiment are shown in Figure 14.

Figure 14 – Experimental results showing the flow patterns in the distribution manifold of a coathanger style die for a two layer polyethylene structure.

Figure 14 shows that the entry section on the left does have a 20% white layer. As the flow moves down the channel, a pool of white material forms in the upper right corner. This material slowly becomes a larger portion of the channel until in the last section on the right the cap layer is approximately 50% of the structure.

The pool of material shown in Figure 14 was caused by secondary flows driven by second normal stress differences in elastic materials (Wheeler and Whissler, 1966, Townsend et al., 1976, Dodson et al., 1987, Dooley and Hilton, 1994, Dooley and Dietsche, 1996, Debbaut et al., 1997, Dooley et al., 1998, Debbaut and Dooley, 1999, Anderson et al., 2006, Yue et al., 2008a, 2008b). Figure 15 shows experimental examples showing how these secondary flows can distort the initial structures shown on the left to the final structures shown on the right. Once again, these structures all contain identical materials so there were no viscosity differences.

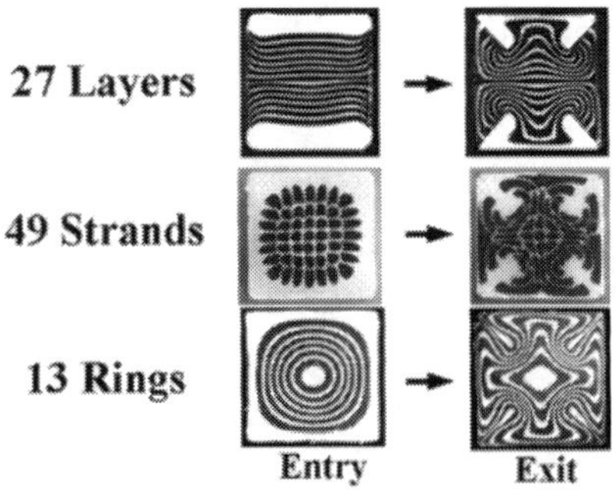

Figure 15 – Elastic layer rearrangement observed experimentally when flowing coextruded structures through a square channel.

A numerical simulation of the 49 strand sample shown in Figure 15 was performed. A viscoelastic flow simulation was performed using a multimode Giesekus model and a mapping method to determine the flow patterns. The results of this simulation are shown in Figure 16.

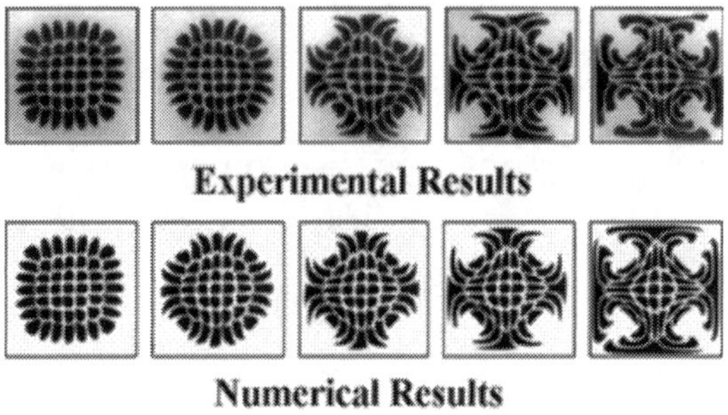

Figure 16 – Comparison of experimental and numerical results for the flow of an elastic polymer through a square channel.

Figure 16 shows a comparison of the experimental results shown in the top portion of the figure and the numerical simulation shown on the bottom of the figure. Even though these interface deformations are very difficult to model, the agreement between the experimental and numerical results is very good.

Conclusions

Experimental and numerical simulations of the coextrusion process have been performed. An experimental technique has been developed that allows the measurement of the rheology of a coextruded structure. In addition, experiments and simulations have been completed that show the importance of feedblock and die geometry on the flow and uniformity of coextruded structures.

Acknowledgements

The author would like to thank Beverly Stout, Chunxia Costeux, Robert Wrisley, Andrew Schlader, Kevin Hughes, Patrick Anderson, and Han Meijer for their contributions to this work.

References

Anderson, P.D., Dooley, J., and Meijer, H.E.H., "Viscoelastic Effects in Multilayer Polymer Extrusion," Appl. Rheol., **16**, 4, 198 (2006).

Debbaut, B., Avalosse, T., Dooley, J. and Hughes, K., "On the Development of Secondary Motions in Straight Channels Induced by the Second Normal Stress Difference: Experiments and Simulations," J. of Non-Newt. Fl. Mech., **69**, 2-3, 255, (1997).

Debbaut, B. and Dooley, J., "Secondary Motions in Straight and Tapered Channels: Experiments and Three-dimensional Finite Element Simulation with a Multimode Differential Viscoelastic Model," J. of Rheol., **43**, 6, 1525 (1999).

Dodson, A. G., Townsend, P. and Walters, K., "Non-Newtonian Flow in Pipes of Non-Circular Cross Section," Comput. Fluids, **2**, 3-4, 121 (1987).

Dooley, J. and Dietsche, L., "Numerical Simulation of Viscoelastic Polymer Flow – Effects of Secondary Flows on Multilayer Coextrusion," Plast. Eng., **52**, 4, 37 (1996).

Dooley, J. and Hilton, B.T., "Layer Rearrangement in Coextrusion," Plast. Eng., **50**, 2, 25 (1994).

Dooley, J., and Rudolph, L., "Viscous and Elastic Effects in Polymer Coextrusion," TAPPI, Polymers, Laminations, Adhesives, Coatings, & Extrusions Conference, Paper 12-5 (2002).

Dooley, J., Hyun, K. S. and Hughes, K.R., "An Experimental Study on the Effect of Polymer Viscoelasticity on Layer Rearrangement in Coextruded Structures," Polym. Eng. and Sci., **38**, 7, 1060 (1998).

Everage, Jr., A. E., "Theory of Stratified Bicomponent Flow of Polymer Melts. I. Equilibrium Newtonian Tube Flow," Trans. Soc. Rheol. 17, 629 (1973).

Han, C. D., "A Study of Bicomponent Coextrusion of Molten Polymers," J. Appl. Polym. Sci., **17**, 1289 (1973).

Han, C. D., Kim, Y. J. and Chin, H. B., "Rheological Investigation of Interfacial Instability in Two-Layer Flat-Film Coextrusion," Polym. Eng. Rev. **4**, 3, 177, (1984).

Khan, A. A. and Han, C. D., "A Study on the Interfacial Instability in the Stratified Flow of Two Viscoelastic Fluids Through a Rectangular Duct," Trans. Soc. Rheol., **21**, 101 (1977).

Khomami, B., "Interfacial Stability and Deformation of Two stratified Power Law Fluids in Plane Poiseuille Flow. 1. Stability Analysis," J. Non-Newt. Fluid Mech., **36**, 289 (1990).

Khomami, B., "Interfacial Stability and Deformation of Two stratified Power Law Fluids in Plane Poiseuille Flow. 2. Interface Deformation," J. Non-Newt. Fluid Mech., **37**, 19 (1990).

Perdikoulias, J., Zatloukal, M., and Birane, T., "Flow Simulation of Coextrusion Feedblocks," SPE-ANTEC Technical Papers, 41, 118 (1995).

Ramanathan, R., Shanker, R., Rehg, T., Jons, S., Headley, D.L. and Schrenk, W.J., "Wave Pattern Instability in Multilayer Coextrusion – An Experimental Investigation," SPE-ANTEC Technical Papers, **42**, 224 (1996).

Schrenk, W. J., and Alfrey, T., Chapter 15 Coextruded Multilayer Polymer Films and Sheets, in Paul, D. R. and Newman, S., (Eds.), Polymer Blends, Academic Press, New York (1978).

Schrenk, W. J., Bradley, N. L., Alfrey, T. Jr. and Maack, H., "Interfacial Flow Instability in Multilayer Coextrusion," Polym. Eng. Sci., **18**, 620 (1978).

Southern, J. H. and Ballman, R. L., "Stratified Bicomponent Flow of Polymer Melts in a Tube," Appl. Polym. Sci., **20**, 175 (1973).

Townsend, P., Walters, K. and Waterhouse, J., "Secondary Flows in Pipes of Square Cross-Section and the Measurement of the Second Normal Stress Difference," J. of Non-Newt. Fl. Mech., **1**, 2, 107 (1976).

Wheeler, J. A. and Whissler, "Steady Flow of Non-Newtonian Fluids in a Square Duct," Trans. Soc. Rheol., **10**, 1, 353 (1966).

White, J. L., Ufford, R. C., Dharod, K. R. and Price, R. L., "Experimental and Theoretical Study of the Extrusion of Two-phase Molten Polymer Systems, J. Appl. Polym. Sci. **16**, 1313 (1972).

Yue, P., Dooley, J., and Feng, J., "A General Criterion for Viscoelastic Secondary Flow in Pipes of Noncircular Cross Section," J. of Rheol., 52, 1, 315 (2008).

Yue, P., Zhou, C., Dooley, J., and Feng, J., "Elastic Encapsulation in Bicomponent Stratified Flow of Viscoelastic Fluids," J. of Rheol., 52, 4, 1027 (2008).

James Lindsay White: A Founder of Polymer Engineering As an Academic Discipline

*Avraam I Isayev**

The University of Akron - Ohio - USA

This talk will provide an overview of life of a prominent scientist and engineer, Professor James Lindsay White, a founder of the polymer processing society. Discussion of Jim's many scientific, engineering and societal contributions along with a description of him as a human being will be given. Various stages of Jim's life including periods of the undergraduate and graduate studies and industrial and academic activities will be described. Jim's numerous contributions to rheology, polymer processing and polymer engineering education will be considered. Jim's activities in writing various historical essays about famous scientists and engineers will also be provided. Finally, some memorable events of Jim's personal life will be presented.

Greener Natural Rubber

*Krisda Suchiva**

Mahidol University - Bangkok - Thailand

Natural rubber (NR) is a strategic material for engineering applications such as tyres, anti-vibration mounts, bearings, fenders, hoses and belts. NR possesses unique properties of high strength and elasticity unmatched by any synthetic rubbers (SR). Unlike its synthetic counterparts, NR is green and renewable material. Therefore, NR will be very important material for the future and for sustainability of mankind. Research and development of NR is essential in order to satisfy the global needs of this very important material. Development of NR should be towards greener NR, that is towards greater reductions in environmental pollution and energy consumption in its production, improvement in its quality which will lead to a greater reduction of waste and energy used for manufacturing of rubber products and enhancement of its properties with the view to increasingly replacing the petroleum-based SR. Intensive researches have been carried out to achieve the above goals, so far with the following results. 1. Greener Production Technology 1.1 Novel Preservative System to Replace the Polluting Ammonia A novel preservative system is being developed to replace the existing, long-serving but polluting, ammonia-based system. Success has been achieved at laboratory level with promising results. The project is being developed at industrial level at present. 1.2 Green Recovering Agent for Skim Rubber New chemical system derived from biopolymers have been developed to replace existing coagulant, concentrated sulphuric acid which is highly toxic and polluting, for recovering of skim rubber during the production of concentrated NR latex. The new recovering agent is being transferred to the concentrated NR latex producing industry. 1.3 Efficient Technology for Drying NR A novel technology has been developed to increase the drying efficiency of NR. Wet NR coagulum can now be dried in a matter of minutes instead of hours for the old technology. An energy saving of around 20% is also achieved. 2. Improvement of Quality of NR 2.1 Improvement of the Consistency of NR A major drawback of NR, despite its excellent properties, is its variations in vulcanization and processing properties. These are caused largely by the nonrubber constituents and also the production technology. Basic studies are being made in order to better understand the causes of inconsistency of NR and the new correction technologies developed. Innovation in production equipment is also being developed which will allow continuous and better controlled production of NR. 2.2 Improvement of the Processability of NR The other problem with NR is that it is tough and highly elastic, making it dufficult to process. A new technology is being developed for the production of low viscosity and elasticity NR. The new grades of NR prove to be easier to process. 3. Improvement of Properties of NR If the green and renewable NR is to replace the petroleum-based SR in the future, its poor gas-barrier, oil-, heat-, oxygen- and ozone-resistances first have to be improved. Chemical modifications of NR by epoxidation reaction to give epoxidized NR is an example of established technology for improving the oil-resistant and gas-barrier properties of NR. Hydrogenation of NR can produce improved heat-, oxygen-and ozone resistant NR but both methods change the properties of NR and are expensive to make. By using reactive blending technology, oil-, heat-, oxygen-, and ozone-resistant NRs have been developed which contain 70-80% NR. Thermoplastic NR has also been developed. The rubber industry will be greener if these specialty NRs can increasingly replace SRs.

ADVANCED CHARACTERIZATION OF POLYMERIC NANOFIBERS

P. SAHA[1], W. SAMBAER[1], M. ZATLOUKAL[1], D. KIMMER[2]

[1]Polymer Centre, Faculty of Technology, Tomas Bata University in Zlín, Zlín, Czech Republic
[2]SPUR a.s., Zlín, Czech Republic

INTRODUCTION

In the last decade, there has been growing interest in nanofibers because the range of applications [1], like filter media, wound dressings [2,3], sensor, catalysis, clothing, etc. is growing. One possible method to produce nanofibers is electrospinning, where a high voltages is applied to produce fibers out of a polymer solution on a grounded collector [4] covered by a support textile. The knowledge about the processing parameters for electrospinning has a great importance for the properties of the nanofiber mats and is already investigated. An important kind of these parameters is material-related, like polymer type, viscosity, electrical conductivity [5], surface tension of the solvent, etc., and equipment-related, such as applied voltage [6], distance between collector and electrode, humidity [7,8], pressure [9] and temperature [9] in the chamber. Several studies have already been done to measure the mechanical properties of single fibers [10-13] and nanofiber mats with different methods [7,14], such as conventional tensile machines or more sophisticated equipment (e.g. atomic force microscope cantilever (AFM) or bending tests). Some investigators have recently turned to digital image analysis to characterize structures and predict some characteristics like porosity [15] and cell infiltration [16,17]. Nevertheless, the influence of the support textile type on the properties is still questionable and the use of digital image analysis can be expanded to other areas where it is possible to predict more nanofiber mat properties. In this work the attention is paid to the difference in mechanical and structural properties of electrospun nanofiber mats produced under the same conditions but with different supporting textiles, where the nanofibers are collecting.

EXPERIMENTAL

Material

A polyurethane (PU) solution based on 4,4′methylenebisphenylisocyanate, poly(3-methyl-1,5-pentanediol)-alt-(adipic,isophtalic acid) and 1,4 butanediol (molar ratio 6:1:5) solved in dimethylformamide (DMF) was synthesized. The prepared solution suitable for electrospinning had a PU concentration of 11.5 wt%, a viscosity of 0.99 Pa.s and a conductivity of 151 μS/cm (adjusted by tetraethylamonium bromide). Two different support textiles were used to collect the nanofibers. One textile was created by 35% poly(m-aramid), 63.8% poly(p-aramid) and 1.2% stainless steel, named as 'Sample 1', and the second support textile was non woven melt-blown polypropylene freshened with Polyfix N, named as 'Sample 2'.

Electospinning process

The produced nanofiber textiles consisted out of three layers of nanofibers produced by electrospinning as described in Figure 1 on the above given support textiles. A voltage of 75kV was applied between the specially designed wire electrode and the grounded collector which was placed 180 mm from the electrode. The electrode rotated at the speed of 7 rpm, and the roll-up speed from the support textile was 0.16 m/min. The environmental conditions in the electrospinning-chamber were the temperature of 27.5°C and relative humidity 28%. Samples for further analyses were taken from the middle part of the produced nanofiber textiles having the fiber diameter distribution as depicted in Figure 2.

Mechanical Test Method

To test mechanical properties, Sentmanat Extensional Rheometer (SER) Universal Testing Platform was used on the rotational rheometer, Advanced Rheometric Expansion System (ARES) [18-20]. The tensile tests were carried out at a constant extensional strain rate of 0.01 s^{-1}. The extensional stress was derived from the torque measured by a 2K FRTN1 transducer (provided by TA Instruments) and plotted as a function of extensional strain.

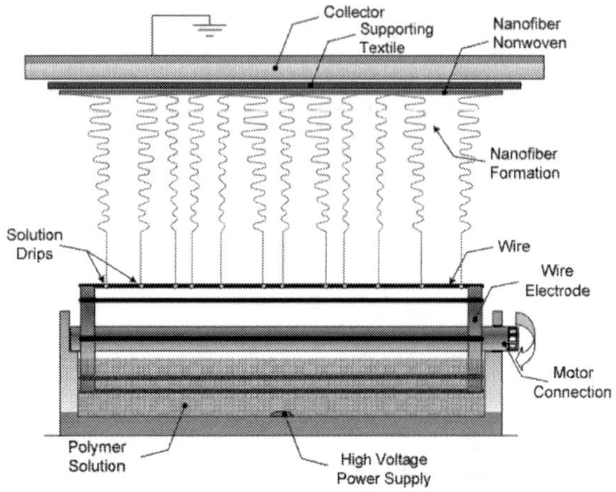

Figure 1: Scheme of electrospinning process.

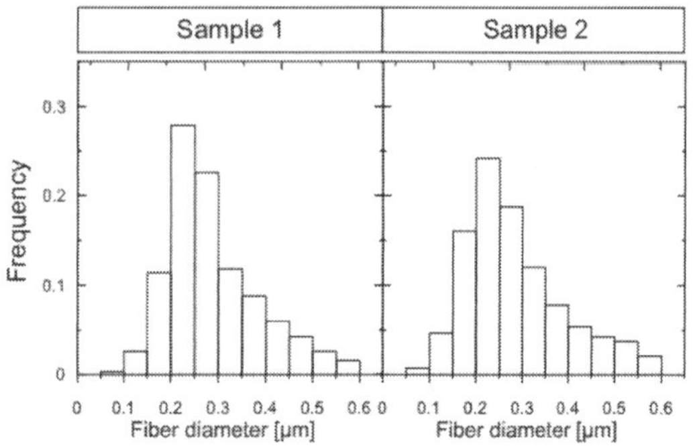

Figure 2: Fiber diameter distributions for both investigated nanofiber nonwovens.

RESULTS AND DISCUSSION

Tensile Tests

Figure 3 shows the results of the tensile tests with included error-bars. As can be seen, there are a big differences between the mechanical properties of the nanofiber mats due to the different support textiles. This graph shows that the fibers from sample 2, spun on the melt blow support textile has higher Young modulus and the rupture stress in comparison with fibers produced on aramid based supporting textile.

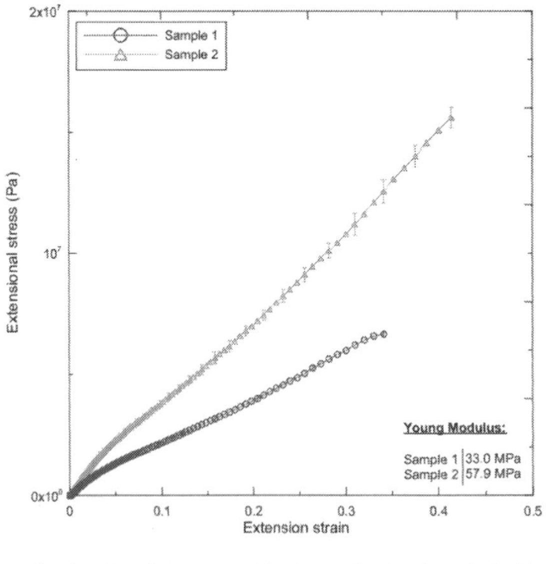

Figure 3: Measured extensional stress as a function of extension strain for sample 1 and 2.

Nanofiber based Nonwovens Structure Analysis

In Figure 4, nanoscale as well as macro-scale view of both, supporting textiles and corresponding PU nanofiber based nonwovens are provided. It is clearly visible that the specific 3D character of the supporting textile texture is 'copied' into PU nanofiber based nonwovens as it is visible on the macro-scale pictures depicted in this figure. In more detail, it is clearly visible that the use of the aramid based supporting textile (Sample 1) during electrospinning process causes creation of highly 3D textured nonwovens whereas the use of melt blown supporting textile (Sample 2) leads to generation of more 2D textures where the thickness profile of PU nonwoven is more even in this case. Note that both SEM pictures used in this work were obtained by using field emission scanning electron microscopy (Vega II LSU, Tescan, Czech republic). The analyzed nanoscale pictures had a resolution of 9009 by 9009 px where one pixel represents a square of 10 by 10 nm.

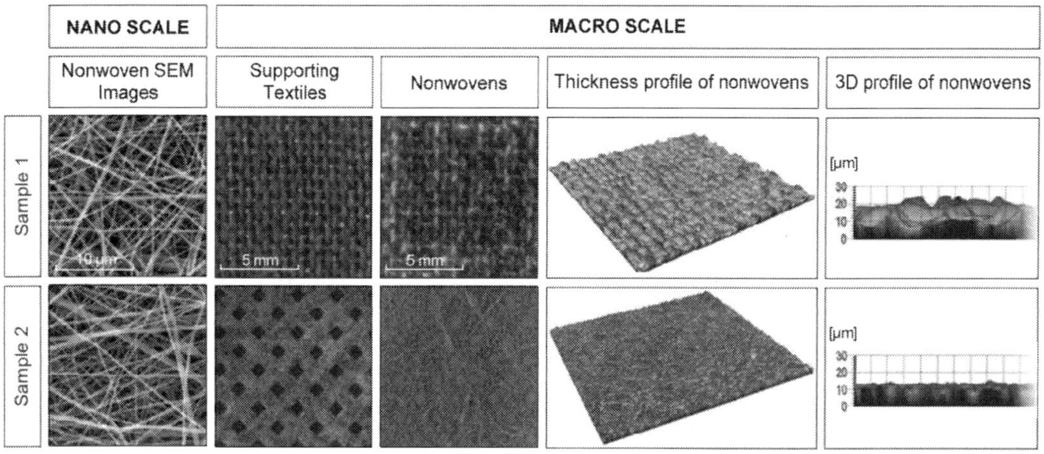

Figure 4: Visualization of the utilized supporting textiles and obtained nanofiber nonwovens through electrospinning process.

Pore Size Distribution Analysis

In this part, novel Scanning Electron Microscope (SEM) picture digital image analysis technique for determination of pore size distribution for nanofiber based nonwovens is introduced. This method is based on the recent work of Ghasemi-Mobarakeh and Semnani et al. [16, 17], who analyze the pores by using different cells with three different characteristic sizes (10, 20 and 30 μm) that could infiltrate the fiber based product. In this work, we have generalized this approach considering virtually unlimited number of cells (depends on particular computer memory allocation capability of the used PC) having sizes from units of nanometers to hundreds of μm which can pass through the fiber nonwoven. In Figure 5, a part of the graphical results are shown to demonstrate how the methodology works. After this procedure is done, the number of circles has been counted and normalized for each sample and plotted versus circle diameter, which represents the pore/hole size in this case (see Figure 6). In order to calculate pore/hole size distribution curve, the derivatives from the Figure 6 has to be calculated. To do that effectively and precisely, the data in Figure 6 were fitted by a combination of two stretched exponentional functions (Kohlrausch-Williams-Watts, KWW, function) having the following form:

$$N(D) = A_1 \exp\left[-\left(\frac{D}{\tau_1}\right)^{\beta_1}\right] + A_2 \exp\left[-\left(\frac{D}{\tau_2}\right)^{\beta_2}\right] \tag{1}$$

where N(D) is normalized number of holes, D is the hole size in nm and A_1, A_2, β_1, β_2, τ_1 and τ_2 are fitting constants. As can be seen in Figure 6, the chosen model (solid line in this graph) describes the discrete data very well and thus the normalized derivative dN(D)=d(normalized number of holes)/d(hole size) can be easily calculated from Eq.1 by using the following expression:

$$dN(D) = \frac{A_1 \exp\left[-\left(\frac{D}{\tau_1}\right)^{\beta_1}\right]\frac{\beta_1}{\tau_1}\left(\frac{D}{\tau_1}\right)^{\beta_1-1} + A_2 \exp\left[-\left(\frac{D}{\tau_2}\right)^{\beta_2}\right]\frac{\beta_2}{\tau_2}\left(\frac{D}{\tau_2}\right)^{\beta_2-1}}{A_1 \exp\left[-\left(\frac{D_s}{\tau_1}\right)^{\beta_1}\right] + A_2 \exp\left[-\left(\frac{D_s}{\tau_2}\right)^{\beta_2}\right]} \tag{2}$$

where D_s is a constant equal to the smallest (circle/hole) diameter that can be detected in the picture. The obtained pore size distribution curves for both investigated samples are depicted in Figure 7. For the pore size distribution curves evaluation, D_{z+1} hole size average (which is sensitive to the highest pores in the analyzed structure), which is defined below has been used

$$D_{z+1} = \frac{\int_{D_S}^{\infty} D^3 \, dN(D) \, dD}{\int_{D_S}^{\infty} D^2 \, dN(D) \, dD} \tag{3}$$

Based on the pore size distribution curves, depicted in Figure 7, can be concluded that the use of the melt blown supporting textile during electrospinning process leads to nanofiber based structure having much smaller pores than structures prepared by the use of the aramid based supporting textile.

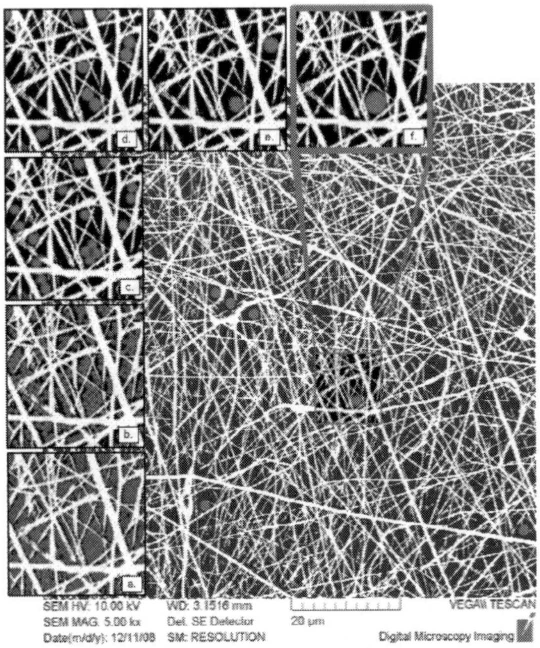

Figure 5: Visualization of pore size distribution analysis for PPNW nonwoven sample for different circle diameters: DA = 250 nm, DB = 450 nm, DC = 850nm, DD = 1050nm, DE = 1250 nm, DF = 1850 nm.

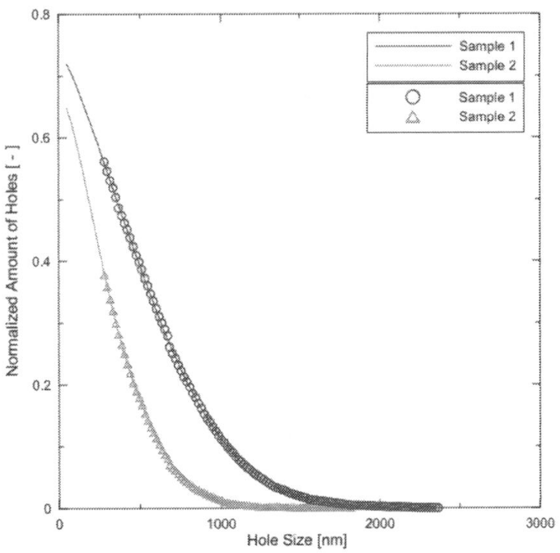

Figure 6: Raw and fitted data for the normalized amount of holes as a function of the hole size for both samples.

Airfactor

For experimental determination of the amount of air in the mats, the air or porosity factor $\bar{\alpha} = 1 - \frac{\rho_f}{\rho_{PU}}$ has been utilized, where ρ_f and ρ_{PU} are densities of fibers and bulk PU, respectively. The density of the fibers was determined from 10 different measurements of the weight and the volume for each sample. Volume has been calculated from the rectangle sample area (width=12.7 mm, length=50 mm) and sample thickness. Note that percentage amount of the air has also been theoretically evaluated based on the black white ratio of SEM images with were converted to binary pictures with a proper threshold level. Theoretically as well as experimentally determined amount of air is summarized for both samples in Table 1. Clearly, the sample having the smallest pores (Sample 2) contain less air than sample having the higher pores (Sample 1). It can also be seen that the error of measured amount of air is much higher than the theoretical one due to thickness measurements difficulties (small sample thicknesses and high elasticity). The differences in airfactor for both samples can explain the big differences in mechanical properties of the fibermats with the same polymer solution and processing conditions spun on different support textiles because the amount of PUR material is different. This is visible in Figure 8 where the maximum stress at break together with E-modulus is plotted as a function of D_{z+1} hole size average and percentage amount for the measured and theoretically determined airfactor.

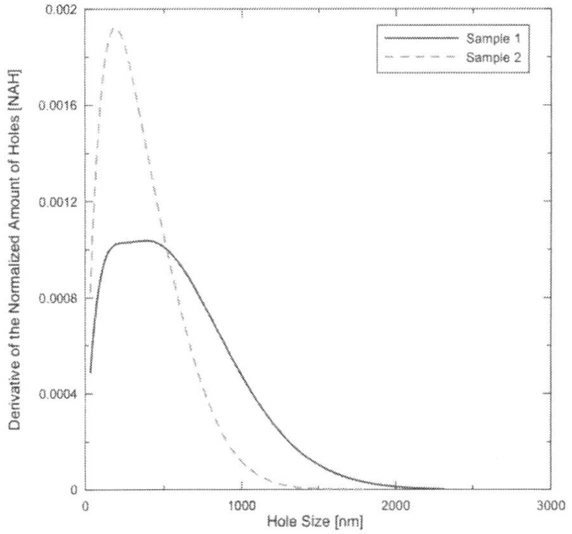

Figure 7: Normalized derivative of the amount of holes as a function of the hole size for both samples.

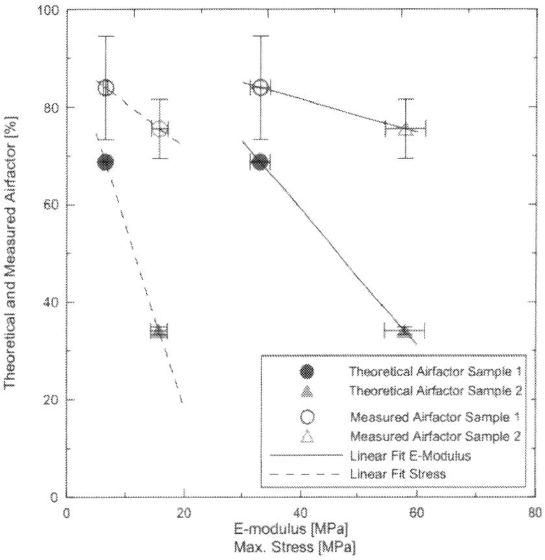

Figure 8: Measured and theoretical airfactor as a function of E-modulus and maximal stress for both samples.

Table 1: Theoretically as well as experimentally determined air factor for both investigated nonwoven samples.

Nonwoven	Experimental Airfactor		Theoretical Airfactor	
Sample 1	83.9%	±4.1%	68.7%	±0.2%
Sample 2	75.5%	±2.9%	34.1%	±0.9%

CONCLUSION

Based on the performed experimental work and proposed novel digital image analysis technique for nanoscale SEM pictures (taking macroscopic structure features into account) has been found that supporting textiles have very high effect on mechanical properties of polyurethane nanofiber nonwovens prepared by electrospinning process (even if their fiber diameter distributions were similar), which can be explained by different porosity of the prepared samples. This study has also shown that there is no direct correlation between the fiber-thickness-distribution and mechanical properties.

ACKNOWLEDGEMENTS

The authors wish to acknowledge Grant Agency of the Czech Republic, grant No. P108/10/1325 (Applied Rheology for advanced polymer processing) and the Ministry of Education CR, grant no. MSM 7088352101, for the financial support.

REFERENCES

1. 1. Huang, Z.M., Y.Z. Zhang, M. Kotiki and S. Ramakrishna, *Composites Science and Technology* **63**, 2223-2253 (2003).
2. Khil MS, Cha Di, Kim HY, Kim IS, Bhattarai N., *J. Biomed. Mater.* **67B**, 675–679 (2003).
3. Kim SH, Nam YS, Lee TS, Park WH. *Silk, Polym. J.* **35**, 185–90 (2003).
4. D. Li and Y.Xia, *Adv. Mater.* **16**, 1151-1170 (2004).
5. Tamer Uyar, Flemming Besenbacher, *Polymer* **49**, 5336–5343 (2008).
6. Delaram Fallahi, Mehdi Rafizadeh, Naser Mohammadi and Behrooz Vahidi, *Polym. Int.* **57**, 1363–1368 (2008).

7. Tan, E.P.S. and C.T. Lim, *Composites Science and Technology* ***66***, 1102-1111 (2006).
8. S. Megelski, J.S. Stephens, D.B. Chase, J.F. Rabolt, *Macromolecules* **35,** 8456 (2002).
9. S. de Vrieze, T. van Camp, A. Nelvig, B. Hagstrom, P. Westbroek, K. de Clerck, *J. Mater. Sci.* **44**, 1357–1362 (2009).
10. Xinnah Wang, Yongda Yan, Michael J. Yost, Stephen A. Fann, Shen Dong and Xiaodong Li, *Journal of Biomedical Materials Research Part* A, 130-135 (2006).
11. Lu-Qi Liu, Dimitrios Tasis, Maurizio Prato and Hanoch Daniel Wagner, *Adv. Mater.* **19**, 1228-1233 (2007).
12. Tan, E.P.S., S.Y. NG and C.T. Lim, *Biomaterials* **26**, 1453-1456 (2005).
13. C.L. Casper, J.S. Stehens, N.G. Tassi, D.B. Chase, J.F. Rabolt, *Macromolecules* **37**, 573-578 (2004).
14. Wang, X.F., X.M. Chen, K.H. Yoon, D.F. Fang, B.S. Hsiao and B. Chu, *Environmental Science & Technology* **39**, 7684-7691 (2005).
15. Mohammed Ziabari, Vahid Mottahitalab, and Akbar Khodaparast Haghi, *Korean J. Chem. Eng.***25**, 923-932 (2008).
16. L. Ghasemi-Mobarakeh, D. Semnani, M. Morshed, *Journal of Applied Polymer Science* **106**, 2536–2542 (2007).
17. Dariush Semnani, Laleh Ghasemi-Mobarakeh, Mohammad Morshed, Mohammad-Hossein Nasr-Esfahani, *Journal of Applied Polymer Science* **111**, 317-322 (2009).
18. M.L. Sentmanat, B.N. Wang, G.H. McKinley, *J. Rheol.* **49**, 585 (2005).
19. M.L. Sentmanat, *Rheol. Acta* **43**, 657 (2004).
20. M.L. Sentmanat, US Patent No.6, 578, 413 (2003).

STRUCTURE DEVELOPMENT IN POLYMER PROCESSING

Takeshi Kikutani

Department of Organic and Polymeric Materials, Graduate School of Science and Engineering, Tokyo Institute of Technology,2-12-1, O-okayama, Meguro-ku, Tokyo 152-8552, Japan – kikutani.t.aa@m.titech.ac.jp

Abstract - Deep understanding of the mechanism of structure development in polymer processing is indispensable for the improvement of the properties of final products through the designing of materials and processes. Melt spinning and drawing processes are probably the most fundamental procedures of polymer processing. These processes are useful for the understanding of the characteristic behavior of various types of polymers because of the uniaxial-nature of both the process and the developed structure. In-situ measurements of process parameters and structure development in the melt spinning and drawing processes are also an effective way for the clarification of the polymer behavior. Utilization of the steady-state process and conversion between Eulerean and Lagrangean coordinates enables us to obtain the information with high time-resolution. It should be noted that the crystallization of polymers under the effect of stress proceeds much faster than that in the quiescent state. Sequential and simultaneous biaxial stretching of films are other continuous industrial processes in which three-dimensional analyses of the structure is necessary. On the other hand, processes such as blow molding and injection molding are the so-called batch-type process in which analyses of not only the structural hierarchy but also structural variation in the products are important. In terms of the theoretical analyses on the mechanism of structure development, stress-optical rule for the molecular orientation, and non-isothermal crystallization kinetics under the effect of flow have been the two main subjects, however, effect of the molecular entanglement on structure development attracts much attention these days. Collaboration of people in the fields of rheology and structure development seems to be required in this field. These subjects will be discussed consulting numerous publications by Prof. James L. White.

AUTHOR INDEX

AUTHOR INDEX

AUTHOR INDEX

AUTHOR INDEX

AUTHOR INDEX

AUTHOR INDEX

AUTHOR INDEX

AUTHOR INDEX

AUTHOR INDEX